2011 GRADUATE PROGRAMS

in Physics, Astronomy, and Related Fields

AIP
American Institute of Physics

American Institute of Physics

Melville, New York

Suite 1NO1
2 Huntington Quadrangle
Melville, NY 11747
Tel. (516) 576-2460
E-mail: mflikop@aip.org
http://www.aip.org

Cover figure originally published in *Applied Physics Letters*, November 24, 2003 issue. Reproduced with permission of Dr. Lee, NTT Basic Research Laboratories, Atsugi, Japan.

International Standard Book Number: 978-0-7354-0840-1
International Standard Serials Number: 0147-1821
AIP Publication R-205.35

Printed in the United States of America

CONTENTS

FOREWORD

The *2011 Graduate Programs in Physics, Astronomy, and Related Fields* provides information on graduate programs in North America. This is the thirty-fourth annual edition.

The great majority of U.S. physics and astronomy doctoral programs and most master's programs in both fields are featured. A substantial number of physics-related fields are listed, including nuclear engineering, electrical engineering, chemical physics, materials science, meteorology, geophysics, medical physics, oceanography, and acoustics departments. The astronomy-related fields, such as astrophysics, atmospheric, space physics, cosmic rays, and others, are covered as well.

I thank the department chairs and administrative assistants for supplying information on their graduate and research programs. Without such cooperation and assistance, this publication and other AIP information-gathering projects that benefit the physics and astronomy communities could not be accomplished. I also thank the AIP staff responsible for this book's production.

The *2011 Graduate Programs in Physics, Astronomy, and Related Fields* has an online component: GradschoolShopper.com, a one-stop site for researching graduate programs in physics, astronomy, and related fields. GradschoolShopper.com adds a new dimension to the print publication: searchability, greater accessibility, and convenience. In addition to a searchable online version, the Web site features helpful links to resources for students and academics. I invite you to visit www.GradschoolShopper.com.

I hope that students, their advisers, and others interested in graduate science education will find this publication useful. I welcome your suggestions for improvements in the format or content of this publication. We are committed to making every version a better product.

H. Frederick Dylla
Executive Director and CEO
American Institute of Physics
October 2010

INTRODUCTION

The *2011 Graduate Programs in Physics, Astronomy, and Related Fields* is designed to provide easily accessible, comparative information on graduate programs and research in physics and in fields based upon the principles of physics. Students planning graduate study, faculty advisers, and others interested in comparative information on graduate programs and physics research will find this information valuable. This is the twenty-fifth annual edition of the book.

The content and format of this edition have remained the same as the previous edition. Two features have appeared in all editions. First, the information on each department is presented, as much as possible, in a tabular format to make it easier to compare information from different departments and to make the listings more compact. Second, for each department information is presented concerning its research expenditures and sources of support. Care should be taken in using this information. While it does give some indication of the level of research activity in a particular research specialty, there certainly is not a one-to-one correspondence between research expenditures and the quality of the research program.

It should be noted that the same approach was used to fund the preparation and distribution of this edition as for the previous editions. Listed graduate departments paid a listing charge to cover the cost of preparing the book and distributing a copy to all departments in North America offering at least a bachelor's degree in physics, astronomy, electrical engineering, or nuclear engineering, and to all engineering schools. Accordingly, AIP was able to prepare a new edition which it otherwise could not have afforded; the book is receiving much wider distribution and hence will have greater use by students than would otherwise be the case. Almost all physics and astronomy doctoral programs in the United States and most master's programs are included. AIP anticipates that the number of listed departments will remain essentially constant in future editions.

Organization of the Book

Each entry in the book describes the graduate programs offered by an academic department at an institution of higher learning in North America. Entries are organized alphabetically by state or province and within each state of province alphabetically by the name of the institution. If more than one department at an institution is listed, the astronomy department is listed second, and the departments in related fields follows.

There are three mechanisms by which a user can locate the listing of a department at a particular institution. First, if the state or province in which the institution is located is known, the entry can be found relatively quickly from Appendix I. Second, Appendix II provides an alphabetical listing of institutions and departments. Third, Appendix III lists institutions and departments by field and highest degree offered.

Both Appendices IV and V are geographically arranged lists that give the reader a synopsis of physics and related-field programs, and the subjects that they offer. These lists include most of the programs featured in the main listings. The main difference between them is that Appendix IV, "Research Specialities of Doctoral Programs," covers the course offerings for Ph.D. programs in physics and related fields, while Appendix V, "Areas of Concentration of Master's Programs," covers master's offerings for almost all of the same programs.

Departments Included

All known departments in North America that had programs leading to a Ph.D. or Master's degree in physics, astronomy, or a physics-related field were invited to submit entries. Additional departments, including medical physics, geophysics, chemical physics, materials science, electrical engineering, nuclear engineering, meteorology, and oceanography department that have physics-oriented research programs, were also invited to submit entries.

The response was excellent. Almost all major U.S. physics doctoral programs and most astronomy programs are listed. There are 228 departments from 202 institutions included in this book. Many of these departments offer graduate degrees in physics-related fields such as those mentioned above.

A few departments with new graduate programs or with graduate programs in a related field may have been omitted because the AIP staff was unaware of them. If so, we would appreciate receiving information about them so they can be included in the 2012 edition.

PART I

UNITED STATES

Geographic Listing of Graduate Programs

ALABAMA AGRICULTURAL AND MECHANICAL UNIVERSITY

DEPARTMENT OF PHYSICS

P.O. Box 1268 Huntsville, Alabama 35762

Web page: http://www.physics.aamu.edu

Students Accepted For Degree	FIELDS		
	Physics	Astronomy	Related Fields
Doctorate	X		
Master's	X		

1. General

President: Dr. Andrew Hugine, Jr.
Dean of Graduate School: Dr. Michael Orok
Department Chairman: Dr. M. D. Aggarwal
Department Telephone Number: (256) 372-5305
Type of Institution: University
Control: Public
Setting: Urban
Total Faculty: 313
Total Graduate Faculty: 240
Total Students: 5,327
Total Graduate Students: 831
Annual Graduate Tuition:
In-state residents: $261*/cr.hr.
Out-of-state residents: $502**/cr.hr.
Tuition rates for: 2010–11
Deferred tuition plan: No
Annual Other Fees: $650
Term: Semester

*$261 is per semester hour for In-state residents.
**$502 is per semester hour for Out-of-state residents.

2. Number of Faculty in Department

The combined total of full-time faculty in the three professorial ranks is 13. The combined total of full-time, part-time, and other faculty at all ranks is 28.

3. Admission, Financial Aid, and Housing

Address admission inquiries to: Dean, School of Graduate Studies
Graduate application fee required: $25
Admission deadline (Fall admission): 7/1
Admission information: For fall admission, 2008–2009, 8 students were accepted from 20 applicants.
Admission requirements: For admission to the graduate programs, a Bachelor's degree in Physical Science/Electrical Engineering Materials Science/optics is required with a minimum undergraduate GPA of 3.00* specified. The GRE is required. The minimum acceptable score for admission is total—850 combined for M.S. The GRE Advanced is required. The minimum acceptable score for admission is 1,000 for Ph.D. The average GRE scores for admissions were total—900. The average GRE Advanced score for admissions was 900. Students from non-English speaking countries are required to demonstrate proficiency in English via the TOEFL exam. Minimum acceptable score for admission is 550.
Undergraduate preparation assumed: Physics: Halliday, Resnick, and Krane, *Physics Part I & II*; Modern Physics: Beiser, *Concepts of Modern Physics*; Mechanics: Arya, *Introduction to Classical Mechanics*; Methods in Mathematical Physics: Boas, *Mathematical Methods in the Physical Sciences*; Electricity and Magnetism: Lorrain, Corson, and Lorrain, *Electric Fields and Waves.*
Address financial aid inquiries to: Dean, School of Graduate Studies, Alabama A&M University, Normal, AL 35762
GAPSFAS application required: Yes
Financial aid deadline: 8/1
Loans available: Yes
On-campus, graduate student housing available: Yes
On-campus, married student housing available: No

*3.00 on 4.00 scale. Students with Bachelor's degrees in optical science or optical engineering programs will be eligible for admission in optics program and with materials science, or materials engineering in materials program.

4. Graduate Degree Requirements

Master's: For admission to the Master of Science program in physics, applicants must have received a Bachelor's degree from a recognized university with a major in any of the physical sciences or electrical engineering or materials science or optics and must have an overall GPA of 3.00 (based on a 4.00 system). Also, students with Bachelor's degrees in optical science, optical engineering programs will be eligible for admission into the optics program and with materials science, or materials engineering programs into materials program. Students with a degree in an area other than physics may be required to take prerequisite undergraduate physics courses.
Thesis Option: The students must complete at least 24 semester hours of course work with a minimum of 12 hours in area of concentration, and write a thesis (6 semester hours credit) on an approved topic under the supervision of a thesis advisor, and satisfactorily defend the finding of the thesis before a committee of faculty appointed by the department and appointed by the Dean of Graduate Studies.
Non-Thesis Option: The students must complete at least 30 semester hours of course work with at least 15 of these being in the area of concentration and pass a comprehensive examination given by the department.
Doctorate: The program is open for admission to students who satisfy the general criteria for admission to graduate-school and who also meet the departmental requirements for admission to the graduate program in the specialization of choice. The applicants with a B.S. in Physics must have an overall GPA of 3.30 (based on a 4.00 system) in the area of concentration and also must have a GRE score of 50 percent in the applicant's major area. These applicants, as well as applicants with Master's degrees, must pass the various examinations described later. Graduates with a major in any of the physical sciences and a minor in physics, as well as graduates in electrical engineering, are eligible for conditional admission. Such students may be required to take additional courses in physics to attain regular status. Students from non-English speaking countries are required to demonstrate proficiency in English via the University's English Competency test for graduate students. A minimum score of 550 on the Test for English as a Foreign Language (TOEFL) WILL BE RE-

QUIRED FOR ADMISSION. Applicants who hold an M.S. Degree in the particular specialization, namely optics or materials science, will be granted provisional admission based on their performance at the Master's level as evidenced by the corresponding transcripts and also based on letters of recommendation from the departmental faculty where they graduated. Such applicants also must have a minimum GPA of 3.30 (based on a 4.00 system) in the major area. Persons holding the M.S. in traditional physics or electrical engineering or chemistry may be eligible for admission subject to the condition given above in this paragraph. Those students may be required to complete some Master's level courses. However, credit will be given only for courses which are in the list of required or optional courses for the specialization to which the applicant will be admitted. In order to earn the Ph.D. degree, a graduate student must earn a total of at least 60 semester hours of credit, with 45 hours in the area of specialization (optics or materials science) and 15 in the general area of physics. In addition to this, a student must pass a departmental qualifying examination before completing 24 semester hours of graduate credits and must also pass a departmental comprehensive examination before being admitted to the Ph.D. candidacy. Also, the student must do research on an approved topic, earn 12 semester hours of credit for the dissertation, and defend the findings of research before a committee of faculty members. A student cannot register for more than 6 credit hours of dissertation during a given semester. A student may skip the M.S. and proceed to the Ph.D. program. There is no foreign language requirement for the degree, but all students will be required to show proficiency in the use of computers. A student must pass three examinations in the following sequence before the degree is awarded:

1. All students seeking a Ph.D. must pass a qualifying examination before completing 24 semester hours of graduate credits. A person who has been admitted on the basis of a Master's degree may take the qualifying examination after the first semester in the program.
2. All students must take a written departmental candidacy examination in the area of specialization before filing for candidacy. This examination must be passed at least nine months before the expected graduation date. A student is considered as a Ph.D. candidate only after passing the departmental examination.

Thesis: Theses may be written *in absentia*.

Table B—Appointments to Graduate Students, 2009–10

Title of Appointee	Appointments Total	Appointments First year	Academic Load Allowed in Credit Hours	Hours of Service Per Week	Stipend for Academic Year ($)
			Semester		
Teaching Assistant	5	1	9	20	12,000
Research Assistant	16	1	9	20	15,000
Fellowships	2	0	9	0	18,000
Total	23	2			

5. Personnel Engaged in Separately Budgeted Research,

Professorial faculty	11
Other faculty	5
Graduate students	21
Undergraduate students	18
Nonteaching research personnel	5
Total	60

6. Separately Budgeted Research Expenditures by Source of Support

	Departmental Research	Physics-related Research Outside Department
Federal government	$1,950,000	$
Private, non-profit organizations	10,000	
Total	$1,960,000	$

7.Separately Funded and Managed Laboratories

Center for Irradiation of Materials	$1,200,000
Total	$1,200,000

Table C—Separately Budgeted Research Expenditures

Research Specialty	No. of Grants	Expenditures ($)
Materials Sci./Metallurgy	4	1,200,000
Optics	3	600,000
Space Science	2	150,000
Total	9	1,950,000

FACULTY

Professor Emeriti

Lal, R. B., Ph.D., Agra, 1963. Solid state physics; materials science; crystal growth.

Lee, C. T., Ph.D., Rice, 1967. Quantum optics.

Tan, A., Ph.D., Univ. of Alabama, Huntsville, 1979. Space science.

Professors

Aggarwal, M. D., Ph.D., Calcutta, 1974. Crystal growth and characterization.

Dokhanian, Mostafa, Optics, Applied Physics, Ph.D., 1999. Alabama A&M University. Optics.

Edwards, Matthew E., Ph.D., Howard Univ., 1977. Materials science/condensed matter, laser optics.

Ila, Daryush, Ph.D., Lowell, 1987. Condensed matter.

Reddy, B. R., Ph.D., Indian Inst. of Tech., Kanpur, 1981. Laser spectroscopy.

Sharma, A., Ph.D., Columbia Univ., 1982. Optics.

Wang, J. C., Ph.D., Massachusetts, 1976. Solid state physics; materials science.

Assistant Professors

Batra, A. K . Ph.D., I.I.T., Delhi, India, 1981. Materials science, physics.

Guggilla, Padmaja, Ph.D., AAMU, 2007. Applied Physics.

Edwards, Vernessa, Ph.D., AAMU, 2004. Normal Alabama. Applied Physics.
Schamschula, M., Ph.D., Univ. of Alabama, Huntsville, 1994. Optics.
Winebarger, Amy, Ph.D., Univ. of Alabama, 1999. Space science, physics.
Zhang, T. X., Nagoya Univ., Japan, 1995. Space science, physics.

Adjunct Professors

Hathaway, David H., Ph.D., Univ. of Colorado, 1979. Astrophysics.
Koshak, William J., Ph.D., Univ. of Arizona. Atmospheric physics.
Nash-Stevenson, Dr. Shelia, Ph.D., AAMU, 1993. Applied physics.
Phanord, Diendonne D., Ph.D., Univ. of Illinois, 1988. Mathematical physics.
Ruffin, Dr. Paul, Ph.D., Univ. of Alabama, 1986. Optics, physics.
Wu, Shi T., Ph.D., Univ. of Colorado, 1967. Aerospace Engineering Science.

Visiting Professors

Bhatnagar, V. P., Space Physics, Ph.D., (Space Physics), Delhi University, Delhi, 1968, P. Eng. (Electrical Engineering), Toronto, 1991.
Johnson, R. Barry, Optical Systems, D.Sc, Southeastern Institute of Technology.
Volz, M., Material Sciences, Ph.D./NASA Administrators Fellow.

Research Associates

Curley, Michael, Ph.D., AAMU, 1997. Applied physics.
Evelyn, A. L., Ph.D., AAMU, 1998.
Kukhtareva, Tanya, M.S. Kiev Univ., Ukraine, 1972.
Wladislaw, Lyatsky, Ph.D., Univ. of St. Petersburg, Russia, 1968.

Research Professors

Bhatnagar, V. P., Ph.D., Delphi Univ., Delhi, 1968, P. Eng. (Elec. Eng.) Toronto, 1991. Space physics.
Kukhtarev, N., Ph.D., Institute of Physics, Kiev, Ukraine, 1973.
Zimmerman, R., Ph.D., Massachusetts Institute of Tech., 1952. Physics.

RESEARCH SPECIALTIES AND STAFF

Theoretical

Atmospheric Physics and solar physics. Bhatnagar, Tan, Winebarger, Wladislaw, Zhang.
Materials Science. J. C. Wang.
Optics. Edwards, Lee.
Nonlinear Adaptive Optics. N. Kukhtarev.

Experimental

Materials Science. Aggarwal, Batra, Evelyn, Guggilla, Ila, Lal, Zimmerman.
Optics. Curley, Dokhanian, Edwards, V. Edwards, Kukhtarev, Kukhtareva, Reddy, Schamschula, Sharma.
Space Science. Bhatnagar, Lyatsky, Tan, Winebarger, Zhang.

FACULTY PUBLICATIONS

Aggarwal, M. D.

M. D. Aggarwal, A. K. Batra, P. Guggilla, M. E. Edwards, B. G. Penn, and J. R. Currie, Jr. "Pyroelctric materials for uncooled infrared detectors: processing, properties, and applications," NASA Technical Memorandum NASA/TM-2010-216373, 88 pages (2010).
A. K. Batra, J. R. Currie, M. A. Alim, and M. D. Aggarwal, "Impedance response of polycrystalline tungsten oxide," J. Phys. Chem. Solids **70**, 1142–1145 (2009).
M. D. Aggarwal, A. K. Batra, R. B. Lal, B. G. Penn, and D. O. Frazier, "Bulk single crystals grown from solution on Earth and in microgravity," to be published in Springer Handbook of Crystal Growth, Springer-Verlag, 2010.
A. K. Batra, J. R. Currie, M. D. Aggarwal, R. B. Lal, M. E. Edwards, and A. Vaseashta, Optoelectronics and Advanced Materials **3**, 124–126 (2009).
A. K. Batra, M. A. Alim, J. R. Currie, and M. D. Aggarwal, "The electrical response of the modified lead titanate-based thick films," Physica B **404**, 1905–1911 (2009).
M. P. Volz, K. Mazuruk, M. D. Aggarwal, and A. Croll, "Interface shape control using localized heating during Bridgman growth," J. Crystal Growth **311**, 2321–2326 (2009).

Batra, A. K.

A.K. Batra, John Corda, Padmaja Guggilla, M. D. Aggarwal, M. E. Edwards, "Electrical properties of Silver Nanoparticles Reinforced $LiTaO_3$:P (VDF-TrFE) Composite Films," SPIE Proceedings, Vol. 7419, Infrared Systems and Photoelectronic Technology IV, Eustace L. Dereniak, John P. Hartke, Paul D. LeVan, Randolph E. Longshore, Ashok K. Sood, Editors, 27 August (2009). **2**, 578–581 (2008).
A. K. Batra, J. R. Currie, M. A. Alim, M. D. Aggarwal, "Impedance studies of polycrystalline tungsten oxide," Journal of Physics and Chemistry of Solids, **70**, 1142–1145 (2009).
Padmaja Guggilla, A. K. Batra, M. E. Edwards, "Electrical characterization of $LiTaO_3$:P (VDF-TrFE) composites," J. Materials Science **44**, 5469–5474 (2009).
A. K. Batra, M. A. Alim, M. D. Aggarwal, J. R. Currie, "The electrical response of modified lead titanate-based thick films," Physica B **404**, 1905–1911 (2009).
A. K. Batra, J. R. Currie, M. D. Aggarwal, R. B. Lal, M. E. Edwards, A. Vaseashta, "Studies on the thick-film organic vapor sensors based on binary metal oxides," Optoelectronics and Advanced Materials-Rapid Comm. **3** 124–126 (2009).

Bommareddi, Rami Reddi

Energy upconversion in holmium doped lead germane tellurite glass, I. Kamma and B. Rami Reddy, J. Appl. Phys. (in press).
Swaroop Bommareddi, M. Dokhanian, and B. Rami Reddy, "Energy upconversion in erbium doped sodium lead germano tellurite glass," Phy. Sta. Sol. **C6** S67–S70 (2009).
I. Kamma, P. Kommidi, and B. Rami Reddy, "High temperature measurement using luminescence of Pr^{3+} doped YAG and Ho^{3+} doped CaF2," Phy. Sta. Sol. **C6**S187–S190 (2009).
I. Kamma, P. Kommidi, and B. Rami Reddy, "Design of a high temperature sensing system using luminescence lifetime measurement," Rev. Sci. Instrum. **79** 096104 (2008).
P. Kommidi and B. Rami Reddy, "Two-hoton excitation stu-

ides in terbium doped yttrium aluminum oxide," J. Appl. Phys. **102** 076105 (2007).

Curley, Michael

H. Jaenisch, J. Handley, M. Curley, M. Edwards, and J.-C. Wang, "Deriving predictive turbulence data models," Proceeding of the SPIE **6971**, 69710H (2008).

N. Kukhtarev, T. Kukhtareva, M. Curley, G. Stargel, S. Sarkisov, "Single-beam phase conjugation for lasers phase locking and image formation," Proceedings of SPIE **7056** (2008).

L. Huey, M. Curley, S. Sarkisov, and J. C. Wang, "Light driven Polymer autho-oscillators," Proceeding of the SPIE **6715**, 671501 (2007).

B. H. Peterson, S. S. Sarkisov, V. N. Nesterov, M. J. Curley, A. Urbas, D. Patel, and J.-C. Wang, "Comparative study of two-photon fluorescent biomarkers at nanosecond and femtosecond pulsed excitation," Proc. SPIE **6442** (2007).

N. Kukhtarev, T. Kukhtareva, M. Curley, H. M. Jaenisch, M. E. Edwards, M. Gu, Z. Zhou, and R. Guo, "Nanosecond electrical and optical pulses and self phase conjugation from photorefractive lithium niobate fibers and crystals," Proc. SPIE **6698** (2007).

Edwards, Matthew

M. D. Aggarwal, A. K. Batra, P. Guggila, M. E. Edwards, B. G. Penn, and J. R. Currie, Jr., "Pyroelectric materials for cooled infrared detectors: Processing, properties, and applications," NASA/TM - 2010-216373, March 2010.

A. K. Batra, J. Corda, P. Gugilla, M. D. Aggarwal, M. E. Edwards, "Electrical properties of silver nanoparticles reinforced $LiTaO_3$:P (VDF-TrFE) composite films," SPIE **7419A**, 2009.

Keka C. Biswas, Jacquelyn Gillespie, Fayequa Majid, Matthew Edwards, and Anjan Biswas, "Dynamics of davydov solitons in a-Helix Protein," Advanced Studies in Biology, **1** (2009).

A. K. Batra, J. R. Currie, M. D. Aggarwal, and M. E. Edwards, "Studies on the thick-film organic vapor sensors based on binary metal oxides," Optoelectronics and Advanced Materials **3**(2), 124–126 (2009).

John Corda, Padmaja Guggilla, A. K. Batra, M. D. Aggarwal, and M. E. Edwards, "Dielectric and Pyroelectric properties of $LiTaO_3$:P (VDF-TrFE) composite films," submitted for publication (2008).

Edwards, Vernessa

N. Kukhtarev, T. Kukhtareva, P. Banarjee, P. Buchhave, J. Wang, and V. Edwards, "Visualization of the refractive index modulation by optical channeling," SPIE Proc. **3793**, 148-156 (1999).

N. Kukhtarev, T. Kukhtareva, V. Edwards, S. Sarkisov, M. Curley, J. C. Wang, V. Rotaru,and O. Korshack, "Optical channeling and diffraction in photorefractive crystals, and chalcogenide films," SPIE Proc. **4803**, Seattle, WA (2002).

N. Kukhtarev, T. Kukhtareva, J. Jones, V. Edwards, M. Bayssie, and J. C. Wang, "Manipulation of microorganisms with optically induced gratings and photovoltaic pulses, OSA conference, October, 2003, Tucson, AZ.

V. Edwards, N. Kukhtarev, T. Kukhtareva, J. Jones, F. Okafor, "Optical trapping using a moving interference pattern for decontamination, cleaning, and separation of liquid solutions, First International Environmental Research Symposium, Jackson State University, October, 2004.

N. Kukhtarev, T. Kukhtareva, V. Rotaru, V. Edwards, F. Buchhave, J. Wang, and S. Bairavarasu, "Optical channeling for radial holographic grating recording in chalcogenide glassy semiconductor films and photo-thermoplastic films, SPIE Proceedings, **5912**, 591209-11 (2005).

Guggilla, Padmaja

A. K. Batra, M. D. Aggarwal, P. Guggilla, M. E. Edwards, B. G. Penn, and J. R. Currie, Jr., "Pyroelectric materials for uncooled infrared detectors: Processing, properties, and applications," NASA/TM-2010-216373, Marshall Space Flight Center, MSFC, AL (March, 2010).

R. Hawrami, P. Guggilla, A. K. Batra, M. D. Aggarwal, and A. Burger, "Advanced scintillator materials based on cesium halides for nuclear radiation detection," communicated with Journal of Crystal Growth.

Padmaja Guggilla, A. K. Batra, M. Dokhanian, and M. D. Aggarwal, "Computation of current responsivity of a bimorph pyroelectric infrared detector," accedpted for publication in Physica B (2010).

Padmaja Guggilla, A. K. Batra, and M. E. Edwards, communicated with the paper "Electrical characterization of $LiTaO_3$:P (VDF-TrFE) composites," Journal of Materials Science **44**(20), 5469 (2009).

A. K. Batra, Padmaja Guggilla, M. D. Aggarawal, Sudip Bhattacharjee, and M. E. Edwards, "Energy harvesting using temporal temperature variations via pyroelectic effect,"Materials for Energy 2010 conference to be held in Karlsruhe/Germany from July 4-8, 2010.

Kukhtarev, Nickolai

N. Kukhtarev, T. Kukhtareva, and F. Okafor, "Optical trapping of nano-(micro) particles by gradient and photorefractive forces," Journal of Holography and Speckle **5**, 1-7 (2009).

N. Kukhtarev, T. Kukhtareva, G. Stargell, and J. Wang, "Pyroelectric and photogalvanic crystal accelerators," J. Appl. Phys. **106**, 014111 (2009).

N. Kukhtarev, T. Kukhtareva, G. Stargell, and J. Wang, "Combined optical and electrical effects in ferroelectric crystal for high laser intensities," Proc. of SPIE **7420** (2009).

N. Kukhtarev and T. Kukhtareva, "Dynamic holography in material science and microbilology," Chapter 12 in New Direction in Holography nd Species," ASP, 2008.

N. Kukhtarev, T. Kukhtareva, and P. Land, "Optical and electrical effects in photorefractive semiconductor and ferroelectric crystals," with graduate student participation has been published in special review book, Nonlinear Optics and Applications, Chapter 10, in Book Nonlinear Optics and Applications, Research Signpost, 2007.

Kukhtareva, Tatiana

N. Kukhtarev, T. Kukhtareva, and F. okafor, "Optical trapping of nano-(micro) particles by gradient and photorefractive forces," Journal of Holography and Speckle **5**, 1-7 (2009).

N. Kukhtarev, T. Kukhtareva, G. Stargell, and J. Wang, "Pyroelectric and photogalvanic crystal accelerators," J. Appl. Phys. **106**, 014111 (2009).

German Telbiz, Yarosalv Kyshenia, Vitaly Shvalagyn, Tatiana Kukhtareva and Florence Okafor, "The ways of formation and spatial stabilization of Ag clusters and nanostructures in mesoporous titania thin film," presented in meeting International School-Seminar, Spectroscopy of Molecules and Crystals *ISSSMC), Ukraine (2009).

N. Kukhtarev, T. Kukhtareva, G. Stargell, J. Wang, Combined optical and electrical effects in ferroelectric crystal for high laser intensities, Proc of SPIE **7420** (2009).

N. Kukhtarev and T. Kukhtareva, "Dynamic holography in material science and microbilogy," Chapter 12 in book

New Direction in Holography and Species, ASP, 2008.

Schamschula, Marius

W. L. Crosson, C. a. Laymon, R. Inguva, and M. Schamschula, "Assimilating remote sensing data in a surface flux-soil moisture model," Hydrol. Proc. **16**, 1645–1662 (2002).

M. Schamschula, W. L. Crosson, C. a. Laymon, and R. Inguva, "Disaggregation of Remotely Sensed Soil Moisture using Neutral Networks," World Automationa Congress, Orlando, FL (2002).

W. Crosson, C. Laymon, A. Limaye, W. Khairy, M. Schamschula, T. Coleman, and R. Ingva, "Assimilation of remote sensing data in a hydrologic model to improve estimates of spatially distributed soil moisture," Proceedings of IGARSS, International Geoscience and Remote Sensing Symposium (June 24-28, Toronto, Canada), 1168–1170 (2002).

T. D. Tsegaye, W. L. Crosson, C. A. Laymon, M. P. Schamschula, and a. B. Johnson, "Application of a Neural network-Based Spatial Disaggregation Scheme for Addressing Scaling of Soil Mositure," in Scaling Methods in Soil Physics, Y. Pachepsky, D. E. Radcliffe, and H. M. Selim (Eds.), CRC Press, Boca Raton, 261–278 (2003).

T. D. Tsegaye, R. Metzl, X. Wang, M. Schamschula, W. Tadesse, D. Clendenon, K. Golson, T. L. Coleman, and G. Schaefer, "A long-term near real time database of metrorological/soil profile data: The Alabama Mesonet (ALMNet)," International Symposium on Remote Sensing of Environment, Saint Petersburg, Russia (2005).

Sharma, Anup

F. A. Calzzani, Jr., R. Sileshi, A. Kassu, J. M. Taguenang, A. Chowdhury, A. Sharma, P. B. Ruffin, C. Brantley, and E. Edwards, "Detection of residual traces of explosives by surface enhanced Raman scattering using gold coated substrates produced by nanospheres imprint technique," Proc. SPIE **6945**, 69451O (2008).

J. M. Taguenaug, A. Kassau, P. B. Ruffin, C. Brantley, E. Edwards, and A. Sharma, "Reversible UV degradation of PMMA plastic optical fibers," Optics Communications **281**, 2089 (2008).

K. Kassu, J. M. Taguenang, and A. Sharma, "Photochemically deposited surface relief gratings of an azo-dye-labeled phospholipid from the aqueous phase," Opt. Lett. **33**, 1656 (2008).

A. Kassu, J. M. Taguenang and A. Sharma, "Photopatterning of polybutadiene substrate by interferometric UV lithography: fabrication of phospholipid microarrays," Applied Optics **46**, 489 (2007).

A. Kassu, J.-M. Taguenang, and A. Sharma, "Nanoscale patterning of phospholipids thin films by interferometric UV lithography," Proc. SPIE **6645**, 664509 (2007).

Wang, J. C.

Michael J. Curley, Sergey S. Sarkisov, Jai-Ching Wang, and Courtney Boykin, "Prototype wireless gas sensor based on chemical sensor using a thin film," SPIE Optic+, photonics Conference, 1–5 August 2010, San Diego Conference Center, San Diego, CA.

N. V. Kukhtarev, T. Y. Kukhtareva, G. Stargell, and J. C. Wang, "Pyroelectric and photogalvanic crystal accelerators," Journal of Applied Physics **106**, 1 (2009).

A. Grabar, "Two channel IR virbration sensor based on dynamic grating in semiconductors and pyroelectrics," Infraed Technology and Applications XXXIV, edited by Bjon F. Andreson, Gabor F. Fulop, and Paul R. Norton, Proc. of SPIE **6940**, 694035 (2008).

N. Kukhtarev, T. V. Kukhtareva, Michael Curley, Gregory Stargell, and J. C. Wang, "Dynamic optical grating in two layer system: ferroelectric and nanostructured metal film," Proc. SPIE **16989**, 698913-1–698913-11 (2008).

LaQuieta Huey, Michael Curley, Sergey S. Sarkisov, and J. C. Wang, "Light-driven polymer autho-oscillators," International Symposium on Optomachatron Technologies ISOT 2007, 8-10 October 2007, Lausanne, Switzerland.

Winebarger, Amy

H. P. Warren, A. R. Winebarger, J. T. Mariska, G. A. Doschek, and H. Hara, "Observation and Modeling of Coronal Moss with the EUV Imaging Spectrometer on Hinode," ApJ **676**, 672W (2008).

A. R. Winebarger, H. P. Warren, and D. A. Falconer, "Modeling X-Ray Loops and EUV "Moss" in an Active Region Core," ApJ **666**, 1245W (2007).

H. P. Warren and A. R. Winebarger, "Static and Dynamic Modeling of a Solar Active Region," ApJ **662**, 1293U (2007).

I. Ugarte-Urra, H. P. Warren, and A. R. Winebarger, "The Magnetic Topology of Coronal Mass Ejection Sources," ApJ **659**, 1673A (2007).

M. J. Aschwanden, A. Winebarger, D. Tsiklauri, and P. Hardi, "The Coronal Heating Paradox," 04/(2007).

Zhang, T. X.

Zhang, T. X., "Fundamental Elements and Interactions of Nature: A Classical Unification Theory," Progress in Phys. **2**, 36–42 (2010).

Zhang, T. X., and Wu, S. T., MHD "Simulation of Non-Flux-Robe Coronal Mass Ejections, J. Geophys. Rev. **114**, A05107 (2009).

Zhang, T. X., "A New Cosmological Model: Black Hole Universe," Progress in Physics, **3**, 3-11, 2009.

Zhang, T. X., MHD "Simulation for the Origin and Magnetic Topology of Solar ^{3}He Events, Astrophys. J., **677**, 692-698 (2008).

Zhang, T. X., "Electric Charge as a Form of Imaginary Energy," Progress in Physics **2**, 79-83 (2008).

AUBURN UNIVERSITY

DEPARTMENT OF PHYSICS

Auburn, Alabama 36849

Students Accepted For Degree	FIELDS		
	Physics	Astronomy	Related Fields
Doctorate	X		
Master's	X		

1. General

President: Jay Gogue
Dean of Graduate School: George Flowers
Department Chairman: Joseph D. Perez
Department Telephone Number: (334) 844-4264
Type of Institution: University
Control: Public
Setting: Small town
Total Faculty: 1,176
Total Graduate Faculty: 1,000
Total Students: 24,530
Total Graduate Students: 3,519
Annual Graduate Tuition:
In-state residents: Full-time—$6,366
Out-of-state residents: Full-time—$16,866
Tuition rates for: 2009–10
Deferred tuition plan: No
Other Fees:
Term: Semester

*Students with assistantships pay no tuition.

2. Number of Faculty in Department

The combined total of full-time faculty in the three professorial ranks is 19.

3. Admission, Financial Aid, and Housing

Address admission inquiries to: Chairman, Physics Graduate Admissions Committee
Graduate application fee required: Apply directly to Physics Dept. and no fee is required
Admission deadline (Fall admission): 8/1
Admission information: For fall admission, 2009–10
15 students were accepted from ~100 applicants.
Admission requirements: A Bachelor's degree in physics or related major is required. A GPA of 3.0 or higher is desired. The verbal and quantitative GRE is required. A combined total greater than 1200 is desired. Students from non-English speaking countries must have TOEFL score of 550 or 213.
Undergraduate preparation assumed: Upper division mechanics, electricity and magnetism, quantum mechanics, and thermal physics. Students sometimes take these courses during first year as graduate students.
Address financial aid inquiries to: Chairman, Physics Graduate Admissions Committee
GAPSFAS application required: No (not required but available)
Financial aid deadline: 8/1
Loans available: No
Address housing inquiries to: Director of Housing, Burton Hall
On-campus, single student housing available: Yes
Cost/semester: $2500*
On-campus, married student housing available: No*
*Adequate off-campus housing is available.

Table A—Faculty, Enrollments, and Degrees Granted

Research Specialty	2009–10 Faculty	Enrollment[1] Fall 2009		No. of Degrees Granted[2] 2009–10 (2005–09)			Median No. of Years for 2009–10 Ph.D.'s
		Master's	Doctorate	Master's	Terminal Master's	Doctorate	
Astrophysics	1						
Space Phys.	3						
Atomic, Molecular & Optical Physics	6						
Condensed Matter Physics	6						
Physics Education	1			Data not available			
Plasma Physics & Fusion	4						
Total							
Full-time Grad. Stud.		48					
Part-time Grad. Stud.							
First-year Grad. Stud.		10					
Median Years in Grad. Study (2008–09 Degrees)							

Undergraduate Degrees, 2008–09 (2004–09): 8(77)

[1] Students not yet committed to a research specialty are entered under non-specialized.
[2] Five-year totals in parentheses.

4. Graduate Degree Requirements

Master's: Each student must complete 30 hours of course work with a minimum grade average of 3.0/4.0. No specific residency period is stipulated. A thesis and nonthesis option is available.
Doctorate: Each student must complete 60 hours of course work including dissertation and research hours. Each candidate must maintain a grade point average of 3.0/4.0 or better. Passage of a general doctoral examination is required of all students. A dissertation based upon original research must be completed and defended in a final oral examination.
Special Equipment, Facilities, or Programs: Epitoxial growth laboratories including MBE; materials characterization facilities including low energy ion accelerator; surface science lab; magnetic fusion laboratory with Compact Toroidal Hybrid device; laboratory plasma devices; dusty plasma laboratory; atomic physics laboratory; Beowulf computer cluster with 96 nodes.

Table B—Appointments to Graduate Students, 2009–10

Title of Appointee	Appointments Total	Appointments First year	Academic Load Allowed in Credit Hours	Hours of Service Per Week	Stipend for 12 mos. Year ($)
			Semester		
Teaching Assistant	25	–		12	19,650
Research Assistant	23	–		Varies	19,650
Total	48	–			

5. Personnel Engaged in Separately Budgeted Research, 2009–10

Professorial faculty	22
Other faculty	4
Postdoctoral appointments	3
Graduate students	41
Undergraduate students	27
Total	97

6. Separately Budgeted Research Expenditures by Source of Support

	Departmental Research	Physics-related Research Outside Department
Federal government	$3,500,000	$
State and local government	$1,500,000	
Total	$5,000,000	$

Table C—Separately Budgeted Research Expenditures

Research Specialty	No. of Grants	Expenditures ($)
	Data not available	

FACULTY

Professors

Bozack, Michael J., Ph.D., Oregon, 1985. Surface physics.
Hanson, James D., Ph.D., Maryland, 1982. Plasma physics.
Hinata, S., Ph.D., Illinois, 1973. Space physics.
Knowlton, Stephen F., Ph.D., MIT, 1984. Plasma physics; space physics.
Lin, Yu, Ph.D., University of Alaska, 1993. Space physics.
Oks, Eugene, Ph.D., Moscow Physical Technological Institute, Moscow, 1975. Plasma physics; atomic and molecular physics; nonlinear dynamics.
Perez, Joseph D., Ph.D., Maryland, 1968. Department head. Space and plasma physics.
Pindzola, Michael S., Ph.D., Virginia, 1975. Atomic and molecular physics.
Robicheaux, Francis J., Ph.D., U. Chicago, 1991. Atomic and molecular physics.
Thomas, Edward, Jr., Ph.D., Auburn University, 1996. Experimental plasma physics.
Tin, Chin-Che, Ph.D., University of Alberta, 1987. Solid state physics.
Williams, John R., Ph.D., North Carolina State, 1974. Nuclear physics; condensed matter physics.

Associate Professors

Dong, Jianjun, Ph.D., Ohio University, 1998. Condensed matter theory, computational physics.
Landers, Allen, Ph.D., Kansas State, 1999. Atomic and molecular physics.
Park, Minseo, Ph.D., North Carolina State University, 1998. Solid state physics.
Simon, Marllin L., Ph.D., Missouri, 1972. Physics education.
Wersinger, Jean-Marie P., Ph.D., Lausanne, 1977. Remote sensing.

Assistant Professors

Fogle, Michael R., Jr., Ph.D., Stockholm University, 2004. Experimental atomic physics.
Loch, Stuart, Ph.D., Univ. of Strathclyde, 2001. Atomic and molecular physics.

RESEARCH SPECIALTIES AND STAFF

Theoretical

Astrophysics. Magnetic fields in stars and galaxies; structure of solar atmosphere. Hinata.
Atomic, Molecular, and Radiative Physics. Photon interactions with atoms, electron scattering by atoms and molecules. Stark Broadening, line shape analysis. Oks, Pindzola, Robicheaux. 5 postdoctoral fellows.
Plasma. Plasma waves and instabilities; magnetic field configurations; particle dynamics. Hanson.
Solid State. Electronic structure; charge transport; dielectric breakdown. Surface dynamics. Chen, Dong, Fromhold.
Space Physics. Magnetospheric plasma physics. Lin, Perez.

Experimental

Atomic Physics. Synchotron studies using imaging techniques; electron spectroscopy; momentum imaging. Landers.
Plasma. Vacuum spark plasmas; magnetic confinement of fusion plasma, plasma diagnostics; plasma spectroscopy; plasma-heating; laboratory simulation of space plasmas. Boivin, Knowlton, Thomas.
Solid State. Wide bandgap semiconductors; epitaxy; solid state switches; surface properties of solids. Bozack, Fromhold, Park, Tin, Williams.
Surface Physics. Surface kinetics; gas-surface interactions; thermal desorption; electron-stimulated desorption; low-work function surfaces; wetting and adhesion; corrosion phenomena. Bozack.

THE UNIVERSITY OF ALABAMA

DEPARTMENT OF PHYSICS AND ASTRONOMY

Tuscaloosa, Alabama 35487

Students Accepted For Degree	FIELDS: Physics	Astronomy	Related Fields
Doctorate	X	*	
Master's	X	*	

*Research in astronomy or astrophysics may be used for a graduate degree in physics.

1. General

President: Robert E. Witt
Dean of Graduate School: David Francko
Department Chairman: Raymond E. White III
Department Telephone Number: (205) 348-5050
Type of Institution: University
Control: Public
Setting: Urban
Total Faculty: 1,211
Total Graduate Faculty: 736
Total Students: 28,807
Total Graduate Students: 4,314
Graduate Tuition:
In-state residents: Full-time—$3,500/semester
Out-of-state residents: Full-time—$9,600/semester
Tuition rates for: 2009–10
Deferred tuition plan: Yes
Term: Semester

2. Number of Faculty in Department

The combined total of full-time faculty in the three professorial ranks is 24. The combined total of full-time, part-time, and other faculty at all ranks is 29.

3. Admission, Financial Aid, and Housing

Address admission inquiries to: Graduate School Office, Box 870118
Graduate application fee required: $50/60 domestic/international
Admission deadline (Fall admission): 4/1
Admission information: For fall admission, 2009–10, 20 students were accepted from 42 applicants.
Admission requirements: For admission to the graduate programs, a Bachelor's degree in physics is required with a minimum undergraduate GPA of 3.0/4.0 specified. The GRE is required. The GRE Advanced may be substituted for the general GRE. Students from non-English speaking countries are required to demonstrate proficiency in English via the TOEFL exam. Minimum acceptable score for admission is 550.
Undergraduate preparation assumed: Halliday and Resnick, *Fundamentals of Physics*; Serway, Moses, and Moyer, *Modern Physics*; Symon, *Mechanics*; Reitz, Milford, *Foundation of Electromagnetic Theory*; Eisberg, Resnick, *Quantum Physics of Atoms*; etc.
Address financial aid inquiries to: Office of Financial Aid, Box 870162
GAPSFAS application required: No
Financial aid deadline: None (Priority Deadline 2/1)
Loans available: Yes
Address housing inquiries to: Director of Housing, Residential Life, Box 870399, Tuscaloosa, AL 35487
On-campus, single student housing available: Yes
Cost/month: $475–$700—1 BR/furnished
On-campus, married student housing available: Yes
Cost/month: $575–$850—2 BR/furnished

Table A—Faculty, Enrollments, and Degrees Granted

Research Specialty	2009–10 Faculty	Enrollment[1] Fall 2009: Master's	Enrollment[1] Fall 2009: Doctorate	No. of Degrees Granted[2] 2009–10 (2005–10): Master's	Terminal Master's	Doctorate	Median No. of Years for 2009–10 Ph.D.'s
Astronomy	3	4	5	1(4)	1(3)	0(5)	–
Astrophysics	3	0	0	0(0)	–	0(1)	–
Atomic, Molecular, & Optical Physics	1	0	1	0(0)	–	0(0)	–
Condensed Matter Physics	9	3	13	3(7)	1(7)	3(14)	5.6
Particles & Fields	8	1	10	2(7)	–	2(5)	6.1
Non-specialized	0	3	1	0(0)	1(1)	0(0)	–
Total		11	30	6(18)	3(11)	5(25)	
Full-time Grad. Stud.		11	30				
Part-time Grad. Stud.		0	0				
First-year Grad. Stud.		4	7				
Median Years in Grad. Study (2009–10 Degrees)				4.2	2.0	5.8	5.8

Undergraduate Degrees, 2009–10 (2005–10): 14(42)

[1]Students not yet committed to a research specialty are entered under non-specialized.
[2]Five-year totals in parentheses.

4. Graduate Degree Requirements

Master's: PLAN I-24 graduate semester hours in an approved program with satisfactory performance required; "B" average; one semester in residence; Master's examination required; thesis required; no language requirement. PLAN II-30 graduate semester hours in an approved program with satisfactory performance required; Master's examination required; thesis not required; no language requirement.
Doctorate: A minimum of 48 graduate semester hours required in an approved program with satisfactory performance; one academic year in residence; oral preliminary examination; dissertation and dissertation examination required.
Thesis: Thesis may be written *in absentia*.
Special Equipment, Facilities, or Programs: Facilities include well-equipped laboratories for research in condensed matter physics, high-energy physics, and image processing. Supporting facilities include a machine shop, electronics shop, computer workstations, and direct access to the campus mainframe computer and the Alabama supercomputer. Faculty and students participate in the Center for Materials for Information Technology and the Tri-Campus Material Science Ph.D. Program.

Table B—Appointments to Graduate Students, 2009–10

Title of Appointee	Appointments Total	Appointments First year	Academic Load Allowed in Credit Hours	Hours of Service Per Week	Stipend for Academic Year ($)
			Semester		
Teaching Assistant	25	11	9	15	13,905[1,2]
Research Assistant	13	0	9	20	14,000[1,2]
University Fellow	2	1	12	0	15,000[1,2]
NASA Fellow	1	0	12	0	18,000
Total	41	12			

[1]Tuition waived.
[2]Summer support mostly available at $3,000–$3,300.

5. Personnel Engaged in Separately Budgeted Research, 7/09–6/10

Professorial Faculty	18
Other Faculty	2
Postdoctoral appointments	5
Graduate students	28
Undergraduate students	17
Total	70

6. Separately Budgeted Research Expenditures by Source of Support

	Departmental Research	Physics-related Research Outside Department
Federal government	$3,200,000	$
Total	$3,200,000	$

7. Separately Funded and Managed Laboratories

Center for Materials for Information Technology (MINT)	$1,000,000
Total	$1,000,000

Table C—Separately Budgeted Research Expenditures

Research Specialty	No. of Grants	Expenditures ($)
Astrophysics	9	500,000
Atomic, Molecular, & Optical Physics	1	200,000
Condensed Matter Physics	14	1,000,000
Particles & Fields	15	1,500,000
Total	39	3,200,000

FACULTY

Professors

Busenitz, Jerome K., Ph.D., Illinois, 1985. Experimental elementary particle physics.

Buta, Ronald J., Ph.D., Texas, Austin, 1984. Galaxy morphology and catalogs.

Butler, William H., Ph.D., U.C. San Diego, 1969. Theoretical condensed matter physics.

Clavelli, Louis J., Ph.D., Chicago, 1967. Theoretical particle physics.

Hardee, Philip E., Ph.D., Maryland, 1976. Theoretical and observational astrophysics.

Harms, Benjamin C., Ph.D., Florida State, 1969. Theoretical particle physics.

Harrell, J. W., Jr., Ph.D., North Carolina, Chapel Hill, 1969. Experimental condensed matter physics.

Keel, William C., Ph.D., California, Santa Cruz, 1982. Galactic nuclei, jets, and galaxy interactions.

Mankey, Gary J., Ph.D., Penn State, 1992. Experimental condensed matter physics.

Piepke, Andreas, Ph.D., Heidelberg, 1990. Experimental elementary particle physics.

Sarker, S. K., Ph.D., Cornell, 1980. Theoretical condensed matter physics.

Schad, Rainer, Ph.D., Hannover, 1991. Experimental condensed matter physics.

Stern, Allen, Ph.D., Syracuse, 1980. Theoretical particle physics.

Tipping, Richard H., Ph.D., Penn State, 1969. Theoretical physics; molecular spectroscopy.

Visscher, Pieter B., Ph.D., California, Berkeley, 1971. Theoretical condensed matter physics; computer simulation.

White, Raymond E. III, Ph.D., Virginia, 1986. Dynamics and hydrodynamics in galaxies and galaxy clusters.

Associate Professors

LeClair, Patrick R., Ph.D., Eindhoven, 2002. Experimental condensed matter physics.

Mewes, Tim, Ph.D., Kaisersleutern, 2002. Experimental condensed matter physics.

Mryasov, Oleg, Ph.D., Russian Academy of Sciences, 1993. Theoretical condensed matter physics.

Stancu, Ion, Ph.D, Rice, 1990. Experimental elementary particle physics.

Assistant Professors

Irwin, Jimmy, Ph.D., Virginia, 1997. Accreting black holes and neutron stars.

Okada, Nobuchika, Ph.D., Tokyo Metropolitan, 1998. Physics beyond the standard model.

Townsley, Dean M., Ph.D., UC Santa Barbara, 2004. White dwarf supernovae.

Williams, Dawn R., Ph.D., UCLA, 2004. Experimental particle astrophysics.

Adjunct Professors

Biermann, Peter L., Ph.D., Gottingen, 1971. Theoretical astrophysics.

Fujiwara, Hideo, D.E., Tokyo, 1969. Experimental condensed matter physics.

Gupta, Arunava, Ph.D., Stanford. Experimental condensed matter physics.

Pandey, Raghvendra K., Ph.D., Cologne. Experimental condensed matter physics.

Adjunct Associate Professor

Crocker, Deborah A., Ph.D., Virginia, 1987. Observational astrophysics.

Professors Emeriti

Alexander, Chester, Jr., Ph.D., Duke, 1968. Experimental condensed matter and chemical physics.

Byrd, Gene G., Ph.D., Texas, Austin, 1974. Theoretical astrophysics.
Coulter, Philip W., Ph.D., Stanford, 1965. Theoretical particle physics.
Izatt, Jerald R., Ph.D., Johns Hopkins, 1960. Experimental nonlinear optics and lasers.
Jones, Stanley T., Ph.D., Illinois, 1970. Physics Education.
Sulentic, Jack W., Ph.D., SUNY, Albany, 1975. Observational astrophysics.

RESEARCH SPECIALTIES AND STAFF

Theoretical

Astrophysics. Extragalactic radio sources; galactic dynamics; three-body problem; solar system dynamics; galactic structure; theory of stellar orbits; extragalactic astronomy; astrodynamics; plasma astrophysics, high energy astrophysics. Biermann, Byrd, Hardee, Townsley, White.

Atomic and Molecular Physics. Scattering theory; molecular spectroscopy; spectral line shapes and intensities; atmospheric applications. Tipping.

Condensed Matter Physics. Electronic structure of solids; magnetic properties; hierarchical and renormalization-group methods; magnetic lattice models. Butler, Mryasov, Sarker, Visscher.

Elementary Particles and Fields. Supersymmetry phenomenology; field theory; quantum black holes; particle astrophysics. Biermann, Clavelli, Harms, Okada, Stern.

Experimental

Astrophysics. Extragalactic radio sources. Interacting pairs and groups. Galaxy morphology. Spectroscopy of AGN. Globular clusters. Buta, Byrd, Irwin, Keel, White.

Condensed Matter Physics. Magnetic materials and thin films; nanoparticles spintronics. Gupta, Harrell, LeClair, Mankey, Mewes, Pandey, Schad.

High-Energy Physics. Detector R&D, Neutrino physics. Busenitz, Piepke, Stancu, Williams. 5 postdoctoral fellows.

THE UNIVERSITY OF ALABAMA AT BIRMINGHAM

DEPARTMENT OF PHYSICS

Birmingham, Alabama 35294

Students Accepted For Degree	FIELDS		
	Physics	Astronomy	Related Fields
Doctorate	X		
Master's	X		

1. General

President: Carol Z. Garrison
Dean of Graduate School: Brian D. Noe
Department Chairman: David L. Shealy
Graduate Program Director: Yogesh K. Vohra
Department Telephone Number: (205) 934-4736
Type of Institution: University
Control: Public
Setting: Urban
Total Faculty: 2,224
Total Graduate Faculty: 1,336
Total Students: 16,874
Total Graduate Students: 5,193
Annual Graduate Tuition:
In-state residents: Full-time—$227/cr. hr. (tuition)
Out-of-state residents: Full time—$568/cr. hr
Tuition rates for: 2009–10
Deferred tuition plan: No
Term: Semester
Note: Semester hours credit awarded

2. Number of Faculty in Department

The combined total of full-time faculty in the five professorial ranks is 20. The combined total of full-time, part-time, and other faculty at all ranks is 24.

3. Admission, Financial Aid, and Housing

Address admission inquiries to: Graduate School Office, HUC 511, 1530 3rd Avenue South, Birmingham, AL 35294-1150
Graduate application fee required: $35 (U.S. citizen); $60 (foreign applicant)
Admission deadline (Fall admission): 7/7
Admission information: For fall admission, 2009–10, 13 students were accepted.
Admission requirements: For admission to the graduate programs, a Bachelor's degree in physics is required with a minimum GPA of B specified. The GRE is required. The GRE Advanced is strongly urged. The average GRE score for 2009–2010 admissions was 1,130 (total). Students from non-English speaking countries are required to demonstrate proficiency in English via the TOEFL exam. Minimum acceptable score for admission is 550.
Undergraduate preparation assumed: Halliday and Resnick & Walker, *Fundamentals of Physics*; Thornton & Rex, *Modern Physics*; Morin, *Introduction to Classical Mechanics*; Griffiths, *Introduction to Electrodynamics*; Reif, *Fundamentals of Statistical and Thermal Physics, Berkeley Course Vol. 5*; Liboff, *Introductory Quantum Mechanics*.

Address financial aid inquiries to: David L. Shealy, Chairman, Department of Physics
GAPSFAS application required: No
Financial aid deadline: Priority deadline 4/1
Loans available: Yes
Address housing inquiries to: Housing Office.
On-campus, single student housing available: Yes
Cost/month: $300–637
On-campus, family housing available: No

Table A—Faculty, Enrollments, and Degrees Granted

Research Specialty	2009–10 Faculty	Enrollment[1] Fall 2009		No. of Degrees Granted[2] 2009–10 (2005–10)			Median No. of Years for 2005–10 Ph.D.'s
		Master's	Doctorate	Master's	Terminal Master's	Doctorate	
Astrophysics	3	0	3	0(2)	1(3)	1(0)	0
Biophysics	3	0	1	0(0)	0(0)	0(0)	0
Condensed Matter Physics	8	3	16	0(5)	1(7)	1(9)	5
Optics	4	0	11	1(4)	0(2)	0(3)	3
Physics Education	2	0	0	0(0)	0(0)	0(0)	–
Total		3	31	1(11)	2(12)	2(12)	
Full-time Grad. Stud.		3	31				
Part-time Grad. Stud.		0	0				
First-year Grad. Stud.		2	11				
Median Years in Grad. Study (2009–10 Degrees)						6	

Undergraduate Degrees, 2009–10 (2005–10): 5(17)

[1]Students not yet committed to a research specialty are entered under non-specialized.
[2]Five-year totals in parentheses.

4. Graduate Degree Requirements

Masters: 30 semester-hours of credit with thesis; minimum B (3.0 average); no residency requirements. Thesis is optional with approval of faculty. An additional "Interdisciplinary Track" for an M.S. degree with thesis option is also offered to non-physics majors and requires a minimum of 12 hours of graduate-level courses offered by other departments.
Doctorate: Minimum residence of three full-time academic years or equivalent periods of part-time enrollment with minimum GPA of B (3.0). Pass: oral placement exam on basic physics concepts; comprehensive exam covering the areas of classical mechanics, quantum mechanics, electromagnetic theory and two selected topics from thermodynamics/statistical mechanics, optics, or solid state physics in no more than two attempts; oral exam on area of research specialization; oral defense of written dissertation proposal; and oral final defense of dissertation. In addition, there is an "Applied Physics Track" for the Ph.D. degree which requires students to complete successfully a sequence of core graduate physics classes in classical mechanics, electromagnetism, quantum mechanics, statistical mechanics, and scientific communication seminars totaling 14 credit hours, 12 credit hours of elective courses in applied physics, 3 credit hours of an Ap-

plied Physics Training course, and Dissertation Research hours.

Thesis: Thesis may be written *in absentia*.

Special Equipment, Facilities, and Programs: The department has active research programs in applied and theoretical astrophysics, biophysics, condensed matter physics, materials science, nanophysics, optics, lasers and laser spectroscopy. Opportunities exist for interaction with major government laboratories including NASA AMES Research Center; Jet Propulsion Lab; NASA Goddard Space Flight Center; NASA Marshall Space Flight Center; the Advanced Photon Source (APS) at Argonne National Lab; the National Synchrotron Light Source (NSLS) at Brookhaven National Lab; the Lawrence Livermore National Lab; Oak Ridge National Lab; Sandia National Lab; the Naval Research Lab; Wright Patterson Air Force Base–Air Force Research Lab; the National Cancer Institute at NIH; National High Magnetic Field Lab, Tallahassee Florida and the Center for Integrated Nanotechnology, Los Alamos National Laboratory.

Graduate Students and faculty have collaborative research Programs with several universities (Auburn University, University of Alabama at Huntsville, Arizona State University, North Carolina State University, Stanford University, University of California at San Diego, University of California at Los Angeles, Technical University of Lodz (Poland), and the General Physics Institute of the Russian Academy of Sciences), as well as with the UAB Medical Center. The department is part of a tri-campus interdisciplinary Materials Science Program.

Astrophysics and Solar-System Physics Labs have capabilities for Raman Imagery using Dilor XY 0.8-m 3-stage Raman spectrometer for acquisition of point spectra and Raman images displaying the distribution of molecular components, with Coherent krypton ion multi-λ laser (blue to IR) and Olympus BX40 microscope enabling common focusing for visible and laser illumination of the sample to 0.5 μm resolution (Lab for Paleobiological Chemical Imagery); Raman Spectroscopy using field-tested mini-Raman spectrometer with Control Development optics, operating at 785 nm or 670 nm, including SDL 8530 (785 nm, 300 mW) and Process Instruments PI-ECL-670-150 (670 nm, 150 mW) lasers, with EIC Raman probes for 785 nm and 670 nm: Transmission Spectroscopy (UV, Vis, IR) using Hewlett-Packard UV/Vis Diode Array Spectrophotometer for spectra from 190–400 nm; Vacuum-UV Spectrometer, for spectra from 100–250 nm; ThermoMattson InfinityGold FTIR Spectrometer for IR spectra in the range from 1–25 μm, resolutions to 0.5 cm^{-1}; Mattson Pollaris FTIR Spectrometer for spectra in the mid-IR (2.5–25 μm), resolutions to 1 cm^{-1}; far-IR to 400 μm with attachments; Mössbauer Spectroscopy with MS-1200 Ranger Scientific spectrometer at 0.1% linearity for transmission and backscatter; 50 mCi Co-57 source; Sample temperatures to 10K with ARS Displex Cryostat to 1000K with furnace; GMW water-cooled magnet for sample fields to 2.5 T; Chemical Analysis using Buck Scientific Gas Chromatograph, Model 910, with FID and PID detectors; D-Star Instruments Isocratic Liquid Chromatograph, model DLC-20; Dycor Quadrupole Mass Spectrometer; Sample Preparation using two ARS closed-cycle helium expansion systems and temperature controllers for T=6.5–300K; UV Photolysis System (Opthos) with microwave generator for 100–250 nm photons; Parr Instruments Reactor, D3141-1, for processing with H_2O at high T and high P; 2" diameter Lindberg Tube Furnace to 1500C with digital multi-step controller.

Biophysics & Biophotonics Lab has capabilities for time-correlated single photon counting and laser tweezers operations using the following equipment: Coherent Verdi (10W)/ Mira 900 D Ti:sapphire laser system, with frequency doubler/ tripler, ps/fs operation, "pulse picker" (for lower excitation rate), broad-tuning optics set. Other capabilities include Fluoromax2 fluorometer, PC-controlled, 180–650 nm excitation, 180–700 nm emission, high sensitivity photon-counting electronics, full spectral corrections, spectral analysis software, Bioelectrospec TIRF system. TIRF bio-sensing setup; fiberoptic bio and chemical sensing setup; blue and infrared diode laser system; diode laser drivers; CO_2 laser based fiber and pipette puller to make fiber probes, CCD based microscope. Absorption Spectrophotometer using JASCO V-530 with programmable thermoelectric temperature controller, 180–1100 nm; Fluorescence Decay Systems using PicoQuant TimeHarp 200 TCSPC PC computer-board, two 50-ps PMTs (Hamamatsu R3809U-50 UV/VIS and R3809U-59 NIR) emission monochromator or wavelength filters, Globals multicomponent global convolution fitting software; Olympus IX-70 inverted microscope: Total Internal Reflection Fluorescence (TIRF)/Laser Tweezers input optics (including TIRF objective), Prior ProScan computerized stage, SIS F-View CCD camera with Microsuite Biological Suite/Scopeview software, 5-mm diam/4-m long fiber optic to relay fluorescence to SPEX fluorometer, multiphoton imaging & spectroscopy with Ti:sapphire laser system; Zeiss IM inverted microscope: with MTI SIT camera plus intensifier, CDS motorized stage. Brookhaven BI-200 SM light scattering system.

Computational Physics Lab has switched 10/100 Mbps LAN with gigabit ethernet connectivity to campus backbone (10GE) and 10GE links to the SouthernCross Roads (SoX) Internet2/NLR Gigapop in Atlanta. This departmental network supports UNIX and Microsoft operating systems running on a range of workstations in faculty, staff, and graduate student offices and labs. The department has a 30 seat pc-cluster for use by general physics students with WebAssign, Activ-Physics, MATLAB, Photoshop7, SigmaPlot10, Scientific Word, MikTeX, comsol multiphysics and other software for research and educational activities. Physics participates in the operation and use of several Beowulf clusters of parallel computing systems on campus, described at UAB IT link for Researchers (http://main.uab.edu/Sites/it/internal/ researchers/). The department has access to two systems housed at the Alabama Supercomputer Center in Huntsville. One system consists of SGI Altix 350 and Altix 450 processors for a total of 228 CPU's with 1510 GB of shared memory and 12 terabytes of storage. The entire system has a floating point performance of 1263 GigaFlops. The second system is a Dense Memory Cluster with 1256 AMD Opteron and Intel Nehelem processors and 6176 GB of distributed memory operating at 2.2 GHz for an overall capacity of 5.97 TFLOPS. The system has 15 terabytes of high performance storage. Physics faculty and students have Internet2 access to national supercomputer centers.

Materials Research Labs and Facilities include Materials Growth and Processing with 1.2 kW and 6 kW Microwave Plasma Chemical Vapor deposition systems for growth of homoepitaxial diamond and nanostructured diamond coatings on metals. Pulsed Laser Deposition Facility for growth of thin films including a Lamda Physik LPX305i Excimer laser, a unique custom-made deposition system developed in collaboration with Neocera, Inc., and a Novel Nanoparticle Beam Pulsed Laser Deposition. Aerosol Reactor for Synthesis and Processing of Nanostructured Materials, Denton Discovery-24 sputtering system with 3-RF/DC capable magnetron sources, as well as spin-coating and dip-coating equipment. Annealing furnaces and thermal evaporators. Materials Characterization Facilities with Bruker EMX EPR spectrometer with 10 GHz microwave bridge. Oxford Instruments ESR900 Continuous Flow Cryostat APD Closed Cycle Low Temperature System. Senteck 400 single-wavelength Ellipsometer. Philips Thin film X-ray diffractometer. Micro-Raman and Photoluminescence spectroscopy for thin films. Nanoindentation facility with Atomic Force Microscopy. Veeco Explorer Atomic Force Microscope with liquid scanning capability. Keithley Model 82 CV measurement system; HP4284 LCR Meter Micro-scratch tester for thin film adhesion measurements. Romulus IV micro-scratch tester for thin film adhesion measurements. Quantum Design model MPMS-5S SQUID magnetometer, Lake Shore model 7000 ac susceptometers. CSM Instruments Nanotribometer and Fogale Microsurf 3D optical profilometer, Bruker Optics Hyperion 3000 infrared microscope with Vertex 70 FTIR spectrometer.

Other Equipment: Diamond anvil cells capable of generating multi-megabar pressures. Electric Discharge Machining of small holes down to 10 microns in diameter in metallic gaskets. Laser heating facility for samples in diamond anvil cells.

Laser and Nonlinear Optics Labs include Absorption spectroscopy performed with a Shimadzu UV-VIS-NIR double beam spectrophotometer UV 3101PC and with a cavity ring down spectrometer coupled to a tunable (200–1200 nm) alexandrite-LiF:F_2^+ color center laser (CCL) combination; Fluorescence and Raman spectroscopies centered around CCS-450 (Janis) closed cycle refrigerator system. Numerous pulsed and CW lasers can be configured for samples excitation. Among them, a Spectra-Physics model PA0270 injection seeded Nd:YAG laser with frequency doubling, tripling, and quadrupling coupled to MOPO System tunable over 400-2500 nm and two Light-Age Raman shifters (H_2 and D_2), and tunable (1100–1250, 550–600, 280–300 nm) LiF:F_2^- CCL; Light Age Alexandrite Laser System PAL101 with variable temporal and spectral outputs coupled to LiF:F_2^+ CCL (800–1200 nm) with frequency doubling, tripling, quadrupling and difference frequency generation (200–8000 nm); EKSPLA PL2143A high energy picosecond Nd:YAG; laser coupled to PG401 OPO tunable over 400–2100 nm 500 Hz repetition rate 1 mJ diode pumped Nd:YAG laser "PULSAR 200" with a pulse duration of 1.5–2 ns with frequency doubling and tripling option; a 35-fs Coherent Legend Elite amplified titanium-sapphire laser, and a Coherent Mira titanium-sapphire 80 MHz laser oscillator. CW lasers include, Coherent Argon and Krypton lasers, Lexel Ti-sapphire laser, SDL824 tunable diode laser, several external cavity multi-wavelength diode lasers, Er-fiber 1550 nm, 10 W linearly polarized ELD laser (IPG Photonics), and several home made Er fiber laser pumped microchip and external cavity mid-IR (2–3 μm) tunable lasers based on Cr_2^+ :ZnS and ZnSe lasers with record (up to 6W) continuous wave output power. The Labs are equipped with state-of-the-art Raman/AFM/NSSOM system for fundamental studies and characterization of materials at nanoscale using Raman, Fluorescence, chemical imaging, as well as topographic, electrical and thermal platforms integrated in one instrument. we also equipped with several spectrometers/spectrographs for measuring fluorescence, excitation, and Raman spectra and kinetics of fluorescence. These include the portable Ocean Optics R2000 fiber coupled Raman system and several Acton Research Corp. SpectraPro scanning monochromators/imaging spectrographs-(SpectraPro 750, 500, 150) with gratings covering UV (200 nm)-middle IR (14000 nm) spectral range coupled to two Princeton Instruments ICCD. Other detectors include numerous PMTs, TE-cooled PbS and InGaAS detector for the 0.7–3 μm range, and a LN-cooled HgCdTe and fast InSb detectors for the 2–14 and 2–5 μm range, respectively. Data acquisition is performed with ARC NCL Spectral Management System, two Standford Instrument boxcar-averages and EGG Instruments 7265 lock-in amplifier interfaced with PC. There is extensive equipment for Z-scan and DFWM characterization of nonlinear optical materials. Spiricon LBA 100 beam profiler is available for beam diagnostics and Wavemeter W-4500 (Burleigh) system for wavelength measurements.

Table B—Appointments to Graduate Students, 2009–10

Title of Appointee	Appointments Total	Appointments First year	Academic Load Allowed in Credit Hours	Hours of Service Per Week	Stipend for Academic Year ($)
			Semester		
Teaching Assistant	10	5	9	20	20,500[1]
Research Assistant	10	3	6–9		20,500–25,000
Fellowship	14	5	12	0	20,776–30,000[2]
Total	34	13			

[1]The department pays tuition, fees, and health insurance for Teaching Assistants.
[2]Fellowships pay full tuition, fees, health insurance, and provide an education budget.

5. Personnel Engaged in Separately Budgeted Research, 10/08–9/09

Professorial faculty	15
Postdoctoral Associates	3
Graduate students	9
Nonteaching Research Personnel	4
Total	31

6. Separately Budgeted Research Expenditures by Source of Support*

	Departmental Research
Federal government	$2,398,058
State and local government	408,375
Private/Industry	276,569
Total	$3,083,002

*For fiscal year Oct. 1, 2008 through Sept. 30, 2009.

Table C—Separately Budgeted Research Expenditures

Research Specialty	No. of Grants	Expenditures ($)*
Astrophysics	4	115,396
Biophysics	5	272,259
Condensed Matter Physics	18	1,187,697
Optics	11	799,558
Physics Education	10	708,092
Total	48	3,083,002

*For fiscal year Oct. 1, 2008 through Sept. 30, 2009.

FACULTY

Professors

Lawson, Chris M., Ph.D., Oklahoma State, 1981. Nonlinear optics; fiber optics; optical sensors; optical coherence imaging and tomography.

Mirov, Sergey B., Ph.D., Lebedev Physical Institute, Moscow, 1983. Experimental quantum electronics; solid state lasers; physics of color centers; laser spectroscopy.

Shealy, David L., Ph.D., Georgia, 1973. Chairman of the Department. Geometrical optics; laser beam shaping optics; radiative transfer; caustic and optical aberration theory.

Vohra, Yogesh K., Ph.D., Bombay, 1980. High pressure physics; synthesis and characterization of diamond crystals and thin films; nanostructured ceramic and polymeric biomaterials.

Wenger, Lowell E., Ph.D., Purdue 1975. Synthesis and characterization of magnetic materials and nanostructures; superconductivity.

Zvanut, Mary Ellen, Ph.D., Lehigh, 1988. Electrical studies and EPR studies of insulators and semiconductors; microelectronics and optoelectronics.

Associate Professors

Camata, Renato P., Ph.D., California Institute of Technology, 1998. Synthesis and properties of metal and semiconductor nanoparticles; nanostructured materials; aerosol strategies in nanomaterials fabrication; pulsed laser deposition of thin films and nanostructured materials.

Gerakines, Perry A., Ph.D., Rensselaer Polytechnic Institute, 1998. Astrophysics; interstellar molecules; interstellar dust; laboratory astrophysics; infrared astrophysics; comets; planetary science; origin of life; observational astronomy.

Harrison, Joseph G., Ph.D., Wisconsin, 1981. Solid state theory; atomic and molecular physics; MRI modeling; chemical kinetics. Simulation of nanoparticle-facilitated hyperthermia.

Kawai, Ryoichi, Ph.D., Waseda, 1985. Condensed matter theory; biophysics theory; materials physics theory; computational physics; complex systems.

Martin, James C., Ph.D., Georgia Tech., 1978. Conformations of biological macromolecules; laser light scattering; optical pattern recognition; Raman spectroscopy.

Nordlund, Thomas M., Ph.D., Illinois, 1977. Structural dynamics of DNA and proteins; protein-DNA recognition; picosecond fluorescence; laser tweezers; biomolecule-nanoparticle interactions.

Stanishevsky, Andrei V., Ph.D., Belarus Acad. of Sciences, 1996. Focused ion beam micro- and nano-fabrication; PVD thin films deposition, characterization, and application; Nanoparticle research.

Wang, Xujing, Ph.D., Texas A&M University, 1995. Theoretical Physics; network theory; biophysics; theoretical and mathematical biology; genetics.

Assistant Professors

Hilton, David J., Ph.D., Cornell University, 2002. Ultrafast spectroscopy, ultrashort pulse generations; ultrafast terahertz spectroscopy; correlated electron materials; superconductvity; high magnetic field spectroscopy; magnetic semiconductors; complex functional nanomaterials; materials in extreme environments.

Kapoor, Rakesh, Ph.D., Bombay, 1989. Biophotonics, biophysics; bioimaging; upconverting materials; biosensors and nanostructured biomaterials.

Research Assistant Professors

Catledge, S. Aaron, Ph.D., Alabama at Birmingham, 1999. Materials Science. Synthesis and properties of nanostructured super-hard materials; Chemical vapor Deposition (CVD) of diamond films and novel nanostructured coatings for biomedical implants; composite scaffolds for tissue engineering; mechanical properties.

Tsoi, Georgiy, Ph.D., Ukraine Acad. of Sciences, 1984. Physics and mathematics. Physical and quantum electronics.

Instructor

DeVore, Todd, Ph.D., Alabama at Birmingham, 1999. Computational physics.

Mohr, Robert, Ph.D., Alabama at Tuscaloosa, 2001. Computational applications to theoretical astrophysical problems.

Professor Emeritus

Agresti, David G., Ph.D., Cal. Tech., 1967.
Bauman, Robert P., Ph.D., Pittsburgh, 1954.
Young, John H., Ph.D., Clark, 1969.

Associate Professor Emeritus

Wdowiak, Thomas J., Ph.D., Case Western Reserve, 1971.

Research Associate Professor Emeritus

Wills, Edward L. Ph.D., Virginia, 1968.

RESEARCH SPECIALTIES AND STAFF

Theoretical

Astrophysics. Computer modeling of astrochemical processes and planetary data from Mars and the outer solar system. Origin of the solar system, impact. Agresti, Gerakines, Mohr.

Biophysics. Macromolecular structure, assembly and dynamics by computer modeling. Kawai, Martin, Nordlund, Wang.

Condensed Matter Physics. Low-dimensional systems; defects in insulators and semiconductors; positron states in condensed-matter; simulation of chemical vapor deposition processes; computational electromagnetics; surface adsorption; *ab initio* molecular dynamics simulations; computational algorithms applicable to massively parallel computers; quantum monte carlo simulations; non-equilibrium statistical mechanics, stochastic processes. Camata, Harrison, Kawai, Mirov, Stanishevsky, Vohra, Zvanut.

Optics. Laser physics, laser spectroscopy, fiber, laser, soft x-ray/UV optics; geometrical optics; nonlinear optics; laser beam shaping; optical design; caustic and optical aberration theory. Hilton, Kapoor, Lawson, Mirov, Shealy.

Experimental

Astrophysics. Astrochemistry of cosmic ices and complex interstellar molecules, molecular evolution and precursors of life, hydrothermal systems, instruments for in-situ planetary science and life search, participation in the Mars Exploration Rover missions, mass extinctions, and Precambrain paleontology, bringing to bear tools such as Mössbauer, uv/vis/ir, Raman, and mass spectroscopies, XRD, and chemical analysis. Agresti, Gerakines, Mohr.

Biophysics & Biophotonics. DNA and protein structure and function via contiuous and time-resolved fluorescence spectroscopy and molecular calculations; fiber-optic biosensors;

TIRF; FRET; transient kinetics of molecular interactions; energy transfer and photophysics of sunscreens; spectroscopy and imaging of assembly and interactions between biomolecules and nanoparticles. Kapoor, Nordlund, Wang.

Condensed Matter Physics. EPR studies of bulk crystals and thin films; optical Mössbauer effect; design and construction of portable Mössbauer spectrometer for use in extraterrestrial studies; high pressure physics; electrical studies of semiconducting and insulating materials; electrical, and optical properties of bulk synthetic diamond and diamond thin films, radiational defects in crystals, optical properties of laser crystals, time-resolved laser spectroscopy; synthesis and characterization of metallic, semiconducting and magnetic materials/nanostructures; superconductivity; aerosol strategies. Camata, Kapoor, Mirov, Stanishevsky, Tsoi, Vohra, Wenger, Zvanut. Three postdoctural associates.

Materials Science/Nanophysics. Nanostructured materials, carbon nanotube synthesis and properties, nanoscale direct writing and patterning, Nanocomposite Biomaterials. Camata, Catledge, Kapoor, Stanishevsky, Vohra, Wenger.

Optics. Laser optics; Laser resonators; Solid state laser materials, Tunable lasers, Laser spectroscopy; UV, holographic projection processing of materials; physiological optics; non-linear optics and nonlinear optical materials; diamond windows for optical spectroscopy; fiber optics; optical sensors; optical imaging; optical coherence; tomography. Hilton, Kapoor, Lawson, Mirov, Shealy. One postdoctoral associate.

Physics Education. Bauman, DeVore, Martin, Wills, Young.

UNIVERSITY OF ALABAMA, HUNTSVILLE

DEPARTMENT OF PHYSICS

Huntsville, Alabama 35899

Students Accepted For Degree	FIELDS		
	Physics	Astronomy	Related Fields
Doctorate	X		X
Master's	X		X

1. General

President: David B. Williams
Dean of Graduate School: Debra M. Moriarity
Department Chair: James A. Miller
Department Telephone Number: (256) 824-2482
Department Website: http://physics.uah.edu
Type of Institution: University
Control: Public
Setting: Urban
Total Faculty: 303 tenured and tenure track; 10 research
Total Graduate Faculty: 385
Total Students: 7,400
Total Graduate Students: 1,500
Annual Graduate Tuition:
In-state residents: Full-time—$3,437 (10 sem. hrs.)
Part-time—$446/cr. hr. (tuition and fees)
Out-of-state residents: Full-time—$7,539 (9 sem. hrs)
Part-time—$1,052/cr. hr. (tuition and fees)
Tuition rates for: 2009–10
Deferred tuition plan: No
Term: Semester

2. Number of Faculty in Department

We have 16 tenured and tenure-track faculty, 3 research faculty, 2 lecturers, and 7 adjunct faculty.

3. Admission, Financial Aid, and Housing

Address admission inquiries to: Chair, Dept. of Physics
Graduate application fee required: $50
Application deadline (for Assistantships; Fall admission):
Foreign nationals 3/1
All others 3/1
Admission information: For fall admission, 2009–10, 24 full-time students were accepted from 33 applicants.
Admission requirements: For admission to the graduate programs, a Bachelor's degree is required with a minimum undergraduate GPA of 3.0 (A=4.0) overall, or 3.0 for last 60 hours of work. Applicants are assumed to have fullfilled the minimum requirements for a major in physics, or its equivalent. Individuals who do not meet this requirement may, on a case-by-case basis, be admitted pending completion of selected undergraduate courses. The GRE is required. The minimum acceptable score required for admission is 1,500 total (Verbal+Quantitative+Analytical) for unconditional admission. Requirements are lower for conditional admission. Students from non-English speaking countries are required to demonstrate proficiency in English via the TOEFL exam. Minimum acceptable score for admission is 500/667, 62/120, or 173/300. Other general requirements of the Graduate School of the University of Alabama in Huntsville may be found in the graduate catalog, which may be found on the UAH website at http://www.uah.edu.
Undergraduate preparation assumed: Symon, *Mechanics*; Griffiths, *Quantum Mechanics*; Griffiths, *Electricity and Magnetism*; Carter, *Classical and Statistical Thermodynamics*, and at least one upper level laboratory class.
Address financial aid inquiries directly to: Financial Aids and Placement Office
GAPSFAS application required: No
Financial aid deadline: 7/1
Loans available: Yes
Address housing inquiries to: Office of University Housing, (256) 824-6108
North Campus Housing:
Cost/Academic year: $5,200, per student (sgl. occup.)
Graduate & Married Student Housing:
Cost/Calendar year: 1 bdrm. apt.—$6,300 unfurnished

Table A—Faculty, Enrollments, and Degrees Granted

Research Specialty	2009–10 Faculty	Enrollment[1] Fall 2009		No. of Degrees Granted[2] 2009–10 (2005–10)			Median No. of Years for 2005–10 Ph.D.'s
		Master's	Doctorate	Master's	Terminal Master's	Doctorate	
Theoretical Astrophysics	5	2	4	0(0)	0(0)	0(2)	6
Experimental Astrophysics	2	1	2	0(0)	0(1)	0(2)	6
Gravitation	1	0	1	0(0)	0(1)	0(0)	–
Optics	3	6	4	0(2)	2(8)	0(5)	5
Solar/Space Physics	8	1	8	0(0)	0(5)	0(1)	5
Other	1	1	1	0(1)	2(5)	0(2)	5
MS without thesis				1(9)	1(6)	–	–
Total	20	11	20	1(12)	5(26)	0(12)	
Full-time Grad. Stud.		25					
Part-time Grad. Stud.		12					
First-year Grad. Stud.		9					
Median Years in Grad. Study (2005–10 Degrees)				2	3	5	–

Undergraduate Degrees, 2009–10 (2005–10): 20(65)

[1]Students not yet committed to a research speciality are entered under "other."
[2]Five-year totals in parentheses.

4. Graduate Degree Requirements

Master's: Thesis Option: 24 semester hours graduate course work, six hours thesis, maintain B average or better; no language requirement. Non-Thesis Option: 33 semester hours graduate course work; maintain B average or better; comprehensive exam; no language requirement.
Doctorate: Minimum 48 semester hours of graduate credits required, plus satisfactory performance in an approved course program. Admission to Ph.D. program dependent upon performance on Comprehensive Exam. Admission to candidacy

on passing Qualifying Exam. Dissertation and final oral exam required; one year residency; no language requirement.

Dissertation: Dissertation may be written *in absentia*.

Special Equipment, Facilities, or Programs: The Department offers both traditional degree programs leading to the M.S. and Ph.D. degrees, and also more specialized "professional" M.S. programs in areas of applied research, such as Optics and Photonics Technology. The traditional degree programs contain both core courses and a wide variety of specialized courses in areas related to contemporary research, while the professional programs also include courses more directly related to workplace applications. The core courses include classical mechanics, electricity and magnetism, thermal and statistical physics, and quantum mechanics. Specialized courses include

(a) Optics—geometrical optics, physical optics, radiometry, polarized light, optical systems design, lasers, optomechanical design and manufacturing, optical fabrication and testing, Fourier optics, and quantum electronics;

(b) Astrophysics/Space Physics—plasma dynamics, stellar atmospheres and interiors, cosmology, high-energy astrophysics, space plasma physics, solar flare physics, and advanced plasma theory;

(c) Solid-State Physics/Materials Science—solid-state physics, quantum theory of solids, structure and properties of materials, materials processing in space, principles of liquid and solid interfaces, and solid-state materials preparation.

The department has state-of-the-art research facilities housed in the modern Optics Building, a 100,000 square foot facility. These facilities include equipment for thin film optics, integrated optics, optical design, space optics, fiber optic sensors, optical surface properties, high-energy lasers, polarimetry, infrared science, biophotonics, quantum optics, nonlinear optics, holography, optical image processing, optical computing, optical signal processing, and data analysis and simulation. Students have also carried out research involving balloon, rocket, and space shuttle payloads. Much student research is performed in collaboration with the staff of numerous research centers, including the Center for Applied Optics (CAO), the Center for Space Plasma and Aeronomic Research (CSPAR), and the NASA-UAH National Space Science and Technology Center (NSSTC). Close collaborations exist with the U.S. Army Missile Command and with the nearby NASA Marshall Space Flight Center. Graduate students regularly present results at professional meetings and have received numerous awards for doing so. All students have access to the department computer room (Apple Macs). Astrophysics students use a 64-node Mac cluster and associated terminals for their research. More computationally-intensive projects may be performed using the Alabama Supercomputer Network.

Incoming graduate students typically hold a teaching assistantship for the first year, funded through the department. In later years, they are encouraged to secure a research assistantship funded through one of the many external contracts and grants held by faculty members (see Table C), or through a national fellowship program such as the NASA Graduate Student Researchers Program or the NSF Graduate Fellowship Program. A normal semester's courseload is nine hours of graduate coursework, plus a one-credit-hour weekly seminar.

Table B—Appointments to Graduate Students, 2009–10

Title of Appointee	Appointments: Total	Appointments: First year	Academic Load Allowed in Credit Hours	Hours of Service Per Week	Stipend for Academic Year ($)
			Semester		
Teaching Assistant	10	5	10	20	12,600
Research Assistant	16	5	10	20	14,000
Total	26	10			

5. Personnel Engaged in Separately Budgeted Research, 2009-10

Professorial faculty	18
Graduate students	16
Undergraduate students	2
Total	36

6. Separately Budgeted Research Expenditures by Source of Support (2009-10)

	Departmental Research	Research Outside Department
Federal government	$2,915,800	0
Private Sector	2,000	0
State government	386,200	$4,340,000
Total	$3,304,000	$4,340,000

Table C—Separately Budgeted Research Expenditures, (2009-10)

Research Specialty	No. of Grants	Expenditures ($)
Theoretical Astrophysics	12	347,900
Experimental Astrophysics	11	916,000
Gravitation	3	51,900
Space/Solar Physics	36	1,496,800
Optics	2	105,200
Education	3	386,200
Total	67	3,304,000

Table D—Physics-related Research Outside Department

Field and Unit Outside Department	No. of Grants	Expenditures ($)
Science Education (in Inst. for Sci. Ed.)	3	4,340,000
Total	3	4,340,000

FACULTY

Professors Emeriti

Comfort, R. Hugh, Ph.D., Univ. of Alabama, Huntsville, 1977. Space Physics.

Dimmock, John, Ph.D., Yale Univ., 1962. Solid-state.

Emslie, A. G., Ph.D., Univ. of Glasgow, 1979. Astronomy.

Franz,, Frank, Ph.D., Univ. of Illinois, 1964. President Emeritus. Atomic physics.

Franz, Judy, Ph.D., Univ. of Illinois, 1965. Condensed matter, atomic physics.

Professors

Fix, John, Ph.D., Univ. of Indiana, 1969. Dean of College of Science. Astrophysics.
Gregory, Don, Ph.D., Univ. of Alabama, Huntsville, 1984. Optics.
Lieu, Richard, Ph.D., Imperial College, London, 1981. Astrophysics.
Miller, James, Ph. D., Univ. of Maryland, 1990. Solar physics.
Zank, Gary, Ph.D., Univ. of Natal, 1987. Pei-Ling Chan Chair of Physics. Space and solar physics.

Research Professors

Paciesas, William, Ph.D., Univ. of California, San Diego, 1978. Experimental X-ray and gamma-ray astrophysics.
Zhang, S. N., Ph.D., Univ. of Southampton, 1990. High-energy astrophysics, black holes.

Associate Professors

Bonamente, Max, Ph.D., Univ. of Alabama, Huntsville, 1999. Astrophysics.
leRoux, Jakobus, Ph.D., Potchefstroom Univ. for C.H.E., 1990. Space and solar physics.
Miller, Richard, Ph.D., Univ. of New Hampshire, 1995. High-energy astrophysics, neutrinos, astrophysics instrumentation.
Pogorelov, Nikolai, Ph.D., Russian Academy of Sciences, 1984. Space and solar physics.

Associate Research Professor

Preece, Robert, Ph.D., Univ. of Maryland, 1990. Astrophysics.

Assistant Professors

Burko, Lior, Ph.D., Israel Institute of Technology, 1998. General relativity, black holes.
Duan, Lingze, Ph.D., Univ. of Maryland, 2002. Ultra-fast laser physics, precision laser frequency control.
Florinski, Vladimir, Ph.D., Univ. of Arizona, Tucson, 2001. Space and solar physics.
Heerikhuisen, Jacob, Ph.D., Univ. of Waikato, 2001. Space and solar physics.
Li, Gang, Ph.D., Indiana Univ., 2000. Space and solar physics.
Sadeghi, Seyed, Ph.D., Univ. of British Columbia, 1999. Nanophotonics, quantum optics, semiconductor laser systems.
Shaikh, Dastgeer, Ph.D., Institute for Plasma Research, 2001. Space and solar physics.

Lecturers

Elsamadicy, Abdalla, Ph.D., A & M Univ., Huntsville, AL, 2002. Surface science.
Strong, Carol, Ph.D., Univ. of Alabama, Huntsville, 1993. Optics, science eduation.

Adjunct Faculty

Abbas, Mian. NASA MSFC.
Baird, James. Professor, Department of Chemistry.
Biermann, Peter. Max Planck Institute for Radioastronomy (Bonn).
Colley, Wes. Senior Research Scientist, Center for Modeling, Simulations, and Analysis.
McGrath, Melissa. Chief Scientist, Science and Missions Systems Directorate, NASA MSFC.
Rice, Rob. Chief Medical Physicist, Clearview Cancer Center.
Sanghadasa, Mohan. Research Scientist, Weapons Sciences Directorate, US Army Research Engineering and Development Command.

RESEARCH SPECIALTIES AND STAFF

Astrophysics/Cosmic Rays. Bonamente, Fix, Lieu, R. Miller, Paciesas, Preece, Zhang.

Space Plasma/Solar Physics. Florinski, Heerikhuisen, leRoux, Li, J. Miller, Pogorelov, Shaikh, Zank.

Space Instrumentation R. Miller.

Advanced Propulsion. Gregory.

Optics. Gregory, Duan, Sadeghi.

Solid State/Materials. Elsamadicy.

Gravitation/Cosmology. Bonamente, Burko, Lieu.

ARIZONA STATE UNIVERSITY

DEPARTMENT OF PHYSICS

Tempe, Arizona 85287-1504

Students Accepted For Degree	FIELDS		
	Physics	Astrophysics	Related Fields
Doctorate	X		X
Master's	X		X

1. General

President: Michael M. Crow
Dean of Graduate School: Maria T. Allison
Department Chair: Robert J. Nemanich
Department Telephone Number: (480) 965-3561
FAX: (480) 965-7954
E mail: physics.grad@asu.edu
Type of Institution: University
Control: Public
Setting: Urban
Total Faculty: 2,641 (Full-time)
Total Students: 68,064
Total Graduate Students: 13,787
Annual Graduate Tuition: †
In-state residents: Full-time—$6,844
Out-of-state residents: Full-time—$20,596*
Part-time rates vary.
Please call for information
Tuition rates for: 2009–10
Deferred tuition plan: No
Term: Semester

†See Graduate Catalog for more detailed information.
*Nonresident tuition is waived for students with Graduate Assistantships.

2. Number of Faculty in Department

The combined total of full-time faculty in the four professorial ranks is 42. The combined total of full-time, part-time, and other faculty at all ranks is 42.

3. Admission, Financial Aid, and Housing

Address admission inquiries to: Director of Graduate Admissions
Graduate application fee required: $70
Admission deadline (Fall admission): 3/1
Admission information: For fall admission, 2009–10, 15 students were accepted from 88 physics applicants. For fall admission, 2009–2010, 14 students were accepted into the Professional Science Master's program (PSM) from 20 applicants.
Admission requirements: For admission to the graduate programs, a Bachelor's degree in physics or a closely related program is required with a minimum undergraduate GPA of 3.0 specified and a minimum Junior-Senior GPA average of 3.0. The GRE is required. Minimum expected GRE scores: 500 verbal, 650 quantitative, 500 analytic. The GRE Physics subject exam is highly recommended, although not mandatory. No minimum score is specified. Students from non-English speaking countries are required to demonstrate proficiency in English via the TOEFL exam. Minimum acceptable score for admission is 550 for paper-based total, 213 minimum for computer-based total.
Undergraduate preparation assumed: Kittel and Kroemer, *Thermal Physics*; Reitz *et al.*, *Electromagnetic Theory*; Jenkins and White, *Optics*; Boas, *Mathematical Methods*; Kittel *et al.*, *Mechanics*, Vol. I; Gasiorowicz, *Quantum Physics*; Eisberg and Resnick, *Modern Physics*.
Address financial aid inquiries to: Graduate Admissions, ADMIN B-285
GAPSFAS application required: No
Financial aid deadline: 6/1
Loans available: No
Address housing inquiries to: Residence Life, Student Services Bldg.
On-campus, single student housing available: Yes
Cost/term: $3,000–5,000 (subject to change)
On-campus, married student housing available: No

Table A—Faculty, Enrollments, and Degrees Granted

Research Specialty	2009–10 Faculty	Enrollment[1] Fall 2009		No. of Degrees Granted[2] 2009–10			Median No. of Years for 2009–10 Ph.D.'s
		Master's	Doctorate	Master's	Terminal Master's	Doctorate	
Biophysics & Biological Physics	11	0	28	0	0	2	
Particle Astrophysics & Cosmology	20	16	60	3	1	0	
Physics and Society	9	0	29	1	0	3	
Physics Nanoscience & Materials Physics	2	19	4	0	1	0	
Total	42	35	121	4	2	5	
Full-time Grad. Stud.		35	121				
Part-time Grad. Stud.		0	0				
First-year Grad. Stud.		0	15				
Median Years in Grad. Study		3	6				

Undergraduate Degrees, 2009–2010: 22

[1]Students not yet committed to a research specialty are entered under non-specialized.
[2]Five-year totals in parentheses.

4. Graduate Degree Requirements

Master's: 30 semester hours required with 3.0 average; at least 18 semester hours resident credit; no language requirement. Final oral examination; thesis required.
Doctorate: 84 semester hours with 3.0 average; at least two semesters continuous full time residence after first year (30 semester hours); at least 30 semester hours earned in residence, with no language requirement. Written, oral comprehensive examination; dissertation required; final oral examination required.
Master of Natural Science: Interdepartmental program in science education, natural sciences, and mathematics for teachers. Does not replace usual teacher certification requirements. 30 semester hours required with 3.0 average; at least 18 semester hours resident credit; no language requirement; not more than 21 semester hours credit (including thesis or project credit) from any one area of natural sciences or mathematics may be applied toward this degree. Thesis or project required; final examination, written, oral, or both is required.
Professional Science Masters in Nanoscience: Interdisciplinary program spanning physics, chemistry, and biochemistry, ma-

terials, and electrical engineering. 30 credit hours either on accelerated 12-month track or part-time 24-month track. 15 hours of core courses (including applied project) and 15 hours elective courses. No thesis requirement.

Thesis: Thesis may be written *in absentia*.

Special Equipment, Facilities, or Programs: Optional M.S. programs in interdisciplinary physics, technical physics, and physics teaching. Optional Ph.D. program in applied physics. Experimental and theoretical diffraction physics and electron microscopy; experimental and theoretical sub-atomic physics; experimental and theoretical condensed matter and materials physics; observational and theoretical astrophysics; science education; interdisciplinary Center for Solid State Science; Center for High-Resolution Electron Microscopy; Surface Science and Ion Beam Analysis facilities; access to University of Arizona and other national and international facilities.

Table B—Appointments to Graduate Students, 2009–10

Title of Appointee	Appointments Total	First year	Academic Load Allowed in Credit Hours	Hours of Service Per Week	Stipend for Academic Year ($)
			Semester		
Teaching Assistant	45	11	12 credit hours	20	15,631
Research Associate	75	4	12 credit hours	20	17,160
Self-Supported	36	0			
Total	156	15			

5. Personnel Engaged in Separately Budgeted Research, 2009–2010

Professorial faculty	40
Postdoctoral appointments	11
Graduate students	75
Undergraduate students	22
Research Scientists	8
Total	156

6. Separately Budgeted Research Expenditures by Source of Support

	Departmental Research	Physics-related Research Outside Department
Federal government	$4,994,000	$929,000
Other government	$85,000	$470,000
Business and industry	$367,000	$15,000
Other	$100,000	62,000
Total	$5,546,000	$1,476,000

Table C—Separately Budgeted Research Expenditures

Research Specialty	No. of Grants	Expenditures ($)
Biophy. & Biological Phys.	43	$3,196,000
Nanoscience & Mat. Phys.	49	$2,473,000
Particle Astrophysics & Cosmology	22	$1,127,000
Physics & Society	12	$238,000
Total	126	$7,034,000

FACULTY

Professors

Alarcon, Ricardo O., Ph.D., Ohio, 1985. Experimental intermediate-energy nuclear physics.

Bauer, Ernst, Ph.D., Universitat Munchen, Germany, 1955. Distinguished Research Professor. Surface and thin film physics, in particular surface electron microscopy.

Bennett, Peter, Ph.D., Wisconsin, 1980. Experimental surface physics; epitaxial growth.

Chamberlin, Ralph V., Ph.D., UCLA, 1984. Experimental condensed matter physics; dynamics of complex systems; nanothermodynamics.

Comfort, Joseph R., Ph.D., Yale, 1968. Experimental nuclear physics; low- and medium-energy nuclear reactions and spectroscopy; reaction mechanisms; nuclear structure models.

Davies, Paul, C. W., Ph.D., UC London, 1970. Astrophysics and cosmology.

Doak, R. Bruce, Ph.D., MIT, 1981. Helium scattering studies of surface phonon physics; time-resolved measurements of dynamical processes at surfaces; tunable ultracold He beams; novel He beam scattering experiments.

Dow, John D., Ph.D., Rochester, 1967. Theoretical solid state physics; theory of materials; experimental scanning tunneling microscopy and surface physics.

Drucker, Jeff, Ph.D., UCSB, 1986. Synthesis and characterization of nanostructured electronic materials for novel opto electronic and photonic applications.

Krauss, Lawrence M., Ph.D., MIT, 1982. Elementary particle physics and cosmology.

Lindsay, Stuart M., Ph.D., Manchester, 1976. Biophysics and nanoscale physics; scanning probe microscopy and nanofabrication; molecular electronics.

McCartney, Martha, Ph.D., Arizona State University, 1989. Electron microscopy techniques, leading expert on electron holography.

Menendez, José, Ph.D., Stüttgart Univ., 1985. Experimental condensed matter physics.

Nelson, Alan C., U.C. Berkeley, 1980. Biophysics in cancer early detection, imaging devices, image reconstructions and diagnostic knowledge generation.

Nemanich, Robert J., Ph.D., Chicago, 1976. Experimental surface science.

Newman, Timothy J., Ph.D., Manchester, 1991. Theory of non-equilibrium systems, statistical physics, stochastic processes; applied to biological systems: especially population dynamics, morphogenesis, tumor growth.

Ponce, Fernando A., Ph.D., Stanford, 1997. Microscopic properties of electronic materials for applications in microelectronics, photonics and optoelectronics.

Rez, Peter, Ph.D., Oxford, 1976. Electron diffraction and microscopy; medical physics; biophysics; solid state theory.

Ritchie, Barry, Ph.D., South Carolina, 1979. Experimental medium-energy nuclear physics.

Sankey, Otto F., Ph.D., Washington (St. Louis), 1979. Theoretical solid state physics; molecular electronics; biophysics.

Schmidt, Kevin E., Ph.D., Illinois, Urbana, 1979. Theoretical solid state physics; computational many-body theory.

Smith, David J., Ph.D., Melbourne, 1978. Regents' Professor. Electron diffraction and high-resolution electron microscopy; electron holography; magnetic materials, and semiconductors.

Spence, John C. H., Ph.D., Melbourne, 1974. Regents' Professor. Ultrahigh resolution electron microscopy; electron channeling; electron microscope contrast theory; excitations in solids by inelastic electron scattering; surface physics; STM; biophysics; x-ray imaging; nanolithography by stem.

Thorpe, Michael F., D. Phil., Oxford University, 1968. Theoretical molecular biophysics and soft condensed matter physics.

Tsen, Kong-Thon, Ph.D., Purdue, 1983. Experimental condensed matter physics.

Tsong, Ignatius S. T., Ph.D., London, 1970. Surface physics; epitaxial thin-film growth processes; wide band gap semiconductors.

Treacy, Mike, Ph.D., Cambridge University, 1980. Diffraction physics, Complex materials. Fluctuation microscopy of disordered materials; Enumeration of hypothetical framework materials; Diffraction phenomena.

Vachaspati, Tanmay, Ph.D., Tufts, 1985. Theoretical cosmology, particle physics, and gravitational physics.

Venables, John A., Ph.D., Cambridge, 1961. Professor of Practice. Electron microscopy and surface physics; atomic processes in adsorption and crystal growth; modeling and graduate education, using the Internet.

Weierstall, Uwe, Ph.D., Research Professor. Eberhard Karls University Tübingen, 1994. Multidisciplinary Research based on electron and X-ray diffraction.

Associate Professors

Belitsky, Andrei V., Ph.D., Bogoliubov Laboratory of Theoretical Physics, Joint Institute for Nuclear Research, Dubna, Russia, 1996. Elementary particle physics, field theory, string theory.

Culbertson, Robert J., Ph.D., Penn State, 1979. Studies of surface modification and characterization of materials, crystal surfaces, and interfaces using ion beams.

Lebed, Richard F., Ph.D., California, Berkeley, 1994. Elementary particle theory. Hadronic physics, fundamental symmetries.

Marzke, Robert, Ph.D., Columbia, 1966. Experimental solid state and chemical physics; NMR, studies of molten ceramics, electrolytes, catalysts and collagen. Biomechanics of hand function.

Matyushov, Dmitry, Ph.D., Kiev State University, Ukraine, 1987. Biophysics, electron transfer, theoretical chemistry.

Maulik, Parikh K., Ph.D., Princeton, 1998. Theoretical physics; black holes; cosmology; classical and quantum gravity; string theory.

Ros, Robert, Ph.D., Basel, 2000. Nanobiophysics, structural biology, molecular recognition.

Shumway, John B., Ph.D., University of Illinois at Urbana-Champaign, 1999. Theoretical condensed matter physics; quantum dots; semiconductors; path-integrals.

Assistant Professors

Chen, Tingyong, Ph.D., Johns Hopkins, 2006. Experimental condensed matter physics; magnetism, superconductivity, nanostructures and nanomaterials.

Easson, Damien A., Ph.D., Brown University, 2002. Particle cosmology; cosmology of the early universe; quantum aspects of gravity.

Lunardini, Cecilia, Ph.D., SISSA, Trieste, 2001. Supernovae neutrinos, nutrido-matter interactions, particle astrophysics.

Ozkan, Banu, Ph.D., Bogaziei University, Istanbul, 2001. Biophysics, modeling, protein folding dynamics.

Vaiana, Sara M., Ph.D., Palermo, 2004. Biophysics, protein self-assembly.

Lecturers

Adams, Gary, Ph.D., Arizona State University, 1992. First principles simulations of semiconductor surfaces and fullerenes, fullerene derivatives and carbon nanotubes.

Covatto, Carl, Ph.D., Arizona State University, 2002. Focused on calculating the evolution of grain size distributions in the outflows of cool stars.

Emeriti Faculty

Herbots, Nicole, Ph.D., U. Catholique de Louvain, Belgium, 1984. Synthesis of new thin-film heterostructures by combined ion and molecular deposition and characterization by a wide variety of modern analysis techniques.

Hestenes, David O., Ph.D., UCLA, 1963. Theoretical foundations of physics; relativistic electron theory; physics education research and development.

Jacob, Richard J., Ph.D., Utah, 1963. Elementary particle theory; intermediate-energy hadronic interactions.

Kaufmann, William B., Ph.D., California, Berkeley, 1968. Theoretical intermediate-energy; nuclear and elementary particle physics.

Mayer, James, Ph.D., Purdue University, 1960. Ion beam analysis; Ion implantation in silicon thin film reactions.

Page, John B., Ph.D., Utah, 1966. Condensed matter theory; dynamical localization in nonlinear lattices, first-principles studies of fullerenes and fullerene polymers; resonance Raman scattering; phonons and electron-phonon interactions in solids.

FACULTY RESEARCH SPECIALTIES

Biophysics & Biological Physics

Stuart Lindsay, Dmitry Matyushov, Alan Nelson, Timothy Newman, Banu Ozkan, Peter Rez, Robert Ros, Otto Sankey, John Spence, Michael Thorpe, Sara Vaiana.

Nanoscale Science & Materials Physics

Gary Adams, Ernst Bauer, Peter Bennett, Ralph Chamberlin, Tingyon Chen, Robert Culbertson, Bruce Doak, John Dow, Jeff Drucker, Nicole Herbots, Robert Marzke, Martha McCartney, Jose Menendez, Robert Nemanich, John Page, Fernando Ponce, Peter Rez, Robert Ros, Kevin Schmidt, John Shumway, David Smith, John Spence, Michael Thorpe, Mike Treacy, Kong-Thon Tsen, Ignatius Tsong, John Venables.

Particle Astrophysics & Cosmology

Ricardo Alarcon, Andrei Belitsky, Joseph Comfort, Paul Davies, Damien Easson, Lawrence Krauss, Richard Lebed, Cecilia Lunardini, Barry G. Ritchie, Richard Jacob, William Kaufmann, Tammay Vachaspati.

Physics and Society

Carl Covatto, Robert Culbertson, David Hestenes, John Venables.

NORTHERN ARIZONA UNIVERSITY

PHYSICS & ASTRONOMY

Flagstaff, Arizona 86011

Students Accepted For Degree	FIELDS Physics	Astronomy	Related Fields
Doctorate			
Master's	X	X	X

1. General

President: John Haeger
Dean of Graduate School: Ramona Mellott
Department Chairman: Kathleen DeGioia Eastwood
Department Telephone Number: (928) 523-2661
Type of Institution: University
Control: Public
Setting: Small town
Total Faculty: 800 (No separate Graduate faculty)
Total Students: 23,600
Total Graduate Students: 5,300
Annual Graduate Tuition:
In-state residents: Full—time—$7,398
Part-time—Based on total number of credits
Out-of-state residents: Full-time—$18,172
Part-time—Based on total number of credits
Tuition rates for: 2010–11
Term: Year

2. Number of Faculty in Department

The combined total of full-time faculty in the three professorial ranks is 9. The combined total of full-time, part-time, and other faculty at all ranks is 13.

3. Admission, Financial Aid, and Housing

Address admission inquiries to: Dept. of Physics and Astronomy, Box 6010, Flagstaff, AZ 86011 or by email graduate.college@nau.edu
Graduate application fee required: $50
Admission deadline (Fall admission): Jan. 15
Admission information: For fall admission, 2009, 8 students were accepted from 15 applicants.
Admission requirements: For admission to the graduate programs, a Bachelors degree in Physics, Astronomy, Chemistry, or a related field is required with a minimum undergraduate GPA of 3.0 specified. The GRE is required. The minimum acceptable score suggested for admission is quantitative 625. The GRE Advanced is required. No minimum score is specified The average GRE scores for admissions were not specified. The average GRE Advanced for admissions was not specified. Students from non-English speaking countries are required to demonstrate proficiency in English via the TOEFL exam with a minimum score of 80.
Undergraduate preparation assumed: Although preparation will vary, we expect at least one semester each of upper-division mechanics, electricity and magnetism, quantum mechanics, and advanced laboratory.
Address financial aid inquiries to: Office of Student Financial Aid (928) 523-4951, Financial.Aid@nau.edu
Financial aid deadline: 3/01
Loans available: Yes
Address housing inquiries to: Office of Residence Life (928) 523-3978, Residence.Life@nau.edu
On-campus, graduate student housing available: Yes
Cost/year: $3,824-$4,736
On-campus, married student housing available: Yes
Cost/month: $712–870

Table A—Faculty, Enrollments, and Degrees Granted

Research Specialty	2009–10 Faculty	Enrollment[1] Fall 2009 Master's	Doctorate	No. of Degrees Granted[2] 2009–10 (2005–09) Master's	Terminal Master's	Doctorate	Median No. of Years for 2009–10 Ph.D.'s
Astrophysics	5	5	–	4(13)	–	–	–
Chemical Physics	0	0	–	0(1)	–	–	–
Condensed Matter Physics	4	4	–	1(8)	–	–	–
Physics Education	2	0	–	0(3)	–	–	–
Other Theoretical/ Math.	1	2	–	0(2)	–	–	–
Total	12	11	–	5(27)	–	–	–
Full-time Grad. Stud.		11	–				
Part-time Grad. Stud.		0	–				
First-year Grad. Stud.		7	–				
Median Years in Grad. Study (2009–10 Degrees)		2		–	–	–	
Undergraduate Degrees, 2009–10 (2005–09): 20(98)							

[1]Students not yet committed to a research specialty are entered under non-specialized.
[2]Five-year totals in parentheses.

4. Graduate Degree Requirements

Master's: 36 hours of graduate courses, including both thesis and non-thesis options. There is no foreign language requirement. Qualifying and comprehensive exams are not required. This program can be interdisciplinary, integrating a broad range of subject areas to enhance student opportunities in the industrial or research world. Individual programs may be customized to meet specific student needs.
Thesis: Thesis may be written *in absentia.*
Special Equipment, Facilities, or Programs: Ultra-high vacuum laboratories, PHI x-ray photoelectron spectroscopy instrumentation, scanning tunneling microscopy, scanning force microscopy, low energy electron diffraction, infrared spectroscopy (Nicolet), mass spectroscopy, Unix workstation laboratory, computational physics laboratory. Local astrophysical facilities include Lowell Observatory, U.S. Naval Observatory, NAU has access to the University of Arizona telescopes.

Table B—Appointments to Graduate Students, 2009–10

Title of Appointee	Appointments Total	First year	Academic Load Allowed in Credit Hours	Hours of Service Per Week	Stipend for Academic Year ($)[1]
			Semester		
Teaching Assistant	9	7	16	20	12,390
Research Assistant	2	0			
Total	11	7			

[1]Out of state tuition waived for TA recipients.

5. Personnel Engaged in Separately Budgeted Research, 7/09–6/10

Professorial faculty	8
Postdoctoral Appointments	1
Graduate Students	4
Undergraduate Student	3
Total	16

6. Separately Budgeted Research Expenditures by Source of Support, 2009–10

	Departmental Research	Physics-related Research Outside Department
Federal government	542,570	$0
State and local government	3,458	0
Other	43,103	0
Total	$589,131	0

Table C—Separately Budgeted Research Expenditures

Research Specialty	No. of Grants	Expenditures ($)
Astrophysics	14	515,465
Condensed Matter Physics	0	3,458
Physics Education	1	70,208
Total	15	589,131

FACULTY

Professors

Delinger, William G., Ph.D., University of Iowa, 1972. Solid state, electronic instrumentation, solar energy, computers.

Dillingham, T. Randall, Ph.D., Kansas State University, 1983. Surface physics, x-ray photoelectron spectroscopy, surface chemistry of ices.

Eastwood, Kathleen DeGioia, Ph.D., University of Wyoming, 1982. Optical and infrared astronomy; star formation, interstellar medium, stellar populations.

Tegler, Stephen C., Ph.D., Arizona State University, 1989. Visible and infrared astronomy, origins of solar systems, comets.

Associate Professors

Barlow, Nadine, Ph.D., University of Arizona, 1987. Planetary astronomy.

Bowman, Gary E., Ph.D., University of Notre Dame, 2000. Foundations of quantum mechanics.

James, Mark C., Ph.D., Kansas State University, 2003. Science Education.

Koerner, David, Ph.D., California Institute of Technology, 1994. Origin of Planetary systems.

Assistant Professor

Trilling, David, Ph.D., University of Arizona, 1999. Planetary science.

Lecturer

Cole, David M., Ph.D., Texas A&M, 1997. Magnetic resonance imaging, science education.

RESEARCH SPECIALTIES AND STAFF

Theoretical

Foundations of Quantum Mechanics. Bowman.

Experimental

Astrophysics. Planetary astrophysics, stellar evolution, origins of solar systems, visible and infrared astronomy, unmanned space missions, star formation, planetary atmospheres. Barlow, Eastwood, Koerner, Tegler, Trilling.

Condensed Matter Physics. Inorganic/Organic materials; semiconductor physics, microsensor development, polymer physics. Delinger, Dillingham.

Physics Education. Science education, Action-based research. Cole, James.

UNIVERSITY OF ARIZONA

DEPARTMENT OF PHYSICS

Tucson, Arizona 85721

Students Accepted For Degree	FIELDS		
	Physics	Astronomy	Related Fields
Doctorate	X	X	X
Master's	X	X	X
Professional Science Masters (PSM)	Medical Physics		

1. General

President: Robert N. Shelton
Dean of Graduate School: Andrew Comrie
Department Chairman: Sumit Mazumdar
Department Telephone Number: (520) 621-6820 (C)
Type of Institution: University
Control: Public
Setting: Urban
Total Faculty: 1,650
Total Graduate Faculty: 1,594
Total Students: 37,036
Total Graduate Students: 8,215
Annual Graduate Tuition: †

6 units or less:
Grad. Assistant Annual Rate: $122
7 units or more:
Grad. Assitant Annual Rate: $134

Tuition rates for: 2009–10
Deferred tuition plan: Yes (if on assistantship)
Term: Semester

†See Graduate Catalog for more detailed information.

2. Number of Faculty in Department

The combined total of full-time faculty in the three professorial ranks is 32. The combined total of full-time, part-time, and other faculty at all ranks is 79.

3. Admission, Financial Aid, and Housing

Address admission inquiries to: Univ. of Arizona, Dept. of Physics, Graduate Coordinator, 1118 E. 4th St., Tucson, AZ 85721.

Graduate application fee required: $75

Admission deadline (Fall semester): Domestic: 1/1; International: 12/1 with assistantship

Admission information: For fall admission, 2010, 47 students were offered admission from 193 applicants.

Admission requirements: For admission to the graduate programs, a Bachelor's degree in physics or a related field is required with an undergraduate GPA of 3.0 in the last 60 units. The GRE, including the physics subject section, are required. The average physics subject score of the 2009 entering class was above the 70th percentile; generally students with scores below the 50th percentile are not admitted. Students from non-English speaking countries are required to demonstrate proficiency in English via the TOEFL and TSE exams. Minimum acceptable score for admission is TOEFL-550, TSE-45.

Table A—Faculty, Enrollments, and Degrees Granted

Research Specialty	2009–10 Faculty	Enrollment[1] Fall 2009		No. of Degrees Granted 2009–10			Median No. of Years for 2009–10 Ph.D.'s
		Mas-ter's	Doc-torate	Mas-ter's	Terminal Master's	Doc-torate	
Applied/Industrial Physics	0	1	0	2	0	0	–
Astrophysics/ Cosmology	2	0	10	1	0	0	–
Atmos/Space Phys. Cosmic Rays	1	0	1	0	0	0	–
Atomic Mass Spec-troscopy Lab	2	0	0	0	0	0	–
Atomic, Molecular & Optical Physics	4	0	20	1	0	1	–
Biophysics	1	0	4	0	0	0	–
Condensed Matter Physics	7	0	10	0	0	2	–
Engineering Physics	0	0	0	0	0	0	–
High Energy/Particles & Fields	9	0	13	1	0	2	–
Mathematical Physics	1	0	0	0	0	0	
Med. Physics	4	5	0	0	0	0	–
Nuclear Physics	1	0	7	0	0	0	–
Physics Education	0	0	1	0	0	0	
Undecided	0	0	0	0	2	0	
Total		6	71	3	2	5	
Full-time Grad. Stud.		6	71	3	2	5	
Part-time Grad. Stud.		0	0	0	0	0	
First-year Grad. Stud.		4	15				

Undergraduate Degrees, 2009–10: Physics: 23 Engineering Physics: 4

[1]Students not yet committed to a research specialty are entered under non-specialized.

Undergraduate preparation assumed: Tipler, *Modern Physics* (Phys. 242); Symon, *Mechanics* (Phys. 321); Wangsness, Griffiths, *Electromagnetic Fields* (Phys. 331/332); Hecht and Zajac, *Optics* (Phys. 320); Fermi, *Thermodynamics*; Callen, *Thermodynamics* (Phys. 325); Liboff, Griffiths, *Quantum Physics* (Phys. 371); Gasiorowicz, *Quantum Physics* (Phys. 472); Herzberg, Kodak, Beveridge, *Atomic Spectra and Atomic Structure* (Phys. 473/474); Kittel, *Introduction to Solid State Physics* (Phys. 460); Arfken, *Mathematical Methods for Physics* (Phys. 475); Melissinos, *Experiments in Modern Physics*; Young, *Statistical Treatment of Experimental Data*; Taylor, *An Introduction to Error Analysis* (all 3 optional); (Phys. 381, 481).

Address financial aid inquiries to: Office of Student Financial Aid, 203 Admin. Building, University of Arizona

GAPSFAS application required: No

Financial aid deadline: 1/1 for assistantships (information available through Dept. of Physics)

Loans available: Yes[1]

Address housing inquiries to: Department of Residence Life, Administration Building

On-campus, single student housing available: Yes
Cost/term: Varies[2]

On-campus, married student housing available: Yes

[1]Inquire at Scholarships and Financial Aids Office.
[2]For rates, check with Department of Residence Life.

4. Graduate Degree Requirements

Master's: In order to qualify for the M.S. degree, the student must complete the qualifying examination. In addition, the student must complete at least 30 units of graduate work, at least 15 of which must be in physics. The student must maintain a 3.0 average. No foreign language required. The student must also satisfy one of the following options: 1. Write a thesis (for which up to six units may be allowed) and pass an oral examination. 2. Take six additional graduate credits in physics and pass a final oral examination. Under this option, the student completes a total of at least 21 units in physics out of the total of 30 units required. 3. Pass the written and oral parts of the comprehensive examination for the Ph.D. Students are not normally admitted into the MS program; students with bachelors degrees who seek the Ph.D. as the terminal degree should apply directly for admission into the Ph.D. program. Such students may then earn the M.S. degree en route to the Ph.D.

Doctorate: The Ph.D. degree requires completion of 36 units of graduate work in physics, with 12 additional units in the minor, with a 3.0/4.0 grade average (exclusive of dissertation credits); passing the qualifying and comprehensive examinations; submitting a dissertation based on independent research; and defending this dissertation in a final examination. Note that independent study can be used to earn a portion of the required graduate units.

Professional Science Masters in Medical Physics: This degree requires completion of at least 36 units of graduate work including 6 units of internship, 12 units of physics, up to 6 units of business and 9 units of other specialty courses.

Other Programs: Interdisciplinary work with other departments is possible, especially in the areas of materials and surface physics, chemical physics, astrophysics, space physics, optical science, medical physics, biophysics, applied math, and mathematical physics.

Thesis: Thesis may be written *in absentia*.

Special Equipment, Facilities, or Programs: Our graduate program is dedicated to research at the cutting edge of discovery. In the past decade alone, we have recruited nearly 20 new faculty members in all major research areas. We also have strong interdisciplinary initiatives, with campus-wide programs in theoretical astrophysics, lunar and planetary sciences, geosciences, optical sciences, biophysics, and applied mathematics. Additional connections to biology, chemistry and medicine provide additional research opportunities. Moreover, in addition to a traditional Ph.D. program we are also one of the few institutions nationwide to have a Professional Science Masters (PSM) program in medical physics. Graduate students have excellent resources at their disposal. General facilities include a shop for students and staff. An extensive collection of physics journals and more than 5,000 physics books are located in the science library building. The University is connected to Internet2 for high speed connectivity to the NSF supercomputer centers, other universities and other supercomputer centers. The University also operates a large SGI supercomputer, numerous UNIX servers, Windows PC & Macintosh public access sites, and a computerized library information system (sabio.arizona.edu). The Physics Department has a large number of Linux PC's, Windows PC's and laser printers (color and b/w) available for general department use. Facilities and equipment for particular research areas in physics include a 3-MV tandem accelerator, spectrographs, spectrometers, superconducting magnets for solid state physics, electron microscopes, electron probes (including Auger, RHEED, and LEED), a materials-processing laboratory, thin-film molecular beam expitaxy (MBE) and sputter epitaxy equipment, thin-film x-ray diffraction facilities, an atomic force microscope, optical tweezers and single molecule detection for biophysics, cryogenic systems covering the entire temperature range of 300 K down to 0.02 K, a computerized Mössbauer spectroscopy system, ultrahigh-resolution infrared and visible lasers, atomic-beam machines, and an observatory for work in experimental relativity and solar seismology. The high-energy experimental group does work at Fermilab and CERN. There is an active program in theoretical physics covering atomic, nuclear, condensed matter, astrophysics, and high-energy theory. The Departments of Physics and Geosciences jointly operate a facility, sponsored by the National Science Foundation, that uses accelerators for radioisotope dating.

Table B—Appointments to Graduate Students, 2009–10

Title of Appointee	Appointments Total	Appointments First year	Academic Load Allowed in Credit Hours	Hours of Service Per Week	Stipend for Academic Year ($)
			Semester		
Teaching Assistant	38	17	–	20	15,042-16,645
Research Assistant	33	0	–	20	15,042-16,645
NASA	0	0	–	0	16,000-22,000
NSF	0	0	–	0	15,000
Self Support	5	1	–		
Other FLW	3	0			
Total	79	18			

5. Personnel Engaged in Separately Budgeted Research, 8/05–5/06

Professorial faculty	33
Graduate students	33
Nonteaching research personnel	31
Total	97

6. Separately Budgeted Research Expenditures by Source of Support

	Departmental Research
Federal government	$3,445,944
Total	$3,445,944

Table C—Separately Budgeted Research Expenditures

Research Specialty	No. of Grants	Expenditures ($)
Accelerator Mass Spectrometry	6	197,138
Astrophysics	5	259,342
Biophysics	1	58,024
Condensed Matter	11	583,400
Nuclear Physics	9	436,237
Optical Physics	5	270,577
Particles & Fields	14	1,658,233
Physics Education Research	1	116,924
Total	52	3,579,875

FACULTY

Professors

Barrett, Bruce R., Ph.D., Stanford U., 1967. Nuclear many-body theory; microscopic shell model calculations; three-nucleon interactions; nuclear collective motion.

Cheu, Elliott, Ph.D., Cornell U., 1991. Experimental high energy.

Dienes, Keith R., Ph.D., Cornell U., 1991. Theoretical high energy physics.

Fang, Li-Zhi, Ph.D., Beijing U., 1956. Theoretical astrophysics and cosmology.

Hsieh, Ke Chiang, Ph.D., U. Chicago, 1969. Experimental cosmic-ray and space physics.

Johns, Ken, Ph.D., Rice U., 1986. Experimental high energy.

Lebed, Andrei, Ph.D., Landau Institute for Theoretical Physics, 1986. Theoretical condensed matter.

Mazumdar, Sumit, Ph.D., Princeton U., 1980. Condensed matter theory, Department Head.

Melia, Fulvio, Ph.D., MIT, 1985. Theoretical astrophysics.

Meystre, Pierre, Ph.D., Swiss Federal Institute of Technology, 1974. Theoretical quantum optics.

Rafelski, Johann, Ph.D., U. Frankfurt, 1973. Relativistic nuclear theory; laser particle physics; nuclear cosmology.

Rutherfoord, John P., Ph.D., Cornell U., 1968. Experimental high-energy physics.

Sarcevic, Ina, Ph.D., U. Minnesota, 1986. High-energy theory, nuclear theory.

Shupe, Michael A., Ph.D., Tufts U., 1976. Experimental high-energy physics.

Toussaint, Douglas, Ph.D., Princeton U., 1978. High-energy theory.

van Kolck, Ubirajara, Ph.D., U. Texas-Austin, 1993. Effective field theories in particle, nuclear, atomic, and astro physics.

Wing, William H., Ph.D., U. Michigan, 1968. Experimental atomic physics and quantum optics.

Zhang, Shufeng, Ph.D., N.Y. Univ. 1991. Condensed matter theory.

Associate Professors

Cronin, Alexander D., Ph.D., U. Washington, 1999. Experimental atomic physics.

Fleming, Sean, Ph.D., Northwestern U., 1995. Nuclear theory, high energy theory.

Jacquod, Philippe, Ph.D., Neuchatel, 1997. Condensed matter theory.

Manne, Srinivas, Ph.D., U. California-Santa Barbara, 1994. Experimental condensed matter.

Stafford, Charles A., Ph.D., Princeton U., 1992. Condensed matter theory.

Su, Shufang, Ph.D., MIT., 2000. Theoretical high energy physics.

Varnes, Erich W., Ph.D., U. California-Berkeley, 1997. Experimental high energy physics.

Visscher, Koen, Ph.D., U. Amsterdam, 1993. Biophysics. Director of Graduate studies.

Assistant Professors

LeRoy, Brian, Ph.D., Harvard U., 2003. Experimental nanoscience.

Sandhu, Arvinder, Ph.D., Tata Institute of Fundamental Research (TIFR), 2004. Experimental atomic, molecular, and optical physics.

Joint Professors

Adamowicz, Ludwik, Professor, Chemistry Physics; Ph.D., Inst. of Physical Chemistry of the Polish Academy of Sciences, 1977. Theoretical chemistry, chemical physics.

Anderson, Brian, Assistant Professor, Optical Science; Ph.D., Stanford U., 1999. Experimental Bose-Einstein condensation.

Binder, Rudolf, Professor, Optical Science; Ph.D., U. Dortmund, Germany, 1988. AMO physics.

Brown, Michael, Professor, Biochemistry, Physics; Ph.D., U. California-Santa Cruz, 1975. Physical chemistry and biochemistry, nuclear magnetic resonance spectroscopy.

Falco, Charles M., Ph.D., California, Irvine 1974, Chair of Condensed matter physics, U of A.

Jessen, Poul, Professor, Optical Science; Ph.D., U. Aarhus, 1993. Atomic and optical physics.

Jull, Timothy, Professor; Ph.D., U. Bristol, 1976. Geosciences.

Kennedy, Thomas G., Professor, Mathematics; Ph.D., U. Virginia, 1984. Mathematical physics.

Lunine, Jonathan L, Professor, Lunar & Planetary Sciences, Physics; Ph.D., Caltech, 1985. Theoretical astrophysics.

Maier, Robert S., Professor, Physics, mathematics; Ph.D., Rutgers State U. New Jersey, 1983. Applied Mathematics, mathematical physics.

Miyashita, Osamu, Assistant Professor, Ph.D. Kyoto University, 2000, Biochemistry and molecular biophysics

Özel, Feryal, Assistant Professor, Ph.D., Harvard U., 2002. Theoretical astrophysics.

Pinto, Philip A., Associate Professor, Computational astrophysics; Ph.D., U. California-Santa Cruz, 1988. Astronomy.

Psaltis, Dimitrios, Associate Professor, Ph.D., U. Illinois, Urbana-Champaign, 1998. Theoretical astrophysics.

Restrepo, Juan, Professor, Applied mathematics, nonlinear dynamics; Ph.D., Pennsylvania State U., 1992. Mathematics, mathematical physics.

Shelton, Robert N., Ph.D., Physics, U. California-San Diego, 1975. President of the U. Arizona.

Tabor, Michael, Professor, Applied Math; Ph.D., D.Sc., Bristol U., 1990. Head, Applied Math, mathematical physics.

Wright, Ewan, Professor of Optical Sciences, Physics; Ph.D., Heriot-Watt U., 1983. Theory and stimulation of light string propagation in air.

Xin, Hao, Associate Professor, Electric and Computer Engineering; Ph.D., MIT, 2000. Microwave and millimeter wave technology.

Adjunct Faculty

Barker, Delmar. Raytheon Corporation.

Chacko, Zackaria, Univ. of Maryland.

Coon, Sidney, Adjunct Professor of Physics, Nuclear Theory.

Denison, Arthur B., Research Coorporation Technologies.

Gallardo, Juan C., Ret., Experimental High Energy.

Pattison, John.

Radziemski, Leon. Research Corporation Technologies.

Timmes, Francis. Group Leader, Theoretical Astrophysics, Los Arizona State Univ.

Weekes, Trevor. Adjunct Professor of Physics, Fred Lawrence Whipple Observatory.

Wiener, Richard J., Research Corporation Technologies.

Research Scientists

Beck, Warren, Senior, Research Scientist; Ph.D., U. Minnesota, 1988. Co-Director, AMS Facility.

Burr, George S., Senior, Research Associate; Ph.D., U. Chicago, 1990. Geophysical sciences.

Curtis, Charles, Research Associate Professor of Physics; Ph.D., U. Arizona, 1978.

Hodgins, Gregory, Asst. Research Scientist; Ph.D., U. Oxford, 1999.

Physics Teacher Preparation Program

Novodvorsky, Ingrid, Ph.D., U. Arizona, 1993. Director, Physics education research and science teacher preparation program.

Lecturers

Jackson, Shawn S., M. A., Washington University in St. Louis, 1988. Astrophysics.

Milsom, John A., Ph.D., Northwestern U., 1996. Astrophysics.

Professors Emeriti

Bickel, William S., Ph.D., Penn State U., 1965. Optical biophysics; atomic physics; physics of music.

Bowen, Theodore, Ph.D., U. Chicago, 1954. Experimental elementary particle and cosmic-ray physics; medical physics.

Chambers, Robert H., Ph.D., Carnegie-Mellon U., 1957. Science education; dislocation dynamics; internal friction.

Donahue, Douglas J., Ph.D., U. Wisconsin, 1952. Atomic and nuclear physics; accelerator mass spectrometry.

Emrick, Roy M., Ph.D., U. Illinois, 1960. Experimental solid state physics.

Garcia, Jose D., Ph.D., U. Wisconsin, 1966. Collision theory; atomic physics; physics education.

Hill, Henry A., Ph.D., U. Minnesota, 1957. Astrophysics and experimental general relativity.

Huffman, Donald R., Ph.D., U. California-Riverside, 1966. Optical properties of solids; astrophysics.

Jenkins, Edgar W., Ph.D., Columbia U., 1962. Experimental elementary particle physics.

Just, Kurt W., Dr.rer.nat., Berlin U., 1954. Gauge theories of gravity and particles.

Kessler, John O., Ph.D., Columbia U., 1953. Applied physics; biophysics and fluids.

Kilkson, Rein, Ph.D., Yale U., 1956. Molecular biophysics.

Kohler, Sigurd, D.Sc., Uppsala, U. Sweden, 1959. Nuclear many-body theory; heavy-ion collisions.

Mahmoud, Hormoz M., Ph.D., U. Indiana, 1953. Field theory.

McIntyre, Laurence C., Jr., Ph.D., U. Wisconsin, 1965. Ion-beam analysis.

Scadron, Michael D., Ph.D., U. California-Berkeley, 1964. Theoretical elementary particle physics.

Stark, Royal W., Ph.D., Case Western Reserve U., 1962. Experimental solid state physics.

Stoner, John O., Jr., Ph.D., Princeton U., 1964. Experimental atomic spectroscopy; thin-film optics.

Thews, Robert L., Ph.D., MIT, 1966. High energy and nuclear theory.

Tomizuka, Carl T., Ph.D, U. Illinois, 1954. Experimental solid state physics.

Vuillemin, Joseph J., Ph.D., U. Chicago, 1965. Experimental solid state physics.

Wangsness, Roald K., Ph.D., Stanford U., 1950. Science education; statistical mechanics; electromagnetic theory.

Weaver, Albert B., Ph.D., U. Chicago, 1952. Cosmic rays.

RESEARCH SPECIALTIES AND STAFF

Theoretical

Astrophysics: Black holes, compact objects, AGN's particle astrophysics, cosmology. Fang, Melia, Milsom, Özel, Psaltis, Rafelski, Sarcevic.

Atomic and Molecular Physics: Ultracold atoms, Bose condensation and atom lasers, cavity QED, ion-atom collision theory, atomic structure calculations. Cronin, Garcia, Jessen, Meystre, van Kolck, Wing.

Condensed Matter: Strongly correlated electron systems, mesoscopic and nanoscopic materials, disordered systems, nonlinear dynamics, quantum chaos, quantum spin systems, superconductivity. Jacquod, Lebed, Mazumdar, Stafford, Zhang.

High Energy: Theory and phenomenology of strong and electroweak interactions, lattice QCD, neutrino physics, Higgs physics, physics beyond the standard model, supersymmetry, grand unification, extra dimensions, and string theory. Dienes, Fleming, Sarcevic, Su, Thews, Toussaint, van Kolck.

Mathematical Physics: Quantum spin systems, stochastic processes, rigorous statistical mechanics, applied probability theory, string theory. Dienes, Kennedy, Maier.

Nuclear Physics: Nuclear many-body theory, heavy ion collisions, quark-gluon substructure of nuclear matter, effective field theories of nuclear forces, chiral perturbation theory, Soft-collinear effective theory, heavy-quark systems, laser-induced particle phenomena. Barrett, Fleming, Rafelski, Sarcevic, Thews, van Kolck.

Physics Education: Secondary school teacher preparation, physics education theory and modeling. Garcia, Novodvorsky.

Science & Technology Policy: Wing.

Experimental

Accelerator Mass Spectrometry and Radioisotope Dating: Radiocarbon dating, cosmic ray effects in terrestrial materials, climate change, geophysical studies. Beck, Burr, Jull.

Atomic & Molecular Physics: Cold atom trapping and interferometry, Bose-Einstein condensates. Anderson, Cronin, Sandhu, Wing.

Biological Physics: Chemotaxis, pattern formation of motile cells, molecular motors, structural and elastic properties of nucleic acids and proteins, membranes, protein electron transfer, protein allostery. Brown, Kessler, Miyashita, Visscher.

Condensed Matter: Self-assembly at solid/liquid interfaces, nonlinear dynamics, pattern formation, hydrodynamics, fullerenes, optical trapping and imaging of biological systems. Falco, LeRoy, Manne, Visscher.

High Energy: Accelerator-based experiments at Fermilab and CERN, including: direct CP violation in kaon decay, search for Higgs, top quark physics, jets, search for extra dimensions, electroweak physics. Cheu, Johns, Rutherfoord, Shupe, Varnes.

Optics: Theoretical and experimental research efforts include investigations of cold trapped atoms, Bose-Einstein condensates, interference of atomic beams, the use of optical tweezers in biophysical measurement, the interaction of light with solids for understanding of energy levels and much more. Bickel, Cronin, Huffman, Mazumdar, Meystre, Visscher, Wing, Wright.

Space Physics & Experimental Astrophysics: Low energy charged particle measurement in the planetary and interplanetary environment, capture and analysis of interplanetary dust particles, application and development of a neutral particle detector for interplanetary and planetary measurements, gamma ray astrophysics, small satellite design, optical communications. Hsieh, Weekes, Wing.

UNIVERSITY OF ARIZONA

DEPARTMENT OF ASTRONOMY

Tucson, Arizona 85721

Students Accepted For Degree	FIELDS		
	Physics	Astronomy	Related Fields
Doctorate		X	
Master's			

1. General

President: Robert N. Shelton
Dean of Graduate School: Andrew Comrie
Department Chairman: Peter A. Strittmatter
Department url: www.as.arizona.edu
Department Telephone Number: (520) 621-2288
Department FAX number: (520) 621-1532
Department Address: 933 North Cherry Avenue
Type of Institution: University
Control: Public
Setting: Urban
Total Faculty: 2,756
Total Graduate Faculty: No formal graduate faculty
Total Students: 38,767
Total Graduate Students/First Professional Headcount: 8,421
Annual Graduate Tuition and Mandatory Fees: †
In-state residents: $7,632
Out-of-state residents: Full-time—$22,544/yr*
Part-time—$1,251/credit
Tuition rates for: 2009–10
Deferred tuition plan: No
Other Fees: Registration: Average total cost with fees
(*In-State*) $3,807
(*Out-of-State*) $11,986
Term: Semester

†See graduate catalog for more detailed information.
*Nonresident tuition is waived for students with graduate assistantships.

2. Number of Faculty in Department

The combined total of full-time faculty in the three professorial ranks is 30. The combined total of faculty and research scientists (including adjunct and joint appointments) is 112.

3. Admission, Financial Aid, and Housing

Web site: http://www.as.arizona.edu
Address admission inquiries to: Chairman, Admissions Committee, Steward Observatory, University of Arizona
Graduate application fee required: $75
Admission deadline (Fall admission): 1/14/2011
Admission information: For fall admission 2010, 9 students were accepted from 120 applicants.
Admission requirements: For admission to the graduate programs, a Bachelor's degree in astronomy or physics is preferred with no minimum undergraduate GPA specified. The GRE is required. No minimum acceptable score is specified. The GRE Advanced Physics is required with no minimum acceptable score specified. The average GRE scores for 2010 admissions were: verbal–561; quantitative–743; total–1,304. The average GRE Advanced score for 2010 admissions was 668. Students from non-English speaking countries are required to demonstrate proficiency in English via the TOEFL exam.
Undergraduate preparation assumed: Normal preparation in mathematics through advanced calculus; physics, including classical mechanics, electromagnetism, quantum mechanics, and thermodynamics.
Address financial aid inquiries to: Admissions Chairman
GAPSFAS application required: No
Financial aid deadline: 1/14/2011
Loans available: Yes
Address housing inquiries to: Department of Student Housing
On-campus, single student housing available: Yes
Cost/term: Varies

Table A—Faculty, Enrollments, and Degrees Granted

Research Specialty	2009–10 Faculty	Enrollment[1] Fall 2009		No. of Degrees Granted[2] 2009–10 (2005–10)			Median No. of Years for 2009–10 Ph.D.'s
		Master's	Doctorate	Master's	Terminal Master's	Doctorate	
Astronomy	30	0	39	0(0)	0(4)	7(32)	6.0
Total		0	39	0(0)	0(4)	7(32)	
Full-time Grad. Stud.		0	39				
Part-time Grad. Stud.		0	0				
First-year Grad. Stud.		0	4				
Median Years in Grad. Study (2009–10 Degrees)		0	6.0	0	0	6.0	6.0

Undergraduate Degrees, 2009–10 (2005–10): 12(68)

[1] Students not yet committed to a research specialty are entered under non-specialized.
[2] Five-year totals in parentheses.

4. Graduate Degree Requirements

Master's: 30 semester hours credit. Thesis with final oral exam required. A GPA of 3.0 (A=4) must be maintained. There is a 30-week residence requirement. No language requirement or written comprehensive or qualifying examination. Normal admission is only to the Ph.D. Program.
Doctorate: Successful completion of course requirements; second year Independent Research; Ph.D. qualifying exam; and dissertation.
Thesis: Thesis may be written *in absentia*.
Special Equipment, Facilities, or Programs: The primary research telescopes of the Steward Observatory include the 6.5-m MMT located on the summit of Mt. Hopkins in the Santa Rita Mountains, the 90-in. (2.3-m) Ritchey-Chrétien reflector on Kitt Peak, the 61-in. (1.55-m) Cassegrain reflector on Mt. Bigelow in the Santa Catalina Mountains, and the Vatican Advanced Technology Telescope (VATT) on Mt. Graham. The VATT has a 1.8-m f/1.0 primary mirror fabricated at the Steward Observatory Mirror Lab; Steward Observatory staff receive 25% of the scheduled observing time on this telescope. The MMT, operated jointly with the Smithsonian Astrophysical Observatory, was upgraded in 2000 to a single 6.5-m mirror constructed at the Mirror Lab. Steward Observatory is a partner in the twin 6.5-m Magellan telescopes at Las Campanas Observatory in Chile (built with Mirror Lab mirrors), giving students and researchers access to the southern skies. The observatory is currently involved in

developing several large (8.4-m) optical telescopes using spin cast mirror technology. The first of these, the Large Binocular Telescope (LBT) on Mt. Graham (two 8.4-m mirrors on a common mount) started normal operation in 2009. The first two 8.4-m mirrors have already been cast for the Giant Magellan Telescope (GMT) and is currently being polished. The GMT will consist of seven 8.4-m mirrors with an equivalent circular aperture of 21.5-m to be located at Las Campanas Observatory. A fourth 8.4-m mirror was cast in 2008 for the Large Synopic Survey Telescope (LSST). The GMT would be, by far, the world's largest. These major telescopes are equipped with a wide variety of instrumentation and detectors and are supported by several smaller instruments used for teaching or special research projects.
A 6.5-m mirror was cast in 2009 for a joint telescope project to be located on Cerro San Pedro Martir in Baja California in collaboration with Mexico, UCB, and UCSC. The main areas of research at the observatory include extragalactic and galactic astronomy, with major specializations in the areas of quasars, active galactic nuclei, degenerate stars, infrared sources, formation of stars and planetary systems, interstellar medium, novae, and radio galaxies. Observational work is concentrated in the optical and infrared, but also includes work at radio, ultraviolet, and x-ray wavelengths using other facilities. The observatory operates the Heinrich Hertz 10-m Submillimeter Telescope (SMT) on Mt. Graham for work at mm and sub-mm wavelengths, and a 12-m radio telescope on Kitt Peak. The research programs also include a wide range of theoretical studies in astrophysics, and an involvement in astronomy in space, such as the Spitzer Space Telescope, the Hubble Space Telescope (HST), and the James Webb Space Telescope (JWST). Laboratory work includes astrobiology, astrochemistry, development of special instruments, CCD and CMOS imaging systems, high frequency receivers for work at submillimeter wavelengths, including observations at the South Pole (Antarctica). The Center for Astronomical Adaptive Optics (CAAO) is developing diffraction limited imaging systems for use with the LBT, Magellan, and GMT. An AO system is in regular operation on the MMT.

Table B—Appointments to Graduate Students, 2008–09

Title of Appointee	Appointments Total	Appointments First year	Academic Load Allowed in Credit Hours	Hours of Service Per Week	Stipend for Academic Year ($)
			Semester		
Teaching Assistant	16	0	12	20	17,227/9 mos.
Research Assistant	39	3	12	20	17,227/9 mos.
NSF Fellowship	3	1	12	20	16,200/12 mos.
Total	58	4			

NSF Fellow = BRUTLAG, SHEVCHUK, TESKE

5. Personnel Engaged in Separately Budgeted Research, 7/08–6/09

Professorial faculty	30
Other faculty	30
Postdoctoral appointments	18
Graduate students	39
Undergraduate students	98
Nonteaching research personnel	113
Total	328

6. Separately Budgeted Research Expenditures by Source of Support

	Departmental Research	Physics-related Research Outside Department
Federal government	$57,589,703	$
Non-Federal	5,112,330	$
Total	$62,702,033	$

Table C—Separately Budgeted Research Expenditures

Research Specialty[1]	No. of Grants[2]	Expenditures ($K)[3]
Astronomy	244	62,702,033
Total	244	62,702,033

FACULTY

Professors

Angel, J. Roger P., Ph.D., Oxford, 1967. High-energy astrophysics; instrumentation; polarization, solar energy.

Arnett, David, Ph.D., Yale, 1965. Nuclear relativistic and computational astrophysics; stellar and galactic evolution.

Bechtold, Jill, Ph.D., Arizona, 1985. Quasars; the intergalactic medium.

Bieging, John H., Ph.D., Caltech, 1974. Interstellar medium; radio astronomy; stellar evolution.

Brown, Robert H., Ph.D., Hawaii, 1982. Kuiper belt objects; brown dwarfs; extra solar planets; IR instrumentation.

Burge, James, Ph.D., Arizona, 1993. Astronomical instrumentation, space optics, adaptive optics.

Cocke, William J., Ph.D., Emeritus, Cornell, 1964. Relativity; turbulence; pulsating radio sources; speckle interferometry.

Coyne, George V., Ph.D., Georgetown, 1962. Magnetic stars; instrumentation.

Eisenstein, Daniel, Ph.D., Harvard, 1996. Cosmology; large-scale structure.

Fan, Xiaohui, Ph.D., Princeton, 2000. Cosmology; quasars; galaxy formation; brown dwarfs.

Fang, Li-Zhi, Ph.D., Beijing, China, 1956. Cosmology clustering at high redshift.

Hoffmann, William F., Ph.D., Emeritus, Princeton, 1962. Infrared astronomy from balloons; instrumentation, nulling interferometry.

Impey, Christopher, Ph.D., Edinburgh, 1981. Quasars; observational cosmology.

Jokipii, J. R., Ph.D., Caltech, 1965. Theoretical astrophysics; plasma physics.

Kennicutt, Robert C., Ph.D., Washington, 1978. Observational extragalactic astronomy; structure and evolution of normal galaxies.

Liebert, James, Ph.D., Caltech, 1965. Theoretical astrophysics; plasma physics, dwarf novae.

Melia, Fulvio, Ph.D., MIT, 1985. Galactic center; relativistic jets and active galactic nuclei; accretion-disk coronae; cataclysmic variables.

Pacholczyk, Andrzej G., D.Sc., Emeritus, Warsaw, 1961. Radioastrophysics; Seyfert galaxies.

Rieke, George H., Ph.D., Harvard, 1969. Stellar and nonstellar infrared observational studies.

Rieke, Marcia, Ph.D., MIT, 1976. Infrared astronomy; instrumentation.

Sarcevic, Ina, Ph.D., Minnesota, 1986. Theoretical particle and nuclear astrophysics.

Sasian, José, Ph.D., Arizona, 1988. Optical design, optical fabrication and testing, optical design for astronomical instrumentation.

Schmidt, Gary D., Ph.D., Arizona, 1978. Evolved stars; active galactic nuclei; optical instrumentation.

Strittmatter, Peter A., Ph.D., Cambridge, 1966. Chairman of the Department. Extragalactic radio resources; speckle interferometry; quasars, accretion disks.

Thompson, Rodger I., Ph.D., MIT, 1970. Infrared spectroscopy; molecular astrophysics; stellar evolution, nucleosynthesis, star formation, galaxy evolution, and cosmology.

Tifft, William G., Ph.D., Emeritus, Caltech, 1958. Space astronomy; galaxies.

Walker, Christopher, Ph.D., Arizona, 1988. Protostellar objects; submillimeter wave instrumentation.

Woolf, Neville J., Ph.D., Emeritus, Manchester, 1959. Circumstellar matter; infrared astronomy, astrobiology.

Zaritsky, Dennis, Ph.D., Arizona, 1991. Magellanic clouds; the distribution and nature of dark matter; large scale structure; interstellar and intergalactic dust; galaxy evolution.

Ziurys, Lucy, Ph.D., UC, Berkeley, 1984. Laboratory microwave spectroscopy; astrochemistry, astrobiology.

Associate Professors

Close, Laird, Ph.D., Arizona, 1995. Adaptive optics; star/planet formation; IR instrumentation.

Davé, Romeel, Ph.D., UC, Santa Cruz, 1998. Galaxy formation; cosmology; intergalactic medium; numerical hydrodynamics.

Hinz, Philip, Ph.D., Arizona, 2001. Instrumentation; adaptive optics; nulling interferometry; exosolar planets.

Meyer, Michael R., Ph.D., Massachusetts, 1996. Star formation; infrared astronomy; initial mass function, astrobiology.

Pinto, Philip A., Ph.D., UC, Santa Cruz, 1988. Supernovae, radiative transfer, cosmic distance scale.

Poss, Richard L., Ph.D., Georgia, 1986. History of astronomy; astronomy and the arts.

Prather, Edward E., Ph.D., Maine, 2000. Research on college teaching and learning in astronomy, physics, and space science; professional development of college faculty.

Psaltis, Dimitrios, Ph.D., Illinois, 1998. Theoretical astrophysics and gravitation.

Zabludoff, Ann, Ph.D., Harvard, 1993. Dependence of galaxy evolution on environment, type, merger history, and mass; structure and evolution of galaxy clusters and groups.

Assistant Professors

Eisner, Joshua, Ph.D., Caltech, 2005. Galactic astronomy and star formation; extrasolar planets.

Guyon, Oliver, Ph.D., University of Paris VI, 2002. Instrumentation; extrasolar planets.

Ozel, Feryal, Ph.D., Harvard, 2002. Theoretical astrophysics.

Shirley, Yancy, Ph.D., Texas at Austin, 2002. Galactic astronomy and star formation; extragalactic astronomy; stellar astronomy; astrobiology.

Astronomers

Axelrod, Timothy, Ph.D., UC, Santa Cruz, 1980. Supernovae, microlensing.

Green, Richard, Ph.D., Caltech, 1977. Quasars; black holes.

Hart, Michael, Ph.D., Arizona, 1991. Astronomical instrumentation; adaptive optics; infrared detectors.

Hill, John M., Ph.D., Arizona, 1984. Large telescope design; clusters of galaxies, instrumentation.

McCarthy, Donald W., Ph.D., Arizona, 1976. Low mass stars; brown dwarfs; infrared speckle interferometry.

Olszewski, Edward, Ph.D., Washington, 1982. Dwarf galaxies; Magellanic clouds.

Schneider, Glenn, Ph.D., Florida, 1985. Infrared astronomy; high contrast imaging; brown dwarfs and exosolar planets; star formation.

Smith, Paul S., Ph.D., New Mexico, 1986. Active galactic nuclei; observing techniques; polarimetry.

Sykes, Mark, Ph.D., Arizona, 1986. Interplanetary dust; asteroids; thermal modeling.

Vilas, Faith, Ph.D., Arizona, 1984. Mineralogy of asteroids and planets.

Associate Astronomers

Egami, Eiichi, Ph.D., Hawaii, 1995. Infrared astronomy, extragalactic astronomy; cosmology.

Engelbracht, Charles, Ph.D., Arizona, 1997. Infrared astronomy, extragalactic astronomy; infrared instrumentation.

Fleming, Thomas, Ph.D., Arizona, 1988. Low mass stars; white dwarfs; x-ray astronomy; nearby stars.

Green, Elizabeth M., Ph.D., Texas, 1981. Stellar evolution; binaries; subdwarf B stars.

Hege, E. Keith, Ph.D., Rensselaer Polytechnic Institute, 1965. Speckle interferometry; optical instrumentation.

Stansberry, John A., Ph.D., Arizona, 1995. Brown dwarfs; debris disks; planetary science.

Assistant Astronomers

Cunha, Katia, Ph.D., Observatorio Nacional/MCT Rio de Janeiro, 1993. High-resolution spectroscopy; stellar abundances; metallicity gradients and chemical evolution.

Halfen, DeWayne T., Ph.D., Arizona, 2006. Astrochemistry, laboratory astrophysics and radio astronomy.

Hinz, Joannah L., Ph.D., Arizona, 2003. Extragalactic astronomy, infrared astronomy.

Kim, Serena, Ph.D., SUNY, 2002. Star and planet formation.

Kulesa, Craig A., Ph.D., Arizona, 2002. Submillimeter wave instrumentation; life cycle of the interstellar medium; star formation.

Su, Kate, Ph.D., Calgary, 2000. Space-based optical and infrared observations on dusty objects.

Weiner, Benjamin, Ph.D., Rutgers, 1998. Galaxy structure and evolution, kinematics, star formation.

Willmer, Christopher N. A., Ph.D., Observatorio Nacional, Brazil, 1990. Extragalactic astrophysics; galaxy evolution; dwarf glaxies; infrared astronomy.

Research Scientists

Codona, Johanan, Ph.D., UC, San Diego, 1985. Adaptive optics; mathematical modeling in wave propagation.

Hubeny, Ivan, Ph.D., Charles University, Prague, 1977. Theory of stellar and planetary atmospheres; radiation transfer; accretion disks.

Lesser, Michael, Ph.D., Arizona, 1988. Astronomical instrumentation; CCD development.

Other Faculty

Staff astronomers and scientists in residence include Richard Allen, Hubert Martin, Blain Olbert, Steve West, and Peter Wehinger.

Adjunct faculty include Willy Benz (University of Bern, Switzerland), Adam Burrows (Princeton University), Wolfgang Duschl (University of Kiel, Germany), William Hartmann (Planetary Science Institute), Gerry Neugebauer, Mark Wagner (Ohio State), and astronomers at the Vatican Observatory, Arizona State University, and Northern Arizona University. The scientific staff is augmented by visiting professors and postdoctoral fellows.

Joint colloquia are held with the National Optical Astronomy Observatory and the National Radio Astronomy Observatory.

COMBINED PARTNERSHIP STAFF

The main building of the Steward Observatory houses the growing research and technical staff, the computer facilities, the MMTO staff and the Tucson offices of the Vatican Observatory, the Smithsonian Astrophysical Observatory, the National Radio Astronomy Observatory, and the Large Synoptic Survey Telescope (LSST). A program in Theoretical Astrophysics (TAP) is jointly run by Steward Observatory, the Physics Department, and the Lunar and Planetary Laboratory (LPL).

RESEARCH SPECIALTIES AND STAFF

Active Galaxies and QSOs. Angel, Bechtold, Impey, Liebert, Melia, Pacholczyk, G. Rieke, M. Rieke, Schmidt, Strittmatter, Wehinger, Woolf.

Astrobiology, Astrochemistry and Origin of Life. Eisner, Halfen, Meyer, Woolf, Ziurys.

Astronomy Education and Outreach. Eisner, Fleming, Impey, McCarthy, Poss, Prather.

Atomic and Molecular Astrophysics. Bieging, Thompson, Walker, Ziurys.

Cosmology. Cocke, Davé, Egami, Eisenstein, Fan, Fang, Impey, Kennicutt, Thompson, Tifft, Zabludoff, Zaritsky.

Galaxies and Galactic Evolution. Arnett, Bechtold, Davé, Eisenstein, Fan, Hill, Impey, Jokipii, Kennicutt, Olszewski, G. Rieke, M. Rieke, Thompson, Tifft, Willmer, Zabludoff, Zaritsky.

Instrumentation. Angel, Bechtold, Brown, Burge, Close, Eisner, Engelbracht, Hart, Hege, Hinz, Hoffmann, Lesser, McCarthy, G. Rieke, M. Rieke, Schmidt, Strittmater, Thompson, Walker, Wehinger, Zaritsky.

Large Telescopes. Angel, Eisner, Hart, Hill, Wehinger, Woolf.

Solar System and Planetary Astronomy. Brown, Close, Jokipii, Sykes, Vilas, Wehinger.

Star Formation. Close, Eisner, Hoffmann, Meyer, Schneider, Thompson, Walker.

Stellar and Galactic Dynamics. Olszewski, Zaritsky.

Stellar Astronomy. Arnett, Bieging, Brown, Close, Fleming, Liebert, McCarthy, Melia, Meyer, Olszewski, G. Rieke, M. Rieke, Schmidt, Strittmatter.

Submillimeter (AKA TERAHERTZ). Walker.

Supernovae and Nucleosynthesis. Arnett, Axelrod, Liebert, Pinto.

UNIVERSITY OF ARIZONA

COLLEGE OF OPTICAL SCIENCES

Tucson, Arizona 85721
www.optics.arizona.edu

Students Accepted For Degree	FIELDS		
	Physics	Astronomy	Related Fields
Doctorate			X*
Master's			X*

*Optical Sciences.

1. General

President: Robert Shelton
Dean of Graduate School: Andrew Comrie
Dean of College of Optical Sciences: James C. Wyant
Department Telephone Number: (520) 621-4111
Department Web Site: www.optics.arizona.edu
Type of Institution: University
Control: Public
Setting: Urban
Total UA Faculty: 2,619
Total UA Graduate Faculty: No formal graduate faculty
Total UA Students: 36,805
Total UA Graduate Students: 7,105
Annual Graduate Tuition and Fees for 2009-2010:
In-state residents: Full-time $4,520
Out-of-state residents: Full-time $12,451*
Deferred tuition plan: Yes (for Graduate Assistants)
Term: Semester

*Nonresident tuition is waived for students with graduate assistantships.

2. Number of Faculty in Department

The combined total of full-time faculty in the three professorial ranks is 62. The combined total of optical sciences and joint faculty at all ranks is 91.

3. Admission, Financial Aid, and Housing

Address admission inquiries to: Dr. Carl F. Maes, Associate Dean, Academic Programs, College of Optical Sciences
Graduate application fee required: $75
Admission deadline (Fall admission): January 1
Admission information: For fall admission, 2009–10, 64 students were accepted from 214 applicants.
Admission requirements: For admission to the graduate programs, a Bachelor's degree in engineering, physics, mathematics, or optics is required with minimum undergraduate GPA of 3.0/4.0 specified. The GRE is normally required. The minimum acceptable score suggested for admission is analytical-65%; quantitative-75%. The GRE Subject Test is recommended but not required. The average GRE scores for 2008-09 admissions were verbal-59%; quantitative-81%; analytical-39%.
Students from non-English speaking countries are required to demonstrate proficiency in English via the TOEFL exam. Minimum acceptable score for admission is 600.
Address financial aid inquiries to: Dr. Carl F. Maes, Associate Dean, Academic Programs, College of Optical Sciences, P.O. Box 210094, Tucson, AZ 85721-0094
GAPSFAS application required: Yes, for certain types of aid
Financial aid deadline: 2/15—early application advisable
Loans available: Yes
Address housing inquiries to: Department of Residence Life
On-campus, single student housing available: Yes
Cost/term: Varies
On-campus, married student housing available: Yes
Cost/month: Varies

Table A—Faculty, Enrollments, and Degrees Granted

Research Specialty	2009 Faculty	Enrollment Fall 2009		No. of Degrees Granted[1] 2009-10 (2005-10)			Median No. of Years for Ph.D.'s
		Master's	Doctorate	Master's	Terminal Master's	Doctorate	
Full-time grads		31	133	10(232)	18(123)	12(152)	5
Part-time grads		32	1				
First-year grads		30	29				
Median years in grad study				2	2	5	
Undergraduate degrees 2005-10	199						
Total	199	93	163	232	123	152	

[1]Five-year totals in parentheses.

4. Graduate Degree Requirements

Master's: Thesis option. A minimum of 32 units of graduate credit in optics or optics-related courses, including 8 units of 910 (thesis) and at least 2 optics laboratory courses, and a final oral examination based primarily on the thesis.
Non-thesis option. A minimum of 35 units of graduate credit in optics or optics-related courses, including at least 2 units of optics laboratory courses; 3 units credit for demonstrated competence in written communication (either by writing an acceptable Master's Report or successfully completing an appropriate course in technical writing); and a final oral examination, based primarily on the subject matter of the courses taken. A cumulative GPA of 3.0 is required for a M.S. degree to be awarded. MS Degree can be completed by distance with one visit to campus.
Doctorate: The equivalent of six semesters of full-time graduate coursework is required (45 units) including at least 2 optics laboratory courses. The equivalent of two semesters of full-time study must be spent in actual residence, and 30 semester units of graduate credit must be earned at the University of Arizona. A dissertation is required and 18 additional units of graduate credit are earned for it. A cumulative GPA of 3.0 is required for a Ph.D. degree to be awarded. A foreign language is not required. Required exams are the written and oral comprehensive examination (usually during the fifth or sixth semester), a disseration proposal examination and the final oral examination.
Theses: Theses and dissertations may be written *in absentia*.
Special Equipment, Facilities, or Programs: The College of Optical Sciences is recognized internationally for its innovative and unusually comprehensive research programs. Research encompasses a broad set of technologies and techniques for exploiting the properties and applications of light and touches virtually every field of science and all modern industries. The

extensive research facilities at the College of Optical Sciences provide the resources for both theoretical and applied research programs in all areas related to optics and the optical sciences. Optical Sciences continually refines and upgrades research facilities. A new 47,000 square-foot facility was completed in early 2006 and enables the College to expand its research capabilities and programs exponentially.

FACULTY

Nobel Laureates

Bloembergen, Nicolaas, Nobel Laureate in Physics, 1981. Professor, Ph.D., University of Leiden, 1948. Nuclear and electronic magnetic resonance. Solid state masers and lasers. Nonlinear optics. Spectroscopy.

Glauber, Roy J., Nobel Laureate in Physics, 2005. Adjunct Professor, Ph.D., 1949, Harvard University. Quantum electrodynamical interactions of light and matter. High-energy collision theory. Statistical correlation of particles produced in high-energy reactions.

Associate Dean, Academic Programs

Maes, Carl F., Ph.D., University of Arizona, 2003. Theory and simulations of lasers resonators, mechanical effects of light on atoms and molecules in high Q cavities, electromagnetic wave propagation, aberrations, and adaptive optics. Experimental work with alkali and solid state lasers, laser linewidth measurement, and adaptive optics operations.

Professors

Angel, J. Roger, P., D. Phil., Oxford University, 1967. Adaptive optics. Instrumentation. Extrasolar planets. Telescope design and optical fabrication. Interferometry.

Armstrong, Neal, Ph.D., University of New Mexico, 1974. New molecular materials through self-assembly and patterning. Interface characterization through surface analysis and scanning probe microscopies. Electrochemistry. Chemical sensors. Analytical chemistry.

Barrett, Harrison H., Ph.D., Harvard University, 1969. Inverse problems in medicine. Applications of statistical decision theory. Medical imaging. Three-dimensional reconstruction. Nuclear medicine.

Binder, Rolf, Ph.D., Universität Dortmund, 1988. Theoretical investigations of the optical properties of semiconductor structures and modeling of semiconductor lasers.

Burge, James H., Ph.D., University of Arizona, 1993. Optical system engineering. Optical design. Opto-mechanics. Pointing and tracking. Detectors. Cryogenic systems. Optical testing and precision metrology. Aspheric surfaces. Ultra-lightweight mirrors for space. Diffractive optics. Stellar interferometry. Astronomical instrumentation.

Chipman, Russell, Ph.D., University of Arizona, 1987. Optical polarization and ophthalmic optics.

Clarkson, Eric, Ph.D., Arizona State University, 1985. Image science, Mathematical optics, Medical imaging, Inverse problems, Image quality assessment.

Dallas, William John, Ph.D., University of California at San Diego, 1973. Picture archiving and communications systems. Electrical current imaging from biomagnetic field measurements. Medical image processing. Image display. Cardiac imaging.

Dereniak, Eustace L., Ph.D., University of Arizona, 1976. Infrared-radiation detection. Imaging spectrometers. Cryogenically cooled detector/electronics technology. CCD and CMOS devices. Infrared detectors using charge transfer concepts. Image processing of infrared sensors.

Falco, Charles, Ph.D., University of California, Irvine, 1974. Metallic superlattices. X-ray optics. Magnetism. Magneto-optics. Far-IR detector materials. Superconductivity. Nucleation and epitaxy of thin films. Multilayered materials and superlattices.

Fallahi, Mahmoud, Ph.D., Toulouse University/CNRS, 1988. High power semiconductor lasers. DFB/DBR lasers. Grating-assisted integrated optics. Photonic integrated circuits. Optical communication. Wavelength multiplexers and demultiplexers. Wavelength filters. Solgel-semiconductor integration. WDM components. Sensors, design and microfabrication. Nanofabrication and nanostructures. Solgel PIC.

Furenlid, Lars, Ph.D., Georgia Institute of Technology, 1988. Development and application of novel detectors, optical configurations, readout electronics, and data-processing methods for biomedical imaging systems, with special emphasis on biological questions related to cancer, cardiovascular, and neurodegenerative diseases.

Gibbs, Hyatt M., Ph.D., University of California, 1965. Fundamental physics of semiconductor quantum-confined structures. Nonlinear optical phenomena. Molecular-beam-epitaxial growth of Ga, Al, In, As heterostructures. Vertical-cavity surface-emitting lasers. Optical instabilities. Nonlinear etalons and waveguides. Optical nonlinearities.

Gmitro, Arthur, Ph.D., University of Arizona, 1982. Magnetic resonance imaging. Optics in medicine. Optical computing.

Greivenkamp, John E., Ph.D., University of Arizona, 1980. Ophthalmic and visual optics. Ophthalmic instrumentation and measurements. Interferometry and optical testing of aspheric surfaces. Optical fabrication. Optical system design. Optical metrology systems. Distance measurement systems. Sampled imaging theory. Optics of electronic imaging systems.

Jessen, Poul S., Ph.D., University of Aarhus, 1993. Quantum state preparation. Coherent control. Quantum tunneling and transport phenomena in optical lattices. Quantum computation. Macroscopic superposition states and coherent evolution in dissipative quantum systems. Few-body quantum state preparation in optical lattices. Matter-wave equivalent of the laser/micromaser. Atom-optics in nano-fabrication. Laser cooling, trapping and manipulation of atoms and ions.

Khitrova, Galina, Ph.D., New York University, 1986. Nonlinear and quantum optics of semiconductor microcavities. Fundamental studies of quantum confined semiconductors.

Kostuk, Raymond K., Ph.D., Stanford University, 1986. Holographic techniques, systems, and materials. Ion exchange waveguide devices-interfacing to polymer and PBG layers. Fiber optic systems including OCDMA, error-correction codes, all-optical network issues. Medical imaging sensors including OCT and holographic filtering of coherent image data.

Mansuripur, Masud, Ph.D., Stanford University, 1981. Optical data storage. Magneto-optics. Optics of polarized light in systems of high numerical aperture. Magnetic and magneto-optical properties of thin solid films. Magnetization dynamics. Integrated optics for optical heads (data storage systems). Information theory. Optical signal processing. Biological data storage. Erbium-doped fiber amplifiers and lasers.

Marcellin, Michael W., Ph.D., Texas A&M, 1987. Digital communication and data storage systems, image and video compression, image processing, digital signal processing.

Mazumdar, Sumit, Ph.D., Princeton University, 1980. Linear and nonlinear optical properties of organic conjugated molecules and polymers. Effects of strong Coulomb correlations, excitons and multiexcitons in organic systems. Organic opto-electronic devices. Strong Coulomb interactions and broken symmetries. Charge and spin density waves and superconductivity in organic and inorganic materials with emphasis on organic charge-transfer solids and transition metal oxides.

Meystre, Pierre, Ph.D., Ecole Polytechnique Federale, 1974. Theoretical quantum optics. Statistical properties of radiation. Laser theory. Nonlinear optics. Atomic physics. Ultracold atoms. Bose condensation and atom lasers. Cavity quantum electrodynamics. Atom optics.

Miller, Joseph M., M.D., Northeastern Ohio Universities College of Medicine, 1985. The effect of astigmatism on visual development. Noninvasive assessment of buried optical elements.

Milster, Thomas, Ph.d., University of Arizona, 1987. Improving data density and signal readout quality in Optical storage systems. Design and implementation of Novel storage modalities, signal detection and components.

Moloney, Jerome, Ph.D., University of Western Ontario, 1977. Mathematical modeling and simulation of photonics systems. Fundamental theory of semiconductor lasers. Modeling high power femtosecond atmospheric light strings. Nonlinear theory of partial differential equations and chaos synchronization in extended complex spatiotemporal interacting systems. Algorithm development for large scale computational photonics systems simulations.

Neifeld, Mark A., Ph.D., California Institute of Technology 1991. Nontraditional imaging. Pattern recognition and neural networks. Parallel coding and signal processing. Volume optical storage. Multiple-quantum-well photonics.

Norwood, Robert, Ph.D., University of Pennsylvania, 1988. Electro-optic polymers and devices, photorefractive polymers, Sol-gels, materials for linear and nonlinear photonic crystals, Organic light emitting diodes, solar cells, and sensors.

Peyghambarian, Nasser, Ph.D., Indiana University, 1982. Optical telecommunication. Fiber optics. Fiber amplifiers and fiber lasers. Integrated optics. Femtosecond laser spectroscopy and dynamics of optical phenomena in semiconductors and organic materials. Nonlinear photonics and high speed optical switching. Characterization of optical materials in terms of speed and nonlinearities. Polymer optoelectronics, photorefractive polymers, organic light emitting diodes and lasers.

Sasian, José M., Ph.D., University of Arizona, 1988. Lens and mirror design, optical fabrication. Optomechanics. Illumination optics. Optical instrumentation for astronomy and biomedical sciences. Conformal optics. Microlithography. Novel optical systems.

Simmons, Joseph H., Ph.D., Catholic University of America, 1969. Quantum-size effects in the optical properties of semiconductor clusters. Optical properties and carrier dynamics in wide-gap semiconductors. Photosensitivity in glass films. Non-linear optical behavior of materials and glasses. Optical spectroscopy of materials at the nano-scale level. Molecular dynamics simulations. Non-linear viscous flow and rheological behavior of molten glasses.

Strickland, Robin, Ph.D., Sheffield University, 1979. Digital image processing, computer vision and signal processing.

Wright, Ewan M., Ph.D., Heriot-Watt University, 1983. Theory and simulation of light string propagation in air. Electromagnetic pulse emission from light string induced plasmas. Supercontinuum generation and self-guiding in condensed media. Bose-Einstein condensation (BEC) in atomic vapors. Mean-field theory and beyond for small BECs. Quantum theory of 1D gases in the Tonks-Girardeau regime. Theory of 1D atom waveguides and interferometers. Theory of light induced waveguides for cold atoms. Optically bound matter. Theory of anyon matter in planar chiral nanostructures.

Wyant, James C., Ph.D., University of Rochester, 1968. Implementation of microcomputers and software to interferometric techniques for optical measurements, in particular for optical testing. Testing of supersmooth optical surfaces. Measurement of optically rough surfaces. Testing of complex aspheric surfaces. Development of commercial optical test equipment based on phase-shifting interferometry.

Ziolkowski, Richard, Ph.D., University of Illinois at Urbana-Champaign, 1980. The application of new mathematical and numerical methods to linear and nonlinear problems dealing with the interaction of acoustic and electromagnetic waves with complex media, metamaterials, and realistic structures.

Associate Professors

Anderson, Brian P., Ph.D., Stanford University, 1999. Experimental Bose-Einstein condensation of dilute gases. Atom lasers. Atom optics. Superfluidity of Bose-Einstein-condensed gases. Vortices, solitons, and vortex rings in BEC. Josephson effects in BEC.

Barton, Jennifer, Ph.D., University of Texas at Austin, 1998. Optical imaging (optical coherence tomography). Laser-tissue Interaction and bioinstrumentation.

Hua, Hong, Ph.D., Beijing Institute of Technology, 1999. Development of 2D and 3D display systems, imaging systems, tracking systems, and interaction methods. Stereoscopic displays. Human-computer interface techniques. Virtual and augmented environments.

Kueppers, Franko, Ph.D., Dr-Ing University Kaiserlautern, 2002. High-speed TDM and high-capacity WDM fiber-optic transmission technologies. Photonic telecommunication systems. Optical transport networks.

Kupinski, Matthew, Ph.D., University of Chicago, 2000. Task based assessment of image quality for both tumor detection and parameter estimation tasks. Statistical characteristics of images and the the objects being imaged. Imaging hardware optimization. Human-observer models for image analysis.

Pau, Stanley, Ph.D., Stanford University, 1996. Micro-optics, MEMS/NEMS for imaging and sensing applications. Optical lithography and novel techniques for nanofabrication. Microfabricated neutral atom trap and ion trap for mass spectrometry and quantum computing. Microfluidic and microfabricated chemical reactor.

Potter, Kelly Simmons, Ph.D., University of Arizona, 1994. Elements for integrated optical systems using both optically active and passive novel photowritable materials. Examination of single and multi-photon processes leading to both linear and non-linear response in optical materials as a result of exposure to either ionizing or non-ionizing radiation. Research into the impact of defect physics on material optical behavior, waveguide device design and optical device performance is particularly emphasized.

Tyo, J. Scott, Ph.D., University of Pennsylvania, 1997. Enabling technologies for optical and microwave remote sensing: Processing of high-dimensional spectropolarimetric data, investigation of statistical properties of hyperspectral imagery, optimization of polarimetric sensors for remote sensing applications, integration of polarimetric and spectral sensors and information, use of spectropolarimetry to improve imaging in scattering media, fusion of multi-dimensional data into intelligible images, design of UWB antennas and antenna arrays, development of sensors, measurement techniques and processing strategies for UWB EM measurements, generation and radiation of electromagnetic transients, utilization of UWB radar for target detection and indentification.

Utzinger, Urs, Ph.D, Federal Institute of Technology, 1995. Optical tissue spectroscopy, optical biosignatures, and bioinstrumentation.

Assistant Professors

Jones, Ronald Jason, Ph.D., University of New Mexico, 2001. Ultrafast laser science, femtosecond frequency combs, extreme nonlinear light/matter interactions and generation, optical frequency metrology, high-resolution spectroscopy.

Peng, Leilei, Ph.D., Purdue University, 2005. High-speed Fourier transform fluorescence spectrometer for hyperspectral imaging.

Research Professors

Biggar, Stuart F., Ph.D., University of Arizona, 1990. Spacecraft and aircraft optical sensors. Optical system design, evaluation and absolute radiometric calibration. Reflectance measurement.

Creath, Katherine, Ph.D., University of Arizona, 1985 and, 2002. Interference microscopy. Interferometric metrology. Optical testing. Optical instrumentation. Image and signal processing. Non-destructive testing and speckle.

Howe, Dennis, Ph.D., University of Rochester, 1975. Application of multi-level modulation to digital optical recording. Finite-element time-domain modeling of super-resolution phenomena in phase change optical discs (Super-RENS). Optical data storage. Estimation/prediction of recovered data reliability and design of new techniques for error control. Design of new recording data formats. Optical (Fourier) filtering and super-resolution.

Associate Research Professor

Dubin, Matthew, Ph.D., University of Arizona, 2002. Ooptical Systems and Engineering, Innovation, design, prototyping and use of systems involving interferometry, alignment, imaging, opto-mechanical design, display design and illumination combining conceptual and analytical skills with system design and practical implementation. Inventor on 21 patents.

Tyler, David, Ph.D., University of New Mexico, 2000. Inverse problems in imaging astronomical imaging techniques and instrumentation, including adaptive optics, imaging and non-imaging measurements of Earth-orbiting satellites.

Assistant Research Professor

Czapla-Myers, Jeffrey, Ph.D., University of Arizona, 2006. Pre-flight and post-launch radiometric calibration of airborne and satellite sensors. Design and testing of field and laboratory radiometers.

RESEARCH SPECIALTIES AND STAFF

Fiber Optics. Chipman, Kueppers, Peyghambarian.

Lasers and Advanced Optical Materials. Armstrong, Fallahi, Mathine, Mazumdar, Peyghambarian.

Medical Optics and Image Science. Barrett, Barton, Chipman. Dallas, Descour, Gmitro, Kupinski, Miller, Peyghambarian, Schwiegerling, Shoemaker.

Nanotechnology. Fallahi, Mansuripur, Peyghambarian, Sarid.

Optical Data Storage. Howe, Kost, Mansuripur, Milster, Neifeld.

Optical Engineering and Testing. Chipman, Dereniak, Greivenkamp, Hua, Jacobs, Sasian, Wyant.

Optical Design and Fabrication. Angel, Burge, Descour, Greivenkamp, Sasian.

Optoelectronic Devices. Fallahi, Peyghambarian.

Quantum Nano-Optics of Semiconductors. Gibbs, Khitrova, Simmons.

Quantum Optics. Anderson, Jessen, Meystre, Wright.

Remote Sensing. Biggar.

Semiconductor Optical Physics. Binder, Moloney, Schulzgen.

Telecommunications. Honkanen, Kostuk, Kueppers, Peyghambarian.

Thin Films. Falco, Peyghambarian.

FACULTY PUBLICATIONS

For a complete listing of College of Optical Sciences faculty publications, please visit the Web site at http://www.optics.arizona.edu

UNIVERSITY OF ARIZONA

DEPARTMENT OF PLANETARY SCIENCES/ LUNAR AND PLANETARY LABORATORY

Tucson, Arizona 85721

Students Accepted For Degree	FIELDS		
	Physics	Astronomy	Related Fields
Doctorate			X
Master's			X

1. General

President: Robert N. Shelton
Dean of Graduate School: Andrew Comrie
Department Head: Michael J. Drake
Department Telephone Number: (520) 621-6963 (C)
Type of Institution: University
Control: Public
Setting: Urban
Total Faculty: 2,147
Total Graduate Faculty: No formal graduate faculty
Total Students: 38,765
Total Graduate Students: 8,419
Annual Graduate Tuition: †
In-state residents: None
**Out-of-state residents*: Full-time—$22,557.12/yr.
Part-time—$839.00/credit hr.
Tuition rates for: 2009–10
Deferred tuition plan: No
***Other Fees:* Registration: Full-time—$6,564.00/yr.
Part-time—$343.00/credit hr.
Arizona Financial Aid Trust Fund: 1–6 units—$15.00/sem.
7 or more units—$28.50/sem.
Recreation Center Fee: 1–3 units—optional
4 or more units—$25/sem.
Term: Semester

†See Graduate Catalog for more detailed information.
*Nonresident tuition is waived for students with Graduate Assistantships.
**In-state and out-of-state residents.

2. Number of Faculty in Department

The combined total of full-time faculty in the three professorial ranks is 32. The combined total of full-time, part-time, and other faculty at all ranks is 37.

3. Admission, Financial Aid, and Housing

Address admission inquiries to: Chairman, Admissions Committee, Department of Planetary Sciences, Kuiper Space Sciences Building, Room 325
email: acad_info@lpl.arizona.edu
URL http://www.lpl.arizona.edu
Graduate application fee required: $75
Admission deadline (Fall admission): 1/15
Admission information: For fall admission, 2010–11, 12 students were accepted from 86 applicants.
Admission requirements: For admission to the graduate programs, a Bachelor's degree in Chemistry, Astronomy, Earth Sciences, Mathematics, Atmospheric Science, Physics or related area, is normally required with a GPA of at least 3.0. A strong background in the physical sciences is essential. The GRE is required of all applicants. No minimum acceptable score is specified. The GRE Advanced is not required in a physical science or other relevant area. No minimum acceptable score is specified. Students from non-English speaking countries are required to demonstrate proficiency in English via the TOEFL exam. Minimum acceptable score for admission is 550.
Address financial aid inquiries to: Admissions Chairman, Dept. of Planetary Sciences, Kuiper Space Sciences Bldg., Room 325.
GAPSFAS application required: No
Financial aid deadline: 1/15
Loans available: Yes
Address housing inquiries to: Dept. of Student Housing, Slonaker House (single graduate students).
On-campus, single student housing available: Yes
Cost/term: Varies*
On-campus, married student housing available: No

*For rates, check with Dept. of Student Housing.

Table A—Faculty, Enrollments, and Degrees Granted

Research Specialty	2009–2010 Faculty	Enrollment[1] Fall 2009		No. of Degrees Granted[2] 2009–2010 (2004–09)			Median No. of Years for 2009–10 Ph.D.'s
		Master's	Doctorate	Master's	Terminal Master's	Doctorate	
Astronomy	1	0	0	0	0	(0)	–
Astrophysics	4	0	2	0	0	(1)	–
Astrobiology	0	1	1	0	0	1	5.75
Atmospheric Physics	1	0	0	0	0	0	–
Atmospheric Sciences	1	0	0	0	0	0	–
Celestial Mechanics	2	0	1	0	0	0	–
Cosmochemistry	6	0	4	0	(1)	(2)	–
Geochemistry	0	0	0	0	0	(1)	–
Geology	4	0	0	0	0	1	4.97
Geophysics	0	0	1	0	0	(2)	–
Kuiper Belt Objects	0	0	0	0	0	0	–
Planetary Atmospheres	6	0	3	0	0	1(1)	5.4
Planetary Geophysics	1	0	2	0	0	0	–
Planetary Physics	3	0	0	0	0	0	–
Planetary Sciences	6	1	12	0	(4)	3(13)	4.74
Planetary Spectroscopy	0	0	0	0	0	(1)	–
Remote Sensing	0	0	1	0	0	(1)	–
Science Education	0	0	1	(1)	0	(1)	–
Space Physics	1	0	0	0	0	(1)	–
Solar Physics	1	0	1	1	0	0	–
Total	37	2	29	(1)	(5)	6(24)	–
Full-time Grad. Stud.		1	29				
Part-time Grad. Stud.		1	0				
First-year Grad. Stud.		0	8				
Median Years in Grad. Study (2009–2010 Degrees)							5.05

Undergraduate Degrees, 2009–10 (2004–09): N/A

[1]Students not yet committed to a research specialty are entered under non-specialized.
[2]Five-year totals in parentheses.
AIP '08-A

4. Graduate Degree Requirements

Master's: A minimum of 30 hours of graduate credit is required for the Master's degree, including a minimum of 15 credit hours selected from the Planetary Sciences core curriculum. Selected courses are to include at least 1 course in each of the core areas—physics, chemistry, and geosciences. The remaining 15 hours are to be selected by agreement between the student and his or her advisor. The student must maintain a B average in all core courses. See graduate catalog for additional information.

Doctorate: All students must complete the 21-unit core program consisting of 505a-505b, 510A, 510B, 512, 517, and 554 (though exceptionally well-prepared students may have parts of this requirement waved). An additional minimum of 18 units must be completed in a specialized area of planetary sciences. A specified reading competence in a modern foreign language is required. Students are expected to complete all requirements for the degree within three or four years following successful completion of the comprehensive examination, which should be taken within three years after enrolling in the program.

The Department of Planetary Sciences' degree programs are conducted in collaboration with the research programs of the Lunar and Planetary Laboratory (LPL). Together, the department and laboratory form an institute uncommonly broad and complete in its approach to planetary science education and research. The department and laboratory participate in many NASA space science missions. Among the missions in which the faculty have participated or are participating in are the Voyager Mission, the Galileo Mission to Jupiter, the Cassini/Huygens Mission to Saturn, the Mars Pathfinder, Near Earth Asteroid Rendezvous, Discovery Missions, NASA Space Shuttle Missions, HiRISE and Phoenix Missions to Mars, and the Ulysses Heliospheric Probe. In addition, LPL scientists make use of Earth orbiting observatories, including the Hubble Space Telescope, the Infrared Space Observatory, and the Ultraviolet Explorer. The laboratory's Space Imagery Center contains one of the most extensive collections of planetary images in the word, beginning with those obtained from the earliest space projects and continuing to most current missions.

LPL's Planetary Imaging Research Laboratory is a modern image processing facility for the analysis of planetary and astronomical data. Also available for student research are cosmochemistry and geochemistry laboratories, including a scanning electron microscope and microprobe facility, an experimental petrology laboratory, a radiochemistry separation and neutron activation laboratory, and a noble gas mass spectrometry laboratory. The numerous telescopes of The University of Arizona Observatories are available for research projects, including instruments on Kitt Peak and in the Santa Catalina Mountains, as well as the Multiple Mirror Telescope on Mt. Hopkins; all are within easy reach of the University campus. Laboratory staff and students also make use of major observatories around the world, including the NASA Infrared Telescope Facility on Mauna Kea, Hawaii. The University is continuing to develop a new observatory site on Mt. Graham, northeast of Tucson. The department participates in interdepartmental programs in theoretical astrophysics and in applied mathematics.

The University's computer center is available to support educational and research activities. The Lunar and Planetary Laboratory maintains a variety of networked computers and workstations in support of the research and educational programs.

Special Equipment, Facilities, or Programs: The Lunar and Planetary Laboratory (LPL) and the Department of Planetary Sciences function as a single unit to carry out solar system research and education. The department and laboratory are housed in the Gerard P. Kuiper Space Sciences Building, the C.P. Sonett Building, and the Phoenix Mission Building. Neighboring facilities include the Tucson headquarters of the National Optical Astronomy Observatory, the National Radio Astronomy Observatory, Steward Observatory, the Optical Sciences Center, the Flandrau Planetarium, and the Planetary Sciences Institute.

The facilities of the University of Arizona Observatories are available to researchers in the LPL. These instruments include the multiple-mirror telescope (six 1.8-m mirrors) on Mt. Hopkins, the 2.3-m telescope on Kitt Peak, two 1.5-m telescopes, and a 1-m telescope on Mt. Lemmon, and several smaller telescopes. For cosmochemical research, the LPL operates a scanning electron microscope and a Cameca XF50 microprobe, high-temperature furnaces for rock melting experiments, and a radiochemistry separation facility for neutron activation analysis. These facilities are used for studying meteorites, lunar samples, and terrestrial analogs. Also available at LPL are a well-equipped electronics shop, a machine-shop, and darkroom facilities.

The Space Imagery Center at the LPL is one of several regional facilities supported by NASA as a repository for images of planets and satellites obtained by spacecraft, as well as for topographical and geologic maps produced from such imagery. The Laboratory maintains an extensive computer network, including central disk servers and numerous workstations. In addition, the various research groups maintain specialized computer systems for data taking and analysis and for theoretical computations. University central computing facilities include a wide variety of systems and network facilities as well as several Convex superminicomputers.

Thesis: Thesis may be written *in absentia*.

Table B—Appointments to Graduate Students, 2009–20010

Title of Appointee	Appointments: Total	Appointments: First year	Academic Load Allowed in Credit Hours	Hours of Service Per Week	Stipend for Academic Year ($)
			Semester		
Teaching Assistant	2	1		20	26,003.00
Research Assistant	3	1		20	26,003.00
Research Associate	8	0		20	28,736.00
NASA	13	2		0	24,000.00
NSF	1	0		0	21,500.00
Space Grant Fellowship	1	1		0	28,000.00
Carson Fellowship	1	1		0	26,005.00
Graduate College Fellow	1	1		0	10,833.00
Other	1	0		0	0
Total	31	7			

5. Personnel Engaged in Separately Budgeted Research, 7/08–6/09

Professorial faculty	32
Other faculty	5
Postdoctoral appointments	13
Graduate students	31
Undergraduate students	0
Nonteaching research personnel	7
Total	88

6. Separately Budgeted Research Expenditures by Source of Support

	Departmental Research	Physics-related Research Outside Department
Federal government	$20,207,489	$
Total	$20,207,489	$

Table C—Separately Budgeted Research Expenditures

Research Specialty	No. of Grants	Expenditures ($)
Astronomy	15	732,470
Atmos./Space Phys., Cosmic Rays	84	18,252,188
Science Education	1	575,000
Planetary Geology	7	647,831
Total	107	20,207,489

FACULTY

Professors

Baker, Victor, Ph.D., Colorado, 1971. Regents'. Planetary surfaces; geomorphology.
Boynton, William V., Ph.D., Carnegie-Mellon, 1971. Cosmochemistry, geochemistry.
Brown, Robert H., Ph.D., U. Hawaii, 1982. Planetary surface processes.
Drake, Michael J., Ph.D., Oregon, 1972. Regents' Professor, Head and Director of the Laboratory. Cosmochemistry, Geochemistry.
Fink, Uwe, Ph.D., Penn State, 1965. Emeritus. Planetary atmospheres, spectroscopy.
Gehrels, Tom, Ph.D., Chicago, 1956. Universal evolution.
Greenberg, Richard J., Ph.D., MIT, 1972. Planetary sciences.
Griffith, Caitlin, Ph.D., U. of New York-Stony Brook, 1991. Planetary atmospheres.
Hubbard, William B., Ph.D., California, Berkeley, 1967. Planetary interiors & atmospheres.
Hunten, Donald M., Ph.D., McGill, 1950. Regents', Emeritus. Earth and planetary atmospheres.
Jokipii, J. R., Ph.D., Cal. Tech., 1965. Regents'. Theoretical astrophysics; space physics.
Larson, Harold P., Ph.D., Purdue, 1967. Distinguished Professor, Emeritus. Planetary sciences.
Lewis, John L., Ph.D., California, San Diego, 1968. Emeritus. Cosmochemistry, Planetary atmospheres.
Lunine, Jonathan I., Ph.D., Cal. Tech., 1985. Planetary sciences and physics.
Malhotra, Renu, Ph.D., Cornell University, 1988. Solar system dynamics.
McEwen, Alfred, Ph.D., Arizona State University, 1988. Planetary Geology.
Melosh, H. Jay, Ph.D., Cal. Tech., 1973. Regents'. Theoretical geophysics; planetary surfaces.
Rieke, George H., Ph.D., Harvard, 1969. Gamma-ray, IR astronomy, cosmic radiation.
Roemer, Elizabeth, Ph.D., California, Berkeley, 1955. Emerita. Comets; minor planets; astrometry.
Smith, Mark, Ph.D., University of Colorado, 1982. Atmospheric chemistry; astrobiology.
Smith, Peter H., M.S., University of Arizona, 1977. Mars photography; planetary atmospheres.
Sonett, Charles P., Ph.D., UCLA, 1954. Regents'. Emeritus. Planetary physics.
Strom, Robert G., M.S., Stanford, 1957. Emeritus. Lunar and planetary surfaces..
Swindle, Timothy D., Ph.D., Washington Univ., 1986. Cosmochemistry.
Yelle, Roger, Ph.D., U. of Wisconsin-Madison, 1984. Planetary atmospheres.

Associate Professors

Giacalone, Joe, Ph.D., Kansas, 1991. Solar and heliospheric physics, astrophysics.
Kursinski, Jr., Emil R., Ph.D., Caltech, 1997. Atmosphysic sciences and planetary sciences.
Lauretta, Dante, Ph.D., Washington University, 1997. Cosmochemistry and planet formation.
Pelletier, Jon, Ph.D., Cornell, 1997. Geosciences, geography, regional development, and planetary sciences.
Showman, Adam, Ph.D., Cal. Tech, 1998. Dynamics and evolution of planetary atmospheres.

Assistant Professors

Byrne, Shane, Ph.D., Cal Tech. 2003. Planetary surfaces.
Rogers, Tamara, Ph.D., University of CA at Santa Cruz, 2006. Solar physics.

Senior Research Scientists

Holberg, Jay B., Ph.D., California, Berkeley, 1974. Planetary rings and atmospheres.
Hood, Lon L., Ph.D., UCLA, 1979. Stratospheric physics and planetary physics.
Kota, Jozsef, Ph.D., Budapest, Hungary, 1980. Theoretical space physics, space weather.
Sandel, Bill R., Ph.D., Rice, 1972. Space plasmas; planetary atmospheres.

Associate Research Scientist

McMillan, Robert S., Ph.D., Texas, 1977. Survey of the solar system for asteroids and comets.

FACULTY

Baker, Victor, R., Regents Professor. Planetary surfaces; geomorphology.
Boynton, William V., Professor. Cosmochemistry; geochemistry.
Brown, Robert H., Professor. Planetary surface processes.
Byrne, Shane, Assistant Professor. Planetary surfaces.
Drake, Michael J., Regents' Professor Head and Director. Cosmochemistry; geochemistry.
Fink, Uwe, Professor Emeritus. Planetary atmospheres; spectroscopy.

Gehrels, Tom, Professor. Universal evolution.
Giacalone, Joe, Associate Professor. Solar and heliospheric physics, astrophysics.
Greenberg, Richard J., Professor. Planetary sciences.
Griffith, Caitlin, Professor. Planetary atmospheres.
Holberg, Jay B., Senior Research Scientist. Planetary rings and atmospheres.
Hood, Lon L., Senior Research Scientist. Stratospheric physics & planetary physics.
Hubbard, William B., Professor. Planetary interiors & atmospheres.
Hunten, Donald M., Regents' Professor Emeritus, Earth and planetary atmospheres.
Jokipii, J. R., Regents' Professor. Theoretical astrophysics; space physics.
Kota, Jozsef, Senior Research Scientist. Theoretical space physic, space weather.
Kursinski, Jr., Emil R., Associate Professor. Atmospheric sciences and planetary sciences.
Larson, Harold P., University Distinguished Professor, Emeritus. Planetary sciences.
Lauretta, Dante, Associate Professor. Cosmochemistry and planet formation.
Lewis, John L., Professor Emeritus. Cosmochemistry planetary atmospheres.
Lunine, Jonathan I., Professor. Planetary sciences and physics.
Malhotra, Renu, Professor. Solar system dynamics.
McEwen, Alfred, Professor. Planetary geology.
McMillan, Robert S., Associate Research Scientist. Survey of the solar system for asteroids and comets.
Melosh, H. Jay, Regents Professor. Theoretical geophysics; planetary surfaces.
Pelletier, Jon, Associate Professor. Geosciences, geography, regional development, and planetary sciences.
Rieke, George H., Professor. Gamma-ray, IR astronomy; cosmic radiation.
Roemer, Elizabeth, Professor Emerita. Comets, minor planets; astrometry.
Rogers, Tamara, Assistant Professor. Solar physics.
Sandel, Bill R., Senior Research Scientist. Space plasmas, planetary astmospheres.
Showman, Adam, Associate Professor. Dynamics and evolution of planetary atmospheres.
Smith, Mark, Professor. Atmospheric chemistry; astrobiology.
Smith, Peter, Professor. Mars photography, planetary atmospheres.
Sonett, Charles P., Regents Professor Emeritus. Planetary physics.
Strom, Robert G., Professor Emeritus. Lunar and planetary surfaces.
Swindle, Timothy D., Professor. Cosmochemistry.
Yelle, Roger, Professor. Planetary Atmospheres.

RESEARCH SPECIALTIES AND STAFF

Theoretical

Astrophysics. Gehrels, Giacalone, Jokipii (2 postdoc).
Atmospheric Sciences. Kursinski.
Celestial Mechanics. Greenberg, Malhotra.
Cosmochemistry. Lewis.
Planetary Atmospheres. Showman (1 postdoc).
Planetary Geophysics. Melosh.
Planetary Sciences. Burne.
Solar Physics. Rogers.
Space Physics. Kota.

Experimental

Astronomy. Roemer.
Astrophyics. Rieke.
Cosmochemistry. M. Smith.
Geology. McEwen, Pelletier (3 postdoc).
Planetary Atmospheres. Fink, Holberg, Hunten.
Planetary Physics. Sandel, Sonett.
Planetary Science. Larson, McMillan, P. Smith.

Theoretical and Experimental

Atmospheric Physics. Hood.
Cosmochemistry. Boynton, Drake, Lauretta, Swindle (2 postdocs).
Geology. Baker, Strom.
Planetary Atmospheres. Griffith, Yelle (2 postdoc).
Planetary Physics. Hubbard.
Planetary Sciences. Brown, Lunine (3 postdocs).

FACULTY PUBLICATIONS

Baker, Victor

J. M. Dohm, H. Miyamoto, G. G. Ori., G. Komatsu, M. Pondrelli, K. J. Kim, R. C. Anderson, A. G. Fairen, T. M. Hare, P. Williams, and 15 coauthors (2010). Linkage Among Geology, Hydrology, Climate, and Life on Earch Point to Possible Life-Containing Environments on Mars.

G. Komatsu, G. G. Ori, M. Cardinale, J. M. Dohn, V. R. Baker, D. A. Vaz, R. Ishimaru, N. Namiki, and T. Matsui (2010). The Search for Methane Gas Emissions Features on Mars.

M. E. Banks, Nicholas P., Lang, Jeffrey S. Kargel, Alfred S. McEwen, Victor R. Baker, John A. Grant, Jon D. Pelletier, and Robert G. Strom (2009). An Analysis of Sinuous Ridges in the Southern Argyre Planitia, Mars Using HiRISE and CTS Images and MOLA Data.

James M. Dohm, Robert C. Anderson, Jean-Piere Williams, Javier Ruiz, Patrick C. McGuire, Debra L. Buczkowski, Ruye Wang, Lucas Scharenbroich, Trent M. Hare, J. E. P. Connerney, and 4 coauthors (2009). Claritas Rise, mars: Pre-Tharsis Magmatism?

James M. Dohm, Jean-Pierre Williams, Robert C. Anderson, Javier Ruiz, Patrick C. McGuire, Goro Komatsu, Alfonso F. Davila, Justin C. Ferris, Dirk Schulze-Makuch, Victor R. Baker, and 6 coauthors (2009). New Evidence for a Magmatic Influence on the Origin of Valles Marineris Mars.

Boynton, William V.

K. Hurley, A. Rowlinson, E. Bellm, D. Perley, I. G. Mitrofanov, D. V. Golovin, A. S. Kozyrev, M. L. Litvak, A. B. Sanin, W. Boynton, and 27 coauthors (2010). A New Analysis of the Short-duration, Hard-spectrum GRB 151103, a Possible Extragalactic Soft Gamma Repeater Giant Flare.

I. G. Mitrofanov, A. Bartels, Y. I. Bobrovnsitsky, W. Boynton, G. Chin, H. Enos, L. Evans, S. Floyd, J. Garvin, D. V. Golovin, and 26 coauthors (2009). Lunar Exploration Neutron Detector for the NASA Lunar Reconnaissance Orbiter.

W. V. Boynton, J. Taylor, S. M. McLennan, A. L. Sprague, and H. E. Newsom (2009). Elemental Compositions of the Martian Surface and Atmosphere and Their Influence on Future Mars Sciences.

M. R. El Maarry, O. Gasnault, M. J. Toplis, D. Baratoux, J. M. Dohm, H. E. Newsom, W. V. Boynton, and S. Karunatillake (2009). Gamma-ray Constraints on the Chemical

Composition of the Martian Surface in the Tharsis Region: A Signature of Partial Melting of the Mantle?

James M. Dohm, Jean-Pierre Williams, Robert C. Anderson, Javier Ruiz, Patrick C. McGuire, Goro Komatsu, Alfonso F. Davila, Justin C. Ferris, Dirk Schulze-Makuch, Victor R. Baker, and 6 coauthors (2009). New Evidence of a Magmatic Influence on the Origin of Valles Marineris, Mars. Robert Brown

M. M. Hedman, P. D. Nicholson, K. H. Baines, B. J. Buratti, C. Sotin, R. N. Clark, R. H. Brown, R. G. French, and E. A. Marouf (2010). The Architecture of the Cassini Division.

Marie Levine, D. Lisman, S. Shaklan, J. Kasting, W. Traub, J. Alexander, R. Angel, C. Blaurock, M. Brown, R. Brown, and 45 coauthors (2009). Terrestrial Planet Finder Coronagraph (TPF-C) light Baseline Concept.

Sebastien Rodriguez, Stephane Le Mouelic, Pascal Rannou, Gabriel Tobie, Kevin H. Baines, Jason W. Barnes, Caitlin A. Griffith, Mathieu Hirtzig, Karly M. Pitman, M. Karly, Christophe Sotin, and 4 coauthors (2009). Global Circulation as the Main Source of Cloud Activity of Titan.

Jason W. Barnes, Robert H. Brown, Jason M. Soderblom, Jason M. Soderblom, Laurence A. Solderblom, Ralf Jaumann, Brian Jackson, Stephanie Le Mouelic, Christophe Sotin, Bonnie J. Buratti, Karley M. Pitman, and 5 coauthors (2009). Shoreline Features of Titan's Ontario Lacus from Cassini/VIMS Observations.

D. S. Choi, A. P. Showman, and R. H. Brown (2009). Cloud Features and Zonal Wind Measurements of Saturn's as Observed by Cassini/VIMS.

A. Coustenis, S. K. Atreya, T. Balint, R. H. Brown, M. K. Dougherty, F. Ferri, M. Fulchignoni, D. Gautier, R. A. Gowen, C. A. Griffith, and 146 coauthors (2009). TandEM: Titan and Enceladus Mission.

Brown, Robert

M. M. Hedman, P. D. Nicholson, K. H. Baines, B. J. Buratti, C. Sotin, R. N. Clark, R. H. Brown, R. G. French, and E. A. Marouf (2010). The architecture of the Cassini Division.

Byrne, Shayne

Colin M. Dundas and Shane Byrne (2010). Modeling Sublimation of Ice Exposed by New Impacts in the Martian Mid-latitudes.

Colin M. Dundas, Alfred McEwen, Serina, Diniega, Shane Byrne, and Sara Martinez-Alonso (2010). New and Recent Gully Activity of Mars as Seen by HiRise.

Kathryn E. Fishbauch, Christine S. Hvidberg, Shane Byrne, Patrick S. Russell, Kenneth E. Herkenhoff, Mai Winstrup, and Randolph Kirk (2010). First High-Resolution Stratigraphic Column of the Martian North Polar Layered Deposits.

P. S. Russell, S. Byrne, and C. J. Hansen (2010). Active Mass Wasting of Ice Layers and Seasonal CO2 Frost in the North Polar Region of Mars.

J. W. Holt, S. Byrne, K. Fishbaugh, S. Christian, N. E. Putzig, R. J. Phillips, and K. Tanaka (2010). Chasma Boreale, mars: A Product of Non-Uniform Polar Accumulation Influenced by Basal Topography.

M. R. Koutnik, D. P. Winebrenner, E. D. Waddington, A. V. Pathare, and S. Byrne (2010). Equilibration Timescales of Ice Flow on Gemina Lingula Indicate Enhanced Flow at Low Temperatures.

Drake, Michael

M. Stimpfl, K. Muralidharan, N. H. de Leeuw, K. Runge, P. A. Deymier, and M. J. Drake (2010). Atomistic Simulations of Adsorption of water onto Foresterite and Fayalite Planar Surfaces: Implication ofr the Origin of water in the Inner Solar System.

J. P. Emery, Y. R. Fernandez, . S. Kelley, C. Hergenrother, J. Ziffer, D. S. Lauretta, M. J. Drake, and H. Campins (2010). Thermophysical Chracterization of Potential Spacecraft Target (101955) 1999 RQ36.

E. Hill, K. Domanik, and M. J. Drake (2010). Metal/Silicate Partitioning of the Moderately Siderophile Elements: The Effect of Temperature and C Concentration.

E. Hill, G. R. Huss, K. Domanik, and M. J. Drake (2010). Subsolids Metal-Olivine Trace Element Partitioning.

E. Hill, K. Domanik, and M. J. Drake (2010). Metal/Silicate Partitioning of Mo, W and V: Effects of T, P, X and fO2.

Fink, Uwe

A. Coradini, D. Grassi, F. Cappaccioni, G. Filacchione, F. Tosi, E. Ammannito, M. C. De Sanctis, V. Formisano, P. Wolkenberg, G. Rinaldi, and 39 coauthors (2009). Martian Atmosphere as Observed by VIRTIS-M on Rosetta Spacecraft.

U. Fink (2009). A Taxonomic Survey of Comet Composition 1985-2004 using CCD spectroscopy.

U. Fink (2009). How Tempel 1 Fits into the Ensemble of Comets: A Spectroscopic Perspective.

R. Linser, U. Fink, B. Reif (2008). Proton-Detected Scalar Coupling Based Assignment Strategies in MAS Solid-State NMR Spectroscopy Applied to Perdeuterated Proteins.

Gehrels, Tom

Tom Gehrels (2009). The Multiverse Origin of Our Physics Does Without Strings, Big Band, Inflation, or Parallel Universes.

Tom Gehrels (2009). The Cosmological Foundation of Our World, See in a Revised History of Our Universe.

Tom Gehrels (2009). 2009 HB=1994 VVI.

Tom Gehrels (2007). The Multiverse and the Origin of Our Universe.

Tom Gehrels (2007). Universes Seen by a Chandrasekhar Equation in Stellar Physics.

Giacalone, Joe

Fan Guo and Joe Giacalone (2010). The Effect of Large Scale Magnetic Turbulence on the Acceleration of Electrons by Perpendicular Collisionless Shocks.

Eileen Chollet, Ruth Skoug, John Steinberg, Nancy Crooker, and Joe Giacalone (2010). Reconnection and Disconnection: Observations of Suprathermal Electron Heat Flux Dropouts.

Marcia Neugebauer and Joe Giacalone (2010). Progress int eh Study of Interplanetary Discontinuities.

Joe Giacalone and Rob Decker (2010). The Origin of Low-energy Anomalous Cosmic Rays at the Solar-wind Termination Shock.

Fan Guo and Joe Giacalone (2009). The Effect of Large Scale Magnetic Turbulence on the Acceleration of Electrons by Perpendicular Collisionless Shocks.

Joe Giacalone and R. B. Decker (2009). The Origin of Low Energy Anomalous Cosmic Rays at the Solar-Wind Termination Shock.

Greenberg, Richard J.

T. A. Hurford, B. G. Bills, P. Helfenstein, R. Greenberg, G. V. Hoppa, and D. P. Hamilton (2009). Geological Implications of a Physical Libration on Enceladus.

Rory Barnes, Brian Jackson, Richard Greenberg, and Sean N. Raymond (2000). Tidal Limits to Planetary Habitability.

Richard Greenberg (2009). Frequency Dependence of Tidal q.

Brian Jackson, Rory Barnes, and Richard Greenberg (2009). Observational Evidence for Tidal Destruction of Exoplanets.

S. N. Raymond, R. Barnes, D. Veras, P. J. Armitage, N. Gorelick, R. Greenberg (2009). Planet-Planet Scattering Leads to Tightly Packed Planetary Systems.

Griffith, Caitlin

Nancy Elias-Rosa, Schuyler Van Dyk, Weidong Li, Adam A. Miller, Jeffrey M. Silverman, Mohan, Ganeshalingam, Andrew F. Boden, Mansi M. Kasliwal, Jozsef Vinko, Jean-Charles Cuillandre, and 6 co-authors (2010). The Massince Progenitor of the Type II-linear Supernova 2009kr.

Katrin Stephan, Ralf Jaumann, Robert H. Brown, Jason M. Soderblom, Lawrence A. Soderblom, Jason W. Barnes, Christophe Sotin, Caitlin A. Griffith, Randolph L. Kirk, Kevin H. Baines and 5 coauthors (2010). Specular Reflection on Tital: Liquids in Kraken Mare.

J. M. Silverman, I. K. W. Kleiser, C. V. Griffith, and A. V. Filippenko (2010). Spectroscopic Identification of CSS100409:154258+021653.

G. Tinetti, P. Deroo, M. R. Swain, C. A. Griffith, G. Vasisht, L. R. Brown, C. Burke, and P. McCuulough (2010). Probing the Terminator Region Atmosphere of the Hot-Jupiter XO-1b with Transmission Spectroscopy.

K. Stephan, R. Jaumann, R. H. Brown, J. M. Soderblom, L. A. Soderblom, J. W. Barners, C. Sotin, C. A. Griffith, R. L. Kirk, K. H. Baines, and 5 coauthors (2010). Detection of a Specular Reflection of Titan by Cassini-VIMS.

Holberg, Jay B.

J. B. Holberg (2010). Sirius B and the Measurement of the Graviational Redshift.

Tuguldur Sukhbold, J. Holberg, and S. Howell (2010). Identifying White Dwarfs in the KELPER Field.

Kenneth J. Mighell, S. Howell, J. Holberg, and W. Sherry (2010). A Calibration Study of Variable Stars in the Kepler Field.

Kevin R. Covey, A. Saha, T. C. Beers, J. J. Bochanski, P. Boeshaar, A. Burgasser, P. Cargile, Y. Chu, C. Claver, K. Cook, and 23 coauthors (2010). Stellar Population Science with LSST.

Jay B. Holberg (2010). Open Questions Regarding the 1925 Measurement of the Gravitational Redshift of Sirius B.

Bradley M. Hansen, W. Hartkopf, J. Holberg, W. C. Jao and S. Ridgway (2010). Stellar Astrophysics with SIM.

Hood, Lon L.

L. L. Hood, W. S. Kiefer, and B. Langlais (2010). Parallel Modeling of the Apollinaris Patera Magnetic and Gravity Anomalies.

L. L. Hood (2010). Central Magnetic Anomalies of Nectarian-ages Lunar Impact Basins: Possible Evidence for an Early Core Dynamo.

L. L. Hood and J. S. Halekas (2010). Lunar Surfaces Magnetometer for Crustal Paleomagnetism and Deep Electromagnetic Sounding Applications.

L. L. Hood (2009). Lunar Magnetism.

L. L. Hood, F. J. Ciesla, N. A. Artemieva, F. Marzari, and S. J. Weidenschilling (2009). Nebular Shock Waves Generated by Planetesimals Passing Through Jovian Resonances: Possible Sites for Chondrule Formation.

Hubbard, William B.

Yohai Kaspi, William B. Hubbard, Adam P. Showman, and Glenn R. Flierl (2010).Gravitational Signature of Juipiter's Internal Dynamics.

W. B. Hubbard, D. W. McCarthy, C. A. Kulesa, S. D. Benecchi, M. J. Person, J. L. Elliot, and A. A. S. Gulbis (2009). Buoyancy Waves in Pluto's High Atmosphere: Implications for Stellar Occultations.

Jonathan J. Fortney, Kevi, Zahnle, Isabelle Baraffe, Adam Burrows, Sarah E. Dodson-Robinson, Gilles Chabrier, Tristan Guillot, Ravit Helled, Franck Hersant, William B. Hubbard, and 2 coauthors (2009). Planetary Formation and Evolution Revealed with a Saturn Entry Probe: The Importance of Noble Gases.

Jason W. Barnes, Curtis S. Cooper, Adam P. Showman, and William B. Hubbard (2009). The Gravitational Signature of Jupiter's Internal Dynamics.

B. Militzer and W. B. Hubbard (2009). Comparison of Jupiter Interior Models Derived from First - principles Simulations.

Hunten, Donald M.

E. Chassefiere, J.-L. Maria, J.-P. Goutail, E. Quemerais, F. LeBlanc, S. Okano, I. Yoshikawa, O. Korablev, V. Gnedykh, G. Naletto, and 42 coauthors (2010). PHEBUS: A Double Ultraviolet Spectrometer of Observe Mercury's Exosphere.

M. H. Wong (2009). Comment on "Transport of Nonmethane Hydrocarbons to Jupiter's Troposphere by descent of smog particles" by Donald M. Hunten (Icarus 194 (2008) 616 622).

D. M. Hunten (2008). Nelson Spencer.

D. M. Hunten (2008). Transport of Nonmethane Hydrocarbons to Jupiter's Troposphere by Descent of Smog Particles.

Jokipii, J. R.

V. Florinski, A. Balogh, J. R. Jokipii, D. J. McComas, M. Opher, N. V. Pogorelov, J. D. Richardson, E. C. Stone, and B. E. Wood (2009). The Dynamic Heliosphere, Outstanding Issues. Report of Working Groups 4 and 6.

K. C. Hsieh, P. C. Frisch, J. Giacalone, J. R. Jokippi, J. Kota, D. E. Larson R. P. Lin, J. G. Luhmann, and Linghau Wang (2009). A Re-Pinterpretation of STEREO/STE Observations and Its Consequences.

J. R. Jokipii and J. Kota (2008). Particle Acceleration at Shocks: Effects of Spatial Variations Along the Shock Face.

J. R. Jokipii (2008). Solar System: A Shock for Voyager 2

J. R. Jokipii (2008). Acceleration and Transport of Energetic Particles Observed in the Inner Heliosphere.

Kota, Jozsef

K. Munakata, Y. Mizoguchi, C. Kato, S. yasue, S. Mori, M. Takita, and J. Kota (2010). Solar Cycle Dependence of the Diurnal Anisotropy of 0.6 TeV Cosmic-ray Intensity Observed with the Matshushiro Underground Muon Detector.

A. Fushishta, Y. Okazaki, T. Naumi, C. Kato, S. Yasue, T. Kuwabara, J. W. Bieber, P. Evenson, M. R. Da Silva, A. Dal Lago, and 6 coauthors (2009). Drift Effects and the Average Features of Cosmic Ray Density Gradient in CIRs During Successive Two Solar Minimum Periods.

J. Kota, J. R. Jokipii, and J. Giacalone (2009). Implications of Voyager ACR Observations on the Injection Rate Along the Termination Shock.

K. Hsieh, A. Czechowski, M. Hilchenback, S. Grzedzielski, and J. Kota (2009). Further Estimation of the Thickness of the Heliosheath Using Updated SOHO Data.

Y. Mizoguchi, K. Munakata, M. Takita, and J. Kota (2009).

The Sidereal Anistoropy of Multi-TeV Cosmic Rays in an Expanding local Interstellar Cloud.

Kursinski, Emil R.

E. R. Kursinski, J. Lyons, C. Newman, M. I. Richardson, D. Ward, and A. C. Otarola (2009). A Global Observing System for Mars: The Dual Satellite Mars Astrobiology and Climate Observatory (MACO).

D. Ward, E. R. Kursinski, A. C. Otarola, K. G. Sammler, and M. Stovern (2009).Development of the ATOMMS Next Generation Satellite to Satellite Occultation for Global Monitoring of Climate.

A. L. Kursinksi, E. R. Kursinski, and C. O. Ao (2009). Water Vapor and OLR Feedbacks over the Recent ENSO Cycle.

K. G. Sammler, E. Kursinski, and D. Ward (2009). The Accuracy of Profiling Ozone via Radio Occultation.

F. Xie, D. L. Wu, C. O. Ao, E. Kursinski, A. Mannucci, and S. Syndergaard (2009). Profiling Stratocumulus-topped Boundary Layers with GPS Radio Occultation.

Larson, Harold P.

D. W. Koerner, S. Kim, D. E. Trilling, H. Larson, A. Cotera, K. R. Stapelfeldt, Z. Wahhaj, S. Fajardo-Acosta, D. Padgett, and D. Backman (2010). New Debris Disk Candidates Around 49 Nearby Stars.

E. F. Erickson, L. J. Allamandola, J.-P. Baluteaun, E. E. Becklin, G. Bjoraker, B. Gordon, C. Michael, J. Lawrence, C. Ceccarelli, E. B. Churchwell, D. P. Clemens, and 40 coauthors (2009). Training of Instrumentalists and Development of New Technologies on SOFIA.

Lauretta, Dante

Jade C. Bond, David P. O'Brien, and Dante S. Lauretta (2010). The Compositional Diversity of Extrasolar Terrestrial Planets: In-Situ Simulations.

J. P. Emery, Y. R. Fernandez, M. S. Kelley, C. Hergenrother, J. Ziffer, D. S. Lauretta, M. Drake, and H. Campins (2010). Thermophysical Characterization of Potential Spacecraft Target (101955) 1999 RQ36.

E. E. Palmer and D. S. Lauretta (2010). A Kamacite Alteration Index for CM Chondrites.

D. S. Lauretta. Trace Element Distributions in the Fukang Pallasite.

D. L. Schrader, H. C. Connolly, and D. S. Lauretta. On the Nebular and Aqueous Signatures in the CR Chondrites.

E. L. Berger, T. J. Zega, and D. S. Lauretta (2010). Microstructures of CI-Chondrite Pyrrhotite and Cubanite.

Lewis, John S.

R. Acquafredda, T. Adam, N. Agafonova, Sanchez P. Alvarez, M. Ambrosio, A. Anokhina, S. Aoki, A. Ariga, T. Ariga, L. Arrabiot, and 252 coauthors (2009). The OPERA Experiment in the CERN to Gran Sasso Neutrino Beam.

John Lewis, Kaj Nystrom, and Pietro Poggi-Corradini (2009). p Harmonic Measure in Simply Connected Domains.

Lunine, Jonathan I.

Jean Schnedier, Alain, Leger, Malcolm Fridlund, Glen J. White, Carlos Eiroa, Thomas Henning, Tom Herbst, Helmut Lammer, Rene Liseau, Fancesco, and 10 coauthors (2010). The Far Future of Exoplanet Direct Characterization.

Lisa Katltenegger, Carlos Eiroa, Ignasi Ribas, Francesco Paresce, Martin Leitzinger, Petra Odert, Arnondl Hanslmeier, Malcolm Fridlund, Helmut Lammer, Charles Beichman, and 14 coauthors (2010). Stellar Aspects of Habitability - Characterizing Target Stars for Terrestrial Planet-Finding Missions.

Lisa Kaltenegger, Frank Selsis, Malcolm Fridlund, Helmut Lammer, Charles Beichman, William Danchi, Carlos Eiroa, Thomas Henning, Tom Herbst, Alain Leger, and 10 coauthors (2010). Deciphering Spectral Fingerprints of Habitable Exoplanets.

O. Aharonson, A. G. Hayes, J. I. Luinine, R. D. Lorenz, M. D. Allison, and C. Elachi (2009). An Asymmetric Distribution of Lakes on Titan As a Possible Concequence of Oribital Forcing.

Julie Castillo-Rogez, Torrence V. Johnson, Man Hoi Lee, Neal J. Turner, Dennis L. Matson, and Jonathan Luinine (2009). Al Decay: Heat Production and a Revised Age for Iapetus.

Malhotra, Renu

G. Vlahovic and R. Malhotra (2009). Student Focused Geospatial Cirriculum Initiatives: Internships and Certificate Programs at NCCU.

K. Y. L. Su, G. H. Rieke, K. R. Stapelfeldt, R. Malhotra, G. Bryden, P. S. Smith, K. A. Misselt, A. Moro-Martin, and J. P. Williams (2009). The Debris Disk Around HR 8799.

Brenae L. Bailey, and Renu Malhotra (2009). Two Dynamical Classes of Centaurs.

Kathryn Volk and R. Malhotra (2009). Resonant Channels to the Distant Kuiper Belt.

David A. Minton and R. Malhotra (2009). Dynamical Erosion of the Asteroid Belt and Implications for the Rate of Large Impacts of the Terrestrial Planets.

Renu Malhotra and T. Ito (2009). Asymmetric Impacts of Near-earth Aseroids on the Moon.

McEwen, Alfred

M. S. Robinson, S. M. Brylow, M. Tschimmel, D. Humm, S. J. Lawrence, P. C. Thomas, B. W. Denevi, E. Bowman-Cisneros, J. Zerr, M. A. Ravine and 13 coauthors (2010). Lunar Reconnaissance Orbiter Camera (LROC) Instrument Overview

C. J. Hansen, N. Thomas, G. Portyankina, A. McEwen, T. Becker, S. Byrne, K. Herkenhoff, H. Kieffer, and M. Mellon (2010). HiRISE Observations of Gas Sublimiation-Driven Activity in Mars' Southern Polar Regions: I. Erosion of the Surface.

Kathryn E. Fishbaugh, Shane Byrne, Kenneth E. Herkenhoff, Randolph L. Kirk, Corey Fortezzo, Patrick S. Russell, and Alfred McEwen (2010). Evaluating the Meaning of "Layer" in the Martian North Polar Layered Deposits and the Impact on the Climate Connection.

Colin M. Dundas and Alfred S. McEwen (2010). an Assessment of Evidence for Pingoson Mars Using HiRISE.

Kelly Jean Kolb, Jon D. Pelletier, and Alfred S. McEwen (2010). Modeling the Formation of Bright Slope Deposits Associated with Gullies in Hale Crater, Mars: Implications for Recent Liquid Water.

C. M. Weitz, R. E. Milliken, J. A. Grant, A. S. McEwen, R. M. E. Williams, J. L. Bishop, and B. J. Thomas (2010). Mars Reconnaissance Orbiter of Light-toned Layered Deposits and Associated Fluvial Landforms on The Plateaus Adjacent to Valles Marineris.

McMillan, Robert S.

R. S. McMillan, P. R. Holvorcem, M. Schwartz, R. Holmes, S. Foglia, G. Hug, and T. B. Spahr (2010). 2010 GZ5.

R. S. McMillan, G. T. Elliott, D. J. Tholen, A. Mainzer, E. Wright, J. Bauer, T. Grav, J. Dailey, J. Masiero, R. Cutri, and 2 coauthors (2010). 2010 FC81.

R. S. McMillan, G. T. Elliott, D. J. Tholen, A. Mainzer, E. Wright, J. Bauer, T. Grav, J. Dailey, J. Mawiero, R. Curti, and 2 coauthors (2010). 2010 FY80.

R. S. McMillan, R. E. Hill, R. A. Kowalski, J. D. Ahern, E. C. Beshore, A. Boattini, G. J. Garradd, A. R. Gibbs, A. D.

Grauer, S. M. Larson, and 8 coauthors (2010). 2010 CV180.

R. S. McMillan, G. Hug, A. Mainzer, E. Wright, J. Bauer, T. Grav, J. Dailey, J. Masiero, R. Curti, R. Walker, and T. B. Spaher (2010). 2010 DJ56.

Melosh, H. Jay

H. Jay Melosh (2009). Airbursts in the Sky With Diamonds? Shock Limits to a younger Dryas Impact.

H. Jay Melosh (2009). An Isotopic Crisis for the Giant Impact Origin of the Moon?

H. J. Melosh, F. Nimmo (2009). An Intrusive Dike Origin for Iapetus' Enigmatic Ridge?

M. J. Drake, H. J. Melosh (2009). Melosh Received 2008 Harry H. Hess Medal.

Pelletier, Jon

Jon. D. Pelletier, Darin Comeau, and Jeff Kargel (2010). Erratum to "Controls of Glacial Valley spacing on Earth and Mars" (Geomorphology 116 (2010) 189-201).

Jon D. Pelletier, Darin Comeau, and Jeff Kargel (2010). Controls of Glacial Valley Spacing on Earth and Mars.

Kelly Jean Kolb, Jon D. Pelletier, and Alfred S. McEwen (2010). Modeling the Formation of Bright Slope Deposits Associated with Gullies in Hale Crater, Mars: Implications for Recent Liquid Water.

B. Lomenick, R. Hao, N. Jonai, R. M. Chin, M. Aghajan, S. Warburton, J. Wang, R. P. Wu, F. Gomez, J. A. Loo, and 7 coauthors (2009). Target Identification Using Drug Affinity Responsive Target Stability (DARTS).

J. D. Pelletier (2009). Alluvial Fan Response to Climatic Change: Insights from Numerical Modeling.

J. D. Pelletier (2009). How do Pediments form? A Numberical Modeling Investigation with Comparison to Pediments in Southern Arizona, USA.

Rieke, George

J. E. Austerman, J. S. Dunlop, T. A. Perera, K. S. Scott, G. W. Wilson, I. Aretxaga, D. H. Hughes, O. Almaini, E. L. Champin, and S. C. Chapman, and 38 coauthors (2010). AzTEC Half Square Degree Survey of the SHADES Fileds-I. Maps, Catalogues and Source Counts (2010).

K. Y. L. Su, G. H. Rieke, K. R. Stapelfeldt, R. Malhotra, G. Bryden, P. S. Smith, K. A. Misselt, A. Moro-Martin, and J. P. Williams (2009). The Debris Disk Around HR 8799.

James Muzerolle, Kevin Flaherty, Zoltan Balog, Elise Furlan, Paul S. Smith, Lori Allen, Nuria Calvet, Paola D'Alessio, S. Thomas Megeath, August Muench, and 2 coauthors (2009). Evidence for Dynamical Changes in the Transitional Protoplanetary Disk with Mid-Infrared Variability.

James Muzerolle, Kevin Flaherty, Zoltan Balog, Elise Furlan, Paul S. Smith, Lori Allen, Nuria Calvet, Paola D'Alessio, S. Thomas Megeath, August Muench, and 2 coauthors (2009). Evidence for Dynamical Changes in a Transitional Protoplanetary Disk with Mid-Infrared Variability.

N. Seymour, M. Huynh, T. Dwelly, M. Symeonidis, A. Hopkins, I. M. McHardy, M. J. Page, and G. Rieke (2009). Investigating the Far-IRRadio Correlation of Start-forming Galaxies to z=3.

Young Shi, George H. Riek, Patrick Ogle, Linhua Jiang, Aleksander M. Diamond-Stanic (2009). Cosmic Evolution of Star Formation in Type-1 Quasar Hosts Since z=1.

Rogers, Tamara M.

T. M. Rogers and K. B. MacGregor (2010). On the Interacation of Internal Gravity Waves with a Magnetic Field−I. Artificial Wave Forcing.

Keith B. McGregor and T. M. Rogers (2010). Reflection and Ducting of Gravity Waves Inside the Sun.

T. Rogers (2009). The Interaction of Internal Gravity Waves and Magnetic Fields in Stellar Interiors.

Tamara Rogers, Amir Shirkhodaie, K. Atindra, Fred Johnson, Chico Foxx, Sean Young, Lamar Westbrook, Tony Marrs, Tom Lewis, Saleh Zein-Sabatto, and 5 coauthors (2009). Distributed Sensor Concepts for Perimeter Surveillance and Vehicle Classification.

Gary Glatzmaier, Martha Evonuk, and Tamara Rogers (2009). Differential Rotation in Giant Planets Maintained by Density-Stratifield Turbulent Convection.

Sandel, Bill R

A. McEwen, E. Turtle, L. Keszthelyi, J. Spencer, N. Thomas, P. Wurz, P. Christensen, K. Khurana, K. H. Glassmeier, U. Auster, and 13 coauthors (2010). Science Rational for an Io Volcano Observer (IVO) Mission.

E. Quemerais, R. Lallement, J. L. Bertauz, B. R. Sandel, V. Izmodenov, and Y. Malama (2010). Ultraviolet Glow from the Hydrogen Wall.

David A. Galvan, Mark B. Moldwin, Bill R. Sandel, and Geoff Crowley (2010). On the Causes of Plasmaspheric Rotation Variability: IMAGE EUV Observations.

M. Spasojevic and B. R. Sandel (2010). Global Estimates of Plasmaspheric Losses During Moderate Distrubance Intervals.

E. Quemerais, B. R. Sandel, and R. Lallement (2009). Voyager UVS Observations of the Hydrogen Wall.

R. Lallement, E. Quemerais, J. Bertaux, B. R. Sandel, J. T. Clarke, and W. Schmidt (2009). Synthetic Study of Heliospheric Ly-alpha Data.

Showman, Adam

Yuan Lian and Adam P. Showman (2010). Generation of Equatorial Jets by Large-scale Latent Heating on the Giant Planets.

David S. Choi, Adam P. Showman, and Ashwin R. Vasavada (2010). The Evolving Flow of Jupiter's White Ovals and Adjacent Cyclones.

A. McEwen, E. Turtle, L. Keszthelyi, J. Spencer, N. Thomas, P. Wurz, P. Christensen, K. Khurana, K. H. Glassmeier, U. Auster, and 13 coauthors (2010). Science Rational for an Io Volcano Observer (IVO) Mission.

L. Han and A. P. Showman (2010). Coupled Convection and Tidal Dissipation in Europa's Ice Shell: 2. Non-Newtonian Viscosity.

A. B. Penny, A. P. Showman, and D. A. Choi (2010). Suppression of the Rhines Effect and the Location of Vortices on Saturn.

J. J. Fortney, M. Shabram, A. P. Showman, Y. Lian, R. S. Freedman, M. S. Marley, and N. K. Lewis (2010). Transmission Spectra of Three-Dimensional Hot Jupiter Model Atmospheres.

Smith Mark A.

B. P. Abbott, R. Abbott, F. Acernese, R. Adhikari, P. Ajith, B. Allen, G. Allen, M. Alshourbagy, R. S. Amin, S. b. Anderson, and 671 coauthors (2010). Searches for Gravitational Waves from Known Pulsars with Science Run 5 LIGO Data.

LIGO Scientific Collaboration, Virgo Callaboration, J. Abadie, B. P. Abbott, R. Abbott, M. Abernathy, T. Accadia, F. Acernese, C. Adams, R. Adhikari, P. Ajith, and 701 coauthors (2010). Predictions for the Rates of Compact Binary Coalescences Observable by Ground-based Gravitational-wave Detectors.

H. Lampeitl, R. C. Nichol, H.-J. Seo, T. Giannantonio, C. Shapiro, B. Bassett, W. J. Perceval, T. M. Davis, B. Dilday, J. Friedman, and 21 coauthors (2010). First-year

Sloan Digital Sky Survey-II Supernova Results: Consistaency and Constraints with Other Intermediate-redshift Data Shets.

C. J. Davis, R. Gell, T. Khanzadyan, M. D. Smith, T. Jenness (2010). A General Catalogue of Molecular Hydrogen Emission-line Objects (MHOs) in Outflows from Young Stars.

The LIGO Scientific Collaboration, The Virgo Collaboration, J. Abadie, B. P. Abbott, R. Abbott, T. Accadia, F. Acernese, R. Adhikari, P. Ajith, B. Allen, G. Allen, and 657 coauthors (2010). Search for Gravitational-wave Inspiral Signals Associated with Short Gamma-Ray Burst During LIGO's Fifth and Virgo's First Science Run.

Smith, Peter H.

John E. Moores, Mark T. Lemmon, Peter H. Smth, Leonce Komguem, and James A. Whiteway (2010). Atmospheric Dynamics at the Phoenix Landing Site as Seen by the Surface Stereo Imager.

Nilton O. Renno, Brent J. Bos, David Catling, Benton C. Clark, Line Drube, David Fisher, Walter Goetz, Stubbe F. Hviid, Horst Uwe Keller, Jasper F. Kok, and 13 coauthors (2009). Possible Physical and Thermodynamical Evidence for Liquid Water at the Phoenix Landing Site.

J. A. Whiteway, L. Komeguiem, C. Dickson, C. Cooks, M. Illnicki, J. Seabrook, V. Popovici, T. J. Duck, R. Davy, P. A. Tayolor, and 14 coauthors. Mars Water-Ice Clouds and Precipitation.

M. H. Hecht, S. P. Kounaves, R. C. Quinn, S. J. West, S. M. Young, D. W. Ming, D. C. Catling, B. C. Clark, W. V. Boynton (2009). Detection of Perchlorate and the Soluble Chemistry of Martian Soil at the Phoenix Lander Site.

W. V. Boynton, D. W. Ming, S. P. Kounaves, S. M. M. Young, R. E. Arvidson, M. H. Hecht, J. Hoffman, P. B. Niles, D. K. Hamara, R. C. Quinn, and 4 coauthors (2009). Evidence for Calcium Carbonate at the Mars Phoenix Landing Site.

P. H. Smith, L. K. Tamppari, R. E. Arvidson, D. Bass, D. Blaney, W. V. Boynton, A. Carswell, D. C. Catling, B. C. Clark, T. Duck, and 26 coauthors (2009). H2O at the Phoenix Landing Site.

Strom, Robert G.

James W. Head, Scott L. Murchie, Louise M. Prockter, Sean C. Solomon, Robert G. Strom, Clark R. Champman, Thomas R. Watters, David T. Blewett, J. J. Gillis-Davis, and Caleb I. Fassett (2009). Evidence for Instrusive Activity on Mercury from the first MESSENERG Flyby.

James W. Head, Scott L. Murchie, Louise M. Prockter, Sean C. Solomon, Clark R. Chapman, Robert G. Strom, Thomas R. Watters, David T. Blewett, Jeffrey J. Gillis-Davis, Caleb I. Fasset , and 3 coauthors (2009). Volcanism on Mercury: Evidence From the First MESSENGER Flyby for Extrusive and Explosive Activity and the Volcanic Origin of Plains.

O. Scholten, S. Buitink, J. Bacelar, R. Braun, A. G. de Bruyn, H. Flacke, K. Singh, B. Stappers, R. G. Strom, and R. Al Yahyaoui (2009). First Results of the NuMoon Experiment.

C. I. Fassett, J. W. Head, D. T. Blewett, C. R. Chapman, J. L. Dickson, S. L. Murchie, R. G. Strom, T. R. Watters (2009). Caloris Impact Basin: Exterior Geomorphhology, Stratigraphy, Morphometry, Radial Sculpture, and Smooth Plains Deposits.

J. L. Rascusin, S. V. Karpov, M. Sokolowski, J. Granot, X. F. Wu, V. Pal'Shin, S. Covino, A. J. van der Horst, S. R. Oates, P. Schady, and 83 coauthors (2008). Broadband Observations of the Naked-eye Y-ray burst GRB080319B.

Swindle, Timothy D.

J. R. Weirich, C. . Isachsen, J. R. Johnson, and T. D. Swindle (2010). Argon Diffusion in Pyroxene and Albite.

A. Wittmann, D. A. Kring, J. M. Friedrich, J. Troiano, R. J. Macke, D. T. Britt, T. D. Swindle, J. R. Weirich, and D. Rumble (2010). Highly Porous and Compositionally Intermediate Ordinary Chondrite LAP 031047.

J. R. Whittman, T. D. Swindle, and D. A. Kring (2009). Clast-rich H-Chondrite Impact Melts.

J. R. Wittman, C. Isachsen, T. D. Swindle, and D. A. Kring (2009). Ar-Ar Impact Ages of Shocked L1 Chondrites.

T. D. Swindle, C. Thomas, O. Mousis, J. I. Luinine, and S. Picaud (2009). Evidence from Ar/39Ar Ages of Lunar Impact Glasses for an Increase in the Impact Rate ???800 Ma Ago.

N. E. B. Zellner, J. W. Delano, T. D. Swindle, F. Barra, E. Olsen, and D. C. B. Whitten (2009). Apollo 17 Regolith, 71501,262: A Record of Impact Events and Mare Volcanism in Lunar Glasses.

Yelle, Roger

T. T. Koskinen, R. V. Yelle, P. Lavvas, and N. K. Lewis (2010). Characterizing the Thermosphere of HD209458b with UV Transit Observations.

J.-Y. Bonnet, R. Thissen, M. Frisari, V. Vuitton, E. Quirico, L. LeRoy, N. Fran, H. Cottin, S. M. Horst, and R. Yelle (2010). HCN Polymers: Composition and Structure Revisited by High Resolution Mass Sectrometry.

J.-E. Wahlund, M. Galand, I. Muller-Wodarg, J. Cui, R. V. Yelle, F. J. Crary, K. Mandt, B. Magee, J. H. Waite, D. T. Young, and 9 coauthors. On the Amount of Heavy Molecular Ions in Titan's Ioonosphere.

I. P. Robertson, T. E. Cravens, J. H. Waite, R. V. Yelle, V. Vuitton, A. J. Coates, J. E. Wahlund, K. Agren, K. Mandt, B. Magee, and 2 coauthors (2009). Structure of Titan's Ionosphere: Model Comparisons with Cassisni Data.

V. Vuitton, P. Lavvas, R. V. Yelle, M. Galand, A. Wellbrock, G. R. Lewis, A. J. Coates, and J.-E. Wahlund (2009). Negative Ion Chemistry in Titan's Upper Atmosphere.

UNIVERSITY OF ARKANSAS

DEPARTMENT OF PHYSICS

Fayetteville, Arkansas 72701

Students Accepted For Degree	FIELDS		
	Physics	Astronomy	Related Fields
Doctorate	X	X	
Master's	X	X	

1. General

President: Alan Sugg
Dean of Graduate School: Collis Geren
Department Chairman: Surendra P. Singh
Department Telephone Number: (479) 575-2506
Type of Institution: University
Control: Public
Setting: Small town
Total Faculty: 983
Total Graduate Faculty: 328
Total Students: 19,849
Total Graduate Students: 3,616
Annual Graduate Tuition:
In-state residents: Full-time—$1,838.82/6 hrs
$2,758.23/9 hrs
Part-time—$294.68/credit
Out-of-state residents: Full-time—$4,350.18/6 hrs
$6,525.27/9 hrs
Part-time—$725.03/credit
Tuition rates for: Fall 2010
Deferred tuition plan: No
Other Fees: $40 Application fee, $50 for international
$498.55 for international students
$39.94 University Fees
$10.19 College Fees
$581 Books, supplies + Lab fees
Term: Semester

2. Number of Faculty in Department

The combined total of full-time faculty in the three professorial ranks is 21. The combined total of full-time, part-time, and other faculty at all ranks is 23.

3. Admission, Financial Aid, and Housing

Address admission inquiries to: Department of Physics
Graduate application fee required: $40, $50 for international students. Fee waived if admitted by department.
Admission deadline (Fall admission): None
Admission information: For fall admission, AY 2009–10, 13 students were accepted from ~100 applicants.
Admission requirements: For admission to the graduate programs, a Bachelor's degree in physics is preferred with a minimum undergraduate GPA of 3.0 (A=4) or 3.2 on last 60 hours credit. The GRE is recommended. The GRE Physics Subject Test is strongly urged. Students from non-English speaking countries are required to demonstrate proficiency in English via the iBT TOEFL exam. Minimum acceptable score for admission is 80. A minimum score of 26 on the speaking section is required for a graduate teaching assistantship.
Undergraduate preparation assumed: Candidates should have an undergraduate degree with the equivalent of a 30-hour major in physics including intermediate level courses in mechanics, electricity and magnetism, thermal physics, quantum mechanics, and mathematics through differential equations.
Address financial aid inquiries to: Graduate Admissions Committee, Department of Physics
GAPSFAS application required: No
Financial aid deadline: None
Loans available: No
Address housing inquiries to: University Housing, 900 Hotz Hall
On-campus, single student housing available: Yes
Cost/academic year: $4,021-single (a/c room and board)
On-campus, married student housing available: No

Table A—Faculty, Enrollments, and Degrees Granted

Research Specialty	2009–10 Faculty	Enrollment[1] Fall 2009		No. of Degrees Granted[2] 2009–10 (2005–09)		
		Master's	Doctorate	Master's	Terminal Master's	Doctorate
Astronomy	3	0	2	1(1)	0(4)	0(1)
Atomic, Molecular, & Optical Physics	8	0	14	0(1)	0(1)	2(5)
Biophyics	2	0	10	0(3)	1(0)	0(0)
Condensed Matter Nano Science	5	0	16	2(1)	2(4)	0(3)
Non-specialized	1	0	1	0(0)	0(0)	0(0)
Physics Education	2	3	0	0(0)	0(12)	0(0)
Total	21	3	43	3(6)	3(21)	2(9)
Full-time Grad. Stud.		2	13			
Part-time Grad. Stud.		3	30			
First-year Grad. Stud.		1	13			

Undergraduate Degrees, 2009–10: 25

[1]Students not yet committed to a research specialty are entered under non-specialized.
[2]Five-year totals in parentheses.

4. Graduate Degree Requirements

Master's: Choose either a 30 credit thesis path or a 36 credit non-thesis path. A GPA of 3.0 (A=4) must be maintained. There is a 30-week residency requirement. Core courses plus physics electives. No language requirement or written comprehensive or qualifying examination.
Doctorate: 40 semester hours of coursework at the graduate level and 18 hours of doctoral dissertations. A GPA of 3.0 (A=4) must be maintained. Residency of two consecutive semesters required after admission to candidacy. No language requirement. Written and oral candidacy examination required by third semester of graduate work. Dissertation with final oral examination.
Other Programs: M.A. degree. Education track. 30 semester hours credit. A GPA of 3.0 (A=4) must be maintained. There is a 30-week residency requirement. No language requirement or written comprehensive or qualifying examination. No thesis. Written report with oral final examination.
Thesis: Thesis may be written *in absentia*.
Special Equipment, Facilities, or Programs: Research facilities include well equipped research laboratories in quantum optics, laser spectroscopy, nonlinear optics, high pressure physics, surface physics, nanoscience, biophysics and computer graphics. The laboratories possess a complete range of equip-

ment including ultra-violet, visible and infrared gas lasers, ultra-high stability CW dye and solid-state laser systems, femtosecond and ultra-high power pulsed solid state laser systems, and a SQUID magnetometer. A $1.5M molecular beam epitaxy (MBE) facility is available in the department for fabricating and characterizing semiconductor heterostructures. Excellent sample characterization and research facilities equipped with X-ray diffractometer (XRD), scanning electron microscope (SEM), atomic force microscope (AFM), Raman spectrometer, etc. are available at the High Density Electronics Center HiDEC at the University.

Table B—Appointments to Graduate Students, 2009–10

Title of Appointee	Appointments Total	Appointments First year	Academic Load Allowed in Credit Hours	Hours of Service Per Week	Stipend for Academic Year ($)
			Semester		
Teaching Assistant	31	8	6–10	20	14,700[1,2]
Research Assistant	13		6–10	20	14,500[1,2] or 14,700
Total	44	8			

[1]Tuition is paid by the University for students having at least a 50% appointment.
[2]Summer teaching and research appointments at 2/9 AY stipend plus tuition waiver are not guaranteed but have been available for all students requesting them in recent years.

5. Personnel Engaged in Separately Budgeted Research, 2009–2010

Professorial faculty	15
Graduate students	13
Postdoctoral appointments	16
Total	44

6. Separately Budgeted Research Grants by Source of Support

	Departmental Research	Physics-related Research Outside Department
Federal government	$9,082,875	
State government	405,875	
Other	11,250	
Total	$9,500,000	

Table C—Separately Budgeted Research Grants

Research Specialty	No. of Grants	Dollar Amount
Atomic, Molecular, & Optical Physics	9	1,417,161
Astronomy	3	779,613
Biophysics	6	2,655,650
Condensed Matter Physics/ Nano Science	4	1,858,000
Physics Education	6	3,786,468
Total	28	10,496,892

FACULTY

Professors

Bellaiche, Laurent, Ph.D., University of Paris XI, 1994. Condensed matter physics.

Gea-Banacloche, Julio, Ph.D., New Mexico, 1985. Theoretical quantum optics.

Gupta, Rajendra, Ph.D., Boston, 1970. Professor Emeritus. Experimental atomic and molecular physics; laser spectroscopy.

Harter, William G., Ph.D., California, Irvine, 1967. Theoretical physics; molecular dynamics; computer graphics.

Hobson, Arthur, Ph.D., Kansas State, 1964. Professor Emeritus. Physics and society; nuclear arms control; physics education; public understanding of science; nuclear war education.

Lacy, Claud H., Ph.D., Texas, Austin, 1978. Astronomy; eclipsing binaries; binary supermassive black holes.

Lieber, Michael, Ph.D., Harvard, 1967. Theoretical physics; scattering theory; quantum field theory.

Pederson, Donald O., Ph.D., Rice, 1971. Vice-Chancellor.

Salamo, Gregory J., Ph.D., CUNY, 1974. Lasers; quantum optics.

Singh, Surendra P., Ph.D., Rochester, 1982. Chair of the Department. Quantum optics; lasers.

Thibado, Paul M., Ph.D., University of Pennsylvania, 1994. Condensed matter physics.

Vickers, Ken, M.S., Arkansas, 1978. Research Professor, Director of micro-electronics photonics program.

Vyas, Reeta, Ph.D., SUNY, Buffalo, 1984. Nuclear theory; quantum optics.

Xiao, Min, Ph.D., Texas, Austin, 1988. Quantum optics.

Associate Professors

Fu, Huaxiang, Ph.D., Fudan University, China, 1994. Condensed matter physics.

Li, Jiali, Ph.D., The City University of New York, 1999. Condensed matter physics; Bio/nano physics.

Oliver, William F., Ph.D., University of Colorado, 1987. Condensed matter physics.

Stewart, Gay, Ph.D., University of Illinois, 1994. Physics education.

Assistant Professors

Gross, Eitan, Ph.D., Bar Ilan University, Israel, 1993. Photochemistry and biophysics.

Kennefick, Daniel, Ph.D., Caltech, 1997. Astrophysics.

Kennefick, Julia, Ph.D., Caltech, 1995. Astronomy; observational cosmology, quasars.

Stewart, John, Ph.D., University of Illinois, 1994. Visiting Assistant Professor. Condensed matter physics, physics education.

Tchakhalian, Jak, Ph.D., University of British Columbia, Canada, 2002. Condensed matter physics.

Adjunct Faculty

Naseem, Hameed A., Ph.D., Virginia Polytechnic Institute & State University, 1984. Adjunct Professor. Condensed matter physics and material science.

Shultz, John, Ph.D., University of Arkansas, 1996. Adjunct Assistant Professor. Lasers and Optics.

RESEARCH SPECIALTIES AND STAFF

Theoretical

Atomic and Molecular Physics. Harter, Lieber.
Condensed Matter. Bellaiche, Fu, Tchakhalian.
Nonlinear Dynamics. Gea-Banacloche, Harter, Singh, Vyas.
Physics Education. Lacy, Stewart, G., Stewart, J.
Quantum Optics and Electronics. Laser theory. Gea-Banacloche, Singh, Vyas, Xiao.

Experimental

Astronomy. Eclipsing binaries; supermassive black holes. D. Kennefick, J. Kennefick, Lacy.
Optics. Self-induced transparency; coherence and fluctuation in lasers; nonlinear optics; photorefraction; photon statistics; spectroscopy, Raman scattering, Brillouin scattering. Oliver, Salamo, Singh, Xiao. 3 postdoctoral research associates.
Condensed Matter and nanoscience. Optical properties of solids; thin films; electronic transport; quantum well structures; surface physics; molecular beam epitaxy (MBE); scanning tunneling spectroscopy; Raman scattering, Brillouin scattering, high-pressure physics; Li, Oliver, Salamo, Tchakhalian, Thibado, and 7 postdoctoral research associates.
Biophysics. Nanopore physics; surfaces, biopolymers; optical tweezers; bioptical and electrical phenomena, neural network, membranes. Gross, Li, Oliver, Salamo, Singh.

CALIFORNIA STATE UNIVERSITY, FULLERTON

DEPARTMENT OF PHYSICS

Fullerton, California 92834-6866

Students Accepted For Degree	FIELDS		
	Physics	Astronomy	Related Fields
Doctorate			
Master's	X		

1. General

President: Dr. Milton A. Gordon
Dean of College of Natural Science & Mathematics: Dr. Steven Murray
Department Chair: Dr. Jim Feagin
Department Telephone Number: (657) 278-3366
Type of Institution: University
Control: Public
Setting: Urban
Total Faculty: 2,265
Total Graduate Faculty: 650
Total Students: 37,000
Total Graduate Students: 3,000
Annual Graduate Tuition:
In-state residents: Full-time—$2,183
6 or fewer units/semester—$1,394
Out-of-state residents: Full-time—$339 per unit
Part-time—Plus in state (6 or fewer units/semester) + $339/unit
Tuition rates for: 2008–09
Term: Semester

2. Number of Faculty in Department

The combined total of full-time faculty in the three professorial ranks is 9. The combined total of full-time, part-time, and other faculty at all ranks is 21.

3. Admission, Financial Aid, and Housing

Address admission inquiries to: CSUF, 800 N. State College Blvd., Fullerton, CA 92834
Graduate application fee required: $55
Admission deadline (Fall admission): March 1; (Spring admission): October 1
Admission information: For fall admission, 2008–09, 6 students were accepted from 10 applicants.
Admission requirements: Students seeking admission to the master's program in physics must have (1) a grade-point average of 2.5 in the last 60 semester units (or the last 90 quarter units), (2) a degree from an accredited college or university with a major in physics or a closely related field (students with majors in fields closely related to Physics must have passed 9 units of upper division physics with a "C" grade or better), (3) a grade-point average of 2.75 for upper-division courses in the physics major. For students with undergraduate degrees in engineering, mathematics or other physical sciences, a GPA of 3.0 in upper-division major courses is required. In addition to the GPA requirements, all applicants must submit (1) a score on the physics subject of the Graduate Record Exam (GRE); students with Physics Bachelor degree may be allowed to take the Physics GRE within the first year of enrollment with Department permission (2) a one-page, 500-word maximum, typed statement of purpose, explaining the students interest in taking a higher degree in physics, and (3) three letters of recommendation. Students with a physics degree will have their physics GRE waived if their GPA is above 3.6 out of 4. International student applicants are required to pass the Test of English as a Foreign Language (TOEFL) with a score of 550 or higher for the computer test and 213 for the paper test. See website: http://physics.fullerton.edu/grad-adm.html
Undergraduate preparation assumed: Upper-division courses in: Mechanics, E&M, Modern Physics, Thermodynamics.
Address financial aid inquiries to: Financial Aid, CSUF, Fullerton, CA 92834
Financial aid deadline: March 2
Address housing inquiries to: Housing, CSUF, Fullerton, CA 92834
On-campus, graduate student housing available: No
On-campus, married student housing available: No

Table A—Faculty, Enrollments, and Degrees Granted

Research Specialty	2007–09 Faculty	Enrollment[1] Fall 2008		No. of Degrees Granted[2] 2008–09 (2004–09)			Median No. of Years for 2008–09 Ph.D.'s
		Master's	Doctorate	Master's	Terminal Master's	Doctorate	
Total		–	–	0(5)	0(5)	–	–
Full-time Grad. Stud.		7	–				
Part-time Grad. Stud.		–	–				
First-year Grad. Stud.		6	–				
Median Years in Grad. Study (2008–09 Degrees)				2	–	–	–
Undergraduate Degrees, 2008–09 (2004–09): 7(72)							

[1] Students not yet committed to a research specialty are entered under non-specialized.
[2] Five-year totals in parentheses.

4. Graduate Degree Requirements

Master's: A minimum of 30 semester units of course work approved by the graduate committee with a grade point average of 3.0. A passing grade on the comprehensive final exam, or a Master's thesis and oral defense of the thesis or completion and presentation of a research project.
Thesis: Theses may not be written *in absentia*.
Special Equipment, Facilities, or Programs: The department has major laboratory facilities in atomic, condensed matter, and optical physics.

Table B—Appointments to Graduate Students, 2008–09

Title of Appointee	Appointments Total	Appointments First year	Academic Load Allowed in Credit Hours	Hours of Service Per Week	Stipend for Academic Year ($)
			Semester		
Teaching Assistant	3	2	–	10–20	3,035–6,070
Research Assistant	0	0	–	–	–
Total	3	0			

5. Personnel Engaged in Separately Budgeted Research, 7/08–6/09

Professorial faculty	6
Undergraduate students	7
Graduate students	1
Total	14

6. Separately Budgeted Research Expenditures by Source of Support

	Departmental Research	Physics-related Research Outside Department
Federal government	$1,069,850	
Total	$1,069,850	

FACULTY

Professors

Cheng, Kwang-Ping, Ph.D., The Catholic University of America, 1990. Astrophysics.

Feagin, Jim, Ph.D., University of North Carolina, 1980. Theoretical Atomic Physics.

Fearn, Heidi, Ph.D., University of Essex, Colchester, England, 1989. Theoretical Quantum Optics and Electrodynamics.

Khakoo, Murtadha, Ph.D., University College, London, England, 1980. Experimental Atomic Physics.

Wanser, Keith, Ph.D., University of California, Irvine, 1982. Experimental Fiber Optics and Theoretical Condensed Matter.

Associate Professor

Childers, J. Gregory, Ph. D., University of Kentucky, 2001. Experimental Atomic Physics.

Loverude, Michael, Ph.D., University of Washington, 1999. Physics Education Research.

Assistant Professors

Smith, Joshua, B.Sc., Syracuse University, Syracuse, New York; Ph.D., Leibniz Universitat, Hannover. Gravitational-Wave Physics.

Tifrea, Ionel, Ph.D., Babes-Bolyai University, Cluj, Romania, 1998. Theoretical Condensed Matter.

Adjunct Professors

Goode, Stephen, Ph.D., University of Waterloo, 1983. Applied Mathematics.

Woodward, James, Ph.D., University of Denver, 1972. History of Science.

CALIFORNIA STATE UNIVERSITY, NORTHRIDGE

DEPARTMENT OF PHYSICS AND ASTRONOMY

Northridge, California 91330-8268
http://www.csun.edu/physics

Students Accepted For Degree	FIELDS		
	Physics	Astronomy	Related Fields
Doctorate			
Master's	X		

1. General

President: Dr. Jolene Koester
Assoc. Vice-President of Graduate Studies, Research, and International Programs: Dr. Mack Johnson
Department Chairman: Ana Cristina Cadavid
Department Telephone Number: (818) 677-2775 (C), *Website*: http://www.csun.edu/physics
Email: physics@csun.edu
Type of Institution: University
Control: Public
Setting: Suburban
Total Faculty: 2,105
Total Students: 35,198
Total Graduate Students: 5,923
Annual Graduate Tuition for 2009–10:
In-state residents: Full-time—$4,801
Out-of-state residents: —$5,961 plus $373 per unit
Deferred tuition plan: No
Other Fees: $200–400
Term: Semester

2. Number of Faculty in Department

The combined total of full-time faculty in the three professorial ranks is 17. The combined total of full-time, part-time, and other faculty at all ranks is 31.

3. Admission, Financial Aid, and Housing

Address admission inquiries to: Office of Admissions & Records
Graduate application fee required: $55
Admission deadline (Fall admission): to be determined*
Admission information: For fall admission, 2009–10, 12 students were accepted from 22 applicants.
Admission requirements: For admission to the graduate programs, a Bachelor's degree in physics or related field is required with a minimum undergraduate GPA of 2.5 in last 60 units and in major. The GRE is required for students whose undergraduate GPA overall is below 3.0. The GRE Advanced is not required. Students from non-English speaking countries are required to demonstrate proficiency in English via the TOEFL. Minimum acceptable score for admission is 550. A writing proficiency test must be passed during the first year.
Undergraduate preparation assumed: Minimum Junior-level-preparation: Symon, *Mechanics*; Lorrain and Corson, *Electromagnetic Fields and Waves*; Sears, *Thermodynamics*; Eisberg, *Modern Physics*; Boas, *Mathematical Methods in Physical Science.*
Address financial aid inquiries to: Financial Aid Office. (Address request for Teaching Assistantships, approximately $19,668/year, to Graduate Advisor)
GAPSFAS application required: No
Financial aid deadline: March 02 of each year
Loans available: Yes
Address housing inquiries to: Housing Office
On-campus, single student housing available: Yes
Cost/month: $517–1,069 per person
On-campus, married student housing available: Yes
Cost/month: $842–1,152 per unit

*Deadlines as early as Feb. 28 may be set by the University for nonresidents and foreign students, and as early as April 30 for residents. For additional details, please write to the Graduate Advisor.

Table A—Faculty, Enrollments, and Degrees Granted

Research Specialty	2009–10 Faculty	Enrollment[1] Fall 2009 Mas-ter's	Enrollment[1] Fall 2009 Doc-torate	No. of Degrees Granted 2009–10 (2004–09) Mas-ter's	Terminal Master's	Doc-torate
Applied Physics	0	0	–	–	0(0)	–
Astrophysics	5	0	–	–	1(4)	–
Bio-Chemical Physics	3	0	–	–	2(2)	–
Condensed Matter Physics	6	0	–	–	0(2)	–
Math Physics	1		–	–		–
Nuclear Physics	3	0	–	–	0(1)	–
Optics	0	0	–	–	6(2)	–
Non-specialized	0	55	–	–	4(9)	–
Total	18	35	–	–	7(20)	–
Full-time Grad. Stud.		8	–			
Part-time Grad. Stud.		35	–			
First-year Grad. Stud.		11	–			
Median Years in Grad. Study (2009–10 Degrees)	2.5			–	–	

Undergraduate Degrees, 2009–10 (2004–09): 16(64)

[1]Students not yet committed to a research specialty are entered under non-specialized.

4. Graduate Degree Requirements

Master's: 30 semester units, including a 14 unit core, 16 units of electives, which may include a six-unit thesis. Minimum GPA of 3.0. Written examination on core for students electing not to do a thesis, or defense of thesis. There are no foreign language or residence requirements specified.
Thesis: Thesis may be written *in absentia.*
Special Equipment, Facilities, or Programs: The department hosts four main facilities:
(1) The Nanotechnology lab conducts interdisciplinary research in nanoscience and nanotechnology. The project consists of building the control electronics, constructing the piezo scanner, sample and tip preparation and writing the control software. The control electronics consist of a simple opamp circuit. The scanner is made of a modification to a PC speaker. The tools used are Atomic Force Microscopy, Scanning Tunneling Microscopy, electrical transport measurements, and nanolithography as the eyes and ears to tell what happens to these small scales, and as hands to change their behavior to realize new devies.
(2) The Computational Materials Theory Center is housed in a 4000 square foot facility that maintains an impressive computational capacity consisting of many multiple processing

computers. The most recent acquisition being an 80 processor Beowulk cluster and a 48 processor SGI Origin 3800. Individual stations allow student access to the state-of-the-art code to study strongly correlated electrons, mechanical properties, surfaces and interfaces, transport mechanisms, nanotechnology and other subfields.
(3) The Center for Supramolecular Studies is an interdisciplinary program in experimental and computational biophysics, bio-chemical physics, and biochemistry. The center operates two Electron Spin Resonance (ESR) spectrometers, a cryogenic IR spectrometer, a Time Resolved Fluorescence Quenching (TRFQ) system, a Nuclear Magnetic Resonance (NMR) unit, high precision instruments for measuring surface tension, viscosity and density, facilities for sample preparation and a computational laboratory to study protein flexibility and stability.
(4) The San Fernando Observatory (SFO) is one of the world's major facilities for solar research locations nine miles from campus. The SFO maintains a 24-inch vacuum solar telescope, a vacuum spectroheliograph, two photometric telescopes, a centimeter-hand radio telescope, computer facilities and smaller instruments.

Table B—Appointments to Graduate Students, 2009–10

Title of Appointee	Appointments Total	Appointments First year	Academic Load Allowed in Credit Hours	Hours of Service Per Week	Stipend for Academic Year ($)
Graduate Assistant	2	0		5-10	14,000
Teaching Asst.	15	5		15-16	16,363
Research Assistant	9	3		4-16	-
Total	26	8			

5. Personnel Engaged in Separately Budgeted Research,

Professorial	13
Postdoctoral appointments	9
Graduate students	17
Professional Research Astronomers	2
Total	41

6. Separately Budgeted Research Expenditures by Source of Support

	Departmental Research
Federal Government	$20,475.78
Private	$3,400.00
Total	$20,509.78

7. Separately Funded Managed Laboratories

San Fernando Observatory	$ 80,000
Computational Materials Theory Center	$ 60,000
The Center for Supramolecular Studies	
Nanotech Lab	
Total	$140,000

Table C—Separately Budgeted Research Expenditures, 7/09-6/10

Research Specialty	No. of Grants	Expenditures ($)
Astrophysics	7	250,000
Condensed Matter Physics	13	1,027,406
Nuclear Physics	1	38,545
Total	21	1,315,351

FACULTY

Professors

Cadavid, Ana C., Ph.D., UCLA, 1989. Sdar physics: plasma physics.

Chapman, Gary, Ph.D., Arizona, 1968. Astronomy; solar physics; spectroscopic study of solar magnetic fields.

Doty, Duane R, Ph.D., UCLA, 1966. Beta decay; neutron detection and interaction of radiation in matter and in the environment.

Kioussis, Nicholas G., Ph.D., Illinois, Chicago, 1984. Solid state theory; magnetic and electronic properties of metals and alloys; rare earth and actinide magnetism; computational materials science.

Lee, Paul L., Ph.D., Cal. Tech., 1971. X-ray spectroscopy; nuclear magnetic resonance and ultrasound imaging.

Lim, Say-Peng, Ph.D., Wisconsin, 1988. Solid state physics; optical properties of materials; magneto optics.

Lu, Gang, Ph.D., China, 1998. Computational materials theory, electronic structure calculations, mechanical properties of materials, multiscale modeling.

Park, Robert T., Ph.D., California, Riverside, 1970. Theoretical high-energy physics.

Peric, Miroslav, Ph.D., U. of Zagreb, Croatia, 1987. Dynamics of spin systems using double modulation method in electron spin resonance.

Ranganathan, Radha, Ph.D., U. Utah, Salt Lake City, 1988. Experimental solid state physics; optical properties of semiconductors.

Seki, Ryoichi, Ph.D., Northeastern, 1968. Theoretical nuclear physics: high energy electron scattering; lattice gauge theory; QCD effective lagrangians; computational biology.

Sheng, Donna, Ph.D., Nanjing University, 1989. Condensed matter.

Assistant Professors

Beloborodov, Igor, Ph.D., Ruhr University, 2000. Granular electronic systems, quantum nonodevices, nanoscale superconductors and magnets, spin-related electronics, strongly correlated electron systems.

Choudhary, Debi Prasad, Ph.D., Gujarat University/PRL, 1982. Optical astronomical instrumentation, solar physics, solar system astronomy and interstellar medium.

Christian, Damian, Ph.D., University of Maryland, 1993. Star formation, extra-solar planets and coments, late type and active stars, stellar abundances.

Postma, Henk, Ph.D., Delft University of Technology, 2001. Nanoscience and technology, carbon nanotubes, one-dimensional electronic systems, single electron-single phonon mesoscopic interactions, quantum information, NEMS and self-assembly of nano circuits.

Ren, Deqing, Ph.D., University of Durham, England, 2001. Astronomical instrumentation, adaptive optics, solar physics.

Shiferaw, Yohannes, Ph.D., University of Pittsburgh, 2001. Computational biology, cardiac electrophysiology and arrhythmias, nonlinear dynamics and pattern formation, statistical mechanics of disordered systems, stochastic processes in biological systems.

Emeritus Professors

Bales, Barney L., Ph.D., Colorado, 1968. Electron spin resonance; anti-tumor antibiotics; micelles; liquid dynamics.
Chow, Paul, Ph.D., Northwestern, 1965. Physics education.
Collas, Peter, Ph.D., UCLA, 1966. Dynamical systems; general relativity.
Lawrence, John, Ph.D., Northeastern, 1968. General relativity; cosmology; theoretical astrophysics; solar astronomy.
Moore, Mortimer, Ph.D., London, 1955. Forensic physics.
Natale, Giovan, Ph.D., UCLA, 1967. Physical acoustics.
Olson, Roy, Ph.D., California, Berkeley, 1958. Electron mean-free path in metals; low-pressure gas electical conduction.
Park, Robert T., Ph.D., California, Riverside, 1970. Theoretical high-energy physics.
Richter, Paul, Ph.D., Arizona State, 1968. Astronomical image enhancement; Fourier optics; digital image processing.
Romagnoli, Robert, Ph.D., Illinois Tech., 1957. Magneto-optics; surface plasma waves.
Sandhu, Harbhajan, Ph.D., Penn State, 1961. Nuclear Physics.
Seki, Ryoichi, Ph.D., Northeastern, 1968. Theoretical nuclear physics: high energy electron scattering; lattice gauge theory; QCD effective lagrangians; computational biology.

Lecturers

Collins, Eric
Dobias, Jan

Adjunct Professors

Penn, Mathew, Ph.D., Honolulu, HI, 1988.
Ruzmaikin, Alexander A., Ph.D., Moscow Ph.-Th. Inst., 1969. Astrophysical and geophysical magnetism; solar dynamo theory.

SUPPORT STAFF

Arciero, Carole
Daskalov, Konstantin
Klevens, Debbie
Ransom, Rideout
Sarca, Victor

RESEARCH SPECIALTIES AND STAFF

Theoretical

Astrophysics and Cosmology. Lawrence.
Computational Materials Science. Kioussis, Lim, Lu, Sheng.
Computational/Statistical Physics. Shiferaw.
Condensed Matter Physics. Beloborodov, Kioussis, Lim, Lu, Sheng. Belobordov.
Medical Physics. Shiferaw.
Mathematical Physics. Cadavid, Collas.
Nanotechnology. Beloborodov.
Nuclear and Hadron Physics. Seki.
Plasma Physics. Ruzmaikin.
Relativity. Collas.
Solar Physics. Cadavid, Lawrence, Penn, Ruzmaikin.
Solid State Physics. Beloborodov, Kioussis, Lim, Lu, Sheng.

Experimental

Adaptive optics. Ren.
Astrophysics. Chapman, Choudhary, Christian, Penn, Ren.
Biophysics. Electron spin resonance. Bates, Peric.
Biophysics, Membranes and Proteins. Ranganathan.
Extra-solar plants. Christian, Ren.
Medical Physics. Lee.
Nanotechnology. Postma.
Nuclear Physics. Doty.
Plasma Physics. Choudhary.
Solar Physics. Chapman, Choudhary, Christian, Penn, Ren.

Directorships

Nicholas Kioussis, Director, Computational Materials Theory Center
Gary Chapman, Director, San Fernando Observatory
Barney Bales, Director, The Center for Supramolecular Studies

SAN DIEGO STATE UNIVERSITY

DEPARTMENT OF PHYSICS

San Diego, California 92182

Students Accepted For Degree	FIELDS		
	Physics	Astronomy	Related Fields
Doctorate			
Master's	X		

1. General

President: Stephen L. Weber
Vice President of Graduate and Research Affairs: Thomas Scott
Department Chairman: Usha Sinha
Department Telephone Number: (619) 594–6240 (C)
Type of Institution: University
Control: Public
Setting: Urban
Total Faculty: 1,795
Total Graduate Faculty: 894
Total Students: 32,817
Total Graduate Students: 6,021
Annual Graduate Tuition:
In-state residents: None
Out-of-state residents: $372/unit
Tuition rates for: 2010–11
Deferred tuition plan: Yes
Other Fees: 0–6 units $2,000
6.1 or more units $3,095
Term: Semester
Webpage: www.physics.sdsu.edu

2. Number of Faculty in Department

The combined total of full-time faculty in the three professorial ranks is 13. The combined total of full-time, part-time, and other faculty at all ranks is 16.

3. Admission, Financial Aid, and Housing

Address admission inquiries to: Admission Office
Graduate application fee required: $55
Admission deadline (Fall admission): 2/1
Admission information: For fall admission, 2010–11, 19 students were accepted from ~33 applicants.
Admission requirements: For admission to the graduate programs, a Bachelor's degree in physics, engineering, or mathematics is required with a minimum undergraduate GPA of 2.85 specified. The GRE is required. The minimum acceptable score required for admission is total–950. Students from non-English speaking countries are required to demonstrate proficiency in English via the TOEFL exam. Minimum score of 550 (213 computer based) required.
Address financial aid inquiries to: www.sa.sdsu.edu/fao
GAPSFAS application required: Yes
Financial aid deadline: 5/31
Loans available: Yes
Address housing inquiries to: www.sdsu.edu/housing
On-campus, single student housing available: Yes
Cost/year: $7,500–$10,500 (room and board)
On-campus, married student housing available: No

Table A—Faculty, Enrollments, and Degrees Granted

Research Specialty	2009–10 Faculty	Enrollment[1] Fall 2009		No. of Degrees Granted[2] 2009–10			Median No. of Years for 2009--10 Ph.D.'s
		Mas-ter's	Doc-torate	Mas-ter's	Terminal Master's	Doc-torate	
Atomic/Nuclear/ Astrophys./Comp. Phys.	–	–	–	4	–	–	–
Condensed Matter Physics	2	4	–	–	1(6)	–	–
Energy Sources & Environ.	1	0	–	–	0(1)	–	–
Medical Physics	2	8	–	–	4(10)	–	–
Nuclear/Astrophysics	4	6	–	–	4(0)	–	–
Optics & Laser Phys.	3	6	–	–	3(7)	–	–
Non-Specialized	–	2	–	–	2(0)	–	–
Physics Education	1	1	–	–	0(2)	–	–
Total	13	41	–	–	14(16)	–	
Full-time Grad. Stud.		10	–	–			
Part-time Grad. Stud.		17	–	–			
First-year Grad. Stud.		9	–	–			
Median Years in Grad. Study (2009–10 Degrees)				–	2.6	–	–

Undergraduate Degrees, 2009–10 (2005--10): 10(37)

[1]Students not yet committed to a research specialty are entered under non-specialized.
[2]Five-year totals in parentheses.

4. Graduate Degree Requirements

Master of Science: *(Radiological Health Physics)*. The student must complete a graduate program in which 18 of the 30 units must include graduate course work in radiological physics, nuclear medicine physics, nuclear instrumentation and radiation biology, and thesis. Other requirements are as for the graduate MS degree in Physics.

Master of Science (Physics): The student must complete a graduate program in which 18 of the 30 units must include graduate course work in quantum mechanics, statistical mechanics, classical mechanics, electricity and magnetism, and thesis. The remaining units must be approved by the graduate advisor. There is a minimum of 24 units in residence. The student must pass a final oral examination on his/her thesis, and maintain a 3.0 GPA. There is no foreign language requirement.

Master of Art: *(Radiological Health Physics)*. The student must complete a graduate program in which 12 of the 30 units must include graduate course work in radiological physics, nuclear medicine physics, nuclear instrumentation and radiation biology. Other requirements are as for the graduate MA degree in Physics.

Master of Art (Physics): The student must complete a graduate program in which 21 of the 30 units include graduate course work in electricity and magnetism, quantum mechanics, statistical mechanics, and classical mechanics. The remaining units must be approved by the students' graduate committee. The student must pass a comprehensive examination and maintain a 3.0 GPA. There is a minimum of 24 units in residence.

Thesis: Thesis may be written *in absentia*, however, residency is required.

Special Equipment, Facilities, or Programs: Liquid helium facility; EPR facility; whole-body counter; nuclear radiation facility, nuclear instrumentation laboratory, laser optics, holography facility; electro-optics measurements laboratory; image processing facility; materials laboratory, ultrafast laser; computational physics lab; Beowulf cluster.

Table B—Appointments to Graduate Students, 2010–11

Title of Appointee	Appointments Total	Appointments First year	Academic Load Allowed in Credit Hours	Hours of Service Per Week	Stipend for Academic Year ($)
			Semester		
Teaching Assistant	15	8	12	18	15,900
Total	15	8			

5. Personnel Engaged in Separately Budgeted Research, 7/09–6/10

Professorial faculty	8
Other faculty	3
Graduate students	10
Undergraduate students	8
Total	29

6. Separately Budgeted Research Expenditures by Source of Support

	Departmental Research	Physics-related Research Outside Department
Federal government	$4,745,000	$
Private, non-profit organizations	15,000	
Business and industry	30,000	
Total	$4,790,000	$

Table C—Separately Budgeted Research Expenditures

Research Specialty	No. of Grants	Expenditures ($)
Condensed Matter Physics	5	270,000
Medical Physics	1	900,000
Nuclear and astrophysics	2	160,000
Optics	2	50,000
Physics Education	3	4,500,000
Total	13	5,880,000

FACULTY

Professors

Burnett, Lowell J., Ph.D., Wyoming, 1972. Emeritus. Nuclear magnetic resonance; artificial membranes; scientific instrumentation.

Cottrell, Don M., Ph.D., Washington, 1967. Emeritus. Cosmic rays; theoretical physics; gravity theory; elementary particles; optical pattern recognition.

Davis, Jeffrey A., Ph.D., Cornell, 1970. Optics; optical pattern recognition; computer generated holograms.

Goldberg, Fred M., Ph.D., Michigan, 1971. Physics education.

Johnson, Calvin, W. Ph.D., Univ. of Washington, 1989. Theoretical nuclear physics, Astrophysics.

Lilly, Roger A., Ph.D., Hawaii, 1968. Emeritus. Atomic spectroscopy, experimental and theoretical; optics.

Morris, Richard H., Ph.D., California, Berkeley, 1957. Electromagnetic theory; laser physics.

Oseroff, Saul B., Ph.D., Nat. Univ. of CUYO, 1972. Emeritus. Experimental condensed matter physics.

Papin, Patrick J., Ph.D., UCLA, 1985. Associate Dean for Academic Affairs. Radiation dosimetry; medical imaging.

Piserchio, Robert J., Ph.D., Arizona, 1966. Emeritus. Musical acoustics.

Rehfuss, Donald E., Ph.D., Oregon, 1962. Emeritus. Physics Education.

Roeder, Stephen B.W., Ph.D., Wisconsin, 1968. Emeritus. Imperial Valley Campus. Magnetic resonance. Scientific instrumentation.

Shore, Herbert B., Ph.D., California, Berkeley, 1965. Emeritus. Theoretical condensed matter physics; computational physics.

Sinha, Usha, Ph.D., IIS Bangalore, 1985. Department Chair and Director of Medical Physics Program. Medical physics.

Sweedler, Alan R., Ph.D., California, San Diego, 1970. Assistant Vice President, International Programs. Physical and environmental science; computer modeling of energy systems.

Templin, Jacques D., Ph.D., UCLA, 1962. Emeritus. Theoretical physics.

Torikachvili, Milton S., Ph.D., University of Campinas, 1978. Experimental condensed matter physics.

Weber, Fridolin, Ph.D., Univ. of Munich, Germany, 1992. Theoretical Nuclear Physics, Astrophysics.

Associate Professors

Anderson, Matt, Ph.D., U. of Oregon, 1998. Ultrafast laser physics.

Baljon, Arlette R.C., Ph.D., U. of Chicago, 1993. Computational soft condensed matter physics.

Bromley, Michael J., Ph.D., Northern Territory University (NTU), Darwin, 2002. Atomic physics.

Wallace, William J., Ph.D., Oregon State University, 1969. Emeritus. Oceanography, History of Science.

RESEARCH SPECIALTIES AND STAFF

Theoretical

Atomic. Bromley.

Computational Physics. Baljon, Bromley, Johnson, Papin, Sweedler, Weber.

Environmental Sciences. Sweedler.

Public Policy. Sweedler.

Experimental

Condensed Matter. Oseroff, Torikachvili.

Education. Goldberg.

Laser Physics. Anderson, Davis, Morris.

Medical Imaging. Papin, Sinha.

Optical Information Processing. Davis.

Radiation Dosimetry. Papin.

SAN DIEGO STATE UNIVERSITY

DEPARTMENT OF ASTRONOMY

San Diego, California 92182

Students Accepted For Degree	FIELDS		
	Physics	Astronomy	Related Fields
Doctorate			
Master's		X	

1. General

President: Stephen L. Weber
Dean of Graduate School: Thomas Scott
Department Chairman: Allen W. Shafter
Department Telephone Number: (619) 594-6182
Type of Institution: University
Control: Public
Setting: Urban
Total Faculty: 1,795
Total Graduate Faculty: 894
Total Students: 32,817
Total Graduate Students: 6,021
Annual Graduate Tuition:
In-state residents: None
Out-of-state residents: $372/credit (waiver possible)
Tuition rates for: 2009–10
Deferred tuition plan: Yes
Other Fees: 0–6 units: $2,000
66.1 units or more: $3,095
Term: Semester

2. Number of Faculty in Department

The combined total of full-time faculty in the three professorial ranks is 6. The combined total of full-time, part-time, and other faculty at all ranks is 12.

3. Admission, Financial Aid, and Housing

Address admission inquiries to: Graduate Admissions. Enrollment Services. SDSU, San Diego, CA 92182-7416
Graduate application fee required: $55
Admission deadline (Fall admission): 2/1
Admission information: For fall admission, 2008–09, 10 students were accepted from 22 applicants.
Admission requirements: For admission to the graduate programs, a Bachelor's degree in a physical science is required with a minimum undergraduate upper division GPA of 2.85 specified. The GRE is required. No minimum score is required. The GRE Physics exam is strongly recommended but not required. Students from non-English speaking countries are required to demonstrate proficiency in English via the TOEFL exam. The minimum score required is 213 on the computer-based TOEFL exam.
Address financial aid inquiries to: The Office of Financial Aid and Scholarships, SDSU, 5500 Campanile Drive, San Diego, CA 92182-7436
GAPSFAS application required: Yes
Financial aid deadline: 5/31
Loans available: Yes
Address housing inquiries to: Office of Housing Administration and Residential Education Office, SDSU, 5500 Campanile Drive, San Diego, CA 92182-1802
On-campus, single student housing available: yes
Cost (room and board)/year: $9,352–10,093
On-campus, married student housing available: Yes

Table A—Faculty, Enrollments, and Degrees Granted

Research Specialty	2008–09 Faculty	Enrollment[1] Fall 2009 Master's	Enrollment[1] Fall 2009 Doctorate	No. of Degrees Granted[2] 2008–09 (2004–09) Master's	Terminal Master's	Doctorate	Median No. of Years for 2008–09 Ph.D.'s
Astronomy	11	19	–	–	4(14)	–	–
Total		19	–	–	4(14)	–	
Full-time Grad. Stud.		13	–				
Part-time Grad. Stud.		4	–				
First-year Grad. Stud.		5	–				
Median Years in Grad. Study (2008–09 Degrees)				–	2.5	–	–

Undergraduate Degrees, 2008–09 (2004–09): 3(15)

[1]Students not yet committed to a research specialty are entered under non-specialized.
[2]Five-year totals in parentheses.

4. Graduate Degree Requirements

Master's: Students must complete a graduate program of which 18 units must be graduate course work in astronomy. An additional 12 units must be in astronomy research or related fields as approved by the department graduate advisor. There is a minimum of 24 units in residence. Students must pass a final oral exam on the thesis or published paper and maintain a 3.0 GPA.
Special Equipment, Facilities, or Programs: Three optical research telescopes of 1.0 meter (two) and 0.6 meter apertures are equipped for photometry, direct imaging, and spectroscopy at SDSU's Mount Laguna Observatory 50 miles east of campus, CCD cameras for use in direct and spectrographic modes automated photometers, and near-infrared imaging camera.

Table B—Appointments to Graduate Students, 2009–10

Title of Appointee	Appointments Total	Appointments First year	Academic Load Allowed in Credit Hours	Hours of Service Per Week	Stipend for Academic Year ($)
			Semester		
Teaching Assistant	8	5	-	10-20	6,000-12,000
Research Assistant	3	2	-	10-20	6,000-12,000
Research Grants	6	0	-	10-20	6,000-12,000
Total	11	7			

5. Personnel Engaged in Separately Budgeted Research, 7/08–6/09

Professorial faculty	6
Graduate students	5
Total	11

6. Separately Budgeted Research Expenditures by Source of Support

	Departmental Research	Physics-related Research Outside Department
Federal Government	$504,580	
Private, nonprofit organizations	$698,593	
Total	$1,203,173	

Table C—Separately Budgeted Research Expenditures

Research Specialty	No. of Grants	Expenditures ($)
Astronomy	15	
Total	15	

FACULTY

Professors

Angione, Ron J., Ph.D., Texas, 1970, Emeritus. Galaxies, QSO's, and atmospheric physics.
Daub, C. T., Ph.D., Wisconsin, 1962, Emeritus. Gaseous nebulae.
Etzel, Paul B., Ph.D., UCLA, 1986. Binary stars; photometry; spectroscopy.
Leach, Robert L., Ph.D., Harvard, 1980. Instrumentation, optical and near-IR detectors, optical observations of star-forming regions, galaxies.
Sandquist, Eric L., Ph.D., U.C. Santa Cruz, 1996. Stellar astrophysics.
Shafter, Allen, Ph.D., UCLA, 1983. Chairman of the Department. Cataclysmic Variable stars; galaxies.
Young, Arthur, Ph.D., Indiana, 1967, Emeritus. Stellar spectroscopy; binary stars.

Associate Professors

Leonard, Douglas C., Ph.D., UC Berkeley 2000. Supernovae; extragalactic distance scale.
Orosz, Jerome A., Ph.D., Yale University, 1996. Black hole binary stars, extrasolar planets, binary stars.
Talbert, Freddie D., Ph.D., Texas, 1968, Emeritus. Galactic clusters.
Welsh, William F. Ph.D., Ohio State, 1993. AGN, cataclysmic variables, extra solar planets.

RESEARCH SPECIALTIES AND STAFF

Theoretical

Astrophysics. Sandquist. Gaseous nebulae. Daub.

Experimental

Atmosphere/Space Physics: Angione, Orosz.
Binary Stars: Angione, Etzel, Orosz, Shafter, Welsh, A. Young.
Extragalactic: Angione, Leach, Leonard, Shafter, Welsh.
Extrasolar Planets: Orosz, Sandquist, Welsh.
Galactic Globular Clusters: Talbert, Sandquist.
Photometry: Angione, Etzel, Leonard, Orosz, Sandquist, Welsh.
Planetary Nebulae: Daub.
Stellar Spectroscopy: Etzel, Leonard, Orosz, A. Young.

SAN FRANCISCO STATE UNIVERSITY

DEPARTMENT OF PHYSICS AND ASTRONOMY

San Francisco, California 94132

Students Accepted For Degree	FIELDS		
	Physics	Astronomy	Related Fields
Doctorate			
Master's	X	X	

1. General

President: Robert A. Corrigan
Dean of Graduate School: Ann Hallum
Department Chair: Susan M. Lea
Department Telephone Number: (415) 338-1659 (C)
Department Fax Number: (415) 338-2178
Type of Institution: University
Control: Public
Setting: Urban
Total Faculty: 1,180
Total Graduate Faculty: 820
Total Students: 26,000
Total Graduate Students: 5,300
Annual Graduate Tuition:
In-state residents: —None
Out-of-state residents: Full-time—$372/credit
Tuition rates for: 2010–11
Deferred tuition plan: No
Annual Other Fees: $5,746(6.1 units or more); $3,664 (6.0 units or less)
Term: Semester

2. Number of Faculty in Department

The combined total of full-time faculty in the three professorial ranks is 12. The combined total of full-time, part-time, and other faculty at all ranks is 20.

3. Admission, Financial Aid, and Housing

Address admission inquiries to: Registrar's Office and/or Department of Physics and Astromony
Graduate application fee required: $55
Admission deadline (Fall admission): 5/1, 4/1 For international students
Admission information: For fall admission, 2009–10, 8 students were accepted from 24 applicants.
Admission requirements: For admission to the graduate programs, a Bachelors' degree in physics, engineering, or mathematics is required with a minimum undergraduate GPA of 3.0 specified. Bachelor's degrees in other disciplines may be acceptable. Consult department. The GRE is required. The GRE Advanced is recommended. Students from non-English speaking countries are required to demonstrate proficiency in English via the TOEFL exam. Minimum acceptable score for admission is 550/213.
Undergraduate preparation assumed: Marion, *Classical Dynamics*; Griffiths, *Electrodynamics*; Zemansky and Dittman, *Heat and Thermodynamics*; Eisberg, *Quantum Physics for Atoms, Molecules, Solids, Nuclei, and Particles*; Boyce and DiPrima, *Ordinary Differential Equations*.
Address financial aid inquiries to: S. M. Lea, Chair, Department of Physics and Astronomy
GAPSFAS application required: No
Financial aid deadline: 4/1
Loans available: Yes
Address housing inquiries to: Housing Office
On-campus, single student housing available: Yes
Cost/term: $4,770
On-campus, married student housing available: No

Table A—Faculty, Enrollments, and Degrees Granted

Research Specialty	2009–10 Faculty	Enrollment[1] Fall 2009 Master's	Enrollment[1] Fall 2009 Doctorate	No. of Degrees Granted[2] 2009–10 (2005–10) Master's	Terminal Master's	Doctorate	Median No. of Years for 2009–10 Ph.D.'s
Astronomy	3	7	–	–	3(12)	–	–
Astrophysics	2	3	–	–	1(9)	–	–
Biophysics	0	0	–	–	0(2)	–	–
Condensed Matter Physics	3	4	–	–	1(9)	–	–
Nuclear Physics	0	0	–	–	0(0)	–	–
Optics	2	6	–	–	1(8)	–	–
Particles & Fields	3	2	–	–	0(3)	–	–
Physics Education	2	1	–	–	1(2)	–	–
Relativity & Gravitation	0	0	–	–	0(2)	–	–
Non-specialized	0	22	–	–	3(8)	–	–
Total		45	–	–	10(53)	–	
Full-time Grad. Stud.		32	–				
Part-time Grad. Stud.		13	–				
First-year Grad. Stud.		10	–				
Median Years in Grad. Study (2009–10 Degrees)			–	2.5		–	

Undergraduate Degrees, 2009–10 (2005–10): 14(53)

[1]Students not yet committed to a research specialty are entered under non-specialized.
[2]Five-year totals in parentheses.

4. Graduate Degree Requirements

Master's: A total of 30 semester units with a "B" average is required. Fifteen semester units in dynamics, electromagnetic theory, mathematical physics, statistical mechanics, and quantum mechanics and six semester units of other graduate physics courses must be included. Nine units of upper division and graduate level courses in mathematics, science, engineering, or other appropriate fields, selected with the approval of the graduate advisor, complete the course requirements. 24 semester units must be completed in residence. Either a Master's Comprehensive Oral Examination or a Master's Thesis with oral defense is required.
Thesis: Thesis may be written *in absentia*.
Special Equipment, Facilities, or Programs: Instrumentation and computational facilities for elementary particle physics; nuclear sources and counters, Mössbauer spectrometer; optics laboratory with spectrometers, laser optics, and holography; vacuum deposition equipment with cleanroom and photolithography facilities; solid state laboratory with NMR and ESR apparatus, ultrahigh vacuum capability, and advanced instrumentation; low-temperature laboratory with temperatures to 20 mK, SQUID systems, high-field superconducting magnets; astronomical observatory, high-quality planetarium.

Table B—Appointments to Graduate Students, 2009–10

Title of Appointee	Appointments Total	Appointments First year	Academic Load Allowed in Credit Hours	Hours of Service Per Week	Stipend for Academic Year ($)
			Semester		
Teaching Assistant	16	8	6–9	16	10,500
Research Assistant	10	2	6–9	10–20	9,000–15,000
Total	26	10			

5. Personnel Engaged in Separately Budgeted Research, 8/09–6/10

Professorial faculty	9
Other faculty	1
Graduate students	12
Undergraduate students	5
Total	27

6. Separately Budgeted Research Expenditures by Source of Support

	Departmental Research	Physics-related Research Outside Department
Federal government	$620,000	
Total	$620,000	

Table C—Separately Budgeted Research Expenditures

Research Specialty	No. of Grants	Expenditures ($)
Astrophysics	7	250,000
Condensed Matter Physics	3	110,000
Optical Physics	3	150,000
Particles & Fields	1	110,000
Total	14	620,000

FACULTY

Professors

Chen, Zhigang, Ph.D., Bryn Mawr, 1995. Lasers and non-linear optics, photorefractive materials; solitons.

Cool, Andrienne, Ph.D., Harvard, 1994. Observational astronomy; stellar astrophysics.

Greensite, Jeffrey P., Ph.D., California, Santa Cruz, 1981. Theoretical elementary particle physics.

Golterman, Maarten, Ph.D., Amsterdam, 1986. Theoretical physics; particle physics.

Lea, Susan M., Ph.D., California, Berkeley, 1974. Department Chair. Theoretical astrophysics: accretion dynamics, x-ray astronomy, plasma astrophysics.

Lockhart, James M., Ph.D., Stanford, 1976. Experimental solid state physics; low-temperature physics; SQUID detectors; instrumentation.

Marzke, Ronald, Ph.D., Harvard, 1994. Observational cosmology; galaxy formation and evolution; universe structure.

Neuhauser, Barbara J., Ph.D., Stanford, 1985. Low-temperature physics; ultralow temperatures; superfluid ^{3}He; neutrino detectors.

Assistant Professors

Barranco, Joseph A., Ph.D., California, Berkeley, 2004. Theoretical and Computational Astrophysics; astrophysical and geophysical fluid dynamics; accretion disks; Star and Planet formation.

Lepeshkin, Nick, Ph.D., New Mexico State, 2001. Non-linear optics, integrated optics, photonic crystals, plasmonic materials.

Mahdavi, Andisheh, Ph.D., Harvard, 2001. Observational and computational astrophysics; groups and clusters of galaxies; dark matter; plasma astrophysics; X-ray astronomy; galactic dynamics; gravitational lensing.

Man, Weining, Ph.D., Princeton, 2005. Experimental and numerical soft condensed matter physics; colloid physics; colloidal suspension thin films; photonic band gap materials; quasicrystals; granular materials.

Adjunct Professors

Adler, Ronald J., Ph.D., Stanford, 1965. Relativity theory; superconductivity.

Barsony, Mary, Ph.D., Cal. Tech., 1989. Observational astronomy; infrared and x-ray astronomy, star formation.

Bland, Roger W., Ph.D., California, Berkeley, 1968. Experimental particle physics; underwater acoustics.

Fischer, Debra, Ph.D., California, Santa Cruz, 1998, Observational astronomy; extrasolar planet searches, orbital dynamics, spectroscopy, spectral synthesis, interferometry.

Lipschultz, Fred, Ph.D., Cornell, 1966. Low temperature physics; physics education.

Marcy, Geoffrey W., Ph.D., California, Santa Cruz, 1982. Stellar astrophysics; observational astronomy.

McCarthy, Chris, Ph.D., California Los Angeles, 2001. Brown dwarf stars; extra-solar planets; stellar spectroscopy.

RESEARCH SPECIALTIES AND STAFF

Theoretical

Astrophysics. Barranco, Barsony, Cool, Fischer, Lea, Mahdavi, Marcy, Marzke, McCarthy.
Elementary Particles. Golterman, Greensite.
General Relativity. Adler.
Mathematical Physics. Lea.
Physics Education. Barranco, Cool, Lea.

Experimental

Astronomy. Barranco, Barsony, Cool, Fischer, Mahdavi, Marcy, Marzke, McCarthy.
Elementary Particles. Bland.
Low Temperature. Bland, Lipschultz, Lockhart, Neuhauser.
Optics. Chen, Lepeshkin, Man.
Solid State. Bland, Chen, Lockhart, Man, Neuhauser.

SAN JOSE STATE UNIVERSITY

DEPARTMENT OF PHYSICS

San Jose, California 95192-0106

Students Accepted For Degree	FIELDS		
	Physics	Astronomy	Related Fields
Doctorate			
Master's	X		

1. General

President: Dr. Jon Whitmore
Associate Vice President Graduate Studies & Research: Dr. Pamela Stacks
Department Chairman: Dr. Kiumars Parvin
Department Telephone Number: (408) 924-5210
Type of Institution: University
Control: Public
Setting: Urban
Total Faculty: 1,740
Total Students: 28,007
Total Graduate Students: 5,766
Estimated Annual Graduate Tuition:
Tuition rates for: 2009–10
Other Estimated Fees: $5,162/yr. (6.1 units & above)
$3,428/yr. (1–6 units)
Term: Semester

2. Number of Faculty in Department

The combined total of full-time faculty in the three professorial ranks is 11. The combined total of full-time, part-time, and other faculty at all ranks is 31.

3. Admission, Financial Aid, and Housing

Address admission inquiries to: Admissions Office
Graduate application fee required: $55
Admission deadline (Fall admission): U.S. Residents-5/1
Foreign Residents-3/1
Admission information: For fall admission, 2008–09, 16 students were accepted from 28 applicants.
Admission requirements: For admission to the graduate programs, a Bachelor's degree is required with a minimum undergraduate GPA of 2.5 specified for the last 60 semester (90 quarter) units attempted at an accredited institution. The GRE Physics is not required for admission to the program, but it must be taken with a score of 550 no later than two semesters prior to graduation. Students from non-English speaking countries are required to demonstrate proficiency in English via the TOEFL exam. Minimum acceptable score for admission is 550.
Undergraduate preparation assumed: Young and Freedman, *University Physics*; Marion, *Classical Dynamics of Particles & Systems*; Beiser, *Concepts of Modern Physics*; Griffiths, *Introduction to Electrodynamics*; Griffiths, *Introduction to Quantum Mechanics*.
Address financial aid inquiries to: Financial Aid Office
GAPSFAS application required: No
Financial aid available only to U.S. residents; deadline: 3/2
Loans available to U.S. residents: Yes
Address housing inquiries to: Attn.: Housing Office
On-campus, single student housing available: Yes
Cost/year: $6,472
On-campus, married student housing available: Yes

Table A—Faculty, Enrollments, and Degrees Granted

Research Specialty	2007–08 Faculty	Enrollment[1] Fall 2007		No. of Degrees Granted[2] 2007–08 (2000–07)			Median No. of Years for 2007–08 Ph.D.'s
		Mas-ter's	Doc-torate	Mas-ter's	Terminal Master's	Doc-torate	
Acoustics	1	0	–	–	0(0)	–	–
Applied Physics	1	0	–	–	0(0)	–	–
Astronomy	4	0	–	–	0(0)	–	–
Astrophysics	4	0	–	–	0(0)	–	–
Atmos./Space Phys., Cosmic Rays	1	0	–	–	0(0)	–	–
Biophysics	1	0	–	–	0(0)	–	–
Celestial Mechanics	1	0	–	–	0(0)	–	–
Computational Physics	4	2	–	–	0(0)	–	–
Condensed Matter Physics	3	0	–	–	0(0)	–	–
Fluids & Rheology	1	0	–	–	0(0)	–	–
Gas Dynamics	1	0	–	–	0(0)	–	–
Holography	2	0	–	–	0(0)	–	–
Laser Spectroscopy	1	0	–	–	0(0)	–	–
Non-linear Physics	1	0	–	–	0(0)	–	–
Optics	3	0	–	–	0(0)	–	–
Particles & Fields	1	0	–	–	0(0)	–	–
Physics Education	2	0	–	–	0(0)	–	–
Plasma Physics & Fusion	1	0	–	–	0(0)	–	–
Quantum Physics	3	0	–	–	0(0)	–	–
Solid State Physics	3	0	–	–	0(0)	–	–
Non-specialized	0	28	–	–	7(36)	–	–
Total		30	–	–	7(36)	–	–
Full-time Grad. Stud.		0	–		5(81)		
Part-time Grad. Stud.		0	–				
First-year Grad. Stud.		11	–				
Median Years in Grad. Study (2006–07 Degrees)				–	–	–	–

Undergraduate Degrees, 2007–08 (2000–08): 6(52)

[1] Students not yet committed to a research specialty are entered under non-specialized.
[2] Five-year totals in parentheses.

4. Graduate Degree Requirements

Master's: A total of 30 semester units with a B average is required. Fifteen semester units in mathematical physics, advanced dynamics, electromagnetic theory, statistical physics, and quantum mechanics and at least six semester units of other graduate physics courses must be included. Twelve units of graduate level and/or upper division courses in mathematics, science, engineering, or other appropriate fields, selected with the approval of the graduate advisor, complete the course requirements. 24 semester units must be completed in residence. All graduate students must attend weekly department seminars in at least one semester, must achieve a satisfactory score on the Physics portion of the GRE no later than two semesters prior to graduation and must satisfy the English writing requirement prior to the semester of graduation. A comprehensive oral examination, literature review, or a thesis presentation is required.

Thesis: Optional.

Special Equipment, Facilities, or Programs: The department has academic options in optics and condensed matter, and offers a concentration in computational physics. The Institute for Modern Optics coordinates optics research and collaboration with local industries. A 4,000-sf instruction and research facility contains equipment for laser spectroscopy, non-linear optics, Fourier optics, holography, and optical metrology. A world-class center for novel laser materials research and modeling of laser systems is located here. The solid state laboratory includes equipment for measuring magnetic susceptibility over a 4–1,000 K temperature range. Computing equipment consists of on-campus workstations and internet access to remote facilities and databases. A 10,000-sf Nuclear Science Facility is used by all science departments. Equipment includes neutron and gamma irradiators; x-ray fluorescence, magnetic resonance, and Mössbauer spectrometers; and a range of radiation analyzing equipment. Several faculty carry out research with colleagues at NASA Ames Research Center. Located in "Silicon Valley," we have research and instructional collaborations with many local instrumentation, materials, semiconductor, and optics companies.

5. Personnel Engaged in Separately Budgeted Research, 7/07–6/08

Professorial faculty	11
Other faculty	1
Graduate students	6
Undergraduate students	10
Total	28

6. Separately Budgeted Research Expenditures by Source of Support

	Departmental Research	Physics-related Research Outside Department
Federal and state government	4,843,000	720,000
Business and industry	840,000	820,000
Total	5,683,000	1,540,000

Table C—Separately Budgeted Research Expenditures

Research Specialty	No. of Grants	Expenditures ($)
Astronomy	2	100,000
Atmos./Space Phys.	4	200,000
Astronomy Education	2	10,000
Novel Laser Materials	2	490,000
Physics Education	1	75,000
Q-Switch Modeling of Lasers	6	4,570,000
Solid State	1	188,000
Total	18	5,683,000

Table D—Physics-related Research Outside Department

Field and Unit Outside Department	No. of Grants	Expenditures ($)
Astrophysics	4	365,000
Optical Spectroscopy	3	720,000
Laser Optics	4	455,000
Total	11	1,540,000

FACULTY

Professors

Bahuguna, Ramendra D., Ph.D., Indian Inst. of Tech., Delhi, 1979. Holographic interferometry; laser speckle metrology; display holography; Fourier optics; fingerprint verification.

Boekema, Carolus, Ph.D., Groningen (NL), 1977. Magnetism and superconductivity in solids; computational condensed matter physics; muon spin research: Mössbauer spectroscopy.

Garcia, Alejandro, Ph.D., Texas, Austin, 1984. Computational fluid mechanics; statistical mechanics.

Holmes, Brian W., Ph.D., Boston, 1980. Musical acoustics; sports physics; physics education.

Kaufman, Michael J., Ph.D., Johns Hopkins University, 1995. Astrophysics: interstellar medium, interactions of young stars with molecular clouds, dynamics and chemistry of molecular shocks, infrared/submillimeter observations.

Lam, Lui, Ph.D., Columbia, 1973. Histophysics; Nonlinear physics; liquid crystals; pattern formation; complex systems.

Parvin, Kiumars, Ph.D., California, Riverside, 1978. Experimental solid state physics; magnetic materials.

Associate Professors

Batalha, Natalie M., Ph.D., University of California at Santa Cruz, 1997. Stellar astrophysics: variable stars, magnetic activity, T Tauri stars; Extra-solar planet detection.

Wharton, Kenneth B., Ph.D., University of California at Los Angeles, 1998. Plasma physics, laser-plasma interactions; sub-picsecond x-ray sources, foundations of quantum mechanics, relativistic quantum mechanics.

Assistant Professors

Beyersdorf, Peter T., Ph.D., Stanford University, 2001. Gravitational wave detection, precision measurements and optical interferometry.

Kress, Monika, Ph.D., Rensselaer, 1997. Astrophysics and planetary science. Computer modeling of protoplanetary disks, comet impacts, and planetary environments; meteorites; astrobiology.

Lecturers

Berman, Irina, Ph.D., Moscow State University, 1989. Superconductivity of conventional, high-temperature and disordered superconductors; localization effects and superconductivity.

Hubickyj, Olenka, Ph.D., CUNY, 1983. RR Lyrae stars' instability strip; primordial atmosphere of Earth; convection turbulence in the Solar Nebula; formation and evolution of gas giant planets.

Kwok, Ray, Ph.D., UCLA, 1990. Condensed Matter Physics, Solid State Physics, Phase Transitions, Electrical and Thermal Transport, Muon Spin Rotation, Applied Superconductivity RF and Microwave Communications, Applied Physics.

Mosqueira, Ignacio, Ph.D., 1995. Cornell University. Planetary Formation.

Sauke, Todd B. Ph.D., University of Illinois, Urbana-Champaign, 1989. Tunable diode laser technology; development of planetary exploration applications for the measurement of isotopic ratios in planetary surface and atmospheric samples.

Sherman, Douglas, Ph.D., University of California at Berkeley, 1987. Hydrogen Solubility and diffusivity in refractory ceramics.

Adjunct Professors

Bolton, Paul R., Ph.D. Yale, 1982. Laser design and laser spectroscopy; laser ablation; strong optical field atomic and plasma physics; ultrafast laser-driven phenomena and diagnostics; laser applications to accelerator development; xuv/x-ray spectroscopy; electron photoinjectors.

Castellano, Timothy, Ph.D., University of Califorina at Santa Cruz, 2001. Detection of extrasolar planets by transit method; stellar main sequence variabilty.

Freund, Friedemann, Ph.D., Marburg Univ., 1959. Defects in crystals; proton conductivity.

Professors Emeriti

Anderson, Merlin F., Ph.D., Oregon State, 1966. Gas dynamics; computer applications.

Becker, Joseph F., Ph.D., NYU, 1976. Spectroscopy; biophysics; optics, optoelectronic devices.

Bloomer, Iris L., Ph.D., London, 1976. General relativity; optical properties of materials.

Finkelstein, Jerome, Ph.D., California, Berkeley, 1967. Theoretical physics; elementary particles.

Gruber, John B., Ph.D., California, Berkeley, 1961. Engineering physics; solid state spectroscopy; rare-earth/transition-metal ion solid state lasers; chemical physics; quantum electronics.

Hamill, Patrick, Ph.D., Arizona, 1971. Atmospheric physics; aerosol physics; celestial mechanics.

Morris, Marvin L., Ph.D., Utah, 1966. High-energy cosmic rays; computer-aided instruction; physics education.

Muirhead, Franklin R., Ph.D., Sheffield, 1961. Solid state physics; magnetism.

Strandburg, Donald L., Ph.D., Iowa State, 1961. Low-temperature physics; magnetism.

Tomley, Leslie J., Ph.D., Washington, 1968. Astrophysics.

Tucker, Allen B., Ph.D., Stanford, 1965. Nuclear physics; cosmic rays; accelerator mass spectrometry; health physics.

Williams, Gareth T., Ph.D., University of Wales, 1960. Optics; holography.

RESEARCH SPECIALTIES AND STAFF

Theoretical

Astrophysics. Batalha, Kaufman, Kress, Tomley.
Atmospheric Physics. Hamill.
Celestial Mechanics. Hamill.
Complex Systems. Lam.
Computational Physics. Boekema, Garcia, Kress, Lam.
Condensed Matter. Boekema, Lam.
Elementary Particles. Finkelstein.
Fluid Mechanics. Garcia.
Gas Dynamics. Garcia.
Nonlinear Physics. Lam.
Physics Education. Becker, Holmes.
Planetary Sciences. Kress.
Quantum Mechanics. Finkelstein, Wharton.
Solid State Physics. Boekema.
The Physics of Musical Instruments. Holmes

Experimental

Acoustics. Holmes.
Applied Physics. Bahuguna.
Astrophysics. Batalha, Kaufman, Kress, Tomley.
Biophysics. Becker.
Holography. Bahuguna, Holmes.
Laser Interactions with Matter. Beyersdorf, Wharton.
Laser Spectroscopy. Becker.
Nonlinear Physics. Lam.
Optics. Bahuguna, Becker, Beyersdorf.
Physics Education. Becker, Holmes.
Plasma Physics. Wharton.
Quantum Optics. Beyersdorf.
Solid State Physics. Boekema, Parvin.
Ultrafast Phenomena. Wharton.

FACULTY PUBLICATIONS

The following bibliography is a representative selection of some of the articles published by the faculty of the Physics Department. A comprehensive list of faculty publications is available by writing to the department.

Bahuguna, Ramendra D.

L. E. Jusinski, R. D. Bahuguna, A. Das, and K. Arya, "Surface Enhanced Raman Spectroscopy from a molecule absorbed on a nanoscale silver Particle cluster in a Holographic Plate," Proceedings of SPIE, vol. 6099, 1 (2006).

R. D. Bahuguna and T. Cosboline, "Jitter mechanism for a real time fingerprint sensor and verification system," U.S. Patent No. 6, 341, 028 (Jan. 28, 2002).

R. D. Bahuguna, K. Arya, J. F. Becker, J. B. Gruber, H. S. Lakkaraju, K. B. Wharton, and G. T. Williams, "The Optics Program in the Physics Department at San José State University," SPIE 7th International Conference on Education in Optics and Photonics, November 2001, Singapore.

Batalha, Natalie M.

W. J. Borucki, D. G. Kock, J. Lissauer, G. Basri, N. Batalha, T. Brown, D. A. Caldwell, J. M. Jenkins, J. J. Caldwell, J. Chritensen-Dalsgaard, W. D. Cochran, E. W. Dunham, T. N. Gautier, J. C. Geary, D. Latham, D. Sasselov, R. L. Gilliland, S. Howell, D. G. Monet, "Kepler Mission Status," ASPC Series **366**, 309 (2007).

N. Batalha, "Selection and Prioritization of Targets for the Kepler Mission," Kepler Project Office Technical Report, KADN-26182. Abstract: AAS, **211**, 135.16 (2007).

N. Batalha, W. Borucki, D. Caldwell, H. Chandrasekharan, T. N. Gautier, J. Jenkins, D. Kock, "Optimization of the Kepler Field of View," Kepler Project Office Technical Report, KADN-26013, Abstract: BAAS **38**, 1118 (2006).

C. Batalha, N. Batalha, S. Alencar, D. Lopes, E. Duarte, Variability of Southern T Tauri Stars (VASTT) III. The Continuum Flux Changes of the TW Hydrae Bright Spot, 2002, Astrophysical Journal **580**, 343.

Becker, Joseph F.

T. B. Sauke and J. F. Becker, "Modeling, measuring, and minimizing the instrument response function of a tunable diode laser spectrometer," Journal of Quantitative Spectroscopy & Radiative Transfer **91**, 453–484 (2005).

T. B. Sauke and J. F. Becker, "An Ultrasensitive Near-Infrared Spectrometer Using an Active Frequency-Locked Continuous-Wave (cw) Ringdown Cavity," NASA Ames Director's Discretionary Fund Report for 2001 (NASA/TM-2002-211398) March 2002, pp. 40–41.

T. B. Sauke and J. F. Becker, "Stable Isotope Laser Spectrometer for Exploration of Mars," Planetary and Space Science **46**, 805–812 (1998).

Berman, Irina V.

Minina N. Ya., Bogdanov E. V., Ilievsky A. A., Polyanskiy A. V., Kraak W., Savin A. M., Berman I. V. Two-dimensional electrons at n-GaAs/AlGaAs heterointerface under uniaxial compression-phys. Stat. Sol. (B) 244, No. 1, 0-65-69, 2007.3.

A. V. Kobelev, R. M. Kobeleva, Yu. L. Protsenko, O. A. Kobelev, I. V. Berman, "Viscoelastic models describing stress relaxation and creep in soft tissues," Mater. Res. Soc. Symp. Proc., V.874@2005 Material Research Society.

I. V. Berman, E. V. Bogdanov, A. A. Ilievsky, N. Ya. Minina, W. Kraak, "Pressure dependence of 2D hole mobility in thermoactivated photoconductivity effect observed in p-GaAs/$Al_{0.5}Ga_{0.5}As$ heterostructures," Phys. Stat. Sol.(b) **241**, N14, 3410–3415 (2004).

Beyersdorf, Peter

J. Xia, P. T. Beyersdorf, M. M. Fejer, and A. Kapitulnik, "Modified Sagnac interferometer for high-sensitivity magneto-optic measurements at cryogenic temperatures," submitted to Applied Physics Letters, March 2006.

P. T. Beyersdorf, S. Kawamura, K. Somiya, F. Kawazoe, and M. Agueros, "Powere-recycled resonant sideband extraction interferometer with polarization detection," Applied Optics, June 2005, **44**, No. 17, 3413–3424.

K. Somiya, P. T. Beyersdorf, K. Arai, S. Sato, S. Kawamura, O. Miyakawa, F. Kawazoe, S. Sakata, A. Sekido, and N. Mio, "Development of a frequency-detuned interferometer as a prototype experiment for next-generation gravitational-wave detectors," Applied Optics, June 2005, **44**, No. 17, 3179–3191.

Boekema, Carolus

C. Boekema, M. C. Browne, and C. Teichgraeber, "The Field (Direction) dependence of observed Magnetism in $YBa_2Cu_3O_7$ Vortex states," Proc. 25th Int. Conf. Low Temp. Phys. (Aug. 2008), N. L. A'dam, J. Phys. Conf. Series 150, 052022 (2009). http://jpcs.iop.org/LT25

C. Boekema and M. C. Brown, "MaxEnt-Burg Application to Muon-Spin Resonance," (MaxEnt 2008, Sao Paulo, Brazil, July 2008). AIP Conf. Proc. #1073, pp. 260–267.

T. Songatikamas, C. Boekema, J. Wong, H. Ngo, and M. C. Browne, "Predicted magnetic fields of loop currents for cuprate superconductivity: A MaxEnt-mSR GdBCO study," Proc. 7th Int. New 3SC Conf (May 2009 Beijing) J. Supercond. Nov. Magn (January 2010) online & ISSN 1557-1939.

Bolton, Paul R.

J. S. T. Ng *et al.*, "Observation of Plasma Focusing of a 28.5 GeV Positron Beam," Phys. Rev. Lett. **87**(24), 244801–1 (12/10/01).

P. R. Bolton *et al.*, "Photoinjector Design for the LCLS," SLAC PUB-8962 and LCLS-TN-01-5 (November 2001).

Castellano, Timothy

T. Castellano, L. Doyle, D. McIntosh, "The visibility of earth transits," PASP Symposium #202, pp. 445–447 (2004).

T. Castellano, "A search for planetary transits of the Star HD 187123 by spot filter CCD differential photometry," Publications of the Astronomical Society of the Pacific **112** (June), 821 (2000).

T. Castellano, J. Jenkins, D. Trilling, L Doyle, and D. Koch, "Detection of planetary transits of the Star HD 209458 in the Hipparcos data set," Astrophysical Journal **20** (March), 532 (2000).

Finkelstein, Jerome

J. Finkelstein "Pure-state informationally complete and 'really' complete measurements," Phys. Rev. A**70**, 052107 (2004).

J. Finkelstein, "Comment on 'How macroscopic properties dictate microscopic probabilities,'" Phys. Rev. A **67**, 026101 (2003).

T. A. Brun, J. Finkelstein, and N. D. Mermin, "How much state assignments can differ," Phys. Rev. A **65**, 032315 (2002).

Freund, Friedemann

F. T. Freund, On the electrical conductivity structure of the stable continental crust, J. Geodynamics **35**, 353–388, 2003.

F. Freund, Charge generation and propagation in rocks, J. Geodynamics **33**, 545–572, 2002.

F. T. Freund, Rocks that crackle and sparkle and glow-Strange pre-earthquake phenomena, J. Sci. Exploration **17**, 37–71, 2003.

García, Alejandro

A. Garcia, *Numerical Methods for Physics*, 2nd Ed., Prentice-Hall, Upper Saddle River, NJ (2000).

A. Donev, J. B. Bell, A. Garcia, and B. Alder, "A hybrid particle-continuum method for hdrodynamics of complex fluids," SIAM Multiscale Modeling and Simulation **8**, 971–911 (2010).

A Donev, B. Alder, and A. Garcia, "A Thermodynamically-Consistent Non-Ideal Stochastic Hard-Sphere Fluid," J. Statistical Physics, P11008 (2009).

Hamill, Patrick

P. Hamill and L. Covey, "Using Lidar Data from Space with Correlative Measurements of Gas Species and a Microphysical Model to Quantify Stratospheric Processes," Opt. Pura. Appl. **41**, 2 (2008).

P. Hamill, Intermediate Dynamic, Jones and bartlett, Boston, MA (2010).

J. L. Alvarellow, K. Zahnle, R. Dobrovolskis, and P. Hamill, "Transfer of mass from I_o and beyond due to cometary impacts," Icarus **194**, 636 (2008).

Hubickyj, Olenka

O. Hubickyj, 2010. Core Accretion Model, in Formation and Evolution of Exoplanets (ed. Rory Barnes), Wiley-Vch, Weinbeim, pp. 101–122.

J. J. Lissauer, O. Hubickyj, G. Gennaro, and P. Bodenheimer, Models of Jupiter's growth incorporating thermal and hydrodynamic constraints, Icarus **199**, 338–350.

O. Hubickyj, J. J. Lissauer, P. Bodenheimer, and G. D'Angelo, 2010. Using Animations to Study the Formation of Gas Giant Planets via the Core Accretion Model. American Geophysical Union, Fall Meeting 2009, abstract #P12B.

Kaufman, Michael J.

M. J. Kaufman, M. G. Wolfir, and D. J. Hollenbach, "[Si II], [Fe II], [C II], and H_2 Emission from Massive Star Forming Regions," Astrophysical Journal **644**, 283 (2006).

J. Franklin, R. L. Snell, M. J. Kaufman, G. J. Melnick, D. A. Neufeld, D. J. Hollenbach, and E. A. Bergin, "SWAS

Observations of Water in Molecular Outflows," Astrophysical Journal **674**, 283 (2008).

M. J. Kaufman, M. G. Wolfire, and D. J. Hollenbach, "The Dense Interstellar Medium in Low and High-Redshift Starbursts: PDR Models," in 'From Z-Machines to ALMA: (Sub)Millimeter Spectroscopy of Galaxies,' ed. A. J. Baker, J. Glenn, A. I. Harris, J. G. Mangum & M. S. Yun, ASP Conference Series, **375**, 43 (2007).

Kress, Monika

M. E. Kress and D. Brownlee, "Origin of Habitable Planets," in Astrobiology: The University of Washington Lectures, eds. W. Sullivan and J. Baross, Cambridge University Press (2007).

M. E. Kress, A. G. G. M. Tielens, and M. Frenklach, "The 'soot line': Polycyclic aromatic hydrocarbons in primitive chondrites as tracers of nebular conditions." Meteoritics and Planetary Science Supplement, July 2007.

M. E. Kress, G. K. Benedix, J. Schutt, and R. P. Harvey. "An unusual strewn field at the Otway Massif, Grosvenor Mountains, Antarctica," Meteoritics and Planetary Science Supplement, July 2007.

Lam, Lui

L. Lam, "Active Walks: The first Twelve Years (Part II)," Int. J. Bifurcation and Chaos, **16**, 239 (2006).

L. Lam, "Active Walks: The First Twelve Years (Part I)," Int J. Bifurication and Chaos, **15**, 2317, (2005).

L. Lam, "The Origin of the International Liquid Crystal Society and Active Walks," Physics (Beijing) **34**, 528 (2005).

Parvin, Kiumars

K. Parvin, X. C. Sun, J. Ma, J. Ly, D. E. Nikles, K. Sun, and L. M. Wang, "Synthesis and Magnetic Properties of Monodisperse Fe_3O_4 Nanoparticles," accepted for publication in J. Appl. Phys., **95**(11), 7121 (2004).

C. Boekema, A. M. Krupski, M. Varasteh, K. Parvin, F. van Til, F. van der Woude, and G. A. Sawatzky, "Cu and Fe Valence States In CuFeS2," accepted for publication in J. Mag. and Mag., **559**, 272 (2004).

X. C. Sun, K. Parvin, J. Ly, and D. E. Nikles, "Magnetic Properties of a Mixture of Two Nanosized Co-S Powders Produced by Hydrothermal Reduction," IEEE Transactions on Magnetics, **30**, No. 5, 2679 (2003).

Sauke, Todd B.

T. B. Sauke and J. F. Becker, "Modeling, measuring, and minimizing the instrument response function of a tunable diode laser spectrometer," Journal of Quantitative Spectroscopy & Radiative Transfer **91**, 453–484 (2005).

T. B. Sauke and J. F. Becker, (1998). "Stable Isotope Laser Spectrometer for Exploration of Mars. Planet." Space Sci. **46**, 805–812.

T. B. Sauke, J. F. Becker, M. Loewenstein, T. D. Gutierrez, and C. G. Bratton, (1994). "An Overview of Isotopic Analysis using Tunable Diode Laser Spectrometry." Spectroscopy, **9**, No. 8, 34–40.

Wharton, Kenneth B.

K. B. Wharton, "Time-Symmetric Quantum Mechanics," Foundations of Physics, **37**, 159 (2007).

STANFORD UNIVERSITY

DEPARTMENT OF PHYSICS

Stanford, California 94305

Students Accepted For Degree	FIELDS		
	Physics	Astronomy	Related Fields
Doctorate	X		
Master's			

1. General

President: John Hennessy
Dean of Research: Ann Arvin
Department Chairman: Steven Kahn
Department Telephone Number: (650) 723-4344
Type of Institution: University
Control: Private
Setting: Suburban
Total Faculty: 1,910
Total Students: 18,498
Total Graduate Students: 11,896
Annual Graduate Tuition:
All Graduate Students: $33,560/$8,390 qtr.
Tuition rates for: 2010–11
Other Fees: Varies (Student health insurance is required)
Term: Quarter

2. Number of Faculty in Department

The combined total of full-time faculty in the three professorial ranks is 36. The combined total of full-time, part-time, and other faculty at all ranks is 43.

3. Admission, Financial Aid, and Housing

Address admission inquiries to: Graduate Admissions Support Section of the Registrar's Office, Stanford University
Graduate application fee required: $125
Admission deadline (Fall admission): 12/14
Admission information: For fall admission, 2010–11, 70 students were admitted from 471 applicants.
Admission requirements: For admission to the graduate programs, a Bachelor's degree in physics (or a related field) is required with no minimum undergraduate GPA specified. The GRE is required. No minimum scores specified. The GRE Subject test is required. No minimum score specified. The average GRE scores reported for 2009–10 were for all applicants. Average scores for 2010–11 are for admitted students only. Average GRE Scores for admitted students were verbal 635, quantitative–795; analytical–4.63; physics subject–785. Students from non-English speaking countries are required to demonstrate proficiency in English via the TOEFL exam. Minimum score required is 600, paper based test; computer based test score must be within 250–300 range.
Address financial aid inquiries to: Graduate Admissions Office, Stanford Univ., Stanford, CA 94305-3005
GAPSFAS application required: No
Financial aid deadline: Check with financial aid office
Loans available: Yes, for U.S. citizens
Address housing inquiries to: Housing assignments, 630 Serra St., Ste. 110, Stanford, CA 94305-6034
On-campus, single student housing available: Yes
Cost/term: Varies
On-campus, married student housing available: Yes
Cost/quarter: Varies
Further information: http://www.stanford.edu/dept/physics

Table A—Faculty, Enrollments, and Degrees Granted

Research Specialty	2009–10 Faculty	Enrollment[1] Fall 2009 Master's	Enrollment[1] Fall 2009 Doctorate	No. of Degrees Granted 2009–10 Master's	No. of Degrees Granted 2009–10 Terminal Master's	No. of Degrees Granted 2009–10 Doctorate
Astrophysics	12	–	28	2	–	4
Atomic & Molecular Physics/Lasers	3	–	15		–	1
Biophysics	1	–	11	1	–	1
Condensed Matter-Physics	11	–	19	1		8
Low Temperature-Physics	2	–	5			1
Particles & Fields	11	–	56	1	1	10
Photon Sci.	–	–	4			2
Relativity & Gravitation	1	–	1		–	1
Non-specialized	–	–	27	0(0)	–	1
Total		–	166		1	29
Full-time Grad. Stud.		–	146			
Part-time Grad. Stud.		–	0			
First-year Grad. Stud.		–	20			
Median Years in Grad. Study (2009–10 Degrees)				2	–	5

[1]Students not yet committed to a research specialty are entered under non-specialized.

4. Graduate Degree Requirements

Master's: The physics department does not offer a separate program for the Master of Science degree, but this degree may be awarded for a portion of the Doctor's degree; one-year residency is required.
Doctorate: A minimum of 10 courses exclusive of dissertation and directed research courses, taken at this University and elsewhere with a minimum GPA of 3.0; comprehensive exam, dissertation, and dissertation exam required; three-year residency required; no language exam required.
Thesis: Thesis may be written *in absentia*.
Special Equipment, Facilities, or Programs: Access to SLAC National Accelerator Laboratory; Teaching Center-Science and Engineering Quad; Hansen Experimental Physics Laboratory; Edward L. Ginzton Laboratory; Kavli Institute for Particle Astrophysics and Cosmology, Center for Space Science and Astrophysics, Laboratory for Advanced Materials.

Table B—Appointments to Graduate Students, 2009–10

Title of Appointee	Appointments Total	Appointments First year	Academic Load Allowed in Credit Hours	Hours of Service Per Week	Stipend for Academic Year ($)
			Quarter		
Research Assistant	133	12	10	20	-
20% RA and 30%				RA8-10	-
TA Combination	36	0	10	TA10-12	
Fellow	33	8	10–18	-	-
Total	166	20			

FACULTY

Professors Emeriti

Chu, Steven, Ph.D., California, Berkeley, 1976. Spectroscopy and quantum electronics.

Fetter, Alexander L., Ph.D., Harvard, 1963. Condensed matter theory.

Hanna, Stanley S., Ph.D., Johns Hopkins, 1947. Nuclear physics.

Lipa, John A., Ph.D., Univ. of Western Australia, 1969. Low-temperature physics.

Little, William A., Ph.D., Rhodes, 1955. Low-temperature and molecular physics.

Ritson, David M., Ph.D., Oxford, 1948. High-energy physics.

Schwettman, H. Alan, Ph.D., Rice, 1962. Low-temperature physics.

Smith, Todd, Ph.D., Rice, 1965. Free-electron laser physics. Emeritus.

Sturrock, Peter (*courtesy appointment*), Ph.D., Cambridge, 1951.

Taylor, Richard E., (*courtesy appointment*), Ph.D., Stanford, 1962.

Turneaure, John, (Research), Ph.D., Stanford, 1967. Experimental general relativity.

Wagoner, Robert, Ph.D., Stanford, 1965. Theoretical and astophysics.

Walecka, J. Dirk, Ph.D., MIT, 1958. Theoretical physics.

Yearian, Mason R., Ph.D., Stanford, 1959. High energy physics.

Wojcicki, Stanley G., Ph.D., California, Berkeley, 1961. High-energy physics.

Professors

Blandford, Roger, Ph.D., Magdalene College, 1974. Theoretical astrophysics. (Joint appointment with SLAC)

Bucksbaum, Philip, Ph.D., UC Berkeley, 1980. Atomic, molecular and optical physics. (Joint appointment: 1/2 SSRL, 1/4 Applied Physics).

Burchat, Patricia, Ph.D., Stanford, 1986. High energy experimental physics.

Cabrera, Blas, Ph.D., Stanford, 1975. Experimental particle astrophysics and condensed matter physics.

Dimopoulos, Savas, Ph.D., Chicago, 1978. Theoretical physics.

Doniach, Sebastian, Ph.D., Univ. of Liverpool, 1958. Theoretical condensed matter physics. (Joint appointment: 1/3 in Physics, 2/3 in Applied Physics.)

Gratta, Giorgio, Laurea, University of Rome, 1986. High energy experimental physics.

Kachru, Shamit, Ph.D., Princeton Univ., 1994. Theoretical high energy physics. (Joint with SLAC)

Kahn, Steven M., Ph.D., UC Berkeley, 1980. Experimental astrophysics. (Joint with SLAC)

Kallosh, Renata, Ph.D., Lebedev Physical Institute, Moscow, 1968. Theoretical physics.

Kapitulnik, Aharon, Ph.D., Tel Aviv University, 1984. (Joint appointment: 1/3 in Physics, 2/3's in Applied Physics).

Kasevich, Mark, Ph.D., Stanford, 1992. Atomic, Molecular and optical physics. (Joint with Applied Physics)

Kivelson, Steven A., Ph.D., Harvard, 1979. Condensed matter theory.

Laughlin, Robert, Ph.D., MIT, 1979. Condensed matter theory.

Levin, Craig, Ph.D., Yale, 1993. Radiology Professor Research; (Courtesy appointment).

Linde, Andrei, Ph.D., Lebedev Physical Institute, Moscow, 1974. Theoretical astrophysics.

Michelson, Peter, Ph.D., Stanford, 1979. Low-temperature physics.

Osheroff, Douglas, Ph.D., Cornell, 1973. Condensed matter physics.

Petrosian, Vahe, Ph.D., Cornell Univ., 1967 (joint appointment: 2/3's in Physics, 1/3 in Applied Physics).

Quake, Stephen, Ph.D., Oxford University, 1999. Bioengineering (Courtesy appointment).

Romani, Roger, Ph.D., CalTech, 1987. Theoretical astrophysics.

Shen, Zhi-Xun, Ph.D., Stanford, 1989. Condensed matter physics (Joint with Applied Physics and SSRL).

Shenker, Stephen H., Ph.D., Cornell, 1980. Theoretical particle physics.

Silverstein, Eva, Ph.D., Princeton, 1996. Theoretical physics and cosmology. (Joint with SLAC)

Susskind, Leonard, Ph.D., Cornell, 1965. Theoretical physics.

Zare, Richard N., Ph.D., Harvard, 1964. (courtesy appointment).

Zhang, Shoucheng, Ph.D., State Univ. of New York, 1987. Theoretical condensed matter physics.

Associate Professors

Abel, Thomas, Ph.D., L. Maxemillian Univ. Munich, 2000. Theoretical astrophysics. (Joint with SLAC)

Allen, Steven, Ph.D., University of Cambridge, 1994. Experimental astrophysics. (Joint with SLAC)

Church, Sarah E., Ph.D., Cambridge, 1991. Observational/experimental astophysics.

Goldhaber-Gordon, David, Ph.D., MIT, 1999. Experimental condensed matter physics.

Manoharan, Hari, Ph.D., Princeton Univ., 1997. Experimental condensed matter.

Moler, Kathryn A., Ph.D., Stanford, 1995. Condensed matter physics. (1/3 Physics, 2/3's Applied Physics).

Assistant Professors

Funk, Stefan, Ph.D., Max Planck-Inst., 2005. Experimental astrophysics.

Graham, Peter, Ph.D., Stanford Univ., 2007. Particle theory.

Kuo, Chao-Lin, Ph.D., UC Berkeley, 2003. Experimental astrophysics.

Qi, Xiaoliang, Ph.D., Tsinghua University, 2007. Theoretical condensed-matter physics.

Senatore, Leonardo, Ph.D., MIT, 2006. Particle theory (joint with SLAC).

Wechsler, Risa, Ph.D., Univ. of Calif. at Santa Cruz, 2001. Theoretical astrophysics (joint with SLAC).

Research Professors

Scherrer, Philip H., Ph.D., Univ. of Calif., Berkeley, 1973. Solar physics.

RESEARCH SPECIALTIES AND STAFF

Theoretical

Astrophysics. Abel, Blandford, Petrosian, Romani, Wagoner, Wechsler.

Biophysics. Doniach.

Condensed Matter. Fetter, Kapitulnik, Kivelson, Laughlin, Zhang.

Cosmology and Particle Physics. Dimopoulos, Kachru, Kallosh, Linde, Senatore, Shenker, Silverstein, Susskind.

Experimental

Astrophysics. Allen, Church, Funk, Kahn, Kuo, Michelson.

Atomic Spectroscopy. Electronic states of atoms in gases and solids; quantum electronics; biophysics and polymer physics; lasers. Bucksbaum, Chu, Kasevich.

Condensed Matter. Goldhaber-Gordon, Kapitulnik, Laughlin, Manoharan, Moler, Shen.

Gravitational Physics. Lipa.

High-Energy Physics. Elementary particle reaction and properties. Burchat, Gratta, Wojcicki.

Low-Temperature Physics. Liquid He; magnetism in solid ^{3}He; low-temperature properties of solids; application of low-temperature techniques; superconducting accelerator; free-electron laser. Cabrera, Osheroff.

STANFORD UNIVERSITY

DEPARTMENT OF APPLIED PHYSICS

348 Via Pueblo Mall
Stanford, California 94305–4090

Students Accepted For Degree	FIELDS		
	Physics	Astronomy	Related Fields
Doctorate			X
Master's			X

1. General

President: John Hennessy
Department Chairman: Hideo Mabuchi
Department Telephone Number: (650) 723–4028
Web Site: http://appliedphysics.stanford.edu
Type of Institution: University
Control: Private
Setting: Suburban
Total Faculty: 1,910
Total Students: 15,319
Total Graduate Students: 8,441
Graduate Tuition per quarter:
All Graduate Students: Full-time—$8,390–$12,900.
Tuition rates for: 2010–11
Other Fees: Varies. Student health insurance required.
Term: Quarter

2. Number of Faculty in Department

The combined total of full-time faculty in the three professorial ranks is 17. The combined total of full-time, part-time, and other faculty at all ranks is 36. In addition, we have 9 Consulting Professors.

3. Admission, Financial Aid, and Housing

Address admission inquiries to: Graduate Admissions Office, 630 Serra Street, Suite 120, Stanford, CA 94305-6032. General information and application process: http://gradadmissions.stanford.edu
Graduate application fee required: $125
Admission deadline (Fall admission): 1/04/11
Admission information: For fall admission, 2010–11, 41 students were accepted from 183 applicants.
Admission requirements: For admission to the graduate programs, a Bachelor's degree in Physics, Mathematics, Chemistry, or Electrical Engineering is required with no minimum undergraduate GPA. The GRE is required. No minimum scores specified. The GRE Advanced is required for the Ph.D. No minimum scores specified. The average GRE percentiles for 2010–11 admissions were verbal–90%; quantitative–92%. The average GRE Advanced percentile for admissions was 86%. Students from non-English speaking countries are required to demonstrate proficiency in English via the TOEFL exam.
Undergraduate preparation assumed: Intermediate undergraduate (junior and senior) courses in mechanics, electricity and magnetism, modern physics, statistical mechanics, and thermodynamics.
Address financial aid inquiries to: Paula P. Perron, Department Administrator, Department of Applied Physics (pperron@stanford.edu)
GAPSFAS application required: No
Financial aid deadline: 1/04/11
Loans available: Yes
Address housing inquiries to: studenthousing@stanford.edu. Web: http://housing.stanford.edu
On-campus, single student housing available: Yes
Cost/term: Varies
On-campus, married student housing available: Yes
Cost/month: Varies

Table A—Faculty, Enrollments, and Degrees Granted

Research Specialty	2009–10 Faculty[3]	Enrollment[1] Fall 2009		No. of Degrees Granted[2] 2009–10 (2005–10)			Median No. of Years for 2009–10 Ph.D.'s
		Master's	Doctorate	Master's	Terminal Master's	Doctorate	
Applied Physics	11	2	187	5(53)	1(6)	2(12)	6.00
Astrophysics	6	0	–	0(0)	0(0)	3(6)	5.75
Atomic, Molecular, & Optical Physics	4		–	0(0)	0(0)	1(2)	4.25
Biophysics	15	0	–	0(0)	0(0)	2(10)	4.75
Chemical Physics	1	0	–	0(0)	0(0)	0(4)	–
Condensed Matter Physics	20	0	–	0(0)	0(0)	5(23)	6.75
Engineering Physics/ Science	1	0	–	0(0)	0(0)	1(2)	7.50
Materials Science	6	0	–	0(0)	0(0)	2(5)	6.50
Optics	3	0	–	0(0)	0(0)	0(3)	–
Physics of Beams	0	0	–	0(0)	0(0)	0(1)	–
Relativity & Gravitation	1	0	–	0(0)	0(0)	0(0)	–
Other (specify) Photonics	10	0	–	0(0)	0(0)	4(31)	5.50
Total		2	137	5(53)	1(6)	20(99)	
Full-time Grad. Stud.		0	139				
Part-time Grad. Stud.		0	0				
First-year Grad. Stud.		2	15				
Median Years in Grad. Study (2009–10 Degrees)		1.5	5.75	–	–	–	–

Undergraduate Degrees, 2009–10 (2005–10): (0) (Graduate Department Only)

[1]Breakdown unavailable for the doctorate enrollment since some students are doing research in various areas. Not possible to pinpoint each student to only one particular area listed.
[2]Five-year totals in parentheses.
[3]Includes Applied Physics faculty as well as faculty from other departments who are supervising the research of Applied Physics graduate students.

4. Graduate Degree Requirements

Master's: Subject Matter: Advanced Mechanics, Electrodynamics, Quantum Mechanics. "B" average required. Total number of course units required-45. No foreign language. No comprehensive and/or qualification examination. No thesis. Terminal M.S. Degree Program offered.
Doctorate: Subject Matter: Advanced Mechanics, Electrodynamics, Quantum Mechanics, Statistical Physics, 1 advanced laboratory, remaining required courses to be distributed between major and minor fields. Total number of course units required-135. "B" average required. Departmental qualification examination. 4th year research progress report. Dissertation. Oral defense of dissertation. No foreign language requirement.

Dissertation: May be written *in absentia*.*

Special Equipment, Facilities, or Programs: The Applied Physics Department participates in the Honors Cooperative Program which offers the opportunity to qualified engineers and scientists employed by companies in the general vicinity of the University to pursue graduate work on a part-time basis, leading to the M.S. degree.

*With the approval of the research advisor, Departmental Graduate Study Committee, and the Department Chair.

Table B—Appointments to Graduate Students, 2009–10

Title of Appointee	Appointments Total	Appointments First year	Academic Load Allowed in Credit Hours	Hours of Service Per Week	Stipend for Academic Year ($)[1]
			Quarter		
Research Assistant	103	12	8–10 units	–	22,500–23,400[3]
Teaching Assistant[2]	–	–	–	–	–
Self-Supporting	1	1	Varies	–	–
Various Fellowships	59	5	Varies	Varies	Varies
Total	163[4]	18[4]			

[1]It is assumed that the academic year is 9 months (3 academic quarters). Stipends stated are for that period of time. Summer quarter awards/stipends are additional.
[2]Applied Physics does not have teaching assistantships (TA) per se, but our students have opportunities to arrange for a TA through other departments. Usually joint RA/TA appointments are made. In this table, the joint appointments are listed under Research Assistant. There have been 20 joint RA/TAs in 2009–10 at various times throughout the academic year.
[3]The stipend level depends upon whether the student is pre- or post-qualified for the PhD. Summer appointments are available usually at a higher level depending upon the arrangement made between the student and research advisor. Students do not have to be registered during the summer quarter. Stipend quoted for 3 quarters.
[4]Several students have dual fellowships, or a fellowship and a RA, and are double-counted. Actual number of students: total 139; first years 17.

5. Personnel Engaged in Separately Budgeted Research, 7/09–6/10

Other faculty	9
Graduate students	137
Total	146

6. Extension Centers and Summer Programs

Summer program is a continuation of the academic year—courses offered, research ongoing in the laboratories.

FACULTY

Professors Emeriti

Beasley, Malcolm R., Ph.D., Cornell, 1968. Low-temperature and condensed matter physics; superconductivity.

Bienenstock, Arthur I., Ph.D., Harvard, 1962. Synchrotron radiation.

Chu, Steven, Ph.D., U. C., Berkeley, 1976. Spectroscopy and quantum electronics, biophysics.

Fetter, Alexander L., Ph.D., Harvard, 1963. Condensed matter theory.

Geballe, Theodore H., Ph.D., California, Berkeley, 1949. Low-temperature and condensed matter physics; superconductivity.

Harris, Stephen E., Ph.D., Stanford, 1963. Quantum electronics and XUV lasers.

Harrison, Walter A., Ph.D., Illinois, 1956. Theoretical condensed matter: electronic structure.

Quate, Calvin F., Ph.D., Stanford University, 1950. Professor of Research. Scanning tunneling microscopy; imaging.

Sturrock, Peter A., Ph.D., Cambridge, 1951. Theoretical solar physics and astrophysics.

Wiedemann, Helmut, Ph.D., Univ. Hamburg, 1971. Professor of Research. Accelerator physics; electron storage rings.

Winick, Herman, Ph.D., Columbia, 1957. Professor of Research. Synchrotron radiation.

Professors

Block, Steven M., Ph.D., California Institute of Technology, 1983. Biophysics.

Bucksbaum, Philip H., Ph.D., U. C. Berkeley, 1980. Atomic, molecular and optical physics; ultrafast science.

Byer, Robert L., Ph.D., Stanford, 1969. Nonlinear optics.

Doniach, Sebastian, Ph.D., Liverpool, 1958. Theory of cooperative phenomena; biophysics.

Fejer, Martin M., Ph.D., Stanford, 1986. Quantum electronics, guided-wave optics, and optical materials.

Fisher, Daniel S., Ph.D., Harvard University, 1979. Theoretical condensed matter physics, biophysics, and evolutionary dynamics.

Hwang, Harold Y., Ph.D., Princeton, 1997. Materials physics; emergent phenomena in oxide heterostructures; devices.

Kapitulnik, Aharon, Ph.D., Tel Aviv, 1984. Theoretical and experimental low-temperature and condensed matter physics; superconductivity.

Kasevich, Mark A., Ph.D., Stanford, 1992. Atomic, molecular and optical physics.

Mabuchi, Hideo, Ph.D., California Institute of Technology, 1998. Chair of the Department. Quantum optics; quantum information and control.

Petrosian, Vahé, Ph.D., Cornell, 1967. Theoretical astrophysics and cosmology.

Shen, Zhi-Xun, Ph.D., Stanford, 1989. Condensed matter physics; electronic structure; photoelectron spectroscopy; synchrotron radiation.

Yamamoto, Yoshihisa, Ph.D., University of Tokyo, 1978. Quantum optics; mesoscopic physics.

Associate Professors

Fisher, Ian R., Ph.D., University of Cambridge, UK. 1996. Condensed matter; materials physics.

Moler, Kathryn A., Ph.D., Stanford University, 1995. Condensed matter; materials physics; physics of small structures and of novel materials.

Reis, David A., University of Rochester, 1999. Condensed matter physics; ultrafast science.

Schnitzer, Mark J. Ph.D., Princeton University, 1999. Biophysics.

Professor of Research

Digonnet, Michel, Ph.D., Stanford, 1984. Fiber optics; sensors; slow and fast light.

Courtesy Professors

Clemens, Bruce M., Ph.D., California Institute of Technology, 1983. Metal-metal multilayers; interfaces and interface reactions; magnetic thin films; x-ray diffraction.

Harris, James S., Ph.D., Stanford University, 1969. Optoelectronic device structures; quantum electronics.

Hesselink, Lambertus, Ph.D., California Institute of Technology, 1977. Nonlinear optics.

Miller, David A. B., Ph.D., Heriot-Watt University, 1979. Electro-optic wave devices; engineering physics.

Moerner, W. E., Ph.D., Cornell University, 1982. Quantum optics.

Osheroff, Douglas D., Ph.D., Cornell University, 1973. Low-temperature and condensed matter physics.

Quake, Stephen R., Ph.D., Oxford University, 1994. Biophysics.

Zhang, Shoucheng, Ph.D., SUNY at Stony Brook, 1987. Theoretical condensed matter.

RESEARCH SPECIALTIES AND STAFF[1]

Theoretical

Biophysics. Doniach. D. S. Fisher

Condensed Matter Physics. Doniach, Fetter, D. S. Fisher, Harrison, Kapitulnik, Zhang.

Solar Physics and Astrophysics. Petrosian, Sturrock.

Experimental

Atomic, Molecular, and Optical Physics. Photonics. Bucksbaum, Byer, Digonnet, Fejer, J. S. Harris, S. E. Harris, Hesselink, Kasevich, Mabuchi, Miller, Moerner, Yamamoto.

Biophysics. Block, Mabuchi, Schnitzer.

Condensed Matter Physics. Materials Research. Beasley, Clemens, I. R. Fisher, Geballe, Hwang, Kapitulnik, Moler, Osheroff, Reis, Shen.

Materials Research. Hwang.

Novel Microscopy and Imaging. Moler, Quate.

Optical Devices. Digonnet, Miller.

Relativity and Gravitation. Byer.

Synchrotron Radiation. Bienenstock, Reis, Shen, Wiedemann, Winick.

[1]Postdoctoral appointments are made through the individual faculty members and independent laboratories. Data not maintained by the reporting unit.

UNIVERSITY OF CALIFORNIA, BERKELEY

DEPARTMENT OF PHYSICS

Berkeley, California 94720-7300

Students Accepted For Degree	FIELDS		
	Physics	Astronomy	Related Fields
Doctorate	X		
Master's	X		

1. General

President: Mark G. Yudof
Chancellor: Robert J. Birgenau
Dean of Graduate Division: Andrew J. Szeri
Department Chair: Frances Hellman
Department Telephone Number: (510) 642-7166 (C)
Website: http://www.physics.berkeley.edu
Type of Institution: University
Control: Public
Setting: Urban
Total Faculty: 2,047
Total Graduate Faculty: Not separated
Total Students: 35,409
Total Graduate Students: 10,258
Annual Graduate Tuition and Fees:
In-state residents: Full-time—None; $12,949.50 (fees including health insurance), $10,939.50
Out-of-state residents: Full-time—$15,312 non-resident tuition; $13,357.50 (fees including health insurance)
Tuition and fees for: 2010–11
All fees subject to change
Deferred tuition plan: No
Term: Semester

2. Number of Faculty in Department

The combined total of full-time faculty in the three professorial ranks is 49. The combined total of full-time, part-time, and other faculty at all ranks is 61 (excludes 43 Emeriti).

3. Admission, Financial Aid, and Housing

Address admission inquiries to: Graduate Student Affairs Officer, Physics Student Services, 370 LeConte Hall #7300, University of California, Berkeley, CA 94720-7300.
Graduate application fee required: U.S. Citizens and current U.S. Residents $70, International applicants $90
Admission deadline: (Fall admission only): Financial Aid and Fellowships—12/17
Admission information: For fall admission, 2010–11, 103 students were accepted from 708 applicants.
Admission requirements: For admission to the graduate programs, a Bachelor's degree in physics is required with a minimum undergraduate GPA of 3.0 specified. Both the General and Advanced Physics GRE are required. The suggested minimum General GRE scores for admission are: verbal–540; quantitative–580; and analytical writing–4.5. The average Advanced Physics GRE score for 2009–10 admissions was 920. Students from non-English speaking countries are required to demonstrate proficiency in English via the TOEFL exam. Minimum acceptable score for admission is 570 (paper), 230(CBT), or 68 (iBT).
Undergraduate preparation assumed: 3 semesters-*General Physics*, Giancoli; 1 semester-*Mechanics*, Taylor; 2 semesters-*Electromagnetism and Optics*, Griffiths; 1 semester-*Thermal/Statistical*, Kittel & Kroemer; 2 semesters-*Atomic Physics and Quantum Mechanics*, Bransden and Joachain, Griffiths; 2 semesters-Advanced Undergraduate Laboratory. Plus mathematics courses in vector calculus, linear algebra, ordinary and partial differential equations, complex variable. (Berkeley undergraduates have in addition 1 semester of physics electives; for example, solid state physics, plasma physics, nuclear and particle physics, relativity.)
Address financial aid inquiries to: Financial Aid, Graduate and Professional Unit, 201 Sproul Hall #1960, University of California, Berkeley, CA 94720-1960, email: fao_grad@berkeley.edu (for Federal Student Direct Loans, other smaller federal loan programs, Federal Work Study Program, and U.C. Parent Grant Program). Only U.S. citizens and eligible non-U.S. citizens may apply for funds administered by the Financial Aid Office. *Inquiries regarding fellowships, teaching and research appointments, or departmental aid should be addressed to the Physics Department.*
FAFSA application required: Yes
Loans available: Yes
Address housing inquiries to: Cal Housing, 2610 Channing Way #2272, University of California, Berkeley, CA 94720-2272, e-mail: reshall@berkeley.edu (single); apt@berkeley.edu (married).
University operated, single student housing available: Yes
Cost/year: $11,110–18,215 (approximate)
University operated, married student housing available: Yes
Cost/month: $980–1,850/mo. (approximate)

Table A—Faculty, Enrollments, and Degrees Granted

Research Specialty	2008–09 Faculty	Enrollment[1] Fall 2008		No. of Degrees Granted[2] 2009–10 (2005–10)			Median No. of Years for 2009–10 Ph.D.'s
		Master's	Doctorate	Master's	Terminal Master's	Doctorate	
Astrophysics	16		49	–	–	11(30)	–
Atomic, Molecular, & Optical Physics	7		17	–	–	2(15)	–
Biophysics	4		14	–	–	3(12)	–
Condensed Matter Physics	19		77	–	–	14(43)	–
Nuclear Physics	1		2	–	–	0(12)	–
Particles & Fields	15		24	–	–	10(32)	–
Plasma Physics and Nonlinear Dynamics	4	–	12	–	–	0(6)	–
Non-specialized			43				
Total		–	258	–	–	40(150)	
Full-time Grad. Stud.		–	258				
Part-time Grad. Stud.		–					
First-year Grad. Stud.		–	47				
Median Years in Grad. Study (2005–06 Degrees)					–		
Undergraduate Degrees, 2009–10 (2004–09): 79(498)							

[1] Students not yet committed to a research specialty are entered under non-specialized.
[2] Five-year totals in parentheses.

4. Graduate Degree Requirements

Master's: 35 semester units in approved program with satisfactory performance; comprehensive exam required; thesis not required; two semester residence requirement; no language requirement.

Doctorate: 38 graduate units in approved program with satisfactory performance, preliminary examination, candidacy qualifying examination, and dissertation required; four semester residency requirement, no language requirement.

Other Programs: Inter-departmental research: Some graduate students are engaged in research problems involving interdepartmental collaboration of which the following are examples: (1) nuclear physics, in programs with the Chemistry Department or the Lawrence Berkeley National Laboratory; (2) astrophysics, with the Department of Astronomy, the Berkeley Center for Cosmological Physics, or the Space Sciences Laboratory; (3) solid state physics, with the Departments of Electrical Engineering and Computer Sciences, and Materials Science and Engineering; (4) plasma physics, with the Departments of Electrical Engineering and Computer Sciences and Nuclear Engineering or the Lawrence Berkeley National Laboratory; (5) Biophysics and medical physics. *Interdisciplinary groups*: There are a number of graduate Interdisciplinary Groups with Ph.D. programs separate from the Ph.D. in Physics, particle physics, with the Berkeley Center for Theoretical Physics.

Dissertation: Dissertation may be written *in absentia*.

Special Equipment, Facilities, or Programs: Research facilities are available at the following laboratories: Lick Observatory, Space Sciences Laboratory, Lawrence Berkeley National Laboratory, Advanced Light Source, National Center for Electron Microscopy, The Molecular Foundry (a Nanoscience User Research Facility); Lawrence Livermore National Laboratory. The following laboratories also make facilities available for graduate student research: Brookhaven National Laboratory, Fermi National Accelerator Laboratory, NASA Ames Center, Kitt Peak Observatory, Stanford Linear Accelerator Center. Berkeley Micro/Nanofabrication Fcility, The Radio Astronomy Laboratory, Argonne National Laboratory, Keck Observatory, CERN, Gran Sasso Underground Laboratory, Assergi, Italy, Kamioka Observatory, Japan, Institute for the Physics and Mathematics of the Universe (IPMU), Tokyo, PSI Switzerland, TRIUMF, Vancouver, BC.

5. Personnel Engaged in Separately Budgeted Research, 7/08–6/09

Professorial faculty	61
Postdoctoral appointments	49
Graduate students	258
Undergraduate students	20
Nonteaching research personnel	12
Total	400

6. Separately Budgeted Research Expenditures by Source of Support

	Departmental Research
Federal government	$14,850,594
Nonprofit Organization	4,429,929
Private Industry	796,144
Other	450,200
Total	$20,526,867

Table B—Appointments to Graduate Students, 2008–09

Title of Appointee	Appointments: Total	Appointments: First year	Academic Load Allowed in Credit Hours	Hours of Service Per Week	Stipend for Academic Year ($)
			Semester		
Graduate Student Instructor	121	35	12 units	10–20	8,318–16,636
Graduate Student Researcher	215	5	12 units	20	25,496.00
Fellowships					
Hertz	2	1	12 units	–	NA
NDSEG (DOD)	1	1	12 units	–	NA
NSF	6	1	12 units	–	NA
U.C. Fellowship	7	4	12 units	–	NA
Samsung	2	0	12 units	–	NA
Total	354	47			

7. Separately Funded and Managed Laboratories

The University of California operates the Lawrence Berkeley National Laboratory separately from the Berkeley campus. This multidisciplinary laboratory supports much faculty and student research in physics and other fields. The number of Lawrence Berkeley National Laboratory staff involved in thesis supervision is included in Table C, but the physics research support at Lawrence Berkeley National Laboratory is displayed separately in Table D. The Space Sciences Laboratory is operated as an Organized Research Unit of the Berkeley campus.

Table C—Separately Budgeted Research Expenditures

Research Specialty	No. of Grants	Expenditures ($)
Astrophysics	37	7,216,319
Atomic & Molecular Physics	45	3,285,939
Biophysics	11	464,779
Condensed Matter Physics—Experiment	76	3,271,233
Condensed Matter Physics—Theory	20	937,599
Cosmological Physics	1	549,837
Nuclear Physics	8	450,758
Particle—Experiment	6	2,157,140
Particle—Theory	17	1,199,984
Plasma Physics & Fluids	10	778,999
Theoretical Physics	1	208,548
Other	9	49,557
Total	241	20,570,688

Table D—Physics-related Research Outside Department

Field and Unit Outside Department[1]	No. of Grants	Expenditures ($)[2]
Lawrence Berkeley National Laboratory Divisions		
Accelerator and Fusion Research (Plasma and Accelerators)	5	6,250,000
Chemistry, Chem. Eng.	4	3,000,000
Experimental Nuclear Science	5	2,184,235
Elementary Particle Physics & Astrophysics		
Experimental	1	22,319,000
Theoretical	1	2,164,000
Materials Sciences (Condensed Matter, Low Temperature, Atomic)	15	4,710,000
MCB, Helen Willis, QB3, HHM1, Cancer Center	50	50,000,000
Campus Departments		
Astronomy Department; Radio Astronomy Laboratory, Theoretical Astrophysics Center, and Center for Integrative Planetary Sciences	127	22,927,301
Materials Science and Engineering	5	1,100,000
Space Sciences Laboratory	200	40,000,000
Total	361	99,654,530

[1] Includes programs which presently include Physics graduate students.
[2] Figures are approximated.

FACULTY

Professors

Anderson, Kinsey A., Ph.D., Minnesota, 1955. Emeritus. Experimental high altitude and space physics.

Arons, Jonathan, Ph.D., Harvard, 1970. Theoretical high-energy astrophysics; x-ray sources; plasma physics.

Bale, Stuart, Ph.D., University of Minnesota, 1994. Experimental astrophysics.

Bardakci, Korkut, Ph.D., Rochester, 1962. Emeritus. Theory of elementary particles.

Birgeneau, Robert J., Ph.D., Yale University, 1966. Chancellor of the University of California at Berkeley. Experimental condensed matter physics.

Budker, Dmitry, Ph.D., California, Berkeley, 1993. Atomic physics.

Bustamante, Carlos J., Ph.D., California, Berkeley, 1981. Biophysics.

Chew, Geoffrey F., Ph.D., Chicago, 1948. Emeritus. Theory of elementary particles.

Chinowsky, William, Ph.D., Columbia, 1955. Emeritus.

Clarke, John, Ph.D., Cambridge, 1968. Experimental superconductivity and low-temperature physics.

Cohen, Marvin L., Ph.D., Chicago, 1964. Theory of condensed matter.

Commins, Eugene D., Ph.D., Columbia, 1958. Emeritus. Experimental atomic and nuclear physics.

Crommie, Michael, Ph.D., California, Berkeley, 1991. Experimental condensed matter physics.

Crowe, Kenneth M., Ph.D., California, Berkeley, 1952. Emeritus. Muon and medium-energy physics.

Davis, Marc, Ph.D., Princeton, 1973. Theoretical astrophysics; extragalactic astronomy; cosmology.

Dynes, Robert C., Ph.D., McMaster University, 1968. Experimental condensed matter physics.

Ely, Robert P., Ph.D., MIT, 1959. Emeritus.

Fajans, Joel, Ph.D., MIT, 1985. Experimental plasma physics.

Falcone, Roger W., Ph.D. Stanford, 1979. Quantum electronics and experimental atomic physics.

Frazer, William R., Ph.D., California, Berkeley, 1959. Emeritus. Elementary particle theory; cosmology.

Freedman, Stuart, J., Ph.D., California, Berkeley, 1972. Experimental nuclear physics; weak interactions; neutron decay; neutrino oscillations.

Gaillard, Mary K., Ph.D., University of Paris, 1968. Emeritus. Theory of elementary particles.

Genzel, Reinhard L., Ph.D., Bonn, Germany, 1978. Experimental astrophysics; infrared and microwave astronomy.

Glaser, Donald A., Ph.D., Cal. Tech., 1950. Emeritus. Professor of the Graduate School. Biophysics and molecular biology; psychophysics and theoretical neuroscience.

Goldhaber, Gerson, Ph.D., Wisconsin, 1950. Emeritus. Experimental elementary particle physics.

Hahn, Erwin L., Ph.D., Illinois, 1949. Emeritus. Experimental condensed matter physics and modern optics.

Hall, Lawrence J., Ph.D., Harvard, 1981. Theory of elementary particles.

Halpern, Martin B., Ph.D., Harvard, 1964. Emeritus. Theory of elementary particles.

Haxton, Wick, Ph.D., UC Santa Cruz, 1971. Theoretical particle physics and astrophysics.

Hellman, Frances D., Ph.D., Stanford University, 1985. Experimental condensed matter physics.

Holzapfel, William L., Ph.D., California, Berkeley, 1996. Experimental astrophysics.

Hořava, Petr, Ph.D., Czech. Academy of Sciences, 1981. theoretical particle physics.

Jackson, J. D., Ph.D., MIT, 1949. Emeritus. Theory of elementary particles.

Jacobsen, Robert G., Ph.D., Stanford, 1991. Experimental elementary particle physics.

Kaufman, Allan N., Ph.D., Chicago, 1953. Emeritus. Plasma theory and nonlinear dynamics.

Kerth, Leroy T., Ph.D., California, Berkeley, 1957. Emeritus. Experimental elementary particle physics.

Kittel, Charles, Ph.D., Wisconsin, 1941. Emeritus.

Knobloch, Edgar, Ph.D., Harvard, 1978. Theoretical astrophysics; fluid dynamics; nonlinear dynamics.

Kunkel, Wulf B., Ph.D., California, Berkeley, 1951. Emeritus. Experimental plasma physics.

Lee, Dung-Hai, Ph.D., MIT, 1982. Theory of condensed matter; quantum phase transitions; strongly correlating electronic systems.

Leone, Stephen R., Ph.D., California, Berkeley, 1974. Gas phase laser spectroscopy.

Lin, Robert P., Ph.D., California, Berkeley, 1967. Experimental high-energy astrophysics; earth magnetospheric, lunar, solar, and interplanetary physics.

Littlejohn, Robert G., Ph.D., California, Berkeley, 1980. Theoretical plasma physics and nonlinear dynamics.

Louie, Steven G., Ph.D., California, Berkeley, 1976. Theory of condensed matter.

Luk, Kam-Biu, Ph.D., Rutgers, 1983. Experimental elementary particle physics.

McKee, Christopher F., Ph.D., California, Berkeley, 1970. Theoretical astrophysics; interstellar medium; high-energy astrophysics.

Mandelstam, Stanley, Ph.D., Birmingham, England, 1956. Emeritus. Theory of elementary particles.

Marrus, Richard, Ph.D., California, Berkeley, 1959. Emeritus. Experimental atomic physics; beam foil spectroscopy.

Mozer, Forrest S., Ph.D., Cal. Tech., 1956. Emeritus.

Muller, R. A., Ph.D., California, Berkeley, 1969. Astrophysics experiment; experimental physics.

Murayama, Hitoshi, Ph.D., University of Tokyo, 1991. Theory of elementary particles.

Orenstein, Joseph W., Ph.D., MIT, 1980. Experimental condensed matter physics.

Packard, Richard E., Ph.D., Michigan, 1969. Emeritus-Experimental condensed matter physics; experimental low-temperature physics.

Perlmutter, Saul, Ph.D., Harvard University, 1986. Experimental astrophysics.

Portis, Alan M., Ph.D., California, Berkeley, 1949. Emeritus. Experimental condensed matter physics; magnetic properties; biophysics.

Price, P. Buford, Ph.D., Virginia, 1958. Emeritus. Professor of the Graduate School. Astrophysics experiment; cosmic radiation and relativistic nuclear physics.

Qiu, Zi Qiang, Ph.D., John Hopkins University, 1990. Experimental condensed matter physics.

Ramesh, Ramamoorthy, Ph.D., University of California at Berkeley, 1987. Experimental condensed matter physics.

Reif, Frederick, Ph.D., Harvard, 1953. Emeritus.

Richards, Paul L., Ph.D., California, Berkeley, 1960. Emeritus. Experimental condensed matter physics; infrared spectroscopy; infrared astrophysics.

Rokhsar, Daniel S., Ph.D., Cornell, 1987. Statistical and many-body physics; biophysics.

Rosenfeld, Arthur H., Ph.D., Chicago, 1954. Emeritus. Physics related to environmental problems; energy and conservation.

Sachs, Rainer K., Ph.D., Syracuse, 1958. Emeritus.

Sadoulet, Bernard, Doctorat d'Etude es Sciences, Orsay, France, 1971. Experimental cosmology.

Schwartz, Charles L., Ph.D., MIT, 1954. Emeritus.

Seljak, Uroš, Ph.D., Massachusetts Institute of Technology, 1995, Cosmology and theoretical astrophysics.

Shank, Charles V., Ph.D., California, Berkeley, 1969. Experimental condensed matter physics. Emeritus.

Shapiro, Marjorie, Ph.D., California, Berkeley, 1984. Experimental elementary particle physics. Chairman, Department of Physics.

Shen, Yuen-Ron, Ph.D., Harvard, 1963. Emeritus. Professor of the Graduate School. Experimental condensed matter physics; quantum and nonlinear optics.

Shugart, Howard A., Ph.D., California, Berkeley, 1957. Emeritus. Experimental atomic physics and atomic beam studies; nuclear spins and moments.

Siegrist, James L., Ph.D., Stanford, 1979. Experimental elementary particle physics.

Smoot, George F., Ph.D., MIT, 1970. Experimental astrophysics.

Stamper-Kurn, Dan, Ph.D., Massachusetts Insitute of Technology, 1999. Bose-Einstein Condensation (BEC) of atoms.

Steiner, Herbert M., Ph.D., California, Berkeley, 1956. Emeritus. Experimental elementary particle physics.

Stevenson, M. Lynn, Ph.D., California, Berkeley, 1953. Emeritus. Experimental elementary particle physics.

Strovink, Mark W., Ph.D., Princeton, 1970. Emeritus. Experimental elementary particle physics.

Suzuki, Mahiko, Ph.D., Tokyo, 1965. Emeritus. Theory of elementary particles.

Townes, Charles H., Ph.D., Cal. Tech., 1939. Emeritus. Professor of the Graduate School. Experimental astrophysics; infrared and microwave astronomy.

Trilling, George H., Ph.D., Cal. Tech., 1955. Emeritus. Experimental elementary particle physics.

Tripp, Robert D., Ph.D., California, Berkeley, 1955. Emeritus. Experimental elementary particle physics.

White, Martin, Ph.D., Yale University, 1992. Theoretical Astrophysics.

Wichmann, Eyvind H., Ph.D., Columbia, 1956. Emeritus. Theory of elementary particles; mathematical physics.

Wurtele, Jonathan S., Ph.D., California, Berkeley, 1979. Theoretical plasma physics.

Yu, Peter, Ph.D., Brown. 1972. Emeritus. Professor of the Graduate School. Experimental condensed matter physics; semiconductor physics.

Zettl, Alex, Ph.D., California, Los Angeles, 1983. Experimental condensed matter physics.

Zumino, Bruno, Ph.D., Rome, 1945. Emeritus. Theory of elementary particles.

Associate Professors

Aganagic, Mina, Ph.D., Cal. Tech., 1999. Theoretical particle physics.

Boggs, Steven E., Ph.D., California, Berkeley, 1998. Gamma ray spectroscopy.

Bousso, Raphael, Ph.D., University of Cambridge, UK, 1997. Theoretical particle physics.

Ganor, Ori, Ph.D., Tel Aviv University, 1996. Theoretical particle physics.

Heinemann, Beate, Ph.D., University of Hamburg, 1999, Experimental particle physics.

Kolomensky, Yury, Ph.D., University of Massachusetts, 1997. BaBar, B hadron decays, with particular emphasis on understanding the origin of CD noninvariance.

Lanzara, Alessandra, Ph.D., University of Rome "La Sapienza," 1999. Experimental condensed matter physics.

Lee, Adrian, Ph.D., Stanford, 1993. Cryogenic far-infared and mm-wave detector development.

Liphardt, Jan T., Ph.D., Cambridge University, 1999. Experimental biophysics.

Moore, Joel E., Ph.D., MIT, 2000. Condensed matter and statistical physics-theory/biophysics theory.

Nomura, Yasunori, Ph.D., University of Tokyo, 2000. Theoretical particle physics.

Vishwanath, Ashvin, Ph.D., Princeton University, 2001. Theoretical condensed matter physics.

Assistant Professors

Battaglia, Marco, Ph.D., University of Helsinki, Finland, 1999. Experimental condensed matter physics.

DeWeese, Michael, Ph.D., Princeton University, 1995, Experimental biophysics.

Häffner, Hartmut, Ph.D., University of Mainz, 2000. Experimental atomic molecular, and optical physics.

Müller, Holger, Ph.D., Humboldt University, Berlin, 2004. Experimental atomic, molecular, and optical physics.

Siddiqi, Irfan, Ph.D., Yale, 2002. Experimental condensed matter physics.

Souza, Ivo, Ph.D., University of Illinois at Urbana-Champaign, 2000. Theoretical condensed matter physics.

Yildiz, Ahmet, Ph.D., University of Illinois, 2004. Experimental biophyiscs.

RESEARCH SPECIALTIES AND STAFF

Theoretical

Astrophysics. Interstellar medium; star formation; binary stars; stellar convection; pulsars; x-ray sources; active galactic nuclei and quasars; galaxy formation; cosmology. Arons, M. Davis, Haxton, McKee, Seljak, White. Graduate-student research assistant positions and postdoctoral positions with the Theoretical Astrophysics Center.

Condensed Matter Physics. Theoretical studies of the properties of semiconductors, metals, and insulators. These include: electronic, vibrational, optical, thermal, superconducting, quantum phase transitions, strongly correlating electronic systems; magnetic, surface, interfacial, phase transitions, and alloy properties of solids. Cohen, D.-H. Lee, Louie, Moore, Souza, Vishwanath. Graduate-student research assistant and postdoctoral positions.

Elementary Particles and Fields. Gauge theories of strong interactions and confinement of color quantum numbers; spectroscopy of new particles; gauge theory of weak and electromagnetic interactions and perturbative QCD; quantum electrodynamics and general field theory; *S*-matrix theory, grand unification theories; supersymmetry; supergravity; superstring theories; quantum field theory and string theory. Aganagic, Bardakci, Bousso, Chew, Gaillard, Ganor, Hall, Halpern, Haxton, Horava, Jackson, Mandelstam, Murayama, Nomura, Suzuki, Wichmann, Zumino. Graduate-student research assistant positions and postdoctoral positions through Department and through Lawrence Berkeley National Laboratory.

Fusion and Plasma Physics. Dynamics of ionized gases far from thermal equilibrium and dominated by long-range electromagnetic fields with applications to controlled fusion, astrophysics, and space science. Kaufman, Littlejohn, Wurtele. Graduate-student research assistant positions and postdoctoral positions.

General Relativity and Mathematical Physics. Algebraic quantum field theory; the physics and mathematics of gauge theories and general relativity. Sachs, Wichmann, Zumino. Graduate-student research assistant positions.

Nonlinear Dynamics. Chaos and approach to chaos; bifurcation theory; fluid dynamics; semiclassical mechanics. Knobloch, Littlejohn. Graduate-student research assistant positions.

Statistical Mechanics. Connection between dimensionality and off-diagonal long-range order in quantum liquids; general features of elementary excitation spectrum of Bose liquid; macroscopic quantum theory and quantum thermodynamics; oxide superconductivity. Rokhsar. Graduate-student research assistant positions.

Experimental

Astrophysics/Cosmic Rays. Magnetospheric physics: space plasmas and fields; auroras; isotopic and elemental composition of cosmic rays; search for new particles and antimatter in cosmic rays; spectrum and anisotropy of the universal microwave radiation; infrared astronomical spectroscopy and spatial interferometry; millimeter and submillimeter spectra; the galactic center; star formation; new astronomical detectors; automated supernova search; x-ray spectroscopy and laboratory astrophysics; high-energy gamma-ray astrophysics; experimental cosmology including particle astrophysics. Anderson, Bale, Boggs, M. Davis, Genzel, Holzapfel, A. Lee, Lin, Mozer, Muller, Perlmutter, Price, Richards, Sadoulet, Smoot, Townes. Graduate-student research assistant positions and postdoctoral positions through the Department; others through the Center for Particle Astrophysics, the Space Sciences Laboratory, and Lawrence Berkeley National Laboratory.

Atomic Physics/Optics. Precision measurements of parity violating effects in atoms; search for electric dipole moment of the electron; laser cooling of atoms; atom trapping; spectroscopy of atoms and molecules with emphasis on problems relevant to astrophysics; spectroscopy and collision processes of highly ionized heavy ions; lifetime and frequency measurements on metastable states of ions and atoms; production and study of laser-induced plasmas; very short wavelength laser transitions. Budker, Chu, Commins, Falcone, Häffner, Leone, Marrus, Müller, Shugart, Stamper-Kurn. Graduate-student research assistant and postdoctoral positions through the Department and through Lawrence Berkeley National Laboratory.

Biophysics. Molecular biophysics and structural biology. Application of atomic force spectroscopy and optical tweezers to biological problems. Bustamante, DeWeese, Liphardt, Rokhsar, Yildiz. Graduate-student research assistant positions.

Building Science. The Center for Building Science, in the Applied Science Division at Lawrence Berkeley National Laboratory, develops efficient lamps and lighting controls, windows, and other equipment. Other major groups measure and improve indoor air quality and develop computer programs to design energy-efficient buildings. Rosenfeld. Graduate-student research positions and postdoctoral positions through Lawrence Berkeley National Laboratory.

Condensed Matter Physics/Low Temperature. Fabrication and study of high-T_c superconductors; development and application of superconducting devices; macroscopic quantum tunneling; theoretical and experimental investigation of 1/f noise. Detection of low signal phenomena by magnetic resonance using the SQUID superconducting quantum interference device; phonon generation by electron spins; study of coherent radiation; nonlinear response of molecules ordered on the surface of a metal. Electronic properties of metal microclusters. Ultralow temperature research on superfluid ^{3}He and ^{4}He; macroscopic quantum interference in superfluid ^{3}He; electrical properties of conductors of reduced dimensionality; charge density wave formation and excitation. Optical properties of metal and semiconductor films with application to solar energy utilization. Infrared spectra of molecules bound to metal surfaces; infrared measurements of superconducting energy gaps; superconducting millimeter wave mixers; far infrared studies of impurities in semiconductors; and far infrared photoconductive detectors. Nonlinear optics; laser spectroscopy; optical properties of solids and liquid crystals and light scattering experiments. Study of Raman and Brillouin spectroscopies in semiconductors. Properties of defects in semiconductors studied by optical and electrical techniques; optical, electrical, and superconducting properties of matter under high pressure; ultrafast processes in solids, scanning tunneling microscopy; nonlinear dynamics in solids; surface physics; interface formation. Birgeneau, Clarke, Crommie, Dynes, Hahn, Hellman, Lanzara, Orenstein, Packard, Portis, Qiu, Ramesh, Richards, Shank, Shen, Siddiqi, Yu, Zettl. Graduate-student research assistant positions through the Department and postdoctoral positions through Lawrence Berkeley National Laboratory.

Elementary Particles. Experiments utilizing particle accelerators such as electron-positron and proton-antiproton colliders as well as fixed target machines to test and extend existing theories of electroweak and strong interactions, and to search for new quarks, new leptons, and other new particles; nonaccelerator experiments to search for double beta decay and other rare processes; development and fabrication of detectors appropriately matched to these goals. Battaglia, Crowe, Goldhaber, Heinemann, Jacobsen, Kerth, Kolomensky, Luk, Shapiro, Siegrist, Steiner, Stevenson, Strovink, Trilling, Tripp. Graduate-student research assistant and postdoctoral positions through the Department and through Lawrence Berkeley National Laboratory.

Fusion and Plasma Physics. Plasma production and heating; magnetic confinement of high-temperature plasma; development and application of plasma diagnostic methods; atomic physics problems related to controlled fusion, accelerator research for heavy-ion driven pellet fusion; single species plasma. Fajans, Kunkel. Graduate-student research assistant and postdoctoral positions through the Department and through Lawrence Berkeley National Laboratory.

Medium Energy and Muon Physics. Studies of nuclear structure and dynamics using particle beam accelerators with momenta up to 1 GeV/c; muon induced fusion, muon spin resonances. Crowe. Graduate-student research assistant positions through the Department and postdoctoral positions through Lawrence Berkeley National Laboratory.

Nuclear Physics. Precision tests of the Standard Electroweak Model and searches for new physics using low energy techniques from nuclear and atomic physics. Nuclear astrophysics. Freedman. Several graduate-assistantships and postdoctoral positions available through the department and Lawrence Berkeley National Laboratory.

UNIVERSITY OF CALIFORNIA, DAVIS

DEPARTMENT OF PHYSICS

Davis, California 95616

Students Accepted For Degree	FIELDS		
	Physics	Astronomy	Related Fields
Doctorate	X		
Master's	X		

1. General

Chancellor: Linda Katehi
Dean of Graduate School: Jeffery C. Gibeling
Department Chairman: Warren Pickett
Department Telephone Number: (530) 752-1501 (C) 5989
Type of Institution: University
Control: Public
Setting: Small town
Total Faculty: 2,558
Total Graduate Faculty: Not separate 2,558
Total Students: 32,153
Total Graduate Students: 4,215
Annual Graduate Tuition:
In-state residents: Full-time—$4,419/quarter
Out-of-state residents: Full-time—$9,453/quarter
Tuition rates for: 2009–10
Deferred tuition plan: Yes (Fee fellowships available)
Other Fees: None
Term: Quarter

2. Number of Faculty in Department

The combined total of full-time faculty in the three professorial ranks is 40. The combined total of full-time, part-time, and other faculty at all ranks is 53.

3. Admission, Financial Aid, and Housing

Address admission inquiries to: Graduate Program, Department of Physics, One Shields Ave., Davis, CA 95616
Graduate application fee required: $70 (US) $90 (International)
Admission deadline (Fall admission only): January 15
Admission information: For fall admission, 2010–11, 64 students were offered admission from 251 applicants.
Admission requirements: An applicant for admission must have a Bachelor's degree from an accredited college or university with a grade-point average of 3.0 or better (on a scale where A=4.0) in upper division coursework and in any graduate courses taken. The GRE general exam is required, and the subject GRE exam in Physics is strongly recommended. No minimum score specified. Students from non-English speaking countries are required to demonstrate proficiency in English via the TOEFL exam. Minimum acceptable score is 550 (paper exam), 213 (computer-based exam), 80 (internet-based exam).
Undergraduate preparation assumed: Typical texts: Tipler, *Physics for Scientists and Engineers*; Serway, Moses, Moyer, *Modern Physics*; Boas, *Mathematical Methods in the Physical Sciences*; Marion and Thornton, *Classical Dynamics*; Griffiths, *Introduction to Electrodynamics*; Reif, *Fundamentals of Statistical and Thermal Physics*; Griffiths, *Quantum Mechanics*.
Address financial aid inquiries to: Graduate Financial Aid Office, One Shields Ave., Davis, CA 95616
GAPSFAS application required: Yes
Financial aid deadline: January 15
Fellowship application due: January 15
Loans available: Yes
Address housing inquiries to: Student Housing Office, One Shields Ave., Davis, CA 95616
On-campus, single student housing available: Yes
Cost/month: $595–750
On-campus, married student housing available: Yes
Cost/month: 1 bdrm., unfurnished-$595[1]
2 bdrms., unfurnished-$678–710[1]

[1]Utilities paid by occupant

Table A—Faculty, Enrollments, and Degrees Granted

Research Specialty	2009–10 Faculty	Enrollment[1] Fall 2008		No. of Degrees Granted[2] 2009–10 (2005–10)			Median No. of Years for 2009–10 Ph.D.'s
		Master's	Doctorate	Master's	Terminal Master's	Doctorate	
Biological Physics	1	0	6	0(0)	0(0)	0(2)	–
Condensed Matter Physics	18	0	41	0(0)	1(1)	5(31)	6
Complex Systems	2	0	6	0(0)	0(0)	1(2)	6
Cosmology/ Astrophysics	10	0	14	0(0)	0(1)	1(11)	6
Nuclear Physics	3	1	5	0(0)	0(0)	1(3)	8
Particles & Fields	12	0	32	0(0)	0(1)	1(16)	6
Physics Education	1	0	4	0(0)	0(0)	1(2)	5
Relativity & Gravitation	1	0	8	0(0)	0(0)	0(0)	–
Non-specialized	0	0	29	13(50)	0(19)	0(0)	–
Total	48	1	145	13(50)	1(22)	10(63)	6
Full-time Grad. Stud.		0	146				
Part-time Grad. Stud.		0	1				
First-year Grad. Stud.		0	29				
Median Years in Grad. Study (2007–08 Degrees)					3	6	–
Undergraduate Degrees, 2009–10 (2005–10): 4(148)							

[1] Students not yet committed to a research specialty are entered under non-specialized.
[2] Five-year totals in parentheses.

4. Graduate Degree Requirements

Master's: Two Master's programs are offered. Plan I requires 32 quarter hours of graduate and upper division courses and a Master's thesis. Plan II requires 36 quarter hours of graduate and upper division course work of which at least 18 hours must be at the graduate level and passing the preliminary exam at the Master's level. The preliminary exam covers senior undergraduate and first-year graduate level physics. Both plans require course work in classical physics, quantum mechanics, and mathematical methods.
Doctorate: The Doctor of Philosophy degree requires a thorough understanding of the foundations of physics and mathematical methods as evidenced by performance on the preliminary exam and an oral exam and submission of a dissertation which must include an original contribution to fundamental physics. Ph.D. students must also complete the graduate core courses in classical physics, statistical physics, quantum mechanics, and mathematical methods. The required curriculum

can be tailored to fit the individual student's preparation and needs. Each graduate student selects a course of study in consultation with a graduate advisor. A student with weaknesses in preparation may be advised to audit or take for credit specific advanced undergraduate courses. A student entering with advanced preparation may skip the core courses by passing the written examination upon entrance to the program. Students are to take a cluster of advanced graduate courses determined by their field of specialization. For more information please visit http://www.physics.ucdavis.edu

Thesis may be written in absentia.

Table B—Appointments to Graduate Students, 2009–10

Title of Appointee	Appointments Total	Appointments First year	Academic Load Allowed in Credit Hours	Hours of Service Per Week	Annual Stipend ($)
			Quarter		
Teaching Assistant	87	29	12	10–20	16,637[1,2,4]
Research Assistant	75	0	12	10–20	15,767–17,030[1,2,3,4]
Unique Davis Fellowships	16	0	12	10–20	924–16,478[1]
Tuition and fee fellowship	24	19	12	10–20	11,632–26,674
Total	202	48			

[1]Full fees paid.
[2]Amount is for academic year.
[3]Theses work is often included which makes the number of hours of research higher.
[4]Students may hold combination of appointments and fellowship.

5. Personnel Engaged in Separately Budgeted Research, 7/09–5/10

Researcher & Postdoctoral appointments	37
Graduate Student Researcher	63
Total	100

6. Separately Budgeted Research Expenditures by Source of Support

	Department
Federal	$4,433,011
Private	997,966
University	340,035
Total	$5,771,012

Table C—Separately Budgeted Research Expenditures

Research Specialty	No. of Grants	Expenditures ($)
Astrophysics/Cosmology	55	1,122,583
Biophysics	0	0
Condensed Matter Physics	40	1,309,646
Computational Physics	11	380,372
Nuclear	5	479,454
Particles & Fields	26	2,478,957
Total	137	5,771,012

FACULTY

Professors

Albrecht, Andreas, Ph.D., Pennsylvania, 1983. Cosmology.

Becker, Robert, Ph.D., Maryland, 1975. Observational astrophysics.

Carlip, Steven, Ph.D., Texas, Austin, 1987. Theoretical high-energy physics; quantum gravity.

Cebra, Daniel A., Ph.D., Michigan State, 1990. Experimental nuclear physics.

Chertok, Maxwell B., Ph.D., Boston, 1997. Experimental high energy physics; hadron collider physics.

Chiang, Shirley, Ph.D., California, Berkeley, 1983. Experimental condensed matter physics and surface physics.

Coleman, Lawrence B., Ph.D., Pennsylvania, 1975. Experimental condensed matter physics; far infrared spectroscopy; phase transitions in solids; structure-property relationships.

Conway, John, Ph.D., Chicago, 1987. Experimental high energy physics.

Corruccini, Linton R., Ph.D., Cornell, 1972. Experimental low-temperature and condensed matter physics; liquid helium.

Cox, Daniel L., Ph.D., Cornell, 1985. Theoretical condensed matter physics.

Crutchfield, James, Ph.D., California, Santa Cruz, 1983. Nonlinear dynamics and complex systems.

Fadley, Charles S., Ph.D., California, Berkeley, 1970. Advanced light source; surfaces and interfaces via angle-resolved photoelectron spectroscopy.

Ferenc, Daniel, Ph.D., University of Zagreb, Croatia, 1992. Experimental nuclear physics; relativistic heavy-ion physics.

Fong, Ching-Yao, Ph.D., California, Berkeley, 1968. Theoretical condensed matter physics.

Galli, Giulia, Ph.D., International School for Advanced studies, Trieste, Italy, 1986. Computational materials theory.

Gunion, John F., Ph.D., California, San Diego, 1970. Theoretical particle physics.

Kaloper, Nemanja, Ph.D., Minnesota, Minneapolis, 1992. Cosmology.

Kiskis, Joseph E., Ph.D., Stanford, 1974. Theoretical particle physics.

Klein, Barry M., Ph.D., NYU, 1969. Theoretical condensed matter.

Knox, Lloyd D., Ph.D., Chicago, 1995. Cosmology.

Ko, Winston T., Ph.D., Pennsylvania, 1971. Experimental particle physics.

Liu, Kai, Ph.D., Johns Hopkins, 1998. Experimental condensed matter.

Lubin, Lori M., Ph.D., Princeton, 1995. Vice Chairperson of the Department. Cosmology.

Luty, Markus, Ph.D., Univ. of Chicago, 1992. Theoretical particle physics; theoretical gravity.

Pellett, David E., Ph.D., Michigan, 1966. Experimental particle physics.

Pickett, Warren E., Ph.D., SUNY, Stony Brook, 1975. Department Chair. Theoretical condensed matter physics; metals and superconductors.

Pines, David, Ph.D., Princeton University, 1950. National Academy of Science Member. Theoretical condensed matter physics.

Rundle, John B., Ph.D., California, Los Angeles, 1976. Condensed matter physics; complex nonlinear systems.

Savrasov, Sergey Y., Ph.D., Lebedev Physical Institute, Moscow, 1994. Theoretical condensed matter physics; electronic structure of solids; computational approaches to strongly-correlated systems.

Scalettar, Richard T., Ph.D., California, Santa Barbara, 1986. Vice Chairperson of the Department. Theoretical condensed matter physics; statistical mechanics; many-body theory; highly correlated systems.

Singh, Rajiv R. P., Ph.D., SUNY, Stony Brook, 1986. Theoretical condensed matter physics; spin glasses; critical phenomena; low-dimensional systems, biological physics.

Svoboda, Robert, Ph.D., Univ. of Hawaii, 1985. Neutrino physics.

Terning, John, Ph.D. University of Toronto, 1990. Theoretical particle physics, electroweak symmetry breaking, supersymmetry, cosmology, extradimensions, AdS/CFT correspondence.

Tripathi, S. Mani, Ph.D., Pittsburgh, 1986. Experimental high-energy physics; productions of direct photons in hadronic interactions.

Tyson, J. Anthony, Ph.D., University of Wisconsin, 1967. National Academy of Science Member. Observational cosmology.

Zhu, Xiangdong, Ph.D., California, Berkeley, 1989. Experimental condensed matter physics; nonlinear optics; laser studies of surfaces.

Zieve, Rena, Ph.D., California, Berkeley, 1992. Experimental condensed matter; low temperature.

Zimanyi, Gergely, Ph.D., Central Research Institute for Physics, Budapest, 1985. Theoretical condensed matter physics; localization; electron gas; low-dimensional system; superconductivity.

Associate Professors

Calderón de la Barca Sánchez, Manuel, Ph.D., Yale University, 2001. Relativistic heavy ion physics.

Cheng, Hsin-Chia, Ph.D., UC Berkeley, 1996. Theoretical particle physics.

Curro, Nicholas, Ph.D., University of Illinois at Urbana-Champaign, 1998. Experimental condensed matter physics.

Erbacher, Robin, Ph.D., Stanford, 1998. Experimental high energy particle physics: top quark physics, searches for new particles and phenomena, particle detector development.

Fassnacht, Christopher D., Ph.D., CalTech, 1999. Cosmology, galaxy structure, Cosmological parameter measurement through gravitational lensing.

Wittman, David M., Ph.D., Univ. of Arizona, 1997. Observational cosmology.

Assistant Professors

Bradac, Marusa, Ph.D., University of Bonn, 2004. Cosmology gravitational lensing, first galaxies, dark matter.

Yu, Dong, Ph.D., University of Chicago, 2005. Experimental condensed matter physics.

Senior Lecturers, SOE

Boeshaar, Patricia, Ph.D., Ohio State University, 1976. Physics and astronomy education.

Webb, David J., Ph.D., Maryland, 1983. Physics education.

Research Physicists

Breedon, Richard, Ph.D., The Rockefeller University, 1988. High energy experimental physics.

Cox, Peter Timothy, Ph.D., University of Michigan, 1980. High energy experimental physics.

Gee, Perry A., M.S., Univ. of Calif. Davis, 2004. Cosmology.

Gregg, Michael D., Ph.D., Yale University, 1985. Cosmology.

Klavins, Peter, M.S., Iowa State University, 1987. Experimental materials physics.

Morein, Gleb, Ph.D., Cornell University, 2000. Computational Geophysics.

Richter, Matthew J., Ph.D., California, Berkeley, 1995. Cosmology.

Scranton, Ryan, E., Ph.D., University of Chicago, 2002. Cosmology.

Smith, John R., Ph.D., California, Davis, 1982. High energy experimental physics.

Stanford, Spencer Adam, Ph.D., University of Wisconsin, Madison, 1990. Cosmology.

Adjunct Professors

de Roeck, Albert, Ph.D., University of Antwerp, Belgium. Experimental particle physics.

Lorenz, Eckart, Ph.D., University of Munich, 1970. Particle astrophysics.

Radousky, Harry B., Ph.D., Illinois, 1982. Experimental condensed matter physics.

Vogt, Ramona, Ph.D., SUNY, Stony Brook, 1989. Theoretical nuclear physics.

Lecturers

Cole, Rodney W., Ph.D., Wyoming, 1978.

Harris, Randy R., Ph.D., California, Davis, 1985.

Professors Emeriti

Brady, F. Paul, Ph.D., Princeton, 1960. Experimental nuclear physics; neutron physics and relativistic nuclear collisions.

Cahill, Thomas A., Ph.D., California, Los Angeles, 1965. Experimental nuclear and atomic physics; analytical applications of accelerator beams; atmospheric physics.

Chau, Ling-Lie, Ph.D., California, Berkeley, 1966. Theoretical particle physics.

Draper, James E., Ph.D., Cornell, 1952. Experimental nuclear physics; particle-gamma and ion-beam atomic spectroscopy.

Erickson, Glen W., Ph.D., Minnesota, 1960. Theoretical physics; quantum field theory.

Garrod, Claude, Ph.D., NYU, 1963. Theoretical physics; quantum-mechanical many-particle systems; statistical mechanics.

Jungerman, John A., Ph.D., California, Berkeley, 1949. Experimental nuclear physics; nuclear arms control.

Knox, William J., Ph.D., California, Davis, 1951. Experimental nuclear physics.

Lander, Richard L., Ph.D., California, Berkeley, 1958. Director, Intercampus Inst. for Research at Particle Accelerators. Experimental particle physics.

McColm, Douglas W., Ph.D., Yale, 1961. Experimental atomic physics; molecular beam studies.

Peek, Neal F., Ph.D., California, Davis 1966. Experimental nuclear physics.

Potter, Wendell H., Ph.D., Illinois, 1970. Science education.

Reid, Roderick V., Jr., Ph.D., Cornell, 1968. Theoretical physics; nucleon-nucleon potentials; nuclear structure.

Yager, Philip M., Ph.D., California, San Diego, 1973. Experimental particle physics.

RESEARCH SPECIALTIES AND STAFF

Theoretical

Complex systems. Computational simulations of complex physical and biological systems; datamining of massive and multidimensional data sets; understanding emergent patterns and coherent structures in nonlinear systems; prediction and forecasting; scaling; network and cluster formation and dynamics. Crutchfield, Rundle. Graduate student research assistant and post-doctoral positions are available.

Condensed Matter Physics. Studies of macroscopic phases of matter, including metals, insulators, superconductors, superfluids, magnets, spin-glasses, supersolids, etc. Thermal and quantum phase transitions and critical phenomena. Microscopic theory and phenomenology of high temperature superconductivity and of superconductivity coexisting with magnetism. Microscopic studies of strongly correlated electron systems. Hubbard, t-J and Heisenberg models. Mesoscopic physics and nanostructures. Electronic structure calculations, quantum Monte Carlo and other advanced numerical methods, are used to study alloy phase stability, semiconductor nanostructures, spintronic materials, and surface physics properties from first principles. Cox, Fong, Galli, Klein, Pickett, Pines, Savrasov, Scalettar, Singh, Zimanyi. Graduate student research assistant and postdoctoral positions are available.

Cosmology/Astrophysics. Physics of the early universe and the formation of cosmic structure, cosmic inflation, dark energy models and probes, cosmic microwave background, precision cosmology. Albrecht, Kaloper, Knox. Graduate student research assistant and postdoctoral positions are available.

Particle Physics. Gauge theories of the electromagnetic, weak, and strong interactions, perturbative and nonperturbative analysis, lattice gauge theory, unification, supersymmetry, extra dimension, symmetry breaking, phenomenology, heavy quark systems, soluble models, quantum gravity. Carlip, Cheng, Gunion, Kiskis, Luty, Terning. Graduate student research assistant and post-doctoral positions are available.

Relativity & Gravitation. Conceptual issues in quantum gravity; low-dimensional models for quantum gravity; black hole thermodynamics; physical and mathematical foundations of general relativity. Carlip. Graduate student research assistant and postdoctoral positions are available.

Experimental

Condensed Matter Physics. Surfaces and interfaces, magnetism, low temperature physics, lattice dynamics, quantum fluctuations, transport properties, high temperature superconductivity, granular materials, light scattering on biological materials and nanostructured materials. Systems under study include atoms, molecules and nanostructures on surfaces, nanoparticles and nanowires, epitaxial multilayer thin films, patterned nanostructures, organic and biological nanostrucutres, ^{4}He vortices, dipolar magnets and frustrated antiferromagnets, exotic superconductors, complex magnetic and ferroelectric materials, and disordered media. Sample synthesis and processing by magnetron sputtering, thermal and e-beam evaporation, electrodeposition, high temperature sintering and single-crystal growth, photo- and e-beam lithography. Characterizations by photoelectron spectroscopy, diffraction, and holography using laboratory sources and high brightness synchrotron radiation at the nearby Lawrence Berkeley National Laboratory, x-ray diffraction, scanning electron microscopy, ultrahigh vacuum scanning tunneling and atomic force microscopy, low energy electron microscopy and diffraction, surface magneto-optical Kerr effect, Raman spectroscopy, far infrared spectroscopy, SQUID and local Hall probe magnetometry, metallographic and thermal analyses, semiadiabatic calorimetry from dilution refrigerator range to high temperatures, transport measurements, and vibrating wire studies. Properties of materials under high pressure in collaboration with Lawrence Livermore National Laboratory. Chiang, Coleman, Corruccini, Curro, Fadley, Liu, Yu, Zhu, Zieve. Graduate student research assistant and postdoctoral positions are available.

Cosmology/Astrophysics. Physics of the early universe, evolution of Active Galactic Nucleii (AGN), quasar absorption line systems, clusters of galaxies, gravitational lenses, large scale surveys. Becker, Bradac, Fassnacht, Gregg, Lubin, Richter, Stanford, Scranton, Tyson, Wittman. Graduate student research assistant and postdoctoral positions are available.

Nuclear Physics. Study of relativistic heavy-ion collisions, creation of quark-antiquark pairs, discovery and study of quark matter or quark-gluon plasma as it is often called. Calderon, Cebra, Ferenc. Graduate student research assistant and postdoctoral positions are available.

Particle Physics. Experiments using particle accelerators and colliders around the world to probe the structure of matter at the deepest possible level. Design and construction of neutrino and WIMP dark matter detectors to search for dark matter and make precision measurements of neutrino properties. Studies of the properties of and the forces between quarks and leptons and searches for new particles. Extensive computer utilization for data analysis and design modeling. Development of new detectors such as micron-size silicon pixel devices. Design and fabricate read-out electronics using new techniques such as ASIC (application specific integrated circuit) and flip chip bump bonding. Breedon, Chertok, Conway, Erbacher, Ko, Lander, Pellett, Svoboda, Tripathi. Graduate student research assistant and postdoctoral positions are available.

Physics Education. The process of how students come to an understanding of physics/science concepts; instructional models; use of visual models in understanding atomic phenomena. Boeshaar, Potter, Webb. Graduate student research assistant positions are available.

UNIVERSITY OF CALIFORNIA, DAVIS

DEPARTMENT OF APPLIED SCIENCE

Davis, California 95616

Students Accepted For Degree	FIELDS		
	Physics	Astronomy	Related Fields
Doctorate			X
Master's			X

1. General

Chancellor: Linda Katehi
Dean of Graduate School: Jeffery Gibeling
Department Chairman: Yin Yeh
Department Telephone Number: (530) 754-8858
Type of Institution: University
Control: Public
Setting: Small town
Total Faculty: 1,493
Total Graduate Faculty: Not separate
Total Students: 30,685
Total Graduate Students: 4,094
Annual Graduate Tuition:
In-state residents: Full-time—$10,617.90
Out-of-state residents: Full-time—$25,623.90
Other Fees: None
Term: Quarter

2. Number of Faculty in Department

Full-time: 18
Adjunct: 4

3. Admission, Financial Aid, and Housing

Address admission inquiries to: Graduate Advisor, Department of Applied Science, University of California, One Shields Avenue, Davis, CA 95616-8254 or EAD advising@ucdavis.edu

Graduate application fee required: $70 domestic/$90 international

Admission deadline (Fall admission only): 4/1 Domestic applications (or until full); 3/1 International applications; 1/15 for all fellowships.

Admission Information: For fall admission, 2009–10, 11 students were accepted from 69 applicants.

Admission requirements: For admission to the graduate programs, a Bachelor's degree in mathematics, physics, chemistry, geology, or engineering is required from an accredited college or university. The GRE is required. Students from non-English speaking countries are required to demonstrate proficiency in English via the TOEFL exam. Minimum acceptable score for admission is 550 (paper)/213 (computer).

Address financial aid inquiries to: Financial Aid Office, 1100 Dutton Hall (530) 752-2390

Financial aid deadline: 1/15

Loans available: Yes

Address housing inquiries to: Student Housing Office. E-mail: http://housing.ucdavis.edu/

On-campus, single student housing available: Yes

On-campus, married student housing available: Yes

Table A—Enrollments, and Degrees Granted

	Enrollment Fall 2009		No. of Degrees Granted 2008–09	
	Master's	Doctorate	Master's	Doctorate
Total			7	6
Full-time Grad. Stud.	8	48		
Part-time Grad. Stud.	–	0		
First-year Grad. Stud.	2	8		
Median Years in Grad. Study			3.5	5
Undergraduate Degrees		**6**		

4. Graduate Degree Requirements

Master's: The Master's degree program provides a broad background in the physical sciences. Requirements for the degree are 36 units of credit in graduate and upper division undergraduate courses with a minimum GPA of 3.0, a comprehensive written examination, and three quarters of academic residence. At least 18 of the 36 units must be graduate courses in the areas specified in the Applied Science program. The comprehensive written examination covers course material specified for study at the Master's level. The same examination also serves as a preliminary examination for the Ph.D. degree. Students with Masters' degrees in areas other than Applied Science should, in general, plan to take some of the course work specified for the Master's program before taking the comprehensive examination. No foreign language requirement. Option MS I: thesis.

Doctorate: The Doctoral program provides specialized training in a particular area of the physical sciences. It consists of both formal course work and independent study and research. No foreign language requirement.

Thesis: Thesis may be written *in absentia.*

Special Equipment, Facilities, or Programs: A competitive feature of the Department is its collaborations with several national labs including Los Alamos, Brookhaven, Lawrence Berkeley and Lawrence Livermore, including the National Ignition Facility for fusion research and the Atlas 44 tern-FLOP supercomputer (www.llnl.gov).

The Department is a partner in the National Center for the design of Biomimetic Nanoconductors (www.nanoconductor-.org), and maintains close ties with the NSF Center for Biophotonics Science and Technology, which conducts cutting-edge, interdisciplinary connecting basic physics to medical applications (cbst.ucdavis.edu).

Table B—Appointments to Graduate Students, 2009–10

Title of Appointee	Appointments Total	Appointments First year	Academic Load Allowed in Credit Hours	Hours of Service Per Week	Stipend for Academic Year ($)
			Quarter		
Teaching Assistant	5	5	-	20	10,132
Research Assistant	31	4	-	20	19,152
Lawrence Livermore National Security Student Employees	2	0	-	20/40[1]	29,294–30,044[2]
Reader	0	0	-	Varies	10.40/hr.
Fellow	8	2	-	Varies	15,000+[2]
Total	46	11			

[1]50% time during the academic year and 100% time during the summer session.
[2]Plus tuition and fees for the academic year.

FACULTY

Professors

Baldis, Hector A., Ph.D., University of British Columbia, 1971. Laser-plasma physics under ICF conditions; high intensity laser-plasma interactions; particle-photon interaction physics; advanced laser-plasma diagnostics.

Cramer, Stephen P., Ph.D., Stanford, 1977. X-ray spectroscopy; bioinorganic chemistry; nitrogen fixation; enzyme kinetics and synchrotron radiation; photosynthesis.

Duan, Yong, Ph.D., University of Pittsburgh, 1996. Computational modeling of biomolecular systems; computational biophysics; computational biology.

Gygi, Francois, Ph.D., Swiss Federal Institute of Technology, 1988. Computational Materials Science, First-Principles Molecular Dynamics, Electronic structure theory, Parallel numerical algorithms.

Harris, Walter, Ph.D., University of Michigan, 1993. Development of optical interferometers for UV-Visible ground/space based remote sensing. Observational studies of chemical, dynamic, and evolutionary processes in the coma and ion tail of comets; photochemsitry and energetic processes in planetary upper atmospheres and near space environments; structure of the heliosphere and local interstellar medium.

Hwang, David Q., Ph.D., Cal. Tech., 1978. Tokamak plasma physics; interaction between plasma waves and plasma particles; tokamak fueling with accelerated compact toroids.

Jensen, Niels Gronbech, Ph.D., Technical University of Denmark, 1991. Modeling of dynamical and statistical systems. Atomic scale modeling of materials, biomaterials, complex fluids. Algorithm development for high performance computing. Interlink between theory, modeling, and experimental characterization.

Kolner, Brian H., Ph.D., Stanford, 1985. Lasers; ultrafast optical phenomena; optical waveform generation and measurement; space-time analogies and temporal imaging; RF/microwave measurement and design; terahertz radiation and spectroscopy.

Krol, Denise M., Ph.D., University of Utrecht, the Netherlands, 1980. Lasers; nonlinear optics; laser spectroscopy; optical materials; wave guides and fibers; properties of glasses.

Luhmann, Jr., N. C., Ph.D., University of Maryland, College Park, 1972. Millimeter wave solid state devices and systems; ultra short pulse electronics; gyrotrons; free electron lasers; plasma physics; phased antenna arrays; advanced accelerators; Compton X-ray light sources for cancer diagnostics and treatment; millimeter wave imaging; MEMS; tokamak diagnostics and far-infrared lasers; microwave/plasma interactions; vacuum microelectronics.

Max, Nelson, Ph.D., Harvard, 1967. Scientific visualization; molecular graphics; photo-realistic image synthesis; volume and flow visualization; computer animation.

McCurdy, C. William, Ph.D., Cal. Tech., 1975. Molecular physics; electron molecule scattering; atoms in intense laser fields; molecule-solid surface scattering.

Miller, Gregory H., Ph.D., California Institute of Technology, 1990. Computational fluid dynamics, shock physics, thermodynamic and constitutive modeling.

Orel, Ann E., Ph.D., UCB, 1981. Theoretical atomic and molecular physics; electron scattering; photoionization; heavy particle dynamics.

Parikh, Atul N., Ph.D., Penn State, 1994. Principles of soft matter assembly, structure, and dynamics; spectroscopy and imaging; materials in biotechnology.

Rocke, David Ph.D., University of Illinois, Chicago, 1972. Statistical analysis of gene expression data, proteomics data by mass spectrometry, and of other high throughout biological assay data; radiation biology; robust statistical methods; formal models in international relations; applicationof statistics in medicine, biology and environmental science.

Vemuri, Rao, Ph.D., UCLA, 1968. Artificial intelligence and soft computing, neural networks, genetic algorithms, software agents. Applications to signal and image processing, multimedia.

Yeh, Yin, Ph.D., Columbia, 1965. Laser diagnostics of materials; laser physics; optics; physical chemistry of molecules; biological macromolecules and cells; optical spectroscopy.

Adjunct Professors

Balhorn, Rodney, Ph.D., University of Iowa, Iowa City, 1972. Toxin and repair protein structure and function, biophysical analysis of DNA-protein complexes, sperm chromatin structure and function, development and application of single molecule analytical techniques.

Canning, Andrew, Ph.D., Edinburgh University, 1988. Material science, scientific computation, parallel computing.

Krishnan, Viswanathan, Ph.D., Indian Institute of Science, 1991. Molecular biophysics; magnetic resonance spectroscopy, structural biology, computational biology, quantum computing, biostatistics and bioinformatics.

Kruer, William, Ph.D., Princeton University, 1969. Plasma physics.

RESEARCH SPECIALTIES AND FACULTY

Computational Science: Canning, Duan, Miller, Jensen, Max, McCurdy, Orel, Rocke, Vemuri, Gyal.

Condensed Matter and Materials Modeling: Jensen, Krol, Miller.

Molecular and Cellular Biosciences: Balhorn, Cramer, Duan, Krishnan, Parikh, Yeh.

Optical, Atomic, and Molecular Science: Cramer, Harris, Kolner, Krol, McCurdy, Orel, Yeh.

Plasma Science: Baldis, Hwang, Kruer, Luhmann.

Space Physics: Harris.

UNIVERSITY OF CALIFORNIA, IRVINE

DEPARTMENT OF PHYSICS AND ASTRONOMY

Irvine, California 92697-4575

Students Accepted For Degree	FIELDS		
	Physics	Astronomy	Related Fields
Doctorate	X		
Master's	X		

1. General

Chancellor: Michael V. Drake
Dean of Graduate School: Frances Leslie
Department Chairman: William H. Parker
Department Telephone Number: (949) 824-6911
Type of Institution: University
Control: Public
Setting: Suburban
Total Faculty: 1,594
Total Students: 27,792
Total Graduate Students: 5,566
Annual Graduate Tuition:
In-state residents: Full-time—None
Out-of-state residents: Full-time—$14,694
Tuition rates for: 2010–11
Deferred tuition plan: No
Other Fees: $13,415
Term: Quarter

2. Number of Faculty in Department

The combined total of full-time faculty in the three professorial ranks is 44. The combined total of full-time, part-time, and other faculty at all ranks, including professional research staff is 82.

3. Admission, Financial Aid, and Housing

Address admission inquiries to: Graduate Admissions, Physics and Astronomy Department
Graduate application fee required: $70 (Domestic), $90 (International)
Admission deadline (Fall admission): 1/15
Admission information: For fall admission, 2010–11, 75 students were accepted from 293 applicants.
Admission requirements: For admission to the graduate programs, a Bachelor's degree in physics is required with a minimum undergraduate GPA of 3.0 specified. The GRE General and Subject tests are required. No minimum scores specified. Students from non-English speaking countries are required to demonstrate proficiency in English via the TOEFL exam. Minimum acceptable score for admission is 550 (paper). In addition, the Test of Spoken English (TSE) is recommended. Minimum acceptable score is 80.
Undergraduate preparation assumed: General Physics: Halliday & Resnick or equiv.; *Mechanics*: Marion or equiv.; *E and M*: Griffiths or equiv.; *Quantum Mechanics*: Liboff or equiv.; *Thermal and Statistical Physics*: Reif or equiv.; *Mathematics for Physics*: Arfken or equiv.; and advanced undergraduate laboratory.
Address financial aid inquiries to: www.grad.uci.edu/prospective/finance_edu.html
GAPSFAS application required: No
Loans available: Yes
Address housing inquiries to: www.housing.uci.edu
On-campus, single student housing with meals available: Yes
Cost/month: $3,078–4,188
On-campus, married student housing available: Yes
Cost/month: $743–1,791

Table A—Faculty, Enrollments, and Degrees Granted

Research Specialty	2009–10 Faculty	Enrollment[1] Fall 2009		No. of Degrees Granted[2] 2009–10 (2005–10)			Median No. of Years for 2009–10 Ph.D.'s
		Master's	Doctorate	Master's	Terminal Master's	Doctorate	
Astrophysics	5	2	18	0(6)	2(5)	5(9)	6
Biophysics	2	1	7	1(4)	0(3)	2(5)	5
Condensed Matter Physics	14	2	57	4(25)	1(12)	10(29)	5.5
Cosmology	4	0	0	0(1)	0(3)	0(1)	
Low Temperature Physics		2	2	0(0)	0(0)	0(0)	
Medical & Health Physics	1	0	2	0(3)	0(2)	1(4)	5
Particles & Fields	13	0	30	2(7)	1(5)	2(14)	6.5
Plasma Physics & Fusion	4	1	12	2(8)	0(1)	3(8)	6
Relativity & Gravitation	1	0	4	0(1)	0(0)	0(0)	
Non-specialized		0	0	0(0)	0(0)	0(0)	
Total	44	8	132	9(55)	4(31)	23(70)	
Full-time Grad. Stud.		8	132				
Part-time Grad. Stud.		0	0				
First-year Grad. Stud.		0	25				

Undergraduate Degrees, 2009–10 (2005–10): 39(137)

[1]Students not yet committed to a research specialty are entered under non-specialized.
[2]Five-year totals in parentheses.

4. Graduate Degree Requirements

Master's: *Master of Science in Physics*. The requirements for the M.S. degree are (1) at least three quarters of residence; (2) mastery of graduate course material, which must be demonstrated by passing, with a grade of B or better, a minimum of eight quarter courses including 211, 213AB, 215A, 223, at least one other course numbered between 200 and 259, and two other courses approved by the graduate advisor; and (3) either [Option A] a research project and written thesis or [Option B] a comprehensive written examination. There is no foreign language requirement.

Option A: The thesis need be of no specified length or format, but must report significant results in readable, meaningful form, at the same time revealing the student's general grasp of the field and awareness of related work.

Option B: The comprehensive exam for the M. S. degree is identical to that for the Ph.D. degree. The level of performance required for the M.S. degree by examination is identical to the required level of performance for the Ph.D. degree.

Doctorate: The principal requirements for the Ph.D. degree are six quarters of residence, passage of a written examination, and successful completion and defense of a dissertation reporting results of original research. In addition, the Ph.D. candidate must complete certain graduate course requirements. Experience in teaching is an integral part of the graduate program, and all Ph.D. students are required to participate in the teaching program for at least three quarters during their graduate careers. Foreign students must pass a campus approved spoken English proficiency exam (required in order to TA) by the time they advance to candidacy. There is no foreign language requirement for the Ph.D. degree.

Thesis: Thesis may not be written *in absentia*.

Table B—Appointments to Graduate Students, 2009–10

Title of Appointee	Appointments Total	Appointments First year	Academic Load Allowed in Credit Hours	Hours of Service Per Week	Stipend for Academic Year ($)
			Quarter		
Teaching Assistant	35	19	12	20	16,637[2]
Research Assistant	63	3	12	20	16,740[2]
Grader	6	2	12	20	8,395[2]
Total	104	24			

[1]Partial fee reimbursement only.
[2]Fees and Non-Resident Tuition (if applicable) paid

5. Personnel Engaged in Separately Budgeted Research, 07/09–06/10

Professorial faculty	59
Postdoctoral appointments	30
Graduate students	132
Non-teaching Researchers	27
Total	248

Table C—Separately Budgeted Research Expenditures

Research Specialty	No. of Grants	Expenditures ($)
Astrophysics	86	2,371,650
Biological & Medical Physics	24	1,063,115
Condensed Matter Physics	62	3,222,046
Particles & Fields	69	7,032,118
Plasma Physics & Fusion	27	3,124,183
Total	268	16,813,112

FACULTY

Professors

Bander, Myron, Ph.D., Columbia, 1962. Emeritus. Elementary particle theory.

Barwick, Steven, Ph.D., California, Berkeley, 1986. Particle astrophysics.

Burke, Kieron, Ph.D., California, Santa Barbara, 1989. Theoretical condensed matter physics.

Chanan, Gary A., Ph.D., California, Berkeley, 1978. Observational astrophysics.

Chen, Liu, Ph.D., California, Berkeley, 1972. Theoretical plasma physics.

Dennin, Michael, Ph.D., California, Santa Barbara, 1995. Experimental condensed matter physics and biological physics.

Feng, Jonathan L., Ph.D., Stanford, 1995. Elementary particle theory, and cosmology.

Fisk, Zachary, Ph.D., UC San Diego, 1969. Experimental condensed matter physics.

Gratton, Enrico, Ph.D., University of Rome, 1969. Biomedical physics.

Hamber, Herbert, Ph.D., California, Santa Barbara, 1980. Elementary particle theory and general relativity.

Heidbrink, William, Ph.D., Princeton, 1984. Experimental plasma physics.

Ho, Wilson, Ph.D., Pennsylvania, 1979. Experimental condensed matter physics and chemistry.

Hopster, Herbert J., Ph.D., RWTH, Aachen, 1977. Experimental condensed matter physics.

Kirkby, David P., Ph.D., Cal. Tech., 1995. Experimental particle physics.

Lankford, Andrew, Ph.D., Yale, 1978. Chairman of the Department. Experimental particle physics.

Lawrence, Jon M., Ph.D., Rochester, 1976 Emeritus. Experimental condensed matter physics.

Lin, Zhihong, Ph.D., Princeton, 1996. Theoretical plasma physics.

Mandelkern, Mark A., Ph.D., California, Berkeley, 1967; M.D., Miami, 1975. Experimental particle physics and medical physics.

Maradudin, Alexei A., Ph.D., Bristol, 1957. Emeritus. Condensed matter theory.

McWilliams, Roger D., Ph.D., Princeton, 1980. Experimental plasma physics.

Mills, Douglas L., Ph.D., California, Berkeley, 1965. Emeritus. Condensed matter theory.

Molzon, William, Ph.D., Chicago, 1979. Experimental particle physics.

Nalcioglu, Orhan, Ph.D., Oregon, 1970. Radiological Sciences and medical physics.

Newman, Riley, Ph.D., California, Berkeley, 1966. Emeritus. Experimental particle physics and gravitational physics.

Parker, William H., Ph.D., Pennsylvania, 1967. Experimental condensed matter physics.

Rostoker, Norman, D.Sc., Carnegie Inst. of Tech., 1950. Emeritus. Plasma physics.

Rutledge, James, Ph.D., Illinois, 1978. Experimental condensed matter physics.

Schultz, Jonas, Ph.D., Columbia, 1962. Emeritus. Experimental particle physics.

Sobel, Henry, Ph.D., Case Inst. of Tech., 1968. Experimental particle physics.

Taborek, Peter, Ph.D., Cal. Tech., 1980. Experimental condensed matter physics.

Trimble, Virginia L., Ph.D., Cal. Tech., 1968. Theoretical astronomy.

White, Steven, Ph.D., Cornell, 1988. Condensed matter theory.

Wu, Ruqian, Ph.D., Beijing, China, 1989. Condensed matter theory.

Yodh, Gaurang, Ph.D., Chicago, 1955 Emeritus. Experimental particle astrophysics.

Yu, Clare, Ph.D., Princeton, 1984. Condensed matter theory and biological physics.

Associate Professors

Barth, Aaron, Ph.D., California, Berkeley, 1998. Observational astrophysics.

Buote, David, Ph.D., Massachusetts Inst. of Tech., 1995. Observational astrophysics and cosmology.

Bullock, James, Ph.D., California, Santa Cruz, 1999. Theoretical cosmology, astrophysics, astronomy.

Casper, David W., Ph.D., Michigan, Ann Arbor, 1990. Experimental particle physics.

Chernyshev, Alexander L., Ph.D., Novosibirsk, Russia, 1995. Condensed matter theory.

Collins, Philip G., Ph.D., California, Berkeley, 1998. Experimental condensed matter physics.

Cooray, Asantha, Ph.D., Chicago, 2001. Theoretical cosmology, astrophysics, and planetary science.

Gross, Steven, Ph.D., Texas, Austin, 1995. Experimental biological physics.

Kaplinghat, Manoj, Ph.D., Ohio State, 1999. Theoretical cosmology and astrophysics.

Rajaraman, Arvind, Ph.D., Stanford, 1998. Elementary particle theory.

Ritz, Thorsten, Ph.D., Univ. of Ulm, 2001. Theoretical biological physics and condensed matter theory.

Shirman, Yuri, Ph.D., California, Santa Cruz, 1997. Theoretical particle physics.

Siwy, Zuzana, Ph.D., Silesian Univ. of Tech., 1997. Experimental biological physics and condensed matter.

Smecker-Hane, Tammy, Ph.D., Johns Hopkins, 1993. Observational astrophysics/astronomy.

Assistant Professors

Alicea, Jason, Ph.D., California, Santa Barbara, 2007. Theoretical Condensed Matter.

Barton, Elizabeth, Ph.D., Harvard, 1999. Experimental cosmology, astrophysics, and astronomy.

Chen, Mu-Chun, Ph.D., Colorado, Boulder, 2002. Theoretical particle physics.

Gulsen, Gultekin, Ph.D., Bogazici Univ. Turkey, 1999. Experimental biological physics.

Krivorotov, Ilya, Ph.D., Minnesota, 2002. Experimental condensed matter physics.

Muftuler, Lufti Tugan, Ph.D., Orta Dogu Teknik Univ., Turkey, 1996. Experimental biological physics.

Su, Lydia, Ph.D., California, Irvine, 1993. Radiological sciences and medical physics.

Taffard, Anyes, Ph.D., Univ. of Liverpool, U.K., 2002. Experimental particle physics.

Whiteson, Daniel, Ph.D., California, Berkeley, 2003. Experimental particle physics.

Research Physicists

Boehmer, Heinrich, Ph.D., Muenster, Germany, 1961. Experimental plasma physics.

Bystritskii, Vitaly, Ph.D., Institute of High Current Electronics, 1977. Experimental plasma physics.

Fang, Taotao, Ph.D., MIT, 2001. Astrophysics.

Garate, Eusebio, Ph.D., Darmouth, 1984. Experimental plasma physics.

Kropp, William, Ph.D., Case Institute of Technology, 1964. Experimental particle physics.

Leskova, Tamara, Ph.D., Inst. of Spectroscopy Russian Academy of Sciences, 1978. Theoretical condensed matter.

Mine, Shunichi, Ph.D., Univ. of Tokyo, 1996. Experimental particle physics.

Smy, Michael, Ph.D., Colorado State Univ., 1997 Experimental particle physics.

Vagins, Mark, Ph.D., Yale, 1994. Experimental particle physics.

RESEARCH SPECIALTIES AND STAFF

Theoretical

Astrophysics. Solar activity; plasma astrophysics; structure and evolution of stars and galaxies. Bullock, Cooray, Kaplinghat, Trimble.

Biological physics. Models of memory; quantum biology; protein modeling. Ritz, Yu.

Condensed matter Physics. Surfaces, superlattices, and ultrathin films; electromagnetic interactions with solids; lattice dynamics; semiconductors. Alicea, Burke, Chernyshev, Dzyaloshinskii, Maradudin, Mills, Ritz, Wallis, White, Wu, Yu.

Cosmlogy. Bullock, Cooray, Feng, Kaplinghat.

Elementary Particles and Fields. Particle phenomenology; astroparticle physics; gauge theories; string theories. Bander, Chen, Feng, Hamber, Mayer, Rajaraman, Shirman.

Plasma Physics. Wave and particle dynamics in plasmas; nonlinear theories and large-scale numerical simulations. Chen, Lin, Rostoker.

Relativity Gravatation. Hamber, Rajaraman

Experimental

Astrophysics. Observational high-energy astrophysics; optical and X-ray astronomy. Barth, Barton, Barwick, Buote, Chanan, Kirkby, Smecker-Hane, Yodh.

Biological physics. Laser tweezers; molecular motors. Collins, Dennin, Gratton, Gross, Gulsen, Muftuler, Siwy.

Condensed Matter Physics. Optical properties of solids and liquids; valence transitions; surface physics; materials science. Collins, Dennin, Fisk, Ho, Hopster, Krivorotov, Lawrence, Parker, Rutledge, Taborek. Superconductivity; tunnel junctions. Parker, Rutledge, Taborek.

Cosmology. Barton, Buote.

Elementary Particles and Fields. Neutrino physics; muon number conservation, proton decay, CP violation, hadron collider physics, astroparticle physics. Barwick, Casper, Lankford, Mandelkern, Molzon, Newman, Schultz, Sobel, Taffard, Whiteson, Yodh.

Medical Physics. Mandelkern, Nalcioglu, Su.

Plasma Physics. Relativistic electron beams; fast ion dynamics; plasma waves and turbulence; collective accelerators. Heidbrink, McWilliams, Rostoker.

Relativity. Gravitation. Newman.

UNIVERSITY OF CALIFORNIA, LOS ANGELES

DEPARTMENT OF PHYSICS AND ASTRONOMY

Los Angeles, California 90095-1547

Students Accepted For Degree	FIELDS		
	Physics	Astronomy	Related Fields
Doctorate	X	X	
Master's			

*In teaching.

1. General

Chancellor, Los Angeles Campus: Gene D. Block
Dean of Graduate School: Claudia Mitchell-Kernan
Department Chair: James Resenzweig
Department Telephone Number: (310) 825-3440 (C)
Type of Institution: University
Control: Public
Setting: Urban
Total Faculty and other teaching staff: 2,711*
Total Graduate Faculty: Not separated
Total Students: 39,984
Total Graduate Students: 13,297**
Annual Graduate Tuition:
In-state residents: Full-time—$4,193/quarter
Out-of-state residents: Full-time—$9,227/quarter
Tuition rates for: 2009–10
Deferred tuition plan: Yes
Other Fees: None
Term: Quarter

*Full Time
**Including interns and residents of the UCLA Center for Health Sciences

2. Number of Faculty in Department

The combined total of full-time faculty in the three professorial ranks is 64. The combined total of full-time, part-time, and other faculty at all ranks is 136 (includes postdoctorals).

3. Admission, Financial Aid, and Housing

Address admission inquiries to: Physics and Astronomy Department, Graduate Affairs Office, 1-707B PAB, Los Angeles, CA 90095-1547 or e-mail: APPLY@Physics.ucla.edu
Graduate application fee required: $70
Admission deadline (Fall admission): 12/15
Admission information: For fall admission, 2009–10, 71 students were accepted from 324 applicants.
Admission requirements: For admission to the graduate programs, a Bachelor's degree in physics is required with a minimum undergraduate GPA of 3.0/4.0 specified. The GRE is required. The GRE Advanced test in Physics is required. There is no minimum established for GRE scores. Students from non-English speaking countries are required to demonstrate proficiency in the TOEFL and UCLA entrance exams in English. Minimum acceptable score for admission is 570 paper or 230 computer. Depending on score, student may be required to take one or more ESL (English as a second language) courses. TSE (test of spoken english) is strongly recommended.
Undergraduate preparation assumed: Marion & Thornton, *Classical Dynamics of Particles and Systems*; Griffiths, *Introduction to Quantum Mechanics* and/or Gasiorowicz, *Quantum Physics*; Arfken, *Mathematical Methods for Physicists*; Wong, *Introduction to Mathematical Physics*; Griffiths, *Introduction to Electrodynamics*; Wangness, *Electromagnetic Fields*; Kittel, *Thermal Physics. (texts in current use).*
Address financial aid inquiries to: www.fao.ucla.edu
GAPSFAS application required: No
Financial aid deadline: 9/1.
Loans available: Yes
Address housing inquiries to: www.housing.ucla.edu
On-campus, single student housing available: Yes
Off-campus, married student housing available: Yes

Table A—Faculty, Enrollments, and Degrees Granted

Research Specialty	2009–10 Faculty	Enrollment[1] Fall 2009		No. of Degrees Granted[2] 2009–10 (2005–10)			Median No. of Years for 2009–10 Ph.D.'s
		Master's	Doctorate	Master's	Terminal Master's	Doctorate	
Accelerator	3	–	8	0(0)	0(0)	0(5)	5.3
Astronomy	0	–	22	0(0)	0(0)	6(22)	5.8
Astrophysics	19	–	6	0(0)	0(0)	2(12)	6.7
Atomic, Molecular, & Optical Phys.	1	–	2	0(0)	0(0)	0(0)	–
Biophysics	6	–	5	0(0)	0(0)	2(11)	4.3
Condensed Matter Physics	13	–	35	0(0)	0(0)	6(29)	7.2
High Energy Physics	8	–	0	0(0)	0(0)	0(0)	–
Low Temperature Physics	2	–	0	0(0)	0(0)	0(0)	
Nuclear Physics	2	–	7	0(0)	0(0)	2(9)	6.5
Particles & Fields	10	–	33	0(0)	0(0)	2(20)	6
Plasma Physics & Fusion	8	–	21	0(0)	0(0)	6(17)	6.5
Non-specialized	0	–	25	12(88)	0(4)	–	–
Other Experimental	0	–	2	0(0)	0(0)	0(0)	–
Total		–	166	12(88)	0(4)	26(126)	
Full-time Grad. Stud.		–	166				
Part-time Grad. Stud.		–	0				
First-year Grad. Stud.		–	23				
Median Years in Grad. Study (2009–10 Degrees)					–	–	6.3

Undergraduate Degrees, 2009–10 (2005–10): 43 physics; 13 astro 6 biophysics

[1]Students not yet committed to a research specialty are entered under non-specialized.
[2]Five-year totals in parentheses.

4. Graduate Degree Requirements

Master's: The Department does not offer a terminal Master's program. The M.S. degree is awarded to students in the Ph.D. program after satisfying a minimum course requirement of 9–11 courses, 6 of which are graduate level physics or astronomy courses, and passing the comprehensive examination at the Master's level of achievement. The residence requirement for the M.S. degree is as follows: minimum period of residence for one academic year of which at least two quarters must be spent on the Los Angeles campus. There is no foreign language requirement. A "B" average in physics or astronomy and an overall "B" average in all courses taken in graduate status are required.
Physics Doctorate: A written comprehensive examination is required and must be passed at the Ph.D. level of performance. It is based on the contents of the five core courses*, 210A, 210B, 215A, 221A, 221B. Students must also take one of the following courses*: 221C, 220, 231A, and pass with "B" or

better. A qualifying dissertation oral and final oral are required. The minimum residence requirement for the Ph.D. is two academic years (six quarters) at UCLA, one of which, ordinarily the second, must be spent in continuous residence. A "B" average in physics and an overall "B" average in all courses taken in graduate status are required. No language requirement.

Thesis: Thesis may not be written *in absentia*.

Astronomy Doctorate: Students are expected to fulfill the usual university requirements for a dissertation, and to pass oral, preliminary and final exams. Students are required to take 10 core courses (including 3 physics courses), one 2-quarter research project during the 2nd year and at least 3 special topics courses. The comprehensive exam occurs during spring of the second year. One year working as a teaching assistant is required.

Thesis: Thesis may be written *in absentia*.

Special Equipment, Facilities, or Programs: The Astronomy Division supports a major program to develop astronomical instrumentation using infrared techniques and encourages graduate student participation. The main off-campus facilities are the Lick Observatory, the Keck Observatory and solar facilities at Mt. Wilson. Graduate students have also used the national observatories, (Kitt Peak, CTIO, NRAO), other major observatories (OVRO, CSO) and satellites (IRAS, HST, ISO, COBE, and ROAST) to obtain data.

*Title of Physics Graduate Courses: 210A,B Electromagnetic Theory; 215A Statistical Mechanics. 220 Classical Mechanics. 221A,B,C Quantum Mechanics, 231A Mathematical Physics.

Table B—Appointments to Graduate Students, 2009–10

Title of Appointee	Appointments Total	Appointments First year	Academic Load Allowed in Credit Hours	Hours of Service Per Week	Stipend for Academic Year ($)
			Quarter		
Teaching Assistant	69	16	N/A	20	16,637–19,506
Research Assistant	76	4	N/A	20	20,928–23,358
Readers	3	3		20	12.08–12.72 H
Tutors	3	–		20	16.74–19.10 H
Total	151	23			

5. Personnel Engaged in Separately Budgeted Research, Fall 2009

Professorial faculty	64
Other faculty	29
Postdoctoral appointments	43
Graduate students	76[1]
Undergraduate students	24
Total	236

[1]Includes 2 employed by other departments

6. Separately Budgeted Research Expenditures by Source of Support

	Departmental Research Physics Division	Astronomy Division
Federal government	$19,458,095	$
Private, nonprofit organizations	4,670,869	
Business and industry	52,354	
Institution's own separately budgeted accounts	652,751	
Total	$24,834,069	$

8. Extension Courses and Summer Programs

UCLA Extension offers various special courses. Inquiries may be directed to UCLA Extension Programs, 113 Unex Bldg., Los Angeles, CA 90095-1349. There is no summer graduate program available.

Table C—Separately Budgeted Research Expenditures

Research Specialty	No. of Grants	Expenditures ($)
Astronomy & Astrophysics	120	5,440,482
Condensed Matter Physics	51	2,000,114
Energy Sources & Evniron.	20	8,492,992
Particles & Fields	38	3,703,394
Plasma Physics & Fusion	48	5,197,087
Total	277	24,834,069

FACULTY

Professors

Arisaka, Katsushi, Ph.D., U. of Tokyo, 1985. High-energy experiment.

Ashour-Abdalla, Maha, Ph.D., Imperial College, 1971. Plasma theory.

Bern, Zvi, Ph.D., California, Berkeley, 1986. Elementary particle theory.

Brown, Stuart, Ph.D., California, Los Angeles, 1988. Condensed matter experiment.

Bruinsma, Robijn, Ph.D., Southern California, 1979. Condensed matter theory.

Chakravarty, Sudip, Ph.D., Northwestern, 1976. Condensed matter theory.

Cline, David, Ph.D., Wisconsin, Madison, 1965. Elementary particle experiment.

Coroniti, Ferdinand, Ph.D., California, Berkeley, 1969.

Cousins, Robert, Ph.D., Stanford, 1981. High-energy experiment.

D'Hoker, Eric, Ph.D., Princeton, 1981. Elementary particle theory.

Ferrara, Sergio, Ph.D., Univ. of Rome, 1968. Elementary particle theory.

Fronsdal, Christian, Ph.D., UCLA, 1957. Field theory.

Gekelman, Walter, Ph.D., Stevens Inst. of Tech., 1972. Plasma experiment.

Gelmini, Graciela, Ph.D., U. Nacional de La Plata, 1981. Elementary particle theory.

Ghez, Andrea, Ph.D., Cal. Tech., 1992. Astrophysics.
Grüner, George, Ph.D., Hungarian Acad. of Science, 1977. Condensed matter experiment.
Hansen, Bradley, Ph.D., Caltech, 1996. Theoretical astrophysics.
Hauser, Jay, Ph.D., Caltech, 1985. High energy experiment.
Holczer, Karoly, Ph.D., Eotvos Lorand U., 1977. Condensed matter experiment.
Huang, Huan Z., Ph.D., MIT, 1990. Nuclear experiment.
Jiang, Hong-Wen, Ph.D., Case Western Reserve, 1989. Condensed matter experiment.
Jura, Michael A., Ph.D., Harvard, 1971. Physics of interstellar medium; mass loss. Vice Chair, Academic Affairs.
Kraus, Per, Ph.D., Princeton, 1995. Elementary particle theory.
Kusenko, Alexander, Ph.D., SUNY, 1994. Elementary particle theory.
Larkin, James, Ph.D., Cal. Tech., 1995. Astrophysics.
Malkan, Matthew A., Ph.D., Cal. Tech., 1983. Quasars and active galaxies: infrared, optical, ultraviolet, and x-ray observations of their line and continuum emission: models of accretion disks around massive black holes.
Mason, Thomas, Ph.D., Princeton, 1995. Soft condensed matter.
McLean, Ian, Ph.D., Glasgow, 1974. Astrophysics.
Miao, Jianwei, Ph.D., Stony Brook, 1999. Soft condensed matter.
Morales, George, Ph.D., California, San Diego, 1973. Plasma theory.
Mori, Warren, Ph.D., UCLA, 1987. Plasma theory.
Morris, Mark R., Ph.D., Chicago, 1975. Vice Chair, Astronomy & Astrophysics. Mass loss envelopes around red giants; The Galactic Center.
Newman, William, Ph.D., Cornell, 1979. Astrophysics.
Ong, Rene, Ph.D., Stanford, 1987. Astroparticle physics.
Patel, C. Kumar N., Ph.D., Stanford, 1961. Condensed matter experiment.
Peccei, Roberto, Ph.D., MIT, 1969. Vice Chancellor for Research. Elementary particle theory.
Pellegrini, Claudio, Ph.D., U. of Rome, 1965. Elementary particle experiment/accelerator.
Putterman, Seth, Ph.D., Rockefeller, 1970. Condensed matter experiment.
Rosenzweig, James, Ph.D., Wisconsin, 1988. Chair of the Department.
Rudnick, Joseph, Ph.D., California, San Diego, 1970. Dean, Division of Physical Sciences.
Saltzberg, David, Ph.D., Chicago, 1994. Elementary particle experiment.
Stenzel, Reiner, Ph.D., Cal. Tech, 1970. Experimental plasma physics.
Tomboulis, E. Terry, Ph.D., MIT, 1976. Elementary particle theory.
Turner, Jean L., Ph.D., California, Berkeley, 1984. Star formation; radio emission from starburst galaxies; molecular line studies of young stellar objects.
Wallny, Rainer, Ph.D., University of Zurich, 2001. Experimental high energy physics.
Whitten, Charles, Ph.D., Princeton, 1966. Nuclear experiment.
Williams, Gary, Ph.D., California, Berkeley, 1974. Low-temperature experiment.
Wright, Edward L., Ph.D., Harvard, 1976. Cosmic background radiation; infrared astronomy; relativity.
Zocchi, Giovanni, Ph.D., Chicago, 1990. Biophysics

Associate Professors

Carter, Troy, Ph.D., Princeton, 2001. Experimental plasma physics.
Furlanetto, Steven, Ph.D., Harvard, 2003. Theoretical Astrophysics.
Gutperle, Michael, Ph.D., Cambridge University, 1997. Elementary particle theory.
Mehta, Mayank, Ph.D., Indian Inst. of Science, Bangalore, 1993. Neurophysics.
Shapley, Alice, Ph.D., Cal. Tech., 2003. Experimental Astrophysics.
Vassiliev, Vladimir, Ph.D., Minnesota, 1997. Astroparticle physics.
Tserkovnyak, Yaroslav, Ph.D., Harvard, 2004. Condensed matter-theory. .

Assistant Professors

Bozovic, Dolores, Ph.D., Harvard, 2001. Soft condensed matter.
Fitzgerald, Michael, Ph.D., California, Berkeley, 2007. Astrophysics.
Hudson, Eric, Ph.D., Univ. of Colorado, 2006. Atomic, Molecular and Optical Physics.
Musumeci, Pietro, Ph.D., UCLA 2004, Elementary Particle Experiment/Accelerator.
Niemann, Christoph, Ph.D., Univ. of Technology, Darmstadt, 2002. Plasma physics.
Regan, B. Chris, Ph.D., California, Berkeley, 2001. Condensed matter

RESEARCH SPECIALTIES AND STAFF

Theoretical

Astrophysics. Stellar structure; stellar atmospheres; plasma astrophysics; binary star evolution; high-energy astrophysics; interstellar and circumstellar processes. Coroniti, Furlanetto, Hansen, Jura, Malkan, Newman, Wright. 2 senior researchers, 2 postdoctoral scholars.
Biophysics. Bruinsma, Rudnick.
Condensed Matter, Solid State, and Statistical Mechanics. Bruinsma, Chakravarty, Rudnick, Tserkovnyak. 1 senior researcher, 2 postdoctoral scholars.
Low-Temperature Physics. Putterman.
Particle Physics, Field Theory, and Source Theory. Bern, D'Hoker, Ferrara, Fronsdal, Gelmini, Gutperle, Kraus, Kusenko, Peccei, Tomboulis. 3 postdoctoral scholars.
Plasma Physics, Space Physics, Astrophysics. Ashour-Abdalla, Coroniti, Morales, Mori. 6 senior researchers, 3 postdoctoral scholars.

Experimental

Atomic, Molecular, & Optical Physics. Hudson. 1 Postdoctoral Scholar.
Accelerator Physics. Musumeci, Pellegrini, Rosenzweig. 4 postdoctoral scholars, 2 senior researchers.
Astrophysics. External galaxies and quasars; galactic nuclei; binary stars; star forming regions; nebulae; the Sun; x-rays; ultraviolet; optical; infrared; radio astronomical instrumentation; polarization. Cline, Fitzgerald, Ghez, Jura, Larkin, Malkan, McLean, Morris, Ong, Shapley, Turner, Vassiliev. 7 postdoctoral scholars, 6 senior researchers.
Biophysics. Bozovic, Gruner, Mehta, Zocchi.
Condensed Matter, Solid State Physics, and Statistical Mechan-

ics. Brown, Gruner, Holczer, Jiang, Mason, Miao, Patel, Putterman, Regan. 4 postdoctoral scholars, 1 senior researcher.

High-Energy and Particle Physics. Arisaka, Cline, Cousins, Hauser, Ong, Saltzberg, Slater, Wallny. 7 senior researchers, 11 postdoctoral scholars.

Low-Temperature Physics. Williams. 1 postdoctoral scholar.

Nuclear and Intermediate-Energy Physics. Huang, Whitten. 2 senior researchers, 3 postdoctoral scholars.

Plasma Physics and Tokamak. Carter, Gekelman, Niemann, Stenzel. 3 postdoctoral scholars, 10 senior researchers.

UNIVERSITY OF CALIFORNIA, RIVERSIDE

DEPARTMENT OF PHYSICS AND ASTRONOMY

Riverside, California 92521
www.physics.ucr.edu

Students Accepted For Degree	FIELDS		
	Physics	Astronomy	Related Fields
Doctorate	X		
Master's	X		

1. General

Chancellor: Timothy White
Dean of Graduate Division: Joseph Childers
Department Chairman: Harry Tom
Department Telephone Number: (951) 827-5332
Type of Institution: University
Control: Public
Setting: Suburban
Total Faculty: 717
Total Graduate Faculty: 717
Total Students: 18,079
Total Graduate Students: 2,371
Annual Graduate Fees and Tuition:
In-state residents: Full-time—$11,254
Out-of-state residents: Full-time—$26,296
Tuition rates for: 2009–10
Deferred tuition plan: Yes
Other Fees: None
Term: Quarter

2. Number of Faculty in Department

The combined total of full-time faculty in the three professorial ranks is 29. The combined total of full-time, part-time, and other faculty at all ranks is 32.

3. Admission, Financial Aid, and Housing

Address admission inquiries to: Graduate Advisor, Physics Department. GOPHYSICS@UCR.EDU
Graduate application fee required: $80 (domestic), $100 (international)
Admission deadline (Fall admission): 1/4 (waived for domestic students) post-deadline applications will be considered on a case-by-case basis.
Admission information: For fall admission, 2009–2010, 66 students were made offers and 25 accepted from 263 applicants.
Admission requirements: For admission to the graduate programs, a Bachelor's degree in physics or equivalent is required with a minimum undergraduate GPA of 3.25 specified. The general and subject GRE is required. The desirable departmental minimum acceptable score is 1,200 combined verbal and quantitative. Students from non-English speaking countries are required to demonstrate proficiency in English via the TOEFL exam. Minimum acceptable score for admission and teaching assistant applications is 550.
Undergraduate preparation assumed: Kittel, *Thermal Physics*; Fowles, *Analytical Mechanics*; Lorrain and Corson, *Electromagnetic Fields and Waves*; Liboff, *Introductory Quantum Mechanics*; Liboff, *Introductory Quantum Mechanics*.

Address financial aid inquiries to: Graduate Advisor, Physics Department
GAPSFAS application required: No
Financial aid deadline: 1/4
Loans available: Yes
Address housing inquiries to: www.housing.ucr.edu
On-campus, single student housing available: Yes
Cost/quarter: $2,820 (average)
On-campus, married student housing available: Yes
Cost/month: $540 to $580 (approx.)

Table A—Faculty, Enrollments, and Degrees Granted

Research Specialty	2005–06 Faculty	Enrollment[1] Fall 2009		No. of Degrees Granted[2] 2008–09 (2006–09)			Median No. of Years for 2008–09 Ph.D.'s
		Master's	Doctorate	Master's	Terminal Master's	Doctorate	
Astrophysics	4	0	10	–	0(3)	0(2)	6
Biophysics[3]	5	0	5	–	0(2)	1(2)	6
Condensed Matter Physics	16	0	56	–	0(7)	10(21)	6
Environmental Physics[3]	2	0	0	–	0(0)	0(0)	6
High-Energy/Particles	12	0	20	–	0(2)	2(15)	6
Non-specialized	0	0	23	–	2(0)	–	–
Total		0	114	–	2(14)	13(40)	
Full-time Grad. Stud.		0	114				
Part-time Grad. Stud.		0	0				
First-year Grad. Stud.		0	25				

Undergraduate Degrees, 2008–09 (2004–08): 13(31)

[1]Students not yet committed to a research specialty are entered under non-specialized.
[2]Five-year totals in parentheses.
[3]Specialty taught by Condensed Matter Physics faculty.

4. Graduate Degree Requirements

Master's Degree: A student is recommended for the degree of M.A. or M.S. in physics upon completion of the following requirements:

1. Satisfactory completion of a minimum of 36 quarter units of approved physics courses taken for a letter grade after admission to graduate study. Of these, at least 24 quarter units must be in the 200 series. Each course must be passed with a grade of "B−" or better. Each student must maintain an average for all courses of "B" or better.
2. Either of the following two plans:

Plan I (Thesis) Satisfactory completion of a thesis in a field of physics to be chosen in consultation with a faculty supervisor. This thesis is approved by a committee designated by the department. In addition, PHYS 401 is required.

Plan II (Comprehensive Examination) Satisfactory performance on the comprehensive examination.

Under either plan, all requirements for the Master's degree must be completed not later than the end of the sixth quarter. Normative Time to Degree Six quarters

Doctorate Degree: The Department of Physics and Astronomy offers the Ph.D. degree in Physics.

It is recommended that students in the Ph.D. program become

associated with a research advisor by the end of their first year.

A student will be recommended for advancement to candidacy for the Ph.D. degree in physics upon completion of requirements (1), (2), and (3) below. The student is recommended for the Ph.D. degree upon completion of requirements (4) and (5) below.

1. **Course Work** Each course must be passed with a grade of "B−" or better. Each student must maintain an average of "B" or better for all courses.

1A. Core courses for students pursuing a program in Physics (other than Astronomy):
PHYS 205 (Classical Mechanics)
PHYS 210A, PHYS 210B, PHYS 210C (Electromagnetic Theory)
PHYS 212A, PHYS 212B (Thermodynamics and Statistical Mechanics)
PHYS 221A, PHY S221B, PHYS 221C (Quantum Mechanics)
PHYS 296 (Summer Research in Physics and Astronomy)
PHYS 401 (Scientific Writing and Illustration)

1B. Core courses for students pursuing a specialization in Astronomy:
PHYS 205 (Classical Mechanics)
PHYS 210A, PHYS 210B, PHYS 210C (Electromagnetic Theory)
PHYS 212A (Thermodynamics and Statistical Mechanics, Part A)
PHYS 213 (Astrophysics of the Interstellar Medium
PHYS 214 (Techniques of Observational Astrophysics
PHYS 218 (Fundamentals of Astrophysics)
PHYS 219 (Cosmology and Galaxy Formation
PHYS 296 (Summer Research in Physics and Astronomy)
PHYS 401 (Scientific Writing and Illustration)

In addition, students in both programs must complete at least three additional graduate lecture courses in the area of their specialization. Students pursuing program 1A should choose electives from section "a-f" below. Students pursuing program 1B should choose electives from section "g" below. The program for each student must be approved by the graduate committee and the student's research advisor. Such a program may entail more than the minimum number of courses, and may also involve a mixture of courses from different areas in addition to those in the lists below.

The elective courses include the following:

a) **Nuclear and Particle Physics** PHYS 225A, PHYS 225B (Elementary Particles)
PHYS 230A, PHYS 230B (Advanced Quantum Mechanics and Quantum Theory of Fields)

b) **Condensed Matter, Surface, Bio Physics, and Optical Physics** PHYS 209A, PHYS 209B (Introduction to Quantum Electronics), PHYS 234 (Physics of Nanoscale systems)
PHYS 235 (Spintronics and Nanoscale Systems)
PHYS 236 (Advanced Imaging Techniques)
PHYS 240A*, PHYS 240B*, PHYS 440C (Condensed Matter Physics)
PHYS 241A, PHYS 241B, PHYS 241C (Advanced Statistical Physics and Field Theory)
PHYS 242 (Physics at Surfaces and Interfaces)
PHYS 246 (Biophysics)
*In this track, students are required to take PHYS 240A and PHYS 240B successively as two of their three additional courses for specialization.

c) **Astrophysics**
PHYS 208 (General Relativity)
PHYS 211A (Radiative Processes in Astrophysics)
PHYS 211B (Astrophysical Fluid Dynamics)
PHYS 214 (Techniques of Observational Astrophysics)
PHYS 215 (Galactic Dynamics)
PHYS 216 (Star Formation)
PHYS 217 (Stellar Structure and Evolution)
Additional astrophysics courses may be taken at other UC campuses through the Intercampus Exchange Program.

d) **Cosmology and Astroparticle Physics**
PHYS 208 (General Relativity)
PHYS 225A, PHYS 225B (Elementary Particles)
PHYS 230A (Advanced Quantum Mechanics)
PHYS 226 (Cosmology)
PHYS 227 (Particle Astrophysics)

e) **Environmental Physics** Two courses chosen from track (b) and two courses chosen from below:
SWSC 203 (Surface Chemistry of Soils)
SWSC 213 (Soil Mineralogy)
ENTX 244/CHEM 244 (Airborne Toxic Chemicals) or other approved graduate-level courses in related fields.

f) **Materials and Nanoscale Physics**
Two courses chosen from track (b) and two additional approved courses from the departments of Chemistry, Chemical and Environmental Engineering, Mechanical Engineering, or Electrical Engineering.

g) **Astronomy**
PHYS 208 (General Relativity)
PHYS 211A (Radiation)
PHYS 213 (Astrophysics of the Interstellar Medium)
PHYS 215 (Galactic Dynamics)
PHYS216 (Star Formation)
PHYS 217 (Stellar Structure and Evolution)
PHYS 226 (Cosmology)

2. **Written Comprehensive Examiniations** Students must have satisfactory performance on a comprehensive examination to be taken at the end of the student's first year. in the event of a failure, a make-up exam is offered in the winter quarter of the school year. The comprehensive examination for students pursuing the physics program consists of an exam that covers Mechanics, Statistical and Thermal Physics, Quantum Mechanics, and Electromagnetism. The comprehensive examination for students pursuing the astronomy specialization consists of an exam that covers Mechanics, Statistical and Thermal Physics, Electromagnetism, and Fundamental Astrophysics.
3. **Oral Qualifying Examination in General Area of Proposed Research** Satisfactory performance on an oral examination in the general area of the student's proposed research. This examination is conducted by a doctoral committee, charged with general supervision of the student's research. It is normally taken during the academic year following that in which the comprehensive examination requirement has been successfully completed. A student who fails this examination on the first attempt may, at the discretion of the committee, be permitted to take it a second time.
4. **Dissertation Examination** Students must complete a dissertation containing a review of existing knowledge relevant to the area of the candidate's research, and the results of the candidates's original research. This research must be of sufficiently high quality to constitute a contribution to knowledge in the subject area.
5. **Final Oral Examination** A final oral defense may be required.

Normative Time to Degree For Students pursuing program 1A: 15 quarters for theoretical physics; 18 quarters for experimental physics; 17 quarters for specialization in environmental physics (theory); 20 quarters for specialization in environmental physics (experimental). For students pursuing the astronomy program, 1B; 18 quarters.

Table B—Appointments to Graduate Students, 2009–10

Title of Appointee	Appointments Total	Appointments First year	Academic Load Allowed in Credit Hours	Hours of Service Per Week	Stipend for Academic Year ($)
			Quarter		
Teaching Assistant	50	25	12	20	16,637[1,2]
Research Assistant	50	0	12	20	15,382[1,3]
College Fellowship	19	19	–	–	11,670[3,4,5,6]
Total	119	44			

[1]Nonresident tuition grants are available for a limited number of students.
[2]Twenty hours per week TA duties.
[3]Full-time summer appointments available up to 3 months at up to $3,488 per month.
[4]25% time (10 hrs/wk) teaching assistantship available to augment stipend.
[5]Amounts vary from $2,000 to $18,000.
[6]Additional $4,000 stipends for summer available.

5. Personnel Engaged in Separately Budgeted Research, 7/08–6/09

Professorial faculty	30
Other faculty	0
Postdoctoral appointments	33
Graduate students	50
Undergraduate students	11
Nonteaching research personnel	7
Total	131

6. Separately Budgeted Research Expenditures by Source of Support

	Departmental Research	Physics-related Research Outside Department
Federal government	$6,380,000	$0
State & Local Gov't.	158,389	0
Total	$6,538,389	

Table C—Separately Budgeted Research Expenditures

Research Specialty	No. of Grants	Expenditures ($)
Condensed Matter Physics	41	3,110,000
Heavy Ion	3	748,000
High-Energy Physics	13	1,640,000
Astrophysics	16	891,000
Total	73	6,389,000

FACULTY

Professors

Barish, Kenneth N., Ph.D., Yale, 1996. Experimental particle and nuclear physics.

Clare, Robert B., Ph.D., MIT, 1982. Experimental high energy physics.

Cummings, Frederick W., Ph.D., Stanford, 1960. Emeritus. Theoretical non-linear dynamics; many-body theory; neural networks.

Desai, Bipin R., Ph.D., UC Berkeley, 1961. Theoretical high-energy physics.

Ellison, John A., Ph.D., Imperial College of Science and Technology (England), 1987. Experimental high-energy physics.

Fung, Sun-Yiu, Ph.D., UC Berkeley, 1963. Emeritus. Experimental high-energy physics.

Gary, John W., Ph.D., UC Berkeley, 1985. Experimental high-energy physics.

Hanson, Gail G., Ph.D., MIT, 1973. Experimental high-energy physics.

Kaus, Peter E., Ph.D., UCLA, 1955. Emeritus. Theoretical high-energy physics.

Kernan, Anne, Ph.D., University College, Dublin, 1958. Emeritus. Experimental high-energy physics.

Liu, Nai-Li Huang, Ph.D., UCLA, 1966. Emeritus. Theoretical physics; condensed matter physics; magnetic semiconductors; surface science.

Ma, Ernest, Ph.D., UC Irvine, 1970. Theoretical particle physics.

MacLaughlin, Douglas E., Ph.D., UC Berkeley, 1966. Emeritus. Experimental condensed matter physics: high-temperature superconductivity; heavy-fermion compounds; magnetic resonance techniques.

McCollum, Donald C., Ph.D., UC Berkeley, 1960. Emeritus. Atomic and molecular physics.

Mills, Allen P., Ph.D., Brandeis, 1967. Experimental condensed matter physics.

Mobasher, Braham, Ph.D., University of Durham, UK, 1988. Observational Astrophysics.

Mohideen, Umar, Ph.D., Columbia University, 1992. Fundamental Precision.

Nickel, John C., Ph.D., Cal. Tech., 1964. Emeritus. Atomic and molecular physics; electron atom scattering. Measurements.

Orbach, Raymond L., Ph.D., UC Berkeley, 1960. Emeritus. Experimental and theoretical condensed matter physics.

Pollak, Michael, Ph.D., Pittsburgh, 1958. Emeritus. Physics of disordered systems.

Seto, Richard, Ph.D., Columbia University, 1983. Experimental particle and nuclear physics.

Shi, Jing, Ph.D., Illinois, 1994. Experimental condensed matter physics.

Simanek, Eugen, Ph.D., Czechoslovak Academy of Sciences, 1963. Emeritus. Theoretical solid state physics; statistical physics.

Tom, Harry W. K., Ph.D., UC Berkeley, 1984. Experimental surface science and optical physics.

Varma, Chandra M., Ph.D., Minnesota, 1968. Theoretical condensed matter physics.

White, R. Stephen, Ph.D., UC Berkeley, 1951. Emeritus. Astrophysics and space physics.

Wilson, Gillian, Ph.D., University of Durham, 1996. Observational Astrophysics.

Wimpenny, Stephen J., Ph.D., Sheffield University (England), 1980. Experimental high-energy physics.

Wudka, Jose, Ph.D., MIT, 1986. Theoretical high-energy physics.

Yarmoff, Jory A., Ph.D., UCLA, 1985. Experimental surface sciences.

Zych, Allen D., Ph.D., Case Western Reserve, 1968. Emeritus Experimental space physics; gamma-ray astronomy; atmospheric and solar gamma-rays and neutrons.

Associate Professors

Bockrath, Marc, Ph.D., UC Berkeley 1999. Experimental condensed matter physics.
Beyermann, Ward, Ph.D., UCLA, 1988. Experimental condensed matter.
Canalizo, Gabriela, Ph.D., University of Hawaii, 2000. Observational Astrophysics.
Kawakami, Roland, Ph.D., UC Berkeley, 1999. Experimental condensed matter physics.
Lau, Chun Ning (Jeanie), Ph.D., Harvard 2001. Experimental condensed matter physics.
Long, Owen, Ph.D., Univ. of Pennsylvania, 1997. Experimental high-energy physics.
Pryadko, Leonid P., Ph.D., Stanford, 1996. Theoretical condensed matter physics.
Shtengel, Kirill, Ph.D., UCLA, 1999. Theoretical condensed matter physics.

Assistant Professors

Aji, Vivek, Ph.D., Univ. of Illinois at Urbana-Champaign, 2002. Theoretical condensed matter physics.
Tsai, Shan-Wen, Ph.D., Brown, 2000. Theoretical condensed matter physics.
Zandi, Roya, Ph.D., UCLA, 2001. Theoretical condensed matter physics.

Adjunct Professor

Sroubek, Zdenek, Ph.D., Czechoslovak Academy of Sciences (Czech Republic), 1961. Solid state physics quantum electronics.

Research Physicists

Heinson, Ann, Ph.D., Imperical College of Science and Technology, University of London (Great Britain), 1988. Experimental high-energy physics.
Nagamine, Kanetada, Ph.D., Univ. of Tokyo, 1969. Experimental muon physics.

RESEARCH SPECIALTIES AND STAFF

Theoretical

Biophysics. Theoretical research on the physics of coiled DNA, the translocution of viral DNA into cells and the properties of viral capsids. Zandi.

Condensed Matter Physics. Research in quantum and statistical mechanics of many body systems including studies of quantum critical phenomena, novel phases and phases transitions in strongly correlated matter, superconductivity, singular Fermi liquids, low-dimensional systems including graphene, quantum computation, Casimir interactions. Aji, Pryako, Shtengel, Tsai, Varma, Zandi. 7 postdoctoral physicists.

Elementary Particles and Fields. Gauge theories; extensions of the standard model of strong, weak and electromagnetic interactions; effective theories of electroweak interactions and their potential impact on future LHC and ILC data; neutrinos and related physics beyond the standard model; the problem of mass, specifically the possible connection between quark and lepton masses and their respective mixing matrices; higher symmetries such as SU(5) and SO(10), as well as models of extra spacetime dimensions. Desai, Ma, Wudka. 1 postdoctoral physicist.

Experimental

Astronomy/Astrophysics/Cosmology: Studies of active galactic nuclei and quasars; formation and evolution of galaxies; high redshift galaxies and galaxy clusters; extremely red objects; weak gravitational lensing; large scale structure formation; dark energy; multi-waveband galaxy surveys. These studies utilize space-based and ground-based telescopes including Chandra, Spitzer, Hubble, Keck, Gemini and CTIO. Canalizo, Mobasher, Wilson. 3 postdoctoral physicists.

Atomic Physics. Positronium atom physics, Bose-Einstein condensation of positronium toward a positronium annihilation gamma-ray laser. Mills. 1 postdoctoral physicist.

Biophysics. Experimental research in single biomolecule force spectroscopy, modeling structure, kinetics and energetics of the assembly of capsid proteins into the complete viral shell, bilayer lipid membrane interferometry, terahertz spectroscopy of water and fully hydrated biomolecules, and DNA molecular motors and DNA computing. Beyermann, Mills, Mohideen, Tom, Zandi. 2 postdoctoral physicists.

Condensed Matter Physics. Strongly correlated electron systems such as heavy-fermion materials, biological materials, high-temperature superconductors, non-Fermi liquids, Bose-Einstein condensates, quantum magnets, magnetic superconductors, spin glasses, fullerenes, and impurity systems. Experimental techniques include nuclear magnetic resonance (NMR) and muon spin rotation (muSR), terahertz spectroscopy, and both elastic and inelastic neutron scattering. NMR experiments are carried out at UCR, and the group travels to "meson factory" facilities such as TRIUMF, Vancouver, Canada, to perform muSR experiments using accelerator-produced beams of muons. Transport and thermodynamic properties are also investigated as functions of temperature and magnetic fields. In some cases, experiments are conducted down to mK temperatures or in pulsed magnetic fields up to 60T. Some of this research is performed off site at user facilities such as Paul Scherrer Institute, LANSCE, and NHMFL. Research is also being conducted on devising an analog neural network computer based on the interactions of DNA molecules, on observing the Bose-Einstein condensation of a dense collection of positronium atoms, on measuring the energy structure of the positronium atom using a BEC positronium atom laser, and on making a Bose-Einstein condensed positronium annihilation gamma ray laser. Beyermann, MacLaughlin, Mills. 3 postdoctoral physicists.

Environmental Physics. Surface physics of the air/water/solid interface and complex materials such as zeolites with application to atmospheric and soil sciences, chemical and bioremediation, and to chemical and biological sensor. Tom, Yarmoff.

Fundamental Precision Measurements. Precision measurements of the Casimir force and other effects of the zero point fields are made using scanning microscopy techniques such as atomic force microscopy and scanning tunneling microscopy. Mohideen. 6 postdoctoral physicists.

High-Energy Astrophysics. Gamma-ray observations in the low and medium energy range (100 keV to 100 MeV) with sufficiently sensitive telescopes will provide unique insights into many outstanding high-energy astrophysics questions. The UCR group is developing a robust Compton telescope as a prototype balloon-borne telescope. This instrument is called the Tracking and Imaging Gamma Ray Experiment (TIGRE) and is being planned to fly in Fall 2005. Zych.

High-Energy Physics. Research efforts are ongoing in the study of proton-antiproton collisions at Fermilab, proton-proton collisions at CERN and electron-positron collisions at SLAC. At Fermilab, high energy proton-antiproton collisions in the CZero detector at the Tevatron Collider are used to study the top and bottom quark production and decay, electroweak physics, and the search for the Higgs boson and other new phenomena. The CMA experiment is now taking data at the CERN Large Hadron Collider, the highest energy collider in the world. Our detector efforts focus on the endcap muon chambers, the silicon strip tracker, and the hadron calorimeter, as well as upgrades for future high-luminosity running. Physics interests include tracking software, top quark physics, search for the Higgs boson, and new physics beyond the Standard Model, such as sypersymmetry. At SLAC the data-taking phase of the BaBar experiment is now complete. The UCR group analysis efforts include measurements of charmless B decays relevant for CP violation studies, searches for new physics in flavor changing neutral current processes, and bottomonium spectroscopy. Clare, Ellison, Gary, Hanson, Heinson, Long, Wimpenny. 6 postdoctoral physicists.

Joint Programs. A joint Ph.D. in physics and M.S. in Soil Physics, joint Ph.D. in Soil Physics and M.S. in Physics, and a Ph.D. in Physics with emphasis on Environmental Physics are offered in conjunction with the Department of Soil and Environmental Sciences. Tom. Interdisciplinary training is offered in the area of chemical physics through collaboration with faculty members in the Department of Chemistry. Tom. The Institute of Geophysics and Planetary Physics (IGPP) offers research opportunities in high-energy astrophysics and plasma astrophysics. Zank, Zych.

Laser Spectroscopy and Nonlinear Optics. Femtosecond and picosecond laser pulses are used to time-resolve chemical processes at surfaces, the rotational dynamics of liquids, electron-phonon coupling in solids, and the orientation of molecules at environmentally significant interfaces. Novel laser sources in the far infrared are being developed. Picosecond time-resolved nanoscale microscopy is used to study the electronic properties of quantum dot structures and the mobility of ferroelectric and ferromagnetic domain walls. Mohideen, Tom. 1 postdoctoral physicist.

Neutrino Factory and Nyon Collider R & D. Muon beam R&D is being carried out as part of the Neutrino Factory and Nyon Collider Collaboration and the Muon Accelerator Program at Fermilab with the goals of developing design for a future neutrino factory and nuon collider. in a neutrino factory, an intense well-controlled beam of high-energy neutrinos is produced fromthe decays of circulating nuons, which can be used to study neutrino oscillations and possible CP violation in neutrinos. In a muon collider, muons and antimyons collide, providing high-energy collisions of fundamental particles in a relatively small accelerator complex, with the potential to advance our fundamental understanding of matter and energy in unique ways. The Muon Ionization Cooling Experiment at the Rutherford Lab will provide a demonstration of muon cooling through emittance exchange. Hanson. 2 postdoctoral physicists.

Neutrino Factory and Muon Collider R & D. Research and development on muon beams is being carried out, as part of the Neutrino Factory and Muon Collider Collaboration, with the goals of producing an intense, well-controlled high-energy beam of the muon decay neutrinos, which can be used to study neutrino oscillations and possible cp violation in neutrinos, and in the further future, colliding beams of high-energy muons and antimuons, which have the potential to advance our fundamental understanding of matter and energy in unique ways. Hanson. 1 postdoctoral physicist.

Physics of nanoscale materials and devices. Porperties of graphene and carbon nanotubes including quantum transport, strain engineering, spin transport, thermoelectric and nanomechanical properties; synthesis of oxide heterostructures, and semiconductors, topological insulators, ultrathin magnetic films, large area graphene, carbon nanotubes, optical probes of interface dynamics, magnetism, and spin coherence; spintronics in semiconductor, metal, and carbon-based materials; thermoelectric and thermo-spintronic phenomena; nanoscale superconductivity; physics of information storage devices. Techniques include nanofabrication by electron beam lithography and focused ion beam milling, electron transport in He^3 and dilution refrigerators, pulsed laser and non-linear optical spectroscopy, scanning probe microscopy, molecular beam epitaxy (MBE), laser-MBE, muon spin relaxation, and SQUIB magnetometry. Bockrath, Kawakami, Lau, Nagamini, Shi, Tom. 4 postdoctoral phsicists.

Proton Spin. What makes up the spin of the proton? It was thought to come from the combined spins of its constituent quarks. However, the "spin crisis" emerged when experiments established that the quarks actually contribute very little. The favored solution to this puzzle is for the spin to be carried by the gluons, which hold the proton together. Experimental studies aimed at solving this puzzle are now under way using the world's first polarized proton collider at Brookhaven National Lab. Barish. 1 postdoctoral physicist.

Relativistic Heavy-Ion Physics. The experimental group studies relativistic nucleus-nucleus collisions at high energies to explore the behavior of extended nuclear matter under extreme conditions of density and temperature. The main goal is to reach the phase transition from ordinary matter to the "quark-gluon plasma." These studies use the PHENIX detector at the Brookhaven Relativistic Heavy Ion Collider (RHIC) where energies of 100 GeV per nucleon are available. Barish, Seto. 3 postdoctoral physicists.

Surface Physics. Faculty members study the geometrical, electronic, and chemical properties of solid surfaces with a wide variety of modern surface spectroscopies, scanning microscopies, and laser-surface interactions. Some experiments are performed at the Brookhaven National Laboratory National Synchrotron Light Source. Tom, Yarmoff. 2 postdoctoral physicists.-

UNIVERSITY OF CALIFORNIA, SAN DIEGO

DEPARTMENT OF PHYSICS

La Jolla, California 92093-0319
E-mail Address: apply@physics.ucsd.edu

Students Accepted For Degree	FIELDS		
	Physics	Astronomy	Related Fields
Doctorate	X		
Master's	X		X

1. General

President: Mark G. Yudof
Chancellor: Marye Anne Fox
Dean of Graduate Studies and Research: Kim E. Barrett
Department Chairman: M. Brian Maple
Department Telephone Number: (858) 534-3293 (Grad. Office)
Type of Institution: University
Control: Public
Setting: Suburban
Total Faculty: 1,205
Total Graduate Faculty: 1,205
Total Students: 28,200
Total Graduate Students: 4,231
Annual Graduate Tuition:
In-state residents: Full-time—None
Out-of-state residents: Full-time—$9,244,50/quarter
Tuition rates for: 2010–11
Deferred tuition plan: No
Other Fees: $4,210.50 (includes health insurance)/quarter
Term: Quarter

2. Number of Faculty in Department

The combined total of full-time faculty in the three professorial ranks is 57; 11 Research Professors; and 14 Professor Emeriti. The combined total of full-time, part-time, and other faculty at all ranks is 78.

3. Admission, Financial Aid, and Housing

Address admission inquiries to: Graduate Admissions Office, Department of Physics (0319), 9500 Gilman Dr., La Jolla, CA 92093-0319
Graduate application fee required: $80 (domestic); $100 (international)
Admission deadline (Fall admission): 12/15/10
Admission information: For fall admission, 2010–11, 116 students were accepted from 487 applicants.
Admission requirements: For admission to the graduate programs, a Bachelor's degree in physics is required with a minimum undergraduate GPA of 3.0 specified. The GRE is required, but no minimum scores are specified. The GRE Advanced is required, but no minimum scores are specified. The average GRE scores for admitted students for 2009–10 were verbal–537; quantitative–775; advanced–811. Students from non-English speaking countries are required to demonstrate proficiency in English via the TOEFL exam. Minimum acceptable TOEFL score for admission is 550 paper, 213 computer.
Undergraduate preparation assumed: David J. Griffiths, *Introduction to Electrodynamics, Prentice Hall*, Daniel Dubin, *Numerical and Analytical Methods for Scientists and Engineers, Wiley*, Thornton and Marion, *Classical Dynamics, Thomson/Brooks Cole*, Dennis Barnaal, *Analog Electronics for Scientific Application, Waveland Press,* Paul Horowitz and Winfield Hill, *The Art of Electronics, Cambridge University Press,* David J. Griffiths, *Introduction to Quantum Mechanics, Prentice Hall*, Stephen Gasiorowicz, *Quantum Physics, Wiley*, J. J. J. Sakurai, *Advanced Quantum Mechanics, Addison Wesley*, Ashley Carter, *Classical and Statistical Thermodynamics, Prentice Hall,* Charles Kittel and Alex Zetti, *Introduction to Solid State Physics, Wiley*, Hans Luth and Harald Ibach, *Solid-State Physics: An Introduction to Principles of Materials Sci., Springer Verlag*.
Address financial aid inquiries to: Student Financial Services, 0013 Graduate Division, 9500 Gilman Dr., La Jolla, CA 92093-0013
FAFSA application required: Yes
Financial aid deadline: 3/1
Loans available: Yes
Address housing inquiries to: Housing Service (0907)
On-campus, single student housing available: Yes*
Cost/month: $405–1100
On-campus, married student housing available: Yes*
Cost/month: $876–1101

*waiting list, limited spaces available

4. Graduate Degree Requirements

Master's: B average in 36 units of graduate work, and comprehensive written exam required; thesis not required; no language requirement; three quarters residency required. No terminal master's program.
Doctorate: B average must be maintained in all course work; comprehensive departmental exam at beginning of second year, completion of five advanced courses, completion of teaching requirement, followed by oral qualifying exam for advancement to candidacy, dissertation, and successful oral defense of dissertation; no language requirement; six quarters residency required.
Other Programs: Ph.D. in Physics (Biophysics) is also available with same requirements as regular Ph.D., except that the departmental exam can be taken at beginning of third year and five courses related to the life sciences are required.
Thesis: Thesis may not be written *in absentia*.

Table A—Faculty, Enrollments, and Degrees Granted

Research Specialty	2008–09 Faculty	Enrollment[1] Fall 2009		No. of Degrees Granted[2] 2008–09 (2004–09)			Median No. of Years for 2007–08 Ph.D.'s
		Master's	Doctorate	Master's	Terminal Master's	Doctorate	
Acoustics	1	–	0	0(0)	0(0)	0(1)	
Applied Physics	7	–	0	0(0)	0(0)	0(1)	–
Astronomy/ Astrophyics	21	–	28	0(0)	0(0)	4(18)	6.16
Atomic, Molecular Physics	1	–	3	0(0)	0(0)	0(1)	–
Biophysics	13	–	27	0(0)	0(0)	5(20)	6.16
Condensed Matter Physics	23	–	42	0(0)	0(0)	6(32)	5.41
Energy Sources & Environ.	0	–	0	0(0)	0(0)	0(0)	–
Fluids Dynamics	8	–	2	0(0)	0(0)	0(0)	–
History & Philosphy	1	–	0	0(0)	0(0)	0(0)	–
Low Temperature Physics	4	–	0	0(0)	0(0)	0(0)	–
Materials Sci.	3	–	1	0(0)	1(3)	0(0)	–
Mathematical Physics	2	–	2				
Nonlinear Dynamics	8	–	3	0(0)	0(0)	0(4)	–
Nuclear Physics	5	–	0	0(0)	0(0)	0(0)	–
Particles & Fields-Elem.	12	–	6	0(0)	0(0)	1(8)	6.11
Physics Education	1	–	0	0(0)	0(0)	0(0)	–
Physics of Beams	4	–	0	0(0)	0(0)	0(0)	–
Plasma Physics & Fusion	10	–	13	0(0)	0(0)	0(10)	–
Polymer Physics/ Science	1	–	0	0(0)	0(0)	0(0)	–
Public Policy	0	–	0	0(0)	0(0)	0(0)	–
Statistical/Thermal	7	–	0	0(0)	0(0)	0(0)	–
Other Theoretical	6	–	5	0(0)	1(0)	0(0)	–
Non-specialized	1	–	45	0(0)	1(0)	6(0)	–
Total		3	175			14(61)	
Full-time Grad. Stud.		0					
Part-time Grad. Stud.		0	0				
First-year Grad. Stud.		1	25				
Median Years in Grad. Study (2008–09 Degrees)							5.83
Undergraduate Degrees, 2008–09: 38(58)							

[1]Students not yet committed to a research specialty are entered under non-specialized.
[2]Five-year totals in parentheses.

Table B—Appointments to Graduate Students, 2008–09

Title of Appointee	Appointments Total	Appointments First year	Academic Load Allowed in Credit Hours	Hours of Service Per Week	Stipend for Academic Year ($)
			Quarter		
Teaching Assistant	63	23	12 Units	20 hrs	15,610.50[1,4]
Research Assistant	96	8	12 Units	20 hrs	18,187.50[2,3]
Dept. Education	4	0	12 Units	20 hrs	17,917.00[3]
Cota Robles Fellowship	2	0	12 Units	20 hrs	15,000.00[3]
San Diego Fellowships	2	1	12 Units	20 hrs	12,000.00[3]
NSF Fellowships	3	0	12 Units	20 hrs	22,500.00[3]
Dept. Energy Computational Science Fellowship	1	0	12 Units	20 hrs	$21,000.00[3]
Training Grant Funding	10	7	12 Units	20 hrs	15,579.00[3]
CALIT[2] Fellows	3	0	12 units	20 hrs	15,000.00[5]
General Atomic Research	3	0	12 Units	20 hrs	18,187.50[3]
Total	175	39			

[1]Usually summer employment available.
[2]Tuition and fees usually included, particularly for first-year students. Totals reflect some overlap; some students are Fellows and also hold appointments as Teaching Assistants.
[3]Plus tuition and fees.
[4]Amount is for the nine-month academic year.
[5]Funds allocated from the California Institute for Telecommunications and Information Technology at UCSD.

5. Personnel Engaged in Separately Budgeted Research, Fiscal 2007–08

Professorial faculty	50
Other faculty	5
Postdoctoral appointments	46
Graduate students	85
Undergraduate students	8
Nonteaching research personnel	10
Total	204

6. Separately Budgeted Research Expenditures by Source of Support

	Departmental Research
Federal government	$12,228,762
State and local government	1,928,475
Private, nonprofit organizations	541,116
Total	$14,698,353

7. Separately Funded and Managed Laboratories

Approximately $5,586,953 million per year of Physics-related research is funded in other research units as:

Center for Astrophysics and Space Sciences
Center for Magnetic Recording Research
Center for Theoretical Biological Physics
Institute of Nonlinear Science
Institute for Pure and Applied Physical Sciences

Table C—Separately Budgeted Research Expenditures

Research Specialty	No. of Grants	Expenditures ($)
Astrophysics/Atmos./Space Phys., Cosmic Rays	7	652,403
Atomic, Molecular, & Optical Physics	1	362,437
Biophysics	23	7,500,854
Condensed Matter Physics	18	3,342,618
High Energy	4	2,308,603
Plasma Physics & Fusion	2	531,438
Total	55	14,698,353

Table D—Physics-related Research Outside Department

Field and Unit Outside Department	No. of Grants	Expenditures ($)
Institute for Neural Computational Science	34	5,822,000
Total	34	5,822,000

FACULTY

Professors

Abarbanel, Henry D. I., Ph.D., Princeton, 1966. Nonlinear dynamics of fluids; optical systems and neural assemblies; geophysical fluid dynamics, biophysics, and physical oceanography.

Arovas, Daniel P., Ph.D., California, Santa Barbara, 1986. Condensed matter theory; statistical mechanics.

Basov, Dmitri N., Ph.D., Lebedev Institute, USSR, 1991. Experimental condensed matter.

Berkowitz, Ami E., Ph.D., Pennsylvania, 1953 (Research Professor). Magnetic materials investigations; correlation of microstructures with magnetic behavior; surface effects; relaxation phenomena.

Branson, James G., Ph.D., Princeton, 1977. Experimental elementary particle physics.

Burbidge, E. Margaret, Ph.D., London Observ., 1943 (Univ. Professor Emeritus). Extragalactic studies, spectrophotometric and imaging; observational work on normal galaxies; galaxies with active nuclei, especially radio galaxies; quasars using Lick Observatory 3-M telescope and Keck Observatory 10-M telescope.

Butov, Leonid V., Ph.D., 1991. Experimental condensed matter physics; semiconductor nanostructures; optics; transport.

Diamond, Patrick H., Ph.D., MIT, 1979. Theoretical plasma physics and astrophysics; nonlinear dynamics.

Di Ventra, Massimiliano, Ph.D., EPFL, 1997. Theoretical condensed matter.

Driscoll, C. Fred, Ph.D., California, 1976. Experimental plasma physics; waves and transport in pure electron and pure ion plasmas; 2D fluid dynamics and turbulence.

Dynes, Robert C., Ph.D., McMaster, 1968. Experimental COndensed Matter, Solid State Physics.

Dubin, Daniel H. E., Ph.D., Princeton, 1984. Theoretical plasma physics; computational; statistical mechanics fluid dynamics.

Feher, George, Ph.D., California, Berkeley, 1954 (Research Professor). Biophysics; photosynthesis; magnetic resonance; mechanisms of crystallization of macromolecules.

Fuller, George M., Ph.D., Caltech., 1981. Theoretical astrophysics; nuclear and elementary particle physics.

Griest, Kim, Ph.D., California, Santa Cruz, 1987. Theoretical and observational astrophysics, theoretical elementary particle physics; dark matter.

Grinstein, Benjamin, Ph.D., Harvard, 1984. Elementary particle theory; quantum field theory; cosmology.

Hirsch, Jorge E., Ph.D., Chicago, 1980. Condensed matter theory.

Hwa, Terence T.-L., Ph.D., MIT, 1990. Statistical mechanics; biological physics; systems biology; molecular evolution; genomics; condensed matter physics; and dynamics of complex systems; polymer physics.

Intriligator, Kenneth A., Ph.D., Harvard, 1992. Theoretical high-energy physics.

Jenkins, Elizabeth, Ph.D., Harvard, 1989. Thermal particle physics, particle astrophysics, nuclear physics.

Kleinfeld, David, Ph.D., California, San Diego, 1984. Computational neuroscience, sensormeter control, optimal imaging, ultra fast optics.

Kuti, Julius, Ph.D., Hungary, 1967. Elementary particles and fields.

Levine, Herbert, Ph.D., Princeton, 1979. Theoretical nonlinear dynamics; biophysics; bioinformatics, condensed matter physics.

Manohar, Aneesh V., Ph.D., Harvard, 1983. Elementary particle physics.

Maple, M. Brian, Ph.D., California, San Diego, 1969. Superconductivity; magnetism, strongly correlated electron phenomena; high-pressure physics; surface science.

McIlwain, Carl E., Ph.D., Iowa, 1960 (Research Professor). Space physics; experimental and theoretical studies of planetary magnetospheres; observational and instrumental astrophysics.

Nguyen-Huu, Xuong, Ph.D., California, Berkeley, 1962 (Professor Emeritus). Biophysics; protein crystallography; and electron microscopy, detectors for x-rays and electrons.

Norman, Michael L. Ph.D., California, Davis, 1980. Computational astrophysics and cosmology.

Okamura, Melvin Y., Ph.D., Northwestern, 1970. Biophysical (optical and magnetic resonance) studies of photosynthetic reaction centers.

O'Neil, Thomas M., Ph.D., California, San Diego, 1965. Theoretical plasma physics.

Onuchic, José, Ph.D., Caltech., 1987. Theoretical biophysics and chemical physics; theoretical studies in electron transfer reactions in chemical and biological systems and in the protein folding problem, bioinformatics.

Paar, Hans P., Ph.D., Columbia, 1974. Experimental high-energy physics.

Peterson, Laurence E., Ph.D., Minnesota, 1960 (Research Professor). X- and gamma-ray astronomy; cosmic rays; space physics; balloon and satellite instrumentation.

Schuller, Ivan K., Ph.D., Northwestern, 1976. Experimental condensed matter physics and materials science (thin films, heterostructures, magnetism, nanostructures, superconductivity).

Schultz, Sheldon, Ph.D., Columbia, 1960 (Research Professor). Negative index of refraction, meta-materials, photonic band gap structures; plasmon resonant particles; advanced instrumentation in biotechnology.

Sham, Lu Jeu, Ph.D., Cambridge, 1963. Condensed matter theory.

Sharma, Vivek A., Ph.D., Syracuse, 1990. Experimental particle physics.

Sinha, Sunil K., Ph.D., Cambridge, 1964. Neutron and X-ray scattering studies of condensed matter.

Suhl, Harry, Ph.D., Oxford, 1948 (Research Professor). Theoretical solid state physics, particularly superconductivity, magnetism, surface kinetics; nonlinear dynamics.

Surko, Clifford M., Ph.D., California, Berkeley, 1968. Experimental studies using positrons and positron-matter interactions, and study of plasma physics using positron.

Tytler, David, Ph.D., London, 1982. Observational cosmology; quasars; ultraviolet and optical observations; statistics, telescopes and astronomical instrumentation.

Vernon, Wayne, Ph.D., Princeton, 1965 (Research Professor). Properties of elementary particles and their interactions; neutrino physics and astrophysics; particle detectors and acceleration techniques; free-electron lasers; Compton backscattered x-ray production.

Wolfe, Arthur M., Ph.D., Texas, 1967. Observational cosmology; galaxy formation; star formation.

Wolynes, Peter G., Ph.D., Harvard, 1976. Theoretical condensed matter, biological and chemical physics.

Wuerthwein, Frank, Ph.D., Cornell, 1995. Experimental elementary particle physics.

Yagil, Avraham, Ph.D., Weizmann Inst., 1988. Experimental elementary particle physics.

Associate Professors

Fogler, Michael M., Ph.D., U. Minnesota, 1997. Theoretical condensed matter.

Groisman, Alexander, Ph.D., Weizmann Inst., 2001. Fluid Dynamics, microfluidics, polymer liquids, biophysics.

Keating, Brian, Ph.D., Brown Univ., 2000. Observational cosmology; cosmic microwave background-experimental observational; cosmic infrared background; centimeter, millimeter, sub-millimeter, and infrared low noise, low temperature instrumentation, detectors and optics.

Murphy, Thomas M., Jr., Ph.D., Caltech, 2000. Experimental astrophysics.

Smith, Douglas E., Ph.D., Stanford, 1999. Single molecule biophysics, polymer physics, optical tweezers, and fluorescence microscopy.

Assistant Professors

Burgasser, Adam, Ph.D., Caltech, 2001. Observational Astrophysics.

Coil, Alison, Ph.D., California, Berkeley, 2004. Observational Astrophysicist.

Dudko, Olga K., Ph.D., Ukraine, 2001. Biophysics.

Shpyrko, Oleg, Ph.D., Harvard U., 2004. Experimental condensed matter.

Wu, Congjun, Ph.D., Stanford University, 2005. Theoretical condensed matter.

Lecturer (PSOE)

Anderson, Michael G., Ph.D., California, Davis, 2006. Physics education.

Faculty Emeriti

Brueckner, Keith A., Ph.D., California, Berkeley, 1950 (Professor Emeritus). Theoretical nuclear physics; statistical mechanics; plasma physics; interaction of lasers with matter; magnetohydrodynamics; theory of metals.

Chen, Joseph C.Y., Ph.D., Notre Dame, 1961 (Professor Emeritus). Theory of atomic and molecular structure and processes; history and philosophy of science.

Fisk, Zachary, Ph.D., California, San Diego, 1969 (Professor Emeritus). Experimental condensed matter physics.

Fredkin, Donald R., Ph.D., Princeton, 1961. Solid state theory; applied magnetics; biophysics.

Goldberger, Marvin L., Ph.D., Chicago, 1948 (Professor Emeritus). Elementary particle physics; quantum field theory; collision theory.

Goodkind, John M., Ph.D., Duke, 1960. Low-temperature experimental research; 2D electrons; solid He; geophysical and fundamental gravity; quantum computing.

Jones, Barbara, Ph.D., London, 1976 (Professor Emeritus). Infrared astrophysics; galactic and extragalactic astronomy; astronomical instrumentation; and research in physics education.

Liebermann, Leonard N., Ph.D., Chicago, 1940 (Professor Emeritus). Magnetism; propagation of underwater sound; molecular and chemical physics; extremely low-frequency electromagnetic waves.

Lovberg, Ralph H., Ph.D., Minnesota, 1955 (Professor Emeritus). Experimental plasma physics; geophysics.

Ride, Sally K., Ph.D., Stanford, 1978. Beam wave interactions; free-electron lasers; space plasma physics.

Shapiro, Vitali, Dr., Sc., Novosibirsk, 1967. Space plasma physics: nonlinear plasma theory, fluid turbulence.

Swanson, Robert A., Ph.D., Chicago, 1958 (Professor Emeritus). Experiments involving properties and interactions of elementary particles; interference and decay of neutral K-mesons; deep inelastic muon scattering; nucleon structure and fragmentation; rare kaon decays and CP violation.

Ticho, Harold, Ph.D., Chicago, 1949 (Professor Emeritus). Experimental elementary particle physics.

Wong, David Y., Ph.D., Maryland, 1958. (Professor Emeritus). Theoretical high-energy physics.

Adjunct Professors

Kobrak, Hans, Ph.D., Chicago, 1961 (Professor Emeritus). Experimental high-energy physics.

Mezei, Ferenc, D. Sc., Hungarian Academy of Science, 1982. Neutron scattering: advanced instrumentation and studies of dynamic phenomena in condensed matter.

Ohkawa, Tihiro, Ph.D., Tokyo, 1955. Experimental plasma physics and controlled fusion.

Pathria, Raj K., Ph.D., Univ. Delhi, 1957. Statistical physics, quantum fluids, and low-temperature physics.

Waltz, Ronald, Ph.D., Chicago, 1970. Theoretical plasma physics; numerical simulation of turbulence in plasma.

RESEARCH SPECIALTIES AND STAFF

Theoretical

Acoustics. Abarbanel

Astrophysics. Diamond, Fuller, Gould, Griest, Norman, Shapiro, Shu, Wolfe. 3 postdoctoral fellows.

Atomic Scattering and Structure. Brueckner, Chen, Gould.

Biophysics. Abarbanel, Dudko, Fredkin, Hwa, Levine, Onuchic. 8 postdoctoral fellows.

Bioinformatics. Hwa, Levine, Onuchic.

Condensed Matter/Solid State Physics. Arovas, DiVentra, Fogler, Fredkin, Hirsch, Hwa, Levine, Sham, Suhl. 3 postdoctoral fellows. Wu, Wolynes.

Elementary Particles and Quantum Field Theory. Fuller, Goldberger, Griest, Grinstein, Intriligator, Jenkins, Kuti, Manohar, Wong. 4 postdoctoral fellows.

Fluid Dynamics. Abarbanel, Diamond, Dubin, Hwa, Levine, Shapiro.

History and Philosophy. Chen.

Mathematical Physics. Diamond, Manohar, Suhl.

Nonlinear Dynamics. Abarbanel, Diamond, Hwa, Levine, Suhl. 1 postdoctoral fellow.

Nuclear Physics and the Quantum Mechanical Many-Body Systems. Brueckner, Fuller, Jenkins, Manohar.

Particle Beams. Ride.

Plasma Physics. Brueckner, Diamond, Dubin, O'Neil, Shapiro, Waltz. 2 postdoctoral fellows.

Statistical and Thermal Physics. Arovas, Brueckner, Diamond, Dubin, Gould, Hwa, Suhl.

Experimental

Astronomy. M. Burbidge, Griest, Jones, Keating, Masek, McIlwain, Murphy, Peterson, Tytler, Vernon, Wolfe. 8 post-doctoral fellows. Coil, Burgasser.

Atomic and Molecular Physics. Surko. 1 postdoctoral fellow.

Biophysics. Feher, Groisman, Kleinfeld, Nguyen-Huu, Okamura, Smith, Wolynes. 1 postdoctoral fellow.

Condensed Matter/Materials Science/Solid State Physics. Basov, Berkowitz, Butov, Fisk, Goodkind, Liebermann, Maple,

Mezei, Schuller, Schultz, Sinha, D.R. Smith. 9 postdoctoral fellows. Dynes.
Electron Microscopy. Nguyen-Huu.
Fluid Dynamics. Driscoll, Surko.
High-Energy Physics. Branson. Kobrak, Liebermann, Masek, Yagil, Paar, Sharma, Swanson, Ticho, Vernon, Wuerthwein. 5 postdoctoral fellows.
Low-Temperature Physics/Magnetism/Superconductivity. Goodkind, Maple, Schuller. 1 postdoctoral fellow.
Nonlinear Dynamics. Driscoll, Maple, Surko. 1 postdoctoral fellow.
Nuclear Physics. Vernon.
Particle Beams. Schultz, Vernon, Ticho.
Physics Education. Anderson.
Plasma Physics. Driscoll, Lovberg, Ohkawa, Surko. 3 postdoctoral fellows.
Polymer Physics. Hwa.
Space Plasma Physics. McIlwain.

UNIVERSITY OF CALIFORNIA, SANTA BARBARA

DEPARTMENT OF PHYSICS

Santa Barbara, California 93106-9530

Students Accepted For Degree	FIELDS		
	Physics	Astronomy	Related Fields
Doctorate	X		
Master's			

1. General

Chancellor: Henry T. Yang
Acting Dean of Graduate School: Gale Morrison
Department Chairman: Mark Srednicki
Department Telephone Number: (805) 893-3888 (C)
Type of Institution: University
Control: Public
Setting: Suburban
Total Faculty: 1,128
Total Graduate Faculty: 1,128
Total Students: 21,108
Total Graduate Students: 2,995
Annual Graduate Tuition: 2008–09
In-state residents: Full-time—$3,186.18/qtr.
Out-of-state residents: Full-time—$8,201.18/qtr.
Deferred tuition plan: No
Other Fees: Health Insurance: $721.25/qtr.
Term: Quarter

2. Number of Faculty in Department

The combined total of faculty in the three professorial ranks is 52. The combined total of other faculty at all ranks is 74.

3. Admission, Financial Aid, and Housing

Address admission inquires to: The Graduate Assistant, Physics Department
Graduate application fee required: $70 for domestic, $90 for international
Admission deadline (Fall admission only): 12/15
Admission information: For fall admission only, 2009–10, 89 students were accepted from over 500 Ph.D. applicants.
Admission requirements: For admission to the graduate programs (only Ph.D. offered), a Bachelor's degree in physics, or related field is required with a 3.0 minimum GPA specified. The GRE Advanced test in physics is also required. No minimum scores are specified. [The average GRE percentage scores for 2009–10 admission were verbal–85.32; quantitative–85.3%; writing–66.2%. The average GRE Advanced percentage score for 2008–09 admission was 79.8%.] Students from non-English speaking countries are required to demonstrate proficiency in English via the TOEFL exam. Minimum acceptable score for admission is 550.
Address financial aid inquiries to: Office of Financial Aid
GAPSFAS application required: No. SAAC required.
Financial aid deadline: 12/15 for fellowships; 12/15 for other aid
Loans available: Yes
Address housing inquiries to: Community Housing Office
University-owned, graduate student housing available:
Cost/month: $762–875 (limited)
University-owned, graduate family student housing available:
Cost/month: $761–1119/mo

Table A—Faculty, Enrollments, and Degrees Granted

Research Specialty	2009–10 Faculty	Enrollment[1] Fall 2009 Master's	Enrollment[1] Fall 2009 Doctorate	No. of Degrees Granted[2] 2009–10 Master's	No. of Degrees Granted[2] 2009–10 Terminal Master's	No. of Degrees Granted[2] 2009–10 Doctorate	Median No. of Years for 2009–10 Ph.D.'s
Astrophysics	–	–	15	2	0	2	9.5
Biophysics	–	–	11	0	–	2	6
Condensed Matter Physics	–	–	68	7	0	7	6
Particles & Fields	–	–	31	1	0	2	6
Relativity & Gravitation	–	–	6	1	0	1	6
Statistical & Thermal	–	–	0	–		0	6
Non-specialized	–	–	1	0	0	–	6
Total		0	132	11	0	14	
Full-time Grad. Stud.		0	132	11	0	10	
First-year Grad. Stud.		0	19	0	0	0	
Median Years in Grad. Study (1999–00 Degrees)				–	–	–	–
Undergraduate Degrees, 2008–09: 56							

[1]Students not yet committed to a research specialty are entered under non-specialized.

4. Graduate Degree Requirements

Master's: The Department of Physics does not offer a terminal M.A. program. Admission is to the Ph.D. program only. Master's degrees may be awarded only in the case of students who leave the Ph.D. program or for continuing students advanced to candidacy for the Ph.D. program who request the M.A. degree. The requirements for the M.A. are (1) completion of 36 quarter-units of work, with a minimum of 32 units of graduate-level courses and the rest approved by the student's academic advisory committee; and (2) successful completion of an M.A. examination administered by the student's graduate advisory committee. (Successful completion of the advancement to candidacy exam fulfills this requirement.)
Doctorate: Requirements include, but are not limited to, directed-preparatory course work, an oral Advancement to Candidacy exam, and the completion and successful defense of a doctoral dissertation. At least 6 regular academic quarters in residence, 3 of which must be completed consecutively prior to advancement to candidacy, required. 3.0 minimum acceptable GPA.
Thesis: Thesis may be written *in absentia*.
Special Equipment, Facilities, or Programs: The Physics building is designed to facilitate experimentation. There are unusually well equipped laboratories in magnetic resonance, laser light scattering, nonlinear fluid mechanics, low-temperature physics, scanning probe microscopy, and the comprehensive experimental program in polymers and organic solids. A unique high-power, tunable, far-infrared free-electron laser is now in operation. Off-campus research is being conducted at CERN (Geneva), Fermilab, The Stanford Linear Accelerator Center, and in the Soudan laboratory. The Kavali Institute for Theoretical Physics, supported by the National Science Foundation, is located in a nearby building within the campus. The Institute conducts research programs of variable duration in sub-areas of the general areas of condensed matter

physics, fundamental particle physics, nuclear physics, astrophysics, and relativity. The Institute invites eminent physicists from around the world to participate in the research. As a center for theoretical study, the Institute has a strong influence on the Department's graduate program.

Table B—Appointments to Graduate Students, 2009–10

Title of Appointee	Appointments Total	Appointments First year	Academic Load Allowed in Credit Hours	Hours of Service Per Week	Stipend for Academic Year ($)
			Quarter		
Teaching Assistant	35	9	12 units (min.)	10–20	16,636
Research Assistant	68	2	12 units (min.)	10–20	*15,696–16,740
Fellowships	19	7	12 units (min.)	-	18,500–31,000
Total	122	18			

5. Personnel Engaged in Separately Budgeted Research, 4/09–3/10

Professorial faculty	52
Other faculty	22
Postdoctoral appointments	35
Graduate students	73
Nonteaching research personnel	21
Total	219

6. Separately Budgeted Research Expenditures by Source of Support

	Departmental Research	Physics-related Research Outside Department
Federal government	$7,388,347	
Other	385,282	$13,597,189
Total	$7,773,629	$13,597,189

Table C—Separately Budgeted Research Expenditures

Research Specialty	No. of Grants	Expenditures ($)
Astrophysics & Cosmology	50	3,110,557.53
Biophysics	4	368,152.26
Condensed Matter Experimental Physics	8	355,741.30
Condensed Matter Theory	2	79,419.60
High Energy Physics	13	2,750,760.80
Theoretical Physics	3	627,619.15
Gravity and Relativity	3	409,681.19
Mathematical Physics	1	71,697.55
Total	84	7,773,629.78

Table D—Physics-related Research Outside Department

Field and Unit Outside Department	No. of Grants	Expenditures ($)
California Nano Systems Inst.	32	5,846,000
Kalvi Institute For Theoretical Physics	6	4,953,537
ITST-Institute for Terahertz Science and Technology	20	2,151,695
ICB-Institute for Collaborative Biotechnologies	2	645,957
Total	60	13,597,189

FACULTY

Professors

Antonucci, Robert, Ph.D., UC Santa Cruz. Observational astrophysics-experimental.

Awschalom, David D., Ph.D., Cornell University. Condensed matter physics-experimental; joint appointment with the Department of Electrical and Computer Engineering.

Balents, Leon, Ph.D., Harvard. Condensed matter physics-theoretical.

Bildsten, Lars, Ph.D., Cornell. Astrophysics-theoretical.

Blaes, Omer M., Ph.D., International School for Advanced Studies, Trieste, Italy. Astrophysics-theoretical.

Bouwmeester, Dik, Ph.D, University of Leiden, NL. Condensed matter-experimental.

Brown, Frank L., Ph.D., Massachusetts Institute of Technology. Theoretical biophysical chemistry.

Campagnari, Claudio F., Ph.D., Yale University. High-energy physics-experimental.

Cannell, David S., Ph.D., Massachusetts Institute of Technology. Experimental study of fluctuations in non-equilibrium fluids.

Carlson, Jean M., Ph.D., Cornell University. Condensed matter physics-theoretical, complex systems, Materials, Geophysics, Biophysics, Neuroscience Ecology Networks.

Cleland, Andrew N., Ph.D., UC Berkeley. Condensed matter physics-experimental.

Eardley, Douglas, Ph.D., UC Berkeley. Relativistic astrophysics.

Fisher, Matthew P. A., Ph.D., University of Illinois. Condensed matter-theoretical.

Giddings, Steve, Ph.D., Princeton University. High energy physics-theoretical.

Gross, David J., Ph.D., UC Berkeley. Particle physics-theoretical. 2004 Physics Nobel Laureate.

Gukov, Sergei, Ph.D., Princeton University. Mathematical Physics.

Gwinn, Carl, Ph.D., Princeton University. Astrophysics-experimental.

Gwinn, Elisabeth G., Ph.D., Harvard University. Condensed matter physics-experimental.

Hansma, Paul K., Ph.D., UC Berkeley. Scanning probe microscopy-experimental.

Heeger, Alan J., Ph.D., UC Berkeley. Director of Institute for Polymers and Organic Solids, 2000 Chemistry Nobel Laureate. Condensed matter physics-experimental. Joint appointment with the Department of Materials.

Horowitz, Gary, Ph.D., University of Chicago. Gravitational physics.

Incandela, Joseph, University of Chicago. High energy physics-experimental.

Kachru, Shamit, Ph.D. Princeton. String theory, Quantum field theory.

Lubin, Philip M., Ph.D., UC Berkeley. Astrophysics and cosmology-experimental.

Ludwig, Andreas W. W., Ph.D., UC Santa Barbara. Condensed matter physics-theoretical.

Marolf, Donald, Ph.D., University of Texas. Gravitational physics.

Martin, Crystal, Ph.D., University of Arizona. Observational astrophysics.

Martinis, John, Ph.D., UC Berkeley. Condensed matter physics-Physical chemistry-theoretical.

Morrison, David, Ph.D., Harvard University, Mathematical Physics.

Nayak, Chetan, Ph.D. Princeton. Theoretical condensed matter physics.

Nelson, Harry N., Ph.D., Stanford University. Elementary particle physics-experimental.

Pincus, Philip A., Ph.D., UC Berkeley. Theoretical Soft Condensed Matter: Biomolecules, self assembly, hydrogen bond networks, highly charged surfaces; joint appointment with the Department of Materials; Biomolecular Science and Engineering.

Polchinski, Joseph, Ph.D., UC Berkeley. Elementary particle physics-theoretical.

Richman, Jeffrey, Ph.D., California Institute of Technology. High energy physics-experimental.

Shea, Joan-Emma, Ph.D., Massachusetts Institute of Technology. Theoretical Biophysical chemistry.

Sherwin, Mark, Ph.D., UC Berkeley. Condensed matter-experimental.

Shraiman, Boris, Ph.D., Harvard University. Biophysics-theoretical.

Silverstein, Eva, Ph.D. Princeton. String theory.

Srednicki, Mark, Ph.D., Stanford University. Particle physics-theoretical.

Stuart, David, Ph.D., UC Davis. Particle physics, high-energy physics-experimental.

Zee, Anthony, Ph.D., Harvard University. Particle physics-theoretical.

Associate Professors

Berenstein, David, Ph.D., University of Texas. High energy physics-theoretical.

Fygenson, Deborah K., Ph.D., Princeton University. Biophysics-experimental.

Oh, Siang-Peng, Ph.D., Princeton University. Astrophysics-theoretical.

Treu, Tommaso L., Ph.D., Scuola Normale Superiore, Pisa, Italy. Observational astrophysics.

Van Dam, Wim, Ph.D., University of Amsterdam. Computer Science.

Assistant Professors

Bleszynski Jayich, Ania, Ph.d. Harvard University. Condensed matter-experimental.

Lipman, Everett, Ph.D., UC Berkeley. Experimental biological physics, single molecule spectroscopy.

Mazin, Ben, Ph.D., California Institute of Technology. Astrophysics.

Monreal, Benjamin, Ph.D., Massachusetts Institute of Technology. High-energy physics.

Xu, Cenke, Ph.D. University of California, Berkeley. Condensed matter-theoretical.

Lecturers

Freedman, Roger, Ph.D., Stanford University. Lecturer with Security of Employment.

Geller, Robert, Ph.D., University of California, Santa Barbara. Continuing Lecturer.

Guruswamy, Sathya, Ph.D. University of Rochester, N.Y. Lecturer. Joint appointment with the College.

Roig, Francesc, Ph.D., University of Massachusetts. Senior Lecturer with Security of Employment. Joint appointment with the College of Creative Studies.

Faculty Emeriti

Allen, S. James, Ph.D., Massachusetts Institute of Technology. Condensed matter physics-experimental.

Ahlers, Guenter, Ph.D., UC Berkeley. Statistical mechanics-experimental.

Barrett, Paul H., Ph.D., UC Berkeley. Professor Emeritus.

Caldwell, David O., Ph.D., UC Los Angeles, Professor Emeritus.

Eisberg, Robert, Ph.D., UC Berkeley. Professor Emeritus.

Fulco, José R., University of Buenos Aires. Professor Emeritus.

Hartle, James B., California Institute of Technology. Professor Emeritus.

Hone, Daniel W., Ph.D., University of Illinois. Professor Emeritus.

Jaccarino, Vincent, Ph.D., Massachusetts Institute of Technology. Professor Emetitus.

Kohn, Walter, Ph.D., Harvard University. Professor Emeritus. 1998 Chemistry Nobel Laureate.

Langer, James S., Ph.D., University of Birmingham. Professor Emeritus.

Lewis, Harold W., Ph.D., UC Berkeley. Professor Emeritus.

Morrison, Rollin J., Ph.D., University of Illinois. Professor Emeritus.

Peale, Stanton J., Ph.D., Cornell University. Research Professor.

Sawyer, Raymond F., Ph.D., Harvard University. Professor Emeritus.

Scalapino, Douglas J., Ph.D., Stanford University, Professor. Condensed matter physics-theoretical.

Schrank, Glen E., Ph.D., UC Los Angeles, Associate Professor Emeritus.

Schrieffer, Robert, Ph.D., University of Illinois. Professor Emeritus.

Sugar, Robert L., Ph.D., Princeton University, Professor Emeritus.

Walker, William C., Ph.D., University of Southern California, Professor Emeritus.

Adjunct Faculty

Brown, Timothy, Ph.D., University of Colorado. Astrophysics-Observational.

Einhorn, Martin, Ph.D., Princeton University. Quantum field theory.

Howell, Dale Andrew, Ph.D., University of Texas at Austin. Astrophysics.

Affiliated Faculty

Safinya, Cyrus R., Ph.D., Massachusetts Institute of Technology. Materials.

Wiltzius, Pierre, Ph.D. E.T.H. Zurich. Soft Condensed Matter and Complex Fluids.

Witherell, Michael, Ph.D., University of Wisconsin. High energy physics-experimental.

RESEARCH SPECIALTIES AND STAFF

Theoretical

Condensed Matter Physics. Superconductivity; magnetism; phase transitions; systems of reduced dimensionality; disordered materials; surface physics; polymers and colloids, Complex Systems, Materials, Geophysics, Biophysics, Neuroscience, Ecology networks. Balents, Carlson, Fisher, Ludwig, Nayak Pincus.

Theoretical High Energy Physics. Quantum field theory; gauge theories; unified theories; string theory, quantum gravity. Berenstein, Giddings, Gross, Horowitz, Marolf, Polchinski, Srednicki, Zee.

Theoretical Astrophysics. Cosmology; galaxy formation; accretion disks and magnetic turbulence; structure of evolved stars and stellar systems; superdense nuclear matter. Bildsten, Blaes, Eardley, Oh, Sawyer.

Theoretical Biophysics. Biomolecular physics of polymers, charged surfaces, polyelectrolytes and lamellar phases. Pincus, Shraiman.

Experimental

Condensed Matter Physics. Basic properties of novel materials, studies of complex fluids and nonlinear phenomena, and the development and study of nanometer-scale electronic, spintronic, superconducting and optical systems. The advanced technology required to pursue this research is provided by several unique, shared research facilities, including molecular beam epitaxy chambers for atomically-precise sample fabrication, a world-class clean room and nanofabrication lab for turning wafers into devices, research and student machine shops, and a free-electron laser facility for exploring terahertz science and technology. In addition, our students have access to the wide range of shared experimental facilities within the Materials Research Laboratory and the new California Nanosystems Institute. Allen, Awschalom, Bouwmeester, Cleland, Cannell, E. Gwinn, Heeger, Martinis, Sherwin.

Experimental Biophysics. Mechanics and dynamics of biological assemblies; scanning probe microscopy for biophysical applications; x-ray diffraction and video microscopy of mixed systems with lipids and biomolecules. Fygenson, Hansma, Lipman.

Experimental Elementary Particle Physics. Experiments at the CERN Large Hadron Collider (CMS), Fermilab (CDF), Stanford Linear Accelerator Center (BaBar), and non-accelerator searches for dark matter (CDMS). Design and construction of particle detectors, especially silicon tracking systems and high-sensitivity, low-background experiments. Campagnari, Incandela, Nelson, Richman, Stuart, Witherell.

Experimental Statistical Mechanics and Biophysics. Studies of critical phenomena in classical fluids and liquid helium; the behavior of enzymes which undergo structural changes. Ahlers, Cannell.

Observational and Experimental Astrophysics and Cosmology. Evolution of galaxies; galactic structure and mass distribution; multiwavelength studies of active galaxies and active galactic nuclei; studies of pulsars and the interstellar medium; ground-based and balloon-borne observations of the cosmic microwave background; detector development. Antonucci, C. Gwinn, Lubin, Martin, Treu.

Polymers and Organic Solids. Fundamental physics of one-dimensional systems; electronic and magnetic properties of conducting polymers and organic compounds. Heeger, Pincus.

UNIVERSITY OF CALIFORNIA, SANTA CRUZ

DEPARTMENT OF PHYSICS

Santa Cruz, California 95064
http://physics.ucsc.edu

Students Accepted For Degree	FIELDS		
	Physics	Astronomy	Related Fields
Doctorate	X		
Master's	X		

1. General

Dean of Graduate Division: Tyrus Miller
Chairman, Department of Physics: David Belanger
Department Telephone Number: (831) 459-4122 (C)
E-mail: gradadvisor@physics.ucsc.edu
Type of Institution: University
Control: Public
Setting: Small town
Total Faculty: 769
Total Graduate Faculty: 769
Total Students: 16,087
Total Graduate Students: 1,425
Annual Graduate Tuition:
In-state residents: Full-time—$4,250 quarterly (estimate)
Out-of-state: Full-time—$9,265 quarterly (estimate)
Tuition rates for: 2009–10
Deferred tuition plan: No
Term: Quarter

2. Number of Faculty in Department

The combined total of full-time faculty in the three professorial ranks is 21. The combined total of full-time, part-time, and other faculty at all ranks is 37.

3. Admission, Financial Aid, and Housing

Address admission inquiries to: UC Santa Cruz, Graduate Application Processing, 1156 High Street, Santa Cruz, CA 95064 http://www.graddiv.ucsc.edu
Graduate application fee required: $70
Admission deadline (Fall admission): 1/15
Admission information: For fall admission, 2010–11, 22 students were offered places from 149 applicants.
Admission requirements: For admission to the graduate programs, a Bachelor's degree in physics is required. 3.0 minimum GPA. The GRE is required. The GRE Advanced Test in Physics is required. Students from non-English speaking countries are required to demonstrate proficiency in English via the TOEFL exam. Minimum acceptable score for admission is 550 on paper based, 220 on computer based, or 83 on new internet based test.
Address financial aid inquiries to: UC Santa Cruz, Financial Aid Office, 1156 High Street, Santa Cruz, CA 95064 http://www2.ucsc.edu/fin-aid
GAPSFAS application required: No
Financial aid deadline: 1/15
Loans available: Yes
Address housing inquiries to: UCSC, Campus Housing Office, http://www.housing.ucsc.edu, 245 Hahn Student Services
On-campus, single student housing available: Yes
Cost/quarter: 12 month contract $943/per month
On-campus married student housing available: Yes
Cost/month: $1,366/2 bedroom apt. (effective July 1, 2010)

4. Graduate Degree Requirements

Master's: Can be done by coursework plus thesis or exams. See our website at http://physics.ucsc.edu for details.
Doctorate: Is completed by coursework plus exams and dissertation. See our website at http://physics.ucsc.edu for details.

Professors

Banks, Tom, Ph.D., M.I.T. Particle theory.
Belanger, David P., Ph.D., UC Santa Barbara. Condensed matter experimental.
Carter, Sue A., Ph.D., University of Chicago. Condensed matter experimental.
Deutsch, Joshua, Ph.D., Cambridge, England. Condensed matter theory.
Dine, Michael, Ph.D., Yale. Particle theory.
Haber, Howard, Ph.D., Michigan. Particle theory.
Johnson, Robert P., Ph.D., Stanford. Particle experimental.
Narayan, Onuttom, Ph.D., Princeton. Condensed matter theory.
Primack, Joel, Ph.D., Stanford. Cosmology.
Ritz, Steven, Ph.D., University of Wisconsin, Madison. Particle physics and astrophysics.
Schlesinger, Zack, Ph.D., Cornell. Condensed matter experimental.
Schumm, Bruce, Ph.D., University of Chicago. Particle experimental.
Seiden, Abraham, Ph.D., UC Santa Cruz. Particle experimental.
Shastry, B. Sriram, Ph.D., Tata Institute. Condensed matter theory.
Young, A. Peter, D. Phil, Oxford. Condensed matter theory.

Associate Professors

Aguirre, Anthony, Ph.D., Harvard. Particle theory.
Smith, David M., Ph.D. UC Berkeley. Particle experiment.

Assistant Professors

Gweon, Gey-Hong, Ph.D., University of Michigan. Condensed matter experimental.
Nielsen, Jason A., Ph.D., University of Wisconsin. Particle experimental.
Profumo, Stefano, Ph.D., SISSA-ISAS. Particle theory.
Sher, Alexander, Ph.D., University of Pittsburg. Biophysics.

Current research interests are described on our website.

UNIVERSITY OF CALIFORNIA, SANTA CRUZ

DEPARTMENT OF ASTRONOMY AND ASTROPHYSICS

Santa Cruz, California 95064
http://astro.ucsc.edu

Students Accepted For Degree	FIELDS		
	Physics	Astronomy	Related Fields
Doctorate		X	
Master's			

1. General

Chancellor: George Blumenthal
Dean of Graduate Division: Tyrus Miller
Chairman: Department of Astronomy & Astrophysics: Sandra M. Faber
Department Telephone Number: (831) 459-2844
E-mail: sliwinsk@ucsc.edu
Type of Institution: University
Control: Public
Setting: Small Town
Total Faculty: 769
Total Graduate Faculty: 769
Total Students: 16,087
Total Graduate Students: 1,425
Annual Graduate Tuition: In-state residents: Full-time—$4,250 quarterly
Out-of-state: Full-time—$9,265 quarterly
Tuition rates for: 2009–10
Deferred tuition plan: No

2. Number of Faculty in Department

The combined total of full-time faculty in the three professorial ranks is 24.

3. Admission, Financial Aid, and Housing

Website: http://graddiv.ucsc.edu/admissions/
Address admission inquiries to: UC Santa Cruz, Graduate Application Processing, 1156 High Street, Santa Cruz, CA 95064
Graduate application fee required: $70
Admission information: For fall admission, 2009–10, 24 students were offered places from 129 applicants
Admission requirements: The Astronomy and Astrophysics Department uses the following data in determining whether a student should be admitted to the graduate program: (1) the number and content of undergraduate courses in physics, math, and astronomy; (2) the grades received in undergraduate courses, particularly in junior and senior level physics, math and astronomy courses. (3) letters of recommendation; (4) GRE achievement scores; (5) GRE General Test scores; the GRE advanced test in Physics is strongly advised; (6) evidence of prior research activity; (7) any other relevant information.
Address financial aid inquiries to: UC Santa Cruz, Financial Aid Office, 1156 High Street, Santa Cruz, CA 95064 http://www2.ucsc.edu/fin-aid
Financial aid deadline: 1/15
Loans available: Yes
Address housing inquiries to: UC Santa Cruz, Campus Housing Office, http://www.housing.ucsc.edu
On-campus, single student housing available: Yes
Cost/quarter: 12 month contract $943/mo.
On-campus married student housing available: Yes
Cost/month: $1,366/2 bedroom apt. (2010)

Table A—Faculty, Enrollments, and Degrees Granted

Research Specialty	2006–07 Faculty	Enrollment Fall 2008 Mas-ter's	Enrollment Fall 2008 Doc-torate	No. of Degrees Granted 2008–2009 Mas-ter's	Terminal Master's	Doc-torate	Median No. of Years for 2008–10 Ph.D.'s
Astronomy	24	0	16	5	0	5	5.8
Total							
Full time Grad. Stud.	0	36					
Part-time Grad. Stud.	0	0					
First-year Grad. Stud.	0	5					
Median Years in Grad. Study (2000–09 Degrees)		6.3			6.3	6.3	

4. Graduate Degree Requirements

Doctorate: Successful completion of course requirements, first year independent research project and presentation at departmental seminar, preliminary examination, and overall performance review during first and second years, Ph.D. qualifying examination; and dissertation. See our website at http://astro.ucsc.edu for details.

5. Personnel Engaged in Separately Budgeted Research

Professorial Faculty	24
Other faculty	2
Postdoctoral appointments	17
Graduate students	36
Non-teaching research personnel	35
Total	114

FACULTY

Professors/Astronomers

Bolte, Michael J., Ph.D., University of Washington. Dynamics of star clusters, ages of star clusters, chemical enrichment history of the Galaxy, observations of interacting galaxies.

Brodie, Jean P., Ph.D., Cambridge University. Galaxies, instrumentation.

Epps, Harland W., Ph.D., University of Wisconsin. Astronomical optics and instrumentation.

Faber, Sandra M., Ph.D., Harvard University. Galaxies, stellar populations, cosmology, instrumentation.

GuhaThakurta, Puragra (Raja), Ph.D., Princeton University. Faint blue galaxies, study of faint stars using multicolor CCD data, search for Kuiper belt comets, gravitational lensing by galaxy clusters, HST studies of dense globular cluster cores, near infrared Tully-Fisher diagram, galactic "cirrus" clouds, interacting Galaxies, dwarf galaxies.

Illingworth, Garth D., Ph.D., Australian National University. Stellar and galaxy dynamics, instrumentation.

Koo, David C., Ph.D., University of California, Berkeley. Cosmology, birth and evolution of galaxies and quasars.

Max, Claire, Ph.D., Princeton University. Adaptive optics, planetary science.

Nelson, Jerry E., Ph.D., University of California, Berkeley. Design and construction of large telescopes; Project Scientist for Keck telescope and Thirty Meter telescope.

Prochaska, Jason, Ph.D., University of California, San Diego. Damped Lya systems in quasars, Lyman limit systems, stellar abundances, thick disk imaging of our galaxy.

Smith, Graeme H., Ph.D., Australian National University. Stellar populations, chromospheric activity among late-type stars.

Vogt, Steven S., Ph.D., University of Texas, Austin. Stellar spectroscopy, instrumentation.

Associate Professors/Associate Astronomers

Bernstein, Rebecca, Ph.D., California Institute of Technology. Formation an evolution of galaxies and stellar populations; astronomical instrumentation and optical design.

Ramirez-Ruiz, Enrico, Ph.D., Cambridge University. Stellar explosions, gamma-ray bursts, accretion physics near compact stars.

Rockosi, Constance, Ph.D., University of Chicago. Galactic structure, stellar populations, CCD detectors, astronomical instrumentation.

Professors

Blumenthal, George R., Ph.D., University of California, San Diego. Cosmology, galaxy formation, high-energy astrophysics.

Laughlin, Gregory, Ph.D., University of California, Santa Cruz. Extra-solar planets, numerical astrophysics, astrophysical phenomena of the extremely distant future.

Lin, Douglas N. C., Ph.D., Cambridge University. Fluid dynamics, star formation, galactic structure, planetary systems, accretion disks.

Madau, Piero, Ph.D., International School for Advanced Studies, (Trieste). Cosmology, high-energy astrophysics.

Margon, Bruce, Ph.D., UC Berkeley. High energy astrophysics, x-ray astronomy, close binary systems, digital sky surveys.

Thorsett, Stephen E., Ph.D., Princeton University. Radio astronomy, high-energy astrophysics, compact objects, relativity.

Woosley, Stanford E., Ph.D., Rice University. Nuclear astrophysics, stellar structure.

Assistant Professors

Fortney, Jonathan, J., Ph.D., University of Arizona. The physics of giant planet atmospheres and interiors, with a focus on exoplanets.

Krumholz, Mark R., Ph.D., University of California, Berkeley. Studies star formation and the interstellar medium using both analytic and numerical techniques, with particular focus on how massive stars and star clusters form, the origin of the stellar initial mass function, the life cycles of molecular clouds, and regulation of the star formation rate on galactic scales.

Associate Adjunct Professors

Dewey, Rachel J., Ph.D., Harvard University. Radio astronomy, pulsar astrophysics, VLBI astrometry.

Steinacker, Adriane, Ph.D., University of Bonn. Magnetohydrodynamical (MHD) simulations of protoplanetary accretion disks and the interaction between turbulent accretion disks and planetary cores.

UNIVERSITY OF SOUTHERN CALIFORNIA

DEPARTMENT OF PHYSICS AND ASTRONOMY

Los Angeles, California 90089-0484

Students Accepted For Degree	FIELDS		
	Physics	Astronomy	Related Fields
Doctorate	X		
Master's			

1. General

President: Steven B. Sample
Associate Vice Provost for Graduate Affairs: Jean Morrison
Department Chairman: Werner Däppen
Department Telephone Number: (213) 740-0848 (C)
Type of Institution: University
Control: Private
Setting: Urban
Total Faculty: 3,200
Total Graduate Faculty: Figure not available
Total Students: 35,000
Total Graduate Students: 18,000
Annual Graduate Tuition:
All Graduate Students: Full-time—$31,176/academic year
Part-time—$1,299 per unit
Tuition rates for: 2009–10
Deferred tuition plan: Yes
Annual Other Fees: $958
Term: Semester

2. Number of Faculty in Department

The combined total of full-time faculty in the three professorial ranks is 26. The combined total of full-time, part-time, and other faculty at all ranks is 42.

3. Admission, Financial Aid, and Housing

Address admission inquiries to: Department of Physics and Astronomy
Graduate application fee required: $85
Admission deadline (Fall admission): 1/1
Admission information: For fall admission, 2009–10, 22 students were accepted from approximately 130 applicants.
Admission requirements: For admission to the graduate programs, a Bachelor's degree in physics is required with a minimum undergraduate GPA of 3.0 specified. The GRE is required. The minimum acceptable score suggested for admission is verbal—500; quantitative—700; total—1,200. The GRE Advanced is required. The minimum acceptable score suggested for admission is 650. The TOEFL exam is recommended for students whose first language is not English. No minimum acceptable score is specified for admission. A minimum TOEFL score of 100 or higher with no less than 20 on each of the four individual sections of the Internet-based TOEFL (iBT) or (600 on paper-based/250 or higher on the computer-based TOEFL) is required for Teaching Assistants.
Undergraduate preparation assumed: Reitz and Milford, *Foundations of Electromagnetic Theory*; Eisberg, *Quantum Physics for Atoms, Molecules, Solids, Nuclei, and Particles*; Saxon, *Elementary Quantum Mechanics*; Reif, *Foundation of Statistical and Thermal Physics*; Boyce and DiPrima, *Elementary Differential Equations and Boundary Value Problems*.
Address financial aid inquiries to: Department of Physics and Astronomy
GAPSFAS application required: No
Financial aid deadline: 1/1
Loans available: Yes
Address housing inquiries to: Housing Services Office, USC, Los Angeles, CA 90089-1332
On-campus, single student housing available: Yes
Cost per term: $3,210–4,435
On-campus, married student housing available: Yes
Cost per month: $1,050–1,225

Table A—Faculty, Enrollments, and Degrees Granted

Research Specialty	2009–10 Faculty	Enrollment[1] Fall 2009		No. of Degrees Granted[2] 2009–10 (2006–10)			Median No. of Years for 2009–10 Ph.D.'s
		Master's	Doctorate	Master's	Terminal Master's	Doctorate	
Astronomy	2	–	1	0(0)	0(0)	0(1)	–
Astrophysics	2	–	4	0(0)	0(0)	0(1)	–
Atmos./Space Phys., Cosmic Rays	4	–	0	0(0)	0(0)	0(0)	–
Atomic, Molecular, & Optical Physics	8	–	2	0(0)	0(0)	0(1)	–
Biophysics	3	–	1	–	–	–	–
Condensed Matter Physics	9	–	17	0(0)	0(1)	3(20)	6
Computational Physics	1	–	6	0(0)	0(0)	1(5)	6
Electro-Physics	0	–	0	0(0)	0(0)	0(2)	0
Laser Physics	3	–	3	0(0)	0()	0(0)	–
Low Temperature Physics	3	–	1	0(0)	0(0)	0(1)	–
Particles & Fields	8	–	13	0(0)	–	1(5)	5
Quantum Information	4	–	9	0	0	1(5)	6
Statistical & Thermal	1	–	0	0(0)	0(0)	0(0)	–
Other Theoretical/ Math.	1	–	3	0(0)	–	0(0)	–
AMO/Chemical Physics	3	–	3	0(0)	–	0(3)	–
Non-specialized	0	–	8	0(13)	2(9)	0(0)	–
Total		0	71	0(13)	2(10)	6(44)	–
Full-time Grad. Stud.		0	71				
Part-time Grad. Stud.		0	0				
First-year Grad. Stud.		0	8				
Median Years in Grad. Study (2009–10 Degrees)				–	–	7	–

Undergraduate Degrees, 2008–09 (2004–08): 11(71)

[1] Students not yet committed to a research specialty are entered under non-specialized.
[2] Five-year totals in parentheses.

4. Graduate Degree Requirements

Master's: The M.S. Physics degree requires satisfactory completion of seven courses of which no more than one course may be directed research. The M.A. Physics degree requires satisfactory completion of eight courses (exclusive of directed research) plus a comprehensive exam. For all master's degrees, a GPA of 3.0 and one-year residency is required; there is no language requirement.

Doctorate: A minimum of 11 courses exclusive of dissertation and directed research courses, taken at this university and elsewhere with a minimum GPA of 3.0; comprehensive exam, qualifying exam, dissertation, and dissertation exam required; one-year residency required; there is no language requirement.

Thesis: Thesis may not be written *in absentia*.

Table B—Appointments to Graduate Students, 2008–09

Title of Appointee	Appointments: Total	Appointments: First year	Academic Load Allowed in Credit Hours	Hours of Service Per Week	Stipend for Academic Year ($)
			Semester		
Teaching Assistant	40	9	12	20	18,000
Research Assistant	36	2	12	20	18,000
Dean's Fellowship	2	0	12	0	18,000
External Fellowship	2	1			
Total	80	11			

5. Personnel Engaged in Separately Budgeted Research, 7/08–7/09

Professorial faculty	24
Postdoctoral appointments	4
Graduate students	16
Undergraduate students	7
Nonteaching research personnel	3
Total	54

6. Separately Budgeted Research Expenditures by Source of Support

	Departmental Research	Physics-related Research Outside Department
Federal government	$2,262,852	$0
Total	$2,262,852	$0

Table C—Separately Budgeted Research Expenditures

Research Specialty	No. of Grants	Expenditures ($)
Astrophysics	10	232,131
Atmos./Space Phys., Cosmic Rays	13	975,423
Atomic, Molecular, & Optical Physics	1	178,260
Condensed Matter Physics	6	534,325
Low Temperature Physics	1	33,053
Particles & Fields	7	151,582
AMO Chemical	7	160,078
Total	45	2,262,852

FACULTY

Professors

Armstrong, Lloyd, Jr., Ph.D., 1996.

Bars, Itzhak, Ph.D., Yale, 1971. Theoretical elementary particle physics.

Bergmann, Gerd, Ph.D., Göttingen, 1963. Experimental condensed matter physics.

Bickers, Nelson E., Ph.D., Cornell, 1986. Theoretical condensed matter physics.

Bozler, Hans, Ph.D., SUNY, Stony Brook, 1972. Experimental low-temperature and condensed matter physics.

Chang, Tu-nan, Ph.D., California, Riverside, 1972. Theoretical atomic physics.

Dappen, Werner, D.Sc., ETH, Zurich, 1978. Theoretical atomic physics; astronomy.

Feinberg, Jack, Ph.D., California, Berkeley, 1977. Experimental laser physics; nonlinear optics.

Gould, Christopher, Ph.D., Cornell, 1978. Experimental low-temperature and condensed matter physics.

Haas, Stephan, Ph.D., Florida State University, 1995. Theoretical condensed matter physics.

Johnson, Clifford, V., Ph.D., Southampton. Theoretical quantum gravity; string theory.

Judge, Darrell L., Ph.D., USC, 1965. Experimental atomic and molecular physics; space physics.

Kalia, Rajiv, Ph.D., Northwestern University, 1976. Computational Condensed Matter Physics.

Kresin, Vitaly, Ph.D., UC Berkeley, 1992. Physics of small clusters.

Nemeschansky, Dennis, Ph.D., Princeton, 1984. Theoretical high-energy physics.

Pilch, Krzysztof, Ph.D., Wroclaw, 1979. Theoretical elementary particle physics.

Rhodes, Edward J., Ph.D., UCLA, 1977. Astronomy.

Saleur, Hubert, Ph.D., Universite Paris 6, 1987. Theoretical elementary particle physics.

Shakeshaft, Robin, Ph.D., Nebraska, 1972. Theoretical atomic physics.

Wagner, William G., Ph.D., Cal. Tech., 1962. Theoretical quantum electronics; elementary particle physics.

Warner, Nicholas P., Ph.D., Cambridge, 1982. Theoretical elementary particle physics.

Associate Professors

Lu, Jia Grace, Ph.D., Harvard University, 1997. Experimental Condensed Matter Physics; Nanoscale Materials and Electronics.

Pierpaoli, Elena, Ph.D., SISSA - ISAS, Trieste, Italy, 1998. Cosmology and Theoretical Astrophysics.

Thompson, Richard S., Ph.D., Harvard, 1965. Theoretical superconductivity and low-temperature physics.

Zanardi, Paolo, Ph.D., Universitá di Roma, 1995. Theoretical Physics and Quantum Information Science.

Assistant Professor

El-Naggar, Moh, Ph.D., Cal.Tech., 2006. Biological Nanostructure.

Joint Appointments

Brun, Todd, Ph.D., Cal. Tech., 1994. Quantum information theory.

Dapkus, Daniel, Ph.D., University of Illinois, 1970. Electrophysics and photonics.

Gundersen, Martin A., Ph.D., USC, 1972. Experimental laser physics.

Hellwarth, Robert W., Ph.D., Oxford, 1955. Theoretical and experimental laser physics; solid state and statistical physics.

Kunc, Joseph, Ph.D., Warsaw Technical Univ., 1974. Plasma physics.

Levi, Anthony F. J., Ph.D., Cambridge University, 1983. Photonic devices.

Lidar, Daniel, Ph.D., Hebrew Univ. of Jerusalem, 1997. Physical theoretical chemistry.

Madhukar, Anupam, Ph.D., Cal. Tech., 1971. Experimental and theoretical quantum-well physics.

Nakano, Aiichiro, Ph.D., University of Tokyo, 1989. Computational Physics.

Penner, Robert, Ph.D., Massachusetts Inst. of Tech., 1982. High energy physics.

Tanguay, Armand, Ph.D., Yale University, 1977. Optics and Photonics.

Vashishta, Priya, Ph.D., Indian Institute of Technology, Kampur, 1967. Computational Physics.

Vilesov, Andrey, Ph.D., St. Petersburg St. Univ., 1985. Physical chemistry.

Zhou, Chongwu, Ph.D., Yale University, 1999. Nanoelectric and nanotechnology molecular electronics.

Research Professors

Peters, Geraldine, Ph.D., UCLA, 1974. Stellar Astrophysics.
Colombo, Loris, Ph.D., University of Milano, 2005. Cosmology.
Nomura, Ken-ichi, Ph.D., USC, 2008. Computational Physics.
Lu, Siyuan, Ph.D., USC, 2006. Physics and Opthalmology.
Wu, Robert, Ph.D., Illinois, 1973. Chemical physics.

RESEARCH SPECIALTIES AND STAFF

Theoretical

Atomic, Molecular and Optical Physics. Interactions of strong electromagnetic radiation (lasers) with matter; multiphoton processes; energy development related basic atomic transitions; many-body approach to atomic transitions; multiply excited atomic resonances; atomic optics; collective properties in atomic traps and confined Bose-Einstein condensates; atomic lithography; atoms and ions in dense stellar plasmas; interactions of high-power laser beams with matter leading to nonlinear optical phenomena; propagation of light in dense inhomogeneous plasmas; free-electron lasers; high-power unstable laser oscillators. Chang, Dappen, Hellwarth, Shakeshaft.

Condensed Matter and Solid State Physics. Superconductivity, superfluidity in ^{3}He; nonlinear transport phenomena in reduced dimensionality; electronic structures, strongly correlated metals; nonlinear phenomena in metals; high-Tc superconductors; quantum magnetism. Bickers, Haas, Kalia, Madhukar, Nakano, Thompson, Vashishta.

Computational Physics. Large-scale simulations of quantum spin liquids, atomic spectra, time-dependent atomic processes in intense fields, multi-scale hybrid simulations of materials, algorithm design, high performance programming environments for massively parallel machines, interactive three-dimensional scientific visualization. Bickers, Chang, Haas, Kalia, Nakano, Shakeshaft, Vashishta.

Elementary Particle Physics. Quantum field theory and unification of fundamental interactions; cosmology; superstring theory; M-theory; gauge theories; supersymmetry and supergravity; conformal field theory; statistical mechanics; mathematical physics; integrable models. Bars, Johnson, Nemeschansky, Pilch, Saleur, Warner.

Quantum Information and Computation. Use of quantum mechanical resources for computation, communication, and other information-processing tasks. Effect of noise and decoherence; quantum error correction and suppression, dynamical decoupling, and decoherence-free subspaces and subsystems; weak and continuous measurements and quantum trajectories; quantum random walks; quantification of entanglement; quantum information theory; algebraic description of quantum states and observables; quantum information and many-body physics; entanglement and quantum phase transitions; quantum algorithms for classical statistical physics; quantum process tomography; geometric phases; adiabatic, holonomic, and topological quantum computation; quantum computing implementations, using quantum dots, linear optics, and magnetic resonance force microscopy; quantum information. Brun, Haas, Lidar, Zanardi.

Experimental

Astronomy, Cosmology, and Helioseismology. Study of the overall structure, composition, origins and evolution of the Universe; analysis of cosmic microwave background data and the study of dark matter and galaxy clusters. Study of the structure and dynamics of the solar atmosphere and interior using observations and theory of solar local and global oscillations; use of helioseismology to probe properties of dense plasmas. Dappen, Pierpaoli, Rhodes.

Atomic, Molecular, and Space Physics. Laser spectroscopy, highly excited atomic states, study of planetary atmospheres from space flight experiments; photoabsorption and emission in planetary atmospheres; vacuum ultraviolet radiation interacting with gaseous plasmas. Judge, Peters, Wu. Space Sciences Center Staff. 3 research scientists.

Biophysics. Research at the interface between biological and inorganic systems. Cell-surface interactions. Microbe-inorganic interactions. Bionenergy production in microbial fuel cells. Electronic and enzymatic activity of extracellular nanostructures. Intersection of information and biological sciences. Biochemical sensors and intelligent bio-mimetic coatings for neural prostheses. Optical imaging and spectroscopic studies of intra-cellular biochemical processes. El-Naggar, Madhukar, S. Lu.

Condensed Matter and Low-Temperature Solid State Physics. Electronic transport and quantum-interference in two-dimensional metals, superconducting films, magnetic surface impurities, anomalous Hall effect and spin-orbit scattering, quantum wires; quasi-one-dimensional conductors; localization; magnetoresistance. Superfluidity in ^{3}He; transport properties of anisotropic conductors and thin alkali metal films; phase transitions in metals of low dimensionality; two-dimensional magnetism of ^{3}He films; development of primary thermometry at ultralow temperatures. Electronic and thermal properties of size-quantized metal nanoclusters; photodissociation, transport, and particle growth in ultra-cold liquid helium clusters. Bergmann, Bozler, Gould, Kresin, J. G. Lu, Madhukar.

Laser Physics and Optics. Laser spectroscopy; nonlinear optical mixing; optical fibers and devices; phase-conjugation, photorefractive effect; Raman-induced Kerr effect spectroscopy; spectroscopy of glassy solids; laser plasma studies; photochemistry of simple molecular systems; interaction in wave guides. Feinberg, Gundersen, Hellwarth, Tanguay. 1 research scientist.

COLORADO SCHOOL OF MINES

DEPARTMENT OF PHYSICS

Golden, Colorado 80401
World Wide Web Home page: http://physics.mines.edu

Students Accepted For Degree	FIELDS		
	Physics	Astronomy	Related Fields
Doctorate	X		X
Master's	X		X

1. General

President: Myles W. Scoggins
Dean of Graduate School: Thomas M. Boyd
Department Chairman: Thomas Furtak
Department Telephone Number: (303) 273-3830
Type of Institution: University
Control: Public
Setting: Small town
Total Faculty: 338
Total Students: 4,858
Total Graduate Students: 1139
Annual Graduate Tuition:
In-state residents: Full-time—$11,540
Out-of-state residents: Full-time—$25,990
Tuition rates for: 2010–11
Deferred tuition plan: Yes
Other Fees: $1,654
Term: Semester

2. Number of Faculty in Department

The combined total of full-time faculty in the three professorial ranks is 15. The combined total of full-time, part-time, and other faculty at all ranks is 34.

3. Admission, Financial Aid, and Housing

Address admission inquiries to: Tim Ohno
Graduate application fee required: $40 ($20 online)
Admission deadline (Fall admission): 3/12
Admission information: For Fall admission, 2009–10, 20 students were accepted from 81 applicants.
Admission requirements: For admission to the graduate programs, a Bachelor's degree in physics is required with a minimum GPA of 3.0 specified. The GRE is required. The minimum acceptable score for admission is 1,100 (verbal +quantitative). The GRE physics subject exam is strongly urged, and is required for financial aid. The average GRE scores are not available. Students from non-English speaking countries are required to demonstrate proficiency in English via the TOEFL exam. The minimum acceptable score for admission is 600 (written) or 215 (computer based).
Undergraduate preparation assumed: One semester of classical mechanics at the level of Marion, two semesters of electromagnetism at the level of Griffiths, one year of modern physics, one semester each of thermodynamics, optics, mathematical physics, and electronics.
Address financial aid inquiries to: Tim Ohno
GAPSFAS application required: No
Financial aid deadline: 3/12
Loans available: Yes
Address housing inquiries to: Housing Office, Colorado School of Mines, (303) 273-3351
On-campus, single student housing available: Yes
Approximate living expenses: $1,250/month

Table A—Faculty, Enrollments, and Degrees Granted

Research Specialty	2008–09 Faculty	Enrollment[1] Spring 2010 Master's	Enrollment[1] Spring 2010 Doctorate	No. of Degrees Granted[2] 2010 (2006–10) Master's	Terminal Master's	Doctorate	Median No. of Years for 2010 Ph.D.'s
Condensed Matter	8	9	23	0(1)	2(9)	2(17)	5
Sub-Atomic	3	13	11	0(1)	5(6)	2(10)	5
Optical	3	1	7	0(1)	1(5)	1(7)	5
Other	7	0	1	0(1)	0(1)	0(0)	5
Total	21	23	42	0(4)	8(21)	5(34)	
Full-time Grad. Stud.		23	40				
Part-time Grad. Stud.		0	2				
First-year Grad. Stud.		12	11				
Median Years in Grad. Study (2010 Degrees)					–	–	5

Undergraduate Degrees, 2010 (2006–10): 62(293)

[1] Students not yet committed to a research specialty are entered as first-year.
[2] Five-year totals in parentheses.

4. Graduate Degree Requirements

Master's: 36 semester hours in approved program with 3.0 GPA; thesis; no foreign language. 27 credit hours including thesis must be taken in residence.
Doctorate: 34 semester hours coursework, 38 of research credit. Coursework includes 12 hours in a specialty area, which include programs in Optical Science and Engineering, Photovoltaics and Electronic Materials, and Nuclear Physics and Astrophysics in addition to topic areas in the other degree programs at CSM. Ph.D. candidacy established by grades or oral examination. Two semesters in residence are required.
Other Programs: Interdisciplinary research is organized under centers: Renewable Energy Materials Research Science and Engineering Center (REMRSEC), Center for Microintegrated Optics for Advanced Bioimaging and Control (MOABC), Nuclear Science and Engineering Center (NUSEC), Golden Energy Computing Organization (GECO). Special solar energy related research programs are available in conjunction with the nearby National Renewable Energy Laboratory (NREL). Interdisciplinary M.S. and Ph.D. degrees are granted in Materials Science and in Nuclear Engineering.
Thesis: Thesis may be written *in absentia*.
Special Equipment, Facilities, or Programs: The department specializes in applied physics. Available materials processing and lithography facilities provide extensive capabilities for making and characterizing nanocrystalline and amorphous semiconductors, patterned nanostructures, and self assembled nanostructures. Capabilities include growth systems (e.g., PECVD, low pressure CVD, MOCVD, sputtering, and elec-

trochemical deposition), transmission electron microscopy and field emission scanning electron microscopy, reactive ion etching, ion implantation, wet etching and cleaning stations, oxide and nitride growth and deposition, dopant diffusion, annealing furnaces, and a clean room with optical lithography. Materials characterization capabilities include surface profilometers, visible and IR spectrometers, visible and FTIR spectroscopic ellipsometers, a complete x-ray analysis laboratory, an imaging x-ray photoelectron spectroscopy system, scanning Auger electron spectroscopy system, temperature dependent Seebeck and Hall effect systems, and characterization of photovoltaic devices and integrated circuits. Also included are an electron microprobe, an atomic force microscope, near-field microscopy, confocal and conventional Raman spectroscopy, and various chemical techniques for determining elemental compositions. Laboratories are also equipped with electron paramagnetic resonance (EPR) and nuclear magnetic resonance (NMR) as well as instruments for temperature dependent electroluminescence and photoluminescence (PL), PL excitation spectroscopy, and microwave modulated PL. The applied optics group maintains a state-of-the-art ultrafast (femtosecond) spectroscopy laboratory, including a 1 terawatt, 1 kHz, Ti:sapphire laser system, a 6 terawatt, 20 Hz, Ti:sapphire laser system used for development of novel x-ray sources, an extended cavity Ti:sapphire oscillator and a diode-pumped Nd:glass laser used for multiphoton imaging, and a fiber oscillator used for ultrafast spectroscopy and nonlinear imaging. The department's microwave and millimeter wave laboratory includes a vector network analyzer with imaging capability. The subatomic physics laboratory includes a 180 keV ion accelerator and facilities for detector development as well as laser-based atmospheric monitoring instrumentation. Faculty participate in experiments at nuclear facilities in Oak Ridge (HRIBF), Argonne (ATLAS), and at TRIMF in Vancouver. Astroparticle physics of ultra-high energy cosmic rays is studied at the Pierre Auger Observatory in Argentina. Computational physics is conducted on a massively 2144-core 268-node supercomputer with 16 Gb of RAM, capable of a peak speed of 23 Tflop.

Table B—Appointments to Graduate Students, 2009–10

Title of Appointee	Appointments Total	Appointments First year	Academic Load Allowed in Credit Hours	Hours of Service Per Week	Stipend for Academic Year ($)
			Semester		
Teaching Assistant	16	9	15	20	20,800
Research Assistant[1]	35	7	15	20	21,632
Total	51	16			

[1] Academic load includes work toward thesis.

5. Personnel Engaged in Separately Budgeted Research, 7/07–6/08

Professional faculty	15
Other faculty	16
Total	31

6. Separately Budgeted Research Expenditures by Source of Support

	Departmental Research	Physics-based Centers
Federal government	$644,317	$5,184,816
Private, nonprofit organizations	144,490	401,074
Total	$788,807	$5,585,890

Table C—Separately Budgeted Research Expenditures

Research Specialty	No. of Grants	Expenditures ($)
Condensed Matter	55	2,847,553
Sub-Atomic Physics	26	2,155,189
Applied Optics	10	621,868
Other	9	750,087
Total	100	6,374,697

FACULTY

Professors

Collins, Reuben T., Ph.D., Cal Tech, 1984. Electronic materials and devices.

Furtak, Thomas E., Ph.D., Iowa State, 1975. Head of Department. Raman spectroscopy; linear and nonlinear optical properties of materials.

Greife, Uwe, Ph.D., University of Bochum, 1994. Experimental nuclear physics and astrophysics.

Kowalski, Frank V., Ph.D., Stanford, 1978. Experimental laser physics and high-resolution spectroscopy.

Lusk, Mark T., Ph.D., California Institute of Technology, 1992. Theoretical and computational condensed matter.

Scales, John A., Ph.D., Colorado, 1984. Mesoscopic materials.

Squier, Jeff, Ph.D., Rochester, 1992. Nonlinear microscopy, micromachining with ultrafast optics, ultrafast x-ray diffraction and absorption.

Taylor, P. Craig, Ph.D., Brown, 1969. Optical, electrical, and structural properties of crystalline and amorphous semiconductors.

Associate Professors

Carr, Lincoln, Washington, 2001. Theoretical condensed matter physics.

Durfee, Charles G. III, Ph.D., Maryland, 1994. Generation, characterization of ultrashort laser pulses, interactions with atoms and plasmas.

Ohno, Timothy, R. Ph.D., Maryland, 1989. Experimental solid state; surfaces.

Sarazin, Frederic, Ph.D., University of Caen, 1999. Experimental nuclear physics and astrophysics.

Wiencke, Lawrence, Ph.D., Columbia, 1992. Experimental particle astrophysics.

Wood, David M., Ph.D., Cornell, 1981. Solid state theory.

Assistant Professor

Wu, Zhigang, Ph.D., College of William and Mary, 2002. Theoretical and computational condensed matter; first principles electronic structure, atomistic model simulations.

Other Faculty

Beach, Joseph, Ph.D., Colorado School of Mines, 2002. Experimental materials physics, photovoltaics. Research Associate Professor.

Bernard, James E., Ph.D., Delaware, 1984. Electronic materials theory. Research Associate Professor.

Cecil, Edward, Ph.D., Princeton, 1972. Experimental nuclear physics; high-temperature fusion plasma diagnostics, University Emeritus Professor.

Coffey, Mark W., Ph.D., Iowa State, 1991. Quantum computing algorithms, superconducting and magnetic systems. Research Professor.

Flammer, P. David, M.S., Colorado School of Mines, 2001. Experimental and theoretical condensed matter physics; plasmonics. Research Assistant Professor.

Flournoy, Alex T., Ph.D., Colorado, 2003. Theoretical particle physics, string theory, and cosmology. Lecturer.

Franceschetti, Alberto, Ph.D., Scuola Internazionale Superiore di Studi Avanzati (SISSA), Trieste, Italy, 1993. Theoretical condensed matter physics. Research Professor.

Ginley, David S., Ph.D., MIT, 1976. Experimental materials physics. Research Professor.

Gray, Frederick, Ph.D., Illinois, 2003. Research Assistant Professor.

Hollingsworth, Russell, Ph.D., Colorado State University, 1985. Senior Scientist, ITN Energy. Experimental device physics. Research Professor.

Kohl, Patrick, Ph.D., University of Colorado-Boulder, 2007. Lecturer.

Kuo, Hsia-Po Vincent, Ph.D., University of Minnesota, 2004. Physics education research. Lecturer.

McNeil, James A., Ph.D., Maryland, 1979. Theoretical nuclear physics. Professor Emeritus.

Olson, Dana C., Ph.D., Colorado School of Mines, 2006. Experimental organic photovoltaics. Research Assistant Professor.

Popescu, Voicu, Ph.D., Ludwig Maximilian University, Munich, Germany, 1999. Theoretical condensed matter physics. Research Assistant Professor.

Ruskell, Todd G., Ph.D., Arizona, 1996. Semiconductor surface physics. Senior Lecturer.

Shayer, Zeev, Ph.D., Tel-Aviv University, 1985. Nuclear physics and engineering. Research Professor.

Smith, Steve J., Ph.D., Michigan, 1996. Condensed matter. Research Associate Professor.

Stone, Charles A., Ph.D., UCLA, 1990. Senior Lecturer.

Stradins, Paul, Ph.D., Latvian Academy of Sciences, 1990. Experimental semiconductor physics; photovoltaics. Research Professor.

Williamson, Don L., Ph.D., Washington, 1971. Experimental solid state; Mössbauer effect, x-ray scattering. Emeritus Professor; Research Professor.

Yarbrough, John M., Ph.D., Colorado School of Mines, 2007. Experimental biophysics. Research Assistant Professor.

Young, Matt, Ph.D., Rochester, 1967. Optics, metrology. Senior Lecturer.

RESEARCH SPECIALTIES AND FACULTY

Theoretical

Condensed Matter. Theoretical many body quantum and classical mechanics in application to ultracold quantum gases: quantum phase transitions; atomic and molecular superfluidity and superconductivity; atom lasers; nonlinear waves; fractals, solitons, and vortices; quantum information science; mathematical physics and inverse problems; novel semiconductor materials and structures; semiconductor alloys; phonon properties; surfaces and interfaces; nanostructures; plasmonic phenomena. Carr, Lusk, Wood, Wu.

Nuclear. Relativistic approaches to nucleon and nuclear structure and scattering. McNeil.

Experimental

Condensed Matter and Advanced Materials. Semiconductor science, electronic devices, optical properties of materials and interfaces, soft condensed matter and liquid crystals, self-assembled monolayers, bio-inorganic composites, quantum nanostructures, surface physics and catalysis, transport phenomena, nonlinear optical properties of surfaces, amorphous materials. Collins, Furtak, Ohno, Scales, Taylor.

Optics, Quasi-Optics and Quantum Electronics. Laser physics and ultrafast optical phenomena, plasmonic electronic systems, nonlinear microscopy and micromachining, ultra-high intensity lasers, ultrafast x-ray diffraction, frequency shifted feedback lasers, precision measurement, radio-frequency wave propagation in random media, ultrasonics, mesoscopic phenomena, quantum chaos. Collins, Durfee, Furtak, Kowalski, Scales, Squier.

Renewable Energy. Solar photovoltaics, third-generation photoconversion, nanostructures, artificial photosynthesis, organic photovoltaics, photoelectrochemistry, optical and electronic properties of crystalline and amorphous semiconductors, thin film photovoltaic materials, photoexcitation and relaxation in nanostructures. Collins, Durfee, Furtak, Ohno, Squier, Taylor, Wu.

Subatomic Physics. Nuclear astrophysics, low energy nuclear physics, astrophysics with radioactive beams, ultra-high energy cosmic ray physics, astroparticle physics, low-energy nuclear reactions, fusion diagnostics, nuclear engineering. Greife, Sarazin, Weincke.

COLORADO STATE UNIVERSITY

DEPARTMENT OF PHYSICS

Fort Collins, Colorado 80523

Students Accepted For Degree	FIELDS		
	Physics	Astronomy	Related Fields
Doctorate	B		
Master's	B	B	

1. General

President: Tony Frank
Dean of Graduate School: Peter Dorhout
Department Chairman: John L. Harton
Department Telephone Number: (970) 491-6206
Type of Institution: University
Control: Public
Setting: Small city
Total Faculty: 1,450
Total Graduate Faculty: No separate graduate faculty
Total Students: 26,413
Total Graduate Students: 3,655
Annual Graduate Tuition for: 2009–10
In-state residents: Full-time—$7,960
Part-time—$359.10/credit
Out-of-state residents: Full-time—$19,612
Part-time—$1006.45/credit
Deferred tuition plan: No
Term: Semester

2. Number of Faculty in Department

The total number of full-time faculty in the three professorial ranks is 19. The total number of full-time, part-time, and other faculty at all ranks is 23.

3. Admission, Financial Aid, and Housing

Address admission inquiries to: Graduate Admissions Committee, Department of Physics, 1875 Campus Delivery, Colorado State University, Fort Collins, CO 80523
Graduate application fee required: $50
Admission deadline: 2/15 (Fall), 10/1 (Spring): late applications can be considered
Admission information: For fall admission, 2010, 23 students were admitted from 114 applicants and 10 have accepted.
Admission requirements: For admission to the graduate programs, a Bachelor's degree in physics or a related field is required with a minimum undergraduate GPA of 3.0/4.0 specified. The GRE Advanced test and the GRE standard test is required. Scores better than 700-quantitative and 600-advanced are desirable. Students from non-English speaking countries are required to demonstrate proficiency in English via the TOEFL exam. Minimum acceptable score for admission is 600.
Undergraduate preparation assumed: Mechanics: Marion, *Classical Dynamics*; Electromagnetism: Reitz and Milford, *Foundations of Electromagnetic Theory*; Thermal Physics: Kittel and Kroemer, *Thermal Physics*; Quantum Mechanics: Griffth, *Introduction to Quantum Mechanics*
Address financial aid inquiries to: Department of Physics
GAPSFAS application required: No
Financial aid deadline: Same as for admission
Loans available: Yes
Address housing inquiries to: Office of Housing and Residence Education, 8032 Campus Delivery, Colorado State University, Fort Collins, CO 80523
On-campus, single student housing available: Yes
Cost/month: $605 (1 bdrm.); $375–440 per bdrm.
On-campus, married student housing available: Yes
Cost/month: $615–760 (2 bdrm.); $728–835 (3 bdrm.)

Table A—Faculty, Enrollments, and Degrees Granted

Research Specialty	2009–10 Faculty	Enrollment[1] Fall 2009 Master's	Enrollment[1] Fall 2009 Doctorate	No. of Degrees Granted[2] 2009–10 (2005–10) Master's	Terminal Master's	Doctorate	Median No. of Years for 2008–09 Ph.D.'s
Acoustics	1	0	0	0(0)	0(1)	0(1)	5
Astronomy	1	0	0	0(0)	0(1)	0(0)	–
Atomic, Molecular, & Optical Physics	6	2	13	2(11)	1(4)	2(2)	5
Condensed Matter Physics	6	4	5	4(3)	0(5)	3(13)	5
Particles & Fields	6	6	4	6(4)	0(3)	0(2)	5
Statistical/Thermal	3	1	2	1(1)	0(3)	0(0)	5
Other	0	10	5	1(3)	0(0)	1(1)	–
Total		23	29	13(22)	1(17)	6(19)	
Full-time Grad. Stud.		23	26				
Part-time Grad. Stud.		0	3				
First-year Grad. Stud.		10	0				
Median Years in Grad. Study (2008–09 Degrees)				2	2	5	–
Undergraduate Degrees, 2008–09 (2003–09): 11(68)							

[1]Students not yet committed to a research specialty are entered under other.
[2]Five-year totals in parentheses.

4. Graduate Degree Requirements

Master's: 2 options: Thesis Option: 30 credits of course work and research are required, 12 credits in core physics courses with another 6 in 500-level physics courses or above; a minimum of 24 credits must be earned at Colorado State University; no language requirement.
Non-Thesis Option: 32 credits of course work are required, 15 credits in core physics courses with another 6 in 500-level physics courses or above; a minimum of 24 credits must be earned at Colorado State University; no language requirement. Journal presentation required.
Doctorate: 72 credits in course work and research in an approved program, 18 credits in core physics courses with another 6 in 500-level physics courses or above; a minimum of 32 credits must be earned at Colorado State University; one-year residency required; no language requirement. Oral examination to determine mastery of specialized field of proposed dissertation required. Dissertation and dissertation defense required.
Thesis: Thesis may be written *in absentia*.
Special Equipment, Facilities, or Programs: Major research facilities in the department include conventional and superconducting magnets, microwave spectrometer for ferromagnetic-resonance studies, esr spectrometers, vibrating-sample magnetometer, ultrasonic spectrometers, x-ray-diffraction instrumentation, several pulsed and cw lasers, tunable dye la-

sers along with detection and signal-processing equipment for quantum-electronics research, sputter-induced resonant-ionization spectroscopy, facilities for semiconductor fabrication and analysis, very-high-pressure equipment, and high-speed work-stations. Many facilities are available within other departments, such as electron microscopes, microelectronic fabrication, molecular beam epitaxy, and nuclear magnetic resonance.

Table B—Appointments to Graduate Students, 2009–10

Title of Appointee	Appointments Total	Appointments First year	Academic Load Allowed in Credit Hours	Hours of Service Per Week	Stipend for Academic Year ($)
			Semester		
Teaching Assistant	22	10	15	20	20,700
Research Assistant	25	0	15	20	20,700
Total	47	10			

5. Personnel Engaged in Separately Budgeted Research, 7/09–6/10

Professorial faculty	12
Postdoctoral appointments	8
Graduate students	45
Total	65

6. Separately Budgeted Research Expenditures by Source of Support

	Departmental Research	Physics-related Research Outside Department
Federal government	$2,132,000	–
Total	$2,132,000	–

Table C—Separately Budgeted Research Expenditures

Research Specialty	No. of Grants	Expenditures ($)
Atomic, Molecular, & Optical Physics	12	810,160
Condensed Matter Physics	16	426,400
Particles & Fields	15	895,440
Total	43	2,132,000

FACULTY

Professors

Bradley, R. Mark, Ph.D., Stanford, 1985. Condensed matter theory; pattern formation in nonequilibrium systems.

Culver, Roger B., Ph.D., Ohio State, 1971. Astronomy; experimental astrophysics.

Fairbank, William M., Jr., Ph.D., Stanford, 1974. Tunable laser spectroscopy; ultrasensitive laser mass spectrometry.

Harton, John L., Ph.D., MIT, 1988. Experimental particle physics.

Krueger, David A., Ph.D., Washington, 1967. Fluids; lidar studies of the atmosphere.

Lee, Siu Au, Ph.D., Stanford, 1976. Laser spectroscopy; laser manipulation of atoms.

Leisure, Robert G., Ph.D., Washington (St. Louis), 1967. Ultrasonics.

Lundeen, Stephen R., Ph.D., Harvard, 1975. Atomic, molecular, and optical physics.

Patton, Carl E., Ph.D., Cal. Tech., 1967. Magnetism and magnetic materials.

Rocca, Jorge J (joint with Electrical & Computer Engineering), Colorado State University, 1983, Lasers, Plasmas, and Quantum Electronics.

She, Chiao-Yao, Ph.D., Stanford, 1964. Quantum electronics and lidar applications.

Toki, Walter H., Ph.D., MIT, 1976. Experimental particle physics.

Wilson, Robert J., Ph.D., Purdue, 1983. Experimental particle physics.

Associate Professors

Eykholt, Richard E., Ph.D., California, Irvine, 1984. Nonlinear dynamics; chaos; mathematical physics.

Field, Stuart B., Ph.D., Chicago, 1986. Vortices in superconductors; nonlinear dynamics.

Gelfand, Martin P., Ph.D., Cornell, 1990. Condensed matter theory.

Roberts, Jacob L., Ph.D., Univ. Colorado-Boulder, 2001. Laser spectroscopy; ultracold plasmas.

Robinson, Raymond S., Ph.D., Colorado State, 1979. Low-density plasmas; space electric propulsion; ion-beam applications, including surface microtexturing.

Assistant Professors

Berger, Bruce E., Ph.D., Cornell Univ., 2000. Experimental particle physics.

Buchanan, Kristen, Ph.D., University of Alberta, 2000. Magnetics.

Buchanan, Norman, Ph.D., University of Alberta, 2003. Particle Astrophysics.

Mostafa, Miguel, Ph.D., Universidada Nacional de Cuyo, 2001, Particle Astrophysics.

Wu, Mingzhong, Ph.D., Huazhong University of Science and Technology, 1999. Magnetics.

Affiliate Faculty

Camley, Robert, Ph.D., U. California-Irvine, 1979. Magnetism.

Craine, Eric, Ph.D., Ohio State U., 1973. Astronomy.

Kabos, Pavel, Ph.D., Slovak Tech. U., 1979. Theoretical Electrotechnique.

Lindsay, Mark D., Ph.D., Harvard U., 1990. Atomic Physics.

Rodriguez, Fernando, Ph.D., U. Cantabria, 1984. Physics.

Sturris, W. Greg, Ph.D., Notre Dame, 1988. Atomic physics.

Wen, Tingdun, Ph.D., National University of Athens, 1996. Solid State Physics.

Yarger, Jeffery, Ph.D., Arizona State U., 1976. Physical Chemistry.

RESEARCH SPECIALTIES AND STAFF

Theoretical

Statistical Physics, Condensed Matter Physics, and Materials Science. Nonlinear dynamical systems; chaos and fractals; mathematical physics; phase transitions; far-from-equilibrium behavior; fractal aggregates; highly disordered materials; percolation theory; polymers; lattice dynamics of molecular crystals and crystallites using phonon theories;

equations-of-state of condensed phases; phase transitions of bulk solids and adsorbed molecular surfaces; computational physics. Bradley, Eykholt, Gelfand.

Experimental

Atomic, Molecular, Optical Physics. Recent highlights include Presidential Early Career Award for cold atom research, and W. M. Keck Foundation grant for Si quantum computer research. Research topics include efficient creation of BEC with new non-evaporative cooling technique and tailored optical traps; study of quantum fluids as a function of confinement geometry; development of Si based quantum computer; laser cooled single atom on demand source for deterministic ion deposition; new techniques for quantum information; investigation of nanostructures fabricated by laser manipulation of cold atoms; Na fluorescence Lidar for both day and night measurements of temperature, zonal wind and meridional wind in the upper atmosphere; studies of atmospheric wave and global changes; fast beam high precision laser-RF spectroscopy of excited states of atoms and molecules; Rydberg atom studies; ultra-sensitive single-atom detection for tagging Ba+ ion in the neutrinoless double-beta decay (EXO collaboration); extreme ultraviolet laser development, nanolithography and imaging. Fairbank, Krueger, Lee, Lundeen, Roberts, Rocca, She

Condensed Matter Physics and Materials Science. Thin-film semiconductors; semiconductor surfaces; solar cells; magnetic thin films, multiferroics; nanomagnetism, magnetic vortex physics in nanodots; magnetic recording physics and materials, ferrite materials, magnetodynamics and magnetic relaxation; magnon Brillouin light scattering; magnetism in rare earth and actinide compounds; chaos and solitons in thin magnetic films; optical spectroscopy at high pressure; structure-physical property relations of novel low-dimensional materials; ultrasonic spectroscopy; metal-hydrogen systems; phase transitions; crystal growth dynamics; conducting polymers; superconducting vortex dynamics; scanning Hall probe microscopy. Buchanan, K., Field, Leisure, Patton, Sites, Wu.

High Energy Physics and Particle Astrophysics. The High Energy Physics members work in the T2K, the KamLAND, and the BaBar experiments and the Particle Astrophysics members are involved with the Auger (Argentina) and the proposed Auger North (Colorado) cosmic ray experiments. The T2K and KamLAND experiments both study neutrino oscillations and the BaBar experiment investigates the weak decays of bottom and charm mesons. T2K P0D detector parts are currently being constructed at CSU. The KamLAND and BaBar experiments are currently analyzing data. The Auger experiment in Argentina is taking data on high energy cosmic rays and is analyzing evidence for the high energy cosmic ray production from active galactic nuclei. Berger, Buchanan, N., Harton, Mostafa, Toki, Wilson.

FACULTY PUBLICATIONS

Berger, Bruce E.

S. Abe *et al.* [KamLAND Collaboration], "Study of the Production of Radioactive Isotopes through Cosmic Muon Spallation in KamLAND," Phys. Rev. C **81**, 025807 (2010).

B. Berger is the first author of the KamLAND Collaboration publication entitled "The KamLAND Full-Volume Calibration Systems," arXiv:0903.0441v1 [nucl-ex] (2009).

Bradley, Mark R.

R. M. Bradley, J. E. Bernard, and L. D. Carr, "Exact Dynamics of Multicomponent Bose-Einstein Condensates in Optical Lattices in One, Two and Three Dimensions," Phys. Rev. A **77**, 033622 (2008).

R. M. Bradley, B. Deconinck, and J. N. Kutz, "Exact Nonstationary Solutions to the Mean-Field Equations of Motion for Two-Component Bose-Einstein Condensates in Periodic Potentials," J. Phys. A **38**, 1901 (2005).

Buchanan, Kristen S.

S. S. Kalarical, M. Betz, A. Bhattarcharya, and K. Buchanan, "Temperature and angular dependence of ferromagnetic resonance linewidth in $La_{0.67}Sr_{0.33}MnO_3$ alloy and superlattice thin films," Technical Digest, 2010 Joint MMM/Intermag Conference, Washington, D.C., AB-02 (2010.

D. J. Keavney, X. M. Cheng, and K. S. Buchanan, "*Polarity reversal of a magnetic vortex core by a unipolar, nonresonant in-plane pulsed magnetic field*," Applied Physics Letters **94**, 1 (2009).

Buchanan, Norman J.

N. J. Buchanan *et al.*, "Design and implementation of the Front End Board for the readout of the ATLAS liquid argon calorimeters," J. Inst **3**, P03004 (2008).V. M. Abazov *et al.* (D0 Collaboration), "Search for Squark and Gluinos in Events with Jets and Missing Transverse Energy using 1.‘ fb-‘ of pp (bar) Collision Data at sqrts(s) =1.96 TeV," Phys. Lett. B **660**, 449 (2008).

Eykholt, Richard E.

M. W. A. M. Hagerstrom, R. Eykholt, A. Kondrashov, and B. A. Kalinikos, "Excitation of Chotic Spin Waves Through Modulation Instability," Phys. Rev. Lett. **102**, 237203 (2009).

A. M. Hagerstrom, W. Tong, M. Wu, B. A. Kalinikos, and R. Eykhold, "Excitation of Chaotic Spin Waves in Magnetic Film Feedback Rings Through Three-Wave Nonlinear Interactions," Phys. Rev. Lett. **102**, 207202 (2009).

Fairbank, William

S.-C. Jeng, W. M. Fairbank, Jr., and M. Miyajima, "Measurements of the Mobility of Alkaline Earch Ions in Liquid Xenon," J. Phys. D: Appl. Phys. **42**, 035302 (2009).

R. Neilson *et al.*, "Characterization of large area APD's for the EXO-200 detector," Nuclear Inst. and Methods in Physics Research A **608**(1), 68–75 (2009).

Field, Stuart B.

D. Dinulovic, H. Saalfeld, Z. Celinski, S. B. Field, and H. H. Gatzen, "Evaluation of an Electromagnetic Microactuator Using Scanning Hall Probe Microscopy Measurements," Appl. Phys. **105**, 07F119 (2009).

D. Dinulovic, H. Saalfeld, Z. Celinski, S. B. Field, and H. H. Gatzen, "Integrated Electromagnetic Second Stage Microactuator for a Hard Disk Recording Head," IEEE Trans. Magn. **44**, 3370 (2008).

Gelfand, Martin

E. Bronson, S. Field, and M. Gelfand, "Equilibrium configurations of Pearl vortices in narrow strips," Phys. Rev. B **73**, 144501 (2006).

D. Steffen and M. Gelfand, "Longitudinal and Hall conductances in model alkali fullerides A_3C_{60}," Phys. Rev. B **69**, 115109 (2004).

Harton, John L.

Upper limit on the cosmis-ray photon fraction at EeV energies from the Pierre Auger Observatory. By the Pierre Auger Collaboration (J. Abraham *et al.*) FERMILAB-PUB-09-668-A, Mar. 2009, 20pp. Published in Astrophar. Phys. **31**, 399–406 (2009); e-print: arXiv:0903:1127 [astro-ph.HE]

Limit on the diffuse flux of ultra-high energy tau neutrinos with the surface detector of the Pierre Auger Observatory. By Pierre Auger Collaboration (J. Abraham *et al.*) FERMILAB-PUB-09-667-A, Mar. 2009 (Received Mar 2009). 19pp. Published in PHys. Rev. D **79**, 102001 (2009); e-print: arXiv:0903:3385 [astro-pph.HE].

Krueger, David A.

D. A. Krueger and C.-Y. She, "Seasonal variability in mesopause region temperatures over Fort Collins, CO (41°N, 105°W) based on lidar observations, 1991 through 2007," J. Atmos. and SOlar Terr. Phys. **71**, 1565–1570 (2009).

C.-Y. She, D. A. Krueger, R. Akmaev, H. Schmidt, E. Talaat, and S. Yee, "Long-term variability in mesopause region temperatures over Fort Collins, CO (41°N, 105°W) based on lidar observations from 1990 through 2007," J. Atmos. and Solar-Terr. Phys. **71**, 1558–1565 (2009).

Lee, Siu Au

A. J. Berglund, S. A. Lee, and J. J. McClelland, "Sub-Doppler laser cooling and magnetic trapping of erbium," Phys. Rev. A **76**, 053418 (2007).

S. J. Rehse, K. M. Bockel, and S. A. Lee, "Laser collimation of an atomic gallium beam," Phys. Rev. A **69**, 063404 (2004).

Leisure, Robert G.

A. Migliori, H. Ledbetter, R. G. Leisure, C. Pantea, and J. B. Betts, "Diamond's elastic stiffnesses from 322K to 10K," J. Appl. Phys. **104**, 053512 (2008).

D. S. Agosta, R. G. Leisure, J. J. Adams, Y. T. Shen, and K. F. Kelton, "Elastic Moduli of a Ti-Zr-Ni i-phase Quasicrystal as a Function of Temperature," Phil. Mag. **87**, 1 (2007).

Lundeen, Stephen R.

J. A. Keele, S. L. Woods, M. E. Hanni, S. R. Lundeen, and W. G. Sturrus, "Optical spectroscopy of high-L Rydberg states of nickel," Phys. Rev. A **81**, 022506 (2010).

M. E. Hanni, J. A. Keele, S. R. Lundeen, C. W. Fehrenbach, and W. G. Sturrus, "Polarizabilities of Pb2+ and Pb4+ from Ionization Energies of Pb+ and Pb3+ from spectroscopy of high-L Rydberg states of Pb+ and Pb3+," Phys. Rev. A **81**, 042512.

Mostafa, Miguel

M. Mostafa, "Upper limit on the cosmic-ray photon fraction at EeV energies from the Pierre Auger Observatory," (arXiv:0903.1127), Astroparticle Physics **31**, Issue 6, 399–401 (2009).

J. Abraham *et al.* [Pierre Auger Collaboration], "Atmospheric effects on extensive air showers observed with the Surface Detector of the Pierre Auger Observatory," Astroparticle Physics **32**, 89 (2009).

Patton, Carl E.

P. Krivosik, S. S. Kalarickal, N. MO, S. Wu, and C. E. Patton, "Ferromagnetic resonance with damping in granular Co-Cr films with perpendiuclar anisotropy," Appl. Phys. Lett. **95**, 52509 (2009).

V. G. Harris, A. Geiler, Y. Chen, S. D. Yoon, M. Wu, A. Yang, Z. Chen, P. He, P. V. Parimi, X. Zuo, C. E. Patton, M. Abe, O. Acher, and C. Vittoria, "Recent advances in processing and aplications of microwave ferrites," J. Magn. Magn. Mat. **321**, 2035–2047 (2009).

Roberts, Jacob L.

M. S. Hamilton, A. R. Georges, and J. L. Roberts, "Influence of Optical Molasses in Loading a Shallow Optical Trap," Phys. Rev. A **79**, 013418 (2009).

A. Gorges, N. Bingham, M. DeAngelo, M. Hamilton, and J. L. Roberts, "Light-assisted collional loss in an $^{85/87}$RB Optial Trap," Phys. Rev. A **78**, 033420 (2008).

She, Chiao-Yao

J. Yue, C.-Y. She, T. Nakamura, S. Harrell, and T. Yuan, "Mesospheric bore formation from large-scale gravity wave perturbations observed by collocated all-sky OH imager and soldium lidar," J. Atomospheric and Solar-Terrestrial Physics **72**, 7–18 (2010).

Z. Liu, D.-C. Bi, X. Song, J.-B. Xia, R. Li, Z. Wang, and C.-Y. She, "Iodine-filter-based hight spectral resolution lidar for atmospheric temperature measurements," Optics Letters **34**, 2712–2714 (2009).

Sites, James R.

G. T. Koishiyev and J. R. Sites, "Impact of Sheet Resistance on 2-D Modelig of Thin-Film Solar Cells," Solar Energy Mat. and Solar Cells **93**, 350–354 (2009).

S. H. Demtsu, D. S. Albin, W. K. Metzger, A. Duda, and J. R. Sites, "Cu-Related Recomination in CdTe Solar Cells," Thin Solid Films **516**, 2251–2254 (2008).

Toki, Walter

W. Dunwoodie *et al.* (BES Collaboration), "Measurement of inclusive momentum spectra and multiplicity distributions of charged particles at S**1/2 sim 2-5 GeV," SLAC-PUB-10095, June 2003, 9pp. Published in Phys. Rev. D **69**, 072002 (2004). E-print: hep-ex/0306055.

J. Z. Bai *et al.* (BES Collaboration), "Observation of a near threshold enhancement in the p anti-p mass spectrum from radiative J/psi→gamma p anti-p decays," SLAC-PUB-9709, UH-511-1019-03, Mar 2003, 5 pp. Published in Phys. Rev. Lett. **91**, 022001 (2003). E-Print: hep-ex/0303006.

Wu, Mingzhong

W. Tong, M. Wu, L. D. Carr, and B. Kalinikos, "Formation of random dark envelope solitons from incoherent waves," Phys. Rev. Letters **014**, 037307 (2010).

Z. Wang, K. Sun, W. Tong, M. Wu, M. Liu, and N. X. Sun, "Microwave-assisted magnetization reversal in large-damping magnetic films: competition between pumping and damping," Phys. Rev. B **81**, 064402 (2010).

UNIVERSITY OF COLORADO, BOULDER

DEPARTMENT OF PHYSICS

Boulder, Colorado 80309

Students Accepted For Degree	FIELDS		
	Physics	Astronomy	Related Fields
Doctorate	X		
Master's	X		

1. General

President: Bruce Benson
Dean of Graduate School: John A. Stevenson
Department Chairman: Paul D. Beale
Department Telephone Number: (303) 492-6952 (C)
World Wide Web Home Page: http://www.colorado.edu/physics/web
Type of Institution: University
Control: Public
Setting: Small city
Total Faculty: 1,517
Total Students: 30,659
Total Graduate Students: 4,900
Annual Graduate Tuition:
In-state residents: Full-time—$5,952.00
Out-of-state residents: Full-time—$16,032.00
Tuition rates for: 2010–11
Deferred tuition plan: Yes
Other Fees: $742.00
Term: Semester

2. Number of Faculty in Department

The combined total of full-time faculty in the three professorial ranks is 56. The combined total of full-time, part-time, and other faculty at all ranks is 77.

3. Admission, Financial Aid, and Housing

Address admission inquiries to: http://www.colorado.edu/physics/Web/education/grad/applying/index.html
Graduate application fee required: $50—U.S. citizens and permanent residents
$70—Foreign students
Admission deadline (Fall admission): 1/1
Admission information: For fall admission, 2010–11, 114 students were admitted from 506 applicants, 37 accepted.
Admission requirements: For admission to the graduate programs, a Bachelor's degree is required with a minimum undergraduate GPA of 3.0 specified. The GRE is required. The GRE Advanced Physics is required, with a minimum score of 500. The average GRE scores for 2010–11 admissions were verbal-584; quantitative-786. The average GRE advanced Physics scores for 2010–11 US admissions were 806. The average GRE scores for 2010–11 international admissions were verbal-532; quantitative-794. The average GRE advanced Physics scores for 2010–11 international admissions were 923. Students from non-English speaking countries are required to demonstrate proficiency in English via the TOEFL exam. Minimum acceptable score for admission is 85-internet based. If you have completed at least one year of full-time academic study at a U.S. institution at the time you apply, you are exempt from this requirement.
An undergraduate program for students entering graduate study in physics should typically include the following:
3 semesters of Introductory Physics,
1 semester of Advanced Classical Mechanics,
2 semesters of Quantum Mechanics,
1 semester of Statistical Mechanics,
2 semesters of Advanced Electricity and Magnetism,
2 semesters of an advanced lab course or project work,
1 semester of an advanced course in Modern Physics (e.g.), Condensed Matter, Geophysics, Atomic, Nuclear or Particle Physics.
3 semesters Calculus
1 semester Linear Algebra
1 semester Differential Equations
General Computing Knowledge

Table A—Faculty, Enrollments, and Degrees Granted

Research Specialty	2008–09 Faculty	Enrollment[1] Fall 2009 Mas-ter's	Enrollment[1] Fall 2009 Doc-torate	No. of Degrees Granted[2] 2009–10 Mas-ter's	No. of Degrees Granted[2] 2009–10 Terminal Master's	No. of Degrees Granted[2] 2009–10 Doc-torate	Median No. of Years for 2009–10 Ph.D.'s
Full-time Grad. Stud.		0	209	19	1	26	
Part-time Grad. Stud.		0	0	0			
First-year Grad. Stud.		0	29	0			
Median Years in Grad. Study (2007–08 Degrees)							7yrs
Undergraduate Degrees, 2008–09: 54							

[1]Students not yet committed to a research specialty are entered under non-specialized.
[2]Five-year totals in parentheses.

Address financial aid inquiries to: University of Colorado at Boulder, Dept. of Physics, 390 UCB, Attn: Graduate Program Asst., Boulder, CO 80309-0390
GAPSFAS application required: No
Financial aid deadline: 1/15
Loans available: Yes
Address housing inquiries to: University of Colorado Housing and Dining Services, 159 UCB, Boulder, CO 80309-0159
On-campus, student housing available: Yes
Cost/month $600–1,200[2] (incl. phone and utilities)

[1]Varying rates depending on number of bedrooms and whether furnished or unfurnished

4. Graduate Degree Requirements

Master's: 30 semester hours, thesis and non-thesis; minimum GPA, 3.0; two semesters or three summers residence minimum; no foreign language requirement; no written comprehensive exam required. Oral presentation required for thesis and non-thesis M.S.; thesis requirements in accordance with Graduate School specifications.
Doctorate: 30 semester hours course work (joint programs have varying course requirements which may be different from the straight physics curriculum) plus 30 doctoral thesis hours; 3.0 minimum GPA, no grades lower than B; six semesters residence minimum; comprehensive exam; oral presentation required; thesis requirements in accordance with our Graduate School specifications.
Thesis: Thesis may be written *in absentia.*
Special Equipment, Facilities, or Programs: Liquid Crystal Ma-

terial Research Center, Extreme Ultraviolet Engineering Research Center, Colorado Center for Lunar Dust and Atmospheric Studies. Programs at the meson facility (LAMPF) at Los Alamos, SLAC, and Fermi National Accelerator Laboratory are available. Experimental and theoretical research programs at JILA, Cooperative Institute for Research in Environmental Sciences, the National Institute for Standards & Technology (NIST), the Laboratory for Atmospheric and Space Physics (LASP), the High Altitude Observatory (HAO), the National Center for Atmospheric Research (NCAR), the National Renewable Energy Laboratory (NREL), the National Oceanic and Atmospheric Administration (NOAA), the U.S. Geological Survey (USGS), and the Center for Opto-Electronic Computing Systems (COCS), Center for Integrated Plasma Studies (CIPS) in astrophysics, atomic, molecular and optical physics, as well as geophysics are available.

Table B—Appointments to Graduate Students, 2009–10

Title of Appointee	Appointments Total	Appointments First year	Academic Load Allowed in Credit Hours	Hours of Service Per Week	Stipend for Academic Year ($)
			Semester		
Teaching Assistant	45	18	9	20	15,284[1,2]
Research Assistant	164	11	9	20	17,604[3,4]
Total	209	29			

[1]Plus tuition waiver and contributions to health insurance.
[2]With summer TA. annual stipend was $16,792.
[3]With summer RA. annual stipend was ranged from $18,480–23,170.
[4]Summer RA's are available with additional stipend.

5. Personnel Engaged in Separately Budgeted Research, 7/09–6/10

Professorial faculty	56[1]
Other faculty	21[1]
Postdoctoral appointments	48[2]
Graduate students	180
Undergraduate students	50
Nonteaching research personnel	25
Total	373

[1]Some joint appointments and adjuncts
[2]Excluding research institutes

7. Separately Funded and Managed Laboratories

CIPS	$1,596,000
JILA	$22,298,319
Lab. for Atmospheric & Space Physics	$40,187,269
Total	$64,081,588

Table C—Separately Budgeted Research Expenditures

Research Specialty	No. of Grants	Expenditures ($)
Atmos./Space Phys., Cosmic Rays	4	290,000
Atomic, Molecular, & Optical Physics	42	4,382,995
Biophysics	5	153,908
Condensed Matter Physics	52	3,820,610
Geophysics	16	977,933
Nuclear Physics	4	552,065
Particles & Fields	5	1,947,609
Physics and Science Education	12	442,020
Physics of Beams	1	212,000
Plasma Physics & Fusion	21	964,000
Women in Science	1	469,854
Total	163	14,212,994

FACULTY

Professors

Anderson, Dana Z., Ph.D., Arizona, 1981. Experiment quantum optics.

Baker, Daniel N., Ph.D., University of Iowa, 1974. Joint appointment with Department of Atmospheric and Space Physics.

Beale, Paul D., Ph.D., Cornell, 1982. Theoretical condensed matter physics.

Cary, John, Ph.D., California, Berkeley, 1979. Theoretical plasma physics.

Clark, Noel A., Ph.D., MIT, 1970. Experimental condensed matter physics.

Cumalat, John P., Ph.D., California, Santa Barbara, 1977. Department Chair. Experimental particle physics.

de Alwis, Senarath P., Ph.D., Cambridge, 1969. Theoretical elementary particles.

DeGrand, Thomas A., Ph.D., MIT, 1976. Theoretical particle physics.

Dessau, Daniel, Ph.D., Stanford, 1992. Experimental condensed matter physics.

Ford, William T., Ph.D., Princeton, 1967. Experimental elementary particles.

Franklin, Allan D., Ph.D., Cornell, 1965. History and philosophy of science.

Goldman, Martin, Ph.D., Harvard, 1965. Plasma physics.

Greene, Chris H., Ph.D., Chicago, 1980. Theoretical atomic physics.

Hasenfratz, Anna, Ph.D., Lorand Eotvos University, 1982. Theoretical high-energy physics.

Horanyi, Mihaly, Ph.D., Lorand Eotvos University, Budapest, 1982. Dusty plasmas.

Kapteyn, Henry, Ph.D., Univ. Calif. Berkeley, 1989. Ultrafast nonlinear optics to generate coherent x-rays; XUV and x-ray lasers, and the use of ultrashort x-ray pulses to probe dynamic processes in chemical and material systems.

Kinney, Edward R., Ph.D., MIT, 1988. Experimental nuclear physics.

Mahanthappa, Kalyana T., Ph.D., Harvard, 1961. Theoretical elementary particles.

Murnane, Margaret, Ph.D., Univ. Calif. Berkeley, 1989. The development and application of novel ultrafast coherent light sources in the visible and x-ray regions of the spectrum.

Nagle, Jamie, Ph.D., Yale, 1996. Experimental nuclear physics.

Nauenberg, Uriel, Ph.D., Columbia, 1963. Experimental elementary particles.
Parker, Scott, Ph.D., U.C.-Berkeley, 1990. Computational plasma physics.
Peterson, Roy J., Ph.D., Washington, 1966. Experimental nuclear physics; intermediate energy physics.
Pollock, Steven, Ph.D., Stanford, 1987. Theoretical nuclear physics.
Price, John C., Ph.D., Stanford, 1986. Experimental low-temperature physics.
Radzihovsky, Leo, Ph.D., Harvard, 1993. Theoretical condensed matter physics.
Rankin, Patricia, Ph.D., U. of London, 1982. Experimental elementary particles.
Ritzwoller, Michael H., Ph.D., U.C., San Diego, 1987. Theoretical geophysics.
Robertson, Scott, Ph.D., Cornell, 1972. Experimental plasma physics; fusion.
Rogers, Charles T., Ph.D., Cornell, 1987. Experimental condensed matter physics.
Shepard, James R., Ph.D., Colorado, 1972. Theoretical and experimental nuclear physics.
Wahr, John, Ph.D., Colorado, 1979. Theoretical geophysics.
Wieman, Carl E., Ph.D., Stanford, 1977. Bose-Einstein condensation; AMO physics.
Zhong, Shijie, Ph.D., University of Michigan, 1994. Theoretical geophysics.

Associate Professors

Becker, Andreas, Ph.D., University Beilefeld, 1997. Theoretical atomic physics.
Betterton, Meredith D., Ph.D., Harvard, 2000. Biophysics; systems biology; bioinformatics; pattern formation.
Finkelstein, Noah, Ph.D., Princeton, 1998. Physics education research.
Gurarie, Victor, Ph.D., Princeton, 1996. Theoretical condensed matter.
Holland, Murray, Ph.D., Oxford, 1994. Theoretical quantum optics; collision studies; Bose-Einstein condensation.
Raschke, Markus B., Ph.D., Technical University of Munich, 1999. Science of nano-optics.
Reznik, Dmitry, Ph.D., University of Illinois, Urbana-Champaign, 1993. Experimental condensed matter and materials physics.
Zimmerman, Eric, Ph.D., Univ. of Chicago, 1998. Experimental particle physics.

Assistant Professors

DeWolfe, Oliver, Ph.D., MIT, 2000. Theoretical elementary particles and fields.
Gopinath, Juliet, Ph.D., MIT, 2005. Joint appointment with Electrial, Computer and Energy Engineering.
Hermele, Michael, Ph.D., U.C. Santa Barbara, 2005. Theoretical condensed matter physics.
Lee, Minhyea, Ph.D., University of Chicago, 2004. Experimental condensed matter and materials physics.
Lewandowski, Heather, Ph.D., University of Colorado, 2002. Experimental atomic and molecular physics.
Marino, Alysia, Ph.D., University of California, Berkeley, 2004. Experimental particle physics (neutrinos).
McElroy, Kyle P., Ph.D., University of California, Berkeley, 2005. Experimental condensed matter physics.
Munsat, Tobin, Ph.D., Princeton, 2001. Experimental plasma physics.
Popovic, Milos, Ph.D., MIT, 2007. Joint appointment with Electrical, Computer and Energy Engineering.
Regal, Cindy, Ph.D., Univ. of Colorado, 2006. Atomic, Molecular and Optical Physics.
Schibli, Thomas, Ph.D., University of Karlsruhe, 2001. Experimental Optics.
Smalyukh, Ivan, Ph.D., Kent State, 2003. Experimental condensed matter physics.
Stenson, Kevin, Ph.D., University of Wisconsin, 1998. Experimental elementary particles and fields.
Uzdensky, Dmitri, A., Ph.d., Princeton, 1998. Theoretical plasma physics and plasma astrophysics.

Professors Adjoint

Cornell, Eric, Ph.D., MIT, 1990. Bose-Einstein condensation, laser and magnetic trapping of neutral cesium using diode lasers and a vapor cell.
Cundiff, Steven, Ph.D., University of Michigan, 1992. Telecommunications, fiber optics, ultrafast optical studies of semiconductors.
Faller, James E., Ph.D., Princeton, 1963. Geophysics, experimental relativity; precision measurement; null experiments.
Hall, John, Ph.D., Carnegie Institute of Technology, 1961. Development of laser stabilization and measurement techniques that lead toward the creation of phase-stable optical frequency sources and their application to precision tests of fundamental principles.
Jin, Deborah, Ph.D., The University of Chicago, 1995. Bose-Einstein condensation in a dilute atomic gas: collective excitations, thermodynamics at the phase transition, interaction effects, vortices and other analogs of superfluidity.
Levine, Judah, Ph.D., New York University, 1966. Application of precision measurement techniques to problem of geophysical interest.
Ye, Jun, Ph.D., Univ. of Colorado, 1997. Atomic and optical physics.

Associate Professor Adjoint

Lehnert, Konrad, Ph.D., University of California, Santa Barbara, 1999. Physics.

Assistant Professor Adjoint

Thompson, James K., Ph.D., MIT, 2003. Precision Measurement.

Professors Attendant Rank

Glaser, Matthew A., Ph.D., Colorado, 1991. Computer simulation techniques for problems in condensed matter physics and statistical physics.
Maclennan, Joseph, Ph.D., Colorado, 1988. Ferroelectric liquid crystals; freely-suspended liquid crystal films; instrumentation.
Smith, James G., Ph.D., University of California, San Diego, 1975. Experimental high-energy physics.

Associate Professor Attendant Rank

Wagner, Stephen, Ph.D., Tthe Johns Hopkins University, 1983. High energy physics.

Assistant Professor Attendant Rank

Perkins, Katherine, Ph.D., Harvard, 2000. Physics education research with a focus on: the use of interactive simulations for teaching and learning physics; students' beliefs about physics (and chemistry); and sustainable course reform.

Research Professor

Bohn, John, Ph.D., University of Chicago, 1995. Cold collisions, semiconductor devices, few-body physics.

Assistant Research Professor

Rey, Ana Maria, Ph.D., University of Maryland 2004. Optical lattices, quantum degenerate Fermi gases, and ultracold Boson-Fermion mixtures.

Senior Instructor

Dubson, Michael, Ph.D., Cornell, 1984. Physics education research.

Lecturers

Diddams, Scott, Ph.D., University of New Mexico, 1996. Experimental laser physics, femtosecond lasers and ultrafast phenomena, nonlinear optics, precision spectroscopy, and optical frequency combs and metrology.

Knill, Emanuel, Ph.D., University of Colorado, 1991. Quantum information science.

Wineland, David, Ph.D., Harvard, 1970. Laser-cooled trapped ions, in the areas of high-resolution spectroscopy, basic plasma physics, and quantum information.

UNIVERSITY OF COLORADO, BOULDER

DEPARTMENT OF ASTROPHYSICAL AND PLANETARY SCIENCES

Boulder, Colorado 80309-0391

Students Accepted For Degree	FIELDS		
	Physics	Astronomy	Related Fields
Doctorate		X	X*
Master's		X	X

*Meterology, Atmospheric/Space Physics, Astrophysics.

1. General

President: Bruce Benson
Dean of Graduate School: John Stevenson
Department Chair: Mitchell Begelman
Department Telephone Number: (303) 492-8915 (C)
World Wide Web Address: http://aps.colorado.edu
Type of Institution: University
Control: Public
Setting: Small city
Total Graduate Faculty: 2,217
Total Students: 30,659
Total Graduate Students: 4,900
Annual Graduate Tuition:
In-state residents: Full-time—$8,928
Out-of-state residents: Full-time—$23,346
Tuition rates for: 2009–10
Deferred tuition plan: Yes
Other Fees: $371.57 plus $30.00 per course
Term: Semester

2. Number of Faculty in Department

The combined total of full-time faculty in the three professorial ranks is 22. The combined total of full-time, part-time, and other faculty at all ranks is 38.

3. Admission, Financial Aid, and Housing

Applications accepted online at: www.colorado.edu/prospective/graduate/apply/process.html
Graduate application fee required: $50
Admission deadline (Fall admission): 1/15
Admission information: For fall admission, 2009–10, 29 students were admitted and 2 accepted from 130 applicants.
Admission requirements: For admission to the graduate programs, a Bachelor's degree in mathematics or physics is recommended with a minimum undergraduate GPA of 3.0 specified. The GRE is required. The minimum acceptable score-suggested for admission is verbal–75%; quantitative–85%; analytical–75%. The GRE Physics Advanced is required. The average GRE scores for 2009–2010 admissions were verbal–88%; quantitative–89%; analytical–63% advanced physics–60%. Students from non-English speaking countries are required to demonstrate proficiency in English via the TOEFL exam. Minimum acceptable score for admission is 500.
Address financial aid inquiries to: Admissions Chair, Department of Astrophysical and Planetary Sciences
GAPSFAS application required: No
Financial aid deadline: 4/1
Loans available: Yes
Address housing inquiries to: Director, Student Housing Office, Campus Box 159
On-campus, single student housing available: Yes
Cost/term: $5,980
On-campus, family student housing available: Yes
Cost/month: $653–259[1]

[1]Varies according to number of rooms and whether furnished or unfurnished.

Table A—Faculty, Enrollments, and Degrees Granted

Research Specialty	2008–09 Faculty	Enrollment[1] Fall 2009 Master's	Enrollment[1] Fall 2009 Doctorate	No. of Degrees Granted[2] 2009–10 Master's	Terminal Master's	Doctorate	Median No. of Years for 2009–10 Ph.D.'s
Astrophysics	17	0	37	7	0	4	6
Planetary Science	5	0	11	2	3	2	6
Total		–	48	9	3	6	6
Full-time Grad. Stud.		–	48				
Part-time Grad. Stud.		–	0				
First-year Grad. Stud.		–	7				
Median Years in Grad. Study (2009–10 Degrees)		2	6				

Undergraduate Degrees, 2009–10 (2009–2010: 20

4. Graduate Degree Requirements

Master's: PLAN I—30 graduate credits with a B average or above plus a thesis with oral defense. Thesis must represent the equivalent of 4 to 6 semester hours of work (included in the 30 credits listed above). PLAN II—30 graduate credits with a B average or above, plus a comprehensive examination. There are no residence or foreign language requirements.
Doctorate: 39 graduate course credits including 4 hours of graduate seminar), 30 hours of thesis credit (all with grades B or above), plus a demonstrated ability to do original research. Comprehensive examination of math, physics, and scientific judgment, and grasp of his/her particular area of research. There is a six-semester residency requirement. Dissertation and satisfactory defense are required.
Other Programs: Ph.D. in plasma physics is available; emphasis can be on astrophysical, laboratory, or theoretical plasma physics.
Thesis: Thesis may be written *in absentia*.

Table B—Appointments to Graduate Students, 2009–10

Title of Appointee	Appointments Total	Appointments First year	Academic Load Allowed in Credit Hours	Hours of Service Per Week	Stipend for Academic Year ($)
			Semester		
Teaching Assistant	8	2	9–10	20	17,604[1,2]
Research Assistant	38	5	9–10	20	17,604[1,3,4]
NASA GSRP	2	0	9–10		18,000
NSF	0	0	9–10		30,000
Total	48	7			

[1]No charge for tuition, Fellowship Supplements for TA's.
[2]Some summer TA's are available with additional stipend.
[3]Stipend increases for approved doctoral candidates.
[4]Summer RA's are available with additional stipend.

5. Personnel Engaged in Separately Budgeted Research, 7/08–6/09

Professorial faculty	22
Other faculty	6
Postdoctoral appointments	33
Graduate students	48
Nonteaching research personnel	37
Total	146

6. Separately Budgeted Research Expenditures by Source of Support

	Departmental Research	Physics-related Research Outside Department
Federal government	$9,498,492	$12,675,897
Total	$9,498,492	$12,675,897

7. Separately Funded and Managed Laboratories

Joint Institute for Laboratory Astrophysics	$18,674,236
Laboratory for Atmospheric and Space Physics	66,867,740
Total	$85,541,976

Table C—Separately Budgeted Research Expenditures

Research Specialty	No. of Grants	Expenditures ($)
Astrophysics	81	9,498,492
Total	81	9,498,492

FACULTY

Professors

Bagenal, Frances, Ph.D., MIT, 1981. Space plasma physics.

Baker, Daniel N., Ph.D., Iowa, 1974. Space plasma physics; planetary magnetospheres; solar terrestrial coupling.

Bally, John, Ph.D., U. Mass., 1980. IR/mm-radio astronomy; star formation; molecular clouds.

Barth, Charles A., Ph.D., UCLA, 1958. Emeritus. Structure and composition of planetary atmospheres, especially Earth, Mars, and Venus.

Begelman, Mitchell C., Ph.D., Cambridge, 1978. Theoretical astrophysics, black holes, high energy.

Burns, Jack, Ph.D., Indiana, 1978. Galaxy clusters, x-ray and radio astronomy.

Cash, Webster C., Ph.D., California, Berkeley, 1978. High-energy astrophysics, X-ray instrumentation.

Conti, Peter S., Ph.D., California, Berkeley, 1963. Emeritus. Stellar spectroscopy; atmospheres and winds of hot stars.

Ergun, Robert E., Ph.D., University of California, Berkeley, 1989. Space physics, magnetospheres, and plasma physics.

Esposito, Larry W., Ph.D., Massachusetts, 1977. Planetary atmospheres; Saturn's rings.

Green, James C., Ph.D., Chair of Department, University of California, Berkeley, 1989. Experimental astrophysics, ultraviolet instrumentation.

Hamilton, Andrew J. S., Ph.D., Virginia, 1983. Theoretical astrophysics, cosmology.

Hansen, Carl J., Ph.D., Yale, 1966. Emeritus. Theoretical astrophysics, stellar structure.

Malville, J. McKim, Ph.D., Colorado, 1961. Emeritus. Solar physics; magnetic fields in flares, prominences, corona.

McCray, Richard A., Ph.D., UCLA, 1967. Emeritus. Theoretical astrophysics.

Shull, J. Michael, Ph.D., Princeton, 1976. Theoretical astrophysics; UV/X-ray space astronomy.

Snow, Theodore P., Jr., Ph.D., Washington, 1973. Optical and ultraviolet astronomy.

Speiser, Theodore W., Ph.D., Penn State, 1964. Emeritus. Observational and theoretical studies of solar-terrestrial physics.

Stocke, John T., Ph.D., Arizona, 1977. Observational extragalactic astronomy.

Thomas, Gary E., Ph.D., Pittsburgh, 1963. Emeritus. Radiative transfer and photochemistry in planetary atmospheres.

Toomre, Juri, Ph.D., Cambridge, 1967. Astrophysical and geophysical fluid dynamics; magnetohydrodynamics; stellar convection.

Associate Professors

Armitage, Philip, Ph.D., Cambridge University, 1996. Star formation, accretion disks extrasolor planets.

Ellingson, Erica, Ph.D., Arizona, 1989. Astrophysics; extragalactic/cosmology galaxy clusters.

Glenn, Jason, Ph.D., Arizona, 1997. Submillimeter instrumentation, interstellar medium, and cosmology, high-Z galaxies.

Rast, Mark, Ph.D., Colorado 1992. Solar Physics convection flows.

Schneider, Nicholas M., Ph.D., Arizona, 1988. Planetary astronomy, Io plasma torms.

Assistant Professors

Darling, Jeremy, Ph.D., Cornell, 2002. Radio astronomy, high-Z molecules, galaxies.

Halverson, Nils, Ph.D., Caltech, 2002. Radio astronomy, CMB cosmology, mm-wave instrumentation, clusters.

Perna, Rosalba, Ph.D., Harvard, 1999. High energy astrophysics gamma ray bursts, accretion physics.

Professors Adjoint

Canup, Robin M., Ph.D., Colorado, 1995. Planetary sciences.

Chapman, Clark, Ph.D., MIT, 1972. Planetary Science.

DeForest, Craig, Ph.D., Stanford, 1995. Solar physics, instrumentation.

Gilman, Peter A., Ph.D., MIT, 1966. Geophysical and astrophysical fluid dynamics; magnetofluid dynamics; differential rotation.

Levison, Harold F., Ph.D., Michigan, 1986. Planetary dynamics.

Porco, Carolyn, Ph.D., Caltech, 1983. Planetary Science.

Ward, William R., Ph.D., Caltech, 1972. Planetary dynamics.

Professors Attendant

Stewart, A. Ian, Ph.D., Belfast, 1965. Planetary atmospheres; airglow and auroral emissions of Earth, Mars, and Venus.

Associate Professor Attendant

Duncan, Douglas, Ph.D., Santa Cruz, 1980. Stellar Astronomy and nucleosynthesis, Director of Observatories.

Research Professors

Ayres, Thomas R., Ph.D., Colorado, 1975. Stellar atmospheres.

Froning, Cynthia, Ph.D., Texas, 1999. Accetion disks, compact binaries instrumentation (O/IR, UV).

Hindman, Brad, Ph.D., Colorado 1995. Solar Physics convection.

Linsky, Jeffrey L., Ph.D., Harvard, 1968. Astrophysics; formation of spectral lines in solar and stellar atmospheres.

Associate Professors Adjoint

Bogdan, Thomas J., Ph.D., Chicago, 1984. Astrophysics.

Grinspoon, David H., Ph.D., Arizona, 1988. Evolution of planetary atmospheres.

Assistant Professor Adjoint

Charbonneau, Paul, Ph.D., Montreal, 1990. Astrophysics, solar physics, planetary interiors, and computational physics.

Instructor

Hornstein, Seth, Ph.D., UCLA, 2006. Astromony and Astrophysics.

Lecturers

Bennett, Jeffrey O., Ph.D., Colorado, 1987. Astrophysics; science education.

Betz, Albert L., Ph.D., University of California, Berkeley, 1977. Infrared spectroscopy, submillimeter instrumentation, and interstellar matter.

Brown, Alexander, Ph.D., St. Andrews, 1978. Astronomy.

Ebbets, Dennis C., Ph.D., Colorado, 1977. Astrophysics.

MacGregor, Keith B., Ph.D., MIT, 1977. Astronomy.

McClintock, William E., Ph.D., Johns Hopkins, 1977. Spectroscopy.

Stewart, Glen R., Ph.D., UCLA, 1982. Planetary science.

Woods, Thomas N., Ph.D., Johns Hopkins, 1985. Solar physics.

RESEARCH SPECIALTIES AND STAFF

(Full-time faculty only)

Theoretical

Astrophysics. Armitage, Begelman, Gnedin, Hamilton, McCray, Perna, Rast, Shull, Toomre, Zweibel.

Atmospheric/Space Physics, Cosmic Rays. Bagenal, Baker, Ergun.

Fluids and Rheology. Toomre.

Planetary Science. Bagenal, Esposito, Pappalardo.

Experimental

Astrophysics. Bally, Burns, Cash, Darling, Duncan, Ellingson, Glenn, Green, Halverson, Schneider, Shull, Snow, Stocke.

Atmospheric/Space Physics, Cosmic Rays. Baker, Barth, Ergun, Esposito, Schneider.

UNIVERSITY OF DENVER

DEPARTMENT OF PHYSICS AND ASTRONOMY

Denver, Colorado 80208-2238
http://www.physics.du.edu

Students Accepted For Degree	FIELDS Physics	Astronomy*	Related Fields
Doctorate	X	X	
Master's	X	X	

*M.S. and Ph.D. in Physics, subfield Astronomy, awarded.

1. General

Chancellor: Robert D. Coombe
Associate Provost for Graduate Studies: Barbara Wilcots
Associate Provost for Research: Cathryn Potter
Department Chairman: Davor Balzar
Department Telephone Number: (303) 871-2238
Web Address: http://www.physics.du.edu
Type of Institution: University
Control: Private
Setting: Suburban
Total Faculty: 612*
Total Graduate Faculty: n/a
Total Students: 11,770
Total Graduate Students: 5,376
Annual Graduate Tuition: **
All Graduate Students: Full-time—$23,736–26,703
Part-time—$989/credit
Tuition rates for: 2010–11
Deferred tuition plan: Yes
Other Fees:
Health Fee: $130/quarter (not mandatory)
Student technology fee: $4/credit hour
Graduate Activity Fee: $50/qtr.
Term: Quarter

*Full time equivalent faculty
**Teaching and research assistants receive full tuition waivers for normal academic loads; the health fee is paid by the university for full-time graduate assistants.

2. Number of Faculty in Department

The combined total of full-time faculty (Prof., Assoc. & Asst.) in the three professorial ranks is 10. The combined total of full-time, part-time, and other faculty at all ranks is 21.

3. Admission, Financial Aid, and Housing

Address admissions inquiries to: Davor Balzar, Chairman, Dept. of Physics and Astronomy.
Graduate application fee required: $60
Admission deadline (Fall admission): 3/1
Admission information: For fall admission, 2010–11, 10 students were accepted from 24 applicants.
Admission requirements: For admission to the graduate programs, a Bachelor's degree in physics (or a physics and mathematics background equivalent to that required for a B.S. degree in physics) is required with a minimum undergraduate GPA of 3.0 specified. Applicants with GPA less than 3.0 may be considered as individual cases. The GRE is required, with a minimum Quantitative Score of 550. Admission preference will be given to those submitting strong GRE Advanced Physics scores. Students from non-English speaking countries are required to demonstrate proficiency in English via the TOEFL exam (minimum acceptable score-550 paper-based; 213 computer-based), or ibT (minimum acceptable score-80); all those who wish to become teaching assistants must demonstrate oral English proficiencies via the ibT (minimum acceptable score-26 on the speaking section) or the TSE exam (minimum acceptable score-50).
Undergraduate preparation assumed: Taylor, *Classical Mechanics*; Griffiths, *Electromagnetism*; Liboff, *Introductory Quantum Mechanics*; advanced undergraduate laboratory; mathematics including differential equations, vector calculus, and linear algebra.
Address financial aid inquiries to: Davor Balzar (address as above). Fax: (303) 871-4405. Internet: physgradinfo@du.edu
GAPSFAS application required: No
Financial aid deadline: 3/1
Loans available: Yes
Address housing inquiries to: Department of Housing, University of Denver (housing@du.edu)
On-campus, single student housing available: Yes
Cost/qtr.: $2,581–2,972
On-campus, married student housing available: Yes
Cost/qtr.: $1,936–2,778

Table A—Faculty, Enrollments, and Degrees Granted

Research Specialty	2009–10 Faculty	Enrollment Fall 2009 Master's	Doctorate	No. of Degrees Granted 2009–10 (2005–10) Master's	Terminal Master's	Doctorate	Median No. of Years for 2009–10 Ph.D.'s
Astrophysics	5	1	5	0(0)	0(1)	0(1)	–
Atmos./Space Phys., Cosmic Rays	4	0	1	0(0)	0(0)	0(0)	–
Condensed Matter Physics	6	0	9	0(0)	1(4)	0(0)	–
Other	3	0	0	0(0)	0(0)	0(0)	–
Biophysics	3	0	2	0(0)	0(0)	0(0)	–
Total		1	17	0(0)	1(5)	0(1)	
Full-time Grad. Stud.		1	17				
Part-time Grad. Stud.		0	0				
First-year Grad. Stud.		1	3				
Median Years in Grad. Study (2009–10 Degrees)				–	–	4	–

Undergraduate Degrees, 2009–10 (2005–10): 5(15)

4. Graduate Degree Requirements

Master's: Option I (Research Thesis): 45 quarter hours in an approved course of study, up to 10 hours of which may be in thesis research; an acceptable thesis; oral final examination (primarily a thesis defense). Option II (No thesis): 45 quarter hours in an approved course of study; oral final examination covering course work. COMMON REQUIREMENTS: Comprehensive Examination. Residence: Enrollment as a graduate student in the University for at least three quarters. There is no departmental foreign language requirement.

Doctorate: Minimum of three years of full-time study beyond the baccalaureate degree, with at least 90 quarter hours of approved graduate credit; acceptable dissertation; comprehensive examination. There is no departmental foreign language requirement. Residence requirement: Enrollment as a graduate student in the University for at least six quarters including at least two consecutive quarters of full-time attendance.

Other Programs: M.S. (Applied Physics) for baccalaureate engineers. For those currently employed, research projects will be matched whenever possible to the employers' programs. Requirements similar to conventional M.S.

Thesis: Thesis may be written *in absentia*.

Special Equipment, Facilities, or Programs: The department has major research programs in: (a) circumstellar material surrounding particularly massive stars; (b) optical and infrared astronomy, using the department's observatory at the summit of Mt. Evans (4,313 m) and other facilities and data sources; (c) condensed matter and materials physics, focusing on using nanofabrication techniques to control and measure thermal, magnetic and electronic properties of thin films and nanostructures down to 300 mK; studies of high resolution microcalorimeter x-ray and gamma-ray detectors; research on transport and recombination effects of carriers in semiconductors, photoconductors, and low-dimensional structures, such as nanoparticles and nanotubes; studies of organic semiconductors, in particular the development of organic photovoltaic devices or "plastic solar cells" for low-cost solar energy harvesting; nanoscale research on ferroelectrics, nano- and biomaterials using x-ray and neutron diffraction and AFM techniques; studies of mechanical properties of environmentally interesting materials. Some graduate research assistantships may be available in cooperation with several nearby Federal laboratories, in particular the National Institute of Standards and Technology (NIST) and the National Renewable Energy Laboratory (NREL).

Table B—Appointments to Graduate Students, 2009–10

Title of Appointee	Appointments Total	Appointments First year	Academic Load Allowed in Credit Hours	Hours of Service Per Week	Stipend for Academic Year ($)
			Quarter		
Teaching Assistant	6	4	27 per AY[1]	20 (max)	18,000[2,3]
Research Assistant	12	0	27 per AY[1]	20 (max)	18,000[2,3]
Total	18	4			

[1]Tuition waived.
[2]Summer research appointments.
[3]Stipend for Academic year 2010-11 will be $19,300.

5. Personnel Engaged in Separately Budgeted Research, 5/09–4/10

Professorial faculty	10
Other faculty	2
Graduate students	18
Undergraduate students	14
Nonteaching research personnel	3
Total	47

6. Separately Budgeted Research Expenditures by Source of Support

	Departmental Research
Federal government	$1,021,697
Total	$1,021,697

Table C—Separately Budgeted Research Expenditures

Research Specialty	No. of Grants	Expenditures ($)
Astronomy	18[1]	392,200
Atmos./Space Phys., Cosmic Rays	5[2]	157,998
Condensed Matter, Materials Science	10	425,412
Other	1	46,087
Total	34	1,021,697

[1]Research supported in part by proceeds from a major bequest, not included in total.
[2]Excludes grants supporting physics research in other University unit(s).

FACULTY

Professor

Stencel, Robert E., Ph.D., Michigan, 1977. Astronomy and astrophysics. Womble Chair in Astrophysics.

Associate Professors

Balzar, Davor, Ph.D., Zagreb, Croatia, 1993. Condensed matter physics, materials science, Chair of the Department.
Calbi, M. Mercedes, Ph.D., Buenos Aires, Argentina, 2000. Condensed matter physics.
Shaheen, Sean, Ph.D., Arizona, 1999. Condensed matter physics; biophysics.

Assistant Professors

Ghosh, Kingshuk, Ph.D., Massachusetts, Amherst, 2003. Biophysics.
Hoffman, Jennifer, Ph.D., Wisconsin-Madison, 2002. Astrophysics.
Loerke, Dinah, Ph.D., Göttingen, Germany, 2004. Biophysics.
Siemens, Mark, Ph.D., Colorado, 2009. Condensed matter physics.
Ueta, Toshiya, Ph.D., Illinois Urbana-Champaign, 2002. Astronomy and astrophysics.
Zink, Barry, Ph.D., UC San Diego, 2002. Condensed matter physics; materials science.

Adjunct Professor

Trott, David A., B.S., Colorado. Astronomy.

Research Professors

Amme, Robert C., Ph.D., Iowa State, 1958. Atomic and molecular physics; environmental materials; pure and applied mechanics, condensed matter physics.
Norris, Jay, Ph.D., Maryland, 1983. Astrophysics.
Ormes, Jonathan F., Ph.D., Minnesota, 1967. High-energy astrophysics.

Lecturer

Iona, Steven, Ph.D., Denver, 1994. Physics Education.

Professors Emeriti

Goldman, Aaron, D.Sc., Technion-Israel Inst. of Tech., 1965. John Evans Professor. Atmospheric physics; atomic and molecular spectroscopy.
Neumann, Herschel, Ph.D., Nebraska, 1965. Physics education.
Olson, John R., Ph.D., Iowa State, 1963. Atmospheric physics; acoustics; electronics.
van der Merwe, Alwyn J., Ph.D., Bern, Switzerland, 1971. Intermolecular forces; history and foundations of physics.
Van Zyl, Bert, Ph.D., Washington, 1963. Atomic and molecular physics.
Williams, Walter John, M.S., Denver, 1963. Upper atmospheric physics.

RESEARCH SPECIALTIES AND STAFF

Theoretical/Computational

Atmospheric molecular spectroscopy. Goldman.
Biophysics. Computational studies of protein-protein interactions. Ghosh.
Condensed Matter Physics. Gas adsorption on nanostructures. Calbi.

Experimental

Astronomy and Astrophysics. Hoffman, Norris, Ormes, Stencel, Ueta.
Biophysics. Complex systems and nonlinear behavior of dynamical networks; molecular dynamics of actin. Loerke, Shaheen.
Condensed Matter Physics. Materials physics and science; applied physics; low-temperature physics; thin films, and nanostructures; measurements of thermal, magnetic, electronic and mechanical properties; organic semiconductors; spintronics; strains and defects; x-ray and neutron diffraction; materials structure; compaction of granular materials, ultra-fast laser optics. Amme, Balzar, Shaheen, Siemens, Zink.
Cosmic Rays. Ormes.

UNIVERSITY OF CONNECTICUT

DEPARTMENT OF PHYSICS

Storrs, Connecticut 06269-3046

Students Accepted For Degree	FIELDS		
	Physics	Astronomy	Related Fields
Doctorate	X		
Master's	X		

1. General

President: Phillip Austin
Vice Provost for Research and Graduate Education: Suman Singha
Department Head: William C. Stwalley
Department Telephone Number: (860) 486-4924
Type of Institution: University
Control: Public
Setting: Small town
Total Faculty: 1,294
Total Graduate Faculty: 1,100
Total Students: 28,677
Total Graduate Students: 6,425
Annual Graduate Tuition:
In-state residents: Full-time—$9,972
Part-time—$554/credit
New England Regional: Full-time—$17,451
Part-time—$970/credit
Out-of-state residents: Full-time—$25,884
Part-time—$1,438/credit
Tuition rates for: 2010–11
Deferred tuition plan: No
Other Fees: General university fee, $204/part-time, $408/half-time, $612/full-time; Activity fee, $13/part-time, $61/half-time and full-time; Matriculation fee, $42; Maintenance fee, $54/part-time, $107/half-time, $213/full-time
Term: Semester

2. Number of Faculty in Department

The combined total of full-time graduate faculty in the three professorial ranks is 25. The combined total of full-time, part-time, and other faculty at all ranks is 67.

3. Admission, Financial Aid, and Housing

Address admission inquiries to:
GRADPHYSICS@UCONN.EDU
Graduate application fee required: $55 online application; $75 paper application
Domestic admission deadline (Fall admission): 6/1
International admission deadline (Fall admission): 6/1
Admission information: For fall admission, 2010–11, 27 students were accepted from 150 applicants.
Admission requirements: For admission to the graduate programs, a Bachelor's degree in physics or a closely allied field with a sufficient concentration in physics is required with a minimum undergraduate GPA of 3.0 specified. Non-degree students may register under the Extended Education Program. Transfer to the Graduate Program may be possible for those with adequate academic performance. The GRE General Test is required for any student who intends to apply for a fellowship and/or assistantship. Three letters of recommendation are also required. No minimum acceptable score is specified. The GRE Advanced is recommended. No minimum acceptable score for admission specified. Students from non-English speaking countries are required to demonstrate proficiency in English via the TOEFL exam. The Graduate School minimum acceptable score for admissions is 550; the normal Department minimum is 600. Foreign students must apply by April 1 and provide an affidavit that they will be able to finance all costs. In exceptional circumstances a later application can be considered.
Address financial aid inquiries to: William C. Stwalley, Head, U-3046
GAPSFAS application required: No
Financial aid deadline: 5/1, need-based aid
4/15, academic merit awards
Loans available: Yes
Address housing inquiries to: Office of the Graduate School, U-1006
On-campus, single student housing available: Yes
Cost/term: $3,401.00
On-campus, married student housing available: Yes

Table A—Faculty, Enrollments, and Degrees Granted

Research Specialty	2008–09 Faculty	Enrollment[1] Fall 2008		No. of Degrees Granted[2] 2009–10 (2005–10)			Median No. of Years for 2009–10 Ph.D.'s
		Master's	Doctorate	Master's	Terminal Master's	Doctorate	
Astrophysics	1	0	2	0(0)	0(0)	0(0)	–
Atomic, Molecular, & Optical Physics	7	0	36	5(12)	1(5)	3(13)	6
Condensed Matter Physics	7	0	19	2(10)	0(4)	0(13)	–
GeoPhysics	1	0	5	2(2)	1(1)	0(0)	–
Nuclear Physics	3	0	14	0(5)	0(0)	0(5)	–
Particles & Fields	4	0	13	2(4)	0(2)	4(5)	7
Polymer Physics/ Science	1	0	1	0(2)	0(0)	1(2)	6
Relativity & Gravitation	1	0	1	0(1)	0(1)	0(1)	–
Non-specialized	0	0	0	0(0)	0(0)	0(0)	–
Total	25	0	91	11(36)	2(13)	8(39)	
Full-time Grad. Stud.			91				
Part-time Grad. Stud.			0				
First-year Grad. Stud.			12				
Median Years in Grad. Study (2009-10 Degrees) (2005–10)							6.5
Undergraduate Degrees, 2009–10 (2005–10): 11(53)							

[1]Students not yet committed to a research specialty are entered under non-specialized.
[2]Five-year totals in parentheses.

4. Graduate Degree Requirements

Master's: Plan A: 15 credits of graduate courses and thesis. Plan B 24 credits of graduate courses; no thesis. Transfer credits not ordinarily accepted. Either of these degrees may (but need not) be part of a Ph.D. program. The courses submitted must be approved in advance by the student's Advisory Committee. An average of "B" or better must be maintained. A final examination is required; it may be written, oral, or both. No foreign language or residency requirement.
Doctorate: The student must complete a plan of study of extent

and quality satisfactory to the student's Advisory Committee and the Dean of the Graduate School. Ordinarily the program will include at least 20 to 24 credits beyond the Master's degree. At least one year must be in residence. The General Examination is usually taken by the end of the fifth semester.

Thesis: Thesis may be written *in absentia*.

Special Equipment, Facilities, or Programs: The Physics Department occupies an 81,350 square foot physics building and 25,000 square feet in the basement of an adjacent Biological Sciences/Physics building which provides research facilities for theoretical and experimental atomic, molecular, and optical physics, condensed-matter physics, and nuclear physics, and theoretical particle physics, astrophysics, and general relativity. In addition, the Institute of Materials Science, adjacent to the Physics Department, provides research facilities for theoretical and experimental condensed-matter physics.

The Physics Department's own computer research network includes three high performance parallel clusters. The Departments Computer Lab provides students and faculty with an access to Linux, Windows, and Mac workstations and a variety of software for numerical analysis, scientific visualization, symbolic processing and program development.

Faculty from the Physics Department have developed joint research programs with Brookhaven National Laboratory, the Thomas Jefferson National Accelerator Facility, the NASA/Caltech jet propulsion laboratory, the Lawrence Livermore National Laboratory, Oak Ridge National Laboratory, the National Institute for Science and Technology, and many other American and foreign universities and research institutions.

Prospective students should consult the brochure "Graduate Education and Research in Physics at the University of Connecticut" which can be requested from the department. It is also available at the website www.phys.uconn.edu.

Table B—Appointments to Graduate Students, 2008–10

Title of Appointee	Appointments Total	Appointments First year	Academic Load Allowed in Credit Hours	Hours of Service Per Week	Stipend for Academic Year ($)
			Semester		
Teaching Assistant	27	6	9	20	19,409–23,343[1]
Research Assistant	21	0	9	20	19,409–23,343[1]
0.5 TA+0.5 RA	16	0	9	20	19,409–23,343[1]
0.5 TA+0.5 Fellow	7	6	12	10	19,409–23,343[1]
0.5 RA+0.5 Fellow	1	0	12	10	19,409–23,343[1]
Total	72	12			

[1]Stipend depends on experience and/or academic level. Tuition is waived and health benefits are in addition to stipend.

5. Personnel Engaged in Separately Budgeted Research, 7/08–6/09

Professorial faculty	24
Research Associates	0
Postdoctoral Fellows	8
Graduate students	91
Total	123

6. Separately Budgeted Research Expenditures by Source of Support

	Departmental Research
Federal government	$2,592,385
State and local government	88,707
Business and industry	86,519
Other (Foundations, Foreign)	37,500
Total	$2,805,111

7. Separately Funded and Managed Laboratories

Institute of Materials Science ~$8,000,000

Table C—Separately Budgeted Research Expenditures

Research Specialty	No. of Grants	Expenditures ($)
Astrophysics	3	300,299
Atomic, Molecular, & Optical Physics	25	1,347,543
Condensed Matter Physics	14	315,774
Geophysics	1	69,475
Nuclear Physics	9	474,917
Particles & Fields	6	259,661
Polymer Physics	1	37,500
Total	59	2,805,111

FACULTY

Professors

Cormier, Vernon F., Ph.D., Columbia, 1976. Professor of geology and geophysics. Wave propagation in deep earth structures.

Côté, Robin, Ph.D., MIT, 1995. Theoretical atomic and molecular physics; ultracold collisions; Bose-Einstein condensation.

Dunne, Gerald V., Ph.D., Imperial College, London, 1988. Theoretical particle physics; quantum field theory; gauge theory.

Dutta, Niloy K., Ph.D., Cornell, 1978. Experimental condensed matter and optical physics; semiconductor laser technology; quantum wires; fiber optic transmission systems.

Eyler, Edward E., Ph.D., Harvard, 1982. Experimental atomic, molecular, and optical physics; precision laser spectroscopy.

Fernando, Gayanath W., Ph.D., Cornell, 1985. Theoretical condensed matter physics; properties of transition metals.

Gibson, George N., Ph.D., U. of Illinois at Chicago, 1990. High-intensity, short-pulse laser physics; laser spectroscopy.

Gould, Phillip L., Ph.D., MIT, 1986. Experimental quantum optics; laser cooling and trapping of atoms.

Hamilton, Douglas S., Ph.D., Wisconsin, 1976. Experimental condensed matter physics; nonlinear optics; light scattering; solid state laser design; dynamics of ions in solids.

Javanainen, Juha, Ph.D., Helsinki, 1981. Theoretical quantum optics; interaction of light with atoms.

Kharchenko, Vasili A., Ph.D., and D. Sc., Ioffe Institute, St. Petersburg, 1977 and 1988. Theoretical physics and x-ray astrophysics; hot atoms in planetary atmospheres; kinetics and theory of atomic collisions.

Kovner, Alex, Ph.D., Tel Aviv, 1985. Theoretical Particle Physics; strongly coupled gauge theories.

Mallett, Ronald L., Ph.D., Penn State, 1973. Theoretical physics; relativity and gravitation; relativistic quantum theory.

Mannheim, Philip D., Ph.D., Weizmann Institute, 1970. Theoretical physics; elementary particle theory; general relativity; astrophysics.

Peterson, Cynthia W., Ph.D., Cornell, 1964. Experimental condensed matter physics; vacuum UV reflection spectroscopy; UV photoemission spectroscopy; biophysics.

Stwalley, William C., Ph.D., Harvard, 1969. Experimental atomic and molecular interactions; laser spectroscopy and dynamics of atoms and molecules; ultracold atoms and molecules.

Wells, Barrett O., Ph.D., Stanford, 1992. Experimental condensed-matter physics; neutron scattering; superconductivity; photoemission.

Associate Professors

Blum, Thomas, Ph.D., University of Arizona, 1995. Theoretical high energy physics, lattice gauge theory, quantum chromodynamics (QCD), electro-weak physics.

Dobrynin, Andrey V., Ph.D., Moscow Institute of Physics and Technology, 1991. Theoretical polymer physics; modeling of self-assembling polymers and polymer/nanoparticle mixtures.

Jones, Richard T., Ph.D., VPI, 1988. Experimental nuclear physics.

Joo, Kyungseon, Ph.D., MIT, 1997. High energy nuclear physics.

Sinkovic, Boris, Ph.D., Hawaii, 1986. Experimental condensed-matter physics; magnetic properties of films, surfaces, and nanostructures.

Yelin, Susanne F., Ph.D., Ludwig-Maximilians University (Munich), 1998. Theoretical quantum optics and condensed matter physics; spin physics in semiconductors; quantum coherence and quantum information; nonlinear optics in atoms and semiconductors.

Assistant Professors

Jain, Menka, Ph.D., University of Puerto Rico, 2004. Experimental condensed matter physics.

Schweitzer, Peter, Ph.D., University of Bochum, 2001. Theoretical nuclear physics.

Faculty Emeriti

Azaroff, Leonid V., Ph.D., MIT, 1954. Experimental condensed matter physics; metal physics; x-ray crystallography.

Bartram, Ralph H., Ph.D., NYU, 1960. Theoretical condensed matter physics; optical and magnetic properties of point imperfections in solids.

Best, Philip E., Ph.D., Western Australia, 1962. Experimental surface physics; electron scattering.

Budnick, Joseph I., Ph.D., Rutgers, 1955. Experimental condensed matter physics; nuclear magnetic resonance; critical phenomena.

Gilliam, O. R., Ph.D., Duke, 1950. Experimental condensed matter physics; electron spin resonance.

Hahn, Yukap, Ph.D., Yale, 1962. Theoretical atomic, molecular and optical physics.

Hayden, Howard, Ph.D., University of Denver, 1967. Experimental condensed matter physics.

Hines, William A., Ph.D., California, Berkeley, 1967. Experimental condensed matter physics; nuclear magnetic resonance and magnetization studies of metals and alloys.

Islam, M. M., Ph.D., Imperial College, London, 1961. Theoretical physics; high-energy scattering; nucleon structure.

Kappers, Lawrence A., Ph.D., Missouri, Columbia, 1970. Experimental condensed matter physics; color centers; optical properties; radiation damage.

Kessel, Quentin, Ph.D., Connecticut, 1966. Experimental atomic and molecular physics; ionization; x-rays; Auger electrons; laboratory astrophysics.

Klemens, Paul G., D.Phil., Oxford, 1950. Theoretical condensed matter physics; conduction properties.

Markowitz, David, Ph.D., Illinois, 1963. Theoretical condensed matter physics.

Otter, Fred A., Ph.D., Illinois, 1959. Experimental characterization and modification of metal and semiconductor surfaces and thin films. Structural, mechanical, electrical, and optical properties.

Pease, Douglas M., Ph.D., Connecticut, 1972. Experimental condensed matter physics; x-ray studies of alloys.

Rawitscher, George, Ph.D., Stanford, 1956. Theoretical physics; nuclear reaction and electron-nucleus scattering.

Russek, Arnold, Ph.D., New York University, 1953. Theoretical atomic, molecular and optical physics.

Schor, Robert, Ph.D., Michigan, 1958. Theoretical physics; chemical physics; biophysics; macromolecules.

Smith, Winthrop W., Ph.D., MIT, 1963. Experimental atomic physics; ion-atom collisions; beam-foil spectroscopy; laser spectroscopy; laboratory astrophysics.

Associated UConn Faculty

Birge, Robert R., Ph.D., Wesleyan, 1972. Schwenk Professor of Chemistry. Biomolecular electronics; biomolecular spectroscopy.

Boggs, Steven A., Ph.D., Toronto, 1972. Research Professor of Materials Science. High voltage dielectrics.

Campagnola, Paul, Ph.D., Yale, 1992. Assistant Professor of Physiology, UConn Health Center. Laser Biophysics.

Edson, James, Ph.D., Penn State, 1989. Associate Professor of Marine Sciences. Boundary layer meteorology with a focus on surface layer turbulence and air-sea interaction.

Huber, Greg, Ph.D., Boston University, 1993. Associate Professor of Cell Biology. Biological physics and mechanics, biocomplexity, soft matter physics, nonequilibrium and nonlinear dynamics.

Liu, Lambo, Ph.D., Stanford, 1993. Associate Professor of Civil and Environmental Engineering. Applied and computational geophysics, continental tectonophysics.

Michels, H. Harvey, Ph.D., Delaware, 1960. Research Professor of Physics. Theoretical atomic and molecular physics.

Montgomery, John A., Ph.D., Columbia, 1978. Research Professor of Physics. Computational molecular physics.

O'Donnell, James, Ph.D., University of Delaware, 1986. Professor of Marine Science. Physics of the coastal ocean, environmental fluid dynamics, mathematical models of environmental processes.

Papadimitrakopoulos, Fotios, Ph.D., Massachusetts, 1993. Associate Professor of Chemistry. Self-assembly of organic, inorganic, biological, and hybrid nanostructures, devices and sensors; organic semiconductors; II-VI and Si nanocrystals; carbon nanotubes.

Roychoudhuri, Chandrasekhar, Ph.D., Rochester, 1973. Research Professor of Physics. Experimental optical physics; semiconductor laser technology.

Schweitzer, Jeffrey S., Ph.D., Purdue, 1972. Research Professor of Physics. Experimental nuclear physics; nuclear astrophysics; solar physics.

Wolgemuth, Charles W., Ph.D., Arizona, 2000. Assistant Professor of Physiology, UConn Health Center. Physics of cellular motility; Morphology, growth, and pattern formation.

Adjunct Faculty

Bahns, John T., Ph.D., Iowa, 1983. Experimental atomic and molecular interactions; laser spectroscopy and dynamics of atoms and molecules. Research Scientist, Northern Illinois Univ.

Bates, Stephen C., Sc.D., MIT, 1977. Experimental condensed-matter physics, President, Thoughtventions Unlimited LLC.

Deveney, Edward F., Ph.D., Connecticut, 1993. Experimental atomic and molecular physics. Assistant Professor, Bridgewater State College.

Fenner, David B., Ph.D., Washington, 1976. Experimental condensed matter physics. Vice President and Director of Research, Epion Corp.

Kussow, Adil-Gerai, Ph.D., A.F. Ioffe Institute of Physics and Technology (St. Petersburg) 1977. Solid state theory: electron excitations, magnetic properties of thin films, theory of defects and fracture, first-principle electronic structure calculations. UMass-Amherst.

Lyyra, A. Marjatta, Ph.D., Stockholm, 1979. Experimental atomic, molecular, and optical physics. Professor of Physics, Temple University.

Nesbet, Robert K., Ph.D., Cambridge, 1954. Computational physics; theoretical chemistry; variational theory; electron scattering by atoms and molecules; electronic structure of atoms, molecules and solids; magnetic impurities and magnetoresistance. IBM Research Division, Almaden, CA.

Pichler, Marin, Ph.D., Connecticut, 2001. Experimental atomic, molecular and optical physics. Assistant Professor of Physics, Goucher College, Baltimore, MD.

Ramsey-Musolf, Michael J., Ph.D., Princeton, 1989. Theoretical nuclear physics; electroweak interactions in nuclei; low energy QCD. Professor of Physics, University of Wisconsin.

Wang, Henry, Ph.D., Iowa, 1991. Experimental atomic and molecular interactions; laser spectroscopy and dynamics of atoms and molecules. Research Scientist, Aerospace Corp.

Zemke, Warren, Ph.D., Illinois Institute of Technology, 1969. Theoretical alkali, atom interactions. Professor of Chemistry, Wartburg College, Waverly, Iowa.

RESEARCH SPECIALTIES AND STAFF

Theoretical

Astrophysics. Dark matter, dark energy, and the cosmological constant problem. Interplay of cosmology and particle physics. Applications of standard and alternate gravity theories to astrophysics and cosmology. Astrophysical plasmas and neutral gas. Physics of the Heliosphere. New astrophysical sources of X-ray emission. Applications of quantum collision theory to the physics of planetary atmospheres and interstellar gas. Kharchenko, Mallett, Mannheim.

Atomic, Molecular and Optical Physics. Atomic and molecular structure. Michels, Montgomery. Interaction of light with atoms. Javanainen, Yelin. Bose-Einstein condensation. Javanainen, Côté. Ultracold collisions. Côté. Quantum coherence and quantum information. Côté, Javanainen, Yelin.

Condensed Matter Physics. Electronic properties of point imperfections in ionic solids. Bartram. Thermal and electronic properties of transition metals. Fernando. Thermal conduction by lattice waves and electrons. Klemens. Spin physics in semiconductors. Yelin. Light-matter interactions. Yelin.

Elementary Particles and Fields. Structure and interactions of elementary particles; properties of gauge fields; interplay of particle physics and astrophysics; lattice gauge theory. Blum, Dunne, Kovner, Mannheim.

General Relativity. Physical processes in strong and weak gravitational fields. Black holes. Quantum Gravity. Theory of closed timelike curves. Mallett.

Mathematical Physics. Quantum field theory; zeta functions and spectral theory; Dunne. Numerical solution of the Faddeev integral equations. Rawitscher. Computational Physics. Computerized Modeling in the Sciences. Rawitscher.

Nuclear Physics. Scattering and rearrangement reactions of projectiles on nuclei. Rawitscher. Lattice gauge theory. Blum. Saturation physics. Kovner. Theory and phenomenology of strong interactions. P. Schweitzer.

Theory. Phenomenology of strong interactions. P. Schweitzer.

Polymer Physics. Self-assembling polymers; polymer/nanoparticle mixtures. Dobrynin.

Relativity. Physical processes in strong and weak gravitational fields. Mallett.

Experimental

Astrophysics. X-ray emission spectra from collisions of highly-charged ions found in the solar wind. Kessel, Smith.

Atomic, Molecular and Optical Physics. Cold ion-neutral collisions. Smith. Quasimolecular processes in heavy ion collisions. Kessel. Laser spectroscopy and atomic and molecular interactions. Stwalley. Laser cooling and trapping of atoms. Gould. High-intensity, short-pulse laser physics. Gibson. Precision laser spectroscopy. Eyler. Ultracold molecules and plasmas. Eyler, Gould, Stwalley.

Condensed Matter Physics. Electronic structure of alloys. Pease. Magnetically ordered systems. Budnick, Sinkovic. High-temperature superconductivity. Budnick, Wells. Surface modification. Kessel. Point defects in nonmetallic crystals via ESR. Gilliam, Kappers. NMR and magnetization studies of metals and alloys. Hines. Point defects in ionic crystals via optical methods. Kappers. Laser spectroscopy of solids. Hamilton. Angular dependent electron spectroscopy from surfaces. Best. Semiconductor laser technology and fiber optic transmission systems. Dutta. Neutron scattering. Wells. Photoemission. Sinkovic, Wells. X-ray absorption spectroscopy. Budnick, Pease. Very high peak power ps diodes for material processing; broadly tunable diodes and novel mux/demux devices for DWDM. Roychoudhuri.

Geophysics. Deep earth and planetary structure; elastic wave propagation; earthquake source properties. Cormier.

Nuclear Physics. Nuclear astrophysics, nuclear structure. Joo, Schweitzer. The structure of the nucleon, low-energy QCD. Jones, Joo, Schweitzer. Meson spectroscopy, Jones. Nanoscale structure and chemistry. J. Schweitzer. Planetary science and forensics. J. Schweitzer.

WESLEYAN UNIVERSITY

DEPARTMENT OF PHYSICS

Middletown, Connecticut 06459-0155

Students Accepted For Degree	FIELDS		
	Physics	Astronomy	Related Fields
Doctorate	X		
Master's			

Table A—Faculty, Enrollments, and Degrees Granted

Research Specialty	2008–09 Faculty	Enrollment[1] Fall 2008		No. of Degrees Granted[2] 2008–09 (2004–09)			Median No. of Years for 2008–09 Ph.D.'s
		Mas-ter's	Doc-torate	Mas-ter's	Terminal Master's	Doc-torate	
Atomic, Molecular, & Optical Physics	3	1	5	1(1)	0(2)	0(3)	6
Condensed Matter Physics	4	0	7	1(3)	0(6)	0(3)	5
Nonlinear Theory	2	0	1	0(0)	0(4)	0(3)	5
Total		1	13		0(12)	0(9)	
Full-time Grad. Stud.		1	13				
Part-time Grad. Stud.		0	0				
First-year Grad. Stud.		0	1				
Median Years in Grad. Study (2008–09 Degrees)				–	–	6	–
Undergraduate Degrees, 2008–09 (2004–09): 17(124)							

[1]Students not yet committed to a research specialty are entered under non-specialized.
[2]Five-year totals in parentheses.

1. General

President: Michael S. Roth
Dean of Graduate Studies: Ishita Mukerji
Department Chairman: Brian Stewart
Department Telephone Number: (860) 685-2030
Type of Institution: University
Control: Private
Total Faculty: 368
Total Graduate Faculty: 111
Total Students: 2,709
Total Graduate Students: 180
Annual Graduate Tuition: 2010–11
All Graduate Students: Full-time—Tuition covered by stipend
Deferred tuition plan: No
Other Fees: $682
Term: Semester

2. Number of Faculty in Department

The combined total of full-time faculty in the three professorial ranks is 9. The combined total of full-time, part-time, and other faculty at all ranks is 11.

3. Admission, Financial Aid, and Housing

Address admission inquiries to: Chairman, Dept. of Physics
Graduate application fee required: No
Admission deadline (Fall admission): 1/31
Admission information: For fall admission, 2010–11, 1 student were accepted from 65 applicants.
Admission requirements: For admission to the graduate programs, a Bachelor's degree in physics is required with no minimum undergraduate GPA specified. The GRE is required. The GRE Advanced is recommended. Students from non-English speaking countries are required to demonstrate proficiency in English via the TOEFL exam. Minimum acceptable score for admission is 83 (internet based), 220 (computer based), or 560 (paper based).
Financial Aid: All students accepted receive full support, including a tuition waiver.
GAPSFAS application required: No
Financial aid deadline: None
Loans available: Yes
Address housing inquiries to: Amy Miller, Office of Residential Life, North College, Middletown, CT 06459
On-campus, single student housing available: Yes
Cost/month: $614
On-campus, married student housing available: Yes
Cost/month: $730–946 (utilities included)

4. Graduate Degree Requirements

Master's: A minimum of eight credits (typically six one-semester courses and two credits in research) with grades of B^- or better. A thesis and oral defense are also required. Course selection is flexible and is done in consultation with members of the student's committee.

Doctorate: While there are no specific course requirements for the Ph.D. degree, each student is required to take one course during every semester of residence. Students must have demonstrated proficiency in the main subject areas of physics before completion of the program. These areas embody quantum theory, including atomic, nuclear, and elementary particle physics, electromagnetism and optics, classical dynamics and relativity theory, thermal, statistical, and solid state physics. Three formal examinations serve to define the student's progress toward the degree. The first, taken during September of the second year, is a short written exam on material at an advanced undergraduate level. During the spring of the second year each student presents to his or her graduate advisory committee a description and defense of a specific research topic. Finally, the dissertation, based on original and significant research, must be defended in an oral examination. There are no foreign language requirements. The spirit of the program is to give the student an early opportunity to become a recognized and significant member of the department. Emphasis is placed on having students "do physics" right from the start, rather than spend one or two years solely on course work before getting into research. To this end, graduate students are expected to join in the research activities of the department upon arrival.

Thesis: *Thesis may be written in absentia.*

Table B—Appointments to Graduate Students, 2009–10

Title of Appointee	Appointments Total	Appointments First year	Academic Load Allowed in Credit Hours	Hours of Service Per Week	Stipend for Academic Year ($)
			Semester		
Research Assistant	3	0	3	20	17,600[1]
1/2 T A+1/2 RA	11	3	3	20	17,600[1]
Total	14	3			

[1]All students receive an additional $5,866 summer research stipend and full tuition waiver.

5. Personnel Engaged in Separately Budgeted Research, 7/08–6/09

Professorial faculty	8
Graduate students	13
Undergraduate students	12
Total	33

6. Separately Budgeted Research Expenditures by Source of Support

	Departmental Research
Outside Funding	$750,000
University	225,000
Total	$975,000

Table C—Separately Budgeted Research Expenditures

Research Specialty	No. of Grants	Expenditures ($)
Atomic, Molecular, & Optical Physics	3	100,000
Condensed Matter Physics	5	540,000
Nonlinear Theory	3	335,000
Total	12	975,000

FACULTY

Professors

Blümel, Reinhold, Ph.D., Technical Univ. of Munich, 1983. Chairman. Theoretical atomic, molecular, optical, and nuclear physics.

Ellis, Fred M., Ph.D., University of Massachusetts, Amherst, 1983. Low-temperature physics.

Hüwel, Lutz, Ph.D., Georg-August Univ., Göttingen, 1980. Atomic and molecular physics.

Morgan, Thomas J., Ph.D., California, Berkeley, 1971. Atomic and molecular physics.

Associate Professor

Starr, Francis, Ph.D., Boston U., 1999. Theoretical soft condensed matter physics, biophysics.

Stewart, Brian, Ph.D., MIT, 1987. Molecular physics.

Voth, Greg, Ph.D., Cornell, 2000. Soft condensed matter.

Assistant Professor

Kottos, Tsampikos, Ph.D., Univ. of Crete, Greece, 1997. Mesocopic Physics and Quantum Chaos.

Othon, Chrstina, Ph.D., Nebraska, 2005. Biophysics & Biochemistry.

Visiting Associate Professor

Westling, Lynn, Ph.D., University of Rochester, 1986. Physics. Pedagogy and the Feminist Critique of Physics.

RESEARCH SPECIALTIES AND STAFF

Experimental

Atomic and Molecular Physics. Laser spectroscopy and collisional studies of laser-prepared Rydberg atoms using fast beams and near thermal beams; nonlinear spectroscopy; fragmentation of highly excited molecules; properties of laser generated sparks; dynamics of inelastic atom-diatomic molecule collisions. These studies are supported by a laser facility supported by the NSF. Hüwel, Morgan, Stewart.

Low-Temperature Condensed Matter, Soft Matter Physics. Physical adsorption; surface physics; molecular motions in adsorbed monolayers using NMR spectroscopy; surface phase transitions; quantum fluids; properties of superfluid films. Ellis, Rollefson.

Soft Condensed Matter Physics, Nonlinear Dynamics, Granular and Fluid Flows, Biophysics. Voth, Othon.

Theoretical

Atomic Physics. Quantum chaos; driven Rydberg systems; atom and ion traps; Bose-Einstein condensation, quantum graphs. Blümel.

Mesoscopic Physics and Bose-Einstein Condensation. Random matrix theory, semi-classics, quantum graphs, quantum dissipation, chaotic scattering, disordered systems. Kottos.

Soft Condensed Matter Physics; Biophysics DNA-based nanomaterials; nanocomposite materials; glass transitions; self-assembly; computational physics. Starr.

YALE UNIVERSITY

DEPARTMENT OF PHYSICS

P. O. Box 208120
New Haven, Connecticut 06520-8120

Students Accepted For Degree	FIELDS		
	Physics	Astronomy	Related Fields
Doctorate	X		
Master's			

1. General

President: Richard C. Levin
Dean of the Graduate School: Jon Butler
Department Chairman: Meg Urry
Department Telephone Number: (203) 432-3650
World Wide Web: www.yale.edu/physics
Type of Institution: University
Control: Private
Setting: Urban
Total Faculty: 3,478
Total Graduate Faculty: Not separated
Total Students: 11,454
Total Graduate Students: 6,143 (incl. Professional)
Annual Graduate Tuition:
All Graduate Students are provided with full tuition fellowship.
Term: Semester

2. Number of Faculty in Department

The combined total of full-time faculty in the three professorial ranks is 56, including 10 whose primary appointment is in Applied Physics. The combined total of full-time, part-time, and other faculty at all ranks is 58, including 10 whose primary appointment is in Applied Physics, and 14 with primary appointments in other departments.

3. Admission, Financial Aid, and Housing

Address admission inquiries to: Director of Graduate Studies, 35 Sloane Physics Laboratory, 217 Prospect St., P.O. Box 208120, New Haven, CT 06520-8120
E-mail address: graduatephysics@yale.edu
Graduate application fee required: $95
Admission deadline (Fall admission): 1/2
Admission information: For fall admission, 2008–09, 40 students were offered admission from 273 applicants.
Admission requirements: For admission to the graduate programs, a Bachelor's degree is required. No minimum undergraduate GPA specified. The GRE is required with no minimum acceptable score specified. The GRE Advanced is required. No minimum acceptable score specified. Students from non-English speaking countries are required to demonstrate proficiency in English via the TOEFL exam. Minimum score required is 250 (CBT). The TSE is also recommended.
Address financial aid inquiries to: Admissions, Yale Graduate School, P.O. Box 208323, New Haven, CT 06520-8323.
GAPSFAS application required: No, but is required if requesting a loan.
Financial aid deadline: 1/3
Loans available: Yes
Address housing inquiries to: Graduate Housing Dept., Yale University, 155 Whitney Ave., P.O. Box 208316, New Haven, CT 06520-8316
On-campus, dormitory housing available: Yes
Cost/year: $4,200–7,160
On-campus, apartment housing available: Yes
Cost/month: $794–1,114

Table A—Faculty, Enrollments, and Degrees Granted

Research Specialty	2008–09 Faculty	Enrollment[1] Fall 2008		No. of Degrees Granted[2] 2008–09(2003–08)			Median No. of Years for 2007–08 Ph.D.'s
		Master's	Doctorate	Master's	Terminal Master's	Doctorate	
Astrophysics	8	0	5	2(4)	0(0)	0(2)	6
Atomic, Molecular, & Optical Physics	4	0	28	3(17)	0(0)	2(13)	7.5
Chemical Physics	2	0	0	0	0(1)	0	–
Condensed Matter Physics	23	0	41	942)	0(0)	1(22)	6.1
Engineering Physics/ Science	0	0	0	0(0)	0(0)	0(0)	6.3
Nuclear Physics	7	0	18	2(19)	0(0)	2(17)	6
Particles & Fields	11	0	12	4(10)	0(0)	0(5)	2
Physics of Beams	1	0	0	0(1)	0(0)	0(0)	–
Relativity & Gravitation	1	0	0	0(0)	0(0)	1(1)	8
Other Experimental	9	0	15	6(16)	0(0)	1(10)	7
Other Theoretical/ Math.	2	0	11	0(0)	0(0)	0(0)	–
Non-specialized	0	0	4	0(23)	0(2)	2(0)	5
Scientific Computation	2	0	0	0(0)	0(0)	0(0)	–
Total		–	74	1(64)	0(3)	16(37)	
Full-time Grad. Stud.		–	134				
Part-time Grad. Stud.		–	0				
First-year Grad. Stud.		–	10				
Median Years in Grad. Study (2008–09 Degrees)				–	–	–	4.75

Undergraduate Degrees, 2007–08 (2003–08): 32(109)

[1]Students not yet committed to a research specialty are entered under non-specialized.
[2]Five-year totals in parentheses.

4. Graduate Degree Requirements

Master's: Students who have successfully advanced to candidacy may petition for the award of M.Phil. degree; no admissions for Master's program only.
Doctorate: Course of Study of three semesters (three courses each semester); graduate seminar; Comprehensive Examination; satisfactory grades (determined by faculty); three years residence; written thesis and oral defense of same before a faculty committee of five.
Thesis: Thesis may be written *in absentia*.
Special Equipment, Facilities, or Programs: Research areas include particle physics, nuclear physics, condensed matter physics, quantum information physics, biological physics, atomic and optical physics, astrophysics, cosmology, and other areas in collaboration with the faculties of engineering and applied science, chemistry, molecular biology and biophysics, mathematics, geology and geophysics, and as-

tronomy. The department occupies the Sloane Physics Laboratory, part of the J. W. Gibbs Laboratories, and the Wright Nuclear Structure Laboratory. Sloane has newly constructed laboratories for research in atomic, molecular, optical, and condensed matter physics. Research on condensed-matter physics is also done in the Department of Applied Physics. The Center for Theoretical Physics is located in Sloane. The Wright Laboratory contains an Extended Stretch Transuranium (ESTU) 20 megavolt tandem electrostatic accelerator. The Wright and Gibbs Laboratories house design facilities used in supporting high-energy experiments at Brookhaven National Laboratories, Fermilab, the Stanford Linear Accelerator Center and the Large Hadron Collider. Experiments are also done at European accelerators and observations taken at South American astronomical observatories. In addition to the centralized University computer system, each research group has its own appropriate computing facilities.

Table B—Appointments to Graduate Students, 2008–09

Title of Appointee	Appointments Total	Appointments First year	Academic Load Allowed in Credit Hours	Hours of Service Per Week	Stipend for Academic Year ($)
			Semester		
Teaching Assistant	65	9	9–12	10	20,925[3]
Research Assistant	91[1]	[2]	–	variable	27,900[4]
Other					6,975[5]
Total	156	91			

[1]Second year students are included in this figure but are not usually appointed into the Assistant in Research positions until summer.
[2]First year students are not included in total as they are appointed to Assistant in Research positions for summer only, usually three months.
[3]9 month academic year stipend (includes university & teaching stipend)
[4]Full 12 month salary for Assistant in Research
[5]Salary for 3 month summer Assistant in Research appointments.

5. Personnel Engaged in Separately Budgeted Research, 2008-09

Professorial faculty	57
Other faculty	4
Postdoctoral appointments	24
Graduate students	106
Total	191

6. Separately Budgeted Research Expenditures by Source of Support

	Departmental Research[1]	Physics-related Research Outside Department
Federal government	$12,764,891	0
Other	3,399,785	0
Total	$16,164,676	$0

[1]Expenditures to 6/30/06

8. Extension Centers and Summer Programs

Extension centers for graduate study are located off-campus at Argonne National Laboratory, Argonne, IL; CIDA Astronomical Observatory, Merida, Venezuela; Brookhaven National Laboratory, Upton, New York; European Organization for Nuclear Research (CERN), Geneva, Switzerland; Fermi National Accelerator Laboratory, Batavia, Illinois; Gesellschaft Für Schwerionenforschung, Darmstadt, Germany; Laue-Langevin Center, Grenoble, France; Los Alamos Scientific Laboratory, Los Alamos, New Mexico; Oak Ridge National Laboratory, Oak Ridge, Tennessee; Stanford Linear Accelerator Laboratory, Palo Alto, California.

Table C—Separately Budgeted Research Expenditures

Research Specialty	No. of Grants	Expenditures ($)
Astrophysics	5	688,643
Atomic, Molecular, & Optical Physics	22	4,729,119
Condensed Matter Physics	9	1,056,119
Nuclear Physics	7	4,461,575
Particles & Fields	17	4,680,036
Physics of Beams	2	549,184
Relativity & Gravitation	0	0
Total	62	16,164,676[1]

[1]Expenditures to 6/30/06

FACULTY

Professors

Alhassid, Yoram, Ph.D., Hebrew, 1979. Theoretical nuclear physics.

Ahn, Charles, Ph.D., Stanford Univ. 1996. Experimental condensed matter physics.

Appelquist, Thomas, Ph.D., Cornell, 1968. Eugene Higgins Professor. Theoretical elementary particle physics.

Bailyn, Charles, Ph.D., Harvard, 1987. Astrophysics and Astronomy.

Baker, O. Keith, Ph.D., Stanford, 1987. Experimental particle physics.

Baltay, Charles, Ph.D., Yale, 1963. Eugene Higgins Professor. Experimental elementary particle physics and experimental astrophysics.

Barrett, Sean E., Ph.D., Univ. of Illinois, 1992. Experimental condensed matter physics.

Casten, Richard F., Ph.D., Yale, 1967. Experimental nuclear physics.

Coppi, Paolo, Ph.D., Caltech, 1991. High energy astrophysics.

DeMille, David P., Ph.D., U. California, Berkeley, 1994. Experimental atomic physics.

Devoret, Michel, Ph.D., Univ. D'Orsay, 1982. Experimental condensed matter physics.

Fleury, Paul A., Ph.D., MIT, 1965. Physics and Engineering.

Girvin, Steven, Ph.D., Princeton, 1977. Theoretical condensed matter.

Glazman, Leonid, Ph.D., Ukr. SSR Academy of Sciences, 1982. Theoretical Condensed matter, nanostructure Physics.

Grober, Robert, Ph.D., Univ. of Maryland, 1991. Experimental condensed matter physics.

Harris, John W., Ph.D., SUNY at Stony Brook, 1978. Relativistic heavy ion and experimental nuclear physics.

Henrich, Victor E., Ph.D., Michigan, 1967. Eugene Higgins Professor of Applied Sciences. Experimental condensed matter physics.

Iachello, Francesco, Ph.D., MIT, 1969. Josiah W. Gibbs Professor. Theoretical nuclear physics.

Lamoreaux, Steve K., PhD., Univ. of Washington, 1986. Experimental particle, atomic, nuclear and condensed matter physics.

Mochrie, Simon, Ph.D., M.I.T. 1985. Experimental condensed matter physics.
Moncrief, Vincent E., Ph.D., Maryland, 1972. Professor of Physics and Mathematics. Gravitation and cosmology.
Natarajan, Priyamvada, Ph.D., Univ. of Cambridge, 1998. Astrophysics.
Parker, Peter D.M., Ph.D., Cal. Tech., 1963. Experimental nuclear physics and nuclear astrophysics.
Prober, Daniel E., Ph.D., Harvard, 1975. Experimental condensed matter physics.
Read, Nicholas, Ph.D., London Univ., 1986. Theoretical condensed matter physics.
Rokhlin, Vladimir, Ph.D., Rice, 1983. Scientific computation.
Sandweiss, Jack, Ph.D., California, Berkeley, 1957. Donner Professor. Experimental elementary particle physics.
Schoelkopf, Robert J., Ph.D., California Institute of Technology, 1995. Experimental condensed matter physics.
Shankar, Ramamurti, Ph.D., California, Berkeley, 1974. Chair of the Department. Theoretical condensed matter and statistical physics.
Stone, A. Douglas, Ph.D., MIT, 1983. Theoretical condensed matter physics.
Tipton, Paul L., Ph.D., Univ. of Rochester, 1987. Experimental particle physics.
Tully, John C., Ph.D., U. Chicago, 1968. Chemistry, applied physics and physics.
Urry, C. Megan, Ph.D., Johns Hopkins, 1984. Astrophysics.
van Dokkum, Pieter, Ph.D., Groningen, 1999. Astrophysics.
Wettlaufer, John, Ph.D., U. Washington, 1991. Geophysics.

Associate Professors

Blawzdziewciz, Jerzy, Ph.D., 1996. Univ. of Warsaw. Theoretical Condensed Matter.
Caines, Helen, Ph.D., Univ. of Birmingham, 1996. Nuclear.
Easther, Richard, Ph.D., Univ. Canterbury, New Zealand, 1994. Theoretical Particle Physics.
Fleming, Bonnie, Ph.D., Columbia Univ., 2001. High energy physics.
Goldberger, Walter, Ph.D., California Inst. of Tech., 2001. Theoretical Particle Physics.
Harris, Jack, Ph.D. UC Santa Barbara, 2000. Atomic.
Ismail-Beigi, Sohrab, Ph.D., M.I.T., 2002. Theoretical condensed matter.
Le Hur, Karyn, Ph.D., Orsay, 1998. Theoretical condensed matter physics.
McKinsey, Daniel, Ph.D., Harvard Univ. 2002. Atomic Physics.
O'Hern, Corey, Ph.D., Univ. of Penn, 1999. Soft condensed matter physics.
Skiba, Witold, MIT, 1997. Theoretical particle.

Assistant Professors

Demers, Sarah, Ph.D., Univ. Rochester, 2005. High Enegy Experimental.
Dufresne, Eric, Ph.D., Univ. of Chicago, 2000. Soft condensed matter physics.
Emonet, Thiery, Ph.D., Univ. La Laguan, 1998.
Geha, Marla, Ph.D., Univ. California, Santa Cruz, 2003. Astronomy and Astrophysics.
Golling, Tobiar, Ph.D., Univ. Bonn, 2005. High Energy Experimental.
Nagai, Daisuke, Ph.D., Univ. of Chicago, 2005. Astrophysics.
North, Jill, Ph.D., Rutgers Univ. 2004. Philosophy of Physics.
Padmanabhan, Nikhil, Ph.D., Princeton Univ., 2006
Rhoades, A. Elizabeth, Ph.D., Univ. of Michigan, 2001. Experimental Biophysics.
Werner, Volker, Ph.D., Cologne Univ., 2004. Experimental Nuclear Physics.

Senior Research Scientists

Adair, Robert K., Ph.D.,Wisconsin, 1951. Experimental elementary particles.
Dhawan, Satish, Ph.D., Univ. of Tsukuba, 1984. Experimental particle physics.
Majka, Richard D., Ph.D., Yale, 1974. Experimental particle physics.
Szymkowiak, Andrew, Ph.D., University of Maryland, 1984. Astrophysics. Experimental nuclear physics.

RESEARCH SPECIALTIES AND STAFF

Theoretical

Astrophysics. Mechanisms for x-ray and gamma ray emission from compact objects. Bailyn, Coppi, Natarajan.
Condensed Matter. Theoretical studies in statistical mechanics, phase transitions, quantum transport and many-body problems. Blawzdziewicz, Girvin, Glazman, Ismail-Beigi, Le Hur, O'Hern, Read, Shankar, Stone, Wettlaufer.
Elementary Particles and Fields. Fundamental quantum field theory; quantum chromodynamics; electroweak symmetry breaking; supersymmetry breaking, supergravity and string models, inflationary cosmology. Appelquist, Easther, Goldberger, Skiba.
General Relativity. Global properties of solutions of Einstein's equations; violations of causality in general relativity; classical and quantum studies of the cosmic censorship conjecture. Moncrief.
Nuclear Physics. Nuclear many-body physics: nuclear structure under extreme conditions and hadronic structure; QCD algebraic methods, statistical and quantum Monte Carlo methods; complex molecules, mesoscopic physics and quantum dots, quantum chaos, dynamics of complex systems and non-equilibrium statistical physics. Alhassid, Iachello.
Scientific Computation. Numerical scattering theory and large scale particle simulations. Rokhlin.

Experimental

Astrophysics. High energy astrophysics, experimental cosmology, nuclear astrophysics. Baltay, Geha, Parker, Szymkowiak, Urry, van Dokkum.
Atomic and Molecular Physics. Precise tests of fundamental symmetries, laser cooling and trapping, atom interferometry, Bose-Einstein condensation, ultra-cold collisions, atomic clocks. DeMille, Jack Harris, Lamoreaux, McKinsey.
Condensed Matter Physics. 2-dimensional electron liquids, spin physics in the quantum hall regime. Biological and Soft Matter physics. See Applied Physics for additional research activities in experimental condensed matter physics. Ahn, Barrett, Devoret, Dufresne, Grober, J. Harris, Mochrie, Prober, Rhoades, Schoelkopf.
Elementary Particles and Fields. Searches for lepton flavor violation in K^+ meson decay; investigations of high-energy $p\bar{p}$ collisions (CDF Project); investigations of high-energy e^+e^- collisions (the SLD and NLC projects); measurements of charmed particle production and decay; studies of deep inelastic scattering of muons from nuclei; precision measure-

ment of the $(g-2)$ value of the muon, and CP violation in K^0L and Brs^0 decays. Adair, Baker, Baltay, Dhawan, Fleming, Majka, McKinsey, Sandweiss, Tipton, Zeller.

Nuclear Physics. Collective modes of nuclei; nuclear models; multi-phonon states; phase transitions; dynamical symmetries; nuclear level lifetimes; high spin states; magnetic rotation; superdeformation; neutron rich nuclei; heavy-ion reactions; gamma ray spectroscopy; nuclear astrophysics; nucleosynthesis; energy generation in stellar objects; relativistic heavy ion collisions; search for a quark-gluon plasma; RHIC physics with the STAR detector. Caines, Casten, Fleming, John Harris, Heinz, Parker, Werner.

UNIVERSITY OF DELAWARE

DEPARTMENT OF PHYSICS AND ASTRONOMY AND BARTOL RESEARCH INSTITUTE

Newark, Delaware 19716

Students Accepted For Degree	FIELDS		
	Physics	Astronomy	Related Fields
Doctorate	X	X	
Master's	X	X	

1. General

President: Patrick T. Harker
University Officer for Graduate Studies: Debra Hess Norris
Physics and Astronomy Department Chair: George C. Hadjipanayis
Department Telephone Number: (302) 831-1995 (C)
Director, Bartol Research Institute: Stuart Pittel
Bartol Research Institute Telephone: (302) 831-8111
Type of Institution: University
Control: Both
Setting: Small University town
Total Faculty: 1,144
Total Students: 15,786
Total Graduate Students: 2,671
**Graduate Tuition*: Full-time—$11,120/semester
Part-time—$1,236/credit
Tuition rates for: 2009–10
Deferred tuition plan: Yes
Other Fees: $484 (Health Service)
Term: Semester

*Teaching and Research Assistant tuition is waived.

2. Number of Faculty in Department

The combined total of full-time faculty in the three professorial ranks is 35 in the Department of Physics and Astronomy. The combined total of full-time, part-time, and other faculty at all ranks is 40.

3. Admission, Financial Aid, and Housing

Address admission inquiries to: Chair of Graduate Admissions Committee
Graduate application fee: Fee of $75 is required at the time of formal application to the Graduate School.
Application deadline (Fall admission): April 15
Admission information: For Fall 2010, 14 students were accepted out of 188 applicants.
Admission requirements: Admission to either the MS or Ph.D. program requires a Bachelor's degree in physics or a closely related field with a minimum GPA of 3.2. Students from non-English speaking countries must demonstrate proficiency in English via the TOEFL exam. For financial support a score greater than 600 is required.
Undergraduate preparation assumed: Electricity and Magnetism, Classical Mechanics, Quantum Mechanics, Thermodynamics.
Address financial aid inquiries to: Chair Graduate Admissions Committee
Financial aid deadline: February 15
Loans available: No
Address housing inquiries to: Office of Housing and Residence Life
On-campus, single student housing available: Yes, but limited
Cost/semester: $2,700–3,500
On-campus, married student housing available: Yes, but limited

Table A—Faculty, Enrollments, and Degrees Granted

Research Specialty	2009–10 Faculty	Enrollment Fall 2009 Master's	Enrollment Fall 2009 Ph.D.	No. of Degrees Granted 2009–10 (2004–09) Master's	Terminal Master's	Ph.D.	Median No. of Years for Ph.D.'s
Astronomy/ Astrophysics	7	1	7	0(0)	1(1)	1(0)	–
Atomic & Molecular Physics (Theory)	3	0	3	0(0)	0(0)	1(5)	–
Atomic & Molecular Physics (Exper.)	3	1	6	0(0)	1(3)	0(5)	–
Condensed Matter Physics (Theory)	3	0	4	0(1)	0(0)	0(5)	–
Condensed Matter Physics (Exper.)	7	1	23	1(1)	0(7)	7(16)	–
Nuclear Physics	1	0	1	0(0)	0(0)	1(0)	–
Particle Physics and Cosmology	4	0	12	0(0)	1(2)	1(4)	–
Particle Astrophysics	8	1	3	0(0)	0(0)	1(0)	–
Space/Plasma Phys.	5	0	7	0(0)	0(1)	0(2)	–
Non-specialized	0	2	4	0(0)	0(0)	0(0)	–
Total		6	70	1(2)	3(14)	12(37)	
Full-time Grad. Stud.		6	70				
Part-time Grad. Stud.		0	0				
First-year Grad. Stud.		2	16				

4. Graduate Degree Requirements

Master's: 24 credit hours of classroom courses plus six credits of M.S. thesis.
Ph.D.: 30 credit hours of classroom courses, passing the Ph.D. qualifying exam, Ph.D. thesis. Students entering the program with a Master's degree may follow the *Ph.D. fast track* which has a reduced course requirement of 12 credits.
Thesis: Thesis may be written *in absentia*.
Special Equipment, Facilities, or Programs: The Department of Physics and Astronomy is housed in Sharp Laboratory, which has its own library, machine and electronics shops as well as research and teaching laboratories, classrooms, and office space. The condensed matter and material science programs have in house scanning and transmission microscopes, a variety of magnetometers, X-ray diffractometers, differential scanning calorimeters, thin-film deposition systems and cryogenic facilities, and make use of accelerator based facilities for X-ray and neutron scattering. The atomic and molecular physics laboratories include femtosecond and high-power pulsed lasers for non-linear optical studies and high resolution multiphoton spectroscopy. The astro-particle physics programs include high-altitude balloon flights and high energy cosmic ray and neutrino experiments in Antarctica (ICECUBE and Anita). Space physics programs maintain a world-wide network of neutron monitors and are involved with MMS, the Magnetosphere MultiScale mission and multi spacecraft missions such as Cluster-2, to study the magnetosphere and the solar wind. Opportunities are available for

participation in several NASA missions: ACE, The Spitzer infrared telecope, the Chandra X-ray satellite and the Hubble Space Telescope. UD is also part of the SMARTS consortium which allows use of a group telescope in Chile. Further programs on campus are the Institute for Energy Conversion and the Center for Composite Materials.

Table B—Appointments to Graduate Students, 2009–10

Title of Appointee	Appointments Total	First year	Academic Load Allowed in Credit Hours	Hours of Service Per Week	Stipend for Academic Year ($)
			Semester		
Teaching Assistant	30	13	6–9	15–20	21,000
Research Assistant	32	3	6–9	15–20	21,000
Fellows	10	0	6–9	15–20	23,000
Self-Supporting	4	2	–	–	–
Total	76	18			

[1]Tuition waived.
[2]Summer and winter appointments available with additional stipend.

5. Personnel Engaged in Separately Budgeted Research, 7/09–6/10

Professorial faculty	35
Other faculty	5
Postdoctoral appointments	13
Graduate students	77
Nonteaching research personnel	6
Total	136

6. Separately Budgeted Research Expenditures by Source of Support 07/1/08–6/30/09 (FY2009)

Federal government	$5,277,574.86
Institutional	11,514.57
Other	1,711,495.25
Total	$7,000,584.68

Table C—Separately Budgeted Research Expenditures

Research Specialty	No. of Grants	Expenditures ($)
Astophysics	10	948,017.33
Astronomy	5	308,620.26
Atomic and Molecular, Optical & Plasma Physics	13	589,307.55
Condensed Matter Physics	31	1,566,746.78
Nuclear Physics	1	61,128.73
Particles & Fields	8	1,662,672.05
Space Physics	27	1,864,091.98
Total	95	7,000,584.68

FACULTY

Professors

Barr, Stephen, Ph.D., Princeton, 1978. Elementary particle theory.

Bieber, John W., Ph.D., Maryland, 1977. Space plasma physics; cosmic rays physics.

Chui, Siu-Tat, Ph.D., Princeton, 1972. Condensed matter theory; low-dimensional and amorphous materials.

Evenson, Paul A., Ph.D., Chicago, 1972. Space physics; solar and cosmic-ray studies.

Gaisser, Thomas K., Ph.D., Brown, 1967. Elementary particle theory; high energy astrophysics.

Glyde, Henry, Ph.D., Oxford, 1964. Chair. Condensed matter theory; neutron studies of condensed matter; liquid helium.

Hadjipanayis, George C., Ph.D., University of Manitoba, 1979. Experimental condensed matter physics; magnetism; nanocrystalline materials.

Leung, Chung Ngoc, Ph.D., Minnesota, 1983. Elementary particle theory.

MacDonald, James, Ph.D., Inst. of Astronomy, Cambridge, 1979. Astronomy and astrophysics; white dwarfs; cataclysmic variables.

Matthaeus, William H., Ph.D., William and Mary, 1979. Space physics; plasma physics; turbulence theory; computational physics.

Mulders, Norbert, Ph.D., Delaware, 1991. Quantum fluids and solids.

Mullan, Dermott J., Ph.D., Maryland, 1969. Astrophysics; solar and stellar physics.

Owocki, Stanley P., Ph.D., Colorado, 1982. Astrophysics; stellar winds.

Pittel, Stuart, Ph.D., Minnesota, 1968. Theoretical nuclear physics and nuclear astrophysics.

Seckel, David, Ph.D., Washington, 1983. Particle astrophysics; cosmology.

Shafi, Qaisar, Ph.D., Imperial College, London, 1971. Elementary particle theory; cosmology.

Shipman, Harry L., Ph.D., Cal. Tech., 1971. Astronomy and astrophysics; white dwarfs.

Stanev, Todor, Ph.D., Sofia, Bulgaria, 1977. Cosmic-ray physics, particle astrophysics.

Szalewicz, Krzysztof, Ph.D., Warsaw, 1977. Theoretical atomic and molecular physics.

Unruh, Karl M., Ph.D., Johns Hopkins, 1983. Condensed matter; finite size effects; thin films.

Walker, Barry Ph.D., New York, Stony Brook, 1996. Light-matter interactions, condensed matter, experimental, optical physics.

Watson, George, Ph.D., Delaware, 1984. Laser spectroscopy of condensed matter; photonic band structure.

Xiao, John Q., Ph.D., Johns Hopkins, 1993. Specialty–metallic thin films and multilayers, superconducting materials; granular materials.

Associate Professors

Morgan, John D. III, Ph.D., California, Berkeley, 1978. Theoretical atomic and molecular physics.

Nowak, Edmund R., Ph.D., Minnesota, 1994. Experimental condensed physics, magnetism, superconductivity, granular materials.

Shah, Ismat, Ph.D., Illinois, 1986. Material Science and Engineering. Thin film, surface, interface, and nanostructures.

Shay, Michael, Ph.D., U. of Maryland, College Park, 1998. Plasma physics, space physics and astrophysics.

Williams, Barbara, Ph.D., Maryland, 1981. Astronomy and astrophysics; radio astronomy; groups of galaxies.

Assistant Professors

DeCamp, Matthew F., Ph.D., Michigan, 2002. Atomic and molecular physics.

Gizis, John, Ph.D., California Inst. of Tech., 1998. Astronomy, subdwarfs, brown dwarfs.

Holder, Jamie, Ph.D., University of Durham, UK, 1997. Cosmic ray physics.

Ji, Yi, Ph.D., The Johns Hopkins University, 2003. Condensed matter and materials physics.

Lorenz, Virginia, Ph.D., University of Colorado at Boulder, 2007. Atomic and molecular physics.

Nikolic, Branislav, Ph.D., New York, Stony Brook, 2000, Theoretical and computational condensed matter physics, transport phenomena.

Safronova, Marianna, Ph.D., Notre Dame, 2001. Quantum computing with neutral atoms, Rydberg atoms.

RESEARCH SPECIALTIES AND STAFF

Theoretical

Astronomy and Astrophysics. MacDonald, Mullan, Owocki, Shay.

Atomic and Molecular Physics. Morgan, Safronova, Szalewicz.

Condensed Matter and Materials Physics. Chui, Glyde, Nikolic.

Cosmic Ray Physics. Stanev.

Elementary Particle Theory. Barr, Gaisser, Leung, Seckel, Shafi, Stanev.

Nuclear Physics. Pittel.

Plasma Physics. Bieber, Matthaeus, Shay.

Space Physics. Bieber, Matthaeus, Shay.

Experimental

Astronomy and Astrophysics. Gizis, Shipman, Williams, Holder.

Atomic and Molecular. DeCamp, Lorenz, Morgan, Szalewicz, Walker.

Condensed Matter and Materials Physics. Glyde, Hadjipanayis, Ji, Mulders, Nowak, Shah, Unruh, Watson, Xiao.

Cosmic Ray Physics. Bieber, Evenson, Holder.

Optics. Walker, Watson.

Space Physics. Evenson.

GEORGE WASHINGTON UNIVERSITY

DEPARTMENT OF PHYSICS

Washington, D.C. 20052

Students Accepted For Degree	FIELDS: Physics	Astronomy	Related Fields
Doctorate	X		
Master's	X		

1. General

President: Steven Knapp
Dean of Graduate School: Peg Barratt
Department Chairman: Barry Berman
Department Telephone Number: (202) 994-6275
Type of Institution: University
Control: Private Setting
Urban Total Faculty: 4,673
Total Graduate Faculty: N/A
Total Students: 20,327
Total Graduate Students: 9,935
Annual Graduate Tuition:
All Graduate Students: $1,147/credit hr.
Tuition rates for: 2010–2011
Deferred tuition plan: Yes
Other Fees: Yes
Term: Semester

2. Number of Faculty in Department

The combined total of full-time faculty in the three professorial ranks is 16. The total number of other faculty is 6.

3. Admission, Financial Aid, and Housing

Address admission inquiries to: Director of Graduate Programs, Dept. of Physics
Graduate application fee required: $60
Admission deadline for financial aid (Fall admission): January 15
Admission requirements: For admission to the graduate programs, a Bachelor's degree in physics or equivalent is preferred with an undergraduate GPA of 3.0. The GRE general is required, the Physics exam is recommended. Students from non-English speaking countries are required to demonstrate proficiency in English via the TOEFL exam with a minimum score over 250 in computer-based, 600 in paper-based, or 100 in internet based test required.
Undergraduate preparation: Subjects at the level of Symon, Mechanics; Lorrain and Corson, *Electricity and Magnetism*; Reif, *Statistical and Thermal Physics*; Eisberg and Park, *Modern and Quantum Physics*.
Address financial aid inquiries to: Department of Physics
GAPSFAS application required: No
Loans available: Yes
Address housing inquiries to: Director of Housing
On-campus, single student housing available: Limited
On-campus, married student housing available: No.

Table A—Faculty, Enrollments, and Degrees Granted

Research Specialty	2009–10 Faculty	Enrollment[1] Fall 2009: Master's	Doctorate	No. of Degrees Granted[2] 2009–10 (2004–10): Master's	Terminal Master's	Doctorate	Median No. of Years for 2009–10 Ph.D.'s
Astrophysics	4	–	3	0(0)	–	0(1)	–
Biophysics	4	–	6	0(0)	–	1(1)	5
Condensed Matter Physics	4	–	3	0(0)	–	0(2)	5
Medical Physics	1	–	2	0(0)	–	1(1)	4
Nuclear Physics	12	–	13	0(1)	0(2)	1(7)	5
Physics Education Research	1					1(0)	4
Non-specialized		–		0(1)	0(2)	–	–
Total	21	0	27	0(2)	0(4)	1(9)	
Full-time Grad. Stud.		0	27				
Part-time Grad. Stud.			0				
First-year Grad. Stud.			8				
Median Years in Grad. Study (2001–10) Degrees				2	3	5	–

Undergraduate Degrees, 2005–06 (2001–10): 7(33)

[1]Students not yet committed to a research specialty are entered under non-specialized.
[2]Five-year totals in parentheses.

4. Graduate Degree Requirements

Master's: M.A. degree with thesis or no thesis options: 30 semester-hours of course work in physics plus thesis, or 36 semester-hours of course work in physics and mathematics, including a tool requirement in computer programming. A 3.0 GPA is required.
Doctorate: A minimum of 72 semester-hours of approved courses for students with only a Baccalaureate. For students with a Master's degree, a minimum of 48 semester-hours is required. Tool requirement: completion of a numerical methods course. A 3.0 GPA is required.
Thesis: Thesis may be written *in absentia*.
Special Equipment, Facilities, or Programs: High-end central computing facility; several departmental computing facilities, including five high-end clusters, two CMP/biophysics research labs; machine shop; Virginia campus facilities contains labs for design, construction, and testing of particle and radiation detectors for use at major accelerator laboratories worldwide.

Table B—Appointments to Graduate Students, 2009–10

Title of Appointee	Appointments: Total	First year	Academic Load Allowed in Credit Hours	Hours of Service Per Week	Stipend for Academic Year ($)
			Semester		
Teaching Assistant	8	4	9	20	18,000–23,000
Research Assistant	13	0	9	0	18,000–23,000
Fellows	5	1	9	0	18,000–25,000
Total	26	5			

5. Personnel Engaged in Separately Budgeted Research, 7/09–6/10

Professorial faculty	20
Postdoctoral Appointments	8
Graduate students	14
Undergraduate students	9
Total	42

6. Separately Budgeted Research Expenditures by Source of Support

	Departmental Research	Physics-related Research Outside Department
Federal government	$1,659,000	
Private Foundations	120,000	
Internal University Support	100,000	
Total	$1,879,000	

Table C—Separately Budgeted Research Expenditures

Research Specialty	No. of Grants	Expenditures ($)
Astrophysics	5	90,000
Biophysics	5	522,000
Nuclear and Particle Physics	21	790,000
Medical Physics	3	117,000
Physics Education Research	4	180,000
Total	38	1,879,000

FACULTY

Professors

Berman, Barry L., Ph.D., Illinois, 1963. Experimental nuclear physics; nuclear astrophysics; biophysics.

Briscoe, William, Ph.D., Catholic, 1978. Experimental nuclear physics and particle physics.

Feldman, Gerald, Ph.D., Washington, 1987. Experimental nuclear physics; physics education research.

Lee, Frank X., Ph.D., Ohio, 1993. Theoretical nuclear and particle physics; computational physics.

Lehman, Donald R., Ph.D., George Washington, 1970. Theoretical nuclear physics.

Opper, Allena, Ph.D., Indiana, 1991. Experimental nuclear and particle physics.

Parke, William C., Ph.D., George Washington, 1967. Theoretical nuclear physics; biophysics; astrophysics.

Reeves, Mark E., Ph.D., Illinois, 1989. Experimental condensed matter physics; biophysics; medical physics.

Research Professors

Maximon, Leonard C., Ph.D., Cornell, 1952. Theoretical nuclear physics; mathematical physics; astrophysics.

Strakovsky, Igor, Ph.D., St. Petersburg, 1984. Experimental nuclear physics; phenomenology.

Associate Professors

Dhuga, Kalvir S., Ph.D., Birmingham, 1980. Experimental nuclear physics, astrophysics.

Eskandarian, Ali, Ph.D., George Washington, 1967. Theoretical nuclear physics; astrophysics.

Haberzettl, Helmut, Ph.D., Bonn, 1979. Theoretical nuclear and particle physics.

Peng, Weiqun, Ph.D., Illinois, 2001. Theoretical biophysics.

Zeng, Chen, Ph.D., Cornell Univ., 1994. Theoretical condensed matter physics; biophysics.

Associate Research Professor

Workman, Ron, Ph.D., British Columbia, 1987. Theoretical nuclear physics; phenomenology.

Assistant Professors

Alexandru, Andrei, Ph.D., Louisiana State University, 2001. Theoretical nuclear physics.

Griesshammer, Harald, Ph.D., Erlangen, 1996. Theoretical nuclear and particle physics.

Qui, Xiangyun, Ph.D., Michigan State University, 2004. Experimental condensed-matter physics, biophysics.

Assistant Research Professors

Micherdzinska, Anna, Ph.D., University of Silesia (Poland), 2003. Experimental nuclear physics.

Paris, Mark, Ph.D., Univ. of Illinois at Urbana-Champaign, 2000. Theoretical nuclear physics.

Wang, Guanyu, Ph.D., Germany, 1998. Theoretical biophysics.

RESEARCH SPECIALTIES AND STAFF

Theoretical

Astrophysics. Eskandarian, Parke, Maximon.

Biophysics. Parke, Peng, Wang, Zeng. 1 postdoctoral fellow.

Condensed Matter Physics. Qiu, Reeves. 3 postdoctoral fellows.

Nuclear and Particle Physics. Alexandru, Eskandarian, Griesshammer, Haberzettl, Lee, Maximon, Paris, Parke, Workman. 2 postdoctoral fellows.

Phenomenology

Nuclear and Particle Physics. Briscoe, Strakovsky, Workman.

Experimental

Astrophysics. Dhuga. 1 postdoctoral fellow.

Biophysics. Qiu, Reeves. 1 postdoctoral fellow.

Condensed-Matter Physics. Qiu, Reeves. 3 postdoctoral fellows.

Medical Physics. Berman, Reeves.

Nuclear and Particle Physics. Berman, Briscoe, Feldman, Opper, Strakovsky. 4 postdoctoral fellows.

Physics Education Research. Feldman.

GEORGETOWN UNIVERSITY

DEPARTMENT OF PHYSICS

Washington, D.C. 20057

Students Accepted For Degree	FIELDS: Physics	Astronomy	Related Fields
Doctorate	X		
Master's	X		

1. General

President: John J. DeGioia
Dean of Graduate School: Timothy A. Barbari
Department Chairman: Edward Van Keuren
Department Telephone Number: (202) 687-5984
Type of Institution: University
Control: Private
Setting: Urban
Total Faculty: 1,957
Total Graduate Faculty: N/A
Total Students: 12,856
Total Graduate Students: 4,490
Annual Graduate Tuition:
In-state Residents: $35,160
Out-of-state Residents: $33,408
Tuition rates for: 2010–11
Deferred tuition plan: No
Yates Recreation Fees: $311
Term: Semester

2. Number of Faculty in Department

The combined total of full-time faculty in the three professorial ranks is 13. The combined total of full-time, part-time and other faculty at all ranks is 26.

3. Admission, Financial Aid, and Housing

Address admission inquiries to: GSAS, Office of Graduate Admissions, Box 571004, Washington, DC 20057-1004

Graduate application fee required: On-line/$75 Hard Copy application

Admission deadline (Fall admission): 2/1

Admission requirements: For admission to the graduate programs, a Bachelor's degree in physics or related field is required with a minimum undergraduate GPA of 3.0, a personal statement, 3 letters of recommendation, and a resume/cv. The GRE Exam and Physics Subject Test are required. Students from non-English speaking countries are required to demonstrate proficiency in English via the TOEFL or IBT exam. A minimum score of 250 (computer) is required for TOEFL. A minimum score of 100 is required for IBT.

Undergraduate preparation assumed: Intermediate level courses in Classical Mechanics, Quantum Mechanics, Electricity and Magnetism, and Statistical and Thermal Physics, as well as a working knowledge of an advanced computer language.

Address financial aid inquiries to: Director of the Graduate Program, Department of Physics, 506 Reiss Science Building, Georgetown University, Washington, DC 20057

GAPSFAS application required: No

Financial aid deadline: 2/1 (Priority Date)

Loans available: No

Address housing inquiries to: www.georgetown.edu/home/housing.html

On-campus, single student housing available: No

On-campus, married student housing available: No

Table A—Faculty, Enrollments, and Degrees Granted

Research Specialty	2009–10 Faculty	Enrollment[1] Fall 2009–10: Master's	Doctorate	No. of Degrees Granted[2] 2009–10 (2004–10): Master's	Terminal Master's	Doctorate	Median No. of Years for 2009–10 Ph.D.'s
Biophysics	2	–	1	0(0)	0(0)	0(0)	–
Hard Condensed Matter	4	–	8	0(0)	0(0)	0(0)	–
Micro/Nano Technologies	4	–	3	0(0)	0(1)	2(2)	–
Optics and Imaging	3	–	6	0(0)	0(0)	0(2)	–
Physics Education	1	–	0	0(0)	0(0)	0(0)	–
Soft Condensed Matter	2	–	4	0(0)	0(0)	0(0)	–
Statistical Physics	3	–	1	0(0)	0(0)	0(0)	–
Ultracold Atoms	2	–	0	0	0	0	–
Non-specialized	0	–	8	0(0)	1(4)	0(0)	–
Total		–	31	–	1(5)	2(4)	
Full-time Grad. Stud.		–	24				
Part-time Grad. Stud.		–	1				
First-year Grad. Stud.		–	5				
Median Years in Grad. Study (2008–09 Degrees)				–	–	–	–

Undergraduate Degrees, 2008–09 (2004–09): 25(72)

[1]Students not yet committed to a research specialty are entered under non-specialized.
[2]Five-year totals in parentheses.

4. Graduate Degree Requirements

Master's: The thesis option requires 36 credits of satisfactory graduate coursework plus a thesis; the non-thesis option requires 42 credits of satisfactory graduate coursework.

Doctorate: The Ph.D. requires 40 credits of satisfactory graduate coursework and a dissertation. Graduate exams include comprehensive and qualifying exams, and a dissertation defense.

Special Programs: Both the M.S. and Ph.D. programs offer a traditional physics track and an Industrial Leadership in Physics (ILP) track. The latter is intended for students interested in scientific careers in industry. The curriculum of the ILP track includes business electives and a year-long internship in industry. Both tracks emphasize communication skills and teamwork.

Thesis: Thesis may be written *in absentia*.

Special Equipment, Facilities, or Programs:

Georgetown GNμlab

GNuLab is a micro-fabrication and materials research facility that is wholly managed by the Department of Physics. GNμLab, comprises a 2000 square foot facility including 1200 square feet of clean room space, and the following capabilities: Optical and electron beam lithography, deposition, evaporation and sputtering systems; Etching: RIE/DRIE equipment and wet TMAH etching; high temperature furnaces; Measurement: Stress, film thickness, FESEM; Wire Bonding.

Soft Matter Lab

Equipment for the synthesis, characterization and manipulation of colloidal particles through the use of our wet chemistry facility. Also, functional facilities for the preparation of biological materials such as cells and bio-polymers. Materials are studied with a high spped laser scanning Leica SP5 confocal microscope equipped with 7-laser lines ranging from UV to near IR. Mechanical measurements are performed with a customized Anton Paar MCR-301 rheometer that works in conjunction with the confocal microscope.

Georgetown Laser Laboratory

The Laser and Optical Characterization lab comprise laser systems for characterization of nanoparticle formation, nonlinear optical effects and spectrally resolved imaging. High powered laser systems include a nanosecond pulsed laser (Quanta-Ray GCR03) pumping an Optical Parametric oscillator (GWU), an Argon/Dye continuous wave laser (Coherent Medical Lambda Plus) and an Ar-pumped femtosecond Ti-:sapphire laser system. Nanoparticle formation is studied with apparatus for photon- and fluorescence correlation spectroscopy, based on a hardware autocorrelator (ALV5000) and photon counting avalanche photodiode. Spectrally resolved fluorescence imaging capabilities are provided by a FALCON chemical imaging microscope (ChemImage). These apparatus are augmented by advanced detection and signal processing instrumentation.

Dynamics Imaging Laboratory

High speed and high resolution digital imaging systems. Software for image acquisition, processing and analysis. High-speed confocal microscope. High-powered optical tweezer.

Beowulf Clusters

The computational groups have 2 Beowulf cluster machines for parallel computing. One is an APPLE cluster and the other is an INTEL cluster. The University has additional parallel computation resources.

Superconductivity and Nanoelectronics Laboratory

Room temperature testing station (with four micromanipulators); Low-noise low temperature (to 1.5 K) transport measurement set-up; Variable temperature high vacuum scanning microscope, JEOL JSPM-4210 scanning probe microscope, capable of performing AFM and STM measurements under ambient conditions, as well as in vacuum, controlled gas environments and temperatures from 100 K to 800 K; Carbon nanotube synthesis facility (CVD) located in GNμLab.

5. Personnel Engaged in Separately Budgeted Research, 7/08–6/09

Professorial faculty	10
Other faculty	4
Postdoctoral appointments	6
Graduate Students	18
Total	38

6. Separately Budgeted Research Expenditures by Source of Support

	Departmental Research	Physics-related Research Outside Department
Federal government	778,633	
Private, non-profit organizations	225,728	
Business & Industry	214,624	
Total	1,218,984	

Table C—Separately Budgeted Research Expenditures

Research Specialty	No. of Grants	Expenditures ($)
Condensed Matter Physics	22	1,218,985
Total	22	1,218,985

FACULTY

Professors

Currie, John F., Ph.D., Cornell, 1977. Device microfabrication; biomedical devices; experimental condensed matter physics.

Freericks, James K., Ph.D., Berkeley, 1991. Theoretical condensed matter physics; computational physics.

Urbach, Jeffrey S., Ph.D., Stanford University, 1993. Experimental condensed matter physics, Biophysics.

Associate Professors

Barbara, Paola, Ph.D., Technical University of Denmark, 1995. Experimental condensed matter physics.

Chiao-Yap, Lydia, Ph.D., Berkeley, 1961. Theoretical condensed matter physics.

Egolf, David A., Ph.D., Duke University, 1994. Computational physics; theoretical soft condensed matter physics; QCD.

Liu, Amy Y., Ph.D., Berkeley, 1991. Theoretical condensed matter physics.

Mathews, Wesley N., University of Illinois, Urbana-Champaign, 1966. Theoretical condensed matter physics.

Paranjape, Makarand, Ph.D., University of Alberta, 1993. Microelectromechanical systems; biomedical microdevices.

Van Keuren, Edward, Ph.D., Carnegie Mellon University, 1990. Optics, experimental condensed matter physics.

Assistant Professors

Blair, Daniel L., Ph.D., Clark University, 2003. Experimental Condensed Matter physics, Biophysics.

Dzakpasu, Rhonda, Ph.D., The University of Michigan, 2003. Experimental and Computational Biophysics.

Rigol, Marcos, Ph.D., University of Stuttgart, Germany, 2004. Computational physics: theoretical condensed matter physics.

Research Professors

Esrick, Mark A., Ph.D., Georgetown University, 1981. Biophysics.

Malinin, George I., Ph.D., Catholic University of America, 1972. Biophysics.

de Vincenz, Andre M., Ph.D., Georgetown University, 1987. Materials science.

Professors Emeriti

Mayer, Walter G., Ph.D., Michigan State University, 1958. Emeritus. Acoustics.

McClure, Joseph A., University of North Carolina, Chapel Hill, 1963. Emeritus. Physics education.

Serene, Joseph, W., Ph.D., Cornell, 1974. Theoretical condensed matter physics.

Adjunct Faculty

Clinton, Thomas W., Ph.D., University of Maryland, 1992. Condensed matter physics.

Hess, Daryl W., Ph.D., University of Illinois, Urbana-Champaign, 1987. Theoretical condensed matter physics.

Lavine, James, Ph.D., University of MD, at College Park, 1971. Semi conductor physics.

Quong, Andrew A., Ph.D., Univ. of California, Irvine condensed matter physics.

Schneider, Thomas W., Ph.D., University of Wyoming, 1993. Chemical and biochemical miscrosensors.

Slakey, Francis, Ph.D., University of Illinois at Urbana-Champaign, 1992. Science policy.

Zlatic, Veljko, Ph.D., Imperial College of London, 1974. Theoretical condensed matter physics.

RESEARCH SPECIALTIES AND STAFF

Theoretical

Biophysics. Axonal chemotaxis. Urbach.

Hard Condensed Matter. Superconductivity, magnetism, strongly correlated materials; structural, electronic, and transport properties; interaction of light with matter. Freericks, Liu, Rigol.

Particles and fields. Effective theories. Egolf.

Statistical Physics. Nonequilibrium dynamical systems, both classical and quantum. Egolf, Freericks, Rigol.

Ultracold gases. Freericks and Rigol.

Experimental

Biophysics. Cellular biophysics. Pattern formation in neural systems, biomaterials. Dzakpasu, Urbach.

Hard Condesend Matter. Superconductivity, superconducting devices, transport in nanostructures. Barbara.

Micro/Nano Technologies. Semiconductor technology, sensors and actuarors, nanotube devices, organic photovoltaic; applications to environmental monitoring, bioengineering, medical imaging. Barbara, Curie, Paranjape, Van Keuren.

Optics and Imaging. Nanoparticle synthesis and characterization, imaging of soft materials, biomedical optics. Blair, Urbach, Van Keuren.

Physics Education. Design of research-based curricular materials to improve student learning. Lindsey.

Soft Condensed Matter. Soft glasses, colloidal and polymer physics, biomaterials, granular material, fluids, nonlinear dynamics. Blair, Urbach.

FACULTY PUBLICATIONS

Barbara, Paola

A. Di Bartolomeo, M. Rinzan, A. K. Boyd, Y. F. Yang, L. Guadagno, F. Giubileo, and P. Barbara, "Electrical properties and memory effects of field-effect transistors from networks of single- and double-walled carbon nanotubes," Nanotechnology, **21**, 115204 (2010) Abstract.

G. Fedorov, P. Barbara, D. Smirnov, D. Jimenez, and S. Roche, "Tuning the bandgap of semiconducting carbon nanotubes by an axial magnetic field," Applied Physics Letters **96**, 132101-132103 (2010) Abstract.

J. Zhou, P. Barbara, and M. Paranjape, "Novel in-situ decoration of single-walled carbon nanotube transistors with metal nanoparticles," Journal of Nanoscience and Nanotechnology, **10**, 3890-3894 (2010).

A. Tselev, Y. F. Yang, J. Zhang, P. Barbara, and S. Shafraniuk, "Carbon nanotubes as nanoscale probes of the superconducting proximity effect in Pd-Nb junctions," Physical Review B, **80**, 054504 (2009) Abstract.

D. Tobias, M. Ishigami, A. Tselev, P. Barbara, E. D. Williams, C. J. Lobb, and M. S. Fuhrer, "Origins of 1/f noise in individual semiconducting carbon nanotube field-effect transistors," Physical Review B **77**(3) (2008) Abstract.

Blair, Daniel L.

M. Caggioni, P. T. Spicer, D. L. Blair, S. E. Iuidberg, and D. A. Weitz, "Rheology and microheology of a microstructured fluid: The gellam gum case," Journal of Rheology **51**, 851 (2007).

H. M. Wyss, D. L. Blair, J. F. Morris, H. A. Stone, and D. A. Weitz, "Mechanism for clogging of microchannels," Phys. Rev. E **74**, 061402 (2006).

D. L. Blair and A. Kudrolli, "Geometry of crumpled paper," Phys. Rev. Lett. **94**, 166107 (2005).

Egolf, David A.

M. P. Fishman and D. A. Egolf, "Revealing the building blocks of spatiotemporal chaos: Deviations from extensivity," Phys. Rev. Lett. **96**, 054103 (2006).

Freericks, James K.

J. K. Freericks, "Quenching Bloch oscillations in an strongly correlated material: The dynamical mean-field theory approach," Phys. Rev. B **77**, 075109 (2008).

A. V. Joura, J. K. Freericks, and Th. Pruschke, "Steady state nonequilibrium density of states of driven strongly correlated lattice models in infinite dimensions," Phys. Rev. Lett. **101**, 196401 (2008).

J. K. Freericks, H. R. Krishnamurthy, and Th. Pruschke, "Theoretical description of time-resolved photoemission spectroscopy: Application to pump-probe experiments," Phys. Rev. Lett. **102**, 136401 (2009).

H. Zenia, J. K. Freericks, H. R. Krishnamurthy, and Th. Pruschke, "Appearance of fragile Fermi liquids in finite width Mott insulators sandwiched between metallic leads," Phys. Rev. Lett. **103**, 116401 (2009).

Liu, Amy Y.

R. R. Zope, T. Baruah, K. C. Lau, A. Y. Liu, M. R. Pederson, and B. I. Dunlap, "Boron fullerenes: From B80 to hole doped boron sheets," Phys. Rev. B **79**, 161403(R) (2009).

A. Y. Liu, "Electron-phonon coupling in compressed 1T-TaS2: stability and superconductivity from first principles," Phys. Rev. B **79**, 220515(R) (2009).

A. Y. Liu, R. R. Zope, and M. R. Pederson, "Structural and bonding properties of bcc-based B80 solids," Phys. Rev. B **78**, 155422 (2008).

A. Y. Liu and I. I. Mazin, "Combining the advantages of superconducting MgB_2 and CaC_6 in one material: Suggestions from first-principles calculations," Phys. Rev. B **75**, 064510 (2007).

Paranjape, Makarand

J. Zhou, M. Paranjape, "Novel In-situ Decoration of Single-walled Carbon Nanotube Transistors with Metal Nanoparticles," J. Nanosci. Nanotech. accepted for publication, 2009.

A. H. Monica, S. J. Papadakis, G. L. Coles, R. Osiander, M. Paranjape, "A lateral carbon nanotube based field emission triode," J. Vac. Sci. Tech. B **26**(2), (2008).

A. H. Monica, S. J. Papadakis, R. Osiander, and M. Paranjape, "Wafer-Level Assembly of Carbon Nanotube Networks Using Dielectrophoresis," Nanotech. **19** (2008).

V. Spinella-Mamo and M. Paranjape, "Using Genetic Algorithms to Characterize Ferrofluid Topographies in Externally Applied Magnetic Fields," J. Magnet. Magnetic Mater. **321**(4), (2008).

J. Zhang, A. Boyd, A. Tselev, M. Paranjape, and P. Barbara, "Mechanism of NO_2 Detection in Carbon Nanotube Field Effect Transistor Chemical Sensors," Appl. Phys. Lett. **88**(12), (2006).

Rigol, Marcos

M. Rigol, B. S. Shastry, and S. Haas, "Effects of Strong Correlations and Disorder in d-Wave Superconductors," Phys. Rev. B **79**, 052502 (2009).

M. Rigol, V. Dunjko, and M. Olshanii, "Thermalization and its mechanism for generic isolated quantum systems," Nature **452**, 854 (2008).

M. Rigol and B. S. Shastry, "Drude weight in systems with open boundary conditions," Phys. Rev. B **77**, 161101 (2008).

M. Rigol, "Breakdown of thermalization in finite one-dimensional systems," Phys. Rev. Lett. **103**, 100403 (2009).

M. Rigol, B. S. Shastry, and S. Haas, "Fidelity and superconductivity in two-dimensional t-J models," Phys. Rev. B **80**, 094529 (2009).

Urbach, Jeffrey S.

D. R. Sisan, D. Yarar, C. M. Waterman, J. S. Urbach, "Event ordering in live cell imaging determined from temporal cross correlation asymmetry," Biophysical Journal, in press.

C. E. Graves, R. G. McAllister, W. J. Rosoff, and J. S. Urbach, "Optical neuronal guidance in three-dimensional matrices," Journal of Neuroscience Methods **179**, 278 (2009).

F. Vega Reyes and J. S. Urbach, "Steady base states for Navier-Stokes granular hydrodynamics with boundary heating and shear," Journal of Fluid Mechanics **636**, 279 (2009).

R. G. McAllister, D. R. Sisan, and J. S. Urbach, "Design and optimization of a high-speed, high-sensitivity, spinning disk confocal microscopy system," Journal of Biomedical Optics, Sep-Oct, **13**, 054508 (2008).

F. V. Reyes and J. S. Urbach, "Effect of inelasticity on the phase transitions of a thin vibrated granular layer," Physical Review E **78**, 051301 (2008).

Van Keuren, Edward

J. Mertzman, S. Kar, S. Lofland, T. Fleming, E. Van Keuren, Y. Tong, S. Stoll, "Surface attached manganese-oxo clusters as potential contrast agents," Chem. Commun. **7**, 788 (2009).

R. Ross, C. Cardona, D. Guldi, S. Sankaranarayanan, M. Reese, N. Kopidakis, J. Peet, B. Walker, G. Bazan, E. Van Keuren, B. Holloway, and M. Drees, "Endohedral fullereness for the advancement of organic photovoltaic devices," Nature Mater. **8**, 208 (2009).

C. Ma, Q. Zhang, E. Van Keuren, "Analysis of symmetric and asymmetric nanoscale slab slot waveguides," Opt. Commun. **282**, 324 (2009).

C. Ma, Q. Zhang, and E. Van Keuren, "Right-angle slot waveguide bends with high bending efficiency," Opt. Express **16**, 14330 (2008).

E. Van Keuren, A. Bone, and C. Ma, "Phthalocyanine Nanoparticle Formation in Supersaturated Solutions," Langmuir **24**, 6079 (2008).

HOWARD UNIVERSITY

DEPARTMENT OF PHYSICS AND ASTRONOMY

Washington, D.C. 20059

Students Accepted For Degree	FIELDS		
	Physics	Astronomy	Related Fields
Doctorate	X		
Master's	X		

1. General

President: Sidney A. Ribeau
Dean of Graduate School: Charles L. Betsey
Department Chairman: Prabhakar Misra
Department Telephone Number: (202) 806-6245
Type of Institution: University
Control: Private
Setting: Urban
Total Faculty: 2,000
Total Graduate Faculty: 310
Total Students: 11,000
Total Graduate Students: 1,066
Annual Graduate Tuition:
In-state residents: Full-time—$10,998/sem.
Part-time—$1,222/credit
Out-of-state residents: Full-time—$10,998/sem.
Part-time—$1,222/credit
Tuition rates for: 2010–11
Deferred tuition plan: Yes
Other Fees: $510/sem.
Term: Semester

2. Number of Faculty in Department

Total full time faculty in three professorial ranks is 16. The combined total of full-time, part-time and other faculty at all ranks is 18.

3. Admission, Financial Aid, and Housing

Address admission inquiries to: Admissions: the Graduate School of Arts and Sciences
Graduate application fee required: $45
Admission deadline (Fall admission): 4/1
Admission information: For fall admission, 2010–11, 4 students were accepted from 6 applicants.
Admission requirements: For admission to the graduate programs, a Bachelor's degree in physics or a closely related field is required with a minimum undergraduate GPA of 3.0 specified. The GRE is required. The minimum acceptable score suggested for admission is verbal–450; quantitative–550; total–1,000. The GRE Advanced is recommended. Students from non-English speaking countries are required to demonstrate proficiency in English via the TOEFL exam. The minimum acceptable score is 550 on paper-based exam and 213 on computer-based exam.
Undergraduate preparation assumed: Marion, *Mechanics*; Reitz, Milford and Christy, *Electricity and Magnetism*; Jenkins and White, *Optics*: Zemansky, *Thermodynamics*; Tipler, *Atomic Physics*.
Address financial aid inquiries to: Prabhakar Misra, Chairman, Dept. of Physics and Astronomy
GAPSFAS application required: Yes
Financial aid deadline: 4/1
Loans available: No
Address housing inquiries to: Student Housing, Administration Building
On-campus, triple/double studio apartment: Yes
Cost: $2,575/1,288 per semester
On-campus, married student housing available: No

Table A—Faculty, Enrollments, and Degrees Granted

Research Specialty	2009–10 Faculty	Enrollment[1] Fall 2010		No. of Degrees Granted[2] 2009 (2004–09)			Median No. of Years for 2008–09 Ph.D.'s
		Master's	Doctorate	Master's	Terminal Master's	Doctorate	
Atmos./Space Phys.,	3	0	2	0(1)	–	0(1)	–
Atomic, Molecular, & Optical Physics	3	0	5	0(0)	–	0(1)	–
Biophysics	2	0	2	0(0)	–	0(0)	–
Condensed Matter Physics	6	1	3	0(0)	–	0(0)	–
Particles & Fields	2	0	1	0(0)	–	0(0)	–
Non-specialized	2	2	0	0(0)	–	0(0)	–
Total		3	13	0(1)	–	0(2)	
Full-time Grad. Stud.		3	13				
Part-time Grad. Stud.		2					
First-year Grad. Stud.		4					
Median Years in Grad. Study (2009–10 Degrees)					–	–	4

Undergraduate Degrees, 2009–10:1, (2005–10): 18

[1]Students not yet committed to a research specialty are entered under non-specialized.
[2]Five-year totals in parentheses.

4. Graduate Degree Requirements

Master's: Modern Physics, Mathematical Methods in Physics, and Advanced Laboratory are recommended. Classical Mechanics, Electromagnetic Theory, Quantum Mechanics, and Statistical Mechanics may be required. Both thesis and non-thesis/comprehensive exam options are available.
Doctorate: A total of 72 hours, including the M.S. requirements and Dissertation are required. The student is required to pass a qualifying examination. Other advanced courses, such as Solid State Physics or Astrophysics, are strongly recommended.
Thesis: Thesis may be written *in absentia*.
Special Equipment, Facilities, or Programs: The department conducts extensive Atmospheric studies through the NASAURC and the NOAA Center for Atmospheric Sciences. The Department houses a modern Laser Spectroscopy Laboratory, a String Theory Group, Experimental Condensed Matter facilities, and a Computational Physics Laboratory.

Table B—Appointments to Graduate Students, 2009–10

Title of Appointee	Appointments Total	First year	Academic Load Allowed in Credit Hours	Hours of Service Per Week	Stipend for Academic Year ($)
			Semester		
Teaching Assistant	8	1	12	20	12,000–16,000[1]
Research Assistant	2	0	12	20	15,000–24,000[1]
Fellowship	7	1	12	–	15,000–30,000[1]
Total	17	2			

[1] In addition, remission of tuition is included.

5. Personnel Engaged in Separately Budgeted Research, 7/09–6/10

Professorial faculty	9
Graduate students	10
Undergraduate students	11
Total	30

6. Separately Budgeted Research Expenditures by Source of Support

	Departmental Research	Physics-related Research Outside Department
Federal government	$2,609,200	$
Total	$2,609,200	$

Table C—Separately Budgeted Research Expenditures

Research Specialty	No. of Grants	Expenditures ($)
Atmos./Space Phys.	6	1,154,000
Atomic, Molecular, & Optical Physics	4	397,200
Condensed Mater	4	296,000
Particle & Fields	1	82,000
Physics Education	2	680,000
Total	17	2,609,200

FACULTY

Professors

Batra, Anand P., Ph.D., Rensselaer Polytechnic Institute, 1966. Solid state physics.

Catchings, Robert M., Ph.D., Wayne State University, 1970. Solid state physics.

Hubsch, Tristan, Ph.D., University of Maryland (College Park), 1987. Elementary particle physics field theory; strings.

Kushawaha, Vikram S., Ph.D., Benaras Hindu University (India), 1973. Laser spectroscopy; atomic physics.

Lindesay, James, Ph.D., Stanford University, 1981. Field theory; computational physics.

Lowe, Walter P., Ph.D., Stanford University, 1983. Condensed matter physics; synchrotron radiation studies.

Misra, Prabhakar, Ph.D., Ohio State University, 1986. Chairman. Molecular spectroscopy; chemical physics.

Salu, Yehuda, Ph.D., University of Tel Aviv (Israel), 1973. Biophysics; medical physics.

Thorpe, Arthur N., Ph.D., Howard University , 1964. Magnetism; solid state physics.

Venable, Demetrius D., Ph.D., American University, 1974. Optical physics.

Associate Professors

Alfred, Marcus, Ph.D., Howard University, 2000. Computational physics; particles and fields.

Demoz, Belay, Ph.D., University of Nevada (Reno), 1985. Atmospheric Physics.

Jenkins, Gregory, Ph.D., University of Michigan, 1991. Atmospheric science.

Joseph, Everette, Ph.D., State University of New York at Albany, Albany, 1997. Atmospheric science.

Assistant Professors

Gatica, Silvina, Ph.D., University of Buenos Aires, 1995. Physics.

Stancil, Kimani, Ph.D., Massachusetts Institute of Technology, 2002. Physics.

Lecturers

Finch, Tehani, Ph.D., Howard University, 2008. Physics.

Instiful, Peter, Ph.D., Howard University, 2007.

Research, Visiting, & Adjunct Faculty

Carruthers, George, Ph.D., University Illinois, 1964. Aerospace and Astronautical Engineering.

Pass, Barry, Ph.D., Rutgers University, 1968. Physics.

Romanyukha, Alex, Ph.D., Russian Academy of Sciences, 1985. Metal Physics.

Senftle, Frank, Ph.D., University of Toronto, 1947. Physics

Ting, Antonio, Ph.D., University of Maryland, 1984. Physics.

RESEARCH SPECIALTIES AND STAFF

Atmospheric Physics. Demoz, Jenkins, Joseph, Venable.

Computational Physics. Alfred, Lindesay.

Condensed Matter. Batra, Catchings, Lowe, Thorpe, Gatica, Stancil.

Optical Physics. Kushawaha, Misra, Venable.

Particle Physics. Alfred, Hubsch, Lindesay.

Spectroscopy. Misra.

Staff

Anne E. Cooke, Administrative Assistant
Ronald W. Crutchfield, Technician
Julius R. Grant, Machinist/Research Technician
Frederick C. Marsh, Technician
Paulette Parkinson-Copeland, Graduate Administrative Aid

CATHOLIC UNIVERSITY OF AMERICA

DEPARTMENT OF PHYSICS

Washington, D.C. 20064

Students Accepted For Degree	FIELDS		
	Physics	Astronomy	Related Fields
Doctorate	X		X
Master's	X		X

1. General

President: John H. Garvey
Dean of Arts and Sciences: Lawrence R. Poos
Department Chairman: Daniel I. Sober
Department Telephone Number: (202) 319–5315
Fax No: (202) 319-4448
Type of Institution: University
Control: Private
Setting: Urban
Total Faculty: 344 Full-time, 370 Part-time
Total Graduate Faculty: 342
Total Students: 6,470
Total Graduate Students: 3,160
Annual Graduate Tuition:
All Graduate Students: Full-time—$31,740/yr.
$15,870/semester
Part-time—$1,245/credit
Tuition rates for: 2009–10
Deferred tuition plan: Yes
Other Fees: $1,110
Term: Semester

2. Number of Faculty in Department

The combined total of full-time faculty in the three professorial ranks is 12. The combined total of full-time, part-time, and other faculty at all ranks is 31.

3. Admission, Financial Aid, and Housing

Address admission inquiries to: Chairman, Dept. of Physics
Graduate application fee required: $55
Admission deadline (Fall admission): 8/1
Admission information: For fall admission, 2007–08, 15 students were accepted from 50 applicants.
Admission requirements: Applications are accepted from students with Bachelors' degrees whose background in physics and mathematics is equivalent to a physics major beginning his senior year. A grade point average of 2.5 is required for admission. The GRE is required, and most graduate fellowships require a combined verbal and quantitative score of 1300. Students from non-English speaking countries are required to demonstrate proficiency in English via the TOEFL exam. Minimum acceptable score for admission is 600.
Undergraduate preparation assumed: Taylor, *Classical Mechanics*; Griffiths, *Introduction to Electrodynamics*; Griffiths, *Introduction to Quantum Mechanics*.
Address financial aid inquiries to: Chairman, Department of Physics
GAPSFAS application required: Yes
Financial aid deadline: 2/15
Loans available: Yes
Address housing inquiries to: Housing Director
On-campus, single student housing available: Yes
On-campus, married student housing available: Yes

Table A—Faculty, Enrollments, and Degrees Granted

Research Specialty	2009–10 Faculty	Enrollment[1] Fall 2009		No. of Degrees Granted[2] 2009–10 (2005–10)			Median No. of Years for 2009–10 Ph.D.'s
		Master's	Doctorate	Master's	Terminal Master's	Doctorate	
Astrophysics	17	0	11	0(0)	0(0)	1(8)	–
Biophysics	2	0	1	0(0)	0(0)	0(2)	–
Condensed Matter Physics	1	0	0	0(0)	0(0)	0(0)	–
Energy Sources & Environ.	3	0	0	0(0)	0(0)	0(0)	–
Materials Science	7	0	7	0(0)	0(0)	1(4)	
Nanotechnology	2	0	2	0(0)	0(0)	0(0)	–
Nuclear Environmental Protection	2	0	–	0(0)	0(0)	0(0)	
Nuclear Physics	3	0	3	0(0)	0(0)	0(1)	
Non-specialized	0	0	2	4(18)	1(6)	0(0)	–
Total		0	26	4(18)	1(6)	2(15)	6.0
Full-time Grad. Stud.		0	21				
Part-time Grad. Stud.		0	5				
First-year Grad. Stud.		0	3				
Median Years in Grad. Study (2009–10 Degrees)		2	2	6	0	0	–

Undergraduate Degrees, 2009–10 (2005–10): 1(8)

[1]Students not yet committed to a research specialty are entered under non-specialized.
[2]Five-year totals in parentheses.

4. Graduate Degree Requirements

Master's: 30 graduate credits required with a GPA of 2.5, of which six may be dissertation guidance if the thesis option is selected. A Master's comprehensive, one year of residency, but no language is required.
Doctorate: 53 graduate credits are required of which 35 must be in physics. Written and oral comprehensives are given after completion of 35 credits and must be passed before beginning thesis work. Students must maintain a GPA of 3.0 in order to qualify to take the comprehensives. Three year residency required. No language is required.
Thesis: Thesis may not be written *in absentia*.
Special Equipment, Facilities, or Programs: The department has cooperative degree programs with many government agencies in the Washington area. Examples are the Naval Research Laboratory, the Naval Surface Warfare Center, Goddard Space Flight Center, and the National Institute for Standards and Technology. CUA is a member of an eleven - university consortium in the Washington, DC area. Courses maybe taken at other Consortium universities.

Table B—Appointments to Graduate Students, 2009–10

Title of Appointee	Appointments Total	Appointments First year	Academic Load Allowed in Credit Hours	Hours of Service Per Week	Stipend for Academic Year ($)
			Semester		
Teaching Assistant	4	2	12	20	18,000[1,2]
Research Assistant	16	1	12	20	18,000–21,000[1,3]
Hubbard Fellowship	1	1	12	0	20,000[1,3]
Total	21	3			

[1]Tuition waived.
[2]Summer appointments may be available with additional stipend.
[3]Calendar year.

5. Personnel Engaged in Separately Budgeted Research, 6/09–5/10

Professorial faculty	11
Other faculty	14
Postdoctoral appointments	7
Graduate students	24
Undergraduate students	4
Nonteaching research personnel	57
Total	117

6. Separately Budgeted Research Expenditures by Source of Support

	Departmental Research	Physics-related Research Outside Department
Federal government	$24,500,000	$0
Business and industry	455,000	
Total	$24,955,000	$0

7. Separately funded and managed centers

The Institute for Astrophysics and Computational Science (IACS) employs 4 postdocs and 44 research scientists in addition to three teaching faculty members. The Vitreous State Laboratory (VSL) employs 2 postdocs and 27 research scientists in addition to four teaching faculty members in the areas of materials science, energy sources and environment, and biophysics. (Director: Ian L. Pegg).

8. Extension Centers and Summer Programs

Catholic U. is a member of an eleven-University Consortium in the Washington, D.C. area. Credits are transferrable.

Table C—Separately Budgeted Research Expenditures

Research Specialty	No. of Grants	Expenditures ($)
Astrophysics	105	5,464,000
Biophysics, Energy Sources & Environ., and Materials Science	40	6,097,000
Nuclear Physics	1	210,000
Total	141	11,771,000

FACULTY

Professor Emeritus

Crannell, Hall L., Ph.D., Stanford, 1964. Experimental nuclear and medium energy physics.
Leibowitz, Jack R., Ph.D., Brown, 1962. Low temperature and condensed matter physics.
Meijer, Paul H. E., Ph.D., Leiden, Netherlands, 1950. Theoretical physics; statistical mechanics and solid state.
Überall, Herbert M., Ph.D., Vienna, 1953 and Cornell, 1956. Theoretical physics; nuclear and acoustics.
Werntz, Carl, Ph.D., Minnesota, 1960. Theoretical physics; nuclear astrophysics.

Professors

Bruhweiler, Frederick, Ph.D., Texas, 1977. Theoretical and observational astrophysics.
Kraemer, Steven B., Maryland, 1985. Theoretical and observational astrophysics.
Macedo, Pedro, Ph.D., Catholic, 1963. Experimental materials science.
Pegg, Ian L., Ph.D., Sheffield, England, 1982. Experimental materials science.
Resca, Lorenzo, Ph.D., Scuola Normale, Superiore di Pisa, Italy, 1970. Theoretical solid state.
Sober, Daniel I., Ph.D., Cornell, 1969. Experimental hadronic and intermediate-energy nuclear physics. medium energy physics.

Associate Professors

DeMello, Duilia, Ph.D., University of São Paulo, 1995. Theoretical and observational astrophysics.
Dutta, Biprodas, Ph.D., Vanderbilt University, 1987. Materials science, nanotechnology.
Klein, Franz J., Ph.D., Bonn, 1996. Experimental hadronic and intermediate-energy nuclear physics.

Assistant Professor

Horn, Tania, Ph.D., Maryland, 2006. Experimental hadronic and intermediate-energy nuclear physics.
Philip, John, Ph.D., Indian Institute of Science, 2001. Materials science, nanotechnology.
Sarkar, Abhijit, Ph.D., Univ. of Illinois at Chicago, 2002. Experimental and theoretical Biophysics.

Adjunct Professors

Gopalswamy, Natchimuthukonar, Ph.D., Indian Institute of Science, 1982. Solar physics.
Kondo, Yoji, Ph.D., University of Pennsylvania, 1965. Astrophysics.
Resta, Raffaele, Ph.D., Pisa, Italy, 1969. Theoretical solid state physics.
Rust, David M., Ph.D., University of Colorado, 1966. Solar magnetic fields; astrophysics.

Adjunct Associate Professors

Disanti, Michael, Ph.D., University of Arizona, 1989. Astrophysics and planetary science.
Mehl, Patrick, Ph.D., Grenoble, 1987. Biophysics.

Adjunct Assistant Professor

Muller, Isabelle, Pierre et Marie Curie, 1986. Physical chemistry.

Research Professors

Aikin, Arthur, Ph.D., Pennsylvania State, 1960. Astrophysics.
Krasnopolsky, Vladimir, Ph.D., Moscow, 1972. Planetary and cometary atmospheres.
Michels, Donald J., Ph.D., Catholic Universtiy of America, 1970. Solar physics.
Ofman, Leon, Ph.D., University of Texas, Austin, 1992. Solar physics; astrophysics.

Research Associate Professors

Clark, Pamela, Ph.D., Maryland, 1979. Geochemistry/remote sensing.
Mohr, Robert K., Ph.D., Massachusetts, 1972. Experimental materials science.
Smith, Myron A., Ph.D., University of Arizona, 1971. Stellar astrophysics.
Starr, Richard D., Ph.D., Illinois, 1978. Gamma-ray astronomy.
Wahlgren, Glenn, Ph.D., Ohio State University, 1986. Astrophysics.

Research Assistant Professors

Chen, Peter, Ph.D., Case-Western Reserve, 1979. Astronomy and advanced optics.
Moran, Thomas, Ph.D., MIT, 1988. Solar physics.

RESEARCH SPECIALTIES AND STAFF

Theoretical

Astrophysics. MHD simulations of sun and solar environment (Ofman); photoionization modeling (Kraemer, Verner); dynamic mideling of solar system (Ipatov); radiative transfer in upper atmospheres (Kutepov).
Biophysics. Statistical mechanical modeling of DNA, proteins and single molecule experiments; statistical approaches to systems biology and biological processes. Sarkar.
Condensed Matter. Semiconductors, insulators and superlattices. Resca.

Experimental

Astrophysics. Solar physics (Brosius, Ofman); observational cosmology & AGNs (DeMello, Kreaemer, Bruhweiler); stellar physics (Wahlgren, Smith, Bruhweiler); upper atmosphere (Aikin); exoplanets and young planetary systems (Bruhweiler); planetary science (Krasnopolsky, Kutepov, Bonev, Villanueva).
Biophyscis. Single-molecule DNA-protein interaction studis using magnetic tweezers; interaction of electromagnetic radiation with cells. Mehl, Sarkar.
Energy Sources & Environmental Science. Nuclear waste treatment technologies; energy recovery; thermoelectric ceramic heterostructures; solar cells; low-carbon impact meterials. Dutta. Pegg, Mohr, Muller.
Materials Science. Structure and properties of novel glass-forming systems; thermoelectric ceramic oxides, semiconductor thin films, solar cells, magnetic thin films, thin film hetrostructures, magnetic tunnel junctions, spin valves, fly ash activity and geopolymers. Dutta, Macedo, Mohr, Muller, Pegg, Philip.
Nanotechnology. Nanoelectronics, nanospintronics, growth of semiconducting and metallic nanostructures, nanoscale device physics, nanoscale sensors, MEMS, bioMEMS. Dutta, Philip.
Nuclear Physics. Nuclear and hadronic structure studies with intermediate-energy electron and photon beams. Horn, Klein, Sober.

FLORIDA ATLANTIC UNIVERSITY

DEPARTMENT OF PHYSICS

Boca Raton, Florida 33431

Students Accepted For Degree	FIELDS		
	Physics	Astronomy	Related Fields
Doctorate	X		
Master's	X		

1. General

President: Mary Jane Saunders
Dean of Graduate Studies: Barry T. Rosson, Ph.D.
Department Chairman: Warner A. Miller, Ph.D.
Department Telephone Number: (561) 297-3380
Type of Institution: University
Control: Public
Setting: Suburban
Total Faculty: 792
Total Students: 27,707
Total Graduate Students: 3,648
Annual Graduate Tuition:
In-state residents: Full-time—$319.96/credit hr
Out-of-state residents: Full-time—$926.42/credit hr
Tuition rates for: 2010–11
Deferred tuition plan: No
Term: Semester

2. Number of Faculty in Department

The combined total of full-time faculty in the three professorial ranks is 17. The combined total of full-time, part-time, and other faculty at all ranks is 19.

3. Admission, Financial Aid, and Housing

Address admission inquiries to: Pedro Marronetti, Recruitment Department of Physics
Graduate application fee required: $30
Admission deadline (Fall admission): 7/1
Admission information: For fall admission, 20010–09, 7 students were accepted, for spring admission 2010–09, 4 students were accepted.
Admission requirements: For admission to the graduate programs, a Bachelor's degree in physics is required with a minimum undergraduate GPA of 3.0/4.0 specified. The GRE is required with a minimum acceptable score for admission of a verbal plus quantitative total of 1,100. The GRE Advanced is not required. Students from non-English speaking countries are required to demonstrate proficiency in English via the TOEFL exam. Minimum acceptable score for M.S./Ph.D. admission is 500 written, 173 computer-based, 61 internet-based.
Undergraduate preparation assumed: Reitz and Milford, *Foundations of Electromagnetic Theory*; Symon, *Mechanics*; Saxon, *Quantum Mechanics*;* Boyce and Deprima, *Elementary Differential Equations and Boundary Value Problems*; Reif, *Statistical and Thermal Physics*;* Boas, *Mathematical Methods in the Physical Sciences*.*

*May be taken during first year of graduate study

Address financial aid inquiries to: Office of Student Financial Aid, 561-297-3530
GAPSFAS application required: No
Financial aid deadline: 4/1
Loans available: No
Assistantships Available: Yes
Address housing inquiries to: Director, Student Housing
On-campus, single student housing available: Yes, limited availability
Contact Housing Office: 561-297-2880
On-campus, married student housing available: No

Table A—Faculty, Enrollments, and Degrees Granted

Research Specialty	2009–10 Faculty	Enrollment[1] Fall 2009		No. of Degrees Granted[2] 2009–10 (2004–09)			Median No. of Years for 2008–09 Ph.D.'s
		Master's	Doctorate	Master's	Terminal Master's	Doctorate	
Biophysics, Neurophysics, Material Physics & Condensed Matter	8	1	5	0(9)	0(11)	3(17)	–
General Relativistic Astrophysics & Quantum Gravity	7	0	9	1(4)	1(2)	0(8)	–
Geophysics	0	1		2			
Physics Education	2	0	0	0(1)	0	0(0)	–
Quamtum optics	2	0	1	0(0)	0(0)	0(0)	–
Other			5				
Theoretical/Math.	2	0	5	0(0)	–	0(0)	–
Total	21	2	25	2(13)	1(14)	1(24)	
Full-time Grad. Stud.		2	25				
Part-time Grad. Stud.		0	2				
First-year Grad. Stud.		1	7				
Median Years in Grad. Study (2008–09 Degrees)				2.3	3	4	–
Undergraduate Degrees, 2008–09: 5							

[1]Students not yet committed to a research specialty are entered under non-specialized.
[2]Seven-year totals in parentheses.

4. Graduate Degree Requirements

Master's: 30 credits in approved program with a 3.0 sustained GPA, including seven credits of thesis research. Students must be in residence for two semesters. Final thesis.
Doctorate: 50 credits in approved program with a 3.0 sustained GPA beyond the M.S., including 30 credits of dissertation research; comprehensive written examination covering mechanics, electromagnetism, quantum mechanics, and statistical mechanics. Dissertation and oral examination.
Other Programs: The MST in physics requires 30 credits with a 3.0 GPA which may include six thesis credits. In addition, a six credit internship requirement must be satisfied for students without teaching experience.
The Ed.D. degree in Curriculum and Instruction is offered for junior college teachers with physics as a first or second teaching field.
**Thesis*: Thesis may be written *in absentia*.*

Table B—Appointments to Graduate Students, 2009–10

Title of Appointee	Appointments Total	Appointments First year	Academic Load Allowed in Credit Hours	Hours of Service Per Week	Stipend for Academic Year ($)
			Semester		
Teaching Assistant	28	9	9	20	10,914 M.S.
Research Assistant	0	0	9	20	15,050 Ph.D.
Total	28	9			15,050 Ph.D.

5. Personnel Engaged in Separately Budgeted Research, 7/09–6/10

Professorial faculty	7
Graduate students	6
Total	13

6. Separately Budgeted Research Expenditures by Source of Support

	Departmental Research	Physics-related Research Outside Department
Federal government	$426,490	$
University	600,000	
Total	$1,026,490	$

Table C—Separately Budgeted Research Expenditures

Research Specialty	No. of Grants	Expenditures ($)
General Relativistic Astrophysics, Biophysics	9	426,490
Total	9	426,490

FACULTY

Professors

Leventouri, Theodora, Ph.D., Athens, 1972. Experimental condensed matter physics, biophysics. X-ray diffraction of biocompatible materials.

Miller, Warner A., Ph.D., The University of Texas at Austin, 1986. Chairman of the Department. Classical and quantum gravity; general relativistic astrophysics, numerical relativity, foundations of quantum mechanics.

Qiu, Shen-Li, Ph.D., CUNY, 1985. Experimental condensed matter; photoemission; electronic structure and magnetic behavior of metals and alloys.

Voss, Richard, Ph.D., UC Berkeley, 1975. Scientific visualization, fractal geometry and theory of chaos.

Wille, Luc T., Ph.D., Ghent, 1983. Theoretical condensed matter; alloys, high-T_c superconductivity.

Associate Professors

Beetle, Christopher, Ph.D., The Pennsylvania State University at University Park, 2000. Classical and quantum gravity; numerical relativity.

Fuchs, Armin, Ph.D., Stuttgart, 1990. Nonlinear dynamical systems; complex systems and brain sciences.

Jirsa, Viktor K., Ph.D., Stuttgart, 1996. Nonlinear dynamics of discrete and spatially extended systems; neural oscillators; modeling and analysis of spatio-temporal brain dynamics; neurocomputing; complex systems and brain sciences.

Lau, Andy W. C., Ph.D., University of California Santa Barbara, 2000. Theoretical soft condensed matter physics; biophysics and statistical mechanics.

Marronetti, Pedro, Ph.D., University of Notre Dame, 1999. Numerical relativity, relativistic astrophysics, neutron star binary systems, gravitational wave physics.

Tichy, Wolfgang, Ph.D., Cornell University, 2001. Numerical relativity, binary black hole systems, gravitational wave physics.

Assistant Professors

Sorge, Korey D., Ph.D., University of Tennessee Knoxville, 2002. Condensed matter physics.

Research Professors and Instructors

Chen, De Huai, Ph.D., CUNY, 1987.

Gross, Robert, Ph.D., Florida Atlantic University, 2002.

Guzman, Angela M., Ph.D., Ludwig-Maximilians Univ, Munich, Germany, 1984. Quantum optics.

Kreymerman, Grigoriy, Ph.D., Acad of Sciences, Soviet Union, 1989. Optics.

Martinez, Leonardo, M.S., State University of New York Albany, 1971.

Professors Emeriti

Bruenn, Stephen W., Ph.D., Columbia, 1968. Theoretical astrophysics; supernovae models; radiation transport.

Burnett, Clyde R., Ph.D., Wisconsin, 1951.

Dean, Nathan W., Ph.D., Cambridge, 1968. Theoretical elementary particle physics; mathematical finance.

Faulkner, John S., Ph.D., Ohio State, 1959. Theoretical physics; theory of alloys.

Jordan, Robin G., Ph.D., Sheffield, 1967. Experimental condensed matter; UV photoemission; alloys.

Lamborn, Bjorn N. A., Ph.D., Florida, 1962. Theoretical physics; nonlinear coupling of plasma waves; relativistic plasmas.

McGuire, James B., Ph.D., UCLA, 1963. Mathematical physics; three-body problem; statistical physics; quantum field theory.

Medina, Fernando D., Ph.D., Princeton, 1975. Experimental condensed matter physics; spectroscopic studies of solids.

RESEARCH SPECIALTIES AND STAFF

Theoretical

Astrophysics. Hydrostatic and hydrodynamic evolution of stars; supernovae models; gamma ray burst models; radiation transport; properties of matter at high densities and temperatures; structure of galactic halos; stellar astrophysics and spectroscopy. Bruenn, Marronetti, Miller.

Biophysics, soft condensed matter physics, neuroscience. Fuchs, Jirsa, Lau.

Computer Simulation of Materials. Simulation and modeling of ecosystems. Wille.

Condensed Matter. Theory of alloys; electronic structures; properties and photoemission; high-temperature superconductivity. Wille.

General relativity; gravitational wave astrophysics, canonical quantum gravity; regge calculus, black hole physics. Miller, Beetle, Bruenn, Marronetti, Tichy.

Statistical mechanics. Phase transitions; nonlinear phenomena, complex systems. Fuchs, Jirsa, Voss.

Experimental

Biophysics, x-ray scatter on biocompatible materials, magnetic nanoparticle physics. Leventouri, Sorge.

Condensed Matter. Electrical, magnetic, structural, and optical properties of solids; thin films; surface effects; alloys; high-temperature superconductors; bioceramics. Jordan, Leventouri, Medina, Qiu.

Optics, quantum optics, spectroscopy. Guzman, Kreymerman, Miller.

FLORIDA INSTITUTE OF TECHNOLOGY

DEPARTMENT OF PHYSICS AND SPACE SCIENCES

Melbourne, Florida 32901-6988

Students Accepted For Degree	FIELDS		
	Physics	Astronomy	Related Fields
Doctorate	X	X	X
Master's	X	X	X

1. General

President: Anthony Catanese
Associate Provost of Graduate School: Randall Alford
Department Head: Terry Oswalt
Department Telephone Number: (321) 674-8098 or (321) 674-8795
Type of Institution: University
Control: Private
Setting: Suburban
Total Faculty: 578
Total Graduate Faculty: 332
Total Students: 8,227
Total Graduate Students: 3,205
Annual Graduate Tuition:
All Graduate Students: $1,040/cr. hr.
Tuition rates for: 2010–11
Deferred tuition plan: No
Term: Semester

2. Number of Faculty in Department

The combined total of full-time faculty in the three professorial ranks is 17. The combined total of full-time, part-time, and other faculty at all ranks is 23.

3. Admission, Financial Aid, and Housing

Address admission inquiries to: Thomas Shea, Graduate Admissions
Graduate application fee required: $50 (Masters), $60 (Ph.D.)
Admission information: For fall admission, 2009–10, 22 students were accepted from 52 applicants.
Admission requirements: For admission to the graduate programs, a Bachelor's degree in physics, astronomy or related subjects is required with an undergraduate GPA of 3.0 on a 4.0 scale. The GRE is recommended (general and subject). Students from non-English speaking countries are required to demonstrate proficiency in English via the TOEFL exam. Minimum acceptable score for admission is 550 (213 computer based, 90 iBT) and a score of 600 (250 computer based, 100 iBT) is necessary to qualify for a teaching assistantship.
Undergraduate preparation assumed: Halliday, Resnick, Walker, *Fundamentals of Physics*; Griffiths, *ElectroDynanmics*; Thornton, *Classical*; Zemansky and Dittman, *Heat and Thermodynamics*; Griffith, *Introduction to Quantum Mechanics*.
Address financial aid inquiries to: Terry Oswalt, Physics and Space Sciences
GAPSFAS application required: No
Financial aid deadline: March 1st
Loans available: No
Address housing inquiries to: Graduate Admissions Office
On-campus, single student housing available: Yes
Cost/term: $3,590 (double), $3,880 (single), upper classman; *Cost/semester*: $4,160(apt)
On-campus, married student housing available: No

Table A—Faculty, Enrollments, and Degrees Granted

Research Specialty[3]	2009–10 Faculty	Enrollment[1] Fall 2009		No. of Degrees Granted[2] 2009–10 (2005–10)			Median No. of Years for 2009–10 Ph.D.'s
		Mas-ter's	Doc-torate	Mas-ter's	Terminal Master's	Doc-torate	
Astronomy	5	6	6	2(6)	–	0(0)	–
Astrophysics	5	2	3	0(0)	–	1(1)	–
Atmos./Space Phys., Cosmic Rays	6	6	5	2(7)	–	1(5)	–
Condensed Matter Physics	2	0	1	0(0)	–	0(1)	–
Engineering Physics	1	0	0	0(0)	–	0(0)	–
Geophysics	6	5	2	1(2)	–	0(0)	–
High Energy	3	2	5	1(4)	–	2(4)	
Space Expl. Research	1	0	0	2(1)	–	0(0)	
Optics	0	0	0	0(0)	–	0(0)	–
Physics Education	4	1	0	0(0)	–	0(0)	–
Total	17	23	22	8(20)	–	4(11)	5.0
Full-time Grad. Stud.		17	18				
Part-time Grad. Stud.		6	4				
First-year Grad. Stud.		3	6				
Median Years in Grad. Study (2005–10 Degrees)				1.9(1.9)	–	5.0(5.0)	–

Graduate Degrees 2009–10 (2005–10): 5(24)
Undergraduate Degrees, 2009–10 (2005–10): 25(147)

[1]Students not yet committed to a research specialty are entered under non-specialized.
[2]Five-year totals in parentheses.
[3]Some faculty are in more than one specialty.

4. Graduate Degree Requirements

Master's: 30 semester-credit-hours with a 3.0 GPA minimum. 2 semesters in residence. 2 comprehensives: 1 written in first semester of residence, one oral or written at completion of M.S. program. Thesis optional.
Doctorate: 45 semester-credit hours beyond the Master's or 75 beyond the Bachelor's. 3.2 GPA minimum. At least 2 calendar years, including one year in doctoral research. Comprehensive written exam before formal admission to candidacy, oral dissertation defense. Dissertation must be submitted to a major physics journal.
Other Programs: M.S. and Ph.D. in space sciences: same general requirements as for physics.
Thesis: Thesis may be written *in absentia*.

5. Special Equipment, Facilities, or Programs

Olin Physical Sciences building: The Department is located in a 70,000 sq.ft. building, which includes specialized laboratories, an observatory, a 3,500 sq.ft. high-bay lab and NASA-qualified clean room.
Astronomy and Astrophysics Laboratory: The Astronomy faculty and students work on a wide variety of topics, including the evolution of white dwarf stars, simulations of cataclysmic variable systems, astrophysical fluid dynamics, accretion phenomena, the physics and evolution of active galactic nuclei

and their jets, cosmology, exoplanets, solar and stellar atmospheres. Observations are conducted from the radio to the gamma-rays, including observations with Hubble Space Telescope, Chandra X-ray Observatory, and Spitzer Space Telescope. Members of the group are involved in the development of instrumentation for the SuperNova Acceleration Probe (SNAP) and in the Canari-Cam Science Team, a guaranteed-time effort on the 10.4m Gran Telescopio Canarias. Resources in the lab include Linux computers, astronomical data reduction packages such as IRAF, AIPS, CIAO, Ftools and HEADAS. The lab is also the control center for remote access to the SARA telescopes in Arizona and Chile and the 0.8m Ortega telescopes on campus.

Geospace Physics Laboratory (GPL): The Geospace Physics Laboratory hosts the space physics research activities of Florida Tech's Space Sciences program. GPL operates a 10-site meridional array of magnetometers up the east coast of the United States (MEASURE array). Measurements from various satellites are used for studying the magnetic wave energy propagation within the geospace environment and the dynamics of the earth's plasmasphere and the storm-time radiation belt. Research includes the study of energetic particle acceleration and propagation within the heliosphere and in interstellar space, involving energetic particle measurements from the Ulysses, ACE, Wind and CRRES spacecraft as well as numerical modeling of particle transport processes.

High Bay Physics and Space Sciences Laboratory: This two-story 3500 sq.ft. hall is available to projects that require a large foot-print and/or heavy utility access. Currently housed in this facility is a ground-link station for a planned experiment for the International Space Station, a 400 sq.ft. NASA-qualified clean room, a magnetic levitation track, the first node of the Florida Tech Grid Computing Project, and the starting payload for the first UNESCO satellite.

High-Energy Physics Laboratory (HEP): Presently, the HEP research is focused on the commisioning of the CMS experiment at CERN and data analysis. Since 2001, the Florida Tech group has responsibilities for calibration of the hadron calorimeters and precision alignment of the muon endcap detectors. The physics analyses are initially focused on measurements of the properties of the top and bottom quarks and search for new gauge bosons. With anticipated higher luminosities, our physics program will switch to search for the Higgs boson and more exotic phenomena at multi-TeV energy scale. Another main research area is the development and construction of a muon tomography system for detecting high-Z materials hidden in cargo, based on advanced micro-pattern gas detectors such as Gas Electron multipliers. The HEP lab houses a state-of-the-art Linux-based computing cluster with about 100 CPU cores that is used for muon tomography detector simulation work and serves as a Tier-3 site on the Open Science Grid for CMS data analysis. The group also conducts research and development on advanced particle detector technology for the Super-LHC upgrade programs. In addition, Florida Tech is a member of the PHENIX experiment at BNL's Relativistic Heavy Ion Collider, which is searching for a new state of matter dubbed the "quark-gluon plasma" and the L3 collaboration at the LEP accelerator.

Maglev Facility: The maglev facility houses a unique 43-foot magnetic levitation and propulsion track, the only such device at an academic insitution. It supports research in controls, aerodynamics, mechanical stability, super-conducting technology and electromagnetic acceleration levitation, to study the feasibility of maglev launch assist for future spacecraft.

SARA Observatories: Florida Tech is the administrative institution for the Southeastern Association for Research in Astronomy (SARA). SARA operates an automated 0.9-m telescope at Kitt Peak National Observatory near Tucson, Arizona and a 0.6m telescope at Cerro Tololo Interamerican Observatory in Chile. Both accessible via Internet and used primarily for CCD imaging and photometry. Ten percent of all observing time on these facilities are allocated to Florida Tech .

Ortega Observatory: A 0.8m Ritchey-Chretién telescope was installed on the FIT campus in Nov. 2007. The telescope is equipped with a wide-field CCD and medium resolution spectrograph.

Scanning Probe Microscopy Laboratory: This facility provides researchers with the ability to image the surface structure of a solid, and to probe the electronic surface properties of a material down to the atomic scale, using a scanning tunneling microscope (STM). This laboratory is also used to investigate novel applications of the STM (e.g., in the field of electrochemistry) and in the development of other types of scanning probe microscopes.

Table B—Appointments to Graduate Students, 2009–10

Title of Appointee	Appointments		Academic Load Allowed in Credit Hours	Hours of Service Per Week	Stipend for Academic Year ($)
	Total	First year			
			Semester		
Teaching Assistant	15	4	3–9	up to 20	13,462
Research Assistant	14	4	3–9	up to 20	12,926
Total	29	8			

5. Personnel Engaged in Separately Budgeted Research, 06/09–05/10

Professorial faculty	17
Graduate Students	20
Post Docs	5
Total	42

6. Separately Budgeted Research Expenditures by Source of Support, 06/09–5/10

	Departmental Research
Federal government	$2,320,901
Other	549,638
Total	$2,870,539

Table C—Separately Budgeted Research Expenditures, 07/08–06/09

Research Specialty	No. of Grants	Expenditures ($)
Astronomy/Astrophysics	19	533,334
Atmos. Phys.	8	372,999
Condensed Matter	2	32,944
High Energy	12	358,938
Physics Education	2	26,760
Space Exploration Research	3	562,193
Space Physics	8	141,646
Total	48	2,870,539

FACULTY

Professors

Baarmand, Marc M., Ph.D., University of Wisconsin-Madison, 1987. Experimental high-energy particle physics at CERN (CMS Experiment), hadroproduction of top and bottom quarks in pQCD, Higgs physics, particle detector technology, grid computing.

Baksay, Laszlo, Ph.D., RWTH Aachen, 1978. Experimental high-energy physics. (L3, CMS at CERN) ; Nuclear physics (PHENIX, BNL); Detector development; International physics education; UNESCO Basic Science; Engineering Physics.

Durrance, Samuel, PhD., University of Colorado, 1980. Space science missions and human space exploration.

Dwyer, Joseph, Ph.D., University of Chicago, 1994. Space physics, solar/heliospheric energy particle observations.

McCay, T. Dwayne, Ph.D., Auburn University, 1974. Provost of the University. Materials processing in space, engineering physics.

Oswalt, Terry D., Ph.D., Ohio State, 1981. Observational astrophysics; stellar evolution, stellar luminosity functions, binary stars, stellar chromospheric activity, and minor planets.

Rassoul, Hamid K., Ph.D., University of Texas, Dallas, 1987. Observation and modeling of auroras, photochemistry of the earth's upper atmosphere, and solar wind-magnetosphere interactions.

Wood, Matthew A., Ph.D., The University of Texas, 1990. Astrophysics, theory and observation of white dwarf stars and cataclysmic variables.

Zhang, Ming Ph.D., Massachusetts Institute of Technology, 1991. Cosmic radiation and interactions with the plasma and magnetic fields in the interstellar medium, the heliosphere, and magnetospheres.

Associate Professors

Hohlmann, Marcus, Ph.D., University of Chicago, 1997. Experimental high energy physics at CERN (L3, CMS), supersymmetry and Higgs searches, detector development.

Perlman, Eric S., Ph.D., University of Colorado, 1994. Active galactic nuclei, jets, observational cosmology, High-energy and multiwavelength astrophysics.

Turner, Niescja E., Ph.D., University of Colorado at Boulder, 2000. Space physics; inner magnetosphere; ring current; energetics of magnetic storms; physics and astronomy education research.

Wang, Ke-Gang, Ph.D., Chinese Academy of Sciences, 1992. Theoretical and computational Materials Physics, and Statistical Physics.

Assistant Professors

Batcheldor, Daniel, Ph.D., University of Hertfordshire, 2005. Active galactic nuclei, galactic dynamics, supermassive black holes.

Liu, Ningyu, Ph.D., Penn. St. U. 2006. Atmospheric electricity, space physics, computational electrodynamics, plasma physics, gas discharge phenomena, high performance computing.

Oluseyi, Hakeem M., Ph.D., Stanford University, 1999. Solar/stellar atmospheres, cosmology, history of astronomy, physics education, instrumentation development.

Sawyer, Benjamin M., M.S., Florida Institute of Technology, 1975. Physics education.

Director of Laboratories

Gering, James A., M.S., Indiana University, 1984. Physics education.

Research Professors

Mantovani, James G., Ph.D., Clemson, 1985. NASA Kennedy Space Center. Condensed matter theory and experiment, particularly surface physics and electron microscopy, Mars and Moon environment.

Principe, Edward L., Ph.D., Penn. St. U. 1996. Materials processing, semiconductor physics, metrology, laboratory management.

Distinguished Research Professor

Foing, Bernard, Ph.D., LPSP/ENSET, Paris 1983. Chief scientist of the European Space Agency Science Program. Space mission (SMART 1), Mars Express, Huygens Probe.

Professors Emeriti

Blatt, Joel H., Ph.D., University of Alabama, 1970. Applied optics and machine/human vision.

Patterson, James D., Ph.D., Kansas, 1962. Theoretical solid state; magnetism; narrow gap semiconductors.

Research Associates

Cara, Mihai, Ph.D., Purdue University, 2008. Active Galactic Nuclei.

Gamayunov, Konstantin, Ph.D., Institute of Terrestrial Magnetism, Ionosphere, and Radio Wave Propagation, Moscow, Russia, 1994. Space Plasma Physics, simulation of waves and energetic particles in the Earth's magnetosphere, energetic particle acceleration propagation in the heliosphere and interstellar medium.

Gnanvo, Kondo, Ph.D., E.N.S.T. de Brelagne, France. High energy physics at CERN, detector development.

Vodopiyanov, Igor, Ph.D., St. Petersburg Nuclear Physics Institute, 1995. High-energy physics at LEP and LHC.

Zhao, Jingkun, Ph.D., National Astronomical Obervatories of China, 2007. Effect of age on stellar activity and mass.

Zuo, Pingbing, Ph.D., Chinese Academy of Sciences, 2008. Observational and theoretical study on cosmic ray propagation in the heliosphere, heliospheric physics, space weather.

RESEARCH SPECIALTIES AND STAFF

Theoretical

Astrophysics: White dwarf and cataclysmic variables. (Wood). Age and evolution of the galaxy. Stellar evolution and rotation (Wood).

Atmospheric Sciences. Multiscale modeling of atmospheric electricity, sprites, jets and related phenomena (Dwyer, Liu, Rassoul).

Materials Physics: Multiscale modeling and property prediction; phase transformation; diffusion; soft materials (Wang).

Particle Physics: Modeling of heavy quark pair production in hadron collisions and tests of perburtative QCD (Baarmand, Baksay, Hohlmann).

Space Physics: Thermospheric-ionospheric coupling, space debris (Rassoul).

Experimental

Astronomy and Astrophysics: Spectroscopy and photometry of binary stars. Minor planets and Exoplanets (Durrance, Oluseyi, and Oswalt); Age and evolution of the galaxy (Oswalt); Stellar evolution and rotation (Oswalt, Wood); Active Galactic nuclei and quasars. (Oluseyi, Perlman), supermassive black holes (Batcheldor).

Astronomy: Development of next-generation ground and space-based observatories, e.g. the Large Synoptic Survey Telescope (Oluseyi). Multi-country research in ethnoastronomy in Africa (Oluseyi), Space-based observatories (Batcheldor), high contrast ratio astronomy (Batcheldor).

Condensed Matter Physics: Scanning tunneling microscopy and optical spectroscopy of semiconductors; Dielectric properties of insulators and granular materials; Mars/Lunar research with NASA KSC. (Durrance, Mantovani, Wang).

Condensed Matter: Carrier transport and generation in high-purity silicon; tests of optical, electrical, and scattering properties of two-dimensional nanostructures for next-generation detectors (Oluseyi).

Geospace Physics Laboratory (GPL): Observational and experimental research in space physics including planetary atmospheres, ionospheric and upper atmospheric physics (airglow and aurora, solar/heliospheric energetic particles, solar modulation of galactic and anomalous cosmic rays, cosmic ray interstellar propagation). (Dwyer, Liu, Rassoul, Turner, Zhang).

High-Energy Physics: Study of electron-positron collisions at LEP (L3 Experiment), heavy-ion collisions (PHENIX Experiment) at BNL, Gaseous Detector Development for Muon Radiography, search for Higgs and new phenomena at TeV energy scale, and proton-proton collisions at LHC (CMS Experiment) at CERN. (Baarmand, Baksay, Hohlmann).

Solar physics: Study of the generation of the Sun's corona and the sources and variability of solar EUV radiation using data from the SoHO, Hinode, TRACE, and STEREO satellites (Oluseyi).

Space Reseach

UNESCO IBSP small satellite project (Baksay, Oluseyi).

Physics Education

Introductory course and laboratory development; international physics education. (Baksay, Gering, Oluseyi, Sawyer, Turner).

Faculty Publications

Please see the website http://cos.fit.edu/pss

FLORIDA INTERNATIONAL UNIVERSITY

DEPARTMENT OF PHYSICS

Miami, Florida 33199

Students Accepted For Degree	FIELDS		
	Physics	Astronomy	Related Fields
Doctorate	X	X	
Master's	X	X	

1. General

President: Mark Rosenberg
Dean of Graduate School: Kevin O'Shea
Department Chairman: Walter Van Hamme
Department Telephone Number: (305) 348-2605 (C)
Type of Institution: University
Control: Public
Setting: Suburban
Total Faculty: 1,554
Total Graduate Faculty: 754
Total Students: 39,718
Total Graduate Students: 7,928
Annual Graduate Tuition:
In-state residents: Full-time—$8,176.32
Part-time—$340.68/credit
Out-of-state residents: Full-time—$20,274.24
Part-time—$844.76/credit
Tuition rates for: 2009–10
Deferred tuition plan: No
Other Fees: $241.98/semester
Term: Semester

2. Number of Faculty in Department

The combined total of full-time faculty in the three professorial ranks is 22. The combined total of full-time, part-time, and other faculty at all ranks is 23.

3. Admission, Financial Aid, and Housing

Address admission inquiries to: Graduate Chair, Physics Dept.
Graduate application fee required: $30
Admission deadline (Fall admission): 6/1 domestic 4/1 international
Admission information: For fall admission, 2009–10, 8 students were accepted.
Admission requirements: For admission to the graduate programs, a Bachelor's degree in physics is required with a minimum undergraduate GPA of 3.0 specified. The GRE is required. The minimum acceptable score required is total—1,120. The GRE Advanced is recommended. Students from non-English speaking countries are required to demonstrate proficiency in English via the TOEFL exam. Minimum acceptable score for admission is 80 (iBT).
Undergraduate preparation assumed: Halliday and Resnick, *Physics*; Reitz, Milford, and Christy, *Electromagnetic Theory*; Becker, *Classical Mechanics*; Liboff, *Quantum Mechanics*; Sears and Salinger, *Thermodynamics*.
Address financial aid inquiries to: Financial Aid Office, University Park
GAPSFAS application required: No
Financial aid deadline: 3/1 Late applications considered
Loans available: Yes
Address housing inquiries to: housing@fiu.edu
On-campus, single student housing available: Yes
On-campus, married student housing available: Yes

Table A—Faculty, Enrollments, and Degrees Granted

Research Specialty	2009–10 Faculty	Enrollment[1] Fall 2009		No. of Degrees Granted[2] 2009–10 (2005–10)			Median No. of Years for 2004–10 Ph.D.'s
		Master's	Doctorate	Master's	Terminal Master's	Doctorate	
Astronomy	3	1	2		0	0(1)	7
Atomic, Molecular, & Optical Physics	1	0	1	0(0)	1(0)	0(0)	6
Biophysics	3	0	5	0(1)	0(1)	0(3)	6
Condensed Matter Physics	3	0	8	0(5)	0(5)	0(1)	5
Nuclear/Particle Physics	0	1	11	0(3)	0(3)	1(6)	7
Physics Education Resrach	2	1	4	0	0	0	0
Non-Specialized	0	0	1	0	0	0	0
Total		3	29	0(9)	1(9)	1(11)	
Full-time Grad. Stud.		31					
Part-time Grad. Stud.		31					
First-year Grad. Stud.		8					
Undergraduate Degrees, 2009–10: 13							

[1]Students not yet committed to a research specialty are entered under non-specialized.
[2]Five-year totals in parentheses.

4. Graduate Degree Requirements

Master's: The program requires 45 credit hours including 15 hours of thesis work. Required courses include Mathematical Methods I, Computational Physics I, Advanced Quantum Mechanics I and II, Advanced Electromagnetic Theory I and II, Statistical Physics, and Advanced Classical Mechanics as well as one three-credit elective in the area of research specialization.
Doctorate: The program requires 90 credit hours including at least 24 hours of Dissertation credit. Course requirements are the same as for the Master's with the addition of three, three-credit electives. All doctoral candidates must pass the Ph.D. qualifying exam no later than three years after entering the program. The exam will be given twice per year and requires a detailed knowledge of all areas of undergraduate and first-year graduate physics. Upon completion of course work, the student shall propose a dissertation topic and defend the proposal before their dissertation committee.
Thesis: Thesis may be written *in absentia*.
Special Equipment, Facilities, or Programs: The department operates three solid state laboratories, and a bio-optics research laboratory. Several theoretical medium-energy/nuclear physics research programs are underway. The department is a member of the SARA consortium operating a 1-m telescope on Kitt Peak. The department has experimental nuclear/particle physics programs at the Thomas Jefferson National Accelerator Facility and at CERN's Large Hadron Collider.

Table B—Appointments to Graduate Students, 2009–10

Title of Appointee	Appointments Total	Appointments First year	Academic Load Allowed in Credit Hours	Hours of Service Per Week	Stipend for Academic Year ($)
			Semester		
Teaching Assistant	18	6	9	20	22,660
Research Assistant	11	2	9	20	22,660
Total	29	8			

5. Personnel Engaged in Separately Budgeted Research, 8/08–8/09

Other faculty	1
Postdoctoral appointments	6
Graduate students	15
Total	22

6. Separately Budgeted Research Expenditures Source of Support

	Departmental Research	Physics-related Research Outside Department
Federal government	$ 2.48M	$
Private	$150,000	$
Total	$ 2.63M	$

Table C—Separately Budgeted Research Expenditures

Research Specialty	No. of Grants	Expenditures ($)
Astronomy	1	120,000
Biophysics	2	226,000
Nuclear Physics	2	832,000
Particle Physics	1	516,000
Physics Ed. Research	2	715,000
Quantum Optics	1	110,000
Solid State Physics	2	107,000
Total	11	2,626,000

FACULTY

Professors

Boeglin, Werner, Ph.D., University of Basel, Switzerland, 1986. Experimental medium-energy nuclear physics; electron scattering.

Bone, R. A., Ph.D., University of the West Indies, 1971. Biophysics; human visual system.

Fiebig, H. R., Ph.D., Münster, 1980. Theoretical nuclear physics; structure of the nucleon.

Gerstman, Bernard, Ph.D., Princeton, 1981. Structure and function of biomolecules.

Markowitz, Pete, Ph.D., William and Mary, Virginia, 1992. Experimental medium-energy nuclear physics; electron scattering.

Maxwell, O. V., Ph.D., SUNY, Stony Brook, 1978. Theoretical medium-energy physics.

Raue, Brian, Ph.D., University of Indiana, 1993. Experimental medium-energy nuclear physics; electron scattering.

Van Hamme, Walter, Ph.D., Ghent, 1981. Stellar structure evolution; close binary systems.

Webb, James, Ph.D., Florida, 1987. Quasars. complex materials.

Zhu, Yifu, Ph.D., U. Virginia, 1987. Nonlinear optics.

Associate Professors

Darici, Yesim, Ph.D., Columbia, Missouri, 1985. Solid state physics; surface physics.

Kramer, Laird, Ph.D., Duke University, 1992. Physics education research.

Li, Wenzhi, Ph.D., Experimental condensed matter physics.

Narayanan, Rajamani, Ph.D., University of California, 1990. Theoretical nuclear physics; structure of the nucleon.

Reinhold, Jörg, Ph.D., Technische Universität München, 1995. Experimental medium-energy nuclear physics, electron scattering.

Sargsian, Misak, Ph.D., Yerevan Physics Institute Armenia, 1993. Theoretical nuclear physics.

Simpson, Caroline, Ph.D., Florida, 1995. Extra-galactic radio astronomy.

Wang, Xuewen, Ph.D., Iowa State, 1987. Theoretical solid state physics.

Assistant Professors

Chapagain, Prem, Ph.D., Theoretical Biophysics. Florida International University, 2005.

Brookes, David, Ph.D., Rutgers University, 2006. Physical education research.

Guo, Lei, Ph.D., Nuclear Physics, Vanderbilt University.

Rodríguez, Jorge, Ph.D., Particle Physics. University of Florida.

RESEARCH SPECIALTIES AND STAFF

Theoretical

Biophysics. The structure and function of biomolecules. Gerstman, Chapagain.

Medium-Energy Physics. Weak interactions in nuclei; pion–nucleus interactions; bag models; quark model of nucleons and nuclei; lattice calculations. Fiebig, Maxwell, Sargsian.

Solid State Physics. Electronic structure and total energy calculations; lattice dynamics. Wang.

Experimental

Astronomy. Stellar evolution and structure; quasars; extra galactic radio astronomy. Van Hamme, Webb, Simpson.

Biophysics. Physics of the visual system; macular pigment; temporal processing, protein folding, interactions of laser light with the human visual system. Bone.

Nuclear/particle Physics. Boeglin, Guo, Markowitz, Raue, Reinhold, Rodriguez.

Nonlinear Optics; Quantum Optics. Zhu.

Solid State Physics. Surface physics; LEED, Auger, XPS, UPS, EELS. Darici, Li.

Physics Education Research. Brookes, Kramer.

FLORIDA STATE UNIVERSITY

DEPARTMENT OF PHYSICS

Tallahassee, Florida 32306-4350

Students Accepted For Degree	FIELDS		
	Physics	Astronomy	Related Fields
Doctorate	X		X
Master's	X		X

1. General

President: Dr. Eric J. Barron
Dean of Graduate Studies: Nancy Marcus
Department Chairman: Dr. Mark A. Riley
Department Telephone Number: (850) 644-2868; Graduate office (850) 644-4473: email: graduate@phy.fsu.edu
Type of Institution: University
Control: Public
Setting: Urban
Total Faculty: 2,150
Total Students: 39,136
Total Graduate Students: 8,370
Annual Graduate Tuition:
In-state residents: Full-time—$308.89/credit hr.
Out-of-state residents: Full-time—$940.29/credit hr.
Tuition rates for: 2008–09
Deferred tuition plan: No
Other Fees: Transportation Access Fee—$6.50/per cr. hr.
Term: Semester
Campus Facility Use Fee: $20 per semester

2. Number of Faculty in Department

The combined total of full-time faculty in the three professorial ranks is 45.

3. Admission, Financial Aid, and Housing

Admission: Admission.fsu.edu
Graduate application fee required: $30
Admission deadline (Fall admission): 2/15
Admission information: For fall admission, 2008–09, 70 students received offers from 413 applicants. 39 accepted offers.
Admission requirements: For admission to the graduate programs, a Bachelor's degree in physics or related science is required with a minimum GPA of 3.0/4.0 during the last two years of undergraduate study. The GRE is required. The minimum acceptable score required for admission is total-verbal+quantitative=1,100. The GRE Advanced is not required. The average GRE scores for 2007–08 admissions were total-1,263 for verbal and quantitative. Students from non-English speaking countries are required to demonstrate proficiency in English via the TOEFL exam. Minimum acceptable score for admission is 550 paper based, 213 computer based or 80 IBTOEFL.
Undergraduate preparation assumed: Electricity and magnetism, Griffiths, *Intro Electrodynamics*; Mechanics, Marion, *Classical Dynamics*; Quantum mechanics, McGervey, *Intro to Modern Physics*; Optics, Hecht, *Optics*; Thermodynamics, Greimer, *Thermodynamics to Statistical Mechanics*.
Address financial aid inquiries to: For Graduate Assistantship Applications, Simon Capstick; Graduate Chair, Dept. of Physics.
GAPSFAS application required: No
Financial aid deadline: 2/15
Loans available: Yes
Address housing inquiries to: University Housing, 133 South Wildwood Drive, Tallahassee, FL 32306-4174 http://www.housing.fsu.edu/
On-campus, single student housing available: Yes
Cost/term: $380–575/month
On-campus, married student housing available: Yes
Cost/month: $355–606/month

Table A—Faculty, Enrollments, and Degrees Granted

Research Specialty	2008–09 Faculty	Enrollment[1] Fall 2008		No. of Degrees Granted[2] 2008–09 (2004–08)			Median No. of Years for 2008–09 Ph.D.'s
		Master's	Doctorate	Master's	Terminal Master's	Doctorate	
Astronomy		0	1	0(0)	0(0)	0(0)	–
Astrophysics		0	4	1(1)	1(0)	0(0)	–
Atomic, Molecular, & Optical Physics		0	1	0(2)	0(1)	0(1)	5.5
Biophysics		0	2	0(5)	0(3)	1(3)	5.5
Condensed Matter Physics		0	55	5(25)	1(6)	7(30)	5.5
Nuclear Physics		0	23	3(24)	2(9)	5(17)	5.5
Particles & Fields		0	14	3(8)	0(3)	4(8)	5.5
Non-specialized		2	12	0(0)	0(0)	0(0)	5.5
Total		1	116	14(61)	5(21)	23(46)	
Full-time Grad. Stud.		0	112				
Part-time Grad. Stud.		0	0				
First-year Grad. Stud.		1	27				
Median Years in Grad. Study (2008–09 Degrees)				2.5	2.5	5.5	5.5

Undergraduate Degrees, 2008–09 (2004–08): 29(77)

[1]Students not yet committed to a research specialty are entered under non-specialized.
[2]Five-year totals in parentheses.

4. Graduate Degree Requirements

Master's: To qualify for the M.S. degree students must either (a) complete 21 hours of graduate work of which at least 18 hours must be in courses numbered 5,000 and above, or (b) take a program of couses which is acceptable to their Supervisor Committee, and which includes four courses from a select list of advanced courses, and pass with a cumulative grade point average of no less than B all six core graduate courses, or c) pass a written Qualifying Examination in the areas of mechanics, statistical mechanics, electrodynamics and quantum mechanics. The classroom phase of the graduate program is designed to introduce students to the basic conceptual tools used in physics and to acquaint them with a variety of research areas. The well prepared incoming student will have had advanced undergraduate courses in Mechanics, Electricity and Magnetism, Modern Physics, Quantum Mechanics, Thermodynamics, and Optics, comparable to the following undergraduate courses at Florida State: PHY 3221 (Mechanics), PHY 4323–4324 (Electricity and Magnetism), PHY 3101 (Intermediate Modern Physics), PHY 4604 (Quantum Theory of Matter AB), PHY 4513 (Thermal and Statistical Physics), PHY 3424 (Optics). Students deficient in one or more of these areas should include in their graduate program whatever undergraduate courses are necessary to remedy these deficiencies. The core graduate courses which contain the material with which every research physicist should be familiar are: PHY 5246 (Theoretical Mechanics); PHY 5524 (Statistical Mechanics); PHY 5346 and PHY 5347 (Electrodynamics A and B); PHY 5645, PHY 5646, PHY 5667 and PHY 5670 (Quantum Mechanics A,B) and a third Quantum Mechanics course.

Examinations: To qualify for the M.S. degree students must pass an Oral Qualifying Examination on their graduate work. In the course Master's degree program, the Qualifying Examination focuses on the subjects of mechanics, statistical mechanics, electrodynamics and quantum mechanics. In the thesis program the oral examination will consist primarily of defense of thesis. Residency: A minimum of two semesters or the equivalent must be completed in residence (must be enrolled for a minimum of 12 semester hours per semester). No language requirement. Minimum GPA of 3.0/4.0.

Doctorate: To qualify for a Ph.D. students must (a) make a formal presentation of some explicit research accomplishment satisfactory to their tentative Ph.D. Supervisory Committee; (b) take a program of courses which is acceptable to their Supervisory Committee, and which includes three courses from a select list of advanced courses; (c) pass with a cumulative grade average of no less than B all six graduate cources; (d) teach two elementary laboratory sections for two semesters; (e) pass a written Qualifying Examination in the advanced under graduate areas of mechanics, statistical mechanics, electrodynamics, and quantum mechanics; (f) pass an PHY 8964 Oral Qualifying Examination in the broad area of their particular specialization within the field of physics; (g) carry out research leading to an acceptable dissertation; and (h) pass orally the PHY 8985 Dissertation Defense of their dissertation. Residency: after having finished 30 semester hours of graduate work or being awarded the Master's degree, the student must be continuously enrolled on the Florida State University campus for a minimum of 24 graduate semester hours credit in any period of 12 consecutive months.

Other Programs: Miscellaneous: The Department of Physics at Florida State University emphasizes strong collaborative efforts between the various research areas and research groups in the department and the strong collaboration of experimentalists and theorists. The location of the Dept. of Scientific Computing, *the Center for Materials Research and Technology*, and the National High Magnetic Field Laboratory has greatly enhanced the research in the department.

The National High Magnetic Field Laboratory is the only user facility of its kind in the Western Hemisphere. The laboratory develops and provides a variety of research magnets at the highest fields available in the world. The laboratory hosts roughly 1,000 scientists annually to perform experiments in scientific disciplines as diverse as biology, chemistry, engineering, geochemsitry, materials science, medicine, and physics.

This unique facility supports an extensive in-house research program that advances its scientific and technical capabilities. The in-house research program is built around leading scientists and engineers who concentrate on the study of strongly correlated electron systems, molecular conductors, magnetic materials, magnetic resonance, cryogenics, and new approaches to measuring materials properties in high magnetic fields. Research at the laboratory is opening new frontiers of science at high magnetic fields, which have enormous potential for commercial and industrial applications. The laboratory also has one of the world's foremost magnet and science technology groups, which designs and builds this new generation of magnets. In 1999, the lab brought online a new 45-Tesla hybrid magnet, the most powerful magnet of its kind in the world. In 2004, the laboratory commissioned the world's first ultra-wide bore 900 MHz NMR magnet for chemical and biomedical research. The National High Magnetic Field laboratory has many exciting research opportunities for graduate students who wish to pursue research at the edge of parameter space in any area of science utilizing these world-class resources and instrumentations.

Thesis: Thesis may be written *in absentia*.

Special Equipment, Facilities, or Programs: The department occupies three adjacent buildings: the Keen Building, an eight-story Physics Research Building, The Leroy Collins Research Laboratory Building, and an undergraduate physics classroom and laboratory building. Extensive experimental facilities include a 9.5-MV Super FN Tandem Van de Graaff accelerator with superconducting post accelerator, a precision Penning trap mass spectrometer, a detector development laboratory for high-energy particle detectors, high-resolution Fourier-transform IR spectrometers, an ion implantation facility, instrumentation for research at liquid helium temperature and thin-film preparation, UHV (including surface analysis, molecular beam epitaxy, and atomic cluster facilities), facilities for high- and low-temperature superconductivity, small-angle and standard x-ray diffractometry, crystal growth facilities and ion implantation facility, scanning electron and tunneling microscopy, image analysis, quasielastic light scattering, polarized electron energy loss spectroscopy, a He atom beam-crystal surface scattering apparatus, and a unique aerosol physics-electron irradiation system. In addition to in-house facilities, ongoing experiments use accelerator and other research equipment at Fermilab, Bates, Brookhaven, Oak Ridge, TJNAF, and CERN. Computational facilities at FSU include an IBM Multiprocessor Supercomputer, a state of the art visualization lab, and a 120 CPU Beowulf cluster in the department. Extensive networking facilities provide access to computers on- and off-campus.

Courses required for the M.S.: Both thesis and non-thesis programs are offered leading to the Master of Science degree.

Non-thesis degree.: To qualify for a non-thesis degree the student must complete at least thirty-three (33) hours in courses numbered 4000 or above, eighteen (18) of which must be in courses numbered 5000 or above. At least twenty-one (21) of the thirty-three (33) hours must be taken on a letter grade basis. At least three (3) of the courses must be from the seven core graduate courses listed above, including at least one Quantum Mechanics course.

Thesis degree: To qualify for a thesis degree, the student must submit an acceptable thesis and complete at least thirty (30) hours in courses numbered 4000 or above, eighteen (18) of which must be in courses numbered 5000 or above. At least eighteen (18) of the thirty (30) hours must be on a letter grade basis. No more than three (3) semester hours of PHY 5918 credit and three (3) semester hours of PHY 5940 credit may be counted toward the Master of Science degree. A minimum of six (6) hours of credit must be earned for the thesis.

Table B—Appointments to Graduate Students, 2008–09

Title of Appointee	Appointments Total	Appointments First year	Academic Load Allowed in Credit Hours	Hours of Service Per Week	Stipend for Academic Year ($)
			Semester		
Teaching Assistant	44	13	12[1]	20	18,000[2–4]
Research Assistant	68	0	12[1]	20	18,000[2–4]
Total	117	13			

[1]The academic load allowed for graduate students is 9–12 hours. The normal load for physics graduate students is 12 hours.
[2]Tuition is paid in addition to the above stipend.
[3]Some summer appointments available at $3,261–3,914.
[4]Higher figure for students with partial fellowship.

5. Personnel Engaged in Separately Budgeted Research, 7/08–6/09

Professorial faculty	45
Other faculty	12
Postdoctoral appointments	36
Graduate students	121
Undergraduate students	15
Total	229

6. Separately Budgeted Research Expenditures by Source of Support

	Departmental Research	Physics-related Research Outside Department
Federal government	$4,057,896.50	$28,525,636.27
Private	53,706.44	20,000.00
Total	$4,111,602.94	$28,545,636.27

Table C—Separately Budgeted Research Expenditures

Research Specialty	No. of Grants	Expenditures ($)
Astrophysics	5	115,279.23
Atomic, Molecular, & Optical Physics	3	350,546.00
Biophysics	3	268,854.85
Condensed Matter Physics	29	34,632,250.12
Nuclear Physics	8	1,996,236.20
Particles & Fields	1	360,000.00
Total	38	35,458,075.35

Table D—Physics-related Research Outside Department

Field and Unit Outside Department	No. of Grants	Expenditures ($)
National High Magnetic Field Laboratory	26	33,879,977.00
Center for Materials Research & Technology (MARTECH)	9	510,742.40
Institute of Molecular Biophysics		
School of Computational Science	1	109,673.47
Total	27	28,545,636.27

FACULTY

Professors

Berg, Bernd, Ph.D., Free Univ., Berlin, 1977. Theoretical physics; lattice gauge theory; computational physics.

Blessing, Susan, Ph.D., Indiana, 1989. Experimental high energy physics; elementary particle physics.

Boebinger, Gregory, Ph.D., Massachusetts Inst. of Technology, 1986. Magnetism; experimental condensed matter physics; correlated electron systems.

Bonesteel, Nicholas, Ph.D., Cornell, 1991. Theoretical physics; condensed matter physics; many-body theory; magnetism, Quantum Hall Effect.

Brooks, James, Ph.D., Oregon, 1973. Experimental physics; low temperature; high magnetic field condensed matter; organic conductors; quantum fluid physics.

Capstick, Simon, Ph.D., University of Toronto, 1986. Theoretical nuclear and particle physics; computational physics.

Cottle, Paul, Ph.D., Yale, 1986. Experimental heavy-ion nuclear physics, teacher preparation.

Dobrosavljevic, Vladimir, Ph.D., Brown University, 1988. Theoretical condensed matter physics; disordered systems and glasses; metal-insulator transitions.

Duke, Dennis, Ph.D., Iowa State, 1974. Theoretical physics, elementary particles, computational physics.

Hill, Stephen, Ph.D, University of Oxford, England, 1994. Experimental Condensed Matter.

Kemper, Kirby, Ph.D., Indiana, 1968. Experimental physics; polarization studies in heavy-ion reactions.

Manousakis, Efstratios, Ph.D., Illinois, 1985. Theoretical physics; condensed matter; many-body theory; superconductivity.

Owens, Joseph, Ph.D., Tufts, 1973. Theoretical physics; elementary particles.

Piekarewicz, Jorge, Ph.D., University of Pennsylvania, 1985. Theoretical nuclear physics; collective nuclear modes; equation of state of dense matter; neutron stars.

Prosper, Harrison B., Ph.D., Manchester, 1980. Experimental high-energy physics; particle physics.

Reina, Laura, Ph.D., Trieste, 1992. Theoretical high energy physics; elementary particles.

Rikvold, Per Arne, Ph.D., Temple, 1983. Theoretical condensed matter physics; statistical physics; surface and interface science.

Riley, Mark, Ph.D., Liverpool, 1985. Chair of Department. Experimental physics; nuclear structure.

Roberts, Winston, Ph.D., University of Guelph, 1988. Theoretical hadron physics.

Schlottmann, Pedro, Ph.D., Technical Univ., Munich, 1973. Theoretical physics; high-T_c superconductors; heavy fermions; magnetism.

Tabor, Samuel, Ph.D., Stanford, 1972. Experimental physics; nuclei far from stability; high-spin states in nuclei.

Van Winkle, David, Ph.D., Colorado, 1984. Experimental condensed matter physics; liquid crystal gels.

von Molnár, Stephen, Ph.D., California at Riverside, 1965. Experimental physics; correlation effects in electronic systems; magnetic semiconductors; magnetic nano-structures.

Wahl, Horst, Ph.D., Vienna, 1969. Experimental physics; elementary particles.

Xiong, Peng, Ph.D., Brown University, 1994. Experimental condensed matter physics; nano-biophysics systems.

Yang, Kun, Ph.D., Indiana University, 1994. Theoretical physics; condensed matter, computational physics.

Zhou, Huan-Xiang, Ph.D., Drexel University, 1988. Computational and experimental biophysics; protein stability folding; protein-protein interactions.

Associate Professors

Adams, Todd, Ph.D., University of Notre Dame, 1997. Experimental high energy physics, particle physics, supersymmetry.

Cao, Jianming, Professor, Ph.D., Rochester, 1996. Experimental condensed matter physics, ultrafast dynamics probed by Lasers.

Eugenio, Paul, Ph.D., University of Massachusetts, 1998. Experimental nuclear physics, quark-gluon structure of matter and hadron spectroscopy.

Hoeflich, Peter, Ph.D., University of Heidelberg, 1986. Theoretical astrophysics; dark energy, dark matter and dark ages.

Lind, David, Ph.D., Rice, 1986. Experimental condensed matter physics; magnetic superlattices.

Ng, Hon-Kie, Ph.D., McMaster, 1984. Experimental physics; far-infrared spectroscopy; superconductivity, highly correlated electron systems; spectroscopy in high magnetic fields.

Shaheen, Shahid, Ph.D., Ruhr-Bochum, 1985. Experimental condensed matter physics; permanent magnet materials and materials science.

Wiedenhoever, Ingo, Ph.D., Cologne University, 1995. Experimental nuclear physics; radioactive beams.

Assistant Professors

Askew, Andrew, Ph.D., Rice University, 2004. Experimental High Energy Physics.

Chiorescu, Irinel, Ph.D., CNRS-Grenoble, France, 2000. Experimental condensed matter physics; magnetic flux qubits.

Crede, Volker, Ph.D., Univ. of Bonn, Germany, 2000. Experimental nuclear physics; quark matter.

Fenley, Marcia, Ph.D., Rutgers, The State University of New Jersey, 1991. Computational biophysics; electrostatics in macromolecules.

Gerardy, Christopher, Ph.D., Dartmouth, 2002. Observational astronomy; stellar explosions.

Okui, Takemichi, Ph.D, University of California, Berkeley, 2003. Theoretical High Energy Physics.

Rogachev, Grigory, Ph.D., Moscow, 1999. Experimental nuclear physics; nucleosynthesis.

Vafek, Oskar, Ph.D., Johns Hopkins, 2003. Theoretical condensed matter physics; quantum phase transitions; superconductivity.

Volya, Alexander, Ph.D., Michigan State, 2000. Theoretical nuclear physics; nuclear structure models.

Warusavithana, Maitri, Ph.D., University of Illinois at Urbana-Champaign, 2005. Experimental Condensed Matter.

Staff Physicists

Balicas, Luis, Ph.D., University of Paris XI-Orsay, 1995; NHMFL Associate Scholar/Scientist.

Brunel, Louis-Claude, Ph.D., University of Lyon, France, 1970; NHMFL Scholar/Scientist.

Choi, Eun Sang, Ph.D., Seoul National University, 1998; NHMFL Assistant Scholar/Scientist.

Engle, Lloyd, Ph.D., Princeton University, 1987; NHMFL Scholar/Scientist.

Frawley, Anthony, Ph.D., Australian National University, 1977. Experimental physics; Relativistic heavy ion physics; QCD matter; Quark gluon plasma.

Jaroszynski, Jan, Ph.D., Institute of Physics of the Polish Academy of Sciences, 1991; NHMFL Assistant Scholar/Scientist.

Kuhns, Philip, Ph.D., College of William and Mary, 1983; NHMFL Scholar/Scientist.

McGill, Stephen, Ph.D., University Pennsylvania, 2004; NHMFL Assistant Scholar/Scientist.

Myers, Edmund, Ph.D., University of Oxford, UK, 1982. Experimental atomic physics; precision measurement.

Popovic, Dragana, Ph.D., Brown University, 1989; NHMFL Scholar/Scientist.

Reyes, Arneil, Ph.D., University of California-Riverside, 1990; NHMFL Scholar/Scientist.

Smirnov, Dmitry, Ph.D., A. F. loffe Physico-Technical Institute of Russian Academy of Sciences, 1996; NHMFL Assistant Scholar/Scientist.

Suslov, Alexey, Ph.D., A. F. Ioffe Physico-Technical Institute of the Russian Academy of Sciences, 1995; NHMFL Assistant Scholar/Scientist.

Tozer, Stanley, Ph.D., The Johns Hopkins University, 1986; NHMFL Scholar/Scientist.

Wang, Yong-Jie, Ph.D., State University of New York at Buffalo, 1993; NHMFL Associate in Research and Instrumentation Physicist.

RESEARCH SPECIALTIES AND STAFF

Theoretical

Astrophysics. Advanced stages of stellar evolution; core-collapse and thermonuclear supernovae; compact stellar remnants; thermonuclear astrophysical combustion; radiative transfer (Gerardy, Hoeflich).

Computational Biophysics. Electrostatics and dynamics of biomolecules in aqueous environments (Zhou, Fenley).

Condensed Matter. Many-body theory of magnetism; magnetic properties of solids; high-temperature superconductivity; heavy fermions; adsorption; phase transitions (Bonesteel, Dobrosavljevic, Manousakis, Rikvold, Schlottmann, Vafek, Yang).

Elementary Particles and Fields. Strong and electroweak interaction phenomenology in high-energy particle physics (Owens, Reina). Lattice gauge theory and numerical simulations of various physical systems (Berg, Duke).

Hadron Physics. (Roberts)

Nuclear Physics. Nuclear structure studies emphasizing transitions; collective nuclear modes; structure and electromagnetic interactions of baryons and nuclei; structure and phases of neutron stars (Capstick, Piekarewicz, Volya).

Experimental

Atomic and Molecular Physics. Precision atomic measurements using Penning trap (Myers).

Condensed Matter. Biomolecular ordering; nano/biophysics; liquid crystals; gels; spintronics; hard magnetic materials; surface physics; sub-picosecond spectroscopy; low- and high-temperature superconductivity; highly correlated electron systems; organic crystals; quantum qubits (Balicas, Boebinger, Brooks, Brunel, Cao, Chiorescu, Choi, Engel, Jaroszynski, Kuhns, Lind, McGill, Ng, Popovic, Reyes, Shaheen, Smirnov, Suslov, Tozer, von Molnár, Van Winkle, Wang, Wiebe, Xiong).

Elementary Particles and Fields. Collider physics; strong and electroweak interactions in high-energy particle physics (Adams, Blessing, S. Hagopian, Prosper, Wahl).

Nuclear Physics. Hadron spectroscopy; heavy-ion reactions and radioactive beams. Photoproduction of baryons and mesons; search for exotic and hybrid mesons; search for new strangeonia states; search for missing baryons, particle detector development and computational physics. Heavy-ion fusion and fragmentation studies. Properties of nuclear systems at high angular momentum and far from stability; laser-induced polarization; octupole structure in nuclei. Light-ion nuclear spectroscopy, alpha, beta, and gamma spectroscopy; relativistic heavy-ion reactions. (Cottle, Crede, Eugenio, Frawley, Kemper, Riley, Rogachev, Tabor, Wiedenhoever).

UNIVERSITY OF CENTRAL FLORIDA

DEPARTMENT OF PHYSICS

Orlando, Florida 32816-2385
http://www.physics.cos.ucf.edu/ and http://www.graduatecatalog.ucf.edu/

Students Accepted For Degree	FIELDS		
	Physics	Astronomy	Related Fields
Doctorate	X	X	X
Master's	X	X	X

1. General

President: John C. Hitt
Dean of Graduate Studies: Patricia Bishop
Department Chairperson: Talat S. Rahman
Department Telephone Number: (407) 823–2325
Department Email: physics@mail.ucf.edu
Type of Institution: University
Control: Public
Setting: Urban
Acres: 1,415
Degrees Granted: 171,659 as of Summer 2008
Total staff: 10
Total Faculty: 60
Total Student: 50,000 +
Total Physics Graduate Students: 92
Graduate Tuition rates for: 2009–10
Full Time In-state per year: $7,351.44
Full Time Out-of-state per year: $26,376.24
Deferred tuition plan: No
Term: Semester
Assistantships Available: Yes (Includes Health Insurance)

	In-State Grad (5000–7999)	Out-of-State Grad (5000–7999)
Tuition	$237.56	$237.56
Differential Tuition		
Out-of-State Fee	–	$754.96
Financial Aid Fee	11.87	11.87
Non-Resident Financial Aid Fee	–	37.74
Capital Improvement Trust Fund Fee	2.44	2.44
Building Fee	2.32	2.32
Transportation Access Fee*	7.94	7.94
Activity & Service Fee	10.64	10.64
Athletic Fee	12.68	12.68
Health Fee*	8.99	8.99
Technology Fee	11.87	11.87
Total/Per Credit Hour Fees	$306.31	$1,099.01

2. Number of Faculty in Department

There are 34 regular faculty, 3 active retired faculty and 23 joint, courtesy and affiliated faculty, making a total of 60.

3. Admission, Financial Aid, and Housing

Address admission inquiries to: Prof. H. Heinrich, Graduate Coordinator, Dept. of Physics, Univ. of Central Florida, Orlando, FL 32816-2385
Apply online: www.graduate.ucf.edu
Graduate application fee required: $30
Fall Admission deadline: International Students–January 15 and Domestic Students–July15
Spring Admission deadline: International Students–July 1 and Domestic Students–December 1
Admission information: For Fall 2010, 88 Ph.D. and 16 MS applications were received. 43 Ph.D., and 7 MS applications were accepted.
Admission requirements: For admission to the graduate programs, a Bachelors degree in Physics is required with a minimum undergraduate GPA of 3.0 (A=4.0) grade point average for the last 60 semester hours of credit earned towards the baccalaureate. A GRE score of at least 1,000 on the combined verbal-quantitative sections of the Aptitude Test is recommended for admission to the Ph.D./M.S. program. Students from non-English speaking countries are required to demonstrate proficiency in English via the TOEFL exam and the SPEAK test. Minimum acceptable score for the TOEFL exam for admission is 550 (or 220 on computer based exam). The SPEAK test is administered once the student arrives at the university. Minimum acceptable for the SPEAK test is 50.1 t can be taken up to 3 times in one year.
Undergraduate preparation assumed: Barger & Olsson, *Classical Mechanics*; Griffiths, *Electromagnetic Theory*; Tipler and Llewellyn, *Modern Physics*; Liboff, *Quantum Physics*; Schroeder, *Thermal Physics*.
Address financial aid inquiries to: Graduate Coordinator, Department of Physics
GAPSFAS application required: No
Fellowship deadline: 1/31
Assistantship deadline: 1/15
Loans available: Yes
Address housing inquiries to: Department of Housing & Residence Life, P.O. Box 163222, University of Central Florida, Orlando, FL 32816-3222
On-campus, single student housing available: Yes
On-campus, married student housing available: No

Table A—Faculty, Enrollments, and Degrees Granted

Research Specialty	2007–08 Faculty	Enrollment Fall 2009–10		No. of Degrees Granted[1] 2009–10			Median No. of Years for 2007–08 Ph.D.'s
		Master's	Doctorate	Master's	Terminal Master's	Doctorate	
Astronomy	5	–	–	–	–	–	–
Atmos./Space Phys., Cosmic Rays	3	–	–	–	–	–	–
Atomic, Molecular, & Optical Physics	2	–	–	–	–	–	–
Bio and Soft Matter Physics	4	–	–	–	–	–	–
Cond. Matter Phys.	19	–	–	–	–	–	–
Optics	3	–	–	–	–	–	–
Particles & Fields	1	–	–	–	–	–	–
Physics Education	5	–	–	–	–	–	–
Total		5	73	7(73)	–	10(49)	
Full-time Grad. Stud.		2	73				
First-year Grad. Stud.		1	21				
Median Years in Grad. Study (2007–08 Degrees)				–	–	–	–
Undergraduate Degrees, 2007–08 (2007–08): 0(6)							

[1]Nine-year totals in parentheses.

4. Graduate Degree Requirements

Master's: A total of 33 semester credit hours is required. The student has the option of choosing courses specialized in General Physics, Condensed Matter Physics and Optical Physics with either a thesis or a non-thesis option. All students must take a set of core courses. The thesis option requires additional semester hours of electives plus 6 semester hours of thesis. The non-thesis option requires electives plus a comprehensive exit exam.

The Physics of MS program also offers a Planetary Science track.

Doctorate: Students have the option of choosing from three specializations: General Physics, Condensed Matter Physics, and Optical Physics. A total of 72 semester credit hours, of which 15 are required dissertation hours, are needed for the doctoral degree. The remaining 57 hours are divided into 18 hours of core courses and a combination of specialization specific electives and research. Upon completion of the core, the student must take the written part of the Ph.D. candidacy examination.

The Physics Ph.D. program also offers a Planetary Science track with specific core course requirements geared towards this track.

For more information, please visit our website.

Special Equipment, Facilities, or Programs: The Department operates the Surface Physics Laboratory, which includes a heavy ion backscattering spectrometer, 400-KEV ion implanter, imaging X-ray photoemission spectrometer, imaging secondary ion mass spectrometer, ultra-high vacuum atomic force microscope and scanning tunneling microscope, digital low energy electron diffraction, digital reflection high energy electron diffraction, angle resolved ultraviolet photoemission spectrometer, and equipment for ultra-thin epitaxial film growth. Other Department laboratories provide equipment for Mössbauer spectroscopy, magnetic susceptibility measurements, SQUID magnetometry, x-ray diffraction, Fourier spectroscopy, Raman and FTIR spectroscopy, inductively coupled plasma etching, e-beam thermal thin film evaporation, submicron optical litography, high-sensitivity on-chip magnetometry, ultra-low temperature high magnetic field high frequency EPR and FMR spectroscopy, and high temperature incubation of bacterial cultures. Equipment for performing biophysical research include an autoclave, a probe sonicator, a PCR thermocycler, a BioSafety hood, multiple centrifuges, and a soft-wall clean room. The Department also has two ultrahigh vacuum systems with an e-beam evaporator, a quartz microbalance, a hybrid atom/ion plasma source, an argon sputter gun, a mass spectrometer, XPS and Auger spectrometers, UPS, TPD, and a variable temperature STM. Computation facilities include three Linux Beowulf clusters with a total of 132 nodes.

Department faculty participate in several campus research centers which provide access to additional specialized equipment. Research in advance materials and microelectronics is supported by the Materials Characterization Facility (MCF), located in the campus Research Park. Department faculty and students using MCF have access to Rutherford backscattering spectroscopy, imaging secondary ion mass spectroscopy, transmission and scanning electron microscopes, a focused ion beam system, a scanning Auger microprobe, X-ray photoelectron spectroscopy, and x-ray diffraction. Faculty members also have access to a 128 processor, 64-bit Opteron cluster at the Interdisciplinary Information Science and Technology Laboratory.

The Department runs Robinson Observatory, an on-campus research and education facility that houses a 0.51-m f/8.2 Ritchey-Chretien astronomical telescope. The primary instrumentation on the telescope is a camera with a research-grade 3072-by-2048 pixel CCD that allows for multi-wavelength imaging.

Thesis: Thesis may be written *in absentia.*

Table B—Appointments to Graduate Students, 2010–11

Title of Appointee	Appointments Total	Appointments First year	Academic Load Allowed in Credit Hours	Hours of Service Per Week	Stipend for Academic Year ($)[2]
			Semester		
Teaching Assistant	36	16	9	20	14,700[1]
Research Assistant	56	5	9	20	14,700–18,000[1]
Total	92	21			

[1]Plus tuition waiver.
[2]Excluding summer.

5. Personnel Engaged in Separately Budgeted Research

Professorial faculty	39
Graduate students	47
Undergraduate students	30
Postdoctoral appointments	12
Total	128

6. Separately Budgeted Research Expenditures by Source of Support

Federal government	$2,252,953
State and local government	25,720
Business and industry	449,978
Total	$2,728,651

FACULTY

Professors

Campins, Humberto, Ph.D., Arizona, 1982. Planetary science; comets and asteroids.

Chang, Zhenghu, Ph.D., Chinese Academy of Sciences, 1998. Optics and Precision Mechanics.

Chernyak, Leonid, Ph.D., Weizmann Institute, 1996. Nanostructure device physics.

Chow, Lee, Ph.D., Clark, 1981. Experimental condensed matter; carbon nanotubes and diamond-like carbon films.

Johnson, Michael, D., Ph.D., Virginia, 1986. Theoretical condensed matter; nanostructures.

Luo, Weili, Ph.D., UCLA, 1989. Experimental condensed matter; complex systems.

Mucciolo, Eduardo, Ph.D., MIT, 1994. Theoretical condensed matter; nanoelectronics.

Peale, Robert E., Ph.D., Cornell, 1989. Experimental condensed matter; far-infrared semiconductor lasers.

Rahman, Talat S., Ph.D., Rochester, 1977. Theoretical condensed matter; surface physics.

Saha, Haripada, Ph.D., Calcutta, 1978. Theoretical atomic, molecular, and optical physics.

Schulte, Alfons F., Dr.rer.nat., Munich, 1985. Biophysics; dynamics of proteins and disordered systems.

Tonner, Brian P., Ph.D., Pennsylvania, 1982. Surface physics; x-ray spectroscopy.

Associate Professors

Bhattacharya, Aniket, Ph.D., Maryland, 1992. Theoretical soft condensed matter; nonlinear dynamics.
Britt, Daniel T., Ph.D., Brown, 1991. Planetary science; geophysics.
del Barco, Enrique, Ph.D., Barcelona, 2001. Experimental condensed matter; single-molecule magnets.
Colwell, Joshua, Ph.D., University of Colorado Boulder, 1989. Planetary science; Plateary rings.
Efthimiou, Costas, Ph.D., Cornell, 1996. Mathematical physics; physics education research.
Harrington, Joseph, Ph.D., MIT, 1994. Planetary science; spectroscopy.
Heinrich, Helge H., Dr.sc.nat., ETH Zurich, 1994. Electron microscopy, nanomaterials.
Ishigami, Masahiro, Ph.D., Maryland, 1999. Electronic properties; physics of graphene.
Kara, Abdelkader, Ph.D., Lille, Saclay, 1985. Theoretical condensed matter; Density Functional Theory.
Khondaker, Saiful I., Ph.D., Cambridge, 1999. Experimental condensed matter; nanoelectronics.
Kokoouline, Viatcheslav, Ph.D., St. Petersburg, 1999. Theoretical atomic and molecular physics.
Leuenberger, Michael N., Ph.D., Basel, 2002. Theoretical condensed matter; quantum computing.
Martin, Eduardo L., Ph.D., La Laguna, 1994. Evolution of low-mass stars; planetary science.
Roldán Cuenya, Beatriz, Ph.D., Duisburg, 2001. Surface physics; catalysis.
Schelling, Patrick K., Ph.D., Minnesota, 1999. Theoretical condensed matter and materials physics.
Stolbov, Sergei, Ph.D., Rostov, 1982. Theoretical condensed physics, catalysis.
Tatulain, Suren, Ph.D., St. Petersburg, 1979. Theoretical biophysics; catalysis.

Assistant Professors

Fernández, Yanga R., Ph.D., Maryland, 1999. Planetary science, comets and asteroids; observatory director.
Ishigami, Masahiro, Ph.D., California, 2004. Condensed matter and materials physics.

Research Professors

Klemm, Richard, PH.D., Harvard, 1974. Theoretical condensed matter; superconductivity; nanomagnetism.

Lecturers

Bindell, Jeffrey B., Ph.D., Brooklyn Poly, 1969. Physics education research.
Brueckner, Thomas J., Ph.D., Montana State, 1997. Physics education research; astronomy.
Cooney, James H., Ph.D., Florida, 2004. Physics education research; cosmology.
Dubey, Archana H., Ph.D., Bhavnagar, 1998. Ferromagnetic fluids.
Flitsiyan, Elena S., Ph.D., Moscow, 1975. Nuclear physics.
Montgomery, Michele M., Ph.D., FIT, 2004. Planetary science; astrophysics.
Velissarius, Christos, Ph.D., Rochester, 1995. Experimental high energy physics.
Zhmudsky, Oleksandr O., Ph.D., Kiev State, 1976. Theoretical physics; nonlinear dynamics.

Faculty Emeriti

Bose, Subir K., Ph.D., Allahabad, 1967. Statistical mechanics; quantum optics.
Brennan, J., Ph.D., Georgia Institute of Technology, 1968.
Llewellyn, Ralph A., Ph.D., Purdue, 1962. Low energy nuclear physics, physics education research.

Distinguished Research Professor

Liboff, Richard, New York University, 1961. Distinguished Professor.

AFFILIATED FACULTY

Nanoscience Technology Center

Hickman, James J., Ph.D.
Professor, Chemistry, Burnett College of Biomolecular Sciences, and Electrical Engineering
Masunov, Artëm E., Ph.D., Assistant Professor
Su, Ming, Ph.D., Assistant Professor
Zhai, Lei, Ph.D., Assistant Professor

The College of Optics and Photonics

Bass, Michael
Boreman, Glenn, D.
Delfyett, Peter J.
Dogariu, Aristide
Glebov, Leonid B.
Hagan, David J.
Kar, Aravinda
Kik, Pieter
Kuebler, Stephen
Richardson, Martin C.
Schoenfeld, Winston Vaughan
Silfvast, William
Soileau, M. J.
Stegeman, George I.
Van Stryland, Eric W.
Zeldovich, Boris Y.

RESEARCH SPECIALTIES AND STAFF

Theoretical

Atomic and Molecular Physics. Kokoouline, Saha.
Condensed Matter Physics. Bhattacharya, Johnson, Leuenberger, Kara, Klemm, Mucciolo, Rahman, Schelling, Stolbov.
Nonlinear dynamics. Zhmudsky.
Optical Physics. Saha.
Particles and Fields. Efhtimiou.
Physics Education. Bindell, Brueckner, Cooney, Efthimiou, Llewellyn.

Experimental

Astronomy and Planetary Sciences. Britt, Brueckner, Campins, Colwell, Fernandez, Harrington, Montgomery.
Biological Physics. Luo, Schulte, Tatulian, Tonner.
Condensed Matter Physics. Chernyak, Chow, del Barco, Dubey, Heinrich, Ishigami, Khondaker, Luo, Peale, Roldan, Schulte, Tonner.
Nuclear and Particle Physics. Flitsiyan, Velissaris.
Optical Physics. Chang, Peale, Schulte.
For inquiries please contact the department by e-mail at: physics@mail.ucf.edu

UNIVERSITY OF FLORIDA

DEPARTMENT OF PHYSICS

Gainesville, Florida 32611-8440

Students Accepted For Degree	FIELDS Physics	Astronomy	Related Fields
Doctorate	X		X
Master's	X		X

1. General

President: J. Bernard Machen
Dean of Graduate School: Henry T. Frierson
Department Chairman: John Yelton
Department Telephone Number: (352) 392-0521 (C)
Type of Institution: University
Control: Public
Setting: Small city
Total Faculty: 4,000
Total Graduate Faculty: 2,982
Total Students: approx. 50,000
Total Graduate Students: 16,536
Annual Graduate Tuition:
In-state residents: $8,059
Out-of-state residents: $22,603
Tuition rates for: 2008–09
Deferred tuition plan: No
Term: Semester

2. Number of Faculty in Department

The combined total of full-time faculty in the three professorial ranks is 44. The combined total of full-time, part-time, and other faculty at all ranks is 48.

3. Admission, Financial Aid, and Housing

Address admission inquiries to: Graduate Affairs Office, P.O. Box 118440
Graduate application fee required: $30
Admission deadline (Fall admission): 2/15
Admission information: For fall admission, 2010, 25 students.
Admission requirements: For admission to the graduate programs, a Bachelor's degree in physics is required with a minimum undergraduate GPA of 3.3 specified. The GRE and GRE Advanced are required. The minimum acceptable GRE score required for admission is total–1,200. The minimum acceptable GRE Advanced score suggested for admission is 560. The average GRE scores for Fall 2010 admissions were verbal–470; quantitative–750; total–1,220. The average GRE Advanced score for Fall 2009 admissions was 780. Students from non-English speaking countries are required to demonstrate proficiency in English via the TOEFL and the TSE exams. Minimum acceptable score for admission is 550 (paper-based), 213 (computer-based) or 80 (internet-based) on the TOEFL.
Undergraduate preparation assumed: Hecht, *Optics*; Gasiorowicz, *Quantum Mechanics*; Griffiths, *Electricity and Magnetism*; Marion, *Mechanics*; Zemansky and Kittel, *Thermal Physics*.
Address financial aid inquiries to: Graduate Affairs Office, P.O. Box 118440, Department of Physics, Gainesville, FL 32611-8440.
GAPSFAS application required: No
Financial aid deadline: 1/31
Loans available: Yes
Address housing inquiries to: Director of Housing (on-campus) or Off-Campus Housing Office
On-campus, single/married student housing available: Yes
Monthly: $380-$500 (1 br)
$430-$605 (2 br)

Table A—Faculty, Enrollments, and Degrees Granted

Research Specialty	2009–10 Faculty	Enrollment[1] Fall 2009 Mas-ter's	Doc-torate	No. of Degrees Granted[2] 2008–09 (2000–09) Mas-ter's	Terminal Master's	Doc-torate	Median No. of Years for 2006–09 Ph.D.'s
Astrophysics	8	0	15	1(3)	0(1)	6(13)	5.67
Biological physics	3	0	3	0(0)	0(4)	1(4)	5.67
Chemical Physics	3	0	9	4(7)	0(3)	2(9)	4.33
Condensed Matter Physics	16	0	41	3(25)	1(5)	17(53)	6.67
Low Temperature Physics	4	0	8	1(5)	0(3)	3(5)	8.33
Mathematical	0.5	0	2	0(1)	0(2)	0(0)	–
Particles & Fields	12	0	23	0(9)	0(3)	2(24)	6.00
Non-specialized	0	0	31	0(1)	4(16)	0(0)	–
Total	46.5	0	132	9(49)	4(32)	21(90)	6.11
Full-time Grad. Stud.		0	132				
Part-time Grad. Stud.		0	0				
First-year Grad. Stud.		0	25				
Median Years in Grad. Study (2006–07 Degrees)				2.34	2.73		6.11
Undergraduate Degrees, 2006–07 (2000–07): 31(153)							

[1]Students not yet committed to a research specialty are entered under non-specialized.
[2]Five-year totals in parentheses.

4. Graduate Degree Requirements

Master's: 30 graduate credits in an approved program with satisfactory performance; Master's exam required; non-thesis option available; one year residence required; no foreign language requirement.
Doctorate: Minimum of 90 hours and satisfactory performance in an approved course program is required, full-time residency for two consecutive semesters is required; no foreign language requirement; preliminary exam, dissertation, and dissertation exam required; teaching experience required; residence and a period of concentrated study required; publishable thesis is required.
Other Programs: Master of Science in Teaching requires 36 credits (half in graduate courses) with satisfactory performance in an approved program; six credits internship required; no thesis required. All candidates for advanced degrees must be registered during the semester in which the degree is awarded.
Thesis: Thesis may be written *in absentia*.
Special Equipment, Facilities, or Programs: Helium liquefier and Microkelvin Laboratory located on campus; library and superb computer facilities includes approximately 2000 processors running Linux 20 TB of disk. Graphics research is done on a 4-cpu SGI 4D/340 with video, Evans & Sutherland Freedom 3200. Mathematical Maple, Macsyma, PV-wave,

Wave front, Mat lab, S-Plus and Explorer software packages are available. Teaching Labs include 100 Pentium pcs with data acquisition hardware and software. Eight station electronics Laboratory is used for training in analog and digital circuits, instrumentation design and computer interfacing and programming. Highlights of the Advanced Physics Laboratory suite include FTIR spectrometer, pulsed NMR apparatus; x-ray diffracts meter, gamma ray spectrometer, helium cryostat, and holography room. Many workstations are distributed in faculty and staff offices. An undergraduate computer lab with 12 Linux pcs is provided for all student access. A full service Machine Shop specializing in all forms of CNC and conventional machining, welding, and soldering. An Electronics Shop is also located in the building specializing in design, prototype instrumentation, consultation, and repair. Cooperative research programs are conducted at Los Alamos National Laboratory, Oak Ridge National Laboratory, Florida Atlantic University, Florida International University, AT&T Bell Laboratories, Brookhaven National Laboratory, Fermilab, Cornell University, CERN, and the National High-Magnetic Field Laboratory (NHMFL). The University of Florida is a member of the consortium (UF, FSU, and LANL) responsible for the operation of the National High-Magnetic Field Laboratory.

Table B—Appointments to Graduate Students, 2008–09

Title of Appointee	Appointments Total	Appointments First year	Academic Load Allowed in Credit Hours	Hours of Service Per Week	Stipend for Academic Year ($)
			Semester		
Teaching Assistant	65	18	9	13.3	21,000
Research Assistant	60	0	9	15.6	21,000
Fellowships	25	8	12	0	23,000–24,000
Total	150	26			

5. Personnel Engaged in Separately Budgeted Research, 7/08–6/09

Professorial faculty	48.5
Postdoctoral appointments	34
Graduate students	77
Undergraduate students	40
Total	199.5

6. Separately Budgeted Research Expenditures by Source of Support

	Departmental Research	Physics-related Research Outside Department
Federal Government	$8,996,000	$
Total	$8,996,000	$

Table C—Separately Budgeted Research Expenditures

Research Specialty	No. of Grants	Expenditures ($)
Condensed Matter Experiment	16	1,950,000
Condensed Matter Theory	13	1,554,000
Theoretical Astro	9	567,000
Experimental Astro	16	5,555,000
Theoretical Particles/Fields	6	456,000
Experimental Particles/Fields	23	2,434,000
Total	121	12,520,000

FACULTY

Professors

Acosta, Darin, Ph.D., California, San Diego, 1993. High energy experiment.

Avery, Paul R., Ph.D., Illinois, 1980. Experimental high-energy physics.

Cheng, Hai Ping, Ph.D., Northwestern, 1988. Computational physics, molecules, clusters and nanostructures.

Detweiler, Steven L., Ph.D., Chicago, 1975. Relativistic astrophysics.

Dorsey, Alan T., Ph.D., Illinois, 1987. Condensed matter theory; pattern formation.

Field, Richard D., Jr., Ph.D., California, Berkeley, 1971. Elementary particle theory.

Fry, James N., Ph.D., Princeton, 1979. Theoretical astrophysics; cosmology.

Hebard, Arthur F., Ph.D., Stanford, 1970. Distinguished Professor. Condensed matter experiment; fullerenes.

Hershfield, Selman P., Ph.D., Cornell, 1989. Condensed matter theory.

Hirschfeld, Peter J., Ph.D., Princeton, 1985. Condensed matter theory.

Ihas, Gary G., Ph.D., Michigan, 1971. Experimental low-temperature and mesoscopic physics.

Ingersent, J. Kevin, Ph.D., Pennsylvania, 1990. Condensed matter theory.

Korytov, Andrey, Ph.D., Dubna, 1991. High energy experiment.

Kumar, Pradeep, Ph.D., California, San Diego, 1973. Condensed matter theory.

Maslov, Dmitri, Ph.D., Landau Institute (Moscow), 1989. Condensed matter theory.

Meisel, Mark W., Ph.D., Northwestern, 1983. Low-temperature physics; condensed matter experiment, biomagnetism.

Mitselmakher, Guenakh, Ph.D., Moscow, 1974. Distinguished Professor. Experimental hadron collider physics.

Muttalib, Khandker A., Ph.D., Princeton, 1982. Condensed matter theory.

Obukhov, Sergei, Ph.D., Landau Institute (Moscow), 1979. Condensed matter theory.

Ramond, Pierre, Ph.D., Syracuse, 1969. Distinguished Professor. Elementary particle theory.

Reitze, David, Ph.D., Texas, 1990. Condensed matter ultra fast experiment, experimental astrophysics.

Rinzler, Andrew, Ph.D., Connecticut, 1991. Condensed matter experiment.

Sabin, John R., Ph.D., New Hampshire, 1966. Molecular and solid state quantum mechanics.

Sikivie, Pierre, Ph.D., Yale, 1975. Elementary particle theory; cosmology.

Stanton, Chris J., Ph.D., Cornell, 1986. Condensed matter theory.

Stewart, Gregory R., Ph.D., Stanford, 1975. Condensed matter physics; novel materials.

Sullivan, Neil S., Ph.D., Harvard, 1972. Condensed matter physics; elementary particle experiment.

Takano, Yasumasa, Ph.D., Helsinki, 1978. Ultralow temperature physics.

Tanner, David B., Ph.D., Cornell, 1972. Distinguished Professor. Condensed matter experiment; elementary particle experiment.

Thorn, Charles B., Ph.D., California, Berkeley, 1971. Elementary particle theory.

Woodard, Richard P., Ph.D., Harvard, 1984. Quantum gravity and quantum field theory.

Whiting, Bernard F., Ph.D., Melbourne, 1979. Theoretical astrophysics.

Yelton, John M., D. Phil., Oxford, 1981. Experimental high-energy physics. Department Chair.

Affiliate Professors

Bartlett, Rodney J, Ph.D., Florida, 1971. Many-electron theory of atoms, molecules and solids.

Mareci, Thomas, Ph.D., Oxford, 1982. Magnetic resonance imaging (MRI).

Micha, David A., Ph.D., Uppsala, 1966. Chemical physics; theoretical, molecular and materials sciences.

Ohrn, N. Yngve, Ph.D., Uppsala, 1963. Quantum theory of matter.

Roitberg, Adrian, Ph.D., Chicago, 1992. Chemical physics.

Shabanov, Sergai, Ph.D., St. Petersburg State University, 1988. Mathematical Physics.

Associate Professors

Biswas, Amlan, Ph.D., Bangalore, India, 1999. Condensed matter experiment.

Hagen, Stephen, Ph.D., Princeton, 1989. Condensed matter experiment, molecular biophysics, optical spectroscopy.

Lee, Yoonseok, Ph.D., Northwestern, 1997. Condensed matter experiment.

Matchev, Konstantin, Ph.D., John Hopkins, 1997. High energy theory.

Müller, Guido, Ph.D., Hannover, 1997. Experimental astrophysics.

Qiu, Zongan, Ph.D., Chicago, 1986. High-energy theory.

Assistant Professors

Furic, Ivan, Ph.D., MIT., 2004.

Matcheva, Katia, Ph.D., John Hopkins University, 2000. Theoretical astrophysics.

Petkova, Aneta, Ph.D., Brandeis University, 2000. Biophysics.

Ray, Heather, Ph.D., University of Michigan, 2004.

Saab, Tarek Khaled, Ph.D., Stanford University 2002. Experimental astrophysics

Professors Emeriti

Adams, E. Dwight, Ph.D., Duke, 1960. Low-temperature physics.

Buchler, J. Robert, Ph.D., California, San Diego, 1969. Astrophysics.

Dufty, James W., Ph.D., Lehigh, 1967. Nonequilibrium statistical mechanics.

Dunnam, F. Eugene, Ph.D., Louisiana State, 1958. Nuclear physics.

Garrett, Richard E., Ph.D., Virginia, 1953. Physics education.

Hanson, Harold P., Ph.D., Wisconsin, 1948. Atomic physics; condensed matter.

Ipser, James R., Ph.D., Caltech, 1969. Relativistic astrophysics.

Klauder, John R., Ph.D., Princeton, 1959. Distinguished Professor. Mathematical physics.

Monkhorst, Hendrik J, Ph.D., Groningen, 1968. Theoretical chemical physics; neutron-free fusion reactor development.

Peterson, Lennart R., Ph.D., MIT, 1966. Aeronomy and atmospheric physics; nuclear theory.

Seiberling, L. Elizabeth, Ph.D., Caltech, 1980. Condensed matter experiment.

Tobey, Frank L., Jr., Ph.D., Michigan, 1962.

Trickey, Samuel B., Ph.D., Texas A&M, 1968. Condensed matter theory and computation.

Van Rinsvelt, Henri A., Ph.D., Utrecht, 1965. Nuclear and applied physics; surface physics.

Senior Associate in Physics

Deserio, Robert, Ph.D., Chicago, 1981. Director of Undergraduate Laboratories.

Research Scientists

Andraka, Bohdan, Ph.D., Temple, 1986. Condensed matter experiment.

Klimenko, Sergey, Ph.D., Novosibirsk 1993. Experimental astrophysics.

Konigsberg, Jacobo, Ph.D., UCLA, 1989. High energy physics.

Xia, Jian-Sheng, Ph.D., UST China, 1989. Low temperature experiment.

RESEARCH SPECIALTIES AND STAFF

Theoretical

Astrophysics. Stellar structure and evolution; variable stars; planetary atmospheres nonlinear dynamics and chaos; high-energy astrophysics; solar-neutrinos; general relativity; black holes; gravitational waves; neutron stars; cosmology: early Universe. Detweiler, Fry, Matcheva, Whiting. 2 postdoctoral associates.

Chemical, Solid State, Molecular, and Atomic Physics. Many-body scattering theory; intermolecular forces; radiation-molecule interactions; propagator methods; density functional theory and local density models; semiempirical molecular orbital methods: electronic structure computation; theory of quantum crystals, molecular crystals, biological molecules and thin films; theory and application of new computational methods; neutron-free fusion reactor development. Bartlett, Cheng, Ingersent, Micha, Sabin, Stanton. 8 postdoctoral associates.

Condensed Matter. Theory of highly correlated systems. Liquid and solid ^{3}He; high-T_c superconductivity and heavy fermions; glasses; disorder and localization; optical interactions in semiconductors; nonlinear phenomena in condensed matter, solitons; nuclear spin ordering in metals; pattern formation. Dorsey, Hershfield, Hirschfeld, Ingersent, Kumar, Muttalib, Obukhov, Stanton. 4 postdoctoral associates.

Elementary Particles and Fields. Standard model of strong and electromagnetic interactions; grand unified theories; string unification; superstring theory; particle astrophysics. Field,

Matchev, Qiu, Ramond, Sikivie, Thorn, Woodard. 3 postdoctoral associates.

Mathematical Physics. Quantum theory. Muttalib, Shabanov.

Non-linear Dynamics and Complexity. Time series analysis; reconstruction of attractors, glasses, and complex fluids. Obukhov.

Experimental

Atomic, Molecular, and Optical Physics. Atom-Laser Interactions. Reitze.

Biophysics and Biological Physics. Protein conformational dynamic and folding; microfluids; in vivo and high resolution MRI, and NMR spectroscopy, bio-optically active processes. Hagen, Mareci, Meisel, Petkova.

Cold Dark Matter Search. Saab.

Condensed Matter Physics. NMR of solids, liquids, and polymers; NMR imaging. Optical properties; quantum magnetic excitations; conducting polymers; organic superconductors; composite materials; thermal and magnetic properties of superconductors; high-temperature superconductors. Surface physics; synchrotron radiation. Electronic properties of novel materials, ultrafast spectroscopy of novel materials. Nonlinear nonequilibrium phenomena. Ultrafast spectroscopy of novel materials; high intensity laser-solid interactions. Andraka, Biswas, Hebard, Lee, Meisel, Reitze, Rinzler, Stewart, Sullivan, Tanner. 10 postdoctoral associates.

Elementary Particle Physics. Axion search. Sikivie, Sullivan, Tanner.

Gravitational Waves. Research in Laser-Interferometer Gravitational-Wave Observatory. Coldwell, Klimenko, Mitselmakher, Müller, Reitze, Tanner, Whiting. 4 postdoctoral associates.

Grid Computing for High Energy Physics. Avery, Bourilkov, Kim. 1 postdoctoral associate.

Hadron Collider Physics. Acosta, Avery, Furic, Konigsberg, Korytov, Mitselmakher, Yelton. 6 postdoctoral associates.

Low-Temperature Physics. Properties of macroscopic quantum systems, in particular liquid and solid ^{3}He, solid H_2 and electronic and magnetic systems down to 10 μK. Thermodynamic, hydrodynamic, magnetic, and transport properties of materials are studied using NMR, ultrasound (P,V,T), and other probes. Ihas, Lee, Meisel, Sullivan, Takano, Xia. 2 postdoctoral associates.

Neutrino Physics. Ray. 1 Postdoctoral associate.

UNIVERSITY OF MIAMI

DEPARTMENT OF PHYSICS

Coral Gables, Florida 33124

Students Accepted For Degree	FIELDS		
	Physics	Astronomy	Related Fields
Doctorate	X		X
Master's	X		X

1. General

President: Donna Shalala
Dean of Graduate School: Terri A. Scandura
Department Chairman: Kenneth J. Voss
Department Telephone Number: (305) 284-2323
Type of Institution: University
Control: Private
Setting: Suburban
Total Faculty: 2,913
Total Graduate Faculty: 2,913
Total Students: 15,600
Total Graduate Students: 5,250
Annual Graduate Tuition:
All Graduate Students: Full-time—$26,640/semester
Part-time—$1,480/credit
Tuition rates for: 2009–10
Deferred tuition plan: Yes
Other Fees: $648 (includes student activity and athletic fees)
Term: Semester

2. Number of Faculty in Department

The combined total of full-time faculty in the three professorial ranks is 18. The combined total of full-time, part-time, and other faculty at all ranks is 26.

3. Admission, Financial Aid, and Housing

Address admission inquiries to: Department of Physics, Graduate Program
Graduate application fee required: $50
Application deadline (Fall admission): 2/1
Admission information: For fall admission, 2010–11, 4 students were accepted from 56 applicants.
Admission requirements: For admission to the graduate programs, a Bachelor's degree in physics is required with a minimum undergraduate GPA of B specified. Exceptions can be made at the discretion of the committee on graduate student advising. The GRE is required for domestic applicants. No minimum acceptable score for admission is specified. The GRE Advanced is required with no minimum score specified. Students from non-English speaking countries are required to demonstrate proficiency in English via the TOEFL exam. Minimum acceptable score for admission is 80 (ibt).
Undergraduate preparation assumed: Griffiths, *Introduction to Electrodynamics*; Schroeder, *Thermal Physics*; Symon, *Mechanics*; Griffiths, *Quantum Mechanics*.
Address financial aid inquiries to: Assistantships: Department of Physics, Graduate Program, Grad. School Loans: Financial Aid Services
GAPSFAS application required: No
Financial aid deadline: 2/1
Loans available: Yes
Address housing inquiries to: Dept. of Residence Halls, P.O. Box 248044, Miami, FL 33124
On-campus, single student housing available: No
On-campus, married student housing available: No

Table A—Faculty, Enrollments, and Degrees Granted

Research Specialty	2009–10 Faculty	Enrollment[1] Fall 2009 Master's	Enrollment[1] Fall 2009 Doctorate	No. of Degrees Granted[2] 2009–10 (2005–10) Master's	Terminal Master's	Doctorate	Median No. of Years for 2009–10 Ph.D.'s
Astrophysics	3	–	5	1(0)	1(1)	0(5)	–
Complex Systems	1	–	4	0(0)	0(0)	–	–
Condensed Matter Physics	5	–	1	0(1)	0(1)	0(3)	–
Optics	3	–	6	0(1)	0(1)	0(1)	–
Particles & Fields	4	–	5	0(2)	0(1)	2(4)	6
Plasma Physics & Fusion	1	–	1	0(1)	0(1)	0(2)	–
Non-specialized	1	–	4	0(4)	0(2)	–	–
Total	18	–	26	1(10)	1(7)	2(15)	
Full-time Grad. Stud.		–	26				
Part-time Grad. Stud.		–	0				
First-year Grad. Stud.		–	5				
Median Years in Grad. Study (2009–10 Degrees)				–	–	6	–
Undergraduate Degrees, 2009–10 (2005–10): 7(35)							

[1]Students not yet committed to a research specialty are entered under non-specialized.
[2]Five-year totals in parentheses.

4. Graduate Degree Requirements

Master's: 30 graduate credits in approved program with B average; score of S or better in the comprehensive departmental exam; two semesters of full-time study or equivalent in residence; no language requirement.
Doctorate: A minimum of 24 graduate credits in approved program with B average; score of P in the comprehensive departmental exam by the student's second yearly attempt at the latest; dissertation and dissertation exam required; one year residency required; no foreign language requirement.
Thesis: Thesis may not be written *in absentia*.
Special Equipment, Facilities, or Programs: Cooperative research programs are conducted with the Rosenstiel School of Marine and Atmospheric Sciences of the University of Miami, and with the Atlantic Oceanographic and Meteorological Laboratory of NOAA and various domestic and foreign universities.

Table B—Appointments to Graduate Students, 2007–08

Title of Appointee	Appointments Total	Appointments First year	Academic Load Allowed in Credit Hours	Hours of Service Per Week	Stipend for Academic Year ($)
			Semester		
Teaching Assistant	19	5	9	15–20	22,000
Research Assistant	5	0	9	15–20	22,000
Total	24	5			

5. Personnel Engaged in Separately Budgeted Research, 6/09–6/10

Professorial faculty	14
Graduate students	20
Total	34

6. Separately Budgeted Research Expenditures by Source of Support

	Departmental Research	Physics-related Research Outside Department
Federal government	$6,067,515	$
Other	50,000	
Total	$6,117,515	$

Table C—Separately Budgeted Research Expenditures

Research Specialty	No. of Grants	Expenditures ($)
Astrophysics	8	1,390,689
Complexity	1	114,999
Condensed Matter Physics	2	341,092
Optics	14	3,006,385
Particles & Fields	5	884,350
Plasma	1	380,000
Total	31	6,117,515

FACULTY

Professors

Alexandrakis, George C., Ph.D., Princeton, 1968. Solid state experiment; transmission resonance; magneto-acoustic propagation in ferromagnetic metals.

Alvarez, Orlando, Ph.D., Harvard, 1979. Theory of elementary particles.

Barnes, Stewart, Ph.D., UCLA, 1972. Solid state theory; many-body theory; superconductivity and magnetism.

Cohn, Joshua L., Ph.D., U. of Michigan, 1989. Condensed matter experiment, materials physics, electronic and lattice transport.

Curtright, Thomas, Ph.D., Cal. Tech., 1977. Theory of elementary particles.

Gordon, Howard R., Ph.D., Penn State, 1965. Optical oceanography; light scattering; radiative transfer; remote sensing.

Huerta, Manuel A., Ph.D., Miami, 1970. Statistical mechanics; plasma physics; numerical simulations in MHD.

Johnson, Neil F., Ph.D., Harvard University, 1989. Theoretical condensed matter physics, biological complexity, social complexity, physical complexity.

Mezincescu, Luca, Ph.D., Bucharest, 1978. Theory of elementary particles.

Nepomechie, Rafael I., Ph.D., Chicago, 1982. Theory of elementary particles.

Voss, Kenneth, J., Ph.D., Texas A&M, 1984. Chairman of the Department. Ocean optics; light scattering; atmospheric optics.

Zuo, Fulin, Ph.D., Ohio State U., 1988. Condensed matter experiment.

Associate Professors

Ashkenazi, Josef, Ph.D., Hebrew Univ. of Jerusalem, 1975. Solid state theory; first-principles band structure methods; many-body physics; high-temperature superconductors.

Galeazzi, Massimiliano, Ph.D., University of Genoa, Italy, 1999. X-ray astrophysics; studies of interstellar/intergalactic medium and X-ray sources; development of X-ray detectors

Gundersen, Joshua O., Ph.D., UC Santa Barbara, 1995. Observational cosmology. Experimental cosmology and astrophysics; cosmic microwave and infrared backgrounds.

Nearing, James C., Ph.D., Columbia, 1965. Associate Chairman of the Department. Theoretical physics; bifurcation theory in fully nonlinear plasma systems.

Assistant Professors

Boynton, Chris G., Ph.D., Miami, 1991. Computational MHD; radiative transfer.

Huffenberger, Kevin M., Ph.D., Princeton, 2006. Cosmology and astrophysics, theory and data analysis.

Korotkova, Olga, Ph.D., University of Central Florida, 2003. Theoretical optics; wave propagation and scattering in random media.

Adjunct and Professors Emeriti

Faber, Shepard, Ed.D., Florida, 1959. Emeritus. Teaching techniques.

Ghandour, Ghassan, Ph.D., UC Berkeley, 1974. Elementary particles; mathematical physics.

Hirschberg, Joseph G., Ph.D., Wisconsin, 1952. Emeritus. Physical optics; Fabry-Perot interferometry; plasma spectroscopy.

Moore, William Franklin, Ph.D., Miami, 1980.

Perlmutter, Arnold, Ph.D., NYU, 1955. Nuclear and particle physics.

Robertson, Harry S., Ph.D., Johns Hopkins, 1949. Emeritus. Statistical mechanics; plasma physics; production, confinement, and heating of plasmas.

Van Vliet, Carolyne M. Ph.D., Free University of Amsterdam, 1956. Adjunct and Emerita (Univ. of Montreal). Equilibrium and non-equilibrium statistical mechanics; stochastic processes; quantum transport in solids.

Zane, Sharon, Ph.D., University of Miami, 2003.

RESEARCH SPECIALTIES AND STAFF

Theoretical

Complex Systems and Complexity. Emergent cooperative phenomena in biological and social systems; dynamical evolution; extreme behavior; prediction and soft control. Johnson.

Cosmology and Astrophyiscs. Origin and evolution of the universe; cosmic microwave background; clustering and large scale structure; dark matter and dark energy; intergalactic medium. Huffenberger.

Elementary Particles. Quantum field theory (especially integrable models); supersymmetry; supergravity; superstrings. Alvarez, Curtright, Ghandour, Mezincescu, Nepomechie.

Environmental Optics. Radiative transfer; remote determination of ocean chlorophyll concentrations. Boynton, Gordon. Wave propagation and scattering in random media. Korotkova.

Optics. Transmission, propagation and scattering of electromagnetic waves in deterministic and random media. Korotkova.

Plasma Physics. Numerical simulations in plasmas and other systems. Boynton, Huerta, Nearing.

Solid State. Electronic structure of solids; many-body physics; high-temperature superconductivity; magnetism. Ashkenazi, Barnes, Johnson. Linear and nonlinear quantum transport; reduced-dimensionality systems. Van Vliet.

Experimental

Cosmology and Astrophysics. Studies of the cosmic microwave and infrared background; studies of the interstellar/intergalactic medium and X-ray sources; instrumentation for low-noise RF and mm-wave detectors and telescopes; development of high-resolution cryogenic microcalorimeters and bolometers cross-correlation studies; statistical analysis. Galeazzi, Gundersen.

Ocean Optics. Light scattering and absorption by marine particulates; instrumentation for measurement of optical properties of the ocean and atmosphere. Voss.

Solid State Physics. Ferromagnetic transmission resonance in metals; spin relaxation; exchange energy; phonon excitation and propagation; nonlinear phenomena, Alexandrakis. Transport and magnetic properties of materials at low temperatures; transition metal oxides, high-temperature and organic superconductors, and reduced dimensional systems (e.g., layered systems and thin films); electrical and thermal conduction, thermoelectric effects; vortex dynamics, critical current, quantum tunneling, laser deposition of ferroelectric thin films. Cohn, Zuo.

More information is available through http://www.physics.miami.edu

UNIVERSITY OF SOUTH FLORIDA

DEPARTMENT OF PHYSICS

Tampa, Florida 33620
http://physics.usf.edu/

Students Accepted For Degree	FIELDS		
	Physics	Astronomy	Related Fields
Doctorate	X		X
Master's	X		X

1. General

President: Judy L. Genshaft
Dean of Graduate School: Karen D. Liller
Department Chair: Pritish Mukherjee
Department Telephone Number: (813) 974-2871
Type of Institution: University
Control: Public
Setting: Urban
Total Faculty: 2,296
Total Students: 44,116
Total Graduate Students: 10,462
Graduate Tuition:
In-state: $331.79/credit
Out-of-state: $818.66/credit
Tuition rates for: 2009–10
Deferred tuition plan: No
Term: Semester

2. Number of Faculty in Department

The combined total of full-time faculty in the three professorial ranks is 24. The combined total of full-time, part-time, and other faculty at all ranks is 32.

3. Admission, Financial Aid, and Housing

Address admission inquiries to: Director of Graduate Studies, Dale E. Johnson
Graduate application fee required: $30
Admission deadline (Fall admission): 2/15; 1/1 international students
Admission information: For fall admission, 2009–2010, 17 new students were enrolled.
Admission requirements: For admission to the graduate programs, a Bachelor's degree in physics or related fields is required with a minimum undergraduate GPA of 3.0 specified. The general GRE is required and the GRE physics subject test is strongly recommended. Students from non-English speaking countries are required to demonstrate proficiency in English via the TOEFL exam. Those applying for a teaching assistantship are required to pass the TSE exam with a minimum score of 50.
Address financial aid inquiries to: Director of Financial Aid, SVC 1102
GAPSFAS application required: Yes
Financial aid deadline: 2/1 (for priority only—loans available at all times)
Loans available: Yes
Address housing inquiries to: Director of Housing and Food Service: Thomas Kane, RAR 229
On-campus. single student housing available: Yes
On-campus, married student housing available: Yes

Table A—Faculty, Enrollments, and Degrees Granted

Research Specialty	2009–10 Faculty	Enrollment[1] Fall 2009 Master's	Doctorate	No. of Degrees Granted[2] 2009–10 (2005–10) Master's	Terminal Master's	Doctorate	Median No. of Years for 2009–10 Ph.D.'s
Atomic, Molecular, & Optical Physics	6	2	9	1(7)	2(7)	2(7)	5.0
Biophysics	4	0	14	0(7)	1(3)	2(5)	4.3
Condensed Matter/ Materials Physics	9	0	26	0(6)	1(1)	1(6)	5.3
Computational	5	0	16	2(7)	0(1)	2(2)	5.0
Total		2	65	3(27)	4(12)	7(20)	
Full-time Grad. Stud.		2	61				
Part-time Grad. Stud.		0	3				
First-year Grad. Stud.		0	16				
Median Years in Grad. Study (2009–10 Degrees)				2.0	2.7	5.0	5.0

Undergraduate Degrees, 2009–10 (2005–10): 9(59)

[1]Students not yet committed to a research specialty are entered under non-specialized.
[2]Five-year totals in parentheses.

4. Graduate Degree Requirements

Master's: The Department offers Master's degrees in Physics and Applied Physics. There are 2 options for the M.S. in Physics: *Thesis Option*: minimum 30 semester hours in an approved program, six of which may be for thesis; final oral exam on the thesis required; no language requirement; one academic year of residence required. *Non-Thesis Option*: minimum 30 semester hours of course work in an approved program; no language requirement; one academic year of residence required.
Dual Master's: the 51-semester-hour program culminates in the student receiving an M.S. degree in Physics and an M.S. Degree in Engineering Science; single thesis included; approximately three years to complete; 15 hrs. of course work each in Physics and Electrical Engineering (microelectronics option); 9 hrs. overlap courses; 6 thesis hrs. each in Physics and Electrical Engineering.
Doctorate: The Department offers a Ph.D. program in Applied Physics. This program encompasses the areas of biophysics, biomedical physics, atomic molecular, optical physics, solid state, materials physics, and physics education. An integrated curriculum incorporating interdisciplinary training in applied physics overlapping these areas of emphasis, and research programs to prepare students for the increasingly technological workforce of the future are key components of the program. These aspects of the program and an industrial practicum provide a bridge between theoretical, fundamental concepts and their practical, engineering applications. These programs offer maximum flexibility and are tailored to suit the interests of students and his/her career objectives.
Dissertation: Required.
Special Equipment, Facilities, or Programs: Experimental and

theoretical research in the Department of Physics is conducted in the following laboratories/programs:

Cellular and Molecular Biophysics Research Laboratory: The Center is highly interdisciplinary involving physics, physiology, cell biology and molecular biology. The structural and functional relationship of membrane proteins, such as ion channels, membrane transporters, and electrogenic pump molecules, as well as their interfaces with external electromagnetic field are currently studied. One of the projects is to study direct energy transform from inorganic energy to the living system by electrically activating the membrane electrogenic pump molecules, Na/K ATPases. Other projects include the study of percutaneous and targeted drug delivery, understanding the mechanisms underlying electrical injury and the development of novel techniques for wound healing. The research involves both basic science and its practical application to biology and medicine. The research program is supported by the National Institute of Health (NIH) since 1994 and many other funding agencies. Our research is conducted and focused on cellular and molecular levels in nanoscales by using broad, state-of-arts techniques including whole cell/patch clamps, various microscopic imagining systems such as multiple-laser confocal microscope and near field microscope, full line of cellular and molecular biology technique (wchen@usf.edu).**

Nanomedicine and Bionanotechnology: Members of this interdisciplinary and entrepreneurial laboratory apply principles of physics and other subjects to design, fabricate and characterize nanostructures, polymer-based biomaterials, especially polypeptide-based materials. Materials of particular interest are nanocoatings for cell and tissue culture, nanocoatings for medical implant devices, nanostructures for combating cancer, and nanofibers for coatings and tissue scaffolds. Spectroscopic, acoustic and microscopic methods are used for physical and chemical characterization. Biological characterization is done by cell culture and other methods. Instrumentation in the laboratory includes a Fourier transform infrared spectrometer, an ultrasensitive dual-beam UV-visible wavelenght spectrometer with Peltier temperature control, a fluorescence microscope equipped with a heated stage, gas control and low light level imaging device, a tissue culture suite, and electrospinning apparatus, a fume hood and various small items of equipment. We collaborate with investigators in several USF departments, the Moffitt Cancer Center, other universities and biotechnology companies (dhaynie@usf.edu).**

Laboratory for Laser Remote Sensing: Laboratory to study precision laser spectroscopy for laser remote sensing of the atmosphere and environmental gases, lidar (laser radar), atmospheric laser progation, and tunable laser detection of trace species in the environment. Past activities include development of a high power KTP OPO laser remote sensing system to measure atmospheric aerosols, a tunable Ho:YSGG differential lidar system to detect sources and sinks of CO2 in the atmosphere, and the development of computer simulation programs to calculate the optical transmission spectrum of the atmosphere _HITRAN-PC_ and simulation of laser remote sensing _LIDAR-PC_. Current work involves deep-UV laser and LED induced fluorescence of trace organic and plastic compounds in water, UV laser induced breakdown spectroscopy _LIBS_ for remote detection of surface contaminants, temperature measurements of super heated LIBS plama using two laser pulse inverse bremsstrahlung absorption, and laser induced thermal emission. killinge@usf.edu)**

Digital Holography & Microscopy Laboratory (DHML): The main theme of our research activities is in the development of novel imaging technologies with emphasis in holographic and interferographic microscopy. In digital holography (DH), the hologram is recorded by a CCD camera, instead of photogrphic plates, and the holographic images are calculated numerically using the electromagnetic diffraction theory. This gives direct access to the phase profile of the optical field and leads to a number of powerful imaging techniques that are difficult or impossible in real space holography. Transparent objects, such as many biological cells, thin film structures and MEMS devices, can be imaged that reveal minute thickness variations with nanometer precision. Optical tomography by digital interference holography (DIH) yields cross-sectional images of biological tissues without actually cutting into them. Cellular motility can be studied by imaging the adhesion layers between a crawling cell and the substrate through the DH of total reflection, important in the study of embryogenesis, neuronal growth, and cancer cell metastasis. Furthermore, we are not only able to image cells and their components, but also manipulate them in full three dimensions, using patterns of light produced by holographic optical tweezers (HOT). Cells and organelles can be captured and tracked, coaxed into artifically patterned growth and motion, and operated on with micromanipulation and microsurgery. Students can expect to work on cutting-edge research topics and be trained extensively in advanced optical design and construction, digital image acquisition, computer programming, electronic instrumentation, and cellular and biomedical laboratory procedures. Digital holographys is an emerging technology that has been experiencing exponential growth in the last decade, and has potential applications in wide-ranging areas including cellular microscopy, metrology, manufacturing processes and testing, medical imaging and diagnostics, biometry, environmental research, and food science, just to name a few. (mkkim@usf.edu)**

Laboratory for Advanced Materials Science and Technology (LAMSAT): to explore innovations in pulsed laser ablation and plasma processes for the growth of thin films of technologically significant materials including super hard materials, magnetic materials, superconductors, and compound semiconductors for solar cells. Past NSF and DOE sponsored research projects have focused on the application of a dual-laser ablation process discovered in this laboratory to grow large-area, particulate-free films of $Cu(InGa)Se_2$ and ZnO for solar cell applications, and to fabricate diamond and diamond-like carbon structures for MEMS applications. One of the recently funded NSF projects focuses on an hybrid process where chemical self-assembly and physical vapor deposition techniques are combined to grow vertically aligned nano-grained films of superhard materials. Novel optical techniques for high resolution, in-situ plasma imaging, and development of new laser-assisted plasma growth processes are being researched. The research encompasses thin film growth, nanostructures, dynamic optical process diagnostics, thin film analysis, characterization and process modeling leading to the fabrication of single-layer and hetero-structure devices (pritish@usf.edu, switanac@usf.edu).**

Laser Physics and Single Crystal Fiber Growth Facility: This laboratory is equipped with a variety of cw and pulsed lasers including frequency-doubled Nd:YAG pumped Ti- sapphire laser, Ar+-pumped dye laser and frequency- stabilized ring dye laser to study nonlinear optics and develop novel fiberoptic devices. Recent research includes study of upconversion processes in rare earth doped laser crystals towards the development of infrared solid state lasers and the invention of a new type of thermal source with potential capabilities in the fabrication of micro-mechanical systems and surface modification on a microscopic scale. This laboratory also uses the laser-heated pedestal growth method to grow rare earth doped crystals in small rod form for spectroscopic and energy transfer studies and optical crystalline material in long fiber form for fiber-optic applications. In particular, techniques for growing high optical quality, high purity sapphire fibers have been developed. Examples of potential applications for sapphire fibers include medical laser delivery and fiber-based sensors for hostile environments. The facility has two growth stations, fabrication equipment, x-ray diffractometer, micro-interferometer, and light sources and spectrometer for optical/spectral characterization. (ndjeu@usf.edu)**

Functional Materials Laboratory: This laboratory is equipped with experimental facilities for studying the electrical and magnetic properties of novel materials. Investigation of the material properties are done over a wide range in temperature ($2\ K<T<350\ K$) and applied magnetic fields up to 7 Tesla. In addition, the frequency-dependent electromagnetic response is probed from DC to 6 GHz. A novel resonant radio-frequency (RF) method has been developed to accurately determine the magnetic anisotropy and switching in materials. Current research projects focus on studies of dynamic magnetic response and high frequency impedance in nanoparticles, composites, thin films, magnetic semiconductors and multiferroic systems. These technologically important materials are promising candidates as building blocks for the next generation multifunctional device. Other interests include magnetocaloric effect in nanostructured materials, spin polarization studies and physics of strongly correlated systems. Ongoing research support by NSF, DoD and DOE. (sharihar@usf.edu).**

The Bio-Nano Research Group: The work of this laboratory is the investigation of the structure/function relationship in biological systems ranging from the single molecule to the multicellular level. Molecular level structures determine the materials properties of the system, which in turn determines the macroscopic biological function. Using the expertise in atomic force microscopy, fluorescence microscopy, rheology, and other techniques found in the laboratory, the physical properties of single molecules and macromolecules are measured, and bulk models are developed and experimentally tested. These models are used to help explain the biology or pathology of systems. Example projects within the lab include investigations of cell surface and extracellular matrix glycoproteins and glycosaminoglycans through single molecule imaging and force spectroscopy. The data from these experiments is used to develop models for the viscoelastic properties of solutions of these biopolymers which are then tested experimentally. These rheological properties are important for the function of tissues ranging from joint interstitial fluids to lung epithelium and will be used to understand the behaviors observed in these systems. The outcomes of the lab are geared to make significant contributions to biomedicine, and as such require a close collaboration with the Departments of Biology and Chemistry and with the School of Medicine. The work is inherently multidisciplinary, and students develop a broad range of skills from physics, biology, and chemistry. (gmatthews@usf.edu) **

Spintronics Lab: Facilities include multi-target sputtering system for thin film heterostructure growth, electron transport measurements in high magnetic fields (7T) and cryogenic temperatures (~1K). Current interests include spin-dependent tunneling, spin-injection, and spin-polarization measurements, and combine modeling to complement experiments when appropriate (cmiller@usf.edu).**

Novel Materials Laboratory: The laboratory is designed for the synthesis and characterization (including structural, optical, electrical, thermal and magnetic) of novel materials for technologically significant applications. The emphasis is on an understanding of the structure-property relationships of material systems; that is, how crystal structure variations affect the electrical, thermal, optical, magnetic, and mechanical properties of materials. The laboratory applies this understanding towards the crystal growth and synthesis of new and novel materials for varying technologically significant applications. The Lab's research is based on new materials for energy-related technologies and includes thermoelectric power conversion and refrigeration, photovoltaics power generation, and magnetic refrigeration. Current materials research includes new semiconductors for electronics and optoelectronics applications, transport properties of "open structured" semiconductors, nanocrystal synthesis and assembly approaches, and new magnetic materials. The research is supported by NSF, DOE, ONR, ARO, NASA and industry. Close collaboration with industry is typical in this interdisciplinary Materials Physics research program that encompasses all aspects of physics and materials science. Students typically acquire a large variety of skill sets and apply these foundations to their applied physics research. (gnolas@usf.edu).**

Materials Simulations Laboratory (MSL): Director: Prof. I. I. Oleynik, Associate Director: V. V. Zhakhovsky. The research program at MSL focuses on modeling the atomic, electronic, and chemical properties of systems of fundamental and technological importance using the powerful arsenal of first-principles density functional theory, tight-binding, and classical interatomic molecular dynamics techniques. The research thrusts are (1) energetic materials, shock physics and materials at extreme conditions; (2) materials for information technology including molecular and spin electronics; (3) ultrafast laser-matter interactions in metals and semiconductors; (4) graphene nanomechanics and surface chemistry; (5) development of novel algorithms and computational methods for large-scale materials simulations using Tera- and Peta-Flop computers. The ultimate goal is to understand fundamental mechanisms and establish structure-property relationships that are difficult or sometimes impossible to obtain from experiment. Other areas of research at MSL include the development of analytic bond order potentials for large-scale atomistic simulations, surface and interface science, including surface chemistry of metal-oxide and metal-polymer interfaces in magnetic tunneling junctions. The research program at MSL is supported by NSF, DARPA, ONR, and ARO. MSL is one of the major users of the NSF funded Teragrid network of supercomputers. The computational facilities at MSL include a 400-CPU Beowulf cluster consisting of the latest 2.7 GHz AMD Shanghai quad-core nodes interconnected by an Infiniband switching fabric, a 12 Tb file server, and 20 linux workstations. MSL personnel include five graduate students, two undergraduate students, two postdoctoral associates, and a research associate professor. MSL is running a vibrant REU program supported by NSF that is specifically focused on attracting minority and female students. Further info can be obtained at http://msl.cas.usf.edu.

Condensed Matter Theory Research Group: This group works in condensed-matter theory, with current projects in crystallography, biological data analysis, and magnetic systems. In collaboration with Dr. Benji Fisher, we have reformulated Fourier-space crystallography into the language of cohomology of groups and applied the results to a wider class of structures than previously considered. Continuing work focuses on homological invariants of a new kind and their possible physical implications. In biology, we have been collaborating with Dr. Chun-Min Lo on analysis of electric cell-substrate impedance-sensing experiments. Looking only at statistical signatures of electrical noise, we can distinguish cancerous from non-cancerous cell cultrures of the same type of cell and can detect physiological effects of the toxin cytochalasin-B at lower concentrations than possible with other techniques. Recent work in magnetic systems includes a statistical-mechanical model of helimagnetism in rare-earth heterostructures and a study of the ballistic-to-diffusive crossover in quantum wires. The latter may have applications in quantum computing. (davidra@ewald.cas.usf.edu)**

Laboratory of Optical Biophysics: One focus of this laboratory is on the spatio-temporal dynamics of excitability changes and of calcium elevations inside neuronal tissues. The spatio-temporal patterns of either property are closely related to the ability of neuronal tissues to change their output (neurotransmitters or hormones) in response to repetitive stimulation of variable duration and/or frequency. These short-term changes are one basic mechanism underlying learning and memory formation in the nervous system in general. Another focus is the physical chemistry of phase separation and aggregation of proteins in solution. Depending on the specifics of the protein interactions and solution conditions, proteins can either stay soluble or undergo a variety of phase transitions including crystallization, liquid-liquid phase separation or precipitation. Using optical techniques, we study the thermodynamics and kinetics of these phase separation phenomena of proteins. Our laboratory, among others, has provided evidence that protein phase diagrams have unusual properties and intriguing aggregation kinetics. These have important implications for protein crystal growth and, even more critically the large class of protein aggregation diseases such as sickle cell anemia, Alzheimer and Parkinson. (mmuschol@usf.edu)**

Computational Soft matter Laboratory: Interest of our group lie in the areas of computational bio-physics and mathematical modeling of biological and social systems. In this context, we focus on molecular dynamics simulations, monte-carlo methods, multi-scale modeling based on mean field theory, dynamical systems with an emphasis on Hamiltonian systems, associated numerical methods, and underlying parallel processing issues. Our current focus is on the study of lipid bilayer systems, which form integral components of celluar membranes. These systems are studied using various computational modeling techniques and validated through close interactions with experimentalists. One aspect of our current work involves the study of hetergeneous model membrane systems. We study the interactions between various membrane components such as phospho- and sphingolipids, and cholesterol that give rise to stable structures such as "rafts" and caveolae. We intend to develop a multi-scale mean filed theory base dmodel into a complete simulation methodology for membrane simulations. The models and methods developed will be used to study the structure and stability of membrane structures such as "raft" and caveolae, which are known to play critical roles in the activation of T-cells in immune response. (pandit@usf.edu)**

Computational condensed matter physics & materials science program: Research interest include the area of theoretical condensed matter physics with a focus on computational nanoscience. The materials of interest include semiconductors, ferroelectrics, ferromagnets and multiferroics in both bulk and low-dimensional forms. Examples are nanotubes, nanowires, nanodots and thin films. An exciting feature of such nanoforms is the appearance of new properties and phenomena that do not exist in bulk. The purpose of my research is to identify these novel features, study their fundamental aspects and explore their new functionalities for future applications in nanoscale devices. An example is utilizing a novel vortex structure that is a unique feature of ferroelectric nanodots in ultrahigh density memory that may increase the current memory capacitance by orders of magnitude. Another research focus is the development of computational techiques that will expand their capabilities beyond existing levels. Examples include the development of first-principle-based techniques for new material forms (nanoscale ferroelectrics and multiferroics) and properties (dielectric loss and tunability). The ultimate research objective is the efficient design of new materials conducted in close collaboration with experimental groups. (iponomar@usf.edu)**

Advanced Materials and Devices Theory Group: Our group is engaged in various problems related to theoretical modeling and description of structural, functional, and nanoscale materials and devices. We pursue two complementary routes - analytical and computational. Analytical techniques based on quantum mechanics, quantum electrodynamics, and many-body theory, are being developed. First principle density functional theory and tight binding models on high-performance supercomputers are being utilized. Currently we are pursuing problems related to the Casimir effect in nanostructured materials, thermoelectric properties of materials with enhanced cooling and power generation performances, simulations of nanostructured materials properties, and related devices. The projects are funded by various national funding agencies. Our group maintains strong collaborations with experimental teams as well as other theoretical groups from the University of South Florida, other universities and national research laboratories. We are devoted to conducting leading edge research to advance our understanding of complex materials and devices using analytical and computatinal methods.(lmwoods@usf.edu)**

The Nanophysics and Surface Science Laboratory: In this laboratory, we investigate condensed matter at the atomic scale. The surface of a material is where the action is; at a surface the material interacts with its environment and thus many chemical and physical processes occur at the interface between a solid and a different medium. Our goal is to understand the structural and electronic properties of surfaces and to tune these properties in order for the surface to perform new or improved functions. Currently investigated surface-functional materials are metal oxides for their use as solid state gas sensors and for solar energy conversion. Modification of surfaces with nanoclusters to improve their functionality is one approach to improve and create new functionalities.

Nanoclusters are aggregates of atoms in the realm between molecules and bulk materials. In this size range condensed matter exhibits new properties, which can be conveniently tuned by controlling their size. In our laboratory we assemble clusters atom by atom in the gas phase and subsequently place them on a support material. This allows investigating the cluster-support interaction and the cluster-size properties relationship. Most of the sample preparation and characterization is done under ultra high vacuum conditions to ensure the integrity of the samples under investigation. In addition to the in-house measurements, some supplementing photoemission and X-ray absorption studies are performed at synchrotron facilities.(mbatzill@usf.edu)

Laboratory of Thin Films and Nanostructures: This laboratory is focused on the development of a new generation of materials and nanostructures for electronic and opto-electronic applications. Current research interests include: chalcogenides materials and devices that change their physical properties, such as phase and volume when illuminated, quantum-dot- and superlattice-heterostructures for LEDs, solar cells and quantum cascade lasers. Also of interest is the application of optical techniques, such as photoreflectance, photoluminescence and Raman spectroscopy, in the characterization of semiconductor devices such as hetero-junction bipolar transistors and quantum well lasers. (munozm@usf.edu).

Nanostructure Optoelectronics Lab: This lab conducts basic studies of material properties, device physics and device engineering. Materials of interest include organic semiconductors (polymers, molecular crystals), semiconductor nanocrystals (quantum dots). One focus is to fabricate and characterize nanostructure optoelectronic devices (solar cells, light emitting diodes and thin film transistors). Ongoing research projects include investigation of hybrid solar cell with organic polymers and semiconductor quantum dots, microstructure organic solar cell arrays MEMS devices, and development of semitransparent flexible solar windows. Reserach facilities include optical spectroscopic setup ns range transient transport setup Mbraun integrated glove box system and organic solar cell and LED fabrication facilities; Solar cell characterization facilities; Low temperature apparatus and high vacuum thermal evaporator (xjiang@usf.edu).**

Solid-state quantum optics lab: This laboratory is equipped with experimental facilities for studying quantum optical phenomena in solid-state nanostructures such as quantum dots, nanocrystals and impurity centers. The long-term goal of this research is to realize controlled light-matter interactions for use in quantum communication and quantum information science. Optical techniques employed include high-resolution spectroscopy, interferometry and multi-photon correlation measurements in both ambient and low-temperature (liquid-helium) environments. Novel optical microcavities are being developed to enhance the interactions of light with single quantum emitters and mechanical resonators for harnessing cavity-electrodynamics and cavity-optomechanics phenomena (andi.muller@gmail.com).**

General Support Facilities: Include a machine shop to build custom mechanical and vacuum parts and an electronics shop capable of custom design, repair and fabrication of electronics and computer components.

**points of contact.

Table B—Appointments to Graduate Students, 2009–10

Title of Appointee	Appointments Total	Appointments First-year	Academic Load Allowed in Credit Hours	Hours of Service Per Week	Stipend for 12 mos. ($)
			Semester		
Grad fellow	13	1			20,000–30,000
Teaching Assistant	34	10	12	20	18,000–23,000[1]
Research Assistant	14	2	12	20	18,000–23,000
Total	61	13			

[1]Tuition waived for up to 12 credit hours/semester.

5. Personnel Engaged in Separately Budgeted Research, 7/09–6/10

Professorial faculty	24
Postdoctoral appointments	12
Graduate students	60
Total	96

6. Separately Budgeted Research Expenditures by Source of Support

	Departmental Research
Private	$65,231
State	
Federal government	$2,969,353
Total	$3,034,584

FACULTY

Professors

Chang, Robert S. F., Ph.D., Cornell, 1976. Solid state laser spectroscopy; energy transfer studies; crystal growth; fiberoptics.

Chen, Wei, Ph.D., Temple Univ. 1988, Biophysics & physiology.

Djeu, Nicholas, Ph.D., Cornell, 1970. Laser physics and nonlinear optics; fiberoptics.

Johnson, Dale E., Ph.D., Univ. of Chicago, 1971. Electron microscopy.

Killinger, Dennis K., Ph.D., Michigan, 1978. Laser optics laser physics; laser remote sensing/LIDAR; Optical Transmission of the Atmosphere. physics; quantum electronics; laser spectroscopy.

Kim, Myung K., Ph.D., Univ. of California, Berkeley, 1986. Digital holography, phase contrast microscopy, optical tomography, biomedical imaging, quantum optics, Laser spectroscopy.

Mukherjee, Pritish, Ph.D., SUNY, Buffalo, 1986. Chairman of the Department. Picosecond lasers and applications; laser-assisted materials growth; semi-conductor and superconductor physics.

Nolas, George S., Ph.D., Stevens Institute of Technology, 1994. Experimental Solid-State, Materials and Condensed Matter Physics.

Srikanth, Hariharan, Ph.D., Indian Inst. of Science, 1993. Experimental condensed matter, materials sciences.

Witanachchi, Sarath, Ph.D., SUNY, Buffalo, 1989. Laser ablation; films; high-T_c superconductors; semiconductors.

Associate Professors

Haynie, Donald T., Ph.D., Johns Hopkins University, 1994. Nanomedicine and bionanotechnology.

Oleynik, Ivan I., Ph.D., Russian Academy of Sciences, 1992. Theoretical condensed matter and chemical physics. Computational materials science.

Rabson, David, Ph.D., Cornell, 1991. Condensed matter theory.

Woods, Lilia, Ph.D., University of Tennessee, 1999. Theoretical Condensed Matter Physics: theory and computation of nanostructures, dispersive interactions, thermoelectric transport.

Assistant Professors

Batzill, Matthias, Ph.D., University of Newcastle upon Tyne, UK, 1999. Surface science; gas-surface interactions; structure and electronic properties of metal oxide surfaces; nanoclusters and quantum dots; solid state gas sensors; photocatalysis and photovoltaic for sustainable and renewable energy.

Jiang, Xiaomei, Ph.D., University of Utah, 2004. Organic electronic materials; fabrication and characterization of light emitting diodes and photovoltaic devices for solar cell applications.

Karaiskaj, Denis, Ph.D., Simon Fraser University, 2002. Two-dimensional spectroscopy on nanostructures, and proteins; optical spectroscopic studies of carbon nanotubes; ultrahigh resolution spectroscopy of semiconductors.

Matthews, Garrett, Ph.D., University of North Carolina, 2001. Biological macromolecules and macromolecular, biopolymers.

Miller, Casey W., Ph.D., Texas, 2003. Experimental condensed matter: Spin-dependent transport properties of novel materials and devices.

Muller, Andreas, Ph.D., University of Texas at Austin, 2007. Experimental quantum optics of nanostructures; quantum dots, nanocrystals, impurity centers; cavity quantum electrodynamics and optomechanics for quantum communication and quantum information science.

Munoz, Martin, Ph.D., University of Campinas, Brazil, 1997. Growth and characterization of semiconductor heterostructure devices with applications in electronics and optoelectronics.

Muschol, Martin, Ph.D., City University of New York, 1992. Neuronal Plasticity, advanced optical techniques to probe cellular mechanisms, Protein crystallization.

Pandit, Sagar A., Ph.D., University of Pune, India, 1999. Computational Biophysics and mathematical modeling of biological and social systems.

Ponomareva, Inna, Ph.D., Russian Academy of Sciences. 2004. Condensed matter physics, numerical quantum chemistry, computational physics, nanoscience, developing and implementation of computational techniques.

Eminent Scholar

Giaever, Ivar, Ph.D., Rensselaer Polytechnic Institute, 1964. Theoretical physics. Nobel Laureate, Physics, 1973. Chair, Physics Executive Advisory Board.

Instructors

Chabot, Michelle D., Ph.D., Univ. of Texas, Austin, 2001. Experimental condensed matter physics, undergraduate lab development, physics education.

Criss, Robert, Ph.D., University of Texas at Dallas, 1993. Applied VUV-VIS spectroscopy, physics education.

Mackay, Kevin, Ph.D., Queen's University, Belfast, N. Ireland, 2000. Extra-solar planets, astronomy education and thin-film magenetic materials.

Woods, Gerald T., Ph.D., University of Tennessee at Knoxville, 2001. Experimental condensed matter, Director of General Physics Labs and Physics Education.

Research Assistant Professor

Lisenkov, Sergei, Ph.D., Russian Academy of Sciences, 2005. Finite-temperature properties of multiferroic materials, Perovskite superlattices and nanostructures, electronic and stability properties of nanotubes and fullerenes.

Research Associate Professor

Zhakhovsky, Vasily V., Ph.D., Russian Academy of Science, 1997. Atomic simulations of laser-matter interactions, shock wave physics and materials at extreme conditions.

Courtesy Professors

Garcia-Rubio, Luis H., Ph.D., MacMaster University, Canada, 1981. Sensors, spectroscopy, material sciences.

Morel, Don L., Ph.D., Tulane Univ., 1971; MBA, Rutgers Univ., 1979. Microelectronics, material sciences.

Visiting Professor

Sakmar, Ismail A., Ph.D., Univ. of California, Berkeley, 1963. High energy particle physics; number theory.

RESEARCH SPECIALTIES AND STAFF

Theoretical

Soft Matter & Biological Physics. Pandit.

Condensed Matter/Material Physics. Oleynik, Ponomareva, Rabson, Woods, L.

Experimental

Atomic, Molecular & Optical Physics. Chang, Djeu, Jiang, Killinger, Kim, Mukherjee, Muschol.

Biological physics. Chen, Kim, Matthews, Muschol.

Condensed Matter/Material Physics. Batzill, Djeu, Jiang, Miller, Mukherjee, Munoz, Nolas, Rabson, Srikanth, Witanachchi.

CLARK ATLANTA UNIVERSITY[1]

DEPARTMENT OF PHYSICS

Atlanta, Georgia 30314

[1]Formed through consolidation of Atlanta University and Clark College.

Students Accepted For Degree	FIELDS		
	Physics	Astronomy	Related Fields
Doctorate			
Master's	X		

1. General

President: Carlton E. Brown
Dean of School of Arts and Sciences: Shirley Williams-Kirksey
Department Chairman: Swaraj S. Tayal
Department Telephone Number: (404) 880–8797
Type of Institution: Four plus
Control: Private
Setting: Urban
Total Faculty: 276 (No separate Graduate faculty)
Total Graduate Students: 671
Annual Graduate Tuition:
All Graduate students: $12,960 full-time, $720/credit hr.
Tuition rates for: 2010–2011
Deferred tuition plan: Yes
Other Fees: $710
Term: Semester

2. Number of Faculty in Department

The combined total of full-time faculty in the three professorial ranks is 8. The combined total of full-time, part-time, and other faculty at all ranks is 8.

3. Admission, Financial Aid, and Housing

Address admission inquiries to: Office of Admissions & Articulation (Graduate), Clark Atlanta University
Graduate application fee required: $40
Admission deadline (Fall admission): July 1
Admission information: For fall admission, 2009–10, 2 students were accepted from 3 applicants.
Admission requirements: For unconditional admission to the graduate program, a Bachelor's degree in physics is required with a minimum undergraduate GPA of 3.0. The GRE is required. Students from non-English speaking countries are required to demonstrate proficiency in English via the TOEFL exam. A minimum score of 500 is required.
Undergraduate preparation assumed: Students should have taken courses in the following areas: classical mechanics, electromagnetic theory, thermal physics, mathematical physics, quantum mechanics, calculus, and differential equations.
Address financial aid inquiries to: Graduate Program Coordinator, Department of Physics
GAPSFAS application required: No
Financial aid deadline: 4/1
Loans available: Yes
Address housing inquiries to: Director, Residential Life
On-campus, single student housing available: Yes
On-campus, married student housing available: No

Table A—Faculty, Enrollments, and Degrees Granted

Research Specialty	2009–10 Faculty	Enrollment[1] Fall 2009		No. of Degrees Granted[2] 2009–10 (2003–09)			Median No. of Years for 2009–10 Ph.D.'s
		Master's	Doctorate	Master's	Terminal Master's	Doctorate	
Atmos./Space Phys., Cosmic Rays	1	0	–	0(1)	–	–	–
Atomic, Molecular, & Optical Physics	2	0	–	0(3)	–	–	–
Condensed Matter Physics	1	0	–	0(0)	–	–	–
Other Theoretical/ Math.	2	4	–	0(5)	–	–	–
Total		4	–	0(9)	–	–	
Full-time Grad. Stud.		4	–				
Part-time Grad. Stud.		0	–				
First-year Grad. Stud.		2	–				
Median Years in Grad. Study (2009–10 Degrees)				2	–	–	–
Undergraduate Degrees, 2009–10 (2003–09): 0(4)							

[1]Students not yet committed to a research specialty are entered under non-specialized.
[2]Five-year totals in parentheses.

4. Graduate Degree Requirements

Master's: 30 hours of graduate courses with a 3.0 GPA are required. The program is designed to provide the basic course work and research skills necessary for the pursuit of a Ph.D. degree in physics. A non-thesis terminal Master's degree in applied physics is also offered.
Thesis: Thesis may be written *in absentia.*
Special Equipment, Facilities, or Programs: Theoretical research is supported in part by computers within the Center for Theoretical Studies of Physical Systems (CTSPS). The computers consist of a CRAY J916 Digital Equipment in a cluster; IBM workstations, RISC 6000/560's and 590's; Sun workstations; and Silicon Graphics workstations. Access through the NSF net/Internet to supercomputing centers at Pittsburgh and LBL is routine. Experimental research in surface science and earth system science is supported by facilities in the T. W. Cole, Jr. Center for Research in Science and Technology.

Table B—Appointments to Graduate Students, 2009–10

Title of Appointee	Appointments		Academic Load Allowed in Credit Hours	Hours of Service Per Week	Stipend for Academic Year ($)
	Total	First year			
			Semester		
Teaching Assistant	3	2	9	Variable	14,400
Research Assistant	1	0	9	Variable	14,400
Total	4	2			

5. Personnel Engaged in Separately Budgeted Research

Professorial faculty	6
Research Scientists	2
Graduate students	4
Undergraduate students	3
Total	15

6. Separately Budgeted Research Expenditures by Source of Support

Federal government	$16,486,000
Total	$16,486,000

Table C—Separately Budgeted Research Expenditures

Research Specialty	No. of Grants	Expenditures ($)
Atomic, Molecular, & Optical Physics	3	–
Condensed Matter Physics	1	–
Other Theoretical/Math.	7	–
Total	11	16,486,000

FACULTY

Professors

Mickens, Ronald E., Ph.D., Vanderbilt, 1968. Fuller E. Callaway Professor. Mathematical physics; modeling of biological and engineering systems; numerical integration schemes; nonlinear dynamics.

Msezane, Alfred Z., Ph.D., Western Ontario, 1973. Director of CTSPS. Theoretical physics; atomic and molecular physics; photon interactions with ground and excited atomic states; electron scattering from atoms and ions.

Puri, Om P., Ph.D., Sauger, 1961. Garfield D. Merner Professor of Science. Experimental physics; investigation of polling process and charge trapping in ceramic and other materials; magnetostriction.

Talukder, Niranjan K., Ph.D. Mechanical Engineering (Bio-Fluid Mechanics). Technical University of Aachen, 1974. Engineering statics and dynamics; experimental studies on cardiovascular fluid dynamics; analytic studies of heat transfer problems.

Tayal, Swaraj S., Ph.D., Roorkee (India), 1982. Theoretical physics; theoretical and computational studies of electron and photon interactions with atoms, ions, and molecules; plasma modeling.

Associate Professors

Mandock, Randal L. N., Ph.D., Georgia Institute of Technology, 1998. Atmospheric sciences; atmospheric boundary layers, sodar, 3-D acoustical imaging of lower troposphere energy transfer near air-ground interface; geophysics.

Wang, Xiao-Qian, Ph.D., International School for Advanced Studies (Trieste, Italy), 1989. Condensed matter; first-principle methods; molecular dynamics simulation; strongly-correlated electron systems.

Williams, Michael D., Ph.D., Stanford University, 1987. Experimental solid state physics; characterization of surfaces and interfaces of optoelectronic semiconductors.

Professional Staff

Harrington, Terry L., M.S., Atlanta University, 1987. Experimental solid state physics; thermoelectret formation; study and characterization in single crystal zirconia and PZT.

Research Scientists

Chen, Zhifan, Ph.D., Texas A & M, 1991. Theoretical physics, excitation, ionization and charge transfer processes in ion-atom collisions; lasers, light sources, modeling, fusion charge transfer, electron-atom scattering.

Felfli, Zineb, Ph.D., Georgia State University, 1993. Theoretical Atomic Physics. Theoretical studies of atomic processes applications; laser physics, astrophysics, plasma physics.

Nduwimana, Alexis, Ph.D., Physics, Georgia Institute of Technology, 2007. Semiconducting nanowires, nano-shell, and multishells.

RESEARCH SPECIALTIES AND STAFF

Theoretical

Atomic and Molecular Physics. Electron/photon scattering from atoms and ions; atomic structure calculations. Chen, Felfli, Msezane, Tayal.

Condensed Matter. Many-body theory; transport theory; molecular dynamics. Wang.

Elementary Particles. Electromagnetic structure of elementary particles. Mickens, Talukder.

Mathematical Physics. Singular perturbation techniques; nonlinear oscillations; wavelets and signal processing; molecular dynamics. Mickens, Wang.

Experimental

Atmospheric Physics. The effects of aerosol and atmospheric constituents on the propagation of electromagnetic radiation; solar radiation measurements; aerosol optical properties; photovoltaic devices; acoustic measurements of the radioactive characteristics of ground cover. Mandock.

Condensed Matter. Characterization of polymers; properties of polymers; surface physics: characterization of surfaces and interfaces of optoelectronic semiconductors using ion and electron spectroscopies. Harrington, Puri, Williams.

EMORY UNIVERSITY

DEPARTMENT OF PHYSICS

Atlanta, Georgia 30322

Students Accepted For Degree	FIELDS		
	Physics	Astronomy	Related Fields
Doctorate	X		
Master's			

1. General

President: James W. Wagner
Dean of Graduate School: Lisa Tedesco
Department Chairman: Kurt Warncke
Department Telephone Number: (404) 727-6584
Type of Institution: University
Control: Private
Setting: Suburban
Total Faculty: 3,600
Total Graduate Faculty: 873
Total Students: 12,755
Total Graduate Students: 5,865
Annual Graduate Tuition:
In-state residents: Full-time—$33,800
Out-of-state residents: Full-time—$33,800
Tuition rates for: 2009–10
Deferred tuition plan: Yes
Other Fees: Activity and Athletic
Term: Semester

2. Number of Faculty in Department

The combined total of full-time faculty in the three professorial ranks is 16. The combined total of full-time, part-time, and other faculty at all ranks is 20.

3. Admission, Financial Aid, and Housing

Address admission inquiries to: Director of Graduate Studies, Department of Physics
Graduate application fee required: none
Admission deadline (Fall admission): 1/3
Admission information: For fall admission, 2009–10, 5 students were accepted into physics.
Admission requirements: For admission to the graduate programs, a Bachelor's degree in physics is required with a minimum undergraduate GPA of 3.0 specified. The GRE is required. The minimum acceptable score suggested for admission are verbal–500; quantitative–700. The TOEFL exam is required for foreign students. The minimum acceptable score is 600. Emory University does not discriminate on the basis of race, color, religion, sex, national origin, handicap, age, or veteran status.
Address financial aid inquiries to: Graduate Coordinator, Department of Physics
GAPSFAS application required: No
Financial aid deadline: 1/3
Loans available: Yes
Address housing inquiries to: University Housing Office (404) 727-8830, Fax (404) 727-8835
On-campus, single student housing available: Yes
On-campus, married student housing available: Yes

Table A—Faculty, Enrollments, and Degrees Granted

Research Specialty	2009–10 Faculty	Enrollment[1] Fall 2009 Master's	Enrollment[1] Fall 2009 Doctorate	No. of Degrees Granted[2] 2009–10 (2000–10) Master's	Terminal Master's	Doctorate	Median No. of Years for 2007–08 Ph.D.'s
Biophysics	6	–	16	–	0(1)	3(5)	7
Soft Condensed Matter Physics	2	–	7	–	0(4)	0(2)	–
Solid State	3	–	–	–	0(0)	0(0)	–
Non-specialized	4	–	–	–	0(0)	0(0)	–
Nuclear Physics	1	–	–	–	0(0)	0(0)	–
Particles & Fields	1	–	–	–	0(1)	0(0)	–
Statistical & Thermal	4	–	2	–	0(7)	3(8)	
Total	21	–	25	–	0(7)	3(8)	
Full-time Grad. Stud.		–	25				
Part-time Grad. Stud.		–	0				
First-year Grad. Stud.		–	5				
Median Years in Grad. Study (2007–08 Degrees)				–	–	–	–

Undergraduate Degrees, 2009–10 (2000–09): 14(101)

[1]Students not yet committed to a research specialty are entered under non-specialized.
[2]Five-year totals in parentheses.

4. Graduate Degree Requirements

Master's: 24 semester hrs. including 20 hrs. of course or seminar work with a gpa of B or better. Two semesters of residence are required. A thesis an oral defense of research work are required.
Doctorate: 48 semester hrs. beyond M.S. level; maximum of 36 hours in research or guided study with an average grade of B^-. A minimum of four semesters in residence are required beyond M.S. level. Ph.D. candidacy is determined by a gpa of B in Physics coursework and a written and orally presented Qualifier Proposal (research proposal). A dissertation and oral defense are required.

Table B—Appointments to Graduate Students, 2009–10

Title of Appointee	Appointments Total	Appointments First year	Academic Load Allowed in Credit Hours	Hours of Service Per Week	Stipend for Academic Year ($)
			Semester		
Teaching Assistant	4	4	12	20	21,500
Research Assistant	20	0	12	–	21,500
Dean's Teaching Fellows	0	0	12		21,500
Total	24	4			

5. Personnel Engaged in Separately Budgeted Research, 7/09–6/10

Professorial faculty	11
Postdoctoral appointments	6
Graduate students	14
Total	31

6. Separately Budgeted Research Expenditures by Source of Support

	Departmental Research
Federal government	$1,945,000
Other Include Institution's own separately budgeted accounts	500,000
Total	$2,445,000

Table C—Separately Budgeted Research Expenditures

Research Specialty	No. of Grants	Expenditures ($)
Biophysics	6	1,181,000
Condensed Matter Physics	5	395,000
Statistical and Thermal Physics	2	369,000
Total	13	1,945,000

Includes institutional cost sharing expenditures.

FACULTY

Professors

Bajaj, Krishan K., Ph.D., Purdue, 1966. Charles T. Winship Professor of Physics. Theoretical solid state physics.
Family, Fereydoon, Ph.D., Clark, 1974. Samuel Candler Dobbs Professor of Condensed Matter Physics.
Hentschel, H. George E., Ph.D., Cambridge, 1978. Theoretical and statistical physics.
Huynh, Boi Hanh, Ph.D., Columbia, 1974. Samuel Candler Dobbs Professor of Physics. Experimental biophysics.
Perkowitz, Sid, Ph.D., Pennsylvania, 1967. Charles Howard Candler Professor of Condensed Matter Physics.

Associate Professors

Berland, Keith M., Ph.D., University of Illinois, 1995. Experimental biophysics.
Day, Edmund P., Ph.D., Stanford, 1973. Experimental biophysics.
DuVarney, Raymond C., Ph.D., Clark, 1968. Adaptive optics.
Finzi, Laura, Ph.D., New Mexico, 1990. Director of Graduate Studies. Experimental biophysics.
Nemenman, Ilya, Ph.D., Princeton University 2000. Theoretical and statistical physics.
Rao, P. Venugopala, Ph.D., Oregon, 1964. Experimental nuclear physics.
Warncke, Kurt, Ph.D., Pennsylvania, 1990. Chair. Experimental biophysics.
Weeks, Eric, Ph.D., Texas Austin, 1997. Director of Undergraduate Studies. Experimental soft condensed matter.

Assistant Professors

Boettcher, Stefan, Ph.D., Washington University, 1993. Statistical physics; critical phenomena.
Rasnik, Ivan, Ph.D., Estadual de Campinas, Brasil, 2000. Experimental biophysics.
Roth, Connie, Ph.D., University of Guelph, 2004. Experimental soft condensed matter, polymerphysics.
Segré, Philip, Ph.D., Maryland, 1993. Experimental soft condensed matter.

Lecturers

Bing, Thomas, Ph.D., University of Maryland, 2007. Physics education.
Brody, Jed, Ph.D., Georgia Inst. Tech., 2003. Photovoltaics.
Coleman, Robert N., M.S., Emory, 1974. Radioecology.
Williamon, Richard M., Ph.D., Florida, 1972. Director, Planetarium. Astronomy.

Adjunct Professors

Cheng, Xiaodong, Ph.D., SUNY at Stony Brook, 1989. Protein crystallography.
Eisner, Robert L., Ph.D., Purdue, 1968. Magnetic imaging.
Galt, James R., Ph.D., Emory University, 1988. Radiology.
Garcia, Ernest, V., Ph.D., Miami, 1974. Medical imaging.
Fine, Alan, Ph.D., University of Pennsylvania, 1987. Physiology and Biophysics.
Malko, John A., Ph.D., Ohio, 1970. Magnetic imaging.
Vicsek, Tamás, Ph.D., Eötvös University. Department of Atomic Physics, Budapest, Hungary.

RESEARCH SPECIALTIES AND STAFF

Theoretical

Solid State Physics: Semiconductor heterostructures and optoelectronic devices. Bajaj.
Statistical and Thermal Physics: Self-organized criticality, optimization, statistical mechanics, quantum-field theory. Boettcher.
Nonequilibrium growth phenomenoa, pattern formation, fractals, surface and interface physics. Family.
Neural networks, spin glasses, nonlinear dynamics, chaos, biomorphogenesis, and turbulence. Hentschel.

Experimental

Biophysics: Novel methods in (two-photon) fluorescence imaging and spectroscopy, developed and applied to molecular and cellular biophysics. Berland.
Dynamics and kinetics of protein-induced conformational changes in DNA relevant to transcription regulation, studied by using single molecule, tethered particle methods, including magnetic tweezers. Finzi.
Structural and functional studies of iron-containing proteins by using Mössbauer and electron paramagnetic resonance (EPR) spectroscopies. Huynh.
Development and application of single molecule fluorescence resonant energy transfer (FRET) methods to protein and nucleic acid structure and dynamics. Rasnik.
Pulsed-EPR and transient optical absorption studies of structural and dynamical bases of metallocenter- and radical-mediated enzyme catalysis; amyloid protein structure; artificial photosynthesis. Warncke.
Soft Condensed Matter Physics: Physical properties and dynamics of polymer molecules, glass transition, structural relaxation, visoelastic deformations. Roth.
Complex liquids, fluid dynamics, multiphase flow, gelation and jamming of colloidal solutions. Segré.
Microscopy of colloidal glasses, nonlinear dynamics, complex fluids, and granular media. Weeks.

GEORGIA INSTITUTE OF TECHNOLOGY

SCHOOL OF PHYSICS

Atlanta, Georgia 30332-0430

Students Accepted For Degree	FIELDS		
	Physics	Astronomy	Related Fields
Doctorate	X		
Master's	X		

1. General

President: G. P. "Bud" Peterson
Provost for Research and Graduate Studies: Gary Schuster
School Chair: Mei-Yin Chou
Department Telephone Number: (404) 894-5200
FAX: (404) 894-9958
Electronic Mail Address: grad.recruiter@physics.gatech.edu
Type of Institution: University
Control: Public
Setting: Urban
Total Faculty: 961
Total Graduate Faculty: 920
Total Students: 18,742
Total Graduate Students: 6,177
Annual Graduate Tuition: 12 months*
In-state residents: Full-time—$4,318 (includes mandatory fees)
Part-time—$360/credit, plus $373
Out-of-state residents: Full-time—$13,102(includes mandatory fees)
Part-time—$1,092/credit, plus $373
Tuition rates for: 2010–10
Deferred tuition plan: No*
Term: Semester

*Students employed as Graduate Research or Teaching Assistants pay $25 per semester tuition plus $823 fee per semester.

2. Number of Faculty in Department

The combined total of full-time faculty in the three professorial ranks is 34. The combined total of full-time, part-time, and other faculty at all ranks is 41.

3. Admission, Financial Aid, and Housing

Address admission inquiries to: Graduate Recruiter, School of Physics
Graduate application fee required: None
Admission deadline (Fall admission): January 1st
Admission information: For fall admission, 2009. 169 applicants; 78 offers; 31 acceptances.
Admission requirements: For admission to the graduate programs, a Bachelor's degree in physics is preferred with a minimum undergraduate GPA of 3.0 preferred for the M.S. program and 3.5 for the Ph.D. program. The GRE General and Physics Tests are required. Students from non-English speaking countries are required to demonstrate proficiency in English via the TOEFL exam. Minimum acceptable score for admission is 625.
Undergraduate preparation assumed: Symon, *Classical Mechanics*; Corson and Lorrain, *Classical Electromagnetic Fields and Waves*, Liboff, *Quantum Mechanics*; Callen, *Thermodynamics*.
Address financial aid inquiries to: Graduate Recruiter, School of Physics or grad.recruiter@physics.gatech.edu
GAPSFAS application required: No
Financial aid deadline: May 1st
Loans available: No
Address housing inquiries to: Housing Office.
E-mail: http://www.housing.gatech.edu
On-campus, single student housing available: Yes (limited)
Average Cost/Month: (Semester Rate: $3,564)
On-campus, married student housing available: Yes (limited)
Cost/month: $1,164–1,355 (1-Bedroom; 2 Bedroom)

Table A—Faculty, Enrollments, and Degrees Granted

Research Specialty	2009–10 Faculty	Enrollment[1] Fall 2009		No. of Degrees Granted[2] 2009 (2005–10)			Median No. of Years for 2009–10 Ph.D.'s
		Master's	Doctorate	Master's	Terminal Master's	Doctorate	
Astrophysics/Gravity	4	1	0	1(0)	0	0	
Atomic & Molecular Phys.	9	1	2	1(2)	0	2(35)	5
Biophysics	7	0	1	0(1)	0	1(2)	–
Computational Physics	6	0	0	0	0	0(1)	
Condensed Matter	15	1	4	1(6)	0	4(40)	5
Materials Science	0	0	0	0	0	0(2)	–
Mathematical Phys.	2	0	0	0	0	0	
Nonlinear Dynamics	10	1	1	1(2)	0	1(16)	5
Nuclear Physics	0	0	0	0(1)	0	0(2)	–
Physics Education	1	0	0	0	0	0(0)	5
Non-specialized	0	3	7	3(45)	0(5)	15(106)	–
Total		7	15	7(58)	0(5)	15(106)	
Full-time Grad. Stud.		7	101				
Part-time Grad. Stud.		0	0				
First-year Grad. Stud.		3	28				
Median Years in Grad. Study (2009–10 Degrees)				2	2	5.5	
Undergraduate Degrees, 2009:			36				

[1]Students not yet committed to a research specialty are entered under non-specialized.
[2]Five-year totals in parentheses.

4. Graduate Degree Requirements

Master's: 30 semester hrs. are required. Thesis is optional. 2.7 GPA is required. One year residency required. No language requirement, no final examination.
Doctorate: The number of credit hours is not stipulated except 9 hrs. in minor with 2.9 GPA required. One year residency required. Comprehensive examination, thesis, thesis examination are required.
Thesis: Thesis may be written *in absentia*.
Special Equipment, Facilities, or Programs: Brochure describing research programs available upon request to department, and on the world wide web at http://www.physics.gatech.edu

Table B—Appointments to Graduate Students, 2008–09

Title of Appointee	Appointments Total	First year	Academic Load Allowed in Credit Hours	Hours of Service Per Week	Stipend for Academic Year ($)
			Quarter		
Teaching Assistant	35	12	18–21	13	22,042.00
Research Assistant	59	0	18–21	13	22,042.00
Student Assistants	0	0	18–21	13	22,042.00
Total	94	12			

5. Personnel Engaged in Separately Budgeted Research, 7/08–6/09

Professorial faculty	35
Other faculty	25
Postdoctoral appointments	23
Graduate students	103
Undergraduate students	0
Total	187

6. Separately Budgeted Research Expenditures by Source of Support

	Departmental Research	Physics-related Research Outside Department
Federal government	$4,968,541	$8,000,000
Other	3,567,030	
Total	$8,535,571	$8,000,000

Table C—Separately Budgeted Research Expenditures

Research Specialty	No. of Grants	Expenditures ($)
Astrophysics	17	229,348
Atomic & Molecular Physics	14	850,960
Biophysics	9	850,960
Chemical Physics	3	112,879
Condensed Matter	61	3,617,234
Nonlinear Dynamics	21	1,177,371
Nuclear Physics	1	124,231
Optics	21	641,625
Other	18	1,507,079
Total	130	8,535,571

FACULTY

Professors

Bellissard, Jean, Ph.D., Univ. of Provence, Marseille, 1974. Mathematical physics.

Chapman, Michael S., Ph.D., MIT, 1995. Experimental quantum optics; atomic physics.

Chou, Mei-Yin, Ph.D., California, Berkeley, 1986. Advance Professor. Theoretical condensed matter physics; electronic structure of materials; computational materials physics.

Conrad, Edward H., Ph.D., Wisconsin-Madison, 1983. Experimental surface physics.

Cvitanović, Predrag, Ph.D., Cornell, 1973. Glen Robinson Chair in Nonlinear Sciences. Director, Georgia Tech Center for Nonlinear Science.

de Heer, Walter A., Ph.D., University of California, 1984. Experimental condensed matter physics, magnetic and electronic properties of clusters, carbon nanostructures.

Erbil, Ahmet, Ph.D., MIT, 1983. Experimental condensed matter physics.

First, Phillip, Ph.D., Illinois, Urbana, 1988. Experimental condensed matter physics.

Gole, James L., Ph.D., Rice, 1971. Experimental chemical and condensed matter physics; optics; material science.

Kennedy, T. A. Brian, Ph.D., Queen's Belfast, 1986. Theoretical quantum optics.

Laguna, Pablo, Ph.D., University of Texas at Austin, 1987. Numerical Relativity.

Landman, Uzi, D.Sc., Haifa, 1969. Institute/Regents Professor. Theoretical condensed matter physics.

Trebino, Rick Ph.D., Stanford University, 1983. Georgia Research Alliance Eminent Scholar. Chair of Ultrafast optical physics.

Uzer, Turgay, Ph.D., Harvard, 1979. Theoretical, molecular, and chemical physics; nonlinear dynamics.

Wiesenfeld, Kurt, Ph.D., California, Berkeley, 1985. Theoretical nonlinear dynamics; biophysics.

You, Li, Ph.D., Colorado, 1993. Physics of light/matter interactions.

Zangwill, Andrew, Ph.D., Pennsylvania, 1981. Theoretical condensed matter physics.

Associate Professors

Davidovic, Dragomir, Ph.D., Johns Hopkins University, 1996. Mesoscopics and low temperature physics.

Grigoriev, Roman, Ph.D., California Institute of Technology, 1998. Theoretical non-linear dynamics.

Kuzmich, Alex, Ph.D., Univ. of Rochester, 2000. Experimental atomic, molecular, and optical physics.

Raman, Chandra, Ph.D., University of Michigan, 1997. Experimental atomic physics.

Sá de Melo, Carlos, Ph.D., Stanford University, 1991. Theoretical condensed matter physics.

Schatz, Michael F., Ph.D., University of Texas, 1991. Experimental nonlinear dynamics; fluid dynamics.

Assistant Professors

Ballantyne, David R., Ph.D., University of Cambridge, 2002. Theoretical astrophysics.

Curtis, Jennifer E., Ph.D., University of Chicago, 2002. Experimental biophysics.

Fernandez de las Nieves, Alberto, Ph.D., University of Granada, 2000. Experimental condensed matter physics.

Goldman, Daniel I., Ph.D., University of Texas at Austin, 2002. Experimental biophysics, nonlinear dynamics.

Jiang, Zhigang, Ph.D., Northwestern University, 2005. Experimental Condensed Matter Physics.

Kim, Harold, Ph.D., Standord University, 2004. Experimental biophysics.

Kindermann, Markus, Ph.D., Universiteit Leiden, 2003. Theoretical condensed matter physics.

Nguyen, Toan T., Ph.D., University of Minnesota, 2002. Theoretical biophysics.

Pustilnik, Michael, Ph.D., Bar-Ilan University, 1997. Theoretical condensed matter.

Riedo, Elisa, Ph.D., University of Milan, 2000. Experimental condensed matter, biophysics.

Shoemaker, Deirdre, Ph.D., University of Texas at Austin, 1999. Numerical Relativity.

Taboada, Ignacio, Ph.D., University of Pennsylvania, 2002. Astrophysics.

Adjunct Professors

Bréchignac, Catherine, Ph.D., University of Paris-Sud, Orsay, 1977. Molecular and cluster physics.

Brown, Kenneth, Ph.D., University of California, Berkeley, 2003. Atomic and molecular physics.

Harvey, Stephen C., Ph.D., Dartmouth College, 1971. Biophysics.

Orlando, Thomas, Ph.D., State, Univ. of New York-Stony Brook, 1988. Experimental physical, analytical and materials chemistry.

Wartell, Roger, Ph.D., University of Rochester, 1971. Experimental biophysics.

Weitz, Joshua, Ph.D., Massachusetts Institute of Technology, 2003. Theoretical biophysics.

Whetten, Robert L., Ph.D., Cornell University, 1984. Experimental chemical, solid-state, and surface physics. Nanometer-scale crystallites, clusters, and size effects.

Zhu, Cheng, Ph.D., Columbia University, 1988. Biophysics.

Research Scientists

Barnett, Robert N., Ph.D., Kansas, 1980. Theoretical condensed matter physics.

Berger, Claire, Ph.D., University Joseph Fourier, Grenoble, 1987. Experimental condensed matter physics.

Bogachek, Eduard N., Ph.D., Kharkov, 1977. Theoretical condensed matter physics; mesoscopics.

Gao, Jianping, Ph.D., Brown, 1989. Theoretical condensed matter physics.

Jenkins, Stewart, Ph.D., Georgia Tech, 2006. Atomic and molecular physics.

Kulp, W. David III, Ph.D., Georgia Tech., 2001. Experimental atomic, molecular and chemical physics.

Luedtke, William D., Ph.D., Georgia Tech., 1984. Theoretical condensed matter physics.

Ray, William Richard, Ph.D., University of Maryland, 2006. Nonlinear.

Ruan, Wen-Ying, Ph.D., Zhongshan University, 1992. Theoretical condensed matter.

Yannouleas, C., Ph.D., University of Maryland, 1982. Theoretical condensed matter physics; theoretical nuclear physics.

Yoon, Bokwon, Ph.D., University of Paris-Sud, Orsay, France, 1997. Theoretical condensed matter physics.

Academic Professionals

Jarrio, Marty, Ph.D., Georgia Tech, 1996. Nuclear physics.

Murray, Eric, Ph.D., Cornell, 1992. Materials science.

Scherbakov, Andrew, Ph.D., Georgia Tech, 1997. Mesoscopic physics.

Sowell, James, Ph.D., Michigan, 1986. Astronomy.

RESEARCH SPECIALTIES AND STAFF

Theoretical

Astrophysics: General relativity; gravitational wave patterns; gravitational interactions of compact binaries; theoretical and phenomenological astrophysics; galaxy and black hole evolution; high-energy particle astrophyics; accretion disks; numerical relativity; cosmology; gravitating systems; black holes; galaxy and black hole evolution; high-energy particle astrophysics; accretion disks; gravitational physics. Ballantyne, Laguna, Shoemaker, 2 postdoctoral fellows.

Atomic and Molecular Physics. Three-body recombination; antihydrogen formation; cold collisions; collisional Stark mixing; Rydberg plasmas; classical-quantal correspondences; atomic Fermi gas transport; optical lattices; spin squeezing of atomic ensembles; Bose-Einstein condensate mixtures; quantum fluctuations; spatial solitary waves; nonlinear optical parametric processes; Rydberg atoms; light/matter interactions. Kennedy, Uzer, You, 1 postdoctoral fellow.

Biophysics. Energy transduction; chemosmosis; noise; protein biosynthesis; energy metabolism; ion channel fluctuations; molecular motors; Hodgkin-Huxley equations; chemo-mechanical energy conversion; energy driven rectification of Brownian motion; quantum mutations in DNA. Nguyen.

Computational Physics. Spatially extended non-equilibrium systems; chaotic mixing in fluids; thin liquid films; dynamics of solid surfaces; epitaxial growth processes; solid-liquid interfaces; melting; glasses; surface diffusion; atomic-scale friction and lubrication; confined complex fluids; electron localization; dynamics of small clusters; kinetic Monte Carlo and molecular dynamics simulations; density functional theory; quantum Monte-Carlo techniques; first-principles electronic structure; Landau-Lifshitz-Gilbert simulations; numerical relativity. Chou, Grigoriev, Landman, Laguna, Shoemaker, Zangwill, 4 postdoctoral fellows.

Condensed Matter and Materials Science. Nanoscience; phase transitions; mesoscopic physics; quantum interference effects; superconductors in high magnetic fields; Bose-Einstein superconductivity; macroscopic quantum phenomena; ferroelectrics; Sutherland-Calogero models; ferromagnets; spintronics; semiconductor quantum dots. Bellissard, Chou, Kindermann, Landman, Pustilnik, Sá de Melo, Zangwill, 3 postdoctoral fellows.

Nonlinear Dynamics and Statistical Physics. Molecular fluctuations; chaotic dynamics; quantum chaos; Husimi-Wigner wave packets; Lyapunov exponent; Rydberg states; trajectory analysis; massively coupled oscillators; chemical reaction dynamics; Hamiltonian flows. Bellissard, Cvitanović, Gregoriev, Uzer, Weitz, Wiesenfeld, You. 1 postdoctoral fellow.

Mathematical Physics. Bloch electrons in magnetic fields; quasicrystals; doped semiconductors; Lie algebras; Non-linear field theory. Bellisard, Cvitanović.

Optical: Classical and Quantum. Quantum optics; atomic Fermi gas transport in optical lattices; nonlinear optics and lasers; radiative interactions; squeezed states, quantum computing, cavity QED. Kennedy, You. 1 postdoctoral fellow.

Physics Education. Matter and Interactions curriculum. Schatz.

Sub-Atomic Physics. Quantization of Non-Linear Field Theories. Cvitanović.

Experimental

Astrophysics: Neutrino and gamma-ray astrophysics. Brown, Taboada. 2 postdocs.

Atomic, Molecular and Chemical Physics. Fundamental properties of ultra-cold condensed gases; atom trapping; multi-atom entanglement; cavity QED; laser Raman and Brillouin scattering; chemical biosensors; photovoltaic devices. Chapman, Gole, Kuzmich, Raman, Trebino, 2 postdoctoral fellows.

Biophysics. Morphogenesis, noise; "g-jitter;" thin organic films; nanotribology. Curtis, Goldman, Kim, Riedo, 1 postdoctoral fellows.

Condensed Matter and Materials Physics. Nanoscience; Soft

matter, Electron diffraction; low-temperature physics; scanning tunneling microscopy; ballistic electron emission spectroscopy; high-resolution x-ray scattering; magnetic susceptibility; magnetic heterostructures; graphene; Josephson tunneling; molecular clusters; thin-film magnetism; semiconductor nanostructures; atomic force microscopes; friction, adhesion; elasticity; wear; nanowires; laser fluorescence; chemiluminescence; mass spectroscopy; amorphous carbon thin films; novel soft materials. Conrad, Davidovic, de Heer, Fernández, Fernández-Nieves, First, Gole, Jiang, Riedo, 6 postdoctoral fellows.

Non-linear Dynamics and Statistical Physics. Spatiotemporal chaos; control/exploitation of chaos; pattern formation in fluids; low-gravity fluid physics; weather-in-a-box; spontaneous and manipulated patters; fluid instabilities; coupled mechanical oscillators. Fernández, Goldman, Schatz.

Optical: Classical and Quantum. Bose and Fermi condensed gases; atom and ion trapping; atom optics; multi-atom entanglement; cavity QED, ultrafast optics; frequency-resolved optical gating (FROG); ultrashort laser pulses. Chapman, Kuzmich, Raman, Trebino.

GEORGIA STATE UNIVERSITY

DEPARTMENT OF PHYSICS AND ASTRONOMY

Atlanta, Georgia 30303

Students Accepted For Degree	FIELDS		
	Physics	Astronomy	Related Fields
Doctorate	X	X	
Master's	X	X	

1. General

President: Mark P. Becker
Director of Graduate Studies, Arts and Sciences: Amber Amari
Department Chair: H. Richard Miller
Department Telephone Number: (404) 413-6033
Type of Institution: University
Control: Public
Setting: Urban
Total Faculty: 1,886
Total Graduate Faculty: Not separated
Total Students: over 30,000
Total Graduate Students: 8,047
Annual Graduate Tuition:
In-state residents: Full-time—$
Part-time—$203/credit
Out-of-state residents: Full-time—$15,214
Part-time—$300/credit
Tuition rates for: 2009
Deferred tuition plan: No
Other Fees: $814
Term: Semester

2. Number of Faculty in Department

The combined total of full-time faculty in the three professorial ranks is 24. The combined total of full-time, part-time and other faculty at all ranks is 27.

3. Admission, Financial Aid, and Housing

Address admission inquiries to: A. G. U. Perera, Director of Physics Graduate Program, uperera@gsu.edu or D. R. Gies, Director of Astronomy Graduate Program, Department of Physics and Astronomy, gies@chara.gsu.edu
Graduate application fee required: $50
Admission deadline (Fall admission): Early: 3/15. Regular: 4/15. Late: 7/15. Astronomy Program: 2/15.
Admission information: For admission, 2008–09, 33 students were accepted from 147 applicants. 6 students were accepted into the Astronomy PhD program from 34 applicants.
Admission requirements: For admission to the graduate programs, no specific degree or GPA is required. Admission is based on the applicant's undergraduate record, GRE scores and recommendations. Remedial work equivalent to a B.S. in Physics will be required of students with degrees in other fields. The GRE is required. The average GRE scores for 2009 admissions were verbal–609; quantitative–712; analytical–416. The GRE Advanced is recommended for financial assistance. No minimum score is required. Students from non-English-speaking countries are required to demonstrate proficiency in English via the TOEFL exam. A student will be tested upon arrival, and course work in English may be required based on the results of this test. Georgia State University, a unit of the University System of Georgia, is an equal opportunity/affirmative action educational institution.
Undergraduate preparation assumed: Eisberg and Resnick, *Quantum Physics*: Griffiths, *Introduction to Electrodynamics*; Mandl, *Statistical Physics*; Kreyszig, *Advanced Engineering Mathematics*; Fowles, *Analytic Mechanics*.
Address financial aid inquiries to: A. G. U. Perera, Director of Physics Graduate Program, or D. R. Gies, Director of Astronomy Graduate Program, Department of Physics and Astronomy
GAPSFAS application required: No
Financial aid deadline: 3/1
Loans available: Yes
On-campus, single student housing available: Yes
On-campus, married student housing available: Yes

Table A—Faculty, Enrollments, and Degrees Granted

Research Specialty	2008–09 Faculty	Enrollment[1] Fall 2009		No. of Degrees Granted[2] 2008–09 (1997–07)			Median No. of Years for 2008–09 Ph.D.'s
		Master's	Doctorate	Master's	Terminal Master's	Doctorate	
Astronomy	6	5	18	9(12)	1(6)	5(16)	6
Astrophysics	2	0	0	0(2)	0(1)	0(3)	6
Atomic, Molecular, & Optical Physics	1	1	4	3(2)	1(1)	3(1)	5
Biophysics	3	1	9	2(3)	1(3)	1(1)	5
Condensed Matter Physics	5	1	12	1(6)	0(4)	5(10)	5
Nuclear Physics	1	0	2	2(2)	0(0)	3(0)	5
Non-specialized	1	3	3	0(1)	0(0)	–	–
Total		20	48	12(28)	2(15)	16(31)	
Full-time Grad. Stud.			62				
Part-time Grad. Stud.			6				
First-year Grad. Stud.			18				
Median Years in Grad. Study (2002–03 Degrees)				2	2	5	–

Undergraduate Degrees, 2008–09 (2008–09): 9(41)

[1]Students not yet committed to a research specialty are entered under non-specialized.
[2]Five-year totals in parentheses.

4. Graduate Degree Requirements

Master's: M.S. students must complete a minimum of 24 semester hours of course work, pass a comprehensive exam, and pass a foreign language exam. An alternate research skill such as computer programming may be substituted for the foreign language exam. M.S. students must either complete an acceptable thesis or complete 6 additional hours of course work.
Doctorate: The Ph.D. degrees each require a minimum of 71 semester hours (beyond the B.S.). Students must complete and defend an acceptable dissertation in physics, or astronomy. Written qualifying and oral preliminary exams and reading exams in two foreign languages also are required. Alternate research skills such as computer programming may be substituted for the foreign language exams.
Thesis: Thesis may not be written *in absentia*.
Special Equipment, Facilities, or Programs: Research in the department currently is supported by the National Science Foundation, the Department of Energy, the National Institutes of Health, and the National Aeronautics and Space Adminis-

tration. Research apparatus within the department includes the following: *X* band and *K* band EPR/ENDOR spectrometers, X-ray diffraction apparatus, UV, X, and gamma irradiation facilities, a wide assortment of CAMAC and NIM modules, positron annihilation lifetime, closed-cycle refrigerator, Ge detector, CO_2 Nd:YAB lasers, monochromators, a 32-channel logic analyzer with a dual channel digital scope, al-mil wire bonder, three FTIR spectrometers one for TR spec down to 10 ns, a steady-state and time-resolved photoluminescence spectrometer, a Raman spectrometers, a high-resolution absorption spectrometer, a high-power frequency agile laser system (200 nm–3000 nm) utilized in a variety of gas phase and materials characterization techiques, ultra high vacuum surface science apparatus including high resolution electron energy loss spectrometer, Auger electron spectrometer, and low energy electron diffraction apparatus. Nuclear researchers utilize the particle accelerators at FermiLab and Brookhaven National Laboratories. The numerous workstations in the department are networked through the University Computer Center to provide access to the center's four mainframe and two Silicon Graphics minisupercomputers. The GSU Library subscribes to more than 250 physics and astronomy journals. Astronomy research facilities on campus include a network of unix-based workstations used extensively for data analysis and image processing. GSU's Hard Labor Creek Observatory (HLCO) is located in a state park 80 kilometers east of Atlanta. There is a darkroom, an electronics shop, and living quarters for observers. The principal telescopes at HLCO are a 16-inch telescope, and the Multi-Telescope Telescope (MTT) with equivalent light-collecting power to a 50-inch telescope, exceeding in size any other telescope in Georgia and listed among the largest telescopes in the Southeast. The Center for High Angular Resolution Astronomy (CHARA) research uses high-spatial resolution imaging techniques to attain image detail beyond that normally obtained with large telescopes. CHARA's major project is the operation of the CHARA Array, an optical and infrared interferometer at Mount Wilson Observatory in California. The CHARA Array consists of six telescopes of 1 m aperture that form a Y-shaped figure with baselines from 30 to 330 m. It is currently the premier instrument of its kind in the world for high angular resolution of stars and their environments. GSU astronomers also participate in observing programs with the Small and Medium Aperture Research Telescope System of telescopes at the Cerro Tololo Interamerican Observatory in Chile.

Table B—Appointments to Graduate Students, 2008–09

Title of Appointee	Appointments Total	Appointments First year	Academic Load Allowed in Credit Hours	Hours of Service Per Week	Stipend for Academic Year ($)[1]
			Semester		
Teaching Assistant	22	7	20	20	6,000–17,000
Research Assistant	17	0	20	See note[2]	19,500
Total	49	7			

[1]Tuition waived. 12-month stipend.
[2]No specific number of hours required.

5. Personnel Engaged in Separately Budgeted Research, 7/08–6/09

Professorial faculty	18
Postdoctoral appointments	9
Graduate students	57
Nonteaching research personnel	5
Total	89

6. Separately Budgeted Research Expenditures by Source of Support

	Departmental Research	Physics-related Research Outside Department
Federal government	$3,085,000	
Other	52,000	
Total	$3,137,000	

Table C—Separately Budgeted Research Expenditures

Research Specialty	No. of Grants	Expenditures ($)
Astronomy	9	1,150,000
Astrophysics	1	170,000
Atomic, Molecular, & Optical Physics	3	210,000
Biophysics	1	300,000
Condensed Matter Physics	4	850,000
Nuclear Physics	1	152,000
Total	50	2,555,000

FACULTY

Professors

Crenshaw, D. Michael, Ph.D., The Ohio State University, 1985. Observation astronomy; space- and ground-based spectroscopy; active galaxies and quasars.

Dietz, Nikolaus, Ph.D., Technical University, Berlin, 1991. Solid state physics, OMCVD and High-Pressure CVD thin film growth (III–V and Chalcopyrites), real-time optical growth diagnostics, optical material characterization (linear and nonlinear material properties), Experimental.

Gies, Douglas R., Ph.D., Toronto, 1985. High-resolution stellar spectroscopy; stellar pulsation; stellar winds; binary star evolution.

He, Xiaochun, Ph.D., Tennessee, 1991. Experimental physics; heavy ion collisions.

Henry, Todd, J., Ph.D., University of Arizona, 1991. Young stars, brown dwarfs, initial mass function.

Manson, Steven T., Ph.D., Columbia, 1966. Regent's Professor. Theoretical physics; atomic and molecular collisions and structure.

McAlister, Harold A., Ph.D., Virginia, 1975. Regent's Professor, Director, CHARA. Observational astronomy; binary stars; speckle and long-base-line interferometry.

Miller, H. Richard, Ph.D., Florida, 1970. Chair of Department. High energy astrophysics variability of AGN.

Morrow, Cherilynn, Ph.D., University of Colorado-Boulder, 1988. Solar physics education research, innovative instruction.

Nelson, William H., Ph.D., Duke, 1970. Assoc. Dean, Research. Experimental physics; ENDOR; ESR; irradiated biological molecules.

Perera, A. G. Unil, Ph.D., Pittsburgh, 1987. Assoc. Chair. Experimental condensed matter physics; electrical and optical properties of semiconductors; device applications.

Stockman, Mark I., D.Sc., Novosibirsk, 1989. Theoretical physics; electronic and optical properties of disordered systems; kinetic and non-linear optical effects in semiconductors.

Wiita, Paul J., Ph.D., Princeton, 1976. Theoretical astrophysics; active galaxies; accretion disks; radio galaxies; quasars.

Professors Emeriti

Bagnuolo, William G. Jr., Ph.D., Caltech, 1976. Observational astronomy; high-resolution imaging; high-resolution spectroscopy; hot stars; irregular galaxies.

Hadley, Joseph H. Jr., Ph.D., Duke, 1963. Experimental physics; ESR, positron annihilation; irradiated organic solids.

Henry, Ronald J., Ph.D., (Provost), Queens, Ireland, 1964. Theoretical physics; atomic collisions and structure.

Hsu, Frank H., Ph.D., Columbia, 1967. Experimental physics; positron annihilation studies of solids.

Mallard, William C., Ph.D., North Carolina, 1959. Experimental physics; defects in solids; acoustical properties of solids; positron annihilation.

Nave, Carl R., Ph.D., Georgia Inst. Tech., 1966. Experimental physics; molecular structure; magnetic resonance; acoustics.

Purcell, James E., Ph.D., Case Inst. of Tech., 1966. Theoretical physics; nuclear and atomic scattering; nuclear structure.

Wingert, David W., Ph.D., Princeton, 1974. Observational astronomy; theoretical astrophysics.

Associate Professors

Apalkov, Vadym M., Ph.D., University of Utah, 1995. Condensed Matter Theory; disordered electronic systems; photonic crystals; quantum cascade lasers.

Hastings, Gary, Ph.D., Imperial College, London (U.K.), 1992. Experimental biophysics; static and time resolved infrared spectroscopies applied to biological systems; experimental physics; energy and electron transfer in photosynthetic organisms.

Mani, Ramesh, Ph.D., University of Maryland, 1990. Experimental Condensed Matter.

Thoms, Brian D., Ph.D., Cornell, 1992. Experimental physics; surface science; wide band gap semiconductors; film growth mechanisms.

Assistant Professors

Dhamala, Mukesh, Ph.D., University of Kansas, 2000. Experimental biophysics; neuroimaging.

Sarsour, Murad, Ph.D., University of Houston, 2002. Experimental Nuclear Physics.

White, Russel, Ph.D., UCLA, 1999. Stellar Astronomy.

Lecturers

Doluweera, Sumith, Ph.D., University of Cincinatti, 2008.

Evans, J., Ph.D., 1998. Georgia State University.

Research Staff

Sturmann, Judit, Ph.D., Vanderbilt, 1999. Research Scientist/CHARA. Astronomical Instrumentation, optical long baseline interferometry.

Sturmann, Laszlo, Ph.D., Vanderbilt, 1997. Senior Research Scientist/CHARA. Astronomical instrumentation, optical long baseline interferometry.

ten Brummelaar, Theo A., Ph.D., Sydney (Australia), 1993. Associate Director/CHARA. Optical propagation in a turbulent atmosphere; long baseline optical stellar interferometry; observational astronomy.

Turner, Nils H., Ph.D., Georgia State, 1998. Research Scientist/CHARA. Astronomical instrumentation, optical long baseline interferometry.

Adjunct Professors

Francombe, Maurice H., Ph.D., London (U.K.), 1958. Experimental physics; semiconductor thin films; magnetic alloys; ferroelectric oxides.

Liu, Hui C., Ph.D., Pittsburgh, 1987. Experimental physics; quantum transport in reduced dimensional semiconductor structures; high frequency device applications.

Ridgway, Steven, Ph.D., Stony Brook, 1972. Experimental infrared astronomy; high resolution Fourier transform spectroscopy; imaging.

Tennakone, Kirti, Ph.D., Hawaii, 1972. Experimental physics, dye sensitized semiconductor nanostructure, solar cells.

RESEARCH SPECIALTIES AND STAFF

Theoretical

Astrophysics. Accretion processes; active galaxies; radio galaxies; quasars. Wiita.

Atomic Physics. Atomic photoionization and photoelectron angular distributions; excitation and ionization of atoms by charged particles; generalized oscillator strengths; properties of atomic ions and excited states. R. Henry, Manson 1 postdoctoral fellow.

Condensed Matter. Electronic and optical properties of disordered systems; kinetic, electrical and transport phenomena in semiconductor devices, and optical effects in semiconductors. Apalkov, Stockman. 2 postdoctoral fellows.

Experimental

Astronomy and Astrophysics. Active galaxies; UV, optical, and x-ray spectroscopy; Quasi-stellar objects; galaxies; photometry; cataclysmic variable stars; star formation; stellar pulsation and mass loss; stellar spectroscopy; binary stars; star formation; speckle and long-base-line interferometry. Bagnuolo, Crenshaw, Gies, T. Henry, McAlister, Miller, White, Wingert. 6 postdoctoral fellows.

Biophysics and Chemical Physics. Radiation damage in organic solids; electron spin resonance; electron-nuclear double resonance; X-ray diffraction; positron annihilation; surface science; static and time resolved FTIR electron transfer in photosynthesis systems and DNA; neuroscience. Cymbalyuk, Dhamala, Hastings, Nave, Nelson, Thoms. 1 postdoctoral fellow.

Condensed Matter Physics. Defects in solids, acoustical, linear and nonlinear optical, electrical and thermal properties of semiconductors; growth of III-N (high-pressure CVD) and waveguided, birefringent heterostructures (OMCVD; real-time growth diagnostics; optoelectronic semiconductor device applications; Dietz, Mani, Perera, Thoms. 5 postdoctoral fellows.

Nuclear Physics. Studies of quark-gluon plasma in heavy-ion collisions at Brookhaven National Labs. He, Sarsour. 1 postdoctoral fellow.

Physics Education. Studies of new methods for integrated learning in physics and astronomy. Doluweera, Evans, Morrow, Thoms, Wilson.

UNIVERSITY OF IDAHO

DEPARTMENT OF PHYSICS

Moscow, Idaho 83844-0903

Students Accepted For Degree	FIELDS		
	Physics	Astronomy	Related Fields
Doctorate	X	X	
Master's	X		X

1. General

President: M. Duane Nellis
Dean of Graduate School: Margrit von Braun
Department Chair: David McIlroy
Department Telephone Number: (208) 885-6380
Email Address: physics@uidaho.edu
Web Page: www.phys.uidaho.edu
Type of Institution: University
Control: Public
Setting: Small town
Total Faculty: 928
Total Students: 11,957
Total Graduate Students: 2240
Annual Graduate Tuition:
In-state residents: Full-time—$5556
Part-time—$282/credit
Out-of-state residents: Full-time—$15,636
Part-time—$786/credit
Tuition rates for: 2009–10
Deferred tuition plan: Yes
Term: Semester

2. Number of Faculty in Department

The combined total of full-time faculty in the three professorial ranks is 12.

3. Admission, Financial Aid, and Housing

Address admission inquiries to: Bernhard Stumpf, Chair Graduate Admissions Committee, Physics Department
Graduate application fee required: $55 ($60 for international students can be waived on request)
Admission deadline (Fall admission): February 1 for priority admission for following Fall term, September 1 for priority admission for following Spring term.
Admission information: For fall admission, 2008–09 8 students were accepted from 36 applicants.
Admission requirements: For admission to the graduate programs, a Bachelor's degree in physics is required with no minimum undergraduate GPA specified. No minimum acceptable scores are specified. Students from non-English speaking countries are required to demonstrate proficiency in English via the TOEFL exam. Minimum acceptable score for admission is 550.
Undergraduate preparation assumed: Equivalent of B.S. in physics
Address financial aid inquiries to: David McIlroy Chair, Physics Department
GAPSFAS application required: No
Financial aid deadline: 2/15
Loans available: Yes
Address housing inquiries to: University Housing
On-campus, single student housing available: Yes
Cost/month: $505–$548
On-campus, Family student housing available: Yes
Cost/month: $505–$798

Table A—Faculty, Enrollments, and Degrees Granted

Research Specialty	2009–10 Faculty	Enrollment[1] Fall Master's	Enrollment[1] Fall Doctorate	No. of Degrees Granted[2] 2007–08 Master's	Terminal Master's	Doctorate	Median No. of Years for Ph.D.'s
Astronomy	1	-	2	-	0	0	–
Atomic, Molecular, & Optical Physics	1	-	-	-	0	0	–
Biophysics	1	-	6	-	1	0	–
Condensed Matter Physics	6	1	11	-	3	2	–
Materials Sci./ Metallurgy	1	2	2	-	0	0	–
Nuclear Physics	2	-	2	-	0	1	–
Particles & Fields	2	-	0	-	0	0	–
Non-Specialized	-	2	2	-	-	-	-
Total		6	23	-	4	3	
Full-time Grad. Stud.		0	0				
Part-time Grad. Stud.		-	0				
First-year Grad. Stud.		2	3				
Median Years In Grad. Study					-	3	5

Undergraduate Degrees, 2008-2009: 6(47)

[1]Students not yet committed to a research specialty are entered under non-specialized.
[2]Five-year totals in parentheses

4. Graduate Degree Requirements

Master of Science (Non-thesis Option): General university M.S. non-thesis requirements apply. The requirement is a minimum of 30 credits in coursework and the credits must be distributed as follows: (1) 20 cr in physics courses numbered 500 and higher (including 2 cr for Phys 501); (2) 10 cr in courses numbered 400 and higher (these may be non-physics courses upon the approval of the physics department Academic Standards Committee). Phys 521, 533, 541–542, and 550 are required.

Students must pass a comprehensive examination, which must be taken at the first offering after the student has completed the core courses required for the M.S. degree. Full-time students may not delay the completion of their core course requirements by avoiding the taking of a core course when offered except with the prior written consent of the Academic Standards Committee and the student's major professor. The examination is written and covers all of general graduate-level physics as defined by the required courses for the M.S. degree. Typically, it will be administered on two different days, with a time limit of approximately three hours for each day. The results of the examination will be evaluated by the physics faculty. If the comprehensive examination is failed, it may be repeated only once; the repeat examination must be taken within a period of not less than three nor more than 14 months following the first attempt.

Master of Science (Thesis Option): General university M.S. requirements for a degree with thesis apply. The student must complete a total of at least 30 credits at 400 level or higher, 20 of which must be at the graduate level, including a maximum of 10 credits in research and thesis. Specific departmental graduate course requirements are 2 credits in Phys 501 and Phys 521, 541–542, and 550. If a student's undergraduate preparation is considered deficient (e.g., if it lacks laboratory experience at the upper-division level), then certain undergraduate courses will be required in the study plan. Such remedial credits are not to be counted towards the total required for the degree. No departmental comprehensive exam is required.

A final defense of the M.S. thesis is scheduled upon completion of the thesis. Full-time students have to take this examination no later than two years after passing the comprehensive examination. The candidate is required to defend his or her work and show a satisfactory knowledge of the field in which the thesis research has been performed. The defense is oral and would typically last for one hour. The exam has to be announced to the physics faculty at least one week in advance. All members of the physics faculty are permitted to attend and ask questions. A recommendation of a majority of the student's graduate committee is necessary to pass the defense. If the defense is failed, it may be repeated only once; the repeat defense must be taken within a period of not less than three months nor more than one year following the first attempt.

Doctorate: General university Ph.D. requirements apply. Correspondence concerning the student's specific goals is encouraged in the preliminary planning of the Ph.D. program. Specific departmental course requirements are: Phys 501 (2 cr), 511, 521, 533, 541–542, 550–551, 571, and at least nine additional semester-hours of physics courses at the 500 level. A typical study plan would include 40 to 50 credits of course work at the 500 level in physics and about 30 credits in research and thesis. The study plan also would include at least six units of upper-division or graduate course work outside of physics. The nature and number of these additional units will depend upon the professional goals of the individual student. In planning a program, the student should consult with the departmental Academic Standards Committee for approval of any particular choice of nonphysics course work. The Ph.D. degree in physics is primarily a recognition of ability and accomplishment in research. The purpose of the course work is to provide the factual and theoretical background for research. Successful completion of course work is not in itself considered as completion of the major requirement for the degree.

All Ph.D. graduate students are required to enroll in Phys 501 (Physics Seminar) each semester while in residence, even if not formally registered for credit in this course.

No formal foreign language requirement exists for Ph.D. candidates; however, in individual cases, depending on the research topic, a reading knowledge in one foreign language may be required by the thesis advisor. A two-Part preliminary examination is required. Part I is taken after the student has completed the courses required for the Ph.D. degree. Full-time students have to take this exam no later than 28 months after entering the graduate program. The examination is written and covers all of general graduate-level physics as defined by the required courses for a Ph.D. degree. Typically, it will be administered on two different days, with a time limit of approximately five hours for each day. The results of the examination will be evaluated by the physics faculty. If the preliminary examination, Part I, is failed, it may be repeated only once; the repeat examination must be taken within a period of not less than three months nor more than 14 months following the first attempt.

Part II of the preliminary examination is set by the major professor of the Ph.D. student for a date within six months after Part I has been passed. The student is required to explain the goals of his or her planned Ph.D. research to the thesis committee and show general familiarity with the fields relevant for the research. Part II is oral and would typically last for one hour. The exam is to be announced to the physics faculty at least one week in advance. All members of the physics faculty are permitted to attend and ask questions. The student's committee certifies to the Graduate College the results of the preliminary examinations. Upon passing, the student is advanced to candidacy for the Ph.D. degree. If Part II is failed, it may be repeated only once; the repeat examination must be taken within a period of not less than three months nor more than one year following the first attempt.

A final defense of the Ph.D. thesis is scheduled upon completion of the dissertation. The candidate is required to defend his or her work and show a superior knowledge of the field in which the thesis research has been performed. The defense is oral and would typically last for one hour. The exam is to be announced to the physics faculty at least one week in advance. All members of the physics faculty are permitted to attend and ask questions. A recommendation of a majority of the student's graduate committee is necessary to pass the defense. If the defense is failed, it may be repeated only once; the repeat defense must be taken within a period of not less than three months nor more than one year following the first attempt.

Thesis: Thesis may be written *in absentia*.

Table B—Appointments to Graduate Students, 2007–08

Title of Appointee	Appointments: Total	Appointments: First year	Academic Load Allowed in Credit Hours	Hours of Service Per Week	Stipend for Academic Year ($)
			Semester		
Teaching Assistant	10	4	12	20	14,126[1,2]
Research Assistant	16	0			
Total	26	4			

[1]Out-of-state tuition waived.
[2]Additional summer support may be available.

5. Personnel Engaged in Separately Budgeted Research, 7/08–6/09

Professorial faculty	10
Graduate students	25
Total	35

6. Separately Budgeted Research Expenditures by Source of Support

	Departmental Research
Federal government	1,612,000
Other government	100,000
Business and industry	4,000
Total	1,716,000

Table C—Separately Budgeted Research Expenditures

Research Specialty	No. of Grants	Expenditures ($)
Astronomy	3	76,000
Bio-Physics	3	730,000
Condensed Matter Physics	8	330,000
Materials Sci./Metallurgy	2	60,000
Other	2	520,000
Total	18	1,716,000

FACULTY

Professors

Machleidt, Ruprecht, Ph.D., Bonn, 1973. Theoretical nuclear physics.

McIlroy, David N., Ph.D., Rhode Island, 1993. Experimental condensed matter physics. Department Chair.

McIver, John, Ph.D. 1979. University of Rochester, V. P. Research, Non-Linear Systems, Laser Physics.

Sammarruca, Francesca, Ph.D., Virginia Polytechnic, 1988. Theoretical nuclear physics.

Yeh, Wei Jiang, Ph.D., SUNY, Stony Brook, 1984. Experimental condensed matter physics.

Associate Professors

Bergman, Leah, Ph.D., N. Carolina State Univ., 1995. Solid State Physics, ultraviolet optical semiconductors.

Berven, Christine, Ph.D. University of Oregon, 1995. Condensed Matter Nanodevice Physics.

Qiang, You, Ph.D., 1997. Albert-Ludwigs University, Germany, Experimental Condensed Matter Physics, Nanophysics.

Stumpf, Bernhard J., Ph.D., Saarland, 1981. Atomic, molecular, and laser physics.

Assistant Professors

Barnes, Jason, , Ph.D. 2004. University of Arizona, Astronomy.

Pozhar, Liudmila, Ph.D., 1994. Institute for Low-Temperature Physics and Engineering, Ukraine, Theoretical Condensed Matter Physics.

Ytreberg, F. Marty, Ph.D., 2000. University of Maine, Computational Biophysics.

Research Assistant Professor

Barnes, Gwen, Ph.D., 2007. University of Arizona. Planetary science.

RESEARCH SPECIALTIES, RESOURCES, AND STAFF

Theoretical

Atomic and Molecular Physics. Atomic and molecular spectroscopy: Stumpf.

Biophysics: Ytreberg.

Condensed Matter Physics: Pozhar.

Nuclear and Particle Physics, Meson production: Sammarruca; Nuclear forces; Relativistic nuclear structure: Machleidt.

Other theory: McIver.

Experimental

Astronomy and Planetary Science: Barnes, J., Barnes, G.

Condensed Matter Physics. Superconductivity; Josephson effect; photoemissions of low dimensional materials; reduced dimensional charge and thermal transport; photonic crystals; nanostructured materials; reduced dimensional charge and thermal transport: Bergman, Berven, McIlroy, Qiang, Yeh.

Special Equipment, Facilities, or Programs

Research equipment includes a machine shop and machinist, dedicated research computer clusters, Raman Spectroscopy, SQUID, Electron Beam Lithography, Micro-Nano Technology Cleanroom, Low-temperature Cryostats equipped with Superconducting magnets, AFM, nanomaterials growth chambers, SEMs and TEMs and access to characterization facilities on campus and at Pacific Northwest National Laboratory.

DePAUL UNIVERSITY

DEPARTMENT OF PHYSICS

Chicago, Illinois 60614

Students Accepted For Degree	FIELDS		
	Physics	Astronomy	Related Fields
Doctorate			
Master's	X		X

1. General

President: Reverend Dennis H. Holtschneider C. M.
Dean of Graduate School: Ralph Erber
Department Chairman: Jesús Pando
Department Telephone Number: (773) 325-7330
webpage: las.depaul.edu/physics
Type of Institution: University
Control: Private
Setting: Urban
Total Faculty: 1842
Total Students: 25,072
Total Graduate Students: 7,795
Annual Graduate Tuition:
All Graduate Students: Full-time—$570.00/credit hr.
Tuition rates for: 2010–11
Deferred tuition plan: Yes
Term: Quarter

2. Number of Faculty in Department

The combined total of full-time faculty in the three professorial ranks is 7. The combined total of full-time, part-time, and other faculty at all ranks is 13.

3. Admission, Financial Aid, and Housing

Address admission inquiries to: Dr. Anuj P. Sarma, Dept. of Physics, Byrne 211, De Paul University, 2219 N. Kenmore Ave., Chicago, IL 60614-3504
Graduate application fee required: $40
Admission deadline (Fall admission): 8/1
Admission requirements: For admission to the graduate programs, a Bachelor's degree in physics, mathematics, chemistry, or engineering is required with a minimum undergraduate GPA of 2.5/4.0 specified. Students from non-English speaking countries are required to demonstrate proficiency in English via the TOEFL exam. Minimum acceptable score for admission is 590, 243 for computer TOEFL.
Undergraduate preparation assumed: Tipler or Senway, *General Physics*; Fowles, *Mechanics*; Tipler, *Modern Physics*; Griffiths, *Electricity and Magnetism*; Schroeder, *Thermal Physics*; Boas, *Mathematical Methods in The Physical Sciences*; Hecht, *Optics*; Griffiths, *Quantum Mechanics*.
Address financial aid inquiries to: Financial Aid Office, De Paul Center, Room 9100, 1 E. Jackson Blvd., Chicago, IL 60604-2287, or email finaid1@depaul.edu
GAPSFAS application required: No
Financial aid deadline: 5/1 (earlier submission of financial and materials advised)
Loans available: Yes
Address housing inquiries to: Department of Housing Services, Centennial Hall, Suite 301, 2345 N. Shefield Ave., Chicago, IL 60614. Seminary, Rm. 326, Chicago, IL 60614-3212
On-campus, single student housing available: Yes

Table A—Faculty, Enrollments, and Degrees Granted

Research Specialty	2009–10 Faculty	Enrollment[1] Fall 2009		No. of Degrees Granted[2] 2007–08 (2002–08)			Median No. of Years for 2005–06 Ph.D.'s
		Master's	Doctorate	Master's	Terminal Master's	Doctorate	
Astrophysics	2	6	–	0(0)	1(5)	–	–
Biophysics	0	0	–	0(0)	0(2)	–	–
Condensed Matter	2	3	–	0(0)	0(0)	–	–
Fluids	1	2	–	0(0)	1(1)	–	–
Nuclear Physics	1	1	–	0(0)	0(4)	–	–
Other Theoretical/ Math.	1	0	–	0(0)	1(3)	–	–
Non-specialized	0	0	–	0(0)	0(0)	–	–
Total		12	–	0(0)	3(15)	–	
Full-time Grad. Stud.		12	–				
Part-time Grad. Stud.			–				
First-year Grad. Stud.		4	–				
Median Years in Grad. Study (2006–07 Degrees)				2	2	–	–
Undergraduate Degrees, 2007–08 (2003–08): 8(30)							

[1]Students not yet committed to a research specialty are entered under non-specialized.
[2]Five-year totals in parentheses.

4. Graduate Degree Requirements

Master's in Applied Physics: 11 (4 quarter-hour each) courses, including thesis; minimum 2.75 grade point average on a scale of 4.0; no time residence requirement; no language requirement; oral thesis examination is required. Faculty expertise in nonlinear optics, fourier optics, laser physics, statistical physics, nuclear physics, condensed matter physics, astrophysics, scientific computation, and nonlinear dynamics.
Thesis: Thesis may be written *in absentia*.
Special Equipment, Facilities, or Programs: NMR equipment: TeachSpin 18 MHz pulsed NMR spectrometer. Optics: an optical spectrum analyzer, a 2-m research grade optical bench, a 4×8 foot holographic table with beam handling optics, He-Ne lasers, 0.55-m Czerny-Turner spectrometer, CCD camera imaging system, an 8 GHz spectrum analyzer. Computational: 10 Pentium-class PCs. 12 Macs Miscellaneous equipment: Mössbauer spectrometer and multichannel analyzers for nuclear physics.

Table B—Appointments to Graduate Students, 2009–10

Title of Appointee	Appointments Total	First year	Academic Load Allowed in Credit Hours	Hours of Service Per Week	Stipend for Academic Year ($)
			Quarter		
Teaching Assistant	6	2	8	16	9,000[1]
Total	6	2			

[1]Tuition waived.

5. Personnel Engaged in Separately Budgeted Research,

Professorial faculty	7
Total	7

6. Separately Budgeted Research Expenditures by Source of Support

	Departmental Research	Physics-related Research Outside Department
Federal government	$20,000	
Private Foundation	0	$40,000
Total	$60,000	$

Table C—Separately Budgeted Research Expenditures

Research Specialty	No. of Grants	Expenditures ($)
Astrophysics	1	20,000
Condensed Matter	1	40,000
Total	3	60,000

FACULTY

Professors

El Saffar, Zuhair M., Ph.D., Wales, 1960. Emeritus. Solid state physics.
Goedde, Christopher G., Ph.D., University of California, Berkeley, 1990. Dynamical systems; nonlinear optics.
Schillinger, Edwin J., Ph.D., Notre Dame, 1950. Emeritus. Physics education.
Stinchcomb, Thomas G., Ph.D., Chicago, 1951. Emeritus. Medical physics, radiation physics.
Van Ostenburg, Donald O., Ph.D., Michigan State, 1956. Emeritus. Solid state physics.

Associate Professors

Behof, Anthony F., Ph.D., Emeritus, Notre Dame, 1965. Optics.
Fischer, Susan M., Ph.D., Notre Dame, 1994. Nuclear physics; gamma-ray spectroscopy.
Lietz, Gerard P., Ph.D., Notre Dame, Emeritus, 1964. Physics education.
Pando, Jesús, Ph.D., University of Arizona, 1997. Astrophysics.
Sarma, Anuj, Ph.D., Univ. of Kentucky, 2000. Astrophysics.

Assistant Professors

Matson, Robert, Ph.D., Oklahoma State, 1994. Fluids.
González Avilés, Gabriela, Ph.D., Northwestern, 2003. Materials Science.
Landahl, Eric, Ph.D., U.C. Davis, 2001. Applied Science.

Instructional Associate

Milton, John, M.S., Saint Louis, 1960. Emeritus. Physics education.
Mihalcea, Gabi, M.S., Kansas State.

Lecturer

Corso, George, Ph.D., Northwestern, 1975. Astronomy.

RESEARCH SPECIALTIES AND STAFF

Theoretical

Astrophysics, Cosmology, large scale structure. Pando.
Dynamical Systems, Nonlinear Optics, Computational Physics. Goedde.

Experimental

Complex Fluid Rheology, Soft Condensed Matter Physics. Matson.
Materials Science. González Avilés.
Nuclear Physics. Fischer.
Star Formation, Radio Astronomy. Sarma.
Ultra-fast Physics. Landahl.

FACULTY PUBLICATIONS

Fischer, Susan M.

S. M. Fischer, T. Anderson, P. Kerns, G. Mesoloras, D. Svelnys, C. J. Lister, D. P. Balamuth, P. A. Hausladen, and D. G. Sarantites, "Shape Coexistance in ^{71}Br and the Question of the Groundstate Spin of ^{71}Kr," Physics Review C **72**, 024321 (2005).

S. M. Fischer, C. J. Lister, and D. P. Balamuth, "Unravelling the Band Crossing in ^{68}Se and ^{72}Kr: The Quest for T=0 Pairing," Physics Review C **67**, 064318 (2003).

N. S. Kelsall, S. M. Fischer, D. P. Balamuth, G. C. Ball, M. P. Carpenter, R. M. Clark, J. Durell, P. Fallon, S. J. Freeman, P. A. Hausladen, R.V. F. Janssens, D. G. Jenkins, M. J. Leddy, C. J. Lister, A. O. Macchiavelli, D. G. Sarantites, D. C. Schmidt, D. Seweryniak, C. E. Svensson, B. J. Varley, S. Vincent, R. Waldsworth, A. N. Wilson, A. V. Affanasjev, S. Frauendorf, I. Ragnarsson, and R. Wyss, "Testing Mean-Field Models Near the N=Z Line: Gamma-ray Spectroscopy of the Tz=1/2 Nucleus ^{73}Kr," Phys. Rev. C **65**, 044331 (2002).

D. G. Jenkins, N. S. Kelsall, C. J. Lister, D. P. Balamuth, M. P. Carpenter, T. A. Sienko, S. M. Fischer, R. M. Clark, P. Fallon, A. Gorgen, A. O. Macchiavelli, C. E. Svensson, R. Wadsworth, W. Reviol, D. G. Sarantites, G. C. Ball, J. Rikovska Stone, O. Juillet, P. Van Isacker, A. V. Afanasjev, and S. Frauendorf,: "T=0 and T=1 States in the Odd-Odd N=Z Nucleus, $_{35}^{70}Br_{35}$" Phys. Rev. C **65**, 064307 (2002).

S. M. Fischer, C. J. Lister, D. P. Balamuth, R. Bauer, J. A. Becker, L. A. Bernstein, M. P. Carpenter, J. Durell, N. Fotiades, S. J. Freeman, P. E. Garrett, P. A. Hausladen, R. V. F. Janssens, D. Jenkins, M. Leddy, J. Ressler, J. Schwartz, D. Svelnys, D. G. Sarantites, D. Seweryniak, B. J. Varley, and R. Wyss, "Alignment Delays in the N=Z Nuclei ^{72}Kr, ^{76}Sr, and ^{80}Zr," Phys. Rev. Lett. **87**, 132501 (2001).

Goedde, Christopher G.

E. Huynh, E. Manzano, A. Medrano, T. Thorn, A. Zavala, C. G. Goedde, and J. R. Thompson, "The Interplay of Thermal and Pump Fluctuations in Stimulated brillouin Scattering," Optics Communications **281**, 836–845 (2008).

A. Betlej, P. Schmitt, P. Sidereas, R. Tracy, C.G. Goedde, J.R. Thompson, "Increased Stokes pulse energy variation from amplified classical noise in a fiber Raman generator," Optics Express **13**, 2948–2960 (2005).

A. Betlej, P. Schmitt, P. Sidereas, R. Tracy, C. G. Goedde, and J. R. Thompson, "Increased Stokes pulse energy variation from amplified classical noise in a fiber Raman generator," Optics Express **13**, 2948–2960 (2002).

L. Garcia, J. Jenkins, Y. Lee, N. Poole, K. Salit, P. Sidereas, C. G. Goedde, and J. R. Thompson, "Influence of classical pump noise on long-pulse multi-order stimulated Raman scattering in optical fiber," Journal of the optical Society of America B **19**, 2727–2736 (2002).

L. Garcia, A. Jalili, Y. Lee, N. Poole, K. Salit, P. Sidereas, C. G. Goedde, and J. R. Thompson, "Effects of Pump Pulse

Temporal Structure on Long-Pulse Multi-Order Stimulated Raman Scattering in Optical Fiber," Optics Communications **193**, 289–300 (2001).

Gonzáles Avilés, Gabriela

H.-R. Wenk, N. Barton, M. Bortolotti, S. C. Vogel, M. Voltolini, G. E. Lloyd, and G. B. González, "Dauphine Twinning and Texture Memory in Polycrystalline Quartz. Part 3. Texture Memory During Phase Transformation," Phys. Chem. Minearls (2009).

G. B. González, J. S. Okasinski, T. O. Mason, T. Buslaps, and V. Honkimäki, "*In situ* Studies on the Kinetics of Formation and Crystal Structure of the Phase In4Sn3O12 Using High-Energy, X-ray Diffraction," J. Appl. Phys. **104**(4), 043520 (2008).

J. Terra, E. Rodriguies Dourado, J. G. Eon, D. E. Ellis, G. González, and A. M. Rossi, "The Structure of Strontium-Doped Hydroxyapatite: An Experimental and Theoretical Study," Phys. Chem. Chem. Phys. (2008).

D. E. Ellis, J. Terra, O. Warschkow, M. Jiang, G. González, J. S. Okasinski, M. J. Bedzyk, A. M. Rossi, J. G. Eon, "A Theoretical and Experimental Study of Leand Substitution in Calcium Hydorxy Apatite," Phys. Chem. Chem. Phys. **8**(8), 967–976 (2006).

O. Warschkow, L. Miljacic, D. E. Ellis, G. B. Gonzálex, and T. O. Mason, "Interstitial Oxygen in Tin-Doped Indium Oxide Transparent Conductors," J. Amer. Ceram. Soc. **89**(2), 616–619 (2006).

Landahl, Eric C.

B. Kraessig, R. W. Dunford, E. P. Kanter, E. C. Landahl, S. H. Southworth, and L. Young, "A simple cross-correlation technique between infrared and hard x-ray pulses," Applied Physics Letters **94**, 171113 (2009).

E. P. Kanter, J. Rudati, D. A. Arms, E. M. Dufresne, R. W. Dunford, D. L. Ederer, C. Hoehr, B. Kraessig, E. C. Landahl, E. R. Peterson, R. Santra, S. H. Southworth, and L. Young, "Characterization of the Spatiotemporal Evolution of Laser-generated Plasmas," J. of Applied Physics **104**, 073307 (2008).

A. Grigoriev, R. Sichel, H. N. Lee. E. C. Landahl, B. Adams, E. M. Dufresne, P. G. Evans, "Nonlinear piezoelectricity in epitaxial ferroelectrics at high electric fields," Physical Review Letters **100**, 027604 (2008).

S. H. Southworth, D. A. Arms, E. M. Dufresne, R. W. Dunford, D. L. Ederer, C. Hoehr, E. P. Kanter, B. Kraessig, E. C. Landahl, E. R. Peterson, J. Rudati, R. Santra, D. A. Walko, and L. Young, "K-edge x-ray absorption spectra of laser-generated Kr^{+} and Kr^{++}," Physical Review A **76**, 043421 (2007).

L. Young, D. A. Arms, E. M. Dufresne, R. W. Dunford, D. L. Ederer, C. Hohr, E. P. Kanter, B. Krassig, E. C. Landahl, E. R. Peterson, J. Rudati, R. Santra, and S. H. Southworth, "X-ray microprobe of orbital alignment in strong-field ionized atoms," Physical Review Letters **97**, 083601 (2006).

Matson, W. Robert

M. G. Kalyankar, W. Matson, B. J. Ackerson, and P. Tong, "Pattern Formation in a Rotating Suspension of Non-Brownian Bouyant Particles," Physics of Fluids **20**, 083301 (2008).

W. R. Matson, B. J. Ackerson, and P. Tong, "Measured Scaling Properties of the Transition Boundaries in a Rotating Suspension of Non-Brownian Settling Particles," J. Fluid mechanics **597**, 233 (2008).

W. R. Matson, B. J. Ackerson, P. Tong, "Dynamics of Rotating Suspensions," Solid State Comm. **139**, 2006.

W. R. Matson, M. Kalyankar, B. J. Ackerson, P. Tong, "Concentration & Velocity Patterns in a Horizontal Rotating Suspension of Non-Brownian Settling Particles," PRE **71**, 031401 (2005).

W. R. Matson, B. J. Ackerson, P. Tong, "Pattern Formation in a Rotating Suspension of Non-Brownian Settling Particles," PRE **67**, 05031 (2003).

Pando, Jesús

J. Pando, L. Sands, and S. Shahacn, "Detection of Protein Secondary Structures via the Discrete Wavelet Transform," Phys. Rev. E **80**, 051909 (2009).

J. Pando, L. L. Feng, and L.-Z. Fang, "The Statistical Discrepency between the IGM and Dark Matter Fields: One-Point Statistics," Astrophysical Journal Supplement **154**, 475 (2004).

B. Kim, P. He, J. Pando, L. L. Feng, and L. Z. Fang, "The Velocity Field of Baryonic Gas in the Universe," Astrophysical Journal **625**, 599 (2005).

L. Feng, J. Pando, and L. Z. Fang, "Intermittent Features of the QSO Lyα Transmitted Flux: Results from Hydrodynamic Cosmological Simulations," Astrophysical Journal 587, 487 (2003).

Sarma, Anuj

A. P. Sarma and E. Momjian, "Detection of the Zeeman Effect in the 36 GHz Class I Ch3OH Maser Line with the EVLA," Astrophysics J. **705**, L176 (2009).

A. P. Sarma, T. H. Troland, J. D. Romney, and T. H. Huynh, "VLBA Observations of the Zeeman Effect in H2O masers in OH 43.8−0.1," Astrophysics J. **674**, 295 (2008).

C. L. Brogan, C. J. Chandler, T. R. Hunter, Y. L. Shirley, and A. P. Sarma, "Arcsecond-Scale Kinematic and Chemical Complexity in Cepheus A East," Astrophysical Journal **660**, L133 (2007).

N. P. Abel, A. P. Sarma, T. H. Troland, and G. J. Ferland, "Thick [O I] and [C II] Emission Towards NGC 6334A," Astrophysical Journal **662**, 1024 (2007).

A. P. Sarma, E. Momjian, T. H. Troland, and R. M. Crutcher, "Very Large Array H I Zeeman Observations of NGC 1275 (Perseus A)," Astronomical Journal **130**, 2566 (2005).

ILLINOIS INSTITUTE OF TECHNOLOGY

PHYSICS DIVISION

Chicago, Illinois 60616

Students Accepted For Degree	FIELDS		
	Physics	Astronomy	Related Fields
Doctorate	X		
Master's	X		

1. General

President: John Anderson
Dean of Graduate School: Ali Cinar
Department Chairman: Russell Betts
Department Telephone Number: (312) 567-3480
Type of Institution: University
Control: Private
Setting: Urban
Total Faculty: 370
Total Graduate Faculty: 370
Total Students: 7,613
Total Graduate Students: 4,974
Annual Graduate Tuition:
All Graduate Students: $975/credit hour
Deferred tuition plan: Yes
Term: Semester

2. Number of Faculty in Department

The combined total of full-time faculty in the three professorial ranks is 21. The combined total of full-time, part-time, and other faculty at all ranks is 34.

3. Admission, Financial Aid, and Housing

Address admission inquiries to: Graduate School
Graduate application fee required: $50
Admission deadline: Jan. 5 for spring, May 15 for summer, Aug. 6 for Fall
Admission information: For fall admission, 2010–11, 22 students were accepted from 50 applicants.
Admission requirements: For admission to the graduate programs, a Bachelor's degree in physics is required with a minimum undergraduate GPA of 3.0 specified. Students from non-English speaking countries are required to demonstrate proficiency in English via the TOEFL exam. Minimum acceptable score for admission–550/213.*
Undergraduate preparation assumed: 1–2 years General Physics–Ohanian; 1 year Mechanics–Marion; 1 year Modern Physics–Taylor; 1 year Electricity and Magnetism–Griffiths; Calculus and Differential Equations. 1 year Quantum Mechanics.
Address financial aid inquiries to: Dept. of Biological, Chemical & Physical Sciences
GAPSFAS application required: No
Financial aid deadline: Jan 31
Loans available: Yes
Address housing inquiries to: Housing Office
On-campus, single student housing available: Yes
Cost/year: $6,692–$17,628 including meals
On-campus, married student housing available: Yes
Cost/month: $783–$1,255

*Paper test/computerized test

Table A—Faculty, Enrollments, and Degrees Granted

Research Specialty	2009–10 Faculty	Enrollment[1] Fall 2009 Master's	Enrollment[1] Fall 2009 Doctorate	No. of Degrees Granted[2] 2009–10 (2004–09) Master's	Terminal Master's	Doctorate	Median No. of Years for 2008–09 Ph.D.'s
Accelerator & Beam Physics	2	0	2	0(0)	0(0)	0(0)	–
Biophysics	5	0	2	0(0)	0(0)	0(4)	–
Condensed Matter Physics	7	1	10	2(5)	0(0)	0(13)	4
Health Physics	2	40	0	3(0)	0(0)	0(0)	–
Particles & Fields	4	0	4	0(0)	0(0)	0(2)	4
Nonspecialized		5	4	0(0)	0(0)	0(0)	–
Total		46	22	2(5)	0(0)	5(19)	
Full-time Grad. Stud.		4	2				
Part-time Grad. Stud.		41	14				
First-year Grad. Stud.		11	3				
Median Years in Grad. Study (2009–10 Degrees)				0	0	5	–
Undergraduate Degrees, 2009–10(2004–09): 3(22)							

[1]Students not yet committed to a research specialty are entered under non-specialized.
[2]Five-year totals in parentheses.

4. Graduate Degree Requirements

Master's: 32 semester hours including eight in thesis research (optional); M.S. comprehensive examination: minimum 3.0/4.0 GPA in approved course work; no foreign language requirements; no formal residency requirement.
Doctorate: 84 semester hours including 32 semester hours Ph.D. core courses and at least 32 semester hours in thesis research; passing Ph.D. qualifying and comprehensive examinations and oral thesis defense; minimum of two semesters in full-time resident study.
Other Programs: Molecular Biochemistry and Biophysics: The department offers interdisciplinary programs leading to M.S. and Ph.D. degrees in molecular biochemistry and biophysics. New advances in our understanding of biological function can be expected from a synthesis of molecular genetics, biochemistry and insights gained from molecular structural information. Individuals with a quantitative, physical approach will be best placed to be innovators in the field. MBB programs complement more traditional graduate programs in biology, chemistry and physics by offering an integrated, molecular-based approach to understanding biological problems, taking insights from all three disciplines.
Thesis: Thesis may be written *in absentia*.
Special Equipment, Facilities, or Programs: Laboratories for experimental research in synchrotron radiation research, superconduction, condensed matter physics, thin films, and particle physics; campus facilities include x-ray diffraction facility, IIT Academic Computing Center and Galvin Library. Scanning probe microscopy, scanning electron microscopy, and XPS/Auger spectroscopy. Collaborative programs with Fermi National Accelerator Laboratory and Argonne National Laboratory.

Table B—Appointments to Graduate Students, 2009–10

Title of Appointee	Appointments Total	First year	Academic Load Allowed in Credit Hours*	Hours of Service Per Week	Stipend for Academic Year ($)
			Semester		
Teaching Assistant	10	2	9	20	15,500
Research Assistant	12	1	–	–	–
Self Supported	2	1	–	–	–
Total	24	4			

*Provided by the Department.

5. Personnel Engaged in Separately Budgeted Research, 9/08–8/09

Professorial faculty	13
Postdoctoral appointments	5
Graduate students	11
Total	29

6. Separately Budgeted Research Expenditures by Source of Support

	Departmental Research
Federal government	$1,950,093.10
State government	0
Total	$1,950,093.10

Table C—Separately Budgeted Research Expenditures

Research Specialty	No. of Grants	Expenditures ($)
Biophysics	2	168,618.21
Condensed Matter Physics	5	828,700.96
Particles & Fields	6	952,773.93
Total	13	1,950,093.10

FACULTY

Professors

Betts, Russell, Ph.D., Univ. of Pennsylvania, 1972. Chair. Nuclear Physics.

Bunker, Grant, Ph.D., Washington, 1984. Associate Chair. Synchrotron radiation, x-ray spectroscopy, computational biophysics, condensed matter.

Irving, Thomas, Ph.D., Biophysics, University of Guelph (1989). Biophysics of muscle contraction, non-crystalline x-ray diffraction, synchrotron radiation instrumentation.

Kallend, John Ph.D., Cambridge, 1971. Computational methods of crystallographic texture analysis.

Kaplan, Daniel, Ph.D., SUNY Stony Brook, 1979. Experimental elementary particle physics.

Lederman, Leon, Ph.D. (Pritzker Professor of Science), Columbia, 1951. Experimental high energy physics; science education.

Morrison, Timothy, Ph.D., Illinois, 1980. Solid state physics.

Schieber, Jay, Ph.D., University of Wisconsin (1989). Research interest: physical properties of soft matter.

Scott, H. Larry, Ph.D., Purdue, 1970. Biophysics.

Segre, Carlo U., Ph.D., California, San Diego, 1981. Solid state physics.

White, Chris, Ph.D., Minnesota, 1990. Experimental elementary particle physics.

Zasadzinski, John, Ph.D., Iowa State, 1979. Solid state physics.

Professors Emeriti

Burnstein, Ray A., Ph.D., Michigan, 1960. Experimental elementary particle physics; new educational techniques.

Johnson, Porter W., Ph.D., Princeton, 1967. Theoretical high energy physics; Science education.

Rubin, Howard A., Ph.D., Maryland, 1967. Experimental elementary particle physics.

Spector, Harold N., Ph.D., Chicago, 1961. Solid state theory; electronic processes in strong electric and magnetic fields and in quantum well systems.

Zwicker, Earl, Ph.D., IIT, 1956.

Distinguished Professor Emeritus

Erber, Thomas, Ph.D., Chicago, 1957. Distinguished. Electrodynamics; magnetism, fatigue, complex systems.

Associate Professors

Coffey, Liam, Ph.D., Chicago, 1985. Condensed matter theory.

Howard, Andrew, Ph.D., Physics, University of California, San Diego (1981). Macromolecular crystallography, methods development, synchrotron radiation.

Spentzouris, Linda, Ph.D., Northwestern, 1996. Accelerator and Beam Physics.

Assistant Professors

Gidalevitz, David, Ph.D., Weizmann, 1996. Membrane biophysics, biomaterials, biosensors and biomimetric thin films, and polymer films.

Sullivan, Zack, Ph.D., Illinois, 1998. Elementary particle theory.

Terry, Jeff, Ph.D., Stanford, 1997. Chemical physics with specialization in synchotron radiation techniques.

Torun, Yagmur, Ph.D., SUNY at Stony Brook, 2000. Accelerator and experimental high energy physics.

RESEARCH SPECIALTIES AND STAFF

Theoretical

Biophysics. Computational and theoretical modeling of biomembrane structure. Scott.

Condensed Matter Physics. Transport properties of semiconducting heterostructures and superlattices; transport in semiconductors in strong electric and magnetic fields, high temperature superconductivity. Coffey, Spector.

High-Energy Physics. Analysis of unified electro-weak theory and quantum chromodynamics; exploring the features of new particles and new interactions high energies; accelerator physics. Johnson, Sullivan.

Quantum Mechanics of Single Atoms. Electrodynamics. High-energy processes in intense magnetic fields; pseudo-random processes and deterministic chaos; hysteresis in magnetic and mechanical systems. Erber.

Experimental

Accelerator and Beam Physics. Kaplan, Spentzouris, Torun.

Biological Physics. Protein structural biophysics as studied using synchrotron sources and x-ray laser; photophysics and photobiology at the molecular level; atomic level simulation; membrane structure. Bunker, Gidalevitz, Longworth, Scott.

Condensed Matter Physics. High temperature superconductivity; magnetism, valence phenomena; synchrotron radiation studies of materials; thermal and electrical properties of solids; amorphous alloys; EXAFS and XANES of materials; radiation damage at low temperatures; growth and properties of thin films; Structures of Complex materials using X-ray techniques: diffraction, scattering and spectroscopy. Materials of interest include catalysts, superconductors, oxygen permeable membranes. Morrison, Segre, Terry Zasadzinski.

Elementary Particles and Fields. Studies of quantum chromodynamics and production and decay of heavy quarks in experiments at Fermilab. Neutrino oscillations, charm CP violation, meson-antimeson mixing, and rare decays of charm, search for CP violation in hyperon decay. Investigation of techniques for high energy muon colliders and neutrino factories. Burnstein, Kaplan, Lederman, Rubin, Torun, White.

Magnetic and Mechanical Hysteresis and Fatigue. Erber.

FACULTY PUBLICATIONS

Bunker, Grant

G. Bunker, "Introduction to XAFS Spectroscopy" book, Cambridge University Press, 2010.

N. Dimakis and G. Bunker, "*Ab initio* self-consistent X-ray absorption fine structure analysis for Metalloproteins," Biophysical Journal, **91**(11), 87–89 (2006).

N. Dimakis and G. Bunker, "Group-fitted ab initio single and multiple-scattering EXAFS Debye-Waller factors," Phys. Rev. B Rapid Communications **65**, 201103R (2002).

C. Karanfil, D. Chapman, G. Bunker, and C. Segre, "Initial results on logarithmic spiral bent laue analyzer for fluorescence x-ray absorption spectroscopy," Rev. Sci. Instrum. **73**(3), 1616 (2002).

G. Khelashvili and G. Bunker, "Practical Regularization Methods for Analysis of EXAFS Spectra," J. Synch. Rad. **6**, 271 (1999).

Burnstein, Ray

R. A. Burnstein and L. M. Lederman,"Using wireless keypads in lecture classes," Physics Teach. **39**, 8 (2001).

R. A. Burnstein et al, "A High-rate spectrometer for the study of charged hyperon and kaon decays," HyperCP collaboration, Nucl. Instrum. Meth. **A** 541, 516 (2005).

D. Rajaram *et al.*, "Search for the lepton-number-violating decay Xi-$\rightarrow$ $p\mu\mu$," HyperCP collaboration, Phys. Rev. Lett. **94**, 181801 (2005).

R. A. Burnstein and L. M. Lederman,"The use and evolution of an audience response system," Audience response systems in Higher education, edited by D. Banks, (Idea Group Inc., Hershey, PA, 2006).

R. A. Burnstein and L. M. Lederman, "Wireless keypads: A new classroom technology using enhanced multiple-choice questions," Physics Education, **24**, 89 (2007).

Coffey, Liam

J. F. Zasadzinski, L. Ozyuzer, L. Coffey, K. E. Gray, D. G. Hinks, and C. Kendziora, "Persistance of Strong Electron Coupling to a Narrow Boson Spectrum in Overdoped B2212 Tunneling Data," Phys. Rev. Lett., **96**, 017004 (2006).

J. F. Zasadzinski, L. Coffey, P. Romano, Z. Yusof, "Tunneling spectroscopy of B2212: Eliashberg analysis of dip feature," Phys. Rev. B **68**, 180504(R) (2003).

L. Coffey, "Calculation of Dip Features in the SIS Conductance of D-Wave Superconductors using Eliashberg Formalism," http://arXiv.org/abs/cond-mat/0103518

Yung-mau Nie and L. Coffey, "Elastic and Inelastic Quasiparticle Tunnelling Between Anisotropic Superconductors," Physical Review B **59**, 11982 (1999).

Gidalevitz, David

F. Neville, A. Ivankin, O. Konovalov, and D. Gidalevitz, "A comparative study on the interactions of SMAP-29 with lipid monolayers," Biochimica et Biophysica Acta-Biomembrances, **1798**, 851 (2010).

H. Sarig, Liran Livneh, V. Held-Kuznetzov, F. Zaknoon, S. Rotem, A. Ivankin, D. Gidalevitz, and A. Mor, "A miniature mimic of host defense peptides with systemic antibacterial efficay," FASEB J. **24**, 1904 (2010).

A. Ivankin, I. Kuzmenko, and D. Gidalevitz, "Cholesterol-Phospholipid Interactions: New Insights from Surface X-ray Scattering Data," Phys. Rev. Letts. **104**, 108101 (2010).

F. Zaknoon, H. Sarig, S. Rotem, L. Livneh, A. Ivankin, D. Gidalevitz, and A. Mor, "Antibacterial properties and mode of action of a short acylysyl oligomer," Antimicrobial Agents and Chemotherapy **53**, 3422 (2009).

F. Neville, Y. Ishitsuka, A. Ivankin, C. S. Hodges, O. Konovalov, A. J. Waring, R. Lehrer, K. Y. C. Lee, and D. Gidalevitz, "Protein interaction with lipid monolayers: grazing incidence X-ray diffraction and X-ray reflectivity study," Soft Matter **4**, 1665 (2008).

Howard, Andy

T. Jin, F. Guo, S. Kim, A. Howard, Y.-Z. Zhang, "X-ray crystal structure of TNF ligand family member TL1A at 2.1 Å," Biochem. Biophys. Res. Comm. **364** 1 (2007).

F. Guo, T. Jin, A, Howard, Y.-Z. Zhang, "Purification, crystallization and initial crystallographic characterization of brazil-nut allergen Ber e 2," Acta Crystallographica F **63**, 976 (2007).

A. Teplyakov, K. Lim, P. P. Zhu, G. Kapadia, C. C. H. Chen, J. Schwartz, A. Howard, P. T. Reddy, A. Peterkofsky, and O. Herzberg, "Structure of phosphorylated enzyme I, the phosphoenolpyruvate:sugar phosphotransferase system sugar translocation signal protein," Proc. Natl. Acad. Sci. **103**, 16218 (2006).

A. J. Howard, "Macromolecular crystallography at third-generation synchrotron radiation sources," Third-generation Hard X-ray Synchrotron Radiation Sources: Sources Properties, Optics, and Experiment. D. M. Mills, ed.; John Wiley & Sons, pp. 311 (2002).

A. J. Howard, "Data processing in macromolecular crystallography," Chapter in: Crystallographic Computing 7: *Proceedings from the Macromolecular Crystallographic Computing School, 1996*. P. E. Bourne and K. D. Watenpaugh, eds. Oxford: Oxford University Press (2000).

Irving, Thomas

J. H. Roh, L. Guo, D. Kilburn, B. Duncan, *et al.*, "Multistage collapse of a bacterial ribozyme observed by time-resolved small angel X-ray scattering," JACS (In press).

B. A. Colson, M. R. Locher, T. Bekyarova, J. R. Patel *et al.*, "Differential roles of regulatory light chain and myosin binding protein-C phosphorylations in the modulation of cardiac force development," J. Physiol. **15**, Pt. 6, 588 (2010).

P. P. de Tombe, R. D. Mateja, K. Tachampa, Ya Mou, *et al.*, "Myofilament lenght dependent activation," J. Mol. Cell Card. **48**(5), 851 (2010).

T. Paunesku, S. Vogt, T. C. Irving, B. Lai, *et al.*, " Biological application of X-ray microprobes," Int. J. rad. Biol. **85**(8), 710 (2009).

G. P. Farman, M. S. Miller, M. C. Reedy, F. N. Soto-Adames, *et al.*, "Phosphorylation and the N-terminal extension of the regulatory light chain help orient and align the myosin heads in Drosophila flight muscle," J. Struc. Biol. **168**(2), 240 (2009).

Johnson, Porter

Relativity for the Mind: An Introduction for Scientists, Rinton Press 2007.

Classical mechanics with applications, World Scientific, 2010.

Kallend, John

Texture and Anisotropy. Cambridge University Press 1998. (co-author.) ISBN 0 521 465168.

Kaplan, Daniel

D. M. Kaplan, M. Goodman, Z. Sullivan (Eds.), "Neutrino Factories, Superbeams, and Beta Beams," *Proceedings of the 11th International Workshop on Neutrino Factories, Superbeams, and Beta Beams- NuFact09*, AIP Conf. Proc. 1222 (2010).

L. Y. Zhu *et al.*, " Measurement of Angular Distributions of Drell-Yan Dimouons in p+p Interactions at 800-GeV/c," Phys. Rev. Lett. **102**, 182001 (2009).

M. Apollonio *et al.*," Accelerator design concept for future neutrino facilities," JINST **4**, P07001 (2009).

C. Kurter, K. E. Gray, J. F. Zasadzinski *et al.*, "Thermal Management in Large Bi2212 Mesas Used for Terahertz Sources," IEEE Trans. Appl. Supercond. **19**(3), 428 (2009).

L. Y. Zhu *et al.*, "Measurement of Upsion production for p+p and p+d interactions at 800-GeV," Phys. Rev. Lett. **100**, 062301 (2008).

Lederman, Leon

R. A. Burnstein and L. M. Lederman, "Comparison of Different Commercial Wireless Keypad Systems," The Physics Teacher **41** (5), 272 (2003).

L. Lederman,"Revolution in Science Education: Put Physics First," Physics Today **54** (9), 44 (2001).

A. Burnstein and L. M. Lederman, "Using Wireless Keypads in Lecture Classes," R. The Physics Teacher **39** (1), 8 (2001).

M. J. Leitch *et al.*, "Nuclear Dependence of Neutral D-Meson Production by 800 GeV/c Protons," Physical Review Letters **72**, 2542 (1994).

G. Danby, J. M. Gaillard, K. Goulianos, L. M. Lederman, N. B. Mistry, M. Schwartz, J. Steinberger, "Observation of High-Energy Neutrino Reactions and the Existence of Two Kinds of Neutrinos," Physical Review Letters **9**, 36 (1962).

Rubin, Howard

D. G. Michael *et al.*, "The Magnetized steel and scintillator calorimeters of the MINOS experiment," Nucl. Instrum. Meth. A **596** (2008).

P. Adamson *et al.*, "A Study of Muon Neutrino Disappearance Using the Fermilab Main Injector Neutrino Beam," Phys. Rev. D **77**, 072002 (2008).

P. Adamson *et al.*, "Measurement of neutrino velocity with the MINOS detectors and NuMI neutrino beam," Phys. Rev. D **76**, 072005 (2007).

P. Adamson *et al.*, "Measurement of the atmospheric muon charge ratio at TeV energies with MINOS," Phys. Rev. D **76**, 052003 (2007).

P. Adamson *et al.*, Charge-separated atmospheric neutrino-induced muons in the MINOS far detector," Phys. Rev. D **75**, 092003 (2007).

Scott, H. Larry

S. W. Chui, H. L. Scott, and E. Jakobsson, "A coarse-grained model based on Morse potential for water and n-alkanes," J. Chem. Theo. & Computation **6**, 851 (2010).

P. W. Tumaneng, S. Pandit, G. Zhao, and H. L. Scott, "Lateral Organization of Complex Lipid Mixtures from Multiscale Modeling," J. Chem. Phys. **132**, 065104 (2010).

S. Pandit and H. L. Scott, "Multiscale Simulations of Heterogeneous Model Membranes," Biochim. Biophys. Acta **1788**, 136 (2009).

S. W. Chiu, S. Pandit, H. L. Scott, and E. Jakobsson, "An Improved United Atom Force Field for Simulation of Mixed Lipid Bilayers," J. Phys. Chem. B **113**, 2748 (2009).

H. S. Pandit, S. W. Chiu, E. Jakobsson, A. Grama, and H. L. Scott, "Cholesterol Packing around Lipids with Saturated and Unsaturated Chains: A Simulation Study," Langmuir **24**, 6858 (2008).

Segre, Carlo

S. Liu, D. Olive, J. Terry, and C. U. Segre, "An X-Ray Absorption Spectrosopy Study of Mo Oxidation in Pb at Elevated Temperatures," J. Nucl. Mat. **392**, 259 (2009).

E. A. Lewis, C. U. Segre, and E. S. Smotkin, "Embedded Cluster -XANES Modeling of Adsorption Porcesses on Pt," Electrochim. Acta **54**, 7181 (2009).

V. S. Prakash Srirangam, S. Chattophadhyay, T. Shibata, J. A. Kaduk, J. T. Miller, C. U. Segre, and S. Shankar, "XAFS studies on a modified Al-Si hypoeutectic alloy," J. Phys. Conf. Series **190**, 012068 (2009).

Q. Jia, E. A. Lewis, E. S. Smotkin, and C. U. Segre, "*In situ* XAFS studies of the oxygen reduction reaction on carbon supported Pt and $PtNi_1$ catalysts," J. Phys. Conf. **190**, 012157 (2009).

S. Stoupin, H. Rivera, Z. Li, C. U. Segre, C. Korzeniewski, D. J. Casadonte, H. Inoue, and E. S. Smotkin, "Structural Analysis of Sonochemically Prepared PtRu Versus Johnson Matthey PtRu in Operating Direct Methanol Fuel Cells," Phys. Chem. Chem. Phys. **10**, 6430 (2008).

Spector, Harold

X. Congxin, "Franz-Keldysh efect in the interband optical absorption of semiconducting nanostructures," J. Appl. Phys. **105**, 084313 (2009).

J. Lee, "Stark effect in the optical absorption in cubical quantum boxes," Physica B **393**, 92 (2007).

H. Ham, "Photoionization Cross-section of Hydrogenic Impurities in Cylindrical Quantum Wires: Finite Well Model," Journal of Applied Physics **100**, 02430 (2006).

J. Lee and W. Chou, "Rashba spin-splitting in parabolic quantum dots," Journal of Applied Physics **99**, 113708 (2006).

J. Lee, W. Ching Chou, and Y. Sheng Huang, "Optical absorption coefficient of quantum wires in strain and electric fields with intermixing interfaces," Physical Review B **72**, 125329 (2005).

Spentzouris, Linda

S. Antipov, W. Liu, W. Gai, J. Power and L. Spentzouris, "Wakefield generation in metamaterial-loaded waveguides," J. Appl. Phys. **102**, 034906 (2007).

S. Antipov, W. Liu, W. Gai, J. Power, and L. Spentzouris, "Double-negative metamaterial research for accelerator applications," Nucl. Instr. Meth. A. **579**, issue 3, 915–923 (2007).

S. Antipov, W. Liu, W. Gai, L. Spentzouris, "Numerical studies of ILC positron target and OMD field effects on beam," J. Appl. Phys. **102**, 014910 (2007).

J. Amundson, W. Pellico, L. Spentzouris, P. Spentzouris, and

T. Sullivan, "An experimentally robust technique for halo measurement using the IPM at the Fermilab Booster," Nucl. Instr. Meth. A. **570**, 1 (2007).

J. Amundson, J. lackey, P. Spentzouris, G. Jungman, and L. Spentzouris, "Calibration of the Fermilab Booster ionization profile monitor," Physical Review Special Topics, Accelerators and Beams, **6**, 10 (2003).

Sullivan, Zack

Z. Sullivan and E. L. Berger, "Trilepton production at the CERN LHC: Standard model sources and beyond," Phys. Rev. **78**, 034030 (2008).

Z. Sullivan and E. L. Berger, "Missing heavy flavor backgrounds to Higgs production," Phys. Rev. D **74**, 033008 (2006).

Z. Sullivan, "Angular correlations in single-top quark and W_{jj} production at next-to-leading order," Phys. Rev. D **72**, 094034 (2005).

Z. Sullivan, "Fast Evaluation of CTEQ Parton Distributions in Monte Carlos," Comput. Phys. Commun. **168**, 25 (2005).

Z. Sullivan, "Understanding single-top-quark production and jets at hadron colliders," Phys. Rev. D **70**, 114012 (2004).

Terry, Jeff

Y. Liu, J. Terry, and S. Jurisson, "Pertechnetate immobilization in aqueous media with hydrogen sulfide under anaerobic and aerobic environments," Radiochimica Acta **95**, 717 (2007).

L. Young, N. Westcott, C. Christensen, J. Terry, D. Lydiate, and M. Reaney, "Inferring the geometry of fourth-period metallic elements in Arabidopsis thaliana seeds using synchrotron-based multi-angle x-ray fluorescence mapping," Annals of Botany **100**, 1357 (2007).

J. F. Collingwood, A. Mikhaylova, M. R. Davidson, C. Batich, W. J. Streit, J. Terry, and J. Dobson, "In situ characterization and mapping of iron compounds in Alzheimer's disease tissue," Journal of Alzheimer's Disease **7**, 267 (2005).

X. Liu, A. D. Compaan, and J. Terry, "Cu K-edge x-ray fine structure changes in CdTe with $CdCl_2$ processing," Thin Solid Films **480**, 95 (2005).

J. G. Tobin, B. W. Chung, G. D. Waddill, R. K. Schulze, J. Terry, J. D. Farr, T. Zocco, D. K. Shuh, E. Rotenberg, K. Heinzelman, and G. Van der Laan, "Resonant photoemission in f-electron systems: Pu and Gd," Physical Review B **68**, 155109 (2003).

Torun, Yagmur

Y. Torun *et al.*, "Muon Acceleration R&D," AIP Conf. Proc. **1182**, 710 (2009).

Y. Torun, "Using Cerenkov Light to Detect Field Emission in Superconducting Cavities," Proc. of PAC09 (2009 Particle Accelerator Conf.), May 4-8, 2009, Vancouver. In press.

M. Apollonio *et al.*, "Accelerator design concept for future neutrino facilities ," JINST **4**, P07001 (2009).

I. Chemakin *et al.*, "Pion production by protons on a thin beryllium target at 6.4-Ge, 12.3-GeV/c, and 17.5-GeV/c incident proton momenta," (E910 Collaboration), Phys. Rev. C **77**, 015209 (2008).

A. Hassanein *et al.*, "The Effects of surface damage on rf cavity operation," Phys. Rev. ST Accel. Beams **9**, 062001 (2006).

White, Christopher

P. Adamson *et al.*, "Search for sterile neutrino mixing in the MINOS long baseline experiment. By MINOS Collaboration," Phys. Rev. D **81**, 052004 (2010).

P. Adamson *et al.*, "Neutrino and Antineutrino Inclusive Charges-current Cross Section Measurments with the MINOS Near Detector. By MINOS Collaboration," Phys. Rev. D **81**, 072002 (2010).

P. Adamson *et al.*, "Observation of muon intensity variations by season with the MINOS far detector. By MINOS Collaboration," Phys. Rev. D **81**, 012001 (2010).

P. Adamson *et al.*, "Search for muon-neutrino to electron-neutrino transitions in MINOS. By MINOS Collaboration," Phys. Rev. D **103**, 261802 (2009).

S. Osprey *et al.*, "Sudden stratospheric warmings seen in MINOS deep underground muon data. By MINOS Collaboration," Phys. Rev. Lett. **36**, L05809 (2009).

Zasadzinski, John

C. Kurter *et al.*, "Evidence of Strong-Coupled Superconductivity in CaC6 from Tunneling Spectroscopy," Physical Review Letters (submitted).

J. F. Zasadzinski *et al.*, "Persistance of Strong Electron Coupling to a Narrow Boson Spectrum in Overdoped Bi2212 Tunneling Data," Physical Review Letters **96**, 017004 (2006).

J. F. Zasadzinski, L. Coffey, P. Romano, Z. Yusof, "Tunneling spectroscopy of B2212: Eliashberg analysis of spectral dip feature," Phys. Rev. B **68**, 180504(R) 2003.

J. R. Zasadzinski, L. Ozyuzer, N. Miyakawa, K. E. Gray, G. G. Hinks, C. Kendziora, "Correlation of Tunneling Spectra in Bi2212 with the Resonance Spin Excitation," Physical Review Letters **87**, 067005 (2001)

NORTHERN ILLINOIS UNIVERSITY

DEPARTMENT OF PHYSICS

DeKalb, Illinois 60115

Students Accepted For Degree	FIELDS		
	Physics	Astronomy	Related Fields
Doctorate	X		
Master's	X		

1. General

President: John G. Peters
Dean of Graduate School: Bradley Bond, Acting
Department Chairman: Laurence Lurio
Department Telephone Number: (815) 753-1772
Type of Institution: University
Control: Public
Setting: Small town
Total Faculty: 1,214
Total Graduate Faculty: 843
Total Students: 24,424
Total Graduate Students: 6,147
Annual Graduate Tuition:
In-state residents: Full-time—$2,700/sem.
Out-of-state residents: Full-time—$5,400/sem.
Tuition rates for: 2008–09
Deferred tuition plan: No
Tuition waived if Research or Teaching Assistant
Annual Other Fees: Full-time—$1,352.22/sem.
Graduate assistants—$1,352.22/sem.
Term: Semester

2. Number of Faculty in Department

The combined total of full-time faculty in the three professorial ranks is 20. The combined total of full-time, part-time, and other faculty at all ranks is 29.

3. Admission, Financial Aid, and Housing

Address admission inquiries to: Graduate School
Graduate application fee required: Yes, $40.00
Admission deadline (Fall admission): 6/1
Admission information: For fall admission, 2010–11, 16 students were accepted from 22 applicants.
Admission requirements: For admission to the graduate programs, a Bachelor's degree in physics or a related discipline is required with a minimum undergraduate GPA of 2.75. The GRE is required. No minimum score is required. The GRE Physics is not required, but recommended for international students. Students from non-English speaking countries are required to demonstrate proficiency in English via the TOEFL exam. Minimum acceptable score for admission is 80/120.
Undergraduate preparation assumed: Corson and Lorrain, *Electricity and Magnetism*; Fowles, *Mechanics*; Weidener and Sells, *Modern Physics*; Fowles, *Optics*.
Address financial aid inquiries to: Department of Physics
GAPSFAS application required: No
Financial aid deadline: 3/1
Loans available: Yes
Address housing inquiries to: Student Housing Services
On-campus, single student housing available: Yes
On-campus, married student housing available: Yes
On-campus housing
Cost/month Stevenson Tower: $4,895/Semester with Meal Plan
Cost/month Northern View Community: $3,480.75/Semester Room Only

Table A—Faculty, Enrollments, and Degrees Granted

Research Specialty	2009–10 Faculty	Enrollment Fall 2009 Mas-ter's	Enrollment Fall 2009 Doc-torate	No. of Degrees Granted[1] 2009–10 (2005–10) Mas-ter's	Terminal Master's	Doc-torate	Median No. of Years for Ph.D.'s
Accelerator	3	1	6	1(5)	0(4)	–	–
Applied Physics	0	0	–	–	0(0)	0(2)	–
Astronomy	0	1	1	–	0(2)	–	
Condensed Matter Physics	16	13	15	0(1)	3(13)	2(4)	–
Medical Physics	1	2	–	–	2(2)	–	
Particles & Fields	7	3	12	0(0)	1(7)	2(7)	–
Physics Education	2	4	–	–	0(2)	–	–
Total	29	24	34	1(6)	5(28)	4(13)	
Full-time Grad. Stud.		20	28				
Part-time Grad. Stud.		4	3				
First-year Grad. Stud.		6	5				
Median Years in Grad. Study		2.5	5.5	–	–	–	

Undergraduate Degrees, 2009–10 (2005–10): 18(46)

[1]Five-year totals in parentheses.

4. Graduate Degree Requirements

Master's: 30 hrs. of course work with 24 in physics; thesis required for pure and applied physics specializations.
Doctorate: Students are required to complete 90 semester hours of graduate course work. This includes 15 hours in five out of six core courses covering classical and quantum mechanics, statistical physics, and electromagnetic theory, and twelve hours in two different areas of physics. A minimum of 24 hours dedicated to dissertation research is required. The remaining hours may include additional dissertation work or other graduate course work in physics and related fields. Students entering the program without a master's degree in physics are required to pass a qualifying examination, which is usually taken at the end of the first year. Successful completion of a candidacy examination based on the core courses and other graduate courses is required of all students in the Ph.D. program. Transfer credits for students entering with a master's degree or with graduate coursework from another institution are allowed, pending approval by the Graduate Studies Committee.
Thesis: Thesis may be written *in absentia*.
Special Equipment, Facilities, or Programs: Students may specialize in four principal areas: condensed matter and materials physics, elementary particles and fields, accelerator physics, and physics education. The department makes special efforts to accommodate the needs of students such as employees of nearby industrial government laboratories and teachers employed in the region who wish to gain advanced degrees in physics on either a part-time or full-time basis. On the departmental faculty are eight condensed matter experimental-

ists and two theoreticians with whom graduate students may work on their thesis research. In addition there are joint and adjunct professors from Argonne National Laboratory and Fermi National Accelerator Laboratory. For solid state experimentation, the department has low- and high-temperature Mössbauer spectrometers; a materials synthesis lab with high-temperature and high-pressure furnaces and thermogravimetric analysis equipment for creating transition element oxides; x-ray diffractometers used for crystal structure determinations; high-vacuum systems for the preparation and study of surfaces; and magnetization, resistivity and magnetoresistive measurements, and two high resolution electron microscopes. The Physics Department has a strong collaborative program with the Advanced Photon Source in x-ray crystallography, inelastic scattering, magnetic x-ray dichroism, high energy x-ray scattering, X-ray and light scattering, surface scattering, and anomalous and resonant scattering spectroscopies. Because of the close departmental ties with the Materials Science Division of the Argonne National Laboratory, both faculty and graduate students make frequent use of research facilities there (1 hour by car).

Among the faculty working on elementary particles and accelerator physics are nine experimentalists and two theoreticians, along with a number of graduate students doing thesis work. At present, the experimentalists participate in the D0 proton-antiproton collision experiment at the Fermi National Acceleration Laboratory (45 minutes by car) and the ATLAS proton-proton-collision experiment at CERN. Detector R&D is ongoing with emphasis on use of scintillator detectors at a future linear collider. Accelerator physics R&D are coordinated through the Northern Illinois Center for Accelerator and Detector Development (NICADD). Current areas include studies of intense electron sources at the Fermilab-NICADD Photoinjector Laboratory and beam diagnostics using resources at NIU, Argonne and Fermilab; Argonne Tandem Linear Accelerator System; muon-based accelerators. Both the particles and accelerator groups collaborate closely with nearby Fermilab and Argonne National Lab where they have access to laboratory and computing facilities. They also have their own high-end computing clusters at the university.

A faculty member works closely with graduate students on methods of physics teaching and serves as a supervisor of their student teaching at selected nearby high schools.

Table B—Appointments to Graduate Students, 2008–09

Title of Appointee	Appointments Total	First year	Academic Load Allowed in Credit Hours	Hours of Service Per Week	Stipend for Academic Year ($)
			Semester		
Teaching Assistant	25	13	10	20	14,175–14,625
Research Assistant	12	1	10	20	14,175–14,625
Total	37	14			

Table C—Separately Budgeted Research Expenditures

Research Specialty	No. of Grants	Expenditures ($)
Accelerator Physics	4	250,806
Condensed Matter Physics	18	918,895
Particles & Fields	6	141,346
Public Service	1	21,000
Total	29	1,332,047

FACULTY

Professors

Blazey, Gerald, Ph.D., U of Minnesota, 1984. Elementary Particles, Experiment.

Chakraborty, Dhiman, Ph.D., SUNY Stony Brook, 1994. Director of Graduate Studies. Elementary Particles, Experiment.

Dabrowski, Bogdan M., Ph.D., Northwestern U, 1987. Condensed Matter, Experiment.

Hedin, David, Ph.D., U of Wisconsin, 1980. Elementary Particles, Experiment.

Lurio, Laurence, Ph.D., Harvard U, 1993. Chair, Condensed Matter, Experiment.

Martin, Stephen, Ph.D., UC Santa Barbara, 1988. Elementary Particles, Theory.

Mini, Susan, Ph.D., Southern Illinois U, 1991. Associate Dean, CLAS. Condensed Matter, Experiment.

Thompson, Carol, Ph.D., U of Houston, 1987. Condensed Matter, Experiment.

Van Veenendaal, Michel, Ph.D., Laboratory of Solid State Physics, RUG, Netherlands, 1994. Condensed Matter, Theory.

Willis, Suzanne, Ph.D., Yale U, 1979. Assistant Chair, Physics education.

Xiao, Zhili, Ph.D., U of Konstanz, 1996. Condensed Matter, Experiment.

Associate Professors

Brown, Dennis, Ph.D., Stanford U, 1993. Condensed Matter, Experiment.

Chmaissem, Omar W., Ph.D., U of Grenoble, France, 1992. Condensed Matter, Experiment.

Coutrakon, George, Ph.D., SUNY Stony Brook, 1983. Medical physics.

Erdelyi, Bela, Ph.D., Michigan State U, 2001. Accelerator physics, Experiment and theory.

Fortner, Michael, Ph.D., Brandeis U, 1989. Elementary Particles, Experiment.

Ito, Yasuo, Ph.D., Cambridge U, 1996. Condensed Matter, Experiment.

Piot, Philippe, Ph.D., U of Grenoble, 1999. Accelerator physics, Experiment.

Winkler, Roland, Ph.D., U Regensberg, 1994. Condensed Matter, Theory.

Assistant Professors

Windelborn, Augden, Ed.D., Northern Illinois U, 1988. Science education.

Emeriti

Albright, Carl H., Ph.D., Princeton U, 1960. Elementary Particles, Theory.
Fedro, Arthur J., Ph.D., Northwestern U, 1965. Condensed Matter, Theory.
Kimball, Clyde W., Ph.D., St. Louis U, 1959. Condensed Matter, Experiment.

Adjunct and Research Faculty

Alp, Ercan, Ph.D., Southern Illinois U, 1984. Condensed Matter, Experiment.
Bhat, Pushpa, Ph.D., Bangalore U, 1982. Elementary Particles, Experiment.
Crabtree, George W., Ph.D., U of Illinois, Chicago, 1974. Condensed Matter, Experiment.
Cummings, MaryAnne, Ph.D., U of Michigan, 1990. Elementary Particles and accelerator physcis, Experiment.
Welp, Ulrich, Ph.D., U of Konstanz, 1988. Condensed Matter, Experiment.
Zaluzec, Nestor, Ph.D., U of Illinois, Urbana-Champaign, 1973. Condensed Matter, Experiment.

RESEARCH SPECIALTIES AND STAFF

Theoretical

Accelarator and Beam Physics. Erdelyi.
Condensed Matter and Materials Physics. Liquid metals; magnetism and cooperative phenomena; many-body theory; optical properties of solids; electronic structure. Fedro, VanVeenendaal, Winkler.
Elementary Particles. Weak interactions; gauge theory; phenomenology; super-symmetric theories. Albright, Martin.

Experimental

Accelerator Physics. Simulation and operation of high brightness photoinjectors. Electron beam diagnostics. Muon and heavy nuclei accelerators, Cummings, Piot.
Condensed Matter and Materials Physics Nanophysics; Mössbauer effect; superconductivity; lattice defects; optical and transport properties of amorphous and crystalline solids; synchrotron radiation; surface physics; magnetic properties of solids; low-temperature physics; x-ray crystallography; materials preparation. Alp, Brown, Chmaissem, Crabtree, Dabrowski, Ito, Kimball, Lurio, Mini, Thompson, Welp, Xiao, Zaluzec.
Elementary Particle/High-Energy Physics. Collider studies of heavy quark production and decay, jet production and searches for quark compositeness; searches for new massive states; detector development. Bhat, Blazey, Chakraborty, Fortner, Hedin.
Medical Physics: Development of Proton Tomography. Coutrakon, Erdelyi.
Physics Education. Willis, Windelborn.

NORTHWESTERN UNIVERSITY

DEPARTMENT OF PHYSICS AND ASTRONOMY

Evanston, Illinois 60208

Students Accepted For Degree	FIELDS		
	Physics	Astronomy	Related Fields
Doctorate	X	X	
Master's			

1. General

President: Morton Schapiro
Dean of Graduate School: Andrew Wachtel
Department Chairman: Heidi Schellman
Department Telephone Number: (847) 491-3685 (C)
Type of Institution: University
Control: Private
Setting: Suburban
Total Faculty: 2,500 (2,200 Full-time Faculty)
Total Graduate Faculty: 1,000
Total Students: 13,300
Total Graduate Students: 5,700
Annual Graduate Tuition:
All Graduate Students: Full-time—$39,840
Part-time—$4,726/course
Tuition rates for: 2010–11
Deferred tuition plan: Yes
Other Fees: None
Term: Quarter

2. Number of Faculty in Department

The number of salaried faculty in the three professorial ranks is 26. The number of faculty including joint, adjunct, and lecturer faculty is 45.

3. Admission, Financial Aid, and Housing

Address admission inquiries to: Graduate Admissions Committee, Department of Physics and Astronomy
Graduate application fee required: $75
Admission deadline (Fall admission): 6/30
Admission information: For Fall admission, 2009–10, 49 students were accepted from 215 applicants.
Admission requirements: For admission to the graduate programs, students should have a Bachelor's degree, preferably in physics, astronomy, or mathematics. Students must submit results of the GRE examinations (regular plus physics advanced test) and three letters of recommendation. For admission decisions, equal weight is given to GPA, GRE scores, and letters of recommendation. Students from non-English speaking countries are required to demonstrate proficiency in English via the TOEFL exam.
Undergraduate preparation assumed: Symon, *Mechanics*; Reitz and Milford, *Electricity and Magnetism*; Reif, *Statistical Mechanics*; Zemansky, *Thermodynamics*; McGervy, *Modern Physics*. Math preparation should include: (1) Ordinary Differential Equations, (2) Partial Differential Equations and Boundary Value Problems, (3) Complex Variable Theory, and (4) Linear Algebra.
Address financial aid inquiries to: Graduate Admissions Committee, Department of Physics and Astronomy
FAFSA application required: Yes, for financial aid.
Financial aid deadline: 12/31; late applications can be considered
Loans available: Yes
Address housing inquiries to: Graduate Student Housing Office, 1915 Maple Ave.
On-campus, unmarried student housing available: Yes
Cost/month: $775
On-campus, married student housing available: Yes
Cost/month: $1,325

Table A—Faculty, Enrollments, and Degrees Granted

Research Specialty	2010–11 Faculty	Enrollment Fall 2010 Master's*	Enrollment Fall 2010 Doctorate	No. of Degrees Granted 2009–10 (2004–10) Master's**	Terminal Master's*	Doctorate	Median No. of Years for 2004–10 Ph.D.'s
Astrophysics	8	–	11	–	–	5(12)	6.2
Biophysics & Complex Systems	3	–	8	–	–	3(4)	6.3
Condensed Matter Physics	13	–	29	–	–	5(28)	6.2
Molecular and Optical Physics	6	–	10	–	–	1(2)	7.2
Nuclear and Particle Physics	6	–	19	–	–	2(15)	6.5
Non-specialized	0	–	7	1(9)	–	0(0)	
Total	36	–	84	1(9)	–	16(61)	6.3
Full-time Grad. Stud.		–	81				
Part-time Grad. Stud.		–	3				
First-year Grad. Stud.		–	13				
Median Years in Grad. Study (2008–09 Degrees)				2	–	6.3	6.3

Undergraduate Degrees, 2008–09 (2004–09): 19(66)

*Terminal Master's degree not offered.
**Left program after completing Master's portion of PhD.

4. Graduate Degree Requirements

Master's: Seven required quarter-courses in physics. "B" average required. Minimum of three quarters of full-time study required. Master's exam. No thesis. No foreign language.
Doctorate: Minimum of two years residency required. 13 quarter-courses in physics and/or astronomy. "B" average required. Departmental preliminary exam. No foreign language.
Thesis: Thesis may be written *in absentia*.
Special Equipment, Facilities, or Programs: Many computing clusters with links to national supercomputing networks. Professionally staffed machine shop and student shop. Printed circuit laboratory and nanofabrication facilities. Condensed-matter facilities include equipment for conductivity, magnetic, and thermal measurements at ultra-low temperatures, NMR spectrometers, thin-film evaporators, crystal growth and orientation facilities, clean rooms, and the shared facilities of the Materials Research Center. Our faculty also has extensive access to major government research facilities, including Fermilab, Argonne National Laboratory, the National High-Magnetic Field Laboratory, and both ground-based and space-based astrophysical observatories.

Table B—Appointments to Graduate Students, 2010–11

Title of Appointee	Appointments Total	Appointments First year	Academic Load Allowed in Credit Hours	Hours of Service Per Week	9-month Stipend ($)
			Quarter		
Teaching Assistant	19	0	3–4	10–12	16,542
Research Assistant	34	1	3–4	30–40	18,000
University Fellow	11	10	3–4	0	19,112
DOE Fellow	1	0	3–4	0	22,500
NASA Fellow	3	0	3–4	0	22,500
GAANN Fellow	8	1	3–4	0	22,500
IGERT Fellow	5	0			22,500
Total	81	13			

5. Personnel Engaged in Separately Budgeted Research, 9/08–8/09

Professorial Faculty	30
Other Faculty	5
Lindheimer Fellow	1
Research Associates	7
Postdoctoral Fellows	22
Graduate Students	41
Total	106

6. Separately Budgeted Research Expenditures by Source of Support, 2008–2009

Agency	Award
Department of Defense	$509,440
Department of Education	$128,628
Department of Energy	2,140,680
NASA	$1,095,533
National Science Foundation	$2,652,980
Total	$6,527,261

8. Extension Centers and Summer Programs

Anglo-Australian Observatory (Coonabarabran, Australia), Argonne National Laboratory (Argonne, Illinois), Brookhaven National Laboratory (Upton, New York), Caltech Submillimeter Observatory (Mauna Kea, Hawaii), Center for Astrophysical Research in Antarctica (University of Chicago), CERN (European Laboratory for Particle Physics) (Geneva, Switzerland), DESY (Deutsches Elektronen-Synchrotron) (Hamburg, Germany), Fermi National Accelerator Laboratory (Batavia, Illinois), Indiana University Cyclotron Facility (Indianapolis), Kitt Peak National Observatory (Tucson, Arizona), Los Alamos Meson Production Facility (Los Alamos, New Mexico), National High Magnetic Field Laboratory (Tallahassee, Florida), National Radio Astronomy Observatory (Socorro, New Mexico), Northwestern University Materials Research Center (Evanston, Illinois), Thomas Jefferson National Laboratory (Newport News, Virginia).

Table C—Separately Budgeted Research Expenditures, 2008-2009

Research Specialty	No. of Grants	Expenditures ($)
Astrophysics	21	1,907,080
Biophysics	5	536,292
Condensed Matter Physics	22	2,741,858
Nonspecialized	1	128,628
Particles & Nuclear Physics	15	1,213,403
Total	64	6,527,261

FACULTY

Professors

Chandrasekhar, Venkat, Ph.D., Yale, 1989. Experimental condensed matter physics.

Dutta, Pulak, Ph.D., Chicago, 1980. Experimental condensed matter physics.

Ellis, Donald, Ph.D., MIT, 1966. Theoretical condensed matter physics.

Freeman, Arthur, Ph.D., MIT, 1956. Theoretical condensed matter physics.

Garg, Anupam, Ph.D., Cornell, 1983. Condensed matter theory.

Halperin, William, Ph.D., Cornell, 1974. Experimental condensed matter physics.

Kalogera, Vassiliki, Ph.D., Illinois, 1997. Theoretical astrophysics.

Ketterson, John, Ph.D., Chicago, 1962. Experimental condensed matter physics.

Marko, John, Ph.D., MIT, 1989. Biological physics.

Meyer, David, Ph.D., UCLA, 1984. Observational astrophysics.

Novak, Giles, Ph.D., Chicago, 1988. Observational astrophysics.

Rasio, Frederic, Ph.D., Cornell, 1991. Theoretical astrophysics.

Sauls, James, Ph.D., SUNY, Stony Brook, 1980. Theoretical condensed matter physics.

Schellman, Heidi, Ph.D., California, Berkeley, 1984. Department Chair. Experimental particle physics.

Seth, Kamal, Ph.D., Pittsburgh, 1957. Nuclear physics.

Taam, Ronald, Ph.D., Columbia, 1973. Theoretical astrophysics.

Ulmer, Melville, Ph.D., Wisconsin, 1970. Department Chair. Observational astrophysics.

Yusef-Zadeh, Farhad, Ph.D., Columbia, 1986. Observational astrophysics.

Associate Professors

de Gouvêa, André, Ph.D., California, Berkeley, 1999. Theoretical particle physics.

Motter, Adilson, Ph.D., UNICAMP (Brazil), 2002. Complex systems.

Schmitt, Michael, Ph.D., Harvard, 1991. Experimental particle physics.

Velasco, Mayda, Ph.D., Northwestern, 1995. Experimental particle physics.

Assistant Professors

Koch, Jens, Ph.D., Freie Universität, Berlin, 2006. Quantum theory. .

Lithwick, Yoram, Ph.D., Caltech, 2002. Theoretical astrophysics.

Low, Ian, Ph.D., Carnegie Mellon, 2000. Theoretical particle physics.

Odom, Brian, Ph.D., Harvard, 2004. Experimental atomic and optical physics.

Lecturers

Brown, Deborah, Ph.D., Northwestern, 1983. Astrophysics.
Rivers, Andrew, Ph.D., New Mexico, 2000. Astrophysics.
Schmidt, Arthur, Ph.D., Notre Dame, 1974. Nuclear physics.
Smutko, Michael, Ph.D., Chicago, 1998. Astrophysics.
Taylor, David, Ph.D., Maryland, 1983. General physics. .
Watkins, Byron, Ph.D., Kentucky, 2004. Experimental condensed matter physics.

Research Professors

Anastassov, Anton, Ph.D., Ohio State, 2000. Experimental particle physics.
Nevirkovets, Ivan, Ph.D., Institute of Metal Physics (Ukraine), 1985. Experimental condensed matter physics.
Ramakrishna, Sai, Ph.D. Indian Institute of Science, 1995. Theoretical molecular physics.
Shafranjuk, Serhii, Ph.D., Kiev, 1985. Experimental condensed-matter physics.
Tomaradze, Amiran, Ph.D., Institute of High-Energy Physics (Russia), 1988. Experimental nuclear physics.

Joint Professors

Bedzyk, Michael, Ph.D., SUNY, Albany, 1982. Experimental condensed matter physics.
Jacobsen, Chris, Ph.D., SUNY Stony Brook, 1988. Experimental atomic and optical physics.
Kumar, Prem, Ph.D., SUNY, Buffalo, 1980. Experimental atomic, molecular, and optical physics.
Patashinski, Alexander, Ph.D. Kharkov-Moscow, 1968. Condensed matter physics, chemistry.
Seideman, Tamar, Ph.D., Weizmann Institute of Science, 1990. Theoretical molecular physics and chemistry.
Shahriar, Selim, Ph.D., MIT, 1992. Experimental quantum optics.
Solla, Sara, Ph.D., University of Washington, 1982. Neural networks and complex systems.
Yuen, Horace P., Ph.D., MIT, 1970. Theoretical condensed matter physics.

Adjunct Professors

Bader, Sam, Ph.D., California, Berkeley, 1974. Experimental condensed matter physics.
Boughezal, Radja, Ph.D., Institute fur Theoretische Physik Universität Zurich (Switzerland) 2005. Theoretical particle physics.
Norman, Mike, Ph.D., Tulane, 1983. Theoretical condensed matter physics.
Roberts, Douglas, Ph.D., University of Oklahoma, 1986. Radio astronomy.
Vinokur, Valerii, Ph.D., Institute of Solid State Physics (USSR), 1979. Theoretical condensed matter physics.

RESEARCH SPECIALTIES AND STAFF

Theoretical

Astrophysics. Neutron stars; formation and dynamics of multiple star systems; stellar atmospheres; Type-I supernovae; dynamics of dense stellar systems; hydrodynamic stellar interactions; X-ray binaries; gravitational waves.

Biophysics and Complex Systems. Nonlinear dynamics, chaos, neural networks, physics of biological systems.

Condensed Matter Physics. Electronic structure of molecules and crystals; electronic, optical, and magnetic structure studies using accurate first principles methods; electronic properties of semiconductors and composite structures; electron-hole liquid at metallic densities and the electron-hole liquid in semiconductors; many-body physics including monomolecular films, chemisorption, liquid and solid helium.

Elementary Particle Physics. Fundamental interactions of elementary particles; quantum chromodynamics; electroweak symmetry breaking; phenomenology of weak interactions; neutrino oscillations; astrophysical particle theory.

Experimental

Condensed Matter Physics. Ultra-low temperature physics; properties of superfluid helium; magnetic, structural, superconducting properties of composition modulated alloys; nuclear magnetic resonance; x-ray studies of monolayer and multilayer films; catalysis and properties of small metallic particles; semiconductor superlattices; light scattering.

Elementary Particle Physics. Studies of electroweak interactions, spectroscopy of heavy quark states, and supersymmetry searches at the 2-TeV $p\bar{p}$ collider (TeV II) at Fermilab; searches for charmomium states at Fermilab; rare K decay studies (CERN); and CMS preparation at CERN.

Nuclear Physics. Studies of the strong interaction in many-body systems; electrons, pion, nucleon, and heavy-ion-induced reactions and scattering on nuclei.

Observational Astrophysics. Optical/UV observations of interstellar gas/dust and quasar absorption line systems; sub-mm polarimetry of interstellar magnetic fields; radio/IR/X-ray observations of supernova remnants, star formation regions and the Galactic Center; sub-mm/UV/X-ray astronomical instrumentation.

FACULTY PUBLICATIONS

Bedzyk, Michael

V. Kohli, M. J. Bedzyk, and P. Fenter, "Direct method for imagin elemental distribution profiles with long-period x-ray standing waves,"Phys. Rev. B **81**, 054112 (2010).

M. O. Ramirez, *et al.*, "Spin-charge-lattice coupling through resonant multimagnon excitations in multiferroic $BiFeO_3$," Appl. Phys. Lett. **94**, 161905 (2009).

Chandrasekhar, Venkat

V. Chandrasekhar, "Thermal transport in superconductor/normal-metal structures," Supercond. Sci. & Tech. **22**, 083001 (2009).

U. R. Singh, *et al.*, "Pseudogap formation in the metallic state of $La_{0.7}Sr_{0.3}MnO_3$ thin films," Appl. Phys. Lett. **93**, 212503 (2008).

de Gouvêa, André

A. de Gouvêa, W. C. Huang, and J. Jenkins, "Pseudo-Dirac neutrinos in the new standard model," Phys. Rev. D **80**, 073007 (2009).

S. Chang and A. de Gouvêa, "Neutrino alternatives for missing energy events at colliders," Phys. Rev. D **80**, 015008 (2009).

Dutta, Pulak

S. Chattopadhyay, *et al.*, "Structural signal of a dynamic glass transition," Phys. Rev. Lett. **103**, 175701 (2009).

K. Kim, *et al.*, "Effects of chitosan on the alignment, morphology and shape of calcite cyrstals nucleating under

Langmuir monolayers," CRYSTENGCOMM **11**, 130 (2009).

Ellis, Donald

D. M. Wells, G. H. Chan, D. E. Ellis, and J. A. Ibers, "UTa_2O(S-2)(3)C1-6: A ribbon structure containing a heterrobimetallic 5d-5f M-3 cluster," J. Sol. State Chem. **183**, 285 (2010).

S. X. Yin and D. E. Ellis, *et al.*, "First-principles investigations of Ti-substitutte hydroxyapatite electronic structure," Phys. Chem. Chem. Phys. **12**, 156 (2010).

Freeman, Arthur

T. K. Bera, *et al.*, "Soluble semiconductors $AAsSe_2$ (A=Li, Na) with a direct-band-gap and strong second harmonic generation: A combined experimental and theoretical study," J. Ame. Chem. Soc. **132**, 3484 (2010).

N. Mansourian-Hadavi, *et al.*, "Transport and band structure studies of crystalline $ZnRh_2O_4$," Phys. Rev. B **81**, 075112 (2010).

Garg, Anupam

A. Vijayaraghavan and A. Garg, "Incoherent Landau-Zener-Stuckelberg transitions in single-molecule magnets," Phys. Rev. B **79**(3), 104423 (2009).

A. Garg, "Conductors in quasistatic electric fields," Ame. J. of Phys. **76**, 615 (2008).

Halperin, William

J. D. Strand, D. J. Van Harlingen, J. B. Kycia, and W. P. Halperin, "Evidence for complex superconducting order parameter symmetry in the low-temperature phase of UPt_3," Phys. Rev. Lett. **103**, 197002 (2009).

Kalogera, Vicky

M. van der Sluys, *et al.*, "Parameter estimation for signals from compact binary inspirals injected into LIGO data," Classical and Quantum Gravity **26**, 204010 (2009).

J. F. Sepinsky, B. Willems, V. Kalogera, and F. A. Rasio, "Interacting binaries with eccentric orbits. II. Secular orbital evolution due to non-conservative mass transfer," Astrophys. J. **702**, 1387 (2009).

Ketterson, John

W. Mu, D. B. Buchholz, M. Sukharev, J. I. Jang, R. P. H. Chang, and J. B. Ketterson, "One-dimensional long-range plasmonic-photonic structures," Opt. Lett., **35**, 550 (2010).

J. I. Jang, S. Mani, J. B. Ketterson, P. Lovera, and G. Redmond, "Nonlinear refractive index and three-photon absorption coefficient of poly(9,9-dioctylfluorence) ," Appl. Phys. Lett. **95**, 221906 (2009).

Kumar, Prem

M. Medic, J. B. Altepeter, M. A. Hall, M. Patel, and P. Kumar, "Fiber-based telecommunication-band source of degenerate entangled photons," Opt. Lett. **35**, 802 (2010).

M. Vasilyev, N. Stelmakh, and P. Kumar, "Estimation of the spatial bandwidth of an optical parametric amplifier with plane-wave pump," J. Mod. Opt. **56**, 2029 (2009).

Lithwick, Yoram

Y. Lithwick, "Formation, survival, and destruction of vortices in accretion disks," Astrophys. J. **693**, 85 (2009).

A.N. Youdin and Y. Lithwick, "Particle stirring in turbulent gas disks: Including orbital oscillations," ICARUS **192**, 588 (2007).

Low, Ian

Q. H. Cao, C. B. Jackson, W. Y. Keung, I. Low, and J. Shu, "Higgs mechanism and loop-induced decays of a scalar into two Z bosons," Phys. Rev. D **81**, 015010 (2010).

E. L. Berger, Q. H. Cao, and I. Low, "Model independent constrainsts among the Wtb, Zb(b)over-bar, and Zt(t) over-bar couplings," Phys. Rev. D **80**, 074020 (2009).

Marko, John

R. Kawamura, "Mitotic chromosomes are constrained by topoisomerase II-sensitive DNA entanglements," J. Cell Biol. **188**, 653 (2010).

J. F. Marko, "Linking topology of tethered polymer rings with applications to chromosome segregation and estimation of the knotting lenght," Phys. Rev. E **79**, 051905 (2009).

Meyer, David

S. I. B. Cartledge, *et al.*, "Interstellar krypton abundances: The detection of kiloparsec-scale differences in galactic nucleosynthetic history, Astrophys," J., **687**(1), 1043, (2008).

Motter, Adilson

A. E. Motter, "Nonlinear dynamics: spontaneous synchrony breaking," Nat. Phys. **6**, 164 (2010).

A. E. Motter, *et al.*, "Relativistic invariance of Lyapunov exponents in bounded and unbounded systems," Phys. Rev. Lett. **102**, 184101 (2009).

Novak, Giles

M. Krejny, *et al.*, "Polarimetry of DG Tau at 350 um," Astrophys. J. **705**, 717 (2009).

G. Novak, J. L. Dotson, and H. Li, "Dispersion of observed position angles of submillimeter polarization in molecular clouds," Astrophys. J. **695**, 1362 (2009).

Rasio, Fred

L. A. Saleh and F. A. Rasio, "The stability and dynamics of planets in tight binary systems," Astrophys. J. **694**, 1566 (2009).

J. M. Fregeau, *et al.*, "The dynamical effects of white dwarf birth kicks in globular star clusters," Astrophys. J. Lett. **695**, L20 (2009).

Sauls, James

J. A. Sauls and M. Eschrig, "Vortices in chiral, spin-triplet superconductors and superfluids," New J. Phys. **11**, 075008 (2009).

J. A. Sauls and M. Eschrig, "Charge dynamics of vortex cores in layered chiral triplet superconductors," New J. Phys. **11**, 075009 (2009).

Schellman, Heidi

V. M. Abazov, *et al.*, "Search for a resonance decaying into WZ boson paris in p(p)over-bar collisions," Phys. Rev. Lett., **104**, 061801 (2010).

T. Aaltonen, *et al.*, "Combination of tevatron searches for the standard model Higgs boson in the W^+W^- decay mode," Phys. Rev. Lett. **104**, 061802 (2010).

Schmitt, Michael

T. Aaltonen, *et al.*, "Search for technicolor particles produced in association with a W boson at CDF," Phys. Rev. Lett., **104**, 111802 (2010).

V. Khachatryan *et al.*, "Transverse-momentum and pseudorapidity distributions of charged hadrons in pp collisions at root s=0.9 and 2.36 TeV," J. High Energy Phys. **2**, 041 (2010).

Seideman, Tamar

S. Ramakrishna, *et al.*, "Origin and implication of ellipticity in high-order harmonic generation from aligned molecules," Phys. Rev. A **81**, 021802 (2010).

I. Nevo, *et al.*, "Laser-induced aligned self-assembly on water surfaces," J. Chem. Phys. **130**, 144704 (2009).

Seth, Kamal

G. Bonvicini, *et al.*, "Measurement of the eta b(1S) mass and the branching fraction for gamma(3S) to gamma eta b(1S)," Phys. Rev. D **81**, 031104 (2010).

S. R. Bhari, *et al.*, "Improved measurement of branching fractions for π-π transitions among Y(nS) states," Phys. Rev. D **79**, 011103 (2009).

Shahriar, Selim

P. Pradhan, G. C. Cardoso, and M. S. Shahriar, "Suppression of error in qubit rotations due to Bloch-Siegert oscillation via the use of off-resonant Raman excitation," J. Phys. B. **42**, 065501 (2009).

X. Liu, *et al.*, "High-speed inline holographic Stokesmeter imaging," Applied Opt. **48**, 3803 (2009).

Solla, Sara

N. Weiler, *et al.*, "Top-down laminar organization of the excitatory network in motor cortex," Nature Neurosci. **11**, 360 (2008).

E. A. Pohlmeyer, *et al.*, "Prediction of upper limb muscle activity from motor cortical discharge during reaching," J. Neural Eng. **4**, 369 (2007).

Taam, Ron

M. J. Cai and R. E. Taam, "Equilibrium structure of prolate magnetized molecular cores," Astrophys. J. Lett. **709**, L79 (2010).

C. C. Heinke, *et al.*, "Further constraints on thermal quiescent X-ray emission from SAX J1808.4-3658," Astrophys. J. **691**, 1035 (2009).

Ulmer, Melville

C. Adami, *et al.*, "Galaxy structure searches by photometric redshifts in the CFHTLS," Astrono. & Astrophys. **509** (2010), in press.

C. Adami, *et al.*, "On the nature of faint low surface brightness galaxies in the Coma cluster," Astrono. & Astrophys. **495**, 407 (2009).

Velasco, Mayda

V. Khachatryan, *et al.* "Transverse-momentum and pseudorapidity distributions of charged hadrons in pp collisions at root s=0.9 and 2.36 TeV," J. High Energy Phys. **2**, 041 (2010).

S. Abdullin, *et al.*, "The CMS barrel calorimeter response to particle beams from 2 to 350 GeV/c," European Phys. J. C **60**, 359 (2009).

Yuen, Horace

H. P. Yuen, "Key generation: Foundations and a new quantum approach," IEEE J. Sel. Top. Quantum Elect. **15**, 1630 (2009).

H. P. Yuen and R. Nair, "Classicalization of nonclassical quantum states in loss and noise: Some no-go theorems," Phys. Rev. A **80**, 023816 (2009).

Zadeh, Farhad Y.

F. Yusef-Zadeh, J. W. Hewitt, R. G. Arendt, B. Whitney, G. Rieke, M. Wardle, J. L. Hinz, S. Stolovy, C. C. Lang, M. G. Burton, and S. Ramirez, "Star formation in the central 400 pc of the Milky Way: Evidence for a population of massive young stellar objects," Astrophys. J. **702**, 178 (2009).

J. W. Hewitt and F. Yusef-Zadeh, "Discovery of new interacting supernova remnants in the inner galaxy," Astrophys. J. Lett. **694**, L16 (2009).

SOUTHERN ILLINOIS UNIVERSITY AT CARBONDALE

DEPARTMENT OF PHYSICS

Carbondale, Illinois 62901-4401

Students Accepted For Degree	FIELDS		
	Physics	Astronomy	Related Fields
Doctorate	X		
Master's	X		

1. General

President: Glen Poshard
Chancellor: Rita Cheng
Dean of Graduate School: John A. Koropchak
Department Chairman: Naushad Ali
Department Telephone Number: (618) 453-2643 (C)
Type of Institution: University
Control: Public
Setting: Small town
Total Faculty: 1,642
Total Graduate Faculty: 850
Total Students (on and off campus): 20,350
Total Graduate Students: 4,799
Annual Graduate Tuition:
In-state residents: Full-time—$7,872.00
Part-time—$328.00/credit
Out-of-state residents: Full-time—$19,680
Part-time—$820.00/credit
Tuition rates for: 2009–10
Deferred tuition plan: Yes
Other Fees: $3,115.00
Term: Semester

2. Number of Faculty in Department

The combined total of full-time faculty in the three professorial ranks is 11.

3. Admission, Financial Aid, and Housing

Address admission inquiries to: Chairman of the Department
Graduate application fee required: $50.00
Admission deadline (Fall admission): 2/15
Admission information: For fall admission, 2007–08, 2 students were accepted from approximately 32 applicants.
Admission requirements: For admission to the Master's programs, a Bachelor's degree in Physics or equivalent, with a minimum GPA of 2.7/4.0 in undergraduate classes is required. For direct admission into the doctoral program, a minimum GPA of 3.25/4.0 in Bachelor's degree courses is required; or the completion of a Master's degree in Physics or equivalent with a graduate GPA of 3.25/4.0. Reporting of GRE scores is strongly recommended for applicants to the Master's degree, and it is required for all applicants to the doctoral program. Students from non-English speaking countries are required to demonstrate proficiency in English via the TOEFL (minimum acceptable score is 550 or 220, if taken by computer or 80 if taken on internet). Teaching assistants, in addition, are required to show proficiency in spoken English by passing an ITA oral test administered during their first semester at SIUC.
Undergraduate preparation assumed: Serway *Modern Physics*, Boas *Mathematical Methods in Physics*, Thornton *Classical Dynamics of Particles & Systems*, Griffiths *Introduction to Electrodynamics*; Pedrotti, Pedrotti, Pedrotti *Introduction to Optics*; Griffiths *Introduction to Quantum Mechanics*. Baierlein *Thermal Physics*.
Address financial aid inquiries to: Chair, Department of Physics, or Student Work and Financial Assistance, Woody Hall
GAPSFAS application required: No
Financial aid deadline: 2/1
Loans available: Yes
Address housing inquiries to: University Housing, Bldg. D (on-campus) or Washington Square, Bldg. C (off-campus)
On-campus, single student housing available: Yes
Cost/term: $2,705–2,965
On-campus, married student housing available: Yes
Cost/term: $2,875–3,060

Table A—Faculty, Enrollments, and Degrees Granted

Research Specialty	2008–09 Faculty	Enrollment[1] Fall 2008		No. of Degrees Granted[2] 2008–09 (2004–09)			Median No. of Years for 2006–07 Ph.D.'s
		Master's	Doctorate	Master's	Terminal Master's	Doctorate	
Applied Physics	5	6	1	1(7)	1(2)	0(1)	6.5
Atomic, Molecular, & Optical Physics	0	0	0	0(0)	0(0)	0(0)	6
Condensed Matter Physics	8	7	4	3(12)	0(1)	1(2)	6.5
Low Temperature Physics	2	1	0	0(0)	0(0)	0(0)	–
Materials Sci./ Metallurgy	6	2	6	0(6)	0(2)	1(0)	6.5
Nuclear Physics	0	0	1	0(4)	0(0)	0(0)	7
Statistical & Thermal	4	5	0	0(1)	0(0)	0(0)	6.5
Other Theoretical/ Math.	2	1	3	0(1)	0(1)	1(2)	6
Total		16	12	4(21)	1(5)	1(3)	
Full-time Grad. Stud.		16	12				
Part-time Grad. Stud.		0	0				
First-year Grad. Stud.		4	3				
Median Years in Grad. Study (2004–05 Degrees)				–	–	–	–
Undergraduate Degrees, 2008–09 (2002–07): 4(13)							

[1] Students not yet committed to a research specialty are entered under non-specialized.
[2] Five-year totals in parentheses.

4. Graduate Degree Requirements

Master's: Master of Science. A total of 30 credit hours are required. Of these, 22 credit hours are in required graduate level courses, no more than 8 credit hours can be in senior-level courses, and no less than three nor more than 6 credit hours can be in Thesis. Candidates for the MS degree are required to pass an examination, written or oral, covering all graduate work, including thesis. Each candidate for the MS degree is required to present one lecture in the graduate seminar; this requirement can be met with the oral thesis defense.
Ph.D. in: Applied Physics. A minimum of 58 credit hours beyond completion of the Bachelor's degree are required. These must include: 19 credit hours of required graduate level courses (with a GPA of at least 3.25 in these courses), an additional 9 credit hours of elective senior or graduate level courses; 6 credit hours of guided research; and a minimum of

24 credit hours of dissertation taken in no less than two academic years of full time work. A qualifying exam, a dissertation proposal examination and a dissertation defense are required; there is no language requirement.

Thesis: Thesis may be written *in absentia*.

Table B—Appointments to Graduate Students, 2008–09

Title of Appointee	Appointments Total	Appointments First year	Academic Load Allowed in Credit Hours	Hours of Service Per Week	Stipend for Academic Year ($)
			Semester		
Teaching Assistant	18	3	15	20	29,628.00[1,2]
Research Assistant	6	0	15	20	9,867.00[1]
Total	25	3			

[1] This amount does not include a tuition waiver, which is granted to all graduate assistants.
[2] This amount includes 1.5 months of support during the summer semester.

5. Personnel Engaged in Separately Budgeted Research, 7/08–6/09

Professorial faculty	10
Total	10

6. Separately Budgeted Research Expenditures by Source of Support

	Departmental Research	Physics-related Research Outside Department
Federal government	$2,310,226	$
State government	39,999	
Private Organizations	73,819	
Total	$2,424,044	$

Table C—Separately Budgeted Research Expenditures

Research Specialty	No. of Grants	Expenditures ($)
Applied Physics	8	792,207
Condensed Matter Physics	1	330,000
Condensed Matter Theory	4	1,092,337
Materials	1	40,000
Other Theoretical/Meth	1	160,000
Other	2	9,500
Total	17	2,424,044

FACULTY

Professors

Ali, Naushad, Ph.D., Alberta, 1984. Experimental solid state and low-temperature physics.

Malhotra, Vivak, Ph.D., Kanpur Univ., 1978. Experimental solid state physics; FTIR; applied physics.

Migone, Aldo, Ph.D., Penn State, 1984. Experimental condensed matter; adsorption.

Associate Professors

Aouadi, Samir, University of British Columbia, Vancouver, BC, Canada, 1994. Applied/condensed matter, ellipsometry of thin films and coatings.

Byrd, Mark, Ph.D., Univ. of Texas, Austin, 1999. Theoretical quantum computation and quantum error connection.

Calbi, Mercedes, Ph.D., Univ. of Buenos Aires, 1999. Theoretical condensed matter, surface physics, thermodynamics and dynamics of gas adsorption.

Kolmakov, Andrei, Ph.D. Kurchatov Institute, Moscow, 1996. (Condensed Matter Physics) Surface and bulk electronic properties of individual nanostructures, fabrication-functionalization-characterization of nanodevices, development of scanning probe microscopy.

Masden, Joseph T., Ph.D., Purdue, 1983. Experimental solid state physics.

Tsige, Mesfin, Ph.D., Case Western Reserve Univ., 2001. Modeling polymer systems with molecular dynamics simulations

Assistant Professors

Silbert, Leonardo, Ph.D., Univ. of Cambridge, England, 1998. Theoretical and computational physics, soft condensed matter and granular matter.

Talapatra, Saikat, Ph.D., Southern Illinois Univ., Carbondale, 2002. Experimental materials; carbon nanotube production; properties and application.

Professors Emeriti

Cutnell, John D., Ph.D., Wisconsin, 1967. Molecular motions and interactions via pulsed nuclear magnetic resonance.

Gruber, Bruno, Ph.D., Vienna, 1962. Group theory methods in physics (molecular, atomic, nuclear, and particle physics).

Henneberger, Walter C., Dr.rer.nat., Göttingen, 1959. Electro dynamics, especially the Aharanov-Bohm effect.

Malik, F. Bary, Dr.rer.nat., Göttingen, 1958. Theoretical atomic and nuclear physics.

Johnson, Kenneth W., Ph.D., Ohio State, 1967. Far-infrared spectroscopy; physics education.

Sanders, Frank C., Ph.D., Texas, Austin, 1968. Theoretical atomic and molecular physics.

Saporoschenko, Mykola, Ph.D., Washington (St. Louis), 1958 Gaseous electronics; ion physics; mass spectrometry; Mössbauer spectroscopy; applications in coal, oil shale.

Watson, Richard E., Ph.D., Illinois, 1938. Nuclear magnetism and applied mathematics.

RESEARCH SPECIALTIES AND STAFF

Theoretical

Quantum computing. Byrd, Gaitan.

Condensed Physics. Topological and geometric effects in condensed matter systems; quantum systems coupled to dissipative environments; Berry's phase; vortices. Gaitan. Gas adsorption on Carbon nanostructures; surface physics. Calbi.

Material Science, Polymers; surface physics. Tsige.

Soft matter. Silbert.

Statistical and Thermal Physics. Computer Simulations of Materials. Calbi, Tsige, Silbert.

Experimental

Applied Physics. Mechanical, frictional, and thermal properties of automotive brake materials; advanced composites from coal by-products; composites from agricultural by-products. Malhotra.

Permanent Magnetic materials. Ali.

Solid State Physics. Synchrotron radiation applied to the study of magnetic and superconducting systems; colossal magnetoresistance; electrical, magnetic and thermal properties of rare earth compounds, spin-glasses and re-entrant magnetic transitions. Ali.

Calorimetric and vibrational studies of phase transitions of solid and liquids confined in nanoporous materials. Malhotra.

Surface Science, nanostructures, scanning probe microscopy, nanosensors. Kolmakov.

Electrical properties of thin films and thin wires. Kolmakov, Masden.

Thermodynamic studies of phases and phase transitions in systems of reduced dimensionality; properties of gases adsorbed on carbon nanotubes; thermodynamic measurements on systems adsorbed on uniform planar substrates. Migone.

Experimental materials; carbon nanotube production; properties and application; novel carbon composites. Talapatra.

Applied condensed matter physics, ellipsometry of thin films and coatings. Aouadi.

THE UNIVERSITY OF CHICAGO

DEPARTMENT OF PHYSICS

Chicago, Illinois 60637

Students Accepted For Degree	FIELDS		
	Physics	Astronomy	Related Fields
Doctorate	X		
Master's			

1. General

President: Robert J. Zimmer
Dean, Division of Physical Sciences: Robert A. Fefferman
Department Chairman: Robert M. Wald
Department Telephone Number: (773) 702-7007
Department e-mail: physics@uchicago.edu URL: http://physics.uchicago.edu/
Type of Institution: University
Control: Private
Setting: Urban
Total Faculty: 2,211
Total Students: 15,094
Total Graduate Students: 9,980
Annual Graduate Tuition:
All Graduate Students: $40,062
Tuition rates for: 2010–11
Deferred tuition plan: No
Other Fees: Student Life Fee:
$260/quarter
University student health plan: $740/qtr basic;
$1,030/qtr comprehensive/advantage
Term: Quarter

2. Number of Faculty in Department

The combined total of full-time faculty in the three professorial ranks is 41. The combined total of full-time, part-time, and other faculty at all ranks is 48.

3. Admission, Financial Aid, and Housing

Address admission inquiries to: physics@uchicago.edu or Graduate Admissions, Department of Physics, 5720 S. Ellis Ave. (60637-1434). Apply online at https://grad-application.uchicago.edu
Graduate application fee required: $55
Admission deadline: December 28 for fall admission. No mid-year admissions.
Admission information: For fall admission, 2010–11, 97 students were accepted from 476 applicants.
Admission requirements: For admission to the doctoral program, a bachelor's degree in any physical science or engineering is required. The General GRE and the Advanced Physics GRE tests are both required. The average GRE Advanced score for 2010–11 admissions was 894. Students from non-English-speaking countries are required to demonstrate proficiency in English via the TOEFL or the IELTS.
Undergraduate preparation assumed: Equivalent of Marion and Thornton, *Classical Dynamics of Particles and Systems*; Reif, *Statistical and Thermal Physics*; Wangsness, *Electromagnetic Fields*: Shankar, *Principles of Quantum Mechanics*; Eisberg and Resnick, *Quantum Physics of Atoms, Molecules, Solids, Nuclei and Particles*; Kittel, *Introduction to Solid State Physics*, 8th ed.

Financial aid: All admitted graduate students are appointed as Teaching or Research Assistants with full financial aid. No aid application is necessary.
FAFSA application required: No
Loans available: Yes
Address housing inquiries to: Graduate Student Housing Office, 5316 S. Dorchester Ave., Chicago, IL 60615-5360
On-campus, graduate student housing available: Yes
Approx. cost/month: Furnished—$870–1,061
Unfurnished—$708–1,604

Table A—Faculty, Enrollments, and Degrees Granted

Research Specialty	2009–10 Faculty	Enrollment[1] Fall 2009 Mas-ter's	Enrollment[1] Fall 2009 Doc-torate	No. of Degrees Granted[2] 2009–10 (2005–10) Mas-ter's	Terminal Master's	Doc-torate	Median No. of Years for 2009–10 Ph.D.'s
Astrophysics	7	–	26	0(0)	0(0)	7(23)	6.8
Atomic, Molecular, & Optical Physics	4	–	3	0(0)	0(0)	0(0)	–
Biophysics	2	–	3	0(0)	0(0)	0(3)	–
Chemical Physics	–	–	1	0(0)	0(0)	0(3)	–
Condensed Matter Physics	11	–	36	0(0)	0(0)	7(59)	6.7
Energy Sources & Environ.	0	–	0	0(0)	0(0)	0(1)	–
Nuclear Physics	1	–	3	0(0)	0(0)	0(0)	–
Particles & Fields	15	–	31	0(0)	0(0)	6(26)	6.1
Physics of Beams	1	–	2	0(0)	0(0)	0(0)	–
Relativity & Gravitation	1	–	2	0(0)	0(0)	0(2)	–
Other (specify) Electron/Ion Microscopy	0	–	0	0(0)	0(0)	0(3)	–
Non-specialized	1	–	21	17(57)	1(2)	0(0)	–
Total		–	127	17(57)	1(2)	20(120)	
Full-time Grad. Stud.		–	127				
Part-time Grad. Stud.		–	0				
First-year Grad. Stud.		–	18				
Median Years in Grad. Study (2000–01 Degrees)				1.5	–	–	6.5
Undergraduate Degrees, 2009–10 (2005–10): 49(175)							

[1]Students not yet committed to a research specialty are entered under non-specialized.
[2]Five-year totals in parentheses.

4. Graduate Degree Requirements

Master's: Although students are not admitted to study for a master's, they may receive a master's degree while studying for the Ph.D. For the master's there is a minimum residence requirement of three quarters of full-time registration or the equivalent, nine quarter-length courses. In addition, one must either pass the Candidacy Examination or pass nine approved graduate courses (six of which are the "core" graduate physics courses) and complete the experimental physics requirement, with a GPA of 2.5 or better. There is no thesis or foreign language requirement.

Doctorate: There is a minimum residence requirement of nine quarters of full-time registration. The candidate must pass the advanced physics laboratory course or participate in a first year experimental research experience, and also pass six advanced physics courses. Four of these advanced courses must be selected from course offerings in either three or four general categories associated with active areas of contemporary physics research; the other two must be advanced, seminar-type elective courses. Other requirements include passing the Ph.D. candidacy examination, defending the dissertation before the candidate's Ph.D. committee, and submitting a paper based on the dissertation to a recognized journal.

Special Facilities and Programs: The Department of Physics at the University of Chicago offers Ph.D. programs in many areas of physics. Students' formal classwork takes place in the modern lecture halls, classrooms, and instructional laboratories of the Kersten Physics Teaching Center. This building also houses special equipment and support facilities for student experimental projects, departmental administrative offices, and meeting rooms. The Center is situated on the science quadrangle near the John Crerar Science Library, which holds over 1,000,000 volumes and provides modern literature search and data retrieval systems.

Student participation is crucial to virtually all research projects, and both graduate and undergraduate research and training are given high priority. Most of the experimental and theoretical research of Physics faculty and graduate students is carried out within the Enrico Fermi Institute, the James Franck Institute, and the Institute for Biophysical Dynamics. These research institutes provide close interdisciplinary contact, crossing the traditional boundaries between departments.

In the Enrico Fermi Institute, members of the Department of Physics carry out theoretical research in particle theory, string theory, field theory, general relativity, and theoretical astrophysics and cosmology. There are active experimental groups in high energy physics, nuclear physics, astrophysics and space physics, infrared and optical astronomy, electron and ion microscopy, and atomic physics. Some of this research is conducted at the Fermi National Accelerator Laboratory, at Argonne National Laboratory, and at the European Organization for Nuclear Research (CERN) in Geneva, Switzerland.

Physics faculty in the James Franck Institute study chemical, solid state, condensed matter, and statistical physics. Fields of interest include chaos, chemical kinetics, critical phenomena, high T_c superconductivity, non-linear dynamics, low temperature, disordered and amorphous systems, the dynamics of glasses, fluid dynamics, surface and interface phenomena, non-linear and nanoscale optics, unstable and metastable systems, laser cooling and trapping, and polymer physics. Much of the research utilizes specialized facilities operated by the Institute, including a low temperature laboratory, a materials preparation laboratory, x-ray diffraction and analytical chemistry laboratories, laser equipment, a scanning-tunneling microscope, and extensive shop facilities. Some members of the faculty are involved in research at Argonne National Laboratory.

The Institute for Biophysical Dynamics includes members of both the Physical Sciences and Biological Sciences Divisions, and focuses on the physical basis for molecular and cellular processes. This interface between the physical and biological sciences is an exciting area that we expect to develop rapidly over the next few years, with a bi-directional impact. Initial research topics include the creation of physical materials by biological self-assembly, the molecular basis of macromolecular interactions and cellular signaling, the derivation of sequence-structure-function relationships by computational means, and structure-function relationships in membranes.

In the areas of chemical, atomic, and biophysics, research toward the doctorate may be done in either the Physics or the Chemistry Department. Facilities are available for research in crystal chemistry, degenerate quantum gases, molecular physics, molecular spectra from infrared to far ultraviolet and Raman spectra, both experimental and theoretical, surface physics, statistical mechanics, radio chemistry, and quantum electronics.

Interdisciplinary research leading to a Ph.D. degree in physics may be carried out under the guidance of faculty committees including members of other departments in the Physical Sciences Division, such as Astronomy and Astrophysics, Chemistry, Computer Science, Geophysical Sciences, Mathematics, or related departments in the Biological Sciences Division.

Table B—Appointments to Graduate Students, 2009–10

Title of Appointee[1]	Appointments Total	Appointments First year	Academic Load Allowed in Credit Hours	Hours of Service Per Week	Stipend for Academic Year ($)
			Quarter		
Teaching Assistant	30[4]	12[4]	3 courses	16	21,915[2]
Research Assistant	64	2	3 courses	19	27,120–28,584[3]
Other (specify)					
Argonne Nat'l. Lab. Fellow	4	0	3 courses	–	29,220[3]
DOE Fellow	1	0	3 courses	–	32,400[3]
GAANN Fellow	16	1	3 courses	–	30,000[3]
NSF Fellow	4	0	3 courses	–	30,000[3]
Bloomenthal Graduate Fellow	1	0	3 courses	–	29,220[3]
Chandrasekhar Fellow	0	1	3 courses	–	31,500[3]
Grainger Graduate Fellow	1	0	3 courses	–	29,220[3]
Candian NSERC Fellow	0	1	3 courses	–	19,000[3]
Hertz Fellow	1	1	3 courses	–	36,000[2]
Fulbright Fellow	2	1	3 courses	–	19,000[3]
Michelson Fellow	2	1	3 courses	–	31,500[3]
Mayer Fellow	1	1	3 courses	–	25,770[3]
Total	127	21			

[1]All TAs, RAs, and fellowship holders receive full-tuition scholarships. All other full-time degree candidates making good academic progress receive partial tuition scholarships.
[2]Stipends are for academic year.
[3]Stipends are for calendar year.
[4]Full-time TAs only.

5. Personnel Engaged in Separately Budgeted Research, 7/09–6/10

Full-time faculty	39
Other Physics faculty	10
Postdoctoral appointments	50
Graduate students	72
Undegraduate students	34
Total	205

6. Separately Budgeted Research Expenditures by Source of Support

	Departmental Research	Physics-related Research Outside Department
Federal government	$19,066,576	$2,387,581
Private, non-profit organizations	853,348	
Business and industry	245,167	
Other (including institution's own separately budgeted accounts)	1,900,147	
Total	$22,065,238	$2,387,581

Table C—Separately Budgeted Research Expenditures

Research Specialty	No. of Grants	Expenditures ($)
Astro Particles and Fields	26	8,926,393
Atmos./Space, Cosmic Rays	9	446,547
Atomic, Molecular, & Optical	10	1,074,398
Beam Physics	1	2,100
BioParticles and Fields	8	903,412
Condensed Matter	41	3,519,314
Other Experimental	1	9,401
Particles & Fields	36	6,032,227
Relativity	7	883,511
Solid State Physics	3	86,345
Statistical	2	181,591
Total	144	22,065,238

Table D—Physics-related Research Outside Department

Field and Unit Outside Department	No. of Grants	Expenditures ($)
Condensed Matter Physics/ MRSEC	1	2,387,581
Total	1	2,387,581

FACULTY

Professors

Abella, Isaac D., Ph.D., Columbia, 1963. Experimental physics; quantum optics; atomic physics; laser spectroscopy.

Blucher, Edward C., Ph.D., Cornell, 1988. Experimental physics; particle physics.

Carena, Marcela, Ph.D., University of Hamburg, 1989. Theoretical physics; elementary particles.

Carlstrom, John E., Ph.D., California, Berkeley, 1988. Experimental physics and astrophysics; star formation and cosmology; observation and new instrumentation.

Cronin, James W., Ph.D., Chicago, 1955. Professor Emeritus. Experimental physics; particle physics; ultra-high energy γ-ray astronomy.

Eastman, Dean E., Ph.D., MIT, 1965. Professor Emeritus. Experimental physics; condensed matter physics.

Freund, Peter G. O., Ph.D., Vienna, 1960. Professor Emeritus. Theoretical physics; particle physics; field theory.

Frisch, Henry, Ph.D., California, Berkeley, 1971. Experimental physics; particle physics.

Geroch, Robert P., Ph.D., Princeton, 1967. Professor Emeritus. Theoretical physics; general relativity.

Guyot-Sionnest, Philippe, Ph.D., California, Berkeley, 1987. Experimental physics; surface physics; nonlinear optical spectroscopy.

Harvey, Jeffrey A., Ph.D., Cal. Tech., 1981. Theoretical physics; particle physics; quantum field theory; superstring theory.

Hildebrand, Roger H., Ph.D., California, Berkeley, 1951. Professor Emeritus. Experimental physics; infrared astronomy.

Isaacs, Eric, Ph.D., MIT, 1988. Experimental physics, condensed matter physics.

Jaeger, Heinrich M., Ph.D., Minnesota, 1987. Experimental condensed matter physics; mesoscopic physics; high-temperature superconductivity.

Kadanoff, Leo P., Ph.D., Harvard, 1960. Professor Emeritus. Theoretical physics; hydrodynamics; statistical physics.

Kang, Woowon, Ph.D., Princeton, 1992. Experimental condensed matter physics; fractional quantum Hall effect; semiconductor physics.

Kim, Kwang-Je, Ph.D., Maryland, 1970. Theoretical physics; beam physics.

Kim, Young-Kee, Ph.D., Rochester, 1990. Experimental elementary particle physics.

Kutasov, David, Ph.D., Weizmann Inst., Israel, 1989. Theoretical physics; quantum field theory; string theory.

Levin, Kathryn, Ph.D., Harvard, 1970. Theoretical physics; solid state physics.

Levi-Setti, Riccardo, Ph.D., Pavia, Italy, 1949. Professor Emeritus. Experimental physics; ion microscopy; secondary ion mass spectrometry; ion-solid interaction.

Lu, Zheng-Tian, Ph.D., University of California at Berkeley, 1994. Experimental physics; atomic physics.

Martinec, Emil J., Ph.D., Cornell, 1984. Theoretical physics; string theory; quantum field theory; elementary particles.

Mazenko, Gene F., Ph.D., MIT, 1971. Theoretical physics; statistical physics.

Merritt, Frank S., Ph.D., Cal. Tech., 1976. Experimental physics; particle physics.

Meyer, Stephan S., Ph.D., Princeton, 1979. Experimental astrophysics; infrared astrophysics; observational cosmology.

Müller, Dietrich, Ph.D., Bonn, 1964. Professor Emeritus. Experimental physics; cosmic rays; high-energy astrophysics.

Nagel, Sidney R., Ph.D., Princeton, 1974. Experimental physics; condensed matter physics; non-linear dynamics.

Nambu, Yoichiro, Sc.D., Tokyo, Japan, 1952. Professor Emeritus. Theoretical physics; particle physics; field theory.

Oddone, Pier, Ph.D., Princeton, 1970. Experimental physics.

Oehme, Reinhard, Ph.D., Goettingen, 1951. Professor Emeritus. Theoretical physics; particle physics; field theory.

Oreglia, Mark J., Ph.D., Stanford, 1980. Experimental physics; particle physics.

Parker, Eugene N., Ph.D., Cal Tech., 1951. Professor Emeritus. Theoretical physics, astrophysics, plasma physics, space physics.

Pilcher, James E., Ph.D., Princeton, 1968. Experimental physics; particle physics.

Rosenbaum, Thomas F., Ph.D., Princeton, 1982. Experimental physics; solid state physics; low-temperature physics.

Rosner, Jonathan L., Ph.D., Princeton, 1965. Theoretical physics; particle physics; field theory.

Rosner, Robert, Ph.D. Harvard, 1976. Theoretical physics; fluid and plasma dynamics; solar physics; high-energy astrophysics.

Savard, Guy, Ph.D., McGill, 1988. Experimental physics; nuclear physics.
Schiffer, John P., Ph.D., Yale, 1954. Professor Emeritus. Experimental physics; nuclear physics.
Shochet, Melvyn J., Ph.D., Princeton, 1972. Experimental particle physics.
Swordy, Simon P., Ph.D., Bristol, 1979. Experimental physics; cosmic rays; space physics.
Turner, Michael S., Ph.D., Stanford, 1978. Theoretical astrophysics; particle physics; cosmology.
Wagner, Carlos E. M., Ph.D., Hamburg, 1989. Theoretical physics; elementary particles; supersymmetric theories.
Wah, Yau W., Ph.D., Yale, 1983. Experimental physics; particle physics.
Wald, Robert M., Ph.D., Princeton, 1972. Chairman of the Department. Theoretical physics; general relativity.
Wiegmann, Paul B., Ph.D., Landau Inst., Moscow, 1978. Theoretical physics; condensed matter physics.
Winstein, Bruce, Ph.D., Cal. Tech., 1970. Experimental physics; particle physics; cosmology.
Witten, Thomas A., Ph.D., California, San Diego, 1971. Theoretical physics; weakly-connected matter.

Associate Professors

Collar, Juan I., Ph.D., South Carolina, 1992. Experimental physics; neutrino and astroparticle physics.
Gruzberg, Ilya A., Ph.D., Yale, 1998. Theoretical physics; condensed matter physics.
Santra, Robin, Ph.D., University of Heidelberg, 2001. Theoretical physics; atomic, molecular and optical physics; chemical physics.
Sethi, Savdeep S., Ph.D., Harvard, 1996. Theoretical physics; quantum field theory; string theory; particle physics.
Zhang, Wendy W., Ph.D., Harvard, 2001. Cond. matter theory.

Assistant Professors

Biron, David, Ph.D., Weizmann Institute, Isreal, 2004. Experimental biophysics.
Canelli, Florencia, Ph.D., University of Rochester, 2003. Experimental physics; elementary particles.
Chin, Cheng, Ph.D., Stanford University, 2001. Laser cooling; trapping, degenerage quantum gases.
Gardel, Margaret L., Ph.D., Harvard, 2004. Experimental biophysics.
Hill, Richard, Ph.D., Cornell University, 2002. Theoretical physics; elementary particles.
Wakely, Scott P., Ph.D., Minnesota, 1999. Experimental astroparticle physics, high-energy astrophysics.

Senior Lecturers

Gazes, Stuart B., Ph.D., MIT, 1983. Experimental physics; nuclear physics.
Reid, David D., Ph.D., Wayne State, 1995. Theoretical physics, discrete space-time, electron- and positron-gas scattering; physics pedagogy.

RESEARCH SPECIALTIES AND STAFF

Visit http://physics.uchicago.edu/research for links to descriptions of individual faculty members' research, selected publications, and other information.

Theoretical

Astrophysics & Cosmology. Cosmology and elementary particle physics. Big-bang nucleosynthesis. Tests of the Big Bang model. Ultra-high energy cosmic-ray processes. Baryogenesis and cosmological phase transitions. Topological defects. Inflationary cosmology. Cosmic microwave background radiation. Dark matter. Formation of structure in the universe. The cosmological constant and dark energy. Aspects of string cosmology. Solar and stellar astrophysics. Astrophysical fluid dynamics. Carlstrom, R. Rosner, Turner. Related work by faculty in the Department of Astronomy and Astrophysics.

Atomic Physics. Trapped Fermi and Bose gases. Ionization dynamics. Inner-shell physics of atoms, molecules, and clusters, strong-field and electron-correlation effects. Free-electron lasers. Ultrafast laser-induced phenomena. Electronic many-body theory. Non-Hermiticity in quantum mechanics. Levin, Santra

Condensed Matter. Macroscopic dynamics of materials, interfacial singularities, and non-linear processes. Turbulent, chaotic, and stochastic behavior in hydrodynamic and other dynamical systems. Spatial self-organization in polymers, surfactant monolayers, colloids and cell assemblies. Physics of magnetic and superconducting materials (systems) driven by a strong interaction. Physics in low dimensions. Fermi liquid and non-Fermi liquid states in many body systems. High temperature superconductivity. Quantum phase transitions. Phase ordering kinetics and defect dynamics. Non-perturbative phenomena in electronic systems; strongly correlated electronic systems, magnetism. Transition between jammed and fluid states in granular matter, glass-forming liquids, and magnetic flux lattices. Integrable models of statistical mechanics and quantum field theory. Stochastic processes. Gruzberg, Kadanoff, Levin, Mazenko, Wiegmann, Witten, Zhang.

Elementary Particle Physics. String theory and unification, duality in gauge theory and string theory, solitons and topological structures, precision electroweak measurements, dark matter candidates, effective field theory, electroweak baryogenesis, low-energy supersymmetry, CP violation, heavy quark physics, confinement in QCD, quantum theory of black holes, large extra dimensions, fermion mass hierarchy, integrable systems. Carena, Freund, Harvey, Hill, Kutasov, Martinec, Nambu, Oehme, J. Rosner, Sethi, Wagner.

Relativity. Black holes. Asymptotic structure. Gravitational radiation. Mathematical aspects of general relativity. Quantum field theory in curved space-times. Quantum gravitation. Alternative theories. Geroch, Wald.

Experimental

Astrophysics. Studies of the cosmic microwave background radiation spectrum and anisotropy with ground and space-based detectors. Search for polarization in the cosmic background radiation. Measurements of the Sunyaev-Zelodovich effect for clusters of galaxies. Measurements of intergalactic radiation fields. High energy gamma-ray astrophysics with atmospheric Cherenkov telescopes. Development of giant air shower array (Auger Project) for investigation of the highest energy cosmic rays. Development of large detectors for high energy cosmic rays on space and balloon payloads. Experimental investigations of cosmic ray electrons and of the elemental and isotopic abundances of cosmic-ray nuclei over a wide energy range. Investigations of solar, magnetospheric, and heliospheric phenomena with satellite and deep space missions. Cosmic dust studies. Development of instruments to detect polarization in the far-infrared emission from interstellar clouds. Investigation of the magnetic field structure of dense cloud cores. Airborne and mountain-top polarimetry. Direct searches for non-baryonic dark matter. Accelerator-based nuclear astrophysics experiments. Carlstrom, Collar, Cronin, Hildebrand, Meyer, Müller, Savard, Schiffer, Swordy, Wakely, Winstein.

Atomic Physics. Bose-Einstein condensation of molecules and fermionic superfluids. Laser cooling and trapping of atoms. Scalable quantum manipulation and quantum computation. Testing time-reversal symmetry in atoms and nuclei. Radio-krypton dating. Chin, Lu.

Beam Physics. Investigation of particle and photon beams and their mutual interactions with the goal of developing novel accelerators or radiation devices. Some current topics are production and acceleration of high-brightness electron beams for linear colliders and free electron lasers; beam dynamics in ionization cooling for muon colliders and neutrino factories; self-amplified spontaneous emission for intense, coherent x-rays; miniature IR radiation source via Smith-Purcell process using electron microscope beams. Theoretical and experimental programs at the Enrico Fermi Institute on campus, at the Argonne National Laboratory Advanced Photon Source, and the A0 facility in Fermilab. K. Kim.

Biophysics. Cell migration and division, physical aspects of biological organization, mechanical behavior of cells, regulation of cell physiology, non-linear dynamics, computational biology, time-resolved fluorescence, confocal microscopy, protein-engineering, signal transduction, gene expression, mathematical modeling, large-scale simulations, stochastic and self-assembly processes, elasticity of polymer networks, optical and holographic traps, single-molecule biophysics, nonlinear optics methods, noise and information in intraneuronal pathways and interneuronal communication, homeostatic regulation of single neuronal function and of the function of small neural circuits, design principles of biological networks, biophysics in vivo — quantifying behavior and physiological activity of neurons, High Power Computation (grid and parallel computing), biophysics of sleep. Biron, Gardel.

Condensed Matter. Optical and electronic transport in normal and superconducting nanocrystals and arrays. Collective effects at ultra-low temperatures including (fractional) quantum Hall effect, vortex tunneling, metal-insulator transitions, and magnetic quantum critical points. Symmetry-breaking and fluctuations in heavy fermion, organic, and high-T_c superconductors. Nonlinear dynamics and flow properties of granular materials. Scaling behavior of liquid flow and droplet breakup. Mathematical analysis and computer simulation of singularity formation. Universal scaling behavior of relaxation phenomena in supercooled liquids and glasses. Microscopic kinetics and dynamics of phase transitions in colloidal suspensions. Manipulation by dynamic optical holographic traps. Molecular regulation within living cells. Self-assembly and morphology of ultrathin polymer films. Biological properties of the cytoskeleton of eukaryotic cells. The mechanical behavior of cells. Eastman, Gardel, Guyot-Sionnest, Isaacs, Jaeger, Kang, Nagel, Rosenbaum.

Elementary Particles. Measurements of properties of the top quark. Searches for supersymmetric particles, the Higgs boson, and other new physics. Precision tests of the standard model in W and Z decays. Studies of $p\bar{p}$ interactions at center-of-mass energies of 1800 GeV. High-precision measurement of CP violation parameters in K decays; high-sensitivity search for rare K decays and for CPT violation. High-precision measurements of hyperon rare decays. High-precision measurements of electroweak interactions at LEP, both near the Z^0 and at center-of-mass energies up to 200 GeV. Searches for new physics including the Higgs boson and supersymmetry; precision measurement of M_W. Preparation for the ATLAS experiment at the LHC (high-energy $p\bar{p}$ interactions at 14 TeV). Research and development on muon colliders and neutrino factories. Use of facilities at Fermi National Accelerator Laboratory and at CERN. Blucher, Canelli, Cronin, Frisch, Y. Kim, Merritt, Oreglia, Pilcher, Shochet, Wah, Winstein.

Ion and Electron Microscopy. Development of high resolution-scanning ion and electron microprobes, imaging micro-analysis by secondary ion mass spectrometry with application to advanced ceramics, visualization of dynamic processes and of biological matter. Levi-Setti.

Nuclear Physics. Studies of the nuclear many-body system: Nuclear structure and interactions, nuclear reactions in astrophysics, nuclear matter under extreme conditions, precision measurements of critical information to nucleosynthesis along the r- and rp-process paths. Low-energy experiments in fundamental interactions and symmetries, exotic nuclear structure, double beta decay, coherent nuclear scattering. Production, cooling and trapping of rare isotopes, R&D for the Rare Isotope Accelerator (RIA) project. Non-nucleonic degrees of freedom in nuclei and phenomena requiring a quark description. Lu, Collar, Savard, Schiffer.

THE UNIVERSITY OF CHICAGO

DEPARTMENT OF ASTRONOMY AND ASTROPHYSICS

Chicago, Illinois 60637

Students Accepted For Degree	FIELDS		
	Physics	Astronomy	Related Fields
Doctorate		X	
Master's		X	

1. General

President: Robert J. Zimmer
Dean of Division of Physical Sciences: Robert Fefferman
Department Chairman: Edward Kolb
Department Telephone Number: (773) 702-8203
Type of Institution: University
Control: Private
Setting: Urban
Total Faculty: 2,211
Total Students: 15,626
Total Graduate Students: 10,492
Annual Graduate Tuition:
All Graduate Students: Full-time—$39,276 (3Q's)
Tuition rates for: 2009–10
Deferred tuition plan: No
Other Fee: $675 (University Student Medical Basic Plan, Student life fee 2009–2010, $864/qtr.
Term: Quarter

2. Number of Faculty in Department

The combined total of full-time faculty in the three professorial-ranks is 27. The combined total of full-time, adjunct, and other-faculty at all ranks is 36. This figure includes joint appointments in other departments and institutes: five in physics, one in chemistry, one in history, and fifteen in the Enrico Fermi Institute.

3. Admission, Financial Aid, and Housing

Address admission inquiries to: Director of Admissions, Department of Astronomy and Astrophysics, 5640 S. Ellis Avenue, Chicago, Illinois 60637
Email: lrebeles@oddjob.uchicago.edu
Internet: http://astro.uchicago.edu
Graduate application fee required: $55—U.S. citizens
$55—Foreign students
Admission deadline (Fall admission): 1/7/07
Admission information: For fall admission, 2009–10, 5 students were admitted from 200 applicants.
Admission requirements: For admission to the graduate programs, a Bachelor's degree, preferably in physics or astronomy, is required, but others will be considered. The GRE and GRE Advanced Physics are required. The average GRE scores for 2009–10 matriculants were verbal-550; quantitative-750; analytical-730. The average GRE Advanced score for 2009–10 matriculants with full financial aid was 750. Students from non-English speaking countries are required to demonstrate proficiency in English via the TOEFL exam. Minimum acceptable score for admission is 90 total (IELTS =7).
Address financial aid inquiries to: Director of Admissions, Department of Astronomy and Astrophysics, 5640 S. Ellis Avenue, Chicago, Illinois 60637
GAPSFAS application required: No
Financial aid deadline: 12/28/07
Loans available: Yes
Address housing inquiries to: Neighborhood Student Apartments, 5316 South Dorchester Ave., Chicago, IL 60615-5360
On-campus, graduate student housing available: Yes
Cost/month: Furnished: $500–1,000; Unfurnished: $541–1,158

Table A—Faculty, Enrollments, and Degrees Granted

Research Specialty	2009–10 Faculty	Enrollment[1] Fall 2008		No. of Degrees Granted[2] 2008–09 (2003–08)			Median No. of Years for 2009–10 Ph.D.'s
		Master's	Doctorate	Master's	Terminal Master's	Doctorate	
Astronomy/ Astrophysics	36	–	29	–	–	4(29)	5.5
Total		–	29	–	–	4(29)	
Full-time Grad. Stud.		–	0				
Part-time Grad. Stud.		–	0				
First-year Grad. Stud.		–	5				
Median Years in Grad. Study (2007–08 Degrees)				1	–	6	–

Undergraduate Degrees, 2007–08 (2003–08): none

[1]Students not yet committed to a research specialty are entered under non-specialized.
[2]Five-year totals in parentheses.

4. Graduate Degree Requirements

Master's: There is a minimum residency requirement of three-quarters full-time registration, or the equivalent—nine courses. In addition, one must pass at least six graduate level physics and astronomy courses with a 3.0 average and either receive a grade of P or M on the Ph.D. candidacy examination or submit and defend successfully at an oral examination the thesis sponsored by a member of the faculty.
Doctorate: There is a minimum residency requirement of nine-quarters of full-time registration, or equivalently, 27 courses. The candidate must complete a required sequence of 10 courses with a 3.0 average, pass the Ph.D. candidacy examination, pass 8 elective advanced courses, submit and defend-successfully a thesis, and have the thesis submitted to a recognized journal.
Thesis: Thesis may be written *in absentia*.
Special Equipment, Facilities, or Programs: Research in astronomy and astrophysics at the University of Chicago covers a broad range of topics, including the Sun and solar-like stars, cosmic rays, the chemical origin of meteorites and comets, interstellar matter, the birth of stars, the death of stars and nucleosynthesis, high energy and relativistic astrophysics, the origins and dynamics of galaxies, and cosmology. The activities involve theoretical, experimental, and observational programs among a community of faculty members from the De-

partments of Astronomy and Astrophysics, Chemistry, Geophysical Sciences, Mathematics, and Physics, with connections to Argonne National Laboratory and Fermi National Accelerator Laboratory.

The students and faculty of the University of Chicago enjoy access to a wide range of observational facilities. The Departmental observational facility is the 3.5-meter aperture telescope at Apache Point Observatory in New Mexico. This telescope has been designed to permit routine remote observing and rapid changeover between instruments, and is instrumented to work from 0.35 mm to 2 microns. A very high-resolution Echelle Spectrograph built by a team led by faculty member Roger Hildebrand allows researchers to determine the composition of stars nearby and to probe the Universe at a time before stars and galaxies existed. Adaptive optics are being developed for the telescope, which will enable faint objects to be studied with a resolution of 0.1 arcsecond; this program is part of a larger NSF-funded effort at Chicago to bring a variety of adaptive-optics techniques to bear on improving the image quality of large-aperture reflecting telescopes. The instrumentation and the adaptive optics are being developed both at the Chicago campus and at the Yerkes Observatory in Williams Bay, Wisconsin. Yerkes serves as a laboratory for development of the instruments and techniques to be used on major telescopes, including the 3.5-meter telescope, the Stratospheric Observatory for Infrared Astronomy (SOFIA), and the Infrared Telescope Facility, and also provides a continuing observational program, with its famous 40-inch (1-meter) refractor and is 41-inch (1-meter) and 0.6-meter reflectors.

In addition, Chicago astronomers regularly use telescopes at the national observatories (Kitt Peak National Observatory, Gemini telescopes and the Cerro Tololo Inter-American Observatory), as well as at other observatories such as the CSO, JCMT, Keck I and II and UKIRT facilities on Mauna Kea, telescopes of the McDonald Observatory in Texas, the 200-inch Hale telescope at Palomar, and the Very Large Array (VLA), BIMA and OVRO radio arrays. Various active NASA satellites (and archives) are also used, including IRAS, Einstein, EXOSAT, IUE, HST, the Chandra X-ray Observatory, HETE-2, ROSAT, COBE, Compton GRO, Rossi XTE, and EUVE, as well as high-altitude balloons. Chicago astronomers will soon participate in observatories that are coming into operation, including the SubMillimeter Array (SMA), SOFIA, and the Gemini 8-meter telescopes.

In collaboration with Fermilab, Princeton University, the Institute for Advanced Study, Johns Hopkins University, a consortium of Japanese institutions, the US Naval Observatory, the Max-Planck-Institute for Astronomy and the University of Washington, we have built a 2.5-meter dedicated telescope, a half-billion-pixel CCD camera, and a 600-object spectrograph to study the large-scale structure of the Universe. The main scientific goals of the Sloan Digital Sky Survey are to construct a three-dimensional map of the Universe by obtaining redshifts for a million galaxies and 100,000 QSOs and accurate digital five-color photometry for 200 million objects. The 30-terabyte SDSS database will soon become the largest and most important astronomical database in existence.

Computing facilities for theoretical and numerical work at Chicago have recently been expanded in the area of visualization, as part of a major NASA-supported initiative in high-performance computing and communications. Much of the campus pioneering of computer networks, graphics, workstations, and telecommunications have been done in this department, which enjoys a close working relationship with the Argonne National Laboratory (ANL), managed by the University of Chicago for the Department of Energy, which has a particularly strong program in high-performance parallel computing.

This department is also the principal host of the Center for Astrophysical Thermonuclear Flashes ("Flash Center"), headed by Donald Lamb. This Center is one of five university-based-centers of excellence funded by the DOE Accelerated Strategic Computing Initiative (ASCI) program, and represents a large collaboration between some 35 University scientists, representing almost all of the Physical Sciences Division's departments and institutes, and scientists at Argonne National Laboratory, at Rensselaer Polytechnic Institute, and at the three DOE defense programs laboratories. The primary focus of the Center is to develop a new generation of computational tools for attacking the problem of nuclear burning on the surfaces of neutron stars and white dwarfs, and in the interior of white dwarfs. This development involves creation of new tools for computing on massively parallel computers; new algorithms for following the complex fluid behavior of astrophysical nuclear flames; and new methods for storing and displaying the resulting data.

A number of faculty in this department are active participants in the Kavali Institute for Cosmological Physics (KICP), a Physics Frontier Center funded by the National Science Foundation. Over the past two decades some of the most important discoveries both in physics and astronomy have come at the cosmology/particle physics boundary. These discoveries, as well as theoretical advances, raise a new set of questions that involve astronomy and particle physics in an indivisible way. This Center is devoted to exploiting the connections between physics at the smallest scale—interactions of the quarks and leptons—and at the largest scale—the constitution and birth of the cosmos itself. The key is an integrated approach—astronomers and physicists, theorists and experimentalists working together, using telescopes and accelerators.

The study of astronomical objects by researchers at Chicago begins nearby, with the solar system. Our proximity to the Sun allows detailed studies of this star. Studies of active regions provide clues to the nature and origin of its magnetic field, and numerical simulations of turbulent compressible-convection help us to understand the nature of energy, angular momentum, and magnetic field transport in its outer layers; tools of helioseismology, together with theory, are being used to probe the interior of the Sun. Observations of other, solar-like, stars are then used by Chicago scientists as a means by which ideas developed in the solar context can be tested: such stars thus become our laboratory.

We can now trace the history of the Universe back to within a fraction of a second of the beginning as well as tracing stars from birth to death. We are asking deep questions about how the Universe began, how stars explode, the origin of the chemical elements and the interworkings of black holes. The confluence of advances in our understanding of the Universe and leaps in technological capability have astrophysics poised for many exciting decades as the next millennium dawns. We are well prepared to participate in this most important and exciting endeavor.

Table B—Appointments to Graduate Students, 2009–10

Title of Appointee	Appointments Total	First year	Academic Load Allowed in Credit Hours	Hours of Service Per Week	Stipend for Academic Year ($)
			Quarter		
Teaching Assistant[3]	13	4		19.5	25,764[2]
Research Assistant	29	0	30 semester hours per academic year	19.5	25,764[2]
NASA Fellow	1	0		–	18,000[1]
NSF Fellow	1	0		–	30,000
Total	44	4			

[1]Stipend for 12 months.
[2]Tuition covered.
[3]Appts. by quarter.

FACULTY

Professors

Carlstrom, John E., Ph.D., California, Berkeley, 1988. Star formation and cosmology; observation and new instrumentation.
Cronin, James W., Ph.D., Chicago, 1955. Ultra high-energy gamma-ray astrophysics.
Cudworth, Kyle M., Ph.D., California, Santa Cruz, 1974. Astrometry; star clusters.
Frieman, Joshua A., Ph.D., Chicago, 1985. Cosmology; particle astrophysics.
Harper, Doyal A., Ph.D., Rice, 1971. Infrared astronomy and the structure of active objects.
Hildebrand, Roger H., Ph.D., California, Berkeley, 1951. Far infrared and submillimeter astronomy.
Hobbs, Lewis M., Ph.D., Wisconsin, 1966. Interstellar matter and galactic structure. Emeritus Faculty.
Hogan, Craig, Ph.D., Cambridge, 1980. Dark Energy; particle astrophysics.
Khokhlov, Alexei, Ph.D., Moscow State University, 1984. Fluid dynamics; supernovae jets; nucleosynthesis; experimental astrophysics.
Kibblewhite, E. J., Ph.D., Cambridge, 1971. Adaptive optics; high-resolution imaging.
Kolb, Edward W., Ph.D., Texas, 1978. Particle physics; cosmology; theoretical astrophysics.
Königl, Arieh, Ph.D., Caltech., 1980. Theoretical high-energy astrophysics.
Kron, Richard G., Ph.D., California, Berkeley, 1978. Director of Yerkes Observatory. Observational studies of active galaxies.
Lamb, Donald Q., Jr., Ph.D., Rochester, 1974. Compact objects; high-energy astrophysics.
Meyer, Stephan S., Ph.D., Princeton, 1979. Infrared astrophysics and observational cosmology.
Oka, Takeshi, Ph.D., Tokyo, 1960. Laser spectroscopy and interstellar molecules. Emeritus Faculty.
Olinto, Angela V., Ph.D., MIT, 1987. Chair of Department. Particle and nuclear astrophysics; cosmology.
Palmer, Patrick E., Ph.D., Harvard, 1968. Radio astronomy and interstellar molecules.
Privitera, Paolo, Ph.D., Laurea, 1993. Ultra-high energy cosmic rays; Extragalactic Astronomy; High Energy Astrophysics.
Rosner, Robert, Ph.D., Harvard, 1975. Fluid and plasma dynamics; solar physics; high-energy astrophysics.
Truran, James W., Ph.D., Yale, 1965. Nuclear astrophysics; evolution of stars and galaxies; high-energy astrophysics.
Turner, Michael S., Ph.D., Stanford, 1978. Cosmology and particle physics; relativistic astrophysics.
Vandervoort, Peter O., Ph.D., Chicago, 1960. Analytical dynamics of galaxies. Emeritus Faculty.
York, Donald G., Ph.D., Chicago, 1970. Interstellar and intergalactic matter; observational cosmology.

Associate Professors

Cattaneo, Fausto, Ph.D., Solar system astronomy, high energy and computational astrophysics.
Dodelson, Scott, Ph.D., Columbia, 1988. Cosmology; gravitational lensing.
Gnedin, Nickolay Y., Ph.D. Princeton University, 1996. Cosmology, galaxy formation, supercomputer simulations.
Hu, Wayne, Ph.D., California, Berkeley, 1995. Precision cosmology, CMB, galaxy surveys, weak lensing.
Kent, Stephen M., Ph.D., Caltech., 1980. Observational studies of galaxies.
Kravtsov, Andrey, Ph.D., New Mexico State University, 1999. Cosmology, structure formation in the universe, numerical simulations.
Miller, Richard H., Ph.D., Chicago, 1957. Numerical experiments and the dynamical evolution of galaxies.

Assistant Professors

Chen, Hsiao-Wen, Ph.D., State University of New York at Stony Brook, 1999. Observational extragalactic astronomy.
Gladders, Michael, Ph.D., University of Toronto, 2002. Observational cosmology and instrumentation.
Hooper, Dan, Ph.D., Wisconsin-Madison, 2003. Theoretical Astrophysics.

RESEARCH SPECIALTIES AND STAFF

Theoretical

Classical Particles and Fields. Dodelson, Frieman, Hu, Königl, Lamb, Olinto, Rosner.
Cosmology, Elementary Particle Astrophysics. Carlstrom, Dodelson, Frieman, Hu, Kolb, Kravtsov, Olinto, Pryke, Turner.
Galactic Dynamics. Miller, Vandervoort.
General Relativity. Geroch (from Mathematics and Physics), Turner, Wald (from Physics).
History of Astronomy. Swerdlow.
Stars, Galaxies, Nuclear Astrophysics. Carlstrom, Hu, Königl, Lamb, Olinto, Rosner, Truran.

Experimental

Cosmochemistry. Clayton (from Chemistry and Geophysical Sciences), Grossman (from Geophysical Sciences).
Extragalactic Astronomy. Carlstrom, Kent, Kron, S. Meyer, York.
High-Energy Astrophysics. Cronin, Müller (from physics), Swordy.
High-Resolution Imaging. Kibblewhite.

High-Resolution Interstellar Spectroscopy. Hobbs.
Infrared Astronomy. Harper, Hildebrand, S. Meyer.
Laser Spectroscopy. Oka.
Radio Astronomy. Carlstrom, Palmer.
Stellar Astronomy. Cudworth, Hobbs, Kron.
Ultra High-Energy Gamma-Ray Astrophysics. Cronin, Müller (from Physics).
UV Spectroscopy, Intergalactic Matter. York.

UNIVERSITY OF ILLINOIS AT CHICAGO

DEPARTMENT OF PHYSICS

Chicago, Illinois 60607

Students Accepted For Degree	FIELDS		
	Physics	Astronomy	Related Fields
Doctorate	X		
Master's	X		

1. General

President: Michael J. Hogan
Interim Dean of Graduate School: Henri Gillet
Department Head: Henrik Aratyn
Department Telephone Number: (312) 996-3400
Type of Institution: University
Control: Public
Setting: Urban Chicago
Total Faculty: 2,574
Total Graduate Faculty: 1,359
Total Students: 26,245
Total Graduate Students: 7,721
Annual Graduate Tuition:
In-state residents: Full-time—$5,136/sem.
Out-of-state residents: Full-time—$11,135/sem.
Tuition rates for: 2009–10
Deferred tuition plan: Yes-Pre-Payment Plan
Other Fees: $1,846/sem.
Term: Semester

2. Number of Faculty in Department

The combined total of full-time faculty in the three professorial ranks is 23.

3. Admission, Financial Aid, and Housing

Graduate College website:: http://grad.uic.edu
Physics Department website: http://physicsweb.phy.uic.edu
Address admission inquiries to: Graduate Admissions, Department of Physics, M/C 273, 845 W. Taylor #2236, Chicago, IL 60607-7059 or FAX: 312-996-9016
E-mail: physics@uic.edu
Graduate application fee required: $50 U.S.; $60 International
Admission deadline (Fall admission): 12/15 for consideration for the University Fellowship; 2/15 priority deadline for domestic applicants, final deadline for international applicants; 5/15 final deadline for domestic applicants.
Admission information: For fall admission, 2010, 15 students were admitted from 131 applicants.
Admission requirements: For admission to the graduate programs, a Bachelor's degree is required, including 20 semester hours of physics coursework and a minimum GPA of 2.75 (A=4.0). The GRE general test is required. Students from non-English speaking countries are required to demonstrate proficiency in English via the TOEFL or IELTS. A complete application will include the paper or online application and payment of the application fee (http://www.uic.edu/depts/oar/grad/apply_grad_degree.html); official transcripts for previous post-secondary coursework and proof of any degrees earned, in signed, sealed envelopes (if foreign, an offical translated copy is also required); GRE and TOEFL or IELTS scores (minimum acceptble scores: TOEFL paper based 550; computer based statement 213; iBT-80; IELTS 6.5 total); 3 sealed letters of recommendation; and academic statement of purpose; an application for graduate appointment if applying for financial assistance.
Undergraduate preparation assumed: Becker, *Mechanics*; Reitz and Milford, *Electromagnetism*; Reif, *Statistical Physics*; Liboff, *Introductory Quantum Mechanics*.
Address financial aid inquiries to: Graduate Admissions, Department of Physics, M/C 273, 845 W. Taylor #2236, Chicago, IL 60607-7059; physics@uic.edu
GAPSFAS application required: No
Financial aid deadline: 2/15 of each year
Loans available: Yes
Address housing inquiries to: Univ. of Illinois at Chicago (M/C 579), 818 S. Wolcott St., Chicago, IL 60612; 312-355-6300; email: housing@uic.edu
On-campus, single student housing available: Yes
On-campus, married student housing available: Yes
www.housing.uic.edu

Table A—Faculty, Enrollments, and Degrees Granted

Research Specialty	2009–10 Faculty	Enrollment Fall 2009		No. of Degrees Granted 2009–10 (2005–10)		
		Master's	Doctorate	Master's	Terminal Master's	Doctorate
Atomic/Molecular	1	0	4	0(1)	0(1)	0(0)
Biophysics	2	0	7	0(2)	0(1)	0(3)
Condensed Matter	8	0	26	2(19)	1(6)	3(13)
Nuclear Physics	4	0	4	0(5)	0(0)	0(3)
Particle Physics	7	0	11	1(4)	0(1)	1(4)
Non-Specialized	0	7	15	0(2)	4(18)	0(0)
Total	22	7	67	3(33)	5(27)	4(23)
Full-time Grad. Stud.		74				
Part-time Grad. Stud.		0				
First-year Grad. Stud.		14				

Undergraduate Degrees, 2004–09: 57

[1]Students not yet committed to a research specialty are entered under non-specialized.
[2]Five-year totals in parentheses.

4. Graduate Degree Requirements

Master's (M.S.): The general requirement for the Master of Science is satisfactory completion of 32 semester hours of work in courses approved by the department. At least 20 of these hours must be at the 500 level; they must include Physics 501 and 502 (Electrodynamics) and Physics 511 and 512 (Quantum Mechanics), and may not include more than 4 hours of Physics 596 (Independent Study) or more than 8 hours of Physics 598 (Thesis Research).
Doctorate (Ph.D.): The minimum requirements for the Ph.D. are: (1) The satisfactory completion of 96 semester hours of course work approved by the department, including at least 36 hours of 500-level courses (exclusive of Physics 596 and 599). These 36 hours must include the sequence Physics 501 and 502 (Electrodynamics), Physics 511 and 512 (Quantum

Mechanics), Physics 561 (Statistical Mechanics), at least one complete sequence chosen from among the following: Physics 521 and 522 (Molecular and Laser Physics), Physics 531 and 532 (Solid State Physics), Physics 551 and 552 (Elementary Particle Physics), Physics 513 and 514 (Quantum Field Theory), and five semesters of the Graduate Seminar, Physics 595. (2) Satisfactory performance in a comprehensive qualifying examination consisting of 400-level problems on classical mechanics, electricity and magnetism, quantum mechanics, and thermodynamics and statistical mechanics. This examination may be repeated once but must be passed no later than January of the student's second year in residence. Details on this examination are available from the department office. (3) Satisfactory performance on an oral examination in the general area of the student's doctoral thesis research, which is to be taken within two years after passing the qualifying examination. The examination will normally start with a brief oral report by the student on his or her proposed research. If the performance is only marginally satisfactory, the student may be asked to retake the examination. (4) Satisfactory completion and defense of a doctoral dissertation. (5) Each student is required to serve as a teaching assistant for at least two semesters.

Thesis: Thesis may not be written *in absentia*.

Special Equipment, Facilities, or Programs: Major research laboratories include the Laboratory of Atomic/Molecular Laser Physics and the Microphysics Laboratory. These research laboratories contain many notable resources including the world's highest spectral brightness tunable ultraviolet laser, dye laser systems, ti: sapphire, various excimer lasers, high-resolution spectrometers, and other computer-assisted optical equipment. The department has molecular beam epitaxy (MBE) ultrahigh-vacuum growth chambers and surface analytical facilities, helium-3 and dilution refrigerators, an ultrasonic spectrometer, Foner and SQUID magnetometers, a susceptibility balance, a Bruker NMR spectrometer, an automated adiabatic calorimeter and on-line computer data handling systems with dedicated mini-and microcomputers. Faculty and students are also using the extensive experimental high-energy physics facilities of Fermilab and the nuclear and solid state facilities at Argonne National Laboratory.

Research is supported by computers within the department as well as the UIC Computer Center. The computers within the department include Digital Equipment Microvax cluster, SUN workstations, and Apple Macintosh and IBM PC microcomputers. These are accessible and interconnected through an Ethernet network and individual connections in most offices. This network, via the Computer Center, is connected to Fermilab, Argonne, and the rest of the world. The UIC Computer Center has an IBM 3090 mainframe and ample peripherals, campus-wide networks (ADN &ADN-II) and has wide bandwidth connections to the supercomputer centers at the University of Illinois at Urbana-Champaign and nationwide.

The Department of Physics' experimental program is supported by a four-person machine shop with ample equipment and a two-person electronics shop with complete facilities for printed circuit layout and construction. The department maintains a sample preparation facility with an arc melter, zone refiner, a spark cutter, x-ray equipment, furnaces, dry boxes, polishing equipment, and other metallurgical instruments.

The Science Library is located in the same building as the Department of Physics. The engineering collection is contained within the Main Library. Terminals and staff help are available for carrying out computer-assisted library searches.

Table B—Appointments to Graduate Students, 2009–10

Title of Appointee	Appointments Total	Appointments First year	Academic Load Allowed in Credit Hours	Hours of Service Per Week	Stipend for Academic Year ($)
			Semester		
Teaching Assistant	39	14	8 min.	20	15,120[1,2]
Research Assistant	23	–	8 min.	20	
Tuition/Fee Waiver	3	0	12 min.	0	15,417[1,2]
Univ. Fellow	7	0	12 min.	0	20,000[2,3]
Total	72	14			

[1]Based on 9-month 50% appointment. Additional support available in summer.
[2]In addition, tuiton and some fees are waived ($19,688).
[3]12 months at $1,667 per month.

5. Personnel Engaged in Separately Budgeted Research, 7/06-6/07

Professorial faculty	22
Postdoctoral appointments	12
Graduate students	39
Nonteaching research assistants	5
Total	78

6. Separately Budgeted Research Expenditures by Source of Support, 7/08–6/09

	Departmental Research	Physics-related Research Outside Department
Federal government	$4,410,455	$
Private organizations	105,754	
Total	$4,516,209	$

Table C—Separately Budgeted Research Expenditures

Research Specialty	No. of Grants	Expenditures ($)
Atomic, Molecular, & Laser Physics	1	1,549,802
Medical Physics and Biophysics	4	472,720
Condensed Matter Physics	18	889,930
High Energy Physics	7	925,880
Nuclear Physics	2	677,876
Total	35	4,516,209

FACULTY

Professors

Adams, Mark R., PhD, SUNY, Stony Brook, 1981. Experimental high-energy physics.

Ansari, Anjum, PhD, University of Illinois, Urbana, 1988. Biophysics experiment.

Aratyn, Henrik, PhD, Copenhagen, N. Bohr Inst., 1984. Department Head. High-energy physics.

Campuzano, Juan C., PhD, Wisconsin, Milwaukee, 1978. Experimental solid state physics.

Crabtree, George, Ph.D., University of Illinois, Chicago, 1974. Condensed matter. Energy.

Gerber, Cecilia, PhD, Universidad de Buenos Aires, Argentina, 1995. Experimental high energy.

Grein, Christoph, PhD, Princeton, 1989. Theoretical condensed matter physics.

Halliwell, Clive, PhD, Manchester, 1971. High-energy physics.

Keung, Wai-Yee, PhD, Wisconsin, Madison, 1980. High-energy physics theory.

Ogut, Serdar, PhD, Yale University, 1995. Theoretical condensed matter physics.

Rhodes, Charles K., PhD, MIT, 1969. Laser physics; atomic and molecular physics.

Schlossman, Mark, PhD, Cornell, 1987. Condensed matter physics experiment.

Schroeder, W. Andreas, PhD, University of London, Imperial College, 1987. Condensed matter physics experiment.

Sivananthan, Sivalingham, PhD, University of Illinois, Chicago, 1988. Experimental solid state physics material science. Biophysics theory.

Stephanov, Mikhail, PhD, Oxford Univ., 1994. Theoretical nuclear and high energy particle physics.

Varelas, Nikos, PhD, University of Rochester, 1994. Experimental high energy particle physics.

Associate Professors

Evdokimov, Olga, PhD, Joint Institute for Nuclear Research, Dubna, Russia, 1999. Experimental Nuclear Physics.

Hofman, David, PhD, SUNY at Stony Brook, 1994. Experimental nuclear physics.

Imbo, Tom, PhD, University of Texas, 1988. High-energy theory; mathematical physics.

Morr, Dirk, PhD, University of Wisconsin, Madison, 1997. Theoretical condensed matter.

Assistant Professors

Cavanaugh, Richard, PhD, The Florida State University, 1999. Experimental High Energy.

Klie, Robert F., PhD, University of Illinois at Chicago, 2002. Condensed matter physics experiment.

Pérez-Salas, Ursula, PhD, University of Maryland, 2000. Biophysics experiment.

Research Staff/Adjunct Professors

Abuel-Rub, Khaled, Research Assistant Prof., PhD, University of Illinois at Chicago, 1992.

Apanasevich, Leonard, Research Assistant Professor, PhD, Michigan State University, 2005.

Borisov, Alexey, Research Associate Professor, PhD, Moscow State University, 1985.

Chang, Yong, Research Assistant Professor, PhD, Shangai Institute of Technical Physics, 1996.

Cho, Michael, Adjunct Assistant Professor, PhD, Drexel University, 1991.

Dutta, Mitra, Adjunct Professor, PhD, University of Cincinnatti, 1981.

Espinoza, Randall, Postdoctoral Research Associate, PhD, University of Illinois at Chicago, 2005.

García Solís, Edmundo, Res. Assoc. Prof., PhD, University of Maryland, 1995.

Goeckner, Hans, Lecturer, PhD, University of Illinois at Chicago, 1995.

Hahn, Suk-Ryong, Res. Assoc. Prof., PhD, Oregon Grad. Institute of Sci. & Tech., 1993.

Hohler, Paul, Postdoctoral Research Associate, PhD, University of Maryland, 2008.

Kang, TaeWon, Adjunct Professor, PhD, Dongguk University, 1982.

Kouznetsov, Sergueí, Research Assistant Professor, PhD, Semyonov Institute of Chemical Physics, 1994.

Kunde, Gerd, Adjunct Professor, PhD, University of Frankfurt, 1994.

Lu, Hui, Adjunct Professor, PhD, Beckman Institute, 1999.

Marko, John, Adjunct Professor, PhD, Massachusetts Institute of Technology, 1989. Biophysics.

Mueller, Mark, Adjunct Professor, PhD, Stanford University, 1984.

Narayanan, Ranjani, Research Assistant Professor, University of Florida, Gainesville, 2009.

Norris, James, Adjunct Professor, PhD, Washington University.

Racz, Ervin, Research Assistant Professor, PhD, University of Szeged, 2006.

Sporken, Robert, Visiting Research Associate Professor, PhD, University of Namur, 1988.

Stroscio, Michael, Adjunct Professor, PhD, Yale University, 1974.

Strom, Derek, Postdoctoral Research Associate, PhD, Northwestern University, 2009.

Tillotson, Andrew, Lecturer, PhD, University of Maryland, 2006.

Yang, Guang, Postdoctoral Fellow, PhD, Sheffield, 2007.

Professors Emeriti

Betts, R. Russell, PhD, Univ. of Pennsylvania, 1972. Nuclear experiment.

Boccara, Nino, Professor, PhD, Paris, 1961.

Bodmer, Arnold, PhD, Manchester, 1953. Theoretical nuclear physics.

Carhart, Richard, PhD, Wisconsin, 1965. Theoretical high-energy physics; environmental physics.

Claus, Helmut, PhD, Karlsruhe, 1965. Experimental solid state physics.

Faurie, Jean-Pierre, PhD, Clermont-Ferrand, 1970. Experimental solid state physics.

Garland, James, PhD, Chicago, 1966. Theoretical solid state physics.

Goldberg, Howard, PhD, California, Berkeley, 1964. Experimental high-energy physics.

Licht, Arthur L., PhD, Maryland, 1963. Theoretical high-energy physics; astrophysics; many-body theory.

McLeod, Donald W., PhD, Cornell, 1962. Experimental high-energy physics.

McNeil, Edward, PhD, University of Illinois-Urbana, 1951.

Montano, Pedro, D.Sc., Technion, 1972. Synchrotron radiation science.

Pagnamenta, Antonio, PhD, Maryland, 1965. Theoretical high-energy physics; microdosimetry; radiation physics.

Sharma, Ram R., PhD, California, Riverside, 1965. Theoretical solid state physics; biophysics.

Solomon, Julius, PhD, California, Berkeley, 1963. Experimental high-energy physics.

Sukhatme, Uday, Sc.D., MIT, 1971. Theoretical high-energy particle physics.

RESEARCH SPECIALTIES AND STAFF

Atomic, Molecular, and Laser Physics. X-ray Microimaging and Bioinformatics. Advanced forms of x-ray generation for biological x-ray mcireholography and the application of 3D holographic imaging for studies of melanoma tumors. Multidisciplinary activity in computational physics which incorporates (1) high power density plasmas, (2) cosmology and particle physics, (3) protein structure and dynamics, and (4) the use of cryptographic procedures for the organization of bioinformatic data. **Experimental and Theoretical:** Rhodes. **Research staff:** Borisov, Racz.

Condensed Matter Physics. Metals, semiconductors, and insulators; Density functional electronic structure calculations; cooperative and critical phenomena and phase transitions; Density functional electronic structure calculations; magnetism in disordered systems; structural instabilities; high-T_c superconductivity; surfaces and thin films; ion implantation; thermodynamic and transport properties; optical properties from Raman scattering, ellipsometry, electrore-flectance, and photocapacitance; growth by molecular beam epitaxy of II–VI semiconducting epilayers and microstructures such as superlattices and tunneling structures; electronic properties of two-dimensional systems (Shubnikov-Dehaas; Quantum Hall Effect); processing and physics of electronic devices. **Experimental:** Campuzano, Klie, Schlossman, Schroeder, Sivananthan. **Theoretical:** Grein, Morr, Ogut. **Research staff:** Chang, Hahn, Tanase.

High-Energy Physics. Collider physics at the D0 Experiment at Fermilab, precision measurements of strong and electroweak interactions, higgs boson and new particle searches, top-quark physics, trigger systems development, silicon microstrip tracking detectros; trigger systems and silicon tracker development for CMS detector as CERN LHC; strings and integrable models, standard model phenomenology, strong and electroweak gauge interactions, CP violation, algebraic and topological aspects of quantum field theory, exotic statistics. **Experimental Faculty:** Adams, Cavanaugh, Gerber, Varelas. **Theoretical Faculty:** Aratyn, Imbo, Keung. **Research Staff:** Apanasevich, Strom.

Biophysics. Theoretical study of protein and nucleic acid structure and dynamics; mechanical properties of biopolymers and biomembranes. **Experimental:** Ansari, Perez-Salas, Schlossman. **Research Staff:** Kouznetzov, Narayanan.

Nuclear Physics. Relativistic heavy ion collisions; dense nuclear matter; Research at RHIC (BNL) and LHC (CERN). **Experimental:** Barranikova, Betts, Halliwell, Hofman.**Theoretical:** Stephanov. **Research Staff:** Apanasevich, García Solís, Hohler, Strom, Yang.

UNIVERSITY OF ILLINOIS AT URBANA-CHAMPAIGN

DEPARTMENT OF PHYSICS
LOOMIS LABORATORY OF PHYSICS

1110 West Green Street
Urbana, Illinois 61801-3080
http://physics.illinois.edu

Students Accepted For Degree	FIELDS		
	Physics	Astronomy	Related Fields
Doctorate	X		
Master's	X		

1. General

President: Michael J. Hogan
Head of Physics: Dale J. Van Harlingen
Assoc. Head of Graduate Programs: John D. Stack
Department Telephone Number: (217) 333-3761
Type of Institution: University
Control: Public
Setting: Suburban
Total Faculty: 3,078
Total Graduate Faculty: 2,736
Total Students: 41,918
Total Graduate Students: 10,709
Annual Graduate Tuition:
In-state residents: Full-time—$12,656
Out-of-state residents: Full-time—$25,922
Tuition rates for: 2009–10
Deferred tuition plan: Yes
Other Fees: $2872 (includes health insurance)
Term: Semester

2. Number of Faculty in Department

The combined total of full-time faculty in the three professorial ranks is 58. The combined total of full-time, part-time, and other faculty at all ranks is 60.

3. Admission, Financial Aid, and Housing

Address admission inquiries to: Graduate Records Secretary, Department of Physics, 1110 W. Green St., Urbana, IL 61801-3080 or through our website. General information and application forms are available at http://physics.illinois.edu.
Graduate application fee required: $60 for domestic applicants $75 for foreign applicants
Admission deadline (Fall admission): 1/15 (10/15 for Spring admission)
Admission information: For 2010 fall admission, 42 students were admitted from 580 applicants.
Admission requirements: For admission to the graduate programs, a bachelor's degree in physics or a related field is required with a minimum undergraduate GPA of 3.0/4.0. On the last 60 hours of work, 20 semester hours of physics beyond elementary physics is also required. The GRE is required. No definite minimum score is set. The average GRE Physics subject score for 2010 admissions was 836. Students from non-English speaking countries are required to demonstrate proficiency in English via the TOEFL or ILETS exam. Exams for admission required of non-native-speaking teaching assistants are the Test of Spoken English (TSE) and the internet-based TOEFL speaking section (minimum score of 24).
Undergraduate preparation assumed: Although preparation will vary, we generally expect one year of upper division mechanics, one year of electricity and magnetism, one semester of optics, one semester of statistical and thermal physics, and one year of quantum mechanics. Also, one or two semesters of laboratory courses are expected.
Address financial aid inquiries to: Graduate Records Secretary, Department of Physics, 1110 W. Green St., Urbana, IL 61801-3080.
GAPSFAS application required: No
Financial aid deadline: 1/15
Loans available: Yes
Address housing inquiries to:
http://housing.illinois.edu. For information by mail: Housing Division, 200 Clark Hall, 1203 S. Fourth, Champaign, IL 61820
On-campus graduate student housing available: Yes
Cost/term: $4,838–6,388/acad. yr.
On-campus, married student housing available: Yes
Cost/month: $598–800/mo

4. Graduate Degree Requirements

Master's: See Academic Information at www address. 32 hours[1] of satisfactory (GPA 2.75/4.0) graduate course work required. All hours must be at the 400-level or higher. 16 of the 32 hours must be in Physics, with at least 8 hours of them at the 500-level. At most, 8 hours of individual study may be counted toward the master's degree. At least 16 hours must be in courses meeting on the Urbana-Champaign campus; credit for graduate work taken elsewhere is by petition only. There is no foreign language requirement.
Doctorate: 96 hours[1] of (2.75/4.0 GPA) satisfactory graduate work. Part of these hours must be thesis work. There is no specific residence requirement, but 64 hours must be taken on the Urbana-Champaign campus. A qualifying examination is required, usually at the beginning of the second year; a preliminary examination is required, usually in the fifth or sixth semester. A thesis and a final examination on the thesis are required. There are no foreign language requirements.
Other Programs: The Medical Scholars Program, which allows students to earn joint MD/Ph.D. degrees, combines cutting edge research in physics with individualized clinical training in medicine. All graduate and medical training is done at the Urbana-Champaign campus. Only US citizens and permanent residents are eligible for admission.
Thesis: Theses may be written *in absentia*.
Special Equipment, Facilities, or Programs: The Department of Physics offers world-class research facilities in many research areas. For a complete description of physics facilities, please consult our website.

Table B—Appointments to Graduate Students, 2009–10

Title of Appointee	Appointments Total	Appointments First year	Academic Load Allowed in Credit Hours	Hours of Service Per Week	Stipend for Academic Year[a] (9 months)[b]
			Semester		
Teaching Assistant	119	24	14	20	16,362–17,055
Research Assistant	107	0	14	20	16,362–17,055
Fellowship	30	10	14	0	varies
Self-supporting (foreign government)	2	1	14	0	varies
Combination TA/RA	18	0	14	20	16,362–17,055
Total	276	35			

[a]For beginning students; stipends increase as students advance toward degree. For academic year 2009–10, the starting stipend for a nine-month academic year appointment is $16,362.
[b]Most students also receive a two-month ($3,636 for 2009) summer research or teaching appointment.

5. Personnel Engaged in Separately Budgeted Research, 7/09–6/10

Professorial faculty	57
Other faculty	8
Postdoctoral appointments	71
Graduate students	107
Undergraduate students	16
Nonteaching research personnel	24
Total	283

6. Separately Budgeted Research Expenditures by Source of Support

	Departmental Research	Physics-related Research Outside Department
Federal government	$20,915,000	$2,431,000
State and local government	74,000	–
Other Government	–	–
Private, non-profit organizations	928,000	–
Business and industry	90,000	–
Other	1,116,000	–
Total	$23,123,000	$2,431,000

7. Separately Funded and Managed Laboratories

Physics faculty-related research projects funded through the Materials Research Laboratory are included in Part 6 and Table C.

Table C—Separately Budgeted Research Expenditures

Research Specialty	No. of Grants	Expenditures ($)
Astrophysics	11	1,325,000
Atomic, Molecular, & Optical Physics	15	2,440,000
Biophysics	22	4,615,000
Condensed Matter Physics	46	5,905,000
Low Temperature Physics	13	2,582,000
Nuclear Physics	7	3,578,000
Particles & Fields	9	2,678,000
Total	123	23,123,000

Table D—Physics-related Research Outside Department

Field and Unit Outside Department	No. of Grants	Expenditures ($)
Astrophysics:		
Astronomy	2	88,000
Biophysics:		
Beckman Institute	4	1,136,000
Institute for Genomic Biology	3	629,000
Chemistry	1	46,000
Condensed Matter Physics:		
NCSA	2	392,000
Mechanical Science and Engineering	1	140,000
Total	13	2,431,000

FACULTY

Professors

Baym, Gordon, Ph.D., Harvard, 1960. George and Ann Fisher Distinguished Professor of Engineering and Center for Advanced Study Professor of Physics. Bose–Einstein condensation in trapped atomic systems and excitons, superfluid helium, matter under extreme conditions, neutron stars.

Beck, Douglas H., Ph.D., MIT, 1986. Experimental nuclear and particle physics; nucleon structure; fundamental symmetries; electron dipole moments.

Ceperley, David M., Ph.D., Cornell, 1976. Founder Professor of Engineering and Center for Advanced Study Professor of Physics. Electronic structure, superfluidity, Monte Carlo methods, physics at high pressure.

Chiang, Tai-Chang, Ph.D., California, Berkeley, 1978. Experimental condensed matter physics; atomically uniform films; electronic properties of impurities, surfaces, and quantum structures.

Clegg, Robert M., Ph.D., Cornell, 1974. Experimental biological physics; dynamic, structural, and thermodynamic studies of functional biological systems.

Cooper, S. Lance, Ph.D., Illinois, 1988. Experimental condensed matter physics; optical spectroscopy; strongly correlated systems; superconductivity.

Eckstein, James N., Ph.D., Stanford, 1978. Experimental condensed matter physics; atomic layer-by-layer molecular beam epitaxy; colossal magnetoresistance.

El-Khadra, Aida X., Ph.D., UCLA, 1989. Theoretical high-energy physics; lattice field theory; quantum chromodynamics; phenomenology.

Errede, Steven M., Ph.D., Ohio State, 1981. Experimental high-energy physics; interactions of the electroweak gauge bosons; physics of music.

Flynn, C. Peter, Ph.D., Leeds, 1960. Experimental condensed matter physics; epitaxy; defects and diffusion; magnetism.

Fradkin, Eduardo H., Ph.D., Stanford, 1979. Quantum Hall effects, strongly correlated systems, superconductors, critical phenomena, disordered systems, field theory.

Gammie, Charles F., Ph.D., Princeton, 1992. Theoretical and computational astrophysics; star formation; planet formation; relativistic accretion flows.

Giannetta, Russell W., Ph.D., Cornell, 1980. Experimental condensed matter physics; superconductivity; magnetic resonance; nanostructures; organic superconductors.

Gladding, Gary E., Ph.D., Harvard, 1971. Associate Head for Undergraduate Programs. Experimental high-energy physics; mixing of charmed mesons; physics education research.

Goldbart, Paul M., Ph.D., Imperial College, London, 1985. Theoretical condensed matter physics, random systems (polymer networks and glasses); mesoscopic physics; superconductivity and superfluidity.

Goldenfeld, Nigel D., Ph.D., Cambridge, 1982. Swanlund Chair. Pattern formation; high-temperature superconductivity; statistical physics; biocomplexity, microbial ecology, evolution.

Gollin, George D., Ph.D., Princeton, 1980. Experimental high-energy physics; CP violation in K-decay; rare B decays.

Greene, Laura H., Ph.D., Cornell, 1984. Swanlund Chair and Center for Advanced Study Professor of Physics. Experimental condensed matter physics; thin-film growth and tunneling in novel superconducting materials.

Ha, Taekjip, Ph.D., Berkeley, 1996. Howard Hughes Medical Investigator. Experimental biological physics; single molecule fluorescence microscopy and spectroscopy; DNA protein interactions; molecular biology.

Kwiat, Paul G., Ph.D, California, Berkeley, 1993. Bardeen Chair of Physics. Experimental quantum optics; optical approaches to quantum information; foundations of quantum mechanics.

Lamb, Frederick K., D.Phil., Oxford, 1970. Fortner Chair for Theoretical Astrophysics. Theoretical astrophysics; plasma, magnetohydrodynamic, and high-energy processes.

Leggett, Anthony J., D.Phil., Oxford, 1964. John D. and Catherine T. MacArthur Chair. Center for Advanced Study Professor of Physics. Foundations of quantum mechanics; superfluidity; high-temperature superconductivity; Bose–Einstein condensation.

Leigh, Robert G., Ph.D., Texas, Austin, 1991. Theoretical high-energy physics; quantum field theory, supersymmetric gauge theory; superstring theory.

Liss, Tony M., Ph.D., California, Berkeley, 1984. Experimental high-energy physics; production and decay of the top quark.

Makins, Naomi C.R., Ph.D., MIT, 1994. Experimental nuclear physics; proton and neutron spin.

Mestre, Jose, Ph.D., U. Massachusetts, 1979. Physics education research; cognitive processes in learning; role and interaction of language in problem solving; educational technologies.

Mouschovias, Telemachos, Ph.D., California, Berkeley, 1974. Theoretical astrophysics; dynamics of interstellar clouds; stellar evolution.

Nayfeh, Munir H., Ph.D., Stanford, 1974. Atomic, molecular, and optical physics; laser atomic spectroscopy.

Oono, Yoshitsugu, Ph.D., Kyushu, 1976. Nonequilibrium statistical physics/dynamical systems; system reduction/asymptotic analysis, including reduction of large data sets.

Peng, Jen-Chieh, Ph.D., U. Pittsburgh, 1975. Experimental medium- and high-energy nuclear physics; parton structures of the nucleons and nuclei.

Phillips, Philip W., Ph.D., U. Washington, 1982. Strongly correlated electronic low-dimensional systems; quantum Hall effect; quantum critical phenomena; quantum magnetism.

Pitts, Kevin T., Ph.D., University of Oregon, 1994. Experimental high-energy physics; CP violation in bottom quark decays.

Schulten, Klaus J., Ph.D., Harvard, 1974. Swanlund Chair. Theoretical biophysics; physics of the living cell; computational physics.

Selen, Mats A., Ph.D., Princeton, 1988. Experimental high-energy physics; experimental astrophysics; decays of the charmed quark.

Selvin, Paul R., Ph.D., California, Berkeley, 1990. Experimental biological physics; structure and dynamics of biological macromolecules; fluorescence.

Shapiro, Stuart L., Ph.D., Princeton, 1973. Theoretical astrophysics and general relativity; physics of black holes and neutron stars; gravitational collapse; generation of gravitational waves.

Stack, John D., Ph.D., California, Berkeley, 1965. Associate Head for Graduate Programs. Theoretical physics.

Stone, Michael, Ph.D., Cambridge, 1976. Quantum Hall effect, superconductivity, and superfluidity.

Thaler, Jon J., Ph.D., Columbia, 1972. Observational cosmology, focusing on the properties of dark matter and dark energy, as well as neutrino masses and diverse phenomena.

Van Harlingen, Dale J., Ph.D., The Ohio State University, 1977. Dept. Head, Willett Professor of Engineering, and Center for Advanced Study Professor of Physics. Experimental condensed matter physics; superconductivity; superconductor device physics.

Weaver, Richard L., Ph.D., Cornell University, 1977. Condensed matter physics; stochastic waves, disordered and complex structures, quantum chaos, random matrix theory, ultrasonics, structural acoustics.

Willenbrock, Scott S., Ph.D., Texas, Austin, 1986. Theoretical high-energy physics; phenomenology; electroweak symmetry breaking; top quark physics; Higgs phenomena.

Wiss, James E., Ph.D., California, Berkeley, 1977. Experimental high-energy physics; photoproduction of charmed mesons; precision study of the B meson.

Associate Professors

Abbamonte, Peter, Ph.D., U. of Illinois at Urbana-Champaign, 1999. Experimental condensed matter physics; resonant soft x-ray scattering; electron self-organization; oxide devices; quantum phase transitions; collective excitations.

Bezryadin, Alexey, Ph.D., J. Fourier Université, 1995. Experimental condensed matter physics; nanometer-scale mescopic physics and molecular electronics; quantum phase transitions.

Dahmen, Karin A., Ph.D., Cornell, 1995. Nonequilibrium dynamical systems; hysteresis; avalanches; earthquakes; population biology; disorder-induced critical behavior.

DeMarco, Brian, Ph.D., U. Colorado, 2001. Experimental atomic, molecular, and optical physics; quantum information science; atomic Bose–Einstein condensates and Fermi gases.

Errede, Deborah M., Ph.D., Michigan, 1987. High energy particle physics; precision measurements of the W mass.

Grosse Perdekamp, Matthias, Ph.D., California, Los Angeles, 1995. Experimental high-energy nuclear physics; nucleon structure; spin-dependent structure of the proton; quark transversity and fragmentation functions.

Hubler, Alfred W., Ph.D., Munich, 1987. Nonlinear dynamics and complex systems.

Lamb, Susan A., D. Phil., Oxford University, 1973. Theoretical astrophysics, computational astrophysics; galaxy collisions and star formation.

Assistant Professors

Aksimentiev, Aleksei, Ph.D., Institute of Physical Chemistry, Warsaw, Poland, 1999. Theoretical and computational biophysics; biomolecular modeling, molecular motors, mechanical proteins, silicon biotechnology, membrane transport.

Budakian, Raffi O., Ph.D., California, Los Angeles, 2000. Experimental condensed matter physics; magnetic resonance force microscopy; micro- and nanomechanical devices.

Chemla, Yann, R., Ph.D., California, Berkeley, 2001. Experimental biological physics; molecular motors; nucleic acid and protein translocases.

Lev, Benjamin, Ph.D., Caltech, 2005. Experimental ultracold atomic and molecular physics; quantum optics; quantum information science.

Mason, Nadya, Ph.D., Stanford University, 2001. Experimental condensed matter physics; quantum properties of carbon nanotubes, superconductivity, quantum phase transitions.

Neubauer, Mark, Ph.D., Pennsylvania, 2001. Experimental particle physics; particle astrophysics; neutrino physics; heavy flavor physics; Higgs boson; electroweak diboson physics.

Vishveshwara, Smitha, Ph.D., California, Santa Barbara, 2002. Theoretical condensed matter physics; strongly correlated systems; localization physics; phase transitions; superconductivity.

Emeriti

Martin, Richard M., Ph.D., Chicago, 1969. Electronic structure; density functional theory; dielectric phenomena; simulations using molecular dynamics and Monte Carlo methods.

Nathan, Alan M., Ph.D., Princeton, 1975. Experimental nuclear physics, physics of baseball.

Slichter, Charles P., Ph.D., Harvard, 1949. Center for Advanced Study Professor of Physics and Chemistry; research professor of physics. Experimental condensed matter physics; nuclear magnetic resonance.

Weissman, Michael B., Ph.D., California, San Diego, 1976. Experimental condensed matter physics; $1/f$ noise, spin glasses, amorphous materials.

RESEARCH SPECIALTIES AND STAFF

Theoretical

Astrophysics and General Relativity. Baym, Gammie, F. Lamb, S. Lamb, Mouschovias, Shapiro.

Biophysics. Aksimentiev, Schulten.

Complex Systems and Nonlinear Dynamics. Dahmen, Goldbart, Goldenfeld, Hubler, Oono, R. Weaver.

Condensed Matter Physics. Baym, Ceperley, Dahmen, Fradkin, Goldbart, Goldenfeld, Leggett, Martin, Oono, Phillips, Stone, Vishveshwara, R. Weaver.

Nuclear Physics. Baym.

Particle Physics and Cosmology. El-Khadra, Leigh, Stack, Stelzer, Willenbrock.

Experimental

Astrophysics. Selen and Thaler.

Atomic, Molecular, Optical, and Quantum-Information Physics. DeMarco, Kwiat, Lev, Nayfeh.

Biological Physics. Chemla, Clegg, Ha, Selvin.

Condensed Matter Physics. Abbamonte, Bezryadin, Budakian, Chiang, Cooper, Eckstein, Flynn, Giannetta, Greene, Mason, Nayfeh, Slichter, Van Harlingen, Weissman.

Nuclear Physics. Beck, Grosse Perdekamp, Makins, Nathan, and Peng.

Particle Physics. D. Errede, S. Errede, Gollin, Liss, Neubauer, Pitts, Selen, and Wiss.

Physics Education. Gladding, Mestre, Selen, Stelzer.

Statistical Physics. Goldenfeld, Weissman.

FACULTY PUBLICATIONS

In 2008–2009, faculty in the Physics Department published 387 journal articles, 8 book chapters, 1 book, and 4 videos.

UNIVERSITY OF ILLINOIS, URBANA-CHAMPAIGN

DEPARTMENT OF ASTRONOMY

Urbana, Illinois 61801

Students Accepted For Degree	FIELDS		
	Physics	Astronomy	Related Fields
Doctorate		X	
Master's		X	

1. General

President: Michael Hogan
Dean of Graduate School: Debasish Dutta
Department Chair: You-Hua Chu
Department Telephone Number: (217) 333-3090
Type of Institution: University
Control: Public
Setting: Suburban
Total Faculty: 2,226
Total Graduate Faculty: 2,061
Total Students: 41,918
Total Graduate Students: 9,630
Annual Graduate Tuition:
In-state residents: Full-time—$9,318
Out-of-state residents: Full-time—$22,584
Tuition rates for: 2009–10
Deferred tuition plan: Yes
Other Fees: $3,338
Term: Semester

2. Number of Faculty in Department

The total of permanent faculty in the three professorial ranks is 15. The total of full-time, part-time, visiting, research, and other faculty at all ranks is 24.

3. Admission, Financial Aid, and Housing

Address admission inquiries to: Admissions Officer, Department of Astronomy, 103 Astronomy Building, 1002 W. Green, Urbana, IL 61801, astronomy@illinois.edu
Graduate application fee required: $60 Domestic, $75 Foreign
Admission deadline (Fall admission): 1/15
Admission information: For fall semester, 2009, 5 students were admitted.
Admission requirements: For admission to the graduate programs, a Bachelors degree is required with a minimum undergraduate GPA of 3.0 on a scale of 4.0 equals A letter grade. The GRE is required. The GRE Advanced Test in Physics is required. Students from non-English speaking countries are required to demonstrate proficiency in English via the TOEFL exam. Minimum acceptable score for admission is 550. Other exams for admission required of foreign teaching assistants are the Test of Spoken English (TSE) or the local SPEAK examination. The policy of the University of Illinois is to comply with all applicable Federal and State Nondiscrimination and Equal Opportunity Laws, Orders, and Regulations. The University of Illinois will not discriminate in its programs and activities against any person because of race, color, national origin, religion, age, sex, handicap, or status as disabled veteran or veteran of the Vietnam era. This nondiscrimination policy applies to admissions, employment, and access to and treatment in University programs and activities.
Address financial aid inquiries to: Admissions Officer, Department of Astronomy, astronomy@illinois.edu
GAPSFAS application required: No
Financial aid deadline: 1/15
Loans available: Yes
Address housing inquiries to: University Housing, http://www.housing.illinois.edu/
On-campus, single student housing available: Yes
Cost/school year: $4,446–6,106/acad. yr.
On-campus, married student housing available: Yes
Cost/month: $584–807/mo.

Table A—Faculty, Enrollments, and Degrees Granted

Research Specialty	2009–10 Faculty	Enrollment[1] Fall 2009		No. of Degrees Granted[2] 2009–10 (2005–10)			Median No. of Years for 2009–10 Ph.D.'s
		Master's	Doctorate	Master's	Terminal Master's	Doctorate	
Astronomy & Astrophysics	17	2	32	2(6)	1(3)	4(19)	5.75
Full-time Grad. Stud.		3	32				
Part-time Grad. Stud.		0	0				
First-year Grad. Stud.		2	3				

Undergraduate Degrees, 2008–09 (2004–09): 8(28)

[1] Students not yet committed to a research specialty are entered under Undecided.
[2] Five-year totals in parentheses.

Table B—Appointments to Graduate Students, 2008–09

Title of Appointee	Appointments Total	First year	Academic Load Allowed in Credit Hours	Hours of Service Per Week	Stipend for Academic Year ($)[1,2]
			Semester		
Fellow	2	0	16/semester	–	17,500-22,500[1,2]
Teaching Assistant	9	3	16/semester	~20	16,317-16,803[1,3]
Research Assistant	22	1	16/semester	~20	16,317-16,803[1,3]
Total	33	4			

[1] Tuition and most student fees waived.
[2] 12-month appointment.
[3] 9-month appointment. Additional summer appointments also available.

4. Graduate Degree Requirements

Master's: 32 credit hours of study in graduate courses are required. There are no thesis or foreign language requirements. Minimum GPA is 3.0/4.0. Further information on exact degree requirements may be obtained from the department.
Doctorate: The Ph.D. degree requires completion of 96 graduate hours in courses in astronomy and related fields (at least 32 of which involve individual study and research), satisfactory performance on a general qualifying examination no later than the beginning of the third year of study, and completion of an original research project culminating in a thesis publishable in whole or in part. Minimum GPA is 3.0/4.0. Thesis may be written in absentia. Further information on exact degree requirements may be obtained from the department.
Special Equipment, Facilities, or Programs: Research Facilities:

The Combined Array for Research in Millimeter-wave Astronomy (CARMA) is operated jointly by a consortium of the California Institute of Technology, the University of California at Berkeley, the University of Illinois, and the University of Maryland, with the partner Universities having guaranteed observing time. The array operates at wavelengths of 1 to 4 mm and consists of ten 6.1-m and six 10.4-m antennas located at Cedar Flat, in the Inyo Mountains of California. CARMA may be used in a wide variety of observational areas, including solar flares, comets, molecular clouds, astrochemistry, star formation, structure of the Milky Way and external galaxies, and cosmology.

The University of Illinois has established the National Center for Supercomputing Applications (NCSA), which hosts some of the most powerful computing systems in the world including a new university-supported 100 Teraflop system (approximately half of the time on this system is reserved for Illinois faculty). Access for large users is by peer review; however, development allocations are directly available for new researchers and graduate students. With these resources, the University of Illinois is a major center for the study of stellar formation and evolution, nucleosynthesis, relativistic astrophysics, quasars, structure formation, and cosmology.

FACULTY

Professors

Chu, You-Hua, Ph.D., California, Berkeley, 1981. Chair of the Department. Interactions between stars and the interstellar medium; physical structures of the interstellar medium of the Magellanic Clouds; space observations using HST, Chandra, Spitzer, XMM-Newton, and FUSE.

Crutcher, Richard M., Ph.D., UCLA, 1972. Research Scientist, NCSA. Radio astronomy; star formation; interstellar magnetic fields; supercomputer data processing; advanced scientific visualization.

Gammie, Charles F., Ph.D., Princeton, 1992. Theoretical and computational astrophysics; black hole astrophysics; star formation; planet formation.

Iben, Icko, Jr., Ph.D., Illinois, 1958. Emeritus. Theoretical astrophysics; stellar structure and evolution; stellar pulsation; nucleosynthesis; binary stars; novae.

Kaler, James B., Ph.D., UCLA, 1964. Emeritus. Planetary and diffuse nebulae; nebular spectroscopy and photometry; symbiotic stars; education.

Lamb, Frederick K., D.Phil., Oxford, 1970. Theoretical astrophysics; plasma, magnetohydrodynamic, and high-energy processes; physics of magnetic white dwarfs, neutron stars, black holes, pulsars, x-ray sources, active galaxies, and quasars.

Mouschovias, Telemachos Ch., Ph.D., California, Berkeley, 1975. Theoretical astrophysics; star formation; interstellar medium; interstellar gas dynamics; magnetohydrodynamics; solar corona; numerical astrophysical fluid dynamics.

Olson, Edward C., Ph.D., Indiana, 1961. Emeritus. Photometry and spectroscopy of interacting binary stars; theoretical continuum and line models as diagnostics of the structure of large accretion disks.

Shapiro, Stuart L., Ph.D., Princeton, 1973. Theoretical astrophysics and general relativity; physics of compact objects (black holes, neutron stars); computational astrophysics and relativity; stellar dynamics.

Snyder, Lewis E., Ph.D., Michigan State, 1967. Emeritus. Molecular astronomy; radio detection of new interstellar and circumstellar molecules; observational studies of maser sources and interstellar clouds; detection and observational studies of cometary molecules.

Thaler, Jon J., Ph.D., Columbia, 1972. Particle physics; dark energy, dark matter, and cosmology.

Thompson, Laird A., Ph.D., Arizona, 1974. Extragalactic studies of the large-scale distribution of galaxies and of galaxy evolution; instrumentation in active/adaptive optical systems for high-resolution imaging from ground-based telescopes.

Webbink, Ronald F., Ph.D., Cambridge, 1975. Emeritus Theoretical astrophysics; close binary systems: structure, evolution, and stability; gravitational wave astronomy.

Associate Professors

Brunner, Robert J., Ph.D., Johns Hopkins University, 1998. Cosmology and extragalactic astronomy: structure formation and evolution, quasars, cluster of galaxies; time domain astronomy, diffuse emission; astronomical data management and data mining.

Fields, Brian D., Ph.D., Chicago, 1994. Cosmology; primordial nucleosynthesis; early universe, dark matter; nuclear astrophysics; cosmic rays; gamma-ray astrophysics; galactic chemical evolution; neutron capture processes; supernovae.

Kemball, Athol, J., Ph.D., Rhodes University, South Africa, 1992. Radio astronomy; high-performance computing; interferometry; late-type evolved stars; gravitational lensing.

Lamb, Susan A., D.Phil., Oxford, England, 1974. Extragalactic studies of interacting and merging galaxies: multiwavelength observations including infrared, UV, and radio; global star formation and stellar population synthesis; N-body and hydrodynamic computer modeling of galaxy collisions; dark matter.

Looney, Leslie L., Ph.D., Maryland, 1998. Radio astronomy; millimeter interferometry techniques; star formation: high spatial resolution imaging, radiative transfer of the inner envelope and circumstellar disk; astronomical instrumentation: millimeter (BIMA and CARMA) and far-infrared (SOFIA/ FIFI LS).

Ricker, Paul M., Ph.D., Chicago, 1996. Computational astrophysics and cosmology; structure formation; galaxy cluster evolution; galaxy formation; N-body simulation; computational fluid dynamics; adaptive mesh refinement.

Sutton, Edmund C., Ph.D., California, Berkeley, 1979. Observational molecular astronomy; physics and chemistry of the interstellar medium; molecular clouds and star-forming regions; instrumentation for millimeter and submillimeter wavelengths.

Assistant Professors

McCall, Benjamin J., Ph.D., Chicago, 2001. Interstellar diffuse bands, infrared astronomy, high-resolution molecular spectroscopy, and interstellar chemistry.

Wong, Tony, Ph.D., California, Berkeley, 2000. Observations of the interstellar medium and star formation in the Milky Way and nearby galaxies; structure and dynamics of disk galaxies; single-dish and interferometric methods in radio astronomy.

FACULTY PUBLICATIONS

A list of faculty publications is available at the department website (http://www.astro.illinois.edu/research/publications/).

BALL STATE UNIVERSITY

DEPARTMENT OF PHYSICS AND ASTRONOMY

Muncie, Indiana 47306

Students Accepted For Degree	FIELDS		
	Physics	Astronomy	Related Fields
Doctorate			X
Master's	X		

1. General

President: Joanne Gora
Dean of Graduate School: Robert Morris
Department Chairman: Thomas Robertson
Department Telephone Number: (765) 285-8860
Type of Institution: University
Control: Public
Setting: Suburban
Total Faculty: 940
Total Graduate Faculty: 973
Total Students: 21,401
Total Graduate Students: 3,664
Annual Graduate Tuition:
In-state residents: Full-time—$7,508 (12–24 hrs.)
Part-time—[1]
Out-of-state residents: Full-time—$20,960 (12–18 hrs.)
Part-time—[1]
Tuition rates for: 2010–11
Deferred tuition plan: No
Other Fees: Approximately $786[2], $80[3], $79[4], $184[5], $90[6]/semester
Term: Semester

[1]Depends on number of sem. hrs. taken in a given semester.
[2]General fee; [3]Course fee; [4]Health fee; [5]Technology fee; [6]Recreation Fee.

2. Number of Faculty in Department

The combined total of full-time faculty in the three professorial ranks is 15. The combined total of full-time, part-time, and other faculty at all ranks is 15.

3. Admission, Financial Aid, and Housing

Address admission inquiries to: Chair, Graduate Selection Committee, Department of Physics and Astronomy
Graduate application fee required: $50
Admission deadline (Fall admission): open
Admission information: For fall admission, 2009–10, 17 students were accepted from 17 applicants.
Admission requirements: Hold Bachelor's degree with a physics major or minor or equivalent from accredited college or university; have minimum 2.75 (4.0) cumulative undergraduate GPA or 3.0 (4.0) cumulative GPA for last two years of undergraduate work; submit scores for Graduate Record Exam General Test or minimum score of 550 on TOEFL exam, if international student from non-English speaking country.
Undergraduate preparation assumed: Fowles, *Mechanics*; Krane, *Modern Physics*; Reitz and Milford, *Electricity and Magnetism*; Pedrotti and Pedrotti, *Optics*; Sears, *Thermodynamics*.
Address financial aid inquiries to: Chair, Graduate Selection Committee, Department of Physics and Astronomy
GAPSFAS application required: No
Financial aid deadline: Open
Loans available: Yes
Address housing inquiries to: Housing Office
On-campus, single student housing available: Yes
Cost/term: $8,438–10,894 per academic year (21 meals/wk.)
On-campus, married student housing available: Yes
Cost/month: $471–660

Table A—Faculty, Enrollments, and Degrees Granted

Research Specialty	2009–10 Faculty	Enrollment[1] Fall 2009		No. of Degrees Granted[2] 2009–10 (2005–10)			Median No. of Years for 2009–10 Ph.D.'s
		Master's	Doctorate	Master's	Terminal Master's	Doctorate	
Astronomy	4	5	–	2(7)	–	–	–
Medical Physics	2	3	–	3(13)	–	–	–
Nano-Science	6	4		4(14)	–	–	–
Nuclear Physics	1	0	–	0(0)	–	–	–
Particle Physics	1	1	–	1(1)	–	–	–
Physics Education	1	0	–	0(10)	–	–	–
Non-Specialized	0	0	–	0(0)	–	–	–
Total	15	13	–	10(45)	–	–	–
Full-time Grad. Stud.		13	–				
Part-time Grad. Stud.		0	–				
First-year Grad. Stud.		3	–				
Median Years in Grad. Study (2009–10 Degrees)				2	–	–	–

Undergraduate Degrees, 2009–10 (2005–10): 8(37)

[1]Students not yet committed to a research specialty are entered under non-specialized.
[2]Five-year totals in parentheses.

4. Graduate Degree Requirements

Doctorate: Ed.D. in Science and Ed.D. in Science Education; 90 semester hours of graduate study which may include one's masters degree work and other graduate work; a minimum of 48 hours must be completed at Ball State.
Master's: M.S.-33 sem. hrs.; 3.0 GPA; formal thesis; oral defense of thesis; 22 sem. hrs. in residency; no language requirement. M.A.-Same as above but with 3 sem. hr. research paper substituting for 6 sem. hr. thesis.
Thesis: Thesis may be written *in absentia*.
Special Equipment, Facilities, or Programs: Research with Notre Dame University Accelerator Group. Cooperative research with Fermi Lab. IBM mainframe, mini- and micro-computing systems available. Center for Computatioinal Nanoscience. Computer-based instrumentation design. Radon Study, Survey Lab, and CCD astronomical image processing.

Table B—Appointments to Graduate Students, 2009–10

Title of Appointee	Appointments Total	First year	Academic Load Allowed in Credit Hours	Hours of Service Per Week	Stipend for Academic Year ($)
			Semester		
Teaching Assistant	8	2	12	20	12,162[1]
Research Assistant	2	0	12	20	12,162[1]
Total	10	2			

[1]Plus full tuition waiver.

5. Personnel Engaged in Separately Budgeted Research, 7/09–6/10

Professorial faculty	5
Other faculty	0
Total	5

6. Separately Budgeted Research Expenditures by Source of Support

	Departmental Research	Physics-related Research Outside Department
Federal government	$50,000	$
Other	85,804	
Total	$135,804	$

7. Extension Centers and Summer Programs

New Hampshire Academy of Applied Science	$5,200
Total	$5,200

FACULTY

Professors

Joe, Yong Suk, Ph.D., Ohio University, 1993. Theoretical condensed matter.

Kaitchuck, Ron, Ph.D., Indiana Univ., 1981. Spectroscopy and photometry of interacting binary stars.

Khatun, Mahfuza, Ph.D., Ohio University, 1985. Theoretical condensed matter.

Robertson, Thomas H., Ph.D., Case Western Reserve, 1978. Galactic structure and kinematics; observational stellar astronomy.

Faculty Emeriti

Cosby, Ronald M., Ph.D., Kentucky, 1971. Semiconductor physics; photovoltaics; solar energy.

Errington, Paul R., Ph.D., West Virginia, 1966. Electronics; microprocessor applications.

Koltenbah, David E., Ph.D., Kent State, 1968. Chemical Physics.

Ober, David R., Ph.D., Purdue, 1968. Chairman of the Department. Nuclear spectroscopy; medium-energy nuclear reactions; radon study.

Place, Ralph L., Ph.D., Kentucky, 1968. Microprocessor applications, expert systems, and computor vision.

Thomas, Gerald P., Ph.D., SUNY, Buffalo, 1968. Elementary particle physics.

Associate Professors

Grosnick, David, Ph.D., University of Chicago, 1986. Elementary Particle Physics.

Islam, Saiful, Ph.D., Ohio University, 1986. Experimental nuclear physics.

Jin, Feng, Ph.D., Wayne State, 1998. Electrical engineering.

Jordan, Thomas, Ph.D., Oklahoma State, 1979. Stellar atmospheres; astrophotography.

Assistant Professors

Berrington, Robert, Ph.D., Indiana Univ., 2000. Numerical and observational extragalactic astronomy, structure and evolution of galaxies, cosmology.

Bryan, Joel A., Ph.D., Texas A & M, 2003. Curriculum and Instruction, Science Education.

Cancio, Antonio, Ph.D., U. of Illinois, 1994. Theoretical condensed matter.

Hedin, Eric, Ph.D., U. of Washington, 1986. Plasma physics.

Maqbool, Mukammad, Ph.D., Ohio University. 2005. Experimental Condensed Matter Physics.

Schmidt, Paul, Ph.D., U. of Georgia, 2006. Spectroscopy of Inorganic Phosphors.

Wijesinghe, Ranjith, Ph.D., Vanderbilt, 1988. Medical physics.

RESEARCH SPECIALTIES AND STAFF

Experimental

Astrophysics. Observational stellar astronomy; galactic structure; extragalactic numerical and observational, CCD imaging; astronomy education; interacting binary stars. Berrington, Jordan, Kaitchuck, and Robertson.

High Energy Nuclear/Particle Physics. Elementary particles (Brookhaven, Fermi, and Argonne National Laboratories). Grosnick

Medical Physics. Biomedical Physics; EEG and MEG imaging; magnetic signals in nerves; radiation treatment planning. Maqbool, and Wijesinghe.

Nanomaterials and Devices. Electronic and photonic materials and devices; gas discharge physics and devices. Jin, Maqbool.

Nuclear Physics. Nuclear structure and nuclear reactions for stellar processes. Islam.

Physics Education Research. Woodrow Wilson Fellowship PhysTEC Noyce. Bryan and Grosnick.

Theoretical

Condensed Matter Theory. Optical properties of semiconductors and nanostructures; density functional theory; nanoscale systems; nanoscience/nanotechnology education and training; optical properties and electron transport in coupled quantum dots; electron transport in nanostructures; molecular nanoelectronics, and quantum cellular automata. Cancio, Hedin, Joe, and Khatun.

Statistical Physics. Low dimensional lattice models. Khatun.

INDIANA UNIVERSITY, BLOOMINGTON

DEPARTMENT OF PHYSICS

Bloomington, Indiana 47405-7105
http://www.iub.edu/~iubphys

Students Accepted For Degree	FIELDS		
	Physics	Astronomy	Related Fields
Doctorate	X		X
Master's	X		

1. General

President: Michael A. McRobbie
Vice-President: (Bloomington)
Dean of Graduate School: James Wimbush
Department Chairman: Richard J. Van Kooten
Department Telephone Number: (812) 855-1247
Type of Institution: University
Control: Public
Setting: Small town
Total Faculty: 1,368*
Total Graduate Faculty: 1,557*
Total Students: 42,347*
Total Graduate Students: 9,857*†
Annual Graduate Tuition:
In-state residents: Full-time—$291.97/credit hr.
Out-of-state residents: Full-time—$850.33/credit hr.
Tuition rates for: 2010–2011
Deferred tuition plan: No
Other Fees: $900 per semester
Term: Semester

*Bloomington Campus
†Including professional schools

2. Number of Faculty in Department

The combined total of full-time faculty in the three professorial ranks is 36. The combined total of full-time, part-time, and other faculty at all ranks is 36.

3. Admission, Financial Aid, and Housing

Address admission inquiries to: Chairman, Graduate Admissions, Dept. of Physics
Graduate application fee required: $55 (domestic); $65 (international)
Admission deadline (Fall admission): 1/15 (domestic) 12/1 (international)
Admission information: For fall admission, 2009–10, 57 students were offered admission from 164 applicants.
Admission requirements: For admission to the graduate programs, a Bachelor's degree in physics is required with a minimum undergraduate GPA of 3.0. The GRE required for applicants who wish to be considered and the GRE Physics Subject Exam 15 strongly recommended. Non-English speaking students countries are required to demonstrate proficiency in English via the TOEFL exam. The minimum score for admission is 550 paper, 213 computer based, 80 Internet based.
Address financial aid inquiries to: Chairman, Graduate Admissions, Physics Dept.
Loans available: Yes
Address housing inquiries to: Halls of Residence, 801 N. Jordan
On-campus, single student housing available: Yes
Cost/academic year: $5,026–6,660 (single rm.) $4,370–5,018 (double rm.)
On-campus, apartment student housing available: Yes
Cost/month: $529–1,099 (furnished and unfurnished)

Table A—Faculty, Enrollments, and Degrees Granted

Research Specialty	2009–10 Faculty	Enrollment[1] Fall 2009		No. of Degrees Granted[2] 2009–10 (2005–10)			Median No. of Years for 2009–10 Ph.D.'s
		Master's	Doctorate	Master's	Terminal Master's	Doctorate	
Astrophysics			1	0(0)	0(0)	0(3)	–
Biophysics			18	0(0)	0(0)	2(5)	7
Chemical Physics			0	0(0)	0(0)	0(2)	–
Condensed Matter Physics			7	0(0)	0(0)	1(12)	6
Nuclear Physics			14	0(0)	0(0)	2(14)	9
Particles & Fields			8	0(0)	0(0)	3(14)	6
Physics Education		1	0	0(0)	0(0)	0(0)	–
Physics of Beams		7	4	0(0)	2(2)	3(12)	4
Other Theoretical/ Math.			2	0(0)	0(0)	1(2)	5
Non-specialized			19	7(44)	1(11)	0(0)	–
Total		8	73	7(44)	3(13)	12(64)	6
Full-time Grad. Stud.		1	73	7			
Part-time Grad. Stud.		8	0	0			
First-year Grad. Stud.		2	12	0			
Median Years in Grad. Study (2009–10 Degrees)		4 yr	3.5 yr	2 yr			

Undergraduate Degrees, 2008–09 (2004–09): 20(69)

[1]Students not yet committed to a research specialty are entered under non-specialized.
[2]Five-year totals in parentheses.

4. Graduate Degree Requirements

http://www.iub.edu/~iub.phys/current/degreeinfo.pdf
Master's: 30 semester hours, at least 20 in physics, 14 of which must be in courses numbered P501 and higher, passed with an average grade of "B" or higher. Physics courses numbered below P501, and passed with a grade of "B⁻" or lower do not count toward this degree. (Seminar, research, and reading courses may not be counted toward the 14 hour requirement.) Master's examination.
Master's in Beam Physics and Technology: [A national program in collaboration with the U.S. Particle Accelerator School (USPAS)] 30 credit hours, including the following: P441 (or equivalent at another institution), P506 (or equivalent), P570, one course at the 500 level or above in laboratory techniques or computational methods, and a Master's thesis course (P802). Four advanced courses in beam physics should be chosen from among the Special Topics courses P571, P671, and P672, with topics to be listed in a syllabus prepared jointly by the I.U. Physics Department and the USPAS. A grade point average of 3.0 or better must be maintained in the courses satisfying the 30 credit-hour requirement. In particular, both P441 and P506 (or equivalents) must be passed with a grade B (3.0) or above. Thesis required. Either an oral defense of the thesis or a written final examination is required, and should take place at Indiana University. The writ-

ten examination may be substituted for the oral defense only with the permission of the thesis committee.

Doctorate: 90 semester hours in course, reading, and research credits; a minimum of 9 credit hours per semester at the P501 level or above with an average grade of "B" or higher (first-year students are allowed a minimum of 7 credit hours at the P501 level or above); minor requirement can be met either outside of Physics or within Physics but outside of student's area of thesis research; written qualifying exam; thesis; final oral exam; a minimum of two consecutive semesters in residence. All candidates are required to undertake supervised teaching as an Associate Instructor for at least one semester. All first time teaching Associate Instructors must enroll in a one-hour graduate credit course, "Practicum in Physics Laboratory." Associate Instructors whose native language is not English are required to take an "Associate Instructor English Exam," which they must pass in order to be qualified to teach. This exam must be passed by the end of the second year of study.

Other Programs: Master of Arts for Teachers: 36 credit hours with a minimum of 20 in physics. *Ph.D. in Astrophysics*: If in residence in the Physics Dept., a student must pass specifically designated parts of the qualifying examinations of both departments; thesis; final oral exam. *Ph.D., in Biophysics:* Students must pass specifically designated parts of the regular departmental qualifying exam, and a biophysics qualifying exam; thesis; final oral exam. *Ph.D. in Chemical Physics*: If in residence in the Physics Dept., same qualifying exam as above; minor in chemistry with eight hours in designated courses; thesis; final oral exam. *Ph.D. in Mathematical Physics*: If in residence in the Physics Dept., same qualifying exam as above, and a special qualifying examination in the Mathematics Department; thesis; final oral exam.

Thesis: Thesis may be written *in absentia*.

Special Equipment, Facilities, or Programs: There is a large joint library for astronomy, computer science, math, and physics in the same building. The Indiana University Cyclotron Facility (IUCF) is a multipurpose laboratory that supports basic research in nuclear, particle, accelerator, and condensed matter physics, and applied reseach in proton radiation effects and medical physics. The facility operates two coupled cyclotrons used primarily for radiation effects studies at the Radiation Effects Research program (RERP) and for treatment of cancer at the Midwest Proton Radiotherapy Institute (MPRI). Also at IUCF, the new Low Energy Neutron Source (LENS) is the first pulsed cold neutron source located at a university; it will provide cold neutrons for three beamlines for small-angle neutron scattering, neutron radiography, and neutron spin echo spectroscopy. Specialized shops for scintillator, wire chamber, and target fabrication are available, with capabilities for design, construction, and testing of large or complex detector and electronics systems. Research equipment in other areas includes facilities for construction and testing of instrumentation for high-energy physics experiments. A 192-node parallel PC cluster is available for research computing. The University provides extensive supercomputing support including an IBM SP cluster with 500 CPUs and access to the High Performance Storage System (HPSS). Condensed matter and low-temperature equipment include two x-ray diffraction systems, one with a high temperature (up to 1300°C) sample chamber; a multi-source high vacuum sputtering system; a 14T superconducting solenoid, other low temperature cryostats with 8T solenoids, two dilution refrigerators, a ^{3}He refrigerator; a helium liquefier; two Auger spectrometers; three low-energy electron diffraction apparatus (LEED); three electron energy low spectrometers (EELS); two scanning tunneling microscopes (STM); microwave network analyzer; a squid magnetometer. Facilities for biophysics research include cell culture and incubation labs, cell sorter, one-photon and two-photon scanning confocal microscopes, instrumentation for multielectrical array recording and general neurophysics instrumentation, as well as access to shared core facilities at the Indiana Molecular Biology Institute. An extensive machine shop now includes a programmable (CNC) milling machine and four full-time machinists.

Table B—Appointments to Graduate Students, 2008–09

Title of Appointee	Appointments		Academic Load Allowed in Credit Hours	Hours of Service Per Week	Stipend for Academic Year ($)
	Total	First year			
			Semester		
Teaching Assistant	28	10	12	20	1,550/mo.
Research Assistant	41	2	12	20	1,800/mo.
Other (specify)	1	0			
GAAN	5	0			
Chem	1	0			
Math	1	0			
Informatics	1	0			
Total	78	12			

5. Personnel Engaged in Separately Budgeted Research, 7/09–6/10

Professorial faculty	36
Postdoctoral appointments	20
Graduate students	83
Scientists (nonteaching)	14
Total	153

6. Separately Budgeted Research Expenditures by Source of Support

	Departmental Research	Physics-related Research Outside Department
State		
Federal government	$14,502,745	$3,301,717
Business and industry	860,818	$101,857
Total	$15,363,563	$3,403,574

Table C—Separately Budgeted Research Expenditures

Research Specialty	No. of Grants	Expenditures ($)
Accelerator	4	643,171
Astrophysics	6	1,212,230
Biophysics	5	1,470,300
Condensed Matter Physics	9	538,396
Data Grids	2	359,486
Nuclear Physics	18	9,628,223
Particles & Fields	8	1,226,253
Instruction	1	114,000
Stu Fllwshps	1	171,504
Total	54	15,363,563

Table D—Physics-related Research Outside Department

Field and Unit Outside Department	No. of Grants	Expenditures ($)
Chemical Physics	18	3,403,574
Total	18	3,403,574

FACULTY

Professors

Baxter, David V., Ph.D., Cal. Tech., 1984. Condensed matter (experimental).

Berger, Michael S., Ph.D., California, Berkeley, 1991. Theoretical physics; elementary particles.

de Ruyter van Steveninck, Robert, Ph.D., Groningen, 1986. Biophysics (experimental).

Fertig, Herbert A. Ph.D., Harvard, 1988. Condensed matter theory.

Glazier, James, Ph.D., University of Chicago, 1989. Biophysics (experimental).

Gottlieb, Steven A., Ph.D., Princeton, 1978. Theoretical physics; elementary particles.

Horowitz, Charles J., Ph.D., Stanford, 1981. Nuclear theory.

Kesmodel, Larry L., Ph.D., Texas, 1974. Condensed matter experimental; surfaces.

Kostelecký, V. Alan, Ph.D., Yale, 1982. Theoretical physics; elementary particles.

Lee, Shyh-Yuan, Ph.D., SUNY, Stony Brook, 1972. Accelerator physics.

Londergan, J. Timothy, Ph.D., Oxford, 1969. Theoretical physics; nuclear theory.

Messier, Mark, Ph.D., Boston University, 1999. Astrophysics (experimental).

Musser, James A., Ph.D., California, Berkeley, 1984. Astrophysics (experimental).

Ogren, Harold O., Ph.D., Cornell, 1970. Elementary particle physics (experimental).

Olmer, Catherine, Ph.D., Yale, 1976. Intermediate energy nuclear physics (experimental).

Ortiz, Gerardo, Ph.D., Ecole Polytechnique Fédérale de Lausanne (EPFL), 1992. Condensed matter theory.

Pynn, Roger, Ph.D., Trinity College, University of Cambridge, 1969. Nuclear physics (experimental).

Serot, Brian D., Ph.D., Stanford, 1979. Nuclear theory.

Snow, W. Michael, Ph.D., Harvard, 1990. Nuclear physics (experimental).

Sokol, Paul E., Ph.D., The Ohio State University, 1981. Condensed matter physics (experimental).

Szczepaniak, Adam P., Ph.D., Washington, 1990. Theoretical physics.

Van Kooten, Richard J., Ph.D., Stanford, 1990. Elementary particle physics (experimental).

Wissink, Scott W., Ph.D., Stanford, 1986. Nuclear physics (experimental).

Professors Emeriti

Alyea, Ethan D., Ph.D., Cal. Tech., 1962. Astrophysics (experimental).

Bacher, Andrew D., Ph.D., Cal. Tech., 1967. Intermediate-energy nuclear physics (experimental).

Bent, Robert D., Ph.D., Rice, 1954. Experimental nuclear physics; nuclear structure, reactions; astrophysics.

Brabson, Bennet, Ph.D., MIT Cambridge, 1966. Elementary particle physics (experimental).

Cameron, John M., Ph.D., UCLA, 1967. Nuclear physics (experimental).

Challifour, John L., Ph.D., Cambridge, 1963. Theoretical physics; mathematical physics.

Crittenden, Ray R., Ph.D., Wisconsin, 1960. Elementary particle physics (experimental).

Dzierba, Alex R., Ph.D., Notre Dame, 1969. Elementary particle physics (experimental).

Goodman, Charles, Ph.D., Rochester, 1955. Nuclear physics (experimental).

Hake, Richard R., Ph.D., Illinois, 1955. Physics education.

Heinz, Richard M., Ph.D., Michigan, 1964. Astrophysics (experimental).

Hendry, Archibald W., Ph.D., Glasgow, 1962. Theoretical physics; elementary particles.

Lichtenberg, Don B., Ph.D., Illinois, 1955. Elementary particle physics (theoretical).

Meyer, Hans Otto, Ph.D., Basel, Switzerland, 1970. Nuclear physics (experimental).

Miller, Daniel W., Ph.D., Wisconsin, 1951. Nuclear physics (experimental); nuclear reactions.

Nann, Herman, Ph.D., Goethe Univ., 1967. Intermediate energy nuclear physics (experimental).

Newton, Roger G., Ph.D., Harvard, 1953. Distinguished Professor. Theoretical and mathematical physics; scattering theory.

Pollock, Robert E., Ph.D., Princeton, 1963. Distinguished Professor. Nuclear physics; nuclear reactions; cyclotron design.

Schaich, William L., Ph.D., Cornell, 1970. Condensed matter theory.

Schwandt, Peter, Ph.D., Wisconsin, 1967. Nuclear physics (experimental).

Swihart, James C., Ph.D., Purdue, 1955. Condensed matter theory.

Walker, George E., Ph.D., CaseWestern Reserve, 1966. Theoretical nuclear physics; intermediate energy.

Wills, John G., Ph.D., Washington, 1963. Theoretical physics.

Associate Professors

Beggs, John, Ph.D., Yale, 1998. Biophysics (experimental).

Carini, John P., Ph.D., Chicago, 1988. Condensed matter (experimental).

Evans, Harold G., Ph.D., U.C.L.A., 1991. Elementary particle physics (experimental).

Setayeshgar, Sima, Ph.D., CalTech, 1997. Biophysics (theoretical).

Tayloe, Rex, Ph.D., Illinois, 1995. Nuclear physics (experimental).

Urheim, Jon, Ph.D., Pennsylvania, 1990 Astrophysics (experimental).

Associate Professor Emeritus

Lurie, Fred M., Ph.D., Illinois, 1963. Condensed matter physics; nuclear magnetic resonance.

Assistant Professors

Bossev, Dobrin, Ph.D., Institute for Chemical Research, Kyoto University, 1999. Condensed Matter Physics (experimental).

Dermisek, Radovan, Ph.D., Ohio State University, 2002. Theoretical Physics; Elementary Particles.

Hess, Mark, H., Ph.D., MIT, 2002. Accelerator Physics (theory).

Kaufman, Lisa, Ph.D., Massachusetts, 2007. Nuclear physics (experimental).

Lammers, Sabine, Ph.D., University of Wisconsin-Madison, 2004. Elementary Particle Physics (experimental).

Liu, Chen-Yu, Ph.D., Princeton University, 2002. Nuclear Physics (experimental).

Long, Josh, Ph.D., Johns Hopkins University, 1997. Nuclear Physics (experimental).

Shepherd, Matthew R., Ph.D., Cornell University, 2004. Elementary Particle Physics (experimental).

Other Graduate Faculty

Bower, Charles, Ph.D., Indiana University, 1988. Astrophysics (experimental).

Gagnon, Pauline, Ph.D., California-Santa Cruz, 1993. Elementary Particle Physics (experimental).

Gardner, Jason, Ph.D., University of Warwick, England, 1996. Condensed Matter Physics (experimental).

Jacobs, William, Ph.D., Washington, 1974. Nuclear Physics (experimental).

Jain, Vivek, Ph.D., University of Hawaii, 1988. Elementary Particle Physics (experimental).

Klein, Susan, Ph.D., California, Berkeley, 1986. Medical physics.

Luehring, Frederick, Ph.D., Northwestern, 1986. Elementary Particle Physics (experimental).

Lunghi, Enrico, Ph.D., International School of Advanced Studies, Italy, 2000. Theoretical physics; Elementary Particles.

Mitchell, Ryan, Ph.D., University of Tennessee, 2003. Elementary Particle Physics (experimental).

Sluka, James, Ph.D., CalTech, 1988. Biophysics (experimental).

Stephenson, Edward, Ph.D., Wisconsin, 1975. Nuclear physics (experimental).

Warren, Garfield, Ph.D., Tuskegee University, 1988. Condensed Matter Physics (experimental).

Zieminska, Daria, Ph.D., Warsaw, 1974. Elementary particle physics (experimental).

RESEARCH SPECIALTIES AND STAFF

Theoretical

Accelerator Physics. Nonlinear beam dynamics; beam-beam interactions; transition energy problems; transverse and longitudinal coherent instabilities and Landau damping; bunched beam cooling; electron storage ring physics; spin motion in synchrotrons. Lee, Hess.

Biological Physics. Intracellular signaling networks. Waves in excitable media. Non-equilibrium systems. Biocomplexity. Theoretical neuroscience. Information theory. Setayeshgar.

Chemical Physics. Electronic transport in alloys; electron-phonon interaction in metals; electronic properties of atoms, molecules, and surfaces; photoelectron cross sections; phase transitions and self-organization; infrared photometry for adsorbates and quantum wells, phasmonics on the nanoscale. Schaich, 2 faculty in Chemistry Department.

Condensed Matter. Quantum Hall effect; superconductivity, spin transport and magnetoresistance; mesoscopics; soft matter; colloidal and biological materials; electron–phonon interaction in metals; optical and electrical properties of solids; collective excitations; many-body theory; surface electrodynamics; random alloys; quantum computation; correlated electronic materials; many-body physics, strongly correlated systems: high-T_c, heavy fermions, fermions in high magnetic fields; exotic superconductors; magnetism and spin systems; quantum fluids and solids; ultracold Fermi and Bose gases; topologically quantum ordered systems; quantum statistical mechanics and field theory methods in condensed matter; quantum information and computation; quantum measurement theory. Fertig, Ortiz, Schaich.

Elementary Particles and Fields. Phenomenology of elementary particle properties and interactions; quantum chromodynamics and electroweak interactions; lattice gauge field theory; solar neutrinos; grand-unified theories; supersymmetry; gravity and supergravity; superstring theory; CPT and Lorentz symmetry. Berger, Dermisek, Gottlieb, Kostelecky, Lunghi.

Nuclear Physics. Study of nuclear structure; medium and high energy nuclear reactions; quantum chromodynamics; hadron spectra and structure, gluon dynamics, relativistic quantum hadrodynamics; neutron stars and nuclear astrophysics, stelar evolution and neutrino transport. Research performed at the Indiana University Nuclear Theory Center, Horowitz, Londergan, Serot, Szczepaniak.

Experimental

Accelerator Physics. Nonlinear beam dynamics; electron cooling; properties of cooled beams; damping of transverse and longitudinal instabilities; spin motion in synchrotrons, with spin rotators (snakes); overlapping spin resonances and snake resonances. Lee, Hess, 4 IUCF physicists.

Astrophysics. Magnetic monopoles; antimatter; supernovae; dark matter searches; bigbang cosmology; neutrino oscillations, dark energy, solar neutrino. Facilities include an assortment of computers, particle detectors, electronics development equipment, data acquisition systems, and spectrophotometers. Experiments are being performed at Fermi National Laboratory, Superkamiokande, and at a number of balloon launch facilities. Bacher, Bower, Heinz, Messier, Mufson (Astronomy Dept.), Musser.

Biological Physics. Experimental and computational neuroscience. Multielectrode recordings in vitro; intracellular and extracellular neural recording in vivo. Experimental biocomplexity. Beggs, deRuyter, Glazier, Sluka.

Chemical Physics. Optical properties of solids; low-temperature properties of metallic solids; chemisorption and catalysis; high-energy electron scattering; nuclear chemistry; chemical vapor deposition of ceramic and other materials; high-temperature x-ray diffraction; solid state NMR. Baxter, Carini, Kesmodel, 6 faculty in Chemistry Department.

Condensed Matter. Confined fluids, neutron scattering, surfactant systems, dynamics of membranes, atomic and electronic transport in disordered solids, compositionally modulated thin films; thin film magnetism and magnetoresistance; metastable systems; surface studies: STM, AFM, EELS; ferromagnetic semiconductors; dynamics of electrons in disordered metals and correlated electron systems; low temperature facilities; dilution refrigerators, superconducting solenoids, squid magnetometer, cryogenic microwave system. Thin film growth using sputtering and CVD soft matter and bio-materials. Baxter, Bossev, Carini, Gardner, Kesmodel, Pynn, Sokol, Warren.

Elementary Particles. Searches for new particles (Higgs bosons, supersymmetric particles, exotics, hybrid systems, glueballs), heavy quark physics (top, bottom, charm), light quarks, neutrino oscillation, testing of fundamental symmetries. Detectors used include drift chambers, drift tubes, scintillating fibers, transition radiation detectors, Cerenkov counters, and calorimeters. IU facilities include data acquisition and numerous data analysis computers, detector construction areas including a high-bay area and large class-10000 cleanroom. Work on DO and MINOS at Fermilab, ATLAS at CERN, experiments at Jefferson Laboratory, and preparing for the Linear Collider. Dzierba, Evans, Gagnon, Jain, Lammers, Luehring, Messier, Mitchell, Musser, Ogren, Shepherd, Urheim, Van Kooten, Zieminska.

Nuclear Physics: Nucleon structure studies: gluon spin distributions, anti-quark and sea quark contributions to nucleon properties, using the STAR detector at RHIC. Weak interaction studies with slow neutrons: precision measurements of neutron decay, *In-pI* and *In-4HeI* weak interactions at NIST, LANSCE, and the new Spallation Neutron Source at Oak Ridge; methods for production of ultra-cold neutrons; search for neutrino oscillations and studies of neutrino-nucleon interactions with the MiniBooNE and SciBooNE detectors at Fermilab. Fundamental symmetry tests: searches for time-reversal violation via electric dipole moments of the electron and neutron. Formation and decay of hot nuclei, damped collisions between heavy nuclei, and nuclear fission at MSU, ATLAS and other labs. Kaufman, Liu, Long, Snow, Stephenson, Tayloe, Wissink. One additional faculty (de Souza) in Nuclear Chemistry.

INDIANA UNIVERSITY, BLOOMINGTON

DEPARTMENT OF ASTRONOMY

Bloomington, Indiana 47405-7105
http://www.astro.indiana.edu/

Students Accepted For Degree	FIELDS		
	Physics	Astronomy	Related Fields
Doctorate		X	X
Master's		X	

1. General

President: Michael A. McRobbie
Dean of Graduate School: James Wimbush
Department Chairman: John Salzer
Department Telephone Number: (812) 855-6911 (C)
Type of Institution: University
Control: Public
Setting: Small town
Total Faculty: 1,368
Total Graduate Faculty: 1,557
Total Students: 42,347*†
Total Graduate Students: 9,857
Annual Graduate Tuition:
In-state residents: Full-time—$291.97 per credit hr.
Out-of-state residents: Full-time—$850.33 per credit hr.
Tuition rates for: 2010–11
Deferred tuition plan: No
Other Fees: $900 per semester
Term: Semester

*Bloomington Campus—Tenure track
†Including professional school

2. Number of Faculty in Department

The combined total of full-time faculty in the three professorial ranks is 8. The combined total of full-time, part-time, and other faculty at all ranks is 12.

3. Admission, Financial Aid, and Housing

Address admission inquiries to: Dr. Liese van Zee, Graduate Advisor, Astronomy Department, Swain Hall West 319
Graduate application fee required: $50 (domestic) $60 (foreign)
Admission deadline (Fall admission): 1/15
Admission information: For fall admission, 2010–11, 5 students were offered admission from 40 applicants.
Admission requirements: For admission to the graduate programs, normally a Bachelor's degree in physics, astronomy, or astrophysics is required with no minimum undergraduate GPA specified. GRE scores, including the Advanced Test in physics, are required. Students from non-English speaking countries are required to demonstrate proficiency in English via the TOEFL exam. The minimum acceptable score for admission is 550.
Undergraduate preparation assumed: Physics and math background sufficient to handle the astronomy in the following texts is assumed: Introduction to Modern Stellar Astrophysics by Carroll, B. and Ostlie, D.
Address financial aid inquiries to: Dr. Liese van Zee, Director of Graduate Studies, Astronomy Department, Swain Hall West 319 or to Office of Scholarships and Financial Aids.
GAPSFAS application required: No
Financial aid deadline: 1/15
Loans available: Yes
Address housing inquiries to: Halls of Residence, 801 N. Jordan
On-campus, single student housing available: Yes
Cost/academic yr: $4,654–6,167 (double); $4,046–4,646 (single)
On-campus, apartment student housing available: Yes
Cost/month: $519–1,119 (includes utilities except telephone; furnished or unfurnished)

Table A—Faculty, Enrollments, and Degrees Granted

Research Specialty	2009 Faculty	Enrollment[1] Fall 2009		No. of Degrees Granted[2] 2009–10 (2005–10)			Median No. of Years for 2009–10 Ph.D.'s
		Mas-ter's	Doc-torate	Mas-ter's	Terminal Master's	Doc-torate	
Astronomy	9	0	20	4(13)	0(4)	2(2)	6.25
Astrophysics	9	–	2	0(0)	0(0)	0(2)	6.25
Total		–	20	3(13)	2(9)	2(3)	
Full-time Grad. Stud.		–	20				
Part-time Grad. Stud.		–	0				
First-year Grad. Stud.		–	3				
Median Years in Grad. Study (2008–09 Degrees)				3	–	–	–
Undergraduate Degrees, 2009–10 (2004–09): 3(17)							

[1]Students not yet committed to a research specialty are entered under non-specialized.
[2]Five-year totals in parentheses.

4. Graduate Degree Requirements

Master's: The M.A. degree requires 30 hours with a minimum GPA of 3.0. There is no specific residence requirement. A thesis may be required at the discretion of the faculty. A final oral exam covering work for the degree is also given.
Doctorate: The Ph.D. in Astronomy requires 90 hours with a minimum GPA of 3.0. There is no specific residency requirement but the student must be continuously enrolled after admission to candidacy. The qualifying examination consists of a written examination (normally after the fourth semester). The requirements for the Ph.D. in Astrophysics include at least four physics courses not required by the Ph.D. in Astronomy. Candidacy is attained by passage of a combination of tests administered by the Physics Department and the Astronomy Department.
Thesis: Thesis may be written *in absentia*.
Special Equipment, Facilities, or Programs: Small local telescopes are available for student training. Most data for thesis research is obtained at national facilities which provide telescopes for optical, radio, and space applications. On campus there is an image analysis laboratory and a solar laboratory. Indiana University has superb centralized supercomputing facilities available for student and faculty research, and the Department has its own computational systems for data processing and scientific computing. The Department is currently using the 3.5-m and the 0.9-m WIYN telescopes at Kitt Peak National Observatory near Tuscon, Arizona in collaboration with the University of Wisconsin, Yale University, and the National Optical Astronomy Observatories.

Table B—Appointments to Graduate Students, 2009–10

Title of Appointee	Appointments Total	Appointments First year	Academic Load Allowed in Credit Hours	Hours of Service Per Week	Stipend for Academic Year ($)
			Semester		
Teaching Assistant	11	3	12	20	14,008
Research Assistant	6	0	12	20	0
Fellowship	3	0			15,000–30,000
Total	17	3			

5. Personnel Engaged in Separately Budgeted Research, 7/09–7/10

Professorial faculty	9
Graduate students	9
Total	18

6. Separately Budgeted Research Expenditures by Source of Support

	Departmental Research	Physics-related Research Outside Department
Federal government	$4,718,432	$1,191,534
Total	$4,718,432	$1,191,534

Table C—Separately Budgeted Research Expenditures

Research Specialty	No. of Grants	Expenditures ($)
Astronomy	43	4,718,432
Total	43	4,718,432

Table D—Physics-related Research Outside Department

Field and Unit Outside Department	No. of Grants	Expenditures ($)
Astrophysics	1	1,191,534
Total	1	1,191,534

[1]Joint grants for the Astrophysics Group are divided between Physics and Astronomy. Only one half of the total amount is listed here.

FACULTY

Professors

Cohn, Haldan N., Ph.D., Princeton, 1979. Dynamical evolution of dense stellar systems; high-performance N-body simulations; globular clusters structure and stellar content; x-ray binaries.

Lugger, Phyllis M., Ph.D., Harvard, 1982. Dynamical evolution of globular clusters and other dense stellar systems; x-ray studies of compact binary stars.

Mufson, Stuart L., Ph.D., Chicago, 1974. High-energy astrophysics; underground cosmic ray physics; neutrino physics.

Pilachowski, Catherine A., Ph.D., Hawaii, 1975. Origin of the Elements in the Milky Way; Star Clusters; Stellar Evolution; the compositions of stars; stellar populations; stellar seismology.

Salzer, John J., Ph.D., University of Michigan, 1987. Galaxy evolution, active galactic nuclei, starburst galaxies, chemical evolution in galaxies, multi-wavelength studies of dwarf galaxies, emission-line surveys.

Professors Emeriti

Burkhead, Martin S., Ph.D., Wisconsin, 1964. Photoelectric photometry; star clusters; galaxies.

Durisen, Richard H., Ph.D., Princeton, 1972. Star formation; astrophysical fluid dynamics; stellar rotation; planetary rings, complex plasmas.

Honeycutt, R. Kent, Ph.D., Case Western, 1968. Stellar astronomy, instrumentation, accretion disks in cataclysmic variables and in other interacting binary stars.

Johnson, Hollis R., Ph.D., Colorado, 1960. Model stellar atmospheres: theoretical stellar spectra; ultraviolet spectra of red-giant stars.

Associate Professors

Deliyannis, Constantine P., Ph.D., Yale, 1990. Stellar evolution; galactic evolution; primordial lithium; Big Bang nucleosynthesis.

van Zee, Liese, Ph.D., Cornell, 1996. Galaxy evolution; element enrichment; star formation; extragalactic neutral hydrogen.

Assistant Professor

Rhode, Katherine, Ph.D., Yale 2003. Extragalactic globular clusters systems; galaxy formation, rotation and evolution of solar-type pre-main-sequence stars.

Research Scientists

Salim, Samir, Ph.D., Ohio State University, 2002. Galaxy Evolution; Star Formation Indicators; Galaxy Biomodality; SED Fitting; Galaxy Surveys; Data Mining; UV Astronomy.

Steiman-Cameron, Thomas Y., Ph.D., Indiana University, 1984. Dynamics of nonplanar astrophysics disks, galaxy formation and evolution, structure of galactic halos, sprial structure of the Milky Way, accretion driven compact x-ray binary stars.

Thornburg, Jonathan, Ph.D., University of British Columbia, 1993. Numerical simulations of gravitational radiation from extreme-mass-ratio binary black hole inspirals/mergers, numerical simulations of binary black hole mergers, gravitational-wave astrophysics.

RESEARCH SPECIALTIES AND STAFF

Theoretical

Dynamical evolution of dense stellar systems; globular clusters; stellar rotation; planetary rings; complex plasma; x-ray studies; high-performance N-body simulations. Cohn, Durisen, Lugger.

Observational

Ground-based and space-based optical and infrared astronomy. Imaging and spectroscopy of stars, star clusters, and external galaxies. Studies of steller abundances and evolution, galaxy evolution, and chemical evolution. Accretion disks interacting binaries, X-ray binaries. Dark energy. Cohn, Deliyannis, Honeycutt, Lugger, Mufson, Pilachowski, Rhode, van Zee.

High-energy particle astrophysics. Neutrino and muon astronomy. Mufson.

Instrumentation. CCD systems, spectrography design, telescope automation. Honeycutt, Pilachowski.

INDIANA UNIVERSITY—PURDUE UNIVERSITY INDIANAPOLIS

DEPARTMENT OF PHYSICS

Indianapolis, Indiana 46202-3273

Students Accepted For Degree	FIELDS Physics	Astronomy	Related Fields
Doctorate	X		
Master's	X		

1. General

Chancellor: Charles Bantz
Dean of Graduate School: Sherry F. Queener
Department Chairman: Andrew Gavrin
Department Telephone Number: (317) 274–6900
Type of Institution: University
Control: Public
Setting: Urban
Total Faculty: 3,041 (2,131 full-time)
Total Students: 30,300
Total Graduate Students: 8,200
Annual Graduate Tuition:
In-state residents: $316.10/hr.
Out-of-state residents: $904.00/hr.
Tuition rates for: 2009–10
Deferred tuition plan: No
Annual Other Fees: $400
Term: Semester

2. Number of Faculty in Department

The combined total of full-time faculty in the four professorial ranks is 16. The combined total of full-time, part-time, and other faculty at all ranks is 21.

3. Admission, Financial Aid, and Housing

Address admission inquiries to: Director of Graduate Program
Graduate application fee required: $55
Admission deadline (Fall admission): 3/15
Admission information: For fall admission, 2009–10, 8 students were accepted from 23 applicants.
Admission requirements: For admission to the graduate programs, a Bachelor's degree in physics or related areas is required with no minimum undergraduate GPA specified. The GRE is required for fellowship applicants. The GRE Advanced is recommended. Students from non-English speaking countries are required to demonstrate proficiency in English via the TOEFL exam. Minimum acceptable score for admission is 550 on the paper based and 213 on the computer.
Undergraduate preparation assumed: Symon, *Mechanics*; Corson and Lorrain, *Electricity and Magnetism*; Eisberg and Resnick, *Quantum Physics*; Rief, Thermodynamics.
Address financial aid inquiries to: Department of Physics Graduate Program
GAPSFAS application required: No
Loans available: Yes
Address housing inquiries to: Department of Physics Graduate Program
On-campus, single student housing available: Yes
Cost/month: $450–750
On-campus, married student housing available: Yes
Cost/month: $450–750

Table A—Faculty, Enrollments, and Degrees Granted

Research Specialty	2009–10 Faculty	Enrollment[1] Fall 2009 Master's	Doctorate	No. of Degrees Granted[2] 2009–10 (2005–10) Master's	Terminal Master's	Doctorate	Median No. of Years for 2008–09 Ph.D.'s
Atomic, Molecular, & Optical Physics	1	0	2	0(2)	–	0(1)	–
Biophysics	6	4	5	1(4)	1(6)	1(1)	–
Condensed Matter Physics	3	4	5	0(3)	1(1)	1(2)	–
Optics	1	4	0	0(1)	–	0(1)	–
Particles & Fields	0	0	0	0(0)	–	0(0)	–
Physics Education	1	1	0	0(0)	0(2)	0(0)	–
Total	12	13	12	1(10)	2(9)	2(5)	
Full-time Grad. Stud.		12	12				
Part-time Grad. Stud.		1	1				
First-year Grad. Stud.		8	2				
Median Years in Grad. Study (2009–10 Degrees)				2	2	5	

Undergraduate Degrees, 2009–10 (2005–10): 4(20)

[1]Students not yet committed to a research specialty are entered under non-specialized.
[2]Five-year totals in parentheses.

4. Graduate Degree Requirements

Master's: Both thesis and non-thesis Master's programs are available. For each program the student must complete 30 credit hours and maintain a grade point average of 2.7/4. Twenty-four credit hours must be in physics/biophysics and 6 hours in mathematics. For the Thesis Master's program, six of the 24 hours are satisfied by completing the thesis. All students must pass a qualifying examination early in their program and an oral examination at the completion of their program. The minimum residence requirement is two semesters of full-time work or equivalent in credits.
Doctorate: Qualified students may pursue the Ph.D. degree at IUPUI in areas where a program has been arranged with Purdue, West Lafayette. Students are usually expected to complete an M.S. degree before pursuing the Ph.D. degree. Currently, a Ph.D. program is available in the area of biological physics, optics, and materials science.
Thesis: Thesis may be written *in absentia*.
Special Equipment, Facilities, or Programs: The NMR facilities of the department consist of three multi-nuclear high-resolution FT spectrometers and one solid state spectrometer. Included among the high-resolution instruments, are narrow-bore 500 and 200 MHz spectrometers, and a wide-bore 300 MHz spectrometer. The solid state instrument has broadline and MAS capabilities, operates at 4.2 T, and is home-built. An x-band EPR spectrometer is used for biophysics research. Full facilities for preparation and characterization of biological samples are available. Two optics research laboratories are equipped with argon laser-pumped CW-frequency doubled Ti-sapphire lasers, diode lasers, and He–Ne lasers among others. State-of-the-art data-acquisition equipment includes digital oscilloscopes, spectrum analyzers, and computer interfaces. High-finesse optical cavities and high-vacuum systems are used to study atomic behavior in laser

fields. Thin-film sputter deposition system with 4 high rate magnetron guns, rf and dc excitation, substrate heating to 1000 K, computer controlled substrate and shutter motions. Chamber is cryopumped with background pressure of 10^{-8} Torr. Two home-build Near-field scanning optical microscopes are operational. A state-of-the-art atomic force microscope is accessible. Computing facilities include various workstations and mainframes and a 40-node cluster.

Table B—Appointments to Graduate Students, 2009–10

Title of Appointee	Appointments Total	First year	Academic Load Allowed in Credit Hours	Hours of Service Per Week	Stipend for Academic Year ($)
			Semester		
Teaching Assistant	8	2	9	20	16,000[1,2]
Research Assistant	5	1	9	40	18,000[1,2]
Fellow	3	0	12	40	22,000
Total	18	3			

[1]Tuition and fees total $400 per semester.
[2]Additional support is available during the summer.

5. Personnel Engaged in Separately Budgeted Research, 7/09–6/10

Professorial faculty	13
Postdoctoral appointments	1
Graduate students	24
Undergraduate students	10
Nonteaching research personnel	1
Total	49

6. Separately Budgeted Research Expenditures by Source of Support

	Departmental Research	Physics-related Research Outside Department
Federal government	$240,000	$50,000
University	210,000	
Total	$450,000	$50,000

Table C—Separately Budgeted Research Expenditures

Research Specialty	No. of Grants	Expenditures ($)
Optics	1	$136,000
Biophysics	1	20,000
Fellowships	10	290,000
Materials Science	3	221,000
Physics Education	2	31,000
Total	15	698,000

Table D—Physics-related Research Outside Department

Field and Unit Outside Department	No. of Grants	Expenditures ($)
Biophysics		
NIH	1	40,000
Total	1	40,000

FACULTY

Professors

Kemple, Marvin D., Ph.D., University of Illinois, 1971. Biological physics; magnetic resonance and fluorescence.

Ou, Zhe-Yu Jeff, Ph.D., Rochester, 1990. Quantum optics; nonlinear optics.

Rao, B. D. Nageswara, Ph.D., Aligarh Muslim University, 1961. Biophysics; magnetic resonance.

Sukatme, Uday, Sc.D. Physics, Massachusetts Institute of Technology, 1971. Executive Vice Chancellor and Dean of Faculties.

Vemuri, Gautam, Ph.D., Georgia Tech, 1990. Nonlinear optics; laser physics.

Wassall, Stephen R., Ph.D., Nottingham, England, 1981. Solid state NMR; biophysics.

Associate Professors

Decca, Ricardo S., Ph.D., Instituto Balseiro, Argentina, 1994. Condensed matter; scanning probe microscopy.

Gavrin, Andrew D., Ph.D., Johns Hopkins, 1992. Chairman of the Department. Materials Science; electron microscopy.

Assistant Professors

Betancourt, Marcos R., Ph.D., UCSD, 1995. Computational biophysics.

Cheng, Ruihua, Ph.D., University of Nebraska-Lincoln 2002. Condensed matter; Magnetic nanostructures.

Joglekar, Yogesh, Ph.D., Indiana University, 2001. Condensed matter; noise spectroscopy.

Petrache, Horia, Ph.D., Carnegie Mellon University, 1998. X-ray scattering, membrane biophysics.

Rader, Andrew J., Ph.D., Michigan State University, 2002. Computational biophysics; protein folding.

Lecturers

Rhoads, Edward, Ph.D., University of Minnesota, 2005. Astronomy.

Ross, John B., Ph.D., Boston University, 1993. Physics Education.

Woodahl, Brian A., Ph.D., Purdue, 1999. Physics Education, theoretical particle physics.

Assistant Scientist

Ray, Bruce D., Ph.D., Indiana University, 1983. Biochemistry; isotope labeling; NMR.

RESEARCH SPECIALTIES AND STAFF

Theoretical

Biophysics. Theoretical modeling of proteins. Protein folding dynamics and thermodynamics. *De novo* protein design. Protein structure prediction by computer simulations. Inferring biological functions from simulations of large-scale motions in proteins and supramolecular assemblies. Betancourt, Rader.

Condensed Matter. Strongly correlated systems including granular high-temperature superconductors, excitonic condensates in semiconductors, noise spectroscopy, and quantum Hall systems. Joglekar.

Magnetic Resonance. Theoretical aspects of magnetic resonance methods used in macromolecular structure studies; computer simulations of nuclear magnetic resonance spectra and electron paramagnetic resonance spectra in biological and nonbiological systems; non-linear chemical exchange. Kemple, Landy, Rao, Ray, Wassall.

Membrane Biophysics. Theoretical modeling of membrane transport of water and non-ionic solutes and osmotic behavior. Lipid-protein interactions. Membrane dynamics and electrostatistics. Kleinhans, Petrache.

Optics. Nonlinear dynamics of diode and solid-state lasers; atomic coherence effects; response to fluctuating fields; quantum noise and measurement; quantum fluctuations in nonlinear optical processes; quantum multiphoton interference. Kemple, Ou, Vemuri.

Experimental

Biological Magnetic Resonance. Macromolecular structure-function relationships in enzyme-substrate complexes of ATP-utilizing enzymes and alcohol dehydrogenase; internal motions of peptides, proteins, and their complexes; NMR and computer simulations of protein dynamics; broadline deuterium NMR of molecular order and dynamics in membranes; MAS NMR determination of peptide conformation; EPR membrane and cytoplasmic studies of reactive oxygen species, and molecular order in model membranes. Kemple, Rao, Ray, Wassall. 1 staff scientist.

Condensed Matter. Spatial and time-resolved spectroscopy in quantum systems. Metal-Insulator transition in superconductors. Correlated electronic systems. Quantum Dots and Casimir Force. Spin-dependent transport. Cheng, Decca.

Materials Science. Artificially structured materials. Ferromagnetic domains and giant magnetoresistive effects in granular metals. Domain wall pinning in amorphous alloys. Scanning electron microscopy with polarization analysis. Fabrication of magnetic nanowires and nanodots. Cheng, Gavrin.

Optics. Fiber optic sensors; semiconductor lasers; optical waveguides and couplers; fiber optic microwave transmission and signal processing; non-linear optics; optical feedback; diode laser and amplifier statistical properties; laser instabilities; chaos and communication; cavity QED; nonlinear optical frequency conversion; photon statistics of nonclassical states; test of EPR nonlocality; multiphoton interference. Ou, Vemuri.

Physics Education. Just-in-Time-Teaching (JiTT), modulus course design. Gavrin.

Scanning Probe Microscopy. Near-field scanning optical microscopy. Atomic force microscopy. Probe-sample interaction effect. Image analysis and deconvolution. Single molecule detection and tracking. Decca.

X-ray Scattering. Small-angle scattering of membrane systems in solution to determine structure and molecular interactions relevant to biological functions. Petrache, Wassall.

PURDUE UNIVERSITY

DEPARTMENT OF PHYSICS

West Lafayette, Indiana 47907

Students Accepted For Degree	FIELDS		
	Physics	Astronomy	Related Fields
Doctorate	X		
Master's	X		

1. General

President: France Córdova
Dean of Graduate School: Mark J. T. Smith
Department Chairman: Nicholas J. Giordano
Department Telephone Number: (765) 494-3000
Department Fax Number: (765) 494-0706
Department E-mail address: physcontacts@purdue.edu
http://www.physics.purdue.edu
Type of Institution: University
Control: Public
Setting: Small city
Tenured/Tenure Track Faculty: 1,918
Other Faculty: 845
Total Students: 39,697
Total Graduate Students: 7,639
Annual Graduate Tuition:
In-state residents: Full-time—$8,638
Out-of-state residents: Full-time—$25,118
Tuition rates for: 2009–10
Deferred tuition plan: Yes
Other Fees: None
Term: Semester

*Fees for TAs, and RAs is $471 per semester.

2. Number of Faculty in Department

The combined total of full-time faculty in the three professorial ranks for 2010 is 56. The combined total of full-time, part-time, and other faculty at all ranks is 62.

3. Admission, Financial Aid, and Housing

Address admission inquiries to: Nicholas J. Giordano, Head, Department of Physics
Graduate application fee required: $55
Admission deadline (Fall admission): 1/15
Admission information: For fall 2009–10 admission, 84 students were accepted from 287 applicants, with 31 matriculated.
Admission requirements: For admission to the graduate programs, a Bachelor's degree in physics is required with a minimum undergraduate GPA of 3.0/4.0. GRE required, but no minimum scores set. The GRE physics is also required. Students from non-English speaking countries are required to demonstrate proficiency in English via the TOEFL exam. The minimum acceptable score for admission is 550 (paper) or 213 (computer based). IBT score 77 (minimum individual scores of 18 writing, 18 speaking, 14 listening, and 19 reading).
Undergraduate preparation assumed: A good preparation for entering students includes a sound knowledge of general physics, intermediate level mechanics, electricity and magnetism, optics, statistical and thermal physics, introductory atomic and nuclear physics including some principles of quantum mechanics. A corresponding mathematical background would include vector analysis, advanced calculus, ordinary differential equations, boundary value problems, and some knowledge of introductory complex analysis. Graduate credit courses are offered at two levels in mechanics, electricity and magnetism, thermal physics and modern physics; first-year students can be placed in courses that will supplement the undergraduate program and correct deficiencies. Strong undergraduate preparation would be provided by adequate study of textbooks at the level of: Marion, *Classical Dynamics*; Griffiths, *Introduction to Electrodynamics*; Reif, *Statistical and Thermal Physics*; Jenkins and White, *Fundamentals of Optics*; and Gasiorowicz, *Quantum Physics*.
Address financial aid inquiries to: Sandy Formica, Graduate Secretary, Department of Physics
GAPSFAS application required: No
Financial aid deadline: none
Loans available: Yes
Address housing inquiries to: Graduate Housing-ghapp@purdue.edu, married and family housing-pvapp@purdue.edu
On-campus, single student housing available: Yes
Cost/month: $397–720
On-campus, married student housing available: Yes
Cost/month: $610–745

Table A—Faculty, Enrollments, and Degrees Granted

Research Specialty	2009–10 Faculty	Enrollment[1] Fall 2009		No. of Degrees Granted[2] 2009–10(2004–09)			Median No. of Years for 2004–10 Ph.D.'s
		Master's	Doctorate	Master's	Terminal Master's	Doctorate	
Applied Physics	2	0	2	0(0)	0(0)	1(3)	–
Astrophysics	6	0	14	0(0)	0(1)	1(4)	–
Biophysics	6	0	16	0(0)	0(1)	4(14)	–
Condensed Matter Physics	17	0	44	0(0)	0(2)	4(28)	–
Geophysics	2	0	2	0(0)	0(0)	0(1)	–
Nuclear Physics	7	0	5	0(0)	0(0)	1(4)	–
Particles & Fields	14	0	22	0(0)	0(1)	5(24)	–
Physics Education	2	0	1	0(0)	0(0)	0(0)	–
Non-specialized	0	4	37	4(32)	4(17)	0(0)	–
Total		4	143	4(32)	4(22)	16(78)	6.3
Full-time Grad. Stud.		4	143				
Part-time Grad. Stud.		0	0				
First-year Grad. Stud.		0	31				
Median Years in Grad. Study (2006–07 Degrees)	6.0						
Undergraduate Degrees, 2006–07: 34							

[1]Students not yet committed to a research specialty are entered under non-specialized.
[2]Five-year totals in parentheses.

4. Graduate Degree Requirements

Master's: Non-thesis option: completion of a minimum of 30 credit hours with at least 24 hours of approved 500–600 level courses in physics, including one laboratory course, and 6 credit hours in 500–600 level mathematics courses, which may be replaced in whole or in part by Methods of Theoretical Physics I, II: grade in a 500-level physics course must be A or B, and grade in a 600-level physics or a mathematics course A, B, or C; minimum graduate grade average of 2.8/4.0; qualifying examination must be taken; written and oral final examinations are given or waived at discretion of student's advisory committee. More than half of the Purdue credits must be earned through the Purdue campus where the degree is conferred. Thesis option: thesis replaces 9 credit hours of physics requirement: final oral examination over thesis is required.

Doctorate: At least 90 hours of credit hours are required for the Ph.D. plan of study. Core requirements include statistical physics (one semester), advanced electricity and magnetism (one semester) quantum mechanics (two semesters), and three graduate-level specialty courses. A core course need not be taken at Purdue if its equivalent has been taken previously.

A student entering with a B.S. Degree and holding a teaching assistantship needs about 2 years to complete all courses. A master's degree or professional doctoral degree from any accredited institution may be considered to contribute up to 30 credit hours toward satisfying the 90 credits required for a Ph.D. degree. An average GPA of 3.0 is required in core courses. At the start of first semester, students are required to take a qualifying examination to demonstrate undergraduate knowledge of mechanics at the level of Marion, *Classical Dynamics*; of electricity and magnetism at the level of Griffiths, *Introduction to Electrodynamics*; and of modern physics at the level of Gasiorowicz, *Quantum Physics*. Students are formally admitted to candidacy for the Ph.D. degree only after they have passed the Ph.D. preliminary examination. The student is eligible to attempt this examination when he or she has completed the core courses with at least a B average. The Preliminary Examination Committee of a given student decides on the nature and coverage of that student's examination. The examination may have written and oral portions. There is no department-wide preliminary examination. After passing the preliminary examination, students can devote practically all of their time to the original research that will serve as the basis for their theses. The research must be of fully professional character and publishable quality. Completion of the Ph.D. requirements include the completion of the thesis, passing an oral examination in defense of the thesis, and preparation of the thesis material for publication.

Computational Science and Engineering: The Computational Science and Engineering (CS&E) Program at Purdue provides students with the opportunity to study a specific science or engineering discipline along with computing in a multidisciplinary environment. The aim of the program is to produce a student who has learned how to integrate computing with another scientific or engineering discipline and is able to make original contributions in both disciplines. The Physics Department is one of the original departments since the inception of this program. The participating departments now number 18 spread over 5 Colleges. Physics CSE students must satisfy both the standard Physics departmental degree requirements and those of the CSE Program. Usually some of the math and specialty course requirements of the Physics Department can be met by courses which simultaneously contribute toward the satisfaction of the CSE requirements; however, generally, both the number of courses and grade requirement are higher for the physics students who elect to specialize in the CSE Program.

M.S. graduates should be well prepared to join and make significant contributions to interdisciplinary research teams. Ph.D. graduates are expected to become leaders in research and development at the forefront of their fields, applying advanced computational techniques and theory to solve key problems.

Thesis: Thesis may be written *in absentia* if necessary.

Research: Research in the Physics department is carried out on all the scales of nature. At the smallest sub-atomic scales (10^{-15} meters) Purdue faculty are unravelling the structure of matter using some of the most powerful particle accelerators created by man at Fermilab in Batavia, IL, CERN in Europe, and the Relativistic Heavy Ion Collider (RHIC) in New York. At the nano to micron scale (10^{-9}–10^{-6} meters) current activities involve such diverse areas as the nature of single electron transistors, the properties of small ensembles of matter, and the function of biological molecules to name just a few. At the "everday" scale (10^{-3}-10^{3} meters) faculty are probing the workings of gravity, "flying through" tumors, characterizing percolation in soil and rocks, and determining the ages of geological features. At the largest scales, the astronomical scale (10^{12}–10^{26} meters), researchers are pursuing the answers to questions concerning the death of stars, the workings of galaxies and the supermassive black holes that power them, to the very basics of how the Universe was born and what is its ultimate fate. The Physics department strives to engage its students at all levels and involve them in the pursuits of very fundamental questions about the world around them. By making them active in the discovery of the answers we hope to inspire the next generation of scientists and communicate to the University and the broader community the beauty and mystery of the world that we live in.

Special Equipment, Facilities, or Programs: Among the major facilities is PRIME Lab, a national center for accelerator mass spectrometry (AMS) which is based on an 8-MeV Tandem Van de Graaff accelerator. AMS is an ultra-sensitive analytical technique for measuring low levels of long-lived radio nuclides and rare trace elements, and has wide applications to the earth and space sciences, biological sciences, and materials sciences.

The department has a class 10,000 clean room used for testing and assembling detectors for use in high energy physics experiments. Members of the Purdue High Energy Physics group have recently built a silicon vertex detector for use in experiments at the Cornell University electron-positron collider and are developing silicon sensors for the CMS Collaboration at the Large Hadron Collider at CERN.

The Physics Department Library has a seating capacity for 100 users and occupies approximately 11,000 square feet of space in the Physics Building. Its collection has in excess of 47,000 volumes and subscriptions of over 250 journals. Most of the journals are also available electronically.

The Physics Department has an Instrument Shop, a Faculty Machine Shop, and an Electronics Shop for building scientific apparatus. The Instrument Shop is staffed by professional machinists and features a CNC (Computer Numerical Control) milling machine and CNC lathe. Many undergraduate and graduate students receive training and practical experience in machining and electronics in the Faculty Machine Shop and Electronics Shop. Machining techniques and safety are taught to physics staff by a professional machinist in the Faculty Machine Shop, which is used by physics faculty, staff, and students. Electronics for both research and instruction are designed, built, and repaired in the Electronics Shop, which is staffed by an electrical engineer.

Purdue University has created Discovery Park for interdisciplinary research in both bio- and nano-science. The Birck Nanotechnology Center and the Bindley Bioscience Center are state-of-the-art facilities where Physics faculty, postdoctoral researchers, and graduate students join colleagues from other disciplines in performing ground-breaking research in nano-physics, biophysics, and Sensory Science & Technology.

Condensed matter experimentalists make use of synchrotron radiation sources at Argonne National Laboratory and Brookhaven National Laboratory. Physicists in High Energy Particle and High Energy Nuclear physics are engaged in experiments at Brookhaven National Laboratory, Fermi National Acceleratory Laboratory, the Cornell Electron Storage Ring, and the CERN Laboratory. Astronomy and astrophysics researchers use facilities at the Whipple telescope at Kitt Peak National Observatory in Arizona, the Hubble Space Telescope, and a variety of other space-based instruments.

Table B—Appointments to Graduate Students, 2009–10

Title of Appointee	Appointments Total	Appointments First year	Academic Load Allowed in Credit Hours	Hours of Service Per Week	Stipend for Academic Year ($)
			Semester		
Teaching Assistant	86	25	12 hours	20	16,612–17,833
Research Assistant	51	2	12 hours	20	17,429–17,833
Total	127	30			

5. Personnel Engaged in Separately Budgeted Research, 7/09–06/10

Professorial faculty	47
Postdoctoral appointments	16
Graduate students	57
Nonteaching research personnel	17
Total	137

6. Separately Budgeted Research Expenditures by Source of Support 2006–07

	Departmental Research
Federal government	$7,309,208
Non-Federal Sponsored Programs	795,131
Total	$8,104,339

Table C—Separately Budgeted Research Expenditures, 2008-09

Research Specialty	No. of Grants	Expenditures ($)
Accelerator Mass Spectrometry	20	718,253
Astrophysics	26	817,649
Applied	8	442,782
Biophysics	9	445,793
Condensed Matter Physics	42	2,146,029
Geophysics	5	164,383
Nuclear Physics	8	626,672
Particle Fields	45	2,467,583
Physics Education	29	275,195
Total	192	8,104,339

FACULTY

Professors

Barnes, Virgil E., Ph.D., Cambridge, 1962. Experimental high-energy physics; elementary particle reactions.

Bortoletto, Daniela, Ph.D., Syracuse, 1989. Experimental high energy physics; elementary particle reactions.

Bryan, Lynn, Ph.D., Purdue University, 1997. Physics education.

Caffee, Marc, Ph.D., Washington University, St. Louis, 1986. Accelerator Mass Spectrometry.

Clark, Thomas E., Ph.D., New York Univ., 1974. Theoretical physics; elementary particle theory.

Córdova, France, Ph.D., California Institute of Technology, 1979. High energy astrophysics.

Cui, Wei, Ph.D., University of Wisconsin, Madison, 1994. High energy astrophysics.

Durbin, Stephen M., Ph.D., Illinois, 1983. Experimental condensed matter physics; biophysics.

Finley, John P., Ph.D., University of Wisconsin-Madison, 1990. High energy astrophysics.

Fischbach, Ephraim, Ph.D., Pennsylvania, 1967. Theoretical physics; elementary particle theory.

Garfinkel, Arthur F., Ph.D., Columbia, 1962. Experimental high-energy physics; elementary particle reactions.

Giordano, Nicholas J., Ph.D., Yale, 1977. Experimental condensed matter physics, musical acoustics, computational biophysics.

Giuliani, Gabriele F., Ph.D., (perfezionamento), Scuola Normale Superiore, Pisa, 1983. Theoretical physics; condensed matter theory.

Gutay, Laszlo J., Ph.D., Florida State, 1964. Experimental high-energy physics; electroweak physics; W, Z pair production, Higgs, SS particles.

Hirsch, Andrew S., Ph.D., MIT, 1977. Experimental high energy nuclear physics; physics education.

Khlebnikov, Sergei, Ph.D., Institute for Nuclear Research of the Academy of Sciences, Moscow, 1988. Elementary particle theory.

Kim, Yeong E, Ph.D., California, Berkeley, 1963. Theoretical physics; nuclear theory; applied nuclear physics; nuclear astrophysics; condensed matter theory; quantum statistical mechanics.

Koltick, David S., Ph.D., Michigan, 1978. Experimental high-energy physics, applied nuclear physics.

Kuo, Tzee K., Ph.D., Cornell, 1963. Theoretical physics; elementary particle theory.

Love, Sherwin T., Ph.D., Stanford, 1978. Theoretical physics; elementary particle theory.

Melosh, H. J., Ph.D., California Institute of Technology, 1972. Geophysics; planetary physics.

Miller, David H., Ph.D., Imperial College, 1963. Experimental high energy physics; elementary particle reactions.

Moffett, Thomas J., Ph.D., Texas, Austin, 1973. Optical astronomy.

Muzikar, Paul, Ph.D., Cornell, 1980. Theoretical physics; condensed matter theory, accelerator mass spectrometry.

Nakanishi, Hisao, Ph.D., Harvard, 1980. Theoretical physics; condensed matter theory; statistical mechanics.

Nolte, David D., Ph.D., California, Berkeley, 1988. Experimental condensed matter physics, biophysics.

Prohofsky, Earl W., Ph.D., Cornell, 1963. Condensed matter theory; biological physics.

Pyrak-Nolte, Laura, Ph.D., California, Berkeley, 1988. Geophysics.

Ramdas, Anant K., Ph.D., Raman Research Inst., 1956. Experimental condensed matter physics and materials sciences; laser optics; chemical physics; biophysics.
Reifenberger, Ronald G., Ph.D., Chicago, 1976. Experimental condensed matter physics.
Scharenberg, Rolf P., Ph.D., Michigan, 1955. Experimental high energy nuclear physics.
Shipsey, I. P. J., Ph.D., Edinburgh, 1986. Experimental highenergy physics; particle astrophysics.
Wang, Fuqiang, Ph.D., Columbia University, 1996. High energy nuclear physics.

Courtesy Professor

Elliott, Daniel S., Ph.D., Michigan, 1981. Experimental atomic, molecular and optical physics; coherent and quantum optics.
Kais, Sabre, Ph.D., Hebrew University of Jerusalem, 1989. Theoretical condensed matter physics.

Associate Professors

Carlson, Erica W., Ph.D., UCLA, 2000. Theoretical condensed matter physics.
Haugan, Mark P., Ph.D., Stanford, 1978. Theoretical physics; astrophysics; relativity; physics education.
Hu, Jiangping, Ph.D., Stanford, 2002. Theoretical condensed matter physics.
Jones, Matthew, Ph.D., 1997. Carleton University (Canada). Experimental high energy physics.
Lister, Matthew L., Ph.D., Boston University, 1999. Observational astronomy and astrophysics.
Lyanda-Geller, Yuli, Ph.D., Ioffe Physico-Technical Institute, 1987. Theoretical condensed matter physics, theoretical atomic, molecular, and optical physics, quantum information science.
Manfra, Michael J., Ph.D., Boston University, 1999. Experimental condensed matter physics.
Neumeister, Norbert, Ph.D., Vienna University of Technology, Austria, 1996. High energy particle physics.
Ritchie, Ken, Ph.D., University of British Columbia, 1998. Experimental biphysics.
Rodriguez, Jorge, Ph.D., University of Illinois, 1995. Theoretical biophysics.
Rokhinson, Leonid, Ph.D., SUNY at StonyBrook, 1996. Experimental condensed matter physics.
Savikhin, Sergei, Ph.D., Tartu State Univ., Estonia, 1991. Experimental biophysics.

Assistant Professors

Chen, Yong, Ph.D., Princeton University, 2005. Experimental condensed matter physics, experimental atomic, molecular and optical physics, nanoscience and nanotechnology.
Csathy, Gabor, Ph.D., Pennsylvania State University, 2001. Experimental condensed matter physics.
Kaufmann, Birgit, Ph.D., University of Bonn, Germany, 1999. Theoretical condensed matter physics.
Kruczenski, Martin, Ph.D., University of Buenos Aires, Argentina, 1998. Theoretical high energy physics.
Lyutikov, Maxim, Ph.D., California Institute of Technology, 1998. Theoretical astrophysics.
Malis, Oana, Ph.D., Boston University, 1999. Experimental condensed matter physics.
Molnar, Dénes, Ph.D., Columbia University, 2002. Theoretical high energy nuclear physics and heavy-ion physics.
Peterson, John, Ph.D., Columbia University, 2003. Observational astrophysics.
Pushkar, Yulia, Ph.D., Freie Uni̇versitat Berlin, Germany, 2003. Experimental biophysics.
Todd, Brian, Ph.D., Case Western Reserve University, 2003. Experimental biophysics.
Xie, Wei, Ph.D., Chinese Academy of Sciences, Beijing, China, 1997. Experimental high energy nuclear physics.
Yang, Chen, Ph.D., Harvard, 2006. Experimental condensed matter physics, nanoscience.

Research Assistant Professors

Rhee, Jaehyon, Ph.D., Michigan State University, 2000. Observational astronomy and astrophysics.
Srivistava, Brijesh, Ph.D., Indian Institute of Technology, Kanpur, India, 1975. Experimental high energy nuclear physics.

Courtesy Assistant Professor

Wasserman, Adam, Ph.D., Rutgers University, 2005. Theoretical physics.

RESEARCH SPECIALTIES AND STAFF

Theoretical

Astrophysics and Relativity. Cosmology; cosmic microwave background; extra dimensions; experimental tests of general relativity; gravitation; plasma and high energy astrophysics; Pulsars and Supernova remnants, active galactive nuclei; gamma ray bursts. Fischbach, Haugan, Khlebnikov, Lyutikov.
Biophysics. Exploring the dynamics of large biomolecules (e.g., metalloproteins, DNA) using theoretical methods for normal mode analysis and simulations of Nuclear Resonant Vibrational Spectroscopy data. Theoretical and computational studies of the electronic structure and magnetic properties of metalloproteins by means of density functional theory (DFT), correlated ab-inito methods, and phenomenological spin Hamiltonians. Development of genetic algorithms in conjunction with high-performance supercomputing for the simulation of biological resonant (Moesbauer, EPR) and XANES spectroscopies. Computational neuroscience and related studies of signal propagation in the brain. Giordano, Prohofsky, Rodriguez.
Condensed Matter Physics. Low-dimensional systems: Ground state properties of two-dimensional electron systems, quantum dots and quantum wires, phase diagrams of electron liquids, Quantum Hall effect, Wigner crystal, spin-orbit interations and polarization of electronic states; transport and optical properties of quantum wells, wires and dots, Luttinger liquid, spin and charge density waves, electron-phonon interactions. Mesoscopics: quantum coherent phenomena in electronic transport, Aharonov-Bohm effect and Berry's phases, weak localization, universal conductance fluctuations, localization and metal-insulator transitions, Kondo effect, spin-dependent phenomena, spin relaxation and non-equilibrium spin polarization. Nano and Quantum physics: Quantum computation in quantum dots; Bose-Einstein condensation, optical lattices, decoherence and dissipation in nanostructures, Coulomb and spin blockade, electron spin resonance and nuclear spin resonance in quantum dots, coherent spin control. Superconductivity and Magnetism: transport in superconductors, Josephson effect, Vortices, BCS and unconventional superconductors. Strongly correlated electronic systems, electronic liquid crystals, high temperature super-

conductivity. Ferromagnetism and Antiferromagnetism, dilute magnetic semiconductors. Transport in ferromagnets, effects of domain walls, spin injection. Spintronics. Carlson, Giuliani, Hu, Kim, Lyanda-Geller, Nakanishi, Rodriguez.

Elementary Particles, Fields Theory and String Theory. Theory and phenomenology of the standard model of elementary particle interactions and its possible extensions; aspects of supersymmetry; neutrino oscillations and their application to astro-physical phenomena; dynamical symmetry breaking; renormalization group studies; cosmological phase transitions; inflationary models of the early universe; brane world models; string theory; string/gauge theory duality. Clark, Fischbach, Khlebnikov, Kuo, Kruczenski, Love.

Geophysics. Ramifications of impact cratering; planetary tectonics; physics of earthquakes and landslides. Melosh.

Nuclear Physics. From keV to TeV energies; theory of three-particle bound and scattering states in nuclear and elementary particle physics; strong, weak, and electromagnetic interactions in nuclei; nuclear many-body problem including the structure of finite nuclei; rotational states of deformed nuclei; theory of nuclear fusion reactions; solar neutrino problem; nuclear astrophysics; study of nuclear matter at extreme conditions; dynamics in relativistic nucleus-nucleus collisions; properties of the quark-gluon plasma. Kim, Molnar.

Statistical Physics. Phase transitions and critical phenomena; phenomenology of first order phase transitions; Ising systems: percolation and other clustering phenomena; quantum percolation and other quantum transport; statistical properties of surfaces and interfaces; scaling in linear and branched polymer chains; kinetics of disorderly growth processes; Brownian motion. Giuliani, Nakanishi.

Experimental

Accelerator Mass Spectrometry. Methods of nuclear physics used to operate a tandem Van de Graaff accelerator and develop new techniques for measuring long-lived radionuclides and other rare particles. Applications are in physics (neutron transport, trace impurities, cross sections), earth science (dating and tracing processes and events, global change, environment), and biological science (drug metabolism, toxicity). Caffee, Muzikar.

Atomic, Molecular and Optical Physics. Studies of two-pathway coherent control processes, including control of photoionization branching ratios, precision measurements of weak optical interactions, and control of photoelectron angular distributions and molecular processes. Coherent optical interactions in trapped ultra-cold atoms. Phase conjugate, four-wave mixing in two-level atomic systems. Bose-Einstein condensates. Quantum manipulation of atoms and molecules with lasers. Quantum information and quantum simulation of condensed matter problems with ultracold atoms and molecules. Coherence and decay in Bose-Einstein Condensates. Quantum interference effects in spin or Bose-Einstein Condensates. Chen, Elliott, Lyanda-Geller.

Applied Physics. Applications of nuclear physics to the detection of hazardous materials in commerce and public areas. Work in conjunction with the Center for Sensing Science and Technology and NSWC, Crane, Indiana to reduce terrorist threat from chemical agents, radiation threats, and explosives. Kim, Koltick.

Astronomy and Astrophysics. Studies of black holes, clusters of black holes, clusters of galaxies, dark matter, dark energy, neutron stars, Galactic structures, very metal-poor stars, horizontal-branch stars, stellar spectroscopy and abundance analysis, active galactic nuclei, relativistic jets, diffuse x-ray and infrared background, and interstellar medium; simultaneous photometry and photoelectric radial velocity observation of ultra-short-period pulsating variable stars. Satellite-based x-ray and gamma ray astronomy; ground-based very high energy gamma-ray experiments (Whipple and VERITAS); instrumentation for high energy astrophysics; radio astronomy and interferometry; optical survey telescopes (LSST). Cui, Finley, Lister, Moffett, Peterson, Rhee.

Biophysics. Modeling of real nervous systems containing small numbers of neutrons, and of learning and memory in simple neural systems. Measuring vibrational properties of metalloproteins and other biomolecules using many techniques, to explore how dynamics controls biological function. Nuclear Resonant Vibrational Spectroscopy conducted at the Argonnne x-ray synchrotron focuses on heme proteins (myoglobin, hemoglobin, cytochromes). Resonant Raman scattering and FTIR are applied to cytochromes and various model heme compounds. Terahertz time-delay spectroscopy is a new technique exploring vibrations of macromolecules beyond the far infrared. Single molecule spectroscopy is being developed to study photosynthetic complexes (PS I) using advanced laser spectroscopy methods. Live cell, single molecule imaging of membrane molecule dynamics and interactions. Durbin, Giordano, Nolte, Pushkar, Ramdas, Ritchie, Savikhin, Todd.

Condensed Matter Physics. Optical absorption, Raman and Brillouin scattering, and photoluminescence of semiconductors and their sub-micron heterostructures; effects of uniaxial stress on vibrational and electronic levels; phonons and magnetic excitations in magnetic diluted semiconductors; solid state plasmas; electron spin resonance and electron-nucleus double resonance; graphene and carbon nanotubes; Mössbauer spectroscopy; nanoscience, nanomaterials and nanodevices; non-linear optics of semiconductors and their quantum well structures; metallic surfaces; resistivity of metals; x-ray studies of quasicrystals; x-ray synchrotron physics; phase problem; x-ray standing waves; superconductivity; magneto-optics; magnetic materials; studies of mesoscopic systems; quantum transport in $GaAs/Al_xGa_{1-x}As$ microstructures; transport studies of "quantum chaos" in both open and closed systems; transport in the fractional quantum Hall effect; scanning Hall probe microscopy; studies of Si/SiO_2 interface roughness; electrical transport in one dimensional nanoscale structures and hetero-structures. Chen, Csathy, Durbin, Giordano, Malis, Manfra, Nolte, Ramdas, Reifenberger, Rokhinson, Yang.

Elementary Particles and Fields. Current experiments in particle physics include the Collider Detector at the Fermilab which is searching for the Higgs particle and SUSY particles, and studying properties of the top quark, in proton anti-proton events at the Tevatron colliding beam accelerator; and CLEO, which is performing precision studies of the weak and strong interactions using charm quarks produced in e^+e^- annihilations at the Cornell Electron Storage Ring. Over the past fifteen years the research program wit the Compact Muon Solenoid experiment at the Large Hadron Collider at CERN in Switzerland has been considerably expanded. Extensive apparatus has been designed, constructed, and installe in CMS. The LHC will begin taking data in late 2009, allowing searches for the Higgs particle, dark matter particles, Supersymmetry, and extra dimensions of space and time. Particle astrophysics work includes contributions to the design, simulation and fabrication of the optical survey telescopy, LSST, and the study of dark energy. Local facilities include state of the art laboratories for the design, fabrication and evaluation of silicon and gas microstrip detectors for particle physics and particle astrophysics, and a major computing facility dedicated to the analysis of data from the LHC. Barnes, Bortoletto, Garfinkel, Gutay, Jones, Koltick, Miller, Neumeister, Peterson, Shipsey. (For more information http://www.physics.purdue.edu/particle/)

Geophysics. Rock mechanics and physics of rocks; physical acoustics of heterogeneous materials and discontinuities; volumetric nondestructive imaging of opaque materials; hydrology and percolation physics. Pyrak-Nolte.

High Energy Nuclear and Heavy Ion Physics. Experimental studies of nuclear matter at high energy densities and temperatures, search for evidence of the formation of the quark-gluon plasma in relativistic nucleus-nucleus collisions by examining statistical and dynamical properties of the reaction products created at the Relativistic Heavy Ion Collider and the Large Hadron Collider. Hirsch, Scharenberg, Srivastava, Wang, Xie.

Physics Education Research. Focus on curricular components and pedagogical methods as it affects student understanding of physics concepts and problem solving strategies. Research on the effective means of dissemination of curriculum innovations. Bryan, Haugan, Hirsch.

UNIVERSITY OF NOTRE DAME

DEPARTMENT OF PHYSICS

Notre Dame, Indiana 46556-5670

Students Accepted For Degree	FIELDS		
	Physics	Astronomy	Related Fields
Doctorate	X		
Master's			

1. General

President: Rev. John I. Jenkins, C.S.C.
Dean, Graduate School: Gregory E. Sterling
Department Chair: Mitchell R. Wayne
Department Telephone Number: (574) 631-6386
Type of Institution: University
Control: Private
Setting: Suburban
Total Faculty: 1,236 fulltime; 128 part-time
Total Graduate Faculty: Not separated
Total Students: 11,861
Total Graduate Students: 3,444
Annual Graduate Tuition:
All Graduate Students: Full-time—$39,775
Part-time—$2,184/credit
Tuition rates for: 2010–11
Deferred tuition plan: Yes
Other Fees: $65 student fee per year, $125 Technology fee per semester, $75 health fee per semester, and $95 parking fee per year.
Term: Semester

2. Number of Faculty in Department

The combined total of full-time faculty in the three professorial ranks is 38, plus two concurrent professors. The combined total of full-time, part-time, and other faculty at all ranks is 68.

3. Admission, Financial Aid, and Housing

Address admission inquiries to: Chair, Graduate Admissions Committee, Dept. of Physics
Graduate application fee required: $35 before Dec. 1, $50 after
Admission deadline (Fall admission): Jan. 15
Admission information: For fall admission, 2009–10, 46 students were accepted from 119 applicants.
Admission requirements: For admission to the graduate programs, a Bachelor's degree in physics is required with a minimum undergraduate GPA of 3.0 in undergraduate courses in physics. The GRE is required. No minimum score is specified. The GRE Physics Test is required. No minimum score is specified. The average GRE scores including verbal, quantitative, and Advanced Physics for 2009–10 admissions were total–2,075 (verbal+quantitative+advanced) and 1,301 (verbal+quantitative+analytical). Students from non-English speaking countries are required to demonstrate proficiency in English via the TOEFL exam. No minimum score required.
Undergraduate preparation assumed: Resnick, Halliday and Krane, *Physics, Volumes 1+2*. Taylor, *Classical Mechanics*; Griffiths, *Quantum Mechanics*; Griffiths, *Introduction to Electrodynamics*, Schroeder, *Introduction to Thermal Physics*.
Address financial aid inquiries to: Director, Financial Aid
GAPSFAS application required: Yes-student loans.
Financial aid deadline: None
Loans available: Yes-Student loans.
Address housing inquiries to: Housing Office, Dean of Students
On-campus, single student housing available: Yes
Cost/term: $490–625/mo.+utilities
On-campus, married student housing available: Yes
Cost/month: $490–780/mo.+utilities

Table A—Faculty, Enrollments, and Degrees Granted

Research Specialty	2009–10 Faculty	Enrollment[1] Fall 2009		No. of Degrees Granted[2] 2009–10 (2006–10)			Median No. of Years for 2009–10 Ph.D.'s
		Master's	Doctorate	Master's	Terminal Master's	Doctorate	
Astrophysics	5	–	9	1(3)	0(2)	1(8)	6.4
Atomic, Molecular, & Optical Physics	4	–	1	0(4)	0(0)	0(2)	6.4
Biophysics	1	–	5	1(4)	1(0)	1(3)	6.4
Condensed Matter Physics	11		22	2(14)	1(1)	1(14)	6.4
Nuclear Physics	9		26	3(10)	1(1)	4(8)	6.4
Particles & Fields	9		16	1(8)	0(4)	1(12)	6.4
Undecided			11				
Total			90	8(43)	3(8)	8(47)	
Full-time Grad. Stud.			90				
Part-time Grad. Stud.			0				
First-year Grad. Stud.			11				

Median Years in Grad. Study (2006–10 Degrees)
Undergraduate Degrees, 2008–09 (2003–09): 23(95)

[1]Students not yet committed to a research specialty are entered under non-specialized.
[2]Five-year totals in parentheses.

4. Graduate Degree Requirements

Master's: The Graduate program in the Physics Department is research oriented. For that reason, the Department does not normally accept students who plan to terminate their studies with a Master's Degree. Under certain conditions a non-research Master's program is available. Students must complete 30 credit hours and maintain a grade point average of 3.0. The student must pass a comprehensive oral examination in the major field. Applicants are cautioned that financial aid is normally restricted to students pursuing Ph.D. programs of study. The minimum residence is two successive semesters.
Doctorate: Students must complete 36 credit hours and maintain a grade point average of 3.0. The minimum residency requirement for the Ph.D. degree is full-time status for four consecutive semesters. The student will normally take a sequence of basic courses in a two-year core curriculum, followed by advanced courses and seminars in specialized areas of study. Included in the core curriculum are one semester of mathematical methods in physics, one semester of experimental methods in physics, one semester each of classical mechanics and statistical mechanics, two semesters of quantum mechanics, two semesters of electrodynamics, and one semester of a research-area course (astrophysics, atomic physics, biophysics, condensed matter physics, elementary particle physics, or nuclear physics) and completion of two

semesters of a breath requirement in physics. Incoming students who have already successfully completed courses equivalent to any of those in the core curriculum will not be expected to take the corresponding core curriculum courses. However, all incoming students are required to take and pass a qualifying examination on undergraduate physics. The student is encouraged to become an active participant in research in the second semester of the first year of his graduate work. Prior to admission to candidacy for the Ph.D. degree, the student must pass comprehensive written and oral examinations. There is no foreign language requirement. Approval of the thesis by the research director and three readers and an oral defense of the thesis complete the requirements.

Thesis: Thesis may be written *in absentia*.

Table B—Appointments to Graduate Students, 2009–10

Title of Appointee	Appointments Total	Appointments First year	Academic Load Allowed in Credit Hours	Hours of Service Per Week	Stipend for Academic Year ($)
			Semester		
Teaching Assistant	45	9	12 credits	12–14	17,500[1]
Research Assistant	37	0	12 credits	12–14	17,500[1]
Fellowships	7	1	12 credits	12–14	18,000[1]
TA & RA	0	0	12 credits	12–14	17,500[1]
Total	89	10			

[1]Average.

5. Personnel Engaged in Separately Budgeted Research, 7/08–6/09

Professorial faculty	36
Other faculty	18
Postdoctoral appointments	17
Graduate students	64
Total	135

6. Separately Budgeted Research Expenditures by Source of Support for the Fiscal Year 7/1/07–6/30/08

	Departmental Research
University	$0
Other	$269,678
Federal government	$10,875,322
Total	$11,145,000

7. Extension Centers and Summer Programs

Two Undergraduate General Physics Courses offered; 4 semester credits each. NSF Research Experience for Teachers and Undergraduates Summer Program. NSF Quarknet Research Experience for High School Students Summer Program.

Table C—Separately Budgeted Research Expenditures

Research Specialty	No. of Grants	Expenditures ($)
Astronomy & Astrophysics	8	511,067
Atomic, Molecular, & Optical Physics	1	46,527
Condensed Matter Physics	10	1,721,244
Nuclear Physics	6	6,395,215
Particles & Fields	9	2,470,947
Total	35	11,145,000

*For Fiscal Year 7/1/08–6/30/09.

FACULTY

Professors

Alber, Mark S., Ph.D., University of Pennsylvania, 1990. Notre Dame Chair in Applied Mathematics and Concurrent Professor of Physics. Dynamical systems treatment of nonlinear partial differential equations with applications to biology and nonlinear optics.

Aprahamian, Ani, Ph.D., Clark, 1986. Experimental nuclear physics; gamma-ray spectroscopy, nuclear masses, lifetimes, astrophysics.

Arnold, Gerald B., Ph.D., UCLA, 1977. Theoretical solid-state physics; magnetism and high-temperature superconductivity.

Berry, H. Gordon, Ph.D., Wisconsin, 1967. Experimental atomic physics.

Bigi, Ikaros I., Ph.D., Munich, 1976. Grace-Rupley II Professor of Physics. Theoretical high-energy physics.

Blackstead, Howard A., Ph.D., Rice, 1967. Experimental physics; solid state physics; magnetism and acoustics.

Bunker, Bruce A., Ph.D.,Washington, 1980. Experimental physics; X-ray, UV, and electron spectroscopy of condensed-matter and biological/environmental systems.

Crawford, Gregory, Ph.D., Kent State University, 1991. Dean, College of Science. Liquid crystal and polymer physics, optics and solid state nuclear magnetic resonance.

Dobrowolska-Furdyna, Malgorzata, Ph.D., Polish Academy of Sciences, 1979. Experimental solid state physics.

Frauendorf, Stefan G., Ph.D., Technical University Dresden, 1971. Theoretical nuclear physics, atomic physics, mesoscopic systems.

Furdyna, Jacek K., Ph.D., Northwestern, 1960. Marquez Professor. Experimental solid state; man-made materials.

Garg, Umesh, Ph.D., SUNY, Stony Brook, 1978. Experimental nuclear physics; nuclear structure; giant resonances; gamma-ray spectroscopy; high spin states.

Garnavich, Peter, Ph.D., University of Washington, 1991. Astrophysics/observational cosmology.

Hyder, Anthony K., Ph.D., Air Force Inst. Tech., 1976. Associate Chair and Director of Undergraduate Studies. Experimental physics; space physics; nuclear physics.

Jankó, Boldizsár, Ph.D., Cornell, 1996. Theoretical condensed matter physics.

Kolata, James J., Ph.D., Michigan State, 1969. Experimental physics; nuclear structure; heavy-ion reactions; radioactive beam physics.

Livingston, A. Eugene, Ph.D., Alberta, 1974. Experimental physics; atomic physics; spectroscopy of highly ionized atoms.

LoSecco, John M., Ph.D., Harvard, 1976. Experimental and theoretical physics; high-energy elementary particle physics.

Mathews, Grant J., Ph.D., Maryland, 1977. Theoretical astrophysics/cosmology; general relativity.

Newman, Kathie E., Ph.D., Washington, 1981. Associate Chair and Director of Graduate Studies. Theoretical physics; statistical mechanics; semiconductors.

Rettig, Terrence, Ph.D., Indiana, 1976. Observational astronomy: comets, solar system formation, and T Tauri stars.

Ruchti, Randal C., Ph.D., Michigan State, 1973. Experimental physics; high-energy elementary particle physics.

Ruggiero, Steven T., Ph.D., Stanford, 1981. Experimental physics; condensed matter and low-temperature physics; superconductivity.

Sapirstein, Jonathan R., Ph.D., Stanford, 1979. Theoretical physics; quantum electrodynamics.

Tanner, Carol E., Ph.D., California, Berkeley, 1985. Experimental physics; atomic physics.

Wayne, Mitchell, Ph.D., UCLA, 1985. Chair of Physics Department. Experimental high-energy elementary particle physics.

Wiescher, Michael, Ph.D., Münster, 1980. Freimann Professor of Physics. Experimental nuclear physics; nuclear astrophysics.

Associate Professors

Balsara, Dinshaw, Ph.D., Illinois, 1990. Theoretical and computational astrophysics.

Collon, Philippe, Ph.D., Wien, 1999. Experimental nuclear physics; new techniques, AMS.

Eskildsen, Morten, Ph.D., Copenhagen 1998. Experimental physics; condensed matter.

Hildreth, Michael D., Ph.D., Stanford, 1995. Experimental high-energy elementary particle physics.

Jessop, Colin, Ph.D., Harvard, 1994. Experimental high-energy.

Kolda, Christopher, F., Ph.D., Michigan, 1995. Theoretical high-energy physics.

Toroczkai, Zoltan, Ph.D., Virginia Polytechnic Institute, 1997. Theoretical condensed matter physics; biophysics.

Assistant Professors

Caprio, Mark A., Ph.D., Yale University, 2003. Nuclear structure theory; many-body physics.

Delgado, Antonio, Ph.D., Universidad Autonoma de Madrid, 2001. Particle physics theory.

Howk, J. Christopher, Ph.D., Wisconsin-Madison, 1999. Observational astrophysics, interstellar and intergalactic media.

Lannon, Kevin P., Ph.D., University of Illinois, 2003. High energy physics.

Ptasinska, Sylwia, Ph.D., Leopold-Franzens-University, 2004. Experimental studies on electron interaction with molecules and radiation damage to DNA and its component biomolecules.

Tang, Xiao-dong, Ph.D., Texas A&M University, 2002. Experimental astrophysics and nuclear physics.

RESEARCH SPECIALTIES AND STAFF

Theoretical

Astrophysics/cosmology. Inflationary cosmology, primordial nucleosynthesis, cosmic microwave background, galaxy formation and evolution, large scale structure, neutrino physics, dark matter, stellar evolution and nucleosynthesis, neutron star binaries, gravity waves, gamma-ray bursts, supernovae. Balsara, Kolda, Mathews. 1 visiting associate professor, 5 postdoctoral research associates, 3 research visitors, 1 adjunct professor.

Atomic Physics. Quantum electrodynamics; weak interactions; atomic many-body theory; photoionization and photoexcitation. Sapirstein. 1 adjunct professor.

Biophysics. Bioinformatics, cellular networks, modeling morphogenesis. Alber, Toroczkai.

Condensed Matter. Many-body problem; high temperature superconductivity; superconductivity and magnetism on the nanoscale; tunneling phenomena; metal-metal interfaces; inhomogeneous and layered superconductors; hopping transport; studies of ordering in semiconductors, magnetic semiconductors. Arnold, Frauendorf, Jankó, Newman. 1 research assistant professor, 2 postdoctoral fellows, 2 research visitors.

Elementary Particle Physics. Formal properties of quantum field theories; supersymmetry, grand unification, spontaneous breaking symmetry, phenomenology of strong and weak processes, rare decays, CP violation; supergravity, extra dimensions, and new particles. Bigi, Delgado, Kolda. 1 research associate professor, 1 adjunct associate professor.

General Relativity. Black holes in a magnetic field, charged black holes, neutron stars, numerical relativity, gravity waves. Mathews.

Nuclear Physics. Many-body problem; nuclear reactions, few-body problem; boson expansions, structure of nuclei with momentum high angular momentum and exotic proton and neutron numbers. Caprio, Frauendorf, Mathews. 2 postdoctoral research associates, 1 adjunct professor.

Statistical Mechanics. Complex networks, Phase transitions; critical phenomena in fluids, networks, computer simulations. Newman, Toroczkai.

Experimental

Astrophysics/Astronomy. Spectra and images of comets, stellar nuclear reaction rates, high redshift supernovae, cosmological parameters. Garnavich, Howk, LoSecco, Rettig, Tang, Wiescher. 1 research professor, 2 research assistant professors, 1 adjunct professor, 2 postdoctoral fellows.

Atomic Physics. Atomic structure, parity violation, tests of fundamental symmetries, excitation mechanisms, and radiative decays in neutral and ionized atoms; precision lifetimes. Berry, Livingston, Tanner.

Biological physics and molecular environmental science: structure and function in metalloproteins, metals in biological and geo-environmental systems, biofilms. Bunker.

Condensed Matter Physics. Low-temperature physics, superconducting microwave absorption, metal and semiconductor superlattices, magnetism, magnetic resonance, magnetoelastic effects, high-temperature superconductivity, optical and far-infrared spectroscopy of semiconductors, crystal growth and MBE of semiconductors, magnetostatic effects, layered superconductors, single-electron tunneling, optical and infrared photoresponse, x-ray absorption spectroscopy and x-ray scattering, condensed matter systems. Blackstead, Bunker, Dobrowolska-Furdyna, Eskildsen, Furdyna, Ruggiero. 1 adjunct professor, 1 adjunct assistant professor, 1 research assistant professor.

High-Energy Elementary Particle Physics. Fermilab D0 experiment (study of the top quark, bottom quark, W boson and physics beyond the standard model); BaBar experiment at SLAC (CP violation in the b-quark system); CMS at CERN (search for the Higgs boson). Hildreth, Jessop, Lannon, LoSecco, Ruchti, Wayne. 1 research professor, 1 research assistant professor, 1 adjunct associate professor, 1 adjunct assistant professor, 1 adjunct assistant research professor, 1 professional specialist, 4 postdoctoral fellows, 1 guest assistant professor.

Nuclear Physics. Nuclear structure; reaction energies; electromagnetic transitions; gamma-ray spectroscopy; high spin states; polarized particles; giant resonances; heavy-ion reactions; radioactive beam studies, nuclear astrophysics. Aprahamian, Collon, Garg, Kolata, Tang, Wiescher. 2 research professors, 1 research assistant professor, 3 adjunct professors, 1 research associate, 2 postdoctoral fellows, 1 professional specialist, 1 assistant professional specialist, 1 visiting research professor, 1 visiting scholar.

Radiation Physics. Low energy electrons in condensed media; determination of product yields in the radiolysis of water, aqueous solutions, liquid hydrocarbons, and liquified rare gases; diffusion-kinetic modeling of transient species produced by ionizing radiation. Ptasinska. 1 research professor.

IOWA STATE UNIVERSITY

DEPARTMENT OF PHYSICS AND ASTRONOMY

Ames, Iowa 50011

Students Accepted For Degree	FIELDS		
	Physics	Astronomy	Related Fields
Doctorate	X	X	
Master's	X	X	

1. General

President: Gregory Geoffroy
Provost: Elizabeth Hoffman
Department Chairman: Joseph Shinar
Department Telephone Number: (515) 294-5441
Type of Institution: University
Control: Public
Setting: Small town
Total Faculty: 1,746
Total Graduate Faculty: 1,349
Total Students: 27,945
Total Graduate Students: 5,424
Annual Graduate Tuition:
In-state residents: Full-time—$6,716[1]
Out-of-state residents: Full-time—$17,816[1]
Tuition rates for: 2009–10
Deferred tuition plan: Yes
Other Fees: $333[2], $92[3] per semester
Term: Semester

[1]Graduate students on 1/4 time or greater assistantships are assessed at the resident rate. In addition, a scholarship of 1/4 tuition is given to students with 1/4 time assistantships, and 1/2 tuition for students with 1/2 time assistantships.
[2]Health and Activity fees.
[3]Computer fee.

2. Number of Faculty in Department

The combined total of full-time faculty in the three professorial ranks is 40. The combined total of full-time, emeritus, adjunct faculty and affiliate at all ranks is 73.

3. Admission, Financial Aid, and Housing

Address admission inquiries to: Josheph Shinar, Chairman, Department of Physics and Astronomy
Graduate application fee required: $30 domestic, $70 foreign
Admission deadline (Fall admission): 2/15
Admission information: For fall admission, 2009–10, 15 students were accepted from 198 applicants.
Admission requirements: For admission to the graduate programs, a Bachelor's degree is required with no minimum undergraduate GPA specified. The General and Physics GRE scores are required. The average GRE scores for 2009–10 admissions were verbal-459; quantitative-756; analytical-6.72. The average GRE Physics score for admissions was 722. Students from non-English speaking countries are required to demonstrate proficiency in English via the TOEFL exam. Minimum acceptable score for admission is 550. They are also advised to take the Test of Spoken English.
Undergraduate preparation assumed: Saxon, *Elementary Quantum Mechanics*; Marion, *Classical Dynamics*; Kittel, *Thermal Physics*; Lorrain and Corson, *Electromagnetic Fields and Waves*.
Address financial aid inquiries to: Director of Student Financial Aids, 210 Beardshear
GAPSFAS application required: No
Financial aid deadline: 3/1
Loans available: Yes.
Address housing inquiries to: Director of Residence, 2419 Friley
On-campus, single student housing available: Yes
Cost: Apt. $482–584/mo.
Grad. Dorm $530/mo. single; $408/mo double; meal plan (unlimited) avail. $2,064/sem.
On-campus, married student housing available: Yes
Cost/month: $482–584

Table A—Faculty, Enrollments, and Degrees Granted

Research Specialty	2009–10 Faculty	Enrollment[1] Fall 2009		No. of Degrees Granted[2] 2009–10 (2005–10)			Median No. of Years for 2009–10 Ph.D.'s
		Master's	Doctorate	Master's	Terminal Master's	Doctorate	
Applied Physics	1	0	7	0(0)	0(0)	0(1)	–
Astrophysics	1	0	2	0(0)	0(6)	0(0)	–
Atmos./Space Phys., Cosmic Rays	5	1	11	0(0)	4(1)	0(0)	–
Biophyiscs	2	0	2	0(0)	0(0)	0(0)	–
Condensed Matter Physics	15	0	47	0(1)	1(6)	9(28)	–
Nuclear Physics	7	0	16	1(2)	1(3)	1(4)	–
Particles & Fields	9	1	9	0(0)	0(0)	0(4)	–
Non-specialized	0	0	0	0(0)	0(1)	0(0)	–
Total	40	2	92	1(3)	6(17)	10(37)	
Full-time Grad. Stud.		2	92				
Part-time Grad. Stud.		0	0				
First-year Grad. Stud.		5	14				

Undergraduate Degrees, 2009–10 (2009–10):6(55)

[1]Students not yet committed to a research specialty are entered under non-specialized.
[2]Five-year totals in parentheses.

4. Graduate Degree Requirements

Master's: The Master of Science degree is offered with and without thesis in various areas of physics (applied physics, atmospheric, high-energy, nuclear, condensed matter as examples) and astronomy. The minimum residential requirement is 30 credits, at least 21 of which must be in physics department graduate courses and 6 of which must be outside the major area. A "B" average (3.0 GPA) must be maintained. There is no foreign language requirement.
Doctorate: The Ph.D. degree in the same areas has a basic requirement of 72 credits, at least one-half of which must be earned at Iowa State University, and 12 of which must be outside the major area. A "B" average (3.0 GPA) must be maintained. There is no foreign language requirement. A qualifying examination given at the beginning of the student's second year, a preliminary oral examination, and a final thesis defense are the other major requirements.
Other Programs: Close relationships exist with the Chemistry, Geological and Atmospheric Sciences, Electrical Engineering and Computer Engineering, Materials Science and Engineering, Computer Science, and Mathematics Departments, and joint programs are possible.

Thesis: Thesis may be written *in absentia*.

Special Equipment, Facilities, or Programs: Five automated ultrasensitive SQUID magnetometers; helium dilution refrigerator millikelvin facility; two rotating anodes and several conventional x-ray diffraction facilities in combination with high and low temperature units and a new liquid surface reflectometer; ultra-high vacuum systems for surface physics studies using LEED, RHEED, and STMs; low temperature, high field magneto-optic spectrometer; microelectronics center for thin-film deposition (MBE, *e*-beam, etc.) and characterization (EELS, x-ray, Auger, ...); magnetic resonance spectrometers with superconducting solenoids for both high- and ultra low-temperature studies; precision spectrometers for neutron scattering and photoemission spectrometry (both carried out at national facilities); high-resolution Ge gamma-ray detectors; a campus-wide network of about 800 workstations linked with fiberoptic (FDDI) and Ethernet connections and running modified Athena software, over 60 of these fast computers are located in the Department, there are an additional 50 networked Unix, and SUN workstations, IBM RISC 6000 Models 550–590 and many MacIntoshes, PC's and windowing terminals, the network is connected to all national labs and supercomputing centers through 45 Mbits/s(T3) lines to the NFSnet/Internet. In addition, there are four multi processor Silicon Graphics Computers, and many PC-clusters ranging up to 128 processors. The clusters run with Linux and communicate via fast switches and either fast or gigabit-ethernet. Electronics and machine shop support; scanning Auger (600 Å resolution) and atomic resolution electron microscopes are available. Research facilities are also utilized at the following laboratories: Fermilab (Batavia, IL), CERN (Geneva, Switzerland), BNL (Upton, NY), SRC (Stoughton, WI), ORNL (Oak Ridge, TN), Advanced Photon Source and Intense Pulsed Neutron Source at Argonne National Laboratory (Argonne, IL), and Stanford Linear Accelerator (Palo Alto, CA). High-energy physics programs include collaboration on: the BABAR experiment at the PEPII B-factory detector, the DELPHI experiment at the CERN LEP collider, and the *D*0 experiment at the Fermilab Tevatron Collider, CMS, and ATLAS at the CERN LHC Collider. A facility for testing high-speed electronics components has been used to evaluate electronics for use in high-energy physics experiments and for applications outside high-energy physics, e.g., the human genome project. The Nuclear Physics group constructed the Level-1 trigger for the PHENIX detector used at the RHIC accelerator (Brookhaven) to search for the quark-gluon plasma. It is also using PHENIX to study nuclear matter under extreme conditions of temperature and density and to probe the properties of the QCD vacuum. It constructed the Late-Energy Trigger for the E864 experiment at the AGS accelerator (Brookhaven) which was used in the search for strange quark matter. Studies on electromagnetic dissociation in relativistic heavy-ion reactions is being carried out. Observational astronomy is pursued at all wavelengths at ISU. The gamma-ray astronomy group is collaborating with the Harvard-Smithsonian Center for Astrophysics, the University of Michigan, and several European groups in a program of very high-energy gamma-ray astronomy centered at the Smithsonian's Whipple Observatory. Optical data are obtained at ISU's Erwin W. Fick Observatory, which houses a 24-in. Cassegrain telescope. Observations are made with the coude spectrograph and direct imaging with the 800 × 800 pixel CCD camera. Optical and infrared data are also obtained with the Hubble Space Telescope, and telescopes at KPNO, CTIO, UKIRT and the Anglo Australian Observatory. Far-IR studies utilize the "Infrared Astronomical Satellite" (IRAS) database and new image processing tools. Ultraviolet and x-ray observations are obtained with the IUE satellite. Spectral line and radio continuum studies are performed with the Very Large Array (New Mexico) and the Australia Telescope National Facility (ATNF). Related research projects in planetary science and meteoritics are carried out in the Geological and Atmospheric Sciences Dept.

Table B—Appointments to Graduate Students, 2009–10

Title of Appointee	Appointments: Total	Appointments: First year	Academic Load Allowed in Credit Hours	Hours of Service Per Week	Stipend for Academic Year ($)
			Semester		
Teaching Assistant	40	15	12	20	13,725–15,300[1]
Research Assistant	52	0	12	20	13,725–15,300[1,2]
Fellows	2	0	12	–	
Total	94	15			

[1]The University provides health insurance for all graduate assistants.
[2]Most advanced research assistants and fellows hold 12-month appointments.

5. Personnel Engaged in Separately Budgeted Research, 7/09–6/10

Professorial faculty	40
Adjunct/Affiliate faculty	6
Postdoctoral appointments	13
Graduate students	94[1]
Non-teaching research personnel	5
Total	158

[1]Most first-year students also receive research support during summer.

6. Separately Budgeted Research Expenditures by Source of Support

	Departmental Research	Physics-related Research Outside Department
Federal government	$19,364,542	$
Other	1,104,364	
Total	$20,468,906	$

7. Separately Funded and Managed Laboratories

Iowa State University operates the Ames Laboratory, a national laboratory of the U.S. Department of Energy. Its current annual budget is approximately $29,000,000. This multidisciplinary laboratory supports faculty and student research in physics and other fields. The number of Ames Laboratory staff involved in thesis supervision is included in section 5, and the physics research support at Ames Laboratory is included in Table C. Other research in areas of applied physics is conducted in several centers maintained by the University's Institute for Physical Research and Technology. Current funding for these projects through the Institute is approximately $18,000,000. Iowa State University also operates a microelectronics center to conduct basic and applied research on electronic and photonic materials, devices, and applications. Research at this laboratory is carried out by its own staff and by faculty and graduate students in physics, electrical and chemical engineering.

Table C—Separately Budgeted Research Expenditures

Research Specialty	No. of Grants	Expenditures ($)
Astrophysics	12	483,083
Biophysics	1	241,250
Condensed Matter Physics	18	17,954,859
High-Energy Physics	6	798,312
Nuclear Physics	8	991,402
Total	41	12,707,309

[1]Includes Institutional Funds.

FACULTY

Professors

Anderson, E. Walter, Ph.D., Columbia, 1965. Experimental physics.

Canfield, Paul C., Ph.D., California, Los Angeles, 1990. Distinguished Professor of Liberal Arts & Science. Experimental physics; design, growth, and characterization of new correlated electron materials.

Goldman, Alan, I., Ph.D., SUNY, Stony Brook, 1984. Distinguished Professor of Liberal Arts and Sciences. Experimental physics; x-ray scattering.

Harmon, Bruce N., Ph.D., Northwestern, 1973. Distinguished Professor of Liberal Arts and Sciences. Theoretical physics; superconducting; magnetic; optical; and lattice dynamical properties of solids.

Hauptman, John M., Ph.D., California, Berkeley, 1974. Experimental physics; high-energy.

Hill, John C., Ph.D., Purdue, 1966. Experimental nuclear physics; relativistic heavy ion physics.

Ho, Kai-Ming, Ph.D., California, Berkeley, 1978. Distinguished Professor of Liberal Arts and Science. Theoretical physics; properties of solids and surfaces.

Johnston, David C., Ph.D., California, San Diego, 1975. Distinguished Professor of Liberal Arts and Sciences. Experimental physics; synthesis and properties of new materials; materials science; magnetic thermal and electric property measurement; high- and low-temperature superconductors.

Kawaler, Steven D., Ph.D., Texas, 1986. Theoretical astrophysics; stellar evolution; stellar pulsation; stellar seismology.

Krennrich, Frank, Ph.D. Lud. Max. Univ., Munich, 1996. Experimental particle astrophysics; gamma-ray astronomy.

Lajoie, John, Ph.D. Yale, 1996. Experimental nuclear physics; relativistic heavy-ion physics.

Luban, Marshall, Ph.D., Chicago, 1962. Theoretical physics; condensed matter.

Ogilvie, Craig A., Ph.D., 1987. Experimental nuclear physics; relativistic heavy-ion physics.

Qiu, Jianwei, Ph.D., Columbia, 1987. Theoretical physics; high-energy and nuclear.

Rosati, Marzia, Ph.D., McGill, 1992. Experimental nuclear physics; relativistic heavy-ion physics.

Rosenberg, Eli I., Ph.D., Illinois, 1971. Experimental physics; high-energy.

Schmalian, Jörg, Ph.D., Freie University Berlin, 1993. Theoretical condensed matter physics, correlated fermion systems.

Shinar, Joseph, Ph.D., Hebrew Univ., 1980. Dept. Chair. Experimental physics; magnetic resonance; optically detected magnetic resonance; optical properties of semiconductors and conducting polymers.

Soukoulis, Costas M., Ph.D., Chicago, 1978. Distinguished Professor of Liberal Arts and Sciences. Theoretical physics; disordered systems; electron and optical localization amorphous magnetism, numerical simulations.

Struck, Curtis, Ph.D., Yale, 1981. Theoretical astrophysics; cosmology; galaxy formation and evolution.

Tringides, Michael, Ph.D., Chicago, 1984. Experimental physics: surface science; LEED; RHEED scanning tunneling microscopy.

Valencia, German, Ph.D., Massachusetts, 1988. Theoretical physics; high-energy.

Vary, James P., Ph.D., Yale, 1970. Theoretical physics; nuclear structure and reactions; quark and gluon structure of hadrons and nuclei.

Whisnant, Kerry, L., Ph.D., Wisconsin, 1982. Theoretical physics; particle phenomenology.

Willson, Lee Anne, Ph.D., Michigan, 1973. University Professor. Stellar astrophysics; mass loss processes; atmospheric structure and dynamics; interpretation of observations.

Professors Emeriti

Barnes, Richard G., Ph.D., Harvard, 1952. Emeritus. Experimental physics; nuclear magnetic resonance.

Borsa, Ferdinando, Ph.D., Pavia, 1961. Experimental solid state physics; nuclear magnetic resonance; phase transitions; superconductivity.

Carter-Lewis, David A., Ph.D., Michigan, 1974. Experimental particle astrophysics; gamma-ray astronomy.

Clem, John R., Ph.D., Illinois, 1965. Emeritus; Distinguished Professor of Liberal Arts and Sciences. Theoretical physics; superconductivity.

Crawley, H. Bert, Ph.D., Iowa State, 1966. Experimental physics; high-energy.

Finnemore, Douglas K., Ph.D., Illinois, 1962. Distinguished Professor of Liberal Arts and Sciences. Experimental physics; superconductivity and very low-temperature phenomena.

Firestone, Alexander, Ph.D., Yale, 1966. Emeritus. Experimental physics; high-energy.

Fuchs, Ronald, Ph.D., Illinois, 1957. Emeritus. Theoretical physics; optical properties of solids.

Hodges, Laurent, Ph.D., Harvard, 1966. Energy and environmental physics; physics education.

Kelly, William H., Ph.D., Michigan, 1955. Experimental physics; nuclear structure and reactions; physics education.

Lamb, Richard C., Ph.D., Kentucky, 1963. Emeritus. Experimental astrophysics; gamma-ray astronomy.

Lassila, Kenneth E., Ph.D., Yale, 1961. Theoretical physics; high energy.

Lynch, David W., Ph.D., Illinois, 1958. Distinguished Professor of Liberal Arts and Sciences. Experimental physics; optical properties and photoelectron spectroscopy of solids.

Peterson, Francis C., Ph.D., Cornell, 1968. Physics education.

Pursey, Derek L., Ph.D., Glasgow, 1952. Emeritus. Theoretical physics; super-symmetric quantum mechanics; foundations of quantum theory: nonlinear physics.

Ross, Dennis K., Ph.D., Stanford, 1968. Emeritus. Theoretical physics; general relativity; fundamental constants; geometrical models of physics.

Stanford, John L., Ph.D., Maryland, 1965. Emeritus. Theoretical and experimental atmospheric physics.

Swenson, Clayton A., D.Phil., Oxford, 1949. Emeritus. Distinguished Professor of Liberal Arts and Sciences. Experimental physics: thermodynamic properties at low temperatures and high pressures.

Weber, Thomas A., Ph.D., Notre Dame, 1961. Theoretical physics; ill-posed problems scattering theory.

Williams, Stanley A., Ph.D., Rensselaer, 1962. Theoretical physics; quark structure of the nucleus.
Wohn, Fred K., Ph.D., Indiana, 1967. Experimental physics; relativistic heavy-ion physics.
Wolford, Donald, Ph.D., Illinois, 1978. Experimental physics: semiconductors; electronic structures; optical properties.
Young, Bing-lin, Ph.D., Minnesota, 1966. Theoretical physics; high-energy.

Associate Professors

Cochran, James H., Ph.D. SUNY, Stony Brook, 1993. Experimental physics; high energy.
Furukawa, Yuji, Ph.D., Kobe University, 1995. Experimental physics; condensed matter.
Kaminski, Adam, Ph.D., University of Illinois at Chicago, 2001. Experimental condensed matter physics.
McQueeney, Robert J., Ph.D., University of Pennsylvania, 1996. Experimental Condensed Matter Physics.
Prell, Soren A., Ph.D., University of Hamburg, 1996. Experimental physics; high-energy. Study of time-dependent CP violation in B meson decays.
Prozorov, Ruslan, Ph.D., Bar-Ilan University 1998. Experimental physics in co-existence and mutual influences of superconductivity and magnetism.
Travesset, Alex, Ph.D., Universitat de Barcelona, 1997. Theoretical condensed matter physics, RG: techniques and applications.
Yu, Edward, Ph.D., Michigan, 1997. Study of biological systems.

Assistant Professors

Chen, Chunhui, Ph.D., Univ. of Pennsylvania, 2003. Experimental Physics; high-energy.
Kerton, Charles, Ph.D., 2000. University of Toronto, Observational Astronomer
Marengo, Massimo, Ph.D., International School for Advanced Studies of Trieste (Italy) 2000. Observational Astronomer.
Sánchez, Mayly, Ph.D., Tufts University, 2003. Experimental Physics; high-energy.
Sivansankar, Sanjeevi, Ph.D., University of Illinois at Urbana-Champaign, 2001. Experimental biophysics.
Tuchin, Kirill, Ph.D., Tel Aviv University, 2001. Theoretical nuclear physics; QCD at high energy.
Wang, Jigang, Ph.D., Rice University, 2005. Experimental physics, condensed matter.
Weinstein, Amanda, Ph.D., Stanford University, 2005. Experimental particle astrophysics.

Affiliate Professor of Physics

Rose, James H., Ph.D., California, San Diego, 1976. Theoretical and applied physics; energetics of metals; acoustics and inverse scattering theory.

Affiliate Assistant Professor

Guo, Xiaofeng, Ph.D., Iowa State University, 1996. Theoretical nuclear physics.

Adjunct Research Professors of Physics

Meyer, W. Thomas, Ph.D., Cornell, 1971. Experimental physics; high-energy.
Vaknin, David, Ph.D., Jerusalem, 1987. Experimental physics; condensed matter; neutron and x-ray scattering; thin organic/biological films; magnetic materials; high temperature superconductors.

Adjunct Research Associate Professors of Physics

Antropov, Vladimir, Ph.D., Ural Politechnical Institute (Sverdlovsk), Physical and Technical department (theoretical physics). Condensed Matter Physics, solid state physics; electronic structure and magnetic properties of alloys of transition metals.
Biswas, Rana, Ph.D., Cornell, 1984. Theoretical physics; electronic and structural properties of solids and surfaces.
Bud'ko, Sergey L., Ph.D., Moscow, 1986. Solid state physics, transport, thermodynamics and magnetic measurements of novel materials.
Kogan, Vladimir G., Ph.D., Technion-Israel, 1977. Theoretical physics; anisotropic superconductors.

Lecturers

Atwood, David, Ph.D., McGill, 1989. Theoretical physics; high-energy.
Fretwell, Helen, Ph.D., Bristol, 1992. Experimental condensed matter physics.
Herrera-Sikoldy, Paula, Ph.D., Barcelona, 1999. Theoretical physics: high-energy.

RESEARCH SPECIALTIES AND STAFF

Theoretical

Applied Physics. Photonic devices. Ho, Soukoulis, 1 postdocoral associate.
Astronomy/Astrophysics. Stellar evolution, stellar winds, and mass loss; pulsating and variable stars. Galaxy formation and evolution, galaxy collisions, and star formation and population evolution in galaxies. Kawaler, Pohl, Struck, Willson.
Condensed Matter. Photoemission; surface properties; optical properties; magnetic properties, electronic structure, lattice dynamics, critical phenomena; localization in disordered and quasiperiodic solids; spin glasses; quantum nanostructures; quantum compuling photonic band gaps, many body theory, and simulations. Antropov, Biswas, Harmon, Luban, Ho, Kogan, Schmalian, Soukoulis, Travesset. 5 Ames Lab physicists. 5 postdoctoral associates, 3 long-term visitors.
High Energy Physics. Phenomenology of the standard model and extensions; neutrino mass and oscillations; quantum chromodynamics; dynamical symmetry breaking and chiral perturbation theory. Atwood, Qiu, Valencia, Whisnant, 4 postdoctoral associate.
Low-Temperature Physics. Superconducting vortex pinning; superconducting-normal proximity effects; Josephson junction arrays; electron-phonon interactions; magnetism. Clem, Harmon, Kogan, Schmalian.
Nuclear Physics. Quark and gluon interactions in nuclei; relativistic heavy-ion and intermediate energy interactions; high-energy reactions of leptons and hadrons with nuclei; nuclear structure. Qiu, Vary. 2 postdoctoral associates.

Experimental/Observational

Applied Physics. Superconducting materials. Finnemore. Magneto-optic devices. Lynch. Amorphous material devices. Shinar. Semiconductors.
Astronomy/Astrophysics (Obs.). Radio, infrared, and optical studies of galaxies; image processing techniques; studies of galaxy clusters at high redshift; photoelectric stellar radial

velocity measurements; multi-wavelength spectroscopy and photometry of variable stars; stellar seismology; identification and study of TeV gamma-ray sources. Carter-Lewis, Kawaler, Kerton, Krennrich, 1 observatory manager, 2 postdoctoral associates.

Condensed Matter. Optical properties; photoemission; neutron scattering; x-ray diffraction; magnetism; thermodynamic measurements; electrical properties; nuclear magnetic resonance; surface studies; thin films; transport properties; new materials design and growth; anisotropic superconductors; x-ray scattering studies of surfaces. Borsa, Canfield, Goldman, Johnston, Lynch, McQueeney, Shinar, Tringides, Vaknin. 6 Ames Lab physicists. 5 postdoctoral associates.

Energy Sources and Environment. Energy conservation; solar energy; radon in homes. Hodges.

High-Energy Physics. Z° studies and b-production in $e^{+}e\mathbf{rs}^{-}$ annihilations, CP violation in the b-quark; $\bar{p}p$ interactions at 1.8 TeV, Higgs particles, heavy leptons, exotic quarks, and top quark studies; quark and gluon jets; glueballs; high-energy neutrino astronomy. Experiments at CERN, SLAC, and Fermilab. Anderson, Cochran, Crawley, Hauptman, Meyer, Prell, Rosenberg, 1 staff physicist. 3 postdoctoral associates.

Low-Temperature Physics. Superconductivity; magnetic molecules; new high-T_c compounds. Borsa, Canfield, Finnemore, Johnston.

Nuclear Physics. Search for quark-gluon plasma and strange quark matter with RHIC collider and AGS accelerator. Studies of nuclear matter under extreme conditions of density and temperature using the PHENIX detector. Studies of electromagnetic dissociation at the CERN-SPS. Hill, Lajoie, Ogilvie, Rosati. 1 staff physicist, 2 postdoctoral associates.

THE UNIVERSITY OF IOWA

DEPARTMENT OF PHYSICS AND ASTRONOMY

Iowa City, Iowa 52242

Students Accepted For Degree	FIELDS		
	Physics	Astronomy	Related Fields
Doctorate	X		X
Master's	X	X	X

M.S. degree is not prerequisite to the Ph.D. Ph.D. not offered in astronomy. Students who wish to pursue program in astronomy beyond M.S. level may qualify for Doctor of Philosophy degree in physics with specialization and a dissertation in astronomy or astrophysics.

1. General

President: Sally Mason
Dean, Graduate College: John C. Keller
Department Chair: Mary Hall Reno
Department Telephone Number: (319) 335-1686
Fax: (319) 335-1753
Web Site: http://www.physics.uiowa.edu
Email: admissions@newton.physics.uiowa.edu
Type of Institution: University
Control: Public
Setting: Small city
Total Faculty: 2,214
Total Graduate Faculty: 1,698
Total Students: 30,328
Total Graduate: 5,720
Annual Graduate Tuition:
In-state residents: Full-time—$3,420/semester
Out-of-state residents: Full-time—$10,222/semester
Tuition rates for: 2009–10
Deferred tuition plan: Yes
Other Fees: Computer fee (full-time) $208.50
Health fee (full-time) $109.50
Student activity fee (full-time) $31.50
Student services fee (full-time) $36.50
Student union fee (full-time) $54.00
Building fee (full-time) $59.50
Arts & Cultural Events Fee (full-time) $12.00
Term: Semester

2. Number of Faculty in Department

The combined total of full-time faculty in the three professorial ranks is 28. The combined total of full-time, part-time, and other faculty at all ranks is 31. There are 9 professors emeriti.

3. Admission, Financial Aid, and Housing

Address admission inquiries to: Dean of Admissions, Calvin Hall
Graduate application fee required: $60 for domestic; $85 for foreign
Admission deadline (Fall admission): Foreign and domestic applications for admission with financial support are due by January 15. Highest priority for departmental financial support is given to those applications received by January 15.
Admission information: For fall admission, 2010–11, 15 new students were enrolled from 166 applicants.
Admission requirements: For admission to the graduate programs, a Bachelor's degree in physics and/or astronomy is required with a minimum undergraduate record of B. The GRE is required. The minimum acceptable score is not specifically stated. The GRE Advanced is strongly recommended. The minimum acceptable score is not specifically stated. Students from non-English speaking countries are required to demonstrate proficiency in English via the TOEFL exam. Minimum acceptable score for admission is 550, or 81 on the iBT test.
Undergraduate preparation assumed: Griffiths, *Quantum Mechanics*; Griffiths, *Introduction to Electrodynamics*; Fowles and Cassiday, *Analytical Mechanics*; Reif, *Statistical and Thermal Physics*.
Address financial aid inquiries to: Director of Student Financial Aid, Calvin Hall
GAPSFAS application required: No
Financial aid deadline: No definite deadline
Loans available: Yes
Address housing inquiries to: Housing, Burge Hall. Phone: 319/335-3009
On-campus, single student housing available: Yes
Cost/Academic Year: *
On-campus, married student housing available: Yes
Cost/month: **

*Double Room: $4,997–5,363/Academic Year Room
$1,195–2,665 academic year board plan
Residence hall options include air conditioning, shared bath facilities and shared kitchen facilities. Board plans range from 20 meals a week (full board) to any 14 or any 10 meals a week, with students choosing which meals make up the total.
**One Bedroom Apt.: $435/mo.+gas & electricity
Two Bedroom Apt.: $480/mo.+gas & electricity or $600/mo. +electricity

Table A—Faculty, Enrollments, and Degrees Granted

Research Specialty	2009–10 Faculty	Enrollment[1] Fall 2009		No. of Degrees Granted[2] 2009–10 (2005–10)			Median No. of Years for 2009–10 Ph.D.'s
		Master's	Doctorate	Master's	Terminal Master's	Doctorate	
Astronomy/ Astrophysics	6	1	10	1(3)	0(4)	0(2)	–
Atmos./Space Phys., Cosmic Rays	4	0	3	0(5)	0(2)	0(5)	–
Condensed Matter and Materials Physics	7	0	16	0(6)	0(3)	1(5)	7
Medical Imaging Physics & Positron Emission Tomography (PET)	2	0	1	0(0)	0(2)	0(0)	–
Nuclear Physics	1	0	2	0(2)	0(0)	0(1)	–
Particles & Fields	7	1	18	2(3)	0(1)	3(5)	5.5
Photonics & Quantum Electronics	6	1	2	0(0)	0(0)	0(0)	–
Plasma Physics & Fusion	5	1	7	0(1)	0(0)	0(2)	–
Non-specified	0	1	6	0(0)	0(0)	0(0)	–
Total		5	65	3(20)	0(12)	4(20)	
Full-time Grad. Stud.		5	65				
First-year Grad. Stud.		1	5				
Median Years in Grad. Study (2009–10 Degrees)				2.3	–	6.25	6.25

Undergraduate Degrees, 2009–10 (2005–10): 9(69)

[1] Students not yet committed to a research specialty are entered under non-specialized.
[2] Five-year totals in parentheses.
[3] This number indicates number of years spent at Iowa.

4. Graduate Degree Requirements

Master's: 30 hours of course work and research with a grade point average of at least 3.00 and an oral final examination are required. Thesis or critical essay options are available. No foreign language requirement is specified. The residence requirement may be fulfilled by completing a minimum of 24 semester hours under the auspices of The University of Iowa.

Doctorate: A minimum of 72 hours of course work and research with a grade point average of at least 3.00, a qualifying examination, comprehensive examination, participation in advanced seminars, and original research are required for the Ph.D. A candidate for the degree will not be recommended until he/she has written the dissertation in proper form for formal publication and has submitted it, with the approval of the research advisor, for publication to a standard scientific journal of wide distribution. There is no specific foreign language requirement. Beyond the first 24 semester hours of graduate work, the residence requirement may be fulfilled by either 1) enrollment as a fulltime student (9 semester hours minimum) in each of two semesters, or 2) enrollment for a minimum of 6 semester hours in each of 3 semesters during which the student holds at least a one-third time assistantship certified by the department as contributing to the student's doctoral program.

Thesis: Thesis may be written *in absentia*.

Special Equipment, Facilities, or Programs: Comprehensive facilities for design, construction, and testing of instruments for space flight and for the decoding, analysis, and display of flight data. Automated 37 cm optical telescope at remote dark site, 3.0 and 4.5 m radio telescopes for instrumentation development. Steady-state magnetized plasma devices (gas discharges and Q-machines) with 1–10 kG magnetic fields. A wide array of gas discharges and diagnostics for the study of plasmas containing charged dust grains. Compact medical cyclotron for production of radionuclides. Molecular-beam epitaxy machines for growth of state-of-the-art III–V semiconductor quantum wells, superlattices, quantum dots, and optoelectronic devices. A large computer cluster for analysis of high-energy nuclear data with direct connections to Fermilab and CERN. A state of the art computer controlled photomultiplier test station for high-energy and nuclear physics detectors. X-ray astronomical instrumentation facility for design of x-ray polarimeters and high-resolution timing instruments to be flown on x-ray space missions. High performance computer cluster with 64 nodes for large-scale kinetic simulation of plasmas. A large number of continuous-wave and pulsed (including ultrafast) lasers are available for spectroscopy.

Table B—Appointments to Graduate Students, 2009–10

Title of Appointee	Appointments: Total	Appointments: First year	Academic Load Allowed in Credit Hours	Hours of Service Per Week	Stipend for Academic Year ($)
			Semester		
Teaching Assistant	28	10	15 max.	20	17,850-19,350[1,2]
Research Assistant	38	6	15 max.	20	17,850-19,350[1,2]
Fellowship	0	0	15 max.	0	17,850-19,350[1,2]
Total	66	16			

[1] Teaching and Research Assistants must pay tuition and fees, set at $7,863 for the academic year 2009–10 for full-time graduate students, but these costs are partially offset by a tuition scholarship of $5,230. The tuition scholarship increases to $7,224 in 2010–11.

[2] Summer appointments are available.

5. Personnel Engaged in Separately Budgeted Research, 7/09–6/10

Professorial faculty	28
Other faculty	12
Postdoctoral appointments	16
Graduate students	40
Undergraduate students	34
Nonteaching research personnel	43
Total	176

6. Separately Budgeted Research Expenditures by Source of Support (Fiscal year 2009)

	Departmental Research
Federal government	$19,329,417
State and local government	2,216,053
Other	6,343
Total	$21,551,813

Table C—Separately Budgeted Research Expenditures

Research Specialty	No. of Grants	Expenditures ($)
Astronomy and Astrophysics	26	1,117,384
Atmos./Space Phys., Cosmic Rays	41	13,985,917
Atomic, Molecular & Optical Physics	0	0
Condensed Matter Physics	28	2,785,200
Nuclear Physics	2	232,969
Particles & Fields	16	1,411,436
Plasma Physics & Fusion	9	942,137
Graduate College and General Research Support from State of Iowa	18	1,076,770
Total	140	21,551,813

FACULTY

Professors

Andersen, David R., Ph.D., Purdue, 1986. Nonlinear optics; quantum electronics; solid state; embedded systems.

Boggess, Thomas F., Ph.D., North Texas State, 1982. Nonlinear optics; ultrafast spectroscopy of semiconductor heterostructures.

Carpenter, Raymon T., Ph.D., Northwestern, 1962. Professor Emeritus. Experimental plasma physics.

Flatté, Michael E., Ph.D., California, Santa Barbara, 1992. Condensed matter and materials theory.

Goree, John A., Ph.D., Princeton, 1985. Experimental plasma physics; biomedical applications of plasmas; soft condensed matter.

Gurnett, Donald A., Ph.D., Iowa, 1965. Experimental space plasma physics.

Hichwa, Richard, Ph.D., Wisconsin-Madison, 1981. Medical physics.

Kaaret, Philip E., Ph.D., Princeton, 1989. X-ray astronomy and instrumentation; black hole binaries and jet ejection from black holes.

Kleiber, Paul D., Ph.D., Colorado, 1981. Associate Chair. Atmospheric physics, chemical physics.

Kletzing, Craig A., Ph.D., California, San Diego, 1989. Experimental space plasma physics; laboratory plasma physics.

Klink, William H., Ph.D., Johns Hopkins, 1964. Professor Emeritus. Theoretical nuclear physics; mathematical physics.
Knorr, Georg, Ph.D., Munich, 1963. Professor Emeritus. Theoretical plasma physics.
Lonngren, Karl E., Ph.D., Wisconsin, 1964. Professor Emeritus. Experimental plasma physics.
Madsen, Mark T., Ph.D., Wisconsin, 1979. Medical physics.
Mallik, Usha, Ph.D., CUNY, 1978. Experimental elementary particle physics.
McCliment, Edward R., Ph.D., Illinois, 1962. Professor Emeritus. Elementary particle physics.
Merlino, Robert L., Ph.D., Maryland, 1980. Experimental plasma physics.
Meurice, Yannick, Ph.D., UCL Louvain-la-Neuve, 1985. Theoretical elementary particle physics.
Mutel, Robert L., Ph.D., Colorado, 1975. Radio astronomy. Space physics, plasma astrophysics.
Neff, John S., Ph.D., Wisconsin, 1961. Professor Emeritus. Observational optical astronomy.
Norbeck, Edwin, Ph.D., Chicago, 1956. Professor Emeritus. Experimental nuclear and elementary particle physics.
Onel, Yasar, Ph.D., London, 1975. Experimental elementary particle and nuclear physics.
Payne, Gerald L., Ph.D., California, San Diego, 1967. Professor Emeritus. Theoretical nuclear physics.
Polyzou, Wayne N., Ph.D., Maryland, 1979. Theoretical nuclear physics, mathematical physics.
Reno, Mary Hall, Ph.D., Stanford, 1985. Chair of the Department. Theoretical elementary particle physics.
Rodgers, Vincent G. J., Ph.D., Syracuse, 1985. Theoretical elementary particle physics.
Schweitzer, John W., Ph.D., Cincinnati, 1966. Professor Emeritus. Theoretical and experimental solid state physics.
Scudder, Jack D., Ph.D., Maryland, 1975. Space plasma physics.
Skiff, Frederick N., Ph.D., Princeton, 1985. Laser spectroscopy; plasma physics.
Smirl, Arthur L., Ph.D., Arizona, 1975. Optical properties of semiconductors; ultrafast photonics; nonlinear optics; laser physics.
Spangler, Steven R., Ph.D., Iowa, 1975. Radio astronomy; plasma astrophysics; space plasma physics.

Associate Professors

Gayley, Kenneth G. Ph.D., California, San Diego 1990. Radiative transfer; radiation hydrodynamics; spectral line diagnostics.
Lang, Cornelia C., Ph.D., California, Los Angeles, 2000. Radio astronomy, x-ray astronomy; observational study of interstellar medium and galactic center.
Nachtman, Jane M., Ph.D., Wisconsin, Madison, 1997. Experimental elementary particle physics.
Newsom, Charles R., Ph.D., Texas, 1980. Experimental elementary particle physics.
Prineas, John P., Ph.D., Arizona, 2000. Semiconductor nanostructures; optoelectronics and photonics; III–V MBE growth; nonlinear optics.
Wohlgenannt, Markus, Ph.D., Utah, 2000. Experimental polymer physics.

Assistant Professors

Howes, Gregory G., Ph.D., California, Los Angeles, 2004. Theoretical and computational plasma physics.
McEntaffer, Randall L., Ph.D., Colorado, Boulder, 2007. X-ray astronomy and instrumentation; diffuse hot interstellar medium.
Pryor, Craig, Ph.D., California, Santa Barbara, 1990. Theoretical condensed matter semiconductor nanostructures.

RESEARCH SPECIALTIES AND STAFF

Theoretical

Astrophysics. Radiation-driven winds from hot stars; plasma-waves and turbulence in the interplanetary and interstellar media. Physics of nonthermal radio sources. Gayley, Mutel, Spangler.
Atmospheric/Space Physics. Physics of space magneto-plasmas and their kinetic properties. Analytical and numerical solution of MHD, Vlasov, and Fokker-Planck equations. Magnetic reconnection and plasma turbulence in the magnetotail; the solar corona and the solar wind. Scudder.
Condensed Matter and Materials Physics. Strong correlation problems and magnetic properties of materials. Electrical, magnetic, and optical properties of nanostructures; spintronics; quantum computation. Flatté, Pryor, Schweitzer.
Elementary Particles and Fields. Particle phenomenology of colliders and neutrino detectors. Renormalization group methods. Gauge/gravity duals, superstrings. Meurice, Reno, Rodgers.
Gravitational Physics. Superstring theory, string theory, gauge/gravity duality, supergravity, conformal field theory, cosmology. Klink, Rodgers.
Mathematical Physics. Mathematical methods with emphasis on group theory, operator algebras, infinite dimensional Lie algebras and nonlinear dynamics. Klink, Meurice, Polyzou, Rodgers.
Nuclear Physics. Numerical and theoretical studies of reactions, structure, and electroweak properties of few hadron and few quark systems using relativistic and non-relativistic quantum mechanics and quantum field theory. Klink, Payne, Polyzou.
Photonics/Quantum Electronics. Coherent states in semiconductors, carrier dynamics in semiconductor lasers and detectors; nonlinear propagation phenomena. Andersen, Flatté, Pryor.
Plasma Physics. Basic plasma physics, theoretical and computational studies of space, astrophysical, and laboratory plasmas, turbulence in kinetic plasmas, development and implementation of high-performance gyrokinetic codes for first-principles simulation of kinetic plasmas. Howes.

Experimental

Astronomy. Radio, x-ray and gamma-ray astronomy. Very long baseline radio astronomy; interstellar radio scintillation; studies of the galactic center; radio continuum observations of galactic and extragalactic sources; radio imaging spectroscopy; x-ray and gamma-ray observations of black hole and neutron star binary systems and the interstellar medium; jet ejection from black holes; plasma astrophysics including radiation hydrodynamics, and astrophysical turbulence. X-ray astronomical instrumentation for polarimetry and high-resolution timing experiments; x-ray sounding rocket payloads; high resolution x-ray spectroscopy. Gayley, Kaaret, Lang, Mutel, Spangler, McEntaffer.
Space Plasmas and Atmospheric Physics, Collisionless magnetic reconnection, energetic particles and waves in the radiation belts; electric and magnetic fields, plasma instabilities, wave phenomena, and radio emissions at Earth, in planetary mag-

netospheres and in the interplanetary medium; auroral phenomena and magnetosphere-ionsphere coupling including global imaging; chemistry and physics of atmospheric dust. Gurnett, Kleiber, Kletzing, Scudder.

Condensed Matter and Materials Physics. MBE growth of III–V semiconductor materials and devices; synthesis and applications of organic semiconductors; electrical transport and magnetic properties of layered ternary transition metal sulfides. Ultrafast optical and electronic properties of semiconductor quantum wells and superlattices; optical measurements of high speed carrier dynamics, including transport, recombination, energy relaxation, and scattering in semiconductor structures and devices. Soft condensed matter, including structure and waves in Coulomb crystals, 2D and 3D experiments under laboratory and microgravity conditions. Boggess, Goree, Prineas, Schweitzer, Smirl, Wohlgenannt.

Elementary Particle Physics. Higgs, SUSY and beyond the standard model searches at Fermilab, CMS/LHC and ATLAS/LHC. Study of CP violation, charmed baryons, and charged and neutral hyperon spin physics at the b-factory BABAR at SLAC. Study of charm baryons at Fermilab. Quartz fiber Cherenkov calorimetry development for collider physics at Iowa-HEP lab. Silicon pixel and microstrip detector development at Fermilab and silicon pixel study in ATLAS/LHC. Digital calorimetry, Compton polarimeter and Particle Flow Algorithm development for a high energy e+e– Linear Collider. Mallik, Nachtman, Newson, Norbeck, Onel.

Medical Imaging Physics and Positron Emission Tomography (PET). Design of radiation detector systems for medical applications and nuclear medicine imaging; development of processing methodologies for analysis of medical images; development of high speed electronics, associated hardware and application software for PET imaging devices; and fabrication of nuclear targets and automated radiochemistry synthesis systems to produce PET radiopharmaceuticals. Hichwa, Madsen.

Nuclear Physics. Study of Pb+Pb at 1000 TeV with CMS/LHC at CERN. Breakup of excited nuclei into large fragments using large national accelerators in the United States and Europe. Norbeck, Onel.

Photonics/Quantum Electronics. Development of ultrafast optical sources and their use in spectroscopy of bulk and microstructure semiconductors; nonlinear optical properties of solid-state materials; coherent processes in semiconductors; semiconductor lasers, detectors, and other photonic devices. Andersen, Boggess, Prineas, Smirl, Wohlgenannt.

Plasma Physics. Plasma waves and instabilities, nonlinear particle dynamics, negative ion plasmas, dusty plasma; interdisciplinary study of strongly-coupled Coulomb crystal structure and dynamics; laboratory simulation of space and astrophysical plasmas; microgravity experiments. Technological topics including; biomedical applications of plasmas; laser scattering and laser-induced fluorescence diagnostics of plasmas; plasma source for plasma processing; particulate contamination in plasma processing. Goree, Kletzing, Longren, Merlino, Skiff.

Teaching Techniques. Remotely controlled imaging telescope linked to regent's institutions. Mutel.

PITTSBURG STATE UNIVERSITY

DEPARTMENT OF PHYSICS

Pittsburg, Kansas 66762

Students Accepted For Degree	FIELDS Physics	Astronomy	Related Fields
Doctorate			
Master's	X		

1. General

President: Steve Scott
Dean of Graduate Studies & Research: Peggy Snyder
Department Acting Chairman: Tim Flood
Department Telephone Number: (620) 235-4391 (X)
Type of Institution: University
Control: Public
Setting: Small town/rural
Total Faculty: 419
Total Students: 7,127
Total Graduate Faculty: 222
Total Graduate Students: 1,264
Graduate Tuition per semester:
In-state residents: Full-time—$2,576
Part-time—$219/hr
Out-of-state residents: Full-time—$6,235
Part-time—$523/hr
Tuition rates for: 2009–10
Deferred tuition plan: No
Annual Other Fees: (campus priviledge fees)
Term: Semester $470; Year $940

2. Number of Faculty in Department

The combined total of full-time faculty in the three professorial ranks is 5. The combined total of full-time, part-time, and other faculty at all ranks is 7.

3. Admission, Financial Aid, and Housing

Address admission inquiries to: Pittsburg State University, Office of Admissions
Graduate application fee required: $35 (US citizen)
Admission deadline (Fall admission): None
Admission information: For fall admission, 2009–10, 5 students were accepted from 13 applicants.
Admission requirements: For admission to the graduate programs, a Bachelor's degree in physics is required with a minimum undergraduate GPA of 2.5 specified (2.7 for international students). GRE Advanced (only) is required. TOEFL exam, minimum acceptable score of 520, required for students from non-English speaking countries.
Undergraduate preparation assumed: B.S. in physics or equivalent.
Address financial aid inquiries to: Pittsburg State University, Dept. of Student Financial Aid, 1701 S. Broadway
GAPSFAS application required: No
Financial aid deadline: None
Loans available: Yes
Address housing inquiries to: Pittsburg State University, Dept. of University Housing
On-campus, single student housing available: Yes
Cost/year: $6,656 (single; 7 day access) (14 meals/wk.) application $45

Table A—Faculty, Enrollments, and Degrees Granted

Research Specialty	2009–10 Faculty	Enrollment Fall 2009 Mas-ter's	Doc-torate	No. of Degrees Granted[1] 2009–10 (2005–10) Mas-ter's	Terminal Master's	Doc-torate	Median No. of Years for 2009–10 Ph.D.'s
Astronomy	3	0	–	0(2)	–	–	–
Atomic, Molecular Optical	1	0	–	0(1)	–	–	–
Condensed Matter Physics	1	0	–	0(4)	–	–	–
Nuclear Physics	1	0	–	1(2)	–	–	–
Other (Experimental)	0	2	–	1(3)	–	–	–
Total		2	–	2(12)	–	–	
Full-time Grad. Stud.		4	–				
Part-time Grad. Stud.		0	–				
First-year Grad. Stud.		3	–				
Median Years in Grad. Study (2008–09 Degrees)				2	–	–	

Undergraduate Degrees, 2009–10 (2005–10): 1(16)

[1] Five-year totals in parentheses

4. Graduate Degree Requirements

Master's: Minimum 32 credit hours, or 30 credit hours with thesis. Maintain 3.0 GPA at the time of graduation. No residency requirement. Qualifying examination required.
Thesis: Thesis may be written *in absentia*.
Special Equipment, Facilities, or Programs: Students enjoy personal attention in an informal atmosphere, where the primary goal is quality teaching based on a collegial rapport between students and teachers. Modern laboratory facilities provide invaluable hands-on experience in front-line research projects. Currently, these include observational astronomy at astrophysical observatory, ultracapacitors, gamma spectroscopy, surface analysis for nanoscale systems and materials science, including microscopy (SEM), (AFM), (STM), x-ray scatterometry, surface plasmons, and Auger spectroscopy.

Table B—Appointments to Graduate Students, 2009–10

Title of Appointee	Appointments Total	First year	Academic Load Allowed in Credit Hours	Hours of Service Per Week	Stipend for Academic Year ($)
			Semester		
Teaching Assistant	3	1	12	20	10,000[1]
Total	3	1			

[1] Nine months tuition assessed at in-state rate; 100% waived.

5. Personnel Engaged in Separately Budgeted Research, 7/09–6/10

Professorial faculty	1
Graduate students	1
Total	2

FACULTY

Professors

Blatchley, Charles C., Ph.D., Louisiana State Univ., 1984. Particle accelerator applications.
Kuehn, David M., Ph.D., New Mexico State, 1990. Planetary.

Associate Professor

Uran, Serif, Ph.D., Illinois Institute of Technology, 2000. Condensed matter.

Assistant Professors

Butler, Rebecca, Ohio State University, 2002. Physics.
Konopelko, Alexander, Ph.D., Tomsk University of Technology, Russia, 1990. Astrophysics.

Lecturer

Scarborough, Kyla, M.S., Pittsburg State University, 2005. Astronomy.

RESEARCH SPECIALTIES AND STAFF

Experimental

Astrophysics. Konopelko.
Astronomy, Planetary Atmospheres. Kuehn, Scarborough.
Atomic, molecular optical. Molecular Spectroscopy. Butler.
Cosmogenic nuclides. Blatchley.
Solid State Physics. Magnetic thin film, binary mixtures, oxidation of metals, optical properties of solids. Uran.

UNIVERSITY OF KANSAS

DEPARTMENT OF PHYSICS AND ASTRONOMY

Lawrence, Kansas 66045

Students Accepted For Degree	FIELDS		
	Physics	Astronomy	Related Fields
Doctorate	X		
Master's	X		X

1. General

Chancellor: Bernadette Gray-Little
Dean of Graduate Studies: Sara Rosen
Department Chairman: Stephen J. Sanders
Department Telephone Number: (785) 864-4626 (C)
Type of Institution: University
Control: Public
Setting: Small city
Total Faculty: 1,570
Total Graduate Faculty: 1,458
Total Students: 30,004
Total Graduate Students: 6,412
Annual Graduate Tuition:
In-state residents: $270.50/credit
Out-of-state residents: $646.25/credit
Tuition rates for: 2009–10
Deferred tuition plan: No
Other Fees: $846.70/year
Term: Semester

2. Number of Faculty in Department

The combined total of full-time faculty in the three professorial ranks is 24. The combined total of full-time, part-time, and other faculty at all ranks is 43.

3. Admission, Financial Aid, and Housing

Address admission inquiries to: Graduate Admissions, Department of Physics and Astronomy, 1251 Wescoe Hall Drive, University of Kansas, Lawrence, Kansas 66045-7572
Graduate application fee required: Yes
Admission deadline (Fall admission): 5/1
Admission information: For fall admission, 2009–10, 10 students were accepted from 60 applicants.
Admission requirements: For admission to the graduate programs, a Bachelor's degree in physics, astronomy, or a related field is desired with a minimum undergraduate GPA of 3.0 (B). The GRE and the GRE Advanced Physics are strongly recommended. Students from non-English speaking countries are required to demonstrate proficiency in English by either the TOEFL or TOEFL iBT exams. The KU administered TSE exam is recommended for those applying for a teaching assistantship and who have not taken the TOEFL iBT exam.
Undergraduate preparation assumed: Mechanics (at the level of the textbook by Marion and Thornton), Electrodynamics (level of D. J. Griffiths), Quantum Mechanics (level of Liboff), Laboratory (level of Melissinos or Brophy); at least two courses in mathematics beyond elementary calculus.
Address financial aid inquiries to: Graduate Secretary, Department of Physics and Astronomy, 1251 Wescoe Hall Drive, University of Kansas, Lawrence, Kansas 66045-7572
GAPSFAS application required: No
Financial aid deadline: 5/1 (priority date), 3/1 (College priority date)
Loans available: No
Address housing inquiries to: Director of Housing, 422 W 11th Stl, Suite DSH, Corbin Hall, University of Kansas, Lawrence, Kansas 66045-3312
On-campus, single student housing available: Yes
Cost/year: starts at $3,642/9 mos. (room and board)
On-campus, married student housing available: Yes
Cost/year: $3,852–7,728 (12 mo. lease)[1]

[1]Up to 1 yr. wait for 2 bdrm. housing

Table A—Faculty, Enrollments, and Degrees Granted

Research Specialty	2009–10 Faculty	Enrollment[1] Spring 2009		No. of Degrees Granted[2] 2009–10 (2005–10)			Median No. of Years for 2009–10 Ph.D.'s[3]
		Master's	Doctorate	Master's	Terminal Master's	Doctorate	
Astronomy	4	–	–	–	–	–	–
Astrobiophysics	2	–	–	–	–	–	
Astrophysics	7	–	–	–	–	–	–
Biophysics	2	–	–	–	–	–	–
Computational Physics	5	–	–	–	–	–	–
Condensed Matter Physics	3	–	–	–	–	–	–
Low Temperature Physics	2	–	–	–	–	–	–
Nuclear Physics	2	–	–	–	–	–	–
Particles & Fields	7	–	–	–	–	–	–
Physics of Beams	1	–	–	–	–	–	–
Plasma Physics & Fusion	2	–	–	–	–	–	–
Relativity & Gravitation	2	–	–	–	–	–	–
Space Phys., Cosmic Rays	5	–	–	–	–	–	
Total		4	37	1(6)	1(27)	4(10)	
Full-time Grad. Stud.		4	37				
Part-time Grad. Stud.		0	0				
First-year Grad. Stud.		0	5				
Median Years in Grad. Study (2007–08 Degrees)				–	–	–	–

Undergraduate Degrees, 2009–10 (2005–10): 15(84)

[1] Students not yet committed to a research specialty are entered under non-specialized.
[2] Five-year totals in parentheses.
[3] Median number of years not available.

4. Graduate Degree Requirements

Master's: 30 hours of advanced courses and at least two hours of Master's Research with satisfactory progress; no foreign language requirement; better than a B average required and a general examination in physics required. 30 hours of resident study is required, but up to six of these may be transferred from another accredited university.
Doctorate: No specific number of credit hours is required; course work should average better than a B; students with a cumulative average grade less than B will be placed on probation; three full academic years of residency are required, two semesters normally consecutive and excluding summer session subsequent to the first year of graduate study must be spent at

the University of Kansas; no foreign language requirement; demonstrated skill in computer programming related to the student's field of study is required; undergraduate certification by the graduate committee and a comprehensive exam are required; a dissertation showing the results of original research is required.

Other Programs: Computational Physics and Astronomy (M.S.).

Thesis: Thesis may be written *in absentia*.

Special Equipment, Facilities, or Programs: Nearby campus Computer Center has a small supercomputer system (a 64-processor SGI Origin 2000) and several high-end Compaq Alpha servers (UNIX and Open VMS). The campus computer network has high-speed redundant connections to the Internet and the University is a member of the Internet II Development Consortium (consisting of more than 120 universities), which provides the latest network connection technology. Condensed matter physics facilities include an advanced materials research lab, a quantum electronics lab, and a semiconductor laser optics lab. These labs are well equipped with thin film deposition systems, a new scanning electron microscope, a unique UHV multi-probe scanning microscopy system, an X-ray diffractometer, SQUID magnetometers, a 6-mK dilution refrigerator, microwave synthesizers and a vector network analyzer, a Nd:YAG laser, and an optical parametric oscillator. A clean room with photo- and electron beam lithography as well as wafer processing tools is also available for micro- and nano-fabrication of solid state devices and circuits. Professionally staffed machine shop and "student" shop. High-energy and nuclear physics groups utilize experimental facilities at various universities and national laboratories as part of collaborative experiments. The Astrobiophysics Working Group sponsors collaborations among the departments of Physics & Astronomy, Ecology and Evolutionary Biology, Geology, and the Biodiversity Research Center. The Department shares with San Diego State University access to a new 1.25-m reflecting telescope located at Mt. Laguna Observatory in southern California. This research-quality instrument is located at an excellent site and is equipped with a CCD imager and a variety of filters. Observing is done remotely from on-campus.

Table B—Appointments to Graduate Students, 2008–09

Title of Appointee	Appointments Total	Appointments First year	Academic Load Allowed in Credit Hours	Hours of Service Per Week	Stipend for Academic Year ($)
			Semester		
Teaching Assistant	21	5	6–12	20	9 mos. 18,224–18,790[1]
Research Assistant	16	0	6–12	20	12 mos. 18,200–25,350
Self-supported	4	0	9–12	0	
Total	41	5			

[1]Plus 100% tuition fee waiver.

5. Personnel Engaged in Separately Budgeted Research, 09–10

Professorial faculty	23
Postdoctoral appointments	8
Graduate students	16
Undergraduate students	20
Nonteaching research personnel	4
Total	70

6. Separately Budgeted Research Expenditures by Source of Support

	Departmental Research	Physics-related Research Outside Department
Federal government	$3,217,525	$
Other	$	$
Total	$3,217,525	$

Table C—Separately Budgeted Research Expenditures

Research Specialty	No. of Grants	Expenditures ($)
Astrobiophysics		38,000
Astronomy		36,279
Cosmology		42,061
Space Physics & Astrophysics		676,577
Condensed Matter Physics		561,578
Nuclear Physics		412,166
Particles & Fields		1,136,346
Biophysics		278,004
Physics of Beams		36,514
Total		3,217,525

FACULTY

Professors

Anthony-Twarog, Barbara J., Ph.D., Yale, 1981. Stellar evolution in open star clusters; CCD and photoelectric photometry; globular clusters; high resolution stellar spectroscopy.

Baringer, Philip S., Ph.D., Indiana, 1985. Experimental physics; elementary particle physics.

Bean, Alice L., Ph.D., Carnegie-Mellon, 1987. Experimental physics; elementary particle physics.

Besson, David Z., Ph.D., Rutgers, 1986. Experimental physics; elementary particle physics.

Cravens, Thomas E., Ph.D., Harvard, 1975. Experimental, theoretical physics; astrophysics; space physics; plasma physics.

Han, Siyuan, Ph.D., Iowa State University, 1986. Experimental condensed matter physics; physics and application of Josephson junctions and SQUIDs; mesoscopic physics; quantum computing.

Hawley, Steven A., Ph.D., University of California, Santa Cruz, 1977. Observational astronomy; spectrophotometry of H II regions and planetary nebulae; astrobiology; human spaceflight.

Melott, Adrian L., Ph.D., University of Texas, 1981. Astrobiophysics and geophysics.

Ralston, John P., Ph.D., Oregon, 1980. Theoretical physics; elementary particle physics; particle astrophysics.

Sanders, Stephen J., Ph.D., Yale, 1977. Chairman of the Department. Experimental nuclear physics.

Shandarin, Sergei F., Ph.D., Moscow Physical Technical Institute, 1975. Astrophysics and cosmology; nonlinear dynamics; computational physics.

Shi, Jicong, Ph.D., University of Houston, 1991. Theoretical physics; nonlinear dynamics; beam dynamics; accelerator physics; computational physics.

Twarog, Bruce A., Ph.D., Yale, 1980. Stellar nucleosynthesis; chemical evolution of galaxies; stellar photometry; high resolution stellar spectroscopy.

Wu, Judy Z., Ph.D., University of Houston, 1993. Experimental condensed matter physics; fabrication, characterization, and application of thin films and nanowires.

Associate Professors

Feldman, Hume A., Ph.D., State University of New York, Stony Brook, 1989. Astrophysics and Cosmology.

Marfatia, Danny, Ph.D., University of Wisconsin, Madison, 2001. Theoretical particle physics; particle astrophysics.

Medvedev, Mikhail V., Ph.D., University of California, San Diego, 1996. Theoretical astrophysics; space physics; plasma physics; nonlinear dynamics; astrobiology.

Murray, Michael J., Ph.D., University of Pittsburgh, 1989. Experimental nuclear physics.

Wilson, Graham W., Ph.D., University of Lancaster, 1989. Experimental physics; elementary particle physics.

Assistant Professors

Antonik, Matthew, Ph.D., University of Maine, 1994. Biophysics.

Fischer, Christopher J., Ph.D., University of Michigan, 2000. Biophysics.

Kong, Kyoungchul (K.C.), University of Florida, Gainesville, 2006. Theoretical physics; elementary particle physics.

Rudnick, Gregory H., Assistant Professor, Ph.D., University of Arizona, 2001. Astronomy; galaxy evolution; galaxy formation.

Zhao, Hui, Ph.D., Northern-Jiaotong University, 2000. Experimental condensed matter physics.

Professors Emeriti

Armstrong, Thomas P., Ph.D., Iowa, 1966. Experimental, theoretical physics; astrophysics; space physics; plasma physics.

Bearse, Robert C., Ph.D., Rice, 1964. Nuclear physics; materials control and accounting; nuclear safeguards; computer database applications.

Davis, Robin E. P., Ph.D., Oxford, 1962. Experimental physics; elementary particle physics.

Eagleman, Joe R., Professor Emeritus, Ph.D., Missouri, 1963. Atmospheric science.

Friauf, Robert J., Professor Emeritus, Ph.D., Chicago, 1953. Experimental condensed-matter physics, diffusion and color centers, molecular dynamics and Monte Carlo simulations.

Kwak, Nowhan, Ph.D., Tufts, 1962. Experimental physics; elementary particle physics.

McKay, Douglas W., Ph.D., Northwestern, 1968. Theoretical physics; elementary particle physics; particle astrophysics.

Munczek, Herman J., Ph.D., Buenos Aires, 1958. Theoretical physics; elementary particle physics.

Sapp, Richard C., Ph.D., Ohio State, 1955. Experimental physics; solid state and low-temperature; low-temperature magnetism.

Shawl, Stephen J., Ph.D., Texas, Austin, 1972. Observational astronomy; stellar astronomy; polarization; globular clusters; astronomy education.

Wiseman, Gordon G., Ph.D., Kansas, 1950. Experimental physics; solid state physics; dielectrics; ferroelectricity.

Wong, Kai-Wai, Ph.D., Northwestern, 1962. Theoretical physics; many-body theory; superconductivity; liquid helium.

RESEARCH SPECIALTIES AND STAFF

Theoretical

Astrobiophysics: Melott, Medvedev

Astrophysics and Cosmology. Dark matter; large-scale structure; γ-ray bursts. Feldman, Marfatia, Medvedev, Melott, Ralston, Shandarin.

Chaos and Dynamical Systems. Ralston, Shi, Medvedev.

Computational Physics. Cravens, Melott, Shandarin, Shi.

Elementary Particle Physics. Symmetry properties and dynamics of elementary particles; particle astrophysics. Kong, Marfatia, Ralston.

Geophysics. Melott.

Plasma Physics. Solar wind; radiation belts. Cravens, Medvedev.

Space Physics. Space probes; trapped particles. Cravens, Medvedev.

Experimental

Astrophysics. Stellar astronomy; nebular astrophysics; galaxy evolution; polarization. Anthony-Twarog, Cravens, Rudnick, Hawley.

Biophysics. Kinetics and thermodynamics of protein-protein and protein-nucleic acid interactions. Antonik, Fischer.

Condensed Matter Physics. Quantum tunneling and coherence, superconducting and single electron devices, high-temperature superconductivity, electronic structure, semiconductors. Han, Wu, Zhao. 4 postdoctoral fellows.

Elementary Particle Physics. Study of proton-antiproton collisions with the D0 experiment at the Fermilab Tevatron; study of electron-positron annihilation with CLEO at CESR; astrophysics research with RICE in Antarctica; study of proton-proton collisions with the CMS experiment at the CERN LHC; research and development work on a future linear electron-positron collider. Baringer, Bean, Besson, Wilson. 3 postdoctoral fellows.

Nonlinear dynamics and granular media.

Nuclear Physics. Heavy-ion reactions at RHIC, and nuclear structure. Murray, Sanders. 1 postdoctoral fellow.

FACULTY PUBLICATIONS

Anthony-Twarog, Barbara

B. J. Anthony-Twarog, C. P. Deliyannis, B. A. Twarog, J. D. Cummings, and R. M. Maderak, "Abundances in NGC 6253 from Hydra Spectroscopy of the Li 6708 A Region," Astronomical Journal **139**, 2034 (2010).

B. A. Twarog, B. J. Anthony-Twarog, and F. Edgington-Giordano, 'The Unevolved Main Sequence of Nearby Field Stars and the Open Cluster Distance Scale," Publications of the Astronomical Society of the Pacific **121**, 1312 (2009).

B. J. Anthony-Twarog, C. P. Deliyannis, B. A. Twarog, Kevin V. Croxall, and J. D. Cummings, "Lithium in the Intermediate-Age Open Cluster NGC 3680," Astronomical Journal **138**, 1171 (2009).

Antonik, Matthew

M. Antonik, S. Felekyan, A. Gaiduk, and CAM Seidel, "Separating Structural Heterogeneities from Stochastic Variations in FRET Distributions via Photon Distribution Analysis," J. Phys. Chem. B **110**, 6970 (2006).

J. Widengren, V. Kudryavtsev, M. Antonik, S. Berger, M. Gerken, and CAM Seidel, "Single molecule detection and identification of multiple species by multiparameter fluorescence detection," Analytical Chemistry **78**(6), 2039 (2006).

A. Gaiduk, R. Kuhnemuth, M. Antonik, and CAM Seidel, "Optical Characteristics of Atomic Force Microscopy Tips for Single-Molecule Fluorescence Applications," Chem. Phys. Chem. **6**, 976 (2005).

Baringer, Philip

V. Khackatryan *et al.* [CMS Collaboration], "Transverse momentum and pseudorapidity distributions of charged hadrons in pp collisions at sqrt(s)=0.9 and 2.36 TeV," JHEP **1002**, 041 (2010); arXiv:1002.0621 [hep-ex].

S. Chatrchyan *et al.* [CMS Collaboration], "Commissioning and Performance of the CMS Silicon Strip Tracker with Cosmic Ray Muons," JINST **5**, T03008 (2010); arXiv:0911.4996 [physics.ins-det].

V. M. Abazov *et al.*, [D0 Collaboration], "Observation of Single Top-Quark Production," Phys. Rev. Lett. **103**, 092001 (2009); arXiv:0903.0850 [hep-ex].

Bean, Alice

T. Rohe *et al.*, "Signal height in silicon pixel detectors irradiated with pions and protons," Nucl. Instrum & Meth. A (2009).

T. MacDonald and A. Bean, "Get Quarked!" Science Scope **33**, 43 (2009).

A. Bean, J. P. Ralston, and J. Snow, "Evidence for observation of virtual radio Cherenkov fields," Nucl. Instrum. & Meth. A **596** (2008).

Besson, David

D. Besson *et al.*, "*In Situ* and Laboratory Studies of Radiowave Propagation through ice and Implications for Siting a large-scale Antarctic Neutrino Detector," Astroparticle Physics **31**, 348 (2009).

J. Libby *et al.* (The CLEO Collaboration), "Inclusive Radiative psi(2S) Decays," Phys. Rev. D **80**, 072002 (2009).

D. Besson *et al.*, "*In Situ* Radioglaciological measurements near Taylor Dome, Antarctica, and Implications for UHE Neutrino Astronomy," Astropart. Phys. **29**, 130 (2008).

Cravens, Thomas

T. E. Cravens, I. P. Robertson, J. H. Waite, Jr., R. V. Yelle, V. Vuitton, A. J. Coates, J.-E. Wahlund, K. Agren, M. S. Richard, V. De La Haye, A. Wellbrock, and F. N. Neubauer, "Model-data comparisons for Titan's nightside ionosphere, Icarus **199**, 174 (2009).

T. E. Cravens, R. L. McNutt, Jr., J. H. Waite, Jr., I. P. Robertson, J. G. Luhmann, W. Kasprzak, and W.-H. Ip, "The plume ionosphere of Enceladus as seen by the Cassini Ion and Neutral Mass Spectrometer," Geophys. Res. Lett. **36**, L08106 (2009).

I. P Robertson, T. E. Cravens, J. H. Waite, Jr., R. V. Yelle, V. Vuitton, A. J. Coates, J. E. Wahlund, K. Agren, K. Mandt, B. Magee, and M. S. Richard, "Structure of Titan's ionosphere: Model comparisons with Cassini data," Planet. Space Sci. **57**, 1834 (2009).

Feldman, Hume

H. A. Feldman, R. Watkins, and M. J. Hudson, "Cosmic Flows on 100 Mpc/h Scales: Standardized Minimum Variance Bulk Flow, Shear and Octupole Moments" (2010), MNRAS (arXiv:0911.5516).

R. Juszkiewicz, H. A. Feldman, J. N. Fry, and A. H. Jaffe, "Nonlinear Effects in the Amplitude of Cosmological Density Fluctuations," JCAP **02**, 021 (2010) (ArXiv:0901.0697).

R. Watkins, H. A. Feldman, and M. J. Hudson, "Consistently Large Cosmic Flows on Scales of 100 h^{-1} Mpc: a Challenege for the Standard ???CDM Cosmology," MNRAS **392**, 743 (2009) (arXiv:0809.4041).

Fischer, Christopher

D. L. Matlock, L. Yeruva, A. K. Byrd, S. G. Mackintosh, C. Langston, C. Brown, C. E. Cameron, C. J. Fischer, and K. D. Raney, "Investigation of translocation, DNA unwiding, and protein displacement by NS3h, the helicase domain from the Hepatitis C virus helicase," Biochemistry **49**, 2097 (2010).

C. J. Fischer, K. Yamada, and D. J. Fitzgerald, "Kinetic Mechanism for Single stranded DNA binding and Translocation by S. cerevisiae Isw2," Biochemistry **48**, 2960 (2009).

C. J. Fischer, A. Saha, and B. R. Cairns, "Kinetic Model for the ATP-dependent Translocation of S. cerevisiae RSC along Double-stranded DNA," Biochemistry **46**, 12416 (2007).

Hawley, Steven

S. A. Hawley, "Spectrophotometry of Southern H II Regions and Strong Line Diagnostics," submitted to publications of the Astronomical Society of the Pacific (2010).

S. A. Hawley and D. F. Merriam, "Kansas From Space: A Century of Viewing the Earth From Above," Publications of the Kansas Academy of Sciences, in press (2010).

S. A. Hawley, "Astromaterials Research and Exploration," NASA TM-2007-214769 (2007).

Han, Siyuan

C.-P. Yang and S. Han, "Rotation gate for a three-level superconducting quantum interference device qubit with resonant interaction," Phys. Rev. A **74**, 044302 (2006).

Z. Zhou, S.-I. Chu, and S. Han, "A unified approach to realize universal quantum gates in a coupled two-qubit system with fixed always-on coupling," Phys. Rev. B **73**, 104521 (2006).

C.-P. Yang and S. Han, "Realization of an n-qubit controlled-U gate with superconducting quantum interference devices or atoms in cavity QED," Phys. Lett. A **73**, 032317 (2006).

Marfatia, Danny

L. Anchordoqui, H. Goldberg, D. Hooper, D. Marfatia, and T. Taylor, "Neutralino dark matter annihilation to monoenergetic gamma rays as a signal of low mass superstrings," Phys. Lett. B **683**, 321 (2010).

V. Barger, D. Marfatia, A. Mustafayev, and A. Soleimani, "Supersymmetric dark matter and lepton flavor violation," Phys. Rev. D **80**, 076004 (2009).

V. Barger, Y. Gao, W. Y. Keung, and D. Marfatia, "Generic dark matter signature for gamma-ray telescopes," Phys. Rev. D **80**, 063537 (2009).

Medvedev, Mikhail

M. V. Medvedev, S. Pothapragada, S. Reynolds, "Modeling spectral variabiliy of prompt Gamma-Ray Burst emission within the jitter radiation paradigm," Astrophys. J. Lett. **702**, L91 (2009).

M. V. Medvedev and A. L. Melott, "Do extragalactic cosmic rays induce cycles in fossil diversity?" Astrophys. J. **664**, 879 (2007).

M. V. Medvedev, L. O. Silva, and M. Kamionkowski, "Cluster magnetic fields from Large-Scale Structure and galaxy shocks," Astrophys. J. Lett. **642**, L1 (2006).

Melott, Adrian

D. Atri, A. L. Melott, and B. C. Thomas, "Lookup tables to compute high energy cosmic ray induced atmospheric ionization and changes in atmospheric chemistry," Journal of Cosmology and Astroparticle Physics **JCAP05**, 008 (2010).

A. L. Melott, B. C. Thomas, G. A. Dreschhoff, and C. K. Johnson, "Cometary airbursts and atmospheric chemistry: Tunguska and a candidate Younger Dryas event," Geology **38**, 355 (2010).

A. C. Overholt, A. L. Melott, and M. K. Pohl, "Testing the link between terrestrial climate change and Galactic spiral arm transit," Astrophysical Journal Letters **705**, L101 (2009).

Murray, Michael

CMS Collaboration (V. Khachatryan *et al.*), "Transverse momentum and pseudorapidity distributions of charged hadrons in pp collisions at $\sqrt{S_{NN}}$=0.9 and 2.36 TeV," JHEP 1002:041, 2010; e-Print: arXiv:1002.0621 [hep-ex].

BRAHMS Collaboration (I. Arsene *et al.*), "Kaon and Pion Production in Central Au+Au Collisions at $\sqrt{S_{NN}}$=62.4 GeV," Phys. Lett. B **687**, 36 (2010).

BRAHMS Collaboration (I. Arsene *et al.*), "Rapidity dependence of proton-to-pion ratio in Au+Au and p+p at $\sqrt{S_{NN}}$=62.4 and 200 GeV," Phys. Lett. B **684**, 22 (2010).

Ralston, John

M. Backovic and John P. Ralston, "Limits on Threshold and 'Sommerfeld' Enhancements in Dark Matter Annihilation," Phys. Rev. D **81**, 056002 (2010).

P. K. Samal, R. Saha, P. Jain, and John P. Ralson, "Signals of Statistical Anisotropy in WMAP Foreground-Cleaned Maps," Mon. Not. Roy. Astron. Soc. **396** 511 (2009).

D. P. Hogan, D. Z. Besson, John P. Ralson, I. Kravchenko, and D. Seckel, "Relativisitc Magnetic Monopole Flux Constraints from RICE," Phys. Rev. D **78**, 075031 (2008).

Rudnick, Gregory

G. Rudnick *et al.*, "The Rest-Frame Optical Lumiosity Function of Cluster Galaxies at z<0.8 and the Assembly of the Cluster Red Sequence," The Astrophysical Journal **700**, 1559 (2009).

C. Papovish, G. Rudnick *et al.*, "Paschen-??? Emission in the Gravitationally Lensed Galaxy SMM J163554.2 +661225," The Astrophysical Journal **704**, 1506 (2009).

T. A. Reichard, T. M. Heckman, G. Rudnic, J. Brinchmann, . Kauffmann, and V. Wild, "The Lopsidedness of Present-Day Galaxies: Connections to the Formation of Stars, the Chemical Evolution of Galaxies, and the Growth of Black Holes," ApJ **691**, 1005 (2009).

Sanders, Stephen

BRAHMS Collaboration (I. Arsene *et al.*), "Kaon and Pion Production in Central Au+Au Collisions at $\sqrt{S_{NN}}$=200 GeV," Phys. Lett. B **687**, 36 (2010).

BRAHMS Collaboration (I. Arsene *et al.*), "Radidity dependence of proton-to-pion ration in Au+Au and p+p at $\sqrt{S_{NN}}$=62.4 and 200 GeV," Phys. Lett. B **684**, 22 (2010); arXiv:0910.3328 [nucl-ex].

S. J. Sanders (for the BRAHMS Collaboration), "Forward-rapidity azimuthal and radial flow of identified particles for $\sqrt{S_{NN}}$=200 GeV Au+Au collisions," Nucl. Phys. **A830**, 179c (2009).

Shandarin, Sergei

Shandarin, S., Habib, S., Heitmann, K. "Origin of the cosmic network in LCDM: Nature vs nurture," Physical Review D **81**, 103006 (2010).

Shandarin, S. F. "The Origin of 'Great Walls'," Journalof Cosmology and Astroparticle Physics **02**, 031 (2009).

Colberg, J. M., Pearce, F., Foster, C., Platen, E., Brunino, S., Basikakos, S., Fairall, A., Feldman, H., Gottlober, S., Hahn, O., Hoyle, F., Muller, V., Nelson, L., Neyrinck, M., Plionis, M., Porciani, C., Shandarin, S. F., Vogeley, M. S., and van de Weygaert, R., "The Aspen-Amsterdam Void Finder Comparison Project," Monthly Notices of Royal Astronomical Society **387**, 933 (2008).

Shi, Jicong

J. Shi, L. Jin, and F. Wang, "Study of beam-beam effects in eRHIC with self-consistent beam-beam simulation," Nucl. Inst. and Meth. A **555**, 6 (2005).

L. Jin, J. Shi, and G. H. Hoffstaetter, "Coherent beam-beam tune shift of unsymmetrical beam-beam interactions with large beam-beam parameters," Phys. Rev. E **71**, 036501 (2005).

J. Shi, L. Jin, and O. Kheawpum, "Multipole Compensation of Long-Range Beam-Beam Interactions with Minimization of Nonlinearities in Poincare Maps of a Storage-Ring Collider," Phys. Rev. E **69**, 036502 (2004).

Twarog, Bruce

B. J. Anthony-Twarog, C. P. Deliyannis, B. A. Twarog, J. D. Cummings, and R. M. Maderak, "Abundances in NGC 6253 from Hydra Spectroscopy of the Li 6708 ??? Region," Astronomical Journal **139**, 2034 (2010).

B. A. Twarog, B. J. Anthony-Twarog, and F. Edgington-Giordano, 'The Unevolved Main Sequence of Nearby Field Stars and the Open Cluster Distance Scale," Publications of the Astronomical Society of the Pacific **121**, 1312 (2009).

B. J. Anthony-Twarog, C. P. Deliyannis, B. A. Twarog, Kevin V. Croxall, and J. D. Cummings, "Lithium in the Intermediate-Age Open Cluster NGC 3680," Astronomical Journal **138**, 1171 (2009).

Wilson, Graham

D0 Collaboration (V. M. Abazov *et al.*), "Search for Higgs Boson Production in Dilepton and Missing Energy Final States with 5.4 fb^{-1} of proton anti-proton collisions at sqrt(s)=1.96 TeV," Phys. Rev. Lett. **104**, 061804 (2010).

ILD Concept Group (H. Stoeck *et al.*), "The International Large Detector, Letter of Intent," DESY 2009-87, FERMILAB-PUB-09-682-E, KEK Report 2009-6, Editors: T. Behnke, D. Karlen, Y. Sugimoto, H. Videau, G. W. Wilson, H. Yamamoto, pp. 1–163, February 2010.

OPAL Collaboration (G. Abbiendi *et al.*), "Measurement of the e+e->W+W- cross section and W decay branching fractions at LEP," Eur. Phys. J. **C52**, 767 (2007).

Wu, Judy

J. Dizon, R. T. Lu, X. Wang, and J. Z. Wu, "Detection of low level dissipation and thermal instability in $YBa_2Cu_3O_{7-???}$ microbridge using low-temperature ear field scanning microwave microscopy," Supercond. Sci. Technol. **23**, 055001 (2010).

F. J. Baca, P. N. Barnes, R. L. Emergo, T. J. Haugan, J. N. Reichart, and J. Z. Wu, "Control of $BaZrO_3$ nanorod alignment in $YBa_2Cu_3O_{7-x}$ thin films by microstructural modulation," Appl. Phys. Lett. **94**, 102512 (2009).

Z. Z. Li and J. Z. Wu, "Gold/Boron core-shell nanocables synthesized from gold-boron eutectic droplets," Nanotechnology **19**, 055606 (2008).

Zho, Hui

B. A. Ruzicka, L. K. Werake, Hui Zhao, S. Wang, and K. P.

Loh, "Femtosecond pump-probe studies of reduced graphene oxide thin films," Applied Phys. Lett. **96**, 173106 (2010).

B. A. Ruzicka, K. Higley, L. K. Werake, and Hui Zhao, "All-optical generation and detection of subpicosecond ac spin-current pulses in GaAs," Physical Review B **78**, 045314 (2008).

Hui Zhao, "Temperature dependence of ambipolar diffusion in silicon-on-insulator," Applied Physics Letters **92**, 112104 (2008).

UNIVERSITY OF KENTUCKY

DEPARTMENT OF PHYSICS AND ASTRONOMY

Lexington, Kentucky 40506

Students Accepted For Degree	FIELDS		
	Physics	Astronomy	Related Fields
Doctorate	X		X
Master's	X		X

1. General

President: Lee Todd
Dean of Graduate School: Jeannine Blackwell
Department Chairman: Michael Cavagnero
Department Telephone Number: (859) 257-6722
Fax: (859) 323-2846
Web Page: http://www.pa.uky.edu/
Type of Institution: University
Control: Public
Setting: Urban
Total Faculty: 2,057
Total Graduate Faculty: 1,579
Total Students: 26,439
Total Graduate Students: 5,437
Annual Graduate Tuition:
In-state residents: Full-time—$7,690*
Part-time—$401.43/credit
Out-of-state residents: Full-time—$16,158*
Part-time—$873.43/credit
Tuition rates for: 2009–10
Deferred tuition plan: No
Other Fees: Student Health $160.00
Term: Semester

*Includes Activities and Health Fee of $160/semester

2. Number of Faculty in Department

The combined total of full-time faculty in the three professorial ranks is 33. The combined total of full-time, part-time, and other faculty at all ranks is 37.

3. Admission, Financial Aid, and Housing

Address admission inquiries to: Director of Admissions, Physics & Astronomy Dept., University of Kentucky, Lexington, KY 40506
www.pa.uky.edu
Graduate application fee required: $50—domestic
$55—international
Admission deadline (Fall admission): 7/19—domestic
2/1—international
Admission information: For fall admission, 2009–10, 10 students were accepted from 120.
Admission requirements: For admission to the graduate programs, a Bachelor's degree in physics or related fields is required with a minimum undergraduate GPA of 2.75 specified. The GRE general test is required. The GRE Subject Test is recommended. No minimum acceptable score is specified. Students from non-English speaking countries are required to demonstrate proficiency in English via the TOEFL exam. Minimum acceptable score for admission is 213.
Undergraduate preparation assumed: Substantial variations in preparation can be accommodated; generally the department expects 1 semester of mechanics (Fowles or Symon); 1 year of electricity and magnetism (Griffith); 1 semester of optics (Pedrotti); 1 semester of thermodynamics and statistical physics (Sears and Salinger); 1 year of atomic, nuclear, and quantum physics (Eisberg); 3 semesters of laboratory courses are expected.
Address financial aid inquiries to: Grad. Admissions Committee
GAPSFAS application required: No
Financial aid deadline: 2/1
Loans available: Yes
Address housing inquiries to: Graduate and Family Housing, 700 Woodland Ave.
On-campus, single student housing available: Yes
Cost/month: $510–630
On-campus, married student housing available: Yes
Cost/month: $630–685

Table A—Faculty, Enrollments, and Degrees Granted

Research Specialty	2009–10 Faculty	Enrollment[1] Fall 2009 Master's	Enrollment[1] Fall 2009 Doctorate	No. of Degrees Granted[2] 2009–10 (2005–10) Master's	Terminal Master's	Doctorate	Median No. of Years for Ph.D.'s
Astronomy & Astrophysics	6	2	8	2	0	0	–
Atomic, Molecular, & Optical Physics	3	0	2	0	0	0	–
Condensed Matter Physics	11	3	16	3	0	1	–
Nuclear Physics	7	0	13	1	0	3	–
Particles & Fields	6	0	7	1	0	0	–
Physics Education	3	2	0	0	0	0	–
Non-specialized	0	1	1	0	0	0	–
Total	37	8	47	8(35)	0(4)	4(26)	
Full-time Grad. Stud.		8	47				
First-year Grad. Stud.		0	10				
Median Years in Grad. Study (2008–09 Degrees)		2	6				–
Undergraduate Degrees, 2008–09 (2005–10) 3(47)							

[1] Students not yet committed to a research specialty are entered under non-specialized.
[2] Five-year totals in parentheses.

4. Graduate Degree Requirements

Master's: PLAN A—24 semester-hours of graduate credit with satisfactory performance and a thesis. No residence requirement. Students must pass physics GRE at 50th percentile or above. PLAN B—Same as Plan A except six hours of course work at the advanced level is substituted for the thesis. Minimum GPA of 3.0 for both plans.
Doctorate: Thirty-six hours of graduate credits required. Requirements set forth by Advisory Committee; minimum GPA of 3.0 in approved course program required; students must pass physics GRE at 50th percentile or above and also pass 6 graduate core physics courses with GPA of 3.0 or higher; one-year teaching experience required; 36 credit hours and one-year full-time residency before and one-year full-time residency after qualification exam.
Other Programs: M.S. in Radiological Health Specialty, coop-

erative program with the Departments of Health Radiation Sciences and Radiation Medicine.

Thesis: Thesis may be written *in absentia*.

Special Equipment, Facilities, or Programs: 6.5 MV, Type CN Van de Graaff accelerator with neutron scattering facility; cryogenic magnet laboratory; regularly use NRAO facilities at Green Bank, WV, Socorro, NM. The TRIUMF cyclotron at University of British Columbia; the KEK proton synchrotron in Tsukuba, Japan; the AGS synchrotron at Brookhaven National Laboratory; SURA/Jefferson Laboratory.

Table B—Appointments to Graduate Students, 2008–09

Title of Appointee	Appointments Total	Appointments First year	Academic Load Allowed in Credit Hours	Hours of Service Per Week	Stipend for Academic Year ($)
			Semester		
Other	1	1		0	0
Teaching Assistant	32	12	9–10	20	14,230[1]
Research Assistant	19	0	9–12	20[2]	14,230–16,600[3]
University Fellows	3	2	12	0	9,000–20,000[4]
NASA Traineeship	0	0	12	0	16,000[6]
Total	55	15			

[1] All tuition is waived. Summer appointments also available.
[2] Expected service for students taking courses. Students not taking courses work full-time.
[3] Out-of-state tuition, and often in-state tuition, is waived.
[4] All tuition is waived. May be supplemented with partial TA or RA appointment.
[5] All tuition is waived. Also has book and travel allowance and possibility of $2,400 summer appointment.
[6] Tuition is waived.

5. Personnel Engaged in Separately Budgeted Research, 2008–09

Professorial faculty	24.0
Other faculty	2.0
Postdoctoral appointments	16.0
Graduate students	20.5
Undergraduate students	5.0
Total	67.5

6. Separately Budgeted Research Expenditures by Source of Support

	Departmental Research	Physics-related Research Outside Department
Federal government	$2,268,178	$5,585,613
Total	$2,268,178	$5,585,613

Table D—Physics-related Research Outside Department

Field and Unit Outside Department	No. of Grants	Expenditures ($)
Computational Physics/Center for Computational Sciences	8	5,585,613
Total	8	5,585,613

FACULTY

Professors

Brill, Joseph W., Ph.D., Stanford, 1978. Experimental solid state physics.

Cao, Gang, Ph.D., Temple, 1992. Experimental condensed matter physics.

Cavagnero, Michael, Ph.D., Chicago, 1983. Chairman of the Department. Theoretical atomic physics.

Connolly, John W. D., Ph.D., Florida, 1966. Theoretical solid state physics.

Das, Sumit R., Ph.D., Chicago, 1983. Theoretical particle physics.

DeLong, Lance E., Ph.D., California, San Diego, 1977. Experimental solid state physics.

Draper, Terrence, Ph.D., California, Los Angeles, 1984. Theoretical particle physics.

Eides, Michael I., Ph.D., Leningrad State University, 1977. Quantum electrodynamics.

Elitzur, Moshe, Ph.D., Weizmann Institute, 1971. Theoretical astrophysics.

Ferland, Gary J., Ph.D., Texas, 1978. Theoretical astrophysics.

Gardner, Susan V., Ph.D., Massachusetts Institute of Technology, 1988. Theoretical nuclear and particle physics.

Gorringe, Tim P., Ph.D., Birmingham, 1984. Experimental intermediate energy nuclear physics.

Kovash, Michael A., Ph.D., Ohio State, 1978. Experimental nuclear physics; intermediate energy.

Li, Bing An, Ph.D., Academia Sinica, Beijing, 1968. Theoretical particle physics.

Liu, Keh-Fei, Ph.D., SUNY, Stony Brook, 1975. Theoretical nuclear and particle physics.

MacAdam, Keith B., Ph.D., Harvard, 1971. Emeritus Experimental atomic and molecular physics.

MacKellar, Alan D., Ph.D., Texas A & M, 1966. Emeritus. Theoretical nuclear and atomic physics.

Martin, Nicholas L. S., D.Phil., Oxford, 1977. Experimental atomic and molecular physics.

McEllistrem, Marcus T., Ph.D., Wisconsin, 1956. Emeritus. Experimental nuclear physics.

Murthy, Ganpathy, Ph.D., Yale, 1987. Theoretical solid state physics.

Ng, Kwok-Wai, Ph.D., Iowa, 1986. Experimental solid state physics.

Shapere, Alfred D., Ph.D., California, Santa Barbara, 1988. Theoretical particle physics.

Shlosman, Isaac, Ph.D., Tel-Aviv, Israel, 1985. Theoretical astrophysics.

Straley, Joseph P., Ph.D., Cornell, 1970. Theoretical solid state physics.

Troland, Thomas H., Ph.D., California, Berkeley, 1980. Observational astronomy.

Yates, Steven W. Ph.D., Purdue 1983. Experimental nuclear physics (Joint with chemistry).

Associate Professors

Christopher, John E., Ph.D., Virginia, 1967. Emeritus Experimental solid state physics, physics education.

Korsch, Wolfgang, Ph.D., Univ. of Marburg, 1990. Experimental nuclear physics.

Levenson, Nancy A., Ph.D., California, Berkeley, 1997. Observational astronomy.

Wilhelm, Ronald, Ph.D., Michigan State University, 1995. Observational Astronomy.

Assistant Professors

Crawford, Christopher, Ph.D., Massachusetts Institute of Technology, 2005. Experimental nuclear physics.

Fatemi, Renee, Ph.D., University of Virginia, 2002. Experimental nuclear physics.

Kaul, Ribhu, Ph.D., Duke University, 2006. Theoretical condensed matter physics.

Plaster, Brad, Ph.D., Massachusetts Institute of Technology, 1999. Experimental nuclear physics.

Seo, Sung, Ph.D., Seol, National University, 2007. Experimental condensed matter.

Strachan, Doug, Ph.D., University of Maryland, 2002. Experimental condensed matter physics.

Adjunct Professor

Menon, Madhu, Ph.D., University of Notre Dame, 1986. Experimental solid state physics.

RESEARCH SPECIALTIES AND STAFF

Theoretical

Astronomy and Astrophysics. Interstellar masers; radiation transport; cosmology; stellar dynamics; high-z radio-galaxies; active galactic nuclei; cataclysmic variable stars; accretion disks; red giant atmospheres. Elitzur, Ferland, Shlosman.

Atomic and Molecular. Atomic Rydberg spectra; Stark and Zeeman effects; dielectric recombination; driven highly excited atomic states; muon and anti-proton interactions with matter; semi-classical ion-atom collisions. Cavagnero.

Condensed Matter Physics. Structure and electronic properties of clusters, defects, and surfaces; catalysis; statistical mechanisms of phase transitions; Josephson networks; quantum dots; quantum Hall effect; tunneling; superconductivity; mesoscopic systems, quantum Monte-Carlo, exact diagonalization, quantum criticality, nano-physics. Connolly, Kaul, Menon, Murthy, Straley.

Nuclear and Particle Physics. Nuclear many-body theory; electron scattering; quark and skyrmion models; gauge theory; hadron spectroscopy and structure; heavy-light quark systems; lattice gauge calculations; string theory; black holes. Das, Draper, Eides, Gardner, Horvath, Li, Liu, Shapere.

Experimental

Astronomy. Observational radio astronomy; interstellar material and star formation; interstellar magnetic fields; kinematics of ancient stars. Levenson, Troland, Wilhelm.

Atomic and Molecular Physics. Collisions of ions with laser-excited, spatially oriented, atoms; interaction of highly excited atoms with external fields; electron-electron coincidence spectroscopy of auto-ionizing atomic states. MacAdam, Martin.

Condensed Matter Physics. Physical properties of new and novel materials; organic metals, magnetic oxides, heavy fermion systems, charge density wave systems, superconducting materials; scanningtunneling electron microscopy, patterned thin films. Brill, Cao, DeLong, Ng, Seo, Strachan.

Nuclear and Particle Physics. TOF neutron spectroscopy and scattering; $(n,n'\gamma)$ spectroscopy; nuclei well off the line of nuclear stability (UNISOR); (p,n) reactions; nuclear astrophysics; intermediate energy, polarization studies; pion production; $(p,p'\gamma)$, (p,γ), and (α,γ) reactions; electron scattering; hyperon interactions in nuclei; weak interactions; muon capture fundamental symmetries; nucleon spin structure; QCD. Crawford, Fatemi, Gorringe, Korsch, Kovash, McEllistrem, Plaster, Yates.

Physics Education. Undergraduate teaching; computer assisted education, teacher preparation. Christopher, Straley.

LOUISIANA STATE UNIVERSITY

DEPARTMENT OF PHYSICS AND ASTRONOMY

Baton Rouge, Louisiana 70803

Students Accepted For Degree	FIELDS		
	Physics	Astronomy	Related Fields
Doctorate	X	*	
Master's	X	*	X

* Degree in Physics with concentration in Astronomy.

1. General

Chancellor: Michael V. Martin
Dean of Graduate School: David Constant
Department Chairman: Michael L. Cherry
Department Telephone Number: (225) 578–2261
Type of Institution: University
Control: Public
Setting: Urban
Total Faculty: 1,502 (1,371 Full time/31 Part time)
Total Graduate Faculty: approximately 1,250
Total Students: 28,194
Total Graduate Students: 4,794
Annual Graduate Tuition: (Required fees included)*9 hrs
In-state residents: Full-time—$3,122/sem.
Out-of-state residents: Full-time—$8,526/sem.
Tuition rates for: 2010–11
Deferred tuition plan: Yes
Other Fees: Yes
Term: Semester
* Students on a full-time graduate assistantship will receive a full exemption in tuition.

2. Number of Faculty in Department

The combined total of full-time faculty in the three professorial ranks is 48. The combined total of full-time, part-time, and other faculty at all ranks is 58.

3. Admission, Financial Aid, and Housing

Address admission inquiries to: Assistantship Committee Chair, Dept. of Physics and Astronomy
Graduate application fee required: $75; waived if application sent directly to department.
Admission deadline (fall admission): Jan. 25
Admission information: For fall admission, 2009–10, more than 220 applications were received; 20 accepted.
Admission requirements: For admission to the graduate programs, a bachelor's degree in physics or a related field is required with a minimum undergraduate GPA of 3.0 The GRE is required. The minimum acceptable total score required is 1,000. The average GRE score for 2009–10 admission was verbal–515; quantitative–767; total–1,282. The GRE Physics Test is strongly recommended for all students. Students from non-English speaking countries are required to demonstrate proficiency in English via the TOEFL exam. The minimum TOEFL score required for an assistantship is 600.
Undergraduate preparation assumed: Gasiorowicz, *Quantum Physics*; Griffiths, *Introduction to Electrodynamics;* Zemansky, *Heat and Thermodynamics*; Marion, *Classical Dynamics of Particles and Systems.*
Address financial aid inquiries to: Assistantship Committee Chairman
GAPSFAS application required: No
Financial aid deadline: 1/25
Loans available: Yes
Address housing inquiries to: Dept. of Residential Life (225) 578-8663
On-campus, single student housing available: Yes
On-campus, university apartments available: Yes

*Graduate school requirement

Table A—Faculty, Enrollments, and Degrees Granted

Research Specialty	2008–09 Faculty	Enrollment[1] Fall 2009		No. of Degrees Granted[2] 2009–10 (2005–10)			Median No. of Years for 2009–10 Ph.D.'s
		Master's	Doctorate	Master's	Terminal Master's	Doctorate	
Astronomy	3	0	9	1(5)	0(3)	2(6)	6.0
Astrophysics/ Space Phys., Cosmic Rays	7	–	8	0(0)	1(2)	0(0)	–
Atomic, Molecular, & Optical Physics	4	–	9	0(0)	0(1)	3(8)	4.8
Computational	6	–	0	0(0)	0(0)	0(1)	–
Condensed Matter/ Solid State Physics/Low Temperature Physics	15	0	24	1(3)	0(2)	0(4)	–
Materials	5	–	–	–	–	–	–
Medical & Health Physics	3	22	2	0(0)	5(21)	0(0)	–
Microstructures	0	–	–	–	–	0(3)	–
Nuclear Physics	3	–	1	0(0)	0(0)	0(4)	6.66
Particles & Fields	3	–	2	0(0)	0(0)	1(2)	5.0
Relativity & Gravitation	4	0	12	1(4)	0(4)	2(6)	5.5
Non-Specialized	12	–	–	–	–	–	–
Total		22	84	3(12)	7(32)	6(24)	
Full-time Grad. Stud.		22	79				
Part-time Grad. Stud.		–	4				
First-year Grad. Stud.		8	18				
Median Years in Grad. Study (2009–10 Degrees)		3.2	5.2				
Undergraduate Degrees, 2009–10 (2005–10): 15(74)							

[1] Students not yet committed to a research specialty are entered under non-specialized.
[2] Five-year totals in parentheses.

4. Graduate Degree Requirements

Master's: The minimum course requirement is 36 semester hours without a thesis or 24 hours with a thesis. Minimum GPA is a "B" average. There is no minimum residence time or foreign language requirement. Thesis candidates have an oral thesis defense, and non-thesis candidates must pass a written comprehensive examination.
Doctorate: 12 hours of advanced courses beyond the core courses are required. The minimum GPA is a "B" average and minimum time for residence is one year. Examinations are (a) the general exam, and (b) dissertation defense. A sub-

stantial portion of the dissertation work must be published in an appropriate refereed professional journal.

Master of Natural Science: Requirements are similar to the M.S. degree except a broader scope of scientific courses (including science education) is emphasized. Thesis may be written *in absentia.*

Master of Science in Medical Physics and Health Physics: The concentration in medical physics requires 29 hours of course work, 6 hours (two semesters) of clinical training at Mary Bird Perkins Cancer center, and a minimum of six hours of thesis research. The concentration in health physics requires 33 hours of course work and a minimum of six hours of thesis research. The degree requires a thesis of publishable quality with an oral defense. Minimum GPA is a "B" average. Program is accredited by the Commission on Accreditation of Medical Physics Educational Programs, Inc. (CAMPEP) for the period 2007-2011.

Special Equipment, Facilities, or Programs: LSU holds the distinction of being one of 25 universities in the nation to hold both land-grant and sea-grant status; is a member of LaSPACE, the Louisiana consortium in NASA's Space Grant program; and is designated as a Doctoral/Research Extensive institution by the Carnegie Foundation. The Hearne Institute for Theoretical Physics carries out interdisciplinary research in relativity and quantum theory.

Extensive in-house facilities include the Center for Advanced Microstructures and Devices (CAMD), a 1.5 GeV electron synchrotron light source currently providing x-rays up to 50 keV for microfabrication and for solid state, surface science, condensed matter, and materials science research. Other condensed matter instrumentation includes a low-temperature (5 mK) dilution refrigerator-high magnetic field (17.5 Tesla) facility used for studying high-temperature semiconductors and for materials research. Thin-films and nanostructures are characterized with STM.

The NSF's Laser Interferometer Gravitational-wave Observatory (LIGO) is located 24 miles from campus. Astrophysical observation at LIGO is underway, as is advanced detector technology development.

Nuclear physics experiments are conducted at Oak Ridge TRIUMF and the National Superconducting Cyclotron Laboratory at Michigan State. The high-energy physics group participates in particle physics experiments at FermiLab (BooNE), the Sudbury Neutrino Observatory (SNO), KAMIOKANDE, and at JPARC/KEK in Japan, and is involved in the deep underground experiments at DUSEL.

Members of the Astronomy, Astrophysics, and Space Science groups are presently conducting observations at Kitt Peak, Cerro Tololo, Lowell Observatory, with the Hubble Space Telescope and SPITZER, and with the Swift and Fermi satellite instruments. These groups are designing x-ray (CASTER) and cosmic ray (CALET) experiments for long duration balloon and space missions, and have recently flown a high energy cosmic ray composition experiment (ATIC) on high altitude balloon flights around the Antarctic. Ultra high energy cosmic rays are being measured in Argentina (AUGER).

Highland Road Park Observatory (HRPO): The observatory is located about 8 miles from campus, includes two fully computer controlled reflecting optical telescopes with 20" and 16" diameter primary mirrors plus a small 10' diameter radio telescope. The optical telescopes are equipped with CCD camera, filter wheel and spectrograph and the system is capable of imaging magnitude 19 stars with an exposure of a couple of minutes. The HRPO is used for teaching our undergradute observational astronomy course, graduate observational techniques course and for student research projects. The HRPO is also used for public outreach during weekly events.

Computational facilities include a large cluster of unix workstations and PowerPCs through which students and faculty gain access to a variety of high-performance computing facilities. These facilities include a 5,440-core, 50 TFlops Supercomputer named Queen Bee and several smaller clusters operated by the Louisiana Optical Network Initiative. In addition, LSU's Center for Computation and Technology (CCT) and High Performance Computing Center (HPC) operate a 1440-core 15 TFlop machine named Tezpur. All of these facilities are used extensively for numerical calculations of general relativity, numerical relativity, analysis of high-energy neutrino data, experimental calculations and simulations of star collisions, gravitational waves, calculations of strongly correlated materials, and simulations of biological materials.

Medical physics facilities at Mary Bird Perkins Cancer Center include Varian Clinac electron and x-ray beams, TomoTherapy HI-ART, BrainLab Novalis, GE PET/CT, HDR brachy therapy, a comprehensive dosimetry labs and elekta synergy and multi-vendor treatment planning lab.

Table B—Graduate Student Appointments, 2009–10

Title of Appointee	Appointments		Academic Load Allowed in Credit Hours	Hours of Service Per Week	Stipend for Academic Year ($)[3]
	Total	First year			
			Semester[3]		
Teaching Assistant	22	13	19	20	17,900–22,900[2]
Research Assistant	57	9	19	20	14,900–28,970[2,3]
Service Assistant[1]	10	9	19	20	14,900–21,900[2]
Fellowship[2]	13	3	19	–	19,900–31,000[2]
Self-supported	4	–	–	–	–
Total	106	34			

[1] Non-teaching duties (i.e., grading, tutoring, proctoring).
[2] Includes full tuition waiver.
[3] 12-month figures.

5. Personnel Engaged in Separately Budgeted Research, 7/09–6/10

Professorial faculty	40
Postdoctoral appointments	32
Graduate students	70
Undergraduate students	63
Nonteaching research personnel	4
Total	209

6. Separately Budgeted Research Expenditures by Source of Support

	Departmental Research	Physics-related Research Outside Department
Federal government	$6,403,899	$ 64,857
Private, nonprofit organizations	1,342,851	55,922
Other—State	973,276	394,782
Total	$8,720,026	$515,561

7. Separately Funded and Managed Laboratories

CAMD (Center for Advanced Microstructures and Devices)	5,312,230
CCT (Center for Computation & Technology	9,529,623
Total	$14,841,853

Table C—Separately Budgeted Research Expenditures

Research Specialty	No. of Grants	Expenditures ($)
Astronomy	26	1,028,844
Astrophysics/Space Phys., Cosmic Rays	28	1,934,570
Atomic, Molecular, & Optical Physics	6	352,142
Condensed Matter Physics	19	2,132,685
Medical Health Physics	10	614,072
Nuclear Physics	6	651,126
Particles & Fields	6	708,807
Relativity and Gravitation	21	1,297,780
Total	122	8,720,026

Table D—Physics-related Research Outside Department

Field and Unit Outside Department	No. of Grants	Expenditures ($)
Federal	1	64,857
Total	1	64,857

FACULTY

Professors

Adams, Philip W., Ph.D., Rutgers, 1986. Transport properties in two-dimensional systems; transport properties of superconducting films.

Browne, Dana, Ph.D., Stanford, 1981. Associate Chairman. Phase transitions and self organization in non-equilibrium systems, electronic structure of materials.

Cherry, Michael L., Ph.D., Chicago, 1978. Chair of the Department. Neutrino physics; cosmic rays; high-energy particle astrophysics.

Clayton, Geoffrey C., Ph.D., Toronto, 1983. Astronomy and astrophysics; interstellar and extragalactic dust; circumstellar dust, R Coronae Borealis stars.

DiTusa, John, Ph.D., Cornell University, 1992. Experimental condensed matter physics.

Dowling, Jonathan P., Ph.D., University of Colorado, 1988. Hearne Research Chair and Co-Director, Hearne Institute for Theoretical Physics. Quantum optics, quantum science and technologies, and photonic crystal theory.

Draayer, Jerry P., Ph.D., Iowa, 1968. Theoretical nuclear structure.

Frank, Juhan, Ph.D., University of Cambridge, England, 1978. Accretion in close binaries and active galactic nuclei.

Giaime, Joseph, Ph.D., MIT, 1995. Gravitational-Wave Physics at the Laser Interferometer Gravitational-Wave Observatory (LIGO). Very Low-Noise Instrumentation.

González, Gabriela, Ph.D., Syracuse University, 1995. Gravitational wave physics at LIGO (Laser Interferometer Gravitational-wave Observatory): instrumental commissioning, detector characterization and data analysis.

Hogstrom, Kenneth R., Ph.D., Rice University, 1977. Director of Medical Physics and Health Physics Program, Chief of Physics—Mary Bird Perkins Cancer Center. Electron beam therapy, image guided radiotherapy; x-ray capture therapy.

Jarrell, Mark, Ph.D., University of CA, Santa Barbara, 1987. Massively parallel simulations of strongly correlated electronic systems.

Johnson, Warren W., Ph.D., Rutgers, 1974. Gravitational radiation detectors; Josephson devices, parametric transducers and quantum nondemolition.

Kurtz, Richard L., Ph.D., Yale, 1983. Director, CAMD. Surface science; experimental condensed matter physics.

Matthews, James M., Ph.D., Wisconsin–Madison, 1984. Experimental cosmic-ray research at extreme energy; Pierre Auger Observatory.

O'Connell, Robert F., Ph.D., Notre Dame, 1962. D. Sc, National University of Ireland, 1975. Boyd Professor. Dissipative and fluctuation phenomena in quantum physics, decoherence and entanglement; general relativity.

Plummer, W., Ward, Ph.D., Cornell University, 1968. Member, National Academy of Sciences. Materials Science, condensed matter physics with emphasis on broken symmetry and reduced dimensionality.

Pullin, Jorge, Ph.D., Instituto Balseiro, Bariloche, Argentina, 1989. Hearne Research Chair in Theoretical Physics. Quantum gravity; classical, quantum mechanical and its astrophysical implications.

Rau, A. Ravi P., Ph.D., Chicago, 1970. Theoretical atomic physics.

Schaefer, Bradley, Ph.D., MIT, 1983. Gamma Ray Bursts. Novae. Supernovae.

Schafer, Ken, Ph.D., University of Arizona, Tucson, 1989. Theory of high intensity ultra-fast laser-matter interactions.

Seidel, Edward, Ph.D., Yale University, 1988. General relativity, numerical relativity, black holes, gravitational waves; relativistic astrophysics; computational physics, high performance computing, scientific and grid computing.

Sprunger, Phillip, Ph.D., University of Pennsylvania, 1993. Condensed matter physics.

Tohline, Joel E., Ph.D., California, Santa Cruz, 1978. Astrophysics; star formation; galaxy dynamics.

Wefel, John, Ph.D., Washington, University in St. Louis, 1971. High energy astrophysics; space science; LaSPACE.

Young, David P., Ph.D., Florida State Univ., 1998. Materials Science.

Zhang, Jiandi, Ph.D., Syracuse University, 1994. Condensed matter physics; Exploring novel properties of complex materials like transition-metal oxides manifested by broken symmetry and reduced dimentionality such as at the surface, interfaces and artificially structured multilayers.

Associate Professors

Blackmon, Jeffrey, Ph.D., University of North Carolina at Chapel Hill, 1994. Nuclear astrophysics and neutrino physics.

Gaarde, Mette, Ph.D., University of Copenhagen, Denmark, 1997. Theory of atomic and optical physics; generation of high order harmonics and attosecond pulses.

Hynes, Robert I., Ph.D., Open University, England, 1999. Multiwavelength observational astronomy of accreting objects.

Jin, Rongying, Ph.D., Swiss Federal Institute of Technology, Zurich, 1997. The development of novel complex materials with intriguing physical properties, such as new phases that exist on the edge of instabilities (unconventional superconductivity, quantum critical phenomena, heavy—Fermion behavior, thermoelectricity etc.).

Kutter, Thomas, Ph.D., University of Heidelberg, Germany (University of Chicago), 1999. Experimental neutrino physics.

Matthews II, Kenneth L., Ph.D., The University of Chicago, 1997. Radiological imaging; nuclear medical imaging.

Stacy, J. Gregory, Ph.D., Maryland, 1980. Gamma-ray detector development; galactic structure and star formation regions.

Stadler, Shane, Ph.D., Tulane University, 1998. Magnetocaloric systems; half-metallic spintronic systems; x-ray absorption spectroscopy.

Vekhter, Ilya, Ph.D., Brown University, 1998. Condensed matter theory; unconventional superconductivity and strongly correlated electron systems.

Assistant Professors

Giesel, Kristina, Potsdam, 2007. Theoretical, gravity and relativity.

Lee, Hwang, Ph.D., Texas A&M University, 1998. Quantum optics; quantum information science.

Moreno, Juana, Ph.D., Rutgers University, 1997. Computational approaches to strongly correlated electron systems.

Sheehy, Daniel, Ph.D., University of Illinois at Urbana-Champaign, 2001. High temperature superconductivity; degenerate bosonic and fermionic atomic gases; vortices in condensed-matter and cold-atom contexts.

Shikhaliev, Polad, Ph.D., Ioffe Physico-Technical Institute, 1998. X-ray and CT imaging; photon counting/energy resolving x-ray detectors; applications to cancer and heart disease.

Singh, Parampreet, Pune, 2004. Theoretical relativity and quantum gravity.

Tzanov, Martin M., Pittsburgh, 2005. Experimental neutrino physics.

Research Faculty

Diener, Peter, Ph.D., University of Copenhagen, Denmark, 1997. Numerical relativity, astrophysical applications.

Guzik, T. Gregory, Ph.D., Chicago, 1980. Solar flares; particle interactions; accelerator experiments; cosmic rays.

Nascimento, Von Braun, Ph.D, Universidade Federal de Minas Gerais. Materials Science, experimental condensed matter physics.

Schnetter, Erik, Ph.D., Tübingen, Germany, 2003. Numerical relativity, relativistic astrophysics, computational physics, high performance computing.

Xiong, Yimin, Univ. Sci. Tech., China, 2005. Material Science, experimental condensed matter physics.

Faculty Emeriti

Chan, Lai-Him, Ph.D., Harvard, 1966. Theory of elementary particles and fields.

Goodrich, Roy G., Ph.D., California, Riverside, 1965. Ball Family Professor. Low-temperature solid state; electrons in metals and phase transitions.

Hamilton, William O., Ph.D., Stanford, 1963. Gravitational radiation; cryogenics; infrared detectors.

Haymaker, Richard W., Ph.D., California, Berkeley, 1967. Field theory; symmetries in high-energy physics, lattice gauge theory.

Landolt, Arlo U., Ph.D., Indiana, 1962. Photometry; variable stars; eclipsing binaries.

Metcalf, William J., Ph.D., Cal. Tech., 1974. Experimental high-energy physics.

Zganjar, Edward F., Ph.D., Vanderbilt, 1966. Heavy-ion reactions; nuclei far from stability; nuclear deformation.

RESEARCH SPECIALTIES AND STAFF

Theoretical

Astrophysics. Star formation; galaxy dynamics; cataclysmic variables; white dwarfs and neutron stars; nucleosynthesis. Frank, Tohline.

Atomic, Molecular and Optical. Electron, atom and ion molecule scattering; intense magnetic fields; variational principles; ultra short pulse laser-atom interactions; dissipation and fluctuation in quantum physics. Gaarde, O'Connell, Rau, Schafer.

Condensed Matter: Theoretical and computational approaches to strongly correlated electron systems. Collective phenomena: magnetism, superconductivity, bose-condensation, quantum critical phenomena. Studies of novel materials, band structure and model calculations. Browne, Jarrell, Moreno, Sheehy, Vekhter.

Elementary Particles and Fields. Field theory; symmetries in high-energy physics. Chan, Haymaker.

Medical Physics/Health Physics. Radiation transport, dose calculations aerosol transport, microdosimetry radiation biology. Hogstrom, Matthews, K., Shikhaliev.

Nuclear Physics. Nuclear structure. Draayer.

Quantum Optics and Quantum Information. Cavity quantum electrodynamics; laser theory; nonlinear optics; photonic crystals; quantum information processing, quantum imaging and sensing. Dowling, Lee.

Relativity. General Relativity, Numerical Relativity, Black Hole formations, Gravitational Wave Detection, Gravitational Wave Theory, Numerical Analysis Simulation and Visualization, Scientific Computing for Relativity, Quantum Gravity. Diener, Lehner, O'Connell, Pullin, Seidel, Singh.

Experimental/Observational

Astronomy. Gamma-Ray Bursts; supernovae; x-ray binaries; dust, novae. Photometry; stellar abundances. Clayton, Hynes, Landolt, Schaefer. 2 postdoctoral researchers.

Center for Advanced Microstructures and Devices (CAMD); 1.5 GeV Synchrotron Radiation Light Source. Kurtz, Sprunger, Stadler.

Condensed Matter. Electrons in metals and phase transitions; optical properties of solids; thermal conductivity, surface science, electron spectroscopies, tunneling microscopes. Adams, DiTusa Goodrich, Jin Kurtz, Nascimento, Plummer, Sprunger, Stadler, Young, Zhang. 4 postdoctoral researches.

Elementary Particles and Fields. Experiments at Fermilab, the Sudbury Neutrino Observatory, Pierre Auger Observatory, JPARC, and KEK. Kutter, J. Matthews, Metcalf, Tzanov. 3 postdoctoral researchers.

Gravitational Physics. Search for gravitational waves with LIGO and development of advanced detectors. Giaime, González, Johnson. 1 postdoctoral researcher.

Low-Temperature Physics. Gravitational radiation; cryogenics; superconductivity; mesoscopic physics. Adams, DiTusa, Goodrich, Johnson, Young.

Medical Physics/Health Physics. Helical tomotheraphy, image-guided radiotherapy, cardiovascular imaging, x-ray photon counting imaging, radioisotope imaging, synchrotron x-ray radiotherapy. Hogstrom, Matthews, K., Shikhaliev.

Nuclear Physics. Heavy-ion reactions; nuclei far from stability; radioactive beams; nuclear astrophysics. Blackmon, Zganjar. 1 postdoctoral.

Space Science/High Energy Astrophysics. Cosmic rays; neutrinos; gamma rays; high-energy nucleus–nucleus interactions; satellite, balloon, accelerator, air shower and underground experiments, Pierre Auger Observatory. Cherry, Guzik, J. Matthews, Stacy, Wefel.

TULANE UNIVERSITY

DEPARTMENT OF PHYSICS AND ENGINEERING PHYSICS

New Orleans, Louisiana 70118–5636
www.physics.tulane.edu

Students Accepted For Degree	FIELDS: Physics	Astronomy	Related Fields
Doctorate	X		
Master's	X		

Address housing inquiries to: Director, Residential Life, 27 McAlister Drive
On-campus housing available: Yes
Cost/month: $446–890

1. General

President: Scott S. Cowen
Department Chairman: James H. McGuire
Department Telephone Number: (504) 865-5520 (C)
Type of Institution: University
Control: Private
Setting: Urban
Total Faculty: 1,297
Total Graduate Faculty: 427
Total Students: 11,157
Total Graduate Students: 3,025
Annual Graduate Tuition:
All Graduate Students: Full-time—$40,584
Tuition rates for: 2009–10
Deferred tuition plan: Yes
Other Fees: $1,310
Term: Semester

2. Number of Faculty in Department

The combined total of full-time faculty in the three professorial ranks is 15. The combined total of full-time, part-time, and other faculty at all ranks is 20.

3. Admission, Financial Aid, and Housing

Address admission inquiries to: Graduate Admissions Officer, Physics Department
Graduate application fee required: None
Admission deadline (Fall admission): 7/1
Admission information: For fall admission, 2009–10, 5 students were accepted from 60 applicants.
Admission requirements: For admission to the graduate programs, a Bachelor's degree in physics or a related field is required with no minimum undergraduate GPA specified. The GRE is recommended. The GRE Advanced is recommended. Students from non-English speaking countries are required to demonstrate proficiency in English via the TOEFL exam. Minimum score-usually 600.
Undergraduate preparation assumed: Giancoli, *Physics*; Marion, *Mechanics*; Marion, *Electricity and Magnetism*; Park, *Quantum Theory*.
Address financial aid inquiries to: Graduate Admissions Officer, Physics and Engineering Physics Department
Financial aid deadline: 2/1
Loans available: Yes

Table A—Faculty, Enrollments, and Degrees Granted

Research Specialty	2009–10 Faculty	Enrollment[1] Fall 2009: Master's	Enrollment[1] Fall 2009: Doctorate	No. of Degrees Granted[2] 2009–10 (2004–09): Master's	Terminal Master's	Doctorate	Median No. of Years for 2008–09 Ph.D.'s
Atomic, Molecular, & Optical Physics	2	–	3	0(0)	–	0(1)	–
Biophysics	1	–	2	0(0)	–	0(1)	–
Chemical Physics	2	–	1	0(0)	–	0(1)	–
Condensed Matter Physics	5	–	10	3(0)	–	2(8)	6
History & Philosophy	1	–	0	0(0)	–	0(0)	–
Material Sci./ Metallurgy	4	–	0	0(0)	–	0(0)	–
Nuclear Physics	3	–	2	0(0)	–	0(1)	–
Polymer Physics/ Science	1	–	3	0(0)	–	0(1)	–
Relativity & Gravitation	1	–	0	0(0)	–	0(0)	–
Statistical & Thermal	2	–	0	0(0)	–	0(0)	–
Other Theoretical/ Math.	2	–	0	0(0)	–	0(0)	–
Non-specialized	0	–	0	0(0)	–	0(1)	–
Total		–	20	0(0)	–	1(14)	
Full-time Grad. Stud.		–	20				
Part-time Grad. Stud.		–	0				
First-year Grad. Stud.		–	5				
Median Years in Grad. Study (2008–09 Degrees)					–	–	6

Undergraduate Degrees, 2008–09 (2004–09): 7(28)

[1]Students not yet committed to a research specialty are entered under non-specialized.
[2]Five-year totals in parentheses.

4. Graduate Degree Requirements

Master's: 24 semester hours with thesis, or 30 semester hours without thesis. No specific grade point average required, but only one grade lower than "B" is allowed. Minimum of two semesters of full-time residency is required.
Doctorate: 48 semester hours of designated course work (maximum of 24 hours transferrable from another institution). No specific grade point average, but only one grade below "B" allowed. Minimum of one year full-time residence required. Proficiency in computer programming. Written and oral preliminary examination, usually during the third year of graduate study. Dissertation based on original research suitable for publication in a recognized professional journal. Oral dissertation defense.
Thesis: Thesis may be written *in absentia*.

Table B—Appointments to Graduate Students, 2009–10

Title of Appointee	Appointments Total	Appointments First year	Academic Load Allowed in Credit Hours	Hours of Service Per Week	Stipend for Academic Year ($)
			Semester		
Teaching Assistant	14	5	12	15	17,500
BOR Fellow	1	0	12	0	20,000
Research Assistant	10	0	12	10	18,000
Total	25	5			

5. Personnel Engaged in Separately Budgeted Research, 7/08–6/09

Professorial faculty	7
Total	20

6. Separately Budgeted Research Expenditures by Source of Support

	Departmental Research
Federal government	$2,500,000
State government	400,000
Private, Non-profit	200,000
Total	$3,100,000

Table C—Separately Budgeted Research Expenditures

Research Specialty	No. of Grants	Expenditures ($)
Atomic, Molecular, & Optical Physics	1	50,000
Condensed Matter Physics	9	1,000,000
Nuclear Physics	5	300,000
Polymer Physics & Biophysics	4	400,000
Total	19	1,750,000

FACULTY

Professors

Diebold, Ulrike, Ph.D., Technische Universität Wien, 1990. Chair. Surface Science.

Ederer, , David L. Ph.D., Cornell University, 1963. Emeritus Professor. Experimental Solid State Physics.

MacLaren, James M., Ph.D., Imperial College, London, 1986. Interim Dean-SLA. Theoretical solid state physics.

Mao, Zhiqiang, Ph.D., University of Science and Technology of China, 1992. Low temperature condensed matter physics.

McGuire, James H., Ph.D., Northeastern, 1969. Department Chair (Murchison-Mallory). Theoretical atomic physics and physics instruction.

Perdew, John P., Ph.D., Cornell, 1971. Solid state theory/density functional theory.

Purrington, R. Daniel, Ph.D., Texas A&M, 1966. Nuclear physics and history of physics.

Reed, Wayne F, Ph.D., Clarkson, 1984. Polymer physics and biophysics.

Rosensteel, George, Ph.D., Toronto, 1975. Mathematical physics.

Tipler, Frank, Ph.D., University of Maryland, 1976. Relativity and Cosmology.

Wietfeldt, Fred, Ph.D., Univ. of California, Berkeley, 1994. Experimental nuclear physics.

Associate Professor

Kaplan, Lev, Ph.D., Harvard University, 1996. Quantum chaos. Chaos.

Assistant Professor

Kim, Dae Ho, Ph.D., Seoul National University, Korea, 2003. Thin-film and nanostructure materials.

Professors of Practice

Horwitz, Norman, B.M.E., University of Pittsburgh. Industrial Engineering.

Schuler, Timothy M., Ph.D., Tulane University, 2004. Experimental solid state physics and physics instruction.

Shakov, Khazhgery "Jerry," Ph.D., Tulane University, 2004. Quantum control and physics instruction.

Adjunct

Csonka, Gábor, Ph.D., Budapest University of Technology, 1993. Solid state theory/density functional theory.

Gaver, Don, Ph.D., Northwestern University, 1998. Fluid dynamics and transport mechanisms.

Norton, Guy, Ph.D., Tulane University, 1990. Classical wave theory, rough surface.

Pratt, Lawrence, Ph.D., University of Illinois, 1977. Thermodynamics and statistical thermodynamics.

Ruzsinszky, Adrienn, Ph.D., Budapest University of Technology and Economics, 2004. Solid state theory/density functional theory.

Vijay, John, T., D. Eng. Sci., Columbia University, 1982. Self assembly and nanostructure materials.

Research Assistant Professors

Alb, Alina, Ph.D., Tulane University, 2004. Polymer physics and biophysics.

Uskov, Dmitry, Ph.D., Moscow Institute of Physics and Technology, 1985. Theoretical atomic and molecular physics.

Li, Schneton, Ph.D., Institute of Physics, Chinese Academy of Sciences, 2004. Surface Science.

RESEARCH SPECIALTIES AND STAFF

Theoretical

History and Philosophy of Physics. Purrington, Tipler.

Many-Body Physics. Topics below, and interaction through the Quantum Theory Group with Departments of Chemistry and Mathematics. McGuire, Perdew, Purrington, Rosensteel.

Nuclear Physics. few-nucleon calculations. Purrington, Rosensteel.

Quantum Chaos. Kaplan.

Solid State and Atomic Physics. Bulk and surface properties of solids; electronic structure of metals, insulators and atoms; fundamentals, approximations and applications of density-functional theory; atomic collisions. MacLaren, McGuire, Perdew.

Experimental

Experimental Particle Physics. Ultracold neutrons, nuclear astrophysics. Wietfeldt.
Laser Spectroscopy. Light scattering in Polymer and Material systems. Reed, Kim.
Polymer Physics & Biophysics. Reed.
Superconductivity. Mao
Materials Engineering. Diebold, Mao, Kim.
Surface Physics. Thin metal films. Absorption and electron stimulated desorption of molecules. Ion-surface interaction. Diebold, Kim, Mao.

FACULTY PUBLICATIONS (see www.physics.tulane.edu for information)

Diebold, Ulrike

S. Li, J. Wang, P. Jacobson, X. Gong, A. Selloni, and U. Diebold, "Correlation between bonding geometry and band gap states at organic-inorganic interfaces: catechol on rutile TiO(110)," Journal of the American Chemical Society (2008), in press.

E. Morales, Y. He, U. Diebold, and B. Delley, "Surface structure on Sn-doped $In_2O_3(111)$ thin films by STM," New Journal of Physics **10**, 12503 (2008).

P. Jacobson, S. Li, C. Wang, and U. Diebold, "Decomposition of catechol and carbonaceous residues on $TiO_2(110)$: A model system for cleaning of EUVL optics," Journal of Vacuum Science and Technology B **26**(6), 2236 (2004).

S. Li, O. Dulub, U. Diebold, "Scanning Tunneling Microscopy Study of a Vicinal TiO_2 Anatase Surface," Journal of Physical Chemistry C **112**, 16166 (2008).

K. Katsiev, A. Kolmakiv, M. Fang, and U. Diebold, "Characterization of individual SnO_2 nanobelts with STM," Surface Science Letters **602**, L112 (2008).

Kaplan, Lev

D. Uskov, L. Kaplan, A.M. Smith, S.D. Huver, and J.P. Dowling, "Maximal Success Probabilities of Linear-Optical Quantum Gates," submitted to Phys. Rev. Letts., arXiv:0808.1926.

E. J. Heller, L. Kaplan, and A. Dahlen, "Refraction of a Gaussian Seaway,," J. Geophys. Res. **113**, C09023 (2008), arXiv:0801.0613.

L. Kaplan and Y. Alhassid, "Interaction Matrix Element Fluctuations in Ballistic Quantum Dots: Random Wave Model," Phys. Rev. B **78**, 085302 (2008) (also selected for Virtual Journal of Nanoscale Science and Technology), arXiv:082.2410.

E. J. Heller, L. Kaplan, and F. Pollmann, "Inflationary Dynamics for Matrix Eigenvalue Problems," Proc. Natl. Acad. Sci. (USA) **105**, 7631 (2008), arXiv:0712.4093.

L. Kaplan and Y. Alhassid, "Interaction Matrix Element Fluctuations in Quantum Dots, Workshop on Nucleu and Mesoscopic Physics (WNMP07)," AIP Conference Proceedings 995, 192, ed. by P. Danielewicz, P. Piecuch, and V. Zelevinsky (2008), arXiv:0712-4095.

Daeho, Kim

H. M. Christen, D. H. Kim, and C. M. Rouleau, "Interfaces in perovskite heterostructures," Applied Physics A-Materials Science and Processing **93**, 807 (2008).

H. M. Christen, M. Varela, and D. H. Kim, "The effect of strain and strain symmetry on the charge-order transition in $Bi_{0.4}Ca_{0.6}MnO_3$films," Phase Transitions **81**, 717 (2008).

D. H. Kim, H.N. Lee, M. D. Biegalsk, and H. M. Christen, "Effect epitaxial strain on ferroelectric polarization in multiferroic $BiFeO_3$films," Appl. Phys. Lett. **92**, 012911 (2008).

M. Angst, R. P. Hermann, W. Schweika, J.-W. Kim, P. Khalifah, H. J. Xiang, M.-H. Whangbo, D. H. Kim, B. S. Sales, and D. Mandrus, "Charge ordered Fe_2OBO_3: intermediate temperature phase with incommensurate modulations arising from geometrical charge frustration," Phys. Rev. Lett. **99**, 256402 (2007).

D. H. Kim, H. N. Lee, M. Varela, and H. H. Christen, "Large ferroleectric polarization in antiferromagnetic $BiFe_{0.5}Cr_{0.5}O_3$ epitaxial films," Appl. Phys. Lett. **91**, 042906 (2007).

Mao, Zhiqiang

W. Bao, Z. Q. Mao, Z. Qu, J. W. Lynn, "Spin-valve effect and magnetoresistivity in single crystalline $Ca_3Ru_2O_7$," Phys. Rev. Lett. **100**, 247203 (2008).

X. F. Xu, Z. A. Xu, T. J. Liu, D. Fobes, Z. Q. Mao, J. L. Luo, and Y. Liu, "Band-dependent normal-state coherence in Sr_2RuO_4: Evidence from Nerst and thermopower measurements," Phys. Rev. Lett. **101**, 057002 (2008).

J. Hooper, M. H. Fang, M. Zhou, D. Fobes, N. Dang, Z. Q. Mao, C. M. Feng, Z. A. Xu, M. H. Yu, C. J. O'Connor, G. J. Xu, N. Anderson, and M. Salamon, "Competing magnetic fluctuations in $Sr_3Ru_2O_7$ probed by Ti doping," Phys. Rev. B **75** (Rapid Communications), 060403 (2007).

D. Fobes, M. H. Yu, M. Zhou, J. Hooper, C. J. O'Connor, M. Rosario, and Z. Q. Mao, "Phase diagram of the electronic states of trilayered ruthenate $Sr_4Ru_3O_{10}$," Phys. Rev. B **75**, 094429 (2007).

Y. J. Jo, L. Balicas, N. Kikugawa, E. S. Cho, K. Storr, M. Zhou, and Z. Q. Mao, "Orbital-dependent metamagnetic response in $Sr_4Ru_3O_{10}$," Phys. Rev. B. **75**, 094413 (2007).

McGuire, James H.

Paperback edtion of *Electron Correlation Dynamics in Atomic Collisions*, 2006. Ethical Misconduct, submitted to APS website (2009.

Purrington, Robert

"The Westerfelt equation with viscous attenuation vs. a casual propagation operator: a numerical comparison," 2009, with Guy Norton. Submitted to the Journal of Sound and Vibration, Dec. 2008.

Reed, Wayne F.

A. M. Alb, M. F. Drenski, W. F. Reed, "Automatic continuous online monitoring of polymerization reactions (ACOMP)," Polymer International **57**, 390 (2008) (Invited).

A. M. Alb, A. Serelis, W. F. Reed, "Kinetic trends in RAFT homopolymerization from online monitoring," Macromolecules **41**, 332 (2008).

A. M. Alb and W. F. Reed, "Recent advances in ACOMP," (Modelling, Monitoring and Control of Polymer Properties) Macromolecular Symposia **271**, 15 (2008).

A. M. Alb and W. F. Reed, "Simultaneous monitoring of polymer and particle characteristics during emulsion polymerization," Macromolecules **41**, 2406 (2008).

B. Zdyrko, P. Bar-Yosef Ofir, A. M. Alb, W. F. Reed, M. M. Santore, "Adsorption of Copolymers Micelles and Aggregates: from Kinetics to Adsorbed Layer Structure," J. Colloid & Interface Sci. **322**, 365 (2008).

Rosensteel, George T.

G. Rosensteel, D. J. Rowe, and S. Y. Ho, "Equations of motion for a spectrum generating algebra: Lipkin–

Meshkov–Glick model," Journal of Physics A: Mathematical and Theoretical **41**, 025208 (2008).

R. A. Schachar, G. G. Liao, R. D. Kirby, F. Kamagar, J. H. Savoie, A. Abolmaali, and G. Rosensteel, "Unexpected shape changes on encapsulated oblate spheroids in response to equatorial traction," J. Phys. A: Math. Theor. **41**, 495204 (2008).

G. Rosensteel and D. J. Rowe, "The competition between SU_3 and pair coupling in the many-fermion sd shell and interacting boson models," Nuclea Physics A **797**, 94 (2007).

Tipler, Frank J.

New Axioms for Rigorous Bayesian Probability Theory (with Maurice Dupre), submitted to London Journal of Mathematics (November 2008).

Heisenberg Uncertainty from the Many-Worlds Point of View, submitted to Journal of Mathematical Physics.

Wietfeldt, Fred

M. S. Dewey, K. Coakley, D. M. Gilliam, G. L. Greene, A. Laptev, J. S. Nico, W. M. Snow, F. E. Wietfeldt, and A. Yue, "Prospects for a New Cold Neutron Beam Measurement of the Neutron Lifetime," (submitted to Nuclear Instruments and Methods, 2008).

F. E. Wietfeldt, J. Byrne, B. Collett, M. S. Dewey, G. L. Jones, A. Komives, A. Laptev, J. S. Nico, G. Noid, E. J. Stephenson, I. Stern, C. Trull, and B. G. Yerozolimsky, "ACORN: An Experiment to Measure the Electron-Antineutrino Correlation in Neutron Decay," (submitted to Nuclear Instruments and Methods, 2008).

M. G. Huber, M. Arif, T. C. Black, W. C. Chen, T. R. Gentile, D. Pushin, F. E. Wietfeldt, and L. Yang, "Precision Measurement of the Neutron-^{3} He Spin- Dependent Scattering Length Using neutron Interferometry," (submitted to Nuclear Instruments and Methods, 2008).

M. G. Huber, M. Arif, T. C. Black, W. C. Chen, T. R. Gentile, D. S. Hussey, D. Pushin, F. E. Wietfeldt, and L. Yang, "Precision Measurement of the n-3 He Incoherent Scattering Leghth Using Neutron Interferometry," (submitted to Phys. Rev. Lett., 2008).

T. R. Gentile, M. S. Dewey, H. P. Mumm, J. S. Nico, A. K. Thompson, T. E. Chupp, R. L. Cooper, B. M. Fisher, I. Kremsky, F. E. Wietfeldt, K. G. Kiriluk, E. J. Beise, "Particle and Photon detection for a neutron radiative decay experiment," Nucl. Inst. Meth. A **579**, 447 (2007).

UNIVERSITY OF NEW ORLEANS

DEPARTMENT OF PHYSICS

New Orleans, Louisiana 70148
physics.uno.edu

Students Accepted For Degree	FIELDS		
	Physics	Astronomy	Related Fields
Doctorate			X*
Master's	X		X

*Interdisiplinary Ph.D. program in Engineering and Applied Science.

1. General

Chancellor: Timothy P. Ryan
Dean of Graduate School: Scott Whittenburg
Department Chairman: C. Gregory Seab
Department Telephone Number: (504) 280-6341
Type of Institution: University
Control: Public
Setting: Urban
Total Faculty: 650
Total Graduate Faculty: 550
Total Students: 11,700
Total Graduate Students: 2,980
Graduate Tuition (Academic Year:
In-state residents: Full-time—$4,872
Part-time—$1,498–4,540
Out-of-state residents: Full-time—$14,460
Part-time—$5,140–12,970
Tuition rates for: 2010–11
Deferred tuition plan: Yes
Term: Semester

2. Number of Faculty in Department

The combined total of full-time faculty in the three professorial-ranks is 10. The combined total of full-time, part-time, and other faculty at all ranks is 12.

3. Admission, Financial Aid, and Housing

Address admission inquiries to: Graduate Admissions
Graduate application fee required: $50
Admission deadline (Fall admission): 7/1 (or pay a $30 late fee)
Admission information: For fall admission, 2010, 10 students were accepted.
Admission requirements: For non-probationary admission to the graduate programs, a Bachelor's degree in physics or a related area is required with a minimum undergraduate GPA of 2.5 specified. The GRE is required. The minimum acceptable score suggested for admission is verbal plus quantitative 1,030. The GRE Advanced is not required. Students from non-English speaking countries are required to demonstrate proficiency in English via the TOEFL exam. Minimum acceptable score for admission is 550 (paper-based) or 79 (iBT).
Undergraduate preparation assumed: Symon, *Mechanics*; Griffiths, *Electricity and Magnetism*; Zemansky, *Thermodynamics*; recommended: Griffiths, *Quantum Mechanics*.
Address financial aid inquiries to: C. Gregory Seab, Chair, Department of Physics
GAPSFAS application required: No
Financial aid deadline: None
Loans available: Yes
Address housing inquiries to: Manager of Student Housing UNO, Lake Front, New Orleans, LA 70148
On-campus, single student housing available: Yes
Cost/term: $1,845 (including meal plan)
On-campus, married student housing available: No

Table A—Faculty, Enrollments, and Degrees Granted

Research Specialty	2008–09 Faculty	Enrollment[1] Fall 2009		No. of Degrees Granted[2] 2009–10 (2005–10)			Median No. of Years for Ph.D.'s
		Master's	Doctorate	Master's	Terminal Master's	Doctorate	
Acoustics	2	1	4	–	0(0)	2(6)	3
Applied Physics	5	1	4	–	0(0)	0(0)	–
Astronomy	1	0	–	–	0(0)	0(0)	–
Astrophysics	1	0	–	–	0(0)	0(0)	–
Computer Science	1	0	–	–	0(0)	0(0)	–
Condensed Matter/ Materials Physics	4	4	8	–	0(0)	0(5)	4
Electromagnetism	2	0	–	–	0(0)	0(0)	–
Geophysics	3	0	–	–	0(0)	0(1)	3
Low Temperature Physics	1	0	–	–	0(0)	0(0)	–
Optics	3	0	–	–	0(0)	0(0)	–
Physics Education	2	0	–	–	0(0)	0(0)	–
Other Theoretical/ Math.	3	1	2	–	0(0)	0(0)	–
Non-specialized	0	0	–	–	0(0)	0(0)	–
Total		6	18	–	0(0)	2(12)	
Full-time Grad. Stud.		5	13				
Part-time Grad. Stud.		1	5				
First-year Grad. Stud.		3	3				

[1] Students not yet committed to a research specialty are entered under non-specialized.
[2] Five-year totals in parentheses.

4. Graduate Degree Requirements

Master's: Master of Science in Applied Physics: Completion of 33 credit hours course work (non-thesis option) or 24 credit hours of course work plus thesis (six cr. hrs.). Completion of the program usually requires about two years work, and a final examination or thesis defense is required. The Master of Science in Applied Physics offers flexibility for students with undergraduate preparation in either physics or an allied science or engineering field. Concentrations in geophysics, computational physics, acoustics, condensed matter, and materials physics are particularly active.
Thesis: Thesis may be written *in absentia.* :
Ph.D. in Engineering and Applied Science: Interdisciplinary, integrative degree program requiring 51 credit hours of course work past the B.S. degree plus an original research dissertation. Up to 30 credit hours can be credited with an M.S. degree from another university. Research areas for students choosing physics as the home department are generally those of the physics faculty.

Table B—Appointments to Graduate Students, 2006–07

Title of Appointee	Appointments Total	Appointments First year	Academic Load Allowed in Credit Hours	Hours of Service Per Week	Stipend for Academic Year ($)
			Semester		
Teaching Assistant	6	2	10	20	11,800[1]
Research Assistant	7	2	10	20	18,500[2]
Total	14	4			

[1] 9-month stipend. In addition, tuition is waived for academic year and summer.
[2] 12-month stipend, tuition waiver.

5. Personnel Engaged in Separately Budgeted Research, 2009–10

Professorial faculty	9
Research Associates	6
Total	15

6. Separately Budgeted Research Expenditures by Source of Support

	Departmental Research
Federal government	$864,247
State government	193,000
Private Industry	190,691
Total	$1,247,938

Table C—Separately Budgeted Research Expenditures

Research Specialty	No. of Grants	Expenditures ($)
Acoustics	4	212,996
Applied Physics	4	224,669
Condensed Matter Physics	6	320,000
Electromagnetism	2	75,000
Geophysics	1	340,273
Physics Education	2	75,000
Total	19	1,247,938

FACULTY

University Research Professors

Ioup, George E., Ph.D., Florida, 1968. Deconvolution, mathematical digital filtering, and spectral estimation; acoustic, geophysical, and aerospace signal analysis and processing; computational physics; underwater acoustics; marine mammal acoustics; higher order correlations and spectra.

Puri, Ashok, Ph.D., CCNY, 1982. Optical properties of condensed matter; geophysical electromagnetic modeling.

Professors

Ioup, Juliette W., Ph.D., Connecticut, 1972. *Seraphia D. Leyda University Teaching Fellow*. Ocean acoustics; deconvolution; seismic inverse theory and data processing; image theory in electromagnetics.

Seab, C. Gregory, Ph.D., Colorado, 1982. *Chairman of the Department and Seraphia D. Leyda University Teaching Fellow*. Astrophysics, inter-stellar medium; shock waves and dust.

Stokes, Kevin L., Ph.D., Rensselaer Polytechnic Institute, 1995. Experimental condensed matter physics, optics, nanoscience.

Associate Professors

Griffith, O. Frank, Ph.D., South Carolina, 1967. Forensic applications.

Malkinski, Leszek, Ph.D., Warsaw, 1991. Experimental condensed matter physics, magnetic-tunneling, patterned nanostructures, spring magnets, and smart materials.

Spinu, Leonard, Ph.D., Orsay, 1998. Experimental condensed matter physics, magnetic nanoparticles. High frequency properties of novel materials.

Professor, Research

Yaremchuk, Max, Ph.D., Shirshov Institute of Oceanology, 1984. Geophysical fluid dynamic, aerodynamics and thermodynamics.

Associate Professor, Research

Peggion, Germana, Ph.D., Florida State, 1985. Geophysical fluid dynamics, ocean circulation modeling. Joint appointment with Civil and Environmental Engineering.

Instructor

Robbert, Patrica S., M.S., University of New Orleans, 1999. *College of Sciences Teaching Fellow*. Experimental condensed matter physics, ellipsometry, applied optics.

Smith, George, Ph.D., University of New Orleans, 1972. Particle physics, acoustics.

Adjunct Faculty

Chin-Bing, Stanley A., Ph.D., New Orleans, 1973. Ocean acoustics.

RESEARCH SPECIALTIES AND STAFF

Theoretical

Computational and Mathematical Physics. Wave propagation, inversion, and signal analysis in acoustics and geophysics; complex image theory and electromagnetic modeling; ocean circulation modeling; modeling and simulation of magnetization processes; inter-stellar dust and shock waves; restoration and enhancement of physical data by noise removal, deconvolution, and other techniques of digital filtering. G. Ioup, J. Ioup, Peggion, Puri, Seab, Spinu.

Experimental

Astronomy. Optical and ultraviolet observations of the Interstellar Medium. Seab.

Condensed Matter/Materials Physics. Optical transport, magnetic properties of nano-scale materials. Malkinski, Spinu,and Stokes.

Geophysics. Electromagnetic modeling; active electromagnetic survey methods; fiber optic spectroscopy. G. Ioup, J. Ioup, Puri.

UNIVERSITY OF MAINE

DEPARTMENT OF PHYSICS AND ASTRONOMY

Orono, Maine 04469-5709

Students Accepted For Degree	FIELDS		
	Physics	Astronomy	Related Fields
Doctorate	X		
Master's	X		X

1. General

President: Robert A. Kennedy
Vice President for Research: Michael J. Eckardt
Department Chair: David J. Batuski
Department Telephone Number: (207) 581-1016
Department Fax Number: (207) 581-3410
Department E-mail Address: physics@maine.edu
Home Page Address: http://www.umephy.maine.edu
Type of Institution: University
Control: Public
Setting: Small town
Total Faculty: 811
Total Graduate Faculty: 665
Total Students: 11,867
Total Graduate Students: 2,384
Annual Graduate Tuition:
In-state residents: $379/credit hr.
Out-of-state residents: $1090/credit hr.
Tuition rates for: 2009–10
Deferred tuition plan: Yes
Other Fees: $1000–3,000/year
Term: Semester

2. Number of Faculty in Department

The combined total of full-time faculty in the three professorial ranks is 14. The combined total of full-time, part-time, and other faculty at all ranks is 15.

3. Admission, Financial Aid, and Housing

Address admission inquiries to: Dr. James McClymer, Department of Physics and Astronomy, Bennett Hall
Graduate application fee required: $50 (May be waived upon request).
Admission deadline (Fall admission): None. However, application review begins in mid-February and continues until programs are filled. Early application is recommended if financial assistance is required.
Admission information: For fall admission, 2009–2010, 6 students were enrolled from 20 applicants.
Admission requirements: For admission to the graduate programs, a Bachelor's degree in physics is normally required with no minimum undergraduate GPA specified. The GRE is required. No minimum score for admission is specified. The GRE Advanced is required. No minimum score for admission is required. Students from non-English speaking countries are required to demonstrate proficiency in English via the TOEFL exam. Minimum acceptable score for admission is 550; for teaching assistants, 600.
Undergraduate preparation assumed: Modem Physics: Beiser, *Concepts of Modern Physics*; Mechanics: Fowles and Cassiday, *Analytical Mechanics*; Electricity and Magnetism: Griffiths, *Introduction to Electrodynamics*; Mathematics: Paul, *Differential Equations for Mathematics, Science, and Engineering*.
Address financial aid inquiries to: Dr. James McClymer, Department of Physics and Astronomy, Bennett Hall
GAPSFAS application required: No
Financial aid deadline: None
Loans available: Yes
Address housing inquiries to: Housing Services, Room 103, 5734 Hilltop Commons, Orono, Maine 04469-5734, (207)581-4580
On-campus, single student housing available: Yes
Cost/term: $4,494–6,256 (room and board) (see http://www.umaine.edu/housing)
On-campus, married student housing available: Yes
Cost/month: $580–826 (see http://www.umaine.edu/housing)

Table A—Faculty, Enrollments, and Degrees Granted

Research Specialty	2009–10 Faculty	Enrollment[1] Fall 2009		No. of Degrees Granted[2] 2009–10 (2005–10)			Median No. of Years for 2009–10 Ph.D.'s
		Master's	Doctorate	Master's	Terminal Master's	Doctorate	
Astronomy	2	0	0	0(0)	0(0)	0(0)	–
Astrophysics	2	1	6	0(0)	0(0)	0(0)	–
Biophysics	3	0	2	0(0)	0(0)	1(1)	–
Chemical Physics	4	0	0	0(0)	0(0)	0(1)	–
Condensed Matter Physics	8	0	2	0(0)	0(0)	0(0)	–
Energy Sources & Environ.	1	1	0	0(0)	0(0)	0(2)	–
Engineering Physics/ Science	3	1	0	0(0)	0(2)	0(0)	–
Geophysics	1	0	12	0(0)	0(0)	1(0)	6
Liquid Crystals	1	0	0	0(0)	0(0)	0(0)	–
Marine Science/ Oceanography	1	0	1	0(0)	0(0)	0(0)	–
Materials Sci./ Metallurgy	4	0	5	0(1)	0(2)	0(0)	–
Medical & Health Physics	1	0	1	0(0)	0(0)	0(0)	–
Nuclear Physics	1	0	0	0(0)	0(0)	0(0)	–
Physics Education	2	0	5	0(0)	0(2)	0(3)	–
Polymer Physics/ Science	1	0	0	0(0)	0(0)	0(0)	–
Relativity & Gravitation	1	0	0	0(0)	0(0)	0(0)	–
Statistical & Thermal	2	0	3	0(0)	0(1)	0(1)	–
Surface Science	4	0	1	0(1)	0(4)	1(1)	6
Other Theoretical/ Math.	1	0	0	0(0)	0(0)	0(0)	–
Non-specialized	0	0	4	0(0)	0(0)	–	–
Total		4	33	0(2)	0(11)	3(0)	
Full-time Grad. Stud.		4	33				
Part-time Grad. Stud.		0	0				
First-year Grad. Stud.		0	65				
Median Years in Grad. Study (2009–10 Degrees)				–	–	6	–
Undergraduate Degrees, 2009–10 (2005–10): 6(42)							

[1]Students not yet committed to a research specialty are entered under non-specialized.
[2]Five-year totals in parentheses.

4. Graduate Degree Requirements

Master's: 30 graduate (semester) credits, with 24 devoted to courses in physics and allied fields; courses must be passed with a minimum grade of C, and no more than six hours of C grade may be applied toward the degree; at least 12 hours of course work must be taken while a full-time student in residence; no foreign language or computer language requirements; no comprehensive or qualifying examinations required: thesis required.

Doctorate: 30 graduate (semester) credits in courses required; courses must be passed with a minimum grade of C, and no more than six hours of C grade may be applied toward the degree. Residence requirement satisfied by registering for a full program of study for two consecutive years following the baccalaureate, or for one year following the award of a Master's degree; no foreign language required, no computer language required; comprehensive examination required; thesis required.

Other Programs: Through cooperative efforts of faculty in the Departments of Physics, Chemistry, Earth Sciences, Electrical and Computer Engineering, and Biochemistry students may work for Master's or Ph.D. degree with concentrations in applied physics, materials science, quaternary studies, and biophysics. *Master's of Engineering (Engineering Physics)*: Requirements same as for M.S. in physics except (1) 9 hours of the 30 must be selected from engineering courses and (2) non-thesis option available with 36 course hours required.

Thesis: Thesis may be written *in absentia*, with permission.

Special Equipment, Facilities, or Programs: The Laboratory for Surface Science and Technology (LASST) unites researchers from Chemistry, Physics, Electrical & Computer Engineering, and Chemical & Biological Engineering in many projects spanning aspects of surface and interface science, thin films, sensors, microsystems, and nanotechnology. Current facilities include thin film synthesis, electron and optical spectroscopies, scanning probe microscopies, x-ray and electron diffraction, focused ion beam-scanning electron microscopy, fluorescence microscopy, device fabrication (class 1000 clean room with photolithography, metallization, wet and dry etch, PECVD, sputtering, mask generation, packaging), and sensor testing (gas delivery systems, electrical and microwave test equipment, data acquisition/integrated electronic test suites). The Physics Education Research Laboratory has several facilities and equipment for conducting research on the learning and teaching of physics: a classroom dedicated to curricular activities based on physics education research (PER) containing computer-equipped work areas for ~24 people, including education software, video equipment, and items for hands-on inquiry-based activities; dedicated clinical interview space, to ensure the anonymity and privacy of students participating in our research work (as required by our Institutional Review Board for testing with human subjects); digital video cameras to record individual interviews as well as classroom interactions; individual computers to help with video analysis, documentation, data management, communication, and other relevant activities; and a high-speed scanner for rapid digital storage of written student responses and other documents.

Astrophysics: Linux/PC workstation network dedicated to research in galactic dynamics, and radio and optical observational astronomy.

The Radon Measurement Laboratory: liquid scintillation counter, 3 HP Ge detectors, x-ray fluorescence, Wrenn detectors, and a portable NaI spectrometer interfaced to a Digidart portable Multichannel Analyzer.

Biophysics and Optics: three laboratories including super-resolution localization microscopy facility and four F-PALM microscopes, image processing compuer cluster, tunable femtosecond pulsed Ti:Sapphire laser and optical parametric oscillator (OPO), cell culture facilities, polymerase chain reaction (PCR) thermal cycler and other equipment for molecular bilolgy, a computer-controlled differential scanning calorimeter (DSC), confocal and two-photon laser-scanning microscopes, fluorescence correlation and cross-correlation microscope, fluorimeter, spectrophotometer, Krypton-Argon and Argon ion lasers, numerous diode lasers spanning visible wavelengths from 400–700 nm, and optical tweezer.

Table B—Appointments to Graduate Students, 2009–10

Title of Appointee	Appointments Total	Appointments First year	Academic Load Allowed in Credit Hours	Hours of Service Per Week	Stipend for Academic Year ($)
			Semester		
Teaching Assistant	22	6	9	20	14,250[1,2]
Research Assistant	14	0	9	20	$19,000-30,000[1,3]
Total	36	6			

[1]Tuition is waived for up to 9 credit hours per semester, and 3 credit hours for the summer term.
[2]Summer appointments at academic year rates.
[3]For a 12-month appointment.

5. Personnel Engaged in Separately Budgeted Research, 7/09–6/10

Professorial faculty	8
Other faculty	6
Postdoctoral appointments	6
Graduate students	25
Undergraduate students	6
Nonteaching research personnel	6
Total	57

6. Separately Budgeted Research Expenditures by Source of Support

	Departmental Research	Physics-related Research Outside Department
Federal government	$1,586,272	$903,263
State and local government	$222,956	$610,250
Private, non-profit organizations	$242,500	$573,728
Business and industry	95,000	
Other	26,500	
Total	$2,173,228	$2,087,241

7. Separately Funded and Managed Laboratories

The University of Maine operates the multidisciplinary Laboratory for Surface Science and Technology (LASST) as a separate unit. Research in areas described previously is carried out by five to ten faculty and staff members, as well as by twenty to thirty graduate and undergraduate students, and a number of postdoctoral personnel. Home Page Address: www.umaine.edu/lasst/

The Maine Center for Research in STEM Education (Center) is an interdisciplinary group of science, mathematics, and education faculty focused on research into student learning of science and mathematics and applications of this research. The Center runs a teacher preparation program, the Master of Science Teaching, for secondary science and mathematics teachers. Home Page Address: www.umaine.edu/center/

Table C—Separately Budgeted Research Expenditures

Research Specialty	No. of Grants	Expenditures ($)
Biophysics	3	558,706
Energy & Env	3	31,500
Health Physics	1	25,500
PER	3	269,509
Science & Mathematics Education	1	25,000
Statistical & Thermal Phy	1	70,000
Surface Science	9	1,193,513
Total	21	2,173,228

Table D—Physics-related Research Outside Department

Research Specialty	No. of Grants	Expenditures ($)
Biophysics	1	163,500
Geophysics	1	25,000
PER	1	163,500
Science and Mathematics Education	3	$121,000
STEM	3	175,500
Surface Science	7	1,438,741
Total	16	$2,087,241

FACULTY

Professors

Astumian, R. D., Ph.D., Univ. Texas, Arlington, 1983. Theoretical and experimental condensed matter physics; biophysics of molecular motors and pumps.

Batuski, David J., Ph.D., New Mexico, 1986. Chair of the Department. Astronomy and astrophysics; observational cosmology; large scale structure in the universe, radio sources in galaxy clusters.

Comins, Neil F., Ph.D., University College, Cardiff, 1978. Observational and theoretical astrophysics; galactic evolution and stability; stellar systems; general relativity, astronomy education.

Hess, C. Thomas, Ph.D., Ohio University, 1967. Environmental nuclear physics; health physics; radioactivity studies.

Kleban, Peter H., Ph.D., Brandeis, 1970. Statistical mechanics; phase transitions and kinetics on surfaces; conformal field theory; instrumentation; electron and ion spectrometry.

Lad, Robert J., Ph.D., Cornell, 1986. Surface physics; thin films; sensor technolog; materials science; ceramics; electronic materials; electron spectroscopy and diffraction; scanned probe microscopy.

McKay, Susan R., Ph.D., MIT, 1987. Theoretical condensed matter physics; phase transitions and critical phenomena; spin glasses, amorphous magnetism, quenched disorder; nonlinear systems and transitions to chaos, pattern formation, systems far from equilibrium, applications of network theory.

Unertl, William N., Ph.D., Wisconsin, 1973. Surface physics; surface analysis techniques; materials science; tribology, paper surface science.

Associate Professors

McClymer, James P., Ph.D., Delaware, 1986. Digital imaging and light scattering from equilibrium and nonequilibrium transitions in liquid crystals and complex fluids.

Mountcastle, Donald B., Ph.D., Virginia, 1971. Molecular biophysics; microcalorimetry; chemical thermodynamics; physics education.

Thompson, John R., Ph.D., Brown University, 1998. Physics education, research on teaching and learning, curriculum development and assessment.

Wittmann, Michael C., Ph.D., University of Maryland, 1998. Physics education, lerning theory development, curriculum development and evaluation, survey design and evaluation.

Assistant Professors

Hess, Samuel T., Ph.D., Cornell, 2002. Experimental and theoretical biophysics. Super-resolution fluorescence microscopy and spectroscopy; function and lateral organization of biomembranes; influenza virus infection; single molecule fluorescence photophysics.

Meulenberg, Robert W., Ph.D., University of California, Santa Barbara, 2002. Experimental condensed matter physics: electronic structure of nanoscale materials; surface and interfacial physics of nanostructures; magnetic materials; applications of synchrotron radiation to materials science.

RESEARCH SPECIALTIES AND STAFF

Theoretical

Astrophysics, Astronomy, and General Relativity. Gravitational radiation detection data analysis; black hole theory; and computer models of galactic dynamics. Comins.

Biophysics. Membrane phase transitions; numerical simulations of optical systems; cooperative interactions. Astumian, S. Hess, Mountcastle.

Physics Education. Learning theories, cognitive development in physics, metacognition and epistemology. Wittmann.

Statistical Mechanics. Theory of phase transitions; conformal field theory; population dynamics; spin glasses; amorphous magnetism; non-linear systems; systems far from equilibrium, pattern formation. Kleban, McKay.

Surface Physics. Energy exchange processes; phase transitions; stepped surfaces; adhesion and friction. Kleban, Unertl.

Experimental

Astronomy (Numerical and Observational). Graduate students observe at the Very Large Array in New Mexico, Steward Observatory in Arizona, and the European Southern Observatory in Chile. Batuski, Comins.

Astronomy Education. Identifying and remediating student misconceptions about the cosmos. Comins.

Astrophysics. Simulations of spiral galaxy dynamics. Analysis of large-scale structure in the universe. Batuski, Comins.

Biophysics. Ultra-high resolution fluorescence microscopy, cell membrane biophysics, single molecule photophysics; molecular motors and pumps. Astumian, S. Hess, Mountcastle.

Engineering Physics. Use of microprocessors and microcomputers for data acquisition; development of new analytical techniques for surface analysis; thin film development; sensor technology; heat transfer; nanotribology. Lad, McClymer, Unertl.

General Relativity. Search for gravity waves using LIGO. Comins.

Liquids, Liquid Crystals. McClymer.

Nanophysics. Electronic structure and optical properties of nanoparticles, magnetic materials, doped and alloy materials. Muelenber, S. Hess.

Nuclear Physics. Studies of radon in homes, schools, and ground-water; nuclear fallout; site characterization; radioactive contamination field studies, Lead-210 dating of sediment cores. C. T. Hess.

Physics Education. Empirical studies of conceptual understanding; development and assessment of instructional materials. Identification of specific student difficulties and productive resources while learning physics concepts. Physics topics include quantum physics, thermal physics, mechanics, wave physics, and sound. Data come from free response written items, surveys, individual and group interviews, and classroom video. Emphasis on the use and understanding of mathematics in upper-division physics. Populations under study include introductory, advanced, and general education students, and preservice K-12 teachers. Mountcastle, Thompson, Wittmann.

Polymer Physics. Metal polymer adhesion, surface structure, mechanical properties, wetting and spreading phenomena. Unertl.

Surface Physics. Crystal structure and reactivity; adsorption at surfaces; surface modification; surface structure of nanoscale materials surface spectroscopy; electronic properties; phase transitions; liquid metal and polymer surfaces; ion-beam modification; model membranes; thin-film synthesis; friction and adhesion; mechanical properties of surfaces; sensors; tribology. Lad, Meulenberg, Unertl. 2 postdoctoral fellows.

FACULTY PUBLICATIONS

Astumian, R. Dean

R. D. Astumian, "Chemical Peristalsis," Proc. Natl. Acad. Sci., **102**, 1843–1847 (2005).

R. D. Astumian, "Paradoxical Games and a Minimal Brownian Motor," American Journal of Physics **73**, 178–183 (2005).

A. Goel, R. D. Astumian, and D. Herschbach, "Tuning and Switching the DNA Polymerase Motor with Mechanical Tension," Proc. Natl. Acad. Sci. **100**, 9699–9704 (2003).

R. D. Astumian, "Adiabatic Pumping Mechanism for Ion Motive ATPases," Phys. Rev. Letts. **91**, 118102 (2003).

R. D. Astumian and P. Hanggi, "Brownian Motors," Physics Today **55**(11), 33–39 (Nov. 2002).

Batuski, David J.

T. Wu, D. J. Batuski, and A. Khalil, "Multi-Scale Morphological Analysis of Application of SDSS DR5 Survey Using the Metric Space Technique," Astrophysical Journal **707**, 1160–1167 (2009).

C. J. Miller, K. S. Krughoff, D. J. Batuski, J. M. Hill, "The MX Northern Abell Cluster Survey II: The Abell/ACO Cluster Redshifts and Spatial Analyses," Astronomical Journal **124**, 1918–1933 (2002).

C. Miller and D. Batuski, "The Power Spectrum of Rich Clusters on Near-Gigaparsec Scales," Astrophysical Journal **551**, 635–642 (2001).

C. Miller, R. Nichol, and D. Batuski, "Possible Detection of Baryonic Fluctuations in the Large-Scale Structure Power Spectrum," Astrophysical Journal **555**, 68–73 (2001).

C. Miller, R. Nichol, and D. Batuski, "Evidence for Acoustic Oscillations in the Matter Power Spectrum," Science **292**, 2302–2303 (2001).

Comins, Neil

N. F. Comins, N. Palestini, E. F. Borra, R. Hohlfeld, L. Wholly, "An Investigation of the Effects of Galactic Interactions."

Discovering the Universe, 8th edition, 2008, W. H. Freeman and Co., New ork.

"The Hazards of Space Travel," book, 2007, Villard press, New York.

"Discovering the Essential Universe," 3rd edition, 2006 textbook, W. H. Freeman and Co., New York.

"Discovering the Universe," 7th edition, 2005 textbook, W. H. Freeman and Co., New York.

Hess, C. Thomas

V. E. Guiseppe, T. J. Gould, C. T. Hess, "Spatial Variation of Waterborne Radon and Temporal Variation of Radon in Water at Nine Maine Schools," Health Phys. **92**(4), 358–365 (2007).

L. J. Osher, L. Leclerc, G. B. Wiersma, C. T. Hess, V. E. Guiseppe, "Heavy Metal Contamination from Historic Mining in Upland Soil and Estuarine Sediments of Egypt Bay, Main, USA," Estuarine, Coastal and Shelf Science **70**, 169–179 (2006).

V. E. Guiseppe, R. Stilwell, C. T. Hess, "A Laboratory Intercomparison of Radon in Water Measurements in Maine," Health Phys. **91**(4), 354–360 (2006).

C. T. Hess, "Measuring and Modeling Exposure from Environmental Radiation on Tidal Flats," Nuclear Instruments and Methods in Physics Research A **537**, 658–665 (2005).

C. T. Hess, "Water Borne Radon in Seven Maine Schools," Health Physics **86**(5), 528–535 (2004).

Hess, Samuel T.

T. J. Gould, V. V. Verkhusha, and S. T. Hess, "Imaging Biological Structures with Fluorescence Photoactivation Localization Microscopy," Nature Protocols 4: 291-308 (2009).

S. T. Hess, "Red Lights, Camera, Photoactivaion!" Nature Methods 6: 124-125 (2009).

T. J. Gould and S. T. Hess, "Nanoscale Imaging of Intracellular Fluorescent Proteins: Breaking the Diffraction Barrier," in Biophysical Tools for Biologists, Volume 2: Methods in Vivo, H. W. Detrich editor, Methods in Cell Biology 89: 329-358 (2008).

T. J. Gould, M. S. Gunewardene, M. V. Gudheti, V. V. Verkhausha, S. R. Yin, J. A. Gosse, and S. T. Hess, "Imaging Molecular Positions and Anisotropies," Nature Methods 5: 1027-1030 (2008).

T. J. Gould, J. Bewersdorf, and S. T. Hess, "A Quantitative Comparison of the Photophysical Properties of Select Quantum Dots and Organic Fluorophores," Z. Phys. Chem. 222: 833-849 (2008).

Kleban, Peter H.

P. Kleban, "Factorization of percolation density correlation functions for clusters touching the sides of a rectangle," Jacob J. H. Simmons, Robert M. Ziff, and Peter Kleban, J. Stat. Mech. (2009) P02067 http://arxiv.org/abs/0811.3080.

T. Prellberg, J. Fiala, and P. Kleban, "Cluster Approximation for the Farey Fraction Spin Chain," J. Stat. Phys. **123**, 455–471 (2006).

J. J. H. Simmons, P. Kleban, R. M. Ziff, "Percolation Crossing Formulas and Conformal Field Theory," J. Phys. A: Math. Theory. 40 (2007) F771-F784 [arxiv: 07051933].

Peter Kleban, Jacob J. H. Simmons and Robert M. Ziff, "Anchored critical percolation clusters and 2-d electrostatics,"

Phys. Rev. Letters 97, 115702 (2006) [arxiv: condmat/0605120].

P. Kleban and D. Zagier, "Crossing Probabilities and Modular Forms," J. Stat. Phys. **113**, 431–454 (2003).

Lad, Robert J.

M. M. Steeves, D. Deniz, and R. J. Lad, "Charge Transport in Flat and nanorod Structured Ruthenium Thin Films," Appl. Phys. Lett. **96**, 142103 (2010).

M. M. Steeves and R. J. Lad, "Influence of nanostructure on Charge Transport in RuO2 Thin Films," J. Vac. Sci. Technol. A, in press (2010).

D. Deniz, D. J. Frankel, and R. J. lad, "Nanostructured Tungsten and Tungsten Trioxide Films Prepared by Glancing Angle Deposition," Thin Solid Films **518**, 4095 (2010).

X. Zhang, m. Byrne, and R. J. Lad, "Structure and Optical Properties of Zr1-xSixN Thin Films on Sapphire," Thin Solid Silms **518**, 1522 (2009).

D. J. Frankel, G. Bernhardt, B. Sturtevant, T. Moonlight, M. Pereira da Cunha, R. J. Lad, "Stable Electrodes and Ultrathin Passivation Coatings for High Temperature Sensors in Harsh Environments," Proc. IEEE Sensors Oct. 26-29, pp. 82 (2008).

McClymer, James P.

J. P. McClymer, "Using Vacuum Food Sealers as a Low -Cost Vacuum Pump," The Physics Teacher **48**(3), 202 (2010).

J. P. McClymer and H. M. Shehadeh, "Photon localization in a nematic liquid crystal", Phys, Rev. A. 79, 031802 (2009).

P. N. Segre and J. P. McClymer, "Fluctuations, stratification and stability in a liquid fluidized bed at low Reynolds number," J. Phys.: Condens. Matter **16** (2004).

J. P. McClymer, "Liquid crystal elecrohydrodynamics in the presence of a magnetic field: the dynamic scattering transition," Liquid Crystals **27**, 1305 (2000).

McKay, Susan R.

Thomas E. Stone and Susan R. mcKay, "Correlated Spin Networks in Frustrated Systems," Physica A, published on-line 2/11/10, in press.

R. K. P. Zia, E. F. Redish, and S. R. Mckay, "Making Sense of the Legendre Transform," Am. J. Phys. **77**, 614 (2009).

I. T. Georgiev and S. R. McKay, "A Nonequilibrium Monte Carlo Renormalization-Group approach Based Upon the Microscopic Master Equation Applied to the Three-State Driven Lattice Gas," Physica D **212**, 233 (2005).

I. T. Georgiev and S. R. McKay, "Position-Space Renormalization-Group Approach for Driven Diffusive Systems Applied to the Asymmetric Exclusion Model," Phys. Rev. E **67**, 056103 (2003).

F. M. Ytreberg and S. R. McKay, "A Quasi-Equilibrium Analysis to Predict the Dependence of Ferrofluid Aggregate Properties on Field Ramping Rate," IEEE Transactions on Magnetics **39**, 2648 (2003).

Meulenberg, Robert W.

R. W. Meulenberg, J. R. I. Lee, A. Wolcott, J. Z. Zhang, L. J. Terminello, T. van Buuren, "Determination of the Exciton Binding Energy in CdSe Quantum Dots." ACS Nano, **3**, 325-330 (2009).

A. Lita, X. Ma, R. W. Meulenberg, T. van Buuren, A. E. Stiegman, "Synthesis and characterization of phase-pure Mn(II)- and Mn(III)-silicalite-2." Inorg. Chem., **47**, 7302 (2008).

J. R. I. Lee, T. Y. J. Han, T. M. Willey, D. Wang, R. W. Meulenberg, J. Nilsson, P. M. Dove, L. J. Terminello, T. van Buuren, J. J. De Yoreo, "Structural Development of Mercaptophenol Self-Assembled Monolayers and the Overlying Mineral Phase during Templated $CaCO_3$ Crystallization from a Transient Amorphous Film." J. Am. Chem. Soc., **129**, 10370 (2007).

J. R. I. Lee, R. W Meulenberg, K. M. Hanif, H. Mattoussi, J. E. Klepeis, L. J. Terminello, T. van Buuren, "Experimental Observation of Quantum Confinement in the Conduction Band of CdSe Quantum Dots." Phys. Rev. Lett., **98**, 146803 (2007).

T. M. Willey, C. Bostedt, T. van Buuren, J. E. Dahl, S. G. Liu, R. M. K. Carlson, R. W. Meulenberg, E. J. Nelson, L. J. Terminello, "Quantum Confinement Observed in the Occupied States of Diamond Clusters." Phys Rev. B **74**, 205432 (2006).

Mountcastle, Donald B.

D. B. Mountcastle, B. R. Bucy, and J. R. Thompson, "Student Estimates of Probability and Uncertainty in Advanced Laboratory and Statistical Physics Courses," 2007 Physics Education Research Conference, L. Hsu, C. Henderson, L. McCullough, Ed., AIP Conf. Proc. **951**, 152–155 (2007).

E. B. Pollock, J. R. Thompson, and D. B. Mountcastle, "Student Understanding of the Physics and Mathematics in P-V Diagrams," 2007 Physics Education Research Conference, L. Hsu, C. Henderson, L. McCullough, Eds., AIP Conf. Proc. **951**, 168-171 (2007).

B. R. Bucy, J. R. Thompson, and D. B. Mountcastle, "Student (Mis)application of Partial Differentiation to Material Properties," in 2006 Phys. Educ. Res. Conf., L. McCullough, L. Hsu, P. Heron, Eds., AIP Conference Proceedings **883**, 157–160 (2007).

J. R. Thompson, B. R. Bucy, and D. B. Mountcastle, "Assessing Student Understanding of Partial Derivatives in Thermodynamics," 2005 Physics Education Research Conference, P. Heron, L. McCullough, J. Marx, Ed., AIP Conf. Proc. **818**, 77–80 (2006).

B. R. Bucy, J. R. Thompson, and D. B. Mountcastle, "What Is Entropy? Advanced Undergraduate Performance Comparing Ideal Gas Processes," in 2005 Physics Education Research Conference, P. Heron, L. McCullough, and J. Marx, Ed., AIP Conf. Proc. **818**, 81–84 (2006).

Thompson, John R.

R. P. Springuel, M. C. Wittmann, and J. R. Thompson, "Defining distances between student responses using data mining methods," accepted for publication in Physical Review Special Topics - Physics Education Research (2010).

J. R. Thompson, W. M. Christensen, and m. C. Wittmann, "Preparing future teachers to anticipate student difficulties in physics in a graduate-level course in physics, pedagogy, and education research," accepted for publication in Physical Review Special Topics- Physics Education Research (2010).

M. j. O'Brien and j. R. Thompson, "Effectiveness of ninth-grade physics in Maine: Conceptual understandign," The Physics Teacher **47**(4), 234–239 (2009).

T. I. Smith, W. M. Christensen, and J. R. Thompson, "Addressing Student Difficulties with Concepts Related to Entropy, Heat Engines and the Carnot Cycle," in 2009 Physics Education Research Conference, C. Henderson, M. Sabella, C. Singh, eds., AIP Conference Proceedings **1179**, 277–280 (2009).

M. C. Wittmann and J. R. Thompson, "Integrated approaches in physics education: A graduate level course in physics, pedagogy, and education research," American Journal of Physics **76**(7), 677–683 (2008).

Wittmann, Michael C.

K. McCann, and M. C. Wittmann (2010) "The role of sign in students' modeling signs of scalar equations," The Physics Teacher **48**(4), 246–249.

E. C. Sayre and M. C. Wittmann (2008) "The plasticity of intermediate mechanics students' coordinate system choice," Physical Review Special Topics Physics Education Research 4 020105. Available at http://prst-per.aps.org/abstract/PRSTPER/v4/i2/e020105.

T. I. Smith and M. C. Wittmann (2008) "Toward a more effective use of the Force and Motion Conceptual Evaluatio," Physical Review Special Topics Physics Education Research 4, 020101. Available at http://prst-per.aps.org/abstract/PRSTPER/v4/i2/e020101.

M. C. Wittmann and J. R. Thompson (2008) "Integrated approaches in physics education: A graduate level course in physics, pedagogy, and education research," American Journal of Physics **76**(7), 677–683 (2008).

M. C. Wittmann (2006) "Using resource graphs to represent conceptual change," Physical Review Special Topics Physics Education Research 2, 020105. Available online at http://prst-per.aps.org/abstract/PRSTPER/v2/i2/e020105.

JOHNS HOPKINS UNIVERSITY

THE HENRY A. ROWLAND DEPARTMENT OF PHYSICS AND ASTRONOMY

Baltimore, Maryland 21218

Students Accepted For Degree	FIELDS Physics	Astronomy	Related Fields
Doctorate	X	X	
Master's			

1. General

President: William Brody
Department Chairman: Daniel H. Reich
Department Telephone Number: (410) 516-7344
Type of Institution: University
Control: Both
Setting: Urban
Total Faculty: 492* (No separate Graduate faculty)
Total Students: 6,437
Total Graduate Students: 1,693
Annual Graduate Tuition:
All Graduate Students: Full-time—$40,680
Part-time—$3,915/per graduate course
Tuition rates for: 2010–11
Deferred tuition plan: N/A
Other Fees: $500 (One time matriculation fee for first year graduate students.)
Term: Semester

*Schools of Arts & Sciences and Engineering

2. Number of Faculty in Department

The combined total of full-time faculty in the three professorial ranks is 30. The combined total of full-time, part-time, and other faculty at all ranks is 66.

3. Admission, Financial Aid, and Housing

Address admission inquiries to: The Henry A. Rowland Department of Physics and Astronomy, Bloomberg Center
Graduate application fee required: $75 (temporarily waived for students with either financial need or foreign exchange problems)
Admission deadline (Fall admission only): 1/14/2011
Admission information: For fall admission, 2010–11, 64 students were accepted from 257 applicants.
Admission requirements: For admission to the graduate programs, a Bachelor's degree in physics, astrophysics, or astronomy is required with no minimum undergraduate GPA specified. The General and Subject GRE scores are required. Letters of recommendation, undergraduate transcripts, and GRE scores are considered equally in granting admission. Students from non-English speaking countries are required to demonstrate proficiency in English via the TOEFL exam. Minimum score required is 600.
Address financial aid inquiries to: Office of Student Financial Services
GAPSFAS application required: No
Financial aid deadline: 4/1/2011
Loans available: Yes
Address housing inquiries to: Housing Office, Wolman Hall
On-campus single/married graduate student housing available: No

Table A—Faculty, Enrollments, and Degrees Granted

Research Specialty 2009–10	Faculty	Enrollment[1] Fall Mas-ter's	Doc-torate	No. of Degrees Granted[2] 2009–2010 Mas-ter's	Terminal Master's	Doc-torate	Median No. of Years for 2009–10 (2005–10) Ph.D.'s
Astrophysics		–	44	N/A	0(3)	2(20)	6.5
Condensed Matter Physics		–	40	N/A	2(4)	2(19)	6.5
Particles & Fields		–	25	N/A	0(0)	4(9)	6.5
Plasma Physics & Fusion		–	1	N/A	1(1)	0(1)	6.5
Total		–	110	N/A	3(8)	8(49)	
Full-time Grad. Stud.*		–	110				
Part-time Grad. Stud.		–	0				
First-year Grad. Stud.		–	19				
Median Years in Grad. Study (2007–08 Degrees)							6.5

[1] Students not yet committed to a research specialty are entered under non-specialized.
[2] Five-year totals in parentheses.
* Does not include non-resident.

4. Graduate Degree Requirements

Doctorate: Entering students elect to work towards a Ph.D. in either Physics or Astronomy and Astrophysics. The two programs have somewhat different requirements (see below). A large measure of flexibility characterizes a typical program of study. The underlying philosophy is that with a minimum of guidance the student soon develops a feeling for what advanced courses are needed and the rate at which he or she can take them. The student has several faculty members as advisors. They will aid the student in selecting the program best suited to his or her needs. Students are expected to demonstrate mastery of upper-level undergraduate material in classical mechanics, electricity and magnetism, quantum mechanics, and statistical physics and thermodynamics by passing written exams before the beginning of their fourth semester of graduate school. Exams covering electricity and magnetism and quantum mechanics are given in September and May while exams covering classical mechanics and statistics and thermodynamics are given in January and May. This schedule permits students to enroll in courses appropriate to their level of preparation.

To receive a Ph.D. in physics, students must complete: Analytical/Numerical Methods for Physicists (2 semesters, 171.415–416), Theoretical Mechanics (1 semester, 171.601) or Advanced Statistical Mechanics (1 semester, 171.703), Electromagnetic Theory (2 semesters, 171.603-604), Quantum Mechanics (2 semesters, 171.605-606), Advanced Physics Lab (1 semester 173.308) or Advanced Laboratory (1 semester, 173.607 or 173.712).

To receive a Ph.D. in astronomy and astrophysics, students must complete: Analytical/Numerical Methods for Physicists (2 semesters, 171.415–416), or equivalent, Stellar Structure and Evolution (1 semester, 171.611), Interstellar Medium and Astrophysical Fluid Dynamics (1 semester, 171.612), Radiative Astrophysics (1 semester, 171.613), Galactic Structure and Stellar Dynamics (1 semester, 171.615), Extragalactic Astronomy (1 semester, 171.617). They must also complete two semesters drawn from the following list of optional courses: Quantum Mechanics (171.605–606), Astrophysical Spectroscopy (171.614), Observational Astronomy (171.618), Statistical Methods for Physics and Astronomy (171.626), Introduction to Plasma Physics and Atomic Processes in Hot Plasmas (171.672), Particle Physics and Cosmology (171.743–744), General Relativity (171.746), Cosmology (171.750), Active Galactic Nuclei (171.754), Fourier Optics and Interferometry (171.755), Astrophysics of Compact Objects (171.756), Advanced Laboratory (173.607–608), or Lab of Advanced Instrumentation (173.712), Planetary Atmospheres (270.623), Planetary Fluid Dynamics (270.661). Students planning on doing an observational thesis are strongly encouraged to take Observational Astronomy.

After successfully completing the exams covering upper-level undergraduate physics, students must pass a Preliminary Oral Exam designed to test for a broad knowledge of basic and modern physics, with an emphasis on order of magnitude estimates and physical reasoning. After passing the Preliminary Oral Examination the student takes the official University Graduate Board Oral Examination. The GBO Examination will focus on a thesis proposal presented by the student. The GBO Examination is normally taken by the end of a student's third year, but not until the student has chosen an advisor and an area of thesis research. After passing the GBO examination the student enters into several years of serious research culminating in the writing of a Ph.D. thesis.

At the conclusion of thesis research, the student defends the written dissertation before a faculty committee.

Thesis: Thesis may be written *in absentia*.

Special Equipment, Facilities, or Programs: The high-energy physics group has facilities for constructing the electronics and detectors needed in modern experiments, and also has independent computing capabilities that allow full analyses of massive amounts of data. Nuclear physics equipment includes facilities for relativistic heavy-ion collision studies and studies of nuclear interactions utilizing muons and pions with cryogenic targets. Among the diverse techniques used for studying condensed matter physics are SQUID (Superconducting Quantum Interference Device) magnetometry/susceptometry, vibrating sample magnetometry, atomic force and magnetic force microscopy, X-ray and electron diffraction, Auger spectroscopy, X-ray fluorescence spectroscopy, electron microscopy, and neutron scattering at the nearby NIST Center for Neutron Research and at other leading international facilities. A variety of cryostats, He^3 refrigerators, and He^3-He^4 dilution refrigerators together with high temperature ovens, electromagnets and superconducting magnets allow measurements to be made from 0.05 K to 1100 K and in magnetic fields up to 12 Tesla. Apparatus for the preparation of samples includes single-crystal growth vacuum furnaces, arc furnaces, several high vacuum and ultra-high vacuum chambers for thin film fabrication using evaporation, MBE, pulsed laser deposition, and sputtering. The department maintains a Class-1000 cleanroom for microfabrication and nanofabrication, and supports an instrument design group with five full-time engineers and a machine shop with three full-time machinists.

Work in the Center for Astrophysical Science (CAS) at JHU is now centered in three areas: developing instrumentation for astronomical observations, particularly from space; optical observation astronomy; and theory. Three large astronomy projects are currently active in CAS: the Far Ultraviolet Spectroscopic Explorer, a free-flying satellite launched May 1999 and the Advance Camera for Surveys (ACS) launched in 2002 and installed on the Hubble Space Telescope (HST). Early scientific data has proven to be promising. Hopkins is a member (with Princeton University, the University of Chicago, and Fermilab) in the Sloan Digital Sky Survey (SDSS), and is a member of the Astrophysical Research Consortium that operates a 3.5-meter telescope at the Apache Point Observatory in New Mexico. The SDSS is a collaborative effort with a major JHU role. It has now imaged about 20 million galaxies and stars and obtained spectra for nearly 200,000 galaxies and quasars. The SDSS will provide the "backbone" for the National Virtual Observatory (NVO), with a team at JHU leading the way. Computer facilities in the Department include a cluster of Sun workstations for incoming graduate students and many other workstations and clusters. These machines support a wide range of functions including data reduction, image processing, simulation of physical processes, and visualization. All are networked to universities, national laboratories, and supercomputer facilities throughout the world. The Johns Hopkins University is the home of the Space Telescope Science Institute. Facilities at the following laboratories and observatories are also frequently used: Brookhaven National Laboratory, Argonne National Laboratory, Stanford Linear Accelerator Center, Fermi National Laboratory, CERN, the University's own Applied Physics Laboratory, National Institute of Science and Technology, Lawrence Berkeley Laboratory, Lawrence Livermore National Laboratory, the White Sands Missile Range, Kitt Peak National Observatory, Cerro Tololo Interamerican Observatory, the Very Large Array of the National Radio Astronomy Observatory, the Las Campanas Observatory, and NASA's Goddard Space Flight Center.

Table B—Appointments to Graduate Students, 2009–10

Title of Appointee	Appointments Total	Appointments First year	Academic Load Allowed in Credit Hours	Hours of Service Per Week	Stipend for Academic Year ($)
			Semester		
Teaching Assistant	48	18	N/A	20[3]	18,500[1,2]
Research Assistant	60	0	N/A	20[3]	24,667[4]
Fellowship	2	1			tuition only
Total	110	19			

[1]Full tuition remission.
[2]Summer Research Assistantships available, approximately $6,000.
[3]Nominal figure.
[4]Maximum for 12 months is $24,667. In addition, full tuition remission.

5. Personnel Engaged in Separately Budgeted Research, 7/05–6/06

Professorial Faculty	30
Other Faculty	7
Postdoctoral appointments	30
Graduate students	104
Undergraduate students	24
Nonteaching research personnel	62
Total	257

6. Separately Budgeted Research Expenditures by Source of Support

	Departmental Research	Physics-related Research Outside Department
Federal government	$16,723,028	$
Private, non-profit organizations	2,075,484	
Other	2,097,007[1]	
Total	$20,895,519	$

[1]Includes institutional cost sharing expenditures.

Table C—Separately Budgeted Research Expenditures

Research Specialty	No. of Grants	Expenditures ($)
Astrophysics	193	11,246,818
Condensed Matter Physics	36	4,785,204
High Energy Experimental	16	951,549
Institutional Cost Sharing/ Dept. Res.	38	3,089,889
Particles & Fields	7	821,945
Plasma Physics & Fusion	2	114
Total	292	20,895,519

FACULTY

Professors

Bagger, Jonathan A., Ph.D., Princeton, 1983. Krieger Eisenhower Professor and Vice Provost for Graduate and Postdoctoral Programs and Special Projects. Theoretical high-energy physics.

Barnett, Bruce A., Ph.D., Maryland, 1970. Experimental Particle Physics.

Bennett, Charles L., Ph.D., MIT, 1984. Experimental cosmology.

Blumenfeld, Barry J., Ph.D., Columbia, 1974. Experimental particle physics.

Broholm, Collin, Ph.D., Copenhagen, 1988. Gerhard H. Dieke Professor. Experimental condensed matter physics.

Chen, Shiyi, Ph.D., Beijing University, 1987. Primary appointment in Mechanical Engineering. Statistical theory and computation of fluid turbulence.

Chien, Chia-Ling, Ph.D., Carnegie-Mellon, 1972. Jacob L. Hain Professor and Director, Materials Research Science and Engineering Center. Experimental condensed matter physics; artificially structured solids.

Chien, Chih-Yung, Ph.D., Yale, 1966. Experimental high-energy physics.

Domokos, Gabor, Ph.D., Dubna, 1963. Professor Emeritus. Algebraic approaches to elementary particle physics; symmetries with application to field theories of particles; strong interactions at high energies.

Eyink, Gregory, Ph.D., Ohio State, 1987. Primary appointment in Applied Mathematics and Statistics. Mathematical physics, fluid mechanics, turbulence.

Feldman, Gordon, Ph.D., Birmingham, England, 1953. Professor Emeritus. Quantum field theory; theory of elementary particles.

Feldman, Paul D., Ph.D., Columbia, 1964. Astrophysics; space physics; planetary and cometary atmospheres; spectroscopy.

Ford, Holland, Ph.D., Wisconsin, 1970. Stellar dynamics; evolution of galaxies; active galactic nuclei; astronomical instrumentation.

Fulton, Thomas, Ph.D., Harvard, 1954. Professor Emeritus. Quantum electrodynamics; atomic theory; high-energy particle physics.

Giacconi, Riccardo, Ph.D., Milan, 1954. University Professor. Astrophysics.

Heckman, Timothy, Ph.D., Professor & Director Center for Astrophysical Sciences. Washington (Seattle), 1978. Astrophysics; active galaxies and quasars.

Henry, Richard C., Ph.D., Princeton, 1967. Professor and Director, Maryland Space Grant Consortium. Astronomy; astrophysics.

Judd, Brian R., D.Phil., Oxford, 1955. Gerhard H. Dieke Professor Emeritus. Theoretical atomic and molecular physics; group theory; solid state theory.

Kim, Chung W., Ph.D., Indiana, 1963. Professor Emeritus. Nuclear theory; elementary particle theory; cosmology.

Kövesi-Domokos, Susan, Ph.D., Budapest, 1963. Professor Emeritus. Theoretical high-energy; physics; astroparticle physics.

Krolik, Julian H., Ph.D., California, Berkeley, 1977. Theoretical astrophysics.

Lee, Yung-Keun, Ph.D., Columbia, 1961. Professor Emeritus. Nuclear physics.

Moos, H. Warren, Ph.D., Michigan, 1961. Research Professor. Astrophysics, plasma physics.

Morava, Jack. Primary appointment in Dept. of Mathematics. Algebraic topology; mathematical physics.

Mountain, Charles Mattias, Ph.D., London University, 1983. Director Space Telescope Science Institute. Star formation in galaxies, capabilities of "second-generation telescope".

Neufeld, David A., Ph.D., Harvard, 1987. Molecular astrophysics; submillimeter and infrared astronomy; interstellar medium.

Norman, Colin A., D.Phil., Oxford, 1973. Theoretical astrophysics.

Pevsner, Aihud, Ph.D., Columbia, 1954. Jacob L. Hain Professor Emeritus. Experimental high-energy physics.

Reich, Daniel H., Ph.D., Chicago, 1988. Professor and Chair. Experimental condensed matter.

Riess, Adam, Ph.D., Harvard, 1996. Professor. Astrophysics.

Robbins, Mark O., Ph.D., California, Berkeley, 1983. Theoretical condensed matter physics.

Searson, Peter, Ph.D., University of Manchester, Institute of Sci. and Tech., 1982. Primary appointment in Materials Science and Engineering; Electronic, nanophase, semiconductor materials.

Strobel, Darrell F., Ph.D., Harvard, 1969. Primary appointment in Earth and Planetary Sciences. Planetary atmospheres and astrophysics.

Sundrum, Raman, Ph.D., Yale University, New Haven, Connecticut, 1990. Alumni Centennial Professor and Director. Theoretical Interdisciplinary Physics and Astrophysics Center. Theoretical high-energy physics.

Swartz, Morris L., Ph.D., University of Chicago, 1983. Experimental high energy physics.

Szalay, Alexander S., Ph.D., Budapest, 1975. Alumni Centennial Professor. Theoretical astrophysics; galaxy formation.

Tesanovic, Zlatko, Ph.D., Minnesota, 1985. Theoretical condensed matter physics.

Walker, J. Calvin, Ph.D., Princeton, 1961. Professor Emeritus. Experimental condensed matter physics; thin films and surfaces; nuclear physics.

Wyse, Rosemary, F. G., Ph.D., Cambridge, 1982. Theoretical astrophysics; galaxy formation and evolution.

Associate Professors

Falk, Michael, Ph.D., University of California, 1998. Primary appointment in Materials Science and Engineering. Theoretical and computational research.

Kaplan, David, Ph.D., Univ. of Washington, 1999. Theoretical particle physics.

Leheny, Robert L., Ph.D., University of Chicago, 1997. Experimental soft condensed matter physics.

Maksimovic, Petar, Ph.D., 1997. MIT, Assistant Professor. Experimental high-energy physics.

Markovic, Nina, Ph.D., Minnesota, 1998. Experimental condensed matter physics.

Melnikov, Kirill, Ph.D., Mainz University, Germany, 1996. Theoretical particle physics.

Tchernyshyov, Oleg, Ph.D., Columbia, 1998. Theoretical condensed matter; high-temperature superconductivity.

Assistant Professors

Armitage, N. Peter, Ph.D., Stanford, 2001. Experimental condensed matter physics.

Gritsan, Andrei, Ph.D., Univ. of Colorado, Bolder, 2000. Experimental high energy physics.

McQueen, Tyrel, Ph.D., Princeton, 2009. Primary appointment in Department of Chemistry. Solid State Chemistry/ Condensed Matter Physics.

Research Professors

Bianchi, Luciana, Ph.D., University of Padua, 1978. Experimental astrophysics.

Blair, William, Ph.D., Univ. of Michigan, 1981. Astrophysics; shockwaves; spectroscopy of plasmas.

Finkenthal, Michael, Ph.D., Hebrew University (Jerusalem). Plasma physics.

Adjunct Professors

Allen, Ronald J., Ph.D., M.I.T., 1967. Adjunct Professor. Space Telescope Science Institute. Spiral structure of galaxies; interstellar medium; radio; optical imaging.

Fall, Michael, Ph.D., Cambridge, 1976. Adjunct Professor (Space Telescope Science Institute): astrophysics.

Ferguson, Henry, Ph.D., JHU, 1990. Adjunct Professor, (Space Telescope Science Institute) observational cosmology, space astronomy.

Hauser, Michael, Ph.D., Caltech, 1967. Adjunct Professor, Space Telescope Science Institute. Cosmology, especially infrared background radiation.

Hornschemeier, Ann, Ph.D., 2002. Penn State Univ. Adjunct Asst. Prof. (NASA Goddard). Astronomy and astrophysics.

Kriss, Gerard, Ph.D., MIT, 1982. Adjunct Professor, Space Telescope Science Institute. Observational astrophysics.

Livio, Mario, Ph.D., Tel-Aviv (Israel), 1978. Space Telescope Science Institute. Adjunct Professor. Theoretical astrophysics; accretion onto white dwarfs; neutron stars and black holes; novae and supernovae.

Nota, Antonella, Ph.D., University of Padova, 1983. Adjunct Professor. Space Telescope Science Institute. Astronomy.

Petrovic, Cedomir, Ph.D., Florida State Univ., 2000. Adjunct Asst. Prof. (Brookhaven National Lab). Condensed matter physics.

Schreier, Ethan, Ph.D., MIT, 1970. Adjunct Professor and President, AUI. Astrophysics, active galaxies and jets.

Stiles, Mark, Ph.D., Cornell University, 1986. Adjunct Professor. NIST. Condensed matter theory.

van der Marel, Roeland, Ph.D., Leiden, 1994. Adjunct Associate Professor (Space Telescope Science Institute). Astronomy.

Weaver, Kimberly, Ph.D., University of Maryland, 1993. Adjunct Professor, NASA Goddard Space Flight Center. Observational astrophysics.

Williams, Robert, Ph.D., Harvard, 1986. Adjunct Professor. Space Telescope Science Institute. Observational astronomy.

RESEARCH SPECIALTIES AND STAFF

Theoretical

Astrophysics. Cosmology and large-scale structure; active galaxies; stellar populations; the interstellar medium; astrophysical magnetohydrodynamics; dark matter; mass-transfer binaries; physics of accretion disks. Krolik, Neufeld, Norman, Riess, Szalay, Wyse.

Atomic. Electronic transitions involving lanthanide ions in crystals and solutions; orthogonal operators for the energy levels of free atoms; group theory in atomic structure and icosahedral systems. Judd.

Condensed Matter. Study of high-Tc superconductivity; quantum critical phenomena; superfluidity; magnetotransport and electrons in high magnetic fields; disordered systems; nonequilibrium processes such as growth and friction. Robbins, Tchernyshyov, Tesanovic.

High Energy. Elementary fields and their interactions; gauge theories and superstring theory; experimental tests of the Standard Model and its extensions such as supersymmetry, technicolor, and theories with additional microscopic or mesoscopic spatial dimensions; dynamics of heavy quark systems; implications of particle physics for astrophysics and cosmology. Bagger, Domokos, Falk, G. Feldman, Fulton, Kaplan, Kim, Kovesi-Domokos, Melnikov, Sundrum.

Experimental

Astrophysics. Ultraviolet astronomy; interstellar medium; infrared and radio astronomy; active galactic nuclei; galactic structure and dynamics; cosmology; planetary atmospheres and magnetospheres; stellar atmospheres and chromospheres; physics of comets; laboratory astrophysics. Bennett, Blair, P. Feldman, Ford, Giacconi, Heckman, Henry, Moos, Strobel.

Condensed Matter. Magnetic thin films; nanostructured materials; superconductivity; strongly correlated electron systems; quantum magnetism; glass transitions; soft condensed matter; surface and interfaces. Armitage, Broholm, C.-L. Chien, Leheny, Markovic, Reich, Walker.

High Energy. Program involves studies of strong, electromagnetic, and weak interaction using counter techniques. Research in progress involves 1) the CDF experiment at the proton-antiproton Tevatron Collider at Fermilab, 2) a future experiment, CMS, at the Large Hadron Collider at CERN, and 3) a satellite experiment, AMS, looking for antimatter in cosmic rays. Barnett, Blumenfeld, C.-Y. Chien, Gritsan, Maksimovic, Pevsner, Swartz.

Nuclear. Properties of relativistic heavy ion collisions. Lee.

Plasma. Spectra of high-temperature plasmas used in fusion; highly ionized atoms. Finkenthal, Moos.

Space Optics. Performance of diffraction gratings (polarization, scattering, and efficiency); study of grating optical systems. Moos.

UNIVERSITY OF MARYLAND

DEPARTMENT OF PHYSICS

College Park, Maryland 20742

Students Accepted For Degree	FIELDS		
	Physics	Astronomy	Related Fields
Doctorate	X		X
Master's	X		X

1. General

President: C. D. Mote, Jr.
Dean of Graduate School: Charles Caramello
Department Chairman: Andrew Baden
Department Telephone Number: (301) 405-3401
(301) 405-5982
Web page: http://www.umdphysics.umd.edu
Type of Institution: University
Control: Public
Setting: Suburban
Total Faculty: 3,752
Total Graduate Faculty: Not separated
Total Students: 36,014
Total Graduate Students: 10,157
Annual Graduate Tuition:
In-state residents: Full-time—$427/credit
Out-of-state residents: Full-time—$921/credit
Tuition rates for: 2009–10
Deferred tuition plan: No
Term: Semester

2. Number of Faculty in Department

The combined total of full-time equivalent faculty in the three professorial ranks is 76. The combined total of full-time, part-time, and other faculty at all ranks is approximately 150*.

*Includes research associates

3. Admission, Financial Aid, and Housing

Address admission inquiries to: Ms. Linda O'Hara, Secretary, Graduate Entrance Committee, Department of Physics
Graduate application fee required: $60
Admission deadline (Fall admission): 1/1
Admission information: For fall admission, 2010–11, 140 students were accepted from approximately 660 applicants.
Admission requirements: For admission to the graduate programs, a Bachelor's degree in a relevant field is required with a minimum undergraduate GPA of 3.0 specified. An average grade of "B" or better in previous work in relevant fields or equivalent indication of aptitude for graduate work is required. Strength of letters of recommendation is considered. Except in most unusual circumstances, both the General and the Advanced Physics GRE's are required. The minimum acceptable score suggested for admission is 700. More weight is placed on the Advanced Physics GRE. The median GRE Advanced score for 2010–11 admissions was 840. Students from non-English speaking countries are required to demonstrate proficiency in English via the TOEFL exams. Minimum acceptable score for admission is 575 total on the paper-based TOEFL (with a writing score of 4.0 or above) or a score of 233 on the computer-based test (CBT) with a writing score of 4.0 or above, or a score of 100 on the internet-based test (IBT) with a writing score of 26 or above. All foreign students admitted with teaching responsibilities, or those who may assume teaching responsibilities at a later date, must participate in the Maryland English Institute (MEI) evaluation before beginning their first semester.
Address financial aid inquiries to: Graduate Entrance Committee, Department of Physics
GAPSFAS application required: No
Financial aid deadline: 1/1
Loans available: Yes
Address housing inquiries to: Office of off-campus housing, 1110 Stamp Student Union, University of Maryland, College Park, MD 20742, (301) 314-3645, http://www.umd.och101.com. A typical apartment complex is Graduate Hills and Graduate Gardens, 7704 Adelphi Road, Hyattsville, MD 20783, (888) 230-7368, www.smc-grad-housing.com.
Cost/month: Efficiency—$883; 1 bdrm.—$925;
2 bdrm.—$1,136

Table A—Faculty, Enrollments, and Degrees Granted

Research Specialty	2009–10 Faculty	Enrollment[1] Fall 2008		No. of Degrees Granted[2] 2008–09 (2004–09)			Median No. of Years for 2004–09 Ph.D.'s
		Master's	Doctorate	Master's	Terminal Master's	Doctorate	
Astrophysics	2	–	–	–	–	0(2)	10.5
Atomic, Molecular, & Optical Physics	13	–	–	–	–	3(6)	6
Chemistry & Biophysics	3	–	–	–	–	0(1)	7
Condensed Matter	22	–	–	–	–	7(31)	6
Dynamical Systems	3	–	–	–	–	1(4)	6
Elementary Particles	10	–	–	–	–	0(10)	6
General Relativity Expt	2	–	–	–	–	0(2)	6.5
Gravitation Theory	4	–	–	–	–	0(8)	6
High Energy	4	–	–	–	–	3(9)	6
Laser Fields	2	–	–	–	–	1(1)	6
Nanophysics	5	–	–	–	–	0(7)	5
Nonlinear Dynamics	4	–	–	–	–	–	–
Nuclear Physics	2	–	–	–	–	(9)	5
Particle Astrophysics	4	–	–	–	–	2(9)	6
Physics Ed Research (PERG)	2	–	–	–	–	0(4)	4.5
Plasma	6	–	–	–	–	4(9)	6
Statistical Mechanics	4	–	–	–	–	1(6)	6
Superconducting Quantum Computing	2	–	–	–	–	0(10)	6
Total Ph.Ds						28(146)	
Total		–	–				
Full-time Ph.D. Students		–	236				
Part-time Students		–	1				
First-year Ph.D. Students		–	48				
Median Years in Grad. Study (2007–08 Degrees)						5.8	–
Undergraduate Degrees, 2008–09 (2005–09): 56(158)							

[1]Students not yet committed to a research specialty are entered under non-specialized.
[2]Five-year totals in parentheses.

4. Graduate Degree Requirements

Master's: Master's Degree non-thesis: Minimum of 30 credits with 18 at graduate level. Must include at least four courses in general physics, also graduate lab. Student must pass at Master's level one part of the Ph.D. Qualifying Exam and prepare a scholarly paper. Student must maintain an overall "B" average in all courses taken. Residence for at least two semester is required. A five-year calendar limit is imposed for both full-time and part-time students.

Master's Degree with thesis: Same minimum credits as above, which will include six credits of thesis and research. Passing one part of Ph.D. Qualifying Exam at Master's level and completing scholarly paper are not required. An oral examination in defense of thesis and covering all course material must be passed.

Doctorate: No minimum number of credits is required. However, a student must have at least 12 credits of thesis research, at least two credits of seminar, and six credits outside specialty. An overall B average must be maintained. At least two semesters of full-time residence are required. Ph.D. Qualifying Exam must be passed by end of second year. No foreign language is required. Each student must present a paper as evidence of scholarly writing ability as a condition for admission for candidacy. This paper must be written independently of and in excess of course requirements.

Thesis: Thesis may be written *in absentia*.

Table B—Appointments to Graduate Students, 2008–09

Title of Appointee	Appointments Total	Appointments First year	Academic Load Allowed in Credit Hours	Hours of Service Per Week	Stipend for Academic Year ($)
			Semester		
Teaching Assistant	48	32	10	20	21,000–24,000[1]
Research Assistant	137	6	10	20	25,000–30,000
Fellow	14	7	Varies	0	Varies
Total	221	45			

[1]12 months

5. Personnel Engaged in Separately Budgeted Research

Professorial Faculty	76
Other Faculty	50
Graduate students	171
Undergraduate students	54
Nonteaching research personnel	34
Postdoctoral appointments	40
Total	425

6. Separately Budgeted Research Expenditures by Source of Support

	Physics	Outside
Federal government	$14,385,863.65	$6,344,769.33
Private, nonprofit	2,880,014.61	975,522.18
Total	$17,265,878.26	$7,320,291.51

Table C—Separately Budgeted Research Expenditures

Research Specialty	No. of Grants	Expenditures ($)
Atomic, Molecular & Optical	15	1,665,945.11
Condensed Matter	31	3,057,983.29
Dynamical Systems & Accelerator Theory	1	115,810.17
Elementary Particles	3	261,811.25
High-Energy Physics	7	1,745,581.68
Materials Research Science Eng. Center	11	1,856,961.26
Nuclear Physics	7	525,232.08
Cosmic Ray Physics	11	4,213,018.97
Plasma Physics	46	4,817,093.29
Relativity & Gravitation	17	805,597.02
Research/Innovation Education	5	484,580.48
Space Physics	14	828,218.70
Superconductivity	12	261,908.04
Biophysics	9	271,582.43
Chemical Physics	8	297,618.56
Nonlinear Dynamics	6	363,322.23
Quantum Physics	3	1,059,072.03
Other Research	25	1,955,054.04
Total	231	24,586,400.63

FACULTY

Professors

Anderson, J. Robert, Ph.D., Iowa State, 1963. Experimental condensed matter physics; diluted magnetic semiconductors; electronic structures and Fermi surfaces of metals and semimetals.

Anlage, Steven, Ph.D., Caltech, 1988. Superconductivity-electromagnetic properties, proximity effect; near-field microwave microscopy; experimental chaos.

Antonsen, Thomas, Ph.D., Cornell, 1977. Plasma physics; coherent sources of radiation.

Baden, Andrew R., Ph.D., California, Berkeley, 1986. Experimental high energy physics with accelerators; data acquisition; high performance computing; data analysis.

Beise, Elizabeth J., Ph.D., MIT, 1988. Experimental nulcear physics-intermediate energy, electron scattering, polarization, few-nucleon and subnucleon systems.

Bhagat, Satindar M., Ph.D., Delhi, 1956. Experimental condensed matter physics; magnetic resonance; exotic magnetic phases; high temperature superconductors.

Brill, Dieter R., Ph.D., Princeton, 1959. General relativity and gravitation; Black Holes; cosmology.

Buonanno, Alessandra, Ph.D., University of Pisa, 1996. Gravity theory, gravitational waves.

Chen, Hsing-Hen, Ph.D., Columbia, 1973. Astrophysics; plasma physics; nonlinear dynamical systems.

Cohen, Thomas D., Ph.D., Pennsylvania, 1984. Nuclear theoretical physics; soliton models of baryons; chiral symmetry; effective low energy models for QCD.

Das Sarma, Sankar, Ph.D., Brown, 1979. Distinguished University Professor. Theoretical condensed matter; many body theory; semiconductor nanostructures; nonequilibrium statistical mechanics.

Drake, James F., Ph.D., UCLA, 1975. Plasma physics; magnetic reconnection; Tokamak transport.

Dorland, William, Ph.D., Princeton, 1993. Turbulence in magnetized plasma; computational physics.

Einstein, Theodore L., Ph.D., Univ. of Pennsylvania, 1973. Theoretical condensed matter physics; surface physics; statistical and thermal physics.

Ellis, Richard F., Ph.D., Princeton, 1970. Experimental plasma physics; plasma waves and instabilities; microwave and far infrared diagnostics for fusion plasmas; plasma probes and analyzers.

Eno, Sarah C., Ph.D., University of Rochester, 1990. Experimental high energy physics with accelerators.

Fisher, Michael E., Ph.D., King's College, 1957. Distinguished University Professor; University System of Maryland Regents Professor. Statistical physics; condensed matter theory; theoretical chemistry; phase transitions and critical phenomena; associated mathematics.

Fuhrer, Michael S., Ph.D., University of California at Berkeley, 1998. Experimental condensed matter.

Gates, S. James, Ph.D., MIT, 1977. The John S. Toll Professor of Physics. Elementary particles-supersymmetry, supergravity, and superstrings.

Goodman, Jordan A., Ph.D., Maryland, 1978. Particle astrophysics.

Greenberg, Oscar W., Ph.D., Princeton, 1956. Elementary particles and field theory.

Greene, Richard L., Ph.D., Stanford, 1967. Experimental condensed matter physics.

Hadley, Nicholas J., Ph.D., California, Berkeley, 1983. High-energy physics.

Hamilton, Douglas C., Ph.D., Chicago, 1977. Experimental space physics; magnetospheric physics; solar wind, solar energetic particles; particle acceleration and transport.

Hammer, David, Ph.D., Univeristy of California at Berkeley, 1991. Physics education—learning and teaching at high school and college levels.

Hassam, Adil B., Ph.D., Princeton, 1978. Plasma physics of the sun; thermonuclear fusion.

Hu, Bei-Lok, Ph.D., Princeton, 1972. General relativity; gravitation and cosmology; quantum field theory; statistical field theory.

Jacobson, Theodore A., Ph.D., Texas, Austin, 1983. Gravitation theory; quantum gravity; black hole thermodynamics.

Jawahery, Abolhassan, Ph.D., Tufts, 1981. High-energy physics with accelerators.

Ji, Xiangdong, Ph.D., Drexel, 1987. Theoretical nuclear physics; quantum chromodynamics; quark and gluon structure of hadrons.

Kirkpatrick, Theodore, Ph.D., Rockefeller, 1981. Theoretical statistical mechanics; condensed matter theory.

Lathrop, Daniel P., Ph.D., Univ. of Texas, 1991. Nonlinear dynamics and chaos; turbulence; fluid dynamics.

Liu, Chuan Sheng, Ph.D., California, Berkeley, 1968. Plasma physics; fusion and space science.

Lobb, Christopher J., Ph.D., Harvard, 1980. Experimental superconductivity; superconducting devices; physics and applications of mesoscopic systems; condensed matter physics.

Milchberg, Howard M., Ph.D., Princeton, 1985. Atomic, Molecular and Optical Physics; Nonliner Optics, Laser and Optical Physics.

Mohapatra, Rabindra N., Ph.D., Rochester, 1969. Elementary particles; quantum field theory; cosmology.

Monroe, Christopher, Ph.D., University of Colorado, Boulder, 1992. Cold atomic physics; quantum information science; ultrafast control of cold atoms; interface between atomic and condensed matter physics; foundations of quantum mechanics.

Orozco, Luis, Ph.D., Univ. of Texas at Austin, 1987. Quantum optics; precision measurement; fundamental interactions.

Ott, Edward, Ph.D., Polytech. Univ., Brooklyn, 1967. Distinguished University Professor. Chaotic dynamics; plasmas.

Paik, Ho Jung, Ph.D., Stanford, 1974. Experimental general relativity; gravitational waves; precision tests of laws of gravity.

Papadopoulos, Dennis, Ph.D., Maryland, 1968. Space plasma physics; lightning; photoconducting plasmas.

Phillips, William D., Ph.D., MIT, 1976. Nobel Laureate in Physics, 1997. Laser cooling; atom trapping; atomic clocks; atomic and optical physics; cold collisions, photoassociative spectroscopy.

Redish, Edward F., Ph.D., MIT, 1968. Physics education research and development.

Rolston, Steven L., Ph.D., SUNY Stony Brook, 1986. Laser cooling of neutral atoms; ultra cold plasmas; Bole-Einstein condensations quantum info.

Roy, Rajarshi, Ph.D., Univ. of Rochester, 1981. Nonlinear dynamics in optical systems; laser physics; wave propagation in optical fibers; coherence and stochastic processes.

Sagdeev, Roald Z., DS, Siberian Branch, USSR Acad. of Sciences, 1962; Ph.D., Institute of Physical Problems, Moscow, 1960. Plasma physics; controlled fusion; space physics; planetary research and astrophysics; arms control; science policy; global security and environment.

Seo, Eun-Suk, Ph.D., Louisiana State University, 1991. Cosmic Ray physics.

Skuja, Andris, Ph.D., California, Berkeley, 1972. Experimental high-energy physics with accelerators; experimental particle physics.

Sreenivasen, Katapalli, Ph.D., Indian Institute of Science, 1970. Fluid mechanics and turbulence; nonlinear dynamics.

Sullivan, Gregory W., Ph.D., Illinois, 1990. Electroweak physics; Standard Model; top quark search.

Wellstood, Frederick, Ph.D., Univ. of California Berkeley, 1988. Superconductivity-High Tc (YBCO) superconducting quantum interference devices; magnetic microscopy; Coulomb blockade electrometers.

Williams, Ellen D., Ph.D., Caltech., 1982. Distinguished University Professor. Condensed Matter Physics; surface science; scanning tunneling microscopy; statistical mechanics of surfaces.

Yakovenko, Victor M., Ph.D., Landau Institute for Theoretical Physics, Moscow, 1987. Condensed matter theory; organic and high-T_c superconductors; the quantum Hall effect; effects of high magnetic fields.

Yorke, James A., Ph.D., University of Maryland, 1966. Distinguished University Professor. Chaos & Non-linear dynamics, statistical physics.

Associate Professors

Appelbaum, Ian, Ph.D., MIT, 2003. Condensed matter.

Bedaque, Paulo, Ph.D., Univ. of Rochester, 1994. Nuclear theory.

Chacko, Zacharia, Ph.D., University of Maryland, 1999. Elementary Particles.

Hoffman, Kara, Ph.D., Purdue, 1998. Particle Astrophysics.

Losert, Wolfgang, Ph.D., City College of the City University of New York, 1998. Nonlinear dynamics and chaos.

Roberts, Douglas A., Ph.D., Univ. of CA, 1994. High energy physics with accelerators.

Assistant Professors

Abazajian, Kevork, Ph.D., UC San Diego, 2001. Elementary particles.

Agashe, Kaustubh, Ph.D., UC Berkeley, 1998. Particle phenomenology; collider signals.

Galitski, Victor M., Ph.D., Univ. of Minnesota, 2002. Condensed matter theory.

Girvan, Michelle, Ph.D., Cornell University, 2003. Nonlinear dynamics.

Hall, Carter, Ph.D., Harvard University, 2002. Experimental nuclear.

Kim, Kiyong, Ph.D., Maryland, 2003. Laser and optical physics.

LaPorta, Arthur, Ph.D., University of California-San Diego, 1996. Biophysics, single molecule biophysics.

Levin, Michael, Ph.D., Massachusetts Institute of Technology, 2006. Condensed Matter Theory.

Ouyang, Min, Ph.D., Harvard, 2001. Condensed Matter Experiment, Nanoelectronics.

Paglione, Johnpierre, Ph.D., Univ. of Toronto, 2004. Condensed Matter Experiment.

Shawhan, Peter, Ph.D., Univ. of Chicago, 1999. Gravitation experiment.

Tiglio, Manuel, Ph.D., Univ. of Cordoba (Argentina), 2000. Gravitation Theory.

Upadhyaya, Arpita, Ph.D., University of Notre Dame, 2000. Biophysics, force generation by actin polymerization, cell polarization, surface and elastic forces in developing tissues, mechanical properties of membrane nanotubes, biological springs.

Research Faculty

Ipavich, Fred M., Sr. Res. Sci., Ph.D., Maryland, 1972. Space physics; interplanetary physics; astrophysics; solar physics; magnetospheric physics.

Kane, Bruce, Sr. Res. Sci., Ph.D., Princeton, 1988. Experimental quantum devices; silicon.

Kellogg, Richard G., Sr. Res. Sci., Ph.D., Yale, 1975. Experimental high energy and particle physics.

Moody, Martin Vol, Sr. Res. Sci., Ph.D., Virginia, 1980. Experimental general relativity; gravitational physics.

Venkatesan, T. Venky, Ph.D., CUNY and Bell Laboratories, 1977. Superconductivity; physics and applications of thin films; surface modification.

Professors Emeriti

Alley, Carroll O., Jr., Ph.D., Princeton, 1962. Atomic physics; quantum electronics—precision time keeping, laser range measurement; quantum mechanics; relativistic gravity.

Boyd, Derek A., Ph.D., Stevens Inst. of Tech., 1973. Plasma physics; plasma diagnostics; far infrared spectroscopy; microwave optics.

Chang, Chia-Cheh (George), Ph.D., Southern California, 1968. Experimental nuclear physics-intermediate energy.

Chang, Chung-Yun, Ph.D., Columbia, 1965. Experimental high-energy physics with accelerators.

Chant, Nicholas S., D. Phil., Lincoln College, Oxford, 1966. Experimental nuclear physics—pion reactions with polarized beams; electron beam experiment at Thomas Jefferson National Accelerator Facility.

Currie, Douglas G., Ph.D., Rochester, 1962. Astrophysics; astrophysical instrumentation; dynamical systems.

DeSilva, Alan W., Ph.D., California, Berkeley, 1961. Plasma physics—plasma diagnostics; light scattering; strongly coupled plasmas.

Drew, H. Dennis, Ph.D., Cornell, 1967. Experimental condensed matter physics; statistical and thermal physics; semiconductor heterostructures; infrared properties of superconductors; near-field optical scanning microscopy.

Dorfman, J. Robert, Ph.D., The Johns Hopkins University, 1961. Statistical and thermal physics; dynamical systems theory.

Dragt, Alex J., Ph.D., California, Berkeley, 1963. Elementary particles and field theory; mechanics; dynamical systems and accelerator theory; charged particle and light optics.

Falk, David S., Ph.D., Harvard, 1959. Theoretical condensed matter physics; statistical and thermal physics; vision.

Glick, Arnold, Ph.D., University of Maryland, 1961. Condensed matter theory; statistical and thermal physics.

Gloeckler, George, Ph.D., Chicago, 1965. Distinguished University Professor. Space physics; heliospheric physics.

Goldenbaum, George C., Ph.D., Maryland, 1966. Plasma physics; fluid dynamics; physics of lightning; environmental science.

Griem, Hans R., Sr. Res. Sci., Ph.D., Kiel, 1954. Plasma physics.

Griffin, James, J., Ph.D., Princeton, 1956. Theoretical nuclear physics; nuclear heavy ion physics; quantum electrodynamics.

Holmgren, Harry D., Sr. Res. Sci., Ph.D., Minnesota, 1954. Experimental nuclear physics—intermediate energy.

Kim, Young Suh, Ph.D., Princeton, 1961. Elementary particles and field theory; group theory.

Korenman, Victor, Ph.D., Harvard, 1966. Assistant Vice President for Academic Affairs. Theoretical condensed matter physics; statistical and thermal physics.

Langenberg, Donald N., Ph.D., California, Berkeley, 1959. Chancellor, Emeritus, University of Maryland System. Condensed matter physics; superconductivity; electronic structure of metals and semiconductors.

Layman, John W., Ed.D., Oklahoma State, 1970. Science Teaching Center. Physics education-use of computers in the laboratory for student conceptual development; constructivist mode of teaching/learning.

Mason, Glenn M., Ph.D., Chicago, 1971. Space plasma physics; cosmic rays; heliospheric physics.

Misner, Charles W., Ph.D., Princeton Univ., 1957. General relativity; physics education.

Park, Robert L., Ph.D., Brown, 1964. Experimental condensed matter physics; surface physics; science policy.

Pati, Jogesh C., Ph.D., Maryland, 1960. Theoretical Particle Physic-Grand unification; supersymmetry; superstrings; particle cosmology.

Richard, Jean-Paul, Ph.D., Univ. of Paris, 1963, Doctorat d'Etat, 1965. Experimental general relativity-gravitational waves, quantum optics.

Roos, Philip G., Ph.D., MIT, 1964. Experimental nuclear physics—lectro-weak interactions; Hadron-induced reactions.

Sucher, Joseph, Ph.D., Columbia, 1957. Elementary particle theory; quantum electrodynamics; composite systems in quantum field theory; relativistic atomic physics.

Toll, John S. Ph.D., Princeton, 1952. Elementary particle theory; dispersion relations; quantum field theory; science education.

Wallace, Stephen J., Ph.D., Washington, 1971. Theoretical physics—scattering theory; nucleon-nucleon interactions, relativistic bound states; electron scattering.

Woo, Ching-Hung, Ph.D., California, Berkeley, 1962. Complexity theory; quantum measurement theory; quantum field theory; history and philosophy of physics.

Associate Professor Emeritus

Kacser, Claude, Ph.D., Oxford, 1959. General physics; teaching of physics; special relativity.

Affiliate Professors

Cumings, John, Ph.D., UC Berkeley, 2002. Condensed matter experiment; materials science.
Hill III, Wendell T., Ph.D., Stanford, 1980. Atomic, Molecular and Optical Physics.
Oehrlein, Gottleib, Ph.D., SUNY-Albany, 1981. Condensed Matter Experiment; Nanostructure fabrication, materials science, thin films.
O'Shea, Patrick, Ph.D., Maryland, 1986. Particles beams.
Phaneuf, Raymond, Ph.D., Wisconsin-Madison, 1985. Condensed Matter Experiment.
Takeuchi, Ichiro, Ph.D., University of Maryland, 1996. Center for Nanophysics and Advanced Materials.
Weeks, John D., Ph.D., Chicago, 1969. Condensed Matter Experiment; Nanoelectronics, thin films.

Adjunct Professors

Bryant, Garnett, Ph.D., Indiana University, 1978.
Campbell, Gretchen, Ph.D., Massachusetts Institute of Technology, 2006.
Clark, Charles W., Ph.D., University of Chicago, 1979.
Julienne, Paul S., Ph.D., University of North Carolina, 1969.
Lett, Paul D., Ph.D., University of Rochester, 1986.
Lynn, Jeffrey W., Ph.D., Georgia Inst. of Tech., 1974. Condensed matter physics; neutron scattering; superconductivity; phase transitions and critical phenomena; magnetic materials.
Mather, John C., Ph.D., UC Berkeley, 1974. Cosmology; far IR astronomy and instrumentation; Fourier transform spectroscopy.
McEnery, Julie, Ph.D., University College Dublin, 1997. Particle astrophysics.
Migdall, Alan, Ph.D., Massachusetts Institute of Technology, 1984.
Porto, James (Trey), Ph.D., Cornell University, 1996.
Solomon, Glen, Ph.D., Stanford University, 1997.
Spielman, Ian, Ph.D., California Institute of Technology, 2004.
Sundrum, Raman, Ph.D., Yale University, 1990.
Taylor, Jacob, Ph.D., Harvard University, 2006.
Tiesinga, Eite, Ph.D., Eindhoven University of Technology, 1993.
Tycko, Robert, Ph.D., UC Berkeley, 1984. Solid state nuclear magnetic resonance (NMR) spectroscopy.
Williams, Carl J., Ph.D., University of Chicago, 1987.

RESEARCH SPECIALTIES AND STAFF
http://www.umdphysics.umd.edu/

Theoretical

Astrophysics (see Relativity and Space Physics and Cosmic Rays).
Atomic, Molecular and Optical Physics (see Experimental).
Condensed Matter. Many-body theory relating to superconductivity, ferromagnetism, effects of critical fluctuations near phase transitions; properties of interfaces, surfaces and surface adsorbates and steps on surfaces; solitons; the quantum Hall effect; two-dimensional systems; theory of superlattices; transport and photoconductivity in microstructures; theory of crystal growth; quantum chaos; first principles band structure calculations; properties of clusters strongly correlated systems—quasi-one-dimensional and organic conductors, high-T_c superconductors, physical effects of high magnetic fields, quantum computation, econophysics. Das Sarma, Einstein, Galitski, Levin, Yakovenko.
Dynamical Systems and Accelerator Theory. Research on the use of Lie algebraic methods for the computation of charged particle beam transport including high-order nonlinear effects. Applications are made to many diverse topics (including beam transport systems, recirculating circular and linear accelerators, damping rings, electron microscopes, light optics, and the long-term behavior of Hamiltonian systems) and many diverse machines (including the Fermilab and SLAC storage rings, the Large Hadron Collider, the Next Linear Collider, and synchrotron light sources). Ott, Girvan.
Elementary Particles and Fields. Unified gauge theories of elementary particle interactions; composite models of quarks and leptons; cosmology; supersymmetry; supergravity; superstrings; spontaneous symmetry breaking and mass generation; neutrino masses and mixings; bound-state models of hadrons; confinement of quarks and color in quantum chromodynamics; problems in quantum electrodynamics; quantum groups; small violations of particle statistics; hidden variables and EPR problems; field theory with finite boundary conditions; coherent and squeezed states; and representations of the Lorentz groups. Abazajian, Agashe, Chacko, Gates, Greenberg, Mohapatra.
Gravitation. Gravitation and cosmology; quantum field theory in curved spacetime; black hole thermodynamics; the early universe and quantum cosmology; nonequilibrium quantum fields; global and causal structures of spacetimes; topology change in quantum gravity; discrete models for spacetime. Brill, Buonanno, Hu, Jacobson, Tiglio.
Low-Temperature Physics (see Condensed Matter).
Physics Education. Graduate assistant training program; seminar in college physics teaching; M.S. and Ph.D. in education in cooperation with Science Teaching Center; special interest in computers and their role in teaching physics; collaboration with AAPT. Hammer, Redish.
Plasmas. Research on a wide variety of topics in laboratory, magnetospheric, and solar plasma physics, including anomalous transport, plasma flows, the generation and dissipation of magnetic fields, nonlinear waves, nonlinear dynamics and chaos, and plasma interaction with intense radiation and high-energy beams. Antonsen, Dorland, Drake, Hassam, Liu, Ott, Papadopoulos, Sagdeev.
Statistical and Thermal Physics. Statistical physics. Classical and quantum phase transitions, phase transitions in ionic systems, folding and other properties of proteins and related bio-molecules, molecular motors, theories and experiments on fluctuations in non-equilibrium steady states, theories of surface structures, theories of the liquid state, quantum mechanics of disordered systems, dynamical systems theory and non-equilibrium statistical mechanics. Fisher, Kirkpatrick (see also Theoretical Condensed Matter).
Theoretical Quarks, Hadrons and Nuclei. QCD and its connection to hadrons and nuclei, e.g., perturbative QCD, lattice QCD, large N QCD, effective field theory, structure of nucleon and hadrons and nucleon-nucleon interactions. Bedaque, Cohen, Ji.

Experimental

Atomic, Molecular and Optical Physics. Laser cooling and trapping of neutral atoms; quantum optics; cavity quantum electrodynamics; correlated photons; quantum feedback and control; spectroscopy and tests of discrete symmetries in atoms; ultracold atoms in optical lattices; linear and non-linear atom optics (deBroglie-wave optics); atom interferometry; ultracold plasmas; collisions of atoms at ultra-low-energy; photoassociation of ultracold atoms; quantum-degenerate gases and

Bose-Einstein condensation; quantum gases in low-dimensions; quantum transport of atoms in periodic potentials; quantum chaos; quantum information and quantum computing. Hill, Milchberg, Monroe, Orozco, Phillips, Rolston.

Biophysics. With outstanding scientists from the fields of biology, physics, chemistry and engineering, Maryland's burgeoning biophysics initiative is addressing complex questions in biology, biomedicine and bioengineering. Using novel theoretical and computational methods and next-generation experimental equipment (such as a holographic laser tweezer array and a two-photon confocal microscope), this interdisciplinary team is exploring biophysics at both the cellular and molecular levels. LaPorta, Losert, Upadhyaya.

Condensed Matter. Electrons in metals and semiconductors; surface properties of solids; spin dynamics of randomized systems; magnetization at low temperatures; ferromagnetism; spin glasses; re-entrant magnetism; diluted magnetic semiconductors; deHass-Van Alphen effect; properties of amorphous systems; electronic band structure of metals and alloys; microwave, far infrared and visible optical studies of superconductors and semiconductors in high magnetic fields and low temperatures; electron diffraction studies of 2D phase transitions; surface geometry and electronic structure; properties of epitaxial films; modulated optical studies of materials prepared by molecular beam epitaxy; studies of equilibrium and driven surface structures; scanning tunneling microscopy; low-energy electron microscopy; surface modification by laser annealing; preparation and study of artificially structured semiconductor and metal systems; low temperature physics of mesoscopic structures, universal conductance fluctuations, persistent currents in normal metal rings, single electron devices. Low temperature physics of mesoscopic structures, universal conductance fluctuations, persistent currents in normal metal rings, single electron devices. Anlage, Bhagat, Fuhrer, Greene, Lobb, Ouyang, Phaneuf, Williams, Wellstood.

Cosmic Ray Astrophysics. Experimental investigations of high energy particles from outer space. Search for exotic matter such as antimatter and dark matter; precision measurements of galactic cosmic rays to understand their origin, acceleration, propagation and to explore the limit of supernova shock wave acceleration. Particle detector development. Balloon-borne and space-based experiments. Monte Carlo simulations. Seo.

High Energy Physics with Accelerators. Search for the fundamental constituents of matter and the properties of the interactions between these constituents using the techniques of high-energy physics; studies of electron-positron collisions with the OPAL detector at LEP in CERN, and $\bar{p}p$ interactions with the DO Detector at the Tevatron in Fermilab dominate the program. The Maryland HEP group is also a member of the Solenoidal Detector Collaboration at the SSCL. Baden, Eno, Hadley, Jawahery, Roberts, Skuja.

Low-Temperature Physics (see Experimental Condensed Matter and Superconductivity Center).

Modern Optics and Astrophysical Investigations. Observations and analysis of Hubble Space Telescope data (in particular, eta Carinae, and the Rings and Satellites of Uranus); observation for the satellite dynamics of LAGEOS III for Lense-Thirring measurement; equipment development, observations and data analysis for the impact of SL9 onto Jupiter; amplitude interferometry; application of modern electro-optical techniques to develop instruments for astrophysical observations, especially for array detectors. Currie.

Nuclear Physics. Experiments in medium energy nuclear reactions and nucleon structure. Studies of the quark-gluon properties of nucleons and mesons using electroweak probes, mostly at the 6 GeV CEBAF electron accelerator at Jefferson Laboratory in Newport News, VA. Experiments use polarization techniques, parity violation, strange particle production in various reactions in the energy regime where the strong force, described by Quantum Chromodynamics, is not perturbative. Beise, Hall.

Particle Astrophysics. Experimental study of ultrahigh energy cosmic rays and nonaccelerator high energy physics. Search for point sources of ultrahigh and very high energy gamma rays using the existing MILAGRO detector. Using the cubic kilometer IceCube detector located beneath the South Pole we will open the field of neutrino astronomy. We explore an energy region where the Universe is opaque to high energy gamma rays originating from beyond the edge of our own galaxy, and where cosmic rays do not carry directional information because of their deflection by magnetic fields. Goodman, Hoffman, Sullivan.

Plasmas. Research on high-temperature ionized gases, including problems of their creation, diagnostics, and confinement, transport and wave properties, scattering and radiation. Ellis, Milchberg, Kim.

Relativity. Antennas sensitive to gravitational radiation are being developed. These include multimode wide band antennas using quantum optics instrumentation. Superconducting transducers for millekelvin temperature antennas are also being developed. A sensitive superconducting gravity gradiometer is being developed for precision test of the inverse square law of gravitation; a highly stable laser clock is being developed. Precision tests of general relativity in space are being studied. Paik, Shawhan.

Space Physics. Experiments on satellites such as SOHO and ACE and deep space probes such as Voyager, Ulysses, and Cassini to study the compositions and distribution functions of space plasmas including the solar wind and solar, heliospheric, and magnetospheric energetic ions. Hamilton, Hill.

Others

Center for Nanophysics and Advanced Materials. Theory and Experiment of Quantum computation using Josephson junction qubits, scanning squid microscopy, superconducting devices, quantum gates, macroscopic quantum effects, decoherence. Experimental and theoretical research on all aspects of superconductivity; new materials, thin films, tunnel and Josephson devices, microwave and infrared properties and mechanisms of superconductivity. Both low-T_c and high-T_c systems. Interdisciplinary projects with Chemistry, Materials and EE departments. Anlage, Bhagat, Das Sarma, Fuhrer, Greene, Lobb, Paglione, Venkatesan, Wellstood, Yakovenko.

Charged Particle Beam Research. Experimental and theoretical research of the physics of space-charge dominated high-brightness beams (instabilities, emittance growth, intensity limits, etc.) for advanced accelerator applications. Transport and longitudinal compression of high-perveance electron beams in a solenoidal focusing channel. Experiments with intense relativistic electron beams: generation of high-power microwaves, and free-electron lasers (FELs). Development of a 10 GHz high-power gyroklystron source for e^+e^- linear colliders. Antonsen.

Dynamical Systems and Accelerator Theory. Research on the use of Lie algebraic methods for the computation of charged particle beam transport and on high current instabilities in linear and circular accelerators. Applications are made to many diverse topics (including beam breakup, recirculating linear accelerators, coupling impedances, electron microscopes, and the long-term behavior of Hamiltonian systems) and many diverse machines (including the large hadron collider linear colliders, the continuous electron beam accelerator and synchrotron light sources).

Institute for Physical Science and Technology (IPST). Chemical Physics, Biophysics, Optical Physics. http://ipst.umd.edu/ fosters excellence in interdisciplinary research and education at the University of Maryland. The Institute focuses on areas which, by virtue of their interdisciplinary nature, fall outside the mainstream of departmental interest and strives to provide an environment in which both theoretical and experimental research can flourish. Currently experimental and theoretical research programs are active in the areas of Applied Mathematics, Chemical Physics, Optical Physics, Space and Upper Atmospheric Physics and Statistical Physics. Roy.

Institute for Research in Electronics & Applied Physics (IREAP) http://www.ireap.umd.edu conducts advanced interdisciplinary research into the physics and application of plasmas and charged particle beams. IPR conducts experimental and theoretical research on high-temperature plasma physics, plasma spectroscopy, relativistic microwave electronics, high-brightness charged particle beams, laser-plasma interactions, nonlinear dynamics (chaos), ion beam microfabrication techniques, and microwave sintering of advanced materials. IPR is recognized internationally as a leading university research center in these areas of research. Lathrop.

Institute for Systems Research (ISR) http://www.isr.umd.edu/ ISR was formed in 1985 as a joint venture betwen the University of Maryland and Harvard University. It is one of 21 interdisciplineary National Science Foundation (NSF) Engineering Research Centers that are specifically chartered to enhance the global competitiveness of U.S. industry. It is the only ERC with the specific mission of conducting systems research.

Joint Quantum Institute (JQI) http://www.jqi.umd.edu/ The JQI was formed in 2006, bringing together researchers from the University of Maryland, the National Institute of Standards and Technology and the Laboratory for Physical Sciences. Topics of study include quantum properties of superconducting qubits; quantum entanglement, control, and transport of atoms in cavities and optical lattices; decoherence studies with atoms and condensed matter systems; spin- and charge-based quantum computing; topological quantum computing; quantum coherence and entanglement; the quantum-classical interface; quasi-one-dimensional superconductors as optical lattices; and quantum computing with the fractional quantum Hall effect.

Maryland Center for Fundamental Physics studies particles, string theory, gravity, cosmology and neutrinos. Abazajian, Agashe, Buonanno, Brill, Chacko, Cohen, Gates, Greenberg, Hu, Jacobson, Ji, Mohapatra, Tiglio, Wallace.

Materials Research Science and Engineering Center (MRSEC) http://mrsec.umd.edu/ The MRSEC is part of a network of national Materials Research centers funded by the National Science Foundation (NSF) The Center's activities focus on three general area: 1) Materials Research, 2) Industrial Collaborations and 3) Educational Outreach. Williams.

Nonlinear Dynamics and Chaos. Research on the theoretical and experimental aspects of nonlinear systems with emphasis on chaotic behavior. Recent research includes the development of methods to control chaos, applications of chaotic dynamics to the transmission of information, studies of fractal properties of vorticity in fluids, the use of chaos theory to explain the occurrence of magnetic fields in astrophysical situations, and applications of chaos to biological/medical systems. Antonsen, Girvan, Lathrop, Losert, Ott, Roy, Sagdeev, Yorke. Program is joint with the Mathematics Department, the Institute for Physical Science and Technology, and the Institute for Plasma Research.

Quantum Electronics. Investigations of Relativity and Quantum Mechanics. Use of lasers, atomic clocks, and single photoelectron detection techniques in the experimental study of some fundamental problems in physics, astronomy, and geophysics. Theoretical studies of the new curved space time gravitational theory of Yilmaz are being intensively pursued.

UNIVERSITY OF MARYLAND

DEPARTMENT OF ASTRONOMY

College Park, Maryland 20742

Students Accepted For Degree	FIELDS		
	Physics	Astronomy	Related Fields
Doctorate		X	
Master's		X	

1. General

President: C. D. Mote, Jr.
Dean of Graduate School: Dr. Charles Caramello
Chairman of Astronomy Department: Stuart Vogel
Department Telephone Number: (301) 405-3001 (C)
Type of Institution: University
Control: Public
Setting: Suburban
Total Faculty: 3,996
Total Graduate Faculty: 1,472
Total Students: 37,195
Total Graduate Students: 10,653
Total Undergrad: 26,542
Annual Graduate Tuition:
In-state residents: $9,420 (10 hrs)
Out-of-state residents: $20,320 (10 hrs)
Tuition rates for: 2009–10
Deferred tuition plan: No
Term: Semester

2. Number of Faculty in Department

The combined total of full-time faculty in the three professorial ranks is 22, including 6 faculty emeriti. The combined total of full-time, part-time, and other faculty at all ranks is 60.

3. Admission, Financial Aid, and Housing

Address admission inquiries to: astrgrad@astro.umd.edu
Graduate application fee required: $60
Admission deadline (Fall admission): 1/15
Admission information: For fall admission, 2009–10, 7 students entered from 116 applicants.
Admission requirements: An undergraduate degree in a related field (normally physics) is required with a minimum undergraduate GPA of 3.0. The Dept. of Astronomy relies on a combination of course grades, letters of recommendation, and the GRE scores in deciding on admission. The GRE Physics test is required. Students from non-English speaking countries are required to demonstrate proficiency in English via the TOEFL, TSE, & TWE exams.
Undergraduate preparation assumed: No formal undergraduate course work in Astronomy is required. However, an entering student should have a basic, working knowledge of the subject, which could be obtained from any one of the many elementary textbooks. It is normally expected that the following subjects should have been studied previous to admission to graduate work: General Physics, Heat, Intermediate Mechanics, Optics, Electricity and Magnetism, Modern Physics, Advanced Calculus and Differential Equations. Any deficiencies in such background courses should be made up as soon as possible.
Address housing inquiries to: Off-campus Housing www.och.umd.edu

Table A—Faculty, Enrollments, and Degrees Granted

Research Specialty	2009–10 Faculty	Enrollment[1] Fall 2009 Master's	Enrollment[1] Fall 2009 Doctorate	No. of Degrees Granted[2] 2009–10 (2005–09) Master's	Terminal Master's	Doctorate	Median No. of Years for 2009–10 Ph.D.'s
Astronomy	22	0	35	6(34)	0(9)	4(21)	7.5
Total		–	35	6(34)	0(9)	4(21)	7.5
Full-time Grad. Stud.		–	35				
Part-time Grad. Stud.		–	0				
First-year Grad. Stud.		–	6				
Median Years in Grad. Study (2008–09 Degrees)				2.5	2.5		7.5
Undergraduate Degrees, 2008–09 (2004–09): 7(45)							

[2] Five-year totals in parentheses.

4. Graduate Degree Requirements

Detailed information is available through the Department's home page: www.astro.umd.edu
Master's: 30 credits (including six credits of Master's research) are required. 12 must be in the major area and 12 at an advanced graduate level. Both thesis and non-thesis options are available. Residence of one year of full-time study is required. A minimum GPA of 3.0 is required for graduation. No foreign language requirement. A written comprehensive exam is required.
Doctorate: Six graduate Astronomy courses plus two courses in supporting areas plus a minimum of 12 credits of doctoral research must be completed. Three years of residency required. A comprehensive exam must be taken prior to admission to candidacy, which must occur within four years of admission to the doctoral program. Dissertation and dissertation defense required (no less than one nor more than four years from admission to candidacy). Minimum GPA of 3.0; no foreign language required.
Thesis: Thesis may be written *in absentia*.
Special Equipment, Facilities, or Programs: In collaboration with four other excellent astronomy departments, the University of Maryland operates CARMA (Combined Array for Research in Millimeter-wave Astronomy), the most powerful millimeter-wave telescope in the world. Graduate students also observe with some of the largest telescopes in the US and around the world, as well as a wide range of space telescopes covering the electromagnetic spectrum from gamma-rays to the submillimeter. Our planetary science team is heavily involved with space missions visiting solar system bodies. Complementing its observational program, the department has a strong theory group, and there is also an important emphasis on the design and building of powerful new instruments.
A number of our students conduct research with distinguished scientists at the nearby NASA Goddard Space Flight Center. The university's scientific partnership with Goddard has recently been further strengthened via the creation of the Joint Space Science Institute (JSI). The first component of JSI is a black hole center, a close collaboration between the Departments of Astronomy and Physics and Goddard scientists that is unique in the work in involving all observational and the-

oretical aspects of black hole research.

An extensive department network provides seamless access to software and hardware on a variety of UNIX and LINUX platforms. The computational astrophysics group maintains and upgrades a Beowulf cluster for computation-intensive science projects and has additional access to a larger cluster maintained by the university.

Table B—Appointments to Graduate Students, 2009–10

Title of Appointee	Appointments Total	Appointments First year	Academic Load Allowed in Credit Hours	Hours of Service Per Week	Stipend for Academic Year ($)
			Semester		
Teaching Assistant	9.5	3.5	10	20	18,910-20,060[2]
Research Assistant	21	1	10[1]	20	24,660-26,542[3]
Fellowship	3.5	1.5	10	0	15,700–19,000
Total	34	6			

[1]Research Assistants are allowed up to 10 credits per semester of tuition remission.
[2]Usually supplemented by summer TA or RA positions.
[3]Research Assistantships are normally for 12 months.

5. Personnel Engaged in Separately Budgeted Research

Professorial faculty	16
Research faculty	98
Graduate students	35
Undergraduate students	29
Total	178

6. Separately Budgeted Research Expenditures by Source of Support

	Departmental Research	Physics-related Research Outside Department
Federal government	$22,811,773	$
Total	$22,811,773	$

Table C—Separately Budgeted Research Expenditures

Research Specialty	No. of Grants	Expenditures ($)
Astronomy	190	22,811,773
Total	190	22,811,773

FACULTY

Professors

A'Hearn, Michael F., Ph.D., Wisconsin, 1966. Cometary physics; photometry.

Bell, Roger A., Ph.D., Australian National, 1961 (Emeritus). Stellar interiors and atmospheres.

Earl, James A., Ph.D., M.I.T., 1957 (Emeritus). Cosmic rays.

Erickson, William C., Ph.D., Minnesota, 1956 (Emeritus). Radio astronomy; extragalactic astronomy.

Hamilton, Douglas P., Ph.D., Cornell, 1994. Planetary astronomy.

Harrington, J. Patrick, Ph.D., Ohio State, 1967. Interstellar matter; stellar atmospheres.

Harris, Andrew I., Ph.D., California, Berkeley, 1986. Astrophysics; instrumentation.

Kundu, Mukul R., Ph.D., Paris, 1957. (Emeritus). Radio astronomy; solar studies; supernovae.

Leventhal, Marvin, Ph.D., Brown Univ., 1964. (Emeritus). Gamma-ray astronomy.

Mundy, Lee G., Ph.D., Texas, 1984. IR astronomy; interstellar matter; mm wave astronomy.

Mushotzky, Richard, Ph.D., California, San Diego, 1976. X-ray astronomy; extragalactic astronomy.

Ostriker, Eve C., Ph.D., California, Berkeley, 1993. Extragalactic astronomy.

Papadopoulos, Konstantinos, Ph.D., Maryland, 1968. Plasma astrophysics; high-energy astrophysics.

Rose, William K., Ph.D., Columbia, 1963. (Emeritus). Radio astronomy; stellar astrophysics; jets.

Veilleux, Sylvain, Ph.D., California, Berkeley, 1989. Extragalactic astronomy.

Vogel, Stuart, Ph.D., California, Berkeley, 1983. Chairman. Interstellar medium; mm wave astronomy.

Wentzel, Donat G., Ph.D., Chicago, 1960. (Emeritus). Solar physics.

Associate Professors

McGaugh, Stacy, Ph.D., Michigan, Ann Arbor, 1992. Extragalatic astronomy.

Miller, M. Coleman, Ph.D., California, California Inst. of Tech., Pasadena, 1990. Theoretical astrophysics.

Reynolds, Christopher, Ph.D., Cambridge, England, 1996. High energy astrophysics.

Richardson, Derek, Ph.D., Cambridge, England, 1994. Computational astrophysics.

Assistant Professors

Bolatto, Alberto D., Ph.D., Boston U., 2000. Extragalactic astronomy.

Ricotti, Massimo, Ph.D., Colorado, 2001. Computational astrophysics.

RESEARCH SPECIALTIES AND STAFF

Theoretical

Cosmology and Galaxy Formation. Ricotti.

Galaxy Structure and Evolution. Spiral structure; star formation; interstellar medium. Ostriker.

High Energy Astrophysics. Active galactic nuclei; black holes; neutron stars; gravitational radiation. Reynolds, Miller.

Planetary Nebulae. Harrington.

Solar System Dynamics. Rings; collisions; solar system origins. Hamilton, Richardson.

Space Plasma Physics. Solar-Terrestrial physics. Papadopoulos.

Observational

CARMA Millimeter-Wave Astronomy. Star formation; interstellar medium; galactic structure, dynamics and evolution; protostellar disks; active galactic nuclei; instrumentation. Harris, Mundy, Vogel.

Comets. Small bodies in the solar system; Deep Impact and EPOXI missions. A'Hearn.

Extragalactic Astronomy. (Optical, Infrared, Radio, X-ray.) Active galactic nuclei; jets; starbursts; low surface brightness galaxies; star formation; galactic winds; intergalactic medium; galaxy clusters; dark matter; cosmology. Bolatto, McGaugh, Mushotzky, Reynolds, Veilleux.

UNIVERSITY OF MARYLAND

DEPARTMENT OF CHEMICAL PHYSICS

College Park, Maryland 20742

Students Accepted For Degree	FIELDS		
	Physics	Astronomy	Related Fields
Doctorate			X
Master's			X

1. General

Chancellor: William E. Kirwan
President: C. D. Mote, Jr.
Dean of Graduate School: Dr. Charles Caramello
Director of Chemical Physics Program: Michael A. Coplan
Department Telephone Number: (301) 405-4780
Type of Institution: University
Control: Public
Setting: Suburban
Total Faculty: 3,867
Total Students: 37,195
Total Graduate Students: 10,653
Annual Graduate Tuition:
In-state residents: $471 credit
Out-of-state residents: $1,016.00/credit
Tuition rates for: 2009–10
Deferred tuition plan: No
Other Fees: $593.63 per semester; (Graduate Students only)
Term: Semester

2. Number of Faculty in Department

The combined total of full-time faculty in the three professorial ranks is 61. The combined total of full-time, part-time, and other faculty at all ranks is 61.

3. Admission, Financial Aid, and Housing

Address admission inquiries to: Professor Michael A. Coplan, Chemical Physics Program, Institute for Physical Science and Technology.
Graduate application fee required: $60
Admission deadline (Fall admission): 2/1
Admission information: For fall admission, 2009–10, 27 total applicants, 3 accepted students accepted from 8 offers.
Admission requirements: For admission to the graduate programs, a Bachelor's degree in physics, chemistry, engineering, or mathematics is required with a minimum undergraduate GPA of 3.5 specified. The GRE (both General and Subject) is required. The average GRE scores for admissions were verbal-541; quantitative-776, analytical 3.84. Students from non-English speaking countries are required to demonstrate proficiency in English via the IBT TOEFL exam. Minimum acceptable score for admissions is 84-99. IELTS minimum acceptable score for admission is 6.5.
Address financial aid inquiries to: Professor Michael A. Coplan, Director, Chemical Physics Program, Institute for Physical Science and Technology
GAPSFAS application required: Yes
Financial aid deadline: 2/1
Loans available: Yes
Address housing inquiries to: University of Maryland Off-Campus Housing, www.och.umd.edu, 1110 Stamp Student Union, College Park, MD 20741
och@umd.edu, 301-314-3645

Table A—Faculty, Enrollments, and Degrees Granted

Research Specialty	2008–09 Faculty	Enrollment Fall 2009		No. of Degrees Granted[1] 2008–09			Median No. of Years for 2008–09 Ph.D.'s
		Master's	Doctorate	Master's	Terminal Master's	Doctorate	
Chemical Physics	58	0	34	3	0	5	5.5
Total		–	34	1	3	8	
Full-time Grad. Stud.		–	34			8	
Part-time Grad. Stud.		–	0				
First-year Grad. Stud.		–	5				5.5
Undergraduate Degrees	N/A						

4. Graduate Degree Requirements

Master's: 30 credit hours of advanced courses with a 3.0 average, including CHEM 684 or ENCH 610, CHEM 687 or PHYS 603, CHEM 691, PHYS 622, and PHYS 623 and an advance laboratory. The student must also have at least two seminar credits and must pass the Qualifying Examination at the Master's degree level. A scholarly paper and oral presentation are also required.
Doctorate: A Ph.D. candidate will be expected to have a background in both chemistry and physics. The Ph.D. Qualifying examination must be successfully passed, usually at the beginning of the second year. In addition, students are required to take an advanced laboratory course, one advanced course, give an oral presentation and submit a scholarly paper. A minimum of 12 hours of registered dissertation credits are required along with the dissertation and dissertation defense.

Table B—Appointments to Graduate Students, 2009–2010

Title of Appointee	Appointments		Academic Load Allowed in Credit Hours	Hours of Service Per Week	Stipend for Academic Year ($)
	Total	First year			
			Semester		
Teaching Assistant	29	5		20	16,297
Research Assistant	6	1			18,519
Total	35	6			

FACULTY

Distinguished University Professors

Alexander, Millard H., Ph.D., University of Paris, 1967. Molecular collisions and energy transfer.
Fisher, Michael E., Ph.D., King's College, London, 1957. Statistical mechanics; condensed matter theory; physical chemistry and associated mathematics.
Gupta, Ashwani K., Ph.D., University of Sheffield, U.K., 1973. Combustion, laser probes, and diagnostics.
Lorimer, George H., Ph.D., Michigan State, 1972. Studies of the mechanism of chaperonin-assisted protein folding.

Ott, Edward, Ph.D., Brooklyn Polytechnic Institute of New York, 1967. Chaos in dynamical systems including fundamental aspects of chaotic dynamics and applications.

Phillips, William D., Ph.D., Massachusetts Institute of Technology, 1976, Interaction of atoms and photons.

Thirumalai, Devarajan, Ph.D., University of Minnesota, 1982. Problems in equilibrium and non-equilibrium statistical mechanics, aspects of the transition from liquid to amorphous state, theoretical study of polymer-colloid interactions, dynamics of protein folding.

Weeks, John D., Ph.D., University of Chicago, 1969. State and dynamic properties of interfaces; pattern formation during crystal growth; theories for the structure and dynamics of liquids.

Williams, Ellen D., Ph.D., California Institute of Technology, 1982. Experimental research on the properties of solid surfaces.

Professors

Anisimov, Mikhail, Ph.D., Moscow State University, 1968. Phase transitions and critical phenomena in fluids, fluid mixtures, liquid crystals, surfactant solutions and other "soft" condensed matter systems.

Briber, Robert M., Ph.D., University of Massachusetts, 1984. Thermodynamics and structure of polymers with an emphasis on the study of complex mixtures of polymers by neutron scattering.

Calabrese, Richard V., Ph.D., University of Massachusetts, 1976. Fluid mechanics of single and multiphase systems.

Carton, James, Ph.D., Princeton University, 1983. Physical oceanography, ocean circulation, dynamics of the tropical Atlantic as it responds to meterological forcing on seasonal and longer timescales.

Chellappa, Ramailingam, Ph.D., Purdue University, 1981. Image analysis and computer vision techniques to medical imaging.

Coplan, Michael A., Ph.D., Yale University, 1963. Electron and ion impact spectroscopy, space physics.

Dagenais, Mario, Ph.D., Rochester University, 1978. Nonlinear optical interactions in condensed matter, optical switching.

Davis, Christopher C., Ph.D., Manchester, U.K. 1970. Quantum electronics; molecular energy transfer, atmospheric trace monitoring.

Dickerson, Russell R., Ph.D., University of Michigan, 1980. Analytical techniques in atmospheric chemistry.

Einstein, Theodore L., Ph.D., University of Pennsylvania, 1973. Surface science, phase transitions of absorbates, surface spectroscopy.

Falvey, Daniel L., Ph.D., University of Illinois, 1989. Photochemical and photophysical behavior of biological molecules.

Fourkas, John, Ph.D., Stanford University, 1991. Ultra fast nonlinear optical spectroscopy of liquids, dynamics of nanoconfined liquids, nonlinear optical microscopy, nontraditional approaches to micro and nanofabrication, dynamics of single molecules and single nanoparticles.

Fuhrer, Michael S., Ph.D., University of California, Berkeley, 1998. Fabrication and experimental studies of electronic and mechanical properties of novel nanostructures.

Fushman, David, Ph.D., University of Kazan, 1985. Theoretical and experimental studies of structure, dynamics, and interactions of biomacromolecules.

Gammon, Robert W., Ph.D., Johns Hopkins University, 1967. Laser light scattering: Brillouin, Rayleigh, and Raman scattering; critical phenomena.

Gilson, Michael, Ph.D., Columbia University, 1988. Molecular modeling and computational chemistry.

Hill, Wendell T. III, Ph.D., Stanford, 1980. Atomic and molecular structure, laser spectroscopy.

Kirkpatrick, Theodore R., Ph.D., Rockefeller University, 1981. Nonequilibrium statistical mechanics, quantum fluids, glasses, disordered electronic systems.

Kofinas, Peter, Ph.D., Massachusetts Institute of Technology, 1994. Synthesis and characterization of block polymers.

Lathrop, Daniel P., Ph.D., University of Texas, Austin, 1991. Nonlinear dynamics and chaos: turbulence, fluid dynamics. Singularities in liquid surface waves, the origin and dynamics of the magnetic field of the Earth.

Mignerey, Alice C., Ph.D., Rochester University, 1975. Heavy-ion induced nuclear reactions.

Milchberg, Howard M., Ph.D., Princeton University, 1985. High-intensity laser-matter interactions.

Rolston, Steven, Ph.D., State University of New York at Stony Brook, 1986. Ultracold neutral plasmas, Bose-Einstein condensation, quantum computation, non-linear atom optics, production of antihydrogen, atomic frequency standards, laser cooling.

Roy, Rajarshi, Ph.D., University of Rochester, 1982. Quantum electronics/optics, noise and nonlinear dynamics in optical systems, laser physics, semiconductor and solid state lasers, fiber and integrated optics, optical bistability, control of spatio-temporal systems, experimental statistical physics.

Reutt-Robey, Janice E., Ph.D., University of California, Berkeley, 1986. Transient surface chemical processes—energy transfer, diffusion.

Sita, Lawrence, Ph.D., Massachusetts Institute of Technology, 1985. Transition and main group metal inorganic and organometallic chemistry, new synthetic methodology, catalyst development, polymers, chemically-tailored surfaces and interfaces, molecular and mesoscopic self-assembly, new nanofabrication processes.

Venkatesan, Thirumalai, Ph.D., City University of New York, 1977. Superconductivity, physics of thin films, surface modification, beam-solid interactions.

Zachariah, Michael, Ph.D. University of California, Los Angeles, 1986. Nanoparticles (aerosols) synthesis, characterization, application, and modeling.

Associate Professors

Dimitrakopoulos, Panagiotis, Ph.D., University of Illinois at Urbana, 1998. Fluid mechanics of drops and bubbles in constrained geometries.

Ehrman, Sheryl, Ph.D., University of California, Los Angeles, 1997. Gas-phase synthesis routes to nanostrctured materials, particle-surface interactions in chemical polishing, chemical transformation and transport of particulates, air pollution.

Jarzynski, Christopher, Ph.D., University of California, Berkeley, 1994. Statistical mechanics at the molecular level, foundations of nonequilibrium thermodynamics, application of statistical mechanics to biophysics, development of efficient numerical schemes for estimating thermodynamic properties of complex systems.

Losert, Wolfgang, Ph.D., City College of City University of New York, 1998. Equilibrium dynamics of granular flows, and the nonlinear dynamics of microstructure formation in alloys and biomaterials.

Martínez-Miranda, Luz J., Ph.D., Massachusetts Institute of Technology, 1985. Analysis and characterization of thin films and buried interfaces, using x-ray scattering and glancing incidence scattering techniques.

Mullin, Amy, Ph.D., University of Colorado, Boulder, 1991. Dynamics of collisions and reactions of high energy molecules, high resolution time-resolved optical probing of state-resolved molecular pathways, vibrationally enhanced chemical reactions, controlling molecular rotation using strong optical fields, reaction mechanisms of radicals and other transient species.

Raghavan, Srinivasa, Ph.D., North Carolina State University, 1998. Complex fluids and soft matter structured at the micron-nanoscale. Rheology microscopy and scattering techniques.

Seo, Eun-Suk, Ph.D., Louisiana State University, 1991. Space-based experiments to cosmic-ray H, He, and heavier nuclei energy spectra at energies approaching 10^{15} eV.

Yu, Yihua Bruce, Ph.D., Johns Hopkins University, 1996. Biomaterials engineering focusing on structure function relationships.

Assistant Professors

Cumings, John, Ph.D., University of California, Berkeley, 2002, Experimental condensed matter.

La Porta, Arthur, Ph.D., University of California, San Diego, 1998. Development of optical techniques to be applied to problems in molecular biology. Single molecular techniques to investigate force, torque generation in molecular motors and to characterize proteins that bind and modify the structure of DNA.

Ouyang, Min, Ph.D., Harvard University, 2001. Spin degree of freedom of electrons and nuclei within ordered nano-engineered architectures with an emphaisis on synthetic methodologies for spin-based hybrid organic-inorganic nanostructures. Spin-charge interactions and spin transport within nanostructured systems using femtosecond optical spectroscopy, magnetotransport and low temperature scanning probe microscopy. New functional nanospintronic devices and technologies.

Rabin, Oded, Ph.D., Massachusetts Institute of Technology, 2004. Synthesis and characterization of nanoparticles and nanowires.

Upadhyaya, Arpita, Ph.D., University of Notre Dame, 2000. Mathematical modeling, quantitative imaging and genetic manipulation to uncover signaling networks and the physical properties of the cell and its surroundings.

Adjunct Professors

Clark, Charles W., Ph.D., University of Chicago, 1979. NIST. Measurement for needed by emerging electronic and optical technologies, ultra cold atoms and molecules, Bose-Einstein condensation.

Nossal, Ralph J., Ph.D., University of Michigan, 1963. NIH. Laser inelastic light scattering, statistical mechanics of condensed media; phase changes and rheological properties of polymer networks and biological gels, membrane biophysics, cell motility and chemotaxis.

RESEARCH SPECIALTIES AND STAFF

Theory

Atomic and Molecular Physics. Molecular collisions and energy transfer, molecular spectra, molecular quantum mechanics, photoelectron spectroscopy, many-body theory of quantum mechanical scattering, atomic and molecular processes at surfaces. Alexander, Clark.

Quantum Electronics and Optical Physics. Multiphoton processes, strong-field laser-atom interaction, Bose-Einstein condensation. Clark, Phillips, Rolston.

Statistical Physics. Nonequilibrium statistical mechanics, interfaces and surfaces, molecular hydrodynamics, phase transitions and critical phenomena, polymer science, disordered solids, thermophysical properties. Anisimov, Calabrese, Fisher, Kirkpatrick, Weeks, Jarzynski, Ott.

Experiments

Atmospheric Physics and Chemistry. Atmospheric spectroscopy, earth's magnetosphere, atmospheric radiation, aerosol physics, atmospheric photochemistry, laser sensing of atmospheric constituents. Dagenais, Davis, Dickerson.

Biophysics: Losert, Thirumalai, Weeks, Lorimer, Fushman, La Porta, Gilson, Upadhyaya.

Condensed Matter Science. Raman, Rayleigh, Brillouin scattering, neutron scattering, nuclear magnetic resonance, critical phenomena and phase transitions, surface science. Gammon, Martínez-Miranda, Reutt-Robey, Williams.

Molecular Physics. Atomic and molecular spectroscopy, electron and ion impact spectroscopy, picosecond spectroscopy, infrared and Raman spectroscopy. Coplan, Davis, Fourkas, Mullin, Hill, Phillips, Rolston, Clark, Sita.

Photochemistry. Laser induced atomic and molecular collisions, ultraviolet and infrared laser photochemistry, photolysis rates in the atmosphere. Dickerson, Falvey.

Quantum Electronics and Optical Physics. Gas laser; light scattering, spectroscopy, nonlinear optics, ultrafast phenomena, laser plasmas, dye lasers and intense field effects. Davis, Gammon, Hill, Milchberg, Phillips, Ouyang, Roy.

Space Physics. Solar and planetary atmospheres, solar wind, cosmic ray. Coplan, Seo.

Thermophysics. Energy conversion, thermodynamics properties, transport properties. Gupta, Anisimov, Raghavan, Dimitrakopoulos, Cumings.

UNIVERSITY OF MARYLAND, BALTIMORE COUNTY

DEPARTMENT OF PHYSICS
APPLIED PHYSICS GRADUATE PROGRAM

Baltimore, Maryland 21250

Students Accepted For Degree	FIELDS		
	Physics	Astronomy	Related Fields
Doctorate	X	X	X
Master's	X	X	X

1. General

President: Freeman A. Hrabowski
Dean of the Graduate School: Janet C. Rutledge
Department Chairman: L. Michael Hayden
Department Telephone Number: (410) 455-2513
Type of Institution: University
Control: Public
Setting: Suburban
Total Faculty: 930
Total Graduate Faculty: 450
Total Students: 12,870
Total Graduate Students: 2,923
Annual Graduate Tuition:
In-state residents: $445/credit
Out-of-state residents: $736/credit
Tuition rates for: 2009–10
Deferred tuition plan: No
Other Fees: $104/credit
Term: Semester

2. Number of Faculty in Department

The combined total of full-time faculty in the three professorial ranks is 19. The combined total of full-time, part-time, and other faculty at all ranks is 50.

3. Admission, Financial Aid, and Housing

Address admission inquiries to: Graduate Admissions Coordinator, Applied Physics, Dept. of Physics
Graduate application fee required: $50
Admission deadline (Fall admission): 7/1
Admission information: For fall admission, 2009–10, 12 students were accepted from 27 applicants.
Admission requirements: For admission to the graduate programs, a Bachelor's degree in physics, chemistry, math, or engineering is required with a minimum undergraduate GPA of 3.0 specified. The GRE is required. The GRE Advanced is desired. The average GRE scores for admissions were verbal-530; quantitative-760; total-1,290. Students from non-English speaking countries are required to demonstrate proficiency in English via the TOEFL exam. Minimum acceptable score for admission is 550. Minimum for T.A. is 600.
Undergraduate preparation assumed: Young and Freedman, *University Physics*; Reif, *Thermal Physics*; Marion, *Newtonian Dynamics*; Tippler and Llewellyn, *Introduction to Modern Physics*; Hecht, *Fundamentals of Optics*; Griffiths, *Quantum Mechanics*; Weber and Arfken, *Mathematical Methods for Physicists*; Griffiths, *Introduction to Electrodynamics*.
Address financial aid inquiries to: Graduate Admissions Coordinator, Applied Physics.
GAPSFAS application required: No
Financial aid deadline: 5/1
Loans available: Yes
Address housing inquiries to: Office of Residential Life
On-campus, single student housing available: Yes
Cost/term: $790/month.
On-campus, married student housing available: No

Table A—Faculty, Enrollments, and Degrees Granted

Research Specialty	2008–09 Faculty	Enrollment[1] Fall 2008		No. of Degrees Granted[2] 2008–09 (2004–08)			Median No. of Years for 2005–06 Ph.D.'s
		Master's	Doctorate	Master's	Terminal Master's	Doctorate	
Astrophysics	4	0	7	2(8)	0(0)	1(0)	
Atmos./Space Phys., Cosmic Rays	4	0	13	2(6)	3(2)	2(7)	6
Atomic, Molecular, & Optical Physics	0	0	0	0(2)	0(0)	0(0)	–
Condensed Matter Physics	4	0	4	2(6)	1(3)	1(3)	6
Materials Sci./ Metallurgy	2	0	0	0(2)	0(2)	0(2)	
Optics	5	1	17	2(4)	2(4)	2(6)	6
Polymer Physics/ Science	1	0	0	0(1)	0(1)	0(1)	6
Other Theoretical/ Math.	1	0	1	0(0)	0(0)	0(0)	–
Non-specialized	0	0	4	0(0)	0(0)	0(0)	–
Total		1	46	8(29)	3(8)	6(19)	
Full-time Grad. Stud.		3	25				
Part-time Grad. Stud.		0	5				
First-year Grad. Stud.		0	9				
Median Years in Grad. Study (2004–05 Degrees)				2.1	2	5	
Undergraduate Degrees, 2008–09 (2004–08): 15(43)							

[1]Students not yet committed to a research specialty are entered under non-specialized.
[2]Five-year totals in parentheses.

4. Graduate Degree Requirements

Master's: Completion of 30 hours of course work, including required core courses in quantum mechanics, and mathematical physics. Minimum acceptable GPA 3.0 (A=4.0). Overall competence must be demonstrated by oral thesis defense (thesis option) or written comprehensive examination (nonthesis option). For thesis option, six hours of the required 30 are for thesis research.
Thesis: Thesis may be written *in absentia*.
Doctorate: Minimum requirement is a total of 46 credit hours, with 28 credit hours of lecture courses at the 600 level or higher and 12 credit hours of Doctoral Research. Overall GPA must be at least 3.0 (A=4). All students will be at least required to complete a core curriculum consisting of Quantum Mechanics, Statistical Mechanics, Mathematical Physics, Classical Mechanics, and Electromagnetic Theory. In addition, they are required to take Computational Physics, Quantum Mechanics II, Advanced Electromagnetic Theory, Physics Seminar and at least one elective course. After completion of the Ph.D. core curriculum, prospective Ph.D. students will be required to pass a written examination in order to qualify for candidacy for the Ph.D. degree. Upon completion of the

doctoral research, the student will be required to write and to defend a dissertation before a committee constituted in accordance with the graduate school regulations. Note that the department now offers M.S. and Ph.D. degrees in Atmospheric Physics.

Table B—Appointments to Graduate Students, 2004–05

Title of Appointee	Appointments Total	Appointments First year	Academic Load Allowed in Credit Hours	Hours of Service Per Week	Stipend for Academic Year ($)[1,2]
			Semester		
Teaching Assistant	12	5	10	20	21,000-27,000
Research Assistant	24	1	10	20	24,000-30,000
Total	36	6			

[1]Plus tuition remission and full health benefits.
[2]All assistants have 12-month stipends of at least $20,000.

5. Personnel Engaged in Separately Budgeted Research, 1/06–12/06

Professorial faculty	19
Graduate students	24
Undergraduate students	5
Research faculty	25
Research associates	17
Total	90

6. Separately Budgeted Research Expenditures by Source of Support

	Departmental Research	Physics-related Research Outside Department
Federal government	$9,550,000	$
Total	$9,550,000	$

Table C—Separately Budgeted Research Expenditures

Research Specialty	No. of Grants	Expenditures ($)
Astrophysics	22	3,000,000
Atmos./Space Phys., Cosmic Rays	30	4,500,000
Condensed Matter Physics	4	300,000
Optics	4	750,000
Quantum Optics	4	1,000,000
Total	70	9,550,000

FACULTY

Professors

Franson, James D., Ph.D., California Institute of Technology, 1977. Quantum optics and quantum computing.

Hayden, L. Michael, Ph.D., California, Davis, 1987. Nonlinear optical properties of polymers; electro-optic techniques; photonic devices.

Hoff, Raymond M., Ph.D., Simon Fraser, 1975. Atmospheric physics; lidar.

Rous, Philip J., Ph.D., Imperial College of Science and Technology. Theoretical physics; surfaces, interfaces and nanostructures.

Shih, Yanhua, Ph.D., Maryland, 1987. Quantum optics; laser physics; nonlinear optics.

Summers, Geoffrey P., Ph.D., Oxford, 1970. Chairman of the Department. Radiation effects in semiconductors; defects in solids.

Associate Professors

George, Ian M., Ph.D., University of Leicester, 1988. Astrophysics, X-ray astronomy.

Henriksen, Mark J., Ph.D., Maryland, 1986. Astrophysics; X-ray astronomy.

Kramer, Ivan, Ph.D., California, Berkeley, 1967. Mathematical modeling.

Martins, Vanderlei, Ph.D., University of São Paolo, 1999. Radiative effects of biomass burning and bio-aerosols.

Pittman, Thomas, Ph.D., University of Maryland, Baltimore County, 1996. Quantum optics and quantum computing.

Reno, Robert C., Ph.D., Brandeis, 1970. Hyperfine interactions in solids; electron microscopy; neutron diffraction measurement.

Sparling, Lynn C., Ph.D, Texas, Austin, 1987. Atmospheric physics; modeling.

Takacs, Laszlo, Ph.D., Eotvos University, 1978. Amorphous and metastable crystalline alloys; energy-dispersive X-ray diffraction; magnetic susceptibility.

Turner, T. Jane, Ph.D., University of Leicester, 1988. Extragalactic astrophysics, X-ray astronomy.

Worchesky, Terrance L., Ph.D., Georgetown, 1984. Optical properties of semiconductors; photonics.

Wu, En-Shinn, Ph.D., Cornell, 1972. Laser light scattering; phase transitions; fluorescence spectroscopy; lipid membranes.

Assistant Professors

Georganopoulos, Markos, Ph.D., University of Thessaloniki, 1989. Broad-band synchrotron emission from relativistic flows in active galaxies, galactic microquasars and gamma-ray bursts.

Gougousi, Theodosia, Ph.D., University of Pittsburg, 1996. Nanoscience, Interfaces.

Research Professors

Krueger, Arlin J., Ph.D., Colorado State University, 1984. Atmospheric sounding techniques and instruments.

Strow, L. Larrabee, Ph.D., Maryland, 1981. High-resolution infrared molecular spectroscopy; atmospheric radiative transfer.

Research Associate Professors

Barnet, Christopher, Ph.D., New Mexico State University, 1990. IR and microwave remote sensing.

Davis, David, Ph.D., University of Maryland, College Park, 1994. Galaxy clusters, X-ray astronomy.

Kundu, Prasun, Ph.D., University of Rochester, 1981. Satellite and ground-based remote sensing.

Larry, David, Ph.D., University of Cambridge, 1991. Atmospheric chemistry and dynamics.

Markus, Thorsten, Ph.D., University of Bremen, 1995. Satellite remote sensing.

McCann, Kevin J., Ph.D., Georgia Institute of Technology, 1974. Lidar and atmospheric aerosols.

Olson, William, Ph.D., University of Wisconsin-Madison, 1987. Remote sensing of precipitation.

Oreopoulos, Lazaros, Ph.D., McGill University, 1996. Cloud modeling and remote sensing.

Pavlis, Erricos C., Ph.D., Ohio State University, 1983. Time-varying gravity observations and precise calibration of altimetric missions.

Torres, Omar, Ph.D., Georgia Institute of Technology, 1989. Ultraviolet radiative transfer modeling.

Varnai, Tamas, Ph.D., McGill University, 1996. Cloud physics and radiation transfer.

Research Assistant Professors

Chiu, Christine, Ph.D., Purdue University, 2003. Remote sensing and radiation transfer.

De Souza-Machado, Sergio, Ph.D., University of Maryland, College Park, 1996. Infrared remote sensing and radiation transfer.

Mehta, Amita V., Ph.D., Florida State University, 1991. Seasonal and interannual climate variability.

Professors Emeriti

Melfi, Harvey, Ph.D., College of William and Mary, 1970. Atmospheric lidar, remote sensing.

Rasera, Robert L., Ph.D., Purdue, 1965. Perturbed gamma-ray angular correlation spectroscopy.

Rubin, Morton H., Ph.D., Princeton, 1964. Theoretical physics; quantum optics.

Adjunct Professors

Beckmann, Volker, Ph.D., University of Hamburg, 2001. Extragalactic astrophysics, X-ray astronomy.

Demoz, Belay, Ph.D., University of Nevada, 1992. Atmospheric physics and chemistry.

Finoguenov, Alexis, Ph.D., Moscow Institute of Physics and Technics, 1997. Chemical evolution of galaxies, groups and clusters of galaxies.

Fitelson, Michael, Ph.D., Pennsylvania State University, 1966. Advanced technologies.

Jacobs, Brian, Ph.D., UMBC, 2003. Quantum optics, quantum information.

Krainak, Michael, Ph.D., Johns Hopkins University, 1980. Laser spectroscopy and remote sensing.

Malak, Henryk, Ph.D., University of Poznam, 1985. Quantum bio-imaging.

Perlman, Eric, Ph.D., University of Colorado, 1999. Galactic nuclei, galactic clusters.

Platnick, Steven, Ph.D., University of Arizona, 1980. Satellite, aircraft, and ground-based cloud remote sensing.

Remer, Lorraine, Ph.D., University of California, Davis, 1991. Climate change, remote sensing.

Winsor, Harry, Ph.D., University of California, Davis, 1978. Advanced technologies.

RESEARCH SPECIALTIES AND STAFF

Theoretical

Atmospheric Physics. Sparling.

Atomic and Molecular Physics. Strow.

Computational Physics. Henriksen, Kramer, Rubin, Rous.

Quantum Theory of Measurement. Franson, Rubin.

Relativity and Astrophysics. Henriksen.

Statistical and Thermal. Finite time processes. Rubin.

Surface Physics. Rous.

Experimental

Applied Physics. Crystallographic texture in materials. Reno.

Astrophysics. X-ray Astronomy, Active galaxies, Extragalactic astrophysics, George, Henriksen, Turner, Georganopoulos.

Atmospheric Physics. Strow, McMillan, Hoff, Martins.

Biophysics. Laser photobleaching; biopolymers. Wu.

Materials Science/Metallurgy. Electron microscopy; X-ray diffraction; mechanical alloying; small angle neutron scattering; phase diagrams. Gougousi, Reno, Takacs.

Optics. Quantum optics, Quantum information, Nonlinear optics; light scattering in fluid mixtures; electro-optics. Franson, Hayden, Pittman, Shih, Worchesky, Wu.

Polymer Physics. Liquid crystals. Hayden, Wu.

Solid State. Semiconductors; optical and electrical properties; low-temperature fluorescence and photoconductivity; Mössbauer effect; gamma-ray perturbed angular correlations; positron annihilation; magnetic measurements. Reno, Summers, Takacs, Worchesky.

Statistical and Thermal. Critical phenomena. Reno, Wu.

UNIVERSITY OF MARYLAND, BALTIMORE COUNTY

DEPARTMENT OF PHYSICS, ATMOSPHERIC PHYSICS GRADUATE PROGRAM

Baltimore, Maryland 21250

Students Accepted For Degree	FIELDS		
	Physics	Astronomy	Related Fields
Doctorate	X		X
Master's	X		X

1. General

President: Freeman A. Hrabowski
Dean of the Graduate School: Janet C. Rutledge
Department Chairman: L. Michael Hayden
Department Telephone Number: (410) 455-2513
Type of Institution: University
Control: Public
Setting: Suburban
Total Faculty: 930
Total Graduate Faculty: 450
Total Students: 12,870
Total Graduate Students: 2,923
Annual Graduate Tuition:
In-state residents: $445/credit
Out-of-state residents: $736/credit
Tuition rates for: 2009–10
Deferred tuition plan: No
Other Fees: $104/credit
Term: Semester

2. Number of Faculty in Department

The combined total of full-time faculty in the three professonial ranks is 19. The combined total of full-time, part-time, and other faculty at all ranks is 50.

3. Admission, Financial Aid, and Housing

Address admission inquiries to: Graduate Admissions Coordinator, Applied Physics, Dept. of Physics
Graduate application fee required: $50
Admission deadline (Fall admission): 7/1
Admission information: For fall admission, 2009–10, 5 students were accepted from 6 applicants.
Admission requirements: For admission to the graduate programs, a Bachelor's degree in physics, chemistry, math, or engineering is required with a minimum undergraduate GPA of 3.0 specified. The GRE is required. The GRE Advanced is desired. The average GRE scores for admissions were verbal-530; quantitative-760; total-1290. Students from non-English speaking countries are required to demonstrate proficiency in English via the TOEFL exam. Minimum acceptable score for admission is 550. Minimum for T.A. is 600.
Undergraduate preparation assumed: Young and Freedman. *University Physics*; Reif, *Thermal Physics*; Marion, *Newtonian Dynamics*; Tippler and Llewellyn, *Introduction to Modern Physics*; Hecht, *Fundamentals of Optics*; Griffiths, *Quantum Mechanics*; Weber and Arfken, *Mathematical Methods for Physicists*; Griffiths, *Introduction to Electrodynamics*.
Address financial aid inquiries to: Graduate Admissions Coordinator, Applied Physics.
GAPSFAS application required: No
Financial aid deadline: 5/1
Loans availabe: Yes
Address housing inquiries to: Office of Residential Life
On-campus, single student housing available: Yes
Cost/term: $790/month.
On-campus, married student housing available: No

Table A—Faculty, Enrollments, and Degrees Granted

Research Specialty	2008–09 Faculty	Enrollment[1] Fall 2008		No. of Degrees Granted[2] 2008–09 (2004–09)			Median No. of Years for 2008–09 Ph.D.'s
		Master's	Doctorate	Master's	Terminal Master's	Doctorate	
Astrophysics	4	0	0	0(0)	0(0)	0(0)	–
Atmos./Space Phys., Cosmic Rays	4	0	13	2(6)	3(2)	2(7)	6
Condensed Matter Physics	4	0	0	0(0)	0(0)	0(0)	–
Materials Sci./ Metallurgy	2	0	0	0(0)	0(0)	0(0)	–
Optics	5	0	0	0(0)	0(0)	0(0)	–
Polymer Physics/ Science	1	0	0	0(0)	0(0)	0(0)	–
Other Theoretical/ Math.	1	0	0	0(0)	0(0)	0(0)	–
Total		0	13	2(6)	3(2)	2(7)	
Full-time Grad. Stud.		5	14				
Part-time Grad. Stud.		0	0				
First-year Grad. Stud.		0	4				
Median Years in Grad. Study (2008–09 Degrees)				2.1	2	6	
Undergraduate Degrees, 2008–09 (2004–09):15(43)							

[1]Students not yet committed to a research specialty are entered under non-specialized.
[2]Five-year totals in parentheses.

4. Graduate Degree Requirements

Master's: Completion of 30 hours of course work, including required core courses in quantum mechanics, mathematical physics, and Atmospheric Physics I and II. Minimum acceptable GPA 3.0 (A=4.0). Overall competence must be demonstrated by oral thesis defense (thesis option) or written comprehensive examination (non-thesis option). For thesis option, six hours of the required 30 are for thesis research.
Thesis: Thesis may be written *in absentia*.
Doctorate: Minimum requirement is a total of 46 credit hours, with 28 credit hours of lecture courses at the 600 level or higher and 12 credit hours of Doctoral Research. Overall GPA must be at least 3.0 (A=4). All students will be at least required to complete a core curriculum consisting of Quantum Mechanics, Statistical Mechanics, Mathematical Physics, Atmospheric Physics I and II, and Electromagnetic Theory. In addition, they are required to take Computational Physics, at least two specialized courses in Atmospheric Physics, Professional Techniques, and Physics Seminar. After completion of the Ph.D. core curriculum, prospective Ph.D. students will be required to pass a written examination in order to qualify for candidacy for the Ph.D. degree. Upon completion of the doctoral research, the student will be required to write and to defend a dissertation before a committee constituted in accordance with the graduate school regulations.

Table B—Appointments to Graduate Students, 2008–09

Title of Appointee	Appointments Total	First year	Academic Load Allowed in Credit Hours	Hours of Service Per Week	Stipend for Academic Year ($)[1,2]
			Semester		
Teaching Assistant	4	3	10	20	21,000–27,000
Research Assistant	15	2	10	20	24,000–30,000
Total	17	5			

[1]Plus tuition remission and full health benefits.
[2]All assistants have 12-month stipends of at least $20,000.

5. Personnel Engaged in Separately Budgeted Research,1/08–12/08

Professorial faculty	19
Graduate students	24
Undergraduate students	5
Research faculty	25
Research associates	17
Total	90

6. Separately Budgeted Research Expenditures by Source of Support

	Departmental Research	Physics-related Research Outside Department
Federal government	$9,550,000	$
Total	$9,550,000	$

Table C—Separately Budgeted Research Expenditures

Research Specialty	No. of Grants	Expenditures ($)
Astrophysics	22	3,000,000
Atmos./Space Phys., Cosmic Rays	30	4,500,000
Condensed Matter Physics	4	300,000
Optics	4	7,500,000
Quantum Optics	4	1,000,000
Total	70	9,550,000

FACULTY

Professors

Franson, James D., Ph.D., California Institute of Technology, 1977. Quantum optics and quantum computing.

Hayden, L. Michael, Ph.D., California, Davis, 1987. Nonlinear optical properties of polymers; electro-optic techniques; photonic devices.

Hoff, Raymond M., Ph.D., Simon Fraser, 1975. Atmospheric physics; lidar, air quality, satellite remote sensing.

Rous, Philip J., Ph.D., Imperial College of Science and Technology. Theoretical physics; surfaces, interfaces and nanostructures.

Shih, Yanhua, Ph.D., Maryland, 1987. Quantum optics; laser physics; nonlinear optics.

Summers, Geoffrey P., Ph.D., Oxford, 1970. Chairman of the Department. Radiation effects in semiconductors; defects in solids.

Associate Professors

George, Ian M., Ph.D., University of Leicester, 1988. Astrophysics, X-ray astronomy.

Henriksen, Mark J., Ph.D., Maryland, 1986. Astrophysics; X-ray astronomy.

Kramer, Ivan, Ph.D., California, Berkeley, 1967. Mathematical modeling.

Martins, Vanderlei, Ph.D., University of São Paolo, 1999. Radiative effects of biomass burning and bio-aerosols.

Pittman, Thomas, Ph.D., University of Maryland, Baltimore County, 1996. Quantum optics and quantum computing.

Reno, Robert C., Ph.D., Brandeis, 1970. Hyperfine interactions in solids; electron microscopy; neutron diffraction measurement.

Sparling, Lynn C., Ph.D, Texas, Austin, 1987. Atmospheric physics; modeling.

Takacs, Laszlo, Ph.D., Eotvos University, 1978. Amorphous and metastable crystalline alloys; energy-dispersive X-ray diffraction; magnetic susceptibility.

Turner, T. Jane, Ph.D., University of Leicester, 1988. Extragalactic astrophysics, X-ray astronomy.

Worchesky, Terrance L., Ph.D., Georgetown, 1984. Optical properties of semiconductors; photonics.

Wu, En-Shinn, Ph.D., Cornell, 1972. Laser light scattering; phase transitions; fluorescence spectroscopy; lipid membranes.

Assistant Professors

Georganopoulos, Markos, Ph.D., University of Thessaloniki, 1989. Broad-band synchrotron emission from relativistic flows in active galaxies, galactic microquasars and gamma-ray bursts.

Gougousi, Theodosia, Ph.D., University of Pittsburg, 1996. Nanoscience, Interfaces.

Research Professors

Krueger, Arlin J., Ph.D., Colorado State University, 1984. Atmospheric sounding techniques and instruments.

Strow, L. Larrabee, Ph.D., Maryland. 1981. High-resolution infrared molecular spectroscopy; atmospheric radiative transfer.

Research Associate Professors

Davis, David, Ph.D., University of Maryland, College Park, 1994. Galaxy clusters, X-ray astronomy.

Kundu, Prasun, Ph.D., University of Rochester, 1981. Satellite and ground-based remote sensing.

Lary, David, Ph.D., University of Cambridge, Churchill College. Chemical data assimilation and neural network applications to atmospheric physics.

Larry, David, Ph.D., University of Cambridge, 1991. Atmospheric chemistry and dynamics.

McCann, Kevin J., Ph.D., Georgia Institute of Technology, 1974. Lidar and atmospheric aerosols.

Olson, William, Ph.D., University of Wisconsin-Madison, 1987. Remote sensing of precipitation.

Oreopoulos, Lazaros, Ph.D., McGill University, 1996. Cloud modeling and remote sensing.

Varnal, Tamas, Ph.D., McGill University, 1996. Cloud physics and radiation transfer.

Research Assistant Professors

Chiu, Christine, Ph.D., Purdue University, 2003. Remote sensing and radiation transfer.

De Souza-Machado, Sergio, Ph.D., University of Maryland, College Park, 1996. Infrared remote sensing, radiation transfer, spectroscopy, plasma physics.

Mehta, Anita, Ph.D., McGill University, 1991. Seasonal and interannual climate variability.

Professors Emeriti

Melfi, Harvey, Ph.D., College of William and Mary, 1970. Atmospheric lidar, remote sensing.

Rasera, Robert L., Ph.D., Purdue, 1965. Perturbed gamma-ray angular correlation spectroscopy.

Rubin, Morton H., Ph.D., Princeton, 1964. Theoretical physics, quantum optics.

Adjunct Professors

Barnet, Christopher, Ph.D., New Mexico State University, 1990. IR and microwave remote sensing.

Beckmann, Volker, Ph.D., University of Hamburg, 2001. Extragalactic astrophysics, X-ray astronomy.

Cahalan, Robert F., Ph.D., University of Illinois, 1973. Cloud physics, atmospheric radiative transfer.

Demoz, Belay B., Ph.D. University of Nevada, Reno. Lidar, mesoscale meteorology.

Finoguenov, Alexis, Ph.D., Moscow Institute of Physics and Technics, 1997. Chemical evolution of galaxies, groups and clusters of galaxies.

Malak, Henryk, Ph.D., University of Poznam, 1985. Quantum bio-imaging.

Markus, Thorsten, Ph.D., University of Bremen, 1995. Satellite remote sensing.

Ormes, Jonathan, Ph.D., University of Minnesota. 1967. Particle and nuclear astrophysics.

Platnick, Steven, Ph.D., University of Arizona. 1980. Satellite, aircraft, and ground-based cloud remote sensing.

Remer, Lorraine, Ph.D., University of California at Davis, 1991. Atmospheric aerosols, remote sensing, climate change, clouds, air quality.

Sinsky, Joel A., Ph.D., University of Maryland, 1967. Acoustics.

Winsor, Harry, Ph.D., University of California, Davis, 1978. Advanced technologies.

RESEARCH SPECIALTIES AND STAFF

Theoretical

Atmospheric Physics. Sparling.
Atomic and Molecular Physics. Strow.
Computational Physics. Henriksen, Kramer, Rubin, Rous.
Quantum Theory of Measurement. Franson, Rubin.
Relativity and Astrophysics. Henriksen.
Statistical and Thermal. Finite time processes. Rubin.
Surface Physics. Rous.

Experimental

Applied Physics. Crystallographic texture in materials. Reno.
Astrophysics. X-ray Astronomy, active galaxies, extra galactic astrophysics. George, Henriksen, Turner, Georganopoulos.
Atmospheric Physics. Strow, McMillan, Hoff, Martins.
Biophysics. Laser photobleaching; biopolymers. Wu.
Materials Science/Metallurgy. Electron microscopy; X-ray diffraction; mechanical alloying; small angle neutron scattering; phase diagrams. Gougousi, Reno, Takacs.
Optics, Quantum Optics, Quantum Information. Nonlinear optics; light scattering in fluid mixtures; electro-optics. Franson, Hayden, Pittman, Shih, Worchesky, Wu.
Polymer Physics. Liquid crystals. Hayden, Wu.
Solid State. Semiconductors; optical and electrical properties; low-temperature fluorescence and photoconductivity; Mössbauer effect; gamma-ray perturbed angular correlations; positron annihilation; magnetic measurements. Reno, Summers, Takacs, Worchesky.
Statistical and Thermal. Critical phenomena. Reno, Wu.

BOSTON COLLEGE

DEPARTMENT OF PHYSICS

Chestnut Hill, Massachusetts 02467-3811

Students Accepted For Degree	FIELDS		
	Physics	Astronomy	Related Fields
Doctorate	X		
Master's	X		

1. General

President: Fr. William P. Leahy, S. J.
Dean of the College and Graduate School of Arts & Sciences: David Quigley
Department Chairman: Michael J. Naughton
Department Telephone Number: (617) 552-3575
Type of Institution: University
Control: Private
Setting: Suburban
Total Faculty: 708
Total Graduate Faculty: Not counted separately
Total Students: 14,796
Total Graduate Students: 4,960
Annual Graduate Tuition:
All Graduate Students: $1,206/credit
Tuition rates for: 2010–11
Deferred tuition plan: No
Other Fees: $422 (health services)
Term: Semester
Health Ins.: Paid by College

2. Number of Faculty in Department

The combined total of full-time faculty in the professorial ranks is 18. The combined total of full-time, part-time, and other faculty at all ranks is 19.

3. Admission, Financial Aid, and Housing

Address admission inquiries to: Admissions, GSAS
Graduate application fee required: $75
Admission deadline (Fall admission): 1/15
Admission information: For fall admission, 2010–11, 9 students were accepted from 110 applicants.
Admission requirements: For admission to the graduate programs, a Bachelor's degree in physics is required with no minimum undergraduate GPA specified. The GRE is required. No minimum acceptable score is specified. The GRE Advanced is required. No minimum acceptable score is specified. Students from non-English speaking countries are required to demonstrate proficiency in English via the TOEFL and GRE exam. Minimum acceptable score for admission is 550 on TOEFL exam.
Undergraduate preparation assumed: Reif, *Fundamentals of Statistical and Thermal Physics*; Lorrain and Corson, *Electromagnetic Fields and Waves* (2nd ed.); Symon, *Mechanics*; Rosenberg, *The Solid State*; Eisberg, *Modern Physics*.
Address financial aid inquiries to: Admissions, GSAS
GAPSFAS application required: Yes
Financial aid deadline: 3/15
Loans available: Yes
Address housing inquiries to: Housing Office
On-campus, single student housing available: No
On-campus, married student housing available: No

Table A—Faculty, Enrollments, and Degrees Granted

Research Specialty	2009–10 Faculty	Enrollment Fall 2009		No. of Degrees Granted 2009–10			Median No. of Years for 2008–09 Ph.D.'s
		Master's	Doctorate	Master's	Terminal Master's	Doctorate	
Condensed Matter Physics	18	0	39	0(0)	0(0)	4(0)	6
Total		0	47	0(0)	0(1)	4(0)	
Full-time Grad. Stud.		0	47				
Part-time Grad. Stud.		0	0				
First-year Grad. Stud.		0	8				
Median Years in Grad. Study (2009–10 Degrees)				–	–	6	–

Undergraduate Degrees, 2009–10: 14 majors

4. Graduate Degree Requirements

Master's: Thesis and non-thesis options are available. They are typically two year programs. Also offered: Master of Science in Teaching.
Doctorate: Comprehensive exam: general and special field, both written and oral parts; thesis with oral defense; course requirements: advanced quantum mechanics, electromagnetic theory, and statistical physics; also, a distributional requirement of electives in four distinct areas of the graduate curriculum; one year residency; no language requirement. All requirements must be completed within eight consecutive years.

Table B—Appointments to Graduate Students, 2009–10

Title of Appointee	Appointments Total	Appointments First year	Academic Load Allowed in Credit Hours	Hours of Service Per Week	Stipend for Academic Year ($)
			Semester		
Teaching Assistant	25	8	Up to 12 as needed	12	19,000[1,2]
Research Assistant	23	0	Up to 12 as needed	12[4]	20,000[1,3]
Total	48	8			

[1]Tuition waived.
[2]Summer appointments guaranteed after first year, ~stipend $1,500/mo.
[3]Some summer appointments available, stipend up to $2,500/month.
[4]Nominal figure. Thesis work is often included which makes the number of hours of research higher. *

5. Personnel Engaged in Separately Budgeted Research, 6/09–5/10

Professorial faculty	14
Postdoctoral appointments	12
Graduate students	32
Non-teaching research personnel	6
Total	64

6. Separately Budgeted Research Expenditures by Source of Support

	Departmental Research	Physics-related Research Outside Department
Federal Gov.	$866,183	$
Other	990,889	33,000
Total	$1,857,072	$33,000

Table C—Separately Budgeted Research Expenditures

Research Specialty	No. of Grants	Expenditures ($)
Condensed Matter Physics	25	1,600,000
Plasma Physics	2	220,000
Total	27	1,820,000

FACULTY

Professors

Naughton, Michael J., Ph.D., Boston University, 1986. Chairman of the Department, Ferris Professor. Experimental condensed matter, correlated electron, nanoscale material, and energy physics.

Bedell, Kevin S., Ph.D., SUNY Stony Brook, 1979. Rourke Professor. Theoretical condensed matter physics.

Broido, David A., Ph.D., University of California at San Diego, 1985. Theoretical condensed matter physics.

Di Bartolo, Baldassare, Ph.D., MIT, 1964. Solid state spectroscopy; flash photolysis of gases and liquids.

Graf, Michael J., Ph.D., Brown University, 1987. Experimental condensed matter physics at low temperatures.

Kempa, Krzysztof, Ph.D., University of Wroclaw, 1980. Theoretical condensed matter physics.

Ren, Zhifeng, Ph.D., Chinese Academy of Sciences, 1990. Experimental condensed matter physics, materials physics, energy physics.

Wang, Ziqiang, Ph.D., Columbia University, 1989. Theoretical condensed matter physics.

Associate Professors

Engelbrecht, Jan R., Ph.D., University of Illinois at Urbana, 1993. Theoretical condensed matter physics, biological physics.

Madhavan, Vidya, Ph.D., Boston University, 2000. Experimental condensed matter physics.

Padilla, Willie J., Ph.D., University of California at San Diego, 2004. Experimental condensed matter physics.

Uritam, Rein A., Ph.D., Princeton University, 1968. Particle theory; history and philosophy of science.

Assistant Professors

Opeil, Cyril P., S. J., Ph.D., Boston College, 2004. Experimental condensed matter physics.

Ran, Ying, Ph.D., Massachusetts Institute of Technology, 2007. Theoretical condensed matter physics, correlated electrons.

Wilson, Stephen D., Ph.D., University of Tennessee, 2007. Experimental condensed matter physics, neutron diffraction.

Distinguished Research Professors

Bakshi, Pradip, Ph.D., Harvard, 1962. Mathematical physics; theoretical plasma physics; quantum field theory.

Kalman, Gabor, D.Sc., Israel Inst. of Tech., 1961. Theoretical plasma physics; many-body physics; astrophysics.

Research Associate Professor/Laboratory Director

Herczynski, Andrzej, Ph.D., Lehigh University, 1987. Fluid dynamics.

Instrumentation Facilities

McMahon, Gregory, Ph.D., University of Saarland, 1994. Focused Ion Beam Facility Manager.

Shepard, Stephen, B.S., Framingham State College. Nanofabrication Cleanroom Facility Manager.

Wang, Dezhi, Ph.D., Boston College, 2007. Electron Microscopy Facility Manager. Materials Physics.

RESEARCH SPECIALTIES AND STAFF

Experimental

Condensed Matter. Low temperature physics, superconductivity, high-T_c materials, heavy-fermion systems, molecular organic conductors, physics in strong magnetic fields, neutron scattering, high resolution STM and STS; fluorescence spectroscopy, nanoscale manipulation of light, nanomaterials for energy, metamaterials. Di Bartolo, Graf, Madhavan, Naughton, Opeil, Padilla, Ren, Wilson.

Energy. Solar photovoltaics; thermoelectric energy conversion; metamaterials for energy conversion. Kempa, Naughton, Opeil, Padilla, Ren.

Materials. Synthesis, characterization and application of insulating, semiconducting, conducting and superconducting nanoscale materials (nanotubes, nanowires, nanoparticles, etc.); thermoelectric materials; photovoltaic materials; metamaterials. Naughton, Opeil, Padilla, Ren, Wilson.

Theoretical

Condensed Matter. Strongly correlated electron systems, high temperature superconductivity, heavy fermion systems, Fermi liquid theory, localized Fermi liquids, quantum Hall effect, metal-insulator transition, electronic, optical and transport properties of nanoscale systems, electromagnetic response of metals, surfaces, semiconductor heterostructures, topological insulators, quantum wells, wires and dots; current driven instabilities of solid state plasmas. Bakshi, Bedell, Broido, Engelbrecht, Kalman, Kempa, Ran, Wang.

History and Philosophy. Selected topics in the foundations of physics, and in the history and philosophy of science. Uritam.

Other Theoretical/Math. Vacuum polarization and electron gas in strong magnetic field; relativistic many body systems in compact astrophysical objects; nonperturbative operator techniques for strongly interacting systems; nonlinear fluctuations-dissipation theorems; nonlinear response functions. Bakshi, Herczynski, Kalman.

Plasma Physics. Strongly coupled plasmas; plasma response function; dense plasmas; plasma phase transitions; dusty plasmas; plasma instabilities in inhomogeneous systems. Bakshi, Kalman.

BOSTON UNIVERSITY

DEPARTMENT OF PHYSICS

Boston, Massachusetts 02215

Students Accepted For Degree	FIELDS		
	Physics	Astronomy	Related Fields
Doctorate	X		X
Master's			X

1. General

President: Robert A. Brown
Dean of CAS: Virginia Sapiro
Associate Dean of Graduate School: J. Scott Whitaker
Chairman: Claudio Rebbi
Director of Undergraduate Studies: Martin Schmaltz
Director of Graduate Studies: James Stone
Department Telephone Number: (617) 353-2600
Type of Institution: University
Control: Private
Setting: Urban
Total Faculty: 4,178
Total Students: 32,557
Total Graduate Students: 13,497
Annual Graduate Tuition:
All Graduate Students: Full-time—$39,314
Tuition rates for: 2009–10
Deferred tuition plan: Yes
Other Fees: GSU $194/year ($97/sem
Health fees $208/year ($104/sem)
Term: Semester

2. Number of Faculty in Department

The combined total of full-time faculty in the three professorial ranks is 38. There are also 4 research faculty, 19 faculty with joint appointments and 25 postdoctoral fellows.

3. Admission, Financial Aid, and Housing

Address admission inquiries to: Graduate Admissions, Physics Department
Graduate application fee required: $70
Admission deadline: (Fall admission) 12/15/10
Admission requirements: For admission to the graduate programs, a Bachelor's degree in physics or astronomy is required. Exceptional candidates from other fields will be considered. Official test results of the Graduate Record Examination (GRE) (General Test and Advanced Subject in Physics) are required. The minimum acceptable score for admission is dependent on the applicant's overall record. Official results of the Test of English as a Foreigh Language (TOEFL) are required of all applicants whose native language is not English. The minimum score requirement is 213 (computer-based test), 550 (paper-based test) or 84 (internet-based test).
Undergraduate preparation assumed: Symon, *Classical Mechanics*; Griffiths, *Electromagnetism*; Tittel, *Thermal Physics*; Liboff, *Quantum Physics*. Students are normally expected to have taken junior-senior courses in mechanics, electromagnetism, quantum or modern physics, and thermal physics.
Address financial aid inquiries to: Graduate School, Financial Aid Office, 705 Commonwealth Avenue, Boston, MA 02215
Financial aid deadline: 12/15/10
Loans available: Yes
Address housing inquiries to: Rental Property Mgmt., Boston Univ.
On-campus, single student housing available: No
Cost/year: approx. $8,400

Table A—Faculty, Enrollments, and Degrees Granted

Research Specialty	2009–10 Faculty	Enrollment[1] Fall 2009		No. of Degrees Granted[2] 2009–10 (2006–10)			Median No. of Years for 2009–10 Ph.D.'s
		Master's	Doctorate	Master's	Terminal Master's	Doctorate	
Bio/polymer	5	–	11	0(3)	0(2)	0(3)	–
Computer Science	1	–	1	0(0)	0(0)	0(1)	–
Condensed Matter Physics (CME)	7	–	36	1(13)	0(5)	2(17)	6
Particles/Fields (HEE)	16	–	13	0(7)	0(4)	2(9)	6.5
Stat/Thermal (CMT)	8	–	41	3(12)	0(0)	9(35)	5.5
Other Theoretical/ Math (HET)	0	–	6	0(0)	0(0)	0(1)	–
Education research	1	–	0	0(0)	0(0)	0(0)	–
Total	38		108	4(35)	0(11)	13(66)	5.5
Full-time Grad. Stud.			108				
Part-time Grad. Stud.			0				
First-year Grad. Stud.			18				
Median Years in Grad. Study (2009–10 Degrees)							6

Undergraduate Degrees, 2009–10 (2006–10): 18*(100**)
*8 astronomy & physics majors
**29 astronomy & physics majors

[1]Students not yet committed to a research specialty are entered under non-specialized.
[2]Five-year totals in parentheses.

4. Graduate Degree Requirements

Master's: Eight semester courses (32 credits) required including Advanced Lab, Mathematical Physics, Statistical Physics and Thermodynamics I, Electrodynamics I, Math Methods, Quantum Mechanics I and II, and one elective course. A passing grade on the departmental comprehensive exam or a Master's thesis. All students must complete a "Scholarly Methods in Physics" course. Each student must satisfy a residency requirement of a minimum of two consecutive regular semesters of full-time graduate study at Boston University.
Doctorate: Eight semester courses (32 credits) beyond those used to fulfill the Master's degree requirements. These must include Advanced Lab, if not already taken to fulfill Master's requirements, and at least five lecture courses numbered between 500 and 850. Up to three non-lecture courses may count toward the total of eight courses, but no more than one directed study course and one seminar course. The five lecture courses must include at least two distribution courses

from the category outside the student's area of specialization. (Category I includes elementary particle and mathematical physics and Category II includes biological physics and condensed matter physics.) Passing with an Honors grade on the departmental comprehensive exam; preliminary oral exam; a departmental seminar; a dissertation and a dissertation defense. Each student must satisfy a residency requirement of a minimum of two consecutive regular semesters of full-time graduate study at Boston University.

Other Programs: Interdisciplinary Ph.D. is also available with many other departments, including Astronomy, Mathematics, Biology, Chemistry, and with departments in the College of Engineering.

Thesis: Thesis may be written *in absentia*.

Special Equipment, Facilities, or Programs: A Scientific Instrument Facility employing five experimental machinists, and an experimental machinist/high vacuum welder. The facility houses a variety of CNC machine tools interfaced to a state-of-the-art CAD/CAM system, manual lathes, milling machines, and grinders, high-vacuum welding equipment, leak detection and precision measurement equipment. The facility also has an assembly expert for fabrication and assembly of complex parts. Electronics Design Facility employing four electrical engineers with extensive CAE tools for analog, digital, and mixed-signal circuit design. Nanoscale Research Facility and Optical Processing Facility with e-beam nanolithography and associated thin films deposition and etching equipment.

Research laboratories include: UHV elastic and inelastic HE surface scattering laboratory; electronic structure laboratory using high-resolution electron and photon spectroscopies; time-resolved x-ray scattering laboratory; low-temperature scanning-optical-microscopy laboratory to probe confined electronic systems; MBE installation for creating wide-bandgap semiconductors and engineered immersion surfaces; nano-optics laboratory with near-field solid immesion and subcellular flouroscent microscopies. Submicron UV photolithography and electron-beam lithographic nano-fabrication facilities. Nanoscale transport laboratory with He^3 and dilution fridge; low temperature, high-frequency and high magnetic field facilities; molecular biophysics and polymer labs with x-ray diffraction, electron, atomic force, confocal, and nearfield scanning probe microscopies, and Fourier transform infrared, Raman, UV-visible and ultrafast laser spectroscopies.

Computer Facilities: An extensive network of computational facilities supports the research activities of the Department. There are networked multiprocessor DELL available to departmental faculty, staff, and students. Additional Unix and Linux servers and workstations, as well as many Windows PC's, are available to research groups. For computationally intensive applications, students have access to supercomputing resources supported through the Center for Computational Science and the Office of Information Technology. At the high end, these currently consist of an IBM BlueGene supercomputer with a peck capacity of 5.7 Tflops (5.7 trillion floating point operations per second), a cluster of IBM pSeries parallel, shared-memory computers, an IBM BladeCenter, and an Intel based Linux cluster. These resources are integrated in a well-endowed distributed computing and visualization environment, which includes a high resolution, stereographic display wall, a laboratory for immersive virtual environments, an Access Grid Conference Facility, the Computer Graphics Laboratory, Myrinet, Gigabit Ethernet, and Fast Ethernet networks. A vast and diverse array of optical fiber connections to the NoX, Metro Ring and commercial ISPs provide multiple Gb/s of bandwidth and connectivity to the Internet, Internet2, and international research networks. The Departmental Computer Facility supports a wide range of software applications for physics data collection, analysis, simulation, and visualization.

Table B—Appointments to Graduate Students, 2009–10

Title of Appointee	Appointments Total	Appointments First year	Academic Load Allowed in Credit Hours	Hours of Service Per Week	Stipend for Academic Year ($)
			Semester		
Teaching Assistant	30	18	12	20	18,400
Research Assistant	73	0	12	20	–
Total	103	18			

5. Personnel Engaged in Separately Budgeted Research, 7/09–6/10

Professorial faculty	38
Postdoctoral appointments	25
Graduate students	73
Undergraduate students	28
Nonteaching research personnel	3
Total	167

6. Separately Budgeted Research Expenditures by Source of Support

	Departmental Research
Federal government	$6,529,107.32
Private, non-profit organizations	393,407.25
Total	$6,922,514.57

Table C—Separately Budgeted Research Expenditures

Research Specialty	No. of Grants	Expenditures ($)
Biophysics	7	313,036.88
Condensed Matter Physics	33	2,070,876.30
Particles & Fields	30	3,858,267.49
Science Education	5	680,333.90
Total	75	6,528,517.96

FACULTY

Professors

Ahlen, Steven P., Ph.D., Univ. of California, Berkeley, 1976. Experimental astrophysics; heavy-ion physics; monopole and quark searches.

Bansil, Rama, Ph.D., Univ. of Rochester, 1975. Biological Physics; polymers.

Bigio, Irving, Ph.D., Univ. of Michigan, 1974 (Joint appointment with Engineering). Biophysics.

Booth, Edward C., Emeritus; Ph.D., Johns Hopkins Univ., 1955. Intermediate-energy nuclear physics.

Brecher, Kenneth, Ph.D., Massachusetts Institute of Technology, 1969 (Joint appointment with Astronomy). Theoretical astrophysics; relativity; cosmology.

Butler, John M., Ph.D., Stanford Univ., 1986. Experimental particle physics; hadron collider physics.
Campbell, David, Ph.D., Cambridge Univ., 1970. Provost. (Joint appointment with Engineering). Condensed matter theory.
Castro Neto, Antonio H., Ph.D., Univ. of Illinois at Urbana-Champaign, 1994. Theoretical condensed matter physics.
Chamon, Claudio, Ph.D., Massachusetts Institute of Technology, 1996. Theoretical condensed matter physics..
Chasan, Bernard, Emeritus; Ph.D., Cornell Univ., 1961. Biological Physics.
Cohen, Andrew G., Ph.D., Harvard Univ., 1986. Mathematical physics; high-energy theory.
Cohen, Robert S., Emeritus; Ph.D., Yale Univ., 1948. Philosophical and historical foundations of physics.
Corinaldesi, Ernesto, Emeritus; Ph.D., The University of Manchester, England, 1951. Quantum mechanics.
DeLisi, Charles, Ph.D., New York Univ., 1969 (Joint appointment with Engineering). Biophysics, biomedical engineering.
De Rújula, Alvaro, Ph.D., University of Madrid, 1968 (Joint appointment with CERN). Theoretical particle physics; phenomenology.
Edmonds, Dean S., Jr., Emeritus; Ph.D., Massachusetts Institute of Technology, 1958. Electronics and instrumentation.
El-Batanouny, Maged, Ph.D., Univ. of California at Davis, 1978. Surface physics; solitons.
Erramilli, Shyamsunder, Ph.D., Univ. of Illinois, 1986. Experimental Biological Physics.
Evans, Evan, Ph.D., Univ. of California, San Diego, 1970 (Joint appointment with Engineering). Biological Physics.
Franzen, Wolfgang, Emeritus; Ph.D., Univ. of Pennsylvania, 1949. Atomic physics; surface physics.
Glashow, Sheldon, Ph.D., Harvard Univ., 1958. Arthur G. B. Metcalf Professor of Science (Joint appointment with the University Professors). Theoretical particle physics.
Goldberg, Bennett B., Ph.D., Brown Univ., 1987. Experimental condensed matter physics.
Hellman, William S., Emeritus; Ph.D., Syracuse Univ., 1961. Elementary particle theory; acoustics.
Kearns, Edwards, Ph.D., Harvard Univ., 1990. Experimental particle physics.
Klein, William, Ph.D., Temple Univ., 1972. Condensed matter theory.
Kreimer, Dirk, Ph.D., Johannes Gutenberg University, 1992 (Joint appointment with Mathematics). Mathematical physics.
Lane, Kenneth D., Ph.D., Johns Hopkins Univ., 1970. Theoretical high-energy physics.
Ludwig, Karl, F. Jr., Ph.D., Stanford Univ., 1986. Experimental condensed matter physics.
Miller, James P., Ph.D., Carnegie-Mellon Univ., 1974. Intermediate- and high-energy experimental physics; muon g-2.
Moustakas, Theodore, Ph.D., Columbia Univ., 1974 (Joint appointment with Engineering). Synthetic novel materials.
Pi, So-Young, Ph.D., SUNY (The State University of New York), Stony Brook, 1974. Theoretical field theory; elementary particles; theoretical astrophysics.
Rebbi, Claudio, Ph.D., International University College of Turin, 1967. Chairman. Theoretical physics; lattice quantum chromodynamics; computational physics.
Redner, Sidney, Ph.D., Massachusetts Institute of Technology, 1977. Acting Chairman of the Department. Statistical physics; stochastic processes.
Roberts, B. Lee, Ph.D., College of William and Mary, 1974. Intermediate- and high-energy experimental physics; muon g-2; CP violation.
Rohlf, James, Ph.D., Caltech., 1980. Experimental particle physics; hadron collider physics.
Rothschild, Kenneth, Ph.D., Massachusetts Institute of Technology, 1974. Biophysics; molecular electronics; physics of vision.
Sergienko, Alexander; Ph.D., Moscow State University, 1987 (Joint appointment with Engineering). Quantum Communications and Photonics.
Shimony, Abner, Emeritus; Ph.D., Yale Univ., 1953, Princeton, 1962. Philosophical and historical foundations of physics; theoretical quantum mechanics.
Skocpol, William J., Emeritus; Ph.D., Harvard Univ., 1974. Condensed matter experimental.
Smith, Kevin, Ph.D., Yale Univ., 1988. Experimental condensed matter physics.
Stachel, John J., Emeritus; Ph.D., Stevens Inst. of Tech., 1962. General relativity; foundations of relativistic space-time theories.
Stanley, H. Eugene, Ph.D., Harvard Univ., 1967. University Professor. Biophysics; polymer physics.
Stone, James L., Ph.D., Univ. of Michigan, 1977. Experimental particle physics and astrophysics; neutrinos; proton decay; monopole studies. Director of Graduate Studies.
Sulak, Lawrence, Ph.D., Princeton Univ., 1970. Experimental particle physics; proton decay; monopoles; muon g-2; neutrinos.
Teich, Malvin C., Ph.D., Cornell Univ., 1966 (Joint appointment with Engineering). Photonics/Lasers.
Whitaker, J. Scott, Ph.D., Univ. of California, Berkeley, 1976. Associate Dean. Experimental colliding-beam physics; supersymmetric particle searches.
Willis, Charles R., Emeritus; Ph.D., Syracuse Univ., 1957. Biophysics; nonlinear physics; statistical physics.
Zimmerman, George O., Emeritus; Ph.D., Yale Univ., 1963. Low-temperature physics; magnetism.

Associate Professors

Carey, Robert, Ph.D., Harvard Univ., 1989. Experimental particle and astro physics.
Meller, Amit, Ph.D., Weizmann Institute of Science, 1998. Biological Physics.
Mertz, Jerome, Ph.D., Univ. of Paris VI, Univ. of California, Santa Barbara, 1991 (Joint appointment with Engineering). Biophysics.
Mohanty, Pritiraj, Ph.D., Univ. of Maryland, College Park, 1998. Experimental condensed matter physics. Associate Prof.
Sandvik, Anders, Ph.D., Univ. of California, Santa Barbara, 1993. Condensed matter computational physics.
Tsui, Ophelia, Ph.D., Princeton Univ., 1996. Experimental Condensed matter.
Schmaltz, Martin, Ph.D., Univ. of California, San Diego, 1995. Director of Undergraduate Studies. Theoretical particle physics.
Swan, Anna, Ph.D., Boston Univ., 1993. (Joint Appt. w. ENG) ECE

Assistant Professors

Averitt, Richard, Ph.D., Rice Univ., 1998. CME.
Bose, Tulika, Ph.D., Columbia Univ. High Energy Physics.
Duffy, Andrew, Ph.D., Queen's University, Kingston, Ontario, Canada, 1995.

Katz, Emanuel, Ph.D., Massachusetts Institute of Technology, 1996. Theoretical high-energy physics.

Polkovnikov, Anatoli, Ph.D., Yale Univ., 2003. Condensed matter theory.

Research Faculty

Hong, Mi Kyung, Ph.D., Univ. of Illinois, 1988. Experimental biophysics.

Ivanov, Plamen, Ph.D., Boston Univ., 1998. Polymer physics.

Krapivsky, Paul, Ph.D., Moscow Physical Technical Institute, 1991. Theoretical Condensed matter physics.

Shank, James, Ph.D., Univ. of California, Berkeley, 1988. High-energy physics. CCS (primary appt.) Physics (secondary appt.)

Yousef, Saul, Ph.D., Carnegie Mellon Univ., 1992. High energy physics. 1994.

RESEARCH SPECIALTIES AND STAFF

Theoretical

Condensed Matter Physics. Strongly interacting electron systems. Low dimensional quantum magnetism and quantum antiferromagnets. High-temperature superconductors and heavy electron systems. Fractional Quantum Hall effect. Surface physics; solitons on surfaces. Structural and vibrational properties of adsorbed atomic layers. Equilibrium and non-equalibrium properties of interacting many particle atomic systems. Quantum Monte Carlo algorithms Graphene. Campbell, Castro , Neto, Chamon, Polkovnikov, Sandvik, and 4 postdoctoral fellows.

Elementary Particle Physics and Cosmology. Physical origin of electroweak and flavor symmetry breaking, including theoretical and phenomenological studies of technicolor, little higgs, extra dimensions and supersymmetry. Quantum chromodynamics. Collider phenomenology. Numerical simulations of lattice gauge theories. Fundamental studies of quantum field theory. Theoretical astrophysics and cosmology; dark matter, inflation, baryogenisis, and the formation of large scale structure. Brower, Cohen, Glashow, DeRujula, Katz, Lane, Pi, Rebbi, Schmaltz, and 5 postdoctoral fellows.

Statistical Physics. Kinetics of phase transitions and coarsening processes. Chemical reactions, stochastic processes, and the role of spatial fluctuations. Structure of heterogeneous networks and network dynamics. Applications to population biology models. Dynamics of social systems. Mechanisms of nucleation and spinodal decomposition. Physics of disordered media; percolation models of disordered materials. Structure of diffusion limited aggregation and viscous fingering. Fractals and multifractals. Hydrogen-bonded network formation in liquid water. Acceleration algorithms for Monte Carlo simulations. First-passage and its applications. Theoretical studies of polymers. Klein, Redner, Stanley, and 8 postdoctoral fellows.

Experimental

Astrophysics. Atmospheric and solar neutrino studies and neutrino astrophysics with the Super-K experiment. Dark matter searches. Ahlen, Kearns, Stone, Sulak, and 2 postdoctoral fellows.

Biological Physics. Energy transduction, ion transport, and signal receptor studies of microbial and vertebrate rhodopsins and their complexes by FTIR, resonance Raman and laser flash spectroscopy. Biomembrane technology and molecular electronics. Structure and electrical properties of membranes. Ultrafast vibrational spectroscopy, STM, AFM imaging of macromolecular assemblies, membrane surfaces, and protein-lipid interactions. Gelation of mucin and mucus. Novel Flourescent Imaging for subcellular microscopy. Dynamics of DNA. Signaling and Information Porcessing in Biochemicals networks. Bansil, Erramilli, Mehta, Rothschild, and 3 postdoctoral fellows.

Condensed Matter Physics. Mesoscopic phenomena; quantum transport and quantum coherence phenomena in nanostructures. Advanced electronic materials. Nano-optics and spectroscopy of quantum dots, photonic bandgap systems, and carbon nanotubes. Single molecule spectroscopy and subcellular imaging. Studies of structural phase transitions in thin-film semiconductors. Synchrotron x-ray scattering studies of kinetics of nucleation, spinodal decomposition, and phase transitions. Growth of artificially structured materials using molecular beam epitaxy, sputtering, and chemical vapor deposition. Properties of high-Tc superconductor-normal interfaces. X-ray emission and photoemission studies of wide band gap semiconductors, organic superconductors, and low dimensional transition metal oxides. Terahertz spectroscopy and time-integrated and time-resolved optical spectroscopy of correlated electron materials. Goldberg, Ludwig, Mohanty. Averitt, Smith, 3 postdoctoral fellows.

Elementary Particle Physics. Investigations of electroweak interactions and top quark production, searches for Higgs boson, and physics beyond the standard at the LHC. Study of neutrino properties using long baseline neutrino oscillation. Study of grand unified theories using proton decay. Ahlen, Bose, Butler, Kearns, Rohlf, Stone, Sulak, and 6 postdoctoral fellows.

Medium-Energy Physics. Precision measurements of the anomalous part of the muon magnetic dipole moment. Precision measurement of the muon lifetime and the Fermi constant. New limit on muon to electron conversion. New limit on the electric dipole moment of the muon and deuteron. Carey Miller, Roberts, and 2 postdoctoral fellows.

Polymer Physics. X-ray, light scattering, and microscopy studies of the structure, dynamics and phase separations of gels, block copolymers, and polymer thin films and interfaces. Bansil, Tsui, and 2 postdoctoral fellows.

Surface Physics. Neutral helium beam scattering studies of the dynamical properties of surfaces and structural phase transitions. Surface quantum motion of hydrogen. Metastable He scattering studies of magnetic properties of surfaces. High resolution photoemission and x-ray emission studies of metals, semiconductors, and oxides. El-Batanouny, Smith.

BOSTON UNIVERSITY

DEPARTMENT OF ASTRONOMY

Boston, Massachusetts 02215

Students Accepted For Degree	FIELDS		
	Physics	Astronomy	Related Fields
Doctorate		X	
Master's		X	

1. General

President: Robert A. Brown
Dean of Graduate School: : Virginia Sapiro
Associate Dean of Graduate School: J. Scott Whitaker
Department Chairman: James Jackson
Department Telephone Number: (617) 353-2625
Type of Institution: University
Control: Private
Setting: Urban
Total Faculty: 3,622
Total Students: 31,766
Total Graduate Faculty: 567
Total Graduate Students: 13,232
Annual Graduate Tuition:
All Graduate Students: Full-time—$37,910
Part-time—$1,184/credit
Tuition rates for: 2009–10
Deferred tuition plan: Yes
Annual Other Fees: $190 Student Union fee; $80 registration fee; $196 health insurance fee
Term: Semester

2. Number of Faculty in Department

The combined total of full-time faculty in the three professorial ranks is 16. The combined total of full-time, part-time, and other faculty at all ranks is 22.

3. Admission, Financial Aid, and Housing

Address admission inquiries to: Graduate Admissions Committee, Astronomy Department, 725 Commonwealth Avenue
Graduate application fee required: $70
Admission deadline (Fall admission): 1/15; late applications accepted
Admission information: For fall admission, 2009–10, 19 students were accepted from 64 applicants.
Admission requirements: For admission to the graduate programs a Bachelor's degree in physics or astronomy is required with a minimum cumulative undergraduate GPA equivalent to B or higher. The GRE is required. The minimum acceptable score suggested for admission is not specified. The GRE Subject Test in Physics is required. The minimum acceptable score suggested for admission is dependent on the applicant's overall record. Students from non-English speaking countries are required to demonstrate proficiency in English via the TOEFL exam. Minimum acceptable score for admission is: computer-based 213.
Address financial aid inquiries to: GRS Admissions Room 112, 725 Commonwealth Avenue
GAPSFAS application required: Yes
Financial aid deadline: 1/15
Loans available: Yes
Address housing inquiries to: Housing Director, Boston Univ.

Table A—Faculty, Enrollments, and Degrees Granted

Research Specialty	2009–10 Faculty	Enrollment[1] Fall 2009 Master's	Enrollment[1] Fall 2009 Doctorate	No. of Degrees Granted[2] 2009–10 (2004–10) Master's	Terminal Master's	Doctorate	Median No. of Years for 2009–10 Ph.D.'s
Astronomy	0	0	5	1(1)	1(1)	0(0)	N/A
Astrophysics	9	0	13	3(9)	1(7)	5(15)	N/A
Atmos./Space Phys. Cosmic Rays	13	0	21	3(12)	0(9)	2(11)	N/A
Total		0	39	7(22)	2(17)	7(26)	
Full-time Grad. Stud.		0	39				
Part-time Grad. Stud.		0	0				
First-year Grad. Stud.		0	6				
Median Years in Grad. Study (2009–10 Degrees)				2	2	5	5
Undergraduate Degrees, 2009–10 (2004–10): 11(104)							

[1] Students not yet committed to a research specialty are entered under Astronomy.
[2] Seven-year totals in parentheses. Degrees awarded in Astronomy only: areas of research indicated here.

4. Graduate Degree Requirements

Master's: Eight semester courses passed with grades of B^- or better. A Master's thesis or a passing grade on the Departmental comprehensive exam is required for a degree in Astronomy. The residence requirement is a minimum of two consecutive semesters of full-time graduate study at Boston University.
Doctorate: Eight semester courses beyond the Master's degree passed with grades of B^- or better; a passing grade on the Departmental comprehensive exam; an oral exam; a dissertation and a dissertation defense. Each student must satisfy a residence requirement of two consecutive semesters of full-time graduate study at Boston University.
Thesis: Thesis may be written *in absentia*.
Special Equipment, Facilities, or Programs: The department shares use of the "Perkins 72" telescope with Lowell Observatory. Departmental facilities include an in-house observatory with 14″, 10″, and 8″ telescopes, a library, and a network of over 200 computers. Individual research groups are involved in instrumentation projects to fabricate equipment used in ground and space based observations.

Table B—Appointments to Graduate Students, 2009–10

Title of Appointee	Appointments Total	Appointments First year	Academic Load Allowed in Credit Hours	Hours of Service Per Week	Stipend for Academic Year ($)
			Semester		
Teaching Assistant	6	6	16	15	18,400
Research Assistant	39	1	16	15	18,400
Total	39	7			

5. Personnel Engaged in Separately Budgeted Research, 7/08–6/09

Professorial faculty	16
Graduate students	39
Undergraduate students	53
Nonteaching research personnel	6
Total	114

6. Separately Budgeted Research Expenditures by Source of Support (2009–10)

	Departmental Research
Federal government	$21,271,716
Total	$21,271,716

Table C—Separately Budgeted Research Expenditures

Research Specialty	No. of Grants	Expenditures ($)
Astrophysics	29	1,361,267
Atmos./Space Phys. Cosmic Rays	46	19,910,449
Total	75	21,271,716

FACULTY

Professors

Bania, Thomas M., Ph.D., Virginia, 1977. Radiospectroscopy; galactic structure; interstellar medium.
Brecher, Kenneth, Ph.D., MIT, 1969. Theoretical high-energy astrophysics; relativity and cosmology, neutron stars.
Chakrabarti, Supriya, Ph.D., Berkeley, 1982. Atmospheric and ionospheric physics; experimental astrophysics.
Clarke, John, Ph.D., Johns Hopkins, 1980. Planetary atmospheres; UV astrophysics; FUV instruments for remote observations.
Clemens, Dan P., Ph.D., Massachusetts, 1985. Infrared, and optical astronomy; interstellar medium; galactic structure; star formation.
Fritz, Theodore A., Ph.D., Iowa, 1967. Space plasma physics; magnetospheric physics; solar wind; rocket and satellites.
Hughes, W. Jeffrey, Ph.D., London, 1974. Magnetospheric physics; space physics.
Jackson, James, Ph.D., MIT, 1986. Department Chairman. Radio, infrared, and gamma-ray astronomy; interstellar medium; starburst galaxies; star formation; the Milky Way; Antarctic astronomy.
Janes, Kenneth A., Ph.D., Yale, 1972. Galactic structure; photometry and spectroscopy; star clusters.
Marscher, Alan P., Ph.D., Virginia, 1977. Extragalactic astrophysics; quasars and active galaxies; radio infrared, x-ray, and gamma-ray astronomy.
Mendillo, Michael, Ph.D., Boston, 1971. Space physics; solar system astronomy; planetary atmospheres.

Research Professors

Crooker, Nancy, Ph.D., UCLA, 1972. Space physics.
Goodrich, Charles, Ph.D., MIT, 1978. Space physics.
Lyon, John, Ph.D., University of Maryland, 1972. Space physics.
Siscoe, George, Ph.D., MIT, 1964. Space physics.
Quinn, Jack, Ph.D., UCSD, 1981. Magnetospheric physics.

Associate Professors

Brainerd, Tereasa G., Ph.D., The Ohio State University, 1992. Theoretical astrophysics; cosmology; computational astrophysics; galaxy formation and evolution; gravitational lensing.
Oppenheim, Meers, Ph.D., Cornell, 1995. Computational and theoretical space plasma physics; particle-wave interactions; meteor trails.
Schwadron, Nathan, Ph.D., Michigan, 1996. Solar wind; magnetic fields; energetic particle sources; cometary x-rays.

Research Associate Professor

Cook, Timothy, Ph.D., University of Colorado, 1991. Space physics.

Assistant Professor

Blanton, Elizabeth, Ph.D., Columbia, 2000. Galactic and radio astronomy.
West, Andrew, Ph.D., U. of Washington, 2005. Kinematics and magnetic activity of Mard L dwarfs; stat. formation in nearby galaxies.

RESEARCH SPECIALTIES AND STAFF

Theoretical

Astrophysics. High-energy astrophysics; neutron stars; quasars and active galaxies. Blanton, Brainerd, Brecher, Marscher.
Space Physics. Magnetospheric physics; space plasma physics; hydromagnetic waves. Crooker, Goodrich, Hughes, Lyon, Oppenheim, Siscoe.

Experimental

Atmospheric Physics. Terrestrial, planetary, and cometary atmospheres; aeronomy; Fabry-Perot interferometry. Chakrabarti, Cook, Mendillo, Schwadron.
Extragalactic Astronomy. Quasars and active galaxies; jets (radio, x-ray, gamma ray); starburst galaxies. Brainerd, Jackson, Marscher, Schwadron, West.
Galactic Astronomy. Chemical history of the galaxy; composition of stars; star clusters; interstellar medium; galactic structure; star formation (optical, radio, infrared). Bania, Blanton, Clemens, Jackson, Janes, Marscher, West.
Space Physics. Solar terrestrial relations; geomagnetic storms; plasma waves; ionospheric physics. Aarons, Basu, Mendillo, Hughes, Chakrabarti, Crooker, Fritz, Siscoe, Spence.

BRANDEIS UNIVERSITY

DEPARTMENT OF PHYSICS, MS 057

Waltham, Massachusetts 02454-9110

Students Accepted For Degree	FIELDS		
	Physics	Astronomy	Related Fields
Doctorate	X	X	
Master's			

1. General

President: Jehuda Reinharz
Dean Graduate School of Arts and Sciences: Malcolm Watson
Department Chairman: John Wardle
Department Telephone Number: (781) 736-2800
FAX Number: (781) 736-2915
Type of Institution: University
Control: Private
Setting: Suburban
Total Faculty: 356
Total Students: 3,317
Total Graduate Students: 872
Annual Graduate Tuition:
Full-time—$38,994:
Part-time—$19,497:
Tuition rates for: 2010–11
Deferred tuition plan: No
Other Fees: Student Insurance:
QSHIP: $1,607 Student
$6,735 Student + Spouse
$4,174 Student + Child
$9,302 Family
Tufts with prescription:
$6,912 Student
$13,834 Student + One
$20,736 Family
Term: Semester

2. Number of Faculty in Department

The combined total of full-time faculty in the three professorial ranks is 17.

3. Admission, Financial Aid, and Housing

Address admission inquiries to: Chairman, Graduate Admissions Committee, Physics Department, Mailstop 057
Graduate application fee required: $75 online/ $100 paper
Admission deadline (Fall admission): 1/15
Admission information: For fall admission, 2010–11, 20 students were accepted from 60 applicants.
Admission requirements: For admission to the graduate programs, a Bachelor's degree is required. No minimum undergraduate GPA specified. The GRE is required. No minimum acceptable score for admission is specified. The GRE Advanced is required. No minimum acceptable score for admission is specified. The average GRE scores for 2009–10 admissions were verbal–506; quantitative–677; total–1,183. Students from non-English speaking countries are required to demonstrate proficiency in English via the TOEFL exam (score of 600 or more required).
Undergraduate preparation assumed: Chabay and Sherwood, Matter & Interaction 3rd edition: Kleppner and Kolenkow, An Introduction to Mechanics; Purcell, Electricity & Magnetism; Young and Freedman, Univ. Physics with Modern Physics and Mastering Physics 12th edition; French, Vibrations and Waves; Simpson, Introduction to Electronics; Griffiths, Introduction to Electomagnetism; Griffiths, Introduction to Quantum Mechanics, plus topics from Shankar, Principles of Quantum Mechanics.
Address financial aid inquiries to: Graduate School
GAPSFAS application required: No
Financial aid deadline: N/A
Loans available: Yes
Address housing inquiries to: Associate Director, Office of Campus Life, USDAN/114
On-campus, single student housing available: Yes
On-campus, married student housing available: Very limited

Table A—Faculty, Enrollments, and Degrees Granted

Research Specialty	2009–10 Faculty	Enrollment[1] Fall 2009		No. of Degrees Granted[2] 2009–10 (2005–10)			Median No. of Years for 2009-10 Ph.D.'s
		Master's	Doctorate	Master's	Terminal Master's	Doctorate	
Astrophysics	2	0	3	0(3)	0(1)	0(0)	–
Biological Physics	3	0	9	0(0)	0(1)	0(0)	–
Biophysics	0	0	4	0(1)	0(2)	1(5)	5
Condensed Matter Physics	5	0	15	0(3)	1(2)	3(12)	6
Particles & Fields	7	0	7	0(5)	2(2)	0(2)	–
Non-specialized	0	0	2	0(1)	0(1)	0(0)	–
Total		0	40	0(13)	3(9)	4(19)	
Full-time Grad. Stud.		0	40				
Part-time Grad. Stud.		0	0				
First-year Grad. Stud.		0	8				
Median Years in Grad. Study (2009–10 Degrees)				6		–	–
Undergraduate Degrees, 2009–10 (2005–10): 9(57)							

[1]Students not yet committed to a research specialty are entered under non-specialized.
[2]Five-year totals in parentheses.

4. Graduate Degree Requirements

Master of Science: One year's residence as a full-time student. Six semester courses in physics numbered above 160 with a grade of at least "B-." A thesis on an approved topic may be accepted in place of a semester course. Satisfactory performance on the Qualifying Examination.
Doctorate: Three year's residence as a full-time student. Nine semester courses of advanced work in physics with a grade of at least "B." Outstanding performance on the Qualifying Examination. Passing of an Advanced Examination in topics related to the student's thesis subject. This examination will normally be taken after preparatory studies in the prospective field of research. Doctoral thesis and final oral examination.
Other Programs: Doctoral program is available in biophysics.
Thesis: Thesis may be written *in absentia*.

Table B—Appointments to Graduate Students, 2009–10

Title of Appointee	Appointments Total	First year	Academic Load Allowed in Credit Hours	Hours of Service Per Week	Stipend for Academic Year ($)[1,2]
			Semester		
Teaching Assistant	16	7	9–12	12	18,000[1]
Research Assistant	24	1	–	–	26,6000[2]
Total	40	8			

[1]$6,000 for 3 summer months.
[2]12 month stipend

5. Personnel Engaged in Separately Budgeted Research, 7/08–6/09

Professorial faculty	14
Postdoctoral appointments	6
Graduate students	24
Undergraduate students	17
Total	61

6. Separately Budgeted Research Expenditures by Source of Support

	Departmental Research	Physics-related Research Outside Department
Federal government	$3,103,746	$
Total	$3,103,746	$

Table C—Separately Budgeted Research Expenditures

Research Specialty	No. of Grants	Expenditures ($)
Astrophysics	2	139,649
Biophysics	1	94,772
Condensed Matter Physics	5	240,797
Elementary Particles	4	1,181,800
Materials Sci./Metallury	8	1,349,724
Relativity & Gravitation	3	156,250
Total	23	3,162,992

FACULTY

Professors

Bensinger, James R., Ph.D., Wisconsin, 1970 Experimental high-energy physics.

Blocker, Craig A., Ph.D., California, Berkeley, 1980. Experimental high-energy physics.

Chakraborty, Bulbul, Ph.D., SUNY, Stony Brook, 1979. Condensed matter theory.

Fraden, Seth, Ph.D., Brandeis, 1987. Liquid crystals, complex fluids, and multiple scattering.

Kirsch, Lawrence E., Ph.D., Rutgers, 1964. Experimental high-energy physics.

Kondev, Jané, Ph.D., Cornell, 1995. Condensed matter theory.

Meyer, Robert B., Ph.D., Harvard, 1970. Physics of liquid crystals; polymer and colloid physics.

Roberts, David H., Ph.D., Stanford, 1973. Experimental radio astronomy; theoretical astrophysics; space astronomy.

Schnitzer, Howard J., Ph.D., Rochester, 1960. Elementary particle theory; quantum theory of fields; string theories.

Wardle, John F. C., Ph.D., Manchester, England, 1969. Chairman. Experimental radio astronomy.

Associate Professors

Lawrence, Albion, Ph.D., Univ. Chicago, 1996. String theory.

Wellenstein, Hermann F., Ph.D., Texas, Austin, 1971. Experimental high-energy physics.

Assistant Professors

Baskaran, Aparna, Ph.D., Univ. Florida, 2006. Non-equilibrium statistical mechanics, biophysics.

Dogic, Zvonimir, Ph.D., Brandeis University, 2001. Complex fluids and biological physics.

Hagen, Michael F., Ph.D., University of California, Berkeley, 2003. Computational physics.

Headrick, Matthew, Ph.D., Harvard, 2003. String Theory.

Samadani, Azadeh, Ph.D., Clark University, 2002. Complex fluids and biological physics.

RESEARCH SPECIALTIES AND STAFF

Theoretical

Astrophysics. Roberts, Wardle.

Biophysics. Hagan, Kondev.

Condensed Matter. Baskaran, Chakraborty, Kondev. 2 postdoctoral fellows.

Elementary Particles and Fields. Supersymmetry; quantum gravity; strings; quantum theory of fields. Headrick, Lawrence, Schniter. 1 postdoctoral fellow.

Experimental

Astrophysics. Radio astronomy of galaxies and quasars. Roberts, Wardle.

Biological Physics. Dogic, Samadani. 1 postdoctoral fellow.

Condensed Matter. Dogic, Fraden, Meyer. 2 postdoctoral fellows.

Elementary Particle Physics. Collider experiments. Bensinger, Blocker, Kirsch, Wellenstein.

CLARK UNIVERSITY

DEPARTMENT OF PHYSICS

Worcester, Massachusetts 01610

Students Accepted For Degree	FIELDS		
	Physics	Astronomy	Related Fields
Doctorate	X		
Master's	X		

Address financial aid inquiries to: Department of Physics
GAPSFAS application required: No
Financial aid deadline: None
Loans available: Yes
Address housing inquiries to: Department of Physics
On-campus, single student housing available: Yes
On-campus, married student housing available: No

1. General

President: David Angel
Department Chairman: Arshad Kudrolli
Department Telephone Number: (508) 793-7169
URL: http://physics.clarku.edu
Type of Institution: University
Control: Private
Setting: Urban
Total Faculty: 187, full-time
Total Graduate Faculty: 187
Total Students: 3,120
Total Graduate Students: 906
Annual Graduate Tuition:
All Graduate Students: Full-time—$36,100
Part-time—$4,512.50/course
Tuition rates for: 2010–11
Deferred tuition plan: No
Other Fees: Health and accident insurance required: approximately $1,439 for single student. Waived on evidence of other insurance.
Term: Semester

Table A—Faculty, Enrollments, and Degrees Granted

Research Specialty	2009–10 Faculty	Enrollment[1] Fall 2009		No. of Degrees Granted[2] 2009–10 (2005–10)			Median No. of Years for 2009–10 Ph.D.'s
		Master's	Doctorate	Master's	Terminal Master's	Doctorate	
Condensed Matter Physics	4	1	9	3(6)	0(1)	1(1)	–
Nuclear Physics	0	0	0	0(0)	0(0)	0(1)	–
Physics Education	1	0	0	0(0)	0(0)	0(0)	–
Statistical & Thermal	1	0	1	0(0)	0(0)	0(2)	–
Total		1	10	3(6)	0(1)	1(4)	
Full-time Grad. Stud.		1	10				
Part-time Grad. Stud.		1	1				
First-year Grad. Stud.		1	2				
Median Years in Grad. Study (2009–10 Degrees)				0	4	5	–
Undergraduate Degrees, 2009–10 (2005–10): 3(19)							

[1] Students not yet committed to a research specialty are entered under non-specialized.

[2] Five-year totals in parentheses.

2. Number of Faculty in Department

The combined total of full-time faculty in the three professorial ranks is 4. The combined total of full-time, part-time, and other faculty at all ranks is 11.

3. Admission, Financial Aid, and Housing

Address admission inquiries to: Department of Physics
Graduate application fee required: $50
Admission deadline (Fall admission): February 1 (recommended)
Admission information: For fall admission, 2009–10, 3 students were accepted from 23 applicants.
Admission requirements: For admission to the graduate programs, a Bachelor's degree in physics, chemistry, mathematics, or engineering is required with a minimum undergraduate GPA of B- specified. The GRE is recommended. The GRE Advanced is recommended. There are no set minimum scores for GRE or GRE Advanced; each case is judged individually. Students from non-English speaking countries are required to demonstrate proficiency in English via the TOEFL exam. No minimum score required; each case is judged individually.

4. Graduate Degree Requirements

Master's: Total courses required: 8 (2 may be transferred, 1 may be for thesis); thesis (or Ph.D. candidacy) required. No language requirement; teaching experience required.
Doctorate: Course requirements: one year in residence beyond M.A. degree (8 courses) is required; 4 area qualification examinations and thesis proposal examination, teaching experience and dissertation are required.
Thesis: Thesis may be written *in absentia*.
Special Equipment, Facilities, or Programs:: The Department stresses research experience at the earliest possible time and encourages all students to enroll in a Research Apprenticeship during their first semester. Students work directly with several faculty members on specific research projects and can begin dissertation work substantially earlier than is usually the case. Clark's size affords a uniquely close association between faculty and students.

Table B—Appointments to Graduate Students, 2009–10

Title of Appointee	Appointments Total	Appointments First year	Academic Load Allowed in Credit Hours	Hours of Service Per Week	Stipend for Academic Year ($)[1,2,4]
			Semester		
Teaching Assistant	5	3	3	15[3]	13,710.00
Research Assistant	4	0	3	20[3]	13,710.00
Independently Supported Graduate Students	2	0	3	20[3]	-
Total	11	3			

[1]Tuition waived.
[2]Some summer appointments are available.
[3]Nominal figure. Thesis work is often included, which makes the number of hours of research higher.
[4]Twelve month stipend is $18,280.

5. Personnel Engaged in Separately Budgeted Research, 7/09–6/10

Professorial faculty	6
Graduate students	8
Undergraduate students	7
Total	21

6. Separately Budgeted Research Expenditures by Source of Support

	Departmental Research	Physics-related Research Outside Department
Federal government	$715,500	$
University research funds	$14,000	$
Total	$729,500	$

Table C—Separately Budgeted Research Expenditures

Research Specialty	No. of Grants	Expenditures ($)
Condensed Matter Physics	7	715,500
University research funds	3	14,000
Total	10	729,500

FACULTY

Professors

Agosta, Charles C., Ph.D., Duke, 1986. Properties of organic superconductors and other materials in high, pulsed magnetic fields.

Kudrolli, Arshad, Ph.D., Northeastern, 1995. Experimental nonlinear physics; granular matter and soft condensed matter.

Landee, Christopher P., Ph.D., Michigan, 1975. Experimental condensed matter physics; low-dimensional quantum magnetism; magnetochemistry.

Associate Professor

Mukhopadhyay, Ranjan, Ph.D., Caltech, 1998. Theoretical condensed matter physics; complex fluids; biological physics.

Faculty Emeriti

Andersen, Roy S., Ph.D., Duke University, 1952. Experimental condensed matter physics; molecular spectroscopy; history and philosophy of science.

Blatt, S. Leslie, Ph.D., Stanford, 1965. Experimental nuclear physics; college/pre-college curriculum development and teacher education.

Gould, Harvey A, Ph.D., California, Berkeley, 1966. Theoretical physics; statistical physics; computer simulation; phase transitions.

Hohenemser, Christoph, Ph.D., Washington (St. Louis), 1963. Experimental physics; nuclear risk assessment; technology assessment; energy and environmental issues.

Kohin, Roger P., Ph.D., University of Maryland, 1961. Experimental condensed matter physics, electron spin resonance studies.

RESEARCH SPECIALTIES AND STAFF

Theoretical

Biological Physics. Physical mechanisms underlying molecular and cellular processes. Membranes; biopolymers; mechanical, biomechanical, and genetic networks. Mukhopadhyay.

Condensed Matter Physics. Soft matter: including polymers, liquid crystals, and gels; complex and disordered systems. Mukhopadhyay.

Energy and Environment. Risk assessment; energy policy. Blatt.

Plasma Physics. Intense electron beams; generation of coherent radiation. Davies.

Statistical and Thermal Physics. Computer simulation; phase transitions. Gould.

Experimental

Granular matter, pattern formation, and soft matter. Kudrolli.

Solid State Physics. Magnetism; low dimensional conductors, organic superconductors, pulsed magnetic fields. Agosta, Landee.

Teaching Methods in Science. Blatt.

FACULTY PUBLICATIONS

Agosta, Charles C.

K. Cho, B. E. Smith, W. A. Coniglio, L. E. Winter, C. C. Agosta, and J. A. Schlueter, "Upper critical field study in the organic superconductor β''- $(ET)_2SF_5CH_2CF_2SO_3$: Possibility of the Fulde-Ferrell-Larkin-Ovchinnikov state," Phys. Rev. B **79**, 220507(R) (2009).

I. Mihut, C. C. Agosta, C. Martin, C. H. Mielke, T. Coffey, M. Tokumoto, M. Kurmoo, J. A. Schlueter, P. Goddard, and N. Harrison, "Incoherent Bragg reflection and Fermi surface hot spots in κ-$(BEDT\text{-}TTF)_2Cu(NCS)_2$," Phys. Rev. B **73**, 125118 (2006).

C. Martin, C. C. Agosta, S. W. Tozer, H. A. Radovan, E. C. Palm, T. P. Murphy, and J. L. Sarrao, "Evidence for the FFLO state in $CeCoIn_5$ from penetration depth measurements," Phys. Rev. B **71**, 020503 (2005).

C. Martin, C. C. Agosta, S. W. Tozer, H. A. Radovan, T. Kinoshota, and M. Tokumoto, "Critical field and Shubnikov-de Haas oscillations of κ-$(BEDT\text{-}TTF)_2Cu(NCS)_2$ under pressure," J. Low Temp. Phys. **138**, 1024 (2005).

T. Coffey, Z. Bayindir, J. F. DeCarolis, M. Bennett, G. Esper, and C. C. Agosta, "Measuring Radio Frequency Properties of Materials in Pulsed Magnetic Fields With A Tunnel

Diode Oscillator," Rev. Sci. Instr. **71**, 4600 (2000).

Blatt, S. Leslie

S. L. Blatt, "Complex Instruction in an Adult Context: Hands-On Science for Pre-Service and In-Service Teachers," *Proceedings of the International Conference on Cooperative Learning* (1996).

P. Nakroshis, S. L. Blatt, C. Landee, and H. Gould, "Order, Disorder, and Chaos: A New Course for Non-Science Majors," Bull. Am. Phys. Soc. **40**, 967 (1995).

S. L. Blatt, H. Gould, M. Gould, P. Nakroshis, C. Barton, and C. Landee, "Discovering Physics: A Successful Approach for Non-Science Students," Bull. Am. Phys. Soc. **39**, 1171 (1994).

S. L. Blatt and H. Gould, "Recent Fractal and Chaos Software Releases Hint at Future Educational Potential," (software review), Computers in Physics **6**, 702 (1992).

S. L. Blatt *et al.*, "Radiative Proton Capture to Excited States in ^{16}O," Phys. Rev. C **39**, 340 (1989).

Gould, Harvey A.

Harvey Gould and Jan Tobochnik, Statistical and Thermal Physics with Computer Applications (Princeton University Press, Princeton, NJ, 2010).

J. Xia, H. Gould, W. Klein, and J. B Rundle,"Near-mean-field behavior in the generalized Burridge-Knopoff earthquake model with variable-range stress transfer," Phys. Rev. E **77**, 031132 (2008).

H. Wang, H. Gould, and W. Klein, "Homogeneous and heterogeneous nucleation of Lennard-Jones liquids," Phys. Rev. E **76**, 031604 (2007).

W. Klein, H. Gould, N. Gulbahce, J. B. Rundle, and K. Tiampo, "The structure of fluctuations near mean-field critical points and spinodals and its implication for physical processes," Phys. Rev. E **75**, 031114 (2007).

H. Gould, J. Tobochnik, and W. Christian, *An Introduction to Computer Simulation: Applications to Physical Systems*, third edition, Addison-Wesley (2006).

Kudrolli, Arshad

D. Tam, J. W. M. Bush, M. Robitaille, and A. Kudrolli, "Tumbling dynamics of flexible wings," Phys. Rev. Lett. **104**, 184504 (2010).

A. Kudrolli, "Concentration dependent diffusion of self-propelled rods," Phys. Rev. Lett. **104**, 088001 (2010).

C. H. Rycroft, A. V. Orpe, and A. Kudrolli, "Physical test of a particle simulation model in a sheared granular system," Phys. Rev. E **80**, 031305 (2009).

K. Safford, Y. Kantor, M. Kardar, and A. Kudrolli, "Structure and dynamics of vibrated granular chains: Comparison to equilibrium polymers," Phys. Rev. E **79**, 061304 (2009).

D. M. Abrams, A. E. Lobkovsky, A. P. Petroff, K. M. Straub, B. McElroy, D. C. Mohrig, A. Kudrolli, and D. H. Rothman, "Growth laws for channel networks incised by groundwater flow," Nature Geoscience **2**, 193 (2009).

Landee, Christopher P.

J. L. White, C. Lee, Ö. Gunaydin-Sen, L. C. Tung, H. M. Chisten, Y. J. Wang, M. M. Turnbull, C. P. Landee, R. D. McDonald, S. A. Crooker, J. Singleton, M.-H. Whangbo, and J. L. Musfeldt, "Charge-spin Coupling in a Quantum Heisenberg Spin Ladder," Phys. Rev. B **81**, 104307 (2010).

H. Kuhne, H.-H. Klauss, S. Grossjohann, W. Brenig, F. J. Litterst, A. P. Reyes, P. L. Kuhns, M. M. Turnbull, and C. P. Landee, "Quantum critical dynamics of a S=1/2 antiferromagnetic Heisenberg chain studied by 13C-NMR spectroscopy," Phys. Rev. B **80**, 045110 (2009).

N. Tsyrulin, T. Pardina, R. R. P. Singh, F. Xiao, P. Link, A. Schneidwind, A. Heiss, C. P. Landee, M. M. Turnbull, and M. Kenzelmann, "Quantum effects in a weakly-frustrated S=1/2 two-dimensional Heisenberg antiferromagnet in an applied magnetic field," Phys. Rev. Lett. **102**, 197201 (2009).

F. Xiao, F. M. Woodward, C. P. Landee, M. M. Turnbull, C. Mielke, N. Harrison, T. Lancaster, S. J. Blundell, P. J. Baker, P. Babkevich, and F. L. Platt, "Observation of 2D XY Behavior in quasi-2D Quantum Heisenberg Antiferromagnets," Phys. Rev. B. **79**, 134412 (2009).

F. M. Woodward, P. J. Gibson, G. Jameson, C. P. Landee, M. M. Turnbull and R. D. Willett, "A Family of Two-Dimensional Heisenberg Antiferromagnets: Synthesis, X-ray structures and Magnetic Behavior of $[Cu(pz)_2](C1O_4)_2$, $[Cu(pz)_2](BF_4)_2$ and $[Cu(pz)_2(NO_3)](PF_6)$," *Inorganic Chemistry* **46**, 4256-4266 (2007).

Mukhopadhyay, Ranjan

Ranjan Mukhopadhyay and Ned S. Wingreen, "Curvature and shape determination of growing bacteria," Phys. Rev. E **80**, 062901 (2009).

H. Wang, N. S. Wingreen, and R. Mukhopadhyay, "Self-organized periodic positioning of protein clusters in growing bacteria," Phys. Rev. Lett. **101**, 218101 (2008).

R. Mukhopadhyay, K. C. Huang, and N. S. Wingreen, "Lipid localization in bacterial cells through curvature-mediated microphase separation," Biophys. J. **107**, 126920 (2008).

P. B. Weichman and R. Mukhopadhyay, "Critical dynamics of the dirty boson problem: revisiting the equality z=d," Phys. Rev. Lett. **98**, 24570 (2007).

D. M. Roma, R. A. Flanagan, A. E. Ruckenstein, A. M. Sengupta, and R. Mukhopadhyay, "Optimal path to epigenetic switching," Phys. Rev. E **71**, 011902 (2005).

HARVARD UNIVERSITY

SCHOOL OF ENGINEERING AND APPLIED SCIENCES

Cambridge, Massachusetts 02138

Students Accepted For Degree	FIELDS		
	Physics	Astronomy	Related Fields
Doctorate	X		
Master's	X		

1. General

President: Drew Faust
Dean of Graduate School: Allan Brandt
Dean of Engineering and Applied Sciences: Cherry A. Murray
Department Telephone Number: (617) 496-9184
Type of Institution: University
Control: Private
Setting: Urban
Total Faculty: 930
Total Graduate Faculty: Not separated
Total Students: 21,125
Total Graduate Students: 12,243; 3,066
Annual Graduate Tuition:
All Graduate students:
Full time $33,696/years 1 & 2
$8,760/years 3 & 4
$2,230/after year 4
Tuition rates for: 2009–10
Deferred tuition plan: Yes
Other Fees: $1,714 for health insurance; $1,126 for University health service.
Term: Semester

2. Number of Faculty in Department

The combined total of full-time faculty in the three professorial ranks is 68. The combined total of full-time, part-time, and other faculty at all ranks is 83 (100 including related faculty in other departments).

3. Admission, Financial Aid, and Housing

Address admission inquiries to: Harvard University, Graduate School of Arts and Sciences, Admissions Office, Holyoke Center, 1350 Massachusetts Avenue, 3rd Floor, Cambridge, MA 02138
Graduate application fee required: $105
Admission deadline (Fall admission): December 31, 2010, for full-time students
Admission information: For fall admission, 2009–10, 170 students were accepted from 1,617 applicants.
Admission requirements: The GRE (general test) is required. Information on average GRE scores: Average GRE Verbal Score–589; Average GRE Quantitative Score–740; Average GRE Writing Score–4.5; there is no arbitrary minimum. A student whose native language is not English should include with the application materials evidence of English proficiency such as TOEFL scores. The minimum acceptable score for admission is 550 80 on the paper test and 213 on the computer test. Requirements for the Bachelor's degree or its equivalent must be met before registration. A high undergraduate GPA (3.77 is average for accepted students) is expected, but there is no rigid minimum.
Address financial aid inquiries to: Graduate School of Arts and Sciences, Admissions Office, Holyoke Center, 1350 Massachusetts Avenue, 3rd Floor, Cambridge, MA 02138.
GAPSFAS application required: Yes (but not parental)
Financial aid deadline: 12/31/10
Loans available: Yes
Address housing inquiries to: The GSAS Office of Housing Services, Dudley House, Room B-2
On-campus, single student housing available: Yes
Cost/year: Graduate dormitory accommodations are available to single students. Rates are $5,674–$8,910 per academic year. Meals are available in the graduate dining hall. Dormitory residents are required to enroll in the meal plan. Rate is $2,160 per academic year.
On-campu, married student housing available: no

Table A—Faculty, Enrollments, and Degrees Granted

Research Specialty	2009–10 Faculty	Enrollment[1] Fall 2009		No. of Degrees Granted[2] (2008–09)			Median No. of Years for 2008–09 Ph.D.'s
		Master's	Doctorate	Master's	Terminal Master's	Doctorate	
Applied Math.	5	5	15	15	4	4	5.5
Applied Physics	9	1	117	12	1	18	6
Computer Science	18	11	59	14	7	10	6
Engineering Sciences	25	12	138	25	10	15	5
Total		29	329	66	29	47	
Full-time Grad. Stud.		26	329				
Part-time Grad. Stud.		3	0				
First-year Grad. Stud.		23	60				
Median Years in Grad. Study (2008–09 Degrees)					6		
Undergraduate Degrees, 2009–10							

[1]Students not yet committed to a research specialty are entered under non-specialized.
[2]Five-year totals in parentheses.

4. Graduate Degree Requirements

Master's: At least 24 hours of eight graduate courses are required for the Master of Science and 16 courses are required for the Master of Engineeirng. The Master's degree program in applied physics at University is designed to provide the basic course work and research skills necessary for the pursuit of a Ph.D. degree in applied physics. A terminal Master's degree in applied physics is also offered.
Thesis: Thesis may be written *in absentia.* Students may also elect the nonthesis option.
Special Equipment, Facilities, or Programs: The program has two elements: course work and research. Theoretical research is supported by computers within the department (CTSPS). Within CTSPS the computers consist of a CRAY J916 Digital Equipment in a cluster; IBM workstations, RISC 6000/560's and 590's; Sun workstations; and Silicon Graphics workstations. Access through the NSF net/Internet to supercomputing centers at Pittsburgh and LBL is routine. Experimental research is supported by facilities in the Research Center for Science and Technology.
A. *Course Work*: The basic core courses consist of the standard courses in classical mechanics, electrodynamics, quantum mechanics, statistical mechanics, and mathematical physics.

In addition, students are required to take at least nine hours of electives in solid state physics, atomic and nuclear physics, physics of fluids, physics of surfaces, applied quantum mechanics, applied mathematics, and computer science. Students must take at least 10 courses for the Ph.D. These courses can be from the departments of Applied Physics, Applied Mathematics, Computer Science, Engineering Sciences, Physics, Mathematics, Statistics, Chemistry, or the life sciences.

B. *Research*: Research training of students emphasizes development of analytical skills in mathematics and theoretical physics. This is attained by each student working on research projects of faculty members. Six hours of graduate credit may be awarded for the completion of a thesis.

Table B—Appointments to Graduate Students, 2009–10

Title of Appointee	Appointments Total	Appointments First year	Academic Load Allowed in Credit Hours	Hours of Service Per Week	Stipend for Academic Year ($)
			Semester		
Teaching Assisant	46	4	9–12	12.5	10,400–11,850
Research Assistant	228	0	9–12	12.5	28,500
Total	274	4			

5. Personnel Engaged in Separately Budgeted Research, 2009–10

Data not available.

6. Separately Budgeted Research Expenditures by Source of Support, 2009–10

Data not available.

Table C—Separately Budgeted Research Expenditures

All Areas	No. of Grants	Expenditures ($)
Applied Math and Mechanical Engineering		$3,913,669.36
Applied Physics and Materials		$15,563,057.61
Computer Science		$3,999,521.79
BioEngineering		$5,740,572.52
Electrical Engineering		$3,558,204.72
Environmental Sciences		$8,359,991.75
IIC (Initiative in Innovating Computing)		%27,955.22
Total		41,162,972.97

FACULTY

Professors

Aizenberg, Joanna, Ph.D., Weizmann, 1996. Biophysics, materials science; soft condensed matter; surface and interface science; biomechanics; cell and tissue engineering.

Anderson, James G., Ph.D., Colorado, 1970. Atmospheric chemistry; gas phase kinetics of free radicals; photochemistry; ozone catalytic loss.

Aziz, Michael J., Ph.D., Harvard, 1983. Physical metallurgy.

Bertoldi, Katia, Ph.D., University of Trento in Italy, 2006. Solid Mechanics.

Brenner, Michael P., Ph.D., Chicago, 1994. Physical mathematics.

Capasso, Federico, Ph.D., University of Rome, 1973. Photonics and quantum optics; materials science and condensed matter physics.

Crozier, Kenneth B., Ph.D., Stanford University. Optics, electromagnetics, and light-matter interactions; photonics and optical devices.

Farrell, Brian F., Ph.D., Harvard, 1981. Dynamic meterology.

Friend, Cynthia M., Ph.D., California, Berkely, 1981. Experimental studies of interfaces and surfaces.

Golovchenko, Jene A., Ph.D., RPI, 1972. Experimental condensed matter; x-ray physics; atomic physics.

Ham, Donhee, Ph.D., Caltech. Radio-frequency and high-speed integrated circuits.

Hau, Lene V., Ph.D., Aahrus, 1991. Experimental atomic physics; Bose-Einstein condensation and quantum optics.

Horowitz, Paul, Ph.D., Harvard, 1970. Experimental astrophysics.

Hu, Eveyln, Ph.D., Columbia University, 1975. Photonics and Optical Devices.

Jacob, Daniel J., Ph.D., Caltech, 1985. Atmospheric chemistry.

Kaxiras, Efthimios, Ph.D., MIT, 1987. Condensed matter theory.

Kuang, Zhiming, Ph.D., Caltech, 2003. Tropical convection and large-scale dynamics; remote sensing.

Lieber, Charles, Ph.D., Stanford University. Materials science and applied physics.

Loncar, Marko, Ph.D., California Institute of Technology, 2003. Electronic and magnetic systems and devices; optics, electromagnetics, and light-matter interactions; photonics and optical devices.

Mahadaevan, L., Ph.D., Stanford University, 1995. Applied mathematics; mechanics.

Manoharan, Vinothan, Ph.D., University of California-Santa Barbara, 2004. Biophysics; materials science; soft condensed matter (with an emphasis on colloids; surface and interface science).

Martin, Paul C., Ph.D., Harvard, 1954. Statistical physics and solid state theory.

Martin, Scot T., Ph.D., California Institute of Technology, 1995. Physical chemistry; environmental chemistry.

Mazur, Eric, Ph.D., Leiden, 1981. Quantum optics and molecular physics.

McElroy, Michael B., Ph.D., Queen's Univ., 1962. Atmospheric physics and chemistry.

Murray, Cherry A., Ph.D., MIT, 1978. Optics, electromagnetics, and light-matter interactions; soft condensed matter; surface and interface science.

Ramanathan, Shriram, Ph.D., Stanford University, 2002. Electronic and magnetic systems and devices, materials science, surface and interface science.

Narayanamurti, Venkatesh, Ph.D., Cornell, 1965. Dean of Engineering and Applied Sciences. Experimental condensed matter and material physics.

Nelson, David R., Ph.D., Cornell, 1975. Statistical physics and condensed matter theory.

Pershan, Peter S., Ph.D., Harvard, 1960. Experimental condensed matter physics; synchrotron radiation studies of properties of matter at interfaces and surfaces.

Rice, James R., Ph.D., Lehigh, 1964. Engineering sciences and geophysics; fracture and faulting dynamics.

Spaepen, Frans, Ph.D., Harvard, 1975. Materials science.

Suo, Zhigang, Ph.D., Harvard University, 1989. The Mechanics of Small Structures for Technologies of Our Time.

Thaddeus, Patrick, Ph.D., Columbia, 1960. Electromagnetic theory and molecular spectroscopy.

Vlassak, Joost J., Ph.D., Stanford University, 1994. Materials engineering.

Weitz, David A., Ph.D., Harvard, 1978. Experimental condensed matter physics.

Westervelt, Robert M., Ph.D., California, Berkeley, 1977. Experimental semiconductor and solid state physics.

Wofsy, Steven C., Ph.D., Harvard, 1971. Atmospheric and environmental sciences.

RESEARCH SPECIALTIES AND STAFF

Applied Mathematics

The dynamics of El Nino • the mechanics of earthquakes • geophysical fluid mechanics • atmospheric turbulence and instability • The Brownian motion of polymers in solution • the kinetics of phase transition • the prediction of material microstructures • the development of novel methods for simulating earthquake mechanics • Motion of cracks in solids • mixing in small geometries • the crumpling of paper • lattice packings in curved geometries • the analysis of materials processes • predicting optimal mechanics for device applications. Brenner, Kaxiras, Mahadevan, Martin, Nelson, Rice, Suo.

Biophysics

The study of complex materials found in tissues and cells using microheological techniques • rheological measurements on cross-linked polymer gels that resemble biopolymer networks • mechanical testing of entangled actin networks • the use of nanopores in solid-state membranes as single-molecule detectors for DNA and other biopolymers • theoretical research in the statistical mechanics of biopolymers • pattern formation during the growth of tissue. Brenner, Golovchenko, Mahadevan, Mazur, Nelson, Weitz.

Electronic and Magnetic Systems and Devices

Investigation of electron and hole transport in novel materials, nanostructures (dots and wires), devices using microscopy techniques • nanowire synthesis techniques and nanoscale assembly/characterization of matter • superconducting phenomena • ultra-sensitive microwave detectors • arrays of magnetic bubbles • Casmir forces and their control in nanoelectromechanical systems (NEMS) • molecular scale sensors capable of detecting single atoms • rapid single molecule DNA sequencing. Aizenberg, Capasso, Golovchenko, Ham, Hu, Kaxiras, Lieber, Loncar, Murray, Narayanamurti, Ramanathan, Westervelt.

Fluid Dynamics

Complex fluids • suspension mechanics • low-Reynolds-number hydrodynamics • microfluidic systems • drainage in foams • flows manipulated by electric fields (e.g., electrokinetics, electrospinning) • coating flows • multiphase flows • systems combining viscous and elastic elements. Brenner, Bertoldi, Mahadevan.

Materials Science

Creation of materials and assemblies and new devices using new materials • phase-, composition-, and morphology selection in materials formation (and how this differs from nanophase materials) • properties of materials in thin film form compared with those in the bulk • the principles and mechanisms involved in the development of stress in thin-film deposition. Aizenberg, Aziz, Clarke, Friend, Kaxiras, Lieber, Mahadevan, Manoharan, Muzur, Narayanmurti, Ramanathan, Spaepen, Vlassak, Weitz.

Oceans, Atmospheres, and Geophysics

The kinetics and thermodynamics of crystal nucleation in small aqueous aerosol particles • the development of new lasers for atmospheric spectroscopy • the material science and theoretical mechanics of crack propagation in geophysics • physics and dynamics of the oceans, atmosphere, climate and climate change. Anderson, Farrell, Jacob, Kuang, Martin, McElroy, Rice, Tzierman, Wofsy.

Optics,Electromagnetics,andLight-MatterInteractions

New coherent light sources based on the quantum cascade laser (compact injection parametric oscillators, Raman injection lasers, and nanostructured semiconductor lasers designed using band-gap engineering, and ultra-short pulses in mid-infrared range) • photonic crystals (injection surface plasmon micro-lasers based on artificial defects and nonlinear optical effects). Capasso, Crozier, Golovchenko, Hau, Hu, Loncar, Mazur, Murray.

Soft Condensed Matter

Study of properties of materials highly deformable by externally applied stresses, electric or magnetic fields, or thermal fluctuations, including polymers, liquid crystals, fluids and complex fluids, surfactants, colloids, foams, and emulsions • fluid interfaces • avalanches in granular materials and the non-equilibrium dynamics of disordered systems. Aizenberg, Brenner, Golovchenko, Mahdevan, Manoharan, Nelson, Pershan, Spaepen, Weitz.

Solid Mechanics

The mechanics and thermdynamics critical to understanding the design and manufacture of engineered materials and structures for a variety of applications including electronic packages, thermal barrier coatings, thin films, and multilayers with a focus on interfaces, high-temperature composites, micromechanical devices, and plate and shell structures at both meso- and macro-scales • mechanics at the cellular and molecular levels • earthquake faulting research, with special emphasis on topics such as the mechanics of frictional sliding and dynamic fracture. Bertoldi, Mahadevan, Rice, Suo, Vlassak.

Surface and Interface Science

Investigation of the atomic structure of interfaces by diffraction and imaging techniques • interfacial thermodynamics quantities and their related phenomena of crystal nucleation and growth • study of mechanical strength • development of new nanostructures, sensors, and catalysts. Aizenberg, Aziz, Friend, Golovchenko, Kaxiras, Mahadevan, Manoharan, Narayanamurti, Pershan, Ramanathan, Spaepen, Vlassak, Weitz.

Theory and Simulation

Electronic and band-structure theory research on III-IV and II-VI compounds, magnetic semiconductors, superlattices and nanostructures, and first principles calculations of surfaces, interfaces, defects, and high-pressure phases • statistical mechanics of collective phenomena, including the study of flux line motion in high-temperature superconductors and phase transitions (in particular, the melting, the behavior of sheet polymers, charge density waves and biophysics) • investigation on electronics in semiconductor heterostructures, low-dimensional antiferromagnets, fractional statistics, inhomogeneous systems, and particles interacting with electromagnetic fields and turbulent fluids and in suspensions. Brenner, Kaxiras, Mahadevan, Nelso, Rice, Suo Wu.

MASSACHUSETTS INSTITUTE OF TECHNOLOGY

DEPARTMENT OF PHYSICS

Cambridge, Massachusetts 02139

Students Accepted For Degree	FIELDS		
	Physics	Astronomy	Related Fields
Doctorate	X		
Master's	X		

Loans available: Yes
Address housing inquiries to: Campus Housing Information W59-200 (617)-253-5148
On-campus, single student housing available: Yes
Cost/month: \$685–1,155
On-campus, married student housing available: Yes
Cost/month: \$1,113–1,647

1. General

President: Susan Hockfield
Dean of Graduate School: Steven R. Lerman
Department Chairman: Edmund Bertschinger
Department Telephone Number: (617) 253-4800 (C)
Type of Institution: University
Control: Private
Setting: Urban
Total Faculty: 1,025
Total Graduate Faculty: 1,025
Total Students: 10,384
Total Graduate Students: 6,152
Annual Graduate Tuition:
All Graduate Students: Full-time—\$38,940
Part-time-\$605*/credit
Tuition rates for: 2010–11
Deferred tuition plan: No
Other Fees: Medical Insurance—\$1,740
Student Life Fee—\$272
Term: Semester

*(9–12 units/course)

2. Number of Faculty in Department

The combined total of full-time faculty in the four professorial ranks is 81. The combined total of full-time, part-time, and other faculty at all ranks is 81.

3. Admission, Financial Aid, and Housing

Address admission inquiries to: Graduate Admissions Officer, Dept. of Physics, Room 4-315
Graduate application fee required: \$75
Admission deadline (Fall admission): 12/15
Admission information: For fall admission, 2010, 88 students were accepted from 628 applicants.
Admission requirements: For admission to the graduate programs a Bachelor's degree in physics is required with no minimum undergraduate GPA specified. The GRE Advanced is required. Students from non-English speaking countries are required to demonstrate proficiency in English via the IELTS exam.
Undergraduate preparation assumed: Reif, *Fundamentals of Statistical and Thermal Physics*: Marion, *Classical Dynamics*, and *Classical Electromagnetic Radiation*; Eisberg, *Fundamentals of Modern Physics.*
Address financial aid inquiries to: Graduate Admissions Officer, Dept. of Physics, Room 4-315
GAPSFAS application required: No
Financial aid deadline: 1/01

Table A—Faculty, Enrollments, and Degrees Granted

Research Specialty	2009–10 Faculty	Enrollment[1] Fall 2009 Master's	Enrollment[1] Fall 2009 Doctorate	No. of Degrees Granted[2] 2009–10 (2005–10) Master's	Terminal Master's	Doctorate	Median No. of Years for 2009–10 Ph.D.'s
Astrophysics	17	0	27	0(1)	0(3)	7(23)	5.0
Atomic, Molecular, & Optical Physics	7	1	28	0(1)	0(2)	3(27)	5.7
Biophysics	4	0	10	0(1)	0(3)	7(15)	6.0
Condensed Matter Physics	15	0	64	1(3)	0(1)	7(40)	6.0
Medical & Health Physics	0	0	0	–	0(0)	0(0)	–
High Energy & Nuclear	34	0	85	–	1(1)	14(66)	5.75
Plasma Physics & Fusion	4	0	24	–	0(0)	6(19)	6.7
Total		1	239	1(6)	1(10)	44(190)	
Full-time Grad. Stud.		1	239				
Part-time Grad. Stud.		0	0				
First-year Grad. Stud.		0	44				
Median Years in Grad. study (2008–09 Degrees)					4.1	5.9	–
Undergraduate Degrees, 2009–10 (2005–10): 80(418)							

[1]Students not yet committed to a research specialty are entered under non-specialized.
[2]Five-year totals in parentheses.

4. Graduate Degree Requirements

Master's: Approximately six graduate level courses in physics are required. A B^- average must be maintained. Thesis required. Residence: one semester. There are no foreign language or comprehensive exam requirements.
Doctorate: Two academic years of full-time graduate work (including thesis) are required for the Ph.D. degree. Two courses are required inside, and two outside, the candidate's doctoral specialty. A B^- average must be maintained. A general Doctoral examination consisting of two written parts and one oral part must be passed in the second or third year of graduate work. Original research, demonstrated through a thesis is required. The thesis and oral defense of the thesis complete the requirements for the doctorate.
Thesis: Thesis may be written *in absentia* (with special permission only).

Table B—Appointments to Graduate Students, 2009–10

Title of Appointee	Appointments Total	First year	Academic Load Allowed in Credit Hours	Hours of Service Per Week	Stipend for Academic Year ($)
			Semester		
Teaching Assistant	30	0	Varies (about 3/4 full level)		28,896
Research Assistant	161	4	Varies (about 3/4 full level)		28,896
University Fellow	49	40	–	–	28,896
Total	240	44			

5. Personnel Engaged in Separately Budgeted Research, 7/09–6/10

Professorial faculty	81
Postdoctoral appointments	21
Graduate students	240
Nonteaching research personnel	5
Total	347

6. Separately Budgeted Research Expenditures by Source of Support

	Departmental Research	Physics-related Research Outside Department
Federal government	$66,728,497	n/a
Private, non-profit organizations	3,741,982	n/a
Total	$70,470,479	–

7. Separately Funded and Managed Laboratories

Center for Materials Science and Engineering	$3,936,091
Kavli Center for Space Research	17,230,150
Laboratory for Nuclear Science	20,663,539
Materials Processing Center	7,798,524
Magnet Laboratory	8,633,912
Plasma Science and Fusion Center	32,134,015
Research Laboratory for Electronics	34,049,607
Spectroscopy Laboratory	2,244,831
Total	$126,690,669

Table C—Separately Budgeted Research Expenditures

Research Specialty	No. of Grants	Expenditures ($)
Astronomy & Astrophysics	–	7,981,684
Atomic, Molecular, & Optical Physics	–	6,559,183
Condensed Matter Physics	–	4,248,949
Biophysics & Health Physics	–	2,242,856
High Energy & Nuclear	–	20,187,975
Plasma & Fusion	–	23,161,760
Physics Education & Other	–	6,088,072
Total	–	70,470,479

FACULTY

Professors

Ashoori, Raymond, Ph.D., Cornell, 1990. Experimental and condensed matter physics.

Becker, Ulrich J., Ph.D., Hamburg, 1964. Experimental physics; high energy.

Belcher, John W., Ph.D., Cal. Tech., 1970. Theoretical physics; solar plasma.

Benedek, George, Ph.D., Harvard, 1953. Experimental physics; light scattering and biophysics.

Bertschinger, Edmund W., Ph.D., Princeton, 1984. Head of the Department. Theoretical physics; astrophysics.

Busza, Wit, Ph.D., University College, 1964. Experimental Physics. Relativistic Heavy Ion.

Canizares, Claude, Ph.D., Harvard, 1972. Experimental physics; X-ray astronomy.

Chen, Min, Ph.D., U.C. Berkeley, 1969. Experimental physics; high energy.

Chakrabarty, Deepto, Ph.D., Cal. Tech., 1996. Experimental Physics; compact objects.

Chuang, Isaac, Ph.D., Stanford, 1997. Experimental physics. Quantum Computation.

Conrad, Janet, Ph.D., Harvard, 1993. Experimental physics; high energy.

Coppi, Bruno, Ph.D., Milan, 1961. Theoretical physics; plasma physics and astrophysics.

Elliot, James, Ph.D., Harvard, 1972. Experimental physics; astronomy.

Farhi, Edward H., Ph.D., Harvard, 1978. Theoretical physics; elementary particle, and quantum computation.

Fisher, Peter H., Ph.D., Cal. Tech., 1988. Experimental physics; high energy.

Freedman, Daniel, Ph.D., University of Wisconsin, 1964. Theoretical physics.

Greytak, Thomas J., Ph.D., MIT, 1967. Experimental physics; low temperature physics.

Guth, Alan J., Ph.D., MIT, 1972. Theoretical physics; elementary particle physics; cosmology.

Hewitt, Jacqueline, Ph.D., MIT, 1986. Experimental physics; gravitational lenses.

Ippen, Erich P., Ph.D., U.C. Berkeley, 1968. Electrical engineering; applied physics; optics.

Jackiw, Roman W., Ph.D., Cornell, 1966. Theoretical physics; elementary particles.

Jaffe, Robert, Ph.D., Stanford, 1972. Theoretical physics; elementary particles.

Joannopoulos, John, Ph.D., U.C. Berkeley, 1974. Theoretical physics; solid state, nanotechnology.

Joss, Paul C., Ph.D., Cornell, 1971. Theoretical physics; astrophysics.

Kardar, Mehran, Ph.D., MIT, 1986. Condensed matter theory, biophysics.

Kastner, Marc, Ph.D., Chicago, 1972. Experimental physics; semi-conductors.

Ketterle, Wolfgang, Ph.D., Ludwig-Maximilian University of Munich, 1986. Experimental physics; atomic resonance and scattering.

Kowalski, Stanley B., Ph.D., MIT, 1963. Experimental nuclear and particle physics.

Lee, Patrick, Ph.D., MIT, 1970. Condensed matter theory.

Levitov, Leonid, Ph.D., Moscow, Physical Technical Institute, 1989. Condensed matter theory.

Litster, J. David, Ph.D., MIT, 1965. Experimental physics; light scattering.

Matthews, June, Ph.D., MIT, 1967. Experimental physics; nuclear structure.
Mavalvala, Nergis, Ph.D., MIT, 1997. Astrophysics, gravity.
Milner, Richard, Ph.D., Cal. Tech., 1984. Experimental nuclear physics.
Moniz, Ernest, Ph.D., Stanford, 1971. Theoretical physics; nuclear structure.
Negele, John, Ph.D., Cornell, 1969. Theoretical physics; nuclear structure.
Porkolab, Miklos, Ph.D., Stanford, 1967. Experimental physics; plasma physics.
Pritchard, David E., Ph.D., Harvard, 1968. Experimental physics; atomic resonance and scattering.
Rajagopal, Krishna, Ph.D., Princeton, 1993. Theoretical nuclear and particle physics.
Rappaport, Saul A., Ph.D., MIT, 1968. Theoretical physics; compact objects.
Redwine, Robert, Ph.D., Northwestern, 1973. Experimental physics; nuclear structure.
Schechter, Paul, Ph.D., Cal. Tech., 1974. Experimental physics; extragalactic astronomy.
Seung, Sebastian, Ph.D., Harvard, 1990. Computational neuroscience.
Taylor, Washington, IV., Ph.D., U.C. Berkeley, 1993. Theoretical physics; string theory.
Tegmark, Max, Ph.D., U.C. Berkeley, 1994. Theoretical physics; cosmology.
Ting, Samuel C. C., Ph.D., Michigan, 1962. Experimental physics; high energy.
van Oudenaarden, Alexander, Ph.D., Delft, 1997. Biological physics.
Wen, Xiao-Gang, Ph.D., Princeton, 1987. Condensed matter theory.
Wilczek, Frank, Ph.D., Princeton, 1974. Theoretical particle physics.
Wyslouch, Boleslaw, Ph.D., MIT, 1987. Experimental physics; high energy.
Zwiebach, Barton, Ph.D., Cal. Tech., 1983. Theoretical; elementary particle theory.

Associate Professors

Burgasser, Adam J., Ph.D., Cal. Tech., 2001. Astrophysics, optical.
Hudson, Eric, Ph.D., U.C. Berkeley, 1999. Experimental and condensed matter physics.
Hughes, Scott, Ph.D., Cal. Tech., 1998. Theoretical physics; gravitational physics.
Lee, Young, Ph.D., MIT, 2000, Experimental physics, X-ray scattering and neutron scattering.
Liu, Hong, Ph.D., Case Western, 1997. Theoretical particle physics; string theory.
Mirny, Leonid, Ph.D., Harvard, 1998. Health Sciences.
Paus, Christoph M.E., Ph.D., III Phys. Institut RWTH Aachen, 1996. Experimental physics, high energy.
Roland, Gunther, Ph.D., Frankfurt, 1993. Experimental physics, relativistic heavy ion.
Sciolla, Gabriella, Ph.D., Torino, 1996. Experimental physics. High energy.
Seager, Sara, Ph.D., Harvard, 1999. Theoretical physics; astrophysics.
Stewart, Iain, Ph.D., Cal Tech, 1999. Theoretical nuclear physics.
Surrow, Bernd, Ph.D., Univ. of Hamburg, 1998. Experimental physics; high energy.
Todadri, Senthil, Ph.D., Yale, 1997. Condensed matter theory.
Vuletic, Vladan, Ph.D., Univ. Munich, 1997. Experimental physics; atomic.

Assistant Professors

Adams, Allen, Ph.D., Stanford, 2000. Theoretical physics, string theory.
Egedal-Pederson, Jan, Ph.D., Oxford, 1998. Experimental physics; plasma.
Figueroa-Feliciano, Enectali, Ph.D., Stanford, 2001. Experimental physics; X-ray, astronomy; cosmology.
Formaggio, Joseph, Ph.D., Columbia, 2001. Experimental physics; high energy.
Gedik, Nuh, Ph.D., U.C. Berkeley, 2004. Experimental condensed matter physics.
Gore, Jeff, Ph.D., U.C. Berkeley, 2005. Biological physics.
Jarillo-Herrero, Pablo D., Ph.D., Delft University of Tech., 2005. Experimental condensed matter physics.
Klute, Markus, Ph.D., Bonn, 2004. Experimental physics, high energy.
McGreevy, John, Ph.D., Stanford, 2002. Theoretical particle physics, string theory.
Monroe, Jocelyn, Ph.D., Columbia, 2006. Experimental physics, high energy.
Nahn, Steven, Ph.D., MIT, 1998. Experimental physics; high energy.
Simcoe, Robert, Ph.D., Cal. Tech., 2003. Experimental physics; optical.
Soljacic, Marin, Ph.D., Princeton, 2000. Theoretical physics; nonlinear optics.
Thaler, Jesse, Ph.D., Harvard, 2004. Theoretical physics, elementary particles.
Winn, Joshua, Ph.D., MIT, 2001. Astrophysics; optical astronomy.
Zwierlein, Martin, Ph.D., MIT, 2007. Experimental Physics; atomic.

Adjunct Professors

Barletta, William, Ph.D., Chicago. Intermediate energy physics.
Moncton, David, Ph.D., MIT, 1975. Experimental condensed matter physics.

RESEARCH SPECIALTIES AND STAFF

Theoretical

Biophysics & Soft Condensed Matter. Kardar, Mirny, Seung.
Cosmology: Guth, Bertschinger, Seager, Tegmark.
Hard Condensed Matter: Joannopoulos, P. Lee, Levitov, Soljacic, Todadri, Wen.
Neutron Stars & Black Holes: Hughes, Joss.
String: Adams, Freedman, Liu, Taylor, Zwiebach, McGreevy.
Particles/Fields: Farhi, Jackiw, Jaffe, Thaler, Wilczek.
Quantum Chromodynamics: Moniz, Negele, Rajagopal, Stewart.

Experimental

Atomic/Optical. Chuang, Greytak, Ippen, Ketterle, Pritchard, Vuletic, Zwierlein.
Biophysics & Soft Condensed Matter. Benedek, Gore, Litster, van Oudenaarden.
Cosmology. Canizares, Elliot, Figueroa-Feliciano, Hewitt, Schechter, Simcoe.
Hard Condensed Matter. Ashoori, Gedik, Hudson, Jarillo-Herrero, Kastner, Y. Lee.

Neutron Stars & Black Holes. Chakrabarty, Mavalvala, Rappaport.
Plasma. Belcher, Coppi, Porkolab, Egedal.
Particles & Fields. Chen, Becker, Conrad, Fisher, Formaggio, Klute, Monroe, Nahn, Paus, Sciolla, Ting.
Quantum Chromodynamics. Busza, Kowalski, Matthews, Milner, Redwine, Roland, Surrow, Wyslouch.
Substellar. A. Burgasser, Winn.

MASSACHUSETTS INSTITUTE OF TECHNOLOGY

DEPARTMENT OF EARTH, ATMOSPHERIC, AND PLANETARY SCIENCES

Cambridge, Massachusetts 02139
http://eapsweb.mit.edu

Students Accepted For Degree	FIELDS		
	Physics	Astronomy	Related Fields
Doctorate			X
Master's			X

1. General

President: Susan Hockfield
Dean of Graduate School: Christine Ortiz
Department Head: Maria T. Zuber
Department Telephone Number: (617) 253–2127
Type of Institution: University
Control: Private
Setting: Urban
Total Faculty: 1,000
Total Graduate Faculty: 1,000
Annual Graduate Tuition:
All students: Full time—$38,940
Part time—$565/unit
Tuition rates for: 2010–11
Deferred tuition plan: No
Other Fees: $272 Student life fee
Term: Semester

2. Number of Faculty in Department

The combined total of full-time faculty in the three professorial ranks is 38. The combined total of full-time, part-time, and other faculty at all ranks is 40.

3. Admission, Financial Aid, and Housing

Address admission inquiries to: Education Office, Department of Earth, Atmospheric, and Planetary Sciences, Room 54-912; (617) 253–3381.
Graduate application fee required: $75
Admission deadline (Fall admission): 1/5
Admission information: For fall admission, 2010–11, 45 students accepted from 196 applicants
Admission requirements: Strong undergraduate emphasis in math and science is necessary. The GRE general test is required. No minimum scores are required. The GRE subject test in chemistry or physics is required for the Planetary Science program. No minimum scores are required. The average GRE scores for admissions were verbal—574; quantitative—723; analytical—689. All non-native English speaking applicants are required to demonstrate proficiency in English via the TOEFL exam or IELTS exam. Minimum acceptable score for admission with TOEFL is 577 (233 for computer-based test, 90 for internet-based). IELTS is the preferred exam, and 7.0 is the minimum acceptable score.
Undergraduate preparation assumed: An undergraduate degree should have strong emphasis on math and science. Specific preparation will depend on the area of study chosen.
Address financial aid inquiries to: EAPS Education Office, Department of Earth, Atmospheric, and Planetary Sciences, MIT Room 54-912, Cambridge, MA 02139. Tel. (617) 253–3381
GAPSFAS application required: No
Financial aid deadline: 1/8
Loans available: Yes
Address housing inquiries to: Graduate Housing Office, MIT Room W59-200, Cambridge, MA 02139. Tel. (617) 253–5148 or graduatehousing@mit.edu
On-campus, graduate student housing available: Yes
Cost: Range (housing only): $662–1,478/month
On-campus, family student housing available: Yes
Cost: Range (housing only): $1,075–1,591/month

Table A—Faculty, Enrollments, and Degrees Granted

Research Specialty	2008–09 Faculty[1]	Enrollment Fall 2009		No. of Degrees Granted 2008–09 (2004–08)			Median No. of Years for 2008–09 Ph.D.'s
		Mas-ter's	Doc-torate	Mas-ter's	Terminal Master's	Doc-torate	
Atmospheres, Oceans, and Climate	13	1	26	0(2)	3(9)	5(15)	5.5
Geophysics	13	2	34	0(0)	1(7)	6(12)	5.5
Geology/ Geochemistry	12	1	17	0(0)	0(19)	2(20)	5.5
Planetary Science	10	–	8	0(1)	1(5)	3(3)	5.5
MIT/WHOI Joint Prog.-EAPS	0	0	69	0(0)	4(19)	10(38)	5.5
Total		4	158	0(3)	9(59)	26(88)	5.5
Full-time Grad. Stud.		4	158				
Part-time Grad. Stud.		0	0				
First-year Grad. Stud.		2	24				
Median Years in Grad.				2	2	5.5	5.5

Undergraduate Degrees, 2008–09 (2004–08): 12(51)

[1]Some faculty are counted under more than one research specialty.

4. Graduate Degree Requirements

Master's: Research Master's—completable in two academic years. Coursework (approximately six graduate level courses) and reasearch leading to thesis. Thesis required.
No foreign language or comprehensive exam requirements for either Master's program. A B-average must be maintained.
Doctorate: Two academic years of full-time graduate work (including thesis) are required for the Ph.D. degree. There are no formal course requirements. A B-average must be maintained. A general Doctoral examination consisting of written and oral parts must be passed in the second year of graduate work. Original research, demonstrated through a thesis, is required. The thesis and oral defense of the thesis complete the requirements for the doctorate.
Other Programs: Earth Resources Laboratory, Center for Global Change Science, and Joint Program with the Woods Hole Oceanographic Institution. See department web page at http://eapsweb.mit.edu or contact the EAPS Education office for more information.
Thesis: Theses may not be written *in absentia*
Special Equipment, Facilities, or Programs: Students have access to the Magellan telescopes at Las Campanas Observatory. Students in the MIT/WHOI Joint Program have access to the extensive oceanographic research facilities of the Woods Hole Oceanographic Institution. The department gives

students access to excellent computer and laboratory facilities. Students may also participate in a variety of field camps—geological, geophysical, astronomical, or oceanographic.

Table B—Appointments to Graduate Students, 2009–10

Title of Appointee	Appointments Total	Appointments First year	Academic Load Suggested in Credit Hours	Hours of Service Per Week	Stipend for Academic Year ($)
			Semester		
Teaching Assistant	11.5	1	36	20	$29,472
Research Assistant	113.5	11	36	20	$28,764
Predoctoral Fellow	31	14	36	0	varies at least $28,764
Total	156	26			

6. Separately Budgeted Research Expenditures by Source of Support

	Departmental Research
Federal government	$13,816,112
State, local & foreign	15,372
Private, non-profit organizations	1,148,172
Business and industry	2,247,658
Other	15,723
Total	$17,243,037

Table C—Separately Budgeted Research Expenditures

Research Specialty	No. of Grants	Expenditures ($)
Astronomy	7	489,648
Geophysics	33	3,231,084
Geol./Geochemistry	46	2,503,287
Marine Sci./Oceanography	3	169,602
Meteorology/Atmos. Science	70	9,284,107
Planetary Science	25	1,565,350
Total	184	17,243,078

FACULTY

Professors

Binzel, Richard P., Ph.D., University of Texas, 1986. Planetary astronomy; collisional evolution of asteroids; physical parameters and surface features of the Pluto-Charon system.

Bowring, Samuel A., Ph.D., University of Kansas, 1985. Origin and evolution of continental lithosphere using radiogenic isotopes. U-Pb geochronology of orogenic belts. Early history of the Earth. Rift magmatism with emphasis on the Baikal and Rio Grande rifts.

Boyle, Edward A., Ph.D., M.I.T., 1976. Paleoceanography and paleoclimatology; variability of the chemical composition of seawater; trace element chemistry of seawater, rivers, and estuaries.

Burchfiel, B. Clark, Ph.D., Yale University, 1961. Origin, development, and structural evolution of the continental crust; active tectonism; studies of the older, more deeply eroded parts of the continental crust. Current studies focus on China and Eastern Europe.

Elliot, James L., Ph.D., Harvard University, 1972. Small-body atmospheres and their changes over time; Kuiper belt; stellar occulation observations; astronomical instrumentation.

Emanuel, Kerry A., Ph.D., M.I.T., 1981. Relationship between cumulus convection and large-scale circulations; parametric representation of convection in large-scale weather forecast and climate models; the Hadley circulation; mesoscale dynamics of fronts and cyclones, tropical cyclone dynamics.

Entekhabi, Dara, Ph.D., M.I.T. 1990. Remote sensing, land-atmosphere interaction, hydrology and hydroclimatology, climate diagnostics.

Evans, J. Brian, Ph.D., M.I.T., 1978. Strength of rocks; the effect of fluids and impurities on strength; recrystallization and grain growth; microstructures of naturally deformed rocks; applications of rock mechanics to tectonic problems; interrelationships of porosity, permeability, and plastic flow.

Ferrari, Raffaele, Ph.D., Scripps Institution of Oceanography, 2001. Turbulence in the ocean and atmosphere using a combination of theory, models and observations. Dynamics of the ocean surface mixed layer. Internal waves. Mixing processes. The role of the ocean in climate.

Flierl, Glenn R., Ph.D., Harvard University, 1975. Professor Flierl and his students are investigating physical and biological dynamics in the ocean and other more general problems in geophysical fluid dynamics. We have examined instabilities of time-dependent basic states, such as tidal flows or varying baroclinicity and are studying how these develop nonlinearly and create turbulent flows. In particular, time-dependent zonal flows which at no instant satisfy the necessary condition for instability of a steady flow nevertheless can generate and sustain an active eddy field. We are also exploring a simplified model of the Jovian atmosphere and have shown that very weak vertical shears can lead to eddy formation and production of zonal jets. In the ocean, we are examining the aggregation processes for Antarctic Krill to understand the role of physical flows and behavior in creating patches on the scales of meters to kilometers. We anticipate that the size and age structure of the population, as well as food sources, ice geometry, and upwelling/downwelling may all contribute. Spatially intermittent distributions are an important element of Krill ecology, and we believe that a better understanding of how the process works and how it evolved is essential.

Grove, Timothy L., Ph.D., Harvard University, 1976. Igneous petrology; magma generation processes in island arc-continental settings and mid-ocean ridges; crystal growth and nucleation; phase transitions in minerals; diffusion in crystalline solids and silicate melts; thermal histories of geologic materials.

Hager, Bradford H., Ph.D., Harvard University, 1978. The physics of geologic processes; numerical modeling of mantle convection in terrestrial planets; numerical modeling of crustal deformation; GPS geodesy.

Herring, Thomas A., Ph.D., M.I.T., 1983. Techniques of space geodesy, including Very Long Baseline interferometry and the use of the Global Positioning System; surface deformations related to plate tectonics and plate boundary zones; effects of whole-Earth dynamics on the nutation series.

Lindzen, Richard S., Ph.D., Harvard University, 1964. Baroclinic instability; the average distribution of winds; average pole-equator temperature variation; response of planetary waves to stationary forcing; transient planetary waves; climatological advance of monsoon rains; gravity waves in the upper atmosphere; hydrodynamic instability.

Marshall, John C., Ph.D., Imperial College, 1980. Dynamics and causes of the general circulation of the atmosphere and ocean; thermocline theory; geostrophic eddies; global-scale ocean modeling.

Morgan, Dale, Ph.D., M.I.T., 1981. Rock Physics; geoelectromagnetism; inverse methods; applied seismology; environmental geophysics.

Plumb, Alan R., Ph.D., University of Manchester, 1972. Eddy transport processes in the atmosphere and ocean; dynamics of the stratosphere and mesosphere and their interaction with the lower atmosphere; large-scale tropospheric dynamics; transport of chemical constituents.

Prinn, Ronald G., Sc.D., M.I.T., 1971. Chemical-dynamical models of the atmosphere; measurement and modeling of the long-lived gases involved in the greenhouse effect and ozone depletion; atmospheric chemistry of carbon and sulfur compounds; integrated global system modelling that couples atmospheric, oceanic and terrestrial physics, chemistry and biology.

Rizzoli, Paola M., Ph.D., Scripps Institution of Oceanography, 1978. Numerical modeling of the ocean general circulation with specific emphasis to the tropical Atlantic ocean, tropical/subtropical interactions, tropical instability waves and the coupled ocean-atmosphere modes; assimilation of oceanographic data into ocean numerical models through ensemble approaches and optimal design of fixed and adaptive observational arrays; physical-biochemical modeling of the Black Sea ecosystem.

Rothman, Daniel H., Ph.D., Stanford University, 1986. Theoretical geophysics. Models of complex natural systems, tied as closely as possible to experimental and observational data. Problems of interest range from geobiological evolution to the dynamics of fluids, rocks, and sand.

Royden, Leigh H., Ph.D., M.I.T., 1982. Regional geology and geophysics; plate tectonics, thermal effects of continental deformation; mechanics of large-scale continental deformation; lithospheric flexure; continental extensions and sedimentary basin formation; uplift and erosion in mountain belts.

Seager, Sara, Ph.D., Harvard University, 1999. Finding and characterizing Earth-like exoplanets. Theoretical models of atmospheres, interiors, and biosignatures of all kinds of exoplanets. Astrobiology.

Summons, Roger, Ph.D., Univ. of New South Wales, 1972. Lipid chemistry of microbes, early biotic and environmental evolution, extinction and radiation events in Earth history, biogeochemical fossils, petroleum, astrobiology.

Wisdom, Jack, Ph.D., California Institute of Technology, 1981. Solar system dynamics; long-term evolution of the orbits and spins of the planets and natural satellites; qualitative behavior of dynamical systems; chaotic behavior; dynamics of planetary rings.

Wunsch, Carl I., Ph.D., M.I.T., 1967. General circulation of the ocean and its time variations, using observations, theory, and models. Tidal dynamics, mixing processes, internal waves. Time series analysis of modern and paleoceanographic records. Paleoclimate on all time scales.

van der Hilst, Robert D., Ph.D., Utrecht University. 1990. Seismic tomography, studies of the Earth's structure with emphasis on mantle beneath convergent plate boundaries; tectonic evolution of subduction systems; mantle dynamics; structure and evolution of continental lithosphere; field studies with portable seismometers.

Zuber, Maria T., Ph.D., Brown University, 1986. Department Head. Theoretical modeling of geophysical processes; analysis of altimetry, gravity and tectonics to determine the structure and dynamics of the Earth and solid planets; development and implementation of spacecraft laser and radio tracking experiments.

Associate Professor

Rondenay, Stéphane, Ph.D., Univ. of British Columbia, 2001. High resolution, teleseismic imaging of the Earth's subsurface; assembly and tectonic evolution of the continental lithosphere; core-mantle boundary processes.

Shim, Sang-Heon (Dan) Ph.D., Princeton Univ., 2001. The physical properties and crystal structures of materials at high pressures and temperatures, and their applications to global scale structures and dynamics of the Earth and planetary interiors.

Weiss, Benjamin P., Ph.D., California Institute of Technology, 2003. Paleomagnetic studies of rocks from Mars, the Moon and Earth; dynamo evolution, planetary histories and geobiology; use and development of SQUID microscopy for paleomagnetism.

Assistant Professors

Bosak, Tanja, Ph.D., California Institute of Technology, 2004. Microbial sediments throughout geologic time as indicators of biological processes and environmental conditions. Morphological and chemical biosignatures, early Earth, astrobiology.

Elkins-Tanton, Linda, Ph.D., M.I.T., 2002. Theory of planet and exoplanet formation and early evolution. Continental magmatism in the absence of subduction, flood basalts, and coincidences with extinctions. Interactions between magma and crustal rocks.

Jagoutz, Oliver, Ph.D., ETH Zurich, 2004. Field related studies of igneous processes; crust mantle interaction; formation and evolution of the oceanic and continental lithosphere.

Malcolm, Alison, Ph.D., Colorado School of Mines, 2005. Wave propagation in complicated media; seismic imaging in the shallow Earth; locating buried resources (primarily oil and gas); applications of microlocal analysis in imaging; nonlinear wave propagation; exploiting information in multiply scattered waves to infer Earth properties.

O'Gorman, Paul, Ph.D., California Institute of Technology, 2006. Large-scale circulation of the atmosphere; interactions of moisture and baroclinic eddies; effect of climate change on the hydrological cycle; turbulence closure theories.

Ono, Shuhei, Ph.D., Pennsylvania State University, 2001. Isotope biogeochemistry of sulfur and oxygen, water-rock-microbe interactions, seafloor hydrothermal deposits, deep biosphere, global sulfur cycles.

Perron, J. Taylor, Ph.D., University of California, Berkeley, 2006. Measurement and modeling of physical processes that shape the surfaces of planets; river networks; biotic effects on landscape evolution; volatile cycling on Mars and Titan.

Selin, Noelle Eckley, Ph.D., Harrard University, 2007. Atmospheric chemistry modeling; biogeochemical cycling of mercury (Hg); air pollution/climate interactions, air pollution health impacts; science-policy interactions.

RESEARCH SPECIALTIES AND STAFF

Theoretical

Marine Science/Oceanography. Dynamics of thermohaline circulation of the ocean, numerical modeling of ocean-climate interactions, and analysis of oceanic data, modeling of the physics, chemistry, and biology of strongly nonlinear eddies and meandering jets, interactions between waves and vortices. Ferrari, Flierl, Marshall, Rizzoli, Wunsch.

Geophysics. Numerical models of nonlinear dynamical systems, fluid dynamics; theoretical models of rock physics; numerical methods for seismology, mantle dynamics, geodesy. Elkins-Tanton, Hager, Herring, Malcolm, Rothman.

Meteorology/Atmospheric Science. Dynamics of the atmosphere and ocean, climate dynamics and modeling, theory of monsoons, coupled ocean-atmosphere models, chemical dynamical models of the atmosphere; inverse methods applied to global trace gas cycles, climatic effects of changes in carbon dioxide concentrations and solar constant, dynamics of planetary atmospheres, data assimilation and adaptive sampling, predictability and ensemble forecasting, mesoscale dynamics of fronts and cyclones, dynamics of tropical intraseasonal oscillations and tropical cyclones, modeling planetary atmospheres. Emanuel, Lindzen, Marshall, Plumb, Prinn, Seager.

Planetary Geophysics. Zuber.

Planetary Science. Numerical experiments and theoretical studies of geophysical fluid dynamics; origins and evolution of planetary jet-stream wind profiles, solar system dynamics, long-term evolution of orbits and spins of planets and satellites, chaotic behavior, dynamics of planetary rings. Planetary history, planetary gravity and magnetic fields, geochemical and geophysical studies of meteorites, atmospheres and interiors of exoplanets. Wisdom, Seager.

Experimental

Atmospheres, Oceans, and Climate. Ocean general circulation, paleoceanography and paleoclimatology, abrupt climate change. Development and application of trace element, organic geochemical, and stable isotopic techniques in oceanography and paleoclimatology. Decadal-to-millennial scale climate change. The origin of organic-rich sediment sequences in the marine environment. The marine nitrogen cycle. Acoustic tomography. Hydrometeorology and hydroclimatology, global measurements of radiatively and chemically important trace gases, climate diagnostic studies. El Niño Southern Oscillation phenomenon, diagnostic studies of the general circulation, satellite observations of planetary atmospheres. Boyle, Entekhabi, Emanuel, Ferrari, Flierl, Lindzen, Marshall, Plumb, Prinn, Rizzoli, Wunsch.

Oceanography/Marine Science. Numerical modeling of ocean general circulation, paleoceanography and paleoclimatology, trace element geochemistry, acoustic tomography. Boyle, Rizzoli, Wunsch.

Geophysics. Application of rock mechanics to tectonic problems, mantle dynamics, numerical modeling of solid-state convection, space geodesy, plate tectonics, seismology, geodetic observation of surface deformation, thermal structure of oceanic lithosphere in the vicinity of hot-spot volcanoes, rock physics, environmental geophysics, seismic tomogrpahy for characterization of the earth's crust and petroleum reservoirs, tectonic evolution of subduction systems, structure and evolution of continental lithospheres. Elkins-Tanton, Evans, Hager, Herring, Malcolm, Morgan, Toksöz, van der Hilst.

Geology/Geochemistry/Geobiology. Rift magmatism, origin and evolution of continental lithosphere using radiogenic isotopes, earth history, active tectonics, structural geology, metamorphic and igneous petrology, geochronology, and numerical simulations of depositional systems. Magma generation processes in arcs, ocean ridges, ocean islands and large igneous processes. Mineralogy of the mantle and mantle processes controlling mantle geochemistry. Mechanics and thermal effects of continental deformation, sediment transport by currents and waves, interpretation of ancient sedimentary environments, process geomorphology, debris-flow rheology, tectonic geomorphology, environmental monitoring of natural terrestrial and marine ecosystems, and the role of climate in the evolution of orogenic systems. Petroleum systems, lipids of cultured microbes and microbial consortia, molecular signatures of hydrothermal ecosystems and signals of biochemical change through time. Bosak, Bowring, Boyle, Burchfiel, Elkins-Tanton, Frey, Grove, Ono, Perron, Royden, Summons.

Planetary Science. Collisional of asteroids, planetary atmospheric dynamics, Kuiper belt, stellar occultations at optical and infrared wavelengths, geodesy, radar and radio studies of physical properties of planets, the Pluto-Charon system, origins and evolution of eddy features (e.g., Jupiter's Great Red Spot), models of planetary lithosphere deformation and the physics of volcanism, development and implementation of space-based laser ranging systems, planetary paleomagnetism and geomagnetism, astrobiology and planetary history. Extrasolar planets. Binzel, Elliot, Seager, Weiss, Zuber.

NORTHEASTERN UNIVERSITY

DEPARTMENT OF PHYSICS

Boston, Massachusetts 02115

Students Accepted For Degree	FIELDS		
	Physics	Astronomy	Related Fields
Doctorate	X		X
Master's	X		X

1. General

President: Joseph E. Aoun
Dean of Graduate School of Arts and Sciences: Bruce Ronkin
Department Chair: Paul Champion, Interim
Department Telephone Number: (617) 373-4240
Type of Institution: University
Control: Private
Setting: Urban
Total Faculty: TBA
Total Students: 22,091
Total Graduate Students: 6,392
Annual Graduate Tuition:
All Graduate Students: $1,100/credit
Tuition rates for: 2010–11
Deferred tuition plan: Yes
Other Fees: $TBA/sem.
Term: Semester

2. Number of Faculty in Department

The combined total of full-time faculty in the three professorial ranks is 27. The combined total of full-time, part-time, and other faculty at all ranks is 39.

3. Admission, Financial Aid, and Housing

Address admission inquiries to: Graduate Coordinator, Physics Department. e-mail gradphysics@neu.edu http://www.physics.neu.edu
Graduate application fee required: $50
Admission deadlines (Fall admission): Ph.D. programs: Feb. 1. Priority date for Masters programs for those interested in finanical aid: Feb. 1; international masters applicants: May 1; Aug 1 is deadline for all domestic masters applicants.
Admission information: For fall admission, 2009–10, 14 students were accepted from 162 applicants.
Admission requirements: For admission to the graduate programs, a Bachelor's degree in physics or a related field is required with no minimum undergraduate GPA specified. The GRE General and Physics Subject Test are required for admission to the Ph.D. program. The minimum acceptable score suggested for admission is not specified. Students from non-English speaking countries are required to demonstrate proficiency in English via the TOEFL or IELTS exams. Minimum acceptable score is 79-80 for the iBT TOEFL.
Undergraduate preparation assumed: Although preparation will vary, a strong background in differential and integral calculus and differential equations is expected. Courses using *Classical Mechanics* (Marion), *Electromagnetic Theory* (Hayt and Buck) *Modern Physics* (Serway) are assumed. It is also desirable, but not required to have studied complex variables and linear algebra, and to have an undergraduate background in most of the following areas: statistical physics and thermodynamics (Sears), optics (Hecht), solid state physics (Kittel), and quantum mechanics (Griffith). (However, students have the opportunity during the first year of graduate study to remedy deficiencies which may exist in their undergraduate background.)
Address financial aid inquiries to: Graduate Coordinator, Physics Department
GAPSFAS application required: No
Financial aid deadline: 2/1 for PhD programs, 5/1 for masters programs.
Loans available: Yes
Address housing inquiries to: Graduate Coordinator, Physics Department
On-campus, single student housing available: Yes
On-campus, married student housing available: No

Table A—Faculty, Enrollments, and Degrees Granted

Research Specialty	2009–10 Faculty	Enrollment[1] Fall 2009 Master's	Doctorate	No. of Degrees Granted[2] 2009–10 (2005–10) Master's	Terminal Master's	Doctorate	Median No. of Years for 2009–10 Ph.D.'s
Biophysics	4	0	9	0(0)	0(0)	1(9)	8
Condensed Matter Physics[3]	12	0	15	0(0)	0(0)	3(18)	6.3
Medical & Health Physics	0	0	0	0(0)	0(0)	0(1)	5.9
Particles & Fields[4]	8	0	8	0(0)	0(0)	1(8)	5
Non-specialized	2	14	6	8(39)	2(11)	–	–
Total		14	38	8(39)	2(11)	5(36)	
Full-time Grad. Stud.		14	38				
Part-time Grad. Stud.		0	0				
First-year Grad. Stud.		14	0				
Median Years in Grad. Study (2009–10 Degrees)				2	2	5	
Undergraduate Degrees, 2009–10 (2005–10): 15(48)							

[1] Students not yet committed to a research specialty are entered under non-specialized.
[2] Five-year totals in parentheses.
[3] Includes sub-specialties: e.g., low temperature physics, mechanics, materials science, statistical & thermal and nanophysics.
[4] Includes sub-specialties: e.g., astrophysics and relativity & gravitation.

4. Graduate Degree Requirements

Master's: Courses: 32 semester hours, of which 24 are in specific courses. Grade average: "B." Time in residence: not stipulated. Foreign languages: not required. Comprehensive and/or qualifying examination: not required. Options: Standard M.S. with/without an M.S. thesis; M.S. with a concentration in applied physics, engineering physics, chemical physics, biophysics, materials physics, mathematical physics, computational physics.
Doctorate: Courses: 42 semester hours. Grade average: "B." Time in residence: one year after qualifying examination. Foreign languages: not required. Qualifying examination: required after completion of one year of graduate courses (with a full undergraduate preparation, two years are needed with less undergraduate preparation). Thesis: required. M.S. degree may be earned while qualifying for Ph.D. degree.
Northeastern University is located in the Back Bay section of Boston, close to the Museum of Fine Arts, the Conservatory of Music, Symphony Hall, and historic Copley Square. It is

an exciting, vibrant place to pursue graduate studies, as Greater Boston is home to more universities and research facilities than any other area in the world.

Thesis research can be undertaken in any one of the department's research specialties or in interdisciplinary areas such as materials physics, mathematical physics, chemical physics, molecular biophysics, or applied engineering physics. A further option allows cooperative research to be done at high-technology industrial, government, national or international laboratories, and at medical research institutions in the Boston area.

The department is housed in the Dana Research Center, with optics, biological physics, and condensed-matter physics labs in the Egan Research Center. There are ample modern research laboratories, department and student machine shops, an electronics shop, conference and seminar rooms, and faculty and graduate student offices. The Egan Center provides a direct interface with materials researchers in chemistry and engineering and includes extensive meeting space in the Technology Transfer Center. Numerous computational facilities are available on campus, including the Advanced Scientific Computing Center (ASCC) in the Dana Research Center, and the Opportunity Research Cluster. In addition to the research they do at campus facilities, faculty members and graduate students also work at research centers located in the United States and Europe. High-energy physics experiments are under way at Fermilab in Batavia, Illinois, and at CERN, Geneva, Switzerland. Astroparticle physics research is performed at the Pierre Auger Observatory in Argentina. Some groups use the synchrotron facilities at Brookhaven National Laboratory, Long Island, New York, and Argonne National Laboratory, Argonne, Illinois, and many faculty members have flourishing collaborations with scientists in Europe, Asia, and South America.

Table B—Appointments to Graduate Students, 2009–10

Title of Appointee	Appointments Total	Appointments First year	Academic Load Allowed in Credit Hours	Hours of Service Per Week	Stipend for Academic Year ($)
			Semester		
Teaching Assistant	30	14	12	20	18,285[1]
Research Assistant	21	0	12	20	18,285[1]
Tuition Fellows	0	0		10	Note[2]
IGERT Fellows[3]	0	0	12	20	30,000[3]
Total	51	14			

[1] Plus remission of tuition, & health insurance.
[2] Remission of tuition only.
[3] Integrative Graduate Education and Research Traineeship plus tuition remission.

5. Personnel Engaged in Separately Budgeted Research, 7/09–6/10

Professorial faculty	26
Other faculty	13
Postdoctoral appointments	18
Graduate students	51
Undergraduate students	106
Nonteaching research personnel	6
Total	220

6. Separately Budgeted Research Expenditures by Source of Support

	Departmental Research	Physics-related Research Outside Department
Federal government	$5,163,971	$13,826
Total	$5,163,971	$13,826

Table C—Separately Budgeted Research Expenditures[1]

Research Specialty	No. of Grants	Expenditures ($)
Biophysics	13	1,275,509
Nanophysics/Condensed Matter Physics[1]	28	2,791,875
Particles & Fields[2]	14	1,082,760
Total	55	5,150,144

[1] Includes: low temperature physics, materials sci./metallurgy, mechanics, statistical & thermal physics.
[2] Includes: astrophysics, relativity & gravitation.

Table D—Physics-related Research Outside Department

Field and Unit Outside Department	No. of Grants	Expenditures ($)
CERN/FERMILAB/Pierre Auger	1	13,826
Total	1	13,826

FACULTY

Professors

Bansil, Arun, Ph.D., Harvard, 1974. Condensed matter theory.

Barabási, Albert-László, Ph.D., Boston University. Theoretical condensed matter physics/complex networks.

Champion, Paul M., Ph.D., Illinois, 1975. Biological and medical physics.

Goldberg, Haim, Ph.D., MIT, 1963. Elementary particle theory.

Heiman, Donald, Ph.D., Univ. of California, Irvine, 1975. Condensed matter experimental physics.

Karma, Alain S., Ph.D., California, Santa Barbara, 1985. Condensed matter theory.

Kravchenko, Sergey, Ph.D., Institute of Solid State Physics, Chernogolovka, 1988. Condensed matter experimental physics.

Lowndes, Robert P., Ph.D., London, 1966. Condensed matter experimental physics.

Markiewicz, Robert S., Ph.D., California, Berkeley, 1975. Condensed matter physics.

Nath, Pran, Ph.D., Stanford, 1964. Elementary particle theory.

Sokoloff, Jeffrey B., Ph.D., MIT, 1967. Condensed matter theory.

Sridhar, Srinivas, Ph.D., Cal. Tech., 1983. Dept. Chair. Condensed matter experimental physics.

Taylor, Tomasz, Ph.D., Warsaw, Poland, 1981. Elementary particle theory.

Widom, Allan, Ph.D., Cornell, 1967. Condensed matter theory.

Williams, Mark C., Ph.D., Minnesota, 1998. Molecular biophysics.

Wood, Darien, Ph.D., California, Berkeley, 1987. High-energy experimental physics.

Associate Professors

Alverson, George O., Ph.D., Illinois, 1979. High-energy experimental physics.

Barberis, Emanuela, Ph.D., California, Santa Cruz, 1996. High-energy experimental physics.

Israeloff, Nathan, Ph.D., Illinois, 1991. Condensed matter experimental.

Menon, Latika, Ph.D., Tata (Bombay), 1998. Nanoscaled materials.

Sage, J. Timothy, Ph.D., Illinois, 1986. Biological and medical physics.

Stepanyants, Armen, Ph.D., Rhode Island, 1999. Condensed matter theory.

Swain, John D., Ph.D., Toronto, 1990. High-energy experimental physics.

Assistant Professors

Bianconi, Ginestra, Ph.D., Notre Dame, 2002. Theoretical Condensed Matter Physics.

Kar, Swastik, Ph.D., Indian Institute of Science, 2004. Condensed Matter Experiment.

Nelson, Brent, Ph.D., California, Berkeley, 2001. Elementary particle theory.

Adjunct Professors

Anchordoqui, Luis, Ph.D., Universidad Nacional de La Plata, Argentina, 1998. Elementary Astroparticle Physics.

Chen, George Tze Yung, Ph.D., Brown University, 1972. Biomedical physics.

Dova, Maria-Teresa, Ph.D., Universidad Nacional de La Plata, 1989. High-energy experimental physics..

Farmelo, Graham, Ph.D., University of Liverpool, 1977. High-energy experimental physics.

Fenker, Howard, Ph.D., Vanderbilt, 1978. High-energy experimental physics.

Kern, Wolfhard, Ph.D., Bonn, Germany, 1958. High-energy experimental physics and education.

Mijnarends, Peter, Ph.D., Delft University of Technology, 1969. Condensed matter theory.

Morgan, C. Robert, Ph.D., MIT, 1969. Condensed-matter theory.

Sauli, Fabio, Ph.D., University of Trieste, 1963. High-energy experimental physics.

Adjunct Assistant Professor

Taylor, Lucas, Ph.D., Imperial College, 1990. High-energy physics.

Adjunct Senior Research Associates

Gongora-Trevino, Maria Araceli, Ph.D., Oxford University, 1984. Condensed matter physics.

Kaprzyk, Stanislaw, Ph.D., Academy of Metallurgy (Krakow), 1981. Condensed-matter theory.

Lindroos, Matti, Ph.D., Tampere Tech, Finland, 1979. Condensed-matter theory.

Emeritus Professors

Aaron, Ronald, Ph.D., Pennsylvania, 1961. Medical physics.

Argyres, Petros N., Ph.D., California, Berkeley, 1954. Condensed matter theory.

Garelick, David A., Ph.D., MIT, 1963. Medical physics.

Glaubman, Michael J. Ph.D., Illinois, 1953. High energy experimental physics.

José, Jorge V., D.Sc., National Univ. of Mexico, 1976.

Malenka, Bertram, Ph.D., Harvard, 1951. Elementary particle theory.

Perry, Clive H., Ph.D., London, 1960. Condensed matter experimental physics.

Reucroft, Stephen, Ph.D., Liverpool, 1969. High-energy experimental physics.

Saletan, Eugene J., Ph.D., Princeton, 1960. High-energy experimental physics.

Shiffman, Carl A., Ph.D., Oxford, 1956. Medical physics.

Srivastava, Yogendra, Ph.D., Indiana, 1964. Condensed matter theory.

Vaughn, Michael T., Ph.D., Purdue, 1960. Elementary particle theory.

von Goeler, Eberhard, Ph.D., Illinois, 1961. High energy experimental physics.

Wu, Fa-Yueh, Ph.D., Washington (St. Louis), 1963. Condensed matter theory.

RESEARCH SPECIALTIES AND STAFF

Theoretical

Theoretical Elementary Particle Physics. The faculty and students in the elementary particle theory goup are actively exploring questions concerning supersymmetry (SUSY), and more specifically its local extension to supergravity (SUGRA), with a view to understanding the connection between the universe at very large and very small scales. This leads to the study of supersymmetry, and supergravity, possible extra dimensions beyond our usual four, and related exotic phenomena such as mini-black holes, which may be produced at accelerators or by ultra high energy cosmic rays. Our formal investigations in superstring theory and M-theory are also conducted with the purpose of making connections between fundamental theory and experiment. The elementary particle theory group at NU initiated the PASCOS and SUSY series of conferences, which have become major conferences in high energy physics. Further, to mark twenty years since the formulation of supergravity (sugra) models at NU, the Physics Department held an international conference (SUGRA20) at NU during the period March 17-20, 2003. (see http://www.sugra20.neu.edu/). The theoretical elementary particle group consists of Professors Hain Goldberg, Pran Nath, Brent Nelson, Yogendra Srivastava, Tomasz Taylor, Michael Vaughn. Emeritus Professors include: Bertram Malenka, Marie Machacek.

Condensed Matter Physics. The group performs research on diverse topics that span forefront areas of hard/soft condensed matter physics and emerging areas at the intersection of physics and other disciplines. Specific research areas include the electronic structure and spectroscopy of high temperature superconductors and other complex materials, nanotribology (atomic-scale friction) in crystalline and polymeric materials, network science with applications to technological, biological and social networks, theoretical/computational materials science, cardiac nonlinear dynamics, and theoretical/computational neuroscience. Bansil, Barabási, José (Emeritus), Karma, Sokoloff, Stepanyants, Widom, Wu (Emeritus).

Experimental

Experimental Nanophysics. The faculty are actively pursuing research at the frontiers of nanoscience. The thrust areas in nanophysics include: left-handed metamaterials for photonic crystals, nanomedicine, spintronics, mesoscopic physics, low-dimensional electronic systems, nanomagnetism, and quantum chaos. Research is aimed at the synthesis of nanoscale materials and devices, as well as fundamental materials issues. The Experimental Nanophysics faculty include Professors Don Heiman, Nathan Israeloff, Sergey Kravchenko, Latika Menon, Clive Perry (Emeritus), and Srinivas Sridhar.

Experimental Particle Physics. The Experimental Elementary Particle and Astroparticle Physics group concentrates its efforts on the following three activities: DØ, CMS, and the Pierre Auger Observatory. The DØ detector at the Tevatron collider at Fermi National Accelerator Laboratory in Illinois measures the products of 2 TeV head-on collisions of protons and antiprotons. At Northeastern we are participating in improving the muon detector system, in event and data visualization, in the upgrade of the trigger system, the analysis of top-quark properties, and electroweak physics of Ws and Zs. The CMS detector at the LHC (the Large Hadron Collider at CERN) is now complete and operating at energy of 7 TeV, well beyond that available at Fermilab. It is searching for signals of new physics such as the elusive Higgs particle and supersymmetric particles. We are working on the APD-based electromagnetic calorimeter readout, the muon endcap chambers, data visualization, and grid-based computing. The Astroparticle physics group has an active program at the Pierre Auger Cosmic Ray Observatory, nearing completion in Argentina. This project aims to elucidate the origin and nature of the highest-energy cosmic rays. Northeastern takes a leading role in development of software for data analysis at Auger. The Experimental Elementary Particle and Astroparticle Physics faculty includes Professors George Alverson, Emanuela Barberis, John Swain, and Darien Wood. 3 postdoctoral fellows.

Biological and Medical Physics. The group performs research on multiple levels from molecules (DNA and proteins) to cells (regulatory and metabolic protein networks) to tissue (cell-to-cell signaling in heart muscle and brain). Eight faculty have externally funded research programs in specific research areas including single molecule DNA-protein interactions, vibrational dynamics of biomolecules, femtosecond protein dynamics, biological networks (signaling, metabolic, and transcription-regulatory networks), cardiac nonlinear dynamics, and theoretical/computational neuroscience. Biological and Medical Physics faculty includes Professors Ronald Aaron (Emeritus), László Barabási, Paul Champion, Jorge José (Emeritus), Alain Karma, J. Timothy Sage, Carl Shiffman (Emeritus), Armen Stepanyants, and Mark Williams.

TUFTS UNIVERSITY

DEPARTMENT OF PHYSICS AND ASTRONOMY

Medford, Massachusetts 02155

Students Accepted For Degree	FIELDS		
	Physics	Astronomy	Related Fields
Doctorate	X	X	
Master's	X	X	

1. General

President: Lawrence Bacow
Dean of Graduate School: Lynne Pepall
Department Chairman: Roger Tobin
Department Telephone Number: (617) 627-3029
Type of Institution: University
Control: Private
Setting: Suburban
Total Faculty: 1,233
Total Students: 9,273
Total Graduate Students: 4,162
Annual Graduate Tuition:
Tuition is waived for students admitted with Teaching Assistant offers.
Other Fees: $40 activities fee

2. Number of Faculty in Department

The combined total of full-time faculty in the three professorial ranks is 25. The combined total of full-time, part-time, and other faculty at all ranks is 25.

3. Admission, Financial Aid, and Housing

Address admission inquiries to: Graduate Student Committee, Physics Department
Graduate application fee required: $75
Application deadline (Fall admission): 1/15 U.S. applicants
12/30 non-U.S. applicants
Admission information: For fall admission 2010–11, 22 students were accepted from 78 applicants.
Admission requirements: For admission to the graduate programs, a Bachelor's degree in physics is required with no minimum undergraduate GPA specified, although strong GPA and recommendations are crucial. The GRE is required and GRE Advanced is strongly recommended. Students from non-English speaking countries are required to demonstrate proficiency in English via the TOEFL exam. Minimum acceptable score for admission is 550 (or 213 CBT).
Address financial aid inquiries to: Graduate Student Committee, Physics Department
GAPSFAS application required: No
Financial aid deadline: 1/15 U.S. applicants
12/30 non-U.S. applicants
Loans available: No
Address housing inquiries to: Housing Office
On-campus, single student housing available: Yes
On-campus, married student housing available: No

4. Graduate Degree Requirements

Master's: Eight graduate level courses in approved program with grades of B^- or better; thesis optional; two semesters residence required; no language requirement; no examination requirements.
Doctorate: Eight graduate courses required. The student must demonstrate proficiency in classical physics and in quantum mechanics. Preliminary exam, dissertation, and dissertation exam required; three academic years of study with at least one year in residence; no language requirement.
Thesis: Thesis may be written *in absentia*.
Special Equipment, Facilities, or Programs: Cooperative research programs are carried out at the Arecibo Laboratory (National Astronomy and Ionospheric Center), Argonne National Laboratory, Brookhaven National Laboratory, European Center for Nuclear Research CERN, Fermi National Accelerator Laboratory, National Radio Astronomy Observatory (Sorocco, NM), and the Soudan II Underground Laboratory. Bio-medical research in cooperation with local hospitals is also possible.

Table B—Appointments to Graduate Students, 2009–10

Title of Appointee	Appointments Total	Appointments First year	Academic Load Allowed in Credit Hours	Hours of Service Per Week	Stipend for Academic Year ($)
			Semester		
Teaching Assistant	13	4	3	20	20,000
Research Assistant	5	0	3	20	20,000
Burlingame Fellow	1	0	3	0	20,000
Total	19	4			

5. Personnel Engaged in Separately Budgeted Research, 7/10–6/11

Professorial faculty	23
Postdoctoral appointments	4
Graduate students	14
Total	41

6. Separately Budgeted Research Expenditures by Source of Support

	Departmental Research	Physics-related Research Outside Department
Federal government	$1,400,000	$
Total	$1,400,000	$

FACULTY

Professors

Cebe, Peggy, Ph.D., Cornell University, 1984. Experimental condensed matter physics.
Ford, Lawrence H., Ph.D., Princeton, 1974. General relativity and cosmology; quantum field theory.
Goldstein, Gary R., Ph.D., Chicago, 1968. Theoretical particle physics.
Gunther, Leon, Ph.D., MIT, 1964. Theoretical condensed matter physics.

Lang, Kenneth R., Ph.D., Stanford, 1969. Astrophysics; radio astronomy.

Mann, W. Anthony, Ph.D., Massachusetts, 1970. Experimental particle physics.

Napier, Austin, Ph.D., MIT, 1978. Experimental particle physics.

Oliver, William P., Ph.D., California, Berkeley, 1969. Chair of the Department, Experimental particle physics.

Schneps, Jack, Ph.D., Wisconsin, 1956. Experimental particle physics.

Sliwa, Krzysztof, Ph.D., Jagiellonian, 1980. Experimental particle physics.

Tobin, Roger, Ph.D., University of California, Berkeley, 1985. Experimental condensed matter physics.

Vilenkin, Alexander, Ph.D., SUNY, Buffalo, 1977. General relativity and cosmology.

Associate Professor

Gallagher, Hugh, Ph.D., Minnesota, 1996. High energy physics.

Assistant Professors

Blanco-Pillado, José, Ph.D., Tufts, 2001. Cosmology.

Marchesini, Danilo, Ph.D., S.I.S.S.A.–I.S.A.S., 2004. Astrophysics.

Sajina, Anna, Ph.D., Univ. of Brithish Columbia, 2006. Astronomy, Astrophysics.

Staii, Cristian, Ph.D., Pennsylvania, 2005. Biophysics.

Adjunct Professors

Chaisson, Eric, Ph.D., Harvard, 1972. Science education.

Hammer, David, Ph.D., Berkeley, 1991. Science Education.

Omenetto, Fiorenzo, Ph.D., Universita di Pavia, Italy, 1997. Applied Physics/Electrical Engineering.

Schwartz, Judah, Ph.D., New York University. 1963. Science and mathematics education.

Thornton, Ronald, Ph.D., Brown Univ., 1976. Science and mathematics education.

Research Faculty

Kafka, T., Ph.D., SUNY, Stony Brook, 1975. Experimental particle physics.

Olum, Kenneth, Ph.D., MIT, 1997. General relativity and cosmology.

Willson, R., Ph.D., Tufts, 1979. Radio astronomy.

Professors Emeriti

Everett, Allen E., Ph.D., Harvard, 1960. Theoretical particle physics; cosmology.

McCarthy, Kathryn A., Ph.D., Harvard, 1957. Experimental solid state physics.

Milburn, Richard H., Ph.D., Harvard, 1954. Experimental particle physics.

Mumford, George S., Ph.D., Virginia, 1954. Astronomy; astrophysics.

Shapira, Yaacov, Ph.D., MIT, 1964. Experimental solid state physics.

RESEARCH SPECIALTIES AND STAFF

Theoretical

Condensed Matter. Macroscopic quantum tunneling; phase transitions; magnetism; superconductivity; Mössbauer effect. Gunther.

Cosmology and General Relativity. Physical processes in the very early universe; cosmic strings; cosmological phase transitions; inflation, quantum gravity; quantum field theory in curved spacetime. Blanco-Pillado, Ford, Olum, Vilenkin. 2 research associates, 1 visiting scholar.

High Energy. Quarks and quantum chromodynamics; electroweak theory; high-energy phenomenology. Goldstein.

Experimental

Astronomy. Radio interferometry of the sun; x-ray and gamma-ray studies of the sun; radio observations of active stars. Multiwavelength observations of stars, nebulae and galaxies. Lang, Willson, Marchesini, Sajina.

Condensed Matter. surface physics; polymers; biophysics. Cebe, Staii, Tobin.

High Energy. High energy neutrino physics, search for neutrino oscillations; top quark studies and search for Higgs particles; heavy quark spectroscopy. Gallagher, Kafka, Mann, Napier, Oliver, Schneps, Sliwa, 2 research associates.

UNIVERSITY OF MASSACHUSETTS, AMHERST

DEPARTMENT OF PHYSICS

Amherst, Massachusetts 01003

Students Accepted For Degree	FIELDS		
	Physics	Astronomy	Related Fields
Doctorate	X		
Master's			

Financial aid deadline: 1/15 (for fellowship consideration)
2/1 (for assistantship consideration)
Loans available: Yes, with GAPSFAS application
Housing information: www.housing.umass.edu

1. General

President: Jack M. Wilson
Chancellor: Robert C. Holub
Dean of the Graduate School: John R. Mullin
Department Head: Donald Candela
Graduate Program Director: Krishna Kumar
Department Telephone Number: (413) 545-2545
World Wide Web Site: http://www.physics.umass.edu
Type of Institution: University
Control: Public
Setting: Small town
Total Faculty: 1,302
Total Graduate Faculty: 1,059
Total Students: 27,016
Total Graduate Students: 6,143
Annual Graduate Tuition:
In-state residents: $110/credit
Out-of-state residents: $414/credit
**Note*: Most fees and tuition waived with TA/RA appointment. (For updated tuition and fees see: http://umass.edu/bursar)
Deferred tuition plan: No

2. Number of Faculty in Department

The combined total of full-time faculty in the three professorial ranks is 29. The combined total of full-time, part-time, and other faculty at all ranks is 33.

Table A—Faculty, Enrollments, and Degrees Granted

Research Specialty	2009–10 Faculty	Enrollment[1] Fall 2009		No. of Degrees Granted[2] 2009–10 (2005–09)			Median No. of Years for 2009–10 Ph.D.'s
		Mas-ter's	Doc-torate	Mas-ter's	Terminal Master's	Doc-torate	
Biophysics	3	0	6	0(0)	0(0)	0(0)	–
Condensed Matter Physics	8	0	31	0(0)	0(4)	3(9)	8
Low Temperature Physics	1	0	1	0(0)	0(0)	1(2)	9
Medical Physics	1	0	0	0(0)	0(0)	0(2)	–
Nuclear Physics	4	0	10	0(0)	0(2)	0(3)	–
Particles & Fields	7	0	16	0(1)	0(4)	1(5)	5.5
Physics Education	0	0	1	0(0)	0(1)	0(1)	–
Polymer Physics	1	0	12	0(0)	0(0)	1(3)	7
Relativity & Gravitation	3	0	2	0(0)	0(0)	0(1)	–
Statistical	3	0	4	0(0)	0(0)	2(6)	8
Non-specialized	2	0	7	5(28)	1(14)	0(0)	–
Total		0	90	5(29)	1(25)	8(32)	
Full-time Grad. Stud.		0	88				
Part-time Grad. Stud.		0	2				
First-year Grad. Stud.		0	17				
Undergraduate Degrees, 2009–10 (2005–09): 16(113)							

[1]Students not yet committed to a research specialty are entered under non-specialized.
[2]Five-year totals in parentheses.

3. Admission, Financial Aid, and Housing

Address admission inquiries to: gradmiss@physics.umass.edu
Graduate application fee required: $40, in-state or out-of-state; $65, international
Admission deadline (Fall admission): 2/1
Admission information: For fall admission, 2009–10, 42 students were accepted from 167 applicants; 15 enrolled.
Admission requirements: A Bachelor's degree in physics or a related area is required with a minimum undergraduate GPA of 2.75. The GRE and the GRE advanced are required. Minimum acceptable score is not specified. The average GRE scores for 2009–10 admissions were verbal–435; quantitative–775; analytical–3.5; advanced–755. Students from non-English speaking countries are required to demonstrate proficiency in English via the TOEFL exam. Minimum acceptable score for admission is 575/paper based, 213/computer based, and 80/Internet based.
Undergraduate preparation assumed: Marion, *Classical Mechanics*; Marion, *Electricity and Magnetism*; Eisberg, *Fundamentals of Modern Physics*; Jenkins and White, *Fundamentals of Optics*; Reif, *Statistical Mechanics*.
Address financial aid inquiries to: gradmiss@physics.umass.edu
GAPSFAS application required: No

4. Graduate Degree Requirements

Master's: 30 credits in approved program with "B" average; Masters examination required; thesis optional; no language or residency requirement. (No terminal M.S. program)
Doctorate: No specific number of graduate credits required; qualifying exam, dissertation, and dissertation exam required; no language requirement; three research area courses required, at least one must be outside the dissertation research area.
Thesis: Thesis/dissertation may be written *in absentia*.
Special Equipment, Facilities, or Programs: **The Center for Hierarchial Manufacturing**, an NSF Nanoscale Science and Engineering Center that provides expansive facilities for nanotechnology research and supports interdisciplinary graduate research; **Nanotechnology Innovation: From Discovery to Product**, an NSF IGERT program providing graduate fellowships in nanotechnology and educational experiences in innovation and entrepreneurship; A low temperature facility capable of operating at 0.015 K.

Table B—Appointments to Graduate Students, 2009–10

Title of Appointee	Appointments Total	First year	Academic Load Allowed in Credit Hours	Hours of Service Per Week	Stipend for Academic Year ($)[1]
			Semester		
Teaching Assistant	26	11.5	10	20	15,063
Research Assistant	43	4.5	10	20	15,063
Polymer	10	0	10	20	15,709
Fellowships	3	1			
Total	82	17			

[1]Tuition, curriculum fee, and health fees waived.

5. Personnel Engaged in Separately Budgeted Research, 6/09–5/10

Professorial faculty	21
Postdoctoral appointments	15
Graduate students	44
Undergraduate students	18
Nonteaching research personnel	0
Total	98

6. Separately Budgeted Research Expenditures by Source of Support (FX2009)

	Departmental Research
Federal government	$4,284,945
Industry	37,805
Private, nonprofit organizations	45,647
University Endowments	675,461
State of Mass.	135,342
Total	$5,179,200*

*UMASS OIR

Table C—Separately Budgeted Research Expenditures

Research Specialty	No. of Grants	Expenditures ($)
Biological Physics	1	22,380
Condensed Matter Physics	28	3,335,314
Low Temperature Physics	2	142,039
Nuclear Physics	3	486,529
Particles & Fields	4	893,684
Statistical Physics	2	290,580
Education	1	8,674
Total	41	5,179,200

Total grant support 5/1/08–4/30/09.

FACULTY

Professors

Candela, Donald, Ph.D., Harvard, 1983. Experimental low-temperature and condensed matter physics.

Donoghue, John F., Ph.D., Massachusetts, 1976. Theoretical high-energy physics.

Goldner, Lori, Ph.D., Cornell, 1984. Biological physics.

Golowich, Eugene, Ph.D., Cornell, 1965. Theoretical high-energy physics.

Hallock, Robert B., Ph.D., Stanford, 1969. Experimental low-temperature and condensed matter physics.

Kumar, Krishna S., Ph.D., Syracuse, 1990. Experimental nuclear physics.

Machta, Jonathan L., Ph.D., MIT, 1980. Theoretical statistical mechanics and condensed matter physics.

Menon, Narayanan, Ph.D., Chicago, 1995. Experimental condensed matter physics.

Miskimen, Rory A., Ph.D., MIT, 1983. Experimental nuclear physics.

Parsegian, V. Adrian, Ph.D., Harvard, 1969.

Prokofiev, Nikolay, Ph.D., Kurchatov, 1987. Theoretical condensed matter physics and computational physics.

Rabin, Monroe S. Z., Ph.D., Rutgers, 1967. Experimental high-energy physics; medical physics.

Svistunov, Boris V., Ph.D., Kurchatov, 1990. Theoretical condensed matter physics.

Traschen, Jennie, Ph.D., Harvard, 1984. Theoretical high-energy physics, relativity and gravitation.

Tuominen, Mark, Ph.D., Minnesota, 1990. Experimental condensed matter physics.

Associate Professors

Blaylock, Guy, Ph.D., Illinois, 1986. Experimental high-energy physics.

Dallapiccola, Carlo J., Ph.D., Colorado, 1993. Experimental high energy physics.

Dinsmore, Anthony D., Ph.D., Pennsylvania, 1997. Experimental condensed matter physics.

Willocq, Stéphane, Ph.D., Tufts, 1992. Experimental high-energy physics.

Assistant Professors

Babaev, Egor, Ph.D., University of Uppsala (Sweden), 2001. Theoretical condensed matter physics.

Brau, Benjamin, Ph.D., MIT, 2002. Experimental high energy physics.

Cadonati, Laura, Ph.D., Harvard, 2002. Experimental Gravitational physics.

Davidovitch, Benjamin, Ph.D., The Weizmann Institute of Science (Israel), 2001. Theoretical condensed matter physics.

Kawall, David, Ph.D., Stanford, 1996. Experimental nuclear physics.

Pocar, Andrea, Ph.D., Princeton, 2003. Experimental neutrino physics.

Ross, Jennifer, Ph.D., University of California, Santa Barbara, 2004. Experimental biological physics.

Santangelo, Christian, Ph.D., University of California, Santa Barbara, 2004. Theoretical condensed matter physics.

Sorbo, Lorenzo, Ph.D., (SISSA/ISAS) of Trieste, 2001. Theoretical High energy physics.

Lecturers

Hatch, Heath, B. S., Univ. Northern British Columbia, 1998.

Kastor, David, Ph.D., Chicago, 1988. Theoretical high-energy physics, relativity and gravitation.

Papirio, Anthony, B. A. Massachusetts, 1978.

Stevens, Jason, B.S., Massachusetts, 1992.

Adjunct and Research Faculty

Muthukumar, Murugappan, Ph.D., Chicago, 1978. Theoretical polymer physics.

Tewari, Shubha, Ph.D., Univ. of California, Los Angeles, 1993. Statistical Physics.

RESEARCH SPECIALTIES AND STAFF

Theoretical

Condensed Matter and Statistical Physics. Complex and disordered systems: phase transitions, dynamics and transport. Computational methods and computational complexity: classical and quantum Monte Carlo methods. Polymers, liquid crystals and polyelectrolytes, self-assembly and pattern recognition of nanostructures; packaging of chromosonal assemblies; transport of biological macromolecules through membranes. Quantum fluids and solids: Bose-Einstein condensation; kinetic theory. Quantum dissipation and decoherence: tunneling, qubits, nanomagnets and low-dimensional conductors. Babaev, Davidovitch, Machta, Muthukumar, Prokofiev, Santangelo, Svistunov.

Particles and Fields. Gauge theories; CP violation, heavy quark physics; structure of weak interactions; physics beyond the Standard Model, gravitation, string theory and cosmology. Donoghue, Golowich, Kastor, Sorbo, Traschen.

Experimental

Biophysics and Medical Physics. Biomaterials; encapsulation of living cells; model studies of interactions among membrane proteins; hydrodynamics of viscous membranes; out-of-equilibrium dynamics of particle suspensions. Single molecule dynamics of motor proteins. Use of accelerated protons for treatment of cancer; image analysis to improve predictive power of pathological examination in ductal carcinoma in situ (breast cancer); SPECT imaging; improvement in IMRT cancer treatment algorithms. Dinsmore, Goldner, Muthukumar, Rabin, Ross.

Low-Temperature Physics. Quantum fluids and solids: superconductivity, spin polarized systems, ^{3}He-^{4}He mixtures, weak-binding systems, helium films, solid helium. Phase transitions: wetting, 2D effects, restricted geometry, quenched disorder, localization, nanostructures. Third sound, NMR, microbalance, thermal techniques, high field/temperature ratio. Candela, Hallock, Tuominen.

Nuclear Physics. Electromagnetic interaction studies of the structure of hadrons using multi-GeV electron and proton beams. At Jlab: Determination of strange quark electromagnetic distributions of the proton by parity-violating electron scattering, precision measurement of the neutral pion lifetime and the chiral axial anomaly. At the Relativistic Heavy Ion Collider (RHIC): Studies of polarized proton-proton collisions to measure the contribution of gluons to the spin of the proton. Neutrino physics; Neutrinoless double beta decay with EXO experiment; Solar neutrinos with borexino experiment. Kawall, Kumar, Miskimen, Pocar.

Particles and Fields. At SLAC: Weak interaction properties in heavy quark systems; CP violation and matter-antimatter asymmetries at the B Factory; planning for the International Linear Collider. At the LHC at CERN; search for physics beyond the Standard Model. Gravitational physics at LIGO. Cosmic ray physics at VERITAS. Blaylock, Brau, Cadonati, Dallapiccola, Rabin, Willocq.

Soft-Condensed-Matter Physics. Complex and disordered systems: fluids in porous media; diffusion and dispersion in random media; avalanche phenomena; flow and rheology of granular materials; glass transitions; fluid-solid interfacial phenomena; dendritic and fractal growth. Polymers and macromolecules: complex fluids; chemical self-assembly; polymer nanostructures; molecular-scale devices. X-ray imaging; optical microscopy, light scattering and ellipsometry. Candela, Dinsmore, Menon, Tuominen.

Nanoscale Science and Condensed Matter Physics. Thin film and nanostructures: device fabrication, electron-beam lithography, time-resolved optical spectroscopy, single electron devices, superconductivity, mesoscopic quantum phenomena, plasmonics, liquid helium, superfluidity, physics in two dimensions, liquid helium mixtures. Magnetic and transport properties of nanostructure. Electrical and optical properties of semiconductors, functional nanostructures, photonic crystal devices and networks. Candela, Dinsmore, Hallock, Tuominen.

UNIVERSITY OF MASSACHUSETTS, AMHERST

DEPARTMENT OF ASTRONOMY

Amherst, Massachusetts 01003

Students Accepted For Degree	FIELDS		
	Physics	Astronomy	Related Fields
Doctorate		X	
Master's		X	

Address housing inquiries to: Housing Assignments Office, 235 Whitmore Bldg. (individual); Family Housing, Wysocki House, 911 N. Pleasant St., Amherst, MA 01002 (family). Housing information and rates can be found at http://www.housing.umass.edu

Residential student housing available: Yes

Family student housing available: Yes

Table A—Faculty, Enrollments, and Degrees Granted

Research Specialty	2009–10 Faculty[3]	Enrollment[1] Fall 2009		No. of Degrees Granted[2] 2009–10 (2005–10)			Median No. of Years for 2009–10 Ph.D.'s
		Mas-ter's	Doc-torate	Mas-ter's	Terminal Master's	Doc-torate	
Astronomy/ Astrophysics	23	0	23	0(3)	0(2)	3(12)	6
Total		0	23	0(3)	0(2)	3(12)	
Full-time Grad. Stud.		0	23				
Part-time Grad. Stud.		0	0				
First-year Grad. Stud.		0	2				
Median Years in Grad. Study (2008–09 Degrees)		–	–	–	–	–	–

Undergraduate Degrees, 2009–10: (2005–10) 20(77)

[1] Students not yet committed to a research specialty are entered under non-specialized.
[2] Five-year totals in parentheses.
[3] Includes Research Faculty

1. General

Dean of Graduate School: John R. Mullin
Chair of Five College Astronomy Dept.: Ronald L. Snell
Head of UMass Astronomy Department: Stephen E. Schneider
Department Telephone Number: (413) 545-2194
Department Fax Number: (413) 545-4223
World Wide Web Site: http://www.astro.umass.edu
Type of Institution: University
Control: Public
Setting: Small town
Total Faculty: 1,320
Total Graduate Faculty: 1,059
Total Students: 27,016
Total Graduate Students: 6,143
Annual Graduate Tuition:
For 2010–11 rates: see http://www.umass.edu/gradschool/ and click on "Prospective Students".
Deferred tuition plan: No
Term: Semester

2. Admission, Financial Aid, and Housing

Address admission and financial aid inquiries to: Chair, Graduate Admissions Committee, Astronomy, LGRT B, 619E, Univ. of Mass., 710 North Pleasant Street, Amherst, MA 01003-9305

E-Mail Address: grad@astro.umass.edu

Graduate application fee required: $40 for Massachusetts residents, $50 for U.S. citizens and permanent residents; $65 for international students

Admission deadline (Fall admission): February 1

Financial aid deadline: February 1 for fellowship consideration
February 1 for assistantship consideration

Loans available: No

Admission requirements: For admission to the graduate program, a Bachelor's degree in a related field is required with a minimum undergraduate GPA of 3.00 specified. The average GPA for 2009–10 admissions was 3.75. The GRE is required. The average GRE scores for 2009–10 admissions were 550 verbal, 766 quantitative, 758 physics. The GRE Physics is required and a minimum suggested score for admission is 600. Students from non-English speaking countries are required to demonstrate proficiency in English via the TOEFL exam. Applicants taking the paper-based test must score 550 or above; those taking the computer-based test must score 213 or above; those taking the internet-based test must score 80 or above.

3. Graduate Degree Requirements

Master's: 30 graduate credits with a "B" average, with no more than six transferred. Of these, 21 must be in the major field. Degree must be finished in three years. There is no language or thesis requirement.

Doctorate: Admission to candidacy for the doctorate is based on two research projects followed by oral exams and satisfactory completion of graduate astronomy and physics course work. A written dissertation, an oral defense, and one-year residency are required for the degree. There is no foreign language requirement.

Thesis: Thesis may be written *in absentia*.

Special Equipment, Facilities, or Programs: The Department operates a high performance parallel Beowulf type computer with 256 processors, and is the U.S. partner in an international project with Mexican institutes to build a 50-m diameter telescope, which will operate at millimeter wavelengths. The Large Millimeter Telescope (LMT) will be completed in 2010, and will be the world's largest single-dish millimeter-wave telescope. Its enormous collecting area and high angular resolution will enable cosmological studies of forming galaxies in the early universe, as well as detailed mapping of sources within our own Galaxy with unprecedented sensitivity. Further information about our graduate program and the research within the Department can be found on our web site at URL http://www.astro.umass.edu

UMass Astronomy: UMass Astronomy is part of the Five College Astronomy Department with Amherst, Hampshire, Mt. Holyoke, and Smith Colleges, which share faculty and facilities.

Table B—Appointments to Graduate Students, 2009–10

Title of Appointee	Appointments Total	Appointments First year	Academic Load Allowed in Credit Hours	Hours of Service Per Week	Stipend for Academic Year ($)
			Semester		
Teaching Assistant	6	2		20	15,063[1]
Research Assistant	16	0		20	15,063[1,2]
Other (specify)					
Gov't of Thailand Fellowship	1	0			16,077
Total	23	2			

[1] Plus waiver of all tuition and most fees.
[2] Summer appointments available at academic year rates.

Table C—Separately Budgeted Research Expenditures[1]

Research Specialty	No. of Grants	Expenditures ($)
Astronomy/Astrophysics	56	3,169,000
Total	56	3,169,000

[1] Total expenditures, 7/01/08–6/30/09.

FACULTY

Professors

Calzetti, Daniela, Ph.D., U. Rome, Italy, 1992. UV optical and infrared astronomy; star formation and stellar feedback in galaxies; the relation between star formation and gas/dust components in galaxies; dust absorption and emission.

Edwards, Suzan, Ph.D., Hawaii, 1980. Star formation; stellar evolution; circumstellar environment of young stars.

Greenstein, George S., Ph.D., Yale, 1968. Relativistic astrophysics and neutron star physics.

Irvine, William M., Ph.D., Harvard, 1961. Radio astronomy and the interstellar medium; solar system physics.

Katz, Neal S., Ph.D., Princeton, 1989. Numerical simulations of the early universe; galaxy formation; modeling the Lyman alpha forest.

Mo, Houjun, Ph.D., Munich University, 1991. The study of galaxy formation, using numerical simulations, analytic models, and statistical comparison with observations.

Schloerb, F. Peter, Ph.D., Cal. Tech., 1978. Director, Five College Radio Astronomy Observatory and UMass LMT Project Director. Planetary radio astronomy; molecular cloud astrophysics; optical interferometry.

Schneider, Stephen E., Ph.D., Cornell, 1985. Head, UMass Department of Astronomy; Radio astronomy; galaxies; galactic dynamics.

Snell, Ronald L., Ph.D., Texas, 1979. Chairman, Five College Astronomy Department. Radio astronomy; molecular clouds and star formation.

Wang, Daniel Q., Ph.D., Columbia University, 1990. X-ray astronomy, pulsars and supernova remnants, interstellar medium, intergalactic medium, structure and evolution of galaxies and galaxy clusters.

Weinberg, Martin D., Ph.D., MIT, 1984. Stellar dynamics; galactic structure and evolution; globular cluster dynamics; binary star systems.

Young, Judith S., Ph.D., Minnesota, 1979. Radio astronomy; extragalactic astronomy; star formation; galaxy evolution.

Associate Professors

Dyar, Darby M., Ph.D., M.I.T, 1985. Planetary science, vis-, IR, LIBS, and Mössbauer spectroscopies.

Giavalisco, Mauro, Ph.D., U. Rome, Italy, 1992. Observational study of galaxy formation and evolution using space-based and ground-based facilities.

Lowenthal, James D., Ph.D., U. Arizona, 1991. Galaxy formation, galaxy evolution, quasar absorption line systems, high-redshift galaxies, starburst galaxies (sub)millimeter galaxies..

Tripp, Todd M., Ph.D., U. Wisconsin-Madison, 1997. Observational study of the evolution of galaxy and star formation in the universe.

Wilson, Grant W., Ph.D., Brown, 1998. Observational cosmology; cosmic microwave background, cosmic infrared background; galaxy formation and evolution; cluster formation; development of high sensitivity bolometric far-infrared receivers.

Yun, Min S., Ph.D. Harvard, 1992. UMass LMT Project Scientist. Observational radio astronomy; galaxy formation and evolution; ultraluminous galaxies and AGNs.

Assistant Professors

Hameed, Salman, Ph.D., New Mexico U., 2001. Star formation in nearby spirals; role of interactions in the evolution of spiral galaxies; astronomy and public policy; history of astronomy.

Pope, Alexandra, Ph.D., U. British Columbia, 2007. Galaxy formation and evolution, submillimetre and infrared galaxies, star formation rates, active galactic nuclei, spectral energy distributions, and the cosmic infrared background.

Research Faculty (all levels)

Erickson, Neal R., Ph.D., California, Berkeley, 1979. Development of advanced mm and sub-mm receivers; and local oscillator sources.

Heyer, Mark, Ph.D., Massachusetts, 1986. Molecular cloud astrophysics and star formation.

Narayanan, Gopal, Ph.D., Arizona, 1997. Radio astronomy; molecular clouds and star formation; development of advanced mm and sub-mm receivers.

UNIVERSITY OF MASSACHUSETTS DARTMOUTH

DEPARTMENT OF PHYSICS

North Dartmouth, Massachusetts 02747-2300

Students Accepted For Degree	FIELDS		
	Physics	Astronomy	Related Fields
Doctorate			
Master's	X		

1. General

Chancellor: Dr. Jean MacCormack
Assoc. Vice Chancellor for Graduate Studies: Dr. Alex Fowler
Department Chair: Dr. J. P. Hsu
Department Telephone Number: (508) 999–8000 (C)
(508) 999-8354
Type of Institution: University
Control: Public
Setting: Suburban
Total Faculty: 368
Total Graduate Faculty: 313
Total Students: 9,155
Total Graduate Students: 1,173
Annual Graduate Tuition:
In-state residents: Full-time—$8,637
Part-time—$479/credit
Out-of-state residents: Full-time—$15,045
Part-time—$835/credit
Tuition rates for: 2009–10
Deferred tuition plan: No
Term: Semester

2. Number of Faculty in Department

The combined total of full-time faculty in the three professorial ranks is 8. The combined total of full-time, part-time, and other faculty at all ranks is 10.

3. Admission, Financial Aid, and Housing

Address admission inquiries to: Graduate Admissions, UMass Dartmouth, 285 Old Westport Road, N. Dartmouth, MA 02747-2300
Graduate application fee required: $40 in state/$60 other
Admission deadline (Fall admission): 4/20:
Admission information: For fall admission, 2008–09, 15 students were accepted from 32 applicants.
Undergraduate preparation assumed: Halliday, Resnick and Walker, *Fundamentals of Physics*; Marion, *Classical Dynamics*; Wangsness, *Electromagnetic Fields*; Gasiorowicz, *Quantum Physics*.
Admission requirements: For admission to the graduate program, a bachelor's degree in Physics or a closely related field is required (no minimum undergraduate GPA specified). The GRE is recommended. The minimum acceptable score suggested for admission is verbal–450; quantitative–600; total–1050. The GRE Advanced is not required. The average GRE scores for admissions were verbal–420; quantitative–755; total–1225. Students from non-English speaking countries are required to demonstrate proficiency in English via the TOEFL exam. Minimum acceptable score for admission is 500. The minimum score to be considered for a teaching assistantship is 550.
Address financial aid inquiries to: Financial Aid Services, UMass Dartmouth, 285 Old Westport Road, N. Dartmouth, MA 02747.
GAPSFAS application required: Yes
Loans available: Yes
Address housing inquiries to: Office of Housing and Residential Life, UMass Dartmouth, 285 Old Westport Road, N. Dartmouth, MA 02747
On-campus, graduate student housing available: Yes
Cost/term: $2,835
On-campus, married student housing available: No

Table A—Faculty, Enrollments, and Degrees Granted

Research Specialty	2009–10 Faculty	Enrollment[1] Fall 2009		No. of Degrees Granted[2] 2009–10 (2005–10)			Median No. of Years for 2007–08 Ph.D.'s
		Master's	Doctorate	Master's	Terminal Master's	Doctorate	
Astronomy	1	0	0	0(0)	0(7)	0(0)	–
Astrophysics	1	2	0	0(0)	1(1)	0(0)	–
Atomic, Molecular, & Optical Physics	1	2	0	0(0)	1(3)	0(0)	–
Marine Sci./ Oceanography	1	2	0	0(0)	1(5)	0(0)	–
Nuclear Physics	1	2	0	0(0)	1(2)	0(0)	–
Relativity & Gravitation	1	4	0	0(0)	2(6)	0(0)	–
Other Theoretical/ Math.	1	3	0	0(0)	2(9)	0(0)	–
Other							
Traffic Engineering	1	1	0	0(0)	0(1)	0(0)	–
Non-specialized	0	2	0	0(0)	0(6)	–	–
Total		18	0	0(0)	8(40)	0(0)	–
Full-time Grad. Stud.		18	0				
Part-time Grad. Stud.		0	0				
First-year Grad. Stud.		9	0				
Median Years in Grad. Study (2007–08 Degrees)		**2**		–	–		

Undergraduate Degrees, 2009–10 (2006–10) : 3(35)

[1] Students not yet committed to a research specialty are entered under non-specialized.
[2] Five-year totals in parentheses.

4. Graduate Degree Requirements

Master's: Candidates for the M.S. degree in Physics at UMass Dartmouth must complete a minimum of thirty (30) semester hours of coursework. Although the program is designed to meet a variety of professional needs, at least fifteen (15) hours of core Physics courses are required. The remaining credit hours may be drawn from (1) other engineering and science courses with prior approval of the Physics Graduate Program Director, (2) the graduate seminar course, (3) research-based courses such as graduate thesis, or (4) certain high-level undergraduate courses in Physics. At most six (6) credit hours may be drawn from each of the last two categories. In addition to coursework, students must complete either (1) a master's thesis project with defense (recommended), a graduate project, or a comprehensive examination.
Thesis: Thesis may be written *in absentia*.
Special Equipment, Facilities, or Programs: Our department maintains a very close relationship with the School of Marine Science and Technology (SMAST), part of the Intercampus

Graduate School at the University of Massachusetts. One faculty has a significant presence there. Students completing the Master's degree in Physics are excellent candidates to continue on at SMAST, which offers a Ph.D. program in Oceanography. These students may even continue their research with the same thesis advisor in that case. Through faculty research, we also currently have access to large facilities such as the MAX-lab facility in Sweden. The department also has two high-performance computer clusters for computational physics research.

Table B—Appointments to Graduate Students, 2009–10

Title of Appointee	Appointments: Total	Appointments: First year	Academic Load Allowed in Credit Hours	Hours of Service Per Week	Stipend for Academic Year ($)
			Semester		
Teaching Assistant	7	2	9–12	20	12,500[1]
Research Assistant	5	3	9–12	20	variable[2]
Total	12	5			

[1] Teaching and research assistantships are sometimes given in the form of a 1/2-time appointment. In this case, the hours of service per week is 10 and the stipend would be half of that shown. Tuition remission is given for all appointments of 1/2-time and above.
[2] The stipend for a full-time research assistant varies considerably according to the funding situation of individual faculty, but in the 2009–2010 academic year, the average calendar year stipend for a full-time research assistant was $15,000.

5. Personnel Engaged in Separately Budgeted Research, 7/09–6/10

Professorial faculty	6
Graduate students	11
Undergraduate students	3
Total	20

6. Separately Budgeted Research Expenditures by Source of Support

	Departmental Research
Federal government	$140,000
State and local government	172,000
Private, non-profit organizations	21,000
Total	$333,000

Table C—Separately Budgeted Research Expenditures

Research Specialty	No. of Grants	Expenditures ($)
Astrophysics	2	87,000
Marine Sci./Oceanography	5	95,000
Nuclear Physics	1	35,000
Relativity & Gravitation	4	45,000
Other Theoretical/Math.	1	21,000
Other		
Traffic Engineering	7	50,000
Total	20	333,000

FACULTY

Professors

Hirshfeld, Alan, Ph.D., Yale University, 1978. Astrophysics, observational astronomy.

Hsu, Jong-Ping, Ph.D., University of Rochester, 1978. Symmetry principles and gauge field theories.

Associate Professors

Khanna, Gaurav, Ph.D., Penn State University, 2000. Theoretical and computational astrophysics, black hole astrophysics, quantum gravity, high performance computing, control and dynamical system theory.

O'Rielly, Grant, Ph.D., University of Melbourne, 1997. Photonuclear physics at intermeditate energies, few-body systems, pion photoproduction, fundamental nuclear symmetries.

Tandon, Amit, Ph.D., Cornell University, 1992. Fluid dynamics, physical oceanography, environmental and computational physics.

Wang, Jay (Jianyi), Ph.D., University of Tennessee, Knoxville, 1992. Theory and simulation of electronic, atomic and optical processes, ion-solids and ion-surface interactions, computational physics.

Zarrillo, Marguerite, Ph.D., University of Central Florida, 1998. Traffic flow modeling, intelligent transportation systems, highway capacity.

Assistant Professor

Fisher, Robert, Ph.D., U.C. Berkeley, 2002. Computational astrophysics, stellar evolution.

RESEARCH SPECIALTIES AND STAFF

Theoretical

Atomic, Molecular, and Optical Physics. Research focuses on electronic and optical properties of matter in interaction with charged particles, photons, and laser pulses. We study electron correlation effects, exotic properties of Rydberg atoms and molecules, Monte Carlo simulations of transport and energy deposition in gases and solids. Each component is computationally oriented and uses proven techniques of computational physics. A related project involves the study of the Lambert W function and its applications in obtaining exact solutions to problems previously thought solvable only numerically. Wang.

Computational astronomy and astrophysics. Fundamental physics of turbulent flows and its application to the two end points of stellar evolution–star formation and supernovae–using combination of theoretical and computational techniques. Fisher.

Marine Science/Oceanography. Numerical models are used to study many aspects of shallow, coastal, and open ocean circulation, as well as analytical and numerical methods upper ocean conditions and circulation. Current research areas include feature-oriented regional modeling systems, multivariate synthesis of biophysical data in different world oceans, biophysical modeling in the northern Humboldt Current, basin scale simulations of the North Atlantic using parallel computational architecture, upper-ocean mixing in the North Atlantic and Southern Ocean, interaction of mesoscale eddies with mixed layers, ageostrophic circulation near fronts in the upper ocean, mathematical representations of ecosystem dynamics in the ocean. Tandon, 1 postdoctoral fellow.

Other: Theoretical/Math. Lorentz and Poincare invariance and space-time symmetry, space-time transformations for non-inertial reference with limiting 4-dimensional symmetry and field theory of non-inertial frames, translation gauge symmetry for gravity and gravitational radiation, gauge field theories and quark confinement, and the transition between classical and quantum physics. Hsu.

Relativity and Gravitation. Research includes theoretical work on the coalescence of black hole binaries under perturbation theory and estimating the properties of the gravitational waves produced therein. This research is relevant for the various gravity wave observatories being constructed worldwide (e.g. LISA, LIGO), which will soon be operational. There is also an interest in high-performance computing and work in quantum gravity via a non-perturbative approach based on the Ashtekar–Sen connection variables. Khanna.

Experimental

Marine Science/Oceanography. Experimental work in oceanography is mainly in the area of remote sensing of the oceans. Satellite-derived data is used to study a variety of interesting physical and physical-biological interactions concerning the relationship between ocean circulation and plankton. The range of projects is shown in the Oceanography section of the Theoretical work. Tandon.

Nuclear Physics. Research currently involves a series of measurements to investigate pion photoproduction near threshold from the proton and (eventually) the neutron. This process is one of the few low-energy phenomena for which exact quantum chromodynamics calculations can be made. Comparison between the theoretical calculations and these new results will form an important test of our understanding of the underlying dynamics of the nucleon. The work is undertaken using the MAX-lab facility at Lund University in Lund, Sweden. O'Rielly.

Other: Traffic Engineering. Research in this area includes transportation modeling, queuing, optimization and car following theory. This work involves the Florida Department of Transportation and the Center for Advanced Transportation Systems Simulation in Orlando, Florida. We are assisting in the design of lane configuration patterns during rush hour, maintenance projects, and unexpected incidents. Zarrillo.

UNIVERSITY OF MASSACHUSETTS, LOWELL

DEPARTMENT OF PHYSICS AND APPLIED PHYSICS

Lowell, Massachusetts 01854

Students Accepted For Degree	FIELDS		
	Physics	Astronomy	Related Fields
Doctorate	X		X
Master's	X		X

1. General

Chancellor: Martin T. Meehan
Department Chairman: Robert H. Giles
Physics Graduate Coordinator: James J. Egan
Department Telephone Number: (978) 934-3750
Type of Institution: University
Control: Public
Setting: Urban
Total Faculty: 394 (Full time only)
Total Students: 11,087 (plus 2,517 in Continuing Education program)
Total Graduate Students: 3,054
Graduate Tuition and Fees:
In-state residents: $10,000/year
Out-of-state residents: $18,730/year
Deferred tuition plan: Yes
Term: Semester

2. Number of Faculty in Department

The combined total of full-time faculty in the three professorial ranks is 18. The combined total of full-time, part-time, and other faculty at all ranks is 43.

3. Admission, Financial Aid, and Housing

Address admission inquiries to: Prof. James J. Egan, Physics Graduate Coordinator, Dept. of Physics and Applied Physics
Graduate application fee required: $50
Admission deadline (Fall admission): April 1.
Admission information: For fall admission, 2009–10, 18 new students entered Physics graduate programs from 70 applicants.
Admission requirements: For admission to the graduate programs, a Bachelor's degree in physics or related area is required with no minimum undergraduate GPA specified. The GRE General Test is required for Ph.D. applicants. No minimum acceptable score is specified. For admission to the Ph.D., program the GRE Physics subject test is strongly suggested. Students from non-English speaking countries are required to demonstrate proficiency in English via the TOEFL exam. Minimum acceptable score is 500 for paper based; 173 for computer based; or 61 for internet.
Undergraduate preparation assumed: Taylor, *Mechanics*; Wangsness, *Electromagnetism*; Liboff, *Quantum Mechanics*; Mandl, *Statistical Mechanics*.
Address financial aid inquiries to: Prof. James J. Egan, Physics Graduate Coordinator, Dept. of Physics and Applied Physics
GAPSFAS application required: No
Financial aid deadline: 3/15
Loans available: Yes
Address housing inquiries to: University Housing Officer
On-campus, single student housing available: Yes
Cost/year: $8,635 per academic year for room and board in residence hall, single occupancy. Includes meal plan.

Table A—Faculty, Enrollments, and Degrees Granted

Research Specialty	2009–10 Faculty	Enrollment[1] Fall 2009–10		No. of Degrees Granted[2] 2009–10 (2006–10)			Median No. of Years for 2009–10 Ph.D.'s
		Master's	Doctorate	Master's	Terminal Master's	Doctorate	
Applied Physics	9	1	3	0(0)	0(0)	0(0)	–
Astronomy	1	0	0	0(0)	0(0)	0(0)	–
Astrophysics	1	0	0	0(0)	0(0)	0(0)	–
Atmos./Space Phys.,				0(1)	0(1)		
Cosmic Rays	1	0	4	0(0)	0(0)	0(2)	–
Atomic, Molecular, & Optical Physics	8	0	0	0(0)	0(0)	0(0)	–
Biophysics	3	0	3				
Condensed Matter Physics	3	0	0	0(0)	0(0)	0(1)	–
Engineering Physics/ Science	2	0	1	0(0)	0(0)	0(0)	–
Materials Sci./ Metallurgy	7	2	4	0(0)	0(0)	1(1)	–
Mechanics	0	0		0(0)	0(0)	0(1)	–
Medical & Health Physics	4	16	16	10(46)	6(30)	1(9)	–
Nanoscience	3	0	3	0(0)	0(0)	0(0)	–
Non-specialized		2	5				
Nuclear Physics	5	0	6	0(3)	1(1)	0(3)	6
Optics & Photonics	7	2	10	2(18)	0(6)	1(5)	6
Particles & Fields	1	0	0	0(0)	0(0)	0(0)	–
Physics Education	3	0	0	0(0)	0(2)	0(0)	–
Polymer Physics/ Science	1	0	5	0(1)	0(0)	1(1)	–
Statistical & Thermal	1	0	0	0(0)	0(0)	0(0)	–
Total		23	56	12(67)	6(38)	4(21)	
Full-time Grad. Stud.		9	47				
Part-time Grad. Stud.		14	3				
First-year Grad. Stud.		9	11				
Median Years in Grad. Study (2009–10 Degrees)				3	2	6	

Undergraduate Degrees, 2009–10 (2004–09): 9(32)

[1] Students not yet committed to a research specialty are entered under non-specialized.
[2] Five-year totals in parentheses.

4. Graduate Degree Requirements

Master's: 30 graduate credits with a "B" average. No foreign language. No qualifying or comprehensive examination required. One year residence required. Thesis or project required. Master's can be obtained enroute to Ph.D. without thesis or project by passing Ph.D. comprehensive exam and 30 graduate credits.
Doctorate: 60 graduate credits with a "B" average, 21 in specific courses. Proficiency in computer programming. Comprehensive written and oral examination required; doctoral research admission oral examination conducted after completion of two-semester research project. One year residence required. Dissertation required.
Research Programs, Special Equipment, Facilities: (www.uml.edu/college/arts_sciences/physics).

The Center for Advanced Materials: is involved in the design, synthesis, characterization and processing of materials for application in new technologies by bringing together state-of-the-art instrumentation, facilities, and expert personnel.

The Photonics Optical Device Center: forms a core of design and fabrication facilities to support various university initiatives requiring innovative semiconductor-based photonic and electronic device technologies which primarily apply semiconductor, dielectric and metallic nanomaterials for new robust photonic devices for defense, medical, and commerical applications. Equipment includes three molecular beam epitaxy machines and conomitant lithography and epilayer characterization facilities.

The Submillimeter-Wave Technology Laboratory: is a leader in terahertz transmitter and receiver technologies, pioneering the design and fabrication of broadband solid-state multiplier sources, ultra-stable optically pumped lasers and laser/microwave hybrid systems. A 20-member research team, with the aid of graduate and undergraduate students, builds and maintains a variety of high-performance solid-state and laser-based measurement systems to generate terehertz frequency radiation resulting int he development of a wide range of materials characterization techniques and high resolution imaging systems for industry, defense, and medial applications.

The Radiation Laboratory: with a 1-W research reactor, an intense Cobalt-60 gamma source and 5.5-MV Van de Graaff accelerator, is a unique interdisciplinary facility for nuclear science and technology research. Applied nuclear research includes materials studies, fast neutron and fission spectroscopy, radiation damage, dosimetry and aerosol transport. The heavy-ion spectroscopy (HI-SPIN) group carries out fundamental nuclear structure research, with experiments at national heavy-ion facilities with high-resolution detector arrays, as well as detector development with industry, using analog and digital signal processing with multi-parameter data acquisition and analysis, for nuclear science, advanced nuclear energy R&D, medical imaging, and homeland security applications.

The Computational Nano-Materials Group: combines theoretical and experimental physics, supporting investigations in solid nanomaterials, device design optimization, and interpretive material and device characterization stuides.

The Advanced Biophotonics Laboratory: provides fundamental expertise on the structural and functional characterization of pathology for exploratory efforts in medical and bioengineering applications. Integrating multiple optical imaging and spectroscopic approaches, researchers monitor biochemical and physiological processes in real time on a variety of spatially different scales.

The Femtosecond Laser Spectroscopy Group: has developed a regenerative amplified femtosecond Ti:Sapphire laser facility to acquire femtosecond laser light-matter interaction data for investigation of material structures and chemical reactions at the molecular level. The laser technology is also used to facilitate the manufacture of micro- and nano-structure materials. Through the development of ultra fast femtosecond optical spectroscopy measurement systems, nanometer scale spatial and temporal resolution material characterization studies are performed and nanostructures on solid surfaces are fabricated using intense femtosecond laser pulse irradiation.

Table B—Appointments to Graduate Students, 2009–10

Title of Appointee	Appointments Total	Appointments First year	Academic Load Allowed in Credit Hours	Hours of Service Per Week	Stipend for Academic Year ($)[1]
			Semester		
Teaching Assistant	27.5	4	12	18	14,780
Research Assistant	22.5	3	12	18	14,780
Total	47	7			

[1] Tuition waived for TAs and RAs.

5. Personnel Engaged in Separately Budgeted Research, 7/09–6/10

Professorial faculty	12
Adjunct faculty	15
Postdoctoral appointments	11
Graduate students	33
Undergraduate students	44
Total	114

6. Separately Budgeted Research Expenditures by Source of Support

	Departmental Research	Physics-related Research Outside Department
Federal government	$7,630,000	$1,500,000
Business and industry	1,029,000	1,000,000
Total	$8,734,000	$2,500,000

Table C—Separately Budgeted Research Expenditures

Research Specialty	No. of Grants	Expenditures ($)
Applied Physics (Sub Millimeter Wave Technology)		6,500,000
Nuclear Physics & Radiation Services		1,364,000
Optics (Photonics)		480,000
Polymer Physics/Material Science/Nanoscience		300,000
Atmospheric/Space Physics[1]		2,500,000
Total		11,144,000

[1] Joint program with the Department of Environmental, Earth, and Atmospheric Sciences.

FACULTY

Professors

Altman, Albert, Ph.D., Maryland, 1962. (Emeritus) Physics education.

Chowdhury, Partha, Ph.D., SUNY, Stony Brook, 1979. Experimental nuclear physics; materials.

Egan, James J., Ph.D., Kentucky, 1969. Graduate Coordinator. Experimental nuclear physics.

French, Clayton S., Ph.D., U. of Lowell, 1985. Radiological science.

Giles, Robert, Ph.D., U Lowell, 1986. Terakertz laser physics, optics. Department Chair.

Goodhue, William D., Ph.D., Univ. of Lowell, 1982. Photonics and optoelectronics; molecular beam epitaxy.

Hardy, F. Raymond, M.S., Lowell Tech., 1962. (Emeritus). Physics education.

Karakashian, Aram S., Ph.D., Maryland, 1970. Theoretical and experimental solid state physics/optics.

Kegel, Gunter H. R., Ph.D., MIT, 1961. Experimental nuclear physics; materials physics.

Kumar, Jayant, Ph.D., Rutgers, 1983. Optics; materials physics.

Mittler, Arthur, Ph.D., Kentucky, 1970. Experimental nuclear physics; physics education.

Podolskiy, Viktor, Ph.D., New Mexico State, 2002. Photonics, Plasmonics, materials science.

Pullen, David J., D.Phil., Oxford, 1963. (Emeritus). Experimental nuclear physics; physics education.

Sajo, Erno, Ph.D, U. Lowell, 1990. Medical Physics, Health Physics, Aerosol Science.

Schier, Walter, Ph.D., Notre Dame, 1964. Experimental nuclear physics.

Sebastian, Kunnat J., Ph.D., Maryland, 1969. Theoretical elementary particle physics, theoretical atomic physics.

Song, Paul, Ph.D., UCLA, 1991. Space physics.

Stimets, Richard W., Ph.D., MIT, 1969. Image processing; optics; astronomy and astrophysics.

Waldman, Jerry, Ph.D., MIT, 1970. (Emeritus). Experimental laser physics, optics.

Associate Professors

McLeod, Roger D., M.S., Lowell Tech., 1966. Theory of vision.

Shen, Mengyan, Ph.D., Univ. of Sci. and Tech. of China, 1990. Nanoscience, femtosecond laser physics.

Tries, Mark A., Ph.D., Univ. Massachusetts Lowell, 1999. Radiological science.

Yaroslavsky, Anna, Ph.D., Saratov State U., 1999. Medical Imaging, biophysics.

Assistant Professors

Wasserman, Daniel, Ph.D., Princeton, 2004. Photonics.

Adjunct Faculty

Antal, John J., Ph.D., Saint Louis Univ., 1952. Neutron radiography.

Baird, Christopher, S., Ph.D., Mass Lowell, 2007. Electromagnetic theory, terahertz imagery.

Bliss, David F., Ph.D., Stony Brook, 2000. Materials science, substrate engineering, crystal growth.

Bobek, Leo, M.S., Univ. of Lowell, 1989. Radiological sciences.

Coulombe, Michael, B.S., U. Mass Lowell, 1989. Microwave systems. Terahertz physics.

Fox, Herbert L., Ph.D., Univ. Massachusetts, Lowell, 2001. Physics education.

Gatesman, Andrew, Ph.D., U. Mass Lowell, 1993. Radar signatures, IR, submillimeter and millimeter wave optical systems.

Goyette, Thomas M., Ph.D., Duke, 1990. Laser systems. Terahertz spectroscopy.

Li, Lian, Ph.D., Univ. Massachusetts Lowell, 1993. Nonlinear optics.

Medich, David, Ph.D., Umass Lowell, 1997. Medical Physics, Health physics.

Montesalvo, Mary, M.S., Univ. of Lowell, 1985; M.B.A. Univ. of Lowell, 1991. Radiation dosimetry.

Mosurkal, Ravi, Ph.D., University of Hyderabad, 1998. Polymer physics.

Narayan, Chandrika, Ph.D., Univ. of Massachusetts, Lowell, 1992. Materials physics, accelerator applications.

Regan, Thomas, M.S., Univ. of Massachusetts Lowell, 1994. Nuclear engineering, neutron radiography, gamma-ray spectroscopy.

Rivard, Mark, Ph.D., Wayne State U., 1998. Medical Physics.

Salesky, Edward T., Ph.D., U. Lowell, 1978. Theoretical applied physics, physics education.

Squillante, Michael R., Ph.D., Tufts, 1980. Nuclear detectors.

Sullivan, Nancy L. B., Ph.D., Univ. Massachusetts Lowell, 1993. Physics Education.

Vangala, Shivashankar R., Ph.D. Univ. Massachusetts Lowell, 2007. Photonics.

Yang, Ke, Ph.D., U. Mass Lowell, 1999. Optics, materials science.

RESEARCH SPECIALTIES AND STAFF

Experimental

Applied Nuclear Physics. Radiation effects; Rutherford backscattering; PIXE; nuclear instrumentation. Chowdury, Egan, Kegel, Regan, Mittler, Schier, Tries.

Medical Physics. French, Medich, Sajo, Tries, Yaroslavsky.

Nanoscience. Femtosecond laser surface interactions. Shen.

Nuclear Physics. Neutron cross sections; fission reaction studies; inelastic neutron scattering; fission product studies; in-beam gamma-ray spectroscopy; high-spin nuclear structure; heavy-ion fusion reactions. Chowdhury, Egan, Kegel, Schier.

Optics and Solid State. Materials tunable visible infrared and far-infrared lasers; opto-electronic materials and devices; image processing; surface plasmons; polymers and biological materials. Biles, Goodhue, Karakashian, Kumar, Podolskiy, Stimets, Waldman, Yang, Wasserman, Yaroslavsky.

Radiological Sciences. Dosimetry; shielding; biological effects of radiation; radon monitoring studies; radiation safety and control; Antal, French, Sojo, Tries, Medich.

Space physics. Solar wind-magnetosphere-ionosphere interactions. Song.

Terahertz-source/Receiver Technologies. Development of coherent sources, receivers and novel imaging systems for applications at terahertz frequencies. Coulombe, Gatesman, Giles, Goyette, Waldman.

Theoretical

Atomic and Molecular Physics. Radiative transitions mesic atoms. Sebastian.

Elementary Particles and Fields. Sebastian.

Optics and Solid State Physics. Quantum optics; dielectric waveguides; surface plasmons; ultraviolet and far-infrared spectra; electronic and vibrational cluster calculations; Baird, Goyette, Podolskiy.

WORCESTER POLYTECHNIC INSTITUTE

DEPARTMENT OF PHYSICS

Worcester, Massachusetts 01609

Students Accepted For Degree	FIELDS		
	Physics	Astronomy	Related Fields
Doctorate	X		
Master's	X		

1. General

President: Dennis D. Berkey
Dean of Graduate Studies and Research: Richard Sisson
Department Chairman: Germano S. Iannacchione
Department Telephone Number: (508) 831-5258
Type of Institution: College
Control: Private
Setting: Urban
Total Faculty: 365*
Total Graduate Faculty: 250**
Total Students: 3,463
Total Graduate Students: 1,163
Annual Graduate Tuition:
All Graduate Students: Full-time—$1,159/credit
Tuition rates for: 2010–11
Deferred tuition plan: No
Annual Other Fees: $85
Term: Semester

*247 Full-time faculty plus 119 part-time faculty headcount.
**201 Full-time graduate faculty plus 49 part-time graduate faculty headcount.

2. Number of Faculty in Department

The combined total of full-time faculty in the three professorial ranks is 12. The combined total of full-time, part-time, and other faculty at all ranks is 15.

3. Admission, Financial Aid, and Housing

Address admission inquiries to: Graduate Studies and Enrollment, gse@wpi.edu
Graduate application fee required: $70
Admission deadline (Fall admission): 2/15
Admission information: For fall admission, 2009–11, 3 students were accepted from 35 applicants.
Admission requirements: For admission to the graduate programs a Bachelor's degree in physics or a related field is required with no minimum undergraduate GPA specified. The GRE is strongly urged. No minimum score specified. The GRE Advanced is strongly urged. No minimum score specified. Students from non-English speaking countries are required to demonstrate proficiency in English via the TOEFL exam. Minimum acceptable score for admission is 550.
Undergraduate preparation assumed: Taylor, *Mechanics*; Lorrain and Corson, *Electromagnetism*; Griffiths, *Quantum Theory*; Phillies, *Statistical Mechanics*.
Address financial aid inquiries to: Graduate Studies and Enrollment, gse@wpi.edu
GAPSFAS application required: No
Financial aid deadline: 2/15
Loans available: No
Address housing inquiries to: Office of Residential Life and Housing, Ellsworth Apt. 16
On-campus, single student housing available: No
On-campus, married student housing available: No

Table A—Faculty, Enrollments, and Degrees Granted

Research Specialty	2008–09 Faculty	Enrollment[1] Fall 2008		No. of Degrees Granted[2] 2008–09 (2004–09)			Median No. of Years for 2008–09 Ph.D.'s
		Master's	Doctorate	Master's	Terminal Master's	Doctorate	
Applied Physics	1	0	0	–	0(0)	0(0)	–
Atomic, Molecular, & Optical Physics	2	0	2	–	0(0)	1(4)	5
Biophysics	1	0	1	–	0(0)	0(0)	–
Chemical Physics	1	0	1	–	0(0)	0(0)	–
Computer Science	2	0	0	–	0(0)	0(0)	–
Condensed Matter Physics	4	1	5	2	0(1)	3(5)	6
Electromagnetism	1	0	0	–	0(0)	0(0)	–
Low Temperature Physics	1	0	0	–	0(0)	0(0)	–
Mechanics	1	0	0	–	0(0)	0(0)	–
Optics	1	0	1	–	0(0)	0(2)	5
Statistical & Thermal	3	0	2	1	0(0)	0(0)	5
Total		1	12	3	0(0)	4(9)	
Full-time Grad. Stud.		1	12				
Part-time Grad. Stud.		1	0				
First-year Grad. Stud.		0	3				
Median Years in Grad. Study (2008–09 Degrees)				–	–	–	6
Undergraduate Degrees, 2008–09 (2004–09): 8(57)							

[1]Students not yet committed to a research specialty are entered under non-specialized.
[2]Five-year totals in parentheses.

4. Graduate Degree Requirements

Master's: 30 semester-hour credits: 24 by approved courses, and 6 by thesis or directed research. Thesis and non-thesis options available. 15 credits in classical mechanics, quantum mechanics, electromagnetism, and statistical mechanics are required.
Doctorate: 90 semester-hour credits beyond the baccalaureate are required, including a minimum of 42 in approved courses and directed study, 30 of dissertation research, and completion and defense of a Ph.D. dissertation. Passage of a qualifying examination is required. Completion of an M.S. degree is not required.

Table B—Appointments to Graduate Students, 2009–10

Title of Appointee	Appointments		Academic Load Allowed in Credit Hours	Hours of Service Per Week	Stipend for Academic Year ($)
	Total	First year			
			Semester		
Teaching Assistant	10	–	10	20	16,965
Research Assistant	1	–	9	20	19,920
Fellowship	2	–	12	20	22,000
Total	13				

5. Personnel Engaged in Separately Budgeted Research, 7/08–6/09

Professorial faculty	4
Graduate students	3
Total	7

6. Separately Budgeted Research Expenditures by Source of Support

	Departmental Research	Physics-related Research Outside Department
Non Federal	$251,801	$50,000
Federal government	$154.813	$
Total	$406,614	$50,000

Table C—Separately Budgeted Research Expenditures

Research Specialty	No. of Grants	Expenditures ($)
Condensed Matter Physics	5	406,614
Total	5	406,614

FACULTY

Professors

Aravind, Padmanabhan K., Ph.D., Northwestern, 1980. Quantum information theory. Associate Department Head.

Keil, Thomas H., Ph.D., Rochester, 1965. Optical properties of solids.

Phillies, George D. J., D.Sc., MIT, 1973. Light scattering spectroscopy and theory of complex fluids.

Ram-Mohan, L. Ramdas, Ph.D., Purdue, 1971. Many-body theory and optical properties of solids.

Zozulya, Alex A., Ph.D., Lebedev Physics Institute of the Academy of Sciences of the U.S.S.R., 1984. Nonlinear optics.

Associate Professors

Burnham, Nancy A., Ph.D., University of Colorado, Boulder, 1987. Mechanical properties of nanostructures, instrumentation and metrology for nanomechanics.

Iannacchione, Germano S., Ph.D., Kent State University, 1993. Experimental soft-condensed matter. Head of the Department.

Quimby, Richard S., Ph.D., Wisconsin, 1979. Optical properties of materials.

Assistant Professors

Koehler, Stephan, Ph.D., University of Chicago, 1997. Structure and dynamics of colloids and granular systems.

Stroe, Izabela, Ph.D., Clark University, 2005. Experimental Biophysics.

Tüzel, Erkan, Ph.D., University of Minnesota, 2006. Computational physics, statistical mechanics of biology and materials.

Director of Physics Education

Koleci, Carolann, Ph.D., Brown University, 2001. Physics Education Research.

RESEARCH SPECIALTIES AND STAFF

Condensed Matter: *Semiconductors*–optical properties of superlattices, heterostructure laser design, spintronics in diluted magnetic semiconductors and devices (Ram-Mohan), *Nanomechanics*– adhesion of microsensor surfaces, compliance of tissue-growth substrates, analysis of force–curve data, atomic–force microscopy calibration (Burnham).

Complex Fluids *Glasses*–theory and simulation, thermodynamics, relaxations (Phillies). *Liquid Crystals*–thermotropic/lyotropic/colloidal systems, phase transitions and critical phenomena, cooperative behavior and self-assembly, quenched random disorder effects, calorimetry instrumentation (Garcia, Iannacchione). *Liquids*–diffusion and transport properties, light scattering spectroscopy of liquids and polymer solutions, wetting phenomena, phase transitions and critical phenomena, superfluidity (Iannachione); *Polymers*–molecular properties of small sample volumes and single molecules, polymer and bio–macromolecular solutions, surfactants, and colloids (Iannacchione); [*Biophysics*–Proteins, dynamics and structure of self-assemblies of biomaterials, DNA, biomechanics, cellular functions (Stroe and Tüzel)

Colloids and Granules: Micro-rheology of complex fluids, self-propulsion, dynamics of cutting soft-materials (Koehler).

Optics: *Lasers*–development of infrared fiber lasers and materials (Quimby), mid-IR and FIR quantum cascade laser design, THz lasers (Ram–Mohan); *Photonics*–fiber amplifiers and optical communications (Quimby); *Spectroscopy*–laser spectroscopy of impurity ions in glasses and crystals (Quimby).

Physics Education Research: Research in physics education focuses on aspects of teaching and learning physics, spanning a broad range of topics from psychology-in studying student behaviors-to computer science-in studying uses of new interactive technologies in learning (Keil, Koleci).

Quantum Physics: *Cold atom*–Bose-Einstein Condensation of bosons and fermions, atom wave guides and interferometers (Zozulya). *Quantum Information*– Bell's theorem quantum algorithms (Aravind). *Wavefunction Engineering*–nanostructures, finite-element modeling of quantum systems and wells, field theory (Ram-Mohan).

CENTRAL MICHIGAN UNIVERSITY

DEPARTMENT OF PHYSICS

Mt. Pleasant, Michigan 48859

Students Accepted For Degree	FIELDS		
	Physics	Astronomy	Related Fields
Doctorate			‡
Master's	X	*	

*Astronomy concentration within physics master's.
‡Ph.D. in Science of Advanced materials in planning.

1. General

President: George Ross
Dean of Graduate School: Roger Coles
Department Chairman: Koblar Jackson
Department Telephone Number: (989) 774-3321
Type of Institution: University
Control: Public
Setting: Small town
Total Faculty: 1,063
Total Graduate Faculty: 650
Total Students: 27,357
Total Graduate Students: 6,777
Annual Graduate Tuition per term for 2009–10:
In-state residents: $5,080
Out-of-state residents: $8,500
Deferred tuition plan: No
Term: Semester

2. Number of Faculty in Department

The combined total of full-time faculty in the three professorial ranks is 12. The combined total of full-time, part-time, and other faculty is 16.

3. Admission, Financial Aid, and Housing

Address admission inquiries to: Koblar Jackson, jacks1ka@cmich.edu

Graduate application fee required: International $45; Domestic $35

Admission deadline: February 1

Admission information: For fall admission, 2009–10, 5 students were admitted from 21 applicants.

Admission requirements: For admission to the physics M.S. program, a Bachelor's degree in physics or a closely related discipline from an accredited science or engineering program. Minimum GPA is 2.7. The GRE is strongly encouraged for applicants seeking assistantship support; three letters of recommendation are required. TOEFL score of IBT 79; CBT 213; Paper 550 is required for international students.

Address financial aid inquiries to: Applicants seeking financial support must complete an online application (see http://www.phy.cmich.edu/Programs/ms_prg.shtml). International students must submit a financial affidavit if no financial support is offered.

Housing: See http://www.cmich.edu/Admissions/Transfers/Housing.htm or call the offices of Residence Life, Bovee University Center, (989) 774-3111

4. Graduate Degree Requirements

Master's: M.S. in Physics. The requirements for the M.S. in Physics are based on a core of 12 semester hours in advanced mechanics, electricity and magnetism, and quantum mechanics. Three hours of seminar are required and six hours credit is given for the thesis. The student and advisor select 9 additional hours in areas of specific value to the student. An overall GPA of 3.0 must be maintained, and no grade below C^- is accepted. There are no language requirements, qualifying examinations, or residency periods required for the Master's degree.

Astronomy: Students may use electives courses to create a specialization in astronomy. Thesis research opportunities in astronomy or astrophysics are available.

Ph.D.: Several of the Physics Department faculty members participate in the Ph.D. program in the Science of Advanced Materials. Information about this program may be found at www.sam.cmich.edu

Special Equipment, Facilities, or Programs: Laser spectroscopy laboratory, rheology laboratory, x-ray crystallography laboratory, astronomical observatory, polymer physics laboratory, and theory workroom. The department is fully networked and includes numerous UNIX and NT work-stations. The department also houses the Center for High Performance Scientific Computing with several UNIX clusters.

5. RESEARCH AREAS

Physics faculty members conduct research in a number of areas, including:

Astronomy: Stellar cluster; variable stars; occultations; interstellar matter.

Astrophysics: Accretion disk models.

Atomic Physics: Polarized atoms; atom cooling.

Chemical Physics: First-principles studies of molecules and atomic clusters; nanocatalysis.

Condensed Matter Physics: Properties of disordered materials and glasses; dielectric spectroscopy.

Materials Physics: X-ray based structure determination; first-principles computational studies of thermoelectric and piezoelectric materials; computational studies of graphene.

Nuclear Physics: Spectroscopy and structure; shell model calculations.

Optics: Laser systems; spectroscopy.

Physics Education: Design and evaluation of courses, curricula, and teaching methods.

Polymer Physics: Characterization and properties of electret polymers.

Rheology. Properties of electro- and magnetorheological fluids.

For more information about faculty and research areas, please visit our website at http://www.phy.cmich.edu

CENTRAL MICHIGAN UNIVERSITY

SCIENCE OF ADVANCED MATERIALS PHD PROGRAM

Mount Pleasant, MI 48859

Students Accepted For Degree	FIELDS		
	Physics	Astronomy	Related Fields
Doctorate	X		X
Master's			

1. General

President: George Ross
Dean of Graduate School: Roger Coles
Program Director: Koblar Alan Jackson
Program telephone: (989) 774-2221
Type of institution: University
Control: Public
Setting: Rural
Total Faculty: 1,063
Total Graduate Faculty: 650
Total Students: 27,357
Total Graduate Students: 6,777
Annual Graduate Tuition per term for 2009–2010:
In-state residents: $5,080
Out-of-state residents: $8,500
Deferred tuition plan: No
Term: semester

2. Number of Faculty in Program

The combined total of full-time faculty in the three professorial ranks is 23. The combined total of full-time, part-time and other faculty is 27.

3. Admission, Financial Aid, and Housing

Address admission inquiries to: Jessica Lapp, samphd@cmich.edu
Graduate application fee required: International $45; Domestic $35
Admission deadline: February 1
Admission information: For fall admission 2009–2010, 6 students were accepted from 19 applicants
Admissions requirement: For admission to the graduate programs, a Bachelor's degree from an accredited science or engineering program. Minimum GPA is 2.7. The GRE is required. Three letters of recommendation are required. TOEFL score of IBT 79; CBT 213; Paper 550 is required for international students.
Address financial aid inquiries to: Applicants are automatically considered for financial support at the time of application. International students must submit a financial affidavit if no financial support is offered.
Housing: See http://www.cmich.edu/Admissions/Transfers/Housing.htm

4. Graduate Degree Requirements

The SAM program is designed as a research intensive program, with a core of interdisciplinary courses on the science of materials. Electives and courses on specialized topics are available through the graduate programs in Physics, Chemistry, Math and other departments. Students entering with a BS in a related field should expect to spend 5 years in residence to complete the program.

The 90 hours of coursework, directed research and thesis credits are required with B grade or above. Relevant MS coursework and thesis credit may be transferred. Students must pass cumulative exams in order to be recommended for candidacy. A thesis prospectus and public presentation is required before dissertation work commences. Upon completion of a written dissertation, an oral dissertation defense is required.

5. Program Description

The Science of Advanced Materials (SAM) doctoral program provides an interdisciplinary environment to train effective researchers for both academic and industrial careers. Classes are small and students are engaged in research with faculty mentors by the end of the first year. The formal coursework focuses on the scientific framework for studying materials and is organized around the key methodologies employed in materials research: modeling, characterization and synthesis. Students gain hands-on experience exploring chemical, physical and mathematical principles of advanced materials. CMU faculty members maintain active research labs in the areas listed below; they are also involved in collaborative relationships with scientists at other institutions and with industry, providing students with additional contact to experts in a variety of materials areas.

6. Research Areas

CMU faculty work in a variety of materials-related areas:

Biobased materials
Electret Polymers
Electronic Properties of nanomaterials
Electroporation
Interfacial electrodynamics
Inorganic Materials Synthesis
Magnetic properties from first-principles
Microscopy
Molecular simulation and Rheology
NMR spectroscopy
Nano devices
Organic catalysts
Materials for energy applications
Polymers for environmental applications
Polymers for medical applications
Raman Spectroscopy
Reticular chemistry
Theory of Atomic Clusters
Theory of Functional Materials
X-Ray Diffraction (XRD)

For faculty information and detailed research areas, please visit our website at http://www.sam.cmich.edu

EASTERN MICHIGAN UNIVERSITY

DEPARTMENT OF PHYSICS AND ASTRONOMY

Ypsilanti, Michigan 48197

Students Accepted For Degree	FIELDS		
	Physics	Astronomy	Related Fields
Doctorate			
Master's	X		X

1. General

President: Susan W. Martin
Dean of Graduate School: Deb deLaski-Smith
Department Head: James J. Carroll, III
Department Telephone Number: (734) 487–4144
Type of Institution: University
Control: Public
Setting: Suburban
Total Faculty: 957
Total Graduate Faculty: Not separated
Total Students: 24,287 (includes on and off campus and workshops)
Total Graduate Students: 5,627
Annual Graduate Tuition:
In-state residents: $401/credit
Out-of-state residents: $821/credit
Tuition rates for: 2010–11
Deferred tuition plan: Yes
Other Fees: $40/semester + $61/credit
Term: Semester

2. Number of Faculty in Department

The combined total of full-time faculty in the four professorial ranks is 12. The combined total of full-time, part-time, and other faculty at all ranks is 20.

3. Admission, Financial Aid, and Housing

Address admission inquiries to: Graduate School
Graduate application fee required: $35 ($25 online)
Admission deadline (Fall admission): March 15
Admission requirements: For admission to the following graduate programs, the Graduate School requires an undergraduate GPA of 2.50. Physics: a Bachelor's degree in physics. Conditional admission may be granted to those without a standard undergraduate preparation in physics. Physics Education: a minor in physics and status as an in-service or prospective teacher. General science: a minor in a science with not less than 30 semester-hours in science and mathematics, and status as an in-service or prospective teacher. The GRE is not required. The GRE Advanced is not required. Students from non-English speaking countries are required to demonstrate proficiency in English via the TOEFL or ELI exam. Minimum acceptable score for admission is 550 or 85, respectively. Admission with lower scores, but requiring special English courses, may be granted.
Address financial aid inquiries to: Graduate Assistantships: Graduate School. Other: Director of Financial Aid
GAPSFAS application required: No
Financial aid deadline: Varies with aid
Loans available: Yes
Address housing inquiries to: housing@emich.edu
On-campus, single student housing available: Yes
Cost/term: $1,828[1]–2,909[2]
On-campus, married student housing available: Yes
Cost/month: $605–645 (1 bdrm.); $650–740 (2 bdrm.)

[1]double occupancy, meal plan extra
[2]single occupancy, meal plan extra

Table A—Faculty, Enrollments, and Degrees Granted

Research Specialty	2009–10 Faculty	Enrollment[1] Fall 2009		No. of Degrees Granted[2] 2009–10 (2005–10)			Median No. of Years for 2008–09 Ph.D.'s
		Master's	Doctorate	Master's	Terminal Master's	Doctorate	
Physics	10	7	–	4(11)	–	–	–
General Science	2	13	–	6(35)	–	–	–
Physics Education	5	7	–	1(8)	–	–	–
Total		27	–	11(54)	–	–	
Full-time Grad. Stud.		10	–				
Part-time Grad. Stud.		15	–				
First-year Grad. Stud.		11	–				
Median Years in Grad. Study (2008–09 Degrees)				3	–	–	–
Undergraduate Degrees, 2008–09 (2004–09): 7(35)							

[1] Students not yet committed to a research specialty are entered under non-specialized.
[2] Five-year totals in parentheses.

4. Graduate Degree Requirements

Master's: Physics: A "B" average in 30 semester hours of graduate credits on an advisor approved program is required. There are no language requirements. 18 hours of program must be taken on campus. An oral examination and a written research report/thesis are required. Physics Education: The requirements are as specified for the Physics Master's program, except that the program emphasizes courses beneficial to secondary school teachers. General science: A program for middle school science teachers. A "B" average in 30 semester hours of approved graduate credits. 18 hours of the program must be taken on campus. There are no language, oral examination, or thesis requirements.
Thesis: Thesis may be written *in absentia*.
Special Equipment, Facilities, or Programs: The department has among its facilities an observatory, remote (including graphics) terminals, microcomputers, a staffed machine shop, a multigigawatt laser laboratory, a plasma physics laboratory, a scanning tunneling microscope, and a modern optics laboratory.

Table B—Appointments to Graduate Students, 2009–10

Title of Appointee	Appointments Total	Appointments First year	Academic Load Allowed in Credit Hours	Hours of Service Per Week	Stipend for Academic Year ($)[1]
			Semester		
Graduate Assistant	8	4	6–12	20	8,400
Total	8	4			

[1] University waives tuition on the first 18 hours of study per fiscal year.

5. Personnel Engaged in Separately Budgeted Research, 7/09–6/10

Professorial faculty	7
Graduate students	6
Undergraduate students	8
Nonteaching research personnel	1
Total	22

6. Separately Budgeted Research Expenditures by Source of Support

	Departmental Research	Physics-related Research Outside Department
Federal	$109,000	$
Other	$330,000	$
Total	$439,000	$

Table C—Separately Budgeted Research Expenditures

Research Specialty	No. of Grants	Expenditures ($)
Condensed Matter	1	49,000
Laser Physics	1	10,000
Space Science	4	60,000
Surface Science	5	320,000
Total	11	439,000

FACULTY

Professors

Behringer, Ernest, Ph.D., Cornell, 1994. Solid state physics.
Carroll, James, Ph.D., West Virginia, 1997. Plasma physics.
Jacobs, Diane A., Ph.D., Texas, Austin, 1984. Solid state physics.
Oakes, Alexandria, Ph.D., Lehigh, 1986. Theoretical mechanics.
Sharma, Natthi, Ph.D., Ohio, 1982. Solid state physics.
Sheerin, James P., Ph.D., Michigan, 1980. Plasma physics.
Shen, Weidian, Ph.D., Wayne State, 1988. Condensed matter.
Thomsen, Marshall, Ph.D., Michigan State, 1984. Solid state physics.
Wylo, Bonnie, Ed.D., Michigan, 1993. Science education.

Assistant Professors

Koehn, Patrick, Ph.D., Michigan, 2002. Space physics.
Kubitskey, Beth, Ph.D., Michigan, 2006. Teacher Ed.
Pawlowski, David J., Ph.D., Michigan, 2009. Space physics.

RESEARCH SPECIALTIES AND STAFF

Theoretical

Acoustics. Oakes.
Computational Physics. Sheerin, Pawlowski.
Electromagnetism. Sharma.
Ethical Issues. Thomsen.
Mechanics. Gyroscopic motion. Oakes.
Optics. Sharma.
Plasma Physics. Koehn, Sheerin, Pawlowski.
Space Physics. Koehn, Sheerin, Pawlowski.
Solid State. Sharma, Thomsen.

Experimental

Condensed Matter. Shen.
Education Research. Carroll, Kubitskey, Wylo.
Fluid dynamics. Jacobs.
Nano Mechanics. Nano Tribology. Shen.
Nonlinear Phenomena. Jacobs, Sheerin.
Optics. Coherent optics. Sharma, Behringer.
Plasma Physics. Carroll, Sheerin.
Solid State. Behringer, Jacobs, Thomsen.
Space Physics. Koehn, Sheerin.
Strength of Materials. Oakes.
Surface Science. Shen.

MICHIGAN STATE UNIVERSITY

DEPARTMENT OF PHYSICS AND ASTRONOMY

East Lansing, Michigan 48824-2320

Students Accepted For Degree	FIELDS		
	Physics	Astronomy	Related Fields
Doctorate	X	X	X
Master's	X	X	X

1. General

President: LouAnna K. Simon
Dean of Graduate School: Karen Klomparens
Department Chairman: Wolfgang Bauer
Department Telephone Number: (517) 884-5532
Type of Institution: University
Control: Public
Setting: Suburban
Total Faculty: 4,985
Total Graduate Faculty: Not separate
Total Students: 47,278
Total Graduate Students: 10,789
Semester Graduate Tuition:
In-state residents: Full-time—$4,304.25/9 credit hours*
Out-of-state residents: Full-time—$8,698.50/9 credit hours*
Tuition rates for: 2009–10
Deferred tuition plan: For Graduate Assistants and On-Campus Residents
Other Fees: $1,342.50/year*
Term: Semester

*Graduate Assistants receive a full tuition waiver, a waiver of all registration fees, and are provided with health insurance.

2. Number of Faculty in Department

The combined total of full-time faculty in the three professorial ranks is 52. The combined total of full-time, part-time, and other faculty is 101.

3. Admission, Financial Aid, and Housing

Address admission inquiries to: Graduate Secretary, Department of Physics & Astronomy, simmons@pa.msu.edu http://www.pa.msu.edu.
Graduate application fee required: $50
Admission deadline (Fall admission): 1/30
Admission information: For fall admission, 2010, 58 students were offered and 21 accepted from a total of 360 applicants.
Admission requirements: For admission to the graduate programs, a Bachelor's degree in physics or astronomy is required with no minimum undergraduate GPA specified. Applicants with degrees in other fields will be considered and admitted on a provisional basis. The GRE is required. No minimum acceptable score for admission is specified. The GRE Advanced Physics is required. No minimum acceptable score for admission is specified. The average GRE scores for 2009–10 admissions were verbal—590; quantitative—760; analytical—4-6. The average GRE Advanced score for 2009–10 admissions was 730. Students from non-English speaking countries are required to demonstrate proficiency in English via the TOEFL exam. Minimum acceptable score for admission is 80 for the Internet Based Test (IBT).
Undergraduate preparation assumed: Bauer and Westfall, University Physics; Marion, Classical Dynamics; Pollack and Stump, Electromagnetism; Griffith, Introduction to Quantum Mechanics.
Address all inquiries to: Graduate Program Director, Michigan State University, Department of Physics & Astronomy, 1312 Biomedical & Physical Sciences Bldg., East Lansing, MI 48824-2320
GAPSFAS application required: Yes, for eligibility for certain scholarships
Assistantship and Financial aid deadline: 1/1
Loans available: Not through Physics & Astronomy Department.
Address housing inquiries to: Residence Halls Assignment Officer, W-185 Holmes Hall
On-campus, single student housing available: Yes
Cost/semester: $3,272.00
On-campus, single and married student apartments available: Yes
Cost/month: $700–774

Table A—Faculty, Enrollments, and Degrees Granted

Research Specialty	2009–10 Faculty	Enrollment[1] Fall 2009		No. of Degrees Granted[2] 2009–10[3] (2001–2010)			Median No. of Years for 2009 Ph.D.'s
		Master's	Doctorate	Master's	Terminal Master's	Doctorate	
Accelerator Physics	4	0	6	2(3)	0(1)	0(7)	–
Astronomy & Astrophysics	10	0	14	3(9)	0(11)	2(18)	–
Atomic, Molecular, & Optical Physics	3	0	1	0(0)	0(0)	1(1)	–
Beam & Mathematical Physics	1	0	3	0(5)	0(4)	0(6)	–
Biophysics	4	0	6	1(5)	0(2)	2(5)	–
Chemical Physics	4	0	2	0(0)	0(0)	0(0)	–
Condensed Matter Physics	13	0	39	2(7)	1(1)	3(78)	–
Materials Science	7	0	0	0(0)	0(0)	0(1)	–
Nuclear Physics	21	1	47	4(13)	1(11)	7(79)	–
Particles & Fields	17	1	15	1(2)	1(3)	4(27)	–
Physics Education	1	0	2	0(0)	0(0)	0(0)	–
Psychoacoustics	1	0	1	0(0)	0(0)	0(4)	–
Non-specialized	0	0	5	0(84)	0(21)	0(0)	–
Total	86*	2	143	13(128)	3(54)	19(22)	–
Full-time Grad. Stud.		2	143				
First-year Grad. Stud.		2	24				
Median Years in Grad. Study (2008–09 Degrees)				2	2	6	
Total Undergraduate Students, 2009–10				294			
Undergraduate Degrees, 2009–10 (2003–10): 41(216)							

[1] Students not yet committed to a research specialty are entered under non-specialized.
[2] Ten-year totals in parentheses.
[3] Includes Spring, Summer and Fall semesters.
*Several faculty members are in more than one area.

4. Graduate Degree Requirements

Master's: 30 semester credit hours with a 3.0 minimum GPA. 6 credits earned on campus is the minimum University residence requirement. No foreign language requirement. Thesis option with between 4 and 10 credits for thesis research (PHY 899) counted toward degree.

Physics Doctorate: There are 8 specific course requirements and a 3.0 GPA must be maintained. Two consecutive semesters is minimum residency requirement. Qualifying and candidacy exam must be passed at Ph.D. level; a thesis and thesis oral are required.

Astronomy & Astrophysics Doctorate: Must take 8 specific core courses with minimum of 3.375 GPA, complete 2-semester 2nd year research project, pass qualifying and candidacy exam and complete a Ph.D. thesis and thesis oral exam.

Other Programs: Accelerator Physics, Atomic and Molecular Physics, Chemical Physics, Biophysics, Beam and Mathematical Physics.

Special Equipment, Facilities, or Programs: Extensive research facilities in solid state and low-temperature physics, astronomy and astrophysics, nuclear physics and high-energy and particle physics. These include: two superconducting heavy-ion cyclotrons, K500 and K1200, with associated apparatus, a modern A1900 fragment separator allowing efficient production and inflight separation of rare isotopes, high-resolution S800 superconducting magnetic spectrograph, high resolution and high efficiency γ-ray, β-ray, neutron and charged particle arrays, 9-T ion trap, rf-fragment separators and laser spectroscopy system, x-ray diffraction apparatus employing x-ray cameras, and a 12 kW rotating anode x-ray source; class 100 clean room, uv mask/aligner, scanning electron microscope, multi pocket e-beam evaporator; photolithographic and electron-beam lithographic facilities for device fabrication with 50 nm resolution; cryogenic facilities, including three helium-3 refrigerators and two helium-3/helium-4 mixing refrigerators; two (55-kG and 105-kG) superconducting magnets; four electromagnets; ac and dc automated SQUID magnetometer; an ultrahigh-vacuum four-gun sputtering system; a 32 cubic meter anechoic room, 174 cubic meter reverberation room and 32 cubic meter double-walled quiet encolsure; a liquid argon calorimeter test station for the DØ and CDF experiments at Fermilab and detector development for future high-energy accelerators; a high-energy physics laboratory which is a state-of-the-art electronics design facility where detectors for experiments are developed, tested, and constructed; and numerous minicomputers and microcomputers in all research areas. Michigan State University is one of the primary partners (along with Notre Dame and the University of Chicago) in the Joint Institute for Nuclear Astrophysics (JINA), which is a $10M program funded as "Physics Frontier Center" focusing on the intersection of nuclear physics and astrophysics by the National Science Foundation. It addresses open questions related to the origin of the elements, compact stellar objects, stellar evolution, and stellar explosions through nuclear physics experiments, astronomical observations, scientific computing and theory. MSU is a partner in the new SOAR 4.1m telescope, which is located on a superb observing site in Chile, but is operated remotely from East Lansing. SOAR is optimized for very high image quality, and offers a wide range of optical and near-infrared imagers and spectrographs.

Table B—Appointments to Graduate Students, 2009–10

Title of Appointee	Appointments Total	Appointments First year	Academic Load Allowed in Credit Hours	Hours of Service Per Week	Stipend for 12 mos. Year ($)
			Semester		
Teaching Assistant	39	18	9	20	19,200
Research Assistant	92	1	6	–	19,200–24,000
Fellowship	12	9	–	–	–
Self	0	0	–	–	–
Total	143	28			

5. Personnel Engaged in Separately Budgeted Research/Teaching, 4/1/09–3/31/10

Professorial faculty	66
Other Faculty	25
Postdoctoral appointments	35
Graduate students	105
Undergraduate students	85
Non-teaching research personnel	48
Total	364

6. Separately Budgeted Research Expenditures by Source of Support

	Departmental Research	Physics-related Research Outside Department
Federal	$35,735,606	–
Private, non-profit organizations	111,764	–
Business & Industry	5,738	–
Other	741,703	–
Total	$36,594,811	–

Funding for the Physics portions of the National Superconducting Cyclotron Laboratory is included in Table 6 and Table C.

7. Separately Funded and Managed Laboratories

National Superconducting Cyclotron Laboratory (NSCL) -
The National Superconducting Cyclotron Laboratory (NSCL) is the leading rare isotope research facility in the United States. Located on the campus of Michigan State University, NSCL scientists and researchers employ a wide range of tools for conducting advanced research in fundamental nuclear science, nuclear astrophysics, and accelerator physics. Funded primarily by the National Science Foundation and MSU, the NSCL operates two superconducting cyclotrons. The K500 was the first cyclotron to use superconducting magnets, and K1200 is the highest-energy continuous beam accelerator in the world.

Southern Astrophysical Research Telescope (SOAR Telescope)
The SOAR Telescope is a 4.1m diameter optical/near-infrared telescope located at 9000 feet altitude in the Andes Mountains of Chile. This superb new instrument was dedicated in early 2004 and is operated by a consortium including Michigan State, the University of North Carolina at Chapel Hill, the National Optical Astronomy Observatory, and the country of Brazil. Michigan State astronomers received 12% of the observing time to carry out a wide variety of observational projects concerning the nature and history of our own Galaxy and of the Universe on a wider scale.

In addition to these two managed facilities, we also have several centers with funding from federal, state, and local sources. These are Joint Institute of Nuclear Astrophysics (JINA), The Center for the Study of Cosmic Evolution (CSCE), Institute of Quantum Sciences (IQS), Quantitative Biology Initiative (QBI), Nanoscale Interdisciplinary Research Team (NIRT) on Structure of Nanocrystals and Coordinated Theoretical-Experimental Project on QCD (CTEQ), Center for Nanomaterials and Assembly (CNDA).

JINA is a NSF funded Physics Frontier Center at the University of Notre Dame, Michigan State University, the University of Chicago, and Argonne National Laboratory. It provides an intellectual center with the goal to enable swift communication and stimulating collaborations across field boundaries (Nuclear Physics and Astrophysics) and at the same time provide a focus point in rapidly growing and diversifying field of Nuclear Astrophysics.

Michigan State has a 14% share of the new 4m SOAR Telescope, which is located in Chile. Our partners are the country of Brazil, the U.S. National Optical Astronomy Observatory and the University of North Carolina at Chapel Hill. MSU gets 12% of the observing time on SOAR. Many of our garduate students are carrying out Ph.D. thesis projects using SOAR. Michigan State also ia a part of the Sloan Digital Sky Survey consortium and we play a leading role in using SDSS to study the halo population of our galaxy.

The IQS is a multidisciplinary center involving Physics, Chemistry, and Mathematics at Michigan State University, with a focus on the rapidly growing area of quantum computing and quantum information sciences.

QBI is a broad interdisciplinary center involving faculty from several biology and physical science departments at MSU, including Physics. Fellowships for QBI are available and nontraditional study programs leading to interdisciplinary degrees are strongly encouraged.

The center for research excellence in complex materials (CORE-CM) and the center for nanomaterials design and assembly (CNDA) are two multidisciplinary centers focusing on synthesizing, characterizing and modeling of soft and hard materials to discover new science and to develop applications to energy generation and storage, and to nanotoxicology and nanomedicine.

CTEQ is a multi-institutional collaboration devoted to a broad program of research projects and cooperative enterprises in high-energy physics centered on Quantum Chromodynamics (QCD) and its implications in all areas of the Standard Model and beyond.

The ATLAS Great Lakes Tier 2 Center (AGL-T2) is, with it's partner location at the University of Michigan, one of the five Tier 2 computing centers for the 450 user U.S. ATLAS physics community. The facility was constructed with a combination of NSF and university resources and is operated through a 5 year grant with U.S. ATLAS. The center consists of a specially constructed room in the BPS building and consists of approximately 2000 job slots (450 of which are MSU Tier 3 nodes) and approximately 0.5 PB of storage. The facility is connected to the University of Michigan and the rest of the ATLAS grid through routers in Chicago. This is accomplished through a dedicated wavelength in redundant segments of the Michigan Lambda Rail 10Gbps fiber loop. The center is approximately 75% installed (2010).

Table C—Separately Budgeted Research Expenditures

Research Specialty	No. of Grants	Expenditures ($)
Astronomy	35	994,814
Astrophysics	4	281,312
Condensed Matter Physics	35	2,463,647
Nuclear Physics	53	29,726,045
Particles & Fields	26	2,868,013
Physics Education	1	139,170
Psychoacoustics	1	121,810
Total	164	36,594,811

Funding for the Physics portions of the National Superconducting Cyclotron Laboratory is included in Table 6 and Table C.

FACULTY

Professors

Baldwin, Jack A., Ph.D., California, Santa Cruz, 1974. Observational studies of quasars, planetary nebulae and HII regions.

Bauer, Wolfgang, Ph.D., Giessen, 1987. University Distinguished Professor and Chairman of the Department. Theoretical physics; intermediate heavy-ion reactions; nuclear transport theory; computational physics; nuclear astrophysics.

Beers, Timothy C., Ph.D., Harvard, 1983. University Distinguished Professor. Observational studies of first- and second-generation stars; stellar cosmo-chronometry; Galactic structure, kinematics and dynamics; Nuclear astrophysics.

Berz, Martin, Ph.D., Giessen, 1986. Beam physics, Mathematical physics, Computational physics.

Birge, Norman O., Ph.D., Chicago, 1986. Experimental condensed matter physics.

Bollen, Georg, Ph.D., University of Kaiserslautern, 1989. Nuclear physics, Atomic and molecular physics.

Brock, Raymond L., Ph.D., Carnegie-Mellon, 1980. Experimental high-energy physics. Gluon resummation phenomenology and applications.

Bromberg, Carl M., Ph.D., Rochester, 1974. Experimental high-energy physics; neutrino oscillations; particle detectors.

Brown, B. Alex, Ph.D., Stony Brook, 1974. Theoretical nuclear physics; shell model theory.

Chivukula, R. Sekhar, Ph.D., Harvard, 1987. Theoretical high-energy physics.

Danielewicz, Pawel, Ph.D., Warsaw University, 1981. Theoretical nuclear physics; relativistic heavy-ion physics.

Donahue, Megan, Ph.D., Colorado, 1990. Space and ground-based observational studies of galaxy clusters.

Duxbury, Phillip M., Ph.D., New South Wales, 1983. Theoretical condensed matter physics; Statistical mechanics.

Dykman, Mark, Ph.D., Kiev, 1973. Theoretical condensed matter physics.

Gelbke, C. Konrad, Ph.D., Heidelberg, 1973. University Distinguished Professor and Director of the National Superconducting Cyclotron Laboratory and the Facility for Rare Isotope Beams Laboratory. Experimental nuclear physics; Nuclear reactions of heavy ions.

Glasmacher, Thomas, Ph.D., Florida State, 1992. University Distinguished Professor and Project Manager of the Facility for Rare Isotope Beams Laboratory. Experimental nuclear physics.

Golding, Brage, Ph.D., MIT, 1966. Experimental condensed matter physics; mesoscopic physics, fluctuation phenomena.

Hartmann, William M., Ph.D., Oxford, 1965. Theoretical condensded matter physics, experiment and theory.

Huston, Joey W., Ph.D., Rochester, 1982. Experimental high-energy physics; direct photon production; experimental test of QCD.

Kester, Oliver, Ph.D., Frankfurt, 1996. Accelerator physics.

Linnemann, James T., Ph.D., Cornell, 1978. Experimental high-energy physics; searches for physics beyond the standard model of Hadron collisions, Particle astrophysics; statistical techniques.

Lynch, William G., Ph.D., Washington, 1980. Experimental nuclear physics.

Mahanti, S. D., Ph.D., California-Riverside, 1968. Theoretical condensed matter physics.

Mittig, Wolfgang, Ph.D., Paris, 1971. Hannah Professor, Physics.

Pope, Bernard G., Ph.D., Columbia, 1971. Experimental high-energy physics; tests of quantum chromodynamics; Hadron collider physics.

Pratt, Scott, Ph.D., Minnesota, 1985. Associate Chairperson and Graduate Program Director. Theoretical nuclear physics.

Pumplin, Jon, Ph.D., Michigan, 1968. Theoretical high-energy physics; strong interaction theory.

Repko, Wayne W., Ph.D., Wayne State, 1967. Theoretical high-energy physics; quantum electrodynamics and strong interaction theory.

Schatz, Hendrik, Ph.D., Heidelberg, 1997. Experimental and theoretical nuclear astrophysics.

Sherrill, Bradley, Ph.D., Michigan State University, 1985. University Distinguished Professor. Experimental nuclear physics.

Simmons, Elizabeth, Ph.D., Harvard, 1990. Dean of Lyman Briggs College. Theoretical high energy physics.

Smith, Horace A., Ph.D., Yale, 1980. Chemical evolution of galaxies; pulsating variable stars.

Stump, Daniel R., Ph.D., MIT, 1976. Associate Chairperson for Undergraduate Studies. Theoretical physics; strong interaction theory; quantum chromodynamics.

Thoennessen, Michael, Ph.D., Stony Brook, 1988. Experimental nuclear physics.

Tomanek, David, Ph.D., Freie Univ. Berlin, 1983. Theoretical condensed matter physics; computational physics.

Voit, G. Mark, Ph.D., Colorado, 1990. Theoretical astrophysics; evolution of galaxy clusters; AGN.

Westfall, Gary, Ph.D., Texas-Austin, 1975. University Distinguished Professor. Experimental heavy-ion physics, reaction mechanisms.

Yuan, Chien-Peng, Ph.D., Michigan, 1988. Theoretical high energy physics.

Zelevinsky, Vladimir, Ph.D., Budker Institute of Nuclear Physics, 1974. Theoretical nuclear physics.

Zepf, Stephen E., Ph.D., Johns Hopkins, 1992. Observational and theoretical studies of evolution of galaxies.

Adjunct Professors

Auerbach, Naftali, Ph.D., Weizmann Institute of Science, Israel, 1966. Physics.

Berger, Edmond, Ph.D., Princeton, 1965. Physics.

Billinge, Simon, Ph.D., Pennsylvania, 1992. Materials Science.

Capriotti, Eugene, Ph.D., Wisconsin, 1962. Astronomy.

Dantus, Marcos, Ph.D., California Institute of Technology, 1991. Chemistry.

Duguet, Thomas, Ph.D., Paris, 2002. Physics.

Fert, Albert, Ph.D., University of Paris-Sud at Orsay, 1970. Physics.

Fisher, Galen, Ph.D., Stanford, 1966. Physics.

Gruebele, Martin, Ph.D., Berkeley, 1988. Chemistry.

Hartung, , Walter, Ph.D., Cornell, 1996.

Hjorth-Jensen, Morten, Ph.D., Oslo, 1993. Physics.

Hussein, Mahir, Ph.D., Massachusetts Institute of Technology, 1971.

Izrailev, Felix, Ph.D., Budker Institute of Nuclear Physics, Russia, 1991. Physics.

Janssens, Robert, Ph.D., University of Catholique de Louvian-Belgium. 1978. Nuclear physics.

Johnson, Ronald, Ph.D., University of Manchester, 1961. Theoretical physics.

Johnstone, Carol, Ph.D., Texas-Austin, 1984. Experimental Nuclear physics.

Kortemeyer, Gerd, Ph.D., Michigan State University, 1997. Physics education.

Kuhn, Leslie, Ph.D., Pennsylvania, 1989. Biophysics.

Lambert, David, Ph.D., Berkeley, 1979. Condensed matter physics.

Langanke, , Karlheinz, Ph.D., University of Muenster Germany, 1985.

Mackay, Michael, Ph.D., Illinois-Urbana, 1985, Chemical engineering.

Marti, Felix, Ph.D., Michigan State University, 1977. Nuclear physics.

Morelli, Donald, Ph.D., Michigan, 1985. Chemical engineering and material science.

Ormand, William, E., Ph.D., Michigan State University, 1985. Nuclear physics.

Otsuka, Takaharu, Ph.D., Tokyo, 1979. Physics.

Platzman, P., Ph.D., California Institute of Tech., 1960.

Sinnis, Constantine, Ph.D., Hawaii, 1990. Physics.

Tostevin, Jeffrey, Ph.D., Surrey, 1978. Physics.

Tsang, Manyee Betty, Ph.D., Washington-Seattle, 1980. Nuclear chemistry.

Wangler, Thomas, Ph.D., Wisconsin, 1964. Accelerator physics.

Wiescher, Michael, Ph.D., Universität Münster, 1980. Nuclear physics.

York, Richard, Ph.D., Iowa, 1976. Accelerator physics.

Associate Professors

Brown, Edward, Ph.D., Berkeley, 1999. Theoretical astrophysics. Compact objects, nuclear astrophysics.

Gade, Alexandra, Ph.D., Cologne, 2001. Experimental nuclear physics.

Loh, Edwin D., Ph.D., Princeton, 1977. Astrophysics.

Nunes, Filomena, Ph.D., Surrey, 1995. Theoretical nuclear physics; few body structures and reactions.

Piermarocchi, Carlo, Ph.D., Ecole Polytechnique Federale de Lausanne, Switzerland, 1998. Theoretical condensed matter physics.

Ruan, Chong-Yu, Ph.D., Texas-Austin, 2000. Experimental condensed matter physics. Experimental molecular and optical physics, chemical physics.

Schmidt, Carl. Ph.D., Harvard, 1990. Theoretical high energy physics.

Tessmer, Stuart, Ph.D., Illinois-Urbana, 1995. Experimental Condensed matter physics.

Tollefson, Kirsten, Ph.D., Rochester, 1997. Experimental high energy physics.

Zegers, Remco G. T., Ph.D., University of Groningen, The Netherlands, 1999. Experimental nuclear physics.

Research Associate Professors

Ghosh, Ruby, Ph.D., Cornell, 1991. Solid state and optical device physics.
Kundu, Arunav, Ph.D., Maryland, 1999. Astronomy.
Makino, Kyoko, Ph.D., Michigan State University, 1998, Beam physics, computational physics.

Assistant Professors

Bogner, Scott, Ph.D., Stony Brook, 2002. Theoretical nuclear physics.
Doleans, Marc, Ph.D., Paris, 2003. Accelerator physics.
Fisher, Wade C., Ph.D., Princeton, 2004. Experimental high energy physics.
Iwasaki, Hironori, Ph.D., Tokyo, Japan, 2001. Experimental nuclear physics.
Lai, Chih-Wei, Ph.D., Berkeley, 2004. Cowen Chair. Experimental condensed matter physics.
Lapidus, Lisa, Ph.D., Harvard, 1998. Experimental biophysics.
McGuire, John A., Ph.D., Berkeley, 2004. Experimental condensed matter physics.
Moore, Michael, Ph.D., Arizona, 1999. Theoretical atomic and molecular physics.
O'Shea, Brian, Ph.D., Illinios-Urbana, 2005. Computational astrophysics and astronomy.
Schwienhorst, Reinhardt, Ph.D., Minnesota, 2000. Experimental high energy physics.
Spyrou, Artemisia, Ph.D., Athens, Greece, 2007. Experimental nuclear physics, nuclear astrophysics.
Wedemeyer, William J., Ph.D., Cornell, 1998. Theoretical and experimental biophysics.
Zhang, Peng Peng, Ph.D., Wisconsin-Madison, 2006. Experimental condensed matter physics.

Research Assistant Professors

Hauser, Reiner, Ph.D., Heidelberg University, 1994. Experimental high energy physics.

Professors Emeriti

Abolins, Maris A., Ph.D., California San Diego, 1965.
Austin, Sam M., Ph.D., Wisconsin, 1960. University Distinguished Professor.
Bass, Jack, Ph.D., Illinois, 1964.
Benenson, Walter, Ph.D., Wisconsin, 1962. University Distinguished Professor.
Blosser, Henry G., Ph.D., Virginia, 1954. University Distinguished Professor.
Carlson, Edward H., Ph.D., Johns Hopkins, 1959.
Galonsky, Aaron I., Ph.D., Wisconsin, 1954.
Harrison, Michael, Ph.D., Chicago, 1960.
Kaplan, Thomas A., Ph.D., Pennsylvania, 1954.
Kashy, Edwin, Ph.D., Rice, 1959. University Distinguished Professor.
Kovacs, Julius S., Ph.D., Indiana, 1955.
McManus, Hugh, Ph.D., Birmingham, 1947.
Parker, Paul M., Ph.D., Ohio State, 1958.
Pollack, Gerald, Ph.D., Cal Tech., 1962.
Pratt, William P., Ph.D., Minnesota, 1967.
Schriber, Stan, Ph.D., McMaster, 1966.
Schroeder, Peter A., Ph.D., Bristol, 1955.
Signell, Peter S., Ph.D., Rochester, 1958.
Stein, Robert, Ph.D., Columbia, 1966.
Thorpe, Michael F., D. Phil., Oxford, 1968; D. Sc., Oxford, 1993. University Distinguished Professor.
Weerts, Hendrick J., Ph.D., RWTH Aachen, 1981.

RESEARCH SPECIALTIES AND STAFF

Astronomy and Astrophysics

Observational cosmology and the evolution of galaxies and galaxy clusters, including theoretical studies and X-ray, optical and infrared observational studies of globular clusters as tracers of galactic evolution, quasars and active galaxies, and galaxy clusters as cosmological probes. Structure and evolution of our Galaxy, including observational studies of first- and second-generation stars, galactic chemical evolution, pulsating variable stars, the structure and chemical composition of planetary nebulae and HII regions, and the dynamics of old stellar populations. Theoretical nuclear astrophysics, physics of compact objects and of supernova explosions. Theoretical modeling of the solar atmosphere. Computational Astrophysics. Astronomical instrumentation for the SOAR telescope. Baldwin, Bauer, Beers, E. Brown, Donahue, Loh, O'Shea, Smith, Voit, Zepf, 14 graduate students and 6 research associates.

Physics - Theoretical

Atomic and Molecular Physics. Nonlinear and quantum optics, coupled matter-light systems, Ion trap physics. Moore, Piermarocchi, 1 graduate student.

Beam and Mathematical Physics. Nonlinear dynamics, beam physics, computational physics, self-validated numerical methods, nonarchimedian analysis; VU Beam program. Berz, Makino, 3 local graduate students and 6 remote graduate students in VU Beam program.

Biophysics. Control theory and optimization methods applied to systems biology; theoretical models of gene regulatory networks; information theory applied to biological systems; development of new statistical methods for the analysis of high-throughput experiments in cellular systems; combinatorial optimization in drug discovery. Duxbury, Piermarocchi.

Condensed Matter Physics. Electronic properties of surfaces, clusters, nanostructures, nanowires, nanotubes, and novel narrow band gap semiconductors, exciter and polariton condensates. Magnetism in strongly correlated systems. Optical properties of quantum dots, quantum wires, and quantum wells. Transport in correlated two-dimensional electron systems; electrons on helium. Many-electron tunneling, activated processes in nonequilibrium systems. Quantum computing using condensed matter systems. Atomic structure of complex materials, nanoparticles, nanoparticle dispersion and their applications to energy and nanomedicine. Photophysics, combinatorial optimization methods in statistical physics with applications to complex materials. Duxbury, Dykman, Mahanti, Piermarocchi, Tomanek, 13 graduate students and 2 research associates.

Elementary Particle Physics. Fundamental structure and interaction of elementary particles; global analysis of high-energy processes to determine the parton (quarks and gluons) structure of hadrons; stringent tests of quantum chromodynamics; origin of spontaneous symmetry breaking in the unified theory of electro-weak interactions; physics of the Higgs particle and the top quark; new phenomena at super high energies; quantum electrodynamics and quantum chromodynamics of bound systems; Monte Carlo methods in high-energy physics; astro-particle physics and constraints of dark energy. Chivukula, Pumplin, Repko, Schmidt, Simmons, Stump,

Yuan, 2 postdoctoral research associates and 3 graduate students.

Nuclear Theory. Quantum many-body theory for nuclear structure, nuclear reactions, nuclear astrophysics, weak interactions physics and properties of mesoscopic systems. Nuclear structure includes applications of ab-initio, configuration interaction, collective model and energy-density functional methods. Nuclear reactions span energies from low- to ultra-relativistic and include quantum transport and coupled-channel methods. Applications are made to structure and reactions of rare isotopes studies at the NSCL and other facilities. For nuclear astrophysics our work relates to the properties of neutron stars, stellar evolution and nucleosynthesis. Applications for fundamental symmetries involve nuclear structure aspects of parity violations, time-reversal violation and standard-model tests from nuclear observables. Bauer, Bogner, Brown, Danielewicz, Nunes, Pratt, Zelevinsky, 2 postdoctoral research associates, 13 graduate students and several visiting faculty.

Physics - Experimental

Accelerator Physics. Design and construction of superconducting cyclotrons and linacs for heavy ions; ECR ion sources and spectrometers and fragment separators. Research in high current beam optics and superconducting radio frequency accelerating structures. Doleans, Hartung, Kester, Marti, York, Zeller, 2 postdoctoral research associates, 6 graduate students.

Acoustics, Psychoacoustics. Signal processing by the human Brain — physiologically based models and psychoacoustical experiments. Emphasis on binaural processing and sound localization in free field and complex room environments. Hartmann, 1 graduate student.

Atomic Physics. Traps, precision measurements. Applications to nuclear physics. Bollen (for students and research associates see Nuclear Physics below).

Biophysics. Fast folding studies of proteins and nucleic acids using optical spectroscopy and microfluidics. Measurement and modeling of unfolded protein dynamics. Scanning probe methods to study biological nanowire systems. Lapidus, Tessmer, 5 graduate students, 2 postdoctoral research associates.

Condensed Matter & Chemical Physics. Quantum phenomena in solids, including quantum transport, optical control of electron and nuclear spins, and electronic control of magnetization in ferromagnets. Growth, structure and properties of novel materials and nanostructures, including diamond growth, ultrafast phenomena and many-body effects in semiconductor nanostructures. Scanning probe microscopy, single-electron capacilance spectroscopy of semiconductor nanostructures, quantum Hall effect. Study of fundamental properties of electronic and photovoltaic nanomaterials using scanning probe microscopy and manipulation of properties of these nanomaterials via surface and interface engineering. Characterization of devices made out of these nanomaterials. Quantum dots: enhanced charge-charge and charge-spin interactions, study properties of single spins in quantum dots. Dye sensitized solar cells, charge transport and carrier enhancement. Surface nonlinear optics. Optical tweezers. Ultrafast phenomena in solids, clusters and surfaces. Material processes far from equilibrium. Ultrafast electron imaging and diffraction. Chemical sensing techniques using fiber optics and solid state devices. Electronic properties of wide band gap semiconductors at high temperature. Semiconductor quantum optics. Transport in metals and alloys, including magnetic multilayers and the interplay between superconductivity and magnetism. Bass, Birge, Ghosh, Golding, Lai, McGuire, Pratt, Ruan, Tessmer, Zhang, 2 postdoctoral research associates, 28 graduate students.

Elementary Particle and High-Energy Physics. Hadron-induced high-energy and super-high-energy interactions; the inclusive production and study of jets and other hadronic final states; the inclusive production of leptons and photons; experimental tests of quantum chromodynamics; measurement of the quark and gluon structure of hadrons; studies of heavy flavor production (including studies of the top quark); experimental tests of electro-weak interactions including the production and study of top quarks; searches for supersymmetric particles and other new phenomena beyond the standard model; Neutrino properties, including interaction cross-sections and oscillation properties: mass hierarchy, mixing angles, a search for CP violation; development of the liquid Argon time projection chambers for neutrino and proton-decay physics; particle astrophysics including cosmic ray physics; design principles for fast digitization, pipelined processing techniques, and computer software for high-energy event selection algorithms; design (and R&D) for experiments to be performed at the Fermi National Accelerator Laboratory (CDF, D0 and NOVA experiments) and the CERN Large Hadron Collider (Atlas experiment). Brock, Bromberg, Fisher, Hauser, Huston, Linnemann, Pope, Schwienhorst, Tollefson, 6 postdoctoral research associates, 13 graduate students.

Nuclear Physics. Test of nuclear models and studies of fundamental interactions; ion trapping for high precision mass measurements; study of rare isotopes with a large excess of neutrons and protons; search for new isotopes and exotic decay modes; exploration of structural changes of rare isotopes, resonance properties of nuclei; isospin dependence of nuclear reactions; nuclear astrophysics, stellar evolution, nova and supernova explosions, galactic chemical evolution, properties of neutron stars and properties of the quark gluon plasma. Bollen, Gade, Gelbke, Glasmacher, Iwasaki, Lynch, Schatz, Sherrill, Spyrou, Thoennessen, Tsang, Westfall, Zegers, 16 postdoctoral research associatesand 35 graduate students, and many short-term visitors.

MICHIGAN TECHNOLOGICAL UNIVERSITY

DEPARTMENT OF PHYSICS

Houghton, Michigan 49931

Students Accepted For Degree	FIELDS		
	Physics	Astronomy	Related Fields
Doctorate	X	X	X
Master's	X	X	X

1. General

President: Glenn D. Mroz
Dean of Graduate School: Jacqueline E. Huntoon
Department Chair: Ravindra Pandey
Department Telephone Number: (906) 487-2086 (C)
Fax: (906) 487-2933
email: physics@mtu.edu
Website: http://www.phy.mtu.edu
Type of Institution: University
Control: Public
Setting: Small town
Total Faculty: 437
Total Students: 7,014
Total Graduate Faculty: 590
Total Graduate Students: 981
Annual Graduate Tuition:
In-state residents: Full-time—$5,355/$595 per credit hour
Out-of-state residents: Full-time—$5,355/$595 per credit hour
Tuition rates for: Fall, 2010, estimated
Deferred tuition plan: No
Annual Other Fees: Variable*
Term: Semester

*$20 Grad. fee for binding of thesis. Any lab fees required in curriculum. $234 per year for health insurance (supported student), if not covered by a policy. Consolidated fee $136/yr.

2. Number of Faculty in Department

The combined total of full-time faculty in the three professorial ranks is 19. The combined total of full-time, part-time, and other faculty at all ranks is 22.

3. Admission, Financial Aid, and Housing

Address admission inquiries to: Dean of Graduate School
Graduate application fee: none
Admission deadline (Fall admission): February 1.
Admission information: For fall admission, 2009–10, 13 students were accepted from 96 applicants. All admitted students are typically offered assistantships or fellowships that supply competitive stipends, full tuition remission, and a health insurance contribution.
Admission requirements: For admission to the graduate programs, a Bachelor's degree in physics is required with a minimum undergraduate GPA of 3.0 specified. However, degree recipients in related areas often apply and are accepted. The GRE is required. The GRE Advanced is not required. Students from non-English speaking countries are required to demonstrate proficiency in English via the TOEFL exam. Minimum acceptable score for admission is 550; (paper-based), 230 (computer-based), 79 (internet-based) .
Undergraduate preparation assumed: Taylor Classical Mechanics; Eisberg and Resnick, *Quantum Physics*; Griffiths, *Introduction to Electrodynamics.*
Address financial aid inquiries to: Graduate Studies Chair, Physics Department/MTU, 1400 Townsend Dr., Houghton, MI 49931-1295 or visit our web site.
GAPSFAS application required: No
Financial aid deadline: None
Loans available: No
Address housing inquiries to: Director of Housing
On-campus, single student housing available: Yes
Cost/year average: $7,732 (double occupancy, with board);
On-campus, married student housing available: Yes
Cost/month: $540/1 bdrm.
$600/2 bdrms.
$775/3 bdrms.

Table A—Faculty, Enrollments, and Degrees Granted

Research Specialty	2009–10 Faculty	Enrollment[1] Fall 2009 Master's	Enrollment[1] Fall 2009 Doctorate	No. of Degrees Granted[2] 2009–10 (2005–10) Master's	Terminal Master's	Doctorate	Median No. of Years for 2008–09 Ph.D.'s
Astronomy	1	2	1	0(0)	1(0)	0(1)	–
Astrophysics	3	1	2	0(0)	0(0)	0(2)	–
Atmos./Space Phys., Cosmic Rays	4	1	7	0(0)	1(1)	1(3)	–
Atomic, Molecular, & Optical Physics	3	2	1	0(0)	0(2)	0(1)	–
Biophysics	1	0	2	0(1)	0(1)	1(1)	–
Condensed Matter Physics	7	1	9	0(1)	0(2)	2(5)	–
Materials Sci./ Metallurgy	3	2	11	0(1)	2(4)	1(5)	–
Non-specialized	0	0	0	0(0)	0(0)	–	–
Total		9	33	0(3)	4(10)	5(18)	
Full-time Grad. Stud.		9	33				
Part-time Grad. Stud.		0	0				
First-year Grad. Stud.		3	9				
Median Years in Grad. Study (2007–08 Degrees)							–

Undergraduate Degrees, 2008–09 (2004–09): 15(45)

[1] Students not yet committed to a research specialty are entered under non-specialized.
[2] Five-year totals in parentheses.

4. Graduate Degree Requirements

Please see requirements at: http://www.phy.mtu.edu describing our Ph.D. programs in Physics and also Engineering Physics.

Table B—Appointments to Graduate Students, 2009–10

Title of Appointee	Appointments Total	Appointments First year	Academic Load Allowed in Credit Hours	Hours of Service Per Week	Stipend for Academic Year ($)
			Semester		
Fellow	0	0	9		30,000
Teaching Assistant	21	11	9	20[1]	16,641–18,837[2]
Research Assistant	19	0	9	20	16,641–18,837[2]
Grad. Assistant	0	0	9	20	16,641–18,837
Total	40	11			

[1]Includes 6–10 contact hours with students and time for preparation, grading, and consulting with students.
[2]Summer appointments with additional stipend and tuition may be available, funds permitting.

5. Personnel Engaged in Separately Budgeted Research, 6/09–7/10

Professorial faculty	14
Graduate students	19
Nonteaching research personnel	5
Total	38

6. Separately Budgeted Research Expenditures by Source of Support*

	Departmental Research	Physics-related Research Outside Department
Federal government	$1,368,816	$
Internal Sources	1,166,434	
Total*	2,535,250	$

FACULTY

Professors

Beck, Donald R., Ph.D., Lehigh, 1968. Computational atomic physics.

Borysow, Jacek I., Ph.D., Texas, Austin, 1986. Laser spectroscopy.

Fick, Brian E. Ph.D., Virginia Polytechnic Institute and State University, 1985. Experimental astro-particle physics.

Hansmann, Ulrich H. E., Freie Universitat, Germany, 1990. Biomolecular modeling.

Jaszczak, John A., Ph.D., Ohio State, 1989. Computational solid state physics; nanotechnology education.

Kostinski, Alexander, Ph.D., Illinois, Chicago, 1984. Radar meteorology, polarization optics, and atmospheric physics.

Levy, Miguel, Ph.D., City Univ. of New York, 1988. Surface physics.

Nemiroff, Robert J., Ph.D., Univ. of Pennsylvania, 1987. Astronomy; astrophysics.

Nitz, David, Ph.D., Rochester, 1978. High energy astrophysics.

Pandey, Ravindra, Ph.D., Univ. of Manitoba, Canada, 1987. Department Chair. Materials theory.

Perger, Warren F., Ph.D., Colorado State, 1986 (part-time). Computational atomic and condensed matter physics.

Seel, Maximillian J., Ph.D., University of Erlangen, West Germany, 1978. Provost. Theoretical chemistry.

Shaw, Raymond, Ph.D., Pennsylvania State University, 1998. Atmospheric sciences.

Suits, Bryan, Ph.D., Illinois, 1981. Condensed matter experiment; NMR.

Associate Professors

Cantrell, Will, Ph.D., University of Alaska, 1999. Atmospheric sciences.

Moran, Peter D., Ph.D. Materials physics; device fabrication and characterization.

Pati, Ranjit, Ph.D., University at Albany, State University of New York, 1998. Condensed matter theory and materials science.

Weidman, Robert S., Ph.D., Illinois, 1980. Condensed matter theory; electronic structure.

Yap, Yoke Khin, Ph.D., Osaka University, 1999. Materials and Laser physics.

Assistant Professors

Huentemeyer, Petra, Ph.D., University of Hamburg, Germany, 2001. Particle astrophysics.

Lee, Kim Fook, Ph.D., Duke University, 2002. Experimental quantum optics.

Mazzoleni, Claudio, Ph.D., University of Nevada, 2003. Atmospheric physics.

RESEARCH SPECIALTIES AND STAFF

More detailed descriptions of faculty, research and facilities can be found at http://www.phy.mtu.edu

Theoretical

Astrophysics, gravitational lensing, high energy astrophysics, close binary stars, and cosmology. Fick, Nemiroff, Nitz.

Atomic, Molecular, and Condensed Matter Physics. Electronic structure of metals and oxides, polymers and biological molecules, point defects, surfaces, metal–insulator transitions, atomistic simulations of materials, relativistic and correlation effects in atomic structure. Beck, Hansmann, Jaszczak, Pandey, Pati, Perger, Seel.

Atmospheric Physics. Application in satellite meteorology, optics and digital image processing. Cloud precipitation and nucleation. Kostinski.

Astrophysics & Astronomy. All-sky monitoring and gamma ray burst detection. Nemiroff.

Experimental

Atmospheric Physics. Application in optics, digital image processing, cloud precipitation and nucleation, climate, and air quality. Cantrell, Mazzoleni, Shaw.

Atomic, Molecular, Optical, and Condensed Matter Physics. X-ray diffraction; laser spectroscopy; magnetic resonance, cosmic rays. Borysow, Fick, Huentemeyer, Lee, Levy, Moran, Nitz, Suits, Yap.

OAKLAND UNIVERSITY

DEPARTMENT OF PHYSICS

Rochester, Michigan 48309-4487

Students Accepted For Degree	FIELDS		
	Physics	Astronomy	Related Fields
Doctorate			X[1]
Master's	X		

[1]Ph.D. in Biomedical Sciences-Medical Physics.

1. General

President: Gary Russi
Executive Director of Graduate Study: Claire Rammel
Department Chairman: Andrei Slavin
Department Telephone Number: (248) 370-3416
Type of Institution: University
Control: Public
Setting: Suburban
Total Faculty: 482
Total Graduate Faculty: 482
Total Students: 18,920
Total Graduate Students: 3,645
Annual Graduate Tuition:
In-state residents: $511/credit
Out-of-state residents: $881.50/credit
Tuition rates for: 2009–10
Deferred tuition plan: No
Other Fees: none
Term: Semester

2. Number of Faculty in Department

The combined total of full-time faculty in the three professorial ranks is 12. The combined total of full-time, part-time, and other faculty at all ranks is 39.

3. Admission, Financial Aid, and Housing

Address admission inquiries to: Department of Physics
Graduate application fee required: None
Admission deadline (Fall admission): August 1
Admission information: For fall admission, 2009–10, 3 students were accepted from 46 applicants.
Admission requirements: For admission to the graduate programs, a Bachelor's degree in physics is required with no minimum undergraduate GPA specified. The GRE and GRE Advanced are required for Ph.D. program only. The GRE Advanced is required for foreign students. Students from non-English speaking countries are required to demonstrate proficiency in English via the TOEFL exam. Minimum acceptable score for admission is 550, 213 computer-based, 79 internet-based, or 77 on the MELab or having a baccalaureate or more advanced degree from an institution in the U.S.
Undergraduate preparation assumed: Fowles, *Analytical Mechanics*; Reitz and Milford, *Electromagnetic Theory*; Tipler, *Modern Physics*; Jenkins and White, *Optics*; Saxon, *Elementary Quantum Mechanics*.
Address financial aid inquiries to: Financial Aid Office, 120 North Foundation Hall
GAPSFAS application required: No
Financial aid deadline: As soon as possible after January 1
Loans available: Yes
Address housing inquiries to: Housing Office, 448 Hamlin Hall
On-campus, single student housing available: Yes
Cost/term: $3,995 (room and board)
On-campus, married student housing available: Yes
Cost/month: $700 (plus utilities)

Table A—Faculty, Enrollments, and Degrees Granted

Research Specialty	2009–10 Faculty	Enrollment[1] Fall 2009		No. of Degrees Granted[2] 2008–09 (2004–09)			Median No. of Years for 2008–09 Ph.D.'s
		Master's	Doctorate	Master's	Terminal Master's	Doctorate	
Applied Physics	0	8	0	0(0)	2(8)	0(0)	–
Astrophysics	1	–	0	0(0)	0(0)	0(0)	–
Condensed Matter Physics	5	–	0	0(0)	0(0)	0(0)	–
Medical & Health Physics	4	–	18	0(0)	0(0)	3(12)	4
Statistical & Thermal	1	–	0	0(0)	0(0)	0(0)	–
Total		8	18	0(0)	2(8)	3(12)	
Full-time Grad. Stud.		8	18				
Part-time Grad. Stud.		0	0				
First-year Grad. Stud.		2	3				
Median Years in Grad. Study (2008–09 Degrees)				2.5	–	–	–
Undergraduate Degrees, 2008–09 (2004–09): 10(41)							

[1] Students not yet committed to a research specialty are entered under non-specialized.
[2] Five-year totals in parentheses.

4. Graduate Degree Requirements

Master's: 36 credits of graduate courses including four credits of PHY 673 (Quantum Mechanics), 24 credits of additional 400–500–600-level courses approved by the department, eight credits of research, including a thesis or a critical essay. No foreign language requirements.
Doctorate: Biomedical Sciences-Medical Physics. 90 semester hours of graduate credit including at least 30 hours of thesis research, grade point average of 3.0 or higher, three full-time equivalent semesters (at least 12 credits/semester) in residence, qualifying examination, dissertation.
Thesis: Thesis may be written *in absentia*.
Special Equipment, Facilities, or Programs: Research facilities in the high pressure optics laboratory include Raman spectrometers with a single or multi-channel detectors, a single grating spectrometer for photoluminescence studies in the visible and infrared regions, argon ion and Ti: sapphire lasers, high pressure cells capable of generating 10 GPa and closed cycle helium refrigerators. Research facilities in the condensed matter physics laboratories include a Faraday magnetometer, an AC susceptometer, a ferromagnetic resonance spectrometer at x-band, systems for wide-band magnetoelectric characterization, 1–40 GHz vector network analyzers, a Philips x-ray diffractometer, one and two kilowatt RF power supplies with 50-W matching networks for silicon ribbon growth, and vacuum facilities for thin film evaporation and fullerene preparation. Biomagnetic research facilities include the Kettering Magnetics Laboratory that provides a non-magnetic environment. Research facilities in the NMR microscopy laboratory include a Bruker AMX 300 NMR spectrometer with a 7-Tesla/89-mm vertical bore superconducting

magnet and micro-imaging accessories. Among research facilities in neighboring hospitals available to medical physics students and faculty are a 3.0-Tesla whole-body NMR system and a 7.0-Tesla/20-cm horizontal bore magnet NMR system for imaging and in vivo spectroscopy, a megawatt-pulsed tunable dye laser and argon ion laser for photodynamic therapy research, a 148-channel whole head SQUID neuromagnetometer, a nuclear medicine laboratory, radiology and CT scanning facilities, an advanced modalities cancer therapy laboratory including radiotherapy and hyperthermia, diagnostic ultrasonic equipment, a laser surgery laboratory and two major hospital medical libraries.

Table B—Appointments to Graduate Students, 2009–10

Title of Appointee	Appointments Total	Appointments First year	Academic Load Allowed in Credit Hours	Hours of Service Per Week	Stipend for Academic Year ($)
			Semester		
Teaching Assistant	3	1	8 (M.S.)	20	6,500
Research Assistant	9	1	12 (Ph.D.)	Variable	14,000–21,000
Graduate Fellowship	5	2	12 (Ph.D.)	Variable	14,000
Total	17	4			

5. Personnel Engaged in Separately Budgeted Research, 7/09–6/10

Professorial faculty	7
Graduate students	11
Total	18

6. Separately Budgeted Research Expenditures by Source of Support

	Departmental Research
Federal government	$7,211,272
Total	$7,211,272

Table C—Separately Budgeted Research Expenditures

Research Specialty	No. of Grants	Expenditures ($)
Condensed Matter Physics	11	2,170,130
Gravitation	1	132,683
Medical & Health Physics	6	4,908,459
Total	18	7,211,272

FACULTY

Professors

Chopp, Michael J., Ph.D., NYU, 1975. Medical physics.
Elder, Ken, Ph.D., University of Toronto, 1989. Statistical physics.
Garfinkle, David, Ph.D., Chicago, 1985. Astrophysics; General relativity.
Roth, Bradley, Ph.D., Vanderbilt University, 1987. Medical physics.
Slavin, Andrei, Ph.D., Leningrad Technical University, 1977. Condensed matter theory; solitons. Chair of the Department.
Srinivasan, Gopalan, Ph.D., Indian Inst. of Technology, 1980. Superconductivity; ferromagnetism.
Venkateswaran, Uma Devi, Ph.D., Missouri, 1985. Optical properties of solids; high pressures.
Xia, Yang, Ph.D., Massey University (New Zealand), 1992. NMR Physics.

Associate Professors

Martins, George, Ph.D., Campinas State University (Brazil), 1994. Condensed matter theory.
Rojo, Alberto, Ph.D., Instituto Balseiro, Bariloche (Argentina), 1990. Condensed matter theory.

Assistant Professor

Khain, Evgeniy, Ph.D., Hebrew University of Jerusalem, 2005. Medical physics and statistical physics.

Visiting Assistant Professor

Surdutovich, Eugene, Ph.D., Wayne State University, 1998. Medical physics.

Professor Emeritus

Liboff, Abraham R., Ph.D., NYU, 1964. Biophysics; medical physics.
Tepley, Norman, Ph.D., MIT, 1963. Solid state and low-temperature physics; biomagnetism; medical physics.

Adjunct Professors

Bleil, Carl, Ph.D., Oklahoma, 1953. Superconductivity.
Brown, Stephen L., Ph.D., Toronto, 1991. Medical biophysics.
Dworkin, Howard J., M.D., Albany Medical College, 1959. Medical physics.
Gerhart, Grant R., Ph.D., Wayne State University, 2000. Condensed matter theory.
Kim, Jae Ho, M.D., Kyunpook University, 1959; Ph.D., Iowa, 1963. Radiobiology and biophysics.
Mantese, Joseph, Ph.D., Cornell University, 1986. Materials science.
Portnoy, Harold, M.D., Wayne State, 1956. Medical physics.
Sabbah, Hani, Ph.D., Oakland University, 1988. Medical physics.
Venkatesan, Srinivasan, Ph.D., University of London, 1974. Fuel cells; renewable energy.
Yan, Di, Ph.D., Washington University, 1990. Medical physics.

Adjunct Associate Professors

Castoldi, Kapila Clara, Ph.D., University of Milan, Italy, 1976. Physics education.
Demetropoulos, Constantine, Ph.D., Wayne State University, 2001. Medical physics.
Ewing, James R., Ph.D., Oakland University, 1992. Medical physics.
Hammond, Robert L., Ph.D., Wayne State University, 2000. Biological physics.
Jiang, Quan, Ph.D., Oakland University, 1991. Medical physics.
Knight, Robert A., Ph.D., Oakland University, 1991. Medical physics.
Liang, Jian, Ph.D., Zhejiang University, China, 1994. Medical physics.
McDermott, Patrick N., Ph.D., University of Rochester, 1985. Medical physics.

Nathanson, S. David, M.D., Witwatersrand, 1966. Medical physics.
Robinson, Stephen E., Ph.D., Hahnemann Medical College, 1973. Medical physics.

Adjunct Assistant Professors

Bowyer, Susan, Ph.D., Oakland University, 1998. Medical physics.
Ionascu, Dan, Ph.D., Northeastern University, 2005. Medical physics.
Jenrow, Kenneth, Ph.D., Oakland University, 1995. Medical physics.
Matuszak, Martha, Ph.D., University of Michigan, 2007. Medical physics.
Zhang, Tiezhi, Ph.D., University of Wiscosin–Madison, 1999. Medical physics.
Zhang, Zheng-Gang, Ph.D., Oakland University, 1995. Medical physics.

Clinical Instructor of Medical Physics

Carlson, Ray, M.S., Wayne State University, 1986. Medical physics.

RESEARCH SPECIALTIES AND STAFF

Theoretical

Astrophysics. General relativity theory. Garfinkle.
Condensed Matter Physics. Elder, Martins, Rojo, Slavin.
Medical physics. Khain, Roth.

Experimental

Medical Physics. Biomagnetism; magnetoencephalography; magnetocardiography; magnetic detection of blood flow; cardiovascular physics; environmental radiation studies; ultrasonic imaging for medical diagnosis; fluid dynamics of cerebral spinal fluids; *in vivo* nuclear magnetic resonance spectroscopy, nuclear magnetic resonance imaging, and radiotherapy, photodynamic therapy, hyperthermia; radiation protection and dosimetry in diagnostic radiology and image reconstruction in computerized tomography. Chopp, Liboff, Tepley, Xia.

Solid State Physics. Elastic properties; electronic structure; Fermi surfaces; gyromagnetism in gases and solids; magnetic phase transitions; superconductivity; semiconductors, ferromagnetism; solid state electronics; materials science; optical properties of solids; physics of high pressures. Slavin, Srinivasan, Tepley, Venkateswaran.

UNIVERSITY OF MICHIGAN

DEPARTMENT OF ASTRONOMY

Ann Arbor, Michigan 48109-1042
http://www.astro.lsa.umich.edu

Students Accepted For Degree	FIELDS		
	Physics	Astronomy*	Related Fields
Doctorate		X	
Master's			

1. General

President: Mary Sue Coleman
Dean of Graduate School: Janet Weiss
Department Chairman: Joel Bregman
Department Telephone Number: (734) 764-3440
Type of Institution: University
Control: Public
Setting: Urban
Total Faculty: 2,987
Total Undergraduate Students: 25,467
Total Graduate Students: 14,526
Annual Graduate Tuition: *
In-state residents: $15,558
Out-of-state residents: $31,468
Tuition rates for: 2007–08
Deferred tuition plan: No
Other Fees: $190
Term: Trimester

*Registration fee: $80 each term

2. Number of Faculty in Department

The combined total of full-time faculty in the three professorial ranks is 16. The combined total of full-time, part-time, and other faculty at all ranks is 16.

3. Admission, Financial Aid, and Housing

Address admission inquiries to: Prof. Douglas O. Richstone, 830 Dennison Bldg., 500 Church St., Ann Arbor, MI 48109-1042
Graduate application fee required: $75
International Student application fee: $75
Admission deadline (Fall admission): 1/12
Admission information: For fall admission, 2008–09 16 students were accepted from 77 applicants.
Admission requirements: For admission to the graduate programs, a Bachelor's degree is required with a minimum undergraduate GPA of 3.0 specified. An undergraduate major in physics, mathematics, or astronomy is assumed. The GRE is required. No minimum acceptable score specified. The GRE Advanced is required. No minimum score specified. Students from non-English speaking countries are required to demonstrate proficiency in English via the TOEFL exam. Candidates for financial assistance must in addition demonstrate English language competence by passing a University of Michigan test.
Address financial aid inquiries to: Chairman, Dept. of Astronomy, 830 Dennison Bldg., 500 Church St., Ann Arbor, MI 48109-1042
GAPSFAS application required: No
Financial aid deadline: 1/12
Loans available: Yes
Address housing inquiries to: University Housing, Univ. of Michigan, 1500 Student Activities Bldg., Ann Arbor, MI 48109-1316
On-campus, single student housing available: Yes
Cost/term: Contact University Housing
On-campus, married student housing available: Yes
Cost/month: Contact University Housing

Table A—Faculty, Enrollments, and Degrees Granted

Research Specialty	20010–11 Faculty	Enrollment[1] Fall 2010 Master's	Enrollment[1] Fall 2010 Doctorate	No. of Degrees Granted[2] 2009–10 (2004–08) Master's	No. of Degrees Granted[2] 2009–10 (2004–08) Terminal Master's	No. of Degrees Granted[2] 2009–10 (2004–08) Doctorate	Median No. of Years for (2007-08) Ph.D.'s
Astronomy	16	–	–	4(14)	1(2)	2(11)	7
Total		–	28	–	–	–	
Full-time Grad. Stud.		–	–				
Part-time Grad. Stud.		–	0				
First-year Grad. Stud.		–	4				
Median Years in Grad. Study (2006–07 Degrees)				–	–	7	–
Undergraduate Degrees, 2007–08 (2003–07): 5(36)							

[1]Students not yet committed to a research specialty are entered under non-specialized.
[2]Five-year totals in parentheses.

4. Graduate Degree Requirements

Master's: 24 semester hours in astronomy and approved cognate courses; a research course; average of "B" or better in 12 hours of 500- and 600-level astronomy courses. *Doctorate*: Requirements for a Master's degree; seven core courses in astronomy; one term equivalent of work experience; doctoral preliminary examination; dissertation; oral defense.
Thesis: Thesis may be written *in absentia*.
Special Equipment, Facilities, or Programs: The Department has a 10% share of the twin 6.5-m Magellan telescopes on Las Campanas in Chile. It also has a share in the 2.4-m and 1.3-m telescopes of the MDM Observatory on Kitt Peak, Arizona. All telescopes have a wide selection of imaging and spectroscopic instruments. The 0.6/0.9-m Curtis Schmidt telescope on Cerro Tololo in Chile is used for wide field CCD imaging. The Radio Astronomy Observatory has a 26-meter paraboloid used at GHz frequencies. For instruction, there is a 0.4-m computer controlled Ritchey-Chretien telescope on central campus with a CCD camera. The Department also operates its own image processing system consisting of a cluster of Sun and PC computers. Graduate students have access to all telescopes and instruments.

Table B—Appointments to Graduate Students, 2009–10

Title of Appointee	Appointments Total	Appointments First year	Academic Load Allowed in Credit Hours	Hours of Service Per Week	Stipend for Academic Year ($)
Teaching Assistant	10	6	-	20	18,780
Research Assistant	15	1	-	20	18,780
Fellowship	9	7	-	0	18,780
Total	34	14			

5. Personnel Engaged in Separately Budgeted Research, 7/09–6/10

Professorial faculty	15
Postdoctoral appointments	16
Nonteaching research personnel	5
Total	36

6. Separately Budgeted Research Expenditures by Source of Support, 2009

	Departmental Research	Physics-related Research Outside Department
Federal government	$3,919,891	$
University	574,135	
Total	$4,494,026	$

Table C—Separately Budgeted Research Expenditures

Research Specialty	No. of Grants	Expenditures ($)
Astronomy	118	2,431,858
Institution	13	345,553
Total	131	2,777,411

FACULTY

Professors

Aller, Hugh D., Ph.D., Michigan, 1968. Radio astronomy; radio polarization; galactic and extragalactic astronomy.

Bregman, Joel, Ph.D., California, Santa Cruz, 1977. Interstellar medium; particle acceleration; gas dynamics; variable extragalactic sources.

Calvet, Nuria, Ph.D., U. California, Berkeley, 1981. Star formation; protostars; protoplanetary disks.

Hartmann, Lee, Ph.D., Wisconsin, 1976. Star formation and evolution of protoplanetary disks.

Mateo, Mario, Ph.D., Washington, 1987. Stellar evolution; stellar populations in the Milky Way and nearby galaxies; kinematics of galaxies, variable stars, and the local extragalactic distance scale.

Richstone, Douglas O., Ph.D., Princeton, 1975. Chairman of the Department. Dynamics of stellar systems; clusters of galaxies; evolution of galaxies; quasars; cosmology.

Associate Professors

Bell, Eric, Ph.D., University of Durham, UK, 1999. Star formation.

Bergin, Edwin, Ph.D., Massachusetts, 1995. Chemistry and physics of molecular clouds and protostellar environments.

Miller, Jon, Ph.D., MIT, 2002. Accretion onto compact objects and relativistic astrophysics.

Monnier, John D., Ph.D., California, Berkeley, 1999. Star formation, stellar evolution, high-resolution imaging, optical interferometry.

Oey, Sally, Ph.D., Arizona, 1995. Massive star feedback processes, multiphase interstellar medium.

Volonteri, Marta, Ph.D., Milan (Italy), 2003. Cosmology, black holes, and galaxy evolution.

Assistant Professors

Gnedin, Oleg, Ph.D., Princeton, 1999. Theoretical Astrophysics, formation and evolution of galaxies.

Ruszkowski, Mateusz, Ph.D., University of Cambridge, 2000. High-energy astrophysics and cosmology, numerical simulations.

Research Scientists

Aller, Margo F., Ph.D., Michigan, 1969. Extragalactic radio astronomy; active compact radio sources.

Hughes, Philip A., Ph.D., Sussex, 1978. Plasma astrophysics and radio sources. Emeritus Research Scientist.

Associate Research Scientist

Seitzer, Patrick, Ph.D., Virginia, 1982. Dynamics and formation of star clusters, dwarf galaxies, CCD detectors and instrumentation.

Assistant Research Scientist

Gultekin, Kayhan, Ph.D., University of Maryland, College Park, 2006. Dynamics of black holes, globular clusters, gravitational waves.

Valluri, Monica, Ph.D., 1993, Indian Institute of Science, galactic dynamics, galaxy evolution.

Professor Emeriti

Cowley, Charles R., Ph.D., Michigan, 1963. Stellar atmospheres; stellar chemistry; peculiar A stars; abundances. Emeritus Professor.

Postdoctoral Research Fellows

Cackett, Edward, Ph.D., University of St. Andrews, UK, 2006. X-ray binaries, AGN, and cooliing neutron stars.

RESEARCH SPECIALTIES AND STAFF

Theoretical

Extragalactic Astronomy. Dynamics of stellar systems; stellar spectra; interacting galaxies; evolution of galaxies; quasars and active galaxies; cosmology. Bregman, Gnedin, Gultekin, Hughes, Irwin, Richstone, Ruszkowski, Volonteri.

Galactic Astronomy. Accretion disks; stellar dynamics; stellar interiors; interstellar matter. Bregman, Gnedin, Mateo, Monnier, Oey, Richstone.

High Energy. Gultekin, Volonteri.

Star Formation and Planet formation. Bergin, Calvet, Hartmann, Heitsch.

Observational

Extragalactic Astronomy. Discovery and analysis of active galaxies and quasars; supernova spectroscopy; dynamics; of galaxies; clustering of galaxies; stellar populations in galaxies; measurement of total flux and its variability of radio sources; linear and circular polarization of radio sources; cosmology. H. Aller, M. Aller, Bregman, Cackett, Mateo, Miller, Oey, Richstone, Seitzer.

Galactic Astronomy. Abundances in chemically peculiar stars; star formation; stellar atmospheres; optical counterparts of x-ray sources; supernova remnants; cataclysmic variables; spectral classification. Cowley, Miller, Monnier, Oey.

Star and Planet Formation. Bergin, Calvet, Hartmann, Monnier.

High Energy. Cackett, Miller.

WAYNE STATE UNIVERSITY

DEPARTMENT OF PHYSICS AND ASTRONOMY

Detroit, Michigan 48201

Students Accepted For Degree	FIELDS		
	Physics	Astronomy	Related Fields
Doctorate	X		
Master's	X		

1. General

President: Jay Noren, M.D.
Department Chair: Ratna Naik, Ph.D
Department Telephone Number: (313) 577-2721 (C)
Type of Institution: University
Control: Public
Setting: Urban
Total Faculty: 2,043
Total Graduate Faculty: 1,721
Total Students: 31,786
Total Graduate Students: 11,021
Annual Graduate Tuition:
In-state residents: $456.50/cr
Out-of-state residents: $1,008.50/cr.
Tuition rates for: 2009–10
Deferred tuition plan: No
Other Fees: Registration fee $155.45
Omnibus Fee: : $34.05/cr.
Term: Semester

2. Number of Faculty in Department

The combined total of full-time faculty in the three professorial ranks is 32. The combined total of full-time, part-time, and other faculty at all ranks is 34.

3. Admission, Financial Aid, and Housing

Address admission inquiries to: Graduate Admissions Committee, Dept. of Physics, Wayne State University, Detroit, MI 48201
website: http://www.physics.wayne.edu;
e-mail: gradadmin@physics.wayne.edu
Graduate application fee required: $50 (U.S. students)
$50 (foreign applicants)
Admission deadline (Fall admission): July 1; For Canadian International Students: May 1
Admission information: For fall admission, 2008–09, 22 students were accepted from 89 applicants.
Admission requirements: For admission to the graduate programs, a Bachelor's degree in physics or related fields is required with a minimum undergraduate GPA of 3.0. A GPA below 3.0 would require a probationary admission. The GRE is strongly urged. No minimum acceptable- score is specified. The GRE Advanced is strongly urged. No minimum acceptable score is specified. Students from non-English speaking countries are required to demonstrate proficiency in English via the TOEFL exam and Test of Spoken English. Minimum acceptable score on TOEFL is 550. Minimum acceptable score on TSE is 50. TSE is absolutely required for Teaching Assistantship consideration.
Undergraduate preparation assumed: K.R. Symon, *Mechanics*; Eisberg, *Fundamentals of Modern Physics*; Reitz, Milford and Christy *Foundations of Electromagnetic Theory*.
Address financial aid inquiries to: Graduate Admissions Committee, Department of Physics and Astronomy
GAPSFAS application required: No
Financial aid deadline: January 5
Loans available: No
Address housing inquiries to: Director, Housing & Residential Life, 598 Student Center Bldg., Detroit, MI 48202; www.housing.wayne.edu
On-campus, single student housing available: Yes
Cost/month: $419 to $1063
On-campus, married student housing available: Yes
Cost/month: $593 to $1063

Table A—Faculty, Enrollments, and Degrees Granted

Research Specialty	2009–10 Faculty	Enrollment[1] Fall 2009		No. of Degrees Granted[2] 2009–10 (2005–10)			Median No. of Years for 2009–10 Ph.D.'s
		Master's	Doctorate	Master's	Terminal Master's	Doctorate	
Applied Physics	1	0	0	0(0)	0(2)	0(2)	–
Astrophysics	2	3	5	0(0)	0(0)	0(0)	–
Atomic, Molecular, & Optical Physics	3	1	2	0(1)	0(7)	0(1)	–
Biophysics/ Biomedical Phys.	1	0	1	0(0)	0(0)	0(0)	–
Condensed Matter Physics	14	1	18	5(6)	3(14)	2(11)	–
Low Temperature Physics	0	0	0	0(0)	0(1)	0(2)	–
Materials Sci.	6	0	0	0(0)	0(1)	0(0)	–
Nuclear Physics	5	0	2	0(0)	0(1)	0(2)	–
Optics	2	0	0	0(0)	0(0)	0(1)	–
Particles & Fields	7	1	10	0(0)	3(7)	1(6)	–
Relativistic Heavy Ions	4	2	9	1(1)	0(3)	0(7)	–
Surface Physics	3	0	0	0(0)	0(1)	0(0)	–
Other Theoretical Math.	9	0	0	0(0)	0(0)	0(0)	–
Non-specialized	1	8	1	0(0)	0(0)	–	–
Total		16	47	6(7)	6(37)	7(32)	
Full-time Grad. Stud.		8	46				
Part-time Grad. Stud.		7	1				
First-year Grad. Stud.		1	12				
Median Years in Grad. Study (2007–08 Degrees)				–	3	6	6
Undergraduate Degrees, 2008–09 (2004–09): 4(37)							

[1] Students not yet committed to a research specialty are entered under non-specialized.
[2] Five-year totals in parentheses.

4. Graduate Degree Requirements

Master's: The Master degree is offered with (M.S.) and without (M.A.) thesis in various areas of physics. Requirements for the M.S. degree are 24 credits of course work at the 5000 level or above plus an eight-credit thesis while the M.A. degree requires 29 credits of course work at the 5000 level or above plus a three-credit essay. Both degrees require at least nine credits at the 7000 level or above with at least half of the course work in physics. Students must maintain a 3.0 GPA and must complete their degree within six years. A final oral exam over the thesis or essay is required of all students.
Doctorate: The Ph.D. degree has a basic requirement of 90 cred-

its which include 30 dissertation credits. Courses at the graduate level in mathematical physics, mechanics and dynamics, quantum mechanics, electromagnetic theory, and statistical mechanics are required for all students as well as certain other courses depending on the area of concentration. A written Ph.D. qualifying exam usually taken after the end of the student's first year, a preliminary oral exam, and a final dissertation defense are the other major requirements. A 3.0 GPA must also be maintained. There is a seven year time limit for completion of the degree.

Other Programs: Smart Sensors & Integrated Microsystems: The Center for Smart Sensors & Integrated Microsystems at Wayne State is an interdisciplinary educational and research program involving faculty from physics, chemistry, engineering, and medicine. Although degrees are awarded in a specific discipline, students register for interdisciplinary courses associated with this program and select research problems ranging from basic science to the development and testing of actual sensors and devices. Interested students should consult the website at http://www.ssim.eng.wayne.edu for more information.

Thesis: Thesis may be written *in absentia*.

Special Equipment, Facilities, or Programs: The department has several well-equipped research laboratories with concentrated efforts in the areas of high-energy nuclear and particle physics, atomic and molecular, applied and condensed matter physics. The relativistic heavy-ion faculty participate in major international collaborations at Brookhaven National Laboratory Alternating Gradient Synchrotron and work on the STAR experiment at the Relativistic Heavy-Ion Collider (RHIC) and the ALICE experiment at the Large Hadron Collider (LHC) at CERN. On campus, the group operates laboratory facilities for the design, construction and testing of precision electromagnetic and hadronic calorimeters. Facilities for testing and development of high-density silicon drift detector arrays for use in nuclear physics experiments are also available. The particle physics groups are part of the BTeV (based at Fermilab), CLEO (at Cornell) and the CDF (at Fermilab) collaborations and have set up facilities on campus for the design and development of electronic systems for the particle detectors. They also have a leadership role in a future high energy electron-positron linear collider. In the low energy positron scattering program, positron beams with energies of 1 to a few 100 eV are generated from sodium-22 sources, that are used in high vacuum systems along with gamma-ray and particle detectors, and target gas production and monitoring instrumentation to measure total, differential elastic, and positronium formation cross sections for positrons scattered by various atoms and molecules. Research programs in condensed matter and material physics have appropriate cryogenic facilities, complete electrical transport, optical, thermal, and magnetic characterization equipment (including two fully automated commercial SQUID magnetometers usable in the 4–800 K temperature range, two ac susceptometers, and a vibrating sample magnetometer), material preparation equipment, x-ray diffractometer, scanning electron microscope, atomic force and scanning tunneling microscopes, and thin-film deposition equipment (including a multi-source electron-beam, a triple-target magnetron sputtering, and a MBE system). Research in low-temperature physics has the capability of attaining temperatures down to 300 mK with a ^{3}He cryostat for electrical, thermal, and magnetic measurements. In addition, access to the film and device fabrication and characterization facilities associated with the Center for Smart Sensors and Integrated Microsystems are available including a class 100 cleanroom. Three microwave spectrometers operating at frequencies of 12, 18, and 80 GHz (in magnetic fields up to 6 tesla) are used to study conduction electrons in normal metals, superconductors, and magnetic systems over the temperature range from 1.5 to 300 K. The thermal wave imaging group uses several visible and infrared lasers, two infrared video cameras, and three image processing systems to investigate thermal properties and defects in materials. The program in surface physics uses a 200-kV ion implanter, and rf and dc sputtering equipment to modify surfaces, while scanning Auger spectrometry and low-energy electron diffraction (LEED) are used to investigate surface properties of materials. Argon ion and helium-neon lasers, a photon correlation spectrometer, and a polarizing microscope with an attached spectrometer and diode array detector are used to investigate properties of liquid crystals. Several research groups possess extensive computational facilities consisting of high performance UNIX and PC workstations that are used in their research. High-speed network links these machines with the rest of the University and Internet. In addition, the faculty members have access to the High Performance Computing Facility which houses a new IBM RS/6000 SP supercomputer and to the Advanced Computing Facility which maintains nearly 100 computers and workstations for scientific computation. Information about the graduate program is also available on the World Wide Web at URL http://www.physics.wayne.edu.

Table B—Appointments to Graduate Students, 2009–10

Title of Appointee	Appointments: Total	Appointments: First year	Academic Load Allowed in Credit Hours	Hours of Service Per Week	Stipend for Academic Year ($)
			Semester		
Teaching Assistant	23	11	10	20	16,586[3,4,6,7]
Research Assistant	23	0	10	20	16,586[3,4,6,7]
Rumble Fellow	4	0	12	0	16,586[1,2,5,6]
Total	50	11			

[1] Dependent allowance—max. $900.
[2] Housing subsidy in University housing—max. $180/month.
[3] Selected student housing subsidy in University housing.
[4] Tuition waived up to 10 credits/semester.
[5] Tuition waived up to 12 credits/semester.
[6] Medical insurance paid by University.
[7] Reduced tuition for dependents.

5. Personnel Engaged in Separately Budgeted Research, 7/09–6/10

Professorial faculty	32
Postdoctoral appointments	4
Graduate students	23
Undergraduate students	10
Nonteaching research personnel	21
Total	90

6. Separately Budgeted Research Expenditures Source of Support

	Departmental Research	Physics-related Research Outside Department
Federal government	$3,704,738	$818,863
State and local government	164,431	1,085,419
Other	50,187	
Total	$3,919,356	$1,904,282

Table C—Separately Budgeted Research Expenditures

Research Specialty	No. of Grants	Expenditures ($)
Applied Physics	1	40,325
Atomic, Molecular, & Optical Physics	0	0
Condensed Matter Physics	8	256,624
Particles & Fields	12	483,563
Relativistic Heavy Ions	3	1,497,936
University Research Support[1]	–	69,866
Total	30	1,850,637

[1] University Research Support distributed as needed to all research projects.

Table D—Physics-related Research Outside Department

Field and Unit Outside Department	No. of Grants	Expenditures ($)
Center for Smart Sensors & Integrated Microsystems	7	818,863
Total	7	818,863

FACULTY

Professors

Bellwied, Rene, Ph.D., Johannes Gutenberg University, Mainz, 1989. Experimental nuclear physics; relativistic heavy-ion collisions; studies related to monoenergetic e^+e^- pair production in super-heavy nucleus-nucleus collisions.

Chang, Jhy-Jiun, Ph.D., Rutgers, 1973. Theory of superconductivity and many-body theories.

Cinabro, David A., Ph.D., Wisconsin, 1991. Experimental high energy physics; charm physics in e^+e^- collisions at CLEO; luminosity spectrum at linear collider.

Cormier, Thomas M., Ph.D., MIT, 1974. Experimental nuclear physics; relativistic heavy-ion collisions; properties of quark matter and the QCD phase transition.

Karchin, Paul E., Ph.D., Cornell, 1982. Experimental high energy physics; heavy quark production in strong interactions; B_s mixing, CP violation, rare decays and Higgs particles at CDF.

Kauppila, Walter E., Ph.D., Pittsburgh, 1969. Experimental atomic and molecular physics; low energy positron scattering (total, differential elastic and positronium formation) by atoms and molecules.

Keyes, Paul H., Ph.D., Maryland, 1972. Experimental condensed matter physics; liquid crystals; phase transitions.

Kuo, Pao-Kuang, Ph.D., Minnesota, 1964. Optics; material physics.

Morgan, Caroline G., Ph.D., Princeton, 1980. Theoretical condensed matter physics; studies of electronic structure and defects in semiconductors.

Naik, Ratna, Ph.D., West Virginia, 1982. Chair of the Department. Experimental condensed matter physics; studies of novel magnetic properties of epitaxially grown superlattices and thin films; sensors and device development.

Pruneau, Claude A., Ph.D., Universite Laval (Quebec), 1987. Experimental nuclear physics; relativistic heavy-ion collisions.

Rolnick, William B., Ph.D., Columbia, 1963. Theoretical elementary particle physics.

Saperstein, Alvin M., Ph.D., Yale, 1956. Theoretical physics; nuclear and elementary particle physics; scattering theory; chaos and mathematical models of arms races; technology and international security policies.

Thomas, Robert L., Ph.D., Brown, 1965. Experimental applied physics; thermal wave imaging. Dean of College of Liberal Arts and Sciences.

Voloshin, Sergei A., Ph.D., Moscow Engineering & Physics Institute, 1980. Experimental nuclear physics; relativistic heavy ion collisions; physics of multiparticle production.

Wadehra, Jogindra M., Ph.D., NYU, 1977. Theoretical atomic and molecular physics; plasma physics; astrophysics. Associate Chair of the Department.

Associate Professors

Bonvicini, Giovanni, Doctor of Physics, Universita di Bologna, 1981. Experimental high energy physics; e^+e^- collisions at CLEO.

Bowen, David, Ph.D., University of Pennsylvania. 1966. Computers, Science, Internet, Creativity.

Gavin, Sean, Ph.D., Illinois, 1987. Theoretical nuclear physics; relativistic heavy ion physics, quark gluon plasma theory, QCD phenomenology.

Harr, Robert F., Ph.D., California, 1990. Experimental high energy physics; CP violation, rare decays and Higgs particles at CDF.

Hoffmann, Peter M., Ph.D., Johns Hopkins, 1999. Experimental condensed matter physics; atomic force microscopy studies of interatomic and intermolecular forces.

Lawes, Gavin, Ph.D., Cornell University, 2001. Experimental condensed matter physics, magnetic ordering magnetic materials and properties, Spin fluctuations.

Mukhopadhyay, Ashis, Ph.D., Kansas State University. 2000. Experimental soft condensed matter physics.

Nadgorny, Boris E., Ph.D., Stony Brook, 1996. Experimental condensed matter physics; spin transport and spin polarization studies, spintronics.

Padmanabhan, K.R., Ph.D., Poona, India, 1975. Experimental condensed matter physics; ion-solid interaction physics; ion channeling.

Petrov, Alexey A., Ph.D., Massachusetts (Amherst), 1997. Theoretical particle physics; heavy quark physics, CP violation, and QCD phenomenology.

Assistant Professors

Huang, Jian, Ph.D., Michigan State University, 2001. Experimental condensed matter physics, quantum charge and spin transport in mesoscopic and nano electron systems, critical phenomena in strongly correlated 1D and 2D systems.

Huang, Zhi-Feng, Ph.D., Tsinghua University, 1999. Theoretical condensed matter physics, non-equilibrium dynamics of nanoscale phases in complex material systems, mechanisms of stained thin films growth, biological aging.

Rehse, Steven J., Ph.D., Colorado State University, 2002. Experimental atomic, molecular and optical physics.

Sakamoto, Takeshi, Ph.D., Kanazawa University, Japan, 2001. Mechanisms of myosin-dependent motility, protein-protein iteractions, actin-myosin interactions and visualization using single molecule imaging techniques *in vitro* and *in vivo*.

Zhou, Zhixian, Ph.D., Florida State University, 2004. Individual nanoscale materials and single organic molecules: synthesis and characterization, nanoscale device fabrication, electrical transport measurements.

RESEARCH SPECIALTIES AND STAFF

Theoretical

Astrophysics. Behavior of atoms and molecules in intense magnetic fields. Wadehra.

Atomic Physics. Electron and positron scattering. Wadehra.

Condensed Matter. Electronic structure and defects in semi-conductors; non-equilibrium superconductivity. Chang, Huang, Morgan.

Elementary Particles and Fields. Heavy quark physics, CP violation, electroweak physics, and QCD phenomenology; Higgs particles and W boson processes. Petrov, Rolnick.

Nuclear Physics. QCD theory and phenomenology; relativistic heavy ion collisions; quark gluon plasma; many-body theory. Gavin.

Other Theoretical/Math. Non-linear and dissipative systems; symmetries and dynamics of classical and quantum systems; specular reflection processes; quantum electrodynamics of many-body systems. Kuo, Rolnick, Saperstein.

Plasma Physics. Production and diagnostics of negative ion beams. Wadehra.

Experimental

Applied Physics. Thermal-wave imaging studies, including photo-acoustics, laser-induced ultrasound, and laser interferometry. Thomas.

Atomic and Molecular Physics. Positron beam interactions with atoms and molecules. Kauppila, Rehse.

Biophysics. Molecular Motor. Single molecule imaging studies with TOtal Internal Reflection Fluorescent (TIRF) microscopy. Sakamoto.

Condensed Matter/Material Physics. Atomic force and scanning tunneling microscopy of surfaces; magnetic materials and device applications; conventional and high-temperature superconductivity; Andreev reflection; electron and Josephson tunneling; spin transport and spin polarization; spintronics; high-frequency behavior of conduction electrons; ion channeling; thin-film and materials research; surface studies and modification; liquid crystals; calorimetric and ultrasonic properties; magnetic resonance spectroscopy; Raman spectroscopy. Hoffmann, Keyes, Kuo, Lawes, Mukhopadhyay, Nadgorny, Naik, Padmanabhan, Sakamoto, Zhou.

Nuclear Physics. Nuclear structure; nuclear lifetime determinators; relativistic heavy-ion collisions; monoenergetic e^+e^- pair production in super-heavy nucleus-nucleus collisions; quark gluon plasma physics. Bellwied, Cormier, Pruneau, Voloshin.

Particle Physics. B_s mixing, heavy quark physics, CP violation, and Higgs particles at CDF (Fermilab); charm physics at Cornell (CLEO); design studies for linear collider. Bonvicini, Cinabro, Harr, Karchin.

Other Experimental Physics: Computers, Networks, Science, Society and Creativity; Bowen.

WESTERN MICHIGAN UNIVERSITY

DEPARTMENT OF PHYSICS

Kalamazoo, Michigan 49008-5252

Students Accepted For Degree	FIELDS		
	Physics	Astronomy	Related Fields
Doctorate	X		
Master's	X		

1. General

President: John M. Dunn
Dean of Graduate College: Lewis Pyenson
Department Chairman: Paul V. Pancella
Department Telephone Number: (269) 387-4940
Type of Institution: University
Control: Public
Setting: Suburban
Total Faculty: 855
Total Graduate Faculty: 631
Total Students: 24,576
Total Graduate Students: 5,029
Annual Graduate Tuition:
In-state residents: —$401.61*/credit
Out-of-state residents: —$850.62*/credit
Tuition rates for: 2009–10
Deferred tuition plan: Yes**
Annual Other Fees: $364/semester (5 or more credit hours); $25 additional foreign student fee
Term: Semester

*Master Graduate Assistants receive up to 18 credit hours of tuition waiver.
Doctoral Associates and Doctoral Graduate Assistants receive up to 24 credit hours of tuition waiver.
**Deferred for Graduate Assistants or Graduate Fellows.

2. Number of Faculty in Department

The combined total of full-time faculty in the three professorial ranks is 17. The combined total of full-time, part-time, and other faculty at all ranks is 21.

3. Admission, Financial Aid, and Housing

Address admission inquiries to: Graduate Advisor, Department of Physics
Graduate application fee required: $40 (Foreign $100)
Admission deadline (Fall admission): 4/1
Admission information: For fall admission, 2009–10, 5 students were accepted from 19 applicants.
Admission requirements: For admission to the graduate programs, a Bachelor's degree in physics or related disciplines is required with a minimum undergraduate GPA of 3.0 specified. The GRE is required for Ph.D. candidates. The GRE Advanced is not required. Students from non-English speaking countries are required to demonstrate proficiency in English via the TOEFL exam. Minimum acceptable score for admission is 500 PBT/173 CBT/61 iBT.
Undergraduate preparation assumed: Halliday and Resnick, *Fundamentals of Physics*; Sprott, *Introduction to Modern Electronics*; Fowles, *Analytical Mechanics*; Christy, Reitz, and Milford, *Electricity and Magnetism*; Eisberg and Resnick, *Quantum Physics*; Meyer-Arendt, *Introduction to Classical and Modern Optics*; Sears, *Thermodynamics, Kinetic Theory and Statistical Thermodynamics.*
Address financial aid inquiries to: Graduate Advisor, Department of Physics
GAPSFAS application required: No
Financial aid deadline: 2/15
Loans available: Yes
Address housing inquiries to: WMU Campus Apartments, 3510 Faunce Student Services Building, Western Michigan University, Kalamazoo, MI 49008-5312
On-campus, single student housing available: Yes**
Cost/month: $315 (single, furnished rooms)*
On-campus, married student housing available: Yes
Cost/month: $586–777 (1–2 bdrm., unfurnished)*
$617–859 (1–2 bdrm., furnished)*

*(2010–2011 rates)
**Apartments may also be available

Table A—Faculty, Enrollments, and Degrees Granted

Research Specialty	2009–10 Faculty	Enrollment[1] Fall 2009		No. of Degrees Granted[2] 2009–10 (2005–09)			Median No. of Years for 2009–10 Ph.D.'s
		Mas-ter's	Doc-torate	Mas-ter's	Terminal Master's	Doc-torate	
Astronomy	2	1	0	0(0)	2(3)	0(0)	–
Atomic, Molecular, & Optical Physics	5	2	8	0(3)	0(2)	3(7)	6
Condensed Matter Physics	5	0	9	3(5)	0(2)	0(4)	–
Nuclear Physics	4	0	7	0(1)	0(2)	0(0)	–
Physics Education	2	0	0	0(0)	0(0)	0(0)	–
Non-specialized	0	3	3	0(0)	0(3)	–	–
Total		6	27	3(9)	2(12)	3(11)	
Full-time Grad. Stud.		5	27				
Part-time Grad. Stud.		1	0				
First-year Grad. Stud.		2	3				
Median Years in Grad. Study (2009–10 Degrees)				2	5	6	6
Undergraduate Degrees, 2009–10 (2005–09): 3(23)							

[1]Students not yet committed to a research specialty are entered under non-specialized.
[2]Five-year totals in parentheses.

4. Graduate Degree Requirements

Master's: 30 semester hours of graduate credit with a GPA of 3.0/4.0 or better. May only transfer 6 hours from another institution. No residency requirement. 15 hours required in Classical Mechanics, Electricity and Magnetism, Statistical Mechanics, Quantum Mechanics, and Research Seminar. In addition, 9 hours in physics, mathematics, or other departments chosen with the consent of the graduate advisor. Pass Qualifying Examination at the master's degree level or satisfactory completion of a Master's Thesis.
Doctorate: 60 semester hours of graduate credit with a GPA of 3.0 or better.
Thesis: Thesis may not be written *in absentia.*
Special Equipment, Facilities, or Programs: A 12-MV tandem Van de Graaff accelerator with associated equipment and electronics together with support staff is used for atomic, nuclear, and applied research. A well-equipped instrument shop and electronic shops with technical support staff are

available. A computer lab is reserved for physics graduate students. The department has ready access to the University alpha-cluster and sun systems. At the University a superconducting NMR spectrometer and scanning electron microscopes are also available.

Table B—Appointments to Graduate Students, 2009–10

Title of Appointee	Appointments Total	First year	Academic Load Allowed in Credit Hours	Hours of Service Per Week	Stipend for Academic Year ($)
			Semester		
Teaching Assistant	11	3	6–9	20	12,804[1,2]
Doctoral Associate	8	0	6–9	20	16,620[3,4]
Total	19	3			

[1]Tuition waived up to 12 credit hours for master students and 18 credit hours for doctoral students.
[2]Summer I/Summer II appointments available.
[3]Tuition waived up to 21 credit hours.
[4]Includes either a Summer I or Summer II appointment.

5. Personnel Engaged in Separately Budgeted Research, 7/09–6/10

Professorial faculty	10
Graduate students	10
Undergraduate students	5
Nonteaching research personnel	3
Postdoctoral appointments	7
Total	35

6. Separately Budgeted Research Expenditures by Source of Support

	Departmental Research	Physics-related Research Outside Department
Federal government	$2,748,783	$
University	162,738	
Total	$2,911,521	$

Table C—Separately Budgeted Research Expenditures

Research Specialty	No. of Grants	Expenditures ($)
Astronomy	2	147,612
Atomic, Molecular, & Optical Physics	3	1,861,820
Condensed Matter Physics	3	556,508
Nuclear Physics	4	314,430
Physics Education	2	178,763
Total	14	2,911,521

FACULTY

Professors

Berrah, Nora, Ph.D., Virginia, 1987. Experimental atomic physics.
Burns, Clement, Ph.D., California, San Diego, 1993. Experimental condensed matter physics.
Chung, Sung G., Ph.D., Tokyo, 1981. Condensed matter theory.
Gorczyca, Thomas, Ph.D., Colorado, 1990. Theoretical atomic physics.
Halderson, Dean W., Ph.D., Kansas, 1974. Intermediate energy nuclear theory.
Kamber, Emanuel Y., Ph.D., London, 1983. Experimental atomic physics.
Korista, Kirk, Ph.D., Ohio State, 1990. Observational astronomy; numerical simulation of gaseous nebulae.
McGurn, Arthur R., Ph.D., California, Santa Barbara, 1975. Condensed matter theory.
Pancella, Paul V., Ph.D., Rice, 1987. Department Chair. Experimental nuclear physics.
Paulius, Lisa, Ph.D., California, San Diego, 1993. Experimental condensed matter physics.
Tanis, John A., Ph.D., NYU, 1976. Experimental atomic physics.
Wuosmaa, Alan H., Ph.D., Pennsylvania, 1988. Experimental Nuclear Physics.

Associate Professors

Henderson, Charles, Ph.D., Minnesota, 2002. Physics education.
Rosenthal, Alvin S., Ph.D., Colorado, 1978. Theoretical nonlinear optics.
Schuster, David Ph.D., Witwatersrand, 1972. Physics education.

Assistant Professors

Bautista, Manuel, Ph.D., Ohio State, 1997. Astronomy.
Famiano, Michael, Ph.D., Ohio State, 2001. Nuclear astrophysics.

Faculty Specialist

Kayani, Asghar, Ph.D., Ohio, 2003. Experimental condensed matter physics.

RESEARCH SPECIALTIES AND STAFF

Theoretical

Astronomy. Atomic data and spectral models for astrophysics; numerical simulations of emissions from gaseous nebulae. Bautista, Korista. 1 postdoctoral associate.
Atomic, Molecular and Optical Physics. Electron-ion collisions; ion-atom collisions; many body theory; R-matrix theory; atomic photoionization; photodetachment; dielectronic recombinations; nonlinear optics. Gorczyca, Rosenthal.
Condensed Matter Physics. Photonic crystals; Anderson localization; scattering of light by the localized surface polaritons of disordered media; inelastic neutron scattering from mixed Ising systems; computer simulations of the dynamics of Heisenberg magnetic alloys; fractons; nano-electronics, spintronics, quantum computing, superconductor-insulator transition, metal-insulator transition, novel many-body techniques. Chung, McGurn.
Nuclear Physics. The nuclear many-body problem; hypernuclear structure; reaction theory; strangeness-exchange reactions; high-energy photo-nuclear reactions; relativistic nuclear structure. Halderson.

Experimental

Astronomy. Acquisition and interpretation of spectroscopic data from active galactic nuclei, high-redshift quasars, and galactic line sources. Korista.

Atomic, Molecular and Clusters. Studies of strongly correlated electron systems—photon-atom, photon-molecule, photon-cluster, and photonnegative ion interactions. Spectroscopy is performed with lasers and third generation light sources. Mechanisms of electronic excitation and charge-changing are investigated for collisions of ions with atomic and molecular targets. Berrah, Kamber, Tanis. 5 postdoctoral associates.

Condensed Matter. Studies of highly correlated electron systems including metal ammonia compounds and high temperature superconductor parent compounds; development of inelastic x-ray scattering using synchrotron radiation; research on the properties of organic semiconductors. Electrical and magnetic properties of high-temperature superconductors, flux vortex dynamics, metal-insulator transitions in rare-earth nickel oxides, electrical transport under high pressure. Ion beam analysis of materials, development of solid oxide fuel cells and hydrogen storage materials. Burns, Kayani, Paulius.

Nuclear Physics. Two-nucleon problem; nucleon-nucleus phenomenology; direct capture of protons by light nuclei; experiments relevant to determination of spectroscopic factors; in-beam gamma ray spectroscopy; pion production; structure of exotic light nuclei, clustering phenomena in light nuclei, few nucleon transfer reactions for nuclear structure and astrophysics, relativistic heavy-ion collisions. Famiano, Pancella, Wuosmaa, 1 postdoctoral associate.

Physics Education. Curriculum design, evaluation, and assessment; cognitive aspects of the teaching and learning of science such as conceptual understanding, problem solving, and epistemology; teacher beliefs and teacher professional development. Henderson, Schuster.

MINNESOTA STATE UNIVERSITY, MANKATO

DEPARTMENT OF PHYSICS AND ASTRONOMY

Mankato, Minnesota 56001

Students Accepted For Degree	FIELDS		
	Physics	Astronomy	Related Fields
Doctorate			
Master's	X		

1. General

President: Richard Davenport
Dean of Graduate School: Anne Blackhurst
Department Chair: Youwen Xu
Department Telephone Number: (507) 389-5743
Type of Institution: University
Control: Public
Setting: Small town
Total Faculty: 731
Total Graduate Faculty: 460
Total Students: 14,954
Total Graduate Students: 1,919
Annual Graduate Tuition:
In-state residents: Full-time—$298/credit
Part-time—$298/credit
Out-of-state residents: Full-time—$490.40/credit
Part-time—$490.40/credit
Tuition rates for: 2009–10
Tuition & Fees to be determined for 2010–11
Deferred tuition plan: No
Other Fees: $33.14/credit
Term: Semester

2. Number of Faculty in Department

The combined total of full-time physics faculty in the three professorial ranks is 12. The combined total of full-time, part-time, and other faculty at all ranks is 12.

3. Admission, Financial Aid, and Housing

Address admission inquiries to: College of Graduate Studies
Graduate application fee required: $40
Admission deadline (Fall admission): July 1
Admission information: For fall admission, 2009–10, 3 students were accepted from 6 applicants.
Admission requirements: For admission to the graduate programs, a Bachelor's degree is required. At least an undergraduate physics minor is required for the M.S. degree. The GRE is recommended. Students from non-English speaking countries are required to demonstrate proficiency in English via the TOEFL exam. A minimum score of 82 (IBT) is required for teaching assistantship. A minimum score of 72 (IBT) is required for admission.
Address financial aid inquiries to: Graduate Studies Office
GAPSFAS application required: F.F.S. required
Financial aid deadline: 3/15 or until all positions are filled
Loans available: Yes
Address housing inquiries to: www.mnsu.edu/reslife
On-campus, single student housing available: Yes
On-campus, married student housing available: No

Table A—Faculty, Enrollments, and Degrees Granted

Research Specialty	2009–10 Faculty	Enrollment[1] Fall 2009		No. of Degrees Granted[2] 2009–10 (2005–10)			Median No. of Years for 2009–10 Ph.D.'s
		Master's	Doctorate	Master's	Terminal Master's	Doctorate	
Astronomy	1	2	–	–	0(0)	–	–
Astrophysics	2	0	–	–	0(0)	–	–
Condensed Matter Physics	3	3	–	–	1(6)	–	–
Medical Physics	1	0	–	–	0(1)	–	–
Nuclear Physics	2	0	–	–	0(0)	–	–
Physics Education	1	1	–	–	2(4)	–	–
Space Physics	1	0	–	–	0(1)	–	–
Other Theoretical/ Math.	1	0	–	–	0(1)	–	–
Non-specialized	0	1	–	–	0(0)	–	–
Total		7	–	–	3(13)	–	
Full-time Grad. Stud.		7	–				
Part-time Grad. Stud.		0	–				
First-year Grad. Stud.		3	–				
Median Years in Grad. Study (2009–10 Degrees)				–	2	–	–
Undergraduate Degrees, 2009–10 (2005–10): 3(24)							

[1] Students not yet committed to a research specialty are entered under non-specialized.
[2] Five-year totals in parentheses .

4. Graduate Degree Requirements

Master's: The Physics MS is the professional degree in physics. The thesis plan requires 30 semester hrs. with a minimum graduate GPA of 3.0 on a 4.0 scale. The alternate plan requires 34 semester hrs. 15 semester hrs. must be earned in residency. Comprehensive written and oral exams are required with the thesis plan. The alternate plan requires a comprehensive written exam.
The Physics Education MS is designed for physics education (student must qualify for licensure). 30 semester hrs. are required with thesis or 34 semester hrs. on the alternate plan, with a GPA of 3.0 on a 4.0 scale. No foreign language requirement. 15 semester hrs. residency requirement. Comprehensive written and oral exams are required with the thesis plan; the written exam is required on the alternate plan. This degree may be earned in Science by completing 28 semester hrs. in science and 6 hrs. of professional education.
Other Programs: Master of Arts in Teaching (MAT). This degree requires a minimum of 34 semester hrs. which must include the 19 hrs. required for Minnesota licensure, a thesis, and internship. No foreign language is required. Graduate GPA of 3.0 on a 4.0 scale required. 15 semester hrs. must be earned as a resident student. Comprehensive written and oral exams required.
Thesis: Thesis may be written *in absentia*.
Special Equipment, Facilities, or Programs: Some of the research equipment available includes an Auger spectrometer, scanning electron microscope, x-ray diffractometer, and facilities for microcomputer and materials research. A 0.5-meter telescope with photometers and a CCD camera is available. Facilities for fabrication and characterization of high-critical-temperature superconductors are available. A 400 kV Van de Graaff particle accelerator is available. A high perfor-

mance computer with 140 64-bit parallel processors is available.

Table B—Appointments to Graduate Students, 2009–10

Title of Appointee	Appointments Total	Appointments First year	Academic Load Allowed in Credit Hours	Hours of Service Per Week	Stipend for Academic Year ($)
			Semester		
Teaching Assistant	5	1	12	15–20	9,000[1]
Research Assistant	0	0	12	15–20	9,000[1]
Total	5	1			

[1] Full tuition waiver.

5. Personnel Engaged in Separately Budgeted Research, 7/09–6/10

Professorial faculty	3
Total	3

6. Separately Budgeted Research Expenditures by Source of Support

	Departmental Research	Physics-related Research Outside Department
Federal government	–	$620,000
University	$5,000	–
Private, non-profit organization	$266,735	–
Total	$271,735	$620,000

Table C—Separately Budgeted Research Expenditures

Research Specialty	No. of Grants	Expenditures ($)
Energy Sources and Environment	2	$266,735
Space Physics	2	$620,000
Medical Physics	1	$5,000
Total	5	$891,735

FACULTY

Professors

Herickhoff, Robert J., Ph.D., Vanderbilt, 1965. Nuclear.
Kipp, Steven L., Ph.D., Pittsburgh, 1980. Astrometry; galactic structure.
Kogoutiouk, Igor, Ph.D., Chernovtsy State University, Ukraine, 1981. Condensed matter, theory.
Palma, Russell L., Ph.D., Rice, 1981. Space physics.
Pickar, Mark A., Ph.D., Indiana University, 1982. Nuclear.
Pierce, James N., Ph.D., Iowa State, 1980. Stellar astrophysics.
Schwartzkopf, Louis, Ph.D., California, Berkeley, 1974. High-temperature superconductors; renewable energy.
Wu, Hai-Sheng, Ph.D., Iowa State, 1988. Condensed matter; semiconductors.
Xu, Youwen, Ph.D., Iowa State, 1987. Condensed matter, superconductivity, magnetism.

Associate Professor

Eskridge, Paul, Ph.D., University of Washington, 1987. Extragalactic astronomy.

Assistant Professors

Brown, Thomas, Ph.D., Montana State University, 2003. Physics education.
Roberts, Andrew D., Ph.D., University of Wisconsin-Madison, 1995. Medical physics.

RESEARCH SPECIALTIES AND STAFF

Theoretical

Stellar Astrophysics. Pierce.
Condensed Matter. Kogoutiouk.
Renewable Energy. Schwartzkopf.

Experimental

Astrometry. Kipp.
Condensed Matter. Wu, Xu.
Galactic Structure. Kipp.
Magnetism. Xu.
Medical Physics. Roberts.
Nuclear, Low Energy. Pickar, Herickhoff, Roberts.
Nuclear, Medium Energy. Pickar.
Nuclear Engineering. Roberts.
Physics Education. Brown.
Surface Physics. Wu.
Superconductivity. Xu.
Extra-Galactic Observational Astronomy. Eskridge.
Space Physics. Palma.

UNIVERSITY OF MINNESOTA, DULUTH

DEPARTMENT OF PHYSICS

Duluth, Minnesota 55812

Students Accepted For Degree	FIELDS		
	Physics	Astronomy	Related Fields
Doctorate			X
Master's	X		X

Loans available: Yes
Address housing inquiries to: Housing Office, 149 Lake Superior Hall
On-campus, single student housing available: Yes
Cost/semester: $3,088 (double occupancy, all meals included)
On-campus, married student housing available: No

1. General

President: R. Bruininks
Dean of Graduate School: Henning Schroeder
Department Head: Jonathan Maps
Department Telephone Number: (218) 726-7124 (C)
Type of Institution: University
Control: Public
Setting: Urban
Total Faculty: 555
Total Graduate Faculty: 428
Total Students: 11,664
Total Graduate Students: 2,242
Annual Graduate Tuition:
In-state residents: Full-time—$11,212*
Part-time—$934*/credit (approx.)
Out-of-state residents: Full-time—$18,310*
Part-time—$1,526*/credit (approx.)
Tuition rates for: 2009–10
Deferred tuition plan: No
Other Fees: $279 student services fee, if registered for more than 6 credits. $5.73/credit computer network access fee, $150 technology fee.
Term: Semester

*Tuition is waived for half-time assistants, up to 14 credits.

2. Number of Faculty in Department

The combined total of full-time faculty in the three professorial ranks is 7. The combined total of full-time, part-time, and other faculty at all ranks is 10.

3. Admission, Financial Aid, and Housing

Address admission inquiries to: Director of Graduate Studies, Physics Department
Graduate application fee required: $75 domestic, $95 international
Admission deadline (Fall admission): 7/15
Admission information: For fall admission, 20010–11, 6 students were accepted from 15 applicants.
Admission requirements: For admission to the graduate programs, a Bachelor's degree in physics or the equivalent is required with minimum GPA of 3.0. The GRE Advanced is recommended. Students from non-English speaking countries are required to demonstrate proficiency in English via the TOEFL exam. Minimum acceptable score for admission is 79.
Undergraduate preparation assumed: Taylor, Classical Mechanics; Griffiths, Introduction to Electrodynamics; Reif, Thermal Physics; Griffiths, Quantum Mechanics.
Address financial aid inquiries to: Director of Graduate Studies, Physics Department
GAPSFAS application required: No
Financial aid deadline: 3/15 preferred.

Table A—Faculty, Enrollments, and Degrees Granted

Research Specialty	2009–10 Faculty	Enrollment[1] Fall 2009 Master's	Enrollment[1] Fall 2009 Doctorate	No. of Degrees Granted[2] 2009–10 (2005–10) Master's	Terminal Master's	Doctorate	Median No. of Years for Ph.D.'s
Atomic, Molecular, & Optical Physics	1	0	–	0(2)	–	–	
Condensed Matter Physics	1	2	–	1(4)	–	–	
Marine Science/ Oceanography	2	0	–	2(3)	–	–	
Particles & Fields	3	6	–	0(3)	–	–	
Physics Education	0	0	–	0(0)	–	–	
Relativity & Gravitation	1	0	–	0(0)	–	–	
Non-specialized	0	2	–	0(0)	–	–	
Total		10	–	3(14)	–	–	
Full-time Grad. Stud.		10	–				
Part-time Grad. Stud.		0	–				
First-year Grad. Stud.		7	–				
Median Years in Grad. Study (2009–10 Degrees)				2	–	–	
Undergraduate Degrees, 2009–10 (2004–10): 4(18)							

[1]Students not yet committed to a research specialty are entered under non-specialized.
[2]Five-year totals in parentheses.

4. Graduate Degree Requirements

Master's: There are two programs of study. The programs are planned with faculty advisors to suit the needs and interests of students. A grade point average of 2.8 must be maintained (on a scale of 4). All students complete a common 14 semester credit core in classical and quantum physics, 6 credits in related fields and a final examination. Plan A requires a Master's thesis. Plan B requires 10 additional credits in approved electives and preparation of one or more papers.
Thesis: Thesis may be written *in absentia*.
Special Equipment, Facilities, or Programs: Lasers and vacuum UV facilities; scanning probe microscopes; low-temperature facility; computing facilities; facilities for vacuum deposition of materials; opportunities in physical limnology and oceanography through Large Lakes Obsevatory; well-funded multidisciplinary Natural Resources Research Institute. Participation in MINOS Nova, and Minerva neutrino experiments.

Table B—Appointments to Graduate Students, 2009–10

Title of Appointee	Appointments Total	Appointments First year	Academic Load Allowed in Credit Hours	Hours of Service Per Week	Stipend for Academic Year ($)
			Semester		
Teaching Assistant	9	7	14	20	13,100[1]
Research Assistant	1	0	14	20	13,100[1]
Total	10	7			

[1]No tuition due from half-time assistants.

5. Personnel Engaged in Separately Budgeted Research, 7/08–6/09

Professorial	6
Graduate students	4
Undergraduate students	7
Total	17

6. Separately Budgeted Research Expenditures by Source of Support

	Departmental Research	Physics-related Research Outside Department
Federal government	$281,000	$256,000
State and local government	$10,000	$50,000
Total	$291,000	$306,000

Table C—Separately Budgeted Research Expenditures

Research Specialty	No. of Grants	Expenditures ($)
Atomic, Molecular, & Optics Physics	2	10,000
Particles & Fields	7	281,000
Total	9	291,000

Table D—Physics-related Research Outside Department

Field and Unit Outside Department	No. of Grants	Expenditures ($)
Oceanography		
Large Lakes Observatory	4	306,000
Total	4	306,000

FACULTY

Professors

Hiller, John R., Ph.D., Maryland, 1980. Theoretical particle physics; computational physics.

Jordan, Thomas F., Ph.D., Rochester, 1962. Theoretical physics; general relativity and cosmology. (Emeritus)

Sydor, Michael, Ph.D., New Mexico, 1964. Condensed matter; optical characterization of semiconductor properties, remote sensing of particulates in water.

Associate Professors

Austin, Jay, Ph.D., MIT/WHOI, 1999. Physical oceanography.

Casserberg, Bo R., Ph.D., Princeton, 1968. Magnetic resonance. (Emeritus).

Habig, Alec, Ph.D., Indiana, 1996. High-energy neutrinos, astrophysics.

Assistant Professors

Gran, Richard, Ph.D., Minnesota, 2002, High-energy physics.

Katsev, Serguei, Ph.D., Ottawa, 2002. Physical Oceanography.

Maps, Jonathan, Ph.D., Massachusetts, 1982. Condensed matter.

RESEARCH SPECIALTIES AND STAFF

Theoretical

Elementary Particles. Quark model calculations. Nonperturbative quantum field theory. Hiller.

General Relativity and Quantum Mechanics. Cosmology. Jordan.

Oceanography. Dynamics of large and mesoscale circulation. Numerical modeling of coastal shelves, estuaries, and large lakes. Coupling between sediment and water column. Sediment early diagenesis. Austin, Katsev.

Experimental

Atomic, Molecular and Optical Physics. Optical characterization of suspended particles. Sydor.

Condensed Matter Physics. Scanning probe microscopy; surface states and excitons in alkali halides; resonance Raman spectroscopy; opto-electronic materials; device physics. Maps, Sydor.

High-energy Neutrinos. Habig, Gran.

Oceanography. Observations of the circulation dynamics of large lakes, estuaries, and coastal shelves. Austin.

UNIVERSITY OF MINNESOTA

SCHOOL OF PHYSICS AND ASTRONOMY
GRADUATE PROGRAM IN PHYSICS

Minneapolis, Minnesota 55455

Students Accepted For Degree	FIELDS Physics	Astronomy	Related Fields
Doctorate	x		x
Master's	x		x

1. General

President: Robert H. Bruininks
Vice Provost and Dean of Graduate Education: Henning Schroeder
School Head: Ronald A. Poling
Department Telephone Number: (612) 624-7375
Type of Institution: University
Control: Public
Setting: Urban
Total Faculty: 4,105
Total Graduate Faculty: 3,600
Total Students: 51,659
Total Graduate Students: 13,170
*Per Semester Graduate Tuition**:
In-state residents: Full-time—$6,022
Out-of-state residents: Full-time—$9571
Tuition rates for: 2010–11
Deferred tuition plan: Yes
*Other Fees***: $783.50 per semester
Term: Semester

*Tuition waived for graduate assistants with 50%-time support. In general, tuition waived at a rate equal to twice the percentage of appointment. Students with graduate assistantships of 25% or greater and their family members are eligible for resident tuition.
**For students with 50%-time graduate assistantships.

2. Number of Faculty in Department

The combined total of full-time professors, associate professors and assistant professors is 43. The total number of active graduate program faculty in physics at all three professorial ranks is 66, including professors emeriti and affiliate members from related university departments.

3. Admission, Financial Aid, and Housing

Address admission inquiries to: Graduate School, 309 Johnston Hall, 101 Pleasant St. S.E., Minneapolis, MN 55455. Web address: www.grad.umn.edu
Graduate application fee required: $75 domestic; $95 foreign
Admission deadline (fall admission): 6/15
Admission information: For fall admission, 2010–11, there were approximately 529 applicants to whom 83 financial aid offers were made.
Admission requirements: For admission to the graduate programs, a Bachelor's degree is required with no minimum undergraduate GPA specified. Undergraduate courses in differential and integral calculus as well as two years of physics courses are required. The GRE and GRE Advanced may be beneficial, but are not required. The average GRE percentiles for 2010–11 admissions were verbal—71%; quantitative—90%; analytical—46. The average advanced GRE percentile is 66%. Students from non-English speaking countries are required to demonstrate proficiency in English via the TOEFL exam. Minimum acceptable score for admission is 550 for the paper-based test; 79 for the internet-based test (21 on writing, 19 on reading); IELTS (6.5); MELAB (80) are also acceptable.
Undergraduate preparation assumed: Applicants are expected to have an undergraduate education with a major in physics, chemistry, mathematics, or engineering, with a significant component of physics. Acceptable preparation would include study in the basic areas of classical physics at the junior level or higher, modern physics, elementary quantum mechanics and mathematics including advanced calculus and differential equations. Exceptions will be considered on an individual basis. Admission and financial aid decisions are based on the overall strength of preparation as demonstrated by the course record, letters of recommendation and GRE scores.
Address financial aid inquiries to: Professor Joseph Kapusta, Director of Graduate Studies, School of Physics and Astronomy, 116 Church Street SE, Minneapolis MN 55455; E-mail:grad@physics.umn.edu
Web address: www.physics.umn.edu
GAPSFAS application required: No
Financial aid deadline: 12/15 for international applicants, 1/15 for domestic applicants
Loans available: No
Address housing inquiries to: Housing Office, Comstock Hall East, 210 Delaware Street, Minneapolis MN 55455 Web: www.housing.umn.edu; Phone: 612-624-2994
On-campus, single student housing available: Yes
Cost/term: Consult Housing Office
On-campus, married student housing available: Yes
Cost/month: Consult Housing Office

Table A—Faculty, Enrollments, and Degrees Granted

Research Specialty	2010-11 Faculty	Enrollment[1] Fall 2010 Master's	Doctorate	No. of Degrees Granted[2] 2010-11 (2005-10) Master's	Terminal Master's	Doctorate	Median No. of Years for 2005-10 Ph.D.'s
Astrophysics & Cosmology	8	2	21	0(1)	0(0)	1(7)	6
Biological Physics	4	0	8	0(1)	1(6)	3(10)	6
Condensed Matter Physics	14	0	30	0(1)	1(11)	7(37)	6
History & Philosophy	1	0	0	0(0)	0(0)	0(0)	–
Nuclear Physics	3	0	7	0(0)	0(0)	3(7)	7
Particles & Fields	16	0	21	1(12)	1(5)	2(22)	7
Physics Education	1	0	3	0(0)	0(0)	1(1)	5
Space & Planetary Physics	3	0	7	0(0)	1(1)	1(2)	7
Non-specialized	0	0	36	0(0)	1(1)	0(0)	7
Total							
Full-time Grad. Stud.		2	133	1(15)	5(24)	18(86)	
Part-time Grad. Stud.		0	0				
First-year Grad. Stud.		0	26				
Median Years in Grad. Study (2009–10 Degrees)				5	3	6	–
Undergraduate Degrees 2009-10 (2005-10):37(151)							

[1]Students not yet committed to a research specialty are entered under non-specialized.
[2]Five-year totals in parentheses.

4. Graduate Degree Requirements

Master's: There are two options: Plan A requires a thesis plus 20 course credits while the Plan B requires 30 credits including four credits involving research project(s) and associated paper(s). No thesis is required for the latter option. For either option, 60% of the course credits must come from this University, and a minimum GPA of 2.8 should be maintained. No foreign language requirements.

Doctorate: 40 graduate course credits which can include some from other graduate schools. Thesis based on original research is defended in the final oral exam. Students need to pass a comprehensive written examination by the beginning of the second year. After most course work is complete, students are expected to demonstrate their readiness to focus on Ph.D. research in the preliminary oral examination. A minimum GPA of 3.3 should be maintained. No foreign language requirements.

Other Programs: Selected faculty in Astrophysics, Biochemistry, Chemical Engineering/Materials Science, Chemistry, Electrical and Mechanical Engineering may direct the research of physics graduate students. The program in the History of Science, Technology and Medicine offers graduate degrees with emphasis in the history of physics for students who wish to pursue this area in conjunction with their studies in physics.

Thesis: Thesis may be written *in absentia*.

Special Equipment, Facilities, or Programs: Departmental facilities include a large modern R&D Machine Shop; Magnetic Microscopy Center and Soudan Underground Research Laboratory in northern Minnesota. University Centers include Characterization Facility for Materials, operating many electron microscopes, scanning probe microscopes and X-ray machines; Nanofabrication Center, Microtechnology Lab with clean room and electron beam lithography, and Supercomputer Institute. High-energy experiments are carried on at CERN, IHEP (Beijing), Fermilab, and non-accelerator laboratories. Extensive computer facilities are available in the School and elsewhere on campus. The endowed W. I. Fine Theoretical Physics Institute encourages theoretical work by physicists at all levels and sponsors workshops on the forefront of physics research topics.

Table B—Appointments to Graduate Students, 2010-11

Title of Appointee	Appointments: Total	Appointments: First year	Academic Load Allowed in Credit Hours	Hours of Service Per Week	Stipend for Academic Year ($)
Teaching Assistant	68.5	22	No limit	20	16,770
Research Assistant	62	1	No limit	20	16,770
Fellow*	12.5	3	No limit	10	16,770
Total	143	26			

*Plus 1/4-time assistantship.

5. Personnel Engaged in Separately Budgeted Research, 7/09–7/10

Professorial faculty	43
Other faculty	7
Postdoctoral appointments	30
Graduate students	133
Undergraduate students	23
Total	193

6. Separately Budgeted Research Expenditures by Source of Support

	Departmental Research
Federal government	$14,842,300
Other	178,407
Total	15,020,707

Table C—Separately Budgeted Research Expenditures

Research Specialty	No. of Grants	Expenditures ($)
Astrophysics & Cosmology	5	1,336,210
Biological Physics	5	525,570
Condensed Matter Physics	17	1,055,933
Education	3	208,465
Nuclear Physics	3	472,147
Particles & Fields	22	6,350,012
Space/Planetary Physics	20	5,072,370
Total	75	15,020,707

FACULTY

Professors

Broadhurst, John H., Ph.D., Birmingham, England, 1959. Experimental nuclear physics; biophysics; astrophysics.

Campbell, Charles E., Ph.D., Washington (St. Louis), 1969. Condensed matter and many-body theory; quantum fluids; quantum statistical mechanics; magnetism.

Cattell, Cynthia A., Ph.D., California, Berkeley, 1980. Space plasma physics; particle acceleration processes; collisionless shocks; magnetic reconnection; non-linear waves; aurora; magnetic and electric field measurements.

Crowell, Paul A., Ph.D., Cornell, 1994. Experimental condensed matter physics; low-dimensional magnetism; spin dynamics and transport in ferromagnet-semiconductor heterostructures; spin transport in metals; magnetization dynamics in ferromagnets.

Cushman, Priscilla B., Ph.D., Rutgers, 1985. The Cryogenic Dark Matter Search; low background radio purity assay, experimental particle physics, advanced instrumentation techniques; CMS collaboration at LHC.

Dahlberg, E. Dan, Ph.D., UCLA, 1978. Experimental physics; magnetism; magnetic imaging and microscopy; magnetotransport.

Goldman, Allen M., Ph.D., Stanford, 1965. Experimental condensed matter physics; superconducting tunneling; localization and superconductivity in disordered and dimensionally constrained systems; quantum phase transitions and quantum critical behavior; properties of superconducting and magnetic oxide materials; epitaxial growth of oxides and oxide heterostructures; electrostatic modification of novel materials.

Halley, J. Woods, Jr., Ph.D., California, Berkeley, 1965. Theoretical physics; simulation and low temperature experiment; many-body theory; magnetism; classical and quantum liquids; percolation; solid-liquid interfaces; statistical mechanics of polymers and reaction networks; electronic structure of disordered normal and superconducting oxides; molecular dynamics simulations; chemical physics relevant to renewable energy devices; Bose Einstein condensate in superfluid helium 4.

Hanany, Shaul, Ph.D., Columbia U., 1993. Astrophysics; observational cosmology; studies of the cosmic microwave background radiation with the goal of understanding the origin and evolution of the universe.

Heller, Kenneth, Ph.D., Washington, 1973. Experimental physics; high-energy physics; neutrino interactions; neutrino oscillations; physics education research; problem solving.

Huang, Cheng-cher, Ph.D., Pennsylvania, 1975. Experimental condensed matter physics; optical and resonant x-ray scattering studies of various liquid crystals phases; molecular arrangements and optical properties of various layered liquid crystal phases with a net polarization; investigation of liquid crystal free-standing film tension.

Kakalios, James, Ph.D., Chicago, 1985. Experimental condensed matter physics; amorphous and nanocrystalline semiconductors; electrical and optical properties; 1/*f* noise; fluctuation phenomena in neurological systems.

Kamenev, Alex, Ph.D., Weizmann Institute, Israel, 1996. Theoretical condensed matter physics; application of the field-theoretical methods in the equilibrium and non-equilibrium statistical mechanics; solid state physics.

Kapusta, Joseph I., Ph.D., California, Berkeley, 1978. Quantum field theory at finite temperature and density; application to relativistic nuclear collisions, astrophysics and cosmology.

Kubota, Yuichi, Ph.D., Tokyo, 1981. Studies of electroweak symmetry breaking-super symmetry, extra dimension, etc. in Hadron collisions; instrumentation for high-energy physics experiments.

Lysak, Robert L., Ph.D., California, Berkeley, 1980. Theoretical space physics; waves in the magnetosphere and ionosphere; auroral acceleration processes; magnetic reconnection; numerical modeling of wave propagation.

Marshak, Marvin L., Ph.D., Michigan, 1970. Experimental elementary particle physics, including long-baseline neutrino physics; cosmic-ray physics and proton decay.

Olive, Keith, Ph.D., Chicago, 1981. Theoretical physics; interplay of particle physics and cosmology.

Poling, Ronald A., Ph.D., Rochester, 1981. Experimental high-energy physics; electron-positron colliding beam experiments; production and decay of heavy-quark states; neutrino interactions and oscillations.

Qian, Yong-Zhong, Ph.D., UC California, San Diego, 1993. Nuclear/particle astrophysics and cosmology; chemical evolution of galaxies; supernova explosions and heavy element nucleosynthesis; neutrino-nucleus interactions; neutrino oscillations.

Rudaz, Serge, Ph.D., Cornell, 1979. Theoretical physics; unified theories of elementary particle interactions supersymmetry; astroparticle physics; relativistic many-body theory.

Rusack, Roger W., Ph.D., Imperial College, U.K., 1979. Experimental physics; high-energy physics; particle physics at the highest available energies; physics at the CERN Large Hadron Collider; advanced instrumentation.

Shifman, Mikhail A., Ph.D., Moscow, 1976. Theoretical high-energy physics; Yang-Mills theories at strong coupling; quantum chromodynamics; supersymmetry-based methods in gauge theories; super-symmetry solutions.

Shklovskii, Boris I., Ph.D., Leningrad, 1968. Electronic transport phenomena in disordered systems; doped and amorphous semiconductors; the quantum Hall effect; charge inversion and gene delivery; transport in ion channels in nanopores; self-assembly of viruses.

Vainshtein, Arkady, Ph.D., Novosibirsk, 1968. Theoretical high-energy physics; gauge theories of fundamental interactions; supersymmetry; nonperturbative effects.

Valls, Oriol T., Ph.D., Brown, 1975. Unconventional superconductivity in high temperature and other materials; proximity effects; nonequilibrium phenomena.

Vinals, J., Ph.D., University of Barcelona, 1983. Theoretical condensed matter physics; nonlinear dynamics, pattern formation, and chaos in extended systems; soft matter and complex fluids; stochastic processes; biological physics.

Voloshin, Mikhail B., Ph.D., ITEP, Moscow, 1977. Theoretical study of nonperturbative properties of elementary particles and of the physical vacuum in quantum field theory; phenomenology of electroweak and strong interactions.

Walsh, Thomas F., Ph.D., California, Berkeley, 1966. Theoretical physics; quantum electrodynamics; quantum chromodynamics; quarks; grand unification.

Wygant, John R., Ph.D., California, Berkeley, 1983. Experimental space plasma physics.

Associate Professors

Cronin-Hennessy, Daniel P., Ph.D., Duke, 1997. Experimental physics; high-energy physics; colliding beam experiments; heavy boson production and decay; heavy quark state production and decay; and neutrino oscillation physics.

Fortson, Lucy, Ph.D., Ph.D., UCLA, 1991. Experimental High-Energy Astrophysics; gamma-ray astronomy; cosmic-ray astrophysics; active galactic nuclei; extra-galactic astronomy with citizen science.

Ganz, Eric, Ph.D., UC Berkeley, 1988. Experimental condensed matter physics; scanning tunneling microscopy studies of diffusion and growth on semiconductor substrates; nanolithography; and quantum chemistry studies of hydrogen storage.

Greven, Martin, Ph.D., MIT, 1995. Experimental condensed matter physics; properties of correlated-electron materials, especially high-temperature superconductors and low-dimensional model magnets; quantum phase transitions; neutron and X-ray scattering experiments; transport and magnetometry measurements; crystal growth of transition metal oxides.

Heger, Alexander, Ph.D., Technical University of Munich, 1998. Astrophysics.

Janssen, Michel, Ph.D., University of Pittsburgh, 1995. History of physics, especially the relativity and quantum revolutions in the early twentieth century.

Mueller, Joachim D., Ph.D., Munich, 1997. Experimental biophysics, fluorescence fluctuation spectroscopy, single molecule detection, and two-photon microscopy; dynamic processes motion, and assembly of biomolecules and membrane systems in vitro and in vivo.

Peloso, Marco, Ph.D., SISSA Trieste, Italy, 2000. Particle physics and cosmology.

Pryke, Clement L., Ph.D., Leeds UK, 1996. Observational cosmology; cosmic microwave background and its polarization; millimeter-wave instrument and telescope design; data analysis and simulation.

Zudov, Michael A., Ph.D., University of Utah, 1999. Experimental condensed matter physics; two-dimensional semiconductor structures; magnetotransport in high Landau levels of two-dimensional electron systems; non-equilibrium and nonlinear transport phenomena; microwave photoconductivity.

Assistant Professors

Mandic, Vuk, Ph.D., U California, Berkeley, 2004. Gravitational wave physics; dark matter; observational cosmology, early universe physics.

Mans, Jeremiah M., Ph.D., Princeton, 2002. Experimental physics; high energy physics; physics of electroweak symmetry-breaking; advanced digital electronics and networking.

Noireaux, Vincent, Ph.D., Paris, 2000. Experimental biophysics; soft condensed matter physics; in vivo and in vitro gene expression, artificial cell, biomimetic systems.

Professors Emeriti

Bayman, Benjamin, Ph.D., Edinburgh, 1955.
Courant, Hans, Ph.D., MIT, 1954.
Dehnhard, Dietrich, Ph.D., Marburg, Germany, 1964.
Gasiorowicz, Stephen, Ph.D., UCLA, 1952.
Giese, Clayton F., Ph.D., Minnesota, 1957.
Hintz, Norton M., Ph.D., Harvard, 1951.
Hobbie, Russell K., Ph.D., Harvard, 1960.
Johnson, Walter H., Ph.D., Minnesota, 1956.
Jones, Roger S., Ph.D., Illinois, 1961.
Kellogg, Paul J., Ph.D., Cornell, 1955.
Mantis, Homer T., Ph.D., NYU, 1950.
Marquit, Erwin, Ph.D., Warsaw, Poland, 1963.
Pepin, Robert O., Ph.D., California Berkeley, 1964.
Peterson, Earl A., Ph.D., Stanford, 1967.
Ruddick, Keith, Ph.D., Birmingham, England, 1964.
Shapiro, Alan E., Ph.D., Yale, 1970.
Tang, Yau Chien, Ph.D., Illinois, 1958.
Waddington, C. J., Ph.D., Bristol, England, 1955.
Zimmermann, William. Jr., Ph.D., CalTech, 1958.

Adjunct Professors

Chelikowsky, James R., Ph.D., California, Berkeley, 1975. Computational materials science.

Daughton, James M., Ph.D., Iowa State, 1963. Models of magnetoresistance and giant magnetoresistance (GMR); electronic device applications of GMR and structures that combine semiconductors and magnetic materials.

de Forcrand, Philippe, Ph.D., California, Berkeley, 1982. Theoretical elementary particle physics; computational physics, lattice gauge quantum chromodynamics.

Palma, Russell, Ph.D., Rice University, 1981. Experimental mass spectrometry; noble gasses and nitrogen in extraterrestrial materials.

Singovski, Alexander V., Ph.D., IHEP, Provino, Russia, 1989. High energy physics; experimental physics; non-perturbative QCD (glueballs, hybrids, etc.) double-gluon interaction (central production) at high energies; protons at high energies; advanced measurement technique.

Withbroe, George L., Ph.D., Michigan, 1965. Solar physics; variation of total solar irradiance; coronal physics; solar wind.

RESEARCH SPECIALTIES AND STAFF

Theoretical

Astrophysics and Cosmology. Theory and phenomenology at the interface between particle physics and cosmology, including extra dimensions and other physics beyond the Standard Model; inflation; supernovae, nucleosynthesis and chemical evolution of galaxies; gamma-ray and X-ray bursts. (See also Department of Astronomy.) Heger, Kapusta, Olive, Peloso, Qian.

Biological Physics. Screening of biophysical polyelectrolytes; biopolymers; phase transitions in biological systems; DNA folding. Shklovskii

Condensed Matter. Mesoscopic and low dimensional electron systems; superconductivity; quantum chaos; magnetism, classical and quantum fluids; nonequilibrium and glass phenomena; polymer theory; semiconductors; disordered systems. Campbell, Halley, Kamenev, Shklovskii, Valls, Vinals.

Nuclear Physics. Quantum field theory at finite temperature and density; relativistic nuclear collisions; applications to astrophysics and cosmology; nucleosynthesis, and neutrino oscillations. Heger, Kapusta, Qian, 3 research associates.

Particle Physics. Gauge theory; quantum field theory; quantum chromodynamics; supersymmetry; nonperturbative properties of elementary particles; phenomenology, the interplay of article theory with cosmology. Olive, Peloso, Rudaz, Shifman, Vainshtein, Voloshin, Walsh, 3 research associates.

Space Physics. Space plasma physics; electromagnetic phenomena in the magnetosphere and ionosphere. Cattell, Lysak.

Experimental

Astrophysics and Cosmology. Understanding the early universe by looking for remnants of the Big Bang. Measurements of microwave background radiation; particle astrophysics; cryogenic dark matter search; solar physics. (See also Department of Astronomy.) Cushman, Fortson, Hanany, Mandic, Pryke, 3 research associates.

Biological Physics. Applications of optical spectroscopy to biomolecules and membrane systems; biomagnetic measurements. Broadhurst, Mueller, Noireaux, 2 research associates.

Condensed Matter. Magnetism; superconductivity; liquid crystals; superfluid helium; amorphous semiconductors; mesoscopic and low dimensional systems; quantum critical phenomena. Crowell, Dahlberg, Ganz, Goldman, Greven, Huang, Kakalios, Zudov, 3 research associates.

Particle Physics. Experiments conducted with high-energy particle accelerators at Fermilab, CERN and IHEP (Beijing); searches for new particles and interactions at the energy frontier; neutrino interaction and oscillation measurements with deep-underground and surface detectors; precision tests of the Standard Model; searches for proton decay. Cronin-Hennessy, Cushman, Heller, Kubota, Mandic, Mans, Marshak, Poling, Rusack, 9 research associates.

Physics Education Research. Investigations to improve the teaching and learning of physics, primarily at the university level. Research centers on problem solving, cooperative learning, diagnostic tools. Heller, 1 research associate.

Space and Planetary Physics. Studies of the solar wind; magnetospheric and ionospheric phenomena; mass spectroscopic study of extraterrestrial element abundances. Cattell, Wygant, 5 research associates.

Other

History of Science, Technology and Medicine. Research carried out on the history of physics from the 17th century Scientific Revolution through the history of 20th century physics. Janssen.

UNIVERSITY OF MINNESOTA

SCHOOL OF PHYSICS AND ASTRONOMY
DEPARTMENT OF ASTRONOMY

Minneapolis, Minnesota 55455

Students Accepted For Degree	FIELDS		
	Physics	Astrophysics	Related Fields
Doctorate		X	
Master's		X	

1. General

President: Robert Bruininks
Dean of Graduate School: Henning Schroeder
Department Chairman: Robert D. Gehrz
Department Telephone Number: (612) 624-4811 (C)
Type of Institution: University
Control: Public
Setting: Urban
Total Faculty: 4,105
Total Graduate Faculty: 3,682
Total Graduate Students: 14,353 (Full and Part-time)
Annual Graduate Tuition: *
In-state residents: Full-time—$5,605.61
Out-of-state residents: Full-time—$9,105.00**
Tuition rates for: 2009–10
Deferred tuition plan: No
Other Fees: $1,189.42/term
Term: Semester

*Tuition and some of the fees are waived for 50% Graduate Assistant support. In general, tuition waived at a rate equal to twice the percentage of appointment.

**The out-of-state tuition is genrally offset by a non-resident tuition waiver bringing the tuition down to the in-state tuition rate.

2. Number of Faculty in Department

The combined total of full-time faculty in the two professorial ranks is 10. Eight additional faculty members are also associated with the graduate program.

3. Admission, Financial Aid, and Housing

Address admission inquiries to: Director of Graduate Studies, Department of Astronomy, School of Physics and Astronomy 116 Church St. S.E. or by e-mail at grad-req@astro.umn.edu
Graduate application fee required: $75 domestic, $95 foreign
Graduate school admission deadline (Fall admission): 6/15
Admission information: For fall admission, 2009–10, 12 students were admitted from 47 applicants.
Admission requirements: For admission to the graduate programs, a Bachelor's degree in astronomy, physics or related fields is required with no minimum undergraduate GPA specified. The GRE is required. No minimum acceptable score for admission is specified. The GRE Advanced physics test is required. No minimum acceptable score for admission is specified. Median GRE scores for 2009–10 admissions were verbal–570; quantitative–770, analytical–4.5; advanced physics–640.
Address financial aid inquiries to: Director of Graduate Studies, Department of Astronomy, School of Physics and Astronomy, 116 Church St. S.E., Minneapolis, MN 55455
GAPSFAS application required: No
Financial aid deadline: 1/10
Loans available: No
Address housing inquiries to: Housing Office, Comstock Hall, 210 Delaware St.
On-campus, single student housing available: Yes
Cost/term: $2,551

Table A—Faculty, Enrollments, and Degrees Granted

Research Specialty	2009–10 Faculty	Enrollment Fall 2009		No. of Degrees Granted[1] 2009–10 (2005–10)			Median No. of Years for 2009–10 Ph.D.'s
		Mas-ter's	Doc-torate	Mas-ter's	Terminal Master's	Doc-torate	
Astrophysics	10	1	19	0(1)	0(6)	2(12)	5
Total		1	19	0(1)	0(6)	2(12)	
Full-time Grad. Stud.		1	19				
Part-time Grad. Stud.		0	0				
First-year Grad. Stud.		0	3				
Median Years in Grad. Study (2009–10 Degrees)				2.5	3	6	–

Undergraduate Degrees, 20089–10 (2005–10): 10(41)

[1]Five-year totals in parentheses.

4. Graduate Degree Requirements

Master's: Plan A and Plan B
Degree Requirements: The master's degree requires a minimum of 30 credits, including one semester of classical physics (Phys 5011) Additional requirements depend on whether the student chooses the thesis (Plan A) or non-thesis (Plan B) option. Plan A emphasizes preparation of a thesis and Plan B emphasizes coursework. Plan A requires 20 semester credits of coursework and 10 thesis credits. Plan B requires 30 semester credits of coursework. Completion of the degree normally takes two years.
Minor Requirements for students majoring in Other Fields: For the master's minor, 8 credits in astrophysics are required.
Language Requirements: None.
Final Exam: A final oral exam is required.
Ph.D. Degree Requirements: The Ph.D. degree requires a minimum of 40 course credits, including a year of classical physics (Phys 5011-5012), and 12 credits in a minor or supporting program; 24 thesis credits are also required. The graduate written exam, offered during the spring, must be passed on the second "real" attempt (first-year students are given a free trial). A second-year project must be defended by the end of the fall of the third year. The preliminary oral exam must be passed by the end of the third year.
Minor Requirements for Students Majoring in Other Fields: For the Ph.D. minor, 12 credits in astrophysics are required.
Language Requirements: None.
Final Exam: A final oral exam is required.
Special Equipment, Facilities, or Programs: The Department is a partner in the Large Binocular Telescope (LBT) consortium with

the University of Arizona; construction was completed in 2007. The LBT facility on Mt. Graham in southeastern Arizona has two 8.4-meter primary mirrors with the combined light gathering power of a single 11.8-meter mirror and the resolution of a 22.8-meter telescope. The partnership also makes immediately available to departmental users a wide range of other telescopes and instrumentation operated by the University of Arizona, including the 10-meter radio telescope on Mt. Graham, 6.5m the MMT on M6 Hopkins, the 90-inch on Kitt Peak and the Magellan telescopes in the southern hemisphere.

The O'Brien Observatory is located 36 miles from the Minneapolis campus. The primary instrument is a 30″ Cassegrain telescope instrumented for optical and infrared observations including CCD imaging and photometry. Computational facilities are on site both for data acquisition and data analysis.

The Mt. Lemmon Observatory is about 50 miles from downtown Tucson, Arizona on the summit of Mt. Lemmon. The 60″ telescope is equipped for astronomical observations on a wide variety of scientific programs. Presently, auxiliary equipment for visual and/or infrared observations includes CCD and infrared imaging, photometers, and polarimeters.

Development and construction of infrared and optical instrumentation is carried out in the Department and makes use of the excellent shop facilities of the School of Physics and Astronomy.

The University of Minnesota Supercomputing Institute for Digital Simulation and Advanced Computation is located in Walter Library as part of the Digital Technology Center (DTC) and houses massively parallel SGI and IBM supercomputers, along with high end visualization equipment. The Department is connected through a fast local area network and possesses SUN and SGI workstations for data processing and analysis. The Laboratory for Computational Science and Engineering (also part of DTC), which is administered through the Department, exists to work with very large data sets, especially those produced in computational and observational astrophysics. It houses supercomputer-class equipment from SGI and other vendors and operates the Power Wall for visualization of complex data.

Table B—Appointments to Graduate Students, 2009–10

Title of Appointee	Appointments Total	Appointments First year	Academic Load Allowed in Credit Hours	Hours of Service Per Week	Stipend for Academic Year ($)
			Semester		
Teaching Assistant	6	2	16	20	16,419[1]
Research Assistant	12	1	16	20	16,419[1]
Grad. Fellow	2	0	16		22,500[1]
No support	0	0	16	-	-
Total	20	3			

[1]Plus tuition benefit.

5. Personnel Engaged in Separately Budgeted Research, 7/09–6/10

Professorial faculty	10
Postdoctoral appointments	2
Graduate students	20
Total	32

6. Separately Budgeted Research Expenditures by Source of Support

	Departmental Research
Federal government	$1,207,266
Total	$1,207,266

Table C—Separately Budgeted Research Expenditures

Research Specialty	No. of Grants	Expenditures ($)
Astrophysics	42	1,207,266
Total	42	1,207,266

FACULTY

Professors

Davidson, Kris D., Ph.D., Cornell, 1970. Emission-line analyses; quasi-stellar objects; supernova remnants; x-ray sources; massive stars.

Gehrz, Robert D., Ph.D., U. Minnesota, 1971. Chair. Infrared astronomy; novae; circumstellar and interstellar dust; development of novel instrumentation.

Humphreys, Roberta M., Ph.D., U. Michigan, 1969. Dir. of Undergraduate studies. Astronomy and astrophysics; stellar spectroscopy; galactic and extragalactic studies; the cosmic distance scale.

Jones, Terry J., Ph.D., U. Hawaii, 1977. Dir. of Graduate studies. Infrared astronomy; high-resolution spectroscopy of stars: interstellar medium; galaxies.

Jones, Thomas W., Ph.D., U. Minnesota, 1972. Theoretical astrophysics; numerical astrophysics; gas dynamics; magnetohydrodynamics; active galaxies; supernova remnants; cosmic ray acceleration.

Rudnick, Lawrence, Ph.D., Princeton, 1974. Radio astronomy; aperture synthesis interferometry; radio galaxies and QSOs; active nuclei.

Skillman, Evan D., Ph.D., U. Washington, 1984. Dir. of Graduate Studies. Chemical abundances; star formation; evolution of galaxies.

Williams, Liliya L. R., Ph.D., U. Washington, 1995. Cosmology; gravitational lensing; formation and evolution of galaxies; QSO absorption line systems.

Woodward, Charles E., Ph.D., U. Rochester, 1987. Infrared astronomy using ground-based facilities and space-platforms.

Woodward, Paul R., Ph.D., U. California, Berkeley, 1973. Numerical astrophysics; hydrodynamic simulations applied to a wide range of astrophysical problems.

Adjunct Faculty

Venn, Kim, Ph.D., U. Texas, 1994. Stellar spectroscopy and model atmospheres analysis techniques used to determine chemical abundances of massive stars.

Research Associate

Kovacs, Attila, Ph.D., 2006. Caltech. Chasing active galaxies and their evolution through the ages via submillimeter and infrared studies, and devising innovative techniques for detecting them across the Universe.

Krejny, Megan, Ph.D., Northwestern University, 2008. Instrumentation and applications for submillimeter and millimeter polarimetry.

MISSISSIPPI STATE UNIVERSITY

DEPARTMENT OF PHYSICS AND ASTRONOMY

Mississippi State, Mississippi 39762
www.msstate.edu/dept/physics

Students Accepted For Degree	FIELDS: Physics	Astronomy	Related Fields
Doctorate			X
Master's	X		

1. General

President: Mark E. Keenum
Vice President for Research and Economic Development: David R. Shaw
Department Chairman: Mark A. Novotny
Department Telephone Number: (662) 325-2806
Type of Institution: University
Control: Public
Setting: Small town
Total Faculty: 1,359*
Total Graduate Faculty: 998
Total Students: 18,601
Total Graduate Students: 4,066
Annual Graduate Tuition:
In-state residents: Full-time—$5,151**
Part-time—$286.25/credit
Out-of-state residents: Full-time—$13,020.50***
Part-time—$723.50/credit
Tuition rates for: 2009–10
Deferred tuition plan: Yes
Other Fees: $1,381/yr
Term: Semester

*Does not include Adjunct and Visiting Faculty.
**Two semester academic year.
***Out-of-state portion of tuition is waived for assistantship holders.

2. Number of Faculty in Department

The combined total of full-time faculty in the three professorial ranks is 15. The combined total of full-time, part-time, and other faculty at all ranks is 21.

3. Admission, Financial Aid, and Housing

Address admission inquiries to: Office of Graduate Studies
Graduate application fee required: $40
Admission deadline (Fall admission): 7/1
Admission information: For fall admission, 2009–10, 13 students were accepted from 47 applicants. Of the students accepted, 10 were offered graduate assistantships.
Admission requirements: For admission to the graduate programs, a Bachelor's degree in physics is required with a minimum undergraduate GPA of 2.75/4.0 specified. The GRE is not required, but the GRE and the GRE Advanced are recommended. The minimum acceptable score suggested for admission is 500. Students from non-English speaking countries are required to demonstrate proficiency in English via the TOEFL exam or the IELTS exam. Minimum acceptable TOEFL score for admission to the MS program is 525 (or 193 computer based) and 550 (or 213 computer based) for the Applied Physics PhD program. Minimum acceptable IELTS score for admission to the M.S. Program is 6.0 and 6.5 for the Applied Physics program.
Undergraduate preparation assumed: Undergraduate major in physics. Deficiencies are corrected by additional course work.
Address financial aid inquiries to: Department of Physics and Astronomy, Box 5167
GAPSFAS application required: No
Financial aid deadline: 2/1
Loans available: Yes
Address housing inquiries to: Housing and Residence Life, Box 9502
On-campus, single student housing available: Yes
Cost/semester: $1,899–3,253
On-campus, married student housing available: Yes
Cost/month: $367–425

Table A—Faculty, Enrollments, and Degrees Granted

Research Specialty	2009–10 Faculty	Enrollment[1] Fall 2009: Master's	Doctorate	No. of Degrees Granted[2] 2009–10 (2005–10): Master's	Terminal Master's	Doctorate	Median No. of Years for 2009–10 Ph.D.'s
Applied Physics	0	0	0	0(0)	0(0)	0(0)	–
Astrophysics	2	1	2	2(3)	0(0)	0(0)	–
Atomic, Molecular, & Optical Physics	5	4	8	2(10)	0(0)	1(4)	4.5
Computational Physics	5	1	8	2(8)	0(0)	2(7)	6
Condensed Matter Physics	3	1	0	0(0)	0(0)	0(0)	–
Materials Sci./ Metallurgy	1	0	0	0(0)	0(0)	0(0)	–
Nuclear Physics	6	3	10	1(9)	0(0)	2(3)	6.5
Optics	2	2	1	1(3)	0(0)	0(0)	–
Physics Education	3	0	0	0(1)	0(0)	0(0)	–
Statistical & Thermal	1	0	0	0(0)	0(0)	0(0)	–
Total		12	29	8(34)	0(0)	5(14)	
Full-time Grad. Stud.		12	25				
Part-time Grad. Stud.		0	4				
First-year Grad. Stud.		6	2				
Median Years in Grad. Study (2009–10 Degrees)				2.5	–	6	–
Undergraduate Degrees, 2009–10 (2005–10): 4(26)							

[1] Students not yet committed to a research specialty are entered under non-specialized.
[2] Five-year totals in parentheses.

4. Graduate Degree Requirements

Master's (M.S. in Physics): A total of 30 credit hours is required with a minimum average grade of "B". The residence requirement is a minimum of 30 weeks. For the thesis option, six credit hours of thesis research are required as is an oral examination on the thesis. For the non-thesis option, a written qualifying examination on the physics core courses and an oral examination are required.
Doctorate (Ph.D. in Engineering with a concentration in Applied Physics): At least three academic years beyond the Bachelor's degree are necessary. The number of credit hours will vary according to the student's needs. A preliminary exam is required for admission to candidacy, after completion of academic coursework. A minimum of 20 credit hours of research

for the dissertation must be scheduled. An oral defense of the dissertation is required.

Thesis: Thesis may be written *in absentia.*

Table B—Appointments to Graduate Students, 2009–10

Title of Appointee	Appointments Total	Appointments First year	Academic Load Allowed in Credit Hours	Hours of Service Per Week	Stipend for Academic Year ($)
			Semester		
Teaching Assistant	15	4	9–12	20	13,952[1,2]
Research Assistant	20	2	9–12	20	13,952–17,000[1,2]
Postdoctoral Associate	1	1		0	38,000[3]
Total	36	7			

[1] Tuition waived.
[2] Some summer stipends for teaching and research are available.
[3] Twelve month appointment.

5. Personnel Engaged in Separately Budgeted Research, 1/1/09–12/31/09

Professorial faculty	12
Graduate students	32
Postdoctoral Appointment	1
Total	45

6. Separately Budgeted Research Expenditures by Source of Support

	Departmental Research	Physics-related Research Outside Department
Federal government	$742,723	$1,851,071
University	301,441	0
Total	$1,044,164	$1,851,071

Table C—Separately Budgeted Research Expenditures

Research Specialty	No. of Grants	Expenditures ($)
Applied Physics	1	$ 20,017
Astrophysics	1	31,287
Atomic, Molecular, & Optical Physics	2	29,607
Computational Physics	2	76,682
Nuclear Physics	5	816,414
Physics Education	1	70,157
Total	12	$1,044,164

Table D—Physics-related Research Outside Department

Field and Unit Outside Department	No. of Grants	Expenditures ($)
Center for Advanced Vehicular Systems (CAVS)	2	$ 412,509
Center for Science, Math, & Technology (CSMT) (Physics Education)	3	937,724
College of Education	1	189,897
College of Engineering	1	82,662
High Performance Computing Collaboratory (HPC) (Computational Physics)	2	138,472
Institute for Clean Energy Technology (ICET) (Atomic & Molecular Physics)	2	$89,807
Total	11	$1,851,071

FACULTY

Professors

Afanasjevs, Anatolijs, Ph.D., Latvian Academy of Sciences, 1993. Habilitated, Latvian State University (Riga) Latvia, 1999. Nuclear physics; computational physics.

Dunne, James A., Ph.D., The American University, 1995. Nuclear physics.

Harpole, Sandra H., Ed.D., Mississippi State, 1986. Physics education.

Lestrade, John Patrick, Ph.D., Rice, 1978. High-energy astrophysics.

Ma, Wenchao, Ph.D., Vanderbilt University, 1985. Nuclear physics.

Monts, David L., Ph.D., Columbia, 1978. Molecular spectroscopy.

Novotny, Mark A., Ph.D., Stanford University, 1978. Head of the Department. Computational physics, materials physics, and statistical mechanics.

Su, Chun Fu, Ph.D., New Orleans, 1976. Molecular spectroscopy.

Winger, Jeffry A., Ph.D., Iowa State University, 1987. Nuclear physics.

Associate Professors

Arnoldus, Hendrik F., Ph.D., Eindhoven University of Technology, The Netherlands, 1985. Theoretical optics.

Clay, R. Torsten, Ph.D., University of Ilinois, 1999. Materials physics, computational physics.

Kim, Seong-Gon, Ph.D., Michigan State University, 1994. Condensed matter physics and materials physics; computational physics.

Wang, Chuji, Ph.D., University of Science and Technology of China, 1998. Molecular spectroscopy, laser diagnostics, instrumentation.

Assistant Professors

Dutta, Dipangkar, Ph.D., Northwestern University, 1999. Nuclear physics.

Pierce, Donna M., Ph.D., University of Maryland, 2006. Astrophysics.

Rupak, Gautam Lan Tai Moong, University of Washington, 1996. Nuclear physics; computational physics; atomic theory.

Adjunct Professors

Carter, H. Kennon, Ph.D., Vanderbilt, 1969. Nuclear physics.
Lindner, Jeffrey S., Ph.D., Mississippi State, 1985. Molecular spectroscopy; polymer physics.
Luthe, John C., Ph.D., Wisconsin, 1975. Theoretical physics.
Rykaczewski, Krzysztof P., Ph.D., Warsaw University, 1983. Nuclear physics.
Singh, J.P., Ph.D., Banares Hindu University, 1980. Laser spectroscopy.
Su, Yi, Ph.D., Wayne State University, 1996. Applied spectroscopy, environmental sensing.

Instructors

Winter, Joshua B., MS, Mississippi State University, 2002.
Worthy, Mark C., M.S.E., University of Huntsville, 1994.

RESEARCH SPECIALTIES AND STAFF

Computational

Materials Physics. Clay, Kim, Novotny.
Nuclear Physics. Afanasjevs, Rupak
Statistical Mechanics. Novotny.

Theoretical

Astrophysics. Lestrade, Pierce.
Chemical Physics. Molecular structure calculations; statistical mechanics. Luthe.
Nuclear Physics. Afanasjevs, Rupak
Physical Optics/Optical Properties of Solids. Arnoldus, Foley, Luthe.
Physics Education. Dunne, Harpole, Pierce, Winter, Worthy.

Experimental

Atomic Molecular and Optical Physics. Microwave, infrared, far infrared, and Raman spectroscopy. Laser spectroscopy. Lindner, Monts, Singh, C. F. Su, Y. Su, Wang.
Gamma-Ray Astronomy. Lestrade.
Nuclear Physics. Carter, Dunne, Dutta, Ma, Rykaczewski, Winger.

THE UNIVERSITY OF MISSISSIPPI

DEPARTMENT OF PHYSICS AND ASTRONOMY

University, Mississippi 38677

Students Accepted For Degree	FIELDS		
	Physics	Astronomy	Related Fields
Doctorate	X		
Master's	X		

1. General

Chancellor: Daniel W. Jones
Dean of Graduate School: Maurice Eftink
Department Chair: Lucien M. Cremaldi
Department Telephone Number: (662) 915-7046
Type of Institution: University
Control: Public
Setting: Suburban
Total Faculty: 729*
Total Graduate Faculty: 502
Total Students: 13,204
Total Graduate Students: 2,062
Annual Graduate Tuition:
In-state residents: Full-time—$5,436**
Part-time—$302.00/sem. hr.
Out-of-state residents: Full-time—$13,896†
Part-time—$772.00/sem. hr.
Tuition rates for: 2010–11
Deferred tuition plan: Yes
Term: Semester

*Does not include acting, adjunct, or visiting faculty
**Two semester academic year
†Out-of-state portion of tuition is waived for assistantship holders.

2. Number of Faculty in Department

The combined total of full-time faculty in the three professorial ranks is 14. The combined total of full-time, part-time, and other faculty at all ranks is 35.

3. Admission, Financial Aid, and Housing

Address admission inquiries to: Dean of the Graduate School, University of Mississippi
Graduate application fee required: $25 (for out-of-state only)
Admission deadline (Fall admission): 4/1
Admission information: For fall admission, 2008–09, 11 students were accepted from 40 applicants.
Admission requirements: For admission to the graduate programs, a Bachelor's degree in physics is required with a minimum undergraduate GPA of 3.0 specified. The minimum acceptable GRE score required is a combined score of 1,000 on the verbal and quantitative parts of the test. The average GRE score for admissions in 2008 was 1,203 (verbal and quantitative combined). Students from non-English speaking countries are required to demonstrate proficiency in English via the TOEFL exam. Minimum acceptable score for admission is 600.
Address financial aid inquiries to: Chair, Department of Physics and Astronomy
GAPSFAS application required: No
Financial aid deadline: 4/1
Loans available: Yes
Address housing inquiries to: Student Housing Office
On-campus, single student housing available: Yes
Cost/term: $2,750 (single); $2,000 (double)
On-campus, married student housing available: No

Table A—Faculty, Enrollments, and Degrees Granted

Research Specialty	2008–09 Faculty	Enrollment[1] Fall 2008 Master's	Enrollment[1] Fall 2008 Doctorate	No. of Degrees Granted 2008–09 (2004–09) Master's	Terminal Master's	Doctorate	Median No. of Years for 2008–09 Ph.D.'s
Particles & Fields							
Unknown		6	17	5	0	3	
Total		6	17	5	0	3	
Full-time Grad. Stud.		6	17				
Part-time Grad. Stud.							
First-year Grad. Stud.		0	3				
Median Years in Grad. Study (2008–09 degrees)				2	2	6	–
undergraduate degrees, 2008-09 (2004-09): 3(59)							

[1]Students not yet committed to a research specialty are entered under non-specialized.
[2]Five-year titals in parentheses.

4. Graduate Degree Requirements

Master's: A Master of Arts degree requires 30 hours of suitable graduate course work. A Master of Science degree requires 24 hours of suitable graduate course work and 6 hours of thesis. The degree program must be completed within six years.
Doctorate: A Ph.D. degree requires three years of study (54 credit hours), a minimum of two years (36 hours) of graduate study at the University of Mississippi, and a minimum of one year full-time graduate work beyond the Master's degree in continuous residence. An average grade of B or above is required for all course work. Successful completion of a preliminary examination (written) and a comprehensive examination are required. The preparation and defense of a dissertation are required. A foreign language is not required.
Thesis: Thesis may be written *in absentia*.

Table B—Appointments to Graduate Students, 2009–10

Title of Appointee	Appointments Total	Appointments First year	Academic Load Allowed in Credit Hours	Hours of Service Per Week	Stipend for Academic Year ($)
			Semester		
Teaching Assistant	23	0	9	20	13,000[1]–16,000
Research Assistant	18	0	9	20	13,000–17,500[1]
Total	41	0			

[1]Tuition Scholarship provided.

5. Personnel Engaged in Separately Budgeted Research, 7/09–6/10

Professorial faculty	18
Graduate students	12
Total	30

6. Separately Budgeted Research Expenditures by Source of Support

	Departmental Research	Physics-related Research Outside Department
Federal government	$1,006,254	$6,487,054.40
Total	$1,006,254	$6,487,054.40

Table C—Separately Budgeted Research Expenditures

Research Specialty	No. of Grants	Expenditures ($)
Atmospheric Physics	1	104,339
Gravity-LIGO	5	749,092
HEP	6	1,883,687
Physical Acoustics	3	416,366
Physics Education	3	48,220
Total	18	3,201,704

Table D—Physics-related Research Outside Department

Field and Unit Outside Department	No. of Grants	Expenditures ($)
National Center for Physical Acoustics	40	9,997,845
Total	40	9,997,845

FACULTY

Professors

Cremaldi, Lucien, Ph.D., Northwestern U., 1983. Particle physics.
Kroeger, Robert S., Ph.D., University of Pittsburgh, 1986. High-energy physics.
Marshall, Thomas, Ph.D., New Mexico Inst. of Mining and Tech., 1981. Atmospheric physics.
Raspet, Richard, Ph.D., Mississippi, 1975. Acoustics.
Summers, Donald J., Ph.D., California, Santa Barbara, 1984. High-energy physics.

Adjunct Faculty

Aranchuk, Vyacheslav, Ph.D., Minsk, Belarus, 1989. Engr. Science.
Arnold, Roy T., Ph.D., Vanderbilt, 1960. Acoustics and atmospheric physics; physics education.
Arnott, Pat, Ph.D., Washington, 1988. Optics.
Atchley, Anthony, Ph.D., Mississippi, 1984. Acoustics.
Cardoso, Victor, Ph.D., CENTRA/IST, 2003.
Caruthers, Jerald, Ph.D., Texas A&M University, 1968.
Crum, Lawrence A., Ph.D., Ohio University, 1967. Acoustics.
Godang, Romulus, Ph.D., Virginia Tech., 2000. HEP.
Keppens, Veerle, Ph.D., Katholieke Universiteit Leuren, Belgium, 1995. Solid state acoustics.
Korman, Murray, Ph.D., Brown University, 1981. Physics.
Lafleur, L. Dwynn, Ph.D., University of Houston, 1969. Solid state acoustics.
Prather, Wayne, Ph.D., Mississippi, 1999. Acoustics.
Rust, W. David, Ph.D., New Mexico Inst. of Mining and Technol., 1973. Atmospheric physics.

Research Professors

Gilbert, Kenneth, Ph.D., Michigan State Univ., 1976. Physical Acoustics.
Sabatier, James, Ph.D., Mississippi, 1985. Acoustics.

Associate Professors

Bombelli, Luca, Ph.D., Syracuse, 1987. General relativity.
Cavaglià, Marco, Ph.D., International School for Advanced Studies of Trieste, 1996. Astrophysics.
Ostrovskii, Igor, Ph.D., KSU, Kiev, 1982. Acoustics.
Quinn, Breese, Ph.D., University of Chicago, 2000. Physics.

Research Associate Professors

Church, Charles, Ph.D. The University of Rochester, 1983. Radiation Biology
Ostrovskaya, Nataliya, Ph.D., KSU, Kiev, 1994.
Waxler, Roger, Ph.D., Columbia University, 1986. Acoustics.

Assistant Professors

Berti, Emanuelle, Ph.D., Univ. of Rome "La Sapienza," 2001. Astrophysics.
Datta, Alakabha, Ph.D., Univ. of Hawaii, 1995. Particle theory.
Gladden, Joseph, Ph.D., Penn State University.
Mobley, Joel, Ph.D., Washington University. 1998.

Research Assistant Professors

Hickey, Craig, Ph.D., University of Alberta, 1994. Geophysics.
Lu, Zhiqu, Ph.D., Universite depau et das Pays de l'Adour, France, 1998. Acoustic thermal-physics.
Stolzenburg, Maribeth, Ph.D., Univ. of Oklahoma, 1996. Atmospheric physics.
Torma, Tibor, Ph.D., UMass, 1996. Particle Physics.

RESEARCH SPECIALTIES AND STAFF

Theoretical

General Relativity. Bombelli, Cavaglià, Berti.
Particle Physics. Datta, Torma.

Experimental

Acoustics. Sound propagation in the atmosphere; physical acoustics; thermal acoustics, refrigerators, and engines. Church, Gilbert, Gladden, Hickey, Lu, Mobley, Raspet, Sabatier, Waxler.
Atmospheric Physics. Atmospheric electricity-thunderstorm electrification. Marshall, Stolzenburg.
Condensed Matter. Ostrovskii, Ostrovskaya.
High-Energy Physics. Cremaldi, Kroeger, Quinn, Summers.

UNIVERSITY OF MISSOURI

DEPARTMENT OF PHYSICS AND ASTRONOMY

Columbia, Missouri 65211

Students Accepted For Degree	FIELDS		
	Physics	Astronomy	Related Fields
Doctorate	X		
Master's	X	X	

1. General

President: Gary D. Forsee
Chancellor: Brady J. Daton
Department Chairman: Peter Pfeifer
Department Telephone Number: (573) 882–3335
FAX: (573) 882–4195
Type of Institution: University
Control: Public
Setting: Urban
Total Faculty: 1,991
Total Graduate Faculty: 1,826
Total Students: 31,314
Total Graduate Students: 7,445
Annual Graduate Tuition:
In-state residents: Part-time—$298.70/credit
Out-of-state residents: Full-time—$771.20/credit
Tuition rates for: 2010–11
Deferred tuition plan: Yes
Other Fees: Recreational Facility Fee–$133.11 per semester (enrolled 6+hours); Activity Fee–$136.89 per semester (enrolled 8+hours); Health Fee–$92.78 per semester (enrolled 6+hours); Information Technology Fee–$12.20 per credit hour.
Term: Semester

2. Number of Faculty in Department

The combined total of full-time faculty in the three professorial ranks is 30. The combined total of full-time, part-time, and other faculty at all ranks is 30.

3. Admission, Financial Aid, and Housing

Address admission inquiries to: Graduate School, 210 Jesse Hall, Columbia, MO 65211. http://gradschool.missouri.edu
Graduate application fee required: Yes, $45 U.S.; $60 Foreign
Admission deadline (Fall admission): 2/1
Admission information: For fall admission, 2010–11, 16 students were accepted from approximately 90.
Admission requirements: For admission to the graduate programs, a Bachelor's degree in Physics is required (no minimum undergraduate GPA specified). The GRE is required. The minimum acceptable score suggested for admission is verbal–400; quantitative–700. The GRE Advanced is recommended. The average GRE scores for admission are verbal 480; quantitative 750. Students from non-English speaking countries are required to demonstrate proficiency in English via the (TOEFL) exam. (Minimum acceptable score for admission is 550 paper, 213 computer or 80 internet based. The IELTS test with a minimum score of 6.5 is an acceptable alternative to the TOEFL.) You are exempt from this requirement if you have completed a year of full-time academic study in the US.
Address financial aid inquiries to: Student Financial Aid, 11 Jesse Hall, Columbia, MO 65211. https://sfa.missouri.edu/ or Dept. of Physics and Astronomy, Director of Graduate Studies
GAPSFAS application required: Yes
Financial aid deadline: 2/1
Loans available: Yes
Address housing inquiries to: Residential Life, 125 Jesse Hall, Columbia, MO 65211. http://reslife.missouri.edu/
On-campus, single student housing available: Yes
On-campus, married student housing available: Yes

Table A—Faculty, Enrollments, and Degrees Granted

Research Specialty	2009–10 Faculty	Enrollment[1] Fall 2010 Master's	Enrollment[1] Fall 2010 Doctorate	No. of Degrees Granted[2] 2008–09 (2004–09) Master's	Terminal Master's	Doctorate	Median No. of Years for 2008–09 Ph.D.'s
Astronomy	2	1	7	1(2)	1(1)	0(0)	–
Biophysics	7	0	7	0(2)	0(0)	2(4)	5
Condensed Matter Physics	16	2	25	4(6)	0(6)	5(15)	5.5
Medical & Health Physics	2	0	0	0(0)	0(0)	0(1)	–
Optics	1	0	1	0(0)	1(0)	0(0)	–
Physics Education	1	0	0	0(0)	0(0)	0(0)	–
Relativity & Gravitation	2	0	1	0(2)	0(0)	0(4)	–
Non-specialized	0	0	1	0(6)	0(2)	0(0)	–
Total		3	42	4(23)	2(9)	7(24)	
Full-time Grad. Stud.		3	42				
Part-time Grad. Stud.		0	0				
First-year Grad. Stud.		0	10				
Median Years in Grad. Study (2008–09 Degrees)				2	2	5	

Undergraduate Degrees, 2008–09 (2004–09): 5(43)

[1] Students not yet committed to a research specialty are entered under non-specialized.
[2] Five-year totals in parentheses.

4. Graduate Degree Requirements

Master's: Requires completion of 30 credit hours beyond the Bachelor's Degree (at least 15 hours of those in 8000 level courses) with a grade of 3.0 (B) or better. Completion of the Departmental Qualifying Examination at the MS pass level or a written thesis is required. No foreign language or computing requirements.
Doctorate: Requires completion of a minimum of 18 hours beyond the Master's Degree, with a grade of 3.0 (B) or better, and completion of the Department Qualifying Examination at the PhD pass level. The degree candidate must also meet the residency requirements. A student is required to have taken a minimum of three full years of graduate work beyond the Bachelor's Degree. To be an official candidate, the student must pass the Comprehensive Examination for the PhD, which is based upon graduate coursework in the Department. Dissertation is required.
Thesis: Thesis may be written *in absentia*.
Special Equipment, Facilities, or Programs: The University of Missouri Research Reactor at Columbia provides a unique opportunity for neutron scattering research. The thermal neutron flux at the beam port is 2×10^{14} neutrons/cm^2 s, the

highest of any US university. Current projects/programs include studies of neutron and X-ray scattering, critical phenomena, surface and interfaces, lattice and liquid dynamics, magnetic materials, optoelectronics, theoretical, computational and experimental biological physics, bio-medical imaging and optics, theory of gravity and relativistic astrophysics, spintronics, and condensed matter theory.

Table B—Appointments to Graduate Students, 2008–09

Title of Appointee	Appointments Total	Appointments First year	Academic Load Allowed in Credit Hours	Hours of Service Per Week	Stipend for Academic Year ($)
			Semester		
Teaching Assistant	23	6	9 cr. hours	20 h	15,000 (9 months)
Research Assistant	22	4			
Total	45	10			

[1] Tuition waived or paid by grant.

5. Personnel Engaged in Separately Budgeted Research, 7/08–6/09

Professorial faculty	28
Other faculty	2
Postdoctoral appointments	15
Graduate students	30
Undergraduate students	9
Total	84

6. Separately Budgeted Research Expenditures by Source of Support

	Departmental Research	Physics-related Research Outside Department
Federal government	$7,141,295	
State and local government	$1,591,095	
Total	$8,732,390	

Table C—Separately Budgeted Research Expenditures

Research Specialty	No. of Grants	Expenditures ($)
Astronomy	19	405,831
Biophysics	4	3,159,930
Condensed Matter Physics	26	3,575,534
Other	1	1,591,095
Total	50	8,732,390

Table D—Physics-related Research Outside Department

Field and Unit Outside Department	No. of Grants	Expenditures ($)
	Breakdown not available	
Total		4,900,000

FACULTY

Professors

Chandrasekhar, H. R., Ph.D., Purdue University, 1973. Optical properties of solids under pressure.

Chandrasekhar, Meera, Ph.D., Brown University, 1976. Optical spectroscopy of semiconductors and superconductors, with an emphasis on high pressure studies.

Chen, Shi-Jie, Ph.D., University of California San Diego, 1994. Theoretical and computational biophysics.

Duncan, Robert, Ph.D., University of California-Santa Barbara, 1988. Vice Chancellor of Research. Low temperature physics.

Forgacs, Gabor, Ph.D., Eotvos Roland University, Budapest, 1978. Physical mechanisms in cell and developmental biology.

Gangopadhyay, Shubhra, Ph.D, Indian Institute of Technology, Kharagpur, 1982. Professor of Electical Engineering (joint position).

Hawthorne, M. Frederick, Ph. D., University of California at Los Angeles 1953. Member of National Academy of Sciences. Professor of Radiology and Director of Nano Medicine (joint position).

Katti, Kattesh Ph.D., Indian Institute of Sciences, Bangalore, India, 1984. Development of site-specific radiopharmaceuticals, chemotherapeutic agents for cancer therapy, chemical and biomedical aspects of optical materials. Professor of Radiology (joint position).

Mashhoon, Bahram Ph.D., Princeton University, 1972. Theory of gravitation and relativistic astrophysics.

Miceli, Paul, Ph.D., University of Illinois at Urbana Champaign 1987. Surfaces and interfaces of Condensed matter Investigated by X-ray and Neutron Scattering.

Pfeifer, Peter, Ph.D., ETH Zurich Swiss Federal Institute of Technology, 1980. Department Chair. Theoretical studies in surface physics, fractals and quantum dynamics, renewable energy.

Satpathy, Sashi, Ph.D., University of Illinois at Urbana Champaign, 1982. Theoretical condensed matter physics: electronic structure and magnetism in solids.

Taub, Haskell, Ph.D., Cornell University, 1971. Structure, phase transitions and dynamics of absorbed films.

Vignale, Giovanni, Ph.D., Northwestern University, 1984. Condensed matter theory.

Associate Professors

Guha, Suchi, Ph.D., Arizona State University, 1996. Fabrication and optical characterization of organics optoelectronic devices.

Kopeikin, Sergei Ph.D., Space Research Institute of the Academy of Science USSR, Moscow, 1986. Gravity and general relativity.

Kosztin, Ioan, Ph.D., University of Illinois at Urbana Champaign, 1997. Theoretical and computational biological physics.

Li, Aigen, Ph.D., Leiden University, 1998. Theoretical Astrophysics.

Montfrooij, Wouter, Ph.D., University of Delft, 1990. Neutron Scattering.

Speck, Angela, Ph.D., University College London, 1998. Astronomy including stellar evolution, astromineralogy and dust around evolved stars, galactic chemical evolution, and meteoritics.

Ullrich, Carsten, Ph.D., Universitat Wurzburg, 1995. Conduced matter theory, density-functional theory.

Wexler, Carlos, Ph.D., University of Washington, 1997. Condensed matter theory.

Yu, Ping, Ph.D., Hong Kong University of Science and Technology, 1998. Optoelectronics and biomedical imaging.

Assistant Professors

Hanuscin, Deborah, Ph.D., University of Indiana, 2004. Physics Education Research.

King, Gavin, Ph.D., Harvard University, 2004. Single molecule biophysics, atomic force microscopy.

Vajk, Owen, Ph.D., Stanford University, 2003. Neutron scattering.

Zou, Xiaoqin, Ph.D., University of California, San Diego, 1993. Theoretical and Computational biophysics.

Associate Teaching Professor

Koszhin, Dorina, Ph.D., Univesity of Illinois at Urbana-Champaign, 1998. Theoretical biophysics.

Assistant Teaching Professors

King, Karen, Ph.D., Dartmouth College, 2003. Medical imaging and biomechanical modeling.

Zhang, Yun, Ph.D., University of California, San Diego, 1999. Experimental condensed-matter physics.

RESEARCH SPECIALTIES AND STAFF

Theoretical

Biophysics. Molecular modeling transport through membranes, RNA folding and assembly, computational drug design. Chen, I. Kosztin, Zou.

Cosmology and General Relativity. Origin and fate of the Universe, gravitational radiation, post-Newtonian gravity, black holes, Kopeikin, Mashhoon.

Condensed Matter Theory. Electronic structure of materials, magnetic devices and spintronics, quantum many-body theory, density-functional theory, transport and optical excitations in semi-conductors, quantum and classical statistical mechanics, fractals and phase transitions. Pfeifer, Satpathy, Ullrich, Vignale, Wexler.

Experimental

Astronomy and Astrophysics. Cosmic dust, planetary and star formation and evolution, galactic chemical evolution, origin or molecules. Li, Speck.

Biophysics. Cellular biomechanics, physical mechanisms of cell and developmental biology, organ printing and tissue engineering, single-molecule atomic force microscopy.Forgacs, G. King.

Medical and Health Physics. Nanomedicine, drug delivery, cancer research. Hawthorne, Katti. Condensed Matter Physics. Organic displays and photovoltaics, Raman scattering, ZnO-based optoelectronics, dielectrics, high pressure optical spectroscopy, magnetic fractals, quantum phase transitions, organic thin films, neutron and x-ray scattering, epitaxial growth, alternative fuel research, hydrogen storage, surface science, multiferroics. H. R. Chandrasekhar, M. Chandrasekhar, Duncan, Gangopadhyay, Guha, Miceli, Montfrooij, Pfeifer, Taub, Vajk.

Optics. Biomedical imaging, nonlinear optics. Yu.

Physics Education. Writing-to-learn strategies, formative assessment tools, inquiry-based teaching methodologies. Hanuscin.

UNIVERSITY OF MISSOURI, KANSAS CITY

DEPARTMENT OF PHYSICS
http://cas.umkc.edu/physics/

Kansas City, Missouri 64110

Students Accepted For Degree	FIELDS		
	Physics	Astronomy	Related Fields
Doctorate	X		
Master's	X		

1. General

President: Gary Forsee
Chancellor: Leo E. Morton
Dean of Graduate School: R. MacQuarrie
Department Chairman: Michael Kruger
Department Telephone Number: (816) 235-1604
Type of Institution: University
Control: Public
Setting: Urban
Total Faculty: 690 (full-time excluding medical school)
Total Students: 14,818
Total Graduate Students: 5,420
Annual Graduate Tuition:
In-state residents: Full-time—$298.70/credit hr.
Out-of-state residents: Full-time—$771.20/credit hr.
Tuition rates for: 2010
Deferred tuition plan: No
Other Fees: Multipurpose building $30.00
Term: Semester

2. Number of Faculty in Department

The combined total of full-time faculty in the three professorial ranks is 10. The combined total of full-time, part-time, and other faculty at all ranks is 13.

3. Admission, Financial Aid, and Housing

Address admission inquiries to: Department of Physics
Graduate application fee required: $35-US $50-foreign
Admission deadline (Fall admission): 7/15, 6/1 for international-students
Admission requirements: For admission to the graduate programs, a Bachelor's degree in physics from an accredited institution is required with a satisfactory academic record. No minimum undergraduate GPA specified. The GRE is strongly urged. The GRE Advanced is not required. Students from non-English speaking countries are required to demonstrate proficiency in English via the TOEFL exam. A minimum score of 575 is required for teaching assistantship, and 525 for admission.
Undergraduate preparation assumed: Simon, *Mechanics*; Reitz and Milford, *Electromagnetic Theory*; Reif, *Thermal Physics*; Serway, Moses, and Moyer, *Modern Physics*; Ohanian, *Principles of Quantum Mechanics*.
Address financial aid inquiries to: Department of Physics, Univ. of Missouri-Kansas City, Kansas City, MO 64110
GAPSFAS application required: Yes
Financial aid deadline: 4/15
Loans available: Yes
Address housing inquiries to: Residence Hall, Univ. of Missouri-Kansas City, 5100 Rockhill Road, Kansas City, MO 64110.
On-campus, single student housing available: Yes–per academic year from $5,732 to $7,254 Room
On-campus, married student housing available: No
meal plans available also: $2,526 to $2,040

Table A—Faculty, Enrollments, and Degrees Granted

Research Specialty	2010 Faculty	Enrollment[1] Winter 2010		No. of Degrees Granted[2]			Median No. of Years for 2006–07 Ph.D.'s
		Mas-ter's	Doc-torate	Mas-ter's	Terminal Master's	Doc-torate	
Applied Physics	1	–	0	–	0(0)	0(0)	–
Astronomy/ Astrophys	1	1	0	–	1(0)	1(1)	–
Chemical Physics	1	–	0	–	0(1)	0(0)	–
Condensed Matter Physics	3	7	3	1	2(19)	0(9)	–
Geophysics	1	–	0	–	0(0)	1(1)	–
Materials Sci./ Metallurgy	2	–	0	–	0(1)	0(2)	–
Optics	0	–	0	–	0(1)	0(1)	–
Polymer Physics/ Science	1	–	0	–	0(0)	0(0)	–
Statistical & Thermal	1	–	0	–	0(1)	0(0)	–
Other Theoretical/ Math.	0	2	0	1	0(2)	1(0)	–
Nuclear Physics	1						
Physics Education	2	2	0	1	0(2)	1(0)	–
Solid State Phys	5	9	7				
Total		21	10	3	2(25)	3(13)	
Full-time Grad. Stud.		18	9				
Part-time Grad. Stud.		3	1				
First-year Grad. Stud.		3	0				
Median Years in Grad. Study (2006–07 Degrees)				–	2	6	–
Undergraduate Degrees, 2004–05 (2000–05): 7(20)							

[1]Students not yet committed to a research specialty are entered under non-specialized.
[2]Five-year totals in parentheses.

4. Graduate Degree Requirements

Thesis Master's: At least 30 hours of graduate credit. A research project must be planned and a thesis reporting this project is required (up to six hours of credit). A grade-point average of 3.0 or better must be maintained. In addition to presenting a satisfactory thesis, the student is required to pass a departmental written examination and a final oral examination.
Doctorate: Interdisciplinary Ph.D. program. Students majoring in physics must choose another related secondary discipline of study. Specific academic requirement is determined by the student's interdisciplinary advisory committee. Minimal requirements for admission: Combined raw score of at least 1,500 in GRE verbal, quantitative, and analytical; undergraduate GPA>2.75 and graduate GPA>3.50. Ph.D. qualifying, comprehensive, and final oral examination for Dissertation defense are required.
Nonthesis Master's: The program of study must include 33 hours of graduate credit. A departmental written examination is required.
Thesis: Thesis may be written *in absentia*.

Table B—Appointments to Graduate Students, 2009–10

Title of Appointee	Appointments Total	Appointments First year	Academic Load Allowed in Credit Hours	Hours of Service Per Week	Stipend for Academic Year ($)
			Semester		
Teaching Assistant	12	–	12	20	11,000[1,3]
Research Assistant	8	3	12	20	11,000–17,000[2,3]
Total	20	3			

[1]Additional income for summer TA.
[2]10- to 12-month appointment.
[3]6 to 9 credit hour fee remission per semester.

5. Personnel Engaged in Separately Budgeted Research, 7/03–6/05

Professorial faculty	8
Research associates	10
Graduate students	12
Total	30

6. Separately Budgeted Research Expenditures by Source of Support

	Departmental Research	Physics-related Research Outside Department
Federal government	$988,731	$
State and local government	776,735	
Total	$1,765,466	$

7. Extension Centers and Summer Programs

Depending on availability of funds and enrollments the department offers a summer program offering general physics courses and physics for science and engineering majors.

Table C—Separately Budgeted Research Expenditures

Research Specialty	No. of Grants	Expenditures ($)
Condensed Matter Physics	14	1,382,904
Physics Education	2	10,698
Polymer Physics/Science	0	250,000
Surface Physics	0	
Astro Physics	2	71,864
Total	18	1,765,466

FACULTY

Professors

Bryant, Paul J., Ph.D., St. Louis, 1957. Emeritus.

Ching, Wai-Yim, Ph.D., Louisiana State, 1974. Curators' Professor. Theoretical condensed matter and materials science.

Jean, Y. C. Jerry, Ph.D., Marquette, 1974. (Joint with Chemistry.) Experimental chemical physics; solid state chemistry.

Kruger, Michael, Ph.D., California, Berkeley, 1992. Curators' Professor. Experimental solid state physics; geophysics. Chairman.

Murphy, Richard D., Ph.D., Minnesota, 1968. Theory of condensed matter; statistical mechanics.

Querry, Marvin R., Ph.D., Kansas State, 1968. Curators' Professor. Optical properties of materials. Emeritus.

Urani, John R., Ph.D., Missouri, Columbia, 1969. Relativity and quantum theory. Emeritus

Wieliczka, David M., Ph.D., Iowa State, 1982. Distinguished Visiting Professor. Experimental solid state physics; photoemission; Auger spectroscopy.

Wrobel, Jerzy M., Ph.D., Wroclaw, 1984. Experimental solid state physics; semiconductor thin films.

Zhu, Da-Ming, Ph.D., Washington, 1988. Experimental solid state physics.

Associate Professors

Leibsle, Fred M., Ph.D., Illinois, 1991. Experimental solid state physics; surface physics.

Stoddard, Elizabeth, Ph.D., Washington University, 2000. Science education.

Waring, Richard C., M.S., Arkansas, 1961. Emeritus.

Assistant Professors

Caruso, Anthony N., Ph.D., Univ of Nebraska Lincoln. 2004. Experimental condensed matter.

McIntosh, Daniel, Ph.D., Univ of Arizona. 2001. Astro Physics.

Priour, Donald, Ph.D., Princeton University, 2000. Condensed matter theory.

RESEARCH SPECIALTIES AND STAFF

Theoretical

Condensed Matter. Electronic structures of ordered and disordered systems; magnetism; liquid metals; superfluid helium; surface physics; computer simulation. Ching, Murphy, Priour.

Nuclear Physics. Stoddard.

Physics Education. Stoddard.

Experimental

Astrophysics. McIntosh.

Biophysics. Biomembrane. Jean.

Chemical Physics. Interaction of positrons, muons, pions, and protons in condensed matter; Compton profile. Jean.

Geophysics. Kruger.

Solid State Physics. Optical properties of condensed materials; STM; AFM; Raman spectroscopy; photoluminescence; photoemission; positron spectroscopy; Auger spectroscopy. Caruso, Jean, Kruger, Leibsle, Wieliczka, Wrobel, Zhu.

FACULTY PUBLICATIONS

Caruso, Anthony

Wisbey, D. and Ning W., Y. Losovyj, I. Ketsman, A. N. Caruso, D. Feng, J. Belot, E. Vescovo, and P. A. Dowben, "Radiation Induced Decomposition for the Metal-Organic Bis(4-cyano-2,2,6,6-tetramethyl-3,5-heptanedionato)copper(II)," Appl. Surf. Sci. **255**, 3576 (2009).

Wisbey, D. S. and Wu, D. F., A. N. Caruso, J. Belot, Ya. B. Losovyj, E. Vescovo, and P. A. Dowben, "Induced spin polarization of copper spin 1/2 molecular layers," Phys. Lett. A **373**, 484 (2009).

Caruso, A. N. and Konstantin I. Pokhodnya, William W. Shum, W. Y. Ching, Bridger Anderson, M. T. Bremer, E. Vescovo, Paul Rulis, A. J. Epstein, and Joel S. Miller, "Direct Evidence of electron spin polarization from an organic-based magnet: [FeII(TCNE)(NCMe)2]

[FeIIICl4]," Phys. Rev. B **79**, 195202 (2009).

Ching, Wai-Yim

W. Y. Ching, Lei Liang and Paul Rulis, Bart Kahr, "Theoretical study of the large linear dichroism of herapathite," Phys. Rev. B **80**, 235132-1-5 (2009).

R. Podgornik A. Siber, R. F. Rajter, R. H. French, W. Y. Ching, and A. Parsegian, "Dispersion interations between optically anisotropic cylinders at all separations: Retardation effect for insulating and semiconducting single wall carbon nanotubes," Phys. Rev. B **80**, 165414-1-10 (2009).

W. Y. Ching, P. Rulis and A. R. Rupini, S. J. Pennycook, "Spectroscopic imaging of EEIs using ab initio data and function field visualization," Ultramicroscopy **109**, 1472-1478 (2009).

Kruger, Michael

"X-Ray Diffraction of Permalloy Nanoparticles Fabricated by Laser Ablation in Water," Mater. Lett. **63**, 893-895 (2009).

"Fractal character of titania nanoparticles formed by laser ablation," J. Appl. Phys. **106**, 054306 (2009).

"Compressibility Minimum in Nanomaterials of a Specific Particle Size," Phys. Rev. B **79**, 125406 (2009).

Leibsle, Fred M.

Ma, X. D., Takeshi, Nakagawa, Yasumasa, Takagi, Marek, Prysbylski, Fred, Icibsle, Toshihiko, Yokoyam, "Magnetic properties of self-assembled Co nanorods grown on Cu(110)-(2x3)N," Phys. Rev. B **78**, 104420 (2008).

Harrington, S. J., Kilway, K. V., Zhu, D.-M., Phillips, J. M. and Leibsle, F. M., "Formate-induced destabilization of the Cu(110) surface," Surf. Sci., **600**, 1193 (2006).

McIntosh, Daniel

Barazza, F., C. Wolf, M. Gray *et al.*, "Relating basic properties of bright early-type dwarf galaxies to their location in Abell 901/902," Astronomy & Astrophysics, **508**, 665-675 (2009).

Heiderman, A., S. Jogee, I. Marinova *et al.*, "Interacting Galaxies in the A901/902 Supercluster with STAGES," Astrophysical Journal **705**, 1433-1455 (2009).

Robaina, A., and E. Bell, R. Skelton, D. McIntosh, *et al.*, "Less than 10 Percent of Star Formation in Z-0.6 Massive Galaxies is Triggered by Major Interactions," Astrophysics Journal **704**, 324-340 (2009).

Priour Jr., Donald

Biddle, J., B. Wang and S. Das Sarma, "Localization in one-dimensional incommensurate lattices beyond the Aubry-Andre model," Phys. Rev. A. **80**, 012303 (2009).

Stoddard, Elizabeth

Al Odom, E. R. Stoddard and S. M. LaNasa, "The relationship of instructional practice and student attitudes to science achievement," Int. J. Sci. Educ., 2007.

Wrobel, Jerzy

Musaev, O. R., Midgley, A. E., Wrobel, J. M., Yan, J., and Kruger, M. B., Fractal character of titania nanoparticles formed by laser ablation, J. Appl. Phys. **106**, 054306 (2009).

Musaev, O. R., Midgley, A. F. Muthu, D. V. S, Wrobel, J. M., Kruger, M. B., "X-Ray Diffraction of Permalloy Nanoparticles Fabricated by Laser Ablation in Water," Mater. Lett. **63**, 893-895 (2009).

Zhu, Da-Ming

Yihong K. and Da-Ming Zhu, "Conductance Mapping of Propton Exchange Membranes by Current Sensing Atomic Force Microscopy," J. Phys. Chem. B **113**, 15040-15046 (2009).

Wang, P. and Da-Ming Zhu, "Molecular Weight Dependence of Viscosity and Shear Modulus of Polyethylene Glycol (PEG) Solution Boundary Layers," J. Phys. Chem. C **113**, 13793-13800 (2009).

Wang, P. and Da-Ming Zhu, "Determine Viscoelastic Properties of Polymer Solution Layers Near a Solid-Fluid Interface using Quartz Crystal Microbalance with Dissiapation Monitoring," IUTAM Symp. on Recent Advances of Acoustics Waves in Solids (Springer, The Netherlands), pp. 414-424 (2009).-

MISSOURI UNIVERSITY OF SCIENCE & TECHNOLOGY

DEPARTMENT OF PHYSICS

Rolla, Missouri 65409

Students Accepted For Degree	FIELDS: Physics	Astronomy	Related Fields
Doctorate	X		
Master's	X		

1. General

President: Gary Forsee
Chancellor: John F. Carney III
Department Chairman: G. Dan Waddill
Department Telephone Number: (573) 341-4781
Type of Institution: University
Control: Public
Setting: Small city
Total Faculty: 88 Part-time
368 Full-time
Total Graduate Faculty: 329
Total Students: 6,154 (on campus)
Total Graduate Students: 998 (on campus)
Annual Graduate Educational Fee:
In-state residents: Full-time—$5,376.60/academic term (18 hrs.)
Part-time—$298.70/credit
Out-of-state residents: Full-time—$13,881.60/AY*
Part-time—$771.20 per credit
Educational fee rates for: 2009–10
Deferred tuition plan: Yes—Payment Plan Available
Other Fees: $911.96 (18 hrs. per year)
Term: Semester

*Out-of-state fees waived for Graduate Assistants

2. Number of Faculty in Department

The combined total of full-time faculty in the three professorial ranks is 18. The combined total of full-time, part-time, and other faculty at all ranks is 23.

3. Admission, Financial Aid, and Housing

Address admission inquiries to: Director of Admission & Student Financial Assistance, 102 Parker Hall
Graduate application fee required: $50, U.S. citizens
$50, International students
Admission deadline (Fall admission): 7/1
Admission information: For fall admission, 2009–10, 10 students were accepted from 28 applicants.
Admission requirements: For admission to the graduate programs, a Bachelor's degree in physics is required with a minimum undergraduate GPA of 3.0 specified. The GRE is required. The minimum acceptable score suggested for admission is total 1,100 (verbal+quantitative). The GRE Advanced is recommended. No minimum acceptable score is used. Students from non-English speaking countries are required to demonstrate proficiency in English via the TOEFL exam. Minimum acceptable score for admission is 570.
Undergraduate preparation assumed: General Physics—Halliday *Fundamentals of Physics*; Modern Physics—Thornton, *Modern Physics for Scientists and Engineers*; Mechanics—Marion, *Classical Dynamics of Particles and Systems*; Thermodynamics—Reif, *Fundamentals of Statistical and Thermal Physics*; Electricity and Magnetism—Griffiths, *Introduction to Electrodynamics*; Quantum Mechanics—Liboff, *Introduction to Quantum Mechanics.*
Address financial aid inquiries to: Office of Student Financial Assistance, G1 Parker Hall
GAPSFAS application required: Yes
Financial aid deadline: 3/01
Loans available: Yes
Address housing inquiries to: Office of Residential Life, 205 W. 12th St.
On-campus, single student housing available: Yes
[1]*Cost/year*: $8,290 double; $9,575, single (TJ North Renovated)
$8,400, double; $9,700, single (TJ South)
$7,440, double; $8,595, single (Quad)
$9,775, double; $10,460, single (Res. College)
On-campus, student apartment housing available: Yes
Cost/month: $615 unfurnished, $705 furnished
−2 bdrm (Nagogami)

[1]Room and board with cable/ethernet

Table A—Faculty, Enrollments, and Degrees Granted

Research Specialty	2009–10 Faculty	Enrollment[1] Fall 2009: Master's	Doctorate	No. of Degrees Granted[2] 2009–10 (2005–10): Master's	Terminal Master's	Doctorate	Median No. of Years for 2009–10 Ph.D.'s
Astrophysics[3]	0	0	3	0(0)	0(0)	1(1)	2.5
Atmos./Cloud & Aerosol Sciences	5	0	5	0(0)	1(2)	0(0)	–
Atomic, Molecular, & Optical Physics	7	1	6	0(2)	0(1)	3(10)	5.5
Biophysics[3]	0	0	3	0(0)	0(0)	2(4)	7
Condensed Matter Physics	7	0	13	0(3)	0(2)	1(6)	6.5
Optics[3]	0	0	0	0(0)	0(0)	0(1)	–
Particles & Fields[3]	0	0	0	0(0)	0(0)	0(1)	–
Physics Education	2	1	0	0(0)	0(1)	0(0)	–
Polymer Physics/ Science	1	0	0	0(0)	0(0)	0(0)	–
Statistical & Thermal	2	0	0	0(0)	0(0)	0(0)	–
Non-specialized	0	3	6	0(1)	0(4)	–	–
Total		5	36	0(6)	1(10)	7(23)	
Full-time Grad. Stud.		5	25				
Part-time Grad. Stud.		0	11				
First-year Grad. Stud.		0	10				
Median Years in Grad. Study (2009–10 Degrees)					4	6	

Undergraduate Degrees, 2009–10 (2005–10): 4(50)

[1]Students not yet committed to a research specialty are entered under non-specialized.
[2]Five-year totals in parentheses.
[3]UMR-UMSL Co-op program.

4. Graduate Degree Requirements

Master's: 30 graduate credit hours for Masters with thesis, 30 graduate credit hours for non-thesis Masters in approved program with satisfactory performance, thesis exam required, no residence or language requirements, B average required.
Doctorate: A minimum of 72 hours with satisfactory performance. Residency requirement of three years (six semesters;

those with Master's degree from UMR or other institution—two years, four semesters), Ph.D. qualifying exam, dissertation, dissertation exam. Language requirement-pass examination or equivalent of one year collegiate level course work with grade of B or better, overall requirement of B grades or better.

Thesis: Thesis may be written *in absentia*.

Special Equipment, Facilities, or Programs: The Cloud and Aerosol Sciences Laboratory provides a wide range of special instrumentation for the study of atmospheric physics, including an assortment of aerosol generators, direct Aitken nuclei counters, condensation nuclei counters, diffusion cloud chambers, and expansion cloud chambers suitable for low temperature applications. A major mobile laboratory facility is available for the characterization of gas turbine and rocket engine exhaust emissions, and is suitable for ground test, airborne, and altitude chamber venues.

The *Graduate Center for Materials Research* provides accessibility to ESCA, auger spectrometers, scanning electron microscope, automatic x-ray spectrometer, mass spectrometers, etc.

The physics department itself provides access to an ion energy loss spectrometer, an ion implantation accelerator system, ESR, a full range of lasers, and general research equipment.

Table B—Appointments to Graduate Students, 2009–10

Title of Appointee	Appointments Total	Appointments First year	Academic Load Allowed in Credit Hours	Hours of Service Per Week	Stipend for Academic Year ($)
			Semester		
Teaching Assistant	15	4	12	20[1]	16,650
Research Assistant	9	0	12	20[1]	16,650
Total	24	4			

[1]Nominal figure. Preparation and grading time, etc., is included. Normal load is 2 to 3 two-hour laboratory sections.
[2]Summer appointments are normally available for all students desiring summer employment.

5. Personnel Engaged in Separately Budgeted Research, 7/09–7/10

Professorial faculty	9
Graduate students	11
Nonteaching research personnel	1
Total	21

6. Separately Budgeted Research Expenditures by Source of Support

	Departmental Research	Physics-related Research Outside Department
Federal government	1,005,401	430,517
Industry	47,185	
State & local government	28,204	
Total	1,080,790	430,517

Table C—Separately Budgeted Research Expenditures

Research Specialty	No. of Grants	Expenditures ($)
Atmos./Cloud & Aerosol Sciences	12	430,517
Atomic, Molecular, & Optical Physics	4	311,179
Condensed Matter Physics	8	311,390
Physics Education	2	27,704
Total	26	1,080,790

Table D—Physics-related Research Outside Department

Field and Unit Outside Department	No. of Grants	Expenditures ($)
Atmos./Cloud & Aerosol Sciences-Chemistry	12	430,517
Total	12	430,517

FACULTY

Professors

Bieniek, Ronald J., Ph.D., Harvard, 1975. Atomic and molecular collision processes; theoretical.

DuBois, Robert D., Ph.D., Nebraska, 1975. Ion-atom collisions-electron spectra; experimental.

Hagen, Donald E., Ph.D., Purdue, 1970. Cloud simulation models and experiments.

Hale, Barbara N., Ph.D., Purdue, 1967. Nucleation of water and ice; theoretical.

Madison, Don H., Ph.D., Florida State, 1972. Electron-atom, ion-atom scattering; theoretical.

Parris, Paul E., Ph.D., Rochester, 1984. Electron transport in disordered solids; theoretical.

Peacher, Jerry L., Ph.D., Indiana, 1965. Atomic collisions; scattering; theoretical.

Pringle, Allan, Ph.D., Missouri, Columbia, 1981. Neutron scattering; magnetic materials; experimental.

Schulz, Michael, Ph.D., U. of Heidelberg, 1984. Ion-atom collisions and electron correlation; experimental.

Waddill, G. Daniel, Ph.D., Indiana, 1987. Chairman of the Department. Characterization of metallic surfaces and interfaces by XPS and UPS; experimental.

Wilemski, Gerald, Ph.D., Yale, 1972. Nucleation theory; theoretical.

Associate Professors

Schmitt, John L., Ph.D., Michigan, 1968. Nucleation phenomena; experimental.

Story, J. Greg, Ph.D., University of Southern California, 1989. Studies of Rydberg atom properties; experimental.

Vojta, S. Thomas, Ph.D., University of Chemnitz, 1994. Correlated electrons and quantum phase transitions; theoretical.

Assistant Professor

Hor, Yew San, Ph.D., Rutgers University, 2004. Growth and characterization of novel bulk and nanostructured materials; experimental.

Jentschura, Ulrich, Ph.D., University of Technology, Dresden, 1999. QED bound-state calculations, relativistic quantum dynamic processes in laser fields, analysis of high-precision experiments; theoretical.

Medvedeva, Julia, E., Ph.D., Inst. of Metal Physics, Russian Academy of Science, 2002. Density functional theory; theoretical.

Yamilov, Alexey, Ph.D., The City University of New York, 2001. Mesoscopic phenomena in light propagation; photonics; theoretical.

RESEARCH SPECIALTIES AND STAFF

Theoretical

Cloud and Aerosol Sciences Laboratory

Atmospheric (Cloud) Physics. Homo- and heteromolecular nucleation studies; properties of water clusters; surface nucleation; simulation of initial stages of cloud formation; condensational drop growth/evaporation; aerosol dynamics; neutron scattering by aerosols, density functional theory of inhomogeneous fluids. B. Hale, Wilemski.

Physics Department

Atomic, Molecular, and Optical Physics. Atomic collisions; scattering; primary ionization. Bieniek, Madison, Peacher.

Condensed Matter Physics. Electron transport in disordered solids; surface phenomena; electronic polymers; density functional theory; mesoscopic phenomena in light propagation; photonics; correlated electrons and quantum phase transitions. Jentschura, Medvedeva, Parris, Vojta, Yamilov.

Experimental

Cloud and Aerosol Sciences Laboratory

Fundamental studies of aerosol generation, measurement of physical and chemical properties, and evolution in the atmosphere. Current emphasis is on the exhaust emissions from jet and rocket engines, with measurement campaigns conducted in-situ with airborne facilities, in ground based facilities in engine test stands, combustor rigs, altitude chambers, and in laboratory combustion facilities. The laboratory has an interdisciplinary flavor with strong interactions with chemistry and engineering. Hagen.

Fundamental studies of gas and liquid phase nucleation using cloud chamber techniques. Schmitt.

Physics Department

Atomic and Molecular Physics. Ion-atom and electron collisions; heavy-ion impact excitation; energy-loss spectrometry; lifetimes; photoionization; electron impact; recoil ion momentum spectroscopy, electron spectroscopy; quantum electronics; Penning reactions; lasers; crossed beam molecular scattering. DuBois, Schulz, Story.

Condensed Matter Physics. Ion implantation; electron spin resonance; magnetism; ferroelectricity; surfaces; superconductivity; electronic ceramics. Hor, Pringle, Waddill.

Physics Education. Development of traditional and non-traditional learning experiences; production of audio-visual aids. Bieniek, Pringle.

UNIVERSITY OF MISSOURI -ST. LOUIS

DEPARTMENT OF PHYSICS AND ASTRONOMY

St. Louis, Missouri 63121

Students Accepted For Degree	FIELDS		
	Physics	Astronomy	Related Fields
Doctorate	X		
Master's	X		

1. General

Chancellor: Thomas F. George
Associate Provost for Academic Affairs and Dean of Graduate School: Judith Walker de Felix
Department Chairman: Bernard Feldman
Department Telephone Number: (314) 516-5931
Type of Institution: University
Control: Public
Setting: Urban
Total Faculty: 884
Total Graduate Faculty: 422
Total Students: 12,496
Total Graduate Students: 2,717
Annual Graduate Tuition:
In-state residents: Full-time—$4,437.48/sem.
Part-time—$369.99/credit hr.
Out-of-state residents: Full-time—$7,580.61/sem.
Tuition rates for: 2010–11
Deferred tuition plan: No
Term: Semester

2. Number of Faculty in Department

The combined total of full-time faculty in the three professorial ranks is 12. The combined total of full-time, part-time, and other faculty at all ranks is 14. For students in the doctoral program, the faculty resources of the cooperating Missouri University of Science and Technology campus are also available.

3. Admission, Financial Aid, and Housing

Address admission inquiries to: Phil Fraundorf
Graduate application fee required: $35 (domestic)
$40 (international)
Admission deadline (Fall admission): 7/1
Admission information: For fall admission, 2010–11, 2 students were accepted from 10 applicants.
Admission requirements: For admission to the graduate programs, a Bachelor's degree in physics or a related field is required with an undergraduate and major field grade point average of B or better. The Graduate Record General Examination is required. The Subject Exam in Physics is optional. To be admitted to the Ph.D. program a student must have a GRE Quantitative score of at least 600 and an Analytical score of at least 3.0. Students from non-English speaking countries are required to demonstrate proficiency in English via TOEFL test with a minimum test score of 570 (230 CBT).
Undergraduate preparation assumed: Marion, *Classical Dynamics;* Wangsness, *Electromagnetic Fields*; Gasiorowicz, *The Structure of Matter: A Survey of Modern Physics.*
Address financial aid inquiries to: Financial Aid Office, 278 Millennium Student Center
GAPSFAS application required: No
Financial aid deadline: 4/1
Loans available: Yes
Address housing inquiries to: Residential Life, 125 South Campus Residence Hall and University Meadows, 2901 University Meadows Drive
On-campus, single student housing available: Yes
On-campus, married student housing available: Yes

Table A—Faculty, Enrollments, and Degrees Granted

Research Specialty	2009–10 Faculty	Enrollment[1] Fall 2009		No. of Degrees Granted[2] 2009–10 (2005–10)			Median No. of Years for 2009–10 Ph.D.'s
		Master's	Doctorate	Master's	Terminal Master's	Doctorate	
Astronomy	3	1	2	1(2)	2(0)	1(0)	–
Astrophysics	1	0	1	0(1)	0(0)	0(0)	–
Atomic, Molecular, & Optical Physics	1	1	0	0(0)	0(0)	0(0)	–
Biophysics	3	0	3	0(3)	0(1)	0(3)	4.5
Condensed Matter Physics	9	2	7	0(6)	0(0)	1(3)	4
Optics	0	0	0	0(0)	0(0)	0(1)	–
Particles & Fields	0	0	0	0(1)	0(0)	0(0)	–
Statistical & Thermal	0	1	0	0(0)	0(0)	0(0)	–
Non-specialized	4	2	2	0(4)	1(0)	0(0)	–
Total		6	15	1(16)	3(1)	2(7)	
Full-time Grad. Stud.		2	12				
Part-time Grad. Stud.		4	3				
First-year Grad. Stud.		2	7				
Median Years in Grad. Study (2009–10 Degrees)				2	2	8	–
Undergraduate Degrees, 2009–10 (2005–10): 1(32)							

[1]Students not yet committed to a research specialty are entered under non-specialized.
[2]Five-year totals in parentheses.

4. Graduate Degree Requirements

Master's: A student must complete 30 credit hours in graduate physics courses with at least 15 of these at the 5000 or 6000 level. The writing of a thesis is optional. A maximum of 6(3) credit hours of Research, P6490, may be counted toward the minimum 15 hours with (without) the thesis option. A comprehensive oral examination must be passed, which includes the defense of the thesis if the student has chosen to write one. A grade point average of a B or better must be maintained during each academic year. The requirements must be fulfilled within six years from the time of admission. At least 2/3 of required graduate credit must be taken in residence. There is no language requirement.
Doctorate: A minimum of 78 hours with satisfactory performance (i.e., B grade point average or better) is required for a Ph.D. in Physics. Residency requirement of three years/six semesters (for those with Master's degree, two years/four semesters) at UM-St. Louis and/or cooperating Missouri S+T campus. Ph.D. qualifying exam, dissertation, dissertation exam administered in cooperation with Missouri S+T.
Thesis: Thesis may be written *in absentia.*
Special Equipment, Facilities, or Programs: The *William L. Clay Center for Nanoscience*, which opened in 1996, is a facility bringing together both physicists and chemists for research in materials science. A focus of the Center is to foster collabo-

rations between its members and colleagues in industry. The Center houses the Microscope Image and Spectroscopy Tech Lab where research at the forefront of nanotechnology is conducted with transmssion electron, scanning probe, and scanning electron microscopes in a building uniquely designed for such work. The Center is spearheading the formation of the Missouri NanoAlliance, a nano-characterization and synthesis network that will facilitate the sharing of resources across Missouri. The *Center for Neurodynamics*, established in 1995, conducts research at the interface between physics and biology, with a focus on the roles of noise and stochastic synchronization in neural processing. The Center has an on-site high speed (CCD) imaging system for studying the spatial dynamics of neural activity in the mammalian brain. Collaborations with St. Louis University will permit high time-resolution magnetoencephalography (MEG) image analysis, making use of a high-speed Internet 2 connection, UMSL's new high-speed (3.8 GHz) 128-node Beowulf cluster, and Missouri's first MEG machine. Astronomers make use of national facilities at Kitt Peak, Cerro Tololo, and Mauna Kea Observatories. The Department maintains both machine and electronic shops. The University provides email and internet services through numerous student labs equipped with computers, flat-bed document scanners, and color printers. The Department maintains a network of UNIX workstations with standard software packages for word and image processing.

Table B—Appointments to Graduate Students, 2009–10

Title of Appointee	Appointments Total	Appointments First year	Academic Load Allowed in Credit Hours	Hours of Service Per Week	Stipend for Academic Year ($)
			Semester		
Teaching Assistant	12	7		20	14,420
Research Assistant	5	0		20	15,500
Total	17	7			

5. Personnel Engaged in Separately Budgeted Research, 7/09–6/10

Professorial faculty	7
Postdoctoral appointments	4
Graduate students	8
Undergraduate students	7
Total	26

6. Separately Budgeted Research Expenditures by Source of Support

	Departmental Research	Physics-related Research Outside Department
Federal government	$554,000	$
Other	180,000	
Total	$734,000	$

Table C—Separately Budgeted Research Expenditures

Research Specialty	No. of Grants	Expenditures ($)
Astronomy	2	404,000
Solid State Theory	1	150,000
Nanomedicine	1	180,000
Total	4	734,000

FACULTY

Professors

Cheng, Ta-Pei, Ph.D., Rockefeller, 1969. Emeritus. Elementary particle physics.

Feldman, Bernard J., Ph.D., Harvard, 1972. Solid state physics. Chairman of the Department.

Flores, Ricardo A., Ph.D., California, Santa Cruz, 1984. Particle astrophysics.

George, Thomas F., Ph.D., Yale University, 1970. Laser/Materials Physics.

Handel, Peter H., Ph.D., Bucharest, 1965. Electromagnetic theory and solid state theory.

Henson, Bob L., Ph.D., Washington (St. Louis), 1964. Statistical physics.

Leventhal, Jacob J., Ph.D., Florida, 1965. Curators' Professor. Atomic physics.

Liu, Jingyue (Jimmy), Ph.D., Arizona State University, 1990. Nanoscience/Nanotechnology.

Moss, Frank E., Ph.D., Virginia, 1964. Curators' Professor. Emeritus. Nonlinear dynamics and biophysics.

Wilking, Bruce A., Ph.D., Arizona, 1981. Astrophysics.

Associate Professors

Bahar, Sonya, Ph.D., University of Rochester, 1997. Biophysics.

Fraundorf, Phil B., Ph.D., Washington (St. Louis), 1980. Solid state physics.

Gibb, Erika, Ph.D., Rensselaer Polytechnic Institute, 2001, Astrophysics, planetary science.

Li, Guoqiang, Ph.D., Shanghai Institute of Optics and fine mechanics, 1996. Biomedical imaging and sensing system.

Majzoub, Eric, Ph.D., Washington (St. Louis), 2000. Energy storage and conversion, computational physics.

Sorrell, Wilfred H., Ph.D., Emeritus. Wisconsin, 1989. Theoretical astrophysics.

RESEARCH SPECIALTIES AND STAFF

Theoretical

Astrophysics and Models of Dust Grains. Sorrell.

Bayesian Interence and Extreme Physics Content Modernization. Fraundorf.

Biophysics. Computational studies of stochastic phase synchronization, spatioteomporal dynamics and information processing in neural ensembles. Bahar.

Chemical Physics. George, Majzoub.

Computational Physics. Majzoub.

Cosmology Particle Astrophysics, Gravitational Lensing, Galaxy and Galaxy Cluster Dynamics, and Supersymmetic Dark Matter. Flores.

Electromagnetism and l/*f* Noise. Handel.

Elementary Particle Physics. Gauge theories of strong, weak, and electromagnetic interactions. Cheng.

Statistical Physics. Henson.

Experimental

Astronomy. Observations of pre-main sequence stars and protoplanetary disks. Gibb, Wilking.

Atomic Physics. Scattering of ions by atoms and resolution of resulting electromagnetic spectra. Leventhal.

Biophysics. In vivo imaging of neural dynamics in epilepsy; electrophysiological studies of the role of stochastic synchronization. Bahar.

Biomedical imaging and sensing systems. Li, Bahar.

Energy Storage and Conversion. Liu, Majzoub.

Nanoscale Characterization, Extraterrestrial and Electronic Materials. Fraundorf, Liu.

Planetary Science. Composition of cometary volatiles. Gibb.

Nonlinear Dynamics. Stochastic and nonlinear processes. Bahar, Moss.

Physics of Fluids. Ion mobilities and coronas in gases and liquids. Henson.

Solid State Physics. Thin-film wear-resistant coatings. Feldman.

WASHINGTON UNIVERSITY

DEPARTMENT OF PHYSICS

St. Louis, Missouri 63130

Students Accepted For Degree	FIELDS		
	Physics	Astronomy	Related Fields
Doctorate	X		
Master's	X		

1. General

Chancellor: Mark S. Wrighton
Dean of Graduate School: Richard Smith
Type of Institution: Private University
Department Chairman: Kenneth F. Kelton
Department Telephone Number: (314) 935–6276
Web site: physics.wustl.edu
E-mail: jmh@wustl.edu
Type of Institution: University
Control: Private
Setting: Suburban
Total Faculty: 3,297
Total Students: 13,761
Total Graduate Students: 6,651
Annual Graduate Tuition:
All Graduate Students: Full-time—$39,400
Part-time—$1,642/credit
Tuition rates for: 2010–11
Deferred tuition plan: No
Annual Other Fees: None
Term: Semester

2. Number of Faculty in Department

The combined total of full-time faculty in the three professorial ranks is 27. The combined total of full-time, part-time, and other faculty at all ranks is 36.

3. Admission, Financial Aid, and Housing

Address admission inquiries to: Julia M. Hamilton, Washington University, Department of Physics, One Brookings Drive, CB 1105, St. Louis, MO 63130-4899, jmh@wustl.edu
Graduate application fee required: $45
Admission deadline (Fall admission): 12/31
Admission information: For fall admission, 2009–10, 14 students were admitted. There were 139 applicants.
Admission requirements: For admission to the graduate programs, a Bachelor's degree is required with no minimum undergraduate GPA specified. The average GRE scores for those who were offered admission for 2008–09 were: verbal–546; quantitative–772; analytic–N.A.; and advanced–760. Both the GRE and GRE Advanced are required. No minimum acceptable score is specified. Students from non-English speaking countries are required to demonstrate proficiency in English via the TOEFL. The minimum acceptable score is 550.
Undergraduate preparation assumed: Mechanics: Marion, *Classical Dynamics of Particles and Systems*; Electromagnetic Theory: Lorrain and Corson, *Electromagnetic Fields and Waves*; Statistical Physics: Reif, *Statistical and Thermal Physics*; Mathematics, through the level of advanced calculus.
GAPSFAS application required: No
Financial aid deadline: 12/31
Loans available: No
Address housing inquiries to: Off-Campus Housing, Box 1075, Telephone (314) 935–5092
On-campus, student housing available: No

Table A—Faculty, Enrollments, and Degrees Granted

Research Specialty	2008–09 Faculty	Enrollment[1] Fall 2009		No. of Degrees Granted[2] 2009–10 (2005–09)			Median No. of Years for 2009–10 Ph.D.'s
		Master's	Doctorate	Master's	Terminal Master's	Doctorate	
Astrophysics	5	–	8	0(0)	0(0)	1(10)	6.0
Biological	3	–	8	0(0)	0(0)	2(9)	4.6
Biomedical & Health Physics	2	–	13	0(0)	0(0)	1(12)	5.0
Condensed Matter Physics	7	–	15	0(1)	0(1)	2(18)	5.6
Nuclear Physics	2	–	4	0(1)	0(1)	0(2)	5.5
Particles & Fields	5	–	4	0(0)	0(0)	1(6)	5.5
Relativity & Gravitation	2	–	2	0(0)	0(0)	0(7)	6.0
Space Physics	6	–	2	0(0)	0(0)	1(2)	6.0
Non-specialized	0	–	27	7(52)	1(6)	0(0)	–
Total		–	83	10(68)	1(6)	8(62)	
Full-time Grad. Stud.		–	82				
Part-time Grad. Stud.		–	1				
First-year Grad. Stud.		–	14				
Median Years in Grad. Study (2008–09 Degrees)				2	–	5.5	–

Undergraduate Degrees, 2008–09 (2003–08): 14(96)

[1]Students not yet committed to a research specialty are entered under non-specialized.
[2]Five-year totals in parentheses.

4. Graduate Degree Requirements

Master's: 30 semester-hours, with "B" average; one year in residence required. No foreign language requirement. 30 semester hours of satisfactory course credits, at least 24 in graduate-level classroom or seminar courses, and at least 12 hours in core graduate courses. Students must maintain a grade point average of B or better. A thesis is not required, but if a satisfactory thesis is submitted, only 24 semester-hours are required.
Doctorate: 72 graduate semester-hours in physics, mathematics, and other approved subjects, including credit earned on thesis research and in supervised teaching. GPA of "B" required in classroom courses. At least two years full-time residence. Experience and demonstrated competence in the teaching of physics is required. No foreign language requirement. Students must take a total of six core 500-level courses. Students are required to pass an oral examination on advanced physics at a level appropriate for a student beginning research in that area. Submission of an original research dissertation, and an oral examination in defense of the dissertation.
Thesis: Thesis may be written *in absentia*.
Special Equipment, Facilities, or Programs: McDonnell Center for the Space Sciences. NANOSIMS—This first-of-its kind instrument is capable of making precise isotopic measurements at a spatial resolution ≤1000Å. Laboratory for Experimental Astrophysics. Laboratory for Ultrasonics with the as-

sociated Biomedical Physics Program. Cardiovascular Biophysics Laboratory in collaboration with the Washington University School of Medicine. Center for Materials Innovation in collaboration with the departments of Chemistry and Earth and Planetary Sciences, and the School of Engineering and Applied Science, Laboratory for High Pressure Physics. The JEOL 2100F scanning transmission electron microscope measures structure and composition of materials at nm-scale. Laboratory for high precision isotope analyses of noble gases.

Table B—Appointments to Graduate Students, 2009–10

Title of Appointee	Appointments Total	Appointments First year	Academic Load Allowed in Credit Hours	Hours of Service Per Week	Stipend for Academic Year ($)
			Semester		
Teaching Fellow	23	0	11[1]	15	19,110[2,5]
Dean's University Fellow	15	14	12		19,110[2,5]
Compton Fellow	1	1	12		6,644
Research Assistant	27	0	9	variable	19,110[2,5]
GAANN	7	0	12	0	30,000[8]
McDonnell Graduate Fellow	1	0	9	0	25,000[3,10]
McDonnell Astronaut Fellow	0	0	9	0	30,000[3,4]
NASA Graduate Fellow	3	0	9	0	22,000[3,9]
NIH Graduate Fellow	1	0	9	0	N.A.
NSF Graduate Fellow	0	0	9	0	30,000
Dissertation Fellow	1	0	0	0	19,110[2,3]
Academic Hughes Fellow	0	0	11 or 12	0	3,000[6]
Summer Hughes Fellow	5	0	0		
Olin Fellow	0	0	12	0	26,250[3,7]
Total	84	15			

[1]Includes 9 credit hours for physics courses, 2 for supervised teaching.
[2]$19,110 for 2009–10.
[3]Plus tuition remission.
[4]$30,500 for 2008–09.
[5]During the first 5 years of graduate study, a Tuition Remission Scholarship is usually concurrent with a Teaching or Research Assistantship.
[6]This is in addition to the stipend from a concurrent appointment, such as a Teaching Fellowship. It is awarded in recognition of superior academic achievement or promise. $3,000 for 2008–09.
[7]$21,735 for 2008–09. May be supplemented by a Research Assistantship in the summer.
[8]$30,000 for 2008–09. Graduate Assistance in Areas of National Need, appointment from September–August.
[9]Academic-year rate.
[10]$25,000 for 2008–09.

5. Personnel Engaged in Separately Budgeted Research, 7/09–6/10

Professorial faculty	23
Other faculty–Res. Prof.	7
Postdoctoral appointments	11
Graduate students	42
Undergraduate students	16
Nonteaching research personnel	7
Total	106

6. Separately Budgeted Research Expenditures by Source of Support

	Departmental Research
Federal government	$5,983,076
Business and industry	83,300
Total	$6,066,376

Table C—Separately Budgeted Research Expenditures*

Research Specialty	No. of Grants	Expenditures ($)
Astrophysics	18	1,824,412
Space Physics	19	1,887,332
Biophysics	3	406,421
Condensed Matter Physics	14	541,147
Medical & Health Physics	14	735,652
Particles & Fields	4	511,686
Relativity & Gravitation	2	159,726
Total	74	6,066,376

*Entries based on 10 months.

FACULTY

Professors

Bender, Carl M., Ph.D., Harvard, 1969. Konneker Distinguished Professor of Physics. Theoretical physics; mathematical physics; particle physics.

Bernard, Claude W., Ph.D., Harvard, 1976. Theoretical physics; particle physics; computational physics.

Bernatowicz, Thomas, Ph.D., Washington (St. Louis), 1980. Astrophysics; extraterrestrial materials; mass spectrometry and transmission electron microscopy.

Buckley, James H., Ph.D., Chicago, 1994. Gamma-ray astronomy; cosmic ray astrophysics.

Carlsson, Anders E., Ph.D., Harvard, 1981. Theoretical biophysics of cells.

Clark, John W., Ph.D., Washington (St. Louis), 1959. Wayman Crow Professor of Physics. Theoretical physics and astrophysics; many-body theory; neural networks; quantum control theory.

Conradi, Mark S., Ph.D., Washington (St. Louis), 1977. Experimental condensed matter physics; high-pressure systems; hydrogen storage in solids; applications of magnetic resonance; MR in medical imaging; hyperpolarized gases.

Cowsik, Ramanath, Ph.D., Bombay, 1968. National Academy of Sciences. Theoretical astrophysics; dark matter; experimental astroparticle physics; gravitation.

Dickhoff, Willem H., Ph.D., Free Univ. of Amsterdam, 1981. Theoretical physics; many-particle theory, nuclear physics.

Gibbons, Patrick C., Ph.D., Harvard, 1971. Experimental condensed matter physics; electronic and lattice structure determinations by electron scattering.

Hohenberg, Charles M., Ph.D., California, Berkeley, 1968. Experimental space science: astrophysics; rare gas mass spectroscopy.

Israel, Martin H., Ph.D., Caltech, 1968. High-energy astrophysics.

Katz, Jonathan I., Ph.D., Cornell, 1973. Theoretical astrophysics; applied physics.

Kelton, Kenneth F., Ph.D., Harvard, 1983. Arthur Holly Compton Professor of Arts & Sciences. Professor of Materials Science and Chairman of Department of Physics. Experimental solid-state physics and materials science.

Miller, James G., Ph.D., Washington (St. Louis), 1969. Albert Gordon Hill Professor. Ultrasonics; biomedical physics; elastic properties of inhomogeneous media.

Ogilvie, Michael C., Ph.D., Brown, 1980. Quantum field theory and particle physics; theoretical physics; computational physics.

Schilling, James S., Ph.D., Wisconsin, Madison, 1969. Experimental solid state physics; high-pressure physics; magnetism and superconductivity.

Solin, Stuart A., Ph.D., Purdue, 1969. Charles M. Hohenberg Professor of Experimental Physics. Professor of Materials Science, Director of the Center for Materials Innovation. Experimental condensed matter/materials physics.

Suen, Wai-Mo, Ph.D., Caltech, 1985. General relativity; cosmology; theoretical astrophysics.

Will, Clifford M., Ph.D., Caltech, 1971. James S. McDonnell Professor of Physics. National Academy of Sciences. Theoretical astrophysics; general relativity.

Joint Professors

Sastry, Shankar M. L., Ph.D., Toronto, 1974. Materials science; metallurgy.

Sobotka, Lee G., Ph.D., California, Berkeley, 1982. Nuclear physics.

Research Professors

Amari, Sachiko, Ph.D., Kobe University, Kobe, Japan, 1986. Presolar grains, meteorites, noble gas and secondary ion mass spectrometry.

Binns, W. Robert, Ph.D., Colorado State, 1969. Astrophysics.

Meshik, Alex P., Ph.D., Vernadsky Insitute, Moscow, 1988. Space physics; rare-gas mass spectrometry.

Zinner, Ernst, Ph.D., Washington (St. Louis), 1972. Astrophysics; experimental space science; extraterrestrial materials.

Associate Professors

Alford, Mark G., Ph.D., Harvard, 1990. Quantum field theory and particle physics; color superconductivity.

Krawczynski, Henric, Ph.D., Hamburg, 1997. High-energy astrophysics; gamma-ray astronomy.

Wessel, Ralf, Ph.D., Cambridge, 1992. Biophysics of neurons and neural computation.

Research Associate Professors

Floss, Christine, Ph.D., Washington (St. Louis). 1991. Space Physics; cosmochemistry

Holland, Mark R., Ph.D., Washington (St. Louis), 1989. Ultrasonics; biomedical physics; biomedical ultrasound.

Leopold, Daniel J., Ph.D., Washington (St. Louis), 1983. Semiconductor physics; electro-optics; materials science; magnetic resonance.

Assistant Professors

Ferrer, Francesc, Ph.D., Unversitat Autónoma de Barcelona, 2001. Particle cosmology: Composition and evolution of the universe, the nature of dark matter and dark energy; ultra high-energy cosmic rays.

Nussinov, Zohar, Ph.D., UCLA, 2000. Condensed matter and materials theory.

Seidel, Alexander, Ph.D., MIT, 2003. Condensed matter and materials theory.

Wang, Yan Mei, Ph.D., California, Berkeley, 2002. Experimental biophysics, single-molecule imaging

Yang, Li, Ph.D., Georgia Institute of Technology, 2006. Condensed matter and materials theory.

Senior Lecturer

Hynes, Kathryn, Ph.D., Washington University (St. Louis).

Emeritus Faculty

Friedlander, Michael W., Ph.D., Bristol, 1955. Cosmic rays; astrophysics; archaeoastronomy.

Luszczynski, Kazimierz, Ph.D., London, 1959. Solid state and low-temperature physics; magnetic resonance.

Phillips, Peter R., Ph.D., Stanford, 1961. Biomedical physics; general relativity and gravitation; astrophysics.

Scandrett, John H., Ph.D., Wisconsin, 1963. Biomedical physics; computer applications.

Shrauner, J. Ely, Ph.D., Chicago, 1963. Theoretical physics; elementary particle theory; high-energy physics.

Sundfors, Ronald K., Cornell, 1963. Ultrasonic studies of solids; acoustic magnetic resonance.

Adjunct Professors

Anderson, Charles H., Ph.D., Harvard, 1962. Biophysics; signal processing; machine vision.

Dixit, Vijai V., Ph.D., Purdue, 1972. Theoretical physics.

Elson, Elliott L., Ph.D., Stanford, 1966. Molecular biophysics.

Falster, Robert, Ph.D., Stanford, 1983. Electronic materials.

Khodel, Victor A., Ph.D., Moscow Engineering & Physics Institute, 1965. Theoretical physics.

Malik, Fazley Bary, Ph.D., Goettingen, 1958. Theoretical physics.

Mandula, Jeffrey E., Ph.D., Harvard, 1966. Theoretical physics; particle physics; mathematical physics.

Rigden, John S., Ph.D., Johns Hopkins, 1960. History of science; molecular physics.

Ristig, Manfred L., Ph.D., Cologne, 1966. Many-body theory; condensed matter theory.

Wickline, Samuel A., M.D., Hawaii, 1980. Cardiology; ultrasonics; biophysics.

Yablonskiy, Dmitriy A., Ph.D., Ukrainian Academy of Sciences, 1972. Magnetic resonance; medical and health physics.

Adjunct Associate Professors

Christian, Eric R., Ph.D., Caltech, 1989. Cosmic rays; high-energy astrophysics.

Comer, Gregory L., Ph.D., University of North Carolina, 1990. General relativity; astrophysics.

Conturo, Thomas E., Ph.D., M.D., Vanderbilt, 1989. Magnetic resonance imaging.

Fraundorf, Philip B., Ph.D., Washington (St. Louis), 1980. Space physics; solid state physics; statistical physics.

Kalyanaraman, Ramki, Ph.D., North Carolina State, 1998. Experimental materials science and solid state physics.

Kovacs, Sandor J., Jr., Ph.D., M.D., Caltech, 1977. Cardiovascular bio-physics; non-linear dynamics.

Redmount, Ian H., Ph.D., Caltech, 1984. General relativity, astrophysics.

Adjunct Assistant Professors

Culver, Joseph P., Ph.D., University of Pennsylvania, 1997. Biomedical physics; diffuse optical tomography.

Leopold, Mary M., Ph.D., Washington (St. Louis), 1985. Optical response theory; electro-optics; mathematical physics.

Sept, David S., Ph.D., Alberta, 1997. Theoretical biophysics; soft condensed matter physics.

Woods, Jason C., Ph.D., Washington University (St. Louis), 2002. Biophysics; applied physics; hyperpolarized-gas MRI.

RESEARCH SPECIALTIES AND STAFF

Theoretical

Applied Physics. Acoustic radiation, hydrodynamics, Mpemba effects, plasma physics. Katz.

Astrophysics. Ultradense matter, neutron stars and quark stars; superfluidity and color superconductivity in compact stars; high-energy astrophysics and astroparticle physics. Alford, Clark, Cowsik, Dickhoff, Ferer, Katz, Khodel.

Biophysics. Force generation in biological systems; self-assembly of biopolymer networks; protein-protein interactions; neural networks; computational neuroscience; cardiovascular physiology and biophysics; nonlinear dynamics. Anderson, Carlsson, Clark, Kovacs, Sept.

Condensed Matter Physics. Quantum fluids; strongly correlated electron systems; metal-insulator transitions; non-Fermi liquids; quantum critically; superconductivity; spin systems; quantum Hall effect; one-dimensional systems; soft condensed matter; magnetism; topological order; transition metal oxides; amorphous and complex ordered structures; glass transition; electronic structure; orbital order; statistical mechanics; mesoscopic physics; optimization and network problems; cold atom physics, excitonic effects. Clark, Dickhoff, Khodel, Nussinov, Seidel, Yang.

Elementary Particles and Fields. Perturbation theory; quantum chromodynamics; Non-Abelian gauge theories and confinement; quark matter; lattice gauge theory; color superconductivity; PT-symmetric theories; semiclassical approximations. Alford, Bender, Bernard, Ferrer, Ogilvie.

Low Temperature Physics. Many-body theory of helium liquids. Clark, Dickhoff, Khodel.

Materials Science/Metallurgy. Alloys; amorphous and defective solids; theory of fracture; nucleation theory, semiconducting nanostructure, graphane, photovoltaics. Carlsson, Kelton, Nussinov, Yang.

Nuclear Physics. Many-body theory of nuclear matter and finite nuclei. Clark, Dickhoff, Khodel.

Relativity. Gravitational radiation, black holes, tests of general relativity; numerical relativity; galactic dynamics, dark matter, cosmology. Cowsik, Ferrer, Suen, Will.

Statistical and Thermal Physics. Statistical mechanics; phase transitions; statistical mechanics of solitons; renormalization group. Carlsson, Nussinov, Ogilvie, Seidel.

Systems Science. Theory of quantum control systems. Clark.

Other Computational Physics. Parallel computation; simulation; numerical analysis, first-principles calculations. Alford, Bender, Bernard, Carlsson, Clark, Dickhoff, Kelton, Miller, Nussinov, Ogilvie, Suen, Yang.

Experimental

Acoustics. Elastic and viscoelastic properties of media including hard and soft tissue; ultrasonic imaging and quantitative ultrasonic imaging and quantitative ultrasonic investigation of the cardiovascular system; characterization of bone and of osteoporosis. Kovacs, Holland, Miller, Wickline.

Applied Physics. Quantitative analysis of images; nondestructive evaluation of composite materials; magnetic resonance imaging of materials; ultrasonic transducers. Conradi, Holland, Leopold, Miller.

Astrophysics/Extraterrestrial Materials. Solid state, ion microprobe, noble gas mass spectrometric and electron microscope investigations of ancient stardust in meteorites and interplanetary dust; nucleosynthesis, stellar evolution, origin and evolution of the solar system including planetary atmospheres and organic molecules, early chronology from studies of (now) extinct isotopes. Amari, Bernatowicz, Floss, Hohenberg, Meshik, Zinner.

Biophysics. Elastic properties of tissue; cardiac mechanics; mechanisms of ultrasonic propagation in tissue; cardiac Doppler ultrasound; wave-guide properties of retinal cells; mechanics of biomembranes; brain imaging; pulmonary physiology; biophysics of neural computation; physics of single neurons; single-molecule imaging. Holland, Kovacs, Miller, Wang, Wessel, Wickline, Woods.

Dark Matter. Laboratory and astronomical searches for dark matter. Buckley, Cowsik.

High-Energy Astrophysics. Cosmic-ray elemental and isotopic composition and energy spectra, gamma-ray and x-ray astrophysics of galactic and extragalactic sources, observations from spacecraft and high-altitude balloons; astrophysics of the highest-energy galactic and extragalactic gamma-ray sources; observations with ground-based atmospheric Cherenkov detectors; correlated optical observations of high-energy astronomical transients; observationsof very high energy neutrinos from high-altitude balloons over Antarctica. Binns, Buckley, Israel, Krawczynski.

Low-Temperature Physics. Phase transitions; absorbed films; ortho-para conversion in hydrogen; hydrogen on surfaces; pressure-induced superconductivity. Schilling.

Condensed Matter/Materials Physics. Hydrogen in metals and ionic and complex solids; nucleation and phase transitions in liquid and solids; synchrotron x-ray diffraction and thermophysical property measurements of equilibrium and non-equilibrium liquids; x-ray and electron microscopy studies of quasicrystalline phases; superconductivity and magnetism under extreme pressures; high-Tc superconductivity; pressure-induced insulator-to-metal transitions; elastic and viscoelastic properties of composites; nanocrystalline materials; thin film growth and characterization; wide and narrow-gap semiconductors; extraordinary magnetoresistance; physisorbed (two-dimensional) matter; nuclear magnetic resonance; photoluminescence and photoconductivity in conducting polymers; electron microscopy and inelastic electron scattering in solids; plasmons in simple metals and composites; magnetic properties of Kondo lattices and weak itinerant ferromagnets, tensile stress-dependent transport properties of narrow-gap semiconductors; 2D physics and magnetic frustration in layered double hydroxides. Conradi, Gibbons, Kelton, Leopold, Miller, Schilling, Solin.

Medical and Health Physics. Quantitative ultrasonics and ultrasonic imaging; quantitative cardiovascular physiology echocardiography; nuclear magnetic resonance imaging (MRO) and positron emission tomography (PET); MRI of lungs with hyperpolarized gas. Conradi, Conturo, Culver, Holland, Kovacs, Miller, Wickline, Woods, Yablonskiy.

Nuclear Physics. Heavy-ion reaction studies from near barrier to relativistic energies; reaction dynamics studied with particle-particle interferometry and atomic x-ray clocks; fragmentation cross sections of cosmic-ray nuclei on various targets; double beta decay; natural nuclear reactors. Binns, Israel, Krawezynski, Meshik, Sobotka.

Relativity and Gravitation. Tests of the equivalence principle; study of forces in the submillimeter domain, including Casimir forces, axion exchange, and violations of the inverse square law of gravitation. Cowsik.

MONTANA STATE UNIVERSITY

DEPARTMENT OF PHYSICS

Bozeman, Montana 59717

Students Accepted For Degree	FIELDS		
	Physics	Astronomy	Related Fields
Doctorate	X		
Master's	X		

1. General

President: Waded Cruzado
Vice Provost for Graduate Education: Carl Fox
Department Chairman: Richard J. Smith
Department Telephone Number: (406) 994-3614 (C)
Type of Institution: University
Control: Public
Setting: Small town
Total Faculty: 700
Total Graduate Faculty: Not separated
Total Students: 12,001
Total Graduate Students: 1,871
Annual Graduate Tuition:
In-state residents: Full-time—$7,500
Out-of-state residents: Full-time—$17,000
Tuition rates for: 2009–10
Deferred tuition plan: Yes
Other Fees: None
Term: Semester

2. Number of Faculty in Department

The combined total of full-time faculty in the three professorial ranks is 17. The combined total of full-time, part-time, and other faculty at all ranks is 47.

3. Admission, Financial Aid, and Housing

Address admission inquiries to: Dr. Rufus Cone, Physics
Application deadline: 1/31
Graduate application fee required: $50
Admission deadline (Fall admission): 5/1
Admission requirements: For admission to the graduate programs, a Bachelor's degree in physics or a related field is required, with a recommended minimum undergraduate GPA of 3.0 in advanced physics and mathematics courses. A minimum overall undergraduate GPA of 3.0 is required. The GRE general and GRE physics subject exams are required. Successful students in our program usually have scores that exceed: verbal–70%, quantitative–80%, analytical–4.0, and physics–40%. Students for whom English is not their first language must demonstrate proficiency in English through the TOEFL exam. The minimum acceptable score for admission is 240-computerized. The TSE exam, with a minimum score of 50, is also required. In lieu of the computerized TOEFL and TSE exams, the ibt-TOEFL exam may be submitted; the minimum overall score must be 80 and the speaking portion must be 26.
Undergraduate preparation assumed: Classical Mechanics (at the level of Marion or Symon), Electricity and Magnetism (Griffiths), Quantum Mechanics (Libof or Gasiorowicz), Statistical and Thermal Physics (Reif).
Address financial aid inquiries to: Prof. Rufus Cone, Physics
GAPSFAS application required: No
Financial aid deadline: 1/31
Loans available: Yes
Address housing inquiries to: Residence Life Office
On-campus, student housing available: Yes
Cost/month: $293–546
On-campus, married student housing available: Yes
Cost/month: $387–701 (some include utilities)

Table A—Faculty, Enrollments, and Degrees Granted

Research Specialty	2009–10 Faculty	Enrollment[1] Fall 2009 Master's	Enrollment[1] Fall 2009 Doctorate	No. of Degrees Granted[2] 2009–10 (2005–10) Master's	Terminal Master's	Doctorate	Median No. of Years for 2009–10 Ph.D.'s
Astrophysics	2	2	5	1(3)	0(1)	0(1)	–
Atmos/Space Phys. Cosmic Rays	1	0	0	0(0)	0(0)	0(1)	–
Condensed Matter Physics	7	1	17	0(10)	1(3)	0(4)	–
Optics	6	1	8	1(3)	2(5)	1(5)	6
Physics Education	2	0	0	0(0)	1(1)	0(1)	–
Relativity & Gravitation	2	0	10	0(4)	0(1)	1(4)	8
Solar Physics	8	0	17	2(9)	0(1)	0(3)	–
Total		4	57	4(29)	4(12)	2(19)	
Full-time Grad. Stud.		4	57				
Part-time Grad. Stud.		0	0				
First-year Grad. Stud.		1	9				
Median Years in Grad. Study (2009–10 Degrees)				2	2.5	7	–
Undergraduate Degrees, 2009–10 (2005–10): 17(74)							

[1]Students not yet committed to a research specialty are entered under non-specialized.
[2]Five-year totals in parentheses.

4. Graduate Degree Requirements

Master's: 20 credits plus thesis or 30 credits without thesis in approved program with satisfactory performance; M.S. exam required; two semesters residency; no language requirements.
Doctorate: Satisfactory performance in approved course program; dissertation; comprehensive and dissertation exams required; four semesters of residency required; no language requirements.
Thesis: Thesis may be written *in absentia*.
Special Facilities and Programs: Optical physics and laser spectroscopy laboratories, including lasers ranging from the ultrastable (few Hz linewidths) to the ultrafast (femtosecond pulses), for research in spectral hole burning phenomena, materials science, and devices, ultrafast holography, smart pixel sensors, Raman lasers, diode laser frequency control and noise characterization, optical frequency standards, and ultrastable optical lasers and cavities; Spectral Information Technology Laboratory (Spectrum Lab) and Optical Technology Center (OpTeC), fostering collaborations with the Department of Chemistry and Biochemistry, and Electrical and Computer Engineering and with local optics industries, and with several national and international laboratories and companies; millimeter-wave magneto-spectroscopy facility; magnetics laboratory; dielectrics laboratory; active materials development and testing laboratory; Magnetic Nanostructure

Growth and Characterization Facility for synthesis and characterization of magnetic films, particles, and interfaces; ion beam laboratory; Image and Chemical Analysis Laboratory; Montana Space Grant Consortium, a statewide program for research, education, and outreach in space science; Big Sky Institute, promoting innovation and excellence in math and science education; and public outreach programs in astrophysics, solar physics, and Mars exploration.

Table B—Appointments to Graduate Students, 2009–10

Title of Appointee	Appointments Total	Appointments First year	Academic Load Allowed in Credit Hours	Hours of Service Per Week	Stipend for Academic Year ($)
			Semester		
Teaching Assisant	24	8	9–12	20	13,500–14,400[1,2,4]
Research Assistant	32	1	9–12	20	13,500–14,400[1,3,4]
Self Supporting	2	1	–	0	
Fellowships	3	0	–	20	15,000
Total	61	10			

[1]13,500 for 1st year and pre-comp students; 14,400 for students passing the comps at the PhD Level.
[2]Additional income for summer TA.
[3]Additional income for summer RA.
[4]Six- to twelve-credit hour fee remission per semester.

5. Personnel Engaged in Separately Budgeted Research

Professorial faculty	17
Other faculty	12
Postdoctoral appointments	4
Graduate students	32
Undergraduate students	80
Non-teaching research personnel	11
Total	156

6. Separately Budgeted Research Expenditures by Source of Support

	Departmental Research	Physics-related Research Outside Department
Federal government	$6,011,088	$
Total	$6,011,088	$

Table C—Separately Budgeted Research Expenditures

Research Specialty	No. of Grants	Expenditures ($)
Space Physics	6	1,075,558
Condensed Matter Physics	15	1,150,941
Optics	11	1,953,423
Physics Education	0	86,847
Relativity & Gravitation	5	395,497
Solar Physics	22	1,348,822
Total	62	6,011,088

FACULTY

Professors

Babbitt, William R., Ph.D., Harvard, 1987. Optical physics; applied optics.

Carlsten, John L., Ph.D., Harvard, 1974. Non-linear optics; laser spectroscopy; atomic physics.

Cone, Rufus, Ph.D., Yale, 1971. Solid state physics; quantum optics and lasers.

Cornish, Neil, Ph.D., Toronto Univ., 1996. Relativity theory and cosmology.

Francis, Gregory E., Ph.D., MIT, 1987. Physics education.

Idzerda, Yves, Ph.D., University of Maryland, 1986. Magnetic materials.

Link, Bennett, Ph.D., University of Illinois, 1991. Astrophysics.

Longcope, Dana, Ph.D., Cornell Univ., 1993. Plasma physics.

Neumeier, John J., Ph.D., U.C. San Diego, 1990. Condensed matter.

Rebane, Aleksander, Ph.D., Estonia, 1985. Optics; lasers.

Smith, Richard J., Ph.D., Iowa State, 1975. Department Head. Ion beams; surface physics.

Tsuruta, Sachiko, Ph.D., Columbia, 1964. Astrophysics; compact objects.

Associate Professors

Adams, Jeffrey, Ph.D., Queens University, 1991. Assistant Vice Provost for Undergraduate Education. Physics education.

Kankelborg, Charles, Ph.D., Stanford Univ., 1996. Solar physics.

Malovichko, Galina I., Ph.D., Kiev, Ukraine, 1987. Optical materials.

Assistant Professor

Qui, Jiong, Ph.D., Nanjing Univ., 1998. Solar and Astrophysics.

Vorontsov, Anton, Ph.D., Northwestern University, 2004. Condensed matter theory.

Research Faculty/Adjunct Faculty

Acton, Loren W., Ph.D., Univ. of Colorado, 1965. Solar physics.

Avci, Recep, Ph.D., Illinois, 1978. Solid state and surface physics.

Canfield, Richard C., Ph.D., Univ. of Colorado, 1968. Solar physics.

Craig, Alan, Ph.D., Univ. of Arizona, 1982. Coherent optics applications.

Drobijev, Mikhail, Ph.D., Moscow Inst. of Physics and Technology, 1986. Molecular biophysics.

Hellings, Ronald, Ph.D., Montana State Univ., 1972. Relativity theory.

Klumpar, David M., Ph.D., University of New Hampshire, 1972. Experimental space physics.

Leamon, Robert J., Ph.D., Univ. of Delaware, 1999. Solar Physics.

Martens, Petrus C., Ph.D., Utrecht University, 1983. Solar Physics.

McKenzie, David E., Ph.D., Univ. of Delaware, 1997. Solar physics.

Riedel, Carla M., Ph.D., University of Minnesota, 1996. Nuclear physics.

Willoughby, Shannon, Ph.D., Tulane Univ., 2003. Theoretical condensed matter and physics/astronomy education.

Faculty Emeriti

Drumheller, John E., Ph.D., Colorado, 1962. Solid state physics; electron spin resonance; magnetism. (Emeritus).

Hermanson, John C., Ph.D., Chicago, 1966. Surface physics theory. (Emeritus).

Kirkpatrick, Larry, Ph.D., MIT, 1968. AAPT officer. Science education. (Emeritus).

Lapeyre, Gerald J., Ph.D., Missouri, Columbia, 1962. Solid state and surface physics; photoemission; electron energy loss. (Emeritus).

Schmidt, V. Hugo, Ph.D., Washington, 1961. Solid state physics; nuclear magnetic resonance; alternate energy. (Emeritus).

Swenson, Robert, Ph.D., Lehigh, 1961. Statistical physics. (Emeritus).

Wheeler, Gerald, Ph.D., SUNY-Stony Brook, 1972. Experimental nuclear physics. (Emeritus).

RESEARCH SPECIALTIES AND STAFF

Theoretical

Astrophysics. Neutron stars, active galactic nuclei, and gamma ray bursters. Link, Tsuruta.

Condensed Matter Physics. Correlated many-body (collective) effects such as superconductivity and superfluidity; influence of magnetic fields, impurities and fluctuations on superconducting properties. Vorontsov.

General Relativity and Cosmology. Gravitational waves; black holes; quantum theory of gravity; early universe; experimental relativity. Cornish, Hellings.

Science Education. Developing and implementing innovative programs for primary and secondary teacher education; developing techniques in education of non-science majors. Adams, Francis, Willoughby.

Solar Physics. Magnetic reconnection; x-ray emission; coronal mass ejection. Acton, Canfield, Leamon, Longcope, Martens, McKenzie, Qiu. 2 research scientists, 1 postdoctoral fellow.

Experimental

Lasers and Optics. Linear and nonlinear optical laser spectroscopy; coherent optical transients; optical hole burning; Raman scattering; solid state laser material. Babbitt, Carlsten, Cone, Craig, Drobijev, Rebane. 1 research scientist.

Condensed Matter Physics and Surface Science: State-of-the-art facilities at MSU enable measurements of physical properties to temperatures as low as 0.3 K; measurements of thermal expansion use a novel dilatometer developed at MSU that is capable of detecting sub-angstrom length changes of specimens to study phase transitions and critical phenomena with superb resolution; characterization of the structure of thin films and buried interfaces for fuel cells and electronic devices; ceramics for fuel cells are fabricated and tested for their electrical properties; defects in advanced materials at the atomic level using a host of techniques such as EPR, ENDOR and optical spectroscopy; Center of Bio-Inspired Nanomaterials utilizes biological molecules as templates for the synthesis of nanoparticles with unusual physical properties; this interdisciplinary effort thrives on close collaboration among biologiest, chemists, and physicists at MSU. Avci, Idzerda, Malovichko, Neumeier, Schmidt, Smith. 4 research scientists, 3 postdoctoral fellows.

Solar Physics. Space instrumentation, including ultraviolet optics, to investigate the high speed dynamics of magnetic reconnection in solar flares. Kankelborg. 1 research scientist.

Space Physics. Solar magnetic activity. Auroral physics. Space instrumentation. Klumpar. 3 research scientists.

CREIGHTON UNIVERSITY

DEPARTMENT OF PHYSICS

Omaha, Nebraska 68178-0001

Students Accepted For Degree	FIELDS		
	Physics	Astronomy	Related Fields
Doctorate			
Master's	X		

1. General

President: Rev. John P. Schlegel, S. J.
Dean of Graduate School: Gail M. Jensen
Department Chair: Janet E. Seger
Graduate Program Director: Sam J. Cipolla
Department Telephone Number: (402) 280-2835
Type of Institution: University
Control: Private
Setting: Urban
Total Faculty: 735
Total Graduate Faculty: 288
Total Students: 7,385
Total Graduate Students: 865
Annual Graduate Tuition:
In-state residents: Full-time—$650*
Part-time—$650/credit*
Out-of-state residents: Full-time—$650*
Part-time—$650/credit*
Tuition rates for: 2009–10
Deferred tuition plan: No
Other Fees: $458/semester (Full time); $47 (Part time)
Technology: $195 (Full time); $79 Part time)
Term: Semester

*Teaching and Research Assistants do not pay tuition or semester and health insurance fees.

2. Number of Faculty in Department

The combined total of full-time faculty in the three professorial ranks is 8. The combined total of full-time, part-time, and other faculty at all ranks is 13.

3. Admission, Financial Aid, and Housing

Address admission inquiries to: Director, Physics Graduate Program, Department of Physics
Graduate application fee required: $50
Admission deadline (Fall admission): None
Admission information: For fall admission, 2009–10, 7 students were accepted from 16 applicants.
Admission requirements: A Bachelors degree in physics or engineering-related field is required with no minimum undergraduate GPA specified. The GRE is required. GRE scores are only used comparatively. The GRE advanced is not required. Students from non-English speaking countries are required to demonstrate proficiency in English via the TOEFL exam. Minimum acceptable score is 550 (written)/213 (computer)/80(iBT).
Undergraduate preparation assumed: Fowles, *Mechanics*; Griffiths, *Electricity and Magnetism*; Tipler, *Modern Physics*.
Address financial aid inquiries to: Director, Physics Graduate Program, Department of Physics
GAPSFAS application required: No
Financial aid deadline: None
Loans available: Yes
Address housing inquiries to: Department of Residence Life, 136 Swanson Hall (402) 280-3016
On-campus, graduate student housing available: Yes
Cost/month: $347–$900/month
On-campus, married student housing available: Yes
Cost/month: $347–900/month (family)

Table A—Faculty, Enrollments, and Degrees Granted

Research Specialty	2009–10 Faculty	Enrollment[1] Fall 2009		No. of Degrees Granted[2] 2009–10 (2005–10)			Median No. of Years for 2009–10 Ph.D.'s
		Master's	Doctorate	Master's	Terminal Master's	Doctorate	
Astro-Particle	2	3	0	1(6)	0(0)	0(0)	–
Atomic, Molecular, & Optical Physics	1	0	0	1(1)	0(0)	0(0)	–
Biophysics	2	3	0	1(6)	0(0)	0(0)	–
Condensed Matter Physics	1	1	0	1(4)	0(0)	0(0)	–
Nuclear Physics	3	3	0	2(9)	0(0)	0(0)	–
Total		10	0	6(26)	0(0)	0(0)	–
Full-time Grad. Stud.		9	0	0			
Part-time Grad. Stud.		1	0	0			
First-year Grad. Stud.		6	0	0			
Median Years in Grad. Study (2009–10 Degrees)		2	0	0	–	–	

Undergraduate Degrees, 2009–10 (2005–10): 4(36)

[1] Students not yet committed to a research specialty are entered under non-specialized.
[2] Five-year totals in parentheses.

4. Graduate Degree Requirements

Master's: Plan A (thesis option) requires 30 credit hours of graduate-level courses, including six credit hours of Thesis Research. Plan B (non-thesis option) requires 33 credit hours of graduate-level courses, including three credit hours of Directed Independent Research for which a report is required. A minimum of fifteen credit hours must be in physics in either Plan. In both Plans, all students are required to take the core graduate courses in classical mechanics, electromagnetics, quantum mechanics, and statistical mechanics. Full time students must enroll in Graduate Seminar each semester. A minimum grade average of 3.0 (B) is required, with no more than two grades of C. A three-part comprehensive exam offered three times each year on (1) mechanics and heat, (2) electromagnetics and optics, (3) modern physics is required to be passed. Each part can be taken separately. Ordinarily a student must devote two semesters and a summer session to resident graduate study.
Thesis: Theses may be written in absentia.
Special Equipment, Facilities, or Programs: Atomic, Molecular and Optical Physics: 200-kV heavy ion accelerator, X-ray fluorescence analyzers of the x-ray tube and radioactive x-ray source variety. High resolution x-ray detectors include a crystal diffractometer, Si(Li) detectors, and a silicon PIN detector.

Condensed Matter Physics: Muffle furnaces, vacuum oven, capacitance bridge, lock-in amplifier, scanning tunneling microscope, photon correlation spectrometer, 5W CW visible laser, CO_2 laser, x-ray diffractometer, evaporator.

Biophysics: Femtosecond titanium sapphire laser, confocal laser scanning microscope, photon correlation spectrometer, fiber optic fluorescence emission/absorption spectrometer, 10W Nd:YAG fiber laser, fiber splicer and launch systems, inverted tissue culture microscope, Zeiss 510 NLO LSM with FLIM equipment available through the Creighton University Integrated Biomedical Imaging core facility.

Computational Physics/Astro-Particle Physics: Multi-node networked Intel-based PC's for simulations.

Nuclear Physics: Compton-scattering spectrometer, gamma-gamma and beta-gamma angular correlation spectrometers, HPGe gamma-ray detector, NIM electronics.

Teaching Certificate Program: Teaching certification and a M.S. degree in physics can be earned in two years (4 semesters, 2 summers). Graduate courses are taken in both the Education and Physics departments. The program includes financial support and tuition remission for three semesters of work as a teaching assistant. A 50% reduction in tuition is available for the remaining credits.

Table B—Appointments to Graduate Students, 2009–10

Title of Appointee	Appointments Total	Appointments First year	Academic Load Allowed in Credit Hours	Hours of Service Per Week	Stipend for Academic Year ($)
			Semester		
Teaching Assistant	8	6	8–11	20	$11,703[1]
Research Assistant	2	1	8–11	variable	12,800
Total	10	7			

[1]Summer appointments ($1500–$3000) available in addition.

5. Personnel Engaged in Separately Budgeted Research, 7/09–7/10

Professorial faculty	8
Graduate students	13
Undergraduate students	33
Nonteaching research personnel	2
Total	56

6. Separately Budgeted Research Expenditures by Source of Support

	Departmental Research	Physics-related Research Outside Department
Federal government	$526,576	$
Private, non-profit organizations	5,750	$
Total	$532,326	$

Table C—Separately Budgeted Research Expenditures

Research Specialty	No. of Grants	Expenditures ($)
Nuclear Physcis	2	200,000
Biophysics	3	92,691
Condensed Matter	2	200,875
Astro-particle/astrophysics	2	38,760
Total	9	532,326

FACULTY

Professors

Cherney, Michael, G., Ph.D., Wisconsin, 1987. Experimental high-energy nuclear physics (relativistic heavy ion physics); control systems; nuclear science education.

Cipolla, Sam J., Ph.D., Purdue, 1969. Experimental atomic physics; inner-shell ionization; response modeling of radiation detectors.

Seger, Janet E., Ph.D., Wisconsin, 1991. Theoretical and experimental high energy nuclear physics.

Associate Professors

Duda, Gintaras, Ph.D., UCLA, 2003. Astro-particle physics; theoretical elementary particle physics; physics education research.

Nichols, Michael G., Ph.D., Rochester, 1996. Experimental biological physics; tissue spectroscopy and microscopy; photodynamic therapy of cancer; cellular mechanics.

Sidebottom, David L., Ph.D., Kansas State, 1989. Glass science; dynamic light scattering; optical spectroscopy; dielectric spectroscopy.

Assistant Professors

Gabel, Jack R., Ph.D., The Catholic University of America, 2000. Observational astrophysics; UV, optical, IR spectroscopy; photoionization modeling of astrophysical plasmas.

McShane, Thomas S., S.J., M.S., Saint Louis, 1956. Control systems; computer simulation; electronics.

Soto, Patricia, Ph.D., University of Groningen, The Netherlands, 2004. Computational molecular biophysics.

Emerti

Kennedy, Robert E., Ph.D., Notre Dame, 1966. History and philosophy of physics.

Zepf, Thomas H., PhD., Saint Louis, 1963. Condensed matter physics.

Visiting Faculty

Gorbunov, Yuri, Ph.D., Universität Siegen, Germany, 2004. Relativistic heavy ion collisions.

Nielsen, Bjorm, Ph.D., Minnesota, 1994. Relativistic heavy ion collisions.

Instructors

Kriegler, David J., M.S., Nebraska, 1976. Astronomy.

Stuva, David R., M.S., Creighton, 1983. General physics.

RESEARCH SPECIALTIES AND STAFF

Theoretical

Nuclear Physics. Ultra-peripheral heavy ion collisions measured by the STAR collaboration at RHIC. Seger.

Astro-particle Physics. Characterization and detection of dark matter; prompt atmospheric lepton flux in high energy cosmic rays; high-energy cosmic rays beyond the GZK cutoff. Duda.

Computational Molecular Biophysics. Protein structure, dynamics, and self-assembly; protein folding and misfolding. Soto.

Experimental

Atomic Physics. Atomic inner-shell ionization. X-ray fluorescence. Cipolla.

Biophysics. Development and application of novel optical techniques to biology and medicine; Multiphoton and confocal laser scanning fluorescence microscopy; Molecular photophysics. Vis/Near-IR tissue spectroscopy; Photodynamic therapy (PDT) of Cancer. Studies of cellular mechanics use an optical stretcher apparatus, which employs a 10 W Nd-YAG fiber laser and an inverted phase-contrast microscope equipped with a fast ccd camera. Nichols.

Condensed Matter Physics. Ionic motion in glasses; dynamic light scattering of the glass transition. Sidebottom.

Nuclear Physics. High energy nuclear physics (relativistic heavy ion physics) in STAR collaboration at RHIC (Brookhaven National Laboratory) and ALICE collaboration at the LHC (CERN). Cherney, Seger, McShane, 2 postdoctoral fellows.

Observational Astrophysics. Observations and analysis of active galactic nuclei using UV, optical, IR spectra from space-based and large ground-based observations; Photoionization modeling of astrophysical plasmas, Studies of energetic mass outflows from quasars. Gabel.

FACULTY PUBLICATIONS

Cherney, M., Seger, J., McShane, T. S. (Creighton Co-authors)

B. S., Nielsen *et al.*, "A system to monitor possible displacements of the inner tracking system of ALICE," Nucl. Instrum. Meth. A **599**, 176 (2009).

A. J. Baltz, Y. Gorbunov, S. R. Klein, and J. Nystrand, "Two-photon interactions with nuclear breakup in relativistic heavy ion collisions," Phys. Rev. C **80**, 044902 (2009).

B. I. Abelev, *et al.*, "Observation of Two-source Inteference in the Photo-production Reaction Au Au->U Au," Phys. Rev. Lett. **102**, 112301 (2009).

Cipolla, S. J.

S. J. Cipolla, "L x-ray production cross sections from 50–300 keV proton impact on selected 4d transition elements," J. Phys. Conf. Series **194**, 082005 (2009).

S. J. Cipolla, "L x-ray production cross sections for proton impact on Fe, Cu, Zn, and Ge," AIP Conference Proceedings **1099**, 180 (2009).

S. J. Cipolla, "An improved version of ISICS: a program for calculating K-, L-, and M-shell cross sections from PWBA and ECPSSR theory on a personal computer," Comp. Phys. Comm. **176**, 157 (2007).

Duda, G. K.

G. Duda and K. Garrett, "Probing Student Online Discussion Behavior with a Course Blog in Introductory Physics," 2008 Physics Education Research Conference, AIP Conf. Proceedings **1064**, 111 (2008).

G. Duda and K. Garrett, "Blogging in the physics classroom: A research-based approach to shaping students' attitudes towards physics," Am. J. Phys. **76**, 1054 (2008).

G. Duda, A. Kemper, and P. Gondolo"Model Independent Form Factors for Spin Independent Neutralino-Nucleon Scattering from Elastic Electron Scattering Data," J. Cosm. Ast. Part. Phys. **04**, 012 (2007).

Gabel, J. R.

J. E. Scott, N. Arav, J. R. Gabel *et al.*, "Variable Intrinsic Absorption in Mrk 279," ApJ, **694**, 438 (2009).

D. M. Crenshaw, S. B. Kraemer, H. R. Schmitt, J. S. Kaastra, N. Arav, J. R. Gabel, and K. T. Korista, "Mass outflow in the seyfert 1 galaxy NGC 5548," ApJ **698**, 281 (2009).

R. Ganguly *et al.*, "Hubble space telescope ultraviolet spectroscopy of 14 low-redshift quasars," AJ **133**, 479 (2007).

Nichols, M. G.

A. E. Ekpenyong, C. L. Posey, J. L. Chaput, A. K. Burkart, M. M. Marquardt, T. J. Smith and M. G. Nichols, "Determination of cell elasticity through hybrid ray optics and continuum mechanical modeling of cell deformation in the optical stretcher," Appl. Opt. **48**, 6344-6354 (2009).

L. Tiede, P. S. Steyger, M. G. Nichols, R. Hallworth, "Metabolic imaging of the organ of corti—A window on cochlea bioenergetics," Brain Res. **1277**, 37-41 (2009).

H. Jensen-Smith, B. Currall, D. Rosino, L. Tiede, M. Nichols, and R. Hallworth, "Fluorescence microscopy methods in the study of protein structure and function," chapter in *Molecular Techniques in Auditory & Vestibular Research*, B. Sokolowski (ed.), Methods in Molecular Biology Series, Humana Press (2008).

Sidebottom, D. L.

D. L. Sidebottom and M. Bassett, "Nearly constant loss in lithium and LiCl-doped borate glasses," Z. Phys. Chem. **223**, 1141 (2009).

R. Fabian, Jr. and D. L. Sidebottom, "Dynamic light scattering in network-forming sodium ultraphosphate liquids near the glass transition," Phys. Rev. **80**, 064201 (2009).

D. L. Sidebottom, "Understanding ion motion in disordered solids from impedance spectroscopy scaling," Rev. Mod. Phys. **81**, 999 (2009).

Soto, P.

B. D. Armstrong, P. Soto, J. E. Shea *et al.*, "Overhauser dynamic polarization and molecular dynamics simulations using pyrroline and piperidine ring nitroxide radicals," J. Magn. Reson. **200**, issue 1, 137-141 (2009).

Z. Zhuang, A. I. Jewett, P. Soto *et al.*, "The effect of surface tethering on the folding of the src-SH3 protein domain," Phys. Biol. **6**, issue 1, 15004 (2009).

M. G. Krone, L. Hua, P. Soto, R. Zhou, B. J. Berne, and J. E. Shea, "Role of water in mediating the assembly of Alzheimer amyloid-beta Abeta16-22 protofilaments," J. Am. Chem. Soc. **130**, 11066 (2008).-

UNIVERSITY OF NEBRASKA-LINCOLN

DEPARTMENT OF PHYSICS AND ASTRONOMY

Lincoln, Nebraska 68588-0299

Students Accepted For Degree	FIELDS		
	Physics	Astronomy	Related Fields
Doctorate	X		
Master's	X	X	

1. General

Chancellor: Harvey S. Perlman
Acting Dean of Graduate School: Kimberly Andrews Espy
Department Chairman: Daniel R. Claes
Department Telephone Number: (402) 472-2770
Type of Institution: University
Control: Public
Setting: Urban
Total Faculty: 1,556
Total Graduate Faculty: 1,449
Total Students: 24,100
Total Graduate Students: 4,591
Annual Graduate Tuition:
In-state residents: Full-time and part-time—$247.00/cr.*
Out-of-state-residents: Full-time and part-time—$665.75/cr.*
Tuition rates for: 2009–10
Deferred tuition plan: No
Other Fees: $539/semester
Term: Semester

*Teaching and Research Assistants do not pay tuition.

2. Number of Faculty in Department

The combined total of full-time faculty in the three professorial ranks is 26. The combined total of full-time, part-time, and other faculty at all ranks is 39.

3. Admission, Financial Aid, and Housing

Address admission inquiries to: Chairman, Graduate Committee, Department of Physics and Astronomy
Graduate application fee required: $45
Admission deadline (Fall admission): None
Admission information: For fall admission, 15 students were accepted from 129 applicants.
Admission requirements: For admission to the graduate programs, a Bachelor's degree is required with a minimum undergraduate GPA of 3.0 specified. Each case is judged on its own merits. The GRE is required. The GRE Advanced is strongly urged. The average GRE scores for 2008–2009 admissions were verbal–447; quantitative–756; total–1,203. The average GRE Physics score for 2008–2009 admission was 790. Students from non-English speaking countries are required to demonstrate proficiency in English via the TOEFL exam or the Michigan Test of English (MTE). Minimum acceptable score for admission is 550 on the TOEFL or 80 on the MTE.
Undergraduate preparation assumed: Symon, *Mechanics*: Corson and Lorrain, *Electricity and Magnetism*; Eisberg, *Modern Physics*.
Address financial aid inquiries to: Chairman, Graduate Committee, Department of Physics and Astronomy
GAPSFAS application required: No
Financial aid deadline: None, but most awards are made by April 1st for fall semester. Foreign students who accept a teaching assistantship must enroll in a no-cost three-week Teaching Institute Program, which normally begins in late July.
Loans available: Yes
Address housing inquiries to: University Housing, 1115 N. 16th St., Lincoln, NE 68588-0622, or email to housing@unl.edu.
On-campus, single student housing available: Yes
Cost/year: $8,111-Double, $8,711-Single, room and board
On-campus, married student housing available: Yes
Cost/month: $430–580–1&2 bdrm. apartment[1]

[1]Unfurnished, some units include utilities.

Table A—Faculty, Enrollments, and Degrees Granted

Research Specialty	2008–09 Faculty	Enrollment[1] Fall 2009		No. of Degrees Granted[2] 2009–10 (2005–10)			Median No. of Years for 2009–10 Ph.D.'s
		Master's	Doctorate	Master's	Terminal Master's	Doctorate	
Astronomy	2	1	0	0(0)	0(0)	0(0)	–
Atomic, Molecular, & Optical Physics	7	5	10	3(4)	1(3)	1(7)	5.92
Condensed Matter Physics	11	4	26	4(21)	1(4)	4(14)	4.67
High Energy	5	1	3	3(2)	1(2)	1(0)	7.67
Non-specialized	1	11	3	0(6)	0(0)	0(0)	
Total		22	42	10(27)	3(7)	6(21)	
Full-time Grad. Stud.		22	42				
Part-time Grad. Stud.		0	0				
First-year Grad. Stud.		15	0				
Median Years in Grad. Study (2009–10 Degrees)				2.44	3.39	5.43	6.1

Undergraduate Degrees, 2009–10 (2004–09): 15(47)

[1]Students not yet committed to a research specialty are entered under non-specialized.
[2]Five-year totals in parentheses.

4. Graduate Degree Requirements

Master's: Option I (for both terminal degree students as well as those continuing in the Ph.D. program): 30 credit hours course work, with minimum of 15 credits in physics, plus a thesis. Option II (for students wishing an interdisciplinary degree program): 36 credit hours of course work, including a minimum of 18 hours in physics and a minimum of nine hours in each of one or more minor subject areas. No thesis required. Option III (for students continuing in the Ph.D. program only): 36 credit hours of coursework, with a minimum of 18 hours in physics. No thesis required. No foreign language required in any option.
Exams: (a) Elementary Exam on elementary physics (grade of B required) and (b) Comprehensive Exam (written and/or oral) covering the student's program of study.
Doctorate: 90 credit hours, research credits included; maximum of 45 credit hours can be transferred; "B" or better grade

average required. No foreign language required. 27 credit hours and 18 consecutive months must be completed in residence.

Exams: (a) Elementary Exam with a passing grade, (b) written Advanced Exams, (c) Comprehensive Exam based on one week of intensive research on a topic approved by the supervisory committee, and (d) Oral Exam in defense of thesis.

Thesis: Thesis may be written *in absentia*.

Special Equipment, Facilities, or Programs: *Astronomy*: Behlen Observatory has a 30-inch reflector telescope equipped with a CCD camera. Both the telescope and the camera are computer controlled. The Minnich Astronomical Computing Center on campus has several computers. A 16-inch rooftop observatory equipped with a CCD camera is operational for student use. *Atomic, Molecular, and Optical Physics*: Behlen Laboratory houses a 4,200 sq. ft. atomic collisions laboratory with four ion accelerators, which cover the energy range from 500 eV–400 keV. These, in conjunction with fast coincidence systems, electrostatic analyzers, a Slevin hydrogen atom source, quadrupole mass analyzers, and several scattering chambers are used to measure the properties of photons, electrons, ions, and neutral particles resulting from energetic collisions. Behlen Laboratory also houses a complete laser-atomic physics laboratory with a high power pulsed Nd:YAG laser system and a high power, narrow bandwidth pulsed tunable dye laser. A state-of-the-art fs laser facility is used for pump-probe studies of spin precession in nanoscale materials. Electron scattering experiments employ UHV GaAs polarized electron and metastable atom sources as well as high-resolution electron spectrometers having provision for energy loss, angular scattering, and angular correlation measurements. Photon polarization measurements are also made. High resolution polarized light spectroscopy is carried out at the Advanced Light Source. *Condensed Matter Physics*: Behlen Laboratory houses an x-ray materials characterization facility, high-field superconductive solenoids, two SQUID magnetometers, several systems for studying magneto-optic properties of thin films over a wide range of temperatures, angle-integrated photoemission, scanning probe microscopy facility with atomic force microscopy, magnetic force microscopy, and near-field scanning optical microscopy, W. M. Keck Center for Fast Dynamics including a femtosecond laser, angle-resolved photoemission, inverse photoemission, low energy electron diffraction, (LEED), angle-resolved thermal desorption, X-ray photoemission, a pulsed nitrogen laser and electron spin analysis facilities, a Raman laser light-scattering facility, a comprehensive laboratory for the study of magnetic materials, as well as high-T_c superconducting materials, and several work stations for theoretical calculations. Several systems for fabrication of nanoscale materials and devices are available: multiple-gun sputtering; focus-ion beam milling; electrodeposition systems for nanodimensional wires and dot arrays.

High Energy Physics: The HEP group performs its research at the Fermilab Tevatron Collider in Batavia, Illinois, at the Large Hadron Collider at CERN in Geneva, Switzerland, and at the Pierre Auger Observatory in Mendoza Province, Argentina, and southeast Colorado.

All: Excellent Instrument (5 full-time machinists) and Electronics (2 full-time technicians) shops are on site.

Library: Love Library houses an extensive collection of physics and astronomy periodicals, books, and monographs. Most research journals are available on-line, as are powerful journal databases.

Table B—Appointments to Graduate Students, 2009–10

Title of Appointee	Appointments Total	Appointments First year	Academic Load Allowed in Credit Hours	Hours of Service Per Week	Stipend for Academic Year ($)
			Semester		
Teaching Assistant	24	5	12	17–20[1]	16,920[2,3]
Research Assistant	27	3	12	17–20	16,920[2,3]
Graduate Research Trainee	10	5	12	20	30,000[2,4]
University Fellowship	10	2	12	13	23,520[2,4,5]
Department Fellowship	0	0	9	13	19,820[2,4,5]
Total	64	15			

[1]Includes estimate of preparation, grading, and proctoring time; regularly scheduled contact hours are 7–10 hours per week.
[2]Tuition is waived.
[3]Summer appointments ($3,560) available in addition.
[4]Stipend is for the calendar year.
[5]Fellowships include 13 hours per week research or teaching assistant.

5. Personnel Engaged in Separately Budgeted Research, 7/09–6/10

Professorial faculty	27
Other faculty	7
Postdoctoral appointments	21
Graduate students	64
Undergraduate students	45
Nonteaching research personnel	3
Total	177

6. Separately Budgeted Research Expenditures by Source of Support

	Departmental Research	Physics-related Research Outside Department
Federal government	$9,759,497	$9,195,732
State and local government	1,064,289	2,784,898
Private, non-profit organizations	384,539	501,322
Business and Industry	39,738	511,549
Total	$11,248,063	$12,993,501

Table C—Separately Budgeted Research Expenditures

Research Specialty	No. of Grants	Expenditures ($)
Atomic, Molecular, & Optical Physics	19	5,414,939
Condensed Matter Physics	26	3,716,602
High Energy Physics	11	1,674,319
Physics Education	1	372,438
Astronomy	1	69,765
Total	45	11,248,063

Table D—Physics-related Research Outside Department

Field and Unit Outside Department	No. of Grants	Expenditures ($)
Nebraska Center for Materials and Nanoscience	99	12,993,501
Total	99	12,993,501

FACULTY

Professors

Batelaan, Herman, Ph.D., U. Utrecht, Netherlands, 1991. Experimental atomic physics; laser cooling and trapping; matter optics and interferometry; ultrafast electron/photon physics.

*__Burns__, Donal J., Ph.D., Queen's, Belfast, 1965. Experimental physics; electron atom collisions; coherence, alignment, and orientation in collisions; curriculum development.

Claes, Daniel R., Ph.D., Northwestern, 1991. Chair of the Department. Experimental high energy physics; D0 experimental at Fermilab; CMS experiment at CERN's large hadron collider; the cosmic ray observatory project (CROP) in Nebraska; the Underground nuclear decay and Neutrino Observatory (UNO).

Dowben, Peter A., Ph.D., Cambridge, England, 1981. Charles Bessey Professor of Physics. Electronic structure of solids and surfaces; nonmetal to metal transitions; surface magnetism; surface ferroelectricity; organic molecular interfaces.

Ducharme, Stephen, Ph.D., Southern California, 1986. Experimental condensed matter and optical physics; physics and applications of two-dimensional ferroelectric polymers.

*__Eckhardt__, Craig J., Ph.D., Yale, 1967. Experimental chemical physics; electronic and vibrational spectra of molecules and molecular crystals; dielectric properties of molecular crystals; detonation mechanism of energetic materials; optical properties of solids; lattice dynamics of molecular crystals; organized molecular monolayers; nanotribology.

Fabrikant, Ilya I., Ph.D., Riga, USSR, 1974. Theoretical, atomic and molecular physics; atomic processes involving negative ions.

Gay, Timothy J., Ph.D., Chicago, 1980. Experimental atomic, nuclear physics; polarized electrons.

Kirby, Roger D., Ph.D., Cornell, 1969. Experimental condensed matter physics; thin film magnetism; fabrication of nanostructural materials; magneto-optic properties; magneto-optic recording; spin dynamics measured using femtosecond laser pump-probe experiments.

Liou, Sy-Hwang, Ph.D., Johns Hopkins, 1985. Experimental condensed matter physics; magnetic properties of thin films; magnetic domain imaging; magnetic sensors.

Schmidt, Edward G., Ph.D., Australian National, 1970. Astronomy and astrophysics; variable stars; spectroscopy and photometry; stellar interiors and evolution.

Sellmyer, David J., Ph.D., Michigan State, 1965. University Professor and George Holmes Distinguished Professor, Director of the Nebraska Center for Materials and Nanoscience, Experimental studies of nanoscale self- and cluster-assembled magnetic structures; exchange-coupled and high-energy magnetic materials; extremely high-density magnetic recording media, new spintronic materials.

Snow, Gregory R., Ph.D., Rockefeller, 1983. Associate Dean for Research, College of Arts and Sciences. Experimental high energy physics; D0 experiment at Fermilab; CMS Experiment at CERN's large hadron collider; the cosmic ray observatory project (CROP) in Nebraska; the Pierre Auger cosmic ray observatory in Argentina.

Starace, Anthony F., Ph.D., Chicago, 1971. George Holmes University Professor. Theoretical atomic physics with emphasis on intense laser-atom interactions and on processes that elucidate few-body dynamics. Research interests include: attosecond and other ultrafast atomic and molecular processes; strong-field atomic processes, such as above threshold ionization/detachment, high-order harmonic generation, laser-assisted electron-atom scattering, and intense laser acceleration of electrons; photoionization and photodetachment processes; atoms in strong external static fields; coherent control of atomic processes; and entanglement and decoherence of spin-based quantum information systems.

Tsymbal, Evgeny Y., Ph.D., Russian Academy of Sciences, Moscow, 1988. Charles Bessey Professor of Physics. Theory of electronic, magnetic, ferroelectric and transport properties of nanostructures.

Umstadter, Donald P., Ph.D., UCLA, 1987. Leland J. and Dorothy H. Olson Chair of Physics. Nonlinear optics of high-intensity laser light in relativistic plasmas, with applications to compact particle accelerators and ultra-short duration x-ray sources. High-energy density physics and extreme states of matter.

*__Woollam__, John A., Ph.D., Michigan State, 1967. University Distinguished Professor and Director of the Center for Microelectronic and Optical Materials Research. Experimental condensed matter physics; optical and electrical properties of solids.

*__Zeng__, Xiao Cheng, Ph.D., Ohio State, 1989. Charles H. Bessey Professor of Chemistry. Computational and theoretical studies of equilibrium and kinetic properties of liquids, solids and nanomaterials.

Emeriti

Burrow, Paul D., Ph.D., California, Berkeley, 1966. Experimental molecular physics; scattering of electrons from atoms and molecules; temporary negative ions; dissociative attachment.

Fuller, Robert G., Ph.d., Illinois, 1965. Properties of ionic solids; Piagetian-based college instruction; use of multi-media in education; computer intensive physics courses; humanized physics; physics education research.

*__Gallup__, Gordon A., Ph.D., Kansas, 1953. Theoretical atomic and molecular physics; theory of vibrational excitation in polyatomic molecules upon electron impact; electron scattering from chiral molecules; atomic and molecular structure.

Hardy, Robert J., Ph.D., Lehigh, 1962. Theoretical physics; statistical mechanics; condensed matter; computer simulations.

Jaecks, Duane H., Ph.D., Washington, 1964. Experimental study of unusual states of atoms and molecules formed in various dynamical processes including photoionization and molecular dissociation using polarized photon-scattered particle correlation measurements and three-particle coincidence measurements; history of scientific instruments.

Jaswal, S.S., Ph.D., Michigan State, 1964. Theoretical condensed matter physics; electron and magnetic properties of solids, surfaces, interfaces, and nanostructures.

Jones, C. Edward, Ph.D., California, Berkeley. Particle theory.

Leung, Kam-Ching, Ph.D., Pennsylvania, 1967. Astronomy and astrophysics; double stars.

Pearlstein, Edgar A., D.Sc., Carnegie Inst. Tech., 1950. Experimental condensed matter physics; optical and electrical properties.

Rudd, M. Eugene, Ph.D., Nebraska, 1962. D.Sc., Concordia College, Moorhead, MN, 1992. Experimental atomic collision physics; history of science and scientific instruments.

Simon, Norman R., Ph.D., Yeshiva, 1968. Analysis of electric and magnetic signals from the brain; EEG and MEG study of sleep.

Weymouth, John W., Ph.D., California, Berkeley, 1951. Applications of physical methods to archaeology; geophysical techniques applied to surveying archeological sites.

Adjunct Professors

Boag, Neil, Ph.D., U. of Bristol, U.K., 1980. Experimental condensed matter physics.

Fridkin, Vladimir, Ph.D., Russian Academy of Sciences, 1958. Experimental condensed matter physics; ferroelectricity.

Hadjipanayis, George C., Ph.D., Manitoba, 1979. Experimental condensed matter physics; magnetic materials and applications.

Leslie-Pelecky, Diandra L., Ph.D., Michigan State, 1991. Experimental condensed matter physics; cluster-assembled magnetic materials; biological applications of magnetic nanoparticles.

Losovyj, Yaroslav Ph.D., U. L'viv, Ukraine, 1984. Condensed matter physics

Manakov, Nikolai L., Ph.D., Voronezh State U., Russia, 1971. Theoretical atomic physics; mathematical physics; theory of intense laser-atom interactions.

Mei, Wai-Ning, Ph.D., SUNY-Buffalo, 1979. Theoretical condensed matter physics; surface physics and molecular dynamics simulations of molecular crystals.

Associate Professors

Adenwalla, Shireen, Ph.D., Northwestern, 1989. Experimental condensed matter physics.

Binek, Christian, Ph.D., U. Duisburg, Germany, 1995. Experimental condensed matter physics; magnetic heterostructures and model systems.

Bloom, Kenneth, Ph.D., Cornell, 1997. Experimental high energy physics.

Dominguez, Aaron, Ph.D., California-San Diego, 1998. Experimental high energy physics.

Gruverman, Alexei, Ph.D., Ural State U., 1990. Fundamental studies of nanoscale physical phenomena in electronic and polar materials.

Uiterwaal, Kees, Ph.D., U. Utrecht, Netherlands, 1994. Photoexcitation and photoionization in ultrashort and intense laser pulses; molecular photochemistry; spatially resolved time-of-flight ion mass spectrometry; nonlinear optical effects; holographic spatial pulse shaping; optical vortices; optical orbital angular momentum.

Assistant Professors

Belashchenko, Kirill, Ph.D., Kurchatov Institute, Moscow, Russia, 1999. Theoretical condensed matter physics.

Centurion, Marin, Ph.D., Cal-Tech, 2005. Experimental ultrafast atomic, molecular, optical and plasma physics; molecular dynamics; laser-induced plasmas; nonlinear optics.

Enders, Axel, Ph.D., Martin Luther U., Germany, 1998. Experimental research on structure, electronic structure, and magnetism of low-dimensional systems.

Kravchenko, Ilya, Ph.D., Kansas, 1999. Experimental high energy physics.

Shadwick, Bradley A., Ph.D., Texas at Austin, 1995. Intense laser-plasma interactions; advanced accelerator concepts; plasma theory; Hamiltonian systems and methods.

Research Associate Professors

Bettis, Clifford L., Ph.D., Oklahoma, 1976. Lecture Demonstrations Manager. Educational physics.

Lee, Kevin M., Ph.D., Nebraska, 1997. Astronomy Education; Instructional Technology; photometric observations of variable stars.

Liu, Yi, Ph.D., Tohoku Univ., Japan, 1988. High resolution and analytical electron microscopy.

Skomski, Ralph, Ph.D., TU Dresden, 1991. Theoretical condensed matter physics; H^- magnetic nanostructures; micromagnetics, intrinsic properties of permanent magnets.

Research Assistant Professors

Banerjee, Sudeep, Ph.D., U. Mumbai (India), 2000. Experimental atomic, molecular and optical physics.

Eads, Michael, Ph.D., Northern Illinois, 2005. High energy physics.

Korlacki, Rafal, Ph.D., U. Silesia, Poland, 2005. Experimental condensed matter physics.

Malik, Sudhir, Ph.D., U. Delhi (India), 1997. High energy physics.

Sokolov, Andrei, Ph.D., Moscow State, 1996. Experimental condensed matter physics; magnetic and electronic properties of materials.

Visiting Professors

Cipolla, Sam J., Ph.D., Purdue, 1969. Atomic physics.

Flocken, John, Ph.D., Nebraska, 1969. Structural phase transitions and ferroelectrics.

Yakovkin, Ivan, Ph.D., Kiev State U., 1983. Condensed matter physics.

Lecturer

Brucker, Melissa, Ph.D., Oklahoma, 2009.

Yenen, Orhan, Ph.D., Nebraska, 1986.

Research Associates

An, Joonhee, Ph.D., California-Davis, 2001. Electronic structure of materials.

Balasubramanian, Balamurugan, Ph.D., Indian Inst. of Technology, 2003. Condensed matter and materials physics.

Burton, J. D., Ph.D., Nebraska, 2008. Properties of advanced magnetic and ferroelectric nanostructures.

Ghebregziabiher, Isaac A., Ph.d., Delaware, 2008. Experimental atomic, molecular, optical, and plasma physics.

Katsanos, Ioannis, Ph.D., Columbia, 2008. High energy physics.

Ketsman, Ihor V., Ph.D., National U. of L'viv, Ukraine, 2006. Surface and interface of solids; surface processes at atomic level; heterogeneous catalytic reactions; gas sensors; development, operation, and maintenance of LV, HV, and UHV systems for processing and characterization of thin films.

Larson, Paul, Ph.D., Michigan State, 2001. Theoretical condensed matter physics.

Lazo-Flores, Jose A., Ph.D., Florida State, 2006. High energy physics.

Li, Xingzhong, Ph.D., U. Science & Technology, Beijing, 1992.

Maharjan, Chakra Man, Ph.D., Kansas State, 2007. High-energy density physics.

Malbouisson, Helena, Ph.D., State University of Rio de Janeiro, 2007. High energy physics.

Michalski, Steven, Ph.D., Nebraska, 2007. Experimental condensed matter physics.

Moorti, Anand, Ph.D., Devi Ahilya Vishwavidyalaya, Indore, India, 2007. Experimental atomic, molecular, optical and plasma physics.

Ngoko Djiokap, Jean Marcel, Ph.D., Catholic University of Louvain, Belgium, 2010. Theoretical atmoic, molecular, optical, and plasma physics.

Pronin, Evgeny, Ph.D., Voronezh State (Russia), 2004. Theoretical atomic physics.

Reece, Timothy J., Ph.D., Nebraska, 2007. Condensed matter and materials physics.

Srivastava, Alok, Ph.D., Indian Institute of Technology, 2002. Experimental atomic, molecular, optical, and plasma physics.

Wu, Ning, Ph.D., Nebraska, 2009. Experimental condensed matter physics.

Wysocki, Aleksander, Ph.D., Nebraska, 2010. Theoretical condensed matter physics.

*Asterisks denote courtesy appointments of faculty from other departments.

RESEARCH SPECIALTIES AND STAFF

Theoretical

Atomic and Molecular Physics. Theory of electron-atom and electron-molecule collisions, electron attachment to molecules and clusters; Rydberg atom collisions, photodetachment of negative ions in external fields. Fabrikant. Theory of vibrational excitation in polyatomic molecules upon electron impact; electron scattering from chiral molecules; atomic and molecular structure. Gallup. Intense laser-matter inactions; high-energy-density physics; plasma-based light sources. Shadwick. 1 postdoc. Theory of intense laser-atom interactions and on processes that elucidate few-body dynamics, including: attosecond and other ultrafast atomic and molecular processes; strong-field atomic processes, such as above threshold ionization/detachment, high-order harmonic generation, laser-assisted electron-atom scattering, and intense laser acceleration of electrons; photoionization and photodetachement processes; atoms in strong external static fields; coherent control of atomic processes; and entanglement and decoherence of spin-based quantum information systems. Starace. 2 postdocs.

Condensed Matter Theory. Electronic structure and magnetism of solids; microstructure/property relations in magnetic materials; theory of microstructural evolution in alloys. Belashchenko. 3 postdocs. Computer simulations of shockwaves; statistical mechanics; theory of thermodynamic, transport, mechanical, and structural properties of solids. R. Hardy. Electronic, magnetic and ferroelectric properties of solids, surfaces, interfaces, and nanostructures. Jaswal. Theory of electronic, magnetic, ferroelectric and transport properties of nanostructures. Tsymbal. 2 postdocs. Investigation of gas-liquid nucleation; understanding friction and lubrication between two solid surfaces; nanoscale materials. Zeng. 2 postdocs.

Non Linear Dynamics. Chaos in dynamical systems. Jones.

Plasma Physics. Intense short-pulse laser-plasma interactions with applications to advanced accelerators and light sources; kinetic theory of plasmas; Hamiltonian structure and methods; advanced numerical methods; large-scale simulations; computational methods. Shadwick. 1 postdoc.

Experimental

Astronomy and Astrophysics. Photometric and spectroscopic study of binary stars; intrinsic variable stars; history of astronomy. Leung. Spectroscopy and photometry of variable stars; stellar interiors and evolution. Schmidt.

Atomic, Molecular, Optical, and Plasma Physics. Vacuum field interaction with atoms and electrons; decoherence in matter interferometry. Batelaan. Low-energy electron scattering process in gases, in particular, temporary negative ion formation and dissociative attachment in bio-molecules. Burrow. Experimental ultrafast atomic, molecular, optical and plasma physics, molecular dynamics; laser-induced plasmas; nonlinear optics. Centurion. 1 postdoc. Molecular and cluster photoionization. Core to Molecular spectroscopy. Electron beam molecular dissociation. Dowben. Collective interactions of molecules and supramolecules. Mechanical energy in molecular processes. Eckhardt. Polarized electron collisions with atoms and molecules. Development of polarized electron sources and electron polarimeters. Neutrino mass measurements. Polarized photoionization of molecules. Gay. Experimental measurement of electron motion and distributions in atoms formed in ion collisions and photoionization. Study of correlated motion of massive particles. Jaecks. Optical testing of historic scientific instruments. Cross sections for inelastic processes in ion-atom, atom-atom, and electron-atom collisions. Rudd. Photoexcitation and photoionization in ultrashort and intense laser pulses; molecular photochemistry; spatially resolved time-of-flight ion mass spectrometry; nonlinear optical effects; holographic spatial pulse shaping; optical vortices; optical orbital angular momentum. Uiterwaal. Nonlinear optics of high-intensity laser light in relativistic plasmas, with applications to compact particle accelerators and ultra-short duration x-ray sources. High-energy density physics and extreme states of matter. Umstadter. 5 postdoctoral/research associates.

Condensed Matter Physics. Magnetic interactions in thin films, exchange bias, neutron scattering, development of solid state neutron detectors; ferromagnetic/ferroelectric thin film interactions. Piezoelectric strain effects on magnetic thin films. Adenwalla. Molecular beam epitaxial growth of magnetic heterolayer structures, extrinsic control of exchange bias for spintronic applications and fundamental aspects of interface magnetism, magnetic nanoparticles, model systems in statistical physics. Binek. 1 postdoc. Surface science; magnetic films, surface ferroelectric transitions, metal-non-metal transitions, metal-insulator interfaces. Dowben. 2 postdocs. Physics and applications of two-dimensional ferroelectric polymers. Ducharme. 2 postdocts. Nonlinear optical properties of molecular solids; piezomodulated Raman and reflection spectroscopy of solids; phase transitions in molecular crystals. Eckhardt. Self-assembled magnetic and molecular nanostructures, low- and variable temperature scanning tunneling microscopy, spin-polarized STM. Enders. Fundamental studies of nanoscale physical phenomena in multiferroic and polar materials by means of scanning probe microscopy (SPM) techniques; Static and dynamic properties of ferroic domains; scaling behavior of ferroelectric-based devices, electronic properties of polar surfaces, SPM-assisted methods for fabrication of nanostructures; SPM studies of electromechanical and mechanical properties of biocompatible materials and biological systems. Gruverman. Magneto-optical Kerr effect in artificially structured magnetic materials with applications to magneto-optical storage devices. Laser interference and patterning of thin films. Optical pump-probe measurements of spin dynamics in magnetic thin films. Kirby. Electronic and magnetic properties of nanostructural materials; magnetic domain imaging applications of scanning probe microscope; magnetic sensors. Liou. Magnetism in self and cluster-assembled nanostructures; magneto-electronic structures and devices; fundamental limits on magnetic recording density; exchange-coupled and hybrid permanent-magnetic materials. Sellmyer. 5 research faculty/postdocs. Optical, electrical, and thermal properties of semiconductors, dielectrics, and metals. Spectroscopic ellipsometry. Woollam. 2 postdocs.

High Energy Physics. The D0 Experiment at Fermilab, Batavia, Illinois—studies of the production and decay properties of the top quark, bottom quark, and intermediate vector bosons, studies in quantum chromodynamics, and searches for physics beyond the Standard Model. The CMS Experiment at CERN, Geneva, Switzerland—the above studies will continue when the Large Hadron Collider opens a new energy frontier in 2009. The group has hardware, software, and physics analysis responsibilities in both D0 and CMS and hosts a Tier-2 computing facility for CMS. The group is involved with the construction and commissioning of the silicon pixel tracking detector for CMS and R&D for a future pixel detector. The Cosmic Ray Observatory Project (CROP)–an education and outreach project to study extensive cosmic-ray air showers with particle detectors located at high schools throughout Nebraska. The Pierre Auger Observatory—the World's largest cosmic ray experiment in Argentina and Colorado. Bloom, Claes, Dominguez, Kravchenko, Snow. 6 postdocs.

UNIVERSITY OF NEBRASKA-LINCOLN

MATERIALS AND NANOSCIENCE PROGRAM
NEBRASKA CENTER FOR MATERIALS AND NANOSCIENCE

Lincoln, Nebraska 68588-0113

Students Accepted For Degree	FIELDS		
	Physics	Astronomy	Related Fields
Doctorate	X		X
Master's	X		X

1. General

Chancellor: Harvey S. Perlman
Dean of Graduate School: Ellen M. Weissinger
Center Director: David J. Sellmyer
Center Telephone Number: (402) 472-7886
Type of Institution: University
Control: Public
Setting: Urban
Total Faculty: 1,556
Total Graduate Faculty: 1,449
Total Students: 24,100
Total Graduate Students: 4,591
Annual Graduate Tuition: (Estimated at time of print)
In-state residents: Full-time & Part-time—$261.75/credit*
Out-of-state residents: Full-time & Part-time—$705.75/credit*
Tuition rates for: 2010–11
Deferred tuition plan: No
Term: Semester

*Teaching and Research Assistants do not pay tuition.

2. Number of Faculty in Department

The combined total of full-time faculty in the three professorial ranks is 72. The combined total of full-time, part-time, and other faculty at all ranks is 79.

3. Admission, Financial Aid, and Housing

Address admission inquiries to: Admissions, Nebraska Center for Materials and Nanoscience, University of Nebraska, Lincoln, NE 68588-0113 (see admission information below).
Graduate application fee required: $45
Admission deadline (Fall admission): None
Admission information: Students desiring a degree in materials science must be admitted to one of the departments participating in the Program (Physics, Chemistry, Chemical Engineering, Electrical Engineering, Mechanical Engineering (Metallurgy), Engineering Mechanics).
Admission requirements: For admission a Bachelor's degree in physics, chemistry, materials science, engineering, or mathematics is required with a minimum GPA of 3.0. The GRE and GRE Advanced are strongly urged. Students from non-English speaking countries are required to demonstrate proficiency in English via the TOEFL exam or the Michigan Test of English (MTE). Minimum acceptable score for admission is 550 on the TOEFL or 80 on the MTE.
Address financial aid inquiries to: Chairman, Graduate Committee of the appropriate department.
GAPSFAS application required: No
Financial aid deadline: None, but most awards are made by April 1st for fall semester.
Loans available: Yes
Address housing inquiries to: Office of University Housing, 1115 N. 16th, Lincoln, NE 68588-0622.
On-campus, single student housing available: Yes
Cost/year: $8,111—Double, $8,711—Single, room and board
On-campus, married student housing available: Yes
Cost/month: $430–$580—1, 2, & 3 bdrm. apartment[1]

[1]Unfurnished, some units include utilities.

Table A—Faculty, Enrollments, and Degrees Granted

Research Specialty	2009–10 Faculty	Enrollment[1] Fall 2009 Mas-ter's	Enrollment[1] Fall 2009 Doc-torate	No. of Degrees Granted[2] 2009–10 (1995–09) Mas-ter's	No. of Degrees Granted[2] 2009–10 (1995–09) Terminal Master's	No. of Degrees Granted[2] 2009–10 (1995–09) Doc-torate	Median No. of Years for 2009–10 Ph.D.'s
Materials Physics	21	16	30				–
Materials Chemistry	18	23	44		Not Available		–
Materials Engineering	33	10	19				–
Other	7						
Total	79	49	93				
Full-time Grad. Stud.							
Part-time Grad. Stud.		N/A					
First-year Grad. Stud.							
Median Years in Grad. Study (2009–10 Degrees)				–	–	–	–

Undergraduate Degrees, 2009–10 (1995–09): not available

[1] Students not yet committed to a research specialty are entered under non-specialized.
[2] Five-year totals in parentheses.

4. Graduate Degree Requirements

Master's: Several options exist including thesis and nonthesis. Generally 30 or 36 hours of coursework are required, depending on the option chosen. See Graduate Studies Bulletin for details.
Doctorate: 90 credit hours, research credits included; "B" or better average required. No foreign language required. 27 hours and 18 consecutive months must be completed in residence.
Thesis: Thesis may be written *in absentia*.
Special Equipment, Facilities, or Programs: The Nebraska Center for Materials & Nanoscience operates and coordinates the following Central Service Facilities: X-Ray Materials Characterization, Electron Microscopy, Crystallography, Metallurgical and Mechanical Characterization, Scanning Probe Microscopy, Materials Preparation, Nanofabrication and Cryogenics. A brief description of each facility follows:
X-Ray Materials Characterization: This facility contains a Bruker-AXS D8 Discover X-Ray Diffractometer with GADDS Area Detector and a Rigaku D/Max-B diffractometer. The Rigaku x-ray diffractometer has both large-angle and small-angle goniometers.
Electron Microscopy: This facility offers researchers the best tools for atomic-scale and submicron materials analysis. The facility includes the latest 200 kV high-resolution electron microscope (JEOL JEM 2010 TEM), an analytical scanning transmission electron microscope (VG HB501 STEM), a

well-equipped scanning electron microscope (JEOL JSM 840A SEM), and the necessary support instruments.

Crystallography: Services provided include unit cell determinations, complete structure determinations, and collection of single crystal data. A Bruker AXS SMART Apex instrument is used to collect the single crystal data. The facility also owns two microscopes in the facility - a Leica ZOOM 2000 microscope and a Meiji stereo zoom microscope with polarizer attachment.

Cryogenics: This facility maintains a continuous supply of the liquid helium and liquid nitrogen necessary for on-campus use.

Materials Preparation: Included in this facility is various equipment for materials and sample preparation. Instruments available include two sputtering systems to fabricate a variety of thin films, especially nanostructured films including overlayers, multilayers, granular solids, clusters, etc. Each sputtering system has three guns. Furthermore, two tube furnaces (Lindberg 54233 and Lindberg 55332) together with vacuum pump stations are available for sample (or target) annealing, doping, and sintering.

Metallurgical and Mechanical Characterization: This facility contains a large variety of equipment to characterize the mechanical and physical properties of a wide range of materials. Equipment available through this facility include a Zeiss ICM metallograph, several Olympus microscopes, hardness testing equipment such as Rockwell, Knoop, and Vickers testers, as well as tension/compression testing machines (MTS, Instron, Satec, Tinius-Olson) and torsion testing equipment. Available equipment for sample preparation includes several different saws, hot presses, facilities for cold-mounting of materials, a belt grinding station and two hand lapping stations for rough and coarse grinding, as well as four variable speed stations for fine polishing.

Scanning Probe Microscopy: Digital Instruments Nanoscope IIIa Dimension 3100 Scanning Probe Microscopy (SPM) Facility provides nanometer-scale characterization of materials by using Scanning Tunneling Microscopy (STM), Atomic Force Microscopy (AFM), and Magnetic Force Microscopy (MFM). MFM facility can scan samples in applied external magnetic fields, which is useful for high resolution magnetic domain imaging under magnetic field.

Nanofabrication: This facility includes a variety of instruments enabling the fabrication of electronic, magnetic and other nanostructures. Equipment includes: Zeiss field-emission microscope, Raith e-beam lithography, focused ion-beam workstation, Suss mask aligner, Trion reactive ion etcher, AJA e-beam evaporation system, ovens, and other.

Specialized Research Facilities: Many state-of-the-art research facilities are available including 14 TeslaNMR spectrometer equipped for solid-state NMR, NIMA Langmuir-Blodgett Trough for monolayer and multilayer films, atomic force microscopes, high-field superconductive solenoids, a SQUID magnetometer, angle-integrated photoemission and electron spin analysis facilities, Raman and Brillouin laser light-scattering facilities, a comprehensive laboratory for the study of magnetic materials, as well as high-T_c superconducting materials, a number of photoemission and inverse photoemission spectrometers (including spin-polarized inverse photoemission), and dedicated minicomputers for theoretical calculations. In addition pulsed laser facilities, atomic force microscope and a comprehensive ellipsometer laboratory are available.

Table B—Appointments to Graduate Students, 2008–09

Title of Appointee	Appointments		Academic Load Allowed in Credit Hours	Hours of Service Per Week	Stipend for Academic Year ($)
	Total	First year			
Information not available					

5. Personnel Engaged in Separately Budgeted Research, 2009–10

Professorial faculty	77
Postdoctoral appointments	24
Graduate students	142
Undergraduate students	15*
Nonteaching research personnel	12
Total	270

*Hourly employment

6. Separately Budgeted Research Expenditures by Source of Support

	Center Research	Physics-related Research Outside Center
Federal government	$18,746,568	$9,759,497
State and local government	4,872,797	1,064,289
Private, non-profit organization	255,117	384,539
Business and industry	627,593	39,738
Total	$24,502,075	$11,248,063

Table C—Separately Budgeted Research Expenditures

Research Specialty	No. of Grants	Expenditures ($)
Materials Chemistry	20	2,421,030
Materials Physics	19	7,226,774
Materials Engineering	50	14,854,271
Total	89	24,502,075

FACULTY

Professors

Berkowitz, David B., Ph.D., Harvard U., 1990. Organic catalyst development and new methods for catalyst screening.

Brand, Jennifer I., Ph.D., California, 1992. Ceramic thin films from cluster beams; supercritical fluid technology.

Burrow, Paul D., Ph.D., California, Berkeley, 1966. Experimental molecular physics; scattering of electrons from atoms and molecules; temporary negative ions.

Chandra, Namas, Ph.D., Texas A&M University, 1986. Mechanics of nano, bio and structural materials and structures; finite deformation; multiscale modeling and simulation; molecular dynamics; superplasticity; composites.

Di Magno, Stephen G., Ph.D., California, 1991. Synthesis and properties of organic solids.

Dowben, Peter A., Ph.D., Cambridge (England), 1981. Surface science, magnetic and ferroelectric films.

Ducharme, Stephen, Ph.D., USC, 1986. Two-dimensional ferroelectric polymers.

Dussault, Patrick H., Ph.D., California Tech., 1986. Synthesis and properties of organic materials.

Dzenis, Yuris A., Ph.D., Texas (Arlington), 1994. Composite materials and mechanics.

Eckhardt, Craig J., Ph.D., Yale, 1967. Statics and dynamics of molecular crystals; solid state optical spectroscopy; Langmuir-Blodgett films; crystal engineering of organic solids.

Feng, Ruqiang, Ph.D., Johns Hopkins, 1992. Experimental mechanics of materials, including high strain rate, shockwave, impact experiments, and the study of inelastic deformation and failure mechanisms of ceramics.

Gay, Timothy J., Ph.D., Chicago, 1980. Experimental atomic physics; state selected ion-atom collisions; spin-polarized sources.

Hanna, Milford, Ph.D., Pennsylvania State University, 1973. Biopolymers, including nanocomposites; bioproducts, including biofuels; and bioprocess engineering.

Harbison, Gerard S., Ph.D., Harvard, 1984. Solid-state NMR spectroscopy; quantum chemical calculations on endohedral fullerenes and ionic crystalline solids.

Hardy, Robert J., Ph.D., Lehigh, 1962. Computer simulations of shock waves, statistical mechanics, theory of thermodynamic and transport properties of anharmonic solids.

Ianno, Natale (Ned) J., Ph.D., Illinois, 1981. Electronic properties of semiconductors and superconductors.

Jaswal, Sitaram S., Ph.D., Michigan State, 1964. Theory of electronic structure of magnetic compounds and multilayers.

Kirby, Roger D., Ph.D., Cornell, 1969. Magneto-optic properties of thin films; light scattering.

Langell, Marjorie, Ph.D., Princeton, 1979. Surface chemistry; Auger and photoemission studies of transition metal oxides.

Larsen, Gustavo, Ph.D., Yale, 1992. Surface chemistry and catalysis.

Liou, Sy-Hwang, Ph.D., Johns Hopkins, 1985. Ultrafine particle magnetic films; high-temperature superconducting films.

Lu, Yongfeng, Ph.D., Osaka Univ., Japan, 1991. Laser material processing and characterization at micrometer and nanometer scales.

Negahban, Mehrdad, Ph.D., Michigan, 1988. Mechanical effects of phase transition in polymers under large strains.

Parkhurst, Lawrence J., Ph.D., Yale, 1965. Optical properties of DNA; solid-state protein thermodynamics.

Pearlstein, Edgar A., D.Sc., Carnegie Inst., 1950. Defects in solids; electrical noise.

Rajca, Andrzej T., Ph.D., Kentucky, 1985. High-spin organic polyradicals; synthesis of two-dimensional magnets from polyradicals.

Rajurkar, K. P., Ph.D., Michigan Technological University, 1982. Non-conventional machining at macro, micro and nano scales; stochastic modeling of manufacturing processes and systems.

Redepenning, Jody G., Ph.D., Colorado State, 1985. Electrochemistry; biocompatible materials.

Robertson, Brian W., Ph.D., Glasgow, Scotland, 1979. Electron microscopy and techniques; nanofabrication; sensor materials and devices.

Saraf, Ravi, Ph.D., Univ. of Massachusetts, Amherst, 1987. Nanoscale material fabrication and devices, and biophysics.

Sellmyer, David J., Ph.D., Michigan State, 1965. Nanomagnetics and magnetoelectronics including self- and cluster-assembled structures, dilute magnetic semiconductors and oxides, exchange-coupled nanocomposites, and fundamental limits on magnetic storage density.

Shield, Jeffrey E., Ph.D., Iowa State Univ., 1992. Microstructural evolution of materials, rapid solidification.

Soukup, Rodney J., Ph.D., Minnesota, 1969. Semiconductors; solar energy studies.

Subramanian, Anuradha, Ph.D., VT, 1995. Biomaterial development and further use of these biomaterials in bioseparations and biomedical applications.

Takacs, James M., Ph.D., California Inst. Tech., 1981. Synthesis of novel polymers.

Timm, Delmar C., Ph.D., Iowa State, 1967. Engineering properties of polymers.

Tsymbal, Evgeny Y., Ph.D., Russian Academy of Sciences, Moscow, 1988. Theory of electronic structure and spin-dependent transport in nanoscale magnetic systems.

Turner, Joseph A., Ph.D., Univ. of Illinois, Urbana-Champaign, 1994. Wave propagation and vibrations, including ultrasonics, NDE, materials characterization, AFM dynamics, and nanoindentation.

Viljoen, Hendrik J., Ph.D., Pretoria (South Africa), 1985. Synthesis and processing of materials with CVD.

Watkins, David K., Ph.D., Florida State, 1984. Electron microscopy studies of minerals.

Williams, P. Frazer, Ph.D., So. California, 1973. Electrical breakdown of semiconductors.

Woollam, John A., Ph.D., Michigan State, 1967. Ellipsometric studies of oxide surfaces; magneto-optic films; optical coatings and electrochromics.

Yang, Yiqi, Ph.D., Purdue University, 1991. Biopolymeric materials; biotextiles; industrial applications of agricultural byproducts.

Zeng, Xiao Cheng, Ph.D., Ohio State Univ., 1989. Statistical mechanics of liquids and solids, phase transition, computer simulations.

Associate Professors

Adenwalla, Shireen, Ph.D., Northwestern, 1989. Experimental condensed matter physics.

Baesu, Eveline, Ph.D., Univ. of California at Berkeley, 1998. Continuum mechanics, plasticity, electrodynamics of continuum media.

Binek, Christian, Ph.D., University of Duisburg, 1995. Magnetic heterostructures in basic research and spintronic applications.

Bobaru, Florin, Ph.D., Cornell University, 2001. Computational mechanics, numerical optimization of advanced materials and systems.

Gruverman, Alexei, Ph.D., Ural State University, 1990. Ferroelectrics, piezoelectrics, scanning probe microscopy.

Hallbeck, Susan M., Ph.D., Virginia Tech, 1990. Ergonomics, Biomedical and Systems Engineering.

Schubert, Mathias, Dr. rer. nat., Universität Leipzig, 1997, Dr. rer. nat. habil., Universität Leipzig, 2003. Condensed matter spectroscopy; ferroic semiconductor thin films and nanostructures.

Tan, Li, Ph.D., University of Michigan, 2002. Unconventional nanolithography; nanoimprint lithography; polymer thin films and devices/sensors.

Yang, Jiashi, Ph.D., Princeton, 1994. Electromechanical materials and devices.

Zhang, Zhaoyan, Ph.D., Penn State U., 2000. Experimental and theoretical study of laser material interactions.

Assistant Professors

Belashchenko, Kirill, Ph.D., Kurchatov Institute, 1999. Electronic structure theory; magnetic, transport, and structural properties of materials and nanostructures.

Cheung, Chin Li (Barry), Ph.D., Harvard, 2002. Synthesis and characterization of materials at the nanoscale.

Choe, Wonyoung, Ph.D., U. Michigan, 1998. Synthesis and properties of zeolites and nanoparticles; magnetorestrictive and hydrogen storage materials.

Enders, Axel, Ph.D., Max-Planck, Halle, Germany, 1999. Scanning probe studies of magnetic nanostructures.

Gu, Linxia, Ph.D., University of Florida, 2004. Soft tissue mechanics; multiscale and multiphysics modeling.

Huang, Jinsong, Ph.D., Unversity of California Los Angeles, 2007. Organic solar cell, organic spintronics, ferroelectric polymers.

Lai, Rebecca Y., Ph.D., University of Texas at Austin, 2003. Ligand induced folding in biopolymers; electrochemical biosensors; surface plasmon resonance biosensors.

Othman, Shadi F., Ph.D., University of Illinois at Chicago, 2004. Microscopic Magnetic Resonance Elastography; MRI.

Pannier, Angela K., Ph.D., Northwestern University, 2007. Biomaterials for nonviral gene delivery and tissue engineering applications.

Schubert, Eva Franke, Ph.D., University of Leipzig, Germany, 1998. Ion beam processing; nanostructured thin film fabrication for optical, electromechanical, and magnetic device applications.

Research Professor

Diestler, Dennis J., Ph.D., Caltech, 1967. Theoretical studies of the fluid solid interface.

Research Associate Professors

Liu, Yi, Ph.D., Tohoku Univ., Japan, 1988. High resolution and analytical electron microscopy.

Skomski, Ralph, Ph.D., Dresden (TU), 1991. Theoretical solid-state magnetism; micromagnetics; permanent magnetism.

Research Assistant Professors

Sokolov, Andrei, Ph.D., Lomonosov Moscow State Univ., Russia, 1996. Magnetoelectronics.

Adjunct Professors

Darveau, Scott A., Ph.D., University of Chicago, 1998. Thin-film photovoltaic materials; Raman microscopy/laser spectroscopy.

Exstrom, Christopher, Ph.D., University of Minnesota, 1995. Characterization and development of novel solar cell film materials; preparation and characterization of novel solvatochromic transition-metal based materials.

Fridkin, Vladimir, Ph.D., Institute of Crystallography, Academy of Sciences of the USSR, 1965. Ferroelectricity; ferroelectric polymers; phase transitions.

Hadjipanayis, George C., Ph.D., Manitoba, 1979. Experimental condensed matter physics; magnetic materials and applications.

Liu, J. Ping, Ph.D., Amsterdam, 1994. Magnetic materials.

Mei, Wai Ning, Ph.D., State Univ. of New York, Buffalo, 1979. Surface structure determination using multiple-scattering theory, molecular dynamics simulations of alkali halides, and molecular solids.

Namavar, Fereydoon, Ph.D., Katholieke Universiteit Leuven, Belgium, 1978. Nanotechnology; alternative surfaces for medical implants; smart surfaces; smart drugs and sensors for medical application.

Palencia, Hector, Ph.D., UNL/UNAM, 2005. Catalysis/nanocatalysts; biofuels, organic synthesis.

Sabirianov, Renat, Ph.D., UTU, 1993. Defects and impurities in magnetic materials.

Smith, Robert W., Ph.D., Oregon State, 1989. Synthesis, crystal growth, and structural characterization of new materials through x-ray diffraction techniques.

Visiting Professor

Boag, Neil, Ph.D., Bristol, 1981. Materials science and synthesis.

RESEARCH SPECIALTIES AND STAFF

Theoretical

Condensed Matter Physics. Structural and vibrational properties of crystalline and amorphous solids using the methods of statistical mechanics (R. Hardy, Zeng). Giant magnetoresistance in magnetic metallic multilayers, spin-dependent tunneling in magnetic tunnel junctions, spin injection into semiconductors (Jaswal, Tsymbal). Permanent-magnet materials, magnetic nanostructures, time-dependent magnetizationprocesses, spin structure of half-metallic ferromagnets, quantum entanglement in nanodots (Skomski).

Mechanics and Materials. Large deformation thermomechanical and mechanical response characterization, continuum thermodynamic modeling (Baesu, Bobaru, Negahban). Fluid-solid interfacial phenomena (Diestler).

Experimental

Condensed Matter Physics. Structural characterization of materials, magnetic systems, polymers, development of solid state neutron detectors (Adenwalla). Exchange bias in magnetic metal/insulator heterosystems, matrix insulated magnetic nanoparticles (Binek). Interplay between magnetic and electric properties of heterogeneous metallic magnetic systems (Sokolov). Electronic band structure and the influence of electronics structure on various phase transitions (Dowben). Ferroelectric nanostructures and polymers (Ducharme, Gruverman). Polarized electron physics (Gay). Magnetism dynamics in thin films, superlattices (Kirby). Magnetic interactions in patterned nanostructures, quantum conductance and magnetoresistance in nanocontacts, nanofabrications (Liou). High-anisotropy magnetic nanocluster-assembled films, nanotube magnetism, spin-logic nanostructures, exchange-coupled nanocomposites (Enders, Liu, Sellmyer).

Materials Chemistry. Synthesis and study of unnatural analogues of biological molecules, catalyst development (Berkowitz). Design and synthesis of inorganic/bio-organic nanoscaled components (Cheung). Development of new robust macrocyclic ligands (DiMagno). Synthesis of peroxide-containing natural products, new methods for organic oxidations based upon ozone, singlet oxygen, and hydrogen peroxide (Dussault). Nanotribology, mechanochemistry, energetic materials supramolecules, and organic ferroelectrics (Eckhardt). Influence of oxidation state on chemical and physical properties of metal-containing materials (George). Structure and properties of solids, NMR (Harbison). Surface properties of transition metal oxides and other metal compounds (Langell). Interactions between DNA, RNA, and proteins (Parkhurst). Design and synthesis of stable high-spin polyradicals (Rajca). Electrochemistry, electrochemical deposition (Redepenning). Structure function studies in supramolecular systems (Choe). Organic synthesis and chemistry, piezoelectric devices (Takacs).

Materials Engineering. Developing new materials and more efficient materials production processes for deposition of thin films, microfibers and microparticles (Brand). Novel devices (Hallbeck). Thin film deposition, high density plasma processing, nanoscale processing (Ianno). Catalysis, adsorption, and materials design (Larsen). Microscale and nanoscale laser material processing and characterization (Lu). Magnetic and electronic thin films, nanoscale wires and devices, and electron probe-based characterization (Robertson). Deposition and characterization of thin films of wear-resistant materials, piezoelectric oxide ceramics (Rohde). Optical properties of semiconductors and nanoscale materials, ellipsometry (Snyder). Deposition and study of semiconductor films (Soukup). Synthesis and development of novel biofunctional materials (Subramanian). Microstructural evolution of materials during processing (Shield). Composite materials comprised of a polymeric matrix (Timm). Piezoelectric materials-(Viljoen). Electrical breakdown of gases and semiconductors, plasma processing of semiconductors (Williams). Laser-materials interactions (Zhang).

Mechanics and Materials. Electromechanical effects, fiber networks, and biomechanics (Baesu). Meshfree methods, structural and multidisciplinary optimization of solids (Bobaru). Functional nanomaterials and nanomanufacturing (Dzenis). Experimental and computational mechanics of materials (Feng). Active materials, composites, micromechanics, and microstructure mechanics (Li). Stochastic wave propagation, experimental ultrasonics, structural acoustics (Turner). Frequency stability of piezoelectric crystal resonators (Yang).

UNIVERSITY OF NEVADA, LAS VEGAS

DEPARTMENT OF PHYSICS AND ASTRONOMY

Las Vegas, Nevada 89154-4002

Students Accepted For Degree	FIELDS		
	Physics	Astronomy	Related Fields
Doctorate	X	X	
Master's	X	X	

1. General

President: Neal J. Smatresk
Dean of Graduate College: Ron Smith
Department Chairman: Tao Pang
Department Telephone Number: (702) 895-3563
Type of Institution: University
Control: Public
Setting: Urban
Total Faculty: 957
Total Graduate Faculty: 690
Total Students: 29,086 (Fall 2009)
Total Graduate Students: 6,378 (Fall 2009)
Annual Graduate Tuition:
In-state residents: Full-time—$239.50/cr.
Part-time—$239.50/cr.
Out-of-state residents: Full-time—$239.50/cr. + $6,170 sem. if registered for 7 or more credits
Part-time—$457.25/cr.
Tuition rates for: 2010
Deferred tuition plan: Yes
Term: Semester

2. Number of Faculty in Department

The combined total of full-time faculty in the three professorial ranks is 16. The combined total of full-time, part-time, and other faculty at all ranks is 21.

3. Admission, Financial Aid, and Housing

Address admission inquiries to: Graduate Program Admissions, Department of Physics and Astronomy
Graduate application fee required: $60/domestic; $95/international
Admission deadline (Fall admission): 6/15
Admission information: For fall admission, 2009–10, 7 students were accepted from 19 applicants.
Admission requirements: For admission to the graduate programs, a Bachelor's degree is required with a minimum undergraduate GPA of 2.75 overall or 3.0 last 2 years. The applicant must have completed 18 semester credits of upper division physics. The GRE General and Advanced Physics are required. Students from non-English speaking countries are required to demonstrate proficiency in English via the TOEFL. Minimum acceptable score for admission is 550.
Undergraduate preparation assumed: Mechanics level—Marion, *Classical Dynamics*; Electricity and Magnetism level—Wangsness, *Electromagnetic Fields*; Quantum Mechanics level—Griffiths, *Quantum Mechanics*; Mathematics level—through Advanced Calculus.
Address financial aid inquiries to: Graduate Program Admissions, Department of Physics and Astronomy
GAPSFAS application required: No
Financial aid deadline: 3/1
Loans available: Yes
Address housing inquiries to: Division of Student Services
On-campus, single student housing available: Yes
Cost/term: $5,228 per semester, room and meal plan.
On-campus, married student housing available: No

Table A—Faculty, Enrollments, and Degrees Granted

Research Specialty	2009–10 Faculty	Enrollment[1] Fall 2009 Master's	Enrollment[1] Fall 2009 Doctorate	No. of Degrees Granted[2] 2009–10 (2005–09) Master's	Terminal Master's	Doctorate	Median No. of Years for 2009–10 Ph.D.'s
Astronomy	5	3	3	0(1)	0(1)	0(3)	–
Astrophysics	2	2	4	0(1)	0(0)	0(0)	–
Atomic, Molecular, & Optical Physics	2	2	1	0(1)	0(1)	1(1)	–
Chemical Physics	1	1	0	0(0)	0(0)	0(0)	–
Condensed Matter Physics	4	8	6	0(2)	0(0)	0(0)	–
Plasma Physics & Fusion	2	0	0	0(0)	0(0)	0(0)	–
Polymer Physical/ Science	1	0	0	0(0)	0(0)	0(1)	–
Total		16	14	0(5)	0(2)	1(5)	
Full-time Grad. Stud.		16	14				
Part-time Grad. Stud.		0	0				
First-year Grad. Stud.		5	2				
Median Years in Grad. Study (2009–10 Degrees)				2	2.5	4	

Undergraduate Degrees, 2009–10 (2005–09): 5(17)

[1]Students not yet committed to a research specialty are entered under non-specialized.
[2]Five-year totals in parentheses.

4. Graduate Degree Requirements

Master's: A minimum of 30 graduate credits is required for the Master of Science degree in Physics or Astronomy including a minimum of 15 credits (excluding thesis) in 700-level courses and six semester hours of research for thesis credit. A final oral exam is required on course work and thesis except for the satisfactory performance on an Astronomy Qualifying Exam for the Astronomy Non-Thesis Option. A grade point average of at least 3.0 is required in all course work which is part of the degree program.
Doctorate: A minimum of 60 semester credits past the Bachelor's degree, including at least 36 graduate level semester

credits in classroom courses in physics, astronomy, or related fields and specified core courses. Course work used to satisfy the requirement for a Master's degree may be included. A minimum grade of B (3.00) is required in each course which is used in the degree program. A minimum of 18 semester credits of dissertation. Qualifying exam and oral defense of the dissertation.

Thesis: Thesis may be written *in absentia* with special approval only.

Programs:

Atomic, Molecular, and Optical Physics: The AMO group consists of six faculty members: four experimentalists and two theorists. The research projects include nonlinear optics, studies of macromolecules, photon correlation spectroscopy, spectroscopy of molecular ions, studies of laser-produced low energy plasmas and trapped ions, atomic and molecular collisions, and modeling of molecular clouds in the interstellar medium. All four research laboratories are equipped with lasers, ultrahigh vacuums, and spectroscopy facilities. Modeling and calculations are conducted in the Computational Physics Laboratory which is partially funded by the W. M. Keck Foundation.

Astronomy/Astrophysics: Faculty research interests include star formation in galaxies, active galactic nuclei, clusters of galaxies, gamma ray bursts, quasars, large scale structure of the universe, and variable stars. Faculty members have guaranteed access to SWIFT, and successfully compete for time on other national facilities. Department facilities include an automated telescope in southern Arizona and access to the Lowell Observatory's 31-inch telescope in Flagstaff, Arizona through UNLV's participation in the National Undergraduate Research Observatory.

Condensed Matter and High Pressure Physics: The UNLV High Pressure Science and Engineering Center (HiPSEC), recently established with support from the U.S. Department of Energy, brings together physicists, chemists, geoscientists, and mechanical engineers to consider fundamental experimental, computational, and engineering problems of materials under high pressures. Faculty and research staff study the equilibrium thermochemical properties, mechanical properties, reaction kinetics, and reaction products at static pressures using in situ x-ray diffraction, absorption, emission, and light-scattering spectroscopy from infrared to x-ray wavelengths, and other chemical and physical methods. Experiments are conducted at three in-house laboratories and in national laboratories including an x-ray beam line in the Advanced Photon Source at the Argonne National Laboratory. In addition, along with experimental studies of complex fluids, state-of-the-art computational techniques are used to study highly correlated electron systems, including d- and f-band metals, clusters, thin films, quantum dots and novel materials.

Department Facilities and Funding Sources: The Department resides in the Robert L. Bigelow Physics Building completed in 1994. Inside this 70,000 square-foot building, there are seven laboratories, other teaching and research laboratories, and supporting facilities including two modern machine shops, a glass shop, and an electronics shop. The Computational Physics Laboratory, partially funded by the W. M. Keck Foundation, has a parallel/distributed computing system with a peak performance of about 8 GFlops. Research in the department is supported by NSF, DOD, DOE, NASA, EPA, the W. M. Keck Foundation, the Bigelow Foundation and the UNLV Foundation.

Table B—Appointments to Graduate Students, 2009–10

Title of Appointee	Appointments Total	Appointments First year	Academic Load Allowed in Credit Hours	Hours of Service Per Week	Stipend for Academic Year ($)
			Semester		
Graduate Assistant	16	4	15	20	12,000[1,2,3]
Research Assistant	10	3	15	20	28,000[1,3]
Total	26	7			

[1]Out-of-state tuition waived (100%). 75% or more of per-credit-hour fees waived.
[2]$12,000/Ph.D.; $10,000/M.S. with additional stipends from department.
[3]Summer research appointments available with additional stipends from department.
[4]Fellowship: does not pay out-of-state tuition and fees.

5. Personnel Engaged in Separately Budgeted Research, 7/08–6/09

Professorial faculty	17
Research faculty	4
Postdoctoral appointments	3
Graduate students	30
Undergraduate students	18
Total	72

6. Separately Budgeted Research Expenditures by Source of Support

	Departmental Research	Physics-related Research Outside Department
Federal government	$14,170,003	$0
State and local government	809,220	$0
Total	$14,979,223	$0

Table C—Separately Budgeted Research Expenditures

Research Specialty	No. of Grants	Expenditures ($)
Astronomy	27	2,351,091
Atomic, Molecular, & Optical Physics	6	706,381
Condensed Matter Physics	16	11,921,751
Total	49	14,979,223

FACULTY

Professors

Chen, Changfeng, Ph.D., Beijing, 1987. Condensed matter theory.

Farley, John W., Ph.D., Columbia, 1977. Laser spectroscopy of molecular ions.

Kwong, Victor H. S., Ph.D., Toronto, 1979. Department Chair. Laser induced plasmas; ion charge transfer processes.

Lepp, Stephen H., Ph.D., Colorado, Boulder, 1984. Atomic and molecular theory.

Pang, Tao, Ph.D., Minnesota, 1989. Condensed matter theory.

Selser, James C., Ph.D. (Applied Science), California, Davis, 1975. Static and dynamic light scattering from macromolecular systems.

Shelton, David P., Ph.D., Manitoba, 1979. Nonlinear optical properties of atoms and molecules.

Zane, Leonard I., Ph.D., Duke, 1970.

Zygelman, Bernard, Ph.D., CUNY, 1983. Atomic and molecular theory.

Associate Professors

Cornelius, Andrew L., Ph.D., Physics, Washington Univ., St. Louis, 1996. Condensed matter experimental.

Pravica, Michael G., Ph.D., Experimental Condensed Matter Physics, Harvard University, 1998.

Proga, , Daniel, Ph.D., Astronomy, Nicolaus Copernicus Astronomical Center, Warsaw, Poland. High energy astrophysics.

Rhee, George, Ph.D., Astronomy, Leiden, 1989. Extragalactic astronomy.

Smith, Diane Pyper, Ph.D. (Astronomy), California, Santa Cruz, 1968. Stellar photometry and spectroscopy.

Spight, Lon D., Ph.D., Nevada, Reno, 1969. Cosmology; interacting galaxies.

Zhang, Bing, Ph.D., Astrophysics, Peking University, 1997.

Assistant Professor

Nagamine, Kentaro, Ph.D., Physics, Princeton University. Cosmology, galaxy formation.

Adjunct Professors

Kernan, Warnick, Ph.D., Nuclear Physics, University of Rochester, 1989.

Kukla, Maija, Ph.D., Chemical Physics, University of Latvia, Riga, Latvia. Materials research.

Perry, Dale L., Ph.D., Chemistry, University of Houston, 1974.

Shaffer, David B., Ph.D., Cal. Tech., 1974. Radio astronomy.

Adjunct Associate Professor

Naduvalath, Balakrishnan, Ph.D., Chemistry, Indian Institute of Technology, 1993.

Faculty Emeriti

Cloud, Stanley D., Ph.D., Oregon, 1968. Nuclear physics; teaching instrumentation.

Weistrop, Donna E., Ph.D. (Astronomy), Cal. Tech., 1971. Extragalactic astronomy.

RESEARCH SPECIALTIES AND STAFF

Theoretical

Astrophysics. Interacting galaxies; digital image processing, gamma ray bursts. Spight, Zhang.

Atomic and Molecular. Astronomical phenomena of atomic and molecular systems; atomic and molecular collisions. Lepp, Zygelman.

Condensed Matter. High-temperature superconductivity; quantum liquids; disordered systems; correlated electron systems; spin systems. Chen, Pang.

Experimental

Astronomy. Galactic clustering and evolution. Photoelectric photometry, spectrophotometry, and spectroscopy of peculiar, upper main-sequence stars; observation and photometry of interacting galaxies. Rhee, Smith, Spight, Zhang.

Atomic and Molecular Physics. Multiply charged ions; charge transfer; molecular ions; cluster ions. Farley, Kwong.

Condensed Matter. Material characterization in high pressure. Cornelius, Pravica.

Chemical Physics. Dynamic behavior of macromolecular systems; nonlinear optics of atoms and molecules. Selser, Shelton.

FACULTY PUBLICATIONS

Chen, Changfeng

Zhuhua Zhang, Changfeng Chen, and Wanlin Guo, "Magnetoelectric Effect in Graphene Nanoribbons on Substrates via Electric Bias Control of Exchange Splitting," Phys. Rev. Lett. **103**, 187204 (2009).

Xuezhi Ke, Changfeng Chen, Jihui Yang, Lijun Wu, Juan Zhou, Qiang Li, Yimei Shu, and P. R. C. Kent, "Microstructure and a Nucleation Mechanism for Naoprecipitates in PbTe-$AgSbTe_2$," Phys. Rev. Lett. **103**, 145502 (2009).

Yufeng Guo, Wanlin Guo, and Changfeng Chen, "Bias voltage induced n-to p-type transition in epitaxial bilayer graphene on SiC," Phys. Rev. Lett. **103**, 039602 (2009).

Yi Zhang, Xuezhi Ke, Changfeng Chen, Jihui Yang, and P. R. C. Kent, "Thermodynamic properties of PbTe, PbSe and PbS: First-principles study," Phys. Rev. B **80**, 024304 (2009).

Cornelius, Andrew L.

R. S. Kumar and A. L. Cornelius, "Structural phase transitions in the potential hydrogen storage compound $RbBH_4$ under compression," to appear in J. Alloys Compd. (2008).

R. S. Kumar, A. Svane, G. Vaitheeswaran, V. Kanchana, E. D. Bauer, M. Hu, M. F. Nicol, and A. L. Cornelious, "Pressure-induced valence change in $YbAl_3$: A combined high-pressure x-ray scattering and theoretical investigation," Phys. Rev. B **70**, 075117 (2008).

R. S. Kumar, X. Ke, A. L. Cornelius, and C. Chen, "Effect of Pressure and Temperature on Structural Stability of Potential Hydrogen Storage Compound Li_3AlH_6," Chem. Phys. Lett. **460**, 442 (2008).

Ravhi S. Kumar, Sandeep Rekhi, D. Prabhakaran, Eunja Kim, M. Somayazulu, Jeremy D. Cook, Timothy Stemmler, Andrew Boothrhoyd, Mark R. Chance, and Andrew L. Cornelius, "Structural studies on $Na_{0.75}CoO_2$ thermoelectric materials at high pressures," Solid State Commun **149**, 1712 (2009).

Ravhi S. Kumar and A. L. Cornelius, "Structural phase transistions in $RbBH_4$ under compression," J. Alloys and Compounds **476**, 5 (2009).

Farley, John W.

A. L. Johnson, D. Koury, J. Welch, T. Ho, S. Sidle, C. Harland, B. Hosterman, U. Younas, L. Ma, and J. W. Farley, "Spectroscopic and microscopic investigation of the corrosion of D-9 stainless steel by lead-bismuth eutectic (LBE) at elevated temperatures. Initiation of thick oxide formation," J. Nucl. Mat. **376**, 265 (2008).

Lepp, Stephen H.

S. Thanki, G. Rhee, and S. Leep, "Fractal Dimension of Galaxy Isophotes," AJ **138**, 941 (2009).

P. F. Weck, E. Kim, S. Lepp, N. Balakrishnan, and H. R. Sadeghpour, "Dimer-induced stabilization of H adsorbate cluster on BN(0001) surface," Phys. Chem. Chem. Phys. **10**, 5165 (2008).

Nagamine, Kentaro

R. Kurosawa, D. Proga, and K. Nagamine, "On the Feedback Efficiency of Active Galactic Nuclei," Astrophysical J. **707**, 823 (2009).

J.-H. Choi and K. Nagamine, "Effects of metal enrichment

and metal cooling in galaxy growth and cosmic star formation history," MNRAS **393**, 1595 (2009).

C. Carilli *et al.*, "Imaging the cool gas, dust, star formation, and AGN in the first galaxies," Astro2010: The Astronomy and Astrophysics Decadal Survey, Science White Papers, no. 37.

Gott, J. R., Hambrick, D. C., Vogeley, M. S., Kim, J., Park, C., Choi, Y-Y., Cen, R., Ostriker, J. P., Nagamine, K., "Genus Topology of Structrure in the Sloan Digital Sky Survey: Model Testing," ApJ **675**, 16 (2008).

K., Zhang, B., Hernquist, L., "Incidence Rate of GRB-host-DLAs at High Redshift," Nagamine, ApJ **686**, L57 (2008).

Pang, Tao

T. Pang, "A Math Makeover: Closing the Gap Between High School and College Math" APS News **18**(11), 6 (2009).

Pravica, Michael G.

P. F. Weck, E. Kim, C. Gobin, and M. G. Pravica, "Organic Cyclic Difluoramino-nitramines: Infrared and Raman Spectroscopy of 3,3,7,7 tetrakis(difluoramino)octahydro 1,5-dinitro-1,5-diazocine (HNFX)," J. Raman. Spectros. **40**, 964 (2009).

S. Tkachev, M. Pravica, E. Kim, and P. Weck, "Raman spectroscopic study of cyclopentane at high pressure," J. Chem. Phys. **130**, 204505 (2009). This work was featured in *Science Letter* in their July 7, 2009 issue under the headline: "Chemical Physics; Research on chemical physics detailed by scientists at University of Nevada."

M. Pravica, B. Yulga, S. Tkachev, and Z. Liu, "High-pressure far- and mid-infrared study of 1,3,5,7 Triaminotrinitrobenzene," J. Phys. Chem. A **113**(32), 9133 (2009). This work was featured in *Science Letter* in their September 8, 2009 issue under the headline: "Physical Chemistry; New findings from University of nevada in the area of physical chemistry described."

S. Tkachev, M. Pravica, E. Kim, E. Romano, P. Weck, "High Pressure Studies of 1, 3, 5, 7 Cyclooctatetraene: Experiment and Theory," J. Phys. Chem. A **112**, 11501 (2008). This work was featured in *Science Letter* in their December 30, 2008 issue under the headline: "Physical Chemistry; Findings from University of Nevada provide new insights into physical chemistry."

H. Giefers and M. Pravica, "Radiation-induced decomposition of PETN and TATB under extreme conditions," J. Phys. Chem. A **10**, 1021 (2008).

Proga, Daniel

D. Proga, "Astrophysics: Quiet is the new loud," Nature **938**, 414 (2009).

R. Kurosawa, D. Proga, and K. Nagamine, "On the Feedback Efficiency of Active Galactic Nuclei," ApJ **707**, 823 (2009).

L. Ciotti, J. P. Ostriker, and D. Proga, "Feedback from Central Black Holes in Elliptical Galaxies. I. Models with Either Radiative or Mechanical Feedback but not Both," ApJ, **699**, 89 (2009).

J. M. Stone and D. Proga, "Anisotropic Wind from Close-in Extra-Solar Planets," ApJ **694**, 205 (2009).

R. Kursosawa and D. Proga, "Three Dimensional Simulations of Dynamics of Accretion Flows Irradiated by a Quasar," ApJ **693**, 1929–1945 (2009).

Rhee, George

G. Rhee, S. Thanki, and S. Lepp, "Fractal dimension of Galaxy Isophotes," Astronomical Journal **138**, 941 (2009).

Shelton, David P.

D. P. Shelton, "Polar domain fluctuations in doped liquid nitrobenzene," J. Chem. Phys. **129**, 134501 (2008).

D. P. Shelton, Electric field of ions in solution probed by hyper-Rayleigh scattering, J. Chem. Phys. **130**, 114501 (2009).

Zhang, Bing

A. Maxham and B. Zhang, "Modeling Gamma-Ray Burst X-Ray Flares within the Internal Shock Model," The Astrophysics Journal **707**, 1623 (2009).

B. Zhang, "Astrophysics: Most distant cosmic blast seen," Nature **461**, 1221 (2009).

B. Zhang, B.-B. Zhang, F. J. Virgili, E.-W. Liang, D. A. Kann, Z.-F. Wu, D. Proga, H.-J. Lv, K. Toma, P. Meszaros, D. B. Burrows, P. W. A. Roming, and N. Gehrels, "Discerning the physical origins of cosmological gamma-ray bursts based on multiple observational criteria: the cases of z=6.7 GRB 080913, z=8.3 GRB 090423, and some short/hard GRBs," The Astrophysical Journal **703**, 1696 (2009).

B.-B. Zhang, B. Zhang, E.-W. Liang, X.-Y. Wang, "Curvature Effect of a Non-Power-Law Spectrum and Spectral Evolution of GRB X-Ray Tails," The Astrophysical J. **690**, L10 (2009).

F. Virgili, E. Liang, and B. Zhang, "Low-Luminosity Gamma-Ray Bursts as a Distinct GRB Population: A Firmer Case from Multiple Criteria Constraints," Montly Notices of the Royal Astronomical Society **392**, 91 (2009).

Zygelman, Bernard

B. Zygelman, "An Elastic Approximation for Spin Flipping Collisions of Hydrogen Atoms," in *Proceeding of the Dalgarno Celebratory Simposium*, Edited by J. Babb *et al.*, World Scientific (2009).

B. Zygelman and J. D. Weinstein, "Theoretical and laboratory study of suppresion effects in fine-structure-changing collisions of Ti with He," Phys. Rev. A **78**, 012705 (2008).

Mei-Ju Lu, Kyle S. Hardman, Jonathan D. Weinstein, and Bernard Zygelman, "Fine-structure-changing collisions in atomic titanium, "Phys Rev. A **77**, 060701(R) (2008).

UNIVERSITY OF NEVADA, RENO

DEPARTMENT OF PHYSICS MS 220

Reno, Nevada 89557–0058

Students Accepted For Degree	FIELDS		
	Physics	Astronomy	Related Fields
Doctorate	X		X
Master's	X		X

1. General

President: Milton Glick
Department Chairman: Roberto Mancini
Department Telephone Number: (775) 784-6792
Web Page: www.physics.unr.edu
Type of Institution: University
Control: Public
Setting: Urban
Total Faculty: 1,662
Total Graduate Faculty: 540
Total Students: 16,867
Total Graduate Students: 3,264
Term: Semester
Annual Graduate Tuition:
In-state residents: Full-time—$176.50/credit
Out-of-state residents: Full-time—$176.50/credit*
Tuition rates for: 2009–10
Deferred tuition plan: Yes

*Plus $5,405/semester if registered for 7 or more credits.

2. Number of Faculty in Department

The combined total of full-time academic faculty in the three professorial ranks is 31. The combined total of full-time, part-time, and other research faculty at all ranks is 49.

3. Admission, Financial Aid, and Housing

Address admission inquiries to: Director of Graduate Studies, Department of Physics
Graduate application fee required: $60 Domestic; $95 International
Admission deadline : Fall, April 30; Spring, October 15
Admission information: For admission, 2009–10, 14 students were accepted from 60 applicants.
Admission requirements: For admission to the graduate programs, a Bachelor's degree in physics is required with a minimum undergraduate GPA of 3.00. Students with related majors are considered, but may be required to take some undergraduate courses based on performance on a qualifying exam administered upon admission. The GRE is recommended but not required. Students from non-English speaking countries are required to demonstrate proficiency in English via the TOEFL exam. Minimum acceptable score for admission is 550 (paper version) or 213 (computer version). Teaching assistants must achieve an acceptable score on an English speaking and comprehension test.*
Undergraduate preparation assumed: Boas. *Mathematical Methods in the Physical Sciences*; Marion and Thornton, *Classical Dynamics*; Griffiths, *Introduction to Electrodynamics*; Griffiths, *Introduction to Quantum Mechanics*; Hecht, *Optics*; Kittel, *Thermal physics*.
Address financial aid inquiries to: Chair, Dept. of Physics
GAPSFAS application required: No
Financial aid deadline: None
Loans available: Yes
Address housing inquiries to: Director of Housing
On-campus, single student housing available: Yes
Cost/term: $4,570–5,090/year—double occupancy
On-campus, married student housing available: Yes
Cost/month: $395 + utilities; waiting list—8 months

*Applicants whose records indicate a deficiency in any of the requirements listed above may be admitted on a probationary basis and may be required to take certain undergraduate courses (which do not carry graduate credit).

Table A—Faculty, Enrollments, and Degrees Granted

Research Specialty	Faculty	Enrollment[1] Fall 2009–10		No. of Degrees Granted[2] 2009–10 (2005–10)			Median No. of Years for Ph.D.'s
		Master's	Doctorate	Master's	Terminal Master's	Doctorate	
Atmospheric Sciences	23	11	15	3(12)	3(12)	1(8)	6
Atomic, Molecular, & Chemical Physics	7	3	18	2(13)	2(9)	2(11)	6
Plasma Physics & Fusion	16	6	18	5(2)	3(3)	2(5)	5
Materials Physics	3	2	2	1(2)	1(3)	0(0)	–
Non-specialized	–	1	0	0(0)	–	–	–
Total		23	53	9(32)	8(26)	6(26)	
Full-time Grad. Stud.		23	32				
Part-time Grad. Stud.		0	1				
First-year Grad. Stud.		2	6				
Median Years in Grad. Study (2009–10 Degrees)				2	2	5.5	–

Undergraduate Degrees, 2009–10 (2005–10): 9(23)

[1]Students not yet committed to a research specialty are entered under non-specialized.
[2]Five-year totals in parentheses.

4. Graduate Degree Requirements

Master's: Plan A requires 30 graduate credits including six thesis credits. Plan B requires 32 graduate course credits. B average required. No language requirement. Qualifying exam on undergraduate physics; final oral exam on thesis work under Plan A; under Plan B final written and oral exam on course work. 24 credits under Plan A and 26 credits under Plan B must be earned in residence.
Doctorate: 48 course credits and 24 dissertation credits. B average required. Written and oral comprehensive exam; final oral defense of dissertation. Minimum of six semesters beyond Bachelor's degree in residence including at least two in succession.
Other Programs: A Master's and Ph.D. program in Atmospheric Sciences is offered in association with the Desert Research Institute with courses and research topics in atmospheric physics.
Thesis: Thesis may be written *in absentia*.
Special Equipment, Facilities, or Programs: Accelerator lab with 150 keV accelerator, 14-GHz ECR multicharged ion source, photon-ion endstation at Advanced Light Source; negative ion beam facility; low temperature laser cooling system; 2-TW Z-pinch pulsed-power device for plasma/X-ray physics

and two 100 TW class lasers at the Nevada Terawatt Facility; Well equipped atmospheric physics labs on campus and at the Desert Research Institute.

Table B—Appointments to Graduate Students, 2008–09

Title of Appointee	Appointments Total	Appointments First year	Academic Load Allowed in Credit Hours	Hours of Service Per Week	Stipend for Academic Year ($)[1]
			Semester		
Teaching Assistant	16	7	9	20	15,000
Research Assistant	37	4	9	20	21,000*
Total	53	11			

[1]Plus full tuition fee waiver. *12 months contracts.

5. Personnel Engaged in Separately Budgeted Research, 7/07–6/08

Professorial faculty	49
Graduate students	76
Total	125

6. Separately Budgeted Research Expenditures by Source of Support

	Departmental Research	Physics-related Research Outside Department
Federal government	$5,200,000	n/a
University	50,000	
Total	$5,250,000	

FACULTY

Professors

Cathey, W. N., Ph.D., Tennessee, 1966. Mössbauer measurements and impurity diffusion studies in solids.

Cowan, T., Ph.D., Yale Univ., 1988. Intense laser interactions with matter, plasma physics and fusion energy research.

Mancini, R., Ph.D., Univ. of Buenos Aires, 1983. Department Chair. Thermodynamics and hydrodynamics of hot, dense matter.

Neill, P. A., Ph.D., Queen's Univ., Belfast, 1984. Alignment and orientation studies in electron atom collisions; ionization and charge transfer in ion-atom collisions.

Phaneuf, R. A., Ph.D., Univ. of Windsor, Canada, 1973. Experimental studies on collisions of multiply charged ions with photons, electrons, atoms, and molecules; atomic processes in plasmas.

Thompson, J. S., Ph.D., Univ. of Tennessee, 1989. Interim Dean. College of Science. Photodetachment of negative ions.

Winkler, P., Dr.rer.nat., Erlangen, 1969; Dr.rer.nat.habil., Erlangen, 1977. Theoretical description of atomic and molecular processes; many-body physics; quantum chemistry.

Winterberg, F., Dr.rer.nat., Goettinger, 1955. Theoretical physics. Elementary particle physics, in particular Planck scale physics; controlled nuclear fusion, in particular inertial confinement fusion; nuclear rocket propulsion.

Research Professors

Barber, P., Ph.D., 1973. Light scattering analyses for dielectric particles.

Chow, J., Sc.D., 1985. Air quality measurement and modeling.

Darling, T., Ph.D., Univ. of Melbourn, 1989. Experimental studies of elasticity and stability in solid state systems.

Gertler, A., Ph.D., UCLA, 1979. Atmospheric chemistry.

Hallett, J., Ph.D., Imperial College, 1958. Airborne and laboratory cloud microphysics; solar energy.

Hudson, J. G., Ph.D., Nevada, 1976. Aerosol and cloud physics.

Kantsyrev, V., D.Sc., Inst. of Analytical Instrumentation, Moscow, Russia, 1992. X-ray spectroscopy of laser-produced plasmas and Z-pinch plasmas.

Safranova, U., Dr. Hab., Vilnus Univ., 1975, Ph.D. 1964, High precision relativistic electronic structure calculations, many-body theory.

Watson, J., Ph.D., Oregon Graduate Center, 1979. Source receptor modeling and observations.

Zielinska, B., Ph.D., 1979. Kinetics and mechanics of gas phase reactions of organic compounds and reactive atmospheric species.

Associate Professors

Arnott, P., Ph.D., Washington State Univ., 1987. Acoustic and optical sensing of the Earth's atmosphere.

Bauer, B., Ph.D., UCLA, 1992. Experimental research involving high-power lasers, Z-pinches linear and nonlinear plasma waves and instabilities.

Derevianko, A., Ph.D., Auburn Univ., 1996. Theoretical research ion atomic and molecular physics. Many-body methods; tests of fundamental symmetries; cold atoms.

McCall, K., Ph.D., Univ. of Mass., Amherst, 1992. Theoretical research on properties of hysteretic nonlinear materials and transport in disordered systems; rock physics.

Sentoku, Y., Ph.D, Osaka Univ. 1999. Modeling the physics of high-energy density matter created by short laser pulses.

Associate Research Professors

Chai, S. K., Ph.D., Univ. of Nevada, Reno, 1978. Planetary boundary layer, cloud dynamics; microphysics; fog formation and atmospheric convection.

Borys, R. D., Ph.D., Colorado State Univ., 1983. Polar meteorology; cloud, snow, and aerosol chemistry.

Bowen, J. L., Ph.D., Univ. of Nevada, Reno, 1976. Air quality meteorology, measurements for air quality.

Fujita, E. M., Ph.D., UCLA, 1992. Source contributions to air pollution and visibility impairment; ozone-precursor relationships; trends in air pollution concentrations and emissions.

Ivanov, V., Ph.D., Lebedev Phys. Inst., 1987. Experimental studies of Z-pinch and laser produced plasmas.

Koracin, D., Ph.D., Univ. of Nevada, Reno, 1989. Numerical modeling; marine boundary layer; topographic flows and air quality; aircraft measurements.

Lowenthal, D. H., Ph.D., Univ. of Rhode Island, 1986. Aerosol sources transport and climate effects.

Moosmüller, H., Ph.D., Colorado State Univ., 1988. Experimental and theoretical research in optical spectroscopy and its application to atmospheric and aerosol physics.

Presura, R., Ph.D., Univ. of Bucharest, 1999. Laboratory studies of astrophysical plasmas.

Rambach, G. D., MSE, Princeton Univ., 1976. Advanced energy systems, energy conversion technology, energy and environmental strategic planning, combustion and propulsion.

Redmond, K. T., Ph.D., Univ. of Wisconsin, Madison, 1982. Climate variability; data management; climate and agriculture.

Rogers, C. F., Ph.D., Univ. of Nevada, Reno, 1977. Aerosol physics; aerosols and climate change; aerosol sampling technology; chemical mechanisms of cloud condensation nuclei.

Shlyaptseva, A., Ph.D., Lebedev Physical Institute, Moscow, 1986. Spectroscopy and modeling of hot, dense plasmas.

Sotnikov, V., Ph.D., Space Research Institute, Academy of Science, Russia, 1978. Modeling of nonlinear phenomena in laboratory and space plasmas.

Assistant Professor

Weinstein, J., Ph.D., Harvard, 2002. Experimental research involving ultracold atoms and molecules.

Assistant Research Professors

Covington, A., Ph.D., Univ. of Nevada, Reno, 1997. Experimental studies of photon interactions with atoms, molecules and solids.

Esaulov, A., Ph.D., Lebedev Inst. 2000. Theoretical modeling of plasmas.

Fuelling, S., Ph.D., Univ. of Nevada, Reno 1991. Experimental atomic and plasma physics.

Kingsmill, D. E., Ph.D., UCLA, 1991. Radar meteorology; convective initation and evolution; quantitative precipitation forecasting; mesoscale processes and complex terrain.

Le Galloudec, N. R., Ph.D., Univ. of Paris VI, 1995. Laser scattering from plasmas.

Mitchell, D. L., Ph.D., Univ. of Nevada, Reno, 1995. Cloud microphysics and radiation transfer.

Sagebiel, J., Ph.D., Univ. of California, Davis. Trace organic species in the atmosphere.

Lecturers

Bach, B., Ph.D., College of William and Mary, 1995. Design and fabrication of optical systems and physics education research.

Bennum, D., Ph.D., Univ. of Nevada, Reno, 1973. Observational investigation of superplanets and applications of astronomy research to education.

Rodríguez, M., Ph.D., Univ. of Nevada, Reno, 1998. Physics and astronomy education research.

RESEARCH SPECIALTIES AND STAFF

Theoretical

Atmospheric Science. Theory of cloud droplet growth by condensation and coalescence; studies of convective processes in plumes and clouds: mixing and radiation transfer processes; mesoscale meteorology and modeling. Arnott, Chai, Koracin, Mitchell.

Atomic and Molecular Physics. Electron impact ionization; dielectronic recombination; resonant states. Winkler.

Atomic Physics: Electronic structure and interactions of atoms; QED effects and parity non-conservation. Derevianko.

Condensed Matter Physics. Properties of non-linear hysteretic materials and transport in disordered systems. McCall.

Magnetohydrodynamics and non-linear phenomena in plasmas. Makhin, Sotnikov.

Many-body methods, fundamental symmetries, cold atoms. Derevianko.

Theory, Modeling and Radiative Properties of Hot, Dense Plasmas. Mancini, Shlyaptseva.

Theory of Elementary Particle Physics on the Planck Scale; Inertial Confinement Fusion; Nuclear Propulsion. Winterberg.

Experimental

Atmospheric Science. Gas particle conversion, lab studies of nucleation and growth of particulates; cloud condensation nuclei; aerosol removal by scavenging; trace elements in snow; atmospheric remote sensing. Arnott, Borys, Chai, Gertler, Hallett, Hudson, Lowenthal, Moosmüller, Wetzel.

Atomic and Molecular Physics. Multi-electron processes in low energy collisions of highly charged ions with atoms. Ali. Alignment and orientation studies in electron-atom collisions. Neill. Interactions of multiply charged ions with atoms, photons, and electrons. Phaneuf. Collisional and photodetachment of negative ions. Covington, Thompson.

High Resolution Spectroscopy. Fusion and x-ray lasers; highly excited states; electron correlation and relativistic effects in atoms and ions; photoionization; application of ion accelerators and synchrotron radiation in research. Bruch.

Plasma Physics. Studies of waves and instabilities in pulsedlaser-produced and Z-pinch plasmas; x-ray diagnostics and instrumentation. Bauer, Kantsyrev.

Observational

Atmospheric Science. Aircraft studies of atmospheric electricity; snow crystal and hail growth; convective plumes; mixing and transfer processes in the Earth's boundary layer; synoptic meteorology; source receptor techniques for atmospheric pollution. Borys, Brown, Chow, Fujita, Gertler, Gillies, Hallett, Hudson, Kingsmill, Watson, Zielinska.

DARTMOUTH COLLEGE

DEPARTMENT OF PHYSICS AND ASTRONOMY

Hanover, New Hampshire 03755–3528

Students Accepted For Degree	FIELDS		
	Physics	Astronomy	Related Fields
Doctorate	X	X*	
Master's	X	X*	

*In Physics with Astronomy dissertation/thesis.

1. General

President: Jim Yong Kim
Dean of Graduate School: Brian Pogue
Department Chair: Miles Blencowe
Department Telephone Number: (603) 646-2854 (C)
E-mail: physics.department@dartmouth.edu
Web Address: http://www.dartmouth.edu/~physics/
FAX: (603) 646-1446
Type of Institution: College
Control: Private
Setting: Small town
Total Faculty: 980
Total Students: 5,700
Total Graduate Students: 1,600
Annual Graduate Tuition (12 months):
Arts and Sciences Graduate Students: Full-time—$51,260*
Tuition rates for: 2009–10
Deferred tuition plan: No
Other Fees: $45
Term: Quarter
*Stipends carry full tuition scholarship

2. Number of Faculty in Department

The combined total of full-time faculty in the three professorial ranks is 17. The combined total of full-time, part-time, and other faculty at all ranks is 24.

3. Admission, Financial Aid, and Housing

Address admission inquiries to: Chair of Graduate Admissions, Dept. of Physics and Astronomy, 6127 Wilder Laboratory. Apply on-line at: www.dartmouth.edu/~physics/academics/apply.html. Only electronic applications and letters of recommendation will be accepted.
Graduate application fee required: $45
Application deadline (Fall admission): 1/15
Admission information: For fall admission, 2010–11, 8 students were accepted from 77 applicants.
Admission requirements: For admission to the graduate programs, a Bachelor's degree in physics, astrophysics, or astronomy is normally required. Mathematics majors with some physics training are considered in some cases. The GRE Aptitude is required. Minimum combined quantitative and verbal score of 1200 is required for admission. The GRE Advanced is required. Letters of recommendation, undergraduate transcripts, and GRE scores are considered equally in granting admission. Students from non-English speaking countries are required to demonstrate proficiency in written and spoken English. Applicants may test with either ETS or the International English Language Testing System (IELTS). Minimum acceptable scores: ILELTS Band score of 7.0; TWE score of 4.5; and a TOEFL score of 600 [paper-based], 250 [computer-based], or 100 [internet-based]. Official test scores must be submitted by the testing agency, no photocopies or reproductions.
Address financial aid inquiries to: Office of Graduate Studies, 6062 Wentworth Hall, Room 305
GAPSFAS application required: No
Financial aid deadline: None
Loans available: Yes
Address housing inquiries to: Housing Programs Office, 7 Lebanon St., Suite 303, Hanover, NH 03755 or call 603-646-2446 or visit www.dartmouthre.com to apply for housing.
On-campus, single student housing available: Yes
Cost/Month: $900-1100
On-campus, married student housing available: No

Table A—Faculty, Enrollments, and Degrees Granted

Research Specialty	2008–09 Faculty	Enrollment[1] Fall 2008		No. of Degrees Granted[2] 2008–09 (2004–09)			Median No. of Years for 2008–09 Ph.D.'s
		Master's	Doctorate	Master's	Terminal Master's	Doctorate	
Applied Physics	2	1	6	0(1)	0(0)	1(3)	5
Astronomy	3	0	9	0(2)	0(1)	1(5)	5
Astrophysics	3	0	1	0(1)	0(0)	0(3)	5
Atmos./Space Phys., Cosmic Rays	3	3	1	0(0)	0(3)	0(7)	5
Atomic, Molecular, & Optical Physics	1	0	2	0(0)	0(0)	0(0)	5
Biophysics	1	0	2	0(1)	0(0)	1(2)	5
Condensed Matter Physics	5	0	3	0(0)	0(3)	0(1)	5
Cosmology	1	0	0	0(0)	0(0)	0(0)	0
Electromagnetism	1	0	0	0(0)	0(0)	0(0)	5
Fluids & Rheology	1	0	0	0(0)	0(0)	0(0)	5
History & Philosophy	1	0	0	0(0)	0(0)	0(0)	5
Particles & Fields	1	0	1	0(0)	0(0)	0(1)	5
Plasma Physics & Fusion	6	0	4	0(0)	0(0)	2(2)	5
Radio Physics	1	0	0	0(0)	0(0)	0(0)	5
Relativity & Gravitation	1	0	1	0(0)	0(0)	0(2)	5
Statistical & Thermal	3	0	0	0(0)	0(0)	0(2)	5
Other Theor/Math	1	0	0	0(0)	0(0)	0(0)	
Non-specialized	0	0	1	0(1)	0(3)	0(0)	5
Quantum Information Science	2	0	1	0(0)	0(1)	0(0)	0
Plasma and Space Physics	6	2	12	0(0)	0(2)	2(3)	0
Total		2	50	0(5)	0(11)	9(35)	
Full-time Grad. Stud.		2	50				
Part-time Grad. Stud.		0	0				
First-year Grad. Stud.		1	15				
Median Years in Grad. Study (2008–09 Degrees)		2	5	–	–	–	–

Undergraduate Degrees, 2009–10 (2005–10): 17(81)

[1]Students not yet committed to a research specialty are entered under non-specialized.
[2]Five-year totals in parentheses.

4. Graduate Degree Requirements

Master's: Minimum of three consecutive terms in residence. Degree credit for eight graduate courses. Two of the eight may be Graduate Research. At least six of the eight should be in Physics and Astronomy. A satisfactory thesis must be completed and defended; or significant co-author of a publication submitted to a referred journal or referred conference proceedings, defended publicly; or passing the Ph.D. qualifying exam. Some teaching experience must be acquired. No foreign language requirement.

Doctorate: Minimum of six terms in residence. Degree credit for twelve graduate courses, exclusive of Graduate Research and teaching courses. Degree credit for at least two terms of supervised undergraduate teaching. A written qualifying examination must be passed. A thesis proposal must be presented and defended. A dissertation of substantial significance and publishable quality must be completed and defended before the faculty. No formal language requirement.

Thesis: Thesis may not be written *in absentia*.

Special Equipment, Facilities, or Programs: Observational facilities including 1.3-m and 2.4-m reflecting telescopes at the MDM Observatory, Kitt Peak, Arizona, and the 10-m SALT telescope at Sutherland, South Africa.

Table B—Appointments to Graduate Students, 2009–10

Title of Appointee	Appointments Total	Appointments First year	Academic Load Allowed in Credit Hours	Hours of Service Per Week	Stipend for Academic Year ($)
			Quarter		
Teaching Assistant	20	12	8	18	23,832[1]
Research Assistant	26	4	8	6	23,832[1]
Own Support	0	0	8	6	0[2]
Other	6	0	8	6	23,832[1]
Total	52	16			

[1] All stipends are for 12 months, and carry a full tuition scholarship; additional supplement for student health insurance

[2] Full tuition scholarship awarded.

5. Personnel Engaged in Separately Budgeted Research, 7/09–6/10

Professorial faculty	14
Postdoctoral appointments	4
Graduate students	36
Undergraduate students	33
Nonteaching research personnel	10
Total	97

6. Separately Budgeted Research Expenditures by Source of Support

	Departmental Research	Physics-Related Research Outside Department
Federal government	$3,448,979	$
Other	70,000	
Total	$3,518,979	$

7. Separately Funded and Managed Laboratories

MDM Observatory (Kitt Peak, Arizona)	$93,000
Total	$93,000

Table C—Separately Budgeted Research Expenditures

Research Specialty	No. of Grants	Expenditures ($)
Astronomy	6	165,600
Atmos./Space Phys., Cosmic Rays	21	2,127,000
Condensed Matter Physics	4	586,379
Particles and Fields	1	10,000
Plasma Physics & Fusion	3	350,000
Relativity and Gravitation	2	100,000
Other (Quantum Information Science)	2	60,000
Systems Science/Engineering	3	120,000
Total	43	3,518,979

FACULTY

Professors

Blencowe, Miles P., Ph.D., London, 1989. Condensed matter theory; mesoscopic physics; open quantum systems; quantum chaos.

Caldwell, Robert, Ph.D., Wisconsin, Milwaukee, 1992. Theoretical cosmology; gravitation, particles, fields; the early universe; large scale structure.

Chaboyer, Brian C., Ph.D., Yale, 1993. Theoretical astrophysics; stellar evolution, galaxy formation.

Fesen, Robert A., Ph.D., Michigan, 1981. Observational astrophysics; supernovae and supernova remnants.

Gleiser, Marcelo, Ph.D., King's College, London, 1986. Field theory and cosmology; the early universe; nonequilibrium phenomena; astrobiology.

Hudson, Mary K., Ph.D., UCLA, 1974. Theoretical space plasma physics.

LaBelle, James W., Ph.D., Cornell, 1985. Experimental space plasma physics.

Lawrence, Walter E., Ph.D., Cornell, 1970. Theoretical condensed matter physics; interacting electron and phonon systems, quantum information theory.

Rogers, Barrett N., Ph.D., MIT, 1991. Laboratory and space plasma theory and simulation; magnetic reconnection; plasma turbulence; magnetic fusion.

Thorstensen, John R., Ph.D., California, Berkeley, 1980. Observational astronomy; Interacting binary stars. Astrometry.

Wegner, Gary A., Ph.D., Washington, 1971. Observational astrophysics.

Wybourne, Martin, Ph.D., Nottingham, 1980. Experimental physics; condensed matter.

Associate Professors

Lynch, Kristina A., Ph.D., UNH, 1992. Experimental space plasma physics; auroral and mesospheric sounding rockets; thermal plasma laboratory experiments.

Rimberg, Alexander, Ph.D., Harvard, 1992. Experimental condensed matter physics; nanoscale physics; quantum phenomena and measurements; microwave and radio-frequency techniques; low-noise electrical measurements.

Viola, Lorenza, Ph.D., Padova, 1996. Theoretical quantum information science and quantum statistical mechanics; open quantum systems; quantum entanglement, quantum chaos.

Assistant Professor

Millan, Robyn, Ph.D., Berkeley, 2002. Experimental space physics; radiation belt losses, x-ray instrumentation for space physics and atmospheric sciences.

Ramanathan, Chandraselchar, Sc.D., MIT, 1996. Experimental quantum information processing; magnetic resonance; many-body physics.

Research Faculty

Crane, Philippe, Ph.D., Yale, 1969. Astrophysics; extra-solar planets.

Denton, Richard E., Ph.D., Maryland, 1986. Computational plasma physics, wave phenomena, reconnection, magnetospheric density.

Kress, Brian, Ph.D., Dartmouth, 2002. Space plasma physics, radiation belt dynamics, fluid and magnetofluid dynamics.

Lyon, John G., Ph.D., Maryland, 1972. Space plasma physics and magnetospheric physics; numerical simulation and computational physics.

Montgomery, David C., Ph.D., Princeton, 1959. Theoretical plasma physics; fluid mechanics, turbulence theory.

Müller, Hans, Ph.D., Dartmouth College, 1997. Space plasma physics and heliospheric physics; cool stellar winds; numerical simulations.

Smith, Timothy, Ph.D., UMass-Lowell, 1990. Experimental nuclear physics studying the quark structure of neutrons and protons.

Adjunct Faculty

Brizard, Alain, Ph.D., Princeton University, 1990. Low-frequency nonlinear gyrokinetic theory, relativistic quasilinear transport driven by arbitrary-frequency electromagnetic fluctuations, variational formulations of exact and reduced, kinetic and fluid plasma equations, applications of Lie-transform methods in plasma physics.

Fulton, Theodore, Ph.D., Cornell University, 1966. Condensed matter physics, particularly single-electron devices and Josephson effects.

Levey, Christopher G., Ph.D., Wisconsin, Madison, 1984. Laser spectroscopy.

Lotko, William, Ph.D., UCLA, 1981. Geospace environment; space plasma physics, modeling, simulation; electromagnetic fields and waves.

Naumann, Robert A., Ph.D., Princeton, 1953. Nuclear spectroscopy.

Pogue, Brian W., Ph.D., McMaster University, Canada, 1996. Medical physics, medical imaging systems software and hardware, laser spectroscopy of tissue, biomedical optics.

RESEARCH SPECIALTIES AND STAFF

Theoretical

Astrobiology. Gleiser.

Astrophysics. Stellar structure and evolution; globular cluster ages; formation of the Milky Way and its galaxies. Chaboyer.

Computational Physics. Caldwell, Chaboyer, Denton, Gleiser, Kress, Lyon, Montgomery, Müller, Rogers.

Condensed Matter and Low-Temperature Physics. Quantum many-body systems, electronic and mechanical properties of mesoscopic systems, quantum measurement theory and decoherence. Blencowe, Lawrence, Viola.

Fluid and Magnetofluid Dynamics. Montgomery, Rogers.

Fundamental Quantum Physics and Quantum Information Science. Measurement, control, information processing at the quantum scale. Blencowe, Lawrence, Viola.

General Relativity and Cosmology. Caldwell, Gleiser.

Particle Physics and Cosmology. Caldwell, Chu, Gleiser.

Plasma and Space Physics. MHD and kinetic theory; computer simulations; waves and instabilities; radiation; transport properties; particle acceleration and heating; applications to space and laboratory plasmas. Denton, Hudson, Kress, Lyon, Montgomery, Müller, Rogers.

Statistical and Quantum Statistical Mechanics. Gleiser, Blencowe, Lawrence, Viola.

Experimental

Astrophysics. Cataclysmic binaries; supernova remnants; white dwarfs, close binary stars, and clusters of galaxies; globular clusters. Chaboyer, Fesen, Thorstensen, Wegner.

Condensed Matter Physics. Phonon and electron transport in low-dimensional systems; cooling techniques for mechanical resonators; nanostructures; mesoscopic physics. Rimberg.

Fundamental Quantum Physics and Quantum Information Science: Measurement, control, information processing at the quantum scale. Ramanathan, Rimberg.

Nuclear Physics. Using electron accelerators to measure the distribution of quarks in neutrons and protons. Smith.

Plasma and Space Physics. Plasma kinetics; microwave plasma interactions; plasma diagnostics; ionospheric and magnetospheric physics; plasma measurement in space; remote sensing of ionospheric plasma processes; auroral particle measurements; mesospheric dust particles; ionospheric and mesospheric sounding rockets; stratospheric balloons, radiation belt losses. LaBelle, Lynch, Millan.

UNIVERSITY OF NEW HAMPSHIRE

DEPARTMENT OF PHYSICS

Durham, New Hampshire 03824

Students Accepted For Degree	FIELDS		
	Physics	Astronomy	Related Fields
Doctorate	X		
Master's	X		

1. General

President: Mark Huddleston
Dean of Graduate School: Harry J. Richards
Department Chairman: Eberhard Möbius
Department Telephone Number: (603) 862-1950
Type of Institution: University
Control: Public
Setting: Small town
Total Faculty: 1,059
Total Graduate Faculty: 600
Total Students: 14,204
Total Graduate Students: 2,359
Annual Graduate Tuition: 09/10 Rates
In-state residents: Full-time—$10,380*
Part-time—$577/credit**
Out-of-state residents: Full-time—$24,350*
Part-time—$1002/credit*
Tuition rates for: 2009–10 (2010-11 rates not yet available)
Deferred tuition plan: No
Other Fees: $2,343/yr.*
Term: Semester

2. Number of Faculty in Department

The combined total of full-time faculty in the three professorial ranks is 36. The combined total of full-time, part-time, and other faculty at all ranks is 36.

3. Admission, Financial Aid, and Housing

Address admission inquiries to: Chair Graduate Admissions Committee, Graduate Admission Coordinator, Physics Department, E-mail address:
physics.grad.info@unh.edu
Graduate application fee required: $65
Admission deadline (Fall admission): 2/15
Admission information: For fall admission, 2010, 42 students were accepted from 115 applicants.
Admission requirements: For admission to the graduate programs, a Bachelor's degree in physics is required with no minimum undergraduate GPA specified. The general GRE test is required. Students from non-English speaking countries are required to demonstrate proficiency in English via a TOEFL exam.
Undergraduate preparation assumed: Barger and Olsson, *Classical Mechanics–A Modern Perspective*; Liboff, *Introductory Quantum Mechanics*; Griffiths, Reitz, and Milford, *Foundations of Electromagnetic Theory.*
Address financial aid inquiries to: Financial Aid Office, Stoke Hall
GAPSFAS application required: No
Financial aid deadline: 2/15
Loans available: Yes
Address housing inquiries to: Residence Office, Pettee House
On-campus, single student housing available: Yes
Cost/term: $6,198/academic year* (09/10 Rate)
On-campus, married student housing available: Yes
Cost/month: $816/1 bdrm.*; $1001/2 bdrm.*
Studio $721

*Includes heat and electricity

Table A—Faculty, Enrollments, and Degrees Granted

Research Specialty	2007–08 Faculty	Enrollment[1] Fall 2009		No. of Degrees Granted[2] 2007–08 (2002–07)			Median No. of Years for 1999–04 Ph.D.'s
		Master's	Doctorate	Master's	Terminal Master's	Doctorate	
Astrophysics/Spac Phys., Cosmic Rays[3]	26	3	37	0(1)	0(7)	2(8)	6
Condensed Matter Phys./Mat. Science	4	0	2	0(0)	0(0)	0(3)	6
Medical Imaging	1	0	1	0(0)	0(0)	0(0)	0(2)
Nuclear Physics	7	0	1	0(1)	0(0)	1(12)	6
Particles & Fields	3	0	2	0(0)	0(1)	0(0)	–
Physics Education	1	0	2	0(0)	0(1)	0(0)	–
Plasma Physics & Fusion	4	0	0	0(0)	0(0)	0(0)	–
Non-speciialized		0	7	0(0)	0(4)	0(0)	–
Total		3	51	0(3)	0(13)	3(23)	
Full-time Grad. Stud.		0	51		3		
Part-time Grad. Stud.		0	0				
Median Years in Grad. Study (2000–01 Degrees)				3	2	6	–
Undergraduate Degrees, 2001–02 (1996–01): 10							

[1] Students not yet committed to a research specialty are entered under non-specialized.
[2] Five-year totals in parentheses.
[3] Combined Astrophysics and Atomic Physics with Space Physics.

4. Graduate Degree Requirements

Master's: Students are required to complete satisfactorily an approved course program of five required courses plus (a) nine additional credits of graduate coursework plus a Master's Thesis and an Oral Thesis Defense **or** (b) twelve additional credits of graduate coursework plus a research project and an oral exam in the form of a seminar. There are no residency or foreign language requirements.
Doctorate: Students are required to complete satisfactorily an approved course program of eight required courses plus four additional courses at the graduate level, pass a written Comprehensive Exam on their undergraduate physics topics by the middle of their second year in the program, pass an oral Qualifying Exam in which a Thesis proposal is presented and discussed, demonstrate proficiency in teaching, complete a Dissertation and pass an oral Dissertation Exam. Students must satisfy a one year residence requirement. Students can earn a Master's Degree after completing thirty graduate credits and passing the written Comprehensive Exam and the Oral Qualifying Exam. There is no foreign language requirement.
Thesis: Thesis may be written *in absentia.*
Special Equipment, Facilities, or Programs: Interdisciplinary Research: The department encourages research in areas related to physics or applied physics. Should a student desire to do research in a field related to physics, special provisions-

may be made. Contact the department chairperson or graduate advisor for details.

Table B—Appointments to Graduate Students, 2009–10

Title of Appointee	Appointments Total	Appointments First year	Academic Load Allowed in Credit Hours	Hours of Service Per Week	Stipend for Academic Year ($)
			Semester		
Teaching Assistant[1]	17	9	6–9	20	14,100
Research Assistant[2]	32	2	6–9	20	16,000
Total	49	8			

[1] The current stipend for a teaching assistant is $13,200 with a full waiver of tuition.
[2] Most students also find support for the summer as research assistants. The stipend is normally 4/9–6/9 of the academic year stipend.

5. Personnel Engaged in Separately Budgeted Research, 7/09–6/10

Professorial faculty	36
Other faculty	0
Graduate students	38
Undergraduate students	22
Nonteaching research personnel	40
Total	118

6. Separately Budgeted Research Expenditures by Source of Support 06/07

	Departmental Research	Physics-related Research Outside Department
Federal government	$14,109,323	$
Non government	210,000	
Total	$7,766,132	$

Table C—Separately Budgeted Research Expenditures

Research Specialty	No. of Grants	Expenditures ($)
Astrophysics/Space Phys., Cosmic Rays	121	12,931,388
Condensed Matter Physics	3	205,867
Medical Imaging	9	725,623
Subatomic Physics	6	456,445
Total	139	14,319,323

FACULTY

Professors

Balling, Ludwig C., Ph.D., Harvard, 1965. Atomic and molecular physics.
Bhattacharjee, Amitava, Ph.D., Princeton Univ., 1981. Peter Paul Chair. Plasma physics.
Calarco, John R., Ph.D., Illinois, 1969. Experimental nuclear physics.
Echt, Olof, Ph.D., Univ. Konstanz (FRG), 1979. Condensed matter physics and materials science.
Forbes, Terry, Ph.D., Colorado, 1978. Astrophysics/solar-terrestrial physics.
Harper, James M. E., Ph.D., Stanford University, 1975. Condensed matter physics and materials science.
Hersman, F. William, Ph.D., MIT, 1982. Experimental nuclear physics and medical imaging with polarized gases.
Isenberg, Philip A., Ph.D., Univ. of Chicago, 1977. Theoretical interplanetary physics, fusion, and plasmas; solar atmosphere.
Kistler, Lynn M., Ph.D., Univ. of Maryland, 1987. Plasma and magnetospheric physics.
Lee, Martin A., Ph.D., Univ. of Chicago, 1971. Theoretical space plasma physics; solar-terrestrial physics.
Maynard, Nelson, Ph.D., Univ. of New Hampshire, 1966. Magnetospheric physics.
McKibben, R. Bruce, Ph.D., Univ. of Chicago, 1972. Space physics.
Möbius, Eberhard, Ph.D., Ruhr Univ., Bochum, Germany, 1977. Plasma and solar terrestrial physics.
Raeder, Joachim, Ph.D., Universität zu Köln, Germany, 1989. Space physics, numerical simulations.
Ryan, James M., Ph.D., California, Riverside, 1978. Astro-physics; solar-terrestrial/atmospheric physics; radiation detectors.
Smith, Charles, Ph.D., College of William and Mary, 1981. Space plasma physics.
Torbert, Roy B., Ph.D., California, Berkeley, 1979. Magnetospheric/ionospheric physics.

Associate Professors

Beane, Silas R. III, Ph.D., Univ. of Texas, Austin, 1994. Theoretical Nuclear and Particle Physics.
Berglund, Per, Ph.D., University of Texas, Austin, 1993. String theory.
Chandran, Benjamin D. G., Ph.D., Princeton University, 1997. Astrophysical plasma physics.
Connell, James, Ph.D., Washington University, 1988. Space physics.
Farrugia, Charles, Ph.D., University of Bern, Switzerland, 1984. Space physics.
Galvin, Antoinette, Ph.D., Univ. of Maryland, 1982. Space physics.
Holtrop, Maurik, Ph.D., Massachusetts Institute of Technology, 1995. Experimental Nuclear & Particle Physics.
Kucharek, Harald, Ph.D., Technical Univ. of Munich, 1989. Magnetospheric and heliospheric physics.
Lessard, Mark, Ph.D., Dartmouth College, 1997. Space and magnetospheric physics.
Lopate, Clifford, Ph.D., Univ. of Chicago, 1989. Space physics.
McConnell, Mark L., Ph.D., Univ. of New Hampshire, 1987. Experimental high energy astrophysics.
Meredith, Dawn C., Ph.D., Cal. Tech., 1987. Physics education.
Pohl, Karsten, Ph.D., University of Pennsylvania, 1997. Experimental Condensed Matter Physics and Materials Science.
Vasquez, Bernard, Ph.D., Univ. of Maryland, 1992. Solar-terrestrial physics.

Assistant Professors

Bloser, Peter F., Ph.D. in Astronomy, Harvard Univ., 2000. Experimental High Energy Astrophysics.
Bravar, Ulisse, Ph.D., New Mexico State Univ., 2001. Experimental High Energy Astrophysics
Chen, Li-Jen, Ph.D., Univ. of Washington, 2002. Computational Space Plasma Physics.
Ebraimi, Fatima, Ph.D., Univ. of Wisconsin, Madison. Computational Space Plasma Physics.
Germaschewski, Kai, Ph.D., Heinerich-Hein-Universitat,

Germany, 2001. Space Physics, Plasma physics, numerical simulations.

Slifer, Karl, Ph.D., Temple University, 2004. Experimental Nuclear Physics.

Tang, Jian-Ming, Ph.D., Univ. of Washington, 2001. Theoretical condensed matter physics.

Affiliate

Patz, Samuel, Ph.D., Brandeis Univ., 1979. Medical Imaging with Polarized Gasses.

Ruset, Julian, Ph.D., UNH 2005. Medical Imaging with Polarized Gasses.

Solvignon, Patricia H., Ph.D., Temple Univ. 2006. Experimental Nuclear Physics.

RESEARCH SPECIALTIES AND STAFF

Theoretical

Astrophysics. Solar, space and magnetospheric physics. Chandran, Forbes, Isenberg, Lee, Raeder, Vásquez.

Condensed matter. Tang.

Laboratory and space plasma physics. Solar, space and magnetospheric physics. Bhattacharjee, Chandran, Raeder.

Nuclear and particle physics. Field Theory. Beane, Berglund.

Theoretical mathematical physics. Berglund, Beane, Raeder.

Experimental

Atmospheric/space plasma physics. Magnetospheric, ionospheric and aurora physics. Farrugia, Galvin, Kistler, Lessard, Maynard, Möbius, Smith, Torbert.

Condensed matter physics. Surface physics; clusters and thin films. Echt, Harper, Pohl.

Cosmic-rays and space radiation physics. Solar neutron physics. Connell, Lopate, McKibben.

High energy astrophysics. Gamma-ray and neutron solar physics. McConnell, Ryan.

Medical imaging. Hersman, Patz, Ruset.

Nuclear and particle physics. Calarco, Hersman, Holtrop, Slifer.

Physics Education. Meredith.

NEW JERSEY INSTITUTE OF TECHNOLOGY

Newark, New Jersey 07102

Dual Diploma NJIT-Rutgers (Newark) Joint Applied Physics Program

Students Accepted For Degree	FIELDS		
	Physics	Astronomy	Related Fields
Doctorate	X		
Master's	X		

1. General

President: Robert A. Altenkirch
Dean of Graduate School: Ronald Kane
Chair of Federated Physics Department: N. M. Ravindra (Ravi)
Direct Number: (973) 596-5742
E-mail: ravindra@mjit.edu; nmravindra@gmail.com
Department Telephone Number: (973) 596–3562 (C)
Type of Institution: University
Control: Public
Setting: Urban
Total Faculty: 416
Total Graduate Faculty: 170
Total Students: 8,840
Total Graduate Students: 2,916
Annual Graduate Tuition: Subject to change
In-state residents: Full-time—$14,200
Part-time—$772/credit
Out-of-state residents: Full-time—$20,168
Part-time—$1,064/credit
Tuition rates for: 2009–10
Deferred tuition plan: Yes
Other Fees: $1005
Term: Semester

2. Number of Faculty in Department

Total number of graduate faculty of the program is 22. 18 from NJIT and 4 from Rutgers (Newark). 13 other graduate faculty are associated with the NJIT-Rutgers (Newark) Joint Applied Physics Graduate Programs.

3. Admission, Financial Aid, and Housing

Address admission inquiries to: Admissions, New Jersey Institute of Technology, Newark, New Jersey 07102-1982
Graduate application fee required: $60
Admission deadline (Fall admission): None
Admission information: For fall admission, 2009–10, 5 students were accepted from 25 applicants.
Admission requirements: For admission to the graduate programs, a Bachelor's degree in Physics with a minimum undergraduate GPA of 3.0. The GRE is required. The minimum acceptable score suggested for admission is verbal–550; quantitative–700; total–1,250. The GRE Advanced is not required. The average GRE scores for admissions were total–1,250. Students from non-English speaking countries are required to demonstrate proficiency in English via the TOEFL exam. Minimum acceptable score for admission is 550 (paper-based) or 213 (computer based).
Undergraduate preparation assumed: Symon, *Mechanics*; Scott, *Electromagnetism*; Mandl, *Quantum Mechanics*; Reif, *Thermodynamics*; Kittel, *Solid State Physics*.
Address financial aid inquiries to: Physics Department, New Jersey Institute of Technology, Newark, New Jersey 07102-1982
GAPSFAS application required: No
Financial aid deadline: 1/15
Loans available: Yes
Address housing inquiries to: Admissions Office, New Jersey Institute of Technology, Newark, New Jersey 07102-1982
On-campus, single student housing available: Yes
On-campus, married student housing available: No

Table A—Faculty, Enrollments, and Degrees Granted

Research Specialty	2008–09 Faculty	Enrollment[1] Fall 2009		No. of Degrees Granted[2] 2009–10 (2002–07)			Median No. of Years for 2008–09 Ph.D.'s
		Master's	Doctorate	Master's	Terminal Master's	Doctorate	
Astrophysics	5	1	6	0(0)	0(0)	2(7)	–
Atomic, Molecular	1	0	0	0(0)	0(0)	0(0)	–
Biophysics	2	0	3	0(0)	0(0)	0(0)	–
Computational	1	0	0	0(0)	0(0)	0(0)	–
Condensed Matter	10	6	13	0(0)	0(0)	3(8)	–
Optical Phenomena	3	0	6	0(0)	0(0)	1(8)	–
Total		7	28	0(0)	0(0)	6(31)	
Full-time Grad. Stud.			28				
Part-time Grad. Stud.			1				
First-year Grad. Stud.			5				
Median Years in Grad. Study (2000–01 Degrees)				5.5	–	–	–
Undergraduate Degrees, 2006–07 (2002–07): 3(3)							

[1]Students not yet committed to a research specialty are entered under non-specialized.
[2]Five-year totals in parentheses.

4. Graduate Degree Requirements

Master's: The interdisciplinary NJIT-Rutgers (Newark) Joint M.S. Degree in Applied Physics requires 30 credits. 24 credits are course work, of which 18 credits are physics courses (including mathematical physics or applied mathematics), and 6 credits are electives. Four graduate physics courses: Classical Mechanics, Classical Electrodynamics I, Quantum Mechanics I, and Statistical Mechanics, are mandatory. 6 credits are thesis research. With the approval of the academic advisor, the student can choose a 3 credit project plus an additional 3 credit course to replace the 6 credit thesis.
Doctorate: For entering students with B.S. or B.A. degrees, the interdisciplinary NJIT-Rutgers (Newark) Joint Ph.D. Degree in Applied Physics requires 75 (above 600 level) credits. 39 credits are course work, and 36 credits are dissertation research. Among the course work 24 credits are physics courses (including mathematical physics or applied mathematics), and 15 credits are electives. Among the 24 credits of physics courses, Classical Mechanics, 621 Classical Electrodynamics I and II, 631 Quantum Mechanics I and II, and Statistical Mechanics, are mandatory. For entering students with M.S. or M.A. degrees, the Joint NJIT-Rutgers (Newark) Ph.D. Degree in Applied Physics requires 54 (above 600 level) credits. 18 credits are course work, and 36 credits are dissertation research. Among the course work 9 credits are physics courses (including mathematical physics or applied mathematics), and 9 credits are electives.
Special Equipment, Facilities, or Programs: Center for Solar-Terrestrial Research, Big Bear Solar Observatory (BBSO), Microelectronics Research Center with class 10 clean room

silicon IC process facility, chemical vapor deposition (CVD) and physical vapor deposition (PVD), and device research labs. Ultrafast optical and optoelectronic phenomena lab and program. Electronic Imaging Center. Owens Valley Solar Array (OVSA). THz Spectrocopy lab, Laser spectroscopy lab. Surface science lab, and Bio-sensor lab associated with Howard Hughes Neuroscience Center. Interdisciplinary applied physics research is carried out in collaboration with Electrical Engineering. Chemistry, and Biological Sciences faculty as well as with the University of Medicine and Dentistry of New Jersey (UMDNJ). Extensive cooperative research with National Solar Observatory, Brookhaven National Laboratory, Bell Labs, National Renewable Energy Laboratory, New Jersey Nano Consortium, US Army Research Lab, and other industrial and federal research laboratories.

Table B—Appointments to Graduate Students, 2008–09

Title of Appointee	Appointments		Academic Load Allowed in Credit Hours	Hours of Service Per Week	Stipend for Academic Year ($)
	Total	First year			
			Semester		
Teaching Assistant	17	4	12	15–20	19,500
Research Assistant	14	1	12	–	18,000–22,000
Total	31	5			

5. Personnel Engaged in Separately Budgeted Research, 2009–10

Professorial faculty	16
Associate professorial faculty	10
Postdoctoral appointments	8
Instructional Staff	3
Graduate students	30
Undergraduate students	5
Total	72

6. Separately Budgeted Research Expenditures by Source of Support

	Departmental Research	Physics-related Research Outside Department
Federal government	$4,116,725	300,000
Private, non-profit organizations	–	–
Business and industry	500,000	$
Total	$4,616,725	$300,000

Table C—Separately Budgeted Research Expenditures

Research Specialty	No. of Grants	Expenditures ($)
Applied Physics	3	800,000
Astrophysics	7	-
Biophysics	1	60,000
Condensed Matter Physics	4	450,000
Optics/Photonics	7	900,000
Total	22	6,383,811

FACULTY

GRADUATE FACULTY OF THE FEDERATED PHYSICS DEPARTMENT

Distinguished Professors/Professors Rank II:

Federici, John, Ph.D., Princeton Univ., 1989. THz and laser spectroscopy.
Gary, Dale E., Ph.D., Univ. of Colorado, 1982. Radio solar physics.
Goode, Philip R., Ph.D., Rutgers Univ., 1968. Astrophysics.
Levy, Roland, Ph.D., Columbia Univ, 1973. CVD, PVD, materials synthesis.
Murnick, Daniel E., Ph.D., MIT, 1966. Laser spectroscopy and applied physics.
Wang, Haimin, Ph.D., Caltech, 1988. Solar physics.

Professors

Chin, Ken K., Ph.D., Stanford Univ., 1986. Applied physics.
Ravindra, N.M., Chair, NJIT, Ph.D., Roorkee, India, 1982. Microelectronics and solid state physics.
Thomas, Gordon, Ph.D., University of Rochester. Optics, biophysics.
Tyson, Trevor, Ph.D., Stanford Univ., 1993. Condensed matter physics.
Wu, Zhen, Ph.D., Columbia Univ., 1984. Atomic and molecular physics; laser spectroscopy and surface science.

Associate Professors

Gerrard, Andrew, Ph.D., Pennsylvania State Univ., 2002. Upper atmospheric research.
Russo, O. L., Dr. Engineering Science, NJIT, 1975. Solid state physics.
Schaden, Martin, Ph.D., University of Vienna, Austria, 1982. Quantum field theory.
Sirenko, Andrei, Ph.D., A. E. Ioffe Physical Technical Institute, Russia, 1993. Optics, materials and device physics.

Assistant Professors

Ahn, Keun, Ph.D., Johns-Hopkins Univ., 2000. Condensed matter physics.
Cao, Wenda, Ph.D., Goettingen (Göttingen), Germany, 2001. Solar Physics.
Prodan, Camelia, Ph.D., University of Houston, 2005, Biophysics.
Zhou, Tao, Ph.D., Max Plank Institute for Solid State Research, 2004. Optical spectroscopy.

Professor Emeritus

Buteau, Leon, Ph.D., University of Florida, Materials Science.

Distinguished Research Professors

Lanzerotti, Louis J., Ph.D., Harvard Univ., 1965. Geophysics and space plasma physics.

Visiting Research Professor

Deng, Na, Ph.D., NJIT, 2007. Space Weather, Solar-Terrestrial Research
Sufian, Abedrabbo, M., Ph.D., NJIT, 1998, Solid state physics, Semiconductors.

Associate Research Professor

Fleishman, Gregory, Ph.D., Ioffe Institute for Physics and Technology, St. Petersburg, Russia, 1998. Nonthermal Electromagnetic Emission in Structural Astrophysical Medium.
Lee, Jeongwoo, Ph.D., California Institute of Technology, 1994. Solar microwave radiation.

Assistant Research Professor

Nita, Gelu, Ph.D., NJIT, 2004. Solar Microwave Radiation.

Research Professor

Abramenko, Valentyna, Ph.D., Ioffe Institute for Physics and Technology, St. Petersburg, Russia, 1990. Solar Magnetic Fields.
Jing, Ju, Ph.D., NJIT, 2005. Solar Magnetic Fields; Solar Activity.
Liu, Chang, Ph.D., NJIT, 2007. Solar Flares and Coronal Mass Ejections; Solar magnetic Fields.
Rimmele, Thomas, Ph.D., University of Freiburg, Germany, 1993.
Varsik, John, Ph.D., University of Hawaii, 1987. Solor Polar Magnetic Fields.
Yurchyshyn, Vasyl, Ph.D., Astronomical Observatory, Kiev, Ukraine, 1998. Solar Flares; Solar Coronal Mass Ejecta.

OTHER GRADUATE FACULTY IN APPLIED PHYSICS

Graduate Faculty (in Other Departments of NJIT and Rutgers) of the Joint Applied Physics Program

Barat, Robert B., MIT, 1990. Applied optics.
Bonder, Edward M., Ph.D., University of Pennsylvania, 1983. Electron microscopy facility.
Grebel, Heim, Ph.D., The Weismann Institute of Science, Israel, 1985. Optoelectronics.
Krasnoperov, Lev N., D.Sc., Institute of Chemical Physics, Moscow, Russia, 1991. Laser photochemistry, laser spectroscopy.
Mendelsohn, Richard, Ph.D., MIT, 1972. Biophysical chemistry.

Associate Graduate Faculty of the Joint Applied Physics Program

Farrow, Reginald, Ph.D., CUNY. Solid state Semiconductors.
Fiory, Anthony, Ph.D., Rugers University, 1968. Rapid thermal annealing and processing. Solid state physics, semiconductors.
Piatek, Slawomir, Ph.D., Rutgers University, 1994. Solar physics.
Shneidman, Vitaly, Ph.D., Academy of Sciences, Ukraine. Computational physics.

RESEARCH SPECIALTIES AND STAFF

Theoretical

Condensed matter physics. Ahn.

Experimental

Atomic, Molecular, and Applied Laser Physics. Federici, Murnick, Sirenko, Wu, Zhou. 2 postdoctoral fellows.
Biophysics-Diagnostics and Sensors. Federici, Murnick, Prodan, Thomas. 4 postdoctoral fellows.
Microelectronics and Micromechanical Devices. Chin, Levy, Ravindra. 3 postdoctoral fellows.
Solar-Terrestrial Physics. CAD, Gary, Gerrard, Goode, Wang. 8 postdoctoral fellows.
Solid State Physics, Materials Science. Chin, Federici, Levy, Ravindra, Russo, Tyson, Wu. 6 postdoctoral fellows.

PRINCETON UNIVERSITY

DEPARTMENT OF PHYSICS

Princeton, New Jersey 08544

Students Accepted For Degree	FIELDS: Physics	Astronomy	Related Fields
Doctorate	X		
Master's			

1. General

President: Shirley Tilghman
Dean of Graduate School: William B. Russel
Department Chairman: Curtis G. Callan
Department Telephone Number: (609) 258-0757 (C)
Type of Institution: University
Control: Private
Setting: Small town
Total Faculty: 1,172
Total Graduate Faculty: Not separated
Total Students: 7,567
Total Graduate Students: 2,520
Annual Graduate Tuition:
Full-time: $38,090
Tuition rates for: 2010–11—$38,090
Deferred tuition plan: No
Other Fees: None
Term: Semester

2. Number of Faculty in Department

The combined total of full-time faculty in the three professorial ranks is 42. The combined total of full-time, part-time, and other faculty at all ranks is 47.

3. Admission, Financial Aid, and Housing

Address admission inquiries to: Graduate Admissions Office, Clio Hall, Princeton University, Princeton, NJ 08544
Graduate application fee required: $65–$105
Admission deadline (Fall admission): 12/15/2010
Admission information: For fall admission, 2010, 53 students were accepted from 489 applicants.
Admission requirements: For admission to the graduate programs, a Bachelor's degree in physics is required with no minimum undergraduate GPA specified. The GRE is required. No minimum acceptable score specified. The GRE Advanced is required. No minimum acceptable score specified. Students from non-English speaking countries are required to demonstrate proficiency in English via the TOEFL exam. No minimum score specified.
Address financial aid inquiries to: Asst. Dean of Financial Affairs, Graduate School, 204 Nassau Hall, Princeton, NJ 08544
GAPSFAS application required: Yes
Financial aid deadline: 1/2
Loans available: N/A
Address housing inquiries to: Graduate Housing Department, MacMillan Bldg., Princeton Univ., Princeton, NJ 08544
On-campus, single student housing available: Yes
Cost per year: $4,092–7,135
On-campus, married student housing available: Yes
Cost/month: $784–1,368/month

Table A—Faculty, Enrollments, and Degrees Granted

Research Specialty	2009–10 Faculty	Enrollment[1] Fall 2009: Master's	Enrollment[1] Fall 2009: Doctorate	No. of Degrees Granted[2] 2008–09: Master's	Terminal Master's	Doctorate	Median No. of Years for 2008–09 Ph.D.'s
Astronomy	2	–	5	–	–	3	5
Astrophysics	5		8	–	–	1	5
Atomic, Molecular, & Optical Physics	2	–	7	–	–	2	5
Biophysics	5	–	14	–	–	5	5
Condensed Matter Physics	10	–	16	–	–	5	5
Mathematical Physics	3	–	0	–	–	–	5
Nuclear & Particle Astrophysics	2	–	2	–	–	1	5
Particles & Fields	15	–	27	–	–	5	5
Other Theoretical/ Math.	3	–	0	–	–	–	5
Non-specialized		–	13	–	–	–	5
Total	47	–	107	–	–	22	
Full-time Grad. Stud.		–	110				
Part-time Grad. Stud.		–	0				
First-year Grad. Stud.		–	13				
Median Years in Grad. Study (2007–2010 Degrees)			5	–	–		

Undergraduate Degrees, 2008–09 (2004–09): 17

[1]Students not yet committed to a research specialty are entered under non-specialized.
[2]Five-year totals in parentheses.

4. Graduate Degree Requirements

Master's: The Master's Degree is conferred only upon passing a General Examination. (Students who want to work toward a Master's degree only will not be admitted.)

Doctorate: The formal course requirements include three core courses to be taken between the beginning of the first year of study and the end of the second year. (This requirement is part of the general examination.) Students taking these courses must achieve a grade of B or higher.

In addition, one year residency is required; general examination, and the dissertation are required.

Thesis: Thesis may be written *in absentia*.

Special Equipment, Facilities, or Programs: Theoretical research spans most of the central topics of modern physics. The department has decades-old traditions of excellence and leadership in these core areas of fundamental physics and it is also rapidly building strength in newer areas, such as theoretical biology. In the newer areas, interaction between Physics and other departments is critical and major university-supported interdisciplinary initiatives provide a strong framework for this cooperation. There is also productive interaction between theorists in the department and those at the nearby Institute for Advanced Studies, even though there is no formal connection between these institutions.

The high energy theory group works on quantum field theory, particle phenomenology and cosmology, string theory and quantum gravity models in various dimensions, and dualities between gauge theories and strings. Some members of the group are also interested in applications of quantum field theory to problems in statistical mechanics and the theory of turbulence.

The cosmology theory group uses astrophysical, particle physics and superstring theory combined with observations to study gravitation and the origin, composition and evolution of our universe.

The theoretical condensed matter group works on quantum many-body theory of systems involving strong correlations and/or disorder, statistical mechanics, biological systems and systems far from equilibrium.

The mathematical physics group is concerned with problems in statistical mechanics, atomic and molecular physics, quantum field theory, and, in general, with the mathematical foundations of theoretical physics.

The theoretical biophysicists work on problems in statistical mechanics and information theory that arise in studying nervous systems, gene expression networks, the organization of genomes and the mechanisms of evolution.

Experiments in High-Energy Particle Physics are directed towards understanding the fundamental interactions and particle structures at extremely small distances. The apparatus is designed and constructed in the physics shops in Jadwin Hall or at the Elementary Particles Laboratory a block away, which contains special facilities for the fabrication of detectors. The experiments are performed at large national and international laboratories, which currently include CERN, Switzerland; Fermilab, Illinois; KEK, Japan and SLAC, California. The data are then analyzed at Princeton.

The Nuclear and Particle Astrophysics group is active in experimental studies of solar neutrinos and dark matter. The goal of the solar neutrino program is to explore neutrino oscillations and solar processes through a measurement of the low energy 7Be neutrino. Neutrinos will be detected with the Borexino liquid scintillation detector located in the Gran Sasso underground laboratory in Italy.

The Dark Matter group is designed to detect WIMPs in our galaxy by their collisions with either xenon or argon nuclei in a scintillation-ionization detector made of the rare gas atoms. Experiments are under development to provide a definitive search for rare WIMP collisions by combining the unique scintillation properties of the rare gas atoms with the low background methods developed for the Borexino solar neutrino experiment.

Research in the Condensed-Matter Physics group seeks to understand electronic behavior in novel low-dimensional solids in which interaction and correlation effects are dominant. Problems investigated have included the Fractional and Integer Quantum Hall effects, high-temperature superconductivity, Kondo effect in quantum dots, spin-density-wave states in organic conductors, highly-frustrated quantum-spin systems, and novel excitations in low-dimensional magnetic systems. The research involves close collaborations between experimentalists and theorists, as well as with faculty in the Chemistry and Electrical Engineering Departments. Experimental groups are also engaged in researching novel patterning techniques using diblock copolymers (with faculty in Chemical Engineering), and techniques for single-molecule detection and separation of biological molecules (with Molecular Biology and the Genomics Center).

In the experimental cosmology group, students often design and build specialized instrumentation to make unique and precise measurements, or analyze cosmological data. In recent years, experimental work has emphasized measurements of the anisotropy and polarization of the cosmic microwave background. Among other projects, Princeton is actively involved in all aspects of the WMAP satellite, is the lead institution for the ACT project, and is a collaborator on the QUIET experiment.

Research in Atomic physics is primarily focused on spin-polarized gases, liquids, and solids, on their properties, interactions, and a wide range of applications. Among applications currently being developed are searches for violation of CP symmetry beyond the Standard Model, tests of Lorentz invariance, development of miniature atomic clocks, ultra-sensitive atomic magnetometers, and new biomedical techniques, such as lung imaging and mapping of the magnetic fields generated by the brain.

Biological physics spans a huge range of subjects, from neurobiology to genomics to fundamentals of protein action. Princeton has strengths in nearly all areas of modern biological physics. Many faculty with a strong physics background who are involved in biological physics are not solely in the Physics Department but have joint appointments with other departments or are completely in other Departments. There is a strong community spirit to biological physics amongst these departments in spite of the vast range of subjects being studied.

Table B—Appointments to Graduate Students, 2009–10

Title of Appointee	Appointments: Total	Appointments: First year	Academic Load Allowed in Credit Hours	Hours of Service Per Week	Stipend for Academic Year ($)
			Semester		
Teaching Assistant	31	0	–	20	24,050–25,750
Research Assistant	36	0	–	20	22,250–23,250
NSF	10	1			
NSERC	3	0			
University Fellowship	27	12			
NDSEG Fellowship	2	0			
Foreign Fellowship	1	0			
Total	110	13			

5. Personnel Engaged in Separately Budgeted Research, 7/08–6/09

Professorial faculty	22
Other faculty	13
Postdoctoral appointments	15
Nonteaching research personnel	39
Total	89

6. Separately Budgeted Research Expenditures by Source of Support

	Departmental Research	Physics-related Research Outside Department
Federal government	11,717,665	$2,530,621
Private, non-profit organizations	453,119	272,526
Business and industry	–	16,677
Total	$12,170,783	$2,819,823

Table C—Separately Budgeted Research Expenditures

Research Specialty	No. of Grants	Expenditures ($)
Astrophysics	15	2,365,048
Atomic & Molecular Phys.	7	1,085,110
Biophysics	4	639,899
Condensed Matter	20	3,630,287
Elem. Particles & Fields	24	4,943,909
Nuclear & Particle Astrophysics	9	2,112,338
Other Theoretical/Math.	4	214,015
Total	83	14,990,606

FACULTY

Professors

Aizenman, Michael, Ph.D., Belfer Graduate School of Science, Yeshiva Univ., 1975. Mathematical physics.
Austin, Robert, Ph.D., Illinois, 1975. Biophysics.
Bialek, William, Ph.D., U.C. Berkeley, 1983. Biophysics.
Calaprice, Frank, Ph.D., California, Berkeley, 1967. Nuclear physics.
Callan, Curtis G., Ph.D., Princeton, 1964. Theoretical physics.
Groth, Edward, Ph.D., Princeton, 1971. Cosmology; gravitation and relativity.
Gubser, Steven, Ph.D., Princeton, 1998. Particle theory.
Haldane, F. Duncan M., Ph.D., Cambridge, 1978. Condensed matter.
Happer, William, Ph.D., Princeton, 1964. Atomic physics.
Huse, David A., Ph.D., Cornell, 1983. Condensed matter physics.
Klebanov, Igor, Ph.D., Princeton, 1986. Theoretical physics.
Lieb, Elliott, Ph.D., MIT, 1956. Mathematical physics.
Marlow, Daniel R., Ph.D., Carnegie-Mellon, 1981. High-energy physics.
McDonald, Kirk, Ph.D., Cal. Tech., 1972. High-energy physics.
Meyers, Peter, Ph.D., California, Berkeley, 1983. High-energy physics.
Nappi, Chiara, Ph.D., Naples, 1976. Theoretical physics.
Ong, Nai-Phuan, Ph.D., California, 1976. Condensed matter physics.
Page, Lyman, Ph.D., MIT, 1989. Cosmology; gravitation; relativity.
Polyakov, Alexandre, Ph.D., Landau Inst., USSR, 1969. Theoretical physics.
Smith, A. J. Stewart, Ph.D., Princeton, 1966. High-energy physics.
Sondhi, Shivaji Lal, Ph.D., UCLA, 1992. Condensed matter physics.
Staggs, Suzanne, Ph.D., Princeton, 1993. Cosmology; gravitation; relativity.
Steinhardt, Paul, Ph.D., Harvard, 1978. Cosmology.
Verlinde, Herman, Ph.D., Utrecht, 1988. Particle theory.
Yazdani, Ali, Ph.D., Stanford University, 1995. Condensed matter.

Associate Professors

Hasan, M. Zahid, Ph.D., Stanford, 2001. Condensed Matter Physics.
Olsen, James, Ph.D., University of Wisconsin-Madison, 1998. High-energy physics.
Romalis, Michael, Ph.D., Princeton, 1997. Atomic physics.
Tully, Christopher, Ph.D., Princeton, 1998. High-energy physics.

Assistant Professors

Bernevig, Bogdan an Andrei, Ph.D., Stanford U. COndensed Matter.
Fowler, Joseph, Ph.D., Chicago, 1999. Cosmology.
Galbiati, Cristiano, Ph.D., Milan, 1999. Nuclear physics.
Gregor, Thomas, PhD, Princeton University, 2005. Biophysics.
Halyo, Valerie, Ph.D., Stanford University, 2000. High-energy physics.
Herzog, Christopher, Ph.D., Harvard, 2002. Particle theory.
Jones, William C., Ph.D., California Institute of Technology, 2005. Cosmology.
Petta, Jason, Ph.D., Cornell, 2003. Condensed matter.
Pretorius, Frans, Ph.D., University of British Columbia, 2002. Theoretical Cosmology.
Seiringer, Robert, Ph.D., Vienna, 2000. Mathematical physics.
Shaevitz, Joshua, Ph.D. Stanford University, 2004, Molecular Biology.
Wang, Liantao, Ph.D., University of Michigan, 2002. Particle theory.

Instructor

Visiting Lecturers with Rank of Professor

Adler, Stephen L., Ph.D., Princeton, 1964. Theoretical physics.
Arkani-Harned, Nima, Ph.D., UC at Berkeley, 1997. Particle theory.
Maldacena, Juan, Ph.D., Princeton, 1996. Particle theory.
Seiberg, Nathan, Ph.D., Tel-Aviv, 1982. Particle theory.
Witten, Edward, Ph.D., Visiting Lecturer with rank of Professor, Princeton, 1976. Theoretical physics.

Associated Faculty

Bhatt, Ravindra, Ph.D., Illinois, 1976. Electrical engineering and computer science.
Car, Roberto, Laurea, Milan, 1971. Chemistry.
Davidson, Ronald C., Ph.D., Princeton, 1966. Plasma physics.
Shayegan, Mansour, Ph.D., MIT, 1983. Electrical Engineering.
Sinai, Ya. G., Ph.D., Moscow State University, 1963. Mathematics.
Spergel, David N., Ph.D., Harvard 1985. Astrophysical Sciences.
Tank, David, Ph.D., Cornell. Biophysics.
Torquato, Salvatore, Ph.D., SUNY, 1980. Chemistry.
Tsui, Dan, Ph.D., Chicago, 1967. Electrical engineering and computer science.
Wingreen, Ned, Ph.D., Cornell, 1989. Molecular Biology.

Research Staff

Adam, Nadia, Ph.D., Cornell, 2006. High Energy Physics.
Allison, Giles Daniel, Ph.D., International School for Advanced Studies. Particle Theory.
Beidenkopf, Haim, Ph.D., Weizmann Institute of Science, 2009. Condensed Matter Experiment.
Chiang, Hsin Cynthia, Ph.D., California Institute of Technology, 2008. Cosmology.
Estevez-Torres, Andre, Ph.D., Université Pierre et Marie Curie. Chemistry.
Fan, Jiji, Ph.D., Yale University, 2009. Particle Theory.
Florescu, Marian, Ph.D., University of Toronto, 2003. Cosmology Theory.
Iyer Biswas, Srividya, Ph.D., Ohio State University, 2009. Biophysics.

Jarosik, Norman, Ph.D., SUNY, Buffalo, 1985. Gravity, cosmology.

Jones, John, Ph.D., Imperial College, London, 2007. High Energy Physics.

Jung, Minkyung, Ph.D., University of Tokyo, 2005. Electrical Engineering.

Kiermaier, Michael, Ph.D., Massachusetts Institute of Technology, 2009. Particl Theory.

Koppinen, Panu, Ph.D., University of Jyväskylä, 2009. Condensed Matter Experiment.

Lopes, Pegna, David, Ph.D., Universitá degli Studi di Pavia. Nuclear Physics.

Stickland, David, Ph.D., Bristol Univ., 1980. High-energy physics.

Wright, Alexander, Ph.D., Queen's University, 2009. Nuclear Physics.

Xie, Zhen, 2001, Scuola Normale Superiore of Pisa. High energy physics.

RESEARCH SPECIALTIES AND STAFF

Theoretical

Condensed Matter. Haldane, Huse, Sondhi.

Mathematical Physics. Aizenman, Seiringer, Lieb.

Particle Theory. Callan, Gubser, Herzog, Klebanov, Polyakov, Verlinde, Wang. 3 postdoctoral fellows.

Relativity and Cosmology. Pretorius, Steinhardt. 1 postdoctoral fellow.

Experimental

Atomic Physics. Happer, Romalis, 2 postdoctoral fellows.

Biophysics. Austin, Bialek. 5 postdoctoral fellow.

Condensed Matter. Hasan, Ong, Petta, Yazdani. 6 postdoctoral fellows.

Gravitation, Relativity, Cosmology. Fowler, Groth, Jones, Page, Staggs. 2 postdoctoral fellows.

High-Energy Particles. Halyo, Jones, Marlow, McDonald, Meyers, Olsen, Smith, Stickland, Tully, Xie, 2 postdoctoral fellows.

Nuclear Physics. Calaprice, Galbiati. 2 postdoctoral fellows.

PRINCETON UNIVERSITY

DEPARTMENT OF ASTROPHYSICAL SCIENCES
PLASMA PHYSICS SECTION

Princeton, New Jersey 08543

Students Accepted For Degree	FIELDS		
	Physics	Astronomy	Related Fields
Doctorate			X*
Master's			

*Ph.D. in plasma physics.

1. General

President: Shirley Tilghman
Dean of Graduate School: William B. Russel
Director of Program in Plasma Physics: Nathaniel J. Fisch
Department Telephone Number: (609) 243-2489 (C)
Type of Institution: University
Control: Private
Setting: Small town
Total Faculty: 1,172
Total Graduate Faculty: Not separated
Total Students: 7,523
Total Graduate Students: 2,479
Annual Graduate Tuition:
All Graduate Students: Full-time—$38,090
Tuition rates for: 2010–11
Deferred tuition plan: No
Other Fees: None
Term: Semester

2. Number of Faculty in Department

The combined total of full-time faculty in the three professorial ranks is 17 (2009–10)

3. Admission, Financial Aid, and Housing

Address admission inquiries to: Office of Graduate Admission, Princeton University, P.O. Box 270, Princeton, NJ 08544-0270
Graduate application fee required: $90
Admission deadline (Fall admission): 12/15/10–All applicants.
Admission information: For fall admission, 2010–11, 9 students were accepted from 41 applicants.
Admission requirements: For admission to the graduate programs, a Bachelor's degree, preferably in physics, mathematics, or engineering is required with no minimum undergraduate GPA specified. The GRE is required. The GRE Advanced in physics or mathematics is required. The average GRE scores (in %) for admissions were Verbal–81.8%, Quantitative–93.2%, writing=4.61/64.9%. The average GRE Advanced score for admissions was 85.6% in physics. Studentsfrom non-English speaking countries are required to demonstrate proficiency in English via the TOEFL exam. No minimum score required.
Undergraduate preparation assumed: A sound undergraduate education in physics and mathematics is assumed, including courses in electricity and magnetism, classical mechanics, thermodynamics and statistical mechanics, quantum mechanics, differential equations, and complex variables.
Address financial aid inquiries to: Office of Graduate Admission, Princeton University, P.O. Box 270, Princeton, NJ 08544-0270.
GAPSFAS application required: No
Financial aid deadline: 12/15/10
Loans available: Yes
Address housing inquiries to: Office of Graduate Admission, Princeton University, P.O. Box 270, Princeton, NJ 08544-0270
On-campus, single student housing available: Yes
Cost/acad. year: $4,092–13,788
On-campus, married student housing available: Yes
Cost/month: $860–1,149

Table A—Faculty, Enrollments, and Degrees Granted

Research Specialty	2009–10 Faculty	Enrollment[1] Fall 2009 Master's	Enrollment[1] Fall 2009 Doctorate	No. of Degrees Granted[2] 2009–10 (2004–09) Master's	Terminal Master's	Doctorate	Median No. of Years for 2008–09 Ph.D.'s
Plasma Physics & Fusion	17	–	35	–	–	7(36)	5.9
Total		–	35	–	–		
Full-time Grad. Stud.		–	35				
Part-time Grad. Stud.		–	0				
First-year Grad. Stud.		–	7				
Median Years in Grad. Study (2008–09 Degrees)				–	–	5.9	
Undergraduate Degrees							

[1]Students not yet committed to a research specialty are entered under non-specialized.
[2]Five-year totals in parentheses (2003–08).

4. Graduate Degree Requirements

Master's: The Department does not sponsor a specific Master's Degree program. However, a Master of Arts degree can be awarded as an incidental degree upon passing the General Examination.
Doctorate: There are no formal curriculum-based requirements in the doctoral program, and no course grades are given. Facility in a foreign language is encouraged, but a readingexamination is no longer required. Students generally take the "Physics Prelims," in the Department of Physics, at the end of their first year, although these may be taken even at the beginning of the year. However, a minimum of one year in residence is required before taking the plasma physics portion (written and oral) of the General Examination.
Thesis: Thesis may be written *in absentia*.
Special Equipment, Facilities, or Programs: Students normally request admission to either the astronomy section or the plasma physics section. The present report gives information only on the plasma physics program. As the name implies, this program has a strong physics orientation. First-year studies typically include graduate courses given by the Department of Physics–quantum mechanics, electricity and magnetism, and statistical mechanics, together with the full-yearintroductory plasma physics course. Course offerings in later years cover plasma waves and instabilities, equilibrium and nonequilibrium statistical mechanics, theories of trans-

port, structure and stability of plasma equilibria, magnetohydrodynamics, kinetic theory, nonlinear processes, and computational methods in plasma physics. Applications of theory are made to astrophysical and geophysical plasmas as well as to the plasmas in controlled fusion research and in other areas of physics and engineering science. Studies of basic plasma physics and of the possibilities of controlled fusion power were initiated in Princeton in 1951. The principal fusion research devices at the Princeton Plasma Physics Laboratory currently include the National Spherical Torus Experiment (NSTX), the Current Drive Experiment-Upgrade (CDX-U), and the Magnetic Reconnection Experiment (MRX). Laboratory scientists and engineers are also involved in the development of innovative confinement concepts such as compact stellarators. In addition, Princeton Plasma Physics Laboratory scientists are collaborating with researchers on fusion science and technology at other facilities, both domestic and foreign. Through its efforts to build and operate magnetic fusion devices, PPPL has gained extensive capabilities in a host of disciplines including advanced computational simulations, lasers, microwave technology, vacuum technology, mechanics, materials science, electronics, computer technology, and high-voltage power systems. In addition, PPPL scientists and engineers are applying knowledge gained in fusion research to other theoretical and experimental areas including plasma thrusters for space propulsion, propagation of intense beams of ions, materials science, solar physics and manufacturing. There are also experimental research opportunities on plasma thrusters, x-ray lasers, nonneutral plasmas, and plasma processing, as well as for theoretical and computational research in diverse areas of plasma physics, including magnetic confinement, plasma heating, transport and stability, plasma astrophysics, laser-plasma interactions, magneto-spheric physics and basic studies in plasma physics. This involves the innovative development of new calculation capabilities as well as the application of state of the art theoretical and computational tools to the interpretation of experimental results. Local computation is carried out on workstations and personal computers with access to Cray Supercomputers and massively parallel computers. Graduate students from their first week participate in both the experimental and theoretical research programs at the Laboratory.

Table B—Appointments to Graduate Students, 2008–09

Title of Appointee	Appointments: Total	Appointments: First year	Academic Load Allowed in Credit Hours	Hours of Service Per Week	Stipend for Academic Year ($)
			Semester		
Research Assistant	25	6	–	20	22,000
DOE FES[1]	3	0	–	–	23,000
NSF[2]	2	0			
DOESSGF[3]	1	0			
NDSEG[4]	3	1			
Fulbright[5]	1	0			
Total	35	7			

[1] Department of Energy Fusion Energy Sciences.
[2] National Science Foundation.
[3] Department of Energy Science Stewardship Graduate Fellowship.
[4] National Defense Science & Engineering Graduate Fellowship
[5] Fulbright International Science & Technology Fellowship

5. Personnel Engaged in Separately Budgeted Research, 10/01/09–4/30/10

Professorial faculty	17
Graduate students	35
Nonteaching research personnel	69
Total	121

6. Separately Budgeted Research Expenditures by Source of Support*

*Funding for the plasma physics graduate program comes from the Princeton Plasma Physics Laboratory's budget. The PPPL is funded by the U.S. Department of Energy through contract with Princeton University.

7. Extension Centers and Summer Programs

Research assistantships which include summer support are available to all graduate students.

FACULTY

Professors

Davidson, Ronald C., Ph.D., Princeton, 1966. Associate Chairman, Astrophysical Sciences, Plasma Physics.

Fisch, Nathaniel J., Ph.D., MIT, 1978. Director of Program in Plasma Physics. Plasma waves, current drive, laser/plasma interaction.

Goldston, Robert J., Ph.D., Princeton, 1977. Experimental plasma physics, plasma heating, transport.

Prager, Stewart C., Ph.D., Columbia U., 1975. Director, Princeton Plasma Physics Laboratory, Experimental plasma physics, stability transport.

Lecturers with Rank of Professor

Cohen, Samuel A., Ph.D., MIT, 1973. Experimental plasma physics.

Hammett, Gregory W., Ph.D., Princeton, 1986. Computational and analytical studies of plasma turbulence and rf heating.

Jardin, Stephen C., Ph.D., Princeton, 1976. Computational magnetohydrodynamics; equilibrium and stability; transport.

Krommes, John A., Ph.D., Princeton, 1975. Plasma stochasticity and turbulence.

Phillips, Cynthia K., Ph.D., Wisconsin, 1982. rf heating.

Reiman, Allan H., Ph.D., Princeton, 1977. Theoretical plasma physics, non-linear dynamics, MHD.

Tang, William M., Ph.D., California, Davis, 1972. Kinetic stability theory.

White, Roscoe B., Ph.D., Princeton, 1963. Resistive magnetohydrodynamic instabilities; parametric instabilities; transport.

Lecturers

Efthimion, Philip C., Ph.D., Columbia U., 1977. Plasma physics, E&M theory and applications, instrumentation.

Ji, Hantao, Ph.D., U. of Tokyo, 1990. Magnetic reconnection; magnetorotational instability (MRI).

Majeski, Richard P., Ph.D., Dartmouth College, 1979. Experimental plasma physics.

Qin, Hong, Ph.D., Princeton University, 1998. Beam physics; gyrokinetic theory; particle simulation.

Associated Faculty

Choueiri, Edgar, Ph.D., Princeton, 1991. Assistant Professor, Mechanical and Aerospace Engineering. Plasma propulsion; space plasma physics; turbulence in collisional plasmas; plasma dynamics and astronautics.

Szymon, Suckewer, Ph.D., Warsaw U., 1966; D.Sc., 1971. Professor, Mechanical and Aerospace Engineering. X-ray lasers development and applications; X-ray microscopy; powerful sub-picosecond lasers and their applications; applications of lasers to biology and medicine; plasma diagnostics; development of compact tokamak as a source of VUV and soft X-ray radiation.

RUTGERS—THE STATE UNIVERSITY OF NEW JERSEY

DEPARTMENT OF PHYSICS AND ASTRONOMY

Piscataway, New Jersey 08854

Students Accepted For Degree	FIELDS		
	Physics	Astronomy	Related Fields
Doctorate	X	X	
Master's	X	X	

1. General

President: Richard McCormick
Dean of Graduate School: Jerome J. Kukor
Department Chair: Ronald Ransome
Department Telephone Number: (732) 445-5500 x2503
FAX: (732) 445-4343
Type of Institution: University
Control: Public
Setting: Suburban
Total Faculty: 2,000 (New Brunswick)
Total Graduate Faculty: 1,900 (New Brunswick)
Total Students: 29,000 (New Branswick)
Total Graduate Students: 8,000 (New Brunswick)
Annual Graduate Tuition:
In-state residents: Full-time—$13,848
Part-time—$577.00/credit
Out-of-state residents: Full-time—$20,100
Part-time—$875.00/credit
Tuition rates for: 2009–10
Deferred tuition plan: Yes
Other Fees: $750–1,500
Term: Semester

2. Number of Faculty in Department

There are approximately 63 full-time faculty in all ranks.

3. Admission, Financial Aid, and Housing

Address admission inquiries to: Dr. Ronald Ransome, Graduate Program Director, Department of Physics and Astronomy, 136 Frelinghuysen Rd., Piscataway, NJ 08854-8019
E-mail: graduate@physics.rutgers.edu
Graduate application fee required: $60
Admission deadline (Fall admission): 1/1 is recommended. Later applications will be considered until 7/15, depending on availability of positions.
Admission information: For fall admission, 2010–11, 45 students were accepted from 260 applicants.
Admission requirements: For admission to the graduate programs, a Bachelor's degree in physics or related field is required with no minimum undergraduate GPA specified. The GRE and the GRE Advanced in Physics are required. No minimum scores specified. The Advanced Physics average was 75% for students admitted. Students from non-English speaking countries are required to demonstrate proficiency in English via the TOEFL or IELTs exam. Minimum acceptable score for admission is 560, 600 for a TA.
Address financial aid inquiries to: Prof. T. Williams, Graduate Program Director, Rutgers University, Department of Physics and Astronomy, 136 Frelinghuysen Rd., Piscataway, NJ 08854-8019, ransome@physics.rutgers.edu
GAPSFAS application required: No
Financial aid deadline: 1/1 (later applications subject to availability)
Loans available: Yes
Address housing inquiries to: Graduate Student Housing, 581 Taylor Rd., Piscataway, NJ 08854
On-campus, single student housing available: Yes
**Cost/calendar yr.*: $8,452–8,906
On-campus, married student housing available: Yes
**Cost/month*: $929–1,176

*Costs for 2009–10

Table A—Faculty, Enrollments, and Degrees Granted

Research Specialty	2009–10 Faculty	Enrollment[1] Fall 2009 Master's	Enrollment[1] Fall 2009 Doctorate	No. of Degrees Granted[2] 2009–10 Master's	No. of Degrees Granted[2] 2009–10 Terminal Master's	No. of Degrees Granted[2] 2009–10 Doctorate	Median No. of Years for 2009–10 Ph.D.'s
Astronomy	9	–	–	0(0)	0(1)	2(7)	–
Biophysics	2	–	–	0(0)	0(0)	0(5)	–
Condensed Matter Physics	21	–	–	0(0)	3(3)	2(20)	–
Nuclear Physics	7	–	–	0(0)	0(0)	0(2)	–
Particles & Fields	16	–	–	0(0)	0(2)	3(18)	–
Physics Education	2	–	–	0(0)	0(0)	0(3)	–
Statistical & Thermal	1	–	–	0(0)	0(0)	0(2)	–
Surface Science	4	–	–	0(0)	0(11)	1(15)	–
Non-specialized	0	–	–	0(3)	1(22)		
Total		–	–	0(3)	4(39)	8(72)	
Full-time Grad. Stud.		–	108				
Part-time Grad. Stud.		–	2				
First-year Grad. Stud.		–	25				
Median Years in Grad. Study (2009–10 Degrees)			–	–	–	–	6.5

[1]Students not yet committed to a research specialty are entered under non-specialized.
[2]Five-year totals in parentheses.

4. Graduate Degree Requirements

Master's: The M.S. degree program is designed for part-time as well as full-time students. A comprehensive oral examination is required of all M.S. candidates. The M.S. degree requires 30 credits of which up to 12 credits may be in upperclass undergraduate (300–400 series) courses. The candidate may choose to write a thesis (in which case, six of the 30 required credits may be devoted to this thesis research) or to submit an essay (which is to be based on material from a course he or she has taken); the thesis must be defended in the oral examination. There is no formal GPA requirement, but no more than three courses with grades of "C" can be counted toward the degree. There is no foreign language requirement.
Doctorate: The candidacy exam consists of an oral exam on a current topic in research. Candidates must present a written report on a current area of research, followed by an oral

presentation and exam. The exam tests the candidate's ability to grasp the relevance, goals, techniques and underlying physics of a current area of research. The exam is normally taken at the start of the second year. In addition, candidates are required to complete a set of core courses with grades of B or better, or pass an exam if exemption is requested based on previous course work. A thesis of original research is required. There is a residence requirement of one year, but no foreign language requirement.

Other Programs: Master of Science for Teachers (MST) Degree: The MST degree is primarily a subject-matter oriented degree for practicing teachers, although others may be accepted. The requirements for the MST degree in physics consist of 30 credits, a comprehensive examination, and an essay or thesis. The courses are chosen in consultation with the departmental advisor to fit the needs of each individual student. Their first aim is to give each candidate the opportunity to further his or her knowledge of physics. Both undergraduate and graduate courses may be used, depending on each person's previous experience.

Thesis: Thesis may be written *in absentia*.

Special Equipment, Facilities, or Programs: The department has 64 faculty members. An additional 9 faculty members from other departments, primarily chemistry and mathematics, are members of the graduate program. This includes new faculty members in high-energy and condensed-matter physics, biophysics, and astronomy. The department is housed in a modern, fully equipped research laboratory with networks of workstations and PCs that provide easy computer access for all students and faculty members. The astrophysics group is focused on galactic dynamics and cosmology and has developed Fabry-Perot interferometers for observatories in Chile and South Africa. Rutgers astronomers have a 10% share of observing time at the 11 meter Southern African Large Telescope. Condensed-matter theory faculty members study strongly correlated electron systems and electronic properties of materials. The multidisciplinary Laboratory for Surface Modification includes 7 physics faculty members and members of the chemistry, materials science, and engineering departments. Research in condensed matter experiment spans low temperature physics, mesoscopic electronics organic conductors, optical scattering spectroscopies, magnetic and multiferroic materials, and two-dimensional systems (e.g. graphene). New research initiatives focus on the synthesis of novel materials and their characterization using optical, scanning-probe, X-ray/neutron diffraction, and transport techniques. High-energy theory research includes phenomenological studies and abstract approaches such as string theory and conformal field theories. High energy experimentalists do research at Fermi Lab and CERN hadron colliders. They search for supersymmetry, the Higgs particle and dark matter with leptons, photons, and jets and also study the top quark and guage bosons. They are also developing detectors and detection technologies for high radiation environments. Nuclear physics research in both theory and experiment span a broad range of questions, including the structure of the nucleon, the interaction of neutrinos with nuclei, the limits of angular momentum and stability in nuclei, and nucleosynthesis. Experiments are carried out at Oak Ridge national lab, Yale, Jefferson Lab in Virginia, and Fermilab.

Table B—Appointments to Graduate Students, 2009–10

Title of Appointee	Appointments Total	Appointments First year	Academic Load Allowed in Credit Hours	Hours of Service Per Week	Stipend for Academic Year ($)
			Semester		
Teaching Assistant	37	9	10 cr. hrs.	12–15	24,312[1]
Research Assistant	31	2	15 cr. hrs.	12–15	24,312[1]
Graduate Fellow	27	9	16 cr. hrs.	None	30,000
Total	95	20			

[1]Plus tuition remission and health benefits.

5. Personnel Engaged in Separately Budgeted Research, 7/01–6/02

Professorial faculty	
Postdoctoral appointments	
Graduate students	
Undergraduate students	
Nonteaching research personnel	
Total	N/A

6. Separately Budgeted Research Expenditures by Source of Support

	Departmental Research
Other	11,000,000
Total	$11,000,000

FACULTY

Professors

Andrei, Eva Y., Ph.D., Rutgers, 1980. Experimental condensed matter physics.

Andrei, Natan, Ph.D., Princeton, 1979. Theoretical elementary particle/condensed matter physics.

Banks, Thomas, Ph.D., MIT, 1973. Theoretical elementary particle physics.

Bartynski, Robert, Ph.D., Pennsylvania, 1986. Experimental condensed matter physics.

Bhanot, Gyan, Ph.D., Cornell, 1979. Systems biology, cancer and population genetics, Professor of biomedical engineering.

Blumberg, Girsh, Ph.D., Estonian Academy of Sciences, 1987. Experimental condensed matter physics.

Case, David, Ph.D., Harvard Univ., 1977. Professor of Chemistry. Theoretical chemistry of biomolecules.

Chandra, Premala, Ph.D., Univ. of Calif.-SB, 1988. Condensed matter theory.

Cheong, Sang-Wook, Ph.D., UCLA, 1989. Experimental condensed matter physics; material science.

Cizewski, Jolie A., Ph.D., SUNY at Stony Brook, 1978. Experimental nuclear physics.

Coleman, Piers, Ph.D., Princeton, 1984. Theoretical condensed matter physics.

Croft, Mark, Ph.D., Rochester, 1977. Experimental condensed matter physics.

Feldman, Leonard, Ph.D., Rutgers, 1967. Experimental condensed matter physics.

Friedan, Daniel, Ph.D., California, Berkeley, 1980. Theoretical elementary particle physics.

Garfunkel, Eric, Ph.D., California, Berkeley, 1983. Professor of Chemistry. Experimental surface science.

Gershenson, Michael E., Ph.D., Institute of Radio Engineering and Electronics (Moscow), 1982. Experimental condensed matter physics.

Gilman, Ronald, Ph.D., Pennsylvania, 1985. Experimental nuclear physics.

Goldin, Gerald A., Ph.D., Princeton, 1969. Mathematical physics. Professor of Mathematics.

Goldstein, Sheldon, Ph.D., Yeshiva University, 1974. Statistical mechanics, foundations of quantum mechanics. Professor of Mathematics.

Gustafsson, Torgny, D.Sc., Chalmers Univ. of Technology, Sweden, 1973. Experimental condensed matter physics; experimental surface physics. Chair of the Department.

Hughes, John Ph.D., Columbia, 1984. Observational astronomy.

Ioffe, Lev, Ph.D., Landau Inst. for Theoretical Physics, Russia, 1985. Theoretical condensed matter physics.

Kalelkar, Mohan S., Ph.D., Columbia, 1975. Associate Chair & Undergraduate Coordinator. Experimental elementary particle physics.

Kloet, Willem M., Ph.D., Utrecht, Netherlands, 1973. Theoretical nuclear physics.

Kojima, Haruo, Ph.D., UCLA, 1972. Experimental condensed matter physics.

Koller, Noemie B., Ph.D., Columbia, 1958. Experimental nuclear physics.

Kotliar, B. Gabriel, Ph.D., Princeton, 1983. Theoretical condensed matter physics.

Langreth, David C., Ph.D., Illinois, 1964. Theoretical condensed matter physics.

Leath, Paul L., Ph.D., Missouri-Columbia, 1966. Theoretical condensed matter physics.

Lebowitz, Joel, Ph.D., Syracuse, 1956. Theoretical statistical mechanics; math physics. Professor of Mathematics and Physics.

Levy, Ronald, Ph.D., Harvard, 1976. Biological physics theory and simulation. Professor of Chemistry, Director BioMaPS Institute.

Lovelace, Claud, B.S., Cape Town, South Africa, 1954. Theoretical elementary particle physics.

Mekjian, Aram, Ph.D., Maryland, 1968. Theoretical nuclear physics.

Moore, Gregory, Ph.D., Harvard, 1986. Theoretical particle physics.

Murnick, Daniel E., Ph.D., Massachusetts Institute of Technology, 1966. Experimental nuclear and atomic physics. Professor of Physics, Newark.

Neuberger, Herbert, Ph.D., Tel Aviv, 1979. Theoretical elementary particle physics.

Olson, Wilma, Ph.D., Stanford, 1971. Mary I. Bunting Professor of Chemistry. Biological physics theory and simulation.

Pryor, Carlton, Ph.D., Harvard, 1982. Experimental astrophysics.

Rabe, Karin, MIT 1987. Theoretical condensed matter physics; theoretical surface physics.

Ransome, Ronald Ph.D., Texas, Austin, 1981. Experimental nuclear physics. Assoc. Chair, Grad Program Director.

Schnetzer, Stephen R., Ph.D., University of California, Berkeley, 1981. Experimental elementary particle physics.

Sellwood, Jeremy, Ph.D., Manchester Univ., England, 1977. Theoretical astrophysics.

Sengupta, Anirvan, Ph.D., Bombay, 1994. Biological Physics.

Shapiro, Joel, Ph.D., Cornell, 1967. Theoretical elementary particle physics.

Somalwar, Sunil, Ph.D., Chicago, 1988. Exp. elementary particle physics.

Strassler, Matthew, Ph.D., Washington, 1993. Theoretical elementary particle physics.

Vanderbilt, David, Ph.D., MIT, 1981. Theoretical condensed matter physics; theoretical surface physics.

Williams, Theodore B., Ph.D., Cal. Tech., 1974. Experimental astrophysics.

Zamick, Larry, Ph.D., MIT, 1961. Theoretical nuclear physics.

Zamolodchikov, Alexander, Ph.D., Inst. of Theoretical and Exp. Physics, Moscow, 1978. Theoretical particle physics.

Associate Professors

Diaconescu, Duiliu-Emanual, Ph.D., Rutgers. 1998. Theoretical high-energy physics.

Etkina, Eugenia, Ph.D., Moscow State Pedagogical Univ., 1997. Physics education. Professor of Education.

Gershtein, Yuri, Ph.D., Moscow Inst. For Physics and Tech., 1996. Experimental high energy physics.

Haule, Kristjan, Ph.D., Ljubljana, 2002. Theoretical condensed matter physics.

Hinch, B. Jane, Ph.D., Cambridge. 1987. Surface studies using atomic and molecular scattering. Professor of Chemistry.

Keeton, Charles, Ph.D., Harvard, 1998. Astronomy.

Kiryukhin, Valery, Ph.D., Princeton, 1997. Experimental condensed matter physics.

Lath, Amitabh, Ph.D., MIT, 1994. Experimental elementary particle physics.

Lukyanov, Sergei, Ph.D., Landau Institute, 1989. Theoretical high-energy physics.

Matilsky, Terry A., Ph.D., Princeton, 1971. Experimental astrophysics.

Thomas, Scott, Ph.D., Texas at Austin, 1993. Theoretical elementary particle physics.

Yuzbashyan, Emil, Ph.D., Princeton, 2004. Theoretical condensed matter physics.

Zimmermann, Frank M., Ph.D., Cornell, 1995. Experimental surface science physics.

Assistant Professors

Baker, Andrew, Ph.D., CalTech, 2000. Observational physics.

Gawiser, Eric, Ph.D., University of California, Berkeley, 1999. Observational astrophysics, cosmology.

Halkiadakis, Eva, Ph.D., Rutgers, 2001. Experimental particle physics.

Jha, Saurabh, Ph.D., Harvard, 2002. Observational cosmology.

Joseph, Charles L., Ph.D., Colorado, 1985. Observational astronomy and detector development.

Morozov, Alexandre, Ph.D., Washington, 2003. Biophysics.

Oh, Seaongshik, Ph.D., Illinois, 2003. Experimental condensed matter physics.

Podzorov, Vitaly, Ph.D., Rutgers, 2002. Experimental condensed matter physics.

Shih, David, Ph.D., Princeton Univ., 2006. Theoretical high energy physics.

Wu, Weida, Ph.D., Princeton, 2004. Experimental condensed matter physics.

RESEARCH SPECIALTIES AND STAFF

Theoretical

Astrophysics. Evolution, structure and dynamics of galaxies, dark matter, gravitational lensing, gravitational N-body simulations. Keeton, Sellwood.

Biophysics. Bhanot, Morozov, Olson, Sengupta.

Condensed Matter Physics. Strongly correlated electron systems, novel superconductors, quantum phase transitions, quantum computing, electronic and structural properties of solids, dielectric and ferroelectric materials, magnetism and multiferroics, equilibrium and non-equilibrium statistical mechanics. N. Andrei, Coleman, Haule, Ioffe, Kotliar, Langreth, Leath, Rabe, Vanderbilt, Yuzbashyan.

High Energy Physics. Quantum field theory; perturbative and lattice QCD calculations; strings and superstrings; supersymmetry; conformal field theory; quantum gravity; phenomenology. Banks, Diaconescu, Friedan, Lovelace, Lukyanov, Moore, Neuberger, Shapiro, Strassler, Thomas, Zamolodchikov.

Nuclear Physics. Nuclear structure; quark dynamics; few-nucleon problem; relativistic heavy-ion reactions; dibaryon resonances; electron scattering; intermediate-energy, hadron scattering. Kloet, Mekjian, Zamick.

Experimental

Astrophysics. Galaxies and clusters of galaxies, cosmology and dark energy, supernovae, galaxies at high redshift, x-ray sources, imaging spectrophotometry, star clusters and dwarf galaxies. Baker, Gawiser, Hughes, Jha, Joseph, Matilsky, Pryor, Williams.

Condensed Matter Physics. Surface physics: geometric structure, electronic structure, molecular adsorption, thin fills; superconductivity; electrical and thermal transport; superfluidity in helium; 2D electron gas; spin resonance; synchrotron radiation. E. Andrei, Bartynski, Blumberg, Cheong, Feldman, Garfunkel, Gershenson, Gustafsson, Kiryukhin, Kojima, Oh, Podzorov, Wu, Zimmermann.

Elementary Particle Physics. Top-quark, Higgs particle searches and Z° physics using $\bar{p}p$ collisions and High energy cosmic rays. Gershtein, Halkiadakis, Kalelkar, Lath, Schnetzer, Somalwar.

Nuclear Physics. Nuclear structure; superdeformed shapes, magnetic moments, nuclei far from stability; intermediate energy electron and proton scattering; neutrino scattering. Cizewski, Gilman, Koller, Ransome.

Physics Education. Etkina, Kalelkar, Shapiro.

STEVENS INSTITUTE OF TECHNOLOGY

DEPARTMENT OF PHYSICS AND ENGINEERING PHYSICS

Hoboken, New Jersey 07030

Students Accepted For Degree	FIELDS: Physics	Astronomy	Related Fields
Doctorate	X		
Master's	X		

1. General

President: Harold J. Raveche
Dean of University Admissions: Daniel Gallagher
Department Director: Knut Stamnes
Department Telephone Number: (201) 216-5665
Type of Institution: University
Control: Private
Setting: Urban
Total Faculty: 400 includes 126 special faculty
Total Graduate Faculty: Not separated
Total Students: 5,934
Total Graduate Students: 3,700
Annual Graduate Tuition:
In-state residents: $1,100/credit
Out-of-state residents: $1,100/credit
Tuition rates for: 2009–10
Deferred tuition plan: No
Other Fees: $60 (application fee)
Term: Semester
World Wide Online Enrollment: 3,500

2. Number of Faculty in Department

The combined total of full-time faculty in the three professorial ranks is 10. The combined total of full-time, part-time and other faculty at all ranks is 25.

3. Admission, Financial Aid, and Housing

Address admission inquiries to: Daniel Gallagher, Dean of University Admissions
Graduate application fee required: $60
Admission deadline (Fall admission): 3/31/10
Admission requirements: For admission to the graduate programs, a bachelor's degree in science or engineering is required with a minimum undergraduate GPA of 3.0 specified. No minimum acceptable score for admission is specified. The average GRE scores for admissions were not calculated. The average GRE Advanced score for admissions is not available. Students from non-English speaking countries are required to demonstrate proficiency in English via the TOEFL exam. Minimum acceptable score for admission is 530.
Undergraduate preparation assumed: "Physics for Scientists and Engineers: A Strategic Approach with Modern Physics w/ Mastering Physics," 5 volume boxed set. Author Randall D. Knight, Publisher Pearson Addison Wesley.
Address financial aid inquiries to: Daniel Gallagher
GAPSFAS application required: Yes
Financial aid deadline: 2/15
Loans available: Yes
Address housing inquiries to: Student Housing, Trina Ballantyne, Dean of Residence Life, Dining Services & Center Operation.
On-campus, single student housing available: Yes
Cost per month: Contact Office of Student Housing
On-campus, married student housing available: No
Cost per semester: Contact Office of Student Housing

Table A—Faculty, Enrollments, and Degrees Granted

Research Specialty	2009–10 Faculty	Enrollment[1] Fall 2009: Master's	Doctorate	No. of Degrees Granted[2] 2009–10 (2005–10): Master's	Terminal Master's	Doctorate	Median No. of Years for 2009–10 Ph.D.'s
Applied Physics	2	0	3	0(0)	0(1)	1(1)	6.5
Atmos./Space Phys., Cosmic Rays	1	0	8	0(0)	0(0)	1(4)	–
Atomic, Molecular & Optical Phys.	4	0	12	1(2)	0(0)	0(6)	6.5
Condensed Matter Physics	5	0	6	0(0)	0(0)	1(7)	–
Electromagnetism	0	0	0	0(0)	0(0)	0(1)	–
Fluids & Rheology	0	0	0	0(0)	0(0)	0(1)	–
Optics	2	0	7	1(0)	0(2)	0(8)	6.5
Plasma Physics & Fusion	0	0	0	0(1)	0(0)	0(4)	–
Statistical & Thermal	0	0	0	0(0)	0(0)	0(1)	–
Other Experimental	0	0	0	0(0)	0(0)	0(0)	–
Non-specialized	0	0	0	6(15)	0(19)	–	–
Total		0	36	8(18)	0(22)	3(33)	
Full-time Grad. Stud.		0	35				
Part-time Grad. Stud.		0	0				
First-year Grad. Stud.		0	1				
Median Years in Grad. Study (2009–10 Degrees)				2	2	6.5	

Undergraduate Degrees, 2009–10 (2005–10): 11(21)

[1] Students not yet committed to a research specialty are entered under non-specialized.
[2] Five-year totals in parentheses.

4. Graduate Degree Requirements

Master's: 30 semester hour credits; 3.0 GPA in physics courses and overall; no residence requirement; no language or other comprehensive exams; thesis optional.
Doctorate: 90 semester hour credits (including Master's credits) of which 50 minimum to be in courses, and 30 minimum to be dissertation research; 3.0 GPA in physics courses and overall; one year residence; no language requirement; comprehensive/qualifying exam (one combined exam); dissertation.
Special Equipment, Facilities, or Programs: The Department has particular strength in areas of applied physics such as optics, atomic and plasma physics, nanotechnology, condensed matter physics.

Table B—Appointments to Graduate Students, 2008–09

Title of Appointee	Appointments Total	First year	Academic Load Allowed in Credit Hours	Hours of Service Per Week	Stipend for Academic Year ($)
			Semester		
Teaching Assistant	16	1	9	20	18,935-21,557[1,2]
Research Assistant	6	9	9	20	
Robert Crooks Stanley Fellowship	2	–	9	–	18,035-21,557[1,2]
Total	36	1			

[1]Higher salary for Master's.
[2]Ph.D. Candidates.

5. Personnel Engaged in Separately Budgeted Research, 7/09–6/10

Professorial faculty	6
Other faculty	3
Postdoctoral appointments	1
Graduate students	6
Nonteaching research personnel	1
Total	17

6. Separately Budgeted Research Expenditures by Source of Support

	Departmental Research	Physics-related Research Outside Department
Federal Government	$5,903,150	$0
Private		$0
NJ		$0
Total	5,903,150	$0

Table C—Separately Budgeted Research Expenditures

Research Specialty	No. of Grants	Expenditures ($)
Atmos./Space Phys.,	2	750,273
Atomic, Molecular, & Optical Physics	5	1,409,693
Applied Physics		
Condensed Matter Physics	3	773,184
Electromagnetism		
Optics	2	2,970,000
Private		
Plasma		
Total	12	5,903,150

FACULTY

Professors

Brucker, E. Byerly, Ph.D., Johns Hopkins, 1959. Experimental-high-energy physics; optics.

Carr, Wayne E., Ph.D., Illinois, 1967. Plasma physics; electron+ion beams; computational physics.

Horing, Norman J., Ph.D., Harvard, 1964. Quantum many-particle theory; solid state physics; surface physics; high-magnetic field effects.

Stamnes, Knut, Ph.D., University of Colorado, 1978. Electron transport and thermalization; kinetic theory; radiation transport; satellite remote sensing; biophotonics for non-invasive diagnostic of biological tissue.

Whittaker, Edward A., Ph.D., Columbia, 1982. Laser techniques; optical diagnosis of gas phase materials; processing reactors; Brillouin scattering; quantum optics.

Associate Professors

Martini, Rainer, Dr. rer.nat (Ph.D), Rheinisch-Westfaelische Technischen Hochschule Aachen, Germany, 1999. High sensitivity laser spectroscopy.

Malinovskaya, Svetlana, Ph.D., Novosibirsk State University, 1993. Laser-matter interaction, coherent control

Yu, Ting, Ph.D., Imperial College, University of London, UK, 1998. Atomic, Molecular and Optical Physics (AMO); Quantum Information and Quantum Optics.

Assistant Professors

Search, Christopher, Ph.D., Univ. of Michigan, 2002. Bose-Einstein condensation, quantum optics, nonlinear optics.

Strauf, Stefan, Dr. rer. nat (Ph.D.), Universität Bremen, Germany, 2001. Nanophotonics, quantum optics.

Adjunct Associate Professors

Hutt, Marvin, Ph.D., New York University, 1987. Optical engineering.

Lenzing, Harry, M. S., Stevens Institute of Technology, 1962. Fellow of the IEEE. Satellite-tracking systems, passive intermodulation (PIM).

Supplee, James, Ph.D., University of Texas at Dallas, 1979. Spectrascopy and semiclassical optics.

Webb, Robert, M.S., New York University, 1966 (Five years additional course work towards Ph.D.).

Research Professor

Tarnovsky, Vladimir, Ph.D., New York University, 1989. Experimental collisions with atoms, molecules, and free radicals.

RESEARCH SPECIALTIES AND STAFF

Theoretical

Electron Transport. Cui, Horing.
Ion Surface Interactions. Horing.
Laser-matter interactions, Malinovskaya.
Optic-electronics Devices Modeling & Simulation.
Quantum Information & Quantum Optics. Ting, Yu.
Quantum Optics. Search, Malinovskaya.
Semiconductor Solid State Theory. Horing.

Experimental

Atmospheric/Space Research. Radiation transport in planetary media including the coupled atmosphere-snow/ice-ocean system. Stamnes.

High Sensitivity Laser Spectroscopy. Martini, Strauf, Supplee, Whittaker.
Nanophotonics. Strauf.
Optical Control of Quantum Systems. Martini, Strauf.
Optical Diagnosis of Gas Phase Materials. Whittaker.
Optical Fiber Devices and Sensors.
Plasma Physics.
Quantum Optics; Ultra Short Pulse (<100 fs), and cw lasers in the visible, MIR and IR Region. Martini, Strauf.

NEW MEXICO INSTITUTE OF MINING AND TECHNOLOGY

DEPARTMENT OF PHYSICS

Socorro, New Mexico 87801

Students Accepted For Degree	FIELDS		
	Physics	Astronomy	Related Fields
Doctorate	X	X	X
Master's	X	X	X

1. General

President: Daniel Lopez
Dean of Graduate School: David J. Westpfahl
Department Chairman: Kenneth B. Eack
Department Telephone Number: (575) 835–5328
Type of Institution: College
Control: Public
Setting: Small town
Total Faculty: 127
Total Students: 1,897
Total Graduate Students: 562
Annual Graduate Tuition: *
In-state residents: Full-time—$2,078.73/semester
Part-time—$230.97/hr.
Out-of-state residents: Full-time—$6,076.18/semester
Part-time—$764.02/hr.
Tuition rates for: 2009–10
Term: Semester

*Tuition waived for students with assistantships (in most cases).

2. Number of Faculty in Department

The combined total of full-time faculty in the three professorial ranks is 12. The combined total of full-time, part-time, and other faculty at all ranks is 12.

3. Admission, Financial Aid, and Housing

Address admission inquiries to: Dean of Graduate Studies, New Mexico Tech.
Graduate application fee required: $16
Graduate admission fee required: $30
Admission deadline (Fall admission): 7/1
Admission information: For fall admission, 2009–10, 3 students were accepted from 21 applicants.
Admission requirements: For admission to the graduate programs, a Bachelor's degree in physics is required with a minimum undergraduate GPA of 3.0/4.0 specified. The GRE is strongly urged. No minimum acceptable score for admission is specified. Students from non-English speaking countries are required to demonstrate proficiency in English via the TOEFL exam. Minimum acceptable score for admission is 540.
Undergraduate preparation assumed: Traditional undergraduate physics program is assumed.
Address financial aid inquiries to: Graduate Office, New Mexico Tech. (for TA/RA information: Dean of Graduate Studies, New Mexico Tech.)
GAPSFAS application required: No
Financial aid deadline: 5/1
Loans available: Yes
Address housing inquiries to: Auxiliary Services, New Mexico Tech.
On-campus, single student housing available: Yes
Cost/semester: $1,261.00/Grad. Dorm.
On-campus, dependent student housing available: Yes
Cost/semester: $2,436.00

Table A—Faculty, Enrollments, and Degrees Granted

Research Specialty	2009–10 Faculty	Enrollment Fall 2008 Mas-ter's	Enrollment Fall 2008 Doc-torate	No. of Degrees Granted[1] 2009–10 (2005–10) Mas-ter's	Terminal Master's	Doc-torate	Median No. of Years for 2009–10 Ph.D.'s
Astrophysics	18	4	16	4(11)	0(7)	0(4)	6
Atmos.	14	2	7	1(2)	0(1)	1(4)	6
Other	7	1	1	1(0)	0(1)	1(1)	–
Total		7	24	5(14)	0(9)	2(9)	
Full-time Grad. Stud.		7	24				
Part-time Grad. Stud.		0	0				
First-year Grad. Stud.		0	7				
Median Years in Grad. Study (2009–10 Degrees)				–	–	–	6
Undergraduate Degrees, 2009–10 (2005–10): 12(64)							

[1]Five-year totals in parentheses.

4. Graduate Degree Requirements

Master's: With thesis: 24 hrs. course work plus six hrs. thesis; without thesis: 27 hrs. course work plus three hrs. independent study, including a paper. Minimum GPA: 3.0, no grade less than C. Should be in residence, minimum 18 hrs. No language requirement. Physics Preliminary Exam required. Thesis topic chosen after consultation with student's advisory committee.
Doctorate: 50 hrs. course work minimum beyond B.A. Minimum GPA: 3.0/4.0. Should be in residence, minimum three semester hours; degree *in absentia* by petition. No foreign language requirement. Physics Preliminary Exam required. Dissertation, oral defense, and paper submitted for publication to a recognized journal required.
Thesis: Thesis may be written *in absentia*, in part, by petition.
Special Equipment, Facilities, or Programs: The 2.4-meter telescopte at New Mexico Tech's Magdalena Ridge Observatory (MRO) is now in operation. The observatory is in the Magdalena Ranger District of the Cibola National Forest, near South Baldy Peak, at an elevation of 10,600 feet. The observatory's optical interferometer is in its construction phase and is expected to obtain first fringes in 2011. In addition to the MRO our astrophysics program takes advantage of the Very large Array (VLA) and Very Long Baseline Arrary (VLBA) of the National Radio Astronomy Onservatory (NRAO), whose Domenici Science Operations Center is on the Tech campus. Strong ties and frequent collaborations between Tech and NRAO give our students opportunities to work with these state-of-the-art facilities. The Langmuir Laboratory for Atmospheric Research, located on a moun-

taintop less than an hour's drive from campus, offers an unparalleled situation for active student participation in observation and research. Tech also operates a powered sailplane, and a dual-polarization radar.

Table B—Appointments to Graduate Students, 2009–10

Title of Appointee	Appointments Total	Appointments First year	Academic Load Allowed in Credit Hours	Hours of Service Per Week	Stipend for Academic Year ($)
			Semester		
Teaching Assistant	10	4	12[3]	20	16,427–25,700[1]
Research Assistant	21	5	12	20	17,419–30,300[2]
Total	31	9			

[1]Stipend for TA−1 is for 1/2-time academic year. TA-2 slightly higher.
[2]Stipend for RA−1 is for 1/2-time academic year. RA-2 slightly higher.
[3]Academic load: Normal 12 hrs., summer: normal 6.

5. Personnel Engaged in Separately Budgeted Research, 7/08–6/09

Professorial faculty	12
Other faculty	20
Graduate students	26
Nonteaching research personnel	8
Total	66

6. Separately Budgeted Research Awarded by Source of Support

	Departmental Research	Physics-related Research Outside Department
Federal government	$5,259,272	3.5M
State government	$96,500	$0
Other	1,227,381	
Total	$8,234,000	6.5M

7. Separately Funded and Managed Laboratories

Magdalena Ridge Observatory[1]	3.5M
Langmuir Laboratory[2]	$86,790
Total	$6,589,800

[1]Astro.
[2]Atmos.

Table C—Separately Budgeted Research Awards

Research Specialty	No. of Grants	Expenditures ($)
Astrophysics	3	243,000
Atmospheric Physics	25	6,340,153
Total	28	6,583,153

FACULTY

Professors Emeriti

Eilek, Jean, Ph.D., British Columbia, 1975. Plasma astrophysics; quasars; radio galaxies; pulsars.

Hankins, Timothy H., Ph.D., California, San Diego, 1971. Radio astronomy of pulsars; instrumentation; signal processing.

Krehbiel, Paul R., Ph.D., Manchester, 1982. Lightning studies; radar meteorology; thunderstorm electrification.

LeFebre, Vernon G., Ph.D., University of Utah 1963. Statistical physics; thermodynamics.

Schery, Stephen C., Ph.D., University of Colorado, 1973. Environmental radioactivity.

Winn, William P., Ph.D., California, Berkeley, 1966. Atmospheric physics; electrical discharges in gases; instrumentation.

Professors

Minschwaner, Kenneth, Ph.D., Harvard, 1992. Radiative transfer and climate; physics of the upper atmosphere.

Raymond, David J., Ph.D., Stanford, 1970. Cloud physics; geophysical fluid dynamics; clouds and climate.

Romero, Van, Ph.D., SUNY, Albany, 1991. Energetic materials; shock phenomena; high energy physics.

Westpfahl, David, J., Ph.D., Montana State, 1985. Dynamics of spiral and dwarf galaxies.

Associate Professors

Creech-Eakman, Michelle, Ph.D., University of Denver, 1997. Stellar astrophysics, mass-loss, optical/IR interferometry, IR instrumentation.

Eack, Kenneth, Ph.D., Univ. of Oklahoma, 1997. Thunderstorm electrification, atmospheric physics.

Hofner, Peter, Ph.D., University of Wisconsin-Madison, 1995. Star formation, interstellar medium, x-ray astronomy, extragalactic interstellar.

Sonnenfeld, Richard, Ph.D., University of California, Santa Barbara, 1987. Experimental physics.

Young, Lisa, Ph.D., 1987, University of New Mexico, Astrophysics; elliptical galaxies.

Assistant Professors

Arendt, Paul, Ph.D., New Mexico Tech, 2003; Elementary processes in extreme electromagnetic fields of astrophysics.

Meier, David, Ph.D., Ph.D., University of California, Los Angeles (2002). Molecular gas chemistry in star formation galaxies; The earliest phases starburst evolution in nearby galaxies.

Morales Juberias, Raúl, Ph.D., University of the Basque Country, Vizcaya (Spain). (2002). Outer planets observations and atmospheric dynamics.

Sessions, Sharon L., Ph.D., University of Oregon, Oregon, 2002. Atmospheric dynamics; statistical mechanics; many-body theory.

Adjunct Faculty

Avramidi, Ivan, Ph.D., Moscow State University, Russia, 1987. Mathematical physics, analysis on manifolds, quantum field theory.

Bakker, Eric, Ph.D., Astronomical instrumentation, Active galactic nuclei, Circumstellar environments.

Buscher, David, Ph.D., University of Cambridge. Optical/IR interferometry, atmospheric seeing measurement, adaptive optics, early and late stages of stellar evolution.

Colgate, Stirling A., Ph.D., Cornell, 1948. Astrophysics; plasma physics; atmospheric physics.

Elvis, Martin, Ph.D., Leicester (UK) Quasars and active galactic nuclei; X-ray astronomy.

Fuchs, Zeljka, Ph.D., New Mexico Tech, Atmospheric dynamics.

Goss, W. Miller, Ph.D., University of California, Berkeley, 1967. Radio astronomy, interstellar medium.

Haniff, Chris, Ph.D., University of Cambridge. Spatial interferometry at optical and near-infrared wavelengths, atmospheric turbulance, imaging theory, evolved stars.

Jurgenson, Colby, Ph.D., University of Denver. State-of-the-art infrared astronomical instrumentation.

Klinglesmith, Dan, Ph.D., Indiana University. Asteroids, robotic telescope operations.

Lopez-Carrillo, Carlos, Doppler radar and data analysis; Tropical dynamics.

Manney, Gloria, Ph.D., Iowa State University. Atmospheric science, stratospheric dynamics/transport, stratospheric polar processes and ozone loss.

Meason, John, Ph.D., University of Arkansas. Nuclear physics, nuclear and space radiation effects, electromagnetic radiation effects and directed energy.

Mihalas, Dimitri, Ph.D., Caltech, Theoretical astrophysics, stellar atmospheres.

Myers, Steven, Ph.D., California Institute of Technology. Cosmology, extragalactic radio astronomy, interferometric imaging algorithms.

Owen, Frazier, Ph.D., Radio Galaxies.

Rison, William, Ph.D., University of Utah, 1980. Atmospheric electricity, radar meteorology, instrumentation.

Rupen, Michael, Ph.D., Princeton University. Gas and dust in galaxies; radio transients.

Ryan, Eileen, Ph.D., University of Arizona, 1992. Asteroid collisional physics, observational and theoretical studies.

Ryan, William, Ph.D., University of Arizona, 1992. Asteroid astronomy; high energy physics.

Taylor, Gregory, Ph.D., University of California, Los Angeles. Very long baseline radio astronomy, active galactic nuclei.

Teare, Scott, Ph.D., University of Guelph, Canada, 1991. Adaptive optics, instrumentation, astrophysics.

Thomas, Ronald J., Ph.D., Utah State University, 1970. Atmospheric physics, instrumenation.

Wrobel, Joan, Ph.D., VLBI.

RESEARCH SPECIALTIES AND STAFF

Theoretical

Astrophysics. Colgate, Westpfahl, Young.

Atmospheric Phys. Atmospheric dynamics. Raymond.

Fluids and Rheology. Raymond.

Experimental

Astrophysics. Radio and x-ray astronomy; binary and variable stars; pulsars; cometary physics; young stars; evolved stars; astronomical instrumentation. Colgate, Goss, Hankins, Rupen, Taylor, Westpfahl, Creech-Eakman, Jurgenson, Haniff, Buscher, Elvis.

Atmospheric Phys. Clouds and lightning; electrical phenomena; pressure waves and pulses; weather radar; precipitation mechanisms; tornadoes; atmospheric charges in gas; radiative transfer and climate; physics of the upper atmosphere. Colgate, Eack, Krehbiel, Minschwaner, Raymond, Rison, Schery, Thomas, Winn. Ice tribocharging, embedded instrumentation. Sonnenfeld.

Energy Sources and Environment. Geothermal energy. Romero.

Nuclear Physics. Nuclear instrumentation. Romero.

Shock Physics. Romero.

NEW MEXICO STATE UNIVERSITY

DEPARTMENT OF ASTRONOMY

Las Cruces, New Mexico 88003

Students Accepted For Degree	FIELDS		
	Physics	Astronomy	Related Fields
Doctorate		X	
Master's		X	

1. General

President: B. Couture
Dean of Graduate School: Linda Lacey
Department Head: Jim Murphy
Department Telephone Number: (575) 646-4438 (C)
Type of Institution: University
Control: Public
Setting: Small city
Total Faculty: 1,215
Total Graduate Faculty: 665
Total Students: 18,497
Total Graduate Students: 3,377
Annual Graduate Tuition:
In-state residents: Full-time—$242/credit
Part-time—$242/credit
Out-of-state residents: Full-time—$17,088
Part-time—1–6 cr.—$712/credit
Tuition rates for: 2010–2011
Deferred tuition plan: Yes
Annual Other Fees: None
Term: Semester

2. Number of Faculty in Department

The combined total of full-time faculty in the three professorial ranks is 10. The combined total of full-time, part-time, and other faculty at all ranks is 14. We also have ~20 adjunct professors.

3. Admission, Financial Aid, and Housing

Address admission inquiries to: Dr. Chris Churchill, Dept. of Astronomy, Box 30001/MSC 4500
Graduate application fee required: $30
Admission deadline (Fall admission): 2/01
Admission information: For fall admission, 2010–11, 5 students were accepted.
Admission requirements: For admission to the graduate programs, a Bachelor's degree in astronomy, physics, other science, or engineering is required with a minimum undergraduate GPA of 3.0. The GRE and the GRE Advanced Physics scores are required. No minimum score for admission is specified. Typical average scores for GRE verbal–570; quantitative–700. Students from non-English speaking countries are required to demonstrate proficiency in English via the TOEFL exam. Minimum acceptable score for admission is 530.
Undergraduate preparation assumed: Math: differential equations; Physics: mechanics, modern physics, some of optics, electricity and magnetism, statistical mechanics, thermodynamics, etc.
Address financial aid inquiries to: Department of Astronomy, Box 30001/MSC 4500
GAPSFAS application required: No
Financial aid deadline: 2/15
Loans available: Yes
Address housing inquiries to: Housing Department, Box 30001/MSC 3BB
On-campus, single student housing available: Yes
Cost/term: $1,550–2,475
On-campus, married student housing available: Yes
Cost/month: $541–735/month

Table A—Faculty, Enrollments, and Degrees Granted

Research Specialty	2009–10 Faculty	Enrollment[1] Fall 2009 Master's	Enrollment[1] Fall 2009 Doctorate	No. of Degrees Granted[2] 2009–10 (2006–10) Master's	Terminal Master's	Doctorate	Median No. of Years for 2007–08 Ph.D.'s
Astronomy	14	13	17	—	2(4)	2(10)	5.5
Total		13	17	—	2(4)	2(10)	
Full-time Grad. Stud.		13	17				
Part-time Grad. Stud.		0	0				
First-year Grad. Stud.		6					
Median Years in Grad. Study (2008–2009 Degrees)				2.9	2.2	5.4	
Undergraduate Degrees: None							

[1]Students not yet committed to a research specialty are entered under non-specialized.
[2]Five-year totals in parentheses.

4. Graduate Degree Requirements

Master's: The M.S. program is closely geared to the Ph.D. program and prospective students are requested to contact the department.
Doctorate: A minimum of 64 credits of graduate work in astronomy and related fields of which 33 are in formal courses. Qualification is ascertained during the student's third semester and a comprehensive oral given after formal course work is completed. Written exam evaluations are based on monthly cumulative examinations. A dissertation and a final oral examination on the dissertation is also required. The residence requirement is two consecutive semesters of full-time graduate work after the first 30 credits.
Thesis: Thesis may be written *in absentia*.
Special Equipment, Facilities, or Programs: The Department is a member of the Astrophysical Research Consortium (ARC), which operates a state-of-the-art 3.5 m telescope at Apache Point, NM and the Sloan Digital Sky Survey with a dedicated 2.5-m telescope. The Department also operates a 1m telescope at Apache Point. The department is home to NASA's Planetery Data System's Atmospheres Node archive of planetary atmosphere-related mission data.

Table B—Appointments to Graduate Students, 2006–07

Title of Appointee	Appointments Total	First year	Academic Load Allowed in Credit Hours	Hours of Service Per Week	Stipend for Academic Year ($)*
			Semester		
Teaching Assistant	10	5	10	20	15,800–16,000
Research Assistant	16	1	10	20	15,800–16,000
Total	26	6			

*Calendar year; approximately $20,125–$20,770.

5. Personnel Engaged in Separately Budgeted Research, 7/07–6/08

Professorial faculty	14
Graduate students	30
Total	44

6. Separately Budgeted Research Expenditures by Source of Support

	Departmental Research
State and College Appropriation	$382,172
Apache Point Observatory	953,371
Federal government	$1,227,596
Total	$2,563,139

7. Separately Funded and Managed Laboratories

New Mexico State University Observatory operations	$328,096
Apache Point Observatory	3,209,857
Total	$3,119,662

Table C—Separately Budgeted Research Expenditures

Research Specialty	No. of Grants	Expenditures ($)
Astronomy	44	2,563,139
Total	44	2,563,139

FACULTY

Professors

Holtzman, Jon, Ph.D., Santa Cruz, 1989. Stellar population ; spiral galaxies; instrument development.
Klypin, Anatoly, Ph.D., Moscow, 1980. Extragalactic astronomy; cosmology.
McNamara, Bernard J., Ph.D., California, Santa Cruz, 1975. Observational astronomy; photoelectric photometry; star formation; cluster photometry and membership.
Walterbos, Reinirus, Ph.D., Leiden, 1986. Interstellar medium; stellar populations; extragalactic.

Associate Professors

Churchill, Chris, Ph.D., California, Santa Cruz, 1997. Quasar absorption lines galaxies and intergalactic medium.
Murphy, James, Ph.D., Washington, 1991. Head of the Department. Atmospheric Sciences; Planetary Atmospheres, Mars exploration missions.
Vogt, Nicole, Ph.D., Cornell, 1995. Galaxies, Galaxy Evolution.

Assistant Professors

Chanover, Nancy, Ph.D., NMSU, 1997. Planetary atmosphere.
Jackiewicz, Jason, Ph.D. Doston U, 2002. Helioseismology.
McAteer, James, Ph.D., Queens U., Belfast, 2003. Solar Physics.

College Professors

Beebe, Reta F., Ph.D., Indiana, 1969. Planetary atmospheres; planetary physics; radiative transfer; cool star atmospheres; equation of state.
Webber, William R., Ph.D., Iowa, 1957. Interplanetary physics; cosmic rays; delta-ray astronomy.

College Assistant Professor

Harrison, Tom, Ph.D., Minnesota, 1989. Infrared astronomy.

Research Associate

Johnson, Joni, Ph.D., Minnesota, 1990. Cataclysmic variables.

Professor Emeritus

Beebe, Herbert, Ph.D., Indiana, 1969. Atmospheres; spectral line formation.

RESEARCH SPECIALTIES AND STAFF

Theoretical

Interstellar medium. Walterbos.
Galaxies. Models of extragalactic objects; star formation. Vogt, Klypin.
Large Scale Structure. Holtzman, Klypin, Vogt.
Planetary Atmospheres. Murphy, Chanover.
Stars. Holtzman, Harrison, McNamara.
Solar Physics. Jackiewicz.

Experimental

Cluster Membership and Dynamics. Magnitude-color relationships; X-ray binaries. McNamara.
Cosmology. Galaxy clusters; large-scale structures. Klypin.
High-Energy Astrophysics. Harrison, McNamara, Webber.
Normal Galaxies. Interstellar medium; star formation. Anderson, Holtzman, Vogt, Walterbos.
Novae. Harrison, Johnson.
Planetary Studies. Positions and motions of Jupiter's atmospheric features; time variation of scattering properties of Jovian and Saturnian atmospheres. R. Beebe, Chanover.
Quasar Absorption Lines. Churchill.
Mars atmospheric modeling and data analysis. Murphy.
Stellar Populations. Holtzman, Walterbos Distant Galaxies, Vogt.

UNIVERSITY OF NEW MEXICO

DEPARTMENT OF PHYSICS AND ASTRONOMY

Albuquerque, New Mexico 87131

Students Accepted For Degree	FIELDS		
	Physics	Astronomy	Related Fields
Doctorate	X		X*
Master's	X		X*

*Optical Science & Engineering

1. General

President: David Schmidly
Dean of Graduate School: Amy Wohlert
Department Chairman: Bernd Bassalleck
Department Telephone Number: (505) 277-2616 (C)
Type of Institution: University
Control: Public
Setting: Urban
Total Faculty: 3,076
Total Graduate Faculty: Not separated
Total Students: 31,763
Total Graduate Students: 5,948
Annual Graduate Tuition:
In-state residents: Full-time—$5,546
Part-time—$233.20/hr.
Out-of-state residents: Full-time—$17,682
Part-time—$738.85/hr.
Tuition rates for: 2009–10
Deferred tuition plan: No
Other Fees: $50/year
Term: Semester

2. Number of Faculty in Department

The combined total of full-time faculty in the three professorial ranks is 27. The combined total of full-time, part-time, and other faculty at all ranks is over 50.

3. Admission, Financial Aid, and Housing

Downloadable application package and complete information online at: http://panda.unm.edu/acadadv/admissions.html
Address admission inquiries to: Coordinator of Program Advisement, Department of Physics and Astronomy, pandainfo@phys.unm.edu
Graduate application fee required: $50
Application deadlines:
International applicants and students who are seeking financial aid must submit materials no later than:
Fall semester: January 15
Spring semester: August 1
Deadlines for domestic students who are *not* seeking departmental financial aid are:
Fall semester: June 1
Spring semester: October 1
Admission information: Out of 119 applicants for Fall 2009 admission, 49 students were offered admission with financial support, and 2 students were admitted without departmental support.
Admission requirements: Students wishing to enter the M.S. or the Ph.D. program in Physics must have an undergraduate degree in physics or its equivalent. Admission to the Optical Science & Engineering graduate programs requires an undergraduate background that includes physics, optics, optical engineering, and/or optoelectronics. Additional information, specific criteria, application forms and instructions are available online at http://panda.unm.edu
Undergraduate preparation assumed: 1 semester of Thermal Physics; 1 year of Mechanics and of E & M.; 1 year of Quantum Physics strongly recommended.
Address financial aid inquiries to: Coordinator of Program Advisement, Department of Physics and Astronomy, pandainfo@phys.unm.edu
GAPSFAS application required: No
Financial aid deadline: January 15 (fall); August 1 (spring)
Departmental financial aid available: Yes
Address housing inquiries to: Housing Office
On-campus, single student housing available: Yes
Cost/term: approx. $3,750/sem. (includes meals)
On-campus, married student housing available: Yes
Cost/month: $570–770

Table A—Faculty, Enrollments, and Degrees Granted

Research Specialty	2009–10 Faculty	Enrollment[1] Fall 2009–10		No. of Degrees Granted[2] 2009–10 (2005–10)			Median No. of Years for 2009–10 Ph.D.'s
		Master's	Doctorate	Master's	Terminal Master's	Doctorate	
Astronomy/ Astrophysics	7	–	–	–	–	–	–
Biophysics	4	–	–	–	–	–	–
Condensed Matter Physics	2	–	–	–	–	–	–
Quantum Information Science	2	–	–	–	–	–	–
Nuclear and Particle Physics	6	–	–	–	–	–	–
Optics and Photonics	4	–	–	–	–	–	–
Other	1	–	–	–	–	–	–
Total							
Full-time Grad. Stud.		15	94				
Part-time Grad. Stud.		8	0				
First-year Grad. Stud.		5	25				
Median Years in Grad. Study (2008–09 Degrees)				3.5	–	6	–
Undergraduate Degrees, 2009–10: 11							

[1]Students not yet committed to a research specialty are entered under non-specialized.
[2]Five-year totals in parentheses.

4. Graduate Degree Requirements

Master's: (Master of Science in Physics and Master of Science in Optical Science and Engineering). Two plans are available for the M.S. Physics: I-A minimum of 24 hours of coursework with 16 hours of 500-level courses, six hours of thesis (599) credit; II-A minimum of 32 hours of coursework with 16 hours of 500-level courses. In either case a minimum GPA of "B" is expected and only 6 hours may be transferred. No foreign language required. Three options are available for the M.S.-Optical Science and Engineering: thesis, non-thesis, or coursework plus, an industrial internship.
Doctorate: (Ph.D. in Physics and in Optical Science and Engineering). The Ph.D. in Physics requires at least 48 semester hours (52 hours for Optical Sciences) exclusive of thesis and

dissertation with 24 hours of physics in courses above 500, exclusive of problems and research courses. Formal requirements are: coursework; two semesters in residence; departmental qualifying procedure; application for and admission to candidacy; the doctoral comprehensive examination; the dissertation; "B" average.

Thesis: Thesis may be written *in absentia*, when approved.

Special Equipment, Facilities, or Programs: Miscellaneous Research opportunities available locally at Sandia National Laboratories, the Air Force Research Laboratory, the VLA (Very Large Array Radio Telescope), Los Alamos National Laboratory. Other facilities with cooperative agreements: Fermi National Accelerator Laboratory, Brookhaven National Laboratory. Some graduate courses are taught at Los Alamos at UNM's Graduate Center by UNM campus faculty and by Los Alamos staff members.

Table B—Appointments to Graduate Students, 2008–09

Title of Appointee	Appointments Total	Appointments First year	Academic Load Allowed in Credit Hours	Hours of Service Per Week	Stipend for Academic Year ($)
			Semester		
Teaching Assistant	37	20	12	20	12,800
Research Assistant	56	3	12	20	12,000–30,000
Total	93	23			

5. Personnel Engaged in Separately Budgeted Research, 7/07–6/08

Professorial faculty	27
Graduate students	51
Nonteaching research personnel	32
Total	110

6. Separately Budgeted Research Expenditures by Source of Support

	Departmental Research
Total	~$6-8 million

FACULTY

Professors

Ahluwalia, Harjit S., Ph.D., Gujarat, 1960. Cosmic radiation; plasma physics; space physics; solar physics.

Bassalleck, Bernd, Ph.D., Karlsruhe, 1977. Chairman of the Department. Experimental nuclear and particle physics.

Cahill, Kevin, Ph.D., Harvard, 1967. Particle theory; lattice gauge theory; protein folding; medical physics.

Caves, Carlton M., Ph.D., California Institute of Technology, 1979. Information physics; quantum information theory; quantum chaos; quantum optics; quantum measurement theory.

Deutsch, Ivan H., Ph.D., California, Berkeley, 1992. Theoretical quantum optics; atom-optics; photonic devices; quantum information processing.

Diels, Jean-Claude, Ph.D., Brussels, 1973. Ultrashort and ultraintense pulse generation; amplification and measurement; ring lasers as gyros and sensors for small ($\ll$0.1 Angstrom) displacements; interaction of intense laser fields with matter; laser induced discharges.

Dunlap, David H., Ph.D., Rochester, 1987. Theory of charge transport in disordered molecular solids.

Finley, Daniel, Ph.D., California, Berkeley, 1968. General relativity; exact solutions of Einstein's field equations; gravitational radiation; complex manifolds and HH spaces; symmetries of nonlinear partial differential equations; (infinite-dimensional) Lie algebras of symmetries.

Gell-Mann, Murray, Ph.D., MIT, 1951. Complex systems; measures of complexity; decoherent histories in quantum mechanics.

Gold, Michael S., Ph.D., California, Berkeley, 1986. Experimental high-energy physics.

Kenkre, V. M., Ph.D., SUNY, Stony Brook, 1971. Theoretical solid-state physics; nonequilibrium statistical mechanics; nonlinear physics, granular materials, chemical physics.

Malloy, Kevin J., Ph.D., Standford, 1984. Semiconductor physics, device physics.

Matthews, John A. J., Ph.D., Toronto, 1971. High energy collider physics; extremely high energy cosmic ray physics; particle physics instrumentation.

McGraw, John T., Ph.D., Texas-Austin, 1977. Astronomy from the Moon; photometric surveys of stars, galaxies, and galactic structure.

Prasad, Sudhakar, Ph.D., Harvard, 1983. Optical interferometric imaging; information dynamics in image processing; quantum optics; optical fibers.

Rand, Richard J., Ph.D., Cal Tech, 1991. Interstellar matter and star formation in galaxies; disk-halo connection; starbursts; interstellar magnetic fields.

Rudolph, Wolfgang, Ph.D., Jena, 1985. Nonlinear optics; laser physics; optical spectroscopy and imaging; ultrashort light pulses and applications.

Seidel, Sally C., Ph.D., Michigan, 1987. Experimental elementary particle physics and instrumentation.

Sheik-Bahae, Mansoor, Ph.D., SUNY, Buffalo, 1987. Nonlinear optics; ultrafast phenomena; solid state physics.

Associate Professors

Fields, Douglas, Ph.D., Indiana, 1991. High energy spin physics; relativistic heavy-ion physics.

Henning, Patricia A., Ph.D., Maryland, 1990. Neutral hydrogen in galaxies; large-scale structure of the Universe; material content of cosmic voids.

Loomba, Dinesh, Ph.D., Boston, 1998. Gravitational lensing; cosmology.

Taylor, Gregory B., Ph.D., California, Los Angeles, 1991. Radio astronomy; active galaxies; time domain astrophysics.

Thomas, James L., Ph.D., Cornell, 1991. Biophysics.

Assistant Professors

Allahverdi, Rouzbeh, Ph.D., Alberta/Edmonton, 2000. Theoretical particle physics and cosmology.

Duan, Huaiyu, Ph.D., Minnesota, 2004. Neutrino and nuclear astrophysics.

Koch, Steven J. Ph.D., Cornell University. 2003. Biophysics.

Lidke, Keith, Ph.D., Minnesota, 2002. Physics/Biophysics.

Pihlström, Ylva M., Ph.D., Chalmers University of Technology, 2001. Radio astronomy; astrophysical instrumentation.

Jointly Appointed Faculty

Brueck, Steven R. J., Ph.D., MIT, 1971. Non-linear optics; optical lithography; silicon manufacturing methodology; nanoscale fabrication; laser-material interactions. (Electrical and Computer Engineering).
Heintz, Philip H., Ph.D., University of Washington, 1971. Biomedical physics (Radiology).
Jain, Ravinder K., Ph.D., California, Berkeley, 1974. Quantum electronics; optoelectronics; experimental solid state physics (Electrical and Computer Engineering).
Moore, Christopher, Ph.D., Cornell University, 1991. Computational physics (Computer Science).
Osiński, Marek, Ph.D., Polish Acad. of Sciences, 1979. Semiconductor lasers; optoelectronic devices and materials; group-III nitrides; degradation mechanisms and reliability; computer simulation. (Electrical and Computer Engineering).

Research Faculty

Ackerman, Mark R., Ph.D., University of New Mexico, 2002.
Atlas, Susan R., Ph.D., Harvard, 1988.
Boyd, Stephen T., Ph.D., California, Los Angeles, 1991.
Epstein, Richard, Ph.D. Stanford, 1972.
Gorelov, Igor V., Ph.D., IT
Hasselbeck, Michael, Ph.D., University of Central Florida, 1995.
Landahl, Andrew, Ph.D., Caltech, 2002.
Nampoothiri, Vasudevan, Ph.D. Indian Institute of Technology, 1999.
Schwoebel, Paul, Ph.D., Cornell, 1987.
Strologas, John, Ph.D., University of Illinois, 2002.
Thomas, Timothy L., Ph.D., Minnesota, 1995
Younus, Imran, Ph.D., Syracuse, 2003.
Zimmer, Peter, Ph.D., University of New Mexico, 2004.

Lecturers

Odom, Boye, M.S., University of Texas at El Paso, 1981.
Saul, Jeff, Ph.D., University of Maryland, 1998.

FACULTY RESEARCH SPECIALTIES

Astronomy and Astrophysics

Adaptive Optics and Interferometry: Diels, McGraw, Pihlström, Prasad, Taylor.
Cosmic-ray and High-energy Astrophysics: Ahluwalia, Gold, Matthews.
Extragalactic Astronomy: Henning, Loomba, Rand, Taylor.
Galactic Astronomy: McGraw.
Radio Astronomy: Henning, Pihlström, Rand, Taylor.
Solar and Space Physics: Ahluwalia.

Biophysics

Experimental: Koch, Lidke, Thomas
Theoretical: Cahill, Kenkre.

Condensed Matter Physics

Experimental: Malloy, Rudolph, Sheik-Bahae.
Theoretical: Dunlap, Kenkre.

Chemical and Molecular Physics: Kenkre, Rudolph.
Materials Physics: Kenkre, Malloy, Sheik-Bahae.
Transport Phenomena: Dunlap, Kenkre, Rudolph.
Tunneling Phenomena: Dunlap, Kenkre.

Quantum Information Science

Atomic Molecular Optical Physics: Caves, Deutsch.
Theoretical: Caves, Deutsch.

Optics

Experimental: Diels, Rudolph, Sheik-Bahae, Thomas.
Theoretical: Caves, Deutsch, Prasad.

Atomic Physics: Deutsch.
High-resolution Spectroscopy and Imaging: Diels, Prasad, Rudolph.
Laser Physics and Nonlinear Optics: Diels, Rudolph, Sheik-Bahae.
Quantum Optics: Caves, Deutsch, Prasad.
Ultrafast Phenomena: Diels, Rudolph, Sheik-Bahae.

Nuclear and Particle Physics

Experimental: Bassalleck, Fields, Gold, Loomba, Matthews, Seidel.
Theoretical: Cahill, Allahverdi, Duan.

Astro-particle Physics: Allahverdi, Duan, Gold, Loomba, Matthews.
Collider Physics: Gold, Seidel.
Fundamental Interactions and Supersymmetry: Allahverdi, Cahill, Gold.
New Particle Searches: Bassalleck, Gold.
Particle Physics Instrumentation: Bassalleck, Fields, Gold, Loomba, Matthews, Seidel.
QCD: Bassalleck, Cahill, Fields, Seidel.
Strangeness Nuclear Physics: Bassalleck.

Other Research Areas in Theoretical Physics

Computational Physics: Cahill.
General Relativity: Finley.
Nonlinear Science: Caves, Finley, Kenkre.
Statistical Physics: Caves, Kenkre.

UNIVERSITY OF NEW MEXICO

DEPARTMENT OF CHEMICAL AND NUCLEAR ENGINEERING

Albuquerque, New Mexico 87131

Students Accepted For Degree	FIELDS		
	Physics	Astronomy	Related Fields
Doctorate			X
Master's			X

1. General

President: Dr. David J. Schmidly
Dean of Graduate School: Dr. Charles B. Fleddermann, Acting Dean
Department Chair: Timothy L. Ward
Department Telephone Number: (505) 277-5431
Type of Institution: University
Control: Public
Setting: Urban
Total Faculty: 3,707
Total Students: 34,674
Total Graduate Faculty: Not separated
Total Graduate Students: 4,393
Annual Graduate Tuition:
In-state residents: Full-time—$4,197.60
Part-time—$233.20/credit
Out-of-state residents: Full-time—$6,244.20
Part-time—$329.40/credit*
Tuition rates for: 2009–10
Deferred tuition plan: No
Annual Other Fees: $40.00 (GPSA fee)
Term: Semester

*6 hrs., in-state rate

2. Number of Faculty in Department

The combined total of full-time faculty in the three professorial ranks is 17. The combined total of full-time, part-time, and other faculty at all ranks is 57.

3. Admission, Financial Aid, and Housing

Address admission inquiries to: Advisor Chemical and Nuclear Engineering Dept., chne@unm.edu
Graduate application fee required: $50
Admission deadline (Fall admission): 1/15 (priority)
Admission information: For Fall admission, 2009–10, 28 students were accepted from 70 applicants.
Admission requirements: For admission to the graduate programs, a Bachelor's degree in engineering or science is required with a minimum undergraduate GPA of 3.0/4.0 specified. The GRE is required. Students from non-English speaking countries are required to demonstrate proficiency in English via the TOEFL exam.
Undergraduate preparation assumed: For Chemical Engineering: Thermodynamics, Transport Phenomenon, Mass Transfer, Applied Ordinary Differential Equations. For Nuclear Engineering: Thermodynamics, Applied Ordinary Differential Equations, Neutron Diffusion. Recommended for Nuclear Engineering: Fluids and Heat Transfer, Nuclear Reactor Theory. Other coursework may be recommended by the Graduate Advisor.
Address financial aid inquiries to: Advisor, Department of Chemical and Nuclear Engineering, chne@unm.edu
GAPSFAS application required: No
Financial aid deadline: 3/1
Loans available: Yes
Address housing inquiries to: Housing Reservations, The University of New Mexico, Albuquerque, NM 87131
On-campus, single student housing available: Yes
Cost/term: $2,200–2,650/semester (meal plans)
Off-campus, married student housing available: Yes
Cost/month: $505–680

Table A—Faculty, Enrollments, and Degrees Granted

Research Specialty	2008–09 Faculty	Enrollment[1] Fall 2009		No. of Degrees Granted[2] 2009–10(2005–10)			Median No. of Years for 2009–10 Ph.D.'s
		Master's	Doctorate	Master's	Terminal Master's	Doctorate	
Biophysics	7	3	11	(3)	0(1)	1(4)	–
Chemical Physics	0	0	0	0(0)	0(0)	0(0)	–
Energy Sources & Environ.	1	2	2	4(7)	0(2)	0(5)	–
Engineering Physics/ Science	0	0	0	0(0)	0(0)	0(0)	–
Materials Sci./ Metallurgy	7	1	10	0(12)	0(2)	4(20)	5
Medical & Health Physics	2	2	0	0(0)	0(10)	0(0)	2
Nuclear Engineering	6	21	14	4(21)	0(5)	0(12)	6
Plasma Physics & Fusion	0	0	0	0(0)	0(0)	0(0)	–
Statistical & Thermal	0	0	0	0(0)	0(0)	0(0)	–
Systems Science/ Engineering	0	0	0	0(0)	0(0)	0(0)	–
Total		29	38	44	20	41	
Full-time Grad. Stud.		29	37				
Part-time Grad. Stud.		0	1				
First-year Grad. Stud.		4	4				
Median Years in Grad. Study (2008 Degrees)				3	3	5	–

Undergraduate Degrees, 2009–10 (2005–10): 21(126)

[1]Students not yet committed to a research specialty are entered under non-specialized.
[2]Five-year totals in parentheses.

4. Graduate Degree Requirements

Master's: Plan I: A minimum of 24 semester-hours of course work exclusive of thesis; a minimum of 9 hours of 500-level course work; at least 18 hours completed in residence at the University; a minimum of 6 hours of thesis credit; a limit of 3 hours of problems courses. Plan II: A minimum of 33 semester-hours of course work; a minimum of 15 hours of 500-level course work; at least 26 hours completed in residence at the University; a limit of 6 hours of problems courses and 8 hours of workshop credit; and, if a major is declared, a minimum of 18 hours in the major and 12 hours in the minor. A grade-point average of 3.0 minimum ("B") is required.
Doctorate: A minimum of 48 semester hours of course work beyond the Bachelor's degree, exclusive of dissertation is required. Twenty-four (24) hours must be completed at UNM, with a maximum of 6 hours of problems courses beyond Master's degree and a minimum of 18 dissertation hours. A maximum of 30 credit-hours completed for the Master's de-

gree, including 6 hours of thesis, may be counted. (A maximum of 30 hours may be transferred from another accredited graduate school provided: grades of "B" or higher were obtained, the student has completed 12 hours in residence at the University of New Mexico, and the transfer is approved by the Dean of Graduate Studies.) At least 18 hours after the Master's degree, exclusive of dissertation, must be 500-level completed at the University of New Mexico. As a general rule, all work toward the requirements for the doctorate must be completed within a 10-year period. A 3.0 ("B") minimum grade point average is required.

Thesis: Thesis may be written *in absentia*.

Special Equipment, Facilities, or Programs: In addition to the well-equipped laboratories on campus, facilities of the Sandia National Laboratories, the Los Alamos National Laboratory, and the Air Force Research Laboratory are utilized for both instruction and research.

Table B—Appointments to Graduate Students, 2009–10

Title of Appointee	Appointments		Academic Load Allowed in Credit Hours	Hours of Service Per Week	Stipend for Academic Year ($)
	Total	First year			
			Semester		
Research Assistant	48	10	12	20	20,400[1–3]
Total	48	10			

[1]Summer support of nominally $3,900.
[2]Tuition waived, fee of $40 paid by student.
[3]Minimum salary.

FACULTY

Professors

Atanassov, Plamen, Ph.D., Bulgaria, 1992. Electrochemical and sensor systems.

Brinker, C. Jeffrey, Ph.D., Rutgers, 1978. Synthesis and processing of porous and composite nanostructures; sol-gel processing; thin films and membranes.

Cecchi, Joseph L., Ph.D., Harvard, 1972. Electronic materials and thin films; plasma etching and deposition.

Datye, Abhaya K., Ph.D., Michigan, 1984. Associate Chair. Heterogeneous catalysis; electron microscopy; materials characterization.

de Oliveira, Cassiano R. E., Ph.D., Queen Mary College, University of London, 1986. Computational radiation transport, nuclear reactor physics, high-performance computing, mathematical modeling of engineering systems.

El-Genk, Mohamed S., Ph.D., New Mexico, 1978. Thermalhydraulic design; reactor safety; reactor design.

Fulghum, Julia E., Ph.D., North Carolina, 1987. Analytical Chemistry; materials characterization; data visualization and analysis.

Prinja, Anil K., Ph.D., Queen Mary College, University of London, 1980. Associate Chair. Radiation transport methods; transport theory; medical physics.

Ward, Timothy L., Ph.D., Washington, 1989. Department Chair. Aerosol processes; aerosol synthesis of materials; ceramic membrane fabrication and characterization.

Associate Professors

Cooper, Gary W., Ph.D., Illinois, 1976. Plasma and fusion engineering.

Edwards, Jeremy, Ph.D., UCSD, 1999. Bioengineering, DNA sequencing technology, systems biology.

Graves, Steven W., Ph.D., Pennsylvania State University, 1998. Protein engineering, Medical diagnostics, Biomedical engineering.

Han, Sang M., Ph.D., California, 1998. Compound semiconductor quantum devices, heterogeneous surface phenomena.

Assistant Professors

Canavan, Heather, Ph.D., George Washington Univ., 2002. Biomaterials, cell/surface interactions, biomedical engineering.

Chi, Eva Y., Ph.D., University of Colorado, Boulder, 2004. Protein folding, dynamics, & stability; Alzheimer's disease.

Dirk, Elizabeth, Ph.D., Rice Univ., 2004. Biomaterials, tissue engineering, drug delivery, biomedical engineering.

Hecht, Adam A., Ph.D., Yale University, 2004. Radiation detection development, nuclear nonproliferation, radiotherapy simulation.

Petsev, Dimiter, Ph.D., University of Sofia, 1996. Complex fluids, nanoscience, electrokinetic phenomena.

Lecturer III

Busch, Robert D., Ph.D., New Mexico, 1976. Nuclear criticality safety; environmental radiation measurements; reactor physics.

Faculty Emeriti

Anderson, Harold M., Ph.D., Wayne State, 1981. Computer simulation of transport phenomena; radioactive waste disposal; wastewater treatment; plasma processing.

Kauffman, David, Ph.D., Colorado, 1970. Geothermal energy; environmental analysis; process design.

Mead, Richard W., Ph.D., Arizona, 1971. Process analysis; hydrometallurgy; fossil energy.

Roderick, Norman F., Ph.D., Michigan, 1971. Plasma and fusion physics; computer modeling.

Adjunct Professors (Letter of Academic Title)

Apblett, Christopher A., Ph.D., Rensselaer Polytechnic Institute, 1992. Research Associate Professor.

Boyle, Timothy J., Ph.D., University of Kansas, 1990. Research Professor.

Brown, Forrest B., Ph.D., University of Michigan, 1981. Research Professor.

Brown, Lee F., Ph.D., University of Delaware, 1963. Research Professor.

Brozik, Susan M., Ph.D., Washington State University, 1994. Research Associate Professor.

Curro, John G., Ph.D., California Institute of Technology, 1969. UNM/NL Professor.

Fan, Hongyou, Ph.D., University of New Mexico, 2000. Research Assistant Professor.

Frink, Laua J., Ph.D., University of Illinois, 1995. Adjunct Assistant Professor.

Evans, Evan A., Ph.D., University of California, San Diego, 1970. Resarch Professor.

Jackson, Nancy, Ph.D., University of Texas, 1990. Research Associate Professor.

King, R. Barry, M.S., University of Houston, 1973. Adjunct Instructor.

Kuznetsov, Igor R., Ph.D., University of Illinois, Urbana-Champaign, 2005. Research Scholar.

Martineau, Richard C., Ph.D., University of Idaho, 2002. Research Professor.

McDaniel, Patrick J., Ph.D., Purdue, 1977. Research Professor.

Medforth, Craig J., Ph.D., University of Liverpool, UK, 1988. Synthesis, characterization and applications of porphyrins and related molecules.

Miller, Warren "Pete" F., Jr., Ph.D., Northwestern University, 1973. Research Professor.

Ruffner, J. Heidi, Ph.D., University of Arizona, 1993. Research Associate Professor,

Shreve, Andrew P., Ph.D., Cornell University, 1991. Research Professor.

Stein, David, Ph.D., University of New Mexico, 1998. Research Assistant Professor.

Taylor, Glenn A., Ph.D., University of North Carolina, 1973. Adjunct. Adjunct Professor.

Tsai, Chung-Yi, Ph.D., Worcester Polytechnic Institute, 1996. Research Assistant Professor.

van Swol, Frank, Ph.D., University of Amsterdam, 1978. Research Professor.

Warsa, James S., Ph.D., University of New Mexico, 1998. Research Professor.

Research Professors

Arthur, Edward, Ph.D., University of Virginia, 1972. Advanced nuclear fuel cycles, nuclear nonproliferation.

Artyushkova, Kateryna, Ph.D., Kent State University, 2001. Materials characterization, data-fusion and data analysis methods to enhance characterization of complex materials and systems.

Challa, Sivakumar R., Ph.D., University of Pittsburgh, 2001. Molecular modeling: theory, simulations, and algorithms. Fluids: Adsorption, thermodynamics, and nano-hydrodynamics. Materials: thermo-physical properties, design, and characterization.

Dunphy, Darren R., Ph.D., University of Arizona, 1999. Synthesis and characterization of self-assembled materials, hybrid materials, Analytical Chemistry.

Ivnitski, Dmitri, Ph.D., Moscow State University, 1989. Analytical biochemistry, electrochemical biosensors, immunosensors, biofuel cells, direct electron transfer.

Loehman, Ronald E., Ph.D., Purdue, 1969. Ceramics engineering.

Pham, Hien, Ph.D., Univ. of New Mexico, 2001. Chromium oxide products; synthesis and characterization of mesoporous oxides.

Sibbett, Scott S., Ph.D., Oregon Health Sciences University, 1985. Research Professor and Co-Director of Center for Biomedical Engineering. Biophysics and bioseparations.

Tournier, Jean-Michel, Ph.D., University of New Mexico, 1996. Thermal management and power systems.

Whitten, David, Ph.D., The Johns Hopkins Univ., 1963. Research Professor and Co-Director of Center for Biomedical Engineering. Editor-in Chief, Langmuir. Biophysics; chemical physics; materials science/metallurgy. Sensors; photophysics of dyes and conjugated polyelectrolytes; biocidal films.

RESEARCH SPECIALTIES AND STAFF

Theoretical

Biophysics. Canavan, Chi, Dirk, Edwards, Graves, Sibbett.

Energy Sources and Environment. Alternate energy sources; fossil energy sources; environment. Atanassov, Busch, Datye, El-Genk.

Engineering Physics. Mathematical modeling of engineering systems; radiation transport; radiation interaction with matter. Prinja, Roderick, de Oliveira.

Fluids and Rheology. Computational fluid mechanics. Petsev.

Fusion and Plasmas. Accelerator technology; plasma physics; plasma processing. Cecchi, Cooper, El-Genk, Prinja, Roderick.

Materials Science. Electron microscopy; characterization of materials; surface science. Artyushkova, Cecchi, Datye, Ward.

Medical and Health Physics. Busch, Cooper, Graves, Hecht.

Nuclear Engineering, Reactor design; safety/surety; space reactors; thermal hydraulics; nuclear criticality radiation transport. Busch, Cooper, El-Genk, Prinja, Roderick, de Oliveira.

Experimental

Biophysics. Atanassov, Chi, Dirk, Edwards, Graves, Sibbett, Whitten.

Bioseparations. Sibbett.

Energy Sources and Environment. Alternate energy sources; fossil energy sources; environment. Atanassov, Busch, Datye, Ward.

Engineering Physics. Radiation diagnostics. Busch, Cooper.

Fluids and Rheology. Thermal hydraulics; hydrometallurgy. El-Genk.

Fusion and Plasmas. Accelerator technology; plasma and fusion engineering. Cecchi, Cooper, Roderick.

Materials Science. Electron microscopy; surface science; aerosol physics. Brinker, Cecchi, Datye, Fulghum, Ward, Whitten.

Medical and Health Physics. Busch, Cooper, Graves, Hecht.

Nuclear Engineering. Reactor safety; thermal hydraulics; nuclear criticality. Busch, Cooper, El-Genk.

Physics of Beams. Cooper.

Radiation Detection and Measurement. Hecht.

BINGHAMTON UNIVERSITY (STATE UNIVERSITY OF NEW YORK)

DEPARTMENT OF PHYSICS, APPLIED PHYSICS, AND ASTRONOMY

Binghamton, New York 13902-6000

Students Accepted For Degree	FIELDS		
	Physics	Astronomy	Related Fields
Doctorate	X		
Master's	X		X

1. General

President: Lois B. Defleur
Vice-Provost and Dean of Graduate School: Nancy Stamp
Department Chairman: Eric Cotts
Department Telephone Number: (607) 777-2217
Type of Institution: University
Control: Public
Setting: Suburban
Total Faculty: 848
Total Graduate Faculty: 504
Total Students: 14,435
Total Graduate Students: 2,920
Annual Graduate Tuition:
In-state residents: Full-time—$8,370
Part-time—$349/credit hr.
Out-of-state residents: Full-time—$13,250
Part-time—$552/credit hr.
Tuition rates for: 2010–11
Deferred tuition plan: No
Other Fees: $1,791 (maximum)
Term: Semester

2. Number of Faculty in Department

The combined total of full-time faculty in the three professorial ranks is 10. The combined total of full-time, part-time, and other faculty at all ranks is 15.

3. Admission, Financial Aid, and Housing

Address admission inquiries to: E. Cotts, Chairman, Department of Physics, Applied Physics, and Astronomy
Graduate application fee required: $60 if filed online; $75 if paper application
Admission deadline (Fall admission): 2/15 (flexible)
Admission information: For spring admission, 2010, 2 students were accepted.
Admission requirements: For admission to the graduate programs, a Bachelor's degree is required with no minimum undergraduate GPA specified. Specialization in physics on the undergraduate level is desirable, but not essential for admission. Two letters of reference required. The GRE is recommended. No minimum acceptable score for admission is specified. The GRE Advanced is recommended. No minimum acceptable score for admission is specified. Students from non-English speaking countries are required to demonstrate proficiency in English via the TOEFL exam. Minimum acceptable score for admission is 550.
Undergraduate preparation assumed: One year of General Physics; one year of Electromagnetic Theory-Marion and Heald, *Classical Electromagnetic Radiation*; one semester of Dynamics-Barger and Olsson, *Classicial Mechanics: A Modern Perspective*; at least a semester of Quantum Mechanics Cohen-Tannoudji *et al.*, *Quantum Mechanics*; and Mathematics through differential equations. Appropriate laboratory experience at the upper undergraduate levels is desirable.
Address financial aid inquiries to: E. Cotts, Chairman, Department of Physics, Applied Physics, and Astronomy
GAPSFAS application required: No
Financial aid deadline: None
Loans available: Yes
Address housing inquiries to: Director of Graduate Housing
On-campus, single student housing available: Yes
Cost/term: $4,974-$11,406
On-campus, married student housing available: Yes
Cost/term: Double $7,608, Family $11,328

Table A—Faculty, Enrollments, and Degrees Granted

Research Specialty	2009–10 Faculty	Enrollment[1] Fall 2009 Mas-ter's	Enrollment[1] Fall 2009 Doc-torate	No. of Degrees Granted[2] 2009–10 (2004–09) Mas-ter's	Terminal Master's	Doc-torate	Median No. of Years for 2004–09 Ph.D.'s
Applied Physics	5		0(0)	3(5)	0(0)	–	–
Atomic, Molecular, & Optical Physics	1		0(0)	0(0)	0(0)	–	
Biophysics	1	0	0(0)	0(0)	0(0)	–	–
Chemical Physics	1	0	0(0)	0(0)	0(0)	–	–
Condensed Matter Physic	5	4	0(0)	1(2)	0(0)	–	–
Energy Sources & Environ.	1	0	0(0)	0(0)	0(0)	–	–
Engineering Physics/ Science	1	0	0(0)	0(0)	0(0)	–	–
Fluids & Rheology	1	0	0(0)	0(2)	0(0)	–	–
Low Temperature Physics	5	0	0(0)	0(0)	0(0)	–	–
Materials Sci./ Metallurgy	2	0	0(3)	0(2)	0(0)	–	–
Medical & Health Physics	1	0	0(0)	0(0)	0(0)	–	–
Optics	1	0	0(0)	0(1)	0(0)	–	–
Particles & Fields	2	0	0(0)	1(4)	0(0)	–	–
Physics Education	1	4	0(0)	0(0)	0(0)	–	–
Statistical & Thermal	1	0	0(0)	0(2)	0(0)	–	–
Other Theoretical/ Math.	2	2	0(0)	0(0)	0(0)	–	–
Non-specialized		5	0(0)	0(0)	0(0)	–	–
Total		16	3	5(15)	0(0)		
Full-time Grad. Stud.		16	3				
Part-time Grad. Stud.		0	0				
First-year Grad. Stud.		3	0				
Median Years in Grad. Study (2003–09 Degrees)				–	2	–	–
Undergraduate Degrees, 2008–09 (2003–09): 13(71)							

[1] Students not yet committed to a research specialty are entered under non-specialized.
[2] Five-year totals in parentheses.

4. Graduate Degree Requirements

Master's: 30 graduate credit hours with at least a B average. Two-semester residence requirement. Choice of thesis or comprehensive exam.
Doctor of Philosophy: at least 24 credit hours of course study (in residence) and 24 additional credit hours of dissertation work. Passing a written qualifying examination, in three parts, cov-

ering the core areas of physics. Successful defence of dissertation.

Other Programs: Master of Science in Physics with a Specialization in Applied Physics is designed for students seeking careers in applied physics. Emphasis is to provide comprehensive education in fundamental physical principles and their applications to enhance the ability to evolve with changing technology and avoid technical obsolescence. Student may study part-time and complete degree in three years, or full-time graduate assistant and complete degree in two years or less. Thesis topics may be drawn from employment, with consent of department and employer.

M.A.T. and M.S.T. programs are designed for students who wish to teach physics at the secondary level. The M.A.T. program is designed for students with a physics background who need education courses; the M.S.T. program is designed for teachers who want to improve their physics background. The credits in professional education courses required for certification are offered, as well as additional work in physics and allied fields. Certified teachers may enroll in the M.S.T. program, in which almost all the training involves substantive physics course work.

Thesis: Thesis may be written *in absentia*.

Special Equipment, Facilities, or Programs: Helium liquefier, AC and DC magnetic susceptibility bridges, x-ray diffractometers, 100,000 kilogauss superconducting magnet, 15″ iron core magnet, differential scanning calorimeters, sputtering equipment, vacuum deposition stations, dilution refrigerator, raman spectrometer, clean room, splat quencher, resistivity bridges, hydrator, cryo-cooler, dielectric analyzer, thermogravimetric analyzer, dynamic mechanical analyzer, thermomechanical analyzer, scanning electron microscope, high pressure intensifier, and a squid magnetometer.

Table B—Appointments to Graduate Students, 2009–10

Title of Appointee	Appointments: Total	Appointments: First year	Academic Load Allowed in Credit Hours	Hours of Service Per Week	Stipend for Academic Year ($)
			Semester		
Teaching Assistant	11	3	12	20	18,000[1]
Research Assistant	5	0	12	20	18,000[2]
Total	16	3			

[1] Appointment also includes tuition waiver.
[2] Appointment also includes tuition waiver.

5. Personnel Engaged in Separately Budgeted Research, 2007–08

Professorial faculty	2
Total	2

6. Separately Budgeted Research Expenditures by Source of Support

Information not available.

Table C—Separately Budgeted Research Expenditures

Research Specialty	No. of Grants	Expenditures ($)
	Information not available	

FACULTY

Professors

Cotts, Eric J., Ph.D., Illinois, 1983. Chairman of Department. Experimental solid state.

Nelson, Charles A., Ph.D., Maryland, 1968. Theoretical high-energy physics; applied physics.

Suzuki, Masatsugu, Ph.D., Tokyo, 1977. Experimental solid state.

Wu, Tsu-Ming, Ph.D., Pennsylvania, 1966. Theoretical solid state; biophysics; applied physics.

Associate Professors

Pompi, Robert L., Ph.D., Cornell, 1968. Experimental solid state; applied physics.

Venugopalan, Srinivasa, Ph.D., Purdue, 1973. Experimental solid state; laser spectroscopy.

White, Bruce E., Jr., Ph.D., Cornell University, 1995. Solid state physics.

Assistant Professors

DeSilva, Theja, Ph.D., University of Cincinnati, 2004. Condensed matter theory, spin-orbit physics.

Lawler, Michael, Ph.D., University of Illinois, Urbana-Champaign, 2006.

RESEARCH SPECIALTIES AND STAFF

Theoretical

Applied Physics. DeSilva, Wu.

Elementary Particle Physics. Quantum Field Theory, and Particle Astrophysics Cosmology. Nelson.

Solid State Physics. Many-body problems in solid state physics and biophysics. DeSilva, Lawler, Wu.

Experimental

Applied Physics. Cotts, Pompi, Venugopalan.

Condensed Matter Physics. Low-temperature condensed matter; localized magnetic moments in metallic crystals; induced valence changes in impurity doped metals; Raman spectroscopy; properties of disordered materials, amorphous metals, layered materials. Cotts, DeSilva, Pompi, Suzuki, Venugopalan, White.

Laser Spectroscopy. Pompi, Venugopalan.

Liquid Crystals. Venugopalan.

Research and Adjunct Teaching Professors

Krentsel, Tatiana. Adjunct Lecturer, Ph.D., Institute of Crystallography, Russian Academy of Sciences, Moscow, Russia, 1994. Physics education

Poliks, Barbara. Research Associate Professor, PhD, Jagiellonian University, Poland, 1982. Computer simulations of polymeric systems including proteins and materials.

Sileo, Richard. Adjuct Lecturer, Ph.D., Applied and Engineering Physics, Cornell University, 1974. Physics education.

Telesca, Andrew, Adjunct Teaching Professor, M.S., State University of NY at Oneonta, 1976. Astronomy and Physics Education.

Wagoner, Shawn, Research Assistant Professor, MS, Physics, North Carolina State University, 1994. Nanofabrication education.

CITY COLLEGE OF THE CITY UNIVERSITY OF NEW YORK

DEPARTMENT OF PHYSICS

New York, New York 10031

Students Accepted For Degree	FIELDS		
	Physics	Astronomy	Related Fields
Doctorate	X		X
Master's	X		X

1. General

President Interim: Robert Paaswell
Acting Dean of Science: Ruth Stark
Department Chairman: Marilyn Gunner
Department Telephone Number: (212) 650-6832
email address: physdept@sci.ccny.cuny.edu
Type of Institution: University
Control: Public
Setting: Urban
Total Faculty: 574[1]
Total Graduate Faculty: 260[1]
Total Students: 15,507[2]
Total Graduate Students: 3,031[3]
Semester Graduate Tuition:
Master Instate: $3,680
Out of State: $575 per credit
Ph.D.: Instate: $370 per credit; Out-of-state: $645 per credit

	Level I[4]	Level II[4]	Level III[4]
In-state residents:	$3,290	$2,060	$815
Out-of-state residents:	$7,713	$4,580	$1,635

Tuition rates for: 2009–10
Deferred tuition plan: Yes
Other Fees: $125
Term: Semester
Web address: http://www.gc.cuny.edu

[1]Full-time faculty.
[2]Full-time equivalents.
[3]Only students in Master's programs are officially listed as City College students. Students in the Ph.D. programs are listed as students of the City University. Neither figure corresponds to what would normally be listed by other institutions, so no number is included here.
[4]The rates quoted are per semester, and correspond to our latest information. Level I—Students who have not yet completed 45 credits of graduate work, including transfer credit. Level II—Students who have completed 45 credits, and passed the First Exam but have not yet completed all their course work. Level III—Students who have completed their course work, and their Second Exam but have not yet completed their degree.

2. Number of Faculty in Department

The combined total of full-time faculty in the three professorial ranks is 27.

3. Admission, Financial Aid, and Housing

Address admission inquiries to: Marilyn Gunner, Chair, Department of Physics
Graduate application fee required: $125
Admission deadline (Fall admission): None
Admission information: For fall admission, 2009–10, 7 students were accepted from 80 applicants.
Admission requirements: [1]For admission to the graduate programs, a Bachelor's degree in physics or a related field is required with a minimum undergraduate GPA of "B" in physics and mathematics courses. However, other factors, such as letters of recommendation are often more important. The GRE is required. No minimum acceptable score specified. The GRE Advanced is required. No minimum acceptable score specified. The average GRE scores for 1995–96 admissions were verbal–390; quantitative–830: total–1,220. The average GRE Advanced score for 1995–96 admissions was 780. Students from non-English speaking countries are required to demonstrate proficiency in English via the TOEFL exam. No minimum score required.
Address financial aid inquiries to: Marilyn Gunner, Chair, Department of Physics
GAPSFAS application required: No
Financial aid deadline: None
Loans available: Yes
On-campus, single student housing available: Yes
On-campus, married student housing available: No

[1]This corresponds to admission into the Ph.D. program.

Table A—Faculty, Enrollments, and Degrees Granted

Research Specialty	2009–10 Faculty	Enrollment[1] Fall 2009		No. of Degrees Granted[2] 2008–09 (2005–10)			Median No. of Years for 2009–10 Ph.D.'s
		Master's	Doctorate	Master's	Terminal Master's	Doctorate	
Applied Physics	3	–	1	–	–	0(1)	0
Atomic & Molecular Physics	2	–	0	–	–	0(0)	–
Biophysics	3	–	6	–	–	0(3)	–
Chemical Physics	1	–	2	–	–	0(0)	–
Condensed Matter Physics	10	–	14	–	–	2(8)	5
Electromagnetism	2	–	0	–	–	0(0)	–
Elem. Particles & Fields	4	–	2	–	–	0(2)	–
Fluids & rheology	4	–	5	–	–	0(2)	5
Low Temperature Physics	2	–	0	–	–	0(0)	–
Medical & Health Physics	2	–	0	–	–	0(1)	–
Optics	5	–	1	–	–	0(2)	–
Physics Education	2		1			1(1)	7
Relativity	2	–	0	–	–	0(0)	–
Statistical & Thermal	5	–	0	–	–	0(0)	–
Other Theoretical Math.	2	–	0	3(21)	3(21)	0(1)	
Total			30	3(21)	3(21)	3(19)	
Full-time Grad. Stud.		7	30				
Part-time Grad. Stud.		4	0				
First-year Grad. Stud.		3	7				
Median Years in Grad. Study (2009–10 Degrees)				–	2	6	6
Undergraduate Degrees, 2008–09 A(2004–09): N/A							

[1] Students not yet committed to a research specialty are entered under non-specialized.
[2] Five-year totals in parentheses.

4. Graduate Degree Requirements

Masters: Students are required to take whatever course work may be necessary for them to achieve Master's level competence in mathematical physics, quantum mechanics, analytical dynamics, and electromagnetic theory. Ordinarily this will require at least one semester of a course in each area. No more than nine credits may be taken in advanced undergraduate courses. Nine credits may be taken in graduate courses in subjects other than physics, upon approval of the Graduate Committee. A written or oral comprehensive examination is required on all or part of the work counting toward the degree unless waived by the Graduate Committee. No thesis or foreign language is required.

Doctorate: Students must complete 60 credits with a minimum overall "B" average, of which at least 30 credits must be taken at the City University. Candidates must pass a First "Qualifying" examination, and a Second Examination in their prospective field of research. A thesis must then be completed and defended. There are no foreign language requirements.

Thesis: Thesis may be written *in absentia*.

Special Equipment, Facilities, or Programs: The department is strong in particle theory. The department has very substantial laboratory facilities in laser-related physics including picosecond and femtosecond spectroscopy applied to many problems. Raman and Rayleigh scattering applied to solids, liquids, and biological materials, Doppler scattering applied to hydrodynamics and biophysics, and other laser interactions such as nonlinear optics, molecular beam interactions, cross section measurements, etc. The department has extensive low-temperature facilities, including a 130-kG superconducting magnet, cryostats (normal helium, helium-III, and dilution), vacuum evaporation facilities, and sample preparation areas. The New York State Center for Advanced Technology in Ultrafast Photonic Materials and Applications focuses on photonics research with medical and commercial applications.

Table B—Appointments to Graduate Students, 2009–10

Title of Appointee	Appointments		Academic Load Allowed in Credit Hours	Hours of Service Per Week	Stipend for Academic Year ($)
	Total	First year			
			Semester		
Teaching Assistant	20	15	unspecified	6–10	24,000[1]
Research Assistant	28	0	unspecified	–	–
Total	44	15			

[1](plus tuition).

5. Personnel Engaged in Separately Budgeted Research, 7/08–6/09

Professorial faculty	14
Other faculty	8
Postdoctoral appointments	27
Graduate students	30
Undergraduate students	7
Total	86

6. Separately Budgeted Research Expenditures by Source of Support (2008–09)

	Departmental Research	Physics-related Research Outside Department
Federal government	$3,153,074	$
Industries	$59,430	
Total	$3,212,504	$

Table C—Separately Budgeted Research Expenditures

Research Specialty	No. of Grants	Expenditures ($)
Biophysics	2	264,223
Condensed Matter	5	648,972
Fluids & Rheology	6	1,152,275
Photonics	3	886,427
Condensed Matter Theory	1	9,430
Total	17	2,961,327

FACULTY

Professors

Alfano, Robert R., Ph.D., NYU, 1972. Picosecond and femtosecond spectroscopy applied to solids, liquids, and biophysics; laser optics; medical applications of photonics.

Birman, Joseph L., Ph.D., Columbia, 1952. Solid state theory; group theory; quantum Hall effect; High Tc superconductors; theory of optical and low-temperature phenomena.

Boyer, Timothy, Ph.D., Harvard, 1968. Classical and quantum theory; electromagnetic theory.

Chang, Ngee-Pong, Ph.D., Columbia, 1963. Particle theory; unification and dynamical symmetry breaking.

Denn, Morton M., Ph.D., University of Minnesota, 1964. Polymeric materials; complex fluids; rheology.

Falk, Harold, Ph.D., Washington, 1962. Statistical mechanics.

Gayen, Swapan K., Ph.D., Univ. of Connecticut, 1984. Photonics, Ultrafast laser spectroscopy. Optical imaging of biological and turbid media, spectroscopy and microscopy of nanoscale systems, tunable solid state lasers.

Gersten, Joel, Ph.D., Columbia, 1968. Solid state theory; nonlinear optics; atomic theory.

Greenberger, Daniel M., Ph.D., Illinois, 1958. Neutron interferometry; quantum theory; relativity; history and philosophy of science.

Gunner, Marilyn, Ph.D., Univ. of Penn, 1988. Experimental and computational biophysics. Chair of the Department

Kaku, Michio, Ph.D., California, Berkeley, 1972. Superstring theory.

Koplik, Joel, Ph.D., California, Berkeley, 1974. Theoretical fluid mechanics; molecular dynamics of fluid flow.

Lubell, Michael S., Ph.D., Yale, 1969. Experimental atomic and elementary particle physics; science and technology policy.

Makse, Hernan, Ph.D., Boston Univ., 1997. Condensed matter, granular matter; soft condensed matter physics.

Nair, V. Parameswaran, Ph.D., Syracuse, 1983. Quantum field theory.

Petricevic, Vladimir, Ph.D., CUNY, 1990. Solid state lasers and materials; photonics.

Polychronakos, Alexios P., Ph.D., California Institute of Technology, 1987. Quantum Field Theory, Mathematical Physics.

Sarachik, Myriam P., Ph.D., Columbia, 1960. Physics of solids at low temperatures, MI transitions, magnetic materials, molecular magnetism.

Schmeltzer, David, D.Sc., Technion, 1980. Theoretical condensed matter; many body physics.

Smith, Frederick W., Ph.D., Brown, 1969. Experimental solid state and semiconductor physics.

Steinberg, Richard, Ph.D., Yale, 1992. Science, physics education.

Associate Professors

Lenzner, Matthias, Ph.D., Friedrich-Schiller Univ., Jena, Germany, 1986. Habilitation, Univ. of Technology, Vienna. Ultrafast Laser, Condensed matter physics.

Meriles, Carlos, Ph.D., Univ. Nac. de Córdoba, Argentina, 1999. Novel Magnetic resonance, spectroscopy.

Punnoose, Alexander, Ph.D., Indian Institute of Science, Bangalore, India, 1999. Theoretical Condensed matter physics.

Shattuck, Mark, Ph.D., Duke Univ., 1995. Experimental fluid mechanics, transport processes, complex fluids.

Tu, Jiufeng J., Ph.D., Cornell Univ., 2000. Physics, optical studies of correlated- and nano-systems.

Vitkalov, Sergey, Ph.D., Russian Academy of Sciences, 1986. Experimental condensed matter physics.

Assistant Professors

Koder, Ronald, Ph.D., Johns Hopkins University, 2000. Experimental and computational biophysics.

RESEARCH SPECIALTIES AND STAFF

Theoretical

Atomic and Molecular Physics. Scattering theory; interaction of intense lasers with atoms; induced atomic fluorescence; atomic structures; correlation effects. Gersten.

Computer Science. Numerical methods in nonlinear equations. Falk, Makse.

Electromagnetism. Investigation of connections between classical and quantum theories of electromagnetism; ideas of zero-point energy; classical electron theory; electromagnetic wave propagation in linear and nonlinear media. Birman, Boyer. 1 postdoctoral fellow.

Elementary Particles and Fields. Gauge theories, string theories, strong interactions, quark-gluon plasma, mathematical aspects of field theory. Chang, Kaku, Nair, Polychronakos. 3 postdoctoral fellows.

Fluids. Nanoscale fluid mechanics, molecular simulation, transport in porous media, superfluid turbulence, rheology; non-Newtonian flow; granular flow; nonequilibrium thermodynamics of granular matter, emulsions. Denn, Koplik, Makse, Shattuck.

Low-Temperature Physics. Superconductivity; phase transitions; one-dimensional structure. Birman, Punnoose, Schmeltzer. 1 postdoctoral fellow.

Material Science. Crystal structures; stability and phase transition theory; dislocations; radiationless recombinations; random media. Birman, Koplik, Punnoose, Schmeltzer.

Optics. Ultrafast phenomena; resonant light scattering; nonlinear phenomena; many-electron effects; decay processes; coherence and quantum optics; optical communications; high-power laser phenomena; nonlocal phenomena; nonlocal plus nonlinear phenomena, sonoluminescence. Birman, Gersten. 2 postdoctoral fellows.

Relativity. Quantum relativity. Greenberger, Kaku.

Solid State Physics. Many-electron effects; elementary excitations in solids and their interactions; phase transitions; dynamics of transitions; amorphous solid; semiconductor physics; mulriphonon process; surface phenomena; statistical mechanics; symmetry properties; band structures; high T_c superconductivity, quantum hall effect; strongly correlated electronic systems. Non-Fermi liquids in low Dimensions, quantum transport in mesoscopic and nanoscale regimes, spintronics, spin Hall effect, persistent currents. Birman, Falk, Gersten, Punnoose, Schmeltzer. 6 postdoctoral fellows.

Statistical and Thermal Physics. Models of magnetic systems; phase transitions; exact results (e.g. bounds for relevant properties); thermodynamics; quantum communication theory; coherence properties; impure systems; neural networks; Monte Carlo calculations; pattern selection. Falk, Koplik, Makse.

Other Theoretical. Connections between classical and quantum theories; neutron interferometry; mathematical physics. Boyer, Greenberger, Polychronakos.

Experimental

Atomic and Molecular Physics. Laser induced atomic fluorescence; scattering cross-section measurements; laser-atomic beam interactions, trapping of positrons. Lubell.

Biophysics. Primary kinetics of photosynthesis and vision; studies of vision pigments; spectroscopic studies of viruses and viral DNA; motility of micro-organisms; sol-gel transitions; radio-isotope tracer kinetics, optical imaging and optical biopsy. Computational protein design and nuclear magnetic resonance to test our understanding of the fundamental interactions underlying protein folding as well as to create new proteins for use in cancer therapies, explosives biosensing and the bioremediation of hazardous wastes. Alfano, Gunner, Koder. 3 postdoctoral fellows.

Chemical Physics. Fluorescence kinetics of dye molecules; kinetics of chemical reactions. Alfano.

Fluids. Phase transitions; excitations in liquids; order parameter and sound propagation; picosecond transient phenomena; supercooled liquid dynamics and liquid-glass phase transition; quantum fluids. Alfano, Shattuck. 2 postdoctoral fellows.

Infrared and Raman studies of superconductors and nano-systems. Tu.

Low-Temperature Physics. Transport phenomena, magnetic materials, metal-insulator transitions, molecular magnets, superconducting materials, alloy properties. Sarachik, Vitkalov. 1 postdoctoral fellow.

Materials Science. Alloy properties growth and characterization of amorphous and crystalline materials solidification of alloys and pattern formation. Sarachik, Smith.

Optics. Ultrafast phenomena, ultrashort light pulse generation, time-resolved laser spectroscopy, optical imaging, nonlinear optics, new laser development, laser-matter interactions, Raman, Rayleigh and infrared spectroscopy, terahertz spectroscopy. Alfano, Gayen, Lenzner, Meriles, Petricevic, Tu. Institute for Ultrafast Spectroscopy and Lasers (20 senior research scientists, postdoctoral research associates and technical staff).

Physics Education Research. Steinberg.

Solid State Physics. Transport properties; superconducting phenomena; optical properties; magnetic properties; allow properties; transitions; excitations in solids and their decay; growth and characterization of amorphous and crystalline materials, surface studies. Alfano, Gayen, Meriles, Petricevic, Sarachik, Smith, Vitkalov. 2 postdoctoral fellows.

Statistical and Thermal Physics. Phase transitions; order parameter and sound propagation. Shattuck. Tunable Solid State Lasers. Gayen, Petricevic.

CLARKSON UNIVERSITY

DEPARTMENT OF PHYSICS

Potsdam, New York 13699-5820

Students Accepted For Degree	FIELDS		
	Physics	Astronomy	Related Fields
Doctorate	X		X
Master's	X		X

1. General

President: Anthony G. Collins
Dean of Graduate School: Peter Turner
Department Chairman: Phill Christiansen
Department Telephone Number: (315) 268-2396
Type of Institution: University
Control: Private
Setting: Small town
Total Faculty: 191
Total Graduate Faculty: Not Separated
Total Students: 2,593
Total Graduate Students: 452
Annual Graduate Tuition:
All Graduate Students: Full-time—$1,136/credit hr.
Tuition rates for: 2010–11
Deferred tuition plan: No
Other Fees: $440
Term: Semester

2. Number of Faculty in Department

The combined total of full-time faculty in the three professorial ranks is 7. The combined total of full-time, part-time, and other faculty at all ranks is 15.

3. Admission, Financial Aid, and Housing

Address admission inquiries to: Physics Dept. Chair
Graduate application fee required: US students $25; foreign students-$35
Admission deadline: Fall admission, 4/15,
Spring admission, 10/1
Admission information: For fall admission, 2009–10, 9 students were accepted from 19 applicants.
Admission requirements: For admission to the graduate programs, a Bachelor's degree in physics is required with no minimum undergraduate GPA specified. The GRE are required. No minimum score is required. Students from non-English speaking countries are required to demonstrate proficiency in English via the TOEFL exam. A minimum score of 550 is required or 213 computer based exam.
Undergraduate preparation assumed: *Mechanics*: Symon or Becker; *Thermodynamics*: Kittel; *Optics*: Hecht; *Electricity and Magnetism*: Lorrain and Corson; *Quantum Mechanics*: Anderson.
Address financial aid inquiries to: Graduate Studies Office
Financial aid deadline: for TA/RA April 15
Loans available: Yes
Address housing inquiries to: Dean of Student Life
On-campus, single student housing available: Yes
On-campus, married student housing available: Yes

Table A—Faculty, Enrollments, and Degrees Granted

Research Specialty	2009–10 Faculty	Enrollment[1] Fall 2009		No. of Degrees Granted[2] 2009–10 (2005–09)			Median No. of Years for 2009–10 Ph.D.'s
		Master's	Doctorate	Master's	Terminal Master's	Doctorate	
Atomic & Molecular Phys.	3	0	0	0(0)	0(0)	0(0)	5
Biophysics	3	0	2	0(0)	1(0)	1(0)	5
Chemical Physics	3	0	0	0(0)	0(0)	0(0)	5
Fusion & Plasmas	1	0	0	0(0)	0(0)	0(0)	–
Optics	1	0	3	0(2)	0(1)	0(3)	5
Solid State	5	0	1	0(0)	1(0)	0(0)	5
Statistical & Thermal	6	0	2	0(1)	0(0)	0(2)	5
Surface/Interface Phys.	3	0	4	1(0)	0(5)	2(0)	–
Other Theoretical/ Math.	3	0	5	0(1)	0(2)	1(4)	4
Total		0	17	1(4)	2(8)	4(9)	
Full-time Grad. Stud.		0	17				
Part-time Grad. Stud.		0	0				
First-year Grad. Stud.		0	1				
Median Years in Grad. Study (2009–10 Degrees)				2	2	6	–
Undergraduate Degrees, 2009–10 (2005–09): 12(52)							

[1]Students not yet committed to a research specialty are entered under non-specialized.
[2]Five-year totals in parentheses.

4. Graduate Degree Requirements

Master's: 30 semester hours with B average and one year of residence are required. There is no language or comprehensive examination. Non-thesis option available.

Doctorate: 90 semester hours with B average and two years of residence are required. A Master's degree may be accepted in lieu of a maximum of 30 hours. The 90 hours must include 6 hours of seminar and a minimum of: 39 hours course work, 24 in major field, 9 in minor field, 6 out-of-department. There is no language requirement. Examinations consist of: a written comprehensive on undergraduate physics, an oral qualifying examination in the area of proposed research, and a defense of the dissertation.

Thesis: Thesis may be written *in absentia*.

5. Personnel Engaged in Separately Budgeted Research, 7/09–6/10

Professorial faculty	7
Graduate students	6
Total	13

Table B—Appointments to Graduate Students, 2009–10

Title of Appointee	Appointments Total	First year	Academic Load Allowed in Credit Hours	Hours of Service Per Week	Stipend for Academic Year ($)
			Semester		
Teaching Assistant	10	1	12[1]	12	21,190[3]
Research Assistant	6	0	12[1]	12	21,190[2]
Tuition Scholar	0	0	15	0	–
Total	16	1			

[1]Plus 6 credit-hours in summer.
[2]Plus tuition.
[3]Minimum. Some appointments carry additional stipends.

6. Separately Budgeted Research Expenditures by Source of Support

	Departmental Research	Physics-related Research Outside Department
Federal government	$662,858	$ not reported
Business and industry	214,903	not reported
Total	$877,761	$ not reported

Table C—Separately Budgeted Research Expenditures

Research Specialty	No. of Grants	Expenditures ($)
Biophysics	1	not reported
Educational Physics	1	not reported
Nonlinear Physics	1	not reported
Statistical & Thermal	1	not reported
Total	4	not reported

FACULTY

Professors

ben-Avraham, Daniel, Ph.D., Bar-Ilan Univ., 1982. Statistical mechanics.

Privman, Vladimir, D.Sc.,Technion, 1982. Statistical mechanics, nanotechnology, colloid science.

Roy, Dipankar, Ph.D., Rennselaer, 1986. Surface physics, optics, nanomaterials.

Schulman, Lawrence, Ph.D., Princeton, 1967. Statistical mechanics.

Sokolov, Igor, Ph.D., St. Petersburg, Russia, 1991. Nanotechnology, soft condensed matter, SPM.

Associate Professor

Wick, David, Ph.D., Clarkson Univ., 1997. Fluids, physics education.

Assistant Professor

Gracheva, Maria, Ph.D., Moscow State Engineering Physics Institute, 1998. Biophysics, solid state physics, nanotechnology.

Research Associates

Cho, Eun-Bum, Ph.D., Seoul National University, 2002.

Dokukin, Maxim, Ph.D., Moscow State University, 2004.

Dokukina, Irina, Ph.D., Lomonosov Moscow State University, 2007.

Research Faculty

Dorfman, Benjamin, D.Sc., Moscow Inst. of Electronic Controlling Machines and Institute of Fine Chemistry.

Mozyrsky, Dima, Ph.D., Clarkson University, 1998.

Adjunct & Joint Faculty

Bollt, Erik, Ph.D., University of Colorado, 1995.

Cheng, Ming-Cheng, Ph.D., Polytechnic University, 1990.

Melnikov, Dmitriy, Ph.D., Lehigh University, 2001.

Ramsdell, Michael, Ph.D., Clarkson University, 2004.

Faculty Emeriti

Glasser, M. Lawrence, Ph.D., Carnegie-Mellon, 1962. Solid state; mathematical physics.

RESEARCH SPECIALTIES AND STAFF

Theoretical

Biophysics. Modeling of protein dynamics. ben-Avraham, Gracheva.

Chemical Physics. Reaction kinetics; surface adsorption. ben-Avraham,Gracheva, Privman, Roy,

Colloid Science. Synthesis, surface deposition. Privman.

Fluids. Numerical studies of turbulence. Wick.

Nanotechnology. Quantum computing; device physics. Gracheva, Mozyrsky, Privman,

Nonlinear optics. Optical sensors. Roy.

Physics Education. Ramsdell, Wick.

Solid State Physics. Quantum wells; semiconductors; ion mobility. Glasser, Gracheva, Shen.

Statistical Mechanics. Phase transitions; scaling: finite size effects; percolation; self-avoiding walks. ben-Avraham, Glasser, Privman, Schulman, Gracheva.

Surface and Interface Physics. Optical and electronic properties of interfaces. Roy.

Theory and stimulation. Cell motility, cell signaling and communication. Gracheva.

Theory of scanning probe microscopy, self-organization, quantum physics. Sokolov.

Experimental

Atomic and Molecular Physics. Molecular self-assembly, sensors, ionic liquids. Roy.

Biophysics. Physical properties of tissues, cytoplasm, and cytoskeleton. Sokolov.

Scanning probe microscopy, nanomaterials, soft condensed matter. Sokolov.

Surface and Interface Physics. Surface analysis; lithium ion batteries (electrodes); photovoltaic materials; solid-liquid interfaces; electrochemical techniques. Roy.

COLUMBIA UNIVERSITY

DEPARTMENT OF APPLIED PHYSICS AND APPLIED MATHEMATICS

New York, New York 10027

Students Accepted For Degree	FIELDS		
	Physics	Astronomy	Related Fields
Doctorate			X
Master's			X

*Graduate students in the Department of Applied Physics and Applied Mathematics can register in either the Fu Foundation School of Engineering and Applied Science or the Graduate School of Arts and Sciences.

1. General

President: Lee C. Bollinger
Dean of The Fu Foundation School of Engineering and Applied Science: Feniosky Peña-Mora
Dean of Graduate School of Arts and Sciences: Henry Pinkham
Department Chairman: Irving P. Herman
Department Telephone Number: (212) 854-4457
Type of Institution: University
Control: Private
Setting: Urban
Total Faculty: 3,630
Total Students: 26,399
Total Graduate Students: 17,875
Annual Graduate Tuition:
All Graduate Students: Full-time—$36,458
Part-time—$1,370/credit
Tuition rates for: 2010–2011
Deferred tuition plan: No
Annual Other Fees: $3,000
Term: Semester

2. Number of Faculty in Department

The combined total of full-time faculty in the three professorial ranks is 34. The combined total of full-time, part-time, and other faculty at all ranks is 52.

3. Admission, Financial Aid, and Housing

Address admission inquiries to: Admissions Office, Rm. 524, S. W. Mudd Bldg., or see http://www.apam.columbia.edu or http://www.seas.columbia.edu/matsci/.
Graduate application fee required: $70
Admission deadline (Fall admission): 12/1, Ph.D. and Eng.Sc.D.; 2/15, M.S., part-time and non-degree.
Admission information: For fall admission, 2009–2010, 36 students matriculated out of 270 applicants, 11 given financial support.
Admission requirements: For admission to the graduate programs, a Bachelor's degree in sciences or engineering is required with no minimum undergraduate GPA specified. The GRE is required. The minimum acceptable score suggested for admission is quantitative-700. The GRE Advanced is required for applicants to the Applied Physics Doctoral program. The GRE Advanced is strongly urged for doctoral applicants in Applied Mathematics and Materials Science and Engineering. The minimum acceptable score suggested for admission is 680. Students from non-English speaking countries are required to demonstrate proficiency in English via the TOEFL exam. Minimum acceptable scores for admission are: 619 (paper-based), 269 (computer-based) or 104 (Internet-based).
Undergraduate preparation assumed: B.S. or B.A. degree in science or engineering from an accredited university.
Address financial aid inquiries to: Admissions Office, Rm. 524, S. W. Mudd Bldg.
FAFSA application required: Yes
Financial aid deadline: 12/1
Loans available: Yes (for U.S. citizens and permanent residents only)
Address housing inquiries to: University Apartment Housing, 400 W. 119th Street, New York, NY 10027
On-campus, single student housing available: Yes
(Rates are for 2009–10 Academic Year)

	Furnished	Unfurnished
Apartment share (private bedroom):	$787–1,339	$771–1,303
Single suite room:	$692–1,106	N/A
Shared suite room:	$596–767	N/A
Studio apartment (limited availability):	$1,122–1,698	$1,024–1,671

On-campus, married student housing available: Yes
Cost/month: $1,179–2,238

Table A—Faculty, Enrollments, and Degrees Granted

Research Specialty	2009–10 Faculty	Enrollment[1] Fall 2009		No. of Degrees Granted[2] 2009–10 (2005–10)			Median No. of Years for 2008–09 Ph.D.'s
		Master's	Doctorate	Master's[3]	Terminal Master's	Doctorate	
Applied Mathematics	13	12	23	14(56))	8(23)	3(10)	5.5
Applied Physics	34	4	0	5(9)	5(8)	0(0)	
Atmos./Space Phys.	4	3	7	1(7)	0(1)	1(1)	6.5
Biophysics/ Biomathematics	4	1	1	0(2)	0(0)	1(1)	6
Chemical Physics	3	0	0	0(0)	0(0)	0(0)	
Condensed Matter Physics	8	2	15	3(26)	0(2)	2(8)	5.5
Electromagnetism	2	0	0	0(0)	0(0)	0(0)	
Fluids	7	0	0	0(0)	0(0)	0(0)	
Geophysics	7	0	0	0(0)	0(0)	0(0)	
Low Temperature Physics	2	0	0	0(0)	0(0)	0(0)	
Materials Science	11	24	15	15(53)	11(26)	1(14)	7
Medical Physics	1	27	3	11(40)	11(40)	0(1)	
Oceanography	3	0	0	0(0)	0(0)	0(0)	
Optical Physics	4	0	3	2(9)	0(0)	0(3)	
Optics	6	0	0	0(0)	0(0)	0(0)	
Plasma Physics & Fusion	6	3	13	3(26)	0(0)	1(14)	6
Scientific Computing	9	0	0	0(0)	0(0)	0(0)	
Total		76	77	54(228)	35(100)	8(53)	
Full-time Grad. Stud.		63	73				
Part-time Grad. Stud.		13	4				
First-year Grad. Stud.		27	13				

Undergraduate Degrees, 2009–10 (2004–09): 42(252)

[1] Students not yet committed to a research specialty are entered under 1st year Graduate Student (Doctorate).
[2] Five-year totals in parentheses.
[3] M.S. and M.Phil.

4. Graduate Degree Requirements

Master's: The Master of Science (M.S.) degree is given for completion of one or more years of study beyond the Bachelor's degree. A minimum of 30 points (or semester hours) of residence credit of approved graduate course work completed at Columbia is required for the degree. A research report is required for the program in Materials Science and Engineering. A minimum grade point average of 2.5 must be maintained. The following are not required for the M.S. degree: knowledge of a foreign language, a comprehensive and/or qualifying examination, and a thesis. All degree requirements must be completed within five years of the beginning of graduate study. The Master of Philosophy (M.Phil.): all requirements for the Ph.D., except for the dissertation. The Program in Medical Physics, which leads to a 36-point M.S. degree and requires a comprehensive examination, is offered in collaboration with faculty from the College of Physicians and Surgeons and the Mailman School of Public Health. It prepares students for careers in medical physics and provides preparation for the ABR certification examination.

Doctorate: For both the Ph.D. and the Eng. Sc.D., candidates must successfully complete 30 points (or semester hours) or more of approved graduate course work beyond the Master's degree. A minimum grade point average of 3.0 must be maintained. Candidates must successfully pass both written and oral qualifying examinations. They must submit and successfully defend an approved dissertation. A knowledge of a foreign language is not required. A time limit of seven years is allowed for completion of the requirements for doctoral degrees. The residence requirement for the Ph.D. degree is satisfied by six Residence Units in the Graduate School of Arts and Sciences, two of which are granted for the Master's degree. The residence requirement for the Eng. Sc.D. degree is satisfied by 12 additional points of credit in APAM E9800 (Doctoral Research Instruction) in the School of Engineering and Applied Science.

Thesis: Thesis may be written *in absentia*.

Special Equipment, Facilities, or Programs: Columbia's Plasma Physics Laboratory contains experiments to better understand and to control high-temperature magnetized plasma. Experiments include a toroidal high-beta tokamak, HBT-EP, a linear steady-state plasma experiment, a large laboratory terrella used to investigate space plasma physics, the CNT stellarator used to study non-neutral plasmas, and a variety of smaller devices used to produce pulsed and continuous plasmas for scientific investigations. Graduate research opportunities also exist with one of several collaborative research programs between Columbia University and other national research projects. These include collaborations with the Princeton Plasma Physics Laboratory, MIT's Plasma Science and Fusion Center, and the Fusion Research Group at General Atomics. Free-electron laser (FEL) research opportunities also exist using the 40 MeV rf linac at the Brookhaven National Laboratory, and a 6 MeV facility is available at Yale University for advanced accelerator research. Research equipment in the Solid-State Physics and Optical Physics laboratories, and the associated Columbia Center for Integrated Science and Engineering, include extensive laser and spectroscopy facilities, a clean room that includes photolithography and thin film fabrication systems, ultra high-vacuum surface preparation and analysis chambers, direct laser writing stations, a molecular beam epitaxy machine, picosecond and femtosecond lasers, and diamond anvil cells. Facilities, and research opportunities, also exist within the interdisciplinary NSF Materials Research Science and Engineering Center, which focuses on complex films composed of nanoparticles, the DOE Energy Frontier Research Center which focuses on conversion of sunlight into electricity in nanometer-sized thin films, and the NSF Nanoscale Science and Engineering Center, which focuses on electron transport in molecular nanostructures. Materials Science and Engineering facilities include transmission and scanning electron microscopes, x-ray diffraction, ellipsometry, x-ray photoelectron microscopy equipment, a clean room with photolithography, deposition, and etching capabilities, magnetic and electrical measurement equipment, laser spectroscopy, laser processing, and mechanical testing. Faculty members have research collaborations with local research and manufacturing centers including Lucent, Exxon, Philips Electronics, IBM, as well as major international research centers. The National Synchrotron Light Source at Brookhaven National Laboratory is used for high-resolution x-ray diffraction and absorption measurements. There are also research opportunities in medical physics at the Columbia Presbyterian Medical Center, as well as at other medical institutes, employing state-of-the-art medical diagnostic imaging and treatment equipment. The Applied Mathematics division is closely linked with the Lamont Doherty Earth Observatory (LDEO), with five faculty members sharing appointments in the Department of Earth and Environmental Sciences. There are also close ties with the NASA Goddard Institute for Space Studies (GISS), with Columbia's Center for Computational Biology and Bioinformatics, and with Columbia's Center for Computational Learning Systems. Ongoing joint research, instruction and supervision of graduate students with LDEO and NASA/GISS span the areas of atmosphere, ocean and climate modeling, geophysical/geological fluid dynamics, earthquake physics, remote sensing, and large-scale scientific computing. The research of the Plasma Lab is supported by a dedicated data acquisition/data analysis system. Computational researchers have local access to the Department's SiCortex supercomputer with 1458 cores as well as to Columbia's 256-processor Linux cluster, and to supercomputer systems at the National Center for Atmospheric Research and the Lawrence Berkeley and Oak Ridge National Laboratories.

Table B—Appointments to Graduate Students, 2009–10

Title of Appointee	Appointments: Total	Appointments: First year	Academic Load Allowed in Credit Hours	Hours of Service Per Week	Stipend for Academic Year ($)
			Semester		
Teaching Assistant	18	12		15	22,500[2]
Research Assistant	52	1		varies	30,000[3]
Other (specify)			Varies[1]		
Fellowship	4	1		0	varies
Total	74	14			

[1]Tuition allowance covers one Resident Unit (no point limit for students in the Graduate School of Arts and Sciences) or 15 points (for students in the Fu Foundation School of Engineering and Applied Science) per semester.
[2]9 months.
[3]12 months.

5. Personnel Engaged in Separately Budgeted Research, 7/09–6/10

Professorial faculty	34
Other faculty	18
Postdoctoral appointments	16
Graduate students	56
Nonteaching research personnel	53
Total	177

6. Separately Budgeted Research Expenditures by Source of Support, 06/09

	Departmental Research
Federal government	$10,594,791
Business and Industry	1,189,044
Other	307,496
Total	$12,091,331

Table C—Separately Budgeted Research Expenditures

Research Specialty	No. of Grants	Expenditures ($)
Atmos./Space Phys.	42	3,765,058
Biophysics	11	281,455
Condensed Matter Physics	26	1,907,832
Materials Science	55	2,289,388
Medical Physics	0	n/a
Physics of Beams	2	6,313
Plasma Physics & Fusion	25	2,527,110
Other (Applied Math)	32	1,314,175
Total	193	$12,091,331

FACULTY

Professors

Bal, Guillaume, Ph.D., University of Paris, 1997. Applied mathematics, partial differential equations with random coefficients, high frequency waves in random media, and application to time reversal, theory of inverse transport and applications in medical imaging and geophysical imaging.

Bienstock, Daniel, Ph.D., MIT, 1985. Applied mathematics, methodology and high-performance implementation of optimization algorithms, applications of optimization: preventing national-scale blackouts, emergency management, approximate solution of massively large optimization problems, higher-dimensional reformulation techniques for integer programming, robust optimzation.

Billinge, Simon J. L., Ph.D., Pennsylvania, 1992. Nanoscale structure-property relationships in functional nanomaterials studied using novel x-ray and neutron scattering techniques coupled with advanced computing; solving the nanostructure problem.

Boozer, Allen H., Ph.D., Cornell, 1970. Plasma theory; theory of magnetic confinement for fusion energy; nonlinear dynamics.

Cane, Mark A., Ph.D., MIT, 1975. Climate dynamics; physical oceanography; geophysical fluid dynamics; computational fluid dynamics; impacts of climate on society, El Niño forecasting.

Chan, Siu-Wai, Sc.D., MIT, 1985. Nanoparticles, electronic ceramics, grain boundaries and interfaces, oxide thin films.

Friedman, Morton B., D.Sc., NYU, 1953. Applied mathematics and mechanics; numerical analysis; parallel computing.

Herman, Irving P., Ph.D., MIT, 1977. Chairman of the Department. Nanocrystals; laser diagnostics of thin-film processing; optical spectroscopy of nanostructured materials; mechanical properties of nanomaterials.

Im, James, Ph.D., MIT, 1989. Laser-induced crystallization of thin films, phase transformations and nucleation in condensed systems.

Keyes, David E., Ph.D., Harvard, 1984. Applied and computational mathematics for PDEs, computational science, parallel numerical algorithms, parallel performance analysis, PDE-constrained optimization.

Kim, Philip, Ph.D., Harvard, 1999. Experimental condensed matter physics, physical properties and applications of nanoscale low-dimensional materials, quantum thermal transport phenomena in 1-dimensional nanoscaled materials, mesoscopic thermoelectricity and thermoelectric applications of nanoscale materials, quantum transport in novel 2-dimensional materials, mesoscopic electron transport and thermodynamic processes for sensors and electric devices.

Mauel, Michael E., Sc.D., MIT, 1983. Plasma physics, waves and instabilities, fusion and equilibrium control; space physics; plasma processing; international energy policy.

Navratil, Gerald A., Ph.D., Wisconsin, 1976. Plasma physics; plasma diagnostics; fusion energy science.

Neumark, Gertrude, Ph.D., Columbia, 1951. Material science and physics of semiconductors, with emphasis on optical and electrical properties of wide bandgap semiconductors and their light-emitting devices.

Noyan, I. Cevdet, Ph.D., Northwestern University, 1984. Characterization and modeling of mechanical and micromechanical deformation; residual stress analysis and nondestructive testing; x-ray and neutron diffraction, microdiffraction analysis.

Osgood, Richard M., Ph.D., MIT, 1973. Nanoscale optical and electronic phenomena, femtosecond lasers and laser probing, low-dimensional physics, integrated optics, nanofabrication and materials growth.

Pinczuk, Aron, Ph.D., University of Pennsylvania, 1969. Spectroscopy of semiconductors and insulators; quantum structures, systems of reduced dimensions, atomic layers of graphene, electron quantum fluids.

Polvani, Lorenzo, Ph.D., MIT, 1988. Atmospheric and climate dynamics, geophysical fluid dynamics, numerical methods for weather and climate modeling, planetary atmosphers.

Ruderman, Malvin A., Ph.D., Cal. Tech., 1951. Theoretical astrophysics; neutron stars; pulsars; early universe; cosmic gamma rays.

Scholz, Christopher H., Ph.D., MIT, 1967. Experimental and theoretical rock mechanics, especially friction, fracture and hydraulic transport properties, nonlinear systems, mechanics of earthquates and faulting.

Sen, Amiya K., Ph.D., Columbia, 1963. Plasma physics; fluctuation and anomalous transport in plasmas; control of plasma instabilities, plasma transport.

Sobel, Adam H., Ph.D., MIT, 1998. Atmospheric science, geophysical fluid dynamics, tropical meteorology, climate dynamics.

Spiegelman, Marc W., Ph.D., Cambridge, 1989. Coupled fluid/solid mechanics, reactive fluid flow, solid earth and magma dynamics, scientific computation/modeling.

Stormer, Horst, Ph.D., University of Stuttgart, 1977. Semiconductors, electronic transport, lower-dimensional physics, transport in nanostructures.

Wang, Wen I., Ph.D., Cornell, 1981. Heterostructure devices and physics; materials properties; molecular beam epitaxy.

Weinstein, Michael I., Ph.D., Courant Institute-NYU, 1982. Applied mathematics, partial differential equations and analysis, waves in nonlinear, inhomogeneous and random media; dynamical systems; multi-scale phenomena, applications to nonlinear optics, mathematical physics; fluid dynamics; geosciences.

Wuu, Cheng Shie, Ph.D., Kansas, 1985. Microdosimetry, biophysical modeling, dosimetry of brachytherapy, gel dosimetry second cancers induced by radiotherapy, medical physics.

Associate Professors

Bailey, William E., Ph.D., Stanford, 1999. Nanoscale magnetic films and heterostructures, materials issues in spin-polarized transport, materials engineering of magnetic dynamics.

Pedersen, Thomas S., Ph.D., MIT, 2000. Plasma physics, magnetic confinement, fusion energy, plasma turbulence, non-neutral plasmas, positron-electron plasmas.

Wiggins, Chris H., Ph.D., Princeton, 1998. Applied mathematics, mathematical biology, biopolymer dynamics, soft condensed matter, genetic networks and network inference, machine learning.

Assistant Professors

Englund, Dirk, Ph.D., Standford, 2008. Quantum optics in photonic nanostructures; photonic crystal optoelectronic devices and networks; quantum information and metrology; nonlinear optics; electron and nuclear spin-dynamics in solid state systems.

Gentine, Pierre, Ph.D., MIT, 2009. Applied mathematics, land-atmosphere interactions, soil-vegetation-transfer-atmosphere models, applications of stochastic processes to hydrology and atmospheric boundary-layer, stochastic rainfall and soil moisture, data-assimilation (filtering) of remote sensing measurements to estimate land-surface variables.

Marianetti, Chris A., Ph.D., Massachusetts Institute of Technology, 2004. Predicting materials properties from first-principles computations; density-functional theory; dynamical mean-field theory; transition-metal oxides; actinides.

Venkataraman, Latha, Ph.D., Harvard, 1999. Single molecule transport, single molecule force spectroscopy, electron transport in nanowires, scanning tunneling microscopy and spectroscopy.

Professors Emeriti

Beshers, Daniel N., Ph.D., University of Illinois, 1956. Metallurgy and materials science.

Chu, C. K., Ph.D., Courant Inst., 1959. Applied mathematics.

Gross, Robert A., Ph.D., Harvard, 1952. Plasma physics, energy, fluid dynamics.

Lidofsky, Leon, Ph.D., Columbia, 1952. Radiation applications in medicine and technology, radiation shielding.

Marshall, Thomas C., Ph.D., Illinois, 1960. Accelerator concepts, free-electron lasers; relativistic beam dynamics and radiation.

RESEARCH SPECIALTIES AND STAFF

Theoretical

Applied Mathematics. Analytical and numerical analysis of PDEs, scientific computation, geophysical fluid dynamics, biomathematics, dynamical systems and chaos, as well as applications to various fields of physics including solid-state, plasma physics, nonlinear optics, medical imaging and the earth sciences and various fields of biology. In collaboration with the NASA Goddard Institute for Space Studies and the Lamont-Doherty Earth Observatory, atmospheric and oceanic modeling. Bal, Bienstock, Boozer, Cane, Courdurier, Friedman, Keyes, Polvani, Scholz, Sobel, Spiegelman, Weinstein, Wiggins. 30 research scientists.

Materials Science. Diffraction physics, dynamic and kinematic scattering of x-rays and neutrons, forward calculation of diffraction profiles and inverse calculation of material structures, theoretical analysis of heterogeneous and homogeneous nucleation kinetics. Billinge, Im, Marianetti, Noyan, 4 research scientists.

Plasma Physics. Basic plasma theory; analytical and computational magnetohydrodynamics; toroidal high-beta equilibrium and stability; space/solar plasmas; linear and nonlinear interactions among waves and particles; plasma turbulence; plasma transport properties; free-electron laser theory. Boozer, Sen. 3 research scientists.

Experimental

Material Science and Engineering. Polycrystalline semiconductor and metallic films, laser irradiation of materials, electronic ceramics, wide-band-gap semiconductors, plasma processing of materials and optical diagnostics of thin-film processing, ceramic nanocomposites, and magnetic films and spintronics, nondestructive testing of material systems, residual stress analysis. Research opportunities also exist in the NSF Materials Research Science and Engineering Center, which focuses on nanocrystals and complex films composed of nanocrystals. Bailey, Chan, Im, Neumark, Noyan. 14 research scientists.

Optical and Laser Physics. Inelastic light scattering, the free-electron laser, accelerators, optical diagnostics of film processing, nonlinear optics, ultrafast optoelectronics, photonic switching, optical physics of surfaces and nanocrystals, laser-induced crystallization, and photon integrated circuits. Herman, Im, Neumark, Osgood, Pinczuk, Wang. 9 research scientists.

Plasma Physics. High-beta tokamak equilibrium, stability, active feedback mode control and transport; HBT-EP Tokamak (R=0.92 m, a=0.15 m, B=0.5 T); design of improved tokamak configurations; plasma heating at microwave and radio frequencies. Joint programs on advanced and innovative fusion confinement at the NSTX device at the Princeton Plasma Physics Laboratory, the DIII-D device at General Atomics investigating high-beta equilibrium and stability, and the LDX device at M.I.T. Plasma diagnostics; CO_2 laser scattering, multi-point Thomson scattering, and tomographic density reconstruction on high-beta plasmas; ultraviolet, optical, infrared, and microwave radiation and scattering. Waves in plasmas: trapped particle instabilities; space plasma physics, nonlinear and chaotic transport in planetary magnetospheres; feedback control of plasma instabilities; anomalous transport due to plasma instabilities. Intense relativistic electron beams: high-power millimeter wave sources-free electron laser; beam diagnostics; relativistic electron beam microwave production; inverse FEL accelerators; plasma processing of semiconductors using microwave sources, non-neutral plasmas. Mauel, Navratil, Pedersen, Sen. 9 research scientists.

Solid State Physics. Nanoscience and nanoparticles, the optical spectroscopy of semiconductor structures that are subjected to high pressure, electronic transport and inelastic light scattering in low-dimensional correlated electron systems, fractional quantum Hall effect, heterostructure physics and applications, molecular beam epitaxy, grain boundaries and interfaces, nucleation in thin films, and surface physics. Research opportunities also exist within the interdisciplinary NSF Materials Research Science and Engineering Center, which focuses on nanocrystals and complex films composed of nanocrystals, and the NSF Nanoscale Science and Engineering Center, which focuses on electron transport in molecular nanostructures. Chan, Herman, Im, Kim, Neumark, Osgood, Pinczuk, Stormer, Venkataraman, Wang. 10 research scientists.

CORNELL UNIVERSITY

DEPARTMENT OF PHYSICS

Ithaca, New York 14853

Students Accepted For Degree	FIELDS		
	Physics	Astronomy	Related Fields
Doctorate	X		
Master's			

1. General

President: David J. Skorton
Dean of Graduate School: Alison Power
Department Chairman: J. Ritchie Patterson
Department Telephone Number: (607) 255-6016
Department Website: *www.physics.cornell.edu*
Department email: *physics@cornell.edu*
Type of Institution: University
Control: Both
Setting: Small city
Total Faculty: 2,908
Total Graduate Faculty: 1,639
Total Students: 20,273
Total Graduate Students: 4,565
Annual Graduate Tuition:
All Graduate Students: Full-time—$29,500*
Tuition rates for: 2010–11
Deferred tuition plan: No
Other Fees: $76 student activity
Term: Semester

*All regular graduate students in physics are normally appointed to positions which provide a stipend, health insurance and full support for tuition.

2. Number of Faculty in Department

The combined total of full-time faculty in the three professorial ranks is 43. The combined total of full-time, part-time, and other faculty at all ranks is 70.

3. Admission, Financial Aid, and Housing

Address admission inquiries to: *physics@cornell.edu*
Graduate application fee required: $70 (online)
Admission deadline (Fall admission): 12/15
Admission information: Fall admissions only. For 2010–11, 85 students were offered admission from 409 applicants, 31 accepted.
Admission requirements: For admission to the graduate program, a Bachelor's degree from an accredited institution: major in physics is recommended but not required, with no minimum undergraduate GPA specified. The GRE general and subject tests are required with no minimum scores specified. The median GRE scores for 2010-11 admissions were: 620–Verbal; 800–Quant.; 4.5–Anal.; 920–Physics Subject. Students from non-English speaking countries are required to demonstrate proficiency in English via the TOEFL exam. Minimum score for admission is 620 (paper)/260 (computer)/105 (online) section minimums: 20 in Writing, 15 in Listening, 20 in Reading and 22 in Speaking.
Address financial aid inquiries to: *physics@cornell.edu* Please note that all regular graduate students in Physics are normally supported throughout the length of their Ph.D. study while they remain in good academic standing.
GAPSFAS application required: No
Financial aid deadline: 12/15
Loans available: Yes
Address housing inquiries to: Housing & Dining Office; 607-255-5368 or housing@cornell.edu, www.housing.cornell.edu
On-campus, single student housing available: Yes
Cost/month: Starting at $388
On-campus, family housing available: Yes
Cost/month: Starting at $860

Table A—Faculty, Enrollments, and Degrees Granted

Research Specialty	2009–10 Faculty	Enrollment[1] Fall 2009		No. of Degrees Granted 2009–10			Median No. of Years for 2009–10 Ph.D.'s
		Master's	Doctorate	Master's	Terminal Master's	Doctorate	
Accelerator/Beams	10	–	7	0(5)	0	0(5)	–
Applied Physics	11	–	3	2(3)	0	0(1)	–
Astronomy	3	–	1	0(4)	0	0(4)	–
Astrophysics	3	–	9	1(2)	0	1(4)	–
Atomic, Molecular, & Optical Physics	6	–	6	2(2)	0	1(6)	–
Biophysics	11	–	23	2(15)	0(1)	4(17)	–
Chemical Physics	4	–	3	1(1)	0	0	–
Computational	11	–	6	3(6)	1(1)	0(4)	–
Condensed Matter/ Solid State Physics	22	–	32	5(20)	0(1)	3(20)	–
Electromagnetism	5	–	0	0(1)	0	0	–
Electronics	4	–	0	0	0	0(1)	–
Engineering Physics/ Science	6	–	1	0	0	0	–
Experimental Colloids	3	–	1	0(1)	0	0	–
Fluids & Rheology	4	–	3	0(4)	0	2(3)	–
High Energy	17	–	19	9(13)	0	3(3)	–
Low Temperature Physics	10	–	6	1(6)	0	4(8)	–
Materials Sci./ Metallurgy	10	–	4	0(3)	1(2)	1(3)	–
Mechanics	4	–	0	0(1)	0	0(1)	–
Nanophysics	8	–	7	1(8)	0	2(18)	–
Non-Linear Dynamics	12	–	1	0(1)	0(0)	0(1)	–
Optics	8	–	3	0(2)	0	0(4)	–
Particles & Fields	18	–	8	2(13)	0(3)	1(20)	–
Physics Education	4	–	0	0	0	0	–
Polymer Physics/ Science	3	–	0	0	0	0(1)	–
Relativity & Gravitation	6	–	5	1(6)	0	2(8)	–
Statistical & Thermal	12	–	3	0(1)	0	0(1)	–
String Theory	2		7	2(4)	0	1(1)	
Surfaces	8	–	0	0	0	0	–
Non-specialized	0	–	16	0(1)	1(1)	0	–
Total		–	174	32(123)	3(9)	25(134)	
Full-time Grad. Stud.		–	174				
Part-time Grad. Stud.		–	0				
First-year Grad. Stud.		–	23				
Median Years in Grad. Study (2009–10 Degrees)				3.4	–	6.2[2]	
Undergraduate Degrees, 2009–10: (5 yr): 20(136)							

[1]Students not yet committed to a research specialty are entered under non-specialized.
[2]Median Years for Ph.D. overall.

4. Graduate Degree Requirements

Master's: Two semesters residence; qualifying exam after one year; thesis and/or Master's exam; maximum time allowed, four years, minimum one year; no foreign language requirement. Master's degree is earned as part of Admission to Candidacy Exam for Ph.D.

Doctorate: Six semesters residence, at least four full-time at Cornell, and no more than two during summers. Qualifying exam after one year, Admission to Candidacy Exam in third year. No fixed course requirements, except one semester Advanced Lab. No foreign language requirement. Maximum time allowed, seven years. Thesis and oral thesis examination. Three-person Special Committees direct course program and supervise research individually for each student. Research in adjacent fields is permissible.

Thesis: Thesis may be written *in absentia*.

Special Equipment, Facilities, or Programs: 1.9 + 1.9 GeV e^+e^- CESR test accelerator; CESR facility also provides high-intensity, high-energy x-rays for NSF-funded synchrotron radiation facility Cornell High-Energy Synchrotron Source (CHESS), and NIH Research Resource MacCHESS (protein crystallographic studies); Microkelvin Laboratory; Center for Advanced Computing; Cornell Center for Materials Research (CCMR) and National Submicron Facility with specialized workshops and equipment; Electron and Optical Microscopy Facility; Nanobiotechnology Center; National Nanofabrication Facility; Center for Nanoscale Systems; and Kavli Institute at Cornell for Nanoscale Science.

Table B—Appointments to Graduate Students, 2009–10

Title of Appointee	Appointments Total	First year	Academic Load Allowed in Credit Hours	Hours of Service Per Week	Stipend for Academic Year ($)
			Semester		
Teaching Assistant	55.5	16.5		20	23,337[1,2]
Research Assistant	74	1.5	See note[2]	20	21,400[1]-22,150
Fellow	19.5	5		-	20,700-30,000[1]
Total	149	23			

[1]Summer appointments available in addition.
[2]Academic load designed by consultation with individual student's committee, not formally by credit hour.

5. Personnel Engaged in Separately Budgeted Research, 7/09–6/10

Professorial faculty	44
Postdoctoral appointments	26
Nonteaching research personnel	55
Total	125

6. Separately Budgeted Research Expenditures by Source of Support†

†Research funding is received through the two laboratories listed in Section 7 rather than through the Physics Department. Additional physics research in astrophysics, materials science, plasma physics, solid state, optics, biophysics, and statistical and thermal physics is carried out in other departments, but complete information is not available.

7. Separately Funded and Managed Laboratories

Lab. for Elem. Particle Physics (2009–10)	$28,342,000
Lab. of Atomic and Solid State Physics (2009–10)	6,687,000
Total	$35,029,000

Table C—Separately Budgeted Research Expenditures

Research Specialty	No. of Grants	Expenditures ($)
	See Section 7	

FACULTY

Professors

Alexander, James P., Ph.D., Chicago, 1985. Elementary particles.

Arias, Tomás A., Ph.D., MIT, 1992. Condensed matter.

Davis, J. C. Seamus, Ph.D., Berkeley, 1989. Solid-state physics.

Dugan, Gerald F., Ph.D., Columbia, 1973. Accelerator physics.

Elser, Veit, Ph.D., California, Berkeley, 1984. Solid state physics.

Flanagan, Eanna, Ph.D., Caltech, 1993. Astrophysics, relativity, gravitational radiation.

Ginsparg, Paul, Ph.D., Cornell, 1981. Digital knowledge networks.

Gruner, Sol M., Ph.D., Princeton, 1977. Biophysics; soft condensed matter.

Hartill, Donald L., Ph.D., Cal. Tech., 1967. Elementary particles.

Henley, Christopher L., Ph.D., Harvard, 1983. Condensed matter.

Hoffstaetter, Georg, Ph.D., Michigan State, 1994. Accelerator tech., beam physics.

LeClair, Andre, Ph.D., Harvard, 1987. Elementary particles.

Lepage, G. Peter, Ph.D., Stanford, 1978. Dean of College of Arts and Sciences. Elementary particles.

McEuen, Paul, Ph.D., Yale, 1991. Nano-structure physics.

Orlov, Yuri, Yerevan Physics Institute, Armenia, 1958. Elementary particles.

Parpia, Jeevak M., Ph.D., Cornell, 1979. Low-temperature physics.

Patterson, J. Ritchie, Ph.D., Chicago, 1990. Elementary particles. Chair of the Department.

Ralph, Daniel C., Ph.D., Cornell, 1993. Condensed matter; low temperature.

Richardson, Robert C., Ph.D., Duke, 1966. Low temperature; solid state.

Rubin, David L., Ph.D., Michigan, 1983. Elementary particles; accelerator physics.

Sethna, James P., Ph.D., Princeton, 1981. Solid state, statistical physics.
Sievers, Albert J., Ph.D., California, Berkeley, 1962. Solid state.
Teukolsky, Saul, Ph.D., Cal. Tech., 1973. Relativity; astrophysics.
Thorne, Robert E., Ph.D., Illinois, 1987. Condensed matter; biophysics.
Tye, Henry, Ph.D., MIT, 1974. Elementary particles.
Wang, Michelle D., Ph.D., Michigan, 1993. Biophysics.
Wasserman, Ira M., Ph.D., Harvard, 1978. Astrophysics.

Associate Professors

Csaki, Csaba, Ph.D., MIT, 1997. Elementary particle physics, quantum field theory.
Franck, Carl P., Ph.D., Princeton, 1978. Solid state physics; chemical physics.
Gibbons, Lawrence K., Ph.D., Chicago, 1993. Elementary particles.
Grossman, Y., Ph.D., Weizmann Institute of Science, 1996. Elementary particles.
Mueller, Erich, Ph.D., UIUC, 2001. Condensed matter.
Perelstein, Maxim, Ph.D., Stanford, 2000. Elementary particles.
Ryd, Anders, Ph.D., UC Santa Barbara, 1996. Elementary particles.

Assistant Professors

Bazarov, Ivan, Ph.D., Far Eastern State Univ., 2000. Elementary particles.
Cohen, Itai, Ph.D., Univ. Chicago, 2001. Solid state physics.
Kim, Eun-ah, Ph.D., UIUC, 2005. Condensed matter.
Liepe, Matthias, Ph.D., Univ. Hamburg, 2001. Accelerator physics.
McAllister, Liam, Ph.D., Stanford, 2005. Elementary particles.
Shen, Kyle, Ph.D., Stanford University, 2005. Solid State Physics.
Thom, Julia, Ph.D., Univ. Hamburg, 2001. Elementary particles.
Vengalattore, Mukund, Ph.D., MIT, 2005. Condensed matter.
Wittich, Peter, Ph.D., Univ. Penn., 2000. Elementary particles.

RESEARCH SPECIALTIES AND STAFF

Theoretical

Astrophysics and Relativity. Flanagan, Teukolsky, Wasserman.
Elementary Particles. Quantum electrodynamics; gauge theories of strong interactions: spectroscopy of new heavy mesons; hadronic structure of the photon. Grossman, McAllister, LeClair, Lepage, Perelstein, Tye.
Low Temperature Physics. Superfluid ^{3}He and ^{4}He; nonequilibrium superconductivity.
Solid State Physics. Equation of state of simple metals; nonlinear phenomena; phonon scattering theory; transport theory; band structure; surface physics. Arias, Elser, Henley, Kim, Mueller, Sethna.
Statistical and Thermal Physics. Phase transitions; turbulence. Arias, Mueller, Sethna.

Experimental

Biophysics. Franck, Gruner, Thorne, Wang.
Elementary Particles. Accelerator physics; photon and charged-lepton interactions with hadrons; electron-positron collisions; high-energy hadron-induced interactions. Alexander, Bazarov, Dugan, Gibbons, Hartill, Hoffstaetter, Liepe, Orlov, Patterson, Rubin, Ryd, Thom, Wittich.
Chemical Physics. Optical properties of polymeric systems. Franck.
Low-Temperature Physics. Superfluidity in ^{3}He and ^{4}He; phase transitions in liquid ^{3}He, ^{4}He, and mixtures and in solid ^{3}He. Parpia, Ralph, Richardson.
Nanostructure Physics: Davis, McEuen, Parpia, Ralph.
Optics. Far-infrared. Sievers, Rubin, Hartill, LeClair, Hoffstaetter.
Solid State Physics. Ultraviolet, visible, infrared and Raman studies of solids; nuclear and electron spin resonance; electronic and thermal transport properties; phase transitions in one and two dimensions. Cohen, Davis, Franck, Gruner, McEuen, Ralph, Shen, Sievers, Thorne, Vengalattore.
Statistical and Thermal Physics. Phase transitions. Franck.

CORNELL UNIVERSITY

SCHOOL OF APPLIED AND ENGINEERING PHYSICS

Ithaca, New York 14853

Students Accepted For Degree	FIELDS		
	Physics	Astronomy	Related Fields
Doctorate			X
Master's			X

1. General

President: David J. Skorton
Dean of Graduate School: Alison G. Power
Director: Frank W. Wise
Department Telephone Number: (607) 255-5198
Type of Institution: University
Control: Both
Setting: Small town
Total Faculty: 1,605
Total Graduate Faculty: 1,605
Total Students: 21,138
Total Graduate Students: 4,689
Annual Graduate Tuition:
All Graduate Students: Full-time—$29,500–$39,450
Tuition rates for: 2010–11
Deferred tuition plan: No
Other Fees: $68
Term: Semester

2. Number of Faculty in Department

The combined total of full-time faculty in the three professorial ranks is 14. The combined total of full-time, part-time, and other faculty, including those with dual appointments, at all ranks is 18 (plus an additional 19 who are members of the Graduate Field of Applied Physics but are members of other departments).

3. Admission, Financial Aid, and Housing

Address admission inquiries to: Graduate Faculty Representative, Applied Physics, 218 Clark Hall
Graduate application fee required: $65
Admission deadline (Fall admission): 1/15
Admission information: For fall admission, 2010–11, typically, about 30 students are admitted from 150 applicants.
Admission requirements: For admission to the graduate programs, a Bachelor's degree in science is required with no minimum undergraduate GPA specified. The GRE and the GRE Advanced are required. Students from non-English speaking countries are required to demonstrate proficiency in English via the TOEFL exam. Last year, students with undergraduate majors in applied physics, electrical engineering, engineering physics, mathematics, materials science, and engineering, mechanical engineering, and physics were accepted.
Address financial aid inquiries to: Graduate School, Caldwell Hall
GAPSFAS application required: No
Financial aid deadline: 1/15
Loans available: Yes
Address housing inquiries to: Graduate Student Housing, Caldwell Hall
On-campus, single student housing available: Yes
Cost/year: $5,600–8,100
On-campus, married student housing available: Yes
Cost/month: $810–970

Table A—Faculty, Enrollments, and Degrees Granted

Research Specialty	2009–10 Faculty	Enrollment[1] Fall 2009		No. of Degrees Granted[2] 2009–10			Median No. of Years for 2009–10 Ph.D.'s
		Master's[3]	Doctorate	Master's[4]	Terminal Master's[3]	Doctorate	
Applied Physics		8	62	8	–	8	–
Total		8	62	8	–	8	
Full-time Grad. Stud.		8	62				
Part-time Grad. Stud.		0	0				
First-year Grad. Stud.		8	11				

Median Years in Grad. Study (2009–10 Degrees)
Undergraduate Degrees, 2009–10: 45

[1]Students not yet committed to a research specialty are entered under non-specialized.
[2]Six-year totals in parentheses.
[3]Master of Engineering in Engineering Physics Program.
[4]Master of Science Degree in Applied Physics.

4. Graduate Degree Requirements

Master of Science in Applied Physics: Minimum residence two units. Full-time study for one semester with satisfactory accomplishment constitutes one residence unit. Requirements must be completed in four years from date of first registration. A written thesis and an oral presentation (e.g., group seminar) are required at the end of the project. Students must maintain a C average (2.0) or better to remain in good standing. No foreign language requirement.
Master of Engineering in Engineering Physics: The Master of Engineering program is a terminal professional Master's degree. The general requirement for the degree is a total of 30 credits for graduate-level courses or their equivalent, distributed as follows: (1) a design project in applied science or engineering (not less than 6 nor more than 12 credits): (2) an integrated program of graduate-level courses (14 to 20 credits); and (3) a required special topics seminar course (four credits).
Doctorate: Six units of residence are required. Full-time study for one semester with satisfactory accomplishment constitutes one residence unit. All requirements must be completed in seven years from date of first registration. No more than two units may be earned in summer. At least four of six units must be earned as full-time student. Two examinations (by the Special Committee) are required for doctoral degree, a uniform Qualifying Exam (oral and written) is taken no later than after three units of residence credit are accumulated, and a final exam, given after completion of dissertation, covers subject matter of dissertation. Publication of thesis by abstract and microfilm is required. No specific GPA required; no foreign language.
Thesis: Thesis may be written *in absentia*.
Facilities: Students in the Graduate Field of Applied Physics are able to access a diverse range of experimental facilities on the Cornell University campus, including: Atomic and Molecular Physics laboratories, Biophysics laboratories, Electron Mi-

croscopy laboratories, Synchrotron X-ray laboratories, Plasma Physics laboratories, Optical Physics and Photonics laboratories, Geophysics laboratories, and Condensed Matter Physics laboratories (including materials physics, surface physics, device physics, and nanophysics). Due to the interdisciplinary and multi-departmental nature of the Graduate Field of Applied Physics at Cornell, the research facilities available are much more extensive than those which can be provided by a single department. Applied Physics is also closely associated with several major research facilities on the Cornell University campus: the Cornell Center for Materials Research (CCMR), the Center for Nanoscale Systems (CNS), the Nanobiotechnology Center (NBTC), the Cornell High Energy Synchrotron Source (CHESS), and the Cornell NanoScale Science & Technology Facility (CNF). Other facilities available for applied physics research include the radar-radio observatory in Arecibo, Puerto Rico; the unique pulsed-power driven high energy plasma facilities operated by the Laboratory of Plasma Studies (LPS); and the NIH Developmental Resourse in Biophysical Imaging and Opto-electronics (DrBio). Finally, close relationships with adjacent fields (Astronomy, Bio-chemistry, Bio-physics, Molecular and Cell Biology, Chemistry, Electrical Engineering, Materials Science, Neutrobiology, and Physics) enable students to make use of facilities not formally represented in the School of Applied and Engineering Physics.

Table B—Appointments to Graduate Students, 2009–10

Title of Appointee	Appointments		Academic Load Allowed in Credit Hours	Hours of Service Per Week	Stipend for Academic Year ($)
	Total	First year			
			Semester		
Teaching Assistant	0	7		20	21,400
Grad. Research Assistant	60	2		20	21,400
Fellowships	2	2		0	varies
Total	62	11			

5. Personnel Engaged in Separately Budgeted Research, 7/08–6/09

Professorial faculty	16[1]
Postdoctoral appointments	23
Graduate students	60
Undergraduate students	3
Total	102

[1]Plus an additional 23 faculty members who are members of the Graduate Field of Applied Physics but are members of other departments, including those who hold dual appointments.

6. Separately Budgeted Research Expenditures by Source of Support

	Departmental Research
Federal & State government	$11,644,219
Business and industry	723,151
University	1,647,531
Total	$14,014,902[1]

[1]There is an estimated additional $15M in Research Expenditures by the 24 faculty members who are members of the Graduate Field of Applied Physics but are members of other departments, including those who hold dual appointments.

FACULTY

Professors

Brock, Joel D., Ph.D., MIT, 1987. High-resolution X-ray scattering studies of condensed matter.

Buhrman, Robert A., Ph.D., Cornell, 1973. High-T_c superconductivity; nanostructure physics; quantum transport and tunneling studies; scanning tunneling microscopy.

Cool, Terrill A., Ph.D., Cal. Tech., 1965. Molecular spectroscopy; chemical physics; ultrasensitive detection laser spectroscopy.

Craighead, Harold G., Ph.D., Cornell, 1980. Nanofabrication; ultrasmall solid state devices and structures.

Gaeta, Alexander, Ph.D., Rochester, 1990. Quantum and nonlinear optics.

Kusse, Bruce R., Ph.D., MIT, 1969. Plasma physics; intense ion beams; inertial fusion.

Lindau, Manfred, Ph.D., Tech. Univ. Berlin, 1983. Biophysics, optical & electrical studies of fusion membrane transport & exocytosis.

Lovelace, Richard V., Ph.D., Cornell, 1970. Plasma physics theory; astrophysics.

Muller, David, Ph.D., Cornell, 1996. Condensed matter; electronic microscopy.

Webb, Watt W., Sc.D., MIT, 1955. Biological physics; chemical physics; cooperative phenomena; hydrodynamics; physical optics; fluctuation correlation spectroscopy.

Wise, Frank W., Ph.D., Cornell, 1988. Director. Ultrafast optics; semiconductor nanostructures.

Associate Professors

Pollack, Lois, Ph.D., Massachusetts Institute of Technology, 1989. Low temperature physics and biological physics.

Xu, Chris, Ph.D., Cornell, 1996. Optics.

Assistant Professor

Fennie, Craig, Ph.D., Rutgers, 2006. Condensed matter; computational matter physics.

Adjunct Professor

Heinekamp, Scott, Ph.D., Brown University. X-ray diffraction.

Adjunct Associate Professor

Bilderback, Donald H., Ph.D., Purdue, 1975. Synchrotron radiation instrumentation.

Hao, Quan, Ph.D., Chinese Academy of Science, 1988. Structural biology of proteins.

Emeritus

Batterman, Boris W., Ph.D., Emeritus. MIT, 1956. X-ray and neutron diffraction; synchrotron radiation; solid state physics.

Fleischmann, Hans H., Emeritus Dr.rer.nat., Technische Hochschule, Munich, 1962. Plasma physics; thermonuclear fusion.

Kostroun, Vaclav O., Ph.D., Emeritus, Oregon, 1968. Low-energy nuclear and atomic physics. (Joint appointment with Nuclear Science and Engineering.)

Silcox, John, Ph.D., Emeritus, Cambridge, 1961. Inelastic electron scattering; atomic resolution electron microscopy.

ASSOCIATED GRADUATE FIELD FACULTY

Professors

DiSalvo, F. J., Jr., Ph.D., Stanford, 1971. Physical and chemical properties of solid state compounds.
Gruner, Sol., Ph.D., Princeton, 1977. Biophysics, condensed matter, advanced X-ray area detectors.
Hammer, David A., Ph.D., Cornell, 1969. Plasma physics; nuclear fusion; high-power electron- and ion-beam physics.
Hines, Melissa, Ph.D., Stanford, 1992.
Houck, James R., Ph.D., Cornell, 1967. Astrophysics.
Kelley, Michael C., Ph.D., California, Berkeley, 1970. Space plasma physics; rocket and satellite instrumentation.
Kitner, Paul M., Ph.D., Minnesota, 1974. Space plasma physics; digital signal processing.
McEuen, Paul L., Ph.D., Yale, 1990. Nanostructures.
Pollock, Clifford R., Ph.D., Rice, 1981. Lasers; molecular spectroscopy; quantum electronics.
Ralph, Daniel, Ph.D., Cornell, 1993. Nanoelectronics and magnetics.
Seyler, Charles E., Jr., Ph.D., Iowa, 1975. Plasma physics; thermonuclear fusion; high-power beams; space plasmas.
Shalloway, David I., Ph.D., MIT, 1975. Theoretical/Computational analysis of protein structure and dynamics.
Tiwari, Sandip, Ph.D., Cornell, 1980. Optical and electronic properties of semiconductor devices.
van Dover, Robert Bruce, Ph.D., Stanford, 1980. Magnetic, Dielectric, superconducting and optical thin films.

Associate Professors

Lipson, Michal, Ph.D., Israel Institute of Technology, 1998. Photonics.
Thompson, Michael O., Ph.D., Cornell, 1984. Rapid phase transformations; nonequilibrium thermodynamics of semiconductor materials.
Zipfel, Warren, Ph.D., Cornell, 1993. Biophysics, biomedical engineering.

Assistant Professors

Erickson, David, Ph.D., U of Toronto, 2004. Fluid and thermal dynamics and optics, nanophotonics, nanofabrication, chemistry and biology.
Park, Jiwoong, Ph.D., U.C. Berkeley, 2003. Synthesis, assembly and characterization of nanoscale materials and devices.

RESEARCH SPECIALTIES OF FACULTY

Astrophysics

Bioimaging and multiphoton microscopy

Biophysics

Condensed matter theory

High-resolution electron microscopy and spectroscopy

Microfluidics and nanofluidics

Nanobiophysics and nanobiotechnology

Nanophotonics

Nanoscale electronic and magnetic devices

Nanostructure physics and chemistry

Nanostructured materials for energy conversion and energy storage

Plasma physics

Quantum and nonlinear optics

Ultrafast optics and lasers

X-ray diffraction

QUEENS COLLEGE OF THE CITY UNIVERSITY OF NEW YORK

DEPARTMENT OF PHYSICS

Flushing, New York 11367-0904

Students Accepted For Degree	FIELDS		
	Physics	Astronomy	Related Fields
Doctorate	X		
Master's	X		

*Doctoral work in physics at Queens College is conducted as a component of the CUNY Ph.D. program in physics. All information supplied below refers only to that portion of the program conducted at Queens College.

1. General

President: James L. Muyskens
Department Chairman: Alexander A. Lisyansky
Department Telephone Number: (718) 997-3350
Type of Institution: College
Control: Public
Setting: Urban
Total Faculty: 631
Total Graduate Faculty: 200
Total Students: 20,711
Total Graduate Students: 4,652
Annual Graduate Tuition:

*Ph.D. Students**	In-State	Out-of-State
*Level I full-time***	$3,290	$645 per credit
*Level II full-time***	$2,060	$4,580
*Level III full-time***	$815	$1,635

*Rates quoted are per semester and correspond to our latest information.

**Level I: Students who have not yet completed 45 credits of graduate work, including transfer credit. Level II: from completion of 45 credits to completion of course work. Level III: from completion of course work to completion of degree.

Tuition rates for: 2009–10
Deferred tuition plan: Yes
Other Fees: Student activity fee for semester ($57.10)
Term: Semester

2. Number of Faculty in Department

The combined total of full-time faculty in the three professorial ranks is 12. The combined total of full-time, part-time, and other faculty at all ranks is 16.

3. Admission, Financial Aid, and Housing

Address admission inquiries to: Prof. Igor Kuskovsky, Department of Physics
Graduate application fee required: $125
Admission deadline (Fall admission): None
Admission information: Admission into the Ph.D. program for fall admission, 2009–10, 2 students were accepted from 60 applicants.
Admission requirements: For admission to the graduate programs, a Bachelor's degree in physics or related field is required with a minimum undergraduate GPA of 3.0 (B) specified. GRE General test is required. GRE Subject test is not required, but is strongly encouraged. No minimum acceptable score for admissions is specified. The average GRE scores for admissions were verbal 570; quantitative 780; total 1350. Students from non-English speaking countries are required to demonstrate proficiency in English via the TOEFL exam, for the Masters program only. No minimum score required.
Address financial aid inquiries to: Prof. Igor Kuskovsky, Department of Physics, Queens College, Flushing, NY 11367
GAPSFAS application required: No
Financial aid deadline: February 1, 2011
Loans available: Yes—Contact Financial Aid Office
On-campus, single student housing available: Yes
On-campus, married student housing available: Yes

Table A—Faculty, Enrollments, and Degrees Granted

Research Specialty	2009–10 Faculty	Enrollment[1] Fall 2010 Master's	Enrollment[1] Fall 2010 Doctorate	No. of Degrees Granted[2] 2009–10 (2004–09) Master's	No. of Degrees Granted[2] 2009–10 (2004–09) Terminal Master's	No. of Degrees Granted[2] 2009–10 (2004–09) Doctorate	Median No. of Years for 2009–10 Ph.D.'s
Atomic Molecular, & Optical Physics	4	0	3	0(0)	0(0)	1(6)	5
Condensed Matter Physics	5	0	8	0(0)	0(0)	1(8)	5
Electromagnetism	0	0	0	0(0)	0(0)	0(1)	–
Physics Education	1	0	0	0(0)	0(2)	0(0)	–
Polymer Physics/ Science	1	0	0	0(0)	0(0)	0(3)	–
Relativity & Gravitation	1	0	0	0(0)	0(0)	0(0)	–
Non-specialized		5	0	2(10)	2(3)	–	–
Total		5	11	2(10)	2(5)	2(18)	
Full-time Grad. Stud.		1	10				
Part-time Grad. Stud.		0	0				
First-year Grad. Stud.		4	1				

Median Years in Grad. Study (2009–10 Degrees) 5
Undergraduate Degrees, 2009–10 (2004–09): 4(29)

[1]Students not yet committed to a research specialty are entered under non-specialized.
[2]Five-year totals in parentheses.

4. Graduate Degree Requirements

Master's: Master of Arts: 30 hours of approved courses plus either a Master's thesis or a written comprehensive exam; an average of B or higher required; no residence or language requirements.
Doctorate: 60 hours of approved courses (at least 30 hours at the City University) with an average of B or higher required. Candidates must pass a First (Qualifying) Examination and a Second Examination in their prospective field of research. A thesis must be completed and defended. There are no foreign language requirements.
Other programs: Science teacher certification is provided through the Science Teacher Career (STC) program run jointly with the Queens College Department of Education.
Thesis: Thesis may be written *in absentia*.
Special Equipment, Facilities, or Programs: Extensive laboratory facilities are used for the preparation and characterization of magnetic, polymer, and electro-optic materials, with support from excellent machine and electronics shops. State-of-the art equipment includes a clean room for nano- and micro-fabrication, two secondary ion mass spectrometers (SIMS), X-ray photoelectron spectroscopy (XPS), ellipsometer, transmission electron microscope (TEM), scanning electron microscope (SEM) with e-beam lithography, AFM-NSOM system, surface profilometer, rheometer, laser

deposition and sputter deposition systems, continuous wave-, nanosecond-, and picosecond laser systems, CCD cameras, coherent autoscan laser spectrometers, microwave vector analyzers, network analyzers, spectrum analyzers, a variable temperature Hall effect measurement system, a table-top thermal evaporator for metallization, and an optical IT superconducting magnet. As a participant in the NSF MRSEC for Polymers at Engineered Interfaces, the Department has access to facilities at Brookhaven National Laboratory, and at neighboring institutions.

Table B—Appointments to Graduate Students, 2008–09

Title of Appointee	Appointments Total	First year	Academic Load Allowed in Credit Hours	Hours of Service Per Week	Stipend for Academic Year ($)
			Semester		
Teaching Assistant	6	3	12	6	$28,000
Research Assistant	3	0	12	20	26,000
Total	11	3			

5. Personnel Engaged in Separately Budgeted Research,

Professional faculty	10
Other faculty	1
Postdoctoral appointments	2
Graduate students	11
Undergraduate students	4
Total	28

6. Separately Budgeted Research Expenditures by Source of Support

	Departmental Research	Physics-related Research Outside Department
Federal government	$820,000	$
State and local government	16,000	
Other Include institution's own separately budgeted accounts	200,000	
Total	$1,036,000	$

Table C—Separately Budgeted Research Expenditures

Research Specialty	No. of Grants	Expenditures ($)
Atomic, Molecular, & Optical Physics	3	430,000
Condensed Matter Physics	6	606,000
Total	9	1,036,000

FACULTY

Professors

Cadieu, Fred J., Ph.D., Chicago, 1970. Experimental physics of magnetic films.

Deych, Lev I., Ph.D., Russia, 1991. Theoretical condensed matter physics.

Genack, Azriel Z., Ph.D., Columbia, 1973. Experimental solid state and optical physics.

Liebovitch, Larry S., Ph.D., Harvard, 1978. Fractals, artificial neural networks.

Lisyansky, Alexander A., Dr. Sci., Ukraine, 1988. Chairman of the Department. Theoretical condensed matter physics.

Schwarz, Steven A., Ph.D., Stanford U., 1980. Experimental polymer physics.

Associate Professors

Klarfeld, Joseph, Ph.D., Yeshiva, 1969. General relativity theory.

Kuskovsky, Igor L., Ph.D., Columbia, 1998. Experimental solid state and optical physics.

Menon, Vinod M., Ph.D. University of Massachusetts, 2001. Experimental solid state and optical physics.

Assistant Professors

Murokh, Lev, Ph.D., Russia, 1996. Quantum theory of nanostructures.

Saini, Sajan, Ph.D., MIT, 2004. Experimental solid state and optical physics.

Vuong, Luat, Ph.D., Cornell University, 2008. Experimental and theoretical nanophotonics, nonlinear optics, and nano-structured organic solar cells.

Theoretical

Condensed Matter Physics. Wave propagation in random media; localization phenomena; photonic crystals; random superlattices, nanostructures. Deych, Lisyansky, Murokh.

Relativity and Gravitation. Quantization of the Gravitational field. Klarfeld.

Statistical Physics. Critical phenomena at phase transitions. Lisyansky.

Experimental

Condensed Matter Physics. Sputtering and pulsed laser deposition and characterization of magnetic films. Properties of surfaces and thin films. Microwave propagation in disordered dielectric materials. X-ray and neutron diffraction of thin films. Self-assembled and colloidal quantum structures. Wide bandgap semiconductors. Organic-inorganic hybrid heterostructures. Quantum wells and dots, nano- and micro-devices. Organic and inorganic solar cells. Cadieu, Genack, Kuskovsky, Menon, Saini. 1 postdoctoral fellow.

Optics. Non-linear optics. Lasing materials, photonic bandgaps, light scattering in random media. Photonic circuits. Microcavity polaritous. Genack, Kuskovsky, Menon, Saini, Vuong. 1 postdoctoral fellows.

Polymer Physics. Diffusion, adhesion, wetting, rheology. SIMS/XPS characterization of polymer interfaces. Schwarz.

Materials Science. Metal/semiconductor interfaces. Characterization of semiconductors, glasses and ceramics. Thin film depth profiling. Cadieu, Kuskovsky, Schwarz. 1 postdoctoral fellow.

RENSSELAER POLYTECHNIC INSTITUTE

DEPARTMENT OF PHYSICS, APPLIED PHYSICS, AND ASTRONOMY

Troy, New York 12180-3590

Students Accepted For Degree	FIELDS		
	Physics	Astronomy	Related Fields
Doctorate	X		
Master's	X	X	

1. General

President: Shirley A. Jackson
Vice Provost & Dean of Graduate Education, Acting: Stanley Dunn
Acting Department Head: Xi-Cheng Zhang
Department Telephone Number: (518) 276-6310
Type of Institution: University
Control: Private
Setting: Urban
Total Faculty: 475
Total Students: 7,656
Total Graduate Students: 1,166
Annual Graduate Tuition:
All Graduate Students: $39,600/yr
Tuition rates for: 2010–11
Deferred tuition plan: No
Other Fees: $1,938/semester
Term: Semester

2. Number of Faculty in Department

The current number of tenured/tenure track faculty in the three professorial ranks is 23. Additionally, there are 6 research faculty, 1 clinical, 1 adjunct faculty, and 15 faculty emeriti.

3. Admission, Financial Aid, and Housing

Address admission inquiries to: Paul Marthens, vice President, Enrollment
Graduate application fee required: $75
Admission deadline (Fall admission): January 15
Admission information: For fall admission, 2009–10, from 150 applicants, 40 students were offered assistantships, 18 accepted.
Admission requirements: For admission to the graduate programs, a Bachelor's degree is required with courses and grades demonstrating ability and preparation adequate for graduate study in physics. Remedial courses available as needed. The GRE is required. There is no specific minimum score requirement. The GRE Advanced is required. There is no specific minimum score requirement. The average GRE scores for Fall 2009 admissions with assistantships were verbal–549, quantitative–745, analytical–4.0, subject–820. Students from non-English speaking countries are required to demonstrate proficiency in English via the TOEFL exam. Minimum acceptable score for admission and teaching assistantship 600. The test of spoken English (TSE) is required by all foreign applicants.
Undergraduate preparation assumed: Students are normally expected to have taken intermediate-level courses in mechanics, electricity and magnetism, quantum physics, thermodynamics, and experimental physics. Typical texts are Marion and Thornton, Griffiths, Brehm and Mullin, Stowe, and Liboff. However, students may take a limited number of remedial courses after enrollment where inadequate preparation has been available, but where other courses and grade records indicate adequate ability.
Address financial aid inquiries to: Office of Financial Aid.
GAPSFAS application required: No
Financial aid deadline: 2/15
Loans available: Yes
Address housing inquiries to: Residence Life and Hospitality Services
On-campus, single student housing available: Yes
Cost/academic year: $6,102 (avg.)
On-campus, married student housing available: Yes
Cost/month: $720 (avg.)

Table A—Faculty, Enrollments, and Degrees Granted

Research Specialty	2008–09 Faculty	Enrollment[1] Fall 2009		No. of Degrees Granted[2] 2009–10 (2004–09)			Median No. of Years for 2008–09 Ph.D.'s
		Master's	Doctorate	Master's	Terminal Master's	Doctorate	
Astrophysics/ Astronomy	3	0	8	0(1)	0(0)	0(0)	–
Biophysics	1	0	4	0(1)	0(1)	1(2)	6
Condensed Matter Physics	11	2	38	4(17)	0(0)	7(21)	4.50
Optics	2	0	7	0(1)	0(0)	1(4)	5.5
Particles & Fields	4	1	6	1(6)	0(0)	4(18)	5.5
Other (specify) Education	0	0	0	0(0)	0(0)	0(0)	–
Total		1	68	5(8)	1(1)	13(45)	
Full-time Grad. Stud.		1	64				
Part-time Grad. Stud.			2				
First-year Grad. Stud.		1	17				
Median Years in Grad. Study (2009–10 Degrees)			5	2	2.5	5	
Undergraduate Degrees, 2009–10 (2005–10): 58(149)							

[1]Students not yet committed to a research specialty are entered under non-specialized.
[2]Five-year totals in parentheses.

4. Graduate Degree Requirements

Master's: 30 semester credit hours, "B" average; one academic year (two semesters) residence minimum. No foreign language, no exams. Thesis or research projects are required, but may be waived for students who pass Ph.D. candidacy exam. Maximum transfer credit six credit hours. Usually six to nine credit hours for thesis or 3 credit hours for research project.
Doctorate: 72 semester credit hours, "B" average; one academic year (two semesters) residence minimum; no foreign language requirement; qualifying examination (eight hours written) by end of first year (mainly on advanced undergraduate level material); candidacy exam (oral, on physics related to proposed thesis research area); thesis dissertation required. Usually about 30 to 45 credit hours of research.
Thesis: Thesis may be written *in absentia*, with on-campus advisor.

Table B—Appointments to Graduate Students, 2009–10

Title of Appointee	Appointments Total	Appointments First year	Academic Load Allowed in Credit Hours	Hours of Service Per Week	Stipend for Academic Year ($)
			Semester		
Teaching Assistant	25	15	9–15	20	16,500
Research Assistant	34		12–15	20	16,500
NASA Fellow	1				16,500
Harry F. Meiners	1	0	–		16,500
IGERT Fell	1	0	–		30,000
Self Supported	2		–		–
RGF	1	1			up to 22,000
GK Fellow	1	1			up to 30,000
Total	66	17			

5. Personnel Engaged in Separately Budgeted Research,

Professorial faculty	22
Research Faculty	5
Adjunct Faculty	1
Postdoctoral appointments	15
Visiting Research Scientists	8
Total	50

6. Separately Budgeted Research Expenditures by Source of Support

	Departmental Research	Physics-related Research Outside Department
Federal government	$6,153,246	$
State and local government	787,854	
Other government	33,284	
Corporate	313,663	
Total	$7,288,047	$

Table C—Separately Budgeted Research Expenditures

Research Specialty	No. of Grants	Expenditures ($)
Astrophysics	10	700,343
Condensed Matter Physics	61	4,054,357
Physics Particle	36	2,268,839
Optics	7	264,514
Total	114	7,288,047

FACULTY

Professors

Adams, G. S., Ph.D., Indiana, 1977. Experimental particle physics; dark matter; exotic hadrons.

García, A., Ph.D., Cornell University, 1987. Theoretical and computational aspects of biomolecular dynamics.

Jackson, S. A., Ph.D., MIT, 1973. Theoretical elementary particle physics.

Lin, S.-Y., Ph.D., Princeton University, 1992. Design, fabrication, and experimental assessment of photonic and plasmonic nanostructures.

Lu, T.-M., Ph.D., Wisconsin, 1976. Thin films and interfaces.

Napolitano, J. J., Ph.D., Stanford, 1982. Experimental nuclear and particle physics.

Nayak, S., Ph.D., Jawaharlal Nehru University, 1995. Theoretical physics; first principle calculations.

Newberg, H., Ph.D., California, Berkeley, 1992. Astrophysics.

Persans, P. D., Ph.D., Chicago, 1982. Spectroscopy and structure of semiconductors, quantum dots, and nanostructures.

Roberge, W. G., Ph.D., Harvard, 1981. Theoretical astrophysics.

Schroeder, J., Ph.D., Catholic, 1974. High pressure physics and biological physics.

Schubert, E. F., Ph.D., University of Stuttgart, 1986. Physics of semiconductor devices. (Joint appointment with ECSE).

Shur, M., Ph.D., A. F. Ioffe Institute, 1967, Dr. Sc. 1992, A. F. Ioffe Institute. Semiconductor physics and device physics ballistic transport, terahertz radiation, UV LEDs.

Stoler, P., Ph.D., Rutgers, 1966. Particle and nuclear physics; structure of hadrons.

Wang, G.-C., Ph.D., Wisconsin, Madison, 1978. Nanostructure physics.

Washington, M., Ph.D., New York University, 1976. Photonic and electronic devices.

Whittet, D. C. B., Ph.D., St. Andrews, U. K., 1975. Astrophysics; observational astronomy; interstellar dust; origins of life.

Zhang, S. B., Ph.D., University of California at Berkeley, 1989. Computational condensed matter theory, physics and chemistry of nano, optoelectronic, photovoltaic, and hydrogen storage materials.

Zhang, X.-C., Ph.D., Brown University, 1986. Acting Department Head. Ultrafast optics, photonics, optoelectronic and terahertz science and technology.

Associate Professors

Korniss, G., Ph.D., Virginia Tech, 1997. Statistical mechanics; Dynamics in complex networks.

Wetzel, C. M., Ph.D., Technical University, Munich, 1993. III-V nitride semiconductor physics and technology.

Wilke, I., Ph.D., Swiss Federal Institute of Technology, 1993. Ultrafast optics; photonics; optoelectronic and terahertz science and technology.

Yamaguchi, M., Ph.D., Hokkaido University, 1991. Acoustic thermal transport in nanoscale materials, TH(2) wave generation, pulse shaping, and optimization, phonon and electron dynamics in condensed matter.

Assistant Professors

Eah, S.-K., Ph.D., Seoul National University, 2001. Synthesis and 2D self-assembly of gold nanoparticles, single-molecule surface-enhanced Raman spectroscopy, catalysis by gold, nano-drum, free-standing film of nanoparticles.

Giedt, J., Ph.D., University of California, Berkeley. Particle phenomenology, lattice field theory, string theory, high energy mathematical and computational physics.

Lewis, M. K., Ph.D., University of Michigan-Ann Arbor. Molecular electronics and hybrid nanostructures.

Research Faculty

Dai, Jianming, Ph.D., Tianjin University, 1994.

Detchprohm, T., Ph.D., Nagoya University, 1996. III-V nitride semiconductor epitaxy and technology.

Lee, S., Ph.D., Michigan, Thin film and interfaces.

Sun, Y., Ph.D., National University of Singapore, 2004. Computational materials science.

Visiting Professor

Schowalter, L.,Ph.D., Illinois, 1981. Material physics.

Adjunct Professor

Xu, J., Ph.D., Institute of Physics, China. Ultrafast optics, terahertz science and technology.

Professors Emeriti

Breed, H., Ph.D. Optics.
Brown, E., Ph.D., Cornell, 1954. Theoretical solid state physics.
Eppenstein, W., M.S., Rensselaer, 1952. Development of instructional resources.
Giaever, Ivar, Ph.D., Rensselaer, 1964. Biological physics.
Hayes, T. M., Ph.D., Harvard, 1968. Condensed matter physics.
Leitner, A., Ph.D., Yale University. Educational physics.
Leung, C. M., Ph.D., California, Berkeley, 1975. Computational astrophysics; radiation transport; astrochemistry.
Levinger, J. S., Ph.D., Cornell, 1948. Theoretical physics.
Medicus, H. A., Ph.D., Swiss Federal Inst. of Tech., 1949. Experimental nuclear physics.
Meltzer, A. S., Ph.D., Princeton, 1956. Astrophysics.
Min, K., Ph.D., Illinois, 1963. Experimental nuclear physics.
Nettel, S. J., Ph.D., MIT, 1960. Theoretical solid state physics.
Resnick, R., Ph.D., Johns Hopkins, 1949. Edward P. Hamilton Distinguished Professor Emeritus. Theoretical physics; physics education.
Salinger, G., Ph.D., University of Illinois. Low temperature physics.
Winhold, E. J., Ph.D., MIT, 1953. Experimental nuclear physics.
Yergin, P. F., Ph.D., Columbia, 1953. Experimental nuclear physics.

RESEARCH SPECIALITIES AND STAFF

Theoretical

Astrophysics. Protoplanetary disks, interstellar shock waves, multifluid magnetohydrodynamics, physics of interstellar dust; Roberge.

Stochastic Dynamics on Complex Networks. Structure and dynamics of biological, social, and artificial networks. Applications to communication networks, epidemic models, and ecological systems. On-campus collaborations and facilities are at the Center for Pervasive Computing and Networking. Korniss

Theoretical condensed matter physics. Electronic structure of nano structured materials. Models describing the structure and electronic properties of surfaces and interfaces and the binding and mobility of adsorbed atoms on metal surfaces. Molecular Electronics and spintronics. Relationships between noise and fractal, and far-from-equilibrium physics. Diffraction from fractals growth/etch front. Kar, Korniss, Lu, Nayak.

Theoretical Particle Physics. This research program focuses on aspects of physical interanctions beyond the Standard Model. This includes investigations of supersymmetric and composite matter using lattice field theory techniques, using the supercomputer facilities at RPI, the Computational Center for Nanotechnology Innovation. We are also actively developing simulation tools for the graphics processing unit architecture, which has a significant price advantage over traditional computing platforms. Giedt.

Experimental

Astrophysics. Data collected at wavelengths from ultraviolet to radio are used to study dust and molecules in interstellar clouds, particularly those giving birth to new stars and planetary systems. Infrared observations from the Spitzer Space Telescope and ground-based telescopes are used to probe the physical properties and chemical composition of dust in molecular clouds, protostars and protoplanetary disks. Our goal is to study the origin and evolution of organic matter in such regions and assess its relevance to exobiology. The enormous astronomical databases of the Sloan Digital Sky Survey (SDSS) and the Sloan Extension for Galactic Understanding and Evolution (SEGUE) are mined to unfold the physical and chemical structure and evolution of the thick disk and halo of the Milky Way Galaxy. A new spectral survey of 9 million Galactic stars is underway, in collaboration with the Chinese LAMOST project. Streams of stars from tidally disrupted dwarf galaxies in the outer parts of the Milky Way are identified and used to study the Galactic dark matter distribution. Astronomical computational methods are developed to classify and statistically analyze hundreds of gigabytes of catalog data. Newberg, Whittet.

Biophysics. Current research in theoretical and computational aspects related to the dynamics and statistical mechanics of biomolecular systems. The objectives are to understand the folding, dynamics, stability and function of biomolecules from physical principles. Protein folding, binding and dynamics are important for understanding how proteins work and how they interact with other biomolecules. Knowledge gained from this research has applications in biotechnology, drug design, and biomaterials. Highly parallel computer simulations of the folding and thermodynamics of proteins and peptides in aqueous solutions are being performed. New simulation methods are being applied to study protein folding, binding and aggregation. Other research interests include the hydrophobic effect, enzyme catalysis, and peptide interactions with membranes. García, Wilke.

Condensed Matter Physics. This research program concentrates on three areas: surfaces, interfaces, thin films, and nanostructures; optical, electronic and energy materials; and electronic transport. New research concepts, materials, and techniques are developed for high technology applications. Many research projects are interdisciplinary. Experimental and theoretical work on surfaces, interfaces, thin films, and nanostructures involves the deposition, growth, and characterization of metals, semiconductors, and insulators from monolayers to multilayers to three dimensional nanostructures. The phenomena that are studied include homo- and hetero-epitaxy, initial stages of epitaxy, nucleation of thin films, surface phase transitions, and interface (solid-solid and solid-liquid) structure and bonding. Theoretical studies using quantum mechanical calculations include semiconductor defect and impurity physics, nanophysics, surface and interface physics, nanostructures under extreme conditions, and weakly interacting many-body systems. Techniques include Auger electron spectroscopy, X-ray photoelectron spectroscopy, high resolution low-energy electron diffraction, reflection high-energy electron diffraction, atomic force microscopy, scanning tunneling microscopy, X-ray absorptions spectroscopy, X-ray crystallography, and ellipsometry. The department's major facilities include ultrahigh vacuum evaporation, III-V group IV molecular beam epitaxy, group III nitrides growth, chemical vapor deposition, atomic layer deposition, and the extensive facilities of the Microelectronics Clean Room. Theoretical work also includes applications of statistical physics and large-scale simulations to study the dynamics of natural, artificial, and social systems, including ecological systems, agent-based models, and social networks. The optical and electronic materials under study include wide bandgap semiconductors, photonic crystals, polymers, semiconductor nanoparticle composites, dielectrics, and magnetic thin films. Optical characterization facilities include Raman, Brillouin, and Rayleigh scattering, photomodulation spectroscopy, photothermal deflection spectroscopy, magneto-optic Kerr effect, and Faraday rotation. Electron transport in semiconductor and metallic materials are under way. This research is expected to enhance understanding of transport in nanostructures. The experimental work includes studies of ballistic electron transport in ultrathin epitaxial multilayers, electrical resistance of metallic films, and plasma wave electronics in high electron mobility transistors. The electron transport in nanoscale systems (single molecule to atomic wire to carbon nanotube) is studied using the state of the art first principles calculation. The current research includes spin assisted transport (Spintronics) at the nanoscale. The computational facilities in the theory group include in house Linux cluster of about 100 processors and the group has access to National Super Computer facilities. Other experimental facilities used in these programs include those at the Center for Integrated Electronics, the Focus Center for Interconnects, the Center for Advanced Interconnect Systems Technologies, the electron microprobe and electron microscope facilities, accelerators at the University at Albany, the National Synchrotron Light Source at Brookhaven National Laboratory, and the Stanford Synchrotron Radiation Laboratory. Dai, Detchprohm, Eah, Lewis, T.-M. Lu, Persans, Schowalter, Schroeder, Schubert, Shur, Sun, Wang, Wetzel, Wilke, Yamaguchi, S. Zhang, X.-C. Zhang.

Educational Development in Physics. Development of new introductory course models. Development of laboratory and demonstration experiments; use of computers in physics instruction; production of written materials, computational physics activities, and computer based multimedia programs.

Optical Physics. Department research in optical physics is directed towards developing new measurement techniques, new optical materials, and novel devices. Faculty research includes terahertz generation, detection, imaging, and spectroscopy; development of novel light emitting devices; photonic nanostructures and optics of metal nanoparticles; spectroscopy of quantum dots and wells; and ultrafast optical effects in semiconductors, magnetic systems, and biological systems. Major facilities include ultrafast laser and terahertz systems associated with the Center for Terahertz Research and the materials growth and optical characterization facilities of the Constellation Chair in Future Chips. Other facilities allow linear and nonlinear optical absorption, luminescence, Raman and Brillouin scattering and various modulation spectroscopies. Dai, Eah, Lin, Persans, Schroeder, Schubert, Wetzel, Wilke, Xu, Yamaguchi, X.-C. Zhang

Experimental Particle and Nuclear Physics. Particle and Nuclear Physics research focuses on the fundamental constituents of matter, and their interactions, governed by the basic principles of quantum mechanics and special relativity. Our major experimental effort focuses on neutrino oscillations, and the determination of the fundamental parameters of the neutrino mixing matrix. This includes the Daya Bay Neutrino Oscillation Experiment, a high sensitivity search for nonzero mixing between the first and third generations, as well as preparations for a long baseline neutrino experiment at DUSEL. We also have a program of experiments at JEfferson Laboratory, where the CEBAF accelerator provides an intense source of high-energy electrons and photons. Experiments at Jefferson Laboratory focus on the structure of baryons involving exclusive reactions, and searching for exotic five-quark baryons. Adams, Giedt, Napolitano, Stoler.

STATE UNIVERSITY OF NEW YORK AT ALBANY

DEPARTMENT OF PHYSICS

Albany, New York 12222

Students Accepted For Degree	FIELDS Physics	Astronomy	Related Fields
Doctorate	X		
Master's	X		

1. General

President: George M. Philip
Dean of Graduate School: Kevin Williams
Department Chair: Carolyn MacDonald
Department Telephone Number: (518) 442–4501
Type of Institution: University
Control: Public
Setting: Urban
Total Faculty: 1,000
Total Graduate Faculty: Not separated
Total Students: 17,040
Total Graduate Students: 4,936
Annual Graduate Tuition:
In-state residents: Full-time—$8,370/yr. (2 sem.)
Part-time—$349/credit
Out-of-state residents: Full-time—$13,250/yr.
Part-time—$552/credit
Tuition rates for: 2009–10
Other Fees: $75 application fee
Term: Semester

2. Number of Faculty in Department

The combined total of full-time faculty in the three professorial ranks is 13. The combined total of full-time, part-time, and other faculty at all ranks is 16.

3. Admission, Financial Aid, and Housing

Address admission inquiries to: Chair, physics@albany.edu
Graduate application fee required: $75
Admission deadline (Fall admission): 6/1 (2/1 for Assistantship)
Admission information: For fall admission, 2009–10, 9 Ph.D. and 9 M.S. were accepted.
Admission requirements: The GRE is strongly recommended. Students from non-English speaking countries are required to demonstrate proficiency in English via the TOEFL. Minimum acceptable score for admission is 213. A score of 250 is required to be eligible for a teaching assistantship.
Undergraduate preparation assumed: Symon, *Mechanics*; Griffiths, *Introduction to Electrodynamics*; Griffiths, *Quantum Mechanics*.
Address financial aid inquiries to: physics@albany.edu
GAPSFAS application required: No
Financial aid deadline: 2/1
Address housing inquiries to: Graduate Housing Offic

Table A—Faculty, Enrollments, and Degrees Granted

Research Specialty	Faculty	Enrollment[1] Fall 2010–11 Master's	Doctorate	No. of Degrees Granted[2] 2010 Master's	Terminal Master's	Doctorate
BioMedical Physics	2.5	–	5	–	–	2(8)
Computational Physics	2	–	6	–	–	0
Condensed Matter & Materials Sci.	2	–	6	–	–	1(5)
Particles & Fields	2.5	–	6	–	–	0(6)
Physics Education	1	–	–	–	–	0(0)
Quantum Theory	3	–	6	–	–	0(5)
Non Specialized	0		13			
Total	13	11	42			3(24)
Full-time Grad. Stud.		9	37			
Part-time Grad. Stud.		2	5			
First-year Grad. Stud.		5	9			
Median Years in Grad. Study (2008–09 Degrees)						–
Undergraduate Degrees, 2009–10: 29(104)						

[1]Students not yet committed to a research specialty are entered under non-specialized.
[2]Five-year totals in parentheses.

4. Graduate Degree Requirements

Master's: 30 graduate course credits, at least 24 on campus, including core, elective and research courses; Master's thesis or passage of Comprehensive Examination.

Doctorate: 60 credit hours beyond Baccalaureate including core, elective and research courses with at least two full-time semesters required. Transfer credit of up to 30 hours allowed. Students are required to pass written Comprehensive Examination followed by Oral Qualifying Examination. Dissertation and dissertation defense examinations are required. Dissertation research may be conducted off-campus in approved programs.

Thesis: Thesis may be written *in absentia*.

Special Equipment, Facilities, or Programs: A transmission electron microscope laboratory studies defects and other materials properties. An EPR facility is used to study biological physics. X-ray research includes applications to materials and medicine. The experimental particle physics program is part of the ATLAS collaborations at CERN and the BABAR collaboration at the Stanford Linear Accelerator Center. A robotics lab studies intelligent behaviours. Cooperative programs have been established with nearby General Electric Research and Development Center, Watervliet Arsenal, N.Y., State Public Health, IBM Watson Research Laboratories. Our website is www.albany.edu/physics/

Table B—Appointments to Graduate Students, 2008–09

Title of Appointee	Appointments Total	First year	Academic Load Allowed in Credit Hours	Hours of Service Per Week	Stipend for Academic Year ($)
			Semester		
Teaching Assistant	24	8	Up to 9	18–20	15,500+summer
Research Assistant	5	0	Up to 9	18–20	
Total	31	10			

[1]Amount does not include Tuition Scholarship Award.
[2]Summer appointments with additional stipends are available.

5. Personnel Engaged in Separately Budgeted Research, 2008–09

Professorial faculty	5
Postdoctoral appointments	1
Graduate students	20
Undergraduate students	2
Total	28

6. Separately Funded and Managed Laboratories

Particle accelerators (a 140-kV Danfysik, a 1-MV tandom NEC, a 400-kV Exitron, and a 4-MV Dynamitron) provide positive ions and electrons for materials research and analysis.

FACULTY

Professors

Alam, M. Sajjad, Ph.D., Indiana, 1975. Experimental particle physics.

Caticha, Ariel, Ph.D., Caltech, 1985. Information physics. Fundamental problem in quantum, statistical and gravitational physics.

Das, Tara P., Ph.D., Calcutta, India, 1955. Theory of electronic structures and associated properties, including hyperfine interactions of atomic, molecular, biological, and condensed matter systems.

Garg, Jagadish B., D.Sc., Paris, 1958. Experimental nuclear physics; neutron resonances.

Inomata, Akira, Ph.D., Rensselaer Polytech. Inst., 1964. Theoretical particle physics and general relativity.

Kuan, Tung-sheng, Ph.D., Cornell, 1977. Materials science and electron microscopy.

Lanford, William A., Ph.D., Rochester, 1972. Materials physics; applied physics; glasses; thin films; ion beam analysis.

MacDonald, Carolyn A., Ph.D., Harvard, 1986. X-ray optics; materials science; and medical physics.

Associate Professor

Ernst, Jesse, Ph.D., Univ. of Rochester, 1995. Experimental particle physics.

Knuth, Kevin, Ph.D., University of Minnesota, 1995. Biophysics, Space physics, Computational physics.

Assistant Professors

Earle, Keith, Ph.D., Cornell University, 1994. Experimental biophysics.

Goyal, Philip, Ph.D., Cambridge, 2005. Information physics, fundamentals of quantum mechanics.

Lunin, Oleg, Ph.D., Ohio State, 2000. String theory, computational particle physics.

Research Professor

Scholes, Charles P., Ph.D., Yale, 1969. Experimental biophysics.

Emeriti

Benenson, Raymond E., Ph.D., Wisconsin, 1955. Particle–solid interactions; hydrogen in metals; teaching experiments.

Kimball, John C., Ph.D., Chicago, 1969. Theoretical condensed matter physics and statistical mechanics.

Lanni, Robert P., M.A., SUNY, Albany, 1954. Physics teaching.

Marsh, Bruce B., Ph.D., Rochester, 1962. Distinguished Teaching Professor. Experimental particle–solid interactions.

Ratcliff, Keith F., Ph.D., Pittsburgh, 1965. Astrophysics; space physics.

Roth, Laura M., Ph.D., Radcliffe, 1957. Theoretical condensed matter physics; liquids and alloys; quantum mechanics.

Scholz, Wilfried W., Ph.D., Freiburg, 1964. Atomic physics; inner shell phenomena; electron and ion collisions; nuclear physics.

RESEARCH SPECIALTIES AND STAFF

Theoretical

Atomic and Molecular Physics. Many-body theory; hyperfine interactions and other energy and wave-function related properties. Das.

Biophysics. Structure of hemoglobin and associated molecules; electronic structure and properties of chlorophyll and related systems. Das.

Computational & Information Physics. Caticha, Knuth, Earle, Das, Goyal.

Elementary Particles. Exotic particles; geometrical models for particles. Inomata. String Theory, Lunin.

Foundation of Quantum Mechanics. Caticha, Inomata. Goyal, Knuth, Earle, Lunin.

X-Ray Scattering Diffraction and Optics. Caticha, Kimball.

Experimental

Medical Biophysics. ESR, EPR of Bioinorganic molecules. Earle, Scholes. X-ray imaging and therapy, MacDonald.

Elementary Particles. Charm Baryon Production from e^+e^- collision. ATLAS Collaboration of CERN. Alam, Ernst.

Solid State and Materials Physics. Surface/interface studies; X-ray crystallography; ion beam characterization of solids. Electron microscopy. Kuan, Lanford, MacDonald.

X-Ray Optics. Multilayer thin film, crystal, and capillary optics, materials and medical applications. MacDonald.

STONY BROOK UNIVERSITY, STATE UNIVERSITY OF NEW YORK

DEPARTMENT OF PHYSICS AND ASTRONOMY

Stony Brook, New York 11794-3800
http://www.physics.sunysb.edu

Students Accepted For Degree	FIELDS		
	Physics	Astronomy	Related Fields
Doctorate	X	X	
Master's	X	X	

1. General

President: Samuel Stanley
Department Chairman: Laszlo Mihaly
Department Telephone Number: (631) 632-8100 (C)
Type of Institution: University
Control: Public
Setting: Suburban
Total Faculty: more than 1,900*
Total Graduate Faculty: †
Total Students: 23,997*
Total Graduate Students: 7,440*
Annual Graduate Tuition:
In-state residents: Full-time—$8,370
Part-time—$349/credit
Out-of-state residents: Full-time—$13,250
Part-time—$552/credit
Tuition rates for: 2010–11
Deferred tuition plan: Yes
Other Fees: $496 per semester; Health insurance; $507 per semester
Term: Semester

†No differentiation of faculty
*Excluding health sciences

2. Number of Faculty in Department

The combined total of faculty in the three professorial ranks is 60. The combined total of faculty at all ranks is 108.

3. Admission, Financial Aid, and Housing

Address inquiries to: Prof. Jacobus Verbaarschot, Director of Graduate Program, Department of Physics and Astronomy
Graduate application fee required: $70
Admission deadline (Fall admission): 1/15;
(Spring admission, contact the Director of Graduate Program first): 10/1
Admission information: For fall admission, 2010–11, 45 students were admitted from 441 applicants.
Admission requirements: A Bachelor's degree in physics or related area is required with a minimum undergraduate GPA of 3.0 and a minimum GPA of 3.0 in physics and mathematics. The GRE is required with no fixed minimum. The Physics GRE is recommended no minimum score. Students from non-English speaking countries are required to demonstrate proficiency in English via the TOEFL, TSE or IELTS exams.
Undergraduate preparation: Typical undergraduate preparation includes courses in classical mechanics, electrodynamics, quantum mechanics, and statistical mechanics/thermal physics
FAFSA application required: Yes, for all US citizens and permanent residents.
Financial aid deadline: 1/15 for assistantships and fellowships
Loans available: Yes (U.S. citizens and permanent residents)
Address housing inquiries to: Regina Lagrasta, Assistant Director, Residence Life, Room 138, Administration Building
On-campus, single student housing available: Yes
Cost/term: See http://studentaffairs.stonybrook.edu/res
On-campus, married student housing available: Yes
Cost/month: See http://studentaffairs.stonybrook.edu/res

Table A—Faculty, Enrollments, and Degrees Granted

Research Specialty	2009–2010 Faculty	Enrollment Fall 2009		No. of Degrees Granted 2009 (2005–09)			Median No. of Years for 2009 Ph.D.'s[3]
		Master's	Doctorate[1]	Master's[1]	Terminal Master's[1]	Doctorate	
Accelerator Physics	3	–	–	–	–	0(3)	6
Astronomy and Astrophysics	12	–	–	–	–	0(3)	6
Atomic, Molecular and Optical	6	–	–	–	–	3(11)	6
Biophysics and X-ray optics	3	–	–	–	–	2(17)	6
Condensed Matter	26	–	–	–	–	7(26)	6
Nuclear	16	–	–	–	–	6(25)	6
High Energy (expt)	11	–	–	–	–	3(12)	6
Scientific Instrumentation	0	2	–	–	3(7)	–	N/A
Yang Inst. Theo. Phys. & Non-specialized	14	8	–	–	–	4(15)	6
Total	91	10	167	20(91)	6(23)	25(112)	
Full-time Grad. Stud.		8	167				
Part-time Grad. Stud.		0	0				
First-year Grad. Stud. (2009–10)		4	26				

Median Years in Grad. Study 2009 Degrees: 6
B.S. Degrees, Dec. 2005–Aug. 2009: 107

[1]Breakdown into all fields is not available.
[2]Special Master's program for exchange students.
[3]Includes time before advancement to Ph.D. candidacy

4. Graduate Degree Requirements

Master of Science in Instrumentation (M.S.I.): Requires a minimum of 2 years of study; one semester of teaching; minor project and Master's thesis.
Master of Arts (M.A.): 30 graduate credits with average grade of B (GPA of 3.0) in approved program with satisfactory performance; either passing approved courses and exams or thesis is required; no residence or language requirements.
Master of Arts in Teaching (M.A.T.): 41 graduate credits with 15 in physics, 20 in education and 6 in supervised teaching as well as an approved teaching project. See SPD@stonybrook.edu
Doctorate (Ph.D.): Satisfactory performance in approved course program; successful completion of written comprehensive examination, dissertation, and dissertation examination. One year of teaching required. One year of residence required. No language requirements.

Special Equipment, Facilities, or Programs: The Department of Physics and Astronomy participates with other departments in programs in biophysics, chemical physics, medical physics, and mathematical physics. A physics student in one of these programs may do thesis research under the supervision of a faculty member in the Physics and Astronomy Department or in one of the cooperating departments. The C. N. Yang Institute for Theoretical Physics, affiliated with the Physics and Astronomy Department, carries on research in particle physics, nuclear physics, statistical mechanics, string and supergravity theory. Cooperative programs are conducted at the Simons Center for Geometry and Physics, at the nearby Brookhaven National Lab, including research at the National Synchrotron Light Source and in the Center for Functional Nanomaterials. The Center for Accelerator Science and Education (CASE) coordinates efforts in accelerator physics. Experiments in particle physics are carried out at the Relativistic Heavy Ion Collider (BNL), at the Tevatron (Fermilab), at Super-Kamiokande (Japan), and at the Large Hadron Collider (CERN). The astronomy group makes regular use of various large optical telescopes, large millimeter wave telescopes and arrays, and space observatories. It has a partnership with the Caltech Palomar Observatory, and it is a member of the SMARTs consortium operating several telescopes at Cerro Tololo, Chile. Faculty are affiliated with the New York Center for Computational Science NYCCS and use its 100 T flops supercomputer and other supercomputers for computational science.

Table B—Appointments to Graduate Students, 2009–10

Title of Appointee	Appointments: Total	Appointments: First year	Academic Load Allowed in Credit Hours	Hours of Service Per Week	Stipend for Academic Year ($)[3]
			Semester		
Teaching Assistant	41	21		15–18	17,500[3]
Research Assistant	124	0	See note[1]	16–20[2]	21,500–28,000[4]
Fellow	6	3		0	21,500–28,000[4]
Total	171	24			

[1]First-year students take about 12 credit hours, with a somewhat lighter load for assistants. Advanced students generally take 9 credit hours.
[2]Nominal figure.
[3]Tuition waived.
[4]Summer stipends included.

5. Personnel Engaged in Separately Budgeted Research, 2009–10

Professorial faculty	60
Other faculty	11
Postdoctoral Appointments	23
Graduate Students	124
Undergraduate Students	5
Nonteaching research personnel	16
Total	239

6. Separately Budgeted Research Expenditures by Source of Support[1]

	Departmental Research[1]	Physics-related Research Outside Department
Federal government	$13,090,366	Unknown
Total	$13,090,366	

[1]Federal figures only.

Table C—Separately Budgeted Research Expenditures[1]

Research Specialty	No. of Grants	Expenditures ($)
Accelerator Physics		39,916
Astronomy		859,157
Atomic, Molecular and Optical		728,002
Condensed Matter		1,541,363
Education		251,932
Nuclear Physics		1,631,142
High Energy Particles and Fields		3,705,348
X-ray Optics and Biophysics		845,994
YITP		654,535
	Direct	10,247,389
	+ indirect	2,842,977
Total		13,090,366

Table D—Physics-related Research Outside Department[1]

Field and Unit Outside Department	No. of Students	Expenditures ($)
Brookhaven National Lab	7	
Dept. of Chemistry	6	
Marine Sci. Res. Center	1	
Dept. of Applied Mathematics	1	
Dept. of Biomedical Engineering	1	
Total	16	

[1]Figures are not known.

FACULTY

Professors

Allen, Philip B., Ph.D., California, Berkeley, 1969. Theoretical Solid State Physics.

Aronson, Meigan, Ph.D. U. Illinois, 1982. Experimental Solid State Physics.

Averin, Dmitrii V., Ph.D., Moscow State U., 1987. Solid State Theory.

****Douglas**, Michael, Ph.D., Caltech, 1988. String Theory.

Drees, Axel, Ph.D., Heidelberg, 1989. Associate Dean, College of Arts and Sciences. Relativistic Heavy Ions.

Engelmann, Roderich, Ph.D., Heidelberg, 1966. Experimental High-Energy Physics.

***Goldhaber**, Alfred S., Ph.D., Princeton, 1964. Theoretical Physics; Nuclear Theory; Particle Physics.

Goldman, Vladimir J., Ph.D., Maryland, 1985. Experimental Solid State Physics.

Gurvitch, Michael, Ph.D., Stony Brook, 1978. Experimental Solid State Physics.

Hemmick, Thomas, Ph.D., Rochester, 1989. Distinguished Professor. Experimental Relativistic Heavy Ion Physics.

Hobbs, John, Ph.D., Chicago, 1991. Experimental High Energy Physics.
Jacak, Barbara, Ph.D., Michigan State U., 1984. Distinguished Professor. Relativistic Heavy Ion Collisions.
Jacobsen, Chris, Ph.D., Stony Brook, 1988. X-ray Microscopy.
Jung, Chang Kee, Ph.D., Indiana, 1986. Experimental High-Energy Physics.
Kharzeev, Dmitri, Ph.D., Moscow State, 1990. Nuclear Theory.
Koch, Peter M., Ph.D., Yale, 1974. Atomic Physics; Chaos.
***Korepin**, Vladimir, Ph.D., Leningrad, 1977. Theoretical Phys.
Kuo, T. T. S., Ph.D., Pittsburgh, 1964. Nuclear Theory.
Lanzetta, Kenneth, Ph.D., Pittsburgh, 1988. Galactic Formation, Cosmology.
Lattimer, James M., Ph.D., Texas, 1976. Nuclear Astrophysics.
Likharev, Konstantin K., Ph.D., Moscow State University, 1969. Distinguished Professor. Mesoscopic Physics.
Lukens, James, Ph.D., U.C., San Diego, 1968. Condensed Matter Nanodevices.
Marburger, John H. III, Ph.D., Stanford, 1967. University Professor. Nonlinear optics.
Marx, Michael, Ph.D., MIT, 1974. Associate Provost. Experimental High-Energy physics.
McCarthy, Robert L., Ph.D., U.C. Berkeley, 1971. Experimental High-Energy Physics.
***McCoy**, Barry M., Ph.D., Harvard, 1967. Distinguished Professor. Theoretical Statistical Mechanics.
Mendez, Emilio, Ph.D., MIT, 1979. Director, Center for Functional Nanomaterials, Brookhaven National Laboratory. Experimental Solid State Physics.
Metcalf, Harold J., Ph.D., Brown, 1967. Distinguished Teaching Professor. Experimental Atomic Physics; Laser Cooling.
Mihaly, Laszlo, Ph.D., Eotvos University, Budapest, 1977. Department Chair. Experimental Solid State Physics.
***Rastelli**, Leonardo, Ph.D., MIT, 2000, Theoretical Physics.
Rijssenbeek, Michael, Ph.D., Amsterdam, 1974. Experimental High-Energy Physics.
***Roček**, Martin, Ph.D., Harvard, 1979. Theoretical Physics.
***Shrock**, Robert, Ph.D., Princeton, 1975. Theoretical Physics.
Shuryak, Edward, Ph.D., Novosibirsk, USSR, 1974. Distinguished Professor. Nuclear Theory.
***Siegel**, Warren, Ph.D., U.C. Berkeley, 1977. Theoretical Physics.
Sprouse, Gene D., Ph.D., Stanford, 1968. Distinguished Professor. Editor in Chief, American Physical Society. Expt. Nuclear Physics.
Stephens, Peter W., Ph.D., MIT, 1978. Exp. Solid State Physics.
***Sterman**, George, Ph.D., Maryland, 1974. Distinguished Professor. Director, YITP. Theoretical Physics.
***van Nieuwenhuizen**, Peter, Ph.D., Utrecht, 1971. Distinguished Professor. Theoretical Physics.
Verbaarschot, Jac, Ph.D., Utrecht, 1982. Director of the Graduate Program. Theoretical Nuclear Physics.
Walter, Frederick M., Ph.D., U.C. Berkeley, 1981. Star Formation; Chromospheres and Coronae.
Zahed, Ismail, Ph.D., MIT, 1983. Theoretical Nuclear Physics.

Associate Professors

Abanov, Alexandre, Ph.D., Chicago, 1997. Theoretical Condensed Matter.
Deshpande, Abhay, Ph.D., Yale, 1995. Polarized photons, RHIC.
***Gonzalez-García**, Concha, Ph.D., University of Valencia, 1991. Theoretical Physics.
Graf, Erlend H., Ph.D., Cornell, 1968. Curriculum Devel.
McGrew, Clark, Ph.D., Irvine, 1994. Expt. High Energy.
Peterson, Deane M., Ph.D., Harvard, 1968. Director of the Undergraduate Program. Stellar Atmospheres.
Weinacht, Thomas, Ph.D., U. Michigan, 2000. Expt. Atomic.

Assistant Professors

Calder, Alan, Ph.D., Vanderbilt U., 1997. Computational Astrophysics
Dawber, Matthew, Ph.D., Cambridge, 2003. Experimental Solid State Physics.
Du, Xu, Ph.D., U. of Florida, 2004. Condensed Matter Experimental.
Koda, Jin, Ph.D., University of Tokyo, 2002. Astronomy.
***Meade**, Patrick, Ph.D., Cornell University, 2006. Theoretical Physics.
Metchev, Stanimir, Ph.D., Cal Tech, 2006. Astronomy.
Schneble, Dominik, Ph.D., U. Konstanz, 2002. Experimental Atomic Physics.
Teaney, Derek, Ph.D., Stony Brook, 2001. Theoretical Nuclear Physics.
Tsybychev, Dmitri, Ph.D., U. Florida, 2004. High Energy.
Zingale, Michael, Ph.D., Chicago, 2000. Computational Astrophysics.

Research Faculty

de Zafra, Robert L., Ph.D., Maryland, 1958. Experimental Atmospheric Physics.
Grannis, Paul D., Ph.D., U.C. Berkeley, 1965. Experimental High-Energy Physics.
Kirz, Janos, Ph.D., U.C. Berkeley, 1963. X-ray Microscopy.
Patel, Vijay, Ph.D., Stony Brook, 2001. Expt. Condensed Matter.
Paul, Peter, Ph.D., Freiburg, 1959. Exp. Nuclear Physics.
Semenov, Vasili, Ph.D., Moscow State University, 1975. Experimental Condensed Matter.
Swesty, Douglas, Ph.D., Stony Brook, 1993. Computational and Nuclear Astrophysics.
Yanagisawa, Chiaki, Ph.D., U. Tokyo, 1981. Experimental High-Energy Physics.

Emeriti

***Brown**, Gerald E., Ph.D., Yale, 1950. Distinguished Professor. Theoretical Nuclear Physics.
Kahn, Peter B., Ph.D., Northwestern, 1960. Professor. Theoretical Physics.
Lee, Linwood, L., Ph.D., Yale, 1955. Professor. Nuclear physics.
McGrath, Robert L., Ph.D., Iowa, 1965. Professor. Nuclear Physics.
Simon, Michal, Ph.D., Cornell, 1967. Professor. Infrared Astronomy.
***Smith**, John, Ph.D., Edinburgh, 1963. Professor. High Energy Theory.
Swartz, Clifford E., Ph.D., 1951, University of Rochester. Professor. School curriculum. Theoretical Physics.
Weisberger, William I., Ph.D., MIT, 1964. Professor. Theoretical Physics.
Yahil, Amos, Ph.D., Cal. Tech., 1970. Professor. Cosmology.
***Yang**, Chen Ning, Ph.D., U. Chicago, 1984. Einstein Professor, Theoretical Physics.

Adjunct and Affiliated Faculty

Aronson, Sam, Ph.D., Princeton, 1968. Director, BNL. Nuclear Physics.

Ben-Zvi, Ilan, Ph.D., Weizmann Institute of Science, 1970. Accelerator Physics.
Bergeman, Thomas, Ph.D., Harvard, 1971. Theoretical Atomic Physics.
***Creutz**, Michael, Ph.D., Stanford, 1970. Theoretical Particle.
Davenport, James, Ph.D., U. Pennsyvania, 1976. Theoretical Condensed Matter.
***Dawson**, Sally, Ph.D., Harvard, 1981. Theoretical Particle.
Dierker, Steven, Ph.D., U. Illinois, 1983. Associate Director, BNL. Experimental Solid State Physics.
Di Mauro, Louis, Ph.D., U. Connecticut, 1980. Optics.
Forman, Miriam, Ph.D., Stony Brook, 1972. Cosmic Rays.
Geller, Marvin, Ph.D., MIT, 1969. Atmospheric Physics.
Johnson, Peter, Ph.D., Warwick U., 1978. Expt. Condensed Matter.
Kao, Chi-Chang, Ph.D., Cornell, 1988. Expt. Condensed Matter.
Karsch, Frithjof, Ph.D., Bielefeld U., 1982. Lattice Gauge Theory.
Ku, Wei, Ph.D., U. Tennessee, 2000. Theoretical Condensed Matter.
Litvinenko, Vladimir, Ph.D., Novosibirsk, 1989. Accelerator Physics.
Maslov, Sergei, Ph.D., Stony Brook, 1996. Theoretical Condensed Matter.
Mould, Richard, Ph.D., Yale, 1957. Professor Emeritus.
Ocko, Ben, Ph.D., MIT, 1984. Condensed Matter Exp.
Oganov, Artem, Ph.D., University College, London, 2002. Associate Professor of Geosciences. Computational Geosciences.
Peggs, Stephen, Ph.D., Cornell, 1981. Accelerator Physics.
Sayre, David, Ph.D., Oxford, 1951. X-ray Microscopy.
Sivaramakrishnan, Anand, Ph.D., U. Texas, 1983. Astronomy.
Spira, Robert, M.A., Stony Brook, 1996.
Smith, Steven O., Ph.D., U.C. Berkeley, 1985. Professor of Biochemistry and Cell Biology, Director, Center for Structural Biology.
Takai, Helio, Ph.D., University of Rio de Janeiro, 1986, Experimental particle physics.
Tolpygo, Sergey, Ph.D., Inst. for Solid State Physics, Russian Acad. of Sciences, 1984. Cond. Matter Nanodevices
Tsvelik, Alexei, Ph.D., Kurchatov Institute of Atomic Energy Moscow, 1980. Condensed Matter Theory.
Venugopalan, Raju, Ph.D., Stony Brook, 1992. High Energy Theory.
Wang, Jin, Ph.D., U. Illinois, 1991. Professor of Chemistry. Biological Physics.
Zhu, Yimei, Ph.D., Nagoya University, 1987. Microscopy.

*Members of the C. N. Yang Institute for Theoretical Physics.

**Members of the Simons Institute for Geometry in Mathematics and Physics.

RESEARCH SPECIALTIES AND STAFF

Theoretical

Astronomy and Astrophysics. Cosmology, galaxy formation and evolution. Nuclear astrophysics, neutrino astrophysics. Neutron stars; equation of state of dense matter; stellar collapse. Calder, Forman, Koda, Lanzetta, Lattimer, Metchev, Zingale.
Atomic, Molecular and Optical Physics. Theory of ultracold quantum gases, non-linear optics. Bergeman, Marburger.
Elementary Particles and Fields. Quantum field theory; unified gauge theory of weak, electromagnetic, and strong interactions; general gauge theory; lattice gauge theory. Creutz, Dawson, Goldhaber, Gonzalez-García, Meade, Rastelli, Roček, Shrock, Siegel, J. Smith, Sterman, van Nieuwenhuizen.
Nuclear. Nucleon-nucleon interaction; meson exchange currents and other mesonic effects in nuclei; effective interactions in nuclei and nuclear matter; heavy ion reactions; Fermi liquid theory; variational and extended semiclassical models of large systems; studies of dense nuclear matter; Random Matrix Theory. Brown, Kharzeev, Kuo, Shuryak, Teaney, Verbaarschot, Zahed.
Relativity. Quantum theory of gravitation; supergravity; supersymmetry; superstrings. Goldhaber, Rastelli, Roček, Siegel, van Nieuwenhuizen.
Solid State. Superconductivity; electron-phonon interactions; magnetic properties; optical properties; quantum computing, Coulomb blockade, quantum Hall effect, density-functional theory. Abanov, Allen, Averin, Davenport, Fernandez-Serra, Ku, Likharev, Maslov, Organov, Tsvelik.
Statistical Mechanics. Mathematical studies of solvable models; relation between statistical mechanics and field theory; Ising models; lattice gauge fields. Goldhaber, Korepin, McCoy, Shrock, Verbaarschot.

Experimental

Accelerator Physics. Ben-Zvi, Peggs, Litvinenko.
Astronomy and Astrophysics. Cosmology; galactic structure and evolution; interstellar molecular clouds; quasar absorption lines; stellar astronomy; chromospheres; coronae; compact objects; star formation. Pre-main sequence objects; high mass star formation. Peterson, Simon, Sivaramakrishnan, Walter.
Atomic, Molecular and Optical Physics. Coherent control of molecules and atoms with tailored, ultrafast laser pulses. Cooling, trapping, and laser spectroscopy of atoms. Ultracold quantum gases. Quantum chaos studies with microwave systems and driven atoms. Koch, Metcalf, Schneble, Weinacht.
Elementary Particles. Particle interactions in high energy collisions; properties of the top and bottom quarks and electroweak bosons; studies of the strong interaction; search for new particles. Rare kaon decays. Neutrino oscillations. Engelmann, Grannis, Hobbs, Jung, Marx, McCarthy, McGrew, Rijssenbeek, Takai, Tsybychev, Yanagisawa.
Nuclear. Relativistic heavy ion collisions; properties of quark-gluon plasma; polarized protons. Deshpande, Drees, Hemmick, Jacak, Paul, Sprouse.
Physics of Biology. Soft X-ray imaging, holography. Jacobsen, Sayre. (Physics students may also do thesis research in biophysics in a cooperative program with several other departments, Brookhaven National Laboratory and Cold Spring Harbor Laboratory. Maslov, S.O. Smith, Wang.)
Solid State. Superconductivity; Josephson effect; x-ray scattering; single electronics; magnetic flux quantum devices; quantum wells, fractional quantum Hall effect; optical spectroscopy, ferroelectrics, Aronson, Dawber, Goldman, Gurvitch, Johnson, Kao, Likharev, Lukens, Mendez, Mihaly, Patel, Semenov, Stephens, Tolpygo, Zhu.

SYRACUSE UNIVERSITY

DEPARTMENT OF PHYSICS

Syracuse, New York 13244-1130

Students Accepted For Degree	FIELDS		
	Physics	Astronomy	Related Fields
Doctorate	X		X
Master's	X		X

1. General

Chancellor: Nancy Cantor
Dean of the Graduate School: Sandra N. Hurd
Department Chair: Peter R. Saulson
Department Telephone Number: (315) 443-3901 (C)
Type of Institution: University
Control: Private
Setting: Urban
Total Graduate Faculty: Full-time instructional faculty; 955; part-time faculty, 94; adjunct faculty, 440. Of the full-time faculty, approximately 88 percent have earned Ph.D. or professional degrees.
Total Students: Fall 2009 enrollment, 13,040 full-time and 696 part-time undergraduate students; 3,812 full-time and 1,868 part-time graduate and law students. Total University enrollment if 19,638.
Total Graduate Students: 5,680
All Graduate Students: Full-time—$1,162/credit
Part-time-$1,162/credit
Tuition rates for: 2010–2011
Deferred tuition plan: Yes
Annual Other Fees: $1,070
Term: Semester

2. Number of Faculty in Department

The combined total of full-time faculty in the three professorial ranks is 29. The combined total of full-time, part-time, and other faculty at all ranks is 33.

3. Admission, Financial Aid, and Housing

Address admission inquiries to: Graduate Coordinator, Physics Department, 201 Physics Bldg. e-mail graduate@phy.syr.edu, or visit our web site at http://physics.syr.edu/graduate
Graduate application fee: $75 (subject to change)
Admission deadline (Fall admission): January 1 (late applications accepted)
Admission information: For fall admission, 2010–2011, 13 students matriculated from 163 applicants.
Admission requirements: For admission to the graduate programs, a Bachelor's degree in physics is recommended but not required, with a minimum undergraduate GPA of 3.0 recommended. The GRE and AGRE are required. The minimum acceptable score suggested for admission is verbal–75%; quantitative–85%. No minimum acceptable score for admission is specified. Students from non-English speaking countries are required to demonstrate proficiency in English. Minimum acceptable score on the TOEFL exam, for admission is 600(pbt); 100(cbt).
Undergraduate preparation assumed: Symon, *Classical Mechanics*; Reitz and Milford, *Foundations of Electromagnetic Theory*; Eisberg, *Fundamentals of Modern Physics.*
Address financial aid inquiries to: Financial Aid Office, 200 Archbold Gymnasium
GAPSFAS application required: No
Financial aid deadline: January 1 (late applications accepted)
Loans available: Yes
Address housing inquiries to: Office of Residence and Dining Services, 202 Steele Hall
On-campus, single student housing available: Yes
Cost/term: See http://housingmealplans.svr.edu/pdf/graduate.pdfmainhous.htm
On-campus, married student housing available: Yes
Cost/month: http://cwis.syr.edu/HousingMealplans/gradhousing.htm

*Address married student housing inquiries to: South Campus Housing, Goldstein Student Center, Room 206

Table A—Faculty, Enrollments, and Degrees Granted

Research Specialty	2009–10 Faculty	Enrollment Fall 2009		No. of Degrees Granted 2009–10 (2005–10)			Median No. of Years for 2009–10 Ph.D.'s
		Mas-ter's	Doc-torate	Mas-ter's	Terminal Master's	Doc-torate	
Biophysics	3	1	6	0(0)	0(0)	3(4)	8
Chemical Physics							
Computer Science	0	0	0	0(0)	0(0)	0(0)	5
Condensed Matter Physics	10	0	27	0(0)	0(1)	1(9)	5
Particles & Fields	11	1	21	0(0)	0(0)	4(19)	5
Relativity & Gravitation	5	0	11	0(0)	0(0)	0(0)	4.5
Non-specialized		1	4				
Total	29	3	69	0(0)	0(1)	8(32)	
Full-time Grad. Stud.		0	68				
Part-time Grad. Stud.			1				
First-year Grad. Stud.		0	18				
Median Years in Grad.Study (2008–09 Degrees)			5.5		–		–
Undergraduate Degrees, 2009–10 (2005–10): 11(73)							

4. Graduate Degree Requirements

Master's: One year minimum in residence is required. A student admitted to graduate work in the Department must take the Comprehensive Examination. The Degree can be achieved in any one of three ways: (1) 24 hours of course work plus a thesis, (2) 30 hours of course work including a Minor Problem and passing the Qualifying Examination, or (3) 36 hours of course work and passing the Qualifying Examination. A B average in course work must be maintained to be eligible for a degree. There are no foreign language requirements.
Doctorate: Satisfactory performance in a course program approved by the student's research committee, which may include courses taken for the M.S. degree. Students must pass a written qualifying examination, a preliminary oral research examination, and a thesis defense. There is no language requirement. At least a B average is required.
Other Programs:
Thesis: Thesis may be written *in absentia*.

5. Personnel Engaged in Separately Budgeted Research

Professorial faculty	29
Other faculty	19
Postdoctoral appointments	18
Graduate students	69
Undergraduate students	53
Total	188

Table B—Appointments to Graduate Students, 2009–10

Title of Appointee	Appointments Total(first year)	Academic Load Allowed in Credit Hours	Hours of Service Per Week	Stipend for Academic Year ($)
Teaching Assistant	25 (18 new)	24	20	20,010
Teaching/Research Assistant	9	24	20	20,010
Research Assistant	31	24	20	20,010
University Fellow	3	30	0	20,755
No Aid	5		0	
Total	69			

Table C—Separately Budgeted Research Expenditures

Research Specialty	No. of Grants	Expenditures ($)
Exp. Astrophysics & Cosmology	6	162,076
Biophysics	5	132,400
Condensed Matter	8	721,530
Low Temperature Phys.	3	280,366
Particles & Fields	8	1,592,713
Relativity & Gravitation	9	648,780
Total	39	3,537,865

FACULTY

Professors

Artuso, Marina, Ph.D., Northwestern, 1986. Elementary particles and fields, experiment.
Balachandran, A. P., Ph.D., Madras, 1962. Elementary particles and fields theory.
Bowick, Mark, Ph.D., Caltech., 1983. Condensed matter theory.
Catterall, Simon, Ph.D., Oxford, 1988. Elementary particles theory; computational physics.
Foster, Kenneth, Ph.D., Caltech., 1972. Biological physics.
Lipson, Edward D., Ph.D., Caltech., 1971. Biological physics.
Marchetti, M. Cristina, Ph.D., Florida, 1982. Condensed matter theory, Department Chair.
Middleton, Alan, Ph.D., Princeton, 1990. Condensed matter, theory; computational physics.
Rosenzweig, Carl, Ph.D., Harvard, 1972. Elementary particles and fields theory.
Saulson, Peter, Ph.D., Princeton, 1981. Experimental relativity and astrophysics.
Schechter, Joseph, Ph.D., Rochester, 1965. Elementary particles and fields, theory.
Schiff, Eric, Ph.D., Cornell, 1979. Condensed matter, Associate Dean, College of Arts & Sciences.
Skwarnicki, Tomasz, Ph.D., Inst. of Nuclear Physics, Krakow, 1986. Elementary particles experiment.
Souder, Paul A., Ph.D., Princeton, 1971. Elementary particles and fields, experiment; medium energy.
Stone, Sheldon, Ph.D., Rochester, 1972. Experimental particle physics.
Vidali, Gianfranco, Ph.D., Penn. State, 1982. Condensed matter experiment.

Associate Professors

Blusk, Steven, Ph.D., Univ. of Pittsburgh, 1995. Elementary particles, experiment.
Movileanu, Liviu, Ph.D., University of Bucharest, 1997. Biophysics, experiment.

Assistant Professors

Armendariz-Picon, Cristian, Ph.D., Ludwig-Maximilians-Universität, Munich, 2001. Cosmology, relativity, elementary particles, theory.
Ballmer, Stefan, Ph.D., MIT, 2006. Theoretical physics, gravitational waves.
Brown, Duncan, Ph.D., University of Wisconsin, Milwaukee, 2004. Theoretical astrophysics, relativity.
Forstner, Martin, Ph.D., University of Texas, Austin, 2003. Condensed matter, experiment.
Hubisz, Jay, Ph.D., Cornell, 2006. Particle physics, cosmology, theory.
LaHaye, Matthew, Ph.D., University of Maryland, College Park, 2005. Experimental condensed matter physics.
Plourde, Britton, Ph.D., University of Illinois, Urbana-Champaign, 2000. Condensed matter, experiment.
Schnee, Richard, Ph.D., UC Santa Cruz, 1996. Experimental/Observational cosmology.
Schwarz, Jennifer, Ph.D., Harvard, 2002. Condensed matter, theory.
Soderberg, Mitchell, Ph.D., Univ. of Michigan, 2000. Elementary particles, experiment.
Watson, Scott, Ph.D., Brown, 2005. Particle physics, cosmology.

Research Professors

Holmes, Richard, Ph.D., Maryland, 1985. Elementary particles; exp.
Mountain, Raymond, Ph.D., Notre Dame, 1992. Elementary particles; experiment.
Wali, Kameshwar, Emeritus & Research Professor, Fellow, American Physical Society.
Wang, Jianchun, Ph.D., MIT, 1997. Elementary particles, exp.

RESEARCH SPECIALTIES AND STAFF

Also see http://physics.syr.edu/research.

Theoretical

Condensed Matter. Soft condensed matter physics, statistical mechanics, and non-equilibrium dynamics including: two-dimensional matter, collective behavior of biological molecules, jamming in granular materials, superconductors, hysteresis in magnets, colloidal particles, topological defects, glassy materials, networks, relationship between algorithms and physics, elastomers. Bowick, Marchetti, Middleton, Schwarz. 3 postdoctoral fellows.

Computational Physics. Study of condensed matter order and optimal distributions on curved interfaces, analysis of phase transitions and phase structure in disordered systems; simulations of charge density waves and flux arrays in superconductors; study of dynamical systems and chaos; numerical simulations on parallel computers; gravitational-wave data analysis and source modeling; grid computing; connections between algorithms and physical principles. Application of distributed processing to large scale quantum theory problems. Simulations of lattice quantum field theories; technicolor and supersymmetric theories. Holographic models of strings. Models of Beyond Standard Model Physics. Bowick, Brown, Catterall, Middleton. 1 postdoctoral fellows.

Elementary Particles and Fields. Quantum gravity; supersymmetry; renormalization theory; chiral symmetries; monopoles and dyons in curved space-time; noncommutative geometry; random surfaces, electroweak theory; quantum chromodynamics; general quantum field theory; constrained field theories; geometric quantization; phenomenological particle dynamics. Simulations of lattice QCD; study of supersymmetric field theories on space-time lattices. Quark gluon plasma. Particle cosmology. Theories with extra dimensions. Simulations of lattice quantum field theories; technicolor and supersymmetric theories. Holographic models of strings. Models of Beyond Standard Model Physics. Armendariz-Picon, Balachandran, Catterall, Hubisz, Rosenzweig, Schechter, Watson. 2 postdoctoral fellow.

Experimental

Astrophysics of the Interstellar Medium. Laboratory studies of physical and chemical processes occurring in the interstellar medium and in planetary atmospheres, including formation of molecular hydrogen and hydrogenation and oxidation reaction on interstellar and/or planetary dust grain analogues. Vidali. 1 postdoctoral fellow.

Biological and Medical Physics. Single-molecule biophysics; membrane biophysics; bionanotechnology and biosensors; protein design; development of new optical technologies; photosensory transduction in microorganisms; bioinformatics; self-organized beating of cilia; phylogenetics and molecular clocks; technology development for telemedicine and human-computer interfacing; medical tomographic image reconstruction; biomedical molecular imaging; PET, SPECT, CT and MRI; medical image registration; ultrafast laser-based X-ray source for radiological applications. Forstner, Foster, Krol, Lipson, Movileanu. 3 postdoctoral fellow.

Gravitation and Cosmology Group. Gravitational-wave Detection and Astrophysics: Searches for gravitational waves using the Laser Interferometer Gravitational Wave Observatory. Commissioning and technology development for advanced gravitational wave detectors. Gravitational wave source modeling and phenomenology. Developing tests of general relativity using gravitational waves. Dark matter detection and early universe cosmology: Development of improved ultra-low-radioactivity environments and detectors of Weakly Interacting Massive Particles, analysis of data from dark matter searches. Theoretical models of dark and cosmic acceleration. Inflation and alternatives. Origin and evolution of cosmological structures. Armendariz-Picon, Ballmer, Brown, Saulson, Schnee. 3 postdoctoral fellows.

High Energy Experimental Particle Physics. Experimental studies of the fundamental Electroweak and Strong interactions as manifested by the decays of beauty and charm quarks, and search for exotic particles. We study b & c quark decays at the LHCb experiment at the CERN LHC hadron collider Geneva, Switzerland, concentrating on rare and CP violating decays. Searches for exotic particle production, including unusual decays of the Higgs boson are also done using LHCb. We also are doing R&D into advanced silicon micro-pattern detectors, such as pixel sensors, and their related readout electronics. Members of the group have discovered several new particles including the B, Ds, Y(1D) and made the first measurements of several very important decay modes of these objects. The group is also starting an effort in Neutrino physics. Artuso, Blusk, Mountain, Skwarnicki, Stone and Wang. 4 postdoctoral fellows.

Intermediate Energy Particle Physics. Use of spin degrees of freedom to study quantum chromodynamics and the Standard Model at low energies. Experiments are underway at Thomas Jefferson National Accelerator Facility (JLab). Holmes, Souder. 1 postdoctoral fellow.

Low Temperature Quantum and Nanoscale Devices: Quantum coherent superconducting circuits; measurement and coupling of circuits for quantum computing; vortex dynamics in nanofabricated thin-film devices; superconducting microwave resonant circuits; nanoelecromechanical systems (NEMS); quantum dynamics of mechanical systems; sensitive environmental gas and bio-sensors; measurements at millikelvin temperatures. LaHaye, Plourde. 3 postdoctoral fellows.

Semiconductors, Thin Films, and Solar Cells. Electronic and optical properties of unconventional semiconductors (amorphous silicon, porous titania, and silicon). Solar cell device physics. Thin-Film growth (plasma, hot-wire). Hybrid organic-inorganic semiconductor devices. Surface physics (structure, kinetics, dynamics, and reactions). Schiff. 1 postdoctoral fellow.

THE ROCKEFELLER UNIVERSITY

CENTER FOR STUDIES IN PHYSICS AND BIOLOGY

New York, New York 10065

Students Accepted For Degree	FIELDS		
	Physics	Astronomy	Related Fields
Doctorate	X		
Master's			

1. General

President: Paul Nurse
Dean of Graduate Studies: Sidney Strickland
Department Chairman: Mitchell Feigenbaum
Department Telephone Number: (212) 327-8086
Type of Institution: University
Control: Private
Setting: Urban
Total Faculty: 69*
Total Graduate Faculty: 69*
Total Students: 196
Total Graduate Students: 196
Annual Graduate Stipend: Full tuition is remitted for all students admitted plus stipend (currently $31,600 per annum).
Other Fees: None
Term: Annual appointments: 7/1–6/30

*Includes only heads of laboratory

2. Number of Faculty in Department

The combined total of full-time faculty in the three professorial ranks is 10. The combined total of full-time, part-time, and other faculty at all ranks is 16.

3. Admission, Financial Aid, and Housing

Address admission inquiries to: Graduate Admissions Administrator and Registrar, Box 177
Graduate application fee required: $80
Admission deadline: 12/6: Deadline
Admission information: For fall admission, 2009–10, 71 students were accepted from 728 applicants.
Admission requirements: For admission to the graduate programs, a Bachelor's degree in physics is required. The GRE General Exam is required and the GRE Subject Exam is highly recommended for admission. Students from non-English speaking countries should submit TOEFL scores or other evidence of proficiency in English.
Undergraduate preparation: In addition to the undergraduate physics curriculum, proficiency in graduate level classical mechanics and special relativity, classical electrodynamics, quantum mechanics, statistical mechanics, and mathematical physics is expected.
Loans available: Full fellowship program
Address housing inquiries to: Dorian Johnson, Housing Office, Box 229 (admitted students only)
On-campus, single student housing available: Yes
Housing Cost/year: $7,848

Table A—Faculty, Enrollments, and Degrees Granted

Research Specialty	2009–10 Faculty	Enrollment[1] Fall 2009 Master's	Enrollment[1] Fall 2009 Doctorate	No. of Degrees Granted[2] 2009–10 (2005–10) Master's	Terminal Master's	Doctorate	Median No. of Years for 2009–10 Ph.D.'s
Experimental Biophysics	1	–	0	–	–	1(5)	–
Experimental High-Energy Physics	3	–	0	–	–	0(1)	–
Theoretical High-Energy Physics	1	–	0	–	–	0(0)	–
Theoretical Physics	5	–	4	–	–	1(1)	–
Total		–	4	–	–	2(7)	
Full-time Grad. Stud.		–	4				
Part-time Grad. Stud.		–	0				
First-year Grad. Stud.		–	0				
Median Years in Grad. Study (2009–10 Degrees)				–	–	4.5	5
Undergraduate Degrees, 2009–10 (2005–10): No undergraduate program							

[1]Students not yet committed to a research specialty are entered under non-specialized.
[2]Five-year totals in parentheses.

4. Graduate Degree Requirements

Doctorate: Each student affiliates with a laboratory for thesis research by September 1st of the second year. The advisor supervises the student's study program. A Faculty Advisory Committee is formed to review the student's progress. When the Faculty Advisory Committee confirms that the student has completed significant experimental or theoretical research, it will approve the student's readiness to write the Ph.D. thesis. The Ph.D. is awarded upon satisfactory completion of the written thesis, presentation in an open lecture to the university and successful completion of an oral defense to the Faculty Advisory Committee.
Thesis: Thesis may not be written *in absentia*.
Special Equipment, Facilities, or Programs: Each student is asked to undertake intensive review and further study in those areas of physics which are of special interest. In addition, some students arrange for individual tutoring. Students have every opportunity to study to the limits of their abilities. Clearly, this approach to graduate education demands of the student initiative, self-discipline, and a strong sense of personal responsibility. A wide variety of electronic instruments, computer facilities, and a machine shop are available for research in experimental physics.

5. Personnel Engaged in Separately Budgeted Research, 7/09–6/10

Professorial faculty	10
Postdoctoral appointments	19
Graduate students	4
Total	33

FACULTY

Professors

Cohen, E. G. D., Ph.D., Amsterdam, 1957. Emeritus. Theoretical physics.
Feigenbaum, Mitchell, Ph.D., MIT, 1970. Chairman of the Department. Theoretical physics.
Goulianos, Konstantin, Ph.D., Columbia University, 1963. Experimental high-energy physics.
Khuri, Nicola, Ph.D., Princeton, 1957. Emeritus. Theoretical high-energy physics.
Leibler, Stanislas, Ph.D., University of Paris, 1984. Theoretical physics.
Libchaber, Albert J., Ph.D., Ecole Normale Superieure, 1965. Experimental biophysics.
Magnasco, Marcelo, Ph.D., The University of Chicago, 1992. Theoretical physics.
Siggia, Eric D., Ph.D., Harvard University, 1975. Theoretical physics.

Associate Professors

Bhatti, Anwar, Ph.D., University of Washington, 1991. Experimental high-energy physics.
Demortier, Luc, Ph.D., Brandeis University, 1991. Experimental high-energy physics.

Adjunct Faculty

Cecci, Guillermo, Ph.D., The Rockefeller University, 1999. Physics.
Gallinaro, Michele, Ph.D., University of Rome, 1996. Experimental high-energy physics.
Khoze, Valery, Ph.D., Institute of Nuclear Physics, Russia, 1978. Theoretical and mathematical physics.
Khuri, Ramzi, Ph.D., Princeton University, 1991. Theoretical physics.
Ren, Hai-Cang, Ph.D., Columbia University, 1984. Theoretical high-energy physics.
Sena, Giovanni, Ph.D., New York University, 2003. Biology.
Sigman, Mariano, Ph.D., The Rockefeller University, 2002. Neuroscience.
White, Sebastian, Ph.D., Columbia University, 1975. Experimental high-energy physics.

Research Associates

Agapiou, John, Ph.D., University College London, 2007. Neuroscience.
Andor, Daniel, Ph.D., University of Cambridge 2005. Theoretical physics.
Arkus, Natalie, Ph.D., Harvard University, 2009. Applied mathematics.
Baule, Adrian, Ph.D., University of Leeds, 2008. Theoretical Physics.
Ciesielski, Robert, Ph.D., Institute of Experimental Physics, Poland, 2004. Experimental high-energy physics.
Chuang, John, Ph.D., University of California, Berkley, 1997. Experimental Biophysics.
Corson, Francis, Ph.D., Ecole Normale Supérieure, 2008. Theoretical physics.
Feinerman, Ofer, Ph.D., Weizmann Institute of Science, 2006. Experimental biophysics.
Francois, Paul, Ph.D., Ecole Normale Superieure, 2002. Theoretical physics.
Geffen, Maria, Ph.D., Harvard University, 2006. Biophysics.
Hektra, Doeke, Ph.D., The Rockefeller University, 2008. Chemical biology.
Katifori, Eleni, Ph.D., Harvard University, 2008. Theoretical physics.
Kuehn, Seppe, Ph.D., Cornell University, 2007. Experimental biophysics.
Kumar, Pradeep, Ph.D., Boston University, 2007. Physics.
Lungu, Gheorghe, Ph.D., University of Florida, 2007. Experimental high-energy physics.
Maeda, Yusuke T., Ph.D., University of Tokyo, 2008. Physics.
Malik, Sarah, Ph.D., University College London, 2010. High-energy physics.
Merrin, Jack, Ph.D., Princeton University, 2006. Physics.
Mesropian, Christina, Ph.D., Rockefeller University, 2000. Experimental high-energy physics.
Rivoire, Olivier, Ph.D., Laboratorie de Physique Theorique et Modeles Statistiques, 2005. Theoretical physics.
Solovey, Guillermo, Ph.D., Universidad de Buenos Aires, 2009. Theoretical physics.
Sorre, Benoit, Ph.D., Institut Curie, 2009. Experimental biophysics.
Warmflash, Aryeh, Ph.D., The University of Chicago, 2008. Theoretical physics.

Students

Frentz, Zak, Experimental biophysics.
Jordan, David, Experimental biophysics.
Oppenheim, Jacob, Theoretical physics.
Taillefumier, Thibaud, Theoretical physics.

RESEARCH SPECIALTIES AND STAFF

Theoretical

Elementary Particle Physics; quantum field theory; statistical physics; kinetic theory; phase transitions; cellular automata; mathematical physics; dynamical systems; turbulance; chaos; condensed matter physics in biology. Cohen, Feigenbaum, Khuri, Leibler, Magnasco, Siggia. 16 postdoctoral associates.

Experimental

Experiments in high-energy physics with the Collider Dectector at Fermilab (CDF collaboration) and with the Compact Magnetic Solenoid detector (CMS collaboration) at the Large Hadron Collider at CERN. Applications of condensed matter physics to problems in biophysics: brownian motors; protein self assembly; statistical mechanics of membrances; DNA and other biomaterials; experimental and computational aspects of DNA and neuron assemblies. Bhatti, Demortier, Goulianos, Libchaber. 3 postdoctoral associates.

UNIVERSITY AT BUFFALO, THE STATE UNIVERSITY OF NEW YORK

DEPARTMENT OF PHYSICS

Amherst, New York 14260-1500
http://www.physics.buffalo.edu

Students Accepted For Degree	FIELDS		
	Physics	Astronomy	Related Fields
Doctorate	X		
Master's	X		

1. General

President: John B. Simpson
Dean of Graduate Education: Dr. John T. Ho
Department Chairman: Francis M. Gasparini
Department Telephone Number: (716) 645-2017 (C)
Department Fax Number: (716) 645-2507
Department E-mail: ubphysics@buffalo.edu
Type of Institution: University
Control: Public
Setting: Suburban
Total Faculty: 2,428
Total Graduate Faculty: 1,513
Total Students: 28,881
Total Graduate Students: 9,513
Annual Graduate Tuition:
In-state residents: Full-time—$4,185/sem.
Part-time—$349/credit hr.
Out-of-state residents: Full-time—$6,625/sem.
Part-time—$552/credit hr.
Tuition rates for: 2009–10 (subject to change)
Deferred tuition plan: No; monthly installment plan is available, however.
*Other Fees**: $1,512.50 for academic year
Term: Semester

*Does not include health insurance of $1,581/yr. (domestic) or $995/yr. (foreign). Health insurance is paid for students on assistantship.

2. Number of Faculty in Department

The combined total of full-time faculty in the three professorial ranks as of Fall 2009 is 29. The combined total of full-time, part-time, and other faculty at all ranks is 48.

3. Admission, Financial Aid, and Housing

Address admission inquiries to: Director of Graduate Studies, Physics Department, 239 Fronczak Hall, Amherst, NY 14260-1500; or send e-mail to: ubphysics@buffalo.edu
Graduate application fee required: $50
Admission deadline (Fall admission): 1/15 for fellowships and teaching assistantships; 2/1 for foreign students not seeking fellowships/assistantships; 7/15 for U.S. citizens not seeking fellowships/assistantships.
Admission information: For fall admission, 2009–10, 68 students were accepted from 204 applicants.
Admission requirements: For admission to the graduate program, a Bachelor's degree in physics with an average of B or above is required, although exceptions can be made in some circumstances. An applicant must satisfy the department of his/her ability to carry out graduate work in physics. GRE tests for verbal, quantitative, analytical, and advanced physics (subject test) are required. No specific acceptable score has been established. Students from non-English speaking countries are required to demonstrate proficiency in English via the TOEFL exam. Minimum acceptable score for admission is 550 (paper-based), 213 (computer-based), or 79 (internet-based). Foreign students making application for teaching assistantships must have a minimum acceptable TOEFL score of 600 (paper-based), 250 (computer-based), or 90 (internet-based), and must pass a Test of Spoken English (SPEAK) before being assigned teaching duties. In lieu of TOEFL, IELTS is acceptable. The total score must be at least 6.5 with no individual score less than 6.0.
Undergraduate preparation assumed: Fowles and Cassiday, *Analytical Mechanics*; Griffiths, *Introduction to Electrodynamics;* Reif, *Statistical and Thermal Physics*; Bransden and Joachain, *Quantum Mechanics*.
Address financial aid inquiries to: Student Response Center, 232 Capen Hall, Buffalo, NY 14260. Tel: (716) 645-2450. FAX: (716) 645-7760. Toll free: 1-866-838-7257.
GAPSFAS application required: No; FAFSA used instead.
Financial aid deadline: None
Loans available: Yes
On-campus, single student housing available: Yes
Cost/academic year: $6,848–10,212 single; $5,928–12,072 (12 month, furnished townhouse). Includes utilities and furnishings.
On-campus, married student housing available: Yes
Address residence hall and apartment inquiries to: University Residence Halls and Apartments, 106 Spaulding Quad, University at Buffalo, Amherst, NY 14261-0054. Tel: (716) 645-2171. Toll free: (866) 285-8806 (U.S. only). Internet: http://www.ub-housing.buffalo.edu/
Abundant and reasonably priced apartments off-campus can be found at the Off-Campus Housing Office at http://www.subboard.com/sbi-och/

4. Graduate Degree Requirements

Master's: (Master of Science): Two options may be taken: a thesis program or nonthesis program. In the thesis program a minimum of 30 credit hours is required. At least 15 semester hours are to be devoted to formal graduate course work, and the remainder culminating in a thesis. The nonthesis program requires 30 credit hours; 15 hours of specified courses plus 9 hours of courses exclusive of research, supervised teaching, and colloquia. Three credit hours of graduate physics laboratory is also required for both options.

Table A—Faculty, Enrollments, and Degrees Granted

Research Specialty	2009–10 Faculty	Enrollment[1] Fall 2009		No. of Degrees Granted[2] 2009–10 (2005–10)			Median No. of Years for 2009–10 Ph.D.'s
		Master's	Doctorate	Master's	Terminal Master's	Doctorate	
Applied Physics	8		1	0(0)	0(0)	0(0)	
Astrophysics	2		3	0(0)	0(0)	0(0)	
Atmos./Space Phys., Cosmic Rays	1		1	0(0)	0(1)	0(0)	
Atomic Molecules & Optical	2			0(0)	0(0)	0(0)	
Biophysics	5		10	0(0)	0(4)	0(2)	
Computational Physics	5	1	1	0(0)	1(1)	0(0)	
Condensed Matter Physics	18	1	22	0(0)	0(10)	1(19)	4.0
Cosmology	3			0(0)	0(0)	1(2)	6.0
High Energy	6	1	8	1(1)	1(4)	0(2)	
Low Temperature Physics	4		3	0(0)	0(1)	0(0)	
Materials Sci.	11		1	0(0)	0(0)	0(1)	
Mechanics	1			0(0)	0(1)	0(0)	
Medical & Health Physics		1	3	0(2)	4(6)	2(11)	7.0
Non-linear Dynamics	1		2	0(0)	0(0)	2(3)	7.5
Nuclear Physics	1	1		0(0)	0(1)	0(0)	
Particles & Fields	6	1	3	0(0)	1(1)	1(2)	7.0
Physics Education	4			0(0)	0(0)	0(0)	
Solid State	10		12	0(0)	0(2)	1(1)	9.0
Spintronics	10		1	0(0)	0(0)	0(0)	
Statistical & Thermal	4			0(0)	0(0)	0(0)	
Other Experimental	3		1	0(0)	0(0)	0(0)	
Other Theoretical	7	1	1	0(0)	0(0)	0(0)	
Non-specialized			5	0(0)	0(0)	0(0)	
Total		7	78	1(3)	7(32)	8(43)	
Full-time Grad. Stud.		7	78				
Part-time Grad. Stud.		0	0				
First-year Grad. Stud.		3	14				
Median Years in Grad. Study (2009–10 Degrees)					3.0	7.0	

Undergraduate Degrees, 2009–10 (2005–10): 30(82)

[1]Students not yet committed to a research specialty are entered under non-specialized.
[2]Five-year totals in parentheses.

Doctorate: A minimum of 72 semester hours of credit must be earned, with at least 36 in graduate physics lecture courses including three credit hours of graduate physics laboratory. In order to insure breadth in the student's Ph.D. program, the Department will evaluate his/her graduate work and may require the student to take specific courses in related fields. Within 18 months of enrollment, both parts of the Qualifying exam must be passed. After an additional 18 months, a short defense of the proposed Ph.D. project must be presented to the Ph.D. committee. A doctoral dissertation is required as well as an oral exam which consists of a defense of the dissertation and other topics determined by the candidate's committee.

Thesis: Thesis may be written *in absentia*.

Special Equipment, Facilities, or Programs: The department has active research experimental and theoretical programs in condensed matter, biophysics, high-energy and elementary particles and programs in computational physics, photonics/biophotonics, statistical physics and astrophysics and cosmology. The department occupies a modern building with major research facilities on the Amherst campus of the university, such as a helium liquifer, SQUID magnetometer, MBE systems, atomic and magnetic force microscopes, pulsed terahertz spectrometer, Raman spectrometer and a 15 Tesla superconducting magnet. Department members are users of several major national facilities, including the National Synchrotron Light Source at Brookhaven National Laboratories, the Tevatron at the Fermi National Accelerator Laboratory, the LHC at Cern, the National High Magnetic Field Laboratory in Tallahassee Florida, the Center for Free Electron Laser Studies at the University of California, Santa Barbara, the W. M. Keck Foundation Free Electron Laser Center at Vanderbilt University in Nashville, Tennessee and the Cornell Nanofabrication Facility.

The department has a joint B.S./M.S. in Computational Physics in cooperation with the Computer Science department. A graduate student in physics can also pursue an Advanced Certificate in Computational Science or in Professional Science Management in Biophysics as an additional credential.

Table B—Appointments to Graduate Students, 2009–10

Title of Appointee	Appointments Total	Appointments First year	Academic Load Allowed in Credit Hours	Hours of Service Per Week	Stipend for Academic Year ($)
			Semester		
Teaching Assistant[1]	40	9	9–12[2]	15–20	$16,500–17,500[3,4]
Research Assistant[1]	25	0	9–12[2]	–	$15,000–20,000[4,5]
CAMBI Fellowship[6]	1	1	9–12[2]	15–20	$19,000[4]
Total	66	10			

[1]Includes a full tuition scholarship and paid health insurance; TAs also receive a waiver of campus-based parking fees.
[2]Students in the terminal phase of their M.S. or Ph.D. may register for less than 9 credits per semester, typically from 1–3 credits.
[3]Summer teaching assistantships are available and pay an additional stipend.
[4]Some assistantship holders also receive additional $2,000–$4,000 Fellowships.
[5]Based on a calendar year appointment.
[6]Includes full tuition scholarship; Fellows pay their own health insurance.

5. Personnel Engaged in Separately Budgeted Research, 7/09–10

Professorial faculty	28
Postdoctoral appointments	14
Graduate students	45
Undergraduate students	6
Total	93

6. Separately Budgeted Research Expenditures by Source of Support, 7/08–09

	Departmental Research	Physics-related Research Outside Department
Federal government	$2,924,625	
Business and industry	377	
Total	$2,925,022	

8. Extension Centers and Summer Programs

During the summer months graduate students have the opportunity to do independent study and research. More than half the departmental faculty are funded by a variety of grants to carry out research during the summer. In addition, many graduate students receive research appointments and teaching assistantships during the summer months.

Table C—Separately Budgeted Research Expenditures

Research Specialty	No. of Grants	Expenditures ($)
Biophysics	7	$448,664
Condensed Matter Physics	33	$1,555,356
Cosmology	3	$80,004
Instruction	1	$18,572
Low Temperature Physics	3	$124,189
Medical & Health Physics	3	$91,691
Particles & Fields	16	$606,546
Total	65	$2,925,022

[1]July 1, 2008-June 30, 2009 sponsored program expenditures.

FACULTY

SUNY Distinguished Professors

Ho, John T., Ph.D., MIT, 1969. SUNY Distinguished Service Professor. Experimental condensed matter physics.

McCombe, Bruce D., Ph.D., Brown U., 1965. Dean of College of Arts and Sciences, SUNY Distinguished Professor. Experimental condensed matter physics; semiconductor physics; semiconductor nanostructures and nanoparticles; optical and infrared spectroscopy and magnetospectroscopy; electrical magneto-transport; spin effects in semiconductors.

Prasad, Paras, Ph.D., Pennsylvania, 1971 (joint appointment with Chemistry). SUNY Distinguished Professor. Photonics, ultrafast optical processes, non-linear optics.

University at Buffalo Distinguished Professor

Gasparini, Francis M., Ph.D., Minnesota, 1970. Chairman of the Department. Experimental low-temperature physics; phase transitions, quantum fluids.

Moti Lal Rustgi Professor of Physics

Petrou, Athos, Ph.D., Purdue, 1983. Experimental solid state physics.

Professors

Baur, Ulrich, Ph.D., Munich, 1985. Theoretical high-energy physics.

Gonsalves, Richard J., Ph.D., Columbia, 1976. Theoretical high-energy physics, computational physics.

José, Jorge, Ph.D., Universidad Nacional Autónoma de México, 1976. Condensed matter theory; theoretical biophysics.

Lee, Yung Chang, Ph.D., Maryland, 1963. Condensed matter physics; many-body theory; statistical mechanics; superconductivity; physics in confined systems.

Luo, Hong, Ph.D., Purdue, 1988. Experimental condensed matter physics; molecular beam epitaxy; microscopy; spintronics, semiconductor nanostructures.

Ram, Michael, Ph.D., Columbia, 1965. Atmospheric physics; climate change; theoretical physics.

Sen, Surajit, Ph.D., Georgia, 1990. Theoretical non-equilibrium many particle physics; nonlinear dynamics; granular materials; metamaterials; dust flow studies; battle problems; disease modeling; mathematical physics; science and math education in middle school level.

Strasser, Gottfried, Ph.D., Innsbruck, Austria, 1991. Experimental solid state physics; semiconductors; molecular beam epitaxy; quantum structures; photonics and electronics.

Weinstein, Bernard A., Ph.D., Brown, 1974. Experimental condensed matter physics, high-pressure properties of hard and soft solids, semiconductor physics, semiconductor nanostructures, oxides, optical spectroscopy, and Raman scattering.

Associate Professors

Cerne, John, Ph.D., UC, Santa Barbara, 1996. Experimental condensed matter physics; strongly correlated electronic materials, magnetic semiconductors, magnetic oxides, graphene, high temperature superconductors, magneto-polarimetry, experimental biophysics.

Han, Jong, Ph.D., Ohio State, 1997. Theoretical condensed matter physics, quantum transport theory, strongly correlated systems, quantum simulations, nanoscale magnetism.

Hu, Xuedong, Ph.D., Michigan, 1996. Theoretical condensed matter physics; theoretical study of nanostructure physics and solid state quantum information processing.

Kinney, William H., Ph.D., Colorado, 1996. Theoretical cosmology, high-energy physics, astrophysics.

Markelz, Andrea, Ph.D., UC, Santa Barbara, 1995. Experimental protein dynamics using terahertz time domain spectroscopy and UV/Vis/IR ultrafast spectroscopy. Experimental condensed matter physics: nanosystems spectroscopy and device development.

Wackeroth, Doreen, Ph.D., Univ. of Karlsruhe, 1995. Theoretical particle physics; phenomenology of particle physics at present and future colliders; electroweak physics; perturbative quantum chromodynamics; supersymmetry.

Zeng, Hao, Ph.D., Nebraska, 2001. Experimental condensed matter physics, nanoscale magnetism, spintronics, nanomaterial synthesis and self-assembly.

Zutic, Igor, Ph.D., Minnesota, 1998. Theoretical condensed matter physics; spin-polarized transport and spintronics, high temperature and unconventional superconductivity, ferromagnetic semiconductors, quantum dots, theoretical nanoscience, computational physics.

Assistant Professors

Ganapathy, Sambandamurthy, Ph.D., Indian Institute of Science, 2000. Experimental condensed matter physics; quantum transport in nanostructures, nanoelectronics, quantum phase transitions.

Iashvili, Ia, Ph.D., Humboldt University, Berlin, Germany, 2000. Experimental elementary particle physics. Research, development and construction of particle detectors. Searches for Higgs and supersymmetric particles, and precision measurements of particle properties at current and future accelerators.

Kharchilava, Avto, Ph.D., Tbilisi State University, 1990. Experimental elementary particle physics. Research, development and construction of particle detectors. Searches for Higgs and supersymmetric particles, and precision measurements of particle properties at current and future accelerators.

Pralle, Arnd, Ph.D., Ludwig-Maximilians-University, Munich and European Molecular Biology Lab (EMBL), Heidelberg, 1999. Experimental Biophysics, soft condensed matter physics. Molecular and cellular mechanics and forces, spatio-temporal patterning, single molecule spectroscopy.

Stojkovic, Dejan, Ph.D., Case Western Reserve University, 2001. Theoretical cosmology, high-energy physics, gravity, astrophysics.

Zhang, Peihong, Ph.D., Pennsylvania State, 2001. Theoretical condensed matter physics, electronic structure theory, nanostructured materials, dilute magnetic semiconductors, wide gap semiconductors, electron-phonon renormalization in metals, quasi-particle properties in strongly correlated materials, high performance computing.

Zheng, Wenjun, Ph.D., Stanford, 2003. Computational modeling of protein structures and dynamics.

Adjunct Professors

Bird, Jonathan P., Ph.D., University of Sussex, UK, 1990. Experimental condensed matter physics.

Cartwright, Alexander N., Ph.D., University of Iowa, 1995. Experimental condensed matter physics; time-resolved optical spectroscopy and ultrafast optical measurements.

Dimock, Jonathan, Ph.D., Harvard, 1971. Mathematical physics.

Krotscheck, Eckhard, Ph.D., Universität zu Köln, 1974. Quantum many-body theory.

Mitin, Vladimir, Doctor of Physical and Mathematical Science, Ukrainian Academy of Science, 1987. Condensed matter theory, modeling and simulations.

Adjunct Assistant Professor

Wang, John, Ph.D., UC, Berkeley, 2001. Cosmology.

Research Professor

DeMarco, Michael, Ph.D., Cincinnati, 1981. Experimental condensed matter physics; Mossbauer effect.

Research Assistant Professor

Jones, Matthew D., Ph.D., Illinois, 1996. Theoretical condensed matter/computational physics.

Faculty Emeriti

Brink, Gilbert O., Ph.D., California, Berkeley, 1957. Experimental atomic and molecular physics.

Fuda, Michael G., Ph.D., Rensselaer Polytech. Inst., 1967. Relativistic quantum mechanics; pion-nucleon scattering, photo- and electro-production of mesons.

Fujita, Shigeji, Ph.D., Maryland, 1960. Statistical mechanics; many-body problems; quantum transport; superconductivity; quantum Hall effect.

Heberle, Juergen, Ph.D., Columbia, 1955. Classical electrodynamics; Müssbauer effect.

Isihara, Akira, D.Sc., Tokyo, 1952. Theoretical solid state and low-temperature physics; statistical mechanics; many-body theory.

Jain, Piyare L., Ph.D., Michigan, 1954. Experimental elementary particle, relativistic heavy-ion physics.

Kao, Yi-Han, Ph.D., Columbia, 1962. Experimental solid state physics; materials physics; low-temperature physics.

Lin, Duo-Liang, Ph.D., Ohio State, 1961. Theoretical condensed matter and optical physics.

Reichert, Jonathan F., Ph.D., Washington, 1962. Experimental condensed matter physics.

Roalsvig, Jan Per, Ph.D., Saskatchewan, 1959. Experimental nuclear physics.

Sachs, Mendel, Ph.D., UCLA, 1954. General relativity; astrophysics; elementary particles, philosophy of physics.

RESEARCH SPECIALTIES AND STAFF

Theoretical

Acoustics. Sen.
Atmospheric Physics. Ram.
Astrophysics. Kinney, Stojkovic.
Biophysics. José, Sen, Zheng.
Computational Physics. Gonsalves, Han, Sen, Zhang, Zheng.
Condensed Matter. Han, Hu, José, Lee, Sen, Zhang, Zutic.
Cosmology. Kinney, Stojkovic, Wang.
Engineering Physics/Science. Sen.
Gravity. Stojkovic.
High Energy. Baur, Gonsalves, Stojkovic, Wackeroth.
Many-Body Theory. Han, Krotsheck, Lee.
Materials Science. Sen, Zhang.
Mathematical Physics. Dimock, Lee, Sen.
Mechanics. Sen.
Nuclear Physics. Lee.
Particles and Fields. Baur, Gonsalves, Kinney, Wackeroth.
Physics Education. Ram, Sen.
Quantum Computing. Hu.
Spintronics. Hu, Zhang, Zutic.
Statistical and Thermal. Lee, Sen.
Superconductivity. Han, Lee, Zutic.

Experimental

Applied Physics. Ganapathy, Luo, Markelz, McCombe, Pralle, Strasser, Weinstein, Zeng.
Atmospheric Physics. Ram.
Atomic, Molecular and Optical Physics. Markelz, Strasser.
Biophysics. Markelz, Pralle.
Condensed Matter. Cerne, Ganapathy, Gasparini, Ho, Markelz, McCombe, Petrou, Prasad, Strasser, Weinstein, Zeng.
Geophysics. Ram.
High Energy. Iashvili, Kharchilava.
High Pressure Physics. Weinstein
Low-Temperature Physics. Ganapathy, Gasparini, McCombe, Strasser.
Materials Science. Cerne, Ganapathy, Luo, Markelz, McCombe, Prasad, Strasser, Weinstein, Zeng.
Optics. Markelz, Prasad.
Particles and Fields. Iashvili, Kharchilava.
Photonics. Markelz, Prasad.
Physics Education. Iashvili, Kharchilava.
Solid State. Ganapathy, Gasparini, Ho, Luo, Markelz, McCombe, Petrou, Prasad, Strasser, Weinstein.
Spintronics. Cerne, Ganapathy, Luo, McCombe, Petrou, Strasser, Zeng.
Statistical and Thermal. Gasparini, Pralle.

UNIVERSITY OF ROCHESTER

DEPARTMENT OF PHYSICS AND ASTRONOMY

Rochester, New York 14627

Students Accepted For Degree	FIELDS		
	Physics	Astronomy	Related Fields
Doctorate	X	X	
Master's	X		

1. General

President: Joel Seligman
University Dean of Graduate Studies: Bruce Jacobs
Department Chair: Nicholas Bigelow
Department Telephone Number: (585) 275-4344
Type of Institution: University
Control: Private
Setting: Urban
Total Faculty: 1,331
Total Graduate Faculty: Not separated
Total Students: 9,431
Total Full-time Graduate Students: 4,057
Annual Graduate Tuition:
All Graduate Students: Full-time—$39,488
Tuition rates for: 2010–11
Deferred tuition plan: Yes
Other Fees: $1,128
Term: Semester

2. Number of Faculty in Department

The combined total of full-time faculty in the three professorial ranks is 27. The combined total of full-time, part-time, and other faculty at all ranks is 52.

3. Admission, Financial Aid, and Housing

Address admission inquiries to: Chair, Admissions Committee, Dept. of Physics and Astronomy
Admission deadline (Fall admission): 1/15
Admission information: For fall admission, 2010–11, 14 students were accepted from 400 applicants.
Admission requirements: For admission to the graduate programs, a Bachelor's degree in physics and/or astronomy is required with no minimum undergraduate GPA specified. The GRE general and Advanced tests are required. Students from non-English speaking countries are required to demonstrate proficiency in English via the TOEFL exam.
Undergraduate preparation assumed: Classical Mechanics in J.B. Marion, *Classical Dynamics of Particles and Systems*; Electricity and Magnetism, for example, in D. J. Griffiths, *Introduction to Electrodynamics*; Thermodynamics, Kinetic Theory, and Statistical Mechanics, for example, by C. Kittel and H. Kroemer, *Thermal Physics*; Quantum Mechanics for example in R. L. Liboff, *Introductory Quantum Mechanics*. Mathematics, good knowledge of advanced calculus, ordinary differential equations; functions of complex variable, boundary value problems, modern algebra.
Address financial aid inquiries to: Chair, Admissions Committee, Dept. of Physics and Astronomy
GAPSFAS application required: No
Financial aid deadline: 1/15
Loans available: Yes
Address housing inquiries to: University of Rochester, University Apartments Office, 020 Gates Wing, Susan B. Anthony Halls, Rochester, NY 14627
On-campus, single and married student housing available: Yes
Cost/month: $504–889

Table A—Faculty, Enrollments, and Degrees Granted

Research Specialty	2009–10 Faculty	Enrollment[1] Fall 2009		No. of Degrees Granted[2] 2009–10 (2005–10)			Median No. of Years for 2009–10 Ph.D.'s
		Master's	Doctorate	Master's	Terminal Master's	Doctorate	
Astronomy/ Astrophysics[3]	9	–	15	0(0)	0(0)	0(9)	-
Atomic, Molecular, & Optical Physics	7	–	41	5(22)	0(2)	3(17)	6.8
Biophysics	1	–	1	1(1)	0(0)	0(0)	–
Chemical Physics	2	–	1	0(2)	1(1)	0(4)	-
Condensed Matter Physics	5	–	7	1(3)	0(0)	1(9)	6.6
Engineering Physics/ Science	4	–	6	1(6)	0(0)	0(3)	-
Medical Physics	2	–	3	0(1)	0(1)	2(6)	6.3
Nuclear Physics	2	–	0	0(1)	0(1)	0(0)	–
Particles & Fields	11	–	23	3(19)	1(2)	3(20)	5.2
Plasma Physics & Fusion	2	–	3	1(5)	0(0)	3(6)	8.1
Other[3]	0	–	0	4(9)	0(1)	0(0)	–
Non-specialized	0	–	19	3(5)	0(1)	0(0)	–
Total		–		19(74)	1(9)	12(74)	
Full-time Grad. Stud.		–	119				
Part-time Grad. Stud.		–	0				
First-year Grad. Stud.		–	26				
Median Years in Grad. Study (2009–10 Degrees)			6.6	–	–	–	–
Undergraduate Degrees, 2009–10 (2005–10): 26(148)							

[1]Students not yet committed to a research specialty are entered under non-specialized.
[2]Five-year totals in parentheses.
[3]We do not offer a MA/MS in Astronomy or Astrophysics.

4. Graduate Degree Requirements

Master's: 30 graduate credits; no minimum grade average required; no language or residency requirements; Master's exam and thesis required. Degrees awarded in physics only.
Doctorate: 90 credit hours beyond Bachelor's degree in approved program required; no minimum GPA; minimum one year residency and full-time enrollment required; no language or computer language required; written preliminary-oral qualifying exams required; thesis and oral thesis exam required. Degrees available in physics or physics and astronomy.
Thesis: Thesis may be written *in absentia*.

Table B—Appointments to Graduate Students, 2009–10

Title of Appointee	Appointments Total	First year	Academic Load Allowed in Credit Hours	Hours of Service Per Week	Stipend for Academic Year ($)
			Semester		
Teaching Assistant	34	26	24 AY	16	23,520[1]
Research Assistant	83	0	24 AY	16	23,520[1]
GAANN Fellows	35	14	24 AY	16	25,520[1]
Marshak Fellows	1	0	24 AY	16	25,520[1]
Univ. Fellows	1	0	24 AY	16	26,680[1]
Total	154	40			

[1]12 months.

5. Personnel Engaged in Separately Budgeted Research, 7/08–6/09

Professorial faculty	26
Non-tenured track or Emeritus Faculty	9
Research Associates and Postdoctoral appointments	24
Graduate students	117
Undergraduate students	10
Research Experience for Undergraduates	41
Total	241

6. Separately Budgeted Research Expenditures by Source of Support

	Departmental Research
Business and industry	–
Federal government	8,453,844
Other, Foundation	128,114
Total	$8,581,958

Table C—Separately Budgeted Research Expenditures

Research Specialty	No. of Grants	Expenditures ($)
Astrophysics	26	899,202
Condensed Matter Experimental Physics	1	114,678
Condensed Matter Theoretical Physics	2	121,789
Nuclear Physics-Experimental	6	357,374
Particle Physics, Experimental	8	4,096,649
Particle Physics, Theory	2	335,188
Physics Education	5	664,837
Quantum Optics Experimental	8	1,433,839
Quantum Optics Theoretical	8	558,452
Total	66	8,581,958

FACULTY

Professors

Agrawal, G. P., Ph.D., Indian Institute of Technology, New Delhi, 1974. Fiber optics, lasers, optical communications.

Betti, R., Ph.D., MIT, 1992. Theoretical plasma physics. Nuclear and mechanical engineering, computational and plasma physics.

Bigelow, N. P., Ph.D., Cornell, 1989. Lee A. Du Bridge Professor. Experimental and theoretical quantum optics and quantum physics. Studies of BEC, laser-cooled and trapped atoms.

Blackman, E. G., Ph.D., Harvard, 1995. Theoretical astrophysics, astrophysical plasmas and magnetic fields; accretion and ejection phenomena; relativistic and high energy astrophysics.

Bocko, M. F., Ph.D., Rochester, 1984. Superconducting electronics, quantum computing, musical acoustics, digital audio technology, sensors.

Bodek, A., Ph.D., MIT, 1972. George E. Pake Professor. Experimental elementary particle physics, proton-antiproton collisions, QCD and structure functions, neutrino physics, electron scattering, and tile-fiber calorimetric detectors.

Boyd, R. W., Ph.D., University of California at Berkeley, 1977. M. Parker Givens Professor. Nonlinear optics.

Castner, T. G., Ph.D., Illinois, 1958. Emeritus. Experimental condensed matter physics, metal insulator transition.

Cline, D., Ph.D., Manchester (England), 1963. Extreme states of nuclei pairing and shape correlations in nuclei.

Conwell, E., Ph.D., Chicago, 1948. Theoretical chemical physics, condensed matter physics, biological physics.

Das, A., Ph.D., SUNY at Stony Brook, 1977. Theoretical particle physics, finite temperature field theory, integrable systems, phenomenology, non-commutative field theory and string/M theory.

Demina, R., Ph.D., Northeastern, 1994. Experimental particle physics, proton-antiproton collisions, top and electroweak physics.

Douglass, D. H., Ph.D., MIT, 1959. Experimental condensed matter physics. Climate change and pollution.

Duke, Charles B., Ph.D., Princeton, 1963. Theoretical condensed matter physics, geophysics and climate.

Eberly, J. H., Ph.D., Stanford, 1962. Andrew Carnegie Professor. Theoretical quantum optics, quantum entanglement, cavity QED, atoms in strong laser fields, dark-state optical control theory.

Fauchet, P. M., Ph.D., Stanford, 1984. Semiconductor materials and device physics, materials sciences, biomedical engineering, and optics.

Ferbel, T., Ph.D., Yale, 1963. Experimental elementary particle physics. Studies of the top quark in hadronic collisions.

Forrest, W. J., Ph.D., California, San Diego, 1974. Emeritus. Observational astrophysics, infrared astronomy, stellar and planetary formation, low-mass stars and brown dwarfs, development of infrared detector arrays and instrumentation.

Foster, T. H., Ph.D., Rochester, 1990. Biological and medical physics.

Frank, A., Ph.D., Washington, Seattle, 1992. Theoretical astrophysics, astrophysical plasmas, numerical hydrodynamics and magnetohydrodynamics.

Gao, Y., Ph.D., Purdue, 1986. Experimental condensed matter physics. Surface physics.

Hagen, C. R., Ph.D., MIT, 1962. Theoretical elementary particle physics, quantum field theory, particularly 2+1 dimensional theories.

Helfer, H. L., Ph.D., Chicago, 1953. Emeritus. Theoretical astrophysics and plasma physics, high-energy astrophysics, dark matter in galactic haloes.

Huizenga, J. R., Ph.D., Illinois, 1949. Emeritus. Nuclear chemistry, nuclear physics.

Jacobsen, E. H., Ph.D., MIT, 1954, Emeritus, Electron Optics.

Knox, R. S., Ph.D., Rochester, 1958. Emeritus. Theoretical biological physics and condensed matter physics. Energy-balance models of climate.

Knox, W. H., Ph.D., Rochester, 1984. Ultrafast sciences and technology, telecommunications, ultrafast biomedical optics and optics education.

Koltun, D. S., Ph.D., Princeton, 1961. Emeritus. Theoretical nuclear physics, meson interactions with nuclei, many body theory, electron scattering.

Manly, S. L., Ph.D., Columbia, 1989. Experimental relativistic heavy ion physics, experimental elementary particle physics.

McCrory, Robert L., Ph.D., MIT 1973. Director of the Laboratory for Laser Energetics. Nuclear and mechanical engineering, computational hydrodynamics.

McFarland, K. S., Ph.D., Chicago 1994. Experimental elementary particle physics, properties of top quarks, neutrino physics, electroweak unification.

Melissinos, A. C., Ph.D., MIT, 1958. Experimental particle physics, high intensity laser particle interactions, free electron lasers, searches for relic gravitational radiation.

Meyerhofer, D. D., Ph.D., Princeton, 1987. Experimental plasma and laser physics. High energy density physics and inertial confinement fusion. High intensity laser-matter interaction experiments, quantum optics.

Novotny, Lukas, Dr. Sci. Techn., ETH Zurich, 1996. Nano-optics, nanoscale phenomena, biophysics.

Okubo, S., Ph.D., Rochester, 1958. Emeritus. Theoretical particle physics and mathematical physics. Lie and non-associative algebras.

Orr, L., Ph.D., Chicago, 1991. Theoretical elementary particle physics, phenomenology, quantum chromodynamics and electroweak physics.

Pipher, J. L., Ph.D., Cornell, 1971. Emeritus. Observational astrophysics, infrared astronomy, Galactic and extragalactic star formation, low-mass stars and brown dwarfs, development of infrared detector arrays and instrumentation.

Rajeev, S. G., Ph.D., Syracuse, 1984. Theoretical particle physics. Nonperturbative quantum field theory applied to strong interactions.

Rothberg, L., Ph.D., Harvard, 1983. Experimental chemical physics, organic electronics and biomolecular sensing.

Savedoff, M. P., Ph.D., Princeton, 1957. Emeritus. Theoretical astrophysics, stellar interiors, interstellar matter, high energy astrophysics.

Schröder, Wolf-Udo, Ph.D., Darmstadt/Germany, 1971. Experimental nuclear physics, dynamics of complex nuclear reactions, fundamental properties of nuclear matter, nuclear transmutation. Nuclear technology applications.

Shapir, Y., Ph.D., Tel Aviv, 1981. Theoretical condensed matter physics, statistical mechanics. Critical phenomena in ordered and disordered systems, fractal growth.

Sharpless, S. L., Ph.D., Chicago, 1952. Emeritus. Observational astrophysics.

Simon, A., Ph.D., Rochester, 1950. Emeritus. Theoretical plasma physics, controlled thermonuclear fusion.

Slattery, P. F., Ph.D., Yale, 1967. Experimental elementary particle physics, investigation of QCD via direct photon production, top quark studies and searches for new phenomena using high energy colliders.

Sobolewski, R., Ph.D., Polish Academy of Science, Warsaw 1983. Applied superconductivity, ultrafast electronics and optoelectronics.

Sproull, R. L., Ph.D., Cornell, 1943. Emeritus. Experimental condensed matter physics.

Stroud, Carlos R., Jr., Ph.D., Washington University, 1969. Quantum optics, short-pulse excitation of atoms and molecules, quantum information.

Tang, Ching W., Ph.D., Cornell, 1975. Doris Johns Cherry Professor. Chemical and condensed matter physics, organic electronics.

Tarduno, J. A., Professor, Ph.D., Stanford, 1987. Geophysics, geomagnetism and geodynamics, plate tectonics and polar wander, geomagnetic reversals, fine particle magnetism, planetary astrophysics.

Teitel, S. L., Ph.D., Cornell, 1981. Statistical and condensed matter physics.

Thomas, J. H., Ph.D., Purdue, 1966. Theoretical astrophysics, astrophysical plasmas, astrophysical fluid dynamics and magnetohydrodynamics, solar physics.

Thorndike, E. H., Ph.D., Harvard, 1960. Experimental elementary particle physics, weak decays of bottom and charm quarks.

Van Horn, H. M., Ph.D., Cornell, 1965. Emeritus. Theoretical astrophysics, degenerate stars.

Watson, D. M., Ph.D., Berkeley, 1983. Observational astrophysics, infrared astronomy, stellar and planetary formation, low-mass stars and brown dwarfs, development of infrared detector arrays and instrumentation.

Wolf, E., Ph.D., Bristol (England), 1948. Wilson Professor. Theoretical optics. Statistical optics, theory of coherence and polarization, inverse scattering, diffraction tomography.

Wolfs, F. L. H., Ph.D., Chicago, 1987. Experimental high-energy/nuclear physics, Relativistic heavy-ion physics, Dark-matter searches.

Zhong, Jianhui, Ph.D., Brown, 1988. Biological and medical physics. Advanced medical imaging, novel MRI techniques, physiological properties, biological tissues.

Associate Professors

Howell, John, Ph.D., Pennsylvania State University, 2000. Experimental quantum optics and quantum physics, quantum cryptography and quantum computation.

Quillen, Alice, Ph.D., Caltech, 1993. Observational astrophysics, galactic structure and dynamics, active galactic nuclei, dynamics of planetary and protoplanetary systems.

Assistant Professors

Badolato, Antonio, Ph.D., California, Santa Barbara, 2005. Semiconductor quantum heterostructures with emphasis on solid-state cavity quantum electrodynamics.

Dery, H., Ph.D., Technion, Israel, 2004. Theory of semiconductor spin electronics.

Garcia-Bellido, Aran, Ph.D., Royal Holloway, University of London, 2002. Experimental particle physics, with interests in supersymmetry and physics of the top quark, and in particular electroweak production of single top quarks.

Jordan, Andrew N., Ph.D., California, Santa Barbara, 2002. Theoretical quantum optics and condensed matter.

Mamajek, Eric E., Ph.D., Arizona, 2004. Observational astronomy, stellar evolution, young stellar clusters, protoplanetary disks, stellar and planetary-system formation.

Ren, Chuang, Ph.D., Wisconsin-Madison, 1998. Theoretical and computational plasma physics, controlled fusion.

RESEARCH SPECIALTIES AND STAFF

Theoretical

Astrophysics. Astrophysical fluid dynamics and magnetohydrodynamics, astrophysical plasmas, computational astrophysics. Accretion disks and hypersonic outflows associated with young stars and degenerate objects. Evolution of protoplanetary disks. Celestial mechanics. Galactic dynamics. Stellar formation and death; planetary nebulae. High-energy astrophysics. Dark matter in galaxy haloes. Dynamo theory of magnetic-field generation in stars, galaxy disks, and planets. Origin and long-term behavior of Earth's magnetic field; field reversals and geodynamics. Physics of sunspots and solar magnetic flux tubes. Laboratory simulation of high-energy-density astrophysical plasmas. Blackman, Frank, Helfer (Emeritus), Quillen, Thomas. 1 postdoctoral research associate, collaborating faculty and staff scientists at the Laboratory for Laser Energetics.

Condensed Matter. Theory of thermodynamic and transport properties of disordered systems also near phase transitions. Theory of flux phases in type II superconductors and Josephson junctions. Scaling properties in clusters and polymers. Complex fluids, colloids, and biosystems. Theory of mesoscopic physics: electronic transport and noise properties. Quantum physics in the solid state – entanglement, measurement and information. Conwell, Duke, Jordan, Shapir, Teitel.

Particle Physics/Nuclear Physics. String theory; matrix model, integrable models; Lagrangian field theory; thermo-field theory; phenomenology, non-perturbative methods in field theory; structure functions of hadrons; supersymmetry, renormalization in quantum mechanics. Das, Hagen, Koltun (Emeritus), Okubo (Emeritus), Orr, Rajeev,

Plasma Physics. Astrophysical plasmas in extreme environments. Fundamental processes common to laboratory and astrophysical plasmas. Space Plasmas interaction of intense lasers with matter. High energy density physics with intense lasers. Hydrodynamic, magnetohydrodynamic and plasma instabilities; particle acceleration. Interaction of intense lasers with matter. Inertial confinement fusion and high-energy-density physics. Compression and heating of pellets to ignition relevant conditions. Blackman, Betti, Frank, Helfer (Emeritus), McCrory, Meyerhofer, Ren, Simon, Thomas, Van Horn (Emeritus). 2 postdoctoral research associates, collaborating faculty and staff scientists at the Laboratory for Laser Energetics.

Quantum Optics. Coherence phenomena in the interaction of light with matter. Subjects include coherent control, quantum entanglement, Bose-Einstein condensates, quantum dots, nanophotonics, atoms in intense laser fields, wave packet states of atoms and molecules, quantum imaging, single-cycle and half-cycle EM pulses, amplitude-coherent chemistry, correlation-induced spectral changes, solitons and inverse scattering theory, diffraction tomography. Agrawal, Badolato, Bigelow, Boyd, Eberly, Howell, Jordan, Novotny, Stroud, Wolf. 4 postdoctoral research associates.

Experimental

Astrophysics. Observations with the NASA Spitzer Space Telescope, the Hubble Space Telescope and Chandra X-ray Observatory, as well as ground-based telescopes. Formation and evolution of stars, protoplanetary disks, dusty debris disks, planets, and brown dwarfs. Mineralogy of dust and chemistry of gases in protoplanetary and debris disks. Structure, dynamics, star formation histories, and chemical evolution of the Milky Way, other galaxies, and nearby young stellar associations. Evolution of stellar rotation, magnetic activity, solar wind, and interaction between Sun and magnetic filed of early Earth. Starbursts and active galaxy nuclei. High relsolution imaging of young stellar systems using adaptive optics. Development of infrared detector arrays and instruments for infrared astronomy. Forrest, Mamajek, Pipher, Quillen, Tarduno, Watson. 2 postdoctoral research associates.

Biological and Medical Physics. Experimental and theoretical research in single molecule spectroscopy and manipulation, photodynamic therapy, diffusion tensor and functional MRI mechanisms and techniques, tissue optics, light scattering, biomolecular sensing, interactions of nanoparticles with biomolecules, and microscopy. Conwell, Foster, Novotny, Rothberg, Zhong.

Condensed Matter. Semiconductor heterojunctions; surface phenomena; synchrotron radiation photoemission; femtosecond time resolved photoemission; ultrafast dynamics in solids; interfaces in organic semiconductors; scanning tunneling microscopy; superconductivity and superconducting films; electron tunneling spectroscopy; metallic, magnetic, and superconducting nanowires; electron-beam lithography and mesoscopic structures. Bocko, Castner (Emeritus), Douglass, Gao, Novotny, Rothberg, Sobolewski.

Nuclear Physics. Structure of exotic nuclei far from stability, relativistic heavy-ion physics, neutrino physics, electron scattering. Bodek, Cline, Manly, McFarland, Schröder, Wolfs. 1 senior staff scientist, 1 postdoctoral research associate.

Particle Physics. Proton-antiproton colliding beam experiment at the Fermi National Accelerator Laboratory—FNAL (CDF and DZERO), proton-proton colliding beam experiments at the CERN LHC Large Hadron Collider (LHC/CMS) e^+e^- colliding beams at Ithaca (LEPP/CLEO-C) and Beijing (BES-III) Relativistic Heavy Ion Collisions at BNL (RHIC/PHOBOS), Neutrino experiments (Fermilab/MINERvA) and (Jparc/T2K. Electron scattering on nuclear targets (JLAB/JUPITER). R+D for future Linear Colliders. Dark matter search at SUSEL (LUX). Bodek, Demina, Ferbel, Garcia-Bellido, Manly, McFarland, Melissinos, Schröder, Slattery, Thorndike, Wolfs; research faculty Budd, Chung, deBarbaro, Ginther, Park, Sakumoto Zielinski and 11 postdoctoral research associates.

Quantum Optics. Quantum interference effects and non-classical states of light, search for locality violation with photons, Bose-Einstein condensation, laser cooling and trapping of atoms and molecules, atom optics, generation of non-classical states of the atom, ultra-cold collisions, cold molecules, novel light sources. Nonlinear optics, quantum coherence, optical solitons. High intensity laser plasma and laser-atom interactions. Bigelow, Boyd, Howell, Knox, W., Meyerhofer, Stroud, and collaborating faculty at the Institute of Optics.

UNIVERSITY OF ROCHESTER

THE INSTITUTE OF OPTICS

Rochester, New York 14627

Students Accepted For Degree	FIELDS		
	Physics	Astronomy	Related Fields
Doctorate			X
Master's			X

1. General

President: Joel Seligman
Director of The Institute: Wayne H. Knox
Department Telephone Number: (585) 275–7764
Type of Institution: University
Control: Private
Setting: Urban
Total Faculty: 1,258
Total Graduate Faculty: Not separated
Total Students: 7,508
Total Graduate Students: Full-time—2,900
Annual Graduate Tuition:
All Graduate Students: Full-time—$39,488
Tuition rates for: 2010–11
Deferred tuition plan: Yes
Other Fees: $2500
Term: Semester

2. Number of Faculty in Department

The combined total of full-time faculty in the three professorial ranks is 29. The combined total of full-time, part-time, and other faculty at all ranks is 32.

3. Admission, Financial Aid, and Housing

Address admission inquiries to: Administrator, Optics Graduate Admissions Committee
Graduate application fee required: $0
Admission deadline (Fall admission): 2/1
Admission information: For fall admission, 2010–11, 23 students were accepted from 217 applicants.
Admission requirements: For admission to the graduate programs, a Bachelor's degree in physics or engineering is required with a minimum undergraduate GPA of 3.0/4.0 specified. (The average for entering students is 3.7/4.0.) The GRE is required. Average scores for our entering students are 80th percentile in all three categories. An Advanced subject exam is not required. The minimum acceptable score suggested for admission is 650. Students from non-English speaking countries are required to demonstrate proficiency in English via the TOEFL exam. Minimum acceptable score for admission is 105 internet based test.
Address financial aid inquiries to: Administrator, Optics Graduate Admissions Committee
GAPSFAS application required: No
Financial aid deadline: 2/15
Loans available: Yes
Address housing inquiries to: University Apartments Office, 1351 Mt. Hope Ave., Rochester, NY 14620
On-campus, single student housing available: Yes*
Cost/month: $600–804
On-campus, married student housing available: Yes*
Cost/month: $690–889

*University-owned, off-campus housing

Table A—Faculty, Enrollments, and Degrees Granted

Research Specialty	2009–10 Faculty	Enrollment[1] Fall 2010 Master's	Enrollment[1] Fall 2010 Doctorate	No. of Degrees Granted[2] 2009–10 (2005–10) Master's	Terminal Master's	Doctorate	Median No. of Years for 2009–10 Ph.D.'s
Optics	29	6	12	11	8(51)	17(43)	7
Total							
Full-time Grad. Stud.		6	89				
Part-time Grad. Stud.		1	0				
First-year Grad. Stud.		6	12				
Median Years in Grad. Study (Degrees)							

Undergraduate Degrees, 2009–10 (2005–10): 11(85)

[1]Students not yet committed to a research specialty are entered under non-specialized.
[2]Five-year totals in parentheses.

4. Graduate Degree Requirements

Master's: M.S. degrees require 30 hours of coursework, including 16 hours of required core courses. The M.S. degrees are normally completed in 9–12 months. There are no residence or foreign language requirements. In a co-op program, students take the first semester of courses, work full-time for 12 months, and then return to campus for the final semester of classes.
Doctorate: General requirements: one year of full-time residence, 90 hours of graduate work (60 hours beyond the M.S.), a year spent as a teaching assistant, successful completion of a written preliminary examination and an oral qualifying examination and completion and defense of a doctoral dissertation. No language requirement.
Thesis: Thesis may not be written *in absentia*.
Special Equipment, Facilities, or Programs: Instruction is offered in optical instrumentation and design, quantum optics and electronics, laser engineering, optics of thin films, electro-optics, holography, interferometry, and most other areas of optical physics and engineering. Well-equipped laboratories allow student thesis research in such areas as ultra-high resolution dye laser spectroscopy, semiconductor lasers, optical physics, nano-optics, optical communications, fiber optics, imaging, nonlinear optics, diffractive optics, gradient index optics, interferometry, image processing, optical materials, and highpower laser physics. In addition to extensive facilities within The Institute, thesis research may be carried out in the Laboratory for Laser Energetics, the School of Medicine and Dentistry, and the Center for Visual Science. Joint projects applying optical techniques in all of these areas are currently underway.

Table B—Appointments to Graduate Students, 2010–11

Title of Appointee	Appointments Total	Appointments First year	Academic Load Allowed in Credit Hours	Hours of Service Per Week	Stipend for Academic Year ($)
			Semester		
Teaching Assistant	20	9	9–16	-	5,000[1,4]–26,000[2,3]
Research Assistant	73		16	-	26,000[2]
Institute Fellow	0		16	-	26,000[2]
Other[5]	0	0	16	-	See[5,6]
Total	93	9			

[1]MS Students.
[2]Plus tuition.
[3]All Ph.D. students are required, as part of their training, to serve one year (usually the second year) as teaching assistants. During this time they carry only 9 credit hours per semester.
[4]Plus partial tuition.
[5]Industrial and foundation support of Fellows at The Institute of Optics.
[6]Minimum of $26,00 plus tuition.

5. Personnel Engaged in Separately Budgeted Research, 7/15–6/07

Information not available

6. Separately Budgeted Research Expenditures by Source of Support

Information not available

8. Extension Centers and Summer Programs

Each summer The Institute of Optics offers several one or two week, non-credit summer courses on various aspects of Optics. Anyone interested for the summer should inquire in the spring.

Table C—Separately Budgeted Research Expenditures

Research Specialty	No. of Grants	Expenditures ($)
	Information not available	

FACULTY

Professors

Agrawal, Govind, Ph.D., Indian Inst. of Tech., 1974. Semiconductor lasers and amplifiers; nonlinear optical phenomena; optical fiber communications.

Bigelow, Nicholas, Ph.D., Cornell, 1989. Quantum Optics and Quantum Physics.

Boyd, Robert W., Ph.D., California, Berkeley, 1977. Nonlinear optics; infrared detection and generation.

Brown, Thomas, Ph.D., Rochester, 1987. Integrated optics; fiber optics: optical properties of solids; quantum electronics.

Eberly, Joseph H., Ph.D., Stanford, 1962. Multiphoton processes; quantum electrodynamics; resonant interaction of light with atoms and molecules.

Fauchet, Philippe, Ph.D., Stanford, 1984. Optical processes in solids; femtosecond laser spectroscopy; materials science and devise applications of light-emitting silicon.

Fienup, James R., Ph.D., Stanford, 1975. Phase retrieval; unconventional imaging; image processing; wave-front sensing.

Foster, Thomas, Ph.D., Rochester. Medical optics, photodynamic therapy.

George, Nicholas, Ph.D., Cal. Tech., 1959. Optical systems; speckle; pattern recognition.

Jacobs, Stephen D., Rochester, 1976. Optical materials.

Knox, Wayne, Ph.D., Rochester, 1983. Director of the Institute, Ultrafast Science and Technology, Telecommunications, and Optoelectronics.

Moore, Duncan T., Ph.D., Rochester, 1974. Geometrical optics; optical instrumentation; gradient index glass; interferometry; medical optics.

Novotny, Lukas, Ph.D., Swiss Federal Institute of Technology. Nano-optics, Light-matter interactions on the subwavelength scale.

Rolland, Jannick, Ph.D., U. of Arizona, 1990. Optical instrumentation and system engineering.

Stroud, Carlos R., Ph.D., Washington (St. Louis), 1969. Quantum optics; short-pulse excitation of atoms and molecules.

Teegarden, Kenneth J., Ph.D., Illinois, 1954. Optical properties of materials.

Wicks, Gary, Ph.D., Cornell, 1981. III–V semiconductors—epitaxial growth, optical properties, and optical devices.

Williams, David, Ph.D., California, San Diego, 1979. Sensitivity and resolution of the human visual system to patterns that are modulated in wavelength, space, and time.

Wolf, Emil, Ph.D., Bristol, 1948; D. Sc., Edinburgh, 1955. Electromagnetic theory and physical optics; diffraction and theory of partial coherence.

Associate Professors

Alonso, Miguel, Ph.D., Rochester, 1996. Mathematical models for wave propagation. Theory of partial coherence. Connection between the ray and wave models.

Bentley, Julie, Ph.D., Rochester Univ., 1995. Leng Design.

Berger, Andrew, Ph.D., MIT. Biomedical optics, Raman Spectroscopy, optical analysis of blood and tissue.

Guo, Chunlei, Ph.D., U. of Connecticut. High intensity laser interactions, with matter.

Krauss, Todd, Ph.D., Cornell, 1998. Nanoscale materials and devices.

Marciante, John R., Ph.D., University of Rochester. Lasers, waveguide, and fiber optics.

Wolf, Seka, Ph.D., Texas. Lasers.

Yoon, Geunyoung, Ph.D., Osaka University. Biomedical and visual optics, and adaptive optics.

Zavislan, James, Ph.D., Rochester. Optical engineering, medical optical instrumentation.

Assistant Professor

Chi, Wanli, Ph.D., Rochester, 2005. Integrated computer and imagining systems.-

APPALACHIAN STATE UNIVERSITY

DEPARTMENT OF PHYSICS AND ASTRONOMY

Boone, NC 28608

Students Accepted For Degree	FIELDS		
	Physics	Astronomy	Related Fields
Doctorate			
Master's	X	X	X

1. General

Chancellor: Kenneth E. Peacock
Dean of Graduate School: Edelma D. Huntley
Department Chairman: Leon Ginsberg (Interim)
Department Telephone Number: (828) 262-3090
Type of Institution: University
Control: Public
Setting: Small town
Total Faculty: 2,476
Total Graduate Faculty: 632
Total Students: 15,387
Total Graduate Students: 1,030
Annual Graduate Tuition:
In-state residents: Full-time—$4,953
Part-time—$/credit 671.45/2
Out-of-state residents: Full-time—$15,373.00
Part-time—$/credit 1,973.20/2
Tuition rates for: 2009-10
Deferred tuition plan: No
Other Fees: $ (included in above)
Term: Semester

2. Number of Faculty in Department

The combined total of full-time faculty in the three professorial ranks is 16. The combined total of full-time, part-time, and other faculty at all ranks is 24.

3. Admission, Financial Aid, and Housing

Address admission inquiries to: Office of Graduate Admissions, ASU Box 32068, Boone, NC 28608
Graduate application fee required: $ 50.00
Admission deadline (Fall admission): July 1
Admission information: For fall admission, 2009–10, 12 students were accepted from 15 applicants.
Admission requirements: For admission to the graduate programs, a Bachelor's degree in physics or a related discipline is required with a minimum undergraduate GPA of 3.0 specified. Students from non-English speaking countries are required to demonstrate proficiency in English via the TOEFL exam. Minimum acceptable score for admission is 550. Recommended score is 580 or better.
Address financial aid inquiries to: Financial Aid Office, ASU Box 32068, Boone, NC 28608
GAPSFAS application required: No
Financial aid deadline: March 15
Loans available: Yes
Address housing inquiries to: Housing Office, P.O. Box 32111
On-campus, graduate student housing available: Yes
Cost/sem.: $2,750-effic., $3,150-1 BR, $3,550-2 BR
On-campus, married student housing available:
Cost/month: $ (same as above)

Table A—Faculty, Enrollments, and Degrees Granted

Research Specialty	2009–10 Faculty	Enrollment[1] Fall 2009		No. of Degrees Granted[2] 2009–10 (2005-10)			Median No. of Years for 2009–10 Ph.D.'s
		Master's	Doctorate	Master's	Terminal Master's	Doctorate	
Astronomy	4	3	–	1(2)	0(0)	0(0)	–
Astrophysics	2	1	–	0(0)	0(0)	0(0)	–
Atmos./Space Phys., Cosmic Rays	3	3	–	3(4)	0(0)	0(0)	–
Atomic, Molecular, & Optical Physics	2	1	–	1(4)	0(0)	0(0)	–
Biophysics			–		0(0)	0(0)	–
Chemical Physics	1	0	–	0(1)	0(0)	0(0)	–
Computer Science	1	0	–	0(0)	0(0)	0(0)	–
Condensed Matter Physics	3	1	–	0(2)	0(0)	0(0)	–
Electronics	2	1	–	0(1)	0(0)	0(0)	–
Energy Sources & Environ.	2	1	–	1(1)	0(0)	0(0)	–
Engineering Physics	6	4	–	2(8)	0(0)	0(0)	–
Fluids & Rheology Geophysics	1	1	–	1(2)	0(0)	0(0)	–
Medical & Health Physics	1	0	–	0(1)	0(1)	0(0)	–
Nanoscience	3	2	–	2(0)	0(0)	0(0)	
Optics	2	1	–	0(1)	1(0)	0(0)	–
Physics Education	3	1	–	0(0)	0(0)	0(0)	–
Other Theoretical/ Math.	1	0	–	0(0)	0(0)	0(0)	–
Total		20		11(27)	1(1)		
Full-time Grad. Stud.		18					
Part-time Grad. Stud.		2					
First-year Grad. Stud.		8					
Median Years in Grad. Study (2009–10 Degrees)		2					

Undergraduate Degrees, 2009–10: (2005–10) 16(57)

[1]Students not yet committed to a research specialty are entered under non-specialized.
[2]Five-year totals in parentheses.

4. Graduate Degree Requirements

Master's in Engineering Physics: Minimum 36 credit hours, or 30 credit hours with thesis. Comprehensive exam is required. Professional Science Master's: 36 credit hours plus an internship. No comprehensive exam.
Special Equipment, Facilities, or Programs: Students enjoy personal attention in an informal atmosphere, where the primary goal is quality teaching based on a collegial rapport between students and teachers. Modern laboratory facilities provide invaluable hands-on experience in front-line research projects. Currently, these include observational astronomy at astrophysical Dark Sky Observatory, a state-of-the-art optics laboratory, optical spectroscopy equipment, time-of-flight mass spectroscopy, time-of-flight secondary ion mass spectrometer and ion-storage facility, surface analysis for nanoscale systems and materials science, and cryopumped thin film vacuum deposition system including microscopy (SEM), (FIB), (AFM), (STM), X-ray microanalysis, surface plasmons, local plasma torch, and Auger spectroscopy.

Table B—Appointments to Graduate Students, 2009–10

Title of Appointee	Appointments Total	Appointments First year	Academic Load Allowed in Credit Hours	Hours of Service Per Week	Stipend for Academic Year ($)
			Semester		
Teaching Assistant	15	8	12	20	10,000[1]
Research Assistant	2	1			
Other (specify)					
Total	17	9			

[1]Maximum available for 9-month academic year Teaching Assistantship.

5. Personnel Engaged in Separately Budgeted Research, 7/08–6/09

Professorial faculty	10
Graduate students	10
Undergraduate students	4
Total	24

Table C—Separately Budgeted Research Expenditures

Research Specialty	No. of Grants	Expenditures ($)
Astronomy	2	44,000
Astrophysics	2	20,000
Atmos./Space Phys., Cosmic Rays	3	100,000
Atomic, Molecular, & Optical Physics	2	67,000
Chemical Physics	1	22,000
Energy Sources & Environ.		160,000
Engineering Physics	1	10,000
Nano Science	3	400,000
Physics Education	1	4,000
Total	16	727,000

FACULTY

Calamai, Anthony G. (Dean, College of Arts & Sciences), Ph.D. in Physics, North Carolina State University. Experimental atomic, molecular, and optical physics; laboratory astrophysics.

Allen, Patricia E., Ph.D. in Physics, Iowa State University. Physis education pedagogy, surface physics.

Burris, Jennifer L., Ph.D. in Physics, Colorado State University. Raman spectroscopy, luminescence, high pressure spectroscopy.

Caton, Daniel B., Ph.D. in Astronomy, University of Florida. Computer applications to astronomical instrumentation; photoelectric photometry of eclipsing binary stars.

Cecile, Danny J., Ph.D. in Physics. Duke University. Quantum chromodynamics, lattice QCD modeling.

Clements, J. Sid, Ph.D. in Nuclear Physics, Florida State University. Experimental applied electrostatics (aerospace and industrial); electrical discharges; electronic instrumentation.

Coffey, Tonya S., Ph.D. in Physics, North Carolina State University. Nanotribology; tribology; ultra-high vacuum technology; microscopy; microanalysis.

Conrad, Brad R., Ph.D. in Physics. University of Maryland. Organic electronics, photovoltaics, nanoscience.

Daw, Adrian N., A.M., Ph.D. in Physics, Harvard University. Experimental atomic, molecular, and optical physics; laboratory astrophysics; solar physics.

Gray, Richard O., Ph.D. in Astronomy, University of Toronto. Stellar spectroscopy/photometry. Fellow, Royal Astronomical Society.

Hester, Brooke C., Ph.D. in Chemical Physics, University of Maryland. Optical trapping and extinction resonance.

Mamola, Karl C., Ph.D. in Physics, Dartmouth College. Spectroscopy, thin film physics. Editor, *The Physics Teacher* magazine.

Pollock, Joseph T., Ph.D. in Astronomy, University of Florida. Quasars; electronic imaging; asteroids.

Rokoske, Thomas L., Ph.D. in Solid State Physics, Auburn University. Electronics; microcomputer applications; robotics; electrical conduction mechanisms in thin films.

Russell, Phillip E., Ph.D. in Materials Science and Engineering, University of Florida. Nanoscience, ion and electron microscopy.

Saken, Jon M., Ph.D. in Astrophysics, University of Colorado. Physics and astronomy education; supernova remnants; ISM.

Sherman, Leah B., Ph.D. in Electrical Engineering, University of Michigan. Opto-electronics.

Sherman, James P., Ph.D. in Physics, Colorado State University. Optics, laser physics, and applications to environmental physics.

Thaxton, Christopher S., Ph.D. in Physics, North Carolina State University. Geophysics, computational physics, sediment transport and erosion studies, electronics, computer interfacing.

DUKE UNIVERSITY

DEPARTMENT OF PHYSICS

Durham, North Carolina 27708-0305

Students Accepted For Degree	FIELDS		
	Physics	Astronomy	Related Fields
Doctorate	X		
Master's			

1. General

President: Richard H. Brodhead
Dean of Graduate School: Jo Rae Wright
Department Chair: Daniel J. Gauthier
Department Telephone Number: (919) 660-2500
Departmental web page: http://www.phy.duke.edu
Type of Institution: University
Control: Private
Setting: Urban
Total Faculty: 2,178
Total Graduate Faculty: 1,280
Total Students: 13,723
Total Graduate Students: 7,275
Annual Graduate Tuition:
All Graduate Students: Full-time—$39,150
Tuition rates for: 2010–11
Deferred tuition plan: Yes
Other Fees: $40 (One time transcript charge)

2. Number of Faculty in Department

The combined total of full-time faculty (including tenure track and secondary appointments) in the three professorial ranks is 40. The combined total of full-time, part-time, and other faculty at all ranks is 61.

3. Admission, Financial Aid, and Housing

Application for admission: Information about admission including applications can be found online at http://www.gradschool.duke.edu
Graduate application fee required: $75 ($65 before Nov. 15 for fall admission; subject to change w/o notice)
Application deadline (Fall admission): 12/15
Admission information: For fall admission, 2010–11, 54 students were accepted from 194 applicants.
Admission requirements: For admission to the graduate programs, a Bachelor's degree in physics or related subject is required with no minimum undergraduate GPA specified. The GRE and the GRE subject is required. The average GRE scores for 2010–11 admissions were verbal–512; quantitative–786; total–1,298; analytic–3.6. The average GRE subject score for 2010–11 admissions was 850. Students from non-English speaking countries are required to demonstrate proficiency in English. Minimum acceptable score for TOEFL (internet based exam) is 83.
Undergraduate preparation assumed: Marion and Thornton, *Classical Dynamics of Particles and Systems*; Griffiths, *Introduction to Electrodynamics*; Kittel and Kroemer, *Thermal Physics*; Bernstein *et al.*, *Modern Physics*; Shankar, *Principles of Quantum Mechanics*.
Address financial aid inquiries to: Ms. Lisa Alfman, Financial Aid Officer, Graduate School Office, 03 Allen Bldg.
GAPSFAS application required: No
Financial aid deadline: All admitted students usually receive financial aid; deadline for acceptance —April 15.
Loans available: Yes
Address housing inquiries to: Dept. of Housing Management, 218 Alexander, Apt. B, Durham, NC 27708-0451.
On-campus, single student housing available: Yes
Cost/12 mos.: $6,150 per student (3 bdrms.) 3 students
$7.090 per student (2 bdrms.) 2 students
$8,950 per student (1 bdrm.) 1 student
On-campus, married student housing available: Yes

Table A—Faculty, Enrollments, and Degrees Granted

Research Specialty	2009–10 Faculty[1]	Enrollment[2] Fall 2009		No. of Degrees Granted[3] 2009–10 (2005–10)			Median No. of Years for 2009–10 Ph.D.'s
		Master's	Doctorate	Master's	Terminal Master's	Doctorate	
Atomic, Molecular, & Optical Physics	6	–	5	1(6)	0(2)	1(8)	6.5
Biophysics	6	–	8	1(4)	0(0)	0(4)	–
Free Electron Lasers	1	–	4	0(2)	0(1)	1(3)	6.0
High Energy Physics	8	–	4	0(2)	1(2)	0(3)	–
Medical Physics	2	–	1	0(0)	0(1)	0(1)	–
Nanophysics	8	–	10	0(1)	0(1)	2(8)	6.2
Nonlinear & Complex Systems	10	–	7	0(2)	0(1)	2(8)	5.7
Nuclear/Particles Theory	5	–	5	1(5)	0(0)	1(4)	6.0
Nuclear Experiment	7	–	12	0(9)	0(0)	3(5)	5.6
String Theory/ Mathematical Physics	3	–	0	0(1)	0(1)	0(1)	–
Non-specialized (Primary Teaching)	6	–	13	0(0)	0(0)	0(0)	–
Total		–	69	3(32)	1(9)	10(45)	
Full-time Grad. Stud.		–	69				
Part-time Grad. Stud.		–	0				
First-year Grad. Stud.		–	13				
Median Years in Grad. Study (2009–10 Degrees)				–	–	–	6.3
Undergraduate Degrees, 2009–10 (2005–10): 6(49)							

[1] Some faculty members are counted in more than one speciality.
[2] Students not yet committed to a research speciality are under non-specialized.
[3] Five-year totals in parenthesis (9/05 through 5/10).

4. Graduate Degree Requirements

Master of Arts (M.A.): Atleat 9 graduate courses (or sufficient placement exam performance), 3.0 GPA average. Oral final examination.
Master of Science (M.S.): Same as M.A. plus written thesis. Final examination on thesis.
Doctor of Philosophy (Ph.D.): Same as M.A. plus preliminary oral exam before dissertation work. Written dissertation. Final examination on dissertation.

Table B—Appointments to Graduate Students, 2009–10

Title of Appointee	Appointments Total	First year	Academic Load Allowed in Credit Hours	Hours of Service Per Week	Stipend for Academic Year ($)
			Semester		
Teaching Assistant	22	11	12	15	19,450[1,2]
Research Assistant	33	1	12	40	19,450[1,2]
Nanoscience Fellow[3]	1	0	–	–	–
Newson Fellow[3]	6	0	–	–	–
Fritz London Fellow[3]	0	0	–	–	–
SURA Fellow[3]	0	0	–	–	–
James B. Duke Fellow[3]	2	1	–	–	–
TPE Fellow[3]	4	0	–	–	–
Chamber Fellow[3]	1	0	–	–	–
Total	69	13			

[1] Plus tuition grant, plus registration and fees.
[2] Summer appointments including tuition remission for the summer also available.
[3] Many fellows receive a supplement, in addition to a TA or an RA stipend.

5. Personnel Engaged in Separately Budgeted Research, 7/09–6/10

Professorial faculty	38
Other faculty	5
Postdoctoral appointments	23
Graduate students	69
Undergraduate students	7
Nonteaching research personnel	34
Total	176

6. Separately Budgeted Research Expenditures by Source of Support 7/08-6/09

	Departmental Research
Federal government	$10,269,149
Private, nonprofit organizations	$115,918
Total	$10,385,067

Table C—Separately Budgeted Research Expenditures*

Research Specialty	No. of Grants	Expenditures ($)
Atomic, Molecular, & Optical Physics	8	964,934
Biophysics	5	385,577
Free Electron Laser laboratory	11	600,971
Geometry and Strings	1	63,127
High Energy Physics	16	1,671,022
Nanophysics	13	761,395
Nonlinear and Complex Systems	14	614,332
Nuclear & Particle Theory	6	666,776
Nuclear Experiment	19	4,656,933
Total	93	10,385,067

*Does not show expenditures of faculty with primary appointments in other departments.

FACULTY

Professors

Aspinwall, Paul, Ph.D., Oxford, 1988. String theory (Primary appointment: Mathematics).

Baranger, Harold U., Ph.D., Cornell, 1986. Theoretical condensed matter physics; nanophysics.

Behringer, Robert P., Ph.D., Duke, 1975. Experimental condensed matter physics; statistical physics; granular Materials; non-linear dynamics.

Beratan, David N., Ph.D., California Institute of Technology, 1985. Theoretical Chemistry; (Primary appointment: Chemistry).

Chang, Albert Mien-Fu, Ph.D., Princeton, 1983. Experimental condensed matter physics; nanophysics.

Edwards, Glenn S., Ph.D., Maryland, 1984. Biophysics; FEL applications research.

Gauthier, Daniel J., Ph.D., Rochester, 1989. Chair of the Department. Quantum information science; nonlinear and complex systems.

Goshaw, Alfred T., Ph.D., Wisconsin, 1966. Experimental elementary particle physics; instrumentation.

Greenside, Henry S., Ph.D., Princeton, 1981. Theoretical Neuroscience. Director of Undergraduate Studies.

Gao, Haiyan, Ph.D., California Institute of Technology, 1994. Experimental medium energy nuclear physics.

Han, Moo-Young, Ph.D., Rochester, 1963. Theoretical physics; elementary particle physics.

Howell, Calvin, Ph.D., Duke, 1984. Nuclear physics; few-nucleon systems.

Kotwal, Ashutosh V., Ph.D., Harvard, 1995. Experimental elementary particle physics; instrumentation.

Liu, Jian-Guo, Ph.D., UCLA, 1990. Computational physics, nonlinear and complez systems, fluid dynamics.

Johnson, Allan G., Ph.D., Duke, 1974. Imaging physics; magnetic resonance imaging (Primary Appointment: Radiology).

Mueller, Berndt, Ph.D., Frankfurt, 1973. Theoretical nuclear and particle physics.

Oh, Seog, Ph.D., MIT, 1981. Experimental elementary particle physics.

Palmer, Richard G., Ph.D., Cambridge, 1973. Theoretical condensed matter physics; complex systems. Director of Graduate Studies.

Petters, Arlie O., Ph.D., MIT, 1991. General relativity and cosmology; gravitational lensing (primary appointment: mathematics).

Samei, Ehsan, Ph.D., Michigan, 1997. Medical Imaging (Primary Appointment; Radiology).

Smith, David R., Ph.D., UCSD 1994. Quantum Optics; photonic cyrstals; metamaterials. (Primary Appointment: Electrical and Computer Engineering).

Thomas, John E., Ph.D., MIT, 1979. Experimental quantum optics; atomic and molecular collision physics.

Tornow, Werner, Ph.D., Universität zu Tübingen, 1974. Experimental nuclear physics; neutrino physics.

Associate Professors

Bass, Steffen, Ph.D., J. W. Goethe University, 1997. Theoretical nuclear and particle physics; relativistic heavy-ion collisions.

Chandrasekharan, Shailesh, Ph.D., Columbia, 1995. Theoretical nuclear and particle physics; lattice field theory.

Curtarolo, Stefano, Ph.D., MIT 2003. Solid state physics; thermodynamics of materials; computational materials science (Primary Appointment: Mechanical Engineering)

Finkelstein, Gleb, Ph.D., Weizmann Institute of Science, 1998. Experimental condensed matter physics; nanophysics.
Kruse, Mark, Ph.D., Purdue, 1996. Experimental elementary particle physics.
Mehen, Thomas, Ph.D., Johns Hopkins, 1998. Theoretical nuclear and particle physics; effective field theory.
Plesser, M. Ronen, Ph.D., Harvard, 1991. String theory; supersymmetry.
Scholberg, Kate, Ph.D., California Institute of Technology, 1997. Experimental elementary particle physics; neutrino physics.
Socolar, Joshua E. S., Ph.D., University of Pennsylvania, 1987. Theoretical condensed matter physics; nonlinear systems; regulatory networks.
Springer, Roxanne P., Ph.D., California Institute of Technology, 1990. Theoretical nuclear and particle physics.
Teitsworth, Stephen W., Ph.D., Harvard, 1986. Experimental condensed matter physics. Associate Chair.
Wu, Ying, Ph.D., Duke, 1995. Free electron laser physics, beam physics.

Assistant Professors

Arce, Ayana Tamu Holloway, Ph.D., Harvard, 2006. Experimental high energy physics.
Buchler, Nicolas Emile, Ph.D., University of Michigan 2001. Biophysics, non-linear dynamics and complex systems. (Joint Appointment with Biology).
Charbonneau, Patrick, Ph.D., Harvard, 2006. Theoretical condensed matter physics, chemical physics. (Primary Appointment: Chemistry).
Walter, Christopher, Ph.D., California Institute of Technology, 1997. Experimental elementary particle physics; neutrino physics.
Yasuda, Ryohei, Ph.D., Keio University, 1998. Biophysics. (Primary Appointment Neurobiology).

Associate Research Professor

Phillips, Thomas J., Ph.D., Harvard, 1986. Experimental elementary particle physics; instrumentation.

Assistant Research Professors

Ahmed, Mohammad W., Ph.D., University of Houston, 1999. Experimental nuclear physics.
Tonchev, Anton, Ph.D., Joint Institute for Nuclear Research, Russia, 1995. Experimental nuclear physics.

Adjunct Professors

Ciftan, Mikael, Ph.D., Duke, 1968. Theoretical physics; solid state theory; statistical thermodynamics.
Everitt, Henry, Ph.D., Duke, 1990. Experimental condensed matter physics; molecular physics; quantum optoelectronics.
Guenther, Robert D. Ph.D., Missouri, 1968. Applied science; tera-hertz optics.
Lawson, Dewey T., Ph.D., Duke, 1972. Acoustics.
Skatrud, David D., Ph.D., Duke, 1984. Molecular collisions; millimeter and submillimeter spectroscopy.
West, Bruce, Ph.D., Rochester, 1970. Biophysics.

Adjunct Assistant Professor

Daniels, Karen E., Ph.D., Cornell University 2002. Experimental non-linear dynamics.

Dutta, Dipangkar, Ph.D., Northwestern, 1999. Experimental medium energy nuclear physics.

Visiting Professor

Hastings, Matthew B., Ph.D., MIT, 1997. Theoretical condensed matter physics.

Lecturer

Brown, Robert, G., Ph.D., Duke, 1982. Theoretical physics; statistical physics.

Professors Emeriti

Bilpuch, Edward G., Ph.D., North Carolina, Chapel Hill, 1956. Experimental nuclear physics.
Evans, Lawrence E., Ph.D., Johns Hopkins, 1960. Theoretical elementary particle physics.
Fairbank, Henry A., Ph.D., Yale, 1944. Experimental low temperature and solid state physics.
Meyer, Horst, Ph.D., Geneva, 1953. Experimental low-temperature and solid state physics.
Roberson, Russell N., Ph.D., Johns Hopkins, 1960. Experimental nuclear physics.
Robinson, Hugh G., Ph.D., Duke, 1954. Atomic and molecular physics.
Walter, Richard L., Ph.D., Notre Dame, 1959. Experimental nuclear physics.
Weller, Henry R., Ph.D., Duke, 1967. Experimental nuclear physics; nuclear structure; gamma-ray studies.

RESEARCH SPECIALTIES AND STAFF

Theoretical

Biophysics: Generation, storage and learning of temporal sequences in songbirds in terms of experimentally known anatomy, physiology, and connectivity of neurons in a songbird's brain. Structure and function of complex dynamical networks, especially genetic regulatory networks. Information transfer between complex networks. Buchler, Greenside, Socolar.
Computational Physics. Numerical techniques for solving nonlinear partial differential equations; Monte Carlo algorithms in field theory and statistical mechanics; molecular dynamics; networks; large scale computations on vector and parallel computers. Computational methods in fluid dynamics, material sciences, plasma physics, and geophysical flow; Emergent behavior in flocking and swarming; Numerical analysis and scientific computing. Baranger, Bass, Brown, Chandrasekharan, Charbonneau, Greenside, Liu, Palmer.
Geometry and Strings: String theory; geometry of space-time, supersymmetry and duality; mirror symmetry, general relativity. Aspinwall, Petters, Plesser.
Nanophysics: Coherence and correlations in nanoscale systems like quantum dots and carbon nanotubes; coulomb blockade; quantum impurity effects; quantum phase transitions; quantum computing quantum entabglement; quantum information; thermodynamics of materials. Baranger, Beratan, Chandrasekharan, Curtarolo, Hastings.
Nonlinear and Complex Systems: Analytical and computational studies of nonlinear and biological systems including genetic networks, heart and brain dynamics; collective behavior in matter and dynamical systems; spin glasses and glasses; adaptive algorithms; static and dynamic critical behavior in

optics and magnetism; granular materials network dynamics; fractal growth; granular matter; in- and out-of-equilibrium dynamical properties of materials self-assembly; micorphase formation; protein aggregation; glass and gel formation. Brown, Buchler, Charbonneau, Greenside, Hastings, Palmer, Socolar.

Nuclear and Particle Physics: Quantum chromodynamics and weak interactions; heavy quark physics; quark-gluon plasma; heavy-ion collisions; effective field theories of nuclear interactions; lattice field theories and Monte Carlo simulations; chaos in classical field theory. Bass, Chandrasekharan, Mehen, Mueller, Springer.

Experimental

Atomic, Molecular, and Optical Physics: Optical traps; atom cooling; strongly interacting Fermi gases; degenerate quantum gasses; Bose-Einstein condensation of molecules; slow, fast and stored light; quantum optics; single photon switching, quantum information; optical noise, optoelectronics; new technologies for optical communication. Electromagnetic properties of materials, photonic crystals and metamaterials. Everitt, Gauthier, Guenther, Skatrud, Smith, Thomas.

Biophysics: Emergent properties and tissue dynamics; fast thermodynamics in laser-tissue interactions; applications of free-electron lasers to biology and medicine; Characterization and control of heart dynamics; stochastic processes in biological systems; optical analysis of molecular dynamics in single synapses; optical stimulation of single synapses; development of high resolution imaging techiques; evolution of bistable and oscillatory dynamics in gene networks. Buchler, Edwards, West, Yasuda.

Free Electron Lasers: Beam physics; FEL and novel light source development; high intensity gamma ray source; FEL applications. Wu.

High Energy Physics: Precision tests of the Standard Model using the top quark, W and Z boson; searches for the Higgs boson, compositeness, new fundamental symmetries and dimensions; tests of the QCD hadron production models; studies of neutrino properties; searches for proton decay; neutrino astrophysics; research program based at Fermilab, CERN and in Japan; state-of-the-art wire chamber and silicon detector development and construction; electronics design for high energy physics experiments. Arce, Goshaw, Kotwal, Kruse, Oh, Phillips, Scholberg, Walter.

Medical Physics: Biomedical imaging; magnetic resonance imaging; magnetic resonance microscopy, x-ray microscopy, tomography and microPET. Johnson, Samei.

Nanophysics: Electronic properties of carbon nanotubes, nano crystals, semiconductor quantum dots and self-assembled DNA structures; physics of Luttinger liquids; scanning tunneling; capacitance and atomic force microscopy; optoelectronic processes in semiconductor microstructures; sub picosecond optical characterization of nanostructures; nanometer-scale photonic, plasmonic and phononic band engineering. Chang, Finkelstein, Teitsworth.

Nonlinear and Complex Systems: Granular materials; dynamics of granular flow; quantum chaos in classical wave systems; chaotic networks; pattern formation and spatiotemporal chaos in fluids far from equilibrium; Rayleigh-Bernard convection. Behringer, Daniels, Gauthier.

Nuclear Physics: QCD and weak interactions in nuclear physics; nucleon structure and nucleon-nucleon interactions; physics of few nucleon systems; electromagnetic nuclear physics; radiative capture reactions using polarized proton and deuteron beams; testing QCD with high intensity gamma-ray source; fundamental symmetry studies with ultra-cold neutrons and the search for neutron electric dipole moment; neutrino oscillations using detectors at KAMLAND; double beta-decay; nuclear astrophysics. Ahmed, Dutta, Gao, Howell, Tonchev, Tornow.

Quantum Information Science: High data rate quantum key distribution; high brightness hyper-entangled sources; multimode quantum communication; multi-element photon counting detector development. Gauthier.

EAST CAROLINA UNIVERSITY

DEPARTMENT OF PHYSICS

Greenville, North Carolina 27858

Students Accepted For Degree	FIELDS		
	Physics	Astronomy	Related Fields
Doctorate	X		X
Master's	X		X

1. General

Chancellor: Steven C. Ballard
Dean of Graduate School: Paul Gemperline
Department Chairman: John C. Sutherland
Department Telephone Number: (252) 328-6739
Type of Institution: University
Control: Public
Setting: Small town
Total Faculty: 1,442 (including medical school)
Total Graduate Faculty: 827
Total Students: 25,990
Total Graduate Students: 5,932
Annual Graduate Tuition:
In-state residents: $3,195
Out-of-state residents: $13,511
Tuition rates for: 2010–11
Deferred tuition plan: Yes
Other fees: $1,916
Term: Semester

2. Number of Faculty in Department

The combined total of full-time faculty in the three professorial ranks is 18. The combined total of full-time, part-time and other faculty at all ranks is 24.

3. Admission, Financial Aid, and Housing

Address admission inquiries to: Graduate School, East Carolina University, Greenville, North Carolina 27858-4353
Graduate application fee required: $60
Admission deadline (Fall admission): 02/15
Admission information: For fall admission, 2009–10, 17 students were accepted from 65.
Admission requirements: For admission to the graduate programs, a Bachelor's degree in physics or related subjects is required with minimum undergraduate GPA of 2.7 for MS, 3.0 for PhD specified. The GRE is required. Scores are used primarily as a guideline. The GRE Advanced is recommended. The average GRE scores for admission were verbal–482; quantitative–714; total–1196. Students from non-English speaking countries are required to demonstrate profiency in English via the TOEFL exam. Minimum acceptable score for admission is 550.
Undergraduate preparation assumed: Symon, *Mechanics*; Griffiths, *Introduction to Electrodynamics*; Sears and Salinger, *Thermodynamics*; Beiser, *Perspective of Modern Physics*; ONeil, *Advanced Engineering Mathematics*; two semesters of advanced laboratory courses.
Address financial aid inquiries to: Financial Aid Officer, Graduate School, 131 Ragsdale, East Carolina University, Greenville, North Carolina 27858-4353
GAPSFAS application required: No
Financial aid deadline: 2/28
Loans available: Yes
Address housing inquiries to: University Housing Service, Jones Residence Hall, East Carolina University, Greenville, North Carolina 27858-4353
On-campus, single student housing available: Yes
Cost/term: $2,175/semester
On-campus, married student housing available: No

Table A—Faculty, Enrollments, and Degrees Granted

Research Specialty	2007–08 Faculty	Enrollment[1] Fall 2007		No. of Degrees Granted[2] 2009–10 (2005–10)			Median No. of Years for 2007–08 Ph.D.'s
		Master's	Doctorate	Master's	Terminal Master's	Doctorate	
Acoustics	2	0	–	0(3)	–	–	–
Atomic, Collision Physics	5	3	–	0(4)	3	–	1
Biomedical Physics	1	–	20	0(5)	–	–	
Computational Physics	3	0	–	0(1)	–	–	–
Medical Physics	3	15	–	6(42)	9	–	–
Laser & Optical Physics	2	0	–	0(2)	–	–	–
Solar Physics	1	0	–	0(0)	–	–	–
Health Physics							
Total		18	23	6(57)			
Full-time Grad. Stud.		18	23				
Part-time Grad. Stud.		0	2				
First-year Grad. Stud.		9	6				
Median Years in Grad. Study (2007–08 Degrees)				1	–	–	–

Undergraduate Degrees, 2007–08 (2003–08): 6(30)

[1]Students not yet committed to a research specialty are entered under non-specialized.
[2]Five-year totals in parentheses.

4. Graduate Degree Requirements

Master's: A minimum of 34 semester hours are required for the Applied Physics (AP) option, a minimum of 38 semester hours are required for the Health Physics (HP) option, and a minimum of 39 semester hours are required for the Medical Physics (MP) option. Major Field Test administered upon entrance into program. For AP option, candidates must write and defend a thesis based on original research. Thesis is not required for MP students.
Doctorate: Minimum 30 semester hours beyond the master; a Master's degree in physics or related areas is preferred; students entering with a BS in physics or related areas will follow the AP Master's curriculum; doctoral written and oral exams covering biomedical physics curriculum; thesis required.
Thesis: Thesis may be written *in absentia*.
Special Equipment, Facilities, or Programs: The department has extensive research laboratories and operates an electronic shop and machine shop. The laboratories include Ion Physics including a 2MV Accelerator, Acoustic Laboratory, Biophysics Laboratory, Biomedical Laser Lab and Medical Imaging

Lab. 32 processor computer cluster. A PC cluster of a total of 32 processors that consists of 16 processors of Dell PowerEdge 1750 with Intel Xeon 3.06GHz CPU. Each processor is equipped with 1 ~ 2 GB memory and one gigabit Ethernet card, and the processors are connected to two 24-port gigabit switches. The operating system is a version of Linux by Gentoo with a Network File System (NFS). A freely available, portable implementation of MPI (message passing interface) library, MPICH, has been installed on the cluster to support parallel computing. Compilers for C, C^{++} and Fortran 90 from the Portland Group and Intel have been installed on the cluster.

Molecular structure characterization lab–Biophysical instrumentation for biomolecular structure characterization using Circular Dichroism Spectroscopy and Linear Dichroism Spectroscopy. Instruments include two Jasco 810 spectrometers (one set for linear dichroism and fluorescence (LD/CD) and the other with stop flow and titration experiments) as well as a Cary UV/VIS Photospectrometer and a MALDI-TOF Mass Spectrometer (undergoing refurbishment). Research projects include spider silk (in collaboration with Biology, ECU and Zoology, Oxford), Prion protein copper binding (Chemistry, ECU), beta-peptide self assembly (Chemistry, ECU) and amyloid A-Beta disease protein intermediates (Chemistry, ECU).

Table B—Appointments to Graduate Students, 2007–08

Title of Appointee	Appointments Total	Appointments First year	Academic Load Allowed in Credit Hours	Hours of Service Per Week	Stipend for Academic Year ($)
			Semester		
Teaching Assistant	0(7)	2(5)	12(12)	15–20	10,000[a]-MS
Research Assistant	0(4)	3(0)	12(12)	15–20	23,520[b]-Ph.D.
Total	22	10			

[a]9 month stipend-additional 2 month support for summer research is usually available.
[b]12 month stipend-additionally tuition remission and health insurance are provided.

5. Personnel Engaged in Separately Budgeted Research, 7/07–6/08

Professorial	21
Total	21

6. Separately Budgeted Research Expenditures by Source of Support

	Departmental Research	Physics-related Research Outside Department
Federal government	$709,392	$
State and local government	65,850	
Private nonprofit organizations	25,000	
Total	$800,242	$

Table C—Separately Budgeted Research Expenditures

Research Specialty	No. of Grants	Expenditures ($)
Biomedical Physics	5	315,000
Total	5	315,000

FACULTY

Professors

Hu, Xin-Hua, Ph.D., California-Irvine, 1991. Biomedical physics.
Joyce, James M., Ph.D., Pennsylvania, 1967. Biomedical physics.
Kempf, Ruth, Ph.D., Rensselaer Polytechnic Institute.
Lapicki, Gregory, Ph.D., NYU, 1975. Theoretical atomic physics.
Seykora, Edward, Ph.D., North Carolina State, 1968. Solar physics.
Shinpaugh, Jefferson L., Ph.D., Kansas State, 1990. Experimental atomic physics. .
Sutherland, John C., Ph.D., Georgia Institute of Technology, 1967. Biophysics.

Associate Professors

Bier, Martin, Ph.D., Clarkson U. 1990. Mathematics, modeling, computational physics.
Day, Orville W., Jr., Ph.D., Brigham Young, 1973. Quantum mechanics of atoms and molecules.
Dingfelder, Michael, Ph.D., Eberhard-Karls University, Germany, 1995. Theoretical physics and radiation modeling.
Justiniano, Edson L.B., Ph.D., Kansas State, 1982. Experimental atomic physics.
Kenney, John M., Ph.D., University of New York at Stony Brook, 1985. Fibril amyloid-like structures.
Li, Yong-qing. Ph.D. Academia Sinica, 1989. Experimental optical physics and biomedical physics.
Lu, Jun Qing, Ph.D., California-Irvine, 1991. Theoretical condensed matter and biomedical physics.
Sprague, Mark W., Ph.D., Mississippi, 1994. Acoustics.

Assistant Professor

Lin, Zi Wei, Ph.D., Columbia University, 1996. Theoretical physics and radiation modeling.

Adjunct Faculty

Payne, Marvin, Ph.D., Univ. of Kentucky, 1965.
Sabelnikov, Alex, Ph.D.
Wolfe, Melodee, M.S., East Carolina U., 1992. Clinical medical physics.

Faculty Emeriti

Adler, Carl G., PhD
Bissinger, George, PhD
Coulter, Byron, PhD
McEnally, Terence E., PhD
Sayetta, Thomas C., PhD
Varlashkin, Paul, PhD
Dinno, Mumtaz A., PhD
Toburen, Larry H., PhD

RESEARCH SPECIALTIES AND STAFF

Theoretical

Atomic Physics. Atomic collisions; penetration of charged particles through matter; innershell ionization and stopping power . Dingfelder, Lapicki.

Computational Physics. Quantum mechanics of atoms and molecules; density functional theory; light scattering from random media; characterization of inhomogeneity in tissue using noncontact laser speckle techniques. Monte Carlo modeling of charged particle track structure. Day, Dingfelder, Flurchick, Lu.

Experimental

Acoustic Physics. Computer simulation and experimental measurement of acoustic radiation from violin using mode analysis; atmospheric and underwater acoustics; computational studies of convective effects on sound propagation; acoustic characterization of Sciaenid fish calls. Bissinger, Sprague.

Astrophysics/Solar Physics. High resolution optical imaging of the solar disc through the earth's atmosphere. Seykora.

Atomic Physics. Experimental atomic physics; ion-atom collisions; charge transfer; recombination and excitation in electron-ion collisions; laser assisted electron-ion and ion-atom collisions; application of atomic physics to biological physics and trace element analysis; measurement of atomic physics to biological physics; measurement of cross sections for collisions involving ions and neutral particles and application of these data in the study of charged particle track structure. Justiniano, Shinpaugh, Toburen.

Biomedical Physics. Effects of physical agents such as temperature, electric fields, and ultrasound on biological systems and interactions between ultrasonic devices and biological tissues; interaction between short laser pulses and biological tissues and cells; coherent imaging of rough surfaces and random medium; electrical activity of the stomach; biological physics of membranes; electron transport in *tissue like* material; development of medical devices. Bier, Day, Dingfelder, Hu, Joyce, Justiniano, Kempf, Kenney, Lapicki, Li, Lin, Lu, Seykora, Shinpaugh, Sibata, Sprague, Toburen.

NORTH CAROLINA STATE UNIVERSITY, RALEIGH

DEPARTMENT OF PHYSICS

Raleigh, North Carolina 27695-8202

Students Accepted For Degree	FIELDS		
	Physics	Astronomy	Related Fields
Doctorate	X		
Master's	X		

1. General

Chancellor: Randy Woodson
Dean of Graduate School: Duane K. Larick
Department Chairman: M. A. Paesler
Department Telephone Number: (919) 515-2521
Type of Institution: University
Control: Public
Setting: Urban
Total Faculty: 2,078
Total Graduate Faculty: 2,500
Total Students: 33,815
Total Graduate Students: 7,991
Annual Graduate Tuition:
In-state residents: Full-time—$6,295
Part-time—prorated
Out-of-state residents: Full-time—$18,343
Part-time—prorated
Tuition rates for: 2010–11
Deferred tuition plan: Yes
Term: Semester

2. Number of Faculty in Department

The combined total of full-time faculty in the three professorial ranks is 49. The combined total of full-time, part-time, and other faculty at all ranks is 52.

3. Admission, Financial Aid, and Housing

Address admission inquiries to: Prof. Harald Ade, Physics Department, Box 8202
Web address: http://www.physics.ncsu.edu
Graduate application fee required: U.S. $65, Int'l $75
Admission target date (Fall admission): U.S. 1/20, Int'l 1/20
Admission information: For fall admission, 2009–10, 67 students were accepted from 139 applicants.
Admission requirements: For admission to the graduate programs, a Bachelor's degree in physics is required with a minimum undergraduate GPA of 3.0/4.0 specified. The GRE General and Physics subject tests are required. No minimum acceptable score is specified. Students from non-English speaking countries are required to demonstrate proficiency in English via the TOEFL exam. Minimum acceptable score for admission is the following: Listening-20; Reading-20; Writing-20, Speaking-20. For a teaching assistantship, a minimum of 23 on the speaking portion is required.
Undergraduate preparation assumed: Griffiths, *Electromagnetic Fields and Waves*; Hand and Finch, *Analytical Mechanics*; Gasiorowicz, *Quantum Physics*. (or equivalent)
Address financial aid inquiries to: Financial Aid Office, 2005 Harris Hall, Box 7302
GAPSFAS application required: No
Financial aid deadline: 3/15
Address housing inquiries to: Housing Assignments Office, 1112 Pullen Hall, Box 7315
On-campus, single student housing available: Yes
Cost/semester.: $2,600
On-campus, married student housing available: Yes
Cost/month: $560–930

Table A—Faculty, Enrollments, and Degrees Granted

Research Specialty	2009–10 Faculty	Enrollment[1] Fall 2009 Master's	Enrollment[1] Fall 2009 Doctorate	No. of Degrees Granted[2] 2009–10 (2005–10) Master's	Terminal Master's	Doctorate	Median No. of Years for 2009–10 Ph.D.'s
Astrophysics	8	2	2	0(1)	0(0)	1(4)	–
Atomic, Molecular, & Optical Physics	3	0	11	0(1)	0(1)	1(2)	–
Biophysics/Soft Matter	4	1	13	2(3)	0(1)	1(3)	–
Nano-Science Materials	16	2	22	1(6)	2(6)	7(23)	6
Nuclear Physics	12	2	20	1(1)	0(0)	3(7)	–
Physics Education	5	0	7	0(1)	0(0)	2(4)	6
Other Theoretical/ Math.	1	1	31	4(8)	0(2)	1(12)	–
Total	49	8	106	8(21)	2(10)	16(55)	
Full-time Grad. Stud.		8	105				
Part-time Grad. Stud.		0	1				
First-year Grad. Stud.		2	16				
Median Years in Grad. Study (2008–09 Degrees)				2	2	6	–
Undergraduate Degrees, 2009–10 (2005–10): 18(118)							

[1]Students not yet committed to a research specialty are entered under non-specialized.
[2]Five-year totals in parentheses.

4. Graduate Degree Requirements

Master's: 30 semester-hours; 3.0/4.0 overall GPA; two semesters residence; no foreign language; no computer language; Option A-comprehensive oral exam and thesis required, or Option B-comprehensive written exam required.
Doctorate: Six semesters beyond the baccalaureate with a 3.0/4.0 overall GPA; no computer language; comprehensive written and oral exams; thesis required.
Thesis: Thesis may be written *in absentia*.
Special Equipment, Facilities, or Programs: The majority of the department is located in 55,000 sq. ft. of the recently renovated Riddick Hall. It houses a number of modern laboratories and teaching facilities, including several clean rooms and shared experimental user facilities. The contiguous NC State Centennial Campus houses several departmental nanoscience/materials laboratories in several buildings. Major facilities for nuclear physics research are provided by the Triangle Universities Nuclear Laboratory. The optical physics and nanoscience/materials laboratories are well-equipped with various laser systems, spectrometers, electron microscopes, materials preparation systems, and extensive data acquisition equipment. In addition, collaboration exists with other departments and with various industrial and governmental laboratories.

The computer facilities are state-of-the-art. At the department and research group level, powerful graphics workstations serve as graphics and communications nodes. The University ranks highly in high performance computing (HPC) and has been awarded a very high bandwidth (Internet2) connectivity to National Supercomputing Centers and other nationally prominent Research Universities.

Table B—Appointments to Graduate Students, 2009–10

Title of Appointee	Appointments Total	Appointments First year	Academic Load Allowed in Credit Hours	Hours of Service Per Week	Stipend for Academic Year ($)
			Semester		
Teaching Assistant	51	14	9	15–20	16,812[1]
Research Assistant	56	1	9	15–20	21,000–25,000[2]
Fellowships	4	1	12	–	30,000
Total	111	16			

[1]Academic year (9 months).
[2]Calendar year (12 months).

5. Personnel Engaged in Separately Budgeted Research, 7/09–6/10

Professorial faculty	50
Other faculty	3
Postdoctoral appointments	15
Graduate Student appointments	72
Undergraduate students	12
Total	152

6. Separately Budgeted Research Expenditures by Source of Support

	Departmental Research
Federal government	$6,761,328
State and local government	972,599
Miscellaneous	1,321,875
Total	$9,055,803

Table C—Separately Budgeted Research Expenditures

Research Specialty	No. of Grants	Expenditures ($)
Astrophysics	19	961,307
Atomic, & Optical Physics, and Synchrotron Radiation	7	476,721
Biophysics	19	1,025,700
Nano-Science/Materials	34	2,563,200
Computational Materials	16	1,159,633
Nuclear & Particle Physics	16	1,815,382
Physics Education	3	282,281
State & Non-Private Miscellaneous	10	921,579
Total	124	9,055,803

FACULTY

Professors

Ade, Harald, Ph.D., SUNY Stony Brook, 1990. Experimental physics; nano-science/materials.

Aspnes, David E., Ph.D., Illinois, 1965. Experimental physics; optical physics/materials.

Beichner, Robert J., Ph.D., SUNY Buffalo, 1989. Physics education research.

Bernholc, Jerzy, Ph.D., Lund, 1977. Theoretical physics; nanoscience/materials.

Blondin, John M., Ph.D., Chicago, 1987. Theoretical physics; astrophysics.

Brown, J. David, Ph.D., Texas, 1985. Theoretical physics; astrophysics and relativity.

Cotanch, Stephen R., Ph.D., Florida State, 1973. Theoretical physics; nuclear physics.

Ellison, Donald C., Ph.D., Catholic, 1982. Theoretical physics; astrophysics.

Fornes, Raymond E., Ph.D., North Carolina State, 1970. Experimental physics; polymer physics.

Golub, Robert, Ph.D., Massachusetts Institute of Technology, 1968. Experimental physics; ultra-cold neutrons.

Gould, Christopher R., Ph.D., Pennsylvania, 1969. Experimental physics; nuclear physics.

Haase, David G., Ph.D., Duke, 1975. Experimental physics; solid state physics; low-temperature physics.

Hallen, Hans, Ph.D., Cornell, 1991. Experimental physics; optics.

Huffman, Paul, Ph.D., Duke, 1995. Experimental physics; low temperature and nuclear physics.

Ji, Chueng, Ph.D., KAIST, Korea, 1982. Theoretical physics; nuclear physics.

Krim, Jacqueline, Ph.D., Washington, 1984. Experimental physics; nano-science/materials.

Lucovsky, Gerald, Ph.D., Temple, 1960. Experimental physics; nano-science/materials.

McLaughlin, Gail, Ph.D., San Diego, 1996. Theoretical physics; nuclear physics; astrophysics.

Mitas, Lubos, Ph.D., Slovak Acad. Sci., 1989. Theoretical physics; nano-science/materials.

Mowat, J. Richard, Ph.D., California, Berkeley, 1969. Experimental physics; atomic physics.

Paesler, Michael A., Ph.D., Chicago, 1975. Head of the Department. Experimental physics; nano-science/materials.

Reynolds, Stephen P., Ph.D., California, Berkeley, 1980. Theoretical physics; astrophysics.

Risley, John S., Ph.D., Washington, 1973. Physics education research.

Roland, Christopher M., Ph.D., McGill, 1989. Theoretical physics; nano-science/materials.

Sagui, Celeste, Ph.D., Toronto, 1995. Theoretical physics; biophysics.

Schaefer, Thomas, Ph.D., Regensburg, 1992. Theoretical nuclear/particle physics.

Young, Albert, Ph.D., Harvard, 1995. Experimental physics; atomic and nuclear physics.

Associate Professors

Buongiorno Nardelli, Marco, Ph.D., Trieste, 1993. Theoretical physics; nano-science/materials.

Clarke, Laura, Ph.D., Oregon, 1998. Experimental physics; nano-science/materials; molecular physics.

Lee, Dean, Ph.D., Harvard, 1998. Theoretical physics; nuclear physics.

Weninger, Keith, Ph.D., UCLA, 1997. Experimental physics; biophysics.

Assistant Professors

Daniels, Karen E., Ph.D., Cornell, 2002. Experimental physics; non-equilibrium granular and fluid systems.

Dougherty, Daniel, Ph.D., University of Maryland, 2004. Experimental physics; nanoscale science and technology.

Gundogdu, Kenan, Ph.D., University of Iowa, 2004. Experimental physis; nanoscale science and technology.

Lazzati, Davide, Ph.D., Universita degli Studi, Milan, 2001. Theoretical physics; astrophysics.

Riehn, Robert, Ph.D., Cambridge, 2003. Experimental physics; biophysics.

Teaching Associate Professor

Pearl, Thomas P., Chicago, 2000. Experimental physics; nanoscience/materials.

Teaching Assistant Professors

Frohlich, Carla, Ph.D., Univeristy of Basel, 2007. Theoretical Physics, nuclear physics, astrophysics.

Kneller, James, Ph.D., Ohio State University, 2001. Theoretical physics, nuclear physics, astrophysics.

Kohlmyer, Matt, Ph.D., Carnegie Mellon University, 2005. Physics education research.

Research Professors

Fortner, Brand, Ph.D., Illinois, 1993. Theoretical physics; astrophysics.

Rowe, Jack, Ph.D., Brown, 1971. Experimental physics: nanoscience/materials.

York, Jimmy, Ph.D., NC State University, 1966. Theoretical physics; other.

Research Associate Professors

Borkowski, Kazimirez, Ph.D., Colorado, 1988. Experimental physics; astrophysics.

Kelley, John H., Ph.D., Michigan State, 1995. Experimental physics; nuclear physics.

Lu, Wenchang, Ph.D., Fudan Univ., 1994. Theoretical physics, nano-science/materials.

Research Assistant Professors

Back, Henning, Ph.D., Virginia Tech, 2004. Experimental physics: nuclear physics.

Bochinski, Jason, Ph.D., Oregon, 2000. Molecular physics; nano-science/materials.

RESEARCH SPECIALTIES AND STAFF

Theoretical/Computational

Astrophysics and Relativity. Theoretical modeling and numerical simulations of supernovae and remnants, gamma-ray bursts, gravitational radiation from supernovae and colliding black holes, accretion onto compact objects, and planetary nebulae; shock waves, particle acceleration, and cosmic rays; neutrinos and nucleosynthesis in supernovae and gamma-ray bursts. Blondin, Borkowski, Brown, Ellison, Lazzati, McLaughlin, Reynolds.

Nanoscience/Materials and Biomolecular Simulations. Large scale simulations of real materials, bio-molecular processes, semiconductors, nanotubes and related nanoscale structures; quantum Monte Carlo simulations; O(N) and multiscale methods; quantum transport; nanostructured materials; phase separation; ferrofluids liquid-state theory; interfaces; diffusion; neural networks; pattern formation; electronic properties of transition-metal oxides and silicates. Bernholc, Buongiorno-Nardelli, Mitas, Roland, Sagui.

Nuclear/Particle Physics. Electromagnetic structure studies of hadrons; relativistic quark models; light-cone quantization; B-physics; glueball and hybrid meson spectroscopy; application to astrophysics and cosmology; neutrino phenomenology; nonperturbative vacuum effects and mixing; CP violation; extra dimensions and physics beyond the standard model; QCD-based description of hadronic interactions; BCS methods and chiral symmetry breaking; Hartree-Fock techniques; lattice gauge theory; many body phenomena and computational algorithms, instantons, finite density QCD, superconductivity; nuclear lattice simulations; effective field theory. Cotanch, Ji, Lee, McLaughlin, Schaefer.

Experimental

Astronomy. Radio, infrared, optical, and x-ray observations of supernova remnants, pulsar-wind nebulae, and planetary nebulae. Borkowski, Reynolds.

Atomic Physics. Laser polarization of atomic vapors, atom trapping. Young.

Biophysics and Soft-condensed Matter. Dynamics of polymers on surfaces and in constrained thin films; nanoprobe tools for signal pathway investigations in cellular biology; single molecule techniques for dynamic and structural studies of proteins and cells; interfacial instabilities; thin film flows; statistical mechanics of granular materials. Ade, Daniels, Hallen, Riehn, Weninger.

Nuclear Physics. Tests of fundamental symmetries; neutron beta decay and electric dipole moments; neutrinos and neutrino oscillations; ultracold neutrons; studies of few nucleon systems; quantum chaos; statistical properties of nuclei; polarized nuclear targets. Golub, Gould, Haase, Huffman, Kelley, Young.

Optics. Near-field optical microscopy and spectroscopy; nano-Raman spectroscopy; optical characterization of electronic materials; linear and non-linear optical spectroscopy of materials, films, surfaces and interfaces under static and dynamic conditions. Aspnes, Gundogdu, Hallen, Paesler.

Physics Education. Reexamination and redesign of modes of instruction and content for large enrollment courses; assessment of student understanding; role of computers including simulation, visualization, computer-based experiments, and student programming; distance learning. Beichner, Kohlmyer, Risley.

Materials Physics and Nanoscale Science and Technology. Nanotribology, micro/nano electromechanical systems, and liquid wetting phenomena; real time spectroscopy and microscopy of nanostructure growth and dynamics; subwavelength optical probes; correlations of eletronic, optical and materials properties of interfaces, nanostructures and nanodevices; atomic level materials preparation and characterization of ultra thin films and related device structures; remote plasma enhanced chemical vapor deposition of semiconductors and insulators; x-ray spectromicroscopy of soft nanomaterials, molecular motion on surfaces; organic dielectrics; conduction and polarization of molecular and macromolecular assemblies; sound propagation in granular materials. Ade, Aspnes, Clarke, Daniels, Dougherty, Gundogdu, Hallen, Krim, Lucovsky, Pearl.

Synchrotron Radiation Research. Microscopy and spectroscopy of complex systems. Ade, Lucovsky, Paesler.

UNIVERSITY OF NORTH CAROLINA, CHAPEL HILL

DEPARTMENT OF PHYSICS AND ASTRONOMY

Chapel Hill, North Carolina 27599-3255

Students Accepted For Degree	FIELDS		
	Physics	Astronomy	Related Fields
Doctorate	X	*	
Master's	X	*	

*Concentration of studies in Astronomy, but degree granted in Physics.

1. General

Chancellor: Holden Thorp
Dean of the Graduate School: Steve Matson
Department Chair: Art Champagne
Department Telephone Number: (919) 962-2078 (C)
Type of Institution: University
Control: Public
Setting: Small town
Total Faculty: 3,223
Total Graduate Faculty: 3,067
Total Students: 28,916
Total Graduate Students: 8,386
Annual Graduate Tuition:
In-state residents: Full-time—$5,413
Part-time—0–2.9 hrs./$1,353.26
3–5.9 hrs./$2,706.50
6–8.9 hrs./$4,059.76
Out-of-state residents: Full-time—$19,811
Part-time—0–2.9 hrs./$4,952.76
3–5.9 hrs./$9,905.50
6–8.9 hrs./$19,811
Tuition rates for: 2009–10
Deferred tuition plan: No
Other Fees: In-state: $1,748.64
Out-of-state: $1,748.64
Term: Semester

2. Number of Faculty in Department

The combined total of full-time faculty in the three professorial ranks is 29. The combined total of full-time, part-time, and other faculty at all ranks is 43.

3. Admission, Financial Aid, and Housing

Address admission inquiries to: g-admit@physics.unc.edu
Graduate application fee required: $77
Admission deadline (Fall admission): Jan. 1
Admission information: For fall admission, 2010, 50 students were accepted from 179 applicants.
Admission requirements: For admission to the graduate program, a Bachelor's degree in physics is required with a minimum GPA of 3.0 specified. The GRE general test and GRE subject test in advanced physics are required. The average GRE scores for applicants admitted in 2009 were: verbal–610; quantitative–750; advanced physics–760. Students from non-English-speaking countries are required to take the TOEFL exam. A score of 550 or better on the TOEFL is required; a score of 600 or better is strongly preferred.
Undergraduate preparation assumed: Symon, *Mechanics*; Martin, *Elements of Thermodynamics*; Hecht and Zajac, *Optics*; Griffiths, *Introduction to Electrodynamics*; Liboff, *Introductory Quantum Mechanics, or comparable.*
Address financial aid inquiries to: Graduate Admissions Committee, Department of Physics and Astronomy, CB #3255, Phillips Hall, UNC, Chapel Hill, NC 27599-3255, g-admit@physics.unc.edu
GAPSFAS application required: No
Financial aid deadline: 3/1
Loans available: Yes
Address housing inquiries to: UNC Housing Student Academic Services Bldg, 450 Ridge Road, CB5500 Chapel Hill, NC 27599
On-campus, single student housing available: Yes
Cost/term: $2,950/semester
On-campus, married student housing available: Yes
Cost/month: $825–$930 (may not be available due to construction)

Table A—Faculty, Enrollments, and Degrees Granted

Research Specialty	2009–10 Faculty	Enrollment[1] Fall 2009		No. of Degrees Granted[2] 2009–10 (2005–10)			Median No. of Years for 2009–10 Ph.D.'s
		Master's	Doctorate	Master's	Terminal Master's	Doctorate	
Applied Physics	9	0	9	0(0)	0(0)	0(0)	–
Astronomy	5	0	10	0(3)	0(0)	1(7)	6
Astrophysics	6	0	6	0(8)	0(0)	0(8)	–
Biological	10	0	11	0(0)	2(0)	1(1)	6
Condensed Matter Physics	16	0	24	9(20)	0(3)	0(18)	6
Cosmology	2	0	4	1(0)	0(0)	0(0)	–
Nanotechnology	3	0	3	0(0)	0(0)	0(0)	–
Nuclear Physics	15	0	16	5(7)	0(0)	2(5)	6
Numerical Hydrodynamics	1	0	1	0(0)	0(0)	0(0)	–
Particles & Fields	2	0	4	0(0)	0(0)	2(0)	
Gravitation	4	0	5	0(2)	0(1)	0(4)	–
Total		0	93	6(40)	1(4)	5(45)	
Full-time Grad. Stud.		0	90				
Part-time Grad. Stud.		0	0				
First-year Grad. Stud.		0	10				
Median Years in Grad. Study (2008–09 Degrees)				2	2	6	6

Undergraduate Degrees, 2008–09 (2004–09): 11(48)

[1] Students not yet committed to a research specialty are entered under non-specialized.
[2] Five-year totals in parentheses.

4. Graduate Degree Requirements

Master's: 30 semester hours required of which three to six may be for thesis or Master's project (if non-thesis option elected). One comprehensive written examination, which also serves as qualifying examination for Ph.D. Oral examination on thesis, or on a Master's project (if non-thesis option is elected), is required. Residency is required.
Doctorate: No specific graduate course credit requirements, but satisfactory completion of approved sequence of courses; a doctoral written examination covering "core" curriculum; a preliminary oral examination for the dissertation, a dissertation, and oral examination on dissertation. Residency is required.
Thesis: Thesis may be written *in absentia*.
Courses taken as a graduate student before enrolling at Chapel Hill may afford students the opportunity to opt out of one or

more first-year core courses and may be eligible for transfer course credit. However, students are required to pass the doctoral written examination, which is offered at the end of the student's first year and covers core material in quantum mechanics, electromagnetic theory, dynamics, and statistical mechanics.

Special Equipment, Facilities, or Programs: The University of North Carolina, together with NC State and Duke Universities, administers the Triangle Universities Nuclear Laboratory three nuclear accelerators in Durham. Special astronomy facilities of UNC, Chapel Hill, include a 61-cm telescopes and a 68-foot planetarium dome with full-dome digital video. Astronomical facilities: observational facilities include the SOAR 4.1-meter telescope in Chile 61 nights time share, 11-meter Southern African Large Telescope (SALT) in South Africa (10 nights), the University's PROMPT obotic telescope array six 0.4-meter telescopes in Chile, one 0.8-meter telescope in Chile under construction, and four 0.4-meter telescopes in Australia under construction. Laboratory facilities include the Goodman Laboratory for Astronomical Instrumentation.

Table B—Appointments to Graduate Students, 2008–09

Title of Appointee	Appointments Total	Appointments First year	Academic Load Allowed in Credit Hours	Hours of Service Per Week	Stipend for Academic Year ($)
			Semester		
Teaching Assistant	44	18	9	18	16,560[2]
Research Assistant	45	4	9	18	22,080[3]
Merit Assistantship	2	2	9	0	18,000[4]
Pogue	0	0	9	0	20,000
B.O.G. Fellow (Board of Governors)	2	2	9	0	20,000[4]
Total	93	26			

[1]Stipend for academic year 2008–09.
[2]Additional stipend for summer work is available.
[3]Stipend for calendar year 2008–09.
[4]Additional supplement awarded by department.

5. Personnel Engaged in Separately Budgeted Research, 7/09–6/10

Professorial faculty	29
Other faculty	14
Postdoctoral appointments	15
Graduate students	90
Undergraduate students	32
Nonteaching research personnel	5
Total	185

6. Separately Budgeted Research Expenditures by Source of Support

	Departmental Research	Physics-related Research Outside Department
Federal government	$12,229,185	$775,859
State and local government	16,180	0
Private, non-profit organizations	52,619	0
Business and industry	48,541	0
Other	1,500,706	
Total	$13,847,231	$775,859

Table C—Separately Budgeted Research Expenditures

Research Specialty	No. of Grants	Expenditures ($)
Astronomy/Astrophysics	52	975,653
Biophysics/Biological Physics	10	150,324
Condensed Matter Physics	88	4,104,642
Nuclear Physics	31	1,839,246
Particles & Fields	19	478,172
Physics Education	6	12,235
Total	209	13,847,231

Table D—Physics-related Research Outside Department

Field and Unit Outside Department	No. of Grants	Expenditures ($)
Nuclear Physics	2	664,910
Biophysics	1	47,384
Condensed Matter	1	63,565
Total	4	775,859

FACULTY

Professors

Cecil, Gerald N., Ph.D., Hawaii, 1987. Observational astronomy; active galactic nuclei.

Champagne, Arthur E., Ph.D., Yale, 1982. Chair of the Depatment. Experimental nuclear physics.

Clegg, Thomas B., Ph.D., Rice, 1965. Experimental nuclear physics; development of polarized beams.

Clemens, J. Christopher, Ph.D., University of Texas-Austin, 1994. Observational astronomy; astrophysics; astronomical instrumentation.

Dolan, Louise A., Ph.D., MIT, 1976. Particle theory.

Engel, Jonathan H., Ph.D., Yale, 1986. Theoretical nuclear physics; violation of fundamental symmetries in nuclei; the role of nuclear effects in solar neutrino and dark matter experiments; *r*-process nucleosynthesis and quark effects in nuclei.

Evans, Charles R., Ph.D., Texas at Austin, 1984. General relativity; numerical hydrodynamics; astrophysics.

Frampton, Paul H., Ph.D., Oxford, 1968. Particle theory; grand unification; cosmology.

Iliadis, Christian, Ph.D., Notre Dame, 1993. Experimental nuclear astrophysics.

Karwowski, Hugon J., Ph.D., Indiana, 1980. Experimental nuclear physics with polarized nuclei.

Khvechtchenko, Dmitri, Ph.D., Landau Institute for theoretical physics, 1989. Theoretical condensed matter physics; quantum transport in strongly correlated systems.

Lu, Jianping, Ph.D., City University of New York, 1988. High-T_c superconductors; fullerenes and fullerides; strongly correlated electron systems; electronic structure and computational physics; disordered system and quantum transports; complex systems; X-ray imaging.

McNeil, Laurie E., Ph.D., Illinois, 1982. Optical studies of materials.

Ng, Yee Jack, Ph.D., Harvard, 1974. Theoretical particle physics; gravitation.

Rowan, Lawrence G., Ph.D., California, Berkeley, 1963. Solid state experiments using EPR.

Superfine, Richard, Ph.D., University of California-Berkeley,

1991. Scanning probe microscopy of biological structures.

Tsui, Frank, Ph.D., Univ. of Illinois-Urbana, Champaign, 1992. Molecular beam epitaxy of transition metals; scanning tunneling microscopy.

Washburn, Sean, Ph.D., Duke, 1982. Effects of quantum-mechanical coherence in charge transport in small systems; ballistic transport in semiconductors.

Wilkerson, John, Ph.D., UNC-CH, 1982. Neutrino physics and neutrino astrophysics.

Wu, Yue, Ph.D., Catholic University of Leuven, 1987. Nuclear magnetic resonance; electron spin resonance in solids.

Zhou, Otto, Ph.D., University of Pennsylvania, 1992. Experimental materials science; structure/property relationships in solids.

Associate Professors

Mersini-Houghton, Laura, Ph.D., University of Wisconsin, 2000. Theoretical cosmology.

Reichart, Daniel E., Ph.D., University of Chicago, 2000. Gamma ray bursts; early universe; interstellar extinction; galaxy clusters.

Qin, Lu-Chang, Sc.D, Massachusetts Institute of Technology, 1994. Electron microscopy, materials science, nanotechnology.

Tiesinga, Paul H. E., Ph.D., Utrecht University, 1996. Computational and theoretical neuroscience; biophysics.

Assistant Professors

Heitsch, Fabian, Ph.D., University of Heidelberg, 2001. Turbulence and Fragmentation in Molecular Clouds.

Henning, Reyco, Ph.D., Massachusetts Institute of Technology, 2003. Experimental neutrino physics.

Kannapan, Sheila, Ph.D., Harvard, 2001. Observational astronomy; galaxy formation and evolution.

López, René, Ph.D., Vanderbilt University, 2002. Nanotechnology; nano-optics; ultra short laser/matter interaction; thin film science.

Oldenburg, Amy, Ph.D., University of Illinois, 2001. Biophysics.

Lecturers

Churukian, Alice, Ph.D., Kansas State University, 2002. Physics Education.

Deardoff, Duane L., Ph.D., North Carolina State University, 2001. Undergraduate laboratories; physics education.

Adjunct and Joint Faculty

Chaffee, Fred, Ph.D., 1968. Astronomy.

Chang, Sha Xiao, Ph.D., Clark University, 1988. Solid State Physics.

Rohm, Ryan M., Ph.D., Princeton, 1985. Quantum field theory; theoretical elementary particle physics.

Rutland, Jonathan, Ph.D., University of North Carolina, 1988. Physics.

Sen, Pabitra, Ph.D., University of Chicago, 1972, Physics.

Silverstone, Murray, Ph.D., University of California, LA, 2000. Astronomy.

Tang, Jie, Ph.D., Osaka University, 1993. Materials science, nanotechnology.

Tonchev, Anton P., Ph.D., Flerov Laboratory of Nuclear Reactions, Joint Institute for Nuclear Research, 1995. Nuclear resonance; fluorescence; nuclear spectroscopy.

Zhang, Jian, Ph.D., 2005, UNC-CH, Experimental Physics.

Professors Emeriti

Briscoe, Charles V., Ph.D., Rice, 1958.
Choi, Sang-il, Ph.D., Brown, 1961.
Christiansen, Wayne A., Ph.D., California, 1968.
Davis, Morris S., Ph.D., Yale, 1950.
Dy, Kian S., Ph.D., Cornell, 1967.
Hernandez, John P., Ph.D., Rochester, 1967.
Hooke, William M., Ph.D., Princeton, 1958.
Hubbard, Paul S., Ph.D., Harvard, 1958.
Kessemeier, Horst, Ph.D., Washington (St. Louis), 1964.
Ludwig, Edward J., Ph.D., Indiana, 1963.
Macdonald, J. Ross, D.Phil., Oxford, 1950.
Merzbacher, Eugen, Ph.D., Harvard, 1950.
Mitchell, Earl N., Ph.D., Minnesota, 1955.
Rose, James, Ph.D., Yale, 1977.
Schroeer, Dietrich, Ph.D., Ohio State, 1965.
Shafroth, Stephen M., Ph.D., Johns Hopkins, 1953.
Slifkin, Lawrence M., Ph.D., Princeton, 1950.
Thompson, William J., Ph.D., Florida State, 1967.
Van Dam, Hendrik, Examen Doctorale, Amsterdam, 1959.
York, Jr., James W., Ph.D., North Carolina State, 1966.

Research Faculty

Falvo, Michael, Ph.D., Univ of North Carolina, Chapel Hill, 1997. Nanoscience.

Kleinhammes, Alfred, Ph.D., Clark, 1990. Condensed matter physics; nuclear magnetic resonance.

O'Brien, E. Timothy, Ph.D., University of California (Santa Barbara), 1974. Cell biology; microscopy.arikh

Parikh, Nalin R., Ph.D., McMaster University, 1985. Ion beam modification of materials; ion beam analysis, microwave plasma deposition/etching.

RESEARCH SPECIALTIES AND STAFF

Theoretical

Astronomy and Astrophysics. Radio, optical, and x-ray astronomy. Evans.

Biological Physics, Computational and Theoretical Neuroscience. Tiesinga. 1 postdoctoral research associate.

Condensed Matter Physics. Many-body theory; fullerenes; high-temperature superconductors; correlated electron systems. Lu, Khvechtchenko.

Field Theory, Particle Theory, Gravitation, and Relativity. Quantum theory of fields and particles; superstring theory; cosmology relativistic dynamics; general relativity; gravitational fields; quantum and supergravity. Dolan, Evans, Frampton, Mersini, Ng, 2 postdoctoral research associates.

Nuclear Physics. Nucleosynthesis in stars and supernovae; violation of fundamental symmetries in nuclei; the role of nuclear effects in solar neutrino and dark matter experiments; Engel. 2 postdoctoral research associates.

Experimental

Astronomy and Astrophysics. Radio, optical, and x-ray astronomy; stellar spectroscopy and photometry. Carney, Cecil, Champagne, Clemens, Iliadis, Kannapan, Reichart, 3 postdoctoral research associates.

Condensed Matter Physics, Materials Science, Microelectronics. NMR studies of quasicrystals, polymer lubricants, nanocrystals and carbon nanotube; Raman and Brillouin scattering from organic semiconductors; AFM and surface science in virtual reality interfaces to microscopy; low-temperature physics and quantum transport; growth and magnetic susceptibility of magnetic semiconductors; electrical and mechanical interaction of carbon nanotubes with surfaces and other tubes; chaos in biological systems. Lopez, McNeil, Parikh, Qin, Superfine, Tsui, Washburn, Wu, Zhou. 8 postdoctoral research associates.

Nuclear Physics. Nuclear astrophysics; radioactive beams; neutrino physics; spin polarization in nuclear reactions; few body physics. Champagne, Clegg, Henning, Iliadis, Karwowski, Wilkerson. 2 postdoctoral research associates.

Biological physics. Scanning probe microscopies applied to biological systems; X-ray imaging for biomedicine, optical coherence tomography. Superfine, Falvo, Zhou, Lu, Oldenburg.

WAKE FOREST UNIVERSITY

DEPARTMENT OF PHYSICS

Winston-Salem, North Carolina 27109

Students Accepted For Degree	FIELDS		
	Physics	Astronomy	Related Fields
Doctorate	X		
Master's	X		

1. General

President: Nathan O. Hatch
Dean of Graduate School: Lorna Moore
Department Chairman: Keith D. Bonin
Department Telephone Number: (336) 758-3223 or 5337
Type of Institution: University
Control: Private
Setting: Suburban
Total Faculty: 1,712
Total Graduate Faculty: 521
Total Students: 7,079
Total Graduate Students: 686
Annual Graduate Tuition:
All Graduate Students: Full-time—$30,358
Part-time—$1,082/sem. hr.
Tuition rates for: 2010–11
Deferred tuition plan: Yes
Other Fees: None
Term: Semester

2. Number of Faculty in Department

The combined total of full-time faculty in the three professorial ranks is 15.5 The combined total of full-time, part-time, and other faculty at all ranks is 24.

3. Admission, Financial Aid, and Housing

Address admission inquiries to: Ms. Judith G. Swicegood, Dept. of Physics, P.O. Box 7507, Wake Forest University, Winston-Salem, NC 27109
Graduate application fee required: $60 domestic/$60 international
Admission deadline: Jan. 15
Admission information: For fall admission, 2009–10, 15 students were accepted from 66 applications. All incoming graduate students receive a new IBM Thinkpad laptop computer with extensive software.
Admission requirements: For admission to the graduate programs, a Bachelor's degree in physics is required with no minimum undergraduate GPA specified. A GPA of at least 3.0 out of 4.0 is strongly recommended. The GRE General Test is required. Students from non-English speaking countries are required to demonstrate proficiency in English via the TOEFL exam. Minimum acceptable score is 575/232 (computer).
Undergraduate preparation assumed: Mechanics—Symon, *Mechanics*; Electricity and Magnetism—Griffiths, *Introduction to Electrodynamics*; Quantum Mechanics—Gasiorowicz, *Quantum Physics*; Thermodynamics—Kittel and Kroemer, *Thermal Physics*.
Address financial aid inquiries to: Dean, Graduate School
GAPSFAS application required: No
Financial aid deadline: Jan. 15
Loans available: Yes, limited
Address housing inquiries to: Residence Housing, P.O. Box 7749, Wake Forest Univ. Winston-Salem, NC 27109
On-campus, single student housing available: None
On-campus, married student housing available: None

Table A—Faculty, Enrollments, and Degrees Granted

Research Specialty	2009–10 Faculty	Enrollment[1] Fall 2009		No. of Degrees Granted[2] 2009–10 (2005–10)			Median No. of Years for 2009–10 Ph.D.'s
		Master's	Doctorate	Master's	Terminal Master's	Doctorate	
Atomic, Molecular, & Optical Physics	2	1	2	–	0(0)	0(2)	5
Biophysics	5	0	9	–	0(5)	2(8)	5
Condensed Matter Physics	5	3	10	–	0(0)	1(9)	6
Medical & Health Physics	2	0	3	–	1(1)	0(0)	–
Particles & Fields	2	0	2	–	0(0)	1(1)	–
Relativity & Gravitation	2	0	2	–	0(1)	1(2)	6
Statistical & Thermal	0	0	0	–	0(0)	0(1)	5
Total	18	4	28	–	1(2)	4(23)	
Full-time Grad. Stud.		4	28				
Part-time Grad. Stud.		0	0				
First-year Grad. Stud.		3	7				
Median Years in Grad. Study (2009–10 Degrees)				–	2	5	5

Undergraduate Degrees, 2009–10 (2004–09): 11(45)

[1]Students not yet committed to a research specialty are not entered in an area.
[2]Five-year totals in parentheses.

4. Graduate Degree Requirements

Master's: 30 semester hours of graduate credit, including no more than six hours of research. Minimum of 12 months full-time in residence. An oral defense of the thesis and a 3.0 average on courses are required.
Doctorate: A written General Exam at the level of material normally covered in the first year of graduate study serves as the preliminary examination. An oral defense of the thesis and a 3.0 average on courses are required. No language is required.
Thesis: Thesis may be written *in absentia*.

Table B—Appointments to Graduate Students, 2009–10

Title of Appointee	Appointments Total	Appointments First year	Academic Load Allowed in Credit Hours	Hours of Service Per Week	Stipend for Academic Year ($)[2]
			Semester		
NSF Fellowship	1	0	12	0	30,000
AHA[3] Fellowship	1	1	12	0	21,000
Dean's Fellowship	1	0	12	0	21,000
Research Assistant	14	3	12	0	20,000[1]
Scholarship	1	1	12	0	–
Teaching Assistant	14	5	12	12	20,000[1]
Total	31	9			

[1]In addition, Teaching and Research Assistants receive a non-taxable scholarship of $30,358 to cover tuition for the year. Stipend is for 2009–10

[2]All incoming graduate students receive a new IBM Thinkpad laptop computer with extensive software.

[3]AHA – American Heart Association.

5. Personnel Engaged in Separately Budgeted Research, 7/08–09

Professorial faculty	14
Research Faculty	5
Postdoctoral Associates	5
Graduate students	17
Undergraduate students	18
Total	59

6. Separately Budgeted Research Expenditures by Source of Support

Information Not Available

Table C—Separately Budgeted Research Expenditures

Research Specialty	No. of Grants	Expenditures ($)
Biophysics	21	1,197,000
Condensed Matter Physics	4	134,000
Nanotech	10	423,000
Astrophysics	2	47,000
Total	36	2,300,000

FACULTY

Professors

Anderson, Paul R., Ph.D., California, Santa Barbara, 1983. General relativity; quantum field theory in curved space.

Bonin, Keith D., Ph.D., Maryland, 1984. Chairman of Deptment. Atomic physics, nanophysics, biophysics, optics.

Carroll, David L., Ph.D., Wesleyan, 1993. Nanostructures and nanotechnology.

Fetrow, Jacquelyn, Ph.D., Penn State 1986. Reynolds Professor. Computational biophysics, computational drug discovery, cheminformatics, systems biology.

Holzwarth, N. A. W., Ph.D., Chicago, 1975. Theoretical solid state physics; electronic structure of bulk solids, surfaces, and molecules.

Kim-Shapiro, Daniel, Ph.D., California, Berkeley, 1993. Biophysics.

Matthews, Eric G., Ph.D., North Carolina, 1977. Thermally stimulated depolarization of defects in insulators; *ab initio* calculations of defect properties.

Williams, R. T., Ph.D., Princeton, 1974. Reynolds Professor. Femtosecond laser studies of defects and electrons in solids.

Associate Professors

Carlson, Eric, Ph.D., Harvard, 1988. Astrophysics and particle physics.

Cook, Gregory B., Ph.D., North Carolina, 1990. General relativity and relativistic astrophysics.

Guthold, Martin, Ph.D., University of Oregon, 1997. Biophysics.

Macosko, Jed, Ph.D., California, Berkeley, 1999. Biophysics of molecular motors and biopolymers.

Salsbury, Fred, Ph.D., California, Berkeley, 1999. Computational biophysics.

Assistant Professors

Jurchescu, Oana, Ph.D., Groningen, Netherlands, 2007. Nanostructures and Nanotechnology.

Thonhauser, Timo, Ph.D., Karl-Franzens University, Austria. Density functional theory.

RESEARCH PROFESSORS

Professor

Holzwarth, G. M., Ph.D., Harvard, 1964. Biophysics.

Kerr, William C., Ph.D., Cornell, 1967. Emeritus. Theoretical solid state and statistical physics; structural phase transitions.

Shields, Howard, Ph.D. Duke, 1956. Emeritus.

Associate Professor

Ucer, K. Burak, Ph.D., University of Rochester, 1997. Ultrafast lasers and spectroscopy, condensed matter physics.

Assistant Professors

Basu, Swati, Ph.D., University of Illinois-Urbana, 1994. Biophysics.

Berry, Joel, Ph.D., Wake Forest University, 2001. Biomedical engineering.

Adjunct Staff

Bourland, J. D., Ph.D., North Carolina. Radiation oncology.

Miller, Timothy, Ph.D., Vanderbilt, 2002. Adjunct Assistant Professor. High energy nuclear and parallel computing.

Roberson, Mark, Ph.D., Princeton, 1989. Signal recognition.

Santago, Peter, Ph.D., North Carolina State, 1986. Adjunct Associate Professor. Image enhancement.

Hodge, William, Ph.D., Wake Forest University, 2008.

RESEARCH SPECIALTIES AND STAFF

Theoretical

Astronomy & Astrophysics. Gravitational physics, general relativity, numerical relativity, black holes, neutron stars, compact binaries, initial data, gravitational waves, quantum field theory in curved space, particle physics. Anderson, Carlson, Cook.

Computational Condensed Matter. Computational materials physics: simulation and prediction of energy storage materials, development of "first principles" simulation methods, condensed matter theory, semi-classical electron dynamics, Berry phase effects, non-linearity, Computational and theoretical materials science, condensed matter physics, solid state physics, density functional theory, first-principles calculations, NMR, van der Waals forces, magnetization. N. Holzwarth, Mattews (vice provost), Kerr (emeritus), Thonhauser, one Post-doctoral associate.

Biophysics. Computational and theoretical biophysics, computational systems biology, protein structure/function relationships, biological network modeling, signal transduction network modeling, molecular physics, drug discovery. Salsbury, Fetrow, two Post-docs.

Experimental

Biological Physics and Optics. Optics, optical trapping, mechanical effects of light, optogenetics, optical microscopy, motor proteins, kinesin, optical and electron paramagnetic spectroscopy, hemoglobin, nitric oxide, nitrite, sickle cell disease, cardiovascular disease, nanobiotechnology, atomic force and optical microscopy, single molecule experiments, protein-DNA interactions, thrombosis & hemostasis, nanofibers, electrospinning, tissue engineering, drug discovery. Bonin, Guthold, G. Holzwarth (emeritus), Kim-Shapiro, Macosko, Shields (emeritus), Swati (research professor), 2 Post-doctoral associates.

Condensed Matter, Optical, Molecular and Atomic Physics, Nanotechnolgy: Electron spin resonance in irradiated organic solids, transport properties, semiconductor trapping centers, ultrafast spectroscopy, laser materials, nanostructures and nanotechnology, nanomotors, solar cells, meta materials and negative index materials, organic electronics, thin-film transistors, field-effect transistors, organic semiconductors, single crystals, microstructure, ultrafast spectroscopy, excitons, scintillators, energy research. Carroll, Jurchescu, Williams, 5 Post-doctoral associates, several visiting scientists.

Center for Nanotechnology and Molecular Materials. The Center is engaged in a broad range of projects from the development of medical technologies, to green energy technologies, to the understanding of the environmental and ethical implications of such nano-based technologies, material design and synthesis, carbon nanotubes, metal nanoparticles, quantum dots, polymers, cage structures, solar cells, biofuels, batteries, high efficiency organic transistors, new lighting systems, antibiotic resistance, wound healing, tissue regeneration. Carroll, 5 Post-doctoral associates, several visiting scientists.

FACULTY PUBLICATIONS

Anderson, Paul R.

(with C. Molina-Paris, and E. Mottola), "Cosmological Horizon Modes and Linear Response in de Sitter Spacetime," Phys. Rev. D **80**(8), 084005 (2009).

Bonin, Keith D.

(with N. R. Gassman, J. P. Nelli, S. Dutta, A Kuhn, Z. Pianowski, N. Winssinger, M. Guthold, and J. C. Macosko), "Selection of Bead-Displayed, PNA-encoded Chemicals," Journal of Molecular Recognition (2009).

(with E. A. Budygin, E. B. Oleson, J. Salek, and S. R. Jones), "Real-Time Voltammetric Detection of Cocaine-Induced Dopamine Changes in the Striatum of Freely Moving Mice," Neuroscience Letters **467**(2) 144 (2009).

(with E. B. Oleson, S. Talluri, S. R. Childers, A. E. Smith, D. C. Roberts, and E. A. Budygin), "Dopamine Uptake Changes Associated with Cocaine Self-Administration," Neuropsychopharmacology **34**, 1174 (2009).

Carroll, David L.

(with R. Jayakanth, A. Manoj, J. Liu, and I. Manna), "A Novel Polymer Nanotube Composite for Photovoltaic Packaging Applications," Nanotechnology **19**(8), 5 (2009).

(with P. Auragudom, A. A. Tangnan, M. A. G. Namboothiry, R. C. Advincula, S. Phanichphant, T. R. Lee), "Defect-free Poly(9,9-bis(2-ethylhexyl)fluorene-2,7-vinylene) for Polymer Light-Emitting Diode (PLED) Devices," Journal of Polymer Research (2009).

(with M. Reyes-Reyes, R. Sandoval, A. Gorbatchev, and J. Napoles-Duarte), "Encapsulation of the Fullerene Derivative [6,6]-Phenyl-C," Journal of Physical Chemistry C **113**(31), 5 (2009).

(with B. McGuirt, J. E. Kielbasa, J. Park, P. Sisco, J. Zhang, E. D. Peterson, C. Murphy, R. D. Adams, and R. T. Williams), "Light Scattering of Interacting Gold Nanorods," 11th ed., Physica Status Solidi B **246**, 3 (2009).

(with J. Talla, S. Curran, D. Birx, D. Zhang, and S. Dias), "Electrical Transport Measurements of Highly Conductive Carbon Nanotube/Poly(bisphenol A. Carbonate) Composite," Applied Physics Letters **105**(7), 2 (2009).

Guthold, Martin

(with N. R. Gassman, J. P. Nelli, S. Dutta, A. Kuhn, K. D. Bonin, Z. Pianowski, N. Winssinger, and J. C. Macosko), "Selection of Bead-Displayed, PNA-encoded Chemicals," Journal of Molecular Recognition (2009).

(with S. Ghosh, S. Dutta, E. Gomes, D. L. Carroll, R. D'Agostino, J. Olson, and W. H. Gmeiner), "Increased Heating Efficiency and Selective Thermal Ablation of Malignant Tissue with DNA-Encased Multiwalled Carbon Nanotubes," ACS Nano **3**, 2667 (2009).

(with C. R. Carlisle, C. Coulais, M. Namboothiry, D. L. Carroll, and R. R. Hantgan), "The Mechanical Properties of Individual, Electrospun Fibrinogen Fibers," Biomaterials **30**, 1205 (2009).

Holzwarth, George M.

(with J. Gagliano, M. Walb, B. Blaker, and J. C. Macosko), "Kinesin Velocity Increases with the Number of Motors Pulling Against Viscoelastic Drag," Eur Biophys J. **13** (2009).

Kim-Shapiro, Daniel B.

(with J. O. Lundberg, M. T. Gladwin, N. Benjamin, N. Bryan, A. Butler, P. Cabrales, A. Fago, M. Feelisch, P. C. Ford, B. A. Freeman, M. Frenneau, J. Friedman, M. Kelm, C. Kevil, A. Kozlov, J. R. Lancaster, D. Lefer, K. McColl, K. McCurry, R. Patel, J. Petersson, T. Rassaf, V. Reutov, G. Richter-Addo, A. Schechter, S. Shiva, K. Tsuchiya, E. Van Faassen, A. J. Webb, B. S. Zuckerman, J. Zweier, and E. Weitzberg), "Nitrate and Nitrite in Physiology, Nutrition and Therapeutics," Nature Chemical Biology **5**, 865 (2009).

"Structure and Function of Hemoglobin and its Dysfunction in Sickle Cell Disease," Cambridge University Press, 2009.

(with M. Gladwin), "Storage Lesion in Banked Blood Due to Hemolysis-Dependent Disruption of Nitric Oxide Homeostasis," Current Opinion in Hematology **16**, 515 (2009).

(with E. V. Faassen, S. Bahrami, M. Feelisch, N. Hogg, M. Kelm, A. V. Kozlov, H. Li, J. Lundberg, R. Mason, H. Nohl, T. Rassaf, A. Samouilov, A. Slama-Schwok, S. Shiva, A. F. Vanin, E. Weitzberg, J. Zweier, and M. T. Gladwin), "Nitrite as Regulator of Hypoxic Signaling in

Mammalian Physiology," Molecular Medicine Reviews **29**, 683 (2009).

(with A. B. Blood, M. Tiso, S. T. Verma, J. Lo, M. S. Joshi, I. Azarov, L. D. Longo, M. T. Gladwin, and G. G. Power), "Increased Nitrite Reductase Activity of Fetal Versus Adult Ovine Hemoglobin," Amer. J. Physiol (2009).

Macosko, J. C.

(with N. R. Gassman, J. P. Nelli, S. Dutta, A. Kuhn, K. D. Bonin, Z. Pianowski, N. Winssinger, and M. Guthold), "Selection of Bead-Displayed, PNA-Encoded Chemicals," Journal o of Molecular Recognition (2009).

(with J. Gagliano, M. Walb, B. Blaker, and G. M. Holzwarth), "Kinesin Velocity Increases with the Number of Motors Pulling Against Viscoelastic Drag," Eur. Biophys J **13** (2009).

(with D. Berleant), "The Genetic Code-more than just a table," Cell Biochem Biophys **55**(2), 107 (2009).

(with Y. Shtridelman), "In vivo Multimotor Force-Velocity Curves by Tracking and Sizing Sub-Diffraction Limited Vesicles," Cellular and Molecular Bioengineering **2**(2), 190 (2009).

(with O. Turunen, and R. Seelke), "Structural and Sequence Analysis of Fast Evolving Duplicated Yeast Genes Indicates Functional Specialization with Partially Relaxed Purifying Selection," FEMS Yeast Res **9**(1) 16 (2009).

Salsbury, Fred R.

(with M. W. Crowder, S. F. Kingsmore, and J. J. Huntley), "Molecular Dynamic Simulations of the Metallo-beta-lactamase from Bacteroides Fragilis in the Presence and Absence of a Tight-Binding Inhibitor," Journal of Molecular Modeling **15**(2), 133 (2009).

(with A. Vasilyeva, J. E. Clodfelter, B. Rector, T. Hollis, and K. Scarpinato) "Small Molecule Induction of MSH2-Dependent Cell Death Suggests a Vital Role of Mismatch Repair Proteins in Cell Death," DNA Repair **8**(1), 103 (2009).

Thonhauser, Timo

(with D. Ceresoli, A. A. Mostofi, N. Marzari, R. Resta, and D. Vanderbilt), "A Converse Approach to the Calculation of NMR Shielding Tensors," J. Chem. Phys. **131**(10), 101101 (2009).

(with D. C. Langreth, B. I. Lundqvist, S. D. Chakarova-Käck, V. R. Cooper, M. Dion, P. Hyldgaard, A. Kelkkanen, J. Kleis, L. Kong, S. Li, P. G. Moses, E. Murray, A. Puzder, H. Rydberg, and E. Schröder), "A Density Fucntional for Sparse Matter," J. Phys.: Condens. Matter. **21**(8), 084203 (2008).

(with D. Ceresoli and N. Marzari), "NMR Shifts for Polycyclic Aromatic Hyrocarbons from First-principles," Int. J. Quantum Chem. **109**(14) 3336 (2009).

(with S. Li, V. R. Cooper, B. I. Lundqvist, and D. C. Langreth), "Stacking Interactions and DNA Intercalation," J. Phys. Chem. B **113**(32), 11166 (2009).

Williams, Richard T.

(with G. A. Bizarri, W. W. Moses, J. Singh, and A. N. Vasilev), "An Analytical Model of Nonproportional Scintillator Light Yield in Terms of Recombination Rates," Journal of Applied Physics **105**, 044507 (2009).

(with M. Nikl, G. P. Pazzi, P. Fabeni, E. Mihokova, J. Pejchal, D. Ehrentraut, and A. Yoshikawa), "Decay Kinetics of the Defect-based Visible Luminescence in ZnO," Journal of Luminescence **129**, 1564 (2009).

(with G. A. Bizarri, N. J. Cherepy, W. S. Choong, G. Hull, W. W. Moses, S. A. Payne, J. Singh, J. D. Valentine, and A. N. Vasilev), "Progress in Studying Scintillator Non-proportionality: Phenomenological Model and Experiments," IEEE Transaction on Nuclear Science **56**, 2313 (2009).

(with G. Bizarri, W. W. Moses, J. Singh, and A. N. Vasiliev), "Simple Model Relating Recombination Rates and Non-proportional Light Yield in Scintillators," Phys. Status Solidi C 6(1), 97 (2009).

(with D. L. Carroll, B. McGuirt, J. E. Kielbasa, J. Park, P. Sisco, J. Zhang, E. D. Peterson, C. Murphy and R. D. Adams), "Light Scattering of Interacting Gold Nanorods," 11th ed., Physica Status Solidi B **246**, 3 (2009).

UNIVERSITY OF NORTH DAKOTA

DEPARTMENT OF PHYSICS AND ASTROPHYSICS

Grand Forks, North Dakota 58202-7129

Students Accepted For Degree	FIELDS		
	Physics	Astronomy	Related Fields
Doctorate	X		
Master's	X		

1. General

President: Robert Kelley
Dean of Graduate School: Joseph Benoit
Department Chairman: Graeme Dewar
Department Telephone Number: (701) 777–2911
Type of Institution: University
Control: Public
Setting: Urban
Total Faculty: over 937 (464 tenured/tenure-track)
Total Graduate Faculty: over 400
Total Students: 13,172
Total Graduate Students: over 2,732
Annual Graduate Tuition: (9 credits/semester)
In-state residents: Full-time—$5,521.32
Part-time—$306.74/credit
Out-of-state residents: Full-time—$13,135.86
Part-time—$729.77/credit
Tuition rates for: 2010–11
Deferred tuition plan: Yes
Other Fees: included in tuition
Term: Semester

2. Number of Faculty in Department

The combined total of full-time faculty in the three professorial ranks is 9. The combined total of full-time, part-time, and other faculty at all ranks is 18.

3. Admission, Financial Aid, and Housing

Address admission inquiries to: Kanishka Marasinghe, Physics Dept.
Graduate application fee required: $35
Admission deadline (Fall admission): 7/25
Admission information: For fall admission, 2009, 9 students were accepted from 17 applicants.
Admission requirements: For admission to the graduate programs, a Bachelor's degree in physics is required with a minimum overall undergraduate GPA of 2.75 or 3.0 for last two years. The GRE is recommended for applicants to the Ph.D. program. The average GRE scores for 2009-10 admissions were MS: verbal–405; quantitative–695; and Ph.D.: verbal–407; quantitative–747. The GRE Advanced is recommended. Students from non-English speaking countries are required to demonstrate proficiency in English via the TOEFL exam. Minimum acceptable score for admission is 550, 213 for computer test, or IBT: speaking 21, listening 19, reading 19, and writing 17.
Address financial aid inquiries to: Kanishka Marasinghe, Department of Physics and Astrophysics
GAPSFAS application required: No
Financial aid deadline: None
Loans available: Yes
Research/Teaching Assistantship available: Yes
Address housing inquiries to: Housing Office, University of North Dakota
On-campus, single student housing available: Yes
Cost/year: $6,753–9,450 with meal plan
On-campus, student housing available: Yes
Cost/month for family: $440–750

Table A—Faculty, Enrollments, and Degrees Granted

Research Specialty	2009–10 Faculty	Enrollment[1] Fall 2009–10		No. of Degrees Granted[2] 2009–10 (2005–10)			Median No. of Years for 2009–10 Ph.D.'s
		Master's	Doctorate	Master's	Terminal Master's	Doctorate	
Astrophysics	2	1	3	0(2)	–	0(0)	–
Condensed Matter Physics	5	1	5	1(8)	–	1(1)	7
Physics Education	1	0	0	0(0)	–	0(0)	–
Medical & Health Physics	1	1	0	0(1)	–	0(0)	–
Total	9	3	8	1(11)	–	0(0)	
Full-time Grad. Stud.		3	8				
Part-time Grad. Stud.		0	0				
First-year Grad. Stud.		0	5				
Median Years in Grad. Study (2009–10 Degrees)				2	–	7	–
Undergraduate Degrees, 2009–10 (2005–10): 3(12)							

[1]Students not yet committed to a research specialty are entered under non-specialized.
[2]Five-year totals in parentheses.

4. Graduate Degree Requirements

Master's: 30 graduate credits. Thesis is required; no language requirement; minimum of two semesters of full-time study in residence.
Doctorate: 90 graduate credits, including a minimum of 9 credits in a related field, in an approved program are required; qualifying exam, comprehensive exam, dissertation, and final oral examination are required; two year residency is required. No language requirement.

Table B—Appointments to Graduate Students, 2009–10

Title of Appointee	Appointments Total	Academic Load Allowed in Credit Hours	Hours of Service Per Week	Stipend for Academic Year ($)
		Semester		
Teaching Assistant	7	10	20	Ph.D. $16,699[1,2]
Research Assistant	2	12	20	M.S.- $14,096[1,2]
No Support	2	12	–	–
Total	11			

[1]Tuition is waived.
[2]Amount is for academic year.

5. Personnel Engaged in Separately Budgeted Research, 7/09–6/10

Professorial faculty	5
Graduate students	4
Total	9

6. Separately Budgeted Research Expenditures by Source of Support (over grant period)

	Departmental Research	Physics-related Research Outside Department
Federal government	$0	$2,400,590
State government	75,500	500,569
University	78,000	0
Total	$153,500	$2,901,159

Table C—Separately Budgeted Research Expenditures

Research Specialty	No. of Grants	Expenditures ($)
Condensed Matter Physics	3	75,500
Total	3	75,500

Table D—Physics-related Research Outside Department

Field and Unit Outside Department	No. of Grants	Expenditures ($)
Astrophysics	2	440,590
Education	1	500,569
Medicine & Health Physics USDA-HNRC	1	1,960,000
Total	4	2,901,159

FACULTY

Professors

Dewar, Graeme A., Ph.D., Simon Fraser University, 1980. Department chair. Solid state physics; magnetoelasticity; magnetic materials; negative index of refraction.

Kim, Ju H., Ph.D., University of Chicago, 1990. Solid state theory, superconductivity.

Lykken, Glenn I., Ph.D., North Carolina, 1966. Health physics; trace element analysis.

Rao, Sesh B., Ph.D., Penn State, 1963. Emeritus. Optics; spectroscopy; atomic and molecular physics.

Schwalm, William A., Ph.D., Montana State University, 1978. Solid state theory, mathematical physics.

Soonpaa, Henn H., Ph.D., Wayne State University, 1955. Emeritus. Solid state physics.

Associate Professors

Marasinghe, Kanishka, Ph.D., University of Missouri-Rolla, 1993. Atomic structure-property relationships of glasses and magnetic materials.

Young, Timothy R., Ph.D., University of Oklahoma, 1994. Supernovae and supernova remnants; hydrodynamics; numerical simulations of radiation and neutrino transport.

Assistant Professors

Barkhouse, Wayne, Ph.D., University of Toronto, 2003. Astrophysics, galaxy clusters.

Oncel, Nuri, Ph.D., University of Twente, 2007. Solid state physics.

Temporary Assistant Professor

Ammons, Edsel, Ph.D., University of Illinois at Urbana-Champaign, 1992. Nuclear physics.

Adjunct Professors

Delhommelle, Jerome, Ph.D., University of Paris XI-Orsay. Molecular simulation of nonequilibrium processes.

DeMuth, David, Ph.D., University of Minnesota, 1999. Astrophysics, physics education.

Jones, Michael, Ph.D., University of North Dakota, 1978. Multidisciplinary science and engineering research.

Lefevre, Russell, Ph.D., University of California–Santa Barbara, 1976. Engineering and physics education.

Momcilovic, Berislav, Ph.D., University of Zagreb, 1973. Medical physics.

Moritz, Brian, Ph.D., University of North Dakota, 2000. Condensed matter theory.

Schwalm, Mizuho, Ph.D., Montana State University, 1978. Solid state theory, physics education.

RESEARCH SPECIALTIES AND STAFF

Theoretical

Astrophysics. Barkhouse, Young.

Mathematical Physics. Fluid and electronic transport in disordered structures, critical phenomena, application of Lie groups to chaotic dynamical systems. M. Schwalm, W. Schwalm.

Solid State Physics. High-temperature superconductivity, magnetic properties, strongly correlated systems. Kim.

Experimental

Astrophysics. Gamma ray bursters, extra-solar planets. Young.

Biophysics. Trace element analysis in environmental and biological matrices. Lykken.

Chemical Physics. Surface physics, X-ray diffraction, phase transitions, trace elements in lead. Lykken, Marasinghe.

Solid State Physics. Transport phenomena, thin films, superconductivity, magnetism, high pressure physics, negative index of refraction, magnetoelasticity. Dewar, Marasinghe, Oncel.

AIR FORCE INSTITUTE OF TECHNOLOGY

DEPARTMENT OF ENGINEERING PHYSICS

Wright-Patterson Air Force Base, Ohio 45433-6583

Students Accepted For Degree	FIELDS		
	Physics	Astronomy	Related Fields
Doctorate	X		X
Master's	X		X

1. General

Commandant: Brig Gen. Walter D. Givhan
Dean of Graduate School: Marlin U. Thomas
Department Chairman: Nancy C. Giles
Department Telephone Number: (937) 255-3636, x4601
Type of Institution: University
Control: Public
Setting: Urban
Total Faculty: 131
Total Graduate Faculty: 131
Total Students: 773
Total Graduate Students: 764
Annual Graduate Tuition: Full Time: $3,900/quarter*
Other Fees: None
Term: Quarter

*Before 1996, all students at AFIT were full-time government employees, mostly officers in the Air Force, and, in effect, were on a full scholarship. As of 1997, AFIT accepts non-government, U.S. citizens as students.

2. Number of Faculty in Department

The combined total of full-time faculty in the three professorial ranks is 24. The combined total of full-time, part-time, and other faculty at all ranks is 30.

3. Admission, Financial Aid, and Housing

Address admission inquiries to: Office of Admissions, AFIT 2950 Hobson Way, Wright-Patterson AFB, OH 45433-7765, Tel (937) 255-6234, x3184; also, see www.afit.edu/en
Graduate application fee required: None
Admission deadline (Fall admission): 3/15
Admission information: For fall admission, 2009–10, 35 students were accepted.
Admission requirements: For admission to the MS programs, a Bachelor's degree in physics or engineering is required with a minimum undergraduate GPA of 3.0. The GRE is required. The minimum acceptable score suggested for admission is verbal—500; quantitative—600. For admission to the PhD programs, a MS degree and GRE scores of 550–verbal and 650–quantitative are required.
Undergraduate preparation assumed: Eisberg and Resnick, *Quantum Physics*: Symon, *Mechanics*; Griffiths, *Electromagnetic Theory*, Kittel, *Thermal Physics*.
GAPSFAS application required: No
Loans available: No
Address housing inquiries to: Housing Office (645CEG/CEH), Wright-Patterson AFB, OH 45433 (Military only)
On-campus, single student housing available: Yes (military only)
Cost/term: Government housing allowance
On-campus, married student housing available: Yes (military only)
Cost/month: Government housing allowance
Off-campus, single and married housing available: Just off Wright-Patterson AFB in several apartment complexes.

Table A—Faculty, Enrollments, and Degrees Granted

Research Specialty	2009–10 Faculty	Enrollment[1] Fall 2008		No. of Degrees Granted[2] 2009–10 (2005–09)			Median No. of Years for 2009–10 Ph.D.'s
		Master's	Doctorate	Master's	Terminal Master's	Doctorate	
Applied Physics	16	32	6	0(1)	0(7)	0(0)	–
Atmos./Space Phys. Cosmic Rays	3	2	2	1(1)	2(14)	0(0)	–
Atomic, Molecular, & Optical Physics	3	1	5	0(0)	1(2)	1(1)	5.0
Biophysics	1	2	0	0(0)	3(5)	0(0)	
Chemical Physics	2	0	1	0(0)	0(1)	0(0)	–
Condensed Matter Physics	5	0	2	0(0)	0(6)	0(2)	–
Energy Sources & Environ.	3	0	0	0(0)	0(0)	0(0)	–
Lasers	6	1	3	0(0)	1(14)	2(2)	–
Materials Sci./ Metallurgy	3	3	0	0(0)	3(6)	0(2)	–
Nuclear Engineering	7	5	2	0(0)	5(20)	0(1)	–
Nuclear Physics	2	1	2	0(0)	1(12)	0(1)	–
Optics	6	4	0	0(0)	4(16)	1(4)	5.0
Plasma Physics & Fusion	1	0	0	0(0)	0(6)	0(0)	–
Total		51	24	1(2)	23(121)	4(13)	
Full-time Grad. Stud.		51	24				
Part-time Grad. Stud.		3	2				
First-year Grad. Stud.		27	8				
Median Years in Grad. Study (2009–10 Degrees)				1.5	1.5	4	
Undergraduate Degrees: N/A							

[1]Students not yet committed to a research specialty are entered under applied physics.
[2]Five-year totals in parentheses.

4. Graduate Degree Requirements

Master's: A Master's degree requires a minimum of 48 quarter hours with a GPA of at least 3.0 (on 4.0 scale). At least three quarters in residence are required. There are no comprehensive or qualifying examinations. A thesis is required, and counts as 12 of the 48 quarter hours which must be submitted for the degree.
Doctorate: A Ph.D. requires a minimum of 36 quarter hours beyond the M.S., with a GPA of at least 3.5 (on 4 scale). At least three consecutive quarters in residence are required. Satisfactory performance on an examination in a specialty area is required for admission to candidacy. At least 24 quarter hours must be taken in the specialty area. A dissertation is required, as is a successful oral defense of dissertation.
Thesis: Theses and dissertations started in residence may be completed *in absentia*.
Special Equipment, Facilities, or Programs: Thesis, dissertation, and faculty research is usually carried out in AFIT operated facilities; however, on occasion such research is carried out in other U.S. Air Force Laboratories through a cooperative agreement.

Table B—Appointments to Graduate Students, 2009–10

Title of Appointee	Appointments Total	Appointments First year	Academic Load Allowed in Credit Hours	Hours of Service Per Week	Stipend for Academic Year ($)
*Research Assistantships	15	5	9	20 hrs	25,000–30,000
USAF, Navy, US Army	60	30	12	N/A	N/A
Total	75	35			

*Support from department or sponsored research funds.

5. Personnel Engaged in Separately Budgeted Research, 2008–09

Professorial faculty	19
Graduate students	33
Postdoctoral appointments	1
Total	53

6. Separately Budgeted Research Expenditures by Source of Support

	Departmental Research	Physics-related Research Outside Department
Federal government	$4,354,373	$
Total	$4,354,373	$

Table C—Separately Budgeted Research Expenditures

Research Specialty	No. of Grants	Expenditures ($)
Biophysics	1	150,000
Chemical Physics	4	160,666
Condensed Matter Physics	2	125,780
Educational Research	5	221,436
Laser Physics	10	1,360,770
Military Applications	11	1,728,724
Nuclear Engineering	6	491,997
Optical Physics	2	115,000
Space Research	1	5,675
Total	42	4,354,373

FACULTY

Professors

Bridgman, Charles J., Ph.D., North Carolina State, 1963. Emeritus. Nuclear weapons fallout modeling; nuclear weapon effects; neutral particle transport.

Burggraf, Larry W., Ph.D., University of Denver, 1981. Chemistry, nuclear instrumentation; environmental science.

Giles, Nancy C., Ph.D., North Carolina State, 1987. Department Head. Solid state physics, materials physics, optical and laser physics.

Hengehold, Robert L., Ph.D., Cincinnati, 1965. Solid state physics; optical and laser diagnostics.

John, George, Ph.D., Ohio State, 1952. Emeritus. Nuclear radiation detection; nuclear science; Mossbauer spectroscopy.

Mathews, Kirk A., Ph.D., Air Force Inst. of Tech., 1983. Nuclear weapons effects; numerical methods of radiation transport.

Perram, Glen P., Ph.D., Air Force Inst. of Tech., 1986. Chemical kinetics and molecular spectroscopy.

Weeks, David E., Ph.D., University of Arkansas, 1989. Chemical physics; molecular structure.

Wolf, Paul, Ph.D., Air Force Institute of Technology, 1985. Associate Dean for Academic Affairs. Molecular physics; thin film technology; laser effects.

Yeo, Yung Kee, Ph.D., Southern California, 1972. Solid state physics; electrical and optical properties of solids.

Associate Professors

Bailey, William F., Ph.D., Air Force Inst. of Tech., 1978. Theory of gas discharge lasers; reaction kinetics.

Marciniak, Michael A., Ph.D., Air Force Institute of Technology, 1995. Semiconductor lasers, photonics, semiconductor materials, optical systems.

Ries, Heidi R., Ph.D., Old Dominion University, 1987. Dean for Research-Nonlinear optical materials, EPR spectroscopy, laser processing of materials.

Tuttle, Ronald F., Ph.D., University of Missouri, Columbia, 1980. Remote sensing, signature analysis, exploitation of electromagnetic radiation.

Assistant Professors

Acebal, Ariel O., PhD, Utah State University, 2008. Space physics; space weather.

Cusumano, Salvatore L., Ph.D., University of Illinois, 1988. Directed energy, control theory, adaptive optics.

Fiorino, Steven T., Ph.D., Florida State University, 2003. Research Faculty. Atmospheric physics, radar meteorology.

Gross, Kevin C., Ph.D., Air Force Institute of Technology, 2007. Remote sensing, chemical physics, atomic and molecular physics.

Hawks, Michael R., PhD, Air Force Institute of Technology, 2006. Optics and remote sensing.

Holtgrave, Jeremy, Ph.D., Air Force Institute of Technology, 2003. Atomic and molecular physics, laser physics.

Kowash, Benjamin, PhD, University of Michigan, 2008. Nuclear engineering, radiation detection, radiation imaging.

Li, Alex Q., Ph.D., Shanghai Institute of Optics and Fine Mechanics, Chinese Academy of Sciences, 1990. Research faculty. Materials characterization, surface physics, atomic force microscope.

Magnus, Amy, Ph.D., Air Force Institute of Technology, 2003. Electro-optics, laser propagation, optical imaging.

McClory, John W., PhD, Air Force Institute of Technology, 2008. Nuclear engineering, radiation effects on electronics.

Petrosky, James C., Ph.D., Rensseler Polytechnic Inst. Radiation effects on semiconductor devices, radiation hardening, interaction of radiation with matter.

Randall, Robert, Ph.D., Univ. of Arizona, 2007. Atmospheric physics.

Sheely, Eugene, PhD., University of Idaho, 1997. Chemical physics, weapons of mass destruction.

Walli, Karl, Ph.D., Rochester Institute of Technology, 2010. Remote sensing.

RESEARCH SPECIALTIES AND STAFF

Theoretical

Atmospheric and Space Physics. Atmospheric modeling; near space environment; ionospheric physics. Acebal, Bailey, Fiorino, Randall.

Chemical Physics. Molecular structure. Burggraf, Sheely, Weeks.

Laser Physics. Laser-plasma interaction; optimization of pumping schemes; analysis of kinetic modeling of cavity discharge. Bailey, Cusumano, Magnus, Perram, Wolf.

Mathematical and Computational Physics. Bailey, Mathews, Weeks.

Meteorology. Atmospheric modeling; numerical weather prediction; radiative transfer. Fiorino, Randall.

Nuclear Weapons Physics. Nuclear weapons effects modeling; particle transport methods; fallout modeling; weapon systems survivability; nuclear power; space applications; radiation damage to materials. Bridgman, Mathews, Tuttle.

Plasma Physics. Ion source development; electric discharges; neutral and charged particle beams; high-power sources of radiation. Bailey.

Remote Sensing. Atmospheric propagation, target detection, and target recognition. Gross, Hawks, Perram, Tuttle, Walli.

Experimental

Chemical Physics. Shock tube reaction kinetics; excited state chemistry of laser candidates. Burggraf, Holtgrave, Perram, Sheely, Wolf.

Environmental Science. Sensor development; environmental materials; remote sensing; atmospheric transport. Burggraf, Perram, Tuttle.

Nuclear Physics. Nuclear radiation detection and applications. Burggraf, John, Kowash, Li, McClory, Petrosky.

Optical Physics. Laser spectroscopy; optical diagnostics; remote sensing; nonlinear optics; optical imaging; optical processing; thin-film optics. Cusumano, Gross, Hawks, Hengehold, Marciniak, Walli.

Solid State Physics. Optical and electrical characterization; semiconductor device physics; photonics. Giles, Hengehold, Marciniak, Ries, Yeo.

BOWLING GREEN STATE UNIVERSITY

DEPARTMENT OF PHYSICS AND ASTRONOMY

Bowling Green, Ohio 43403

Students Accepted For Degree	FIELDS Physics	Astronomy	Related Fields
Doctorate			
Master's	X		X

Table A—Faculty, Enrollments, and Degrees Granted

Research Specialty	2009–10 Faculty	Enrollment[1] Fall 2009 Mas-ter's	Doc-torate	No. of Degrees Granted[2] 2009–10 (2005–10) Mas-ter's	Terminal Master's	Doc-torate	Median No. of Years for 2005–10 Ph.D.'s
Acoustics	0	1	–	–	0(3)	–	–
Astronomy	0	4	–	–	0(3)	–	–
Astrophysics	0	–	–	–	–	–	–
Materials Sci./ Metallurgy	1	2	2	–	1(4)	–	–
Optics	0	1	–	–	0(7)	1(1)	5
Physics Education	0	8	–	–	5(16)	–	–
Other (Computational Phys.)	2	1	–	–	2(4)	–	–
Non-specialized	–	0	0	–	–	–	–
Total		17	2	–	8(37)	1(1)	–
Full-time Grad. Stud.		9					
Part-time Grad. Stud.		8					
First-year Grad. Stud.		4					
Median Years in Grad. Study (2009–10 Degrees)				–	2	5	

Undergraduate Degrees, 2009–10 (2005–10): 3(21))

[1]Students not yet committed to a research specialty are entered under non-specialized.
[2]Five-year totals in parentheses.

1. General

President: Carol Cartwright
Dean of Graduate School: Deanne Snavely
Department Chairman: Robert I. Boughton
Department Telephone Number: (419) 372-2421
Type of Institution: University
Control: Public
Setting: Small town
Total Faculty: 935
Total Graduate Faculty: 1,090
Total Students: 20,491
Total Graduate Students: 4,228
Annual Graduate Tuition:
In-state residents: Full-time—$17,232
Part-time—$478/credit
Out-of-state residents: Full-time—$28,194
Part-time—$784/credit
Tuition rates for: 2010–11
Deferred tuition plan: Yes
Other Fees: $1,650/semester (Full-time)
$330/semester (Part-time)
Term: Semester

2. Number of Faculty in Department

The combined total of full-time faculty in the three professorial ranks is 7. The combined total of full-time, part-time, and other faculty at all ranks is 14.

3. Admission, Financial Aid, and Housing

Address admission inquiries to: Graduate College Admissions Office
Graduate application fee required: $30
Admission deadline (Fall admission): None.
Admission information: For fall admission, 2010–11, 5 students were accepted from 38 applicants.
Admission requirements: B.S. with major in physics or minor in physics with major in cognate field. The GRE is required. For the GRE Advanced, no minimum acceptable score for admission is specified. Students from non-English speaking countries are required to demonstrate proficiency in English via the TOEFL exam. The minimum suggested score is 550.
Address financial aid inquiries to: Graduate Coordinator, Department of Physics and Astronomy
GAPSFAS application required: No
Financial aid deadline: 3/1
Loans available: Yes
Address housing inquiries to: Off-Campus Student Services (419) 372-2843
On-campus, single student housing available: Yes
Cost/term: Varies
On-campus, married student housing available: No

4. Graduate Degree Requirements

Master's: Plan I: 30 semester hrs.: 3.0 GPA; formal thesis; oral exam on thesis; 24 semester hrs. in residence; no language requirement. Plan II: 32 semester hrs.; 3.0 GPA; no thesis, but comprehensive exam and scholarly paper; 24 semester hrs. in residence; no language requirement.
Other Programs: MAT degree: 22–26 semester hrs. in academic major, 8–14 semester hrs. in education; 3.0 GPA, research paper required.
Thesis: Thesis may be written *in absentia*.
Special Equipment, Facilities, or Programs: Career-oriented terminal degree programs in astrophysics, computational physics, and solid state instrumentation are offered in addition to the Ph.D. preparatory Master's degree. Students can pursue a Ph.D. through a cooperative program with the University of Toledo. There is no charge to graduate students for use of either the time-share mainframe computers or the many workstations that are located in the Department. Remote access to the Ohio Supercomputer Center is also available. Laboratory facilities are available for research in non-linear optics, magnetic resonance, and low temperature transport properties. Additional facilities available for graduate student research include electron microscopes, CCD camera, and 0.5-m telescope.

Table B—Appointments to Graduate Students, 2008–09

Title of Appointee	Appointments Total	Appointments First year	Academic Load Allowed in Credit Hours	Hours of Service Per Week	Stipend for Academic Year ($)
			Semester		
Teaching Assistant	12	5		10	10,075[1,2]
Fellowship	2	–		20	15,231[1,2]
Total	14	5			

[1]Tuition and fees waived.
[2]Does not include 2/9 summer stipend.

5. Personnel Engaged in Separately Budgeted Research, 2008–09

Professorial faculty	6
Graduate students	6
Undergraduate students	7
Total	19

6. Separately Budgeted Research Expenditures by Source of Support

	Departmental Research	Physics-related Research Outside Department
Federal government	0	0
State and local government	$55,000	
Total	$55,000	0

Table C—Separately Budgeted Research Expenditures

Research Specialty	No. of Grants	Expenditures ($)
Materials Science	1	55,000
Total	1	55,000

FACULTY

Professors

Boughton, Robert I., Ph.D., Ohio State, 1968. Department Chair, Low temperature solid state; Physics education.

Duncan, G. Comer, Ph.D., Brandeis, 1971. Emeritus. General relativity and gravitation; computational physics; numerical relativity; acoustics; phonation.

Fulcher, Lewis P., Ph.D., Virginia, 1969. Theoretical nuclear physics; quantum electrodynamics; particle physics; acoustics; phonation.

Laird, John B., Ph.D., Yale, 1983. Stellar astronomy; physics education.

Smith, Dale W., Ph.D., Washington, 1978. Director of the Planetarium. Planetary astronomy.

Associate Professors

Crandall, A. Jared, Ph.D., Michigan State, 1967. Emeritus. Microcomputer interfacing; solar energy research.

Layden, Andrew C., Ph.D., Yale, 1993. Stellar astronomy.

Shirkey, Charles T., Ph.D., Ohio State, 1969. Emeritus.

Xi, Haowen, Ph.D., Lehigh, 1993. Computational physics.

Assistant Professor

Zamkov, Mikhail A., Ph.D., Kansas State, 2003. Nanosystems, thin films, photovoltaic cells.

Lecturer

Tiede, Glenn P., Ph.D., Ohio State, 1997. Stellar populations.

Instructors

Attygalle, Lilani, Ph.D., Univ. Toledo, 2008. Physics.

Blanton, Miles C., Ph.D., Univ. North Carolina, 2008. Astronomy.

Dellenbusch, Kate E., Ph.D., Univ. Wisconsin, 2008. Astronomy.

Mandell, Eric, Ph.D., Missouri, 2007. Physics.

Rogel, Allen B., Ph.D., Indiana, 2005. Astronomy.

RESEARCH SPECIALTIES AND STAFF

Theoretical

Acoustics. Human phonation, modeling the human larynx. Duncan, Fulcher.

Computational Physics. Supercomputer simulations in astrophysics and solid state physics; parallel algorithms on transputer arrays. Laird, Xi.

Materials Science. Non-linear optics; thin films. Xi.

Experimental

Acoustics. Measurements of air flow and pressure in human larynx. Fulcher.

Astronomy. Stellar spectroscopy and abundances; stellar populations; chemical evolution of galaxies; variable stars; cataclysmic variables. Laird, Layden, Rogel, Smith, Tiede.

Materials Science. Non-linear optics and devices; thin film devices; nanosystems; photovoltaic cells. Zamkov.

Physics Education. Active learning strategies and materials; teacher preparation. Boughton, Laird.

CASE WESTERN RESERVE UNIVERSITY

DEPARTMENT OF PHYSICS

Cleveland, Ohio 44106-7079

Students Accepted For Degree	FIELDS		
	Physics	Astronomy	Related Fields
Doctorate	X		
Master's	X		

1. General

Department Chair: Kathleen Kash
Associate Chair: Corbin Covault
Director of Graduate Studies: Corbin Covault
Director of Admissions: Walter Lambrecht
Director of Physics Entrepreneurship Program: Robert Brown
Director of Center for Education and Research in Cosmology and Astrophysics: Glenn Starkman
Department Telephone Number: (216) 368-4000
FAX: (216) 368-4671
E-mail: admissions@phys.case.edu
World Wide Web: http://physics.case.edu
Type of Institution: University
Control: Private
Setting: Urban
Total Faculty: 2,720
Total Students: 9,738*
Total Graduate Students: 5,510**
Annual Graduate Tuition:
All Graduate Students: Full-time—$25,740
Part-time—$1,430/credit
Tuition rates for: 2010–11
Deferred tuition plan: Yes
Other Fees: $1,320
Term: Semester

*Including nondegree
**Including part-time and professional students

2. Number of Faculty in Department

The combined total of full-time faculty in the three professorial ranks is 22.

3. Admission, Financial Aid, and Housing

Outstanding programs and cutting-edge research are offered in particle/astrophysics, cosmology, condensed matter physics, optics, nanoscience and nanotechnology, and medical imaging physics.

The department is ranked in the top 10% of the U.S. Physics Ph.D. programs in terms of faculty scholarly output by Academic Analytics, LLC. Faculty research support is high, averaging more than $300,000 per professor per year.

Strong research interactions exist with the Departments of Astronomy, Biomedical Engineering, Chemical Engineering, Chemistry, Macromolecular Science and Engineering and-Mechanical Engineering, as well as with Case Western Reserve University's School of Medicine and the Great Lakes Energy Institute. Superb on-campus facilities include the Electronics Design Center, the Center for Education and Research in Cosmology and Astrophysics, the Swagelok Center for Surface Analysis of Materials, the Center for Layered Polymeric Systems, and the Great Lakes Institute for Energy Innovation. The average time to Ph.D. is only 5.2 years, with 100% job placement. Approximately one-third of our Ph.D.s go into academic careers, one-third into industry, and one-third into national labs and/or physics-related careers.

The department and campus environments offer support to students through a wide variety of resources, including the Physics Graduate Student Association, the Graduate Student Senate, the Center for Women, and many on-campus cultural organizations. Planning for the establishment of an on-campus child care center is underway. The university is situated in a park-like setting within the city of Cleveland, and is only steps from the world-renowned Cleveland Orchestra and Cleveland Museum of Art. Theaters, concert venues, numerous other museums, major league sports, water recreation, and very affordable housing are all within ten minutes of campus, and outdoor summer and winter activities are a short drive away.

Address admissiion inquiries to: Admissions Director, Physics; (e-mail): admissions@phys.case.edu

Graduate application fee required: Free for paper application and $8 for on-line application (see http://www.phys.case.edu/grad/apply.php for application material)

Admission deadline (Fall admission): Jan. 15. With rare exceptions, first year students are not admitted for the spring semester beginning in January. Transfer students who intend to matriculate in January are treated on a case-by-case basis.

Admission information: For fall admission, 2010–11, 44 students were accepted from 233 applicants.

Admission requirements: For admission to the graduate programs, a Bachelor's degree in physics, mathematics, or related field is required, with a minimum undergraduate GPA of 3.0/4.0 specified. The GRE and physics subject exams are required. No minimum acceptable score is specified. Students from non-English speaking countries are required to demonstrate proficiency in English via the TOEFL exam. Minimum acceptable score for admission is 550 (paper), or 213 (computer), or 79 (internet based).

Undergraduate preparation assumed: Taylor, *Classical Mechanics*; Griffiths, *Electrodynamics*; Kittel, *Thermal Physics*; Griffiths, *Quantum Mechanics*; or equivalent textbooks; one or two years of advanced laboratory courses.

Address financial aid inquiries to: Director of Admissions, admissions@phys.case.edu

Address Physics Entrepreneurship inquiries to: PEP Director, rwb@case.edu

GAPSFAS application required: No
Financial aid deadline: January 15
Loans available: Yes
Address housing inquiries to: Dean, Graduate Studies
On-campus, single student housing available: No
On-campus, married student housing available: No

Table A—Faculty, Enrollments, and Degrees Granted

Research Specialty	2009–10 Faculty	Enrollment[1] Fall 2009 Master's	Doctorate	No. of Degrees Granted[2] 2009–10 (2005–10) Master's	Terminal Master's	Doctorate	Median No. of Years for 2009–10 Ph.D.'s
Applied Physics	6	–	0	0(0)	0(0)	0(2)	5
Astrophysics	7	–	13	3(6)	0(0)	4(12)	5.5
Atomic, Molecular, & Optical Physics	5	–	0	0(0)	0(0)	0(0)	–
Biophysics	2	–	1	0(0)	0(0)	1(1)	–
Condensed Matter Physics	10	–	13	4(20)	1(2)	3(13)	5
Fluids & Rheology	2	–	0	0(0)	0(0)	0(1)	–
Low Temperature Physics	1	–	1	0(2)	0(0)	0(2)	5.5
Imaging Physics	1	–	3	1(5)	0(0)	1(5)	5
Nanoscience	8	–	3	0(0)	0(0)	1(1)	5
Optics	10	–	5	1(8)	0(0)	0(4)	5
Particles & Fields	7	–	3	3(3)	0(0)	0(3)	5
Physics Entrepreneurship	2	8	0	0(0)	4(19)	0(0)	–
Polymer Physics/ Science	4	–	1	0(0)	0(0)	0(1)	5.5
Relativity & Gravitation	4	–	4	0(0)	0(0)	0(2)	5
Statistical & Thermal	4	–	0	0(0)	0(0)	0(0)	–
Surfaces & Interfaces	3	–	1	0(0)	0(0)	0(2)	5
Total		8	45	9(44)	5(21)	10(48)	
Full-time Grad. Stud.		5	45				
Part-time Grad. Stud.		3	0				
First-year Grad. Stud.		4	11				
Median Years in Grad. Study (2009–10 Degrees)				1.5	1.7	5.2	–

Undergraduate Degrees, 2009–10 (2005–10): 36(98)

[1]Students not yet committed to a research specialty are entered under non-specialized.
[2]Five-year totals in parentheses.

4. Graduate Degree Requirements

Master's: 27 graduate credit hours in approved program including six required hours; Master's exam required; thesis option; no residence or language requirement.

Doctorate: Up to 36 hours of coursework is required (may be reduced by graduate coursework done elsewhere); comprehensive and topical exams, dissertation, and dissertation exam required; one year residency; no language exam required. see http://www.phys.cwru.edu/grad/phd.php

Thesis: Thesis may be written *in absentia*.

Special Equipment, Facilities, or Programs: A wide variety of facilities and programs is available within the department, and in addition there are collaborative programs with other departments, including Macromolecular Science, Chemistry, Astronomy, Materials Science, and the Medical School. In astrophysics research, experiments in collaboration with other universities are being performed to determine the nature of elementary particle dark matter in the universe by direct detection in underground detectors, while other experiments are performed to search for high-energy gamma rays and to explore the Cosmic Microwave Background. High-energy physics experiments are undertaken at various National Laboratories. Theoretical work on astrophysics and cosmology, as well as particle, condensed matter physics, and quantum computing, covers a large number of research topics. Condensed matter studies include measurements of dielectric; optical and nonlinear optical properties; thin-film properties; nanoscopic physics; quantum computing; liquid crystal and complex fluid properties; semiconductor crystal growth; quantum wells, wires, and dots; other nanoscopic structures; spintronics; organic electronics; and photovoltaics. A wide range of facilities is available in surface physics and in optics. Among the collaborative programs are experimental and theoretical studies of phase transitions in polymers and of liquid crystals, photovoltaic materials, surface physics, the physics of imaging, fluid physics, dark matter detection, and measurements of fundamental parameters in cosmology. Departmental computing facilities are extensive, and are used in both research and courses. Weekly specialized seminars in particle/astrophysics and condensed matter physics take place, in addition to a weekly departmental colloquium.

The Physics Department has been recognized six times times by the U.S. Dept. of Education as meeting vital national needs. Special graduate fellowships are available.

In addition to a traditional physics program, the Department maintains a Physics Entrepreneurship Masters degree program. The program is designed to empower physicists as entrepreneurs and to enable students and graduates to build on their physics skills to start new high-tech businesses or to launch new product lines in existing companies.

Table B—Appointments to Graduate Students, 2009–10

Title of Appointee	Appointments Total	First year	Academic Load Allowed in Credit Hours	Hours of Service Per Week	Stipend for Academic Year ($)*
			Semester		
Teaching Assistant	14	10	9	14–16	22,380
Research Assistant	21	0	9	14–16	22,380
Other	10	1	9	14–16	22,380
Total	45	11			

*Plus tuition, for 12 month Academic year 2010-11.

5. Personnel Engaged in Separately Budgeted Research, 7/09–7/10

Full Time Professorial faculty	22
Visiting faculty	5
Postdoctoral appointments	17
Graduate students	45
Undergraduate students	98
Total	187

6. Separately Budgeted Research Expenditures by Source of Support

	Departmental Research	Physics-related Research Outside Department
Federal government	$6,771,944	$
Other	723,859	
Total	$7,495,853	$

Table C—Separately Budgeted Research Expenditures

Research Specialty	No. of Grants	Expenditures ($)
Bio and Medical Physics	5	188,188
Condensed Matter and Optics	26	2,397,373
General	6	668,468
Particles, Fields, & Astrophysics	25	4,241,824
Total	61	7,495,853

FACULTY

Professors

Akerib, Daniel S., Ph.D., Princeton, 1991. Experimental particle astrophysics, dark matter, low temperature detectors, particle physics.

Alexander, Iwan, Ph.D., Washington State 1981. Joint appointment with mechanical engineering, fluid physics, microgravity.

Brown, Robert W., Ph.D., MIT, 1968. Institute Professor. Theoretical physics; elementary particles; imaging physics.

Chottiner, Gary S., Ph.D., Maryland, 1980. Experimental condensed matter physics; surface physics.

Covault, Corbin, Ph.D., Harvard, 1991. Experimental high energy astrophysics.

Kash, Kathleen, Ph.D., MIT 1982. Chair of the Department. Experimental condensed matter physics; optics; mesoscopic physics.

Kowalski, Kenneth L., Ph.D., Brown, 1963. Theoretical physics; nuclear and elementary particle physics.

Lambrecht, Walter R. L., Ph.D., Ghent, 1980. Theoretical condensed matter physics; electronic structure of materials.

Luck, Earle, Ph.D., Texas, 1977. Joint appointment with Astronomy. Stellar and galactic chemical evolution; stellar abundance analysis; spectrum synthesis techniques.

Mihos, Christopher, Ph.D., Michigan, 1992. Joint appointment with Astronomy. Observational and computational astrophysics; galactic dynamics; galaxy clusters; galaxy evolution.

Morrison, Heather, Ph.D., Australian National Univ., 1988. Joint appointment with Astronomy. Galaxy structure, formation, and evolution, especially Milky Way and Local Group.

Petschek, Rolfe G., Ph.D., Harvard, 1981. Theoretical physics; statistical physics; condensed matter physics.

Rosenblatt, Charles, Ph.D., Harvard, 1978. Experimental condensed matter physics; liquid crystals and complex fluids; optics; microgravity; fluid physics.

Ruhl, John, Ph.D., Princeton, 1993. Experimental particle astrophysics, cosmic microwave background.

Shutt, Thomas, Ph.D., Agnar Pytte Professor. Univ. of California, Berkeley, 1993. Experimental particle astrophysics, dark matter, neutrino physics.

Sibata, Claudio, Ph.D., Wisconsin, 1984. Joint appointment with medicine. Biophotonics.

Singer, Kenneth D., Ph.D., Pennsylvania, 1981, Ambrose Swaysey Professor. Experimental physics; nonlinear optics; organic electronics; photovoltaics.

Starkman, Glenn, Ph.D., Stanford 1988. Theoretical physics, cosmology, particle physics, and astrophysics.

Tabib-Azar, Massood, Ph.D., RPI, 1986. Joint appointment with Electrical Engineering. Experimental device physics.

Taylor, Cyrus C., Ph.D., MIT, 1984. Dean of the College of Arts and Sciences, Albert A. Michelson Professor of Physics. Theoretical physics, theoretical and experimental elementary particle physics; physics of entrepreneurship.

Taylor, Philip L., Ph.D., Cambridge, 1962. Perkins Professor of Physics. Theoretical condensed matter physics; physics of polymers and liquid crystals.

Tien, Norman, Ph.D., U. California at San Diego, 1993. Microelectromechanical systems.

Associate Professors

Beddar, A. S., Ph.D., Wisconsin, 1990. Joint appointment with medicine. Biophotonics.

Dragowsky, Michael R., Ph.D., Oregon State University, 1999. Experimental particle astrophysics, dark matter.

Jankowsky, Eckhard, Ph.D., Dresden Institute of Technology, 1996. Joint appointment with Biochemistry. Experimental biophysics, single molecule fluorescence, enzyme kinetics.

Mathur, Harsh, Ph.D., Yale, 1994. Theoretical condensed matter physics; localization and mesoscopic physics. Cosmology and particles.

McEnery, Maureen, Ph.D., Johns Hopkins, 1993, Joint appointment with Medical School. Biophysics.

Shan, Jie, Ph.D., Columbia, 2001. Experimental condensed matter physics; ultrafast spectroscopy; terahertz time-domain spectroscopy.

Assistant Professors

Berezovsky, Jesse, Ph.D., University of California at Santa Barbara, 2007. Experimental condensed matter. Transport, quantum coupling of spins and photons, quantum information.

de Rham, Claudia, Ph.D., University of Cambridge, 2005. Theoretical cosmology and particle physics.

Gao, Xuan, Ph.D., Columbia, 2003. Experimental condensed matter physics; Applied physics; electronic properties of low dimensional nanostructures; semiconductor nanowires; nanosensors.

Tolley, Andrew J., Ph.D., University of Cambridge, 2003. Theoretical cosmology and particle physics. Dark energy.

Zehavi, Idit, Ph.D., Hebrew University of Jerusalem, 1999. Joint appointment with Astronomy. Theoretical astrophysics; cosmology; large-scale structure; galaxy and structure formation.

RESEARCH SPECIALTIES AND STAFF

Theoretical

Astrophysics, Cosmology, and Gravitational Physics. Neutrino astrophysics; early universe cosmology; dark matter; darkenergy; large scale structure; gravitational lensing; black hole evaporation; stellar evolution; cosmic strings; cosmic microwave background. de Rham, Mathur, Starkman, C. Taylor, Tolley. 3 postdoctoral fellows.

Condensed Matter Physics. Electronic properties of metals and semiconductors; photovoltaics, crystal growth; transport properties in ordered and disordered materials; band structure; deformation potentials; localization, thermo-electricity; interface and surface physics; lattice vibrations. Lambrecht, Mathur, P. Taylor, Petschek. 3 postdoctoral fellows.

Elementary Particles and Fields. Electroweak theory; standard model; cosmology; black hole physics; superstring theory SSC physics; supersymmetry, field theories at finite temperature; quark-gluon plasma; diffractive excitation mechanisms. de Rham, Kowalski, Mathur, Starkman, C. Taylor, Tolley. 2 postdoctoral fellows.

Imaging Physics. Algorithm development; bio-data acquisition and analysis; rf coil theory; inverse scattering theory; diagnostic imaging. Brown.

Liquid Crystals. Phase transitions, dynamics, symmetry and surface effects, nonlinear behavior. Petschek, P. Taylor. 1 postdoctoral.

Polymer Physics. Equations-of-state; phase transitions; dynamical behavior; piezoelectric effects; polymer liquid crystals. Petschek, P. Taylor.

Statistical Physics. Statics and dynamics of phase transitions; pattern formation and dendritic growth; liquid crystals, polymeric liquid crystals, complex fluids; oscillatory chemical reactions; membrane noise. Petschek, P. Taylor.

Experimental

Astrophysics and Cosmic Rays. Dark matter; low temperature detectors; neutrino experiments; cosmic microwave background; high energy cosmic rays; gamma ray astrophysics. Akerib, Covault, Dragowsky, Ruhl, Shutt, 4 postdoctoral fellows.

Condensed Matter Physics. Electronic structure of metals and alloys; surfaces; crystal growth; thin films; amorphous films; dielectric and cohesive properties; dielectric and mechanical relaxation; organic electronics; transport properties of nanostructures, quantum wells, mesoscopic systems, fuel cells; soft matter. Berezovsky, Chottiner, Gao, Kash, Rosenblatt, Singer, Shan. 5 postdoctoral fellows.

Elementary Particles and Fields. Collider physics. C. Taylor, Akerib, Covault.

Liquid Crystals and Complex Fluids. Phase transitions; optical, magnetic, and electrical properties, microgravity, nanostructured LCs, symmetry effects. Rosenblatt. 2 postdoctoral fellows.

Nanoscopic Physics. Quantum dots, wires, molecular electronics, nanoscopic surface modification, nanowires and sensors. Berezovsky, Gao, Kash, Rosenblatt, Shan, Singer. 3 postdoctoral fellows.

Optics. Linear and nonlinear optical properties of organic, polymeric materials, and mesoscopic systems, photovoltaics, ultrafast spectroscopy. Berezovsky, Kash, Rosenblatt, Shan, Singer. 2 postdoctoral fellows.

Polymer Physics. Phase transformations; dielectric properties; magnetic and electric field effects; optical mechanical properties. Rosenblatt, Shan, Singer.

Surface Physics. Surface magnetization; secondary electron emission; surface analysis; physi- and chemisorption. Chottiner.

Fluid Physics. Interface instabilities, magnetic levitation. Rosenblatt.

Special Programs

Center for Education and Research in Cosmology and Astrophysics: A new center created in collaboration with the Cleveland Museum of Natural History's Shafran Planetarium and CWRU's Astronomy Department to promote research and education in cosmology and astrophysics. http://cerca.case.edu

Institute for Advanced Materials. The Institute for Advanced Materials brings together internationally recognized faculty researchers to engage in multi-disciplinary efforts on a broad range of materials that not only are ubiquitous in everday life, but are cornerstones to many key technology areas. Specifically, IAM focuses on strategic research that impacts national needs in human health, energy, and the environment. The four focus areas are: Fundamental Materials Research, Materials for Human Health, Materials for Energy, and Materials for Sustainability. http://iam.case.edu

The Institute for the Science of Origins. The Institute for the Science of Origins (ISO) is a collaborative team of faculty members and researchers from diverse scientific disciplines seeking to understand how complex systems emerge and evolve, from the universe to the mind, from microbes to humanity. http://www.case.edu/origins

The Michelson Postdoctoral Lectureship is an annual prize sponsored by Case Western Reserve University. It is awarded to an outstanding recent Physics Ph.D. based on an international competition. The winner spends one week in residence in the Department, and delivers several seminars and a departmental colloquium on his/her research.

Physics Entrepreneurship Masters Degree: To empower physicists as entrepreneurs and enable graduate students to build on their physics skills to start new high-tech businesses or to launch new product lines in existing companies.

Workshops and Conferences: The Department regularly holds national and international meetings on a variety of topics. Recent conferences have included: *The Future of Cosmology, Future Physics and Future Facilities, The Cosmic Microwave Background, Great Lakes Cosmology Workshop, the American Vacuum Society Conference, International Workshop on MRI, Einstein's Legacy, and Confronting Gravity.*

Outreach: The Department works with high school teachers and students to improve science education locally and nationally. The Department also hosts a web site of a national program in astronomy education called *Ask an Astronomer*.

Recent Books by Faculty include Magnetic Resonance Imaging: Physical Properties and Sequence Design (Robert Brown), A Quantum Approach to Condensed Matter Physics (Philip L. Taylor).

NSF REU in Physics. A 10-week program hosting ten undergraduate students for physics research projects on campus during the summer.

International Programs. The department spearheaded three university-wide student and faculty exchange programs with the University of Calabria (Italy), Nagaoka University of Science and Technology (Japan) and Université Pierre et Marie Curie (U. Paris 6, France).

CASE WESTERN RESERVE UNIVERSITY

WARNER AND SWASEY OBSERVATORY

Cleveland, Ohio 44106-7215

Students Accepted For Degree	FIELDS		
	Physics	Astronomy	Related Fields
Doctorate		X	

Address housing inquiries to: Housing Bureau
On-campus, single student housing available: No
On-campus, married student housing available: No

1. General

Department Chair and Director: Chris Mihos
Department Telephone Number: (216) 368-3728
FAX: (216) 368-5406
email: wsobs@grendel.astr.cwru.edu
World Wide Web: http://astronomy.case.edu
Type of Institution: University
Control: Private
Setting: Urban
Total Faculty: 2,720
Total Students: 9,738*
Total Graduate Students: 5,510**
Annual Graduate Tuition:
All Graduate Students: Full-time—$25,740
Part-time—$1,430/credit
Tuition rates for: 2010–11
Deferred tuition plan: Yes
Annual Other Fees: $1,320
Term: Semester

*Including non-degrees
**Including part-time, and including professional students

2. Number of Faculty in Department

The combined total of full-time faculty in the three professorial ranks is 4. The combined total of full-time, adjunct and other faculty are 7.

3. Admission, Financial Aid, and Housing

Address admission inquiries to: Chris Mihos, Chair, Dept. of Astronomy, Director, Warner and Swasey Obs.
Graduate application fee required:
Admission deadline (Fall admission): 15 Jan
Admission information: For fall admission, 2010–11, 1 students were accepted from 12 applicants.
Admission requirements: For admission to the graduate programs, a Bachelor's degree in physics or astronomy is required. The Department requires a minimum undergraduate GPA of 3.00 (out of 4.00 or equivalent) from an accredited school. The Graduate Record Exam is required as is the Physics Subject area test. Three letters of recommendation are required. For students whose first language is not English the TOEFL exam is required with a minimum score of 550 (paper) or 213 (computer). The Department reserves the right to modify or waive these requirements as it sees fit within the minimum criteria established by the Graduate School.
Address financial aid inquiries to: C. Mihos, Chair, Dept. of Astronomy, Warner and Swasey Obs.
GAPSFAS application required: No
Financial aid deadline: 15 Jan.
Loans available: Yes

Table A—Faculty, Enrollments, and Degrees Granted

Research Specialty	2009–10 Faculty	Enrollment[1] Fall 2009		No. of Degrees Granted[2] 2009–10 (2005–10)			Median No. of Years for 2009–10 Ph.D.'s
		Mas-ter's	Doc-torate	Mas-ter's	Terminal Master's	Doc-torate	
Astronomy	4	0	2	0(0)	0(0)	0(1)	5
Total		0	2	0(0)	0(0)	1(1)	
Full-time Grad. Stud.		0	2				
Part-time Grad. Stud.		0	0				
First-year Grad. Stud.		0	0				
Median Years in Grad. Study (2009–10 Degrees)				–	–	5	–
Undergraduate Degrees, 2009–10(2005–10): 3(34)							

[1] Students not yet committed to a research specialty are entered under non-specialized.
[2] Five-year totals in parentheses.

4. Graduate Degree Requirements

Doctorate: Approximately two years of courses (with cumulative quality-point average of 3.0 or above); must pass satisfactorily a qualifying exam administered by the department; dissertation and defense. Residency requirement for the Ph.D. is one academic year, full-time. There are no total course hours required.
Thesis: Thesis may be written *in absentia*.
Special Equipment, Facilities, or Programs: The Department owns and operates the Burrell Schmidt telescope, located at Kitt Peak National Observatory in Arizona. This telescope has a 2.6 square degree field and is optimized for deep surface photometry. We are also members of the Sloan Digital Sky Survey III, which provides opportunities for research into Milky Way stellar populations and large scale structure.

Table B—Appointments to Graduate Students, 2009–10

Title of Appointee	Appointments		Academic Load Allowed in Credit Hours	Hours of Service Per Week	Stipend for Academic Year ($)*
	Total	First year			
			Semester		
Teaching Assistant	1	0	9	10	$22,240
Research Assistant	1	0	9	–	$22,368
Total	2	0			

*Plus tuition stipend for academic year.

5. Personnel Engaged in Separately Budgeted Research,

Professorial faculty	4
Posdoctoral appointments	1
Graduate students	2
Undergraduate students	9
Nonteaching research personnel	1
Total	17

6. Separately Budgeted Research Expenditures by Source of Support

	Departmental Research	Physics-related Research Outside Department
NSF	$163,575	0
Private non-profit organizations	$78,609	0
Total	$242,184	0

Table C—Separately Budgeted Research Expenditures

Research Specialty	No. of Grants	Expenditures ($)
Astronomy.	19	242,184
Total	19	242,184

FACULTY

Professors

Luck, R. Earle, Ph.D., Texas, Austin, 1977. Worcester R. and Cornelia B. Warner Professor of Astronomy .

Mihos, J. Christopher, Ph.D., University of Michigan, 1992. Chair and Director of the Warner and Swasey Observatory.

Morrison, Heather L., Ph.D., Australian National University, 1989.

Ruhl, John, Ph.D., Princeton, 1993. Joint appointment with Physics.

Starkman, Glenn D., Ph.D., Stanford, 1988. Joint appointment with physics.

Assistant Professor

Zehavi, Idit, Ph.D., Hebrew University of Jerusalem, 1999.

Adjunct Professor

Kriessler, Jeff, Ph.D., Michigan State, 1997.

Observatory Manager

Harding, Paul, Ph.D., U. Arizona, 2001.

RESEARCH SPECIALTIES AND STAFF

Observational

Galactic Structure and Stellar Populations. Morrison, Harding.

Galaxy Evolution; Galaxy Clusters. Mihos.

Studies of stellar and galactic chemical evolution; high dispersion spectroscopy and abundance analysis. Luck.

CWRU astronomy owns and operates the Burrell Schmidt wide-field telescope at Kitt Peak Observatory, and is also a member of the SDSS-III collaboration.

Theoretical

Cosmology, Large-Scale Structure; Galaxy and Structure Formation. Zehavi, Mihos

Galactic Evolution and Galactic Dynamics. Mihos, Morrison.

KENT STATE UNIVERSITY

DEPARTMENT OF PHYSICS

Kent, Ohio 44242

Students Accepted For Degree	FIELDS		
	Physics	Astronomy	Related Fields
Doctorate	X		X
Master's	X		X

1. General

President: Dr. Lester A. Lefton
Vice President for Research: Dr. John L. West
Department Chair: Prof. Dr. James T. Gleeson
Department Telephone Number: (330) 672-2246
Website: http//:www.kent.edu/cas/physics/
Type of Institution: University
Control: Public
Setting: Suburban
Total Faculty: 669 (Tenure-Track)
Total Students: 25,100
Total Graduate Students: 5,200
Annual Graduate Tuition: 2010–11
In-state residents: Full-time—$437/credit
11 hrs. or more/sem. $4,803
6 hrs. (summer) $2,622
Total for year $12,228
Out-of-state residents: Full-time—$779/credit
11hrs. or more/sem. $8,561
6 hrs. (summer) $4,674
Total for year $21,796
Tuition rates for: Fall 2010–11, effective Fall semester
Deferred tuition plan: Yes
Term: Semester

2. Number of Faculty in Department

The combined total of full-time faculty in the three professorial ranks is 19. The combined total of full-time, part-time, and other faculty at all ranks is 21.12.

3. Admission, Financial Aid, and Housing

Address admission inquiries to: Dr. Declan Keane, Graduate Coordinator, Department of Physics (physgradprogram@kent.edu)
Graduate application fee required: $30 ($60 for international applicants)
Admission deadline (Fall admission): 1/31 (Graduate study may also be initiated during the spring). Early application is recommended for consideration for financial aid.
Admission information: Typical class consists of 8–12 new students.
Admission requirements: For admission to the graduate program, a Bachelor's degree in physics or a related discipline is required with a minimum undergraduate GPA of 3.0 on a scale of 4.0. The subject (physics) GRE test is normally required. Students from non-English speaking countries are required to demonstrate proficiency in English with a minimum TOEFL score of 560 (paper-based); 220 (computer-based); or 83 (Internet-based).
Address financial aid inquiries to: Dr. Declan Keane, Graduate Coordinator, Department of Physics
GAPSFAS application required: No
Financial aid deadline: Same as for Admission
Address on-campus housing inquiries to: Department of Residence Services, Kent State University, P.O. Box 5190, Kent, OH 44242-0001 Email: housing@res.kent.edu
Phone: 330-672-7000 (or 1-800-706-8941)
On-campus, graduate student housing available: Yes
Cost/semester: dorms, 2-person double–$2,588; 2-person suite–$3,428, single-$2,844; single with shared bath–$3,438; single semi-suite with shared bath–$3,770.
On-campus, married student housing available: Yes
Cost/month: $720–1 bedroom
$750–2 bedrooms
All utilities (including cable TV and Internet access), but excluding phone are included in the rates shown.
Off-campus: In addition, a variety of reasonably-priced rental housing can be found in the Kent area. Campus Bus Service provides a transportation network for the Kent campus and links the campus with shopping centers and residential neighborhoods in nearby communities.

Table A—Faculty, Enrollments, and Degrees Granted

Research Specialty	2009–10 Faculty	Enrollment[1] Fall 2009		No. of Degrees Granted[2] 2009–10 (2005–10)			Median No. of Years for 2009–10 Ph.D.'s
		Master's	Doctorate	Master's	Terminal Master's	Doctorate	
Biological Physics	4	0	2	0(1)	0(0)	0(2)	–
Liquid Crystals & Soft Condensed Matter	6	1	16	1(3)	0(1)	5(9)	8.0
Low Temperature Physics & Superconductivity	7	0	1	0(2)	0(1)	4(1)	6.7
Nuclear Physics	7	0	8	2(3)	2(2)	4(7)	7.0
Non-specialized	0	0	17	0(0)	0(1)	0(0)	–
Total		1	44	3(9)	2(5)	13(19)	
Full-time Grad. Stud.		1	42				
Part-time Grad. Stud.		2	0				
First-year Grad. Stud.		0	11				
Median Years in Grad. Study (2009–10 Degrees)				2.3	3.7	7.2	
Undergraduate Degrees, 2009–10 (2005–10):12(48)							

[1]Students not yet committed to a research specialty are entered under non-specialized.
[2]Five-year totals in parentheses.

4. Graduate Degree Requirements

Master's: 32 semester-hours of graduate courses. The Master of Sciences (M.S.) degree requires a thesis whereas the Master of Arts (M.A.) degree does not.
Doctorate: 90 semester-hours of courses, seminar, and research work beyond the Bachelor's degree or 60 hours beyond M.S. required. Passing the candidacy examination, and successful defense of dissertation are required.
Thesis: Thesis may be written *in absentia*.
Special Equipment, Facilities, or Programs-
Condensed Matter: Extensive facilities focused on liquid crystal research, including light scattering, conventional and

synchrotron x-ray, and neutron scattering studies. Millikelvin cryostats, high field magnets, and SQUID magnetometry are employed in condensed matter research. The Department also utilizes the facilities in other KSU units, e.g., microfabrication equipment at the Liquid Crystal Institute.

Nuclear and Particle Research: Experimental research is performed at accelerator facilities at JLAB, RHIC, and MAMI-C (Mainz).

Planetarium: The Department operates a planetarium and an observatory for instructional purposes and community outreach.

Table B—Appointments to Graduate Students, 2009–10

Title of Appointee	Appointments Total	Appointments First year	Academic Load Allowed in Credit Hours	Hours of Service Per Week	Stipend for Academic Year ($)
			Semester		
Teaching Assistant	15.5	11	15	20	18,000[1]
Research Assistant	28.5	0	15	20	18,000[1]
Total	44.0	11			

[1]All tuition and fees (resident and non-resident) are provided, as well as partial health care insurance. Summer stipends vary from $3,000 to $6,000. The stipend for 2010–11 will be $18,000.

5. Personnel Engaged in Separately Budgeted Research, 7/08–6/09

Professorial faculty	18
Postdoctoral appointments	5
Graduate students	16.5
Research Engineer	1
Senior Research Fellow	1
Total	41.5

6. Separately Budgeted Research Expenditures by Source of Support 7/08–6/09

	Departmental Research
Federal government	$1,888,685.53
State government	66,383.86
Business and industry	45,264.42
Institution	149,518.80
Total	$2,149,852.61

Table C—Separately Budgeted Research Expenditures

Research Specialty	No. of Grants	Expenditures ($)
Condensed Matter Physics	24	939,592.70
Nuclear Physics	12	1,206,331.13
Physics Education	–	3,928.78
Total	36	2,149,852.61

FACULTY

Professors

Allender, David W., Ph.D., University of Illinois at Urbana-Champaign, 1975. Condensed matter theory; liquid crystals, biophysics.

Almasan, Carmen C., Ph.D., University of South Carolina, 1989. Experimental condensed matter physics; superconductivity, magnetism.

Finotello, Daniele, Ph.D., SUNY at Buffalo, 1985. Nuclear magnetic resonance and calorimetry studies of liquid crystals.

Gleeson, James T., Ph.D., Kent State University, 1991. Chair of the Department. Experimental liquid crystal physics; dynamics.

Keane, Declan, Ph.D., University College (Dublin), 1981. Relativistic nuclear collisions.

Kumar, Satyendra, Ph.D., University of Illinois at Urbana-Champaign, 1981. Liquid crystal physics and applications; biophysics; nanostructures.

Manley, D. Mark, Ph.D., University of Wyoming, 1981. Experimental/phenomenological hadronic physics; baryon spectroscopy.

Mann, Elizabeth K., Ph.D., Université Pierre et Marie Curie (Paris VI), 1992. Experimental soft-matter physics, surface physics.

Margetis, Spyridon, Ph.D., University of Frankfurt (Germany), 1990. Relativistic nuclear collisions.

Petratos, Gerassimos G., Ph.D., American University, 1988. Experimental nuclear and particle physics.

Quader, Khandker, Ph.D., SUNY at Stony Brook, 1983. Condensed matter theory; normal and pairing state properties in correlated matter, unconventional superconductivity; superfluidity in ultracold atoms.

Sprunt, Samuel N., Jr., Ph.D., Massachusetts Institute of Technology, 1989. Experimental liquid crystal physics, phase transitions.

Tandy, Peter C., Ph.D., Flinders University of South Australia, 1973. Theoretical nuclear and particle physics.

Associate Professors

Ellman, Brett D., Ph.D., University of Chicago, 1992. Experimental condensed matter physics; superconductivity.

Katramatou, A. Mina T., Ph.D., American University, 1988. Physics education; nuclear and particle physics.

Portman, John J., Ph.D., University of Illinois at Urbana-Champaign, 2000. Condensed matter theory; biological physics, protein structure and interactions.

Schroeder, Almut, Ph.D., Universität Karlsruhe (Germany), 1991. Experimental condensed-matter physics; magnetism, neutron scattering.

Assistant Professor

Balci, Hamza, Ph.D., University of Maryland, 2004. Experimental biophysics: DNA repair, single molecule biophysics, biomedical physics and imaging.

Dzero, Maxim, Ph.D., Florida State University, 2003. Condensed matter theory; emergent behavior in complex materials; Kondo/topological insulator; non-equilibrium superfluidity.

Staff

Aldhizer, Wade. Machinist and Shop Supervisor.

Baldwin, Alan R., M.S.E.E., Clarkson College of Technology, 1971, Ph.D., Kent State University, 1994. Research Engineer.

Putman, Gregory W., M.S., Indiana State University, 1997. Academic Laboratory Manager.

Zhang, Wei-ming, Ph.D., The Johns Hopkins University, 1985. Senior Research Fellow.

Faculty Emeriti

Anderson, Byron D., Ph.D., Case Western Reserve University, 1972. Experimental medium-energy nuclear physics.

Christensen, Stanley H., Ph.D., Cornell University, 1963.Physics education.

Doane, J. William, Ph.D., University Missouri, 1965. Nuclear magnetic resonance and liquid crystals.

Fai, George, Ph.D., Eötvös Loránd (Budapest), 1974. Theoretical nuclear physics; relativistic nuclear collisions.

Gelerinter, Edward, Ph.D., Cornell University, 1966. Electron paramagnetic resonance and optical studies of liquid crystals.

Hubin, Wilbert N., Ph.D., University of Illinois at Urbana-Champaign, 1969. Computer hardware and physics education.

Lee, Michael A., Northwestern University, 1977. Condensed-matter theory; biophysics; physics education.

Madey, Richard, Ph.D., University of California, Berkeley, 1952. Experimental medium-energy nuclear physics; transport of fluids through porous media.

Moroi, David S., Ph.D., The Johns Hopkins University, 1959. Theory of particles with internal degrees of freedom; liquid crystal theory.

Spielberg, Nathan, Ph.D., Ohio State, 1952. X-ray physics; structure of liquid crystals.

Uhrich, David L., Ph.D., University of Pittsburgh, 1966. Mössbauer effect.

Watson, John W., Ph.D., University Maryland, 1970. Experimental nuclear and particle physics.

RESEARCH SPECIALTIES AND STAFF

Theoretical

Biophysics. Statistical mechanics of biopolymers and membranes. Theory and computer modeling of biopolymer dynamics: protein folding; protein-protein interactions; RNA dynamics. Computational modeling of complex biological reaction networks. Structure and function of lipid membranes; protein interactions at molecular interfaces. Bioinformatics. Allender, Portman.

Condensed Matter Physics. Temperature, surfaces, and electric, magnetic, or optical fields driven phase transitions. Landau theory and elastic deformations, wetting and biaxiality in liquid crystals. Accurate electronic structure of atoms and molecules. Numerical solutions for quantum systems using Green's Function Monte Carlo Method. Optics of liquid crystals and optical properties of liquid crystal devices. Physics of novel correlated systems including high temperature superconductors, layered materials, heavy fermions, Kondo/topological insulation, ultracold atomic systems, correlated quantum matter in atomic traps. Phenomenological and microscopic approaches employing methods of quantum statistical mechanics, main-body Green's function and functional integral techniques, Fermi liquid theory, transport equations, Allender, Dzero, Portman, Quader.

Liquid Crystal Physics. Phase transitions; diffusion; electrohydrodynamics; electric and magnetic properties, technological applications in LCDs, photonic and other devices. Allender.

Nuclear and Particle Physics. Studies of the equation of state of nuclear matter. Development of methods to analyze information provided by experiment on the quark-gluon plasma and the phase transition leading to this state of matter. Tandy.

Experimental

Biophysics. DNA repair, single molecule biophysics, biomedical physics, and imaging. Static and dynamic properties of amphiphilic materials such as lipids, surfactants and cholesterols. Fluorescence correlation spectroscopy on model membranes. Balci, Kumar, Mann.

Condensed Matter Physics. Transport and magnetic behavior of high temperature superconductors. Exotic superconductors, unusual magnets, novel insulators, quantum phases transitions. Momentum-space geometry of the gap structures of heavy fermions and high-temperature superconductors. Measurements of the magnetic susceptibility, magneto-resistance, Hall effect, neutron scattering. The superfluid transition in helium films and confined to porous media near the lambda point. Almasan, Ellman, Finotello, Schroeder.

Liquid Crystal Physics. Studies of phase transitions in bulk and confined to porous media. Determination of critical exponents, orientational and translational order using calorimetry, x-ray, nuclear magnetic resonance, and optical techniques. Studies of surface-induced liquid crystal properties and alignment layers. Phase separated ferroelectric smectic liquid crystal displays. Elastic properties and dynamics of polymer stabilized liquid crystals. Interfacial phenomena, phase behavior, and hydrodynamics of two dimensional systems. Out of equilibrium behavior of complex fluids. Optical determination of electro-hydrodynamic convection. Pattern formation. Finotello, Gleeson, Kumar, Mann, Sprunt.

Nano-materials. Optical components and devices at sub-micron scale, carbon nanotube polymer composites. Kumar, Sprunt.

Nuclear and Particle Physics. Development of neutron detectors and neutron polarimeters. Pattern recognition software for charged-particle tracking detectors. Measurements of fundamental structure functions of the neutron and proton. Phases of nuclear matter, including the search for quark gluon plasma using heavy ions and 4 pi detectors. Nuclear structure and nuclear reaction mechanisms using electron and hadron beams. Elastic electromagnetic form factors of few-body nuclei. Phenomenological studies of pion-nucleon and antikaon-nucleon scattering to extract information on baryon resonances. Katramatou, Keane, Manley, Margetis, Petratos.

MIAMI UNIVERSITY

DEPARTMENT OF PHYSICS

Oxford, Ohio 45056

Students Accepted For Degree	FIELDS		
	Physics	Astronomy	Related Fields
Doctorate			
Master's	X		

1. General

President: David Hodge
Dean of Graduate School: Bruce Cochrane
Department Chairman: Michael J. Pechan
Department Telephone Number: (513) 529-5625 (C)
Type of Institution: University
Control: Public
Setting: Small town
Total Faculty: 1,529
Total Graduate Faculty: 827
Total Students: 16,884
Total Graduate Students: 2,213
Annual Graduate Tuition:
In-state residents: Full-time—$11,796
Out-of-state residents: Full-time—$25,428
Tuition rates for: 2009–2010
Deferred tuition plan: No
Term: Semester

2. Number of Faculty in Department

The combined total of full-time faculty in the three professorial ranks is 15. The combined total of full-time, part-time, and other faculty at all ranks is 20.

3. Admission, Financial Aid, and Housing

Address admission inquiries to: Graduate School Office, 102 Roudebush Hall
Graduate application fee required: $35
Admission deadline (Fall admission): Review begins 3/1
Admission information: For fall admission, 2009–10, 10 students were accepted from 33 applicants.
Admission requirements: For admission to graduate programs, a Bachelor's degree in physics or related areas (consult our departmental Graduate Student Advisor about appropriateness of related area) is required with a minimum undergraduate GPA of 2.75/4.0 specified. The GRE general and physics tests are strongly encouraged, but not required. A research statement is required. Students from non-English speaking countries are required to demonstrate proficiency in English via the TOEFL exam.
Undergraduate preparation assumed: *Classical Mechanics*, Symon; *Electromagnetism*, Griffiths; *Quantum Mechanics*, Griffiths; *Statistical and Thermal Physics*, Reif. Courses in linear algebra and differential euqations.
Address financial aid inquiries to: Graduate School Office, 102 Roudebush Hall
GAPSFAS application required: No
Financial aid deadline: 3/1
Loans available: Yes
Housing: www.muohio.edu/graduate/student.cfm

Table A—Faculty, Enrollments, and Degrees Granted

Research Specialty	2009–10 Faculty	Enrollment Fall 2009		No. of Degrees Granted[1] 2008–09 (2005–09)			Median No. of Years for Ph.D.'s
		Master's	Doctorate	Master's	Terminal Master's	Doctorate	
Astrophysics	1	0	–	–	0(6)	–	–
Atmospheric	0	0	–	–	0(4)	–	–
Biophysics	2	3	–	–	1(6)	–	–
Computational	1	0	–	–	0(3)	–	–
Condensed Matter	5	7	–	–	4(11)	–	–
Physics Education	2	1	–	–	1(1)	–	–
Optics/Atomic	4	5	–	–	4(13)	–	–
Other		0	–	–	1(4)	–	–
Total	15	16	–	–	11(47)	–	
Full-time Grad. Stud.		16	–				
Part-time Grad. Stud.		0	–				
First-year Grad. Stud.		10	–				
Median Years in Grad. Study		2	–	–	–	–	
Undergraduate Degrees, 2009–10 (2006–10): 12(76)							

[1]Five-year totals in parentheses.

4. Graduate Degree Requirements

Master of Science: 1) Thesis option: A minimum of 30 semester hours of graduate course work, research, and thesis credit is required. You must write a thesis proposal and defend it before your thesis committee. Subsequent completion and defense of the thesis are required. 2) Non-thesis option: A minimum of 36 semester hours of graduate credit is required. A comprehensive examination must also be passed.
The thesis option is strongly recommended. For either the thesis or non-thesis option you are expected to show proficiency in the areas of quantum physics, classical mechanics, electromagnetism, statistical physics, and mathematical methods used in physics. Evidence of proficiency means successful completion of courses at the graduate level. Graduate course work is selected in consultation with the thesis director (thesis option) and graduate program director.
Special Equipment, Facilities, or Programs: Faculty currently qualified to direct graduate student research maintain, or have access to the following:

- Cold atom trap and optical lattice with tunable diode lasers
- Pulsed laser polarization spectrometer
- Nanosecond time-resolved fluorescence spectrometer and sectioning microscope
- Cell culture facilty of laminar flowhood, CO_2 incubators
- Near field scanning optical microscope
- Electron beam lithography system
- Gamma-ray spectrometer
- Quantum Design Physical Properties Measurement System
- Ferromagnetic resonance spectrometer
- Grid cluster for computation
- 12 Tesla superconducting magnet and cryostat
- Class 1000 cleanroom to fabricate nanodevices
- Automated printed circuit board prototyping facility

Table B—Appointments to Graduate Students, 2009–10

Title of Appointee	Appointments: Total	First year	Academic Load Allowed in Credit Hours	Hours of Service Per Week	Stipend for Academic Year ($)
			Semester		
Teaching Assistant	16	10	10 (min) 14 (max)	20	14,262[*]
Total	16	10			

[*]Graduate students receive a duty-free summer stipend of $1,800

5. Personnel Engaged in Separately Budgeted Research, 7/09–6/10

Professorial faculty	6
Graduate students	13
Total	19

6. Separately Budgeted Research Expenditures by Source of Support

	Departmental Research	Physics-related Research Outside Department
Private, Non-profit	96,000	
Federal government	188,000	
State government	627,000	
Total	$911,000	$

Table C—Separately Budgeted Research Expenditures

Research Specialty	No. of Grants	Expenditures ($)
Atomic, Molecular, & Optical	2	65,000
Physics Education	1	627,000
Condensed Matter	3	147,000
Biophysics	1	
Total	7	911,000

FACULTY

Professors

Houk, T. William, Ph.D., Maryland, 1971. Biophysics.

Jaeger, Herbert, Ph.D., Oregon State University, 1987. Experimental solid state.

Pechan, Michael J., Ph.D., Iowa State, 1977. Department Chair. Solid state; experimental physics.

Rice, Perry, Ph.D., Arkansas, 1988. Theoretical quantum optics and quantum information.

Taylor, Beverley A. P., Ph.D., Clemson, 1978. Physics education.

Yarrison-Rice, Jan, Ph.D., Arkansas, 1990. Experimental nonlinear optics, nanotechnology.

Associate Professors

Alexander, Stephen, Ph.D., Pennsylvania State University, 1990. Astrophysics.

Bali, Samir, Ph.D., University of Rochester, 1994. Experimental quantum optics.

Bayram, S. Burcin, Ph.D., Old Dominion Univ., 1998. Experimental atomic, molecular, optical physics.

Urayama, Paul, Ph.D., Princeton University, 2001. Experimental biophysics.

Assistant Professors

Blue, Jennifer, Ph.D., University of Minnesota, 1997. Science education.

Clayhold, Jeffrey, Ph.D., Princeton University, 1989. Experimental condensed matter physics.

Clemens, James, Ph.D., University of Oregon, 2003. Theoretical quantum optics, quantum information.

Eid, Khalid, Ph.D., Michigan State University, 2002. Nanotechnology, magnetism, semiconductors.

RESEARCH SPECIALTIES AND STAFF

Astrophysics. Numerical simulations with gravitational N-body codes. *Theoretical* — Alexander.

Atomic, Molecular, and Optical Physics. Polarization spectroscopy and electron-imaging spectroscopy of alkali-rare gas collisions, line narrowing of high-power broad area diode lasers, dynamics of laser-cooled atoms in magneto-optical traps and optical lattices, electromagnetically induced transparency and absorption in atoms, imaging and optical sensing in turbid media. *Experimental:* Bali, Bayram.

Biophysics. High-pressure studies. Methods of fluorescence-based metabolic sensing and microscopy imaging applied to piezophysiological studies. *Experimental:* Urayama.

Computational Physics. Scientific visualization, image processing and analysis, and the use of computer graphics and animation to enhance comprehension of physical phenomena. *Theoretical:* Clemens, Rice.

Condensed Matter and Nanotechnology. Fabrication and optical characterization of nanoscale materials and devices using a variety of methods including electron beam lithography and photolithography, magnetoresistance in nanodevices at cryogenic temperatures, nanoscale magneto dynamics in reduced dimensional systms, angular correlation spectroscopy of ceramic materials, sub-micron transport of cold atoms in optical lattices, electronic and thermal properties of novel solid state materials. *Experimental:* Bali, Clayhold, Eid, Jaeger, Pechan, Yarrison-Rice.

Physics Education. Strategies for teaching scientific reasoning and problem-solving skills in introductory physics classes, elementary school and K-12 science education. Blue, Taylor.

Quantum Optics and Quantum Information. Photon correlation spectroscopy of cold trapped atoms in optical lattices, cavity QED, theoretical and computational modeling of light and matter for generating nonclassical or entangled states of light and atoms; applications include quantum teleportation, quantum error correction, quantum key distribution and general quantum algorithms. Relativistic quantum information theory. *Theoretical:* Clemens, Rice. *Experimental:* Bali.

OHIO UNIVERSITY

DEPARTMENT OF PHYSICS AND ASTRONOMY

Athens, Ohio 45701

Students Accepted For Degree	FIELDS		
	Physics	Astronomy	Related Fields
Doctorate	X		
Master's	X		

1. General

President: Roderick McDavis
Vice President for Research Programs: Rathindra Bose
Dean of Graduate College: Rathindra Bose
Department Chair: Joseph Shields
Department Telephone Number: (740) 593-1718
Type of Institution: University
Control: Public
Setting: Small town
Total Faculty: 727
Total Graduate Faculty: Not available
Total Students: 20,413
Total Graduate Students: 2,826
Annual Graduate Tuition:
In-state residents: Full-time—$2,613/qt.
Part-time—$234/credit
Out-of-state residents: Full-time—$5,957/qt.
Part-time-$496/credit
Tuition rates for: 2009–10
Deferred tuition plan: Yes
Other Fees: General fee $308/qt.
Health Insurance $1,098/yr.
Technology Fee $65/quarter
Term: Quarter

2. Number of Faculty in Department

The combined total of full-time faculty in the three professorial ranks is 27. The combined total of full-time, part-time, and other faculty at all ranks is 34.

3. Admission, Financial Aid, and Housing

Address admission inquiries to: Graduate Appointments Committee Chair (see http://www.phy.ohiou.edu)
Graduate application fee required: $55.00
Admission deadline (Fall admission): April 1
Admission information: For fall admission, 2009–10, 15 students were accepted from 128 applicants.
Admission requirements: For admission to the graduate programs, a Bachelors degree in science, mathematics, or engineering is required; an undergraduate GPA of 3.0 is required. The degree from an institution outside the USA must be equivalent to a four-year program in the USA. The GRE General and GRE Subject (Physics) examinations are recommended. Students from non-English speaking countries are required to demonstrate proficiency in English via the TOEFL or equivalent. A score of 600 (paper) or 250 (computer) is recommended for full-time graduate study. For teaching associateships, the Test of Spoken English is also required for all international students. An examination in English is given upon arrival, and students may be required to enroll in English language instruction.
Undergraduate preparation assumed: Students entering with a B.S. degree in physics are normally assumed to have completed studies in the following basic subjects to the levels indicated. However, less prepared students, whose record is otherwise good, are routinely allowed to make up deficiencies in their physics preparation. This applies particularly to students with undergraduate degrees in Chemistry, Mathematics, Teaching of Physics, etc. Texts: Mechanics—J. B. Marion, *Classical Dynamics*; Electricity and Magnetism—D. J. Griffiths, *Introduction to Electrodynamics*; E. M. Purcell, *Electricity and Magnetism*; Modern Physics—Weidner and Sells, *Elementary Modern Physics*; Quantum Mechanics—Eisberg and Resnick, *Quantum Physics*; Thermophysics—Sears and Salinger, *Thermodynamics, Kinetic Theory and Statistical Mechanics*; Mathematics—calculus, including differential equations, Fourier series, complex variables, vector operators, basic algebra, including matrices and determinants.
Address financial aid inquiries to: Graduate Appointments Committee Chair
GAPSFAS application required: No
Financial aid deadline: April 1
Loans available: Yes
Address housing inquiries to: Housing Office, Chubb Hall
On-campus, single student housing available: Yes
Cost/term: $1193/quarter
On-campus, married student housing available: Yes
Cost/month: $578–707/month

Table A—Faculty, Enrollments, and Degrees Granted

Research Specialty	2008–09 Faculty	Enrollment[1] Fall 2009		No. of Degrees Granted[2] 2009–10 (2005–10)			Median No. of Years for 2008–09 Ph.D.'s
		Master's	Doctorate	Master's	Terminal Master's	Doctorate	
Astrophysics	4	0	9	2(4)	1(4)	0(3)	6.2
Biophysics	4	0	6	2(4)	1(3)	1(4)	5.5
Condensed Matter Physics	5	0	17	0(3)	0(2)	2(11)	5.8
Nuclear Physics	9	0	18	0(1)	0(1)	1(10)	6.2
Nanoscience	6	0	15	0(0)	0(2)	0(3)	5.5
Non-specialized	0	0	8	0(1)	0(0)	0(0)	–
Total		0	73	4(13)	1(12)	5(31)	
Full-time Grad. Stud.		0	73				
Part-time Grad. Stud.		0	0				
First-year Grad. Stud.		0	15				
Median Years in Grad. Study (2005–10 Degrees)				2.6	2.4	5.6	–
Undergraduate Degrees, 2009–10 (2004–09): 3(14)							

[1]Students not yet committed to a research specialty are entered under non-specialized.
[2]Five-year totals in parentheses.

4. Graduate Degree Requirements

Master's: 45 quarter hours minimum; 3.0 (B) average minimum; thesis optional; residence requirement not specified; no exams except in courses and for thesis; no foreign language required.
Doctorate: There is no specified number of course hours but a series of core courses are required; a minimum 3.0 (B) average must be maintained. A comprehensive review, three quar-

ters continuous residence minimum, and dissertation are required. No foreign language required.

Other Programs: Interdepartmental studies (e.g., communications, education) and other special programs available by arrangement.

Special Equipment, Facilities, or Programs: The Physics Department occupies two wings of Clippinger Research Laboratories, the newly renovated Edwards Accelerator Laboratory building which contains Ohio University's 4.5 MV high-intensity tandem Van de Graaff accelerator; and the Surface-Science Research Laboratory. The department has a well-equipped and staffed machine shop in addition to specialized research equipment. The computer facilities are excellent and include networked PCs, workstations, a Beowulf cluster and the CRAY T-90 and T3E at the Ohio Supercomputer Center. All are available to students over high speed networks (Internet II).

Thesis: Thesis may be written *in absentia*.

Table B—Appointments to Graduate Students, 2009–10

Title of Appointee	Appointments Total	Appointments First year	Academic Load Allowed in Credit Hours	Hours of Service Per Week	Stipend for Academic Year ($)
			Quarter		
Govt. Scharship	1	0	–	–	
Teaching Assistant	44	18	12–18	15	17,754
Research Assistant	28	0	12–18	–	17,754
Total	73	18			

5. Personnel Engaged in Separately Budgeted Research, 7/09–6/10

Professorial faculty	27
Postdoctoral appointments	14
Graduate students	43
Undergraduate students	14
Total	98

6. Separately Budgeted Research Expenditures by Source of Support

	Departmental Research	Physics-related Research Outside Department
Federal government	$3,284,578.00	$
Private	54,000.00	$
Total	$3,338,578.00	

Table C—Separately Budgeted Research Expenditures

Research Specialty	No. of Grants	Expenditures ($)
Astrophysics	6	235,353.00
Biophysics	4	226,495.00
Condensed Matter Physics	4	411,593.00
Nuclear Physics	5	1,466,576.00
Nanoscience	3	998,561.00
Total	22	3,338,578.00

FACULTY

Professors

Brune, Carl R., Ph.D., California Institute of Technology, 1994. Experimental nuclear astrophysics.

Drabold, David A., Ph.D., Washington U, 1989. Distinguished Prof. Theoretical condensed matter; computational methodology for electronic structure; theory of topologically disordered material.

Elster, Charlotte, Bonn (Germany), 1986. Dr.rer.nat. Nuclear and intermediate-energy theory.

Govorov, Alexander O., Ph.D., Novosibirsk (Russia), 1991. Theoretical condensed matter; semiconductor nanostructures; nanoscience.

Grimes, Steven M., Ph.D., Wisconsin, Madison, 1968, Emeritus. Distinguished Prof. Nuclear physics.

Hicks, Kenneth H., Ph.D., Colorado, 1984. Nuclear and intermediate energy physics.

Ingram, David C., Ph.D., Salford (UK), 1980. Thin films; atomic collisions in solids; surface physics.

Jung, Peter, Ph.D., Ulm (Germany), 1985. Distinguished Professor. Non-equilibrium statistical physics, non-linear stochastic processes, pattern formation, biophysics.

Kordesch, Martin E., Ph.D., Case Western, 1984. Surface physics, wide gap materials.

Phillips, Daniel, Ph.D., Flinders (Australia), 1995. Theoretical nuclear and particle physics.

Prakash, Madappa, Ph.D., Bombay, 1979. Theoretical nuclear astrophysics.

Shields, Joseph C., Ph.D., California (Berkeley), 1991. Chair of the Department. Astrophysics; interstellar medium; active galactic nuclei.

Smith, Arthur R., Ph.D., Texas, 1995. Experimental semiconductor physics; thin films.

Statler, Thomas S., Ph.D., Princeton, 1986. Astrophysics; galatic structure and dynamics.

Ulloa, Sergio E., Ph.D., SUNY, Buffalo, 1984. Theoretical condensed matter.

Wright, Louis E., Ph.D., Duke, 1966. Emeritus. Nuclear theory, electrodynamics, intermediate energy theory.

Associate Professors

Boettcher, Markus, Ph.D., Bonn (Germany), 1997. High energy astrophysics.

Braslavsky, Ido, Ph.D., Israel Inst. of Tech., 1998. Biophysics.

Castillo, Horacio E., Ph.D., Illinois, 1998. Theoretical condensed matter.

Hla, Saw-Wai, Ph.D., University of Ljubljana (Slovenia), 1997. Experimental nanoscience.

Lucas, Mark, Ph.D., Illinois, 1995. Experimental nuclear and particle physics.

Neiman, Alexander, Ph.D., Saratov (Russia), 1998. Biophysics and non-linear stochastic processes.

Tees, David F. J., Ph.D., McGill, 1995. Biophysics.

Assistant Professors

Clowe, Douglas, Ph.D., Hawaii, 1998. Observation astrophysics.

Chen, Gang, Ph.D., Lehigh University, 2004. Experimental condensed matter physics.

Frantz, Justin, Ph.D., Columbia, 2004. Experimental Nuclear Physics.

Roche, Julie, Ph.D., Univ. B. Pascal (France), 1998. Nuclear and intermediate energy physics.

Sandler, Nancy, Ph.D., Illinois, 1998. Theoretical condensed matter.

Schiller, Andreas, Ph.D., University of Oslo (Norway), 2000. Experimental nuclear physics.

Stinaff, Eric, Ph.D., Iowa State, 2002. Experimental nanoscience.

RESEARCH SPECIALTIES AND STAFF

Theoretical

Astrophysics. Studies of galaxies and galaxy clusters, with emphasis on galaxy structure and dynamics, cluster central galaxies and intergalactic medium, quasars, and supermassive black holes in galaxy nuclei. High energy astrophysics related to accretion onto compact objects, relativistic jets, and gamma-ray bursts. Nebular astrophysics applied to active galactic nuclei and starbursts. Investigations into these topics employ multiwavelength observations with national facilities (Hubble Space Telescope, Chandra X-Ray Observatory, MMT Observatory) as well as theoretical efforts including analytic calculations and large-scale numerical simulations. Boettcher, Shields, Statler.

Biophysics. Current projects include computational modeling of complex cellular signaling networks, especially intracellular and intercellular calcium signaling, modeling of neural and glial functions in healthy and epileptic tissue, stochastic modeling of electro-receptors in paddle fish, modeling of the neuronal circuitry of the cat's retina, stochastic and coherence resonance in excitable biologic systems, nano-scale ion channel and receptor clusters, modeling slow axonal transport. Jung, Neiman.

Condensed Matter Theory. Statistical mechanics and nonequilibrium dynamics of disordered systems and glassy materials. Some areas of interest include: nanoscale-sized dynamical heterogeneities in glassy materials, slow activation-controlled motion of topological defects (e.g.: vortices, dislocations), and disordered electronic systems. Methodology of first principles simulation: development of local basis density functional methods, time-dependent density functional theory and efficient computation of Wannier functions, and the single particle density matrix. Theory of disordered insulators: Anderson transition, photo-structural response, novel schemes for structural modeling of glasses and studies of pressure-induced polyamorphism. Optical and transport phenomena in nanoscale systems, including quantum dots, rings, and channels. Recent activity covers excitons in quantum rings, spin transport in nanocrystals, and quantum acousto-electric interactions on nanoscale. Other nanoscience problems of interest include electronic transport in complex molecule systems, the role of controlled disorder on the metallic or insulating nature of one- and two-dimensional systems, and the role of collective effects on the optical and transport properties of quantum dot arrays. Studies of low-dimensional strongly correlated electron systems, disordered electronic systems and quantum Hall effect physics. Castillo, Drabold, Govorov, Sandler, Ulloa.

Mathematical and Computational Physics. Quantum simulations, *ab initio* calculations, and visualization of many-body and few-body systems in condensed matter and nuclear physics. Numerical methods and algorithmic development for high performance vector and parallel computers. Analytical and algorithmic studies in differential and integral equations, probability theory, and series expansions. Drabold, Elster, Phillips, Ulloa, Wright.

Nuclear and Intermediate Energy Physics. Research in theoretical nuclear and particle physics at Ohio University has as its major component the modeling of processes involving atomic nuclei with mass numbers 1, 2, 3, and 4. We attempt to reveal aspects of the forces that are at work inside the nucleus by examining data obtained when targets made of hydrogen and helium isotopes are bombarded with photons, electrons, neutrons, and pions. In order to understand the dynamics of the nucleus we build theoretical descriptions of these reactions and compare our predictions to the experimental results. "Light nuclei" (nuclei containing up to four nucleons) are particularly useful in this regard because once the nuclear dynamics is specified the Schrodinger equation for these systems can be solved exactly. A recent focus of our group has been the application of effective field theory techniques to such reactions. Using these, and other, theoretical techniques we have worked on nucleon-nucleon scattering, nucleon-deuteron scattering at intermediate energies, meson-production in nucleon-nucleon collisions at intermediate energies, relativistic effects in nuclear physics, electron-deuteron scattering, Compton scattering from the proton, deuteron and Helium-3, pion photo- and electro-production on the proton and charge-symmetry breaking in nuclear physics. Charge-symmetry breaking is of particular current interest, since here nuclear reactions such as the production of neutral pions in deuterium-deuterium collisions reveal aspects of Quantum Chromodynamics associated with the difference between up and down quarks. Lastly, we have also worked on providing reliable predictions for processes of relevance to astrophysics and cosmology, e.g., neutron interactions that contribute to supernova and neutron-star cooling. Such projects are relevant to the Ohio University's newly-established Research Priority on 'Structure of the Universe: From Quarks to Superclusters.' Elster, Phillips, Prakash, Wright.

Experimental

Astrophysics. Spectroscopic observations of stellar motions and stellar populations in elliptical galaxies, and evidence for dark matter. Ionized gas in galaxies. Gravitational lens in X-ray studies of galaxy clusters. Nuclear physics applied to astrophysics. Brune, Clowe, Shields, Statler.

Biological Physics. Stochastic resonance in psychophysics and animal behavior. Studies of stochastic non-linear dynamics in paddlefish electroreceptors. Experimental determination of the response of single cell adhesion molecules to applied forces using a microcantilever device. Lipid bilayer tether pulling on leukocytes and platelets using micropipette aspiration. Studies of cell adhesion in pressure gradients in micropipettes. Determination of cell membrane mechanical properties. Optical studies of biomolecules at the single molecule level using Total Internal Reflection Fluorescence microscopy and Fluorescence Resonance Energy Transfer. Biomineralization. Studies of ice-modifying antifreeze proteins. Studies of DNA-protein interactions using optical methods. Braslavsky, Jung, Neiman, Tees.

Condensed Matter. Current projects encompass various areas in nanoscale science probe techniques, materials characterization by ion beams, synthesis and characterization of photonic and electronic materials, spin electronics and low-temperature mesoscopic physics. Relevant projects are illustrated by the following list: Thin film growth (by molecular beam epitaxy) and characterization (using scanning probe microscopy techniques, inc. spin-polarized) of the structural, electronic, and magnetic properties of transition metal nitride layers and magnetic-doped nitride semiconductors. Single atom/molecule manipulation using ultra high vacuum low temperature scanning tunneling microscopy, development of single molecule electronics and mechanical devices, molecular and metal thin films, surface science, and Microscopy techniques. Amorphous semiconductors and their photonic properties. MeV ion beam analysis of materials, and measurement of relevant cross-sections. Ion beam and plasma deposition of materials and their characterization. Synthesis of zeolite-related materials with catalytic and novel nanoelectronic properties, and their characterization by X-ray diffraction and electronic measurements at variable temperatures. Synthesis and characterization of nanowires and molecular wires from complex inorganic precursors. Electronic properties of single biological macromolecules, such as DNA, and their assembly into structures with nanoelectronic potential. Organic semiconductor materials, growth, structural and electronic characterization, and organic electronic device fabrication. Mesoscopic physics and transport properties of low-dimensional semiconductor and metallic structures, fabricated by nanoscale lithographic techniques, and measured at very low temperatures and high magnetic fields. Semiconductor heterostructure physics. Spin electronics in low-dimensional structures. Chen, Hla, Ingram, Kordesch, Smith, Stinaff.

Nuclear and Intermediate Energy Physics. Contemporary research in experimental nuclear physics involves collaboration with scientists from many different institutions and heavy use of specialized accelerator facilities around the world. Ohio University nuclear physicists are recognized leaders in a variety of experimental programs spanning a broad energy domain. At higher energies our faculty are leading research programs at Jefferson Laboratory in Virginia. These include: the study of electromagnetic production of strange baryons and the search for new exotic baryons in Hall B; precision measurements of the weak charge of the proton in Hall C; the study of the nature of the gluonic flux tube in Hall D. An active program studying the photo-excitation of the nucleon is also ongoing at the SPring-8 experiment in Japan. At lower energies, our faculty are directing research programs in several distinct areas, including: fundamental symmetries in nuclear reactions via precision tests of charge symmetry breaking at TRIUMF in Canada; exotic nuclei far from the line of stability at GANIL and the Hahn-Meitner Institute in Europe; measurements of neutron cross sections at Los Alamos in New Mexico; studies of nuclear level densities at the Holifield Radioactive Ion Beam Facility in Tennessee; studies of pion photoproduction and QCD sum rules with the LEGS facility at Brookhaven National Laboratory in New York. Our Department also operates the high-intensity Ohio University Tandem Van de Graff accelerator with its unique beam swinger magnet and long flight path for high precision measurements of various nuclear cross sections and projects in medical physics. Brune, Frantz, Grimes, Hicks, Roche, Schiller.

THE OHIO STATE UNIVERSITY

DEPARTMENT OF PHYSICS

Columbus, Ohio 43210

Students Accepted For Degree	FIELDS		
	Physics	Astronomy	Related Fields
Doctorate	X		
Master's	X		

1. General

President: E. Gordon Gee
Dean of Graduate School: Patrick Osmer
Department Chairperson: James Beatty
Department Telephone Number: (614) 292-5713
Type of Institution: University
Control: Public
Setting: Urban
Total Faculty: 5,374
Total Graduate Faculty: 3,026 (regular faculty)
Total Students: 55,014
Total Graduate Students: 10,741
Annual Graduate Tuition:
In-state residents: Full-time—$10,169*
Out-of-state residents: Full-time—$25,029*
Tuition rates for: 2009–10
Deferred tuition plan: No
Other Fees: Recreation ($82), Activity ($25), City Bus ($9)
Term: Quarter

*Academic year, three quarters

2. Number of Faculty in Department

The combined total of full-time faculty in the three professorial ranks is 60 (56 on the main campus and 4 on the regional campuses).

3. Admission, Financial Aid, and Housing

Address admission inquiries to: Graduate Studies Chairperson, Department of Physics, 191 W. Woodruff Avenue
Graduate application fee required: $5 for pre-application
Admission deadline (Fall admission): 1/1
Admission information: For fall admission, 2009–10, 83 students were offered admission from 401 applicants.
Admission requirements: For admission to the graduate programs, a Bachelor's degree in physics or related field is required with a minimum undergraduate GPA of 3.0/4.0 specified. The GRE is required. No minimum score set. The GRE Advanced is required. The minimum acceptable score suggested for admission is 600. The average GRE Advanced score for 2009–10 admissions was 683. Students from non-English speaking countries are required to demonstrate proficiency in English via the TOEFL ibt exam. Minimum acceptable score for admission is 79. Graduate Teaching Associates must demonstrate their fluency in spoken English before they will be allowed to assume classroom teaching duties.
Undergraduate preparation assumed: Symon, Mechanics; Lorrain and Corson, Electromagnetic Fields and Waves; Eisberg and Resnick, *Quantum Physics*: Reif, *Statistical and Thermal Physics*.
Address financial aid inquiries to: Vice-Chair for Graduate Studies, Department of Physics, 191 W. Woodruff Ave. OSU, Columbus, OH 43210.
GAPSFAS application required: No
Financial aid deadline: 1/1
Address housing inquiries to: Director, Graduate Student Housing, 350 Morrill Tower, 1910 Cannon Drive
On-campus, single student housing available: Yes
Cost/month: $715
On-campus, married student housing available: Yes
Cost/month: $535, 1 bdrm.; $670, 2 bdrm. (incl. heat)

Table A—Faculty, Enrollments, and Degrees Granted

Research Specialty	2009–10 Faculty	Enrollment[1] Fall 2009		No. of Degrees Granted[2] 2009 (2005–09)			Median No. of Years for 2009–10 Ph.D.'s
		Master's	Doctorate	Master's	Terminal Master's	Doctorate	
Astrophysics	5	0	3	–	–	1(4)	–
Atomic, Molecular, & Optical Physics	8	0	28	–	–	0(9)	–
Biological Physics	3	0	1	–	–	0(3)	
Condensed Matter Physics	20	0	62	–	–	16(44)	–
Nuclear Physics	9	0	9	–	–	2(5)	–
Particles & Fields	13	0	27	–	–	4(25)	–
Physics Education	2	0	5	–	–	2(7)	–
Non-specialized	–	0	41	–	–	–	–
Total		0	176		–	24(97)	
Full-time Grad. Stud.		0	176				
Part-time Grad. Stud.		0	0				
First-year Grad. Stud.		0	42				
Median Years in Grad. Study (2009–10 Degrees)				–	–	–	6.3

Undergraduate Degrees, 2009 (2005–09): 36(144)

[1]Students not yet committed to a research specialty are entered under non-specialized.
[2]Five-year totals in parentheses.

4. Graduate Degree Requirements

Master's: Plan A (Thesis): Complete 45 credit hours of graduate level course work including work on thesis. Pass departmental qualifying examination and final oral exam on thesis. Plan B (no thesis): Complete 50 credit hours of graduate level course work. Demonstrate competence in individual work, experimental or theoretical. No language requirement for either plan. Master's degree can be earned as part of General Exam for Ph.D.
Doctorate: Complete 120 credit hours of graduate level course work including research on dissertation. Of this amount 45 hours may be transferred from a Master's degree at OSU, or up to 90 hours from acceptable graduate work at another institution. Pass candidacy Examination for Ph.D. candidacy. Pass oral final examination on dissertation. No language requirement.
Other Programs: M.S. and Ph.D. in Chemical Physics.
Thesis: Thesis may be written *in absentia* by special permission.
Special Equipment, Facilities, or Programs: Departmental computer center (VMS cluster and UNIX network), University Supercomputer Center. Helium liquefaction facility, low-temperature cryostats operating down to the sub-milliKelvin

range. Extremely high-resolution grating spectrometers, tunable-diode laser spectrometers and Fourier transform spectrometers for visible, near-infrared, and far-infrared spectroscopy of gases and solids; continuous, pulsed, and picosecond laser sources, low-temperature magnetic resonance spectrometers (NMR and EPR). Facilities for the detection of charged particles, neutrons, and high-energy gamma rays. Among the facilities available in our interdisciplinary Materials Research Laboratory are several specialized types of electron microscopes, high-resolution FT and Raman spectrometers, and numerous sample preparation devices (an ion mill, a controlled-atmosphere furnace, thin-film evaporators, etc.).

Table B—Appointments to Graduate Students, 2009–10

Title of Appointee	Appointments Total	Appointments First year	Academic Load Allowed in Credit Hours	Hours of Service Per Week	Stipend for Academic Year ($)
			Quarter		
Teaching Assistant	66	33		16[1]	17,172-17,838[2]
Research Assistant	81	2		Full time	17,172-17,838[2]
Fellow	29	7		0	18,486[2]
Total	176	42			(Au09)

[1] Includes classroom contact with students of 4–10 hours plus time for preparation, grading, and consulting with students.
[2] In addition to this stipend, all tuition and general fees are waived. 9 months stipend. (summer support is also normally available).

5. Personnel Engaged in Separately Budgeted Research, 7/09–6/10

Professorial faculty	60
Emeritus faculty	2
Postdoctoral appointments	43
Graduate students	176
Undergraduate students	37
Nonteaching research personnel	21
Total	339

6. Separately Budgeted Research Expenditures by Source of Support

	Departmental Research	Physics-related Research Outside Department
Federal government	$12,585,473	$1,581,135
State and local government	570,850	103,142
Private, nonprofit organizations	1,569,446	9,173
Business and industry	638,136	102,974
Other	0	0
Total	$15,363,905	$1,796,424

Table C—Separately Budgeted Research Expenditures

Research Specialty	No. of Grants	Expenditures ($)
Astrophysics	6	735,837
Atomic, Molecular, & Optical Physics	32	4,675,477
Biological Physics	15	906,884
Condensed Matter Physics	57	4,689,786
Nuclear Physics	8	1,067,770
Particles & Fields	16	2,818,229
Physics Education	7	469,922
Total	141	15,363,905

FACULTY

Professors

Agostini, Pierre, Doctorat, Univ. AIX Marseille, 1967. Experimental atomic, molecular, and optical physics.

Andereck, C. David, Ph.D., Rutgers, 1980. Experimental condensed matter physics.

***Aubrecht**, Gordon J., Ph.D., Princeton, 1976. Theoretical high-energy physics.

Beatty, James, Ph.D., Chicago, 1986. Experimental astrophysics.

Braaten, Eric, Ph.D., Univ. of Wisconsin-Madison, 1981. Theoretical high-energy physics.

Brillson, Leonard, Ph.D., University of Pennsylvania, 1972. Experimental condensed matter physics and electrical engineering.

De Lucia, Frank C., Ph.D., Duke, 1969. Experimental atomic, molecular, and optical physics.

Dimauro, Louis F., Ph.D., U. of Connecticut, 1980. Experimental atomic, molecular, and optical physics.

Durkin, L. Stanley, Ph.D., Stanford, 1981. Experimental high-energy physics.

Epstein, Arthur J., Ph.D., Pennsylvania, 1971. Experimental condensed matter physics.

Freeman, Richard R., Ph.D., Harvard, 1973. Experimental atomic, molecular, and optical physics.

Furnstahl, Richard J., Ph.D., Stanford, 1985. Theoretical nuclear physics.

Gan, K. K., Ph.D., Purdue, 1985. Experimental high-energy physics.

Hammel, P. Chris, Ph.D., Cornell, 1984. Experimental condensed matter physics.

Heinz, Ulrich, Ph.D., Goethe Univ., 1980. Nuclear theory.

Herbst, Eric, Ph.D., Harvard, 1972. Theoretical astrophysics; atomic, molecular, and optical physics.

Ho, Tin-Lun (Jason), Ph.D., Cornell, 1977. Theoretical condensed matter physics.

Honscheid, Klaus, Ph.D., Univ. of Bonn, 1988. Experimental high energy physics.

Hughes, Richard E., Ph.D., Pennsylvania, 1992. Experimental high energy physics.

Humanic, Thomas J., Ph.D., Univ. of Pittsburgh, 1979. Experimental nuclear physics.

Jayaprakash, Ciriyam, Ph.D., Illinois, 1978. Theoretical condensed matter physics.

Kagan, Harris P., Ph.D., Minnesota, 1979. Experimental high-energy physics.

Kass, Richard, Ph.D., California, Davis, 1978. Experimental high-energy physics.

Lemberger, Thomas R., Ph.D., Illinois, 1978. Experimental condensed matter physics.

Ling, Ta-Yung, Ph.D., Wisconsin, 1971. Experimental high-energy physics.

Lisa, Michael A., Ph.D., Michigan State, 1993. Experimental nuclear physics.

***Mainland**, G. Bruce, Ph.D., Texas, Austin, 1971. Theoretical high-energy physics.

Mathur, Samir, Ph.D., Univ. of Bombay, 1987. High energy theory.

Patton, Bruce R., Ph.D., Cornell, 1971. Theoretical condensed matter physics.

Pelz, Jonathan P., Ph.D., U. California, Berkeley, 1987. Experimental condensed matter physics.

Perry, Robert J., Ph.D., Maryland, 1984. Theoretical nuclear physics.

***Putikka**, William O., Ph.D., Wisconsin, 1988. Theoretical condensed matter physics.

Raby, Stuart A., Ph.D., Tel Aviv Univ., 1976. Theoretical high-energy physics.

Randeria, Mohit, Ph.D., Cornell, 1987. Theoretical condensed matter physics.

Shigemitsu, Junko, Ph.D., Cornell, 1978. Theoretical high-energy physics.

Sooryakumar, R., Ph.D., Illinois, Urbana-Champaign, 1980. Experimental condensed matter physics.

Steigman, Gary, Ph.D., NYU, 1968. Theoretical astrophysics and cosmology.

Stroud, David G., Ph.D., Harvard, 1969. Theoretical condensed matter physics.

Sugarbaker, Evan R., Ph.D., Michigan, 1976. Experimental nuclear physics.

Trivedi, Nandini, Ph.D., Cornell, 1987. Theoretical condensed matter physics.

Walker, Terrence P., Ph.D., Indiana, 1987. Theoretical astrophysics and cosmology.

Wilkins, John W., Ph.D., Illinois, 1963. Theoretical condensed matter physics.

Winer, Brian L., Ph.D., Berkeley, 1991. Experimental high energy physics.

Associate Professors

Bao, Lei, Ph.D., U. of Maryland, 1999. Physics Education.

Beacom, John, Ph.D., Wisconsin, 1997. Theoretical astrophysics.

Bundschuh, Ralf, Ph.D., Univ. of Porsdam, 1996. Theoretical biophysics.

Gramila, Thomas, Ph.D., Cornell, 1989. Condensed matter experiment.

***Jeschonnek**, Sabine. Ph.D., Bonn Univ., 1996. Nuclear theory.

Kilcup, Gregory P., Ph.D., Harvard, 1986. Theoretical high-energy physics.

Kovechegov, Yuri, Ph.D., Columbia, 1998. Theoretical nuclear physics.

Lafyatis, Gregory P., Ph.D., Harvard, 1982. Experimental atomic, molecular, optical physics.

Schumacher, Douglass, Ph.D., Michigan, 1995. Experimental atomic, molecular, and optical physics.

Van Woerkom, Linn D., Ph.D., University of Southern California, 1987. Experimental atomic, molecular, and optical physics.

Zhong, Dongping, Ph.D., Calif. Institute of Technology, 1999. Experimental biophysics.

Assistant Professors

Gupta, Jay, Ph.D., U.C. Santa Barbara, 2002. Experimental condensed matter physics.

Heckler, Andrew, Ph.D., Washington, 1994. Physics education.

Johnston-Halperin, Ezekiel, Ph.D., U. C. Santa Barbara, 2003. Experimental condensed matter physics.

Meyer, Julia, Ph.D., Koln, 2001. Theoretical condensed matter physics.

Poirier, Michael, Ph.D., Chicago, 2001. Experimental biophysics.

Yang, Fengyuan, Ph.D., Johns Hopkins, 2001. Condensed matter experiment.

Adjunct Professors

Weilhammer, Peter, Ph.D., University of Munich, 1969. CERN. Experimental high energy physics.

Winnewisser, Brenda, Ph.D., Duke, 1965. Professor. Experimental atomic, molecular, and optical physics.

Winnewisser, Manfred, Dr.rer.nat., Technical University of Karlsruhe, 1960. Professor. Experimental atomic, molecular, and optical physics.

*Regional campuses Physics faculty members.

RESEARCH SPECIALTIES AND STAFF

Theoretical

Astrophysics and Cosmology. Early universe theories and big-bang cosmology; primordial and stellar nucleosynthesis; large-scale structure; the particle physics/astrophysics connection; formation of interstellar molecules; star formation. Beacom, Herbst, Steigman, Walker.

Atomic, Molecular, and Optical. Hamiltonians and spectra of molecules with large amplitude motions; reactive and inelastic scattering processes; physisorption. Herbst.

Biological Physics. Genetic network analysis; immunological modeling; ecological modeling; biological sequence analysis; modeling of single molecule experiments; RNA folding. Bundschuh, Jayaprakash.

Condensed Matter. Theory of high-temperature superconductors; Josephson junction arrays; novel approaches to electronic structure theory; statistical mechanics of cellular automata; lattice-gas hydrodynamics; Bose-Einstein condensation of atomic gases; quantum Hall effect; phase transitions at interfaces in microemulsion systems; the wetting transition; equilibrium crystal shapes; properties of liquid metals and semiconductors; *ab initio* molecular dynamics; Ising models; lattice gases and frustrated spin systems; superfluid ^{3}He in disordered media; electrical and nonlinear optical properties of inhomogeneous media; liquid crystals; nonlinear transport; ultrafast dynamics of strongly correlated nanostructures; transport and optical properties of nanostructures. Ho, Jayaprakash, Meyer, Patton, Putikka, Randeria, Stroud, Trivedi, Wilkins.

High Energy. Phenomenology of the strong, and electroweak interactions; nonperturbative methods in quantum field theory; supersymmetry; unified model building; lattice gauge theory; heavy-quark systems; light front quantization. Braaten, Kilcup, Mathur, Pinsky, Raby, Shigemitsu.

Nuclear. Quantum chromodynamics, effective hadronic field theories, and quark models; nonperturbative light-front field theories; nonequilibrium and thermal field theory; relativistic treatments of nuclear reactions, nuclear matter, and structure;

nonlocal potentials and nuclear scattering. Furnstahl, Heinz, Jeschonnek, Kovechegov, Perry.

Experimental

Astrophysics, Cosmic Ray Detection; Supernova Neutrino Detector Development. Beatty.

Atomic, Molecular, and Optical. Ion and neutral trapping; superconductor vortices; interaction of matter with high-intensity lasers; femtosecond laser spectroscopy; high-precision spectroscopic studies of small molecules, with particular interest in molecules of astrophysical significance; non-reactive collisions at very low temperatures. Agostini, De Lucia, Dimauro, Freeman, Lafyatis, Schumacher, Van Woerkom.

Biological Physics. Protein/enzyme dynamics; femtosecond spectroscopy; Biological macromolecular hydration; chromatin structure; DNA repair; DNA mechanics; single molecule experiments. Poirier, Zhong.

Condensed Matter. Chaos and turbulence in classical systems; pattern formation and transitions in convecting classical fluids; effects of noise on nonlinear dynamical systems; noise-induced transitions in liquid crystals; spin waves in normal and superfluid ^{3}He; atomic scattering at the surface of liquid helium; excitations at interfaces in liquid and solid helium; thermodynamic, transport, IR-UV spectroscopy, time-resolved (picosecond) spectroscopy, and magnetic properties of novel materials such as conducting and semiconducting polymers, including light-emitting devices; magnetic properties of molecular ferromagnets; charge and spin transport, magnetic, optical, and other properties of high-T_c oxide superconductors; nonlinear responses and chaotic studies of magnetic materials at high frequencies; vortex unbinding phase transitions in two-dimensional superconducting films; Josephson junction arrays; metal-insulator transitions; optical properties of quantum well structures; optical and superconducting properties of metallic superlattices; Brillouin scattering and surface acoustic excitations; ferromagnetic resonance and chaos, magneto-optic properties and magnetic semiconductor studies in insulating garnet films; exchange coupling in magnetic multilayers; atomic-scale properties of semiconductor surfaces, strain-layer semiconductor growth, and microscopic electronic transport in insulating films; magnetic resonance in solids, including high T_c cuprate, fullerene, and other unusual superconductors. Andereck, Epstein, Gramila, Gupta, Hammel, Johnston-Halperin, Lemberger, Pelz, Sooryakumar, Yang.

High Energy. Electron-positron annihilations using the CLEO detector at CESR; spectroscopy of particles containing beauty and charm quarks; study of tau lepton, B-mixing and CP violation. Electron-proton scattering using the Zeus detector at HERA in Germany; search for extensions to the standard model and new particles up to mass scale of 800 GeV; measurement of structure functions. Proton-anti-proton scattering using the CDF detector at Fermilab; studies of the top quark and new particle production. Proton-proton collisions with the CMS detector at CERN's LHC; measurement of the properties of the Higgs boson; search for extensions beyond the standard model. Instrumentation and detector development for high energy physics experiments. Durkin, Gan, Honscheid, Hughes, Kagan, Kass, Ling, Winer.

Nuclear. Relativistic heavy ion collisions; boson interferometry; LHC Alice Experiment; RHIC Star Experiment; silicon drift detector development; intermediate energy charge exchange reactions. Humanic, Lisa, Sugarbaker.

Physics Education

The improvement of instruction in physics has always been an integral part of the work of the academic physics community. The establishment of research programs in this area have recently brought it formal recognition as one of the subfields of physics. Physics education research approaches its work in the context of the specific problems posed by instruction and learning in physics while drawing on relevant aspects of knowledge from cognitive psychology and pedagogy. Physicists, with their intimate knowledge of physics and the ability to use this knowledge to address complex problems, are in a unique position to make contributions to this field. In recent years, physics education research has ranged widely—from the study of student behaviors and cognitive skills, to the uses of new interactive technologies for learning physics. The research often leads to development of materials that improve student learning and that are used to evaluate that learning. Members of the OSU Physics Education Research Group are especially interested in the following areas: curriculum development; evaluation and assessment; professional development; analysis of student problem solving strategies; construction of conceptual models; the use of technology to enhance learning; and system change research. Bao, Heckler.

Others

The Ohio State University Center for Cosmology and AstroParticle Physics (CCAPP) is supported by a $5M award from the Provost's Targeted Investment in Excellence Program, by a $5M Exploration of Space endowment, and other private endowments. CCAPP's mission is to support collaborative research between the OSU departments of Astronomy and Physics in areas where OSU can make fundamental impact: dark energy, dark matter, and multi-messenger particle astrophysics. Through the Center, OSU has joined some of the world's leading research efforts: DES, GLAST, AUGER, IceCube, and SDDSS-III. Of particular importance is the Center's identity as a collection point for the world's best young researchers in the areas of cosmology and particle-astrophysics. The Center houses approximately 20 faculty, 7 CCAPP Postdocs and other mission specific postdocs, graduate students, and staff. It typically hosts 100 visitors a year along with 10 mini-workshops and several collaboration meetings. CCAPP researchers receive federal support from the Department of Energy, NASA, and the National Science Foundation. Physics faculty: Beacom, Beatty, Honscheid, Hughes, Kass, Steigman, Walker, Winer. (www.ccapp.osu.edu)

The Center for Electronic/Magnetic Nanoscale Composite Multifunctional Materials (ENCOMM) is supported by a $4.1M award from the Provost's Targeted Investment in Excellence Program. ENCOMM builds on broad strength at OSU in electronic, magnetic and organic materials to address cutting edge challenges in understanding and developing complex multicomponent materials. These problems are inherently multidisciplinary and require state-of-the-art facilities. ENCOMM's mission is to build and nurture the teams that can compete effectively for multidisciplinary block funded centers, to create the environment in which these theoretical and experimental teams can form and interact, and to provide the infrastructure needed to perform the research that will define this field. ENCOMM membership extends across the departments of Physics, Chemistry, Mechanical Engineering, Materials Science & Engineering, Electrical & Computing Engineering, and Biomedical Engineering. ENCOMM formed and nurtured the Interdisciplinary Research Groups that successfully competed for our new NSF funded Materials Research Science and Engineering Center (MRSEC). ENCOMM includes 44 faculty from six departments. Physics faculty: Hammel, Johnston-Halperin, Brillson, Epstein, Gramila, Gupta, Jayaprakash, Lemberger, Meyer, Pelz, Poirier, Putikka, Randeria, Sooryakumar, Stroud, Trevedi, Wilkins, Yang. (www.physics.ohio-state.edu/ENCOMM)

Our Materials Research Science and Engineering Center (MRSEC) called the Center for Emergent Materials is supported by an NSF award of $10.8M over six years with an additional $6.2M in institutional support. The CEM is comprised of two interdisciplinary research groups whose activities focus on the general area of magnetoelectronics: the first "Towards Spin-Preserving, Heterogeneous Spin Networks" is focused on broadening the application of spintronics to new materials through collaborative development of innovative characterization and investigational tools and new materials growth and testing approaches. The second IRG titled "Double Perovskite Interfaces and Heterostructures" is creating new functionality in oxide materials by combining novel perovskite compounds into composite structures with controlled interfaces. By tuning magnetic and electronic properties of the constituent materials and exploiting controlled strain at interfaces new properties emerge. The Center includes 27 graduate research fellows, 5 post doctoral researchers, 20 undergraduate students, and 3 staff and 21 faculty members. Physics faculty: Johnston-Halperin, Gupta, Hammel, Heckler, Epstein, Meyer, Pelz, Stroud, Brillson, Lemberger, Randeira, Trivedi, Yang. (cem.osu.edu)

THE OHIO STATE UNIVERSITY

DEPARTMENT OF ASTRONOMY

Columbus, Ohio 43210

Students Accepted For Degree	FIELDS		
	Physics	Astronomy	Related Fields
Doctorate		X	
Master's			

1. General

President: E. Gordon Gee
Dean of Graduate School: Patrick S. Osmer
Department Chairman: Bradley Peterson
Department Telephone Number: (614) 292-1773 (C)
Type of Institution: University
Control: Public
Setting: Urban
Total Faculty: 5,854
Graduate Faculty: 3,076
Total Students: 55,014
Total Graduate Students: 10,741
Annual Graduate Tuition Before Candidacy:
In-state residents: Full-time—$10,713*
Out-of-state residents: Full-time—$25,953*
After candidacy:
In-state residents: Full-time—$3,458*
Out-of-state residents: Full-time—$8,030*
Tuition rates for: 2009–10
Deferred tuition plan: No
Term: Quarter

*Academic year, three quarters

2. Number of Faculty in Department

The combined total of full-time faculty solely in the Astronomy Department in the three professorial ranks is 16. The combined total of full-time and faculty with joint appointments at all ranks is 20.

3. Admission, Financial Aid, and Housing

Address admission inquiries to: Graduate Committee Chairman, Dept. of Astronomy, 140 W. 18th Ave.
Graduate application fee required: $40 domestic; $50 foreign
Admission deadline (Fall admission): 1/4 domestic; 11/30 foreign.
Admission information: For fall admission, 2009–10, 12 students were accepted from 77 applicants.
Admission requirements: For admission to the graduate programs, a Bachelor's degree in astronomy, physics, or math is suggested with a minimum undergraduate GPA of 3.0/4.0 specified. The GRE and the GRE Advanced (Physics) are required. Students from non-English speaking countries are required to demonstrate proficiency in English via the TOEFL exam. Minimum acceptable score for admission is 79. Graduate Teaching Associates must demonstrate their fluency in spoken English before they will be allowed to assume classroom teaching duties.
Address financial aid inquiries to: Graduate Committee Chair, Dept. of Astronomy, 140 W. 18th Ave.
GAPSFAS application required: No
Financial aid application deadline: 1/15
Loans available: Yes
Address housing inquiries to: Housing Services, 350 Morrill Tower, 1910 Cannon Drive.
On-campus, single student housing available: Yes
Cost/month: $730
On-campus, married student housing available: Yes
Cost/month: $610—1 bdrm (includes heat);
$755—2 bdrm. (includes heat)

Table A—Faculty, Enrollments, and Degrees Granted

Research Specialty	2009–10 Faculty	Enrollment Fall 2009		No. of Degrees Granted[1] 2009–10 (2005–10)			Median No. of Years for 2009–10 Ph.D.'s
		Master's	Doctorate	Master's	Terminal Master's	Doctorate	
Astrophysics	21	–	29	2(12)	0(2)	5(24)	6
Full-time Grad. Stud.		–	29				
Part-time Grad. Stud.		–	0				
First-year Grad. Stud.		–	6				
Undergraduate Degrees, 2009–10 (2005–10): 7(37))							

[1] Five-year totals in parentheses.

4. Graduate Degree Requirements

Master's: A minimum of 45 quarter hours at the graduate level is required, with a cumulative point-hour ratio of 3.0. At least 36 hours of graduate credit must be completed in a period of three quarters at this university. A final comprehensive examination or a thesis is required.
Doctorate: A minimum of 135 quarter hours at the graduate level is required, of which 45 hours may be credited for completion of a Master of Science earned elsewhere. An average point-hour ratio of 3.0 is required in all graduate courses. A general examination (which admits a student to candidacy) and a final oral defense of the dissertation are required.
Thesis: Thesis may be written *in absentia*.
Special Equipment, Facilities, or Programs: The Department is a partner in the Large Binocular Telescope (LBT), with twin 8.4-meter mirrors on Mt. Graham in Arizona. The Department also has a one-fourth share of the MDM Observatory with modern 2.4-m and 1.3-m telescopes located on Kitt Peak. Instrumentation available on the MDM telescopes includes (1) TIFKAM, an infrared imager and spectrometer with a Hawaii-2 1024×1024 HgCdTe array, (2) CCD spectrographs for moderate resolution observations of faint sources, and (3) CCD cameras for direct imaging, including $8k^2$ and $4k^2$ wide-field imaging cameras. New advanced instrumentation is built by the Astronomy Department's Imaging Sciences Laboratory. Recent projects include the MODS optical spectrographs, a fixed secondary mirror, and the aluminization system for the LBT. Students interested in astronomical instrumentation are encouraged to participate in designing and constructing modern instruments.
The department has a network of 100 LINUX and Windows workstations. Each graduate student is assigned a workstation for their own research use. The department also has a number of high-performance multi-processor computers for intensive calculations, including Beowulf clusters, and students regularly have access to the Ohio Supercomputing Center.

The Department also provides machine and electronic shops and a complete support staff.

Table B—Appointments to Graduate Students, 2009–10

Title of Appointee	Appointments Total	First year	Academic Load Allowed in Credit Hours[1]	Hours of Service Per Week	Stipend for Academic Year ($)
			Quarter		
Teaching Assistant	9	1	9–15	20	23,592
Research Assistant	9	0	9–15	20	23,592
University Fellowship	7	4	15	0	25,592
NSF Fellowship	2	0	15	0	30,000
Fullbright Fellowship	1	1	15	0	–
Total	28	6			

[1]Before admission to candidacy.

5. Personnel Engaged in Separately Budgeted Research, 7/09–6/10

Professorial faculty	17
Graduate students	19
Total	36

6. Separately Budgeted Research Expenditures by Source of Support 7/08–6/09

	Departmental Research	Physics-related Research Outside Department
Federal government	$2,301,543	$1,139,871[1]

[1]Represents expenditures by faculty who hold joint appointments in Physics and Astronomy. Amount is not included in Departmental Research.

Table C—Separately Budgeted Research Expenditures

Research Specialty	No. of Grants	Expenditures ($)
Astrophysics	84	3,441,414

FACULTY

Professors

Beacom, John F., Ph.D., Wisconsin, 1997. Particle/nuclear astrophysics and cosmology; neutrinos.

Beatty, James J., Ph.D., Chicago, 1986. Experimental particle astrophysics: ultrahigh energy cosmic rays and neutrinos.

Gould, Andrew P., Ph.D., Stanford, 1988. Microlensing, extrasolar planets; identifying and measuring dark matter.

Herbst, Eric, Ph.D., Harvard, 1972. Molecular astrophysics; scattering processes.

Kochanek, Christopher S., Ph.D., Caltech, 1989. Galaxy structure; gravitational lenses; cosmology.

Mathur, Smita, Ph.D., Tata Institute of Fundamental Research, 1991. Active galactic nuclei; X-ray astronomy; high redshift universe.

Osmer, Patrick S., Ph.D., Caltech, 1970. Spectroscopy and evolution of quasars.

Peterson, Bradley M., Ph.D., Arizona, 1978. Chairman. Active galactic nuclei; supermassive black holes; QSO absorption lines.

Pinsonneault, Mark H., Ph.D., Yale, 1988. Theoretical stellar structure and evolution.

Pogge, Richard W., Ph.D., U.C. Santa Cruz, 1988. Active galaxies; supermassive black holes; instrumentation; extrasolar planets.

Pradhan, Anil K., Ph.D., University College, London, 1977. Atomic and molecular physics.

Ryden, Barbara S., Ph.D., Princeton, 1987. Large-scale structure of the Universe; cosmology.

Sellgren, Kristen, Ph.D., Caltech, 1983. Interstellar dust; galactic center; infrared astronomy.

Stanek, Krzysztof Z., Ph.D., Princeton, 1996. Gamma-ray bursts; extragalactic distance scale; extra-solar planets; most massive stars; variability searches.

Steigman, Gary, Ph.D., NYU, 1968. Cosmology; Big Bang nucleosynthesis.

Walker, Terrence P., Ph.D., Indiana, 1987. High-energy nuclear physics and cosmology.

Weinberg, David H., Ph.D., Princeton, 1989. Large-scale structure; cosmology.

Associate Professors

Gaudi, B. Scott, Ph.D., Ohio State, 2000. Extrasolar planets; Kuiper Belt; astrobiology.

Terndrup, Donald M., Ph.D., U.C. Santa Cruz, 1986. Galactic bulge; stellar rotation; stellar populations.

Assistant Professors

Johnson, Jennifer A., Ph.D., U.C. Santa Cruz, 1999. Stellar abundances; stellar evolution.

Martini, Paul, Ph.D., Ohio State, 2000. Active galactic nuclei; clusters of galaxies; instrumentation.

Thompson, Todd, Ph. D., Arizona, 2002. Supernovae, gamma-ray bursts, starburst galaxies, stellar feedback processes.

Research Scientists

Atwood, Bruce, Ph.D., Wesleyan, 1975. Astronomical instrumentation; extragalactic astronomy.

Nahar, Sultana, Ph.D., Wayne State University, 1987. Radiative and collisional atomic processes.

RESEARCH SPECIALTIES AND STAFF

Theoretical

Astronomy and Astrophysics. Gaseous nebulae and interstellar medium; stellar atmospheres and radiative transfer; compact objects, atomic and molecular physics; dynamics of systems; cosmology; extrasolar planetary systems. Beacom, Gaudi, Gould, Pinsonneault, Pradhan, Ryden, Steigman, Thompson, Walker, Weinberg.

Experimental

Astronomy and Astrophysics. Stellar and extragalactic spectroscopy; stellar populations; stellar photometry; interstellar gas and dust; star formation; nuclear astrophysics; active galactic nuclei; observational cosmology; interacting binary systems; extrasolar planetary systems; infrared observations; galactic structure; galactic center. Beatty, Gaudi, Herbst, Johnson, Mathur, Kochanek, Martini, Osmer, Peterson, Pogge, Sellgren, Stanek, Terndrup.

THE UNIVERSITY OF AKRON

DEPARTMENT OF PHYSICS

Akron, Ohio 44325-4001

Students Accepted For Degree	FIELDS: Physics	Astronomy	Related Fields
Doctorate			X
Master's	X		

1. General

President: Luis M. Proenza
Dean of Graduate School: George R. Newkome
Department Chair: Robert Mallik
Department Telephone Number: (330) 972-7078
Type of Institution: University
Control: Public
Setting: Urban
Tota Faculty: 1,907
Total Graduate Faculty: 958
Total Students: 27,911
Total Graduate Students: 4,103
Annual Graduate Tuition:
In-state residents: Full-time—$397.55/credit
Out-of-state residents: Full-time—$407.55/credit
Tuition rates for: 2009–10
Deferred tuition plan: Yes
Annual Other Fees: $200 (approx.)
Term: Semester

2. Number of Faculty in Department

The combined total of full-time faculty in the three professorial ranks is 10. The combined total of full-time, part-time, and other faculty at all ranks is 15.

3. Admission, Financial Aid, and Housing

Address admission inquiries to: Dr. Ben Nu, Department of Physics
Graduate application fee required: $30 domestic, $40 international
Admission deadline: recommended, 6 months prior to intended date of enrollment.
Admission information: For fall admission, 2009–10, 6 students were accepted from 18 applicants.
Admission requirements: For admission to the graduate programs, a Bachelor's degree in physics is preferred, but applicants with Bachelors' degrees in another field of science, engineering, mathematics, or science education will be considered on an individual basis. A minimum undergraduate GPA of 2.75/4.0 is required for full admission. Students from non-English speaking countries are required to demonstrate proficiency in English via the TOEFL exam and in-house test administered by UA. The minimum computer-based TOEFL score is 213.
Undergraduate preparation assumed: Ohanian, *Modern Physics*, Fowles and Cassiday, *Mechanics*, Griffiths, *Introduction to Electrodynamics*, Ohanian, *Principles of Quantum Mechanics*, Kittel and Roemer, *Thermal Physics*, Kittel, *Introduction to Solid State Physics.*
Address financial aid inquiries to: Dr. Ben Yu-Kuang Hu, Department of Physics
GAPSFAS application required: No
Financial aid deadline: 5/1
Loans available: Yes
Address housing inquiries to: Graduate School, The University of Akron
On-campus, single student housing available: No
On-campus, married student housing available: No

Table A—Faculty, Enrollments, and Degrees Granted

Research Specialty	2009–10 Faculty	Enrollment[1] Fall 2009: Master's	Doctorate	No. of Degrees Granted[2] 2009–10 (2005–10): Master's	Terminal Master's	Doctorate	Median No. of Years for 2008–09 Ph.D.'s
Chemical Physics	5	0	1	–	0(0)	0	–
Condensed Matter Physics	6	7	0	–	4(20)	–	–
Electromagnetism	1	0	0	–	0(0)	–	–
Fluids & Rheology	2	0	0	–	0(0)	–	–
History & Philosophy	1	0	0	–	0(0)	–	–
Low Temperature Physics	2	0	0	–	0(0)	–	–
Particles & Fields	1	0	0	–	0(0)	–	–
Physics Education	3	0	0	–	2(3)	–	–
Polymer Physics/ Science	3	3	1	–	1(8)	–	–
Relativity & Gravitation	1	0	0	–	0(0)	–	–
Statistical & Thermal	3	0	0	–	1(3)	–	–
Other Theoretical/ Math.	0	1	0	–	0(0)	–	–
Total		11	2	–	8(34)	0	
Full-time Grad. Stud.		17	4				
Part-time Grad. Stud.		0	0				
First-year Grad. Stud.		5	0				
Median Years in Grad. Study (2008–09 Degrees)				–	2	–	–

Undergraduate Degrees, 2008–09 (2004–09): 41

[1] Students not yet committed to a research specialty are entered under non-specialized.
[2] Five-year totals in parentheses.

4. Graduate Degree Requirements

Master's: 30 semester credits; 3.0 GPA. A thesis in a current area of active interest is required.
Other Programs: The interdisciplinary Ph.D. program in Polymer Science includes an option in Polymer Physics. Three faculty members of the physics department have some affiliation with this internationally recognized program. One Physics faculty has a joint appointment and directs Ph.D. research in Polymer Science and one in Polymer Engineering. Five members of the Physics Department have joint appointments with Chemistry allowing them to supervise Ph.D. research in the degree program of Physical Chemistry.
Thesis: Thesis may be written *in absentia.*
Special Equipment, Facilities, or Programs: The Department of Physics is housed in Ayer Hall with space and facilities for instruction and research. An NMR laboratory provides facilities for research on molecular motions of large polymer molecules in the viscous liquid phase in solutions and in rubbery materials including composites.A class-100 clean room

houses a laminar flow hood with wet/dry capabilities and an in-house modified Atomic Force Microscope for surface characterization and nanolithography. Surface science laboratories are dedicated to the study of surfaced related phenomena. These are equipped with several high vacuum systems for the thermal and sputter deposition of thin films. The morphology and chemical reactivity of these thin films are characterized *ex-situ* by atomic force and scanning tunneling spectroscopy; inelastic electra tuneling spectroscopy and FTIR. ultra-high vacuum systems are also available for the *in-situ* study of surface phenomena. Measurement techniques include scanning tunneling microscopy, Auger electron spectroscopy, X-ray photoelectronic spectroscopy, low energy electron diffraction, temperature programmed desorption and mass spectroscopy. Facilities available for computational research include dual quad core processor workstations, an SGI workstation, a quad core workstation and small clusters. These facilities are used to support instruction and theoretical research in condensed matter, polymer physics, statistical mechanics, critical phenomena, exact enumerations, Monte Carlo simulations, etc. Studies of physical properties of polymeric materials also utilize the extensive facilities of the College of Polymer Science and Polymer Engineering.

Table B—Appointments to Graduate Students, 2008–09

Title of Appointee	Appointments Total	Appointments First year	Academic Load Allowed in Credit Hours	Hours of Service Per Week	Stipend for Academic Year ($)[1]
			Semester		
Teaching Assistant	14	6	10	20	14,000
Total	14	6			

[1]Calendar year, July 1–June 30

5. Personnel Engaged in Separately Budgeted Research,

Professorial faculty	3
Graduate students	4
Total	7

6. Separately Budgeted Research Expenditures by Source of Support

	Departmental Research	Physics-related Research Outside Department
Federal government	$50,000	–
State and local government	274,000	–
Total	$229,257	

Table C—Separately Budgeted Research Expenditures

Research Specialty	No. of Grants	Expenditures ($)
Biophysics	1	50,000
Nanotechnology	2	274,000
Total	3	324,000

FACULTY

Professors

Gujrati, P. D., Ph.D., Columbia, 1979. Theoretical physics; phase transitions and critical phenomena; polymer physics; combinatorics and graph theory; renormalization group and field theory.

Mallik, Robert R., Ph.D., Leicester Polytechnic, 1985. Chair. Low-temperature physics; surface physics; electron tunneling; scanning probe microscopy.

Ramsier, Rex D., Ph.D., Univ. of Pittsburgh, 1994. Surface science; surface functionalization, nanotechnology, nanofibers, physics education.

von Meerwall, Ernst D., Ph.D., Northwestern, 1969. (Emeritus) Distinguished Professor; Associate Dean, Polymer Science. Condensed matter physics, mainly polymers; NMR, diffusion, structure-property relations; numerical methods and computation.

Associate Professors

Buldum, Alper, Ph.D., Bilkent University, 1998. Condensed matter theory, nanoscience and nanotechnology; carbon nanotubes, quantum transport, molecular electronics.

Chen, Ang, Ph.D., Zhejiang University, 1994. Experimental condensed matter physics; materials physics; ferroelectric physics, ferroelectric/piezoelectric physics, ferroelectric/piezoelectric oxides/polymers and devices; nanotechnology; characterization of dielectric/ferroelectric properties in a wide temperature and frequency range.

Hu, Ben Yu-Kuang, Ph.D., Cornell University, 1990. Theoretical condensed matter physics. Frictional drag in coupled electronic bilayers, transport in mesoscopic and nanoscale systems.

Luettmer-Strathmann, Jutta, Ph.D., Univ. of Maryland, College Park, 1994. Statistical mechanics; polymer physics; static and dynamic properties of simple and complex fluids; phase transitions and critical phenomena; mode-coupling, integral-equation and simulation techniques.

Lyuksyutov, Sergei, Ph.D., USSR Acad. Sci., 1991. Experimental surface physics; nanolithography, photorefractive optics, small-angle neutron scattering.

Assistant Professor

Dordevic, Sasa, Ph.D., University of California, San Diego, 2002. Experimental condensed matter physics; high temperrature superconductivity, heavy fermions, low dimensional metals, vortex dynamics, strongly correlated electron systems; infrared, optical and magneto-optical spectroscopy.

RESEARCH SPECIALTIES AND STAFF

Theoretical

Computational Physics. Monte Carlo simulaitons, nanotechnology, atomic scale modeling. Buldum, Hu, Luettmer-Strathmann, Paramonov.

Field Theory. Renormalization; supersymmetry. Gujrati. General Relativity.

Polymer Physics. Glass transitions, branched polymers, collapsed phase. Dynamical and topological properties, phase separation, transport properties. Gujrati, Luettmer-Strathmann.

Solid State Physics. Transport, mesoscopic systems; two-dimensional electron gases; superlattices; electron tunneling. Buldum, Gujrati, Hu, Luettmer-Strathmann.

Statistical Physics. Phase transitions and critical phenomena, static and dynamic properties of simple and complex fluids. Gujrati, Luettmer-Strathmann.

Computational Physics

Monte Carlo simulations, nanotechnology, atomic scale modeling. Buldum Hu, Luettmer-Strathmann, Paramonov.

Experimental

Chemical Physics. Chu, Creel, Dordevic, Henriksen, Mallik, Ramsier, von Meerwall.

Computer-Aided Instruction. Software development. Griffin, Hu, Luettmer-Strathmann.

Computer Physics. Applications of computers to research data accumulation and processing. Griffin, von Meerwall.

Physics Education. Programs for secondary school science teachers and physics teaching. Griffin, Henriksen, Mallik, Ramsier.

Polymer Physics. Deformation and fracture processes; crystallization; adhesion; NMR spectroscopy. von Meerwall; AFM lithography in polymer materials: Lyuksyutov, Paramonov.

Solid State Physics. Hall effect, quantum size effects, pulsed NMR diffusion, high temperature superconductivity. Chu, Creel, Dordevic, Griffin, Henriksen, Mallik, Ramsier, von Meerwall.

Surface Physics. Elastic and inelastic electron tunneling. Fourier Transform infrared spectroscopy. Scanning tunneling microscopy. Atomic force microscopy. Thermal desorption spectroscopy, nanolithography. Henriksen, Lyuksyutov, Mallik, Ramsier.

Photorefractive Optics: Ferroelectric electron emission, Holography: Lyuksyutov.

FACULTY PUBLICATIONS

Buldum, A.

C. B. Clemens, P. Hamrick, J. Heminger, K. L. Kreider, G. W. Young, A. Buldum, E. Evans, and G. Zhang, "Modeling, simulation, and experiments of coating growth on nanofibers," J. Appl. Phys. **103**, 044304 (2008).

J. P. Wilber, C. B. Clemons, G. W. Young, A. Buldum, and D. D. Quinn, "Continuum and atomistic modeling of interacting graphene layers," Phys. Rev. B **75**, 045418 (2007).

A. Patil, T. Ohashi, A. Buldum, and L. Dai, "Controlled and atomistic modeling of interacting graphene layers," Appl. Phys. Lett. **89**, 103103 (2006).

Chen, A.

A. Chen, Zhi Yu, Zhi Jing, Ruyan Guo, A. S. Bhalla, and L. E. Cross, "Piezoelectric and Electrostrictive strain behavior of Ce-doped $BaTiO_3$ ceramics, Appl. Phys. Lett. **80**(18), 3424–26 (2002).

A. Chen and Zhi Yu, "Dielectric relaxor and ferroelectric relaxor: doped paraelectric $SrTiO_3$," J. Appl. Phys. **91**(3), 1487–1494 (2002).

A. Chen, Zhi Yu, L. E. Cross, Ruyan Guo, and A. S. Bhalla, "Dielectric relaxzation and conduction in $SrTiO_3$ thin films under dc bias," Appl. Phys. Lett. **79**(6), 818–820 (2001).

Dordevic, S.

S. V. Dordevic, L. W. Kohlman, N. Stojiloci, R. Hu, and C. Petrovic, Phys. Rev. B **80**, 115114 (2009).

S. V. Dordevic, L. W. Kohlman, L. C. Tung, Y.-J. Wang, A. Gozar, G. Logvenov, and I. Bozovic, Phys. Rev. B.**79**, 134503 (2009).

C. C. Homes, S. V. Dordevic, A. Gozar, G. Blumberg, T. Room, D. Huvonen, U. Nagel, A. D. La Forge, D. N. Basov and H. Kageyama, Phys. Rev. B **79**, 125101 (2009).

Gujrati, P. D.

L. Hong, P. D. Gujrati, V. N. Novikov, and A. P. Sokolov, "Molecular cooperativity in the dynamics of glass-forming systems: A new insight," J. Chem. Phys. **131**, 194511 (2009).

P. D. Gujrati, "Comments on Entropy of polydisperse chains: Solution on the Bethe lattice," J. Chem. Phys. **128**, 184904 (2008).

R. Batman and P. D. Gujarati, "Compressible or incompressible blend of interacting monodisperse star and linear polymers near a surface ," J. Chem. Phys. **128**, 124903 (2008).

Hu, B. Y.-K.

Ben Yu-Kuang Hu, "Relativistic momentum and kinetic energy, and $E=mc^2$," Eur. J. Phys. **30**, 325-330 (2009).

Ben Yu-Kuano Hu, E. H. Hwang, and S. Das Sarma, "Density of States of Disordered Graphene," Phys. Rev. B **78**, 165411 (2008).

Wang-Kong Tse, Ben Yu-Kuang Hu, and S. Das Sarma, "Chirality-Induced Dynamic Kohn Anomalies in Graphene," Phys. Rev. Lett. **101**, 066401 (2008).

Luettmer-Strathmann, J.

F. Rampf, W. Paul, and K. Binder, "Transitions of tethered polymer chains," J. Chem. Phys. **128**, 064903 (2008).

J. Luettmer-Strathmann and M. Mantina, "Local and chain dynamics in miscible blends: A Monte Carlo simulation study," J. Chem. Phys. **124**, 174907 (2006).

B. J. de Gans, R. Kita, S. Wiegand, and J. Luettmer-Strathmann, "Unusual thermal diffusion in polymer solutions," Phys. Rev. Lett. **91**, 245501 (2003).

Lyuksyutov, S.

S. F. Lyuksyutov, P. Buchhave, and M. V. Vasnetsov, "Self-excitation of space charge waves," Phys. Rev. Lett. **79**, 67–70 (1997).

S. F. Lyuksyutov, R. A. Vaia, P. B. Parmonov, S. Juhl, L. Waterhouse, R. M. Ralich, and G. Sigalov, "Electrostatic nanolithography in polymers using atomic force microscopy," Nature Mater. **2**, 468–472 (2003).

S. F. Lyuksyutov, "Nano-patterned in polymeric materials and biological objects using atomic force microscopy electrostatic nanolithography," Current Nanoscience **1**, 245–251 (2005).

Mallik, R. R.

I. Dolog, R. R. Mallik, A. Mozynski, and J. Hu, "Adsorption of 7-Ethynyl-2,4,9-trithia-tricyclo[3.3.1.1^{3,7}]decane on ultra-thin CdS films," Surface Sci. **600**, 2972–2979 (2006).

I. Dolog, R. R. Mallik, and S. F. Lyuksyutov, "Robust functionalization of amorphous cadmium sulfide using z-lift amplitude modulated atomic force microscopy assisted electrostatic nanolithography," Appl. Phys. Lett. **90**, 213111 (2007).

E. Rowicka, D. Kashyn, M. A. Reagan, T. Hitano, P. B. Paramonov, I. Dolog, R. R. Mallik, and S. F. Lyuksyutov, "Influence of water condensation on charge transport and electric breakdown between atomic force microscope tip, polymeric and (semiconductor) CdS surfaces," Current Nanosci. **4**(2), 166–172 (2008).

Ramsier, R. D.

A. F. Lotus, E. T. Bender, E. A. Evans, R. D. Ramsier, D. H. Reneker, and G. G. Chase, "Electrical, Structural, and Chemical Properties of Semiconducting Metal-Oxide Nanofiber Yarns," J. Appl. Phys. **103**, 024910 (2008).

S. J. Park, S. Bhargava, E. T. Bender, G. G. Chase, and R. D. Ramsier, "Palladium Nanoparticles Supported by Alumina Nanofibers Synthesized by Electrospinning," J. Mater. Res. **23**, 1193 (2008).

M. W. Kovacik, R. A. Mostardi, D. R. Neal, T. F. Bear, M. J. Askew, E. T. Bender, J. I. Walker, and R. D. Ramsier, "Differences in the Surface Composition of Seemingly Similar F75 Cobalt-Chromium Micro Sized Particles Can Affect Synovial Fibroblast Viability," Colloids Surfaces B: Biointerfaces **65**, 269 (2008).

von Meerwall, E.

H. Lin, W. L. Mattice, and E. D. von Meerwall, "Chain Dynamics of Bidisperse Polyethylene Melts: A Monte-Carlo Study on a High Coordination Lattice," Macromolecules **40**, 959–966 (2007).

E. D. von Meerwall, H. Lin, and W. L. Mattice, "Trace Diffusion of Alkanes in Polyethylene: Spin-Echo Experiment and Monte-Carlo Simulation," Macromolecules **40**, 2002–2007 (2007).

N. Waheed, W. L. Mattice, and E. D. von Meerwall, "Enhanced Diffusion at Intermediate Stereochemical Composition in Polypropylene by Dynamical Monte Carlo," Macromolecules **40**, 1504–1511 (2007).

UNIVERSITY OF CINCINNATI

DEPARTMENT OF PHYSICS

Cincinnati, Ohio 45221-0011

Students Accepted For Degree	FIELDS		
	Physics	Astronomy	Related Fields
Doctorate	X	*	
Master's	X	**	

*Ph.D. in Physics with Astronomy Dissertation.
**MS in Physics with Astronomy Emphasis.

1. General

President: Gregory H. Williams
Department Head: Kay Kinoshita
Department Telephone Number: (513) 556-0501
Department Fax Number: (513) 556-3425
E-Mail: ucphysics@ucmail.uc.edu
Type of Institution: University
Control: Public
Setting: Urban
Total Faculty: Full-time—2,654
Part-time—3,229
Total Graduate Faculty: 1,702
Total Students: 39,667
Total Graduate Students: 5,561
Quarterly Graduate Fees: [1]
In-state residents: Full-time—$3,888
Part-time—$390/credit
Out-of-state residents: Full-time—$7,471
Part-time—$748/credit
Tuition rates for: 2010–11
Deferred tuition plan: Yes
Annual Other Fees: [2]
Term: Quarter

[1]Excludes General Fee of $412, per quarter.
[2]All full-time students are required to be covered by health insurance. Foreign students must purchase University insurance. ITIE fee $112/quarter.

2. Number of Faculty in Department

The combined total of full-time faculty in the three professorial ranks is 27. The combined total of full-time, part-time, and other faculty at all ranks is 42.

3. Admission, Financial Aid, and Housing

Address admission inquiries to: Graduate Program Director, Department of Physics, P.O. Box 210011 or at physics.grad@uc.edu
Graduate application fee required: $45
Admission deadline (Fall admission): April 15.
Information available on World Wide Web: http://homepages.uc.edu/physics/index.html
Admission information: For fall admission, 2010–10, 14 students were offered admission from 120 applicants.
Admission requirements: For admission to the graduate programs, a Bachelor's degree in physics or a related science or engineering discipline is required. Students applying for graduate stipends should have a GPA of 3.0/4.0 or better. The GRE general exam and the GRE physics subject test are recommended for all applicants. No minimum score is set. Students from non-English speaking countries are required to demonstrate proficiency in English via the TOEFL exam. Minimum acceptable score for admission is 520 paper based; 190 computer based; or 68 internet based. TOEFL exam, and an overall band score of 6.5 for the IELTS exam. An oral English test is given after arrival.
Undergraduate preparation assumed: Symon, *Mechanics*; Lorrain and Corson, *Electromagnetic Fields and Waves*; Kittel, *Thermal Physics*: Fermi, *Thermodynamics*; Anderson, *Modern Physics and Quantum Mechanics*.
Address financial aid inquiries to: Graduate Program Director, Department of Physics, P.O. Box 210011 or at physics.grad@uc.edu
GAPSFAS application required: No
Financial aid deadline: April 15
Loans available: Yes
Address housing inquiries to: Graduate & Family Housing, P.O. Box 210045 or at ucgradfa@uc.edu
On-campus, student dormitory housing available: Yes
On-campus, graduate student apartments available: Yes

Table A—Faculty, Enrollments, and Degrees Granted

Research Specialty	2009–10 Faculty	Enrollment[1] Fall 2009 Master's	Enrollment[1] Fall 2009 Doctorate	No. of Degrees Granted[2] 2009–10 (2005–09) Master's	Terminal Master's	Doctorate	Median No. of Years for 2009–10 Ph.D.'s
Astrophysics	2	1	2	1(1)	1(2)	1(2)	7.21
Condensed Matter Physics	12	0	15	0(0)	0(3)	4(17)	6.78
High Energy Physics	10	0	10	0(0)	0(0)	3(11)	5.95
Medical & Health Physics	0	0	3	0(1)	0(2)	1(1)	–
Other Computational Bio-Physics	0	0	5	0(0)	0(0)	0(0)	–
Physics Education	1	1	0	1(0)	1(0)	0(0)	–
Non-specialized	0	0	17	4(12)	3(8)	0(0)	–
Total	25	2	52	5(14)	5(15)	9(31)	
Full-time Grad. Stud.		1	52				
Part-time Grad. Stud.		0	0				
First-year Grad. Stud.		1	13				
Median Years in Grad. Study (2009–10 Degrees)				1.78	4.10	6.67	

Undergraduate Degrees, 2008–09 (2004–09): 13(26)

[1]Students not yet committed to a research specialty are entered under non-specialized.
[2]Five-year totals in parentheses.

4. Graduate Degree Requirements

Master's: There are two options available. Neither contains a language requirement. A. *Thesis Option*: Completion of 40 quarter credits of formal courses in the department. Participation in Graduate Colloquium and Seminar each resident, full-time term. "B" average course work. Research project based on a minimum of 8 course credits of work, and culminating in a written thesis. Final oral examination on thesis and related topics. B. *Non-Thesis Option*: Satisfactory completion of 48 quarter credits of formal courses. Participation in Graduate Colloquium and Seminar each resident, full-time term. "B" average in course work and satisfactory performance on a written examination.
Doctorate: Satisfactory completion of 135 quarter credits of graduate level course work, including seminars and research work on thesis topic. Pass written Qualifying Examination.

Oral examination covering chosen field of research. Dissertation. Public presentation and defense of dissertation. Residency requirement of 45 credits. Teaching requirement. No language requirement.

Other Programs: Numerous interdisciplinary research opportunities exist. Recent thesis research projects have been in such areas as Medical Physics, Chemistry, or Physics applied to various problems in the Engineering disciplines. The Department participates in the Interdisciplinary Graduate Degree Program of the Division of Graduate Education and Research. The Program allows custom tailoring of interdisciplinary studies to the individual student's interests.

Thesis: Thesis may be written *in absentia*.

Special Equipment, Facilities, or Programs:

A wide variety of facilities and programs are available to us, both within the department and elsewhere on campus. We have a modern research laboratory and office building which house our on-campus research activities. We have strong research ties with the Departments of Chemistry, Electrical Engineering, Chemical and Materials Engineering, and Radiology. In particular physics, work is undertaken at the national and international accelerator laboratories (SLAC, Fermilab, and KEK). Our condensed matter laboratories are well equipped with the normal variety of ultrasensitive measurement, analytical and sample preparation equipment. Special items include dilution refrigerators for the milli-Kelvin range, argon, and carbon dioxide lasers for Brillouin and Raman studies, and an excellent photolithographic and microelectronics laboratory.

Table B—Appointments to Graduate Students, 2009–10

Title of Appointee	Appointments: Total	Appointments: First year	Academic Load Allowed in Credit Hours	Hours of Service Per Week	Stipend for Academic Year ($)
			Quarter		
Teaching Assistant	30	13	12–19	20 max.[1]	13,430–16,470[2,3]
Research Assistant	11	0	12–19	20 max.	13,430–16,470
Total	41	13			

[1]Usually includes 3–8 hours of classroom/laboratory teaching, plus time for preparation, grading, and consulting with students.
[2]Tuition and fees waived, except health insurance, and ITIE fee. 10 month appointment.
[3]Summer assistantships are also available.

5. Personnel Engaged in Separately Budgeted Research, 7/08–06/09

Professorial faculty	18
Postdoctoral appointments	6
Graduate students	23
Undergraduate students	2
Total	49

6. Separately Budgeted Research Expenditures by Source of Support, 07/08–06/09

	Departmental Research
Federal government	$1,729,911
Private industry	$16,500
Total	$1,746,411

Table C—Separately Budgeted Research Expenditures by Research Specialty, 07/08–06/09

Research Specialty	No. of Grants	Expenditures ($)
Astrophysics	2	18,212
Condensed Matter Physics	5	315,169
High Energy Physics	4	852,871
Physics Education	2	5,002
Total	13	1,191,254

FACULTY

Professors

Argyres, Philip C. Ph.D., Princeton, 1989. Theoretical particle physics.

Endorf, Robert J., Ph.D., Carnegie-Mellon, 1971. Physics education.

Esposito, F. Paul, Ph.D., Chicago, 1971. Theoretical physics; general relativity and astrophysics; condensed matter theory.

Hanson, Margaret M., Ph.D., Colorado, 1995. Experimental astronomy and astrophysics.

Jackson, Howard E., Ph.D., Northwestern, 1971. Experimental condensed matter physics; laser light scattering studies.

Johnson, Randy A., Ph.D., California, Berkeley, 1975. Experimental high-energy physics.

Kim, Young H., Ph.D., Florida, 1986. Experimental condensed matter physics.

Kinoshita, Kay, Ph.D., California, Berkeley, 1981. Experimental particle physics.

Ma, Michael, Ph.D., Illinois, 1983. Theoretical condensed matter physics.

Meadows, Brian T., D. Phil., Oxford, 1966. Experimental high-energy physics.

Newrock, Richard S., Ph.D., Rutgers, 1970. Experimental condensed matter physics.

Pinski, Frank J., Ph.D., Minnesota, 1977. Theoretical condensed matter physics.

Scanio, Joseph J. G., Ph.D., California, Berkeley, 1967. Theoretical high-energy physics.

Schwartz, Alan, Ph.D., Harvard, 1988. Experimental particle physics.

Sitko, Michael, Ph.D., Wisconsin, 1980. Experimental astronomy; cosmic dust; protostellar disks; comets.

Smith, Leigh M., Ph.D., Illinois, 1988. Experimental condensed matter physics.

Sokoloff, Michael D., Ph.D. California, Berkeley, 1983. High-energy experimental physics.

Wijewardhana, L. C. R., Ph.D., MIT, 1984. Theoretical particle physics.

Associate Professors

Kagan, Alexander L., Ph.D., Chicago, 1989. Theoretical particle physics.

Mast, David B., Ph.D., Northwestern, 1982. Experimental condensed matter physics.

Serota, Rostislav A., Ph.D., MIT, 1987. Theoretical condensed matter physics.

Wagner, Hans-Peter A., Ph.D., Regensburg (Germany), 1991. Experimental condensed matter physics.

Assistant Professors

Bolech, Carlos J., Ph.D., Rutgers, 2002. Theoretical condensed matter physics.
Kogan, Andrei B., Ph.D., Duke, 2000. Experimental condensed matter physics.
Shah, Nayana B., Ph.D., Rutgers, 2003. Theoretical condensed matter physics.
Zupan, Jure, Ph.d., Univ. of Ljubljana, 2002. Theoretical particle physics.

Professors Emeriti

Chow, William S., Ph.D., Case Inst. of Tech., 1964. Theoretical condensed matter physics; theory of semiconductors.
Fenichel, Henry, Ph.D., Rutgers, 1964. Experimental condensed matter physics; fluids; optics; holography.
Goodman, Bernard, Ph.D., Pennsylvania, 1955. Theoretical condensed matter physics.
Jha, Shacheenatha, Ph.D., Edinburgh Univ., 1950. Experimental condensed matter physics.
Joiner, William C. H., Ph.D., Rutgers, 1962. Experimental condensed matter physics; superconductivity.
Russell, James E., Ph.D., Yale, 1958. Theoretical physics; exotic atoms; chemical physics.
Suranyi, Peter, Ph.D., Joint Inst. for Nucl. Res., Dubna, 1964. Theoretical physics; high-energy theory; statistical mechanics.
Witten, Louis, Ph.D., Johns Hopkins, 1951. Theoretical physics; general relativity.
Zhang, Fu-Chun, Ph.D., Virginia Tech., 1983. Theoretical condensed matter physics.

RESEARCH SPECIALTIES AND STAFF

Theoretical

Astrophysics and Relativity. Gauge theories of gravity and supergravity; quantum gravity; black holes; brane-world cosmology; baryogenesis. Argyres, Esposito, Scanio, Suranyi, Wijiwardhana, Witten.

Atomic and Molecular Physics. Scattering theory; rearrangement collisions; chemical phenomena during atomic cascade when muons and negative mesons are stopped in helium; details of capture mechanisms for negative pions and kaons stopped by high-Z atoms. Russell.

Condensed Matter, Low-Temperature Statistical Physics. Superconductivity; dynamical response properties of metal electrons; disordered systems; electronic structure of metals and alloys; itinerant magnetism; strongly correlated Fermions and boson systems; large scale many body simulations; quantum and classical fluids; quasi-one-dimensional conductors: thermal properties of many-body systems; semiconductor superlattices; transport properties of semiconductors; nonlinear effects in semiconductors; phase transitions and critical phenomena; gas dynamics; high-T_c superconductivity; mesoscopic physics; metal-insulated transitions, random magnets; heavy Fermion systems, quantum chaos, fractional quantum Hall effect. Goodman, Ma, Pinski, Serota, Zhang.

Elementary Particles and Fields. Symmetry and constituent models of hadrons and leptons; grand unified theories; supersymmetric gauge theories; supergravity; superstring theories; M-theory, matrix models; strong interaction phenomenology and dynamics; nonperturbative methods; lattice field theories; renormalization group; phase transitions in quantum field theories; electro-weak symmetry breaking and fermion mass generation; flavor physics; conformal field theories. Argyres, Esposito, Kagan, Scanio, Suranyi, Wijewardhana, Witten.

Quantum signatures of classically chaotic and regular systems; thermodynamical and kinetic properties of small particles; orbital and spin magnetism of Mesoscopic systems. Serota.

Experimental

Astronomy and Astrophysics. Near-IR spectroscopy of massive young stars; star formation and galactic structure; mid-IR spectroscopy of protostellar disks, comets, and solar system formation and evolution. Hanson, Sitko.

Condensed Matter, Materials Science, Low Temperature, and Optics. Laser light scattering including Raman, Brillouin and photoluminescence; femtosecond and picosecond spectroscopy; spatially- and temporally-resolved submicron microscopies; electronic and optical properties of low dimensional semiconductors including quantum dots, nanowires and nanowire heterostructures; near field optical semiconductor spectroscopy of photonic crystals, optical waveguides, and plasmonic nanostructures; phonons in insulator and semiconductor nanostructures; spin properties and dynamics of excitons in semiconductor nanostructures; electron beam lithography of micro- and nano-scale biological and chemical sensors and electronic devices; ultralow temperatures; transport properties of one-dimensional wires; Luttinger liquid effects; Coulomb drag interactions between wires; spin transistors and spintronics; spin-orbit coupling in quantum point contacts; Non-equilibrium dynamics in nanoscale systems: Electronic correlation phenomena in semiconductor quantum dots, non-equilibrium Kondo effect, photon-assisted transport, dynamic conductance spectroscopy, coherent manipulation of spin-based quantum systems, electro-optical phenomena in nanostructured materials. Organic molecular beam deposition (OMBD) and optical characterization of organic and hybrid organic/semiconductor nanostructures and organic light emitting diodes; Study of linear and nonlinear optical properties in organic/semiconductor nanostructures and waveguides: stimulated emission; second harmonic gernation; two-photon absorption; nonlinear refractive index; coherent exciton dynamics in II-VI quantum wells, - wires and - dots; phase coherent photorefractive effect in quantum wells; optical coherence imaging; near-field microwave scanning of HTS materials, composite ceramics, layered semiconductors and biological samples; vibrational thermodynamics of micro and nano particle systems; plasma treatment and surface functionalization of carbon nanotubes and other nano particle materials; near field spectroscopy of single wall carbon nanotubes; far-infrared (FIR) studies of metal-insulator transition in highly correlated layered systems, FIR charge dynamics in superconducting cuprates, and FIR absorption by small metal particles. Electrical and thermal transport studies of pristine and functionalized carbon nanotubes; Microwave electrical transport in carbon nanotubes and other nano-structured materials; Electrical and magnetic properties of functionalized nano-structured materials for use in bio-sensors and bio-medical sensors; Raman spectroscopy investigations of the source of lead contamination in domestic drinking water system components. Fenichel, Jackson, Joiner, Kim, Kogan, Mast, Newrock, Smith, Wagner.

Medical Physics. Development of micro-miniature microwave and surface acoustic wave devices for biochemical sensors; investigation of membrane structures using near-field scanning microwave microscopy and ultrasonic acoustics. Mast.

Particle Physics, High-Energy Physics.

At Fermilab: neutrino and antineutrino experiments; studying high-energy neutrino interactions and neutrino oscillations.

At SLAC: heavy quark physics, especially CP violations in B-meson decays using the Babar detector, and charm meson decays.

At KEK (Japan): studies of CP violation, mixing, and rare processes in B-meson and D-meson decays, using the Belle detector.

Johnson, Kinoshita, Meadows, Schwartz, Sokoloff.

Physics Education. Improving introductory physics courses with the implementation of active learning and inquiry; improving the training of graduate teaching assistants; professional development programs for primary, middle school, and high school teachers; collaboration with other science departments, the College of Education, Criminal Justice, and Human Services and local school districts to improve K-16 science teaching. Endorf.

UNIVERSITY OF DAYTON

ELECTRO-OPTICS PROGRAM

Dayton, Ohio 45469-0245

Students Accepted For Degree	FIELDS Physics	Astronomy	Related Fields
Doctorate			X
Master's			X

1. General

President: Daniel J. Curran
Dean of Graduate School: Paul Vanderburgh
Dean, School of Engineering: Tony Saliba
Director: Joseph W. Haus
Department Telephone Number: (937) 229-2797
Type of Institution: University
Control: Private
Setting: Urban
Total Faculty: 937
Total Students: 10,909
Total Graduate Students: 3,502
Annual Graduate Tuition: $729
Tuition rates for: 2010–11
Deferred tuition plan: Yes
Annual Other Fees: None
Term: Semester

2. Number of Faculty in Department

The combined total of full-time faculty in the three professorial ranks is 16. The combined total of full-time, part-time, and other faculty at all ranks is 22.

3. Admission, Financial Aid, and Housing

Address admission inquiries to: Electro-Optics Program, University of Dayton, 300 College Park, CPC572, Dayton, Ohio 45469-2951
Graduate application fee required: Domestic applications online, no fee. International applications there is a $50.00 application fee. Paper applications are no longer accepted for Domestic or International.
Admission deadline (Fall admission): 8/1
Admission information: For fall admission, 2009–10, 15 students were accepted from 70 applicants.
Admission requirements: For admission to the graduate programs, a Bachelor's degree in physics, optics, or engineering is required. Students who have degrees in chemistry, computer science, applied mathematics, or in related sciences are encouraged to apply, but they may be required to take a limited amount of undergraduate work. International students are required to demonstrate proficiency in English via the TOEFL exam. Minimum acceptable score is 550.
Address financial aid inquiries to: Office of Scholarships and Financial Aid, University of Dayton, 300 College Park, CPC572, Dayton, Ohio 45469-2951
GAPSFAS application required: Yes
Financial aid deadline: 3/1
Loans available: Yes
Address housing inquiries to: Housing Services, Gosiger Hall 212, University of Dayton, Dayton, Ohio 45469-0950
On-campus, single student housing available: Yes
Cost/term: Varies
On-campus, married student housing available: No

Table A—Faculty, Enrollments, and Degrees Granted

Research Specialty	2008–09 Faculty	Enrollment[1] Fall 2009 Master's	Doctorate	No. of Degrees Granted 2009–10 (2000–09) Master's	Terminal Master's	Doctorate	Median No. of Years for 2008–09 Ph.D.'s
Optics	16	27	19	10(50)	–	6(26)	3.5
Total		23	–	13(27)	–	4(15)	
Full-time Grad. Stud.		17	8				
Part-time Grad. Stud.		6	6				
First-year Grad. Stud.		5	3				
Median Years in Grad. Study (1999–00 Degrees)				2	–	3.5	–
Undergraduate Degrees: N/A							

[1]Five-year totals in parentheses.

4. Graduate Degree Requirements

Master's in Electro-Optics: 24 semester hours of courses (including 18 semester hours of core courses, 3 semester hours of elective, and three 1-semester-hour EO laboratory courses) and six semester hours of thesis in the case of thesis option. A nonthesis option is available where thesis is replaced by six semester hours of approved electives.
Doctorate in Electro-Optics: A minimum of 90 semester-hours beyond the Bachelor's degree. This includes 12 semester hours of 600-level EO courses, 30 semester hours of dissertation, and 6 semester hours of mathematics. In addition, comprehensive exam and residency are required. All entering Ph.D. students are expected to have completed an M.S. in electro-optics or its equivalent.
Other Programs: A Ph.D. in Electrical Engineering is available to students who would like to emphasize electro-optics in their dissertation research, but meet all other electrical engineering requirements.
Thesis: Thesis may be written *in absentia*.
Special Equipment, Facilities, or Programs: Electro-optics facilities include a total of 25 research laboratories dedicated to ellipsometry, optical processing, optical metrology, pattern recognition, nonlinear optics, spectroscopy, and nanophotonics fabrication and characterization. A wide range of optical and optical mounting equipment is available including a variety of pulsed lasers and continuous wave lasers. The University is located within minutes of the Wright-Patterson Air Force Base where opportunities are often available to EO students to work at one of the many government laboratories at this facility, including a graduate co-op program.

Table B—Appointments to Graduate Students, 2010–11

Title of Appointee	Appointments Total	Appointments First year	Academic Load Allowed in Credit Hours	Hours of Service Per Week	Stipend for Academic Year ($)
			Semester		
Teaching Assistant	7	2	10	20	15,000[1,2,3]
Research Assistant	19	3	10	20	21,000[1,3]
Minority Scholarship	1	1			
Dissertation Fellowship	1	1			
Total	28	7			

[1] Tuition and fees waived.
[2] Converted to research assistantship during summer.
[3] Research during the summer pays at least $1,200 per month.
[4] Continues through the summer at the same monthly rate.

5. Personnel Engaged in Separately Budgeted Research, 8/10–5/11

Professorial faculty	8
Other faculty	8
Graduate students	46
Total	62

6. Separately Budgeted Research Expenditures by Source of Support

The University of Dayton conducts research and development in a wide range of areas of applied science and engineering. Much of this is through sponsored research of the University of Dayton Research Institute. The annual total dollar volume of this research is over $45 million. Breakdown by source of support is not available.

FACULTY

Professors

Banerjee, Partha, Ph.D., Univ. of Iowa, 1983. Chair of ECE. Nonlinear Optics and Optoelectronic Materials.
Chatterjee, Monish, Ph.D., Univ. of Iowa, 1985. Acousto-optics.
Duncan, Bradley D., Ph.D., Virginia Polytechnic, 1991. Fiber optics; LADAR imaging; EO sensors; Fourier optics.
Hardie, Russell C., Ph.D., Delaware, 1992. IR signal processing.
Haus, Joseph W., Ph.D., Catholic U., 1975. Director. Quantum and nonlinear Optics.
Loomis, John S., Ph.D., Arizona, 1980. Optical design; geometrical and physical optics; image processing.
O'Hare, Michael J., Ph.D., SUNY, Buffalo, 1966. Theoretical physics; solid state; optical properties.
Powers, Peter E., Cornell, 1994. Nonlinear optics.
Taylor, Phil, Ph.D., Penn State, 1984. Laser-induced fluorescence/laser-induced photolysis.
Vorontsov, Mikhail A., Ph.D., Doctor of Science in Physics and Mathematics, Lomonosov Moscow State University, Moscow, Russia (1989). Adaptive optics, laser beam control, imaging through turbulence; wavefront sensing and control; optical communications.

Associate Professors

Sarangan, Andrew, Ph.D., Univ. of Waterloo, 1996. Semiconductor lasers, fiber optics.
Zhan, Qiwen, Ph.D., Univ. of Minnesota, 2002.

Associated Faculty/Scientists

Dierking, Mat, Ph.D., University of Dayton, 2009. Laser Radar.etry and IR Systems.
Grote, James G., Ph.D., University of Dayton, 1994. Nonlinear optics; optoelectronic materials.
McManamon, Paul, Ph.D., Ohio State University, 1977. Sensors.
Polnau, Ernst, Ph.D., University of Bern, Switzerland, 1999. Atmospheric Objects.
Schepler, Ken, Ph.D., The University of Michigan, 1975. Laser Physicist.
Watson, Edward A., Ph.D., University of Rochester, 1991. Active imaging systems; nonmechanical beam steering.
Weyrauch, Thomas, Ph.D., Technische Hochschule Darmstadt, Germany, 1997. Adaptive Optics.

Professor Emeritus

Yaney, Perry, Ph.D., University of Cincinnati, 1963. Non-Linear Optics.

RESEARCH SPECIALTIES AND STAFF

Adaptive Optics. Vorontsov.

Ellipsometry. Zhan.

Guided Wave Optics. Duncan, Grote, Haus, Sarangan.

Laser Diagnostics and Spectroscopy. Taylor, Yaney.

Machine Vision. Loomis.

Nonlinear Optics. Banerjee, Grote, Haus, Powers.

Optical Metrology. Loomis, Zhan.

Optical Processing/Computing. Duncan, Hardie.

Optical Systems and Devices. Duncan, Loomis, Sarangan, Watson, Yaney.

Optoelectronic Materials. Banerjee, Detrio, Grote, Powers.

Quantum Optics. Haus, O'Hare.

Radiometry. Baxley.

FACULTY PUBLICATIONS

Please visit the Electro-Optics Program website, http://www.engineering.udayton.edu/eop/.

UNIVERSITY OF TOLEDO

DEPARTMENT OF PHYSICS AND ASTRONOMY

Toledo, Ohio 43606

Students Accepted For Degree	FIELDS: Physics	Astronomy	Medical Physics
Doctorate	X	X	X
Master's	X	X	

1. General

President: Dr. Lloyd A. Jacobs
Interim Provost and Executive Vice President: William McMillen
Department Chairman: Dr. Karen S. Bjorkman
Department Telephone Number: (419) 530-2241
E-mail: willie.brown@utoledo.edu
Type of Institution: University
Control: Public
Setting: Urban
Total Faculty: 1,151
Total Graduate Faculty: 563
Total Students: 23,064
Total Graduate Students: 4,924
Annual Graduate Tuition: for 2 semesters 2010–11*
In-state residents: Full-time—$11,040
Part-time—$460/credit hour
Out-of-state residents: Full-time—$20,928
Part-time—$872/credit hour
Tuition rates for: 2010–11
Deferred tuition plan: Yes
Other Fees: $575.04/semester for general fee for a full-time student; $287.52 for a part-time grad student.
Term: Semester
Department Website: www.physics.utoledo.edu

2. Number of Faculty in Department

The combined total of full-time faculty in the three professorial ranks is 20. The combined total of full-time, part-time, adjunct, visiting, research, and emeritus, and faculty at all ranks is 40.

3. Admission, Financial Aid, and Housing

Address admissions inquiries to: Graduate Admissions Committee, Department of Physics and Astronomy
Graduate application fee required: $45
Application deadlines (fall): admission without assistantship, 5/31 International, 7/15 Domestic; for assistantship, 1/15 to be considered in the first round.
Admission information: For fall admission, 2010–11, 8 students were accepted from 89 applicants.
Admission requirements: For admission to the graduate programs, a Bachelor's degree in physics is required with a minimum undergraduate GPA of 2.7 specified. All applicants are required to present scores on the General GRE (no minimum score specified). The GRE physics subject test is encouraged but not required. Students from non-English speaking countries are required to demonstrate proficiency in English via a score on the TOEFL or at least 550 (paper) or 213 (computer), or 80 (internet based, preferred)
Undergraduate preparation assumed: Wangsness, *Electricity and Magnetism*; Marion and Thornton, *Mechanics*; Townsend. *A Modern Approach to Quantum Mechanics*; Halliday and Resnick, *Fundamentals of Physics*; Kittel, *Introduction to Solid State Physics*; Eisberg and Resnick, *Quantum Physics of Atoms, Molecules, Solids, Nuclei, Particles.*
Address financial aid inquiries to: Director of Financial Aid Office
GAPSFAS application required: Yes or FAF (financial aid form)
Financial aid deadline: March 1st
Loans available: Yes
Address housing inquiries to:
On campus: Residence Life
100 Dowd Hall
Off campus (Graduate/Undergraduate): Room 2561
Student Union
On-campus, single student housing available: Yes.
Cost/term: (Fall 2010) double occupancy room and board $2,688–3,656/semester
On-campus, married student housing available: No

Table A—Faculty, Enrollments, and Degrees Granted

Research Specialty	2009–10 Faculty	Enrollment[1] Fall 2009: Master's	Doctorate	No. of Degrees Granted[2] 2009–10 (2005–10): Master's	Terminal Master's	Doctorate	Median No. of Years for 2009–10 Ph.D.'s
Astronomy	7	3	8	1(2)	1(5)	2(7)	6
Astrophysics	7	1	2	1(1)	0(1)	1(1)	5
Atomic, Molecular & Optical Physics	4	0	3	0(0)	0(1)	0(1)	–
Biophysics	1	0	1	0(1)	0(0)	0(0)	–
Condensed Matter Physics	8	0	5	1(4)	0(2)	2(4)	7
Materials Sci./ Metallurgy	6	13	16	1(1)	0(3)	3(5)	6.2
Medical/Health	3	0	2	0(0)	0(0)	1(4)	9
Optics	1	0	1	0(0)	0(0)	0(1)	–
Plasma Physics & Fusion	0	0	0	0(0)	0(0)	0(1)	–
Non-Specialized	0	0	0	0(0)	0(0)	0(0)	–
Total		17	38	4(10)	1(12)	9(34)	6
Full-time Grad. Stud.		16	37				
Part-time Grad. Stud.		1	1				
First-year Grad. Stud.		1	2				
Median Years in Grad. Study (2009–10 Degrees)				4.5	2.0		6
Undergraduate Degrees, 2008–09 (2009–10: 8(27)							

[1]Students not yet committed to a research specialty are entered under non-specialized.
[2]Five-year totals in parentheses.

4. Graduate Degree Requirements

Master's: Requires 30 hours including at least 26 in physics, research thesis, and oral final examination.
Doctorate: Requires 90 hours, including 30 to 48 for research. Requires qualifying, comprehensive, and final oral examination. No language requirement.
Other Programs: **Materials Science** option, involving courses from physics, chemistry, and engineering. **Astrophysics Option** (Ph.D.) **Master of Science and Education** offered in cooperation with College of Education. Requires 32 hours, with 18 hours in physics, and a thesis or project in either

physics or education. **Medical Physics** option (Ph.D.), offered jointly with the Medical College, CAMPEP accredited.

Special Equipment, Facilities, or Programs: Graduate study and research in the following specialty areas: Astronomy and Astrophysics, Atomic Physics, Solid State Physics, Materials Science, Optics, Photovoltaics, Theoretical Physics. Positive and negative ion accelerators are used for atomic physics and for ion implantation in semiconductors. Ritter Observatory, located on campus, houses a modern one meter reflector with associated spectrographs and solid state detectors. Facilities include systems for physical and chemical vapor deposition (CVD) of thin films, including magnetron sputtering, molecular beam epitaxy, plasma enhanced CVD, and hot wire CVD. Facilities for materials characterization and device testing include: optical analysis including transmission, reflection, photoluminescence, Raman spectroscopy, spectroscopic ellipsometry, scanning tunneling, atomic force, and scanning electron microscopy; current-voltage capacitance, Hall effect, and quantum efficiency. Other research equipment allows a wide variety of atomic-deposition, film-growth, and materials characterization techniques. Computational facilities include two Beowulf supercomputer clusters.

Table B—Appointments to Graduate Students, 2008–09

Title of Appointee	Appointments Total	Appointments First year	Academic Load Allowed in Credit Hours	Hours of Service Per Week	Stipend for Academic Year ($)
			Semester		
Teaching Assistant	33	14	12	20	9,500–16,000[1,2,3]
Research Assistant	39	0	12	20	9,500–16,000[1,2,3]
University Fellowship	0	0	12	20	
Total	72	14			

[1]Tuition waived.
[2]Amount is for academic year. Summer appointments are available with an additional $3,000–$6,000 stipend.
[3]$9,500 stipend is for students working toward a terminal M.S. degree. $16,000 stipend is for students working toward a PhD.

5. Personnel Engaged in Separately Budgeted Research, 7/08–6/09

Graduate students	39
Post Doctoral Research Professors/ Technicians	22
Professorial faculty	19
Undergraduate students (REU)	15
Total	95

6. Separately Budgeted Research Source of Active Grants

	Departmental Research
Federal government	$6,692,525
State and local gov't.	9,228,137
Private, nonprofit	116,720
Total	$16,037,382

Table C—Separately Budgeted Research Grants (2005–06)

Research Specialty	No. of Grants	Expenditures ($)
Astrophysics	14	1,029,832
Atomic, Molecular, & Optical Physics	2	108,234
Materials Sci./Condensed Matter	19	14,870,716
Physics Education	2	88,600
Total	37	16,037,382

FACULTY

Professors

Amar, Jacques G., Ph.D., Temple U., 1985. Dynamics of thin-film and epitaxial growth; materials science; surface physics.

Anderson-Huang, Lawrence S., Ph.D., U. California, Berkeley, 1977. Astronomy and astrophysics; stellar atmosphere theory.

Bjorkman, Karen S., Ph.D., U. Colorado-Boulder, 1989. Chair. Hot stars; pre-main sequence stars; circumstellar disks.

Collins, Robert W., Ph.D., Harvard U., 1982. Materials science, thin films, optics of solids, ellipsometry and polarimetry.

Deng, Xunming, Ph.D., U. Chicago, 1990. Photovoltaics; thin-film semiconducting materials; applied physics.

Federman, Steven R., Ph.D., NYU, 1979. Astrophysics; interstellar chemistry; laboratory astrophysics.

Gao, Bo, Ph.D., Univ. of Nebraska-Lincoln, 1989. Theoretical atomic physics.

Heben, Michael, Ph.D., Cal. Tech. 1990. Synthesis and materials physics and chemsitry of nanotube materials for energy and hydrogen storage.

Karpov, Victor G., Ph.D., Leningrad Polytechnic Institute, 1979. Dr. of Science. Institute for Nuclear Physics, Russian Academy of Science, 1986. Condensed matter; theoretical physics.

Kvale, Thomas J., Ph.D., U. Missouri-Rolla, 1984. Negative ion spectroscopy, ion-atom collisions; medical physics.

Lee, Scott A., Ph.D., U. Cincinnati, 1983. Ultra-high pressure physics; phase transitions of DNA; biological physics.

Palmer, James F., M.S., U. Florida, 1977. Health physics.

Associate Professors

Bjorkman, Jon E., Ph.D., U. Wisconsin, Madison, 1992. Stellar winds; radiation transfer; young stellar objects.

Cheng, Song, Ph.D., Kansas State U., 1991. Experimental physics; ion-atom/molecule collisions; beam-foil spectroscopy.

Ellingson, Randall J., Ph.D., Cornell U., 1994. Basic physics of nanostructured materials for solar energy conversion; transient laser spectroscopy.

Khare, Sanjay, V. Ph.D., U. Maryland, 1996. Theoretical and computational; materials science, mechanical, strucutural and electronic properties.

Marsillac, Sylvain, Ph.D., U. Nantes, France, 1996. Photovoltaics Materials science, thin films, applied physics.

Megeath, S. Thomas, Ph.D., Cornell U., 1993. Star and planet formation; planet detection, infrared and astronomy millimeter-wave.

Assistant Professors

Chandar, Rupali, Ph.D., Johns Hopkins U., 2000. Extragalactic astronomy; galaxy formation and evolution, star clusters, stellar populations.

Smith, J. D., Ph.D., Cornell U., 2001. Extragalactic astronomy; infrared astrophysics; astrophysical dust.

Active Emeritus Professors

Bagley, Brian G., Ph.D., Harvard U., 1968. Development and analysis of solid-state devices for optical computing and communications.

Bohn, Randy, G., Ph.D., Ohio State, 1969. Thermal properties of solid hydrogen.

Compaan, Alvin D., Ph.D., U. Chicago, 1971. Active Emeritus Prof. Photovoltaics, semiconductor physics; thin-film growth and characterization.

Curtis, Larry J., Ph.D., U. Michigan, 1963. Atomic structure and lifetimes; laboratory astrophysics.

Deck, Robert T., Ph.D., U. Notre Dame, 1961. Nonlinear optics and photonics, design of all-optical light signal processing elements.

Ellis, David G., Ph.D., Cornell U., 1964. Theory of atomic structure and spectra; correlation in multi-electron wavefunctions.

Witt, Adolf N., Ph.D., U. Chicago, 1967. Interstellar dust, radiative transfer, photoluminescence by interstellar nanoparticles.

RESEARCH SPECIALTIES AND STAFF

Theoretical

Astrophysics. Cosmochemistry; interstellar molecular gas; interstellar dust; spectroscopy, polarimetry, and theory of stellar winds, disks and envelopes; star formation; computational methods. Anderson, J. Bjorkman, Federman.

Atomic, Molecular and Optical Physics. Theory of atomic structure and spectra, including high-performance compuational techniques (Ellis, Curtis); quantum theories of two-atom, few-atom, and many-atom systems, including Bose-Einstein condensates (Gao); plasma discharges, atomic structure, Rydberg states, radioactive transitions, and photoionization (Theodosiou); theory and design of optical integrated circuits, components, and devices (Bagley, Deck).

Condensed Matter/Materials Physics: computational study of materials, surfaces and interfaces (Amar, Khare); disordered sytems (Khare, Karpov); non-equilibrium systems and thin-film growth (Amar); phase transition kinetics and thin-film photovoltaics (Karpov); quantum many-atom systems (Gao).

Experimental

Astronomy. Interstellar gas, molecular cloud chemistry, cosmochemistry, laboratory astrophysics (Federman); interstellar dust; light scattering, photoluminescence (Witt); stellar spectroscopy and polarimetry, Be stars, Herbig Ae/Be stars, circumstellar disks and envelopes (K. Bjorkman); extragalactic star clusters, star formation (Chandar); star formation, infrared and millimeter observations; extrasolar planet detection (Megeath).

Atomic, Molecular, and Optical Physics. Beam-foil spectroscopy; atomic lifetime measurements; ion-atom collisions. (Cheng, Curtis, Federman, Kvale); excitation/ionization, charge transfer, secondary emission of elecrons from surfaces, photodetachment (Kvale); semi-empirical techniques for structure of highly excited molecules and atoms (Curtis); integrated optics and non-linear optics (Bagley, Deck); spectroscopic ellipsometry and polarimetry (Collins)

Biological physics. DNA structure and bonding to cancer drugs, phase transitions in hyaluronic acid (Lee).

Condensed Matter/Materials Physics. Photovoltaic materials and devices: amorphous silicon (Deng), CuInSe(2) (Marsillac) and CdTe (Compaan); photoelectrochemical H_2 generation (Deng); real time metrology (Collins).

Medical/Health Physics. Applied accelerator-based physics; applications to radiation therapy. (Kvale, Palmer); radiation oncology (Health Science Campus): hyperbarric medicine (Feldmeier); treatment of gastrointestinal cancer (Dobelbower); radiation beam modeling with Monte Carlo simulation techniques, optimization in IMRT delivered external beam radiotherapy, stereotactic radiosurgery, intra-operative radiation therapy, and three-dimensional dosimetric analysis and quantitative bremsstrahlung SPECT imaging for gamma-emitting radiopharmaceuticals (Parsai) Medical physics—diagnostic Radiology: (Health Science Campus): Tomosynthesis imaging techniques in mammography, perfusion techniques for functional MRI and BOLD functional MRI, MR proton spectroscopy, diagnostic imaging system performance testing (Dennis)

OKLAHOMA STATE UNIVERSITY

DEPARTMENT OF PHYSICS

Stillwater, Oklahoma 74078

Students Accepted For Degree	FIELDS		
	Physics	Astronomy	Related Fields
Doctorate	X		X
Master's	X		X

1. General

President: V. Burns Hargis
Dean of Graduate College: Gordon A. Emslie
Department Chairman: James P. Wicksted
Department Telephone Number: (405) 744-5796 (C)
Type of Institution: University
Control: Public
Setting: Small town
Total Faculty: 1,200
Total Graduate Faculty: 1,100
Total Students: 21,000
Total Graduate Students: 4,670
Annual Graduate Tuition:
In-state residents: $155/credit
Out-of-state residents: $602/credit
Out of state tuition is waived as well as 6 cr. hr in-state tuition per semester for Teaching/Research Assistants. Additional Ph.D. Incentive Awards are available.
Other Fees: $104
Term: Semester

2. Number of Faculty in Department

The combined total of full-time faculty in the three professorial ranks is 25. The combined total of full-time, part-time, and other faculty at all ranks is 28.

3. Admission, Financial Aid, and Housing

Address admission inquiries to: Graduate Coordinator, Department of Physics
Graduate application fee required: $40 US/$75 INT'L
Admission deadline (Fall admission): 6/1
Admission information: For fall admission, 2008–09, 23 students were accepted from over 100 applicants.
Admission requirements: For admission to the graduate programs, a Bachelor's degree in physics or related field is required. The GRE (both general and subject) is recommended. Students from non-English speaking countries are required to demonstrate proficiency in English via the TOEFL or IELTS exam. Minimum acceptable score for admission is 600, 620 for consideration as a teaching assistant.
Address financial aid inquiries to: Office of Admissions
GAPSFAS application required: Yes
Financial aid deadline: 6/1
Loans available: Yes
Address housing inquiries to: Manager, University Apartments, or Residence Halls Housing, Iba Hall.
On-campus, single student housing available: Yes
Cost/term: $550/month with unlimited meals
On-campus, graduate student appartments available: Yes
Cost per month: $573–958 (all bills paid)

Table A—Faculty, Enrollments, and Degrees Granted

Research Specialty	2009–10 Faculty	Enrollment[1] Fall 2009		No. of Degrees Granted[2] 2009–10 (2005–10)			Median No. of Years for 2009–10 Ph.D.'s
		Master's	Doctorate	Master's	Terminal Master's	Doctorate	
Astrophysics	2	0	0	0(0)	0(0)	0(0)	–
Atomic, Molecular, & Optical Physics	2	0	2	1(3)	0(4)	1(4)	–
Biophysics	3	0	2	0(3)	1(2)	2(5)	–
Chemical Physics	2	0	1	0(1)	0(2)	0(2)	–
Condensed Matter Physics	9	0	11	1(3)	0(3)	2(8)	–
Fluids & Rheology	2	0	0	0(0)	0(0)	0(1)	–
Materials Sci./ Metallurgy	4	0	0	0(0)	0(0)	0(0)	–
Optics	3	0	1	0(1)	0(1)	0(3)	–
Particles & Fields	2	0	9	0(0)	0(2)	0(4)	–
Photonics	6	0	9	0(0)	0(0)	0(0)	–
Physics Education	1	0	0	0(0)	0(0)	1(2)	
Polymer Physics/ Science	2	0	0	0(0)	0(0)	0(0)	–
Quantum Optics	1	0	4	0(0)	0(0)	0(0)	–
Radiation Physics	4	2	6	0(0)	0(0)	0(0)	–
Statistical & Thermal	1	0	0	0(0)	0(0)	0(0)	–
Non-specialized	0	1	0	0(0)	0(0)	–	–
Total		3	45	2(11)	1(14)	6(29)	
Full-time Grad. Stud.		3	45				
Part-time Grad. Stud.		0	0				
First-year Grad. Stud.		0	8				
Median Years in Grad. Study		3	5				–

Undergraduate Degrees, 2009–10 (2005–10): 10(22)

[1]Students not yet committed to a research specialty are entered under non-specialized.
[2]Five-year totals in parentheses.

4. Graduate Degree Requirements

Master's: 24 semester hours of approved physics courses plus 6 hours thesis. Options in optics and photonics medical physics. No language requirement. Last eight semester hours and 21 total semester hours must be completed in residence. At least a B average is required. In addition, a "Professional" M.S. in Physics is offered as a 32 credit hour (Report) plan.
Doctorate: 90 hours of approved courses (including thesis research) beyond Bachelor's degree. Departmental preliminary exam required. Minimum 30 semester hours and one of last two years in residence. Qualifying exam and dissertation defense. No language requirement. At least a B average is required.
Special Equipment, Facilities, or Programs: Materials Growth and Characterization Laboratory; MBE lab; Rubidium and sodium BEC labs; 2-MeV Van de Graaff accelerator; ESR; Solid State NMR; femtosecond spectroscopy; optical absorption and fluorescence spectroscopy, Radiation Dosimetry Laboratory, Brillouin Scattering Laboratory. Mendenhall Observatory (0.6 m RC robotic telescope); Multiple faculty Beowolf PC clusters for their research programs. Physics, along with Electrical Engineering, participates in the multidisciplinary Ph.D. Photonics programs. The types of experimental techniques in progress include: photon correlation, Raman scattering, Brillouin scattering, whispering gallery modes,

four-wave mixing, time-resolved site selection, holographic gratings, picosecond pulse-probe, and multiphoton excitation. The instrumentation includes a variety of solid state, liquid, and gas lasers, nonlinear optical crystals autocorrelators, streak cameras, FTIR, optical multichannel analyzers, boxcar integrators, and signal averagers, along with the standard monochromators, spectrum analyzers, detectors, and cryogenic equipment required for conventional spectroscopy.

Table B—Appointments to Graduate Students, 2008–09

Title of Appointee	Appointments Total	Appointments First year	Academic Load Allowed in Credit Hours	Hours of Service Per Week	Stipend for Academic Year ($)
			Semester		
Teaching Assistant	32	8	9	20	16,071
Research Assistant	21	0	9	20	18,000
Total	53	8			

5. Personnel Engaged in Separately Budgeted Research, 7/08–6/09

Professorial faculty	28
Postdoctoral appointments	3
Graduate students	39
Undergraduate students	40
Total	110

6. Separately Budgeted Research Expenditures by Source of Support

	Departmental Research	Physics-related Research Outside Department
Federal government	$680,795	$
State and local government	796,622	
Private, non-profit organizations	143,461	
Total	$1,620,878	$

Table C—Separately Budgeted Research Expenditures

Research Specialty	No. of Grants	Expenditures ($)
Condensed Matter Physics	17	714,731
Optics	1	347,500
Particles & Fields	5	242,307
Biophysics	1	205,000
Statistical & Thermal Physics	1	111,334
Total	25	1,620,878

FACULTY

Professors

Ackerson, Bruce J., Ph.D., Colorado, 1976. Dynamic light scattering, colloids; critical phenomena in fluids.

Agarwal, Girish, Ph.D., Rochester, 1969. Quantum Optics; Nonlinear Optics; Quantum Information Science and Foundations of Quantum Mechanics; Surface Optics-nano photonics.

Babu, K. S., Ph.D., Hawaii, 1986. Grand unification model building, fermion mass mixing; neutrinos.

Bandy, Donna K., Ph.D., Drexel, 1984. Theoretical laser physics; instabilities, nonlinear behavior, optical devices.

Emslie, A. Gordon, Ph.D., D.Sc., Glasgow (Scotland), 1979, 1997. Solar physics.

Harmon, H. James, Ph.D., Purdue, 1974. Biophysics; high resolution high-speed optical spectroscopy; spectroscopy determination of enzyme kinetic intermediates; design of solid state chemical sensors; photochemical reaction of porphyrins.

McKeever, Stephen W. S., Ph.D., Univ. College of N. Wales (Bangor), 1975. Experimental solid state physics; thermoluminescence; thermally stimulated polarization currents; radiation dosimetry; semiconductors.

Mintmire, John W., Ph.D., Florida 1980. Computational materials physics, electronic structure theory, nanostructured materials.

Nandi, Satyanarayan, Ph.D., Chicago, 1975. Theoretical High Energy Physics, grand unification, supersymmetry, extra dimensions, physics at LHC.

Perk, Jacques H. H., Ph.D., Leiden, 1979. Theoretical physics; exactly solvable models in statistical mechanics.

Rosenberger, Albert T., Ph.D., Illinois, 1979. Experimental and theoretical optical physics; microresonator optics and plasmonics.

Sherwood, Peter, M. A., Ph.D., Sc.D., Cambridge University, 1970, 1995. X-ray photoelectron spectroscopy of advanced materials surfaces.

Wicksted, James P., Ph.D., CUNY, 1983. Head of Department. Experimental solid state physics; Raman and Brillouin scattering, nonlinear-optics, rare-earth doped glasses, nanomaterials.

Xie, Aihua, Ph.D., Carnegie Mellon, 1987. Biophysics; structural dynamics of proteins; molecular mechanism of receptor activation in signal transduction; biomedical application of lasers.

Xie, Xincheng, Ph.D., Maryland, 1988. Theoretical solid state; fractional quantum Hall effect; high-T_c superconductivity; transport properties of semiconductors.

Associate Professors

Hauenstein, Robert J., Ph.D., Caltech., 1987. Experimental semiconductor physics, Molecular beam epitaxial growth; heterostructures.

Shull, Peter O., Ph.D., Rice, 1982. Supernova remnants, exoplanets, near-Earth asteroids.

Summy, Gilford, Ph.D., Griffith (Australia) 1995. BEC, quantum chaos, atom optics.

Yukihara, Eduardo, Ph.D., Sao Paulo (Brazil), 2001. Experimental solid state, radiation dosimetry, thermoluminescence.

Assistant Professors

Benton, Eric, Ph.D., Dublin, 2004. Ionizing radiation dosimetry, radiation protection, effects of radiation on living organisms.

Guo, Yin, Ph.D., Maryland, 1992. Computational physics, dynamical process of complex systems, molecular dynam simulations.

Khanov, Alexander, Ph.D., University of Rochester, 2004. Experimental high energy physics.

Liu, Yingmei, Ph.D., University of Pittsburg, 2004. Atom, molecular and optical physics.

Rizatdinova, Flera, Ph.D., Moscow State University, 1994. Experimental high energy physics.mental solid state, radiation dosimetry, thermoluminescence.

Zhou, Donghua, Ph.D., College of William and Mary, 2003. Biophysics.

Assistant Research Professor

DeWitt, Regina, Ph.D., University of Heidelberg, Germany, 2002. Radiation Physics.

Research Scientist

Wang, L., Ph.D., Wuhan Institute of Physics and Mathematics, 2006. Magnetic Resonance and Atomic and Molecular Physics.

Adjunct Faculty

Akselrod, M., Ph.D., Urals State Technical University, 1983. Solid State Physics.
Chen, W., Ph.D., University of Oregon, 1988. Physics.
Lucas, A., Ph.D., Oklahoma State University, 2003. Physics.
Perk, H., Ph.D., SUNY at Stony Brook, 1973. Theoretical Physics.

Emeriti Faculty

Dixon, G., Ph.D., University of Georgia, 1967. Physics
Lange, J., Ph.D., Pennsylvania State University, 1964. Solid State Physics; Acoustics.
Martin, J., Ph.D., Iowa State University, 1967. Physics.
Swamy, N., Ph.D., Florida State University, 1958. Physics.
Westhaus, P., Ph.D., Washington (St. Louis), 1966. Theoretical Physics.
Wilson, T., Ph.D., Florida, 1966. Theoretical physics; electronic structure of point defects.

RESEARCH SPECIALTIES AND STAFF

Theoretical

Astrophysics. Supernova remnants; astrophysical jets; solar physics. Emslie.
Atomic and Molecular. Density functional theory of electronic structure. Guo, Mintmire.
Elementary Particles and Fields. Grand unification, supersymmetry, extra dimension, physics at LHC, fermion masses and mixings, neutrinos. Nandi, Babu. 1 postdoctoral fellow.
Optics and Photonics. Nonlinear behavior of laser systems; modeling of optical instabilities, quantum optics, nonlinear optics. Bandy, Rosenberger.
Physics Education. Various instructional strategies. Ackerson.
Condensed Matter Physics. Electronic structure of disordered systems; density functional theory, low-dimensional materials, dielectric response theory; optical properties of defects; vibronics; high-T_c materials; transport in semiconductors; quantum Hall effect. Guo, Mintmire, X. Xie. 1 postdoctoral fellow.
Quantum Optics and Quantum Information Science. Coherent Control; Ultraslow Light, Super-resolution; Quantum Imaging, Quantum Entanglement; Decoherence; Quantum Optics of Semiconductor Dots; Integrated structures and Nano-mechanical Quantum Devices. Agarwal.
Statistical Mechanics. Ising model, stochastic processes, exactly solvable models; low-dimensional systems, quasicrystals. H. Perk, J. Perk.

Experimental

Astrophysics. Supernova remnants; exoplanets; near-Earth asteroids; solar physics. Shull.
Atomic Molecular Optical BEC. Summy, Liu.
Biophysics. Laser effects on biological materials; high resolution high-speed optical spectroscopy; enzyme kinetics; photochemical reactions of porphyrins; protein dynamics; protein structure; membrane proteins; amyloid proteins; solid-state NMR. Chen, Harmon, A. Xie, Zhou. 1 postdoctoral fellow.
Chemical Physics. Photocatalysis, Photoenergy conversion, monolayer surfaces, solid-state catalysts, photoreductive chemistry. Harmon.
Complex Fluids. Light scattering; phase transitions in colloids; dynamics of flow systems. Ackerson.
Condensed Matter Physics. Optical, electrical, thermal, acoustical, structural and mechanical properties of solids; laser materials; ESR; energy transfer; epitaxial growth, nanoparticles and nanotubes. Harmon, Hauenstein, McKeever, Wicksted, Yukihara. 1 postdoctoral fellow.
Experimental High Energy Physics. Khanov, Rizatdinova.
Optics and Photonics. Response of materials to coherent excitation. Optical gain, frequency mixing, multiphoton effects, dephasing times, nonlinear susceptibility measurements involving free and trapped carriers, optical switching, and storage. Microresonators and plasmonics. Ackerson, Rosenberger, Wicksted.
Quantum Information Science. Spin squeezing. Liu.
Radiation Physics. Dosimetry materials and measurements, radiation effects. Akselrod, Benton, DeWitt, Lucas, McKeever, Yukihara. 1 postdoctoral fellow.
Surface Chemistry and Physics. Electrode surfaces, corrosion systems, material surface properties, carbon fiber surfaces thin films, and plasmonics. Sherwood, Rosenberger, Wang.

UNIVERSITY OF OKLAHOMA

HOMER L. DODGE
DEPARTMENT OF PHYSICS AND ASTRONOMY

Norman, Oklahoma 73019

Students Accepted For Degree	FIELDS		
	Physics	Astronomy	Related Fields
Doctorate	X	X	X
Master's	X	X	X

1. General

President: David L. Boren
Dean of Graduate School: T. H. Lee Williams
Department Chairman: Gregory A. Parker
Department Telephone Number: (405) 325-3961
Type of Institution: University
Control: Public
Setting: Suburban
Total Faculty: 1,822
Total Graduate Faculty: 1,233
Total Students: 26,656
Total Graduate Students: 3,900
Annual Graduate Tuition:
In-state residents: Full-time—$279.10/credit
Out-of-state residents: Full-time—$688.80/credit
Tuition rates for: 2009–10
Deferred tuition plan: No
Other Fees: $216
Term: Semester

2. Number of Faculty in Department

The combined total of full-time faculty in the three professorial ranks is 30.

3. Admission, Financial Aid, and Housing

Address admission inquiries to: Graduate Selection Committee, Homer L. Dodge, Department of Physics and Astronomy, University of Oklahoma, 440 West Brooks-100, Norman, OK 73019;
e-mail: inquiry@physics.ou.edu;
Website: physics.ou.edu
Graduate application fee required: $40 U.S./$90 International
Admission deadline (Fall admission): 2/15
Admission information: For fall admission, 2009–10, 12 students were accepted from 78 applicants.
Admission requirements: For admission to the graduate programs, a Bachelor's degree in physics and/or astronomy is required with a minimum undergraduate GPA of 3.0/4.0 specified. The GRE general and subject exams are required. Students from non-English speaking countries are required to demonstrate proficiency in English via the TOEFL exam. Minimum acceptable score for admission is 250.
Undergraduate preparation assumed: Marion, *Classical Dynamics of Particles and Systems;* French, *Vibrations and Waves*; Lorrain, *Electromagnetic Fields and Waves*; Saxon, *Elementary Quantum Mechanics*.
Address financial aid inquiries to: Director of Financial Aid, University of Oklahoma, 1000 Asp Ave., Norman, OK 73019
GAPSFAS application required: No
Financial aid deadline: 3/1
Loans available: Yes
Address housing inquiries to: Housing Office, 1406 Asp Ave., Norman, OK 73019
On-campus, single student housing available: Yes
Cost/semester: $3,451–5,010 (includes meals)
On-campus, married student housing available: Yes
Cost/month: $480–842 (all utility bills paid)

Table A—Faculty, Enrollments, and Degrees Granted

Research Specialty	2009–10 Faculty	Enrollment[1] Fall 2009		No. of Degrees Granted[2] 2009–10 (2004–100)			Median No. of Years for 2009–10 Ph.D.'s
		Master's	Doctorate	Master's	Terminal Master's	Doctorate	
Applied Physics	1	0	0	0(0)	0(0)	0(0)	0
Astrophysics	7	5	7	0(3)	0(7)	3(7)	0
Atomic Molecular, & Optical Physics	6	2	15	0(3)	0(1)	1(5)	6
Condensed Matter Physics	8	3	11	0(3)	1(2)	3(6)	6
Engineering Physics/ Science	15[3]	0	0	0(2)	0(2)	0(0)	–
Particles & Fields	8	0	13	1(2)	0(0)	2(8)	6
Storms/Meteorology	2	0	0	0(0)	0(0)	0(0)	0
Non-specialized	0	2	5	0(0)	1(2)	0(0)	–
Total		12	51	1(13)	5(14)	9(26)	
Full-time Grad. Stud.		8	59				
Part-time Grad. Stud.		1	0				
First-year Grad. Stud.		0	8				
Median Years in Grad. Study (2009–10 Degrees)				6	2.5	6.0	–

Undergraduate Degrees, 2009–10 (2004–10): 6(53)

[1] Students not yet committed to a research specialty are entered under non-specialized.
[2] Five-year totals in parentheses.
[3] Faculty drawn from Physics, Electrical Engineering and Chemical Engineering departments.

4. Graduate Degree Requirements

Master's: A student must complete 30 hours of course work including preparation of a thesis, or 32 hours of course work (no-thesis program), taken in accordance with the general rules of the Graduate College. The allowable minimum of credits in physics and astronomy is 18 hours, six hours of which must be at the 5000 level or above.

Doctorate: The student must complete a minimum of 36 hours of course work at the 5000 level or above, excluding the credit hours granted for preparation of the thesis or dissertation describing original research. These hours include 21 hours of specific required courses. Another 54 hours of graduate course work is required as appropriate for the student's field of research specialization, including research hours. The qualifying exam is offered semi-annually and is usually taken at the end of the first year of graduate study. The general examination for the Ph.D. degree consists of a written report and an oral exam, including a presentation of a topic related to the field of specialization and a probing of the student's knowledge of general principles, and is taken before the student begins dissertation research. The Ph.D. in physics may include an emphasis in astronomy, or astrophysics.

Other Programs: Advanced degrees in Engineering Physics are also offered, including M.S. in Engineering Physics, and

Ph.D. in Engineering. Specialization areas include Atomic Molecular Physics, Solid State Physics, Physics of Weather, Materials Characterization, and others.

Thesis: Thesis and dissertations may be written *in absentia*.

Special Equipment, Facilities, or Programs: To facilitate theoretical and experimental research, an extensive computer network consisting of over 60 machines (mostly Suns) and 4Tb of disk space is located in the Department. The university has a new super computer center "OSCER". For astronomical research AIPS and IRAF software are available. Well-equipped laboratories for experimental research in atomic and molecular collisions, semiconductor physics, and materials characterization are located in the Department, including an MBE machine, with in situ STM, a semiconductor clean room, high magnetic field, low temperature, far-infrared and cryogenic facilities. Other experimental research programs make use of facilities at Cornell University (CLEO), National Magnet Laboratory, Los Alamos, and Fermi National Laboratories, and Kitt Peak, VLA, and Cerro Tololo Inter-American observatories. An instrument and machine shop within the Department also supports the experimental research efforts. Engineering Physics was established at the University of Oklahoma in 1924 as the first program of its kind in the United States. Due to its interdisciplinary character, the program is able to make use of the extensive research facilities of the College of Engineering and the Department of Physics and Astronomy. At present the program is centered around a dual-chamber MBE system for growth of narrow-gap semiconductor films.

Table B—Appointments to Graduate Students, 2009–10

Title of Appointee	Appointments: Total	Appointments: First year	Academic Load Allowed in Credit Hours	Hours of Service Per Week	Stipend for Academic Year ($)[1]
			Semester		
Teaching Assistant	29	19	9	10–15	14,420–18,540[1]
Research Assistant	31	1	9	20	12,000–23,250[1]
Total	60	10			

[1] Tuition waived.

5. Personnel Engaged in Separately Budgeted Research, 7/09–6/10

Professorial faculty	28
Postdoctoral appointments	9
Graduate students	36
Undergraduate students	10
Total	83

6. Separately Budgeted Research Expenditures by Source of Support

	Departmental Research	Physics-related Research Outside Department
Federal and State government	$4,424,462	$
Total	$4,424,462	$

Table C—Separately Budgeted Research Expenditures

Research Specialty	No. of Grants	Expenditures ($)
Astrophysics	16	446,623
Atomic, Molecular, & Optical Physics	14	726,816
Condensed Matter Physics	26	1,768,368
Particles & Fields	18	1,301,142
Physics Education	1	181,509
Total	75	4,424,462

FACULTY

Professors

Baron, Edward A., Ph.D., SUNY, Stony Brook, 1985. Astrophysics of condensed objects, particularly supernova.

Baer, Howard, Ph.D., Wisconsin, 1984. High energy theory; supersymmetrys phenomenology.

Branch, David, Ph.D., Maryland, 1969. Spectroscopic astrophysics; supernovae.

Cowan, John J., Ph.D., Maryland, 1976. Stellar evolution and nucleosynthesis; supernovae; cosmology.

Doezema, Ryan E., Ph.D., Maryland, 1971. Chairman of the Department. Experimental solid state physics; 2D electron-systems in semiconductors; superconductivity.

Furneaux, John E., Ph.D., California, Berkeley, 1979. Experimental semiconductor physics and low temperature physics.

Gutierrez, Phillip, Ph.D., California, Riverside, 1983. Experimental high-energy physics; Fermilab D0 experiment.

Henry, Richard C., Ph.D., Michigan, 1983. Chemical evolution of galaxies; abundance of planetary nebulae; evolution of intermediate mass stars.

Johnson, Matthew, Ph.D., Cal. Tech., 1989. Experimental semiconductor and surface physics; scanning tunnelling microscopy.

Kantowski, Ronald, Ph.D., Texas, Austin, 1966. Unified classical and quantum field theories.

Milton, Kimball, Ph.D., Harvard, 1971. High-energy theory, particularly the development of nonperturbative methods to be applied to quantum chromodynamics and other field theories.

Morrison, Michael, Ph.D., Rice, 1976. Theoretical atomic and molecular physics, particularly electron and positron collisions and near-threshold excitations.

Mullen, Kieran, Ph.D., Michigan, 1989. Theoretical solid state-physics.

Parker, Gregory, Ph.D., Brigham Young, 1976. Theoretical-atomic physics, particularly reactive scattering.

Romanishin, William, Ph.D., Arizona, 1980. Extragalactic astronomy; clusters of galaxies; active galactic nuclei.

Santos, Michael, Ph.D., Princeton, 1992. Experimental semiconductor and surface physics; MBE growth of narrow gap systems.

Skubic, Patrick, Ph.D., Michigan, 1977. Experimental high-energy physics; Fermilab-D0 Experiment.

Watson, Deborah K., Ph.D., Harvard, 1977. Theoretical atomic and molecular physics; dimensional perturbation theory, Bose-Einstein condensates.

Associate Professors

Abbott, Braden, Ph.D., Purdue, 1994. Experimental high energy physics; DO experiment.

Abraham, Eric, Ph.D., Rice, 1996. Experimental atomic and molecular physics.

Bumm, Lloyd A., Northwestern Univ., 1991. Nanophysics.

Kao, Chung, Ph.D., Texas, Austin, 1990. High energy physics; electroweak symmetry breaking; supersymmetry.

Leighly, Karen, Ph.D., Montana State, 1991. Active Galactic Nuclei (AGN).

Mason, Bruce A., Ph.D., Maryland, 1985. Theoretical solid state and device physics; computational physics.

Murphy, Sheena, Ph.D., Cornell, 1991. Experimental semiconductor and superconductor physics; low temperature physics.

Ryan, Stewart, Ph.D., Michigan, 1971. Applied physics; materials characterization.

Shafer-Ray, Neil, Ph.D., Columbia, 1990. Experimental atoms; molecular physics; laser physics.

Shaffer, James P., Rochester, 1999. Atomic, molecular, chemical physics; optics.

Strauss, Michael, Ph.D., UCLA, 1988. Experimental high energy physics; Fermilab D0 experiment.

Wang, Yun, Ph.D., Carnegie Mellon, 1991. Cosmic microwave background anisotropy and supernovae.

Adjunct Faculty

Beasley, William, Ph.D., Dallas, Texas, 1974. Professor. Meteorology.

Crompton, Robert, Ph.D., Adelaide, 1954. Professor. Electron and ion diffusion.

Feldt, Andrew N., Ph.D., Oklahoma, 1980. Assistant Professor. Atomic and molecular theory.

MacGorman, Donald, Ph.D., Rice, 1978. Associate Professor. Atmospheric electricity.

Rust, David, Ph.D., New Mexico Inst. Tech., 1973. Professor. Atmospheric and plasma physics.

Snow, Joel, Ph.D., Yale, 1983. Assoc. Professor.

RESEARCH SPECIALTIES AND STAFF

Theoretical

Astronomy and Astrophysics. Cosmology; extragalactic astronomy; nucleosynthesis; stellar atmospheres; stellar evolution; supernovae. Baron, Branch, Cowan, Henry, Wang.

Atomic and Molecular Physics. Atomic collisions. Morrison, Parker, Watson.

Elementary Particles and Fields. Quantum field theory; particle physics phenomenology; gravity theory. Kao, Milton, Kantowski, 1 postdoctoral fellow.

Solid State Physics. Mason, Mullen. 1 postdoctoral fellow.

Experimental

Astronomy and Astrophysics. Binary and variable stars; extragalactic astronomy; extragalactic H regions; supernovae. Branch, Cowan, Henry, Leighly, Romanishin.

Atmospheric Physics. Storm electricity; physics of lightning. MacGorman. Rust.

Atomic, Molecular, and Chemical Physics. Atomic and molecular scattering; lasers; laboratory astrophysics; multiphoton ionization; reactive scattering. Abraham, Shaffer, Shafer-Ray.

Elementary Particle Physics, Fermilab, CERN. Abbott, Gutierrez, Skubic, Strauss. 1 postdoctoral fellow and 1 Research Scientist.

Solid State and Applied Physics. Bumm, Doezema, Furneaux, Johnson, Murphy, Ryan, Santos. 2 postdoctoral fellows.

OREGON STATE UNIVERSITY

DEPARTMENT OF PHYSICS

Corvallis, Oregon 97331-6507
(www.physics.oregonstate.edu)

Students Accepted For Degree	FIELDS		
	Physics	Astronomy	Related Fields
Doctorate	X		
Master's	X		

1. General

President: Edward J. Ray
Dean of Graduate School: Sally Francis
Department Chairman: Henri J. F. Jansen
Department Telephone Number: (541) 737-4631
Type of Institution: University
Control: Public
Setting: Small town
Total Faculty: 3,566
Total Graduate Faculty: 1,800
Total Students: 21,969
Total Graduate Students: 3,203
Annual Graduate Tuition:
In-state residents: Full-time—$11,419
Part-time—*
Out-of-state residents: Full-time—$17,487
Part-time—*
Tuition rates for: 2009–10
Deferred tuition plan: Yes
Other Fees: $546/term; $1,638/annual
Term: Quarter

*See Graduate Catalog

2. Number of Faculty in Department

The combined total of full-time faculty in the three professorial ranks is 12. The combined total of full-time, part-time, and other faculty at all ranks is 16.

3. Admission, Financial Aid, and Housing

Address admission inquiries to: Department of Physics, 301 Weniger Hall
Graduate application fee required: $60
Admission deadline (Fall admission): January 15
Admission requirements: For admission to the graduate programs, a Bachelor's degree is required with a minimum undergraduate GPA of 3.0* specified. The GRE is strongly urged. The GRE is strongly recommended for foreign students. No minimum acceptable score is specified. The GRE Advanced is not required. The average GRE scores for 1999–2000 admissions were verbal–481; quantitative–659; analytical–631. Students from non-English-speaking countries are required to demonstrate proficiency in English via the TOEFL exam. Minimum acceptable score for admission is 550 (written), 213 (computer, 80 (iBT).
Undergraduate preparation assumed: Halliday and Resnick, *Fundamentals of Physics*; Krane, *Modern Physics*; Boas, *Mathematical Methods in the Physical Sciences*; Marion, *Classical Dynamics of Particles and Systems*; Brophy, *Basic Electronics for Scientists*; Griffiths, *Introduction to Electrodynamics*; Liboff, *Quantum Physics*; Hecht, *Optics*; Kittel and Kroemer, *Thermal Physics.*
Address financial aid inquiries to: Department of Physics
GAPSFAS application required: No
Financial aid deadline: Jan. 15†
Loans available: Yes
Address housing inquiries to: University Housing and Dining Services, Oregon State University 97331-1317
On-campus, single student housing available: Yes
Cost/year: $8,838, room & board
On-campus, married student housing available: Yes
Cost/month: $474+ (6-9 mo. waiting list)

*May be appealed to Admissions Committee
†Awards are made after 2/1 if funds are still available

Table A—Faculty, Enrollments, and Degrees Granted

Research Specialty	2009–10 Faculty	Enrollment[1] Fall 2009 Master's	Enrollment[1] Fall 2009 Doctorate	No. of Degrees Granted[2] 2009–10 (2004–06) Master's	Terminal Master's	Doctorate	Median No. of Years for 2008–09 Ph.D.'s
Atomic, Molecular, & Optical Physics	5	0	10	0(1)	0	1(7)	5.5
Computational Physics	4	0	4	0(2)	0	0(2)	
Condensed Matter Physics	6	2	9	0(4)	0	0	
Nuclear Physics	0	0	2	0(0)	0(0)	0(0)	
Particles & Fields	0	1	0	0(0)	0(0)	0(0)	
Physics Education	2	0	0	0(1)	0(0)	0(0)	
Relativity & Gravitation	1	0	1	0(0)	0(0)	0(0)	
Non-specialized	0	5	8	0(0)	2(2)	0(1)	
Total		8	30	0(8)	2(2)	1(11)	
Full-time Grad. Stud.		8	30				
Part-time Grad. Stud.		0	0				
First-year Grad. Stud.		4	5				
Median Years in Grad. Study (2008–09 Degrees)					3.0	5.5	
Undergraduate Degrees, 2008–09 (2004–09): 15(88)							

[1]Students not yet committed to a research specialty are entered under non-specialized.
[2]Five-year totals in parentheses.

4. Graduate Degree Requirements

Master's: 45 term hours of credit, with a 3.0 grade point average (minimum), with approximately 2/3 of the credit in the major and the remaining 1/3 in a minor if chosen. The optional minor is ordinarily completed within the physics department; however, students seeking a more flexible program may plan a minor in another discipline. Residence requirements include one academic year full-time load (15 hours per term) or fair equivalent. A maximum of 15 term hours of transferred credit may be applied toward the residence requirement. There is no foreign language or computer language requirement. Thesis is optional. Completion of a project is required if the non-thesis option is chosen. Satisfactory performance is also required in a two-hour oral examination on the major and minor subjects.
Doctorate: The dostor of philosophy degree is granted primarily for creative attainments. Broadly viewed, the departmental

requirements, advancement to candidacy, and completion of a thesis. There is no rigid credit requirement; however, the equivalent of at least three years of full-time graduate work beyond the bachelor's degree (at least 108 graduate credits) is required. All graduate student programs must consist of, at a minimum, 50% graduate stand-alone courses. After admission into the doctoral program, a minimum of one full-time academic year (at least 36 graduate credits) should be devoted to the preparation of the thesis. The equivalent of one full-time academic year of regular non-blanket course work (at least 36 graduate credits) must be included on a doctoral program. Advancement to Ph.D. candidacy requires satisfactory performance on written and oral comprehensive examinations. A thesis and oral defense of the thesis are required.

Thesis: Thesis may be written *in absentia*.

Special Equipment, Facilities, or Programs: A laboratory for preparation and characterization of thin solid films includes facilities for deposition by evaporation, rf sputtering, and pulsed laser techniques. Films are characterized by AC and DC transport, optical transmission and reflection, and Hall effect measurements. Neutron diffraction studies of magnetic semiconductors are conducted at the facilities of NIST. Spectrometers and magnetic fields up to 8 T are used for solid state nuclear magnetic resonance and nuclear quadrupole resonance studies of semiconductors, conducting oxides and related materials. Laser cooling and atom trapping studies are being done using diode lasers. Cold atoms are used in studies of atom interferometry. Nonlinear optical studies of thin films and interfaces are conducted in a picosecond pulsed laser laboratory. A Ti-sapphire femtosecond oscillator and 532 nm diode-pumped solid-state laser are used for terahertz studies of semiconductors.

Computional physics research is conducted in several areas including nuclear and solid state physics. The Oregon State Physics Department is one of the few in the USA with a degree in Computational Physics. The department maintains a 20 node Solaris Beowulf cluster for research and advanced computational projects, as well as a 30 node cluster for CP education. Laboratory courses offer instruction in interfacing computers for laboratory experiments as well as practical experience in computational physics. The department maintains an electronics shop and has access to on-campus machine shops.

Physics eduction research is conducted in the context of the "Paradigms in Physics" program, a unique curriculum developed for upper division physics instruction, and in the implementation of interactive teaching techniques in large, lower division courses. In the "Paradigms" curriculum, course content has been rearranged to better reflect the way professional physicists think about their field and to incorporate pedagogy that assigns to the students more responsibility for their own learning.

Cooperative arrangements permit students to pursue advanced physics degrees through research in other departments, including chemistry and electrical engineering.

Table B—Appointments to Graduate Students, 2008–09

Title of Appointee	Appointments: Total	Appointments: First year	Academic Load Allowed in Credit Hours	Hours of Service Per Week	Stipend for Academic Year ($)
			Quarter		
Teaching Assistant	26	4	12	15	14,000[1,2]
Research Assistant	11	0	12	15	14,000–14,400[1,2]
Total	37	4			

[1]Tuition waived.
[2]Summer appointments available with additional $1,500–2,500 stipend.

5. Personnel Engaged in Separately Budgeted Research, 7/08–6/09

Professorial faculty	8
Graduate students	14
Total	22

6. Separately Budgeted Research Expenditures by Source of Support

	Departmental Research
Private foundations	31,032
Federal government	1,021,947
Total	1,052,979

7. Extension Centers and Summer Programs

There are no extension centers. Summer Term offerings include five undergraduate sequences or courses and research or independent study for graduate students. REU program and other summer research programs for undergraduates are available.

Table C—Separately Budgeted Research Expenditures

Research Specialty	No. of Grants	Expenditures ($)
Comp Physics	2	6,705
Atomic, Molecular, & Optical Physics	18	522,831
Condensed Matter Physics	14	395,735
Physics Education	2	127,707
Total	36	1,052,979

FACULTY

Professors

Jansen, Henri J. F., Ph.D., Groningen, 1981. Chairman of the Department. Theoretical solid state physics.

Krane, Kenneth S., Ph.D., Purdue, 1970. Experimental nuclear and solid state physics. Emeritus.

Landau, Rubin H., Ph.D., Illinois, 1970. Theoretical nuclear and particle physics; computational physics. Emeritus.

Manogue, Corinne A., Ph.D., Texas, 1984. Theoretical particle physics; physics education research.

Stetz, Albert W., Ph.D., California, Berkeley, 1968. Experimental intermediate energy nuclear physics. Emeritus.

Tate, Janet, Ph.D., Stanford, 1988. Experimental solid state physics.

Warren, William W., Ph.D., Washington (St. Louis), 1965. Experimental solid state physics. Emeritus.

Associate Professors

Giebultowicz, Tomasz M., Ph.D., Univ. of Warsaw, 1975. Experimental solid state physics.
Hetherington, William M., III, Ph.D., Stanford, 1977. Experimental optical and chemical physics.
Lee, Yun-Shik, Ph.D., Texas, Austin, 1997. Experimental optical and solid state physics.
McIntyre, David H., Ph.D., Stanford, 1987. Experimental atomic and optical physics.
Ostroverkhova, Oksana G., Ph.D., Case Western Reserve, 2001. Experimental optical and chemical physics.
Podolskiy, Viktor, Ph.D., New Mexico State, 2002. Theoretical optical physics.

Assistant Professors

Demaree, Dedra, Ph.D., Ohio State U., 2006. Physics education research.
Minot, Ethan D., Ph.D., Cornell U., 2004. Experimental condensed matter physics.
Roundy, David, Ph.D., U. Cal. Berkeley, 2001. Theoretical condensed matter physics.
Schneider, Guenter, Ph.D., Oregon State U., 1999. Theoretical condensed matter physics.

RESEARCH SPECIALTIES AND STAFF

Theoretical

Condensed Matter Physics. Electronic structure of solids, magnetism, metal clusters and reduced dimensionality, aqueous interfaces and solutions; density functional theory. Jansen, Roundy, Schneider.
Optical Physics. Electromagnetic properties of nano- and microstructured materials, fundamentals of negative refractive index, nano-photonics and plasmonics, quantum chaos in optical systems. Podolskiy.
Particle Physics. Octonions, field theory in curved space-time. Manogue.

Experimental

Atomic, Molecular, and Optical Physics. Non-linear optical studies of surfaces and interfaces. Hetherington. Ultrafast processes and nonlinear optics in organic materials. Ostroverkhova. Laser cooling and trapping of atoms; atom interferometry; laser spectroscopy; quantum optics. McIntyre. Terahertz spectroscopy and ultrafast carrier dynamics in semiconductors using femtosecond lasers; photonic crystals. Lee.
Condensed Matter Physics. Electrical transport and optical studies of transparent, conducting thin films. Tate. Neutron diffraction studies of magnetic semiconductor heterostructures, "spintronics." Giebultowicz. NMR and NQR studies of electron dynamics in conducting oxides. Warren. Carbon nanotube biosensors. Minot.

Physics Eduction

Computational Physics: Undergraduate education in computational physics. Landau.
Undergraduate Physics. Curriculum reform, studies of student learning problems. Manogue, Demaree.

FACULTY PUBLICATIONS

Demaree, Dedra

S. Li and D. Demaree, "Student Physics Learning Across Communities of Practice," 2010 AERA proceedings: Denver, Colorado.
S. Li and D. Demaree, "Promoting productive communities of practice: an instructor's perspective," 2009 PERC Proceedings: Ann Arbor, Michican. Denver, Colorado.
S. Allie, D. Demaree, J. Taylor, F. Lubben, and A. Buffler, "Invited Paper: Making Sense of Measurements, Making Sense of the Textbook," 2008 PERC conference: Edmonton, Canada.
D. Demaree, S. Alie, M. Low and J. Taylor, "Quantitative and qualitative analysis of student textbook summary writing," 2008 PERC Proceedings: Edmonton, Canada.

Giebultowicz, Tomasz M.

H. Kepa, C. F. Majkrzak, A. Yu. Sipatov, A. G. Fedorov, T. A. Samburskaya, and T. M. Giebultowicz, "Interlayer coupling in EuS/SrS, EuS/PbSe and EuS/PbTe magnetic semiconductor superlattices," Journal of Physics (2009).
E. Przezdziecka, E. Dynowska, W. Paszkowicz, W. Dobrowolski, H. Kepa, T. M. Giebultowicz, E. Janik, and J. Kossut, "MnTe and ZnTe grown on sapphire by MBE," Thin Solid Films **516**, 4813–4818 (2008).

Jensen, Henri

V. Nguyen, J. Meuli, B. Brooks, H. Jansen, J. Westall, and M. Koretsky, "Determining Localized Anode Condition to Maintain Effective Corrosion Protection," Peer Reviewed Final Report to ODOT, August 2009.

Landau, Rubin H.

R. H. Landau and C. C. Bordeianu, *A Survey of Computational Physics* (Princeton University Press, Princeton, 2008).
R. H. Landau, M. J. Paez, and C. C. Bordeianu, *Computational Physics, Problem Solving with Computers*, Second Edition (Wiley-VCH, Berlin, 2007).
R. H. Landau, "Resource Letter CP-2: Computational Physics," Amer. J. of Phys. **76**, 296-306 (2008).
R. H. Landau, *What to Teach? Computational Science as an Improved Model for Science Education*, Microsoft Research Faculty Summit on Computational Education for Scientists, July 2008.

Lee, Yun-Shik

Y.-S. Lee, A. D. Jameson, J. L. Tomaino, J. P. Prineas, J. T. Steiner, M. Kira, and S. W. Koch, "Terahertz and optical frequency mixing in semiconductor quantum-wells," Proc. SPIE **7582**, 75820Y (2010).
A. D. Jameson, J. L. Tomaino, Y.-S. Lee, J. P. Prineas, J. T. Steiner, M. Kira, and S. W. Koch, "Transient optical response of QW excitons to intense narrowband THz pulses," Appl. Phys. Lett. **95**, 201107 (2009).
Y.-S. Lee, J. R. Danielson, J. P. Prineas, J. T. Steiner, M. Kira, and S. W. Koch, "Terahertz-induced extreme nonlinear transients in semiconductor quantum wells," Phys. Stat. Sol. (C) **6**, 457 (2009).
Y.-S. Lee, *Principles of Terahertz Science and Technology* (Springer, New York, 2009).
J. R. Danielson, A. D. Jameson, J. L. Tomaino, H. Hui, J. D. Wetzel, Y.-S. Lee, and K. L. Vodopyanov, "Intense narrow band terahertz generation via type-II difference-frequency generation in ZnTe using chirped optical pulses," J. Appl. Phys. **104**, 033111 (2008).

W. Hurlbut, Y.-S. Lee, K. L. Vodopyanov, P. S. Kuo, and M. M. Fejer, "Multiphoton absorption and nonlinear refraction of GaAs in the mid-infrared," Opt. Lett. **32**, 668 (2007).

J. R. Danielson, Y.-S. Lee, J. P. Prineas, J. T. Steiner, M. Kira, and S. W. Koch, "Interaction of Strong-Cycle Terahertz Pulses with Semiconductor Quantum Wells," Phys. Rev. Lett. **99**, 237401 (2007).

McIntyre, David H.

D. H. McIntyre, J. Tate, and C. A. Manogue, "Integrating Computational Activities into the Upper-Level Paradigms in Physics Curriculum at Oregon State University," Am. J. Phys. **76**, 340–346 (2008).

R. Kykyneshi, D. H. McIntyre, J. Tate, C.-H. Park, and D. A. Keszler, "Electrical and optical properties of epitaxial transparent conductive BaCuTeF thin films deposited by pulsed laser deposition," Solid State Sci. **10**, 921–927 (2008).

K. Jiang, A. Zakautayev, J. Stowers, M. D. Anderson, J. Tate, D. H. McIntyre, D. C. Johnson, and D. A. Keszler, "Low temperature, solution processing of TiO2 thin films and fabrication of multilayer dielectric optical elements," Solid State Sci. **11**, 1692–1699 (2009).

M. J. Kendrick, D. H. McIntyre, and O. Ostroverkhova, "Wavelength dependence of optical tweezer trapping forces on dye=doped polystyrene microspheres," J. Opt. Soc. Am. **26**, 2189-2198 (2009).

A. Zakutayev, R. Kykyneshi, G. Schneider, D. H. McIntyre, and J. Tate, "Electronic structure and excitonic absorption in BaCuChF (Ch=S, Se, Te)," Phys Rev. B **81**, 155103 (2010).

Minot, Ethan D.

L. Prisbrey, G. Schneider, and E. Minot, "Modeling the electrostatic signature of single enzyme activity," J. Phys. Chem. B, **114**, 3330-3333 (2010).

T. DeBorde, M. Leyden, J. C. Joiner, and E. D. Minot, "Identifying individual single-walled and double-walled carbon nanotubes by atomic force microscopy," Nano Letters **8**, 3568–3571 (2008).

I. Heller, A. M. Janssens, J. Mannik, E. D. Minot, S. G. Lemay, and C. Dekker, "Identifying the mechanism of biosensing with carbon nanotube transistors," Nano Letters **8**, 591 (2008).

Ostroverkhova, Oksana

M. J. Kendrick, D. H. McIntyre, and O. Ostroverkhova, "Wavelength dependence of optical trapping forces on dye-doped polystyrene microspheres," J. Opt. Soc. Am. B **26**, 2189-2198 (2009).

A. D. Platt, J. Day, S. Subramanian, J. E. Anthony, and O. Ostroverkhova, "Optical, fluorescent, and (photo)conductive properties of high-performance functionalized pentacene and anthradithiophene derivatives," J. Phys. Chem. C. **113**, 14006-14014 (2009).

J. Day, A. Platt, S. Subramanian, J. E. Anthony, and O. Ostroverkhova, "Influence of organic semiconductor-metal interfaces on the photoresponse of functionalized anthradithiophene thin films," J. Appl. Phys. **105**, 103703 (2009).

J. Day, A. Platt, O. Ostroverkhova, S. Subramanian, and J. E. Anthony, "Organic semiconductor composites: influence of additives on the transient photocurrent," Appl. Phys. Lett. **94**, 013306 (2009).

J. Day, S. Subramanian, J. E. Anthony, Z. Lu, R. J. Twieg, and O. Ostroverkhova, "Photoconductivity in organic thin films: from picoseconds to seconds after excitation," J. Appl. Phys. **103**, 123715 (2008).

Podolskiy, Viktor A.

S. Thongrattanasiri, J. Elser, and V. A. Podolskiy, "Quasi-planar optics: computing light propagation and scattering in planar waveguide arrays," J. Opt. Soc. Am. **26**, B102 (2009).

A. Kabashin, P. evans, S. Pastkovsky, W. Hendren, G. A. Wurtz, R. Atkinson, R. Pollard, V. Podolskiy, and A. V. Zayats, "Plasmonic nanorod metamaterials for biosensing," Nature Materials **8**, 867 (2009).

R. J. Pollard, A. Murphy, W. R. Hendren, P. R. Evans, R. Atkinson, G. A. Wurtz, A. V. Zayats, and V. A. Podolskiy, "Optical nonlocalities and additional waves in epsilon-near-zero metemeterials," Phys. Rev. Lett. **102**, 127405 (2009).

S. Thongrattanasiri and V. A. Podolskiy, "Hyper-gratings: nanophotonics in planar anisotropic metamaterials," Opt. Lett. **34**, 890 (2009).

M. A. Noginov, G. Zhu, M. f. Mayy, B. A. Ritzo, N. Noginova, and V. A. Podolskiy, "Stimulated emission of surface plasmon polaritons," Phys. Rev. Lett. **101**, 226806 (2008).

Roundy, David

A. F. Oskooi, D. roundy, M. Ibanescu, P. Bermel, J. D. Joannopoulos, and S. G. Johnson, "Meep: A flexible free-software package for electromagnetic simulations by the FDTD method," Comp. Phys. Comm. **181**, 687-702 (2010).

D. A. Freedman, D. Roundy, and T. A. Arias, "Elastic effects of vacancies in strontium titanate: Short- and long-range strain fields, elastic dipole tensors, and chemical strain," Phys. Rev. B **80**, 2972-2974 (2009).

Schneider, Guenter

A. Zakutayev, R. Kykyneshi, G. Schneider, D. H. McIntyre, and J. Tate, "Electronic structure and excitonic absorption in BaCuChF (Ch=S, Se, and Te)," Phys. Rev. B **81**, 155103 (2010).

L. Prisbrey, G. Schneider, and E. Minot, "Modeling the Electrostatic Signature of Single Enzyme Activity," J. Phys. Chem. B **114**, 3330-3333 (2010).

A. Bagrets, R. Werner, F. Evers, G. Schneider, D. Schooss, and P. Wolfle, "Lowering of surface melting temperature in atomic clusters with a nearly closed shell structure," Phys. Rev. D **81**, 075435 (2010).

J. A. Spies, R. Schafer, J. F. Wager, P. Hersh, H. Platt, D. A. Keszler, G. Schneider, R. Kykyneshi, J. Tate, X. Liu, A. D. Compaan, and W. N. Schafarman, "pin Double-Heterojunction Thin-Film Solar Cell p-layer Assessment," Solar Energy Materials & Solar Cells **93**, 1296 (2009).

M. Weinert, G. Schneider, R. Podloucky, and J. Redinger, "FLAPW: Applications and implementations," J. Phys.: Condens. Matter **21**, 084201 (2009).

Tate, Janet

A. Zakutayev, R. Kykyneshi, J. Kinney, D. H. McIntyre, G. Schneider, and J. Tate, "Excitonic absorption and emission in thin-film BaCu(Q,Q')FQ,Q'=S,Se,Te," Phys. Rev. B **81**, 155103 (2010).

A. Zakutayev, J. Tate, H. A. S. Platt, D. A. Keszler, A. Ba rati, W. Jaegermann, and A. Klein, "Band alignment at the BaCuSeF/ZnTe interface," Appl. Phys. Lett. **96**, 162110 (2010).

A. Zakutayev, J. Tate, H. A. S. Platt, D. A. Keszler, C. Hein, T. Mayer, A. Klein, and W. Jaegermann, "Electronic properties of BaCuChF (Ch=S,Se,Te) surfaces and BaCuChF/

ZnPc interfaces," J. Appl. Phys. **107**, 103713 (2010).

K. Jiang, A. Zakutayev, J. Stowers, M. D. Anderson, J. Tate, D. H. McIntyre, D. C. Johnson, and D. A. Keszler, "Low-temperature, solution processing of TiO2 thin films and the fabrication multilayer dielectric optical elements," Solid State Sciences **11**, 1692-1699 (2009).

J. Tate, H. L. Ju, J. C. Moon, A. Zakutayev, A. P. Richard, J. Russell, and D. H. McIntyre, "Origin of p-type conduction in single crystal CuAlO2," Phys. Rev. B **80**, 165206 (2009).

PORTLAND STATE UNIVERSITY

DEPARTMENT OF PHYSICS

Portland, Oregon 97207

Students Accepted For Degree	FIELDS		
	Physics	Astronomy	Related Fields
Doctorate	X*		
Master's	X		

*Applied Physics

1. General

President: Wim Wiewel
Vice Provost for Graduate Studies & Research: W. Feyerherm
Department Chairman: Erik Bodegom
Department Telephone Number: (503) 725-3812
Type of Institution: University
Control: Public
Setting: Urban
Total Faculty: 1,457 (full-time and part-time)
Total Graduate Faculty: Not separated
Total Students: 27,972
Total Graduate Students: 6,298
Annual Graduate Tuition:
In-state residents: Full-time—$9,867 (incl. fees)
Part-time—$374/credit*
Out-of-state residents: Full-time—$14,592 (incl. fees)
Part-time—$549/credit*
Tuition rates for: 2009–10
Deferred tuition plan: Yes
Annual Other Fees: 180
Term: Quarter

*Includes fees per credit/term incidental, health, Recreation Center bldg. fee.

2. Number of Faculty in Department

The combined total of full-time faculty in the three professorial ranks is 16. The combined total of full-time, part-time, and other faculty at all ranks is 21.

3. Admission, Financial Aid, and Housing

Address admission inquiries to: Department of Physics
Graduate application fee required: $50
Admission deadline (Fall admission): 3/1 (can be waived)
Admission information: For fall admission, 2008–09, 13 students were accepted from 24 applicants.
Admission requirements: For admission to the graduate programs, a Bachelor's degree is required with a minimum undergraduate GPA of 2.75 specified. The GRE is required. The GRE Advanced is not required. Students from non-English speaking countries are required to demonstrate proficiency in English via the TOEFL exam. Minimum acceptable score for admission is 550, computer based score: 213.
Undergraduate preparation assumed: Resnick and Halliday, *Fundamentals of Physics*; Tipler, *Modern Physics*; Fowles, *Analytical Mechanics*; Hecht, *Modern Optics*; Reif, *Statistical Physics*; Griffiths, *Intro to Electrodynamics*.
Address financial aid inquiries to: Financial Aid Office
GAPSFAS application required: No
Financial aid deadline: 3/1 (priority deadline)
Loans available: Yes
Address housing inquiries to: www.pdx.edu/housing/housing@pdx.edu
Tel.: (503) 725-4333
On-campus, single student housing available: Yes
Cost/month: $400–689 (sleeper-studio)
On-campus, married student housing available: Yes
Cost/month: $758–1,010, 1 bdrm.; $1,100–1,214 2 bdrm.

Table A—Faculty, Enrollments, and Degrees Granted

Research Specialty	2009–10 Faculty	Enrollment[1] Fall 2009		No. of Degrees Granted[2] 2008–09 (2004–09)			Median No. of Years for 2005–06 Ph.D.'s
		Master's	Doctorate	Master's	Terminal Master's	Doctorate	
Atmos. Physics & Chemistry	3	1	3	1(2)	–	0(1)	–
Atomic, Molecular, & Optical Physics	3	1	1	1(3)	–	0(0)	–
Biophysics	2	2	3	0(9)	–	0(1)	6
Condensed Matter Physics	2	5	6	2(12)	–	0(1)	6
Electron Optics	2	0	1	0(2)	–	0(1)	12
Materials Sci.	3	3	6	1(3)	–	0(1)	–
Nanoscience	3	5	8	2(7)	–	0(0)	–
Non-specialized	0	4	0	0(0)	–	0(0)	–
Total		21	28	5(38)	–	0(6)	–
Full-time Grad. Stud.		16	22			1(5)	
Part-time Grad. Stud.		5	6				
First-year Grad. Stud.		7	5				
Median Years in Grad. Study (2006–07 Degrees)				2.1	–	–	–

Undergraduate Degrees, 2008–09 (2004–09): 18(77)

[1]Students not yet committed to a research specialty are entered under non-specialized.
[2]Five-year totals in parentheses.

4. Graduate Degree Requirements

Master's: Program approval required. 45 quarter credit hours required . A GPA of 3.0/4.0 or better continuously is required. 30 quarter credit hours must be taken in residence. A final oral examination is required.
Doctorate: Program approval required. A minimum GPA of 3.0/4.0 continuously . A minimum of 3 consecutive quarters in residence is required. Qualifying examinations and oral defense of thesis required.
Thesis: Thesis may be written *in absentia*.

Table B—Appointments to Graduate Students, 2008–09

Title of Appointee	Appointments Total	Appointments First year	Academic Load Allowed in Credit Hours	Hours of Service Per Week	Stipend for Academic Year ($)
			Quarter		
Teaching Assistant	17	2	12	12	20,000
Research Assistant	12	0	9	12	20,000
Total	29	2			

[1]Plus Tuition.

5. Personnel Engaged in Separately Budgeted Research,

Professorial faculty	17
Other faculty	6
Graduate students	27
Undergraduate students	12
Nonteaching research personnel	8
Total	67

6. Separately Budgeted Research Expenditures by Source of Support

	Departmental Research	Physics-related Research Outside Department
Federal government	$2,200,000	$
Private, nonprofit organizations	100,000	
Business and Industry	450,000	
Total	$2,750,000	$

Table C—Separately Budgeted Research Expenditures

Research Specialty	No. of Grants	Expenditures ($)
Atmos/Phys. Chemistry	2	600,000
Biophysics	2	200,000
Electron Optics	2	50,000
Materials Sci.	2	350,000
Nanotechnology	9	1,480,000
Total	17	2,680,000

FACULTY

Professors

Abramson, Jonathan J., Ph.D., Rochester, 1975. Biophysics.
Bodegom, Erik, Ph.D., Catholic, 1982. Chairman of the Department. Complex systems; charge coupled devices.
Carruthers, John, Ph.D., Toronto, 1967. Materials Science and Engineering.
Freeouf, John L., Ph.D., Univ. Chicago, 1973. Optical Studies of Semiconductors.
Jiao, Jun, Ph.D., Arizona, 1997. Electron microscopy; electron field emission of nanomaterials and carbon nanoclusters.
Khalil, M. Aslam Khan, Ph.D., Texas, 1976. Atmospheric physics and chemistry.
Könenkamp, Rolf, Ph.D., Tulane, 1984. Solid state physics; electron optics.
Leung, Pui-Tak, Ph.D., SUNY, Buffalo, 1982. Atomic, optical, and surface physics.
Mitchell, Drake C., Ph.D., Oregon, 1987. Biophysics.
Solanki, Raj, Ph.D., Colorado State, 1982. Semiconductors.

Professors Emeriti

Casperson, Lee, Ph.D., Caltech, 1971.
Dash, John, Ph.D., Penn State, 1966. Metallurgy.
Howard, Donald G., Ph.D., California, Berkeley, 1964.
Nussbaum, Rudi H., Ph.D., Amsterdam, 1959.
Rempfer, Gertrude F., Ph.D., Washington, 1939. Electron optics.
Semura, Jack, Ph.D., Wisconsin, 1972. Statistical mechanics.
Smejtek, Pavel K., Ph.D., Czechoslovak Academy, 1965.
Takeo, Makoto, Ph.D., Oregon, 1953.

Associate Professors

La Rosa, Andres H., Ph.D., North Carolina State, 1996. Opto/ultrasonic near-field microscopy; optical MEMS.
Möck, Peter, Ph.D., Humboldt-University of Berlin, 1991. Electron microscopy; quantum dots; x-ray diffraction.
Sánchez, Erik J., Ph.D., Portland State, 1999. Microscopy, lasers, nanotechnology.
Straton, Jack, Ph.D., Oregon, 1986. Quantum Scattering Theory.

Assistant Professors

Butenhoff, Chris, Ph.D., Portland State U., 2010. Atmospheric physics.
Rice, Andrew, Ph.D., UC Irvine, 2002. Atmospheric physics and chemistry.
Seipel, Bjoern, University of Tübingen, 1999. Microscopy, x-ray diffraction.
Widenhorn, Ralf, Portland State 2005. Thermally activated processes, charge coupled devices.

RESEARCH SPECIALTIES AND STAFF

Theoretical

Atomic, Molecular, and Optical Physics. Stopping power theory; sum rule calculations; near-field optics; surface effects. La Rosa, Leung, Takeo, Sánchez, Straton.
Computing Nanoarchitectures. Carruthers
Global Emissions Inventories of Greenhouse Gases. Butenhoff, Khalil.
Global Environmental Change. Climate modeling; atmospheric dispersion and chemistry. Butenhoff, Khalil, Rice.
Modelling of Electromagnetic fields. Sánchez.
Surface Physics. Molecular fluorescence; metallic nanoparticles and plasmonics. Leung.

Experimental

AFM and STM. La Rosa, Sánchez.
Atmoshperic science. Rice, Khalil.
Biological imaging. Sánchez.
Biophysics. Abramson, Mitchell, Smejtek.
Carbon nanotubes and nanoclusters. Jiao, Sánchez.
Charge-coupled devices. Bodegom, Widenhorn.
Crystallographic identification of compounds. Möck.
Electron optics. Rempfer, Könenkamp.
Electronic device physics. Freeouf, Könenkamp, Jiao, Solanki.
Global Change Science. Design of field experimentation, instrumental analysis of air and water samples. Butenhoff, Khalil, Rice.
Lasers. Sánchez.
Microscopy. Jiao, La Rosa, Möck, Sánchez, Könenkamp, Seipel.
Nanometrology, Nanoelectronics, Biosensors. Freeouf, Carruthers, Solanki, Jiao, Könenkamp, Straton.
Quantum dots. Möck.
Solid state physics. Freeouf, Möck, Könenkamp, Solanki.

UNIVERSITY OF OREGON

DEPARTMENT OF PHYSICS

Eugene, Oregon 97403-1274
http://physics.uoregon.edu

Students Accepted For Degree	FIELDS		
	Physics	Astronomy	Related Fields
Doctorate	X		
Master's	X		

1. General

President: Richard Lariviere
Dean of Graduate School: Richard Linton
Department Chairman: Stephen Kevan
Department Telephone Number: (541) 346-4751
Type of Institution: University
Control: Public
Setting: Small city
Total Faculty: 1,786
Total Graduate Faculty: No distinction made
Total Students: 22,386
Total Graduate Students: 3,872
Annual Graduate Tuition:
In-state residents: Full-time—$12,697.50 (9–16 credits)
Part-time—reduced rates
Out-of-state residents: Full-time—$17,989.50 (9–16 credits)
Part-time—reduced rates
Tuition rates for: 2009–10
Deferred tuition plan: Yes
Other Fees: included in tuition
Term: Quarter

2. Number of Faculty in Department

The combined total of full-time faculty in the three professorial ranks is 28. The combined total of full-time, part-time, and other faculty at all ranks is 31.

3. Admission, Financial Aid, and Housing

Address admission inquiries to: Department of Physics
Graduate application fee required: $50.00
Admission deadline (Fall admission): 1/15
Admission information: For fall admission, 2010, 46 students were accepted from 228 applicants. All 46 were offered teaching assistantships.
Admission requirements: For admission to the graduate programs, a Bachelor's degree in physics or related subject is required with a minimum undergraduate GPA of 3.0 in advanced physics and mathematics courses. The GRE is required. No minimum acceptable score for admission is specified. The GRE Advanced Test in physics is required. The median GRE scores for fall, 2010 admission with teaching assistantships were verbal–560; quantitative–790; total–2,120. The median physics advanced score was 780. Students from non-English speaking countries are required to demonstrate proficiency in English via the TOEFL exam. Minimum acceptable score is 500 for admission, 600 for teaching assistants. Supplemental English training after arrival may be required for TOEFL scores below 575. Therefore, scores above 600 are given preference.
Undergraduate preparation assumed: Familiarity with material at a level found in the following text books: Classical Mechanics: Chow, *Analytical Mechanics*: Electricity and Magnetism: Griffiths, *Introduction to Electrodynamics*; Statistical and Thermal Physics: Schroeder, *Introduction to Thermal Physics*; Modern Physics: Griffiths, *Intro to Quantum Mechanics*.
Address financial aid inquiries to: Department of Physics
GAPSFAS application required: No
Financial aid deadline: 1/15 complete application file (to assure consideration)
Loans available: Yes
Address housing inquiries to: University Housing, Walton Hall, University of Oregon (htp://housing.uoregon.edu/About/)
On-campus, single student housing available: Yes
Cost/term: $4,133–6,295 (room and board)
On-campus, married student housing available: Yes
Cost/month: $528–985

Table A—Faculty, Enrollments, and Degrees Granted

Research Specialty	2009–10 Faculty	Enrollment[1] Fall 2009		No. of Degrees Granted[2] 2009–10 (2005–09)			Median No. of Years for 2003–08 Ph.D.'s
		Master's	Doctorate	Master's	Terminal Master's	Doctorate	
Applied Physics	0	15	0	3(0)	3(19)	0(0)	–
Astrophysics	3	0	0	0(0)	0(0)	1(0)	8
Atomic, Molecular, & Optical Physics	7	0	0	0(0)	0(0)	4(15)	6
Biophysics	3	0	0	0(0)	0(0)	2(1)	5
Chemical Physics	0	0	0	0(0)	0(0)	0(0)	
Condensed Matter Physics	7	0	0	0(0)	(0)	5(10)	6
Particles & Fields	8	0	0	0(0)	(0)	1(7)	7
Physics Education	3	0	0	0(0)	0(0)	0(0)	–
Non-specialized	0	3	87	4(54)	2(0)	–	–
Total		18	87	7(41)	5(19)	13(35)	
Full-time Grad. Stud.		18	87				
Part-time Grad. Stud.		0	0				
First-year Grad. Stud.		15	18				
Median Years in Grad. Study (2007–08 Degrees)		1.5	7				

Undergraduate Degrees, 2007–08 (2003–08): 20(91)

[1]Students not yet committed to a research specialty are entered under non-specialized.
[2]Five-year totals in parentheses.

4. Graduate Degree Requirements

For more detail, see: http://physics.uoregon.edu/physics/grad/handbook.html
Master's: 45 credit hours of graduate level courses, including 32 credits of physics and at least 24 credits of University of Oregon graded courses, are required. These must include at least one 3-term sequence in physics at the 600-level and an approved sequence in mathematics. A maximum of 15 hours of credits earned in another accredited graduate school with a grade of B or better may be counted. A minimum GPA of 3.0 must be maintained. Command of a foreign language is recommended but not required. A Master's final examination, or a thesis, or a certain course requirement has to be satisfactorily completed.

Applied Physics Master's: The department offers an Applied Physics Masters Program that leads to a professional M.S. degree. This degree is an alternative to the research-based Ph.D. and is more oriented towards the needs of industrial physicists than the traditional masters degree. This program includes an Internship component. The applied Physics Masters Program is offered through the U of O Materials Science Institute (see http://internship.uoregon.edu).

Doctorate: The student must pass a qualifying examination covering advanced undergraduate physics in mechanics, electricity and magnetism, quantum mechanics; statistical mechanics and thermal physics.Students generally must complete core graduate courses in mechanics, statistical physics, electromagnetic theory, and quantum mechanics, though they can be excused based on previous study. In addition, students must take a total of six other one-quarter courses, chosen from the following areas: Condensed Matter Physics; Nuclear and Particle Physics; Atomic Physics and Molecular Physics; Astronomy and Early Universe Physics; Experimental and Theoretical Techniques; Interdisciplinary. An oral comprehensive examination and a thesis are required. Proficiency in a foreign language is recommended but not required. Three years work beyond the Bachelor's degree is required, of which three consecutive terms must be on the Eugene, Oregon campus.

Thesis: Thesis may be written *in absentia*.

Special Equipment, Facilities, or Programs:

Institutes and Centers

The University of Oregon has several interdisciplinary institutes in which many physics faculty participate:

The Materials Science Institute (MSI, http://www.uoregon.edu/~msiuo/) focuses much of its efforts on the creation and study of new materials and devices, but also addresses more abstract questions in experiment and theory. The MSI has a wide range of fabrication and characterization capabilities located both in individual laboratories and in common facilities. An important mission of the MSI is education and in this connection it promotes integrated research between various departments and conducts Summer Industrial Internship programs in semiconductor processing, polymer technology, and (with the OCO), optics and photonics. The MSI is a founding member and partner of the Oregon Nanoscience and Microtechnologies Institute (ONAMI, http://www.onami.us/).

The Oregon Center for Optics (OCO, http://oco.uoregon.edu/) promotes and facilitates research and education in the sciences at the University of Oregon wherever optical science is involved in an essential fashion in either its fundamental aspects or its technological applications. The OCO has a broad range of state-of-the-art lasers and spectroscopy equipment located in individual laboratories and also in common facilities.

The Institute of Theoretical Science (ITS, http://www.uoregon.edu/~its/) is a center for theoretical research in over-lapping areas of physics, chemistry and mathematics. It provides an environment in which theorists can share common themes and mathematical approaches.

The Institute of Molecular Biology (IMB, http://www.molbio.uoregon.edu/) is comprised of biologists, chemists, and physicists pursuing a molecular-level understanding of living systems. It runs a weekly seminar series and operates common facilities to assist with imaging, cell culture, and analytic characterization.

The Center for High Energy Physics (UOCHEP, http://uoregon.edu/~chep/) supports experimental and theoretical high energy physics research activities at the University of Oregon and at various external laboratories, including the CERN, Fermilab and SLAC accelerator facilities and the LIGO gravitational wave observatories.

Departmental and Other Facilities

Pine Mountain Observatory (http://pmo-sun.uoregon.edu/) houses several telescopes and is equipped with CCD cameras for remote data collection.

The Center for Advanced Materials Characterization at Oregon (CAMCOR, http://www.uoregon.edu/)houses capital-intensive equipment for microanalysis, surface analysis, electron microscopy, semiconductor device fabrication, as well as traditional chemical characterization for users from inside and outside the University. The staff members who run the facilities are experienced in sample preparation, data collection and data analysis. A new building was occupied by CAMCOR in Spring, 2008 and expansion into basement space of another new building is planned for 2011.

The Shared Laser Facility (SLF) is a multidisciplinary laboratory available to the university community and others by arrangement. Faculty members may either set up long-term experiments in the SLF or use shared equipment for short-term experiments. SLF personnel also provide expertise in setting up experiments in user laboratories.

The Technical Services Administration (TSA) maintains professional and student machine shops, and an electronics shop.

Table B—Appointments to Graduate Students, 2009–10

Title of Appointee	Appointments: Total	Appointments: First Year	Academic Load Allowed in Credit Hours	Hours of Service Per Week	Monthly Stipend for Academic Year ($)
			Quarter		
Teaching Assistant	35	18	9–16	16	1,663,16
Research Assistant	32	0	9–16	16	1,663,16
Total	67	18			

5. Personnel Engaged in Separately Budgeted Research, 7/07–6/08

Professorial faculty	31
Other faculty	2
Postdoctoral appointments	4
Graduate students	50
Undergraduate students	9
Nonteaching research personnel	13
Total	109

6. Separately Budgeted Research Expenditures by Source of Support

	Departmental Research	Physics-related Research Outside Department
Federal government	$7,017,737	$
State and local government	155,172	
Private, non-profit organizations	149,030	
Other	1,970,287	
Total	$9,292,226	$

7. Separately Funded and Managed Centers

Laboratories

Pine Mountain Observatory	$129,296
Total	$129,296

Table C—Separately Budgeted Research Expenditures

Research Specialty	No. of Grants	Expenditures ($)
Astronomy	0	0
Astrophysics	5	156,053
Biophysics	10	951,427
Condensed Matter Physics	30	1,453,947
Energy Sources & Environ.	11	238,321
Low Temperature Physics	1	48,429
Nuclear Physics	0	0
Optics	24	1,336,594
Particles & Fields	12	1,926,381
Physics Education	4	1,061,876
Total	100	7,172,908

FACULTY

Professors

Belitz, Dietrich, Dr.rer.nat., Munich, 1982. Condensed matter theory.
Bothun, Gregory D., Ph.D., Washington, 1981. Astronomy.
Brau, James E., Ph.D., MIT, 1978. Experimental elementary particle physics.
Cohen, J. David, Ph.D., Princeton, 1976. Experimental solid state physics.
Crasemann, Bernd, Ph.D., California, Berkeley, 1953. Emeritus. Atomic physics.
Csonka, Paul, Ph.D., Johns Hopkins, 1963. Elementary particle theory; accelerators.
Deshpande, Nilendra G., Ph.D., Pennsylvania, 1965. Elementary particle theory.
Donnelly, Russell J., Ph.D., Yale, 1956. Emeritus. Classical and superfluid hydrodynamics.
Frey, Raymond E., Ph.D., California, Riverside, 1984. Experimental elementary particle physics.
Girardeau, Marvin, Ph.D., Syracuse 1958. Emeritus.
Haydock, Roger, Ph.D., Cambridge, 1972; Sc.D., Cantab, 1989. Solid state theory.
Hsu, Stephen D. H., Ph.D., University of California, Berkeley, 1991. Elementary particle theory.
Hwa, Rudolph C., Ph.D., Brown, 1962. Emeritus. Elementary particle theory.
Imamura, James N., Ph.D., Indiana, 1981. Astrophysics.
Kevan, Stephen D., Ph.D., California, Berkeley, 1980. Experimental solid state physics.
Lefevre, Harlan W., Ph.D., Wisconsin, 1961. Emeritus. Experimental nuclear physics.
Matthews, Brian W., Ph.D., Adelaide, 1963. Protein crystallography.
McDaniels, David K., Ph.D., Washington, 1960. Emeritus. Experimental nuclear physics; solar energy.
Moseley, John T., Ph.D>, Grorgin Tech, 1969. Emeritus.
Overley, Jack C., Ph.D., Cal. Tech., 1960. Emeritus. Experimental nuclear physics.
Park, Kwangjai, Ph.D., California, Berkeley, 1965. Emeritus. Fluid mechanics.
Rayfield, George W., Ph.D., California, Berkeley, 1964. Emeritus. Membrane biophysics.
Raymer, Michael G., Ph.D., Colorado, 1979. Quantum optics.
Remington, Stephen James, Ph.D., Oregon, 1977. Protein crystallography.
Schombert, James, Ph.D., Yale, 1984. Astronomy.
Sokoloff, David R., Ph.D., MIT, 1972. Science education.
Soper, Davison E., Ph.D., Stanford, 1971. Elementary particle theory.
Strom, David, Ph.D., Wisconsin, Madison, 1986. Experimental elementary particle physics.
Taylor, Richard, Ph.D., 1988, Univ. of Nottingham, U.K. Experimental solid state physics.
Toner, John, Ph.D., Harvard, 1981. Condensed matter theory.
van Enk, Stephen, Ph.D., Theoretical Optical Physics.
Wang, Hailin, Ph.D., Michigan, 1990. Quantum optics.
Zimmerman, Robert L., Ph.D., Washington, 1963. Emeritus. Theoretical astrophysics; general relativity.

Associate Professors

Deutsch, Miriam, Ph.D., Hebrew Univ., 1996. Quantum optics.
Gregory, Stephen, Ph.D., Waterloo, 1975. Associate Chairman for Graduate Studies. Experimental condensed matter.
Kribs, Graham, Theoretical High Energy Physics.
Nöckel, Jens, Ph.D., Yale, 1997. Optical Physics.
Steck, Daniel, Ph.D., Texas, 2001. Experimental and theoretical optical physics.
Torrence, Eric, Ph.D., MIT, 1997. Experimental High Energy Physics.

Assistant Professors

Chang, Spencer, Ph.D., Harvard, 2004. Theoretical High Energy Physics.
Corwin, Eric, Ph.D., Chicago, 2007. Experimental Soft Condensed Matter and Biophysics.
Parthasarathy, Raghuveer, Ph.D., Univ. Chicago, 2002. Experimental solid state and biophysics.

RESEARCH SPECIALTIES AND STAFF

Theoretical

Astrophysics. Modeling of astrophysical fluid flows and plasmas. Imamura.
Condensed Matter. Electronic energy structure; cohesion; and electronic properties of solids; localization of electronic states in amorphous semiconductors. Belitz, Haydock, Toner.
Elementary Particles and Fields. Theory of elementary particles. Chang, Csonka, Deshpande, Hsu, Hwa, Kribs, Soper. 3 postdoctoral fellows.
Fluids. Superfluid mechanics. Donnelly.
Optics. Theoretical optics. Nöckel, van Enk.

Experimental

Astrophysics. Photometry of binary star x-ray sources, observational cosmology, galaxy formation and evolution. Bothun, Schombert.
Biophysics. X-ray crystallography of proteins; membrane biophysics. Corwin, Matthews, Parthsarathy, Rayfield, Remington.
Condensed Matter. Phonon and electron transport in low-dimensional systems; electronic properties of amorphous semicon-

ductors; surface physics. Cohen, Gregory, Kevan, Corwin, Taylor. 2 postdoctoral fellows.

Elementary Particles. Brau, Frey, Strom, Torrence. 2 postdoctoral fellows. 1 senior research associate.

Energy Sources and environment. Bothun, Cohen, McDaniels.

Fluid Mechanics. Classical and superfluid hydrodynamics. Donnelly.

Optics. Quantum optics. Cold atoms. Deutsch, Gregory, Raymer, Steck, Wang. 1 postdoctoral fellow.

Gravity. Brau, Frey, Strom. 2 postdoctoral fellows.

BRYN MAWR COLLEGE

DEPARTMENT OF PHYSICS

Bryn Mawr, Pennsylvania 19010-2899

Students Accepted For Degree	FIELDS		
	Physics	Astronomy	Related Fields
Doctorate	X		
Master's	X		

1. General

President: Jane Dammen McAuliffe
Dean of Graduate Studies: Elizabeth F. McCormack
Department Chairman: Michael W. Noel
Department Telephone Number: (610) 526-5358
Department web page: www.brynmawr.edu/physics
Type of Institution: College
Control: Private
Setting: Suburban
Total Faculty: 158
Total Students: 1,771
Total Graduate Students: 464†
Annual Graduate Tuition:
All Graduate Students: Full-time—$31,740
Part-time—$5,290/unit*
Tuition rates for: 2009–10
Deferred tuition plan: No
Other Fees: None
Term: Semester

*Full load is 6 units/year.
†144 in the Graduate School of Arts and Sciences

2. Number of Faculty in Department

The combined total of full-time faculty in the three professorial ranks is 5. The combined total of full-time, part-time, and other faculty at all ranks is 7.

3. Admission, Financial Aid, and Housing

Address admission inquiries to: Dean of the Graduate School of Arts and Sciences.
Graduate application fee required: $30
Admission deadline (Fall admission): January 4
Admission information: For fall admission, 2008–09, 4 of 8 applicants were accepted.
Admission requirements: For admission to the graduate programs, a Bachelor's degree in physics or a closely related field is required with no minimum undergraduate GPA specified. The GRE is required. No set minimum scores. The GRE-subject test in physics is required. No set minimum scores. Students from non-English speaking countries are required to demonstrate proficiency in English via the TOEFL or IELTS exam; minimum TORFL score 600 paper based, 250 computer based, or 100 iBT; minimum IELTS score 7.
Address financial aid inquiries to: Dean of Graduate Studies
GAPSFAS application required: Yes
Financial aid deadline: January 4
Loans available: Yes
Address housing inquiries to: Angie Sheets, Director of Residential Life
On-campus, single student housing available: No
On-campus, married student housing available: No

Table A—Faculty, Enrollments, and Degrees Granted

Research Specialty	2009–10 Faculty	Enrollment[1] Fall 2009		No. of Degrees Granted[2] 2008–09 (2004–09)			Median No. of Years for 2008–09 Ph.D.'s
		Master's	Doctorate	Master's	Terminal Master's	Doctorate	
Atomic, Molecular, & Optical Physics	2	0	1	1(1)	0(2)	0(2)	–
Condensed Matter Physics	2	0	0	0(0)	–	0(0)	–
Nonlinear Dynamics	0	0	0	0(0)	–	0(1)	–
Theoretical Physics	1	0	1	0(0)	–	0(0)	–
Non Specialized	0	0	1	0(0)		0(0)	
Total		0	3	1(1)	0(2)	0(3)	
Full-time Grad. Stud.		0	3				
Part-time Grad. Stud.		0	0				
First-year Grad. Stud.		0	0				
Median Years in Grad. Study (2005–06 Degrees)				–	–	–	
Undergraduate Degrees, 2005–06 (2001–06): 6(33)							

[1]Students not yet committed to a research specialty are entered under non-specialized.
[2]Five-year totals in parentheses.

4. Graduate Degree Requirements

Master's: Oral qualifying examination; one full year of work with satisfactory performance done in residence. Master's paper and oral Master's examination.
Doctorate: At least three full years of work in residence with satisfactory performance; oral qualifying examination; preliminary examinations; thesis; final examination.
Thesis: Thesis may be written *in absentia*.
Special Equipment, Facilities, or Programs: Cooperative agreements with the University of Pennsylvania and Drexel University allow Bryn Mawr graduate students to take work in special field areas not available at Bryn Mawr. The physics collection in the new science library contains an excellent collection of monographs and current journal subscriptions. The college and the department have excellent state-of-the-art computing facilities for data acquisition, modeling, and data analysis as well as high-speed computer links to the national and international physics communities.

Table B—Appointments to Graduate Students, 2010–11

Title of Appointee	Appointments Total	Appointments First-year	Academic Load Allowed in Credit Hours	Hours of Service Per Week	Stipend for Calendar Year ($)
			Semester		
Teaching Assistant	1	0	8	17	22,750[1]
Research Assistant	2	0			22,750[1]
Total	3	0			

[1]Plus tuition waiver and $1,500 towards health insurance for U.S. students, $1,064 for international students.

5. Personnel Engaged in Separately Budgeted Research, 2010–11

Professorial faculty	5
Postdoctoral Fellows	1
Graduate students	3
Undergraduate students	7
Total	16

6. Separately Budgeted Research Expenditures by Source of Support, 2009–10

	Departmental Research	Physics-related Research Outside Department
Federal government	$205,000	–
College	110,500	–
Total	$315,500	–

Table C—Separately Budgeted Research Expenditures

Research Specialty	No. of Grants	Expenditures ($)
Atomic, molecular and optical physics	2	155,000
Theoretical physics	1	50,000
Total	1	205,000

FACULTY

Professors

Beckmann, Peter A., Ph.D., British Columbia, 1975. Solid state dynamic nuclear magetic resonance.

McCormack, Elizabeth F., Ph.D., Yale, 1989. Dean of Graduate Studies. Atomic, molecular and optical physics.

Associate Professor

Noel, Michael W., Ph.D., Rochester, 1996. Department Chair. Atomic, molecular and optical physics.

Laboratory Coordinator and Lecturer

Matlin, Mark, D., Ph.D., Maryland, 1991. General relativity.

Assistant Professors

Cheng, Xuemei May, Ph.D., Johns Hopkins, 2006. Condensed matter physics; magnetic materials.

Schulz, Michael B., Ph.D., Stanford, 2002. Theoretical physics with a focus on string theory.

RESEARCH SPECIALTIES AND STAFF

Theoretical

Theoretical Physics. String theory and its applications to quantum field theory, cosmology, and particle physics. Schulz.

Experimental

Atomic, Molecular, and Optical Physics. Laser-based studies of atomic and molecular excited-state structure and decay dynamics, including photoionization, autoionization, predissociation, photodissociation. Nonlinear optical techniques including multiphoton excitation and detection, laser-induced grating spectroscopy, degenerate four wave mixing and vacuum ultraviolet light generation. McCormack.

Atomic, Molecular, and Optical Physics. Current research focuses on understanding and controlling resonant energy transfer in ultracold samples of highly excited atoms. We use laser cooling and trapping techniques to prepare and manipulate the atomic sample and study the extremely long range, many body interactions that result when the atoms are excited to weakly bound states. Noel.

Condensed Matter Physics. Time resolved magetic imaging via photoemission electron microscopy. Magnetic thin films with perpendicular anistropy. Ferromagnetic semiconductors and spintronic materials. Cheng.

Solid State Dynamic Nuclear Magnetic Resonance (NMR). We perform ^{1}H and ^{19}F solid state nuclear magnetic resonance relaxation studies in ionic solids and in organic molecular solids and subsequently model the motion with knowledge of the equilibrium structure. Collaborators are at the University of Delaware (NMR spectroscopy), University of California at San Diego (X-ray diffraction), Villanova University (X-ray diffraction), and Chengdu, China (electronic structure calculations). Beckmann.

FACULTY PUBLICATIONS

Beckmann, Peter A.

P. A. Beckmann, "A Review of Polytopism in Lead Iodide," Cryst. Res. Technol. **45**, 455–460 (2010).

P. A. Beckmann, W. G. Dougherty Jr., and W. S. Kassel, "Methyl and *t*-Butyl Reorientation in an Organic Molecular Solid," Solid State Nuc. Mag. Resonan. **36**, 86–91 (2009).

G. Neue, S. Bai, R. E. Taylor, P. A. Beckmann, A. J. Vega, and C. Dybowski, "^{119}Sn spin-lattice relaxation in α-SnF_2," Phys. Rev. B **79**, 214302 1-5 (2009).

O. Dmitrenko, S. Bai, P. A. Beckmann, S. van Bramer, A. J. Vega, and C. Dybowski, "^{207}Pb Chemical Shielding in $PbMoO_4$ and $PbCl_2$: the Effects of Temperature and Lattice Expansion," J. Phys. Chem. A **112**, 3046–3052 (2008).

X. Wang, A. L. Rheingold, A. G. DiPasquale, F. B. Mallory, C. W. Mallory, and P. A. Beckmann, "The Quenching of Isopropyl Group Rotation in Van der Waals Molecular Solids," J. Chem. Phys. **128**, 124502, 1-3 (2008).

Cheng, Xuemei May

X. M. Cheng, K. S. Buchanan, R. Divan, K. Y. Guslienko, and D. J. Keavney, "Nonlinear vortex domains in ferromagnetic disks," Phys. Rev. B **79**, 172411 (2009).

D. J. Keavney, X. M. Cheng, and K. S. Buchanan, "Polarity reversal of a magnetic vortex core by a unipolar, nonresonant in-plane pulsed magetic field," Appl. Phys. Lett. **94**, 172506 (2009).

D. S. Gianola, C. Eberl, X. M. Cheng, and K. J. Hemker, "Stress-driven surface topography evolution in nancrystalline Al Thin films," Advanced Materials **20**, 303 (2008).

V. Rose, X. M. Cheng, D. J. Keavney, J. W. Freeland, K. S. Buchanan, B. Ilic, and V. Metlushko, "The breakdown of the fingerprinting of vortices by hysteresis loops in circular multilayer ring arrays," Appl. Phys. Lett. (cover) **91**, 132501 (2007).

Y. L. Iunin, Y. P. Kabanov, V. I. Nikitenko, X. M. Cheng, D. Clarke, O. A. Tretiakov, O. Tchemyshyov, A. J. Shapiro, R. D. Shull, and C. L. Chien, "Asymmetric Domain Nucleation and Unusual Magnetization Reversal in Ultrathin Co Films with Perpendicular Anistropy," Phys. Rev. Lett. **98**, 117204 (2007).

McCormack, Elizabeth F.

J. Croman and E. F. McCormack, "Energies and lifetimes of the predissociative v=12 and 13 levels of the $D^1\Pi_u^+$ state in H_2," J. Phys. B: At. Mol. Opt. Phys. **41**, 035103 (2008).

R. C. Ekey, A. Marks, and E. F. McCormack, "Double resonance spectroscopy of the $B''\bar{B}\ '\Sigma_u^+$ state of H_2," Phys. Rev. A **73**, 023412 (2006).

Noel, Michael W.

T. J. Carroll, C. Daniel, L. Hover, T. Sidie, and M. W. Noel, "Simulations of dipole-dipole interaction between two spatially separated groups of Rydberg atoms," Phys. Rev. A **80**, 052712 (2009).

T. J. Carroll, S. Sunder, and M. W. Noel, "Many-body interactions in a sample of ultracold Rydberg atoms with varying dimensions and densities," Phys. Rev. A **73**, 032725 (2006).

Schulz, Michael B.

R. Donagi, P. Gao, and M. B. Schulz, "Abelian Fibrations, String Junctions, and Flux Geometry Duality," JHEP 0904, 119 (2009); arXiv: 0810.5195

A. Lawrence, T. Sander, M. B. Schulz, B. Wecht, "Torsion and Supersymmetry Breaking," JHEP **0807**, 042 (2008); arXiv:0711.4787.

M. Cvetic, T. Liu, M. B. Schulz, "Twisting K3 x T^2 Orbifolds," JHEP **0709**, 092 (2007); hep-th/0701204.-

CARNEGIE MELLON UNIVERSITY

DEPARTMENT OF PHYSICS

Pittsburgh, Pennsylvania 15213

Students Accepted For Degree	FIELDS		
	Physics	Astronomy	Related Fields
Doctorate	X		
Master's	X		

1. General

President: Jared L. Cohon
Department Head: Gregg Franklin
Graduate Admissions Telephone Number: (412) 268-2849
Fax #: (412) 681-0648
E-mail: physgrad@andrew.cmu.edu
www.cmu.edu/physics
Type of Institution: University
Control: Private
Setting: Urban
Total Faculty: 1,368
Total Graduate Faculty: 1,368
Total Students: 11,371
Total Graduate Students: 5,420
Annual Graduate Tuition:
All Graduate Students: Full-time—$36,000/year
Part-time—$500/per unit
Tuition rates for: 2010–11
Deferred tuition plan: No
Other Fees: $400/year
Term: Semester

2. Number of Faculty in Department

The combined total of full-time tenure track faculty in the three professorial ranks is 34. The combined total of full-time and other faculty at all ranks is 41.

3. Admission, Financial Aid, and Housing

Address admission inquiries to: Graduate Studies, Department of Physics, Carnegie Mellon University, Pittsburgh, PA 15213
E-mail: physgrad@andrew.cmu.edu
Graduate application fee required: No
Application deadline (Fall admission): 1/1
Admission deadline (Fall admission): 4/15
Admission information: For fall admission, 2010–11, 19 students were accepted from 301 applicants.
Admission requirements: For admission to the graduate programs, a Bachelor's degree in physics or related field is required. The GRE and GRE Subject Test are required. For students from non-English speaking countries the TOEFL exam is required. No minimum scores specified.
Undergraduate preparation assumed: A typical student will have completed intermediate courses in mechanics (Marion); electricity and magnetism (Griffiths and Wangsness); modern physics (Eisberg and Resnick); wave mechanics (Townsend); thermodynamics and statistical mechanics (Reif); modern physics laboratory (Melissinos).
Address financial aid inquiries to: Graduate Studies, Department of Physics
GAPSFAS application required: No
Financial aid deadline: N/A
Loans available: N/A
Address housing inquiries to: Graduate Studies, Department of Physics
On-campus, single student housing available: Undergraduate only
On-campus, married student housing available: No

Table A—Faculty, Enrollments, and Degrees Granted

Research Specialty	2009–10 Faculty	Enrollment[1] Fall 2009 Master's	Enrollment[1] Fall 2009 Doctorate	No. of Degrees Granted[2] 2009–10 (2005–10) Master's	Terminal Master's	Doctorate	Median No. of Years for 2009–10 Ph.D.'s
Applied Physics	4	–	3	0(0)	–	0(5)	–
Astrophysics	7	–	3	0(0)	–	1(5)	5
Biological Physics	8	–	9	0(0)	–	1(1)	5
Computational Physics	10	–	0	0(0)	–	0(0)	–
Condensed Matter Physics	8	–	13	0(0)	–	2(15)	5
Medium Energy/ Nuclear Physics	6	–	9	0(0)	–	1(9)	5
Particles & Fields	10	–	10	0(0)	–	2(10)	5
Quantum Foundations	1	–	3	0(0)	–	0(0)	–
Statistical & Thermal	3	–	0	0(0)	–	0(0)	–
Non-specialized	0	–	17	7(65)	–	–	–
Total		–	67	7(65)	–	7(45)	
Full-time Grad. Stud.		–	67				
Part-time Grad. Stud.		–	0				
First-year Grad. Stud.		–	14				
Median Years in Grad. Study (2009–10 Degrees)				2	–	5	–

Undergraduate Degrees, 2009–10 (2005–10): 31(183)

[1]Students not yet committed to a research specialty are entered under non-specialized.
[2]Five-year totals in parentheses.

4. Graduate Degree Requirements

Master's: 32 semester hours (96 units) of course work with grade average of B or above. 4 semester hours (12 units) experimental work required. No thesis or foreign language requirements. Written qualifying examination required. Residence, one year.
Doctorate: Satisfactory performance in an approved program. Additional course requirements will depend on level of preparation. Comprehensive oral and written qualifying examinations, annual research reviews, thesis, and final thesis defense required. One year residence as full-time student required. There is a teaching requirement for the Ph.D. degree.
Thesis: Thesis may be written *in absentia*.
Special Equipment, Facilities, or Programs: Astrophysics research is integrated within the Bruce and Astrid McWilliams Center for Cosmology which brings together astrophysicists, particle physicists, computer scientists and statisticians to advance our understanding of dark matter and dark energy which dominate the Universe. Observational astrophysics is carried out using a variety of space-based and ground-based telescopes. Computation for astrophysics research utilizes 3 parallel Linux in-house clusters with 196 cores, 320 Gb RAM total and 40 Tb disk space as well as a 600 core supercomputer.

The Department maintains facilities for condensed matter and biological physics research, including apparatus for: x-ray diffraction and reflection, laser spectroscopies, calorimetry, magnetic and electrical transport measurements, optical characterization of interfaces, scanning tunneling and atomic force microscopies, and sample preparation. Scattering experiments are carried out at an in-house x-ray facility including fix tube and rotating anode sources as well as at national synchrotron and neutron facilities. Computation facilities for these groups include an in-house 128 core cluster. Collaborations with other departments provide access to additional facilities including; clean room facilities, electron microscopies, optical microscopies, magnetic measurements, and fluids and interface characterization.

High-energy research is carried out by faculty utilizing facilities at the Fermi National Accelerator Laboratory in Chicago, at CERN in Geneva, Switzerland, at CESR in Cornell, and at BES in Beijing, China. A data analysis laboratory is maintained on campus, as are laboratories for the development of detection systems.

The medium-energy physics group builds and carries out experiments at the Thomas Jefferson National Accelerator Facility (JLab) in Virginia. Present work centers on the GlueX experiment which is part of the planned upgrade to JLab and is currently being led by members of the CMU group. The group utilizes a 290-core computer cluster for computational studies.

Deparmental facilities include machine shops, numerous computer clusters, and a stock room. The University Computing Center operates an extensive system of networked scientific workstations and micro-computers with central fileservers for research and educational applications. Access to a Cray XT3 MPP supercomputer as well as sets of SMP machines are available through the Pittsburgh Supercomputing Center. The Physics Department is located in Wean Hall, which also houses the science and engineering library.

Table B—Appointments to Graduate Students, 2008–09

Title of Appointee	Appointments: Total	Appointments: First year	Academic Load Allowed in Credit Hours	Hours of Service Per Week	Stipend for Academic Year ($)[1,2]
			Semester		
Teaching Assistant	31	12	Varies	15	18,000
Research Assistant	36	2	Varies	-	18,000
Total	67	14			

[1]Full-time students for Summer 2009 received three additional months stipend.
[2]All tuition and fees are covered in addition to stipend.

5. Personnel Engaged in Separately Budgeted Research, 7/08–6/09

Professorial faculty	34
Postdoctoral appointments	22
Graduate students	36
Undergraduate students	40
Nonteaching research personnel	9
Total	141

6. Separately Budgeted Research Expenditures by Source of Support

	Departmental Research	Physics-related Research Outside Department
Federal government	$4,938,809	$354,000
Private Industry	194,749	
University	1,375,041	
Total	$6,508,599	$354,000

Table C—Separately Budgeted Research Expenditures

Research Specialty	No. of Grants	Expenditures ($)
Astrophysics	8	519,063
Biological Physics	17	1,713,868
Condensed Matter Physics	19	963,254
Medium Energy/Nuclear	9	1,426,458
Particles & Fields	15	1,575,319
Quantum Foundations	4	286,075
Statistical & Thermal Physics	2	6,562
Total	74	6,508,599

Table D—Physics-related Research Outside Department

Field and Unit Outside Department	No. of Grants	Expenditures ($)
Electrical and Computer Engineering Department	6	354,000
Total	6	354,000

FACULTY

Professors

Feenstra, Randall M., Ph.D., California Institute of Technology, 1982. Experimental condensed matter physics; semiconductor surfaces.

Ferguson, Thomas, A., Ph.D., UCLA, 1978. Experimental high-energy physics; CMS at CERN.

Franklin, Gregg B., Ph.D., MIT, 1980. Experimental medium energy/nuclear physics; production and interactions of strange hadrons; strange sea quarks in the nucleon.

Garoff, Stephen, Ph.D., Harvard, 1977. Experimental condensed matter physics; surfaces and interfaces.

Gilman, Frederick, Ph.D., Princeton University, 1965. Buhl Professor of Physics. Theoretical elementary particle physics; CP violation, heavy quarks and leptons.

Griffiths, Richard E., Ph.D., University of Leicester, U.K., 1972. Experimental astrophysics; observational cosmology.

Griffiths, Robert B., Ph.D., Stanford, 1962. Otto Stern Professor of Physics. Theoretical physics; foundations of quantum mechanics.

Holman, Richard F., Ph.D., Johns Hopkins, 1982. Theoretical particle physics and cosmology; inflation, dark energy.

Levine, Michael J., Ph.D., California Institute of Technology, 1963. Theoretical elementary particle physics; Director Pittsburgh Supercomputer Center.

Lösche, Mathias, Ph.D., Technical U. of Munich, Germany, 1986. Experimental biological physics; molecular and membrane biophysics.

Majetich, Sara A., Ph.D., Georgia, 1987. Experimental condensed matter physics; semiconductor and magnetic nanoparticles.

Meyer, Curtis A., Ph.D., University of California, Berkeley, 1987. Experimental medium-energy/nuclear physics; meson and glueball spectroscopy; strangeness physics.

Morningstar, Colin J., Ph.D., University of Toronto, 1991. Theoretical medium-energy physics; nonperturbative phenomena in quantum field theories.

Nagle, John F., Ph.D., Yale, 1965. Experimental and theoretical biological physics; statistical mechanics of phase transitions, biomembranes.

Paulini, Manfred, Ph.D., University of Erlangen, Germany, 1993. Experimental high-energy physics; CDF at Fermilab; CMS at CERN.

Peterson, Jeffrey B., Ph.D., University of California, Berkeley, 1985. Experimental astrophysics; observational cosmology.

Quinn, Brian P., Ph.D., MIT, 1984. Experimental medium energy/nuclear physics; production and interaction of strange hadrons; strange sea quarks in the nucleon.

Rothstein, Ira Z., Ph.D., University of Maryland at College Park, 1992. Theoretical particle physics and cosmology; LHC theory; gravity waves.

Russ, James S., Ph.D., Princeton, 1966. Experimental high-energy physics; CDF at Fermilab; particle astrophysics.

Schumacher, Reinhard A., Ph.D., MIT, 1983. Experimental medium-energy/nuclear physics; production and interactions of strange hadrons; strange sea quarks in the nucleon.

Sekerka, Robert F., Ph.D., Harvard, 1965. University Professor of Physics. Theoretical condensed matter physics; problems in materials science.

Suter, Robert M., Ph.D., Clark, 1978. Experimental condensed matter physics; x-ray and neutron scattering studies.

Swendsen, Robert H., Ph.D., Pennsylvania, 1971. Theoretical condensed matter physics; computer simulations; statistical mechanics of phase transitions and biological molecules.

Tristram-Nagle, Stephanie, Ph.D., University of California, Berkeley, 1981. Experimental biophysics; membrane biophysics.

Vogel, Helmut, Ph.D., University of Erlangen, Germany, 1979. Experimental high-energy physics; CMS at CERN.

Widom, Michael, Ph.D., University of Chicago, 1983. Theoretical condensed matter physics; quasicrystals, ferrofluids, DNA.

Associate Professors

Briere, Roy A., Ph.D., University of Chicago, 1995. Experimental high-energy physics; BES at Beijing.

Croft, Rupert, Ph.D., Oxford University, 1995. Theoretical astrophysics/cosmology; simulations of the evolution of the universe.

Deserno, Markus, Ph.D., University of Mainz, Germany, 2000. Theoretical condensed matter and biophysics; membrane structure and properties.

Di Matteo, Tiziana, Ph.D., University of Cambridge, 1998. Theoretical astrophysics/cosmology; cosmological simulations.

Evilevitch, Alex, Ph.D., Lund University, Sweden, 2001. Experimental biological physics; physics of viruses.

Assistant Professors

Trac, Hy, Ph.D., University of Toronto, 2004. Theoretical astrophysics/cosmology; evolution of the dark matter, baryons, and stars.

Woods, Kristina, Ph.D., Stanford, 2004. Experimental biological physics; terahertz probes of fast molecular conformational dynamics.

Emeriti

Berger, Luc, Ph.D., Lausanne, Switzerland, 1960. Experimental and theoretical condensed matter physics; studies of metallic ferromagnets.

Edelstein, Richard M., Ph.D., Columbia, 1960. Experimental high-energy physics; dynamics of strong interactions.

Engler, Arnold, Ph.D., Berne, Switzerland, 1953. Experimental high-energy physics; colliding beams techniques.

Fetkovich, John G., Ph.D., Carnegie Mellon, 1959. Special Assistant to the President for Academic Affairs.

Kisslinger, Leonard S., Ph.D., Indiana, 1956. Theoretical nuclear and particle physics; nonperturbative QCD.

Kraemer, Robert W., Ph.D., Johns Hopkins, 1962. Experimental high-energy physics; colliding beams techniques.

Rayne, John A., Ph.D., Chicago, 1954. Experimental condensed matter physics; electronic and magnetic properties of metals and alloys; ultrasonic absorption in solids.

Schumacher, Robert T., Ph.D., Illinois, 1955. Musical acoustics; magnetic resonance in solids.

Wolfenstein, Lincoln, Ph.D., Chicago, 1949. University Professor of Physics. Theoretical elementary particle physics; weak interactions and symmetry principles.

Vander Ven, Ned S., Ph.D., Princeton, 1962. Experimental condensed matter physics; electron and nuclear spin resonance in solids.

Young, Hugh D., Ph.D., Carnegie Mellon, 1959. Physics education.

Faculty by Courtesy

Anna, Shelley, Ph.D., Harvard, 2000. Dynamic of soft matter; fluid mechanics.

Greve, David, Ph.D., Lehigh, 1979. Physics and development of novel sensors.

Islam, Mohammad, Ph.D., Lehigh, 2000. Structure, dynamics and self-assembly of soft matter; properties of nanoscale structures.

Maloney, Craig, Ph.D., University of California, Santa Barbara, 2005. Mechanical response of solid-like materials.

Mandal, Maumita, Ph.D., Hyderobad, India, 2004. RNA structure and conformational rearrangements.

McHenry, Michael, Ph.D., MIT, 1988. Magnetic properties of materials.

Zhu, Jian-Gang, Ph.D., University of California, San Diego, 1983 and 1989. Magnetic data storage technologies.

RESEARCH SPECIALTIES AND STAFF

For details of our research activities see www.cmu.edu/physics

Experimental

Astrophysics. Astrophysics research is integrated within the Bruce and Astrid McWilliams Center for Cosmology which brings together astrophysicists, particle physicists, computer scientists statisticians to advance our understanding of dark matter and dark energy which dominate the Universe. Observational cosmology; measurements of anisotropy in the 2.7 K cosmic microwave background radiation at millimeter wavelengths; 21 cm hydrogen radiation cosmology; development of high-sensitivity receivers at millimeter wavelengths; studies of galactic clustering and gravitational lensing; early evolution of galaxies and the universe, using data drom the

Hubble Space Telescope; study of clusters of galaxies, using x-ray and optical observation. CMU is a participating member in the South African Large Telescope, the Sloan Digital Sky Survey and the Large Synoptic Survey Telescope. R. E. Griffiths, J. Peterson.

Biological Physics. Structure and function of biomembranes; NMR studies of the structure of proteins and optical microscopic studies of cell structure; molecular origins of Alzheimer's disease; biofluid mechanics of lung airways; biopolymer dynamics; physics of viruses. A. Evilevitch, S. Garoff, M. Lösche, J. Nagle, S. Tristram-Nagle, K. Woods. Courtesy faculty: M. Mandal.

Condensed Matter and Applied Physics. Properties and applications of nanoparticles and nanostructures; structure of thin organic and metal solid films; structure and properties of liquid/solid interfaces; wetting of fluids on solids; structure of semiconductor and metal surfaces; influence of surface properties on semiconductor devices; magnetic films for data storage; x-ray scattering from thin films and surfaces; x-ray microscopy for characterization of grain structure and growth in metals; microfluidics; interfacial fluid mechanics; properties and application of nanotubes and nanorods. Many of these activites are carried out in active collaboration with other departments, institutes and centers in the science and engineering colleges. R. Feenstra, S. Garoff, S. Majetich, R. Suter, K. Woods. Courtesy faculty: S. Anna, D. Greve, M. Islam, M. McHenry, J. Zhu.

High-Energy Physics. Study of heavy quark production and decay properties with CDF at Fermilab including matter-antimatter oscillations, CP violations and quarkonia spin alignment measurements. Studies of the properties of charm quarks using the CLEO detector at Cornell and the BES facility in Beijing, China. Operation and physics with the CMS detector at the LHC collider at CERN. Ultra high energy cosmic ray neutrinos. R. Briere, T. Ferguson, M. Paulini, J. Russ, H. Vogel.

Medium-Energy/Nuclear Physics. Strong QCD; the spectrum of excited baryons; gluonic excitations of mesons and quark confinement; fundamental form factors of the proton and neutron; strangeness content of the nucleon; electromagnetic interactions with hadronic systems. G. Franklin, C. Meyer, B. Quinn, R. A. Schumacher.

Theoretical

Astrophysics. High reolution, large scale computational studies of structure formation in the universe; evolution of supermassive galaxies and black holes; nature of dark energy and dark matter. Analaysis of data from Sloan Digital Sky Survey and Large Synoptic Survey Telescope. R. Croft, T. Di Matteo, H. Trac.

Biological Physics. Theoretical analysis of biomembranes; Monte Carlo simulations of proteins; elastic continuum theory and differential geometry of fluid membranes, statistical physics and coarse-grained molecular dynamics simulations of membranes and peptides; structure of viruses and nucleic acids. M. Deserno, J. Nagle, R. Swendsen, M. Widom.

Computational Physics. Computational physics at Carnegie Mellon is an umbrella that encompasses a rapidly growing and highly interdisciplinary set of activites that are taking place in all areas of the Department. R. Croft, M. Deserno, T. Di Matteo, M. Levine, C. Meyer, C. Morningstar, M. Paulini, R. Sekerka, R. Swendsen, H. Trac, M. Widom.

Condensed Matter. Nonlinear analysis and simulation of hydrodynamics and crystal growth; Monte Carlo studies of complex fluids, biological molecules, disordered solids and phase transitions; modeling of quasicrystals, ferroelectrics and incommensurate phases. J. Nagle, R. Sekerka, R. Swendsen, M. Widom. Courtesy faculty: C. Maloney, J. Zhu.

High-Energy Physics. Quantum gauge field theories and their applications to experiments; weak interaction phenomenology; CP violation; heavy quark phsyics; inflationary universe dynamics; topological defects and their applications in cosmology; gravity wave physics; LHC phenomenology. F. Gilman, R. Holman, I. Z. Rothstein, L. Wolfenstein.

Medium-Energy/Nuclear Physics. Strong and weak nuclear force; formation of hadrons, confinement, exotic forms of matter; Markov-chain and Monte Carlo computation of QCD; lattice gauge theory; QCD sum rules. L. Kisslinger, C. Morningstar.

Quantum Foundations and Qunatum Information Theory. Reformulation of quantum theory using consistent histories and decoherence and application of quantum mechanics in computing. R. B. Griffiths.

DREXEL UNIVERSITY

DEPARTMENT OF PHYSICS

Philadelphia, Pennsylvania 19104

Students Accepted For Degree	FIELDS		
	Physics	Astronomy	Related Fields
Doctorate	X		
Master's	X		

1. General

President: John A. Fry
Dean of the College of Arts and Sciences: Donna Murasko
Department Head: Michel Vallières
Department Telephone Number: (215) 895-2708
Type of Institution: University
Control: Private
Setting: Urban
Total Faculty: 1,500
Total Students: 22,493
Total Graduate Students: 9,009
Annual Graduate Tuition:
All Graduate Students: $915/cr. hr.
Tuition rates for: 2010–11
Deferred tuition plan: Yes
General Fees: $240 full-time
$120 part-time
Term: Quarter

2. Number of Faculty in Department

The combined total of full-time faculty in the three professorial ranks is 21. The combined total of full-time, part-time, and other faculty at all ranks is 36.

3. Admission, Financial Aid, and Housing

Address admission inquiries to: Office of Graduate Admissions, enroll@drexel.edu
Graduate application fee required: $75 (waived for online application)
Admission deadline (Fall admission): 2/1 for Ph.D. program with assistantship support (TA/RA). 9/1 for MS program only (no financial support available)
Admission information: For fall admission, 2010–11, 7 students were accepted with full financial support from 135 applicants.
Admission requirements: For admission to the graduate programs, a Bachelor's degree in an approved program is required with a minimum undergraduate GPA of 3.0/4.0 specified. The GRE general tests is required for financial aid. Students from non-English speaking countries are required to demonstrate proficiency in English via the TOEFL exam. Minimum acceptable score for admission is 550, (80 IBT, 213 CBT) but opportunities for financial aid are greater for TOEFL scores near or above 600 (100 IBT, 250 CBT). Teaching assistants educated in non-English-speaking countries must complete a special English program.
Undergraduate preparation assumed: Advanced undergraduate coursework in classical mechanics, electromagnetism, statistical physics, and quantum mechanics. Mathematics coursework in differential equations and linear algebra.
Address program questions and financial aid inquiries to: Dr. Michael Vogeley, Director of Graduate Studies, Department of Physics, physics@drexel.edu
GAPSFAS application required: No
Loans available: No
Address housing inquiries to: Office of Residential Living
On-campus, single student housing available: Yes
Cost/quarter: $2,975 (undergraduate priority)
On-campus, married student housing available: No
Off-campus: $695/month (average) $425-1,124/month

Table A—Faculty, Enrollments, and Degrees Granted

Research Specialty	2009–10 Faculty	Enrollment[1] Fall 2009		No. of Degrees Granted[2] 2009–10 (2005–09)			Median No. of Years for 2009–10 Ph.D.'s
		Master's	Doctorate	Master's	Terminal Master's	Doctorate	
Astrophysics	4	0	12	2(3)	0(2)	2(1)	6
Biophysics	5	0	11	0(7)	0(3)	0(5)	–
Condensed Matter Physics	4	0	3	0(1)	0(3)	0(0)	–
Particles & Fields	4	0	5	0(1)	0(1)	0(1)	–
Other Theoretical/ Math.	2	0	7	0(5)	0(0)	1(2)	–
Non-specialized	2	0	0	0(0)	0(0)	0(0)	7
Total		0	38	2(17)	0(9)	3(9)	
Full-time Grad. Stud.			36				
Part-time Grad. Stud.			2				
First-year Grad. Stud.			7				
Median Years in Grad. Study (2009–10 Degrees)				2	2	6	6
Undergraduate Degrees, 2009–10 (2005–09):15(36)							

[1]Students not yet committed to a research specialty are entered under non-specialized.
[2]Five-year totals in parentheses.

4. Graduate Degree Requirements

Master's: The requirement for the Master's degree is 45 quarter credits in an approved program. The student is required to maintain at least a 3.0 GPA. There is no thesis or foreign language requirement for the M.S. degree. There is no specific residence requirement for the M.S. degree. There are no examinations required for the M.S. degree.
Doctorate: In addition to required graduate-level coursework in physics, the successful Ph.D. candidate must (a) pass the Ph.D. candidacy examinations, both written and oral; and (b) perform original research, write a satisfactory thesis describing that research, and defend the thesis in an oral examination. There is no foreign language requirement.
Thesis: Thesis may be written *in absentia*.
Special Equipment, Facilities, or Programs: Students in the Graduate program are able to access a diverse range of experimental facilities including:
(1) Astrophysics Facilities: Numerical Astrophysics Facility, primarily networked LINUX and Mac OS X workstations emphasizes theoretical and numerical studies of stars, star clusters, the early Universe, galaxy distributions, cosmology modeling and gravitational lensing. The facility also employs special purpose high performance computers, such as the Gravity Pipeline Engine (GRAPE), a new Beowulf cluster

(128 processors, 128G RAM, 2TB RAID disk) and a system using Graphics Processing Units to achieve computational speeds of up to a trillion floating point operations per second. The Joseph R. Lynch Observatory houses a 16 inch Mead Schmidt-Cassegrain telescope equipped with SBIG CCD camera. Drexel is a participant in the Sloan Digital Sky Survey, which operates a 2.5m telescope at Apache Point, NM and the Large Synoptic Survey Telescope to be built in Chile (first light 2014).

(2) *Biophysics Facilities*: (a) Modulated excitation kinetics laboratory uses frequency domain techniques to follow internal dynamics of biological molecules. (b) Spatially resolved kinetics laboratory uses simultaneously resolved spatio-temporal data at microscopic resolution to follow biological self-assembly processes, such as polymerization of sickle hemoglobin. (c) Atomic Force Microscope (AFM) facility to study the structure and interaction of macromolecule via imaging, and to investigate the mechanical and kinetic properties of individual protein molecules via nanomanipulation. (d) Computational Biophysics facility including two Beowulf clusters (44-node dual-core Xeon, 43-node dual quad-core Xeon [344 cores]), 24TB RAID disk server, and ten linux workstations connected through a gibabit network. (e) Preparative laboratory provides facilities for biological sample purification and characterization.

(3) *Condensed Matter Facilities*: Ultra-low temperature laboratory has a dilution refrigerator, ^{3}He and ^{4}He cryostats and microwave sources to study quantum phenomena in nano- and microscale devices, superconducting qubits, nanostructrures and quantum fluids and solids.

(a) Magnetic material: laboratory conducts research on amorphous magnetic thin films, fiber optical sensors, (b) Surface science laboratory has scanning probe microscopy to study surface structure interfaces at the atomic level.

(4) *Particle Physics Facilities*: Detector development laboratory provides experimental support for an international research program in nonaccelerator particle and nuclear physics performing tests of invariance principles and conservation laws, and neutrino oscillations.

(5) *General Support Facilities*: Include an electronics shop capable of custom design and fabrication of electronics and computer components, and a machine shop to assist in the design, construction, and repair of mechanical component.

Table B—Appointments to Graduate Students, 2009–10

Title of Appointee	Appointments Total	Appointments First-year	Academic Load Allowed in Credit Hours	Hours of Service Per Week	Stipend for Calendar Year ($)
			Quarter		
Teaching Assistant[1]	24	7	9	20	21,000
Research Assistant[1]	11	0	9	20	21,000–24,000
Self-supported	3	0	9	–	
Total	38	7			

[1]Plus tuition remission and health insurance.

Table C—Separately Budgeted Research Expenditures

Research Specialty	No. of Grants	Expenditures* ($) Fiscal Year 2008–09
Astrophysics	11	1,280,000
Biophysics	4	646,000
Condensed Matter	1	40,000
Particles & Fields	2	214,000
Nonlinear Dynamics	1	13,000
Total	19	2,193,000

*actual expenditures, not indirect costs.

FACULTY

Professors

Bose, Shymalendu, Ph.D., Maryland, 1967. Fullerenes and carbon nanotubes; superconductivity; x-ray and electron spectroscopies of solids

DiNardo, N. John, Ph.D., Pennsylvania, 1982. Studies of surfaces and interfacial phenomena in solids.

Ferrone, Frank, Ph.D., Princeton, 1974. Experimental and theoretical protein dynamics; kinetics of biological self-assembly. Sickle Hemoglobin.

Finegold, Leonard X., Ph.D., London, 1959. Biophysics; granular physics.

Gilmore, Robert, Ph.D., MIT, 1967. Analysis of data from chaotic dynamical systems; applications of group theory to problems in atomic, molecular, nuclear, and solid state physics.

House, Frederick B., Ph.D., Wisconsin, 1965. Satellite meteorology; Earth energy budget.

Lane, Charles C., Ph.D., Cal. Tech., 1987. Nonaccelerator-based particle physics. Solar neutrinos and neutrino oscillations (Projects CHOOZ and KamLAND).

Lim, Tech-Kah, Ph.D., Adelaide, 1968. Physics education.

McMillan, Stephen L. W., Ph.D., Harvard, 1983. Stellar dynamics; large-scale computations of stellar systems.

Steinberg, Richard I., Ph.D., Yale, 1969. Experimental tests of invariance principles and conservation laws; solar neutrinos and neutrino oscillations (Project CHOOZ).

Tyagi, Somdev, Ph.D., Brigham Young, 1976. Physics of high-temperature superconductivity; magnetic properties of thin-sputtered films of amorphous metallic alloys; fiber optical sensors giant magnetoresitive (GMR) materials.

Vallières, Michel, Ph.D., Pennsylvania, 1972. Department Head. Large-scale shell-model calculations; computer architecture for nuclear physics problems.

Vogeley, Michael S., Ph.D., Harvard, 1993. Cosmology; Sloan Digital Sky Survey; formation of structure in the universe.

Yuan, Jian-Min, Ph.D., Chicago, 1973. Theoretical and computational biophysics, biological pathways and networks, protein folding and stability, protein aggregation, systems biology, and nonlinear dynamics.

Associate Professors

Cruz Cruz, Luis, Ph.D., MIT, 1994. Molecular dynamics of proteins; Spatial correlations; Cellular automata.

Goldberg, David M., Ph.D., Princeton, 2000. Gravitational lensing; cosmic microwave background; cosmology; computational physics.

Richards, Gordon, Ph.D., Chicago, 2000. Quasars, Quasars Absorption Lines, Gravitational Lensing, Galaxy Evolution, Sloan Digital Sky Survey.

Urbanc, Brigita, Ph.D., Ljubljana. 1994. Slovenia. Computational biophysics; Protein folding and assembly.

Yang, Guoliang, Ph.D., Southern Illinois, 1992. Atomic force microscope (AFM) study of single protein molecules; elastic properties of individual DNA and protein molecules.

Assistant Professors

Maricic, Jelena, Ph.D., Hawaii, 2005. Neutrino Oscillations, Geoneutrinos, Solar Neutrinos, and Neutrino Applications.

Ramos, Roberto, Ph.D., Washington, 1999. Low temperature condensed matter physics, quantum computing, nanoscience, quantum fluids and solids.

Teaching Faculty

Trout, Joseph, Ph.D., Philadelphia, 1998. In charge of Instructions and Laboratories.

Research Faculty

Allred, Joel, Ph.D., Washington, 2005. Simulations of solar and stellar flares.

Aprelev, Alexey, Ph.D., St. Petersburg, 1995. Experimental Biophysics.

MacNeice, Peter, Ph.D., Cambridge, 1994. Solar Physics; Magnetohydrodynamics; high-performance computing; parallel adaptive mesh refinement techniques.

Olson, Kevin, Ph.D., Massachusetts. Development of parallel, numerical algorithms for Astrophysics applications.

Vesperini, Enrico, Ph.D., Pisa, 1994. Evolution of galaxies and globular clusters, numerical simulations.

Postdoctoral Fellows

Barz, Bogdan, Ph.D., Columbia, 2009. Molecular dynamics of proteins.

Jampani, Srinivasa, Ph.D., Hyderabad, 2010. Computational biophysics.

Rajesh, Deo, Ph.D., Georgia, 2007. Multiwavelength studies of AGNs using the Sloan Digital Sky Survey and Spitzer Space telescopes.

Sereda, Yuri, Ph.D., Kharkiv, 2001. Molecular dynamics of proteins.

Zbiri, Karim, Ph.D., Nantes, 2007. Nuclear Physics, heavy ions collisions and Monte Carlo simulations.

Visiting Faculty

Spicer, Daniel, Ph.D., Maryland, 1976. Space and solar plasma physics, magnetohydrodynamics and numerical 3D.

RESEARCH SPECIALTIES AND FACULTY

Theoretical

Astrophysics. Goldberg, McMillan, MacNiece.
Biophysics. Cruz, Urbanc, Yuan.
Condensed Matter Theory. Bose.
Mathematical Physics. Gilmore.

Experimental

Astrophyics. Richards, Vogeley.
Biophysics. Ferrone, Yang.
Condensed Matter Physics. Ramos, Tyagi.
Nonaccelerator-based Particle Physics. Lane, Maricic.

GRADUATE PROGRAM HIGHLIGHTS

- Begin research in the first year with freedom to explore different areas of physics before choosing a thesis topic.
- Participation by students in major world-wide research collaborations, including KamLAND, Double Chooz, the Sloan Digital Sky Survey, and the Large Synoptic Survey Telescope.
- Collaboration with researchers at nearby institutions including Princeton and U. Pennsylvania.
- All coursework and exam requirements finished by June of second year (with award of the M.S. in Physics along the way).
- An active tightly-knit community of graduate students that enjoys dinners and outings together.
- Physics Graduate Student Association run by our students and funded by the University.
- Weekly graduate student-only research seminars (free lunch!).
- Mentoring program that matches new students with current students.
- Interaction with world-renowned researchers who visit Drexel for our colloquium series and the annual Kaczmarczik Lecture, which has featured several Nobel laureates.

Our graduate program is a very important part of our department. We currently have 38 graduate students working in a wide range of areas of research. Students are strongly encouraged to begin research from the moment they arrive, and first and second year students do a number of small research projects. There are opportunities for research in a variety of specialized areas, and students are encouraged to "shop around" prior to their thesis. We also offer topical courses in areas of current research, including Astrophysics, Biophysics, Nanoscience, Nonlinear Dynamics, Particle Physics, and Solid State.

INDIANA UNIVERSITY OF PENNSYLVANIA

DEPARTMENT OF PHYSICS

Indiana, Pennsylvania 15705-1087

Students Accepted For Degree	FIELDS		
	Physics	Astronomy	Related Fields
Doctorate			
Master's	X		

1. General

President: Dr. Tony Atwater
Dean of Graduate School: Dr. Timothy P. Mack
Department Chairman: Dr. Devki Talwar
Department Telephone Number: (724) 357-2370
Type of Institution: University
Control: Public
Setting: Small town
Total Faculty: 678 (full-time)
Total Graduate Faculty: 492
Total Students: 14,638
Total Graduate Students: 2,347
Annual Graduate Tuition:
In-state residents: Full-time—$6,666
Part-time—$370/credit
Out-of-state residents: Full-time—$10,666
Part-time—$593/credit
Tuition rates for: 2009–10
Deferred tuition plan: No
Other Fees: $1,625.90/year
Term: Semester

2. Number of Faculty in Department

The combined total of full-time faculty in the three professorial ranks is 11. The combined total of full-time, part-time, and other faculty at all ranks is 13.

3. Admission, Financial Aid, and Housing

Address admission inquiries to: Director of Graduate Studies, Department of Physics
Graduate application fee required: $40
Admission deadline (Fall admission): 8/15
Admission information: For fall admission, 2009–10, 4 full-time students were accepted.
Admission requirements: For admission to the graduate programs, a Bachelor's degree in physics or a related field is required from a college or university accredited by the Middle States Assoc. of Colleges and Secondary Schools or an equivalent regional accrediting agency, with a minimum undergraduate GPA of 2.6 specified. The GRE is required by the end of the first semester. The minimum acceptable score suggested for admission is verbal—400; quantitative—500; analytic—500; total—1,400. The GRE Advanced is not required. Students from non-English speaking countries are required to demonstrate proficiency in English via the TOEFL exam. Minimum acceptable score for admission is 540.
Address financial aid inquiries to: Dr. Timothy P. Mack, Dean, Graduate School
GAPSFAS application required: No
Financial aid deadline: 3/15
Loans available: Yes
Address housing inquiries to: Michael W. Lemaster, Director, Housing and Residence Life
On-campus, single student housing available: No
On-campus, married student housing available: No

Table A—Faculty, Enrollments, and Degrees Granted

Research Specialty	2009–10 Faculty	Enrollment[1] Fall 2009		No. of Degrees Granted[2] 2009–10 (2005–10)			Median No. of Years for 2009–10 Ph.D.'s
		Master's	Doctorate	Master's	Terminal Master's	Doctorate	
Applied Physics	3	1	N	1(2)	0(2)	–	–
Condensed Matter Physics	1	3	–	2(1)	0(3)	–	–
Low Temperature Physics	0	0	N	0(0)	0(0)	–	–
Materials Sci./ Metallurgy	3	1	–	0(2)	0(1)	–	–
Nuclear Physics	2	0	–	0(2)	0(1)	–	–
Physics Education	2	1	–	1(1)	0(1)	–	–
Non-specialized	0	0	–	–	–	–	–
Total		6	–	4(8)	0(8)	–	
Full-time Grad. Stud.		6	–				
Part-time Grad. Stud.		0	–				
First-year Grad. Stud.		4	–				
Median Years in Grad. Study (2009–10 Degrees)				2	2	–	–
Undergraduate Degrees, 2009–10 (2005–10): 4(26)							

[1]Students not yet committed to a research specialty are entered under non-specialized.
[2]Five-year totals in parentheses.

4. Graduate Degree Requirements

Master's: A minimum of 24 credit hours of course study is required with a 3.0 average. A thesis, minimum 6 credit hours, is required. Students must take a comprehensive exam. There is a minimum 1 year residency requirement.
Master of Arts: There are two programs possible. One involves 24 or 27 credit hours of course work with a 3 or 6 credit hour thesis. The second involves no thesis with 33 hours course work.
Thesis: Thesis may be written *in absentia*.
Special Equipment, Facilities, or Programs: Available research facilities include a class 100 clean room, a Magnetism Laboratory, and a materials fabrication and characterization facility (e-beam, sputtering and thermal evaporation system; photo lithography; C-V, I-V, DLTS, and Hall measurement equipment; ellipsometer, FTIR, etc.) and a high resolution gamma ray spectrscopy facility. Support facilities include a machine shop and an electronic shop.

Table B—Appointments to Graduate Students, 2009–10

Title of Appointee	Appointments		Academic Load Allowed in Credit Hours	Hours of Service Per Week	Stipend for Academic Year ($)
	Total	First year			
			Semester		
Teaching Assistant	8	5		20	5,910
Total	8	5			

5. Personnel Engaged in Separately Budgeted Research, 7/07–6/08

Faculty	2
Grad Students	0
Total	2

6. Separately Budgeted Research Expenditures by Source of Support

	Departmental Research	Physics-related Research Outside Department
Federal Govt.	$34,000	
State Govt.	$15,000	
Total	$49,000	

Table C—Separately Budgeted Research Expenditures

Research Specialty	No. of Grants	Expenditures ($)
Condensed Matter Physics	1	15,000
Nuclear Physics	1	34,000
Total	2	49,000

FACULTY

Professors

Eck, John S., Ph.D., The John Hopkins University, 1967. Physics.

Freeman, W. Larry, Ph.D., Clemson, 1976. Theoretical solid state physics.

Haija, A. J., Ph.D., Pennsylvania State University, 1977. Condensed matter, exp.

Karimi, Majid, Ph.D., Oklahoma, 1984. Computational physics.

Numan, Muhammad, Ph.D., William and Mary, 1982. Semiconductor device physics.

Talwar, Devki N., Ph.D., Allahabad, 1976. Chairman of the Department. Condensed matter theory.

Wijekumar, V., Ph.D., Ohio State, 1985. Low-energy nuclear physics.

Associate Professors

Kenning, Gregory, Ph.D., Michigan St. 1988. Condensed matter, exp.

Sobolewski, Stanley J., Ph.D., SUNY at Buffalo, 1988. Educational research.

Zhou, Feng, Ph.D., Shanghai Institute of Optics and Fine Mechanics, 1989. Optics.

Assistant Professor

Freda, Ronald, M. S., Case Western Reserve Uni., 1973. Phys. Education.

RESEARCH SPECIALTIES AND STAFF

Experimental

Condensed Matter/Optics. Haija.
Magnetism. Kenning.
Materials Physics. Numan, Zhou.
Nuclear Physics. Low-energy nuclear reactions. Eck, Wijekumar.
Physics Education. Freda, Sobolewski.
Semiconductor Device Physics. Numan.
Solid State. Electro-optical device fabrication. Freeman.

Theoretical

Computational Physics. Karimi.
Heterojunction Device Modeling. Talwar.
Quantum Size Effects. Freeman.
Solid State. Electronic, vibrational, and magnetic properties of semiconductors; transport in and superlattices. Talwar.

LEHIGH UNIVERSITY

DEPARTMENT OF PHYSICS

Bethlehem, Pennsylvania 18015

Students Accepted For Degree	FIELDS		
	Physics	Astronomy	Related Fields
Doctorate	X	X	
Master's	X	X	

1. General

President: Alice P. Gast
Department Chairman: Volkmar Dierolf
Department Telephone Number: (610) 758-3930
Department web page: www.physics.lehigh.edu
Type of Institution: University
Control: Private
Setting: Small city
Total Faculty: 440 full-time
Total Graduate Faculty: 440
Total Students: 7,000
Total Graduate Students: 2,190
Annual Graduate Tuition:
All Graduate Students: $1,185/credit
Tuition rates for: 2010–11
Deferred tuition plan: Yes
Other Fees: None
Term: Semester

2. Number of Faculty in Department

The combined total of full-time faculty in the three professorial ranks is 20. The combined total of full-time, part-time, and other faculty at all ranks is 29.

3. Admission, Financial Aid, and Housing

Address admission inquiries to: Graduate Admissions Officer, Lehigh University, Dept. of Physics, 16 Memorial Drive, East, Bethlehem, PA 18015. E-mail: LG00@lehigh.edu
Graduate application fee required: $75
Admission deadline (Fall admission): March 15
Admission information: For fall admission, 2009–10, 15 students were accepted from 98 applicants.
Admission requirements: For admission to the graduate programs, a Bachelor's degree in physics or a related field is required with a minimum undergraduate GPA of 3.0. The GRE is required. The minimum acceptable score suggested for admission is verbal–550; quantitative–650; total–1,200. The GRE Advanced is required. The minimum acceptable score suggested for admission is 600. Students from non-English speaking countries are required to demonstrate proficiency in English via the TOEFL exam and the University SPEAK test for teaching assistants. Applicants are normally expected to have minimum TOEFL scores as follows: 600 for paper based; 213 for computer based (250 preferred); 85 for iBT composite (25 writing, 24 speaking, 21 reading, 15 listening).
Undergraduate preparation assumed: Intermediate mechanics, electricity and magnetism, atomic and quantum physics, thermodynamics, and laboratory experience. Mathematics through partial differential equations. Typical texts include Symon, Griffiths, Eisberg and Resnick, Merzbacher, Reif, and Van Ness. Able students with inadequate preparation may take a limited number of remedial courses after enrollment.
Address financial aid inquiries to: Graduate Admissions Officer, Lehigh University, Dept. of Physics, 16 Memorial Drive, East, Bethlehem, PA 18015
GAPSFAS application required: No
Financial aid deadline: 3/15
Loans available: No
Address housing inquiries to: Residence Operations Office, Lehigh University, Rathbone Hall, 63 University Drive, Bethlehem, PA 18015
On-campus, single student housing available: Yes
Cost/month: $595*, efficiency $505*
On-campus, married student housing available: Yes
Cost/month: $505*–670*

*Plus utilities

Table A—Faculty, Enrollments, and Degrees Granted

Research Specialty	2009–10 Faculty[3]	Enrollment[1] Fall 2009 Master's	Enrollment[1] Fall 2009 Doctorate	No. of Degrees Granted[2] 2009–10 (2005–10) Master's	Terminal Master's	Doctorate	Median No. of Years for 2009–10 Ph.D.'s
Astrophysics & Astronomy	3	0	4	2(4)	0(0)	0(3)	
Atomic, Molecular, & Optical Physics	2	0	5	0(4)	0(1)	1(5)	5
BioPhysics	1	0	4	3(5)	0(2)	0(3)	
Condensed Matter Physics	6	0	10	5(10)	0(1)	1(8)	5
Nano-science	1	0	2	0(1)	0(0)	0(1)	
Optics	2	0	9	4(9)	0(1)	1(4)	6
Plasma Physics & Fusion	4	0	2	1(2)	0(0)	0(2)	
Polymer Physics/ Science	1	0	3	2(3)	0(0)	1(1)	5
Statistical & Thermal	1	0	1	0(2)	0(1)	0(1)	
Non-specialized	0	1	7	0(0)	0(0)	0(0)	
Total		1	47	17(40)	0(6)	4(28)	
Part-time Grad. Stud.		0	0				
First-year Grad. Stud.		1	8				
Median Years in Grad. Study (2009–10 Degrees)							5.3
Undergraduate Degrees, 2009–10 (2005–10): 8(39)							

[1]Students not yet committed to a research specialty are entered under non-specialized.
[2]Five-year totals in parentheses.
[3]Each faculty member is assigned to only one specialty in this chart. However, since many faculty work in multiple fields, these numbers underrepresent activity in several areas.

4. Graduate Degree Requirements

Master's: 30 credit hours required, including a research project. No minimum grade average, but more than four grades below B cause a student to become ineligible for further graduate work. All work for M.S. must be done in residence at Lehigh. No foreign language requirement, and no requirement for comprehensive/qualifying exam or thesis.
Doctorate: 9 credits of course work beyond M.S. required, but Ph.D. programs usually include 20 or more credits beyond M.S. Minimum time requirement of two years, at least one

year in residence. Qualifying exam, general exam, and thesis defense required. No departmental or university language requirements.

Other Programs: The additional interdepartmental areas of research include materials science, surface science, photonics, and geophysics.

Thesis: Thesis may be written *in absentia*.

Special Equipment, Facilities, or Programs: Research facilities are housed in the Sherman Fairchild Center for the Physical Sciences, containing Lewis Laboratory, the Sherman Fairchild Laboratory for Solid State Studies, and a large connecting research wing. Well-equipped laboratory facilities are available for experimental investigations in research areas at the frontiers of physics. Instruments used for experimental studies include a wide variety of laser systems ranging from femtosecond and picosecond pulsed lasers to stabilized single-mode cw Ti-sapphire and dye lasers. There is also a Fourier-transform spectrometer, cryogenic equipment that achieves temperatures as low as 0.05 K and magnetic fields up to 9 Telsa, a facility for luminescence microscopy, and a laser-tweezers system for studies of complex fluids. A 3MeV van de Graaff accelerator is used to study radiation-produced defects in solids. The Fairchild Laboratory also contains a processing laboratory where advanced Si devices can be fabricated and studied. All laboratories are well furnished with electronic instrumentation for data acquisition and analysis. Several professors are members of the interdisciplinary Center for Optical Technologies that offers a wide range of state-of-the-art facilities including a fiber drawing tower, waveguide and fiber characterization labs, and a new epitaxy facility for the growth of III-V semiconductor structures and devices. Extensive up-to-date computer facilities are available on campus and in the department. All computing resources can be accessed directly from graduate student and faculty offices through a high speed backbone. Researchers have access to the national Research Internet (Internet 2) via a 155 Mbps gateway.

Table B—Appointments to Graduate Students, 2009–10

Title of Appointee	Appointments Total	Appointments First-year	Academic Load Allowed in Credit Hours	Hours of Service Per Week	Stipend for Calendar Year ($)
			Semester		
Teaching Assistant	22	5	9	16	22,860–23,540
Research Assistant	22	0	9	20	22,860–23,540
Fellowship	3	3	12	0	23,300–32,000
Total	47	8			

Stipends are for 12 months

5. Personnel Engaged in Separately Budgeted Research, 7/09–6/10

Professorial faculty	21
Postdoctoral appointments	16
Graduate students	47
Undergraduate students	20
Total	104

6. Separately Budgeted Research Expenditures by Source of Support

	Departmental Research	Physics-related Research Outside Department
Federal government	$3,056,670	$2,205,240
Institutional and State	410,840	873,000
Total	$3,467,510	$3,078,240

Table C—Separately Budgeted Research Expenditures

Research Specialty	No. of Grants	Expenditures ($)
Astronomy & Astrophysics	3	$124,200
Atomic, Molecular, & Optical Physics	2	161,160
Biophysics	6	690,880
Condensed Matter Physics	10	613,680
Plasma Physics & Fusion	10	638,870
Statistical & Thermal Physics	2	202,710
Applied Optics	8	742,280
Nanosciences	5	293,730
Total	46	3,467,510

FACULTY

Professors

Biaggio, Ivan, Ph.D., ETH-Zurich, 1993. Experimental condensed matter; non-linear optics.

DeLeo, Gary G., Ph.D., Connecticut, 1979. Associate Chairperson. Astrophysics.

Dierolf, Volkmar, Ph.D., Utah, 1992. Chairperson. Experimental condensed matter; optical spectroscopy and microscopy.

Folk, Robert T., Ph.D., Lehigh, 1958. Theory of very light nuclei; elastic properties of solids.

Fowler, W. Beall, Ph.D., Rochester, 1963. Emeritus. Theory of electronic and optical properties of nonmetallic solids.

Gunton, James D., Stanford, 1967. Joseph A. Waldschmitt Professor. Condensed matter theory.

Hickman, A. Peet, Ph.D., Rice, 1973. Theoretical atomic, molecular, and optical physics.

Huennekens, John P., Ph.D., Colorado, 1982. Atomic and laser physics; molecular spectroscopy; atomic collisions; nonlinear optics.

Kanofsky, Alvin S., Ph.D., Pennsylvania, 1966. High-energy experimental physics.

Kim, Yong W., Ph.D., Michigan, 1968. Atomic physics; statistical physics.

Koch, Thomas L., Ph.D., Cal-Tech., 1982. Director, Center for Optical Technologies. Optoelectronics; photonics.

Kritz, Arnold H., Ph.D., Yale 1961. Theoretical plasma physics.

McCluskey, George E., Jr., Ph.D., Pennsylvania, 1965. Astronomy, astrophysics.

Ou-Yang, H. Daniel, Ph.D., UCLA, 1985. Soft condensed matter and biophysics.

Rickman, Jeffrey M., Ph.D., Carnegie-Mellon, 1989. Solid state theory.

Stavola, Michael J., Ph.D., Rochester, 1980. Shermain Fairchild Professor and Associate Dean, College of Arts and Sciences. Vibrational spectroscopy; defects in semiconductors.

Toulouse, Jean, Ph.D., Columbia, 1981. Nonlinear fiber optics, photonics; dielectric, Raman, and neutron scattering studies of disordered ferroelectrics.

Associate Professors

Licini, Jerome C., Ph.D., MIT, 1987. Low-temperature quantum transport phenomena in Si and GaAs devices, and nanotechnology.
Shaffer, Russell A., Ph.D., Johns Hopkins, 1962. Emeritus. Theory of elementary particles.

Assistant Professors

McSwain, M. Virginia, Ph.D., Georgia State, 2004. Astrophysics.
Rotkin, Slava, Ph.D., Ioffe Inst., St. Petersburg, 1997. Nanoscience, condensed matter theory.
Vavylonis, Dimitrius, Ph.D., Columbia, 2000. Biophysics.

Research Scientists

Bateman, Glenn, Ph.D., Princeton, 1970. Plasma physics.
Pankin, Alexei, Ph.D., Kiev, 1998. Plasma physics.

Visiting Research Scientists

Abou, Berengere, Ph.D., Universite Paris VI, 1998. Microrheology.
Chakrabarti, Amit, Ph.D., Univ. of Minnesota, 1987. Statistical physics.
Ge, Wenping, Ph.D., Shanghai Jiaotong Univ., 2003. Non-linear optics.
Ha, Chungil, M. S., Pusan National Univ., 2006. Optics.
Huang, Chien-Hua, M. S., National Cheng Kung Univ., 2004. Optics.
Iolin, Eugene, Ph.D., Estonian Acad. Sci., 1978. Condensed matter.
Lin, Aoxiang, Ph.D., Gwangju Inst. of Sci. & Tech., 2008. Condensed matter.
Liu, Ya, Ph.D., Brandeis Univ., 2009. Statistical physics.
Lopez, Antonio Perez, Ph.D., Universitat de les Illes Balears, 2009. Statistical physics.
Narendran, Manikandan, Ph.D., Indian Inst. Sci., 2007. Optics.
Rafiq, Tariq, Ph.D., Chalmers Univ. of Tech., 2004. Plasma physics.
Ryan, Gillian, M.S., Dalhousie Univ., 2006. Biophysics.
Ryasnyanskiy, Alek, Ph.D., Uzbek Acad. Sci., 2002. Non-linear optics.
Veksler, Alex, Ph.D., Ben-Gurion Univ., 2005. Biophysics.
Yusuf, Eddy, Ph.D., Florida State Univ., 2005. Biophysics.
Zhou, Liangcheng, Ph.D., Lehigh Univ., 2010. Nanophotonics.

Adjunt Faculty

Cereghetti, Paola, Ph.D., Swiss Federal Inst. of Tech., 2000. Physics.
Glueckstein, Jon, Ph.D., Wisconsin, 1997. Physics.
Loomis, John, M. S., Univ. of Massachusetts, 1973. Astronomy.
Lucic, Dragan, M. S., Univ. of Colorado, 1997. Physics.
Veltchev, Iavor, Ph.D., Vrije Universiteit Amsterdam, 2001. Physics.

RESEARCH SPECIALTIES AND STAFF

Theoretical

Astrophysics. Ultraviolet spectroscopy and gas dynamics of interacting binary systems; orbits of binary stars; N-body dynamics. DeLeo and McCluskey.
Atomic, Molecular, and Optical Physics. Charge exchange collisions; fine-structure changing collisions; optical processes in gases; molecular hyperfine spectroscopy. Hickman.
Biophysics. Physical and engineering principles involved in the assembly of actin proteins into filaments and larger scale structures. Statistical mechanics and soft matter physics applied to actin protein assemblies and the emergent collective properties. Vavylonis.
Elasticity. Solution of integral equations for mixed boundary value problems. Folk.
Physics of nano- and molecular scale systems. Quantum mechanics of one-dimensional systems, many-body effects in carbon nanotubes, physics of nanotube/nanowire devices, electron transport in molecular systems, modeling of interaction between nano- and biological systems. Rotkin.
Plasma Physics. Integrated modeling codes are developed and used to predict temperature, momentum and density profiles in magnetically confined controlled fusion plasma experiments. Theoretically derived physics models are developed for use in these codes and detailed comparisons are made between our simulations and experimental data in order to understand the physics of transport and confinement in plasmas. There are active collaborations with theory and experimental groups, both nationally and internationally. Bateman, Kritz, Pankin.
Solid State Physics. Electronic and vibrational properties of defects in semiconductors and insulators. Fowler.
Statistical Physics. Pattern formation in nonlinear, non-equilibrium systems. Kinetics of first order phase transitions focusing on crystallization of globular proteins. Cell-cell communication via calcium oscillations. Gunton.

Experimental

Astronomy. Observational studies to understand the formation and evolution of stars. Particular areas of interest are young open clusters, binary stars, X-ray binaries and pulsars, the formation of disks in Be stars, and the origin of magnetic fields in massive stars. Lehigh has a significant amount of telescope access as a partner in the SMARTS Consortium. McSwain.
Atomic Physics. Collisional processes in atomic vapors including excitation transfer and "energy pooling," line-broadening, quenching, diffusion, resonance exchange and velocity-changing collisions; molecular spectroscopy of bound singlet and triplet states of alkali diatomics, photodissociation, predissociation, and bound-free emission. Huennekens.
Complex Fluids; soft condensed matter and biophysics; application of optical imaging, trapping, and manipulation for cell mechanics studies. Ou-Yang.
Electron attachment; short optical pulses; excitation transfer in gas discharges; optically assisted gas phase reactions. Kim.
Nanoscience. Quantum mechanical theory of carbon nanotubes, DNA nanotube hybrids; optics, optoelectronics & electronics of nanotube devices; nanotube-organics complexes. Rotkin.
Nonlinear Optics. Multiple orders of light-matter interactions. Time-resolved spectroscopy of second and third-order nonlinear optical effects in organic and inorganic materials. Optical frequency conversion and all-optical switching. Biaggio.
Photonics, Fiber Optics. Nonlinear effects in optical fibers and waveguides. Dierolf, Toulouse.
Physics of Fluids. Nonlinear dynamics in fluid systems; dynamics of small particle suspensions; light scattering loss spectroscopy; instabilities of interfaces. Kim.
Plasma Physics. Collisional and collisionless phenomena of very dense plasmas in or near a local thermodynamic equilibrium; anomalies in radiation transport properties; lowering of ion-

ization potentials in dense plasmas; laser-produced plasmas. Kim.

Solid State Physics. Quantum transport behavior of electrons, conduction in ultrasmall silicon MOSFETs and gallium-arsenide devices and carbon nanotubes at low temperature and high magnetic field. Licini.

Defects in semiconductors. Current interest is in defect complexes that contain light-element impurities such as H, C, O, and N. Vibrational spectroscopy and uniaxial stress techniques are used to elucidate microscopic properties. Stavola.

Raman and neutron scattering, dielectric and ultrasonic spectroscopies, collective vibrational dynamics of disordered ferroelectrics and glasses. Toulouse.

Point defects in insulating materials with ferroelectric domain walls and other dopants; optical spectroscopy under application of hydrostatic pressure, and magnetic fields; carrier localization in wide band gap semiconductors. Dierolf.

Charge transport in insulators and semiconductors; nonlinear optical spectroscopy. Biaggio.

Statistical Physics. Intrinsic fluctuations in fluids under external forcing, such as Brownian motion; chaotic transitions; light scattering from fractals; 1/f-dynamics of granular avalanches. Kim.

PENNSYLVANIA STATE UNIVERSITY

DEPARTMENT OF PHYSICS

University Park, Pennsylvania 16802

Students Accepted For Degree	FIELDS		
	Physics	Astronomy	Related Fields
Doctorate	X		
Master's	X		

1. General

President: Graham B. Spanier
Dean of Graduate School: Henry Foley
Department Head: Jayanth R. Banavar
Department Telephone Number: (814) 865-7533
Type of Institution: University
Control: Public
Setting: Small town
Total Faculty: 3,159**
Total Students: 44,832*
Total Graduate Students: 6,202*
Annual Graduate Tuition:
In-state residents: Full-time—$16,332/academic year
Part-time—$680/credit
Out-of-state residents: Full-time—$28,426/academic year
Part-time—$1,184/credit
Tuition rates for: 2009–10
Deferred tuition plan: Yes
Other Fees: Thesis fees: M.S.-$25; Ph.D.-$95
Term: Semester

*University Park Campus

2. Number of Faculty in Department

The combined total of full-time faculty in the three professorial ranks is 37. The combined total of full-time, part-time, and other faculty at all ranks is 48.

3. Admission, Financial Aid, and Housing

Address admission inquiries to: Dept. of Physics, 104 Davey Laboratory; graduate-admissions@phys.psu.edu
Graduate application fee required: $65 upon matriculation
Admission deadline (Fall admission): April 15
Admission information: For fall admission, 2009–10, 20 new students were enrolled.
Admission requirements: For admission to the graduate programs, a Bachelor's degree in physics or an allied field is required with no minimum undergraduate GPA specified, but a 3.0 average in junior-senior courses in physics and mathematics is required. The best-qualified applicants will be accepted up to the number of spaces available. The GRE is required. The minimum acceptable score required for admission is not fixed. The GRE Advanced is required. The minimum acceptable score required for admission is not fixed. The average GRE scores for 2009–10 admissions were quantitative–(91%); physics–(75%). Students from non-English speaking countries are required to demonstrate proficiency in English via the TOEFL exam. Minimum score required is 550 (paper) or 80 with a 19 on the speaking section (iBT).
Typical undergraduate preparation Minimum: Marion & Thornton, *Classical Dynamics of Particles and Systems*; Griffiths, *Introduction to Electrodynamics*; Griffiths, *Introduction to Quantum Mechanics*; Reif, *Fundamentals of Statistical and Thermal Physics*.
Address financial aid inquiries to: Graduate Admissions, Dept. of Physics, 104 Davey Laboratory
GAPSFAS application required: No
Financial aid deadline: 1/15
Loans available: Yes
Address housing inquiries to: Assignment Office, 201 Johnston Commons
On-campus, single student housing available: Yes
Cost/month: $720
On-campus, married student housing available: Yes
Cost/month: $900–1,165

Table A—Faculty, Enrollments, and Degrees Granted

Research Specialty	2008–09 Faculty	Enrollment[1] Fall 2009 Master's	Enrollment[1] Fall 2009 Doctorate	No. of Degrees Granted[2] 2009–10 (2005–10) Master's	Terminal Master's	Doctorate	Median No. of Years for 2009–10 Ph.D.'s
Acoustics	1	1	1	1(1)	0(0)	0(1)	5.6
Atomic, Molecular, & Optical Physics	6	0	13	0(0)	0(0)	0(3)	6.6
Biological	5	0	8	1(1)	0(1)	1(3)	6.0
Condensed Matter Physics	18	2	56	0(6)	2(5)	2(40)	5.6
Particles & Fields	12	0	20	0(1)	0(1)	2(9)	6.5
Relativity & Gravitation	5	0	19	0(1)	0(2)	2(15)	5.6
Other	2	1	0	0(0)	0(6)	0(0)	–
Total		4	117	1(9)	2(15)	7(71)	
Median Years in Grad. Study (2009–10 Degrees)				–	–	5.6	–

Undergraduate Degrees, 2009–10 (2005–10): 35(168)

[1]Students not yet committed to a research specialty are entered under other.
[2]Five-year totals in parentheses.

4. Graduate Degree Requirements

Master's: 30 credits (semester-equivalent), at least 18 at the graduate (not dual) level; 3.0 minimum average. No residence requirement. No foreign language. No examinations. Thesis or research paper may be submitted.
Doctorate: No fixed total credits; 21 graduate course credits specified, a minimum of 15 additional course credits required. Overall 3.0 average grade. Minimum of two semesters residence after candidacy. No foreign language. Candidacy examination during first year; comprehensive examination, usually during second year. Thesis required.
Other Programs: M.Ed.
Thesis: Thesis may be written *in absentia*.
Special Equipment, Facilities, or Programs: Several faculty members are affiliated with the Intercollege Materials Research Laboratory.

Table B—Appointments to Graduate Students, 2009–10

Title of Appointee	Appointments Total	Appointments First year	Academic Load Allowed in Credit Hours	Hours of Service Per Week	Stipend for Academic Year ($)
			Semester		
Teaching Assistant	52	19	9–12	20	15,705
Research Assistant	58	1	9–12	20	17,280
Fellowships	2	0	–	–	–
Self-supported	3	0	–	–	–
Traineeships	0	0			
Other	2	0			
Total	117	20			

5. Personnel Engaged in Separately Budgeted Research, 1/09–12/09

Professorial faculty	48
Postdoctoral appointments	29
Graduate students	65
Undergraduate students	40
Nonteaching research personnel	10
Total	192

6. Separately Budgeted Research Expenditures by Source of Support, 7/1/08–5/31/09

	Departmental Research
Federal government	$8,313,160
Private, nonprofit organizations	2,102,577
Total	$10,415,737

Table C—Separately Budgeted Research Expenditures

Research Specialty	No. of Grants	Expenditures ($)
Atomic, Molecular, & Optical Physics	8	1,374,245
Condensed Matter Physics	23	5,673,118
Gravity	5	709,633
HE	16	2,658,741
Total	52	10,415,737

FACULTY

Professors

Albert, Reka, Ph. D., Notre Dame, 2001. Statistical mechanics, network theory, systems biology.

Anderson, James, Ph. D., Princeton, 1963. Evan Pugh Professor. Quantum chemistry by Monte Carlo methods.

Ashtekar, Abhay V., Ph. D., Chicago, 1974. Eberly Professor. Gravity, geometry, and physics.

Banavar, Jayanth R., Ph. D., Pittsburgh, 1978. Distinguished Downsbrough Department Head. Condensed matter physics and statistical mechanics.

Castleman, A. W., Ph. D., Polytechnic Institute of New York, 1969. Eberly Distinguished Chair in Science. Laser photo physics, clusters, quantum confinement effects.

Chan, Moses H. W., Ph. D., Cornell, 1974. Evan Pugh Professor. Low-temperature physics; phase transitions in two and three dimensions; superfluid and liquid-vapor transitions in random media.

Cole, Milton W., Ph. D., Chicago, 1970. Distinguished Professor. Theoretical surface physics; superfluids; statistical mechanics.

Collins, John, Ph. D., Cambridge, 1975. Quantum field theory; perturbation effects in QCD; renormalization theory; neuroscience.

Coutu, Stephane, Ph.D., California Institute of Technology, 1993. High-energy cosmic-ray positrons and antiprotons as a possible sign for annihilating dark matter particles.

Cowen, Douglas, Ph. D., University of Wisconsin-Madison, 1990. Astrophysics, particles and fields.

Crespi, Vincent, Ph.D., University of California, 1994. Theory of superconducting, transport, electronic and structural/mechanical properties of novel materials.

Cutler, Paul H., Ph. D., Penn State, 1958. Emeritus. Cutler Professor. Theoretical surface physics; tunneling theory, electron emission in microstructures.

Diehl, Renee, Ph. D., Washington (Seattle), 1982. Surface physics; high-resolution LEED; surface defects; adsorbate structure and phase transitions.

Fichthorn, Kristen A., Ph. D., Michigan, 1989. Statistical mechanics and computer simulation.

Finn, Lee Samuel, Ph. D., Caltech, 1987. Gravitational wave astronomy; sources of gravitational radiation; relativistic astrophysics; numerical relativity.

Freed, Norman, Ph. D., Western Reserve, 1964. Associate Dean of the Eberly College of Science. Nuclear structure theory; intermediate-energy nuclear theory.

Gibble, Kurt, Ph.D., Colorado, 1990. Experimental atomic physics.

Günaydin, Murat, Ph. D., Yale, 1973. Theoretical physics; super-strings; super-gravity; Kaluza-Klein theories.

Heppelmann, Steven F., Ph. D., Minnesota, 1981, Experimental high-energy physics.

Herman, Roger M., Ph. D., Yale, 1962. Emeritus. Theoretical problems in the interaction of simple atoms and molecules; nonlinear optics; liquid crystals.

Jain, Jainendra K., Ph. D., Stony Brook, 1985. Mueller Professor Theoretical solid state physics.

Larson, Daniel J., Ph.D., Harvard, 1971. Verne M. Willaman Dean of the Eberly College of Science. Atomic, molecules and optical physics.

Li, Qi, Ph. D., Peking, 1989. Various aspects of fundamentals and applications of thin films, ultrathin films, multiplayer and superlattice structures.

Liu, Ying, Ph. D., Minnesota, 1991. Experimental condensed matter physics; low dimensional systems and superconductivity.

Mahan, Gerald, Ph.D., University of California at Berkeley, 1964. Distinguished Professor. Condensed matter theory, transport and optical properties, solid state devices.

Mallouk, Thomas E., Ph.D., Berkeley, 1983. DuPont Professor. Photocatalysis, molecular electronics, fuel cell electrochemistry, environmental remediation, chemical sensing, and motion on the nanoscale.

Maynard, Julian D., Ph. D., Princeton, 1974. Distinguished Professor, Acoustics; liquid and solid helium; interface and two-dimensional phenomena.

Meszaros, Peter, Ph.D., University of California at Berkeley, 1972. Eberly Professor. Theoretical high-energy astrophysics; gamma-ray burst sources; ultra-high energy neutrinos and photons; cosmology; neutron stars and black holes; active galactic nuclei.

Robinett, Richard W., Ph. D., Minnesota, 1981. Associate Department Head. Mathematical physics, time-dependent quantum systems, and pedagogical issues related to quantum mechanics.

Samarth, Nitin, Ph. D., Purdue, 1986. Associate Department Head. Magnetic superlattices; Magneto-optics; semiconductors.

Schiff, Steven, Ph.D., Duke University Medical Center, 1985. Brush Professor. Experimental biological physics.

Schiffer, Peter E., Ph. D., Stanford, 1993. Associate Vice President for Research. Director of Strategic Initiatives. Condensed matter experiments.

Sommers, Paul, Ph.D., University of Texas at Austin, 1973. High-energy cosmic rays, astrophysics, and general relativity.

Strikman, Mark, Ph. D., Leningrad Nuclear Physics Institute, 1978. Applications of high-energy physics to nuclei.

Terrones, Mauricio, Ph. D., Sussex, 1998. Production of nanomaterials, electron microscopy techniques for analysis, and molecular simulations.

Weiss, David, Ph.D., Stanford, 1993. Experimental atomic physics.

Willis, Roy F., Ph. D., Cambridge, 1967. Solid state physics; surface physics; electron spectroscopy of thin films and surfaces; synchrotron source radiation physics.

Associate Professors

Bojowald, Martin, Ph.D., RWTH Aachen, 2000. Quantum gravity; classical and quantum cosmology; classical and quantum aspects of black holes; general aspects of quantization; Poisson geometry and non-commutative geometry.

Owen, Benjamin, Ph.D., California Institute of Technology, 1998. Gravitational waves; neutron stars.

Roiban, Radu, Ph.D., State University of New York at Stony Brook, 2001. String theory, gauge theories, quantum field theory.

Sofo, Jorge, Ph.D., Inst. Balseriro, Bariloche, Argentina, 1991. Director of the Materials Simulation Center. Theoretical and computational methods to link properties and structures of materials.

Assistant Professors

DeYoung, Tyce, Ph.D., University of Wisconsin-Madison, 2001. High energy particle astrophysics, especially neutrino and gamma ray astronomy.

Gemelke, Nathan Ph.D., Stanford University, 2007. Bose-Einstein condensation, quantum gases, laser cooling and trapping, quantum information and computation.

Jin, Dezhe Ph.D., University of California, San Diego, 1999. Theory of biological neural networks and computational models of neurobiological functions.

Kozhevnikov, Alexey A., Ph.D., Yale, 2001. Biological physics, neural computations, novel experimental techniques in neuroscience.

Mocioiu, Irina, Ph.D., State University of New York at Stony Brook, 2002. High energy physics and its connections to astrophysics and cosmology.

O'Hara, Kenneth, Ph.D., Duke, 2000. Downsbrough Professor. Experimental atomic, molecular & optical physics, condensed matter physics.

Stasto, Anna, Ph.D., Institute of Nuclear Physics at Polish Academy of Science, 1999. Elementary particle physics, phenomenology of strong interactions, hadronic interactions at high energies.

Zhu, Jun, Ph.D., Columbia University, 2002. Experimental condensed matter physics, low temperature transport and scanned probe experiments. Mesoscopic and nanoscale systems.

Distinguished Visiting Professor

Penrose, Roger, Ph. D., St. John's, Cambridge (UK), 1958. General relativity; Cosmology; foundations of quantum mechanics.

RESEARCH SPECIALTIES AND STAFF

Theoretical

Biological-Physics and Neuroscience. Computational models for birdsong experiments; modeling of neural mechanisms by which high-level memories are coded; biophysically realistic models of rhythm generation, seizures, and Parkinson's disease; prototype feedback control systems; molecular level interactions among genes and proteins; complex systems; framework for understanding proteins; statics and dynamics of ecological communities; scaling in ecology. Albert, Banavar, Collins, Jin, Schiff.

Condensed Matter Physics. Tunneling phenomena in normal metals and superconductors; electronic properties of surfaces and interfaces; contact and transport phenomena in submicron systems; lattice dynamics of metals; liquid metals, and alloys; theory of the scanning tunneling microscope; field emission of electrons; photoemissions; ion emission from solid and liquid metals; dynamical theory of particle-surface interaction; adsorption of atoms on surfaces; two-dimensional phase transitions; statistical mechanics; Hubbard model; high-T_c superconductivity. Monte Carlo simulation in hydrodynamics, solid-liquid-vapor systems, spin glasses, magnetic systems, and porous media; electronic structure theory; carbon nanotubes and nanostructures. Albert, Anderson, Banavar, Cole, Crespi, Cutler, Jain, Mahan, Sofo.

Elementary Particles and Fields. Application of relativistic quantum theory to internal structure of subnuclear particles; quantum chromodynamics and quantum electrodynamics; quantum chromodynamics in nuclei; collider and supercollider phenomenology; electroweak interactions; supersymmetry; conformal field theory. M/Superstring theory, supergravity. J. Collins, Gunaydin, Meszaros, Mocioiu, Roiban, Stasto, Strikman.

Gravitational Physics. Classical and quantum theories of gravitation; nonperturbative approaches to quantum field theory and quantum gravity; gravitational waves; astrophysical applications of general relativity; numerical relativity; mathematical physics; cosmology. Ashtekar, Bojowald, Finn, Meszaros, Owen, Penrose.

Experimental

Acoustics. Sound propagation in superfluid helium; acoustical holography; nonlinear acoustics in random media. Maynard.

Atomic and Molecular Physics. Bose Einstein condensation; laser-cooled atomic clocks; the structure and dynamics of atomic and molecular clusters; laser cooling and trapping of atoms; tests of fundamental symmetries; quantum computation; quantum scattering of cold atoms; optical lattices. Castleman, Gemelke, Gibble, O'Hara, Weiss.

Biological Physics and Neuroscience. Neural organization of complex motor sequences; neural activity in functioning neural circuits. Kozhevnikov.

Elementary Particle and High-Energy Physics. Particle Astrophysics. Studies of the properties of quarks and gluons in hadronic interactions; studies of nuclear color transparency with hard hadronic elastic collisions; direct measurements of high-energy cosmic rays and cosmic antimatter particles; studies of the highest-energy cosmic rays; neutrino astrophysics; gamma-ray astrophysics. Coutu, Cowen, DeYoung, Heppelmann, Sommers.

Condensed Matter and Material Physics. Sound modes in liquid helium; critical phenomena; phase transitions of adsorbed molecular film; superfluidity in reduced dimensionality and at millikelvin temperatures; wave propagation in random media; momentum distributions in quantum liquids and solids using deep nonelastic neutron scattering; thermodynamic and elastic properties of high-T_c superconductors and quasicrystals. Mechanical and optical properties of metal matrix composites; electrical and optical properties and x-ray diffraction of thin film, surfaces, and interfaces; thin-film sputtering phenomena. Relaxation and amorphous semiconductors; Raman spectroscopy; transport properties; band structure of transition metals; superlattice structures; quantum well heterostructures; low-temperture mesoscopic physics; magneto-optics; luminescence; phonons in clusters, amorphous and liquid semiconductors; investiation of adsorption-desorption phenomena; interaction between surface atoms, random walk of atoms on surfaces, and structural and composition analysis of surface layers with the field ion microscopic and the atom-probe field ion microscope; diffraction of atomic beams by surfaces; photoemission; optical properties; inelastic neutron scattering; high-resolution LEED; surface defects; structural studies of adsorbed overlayers; phaser transitions of adsorbates; liquid metal ionization; laser-solid surface interactions; scanning tunneling microscopy; magnetic and high-T_c superconducting thin films. Frustrated magnets; granular materials; optic magnetic and super conducting geometrics; ferro electric thin films; magnetic semiconductors. Castleman, Chan, Diehl, Li, Liu, Maynard, Samarth, Schiffer, Terrones, Willis, Zhu.

Materials Physics. Mechanical and optical properties of metal matrix composites; electrical and optical properties and x-ray diffraction of thin film, surfaces, and interfaces; thin-film sputtering phenomena. Relaxation and amorphous semiconductors; field ion microscopy. Li, Liu, Schiffer, Willis.

Solid State and Surface Physics. Raman spectroscopy; transport properties; band structure of transition metals; superlattice structures; quantum well heterostructures; low-temperature mesoscopic physics; magneto-optics; luminescence; phonons in clusters, amorphous and liquid semiconductors; investigation of adsorption-description phenomena; interaction between surface atoms, random walk of atoms on surfaces, and structural and composition analysis of surface layers with the field ion microscopic and the atom-probe field ion microscope; diffraction of atomic beams by surfaces; photoemission; optical properties; inelastic neutron scattering; high-resolution LEED; surface defects; structural studies of adsorbed overlayers; phase transitions of adsorbates; liquid metal ionization; laser-solid surface interactions; scanning tunneling microscopy; magnetic and high-T_c superconducting thin films. Castleman, Diehl, Li, Liu, Samarth, Willis, Zhu.

TEMPLE UNIVERSITY OF THE COMMONWEALTH SYSTEM OF HIGHER EDUCATION

DEPARTMENT OF PHYSICS

Philadelphia, Pennsylvania 19122

Students Accepted For Degree	FIELDS		
	Physics	Astronomy	Related Fields
Doctorate	X		
Master's	X		

1. General

President: Ann W. Hart
Acting Dean of Graduate School: Richard M. Englert
Department Chairman: Rongjia Tao
Department Telephone Number: (215) 204-7634 (C)
Type of Institution: University
Control: Both
Setting: Urban
Total Faculty: 3,481
Total Graduate Faculty: 905
Total Students: 38,956
Total Graduate Students: 4,773
Annual Graduate Tuition:
In-state residents: $590/credit
Out-of-state residents: $861/credit
Tuition rates for: 2009–10
Deferred tuition plan: Yes
Other Fees: $250
Term: Semester

2. Number of Faculty in Department

The combined total of full-time faculty in the three professorial ranks is 18. The combined total of full-time, part-time, and other faculty at all ranks is 28.

3. Admission, Financial Aid, and Housing

Address admission inquiries to: Chairman, Graduate Program Committee, Physics Department, Barton Hall
Graduate application fee required: Graduate Application Fee, nonrefundable (required of all applicants) $60
Admission deadline (Fall admission): None (no official deadline, as soon as possible)
Admission information: For fall admission, 2009–10, 8 students were accepted from 34 applicants.
Admission requirements: For admission to the graduate programs, a Bachelor's degree in physics or a related area is required. A minimum GPA of 3.0 in physics and mathematics is required. The GRE is required by the Graduate Admissions Office of the College of Science and Technology. Students from non-English speaking countries are required to demonstrate proficiency in English via the TOEFL exam. Minimum acceptable score for admission is 575 paper based, 230 computer based.
Undergraduate preparation assumed: (For example) Halliday and Resnick, *Fundamentals of Physics*; Tipler, *Modern Physics*; Symon, *Mechanics*; Reitz and Milford, *Electromagnetic Theory*; Zemansky, *Heat and Thermodynamics*; Jenkins and White, *Optics*; Saxon, *Quantum Mechanics*. Less well-prepared students may be accepted, but will be expected to take the appropriate undergraduate courses.
Address financial aid inquiries to: Chairman, Graduate Program Committee, Physics Dept., Barton Hall-009-00, 1900 N. 13th St., Philadelphia, PA 19122-6082. Fax: 215-204-5652. E-mail: physics@temple.edu
GAPSFAS application required: No
Financial aid deadline: None
Loans available: Yes
Address housing inquiries to: Foreign Students: International Services, 203B Vivacqua Hall, University Housing, 1910 Liacouras Walk
On-campus, single student housing available: Yes
Cost/term: Varies
On-campus, married student housing available: Yes
Cost/month: Varies

Table A—Faculty, Enrollments, and Degrees Granted

Research Specialty	2009–10 Faculty	Enrollment[1] Fall 2008		No. of Degrees Granted[2] 2008–09 (2004–09)			Median No. of Years for 2008–09 Ph.D.'s
		Master's	Doctorate	Master's	Terminal Master's	Doctorate	
Atomic, Molecular, & Optical Physics	3	–	8	0(0)	–	3(9)	–
Biophysics	1	–	0	0(0)	–	0(1)	–
Materials Sci./ Metallurgy	2	–	0	0(0)	–	0(0)	–
Optics	3	–	3	0(0)	–	0(0)	–
Particles & Fields	5	–	6	0(0)	–	1(2)	–
Physics Education	3	–	0	0(0)	–	0(0)	–
Relativity & Gravitation	1	–	3	0(0)	–	1(1)	–
Solid State	14	–	14	0(0)	–	2(2)	–
Statistical & Thermal	3	–	0	0(0)	–	0(0)	–
Non-specialized	0	2	10	0(3)	1	0(0)	–
Total		–	44	0(3)	1	6(14)	
Full-time Grad. Stud.		–	44				
Part-time Grad. Stud.		–	–				
First-year Grad. Stud.		–	8				
Median Years in Grad. Study (2009–10 Degrees)				–	–	–	–

Undergraduate Degrees, 20089–10 (2005–10: Information not available

[1]Students not yet committed to a research specialty are entered under non-specialized.
[2]Five-year totals in parentheses.

4. Graduate Degree Requirements

Master's: 24 graduate credits of course work with satisfactory performance; Master's exam required; thesis not required; no residence or language requirement.
Doctorate: No specific number of graduate credits required, but satisfactory performance in an approved course program is required; comprehensive written and oral exam, dissertation and dissertation exam required; one year residency required; no language exam required.
Other Programs: Chemical Physics interdisciplinary program similar to regular doctoral program.
Thesis: Thesis may be written *in absentia*.
Special Equipment, Facilities, or Programs: The department is housed in Barton Hall, which has "smart" lecture theaters, offices, classrooms and laboratories. The Science Library

contains frequently used journals and books; several thousand additional volumes are located in the Paley Library across the street from Barton Hall. A student shop and a materials preparation facility are available. The University computer facilities are based on a UNIX cluster composed Digital Equipment Corporation Alphas including a high-performance numerical compute-server. The departmental computer facilities include a local area network (LAN) of six Silicon Graphics IRIS Indigo R4000 workstations, a number of Windows NT workstations and 8 host LAN of Linux workstations. The deparmental local area networks are connected to a high-performance fiber-optic campus backbone through which all University computer facilities can be reached. High-speed access to the Internet is readily available from all departmental computers. Electronic information retrieval is provided by Temple University library's Scholars Information System, which subscribes to a wide range of on-line databases. The research laboratories are conducting a variety of studies on optical hole-burning and multiple quantum well structures; laser-based molecular spectroscopy; low-temperature properties of alloys and intermetallics, including valence fluctuations and heavy fermion behavior; high temperature superconductivity; Mossbauer spectroscopy; neutrino oscillation; nucleon structure; dark matter detection; soft-condensed matter physics and electro- and magneto-rheology. The department also uses outside facilities, including the Los Alamos Meson Physics Facility, the Brookhaven National Laboratory, the Stanford Linear Accelerator Center, the Thomas Jefferson National Accelerator Facility, and the National High Magnetic Field Laboratory. Theoretical work is being conducted in such areas as elementary particles and their interactions, statistical mechanics, biophysics, general relativity, and condensed-matter theory.

Table B—Appointments to Graduate Students, 2008–09

Title of Appointee	Appointments Total	Appointments First year	Academic Load Allowed in Credit Hours	Hours of Service Per Week	Stipend for Academic Year ($)
			Semester		
Teaching Assistant[1]	23	8	9 credits	20	15,426
Research Assistant[2]	14		(3 credits		
Other	2		per course)		
Fellowship	5				
Total	44	8			

[1]TAs get an additional $stipend for summer, $15,889 total for academic year.
[2]Externally funded RAs get about $21,186/year. College funded RAs get $15,889/yr. plus an additional $stipend for summer.

5. Personnel Engaged in Separately Budgeted Research, 7/08–09

Professorial faculty	11
Postdoctoral appointments	4
Graduate students	14
Total	29

6. Separately Budgeted Research Expenditures by Source of Support, 7/08–7/09

	Departmental Research	Physics-related Research Outside Department
Federal government	$1,860,000	$150,000
Private, nonprofit organizations	250,000	
Total	$2,110,000	$150,000

Table C—Separately Budgeted Research Expenditures

Research Specialty	No. of Grants	Expenditures ($)
Chemical Physics	4	405,000
Particles & Fields	4	570,000
Condenced Matter	6	1,135,000
Total	14	2,110,000

Table D—Physics-related Research Outside Department

Field and Unit Outside Department	No. of Grants	Expenditures ($)
Condensed Matter	1	150,000
Total	1	150,000

FACULTY

Professors

Burkhardt, Theodore, Ph.D., Stanford, 1967. Statistical physics; application of the renormalization group.

Dubeck, Leroy, Ph.D., Rutgers, 1965. Development of pre-college science materials.

Forster, Dieter, Ph.D., Harvard, 1969. Statistical mechanics and many-body theory.

Hasan, Zameer, Ph.D., Australian National University, 1979. Optical and magneto-optical properties of solids; laser materials scanning; tunneling microscopy laser spectroscopy.

Lyyra, Marjatta, Ph.D., Stockholm, 1979. Laser spectroscopy of molecules and atoms.

Martoff, C. J., Ph.D., California, Berkeley, 1980. Experimental particle physics; development of particle detectors using superconductivity.

Meziani, Zein-Eddine, Ph.D., University of Paris VII, 1984. Experimental particle physics; quark structure of the nucleon and nucleon structure functions.

Riseborough, Peter, Ph.D., Imperial College, London, 1977. Theoretical condensed matter physics and statistical mechanics.

Tahir-Kheli, Raza, Ph.D., Oxford, 1962. Theory of magnetism; randomly disordered systems.

Tao, Rongjia, Ph.D., Columbia, 1982. Energy Science, electrorheological and magneto-rheological fluids, superconductivity, photonic crystals, nonlinear optics.

Xi, Xiao Xing, Ph.D., Peking U., 1987. Superconductivity, film, materials physics.

Associate Professors

Dziembowski, Zbigniew, Ph.D., Warsaw, 1975. Theoretical particle physics.

Gawlinski, E., Ph.D., Boston, 1983. Chairman of the Department. Statistical mechanics biophysics and computational physics.

Ivarone, Maria, Ph.D., U. of Napoli, 1996. Experimental condensed matter physics.

Lin, C. L., Ph.D., Temple, 1985. Highly correlated electron systems.

Yuen, T., Ph.D., Temple, 1990. Experimental condensed matter physics; Mössbauer spectroscopy.

Assistant Professors

Metz, Andreas, Ph.D., U. Mainz, 1967. Theoretical nuclear and particle physics.

Santamore, Deborah, Ph.D., California Inst. Tech., 2003. Theoretical condensed matter physics and atomic, molecular, and optical physics.

Adjunct Professors

Ha, Yuan K., Ph.D., Yale U., 1982. General relativity.

Wu, Dong Ho, Ph.D., Tufts, 1991. Experimental condensed matter physics and optical physics.

Research Professor

Kotchigova, S., Ph.D., St. Petersburg Univ., 1981. Relativistic ab initio study of alkaline earths.

Research Associate Professor

Romanov, D., Ph.D., Novosibirsk Inst. for Semiconductor Physics, 1986. Optical control of nanosystems.

Research Assistant Professor

Ahmed, E. Ph.D., Temple U., 2007. Atomic molecular and optica physics.

Chen, Ke, Ph.D., Chinese Acad. Sciences, 2001. Superconductivity, thin film.

Schulte, E., Ph.D., UIUC, 2002. Nuclear and particle physics.

Tang, H., Ph.D., Northwest Polytec U. (China), 2006. Complex fluids.

Xu, Xiaojun, Ph.D., Chinese Acad. Sciences, 1995. Superconductivity, complex fluids.

Associate Professor for Teaching/Instruction

Borovitskaya, Elena, Ph.D., Inst. for Appl. Phys., Russian Academy of Science, Nizhini, Novogorod, 1988. Theoretical condensed matter physics; optics and transport.

Assistant Professors for Teaching/Instruction

Mackie, Matt, Ph.D., U. of Connecticut, 1999. Theoretical atomic, molecular, and optical physics.

Tsenkov, Tsvetlin, Ph.D., Drexel U., 2004. Nonlinear dynamics.

Professors Emeriti

Auerbach, Leonard B., PhD., California, Berkeley, 1962. Experimental particle physics; investigations of the properties of fundamental particles.

Franklin, Jerrold, Ph.D., Illinois, 1956. Theoretical particle physics; quark and parton theory; S-matrix theory.

Intemann, Robert L., Ph.D., Stevens Inst. of Tech., 1964. Atomic physics; inner shell processes.

Larsen, Sigurd Y., Ph.D., Columbia, 1962. Quantum statistical-mechanics; few-body problem; hyperspherical harmonics; molecular physics; chemical reactions.

Mihalisin, Ted W., Ph.D., Rochester, 1967. Condensed matter physics.

Neville, Donald E., Ph.D., Chicago, 1962. Theoretical particle physics; symmetries and quark models; quantum field theory; quantum gravity.

Offenbacher, Elmer L., Ph.D., Pennsylvania, 1951.

Weinberg, Robert B., Ph.D., Columbia, 1963.

RESEARCH SPECIALTIES AND STAFF

Atomic, Molecular and Optical Physics. Ahmed, Hasan, Intemann, Kotchigova, Larsen, Lyyra, Mackie, Romanov, Santamore, Tao, Wu.

Condensed Matter Physics. Chen, Hasan, Ivarone, Lin, Mihalisin, Riseborough, Romanov, Santamore, Tahir-Kheli, Tang, Tao, Wu, Xi, Xu, Yuen.

Elementary Particle Physics. Auerbach, Dziembowski, Franklin, Martoff, Metz, Meziani, Neville, Schulte.

General Relativity. Ha.

UNIVERSITY OF PITTSBURGH

DEPARTMENT OF PHYSICS AND ASTRONOMY

http://www.physicsandastronomy.pitt.edu

Pittsburgh, Pennsylvania 15260

Students Accepted For Degree	FIELDS		
	Physics	Astronomy	Related Fields
Doctorate	X		X
Master's	X		

1. General

Chancellor: Mark A. Nordenberg
Dean: N. John Cooper
Department Chairman: David Turnshek
Department Telephone Number: (412) 624-9000
Web Site: http://www.physicsandastronomy.pitt.edu
Department Fax Number: (412) 624-9163
Type of Institution: University
Control: Both
Setting: Urban
Total Faculty: Main Campus: 4,686
Total Graduate Faculty: 1,387
Total Students: Main Campus: 28,328
Total Graduate Students: 10,297
Term: Trimester
Annual Graduate Tuition:
Tuition rates for: 2009–10
In-state residents: Full-time—$16,402 (2 term)
Part-time—$665/credit
Out-of-state residents: Full-time—$28,694 (2 term)
Part-time—$1,175/credit
Deferred tuition plan: Yes
*Other Fees**:
Health—$85 (full-time)
Activities—$20 (full-time)/$10 (part-time)
Computing & Network Services—$150 (full-time)/ $75 (part-time)
Security, Safety & Transportation—$90 (full-time and part-time)

*Fees are per term and subject to change.

2. Number of Faculty in Department

The combined total of full-time faculty in the three professorial ranks is 35. The combined total of full-time, part-time, and other faculty at all ranks is 52.

3. Admission, Financial Aid, and Housing

Address admission inquiries to: Admissions Officer, Department of Physics and Astronomy, 100 Allen Hall, Univ. of Pittsburgh, Pittsburgh, PA 15260. pagrad@pitt.edu
Admission deadline (Fall admission): 1/31. Late applications are accepted on the basis of space availability.
Admission information: For fall admission, 2009–10, 84 students were accepted from approximately 299 applicants.
Admission requirements: For admission to the graduate programs, a Bachelor's degree in one of the physical sciences, mathematics or engineering is required with a minimum undergraduate GPA of 3.0 in mathematics and physics specified. The GRE and the GRE Advanced are required under normal circumstances. Students from non-English speaking countries are required to demonstrate proficiency in English via the TOEFL (577), iBT (90) with strong speaking component, or IELTS (6.5) exams. (Minimum score).
International Students: See http://www.ois.pitt.edu/
Department Financial Aid: Considered at time of admission review. Teaching and research assistantships, several fellowships for new students.
GAPSFAS application required: No
Student Loan Financial Aid deadline: Recommended 90 days prior to term, inquiries to: Financial Aid Office, 4227 Fifth Ave., Alumni Hall, University of Pittsburgh, Pittsburgh, PA 15260
Loans available: Yes (some federal restrictions)
Address housing inquiries to: Dept. Property Management, 127 N. Bellefield Ave., Pittsburgh, PA 15213
http://www.ocl.pitt.edu and www.pc.pitt.edu/index.html
Near-campus, various housing available: Limited
On-campus, married student housing available: Yes

Table A—Faculty, Enrollments, and Degrees Granted

Research Specialty	2009–10 Faculty	Enrollment[1] Fall 2009–10 Mas-ter's	Doc-torate	No. of Degrees Granted[2] 2009–10 Mas-ter's	Terminal Master's	Doc-torate
Astrophysics/Cosmology	9	0	14	0(0)	0(0)	1(12)
Condensed Matter	12	0	32	0(0)	0(0)	3(26)
Particle Physics	13	0	14	0(0)	0(0)	2(10)
Physics Education Res.	2	1	3	0(0)	1(1)	0(2)
Non-specialized	0	1	15	18(66)	5(23)	–
Total	44	2	78		1(2)	6(46)
Full-time Grad. Stud.		0	78			
Part-time Grad. Stud.		0	1			
First-year Grad. Stud.		0	18			
Undergraduate Degrees[2], 2009–10 (2005–09): 16(99)						

[1]Students not yet committed to a research specialty are entered under non-specialized.
[2]Five-year totals in parentheses.

4. Graduate Degree Requirements

Doctorate: Ph.D. students must successfully complete the following six graduate-level core courses: Dynamical Systems (one term), Statistical Mechanics and Thermodynamics (one term), Classical Electricity and Magnetism (one term), Mathematical Methods (one term), and Non-relativistic Quantum Mechanics (two terms). Exemptions from any of these courses may be granted if a student has successfully completed an equivalent course elsewhere. Students must complete these core courses with a grade point average of at least 3.00, which corresponds to a B average; they must also maintain a GPA of at least 3.00 in all of their graduate courses. In order to satisfy the Ph.D. Comprehensive Examination requirement, students must achieve a score of at least 60% on the final examination in each of the six core courses. This requirement must be fulfilled within the first two years unless an extension is granted. After passing the Ph.D. Comprehensive Examination, the student must find a research advisor and begin the process that leads to Admission to Candidacy and ultimately to the preparation and defense of a satisfactory dissertation. All Ph.D. students are required to serve for two terms as a Teaching Assistant in introductory undergraduate laboratories or recitations. An exemption may be granted if a student has substantial prior teaching experience. There is no foreign language requirement. There is a residence requirement of six full terms, with a total of 72 credit hours. Under some circumstances up to two terms of prior graduate work may be transferred from another institution.

Master's: Candidates for the M.S. degree must satisfy the Preliminary Evaluation, which requires the successful completion of at least one course in each of the following core subjects: Dynamical Systems, Statistical Mechanics and Thermodynamics, Electricity and Magnetism, and Quantum Mechanics, with a final examination score of at least 50% for courses at the graduate level or 60% for courses at the advanced undergraduate level. M.S. candidates may elect one of three alternative options to earn the degree: (1) Submit a thesis and successfully complete at least six courses (at least four must be at the graduate level and the balance at the advanced undergraduate level); (2) Submit no thesis but successfully complete at least eight courses (at least four must be at the graduate level and the balance at the advanced undergraduate level); (3) Submit no thesis but successfully complete at least six courses at the graduate level. M.S. students must maintain a grade point average of at least 3.00, which corresponds to a B average, in the core subjects **and** in all of their courses. There is no foreign language requirement. There is a residence requirement of two full terms with a total of 24 credits.

Other Programs: Interdisciplinary research programs may be arranged on a case-by-case basis. There have been Physics Doctorates awarded for work done in collaboration with the faculty members in the Chemistry Department, the Mathematics Department, the Materials Science Department, the Electrical and Chemical Engineering Departments, the Department of Biological Sciences, the Department of Computational Biology and the Department of Radiology in the School of Medicine, among others.

Special Equipment, Facilities, or Programs: The Department of Physics and Astronomy is located on the main campus in a complex of five interconnecting buildings. The Department facilities include a professionally-staffed maching shop, electronics shop, and glass-blowing shop; a large stockroom; and specialized clean rooms for electronics assembly, as well as extensive Departmental and University computer resources. Departmental students have easy access to the facilities and expertise available at the Gertrude E. and John M. Peterson Institute of NanoScience and Engineering (PINSE) on the University campus and the Pittsburgh Supercomputing Center (PSC). Other local facilities include Allegheny Observatory. Experiments in particle physics are carried out at national and international facilities such as Fermilab in Chicago and CERN in Switzerland and J–PARC in Japan. Similarly, astrophysics/cosmology ground-based programs are conducted at national and international observatories located at, for example: Kitt Peak and Mount Hopkins, in Arizona, Cerro Tololo, Las Campanas, and La Silla in Chile; Mauna Kea in Hawaii; and Apache Point in New Mexico for collection of Sloan Digital Sky Survey data. Pitt faculty also make use of space-based telescopes, for example: the Hubble Space Telescope; the Chandra X-Ray Telescope; and the GALEX UV Telescope. Pitt faculty are also members of several current and/or future large-telescope consortia: the Sloan Digital Sky Survey (SDSS); the Atacama Cosmology Telescope (ACT); the Panaramic Survey Telescope & Rapid Response System (Pan-STARRS); and the Large Synoptic Survey Telescope (LSST).

Table B—Appointments to Graduate Students, 2009–10

Title of Appointee	Appointments Total	Appointments First year	Academic Load Allowed in Credit Hours	Hours of Service Per Week	Stipend for Academic Year ($)
			Semester		
Teaching Assistant	28	12	15	15–20	15,065[1]
Research Assistant	37	2	15	15–20	15.065[1]
Teaching Fellow	2	0	15	15–20	15,675[1]
Part-time Instructors	0	0	15	15–20	n/a
A&S Predoctoral Fellows	2	3		15-20	17,972[2]
A&S Summer Fellows	1	1	3	15–20	17,000[4]
A. W. Mellon Fellow	2	0	15	0	17,500[2]
Zaccheus Daniel Fellow	1	0	15	0	10,000[2]
Mary L. Warga Fellow	1	1	15	0	17,500[2,3]
E. Baranger Fellow (new for Fall 2005)	0	0	15	0	17,500[2]
Total	77	19			

[1]These are two-term salaries. Some third term appointments are available. Most courses are offered during the first two terms. Students are encouraged to engage in research during the third term. Scholarships covering tuition and required fees supplement these appointments. Individual medical insurance is covered.

[2]Two-term stipend available in 8- or 12-month payments. May be supplemented with third term teaching or research from department, if available.

[3]Supplemental funds of $4,000 for travel, research, and other professional expenses during the first year.

[4]Supplements first-year students' summer support.

University fellowships exist for certain groups who continue to be underrepresented in the national pool of earned doctorate degrees as well as within the professorate at the University.

5. Personnel Engaged in Separately Budgeted Research, 7/08–6/09

Professorial faculty	24
Other faculty	6
Postdoctoral appointments	17
Graduate Students	66
Undergraduate Students	48
Nonteaching research personnel	4
Total	165

6. Separately Budgeted Research by Source of Support 7/08–6/09 ($K)

	Departmental Research
Federal government	$5,358
Other	0
Total	$5,358

7. Separately Funded and Managed Laboratories: Allegheny Observatory

Table C—Separately Budgeted Research Expenditures

Research Specialty	No. of Grants	Expenditures ($K)
Astronomy & Astrophysics	24	900
Condensed Matter & Solid State	31	2,281
Particle Physics	18	1,822
Other	6	355
Total	79	5,358

FACULTY

Professors

Boyanovsky, Daniel, Ph.D., California, Santa Barbara, 1982. Theoretical condensed matter physics; particle astrophysics, and astrophysics and cosmology

Coalson, Rob, Ph.D., Harvard, 1984. Chemical physics.

Duncan, H. E. Anthony, Ph.D., MIT, 1975. Theoretical particle physics.

Dytman, Steven A., Ph.D., Carnegie-Mellon, 1978. Experimental particle physics, neutrino physics.

Hillier, D. John, Ph.D., Australian National, 1984. Theoretical and observational astrophysics; computational physics.

Jasnow, David M., Ph.D., Illinois, 1969. Theory of phase transitions; statistical physics, biological physics.

Levy, Jeremy, Ph.D., California, Santa Barbara, 1993. Experimental condensed matter, quantum information.

Maher, James V., Ph.D., Yale, 1969. Experimental solid state physics; critical phenomena; physics of fluids.

Petek, Hrvoje, Ph.D., California, Berkeley, 1985. Experimental condensed matter/AMO, nanoscience, solid-state physics.

Roskies, Ralph Z., Ph.D., Princeton, 1966. Director, Supercomputer Center. Theoretical particle physics; use of computers in theoretical physics.

Schulte-Ladbeck, Regina, Ph.D., Heidelberg, 1985. Astrophysics.

Shepard, Paul, Ph.D., Princeton, 1969. Experimental particle physics.

Snoke, David W., Ph.D., Illinois, 1990. Experimental condensed matter and solid state physics.

Turnshek, David A., Ph.D., Arizona, 1981. Chair of the Department. Astrophysics; extragalactic astronomy; observational cosmology.

Wu, Xiao-Lun, Ph.D., Cornell, 1987. Experimental condensed matter; biological physics.

Yang, Judith, Ph.D., Cornell, 1993. Materials Science & Engineering.

Associate Professors

Boudreau, Joseph, Ph.D., Wisconsin, 1991. Experimental particle physics.

Devaty, Robert P., Ph.D., Cornell, 1983. Experimental solid state physics.

Kosowsky, Arthur, Ph.D., Chicago, 1994. Theoretical and Experimental cosmology and astrophysics.

Leibovich, Adam, Ph.D., California Institute of Technology, 1997. Director of Graduate Studies. Theoretical particle physics.

Liu, W. Vincent, Ph.D., Texas at Austin, 1999. Theoretical condensed matter.

Mueller, James A., Ph.D., Cornell, 1989. Undergraduate Program Director. Experimental particle physics.

Naples, Donna, Ph.D., Maryland, 1993. Experimental neutrino physics.

Paolone, Vittorio, Ph.D., California, 1990. Experimental high-energy particle physics.

Savinov, Vladimir, Ph.D., Minnesota, 1996. Experimental particle physics.

Singh, Chandralekha, Ph.D., California, Santa Barbara, 1993. Polymer physics; physics education research.

Swanson, Eric, Ph.D., Toronto, 1991. Theoretical particle physics.

Assistant Professors

Daley, Andrew, Ph.D., University of Innsbruck, 2005. Theoretical Condensed Matter.

D'Urso, Brian, Ph.D., Harvard, 2003. Experimental condensed matter physics, nanoscience.

Dutt, Gurudev, Ph.D., University of Michigan, 2004. Experimental condensed-matter/AMO science, quantum information.

Freitas, Ayres, Ph.D., University of Hamburg, 2002. Theoretical particle physics.

Newman, Jeffrey, Ph.D., U. C. Berkeley, 2000. Astrophysics, extragalactic astronomy, observational cosmology.

Salman, Hanna, Ph.D., Weizmann Institute of Science, 2002. Experimental biological physics.

Wood-Vasey, Michael, Ph.D., California at Berkeley, 2004. Astrophysics, extragalactic astronomy, observational cosmology.

Zentner, Andrew, Ph.D., The Ohio State University, 2003. Chair, Graudate Student Admission Committee. Theoretical cosmology.

Professors Emeriti in Residence

Baranger, Elizabeth U., Ph.D., Cornell, 1954.

Cleland, Wilfred E., Ph.D., Yale, 1964.

Cohen, Bernard L., Ph.D., Carnegie-Mellon, 1950.

Engels, Jr., Eugene, Ph.D., Princeton, 1962.

Gerjuoy, Edward, Ph.D., California, Berkeley, 1942.

Goldburg, Walter I., Ph.D., Duke, 1955.

Janis, Allen I., Ph.D., Syracuse, 1957.

Johnsen, Rainer, Ph.D., Kiel, Germany, 1966. Experimental atomic and plasma physics.

Koehler, Peter F. M., Ph.D., Rochester, 1967. Experimental high-energy particle physics; physics education research.

Newman, Ezra T., Ph.D., Syracuse, 1956.

Saladin, Juerg X., Ph.D., Eidgenossische Technische Hochschule, Switzerland, 1959.

Vincent, C. Martin, Ph.D., Witwatersrand, South Africa, 1966.

Adjunct Associate Professor

Boyd, Steven, Ph.D., University of Sydney, 1998.

Research Professors

Choyke, W. James, Ph.D., Ohio State, 1952. Experimental Solid state physics; defect states in semiconductors, large-bandgap spectroscopy.

Pratt, Richard H., Ph.D., Chicago, 1959.

Weisheit, Jon, Ph.D., Rice, 1970. Theoretical atomic physics, astrophysics.

Winicour, Jefferey, Ph.D., Syracuse, 1964. General relativity numerical relativity.

Research Associate Professor

Rao, Sandhya, Ph.D., Pittsburgh, 1994. Astrophysics; extragalactic astronomy; observational cosmology.

Research Assistant Professor

Feng, Min, Ph.D., Chinese Academy of Sciences, 2005. Experimental Condensed Matter.

Zhao, Jin, Ph.D., University of Science and Technology of China, 2003. Theoretical condensed matter, nanoscience.

Lecturer/Lab Supervisor

Clark, Russell, Ph.D., Louisiana State, 1997. Physics education research, neutrino physics.

RESEARCH SPECIALTIES

Theoretical

Astrophysics and Cosmology. Early universe physics; statistical cosmology; radiative transfer. Stellar atmospheres and massive stars. Boyanovsky, Hillier, Kosowsky, Weisheit, Zentner.

Atomic Physics. Interactions of electrons and photons in the field of an atom; atomic processes in high temperature/density plasmas; bremsstahlung; pair production. Pratt.

Condensed Matter, Solid State and Statistical Physics. Phase transitions; disordered systems; nonequilibrium behavior; polymer physics; biological physics; atomic cold gases; superconductivity; quantum kinetics. Boyanovsky, Daley, Jasnow, Liu, Zhao.

General and Numerical Relativity and Relativistic Astrophysics. Gravitational lensing; Gravitational radiation and black hole physics. E. Newman, Winicour.

Intermediate Energy Physics. Medium-energy reactions with photons and mesons; heavy quark phenomenology; QCD modeling; spin density matrix physics; mathematical methods for nonlinear dynamics. Swanson.

Particle Physics. Gauge field theories; lattice calculations; nonperturbative effects; weak interaction models and phenomenology; heavy-quark physics; super symmetry. Boyanovsky, Duncan, Freitas, Leibovich.

Experimental

Applied Physics. surface science; polymers; computational physics; optoelectronics; quantum computing; porous semiconductors; large bandgap semiconductors interdisciplinary research. Choyke, Coalson, Devaty, Levy, Snoke, and selected members of the departments of biological science, chemistry, material science, electrical engineering, and chemical engineering.

Astrophysics. Local and distant galaxies; quasars; studies of quasar absorption line systems; statistical analysis of the properties of galaxies; clustering and large-scale structure; dark matter and dark energy; cosmic microwave background; galactic and intergalactic medium; massive stars; model stellar atmospheres. Hillier, Kosowsky, J. Newman, Rao, Schulte-Ladbeck, Wood-Vasey, Turnshek, Zentner.

Atomic and Atmospheric Physics. Experimental low-energy plasma physics; electron-ion recombination and ion-atom/molecule reactions; applications to planetary atmospheres; airglow studies; thermospheric dynamics. Johnsen.

Condensed Matter. Nanoscience; biological physics; solid state physics; phase transitions; large bandgap semiconductors; quantum well phenomena; ferroelectrics; coherence phenomena in condensed gases and solids; semiconductor spintronics and quantum computation; atomic-scale optical spectroscopy; ultrafast phenomena; pattern formation; two-dimensional condensed matter systems; turbulence. Choyke, Devaty, D'Urso, Dutt, Feng, Goldburg, Levy, Petek, Salman, Snoke, Wu.

Particle Physics. Study of the properties of elementary constituents of matter (quarks and leptons) at the Tevatron proton-antiproton collider, located a the Fermi National Accelerator Laboratory. Involved with the European Large Hadron Collider, which may uncover the elusive Higgs boson as well as a spectrum of new particles arising from "supersymmetry." Studies of fundamental properties of neutrinos, such as oscillations, mass differences, and neutrino-nucleus interactions. Boudreau, Cleland, Dytman, Mueller, Naples, Paolone, Savinov, Shepard.

Physics Education Research. Identification of sources of student difficulties in learning concepts in both introductory and advanced-level physics courses; design, implementation, and outcome assessment of changes in curricular offerings/pedagogical methods that are designed to reduce these difficulties. Clark, Koehler, Singh.

BROWN UNIVERSITY

DEPARTMENT OF PHYSICS

Providence, Rhode Island 02912

Students Accepted For Degree	FIELDS		
	Physics	Astronomy	Related Fields
Doctorate	X		
Master's			

1. General

President: Ruth J. Simmons
Department Chair: James Valles, Jr.
Department Telephone Number: (401) 863-2641
Type of Institution: University
Control: Private
Setting: Urban
Total Faculty: 686
Total Graduate Faculty: 686
Total Students: 8,261
Total Graduate Students: 1,832
Annual Graduate Tuition:
All Graduate Students: Full-time—$39,928
Part-time—$4,991/course
Tuition rates for: 2010–11
Deferred tuition plan: Yes
Annual Other Fees: $40 Activity Fee
$650 mandatory health fee
Term: Semester

2. Number of Faculty in Department

The combined total of full-time physics faculty in the three professorial ranks is 27. The combined total of full-time, part-time, and other faculty at all ranks is 43.

3. Admission, Financial Aid, and Housing

Address admission inquiries to: Physics_Admission@Brown.edu.
Graduate application fee required: $70
Admission deadline (Fall admission): 1/1
Admission information: For fall admission, 2010–11, 63 students were accepted from 307 applicants; of these, 19 students joined the Department.
Admission requirements: For admission to the graduate program, a Bachelor's degree in physics or related field is required with no minimum undergraduate GPA specified. The GRE general is required. The GRE subject/Advanced is strongly recommended. Students from non-English speaking countries are required to demonstrate proficiency in English via the TOEFL exam. Minimum acceptable score for admission is 577 paper-based test, 233 computer-based test, 90 internet-based test (Ibt).
Undergraduate preparation assumed: Undergraduate requirements flexible to some extent; preference given for strong upper-class study in mechanics, E&M, wave theory, modern physics, and mathematics through partial differential equations. Purcell, *Electricity and Magnetism*; Schey, *Div, Grad, Curl and All That*; Feynman, *The Feynman Lectures on Physics*, Vol. II recommended; French and Taylor, *An Introduction to Quantum Physics*; Marion, *Classical Electromagnetic Radiation*; Gasiorowicz, *Quantum Physics*; Reif, *Fundamentals of Statistical and Thermal Physics*; French, *Vibrations and Waves*; Kibble, *Classical Mechanics* (2nd. ed. or later).
Information on Graduate support and Financial Aid can be found at: http://gradschool.brown.edu/go/support.
Address housing inquiries to: Office of Auxiliary Housing. Box 1902, Telephone: (401) 863-2541
On-campus, Non-dormitory single student housing available: Yes
Cost/year (9 mos.): $5,958
On-campus, Non-dormitory married student housing available: Yes

Table A—Faculty, Enrollments, and Degrees Granted

Research Specialty	Faculty	Enrollment[1] Fall 2000		No. of Degrees Granted[2] 2008–09		
		Master's	Doctorate	Master's	Terminal Master's	Doctorate
Astrophysics	4	–	12	–	–	0(7)
Biophysics	3	–	5	–	–	4(13)
Condensed Matter Physics	10	–	26	–	–	8(29)
Particles & Fields	9	–	19	–	–	5(16)
Other	–	–	16	–	–	1(5)
Non-specialized	–	–	23	–	–	0(0)
Total	26	0	101	11(63)	0(6)	18(70)
Full-time Grad. Stud.		0	101			
Part-time Grad. Stud.		0	0			
First-year Grad. Stud.		0	18			
Median Years in Grad. Study (2008–09)			6			

[1]Students not yet committed to a research specialty are entered under non-specialized.
[2]Five-year totals in parentheses.

4. Graduate Degree Requirements

Master's: Ordinarily, only applicants for the Ph.D. are accepted. An Sc.M. can be earned *en route*. Approved sequence of eight semester courses (32 credits). At least seven courses at Brown with more "B's" or "A's" than "C's." No foreign language required. Thesis usually not required. There are no requirements for a comprehensive exam.
Doctorate: Equivalent of three full-time years. Six one-semester "core-courses"—quantum mechanics (2), Classical physics (mechanics and electricity and magnetism) (2) experimental techniques (1) and statistical mechanics (1). Also, four advanced courses in area of research specialization. Required exams: qualifying (Sem. 3), preliminary (Sem. 5); also required: thesis defense (oral). No foreign language requirement.
Thesis: Thesis may be written *in absentia*.
Special Equipment, Facilities, or Programs: (a) Brown is a member of Universities Research Association, Inc. (URA), part of the Fermilab Research Alliance which operates the Fermi National Accelerator Laboratory (FNAL) in Batavia, Illinois, and other facilities. Brown physicists are involved in the D0 experiment at the 2 TeV Tevatron collider located at FNAL, as well as in the CMS experiment at the 14 TeV Large Hadron Collider (LHC) located at CERN, in Geneva, Switzerland. Brown leads experimental collaborations which operate rare particle search experiments at international under-

ground laboratories at Gran Sasso, Italy, and Sanford Lab, Homestake Mine, South Dakota. Brown in involved with ground-based, balloon-borne and satellite-based cosmology and astrophysics experiments. Researchers at Brown are collaborating on telescope projects in Arizona and Chile, and balloon flights launched from Texas and the Antarctic. Using equipment designed and built at Brown, as well as the National Laboratories, data are recorded in experimental runs and analyzed at Brown with extensive use of computer systems. (b) The Physics Department is an active participant in Brown's Institute for Molecular and Nanoscale Innovation (IMNI), an umbrella organization which supports centers and collaborative research teams in targeted areas of the molecular and nanosciences. IMNI is a "polydisciplinary" venture with 60 faculty participants representing nine departments across campus. IMNI serves as a focal point for interaction with industry, government, and affiliated hospitals. (c) Physics is also associated with the Institute for Brian and Neural Systems and the Center for Biomedical Engineering at Brown. (d) Extensive computer facilities are available. These include a variety of Windows and Linux/UNIX workstations within the department, all of which are connected via Ethernet. In addition, the department has several powerful Zinux clusters for dealing with problems needing large scale computation. Several high speed network links provide worldwide access to experimental facilities, and enable extensive and efficient use of national supercomputing centers. A department web server provides access to personal home pages of faculty, staff, and students, as well as general departmental information.

Table B—Appointments to Graduate Students, 2009–10

Title of Appointee	Appointments: Total	Appointments: First year	Academic Load Allowed in Credit Hours[1]	Hours of Service Per Week	Stipend for Academic Year ($)[4]
			Semester		
Teaching Assistant	26	15	12	15[2]	19,500
Research Assistant	60	1	12	15[3]	19,500
Fellowships	12	3	16	-	19,500
Total	98	19			

[1]At Brown, one course is equivalent to four semester hours.
[2]One-half teaching and one-half preparation.
[3]Research.
[4]Plus tuition, health services fee, and health insurance.

5. Personnel Engaged in Separately Budgeted Research, 2009–10

Professorial faculty	25
Postdoctoral appointments	21
Total	46

6. Separately Budgeted Research Expenditures by Source of Support

	Departmental Research	Physics-related Research Outside Department
Institution	1,182,730	Information not available to Physics Dept.
Federal government	5,526,780	
Private, non-profit organizations	139,027	
Business and industry	199,517	
Other	1,387,046	
Total	8,435,100	

Table C—Separately Budgeted Research Expenditures

Research Specialty	No. of Grants	Expenditures ($)
Astrophysics	11	1,539,473
Biophysics	13	1,332,417
Condensed Matter Physics	23	1,935,414
Particles & Fields	26	3,627,796
Total	73	

FACULTY

Professors

Baird, James C.,[1] Ph.D., Rice, 1959. Experimental atomic physics.

Cooper, Leon N., Ph.D., Columbia, 1954. Thomas J. Watson, Sr. Professor of Science. Neural studies; theoretical condensed matter physics.

Cutts, David, Ph.D., California, Berkeley, 1968. Experimental high-energy physics.

Gaitskell, Richard, Ph.D., Oxford University, 1993. Experimental high-energy physics.

Guralnik, Gerald S., Ph.D., Harvard, 1964. Theoretical high-energy physics.

Heintz, Ulrich, Ph.D.,SUNY, Stony Brook, 1991. Experimental high-energy physics.

Jevicki, Antal, Ph.D., CUNY, 1976. Theoretical high-energy physics.

Kosterlitz, J. Michael, Ph.D., Oxford, 1969. Theoretical condensed matter physics.

Landsberg, Greg L., Ph.D., State Univ. of New York, Stony Brook, 1994. Experimental high energy physics

Ling, Xinsheng, Ph.D., Univ. of Conn., 1992. Experimental condensed matter physics, biological physics.

Maris, Humphrey J., Ph.D., Imperial College, 1963. Experimental and theoretical condensed matter physics.

Marston, J. Bradley, Ph.D., Princeton, 1989. Theoretical condensed matter physics.

Narain, Meenakshi, Ph.D., State University of New York, Stony Brook, 1991. Experimental high-energy physics.

Nurmikko, Arto W.,[2] Ph.D., California, Berkeley, 1971. Experimental semiconductor physics.

Pelcovits, Robert A., Ph.D., Harvard, 1978. Theoretical condensed matter physics.

Stratt, Richard M.,[1] Ph.D., California, Berkeley, 1979. Theoretical condensed matter physics.

Tan, Chung-I, Ph.D., California, Berkeley, 1968. Theoretical high-energy physics.

Valles, James M., Jr., Ph.D., Univ. of Mass., 1988. Experimental condensed matter physics, biological physics.

Xiao, Gang, Ph.D., Johns Hopkins, 1988. Experimental condensed matter physics.

Xu, Jimmy,[2] Ph.D., Minnesota, 1987. Experimental condensed matter physics.

Ying, See-Chen, Ph.D., Brown, 1968. Theoretical condensed matter physics, biological physics.

Zaslavsky, Alexander,[2] Ph.D., Princeton University, 1991. Experimental condensed matter physics.

Associate Professors

Dell'Antonio, Ian P., Ph.D., Harvard, 1995. Experimental astrophysics.

Lowe, David A., Ph.D., Princeton, 1993. Theoretical high energy physics.

Mitrovic, Vesna, Ph.D., Northwestern University, 2001. Experimental condensed matter.

Tang, Jay X., Ph.D., Brandeis University, 1995. Experimental condensed matter, biological physics.

Tucker, Gregory S., Ph.D., Princeton, 1991. Experimental astrophysics.

Assistant Professors

Feldman, Dmitri, Ph.D., Landau Inst. for Theoretical Phys., 1998. Theoretical condensed matter.

Koushiappas, Savvas, Ph.D., Ohio State University, 2004. Experimental Astrophysics.

Spradlin, Marcus, Ph.D., Harvard University, 2001. Theoretical high-energy physics.

Stein, Derek, Ph.D., Harvard University, 2003. Experimental condensed matter physics, biological physics.

Volovich, Anastasia, Ph.D., Harvard University, 2002. Theoretical high-energy physics.

Professors (Research)

Elbaum, Charles, Ph.D., Toronto, 1954. Neural Studies, experimental condensed matter physics.

Fried, Herbert M., Ph.D., Stanford, 1957. Theoretical high-energy physics.

Gerritsen, Hendrik J., Ph.D., Leiden, 1955. Experimental nonlinear optics; Experimental condensed matter physics.

Lanou, Robert E., Jr., Ph.D., Yale, 1957. Experimental high-energy physics.

Partridge, Richard, Ph.D., California Institute of Technology, 1984. Experimental high-energy physics.

Seidel, George M., Ph.D., Purdue, 1958. Experimental condensed matter physics.

Professor of Physics (MBL)

Oldenbourg, Rudolf, Ph.D., University of Konstanz, Germany, 1981. Condensed matter physics.

Assistant Professor (Research)

Li, Guanglai, Ph.D., Nanjing University, 1999. Experimental condensed matter.

Speer, Thomas, Ph.D., University of Geneva, 2000. Experimental high-energy physics.

Adjunct Professors (Research)

Ala-Nissila, Tapio, Ph.D., Temple, 1986. Theoretical condensed matter physics.

Brower, Richard, Ph.D., University of California-Berkeley, 1996. Theoretical high energy physics.

Granato, Enzo, Ph.D., Brown University., 1986. Condensed matter theory.

Lawandy, Nabil, Ph.D., Johns Hopkins University, 1980. Experimental quantum optics.

Adjunct Associate Professor

Targan, David, Ph.D., Minnesota, 1988. Astronomy.

Adjunct Assistant Professors (Research)

Antonelli, G. Andrew, Ph.D., Brown University, 2001. Experimental condensed matter physics.

Morath, Christopher, Ph.D., Brown University, 1992. Experimental condensed matter physics.

[1]Principal appointment in Chemistry.
[2]Principal appointment in Engineering.

RESEARCH SPECIALTIES AND STAFF

Theoretical

Physics of Elementary Particles. Current activities include studies in quantum field theory, quantum chromodynamics, gauge/gravity duality, nonperturbative methods in field theory, solitons, monopoles, spontaneous symmetry breaking, lattice field theories, renormalization group, field theoretic approaches to condensed matter, gauge theories of weak and electromagnetic interactions, grand unification theory and phenomenology, phenomenology of scattering and production processes, the quantum theory of gravitation, super symmetry, supergravity, superstrings, and cosmology. Guralnik, Jevicki, Lowe, Spradlin, Tan, Volovich.

Physics of Condensed Matter. Research problems currently under investigation include interference and interaction in mesoscopic systems including the quantum Hall effect, quantum wires and quantum phase transitions; strongly correlated electrons in layered materials; development of a non-equilibrium statistical mechanics of planetary climates; non-equilibrium transport in nanostructures; modeling actinide complexes in aqueous solution; liquid crystal physics and its interface with biology; dynamics of biopolymers in nanochannels; strain relaxation and dynamics of heteroepitaxial nanostructures; microscopic theories of friction; out of equilibrium systems in the presence of weak stochastic noise; ultrafast dynamics in liquids. Feldman, Kosterlitz, Marston, Pelcovits, Stratt, Ying.

Astrophysics and Cosmology. Cosmological models for structure formation and particle-physics predictions of the nature of the dark matter are analyzed. Computational and analytic tools are used to predict the distribution of matter on sub- and super-galaxy scales and to aid in the design of the next generation of cosmological experiments. Koushiappas.

Neural Science. The major goal of the research is to elucidate the biological mechanisms that underlay learning and memory: to find principles of organization that can account both for experimental data on the cellular level. Among the detailed objectives are the following: to clarify the dependence of learning on synaptic modification; to elucidate the principles that govern synapse formation or modification; to use principles of organization that can account for observations on a cellular level to construct network models that can learn, associate and reproduce such higher level cognitive acts as abstraction, computation and language acquisition. Cooper.

Experimental

Physics of Elementary Particles. The properties of elementary particles and their interactions are being investigated, with current effort focused on the study of proton-proton collisions at the highest available energy with the CMS experiment at the Large Hadron Collider at CERN and proton-antiproton collisions at the previous energy frontier facility: the DØ experiment at Fermilab Tevatron accelerator. The CMS program is focused on searches for new particles, forces, and properties of space-time, beyond the predictions of the Standard Model of particle physics. That includes searches for supersymmetry and other heavy partners of the known particles, extra spatial dimensions, and new forces. In addition to this avenue to discovery, the CMS and DØ programs include precision measurements of the properties of electroweak and strong interactions, in particular measurement of the top-quark properties. An important component of the current DØ and near-future CMS program is the search for the last missing piece of the Standard Model-the Higgs boson. High-performance LHC Computing Grid networking and videoconferencing facilities provide tight links between Brown, CERN, and Fermilab. Local computer cluster connected to the Grid allows for massive parallel computing support of the DØ and CMS physics program. Cutts, Heintz, Landsberg, Narain.

Physics of Condensed Matter. Studies of magnetoconductive, optical, and mechanical properties of amorphous metals and semiconductors, magnetic solids and high T_c superconductors; mesoscopic superconducting arrays and colloidal model systems; flux lattices in type-II superconductors; low-temperature scanning probe techniques and devices; giant and colossal magnetoresistance effects in magnetic superlattices, granular solids and oxides; electron-electron interactions in two dimensional electron systems at low temperatures, electronic and magnetic properties of artificial superlattices, quantum wires and dots; the quantum Hall effect; nonlinear optical phenomena and plasma dynamics studies in semiconductors using picosecond and femtosecond laser pulses; studies of ultrasonic and thermal properties of solids using picosecond laser pulses; properties of liquid and solid ^{4}He and ^{3}He, including elementary excitations and their interactions cavitation and levitation of superfluid ^{4}He; NMR studies of the structure and bonding of crystalline and glassy solids; nuclear quadrupole resonance (NQR) studies of electronic distributions in molecules of biological importance and in inorganic solids; studies of chemisorption phenomena, surface band structure, reconstruction and two-dimensional phase transitions by low-energy electron diffraction, photoelectron spectroscopy. Baird, Ling, Maris, Mitrovic, Nurmikko, Stein, Tang, Valles, Xiao, Xu, Zaslavsky.

Astrophysics and Cosmology. The origins and evolution of the universe are being measured. Topics of research include the following: Studies of the Cosmic Microwave Background from satellite, balloon-borne and ground-based missions, measurements of the CMB to measure properties of the early Universe. A parallel effort in sub-millimeter cosmology is being carried out using the BLAST balloon-borne observatory to study the epoch of formation of the first galaxies and the dynamics of star formation in our own galaxy. Wide-field optical and near-imaging surveys are being carried out with telescopes in Arizona and Chile to map out the gravitational lensing signal and measure the shear correlation function and the growth of clustering over cosmic time to measure the evolution of the dark energy equation of state. Studies of mass substructure from gravitational lensing maps of clusters of galaxies taken with HST and ground-based telescopes are being used to measure the clustering properties of Dark Matter. Investigations using the next generation of wide-field survey instruments to map the galaxy group and cluster distribution out to high redshift are being planned. Studies of the galaxy interaction and star formation properties through optical photometry, spectroscopy, NIR photometry and radio spectral line observations are being carried out. A large experimental program for the direct detection of particle dark matter is also being coordinated by Brown, including the XENON10, LUX and LZ20 experiments. The experiments are constructed and operated in deep underground laboratories, such as Gran Sasso, Italy and the new national underground facility at the Sanford Lab located in the Homestake Mine, South Dakota. The goal is to identify the rare events associated with exotic particles, that could make up the dark matter halo of the Milky Way, directly interacting within the experiments' detectors. Dell'Antonio, Koushiappas, Gaitskell, Tucker.

Biological Physics. Research problems currently under investigation include: development of single-molecule DNA sequencing technology using solid-state nanoproes; electro-fluidics for single molecule biophysics; electronic DNA barcode sequencing; electrokinetic energy harvesting in the presence of hydrodynamic slip; probing the sequence and dynamics of single DNA molecules using solid-state nanopores, optical tweezers, and binding proteins; DNA sequencing using nanopore mass spectrometry; biophysical mechanism of bacterial swimming and adhesion; biomechanics of actin networks regulated by physical mechanisms; mechanics of intracellular pathogens and biomimetic systems propelled by actin comet tails; neutrophil mechano-sensing using traction microscopy; swimming and force sensing on microorganisms. Ling, Stein, Tang, Valles.

CLEMSON UNIVERSITY

DEPARTMENT OF PHYSICS AND ASTRONOMY

Clemson, South Carolina 29634-0978

Students Accepted For Degree	FIELDS		
	Physics	Astronomy	Related Fields
Doctorate	X	X	X
Master's	X	X	X

1. General

President: James F. Barker
Dean of Graduate School: J. Bruce Rafert
Department Chairman: P. A. Barnes
Department Telephone Number: (864) 656-3416
Type of Institution: University
Control: Public
Setting: Small town
Total Faculty: 1,452
Total Students: 19,111
Total Graduate Students: 3,765
Graduate Tuition:
Residents: Full-time-$4,126/semester
Part-time-$525/credit hr
Non-residents: Full Time-$8,210/semester
Part Time-$1,050/credit hr
Tuition rates for: 2010–11
Deferred tuition plan: Yes
Term: Semester

2. Number of Faculty in Department

The combined total of full-time faculty in the three professorial ranks is 25. The combined total of full-time, part-time, and other faculty at all ranks is 35.

3. Admission, Financial Aid, and Housing

Address admission inquiries to: Dr. Murray Daw, Graduate Student Recruiter, Dept. of Physics and Astronomy. Foreign students should apply directly to the Graduate School for a special self-managed application package.
Graduate application fee required: $70 domestic, $80 international
Admission deadline (Fall admission): 12/15
International students: Jan. 15
Admission information: For fall admission, 2009–10, 28 students were accepted from 71 applicants.
Admission requirements: For admission to the graduate program, a Bachelor's degree is required with no minimum undergraduate GPA specified. Usual preparation is under-graduate major in physics. Students from other fields will have an opportunity to make up deficiencies. The GRE Aptitude and Advanced tests are required. No minimum score is specified for those tests. The average GRE scores are verbal 560; quatitative 715. Students from non-English speaking countries are required to demonstrate proficiency in English via the TOEFL exam. Minimum TOEFL scores of 90 (internet), or 570 (paper) ensure that international students will be considered for admission and assistantships. We will often also conduct a phone interview with the applicant. To be considered for a teaching assistantship, these students must also demonstrate a proficiency in spoken English by an acceptable score on the TSE.
Undergraduate preparation assumed: Courses based upon texts such as Hecht, *Optics*, Griffiths, *Introduction to Electrodynamics*, Marion and Thornton, *Mechanics*, Eisberg, *Modern Physics*, Stowe, *Introduction to Statistical Mechanics and Thermodynamics*, Griffiths and Townsend, *Quantum Physics*. Included with these standard areas of study should be an advanced undergraduate laboratory in Experimental Physics, mathematics including Differential Equations, Complex Variable, Fourier Analysis and Operational Mathematics. Some knowledge of computer programming including standard methods using Mathematica, Maple, MatLab, etc., will also be helpful.
Address financial aid inquiries to: Office Manager, Dept. of Physics and Astronomy
GAPSFAS application required: No
Financial aid deadline: None
Loans available: Yes
Address housing inquiries to: Housing Office, 200 Mell Hall
On-campus, single student housing available: Yes
Cost/semester: $1,730–2,950
On-campus, married student housing available: No

Table A—Faculty, Enrollments, and Degrees Granted

Research Specialty	2009–10 Faculty	Enrollment Fall 2009		No. of Degrees Granted[1] 2009–2010		
		Master's	Doctorate	Master's	Terminal Master's	Doctorate
Astronomy	2	0	0			
Astrophysics	3	1	14		1	3
Atmos./Space Phys.	5	0	3	2		
Biophysics	4	1	4		1	
Condensed Matter/ MaterialPhysics	8	4	11		1	2
Radiation Physics	1					
Solid State/ Surface Physics	1	0	6			
Total	24	6	38	2(38)	3	5(22)
Full-time Grad. Stud.		6	38			
Part-time Grad. Stud.		0	0			
First-year Grad. Stud.			17			
Undergraduate Degrees, 2009–10 (2005–10): 19(83)						

[1] Five-year totals in parentheses.

4. Graduate Degree Requirements

Master's: 30 semester credits of course work are required including six credits of thesis research with an oral defense; for non-thesis degrees 36 credits are needed. A grade point average of 3.0 is required. There are no foreign language requirements.
Doctorate: Students entering the Ph.D program directly have the same course requirements as those entering the master's program. In addition, the student must successfully complete a departmental study plan comprising a suitable number of courses, and complete and defend a dissertation based on original research work. The student must pass a set of Ph.D. qualifying exams. A 3.0 grade point average is required. There are no foreign language requirements.
Thesis: Thesis may be written *in absentia*.
Special Equipment, Facilities, or Programs: The department is

housed in the four-story 64,000 sq. ft. physics and astronomy building. A fully equipped machine shop and computing facilities are available. Office space is provided for graduate students. Extensive research facilities include a scanning tunneling microscope nanomaterial processing laboratory with electric arc discharge, pulsed laser vaporization and CVD synthesis capabilities, bulk and thin film thermoelectric materials growth facilities, Raman scattering, infrared/visible spectroscopy, electron microscopy, atomic force microscopy, and electrical transport measurements used extensively for characterizing carbon nanotubes, nanodiamond, semiconducting oxide nanobelts and nanowires. Access is also available to the SARA and Super Lotis telescopes.

Table B—Appointments to Graduate Students, 2009–10

Title of Appointee	Appointments		Academic Load Allowed in Credit Hours	Hours of Service Per Week	Stipend for Academic Year ($)[1]
	Total	First year			
			Semester		
Teaching Assistant	30	14	12	20	5,334–16,000
Research Assistant	12	3	6	20	4,500–32,659
TA & RA	4				10,334–29,050
Total	42	17			

[1]Stipends are based on the academic year. Graduate student tuition is waived for TA's and RA's; however, graduate students are required to pay their graduate student fees. Teaching assistants are not required to teach during the summer as long as they are working toward their degree.

5. Personnel Engaged in Separately Budgeted Research,

Professorial faculty	24
Postdoctoral appointments	0
Graduate students	45
Undergraduate students	13
Total	82

Table C—Separately Budgeted Research Expenditures

Research Specialty	No. of Grants	Expenditures ($)
Astronomy	15	
Astrophysics	8	
Atmos./Space Phys.	10	
Biophysics	2	
Condensed Matter/ Material Physics	7	
Solid State/Surface Physics	3	
Total	45	

6. Separately Budgeted Research Expenditures by Source of Support

	Departmental Research
Federal government	$2,775,666
State government	$249,498
Private Sources	93,562
Total	$3,118,726

7. Extension Centers and Summer Programs

Graduate and undergraduate courses are offered and research is carried on during two five week summer terms.

FACULTY

Professors

Barnes, Peter A., Ph.D., Simon Fraser University, 1969. Department Chair. Condensed Matter Physics.

Daw, Murray S., Ph.D., Caltech, 1981. Solid state theory; structure and dynamics of defects in solids.

Hartmann, Dieter H., Ph.D., California, Santa Cruz, 1989. Gamma-ray astronomy; nucleosynthesis; galactic structure.

Larsen, M. F., Ph.D., Cornell, 1979. Atmospheric/space physics.

Leising, Mark D., Ph.D., Rice, 1987. Gamma-ray astronomy; supernovae.

McNulty, Peter J., Ph.D., SUNY, Buffalo, 1965. Radiation physics.

Meyer, Bradley S., Ph.D., Chicago, 1989. Nuclear astrophysics; supernova theory; cosmology.

Meriwether, J. W., Jr., Ph.D., Maryland, 1970. Atmospheric space physics; optics.

Rafert, James B., Ph.D., University of Florida, 1978. Graduate Dean. Stellar astrophysics, remote sensing physics, hyperspectral imaging.

Rao, Apparao M., Ph.D., Kentucky, 1989. Condensed Matter Physics; nanomaterial synthesis, mechanical properties, chem-bio sensing, and solid state spectroscopy.

Tritt, Terry M., Ph.D., Clemson University, 1985. Experimentalist in Condensed Matter and Materials Physics: Electrical and Thermal Transport Phenomena.

Associate Professors

Alexov, Emil, Ph.D., in Physics, Sofia University, Bulgaria, 1990. Protein structure prediction and the use of the predicted structure to infer the biological properties of proteins.

Brittain, Sean, Ph.D., University of Notre Dame, 2004. Astrophysics, planet formation, circumstellar disks, spectroscopy.

Flower, Phillip J., Ph.D., Washington, Seattle, 1976. Stellar evolution; star clusters.

Ke, Pu-Chun, Ph.D., Victoria University, Australia, 2000. Molecular biophysics, optical microscopy, laser trapping.

King, J. R., Hawaii, 1993. Stellar abundances, stellar atmospheres, Galactic populations, high resolution spectroscopy.

Lehmacher, Gerald, Ph.D., Bonn University, 1993. Atmospheric physics, processes in the mesosphere and lower thermosphere.

Marinescu, D. C., Ph.D., Purdue, 1996. Condensed matter theory.

Oberheide, Jens, Ph.D., Wuppertal University, 2000. Atmospheric and geospace physics, climate and weather of the Sun-Earth system.

Sosolik, Chad, E., Ph.D., Cornell University, 2001. Surface physics.

Assistant Professors

Drymoitis, Fivos R., Ph.D., Florida State University, 2002. Condensed matter physics, crystal growth.

He, Jian, Ph.D., The University of Tennessee, 2004. Condensed matter physics. Single crystal growth and characterizations, funtional nanocomposite materials.

Kang, Hye Jung, Ph.D., University of Tennessee, 2005. Condensed matter physics; experimentalist using neutron scattering.

Spano, Meredith Newby, Ph.D., Florida State University, 2002. Molecular biophysics.

Tewari, Sumanta, Ph.D., University of California, Los Angeles, 2003. Condensed matter theory, topological quantum computation, high temperature superconductivity, cold atomic physics in optical traps and lattices.

Lecturers

Brown, Jason, Ph.D., Clemson University, 1999. Physics.

Pope, Amy, Ph.D., Clemson University, 2002. Condensed matter physics, physics instruction.

The, Lih-Sin, Ph.D., Arizona, 1989. Gamma-ray astronomy, supernova remnants, stellar nucleosynthesis.

Research Professors

Fesen, Casandra, Ph.D., University of Michigan, 1981. Modeling large scale thermosphere/ionosphere dynamics.

Skove, Malcolm J., Ph.D., Virginia, 1960. Alumni Professor Emeritus of Physics. Superconductivity; transport effects in whiskers.

Research Associates

Lowhorn, Nathan, Ph.D., Clemson University, SC, 2005.

Ponnambalam, Vijayabarathi, Ph.D., Indian Institute of Technology, 1997. Solid state physics.

RESEARCH SPECIALTIES AND STAFF

Theoretical

Astrophysics. Nucleosynthesis; space astrophysics; stellar atmospheres; cosmic rays; gamma-ray bursts; supernova theory; origin of solar system; stellar evolution; cosmology. Brittain, Hartmann, King, Leising, Meyer.

Atmospheric/Space Physics. Studies of atmospheric wave dynamics, propagation and their interaction with chemical and airglow processes; ionospheric electrodynamics, plasma; Climate and Weather of the Sun-Earth System. Larsen, Lehmacher, Meriwether, Oberheide.

Biophysics. DNA repair mechanisms; quantum biology. Ke, Spano.

Computational Biophysics and Bioinformatics. Predicting protein-protein interactions and 3D structures of the protein-protein complexes; computer simulations of protein-protein interactions, the role of conformation changes, pH and salt concentration; sequences alignment, multiple sequence alignment and correlated mutations analysis; investigation of the role of plastocyanin/ferredoxin dimorphism in the photosynthesis; analysis of the molecular basis of Huntington disease. Alexov.

Particle, Nuclear, and Radiation Physics. Radiation in space; effects of ionizing radiation on microelectronic circuits; single event effects in microelectronic and optoelectronic circuits, nuclear spallation reactions, dosimetry and microdosimetry, interaction of radiation with matter; interaction of EM radiation with atoms and molecules. McNulty.

Condensed matter. Surface phenomena, including scattering; anharmonic effects in crystal lattices; magnetic, optic, transport properties of semiconductor mesoscopic structures (quantum wells, superlattices); broken symmetry states; charge and spin density waves; non-equilibrium superconductivity, topological computation, high temperature superconductivity, cold atomic physics in optical traps and lattices. Daw, Marinescu, Tewari.

Experimental

Astronomy. Gamma-ray astronomy; observational astronomy; stellar evolution. Stellar atmospheres, circumstellar evolution, planet formation, stellar accretia, abundance determinations, Galactic chemical evolution, starclusters close binary star systems. Brittain, Hartmann, King, Leising, Flower, Rafert, The.

Atmospheric/Space Physics. Rocket, radar, and spacecraft studies of ionospheric dynamics, electrodynamics, and plasma physics; studies of atmospheric dynamics and composition with lidar and Fabry-Perot systems. Sounding rocket instrumentation, density, temperature, and turbulence in the mesosphere and lower thermosphere. Satellite data analysis of vertical coupling processes. Larsen, Lehmacher, Meriwether, Oberheide.

Biophysics. Biomedical optical imaging; fluorescence tomography; optical spectroscopy; microwave impaging; ultrasound tomography; bioluminescence tomography; x-ray tomosynthesis; and spectroscopy; DNA repair mechanisms; structure of biological molecules; biological effects of radiation damage; mechanisms of carcino-genesis; mechanisms for single event effects in microelectronics, microdosimetry using microelectronic technology, structural biology of RNA, biomolecular structure function relationships, NMR spectroscopy, fluorescence spectroscopy, single molecule biophysics, nanoscience. Ke, Spano, Barnes, McNulty.

Condensed Matter Physics. Synthesis of nanostructured materials using electric arc discharge, pulsed laser vaporization, neutron scattering, and CVD methods. Optical characterization of novel materials by Raman scattering, infrared/visible and fluorescence spectroscopy. Mechanical properties of one-dimensional materials and chem-bio sensing using harmonic detection of resonance technique. Superconducting nanotubes. Rao, Kang.

Materials Physics. Thermoelectric materials and applied physics; thermophysical properties of novel materials, including investigations of low temperature heat capacity and thermal transport; investigations of thermal, magnetic and electronic transport properties of exotic systems, low-dimensional conductors, strongly electron correlated materials and phase transition materials; high temperature thermophysical properties of novel materials; synthesis of thermoelectric nanomaterials and composite thermoelectrics. Tritt, Drymiotis, He.

Surface Physics. Atomic and molecular beam interactions at surfaces; formation and characterization of surface nanostructures with an energetic beam and scanning tunneling microscope. Sosolik.

FACULTY PUBLICATIONS

Alexov, Emil

Teng, S., Madej, T., Panchenko, A., and Alexov, E. "Modeling Effects of Human Single Nucleotide Polymorphisms on Protein-Protein Interactions," Biophys. J. **76**, 2178-2188 (2009).

Talley Kemper, Ng Carmen, Snoppell Michael, Kundrotas Petras and Alexov E. "On the electrostatic component of protein-protein binding free energy" PMC Biophysics 1:2, 1-23 (2008).

Petras J. Kundrotas, Marc F. Lensink, Alexov, E. "Homology-based modeling of 3D structures of protein-protein complexes using alighnments of modified sequence profiles" Int. J. Biol. Macromol. **43**, 198-208 (2008).

Teng S., Michonova-Alexova E and Alexov, E. "Approaches and Resources for prediction of the Effects of Non-Synonymous Single Nucleotide Polymorphism on Protein Function and Interactions" Current Pharmaceutical Biotech. **9**, 123-133 (2008).

Brittain, S. D.

E. L. Gibb, S. D. Brittain, T. W. Rettig, M. Troutman, T. Simon, and C. Kulesa, *CO and* H_3^+ *Toward MWC 1080, MWC 349 and LkH*α*101*, Astrophysical J. 715 (2010).

D. A. Tilley, D. Balsara, S. Brittain, and T. Rettig, *Dust Settling in Magnetorotationally-Driven Turbulent Discs II: The Pervasiveness of the Streaming Instability and its Consequences*, Monthly Notices of the Royal Astronomical Society **403**, 211 (2010).

S. D. Brittain, T. Simon, T. Rittig, E. Gibb, and J. Liskowsky, *The Ongoing Outburst of V1647 Ori*, Astrophysical Journal **708**, 85 (2010).

S. Brittain, J. Najita, and J. Carr, *Tracing the inner edge of the disk around HD 100546 with Rovibrational CO Emission Lines*, Astrophysical Journal **702**, 85 (2009).

C. A. Grady, G. Schneider, M. L. Sitko, G. M. Williger, K. Hamaguchi, S. D. Brittain, K. Ablordeppey, D. Apai, L. Beerman, W. J. Carpenter, K. A. Collins, M. Fukagawa, H. B. Hammel, Th. Henning, D. Hines, R. Kimes, D. K. Lynch, F. Menard, R. Pearson, R. W. Russell, M. Silverstone, P. Smith, M. Troutman, D. Wilner, and B. Woodgate, *The Shadowy Life of SAO 206462's Transitional Disk*, Astrophysical Journal **699**, 1822 (2009).

Daw, M. S.

E. Ertekin, M. Daw, and D. C. Chrzan, "Topological description of the Stone-Wales defect formation energy in carbon nanotubes and graphee," Physical Review B **79**, 155421 (2009). Selected for April 27, 2009 issue of Virtual Journal of Nanoscale Science and Technology.

X. Gao and M. S. Daw, "First-principles super-cell investigation of rattling effect in Li-doped KCl," J. Phys.: Condens Matter **21**, a045401 (2009).

J. ballato, T. Hawkins, P. Foy, B. Kokuoz, R. Stolen, C. McMillen, M. Daw, Z. Su, T. M. Tritt, M. Dubinskii, J. Zhang, T. Sanamyan, and M. J. Matthewson, "On the Fabrication of all-glass optical fibers from crystals," J. Appl. Phys. **105**, 053110 (2009).

V. Ponnambalam*, X. Gao, S. Lindsey, P. Alboni, Z. Su, B. Zhang, F. Drymiotis, M. S. Daw, and T. M. Tritt, "Thermoelectric properties and electronic structure calculations of low thermal conductivity Zintl phase series M16X11 (M=Ca and Yb; Z=Sb and Bi)," J. Alloys and Compounds **484**, 80–85 (2009).

D. Dickel and M. S. Daw, "Improved calculation of vibrational mode lifetimes in anharmonic solids-Part I: Theory," Computational Materials Science **47**, 698–704 (2010).

Drymoitis, Fivos

F. R. Drymiotis, "Resonant Ultrasound Spectroscopy: Overview and Applications," International Journal of Modern Physics B **24**, 1047–65 (2010).

F. R. Drymiotis, T. B. Drye, Y. S. Wang, J. He, D. Rhodes, K. Modic, S. Cawthorne, and Q. R. Zhang, "Structure formation and very low thermal conductivity in Pb:Te:Ag:Se mixtures," Journal of Applied Physics **107** (2010).

F. Drymiotis, T. Drye, D. Rhodes, Q. Zhang, J. C. Lashey, Y. Wang, S. Cawthorne, B. Ma, S. Lindsey, and T. Tritt, "Glassy thermal conductivity in the two-phase $Cu_xAg_{3-x}SbSeTe_2$ alloy and high temperature thermoelectric behavior," Journal of Physics-Condensed Matter **22** (2010).

V. Ponnambalam, X. Gao, S. Lindsey, P. Alboni, Z. Su, B. Zhang, F. Drymiotis, M. S. Daw, and T. M. Tritt, "Thermoelectric properties and electronic structure calculations of low thermal conductivity Zintl phase series M16X11 (M=Ca and Yb; Z=Sb and Bi)," Journal of Alloys and Compounds **484**, 80-5 (2009).

Hartmann, Dieter H.

J. Greiner *et al.*, "GRB 080913 at redshift 6.7," The Astrophysical Journal **693**, 1610 (2009).

Y. Mizuno, B. Zhang, B. Giacomazzo, K. Nishikawa, P. Hardee, S. Nagataki, and D. Hartmann, "Magnetohydrodynamic effects in propagating relativistic jets: Reverse shock and magnetic acceleration," The Astrophysical Journal Letters **690L**, 47 (2009).

D. Hartmann, "Probing cosmic chemical evolution with GRBs," New Astronomy Reviews **52**, 450 (2008).

R. Voss, R. Diehl, D. Hartmann, and K. Kretschmer, "Population synthesis models for ^{26}Al production in starforming regions," New Astronomy Reviews **52**, 436 (2008).

A. Updike *et al.*, "The rapidly flaring afterglow of the very bright and energetic GRB 070125," The Astrophysical Journal **685**, 361 (2008).

He, Jian

X. Xu, A. F. Bangura, J. G. Analytis, J. D. Flecther, M. M. J. French, N. Shannon, J. He, S. Zhang, D. G. Mandrus, R. Jin, and N. E. Hussey, "Directional Field-induced Metallization of Quasi-one-dimensional $Li_{0.9}Mo_6O_{17}$," Phys. Rev. Lett **102**, 206602 (2009).

S. N. Zhang, J. He, X. H. Ji, Z. Su, S. H. Yang, T. J. Zhu, X. B. Zhao, and T. M. Tritt, "Effects of Ball-milling Atmosphere on the Thermoelectric Properties of TAGS-85 Compounds," J. Elec. Mater. **38**, 1141 (2009).

F. Wang, J. V. Alvarez, J. W. Allen, S.-K. Mo, J. He, R. Jin, D. Mandrus, and H. Höchst, "Quantum Critical Scaling in the Single Particle Spectrum of a Novel Anisotropic Metal," Phys. Rev. Lett. **103**, 136401 (2009).

Y. Cui, J. He, G. Amow, and H. Kleink, "Thermoelectric Properties of n-type Double Substituted $SrTiO_3$ Bulk Materials," Dalton Trans. **39**, 1031 (2010).

K. Yang, J. He, Z. Su, J. B. Reppert, M. J. Skove, T. M. Tritt, and A. M. Rao, "Inter-tube Bonding, Graphene Formation and Anisotropic Transport Properties in Spark Plasma Sintered Multi-wall Carbon Nanotube Arrays," Carbon **48**, 76 (2010).

Kang, Hye Jung

M. Kofu, S.-H. Lee, M. Fujita, H.-J. Kang, H. Eisaki, and K. Yamada, "Hidden Quantum Spin-Gap State in the Static Stripe Phase of High-Temperature $La_{2-x}Sr_xCuO_4$ Superconductors," Phys. Rev. Lett. **102**, 047001 (2009).

M. Kofu, J.-H. Kim, S. Ji, S.-H. Lee, H. Ueda, Y. Qiu, H.-J. Kang, M. A. Green, and Y. Ueda, "Weakly Coupled s=1/2 Quantum Spin Singlets in $Ba_3Cr_2O_8$," Phys. Rev. Lett. **102**, 037206 (2009).

J. Wen, Z. Xu, G. Xu, J. M. Tranquada, G. Gu, S. Chang, and H. J. Kang, "Magnetic field induced enhancement of spin-order peak intensity in $La_{1.875}Ba_{0.125}CuO_4$," Phys. Rev. B **78, 212506 (2008).**

C. Stock, C. Broholm, J. Hudis, H. J. Kang, and C. Petrovic, "Spin Resonance in the d-Wave Superconductor $CeCoIn_5$," Phys. Rev. Lett. **100**, 087001 (2008).

S. Li, Z. Yamani, H. J. Kang, K. Segawa, Y. Ando, X. yao, H. A. Mook, and P. Dai, "Quantum spin excitation through the metal-to-insulator crossover in $YBa_2Cu_3O_{6+y}$," Phys. Rev. B **77**, 014523 (2008).

Ke, Pu-Chun

R. Chen, T. A. Ratnikova, M. Stone, S. Lin, M. Lard, G. Huang, J. S. Hudson, and P. C. Ke, "Differential Uptake of Carbon Nanoparticles by Plant and Mammalian Cells," Small **6**, 612 (2010).

L. Monticelli, E. Salonen, P. C. Ke, and I. Vattulainen, "Effects of Carbon Nanoparticles on Lipid Membranes: A Molecular Simulation Perspective," Soft Matter **5**, 4433 (2009).

J. Shang, T. A. Ratnikova, S. Anttalainen, E. Salonen, P. C. Ke, and H. T. Knap, "Experimental and Simulation Studies of Real-Time Polymerase Chain Reaction in the Presence of a Fullerene Derivative," Nanotechnology **20**, 415101 (2009).

S. Lin, P. Bhattacharya, D. E. Brune, N. Rajapakse, and P. C. Ke, "Effects of Quantum Dots Adsorption on Algal Photosynthesis," J. Phys. Chem. C **113**, 10962 (2009).

S. Lin, J. Reppert, Q. Hu, J. S. Hudson, M. L. Reid, T. Ratnikova, A. M. Rao, H. Luo, and P. C. Ke, "Uptake, Translocation and Transmission of Carbon Nanomaterials in Rice Plants," Small **5**, 1128 (2009).

King, Jeremy R.

E. J. Bubar and J. R. King, *Spectroscopic Abundances and Membership in the Wolf 630 Moving Group*, Astron. J., submitted (2010).

S. C. Schuler, A. L. Plunkett, and J. R. King, *FeI and Fe II Abundances of Cool Dwarfs int he Pleiades Open Cluster*, Publ. Astron. Soc. Pacif., submitted (2009).

J. R. King, S. C. Schuler, L. M. Hobbs, and M. H. Pinsonneault, *Li I and K I Scatter in Cool Pleiades Swarfs*, Astrophys. J. **710**, 1610 (2010).

S. C. Schuler, J. R. King, L.-S. The, *Stellar Nucleosynthesis in the Hyades Open Cluster*, Astrophys. J. **701**, 837 (20009).

Y. Chen, J. R. King, and A. M. Boesgaard, *The Origin of the Metal -Poor Common Proper Motion Pair HD 134439/ 134440: Insights from New Elemental Abundances*, Astron. J., submitted (2009).

Larsen, Miguel

D. L. Hysell, G. Michhue, M. Nicolls, C. Heinselman, and M. F. Larsen, Assessing auroral electric field variance with coherent and incoherent scatter radar, J. Atmos, Solar Terr. Phys., 2009. Doi:10.1016/j.jastp.2008.10.013.

L. Sangalli, D. J. Knudsen, M. F. Larsen, T. Zhan, R. F. Pfaff, and D. Rowland, Rocket-based measurements of ion velocity, neutral wind, and electric field in the collisional transistion region of the auroral ionosphere, J. Geophys. Res., **114**, A04306, 2009. Doi:10.1029/2008JA013757.

D. P. Drob, J. T. Emmert, G. Crowley, J. M. Picone, G. G. Shepherd, W. Skinner, P. Hays, R. J. Niciejewski, M. Larsen, C. Y. She, J. W. Meriwether, G. Hernández, M. J. Jarvis, D. P. Sipler, C. A. Tepley, M. S. O'Brien, J. R. Bowman, Q. Wu, Y. Murayama, S. Kawamura, I. M. Reid, and R. A. Vincent, An empirical model of the Earth's horizontal wind fields: HWM07, J. Geophys. Res., **113**, A12304 (2008). Doi:10:1029/2008JA013668.

D. L. Hysell, G. Michhue, M. F. Larsen, R. Pfaff, M. Nicolls, C. Heinselman, and H. Bahcivan, Imaging radar observations of Farley Buneman waves during the JOULE II experiment, Ann. Geophys., 1837-1850 (2008). Doi:www.ann-geophysics.net/26/1837/2008.

Lehmacher, Gerald A.

G. A. Lehmacher, E. Kudeki, A. Akgiray, L. Guo, P. Reyes, and J. Chau, Radar cross sections for mesospheric echoes at Jicamarca, Ann. Geophys. **27**, 2675–2684 (2009).

Moreels, G., J. Clairemidi, M. Faivre, D. Mougin-Sisini, M. N. Kouahla, J. W. Meriwether, G. A. Lehmacher, E. Vidal, and O. Veliz, Steroscopic imaging of the hydroxyl layer at low latitudes, Planet. Space Sci., **56**(11), 1467-1479 (2008).

L. Guo and G. Lehmacher, First meteor radar observations of tidal oscillations over Jicamarca (11.95° S, 76.87° W), Ann. Geophys. **27**, 2575–2583 (2009).

Leising, Mark D.

R. Diehl and M. Leising, "Gamma-Rays from Positron Annihilation. Proceedings of Science, 7th INTEGRAL Workshop, **67**, 001 (2009).

M. Leising and R. Diehl, INTEGRAL Studies of Nucleosynthesis Lines. Proceedings of Science, 7th INTEGRAL Workshop, **67**, 007 (2009).

M. Leising and R. Diehl, Gamma-ray Line Studies of Nuclei in the Cosmos. Proceedings of Science, 10th Symposium on Nuclei in the Cosmos **53**, March 2009.

Marinescu, D. Catalina

C. P. Moca, D. C. Marinescu, and S. Filip, "Spin Hall Effect in a Symmetric Quantum Well by a Random Rashba Field," Phys. Rev. B **77**, 193302 (2008).

Meriwether, John W.

Emmert, J. D. P. Drob, G. Hernández, M. J. Jarvis, J. W. Meriwether, R. J. Niciejewski, G. G. Shepherd, D. P. Sipler, and C. A. Tepley, DWM-07 global empirical model of upper thermospheric storm-induced disturbance winds, J. Geophys. Res., **113**, doi: 10.1029/2008JA013541, 2008.

Makela, Jonathan, J., John W. Meriwether, José P. Lima, Ethan S. Miller, Shaun J. Armstrong, The Remote Equatorial Nighttime Observatory of Ionospheric Regions Project and the International Heliospherical Year, Earth Moon Planet, doi: 10.1007/s11038-008-9289-0, 2009.

Moreels, G., J. Clairemidi, M. Faivre, D. Pautet, F. Rubio da Costa, P. Rousselot, J. W. Meriwether, G. A. Lehmacher, E. Vidal, J. L. Chau, G. Monnet, Near-infrared sky background fluctuations at mid- and low latitudes, Exp. Astron **22**, 87-107 (2008).

Meriwether, J. W. (2009), Thermospheric dynamics at low and mid-latitudes during magnetic storm activity, in Mid-Latitude Ionospheric Distrubances and Dynamics, edited by A. Coster, T. Fuller-Rowell, P. Kintner, A. Mannucci, and M. Medillo.

Meriwether, J., M. Faivre, C. Fesen, P. Sherwood, and O. Veliz (2008), New results on equatorial thermospheric winds and the midnight temperature maximum, Ann. Geophys. **26**, 447-466.

Meyer, Bradley S.

P. Hoppe, J. Leitner, B. S. Meyer, L.-S. The, M. Lugaro, and S. Amari, "An Unusual Presolar Silicon Carbide Grain from a Supernova: Implications for the Production of Silicon-29 in Type II Supernovae," Astrophys. J **691**, L691 (2009).

G. R. Huss, B. S. Meyer, G. Srinivasan, J. N. Goswami, SSahijpal, "Stellar sources of the short-lived radionuclides inthe early solar system," Geochim. Cosmochim. Acta. **73**, 4922 (2009).

M. F. El Eid, L.-S. The, and B. S. Meyer, "Massive Stars: Input Physics and Stellar Models," Space Sci. Rev. **147**, 1 (2009).

A. N. Krot, K. Nagashima, F. J. Ciesla, B. S. Meyer, I. D. Hutcheon, A. M. Davis, G. R. Huss, and E. R. D. Scott, "Oxygen Isotopic Composition of the Sun and Mean Oxygen Isotopic Composition of the protosolar Silicate

Dust: Evidence from Refractory Inclusions," Astrophys. J **713**, 1159 (2010).

A. N. Krot, K. Nagashima, M. Bizzarro, G. Huss, A. M. Davis, B. S. Meyer, and A. A. Ulyanov, "Multiple Generations of Refractory Inclusions in the Metal-Rich Carbonaceous Chondrites Acfer 182/214 and Isheyevo", Astrophys. J. **672**, 713 (2008).

Oberheide, Jens

H.-L. Liu, B. T. Foster, M. E. hagan, J. M. McInerney, A. Maute, L. Qian, A. D. Richmond, R. G. Roble, S. C. Solomon, R. R. Garcia, D. Kinnison, D. R. Marsh, A. K. Smith, J. Richter, F. Sassi, and J. Oberheide, Thermosphere Extension of the Whole Atmosphere Community Climate Model, J. Geophys. Res. submitted (2010).

F.-J. Lubken, J. Austin, U. Langematz, and J. Oberheide, Introduction to the Climate and Weather of the Sun Earth System special section, J. Geophys. Res., in press (2010).

D. Offermann, P. Hoffmann, P. Knieling, R. Koppmann, J. Oberheide, and W. Steinbrecht, Long-term Trends and Solar Cycle Variations of Mesospheric Temperature and Dynamics, J. Geophys. Res., in press (2010).

W. E. Ward, J. Oberheide, L. P. Goncharenko, T. Nakamura, P. Hoffmann, W. Singer, L. C. Chang, J. Du, D.-Y. Wang, P. Batista, B. Clemesha, A. H. Manson, C. Meek, D. M. Riggin, C.-Y. She, T. Tsuda, and T. Yuan, On the consistency of model, ground-based and satellite observations of tidal signatures: Initial results from the CAWSES tidal campaigns, J. Geophys. Res. **115**, D07107 (2010).

J. Oberheide, J. M. Forbes, K. Hausler, Q. Wu, and S. L. Bruinsma, Tropospheric tides from 80-400 km: propagation, inter-annual variability and solar cycle effects, J. Geophys. Res. **114**, D00105 (2009).

Rao, Apparao M.

D. Dickel, M. J. Skove, and A. M. Rao, "An Analytic Characterization of the Harmonic Detection of Resonance Method," J. Appl. Phys. **106**, 044515 (2009).

J. Haruyama, M. Matsudaira, T. Shimizu, J. Nakamura, T. Eguchi, T. Nishio, Y. Hasegawa, H. Sano, Y. Iye, J. Reppert, and A. M. Rao, "Pressure Dependence of Meissner Effect in Films of Ropes of Boron-doped Carbon Nanotubes," Superlattices and Microstructures **46**, 333 (2009).

R. Podila, W. Queen, A. Nath, A. Fazzio, A. Schonalez, J. He, G. Dalpian, M. J. Skove, S. J. Hwu, and A. M. Rao, "Origin of FM in Micro-and Nano Zinc Oxide," Nano Lett. **10**, 1383 (2010).

S. H. Park, P. Theilmann, K. Yang, A. M. Rao, and P. R. Bandaru, "The Influence of Coiled Nanostructure on the Enhance of Dielectric Constants and ELectromagnetic Shielding Efficiency in Polymer Composities," Appl. Phys. Lett. **96**, 043115 (2010).

K. Yang, J. He, Z. Su, J. B. Reppert, M. J. Skove, T. M. Tritt, and A. M. Rao, "Inter-Tube Nonding, Graphene Formation and Anisotropic Transport Properties in Spark Plasma Sintered Multi-Wall Carbon Nanotube Arrays," Carbon **48**, 756 (2010).

Sosolik, C.E.

M. P. Ray, R. E. Lake, L. B. Thomsen, G. Nielson, O. Hansen, I. Chorkendorff, and C. E. Sosolik, "Towards hot electron mediated charge exchange in hyperthermal energy ion-surface interactions," Journal of Physics: Condensed Matter **22**, 084010 (2010).

R. E. Lake, J. M. Pomeroy, and C. E. Sosolik, "Energy dissipation of highly charged ions on AL oxide films," Journal of Physics: Condensed Matter **22**, 084008 (2010).

M. P. Ray, R. E. Lake, and C. E. Sosolik, "Subsurface excitations in a metal," Phys. Rev. B **80**, 161405(R) (2009).

M. P. Ray, R. E. Lake, and C. E. Sosolik, "Energy transfer in quasibinary and collective scattering events at a Ag(001) surface," Phys. Rev. B **79**, 155446 (2009).

M. P. Ray, R. E. Lake, and C. E. Sosolik, "Alkali Ion Scattering from Ag(001) and Ag Thin Films at Low and Hyperthermal Energies," Nucl. Instrum. Meth. B **267**, 615 (2009).

Spano, Meredith Newby

M. N. Lambert, . Vocker, S. Redemann, S. Blumberg, A. Gajraj, J. C. Meiners, and N. G. Walter, "Correction: Mg2+ induced compaction of single RNA molecules monitored by tethered particle microscopy," Biophysical Journal **98**, 2041 (2010).

Tewari, Sumanta

J. D. Sau, R. M. Lutchyn, S. Tewari, and S. Das Sarma, "A generic new platform for topological quantum computation using semiconductor heterostructures," Phys. Rev. Lett. **104**, 040502 (2010).

S. Tewari, J. D. Sau, and S. Das Sarma, "A theorem for the existence of Majorana fermion modes in spin-orbit-coupled semiconductors," Annals Phys. **325**, 219–231 (2010).

C. Zhang, S. Tewari, and S. Chakravarty, "Quasiparticle Nernst effect in the cuprate superconductors from the d-density wave theory of the pseudogap phase," Phys. Rev. B **81**, 104517 (2010).

S. Tewari, R. M. Lutchyn, and S. Das Sarma, "Effects of a dilute gas of fermions on the superfluid-insulator phase diagram of the Bose-Hubbard model," Phys. Rev. B **80**, 054511 (2009).

S. Tewari and C. Zhang, "Effects of quasiparticle ambipolarity on the Nernst effect in underdoped cuprate superconductors," Phys. Rev. Lett. **103**, 077001 (2009).

Li-Sin, The

M. F. El Eid, L.-S. The, and B. S. Meyet, L.-S. The, M. Lugaro, and S. Amari, "An Unusual Presolar Silicon Carbide Grain from a Supernove: Implications for the Production of Silicon-29 in Type II Supernove," The Astrophysical Journal Letters **691**, L20 (2009).

C. Vockenhuber, C. O. Ouellet, L.-S. The, L. Buchmann, J. Caggiano, A. A. CHen, J. M. D'Auria, B. Davids, L. Fogarty, D. Frekers, A. Hussein, D. A. Hutcheon, W. Kutschera, D. Ottewell, M. Paul, M. M. Pavan, J. Pearson, C. Ruiz, G. Ruprecht, M. Trinczek, and A. Wallner, "40Ca(α,γ)44Ti and the production of 44Ti in supernovae," Journal of Physics G: Nuclear and Particle Physics **35**, 014034 (2008).

L.-S. The, M. F. El Eid, and B. S. Meyer, "s-Process Nucleosynthesis in Advanced Burning Phases of Massive Stars," The Astrophysical Journal **655**, 1058 (2007).

Tritt, Terry M.

T. M. Tritt, H. Bottner, and L. D. Chen, "Thermoelectrics: Direct Solar Thermal Energy Conversion," invited article for Special MRS Bulletin: Harnessing Materials for Energy MRS Bulletin, **33**, 366 (2008).

X. Ji, B. Zhang, Z. Su, T. Holgate, J. He. and T. M. Tritt, "Nanoscale Granular Boundaries in Polycrystalline $Pb_{0.75}Sn_{0.25}Te$: An Innovative Approach to Enhance the Thermoelectric Figure of Merit," Phys. Stat. Sol. A **206**, 221 (2009).

J. Ballato, T. Hawkins, P. Foy, B. Kokuoz, C. McMillen, Z. Su, T. M. Tritt, M. Dubinskii, and M. J. Matthewson, "On the Fabrication of Er-doped YAG Optical FIber," J. Appl. Phys. **105**, 053110 (2009).

H. Xu, K. Kleinke, T. Holgate, H. Zhang, Z. Su, T. Tritt, and H. Kleinke, Thermoelectric Performance of $Ni_yMo_3Sb_{7-x}Te_x$ ($y<0.1$, $1.5<x<1.7$)," J. Appl. Phys. **105**, 053703 (2009).

W. Xie, X. Tang, Y. Yan, Q. Zhang, T. M. Tritt, "Unique Low-Dimensional Structure and Enhanced Thermoelectric Performance for P-type $Bi_{0.52}Sb_{1.48}Te_3$ Bulk Materials, Appl. Phys. Lett. **94**, 102111 (2009).

UNIVERSITY OF SOUTH CAROLINA

DEPARTMENT OF PHYSICS AND ASTRONOMY

Columbia, South Carolina 29208

Students Accepted For Degree	FIELDS		
	Physics	Astronomy	Related Fields
Doctorate	X		
Master's	X		

1. General

President: Harris Pastides
Interim Dean of Graduate School: Timothy Mousseau
Department Chairman: Chaden Djalali
Department Telephone Number: (803) 777-8105
Type of Institution: University
Control: Public
Setting: Urban
Total Faculty: 1,560
Total Students: 28,481
Total Graduate Students: 6,527
Graduate Tuition:
Residents: Full-time—$4,894/semester
Part-time—$484/credit hr
Non-residents: Full-time—$10,540/semester
Part-time—$1,028/credit hr
Tuition rates for: 2009–10
Deferred tuition plan: Yes
Other Fees: $500 Foreign Student Enrollment Fee
Term: Semester

2. Number of Faculty in Department

The combined total of full-time faculty in the three professorial ranks is 24. The combined total of full-time, part-time, and other faculty at all ranks is 25.

3. Admission, Financial Aid, and Housing

Address admission inquiries to: Graduate Director, Department of Physics and Astronomy
Graduate application fee required: $50
Admission deadline (Fall admission): 7/1
Admission information: For fall admission, 2009–10, 25 students were accepted from 80 applicants.
Admission requirements: For admission to the graduate programs, a Bachelor's degree is required with no minimum undergraduate GPA specified. Usual preparation is undergraduate major in physics. Students from other fields will have an opportunity to make up deficiencies. The GRE isrequired. Students from non-English speaking countries are required to demonstrate proficiency in English via the TOEFL exam. Minimum acceptable score for admission is 570 (paper-base) or 80 (Internet-based).
Undergraduate preparation assumed: One semester advanced undergraduate courses in Mechanics, Electromagnetism, Modern Physics, Quantum Theory, and Experimental Physics, as well as mathematics through Differential Equations and Advanced Calculus, are assumed. Provision is made for students to make up deficiencies in these areas.
Address financial aid inquiries to: Graduate Director, Dept. of Physics and Astronomy or www.sc.edu/financialaid
GAPSFAS application required: No
Financial aid deadline: see www.sc.edu/financialaid
Loans available: Yes
Address housing inquiries to: University Housing
On-campus, single student housing available: Yes
Cost/month: $625–990
On-campus, married student housing available: Yes
Cost/month: $625–990

Table A—Faculty, Enrollments, and Degrees Granted

Research Specialty	2009–10 Faculty	Enrollment[1] Fall 2009		No. of Degrees Granted[2] 2009–10 (2005–10)			Median No. of Years for 2009–10 Ph.D.'s
		Master's	Doctorate	Master's	Terminal Master's	Doctorate	
Astrophysics	1	0	3	–	0(1)	1(2)	6
Condensed Matter Physics	7	2	6	1(2)	2(11)	0(2)	5.5
Nuclear Physics	7	3	10	0(1)	1(6)	0(6)	5
Particles & Fields	6	1	10	0(1)	0(1)	2(7)	6.5
Other Theo/Math	4	0	3	0(1)	0(0)	0(1)	–
Other (specify)	0	0	–	–	–	–	–
Non-specialized	0	0	7	–	–	–	–
Total	25	5	39	1(5)	3(19)	3(18)	
Full-time Grad. Stud.		5	38				
Part-time Grad. Stud.		0	1				
First-year Grad. Stud.		1	6				
Median Years in Grad. Study (2009–10 Degrees)				4	–	6	–
Undergraduate Degrees, 2009–10 (2005–10): 13(28))							

[1]Students not yet committed to a research specialty are entered under non-specialized.
[2]Five-year totals in parentheses.

4. Graduate Degree Requirements

Master's: The requirements for the Master of Science degree include 30 semester hours of course work; a thesis; and an oral comprehensive examination. The minimum residence is two semesters. There is no foreign language requirement.
Doctorate: The requirements for the degree of Doctor of Philosophy include 60 semester hours of advanced course work (or 30 semester hours beyond the Master's), written and oral examination for admission to candidacy, a reading knowledge of one foreign language, and a dissertation. Three years of residence are required, at least one of which is at the University of South Carolina.
Other Programs: The MAT program includes 30 semester hours of graduate work with a distribution of graduate credit of from 6 to 15 credits in professional education and from 15 to 24 credits in the teaching area.
Thesis: Thesis may be written *in absentia*.
Special Equipment, Facilities, or Programs: The Department is housed in the eight-story Physical Sciences Center, which also contains a machine shop and computer terminal facilities. Individual offices are provided for graduate students. Equipment for experimental research includes high- and low-temperature electron spin resonance spectrometers (cw and pulsed), two Mössbauer spectrometers, and a superconducting quantum interference device susceptometer. In addition to on-campus research facilities, University of South Carolina faculty and graduate students are utilizing the Thomas Jefferson National Accelerator Facility (JLab), Fermi National Ac-

celerator Laboratory (Fermilab), Stanford Linear Accelerator Center (SLAC), and Instituto Nazionale di Fisica Nucleare (INFN) in Italy.

Table B—Appointments to Graduate Students, 2009–10

Title of Appointee	Appointments Total	First year	Academic Load Allowed in Credit Hours	Hours of Service Per Week	Stipend for Academic Year ($)
			Semester		
Teaching Assistant	19	5	12	20	15,250[1,2]
Research Assistant	18	0	6	20	15,625[1,2,3]
Total	37	5			

[1]Graduate assistants are charged the in-state tuition rate of $467 per credit for 1–11 hours or $4,718 for 12–16 credit hours.
[2]Some summer appointments are available.
[3]In general, RAs do not have to pay tuition (paid by grant).

5. Personnel Engaged in Separately Budgeted Research,

Professorial faculty	25
Graduate students	24
Total	49

6. Separately Budgeted Research Expenditures by Source of Support

	Departmental Research	Physics-related Research Outside Department
Federal government	$2,776,909	$
State government	$6,000	
Total	$2,775,909	$

7. Extension Centers and Summer Programs

Graduate and undergraduate courses are offered and research is carried on during two five-week summer terms.

Table C—Separately Budgeted Research Expenditures

Research Specialty	No. of Grants	Expenditures ($)
Astronomy	5	93,016
Computational Physics	0	0
Condensed Matter Physics	5	684,266
Nuclear Physics	7	659,036
Particles & Fields	5	1,186,004
Other Theoretical/Math.	3	153,587
Total	25	2,775,909

FACULTY

Professors=15

Associate Professors=4

Assistant Professors=6

RESEARCH SPECIALTIES AND STAFF

Theoretical

Foundations of Quantum Theory. Generalized renormalization phenomena; effects of potentials and related topological concepts; generalized gauge invariance and pseudoperturbations. Altschul, Mazur.

Mathematical Physics. Lie group applications. Gudkov, Pershin. 1postdoctoral fellow.

Nuclear Physics. Effective field theory, astrophysics, quark models. Kubodera, Myhrer.

Solid State. Monte Carlo calculations of the properties of spin glasses; magnetic resonance; theory of high-T_c superconductivity. Bazaliy, Creswick.

Statistical and Thermal. Symmetry properties and thermal properties of matter and radiation; nonlinear effects; critical phenomena and processes far from equilibrium. Creswick.

Experimental

Astronomy and Astrophysics. Optical, infrared, and ultraviolet studies of quasars and distant galaxies to investigate chemical and morphological evolution of galaxies, interstellar/intergallactic matter, and intergalactic background radiation. Kulkarni.

Nuclear Physics. The intermediate energy nuclear physics group has been playing a leadership role in research addressing the question of the structure and interaction of nucleons and nuclei in terms of quantum chromodynamic and has been making significant contributions. The research program uses multi-GeV photon and electron beams in Halls A and B from the Continuous Electron Beam Accelerator Facility (CEBAF) located at Thomas Jefferson National Laboratory (JLab) and are co-spokespersons of nine JLab experiments pursuing this quest in three interwoven areas: the study of medium modifications of hardrons, the study of the excited states of the nucleon, and the study of the transition from quark-gluon to pion-nucleon degrees of freedom in exclusive processes. Djalali, Gothe, Ilieva, Strauch, Tedeschi. 4 postdoctoral fellows.

Particle Physics. High-energy proton-proton collisions are used to search for new particles indicative of physics beyond the standard model: the Higgs boson, supersymmetry, gravitons, black holes, extra dimensions, etc. Accelerator-based neutrino experiments study neutrino oscillations, masses and angles, electroweak parameters, as well as neutrino-induced reactions at new levels of precision. Avignone, Mishra, Petti, Purohit, Rosenfeld, Wilson. 2 postdoctoral fellows.

Condensed Matter Physics. Theory of critical phenomena. Numerical methods in statistical and quantum physics. Monte Carlo simulation. Bulk and nano-structured thermoelectric materials. Aharanov-Bohm charge spectroscopy. Colossal magneticresistance. Electron-spin resonance. Biophysics. Domain Dynamics in ferroelectrics. Superconductivity - lead inverse opals, high-dissipation phenomena, flux-vortex dynamics, metal-insulator transitions, magnetic penetration depth. Magnetic manoparticles and ultrasonic cavitation. Crawford, Creswick, Crittenden, Datta, Kunchur, Webb. 1 postdoctoral fellow.

SOUTH DAKOTA SCHOOL OF MINES AND TECHNOLOGY

DEPARTMENT OF PHYSICS

Rapid City, South Dakota 57701

Students Accepted For Degree	FIELDS		
	Physics	Astronomy	Related Fields
Doctorate			X
Master's	X		

1. General

President: Robert A. Wharton
Dean of Graduate Education & Research: Anthonete Logar
Department Head: Andre Petukhov
Department Telephone Number: (605) 394–2361
Type of Institution: College
Control: Public
Setting: Small town
Total Faculty: 116
Total Students: 2,070
Total Graduate Students: 232
Annual Graduate Tuition:
In-state residents: $133.70/credit
Out-of-state residents: $394.25/credit
Tuition rates for: 2008–09
Deferred tuition plan: No
Annual Other Fees: Call for amount
Term: Semester

2. Number of Faculty in Department

The combined total of full-time faculty in the three professorial ranks is 5.

3. Admission, Financial Aid, and Housing

Address admission inquiries to: Graduate Admissions
Graduate application fee required: $35.
Admission deadline Fall admission: 2008–09: 7/1, U.S.; 4/1 overseas; spring admission: 11/1, U.S.; 9/1, overseas.
Admission requirements: The Graduate Office encourages applications from qualified students holding bachelor's degrees in relevant areas of engineering and science. The GRE, three letters of recommendation and a GPA of 3.0 or better is required. The GRE Advanced is recommended. Students from non-English speaking countries are required to demonstrate proficiency in English via the TOEFL exam. Minimum acceptable score for admission is 520/190/68, but less than 560/220/83 requires testing.
Undergraduate preparation assumed: Benson, *Calculus-Based University Physics* (Halliday, Resnick, Walker or equivalent); Optics (Hecht or equivalent); *Classical Mechanics* (Thorton, Marion); *Electrodynamics* (Griffiths or equivalent); *Thermodynamics, Statistical Mechanics* (Baierlein or equivalent); *Modern Physics*.
Address financial aid inquiries to: Chairman, Andre Petukov, Department of Physics.
GAPSFAS application required: No
Financial aid deadline: 3/15
Loans available: No
Address housing inquiries to: Maureen C. Wilson, Director of Residence Life
On-campus, single student housing available: Yes
Cost/term: $1,366/semester
On-campus, married student housing available: No

Table A—Faculty, Enrollments, and Degrees Granted

Research Specialty[2000-01] (1998-01)	Faculty	Enrollment[1] Fall		No. of Degrees Granted[2] 2007–08			Median No. of Years for 2007 Ph.D.'s
		Master's	Doctorate	Master's	Terminal Master's	Doctorate	
Mateials Science	1		1				
Nanoscience and Engineering	2		4				
Physics	5	7					
Total		7	5				
Full-time Grad. Stud.		6	4				
Part-time Grad. Stud.		1	0				
First-year Grad. Stud.		2	2				
Median Years in Grad. Study (2007–08 Degrees)				–	–	–	–

Undergraduate Degrees, 2007–08 (2004–07): 4(12)

[1]Students not yet committed to a research specialty are entered under non-specialized.
[2]Five-year totals in parentheses.

4. Graduate Degree Requirements

Master's: M.S. degree in Physics. A program of at least 32 credit hours of course work and research; at least 19 credit hours of graduate course work (700-numbered courses and above); 7 credit hours of thesis research; a satisfactory thesis based upon individual research; meeting or exceeding 3.0 GPA; passing an examination on general knowledge in physics and successfully defending the thesis; a minimum of two semesters in residence; at the discretion of the Physics Department, a non-thesis program may be allowed.
Doctorate: Ph.D. in Materials Engineering and Science. A minimum of 80 semester credit hours beyond the Bachelor's degree including a minimum of 50–60 semester credit hours of course work beyond the Bachelor's degree or 26–36 semester credit hours beyond the Master's degree; meeting or exceeding 3.0 GPA; at least two consecutive semesters of residence as a full-time student; a comprehensive examination, a dissertation representing at least the equivalent of one academic year of full-time research.
Ph.D. in Nano Science and Engineering. Minimum of 80 credit hours, including at least 40 credits (or more) of graduate level coursework, including 14 credits of multidisciplinary Nano Science and Engineering core coursework.
Thesis: Thesis may not be written *in absentia.*

Table B—Appointments to Graduate Students, 2007–08

Title of Appointee	Appointments		Academic Load Allowed in Credit Hours	Hours of Service Per Week	Stipend for Academic Year ($)
	Total	First year			
			Semester		
Teaching Assistant	3	2	9	20/10[1]	$8,000[1]
Research Assistant	5	3			
Total	8	5			

[1]Half-time appointment.

5. Personnel Engaged in Separately Budgeted Research, 7/07–6/08

Professorial faculty	5
Graduate students	5
Total	10

6. Separately Budgeted Research Expenditures by Source of Support

	Departmental Research	Physics-related Research Outside Department
Federal government	$813,800	$
State government	26,000	
Total	$839,800	$

Table C—Separately Budgeted Research Expenditures

Research Specialty	No. of Grants	Expenditures ($)
Condensed Matter Physics	3	$839,000
Total	3	$839,000

FACULTY

Professors

Foygel, Michael, D. Sc., State Technical University, St. Petersburg (Leningrad Polytechnic Institute), Russia, 1986. Condensed Matter Theory; Physics of Disordered Solids; Thermal and electrical conductivity of carbon nanotube composites; Electronic transport in molecular monolayer; Variable-range hopping of spin and lattice polarons; Giant magnetoresistance in magnetic semiconductors and nanostructures; Non-equilibrium electronic processes (tunnel recombination, transport phenomena, light absorption, luminescence, photoconductivity) in disordered semiconductors and delta-doped semiconductor structures. Recent research activities are concentrated on NASA funded projects related to space cryogenic fluids management, such as modeling of liquid hydrogen loading operations.

Petukhov, Andre, G., Ph.D., State Technical University, Russia, 1981. Department Head. Condensed Matter Theory; Quantum information; Electronic structure of materials; Theory of nanoscale semiconductor materials and devices; Spin-dependent transport and magnetism in semiconductors; Single-atom electronics; processing; Quantum transport in magnetic heterostructures and nanostructures; Giant magnetoresistance caused by magnetic polarons. Recent research activities are focused on spintroic devices, magnetic quantum dots, quantum metrology and atom interferometry.

Sobolev, Vladimir, Ph.D., Physics and Technology Institute of the Academy of Sciences of Ukraine, 1974. Condensed Matter Physics; Properties of materials with spontaneous polarization (ferroelectronic magnetic and magnetoelectric substances); Phase transitions in materials with spontaneous polarization; Experimental and theoretical studies of materials for special applications in microelectronics; Structure of magnetic domain walls and their dynamics in ultrathin magnetic films; Magnetic properties of ferromagnets with structural bulk and surface defects, Kinetic and relaxation behavior of systems of interacting quasiparticles in substances subjected to the action of intense external alternating fields (magnetic and electric fields). Recent research activities are concentrated on experimental studies of non-homogeneous states caused by coexistence of ferroelectric and antiferroelectric phases, and galvano-magnetic properties of carbon nanofiber materials.

Associate Professors

Corey, Robert L., Ph.D., Washington University, St. Louis, 1992. Solid-state nuclear magnetic resonance; characterization of electronic materials, hydrogen-metal systems, and structural materials. Current research focuses on nuclear magnetic resonance (NMR) investigations of hydrogen motion in candidate solid-state hydrogen storage materials. The experimental work mainly involves ^{1}H, ^{2}H, ^{23}Na, ^{27}Al, ^{7}Li, and ^{11}BNMR spectroscopy at variable field, temperature, and pressure in binary metal hydride and complex metal hydride crystals.

Assistant Professor

Dr. Bai, Xinhua, Ph.D., Beijing University, Beijing, China, 1996. Experimental astroparticle physics, muon measurements, and dark matter search. Recent research activities are concentrated on search for WIMPs with LUX experiment at DUSEL, studies of muon bundle physics with high-energy neutrino astronomy project IceCube at the South-Pole, underground background studies using IceCube and LUX data and simulation, and a conceptual design of a surface array at DUSEL to eliminate cosmic ray related uncertainties in dark matter search.

RESEARCH SPECIALTIES AND STAFF

Theoretical

Condensed Matter Physics. Theory of electronic phenomena in semiconductors; electrical conductivity; light absorption; photo-luminescence; photo-conductivity; theory of semiconductor materials and devices; magnetic nanostructures; quantum wells, resonant-tunneling devices. Foygel, Petukhov, Sobolev.

Experimental

Condensed Matter Physics. Nuclear magnetic resonance; characterization and investigation of properties of electronic materials. Corey, Sobolev.

Astroparticle physics. Muon measurements, conceptual design of a surface detectors array, and dark matter search. Bai.

UNIVERSITY OF TENNESSEE, KNOXVILLE

DEPARTMENT OF PHYSICS AND ASTRONOMY

Knoxville, Tennessee 37996-1200

Students Accepted For Degree	FIELDS		
	Physics	Astronomy	Related Fields
Doctorate	X		
Master's	X		

1. General

Acting President: Jan Simek
Vice President for Research: Dr. David Millhorn
Department Head: Soren P. Sorensen
Department Telephone Number: (865) 974-3342
Type of Institution: University
Control: Both
Setting: Urban
Total Faculty: 1,400
Total Students: 27,107
Total Graduate Students: 6,101
Annual Graduate Tuition:
In-state residents: Full-time—$3,756/sem. (est.)
Out-of-state residents: Full-time—$10,804/sem (est.).
Tuition rates for: 2010–11
Deferred tuition plan: Yes
Other Fees: in-state: $400/sem.
out-of-state: $611/sem.
Term: Semester

2. Number of Faculty in Department

The combined total of full-time faculty in the three professorial ranks is 32. The combined total of full-time, part-time, and other faculty at all ranks is 112.

3. Admission, Financial Aid, and Housing

Address admission inquiries to: Graduate School, 111 Student Services Bldg.
Graduate application fee required: $35
Admission deadline (Fall admission): 2/1
Admission requirements: For admission to the graduate programs, a Bachelor's degree in physics, mathematics, or engineering is required with a minimum undergraduate GPA of 2.7 specified for American students and 3.0 for international students. The GRE subject test is strongly urged. No minimum acceptable score for admission is specified. Students from non-English speaking countries are required to demonstrate proficiency in English via the TOEFL exam. Minimum acceptable score for admission is 550 (paper based), 213 (computer based), or 80 (ibt).
Undergraduate preparation assumed: Griffiths, *Introduction to Electrodynamics*, Marion and Thornton, *Classical Dynamics*, Griffiths, *Introduction to Quantum Mechanics*.
Address financial aid inquiries to: Financial Aid Office, 115 Student Services Building
GAPSFAS application required: No
Loans available: Yes
Address housing inquiries to: University Housing, 405 Student Services Building and UT Rental Property, 472 South Stadium Hall
On-campus, single student housing available: No
On-campus, married student housing available: No

Table A—Faculty, Enrollments, and Degrees Granted

Research Specialty	2009–10 Faculty	Enrollment[1]		No. of Degrees Granted[2] 2009–10 (2005–10)			Median No. of Years for 2009–10 Ph.D.'s
		Master's	Doctorate	Master's	Terminal Master's	Doctorate	
Accelerator Physics	3	0	0	0(0)	0(0)	0(0)	–
Astrophysics	3	1	7	2(3)	0(3)	3(5)	-6
Atomic, Molecular, & Optical Physics	6	0	3	1(1)	0(0)	1(1)	-5
Biophysics	2	2	3	1(1)	1(1)	2(2)	5
Chemical Physics	2	2	3	1(1)	1(1)	2(2)	5
Condensed Matter Physics	18	4	35	2(4)	1(5)	10(22)	–
Medical Physics	2	0	0	0(0)	0(0)	0(0)	-5
Nuclear Physics	12	0	15	3(4)	1(7)	5(14)	6
Particles & Fields	8	0	12	0(1)	2(1)	2(1)	5
Other Experimental	4	1	4	0(1)	0(0)	0(1)	–
Other Theoretical/ Math.	2	1	9	0(0)	0(0)	0(0)	–
Non-specialized		5	8	0(0)	1(2)	0(2)	–
Total		16	99	10(16)	7(20)	25(50)	
Full-time Grad. Stud.		12	97				
Part-time Grad. Stud.		2	1				
First-year Grad. Stud.		5	23				
Median Years in Grad. Study (2007–08 Degrees)				2.5	2.5	5.5	

Undergraduate Degrees, 2009–10 (2005–10): 7(43)

[1]Students not yet committed to a research specialty are entered under non-specialized.
[2]Five-year totals in parentheses.

4. Graduate Degree Requirements

Masters:
Thesis Option: The course requirements include 24 semester hours of physics courses, of which at least 12 semester hours are taken from Physics 506, 513–514, 521–522, 531, 541, 571, 573. Each candidate must present an acceptable thesis, 6 hours of 500, and pass an oral examination on course material and thesis.
The department offers an M.S. thesis program with a concentration in geophysics. Program requirements are: 12 hours from Physics 506, 513–514, 521–522, 531, 541, 571, 573; a minimum of 12 additional hours in geology, geophysics, and/or physics, as approved by the student's committee; and the presentation of an acceptable thesis, 6 hours of Physics 500, an advanced seminar, and the passing of an oral examination on course material and thesis.
Project Option: The course requirements include a minimum of 30 hours of graduate credit in courses composed of Physics 506, 513–514; 6 hours from Physics 593, 594 for a Project in Lieu of Thesis; 9 hours from general physics: 411–412, 421, 431–432, 461–462, 507, 508, 521–522, 531, 541, 555, 571, 573 (at least 3 hours above the 500-level); and 6 hours from a single minor field outside of the physics department, such as computer science, mathematics, engineering, chemistry, biology, education, business, or law.
The candidate must pass an oral examination on course material and on the Project representing the culmination of an original research project completed by the student. A written report must be approved and accepted by the Physics Gradu-

ate Committee and the Department Head. An electronic version of the written report must also be submitted to the permanent electronic archive of the Physics Department available to the Internet.

Non-Thesis Option: Students seeking the non-thesis option must apply to the department's graduate committee for permission to enroll under this program. The requirements are the satisfactory completion of 30 hours of course work composed of 18 semester hours from Physics 513–514, 521–22, 531–32, 541–42, and 571–72; 6 semester hours in a minor field; and 6 semester hours from other courses numbered above 400 (preferably of advanced laboratory nature). At least 20 hours must be taken at the 500-level or above. In addition, the candidate must pass a written examination administered by his/her committee.

The Doctoral Program: All students are expected to take the graduate core curriculum in physics consisting of the following courses: Physics 521–22, 531, 541, 551, and 571. Students specializing in chemical physics may substitute Chemistry 572 for Physics 551, and should complete at least 6 semester hours from Chemistry 530, 570, 571, 573, 595, 630, 670, and 690. Students must take a minimum of 15 hours of 600-level courses, with 6 of these hours in their area of specialization. Physics 601–02 are normally required of students specializing in atomic physics; Physics 621–22 of students in nuclear physics; Physics 626–27 of students in elementary particle physics (and/or Physics 613–14 for students specializing in theoretical high-energy physics); and Physics 671–72 of students in condensed matter and surface physics. Students concentrating in nanomaterials must take a minimum 15 hours of 600-level courses, of which at least 6 hours are offered by the department and at least 6 hours are from a list of courses offered by several departments which are appropriate for a concentration in nanomaterials.

To be admitted to Ph.D. candidacy students must: a) fulfill all general requirements by the Graduate School, b) pass the qualifying examination, c) have at least a 3.0 GPA on the graduate core curriculum in physics, d) form a doctoral committee, and e) pass the comprehensive examination.

The qualifying examination is designed to test the student's general knowledge of the fundamentals of physics. The performance needed to pass this examination corresponds to a mature command of the material typically included in the undergraduate physics major curriculum. The qualifying examination should be passed after the student's first year of study. Based on the student's performance on a) the qualifying examinations, b) the course work, c) the GRE scores and d) optional research participation, the faculty will decide if the student will be allowed to continue in the Ph.D. program. Students are required to find a research advisor and form a doctoral committee before the end of the second year of study. This committee is responsible for advising the student and monitoring his/her progress towards the doctoral degree.

The comprehensive examination is designed to test the students on a) specific knowledge and skills in the areas essential to the student's research program, b) capability to successfully complete the doctoral dissertation and c) general knowledge of the graduate core curriculum. The most essential component of this examination is the presentation and defense of an original research proposal. The comprehensive examination must be passed before the end of the third year of study. It contains both a written and an oral component and is conducted by the student's doctoral committee and an additional faculty member appointed by the department head.

The dissertation topic will be chosen with reference to one of the fields in which research facilities can be made available either at The University of Tennessee laboratories in Knoxville; The University of Tennessee Space Institute at Tullahoma, Tennessee; the Oak Ridge National Laboratory, Oak Ridge, Tennessee; or at other research facilities used by the University faculty.

Special Equipment, Facilities, or Programs: Special experimental facilities include high power ultrafast tunable lasers; photoelectron and reflectron time-of-flight mass spectrometers; ultrahigh resolution scanning tunneling microscopy systems; low energy electron diffraction systems; soft x-ray spectrometers; a large computer-controlled, high-resolution molecular spectrometer; computer-controlled nuclear data analysis systems; helium dilution refrigerator for ultralow temperature studies; and a complete biophysics laboratory. The department has fully staffed instrument, machine, and electronics shops. Research facilities are also available at Oak Ridge National Laboratory in nuclear, heavy-ion atomic, and condensed matter physics. A Ph.D. program in Chemical Physics is conducted jointly with the Department of Chemistry. The department also conducts a resident Ph.D. program in physics at the University of Tennessee Space Institute in Tullahoma. Facilities there support research in atomic and molecular physics, laser physics, chemical physics, quantum optics, infrared spectroscopy, and laser scattering. A cooperative arrangement permits research appointments for selected students at the nearby Air Force Arnold Engineering Development Center research operation.

Table B—Appointments to Graduate Students, 2009–10

Title of Appointee	Appointments: Total	Appointments: First year	Academic Load Allowed in Credit Hours	Hours of Service Per Week	Stipend for Academic Year ($)
			Semester		
Teaching Assistant	61	26	9–15	20	17,335
Research Assistant	39	0	9–15	20	17,335–30,000
Total	100	26			

5. Personnel Engaged in Separately Budgeted Research, 7/07–6/08

Professorial faculty	48
Postdoctoral appointments	38
Graduate students	64
Non-teaching research personnel	46
Total	196

6. Separately Budgeted Research Expenditures by Source of Support

	Departmental Research
Federal government	$6,526,522
State and local government	1,703,070
Total	$8,729,592

7. Separately Funded and Managed Laboratories

UT Space Institute, Tullahoma, TN.	$27,000,000
Total	$27,000,000

Table C—Separately Budgeted Research Expenditures

Research Specialty	No. of Grants	Expenditures ($)
Applied	12	333,208
Astrophysics	2	77,024
Atomic, Molecular, & Optical Physics	3	452,230
Condensed Matter Physics	42	2,706,507
Nuclear Physics	26	4,045,808
Particles & Fields	8	550,010
Physics Education	3	64,805
Other Theoretical/Math.		
Total	96	8,229,592

FACULTY

Professors

Barnes, Frank E., Ph.D., Caltech, 1977. Theoretical elementary particle physics.
Bingham, Carroll R., Ph.D., Tennessee, 1965. Experimental nuclear physics.
Blass, William E., Ph.D., Michigan State, 1963. Experimental and theoretical molecular spectroscopy. (Emeritus)
Breinig, Marianne, Ph.D., Oregon, 1979. Atomic and molecular physics.
Bugg, William M., Ph.D., Tennessee, 1959. Experimental elementary particle physics. (Emeritus).
Callcott, Thomas A., Ph.D., Purdue, 1965. Experimental solid state (Emeritus).
Cardall, Christian, Ph.D., California-San Diego, 1997, Nuclear-Astrophysics
Childers, Robert W., Ph.D., Vanderbilt, 1963. Theoretical elementary particles. (Emeritus).
Compton, Robert N., Ph.D., Tennessee, 1964. Experimental chemical physics.
Dagotto, Elbio, Ph.D., Bariloche, 1985. Theoretical condensed matter physics.
Dai, P., Ph.D., Missouri, 1993. Experimental condensed-matter physics.
Duckett, Kermit E., Ph.D., Tennessee, 1964. Experimental fiber studies. (Emeritus).
Efremenko, Yuri, Ph.D., ITEP, Moscow, Russia, 1989. Experimental particle physics.
Eguiluz, Adolfo, Ph.D., Brown, 1976. Condensed matter.
Elston, Stuart, Ph.D., Massachusetts, 1975. Atomic and molecular physics.
Georghiou, Solon, Ph.D., Manchester, 1968. Biophysics (Emeritus).
Greene, Geoffrey L., Ph.D., Harvard, 1977. Experimental nuclear physics.
Guidry, Michael W., Ph.D., Tennessee, 1974. Experimental nuclear physics.
Handler, Thomas, Ph.D., Rutgers, 1974. Elementary particles.
Kamychkov, Yuri A., Ph.D., Moscow, 1970. Experimental elementary particle physics.
Levin, Jon C., Ph.D., University of Oregon, 1986. Experimental atomic physics.
Macek, Joseph H., Ph.D., Rensselaer, 1964. (Distinguished Scientist.) Atomic physics.
Mezzacappa, Anthony, Ph.D., Texas at Austin, 1988. Nuclear astrophysics.
Moreo, Adriana, Ph.D., Bariloche, 1985. Theoretical condensed matter physics.
Nazarewicz, Witold, Ph.D., Warsaw, 1981. Nuclear physics.
Painter, Linda R., Ph.D., Tennessee, 1968. Experimental liquid state.
Pegg, David J., Ph.D., New Hampshire, 1970. Experimental heavy-ion atomic physics (Emeritus).
Plummer, E. Ward, Ph.D., Cornell, 1968. (Distinguished Scientist.) Experimental condensed matter physics.
Quinn, John J., Ph.D., Maryland, 1958. Theoretical solid state physics.
Read, Kenneth F., Ph.D., Cornell, 1987. Physics of elementary particles.
Riedinger, Leo L., Ph.D., Vanderbilt, 1969. Experimental nuclear physics.
Shih, Chia C., Ph.D., Cornell, 1967. Theoretical elementary particle and atomic physics; experimental medical physics (Emeritus).
Siopsis, George, Ph.D., Calif. Inst. Tech., 1987. Theoretical elementary particle physics.
Sorensen, Soren, Ph.D., Niels Bohr Inst., 1981. Head of the Department. Experimental nuclear physics.
Thompson, James R., Ph.D., Duke, 1969. Experimental solid state.
Weitering, Harm H., Ph.D., Univ. of Groningen, 1991. Experimental condensed matter physics.
Zhang, Zhenyu, Ph.D., Rutgers University, 1989, Condensed matter physics

Associate Professors

Grzywacz, Robert, Ph.D., Warsaw, 1997. Experimental nuclear physics.
Papenbrock, Thomas, Ph.D., Max-Planck Institute, 1996. Theoretical nuclear physics.
Spanier, Stefan M., Ph.D., Univ. Mainz, 1994. Experimental high energy physics.

Assistant Professors

Jones, Kate, Ph.D., Univ. Surrey, 2000. Experimental Nuclear Physics.
Joo, Jaewook, Ph.D., Rutgers, 2004.
Mannella, Norman, Ph.D., U.C. Davis (CA) 2003. Experimental Condensed Matter.

Lecturers

Abdelrazek, Margie, Ph.D., Florida State University, 2001.
Dandaneau, Debra, Ph.D., Univ. of Colorado, Boulder, 2006.
Daunt, Stephen, Ph.D., Queen's, 1976.

Adjunct Professors

Aytug, Tolga, Ph.D., Kansas, 2000.
Bardayan, Daniel, Ph.D., Yale, 1999.
Burgdoerfer, Joachim, Ph.D., Freie Univ., Berlin, 1982.
Calarco, John, Ph.D., Illinois, Urbana-Champaign, 1969.
Cooke, John, Ph.D., Ga. Inst. of Technology, 1965.
Datskos, Panos, Ph.D., Tennessee, 1988. Molecular and chemical physics.
Dean, David, Ph.D.,Vanderbilt, 1991.
Duckett, Kermit, Ph.D., Tennessee, 1964. (Emeritus)

Egami, Takeshi, Ph.D., Pennsylvania, 1971.
Fernández-Baca, Jaime, Ph.D., Maryland, 1986.
Galindo-Uribarri, Alfredo, Ph.D., Univ. of Toronto, 1991.
Garrett, William R., Ph.D., Alabama, 1963.
Grice, Warren, Ph.D., Rochester, 1997.
Henderson, Stuart, Yale, 1991.
Holmes, Jeffrey, Ph.D., Calif. Inst. of Technology, 1976.
Keppens, Veerle, Katholieke Universiteit Leuven, Belgium, 1995.
Kristic, Predrag, Ph.D., City College of C.U.N.Y., New York, 1982.
Li, An-Ping, Ph.D., Peking Univ., 1997.
Mandrus, David G., Ph.D., Stonybrook, 1992.
Mason, Thom E., Ph.D., McMaster, 1990.
Melnichenko, Yuri B., Ph.D., Kiev State Univ., 1985.
Messer, Otis E., Ph.D., Tennessee, 2000.
Nagler, Stephen, Ph.D., Univ. of Toronto, 1982.
Passian, Ali, Ph.D., Univ. of Tennessee, 2000.
Ramsey, Chester, Ph.D., Tennessee, 2000.
Reinhold, Carlos, Ph.D., Buenos Aires, Argentina, 1988.
Shen, Jian, Ph.D, Institut Mikrostrukturphysik, 1996.
Simpson, John, Ph.D., Univ. Arizona, 1995.
Singh, David, Ph.D., Univ. of Ottawa, Canada, 1985.
Smith, Michael, Ph.D., Yale, 1990.
Sokolov, Alexei, Ph.D., Russian Acad Sciences, 1986.
Stockli, Marti, Ph.D., ETH, Zurich, Switzerland, 1978.
Stone, Jirina, Ph.D., Charles Univ., Prague, 1975.
Sun, Yang, Ph.D., Technische Universität, München, 1991.
Townsend, David, Ph.D., London, 1970.
Van Berkel, Gary, Ph.D., Washington State, 1987.
Zucker, Alexander, Ph.D., Yale, 1950.

Research Professors

Ferrell, Thomas L., Ph.D., Clemson, 1969. Experimental surface physics.
Ovchinnikov, Serguei, Ph.D., Technical Inst. of Lenigrad, 1985. Atomic physics.
Pinnaduwage, Lal, Ph.D., Pittsburgh, 1986. Atomic and molecular physics
Stone, Nicholas J., Ph.D., Oxford University, 1963. Nuclear physics.
Thundat, Thomas, Ph.D., State U. of NY at Albany, 1987. Atomic and molecular physics.
Wong, Cheuk-Yin, Ph.D., Princeton, 1966.

Research Assistant Professors

Hix, W. Raph, Ph.D., Harvard, 1995.
Senesac, Larry, R., Ph.D., University of Tennessee, 1997. Atomic and molecular physics.
Tselev, Alexander, Ph.D., Dresden Univ. of Technology, Gemany, 2000.

Zhu, Wenghang, Ph.D., Inst. of Physics, Chinese Acadamy of Sciences, 2004.

Director of Undergraduate Laboratories

Parks, James E., Ph.D., Kentucky, 1970. Experimental atomic physics.

RESEARCH SPECIALTIES AND STAFF

Theoretical

Astrophysics. Core collapse supernovae, novae, computational modeling, nuclear astrophysics. Cardall, Guidry, Mezzacappa.
Atomic, Molecular, and Optical Physics. Atomic scattering theory, interaction of electromagnetic radiation with atoms and electron correlation with atoms. Macek.
Chemical Physics. Quantum magnets, neutron scattering, superconductivity, magnetism, weakly-bound complexes. Barnes, Dagotto, Macek, Moreo.
Condensed Matter Physics. Strongly correlated electron systems, colossal magnetoresistance, high Tc superconductivity, transport properties, nanostructures, many-body excitations, thin-film growth, surface physics, quantum Hall effect, quantum magnetism. Barnes, Dagotto, Eguiluz, Moreo, Quinn, Zhang.
High Energy Physics. Strings and quantum gravity, scattering by black holes, hadron physics, heavy-ion collisions, quantum spin systems. Barnes, Siopsis.
Nuclear Physics. Nuclear structure, many body problem, physics of open systems, quantum chaos and random matrices, spin chains. Dean, Nazarewicz, Papenbrock.

Experimental

Chemical Physics. Laser spectroscopy, negative ions, chirality, synchrotron spectroscopy of atoms. Compton.
Condensed Matter Physics. Neutron scattering, high Tc superconductivity, magnetism and lattice effects in colossal magnetoresistance, physics of novel materials and complex electron systems, low dimensional materials, nanostructures, surface and interface physics, x-ray spectroscopy, thin-film materials. Dai, Egami, Mannella, Thompson, Weitering.
High Energy Physics. CP violation measurements, meson spectroscopy, electron neutrino detection, neutrino oscillation. Efremenko, Handler, Kamychkov, Spanier.
Nuclear & Nuclear Structure Physics. Decay spectroscopy, experimental nuclear astrophysics, gamma ray spectroscopy, relativistic heavy ion physics, hot and dnese nuclear matter, neutron physics. Bingham, Greene, Grzywacz, Jones, Sorensen, Read, Riedinger.

Other

Physics Education. Breinig, Levin, Elston.
Molecular Spectroscopy. Blass.

VANDERBILT UNIVERSITY

DEPARTMENT OF PHYSICS AND ASTRONOMY

Nashville, Tennessee 37235

Students Accepted For Degree	FIELDS		
	Physics	Astronomy	Related Fields
Doctorate	X*	X°	
Master's	X	X	X+

1. General

Chancellor: Nicholas S. Zeppos
Dean of Graduate School: Dennis Hall
Department Chairman: Robert J. Scherrer
Department Telephone Number: (615) 322-2828
Type of Institution: University
Control: Private
Setting: Suburban
Total Faculty: 2,689*
Total Graduate Faculty: 972*
Total Students: 11,252*
Total Graduate Students: 4,558*
Annual Graduate Tuition:
All Graduate Students: $1,568/semester-hr.
Tuition rates for: 2009–10
Deferred tuition plan: Yes
Other Fees: $2,295
Term: Semester

*Excludes Medical and Law Schools
°Physics degree
+Masters in Medical Physics administered through the Medical School.

2. Number of Faculty in Department

The combined total of full-time faculty in the three professorial ranks is 31. The combined total of full-time, part-time, and other faculty at all ranks is 71.

3. Admission, Financial Aid, and Housing

Address admission inquiries to: Graduate School. 411 Kirkland Hall, Vanderbilt University, Nashville, TN 37240
Graduate application fee required: $40*
Admission deadline (Fall admission): 1/15
Admission information: For fall admission, 2009–10, 35 students were accepted from 167 applicants.
Admission requirements: For admission to the graduate programs, a Bachelor's degree in physics is required with a minimum undergraduate GPA of 3.0/4.0 (B average) specified. The GRE is required. The minimum acceptable score suggested for admission is verbal-550; quantitative-550; total-1,100. Applicants in physics should have quantitative scores greater than 650. The GRE Physics is required. The Graduate School's minimum acceptable score suggested for admission is 650. Students from non-English speaking countries are required to demonstrate proficiency in English via the TOEFL exam. Minimum acceptable score for admission is currently 500; much higher scores (~600) are required of teaching assistants.
Undergraduate preparation assumed: Resnick and Halliday, *Physics*; Eisberg and Resnick, *Quantum Physics of Atoms, Molecules, Solids, Nuclei, and Particles*; Reitz, Milford, Christy, *Foundations of Electromagnetic Theory*; Zemansky, *Heat and Thermodynamics*; Reif, *Fundamentals of Statistical and Thermal Physics*; Symon, *Mechanics*; Saxon, *Elementary Quantum Mechanics*.
Address financial aid inquiries to: Robert J. Scherrer, Chair, Department of Physics and Astronomy
GAPSFAS application required: No
Financial aid deadline: 1/15
Loans available: Yes
Address housing inquiries to: Housing Division, Box 351677, Station B
On-campus, single student housing available: No
On-campus, married student housing available: No

*Waived for online applicants

Table A—Faculty, Enrollments, and Degrees Granted

Research Specialty	2009–10 Faculty	Enrollment[1] Fall 2009		No. of Degrees Granted[2] 2009–10 (2005–09)			Median No. of Years for 2008–09 Ph.D.'s
		Master's	Doctorate	Master's	Terminal Master's	Doctorate	
Applied Physics	1	0(0)	2	0(0)	0(0)	0(3)	–
Astronomy/Astrophysics	4		11	0(4)	0(0)	2	–
Atomic, Molecular, & Optical Physics	2		1	0(4)	0(0)	2(11)	6.5
Biophysics	2	–	7	0(1)	0(1)	0(4)	5.0
Condensed Matter/ Nanoscience	7		15	0(7)	0(1)	1(8)	7.5
Free Electron Laser	–		–	0(0)	0(1)	0(9)	–
Medical & Health Physics	1		13	0(7)	0(0)	0(6)	6.0
Nuclear Physics	8		7	0(7)	0(1)	0(5)	7
Particle & Fields	6		2	0(1)	0(2)	0(6)	6.0
Non-specialized		–	13	0(0)	0(0)	0(0)	–
Total	31	0	71	0(31)	0(6)	5(51)	
Full-time Grad. Stud.			71				
Part-time Grad. Stud.			0				
First-year Grad. Stud.			13				
Median Years in Grad. Study (2009–10 Degrees)				–	–	–	7.3

Undergraduate Degrees, 2009–10:18 (2005–09): 62

[1]Students not yet committed to a research specialty are entered under non-specialized.
[2]Five-year totals in parentheses.

4. Graduate Degree Requirements

Master's: (Physics) 24 semester-hours of course work plus research thesis; B average required; can be done in one year but often requires three semesters; no foreign language; no comprehensive exams. Master's also awarded without thesis on basis of 42 hours, Ph.D. qualifying exam, some research experience. (Astronomy) Non-thesis option not available; normally requires 4 semesters; oral exam required. (Medical Physics) Thesis and non-thesis option; oral exam required.
Doctorate: (Physics) 72 semester-hours of course work—up to 36 of this can be research; B average required in formal course work; completion of core coursework and oral qualifying exam after first or second year establishes Ph.D. candidacy; dissertation in physics or astronomy; one year residency.

Other Programs: Master of Science in Medical Physics; MAT. Interdisciplinary Graduate Program in Materials Science. Interdisciplinary work is encouraged and is tailored to fit the needs of the individual student; Masters of Science at Fisk University, Ph.D. at Vanderbilt University.

Thesis: Thesis may be written *in absentia*.

Special Equipment, Facilities, or Programs: The Vanderbilt Free-Electron Laser Center contains an RF-linac free-electron laser. Scientific collaborations at the Center span several natural science departments, the School of Engineering and the Medical School. A unique Compton scattering, tunable, pulsed, monochromatic x-ray source has recently been constructed and is available for applications in physics, biology, and medicine. The Center houses laser-Raman and FTIR spectrometers, a high resolution time-of-flight mass spectrometer, a time-resolved fluorometer, two nanosecond/picosecond CPA Nd:YAG and tunable dye laser systems, a 10-5 picosecond laser, an excimer laser. Two femto-second optical parametric amplifier systems span a wave-length range from 200 nm to 15 μm. The Nanofabrication Laboratory, a collaborative activity with the School of Engineering, has a focused ion beam, state-of-the-art pulsed laser thin-film deposition system, atomic-force microscope and a variety off diagnostics for studying nanometer-scale structures and materials. This laboratory is a centerpiece of the Vanderbilt Institute of Nanoscale Science and Engineering. Vanderbilt University is a member of the University Radioactive Ion Beam (UNIRIB) Consortium at Oak Ridge National Laboratory that is in charge of a new generation recoil mass spectrometer. The new RMS was initiated by Vanderbilt for use with the new Holifield Radioactive Ion Beam Facility (HRIB). The Joint Institute for Heavy-Ion Research at Oak Ridge, Tennessee is operated by Vanderbilt University, the University of Tennessee, and Oak Ridge National Laboratory to support users of the Holifield HRIB, and sponsors a visitor's program which brings distinguished scientists for research and lecturing. Members of the nuclear structure physics group have additional cooperative research programs at Argonne National Laboratory, Lawrence Berkeley Laboratory, Idaho National Engineering Laboratory, Joint Institute for Nuclear Research in Russia, the Universities of Frankfurt, Tsinghua, and Bucharest, as well as access to a number of supercomputers around the country. The Vanderbilt Relativistic Heavy Ion (RHI) nuclear group plays a senior management role in the PHENIX relativistic heavy-ion experiment approved at RHIC. The RHI group studies nuclear matter under extreme conditions of temperature and density, and searches for exotic new states of matter. Experiments are conducted at the Alternate Gradient Synchrotron at Brookhaven National Laboratory. The Vanderbilt Center for Heavy Ion and Particle Physics (VCHIPP) consists of clean rooms, electronics, and other infrastructure for the development and testing of large scale detectors for nuclear and particle physics. The Atomic, Molecular, Optical and Surface (AMOS) groups study the dynamics of surface and interface processes under a wide variety of conditions and has collaborators in the Departments of Chemistry and in the School of Engineering. The AMOS resources include the Vanderbilt Free-Electron Laser, femosecond Ti-Sapphire lasers, OPAs, linear and ring dye lasers, excimer and excimer-pumped lasers, frequency doublers, ultrafast tunable lasers, low-energy ion electron and atomic beam sources, 300 kV ion implanter, and visible and vacuum-ultraviolet spectrometers. Members of the groups are involved in cooperative research programs at the Max-Planck-Institute in Garching, the Synchrotron Radiation Center in Madison, Wisconsin, the Universities in Vienna, Berlin, and Krakow. Members of the biological physics group investigate cellular electric, magnetic and mechanical phenomena. Included in the laboratories are a magnetic imaging facility with high-resoution SQUID magnetometers and microscopes, scanning stages and magnetic shields, numerous video fluorescence microscopes, and a confocal microscope with a coupled laser microsurgery system. Studies underway are examining the non-linear electrodynamics of cardiac tissue, intracellular and paracrine signaling in cellular biosystems, coupling among genetics, morphologic change and the mechanics of soft condensed matter, and nanoparticle and nanocluster labeling and spectroscopy. The project to Instrument and Control the Single Cell is developing tools and techniques for cellular biophysics and wide-bandwidth metabolic measurements. The group, in conjunction with the Vanderbilt Institute for Integrative Biosystems Research and Educations (VIBRE) operates a BioMEMS Fabrication Facility that includes three class-100 cleanrooms for photolithography, soft lithography, for biomicrofluidics, e-beam and ion-etch fabrication of metal microelectrode arrays, and extensive cellular biophysics instrumentation. The group uses infrared, visible, and ultraviolet spectroscopic techniques for investigating the dynamics of biopolymers and laser-tissue interactions. The high-energy physics group at Vanderbilt University has research projects at Cornell, SLAC, and Fermilab. The astronomy group actively uses national and international ground-based and space-platform observatories. Vanderbilt is a partner in the SMARTS (Small and Medium Aperture Research Telescope System) that operates telescopes at the Cerro Tololo Inter-American Observatory in Chile. Vanderbilt University is a charter member of the Extreme Universe Space Observatory (EUSO) Consortium, formed of three U.S. universities, two national laboratories, and seven European nations. The experiment is funded by NASA to put a lens and electronics on board the International space station in the year 2009. The energy, direction and composition of nature's most energetic particles will be measured. The Department has exceptionally good research computing facilities. The Vanderbilt Multi-Processor Integrated Research Engine (VAMPIRE) cluster of 200 CPU nodes at present and funded to more than double its size in the next three years. The facility is one of the top university super computing resources in the United States. There is also a wide variety of modern computational facilities within each of the research groups. Vanderbilt University is a participant in the Internet II network and thus involved in developing and using the next generation network. The materials and nanoscience physics group focuses on the growth and analysis of thin films with enhanced electronic and optical properties. Resources include a 2.0 MeV Van de Graaff accelerator, 4 Kelvin optical cryostate, a 300 KeV accelerator for ion scattering analysis and ion implantation, a Si based molecular beam epitaxy (MBE) system, a combined focused ion beam-pulsed laser deposition system for nanostructure fabrication, and various apparatus for growth, automated nanocrystalline thin film deposition system, annealing and film measurement. Members of the group collaborate with Oak Ridge National Lab, the Engineering School and other centers for materials science, in an extensive interdisciplinary program. The theoretical condensed matter physics group and the nuclear theory group have joint research activities with Oak Ridge National Laboratory where the computational facilities include several massively parallel computers. The condensed matter theory group has programs in quantum transport, semiconductor physics, nanocatalysis, interaction of light and ion beams with matter, and the physics of complex transition-metal oxides.

Table B—Appointments to Graduate Students, 2009–10

Title of Appointee	Appointments Total	Appointments First year	Academic Load Allowed in Credit Hours	Hours of Service Per Week	Stipend for Academic Year ($)
			Semester		
Teaching Assistant	27	9	12	20	17,500[1-4]
Research Assistant	43	3	–	–	23,250[1-4]
Self-supporting students	1	1			
Total	71	13			

[1]In addition to the preceding stipend amounts, some of the 74 graduate students also receive supplemental awards as follows:

	Amount Per Award	Total
University Graduate Fellowship	$5,000	2
William & Nancy McMinn Graduate Fellowship	$7,500	3
Harold Stirling Vanderbilt Graduate Fellowship	$3,000	1

[2]Stipend amounts for Research Assistants are partially funded by the following:

	Amount Per Award	Total
William & Nancy McMinn Graduate Fellowship	$7,500	2
UGF	$5,000	2
HSV	$3,000	1
A&S Select Fellowship	$3,000	1

[3] Graduate Student Teaching and Research Assistants are compensated for 100% of their tuition through waiver, training grants, etc.

[4] Summer compensation is generally available to Graduate Student Teaching and Research Assistants as follows:

a. One full-time equivalent Teaching Assistant is employed for Summer Semester at $3,250 each.

b. Research Assistantships are generally available during the summer for academic year Teaching Assistants at $3,250 each.

5. Personnel Engaged in Separately Budgeted Research, 7/09–6/10

Professorial faculty	27
Postdoctoral appointments	21
Other faculty	15
Undergraduate students	1
Total	64

6. Separately Budgeted Research Expenditures by Source of Support

	Departmental Research	Physics-related Research Outside Department
Federal government	$4,753,312	$
Total	$4,753,312	$

Table C—Separately Budgeted Research Expenditures

Research Specialty	No. of Grants	Expenditures ($)
Astronomy	11	467,183
Atomic, Molecular, & Optical Physics	9	$1,260,973
Biophysics	4	761,905
Condensed Matter Physics	8	1,224,124
Nuclear Engineering	4	660,184
Particle Physics	10	378,943
Total	46	4,753,312

FACULTY

Professors

Brau, Charles A., Ph.D., Harvard, 1965. Theoretical and experimental physics; atomic and molecular physics; lasers; electron accelerators.

Ernst, David J., Ph.D., MIT, 1970. Nuclear theory; intermediate energy nuclear reactions; neutrino oscillation phenomenology; hadronic structure.

Feldman, Leonard C., Stevenson Professor of Physics, Ph.D., Rutgers-The State University, 1967. Materials physics; surface and thin films science; low dimensional materials.

Gore, John C., Ph.D., U. London, 1976. Development and application of imaging; magnetic resonance imaging and spectroscopic techniques.

Greene, Senta V., Ph.D., Yale, 1992. Experimental physics; relativistic heavy ion collisions.

Haglund, Richard F., Jr., Ph.D., North Carolina, 1975. Experimental physics; nanoscale nonlinear optics and phase transitions; laser modification of surfaces and films; free-electron laser applications including polymer thin-film deposition and biomolecular mass spectroscopy.

Hamilton, Joseph H., Ph.D., Indiana, 1958. Landon C. Garland Distinguished Professor. Experimental nuclear physics; nuclear structure and reactions with heavy ions; fission processes.

Kephart, Thomas W., Ph.D., Northeastern, 1981. Theoretical physics; elementary particles, field theory, and cosmology.

Maguire, Charles F., Ph.D., Yale, 1973. Experimental physics; high-energy and relativistic heavy-ion collisions; studies of the nuclear equation of state and the quark-gluon plasma.

Oberacker, Volker E., Ph.D., Frankfurt, 1977. Theoretical nuclear physics; computational physics; structure of exotic nuclei.

Pantelides, Sokrates T., Ph.D., Illinois, 1973. William A. and Nancy F. McMinn Professor of Physics. Theoretical physics; semiconductor physics; first principles atomic-scale dynamics; mesocopic dynamics in complex solids; interactions of light with matter.

Ramayya, Akunuri V., Ph.D., Indiana, 1964. Experimental nuclear physics; nuclear structure and reactions with heavy ions; fission processes.

Scherrer, Robert J., Ph.D., Chicago, 1986. Astrophysics, theory. The physics of the early universe and the large-scale structure of the universe, including studies of primordial nucleosynthesis, dark energy, the cosmic microwave background, and particle physics in the early universe.

Sheldon, Paul D., Ph.D., California, Berkeley, 1986. Experimental physics; high-energy particles.

Tolk, Norman H., Ph.D., Columbia, 1966. Experimental physics; inelastic interactions with surfaces; particle-solid interactions; quantum-mechanical phase-interference effects; free-electron laser applications.

Umar, Sait A., Ph.D., Yale, 1985. Theoretical computational physics; nuclear theory, heavy-ion nuclear and atomic physics; models of supernovae.

Weintraub, David A., Ph.D., UCLA, 1989. Observational X-ray infrared and submillimeter astronomy; pre-main-sequence stars.

Weiler, Thomas J., Ph.D., Wisconsin, 1976. Theoretical physics; elementary particles, high-energy astrophysics, and cosmology.

Wikswo, John P., Jr., Ph.D., Stanford, 1975. Gordan A. Cain University Professor. Biological physics, biomedical engineering, cardiac and cellular electrophysiology, cellular instrumentation and control, complex matter, electromagnetism, non-destructive evaluation, non-linear dynamics and non-equilibrium behavior, SQUID magnetometry.

Associate Professors

Csorna, Steven E., Ph.D., Columbia, 1974. Experimental physics; high-energy particles; detector R&D.

Johns, Will, Ph.D., Colorado-Boulder, 1995. Experimental physics, high-energy physics.

Stassun, Keivan, Ph.D., Wisconsin, 2000. Observations and modeling of star formation; science pedagogy; diversity issues.

Velkovska, Julia, Ph.D., SUNY, 1997. Experimental physics; relativistic heavy ion collisions.

Weller, Robert A., Ph.D., Caltech., 1978. Joint appointee with Electrical and Electronic Engineering and Computer Science. Experimental physics, electronic and photonic materials; micro and nanotechnology, ion-solid interactions, radiation effects and reliability in microelectronics.

Assistant Professors

Berlind, Andreas, Ph.D., Ohio State, 2001. Large scale structure and galaxy formation; ultra-high cosmic energy cosmic rays.

Bolotin, Kirill I., Ph.D., Cornell, 2006. Experimental condensed matter physics; electron and spin transport in nano-scale systems (graphene).

Dickerson, James H., Ph.D., SUNY@Stony Brook, 2002. Experimental physics optical and electro-optical spectroscopy of nanostructures; fundamental phenomena of semiconductors and insulators; nanostructured thin films.

Holley-Bockelmann, Kelly, Ph.D., Michigan, 1999. Galaxy dynamics; N-body simulations; supermassive black holes; gravitational waves.

Hutson, M. Shane, Ph.D., Virginia, 2000. Experimental biophysics.

Varga, Kalman, Ph.D., U. Debrecen, 1996. Theoretical and computational research on multiscale modeling of materials.

Xu, Yaqiong, Ph.D., Rice University, 2006; Chinese Academy of Sciences, 2002. Interaction between single-walled carbon nanotubes and DNA.

Distinguished Research Professor

O'Dell, Robert C., Ph.D., Wisconsin, 1962. Optical imaging and spectroscopy, protoplanetary disks, planetary nebula.

Research Professor

Webster, Medford S., Ph.D., Washington, 1959. Experimental physics; high-energy physics and astrophysics; photon JETS; high-energy neutrinos.

Research Associate Professors

Hmelo, Anthony B., Ph.D., SUNY at Stony Brook, 1987. Materials physics; surface and thin films science; low dimensional materials.

Idrobo, Juan Carlos, Ph.D., UC-Davis, 2004. Atomic scale structure-property relationships of defects and interfaces in complex materials.

Mendenhall, Marcus H., Ph.D., Cal. Tech., 1983. Experimental; free electron laser.

Oxley, Mark, Ph.D., University of Melbourne, 1999. Theoretical physics; first principles of atomic-scale dynamics; semiconductor physics.

Research Assistant Professors

Avanesyan, Sergey, Ph.D., Moscow State University, 1987. Experimental physics; laser modification of surfaces and films.

Batyrev, Iskander G., Ph.D., Moscow, 1995. Theoretical condensed matter.

Bradshaw, Leonard, Ph.D., Vanderbilt, 1995. Living state physics.

Engh, Daniel, Ph.D., Vanderbilt, 2002. Experimental physics; high energy particles.

Gabella, William E., Ph.D., Colorado, 1991. Experimental; free electron laser.

Huang, Shengli, Ph.D., University of Science and Technology, China, 2004. Experimental physics; relativistic heavy ion collisions.

Issah, Michael, Ph.D., Stony Brook, 2006. Flow analysis in PHENIX (RHIC) and CMS (LHC) collaborations.

Ivanov, Borislav, Ph.D., Univ. Chemical Tech. & Metallurgy, Bulgaria, 1994. Experimental biophysics.

Jarvis, Jonathan, Ph.D., Vanderbilt, 2009. High brightness electron sources for free electron lasers.

Pepper, Joshua, Ph.D., Ohio State University, 2007. Theory and practice of searching for extrasolar planets.

Tackett, Alan, Ph.D., Wake Forest, 1998. Research computing systems administrator.

Velkovsky, Momchil, Ph.D., Stony Brook, 1997. Experimental biological physics.

Xu, Ying, Ph.D., Rochester, 2004. Spin dynamcis of charge carriers in semiconductors.

Zavalin, Andrey I., Ph.D., Moscow Engineering Physics Inst., 1990. Free Electron Laser; nonlinear optics and spectroscopy.

RESEARCH SPECIALTIES AND STAFF

Theoretical

Computational Nuclear and Atomic Physics. Nuclear structure; intermediate-energy and high-energy nuclear reactions; hadronic structure; supernovae modeling, netron oscillation phenomenology. Ernst, Oberacker, Umar. 1 postdoctoral fellow.

Condensed Matter. Equilibrium atomic configurations and atomic-scale dynamics of bulk defects, surfaces and interfaces; growth process; grain boundary dynamics; interaction of radiation with materials; many-body effects in nanostructures, ultrafast dynamics of strongly correlated systems. Pantelides, Varga. 3 research faculty.

Elementary Particles and Fields and Cosmology. Electroweak symmetry breaking; supersymmetry; unification of forces; super-strings; cosmology and particle astrophysics; highest energy cosmic rays; particle and dark matter phenomenology. Kephart, Scherrer, Weiler. 2 research faculty.

Experimental

Applied Physics. Ion-beam and laser modification of optical and electronic materials; ultrasensitive analysis of thin-film growth processes; laser-induced desorption spectroscopy of nonmetallic surfaces using visible, ultraviolet and free-electron lasers; nonlinear optical physics in quantum-dot composites; biomolecular mass spectrometry. Feldman, Haglund, Schriver, Weller. 3 research faculty.

Astronomy/Astrophysics. Observations of star forming nebulae and planetary nebulae; protoplanetary disks around young stars and the formation of planetary systems; rotational evolution of young stars; computational modeling of star formation; measurement of the expansion history of the Universe using high-redshift supernovae; the origin of the magnetospheric plasma. Chappell, O'Dell, Stassun, Weintraub. 2 research faculty. Holley-Bockelmann, Berlind

Atomic, Molecular, and Laser Interactions at Surfaces. Interactions of ions, atoms, electrons, and, synchrotron and laser photons with surfaces, interfaces, and thin films, with emphasis on electronic processes; carrier and spin dynamics; damage processes; desorption induced by electronic transitions, quantum interference effects, modification of surface electronic structure. Tolk. 2 research faculty.

Biophysics. Action potential propagation, shock response, and non-linear dynamics in excitable system such as cardiac and smooth muscle; magnetic imaging of bioelectric currents, magnetic markers, and remanent geomagnetism; laser-tissue interactions; cellular development, differentiation morphogenesis; fluorescence imaging; instrumenting and controlling single cells; membrane transporters and channels. Hutson, Piston, Wikswo. 9 research faculty.

Elementary Particles. Experiments with the FNAL tagged photon beam for very high statistics, strong and weak charm decays, and to study the structure of the photon at high energy. Experiment E831 (FOCUS) with the FNAL wideband photon beam for very high statistics, strong and weak decays. Fermi lab experiment BTEV in the tevatron collider for ultra high statistics, strong and weak decays of beauty and charm particles. Massively parallel computing on campus with the VAMPIRE PC linux cluster. Johns, Sheldon, Webster. 4 research faculty.

Electron-Positron Interactions at CLEO. Heavy quark physics, strong and weak decays of charmed and bottom mesons, CP violation and baryons using data from CLEO II at Cornell. Detector R&D, including robot for drift chamber stringing. Csorna.

Free-Electron Lasers. Generation of tunable, monochromatic, and coherent radiation at IR to x-ray wavelengths; application of radiation to materials science, biophysics, and medical physics. Brau. 3 research faculty.

Medical Physics. Interdisciplinary radiological medical physics, both diagnostic and therapy. Coffey, Duggan, Patton, Price.

Nanoscale Materials Physics. Synthesis and characterization of lower dimensional quantum structures. Growth and characterization of thin films and modified surfaces with enhanced optical, magnetic, and electronic properties. Dickerson, Feldman, Haglund, Hertel, Schriver, Tolk, Weller. 8 research faculty.

Nuclear Physics. Nuclear structure, Coulomb excitation; fission processes, in-beam gamma-ray spectroscopy of heavy-ion reactions; isotope separator and recoil mass spectrometer work to study nuclei far from line of stability; the behavior of nuclear matter at extreme temperatures; the search for quark gluon plasma and the colored glass condensate. Greene, Hamilton, Maguire, Ramayya, Velkovska. 5 research faculty.

BAYLOR UNIVERSITY

DEPARTMENT OF PHYSICS

Waco, Texas 76798

Students Accepted For Degree	FIELDS		
	Physics	Astronomy	Related Fields
Doctorate	X		
Master's	X		

1. General

President: Kenneth W. Starr
Dean of Graduate School: Larry Lyon
Department Chairman: Gregory A. Benesh
Department Telephone Number: (254) 710-2511
Type of Institution: University
Control: Private
Setting: Urban
Total Faculty: 903
Total Graduate Faculty: 619
Total Students: 15,025
Total Graduate Students: 2,332
Annual Graduate Tuition:
All Graduate Students: Full-time—$20,700
Tuition rates for: 2010–2011
Deferred tuition plan: Yes
*Other Fees**: $115/hr.—11 hrs. or less
Term: Semester

*Student service fees/sem.

2. Number of Faculty in Department

The combined total of full-time faculty in the three professorial ranks is 15. The combined total of full-time, part-time, and other faculty at all ranks is 21.

3. Admission, Financial Aid, and Housing

Address admission inquiries to: Graduate Admissions Office
Graduate application fee required: $40
Admission deadline (Fall admission): 2/15
Admission information: For fall admission, 2010–2011, 7 students were accepted from 38 applicants.
Admission requirements: For admission to the graduate programs, a Bachelor's degree in physics is required with a minimum undergraduate GPA of 3.0 specified. The GRE General Test and GRE Subject Test (Physics) are required. A minimum acceptable score of 1,000 is required on the verbal plus quantitative portions of the General Test. Students from non-English speaking countries are required to demonstrate proficiency in English via the TOEFL exam. Minimum acceptable score for admission is 550.
Undergraduate preparation assumed: 32 semester hours of undergraduate physics including 8 semester hours at senior level; 18 semester hours of undergraduate math, including differential equations; 1 semester of undergraduate chemistry.
Address financial aid inquiries to: Dr. Walter Wilcox, Dept. of Physics, One Bear Place #97316; Waco, Texas 76798-7316
GAPSFAS application required: No
Financial aid deadline: 4/1
Loans available: No
Address housing inquiries to: Dean of Students, Baylor Univ.
On-campus. single student housing available: Yes
Cost/month: $490–585 ($756–851/with meals)
On-campus. married student housing available: Yes (available near campus)
Cost/month: $575 and up

Table A—Faculty, Enrollments, and Degrees Granted

Research Specialty	2009–10 Faculty	Enrollment[1] Fall 2009		No. of Degrees Granted[2] 2008–09 (2004–09)			Median No. of Years for 2008–09 Ph.D.'s
		Master's	Doctorate	Master's	Terminal Master's	Doctorate	
Atmos./Space Phys., Cosmic Rays	2	3	4	0(2)	0(0)	0(2)	–
Atomic, Molecular, & Optical Physics	1	0	0	0(2)	1(1)	0(1)	–
Condensed Matter Physics	6	1	1	0(1)	0(0)	0(2)	–
Particles & Fields	6	2	10	1(3)	0(0)	5(7)	5.0
Non-specialized	0	0	8	0(4)	1(1)	0(0)	–
Total	15	6	23	1(12)	2(2)	5(12)	
Full-time Grad. Stud.		6	23				
Part-time Grad. Stud.		0	0				
First-year Grad. Stud.	–	0	8				
Median Years in Grad. Study (2008–09 Degrees)				2.0	5.3	5.0	–

Undergraduate Degrees, 2009–10 (2005–10): 5(19)

[1]Students not yet committed to a research specialty are entered under non-specialized.
[2]Five-year totals in parentheses.

4. Graduate Degree Requirements

Master's: 30 semester hrs. with thesis; 36 semester hrs. without thesis; minimum GPA of 3.0; residence requirements: two full semesters or three full summers; comprehensive oral examination required for thesis and nonthesis degree; thesis requirements: under supervision of thesis director and three graduate faculty members. Completed draft of research thesis must meet approval of all committee members.
Doctorate: 78 semester hrs. with dissertation; minimum GPA of 3.0; residence requirements: at least two consecutive semesters (summer does not count as a full semester); preliminary (qualifying) examination: 12 hours written examination covering quantum mechanics, classical mechanics, electricity and magnetism, mathematical physics, statistical mechanics; and other topics (must be taken prior to admission to candidacy for Ph.D.); final oral exam: given after all course, research, and dissertation requirements have been fulfilled; dissertation requirements: covers program of original research, the results of which reveal scholarly competence and are publishable in AIP journals or equivalent.
Thesis: Thesis may be written *in absentia*.
Special Equipment, Facilities, or Programs: Scanning tunneling microscope (STM) system. Metalorganic chemical vapor deposition (MOCVD) system. Equipment for x-ray diffraction studies. Surface analysis system (XSAM 800) allowing for the characterization of surface and atomic electronic structure(s) using ARXPS, ISS, AES, AM, and LEED. Hypervelocity accelerator lab including dust particle accelerator, light gas gun accelerators, laser gas cell accelerator system and laser hypervelocity impact simulation system, 2 GEC RF/DC Reference Cells used for complex (dusty) plasma and

colloidal plasma physics. Low field MRI. Spectra Nd:YAG laser, optical parametric oscillator FTIR spectrometer.

Table B—Appointments to Graduate Students, 2010–11

Title of Appointee	Appointments Total	Appointments First year	Academic Load Allowed in Credit Hours	Hours of Service Per Week	Stipend for Academic Year ($)
			Semester		
Teaching Assistant	22	6	10	15	20-26,000[1]
Research Assistant	7	1	10	15	20-26,000[1]
Total	29	7			

[1]Tuition waived.

5. Personnel Engaged in Separately Budgeted Research, 6/10–6/11

Professorial faculty	10
Graduate students	10
Total	20

6. Separately Budgeted Research Expenditures by Source of Support

	Departmental Research
Federal government	$4,051,000
Private, non-profit organizations	659,800
Total	$4,710,800

Table C—Separately Budgeted Research Expenditures

Research Specialty	No. of Grants	Expenditures ($)
Astrophysics/Space Physics	4	$422,100
Condensed Matter Physics	5	$1,102,500
NSF Summer Research Program	1	$143,000
Particles & Fields	8	$640,800
Physics & Society	2	$2,408,000
Total	20	4,710,800

FACULTY

Professors

Benesh, Gregory A., Ph.D., Northwestern, 1980. Theoretical condensed matter physics.

Hyde, Truell II, Ph.D., Baylor, 1988. Theoretical and Experimental Complex plasmas, Space physics and interplanetary particles. Astrophysics, magnetoelectrodynamics.

Wang, Anzhong, Ph.D., Ioannina (Greece) 1991. Theoretical gravity and cosmology.

Ward, Bennie, Ph.D., Princeton, 1973. Theoretical particle physics and Quantum General Relativity.

Wilcox, Walter M., Ph.D., UCLA, 1981. Theoretical elementary particle physics.

Associate Professors

Ariyasinghe, Wickramasinghe, Ph.D., Baylor, 1987. Experimental atomic, molecular, and solid state physics.

Cleaver, Gerald, Ph.D., Cal. Tech., 1993. Superstring theory.

Dittmann, Jay R., Ph.D., Duke, 1998. Experimental high energy particle physics.

Olafsen, Jeffrey, Ph.D., Duke, 1994. Experimental nonlinear dynamics.

Olafsen, Linda, Ph.D., Duke, 1997. Experimental condensed matter physics.

Park, Kenneth, Ph.D., University of Rochester, 1993. Experimental surface physics.

Russell, Dwight, Ph.D., Vanderbilt University, 1986. Experimental surface physics.

Assistant Professors

Hatakeyama, Kenichi, Ph.D., Rockefeller, 2003. Experimental high energy particle physics.

Matthews, Lorin, Ph.D., Baylor, 1998. Theoretical space physics.

Zhang, Zhenrong, Ph.D., Chinese Academy of Science, 2002. Experimental condensed matter physics.

Senior Lecturers

Kinslow, Linda, Ph.D., Baylor, 1979. Many body theory.

Schaub, Edward, M.S., Baylor, 1992. Space physics.

Vasut, John, Ph.D., Baylor, 2001. Theoretical dusty plasmas.

RESEARCH SPECIALTIES AND STAFF

Theoretical

Complex Plasmas and Space Physics. Dust dynamics in planetary rings, cometary comas, protostellar/protoplanetary clouds. Charging processes and colloidal crystal formation in dustry plasmas. Hyde, Matthews.

Condensed Matter. Embedding problems. Electronic structure of surfaces. Surface energies, magnetism, and catalysis. Benesh.

Cosmology and Gravitation. Wang.

Elementary Particle Physics. Ward, Wilcox.

Superstring/M Theory. Cleaver.

Experimental

Atomic Physics. Heavy-ion-induced Auger electron studies, chemical binding effects on Auger electrons and energy-loss mechanisms, intermediate and high-energy electron scattering. Ariyasinghe.

Complex Plasmas and Space Physics. Hypervelocity impact phenomena as related to orbital debris and fusion devices. Dusty plasmas, laser physics, shock physics, *in situ* instrumentation. Hyde, Matthews.

Condensed Matter. Optical and electronic properties of III-V semiconductors. Infrared semiconductor lasers. L. Olafsen.

High-Energy Particle Physics. Studies of high-energy hadron collisions with the CDF experiment at the Fermi National Accelerator Laboratory and the CMS experiment at CERN. Dittmann, Hatakeyama.

Nonlinear, Chaotic, and Non-equilibrium Systems. Chemical, granular, and soft condensed matter physics, insect biomechanics, dissipative and dynamical systems. J. Olafsen

Surface Physics. Surface atomic and electronic structure of transition metals and compounds, adsorbate-induced surface modifications, atomically resolved imaging of surface catalytic reactions on model catalytic systems, material characterization under ultrahigh vacuum. Park, Russell, Zhang.

RICE UNIVERSITY

DEPARTMENT OF PHYSICS AND ASTRONOMY

Houston, Texas 77251

Students Accepted For Degree	FIELDS Physics	Astronomy	Related Fields
Doctorate	X	X	
Master's			

*Masters degree may be awarded as a part of the qualification process for a Ph.D.

1. General

President: David W. Leebron
Provost: George McLendon
Department Chairman: F. Barry Dunning
Department Telephone Number: (713) 348-4938
Type of Institution: University
Control: Private
Setting: Urban
Total Faculty: 644
Total Graduate Faculty: 523
Total Students: 5,556
Total Graduate Students: 2,277
Annual Graduate Tuition:
All Graduate Students: Full-time—$33,120
Part-time—$1840/credit[1]
Tuition rates for: 2010–11
Deferred tuition plan: Yes
Other Fees: $500
Term: Semester

[1]Plus $135 registration fee.

2. Number of Faculty in Department

The combined total of full-time faculty in the three professorial ranks is 42. The combined total of full-time, research, adjunct, and other faculty at all ranks is 56.

3. Admission, Financial Aid, and Housing

Address admission inquiries to: Graduate Admissions, Department of Physics and Astronomy, MS61, Rice University, P.O. Box 1892, Houston, Texas 77251-1892
Graduate application fee required: $70
Admission deadline (Fall admission): January 15
Admission information: For fall admission, 2010–11, 51 students were accepted from 229 applicants.
Admission requirements: For admission to the graduate programs, a Bachelor's degree in physics is normally required. The GRE is required. The GRE Advanced is required (except for MST students). The average GRE scores for admissions were verbal—545; quantitative—771. The average GRE advanced score for admissions was 869. Students from non-English speaking countries are required to demonstrate proficiency in English via the TOEFL exam. Minimum acceptable score for admission is 250 on the computer-based version or 90 on the internet-based test.
Undergraduate preparation assumed: Advanced undergraduate courses in mechanics, electricity and magnetism, quantum mechanics and statistical physics. Mathematics at least through partial differential equations and complex analysis, and one year of advanced undergraduate laboratory.
Address financial aid inquiries to: Graduate Admissions, Department of Physics and Astronomy, MS61, Rice University, P.O. Box 1892, Houston, TX 77251-1892
GAPSFAS application required: No
Financial aid deadline: January 15
Loans available: Yes
Address housing inquiries to: Graduate House, 1515 Bissonnet, Houston, TX 77005-1846
On-campus, single student housing available: Yes
**Cost/mo.*: Single $550–825; Double $550–645/person
On-campus, married graduate student housing available: No

*With $350 deposit

Table A—Faculty, Enrollments, and Degrees Granted

Research Specialty	2009–10 Faculty	Enrollment[1] Fall 2009 Master's	Doctorate	No. of Degrees Granted[2] 2009–10 (2006–10) Master's	Terminal Master's	Doctorate	Median No. of Years for 2009–10 Ph.D.'s
Astronomy/ Astrophysics	9	9	4	1(10)	0(1)	2(9)	7.5
Atmos./Space Phys., Cosmic Rays	5	7	5	2(14)	1(2)	2(12)	6
Atomic, Molecular, & Optical Physics	4	11	6	5(12)	0(3)	3(14)	7.5
Biophysics	4	4	4	1(12)	0(4)	3(9)	6
Chemical Physics	1	0	1	0(0)	0(0)	1(3)	6.5
Condensed Matter Physics	10	10	5	3(10)	1(1)	4(14)	6
Particles & Fields	8	2	2	1(4)	0(1)	1(3)	7.5
Non-specialized	0	21	0	0(0)	0(0)	0(0)	–
Total		58	27	13(62)	2(12)	16(64)	
Full-time Grad. Stud.		58	27				
Part-time Grad. Stud.		0	0				
First-year Grad. Stud.		19	0				
Median Years in Grad. Study (2009–10 Degrees)				3.7	4.3	6.7	–

Undergraduate Degrees, 2009–10 (2006–10): 16(87)

[1]Students not yet committed to a research specialty are entered under non-specialized.
[2]Five-year totals in parentheses.

4. Graduate Degree Requirements

Master's: The Master of Science is a research degree, normally undertaken as the first stage of doctoral study. The M.S. requires completion of four core courses chosen from a restricted group, at least two other courses appropriate for the student's area of interest, and research, for a minimum program totaling 30 credit hours. The student must also pass an oral defense of the Master's thesis, prepared under the direction of a departmental faculty member. A paper submitted to a refereed journal may substitute for the M.S. thesis at the discretion of the examining committee. Previous graduate work will be considered on an individual basis.
Doctorate: To be eligible for the Ph.D. degree, graduate students must demonstrate to the department their ability to engage in advanced research. This is normally accomplished by successfully defending the Master's research and passing an oral examination in the student's sub field. To obtain the Ph.D., students must complete two additional courses beyond the

M.S. requirements, complete the teaching practicum, and publicly defend a Ph.D. research thesis done under the direction of a departmental faculty member. The department does not have a foreign language requirement. Students must complete a total of 90 credit hours of approved graduate-level study beyond the bachelor's degree, including the teaching practicum and thesis research. They must be in residence for at least four semesters of full-time graduate study at Rice for the Ph.D.

All graduate students in the doctoral programs share nominal teaching responsibilities in the department for a total of five semesters, beginning in the second semester of residence. Assignments typically include some combination of grading and instruction in the undergraduate laboratories. The total time commitment, including preparation, is expected to be about six hours per week.

Thesis: Theses may not be written *in absentia*.

Table B—Appointments to Graduate Students, 2009–10

Title of Appointee	Appointments Total	Appointments First year	Academic Load Allowed in Credit Hours	Hours of Service Per Week[1]	Stipend for Academic Year ($)
			Semester		
Research Assistant	76	0	15	5	25,700[2]
Rice Fellowship	21	21	15	5	19,275[3]
Total	97	21	–		

[1]All students assist with grading and laboratory instruction for 5 semesters.
[2]12 months. Increases to $26,700 with Ph.D. candidacy.
[3]9 months.

5. Personnel Engaged in Separately Budgeted Research, 7/09–6/10

Professorial faculty	39
Nonteaching research scientists	10
Postdoctoral appointments	14
Graduate students	76
Nonteaching research personnel	3
Total	142

6. Separately Budgeted Research Expenditures by Source of Support

	Departmental Research
Federal government	$7,738,000
State government	32,000
Private, non-profit organizations	1,230,000
Total	$9,000,000

Table C—Separately Budgeted Research Expenditures

Research Specialty	No. of Grants	Expenditures ($)
Astronomy/Astrophysics	30	1,019,000
Atmos./Space Phys. Cosmic Rays	35	2,336,000
Atomic, Molecular, & Optical Physics	20	1,991,000
Biophysics	18	890,000
Condensed Matter Physics	23	1,490,000
Particles & Fields	9	1,274,000
Total	135	9,000,000

FACULTY

Professors

Alexander, David, Ph.D., University of Glasgow, 1988. Solar and solar-terrestrial physics.

Chan, Anthony A., Ph.D., Princeton, 1991. Theoretical plasma physics, space physics.

Corcoran, Marjorie D., Ph.D., Indiana, 1977. Experimental particle physics.

Deem, Michael W., Ph.D.[1], University of California at Berkeley, 1994. Statistical mechanics and bioinformatics.

Du, Rui-Rui, Ph.D., University of Illinois, 1990. Experimental condensed matter physics.

Dufour, Reginald J., Ph.D., University of Wisconsin at Madison, 1974. Observational astrophysics of nebulae and galaxies.

Dunning, F. Barry, Ph.D., University College, London (England), 1969. Chair. Experimental atomic, molecular and optical physics, physics of surfaces, thin film magnetism.

Halas, Naomi J., Ph.D.[2], Bryn Mawr College (1986). Experimental condensed matter physics.

Hannon, James P., Ph.D., Rice, 1967. Theoretical condensed matter physics; coherent gamma-ray optics; cooperative phenomena.

Hartigan, Patrick M., Ph.D., University of Arizona, 1987. Observational astrophysics and protostars and jets.

Hill, Thomas W., Ph.D., Rice University, 1973. Space plasma physics, planetary magnetospheres.

Huang, Huey W., Ph.D., Cornell, 1967. Theoretical statistical physics; theoretical biophysics; X-ray, neutron and optical spectroscopies on membrane physics.

Hulet, Randall G., Ph.D., MIT, 1984. Experimental atomic physics and quantum optics; laser cooling and atom trapping.

Killian, Thomas C., Ph.D., MIT, 1999. Experimental atomic, molecular and optical physics, ultracold plasmas.

Kono, Junichiro, Ph.D.[2], State University of New York at Buffalo (1995). Experimental condensed matter physics.

Lane, Neal F., Ph.D., University of Oklahoma, 1964. Science policy.

Levy, Eugene H., Ph.D., Chicago, 1971. Planetary geophysics, solar and space physics.

Liang, Edison P., Ph.D., University of California at Berkeley, 1971. High-energy astrophysics, cosmology.

Natelson, Douglas A., Ph.D., Stanford, 1998. Experimental condensed matter physics.

Nordlander, Peter, Ph.D., Chalmers (Sweden), 1985. Condensed-matter theory; electronic properties of surfaces.

Padley, B. Paul, Ph.D., University of Toronto, 1987. Experimental particle physics.

Rau, Carl, Ph.D., Munich Technical, 1970, Dr. habil., Munich, 1980. Surface science, magnetism.

Reiff, Patricia H., Ph.D., Rice University, 1975. Space plasma physics and auroral physics.

Roberts, Jabus B., Jr., Ph.D., University of Pennsylvania, 1969. Experimental elementary particle physics.

Scuseria, Gustavo E., Ph.D.[3], University of Buenos Aires (1983). Theoretical quantum chemistry and condensed matter physics.

Si, Qimiao, Ph.D., University of Chicago, 1991. Theoretical condensed matter physics; strongly correlated electron systems.

Stevenson, Paul M., Ph.D., Imperial College (London), 1979. Quantum field theory; elementary particle phyisics.

Toffoletto, Frank R., Ph.D., Rice University, 1987. Magnetospheric physics and numerical simulations.

Associate Professors

Baring, Matthew G., Ph.D., Cambridge, 1989. Theoretical high-energy astrophysics of compact objects.

Dodds, Stanley A., Ph.D., Cornell, 1975.

Hafner, Jason H., PhD., Rice University, 1998. Biophysical applications of nanotechnology.

Johns-Krull, Christopher M., Ph.D., University of California, Berkeley, 1994. Observational astronomy of stellar evolution.

Pu, Han, Ph.D., University of Rochester, 1999. Theoretical atomic and optical physics.

Assistant Professors

Bradshaw, Stephen J., Ph.D., University of Cambridge (2003). Solar physics.

Ecklund, Karl M., Ph.D., Stanford University (1996). Experimental high-energy physics.

Fossati, Giovanni, Ph.D., International School for Advanced Studies, 1998. High energy astrophysics.

Geurts, Frank, Ph.D., Universiteit Utrecht (1998). Experimental high-energy physics.

Imambekov, Adilet, Ph.D., Harvard University (2007). Theoretical atomic and moleclar physics.

Kiang, Ching-Hwa, Ph.D., California Institute of Technology, 1995. Molecular biophysics.

Morosan, Emilia, Ph.D., Iowa State University (2005). Synthesis and characterization of novel electronic and magnetic materials.

Nevidomskyy, Andriy, Ph.D., University of Cambridge (2005). Theoretical condensed matter physics.

Raghu, Srinivas, Ph.D., Princeton University (2006). Theoretical condensed matter physics.

Senior Faculty Fellows

Llope, William J., Ph.D., State University of New York at Stony Brook, 1992. Experimental nuclear and particle physics.

Sazykin, Stanislav, Ph.D., Utah State University, 2000. Space plasma physics.

Smith, Ian A., Ph.D., Washington University, 1990. Observational high-energy astrophysics.

Yepes, Pablo, Ph.D., University of Santiago de Compostela, 1988. Experimental particle physics.

Faculty Fellow

Lapotki, Dmitri, Ph.D., Belarussian State University (1988). Experimental biophysics.

Adjunct Faculty

Aschwanden, Markus J., Ph.D., ETH Zurich (1987). Observational solar physics.

Bolech, Carlos J., Ph.D., Rutgers, 2002. Theoretical condensed matter physics.

Burch, James L., Ph.D., Rice University, 1968. Space plasma physics; magnetospheric and auroral physics.

Chang-Diaz, Franklin R., Ph.D., MIT, 1977. Plasma rocket propulsion; plasma physics and fusion technology.

Kirchner, Stefan, Ph.D., Technical University of Karlsruhe (2002). Theoretical condensed matter physics.

Li, Hui, Ph.D., Rice University, 1995. High energy astrophysics.

Newman, James H., Ph.D., Rice University, 1984. Atomic collision physics.

Rimberg, Alexander J., Ph.D., Harvard University, 1992. Experimental condensed matter physics.

Sumners, Carolyn, Ed.D., University of Houston, 1979. Science teaching.

Young, David T., Ph.D., Rice University (1970). Space plasma physics.

[1]Joint appointment in Bioengineering

[2]Joint appointment in Electrical and Computer Engineering

[3]Joint appointment in Chemistry

RESEARCH SPECIALTIES AND STAFF

Theoretical

Astronomy. High energy electromagnetic processes; dense and/or intensely magnetized plasmas; accretion disk phenomena; gamma-ray bursters; chemical evolution of galaxies. Baring, Fossati, Hartigan, Liang.

Atomic Molecular and Optical Physics. Degenerate quantum gases; atom optics and quantum optics. Pu.

Condensed Matter Physics. Strongly-correlated electron systems in high temperature superconductors and mesoscopic structures; non-Fermi liquid behavior in various systems; particle-surface interactions. Hannon, Imambekov, Nevidomskyy, Nordlander, Raghu, Si.

Nuclear and Particle Physics. Nonperturbative methods in quantum field theory. Stevenson.

Space Physics. Magnetospheric structure and dynamics, both terrestial and planetary; electromagnetic wave-particle interactions. Chan, Hill, Sazykin, Toffoletto.

Experimental

Astronomy. Multiwavelength imagery and spectroscopy of astrophysical plasmas, particularly HII regions, nebulae and star-forming regions; gamma-ray bursters and pulsars, solar corona and solar-terrestrial interactions. Alexander, Bradshaw, Dufour, Hartigan, Johns-Krull, Oberlack, Smith.

Atomic, Molecular, and Optical Physics. Quantum gases and Bose-Einstein condensation; use of Rydberg atoms to probe chaotic dynamics, ion-surface interactions, and electron-molecule interactions. Dunning, Hulet, Killian.

Biophysics. Studies of membrane ion channels; nanoscale probes; biological macromolecular assemblies; statistical mechanics and bioinformatics; photothermal techniques. Deem, Hafner, Huang, Kiang, Lapotki.

Condensed Matter Physics. Electrical transport in lithographically and chemically formed nanostructures; spin-polarized surface spectroscopies; novel electronic materials. Du, Dunning, Kono, Halas, Morosan, Natelson, Rau, Scuseria.

Nuclear and Particle Physics. Detector development, hardware and software, for RHIC and CMS; study of quark-gluon plasma at RHIC; D0 experiment at Fermilab. Corcoran, Ecklund, Geurts, Llope, Padley, Roberts, Yepes.

Space Physics. Mission planning and data analysis for earth and planetary plasma probes. Reiff.

RICE UNIVERSITY

PROFESSIONAL SCIENCE MASTER'S PROGRAM IN NANOSCALE PHYSICS

Houston, Texas 77005

Students Accepted For Degree	FIELDS: Physics	Astronomy	Related Fields
Doctorate			
Master's	X		

Address housing inquiries to: Graduate House
1515 Bissonnet
Houston, TX 77005-1813

On-campus, single student housing available: Yes
Cost/mo.: $380–790
On-campus, married graduate student housing available: No

1. General

President: David Leebron
Dean of Natural Sciences: Dan Carson
Track Director: F. Barry Dunning
Program Telephone Number: (713) 348-3188
E-mail: profms@rice.edu
Website: http://www.profms.rice.edu
Type of Institution: University
Control: Private
Setting: Urban
Total Faculty: 598
Total Graduate Faculty: 486
Total Students: 5,145
Total Graduate Students: 2,144
Annual Graduate Tuition:
All Graduate Students: Full-time—$25,900
Part-time—$1,440/credit
Tuition rates for: 2010–2011
Deferred tuition plan: Yes
Other Fees: $540
Term: Semester

*This program will cost three semesters of graduate tuition plus a nominal fee for the semester when the student is on internship.

2. Number of Faculty in Department

The combined total of full-time faculty in the three professorial ranks is 37. The combined total of full-time, part-time, and other faculty in nanoscale physics is 12.

3. Admission, Financial Aid, and Housing

Address admission inquiries to: Professional Master's Program, MS-103, Rice University, P.O. Box 1892, Houston, TX 77251-1892
Graduate application fee required: No
Admission deadline (Fall admission): 2/25
Admission information: For fall admission, 2009–10, 8 students were accepted from 14 applicants.
Admission requirements: For admission to the graduate programs, a Bachelor's degree in physics, or related field, is required. Minimum undergraduate GPA is 3.0. The General GRE is required. The average GRE scores for admissions were verbal 500; quantitative 700. Students from non-English speaking countries are required to demonstrate proficiency in English via the TOEFL exam. Minimum acceptable score for admission is 600, or 250 computer based.
Undergraduate preparation assumed: Undergraduate courses in quantum mechanics, differential equations, and calculus.
Address financial aid inquiries to: No financial aid available.
GAPSFAS application required: No
Financial aid deadline: N/A. Some corporate scholarships
Loans available: Yes

Table A—Faculty, Enrollments, and Degrees Granted

Research Specialty	2009–10 Faculty	Enrollment[1] Fall 2009: Master's	Doctorate	No. of Degrees Granted[2] 2009–10: Master's	Terminal Master's	Doctorate	Median No. of Years for Ph.D.'s
Atmos./Space Phys., Cosmic Rays	7	–	0	0(0)	0(0)	–	–
Atomic, Molecular, & Optical Physics	4	2	0	0(3)	0(0)	–	–
Biophysics	4	–	0	0(0)	0(0)	–	–
Condensed Matter Physics	1	–	0	0(0)	0(0)	–	–
Nanocrystals	2	–	0	0(0)	0(0)	–	–
Total		2	0	0(4)	0(0)	0(0)	–
Full-time Grad. Stud.		8	–				
Part-time Grad. Stud.		0	–				
First-year Grad. Stud.		2	–				
Median Years in Grad. Study				2			

[1] Students not yet committed to a research specialty are entered under non-specialized.
[2] Five-year totals in parentheses.

4. Graduate Degree Requirements

Master's: The Master of Science degree in Nanoscale Physics will provide students the knowledge to successfully navigate the emerging field of nanoscale science and technology. The 21-month program begins with two semesters of coursework at Rice followed by a three- to six-month industrial internship. After the internship, students will return to Rice for a final semester of coursework. All students are required to complete Nanostructures and Nanotechnology I and II, Methods of Experimental Physics I and II, Characterization and Fabrication at the Nanoscale, and Computational Physics. In addition to these six core physics courses, students will enroll in one management course, one policy course, and a seminar jointly with students involved in the other professional master's tracks. Students will be able to choose four elective courses, for a minimum of 40 total credit hours for the program. At least 24 of these credit hours must be taken at Rice. No thesis is required, however students must present their internship project in both oral and written form.

FACULTY

Professors

Barron, Andrew R., Professor, Ph.D., Imperial College of Science and Technology, University of London, 1986. Applications of inorganic chemistry to the materials science of aluminum, gallium and indium.

Carson, Dan, Dean of Natural Sciences, Ph.D., Temple University. Biochemistry and cell biology.

Colvin, Vicki L., Professor, Ph.D., University of California, Berkeley, 1994. Nanocrystals, confined liquids and glasses, porous solids, and photonic band gap materials.

Dunning, F. Barry, Professor, Ph.D., University College London, 1969. Experimental atomic and molecular physics, surface physics, spin dependent phenomena, surface magnetism, chemical physics, optics and instrumentation.

Hafner, Jason H., Associate Professor, Ph.D., Rice University, 1998. Carbon nanotube synthesis: chemical kinetics and device fabrication. Lipid bilayer substrates for biological atomic force microscopy.

Killian, Thomas C., Associate Professor, Ph.D., Massachusetts Institute of Technology, 1999. Atomic, molecular, and optical physics: cold collisions, Bose-Einstein condensation, fundamental measurements, high-resolution spectroscopy, atom-photon interactions, and low temperature plasmas.

Natelson, Douglas, Associate Professor, Ph.D., Stanford University, 1998. Nanoscale physics, in particular the electrical and magnetic properties of systems with characteristic dimensions approaching the single-nm scale.

Toffoletto, Frank R., Associate Professor, Ph.D., Rice University, 1987. Magnetospheric physics, numerical simulations, space weather.

RESEARCH SPECIALTIES AND STAFF

Experimental

Atmos./Space Physics, Cosmic Rays. Toffoletto.
Atomic, Molecular, and Optical Physics. Dunning and Killian.
Biophysics. Hafner.
Condensed Matter Physics. Dunning, Natelson, and Rimberg.
Nanocrystals. Colvin.

SOUTHERN METHODIST UNIVERSITY

DEPARTMENT OF PHYSICS

Dallas, Texas 75275-0175

Students Accepted For Degree	FIELDS: Physics	Astronomy	Related Fields
Doctorate	X		
Master's	X		

1. General

President: R. Gerald Turner
Dean of Graduate School: James E. Quick
Department Chairman: Fredrick I. Olness
Department Telephone Number: (214) 768-2495 (C)
Type of Institution: University
Control: Private
Setting: Suburban
Total Faculty: 660
Total Students: 10,000+
Total Graduate Students: 4,663
Annual Graduate Tuition:
All Graduate Students: $1,381/credit
Tuition rates for: 2010–11
Deferred tuition plan: No
Other Fees: $176/credit
Term: Semester

2. Number of Faculty in Department

The combined total of full-time faculty in the three professorial ranks is 10. The combined total of full-time, part-time and other faculty at all ranks is 19.

3. Admission, Financial Aid, and Housing

Address admission inquiries to: Southern Methodist University, Graduate Admissions—Physics, P.O. Box 750240, Dallas, TX 75275-0240
Graduate application fee required: $75
Admission deadline (Fall admission): February 1
Admission information: For fall admission, 2010–11, 4 students were accepted from 6 applications.
Admission requirements: For admission to the graduate programs, a Bachelors degree in physics is required with a minimum undergraduate GPA of 3.0 specified. The GRE is required. The minimum acceptable score suggested for admission is 1,300. The GRE Advanced is recommended. Students from non-English speaking countries are required to demonstrate proficiency in English via the TOEFL exam. Minimum acceptable score for admission is 550.
Undergraduate preparation assumed: Serway, *Physics*; Marion and Thornton, *Classical Dynamics*; Griffiths, *Introduction to Electrodynamics*; Reif, *Fundamentals of Statistical and Thermal Physics*; Gasiorowicz, *Quantum Physics*; Bevington, *Data Reduction and Error Analysis*; Arfken/Weber, *Mathematical Methods for Physicists*.
Address financial aid inquiries to: Director of Graduate Studies, P.O. Box 750240, Dallas, TX 75275-0240
GAPSFAS application required: No
Financial aid deadline: 5/15
Loans available: Yes
Address housing inquiries to: Office of Housing and Residence Life, P.O. Box 215, Dallas, TX 75275
On-campus, single student housing available: Yes
Cost/academic year: $7,260 (various meal plans available)
On-campus, married student housing available: Yes
Cost/month: $957 (1 bedroom various meal plans available)

Table A—Faculty, Enrollments, and Degrees Granted

Research Specialty	Faculty	Enrollment[1] Fall 2010: Master's	Doctorate	No. of Degrees Granted[2] 2009–10 (2005–10): Master's	Terminal Master's	Doctorate	Median No. of Years for 2009–10 Ph.D.'s
Astrophysics	1	–	0	0(0)	–	0(0)	–
Atmos/Space Phys., Cosmic Rays	1	–	0	0(0)	–	0(0)	–
Condensed Matter Physics	2	–	0	0(0)	–	0(0)	–
Particles & Fields	9		14	0(5)	0(0)	1(5)	6
Other Theoretical/ Math.	4	–	1	0(0)	–	1(5)	–
Total			15	0(5)	0(0)	2(10)	
Full-time Grad. Stud.			15				
Part-time Grad. Stud.			1				
First-year Grad. Stud.			1				
Median Years in Grad. Study (Degrees)		2	6	–	–	–	–

Undergraduate Degrees, 2009–10 (2005–10): 2(18)

[1]Students not yet committed to a research specialty are entered under non-specialized.
[2]Five-year totals in parentheses.

4. Graduate Degree Requirements

Master's: Students seeking the M.S. degree in physics must complete either 33 semester hours of approved graduate course work or 30 semester hours of courses including a research thesis. Every M.S. degree program must contain at least 18 semester hours of graduate courses in physics, including Classical Mechanics, Electromagnetic Theory, and Quantum Mechanics.
Doctorate: Students seeking the Ph.D. in Physics must complete eight specified 3-credit core courses (Mechanics, Electricity and Magnetism (2), Quantum Mechanics (2), Statistical Mechanics, Mathematical Methods and Quantum Field Theory), four more 3 credit-hour elective graduate courses, 12 credit-hours of research, and a 12 credit-hour dissertation. Additionally, the student must pass a comprehensive examination after completion of the core courses.
Thesis: Thesis may be written *in absentia*.
Special Equipment, Facilities or Programs: The department maintains a fast signal processing opto-electronics laboratory, that is used for R&D for the ATLAS detector, as well as collaborative projects with local hi-tech companies. Students participating in research use the facilities provided by national research laboratories: Fermilab, Brookhaven National Laboratory, and CERN-the European Center for Particle Physics. SMU faculty members participate in the following experiments: ATLAS, DO, NOVA, ebubble, and CTEQ.

Table B—Appointments to Graduate Students, 2008–09

Title of Appointee	Appointments Total	Appointments First year	Academic Load Allowed in Credit Hours	Hours of Service Per Week	Stipend for Academic Year ($)
			Semester		
Teaching Assistant	7	7	9	15	20,400
Research Assistant	8	0	9	20	21,600
Total	15	7			

5. Personnel Engaged in Separately Budgeted Research, 7/09–6/10

Professorial faculty	10
Other faculty	4
Postdoctoral appointments	11
Graduate Students	4
Nonteaching research personnel	3
Total	32

6. Separately Budgeted Research Expenditures by Source of Support

	Departmental Research	Physics-related Research Outside Department
Federal government	$2,000,000	None
Other		
Total	$2,000,000	None

Table C—Separately Budgeted Research Expenditures

Research Specialty	No. of Grants	Expenditures ($)
Particles & Fields	10	2,000,000
Total	10	2,000,000

FACULTY

Professors

Olness, Fredrick I., Ph.D., University of Wisconsin-Madison, 1985. Particle theory and phenomenology.

Stroynowski, Ryszard, Ph.D., University of Geneva, Switzerland, 1973. Chair. Experimental high-energy physics.

Associate Professors

Coan, Thomas E., Ph.D., University of California at Berkeley, 1989. Experimental high-energy physics.

Hornbostel, Kent J., Ph.D., Stanford University, 1988. Field theory, lattice gauge theory.

Vega, Roberto, Ph.D., University of Texas, 1988. Particle theory and phenomenology.

Ye, Jingbo, Ph.D., Swiss Federal Institute of Technology, 1992. Experimental high-energy physics.

Assistant Professors

Cooley, Jodi, Ph.D., Wisconsin (Madison).

Kehoe, Robert L. P., Ph.D., Notre Dame, 1997. Experimental high-energy physics.

Nadolsky, Pavel, Ph.D., Michigan State University.

Sekula, Stephen, Ph.D., Wisconsin (Madison).

Visiting Assistant Professor

McElgin, William, Ph.D., University of Chicago.

Senior Lecturer

Dalley, Simon, Ph.D., University of Southhampton. 1991. Theoretical elementary particle physics.

Scalise, Randall J., Ph.D., Pennsylvania State University, 1994. Theoretical elementary particle physics.

Research Associate Professor

Liu, Tiankuan, Ph.D., University of Science and Technology of China, 1998. Nuclear Physics.

Research Assistant Professors

Gong, Datao, Ph.D., University of Science and Technology of China, 1999. Particle physics.

Xiang, Annie Chu, Ph.D., Department of Electrical and Computer Engineering, Rice University, 2002. Fiber Optics Communication.

Research Scientists

Firan, Ana
Geuzzi, Marco
Hadavand, Haleh
Hoffman, Julia
Joffe, David
Kami, Sami
Liu, Chonghan
Park, Kwangwoo
Randle-Conde, Aidan
Renkel, Peter
Scorza, Silvia

Adjunct Professor

Cotton, John, M.S., Southern Methodist University, 1991. Astronomy.

RESEARCH SPECIALTIES AND STAFF

Theoretical

Field Theory. Lattice gauge theory; QCD. Hornbostel.

Particles and Fields. QCD structure functions; heavy quark approximation; Monte Carlo simulations; Higgs production and scattering backgrounds. Dalley, Nadolsky, Olness, Scalise, Vega. 2 postdoctoral research associates.

Experimental

Particle Physics. Participation in ATLAS experiment at CERN, eBubble (BNL), DO (FNAL), NOVA (FNAL), and LBNE (FNAL). SuperCDMS (Soudan, SNOLAB) Coan, Cooley, Gong, Kehoe, Liu, Sekula, Stroynowski, Xiang, Ye. 8 postdoctoral research associates, 1 machinist, 1 computer professional.

TEXAS CHRISTIAN UNIVERSITY

DEPARTMENT OF PHYSICS AND ASTRONOMY

Fort Worth, Texas 76129

Students Accepted For Degree	FIELDS		
	Physics	Astronomy	Related Fields
Doctorate	X	X	
Master's	X		

1. General

President: Victor Boschini
Dean: D. Kouris
Department Chairman: T. Waldek Zerda
Department Telephone Number: (817) 257-7375 (C)
Department E-mail Address: physics@tcu.edu
World Wide Web: www.phys.tcu.edu
FAX: (817) 257-7742
Type of Institution: University
Control: Private
Setting: Urban
Total Faculty: 523
Total Students: 8,853
Total Graduate Students: 1,213
Annual Graduate Tuition:
All Graduate Students: $980/sem. hr.
Tuition rates for: 2009–10
Deferred tuition plan: Yes
Other Fees: $24/year
Term: Semester

2. Number of Faculty in Department

The combined total of full-time faculty in the three professorial ranks is 8. The combined total of full-time, part-time, and other faculty at all ranks is 9.

3. Admission, Financial Aid, and Housing

Address admission inquiries to: The Physics and Astronomy Department
Graduate application fee required: $50
Admission deadline (Fall admission): 8/15
Admission information: TCU Physics and Astronomy Department accepted four students for the 2009–10 year.
Admission requirements: For admission to the graduate programs, a Bachelor's degree in physics is required with a minimum undergraduate GPA of 3.0/4.0 specified. The GRE is required. The GRE Advanced is recommended. No minimum scores required. Students from non-English speaking countries are required to demonstrate proficiency in English via the TOEFL exam. Minimum acceptable score for admission is 550, better than 600 recommended.
Undergraduate preparation assumed: B.A. or B.S. with a physics major, or 24-semester-hour equivalent including intermediate or advanced undergraduate courses in mechanics, electricity and magnetism, atomic and nuclear or modern physics, or their equivalents. Twelve semester hours must be of junior or senior level. Required are mathematics through differential equations and a course in general chemistry.
Address financial aid inquiries to: The Physics and Astronomy Department
GAPSFAS application required: No
Financial aid deadline: 3/1
Loans available: Yes
Address housing inquiries to: Housing and Residential Living Office
On-campus, single student housing available: No
On-campus, married student housing available: No

Table A—Faculty, Enrollments, and Degrees Granted

Research Specialty	2009–10 Faculty	Enrollment Fall 2009		No. of Degrees Granted[1] 2010–10 (2004–10)			Median No. of Years for 2009–10 Ph.D.'s
		Master's	Doctorate	Master's	Terminal Master's	Doctorate	
Astronomy	1	–	2	–	0(0)	0(2)	5
Atomic, Molecular, & Optical Physics	4	–	6	–	0(3)	2(4)	5
Condensed Matter Physics	1	–	3	–	0(1)	1(6)	5
Statistical & Thermal	1	–	2	–	0(0)	1(0)	5
Total		–	13	–	1(4)	4(12)	
Full-time Grad. Stud.		–	11				
Part-time Grad. Stud.		–	2				
First-year Grad. Stud.		–	4				
Median Years in Grad. Study (2009–10 Degrees)	5	–	–	–	–	–	
Undergraduate Degrees, 2009–10 (2005–10): 4(10))							

[1]Five-year totals in parentheses.

4. Graduate Degree Requirements

Master's:
The M.S. Degree requires 30 approved semester hours with a thesis or 36 semester hours without a thesis. Course requirements for the degree are: Quantum Mechanics I & II, three courses from Classical Mechanics, Electrodynamics I & II, Solid State Physics or Statistical Physics, plus a minimum of 6 additional semester hours in Physics. An oral exam over course work and thesis, if any, is required.
Doctorate: (Ph.D. only) The Ph.D. degree is available on a Physics or Astrophysics track. An M.B.A. degree may also be earned in combination with the Ph.D.
1. Course work: The following core of courses is required and are normally completed during the first four semesters of graduate study. The core courses must be completed with an average grade of 2.75 (out of 4.0) or better.
Physics Track: Quantum Mechanics I & II; 12 semester hours selected from Classical Mechanics, Statistical Physics, Solid State Physics, or Electrodynamics I or II, and 9 semester hours of Research Problems in Physics. Additional course work may be required to ensure adequate preparation for the specified courses.
Astrophysics Track: Astrophysics, Quantum Mechanics I, Electrodynamics I; 12 semester hours selected from Quantum Mechanics II, Classical Mechanics, Statistical Physics or Electrodynamics II; and 9 semester hours of Research Problems in Astronomy. Students may also be required to take Advanced Topics in Astrophysics to ensure an adequate background for their dissertation research.
Each full-time student is required to participate in graduate seminars. The course requirements for any course other than

Research Problems may also be met by satisfactory performance on a written examination or by transfer of credit in an equivalent course from another institution.

2. Pre-dissertation examination: This exam is normally taken during the fourth semester of graduate study, and consists of three parts: first, a written report submitted to the advisory committee on a research project either completed or proposed for a dissertation; second, a colloquium based on the written research report; and third, an oral examination over the research report given by the advisory committee and faculty. Successful completion of the predissertation examination and the required core work constitute admission to candidacy for the Ph.D. degree. Unsuccessful completion of the pre-dissertation exam may result in the student being advised to complete the requirements for a Master of Science degree.
3. Dissertation: Completion of a dissertation consisting of an original research project directed by a faculty member at TCU. A final oral exam in defense of the dissertation is required and a paper based on the dissertation research must be submitted for publication in an appropriate scientific journal.
4. Teaching Requirements: Each full-time graduate student pursuing a degree in physics is required to participate in the undergraduate teaching function of the department. The faculty are committed to effective teaching and believe that experience in the teaching of physics is an integral part of graduate education. This requirement is met by assisting in undergraduate labs, giving laboratory instructions, grading papers, conducting problems sessions, or offering tutorial help. No more than 8 hours per week.

The Ph.D. with an M.B.A. Option

The Ph.D. in Physics is also available with a Business option. Students entering the Ph.D. program with a B.S. degree are normally expected to compete the Ph.D. requirements within five years. At the end of the fourth year of graduate studies, a candidate for the Ph.D. degree who has demonstrated sufficient progress in dissertation research, may apply for the MBA option. During the fifth year the student is expected to continue with the dissertation on a reduced scale, and, if on Departmental Teaching Assistantship, to perform designated departmental teaching duties. Students entering the Ph.D. program with advanced standing (M.S. degree or more) can request an accelerated program. In addition to the course work, qualifying examinations, and dissertation requirements specified above for the Ph.D. degree in Physics or Astrophysics, students electing to take the MBA Option will take 18 hours of MBA course work during two consecutive semesters as outlined in the TCU Graduate Studies Bulletin. Students are required to attend the Team Building and Skills workshop conducted by the School of Business. Students are assessed a fee for the workshop. The results of the GRE will be accepted in lieu of the GMAT. Students who wish to continue their studies in the program after their first year of business courses and pursue the MBA degree will be required to complete such additional course work as required of other MBA students and as outlined in the TCU Graduate Studies Bulletin. The maximum term of fellowship or assistantship support through the Department of Physics is 5 year for the Ph.D. Degree or the Ph.D. with MBA Option. Support for MBA courses from the TCU Physics Department fellowships or assistantship is limited to 18 hours. Financial support for the additional 24 hours required for completion of the MBA degree would be the student's responsibility. However, students would be eligible to apply for financial aid from the School of Business.

Thesis: Thesis may be written *in absentia*.

Table B—Appointments to Graduate Students, 2009–10

Title of Appointee	Appointments Total	Appointments First year	Academic Load Allowed in Credit Hours	Hours of Service Per Week	Stipend for Academic Year ($)
Teaching Assistant	11	4	1–9	8	18,000
Total	11	4			

6. Separately Budgeted Research Expenditures by Source of Support

	Departmental Research
NSF	173,025
NASA	216,744
Private, non-profit organizations	29,500
Total	$419,269

FACULTY

Professors

Graham, W. R. M., Ph.D., York, Toronto, 1971. Molecular physics; solid state physics.

Gryczynski, Z.K., Ph.D., Gdansk, Poland, 1986. Biophysics.

Kouris, Demitris, Ph.D., Northwestern U., 1987. Nanomechanics.

Miller, Bruce Neil, Ph.D., Rice, 1969. Theoretical physics; statistical mechanics.

Rittby, C. Magnus L., Ph.D., Univ. of Stockholm, 1985. Theoretical, molecular, and atomic physics.

Zerda, Tadeusz W., Ph.D., Silesian, Poland, 1978. Molecular physics; high pressure physics. Chair of the Department.

Assistant Professors

Strzhemechny, Yuri, Ph.D., CUNY, 2000. Photoluminescence.

Frinchaboy, Peter, Ph.D., Virginia, 2006. Astronomy.

Instructor

Ingram, Douglas, Ph.D., Univ. Washington, 1996. Astronomy.

RESEARCH SPECIALTIES AND STAFF

Theoretical

Atomic and Molecular Physics. Rittby.

Statistical Physics. Miller.

Experimental

Astronomy. Frinchaboy.

Biophysics. Gryczynski.

Fourier Transform Spectroscopy. Molecular clusters in solids. Graham.

Spectroscopy. High pressure. Materials Science. Zerda.

Photoluminescence, optoelectronics, semiconductors. Strzhemechny.

TEXAS STATE UNIVERSITY–SAN MARCOS

DEPARTMENT OF PHYSICS

San Marcos, Texas 78666

Students Accepted For Degree	FIELDS		
	Physics	Astronomy	Related Fields
Doctorate			
Master's	X		

1. General

President: Denise M. Trauth
Dean of Graduate School: J. Michael Willoughby
Department Chairman: David Donnelly
Department Telephone Number: (512) 245-2131 (C)
Type of Institution: University
Control: Public
Setting: Small town
Total Faculty: 1,371
Total Graduate Faculty: 820
Total Students: 30,000
Total Graduate Students: 4,060
Annual Graduate Tuition:
In-state residents: Full-time—$7,665*
Out-of-state residents: Full-time—$14,610
Tuition rates for: 2009–10
Deferred tuition plan: No
Other Fees: included above
Term: AV

*9 hrs./sem.+6 hrs./summer

2. Number of Faculty in Department

The combined total of full-time faculty in the three professorial ranks is 8. The combined total of full-time, part-time, and other faculty at all ranks is 13.

3. Admission, Financial Aid, and Housing

Address admission inquiries to: David Donnelly, Physics Dept., Graduate Advisor
Graduate application fee required: $25
Admission deadline (Fall admission): 7/15
Admission information: For fall admission, 2009–10, 4 students were accepted from 4 applicants.
Admission requirements: For admission to the graduate programs, a Bachelor's degree in physics is required with a minimum undergraduate GPA of 2.75/4.0 specified. The GRE is required. The minimum acceptable score suggested for admission is verbal-350; quantitative-650; total-1,000. The GRE Advanced is not required. Students from non-English speaking countries are required to demonstrate proficiency in English via the TOEFL exam. Minimum acceptable score for admission is 550.
Undergraduate preparation assumed: Halliday, Resnick, & Walker, *Introductory Physics*; Eisberg & Resnick, *Quantum Physics*; Fowles & Cassiday, *Mechanics*; Kittel, *Thermal Physics*; Saxon, *Elementary Quantum Mechanics*; Reitz, Milford, & Christy, *Foundations of Electromagnetic Theory*; Boas, *Mathematical Methods.*
Address financial aid inquiries to: General: Financial Aid Office, JCK Bldg., TSU-SM, San Marcos, TX 78666; (512) 245-2315; send assistantship inquiries to Physics Chairman.
GAPSFAS application required: No
Financial aid deadline: 4/1
Loans available: Yes
Address housing inquiries to: Residence Life, TSU-SM, San Marcos, TX 78666 (512) 245-2382
On-campus, graduate student housing available: Yes
Cost/term: $1,368–1,677
On-campus, married student housing available: Yes
Cost/month: $310–725

Table A—Faculty, Enrollments, and Degrees Granted

Research Specialty	2009–10 Faculty	Enrollment[1] Fall 2009 Master's	Enrollment[1] Fall 2009 Doctorate	No. of Degrees Granted[2] 2009–10 Master's	Terminal Master's	Doctorate	Median No. of Years for 2006–07 Ph.D.'s
Astronomy	2	0	–	0(0)	–	–	–
Astrophysics	1	0	–	0(0)	–	–	–
Atomic, Molecular, & Optical Physics	1	0	–	0(0)	–	–	–
Computer Science	0	0	–	0(1)	–	–	–
Condensed Matter Physics	7	16	–	7(21)	–	–	–
Nuclear Physics	0	0	–	0(0)	–	–	–
Optics	1	0	–	0(1)	–	–	–
Physics Education	0	0	–	0(0)	–	–	–
Relativity & Gravitation	1	0	–	0(0)	–	–	–
Other Theoretical/ Math.	2	2	–	0(0)	–	–	–
Non-specialized	0	1	–	0(1)	–	–	
Total		19	–	7(22)	–	–	
Full-time Grad. Stud.		17	–		–	–	
Part-time Grad. Stud.		16	–				
First-year Grad. Stud.		3	–				
Median Years in Grad. Study (2006–07 Degrees)				2	–	–	–
Undergraduate Degrees, 2007–08: 11(40)							

[1] Students not yet committed to a research specialty are entered under non-specialized.
[2] Five-year totals in parentheses.

4. Graduate Degree Requirements

Physics Master's: A Masters of Science degree with thesis requires 15–18 semester hours of physics coursework, 9–6 hours of coursework in another science, and a minimum of 6 hours of thesis supported by a thesis oral defense. A Masters of Science degree without a thesis is available which requires 6 additional hours of physics coursework in lieu of the thesis, and 6 hours of additional coursework.
Materials Physics Masters: A Masters of Science degree requiring 35 semester hours of physics coursework. The curriculum emphasizes topics of interest to the microelectronics industry with a required thesis and industrial internship. In all degrees, an oral thesis defense is required. A minimum of 3.0/4.0 GPA is required for graduation, and a minimum of 24 hours of coursework must be completed on campus. There is no foreign language requirement.
Thesis: Thesis may be written *in absentia*.
Special Equipment, Facilities, or Programs: The Department has a special emphasis on experimental materials physics and thin film solid state physics research. This emphasis is fo-

cused toward preparing Master's graduates for professional employment in the semiconductor wafer fabrication industry or thin film magnetic storage industry. Thesis research may utilize thin film sputtering (magnetron and dual ion beam), scanning electron microscopy, energy dispersive spectroscopy, high resolution x-ray diffraction/reflectivity, scanning probe microscopy (AFM and STM), magnetometry, resistivity, high temperature furnace/oven, ellipsometry, and photoluminescence. Competitive opportunities for industry internships are available. In addition, the Department maintains a small astronomical observatory (16″ reflector) for student use.

Table B—Appointments to Graduate Students, 2009–10

Title of Appointee	Appointments Total	Appointments First year	Academic Load Allowed in Credit Hours	Hours of Service Per Week	Stipend for Academic Year ($)
			Semester		
Teaching Assistant	10	2	12 sem. hrs.	20	11,886
Research Assistant	5	1	12 sem. hrs.	20	11,886
Total	15	3			

[1]All tuition and fees (resident and nonresident) are waived; first-year stipend, $8,500; half or full-time summer support also provided, depending on availability of funds.
[2]University fellow carries 20-hr. service for one term.

5. Personnel Engaged in Separately Budgeted Research, 7/09–6/10

Professorial faculty	6
Graduate students	4
Undergraduate students	6
Total	14

6. Separately Budgeted Research Expenditures by Source of Support

	Departmental Research	Physics-related Research Outside Department
Federal government	$500,000	$
Business and industry	25,000	
Total	$525,000	$

Table C—Separately Budgeted Research Expenditures

Research Specialty	No. of Grants	Expenditures ($)
Condensed Matter Physics	8	$525,000
Total	8	$525,000

FACULTY

Professors

Droopad, Ravi, Ph.D., Imperial College, London, Experimental Condensed Matter.

Olson, Donald W., Ph.D., University of California at Berkeley, 1975. Astrophysics; general relativity; computational astronomy.

Associate Professors

Geerts, Wilhelmus J., Ph.D., University of Twente, Enschede, The Netherlands, 1992. Nanostructured magnetic & semiconductor materials: transport properties & optical characterization.

Piner, Edwin, Ph.D., NC State Univ., 1998, GaN Devices.

Spencer, Gregory W., Ph.D., Univ. of Florida, 1986. Advanced electronic devices and materials.

Ventrice, Carl A., Ph.D., Drexel University, 1991. Physics.

Assistant Professors

Lee, Byounghak, Ph.D., Indiana University, 2002. First principles simulations of electronic structure of nanomaterials.

Theodoropoulou, Nikoleta, Ph.D., University of Florida, 2002. Properties of dilute magnetic nanostructures.

Lecturers

Doescher, Russell L., M.S., Southwest Texas State University, 1992. Computational and observational astronomy; physics education.

Mount, Jennifer, M.S., Texas State Univ., 2004.

Scolfaro, Luisa, Ph.D., Univ. of Sao Paulo, 1988.

RESEARCH SPECIALTIES AND STAFF

Theoretical

General relativity. Olson.

Experimental

Artificially Structured Thin Film Materials. Geerts.

Computational Modeling of Historical Events in Astronomy. Doescher, Olson.

Electric Transport Properties of Thin Film Materials. Geerts, Lee.

Fabrication and characterization of dilute magnetic nanostructures. Theodoropoulou.

Giant Magnetoresistance Materials. Geerts, Theodoropoulou.

Materials Physics Education. Novel focused revisions to traditional physics curriculum appropriate for enhanced preparedness for employment in applied thin film physics industries (semiconductor & magnetic storage industries). Spencer.

Optical Characterization of Thin Film Materials. Geerts.

Scanning Electron Microscopy Analysis of Thin Film Materials. Spencer.

Semiconductor Diffusion Barrier and Intermetallic Thin Film Materials. Geerts.

GaN Devices, Piner

Novel CMOS Applications, Droopad

Modeling of Condensed Matter Systems, Lee

FACULTY PUBLICATIONS

Droopad, Ravi

D. L. Feldwinn, J. B. Clemens, J. Shen, S. R. Biship, T. J. Grassman, A. C. Kummel, R. Droopad, and M. Passlack, *Anomalous Hybridization in the In-Rich InAs(001) Reconstruction*, Surface Science **603**, 3321 (2009).

I. G. Thayne, R. J. W. Hill, M. C. Holland, X. Li, H. Zhou, D. S. Macintyre, S. Thomas, K. Kalna, C. R. Stanley, A. Asenov, R. Droopad, and M. Passlack, *Review of Current Status of III-V MOSFETs*, ECS Transactions **19**, 275 (2009).

J. B. Clemens, S. R. Bishop, D. L. Feldwinn, R. Droopad, and

A. C. Kummel, *Initial States of the Autocatalytic Oxidation of the InAs(001)-(4x2/c(8x2) Surface by Molecular Oxygen*, Surface Science **603**, 2230 (2009).

M. Passlack, R. Droopad, P. Feyes, and L. Q. Wang, *Electrical Properties of Ga_2O_3/GaAs Interfaces and GdGaO Dielectrics in GaAs-Based MOSFETs*, IEEE Electron. Device Letters **30**, 2 (2009).

M. Passlack, R. Droopad, . Yu, N. Medendorp, D. Braddock, X. W. Wang, T. P. Ma, T. Buyuklimanli, *Screening of Oxide/GaAs Interfaces for MOSFET Applications*, IEEE Electron. Device Letters **29**, 1181 (2009).

Geerts, Wilhelmus

J. R. K. Pandey, P. Padmini, R. Schad, J. Dou, H. Stern, R. Wilkins, R. Dwivedi, W. J. Geerts, and C. O'Brien, *Novel magnetic Semiconductors in Modified $FeTiO_3$ for Radhard Electronics*, J. Electroceramics **22**, 334 (2009).

Lee, Byounghak

B. Lee and L.-W. Wang, *Electronic Structure of ZnTe:O and its Usability for Intermediate Band Solar Cell*, Appl. Phys. Lett. **96**, 071903 (2010)

C. Yang, J. C. Meza, B. Lee, and L.-W. Wang, *KSSOLV-A Toolbox for Density Functional Theory Calculations*, ACM Trans. Math. Software **36**, 10 (2008).

B. Lee, L.-W. Wang, and A. Canning, *Effects of d-Electrons in Pseudopotential Screened-Exchange Density Functional Calculations*, J. Appl. Phys. **103**, 113713 (2008).

J. Schrier, B. Lee, and L.-W. Wang, *Mechanical and Electronic-Structure Properties of Compressed CdSe Nanocrystals*, J. Nanoscience Nanotechnology **8**, 1994 (2008).

Olson, Don

D. W. Olson and L. E. Jasinski, *Abraham Lincoln's Celestial Connections*, Sky & Telescope **117**, 66 (2009).

D. W. Olson, R. L. Doescher, K. N. Beicker, and A. F. Gregory, *Moon and Tides at Caesar's Invasion of Britain in 55 B.C.*, Sky & Telescope **116**, 18 (2008).

Piner, Edwin

M. R. Armstrong, E. J. Reed, K.-Y. Kim, J. H. Glownia, W. M. Howard, E. L. Piner, and J. C. Roberts, *Observation of Terahertz Radiation Coherently Generated by Acoustic Waves*, nature Physics (2009).

J. W. Chung, E. L. Piner, and T. Palacios, *N-Face GaN/AlGaN HEMTs Fabricated Through Layer Transfer Technology*, IEEE Electron. Device Letters **30**, 113 (2009).W. Johnson, S. Singhal, A. Hanson, R. Therrien, A. Chaudhari, W. Nagy, P. Rajagopal, Q. Martin, T. Nichols, A. Edwards, J. Roberts, E. L. Piner, I. Kizilyalli, and K. J. Linthicum, *GaN-on-Si HEMTs: From Device Technology to Product Insertion*, Mater. Res. Soc. Symp. Proc., **1068**, 3 (2008).

J. C. Roberts, J. W. Cook Jr., P. Rajagopal, E. L. Piner, and K. J. Linthicum, *AlGaN Transition Layers on Si(111) Substrates-Observations of Microstructure and Impact on Material Quality*, Mater. Res. Soc. Symp. Proc. **1068**, 147 (2008).

B. S. Kang, H. T. Wang, F. Ren. B. P. Gila, C. R. Abernathy, S. J. Pearton, D. M. Dennis, J. W. Johnson, P. Rajogopal, J. C. Roberts, E. L. Piner, and K. J. Linthicum, *Exhaled-Breath Detection Using AlGaN/GaN High Electron Mobility Transistors Integrated with a Peltier Element*, Solid State Lett. **11**, J19 (2008).

Theodoropoulou, Nikoleta

A. Sharma, N. Theodoropoulou, S. Wang, K. Xia, W. P. Pratt Jr., and J. Bass, *Sensitivity of Ag/Al Interface Specific Resistances to Interfacial Intermixing*, J. Appl. Phys. **105**, 123920 (2009).

N. Theodoropoulou, A. Sharma, W. P. Pratt Jr., and J. Bass, *Low Frequency Magnetization Excitations in Magnetic Nanopillars Induced by a DC Spin Polarized Current*, J. Appl. Phys., **105**, 07D122 (2009).

Q. Fowler, B. Richard, A. Sharma, N. Theodoropoulou, R. Loloee, W. P. Pratt Jr., and J. Bass, *Spin-Diffusion Lengths in Dilute Cu(Ge) and Ag(Sn) Alloys*, J. Magn. Magn. Mat. **321**, 99 (2009).

N. Theodoropoulou, A. Sharma, W. P. Pratt Jr., J. Bass, M. D. Stiles, and J. Xiao, *Two Simple Tests for Models of Current-Induced Magnetization Switching*, J. Appl. Phys. **103**, 07A705 (2008).

A. Sharma, N. Theodoropoulou, T. Haillard, R. Acharyya, R. Loloee, W. P. Pratt Jr., J. Bass, J. Zhang, and M. A. Crimp, *Current-Perpendicular-to-Plane Magnetoresistance of Ferromagnetic F/Al Interfaces (F=Py, Co, Fe, and $Co_{91}Fe_9$) and Structural Studies of Co/Al and Py/Al*, Phys. Rev. B **77**, 224438 (2008).

Ventrice Jr., Carl A.

I. Jung, D. A. Field, N. J. Clark, Y. Zhu, D. Yang, R. D. Piner, S. Stankovich, D. A. Dikin, H. Geisler, C. A. Ventrice Jr., and R. S. Ruoff, *Reduction Kinetics of Graphene Oxide Determined by ELectrical Transport Measurements and Temperature Programmed Desorption*, J. Phys. Chem. **113**, 18480 (2009).

D. Yang, A. Velamaakanni, G. Bozoklu, A. Park, M. Stoller, R. D. Piner, S. Stankovich, I. Jung, D. A. Field, C. A. Ventrice Jr., and R. S. Ruoff, *Chemical Analysis of Graphene Oxide Films Heat and Chemical Treatments by X-ray Photoelectron and Micro-Raman Spectroscopy*, Carbon **47**, 145 (2009).-

TEXAS TECH UNIVERSITY

DEPARTMENT OF PHYSICS

Lubbock, Texas 79409

Students Accepted For Degree	FIELDS		
	Physics	Astronomy	Related Fields
Doctorate	X		X
Master's	X		X

1. General

Chancellor: Kent Hance
President: Guy Bailey
Dean of Graduate School: Fred Hartmaister
Department Chairman: Roger Lichti
Department Telephone Number: (806) 742-3767
Fax: (806) 742-1182
Department Homepage: http://www.phys.ttu.edu
Type of Institution: University
Control: Public
Setting: Urban
Total Faculty: 2,671
Total Graduate Faculty: 767
Total Students: 30,049
Total Graduate Students: 5,175*
Annual Graduate Tuition: **
In-state residents: Full-time—$7,398/year
Part-time—$212.50/credit
Out-of-state residents: Full-time—$14,046/year
Part-time—$489.50/credit
Tuition rates for: 2009–10
Deferred tuition plan: Yes (3 installments)
Other Fees: $2,298 average for full-time (some can be waived)
Term: Semester

*Excluding Law School
**Minimum fee of $50
***Includes minimum of $30 per hour charged by each college in addition to regular graduate tuition of $50 per hour (in-state) and $325 per hour (out-of-state). (Full-time graduate tuition based on 24 credit hours in academic year).

2. Number of Faculty in Department

The combined total of full-time faculty in the three professorial ranks is 19. The combined total of full-time, part-time, and other faculty at all ranks is 30.

3. Admission, Financial Aid, and Housing

Address admission inquiries to: Graduate Recruiter, Dept. of Physics, E-mail: m.sanati@ttu.edu. For more information see the department world wide web page at http://www.phys.ttu.edu
Graduate application fee required: Foreign Students—$75 (nonrefundable); Domestic students—$50 (nonrefundable)
Admission deadline (Fall admission): 3/1 (for support)
Admission information: For fall admission, 2009–2010, 18 students were accepted from more than 75 applicants.
Admission requirements: A Bachelors degree in physics is required for admission to the graduate programs in physics. For students with a Bachelor's degree in a related field, undergraduate leveling may be required. The GRE General Test is required and GRE Physics subject test is recommended. An undergraduate GPA of 3.0 is preferred. For the past several years, the average General GRE scores were verbal-490; quantitative-710; total-1,200. A minimum GRE total score of 1,150 is required to obtain financial support from the department. Students from non-English speaking countries are required to demonstrate proficiency in English via the TOEFL or IELTS exams. A minimum TOEFL score of 550 (written), 213 (computer), or 79 (internet) is required for admission and financial support. The IELTS score of 6.5 or better is also accepted. All new foreign teaching assistants are required to pass an English workshop administered by the University.
Address financial aid inquiries to: Graduate Recruiter, Dept. of Physics. E-mail: m.sanati@ttu.edu
GAPSFAS application required: No
Loans available: Yes
Address housing inquiries to: Director, Housing Office, Box 41141, Lubbock, TX 79409-1141
On-campus, single student housing available: Yes
Cost/sem: $4,516* fall semester;
$3,010 spring semester (room and board)
On-campus, married student housing available: No

*High rate (includes Diamond meal plan/$3,427-20 meals/wk.). Single rooms available for additional $500/sem. Lower rates available upon request.

Table A—Faculty, Enrollments, and Degrees Granted

Research Specialty	2008–09 Faculty	Enrollment[1] Fall 2009		No. of Degrees Granted[2] 2008–09 (2004–09)			Median No. of Years for 2008–09 Ph.D.'s
		Master's	Doctorate	Master's	Terminal Master's	Doctorate	
Applied Physics	2	2	0	0(3)	3(14)	0(1)	6
Atomic, Molecular, & Optical Physics	3	1	2	0(1)	0(1)	0(1)	6
Biophysics	3	2	9	0(1)	0(0)	0(2)	5.5
Condensed Matter Physics	6	2	17	0(8)	0(2)	1(8)	5.5
Energy Sources & Environ.	1	0	0	0(0)	0(0)	0(0)	–
Particles & Fields	5	0	5	0(0)	0(1)	0(1)	6
Physics Education Research	0	2	2	0(0)	0(0)	0(0)	–
Non-specialized	2	5	0	5(0)	0(0)	0(0)	–
Total		14	35	5(17)	3(18)	1(13)	
Full-time Grad. Stud.		14	35				
Part-time Grad. Stud.		1	0				
First-year Grad. Stud.		17	0				
Median Years in Grad. Study (2007–08 Degrees)				3	2.5	5.8	5.8
Undergraduate Degrees, 2007–08 (2003–08):							

[1]Students not yet committed to a research specialty are entered under non-specialized.
[2]Five-year totals in parentheses.

4. Graduate Degree Requirements

Master's: The M.S. with Thesis requires a minimum of 24 hours of graduate course work and 6 hours of thesis with a minimum GPA of 3.0. A thesis based on a research problem and a final oral exam over the research problem are required. A professional M.S. degree option in applied semiconductor physics is also offered. This option includes two internship semesters and requires departmental reports in lieu of a thesis. Minimum residence time is one academic year. M.S. degrees with either physics or applied physics options are offered.

Doctorate: A minimum of 60 hours beyond the B.S. degree plus 12 hours of dissertation with a minimum GPA of 3.0. A minimum of 3 years of graduate study beyond the B.S. degree with 1 year of residence beyond the M.S. degree or equivalent is required. After completing the core courses—typically after one year—all candidates must pass a written exam over the core curriculum. A dissertation on an original research project and an oral defense of the dissertation are required. Ph.D. degrees with either physics or applied physics options are offered. A specialization in chemical physics, in cooperation with the Department of Chemistry and Biochemistry, is also available.

Dissertation: Dissertation may be written *in absentia*.

Special Equipment, Facilities, or Programs: Program in experimental particle physics; J. Fred Bucy and Odetta Greer Bucy Chair in Physics, Dr. Richard Wigmans. NanoTech Center research program in nanotechnology headed by Dr. Mark Holtz.

Table B—Appointments to Graduate Students, 2008–09

Title of Appointee	Appointments: Total	Appointments: First year	Academic Load Allowed in Credit Hours	Hours of Service Per Week	Stipend for Academic Year ($)
			Semester		
Teaching Assistant	28	13	12	20	15,600[1]
Research Assistant	16	0	12	variable	8,000–15,600[1]
Bucy Scholar	1	1	12	variable	12,500[1]
Welch Fellow	2	1	12	variable	8,000–15,600[1]
Total	47	15			

[1]Average stipend for nine months.

5. Personnel Engaged in Separately Budgeted Research, 9/08–8/09

Professorial faculty	12
Postdoctoral appointments	4
Graduate students	31
Undergraduate students	12
Other faculty	1
Total	60

6. Separately Budgeted Research Expenditures by Source of Support

	Departmental Research	Physics-related Research Outside Department
Federal government	$2,154,954	2,002,875
State and local government		
Other[1]	160,000	
Total	2,314,954	2,002,875

[1]Includes $160,000 institutional matching funds.

Table C—Separately Budgeted Research Expenditures

Research Specialty	No. of Grants	Expenditures ($)
Biophysics	3	245,686
Condensed Matter Physics	7	865,911
Physics Education	3	464,357
Particles & Fields	3	744,000
Total	16	2,314,954

Table D—Physics-related Research Outside Department

Field and Unit Outside Department	No. of Grants	Expenditures ($)
Nanotech	4	2,002,875
Total	4	2,002,875

FACULTY

Professors

Akchurin, Nural, Ph.D., Iowa, 1990. Experimental particle physics; supersymmetry and Higgs searches; polarization; accelerator physics; particle detectors.

Cheng, Kwan Hon, Ph.D., Waterloo, Canada, 1983. Experimental biophysics; time-resolved fluorescence spectroscopy; membranes; nuclear magnetic resonance imaging; biochips.

Estreicher, Stefan K., Ph.D., Zürich, Switzerland, 1982. *Ab initio* calculations of properties of defects in semiconductors; molecular dynamics.

Holtz, Mark W., Ph.D., Virginia Tech., 1987. Materials Physics, nanoscience, optical properties of condensed matter; semiconductors, epitaxy.

Lichti, Roger, L., Ph.D., Illinois, 1972. Department Chairman. Muon spin research; defect properties; hydrogen defect chemsitry; semiconductor physics; magnetism; phase transitions.

Lodhi, M. A. K., Ph.D., London, England, 1963. Nuclear theory and structure; new and renewable sources of energy and environment; advanced space power generation and management.

Myles, Charles W., Ph.D., Washington (St. Louis), 1973. Theoretical and computational materials physics, with emphasis on semiconductor materials. Clathrate materials and thermoelectrics. Electronic properties of defects, electronic banstructures, properties of semiconductor alloys. High electric field transport. Molecular Dynamics and Monte Carlo computer simulations.

Quade, C. Richard, Ph.D., Oklahoma, 1962. Theoretical molecular dynamics; magnetization of impurity ions.

Wigmans, Richard, Ph.D., Vrije Universiteit Amsterdam, 1975. Holder of the J. Fred Bucy and Odetta Greer Bucy Chair in Physics. Experimental high-energy particle physics; particle detectors; calorimetry; astrophysics; cosmelogy.

Associate Professors

Gibson, Thomas L., Ph.D., Oklahoma, 1982. Quantum collision theory; low-energy positron-molecule collisions; concurrent computational techniques; Monte Carlo simulations.

Glab, Wallace L., Ph.D., Illinois, Urbana, 1984. Experimental atomic and molecular physics; laser spectroscopy of excited states of atoms and molecules; multiphoton ionization and photoelectron spectroscopy of small molecules.

Huang, Juyang, Ph.D., SUNY, Buffalo, 1987. Experimental and theoretical membrane biophysics; liposome technology; drug delivery; biochip; fluoresence microscopy; X-ray diffraction; Monte Carlo simulations.

Lamp, C. David, Ph.D., Missouri-Columbia, 1984. Experimental solid state physics; uniaxial stress transient spectroscopy; semiconductor materials; materials science; physics education; science training for secondary school teachers.

Thacker, Beth A., Ph.D., Cornell Univ., 1990. Theoretical particle physics, physics education.

Assistant Professors

Grave de Peralta, Luis, Ph.D., Texas Tech University, 2000. Plasmonics, photonics, nanotechnolgy.

Lee, Sung-Won, Ph.D., University of Glasgow, Glasgow, UK. 2000. Experimental Particle Physics.

Park, Soyeun, Ph.D., University of Texas-Austin, 2003. Experimental biophysics.

Sanati, Mahdi, Ph.D., University of Cincinnati, 1999. Theoretical condensed matter physics; structural phase transformation in solids; solitons in physical systems.

Volobouev, Igor, Ph.D., Southern Methodist University, Dallas, Texas, 1997. Experimental Particle Physics.

Professors Emeriti

Borst, Walter L., Ph.D., Berkeley, 1968.
Hatfield, Lynn L., Ph.D., Arkansas, 1966.
Mires, Raymond W., Ph.D., Oklahoma, 1964.
Thomas, Henry C., Ph.D., Vanderbilt, 1950.

Joint Professors

Kristiansen, Magne, Ph.D., Texas, 1967. Pulsed power; electrical space propulsion; rf wave propagation in magnetized plasmas; high power microwaves. Explosive pulsed power generators. (Electrical Engineering.)

Krompholz, Herman G., Ph.D., Darmstadt, Germany, 1977. Pulsed power; high-speed plasma diagnostics; surface discharges; high power microwave propagation. (Electrical Engineering.)

Quitevis, Edward L., Ph.D., Harvard, 1981. Ultrafast spectroscopy; nonlinear optics; photophysics; molecular aggregates; membranes and micelles; liquids. (Chemistry.)

Sill, Alan F., Ph.D., American University, 1987. Particle physics and high performance computing.

Adjunct Professors

Kubricht, William, MMSc, DABR, Medical Physicist, Diplomat of the American Board of Radiology.

Mark, Rufus, MD. Radiation Oncology, Medical Physics.

Nair, Murali, MD, DABR, Diplomat of the American Board of Radiology. Medical Physics.

Torres, Carlos, MD. Radiation Oncology, Medical Physics.

RESEARCH SPECIALTIES AND STAFF

Theoretical

Applied Physics. New and renewable energy sources; power conversion systems for space use. Lodhi. Photoconductive switch simulations, Myles.

Atomic, Molecular, and Chemical Physics. Theory of vibration-rotation fine structure and intramolecular forces. Quade. Low-energy electron-molecule collisions; Computational techniques. Gibson.

Biophysics. Cell membranes; cholesterol domains; multibody interactions; anomalous diffusion; Monte Carlo and dynamic simulations. Huang.

Condensed Matter Physics. Defects in semiconductors, molecular dynamics and Monte Carlo simulations; breakdown in semiconductors; impurities and complexes in semiconductors; molecular orbital theory. Structural phase transformation; high field transport, clathrates and thermoelectrics, band structure solitons, Estreicher, Myles, Sanati.

High Energy Physics. Short-range correlations; pion and photonuclear reactions; high-energy electron scattering, nuclear structure; hadron-Quark-hybrid model and Quarknia; charge exchange phenomena. Lodhi. Lattice gauge theory. Thacker.

Physics Education Research. Assessment and curriculum development. Thacker, Lamp.

Experimental

Applied Physics. High voltage breakdown of insulators. Krompholz. Aging of insulators in the space environment; electric arcject thrusters; high power microwaves. Kristiansen. High-speed plasma diagnostics. Krompholz. Materials physics and characterization. Holtz, Lamp, Lichti. Medical physics and diagnostic imaging. Cheng.

Atomic, Molecular, and Chemical Physics. Laser spectroscopy; pulsed and cw fluorescence. Cheng, Glab, Quitevis. Laser applications. Cheng, Glab, Quitevis. Electromagnetic interactions. Secondary electron emission from insulators. Gibson.

Biophysics. Molecular spectroscopy of membranes; quantitative magnetic resonance imaging; membrane electrophysiology, Liposome technology; fluorescence microscopy; X-ray diffraction; drug delivery system; biochip conformal radiation dosimetry. Cheng, Huang, Park.

Condensed Matter Physics and Nanotechnology. Magnetism. Muon spin rotation defect characterization. Lichti. Semiconductor materials, nanoscience; optical properties; Raman scattering. Holtz. Photonics; Plasmonics. Grave de Peralta.

Particle Physics. Proton-proton and antiproton-proton scattering at TeV energies (CDF, CMS). New particle searches. Heavy quark spectroscopy. Research and development of high energy particle detectors (calorimeters, high-precision tracking devices). Accelerator physics. Akchurin, Lee, Volobouev, Wigmans.

Texas Tech is committed to the principle that in no aspect of its programs shall there be differences in the treatment of persons because of race, creed, national origin, age, sex, or disability, and that equal opportunity and access to facilities shall be available to all.

THE UNIVERSITY OF TEXAS AT BROWNSVILLE AND TEXAS SOUTHMOST COLLEGE

DEPARTMENT OF PHYSICS AND ASTRONOMY

Brownsville, Texas 78520

Students Accepted For Degree	FIELDS		
	Physics	Astronomy	Related Fields
Doctorate	X(*)		
Master's	x		

(*) The Doctorate Degree is granted and administered by the Department of Physics at the University of Texas at San Antonio. For more details please refer to the description below and to the descritpion of the doctoral program at the Univ. of Texas at San Antonio. Under the memorandum of understanding, students in the program can select to perform their dissertation research under the supervision of a faculty member at the University of Texas at Brownsville.

1. General

President: Juliet V. García
Dean of Graduate School: Charles Lackey
Department Chairman: Soma Mukherjee
Department Telephone Number: (956) 882-6779
Type of Institution: University
Total Faculty: 395
Total Graduate Faculty: 212
Total Students: 17151
Total Graduate Students: 892
Annual graduate Tuition: *
In-state residents: Full-time—$2,096 (9 credits)
*Out-of-State residents**: Full-time—$4,886* (9 credits)
Tuition rates for: Fall 2010
Deferred tuition plan: No
Other Fees: **
Term: Semester

*Teaching or Research Assistants are charged the same rate as in-state residents
**See Graduate Catalog for more detailed breakdown

2. Number of Faculty in Department

The combined total of full-time faculty in the three professorial ranks is 14.

3. Admission, Financial Aid, and Housing

Address admission inquiries to: Graduate Program Coordinator, Dept. of Physics and Astronomy, The University of Texas at Brownsville, 80 Fort Brown, Brownsville, TX 78520; Email: gpcoordinator@phys.utb.edu
Graduate application fee required: $30
Admission deadline: (Fall) July 1, (Spring) Dec 1 for M.S. Physics. International applicants are recommended to apply at least two months in advance of these deadlines. For the doctorate program, the application deadlines are as given by the University of Texas at San Antonio (UTSA).
Admission information: 4 students were admitted for the Fall 2009 period out of 11 applicants.
Admission requirements: For admission to the M.S. program in Physics at the University of Texas at Brownsville (UTB), a Bachelor's degree in Physics or a related field is required with a minimum undergraduate GPA of 3.0 on a 4.0 scale. The GRE general exam is required. Students from non-English speaking countries are required to demonstrate proficiency in English via the TOEFL exam. Two letters of recommendation from people familiar with the applicant's undergraduate scholastic record are required. Applicants who do not meet the above criteria can apply for conditional admission into the program.
Students enrolled in the cooperative Ph.D. program between UTB and UTSA reside at UTB and perform their dissertation research under the supervision of a graduate faculty member of the UTB Physics and Astronomy Department. Admission and graduation requirements are the same as those established for the UTSA Ph.D. program in Physics. Students holding a Bachelor's degree first enroll in the UTB M.S. program in Physics in order to complete coursework requirements and then enroll in the UTSA Ph.D. program in Physics to take research, dissertation, thesis and advanced elective courses. Students already holding a Master's degree who wish to pursue research at UTB should contact the UTB Physics and Astronomy Graduate Program Coordinator.
Address financial aid inquires to: Graduate Program Coordinator, Dept. of Physics and Astronomy, The University of Texas at Brownsville, 80 Fort Brown, Brownsville, TX 78520; Email: gpcoordinator@phys.utb.edu
GAPSFAS application required: No
Loans Available: Yes
Address housing inquiries to: Residential Life Office, 1915 University Blvd., Brownsville, TX 78520
(956) 882-7191, (956) 882-6809 (fax), housing@utb.edu
Assistantships: $15,000–$19,000 with merit-based tuition assistance
On-campus, single student housing available: Yes
On-campus, married student housing available: No

Table A—Faculty, Enrollments, and Degrees Granted

Research Specialty	2008–09 Faculty	Enrollment 2009–10		No. of Degrees Granted 2009			Median No. of Years for Ph.D.'s
		Master's	Doctorate	Master's	Terminal Master's	Doctorate	
Astronomy & Astrophysics	3	3	1	0	2	0	–
Atomic,Molecular, & Optical Physics	2	1	0	0	0	0	–
Biophysics	2	0	0	0	1	0	–
Nanoscience	1	1	0	0	0	0	–
Physics Education	1	0	0	0	0	0	–
Relativity & Gravitation	5	7	0	0	1	0	–
Non-specialized	–	0	0	0	1	0	–
Total	14	12	1	0	5	0	
Full-Time Grad. Stud.		12					
Part-Time Grad. Stud.		2					

[1]Students not yet committed to a research specialty are entered under non-specialized
[2]The cooperative doctorate program with UTSA was started in 2009.

4. Graduate Degree Requirements

Master: Students can choose a thesis or non-thesis option within this program. Both options require 30 semester hours of credit for successful completion. The thesis option requires 18 hours of course work, 6 hours of Graduate Research and 6 hours of Thesis. The non-thesis option requires 27 hours of course work and a 3-hour course, Research Problems in Physics. A M.S. with Emphasis in Applied Physics is available starting from Fall 2010. Students taking this emphasis will specialize in advanced optics, electromagnetics and nanophotonics.

Doctorate: The requirements for the UTSA Ph.D. program in Physics apply.

Special Equipment, Facilities, or Programs: UTB/TSC is a member institution of the LIGO-Virgo Collaboration (LVC) which operates four large scale detectors of gravitational waves (three in the United States and one in Europe). As members of the LVC, faculty members in the Physics department conduct research in the areas of gravitational wave data analysis, detector characterization and instrumentation. Graduate and undergraduate students participating in LVC activities regularly visit the LIGO facilities located in Washington state and Louisiana. Members of the LVC have access to the computing resources of the Open Science Grid for data analysis and detector characterization tasks. ARCC—the Arecibo Remote Command Center is a facility for remote observations using the Arecibo radio telescope at Puerto Rico. The research conducted with this facility includes searches and studies of radio pulsars. The ARCC also enables research based on observations made with the Green Bank and Parkes radio telescopes. For biophysics research, a Nanoscope IV Atomic Force Microscope is available for imaging biological nanostructures and doing force spectroscopy. Optical tweezers with TIRF fluorescence for single molecule studies and a Cell Culture Facility, including a minus 80 freezer and laminar flow fume hoods, for cancer cell studies are available. The department operates two computer clusters, "Funes" and "Futuro", for astrophysical computations, data analysis, and modeling of photonic crystals with FDTD. Futuro is a 120 processor (2.4 GHz, 12 GB RAM/processor) IBM cluster and Funes is the older cluster with 70 processors. The department is also involved in research related to LISA, a future NASA and ESA space-based gravitational wave detector, in collaboration with the NASA Goddard Space Flight Center. The research involves operation of FPGA-based phase meters and realization of the optical transmission and detection setup in the optics lab. The laboratory includes 2 soft-wall clean-room enclosures, a highly stabilized 10-W Nd:YAG MOPA laser (Lightwave) and other infrared and visible-range lasers with optics and optical-beam diagnostic equipment.

Table B—Appointments to Graduate Students, FALL 2009

Title of Appointee	Appointments: Total	Appointments: First year	Academic Load Allowed in Credit Hours	Hours of Service Per Week	Stipend for Academic Year ($)[1,2,3]
			Semester		
Teaching Assistant	1	–	9	20	$10,000–$15,000
TA and RA combination	7	–	9	20	$15,000–$19,000
Research Assistant	4	–	9	20	$15,000–$19,000
Total	12				

[1] During the first year, most students are Teaching Assistants and Research Assistants.
[2] Out-of-state tuition is waived for students working on Campus.
[3] Additional teaching appointments are available in the summer.

5. Personnel Engaged in Separately Budgeted Research, 7/09–6/10

Professorial faculty	14
Other faculty	2
Postdoctoral appointments	1
Graduate Students	12
Undergraduate Students	29
Total	58

6. Separately Budgeted Research Expenditures by Source of Support (FY 2009)

Federal government	$4,264,160
Private, non-profit Organizations	$30,437
Institutional Funds	$191,873
State Government	304,000
Total	$4,790,470

FACULTY

Professors

Díaz, Mario, Ph.D., University of Córdoba, 1987. Gravitational wave astronomy, instrumentation and modeling.

Price, Richard H., Ph.D., California Institute of Technology, 1971. General relativity, astrophysics.

Associate Professors

Benacquista, Matthew, Ph.D., Montana State University, 1988. Astrophysics.

Dukes, Phillip R., Ph.D., Brigham Young University, 1996. Physics education.

Guevara, Natalia V., Ph.D., Moscow State University, 1989. Biophysics.

Mukherjee, Soma, Ph.D., University of Calcutta, 1991. Gravitational wave astronomy, data analysis and detector characterization.

Romano, Joseph, Ph.D., Syracuse University, 1991. Gravitational wave detection, data analysis.

Mohanty, Soumya D., Ph.D., Pune University, 1998. Gravitational wave astronomy, data analysis, computational methods.

Jenet, Fredrick, Ph.D., California Institute of Technology, 2001. Astrophysics. Gravitational wave detection with Pulsar timing.

Assistant Professors

Creighton, Teviet, Ph.D., California Institute of Technology, 2000. Astrophysics

Hanke, Andreas, Ph.D., University of Wuppertal, 1998. Statistical mechanics, soft condensed matter physics, biophysics, nanoscience.

Quetschke, Volker, Ph.D., University of Hannover, 2003. Laser physics, space-based experiments.

Rakhmanov, Malik, Ph.D., California Institute of Technology, 2000. Experimental optics, nanophotonics.

Touhami, Ahmed, Ph.D., University of Pierre & Marie Curie, 1993, Biophysics, experimental optics.

RESEARCH SPECIALTIES AND STAFF

Theoretical

Astronomy and Astrophysics. Relativistic binaries; pulsar searches, Pulsar timing. Benacquista, Creighton, Jenet.

Biophysics. Modeling in molecular cell biology; conformational dynamics of DNA. Hanke.

Nanoscience. Casimir effect in nanomechanical systems. Hanke

Physics Education. Java applets for physics education; visualization of relativistically moving objects. Dukes.

Relativity and Gravitation: relativistic astrophysics, gravitational wave data analysis, computational methods and algorithm development. Diaz, Mohanty, Mukherjee, Price, Romano, 1 postdoctoral fellow.

Experimental

Astronomy and Astrophysics. Radio observations with Arecibo, Green Bank and Parkes radio telescopes. Jenet.

Atomic, Molecular, & Optical Physics. Adaptive control of laser beam; adaptive optics, phasefront sensing; laser frequency stabilization; optical tweezers and wrenches; optical readout of AFM; photonic crystals. Quetschke, Rakhmanov.

Biophysics. Structure and functions of lipoproteins; ESR and fluorescence spectroscopy in studies of lipid-protein interactions; Atomic Force Microscopy. Guevara, Touhami. Graduate-student research assistant and postdoctoral positions.

UNIVERSITY OF NORTH TEXAS

DEPARTMENT OF PHYSICS

Denton, Texas 76203

Students Accepted For Degree	FIELDS		
	Physics	Astronomy	Related Fields
Doctorate	X		
Master's	X		

1. General

President: Dr. V. Lane Rawlins
Dean of Graduate School: Dr. Michael Monticino
Department Chair: Christopher L. Littler
Department Telephone Number: (940) 565-2626
Department e-mail: physics@unt.edu
Department website: http://www.physics.unt.edu
Type of Institution: University
Control: Public
Setting: Small town
Total Faculty: 1,368
Total Graduate Faculty: 965
Total Students: 36,123+
Total Graduate Students: 7,649
Annual Graduate Tuition: Posted 5/1/09
In-state residents: Full-time—$5,395.76
Part-time—$299,76/credit[1]
Out-of-state residents: Full-time—$10,381.76[2]
Part-time—$576.76/credit
Tuition rates for: 2009–10
Deferred tuition plan: Yes
Term: Semester

[1]18 credit hours/calendar year
[2]Out of state tuition waived for employees, e.g. Teaching Assistants and Research Assistants (9 credit hours/long semester)

2. Number of Faculty in Department

The combined total of full-time faculty in the three professorial ranks is 20. The combined total of full-time, part-time, and other faculty at all ranks is 25.

3. Admission, Financial Aid, and Housing

Address admission inquiries to: Admissions Office, University of North Texas, 1155 Union Circle, #311067, Denton, TX 76203
Application fee required: $50 Residents; $75 Nonresidents
Admission deadline (Fall admission): 7/15
Admission information: For fall admission, 2009–10, 25 students were accepted from 57 applicants
Admission requirements: For admission to the graduate programs, a Bachelor's degree in physics is required with a minimum undergraduate GPA of 2.8 specified. The GRE is required. Minimum acceptable score is Q(704)+A(3.07). The GRE Advanced is recommended. The average GRE scores for admissions were Q(710)+A(4.0). Students from non-English speaking countries are required to demonstrate proficiency in English via the TOEFL exam. Minimum score of 79/550/213 is required.
Undergraduate preparation assumed: Cutnell & Johnson *Physics*; Krane *Modern Physics*; Fowles *Analytical Mechanics*; Arfken *Essential Mathematical Methods for Physicists*; Serway & Jewett *Physics for Scientists & Engineers*; Griffith *Introduction to Quantum Mechanics*; Bowley *Introductory Statistical Mechanics*.
Address financial aid inquiries to: Student Financial Aid, 1155 Union Circle, #311370, ESSC Room 228, Denton, TX 76203-5017.
GAPSFAS application required: No
Financial aid deadline: 6/1
Loans available: Yes
Address housing inquiries to: Housing Dept., University of North Texas, P.O. Box 13617, Denton, TX 76203-3617
On-campus, single student housing available: Yes
On-campus, married student housing available: No

4. Graduate Degree Requirements

Master's: Option 1: Master of Arts, 30 semester credit hours. this includes 6 hours of thesis.
Option 2: Master of Science: 33 semester credit hours. This includes 6 hours of Problems in Lieu of Thesis, which are independent though not necessarily original studies that may be experimental, computational, tutorial, bibliographic, pedagogic, or a combination of these. As part of the requirements for each problems course, the student must present a formal written report of the work done in the course, which must be approved by an advisory committee.
Option 3: Master of Science: 36 semester credit hours. Coursework option.
Doctorate: A Ph.D. candidate must pass a qualifying examination over the core material of the graduate curriculum and receive approval of a research proposal from an advisory committee; or complete six (6) core courses with a minimum of 3 A's and 3 B's and receive approval of a research proposal from an advisory committee, to be accepted into candidacy. Course work and research amounting to the equivalent of two academic years beyond the Master's degree or three years beyond the Bachelor's degree may be considered a minimum.
Thesis: Thesis may be written *in absentia*.

Table A—Appointments to Graduate Students, 2009–10

Title of Appointee	Appointments		Academic Load Allowed in Credit Hours	Hours of Service Per Week	Stipend for Academic Year ($)
	Total	First year			
			Semester		
Teaching Assistant	31	5	9 hrs per semester	20	18,567
Research Assistant	12	3	9 hrs per semester	20	18,900
Total	43	8			

5. Personnel Engaged in Separately Budgeted Research, 9/08–8/09

Professorial faculty	19
Other faculty	2
Postdoctoral appointments	2
Graduate students	50
Undergraduate students	0
Nonteaching research personnel	1
Total	72

6. Separately Budgeted Research Expenditures by Source of Support

	Departmental Research	Physics-related Research Outside Department
Federal government	$1,318,395.57	
State and local government	264,637.48	
Private, nonprofit organizations	293,725.66	
Business and industry		$123,612.41
Total	$2,000,371.12	

Table B—Separately Budgeted Research Expenditures

Research Specialty	No. of Grants	Expenditures ($)
Atomic, Molecular, & Optical	2	150,345.05
Condensed Matter	3	633,211.51
Experimental Microwave	0	
Nuclear	0	
Optical	0	
Semi-Conductor	2	106,817.82
Solid State	0	
Theoretical	5	1,063,698.13
Astronomy	3	46,298.61
Total	15	2,000,371.12

Table C—Physics-related Research Outside Department

Field Unit Outside Department	No. of Grants	Expenditures ($)
Phillip Morris	1	123,612,41
Total	1	$123,612.41

FACULTY

Regents Professors

Duggan, Jerome L., Ph.D., Louisiana State, 1961. Materials Analysis with Ion Beams, Study of Low Energy Nuclear Reactions, Innershell Ionization Studies, Nuclear Electronics and Measurement, Nuclear Physics Experiments for Physics Undergraduates.

Hu, Zhibing, Ph.D., McMasters University, Canada, 1988. Soft Condensed Matter Physics, Polymer Chemistry and Physics, Light Scattering Characterization, Colloidal Chemistry and Physics, Polymer Gels and Related Biomaterials, Nanostructured Materials.

McDaniel, Floyd, Ph.D., Georgia, 1971. Semiconductor Materials, Materials Science, Accelerator Mass Spectrometry, Ion-atom Interactions, Nuclear Physics.

Professors

Grigolini, Paolo, Ph.D., University of Pisa, Italy, 1969. Science of Complexity, Joint action of order and randomness as a source of long-range correlation, Self-organization in physics, biology and material science, From dynamics to thermodynamics and from quantum to classical physics: the anomalous versus the ordinary statistical mechanical perspective.

Kobe, Donald H., Ph.D., Minnesota, 1961. Quantum Mechanics, Quantum Field Theory, Electromagnetic Theory, Classical Mechanics.

Krokhin, Arkadii, Ph.D., Kiev State Univ., Ukraine, 1983. Solid state physics, dynamical systems, quantum chaos.

Littler, Chris, Ph.D., North Texas, 1984. Solid State Physics.

Matteson, Samuel E., Ph.D., Baylor University, 1976. Ion-solid interactions, Ion Beam Analysis of Materials, Semiconductor Materials, Acoustics, Ion Optics.

Mueller, Dennis W., Ph.D., University of Nebraska, 1988. Atomic and Molecular Physics, Nano Technology, NonDestructive Inspection (NDI), Instrumentation and Measurement Science, Microelectronics.

Neogi, Arup, Ph.D., Yamagata University, Japan, 1992. Semiconductor Nanostructures/Semiconductor Physics, Lasers and Nonlinear Optics, Ultrafast Optics and Optical Spectroscopy, Optical Communication, Plasmonic Nanoscience.

Ordonez,, Carlos A., Ph.D., UT Austin. Penning Trap Related Research Including Antimatter Studies, Cryogenic Heat Engine Research, Plasm Space-Charge Research including Plasma Sheath Related Studies.

Perez, Jose M., Ph.D., University of California-Berkeley, 1983. Nanotechnology, Carbon Nanotubes, Quantum Dots, Spintronics, Scanning Tunneling Microscopy.

Quintanilla (Ward), Sandra, Ph.D., University of London, 1986. Theoretical Atomic Physics in Particular Three-body (Coulomb) Problems, Low- and High-energy Positron Collisions, Electron Collisions, Photodetachment of Negative Ions, Variational Principles, Modified Effective Range Theories, Formulation of Atomic Theory.

Roberts, James A., Ph.D., Oklahoma, 1967. Microwave interaction with Matter, Molecular and Atomic Spectroscopy, Plasma Physics, Electronic Systems, Astronomy.

Associate Professors

Kowalski, Jacek M., Ph.D., Poland, 1973. Statistical Physics, Quantum Theory of Magnetism, Artificial and Biological Neuronal Networks.

Shiner, David, Ph.D., University of Michigan, 1988. Atomic Physics, Laser Spectroscopy, Precision Measurement.

Weathers, Duncan, Ph.D., Cal Tech, 1989. Ion-solid interactions, with an Emphasis on Sputtering, Accelerator-based Materials Characterization Techniques, Ion Optics, Resonance Ionization Spectroscopy, Instrumentation.

Assistant Professors

Philipose, Usha, Ph.D., University of Toronto, Canada, 2006. Growth and Characterization of Semiconductor Nanowires.

Reinert, Tilo, Ph.D., University of Leipzig, Germany, 2001. Ion bean physics.

Rout, Bibhudutta, Ph.D., Institute of Physics, Bhubaneswar, India, 2001. Growth, Characterization and Thermal Behavior of

Epitaxial Metallic Layers on Semiconductors and their Self-assembled Microstructures.

Rostovtsev, Yuri, Russian Academy of Sciences, 1991. EIT, Quantum Coherence, FELS.

Shemmer, Ohad, Ph.D., Tel Aviv University, Israel, 2004. Black Hole Mass, Accretion Rate, and Metal Abundances in Active Galactic Nuclei.

RESEARCH SPECIALTIES AND STAFF

Theoretical

Atomic and Molecular Physics. Electron & Positron Scattering, Quintanilla (Ward).

Plasma Physics; Penning Trap; Antimatter Studies; Cryogenic Heat Engine Research. Ordonez.

Low-Temperature Physics. Many-body theory; quantum fluids. Kobe.

Nonlinear Dynamics. Biophysics. Grigolini, Kowalski.

Solid State. Optical properties of solids; transport processes. Deering, Kobe, Kowalski, Krokhin, Philipose.

Statistical Physics. Complexity. Deering, Grigolini, Kowalski, Krokhin.

Experimental

Applied Physics. Ion-implantation and trace analysis in semiconductors; trace analysis; accelerator; mass spectrometry. Duggan, Matteson, McDaniel, Weathers, Rout.

Astronomy. Ultraviolet, evolution of galaxy. Atomic and Molecular Physics. Ionization processes for inner shell electrons; ion penetration. Sputtering. Matteson, Shemmer, Weathers.

Microwave Spectroscopy. Collision cross sections. Roberts.

Optics. Laser technology and applications of lasers. Neogi, Shiner.

Solid State. Semiconductor materials growth; Scattering mechanisms and influence of impurities; transport coefficients; laser spectroscopy. Golding, Littler, Neogi, Perez, Rout.

Polymers; Hydrogels; Biomaterials; Light scattering. Hu.

UNIVERSITY OF TEXAS AT ARLINGTON

DEPARTMENT OF PHYSICS

Arlington, TX 76019
WWW Home page: http://www.uta.edu/physics

Students Accepted For Degree	FIELDS: Physics	Astronomy	Related Fields
Doctorate	X	X	X
Master's	X	X	X

1. General

President: James D. Spaniolo
Dean of Graduate School: Philip Cohen
Department Chairman: Alex Weiss
Department Telephone Number: (817) 272-2266
Type of Institution: University
Control: Public
Setting: Urban
Total Faculty: 1,302
Total Graduate Faculty: 775
Total students: 25,084
Total Graduate Students: 6,099
Annual Graduate Tuition: †
Tuition rates for: Fall 2008
In-state residents: Full-time—$3,358 (9 credits)
*Out-of-state residents**: Full-time—$5,887 (9 credits)
Deferred tuition plan: Yes
Other Fees: †
Term: Semester

*Teaching or research assistants employed half-time are charged the same rate as in-state residents
†See Graduate Catalog for more detailed breakdown.

2. Number of Faculty in Department

The combined total of full-time faculty in the three professorial ranks is 23.

3. Admission, Financial Aid, and Housing

Address admission inquiries to: Physics Dept., Box 19059
Graduate application fee required: $40 for U.S. and $70 for foreign students
Admission deadline: Applications are accepted throughout the year.
Admission information: Ten students were admitted out of 45 applicants in Fall 2008.
Admission requirements: For unconditional admission to the graduate program, a Bachelor's degree in physics or in related fields with a minimum 3.0 GPA on a 4.0 scale is required. The applicant should have a score of minimum 1,000 (verbal+quantitative) in GRE. The GRE advanced is recommended. The average total GRE score for 2009–10 admissions in Physics was 1150. Students from non English speaking countries are required to demonstrate proficiency in English via the TOEFL exam. Minimum acceptable TOEFL score for admission and financial aid is 550, respectively.
The Doctor of Philosophy in Physics and Applied Physics program is designed to prepare broadly trained physicists who would be engaged primarily in research, teaching and development in academia or industry. To be admitted to this program, an applicant must satisfy the general requirements of the Graduate School, and have a Masters degree or 30 credit hours of graduate courses in physics or related fields.
Undergraduate preparation assumed: Junior and senior physics at the level suggested by the following textbooks: Carter, *Statistical Physics*; Wangsness, *Electromagnetic Fields*; Gasiorowicz, *Quantum Mechanics*; Marion, *Classical Dynamics of Particle Systems*; Riley, Hobson and Bence, *Mathematical Methods for Physics and Engineering*.
Address financial aid inquiries to: Chairman, Graduate Admissions Committee, Department of Physics
GAPSFAS application required: No
Financial aid deadline: Yes
Loans available: Yes
Address housing inquiries to: Housing Office, 210 University Center
On-campus, graduate student housing available: Yes
On-campus, married student housing available: Yes

Table A—Faculty, Enrollments, and Degrees Granted

Research Specialty	2008–09 Faculty	Enrollment[1] Fall 2009: Master's	Doctorate	No. of Degrees Granted[2] 2008–09 (2004–08): Master's	Terminal Master's	Doctorate	Median No. of Years for 2009–10 Ph.D.'s
Applied Physics	2	1	2	1(1)	–	1(3)	–
Astrophysics	3	2	4	1(5)	–	1(1)	–
Atomic, Molecular, & Optical Physics	1	0	0	–	–	0(0)	–
Condensed Matter	7	6	4	4(13)	–	4(7)	–
Nanobio Physics	2	1	3	2(2)	–	0(0)	–
Particles & Fields	6	4	4	1(3)	–	1(3)	–
Non-specialized		3	0	4(5)	–	–	–
Space Physics	2	3	2	1(1)			
Total	23	20	17	–	–	6(14)	–
Full-time Grad. Stud.		14	17	14(30)	–		
Part-time Grad. Stud.		1	0				
First-year Grad. Stud.		5	5				
Median Years in Grad. Study (2001–02 Degrees)				–	–	–	–

Undergraduate Degrees, 2001–02 (1997–02): 3(28)

[1]Students not yet committed to a research specialty are entered under non-specialized.
[2]Five-year totals in parentheses.

4. Graduate Degree Requirements

Master's: The Master of Science in physics requires a minimum of 30 credit hours, of which 24 hours, including six hours of thesis, will be in physics, and six may be selected from physics, mathematics, chemistry, geology, biology or engineering as approved by the Graduate Advisor. The completion of this degree normally takes two years. Foreign language, comprehensive, and qualifying exams are not required. However, a grade point average of 3.0 (on a scale of 4.0 maximum) must be maintained for all work undertaken as a graduate student. The student must conduct research leading to a thesis which must be defended in an oral exam. Thesis may be written *in absentia*. Non-thesis option is also available.
Doctorate: For the completion of the degree, the student must 1) demonstrate competence in a minimum of 39 credit hours of core courses in physics, chosen under the guidance of the

supervising committee, and approved in advance by the Graduate Studies Committee, 2) complete 9 credit hours of internship or six credit hours of research with a written report and three hours of Applied Physics course, 3) pass qualifying and comprehensive examinations, and 4) conduct research leading to a dissertation which must be defended in an oral exam. Dissertation may be written *in absentia*.

Special Equipment, Facilities, or Programs: angular correlation of annihilation radiation (ACAR) system, *four-probe* electrical conductivity measurement system with closed-cycle liquid helium cryostat, high pressure diamond anvil cell, low energy positron beams, magnetic resonance spectrometer, photoemission spectrometer, photoluminescence spectrometer, position annihilation induced Auger electron spectroscopy, Raman spectrometer, scanning Auger microbe, thin-film deposition systems, variable temperature vibrating sample magnetometers, scanning electron microscopy with polarization analysis (SEMPA), alternating gradient magnetometer (AGM), magnetic properties measurement system (MPMS) SQUID magnetometer and arc melting furnace, thermal particle analyzer, rapid thermal processor (RTP) and high-energy physics detector construction facility. Computational facilities at UTA include high performance computing environment that combines multiple independent systems connected via a private high-speed network. The servers at this facility feature Alpha, Intel IA-32, and Itanium architectures running Compaq's Tru64 UNIX and Redhat Linux. The high-energy physics group of the physics department has a farm of Linux PC's that participate in high level simulation and a distributed and parallel computing cluster facility run jointly with Computer Science and Engineering department. This facility has 160 CPUs and over 70 TB of disk space available. The HEP group is also establishing a 5000 processor grid from which will support analysis of data remote ATLAS experiment at CERN. University-wide parallel computing facilities are also available through Beowulf Cluster. Separately, the physics department has a computing laboratory with multiple PCs, printers and scanners. There is also currently 32 node Beowulf-based computational physics cluster in the Department. In addition, research groups in the department have PCs and workstations.

Table B—Appointments to Graduate Students, 2009–10

Title of Appointee	Appointments Total	Appointments First year	Academic Load Allowed in Credit Hours	Hours of Service Per Week	Stipend for Academic Year ($)
			Semester		
Teaching Assistant	21	5	9	20	18,960
Research Assistant	20	0	9	20	18,960
Total	41	5			

[1]Salary will increase with experience, and out-of-state tuition is waived for 20 hours of service.
[2]Additional teaching and research appointments are available in the summer.

5. Personnel Engaged in Separately Budgeted Research, 6/04–5/05

Professorial faculty	23
Visiting professors	2
Post-doctoral Research Associates	11
Graduate students	35
Undergraduate students	35
Total	106

6. Separately Budgeted Research Expenditures by Source of Support

	Departmental Research	Physics-related Research Outside Department
	Breakdown not available	
Total	$3,101,842	

Table C—Separately Budgeted Research Expenditures

Research Specialty	No. of Grants	Expenditures ($)
Astrophysics	5	97,092
Condensed Matter Physics	9	1,524,000
High Energy Physics	13	1,480,750
Space Physics	4	91,359
Total	31	3,193,201

FACULTY

Professors

Black, Truman D., Ph.D., Rice University, 1964. Electron paramagnetic resonance; ultrasonics; laser optics.

Brandt, Andrew, Ph.D., UCLA, 1992, Experimental high-energy physics.

De, Kaushik, Ph.D., Brown University, 1988. Experimental high-energy physics.

Koymen, Ali R., Ph.D., University of Michigan, 1984. Surface physics; surface magnetism; positron physics.

Liu, Ping, Ph.D., University of Amsterdam, the Netherlands, 1994. Condensed matter physics; magnetic materials; nanomaterials.

López, Ramón E., Ph.D., Rice University, 1986. Space physics; Physics and Science Education.

Musielak, Zdzislaw, Ph.D., University of Gdansk, Poland, 1980. Theoretical astrophysics; cosmology; chaos and nonlinear physics.

Ray, Asok, Ph.D., Texas Tech University, 1977. Condensed matter theory; clusters; electron transport theory.

Rubins, Roy S., D. Phil., Oxford University, England, 1961. Magnetic resonance; microwave absorption studies.

Sharma, Suresh C., Ph.D., Brandeis University, 1976. Positron physics; high pressure physics; surface science; nano-materials.

Weiss, Alex, Ph.D. Brandeis University, 1983. Positron physics; surface physics.

White, Andrew P., Ph.D., London, England, 1972. Experimental high-energy physics.

Associate Professors

Cuntz, Manfred, Ph.D., Heidelberg, Germany, 1988, Theoretical astrophysics; observational astronomy, astrobiology.
Fazleev, Nail, Ph.D., Kazan University, Russia, 1978. Theoretical condensed matter physics.
Yu, Jaehoon, Ph.D., SUNY, Stony Brook, 1993. Experimental high-energy physics.
Zhang, Qiming, Ph.D., SISSA, Trieste, Italy, 1989. Theoretical condensed matter.

Assistant Professors

Chen, Wei, Ph.D., Peking University, China, 1992. Nanobio physics.
Deng, Yue, Ph.D., University of Michigan, 2006. Space physics.
Farbin, Amir, Ph.D., University of Maryland, 2004. Experimental high-energy physics.
Huda, Muhammad, Ph.D., University of Texas at Arlington. Condense matter theory.
Jackson, Christopher, Ph.D., Florida State University, 2005. Theoretical high energy physics.
Mohanty, Samarendra, Ph.D.
Park, Sangwook, Ph.D., Purdue University, 1998. Astrophysics.

RESEARCH SPECIALTIES AND STAFF

Theoretical and/or Data Analysis

Astrophysics: Cuntz, Musielak, Park.
Atomic and molecular clusters and nanostructures: Ray.
Chaos and nonlinear physics: Musielak.
Condensed matter: Fazleev, Huda, Ray, Zhang.
Physics/Science Education: López.
Space physics: Deng, López.
Transport theory, electron transport: Ray.

Experimental

Condensed matter physics: Black, Koymen, Liu, Rubins, Sharma.
High energy physics: Brandt, De, Farbin, White, Yu, Jackson.
Nanobio physics: Chen, Mohanty
Nanomaterials: Koymen, Liu, Sharma, Weiss.
Optics: Black, Sharma.
Positron physics: Koymen, Sharma, Weiss, Jackson.
Surface physics: Koymen, Sharma, Weiss.

UNIVERSITY OF TEXAS AT AUSTIN

DEPARTMENT OF PHYSICS

Austin, Texas 78712

Students Accepted For Degree	FIELDS		
	Physics	Astronomy	Related Fields
Doctorate	X		
Master's	X		

1. General

President: William Powers, Jr.
Dean of Graduate Studies: Victoria Rodríguez
Department Chairman: Richard Hazeltine
Department Telephone Number: (512) 471-1153 (C) 471-1664
Department Web Site: http://www.ph.utexas.edu
Type of Institution: University
Control: Public
Setting: Urban
Total Faculty: 3,887
Total Graduate Faculty: 1,915
Total Students: 50,995
Total Graduate Students: 12,827
Annual Graduate Tuition:
In-state residents: $7,330
Out-of-state residents: $14,568
Tuition rates for: 2009–10
Deferred tuition plan: Yes
Term: Semester

*All required fees included in the above amounts. Other fees may vary.

2. Number of Faculty in Department

The combined total of full-time faculty in the three professorial ranks is 51. The combined total of full-time, part-time, and other faculty at all ranks is 55.

3. Admission, Financial Aid, and Housing

Address admission inquiries to: Graduate Coordinator, Department of Physics, RLM 5.208, University of Texas at Austin, University Station, C1600, Austin, TX 78712-1081
Graduate application fee required: $50 for U.S. applicants; $75 for International applicants
Admission deadline (Fall admission): 12/1
Admission information: For fall admission, 2009–10, 99 students were accepted from 402 applicants.
Admission requirements: For admission to the graduate programs, a Bachelor's degree is required with minimum undergraduate GPA of 3.0 specified. The GRE is required. The GRE Advanced Physics is required for admission. The average GRE scores for 2009–10 admissions were verbal–524; quantitative–773; total–1,297. The average GRE Advanced score for 2009–10 admission was 812. Students from non-English speaking countries are required to demonstrate proficiency in English via the TOEFL exam. Minimum acceptable score recommended for admission is 550. The TOEFL is absolutely required for foreign applicants and cannot be waived, substituted, or delayed. Foreign students who accept teaching assistantships must pass an English language proficiency assessment before any appointment can be made.
Undergraduate preparation assumed: For graduate work in physics, it is assumed that the student has an undergraduate background that includes the following: mechanics at the level of Halliday, Resnick, and Krane, *Physics*, Vol. 1; electricity and magnetism at the level of Halliday, Resnick, and Krane, *Physics*, Vol. 2; thermodynamics at the level of Kittel and Kroemer, *Thermal Physics*; atomic physics at the level of Morrison, Estle, and Lane, *Quantum States of Atoms, Molecules and Solids*; quantum mechanics at the level of Morrison, *Understanding More Quantum Physics*.
Address financial aid inquiries to: Graduate Coordinator, Department of Physics, University of Texas at Austin, RLM 5.208, Austin, TX 78712-1081, or to: graduate@physics.utexas.edu
GAPSFAS application required: No
Financial aid deadline: 12/1
Loans available: Yes
Address housing inquiries to: Division of Housing and Food Service, P.O. Box 7666, University of Texas at Austin, Austin, TX 78712-7666
On-campus, single student housing available: Yes—limited availability
Cost/per year 2010–11 rates: $8,184-double
$11,360 for single room
On-campus, married student housing available: No*
Cost/month: $510–744. Rates subject to change. Currently a waiting list for University housing.

*Not on campus, but University housing for married students is available (about 5 mi. from campus, with shuttle-bus service).

Table A—Faculty, Enrollments, and Degrees Granted

Research Specialty	2009–10 Faculty	Enrollment[1] Fall 2009–10		No. of Degrees Granted[2] 2009–10 (2005–10)			Median No. of Years for 2009–10 Ph.D.'s
		Master's	Doctorate	Master's	Terminal Master's	Doctorate	
Acoustics	–	0	0	0(0)	0(1)	0(0)	–
Atomic, Molecular, & Optical Physics	9	4	29	3(7)	1(12)	6(24)	8
Biophysics	2	3	14	3(4)	0(7)	0(7)	–
Condensed Matter Physics	14	1	65	1(1)	0(7)	9(58)	6
Cosmology & String Theory	5	2	20	1(1)	1(1)	2(9)	8
High Energy Physics	7	0	14	0(0)	0(2)	4(14)	6
Nonlinear Dynamics	2	1	6	1(3)	0(4)	1(7)	5
Nuclear Physics	1	0	0	0(0)	0(0)	0(2)	–
Particles & Fields	5	4	9	1(1)	1(2)	2(15)	7
Plasma Physics & Fusion	7	1	24	1(2)	0(6)	5(19)	6
Relativity & Gravitation	1	1	9	0(1)	1(2)	3(7)	7
Statistical & Thermal	2	0	3	0(0)	0(0)	0(8)	–
Non-specialized	–	2	20	0(0)	0(1)	0(0)	–
Total	55	19	213	11(20)	3(45)	32(170)	
Full-time Grad. Stud.		17	202				
Part-time Grad. Stud.		2	11				
First-year Grad. Stud.		4	50				
Median Years in Grad. Study (2009–10 Degrees)				2	5	7	
Undergraduate Degrees, 2008–09 (2004–09)							

[1] Students not yet committed to a research specialty are entered under non-specialized.

4. Graduate Degree Requirements

Master of Arts: The time required for the degrees will average about one calendar year plus one semester for a student with a strong undergraduate background. Requirements include 30 semester hours with a "B" average. Eighteen to twenty-four semester hours, including the thesis, must be in the major program. The minor, which is obligatory, consists of a minimum of six hours in a supporting subject or subjects outside the major program. Each program must include at least thirty semester hours of graduate work, including the thesis. All completed work included in the degree program at the time of admission to candidacy must have been taken within the previous six years.

The Master of Science in Applied Physics is designed to provide students with a broad background in physics and related fields, with an emphasis on those aspects of the science most used in an industrial setting. The required physics courses include PHY 380N, 387K, and 389K, a course in the physics of sensors, and a technical seminar. A thesis is also required. The supporting work must be in engineering, chemistry or geological sciences.

Doctorate: A student must fulfill the following requirements to be admitted to candidacy for the Ph.D. degree in physics: (1) Fulfill the core course requirement described below; (2) show evidence of exposure to modern methods of experimental physics; this exposure may be gained in a senior-level laboratory course taken by the student as an undergraduate and approved by the graduate adviser and the Chairman of the Graduate Studies Committee, by previous participation in an experimental program or in Physics 380N; and (3) fulfill the oral examination requirement described below.

Core Courses: During the first two years of graduate student, the student must take four core courses; Classical Mechanics (385K), Statistical Mechanics (385L), Electromagnatism I (387K), or Electromagnatism II(387L), and Quantum Mechanics I (389K) or Quantum Mechanics II (389L). The student must earn an official grade of at least "B" in each course and a grade point average of at least 3.30 in the four courses. The student may ask for the grade he or she earns in Physics 380N to be substituted for the grade in one of the core courses when the average is computed. A well-prepared student may seek to fulfill the core course requirement by earning satisfactory grades on the final examinations for some of these courses rather than by registering for them; in this case, the student does not receive graduate credit for these courses and the grade is not counted toward the required average.

The oral qualifying examination: After satisfying the first two requirements above, and within twenty-seven months of entering the program, the student must take an oral qualifying examination. The examination consists of a presentation before a committee of four physics faculty members, one of whom is a member of the Graduate Studies Subcommittee. The presentation is open to all interested parties. It is followed by a question period restricted to the student and the committee. The questions during this session are directed to clarifying the presentation and determining whether the student has a solid grasp of the basic material needed for research in his or her specialization. The student passes the examination by obtaining a positive vote from at least three of the four faculty members on the oral qualifying committee. Each Program of Work for the doctoral degree must include at least four advanced courses in physics; a list of acceptable courses is maintained by the Graduate Studies Subcommittee. The program must also include three courses outside the student's area of specialization. One of these must be an advanced physics course; another must be outside the Department of Physics; the third may be either an advanced physics course or a course outside the Department of Physics. A dissertation is required of every candidate, followed by a final oral examination covering the dissertation and the general field of the dissertation.

Special Equipment, Facilities, or Programs: Modern facilities for graduate study and research include a large-scale cryogenic laboratory; nuclear magnetic and electron paramagnetic resonance laboratories; extensive facilities for tunneling and force microscopy and nanostructure characterization, SQUID magnetometry, and electron spectroscopy; well-equipped laboratories in optical spectroscopy, quantum optics, femtosecond spectroscopy and diagnostics, and electron-atom and surface scattering; and facilities including a table-top 100 terawatt laser for strong field physics studies for turbulent flow and nonlinear dynamics experiments and two petawatt lasers (one Ti-sapphire providing 30J in 30fs and another glass laser at 200J in 150fs). Plasma physics experiments are conducted at the major national tokamaks in Boston and San Diego as well as on the local machine, The Helimak. Experiments in high-energy heavy ion nuclear and particle physics are conducted at large accelerator facilities such as Brookhaven National Laboratory (New York), Fermi National Laboratory (Illinois), and Germany's Deutsches Electron Synchrotron. Theoretical work in plasma physics, condensed matter physics, acoustics, nonlinear dynamics, relativity, astrophysics, statistical mechanics, and particle theory is conducted within the Department of Physics. Students have access to excellent computer and library facilities including Ranger, the 10th fastest computer at 504 Tflops. The department maintains and staffs a machine shop, student workshop, low-temperature and high-vacuum shop, and an electronics design and fabrication shop.

Table B—Appointments to Graduate Students, 2009–10

Title of Appointee	Appointments: Total	Appointments: First year	Academic Load Allowed in Credit Hours	Hours of Service Per Week	Stipend for Academic Year ($)
			Semester		
Teaching Assistants*	92	40	9–12	10–20	8,275–17,610
Research Assistants	109	2	9–12	10–20	Salary proportionate
Assistant Instructors	22	0	9–12	10–20	9,335–18,670
Total	223	42			

*Teaching Assistants include graders.

5. Personnel Engaged in Separately Budgeted Research, 2008–09

Professorial faculty	55
Postdoctoral appointments	38
Graduate students	130
Undergraduate students	48
Nonteaching research personnel	50
Total	321

6. Separately Budgeted Research Expenditures by Source of Support

	Departmental Research	Physics-related Research Outside Department
Federal government	$15,422,544	$
State and local government	2,195,583	
Private, nonprofit organizations	1,696,511	
Business and industry	524,473	
Total	$19,839,111	$

Table C—Separately Budgeted Research Expenditures

Research Specialty	No. of Grants	Expenditures ($)
Atomic, Molecular, & Optical Physics	10	$661,254
Condensed Matter Physics	47	3,679,314
High Energy Physics	13	1,298,115
Nonlinear Dynamics	11	1,076,024
Nuclear Physics	3	458,080
Particles & Fields	5	696,654
Physics Education	3	160,109
Plasma Physics & Fusion	50	11,136,887
Relativity	1	81,473
Statistical & Thermal Physics	7	671,201
Total	150	$19,839,111

FACULTY

Professors

Antoniewicz, P. R., Ph.D., Purdue, 1965. Theoretical investigation of electromagnetic wave propagation and transport properties in solids and liquids; investigation of the properties of atoms and molecules on metal surfaces.

Bengtson, R. D., Ph.D., Maryland, 1968. Experimental plasma physics; atomic reactions in plasmas.

Berk, H. L., Ph.D., Princeton, 1964. Theoretical plasma physics; computer simulation of plasmas.

Böhm, A., Dr.rer.nat., Universität Marburg, 1966. Particle phenomena in terms of algebraic and group-theoretical methods.

Chelikowsky, J., Ph.D., California, Berkeley, 1975. Solid State Physics. Computational Materials Science.

Chiu, C. B., Ph.D., California, Berkeley, 1965. Theoretical particle physics, particularly in quantum chromodynamics; confinement problems; subquark and sublepton models; theories in hadron collisions.

Coker, W. R., Ph.D., Georgia, 1966. Theoretical nuclear physics, with emphasis on scattering and reactions of hadrons and nuclei at medium energies.

de Lozanne, A. L., Ph.D., Stanford, 1982. Low-temperature vacuum-tunneling microscopy.

Dicus, D. A., Ph.D., UCLA, 1968. Field theory of strong, weak, and electromagnetic interactions; astrophysical implications of the weak force.

Distler, J., Ph.D., Harvard, 1987. High energy theory, mathematical physics, and string theory.

Ditmire, T., Ph.D., California, Davis, 1995. Intense ultrafast laser interactions.

Downer, M. C., Ph.D., Harvard, 1983. Atomic and molecular physics; atomic physics; femtosecond spectroscopy; condensed matter surfaces; high-field atomic and plasma physics.

Erskine, J. L., Ph.D., Washington, 1973. Experimental solid state physics; surface physics; magnetism.

Fink, M., Dr.Phil., Technische Hochschule Karlsruhe, 1966. Electron diffraction.

Fischler, W., Ph.D., Université Libre de Bruxelles, 1976. Theoretical physics; particle theory; invisible axion and supersymmetry.

Fitzpatrick, R., Ph.D., Sussex, 1988. Magnetic reconnection and gross plasma instabilities in fusion, terrestrial, and astrophysical contexts.

Frommhold, L. W., Ph.D., Universität Hamburg, 1959. Atomic and molecular physics.

Gentle, K. W., Ph.D., MIT, 1962. Experimental plasma physics.

Gleeson, A. M., Ph.D., Pennsylvania, 1965. Field theory of strong interactions and the physics of superdense matter.

Hazeltine, R., Ph.D., Michigan, 1968. Theoretical plasma physics.

Heinzen, D., Ph.D., MIT, 1988. Atomic and molecular physics; laser-cooling and atom trapping; Bose-Einstein condensation.

Hoffmann, G. W., Ph.D., UCLA, 1971. Experimental nuclear physics and chemistry using medium-energy projectiles.

Horton, C. W., Jr., Ph.D., California, San Diego, 1967. Theoretical plasma physics.

Kaplunovsky, V., Ph.D., Tel Aviv University, 1983. Particle theory; string phenomenology.

Keto, J. W., Ph.D., Wisconsin, Madison, 1972. Reactions and radiative processes of excited atoms and molecules; laser spectroscopy high power lasers.

Kleinman, L., Ph.D., California, Berkeley, 1960. Solid state theory. The electronic structure of solids, surfaces, and clusters; chemisorption.

Lang, K., Ph.D., Rochester, 1985. Rare decay of the K-meson.

MacDonald, A. H., Ph.D., University of Toronto, 1978. Condensed matter theory with emphasis on electron-electron interactions.

Marder, M. P., Ph.D., California, Santa Barbara, 1986. Nonlinear dynamics; statistical physics of solids.

Markert, J. T., Ph.D., Cornell, 1987. Condensed matter experiment; crystal growth; high-T_c materials; magnetic materials; magnetic resonance; magnetic microscopies.

Matzner, R. A., Ph.D., Maryland, 1967. General relativity and cosmology; manifolds with little symmetry; kinetic theory; conservation laws in general relativity; black hole physics and gravitational radiation.

Morrison, P. J., Ph.D., California, San Diego, 1979. Plasma physics.

Niu, Q., Ph.D., Washington, 1985. Field theory of condensed matter; theory of superconductivity; mesoscopic physics; quantum transport and diffusion.

Raizen, M., Ph.D., Texas, Austin, 1989. Atomic, molecular, and optical physics; atom optics; quantum chaos.

Reichl, L. E., Ph.D., Denver, 1969. Nonequilibrium quantum statistical mechanics; Brownian motion; nonlinear dynamics.

Riley, P. J., Ph.D., University of Alberta, 1962. Experimental studies of the nucleon-nucleon interaction at medium energy, actions in decaying plasmas; environmental effects on spectra.

Ritchie, J. L., Ph.D., Rochester, 1983. High-energy/nuclear physics.

Schwitters, R. F., Ph.D., MIT, 1971. Chairman of the Department. Experimental high-energy physics—detector development and B-physics studies.

Shih, C. K., Ph.D., Stanford, 1988. Condensed matter; study of surface properties of microelectronic materials.

Sitz, G. O., Ph.D., Stanford, 1987. Atomic and molecular experiment; oriented molecules; surface scattering.

Sudarshan, E. C. G., Ph.D., Rochester, 1958. Elementary particle physics; quantum optics; quantum field theory; classical mechanics; foundations of physics.

Swinney, H. L., Ph.D., Johns Hopkins, 1968. Equilibrium and nonequilibrium phase transitions; dynamics of nonlinear systems.

Weinberg, S., Ph.D., Princeton, 1957. Theoretical physics.

Associate Professors

Demkov, A., Ph.D., Arizona, 1995. Condensed Matter theory, physics of electronic materials, surfaces, & interfaces; thin films & devices; novel materials; quantum transport.

Florin, E.-L., Technische Universität Munchen, 1990. Experimentalist, non-linear dynamics.

Kopp, S., Ph.D., Chicago, 1994. CP violation, weak decays of heavy quarks, neutrino oscillations.

Paban, S., Ph.D., Universidad de Barcelona, 1988. Quantum mechanics, particle phenomenology, string theory.

Shvets, G., Ph.D., MIT, 1995. Theory & simulations: laser-plasma interactions, plasma based accelerators, photonics, nano-plasmonics. Experiment: phonon-assisted nanolithography, compact surface-wave accelerators.

Tsoi, M., Ph.D., Universität Konstanz, 1998. Condensed matter experiment, nanostructures, spintronics.

Turner, J. S., Ph.D., Indiana, 1969. Nonequilibrium statistical mechanics and thermodynamics; theoretical and experimental studies of chemical instabilities; dynamics of nonlinear systems.

Yao, Z., Ph.D., Harvard, 1997. Nanostructures and mesoscopic physics.

Assistant Professors

Fiete, G. A., Ph.D., Harvard, 2003. Theory of quantum matter and correlated electrons at the nanoscale.

Li, X., Ph.D., Michigan, 2003. Experimental condensed matter physics, femtosecond spectroscopy; phase-sensitive nonlinear optical interactions in semiconductors.

Markert, C., Ph.D., Johann Wolfgang Goethe Universität, 2001. Nuclear physics; experimental heavy-ion physics; the Quark-Gluon Plasma (QGP) phase.

Shubeita, G. T., Ph.D., Université de Lausanne, 2002. Biophysics-integrating biophysical approaches with molecular biology, genetics, and proteomics.

Professors Emeriti

de Wette, F. W., Ph.D., Universiteit Utrecht, 1959. Theoretical study of structural, thermodynamics, and scattering properties of crystal surfaces.

DeWitt-Morette, C., Doctorat d'Etat, Universite de Paris, 1947. Functional integration; differential geometry; mathematical physics.

Drummond, W. E., Ph.D., Stanford, 1958. Theoretical plasma physics.

Gavenda, J. D., Ph.D., Brown, 1959. Study of properties of conduction electrons in metals using ultrasonic and electromagnetic waves.

Griffy, T. A., Ph.D., Rice, 1961. Theoretical medium-energy physics; underwater acoustics.

McCormick, W. D., Ph.D., Duke, 1959. Experimental low-temperature and solid state physics; phase transitions in solids (critical phenomena); instabilities in nonequilibrium systems.

Moore, C. F., Ph.D., Florida State, 1964. Detection and measurement of the interactions and involvement of the nuclear continuum in scattering experiments; atomic interactions in highly ionized atoms.

Nolle, A. W., Ph.D., MIT, 1947. Magnetic resonance and optical studies of solids; viscoelastic phenomena.

Oakes, M. E., Ph.D., Florida, 1964. Theoretical and experimental studies of wave propagation in plasmas.

Schieve, W. C., Ph.D., Lehigh, 1959. Nonequilibrium statistical mechanics; quantum optics; stochastic processes.

Swift, J. B., Ph.D., Illinois, 1968. Studies of nonlinear dynamics; phase transitions.

Thompson, J. C., Ph.D., Rice, 1956. Studies of electronic states in disordered systems (metallic and semiconducting) by galvanomagnetic parameters; optical properties; photoemission; the metal-nonmetal transition.

Udagawa, T., Ph.D., Tokyo Univ. of Education, 1962. Theoretical nuclear physics.

RESEARCH SPECIALTIES AND STAFF

Theoretical

Condensed Matter Physics. *Ab-initio* electronic structure calculations of the physical, electronic, and magnetic (including noncolinear magnetic systems) properties of solids, surfaces, interfaces and liquids; molecular dynamics calculations of properties of solids, liquids, and crack propagation; density functional theory; Berry phases in polarization theory and spinwave theory; Block electrons in magnetic fields, quantum Hall effect; quantum theory of thin film growth and surface diffusion; theory of mesoscopic phenomena, phonon calculations and lattice dynamics for high T_c superconductors; theory of atom surface interactions, physisorption, chemisorption. Antoniewicz, Chelikowsky, Demkov, Kleinman, MacDonald, Marder, Niu, Shvets, Swift. 13 postdoctoral fellows, 1 research scientist.

Elementary Particles and Fields. Phenomenological studies of the properties of matter ranging from medium-energy physics; symmetries in elementary particle physics; field theory of strong interactions and the physics of superdense matter; quantum chromodynamics; confinement problems; subquark and sublepton models; and supersymmetry. In addition, there is work on quantum optics; basic quantum field theory and quantum mechanics; classical mechanics; particle phenomena in terms of algebraic and group-theoretical methods; and electromagnetic interactions. Bohm, Chiu, Dicus, Gleeson, Sudarshan.

Fusion and Plasmas. Kinetic theory and transport theory; turbulent heating; collisionless shock waves; plasma turbulence; computer simulation of plasmas; stability theory controlled fusion; plasma dynamics. Berk, Fitzpatrick, Hazeltine, Horton, Morrison, Shvets. 4 postdoctoral fellows. 16 research scientists (non-faculty).

Nonlinear Dynamics. Dynamics of materials, especially fracture and dislocation dynamics. Instabilities and turbulence in fluids, granular media, liquid crystals, and chemical reaction-diffusion systems; chaos in low dimensional dynamical systems. Marder, Swift. 1 research scientist.

Nuclear Physics. Scattering and reactions of hadrons and nuclei at medium energies; nuclear structure in the low-energy region using neutron scattering techniques; nuclear structure and reaction mechanism. Coker, Udagawa.

Physics Education. Curriculum development and evaluation at the university level, science teacher preparation program, computer-based education. Chiu, Gleeson, Marder, Turner.

Relativity. Quantum theory of space time; techniques of quantization in curved space-time; string theory; path integration; stochastic processes; critical phenomena in gravitational collapse; computational relativity; cosmology; exact solutions in general relativity; conformal properties of space-time; manifolds with little symmetry; kinetic theory; conservation laws in general relativity; black hole physics; black hole interactions; gravitational radiation; interaction of matter with gravitation. C. DeWitt-Morette (Emeritus), Fischler, Matzner, Weinberg.

Statistical Mechanics and Complex Systems. Nonequilibrium statistical physics; thermodynamic processes; nonequilibrium quantum statistical mechanics; quantum chaos; mesoscopic physics; nonlinear dynamics; complex systems theory; Brownian motion. Reichl, Turner. 9 post-doctoral fellows, 4 research scientists.

The Theory Group. Research spans the range from studies of physics at the most fundamental level to exploration of phenomenologically relevant current issues in elementary particle physics. On the more fundamental level, the work continues in gravity and quantum cosmology, conformal field theories, superstring theories, and M theory, with special attention to the links between these topics, and to the implication of superstring and M theory for effective field theories at accessible energies. Such theories offer the hope of uniting all forces including gravitation in a theory of superstrings. So far it seems that these theories allow for the first time a satisfactory elimination of the infinities that have plagued all earlier quantum theories of gravitation. Distler, Fischler, Kaplunovsky, Paban, Weinberg. 4 postdoctoral fellows, 6 research scientists.

Underwater Acoustics. Griffy (Emeritus).

Experimental

Atomic, Molecular, and Optical Physics. Atom optics, quantum transport in optical lattices, quantum chaos with ultra-cold atoms. Ultra-cold collisions, Bose-Einstein condensation, and search for atomic electric dipole moment, state-resolved molecular-surface scattering and gas-surface dynamics. Raman spectroscopy. Electron diffraction. Neutrino rest mass experiments. Laser spectroscopy of nanoparticles, and development of new materials. Molecular collision and sonoluminescence. Femtosecond spectroscopy, high-power lasers. Ditmire, Downer, Fink, Frommhold, Heinzen, Keto, Raizen, Sitz. 4 postdoctoral fellows, 14 research scientists.

Condensed Matter Physics. Surface and thin-film magnetism, dynamics of magnetization reversal, magnetic switching, Barkhansen noise, domain dynamics; magnetic and electronic effects in ultrathin films multilayers and nanostructures, normal and superconducting properties of high-temperature superconductors; nonlinear optical response of solids, femtosecond spectroscopy of solid state systems; nanostructure fabrication and characterization based on scanning tunneling microscopy; intrinsic phenomena at surfaces and interfaces studied by electron diffraction, spectroscopy, atom surface scattering, linear and nonlinear optical spectroscopy, and scanning probe techniques, including near-field optical microscopy; thin-film nucleation and growth; cluster physics, mesoscopic phenomena in solids, materials synthesis including novel magnetic and superconducting materials; transport and magnetic characterization; strongly correlated electron systems; mechanical properties of materials including fracture. de Lozanne, Downer, Erskine, J. Markert, Shih, Tsoi, Yao. 11 postdoctoral fellows, 1 research scientist.

Experimental High-Energy Relativistic Heavy Ion Physics. The research focuses on two experiments: (1) E896 (using the AGS at the Brookhaven National Laboratory)—a *definitive search* for the short-lived *HO* di-baryon, a strangeness $=-2$, 6-quark object predicted by bag models. E896 also searches for other short-lived objects composed of strange hadrons which may be produced in high energy nucleus-nucleus collisions. (2) STAR [Solenoidal Tracker at RHIC (Relativistic Heavy Ion Collider)] at the Brookhaven National Laboratory, to study primordial matter at conditions of extreme temperature and pressure. Such matter is produced through central collisions of circulating beams of Au ions of momenta 100 GeV/c per nucleon (total center-of-momentum energy=40 TeV). STAR searches for evidence of the formation of a *quark-gluon plasma* (a phase of nuclear matter in which quarks and gluons are not confined within nucleons or mesons), and it searches for evidence of the restoration of the fundamental *chiral symmetry* of the strong interaction at high temperature. Both experiments explore the most fundamental physics and chemistry of nature as it may have existed during the early evolution of the Universe (about 10^{-7}–10^{-6} seconds after the *Big Bang*). Hoffmann, Riley. 2 research scientists, 1 postdoctoral fellow.

Fusion and Plasma. Plasma turbulence and transport, plasma heating, plasma propulsion, plasma spectroscopy, plasma diagnostics, plasma processing, atomic reactions in plasmas. Bengtson, Gentle, Oakes (Emeritus).

High-Energy Physics. Study of the properties of elementary particles, particularly kaons, B-mesons, and neutrinos. Study of rare decays of the kaons, tests of conservation laws and CP violation. Study of B-meson decays, information on CP violation. Neutrino oscillation measurements, information on neutrino mass. Experiments at national and international accelerator laboratories. Detector development, applications of particle detectors to medical imaging. Kopp, Lang, Ritchie, Schwitters. 2 research scientists, 1 postdoctoral fellows.

High-Field Physics. Wake-field accelerators, terawatt lasers, optical properties of nanostructured plasmas at high fields. Downer, Keto. 1 postdoctoral fellow, 1 Research Scientist.

Nonlinear Dynamics. Patterns and transport properties of granular media (for example, vibrating sand); dynamics of planetary-type flows and applications of Hamiltonian theory to the jet stream and atmospheric and oceanic transport and turbulence; convection in fluids (driven by buoyancy or by surface tension), chemical reaction-diffusion patterns; control of nonlinear regular and chaotic systems; motility of cells *in vitro*; nonlinear pattern formation in cell membranes; elastic properties of cells, crack propagation in amorphous and crystalline solids; dynamics of dislocation in solids. Florin, McCormick (Emeritus), Raizen, Shubeita, Swinney. 2 postdoctoral fellows, 1 research fellow.

Nuclear Physics. Detection and measurement of the interactions and involvement of the nuclear continuum in scattering experiments; atomic interactions in highly ionized atoms; nucleon-nucleon interaction at medium energy; neutron and triton induced reactions; nuclear structure physics; medium-energy physics; pion-induced nuclear reactions; angular correlations; heavy-ion induced reactions; atomic collisions. Hoffmann, Riley. 1 research fellow.

UNIVERSITY OF TEXAS, DALLAS

PHYSICS DEPARTMENT

Richardson, Texas 75080-3021

Students Accepted For Degree	FIELDS		
	Physics	Astronomy	Related Fields
Doctorate	X		X
Master's	X		X

1. General

President: David E. Daniel
Dean of Graduate Studies: Austin J. Cunningham
Dean, Natural Sciences and Math: Myron B. Salamon
Head of Physics Department: Robert Glosser
Department Telephone Number: (972) 883–2884
Type of Institution: University
Control: Public
Setting: Suburban
Total Faculty: 791
Lecturers: 372
Total Graduate Faculty: 419
Total Students: 15,783
Total Graduate Students: 5,982
Semester Graduate Tuition & Fees (guaranteed tuition plan):
In-state residents: $2,234/3 hrs.; $4,701/9 hrs
Out-of-state residents: $3,0483/3 hrs. $8,506/9 hrs.
Tuition rates for: 2009–10
Deferred tuition plan: Yes
Term: Semester

Tuition and fees are waived for Teaching Assistants and Research Assistants.

2. Number of Faculty in Department

The combined total of full-time faculty in the three professorial ranks is 21. The combined total of full-time, part-time, and other faculty at all ranks is 38.

3. Admission, Financial Aid, and Housing

Address admission inquiries to: Marjorie D. Renfrow, Graduate Counselor, http://www.utdallas.edu/dept/physics (email: margie@utdallas.edu) Physics Programs EC36
Graduate application fee required: $100 (international), $50 (citizens) **Payable at registration.**
Admission deadline: 45 days prior to registration
Admission information: For fall admission, 2009–10, 27 students were accepted from 50 completed applications.
Admission requirements: For admission to the graduate programs, a Bachelor's degree in physics or a related field is required with a minimum undergraduate GPA of 3.0. The GRE is required. Students must have a minimum of 700 on the quantitative and 500 on the verbel. Applicants with lower scores will be considered on an individual basis. The GRE subject test is not required. The average GRE scores for 2009–10 admissions were verbal–504; quantitative–760; total–1,264. Students from non-English speaking countries are required to demonstrate proficiency in English via the TOEFL exam with a minimum score of 213 or atleast 80 on the iBT which is less than two years old or 213 on the computer test.
Address financial aid inquiries to: Marjorie D. Renfrow, Graduate Counselor and Student Coordinator, Mail Station EC36
GAPSFAS application required: No
Financial aid deadline: 3/1
Loans available: Yes
Address housing inquiries to: Student Services, Mail Station GR2.1
On-campus, single student housing available: Yes
On-campus, married student housing available: Yes

Table A—Faculty, Enrollments, and Degrees Granted

Research Specialty	2009–10 Faculty	Enrollment[1] Fall 2008		No. of Degrees Granted[2] 2009–10 (2005–10)			Median No. of Years for 2009–10 Ph.D.'s
		Master's	Doctorate	Master's	Terminal Master's	Doctorate	
Applied Physics		2			1(11)	–	
Astrophysics	1	2	5	1(6)	1(2)	1(2)	6
Atmos./Space Phys., Cosmic Rays	6		6	0(7)		2(8)	6
Condensed Matter Physics/(NANO)	8	2	29	2(9)		3(13)	6
Medical & Health Physics	0		1				
Particles & Fields	2	0	5	0(1)			
Relativity & Gravitation	1						
Other (Physics Education)							
Non-specialized	3	8	6	1(6)	2(32)		
Total		15	52	4(24)	4(45)	6(23)	
Full-time Grad. Stud.	52						
Part-time Grad. Stud.	15						
First-year Grad. Stud.	17						
Median Years in Grad. Study (2009–10 Degrees)	6						

Undergraduate Degrees, 2009–10 (2005–10): 12(42)

[1]Students not yet committed to a research specialty are entered under non-specialized.
[2]Five-year totals in parentheses.

4. Graduate Degree Requirements

Master's: For the M.S. all students must complete at least 26 hours of graduate physics courses, including a 12-hour "core." The degree is completed either by six hours of research, including a thesis, or by six hours of additional graduate courses.
Applied Physics Master of Science (MSAP): Offered for students wishing to emphasize applications encountered in most industrial and high technology environments. The Program is open to anyone who has a scientific or engineering degree. Program requirements consist of 16 credit hours of core courses and additional electives for a total of 32 hours. These hours may include an industrial internship or research project.
Doctorate: The Ph.D. students must complete the 32-hour core, a minimum of 4 elective courses, 2 from within his/her area of specialization and 2 selected from different areas within the department plus whatever his/her committee requires. A Ph.D. candidate must pass a written qualifying exam that is presented twice each academic year. Once a dissertation topic has been selected and a faculty committee formed, the student presents a dissertation proposal to his/her committee for approval, presents a seminar, and is given an oral examination on the dissertation topic and related subjects. The student

must then complete an acceptable dissertation and present a seminar. A successful defense of the dissertation concludes the requirements for the Ph.D. degree.

Thesis: Thesis may be written *in absentia*.

Research: Research in *Atmospheric and Space Sciences* encompasses theory, modeling and experiment, with emphasis on the terrestrial upper atmosphere, planetary atmospheres, the effects of weather and climate, and space instrument development. Much of the research is conducted in the Hanson Center for Space Sciences that provides research opportunities using large satellite databases, rocket instruments and data, and computer models.

The High Energy Physics group collaborates on the ATLAS experiment at CERNS large Hadron collider. ATLAS is searching for the Higgs Boson and evidence for physics beyond the standard model. UTD works on the pixel and inner detector-general subsystems. UTD hosts an ATLAS Tier 3gs GRID computing site, the first such facility in ATLAS.

Research in *Materials Science* is focused on advanced nanostructures such as carbon nanotubes, photonic crystals and organic multilayers and on the non-destructive examination of materials using optical and magnetic (ESR and SQUID) probes. The Nanotechnology Institute houses a wide variety of experimental equipment allowing the production of nanostructures and their examination (SEM AFM/STM, XRD etc) for use in many areas including energy production and harvesting, and organic superconductors. Optical examination of materials (both inorganic and biological) is accompllished using modulation spectroscopy, Raman scattering, phtoluminescence and laser-induced breakdown spectroscopy (LIBS). Optical research also includes using femtosecond pump-probe spectroscopy, laser microscopy, photoluminescence and absorption measurements at picosecond timescales. Optical examination of materials is accomplished using modulation spectroscopy, Raman scattering and photoluminescence.

Computation study of nanomaterials using atomistic and quantum simulations.

Research in *Relativity and Astrophysics* is devoted mainly to theoretical problems in general relativity and cosmology, including the topology of the big bang and how recent observations may constrain general classes of cosmological models.

Table B—Appointments to Graduate Students, 2009–10

Title of Appointee	Appointments Total	First-year	Academic Load Allowed in Credit Hours	Hours of Service Per Week	Starting Stipend for Academic Year ($)
			Semester		
Teaching Assistant	23	8	9	20	18,000*
Research Assistant	20	2	9	20	18,000*
Total	13	10			

*Plus paid insurance, tuition, and fees.

5. Personnel Engaged in Separately Budgeted Research, 7/09–6/10

Professorial faculty	10
Graduate students	20
Nonteaching research personnel	3
Postdoctoral appointments	8
Total	41

6. Separately Budgeted Research Expenditures by Source of Support

	Departmental Research	Physics-related Research Outside Department
Federal government	$3,392,412	
Private, non-profit organizations	50,000	
Total	$3,442,412	

Table C—Separately Budgeted Research Expenditures

Research Specialty	No. of Grants	Expenditures ($)
Atmospheric & Space Science	14	2,252,321
Condensed Matter/Nanotech	10	981,758
Particles & Fields	1	208,333
Total	25	3,442,412

Table D—Physics-related Research Outside Department

Field and Unit Outside Department	No. of Grants	Expenditures ($)
Physics Education	0	0
Medical & Health	0	0
Total	0	0

FACULTY

Professors

Chaney, Roy C., Ph.D., Oklahoma, 1971. Calculation of electronic structure of solids by LCAO; space science instrumentation.

Cunningham, Augustine J., Ph.D., Belfast, 1969. Ion–electron recombination processes; ion–molecule reactions; high-temperature and pressure gas kinetics; ultraviolet spectroscopy; plasma etching, e-beam lithography.

Earle, Gregory D., Ph.D., Cornell, 1988. Space physics; neural networks and instrumentation.

Fenyves, Ervin J., Ph.D., Budapest, 1950. Elementary particles; cosmic rays; gamma-ray astrophysics; gamma-ray and neutrino detectors.

Glosser, Robert, Ph.D., Chicago, 1967. Optical properties of solids.

Heelis, Roderick A., Ph.D., Sheffield, 1973. Plasma processes and electrodynamics in planetary atmospheres and ionospheres; space flight instrumentation.

Hoffman, John H., Ph.D., Minnesota, 1958. Ionospheric composition; planetary atmospheres; mass spectroscopy; stratospheric cluster ion composition.

Izen, Joseph M., Ph.D., Harvard, 1982. Elementary particles, charm, bottom, and τ decay, e^+e^- collider experiments, high-energy physics computing.

Lee, Mark, Ph.D., Stanford, 1991. Pure and applied condensed matter physics, science and engineering of novel electronic and optical materials and electronic and photonic device engineering.

Leslie-Pelecky, Diandra L., Ph.D., Michigan State, 1991. Biomedical applications of magnetic nanomaterials; fundamental properties of magnetic materials.

Lou, Xinchou, Ph.D., State University of New York at Albany, 1989. Elementary particles physics, bottom and charm physics, e^+e^- colliders, offline software and distributed computing.

Rindler, Wolfgang, Ph.D., London, 1956. Special and general relativity; cosmology; spinors.

Salamon, Myron B., Ph.D., Berkeley, 1966. Experimental studies of unconventional superconductors, manganites and layered magnetic materials. Low-temperature physics, neutron and x-ray scattering.

Tinsley, Brian, Ph.D., Canterbury, 1963. Airglow; aurora; theoretical research in aeronomy; instrumentation for atmospheric spectroscopy.

Zakhidov, Anvar, Ph.D., Institute of Spectroscopy, U.S.S.R. Academy of Sciences, 1981. Nanotechnology; photonic crystals; carbon nanotubes; organic molecular crystals.

Associate Professors

Anderson, Phillip C., Ph.D., UTD, 1990. Ionospheric and magnetospheric electrodynamics; space weather; space environment effects on human systems.properties of materials.

Gartstein, Yuri, Ph.D., Institute for Spectroscopy, USSR Academy of Sciences, 1988. Condensed matter physics with emphasis on nanoscience; electronic, optical and transport properties of organic materials.

Lary, David J., Ph.D., Cambridge, 1991. Computational and information systems to facilitate discovery and decision support in earth system Science.

Assistant Professors

Ishak-Boushaki, Mustapha, Ph.D., Queen's University, 2002. Classical and modern cosmology; relativity; gravitational lensing (cosmic shear); cosmological models; computer algebra systems applied to relativity.

Malko, Anton V., Ph.D., New Mexico State/Los Alamos National Labs, 2002. Femtosecond laser spectroscopy of Nanomaterials such as semiconductor quantum dots, wires and wells; photoluminescence spectroscopy and microscopy; quantum optics; photoluminescence spectroscopy of single nanoparticles; solid state physics; laser physics.

Slinker, Jason D., Ph.D., Cornell, 2007. Organic optoelectronic devices and laboratory assays. Devices include light emitting electrochemical cells and electrochemical biosensors with DNA-modified electrodes.

Senior Lecturers

MacAlevey, Paul J., Ph.D., UTD.
Rasmussen, Beatrice, M.S., UTD.

RESEARCH SPECIALTIES AND STAFF

Atmospheric and Space Physics. Aeronomy; thermospheric, ionospheric and magnetospheric physics; planetary atmospheres. Earle, Heelis, Tinsley. Instrumentation and data analysis for various satellites and deep space probes; microphysics of clouds, climate. Anderson, Cunningham, Earle, Heelis, Hoffman. Atmospheric electricity. Thermal properties of airless planetary regoliths, distribution of volatiles in the Martian crust, misconceptions in physics and astronomy education, space science and physics educational outreach programs. Urquhart.

High Energy Physics. Charm, bottom, and τ decays at e^+e^- colliders; simulation of fixed target detectors for b physics. Chaney, Fenyves, Izen, Lou.

Optics. Quantum and nonlinear optics, single and multiphoton emission processes. Ultrafast laser spectroscopy. Malko.

Cosmology, Astrophysics, Relativity. gravitational radiation; exact solutions of Einstein's field equations. Rindler. Classical and modern cosmology; gravitational lensing (cosmic shear); cosmological models; computer algebra systems applied to relativity; Ishak-Boushaki.

Condensed Matter. Calculation of electronic structure by LCAO. Chaney. Raman, photoluminescence, and modulation spectroscopy of solids. Gartstein, Glosser. Unconventional superconductivity. Salamon. Magnetism; disordered and nanoscale magnets. Leslie-Pelecky; manganites and magnetic multilayers. Salamon. Femtosecond laser spectroscopy of materials, photoluminescence, absorption spectroscopy. Malko.

Nanoscience and Nanotechnology; photonic crystals; carbon nanotubes; organic molecular crystals. Gartstein, Zakhidov. Quantum semiconductor nanostructure, optical properties. Malko.

UNIVERSITY OF TEXAS, EL PASO

DEPARTMENT OF PHYSICS

El Paso, Texas 79968

Students Accepted For Degree	FIELDS		
	Physics	Astronomy	Related Fields
Doctorate			X
Master's	X		X

1. General

President: Diana Natalicio
Dean of Graduate School: Patricia Witherspoon
Department Chairman: Vivian Incera
Department Telephone Number: (915) 747-5715
Type of Institution: University
Control: Public
Setting: Urban
Total Faculty: 1,158
Total Students: 20,198
Total Graduate Faculty: Not Available
Total Graduate Students: 3,097
Annual Graduate Tuition:
In-state residents: $2,503/sem.
Out-of-state residents: $5,298/sem.*
Tuition rates for: 2010–11
Deferred tuition plan: Yes
Annual Other Fees: **
Term: Semester

*Teaching assistant pay in-state tuition
**See Graduate Catalog for detailed breakdown

2. Number of Faculty in Department

The combined total of full-time faculty in the three professorial ranks is 14. The combined total of full-time, part-time, and other faculty at all ranks is 22.

3. Admission, Financial Aid, and Housing

Address admission inquiries to: Vivian Incera, Physics Department
Graduate application fee required: $45 ($80 for international students)
Admission deadline (Fall Admission): 8/1
Admission information: For fall admission, 2009–10, 11 students were accepted.
Admission requirements: For admission to the graduate programs, a Bachelor's degree in physics or equivalent is required with a minimum undergraduate GPA of 3.0 specified. The GRE is suggested. Students from non-English speaking countries are required to demonstrate proficiency in English via the TOEFL exam. Minimum acceptable score for admission is 550.
Undergraduate preparation assumed: Symon, *Mechanics*; Pugh and Pugb, *Principles of Electricity and Magnetism*; Park, *Introduction to Quantum Theory.*
Address financial aid inquiries to: Vivian Incera, Physics Department
GAPSFAS application required: No
Financial aid deadline: None
Loans available: Yes
Address housing inquiries to: Director of Housing
On-campus, single student housing available: Yes
Cost/month: $540, single; $375, two-person efficiencies; $515, two-bedroom; $490, four-bedroom. New for fall 2010: $475, on bdrm apt; $450, two bdrm apt. This is per person. All utilities are paid and includes basic cable, internet, and a phone for local calls.
On-campus, married student housing available: Yes
Cost/Month: Varies

Table A—Faculty, Enrollments, and Degrees Granted

Research Specialty	2008-09 Faculty	Enrollment[1] Fall 2009		No. of Degrees Granted[2] 2009-10 (2005-09)			Median No. of Years for 2009-10 Ph.D.'s
		Master's	Doctorate	Master's	Terminal Master's	Doctorate	
Astronomy	0	0	–	1(0)	0(0)	–	–
Biophysics	1						
Environmental Physics	1	1	–	1(0)	0(0)	–	–
Geophysics	0	0	–	1(0)	0(0)	–	–
Gravity	0	0	–	1(0)	0(0)	–	–
Materials	4	2	–	0(0)	0(0)		
Medical Physics	1						
Nuclear Physics	1	2	–	1(0)	0(0)	–	–
Non-Specialized	0	18	–	0(0)	0(0)	–	–
Radars	0	1	–	0(0)	0(0)	–	–
Solid State	3	2	–	0(0)	0(0)	–	1
Phys. Ed	1	2	–	1(0)	0(0)	–	–
Total		28	–	6(0)	0(0)	–	–
Full-time Grad. Stud.		11	–				
Part-time Grad. Stud.		0	–				
First-year Grad. Stud.		6	–				
Median Years in Grad. Study (2009–10 Degrees)				2	2	–	–

Undergraduate Degrees, 2009–10 (2005–09): 4

[1]Students not yet committed to a research specially are entered under non-specialzed.
[2]Five-year totals in parentheses.

4. Graduate Degree Requirements

Master's: The department offers a program of courses and research leading to the degree of M.S. in Physics. Two routes may be taken. Plan 1 requires 30 semester hours of credit: 24 hours of course work plus a 6 hour thesis (Physics 5398 and 5399). Plan 2 requires 36 hours of course work, including a successful completion of a research problem (Physics 5391) being substituted for a thesis. A grade average of B is required. There are no foreign language, comprehensive, or qualifying exam requirements.
Doctorate: The Department of Physics participates in two interdisciplinary doctoral programs in materials science and engineering and in environmental science. Information about this program may be obtained from the Graduate Advisor, Department of Physics.
Other Programs: A Master of Science degree in geophysics is also offered. The requirements are basically the same as for the Master of Science degree in physics; a total of 30 semester hours of credit, 24 hours of course work, plus a six hour thesis.
Thesis: Thesis may be written *in absentia*.

Table B—Appointments to Graduate Students, 2008–09

Title of Appointee	Appointments Total	Appointments First year	Academic Load Allowed in Credit Hours	Hours of Service Per Week	Stipend for Academic Year ($)
			Semester		
Teaching Assistant	11	6	10	20	10,276
Fellow	3	2	10	20	10,000–12,000
Total	14	8			

5. Personnel Engaged in Separately Budgeted Research, 7/08–6/09

Professorial faculty	6
Graduate students	6
Undergraduate students	4
Total	16

FACULTY

Professors

López, Jorge A., Ph.D., Texas A&M, 1986. Theoretical physics; nuclear physics.

Ferrer, Efrain J., Ph.D., P.N. Lebedev Physical Institute, Moscow, USSR. Theoretical physics.

Incera, Vivian, Ph.D., P.N. Lebedev Physical Institute, Moscow, USSR. Theoretical Physics.

Associate Professors

Botez, Cristian E., Ph.D., The University of Missouri, Columbia

Fitzgerald, Rosa M., Ph.D., U.C. Riverside, 1992. Condensed matter theory; environmental studies.

Hagedorn, Eric A., Ph.D., University of Wisconsin-Milwaukee. Testing and measurement in physics education.

Manciu, Felicia, Ph.D., State University of New York at Buffalo

Manciu, Marian, Ph.D., State University of New York at Buffalo

Ravelo, Ramon, Ph.D., Boston University, 1990. Computational physics; condensed matter; nonlinear phenomena.

Assistant Professors

Baruah, Tunna,

Durandurdu, Murat, Ph.D., Ohio University

Li, Chunqiang, Ph.D., Princeton University, Princeton, NJ

RESEARCH SPECIALTIES AND STAFF

Theoretical

Environmental Physics. Fitzgerald.

Materials science. Baruah, Durandurdu, Ravelo.

Nuclear Physics. J. López.

Experimental

Acoustics. Cooper.

Materials Science. Botez, F. Manciu, M. Manciu.

Physics Education. E. Hagedorn

THE UNIVERSITY OF TEXAS AT SAN ANTONIO

DEPARTMENT OF PHYSICS AND ASTRONOMY

San Antonio, Texas 78249

Students Accepted For Degree	FIELDS		
	Physics	Astronomy	Related Fields
Doctorate	X		
Master's	X		

Preamble

The graduate program in Physics at UTSA includes a partnership between three institutions: the Department of Physics and Astronomy at the University of Texas at San Antonio (UTSA), the Department of Physics and Astronomy at the University of Texas at Brownsville (UTB) and the Space Science and Engineering Division at the Southwest Research Institute (SwRI) in San Antonio. The Department of Physics and Astronomy at the University of Texas at San Antonio is the lead institution and the degree granting institution. Selected faculty members from UTB and SwRI are appointed as adjoint faculty members and can Chair dissertation committees as well as serve as Thesis and Dissertation advisors for MS and PhD students, respectively.

1. General

President: Ricardo Romo
Dean of Graduate School: Dorothy Flannagan
Department Telephone Number: (210)-458-5454
Type of Institution: University
Control: Public
Total Faculty: 1,307
Total Graduate Faculty: 648
Total Students: 28,955
Total Graduate Students: 3,678
Annual Graduate Tuition:
In state residents: Full Time 6,985/yr
Part time 5,197/yr
Out of state residents: Full Time 18,619/yr
Part-time 13,507/yr
Tuition rates: 2009–2010
Deferred Tuition plan: yes
Department website: http://physics.uts.edu

2. Number of Faculty in Department

The combined number of faculty at the three professional ranks including adjoint faculty from Univ. of TX at Brownsville and SwRI is 37. The faculty distribution is the following: 14 Tenure and Tenure-Track (T-TT) faculty at UTSA, 14 T-TT adjoint faculty at UTB and 9 adjoint faculty at SwRI. The combined total of full-time, part-time, and other faculty at all ranks is 41.

3. Admission, Financial Aid, and Housing

Official Transcripts and official general GRE scores: are required and have to be submitted before the application deadline. There are no minimum scores to be eligible to apply. The Physics GRE is not required but is strongly recommended. For international students the TOEFL or the IELTS score is required. The minimum required TOEFL score is 61 for Master's and 79 for PhD applicants. The minimum required IELTS score is 5 for Master's and 6.5 for PhD applicants.

Address admission inquiries to: Dr. Lorenzo Brancaleon (lorenzo.brancaleonutsa.edu)
Graduate Application Fee: Online Application Paper Application UTSA Graduates or
Degree Candidates—online $30—paper $35 : Non-UTSA graduates
online $45—paper $50 : International applicants
online $80—paper $85 : Late Fee
For Late Application (this fee is applicable to all) $10

Table A—Faculty, Enrollments, and Degrees Granted

Research Specialty	Faculty	Enrollment[a] Fall		No. of Degrees Granted			Median No. of Years for Ph.D.'s
		Master's	Doctorate	Master's	Terminal Master's	Doctorate	
Astrophysics	1		3	0(0)	0(0)		
Biophysics	4	1	6	0(0)	0(0)		
Condensed Matter and Materials Science	5	3	13	0(2)	0(1)		
Nanoscience	3	2	8	0(0)	1(0)		
Space Physics	10	4	9	2(0)			
Computational Physics	1		2	0(0)	0(0)		
Non specialized		15	7				
Total	24	25	48				

[1]Students not yet committed to a research specialty are entered under non-specialized.
[a]For other specialties please refer to the university of Texas at Brownsville data

4. Graduate Degree Requirements

Master's: The Master of Science program requires the successful completion of a minimum of 30 semester credit hours including 12 hours of core classes, 6 hours of Directed Research and 3 hours of research seminars.

Doctorate: The doctoral degree requires a minimum of 81 semester credit hours beyond the baccalaureate degree. The coursework in the Program of Study includes a Core Curriculum (12 semester credit hours) and Advanced Electives (27 semester credit hours) including graduate courses offered by other departments with the approval of the student's Graduate Advisor. Research hours, including Research Seminar (3 semester credit hours), Directed and Doctoral Research (27 semester credit hours) and Dissertation (12 semester credit hours), totaling at least 42 semester credit hours, complete the Program of Study.

Special equipment facilities or programs: atomic layer deposition, sputterer, two clean rooms, surface profiler, XRD, thin film stress measurement, vector network analyzer, hot embosser, ellipsometer, tensile tester, porosimeter, three atomic force microscopes, FTIR spectrometers, inverted microscopes, micro-Raman spectrometer, two systems for pulsed laser deposition, barrel etcher, LAM etcher, three scanning electron microscopes, two transmission electron microscopes, two mass spectrometers, two facilities for femtosecond spectroscopy and imaging, circular dichroism spectrometer, fluorescence spectrometers, UV-Vis absorption spectrometers, system for picoseconds fluorescence lifetime detection, Raman spectrometer.

Table B—Appointments to Graduate Students, 2009-10

Title of Appointee	Appointments Total	Appointments First year	Academic Load Allowed in Credit Hours	Hours of Service Per Week	Stipend for Academic Year ($)
Teaching Assistant	5		9	20	25,000
Research Assistants	6		9	20	$25,000
Others	1				
Total	12				

5. Personnel Engaged in Separately Budgeted Research,

Professorial faculty	14
Other faculty	12
Postdoctoral appointments	14
Graduate Students	40
Undergraduate Students	10
Visiting Researchers	5
Total	95

6. Separately Budgeted Research Expenditures

	Departmental Research	Research Outside the Department
FY 2008	1,387,312	
Total	1,387,312	

FACULTY

Professors

Ayon, Arturo, PhD, Cornell University, 1996, Sensor arrays, Micro-chemical reactors, The demonstration of miniaturized muon sensors, Micropropulsion employing solid fuels, Negative index of refraction materials for imaging and other photonic applications, The utilization of MEMS actuators on phase array antennas, CMOS-compatible microwave varactors and other radio-frequency projects.

Chen, Chonglin, PhD, Pennsylvania State University, 1994, electronic thin films, surface science and interface phenomena, nanostructures and nanophenomena, advanced materials, nano/micro scale characterizations, and novel device fabrications.

Chen, Liao Y., PhD, Academia Sinica, China, 1988, Biological Physics, Chemical Physics, Condensed Matter Physics

Nash, Patrick L., PhD, University of North Carolina at Chapel Hill, 1981, cosmology (unification of inflation dark matter and dark energy); modeling of nano-materials; mathematical physics; computational physics

Sardar, Dhiraj K., PhD, Oklahoma State University, 1980, optical properties of a variety of technologically important materials using high resolution laser spectroscopy techniques, optical properties of biological tissues as well as studying the laser-tissue interaction, an exciting area of biophysics.

Jose-Yacaman, Miguel, PhD, National University of Mexico, Mexico, 1973, Chairman of the Department, structure and properties of nanoparticles including metals, semiconductors, and magnetic materials. Synthesis and characterization of new materials most of them nanoparticles, surfaces and interfaces, defects in solids, electron diffraction and imagining theory, quasicrystals, archaeological materials, and catalysis.

Associate Professors

Brancaleon, Lorenzo, PhD, University of Parma, Italy, 1996, molecular biophysics, optical spectroscopy, biomaterials, photobiophysics.

Chabanov, Andrey A., PhD, City University of New York, 2002, microwave properties of magnetic photonic crystals and their applications in antennas, propagation and localization of microwaves in random cavities and waveguides, fabrication and optical properties of photonic band gap materials for photonics applications, photon localization and lasing in disordered microstructures

Schlegel, Eric M., PhD, Indiana University, 1983, Associate Professor, X-ray line emission from cataclysmic variables, supernovae

Assistant Professors

Koinov, Zlatko, PhD, St. Petersburg Electrotechnical University, Russia, 1999, Nanophysics: Optical properties of single and couple quantum wells. Bose-Einstein condensation: Formation of Bose-Einstein condensate of excitons; polariton condensation in microcavities: the effects of the symmetry-breaking disorder on the condensate; collective excitations in superconductors Magnetooptics: Quantum-well excitons in high magnetic fields; Bethe-Salpeter equations for magnetoexcitons; Confinement of magnetoexcitons in quantum wells due to inhomogeneous magnetic fields. Strongly correlated systems: Hubbard model; High-temperature superconductivity.

Lopez-Lozano, Xochitl, PhD, Universidad Autonoma de Puebla, Mexico, 2005, structural, electronic, optical and catalytic properties of materials at the nanometer scale through semi-empirical and ab initio density-functional theory calculations

Marucho, Marcelo, PhD, National University of La Plata, Argentina, 2002, theoretical and computational research in chemical physics, biophysics, and polymer physics, mathematical physics, with particular emphasis on the development of new analytical and numerical tools.

Nash, Kelly, PhD, University of Texas at San Antonio, 2009, applications of fluorescent polymer composites, diodes, lighting displays, optical biosensors, drug delivery

Peralta, Xomalin G., PhD, University of California at Santa Barbara, 2002, terahertz spectroscopy, biophysics, nanoparticles.

Adjoint Professors from SwRI (Space Science)

Allegrini, Frédéric, Adjoint Assistant Professor, Ph.D., University of Bern, Switzerland

Boice, Daniel, Adjunct Associat Professor, Ph.D., New Mexico State University

Desai, Mihir I., Lead Adjoint Professor, Ph.D., University of Birmingham, UK

Goldstein, Jerry, Adjoint Assistant Professor, Ph.D., Dartmouth College

Jahn, Jorg-Micha, Adjoint Assistant Professor, Ph.D., Dartmouth College

Livi, Stefano Adjoint Professor, Ph.D., University of Florence (Italy).

McComas, David J., Adjoint Professor, Ph.D., University of California at Los Angeles
Pollock, Craig, Adjoint Professor, Ph.D., University of New Hampshire
Valek, Phillip, Adjoint Assistant Professor, Ph.D., Auburn University
Waite, Jack H., Adjoint Professor, Ph.D.

Senior Lecturer

Konno, Ishiro, Ph.D., Arizona State University.
Koynova, Aeta, Ph.D., Loffe Physico-Technical Institute, St. Petersburg, Russia.
Lopez-Mobilia, Rafael, Ph.D., University of Texas at Austin.

UNIVERSITY OF TEXAS HEALTH SCIENCE CENTER AT SAN ANTONIO

DEPARTMENT OF RADIOLOGY

San Antonio, Texas 78229-3900

Students Accepted For Degree	FIELDS Physics	Astronomy	Related Fields
Doctorate	X		X
Master's	X		X

1. General

President: William L. Henrich, MD
Dean of Graduate School: Robert L. Reddick, MD
Program Director: Geoffrey D. Clarke, Ph.D.
Department Telephone Number: (210) 567-5550
Website: http://radsci.uthscsa.edu
Type of Institution: Health University
Control: Public
Setting: Urban
Total Faculty: 1,562
Total Graduate Faculty: 224
Total Students: 3,060
Total Graduate Students: 338
Annual Graduate Tuition:
In-state residents: Full-time—$128/credit
Part-time—$128/credit
Out-of-state residents: Full-time—$464/credit
Part-time—$464/credit
Tuition rates for: 2010–11
Deferred tuition plan: Yes
Other Fees: *Lab Fees* None
Term: Semester

2. Number of Faculty in Department

The combined total of full-time faculty in the three professorial ranks is 33. The combined total of full-time, part-time, and other faculty at all ranks is 69.

3. Admission, Financial Aid, and Housing

Address admission inquiries to: Admissions & Records, UTHSCSA, MSC 7702, 7703 Floyd Curl Dr., San Antonio, Texas 78229-3900
Graduate application fee required: $10
Admission deadline (Fall admission): March 1, 2011
Admission information: For fall admission, 2010–11, 13 students were accepted from 116 applicants.
Admission requirements: For admission to the graduate programs, a Bachelors degree in natural science or engineering is required with a minimum undergraduate GPA of 3.0 specified. The GRE is required. The average GRE scores for admissions was 1,225. Students from non-English speaking countries are required to demonstrate proficiency in English via the TOEFL exam. Minimum acceptable score for admission is 560 (paper test), 220 (computer test), 68 (iNet test).
Admission Procedure: Decisions on acceptace of applications are made by the Radiological Sciences Committee on Graduate Studies (COGS). The Application Review Committee will present the candidates for acceptance and rejection to the COGS. By majority vote, a student may be recommended for admission to the Dean of the Graduate School of Biomedical Sciences. The applicants are informed, in writing, by the Dean of the action taken on their application and of any contingencies imposed upon acceptance. Applications that are incomplete by the deadline will be rejected but reapplication may be made after all materials are received.

The Graduate Program in Radiological Sciences uses the following guidelines, approved by COGS on June 17, 1992, for selection of applicants: 50% from Texas, 25% outside Texas, 25% foreign.

Applicants must have undergraduate credit for the following courses: Biology: Two semesters of general biology (two years of Radiation Biology). Chemistry: Two semesters of general chemistry (through biochemistry for Radiation Biology). Physics: Two years of general physics (two semesters for Radiation Biology). Students enrolling in a medical physics track should have an undergraduate degree in physics or a related engineering or physical science with coursework equivalent to a minor in physics (includes at least three upper level undergraduate courses). Applicants with deficiencies in their physics background may be conditionally admitted to a medical physics track, but remedial education in physics shall be completed by the student prior to admission into candidacy.
Address financial aid inquiries to: Dir. of Financial Aid, UTHSCSA, Rm. 318L, 7703 Floyd Curl Dr., San Antonio, Texas 78229-3900.
GAPSFAS application required: Yes
Financial aid deadline: April 1
Loans available: Yes
Address housing inquiries to: Private housing available near campus.

Table A—Faculty, Enrollments, and Degrees Granted

Research Specialty	2009–10 Faculty	Enrollment Fall 2009 Master's	Doctorate	No. of Degrees Granted[1] 2008–09 (2007–08) Master's	Terminal Master's	Doctorate	Median No. of Years for 2008–09 Ph.D.'s
Biophysics	9	–	2	–	–	–	–
Human Imaging	8	–	12	–	–	1(0)	.5
Medical & Health Physics	16	9	33	4(5)	–	7(7)	.5
Total		9	47	4(5)	–	8(7)	–
Full-time Grad. Stud.		9	41				
Part-time Grad. Stud.		0	6				
First-year Grad. Stud.		6	4				

Median Years in Grad. Study (2008–09 Degrees) PhD degree: 5 yr; MS degree: 2.5 yr
Undergraduate Degrees, 2001–02 (1997–02): N/A

[1]Five-year totals in parentheses.

4. Graduate Degree Requirements

Master's: Completion of all required courses (GPA greater than 3.0), completion of 30 graduate credits, successful completion of parts I and II of the qualifying examination, presentation and defense of a Masters Thesis.
Doctorate: Same as above plus: successful completion of part III

of Qualifying exam (NIH Style research proposal and defense), presentation and defense of Ph.D. Dissertation.

Thesis: Thesis may be written *in absentia*.

Special Equipment, Facilities, or Programs: Radiology Dept.: PET/SPECT/CT small animal imager; micro-ultrasound; atomic force microscopy, TOF mass spectrometer; x-ray photoelectron spectrometer; clinical MRI and Interventional Radiology (RF ablation).

Research Imaging Center: 3 PET scanners; 3 MRI systems; 2 cyclotrons; electrophysiology lab; transcranial magnetic stimulation.

Cancer Therapy & Research Center: 2 CT scanners; 7 linear accelerators; 3 treatment planning systems; TLD dosimetry; BANG gel system.

Additional research and clinical practice opportunities are available at Wilford Hall Medical Center, Brooke Army Medical Center and the Veterans Administration Hospital.

Table B—Appointments to Graduate Students, 2009–10

Title of Appointee	Appointments Total	Appointments First year	Academic Load Allowed in Credit Hours	Hours of Service Per Week	Stipend for Academic Year ($)
			Semester		
Teaching Assistant	34	6	9	20	24,783
Fellows	5	1	9	20	varies
Total	39	7			

5. Personnel Engaged in Separately Budgeted Research, 7/06–6/07

Professorial faculty	33
Postdoctorial appointments	5
Graduate students	65
Nonteaching research personnel	19
Total	122

6. Separately Budgeted Research Expenditures by Source of Support

	Departmental Research	Physics-related Research Outside Department
Federal government	$638,797	$4,600,433
Private, non-profit organizations	254,519	
Business and industry	266,834	6,560,189
Other Include institution's own separately budgeted accounts	851,500	3,404,239
Total	$3,011,650	$14,564,861

Table C—Separately Budgeted Research Expenditures

Research Specialty	No. of Grants	Expenditures ($)
Biophysics	2	483,853
Chemical Physics	3	403,100
Extramural Student Research	4	250,697
Human Imaging	6	1,799,217
Medical & Health Physics	3	3,566,035
Total	18	6,502,902

FACULTY

Professors

Bower, James, Ph.D., University of Wisconsin, 1981.
Chinitipalli, Kedar, M.D., Guntar Medical College, India, 1975.
Clarke, Geoffrey, Ph.D., UT Southwestern, Dallas, 1984.
Fox, Peter, M.D. Georgetown University, 1979.
Glickman, Randolph D., Ph.D., University of Toronto, 1978.
Goins, Beth A., Ph.D., Univ. of Tennessee, Knoxville, 1988.
Lancaster, Jack L., Ph.D., Univ. TX Health Sci. Center Dallas, 1978.
McDavid, William, Ph.D., Univ. TX Health Sci. Center San Antonio, 1976.
Phillips, William T., M.D., Univ. TX Medical Branch at Galveston, 1980.
Papanikolou, Nikos, Ph.D., Univ. Wisconsin, Madison, 1994.
Robin, Donald, Case Western Univ., 1984
Sprague, Eugene, Ph.D., Univ. TX Health Sci. Center San Antonio, 1979.

Associate Professors

Davis, M. Duff, Ph.D., University of Zurich, 1985.
Hatab, Mustapha, Ph.D., UT Southwestern, Dallas, 1995.
Jerabek, Paul, Ph.D., University of California, Irvine, 1982.
Natarajan, Mohan, Ph.D., University of Madras, India, 1986.
Wiatrowski, Wayne A., Ph.D., Univ. TX Health Sci. Center San Antonio, 1979.
Suri, Rajiu, M.D., Poonjab Univ., 1991.

Assistant Professors

Brewer, Patricia, Ph.D., Uniformed Services Univ. of the Health Sciences, 1984.
Charlton, Michael, Ph.D., Texas A&M Univ. College Station, 2001.
Esquivel, Carlos, Ph.D., Univ. of TX Health Science Ctr., San Antonio, 2005.
Gutiurrez, Alonso, Univ. of Wisconsin, 2007.
Hardies, L. Jean, Ph.D., Univ. TX Health Sci. Center San Antonio, 1986.
Kochunov, Peter, Ph.D., Univ. TX Health Sci Center, San Antonio, 2000.
Laird, Angela, Univ. of Wisconsin, 2002.
Narayana, Shalini, Ph.D., University of Iowa, 1996.
Peng, Qi, Ph.D., U Texas Southwestern, Dallas, 2005.
Shi, Chenyu, Ph.D., Rensselaer Polytechic Institute, 2004.
Stathakis, Sotitios, Ph.D., Univ. of Ratras, Hellas, Greece, 2003.
Swanson, Gregory, MD, Univ. Texas Med. School, Houston, 1991.
Wicha, Nicole, Ph.D., Univ. California, San Diego, 2002.

Adjunct Associate Professors

Bice, William, Ph.D., University of Florida, 1985.
Goff, David L., Ph.D., University of Texas, Austin, 1969
Leidholdt, Edwin, Ph.D., University of Virginia, 1983.
Levy, Louis, Ph.D., UT Health Sci Cntr, San Antonio, 1974.
Marbach, James, Ph.D., Univ. TX Health Science Center, Houston, 1978.
Prestidge, Bradley, M.D., Uniformed Services Univ. of the Health Sci., 1985.
Watts, Ronald, Ph.D., Univ. of TX Health Science Ctr., San Antonio, 1994.

Adjunct Assistant Professors

Blough, Melissa, Ph.D., Univ. TX Health Sci. Center San Antonio, 1999.
Cheng, Chih-Yao, Ph.D., UT Health Sci Center, San Antonio, 2006.
Feng, Ching-mei, Ph.D., Univ. of TX Health Science Ctr., San Antonio, 2003.
Goff, David, Ph.D., University of California, Los Angeles, 1995.
Keener, Carl, Ph.D., Univ. TX Health Sci. Center San Antonio, 1996.
Lee, Nina, Ph.D., Univ. TX Health Sci. Center San Antonio, 1998.
Ozus, Bahadir, Ph.D., UT Health Sci Center at San Antonio, 2004.
Payne, William, Ph.D., UT Health Sci Center at San Antonio, 1977.
Sadeghi, Amir, Ph.D., UT Health Sci Center at San Antonio, 1996.
Prete, James, Ph.D., Univ. TX Health Sci. Center San Antonio, 1998.
Shriver, Christy, M.S., Texas A & M University, 1972.
Tucker, Jonathan, Ph.D., Univ. TX Health Sci. Center San Antonio, 2001.
Vail, Neal, Ph.D., UT Austin, 1994.
Wang, Minghong, New Mexico St. U., 1991.

RESEARCH SPECIALITIES AND STAFF

Experimental

Biological Chemistry. Goins, Hardies, Sprague.
Brain Function Imaging. Fox.
Chemistry of Radiopharmaceuticals. Goins, Phillips.
Laser Physics & Laser Applications in Medicine. Glickman.
Magnetic Resonance Imaging (MRI). Clarke, Fox, Peng, Lancaster.
Medical Electronic Imaging. Lancaster, Tucker.
Nuclear Medicine Imaging Physics. Lancaster, Phillips.
Positron Emission Tomography (PET). Fox, Jerabek.
Radiation Biology and Radiation Toxicology. Natarajan.
Tomotherapy, Brachytherapy, and Radiation Therapy Physics. Bice, Papanikolaou, Shi, Sotirios.
X-ray Beam Analysis & Characterization. Hatab, Tucker, Waggener.

Selected Recent Student Publications

Lin, L., Shi C., Eng T., et al. "Evaluation of inter-fractional setup shifts for site-specific helical tomotherapy treatments," Technol. Cancer Res. Treat. **8**(2): 115 (2009).
Giantsoudi, D., Stathakis S., Liu Y., et al. "Monte Carlo Modeling and Commissioning of a Dual-layer Micro Multileaf Collimator," Technol. Cancer Res. Treat. **8**(2): 105 (2009).
Page, L., Maswadi S., Glickman R., et al. "Optoacoustic imaging: an application to the detection of foreign bodies," Proc. SPIE **27**, 7177 (2009).
Sarkar, V., Shi C., Rassiah-Szegedi P., et al. "Feasibility study on the use of digital tomosynthesis with individually-acquired megavoltage portal images for target localization," J. of B.U. Oncol. **14**(1): 103 (2009).
Soundararajan, A., Bao A., Phillips W. T., et al. " [(186)Re] Lipsomal doxorubicin (Doxil): in vitro stability, pharmacokinetics, imaging and biodistribution in a head and neck squamous cell carcinoma xenograft model," Nucl. Med. & Biol. **36**(5): 515 (2009).
Lin, Al, Fox P.T., Yang Y., et al. " Evaluation of MRI models in the measurement of CMRO2 and its relationship with CBF," Magn. Reson. Med. **60**(2): 380 (2008).
Lin, L., Shi C., Liu Y., et al. "Development of a novel post-processing treatment planning platform for 4D radiotherapy," Technol. Cancer Res. Treat. **7**(2): 125 (2008).
Roland, T. Esquivel C., Stathakis S., Papanikolau N. "On the evaluation of the relative sensitivity of commercial TLD readers using well characterized TLD chips," J. B. U. Oncol. **13**(4): 547 (2008).
Su, FC., Shi C., Crownover R., et al. "Dosimetric impacts of gantry angle misalignment on prostate cancer treatment using helical tomotherapy," Technol. Cancer Res. Treat. **7**(4): 287 (2008).
Zavaleta, CL., Goins B. A., Bao A., et al. "Imaging of 186 Re-liposome therapy in ovarian cancer xenograft model of peritoneal carcinomatosis," J. Drug Targeting **16**(7): 626 (2008).

BRIGHAM YOUNG UNIVERSITY

DEPARTMENT OF PHYSICS AND ASTRONOMY

Provo, Utah 84602

Students Accepted For Degree	FIELDS		
	Physics	Astronomy	Related Fields
Doctorate	X	X	
Master's	X	X	

1. General

President: Cecil O. Samuelson
Academic Vice President: John S. Tanner
Department Chair: Ross L. Spencer
Department Telephone Number: (801) 422-4361 (C)
Department Fax Number: (801) 422-0553
Type of Institution: University
Control: Private
Setting: Small town
Total Faculty: 1,283
Total Graduate Faculty: 1,084
Total Students: 32,955
Total Graduate Students: 3,355
Annual Graduate Tuition:
In-state residents: Full-time—$5,420*
Part-time—$301 per hr.**
Out-of-state residents: Full-time—$5,420*
Part-time—$301 per hr.**
Tuition rates for: 2009–10
Deferred tuition plan: Yes
Other Fees: None
Term: Semester

*$10,840 if non-member of the Church of Jesus Christ of Latter-day Saints.
**$602 if non-member of L.D.S. Church.

2. Number of Faculty in Department

The combined total of full-time faculty in the three professorial ranks is 30. The combined total of full-time, part-time, and other faculty at all ranks is 38.

3. Admission, Financial Aid, and Housing

Address admission inquiries to: Dr. Ross L. Spencer, Chair, Department of Physics and Astronomy
Graduate application fee required: Yes
Admission deadline (Fall admission): 1/15
Admission information: For fall admission, 2009–10, 11 students were accepted from 17 applicants.
Admission requirements: For admission to the graduate programs, a Bachelor's degree in physics or astronomy, is required with a minimum undergraduate GPA of 3.0 in the last 60 hours of coursework. The GRE, subject, is required. Students from non-English speaking countries are required to demonstrate proficiency in English via the TOEFL or equivalent. Minimum acceptable score for admission is 580 (237 computer based).
Undergraduate preparation assumed: Fowles, Analytical Mechanics; Griffiths, Foundations of Electromagnetic Theory; Schroder, Thermal Physics; Griffiths, Quantum Mechanics; Hecht, Optics.
Address financial aid inquiries to: Director of Financial Aid, A-41 ASB
GAPSFAS application required: No
Financial aid deadline: 1/15
Loans available: Yes
Address housing inquiries to: Director of Housing, 100 SASB
On-campus, single student housing available: Yes*
Cost/semester: Room and board $1,015-4,090
On-campus, married student housing available: Yes*
Cost/month: 1 bdrm./$560–588; 2 bdrm./$615; 3 bdrm./$715

*Apply one year prior to arrival.

Table A—Faculty, Enrollments, and Degrees Granted

Research Specialty	2008–09 Faculty	Enrollment[1] Fall 2008 Master's	Enrollment[1] Fall 2008 Doctorate	No. of Degrees Granted[2] 2008–09 (2004–09) Master's	Terminal Master's	Doctorate	Median No. of Years for 2008–09 Ph.D.'s
Acoustics	3	5	3	0(11)	0	1(2)	9
Astrophysics	6	6	2	0(1)	0	0(1)	–
Atomic, Molecular, & Optical Physics	7	4	3	0(7)	0	1(2)	6
Condensed Matter Physics	7	4	3	1(1)	0	0(2)	–
Nuclear Physics	0	0	0	0(2)	0	0(1)	–
Plasma Physics & Fusion	3	0	1	0(3)	0	1(1)	9
Relativity & Gravitation	2	1	0	0(1)	0	0(1)	–
Other Theoretical/ Math.	2	1	3	0(3)	0	0(0)	–
Total	30	21	15	1(29)	0	2(10)	
Full-time Grad. Stud.		21	15				
Part-time Grad. Stud.		0	0				
First-year Grad. Stud.		9	2				
Median Years in Grad. Study (2008–09 Degrees)				2.5		8	–

Undergraduate Degrees, 2008–09 (2004–09): 57(325)

[1]Students not yet committed to a research specialty are entered under non-specialized.
[2]Five-year totals in parentheses.

4. Graduate Degree Requirements

Master's: Minimum of 24 hours course credit (7.5 may be transferred) plus thesis (6 credit hour minimum); thesis required; oral examination in support of thesis required; residence of 20 semester hours taken on Provo campus; minimum GPA: 3.0; no language requirement. Before admission to candidacy, a student must be accepted as a research student by a department faculty member and submit a proposed study list.
Doctorate: Semester hours requirement: 36 hours approved coursework (exclusive of graduate seminars) plus dissertation (18 hour minimum); minimum GPA: 3.0. Ordinarily, two years of full-time coursework must be taken on the Provo campus, of which two full-time semesters must be consecutive. Written qualifying examination in the first year plus oral exam in support of dissertation are required. Before admission to candidacy, a student must be accepted as a research student by a departmental faculty member, and submit a proposed study list.
Other Programs: The Ph.D. in Physics and Astronomy has the same requirements as for the Ph.D. in Physics except the

course requirements are appropriate to Astronomy and Astrophysics.

Thesis: Thesis may be written *in absentia*.

Special Equipment, Facilities, or Programs: Mountain observatory with 24-inch and 0.9-meter telescopes; ultra-short pulse laser system; laser cooling system; XUV reflectometers; on-campus supercomputers; atomic force and scanning tunneling microscopes; anechoic chamber, reverberation chamber.

Table B—Appointments to Graduate Students, 2008–09

Title of Appointee	Appointments: Total	Appointments: First year	Academic Load Allowed in Credit Hours	Hours of Service Per Week	Stipend for Academic Year ($)
			Semester		
Teaching Assistant	14	4	12	20	18,330–19,230[1]
Research Assistant	16	6	12	20	19,080–20,370[1,2]
Total	30	10			

[1]Varies with experience.
[2]An additional 18 assistantships of $5,720–6,210 are given during summer terms.

5. Personnel Engaged in Separately Budgeted Research, 2008–09

Professorial faculty	13
Graduate students	24
Undergraduate students	76
Nonteaching research personnel	1
Total	114

6. Separately Budgeted Research Expenditures by Source of Support (2008) Annual Year

	Departmental Research	Physics-related Research Outside Department
Federal government	1,201,000	$
Other	190,000	
University	936,000	
Total	$2,127,000	$

Table C—Separately Budgeted Research Expenditures

Research Specialty	No. of Grants	Expenditures ($)
Acoustics	4	339,000
Astrophysics	1	172,000
Atomic, Molecular, & Optical Physics	7	560,000
Condensed Matter Physics	5	513,000
Plasma Physics	0	40,000
Relativity	2	163,000
Theoretical Physics	0	340,000
Total	18	2,127,000

FACULTY

Professors

Allred, David D., Ph.D., Princeton, 1977. Lasers; x-rays; thin-film physics.

Bergeson, Scott D., Ph.D., Wisconsin, 1995. Laser cooling.

Berrondo, Manuel, Ph.D., Upsala, Sweden, 1973. Solid state theory.

Migenes, Victor, Ph.D., University of Pennsylvania, 1989. Astrophysics.

Moody, Joseph Ward, Ph.D., Michigan, 1986. Experimental astrophysics.

Peatross, Justin, Ph.D., University of Rochester. High intensity laser physics.

Rees, Lawrence B., Ph.D., Maryland, 1983. Medium and low energy nuclear physics.

Sommerfeldt, Scott D., Ph.D., Pennsylvania State University, 1989. Dean of the College. Acoustics and structural vibrations.

Spencer, Ross L., Ph.D., Wisconsin, 1979. Chair of the Department. Theoretical plasma physics.

Stokes, Harold T., Ph.D., Utah, 1977. Theoretical solid state physics.

Turley, R. Steven, Ph.D., Massachusetts Institute of Technology, 1984. Computational electromagnetics and atomic physics.

Associate Professors

Campbell, Branton J., Ph.D., UC Santa Barbara, 1999. Experimental condensed matter.

Christensen, Clark G., Ph.D., Cal. Tech., 1972. Astrophysics.

Colton , John S., Ph.D. University of California, Berkeley, 2000, Experimental condensed matter.

Davis, Robert C., Ph.D., Utah, 1996. Experimental condensed matter.

Durfee, Dallin S., Ph.D., Massachusetts Institute of Technology, 1999. Atomic Physics.

Hart, Grant W., Ph.D., Maryland, 1983. Plasma physics; lasers.

Hart, Gus, Ph.D., University of California, Davis, 1999. Computational condensed matter.

Hess, Bret C., Ph.D., Iowa State, 1988. Experimental condensed matter.

Hintz, Eric G., Ph.D., Brigham Young, 1995. Observational astrophysics.

Hirschmann, Eric W., Ph.D., University of CA, Santa Barbara, 1996. Theoretical and computational physics.

Leishman, Timothy W., Ph.D., Pennsylvania State University, 2000. Acoustics.

Neilsen, David, Ph.D., University of Texas at Austin, 1999. Theoretical and computational physics.

Peterson, Bryan G., Ph.D., Brigham Young, 1983. Laser physics.

Vanfleet, Richard, Ph.D., University of Illinois, 1999. Experimental condensed matter.

Van Huele, Jean-Francois, Ph.D., Brussels Free University, 1987. Theoretical quantum electrodynamics.

Ware, Michael, Ph.D., Brigham Young University, 2001. Quantum optics.

Assistant Professors

Chesnel, Karine, Ph.D., University of Joseph Fourier, 2002. Experimental condensed matter.

Gee, Kent L., Ph.D., Pennsylvania State University, 2005. Acoustics.

Stephens, Denise, Ph.D., New Mexico State, 2002. Astrophysics.

RESEARCH SPECIALTIES AND STAFF

Theoretical

Acoustics: Active control of sound and vibration; aeroacoustics; architectural acoustics; audio acoustics; sound-structure interaction; acoustical treatments; nonlinear acoustics. Gee, Leishman, Sommerfeldt.

Atomic and Molecular Physics. Computational study of the nonlinear interaction of intense waves with gases; computational studies of the reflection of electromagnetic waves from rough surfaces; quantum electrodynamics. Turley, Peatross, Ware.

Condensed Matter. Material Simulations, first Principles calculations, alloy modeling, electronic structure and excitations in nanosystems, the group-theoretical description of phase transitions, incommensurately-modulated structures. B. Campbell, Gus Hart, B. Hess, H. Stokes.

Plasma Physics. Basic plasma studies focusing on non-neutral plasmas and particle simulations. Spencer.

Relativity. Large-scale computation and numerical evolution of the dynamic behavior of very strong gravitational fields, including the description of gravitational collapse and the formation and evolution of black holes. Hirschmann, Neilsen.

Theoretical/Math. Quantum, relativity, and complexity; quantum foundations and information, spintronics, field theory, parallel computing, chaos, complex behavior, consensus/frustration models. Berrondo, Van Huele.

Experimental

Acoustics. Active control of sound and vibration; aeroacoustics; architectural acoustics; audio acoustics; sound-structure interaction; acoustical treatments; electroacoustics; nonlinear acoustics; passive noise control. Gee, Leishman, Sommerfeldt.

Astrophysics. Narrow-band photometry and spectrophotometry; broad-band photometry; galactic clusters; luminosity function of galaxies; metal abundances; variable stars; interstellar reddening. Christensen, Hintz, Moody, Stephens, Migenes.

Atomic Molecular Optical. Physics of laser-cooled gases and plasmas; laser photochemistry and plasma diagnostics; VUV optics and sources; ultrafast lasers; high harmonic generation; mater interferometry. Allred, Bergeson, Durfee, Peatross, Turley, Ware.

Condensed Matter. Nanofabrication, carbon nanotube microstructures, biomolecule-templated electronics, electron microscopy, semiconductor spin dynamics, optical spectroscopy of nanostructures, nanomagnetism, nanostructure-property relations, magnetometry, magnetic x-ray spectroscopy, x-ray diffuse scattering, XUV optical thin films. Allred, Campbell, Chesnel, Colton, Davis, hess, Vanfleet.

Nuclear Physics. Particle-induced x-ray emission (PIXE) spectroscopy and related ion beam analysis techniques. Rees.

Physics Education. Needs studies; curriculum development; course development; instructional strategies. Rees.

Plasma Physics. Basic plasma studies focusing on non-neutral plasmas. Grant, Hart, Peterson.

UNIVERSITY OF UTAH

DEPARTMENT OF PHYSICS AND ASTRONOMY

Salt Lake City, Utah 84112
http://www.physics.utah.edu

Students Accepted For Degree	FIELDS		
	Physics	Astronomy	Related Fields
Doctorate	X		X
Master's	X		X

1. General

President: Michael Young
Dean of Graduate School: Charles Wight
Department Chairman: David Kieda
Department Telephone Number: (801) 581-6901 (C)
Type of Institution: University
Control: Public
Setting: Urban
Total Faculty: 3,582
Total Tenured Faculty: 1,495
Total Students: 29,284
Total Graduate Students: 7,418
Annual Graduate Tuition: *
In-state residents: Full-time—$2,352 (9 hr/sem, tuition + fees)
Out-of-state residents: Full-time—$7,418 (9 hr/sem, tuition + fees)
Tuition rates for: 2009–10
Deferred tuition plan: No
Other Fees: Some labs require special fees
Term: Semester

*Sliding scale per semester

2. Number of Faculty in Department

The combined total of full-time faculty in the three professorial ranks is 35. The combined total of full-time, part-time, and other faculty at all ranks is 72.

3. Admission, Financial Aid, and Housing

Address admission inquiries to: Graduate Secretary, Department of Physics and Astronomy, 115 South 1400 East #201, Salt Lake City, UT 84112-0830, admissions@physics.utah.edu
Graduate application fee required: $45; foreign students—$65
Admission deadline (Fall admission): February 1
Admission information: For fall admission, 2009, 30 students were accepted from 168 applicants.
Admission requirements: For admission to the graduate programs, a Bachelor's degree in physics is required with a minimum undergraduate GPA of 3.0 specified. The GRE Physics is required. No minimum specified. The GRE Advanced is required. A minimum score of 640 is recommended. Students from non-English speaking countries are required to demonstrate proficiency in English via the TOEFL exam. Minimum acceptable score for admission is 500, recommended—600.
Undergraduate preparation assumed: Marion, *Electricity and Magnetism*; Saxon, *Quantum Mechanics*; Marion, *Classical Dynamics*; Tipler, *Modern Physics*.
Address financial aid inquiries to: Graduate Secretary, Department of Physics and Astronomy, 115 South 1400 East #201, Salt Lake City, UT 84112-0830, admissions@physics.utah.edu
GAPSFAS application required: No
Financial aid deadline: 2/1
Loans available: Yes
Address housing inquiries to: Office of Residential Living, 5 Heritage Center, Salt Lake City, UT 84112-2036 http://www.housing.utah.edu/sage.html
Graduate student single housing varies from $3,096–4,116/academic year includes utilities, internet, cable TV.

Table A—Faculty, Enrollments, and Degrees Granted

Research Specialty	2009–10 Faculty	Enrollment[1] Fall 2009 Mas-ter's	Enrollment[1] Fall 2009 Doc-torate	No. of Degrees Granted[2] 2009–10 (2003–10) Mas-ter's	Terminal Master's	Doc-torate	Median No. of Years for 2009–10 Ph.D.'s
Acoustics	1	0	2	0(3)	0(0)	1(1)	–
Astronomy	5	0	0	0(0)	0(0)	0(0)	–
Astrophysics	6	0	5	0(3)	0(0)	0(2)	–
Atmos./Space Phys., Cosmic Rays	10	0	16	3(6)	0(0)	4(9)	–
Atomic, Molecular, & Optical Physics	1	0	3	0(1)	0(0)	1(4)	–
Biophysics	3	0	4	0(0)	0(0)	0(1)	–
Chemical Physics	1	0	1	0(0)	0(0)	0(1)	–
Condensed Matter Physics	16	0	41	4(17)	0(0)	8(22)	–
Medical & Health Physics	3	0	7	0(0)	0(0)	0(0)	–
Particles & Fields	2	0	0	0(4)	0(0)	1(1)	–
Physics Education	3	0	0	0(0)	0(0)	0(0)	–
Relativity & Gravitation	0	0	2	0(2)	0(0)	0(0)	–
Other Theoretical/ Math.	0	0	2	0(1)	0(0)	0(5)	–
Other (specify) Instrumentation	0	7	0	4(18)	0(0)	0(0)	–
Non-specialized	0	3	30	0(0)	0(0)	0(0)	–
Total		10	111	11(69)	0(0)	15(46)	
Full-time Grad. Stud.			96				
Part-time Grad. Stud.		7	1				
First-year Grad. Stud.		3	14				
Median Years in Grad. Study (2007–08 Degrees)		2	5				

Undergraduate Degrees, 2007–08 (2000–08): 24(108)

[1] Students not yet committed to a research specialty are entered under non-specialized.
[2] Five-year totals in parentheses.

4. Graduate Degree Requirements

Master's: 30 graduate-semester-hours with a 3.0 grade average, in an approved program with satisfactory performance on Departmental Common Exam. Either thesis or nonthesis M.S. available. *Master's of Instrumentation*: 30 graduate semester hours with a 3.0 grade average. Nine to fifteen hours will be related to the instrumentation project. No language required. For admission, a Bachelor's degree in engineering, biology, chemistry, or some related field may be substituted for a degree in physics.
Doctorate: 45 graduate-semester-hours required. Satisfactory

performance on Departmental Common Exam or GRE Physics required for admission to Ph.D. program (no set minimum). Satisfactory performance (3.0 average) in an approved course program is required. Qualifying exam, dissertation, and dissertation exam required. Teaching experience required, and one of last two years must be in residence. No language requirement.

Other Programs: Interdisciplinary studies available in chemical physics, and a variety of other areas by special arrangement. The Ph.D. program in Medical Physics is an inter-disciplined program in which complex medical and biological stystems are studied using physics-based techniques and models. M.A. requirements are the same as M.S., except proficiency in one foreign language is required. M.Phil. requirements the same as Ph.D., except no dissertation required.

Thesis: Thesis may be written *in absentia*.

Special Equipment, Facilities, or Programs: The Department maintains extensive facilities for teaching and research. The INSCC Building (Intermountain Network & Scientific Computation Center) houses 7 state of the art laser labs on the first floor run by several condensed matter and biophysics experimental groups. The research of one group focuses on time-resolved and steady state (cw) investigations of photo-excitations in solids, particularly in semiconductors. A state of the art Ti:sapphire laser gives time resolution of 10 fs. Current efforts in this femtosecond lab include pulse amplification with a Nd:YAG laser and the generation of continuum pulses from the near-infrared to the UV. In the picosecond lab, two tunable synchronously pumped dye lasers are used for further photoexcitation studies. A 2D streak camera is used to measure the photoluminescence spectrum evolution with picosecond resolution. The laser laboratory has also been used to study optical non-linear spectra in electronic polymers, solids and other semiconductors. This includes spectra of two-photon absorption, non-linear refractive index and third harmonic generation. Light absorption is measured from the UV to the far IR using self-contained commercial instruments (Cary 17 DX and Bruker IFS88, both recently upgraded with modern electronics), operated either by reserachers or as a service provided by members of the technical staff. The single molecule spectroscopy group runs a low temperature laser microscopy lab centering around a helium cryostat and a one-box femtosecond laser system with wide (680nm- 1080nm) automated tunability. An FEI NovaNano Field emission Scanning Electron Microscope with 1.0 nm resolution (1.6 nm at 1 keV or low vacuum) is widely used for imaging. EDS analysis and e-beam lithography, a Leo 440i SEM is used for images requiring extremely large depth of field as well as for teaching. The Scanning Probe Microscopy group has many scanning probe microscopes, including several Atomic Force Microscopes, a Scanning Tunneling Microscope, two Near-field Optical Microscopes, a Scanning Capacitance Microscope and an ultra-high vacuum AFM/STM system. Two new biophysics labs are under construction. The first lab is located in the INSCC building. The focus of the group is on single molecule studies of molecular motor activity and other protein interactions. Equipment includes (or will shortly include) a high-resolution optical microscope with optical trapping and fluorescence capabilities, as well as auxiliary biological research equipment (e.g. low-temperature refrigeration facilities and a Beckman TL-100 ultracentrifuge). The second Biophysics lab is under construction in the James Fletcher Building. This lab focuses on understanding the mechanism of enveloped virus budding using single molecule, fluorescence spectroscopy and high resolution live cell imaging technologies. A new iMIC digital microscope would be installed that is capable of confocal, TIRF, live cell imaging and fluorescence correlation spectroscopy. In addition, the Department operated a fully equipped Opto-Electronic Materials Laboratory for chemical synthesis (including organic semiconductors not commercially available), purification, growth of single crystals, vacuum/controlled atmosphere annealing, sample cutting and polishing, thin-film deposition via thermal evaporation of rf sputtering, as well as a wide variety of techniques for chemical and physical characterization. A low-temperature AFM/STM (5 Kelvin) from Omicron Nanotechnology should arrive by the end of 2010.

The Astronomy and Astrophysics Group consists of nine full-time faculty members who are leading research programs at world-class astronomical factilities. As full institutional members of the Third Sloan Digital Sky Survey, the astronomy research group pursues an active research program in the BOSS, APGEE, and SEGUE surveys. The astronomy group also pursues observational research using facilities in Chile, Hawaii, and the Southwestern US. The astronomy group is a key member of the proposed BigBOSS Observatory, a stage IV baryon acoustic oscillation survey designed to elucidate the formation of galaxies in the early universe, and properties of dark energy and dark matter. The Department operates the 32″ Willard L. Eccles Observatory on Frisco Peak, Utah, approximately 200 miles from Salt Lake City. This high altitude (9600 ft a.s.l.) observatory is being developed for IR spectroscopy and imaging surveys. The department operates a pair of 3-meter interferometric telescopes at StarBase Utah, approximately 35 miles west of Salt Lake City. The South Physics observatory on campus houses the department 14″ fully automated telescope and several others with CCD photometers, spectrometers, and other accessories. The gamma-ray research group pursues astrophysics research with the Very Energetic Radiation Imaging Telescope Array System (VERITAS) located near Tucson, Arizona. Its four 10m telescopes make sterescopic measurements of TeV gamma rays from black holes, supernova remnants, pulsars, and active galactic nuclei. Faculty members also pursue cosmic ray and gamma-ray research at the High Altitude Water Cherenkov (HAWC) observatory, located on Sierra Negra, Mexico. The University of Utah is the host institution for the Telescope Array (TA) and Telescope Array Low Energy extension (TALE) Projects, located 125 miles from Salt Lake City in the west-central Utah desert. Its ground array has more than 500 scintillation detectors covering 750 square kilometers, accompanied by three air-fluorescence detectors. Both TA and TALE are designed to study the highest energy particles known, and both experiments make extensive use of the air-fluorescence technique first successfully employed at Utah by the Fly's Eye Experiment (1976-1991) here at University of Utah, Degree programs in astronomy are currently offered.

The department has a robust wired and wireless local network designed with growth and flexibility in mind. The local network is integrated into a cutting edge university network dedicated to providing premier Internet services. Core user and computational services are provided by a dozen Sun Fire and Sun Enterprise servers accompanied by several powerful Linux and Windows servers. Data storage and backup are provided by a growing storage area network (SAN) currently totaling roughly one terabyte of disk and ten terabytes of tape storage. We provide access to a large suite of programs for departmental use including Maple, Matlab, Mathematica, LabVIEW, Microsoft software, and educational software. In addition, the University provides deeply discounted prices on hundreds of software titles through their office of software licensing. There are numerous open access terminals, desktops, and printers in the department library, study areas, and the five open computer labs. Individual workstations are a mixture of Windows, Linux, Macintosh, and UNIX. The department also supports several research groups that have various computational computers and multiple terabyte size data arrays, usually based on UNIX type architectures. Research groups also have access to large computational clusters through the University's Center for High Performance Computing (CHPC), totaling well over 1500 processors in various clusters.

Research is also supported by a professional Research Machine Shop, as well as a Student Shop, the latter open to all faculty, staff, and students who have completed a training course. Both shops are equipped with state-of-the-art CNC lathes and mills, as well as cutting, drilling, and welding equipment. The well-equipped wood shop allows fabrication of non-magnetic supports and shipping containers. The ample stockroom saves time and effort in procuring both common and hard-to-find materials and supplies.

Table B—Appointments to Graduate School, 2009–10

Title of Appointee	Appointments Total	Appointments First year	Academic Load Allowed in Credit Hours Per Semester	Hours of Service Per Week	Stipend for Academic Year ($)
Teaching Assistant	43	13	9–12	20	17,043 level 1 19,891 level 2
Research Assistant	28	2	9–12	20	18,206 level 1 19,808 level 2
Other (specify)	3*	1**	–	–	–
Total	72	17			

* WEST Fellowships

**Own Funding

[1]Includes $6,405 of tuition (out of state rate).

[2]Summer RAs and a limited number of summer TA's are available.

5. Personnel Engaged in Separately Budgeted Research, 7/09–6/10

Professorial faculty	26
Other faculty	8
Postdoctoral appointments	17
Graduate students	27
Undergraduate students	30
Nonteaching research personnel	7
Total	115

6. Separately Budgeted Research Expenditures by Source of Support 7/09–6/10

	Departmental Research	Physics-related Research Outside Department
Federal government		
Business and industry	Breakdown unavailable	
Other		
Total	$3,097,676	$1,449,866

Table C—Separately Budgeted Research Expenditures, 7/09–6/10

Research Specialty	No. of Grants	Expenditures ($)
Astrophysics	5	197,541
Atomic Physics	1	252,261
Condensed Matter	10	1,867,097
High Energy/Cosmic Ray	6	1,565,290
Particle	5	367,718
Medical & Health Physics	1	297,635
Total	28	4,547,542

FACULTY

Distinguished Professors

Efros, Alexei L., Ph.D., Ioffe Physico Technical Institute, 1972. Theoretical condensed matter.

Vardeny, Zeev V., Ph.D., Technion, Israel, 1979. Experimental condensed matter.

Wu, Yong-Shi, Ph.D., Chinese Academy of Science, 1965. High-energy theory.

Professors

Ailion, David C., Ph.D., Illinois, 1964. Experimental condensed matter.

Bromley, Benjamin C., Ph.D., Dartmouth College, 1994. Theoretical astrophysics.

Cassiday, George L., Ph.D., Cornell, 1968. Cosmic rays.

DeFord, John W., Ph.D., Illinois, 1962. Physics education.

DeTar, Carleton E., Ph.D., California, Berkeley, 1970. Elementary particle theory.

Gondolo, Paolo, Ph.D., California, Los Angeles, 1991. Cosmology; Dark matter.

Harris, Frank E., Ph.D., California, Berkeley, 1954. Chemical physics.

Jui, Charles C., Ph.D., Stanford, 1992. Cosmic rays.

Kieda, David, Ph.D., Pennsylvania, 1989. Experimental high-energy astrophysics.

Lupton, John M., Ph.D., Durham, London, 2001. Experimental condensed matter.

Mattis, Daniel C., Ph.D., Illinois, 1957. Theoretical condensed matter.

Mishchenko, Eugene, Ph.D., Landau Institute for Theoretical Physics. Moscow, 1998. Theoretical condensed matter.

Raikh, Mikhail, Ph.D., Ioffe Physico Technical Institute, 1981. Theoretical condensed matter.

Saam, Brian T., Ph.D., Princeton, 1995. Experimental atomic/molecular.

Sokolsky, Pierre V., Ph.D., Illinois, 1973. Cosmic rays; high-energy physics.

Symko, Orest G., Ph.D., Oxford, 1967. Low-temperature physics; thermoacoustics.

Thomson, Gordon, Ph.D., Harvard, 1972. Cosmic rays; high energy physics.

Williams, Clayton C., Ph.D., Stanford, 1984. Experimental condensed matter.

Distinguished Professors Emeriti

Lüty, Fritz W., Ph.D., Stuttgart, 1955. Experimental condensed matter.

Taylor, P. Craig, Ph.D., Brown, 1969. Experimental condensed matter.

Professors Emeriti

Ball, James S., Ph.D., California, Berkeley, 1960. Elementary particle theory.
Bergeson, Haven E., Ph.D., Utah, 1962. Cosmic rays.
Dick, B. Gale, Ph.D., Cornell, 1958. Theoretical condensed matter.
Fowles, Grant R., Ph.D., California, Berkeley, 1950. Optics.
Gibbs, Peter, Ph.D., Utah, 1951. Condensed matter.
Johnson, Owen W., Ph.D., Utah, 1962. Experimental condensed matter.
Kuchar, Karel V., Ph.D., Charles, Prague, 1966. Relativity.
Ohlsen, William D., Ph.D., Cornell, 1961. Experimental condensed matter.
Price, Richard, Ph.D., CalTech, 1971. Relativity.
Sutherland, T. Bill, Ph.D., SUNY, Stony Brook, 1968. Theoretical condensed matter.
Williams, George A., Ph.D., Illinois, 1956. Experimental condensed matter.

Associate Professors

Bergman, Douglas, Ph.D., Yale, 1997. Cosmic rays; high energy physics.
Boehme, Christoph, Ph.D., Philipps, Marburg, 2002. Experimental condensed matter.
LeBohec, Stephan, Ph.D., Paris XI, 1992. Experimental high energy astrophysics.
Springer, Wayne R., Ph.D., Univ. of Maryland, 1991. Experimental astrophysics.
Starykh, Oleg, Ph.D., Russian Academy of Science, 1991. Theoretical condensed matter.
van den Bosch, Frank, Ph.D., Leiden, The Netherlands, 1997. Astronomy.

Assistant Professors

Bolton, Adam, Ph.D., MIT, 2005. Astronomy.
Dawson, Kyle, Ph.D., Berkeley, 2004. Astronomy.
Deemyad, Shanti, Ph.D., Washington Univ., St. Louis, 2004. Experimental condensed matter.
Gerton, Jordan, Ph.D., Rice, 2001. Experimental condensed matter.
Ivans, Inese, Ph.D., UT Autstin, 2002. Astronomy.
Rogachev, Audrey, Ph.D., Nagoya, Japan, 2000. Experimental condensed matter.
Saffarian, Saveez, Ph.D., Washington Univ., St. Louis, 2003. Biophysics.
Vershinin, Michael, Ph.D., Illinois, 2004. Biophysics.

Adjunct Professors (Not in residence)

Bjorken, James D., Ph.D., Stanford, 1959. High energy theory.
Blinc, Robert, Ph.D., Ljubljana, 1959. Experimental condensed matter.
Chubukov, Andrey, Ph.D., Moscow State University, 1985. Condensed matter.
Ehrenfreund, Eitan, Ph.D., Hebrew University, 1970. Experimental condensed matter.
Facelli, Julio C., Ph.D., University of Buenos Aires, Argentina, 1981. Nuclear magnetic resonance.
Gaisser, Thomas K., Ph.D., Brown, 1967. Cosmic rays; elementary particle physics.
Johnson, Christopher R., Ph.D., Utah, 1989. Physics; applied math.
Karl, Gabriel, Ph.D., Toronto, 1964. Theoretical high energy.
Liu, Feng, Ph.D., Virginia Commonwealth University, 1990. Chemical physics.
Madan, Arun, Ph.D., Dundee, Scotland, 1973. Condensed matter.
McCamey, Dane, Ph.D., New South Wales, Sydney, Australia, 2007. Experimental condensed matter.
Ormes, Jonathan F., Ph.D., Minnesota, 1967. Cosmic rays.
Parker, Dennis, Ph.D., Utah, 1978. Medical biophysics; computing.
Shahbazyan, Tigran, Ph.D., Utah, 1995. Theoretical condensed matter.
Shapiro, Boris, Ph.D., USSR Academy of Science, 1970. Theoretical condensed matter.
Stringfellow, Gerald, Ph.D., Stanford, 1967. Semiconductor physics.

Adjunct Associate Professors (Not in residence)

Blair, Steven, Ph.D., Colorado, Boulder, 1998. Nanophotonics.
Jeong, Eun-Kee, Ph.D., Washington University, St. Louis, 1991. Medical physics; MRI.
Martens, Kai, Ph.D., Heidelberg, 1994. Cosmic rays.
Nahata, Ajay, Ph.D., Columbia, 1997. Nanotechnology; Optics.

Adjunct Assistant Professors (Not in residence)

Bartl, Michael, Ph.D., Karl-Franzens-University, Graz, Austria, 2000. Chemical physics; Nanotechnology.
Huentemeyer, Petra, Ph.D., Hamburg, Germany, 2001. High energy astrophysics.
Paul, Prabasaj, Ph.D., Utah, 1995. Theoretical condensed matter.

Research Professors

Gellermann, Werner, Ph.D., Tech. Universitate, Hanover, 1978. Experimental condensed matter.
Worlock, John M., Ph.D., Cornell, 1962. Condensed matter.

Research Associate Professors

Belz, John, Ph.D., Cosmic rays.
Laicher, Gernot, Ph.D., Utah, 1994. Experimental condensed matter.
Matthews, John N., Ph.D., Rutgers, 1995. Cosmic rays.

Research Assistant Professors

AbuZayyad, Tareq, Ph.D., Utah. Cosmic rays.
Cady, Robert, Ph.D., Utah, 1983. Cosmic rays.
Mkhitaryan, Vagharsh, Ph.D., Yerevan Physics Institute, 1997. Theoretical condensed matter.
Nguyen, Tho, Ph.D.,. Iowa, 2008. Experimental condensed matter.

Professor (Lecturer)

Ingebretsen, Richard J., Ph.D., Utah, 1989. M.D., Utah, 1993. Physics education.

Associate Professors (Lecturer)

Pantziris, Anthony, Ph.D., Brown, 1987. Physics education.
Stone, Christopher, Ph.D., Utah, 1992. Physics education.

Instructor (Lecturer)

Higgs, Lynn B., MS., Utah, 1972. Physics education.

Research/Lecturer Emeritus

Rudolph, Sidney, Ph.D., Utah, 1986. Physics education.

Research Associates

Baird, Doug, Ph.D., pending, Utah. Experimental condensed matter.

Bange, Sebastian, Ph.D., Universi ty of Potsdam, 2009. Experimental condensed matter.

Blake, S. Adam, Ph.D., Utah, 2009. Experimental high energy physics.

Brown, Peter, Ph.D., Penn State, 2009. Astronomy.

Browstein, Joel, Ph.D., Perimeter Institute, 2009. Astronomy.

Chaudhuri, Debansha, Ph.D., Arizona, 2003. Theoretical high energy physics.

Dall'Asen, Analia, Ph.D., Buenos Aires, 2005. Experimental condensed matter.

Ermakov, Igor, Ph.D., Russian Academy of Science, Moscow, 1997. Experimental condensed matter.

Foley, Justin, Ph.D., Trinity College-Dublin, 2005. Theoretical high energy physics.

Godambe, Sagar, Ph.D., University of Mumbai, 2008. Experimental high energy physics.

Ivanov, Dmitri, Rutgers. Experimental high energy.

Levkova, Ludmila, Ph.D., Columbia, 2004. Theoretical high energy.

Li, Dongbo, Ph.D., Pennsylvania, 2008. Experimental condensed matter physics.

Oktay, Mehment, Ph.D., Iowa, 2001. Theoretical high energy physics.

Rodriguez, Doug, Ph.D., Utah, 2010. Experiemental high energy physics.

Scott, Lauren, Ph.D., Washington University, 2005. Experimental high energy.

Sharifzadeh, Mohsen, Ph.D., Utah, 2005. Experimental condensed matter.

Simmerer, Jennifer, Ph.D., University of Texas, Austin, 2006. Astrophysics.

Stirling, Spencer, Ph.D., Texas, 2008. Theoretical high energy physics.

Stokes, Benjamin, Ph.D., Utah, 2006. experimental high energy.

Stratton, Sean, Rutgers. Experimental high energy.

Stroman, Thomas, Ph.D., Iowa State, 2010. Experimental high energy.

Vincent, Stephane, Ph.D., University of Nice, 2007. Experimental high energy physics.

FACULTY RESEARCH SPECIALTIES

Theoretical

Astronomy: van den Bosch

Astrophysics, Relativity, Cosmology. Bromley, Gondolo.

Chemical Physics. Harris.

Condensed Matter. Efros, Mattis, Mishchenko, Mkhitaryan, Raikh, Starykh, Worlock, Wu.

Elementary Particle. DeTar, Wu.

Physics Education. DeFord, Higgs, Ingebretsen, Pantziris, Stone.

Experimental

Acoustics. Symko.

Astronomy. Bolton, Dawson, Ivans, Springer

Astrophysics. Kieda, LeBohec

Atomic Physics. Saam.

Biophysics. Gerton, Saffarian, Vershinin.

Condensed Matter. Ailion, Boehme, Deemyad, Laicher, Lupton, Nguyen, Rogachev, Vardeny, C. Williams.

Cosmic Rays. Abu-Zayyad, Belz, Bergman, Cady, Cassiday, Jui, Matthews, Sokolsky, Springer, Thomson.

Medical & Health. Gellerman.

UTAH STATE UNIVERSITY

DEPARTMENT OF PHYSICS

Logan, Utah 84322-4415
www.physics.usu.edu

Students Accepted For Degree	FIELDS		
	Physics	Astronomy	Related Fields
Doctorate	X		
Master's	X		

1. General

President: Stan Albrecht
Dean of Graduate School: Byron Burnham, Dean
Department Chairman: Jan J. Sojka
Department Telephone Number: (435) 797-2857
Type of Institution: University
Control: Public
Setting: Small town
Total Faculty: 870
Total Students: 14,458
Total Graduate Faculty: Not separated
Total Graduate Students: 1,721
Annual Graduate Tuition:
In-state residents: Full-time—$4,805.58[1]
Out-of-state residents: Full-time—$14,982,08[1]
Foreign students—$13,606[2]
Tuition rates for: 2010-11[3]
Deferred tuition plan: Yes
Term: Semester

[1]2 semesters, 10 hrs/semester, fees included.
[2]$100 per semester fee assessed for International students.
[3]Tuition subject to change.

2. Number of Faculty in Department

The combined total of full-time, tenure track faculty in the three professorial ranks is 15. There are 9 full-time research faculty and the combined total of full-time, part-time, and other faculty at all ranks is 16.

3. Admission, Financial Aid, and Housing

Address admission inquiries to: School of Graduate Studies, UMC 0900
Graduate application fee required: $55-U.S. citizens
$55-Foreign
Admission deadline: Fall, 6/15
Admission information: For fall admission, 2009–10, 6 students were accepted from 17 applicants.
Admission requirements: For admission to the graduate programs, a Bachelor's degree in physics or a related field is required with a minimum undergraduate GPA of 3.0 specified. The GRE is required. Students from non-English speaking countries are required to demonstrate proficiency in English via the TOEFL exam. Minimum acceptable score for admission is 550.
Undergraduate preparation assumed: Symon, *Mechanics*; Griffiths, *Introduction to Electrodynamics*; Kittel and Kroemer, *Thermal Physics*; Pendrotti and Pendrotti, *Optics*; Liboff, *Quantum Mechanics*.
Address financial aid inquiries to: Financial Aid Office, UMC 1800
GAPSFAS application required: Yes
Financial aid deadline: 6/30
Loans available: Yes
Resident Tuition Remission for Doctoral Students: A student who is matriculated in a doctoral degree program and is a graduate assistant or graduate fellow receiving at least $600 per month may be awarded a resident (instate) tuition remission. The student must be registered for at least nine credits.
Address housing inquiries to: USU Housing Office, UMC 8600
On-campus, single student housing available: Yes
Cost/semester: $940–1,310⁺ (dorms.); $1,030–1,430 (apts.)*†
On-campus, married student housing available: Yes*
Cost/month: $450–600

*In addition, off-campus apts. are available-$450 per mo. and up; Trailer Court-$195–221 per mo. (rental fee per space)
†Costs subject to change.
⁺Does not include food costs now.

Table A—Faculty, Enrollments, and Degrees Granted

Research Specialty	2008–09 Faculty	Enrollment[1] Fall 2009 Master's	Enrollment[1] Fall 2009 Doctorate	No. of Degrees Granted[2] 2008–09 (2004–09) Master's	Terminal Master's	Doctorate	Median No. of Years for 2008–09 Ph.D.'s
Atmos./Space Phys.,	6	1	8	2(6)	–	2(11)	4
Electromagnetism & Plasma Physics	2	0	5	2(2)	–	2(6)	5
Nonlinear Dynamics	1	0	0	0(0)	–	0(0)	–
Physics Education	0	0	0	0(0)	–	0(1)	–
Relativity & Gravitation	3	2	3	0(0)	–	0(2)	–
Surface Physics	4	3	3	0(0)	–	0(4)	–
Non-specialized		1	2				
Total	16	5	25	4(17)	–	5(21)	
Full-time Grad. Stud.		3	25				
Part-time Grad. Stud.		2	0				
First-year Grad. Stud.		0	2				
Median Years in Grad. Study (2004–09 Degrees)				3	–	6	–
Undergraduate Degrees, 2009–10 (2005–10): (82)							

[1]Students not yet committed to a research specialty are entered under non-specialized.
[2]Five-year totals in parentheses.

4. Graduate Degree Requirements

Master's: The minimum number of credit hours for an M.S. is 30. Of these, 12 credits (four courses) must come from any of the courses listed below for doctoral students. The remaining credits may come from other graduate level courses and credit for research. In consultation with the M.S. student's supervisory committee, the student will submit and orally defend either a Plan A research thesis (worth from 6 to 15 credits) or a Plan B research report (worth 2 or 3 credits).
Doctorate: The minimum number of credits for a Ph.D. is 90 beyond the bachelor's degree or 60 beyond the master's degree. The Ph.D. student will complete the following minimum coursework (9 semester courses): classical mechanics (1), electrodynamics (1), quantum mechanics (1), statistical mechanics (1), mathematical and computational physics (2),

"state of matter" (1), and advanced topics (2). The Ph.D. student must formally Qualify for the Ph.D. program about one year after commencing study and pass a Candidacy Examination, normally about two years after commencing study. The rules for Qualification and the Candidacy Examination are designed to facilitate the student's attainment of the Ph.D. degree. Qualification is conferred by vote of the faculty and is based on whatever relevant information the student wishes to have presented, but must include faculty evaluation of coursework taken at Utah State (normally, at least four courses). Prior to taking the Candidacy Examination the student must have passed at least 7 of the 9 courses mentioned above with a 3.0 GPA and established a graduate supervisory committee. The Candidacy Examination consists of an oral presentation and follow-up questioning pertaining to a research topic set by the student's supervisory committee. The Ph.D. student also presents a written thesis and an oral thesis defense.

Special Equipment, Facilities, or Programs: Research in the Physics Department involving graduate students focuses on upper atmospheric and space physics, surface physics and nanofabrication, general relativity and field theory, plasma physics and electromagnetism, physics education, and nonlinear dynamics. On-campus experimental work is conducted in the following laboratories: (1) a lidar laboratory, in which the physical and chemical properties of the atmosphere from 40 to 90 km are probed, that includes an 18 W, Nd:YAG laser pulsed at 30 Hz, a 44 cm Newtonian telescope, and associated electronics; (2) two nanosystem fabrication and microscopy laboratories, in which novel electronic materials and reactions on semiconductor surfaces are investigated, that includes two scanning tunneling microscopes and an MBE growth facility; (3) a surface modification and characterization laboratory, in which the electronic properties of materials and their relevance to spacecraft charging are examined, that includes multiple UHV chambers in which electron and ion beam bombardment as well as Auger, LEED, and UV and photoemission spectroscopy can be performed; and (4) a femtosecond dynamics laboratory, in which the nature of rapid electron-phonon interactions are explored, that includes an ultrafast Ti-sapphire laser with associated optics and instrumentation. Graduate students also conduct experimental research on atmospheric dynamics at off-campus facilities. Frequent and continuing work is done using radar and optical facilities at Sondestrom (Greenland), Millstone Hill (Massachusetts), Arecibo (Puerto Rico), and Jicamarca (Peru). The University maintains an aeronomy observatory, capable of characterizing the atmosphere from 40 to 500 km, at Bear Lake (30 km away). Bear Lake Observatory is equipped with a magnetometer, ionosonde, and various spectroscopic imaging systems. Students also have direct access to data from a constellation of microsatellites (ION-F). Theoretical physics research involves general relativity gravitation, gravitational waves and relativistic astrophysics, classical and quantum field theory, unified field theories, and various topics in mathematical physics. An internationally renowned atmospheric and space physics modeling and data analysis program, conducted in close collaboration with the Center for Atmospheric and Space Science, employs state-of-the-art, distributed computing systems and large geophysical databases.The plasma fusion group is developing models of hot plasmas that are applied to fusion reactor experiments. Current work on complex systems focuses on processes in granular and fractured materials, and on the nature of computation in biological systems.

Table B—Appointments to Graduate Students, 2009–10

Title of Appointee	Appointments Total	Appointments First year	Academic Load Allowed in Credit Hours	Hours of Service Per Week	Stipend for Academic Year ($)
			Semester		
Teaching Assistant	10.5	2	9–12	20	11,500 (9 months)
Research Assistant	7	0	9–12	20	18,000 (12 months)
Total	17.5	2			

5. Personnel Engaged in Separately Budgeted Research, 7/08–6/09

Professorial faculty	13
Other faculty	6
Graduate students	15
Undergraduate students	28
Total	62

6. Separately Budgeted Research Expenditures by Source of Support

	Departmental Research	Physics-related Research Outside Department
Federal government	$	$
Private, nonprofit organizations	Breakdown not available	
Business and industry		
Other		
Total	$513,047	$735,638

Table C—Separately Budgeted Research Expenditures

Research Specialty	No. of Grants	Expenditures ($)
Atmos./Space Phys., Cosmic Rays	5	735,638
Fundamental Theory	0	–
Nonlinear Dynamics	2	110,000
Plasma Fusion	3	214,001
Surface Physics	1	189,046
Total	11	1,248,685

FACULTY

Professors

Bialkowski, Stephen E., Ph.D., Utah, 1978. (Adjunct). Non-linear optics; laser spectroscopy.

Dennison, J. R., Ph.D., Virginia Tech, 1985. Solid state and surface physics.

DeVito, Raymond, Ph.D., Michigan State, 1979. (Adjunct). Medical physics, patent law.

Edwards, W. Farrell, Ph.D., Cal Tech., 1960. Electromagnetic theory.

Fejer, Bela, Ph.D., Cornell, 1974. Space plasma physics.

Hall, Leonard F., Ph.D., Wisconsin, Madison, 1977. (Adjunct). Structure forming systems.

Hansen, Wilford N., Ph.D., Iowa State, 1956. (Emeritus). Reflection spectroscopy; surface physics.
Hatch, Eastman N., Ph.D., Cal Tech., 1956. (Emeritus). Nuclear physics.
Lind, Don L., Ph.D., California, Berkeley, 1964. (Emeritus). Space physics.
Lind, V. Gordon, Ph.D., Wisconsin, Madison, 1964. (Emeritus). Medium-energy nuclear physics.
Moore, R. Gilbert, B.S., New Mexico State, 1949. (Adjunct). Atmospheric physics.
Peak, David, Ph.D., Albany, 1969. Complex materials and dynamics.
Pendleton, William R., Ph.D., Arkansas, 1964. (Emeritus). Atomic and molecular physics.
Raitt, W. John, Ph.D., King's College, London, 1963. (Emeritus). Space plasma physics.
Rees, David, Ph.D., Univ. College, London, 1967. (Adjunct). Optics; atmospheric phenomenon.
Russell, Ray, Ph.D., California, San Diego, 1978. (Adjunct). Space science.
Schunk, Robert W., Ph.D., Yale, 1970. Director, Center for Atmospheric and Space Science. Space plasma physics.
Shen, T. C., Ph.D., Maryland, 1985. Surface science.
Shinn, Neal, Ph.D., Mass. Inst. of Tech., 1983. (Adjunct). Chemical physics.
Sojka, Jan J., Ph.D., Univ. College, London, 1976. Department Head, Assistant Director, Center for Atmospheric and Space Sciences. Space plasma physics.
Taylor, Michael, Ph.D., Southampton, 1986. Atmospheric physics.
Torre, Charles, Ph.D., North Carolina, 1985. Asst. Dept. Head. Gravitational Physics, Field Theory, Mathematical Physics.
Tucker, John R., Ph.D., Harvard, 1971. (Adjunct). Device physics; superconductivity.
Wickwar, Vincent, Ph.D., Rice, 1971. Atmospheric physics.

Associate Professors

Balasubramaniam, K. S., Ph.D., Indian Institute of Science and Indian Institute of Astrophysics, 1989. (Adjunct). Solar physics.
Davis, I. Lee, Ph.D., Utah State, 1983. (Adjunct). Condensed matter.
DeGaris, Hugo, Ph.D., Brussels, Belgium, 1992. (Adjunct). Artificial intelligence.
Dyer, James S., Ph.D., Utah State, 1988. (Adjunct). Materials chemistry and contamination control.
Held, Eric D., Ph.D., Wisconsin, Madison, 1999. Plasma physics.
Marshall, Jill A., Ph.D., Texas, Austin, 1984. (Adjunct). Science education.
McAdams, Robert E., Ph.D., Wisconsin, Madison, 1964. (Emeritus). Medium-energy nuclear physics.
Riffe, D. Mark, Ph.D., Cornell, 1989. Optical studies of surfaces.
Wheeler, James T., Ph.D., Chicago, 1986. Relativity; Particle physics.
Vieira, David, Ph.D., California, Berkeley, 1978. (Adjunct). Medium-energy nuclear physics.
Zavyalov, Vladimir, Ph.D., Ural State, 1994. (Adjunct). Condensed matter modeling.

Assistant Professors

King, Jeremy R., Ph.D., Hawaii, 1993. (Adjunct). Astrophysics.
Larson, Shane L., Ph.D., Montana State, 1999. Gravitational Physics and Relativistic Astrophysics.
Scherliess, Ludger, Ph.D., Utah State, 1997. Space Physics.
Yang, Haeyeon, Ph.D., Brown, 1996. Surface science.

Research Professors

Berkey, F. Thomas, Ph.D., Alaska, 1971. (Adjunct). Space physics.
Howard, Allen Q., Ph.D., Colorado, 1972. Electromagnetic theory. (Adjunct)
Miller, Kent, Ph.D., Illinois, Urbana-Champaign, 1977. Atmospheric physics.
Wilkerson, Thomas D., Ph.D., Michigan, 1962. (Adjunct). Atmospheric physics.

Research Associate Professors

Barakat, Abdallah, Ph.D., Utah State, 1982. Theoretical plasma physics.
Demars, Howard, Ph.D., Utah State, 1986. Theoretical plasma physics.
Doyle, Timothy E., Ph.D., Utah State, 2004. Complex media.
Hansen, J. Stephen, Ph.D., Durham, 1975. Image processing.
Singh, Ajay Kumar, Ph.D., Institute for Plasma Research, 1993. Experimental Tokamak physicist, Plasma Physics.
Zhu, Lie, Ph.D., Alaska, 1990. Space plasma physics.

Senior Lecturer

Triplett, Tonya, M.S., Utah State, 2003. Physics education.

UNIVERSITY OF VERMONT

DEPARTMENT OF PHYSICS

82 University Place
Burlington, Vermont 05405-0125
http://www.uvm.edu/physics

Students Accepted For Degree	FIELDS		
	Physics	Astronomy	Related Fields
Doctorate			X
Master's	X		X

1. General

President: Daniel M. Fogel
Dean of Graduate School: Domenico Grasso
Department Chairman: Dennis Clougherty
Department Telephone Number: (802) 656-2644
Type of Institution: University
Control: Public
Setting: Urban
Total Faculty: 1,068 full-time, 231 part-time
Total Graduate Faculty: 613
Total Students: 13,391
Total Graduate Students: 1,976
Annual Graduate Tuition:
In-state residents: $5,856 (credit hours, $488/cr. hr., $829 per semester)
Out-of-state residents: $14,784 (12 cr. hr., + $1,232/cr.hr.)
Tuition rates for: 2009–10
Deferred tuition plan: No
Other Fees: $829 comp fees per semester
Term: Semester

2. Number of Faculty in Department

The combined total of full-time faculty in the three professorial ranks is 9. The combined total of full-time, part-time, and other faculty at all ranks is 15.

3. Admission, Financial Aid, and Housing

Address admission inquiries to: Physics Dept., A405, Cook Bldg.
Graduate application fee required: $40 online
Admission deadline (Fall admission): 4/1
Admission information: For fall admission, 2009–10, 5 students were accepted from 18 applicants.
Admission requirements: For admission to the graduate programs, a Bachelor's degree in physics is required with no minimum undergraduate GPA specified. The general GRE is required and cannot be waived. The Advanced GRE is ordinarily required, but can be waived in special cases. No minimum acceptable score is specified. The average GRE scores for 2009–10 admissions were verbal-550; quantitative-728; analytical writing 3.9. Students from non-English speaking countries are required to demonstrate proficiency in English via the TOEFL exam. Minimum acceptable score for admission is 550, for teaching assistant 600.
Undergraduate preparation assumed: John R. Taylor, *Classical Mechanics*; David Griffiths, *Introduction to Electrodynamics;* Kittel & Kroemer, *Thermal Physics*; David Griffiths, *Introduction to Quantum Mechanics*.
Address financial aid inquiries to: Physics Dept., A405, Cook Bldg.
GAPSFAS application required: No
Financial aid deadline: 3/1 for FAFSA
Loans available: Yes
Address housing inquiries to: Office of Family Housing, Fort Ethan Allen, 36 Catamount Lane, Colchester, VT 05446
Single student housing available: Yes
Cost/month: $755 per mo.
Family Housing available: Yes
Cost/month: $990–$1,300 per month for year (varies with # of bdrms)

Table A—Faculty, Enrollments, and Degrees Granted

Research Specialty	2009 Faculty	Enrollment[1] Fall 2009		No. of Degrees Granted[2] 2009–10 (2003–09)			Median No. of Years for 2009–10 Ph.D.'s
		Master's	Doctorate	Master's	Terminal Master's	Doctorate	
Acoustics	1	0	–	0(0)	0(2)	–	–
Astronomy	1	1	–	0(0)	0(0)	–	–
Astrophysics	1	0	–	0(0)	0(0)	–	–
Biophysics	4	2	–	0(0)	0(5)	–	–
Condensed Matter Physics	5	1	–	0(1)	0(6)	–	–
History & Philosophy	1	0	–	0(0)	0(2)	–	–
Low Temperature Physics	2	0	–	0(0)	0(0)	–	–
Materials Sci.	4	0	12	1(0)	0(0)	2(5)	–
Non-Linear Physics	1	0	–	0(0)	0(0)	–	–
Optics	2	0	0	0(0)	0(0)	–	–
Physics Education	1	0	–	0(0)	0(6)	–	–
Polymer Physics/ Science	2	0	0	0(0)	0(0)	–	–
Ultrasonics	1	0	–	0(0)	0(0)	–	–
Total		4	12	1(1)	1(21)	–	
Full-time Grad. Stud.		4	12		1	2	
Part-time Grad. Stud.		0	0				
First-year Grad. Stud.		1	4				

Median Years in Grad. Study (2009–10 Degrees)
Undergraduate Degrees, 2009–10 (2003–08): 3(0)

[1]Students not yet committed to a research specialty are entered under non-specialized.
[2]Five-year totals in parentheses.

4. Graduate Degree Requirements

Master's: A minimum of 30 semester hours of graduate credit is required. Of the 30 hours, 21 must be completed in residence. At least six must be in thesis research and 9 in other courses numbered above 300 (graduate students only). No more than 15 hours of thesis research may be included in the degree program. The candidate must pass a written and oral comprehensive examination, as well as an oral examination on his thesis. The graduate student must maintain a B average. There are no foreign language requirements.
Materials Science Program: Masters of Science and the Doctor of Philosophy degrees are offered in this interdisciplinary program. The faculty are drawn from the departments of Chemistry, Electrical Engineering, Mechanical Engineering and Physics. The program is commited to educating the students in the application of basic sciences and engineering to

promote understanding of the properties of materials, their development and applications, and to carry out advanced and stimulating research in these areas. The research program pursued in Materials Science at the University of Vermont have two areas of specialization: Electronic Materials and Bio-/Polymeric Materials. Each student must meet the general requirements or admission as outlined under the "Regulations of the Graduate College." Students in the program are sponsored by the participating department which best reflects the student's backgound and interest.

Thesis: Thesis may be written *in absentia*.

Special Equipment, Facilities, or Programs: Research is concentrated in areas of astrophysics, biological physics, polymer physics, materials science, and solid state physics. Collaboration is feasible with other departments of science, engineering, and medicine of this geographically small campus. There is especially close cooperation with the Departments of Electrical Engineering, the Medical School, and Chemistry. The Department shares a building with the Department of Chemistry.

Table B—Appointments to Graduate Students, 2009–10

Title of Appointee	Appointments Total	Appointments First year	Academic Load Allowed in Credit Hours	Hours of Service Per Week	Stipend for Academic Year ($)
			Semester		
Teaching Assistant	9	1		18–20	15,000
Research Assistant	3				
Total	12	1			

5. Personnel Engaged in Separately Budgeted Research, 7/08–6/09

Professorial faculty	7
Postdoctoral appointments	1
Graduate students	12
Total	20

6. Separately Budgeted Research Expenditures by Source of Support

	Departmental Research	Physics-related Research Outside Department
Federal government	$838,191	14,146
Total	$838,191	14,146

Table C—Separately Budgeted Research Expenditures

Research Specialty	No. of Grants	Expenditures ($)
Astronomy	2	257,957
Biological Physics	1	93,645
Biophysics	1	14,146
Condensed Matter Physics	5	486,589
Total	9	852,337

FACULTY

Professors

Clougherty, Dennis P., Ph.D., MIT, 1989. Theoretical condensed matter physics.
Rankin, Joanna M., Ph.D., Iowa, 1970. Radio astrophysics; history of science.
Wu, Jun-Ru, Ph.D., California, 1985. Experimental condensed matter physics and ultrasound.

Associate Professors

Chu, Kelvin, Ph.D., University of Illinois, 1995. Experimental biophysics; protein dynamics; low-temperature physics.
Headrick, Randall, Ph.D., University of Pennsylvania, 1988. Molecular beam epitaxy, x-ray scattering surface processing.
Spartalian, Kevork, Ph.D., Carnegie-Mellon, 1974. Mössbauer spectroscopy; biological physics; physics education.
Yang, Jie, Ph.D., Princeton, 1987. Experimental biophysics; atomic force microscopy.

Assistant Professors

Furis, Madalina, Ph.D., Univ. of Buffalo (SUNY), 2004. Ultrafast spectroscopy, time-resolved photoluminescence.
Kotov, Valeri, Ph.D., Clarkson Univ., 1996. Condensed matter theory.

Emeriti

Arns, Robert G., Ph.D., Michigan, 1960. History of science.
Brown, John S., Ph.D., Rutgers, 1967. Solid state theory and liquid metals.
Detenbeck, Robert W., Ph.D., Princeton, 1962. Physics education.
Nyborg, Wesley L., Ph.D., Penn State, 1947. Biological physics and ultrasound.
Smith, David Y., Ph.D., Rochester, 1962. Optical and X-ray properties of matter.

Senior Lecturer

Sanders, Malcolm, Ph.D., Yale, 1984. Applied physics, non linear systems, chaos.

Lecturers

Malghani, Shaheen, Ph.D., Univ. of Vermont, 1991. Solid state and materials physics.
Manley, Don, M.A., Univ. of Oregon, 1954. Physics.
Pepe, Jason, M.S., Univ. of Vermont, 2003. Physics.
Perry, John, Ph.D., Univ. of Rochester, 1992. Astrophysics.

RESEARCH SPECIALTIES AND STAFF

Experimental

Acoustics. Sanders
Astrophysics. Rankin.
Atomic Force Microscopy. Yang.
Biological and Medical Physics. Chu, Nyborg, Spartalian, Wu, Yang.
Condensed Matter and Materials Physics. Furis, Headrick, Wu.
History of Science. Arns.
Mössbauer Spectroscopy. Spartalian.
Protein Dynamics. Chu.
Ultrasonics. Nyborg, Wu.
X-ray Optics. Smith.

Theoretical

Condensed Matter. Electronic and transport properties of metals, random alloys, and liquid metals; lattice dynamics; order-disorder phase transitions in alloys. Superconductivity. Strongly correlated electron systems; electronic properties of graphene. Clougherty, Kotov.

Surface Physics. Electronic surface properties of metals and alloys; properties of adatoms and adlayers on graphite. Clougherty.

Ultrasonics. Physical mechanisms for biological effects of ultrasound. Nyborg, Wu.

COLLEGE OF WILLIAM AND MARY

DEPARTMENT OF PHYSICS

Williamsburg, Virginia 23187-8795

Students Accepted For Degree	FIELDS		
	Physics	Astronomy	Related Fields
Doctorate	X		X
Master's			

1. General

President: W. Taylor Reveley III
Dean of Graduate School: S. Laurie Sanderson
Department Chairman: Keith A. Griffioen
Director, Physics Graduate Admissions: Marc Sher
Department Telephone Number: (757) 221-3500
Type of Institution: University
Control: Public
Setting: Small town
Total Faculty: 796
Total Graduate Faculty: Not applicable
Total Students: 7,874
Total Graduate Students: 2,038
Annual Graduate Tuition:
In-state residents: Full-time—$10,768
Part-time—$345/credit
Out-of-state residents: Full-time—$24,638
Part-time—$920/credit
Tuition rates for: 2010–11
Deferred tuition plan: No
Other Fees: None
Term: Semester

2. Number of Faculty in Department

The combined total of full-time faculty in the three professorial ranks is 30. The combined total of full-time, part-time, and other faculty at all ranks is 55.

3. Admission, Financial Aid, and Housing

Address admission inquiries to: Director, Graduate Admissions, Department of Physics or email grad@physics.wm.edu. Additional information is available on the www at http://www.wm.edu/physics
Graduate application fee required: $45
Admission deadline (Fall admission): February 1
Admission information: For fall admission, 2009–10, 27 students were accepted from 102 applicants.
Admission requirements: For admission to the graduate programs, a Bachelor's degree in physics or a related field is required with a minimum GPA of 2.5/4 specified. The GRE is required. No minimum acceptable score for admissions is specified. The GRE subject test is required. No minimum acceptable score for admissions is specified. Students from non-English speaking countries are required to demonstrate proficiency in English via their GRE and TOEFL scores and letters from referees.
Undergraduate preparation assumed: Marion & Thornton, Mechanics; Griffiths, Quantum Physics; Griffiths, Electricity and Magnetism; Kittel & Kroemer, Thermal Physics.
Address financial aid inquiries to: Director, Graduate Admissions, Department of Physics or email grad@physics.wm.edu
GAPSFAS application required: No
Financial aid deadline: None
Loans available: Yes
Address housing inquiries to: Office of Residence Life
On-campus, graduate student housing available: Yes
Cost/semester: $2,806–$2,856
On-campus, married student housing available: No

Table A—Faculty, Enrollments, and Degrees Granted

Research Specialty	2010–11 Faculty[3]	Enrollment[1] Fall 2009		No. of Degrees Granted[2] 2009–10 (2005–10)			Median No. of Years for 2009–10 Ph.D.'s
		Mas-ter's	Doc-torate	Mas-ter's	Terminal Master's	Doc-torate	
Applied Physics	4	–	2	0(1)	0(0)	1(2)	5
Atomic, Molecular, & Optical Physics	6	–	12	5(11)	0(0)	1(4)	7
Biophysics	2		1	0(2)	0(1)	0(1)	–
Chemical Physics	2	–	0	0(0)	0(1)	0(1)	–
Computational Physics	6	–	1	1(1)	0(0)	0(0)	–
Condensed Matter Physics	9	–	9	1(8)	0(0)	1(10)	6
Cosmology	1	–	0	0(0)	0(0)	0(0)	–
Materials Sci./ Metallurgy	1	–	0	0(0)	0(0)	0(0)	–
Nonlinear Dynamics	2	–	2	0(0)	0(0)	0(1)	–
Nuclear Physics	11	–	12	3(9)	0(0)	2(9)	7
Particles & Fields	13	–	10	3(13)	0(0)	1(7)	5.5
Physics of Beams	1	–	0	0(0)	0(0)	0(1)	–
Plasma Physics & Fusion	2	–	2	0(1)	0(0)	0(2)	–
Non-specialized	0	–	14	1(9)	0(2)	0(0)	–
Total		–	65	14(55)	0(4)	6(38)	
Full-time Grad. Stud.		–	65				
Part-time Grad. Stud.		–	0				
First-year Grad. Stud.		–	14				
Median Years in Grad. Study (2009–10 Degrees)				1.5		6	–

Undergraduate Degrees, 2000–10 (2005–10): 13(90)

[1]Students not yet committed to a research specialty are entered under non-specialized.
[2]Five-year totals in parentheses.
[3]Some faculty have multiple research specialties.

4. Graduate Degree Requirements

Master's: For the M.S. degree, the requirements are taking the PhD qualifying exam and 32 satisfactory credits of graduate work with a B average. A student progressing toward the Ph.D. degree will usually satisfy the M.S. requirements en route. At least one semester must be spent in residence and a minimum of one semester of teaching is required of all candidates. There are no foreign language or thesis requirements.
Doctorate: For the Ph.D., required courses include Classical Mechanics, Mathematical Physics, Quantum Mechanics I & II, Classical Electricity and Magnetism I &II, Field Theory & Relativistic Quantum Mechanics, Statistical Physics & Thermodynamics. In addition, 2 semesters of Colloquium, Teaching Physics, at least one elective from inside and at least one

outside the student's field of study may be required. The candidate must, in addition to passing the qualifying exam, demonstrate a mastery of the material in the first and second year courses, either by doing well in these courses or by individual examinations. A student must maintain a B average for all course work. There is a one year residence minimum for the degree. The research must be a significant original contribution. The dissertation must be approved by the candidate's faculty committee, and must be successfully defended in a public oral examination. A Ph.D. candidate must teach a minimum of two semesters. There are no foreign language requirements.

Thesis: Thesis may be written *in absentia*.

Special Equipment, Facilities, or Programs: The Department is housed in the William Small Physical Laboratory, which contains its own library, machine shop and other support facilities in addition to research and teaching laboratories, classrooms, and offices. The Physics department has many workstations and personal computers; and access to supercomputers is available through national and international networks. Extensive computational resources are available through the Center for Piezoelectric Design and the nuclear/particle group. The new high field solid state Nuclear Magnetic Resonance Laboratory opened in spring 2005. A 6-GeV continuous electron beam accelerator facility, Thomas Jefferson National Accelerator Facility and Applied Research Center (ARC) is located in nearby Newport News. Faculty and graduate students are or have been engaged in experiments at Fermilab, Jefferson Lab, NASA-Langley, TRIUMF (Vancouver, Canada), Brookhaven National Laboratory, Soudan Underground Laboratory, CERN, SLAC, and DUBNA, Russia.

Table B—Appointments to Graduate Students, 2009–10

Title of Appointee	Appointments: Total	Appointments: First year	Academic Load Allowed in Credit Hours	Hours of Service Per Week	Stipend for Academic Year ($)[1,2]
			Semester		
Teaching Assistant	37	14	18	20	16,500
Research Assistant	28	0	18	20	16,500
Self-support	4	1	18	–	–
Total	65	14			

[1]Students normally continue research through the summer, bringing stipend total to $22,000/yr.

[2]Tuition and fees, as well as health insurance are paid by the department.

5. Personnel Engaged in Separately Budgeted Research, 4/09–4/10

Professorial faculty	26
Other faculty	12
Postdoctoral appointments	10
Graduate students	70
Undergraduate students	14
Nonteaching research personnel	4
Total	134

6. Separately Budgeted Research Expenditures by Source of Support, 4/30/09–4/30/10

	Departmental Research	Physics-related Research Outside Department
Federal government	$3,815,757.41	372,576.28
State and local government	734,006.94	
Private, nonprofit orgaizations	93,122.55	
Other	428,719.13	
Total	$5,071,606.03	372,576.28

Table C—Separately Budgeted Research Expenditures

Research Specialty	No. of Grants	Expenditures ($)
Atomic, Molecular, & Optical Physics	13	$401,531
Condensed Matter Physics	13	$1,137,502
Nuclear Physics	12	$1,207,135
Particles & Fields	10	$920,399
Plasma Physics & Fusion	5	$138,508
REU, SURA, JLab, VASpace, Private, State	28	$1,266,532
Total	81	$5,071,607

Table D—Physics-related Research Outside Department

Field and Unit Outside Department	No. of Grants	Expenditures ($)
NASA-Langley Atmospheric/ Space Physcs	7	372,576.28
Total	7	372,576.28

FACULTY

Professors

Armstrong, David S., Ph.D., British Columbia, 1989. Nuclear and particle experiments.

Averett, Todd D., Ph.D., Virginia, 1995. Nuclear experiments.

Carlson, Carl E., Ph.D., Columbia, 1968. Particle and nuclear theory.

Carone, Christopher D., Ph.D., Harvard, 1994. Particle theory.

Cooke, William E., Ph.D., MIT, 1976. Atomic, molecular, and optical physics experiments.

Delos, John B., Ph.D., MIT, 1970. Atomic and molecular theory; nonlinear dynamics and chaos.

Griffioen, Keith A., Ph.D., Stanford, 1984. Chairman of the Department. Nuclear and particle experiments.

Hoatson, Gina L., Ph.D., East Anglia, 1980. Experimental condensed matter; chemical physics; molecular spectroscopy.

Kossler, William J. Ph.D., Princeton, 1964. Nuclear and condensed matter experiments and physics of beams.

Krakauer, Henry, Ph.D., Brandeis, 1975. Condensed matter theory and computational physics.

Manos, Dennis M., Ph.D., Ohio State, 1976. Applied physics experiments; plasma physics; biophysics; nanotechnology.

McKeown, Robert D., Ph.D., Princeton, 1979. Deputy Director, Nuclear and particle experiments.

Pennington, Michael R., Ph.D., London, 1971. Associate Director for Theoretical and Computational Physics, TJNAF. Hadronic theory.
Perdrisat, Charles F., D.Sc., Swiss Fed. Inst. Tech., 1961. Nuclear and particle experiments.
Sher, Marc T., Ph.D., Colorado, 1980. Particle theory.
Tracy, Eugene R., Ph.D., Maryland, 1984. Plasma theory; nonlinear dynamics.
Vahala, George M., Ph.D., Iowa, 1972. Plasma theory; lattice Boltzmann; quantum lattice algorithms and computational physics.
Zhang, Shiwei, Ph.D., Cornell, 1993. Condensed matter theory and computational physics.

Associate Professors

Erlich, Joshua, Ph.D., MIT, 1999. Particle theory and cosmology.
Lukaszew, Rosa A., Ph.D., Wayne State, 1996. Condensed matter experiments.
Nelson, Jeffrey K., Ph.D., Minnesota, 1994. Director, Physics Graduate Program. Particle experiment and computational physics.

Assistant Professors

Aubin, Seth A. M., Ph.D., SUNY at Stony Brook, 2003. Atomic, molecular, and optical physics experiments.
Deconinck, Wouter, Ph.D., Michigan, 2008. Nuclear experiments.
Detmold, William, Ph.D., Adelaide, 2002. Hadronic theory and lattice field theory.
Kordosky, Michael A., Ph.D., Texas-Austin, 2004. Particle experiments.
Novikova, Irina, Ph.D., Texas A&M, 2003. Atomic, molecular, and optical physics experiments.
Orginos, Konstantinos N., Ph.D., Brown, 1998. Lattice gauge theory and computational physics.
Qazilbash, Mumtaz, Ph.D., Maryland, College Park, 2004. Condensed matter experiments.
Rossi, Enrico, Ph.D., Texas-Austin, 2005. Condensed matter theory.
Vahle, Patricia L., Ph.D., Texas-Austin, 2004. Particle experiments.

Adjunct Professors

Bosted, Peter, Ph.D., MIT, 1980. Nuclear experiments.
Carlini, Roger D., Ph.D., New Mexico, 1978. Nuclear and particle experiments.
Danehy, Paul M., Ph.D., Stanford, 1995. Atomic, molecular, and optical physics experiments.
Lung, Allison F., Ph.D., American, 1992. Nuclear and particle experiments.
Majewski, Stanislaw, Ph.D., Warsaw, 1979. Particle detectors.
Osborne, Alfred R., Ph.D., Houston, 1974. Nonlinear phenomena in fluids; physical oceanography.
Reilly, Anne C., Ph.D., Michigan, 1996. Atomic, molecular, and optical physics experiments; condensed matter experiments.
Sasinowski, Maciek, Ph.D., William and Mary, 1995. Biophysics.
Vanderhaeghen, Marc, Ph.D., Ghent, 1995. Nuclear theory.
Wolf, Stuart A., Ph.D., Rutgers, 1969. Condensed matter experiments.

Professors Emeriti

Champion, Roy L., Ph.D., Florida, 1966. Atomic and molecular experiments.
Eckhause, Morton, Ph.D., Carnegie-Mellon, 1962. Nuclear and particle experiments.
Gross, Franz L., Ph.D., Princeton, 1963. Nuclear and particle theory.
Kane, John R. Ph.D., Carnegie-Mellon, 1964. Nuclear and particle experiments.
McKnight, John L., Ph.D., Yale, 1957. Foundations of quantum theory; history and philosophy of science.
Petzinger, Kenneth G., Ph.D., Penn, 1971. Condensed matter theory.
Remler, Edward A., Ph.D., North Carolina, 1963. Nuclear and particle theory.
Schone, Harlan E., Ph.D., California, Berkeley, 1960. Condensed matter experiments; magnetic resonance.
Thomas, Anthony, Ph.D., Flinders, 1974. Director of the Theory Center at Thomas Jefferson National Accelerator Facility. Nuclear theory.
von Baeyer, Hans C., Ph.D., Vanderbilt, 1964. Particle theory; public understanding of science.
Walecka, J. Dirk, Ph.D., MIT, 1958. Nuclear and particle theory.
Welsh, Robert E., Ph.D., Penn State, 1960. Nuclear and particle experiments.

Research Professor

Venkataraman, Malathy D., Ph.D., Kerala, 1968. Spectroscopy; chemical physics.

Research Associate Professor

Benner, D. Chris, Ph.D., Arizona, 1979. Molecular spectroscopy; chemical physics.

Research Assistant Professor

Mikhailov, Eugeniy E., Ph.D., Texas A&M, 2003. Atomic, molecular, and optical physics experiments.

Research Scientist

Purwanto, Wirawan, Ph.D., William and Mary, 2005. Condensed matter theory.

Research Associates

Chang, Chia-chen, Ph.D., Penn State, 2006. Condensed matter theory.
Kuschner, Karl W., Ph.D., William and Mary, 2009. Atomic, molecular, and optical physics experiments.
Lee, Hosik, Ph.D., Texas-Austin, 2008. Atomic, molecular, and optical physics theory.
Ma, Fengjie, Ph.D., Inst. Physics, CAS, 2004. Condensed matter theory.
Mathis, Mark, Ph.D., Johns Hopkins, 2010. Particle experiments.
Meinel, Stefan, Ph.D., Cambridge, 2010. Hadronic theory.
Walding, Joseph J., Ph.D., Imperial College of London, 2010. Particle experiments.
Zhao, Bo, Ph.D., U. Connecticut, 2009. Nuclear experiments.
Zhao, Libo, Ph.D., Inst. Phys., CAS, 1994. Atomic, molecular, and optical theory.

Research Engineers

Bensel, John, Ph.D., U. Pennsylvania, 1973.
Riso, Jose E., B.S., Sociedad Argentina de Grafologia, 1988.
Walter, Eric, Ph.D., Pennsylvania, 2001. Scientific Programmer.

Director of Teaching Labs

Hancock, A. Dayle, Ph.D. Houston, 1981.

RESEARCH SPECIALTIES AND STAFF

Theoretical

Atomic and Molecular Physics. Order and chaos in classical and quantum systems; atoms in strong fields; atomic and molecular collisions. Delos, Lee, Zhao.

Condensed Matter Physics. Electronic properties of materials; positive muons in solids; surface physics; high-temperature superconductivity ferroelectrics; ultra-cold atoms; graphene; two-dimensional systems; strongly correlated electron systems; computational physics. Chang, Krakauer, Ma, Purwanto, Rossi, Walter, Zhang.

Hadronic Physics. Perturbative and nonperturbative QCD; lattice gauge theory; effective field theories for hadrons. Carlson, Detmold, Meinel, Orginos, Pennington, Vanderhaeghen.

High Energy Particle Physics. Electroweak phenomenology and symmetry breaking; extensions of the standard model; supersymmetry, grand unification and extra dimensions; string theory, cosmology. Carone, Erlich, Sher.

Plasma and Fluid Physics. Magnetohydrodynamics; kinetic theory; turbulence; numerical simulation of plasmas; applications to fusion; nonlinear dynamics and chaotic signal process; ocean waves; developing type-II quantum computer algorithms for MHD. Supercomputers are used at DoE-NERSC, DoD-NAVO, DoD-ERDC, and Earth Simulator (Japan). Osborne, Tracy, Vahala.

Experimental

Atomic, Molecular, and Optical Physics. Ion-atom and ion-molecule collisions; collisional detachment; inelastic scattering; collisions of ions with surfaces; interactions of lasers with atoms; ultra-cold quantum gases; quantum optics; studies of nonlinear processes. Aubin, Cooke, Danehy, Kuschner, Mikhailov, Novikova, Reilly.

Chemical Physics. Tunable diode laser and Fourier transform spectroscopy in support of atmospheric studies. Benner, Devi. Pulsed FT NMR studies of structure and dynamics in solids, liquid crystals, and polymers. Hoatson.

Condensed Matter Physics. Nuclear magnetic resonance, muon spin rotation, hydrogen in metals, internal fields in solids, electronic structure of metals, microwave properties of high-T_c superconductors, ferroelectrics, piezoelectrics, metallic thin films, magnetic nanostructures. Mesoscopic systems, correlated electron systems. Hoatson, Kossler, Luepke, Lukaszew, Manos, Qazilbash, Reilly, Wolf.

High Energy Particle Physics. Experiments at Fermilab and the Soudan Underground Laboratory. Neutrino masses and mixing. CP violation in neutrinos. Neutrino interactions on nucleons and nuclei. Structure of the weak current. Particle astrophysics and cosmic ray physics. Reactor and long baseline neutrino oscillation experiments. Kordosky, Mathis, McKeown, Nelson, Vahle, Walding.

Nuclear and Hadronic Physics. Intermediate energy experiments at Jefferson Lab. Measurements of the structure of the nucleons and nuclei via electromagnetic and electroweak interactions. Muon absorption, meson spectroscopy, hyper-polarized nuclear targets. Armstrong, Averett, Bosted, Carlini, Deconinck, Griffioen, Lung, Majewski, McKeon, Perdrisat, Riso, Zhao.

Plasma Physics. Properties of glow discharges; glow discharge effects on surfaces. Manos.

Surface Physics. Physical and chemical properties of surfaces; gas adsorption and desorption phenomena: ultrahigh vacuum technology and instrumentation. Manos.

Computational

Turbulence, macroscopic nonlinear systems, soliton theory, wave propagation, signal processing, Monte Carlo simulations, ab initio condensed matter calculations, and lattice quantum chromodynamics. Detmold, Krakauer, Nelson, Orginos, Vahala, Zhang.

GEORGE MASON UNIVERSITY—PHYSICS PH.D.

DEPARTMENT OF PHYSICS AND ASTRONOMY

Fairfax, Virginia 22030-4444

Students Accepted For Degree	FIELDS		
	Physics	Astronomy	Related Fields
Doctorate	X	X	X
Master's	X	X	X

1. General

President: Alan G. Merten
Dean (College of Science): Vikas Chandhoke
Department Chairperson: Michael E. Summers
Department Telephone Number: 703-993-1280
Type of Institution: University
Control: Public
Setting: Suburban
Total Faculty: 1331
Total Graduate Faculty: Not broken down
Total Students: 32,067
Total Graduate Students: 12,365
Annual Graduate Tuition:
In-state residents: $431.50/credit
Out-of-state residents: $1046.50/credit
Tuition Rate for: 2010-2011
Deferred tuition plan: Yes
Other Fees: Yes
Term: Semester

2. Number of Faculty in Department

The combined total of full-time faculty in the three professional ranks is 27. The combined total of full-time, part-time and other faculty at all ranks is 36.

3. Admission, Financial Aid, and Housing

Address admission inquiries to: Office of Admissions, Physics PhD, George Mason University, 4400 University Drive, Fairfax, Virginia, 22030.
Graduate application fee required: $60
Admission deadline (Fall admission): 15 April
Admission information: For fall admission, 2010-2011, 16 students were accepted from 31 applicants.
Admission requirements: For admission to the graduate programs, a Bachelor's degree in physics, or a related discipline is required with a 3.0 GPA specified. The GRE is required. Students from non-English speaking countries are required to demonstrate proficiency in English via the TOEFL exam. Minimum acceptable for admission is 88.

4. Graduate Degree Requirements

Doctorate: The Ph.D. program in Physics with tracks in physics and astonomy requires 72 credits hours from the following categories: (1) 12 credit hours of common core courses; (2) 6 credit hours of Physics and Astronomy Electives (3) 3 credit hours of Seminar in Physics and Astronomy (4) 27 credit hours of General Elective including preliminary research credits (5) 24 credit hours of Dissertation Research. For those holding a master's degree, the 72 hours may be reduced by up to 30 hours.

Table A—Faculty, Enrollments, and Degrees Granted

Research Specialty	2009–10 Faculty	Enrollment Fall 2009		No. of Degrees Granted 2009–10 (2005–09)			Median No. of Years for 2009–10 Ph.D.'s
		Master's	Doctorate	Master's	Terminal Master's	Doctorate	
Astrobiology	1	–	1	–	–	–	–
Astronomy	2	–	–	–	–	–	–
Astrophysics	4	–	6	–	–	1	–
Atomic Physics	2	–	–	–	–	–	–
Biophysics	3	–	–	–	–	–	–
Condensed Matter	2	–	1	–	–	–	–
Health Physics	1	–	–	–	–	–	–
Nonlinear Physics	2	–	2	–	–	–	–
Nuclear Physics	3	–	–	–	–	–	–
Optics	1	–	–	–	–	–	–
Particles & Fields	3	–	1	–	–	–	–
Physics Education	3	–	–	–	–	–	–
Plasma Physics	1	–	–	–	–	–	–
Space Weather	4	–	3	–	–	–	–
Statistical Physics	1	–	–	–	–	–	–
Total	36	–	71	0	–	1	–
Full-time Grad. Stud.		–	24	–	–	–	–
Part-time Grad. Stud.		–	10	–	–	–	–
First-year Grad. Stud.		–	16	–	–	–	–

Table B—Separately Budgeted Research Expenditures

Research Specialty	No. of Grants	Expenditures ($)
Astronomy & Astrophysics	14	$1,450,000
Atomic Physics	2	$627,000
Biophysics	1	$279,000
Condensed Matter	3	$979,000
Particle Physics	2	$169,700
Space Weather	7	1,483,500
Total	25	$4,989,500

Other Programs: The Department of Physics and Astronomy offers a MS in Applied Physics.
Special Equipment, Facilities or Programs: Research facilities are available, including an observatory and computing facilities, with access to major national laboratories located in the metropolitan area.

5. Separately Funded and Managed Laboratories

Krasnow Institute for Advanced Study

FACULTY

Professors

Duxbury, Thomas, M.S., Purdue University, 1966, Space Sciences.
Dworzecka, Maria, Ph.D., Warsaw University, 1969. Quantum mechanics, physics education. Nuclear physics.
Ehrlich, Robert, Ph.D., Columbia University, 1964. Physics, particle physics, physics education. Climate change.
Ellsworth, Robert, Ph.D., University of Rochester, 1966. Particle physics, cosmic rays.

Lieb, B. Joseph, Ph.D., College of William and Mary, 1971. Nuclear physics, planetary physics, atmospheric physics.

Meier, Robert, Ph.D., University of Pittsburgh, 1966. Upper atmospheric and ionospheric physics, and remote sensing.

Mishin, Yuri, Ph.D., Moscow Institute of Steel and Alloys, 1985. Computational materials science. Solid State.

Poland, Arthur, Ph.D., Indiana University, 1968. Solar Physics, Astrophysics, Space Weather Sciences.

Satija, Indu, Ph.D., Columbia University, 1983. Nonlinear dynamics, condensed matter.

Summers, Michael, E., Ph.D., California Institute of Technology, 1985. Planetary Sciences.—Atmospheric Physics, Astrobiology.

Trefil, James, Ph.D., Stanford University, 1966. Theoretical physics, science education.

Associate Professors

Barreto, Ernest, Ph.D., University of Maryland College Park, 1996. Nonlinear dynamics, neuroscience.

Oerter, Robert Ph.D., University of Maryland, College Park, 1989. Particle physics, string theory.

Opher, Merav, Ph.D., University of São Paulo, 1998. Space weather, plasma physics, coronal mass ejections.

Rubin, Philip, Ph.D., University of California, Los Angeles (UCLA), 1989. Particle physics, physics education.

Satyapal, Shobita, Ph.D., University of Rochester, 1995. Astrophysics, infrared astronomy.

Sauer, Karen, Ph.D., Princeton University, 1998. Atomic and molecular physics, and magnetic resonance.

So, Paul, Ph.D., University of Maryland, College Park, 1995. Nonlinear dynamics, neuroscience.

Weingartner, Joseph, Ph.D., Princeton University, 1999. Theoretical astrophysics, cosmic dust.

Assistant Professors

Cressman, J. Robert, Assistant Professor of Physics and Astronomy. B.S. Union College, 1995; Ph.D., University of Pittsburgh, 2003. Experimental biophysics.

Geller, Harold A., D.A., George Mason University, 2005. Education, observational astronomy, astrobiology.

Gliozzi, Mario, Assistant Professor of Physics, Ph.D., Torino University. Astrophysics.

Jazaeri, Amin, M.S., University of Southern California, 1994. Image processing, hyperspectral imaging.

Nikolic, Predrag, Assistant Professor, Ph.D., MIT, Condensed matter theory.

Rosenberg, Jessica, Assistant Professor of Physics and Astronomy. B.A., Wesleyan University, 1993; Ph.D., University of Massachusetts, 2000. Astronomy.

Tian, Ming, Assistant Professor of Physics and Astronomy. B.S., Nankai University, 1984; M.S., Changchum Institute of Physics, Chinese Academic of Sciences, 1987; Ph.D., Paris-Sud University, 1997. Quantum computing; Quantum optics.

Wyczalkowski, Ania, Ph.D., University of Maryland, College Park, 1998. Condensed matter, critical phenomena in fluids.

Zhao, Erhai, Assistant Professor of Physics, Ph.D., Fudan University, 2005, Condensed matter theory.

Professor Emeritus

Mielczarek, Eugenie, Ph.D., Catholic University, 1963. Physics of biological systems, solid state physics.

Instructors

Ericson, Rebecca, M.S., Creighton University, 1975. Astronomy education.

Ewell, Mary, M.S., George Mason University, 2000. Physics education.

RESEARCH SPECIALTIES

Theoretical

Astrophysics: Satyapal, Summers, Rosenberg, Weingartner.
Planetary Science: Lieb, Summers.
Condensed Matter: Satija, Wyczalkowski, Zhao.
Nonlinear Dynamics: Barreto, Satija, So.
Particle Physics: Ehrlich, Oerter, Rubin.
Quantum Theory: Tian, Nikolic, Satija.
Space Physics: Opher.

Experimental

Atomic Physics: Sauer-Tian.
Astronomy Education: Ericson, Geller.
Biophysics: Barreto, Cressman, Mielczarek, So.
Nonlinear Dynamics: Barreto, So-Cressman.
Particle Physics: Ellsworth, Rubin, Sauer.
Physics Education: Dworzecka, Ehrlich, Ewell, Geller, Trefil.

FACULTY PUBLICATIONS

Please see department web site: http://www/physics.gmu.edu

GEORGE MASON UNIVERSITY
College of Science

DEPARTMENT OF COMPUTATIONAL AND DATA SCIENCES

Fairfax, Virginia 22030-4444

1. General

President: Alan G. Merten
College of Science Dean: Chandhoke
Department Chair: D.A. Papaconstantopoulos
Department Telephone Number: (703) 993-3807
Type of Institution: University
Control: Public
Setting: Suburban
Total Faculty: 760
Total Graduate Faculty: Not broken down
Total Students:
Total Graduate Students: 7,269
Graduate Tuition:
In-state residents: Full-time—$431.50/credit
Part-time—$431.50/credit
Out-of-state residents: Full-time—$1,046.50/credit
Part-time—$1,046.50/credit
Tuition rates for: 2010–11
Deferred tuition plan: Yes
Term: Semester

2. Number of Faculty in Department

The combined total of full-time faculty in the three professorial ranks is 16. The combined total of full-time, part-time, and other faculty at all ranks is 25.

3. Admission, Financial Aid, and Housing

Address admission inquiries to: COS Office of Graduate Admissions, 4400 University Drive, MS 6A3, Fairfax, VA 22030-4444

Graduate application fee required: $60 online and $75 paper

Admission deadline (Fall admission): 3/1

Admission information: For fall admission, 2008–09, 20 students were accepted from 28 applicants.

Admission requirements: For admission to the graduate programs, a Bachelor's degree in one of the physical sciences, math or a related discipline is required with a 3.0 GPA specified. The GRE is required. Students from non-English speaking countries are required to demonstrate proficiency in English via the TOEFL exam. Minimum acceptable score for admission is 575.

Undergraduate preparation assumed: Students applying to the SCS doctoral program should have a solid academic background in computer science and either physics or chemistry or mathematics.

Table A—Faculty, Enrollments, and Degrees Granted

Research Specialty	2009 Faculty	Enrollment Fall 2009–10 Master's	Enrollment Fall 2009–10 Doctorate	No. of Degrees Granted 2009–10 (2004–10) Master's	Terminal Master's	Doctorate	Median No. of Years for Ph.D.'s
Space Weather and Astrophysics	9		12			6	
Materials Science	5		14			1	
Fluid Dynamics	5		8	2		4	
Statistics	2		21			2	
Computational Mathematics	2		8	1		1	
Remote Sensing	2		3	3		2	
Total			66	6		16	

4. Graduate Degree Requirements

Doctorate: The Ph.D. program in Computational Sciences and Informatics (CSI) requires 72 credit hours with the following minimums: (1) 12 hours of common CSI core courses; (2) 15 hours of required concentration courses; (3) 18 hours of electives; (4) 24 hours of dissertation research. For those holding a master's degree, the 72 hours may be reduced by up to 30 hours.

Other Programs: M.S. degree in Computational Science and B.S. degree in Computational and Data Sciences

Special Equipment, Facilities, or Programs: The CDS department houses two dozen high-performance PC workstations. The Department also provides students with access to its own parallel supercomputer. External supercomputing facilities are also available.

Table B—Appointments to Graduate Students, 2010–11

Title of Appointee	Appointments Total	Appointments First year	Academic Load Allowed in Credit Hours	Hours of Service Per Week	Stipend for Academic Year ($)
			Semester		
Research Assistant	30	5	12	20	15,500
Total	30	5			

[1]All tuition and fees (resident and nonresident) are waived; first-year stipend, $15,000; half or full-time summer support also provided, depending on availability of funds.
[2]University fellow carries 20-hr. service for one term.

5. Personnel Engaged in Separately Budgeted Research, 7/09–6/10

Professorial faculty	15
Graduate students	30
Undergraduate students	7
Total	42

6. Separately Budgeted Research Expenditures by Source of Support Per Year

	Departmental Research
Federal government	$4,786,000
Total	$4,786,000

Table C—Separately Budgeted Research Expenditures

Research Specialty	No. of Grants	Expenditures ($)
Space Weather and Astrophysics	7	1,700,000
Materials Science	7	1,643,000
Fluid Dynamics	9	1,393,000
Statistics	2	50,000
Total	25	4,786,000

FACULTY

Professors

Blaisten-Barojas, Estela, Ph.D., Universite de Paris VI (France), 1974. Computational physics, condensed matter physics.

Becker, Peter A., Ph.D., University of Colorado, 1987. Astrophysics, applied mathematics.

Gentle, James E., Ph.D., Texas A&M, 1974. Computational statistics, numerical analysis.

Löhner, Rainald, Ph.D., University College of Swansea (Wales, U.K.), 1984. Computational fluid dynamics.

Papaconstantopoulos, Dimitrios, Ph.D., The University of London, 1967. Solid state physics.

Wegman, Edward J., Ph.D., University of Iowa, 1968. Computational statistics.

Associate Professors

Borne, Kirk, Ph.D., California Institute of Technology, 1983. Astronomy, astrophysics.

Cebral, Juan Raul, Ph.D., George Mason University, 1996. Computational fluid dynamics.

Wallin, John F., Ph.D., Iowa State, 1989. Astronomy, astrophysics.

Yang, Chi, Ph.D., Shanghai Jiao Tong University, 1988. Computational fluid dynamics, numerical ship hydrodynamics.

Zhang, Jie, Ph.D., University of Maryland, 1999. Solar physics, space weather.

Zoltek, Stanley M., Ph.D., State University of New York, 1976. Differential geometry.

Assistant Professors

Camelli, Fernando, Ph.D., George Mason University, 2002. Transport and dispersion of pollutants in the atmosphere, turbulence Modeling for CFD, parallel computing.

Griva, Igor, Ph.D., George Mason University, 2002. Nonlinear optimization and applications.

Sheng, Howard, Ph.D., Shenyang National Laboratory for Materials, Chinese Academy of Sciences, 1997. Materials Science.

Weigel, Robert, Ph.D., University of Texas, Austin, 2000. Magnetospheric physics, space weather, and geomagnetism.

Research Professors

Bilitza, Dieter, Ph.D., Albert Ludwigs University, 1984. Ion aspheric phyisics.

Dere, Ken, Ph.D., Catholic University, 1980. Solar physics, space weather.

Odstreil, Dusan, Ph.D., Comenius University, 1984. Space Weather.

Poland, Arthur, Ph.D., Indiana University, 1969. Solar atmosphere, space weather.

Titarchuk, Lev, Ph.D., Space Research Institute (Moscow), 1972. Astrophysics.

RESEARCH SPECIALTIES

Theoretical

Astrophysics. Becker, Borne, Dere, Poland, Wallin, Zhang.

Condensed Matter. Blaisten-Barojas, Papaconstantopoulos, Sheng.

Fluid Dynamics. Cebral, Lohner, Yang, Camelli.

Mathematics. Griva, Zoltek.

Quantum Computing, Gomez.

Space Sciences. Weigel, Zhang.

Statistics. Gentle, Wegman.

VIRGINIA COMMONWEALTH UNIVERSITY

DEPARTMENT OF PHYSICS

Richmond, Virginia 23284-2000

Students Accepted For Degree	FIELDS		
	Physics	Astronomy	Related Fields
Doctorate			X
Master's	X		

1. General

President: Michael Rao
Dean of Graduate Studies: F. Douglas Boudinot
Department Chairman: Alison Baski
Department Telephone Number: (804) 828-1818 (C) 8295
Department FAX Number: (804) 828-7073
Department WEB address: http://www.vcu.edu/hasweb/phy
Type of Institution: University
Control: Public
Setting: Urban
Total Faculty: 1,919
Total Students: 32,436
Total Graduate Students: 7,644
Annual Graduate Tuition:
In-state residents: Full-time—$8,616
Part-time—$479/credit
Out-of-state residents: Full-time—$17,883
Part-time—$994/credit
Tuition rates for: 2010–11
Deferred tuition plan: Yes
Annual Other Fees: $1,898/2,311.
Term: Semester

2. Number of Faculty in Department

The combined total of full-time faculty in the three professorial ranks is 9. The combined total of full-time, part-time, and other faculty at all ranks is 17.

3. Admission, Financial Aid, and Housing

Address admission inquiries to: School of Graduate Studies, Virginia Commonwealth University, P.O. Box 843051, Richmond, VA 23284-3051
Graduate application fee required: $50
Admission deadline (Fall admission): 8/1 for M.S. in Physics/Applied Physics, 3/1 for Medical Physics
Admission requirements: For admission to the graduate programs, a Bachelor's degree in physics or engineering is recommended with a minimum undergraduate GPA of 2.7 specified. The GRE is required. Students from non-English speaking countries are required to demonstrate proficiency in English via the TOEFL exam. Minimum acceptable score for admissions is 550 (paper based) and 213 (computer-based).
Undergraduate preparation assumed: A typical student will have completed intermediate and/or advanced courses in *Classical Mechanics*, Marion and Thornton; *Electricity and Magnetism*, Reitz, Milford, and Christy; and *Modern Physics*, Eisberg and Resnick. Deficiencies in advanced courses may be made up while a graduate student.
Address financial aid inquiries to: UES/Financial Aid, 901 West Franklin St., Richmond, VA 23284-3026
GAPSFAS application required: No (Use form sent with application)
Financial aid deadline: 3/1, for priority consideration.
Loans available: Yes
Address housing inquiries to: Housing Office, 711 W. Main St.
On-campus, single student housing available: Yes
On-campus, married student housing available: No

Table A—Faculty, Enrollments, and Degrees Granted

Research Specialty	2009–10 Faculty	Enrollment[1] Fall 2009 Mas-ter's	Enrollment[1] Fall 2009 Doc-torate	No. of Degrees Granted[2] 2009–10 (2005–10) Mas-ter's	Terminal Master's	Doc-torate[3]	Median No. of Years for 2009–10 Ph.D.'s
Chemical Physics	8	0	3	–	0(0)	1(3)	–
Condensed Matter Physics	8	7		–	2(17)		–
Nanoscience & Nanotechnology	8	–	2	–	–	0	–
Relativity & Gravitation	1	2	0	–	2(4)	0(0)	–
Total		9	5	–	4(21)	1(3)	
Full-time Grad. Stud.		9	5				
Part-time Grad. Stud.							
First-year Grad. Stud.		5	2				
Median Years in Grad. Study (2009–10 Degrees)				–	2	–	
Undergraduate Degrees, 2009–10 (2005–10): 19(69)							

[1]Students not yet committed to a research specialty are entered under non-specialized.
[2]Five-year totals in parentheses.
[3]Under the Chemical Physics program in collaboration with the Department of Chemistry.

4. Graduate Degree Requirements

Master's in Physics/Applied Physics: Completion of 30 approved graduate credits with at least 15 credits of didactic or laboratory course work, and successful completion of a Master's Thesis. Each student will choose an advisor during the first semester and propose a Plan of Study to fulfill the student's individual career goals.
Thesis: Thesis may not be written *in absentia*.
Special Equipment, Facilities, or Programs:: The department has facilities for surface and material physics research, including an Atomic Force Microscope, and equipment for Raman and photoluminescence. Other analytical equipment (SEM, XPS, TEM) is available in a shared facility.

Table B—Appointments to Graduate Students, 2009–10

Title of Appointee	Appointments Total	Appointments First year	Academic Load Allowed in Credit Hours	Hours of Service Per Week	Stipend for Academic Year ($)
			Semester		
Teaching Assistant	9	5	12	15–20	12,006 + tuition
Research Assistant	4	1	12	15–20	12,006 + tuition
Total	13	6			

5. Personnel Engaged in Separately Budgeted Research, 7/09–6/10

Professorial faculty	5
Nonteaching research personnel	11
Total	16

6. Separately Budgeted Research Expenditures by Source of Support

	Departmental Research	Physics-related Research Outside Department
Federal government	$1,270,000	$
State & Local government		
Private		
University		
Total	1,270,000	$

Table C—Separately Budgeted Research Expenditures

Research Specialty	No. of Grants	Expenditures ($)
Condensed Matter Physics	12	1,270,000
Total	12	1,270,000

FACULTY

Professors

Baski, Alison A., Ph.D., Stanford University, 1991. Semiconductor surface studies.

Jena, Purusottam, Ph.D., California, Riverside, 1970. Electronic structure theory of metals and alloys, semiconductors, intermetallics, and insulators; atomic clusters and cluster assembled materials.

Khanna, Shiv N., Ph.D., Delhi, 1976. Theoretical solid state; electronic structure of amorphous metals and small atomic clusters.

Associate Professors

Bertino, Massimo F., Ph.D., MPI-Germany, 1996. Production of nanostructures by photo-lithographic methods.

Bishop, Marilyn F., Ph.D., California, Irvine, 1976. Charge density waves; superconductivity; biophysics, semiconductor.

Gowdy, Robert H., Ph.D., Yale, 1968. General relativity and cosmology.

Reshchikov, Michael A., Ph.D., Ioffe Physical-Technical Institute, 1989. Defects in semiconductors; photoluminescence.

Assistant Professor

Demchenko, Denis, Ph.D., South Dakota School of Mines & Technology, 2002. Theoretical and computational nanoscience, electronic structure theory of semiconductor nanocrystals.

Ye, Dexian, Ph.D., Rensselaer Polytechnic Institute, 2006. Fabrication and characterization of nanostructured surfaces.

Teaching Assistant Professors

Ameen, David B., Ph.D., Virginia Commonwealth University, 2000.

Full-Time Instructors

Reveles, J. Ulises, Ph.D., Cinvestav, Mexico City, 2004.

McMullen, J. Thomas, Ph.D., Queen's, 1968.

Skrobiszewski, John L., M. S., Virginia Commonwealth University, 2003.

RESEARCH SPECIALTIES AND STAFF

Theoretical

Biophysics. Polymerization kinetics of biological polymers; light scattering from polymers photonic band structure in biological systems. Bishop.

Chemical Physics. Properties of metal clusters and cluster assembled materials. Jena, Khanna.

General Relativity. Dynamical structure of gravitational theories; gravitational waves; exact solutions of Einstein's equations; quantum gravity. Gowdy.

Solid State Physics. Electronic structure of defects and defect complexes; electronic and magnetic properties of multilayer thin films, dilute magnetic semiconductors, metal oxides and hydrogen storage materials in bulk and nanostructured forms; transport theory for simple metals and charge density waves; electronic, structural, elastic properties of semiconductors nanostructures and nanoscale photovoltaics. Bishop, Demchenko, Jena, Khanna, Reshchikov.

Experimental

Surface and Materials Science. Growth and characterization of semiconductor and metal systems using scanning probe microscopy techniques, modulation spectroscopy, Raman, and photoluminescence. Photo-lithographic fabrication of nanocomposites for structural and energetic applications. Baski, Bertino, Reshchikov, Ye.

UNIVERSITY OF VIRGINIA

DEPARTMENT OF PHYSICS

Charlottesville, Virginia 22904-4714

Students Accepted For Degree	FIELDS		
	Physics	Astronomy	Related Fields
Doctorate	X		
Master's	X		

1. General

President: Teresa Sullivan
Associate Dean of Graduate School: Robert Fatton
Department Chairman: Joe Poon
Department Telephone Number: (434) 924-3781
E-mail address: grad-info-request@physics.virginia.edu
Type of Institution: University
Control: Public
Setting: Small city
Total Faculty: 2,159
Total Graduate Faculty: 2,159
Total Students: 20,895
Total Graduate Students: 6,598
Annual Graduate Tuition:
In-state residents: Full-time—$13,826
Out-of-state residents: Full-time—$23,822
Tuition rates for: 2010–11
Deferred tuition plan: Yes
Other Annual Fees: $54
Term: Semester

2. Number of Faculty in Department

The combined total of full-time faculty in the three professorial ranks is 40.

3. Admission, Financial Aid, and Housing

Address admission inquiries to: Graduate Admissions Advisor
Graduate application fee required: $60
Admission deadline (Fall admission): 7/15
Admission information: For fall admission, 2010–11, 11 students were accepted from 196 applicants.
Admission requirements: For admission to the graduate programs, a Bachelor's degree is required. The GRE is required. No minimum acceptable score is specified. The GRE Subject is required. No minimum acceptable score is specified. The average GRE scores for 2010–11 admissions were verbal–544; quantitative–769; total–1,313. The average GRE Subject score for 2010–11 admissions was 813. Students from non-English speaking countries are required to demonstrate proficiency in English via the TOEFL exam. Minimum acceptable internet based score for admission is 90.
Undergraduate preparation assumed: Marion and Thornton, *Classical Dynamics of Particles and Systems* (mechanics); Kittel, *Thermal Physics* (statistical physics); Fermi, *Thermodynamics* (statistical physics); Marion, *Classical Electromagnetic Radiation* (electromagnetism); Gasioriowicz, *Quantum Physics* (quantum mechanics).
Address financial aid inquiries to: Graduate Admissions Advisor
GAPSFAS application required: No
Financial aid deadline: 1/7
Loans available: Yes
Address housing inquiries to: housing@virginia.edu
On-campus, single student housing available: Yes
Cost/year: $5,240–7,310
On-campus, married student housing available: Yes
Cost/month: $675–842

Table A—Faculty, Enrollments, and Degrees Granted

Research Specialty	2008–09 Faculty	Enrollment[1] Fall 2008		No. of Degrees Granted[2] 2008–09 (2004–09)			Median No. of Years for 2008–09 Ph.D.'s
		Master's	Doctorate	Master's	Terminal Master's	Doctorate	
Atomic, Molecular, & Optical Physics	7	1	19	0(1)	0(2)	2(19)	6
Biophysics	2	0	4	0(1)	0(0)	1(1)	8
Chemical Physics	4	0	0	0(0)	0(0)	0(0)	–
Condensed Matter Physics	13	0	19	0(1)	0(0)	3(10)	5.3
Engineering Physics/ Science	3	0	5	0(0)	0(1)	0(1)	8
Material Science/ Metallurgy	1	0	0	0(0)	0(0)	0(2)	7
Nuclear Physics	14	0	6	0(0)	0(4)	0(9)	7
Particles & Fields	8	2	19	0(4)	0(2)	1(16)	9
Physics Education	2	63	0	0(0)	21(46)	0(0)	–
Non-specialized	0	0	29	0(1)	0(0)	0(0)	–
Total		66	101	0(8)	21(55)	7(58)	
Full-time Grad. Stud.		3	101				
Part-time Grad. Stud.		63	0				
First-year Grad. Stud.		0	26				
Median Years in Grad. Study (2007–08 Degrees)				4.3	2.5	7.8	6.9

Undergraduate Degrees, 2008–09 (2004–09): 45(200)

[1]Students not yet committed to a research specialty are entered under non-specialized.
[2]Five-year totals in parentheses.

4. Graduate Degree Requirements

Master's: 24 graduate credits in approved program with satisfactory performance, thesis, and thesis exam; no language requirement.
Doctorate: 72 graduate credits required; satisfactory performance in an approved course program is required; comprehensive exam, dissertation, and dissertation exam; two semesters residency required; no language required.
Other Programs: The Department also offers M.S. and Ph.D. degrees in Engineering Physics in cooperation with the School of Engineering and Applied Science. A Ph.D. in Biophysics is available through an interdisciplinary program associated with the physics department and other science departments of the University. A Master of Arts in Physics Education (MAPE) degree is offered to High School teachers.
Thesis: Thesis may be written *in absentia*.
Special Equipment, Facilities, or Programs: Department facilities include machine and electronics shops, a physics library, and extensive computing resources. Most on-site research is carried out in the J. W. Beams Laboratory building, with additional facilities on campus for work in laser science, materials science, nanoscale fabrication, and particle detector development. The department is active at many off-site facilities as well, including major particle accelerators, neutron sources, and synchrotron sources in the US and abroad.

Table B—Appointments to Graduate Students, 2008–09

Title of Appointee	Appointments: Total	Appointments: First year	Academic Load Allowed in Credit Hours	Hours of Service Per Week	Stipend for Academic Year ($)
			Semester		
Teaching Assistant	39	23	12	20	17,900
Research Assistant	47	0	12	20	17,900
Dept. and/or Grad. School Fellowships	12	3	12	0	17,900–18,000
External Fellowships	6	0	12	0	
Total	104	26			

5. Personnel Engaged in Separately Budgeted Research, 6/08–5/09

Professorial faculty	12
Postdoctoral appointments	22
Graduate students	47
Undergraduate students	17
Total	98

6. Separately Budgeted Research Expenditures by Source of Support

	Departmental Research
Federal government	$5,775,837
State and local government	419,731
Private, nonprofit organizations	
Institute for Nuclear and Particle Physics	502,035
Total	$6,697,602

Table C—Separately Budgeted Research Expenditures

Research Specialty	No. of Grants	Expenditures ($)
Atomic and Molecular Physics	23	1,072,296
Condensed Matter	33	1,588,488
Nuclear Physics	35	2,572,714
Particles & Fields	18	1,359,166
Physics Education	1	104,937
Total	110	6,697,601

FACULTY

Professors

Arnold, Peter B., Ph.D., Stanford, 1986. Theoretical particle physics.

Bloomfield, Louis A., Ph.D., Stanford, 1983. Experimental atomic and solid state physics.

Cates, Gordon D., Jr., Ph.D., Yale, 1987. Experimental nuclear and atomic physics.

Cox, Bradley B., Ph.D., Duke, 1967. Experimental high-energy particle physics.

Dukes, Edmond C., Ph.D., Michigan, 1984. Experimental elementary particle physics.

Fendley, Paul, Ph.D., Harvard, 1990. Theoretical condensed matter and particle physics.

Fowler, Michael, Ph.D., St. John's College, Cambridge, 1962. Theoretical physics; field theory and solid state theory; physics education.

Gallagher, Thomas F., Ph.D., Harvard, 1971. Collisions and spectroscopy of atoms and molecules.

Hess, George B., Ph.D., Stanford, 1967. Experimental solid state physics; liquid helium; physisorption.

Hung, Pham Q., Ph.D., UCLA, 1978. Theoretical particle physics; cosmology.

Jones, Robert R., Jr., Ph.D., Virginia, 1990. Experimental atomic molecular and optical physics.

Lee, Seung-Hun, Ph.D., Johns Hopkins, 1996. Experimental condensed matter physics.

Norum, Blaine E., Ph.D., MIT, 1979. Experimental nuclear and particle physics.

Počanić, Dinko, Ph.D., Zagreb, 1981. Chairman of the Department. Experimental intermediate-energy nuclear and particle physics.

Poon, S. Joseph, Ph.D., Caltech., 1978. Experimental solid state physics; nanostructured materials; quasicrystals; thermoelectric compounds.

Thacker, Harry B., Jr., Ph.D., UCLA, 1973. Elementary particle physics and quantum field theory.

Thornton, Stephen T., Ph.D., Tennessee, 1967. Experimental nuclear physics; physics education.

Associate Professors

Hirosky, Robert J., Ph.D., Rochester, 1994. Experimental particle physics.

Kolomeisky, Eugene B., Ph.D., Academy of Sciences of the USSR, Moscow, 1988. Theoretical condensed matter.

Liyanage, Nilanga, Ph.D., MIT, 1999. Experimental nuclear and particle physics.

Louca, Despina A., Ph.D., Pennsylvania, 1997. Experimental condensed matter.

Pfister, Olivier, Ph.D., University of Paris-North, France, 1993. Experimental atomic, molecular, and optical physics.

Sackett, Charles A., Ph.D., Rice 1998. Experimental atomic, molecular and optical physics.

Shivaram, Bellave S., Ph.D., Northwestern, 1984. Experimental solid state physics.

Yoon, Jongsoo, Ph.D., Penn State, 1997. Experimental condensed matter physics.

Assistant Professors

Baeßler, Stefan, Ph.D., University of Heidelberg, 1996. Experimental nuclear and particle physics.

Dawson, Chrstopher, Ph.D., University of Southampton, 1998. Theoretical nuclear and particle physics.

Group, R. Craig, Ph.D., Florida, 2006. Experimental high energy physics.

Klich, Israel, Ph.D., Israel Institute of Technology, 2004. Theoretical condensed matter; theoretical mathematical physics.

Lamacraft, Austen, Ph.D., University of Cambridge, 2002. Theoretical condensed matter physics.

Neu, Christopher, Ph.D., Ohio State, 2003. Experimental high energy physics.

Paschke, Kent D., Ph.D., Carnegie Mellon, Ph.D., 2001. Experimental nuclear and particle physics.

Vaman, Diana, Ph.D., SUNY, Stony Brook, 2001. Theoretical nuclear and particle physics.

Zheng, Xiaochao, Ph.D., M.I.T., 2002. Experimental nuclear and particle physics.

Research Professors

Crabb, Donald G., Ph.D., Southhampton, 1967. Experimental nuclear and particle physics.
Day, Donal B., Ph.D., Virginia, 1979. Experimental nuclear and particle physics.
Lindgren, Richard A., Ph.D., Yale, 1969. Experimental nuclear and particle physics, physics education.

Research Associate Professor

Liuti, Simonetta, Ph.D., Universitád: Roma, 1989. Theoretical nuclear and particle physics.

Adjunct Faculty

Cardman, Lawrence S., Ph.D., Yale, 1972. Experimental nuclear and particle physics.
Gillies, George T., Ph.D., Virginia, 1980. Engineering Physics.
Imlay, Richard, Ph.D., Princeton, 1967. Experimental high energy.
Lehmann, Kevin, Ph.D., Harvard, 1983. Experimental chemical physics; experimental atomic, molecular, and optical physics.
Williams, Mark B., Ph.D., Virginia, 1990. Medical Physics.

Professors Emeriti

Brill, Arthur S., Ph.D., Pennsylvania, 1956. Experimental biophysics; proteins and transition metal ions.
Celli, Vittorio, Dottore in Fisica, Pavia, 1958. Theoretical solid state physics; surface studies.
Conetti, Sergio, Dottore in Fisica, Trieste, 1967. Experimental high-energy particle physics.
Coopersmith, Michael, Ph.D., Cornell, 1962. Theoretical physics; statistical mechanics; phase transitions.
Deaver, Bascom S., Jr., Ph.D., Stanford, 1962. Experimental solid state physics: superconducting devices.
Fishbane, Paul M., Ph.D., Princeton, 1967. Theoretical physics; elementary particles.
Minehart, Ralph C., Ph.D., Harvard, 1962. Experimental nuclear and particle physics.
Ruvalds, John, Ph.D., Oregon, 1967. Theoretical solid state physics.
Ritter, Rogers C., Ph.D., Tennessee, 1961. Gravitation; precision measurements; medical physics.
Schnatterly, Stephen, Ph.D., Illinois, 1965. Experimental solid state physics; soft x-ray and inelastic electron scattering spectroscopy of solids, atoms, and molecules.
Sobottka, Stanley E., Ph.D., Stanford, 1960. X-ray crystallography; x-ray detector development; experimental nuclear physics.
Weber, Hans J., Ph.D., Frankfurt, 1965. Theoretical nuclear and particle physics.
Ziock, Klaus O. H., Ph.D., Bonn, 1956. Experimental nuclear and particle physics.

RESEARCH SPECIALTIES AND STAFF

Theoretical

Condensed Matter/AMO. Field theoretic models for solid state systems; many-body physics in ultracold atomic gases; quantum Hall effect; Bethe Ansatz systems; topological quantum computation; phase transitions and renormalization group methods in statistical physics; Bose-Einstein condensation; theory of macroscopic quantum phenomena; pattern formation; nonperturbative statistical mechanics. Fendley, Kolomeisky, Lamacraft.

High Energy Physics. Theoretical studies of high-energy physics including properties of quantum chromodynamics; lattice gauge theory; string/gauge duality; high-temperature field thoery; electroweak interactions; grand unified theories; supersymmetry; neutrino physics including models of neutrino masses; dark matter; dark energy; cosmology. Arnold, Hung, Thacker, Vaman.

Nuclear and Hadronic Physics. Lattice gauge theory; inclusive and exclusive deep inelastic electron and neutrino scattering on nucleons and nuclei; the spin composition of quarks and gluons within hadrons; the role of QCD in hadronic structure. Dawson, Liuti.

Experimental

Atomic, Molecular and Optical Physics. Laser manipulation and spectroscopy of atoms, ions, small molecules, and clusters, including Bose-Einstein condensation in dilute vapors; atom interferometry; quantum optics; quantum information; optical interferometry; dipole-dipole interactions between cold Rydberg atoms; ultracold plasmas; observation and control of electronic wavepackets in Rydberg atoms using microwave, THz, and optical fields; dynamics of atoms and molecules in intense femtosecond laser pulses; high-order harmonic generation in gases; spectroscopy of single and doubly excited Rydberg atoms; studies of magnetic properties of clusters; photo-detachment and photoionization; development of new techniques in laser spectroscopy; noble gas hyper polarization via spin exchange with optically pumped alkali atoms, optical control of chemical processes; cavity ring-down spectroscopy; spectroscopy using helium nano-droplet isolation; investigation of highly excited vibrational states; microwave-optical double resonance. Bloomfield, Cates, Gallagher, Jones, Lehmann, Pfister, Sackett.

Condensed Matter Physics. Studies of the electronic, magnetic, optical, acoustic, thermoelectric and nanostructural properties of solids ranging from amorphous to crystalline systems that exhibit unusual properties. Recent activities include synthesis and characterization of colossal magnetoresistive manganites, heavy fermion materials, high temperature superconductors, bulk metallic glasses, and intermetallic compounds; development of solid state receivers and sources using GaAs Schottky diodes and superconducting tunnel junctions; measurements of thermodynamic properties of low dimensional electronic systems including superconductors and nano-structured materials at low temperatures with high magnetic fields; characterization of static and dynamic lattice effects in perovskite oxides, intermetallic alloys, ferromagnets and martensites using the pair density function analysis in the studies of phase transitions, measurement of magnetic and quantum correlation effects in heavy fermion and high-temperature superconductors, studies of electronic excitations in thin films using inelastic electron scattering, and studies of solubility and segregation in binary mixture films using ellipsometry and infrared absorption. The condensed matter community has access to a variety of cryogenic facilities, high-field magnets, different scanning-probe instruments such as scanning tunneling, force, and optical microscopes, various vacuum thin-film deposition and etching systems, and a range of microwave and millimeter-wave analytic instruments, electron-beam microscopy, and photolithography facilities, as well as access to national labs where high magnetic field sources are available. In addition, many research projects are carried out in collaboration with the Electrical Engineering and Materials Science Departments at UVa. The group also collaborates and performs research at national and international neutron and x-ray facilities and carries out high precision measurements on the atomistic properties of materials particularly under high pressure. Deaver, Hess, Lee, Louca, Poon, Shivaram.

High Energy Physics. The experimental particle physics group at the University of Virginia is currently involved in several endeavors around the globe. At the energy frontier Virginia has a significant effort on the CMS experiment at the LHC at CERN, complementing continuing enterprises on both the D0 and CDF experiments at the Fermilab Tevatron. The collider physics groups are interested in supersymmetry, top quark and QCD measurements, and the search for the Higgs and other new phenomena. At the intensity frontier the group is playing a major role on the NOVA experiment at Fermilab, which will be one of the flagship neutrino experiments in the coming decade, and on the Mu2e experiment at Fermilab, which will be searching for charged lepton flavor violation with unprecedented sensitivities. Arenton, Cox, Dukes, Hirosky, New, Paschke, Pocanic Zheng. (1 Senior Scientist).

Nuclear and Particle Physics. Researchers at UVa lead active programs at various accelerator laboratories in the United States and abroad, in the fields of electronuclear physics, neutron physics, fundamental symmetries, nuclear-polarized target development, and accelerator physics. UVa professors perform research with electron accelerator at Thomas Jefferson National Accelerator Facility (TJNAF), using a variety of techniques in a broad range of studies probing low and medium energy quantum chromodynamics (QCD). Topics addressed include the longitudinal and transverse spin structure of the nucleon, nucleon and nuclear form factors, nucleonic excitations, QCD sum rules, tests of effective field theories, structure of few-body systems, and the correspondence between low-energy and high-energy QCD phenomena. Additional tests of QCD sum rules and effective field theories are carried out at the HIGS facility at TUNL in North Carolina. UVa professors lead studies of fundamental electroweak physics at the low-energy precision frontier, including studies of the weak interaction through pion decay at the Paul Scherrer Institut in Switzerland, parity-violating electron scattering studies at TJNAF, and neutron decay studies planned for the Spalation Neutron Source (SNS) at Oak Ridge National Laboratory. The boundary of electroweak and QCD physics will also be studied through hadronic parity violation at the SNS, while a search for new interactions using gravitationally bound neutrons is performed at the ILL in Grenoble, France. Topics in accelerator physics addressed by UVa researchers include projects to improve polarized electron sources and superconducting radio frequency cavities, which will be key technologies in the development of future high energy electron accelerators. Research activities carried out locally at the university include development of nuclear-polarized solid targets for particle scattering experiments. A second laboratory is dedicated to developing polarized noble gas targets for particle physics, as well as exploring the application of polarized noble gases in medical imaging. A detector development laboratory is reserved for large-scale particle-physics detector projects. Baeßler, Cardman, Cates, Crabb, Day, Lindgren, Liyanage, Minehart (Emeritus), Norum, Paschke, Pocanic, Zheng. (4 staff research scientists, 1 staff principle scientist).

Physics Education. The department has an outreach professional development program offering local and distance learning courses to K-12 teachers that include physical/Earth science teachers and high school physics teachers. Courses are offered at sites throughout Virginia, in residential summer workshops in Charlottesville, and distance learning, web-based courses on the Internet. Teachers take these courses for endorsement, recertification, or simply to increase their physics content. The Masters of Arts in Physics Education degree (MAPE) program is active with a rolling application deadline. Lindgren, Thornton.

UNIVERSITY OF VIRGINIA

DEPARTMENT OF ASTRONOMY

Charlottesville, Virginia 22904

Students Accepted For Degree	FIELDS		
	Physics	Astronomy	Related Fields
Doctorate		X	
Master's		X	

1. General

President: : John T. Casteen III
Dean of Graduate School: Meridith J. Woo
Department Chairman: John F. Hawley
Department Telephone Number: (434) 924-7494 (C)
Type of Institution: University
Control: Public
Setting: Small town
Total Faculty: 2,159
Total Graduate Faculty: 2,159
Total Students: 20,895
Total Graduate Students: 6,598
Annual Graduate Tuition and Required Fees:
In-state residents: Full-time—$12,628
Out-of-state residents: Full-time—$22,628
Tuition rates for: 2009–2010
Deferred tuition plan: Yes
Other Fees: $57
Term: Semester

2. Number of Faculty in Department

The combined total of full-time faculty in the three professorial ranks is 16. The combined total of full-time, part-time, and other faculty at all ranks is 62.

3. Admission, Financial Aid, and Housing

Address admission inquiries to: Graduate Admissions, Dept. of Astronomy, Box 400325, Charlottesville, VA 22904-4325. Or send email to gradadm@www.astro.virginia.edu. Or see our home page at http://www.astro.virginia.edu.
Graduate application fee required: $60
Admission deadline (Fall admission): 12/31
Admission information: For fall admission, 2010, 21 students were accepted from 79 applicants. Six enrolled.
Admission requirements: For admission to the graduate programs, a Bachelor's degree in physics or astronomy is required with a minimum undergraduate GPA of 3.0 specified. The GRE is required. No minimums stated. The GRE Advanced Physics is required. No minimums stated. Scores better than 600–verbal, 700–quantitative, and 630–advanced are desirable. Students from non-English speaking countries are required to demonstrate proficiency in English via the TOEFL exam. Minimum acceptable score for admission is 630 on the paper test or 267 on the computer version.
Address financial aid inquiries to: Graduate Admissions, Dept. of Astronomy, Box 400325, Charlottesville, VA 22904-4325. Or send email to gradadm@www.astro.virginia.edu. Or see our home page at http://www.astro.virginia.edu.
GAPSFAS application required: No
Financial aid deadline: 12/3
Loans available: Yes
Undergraduate preparation assumed: Sound undergraduate training in physics is assumed, but no background in astronomy is necessary.
Address housing inquiries to: Director of Housing, Station #1, Emmet House, UVa, Charlottesville, VA 22904
On-campus, single student housing available: Yes
On-campus, married student housing available: Yes

Table A—Faculty, Enrollments, and Degrees Granted

Research Specialty	2009–10 Faculty	Enrollment[1] Fall 2009		No. of Degrees Granted[2] 2009–10 (2005–10)			Median No. of Years for 2009–10 Ph.D.'s
		Master's	Doctorate	Master's	Terminal Master's	Doctorate	
Astronomy/ Astrophysics	16	10	25	3(25)	0(2)	6(14)	7
Total	16	10	25	3(25)	0(2)	6(14)	
Full-time Grad. Stud.		11	21				
Part-time Grad. Stud.		0	0				
First-year Grad. Stud.		5	0				
Median Years in Grad. Study (2009–10 Degrees)				2	–	7	–

Undergraduate Degrees, 2009–10 (2005–10): 7(32)

[1]Students not yet committed to a research specialty are entered under non-specialized.
[2]Five-year totals in parentheses.

4. Graduate Degree Requirements

Master's: 30 hours of credit in approved courses plus demonstration of research competency is required. Student must pass a qualifying exam at the beginning of the second semester. Student must be in residence for two semesters.
Doctorate: In addition to requirements for a Master's degree, the student must take 24 additional credit hours of approved courses and pass a comprehensive exam at the beginning of the fourth semester. Ph.D. dissertation is required, as are two additional years of residency.
Thesis: Thesis may be written *in absentia*.
Special Equipment, Facilities, or Programs: The Department operates two local observing sites, with 66-cm, 76-cm, and 100-cm telescopes equipped with fast-readout, large-format CCD imaging devices, IR imaging cameras as well as a fiber optic spectrograph. A modern IR lab is actively engaged in camera and spectrograph development and detector characterization programs. The Department has a fully equipped machine shop and electronics shop with a machinist and electronics technician on staff. The Department is part of the Large Binocular Telescope (LBT) consortium through the Research Corporation. The department has access to Steward Observatory (University of Arizona) telescopes. The Department is a member of the Astrophysical Research Corporation, the Apache Point Observatory, and the Sloan Sky Survey III project. The National Radio Astronomy Observatory has its headquarters located on campus, and students may work with astonomers there. The Virginia Institute for Theoretical Astronomy, which includes visitors and postdoctoral fellows, is affiliated with the Department. The Department maintains an extensive network of Linux workstations for research, as well as WIN-XP PCs for instructional and administrative purposes. The Department operates a 24 node dual core Linux

Beowulf cluster with Infiniband interconnect for computational research. Department members also have access to University research Beowulf clusters.

Table B—Appointments to Graduate Students, 2009–10

Title of Appointee	Appointments Total	Appointments First year	Academic Load Allowed in Credit Hours	Hours of Service Per Week	Stipend for Academic Year ($)
			Semester		
Teaching Assistant	8	3	12	20	17,600[1]
Research Assistant	11	0	12	20	17,600[1]
Fellowship	16	2	12	20	17,600[1]
Total	35	5			

[1]Tuition is paid or waived for all supported students. Additional summer research or teaching support is available for most students. Health insurance is provided.

5. Personnel Engaged in Separately Budgeted Research, 7/08–6/10

Professorial faculty	16
Postdoctoral faculty	11
Graduate students	35
Undergraduate students	8
Nonteaching research personnel	8
Total	70

6. Separately Budgeted Research Expenditures by Source of Support

	Departmental Research	Physics-related Research Outside Department
Federal government	$3,083,654	$
Other	585,870	
Total	$3,669,524	$

Table C—Separately Budgeted Research Expenditures

Research Specialty	No. of Grants	Expenditures ($)
Astronomy/Astrophysics	76	
Total	76	

FACULTY

Professors

Chevalier, Roger A., Ph.D., Princeton, 1973. Supernovae; gas dynamics.

Hawley, John F., Ph.D., Illinois, 1984. Department Chairman. Computational astrophysics; black hole accretion; magnetohydrodynamics.

Ianna, Philip A., Emeritus, Ph.D., Ohio State, 1968. Astrometry; white dwarfs; star clusters; low mass stars.

Johnson, Robert E., (Dept. of Materials Science) Ph.D., Wisconsin. Atomic and molecular physics; planetary satellites; interplanetary dust.

Li, Z. Y., Ph.D., Colorado, 1993. AGN, star formation, astrophysical jets & winds.

Majewski, Steven R., Ph.D., Chicago, 1991. Galactic structure and dynamics; high redshift galaxies; astrometry.

O'Connell, Robert W., Ph.D., Cal. Tech., 1970. Stellar populations in galaxies; active galaxy nuclei; space astronomy.

Rood, Robert T., Ph.D., MIT, 1969. Stellar evolution; nucleosynthesis.

Sarazin, Craig L., Ph.D., Princeton, 1975. Interstellar medium; high-energy astrophysics; clusters of galaxies.

Saslaw, William C., Ph.D., Cambridge, 1968. Cosmology; radio galaxies; stellar dynamics.

Skrutskie, Michael F., Ph.D., Cornell, 1987. Infrared astronomy, astronomical instrumentation, brown dwarfs, galactic structure.

Thuan, Trinh X., Ph.D., Princeton, 1974. Structure and evolution of galaxies; interstellar material in galaxies.

Tolbert, Charles R., Ph.D., Emeritus, Vanderbilt, 1963. Galactic structure; photometry.

Whittle, Mark, Ph.D., Cambridge, 1982. Active galaxy nuclei.

Associate Professors

Evans, Aaron S., Ph.D., Hawaii, 1996. Extragalactic astronomy; active galactic nuclei; infrared and radio astronomy.

Murphy, Edward M., Ph.D., Virginia, 1996. Interstellar medium; astronomy education.

Verbiscer, Anne J., Ph.D., Cornell, 1991. Planetary astronomy, planetary ices.

Assistant Professors

Arras, Phil, Ph.D., Cornell, 1999. Theoretical astrophysics, stellar and planetary physics, relativistic astrophysics.

Indebetouw, Remy, Ph.D., Colorado, 2001. Infrared and radio astronomy; interstellar medium; star formation.

Johnson, Kelsey E., Ph.D., Colorado, 2001. Starburst galaxies; massive star clusters; star formation; multi-wavelength observations.

Research Professors

Bastian, Timothy S., Ph.D., Colorado, 1987. Plasma astrophysics; solar radio astronomy.

Braatz, James A., Ph.D., Maryland, 1996. Cosmic masers; active galactic nuclei, radio astronomy, analysis software and algorithms.

Bradley, Richard F., Ph.D., Virginia, 1992. Radio astronomy instrumentation.

Bridle, Alan H., Ph.D., Cambridge, 1967. Radio interferometry; active galaxy nuclei.

Brogan, Crystal, Ph.D., Kentucky, 2000. Star formation; interstellar medium.

Burton, W. Butler, Ph.D., Leiden, 1970. Interstellar medium, HI, galactic structure.

Carilli, Christopher, Ph.D., MIT, 1989. Radio astronomy; cosmology.

Condon, James J., Ph.D., Cornell, 1972. Radio interferometry; active galaxy nuclei.

Cotton, William, Ph.D., Texas, 1975. Radio interferometry.

Fisher, J. Richard, Ph.D., Maryland, 1972. Radio astronomy instrumentation.

Fomalont, Edward B., Ph.D., Cal. Tech., 1967. Radio astronomy.

Friel, Eileen, Ph.D., UCSC, 1986. Stellar astronomy; Galactic structure.

Hibbard, John E., Ph.D., Columbia, 1995. H I in interacting galaxies.

Hogg, David E., Ph.D., Toronto, 1962. Radio interferometry; extragalactic radio sources.

Hunt, Gareth, M.S., Manchester, 1970. Astronomical data analysis.
Jewell, Phillip, Ph.D., Illinois, 1982. Interstellar chemistry.
Kellermann, Kenneth I., Ph.D., Cal. Tech., 1963. Radio interferometry; compact radio sources.
Kerr, Anthony R., Ph.D., Melbourne, 1969. Microwave electronics.
Kunkel, William E., Ph.D., Texas, 1967. Galactic dynamics, stellar astronomy.
Liszt, Harvey S., Ph.D., Princeton, 1973. Interstellar molecules.
Lo, K-Y. (Fred), Ph.D., MIT, 1974. Active galaxies, star formation, interstellar molecules.
Lonsdale, Carol J. Ph.D., Edinburgh, 1980. Infrared astronomy, galaxy formation, active galactic nuclei.
Mangum, Jeffrey, Ph.D. Virginia, 1990. Star formation; interstellar chemistry.
Mason, Brian, Ph.D., Penn., 1999. Radio astronomy; cosmology.
Murphy, Patrick, Ph.D., Univ. College (Dublin), 1983. Astronomical data analysis.
Pan, Shing-Kuo, Ph.D., Columbia, 1984. Radio instrumentation.
Pospiezalski, Marian W., Ph.D., Warsaw, 1976. Microwave electronics.
Ransom, Scott M., Ph.D., Harvard, 2001. Pulsars; neutron stars.
Roberts, Mort, Ph.D., Berkeley, 1958. Radio astronomy.
Seidelmann, P. Kenneth, Ph.D., Cincinnati, 1968. Astrometry; dynamical astronomy.
Sheth, Kortik, Ph.D., Maryland, 2001. Galactic evolution; molecular astrophysics.
Uson, Juan, Ph.D., Madrid, 1979. Radio interferometry; clusters of galaxies.
Vanden Bout, Paul A., Ph.D., California, Berkeley, 1966. Interstellar medium; x-ray astronomy.
Webber, John, Ph.D., Caltech, 1970. Radio interferometry.
Wootten, H. Alwyn, Ph.D., Texas, 1968. Interstellar molecules.
Yin, Qi-Feng, Ph.D., Beijing, 1960. Radio astronomy.

Research Scientists

Black, Gregory J., Ph.D., Cornell, 1997. Planetary satellites, radar astronomy.
Hearty, Frederick R., Ph.D., Colorado, 2007. Astronomical instrumentation, extragalactic astronomy.
McDavid, David A., Ph.D., Amsterdam, 2001. Emission line stars, polarimetry.
Nelson, Matthew J., Ph.D., Wisconsin, 1987. Variable stars; compact objects, space astronomy; astronomical instrumentation and telescope engineering.
Patterson, Richard, Ph.D., Virginia, 1994. Dwarf galaxies, galactic structure, astrometry.
Wilson, John C., Ph.D., Cornell, 2002. Infrared astronomy, instrumentation, brown dwarfs.

Postdoctoral Fellows

Chen, Chang-Hui R., Ph.D., Illinois, 2007. Star formation; interstellar medium; supernovae.
Drosback, Meredith M., Ph.D., Colorado, 2006. Star formation; infrared and millimeter astronomy.
Garcia Pẽrez, Ana, Ph.D., Uppsala, 2006. Galactic chemical evolution; stellar astronomy.
Guan, Xiaoyue, Ph.D., Illinois, 2009. Computational astrophysics; high-energy astrophysics.
Kepley, Amanda A., Ph.D., Wisonsin-Madison, 2008. Radio astronomy; irregular galaxies.
Kim, DongCheng, Ph.D., Hawaii, 1995. Infrared astronomy; active galatic nuclei.
Nidever, David, Ph.D., Virginia, 2010. Glaactic structure and dynamcis; radio and infrared astronomy.
Sanna, Nicoletta, Ph.D., Rome, 2010. Stellar populations; starburst galaxies; stellar astronomy.
Sivakoff, Gregory, Ph.D., Virginia, 2006. X-ray astronomy; elliptical galaxies; active galactic nuclei; radio astronomy.
Sun, Ming, Ph.D., Harvard, 2005. X-ray Astronomy; galaxy clusters; galaxy evolution.
Yao, Lihong, Ph.D., Toronto, 2009. Starburst galaxies; molecular astrophysics; starformation.

RESEARCH SPECIALTIES AND STAFF

Theoretical

Computational Astrophysics. Hawley.
Cosmology. Dynamics and evolution of galaxies. Saslaw, Thuan.
High-Energy Astrophysics. X-ray sources in galaxies and clusters of galaxies; supernovae; collapsed stars. Arras, Chevalier, Hawley, Li, Sarazin.
Interstellar Medium. Chevalier, Sarazin.
Star Formation. Li.
Stellar Evolution. Globular clusters; degenerate stars; binary stars. Arras, Li, Rood.
Strong Radio Sources and Active Galaxy Nuclei. Hawley, Saslaw, Whittle.

Observational

Astrometry. Parallaxes and proper motions. Ianna, Majewski, Patterson, Seidelmann.
Infrared Astronomy and Astronomical Instrumentation. Hearty, M. Nelson, Skrutskie, Wilson.
Interstellar medium. Indebetouw, K. Johnson.
Optical, UV, and X-ray Observations of Extragalactic Objects. Stellar and interstellar content of galaxies; galaxy dynamics; clusters of galaxies; active galaxy nuclei. Chevalier, Hearty, K. Johnson, Majewski, O'Connell, Sarazin, Thuan, Whittle.
Planetary astronomy. Arras, Black, R. Johnson, Verbiscer.
Radio and Microwave Astronomy. Interferometry; neutral hydrogen and molecular line observations of galactic and extragalactic sources; pulsars. Hearty, Rood, Thuan, Tolbert.
Space Astronomy. Majewski, O'Connell, Patterson, Sarazin, Whittle.
Star Formation. Indebetouw, K. Johnson.
Starburst Galaxies. K. Johnson, O'Connell.
Stellar Astrophysics. Nelson, O'Connell, McDavid, Rood, Wilson.

UNIVERSITY OF WASHINGTON

DEPARTMENT OF PHYSICS

Seattle, Washington 98195

Students Accepted For Degree	FIELDS		
	Physics	Astronomy	Related Fields
Doctorate	X		
Master's	X		

1. General

President: Mark Emmert
Dean of Graduate School: Gerald Baldasty
Department Chairman: Blayne Hecked
Department Telephone Number: (206) 543-2771
FAX: (206) 685-0635
URL: http://www.phys.washington.edu
Type of Institution: University
Control: Public
Setting: Urban
Total Faculty: 6,022
Total Students: 41,855
Total Graduate Students: 14,130
Annual Graduate Tuition:
In-state residents: Full-time—$11,495*
Part-time—$3,067*/2 credits
($3,535 for summer quarter)
Out-of-state residents: Full-time—$25,836*
Part-time—$6,877*/2 credits
($7,982for summer quarter)
Deferred tuition plan: No (payroll tuition deduction)
Other Fees: None
Term: Quarter

*This tuition rate is for the academic year (9 months).
Students holding assistantships of half-time or more receive a tuition waiver, but pay approximately $250/qtr in student fees.

2. Number of Faculty in Department

The combined total of full-time faculty in the three professorial ranks is 53. The combined total of full-time, part-time, and other faculty at all ranks is 137.

3. Admission, Financial Aid, and Housing

Address admission inquiries to: Graduate Program Assistant, Department of Physics, University of Washington Box 351560, Seattle, WA 98195-1560; (206), 543-2488; grad@phys.washington.edu

Graduate on-line application fee required: $50

Admission deadline (Fall admission): January 15

Admission information (Ph.D. program): For Fall admission, 2009–10, 99 students were admitted from 392 applicants. Of the 99 students who were admitted, 30 accepted.

Admission requirements (Ph.D. program): For admission to the Ph.D. program, a Bachelor's degree in physics is required with a minimum undergraduate GPA of 3.0. Both the GRE Aptitude and Advanced Physics tests are required. There is no minimum score requirement but considerable weight is given to these scores. The average GRE advanced score for admission was 815. Students from non-English speaking countries are required to demonstrate proficiency in English via the TOEFL exam. Minimum acceptable score for admission is 580 on the paper-based TOEFL and 237 on the computer-based TOEFL. A test of English as a spoken language is required.

Table A—Faculty, Enrollments, and Degrees Granted

Research Specialty	2009–10 Faculty	Enrollment[1] Fall 2009		No. of Degrees Granted 2009–10			Median No. of Years for 2009–10 Ph.D.'s
		Master's[2]	Doctorate	Master's	Terminal Master's[2]	Doctorate	
Applied Physics	0	38	–	0	3	0	
Astronomy	8	0	–	0	0	1	6.75
Astrophysics	15	0	–	0	0	0	
Biophysics	10	0	–	0	0	0	
Chemical Physics	3	0	–	0	0	0	
Condensed Matter Physics	37	0	–	0	0	3	4.13
Energy Sources & Environ.	2	0	–	0	0	0	
Geophysics	3	0	–	0	0	0	
Low Temperature Physics	0	0	–	0	0	2	5.5
Nuclear Physics	29	0	–	0	0	3	7.41
Optics	8	0	–	0	0	0	
Particles & Fields	27	0	–	0	0	4	6
Physics Education	5	0	–	0	0	0	
Plasma Physics & Fusion	1	0	–	0	0	1	6.25
Relativity & Gravitation	6	0	–	0	0	0	
Statistical & Thermal	1	0	–	0	0	0	
Non-specialized	3	0	–	19	0	0	
Total	159	38	136	19	3	14	
Full-time Grad. Stud.		6	133				
Part-time Grad. Stud.		32	3				
First-year Grad. Stud.		5	21				
Median Years in Grad. Study (2009–10 Degrees)				1.7	2.2	5.91	
Undergraduate Degrees, 2009–10:75(254)							

[1]Students not yet committed to a research specialty are entered under non-specialized.
[2]Terminal (Evening) Master's Program in Applications of Physics.
[3]Including Adjunct and Affiliate Faculty.

3. Graduate Degree Requirements

Typical undergraduate preparation (Ph.D. program): Upper-division courses in mechanics; electricity and magnetism; statistical physics and thermodynamics; modern physics, including an introduction to quantum mechanics; and advanced laboratory work. Preparation in mathematics to include vector analysis, complex variables, ordinary differential equations, Fourier analysis, boundary-value problems, and special functions.

Examples of physics texts: *Theoretical Mechanics of Particles and Continua*, Fetter & Walecka; *Introduction to Electrodynamics*, Griffiths; *Fundamentals of Statistical & Thermal Physics*, Reif; *Introduction to Quantum Mechanics*, Griffiths.

Evening Master's Degree Program in Applications of Physics: A Master's degree track in applications of physics is available for people who are currently employed and whose background is in physical science or engineering. Courses are offered in the evening. This program is intended for part-time students, so assistantships and fellowships are not offered, nor can student visa applications be supported.

Admission requirements (Evening Master's Degree program): For admission to the Evening Master's Degree program, applicants should have an undergraduate degree in physical science, engineering, mathematics, or computer science, including introductory courses in calculus and physics. An average grade of 3.0 in junior- and senior-level physical science or engineering courses is required. The GRE Aptitude is not required.

Address financial aid inquiries to: Office of Student Financial Aid, University of Washington, Box 355880, Seattle, WA 98195-5880; (206) 543-6101; osfa@u.washington.edu; http://www.washington.edu/students/osfa

GAPSFAS application required: No

Financial aid deadline: 2/15

Loans available: Yes

Address housing inquiries to: Student Services Office, Housing and Food Services, University of Washington, 301 Schmitz Hall, Box 355842, Seattle, WA 98195–5842; (206) 543-4059; hfsinfo@u.washington.edu; http://hfs.washington.edu

On-campus, single student housing available: Yes
Cost/month: $661

On-campus, married student housing available: Yes
Cost/month: $700–900 (3–12 mo. waiting period)

4. Graduate Degree Requirements

Master's: Minimum of 36 approved credits. 18 in courses numbered 500 or above, including a minimum of three credits in Physics 600 research. At least 18 credits must be in graded courses. No thesis required. No foreign language required. Must pass qualifying examination (Ph.D. program). Must submit project report and pass a final oral examination (Evening Master's Degree program). A minimum of three full-time quarters of residency required. Part-time quarters may be accumulated to meet this requirement. Need a grade point average of 3.0.

Doctorate: Grade point average above 3.0 is required; sequence of required courses must be taken; qualifying examination, general examination, and a final examination which is usually a defense of the dissertation are required; 18 graded credits at University of Washington required; minimum of three academic years of resident study; minimum of 27 credits of dissertation over period of at least 3 quarters; some teaching experience required; no language exam required; you must be registered the quarter you receive your degree.

Thesis: Thesis may be written *in absentia*.

Special Equipment, Facilities, or Programs: At our Center for Experimental Nuclear Physics and Astrophysics, an FN tandem Van de Graaff accelerator can be used with negative-ion injection to reach energies of 18 MeV for protons and higher energies for heavier ions. For nuclear astrophysics experiments, with a terminal ion source, the proton (or helium ion) beam energy can be as low as 100 keV, with currents up to 30 microamps. Our High-Energy Laboratory maintains facilities for preparation and analysis of experiments performed at off-campus accelerators. We have a ^{3}He dilution refrigerator for extremely low-temperature research, a laser facility for generating tunable optical radiation at precisely controlled frequencies, and x-ray equipment including a rotating anode x-ray diffraction lab and an extended x-ray absorption fine structure (XAFS) facility for determining the atomic structure of condensed matter. Facilities for research into nanostructure include two atomic force microscopes, one at room temperature and the other at low temperature in extra-high vacuum, and a 14 Tesla superconducting magnet. The Department also houses the Institute for Nuclear Theory (INT), a national facility funded by the Department of Energy to host visitor programs for the exploration of current topics in nuclear theory. The INT is closely integrated with the Physics Department both physically and intellectually, with INT Senior Fellows supervising thesis research in physics, and INT seminars being frequently attended by members of the Physics department. Computing facilities include numerous modern workstations and server machines, plus access to both University mainframes and national supercomputing centers. Since 1994, the Department has been located in a recently-constructed building with state-of-the-art facilities for instruction and research.

Table B—Appointments to Graduate Students

Title of Appointee	Appointments		Academic Load Allowed in Credit Hours	Hours of Service Per Week	Stipend for Academic Year ($)
	Total[1]	First year[1]			
			Quarter		
Teaching Assistant	48	16	10–18	20	$18,931-20,339 (11 mo. appts.)[2]
Research Assistant	88	4	10–18	20	$21,420-24,756 (12 mo. appts.)[3]
Fellowship	6	1	10–18	N.A.	$24,500 (12 mo. appts.)[3]
Total	134	23			

[1] These numbers reflect Autumn quarter 2008.
[2] Based on an 11 month appointment.
[3] Based on a 12 month appointment.

5. Personnel Engaged in Separately Budgeted Research

Professorial faculty	73
Professor	36
Emeritus Prof.	21
Associate Prof.	5
Assistant Prof.	4
Research Prof.	2
Res. Assoc. Prof.	1
Res. Asst. Prof.	4
Postdoctoral Res. Assoc.	26
Graduate students	73
Degreed Professional Staff*	2
Total	174

*Does not include hourly students, classified staff such as technicians, REUs, or support staff on State funds.

6. Separately Budgeted Research Expenditures by Source of Support

	Departmental Research	Physics-related Research Outside Department
Federal government	$12,772,088	–
Non-federal	143,100	–
Total	$12,915,188	–

Table C—Separately Budgeted Research Expenditures

Research Specialty	No. of Grants	Expenditures ($)
AMO	9	1,048,901
Computing	4	101,856
Condensed Matter	26	1,704,221
Nuclear Physics	10	5,793,109
Particles and Fields	13	2,170,504
Physics Education	3	812,790
Relativity and Gravitation	8	1,283,807
Total	73	12,915,188

FACULTY

Professors

Bertsch, George F., Ph.D., Princeton, 1965. Theoretical physics; quantum many-body physics with application to nuclei.

Bulgac, Aurel, Ph.D., Leningrad Nuclear Physics Institute, 1977. Theoretical nuclear physics.

Burnett, Thompson H., Ph.D., California, San Diego, 1968. Experimental elementary particle physics.

Chaloupka, Vladimir, Ph.D., Geneva, 1975. Experimental elementary particle physics.

den Nijs, Marcel, Ph.D., Katholieke University, Netherlands, 1979. Theoretical condensed matter physics.

Ellis, Stephen D., Ph.D., Cal. Tech., 1971. Theoretical elementary particle physics.

Garcia, Alejandro, Ph.D., University of Washington, 1991. Experimental nuclear physics.

Gundlach, Jens, Ph.D., Univeristy of Washington, 1990. Experimental nuclear physics.

Heckel, Blayne, Ph.D., Harvard, 1981. Experimental atomic physics.

Heron, Paula, Ph.D., Western Ontario, 1995. Physics education.

Kaplan, David B., Ph.D., Harvard, 1985. Theoretical nuclear physics.

Lubatti, Henry J., California, Berkeley, 1966. Experimental elementary particle physics; Visual Techniques Laboratory.

McDermott, Lillian C., Ph.D., Columbia, 1959. Physics education.

Miller, Gerald A., Ph.D., MIT, 1972. Theoretical nuclear physics.

Nelson, Ann E., Ph.D. Harvard, 1984. Theoretical particle physics.

Olmstead, Marjorie A., Ph.D., California, Berkeley, 1985. Experimental condensed matter physics.

Rehr, John J., Ph.D., Cornell, 1972. Theoretical condensed matter physics.

Robertson, R. G. Hamish, Ph.D., McMaster University, 1971. Experimental nuclear physics.

Rosenberg, Leslie, Ph.D., Stanford, 1985. Experimental astrophysics.

Rothberg, Joseph E., Ph.D., Columbia, 1963. Experimental elementary particle physics.

Savage, Martin J., Ph.D., CIT, 1990. Theoretical nuclear physics.

Schick, Michael, Ph.D., Stanford, 1967. Theoretical, statistical and condensed matter physics.

Shaffer, Peter S., Ph.D., Washington, 1993. Physics Education.

Sharpe, Stephen, R., Ph.D., California, Berkeley, 1983. Theoretical elementary particle physics.

Son, Dam Thanh, Ph.D., Russian Academy of Science, Moscow, 1994. Theoretical nuclear physics.

Sorensen, Larry B., Ph.D., Illinois, 1979. Experimental condensed matter physics.

Spivak, Boris, Ph.D., Leningrad Politecknical Institute, 1978. Theoretical condensed matter physics.

Van Dyck, Robert S., Ph.D., California, Berkeley, 1971. Experimental atomic physics.

Watts, Gordon T., Ph.D., U. of Rochester, 1994. Experimental elementary particle physics.

Wilkes, Richard J., Ph.D., Wisconsin, 1974. Experimental high-energy physics and space science.

Yaffe, Laurence G., Ph.D., Princeton, 1980. Theoretical elementary particle physics.

Associate Professors

Andreev, Anton, Ph.D., MIT, 1996. Theoretical condensed matter physics.

Cobden, David, H., Ph.D., Cambridge, 1991. Experimental condensed matter physics.

Goussiou, Anna, Ph.D., University of Wisconsin, Madison, 1995. Particle physics.

Karch, Andreas, Ph.D., Humbold, Germany, 1998. Theoretical particle physics.

Raschke, Markus B., Ph.D., Max-Planck Institute for Quantum Optics and Technology University, Munich, 1999. Chemistry and nanoparticles.

Seidler, Gerald T., Ph.D., U. of Chicago, 1993. Experimental condensed matter physics.

Assistant Professors

Blinov, Boris B., Ph.D., University of Michigan, 2000. Experimental atomic physics.

Gupta, Subhadeep, Ph.D., MIT, 2003. Experimental atomic physics.

Morales, Miguel F., Ph.D., California, Santa Cruz, 1992. Experimental astrophysics and cosmology.

Tolich, Nikolai, Ph.D., Stanford, 2005. Experimental nuclear physics.

Research Professors

Doe, Peter J., Ph.D., Durham University, 1977. Experimental nuclear physics.

Trainor, Thomas A., Ph.D., North Carolina, 1973. Experimental nuclear physics.

Research Associate Professor

Zhao, Tianchi, Ph.D., Columbia, 1986. Experimental elementary particle physics.

Research Assistant Professors

Enomoto, Sanshiro, Tohoku University, 2005.

Forbes, Michael, MIT, 2005.

Lin, Huey-Wen, Columbia University, 2006.

Miller, Michael L., Ph.D., Yale, 2004. Experimental nuclear and particle physics.

Romatschke, Paul, Ph.D., Technical University of Vienna, Austria. 2003.

Rosati, Stefano, University of Bonn, 2002.

Schlamminger, Steven, Ph.D., University of Zurich, 2002. Gravitational physics.

Stetzer, Mackenzie, Ph.D., Pennsylvania, 2000. Physics education and experimental condensed matter physics.

Professors Emeriti

Adelberger, Eric G., Ph.D., Cal. Tech., 1967. Experimental nuclear physics; gravitation.
Baker, Marshall, Ph.D., Harvard, 1958. Theoretical elementary particle physics.
Bardeen, James M., Ph.D., Cal. Tech., 1965. Theoretical astrophysics and relativity.
Bodansky, David, Ph.D., Harvard, 1950.
Boulware, David G., Ph.D., Harvard, 1962. Chairman of the Department. Theoretical physics; astrophysics; relativity; elementary particles.
Boynton, Paul E., Ph.D., Princeton, 1967. Astronomy; astrophysics; gravitation.
Brown, Frederick C., Ph.D., Harvard, 1950. Experimental condensed matter physics.
Brown, Lowell S., Ph.D., Harvard, 1961. Theoretical elementary particle physics.
Clark, Kenneth C., Ph.D., Harvard, 1947.
Cook, Victor, Ph.D., California, Berkeley, 1962. Experimental elementary particle physics.
Cramer, John G., Jr., Ph.D., Rice, 1961. Experimental nuclear physics.
Dash, J. Gregory, Ph.D., MIT, 1948.
Dehmelt, Hans G., Ph.D., Gettingen, 1950. Experimental atomic physics.
Forston, E. Norval, Ph.D., Harvard, 1963. Experimental atomic physics.
Halpern, Isaac, Ph.D., MIT, 1948.
Henley, Ernest M., Ph.D., California, Berkeley, 1952.
Ingalls, Robert L., Ph.D., Carnegie-Mellon, 1962. Experimental condensed matter physics.
Puff, Robert D., Ph.D., Harvard, 1960. Theoretical statistical physics.
Stern, Edward A., Ph.D., Cal. Tech., 1955. Experimental condensed matter physics.
Storm, Derek W., Ph.D., Washington, 1970. Director, Nuclear Physics Laboratory. Experimental nuclear physics.
Thouless, David J., Ph.D., Cornell, 1958. Theoretical condensed matter physics.
Vilches, Oscar E., Ph.D., Universidad Nacional de Cuyo, Argentina, 1966. Experimental low-temperature physics.
Wilets, Lawrence, Ph.D., Princeton, 1952.
Williams, Robert W., Ph.D., MIT, 1948.

Affiliate Professors

Alberg, Mary A., Ph.D., Washington, 1974. Theoretical nuclear physics.
Balantekin, A. Baha, Ph.D., Yale, 1982. Theoretical nuclear physics.
Barrett, Bruce R., Ph.D., Stanford, 1967. Nuclear many-body theory.
Bichsel, Hans, Ph.D., University of Basel, 1951. Experimental nuclear physics.
Bowles, Thomas J., Ph.D., Princeton, 1978. Experimental nuclear physics.
Cahn, John, Ph.D., California, Berkeley, 1953. Theoretical condensed matter physics.
Chayes, Jennifer T., Ph.D., Princeton, 1983. Mathematical physics.
Cleveland, Bruce T., Ph.D., Johns Hopkins, 1970. Experimental Neutrino physics.
Elliot, Steven R., Ph.D., California, Irvine, 1987. Experimental nuclear physics.
Friedman, William A., Ph.D., MIT, 1966. Nuclear physics.
Nordtvedt, Kenneth, Ph.D., Stanford, 1964. General Relativity.
Raab, Frederick J., Ph.D., SUNY, Stony Brook, 1980. Experimental gravitational physics.
Riedel, Eberhard K., Ph.D., München, 1966. Theoretical condensed matter and statistical physics.
Strassler, Matthew, Ph.D., Stanford, 1993. Quantum field theory, string theory and particle physics.
Stubbs, Christopher W., Ph.D., Washington, 1988. Experimental and observational cosmology.
Tung, Wu-Ki, Ph.D., Yale, 1966. Theoretical elementary particle physics.
Van Bibber, Karl, Ph.D., MIT, 1976. High energy physics.
Wettlaufer, John S., Ph.D., Washington, 1991. Ice physics.

Affiliate Associate Professor

van Kolck, Ubirijara, Ph.D., U. of Texas, Austin, 1993. Theoretical nuclear physics.

Adjunct Professors

Baker, David, Ph.D., California, Berkeley, 1989. Biological physics.
Buck, Warren W., Ph.D., William and Mary, 1976. Theoretical nuclear and elementary particle physics.
Campbell, Charles, Ph.D., Texas, Austin, 1979. Chemistry.
Drobny, Gary, Ph.D., California, Berkeley, 1981. Chemistry.
Dunham, Scott T., Ph.D., Stanford University, 1985. Modeling and simulation of microfabrication processes and device behavior.
Fine, Arthur I., Ph.D., U. Chicago, 1963. Foundations of physics.
Hawley, Suzanne L., Ph.D., U. Texas, Austin, 1989. Theoretical astrophysics.
Holzworth, Robert, Ph.D., California, Berkeley, 1977. Geophysics.
Jarboe, Thomas R., Ph.D., California, Berkeley, 1974. Plasma physics.
Krishnan, Kannon M., Ph.D., UC Berkeley, 1984. Condensed matter experiment.
Ohuchi, Fumio, Ph.D., University of Florida, 1981. Experimental condensed matter physics.
Quinn, Thomas R., Ph.D., Princeton, 1986. Astrophysics.
Reinhardt, William P., Ph.D., Harvard, 1968. Chemistry.
Rieke, Frederick M., Ph.D., California, Berkeley, 1991. Sensory signal processing and computation.
Winglee, Robert, Ph.D., University of Sidney, 1984. Geophysics.

Adjunct Associate Professors

Dalcanton, Julianne, Ph.D., Princeton, 1995. Astrophysics.
Keller, Sarah L., Ph.D., Princeton U., 1995. Biophysics.
Lin, Lih, Ph.D., UCLA, 1996. Electrical engineering.
Rieke, Frederick M., Ph.D., California, Berkeley, 1991. Sensory signal processing and computation.

Adjunct Assistant Professors

Agol, Eric, Ph.D., UC Santa Barbara, 1997. Theoretical astrophysics.
Doran, Charles, Ph.D., Harvard, 1999. Mathematics.
Fairhall, Adrienne, Ph.D., Weizmann Institute of Science, Israel, 1998. Theoretical physics.
Iqbal, Amer, Ph.D., MIT, 2000. Mathematical physics and string theory.

Ginger, Jr., David S., Ph.D., Cambridge, 2001. Experimental condensed matter physics.

RESEARCH SPECIALTIES AND STAFF

Experimental

Astrophysics. Boynton, Burnett, Chaloupka, Dalcanton, Doe, Robertson, Stubbs, Wilkes.

Atomic and Molecular Physics. Blinov, Dehmelt, Fortson, Heckel, Nagourney, Van Dyck.

Biological Physics. Rieke.

Condensed Matter, Low-Temperature and Surface Physics. Campbell, Chopelas, Cobden, Dash, Drobny, Ginger, Ingalls, Krishnan, Ohuchi, Olmstead, Seidler, Sorensen, Stern, Vilches.

Cosmic Rays. Wilkes.

Elementary Particles and Fields. Burnett, Chaloupka, Cook, Lubatti, Mockett, Rothberg, Watts, Zhao.

Geophysics. Holzworth, Winglee.

Gravitation. Adelberger, Boynton, Raab.

Nuclear Physics. Adelberger, Bowles, Cleveland, Cramer, Doe, Elliott, Gundlach, Robertson, Snover, Storm, Trainor, Weitkamp.

Physics Education. Heron, L. McDermott, Shaffer.

Plasmas. Jarboe.

Theoretical

Astrophysics. Agol, Quinn, Savage.

Astrophysics and Relativity. Bardeen, Boulware, Hogan.

Atomic Physics. Wilets.

Biophysics. D. Baker, Keller.

Condensed Matter and Statistical. Andreev, Ao, den Nijs, Dunham, Puff, Rehr, Riedel, Schick, Spivak, Thouless.

Elementary Particles and Fields. M. Baker, Boulware, L. Brown, Ellis, Henley, Karch, Nelson, Sharpe, Yaffe.

Foundations of Physics. Fine.

General Relativity. Nordtvedt.

Nuclear Physics. Balantekin, Barrett, Bertsch, Buck, Bulgac, Detmold, Henley, D. Kaplan, Miller, Savage, Son, van Kolck, Wilets.

UNIVERSITY OF WASHINGTON

DEPARTMENT OF ASTRONOMY

Seattle, Washington 98195-1580

Students Accepted For Degree	FIELDS		
	Physics	Astronomy	Related Fields
Doctorate		X	
Master's		X	

1. General

President: Dr. Mark Emmert
Acting Dean of Graduate School: Dr. Jerry Baldasty
Department Chairman: Dr. Suzanne Hawley
Department Telephone Number: (206) 543-2888
URL: http://www.astro.washington.edu
Type of Institution: University
Control: Public
Setting: Urban
Total Faculty and Staff: 28,860
Total Graduate Students: 11,070
Graduate Tuition: 2009–10
In-state residents: $10, 727
Out-of-state residents: $24,067
Graduate Students with Appointments: $1,050
Deferred tuition plan: No
Term: Quarter

2. Number of Faculty in Department

The combined total of full-time faculty in the three professorial ranks is 17. The combined total of full-time, part-time, and other faculty at all ranks is 25.

3. Admission, Financial Aid, and Housing

Address admission inquiries to: Graduate Program Assistant, Department of Astronomy Box 351580, Seattle, WA 98195. (206) 543-2888. grad@astro.washington.edu.
Graduate application fee: $65
Admission deadline (Fall admission): December 31.
Admission information: There were a total of 90 applicants for Autumn 2009. Out of 21 admission offers, 8 accepted.
Admission requirements: A Bachelor's degree in astronomy, physics, math, or geology (other field related to astronomy) is required with no minimum GPA specified. Both GRE and GRE Advanced Physics are required. The average GRE Advanced score for 2009–10 admissions was 691. Students from non-English speaking countries are required to demonstrate proficiency in English via the TOEFL exam. Minimum acceptable score is 92 on the internet based TOEFL exam.
Undergraduate preparation assumed: Undergraduate preparation assumed is difficult to specify owing to the great range of backgrounds of incoming graduate students. The equivalent of an undergraduate physics program is typical.
Address financial aid inquiries to: Office of Student Financial Aid, (206) 543-6101. osfa@u.washington. edu. http://www.washington.edu/students/osfa.
GAPSFAS application required: No
Financial aid deadline: 2/15
Loans available: Yes
Address housing inquiries to: Housing Office, (206) 543-4059. http://hfs.washington.edu.
On-campus, single student housing available: Yes
Cost/month: $661–855
Off-campus, married student housing available: Yes
Cost/month: $708–1,000

Table A—Faculty, Enrollments, and Degrees Granted

Research Specialty	2009–10 Faculty	Enrollment Fall 2009		No. of Degrees Granted 2009 (2005–10)			Median No. of Years for 2008–09 Ph.D.'s
		Master's	Doctorate	Master's	Terminal Master's	Doctorate	
Astronomy	17	0	31	0	0	3	5.7
Total		–	31	0	0	3	
Full-time Grad. Stud.		–	31				
Part-time Grad. Stud.		–	0				
First-year Grad. Stud.		–	8				

Undergraduate Degrees, 2009–10 (2004–10): 28(146)

4. Graduate Degree Requirements

Master's: With Thesis: 36 approved credits, of which 18 are in astronomy courses at the 500 level or above and nine are thesis research. Without Thesis: 36 approved credits, of which 18 are in astronomy courses at the 500 level or above.
Doctorate: Admission Requirements: Entering students are expected to have a strong background in physics and mathematics. Graduation Requirements: Passage of the departmental qualifying examinations. Master's degree in astronomy or equivalent knowledge; at least three quarters of teaching experience in astronomy; dissertation and final examination. Students interested in work in theoretical astrophysics may take additional courses in physics and mathematics. Students working on other topics may take certain courses in related fields, such as astrobiology, astronautics, atmospheric sciences, geophysics, or electrical engineering. A knowledge of computer programming is useful.
Thesis: Thesis may be written *in absentia*.
Special Equipment, Facilities, or Programs: The Department owns, in consortium with several other universities, a 3.5-meter telescope at Apache Point, NM, and receives 25% of the observing time on this facility. It is operated largely remotely over the internet, and used heavily for graduate student dissertation research. UW is also a participant in the Sloan Digital Sky Survey, a project to make a digital photometric and spectroscopic map of 25% of the celestial sphere, using a special purpose telescope also on Apache Point. The Department is a founding member of the future Large Synoptic Survey Telescope. The Department also operates a 0.8m telescope in the Cascade Mountains of Washington, for the use of its students. Additional facilities in Seattle include an electron microscopy laboratory for analysis of cosmic dust particles and laboratories for developing astronomical telescopes and instrumentation. Members of the faculty are on teams that supplied instrumentation for the Hubble Space Telescope. Faculty and students are also extensive users of radio and optical facilities at national centers, and of national supercomputing facilities. The Department operates a large network of Linux workstations in support of all of these efforts.

Table B—Appointments to Graduate Students, Autumn 2009

Title of Appointee	Appointments Total	First year	Academic Load Allowed in Credit Hours	Hours of Service Per Week	Stipend for Academic Year ($)
			Quarter		
Teaching Assistant	9	4	10	20	13,725–15,849
Research Assistant	5	2	10	20	13,725–15,849
Fellowship	5	2	10	0	varies
Total	31	8			

5. Personnel Engaged in Separately Budgeted Research, 7/09–10

Professorial faculty	18
Postdoctoral appointments	18
Graduate students	31
Total	67

FACULTY

Professors

Anderson, Scott, Ph.D., Washington, 1985. High-energy astrophysics; compact binaries; quasars.

Balick, Bruce B., Ph.D., Cornell, 1971. Planetary nebulae, their structure and evolution; gas dynamics; active nuclei and their impact on galactic structure.

Brownlee, Donald E., Ph.D., Washington, 1971. Interplanetary dust; comet physics; meteoritics; origin of the solar system.

Dalcanton, Julianne, Ph.D., Princeton, 1995. Galaxy evolution and formation; cosmology; galactic dynamcis.

Hawley, Suzanne, Ph.D., Texas, 1989. Chair of the Department. Low mass stars; variable stars; star clusters; dwarf galaxies; galactic structure.

Quinn, Tom, Ph.D., Princeton, 1986. Astrophysical dynamics on a wide range of scales, from asteroids to clusters of galaxies; solar system studies.

Sullivan, Woodruff T. III, Ph.D., Maryland, 1971. Astrobiology; galaxies; clusters of galaxies; distance scale; history of radio astronomy.

Szkody, Paula, Ph.D., Washington, 1975. Cataclysmic variables; white dwarfs.

Associate Professors

Agol, Eric, Ph.D., University of California, Santa Barbara, 1997. Relativistic astrophysics and gravity; black holes; active galaxies; accretion disks; extrasolar planets.

Connolly, Andrew, Ph.D., Univ. of London, 1993. Formation and evolution of galaxies; cosmology; astronomical surveys.

Ivezic, Zeljko, Ph.D., University of Kentucky, 1995. Deep sky surveys; quasars; stellar populations; asteroids; origin of interstellar dust.

Meadows, Victoria, Ph.D., Univ. of Sydney, 1994. Planetary atmospheres; astrobiology.

Research Professors

Becker, Andrew, Ph.D., Washington, 2000. Assistant Professor. Time domain science; techniques of massive survey astronomy; data mining.

Governato, Fabio, Ph.D., Rome, 1995. Associate Professor. Galaxy & clusters; cosmic structure formation; planet formation.

King, Ivan, Ph.D., Harvard, 1952. Professor. Stellar populations; star clusters; structure & dynamics of globular clusters.

Professors Emeriti

Baum, Bill, Ph.D., Caltech, 1950. External galaxies; solar system; astronomical instrumentation.

Böhm, Karl Heinz, Ph.D., Kiel, 1954. Stellar atmospheres and interiors; nebulae, circumstellar matter.

Böhm-Vitense, Erika, Ph.D., Christian Albrechts, 1951. Stellar atmospheres; convection; stellar chromospheres and coronae; magnetic stars; cepheids.

Hodge, Paul, Ph.D., Harvard, 1960. Galaxies; the Magellanic Clouds.

Lutz, Julie, Ph.D., Illinois, 1972. Planetary nebulae and symbiotic stars; astronomy education.

Wallerstein, George, Ph.D., Cal.Tech., 1958. Spectra of variable stars; chemical composition of stellar atmospheres; interstellar lines.

Lecturers

Larson, Ana, Ph.D., British Columbia, 1996. Senior Lecturer. Modeling stellar atmospheres; precise radial velocities of stars.

Laws, Christopher, Ph.D., Washington, 2004. Extrasolar planetary systems; stellar evolution; chemical evolution.

Smith, Toby, Ph.D., Washington, 1995. Senior Lecturer. Terrestrial impact craters; meteoritics.

RESEARCH SPECIALTIES AND STAFF

Quasars and other Active Galaxies. Nuclear properties; dynamics; ejecta and ejection mechanisms; lensing; accretion; absorption lines. Agol, Anderson, Balick, Ivezic.

Astrobiology and Planetary Studies. Formation of solar system; extrasolar planets; planetary atmospheres; asteriods, meteorites. Agol, Brownlee, Hodge, Ivezic, Meadows, Sullivan, Quinn.

Clusters of Stars. Evolution; abundance determinations; statistical properties. Hawley, Hodge, King, Wallerstein.

Compact Objects. Degenerate stars; black holes; cataclysmic variables and other compact binaries. Agol, Anderson, Bohm, Szkody.

Cosmology. Large-scale clustering; cosmological parameters; big bang nucleosynthesis; cosmic abundances. Becker, Connolly, Quinn.

Galactic Nebulae. Supernova remnants; hot and cool components in the interstellar medium; H II regions, and planetary nebulae; Herbig-Haro objects. Balick, Bohm, Lutz, Wallerstein.

Galaxies. Structure and dynamics; formation; dark matter; gaseous and stellar content; internal motions; extragalactic distance scale; clusters of galaxies; properties of giant H II regions; normal galaxy nuclei. Balick, Connolly, Dalcanton, Governato, Hawley, Hodge, Ivezic, Quinn, Sullivan.

History of Astronomy. Sullivan.

Stars. Chemical composition; magnetic activity, chromospheres, coronae, flares; circumstellar envelopes; theory of convection; equation-of-state; variable stars; low mass stars and brown dwarfs. Bohm, Bohm-Vitense, Hawley, Hodge, Ivezic, Szkody, Wallerstein.

WASHINGTON STATE UNIVERSITY

DEPARTMENT OF PHYSICS AND ASTRONOMY

Pullman, Washington 99164-2814

Students Accepted For Degree	FIELDS		
	Physics	Astronomy	Related Fields
Doctorate	X	X	X
Master's	X	X	X

1. General

President: Elson S. Floyd
Dean of Graduate School: Howard Grimes
Chairman: Steven Tomsovic
Department Telephone Number: (509) 335-1698
Type of Institution: University
Control: Public
Setting: Small town
Total Faculty: 2,228
Total Graduate Faculty: 834
Total Students: 26,101 (statewide)
Total Graduate Students: 3,576
Annual Graduate Tuition:
In-state residents: Full-time—$8,862
Part-time—$443/credit
Out-of-state residents: Full-time—$21,660[1]
Part-time—$1,083/credit
Tuition rates for: 2010–11 estimated
Deferred tuition plan: Yes
Other Fees: $838
Term: Semester

[1]Out-of-state residents with 1/2-time appointments may pay in-state tuition rates.

2. Number of Faculty in Department

The combined total of full-time faculty in the three professorial ranks is 17. The combined total of full-time, part-time, and other faculty at all ranks is 41.

3. Admission, Financial Aid, and Housing

Address admission inquiries to: Chair, Graduate Studies, Department of Physics and Astronomy
www.physics.wsu.edu
Graduate application fee required: $50
Admission deadline (Fall admission): 7/15[1]
Admission information: For fall admission, 2009–10, 12 students were accepted from 124 applicants.
Admission requirements: For admission to the graduate programs, a Bachelor's degree is required with a minimum GPA of 3.0 in the last half of the undergraduate work. The GRE General Test and Subject Test in Physics are required of all applicants. No minimum acceptable scores are specified. Students from non-English speaking countries are required to demonstrate proficiency in English via the TOEFL exam. Minimum acceptable score for admission is 550 (paper based) or 213 (computer-based).
Undergraduate preparation assumed: Symon, *Mechanics*; Reitz, Milford, and Christy, *Foundations of Electromagnetic Theory*; Zemansky, *Heat and Thermodynamics*; Liboff, *Introductory Quantum Mechanics*; Boas, *Mathematical Methods in the Physical Sciences*; Eisberg and Resnick, *Quantum Physics of Atoms, Molecules, Solids, Nuclei, and Particles*. Preparation in optics, solid state physics, nuclear physics, and/or acoustics is encouraged.
Address financial aid inquiries to: Chair, Graduate Studies, Department of Physics and Astronomy
GAPSFAS application required: No
Financial aid deadline: 3/1
Loans available: Yes
Address housing inquiries to: Housing and Dining Financial Services, Washington State University, Pullman, WA 99164-1722
On-campus, single student housing available: Yes
Cost/term: E-mail: housing@wsu.edu
On-campus, married student housing available: Yes
Cost/month: E-mail: housing@wsu.edu

[1]International students should contact WSU Graduate Admissions office for deadlines. E-mail: gradsch@wsu.edu

Table A—Faculty, Enrollments, and Degrees Granted

Research Specialty	2009–10 Faculty	Enrollment[1] Fall 2009 Master's	Enrollment[1] Fall 2009 Doctorate	No. of Degrees Granted[2] 2009–10 (2005–10) Master's	Terminal Master's	Doctorate	Median No. of Years for 2009–10 Ph.D.'s
Acoustics	2	0	5	0(3)	0(1)	1(5)	6
Astronomy/ Astrophysics	4	0	9	0(1)	0(2)	1(2)	5
Atomic, Molecular, & Optical Physics	2	0	4	2(4)	0(0)	1(2)	5.5
Biophysics	3	0	1	0(1)	0(0)	0(0)	
Chemical Physics	5	0	0	0(0)	0(0)	0(0)	
Condensed Matter Physics	9	0	4	1(1)	0(3)	1(1)	7.5
Fluids & Rheology	1	0	0	0(0)	0(0)	0(0)	
High Energy/Particle	0	0	0	0(0)	0(0)	0(0)	
Low Temperature Physics	4	0	0	0(0)	0(0)	0(0)	
Materials Sci.	9	1	4	0(0)	0(0)	0(0)	
Optics	5	0	8	1(1)	1(3)	0(5)	
Polymer Physics	2	0	0	0(0)	0(0)	0(0)	
Shock Physics	2	0	5	0(2)	0(0)	0(1)	
Surface Science	5	0	2	0(0)	0(0)	1(1)	7
Theoretical/ Math.	5	0	3	0(0)	0(0)	0(0)	
Nano Physics	4	0	1	0(0)	0(0)	0(0)	
Non-specialized	0	3	14	0(0)	2(5)	0(0)	
Total		4	60	4(13)	3(14)	5(17)	
Full-time Grad. Stud.		4	60				
Part-time Grad. Stud.		0	0				
First-year Grad. Stud.		0	13				
Median Years in Grad. Study (2009–10 Degrees)				3.5	3.7	6.2	

Undergraduate Degrees, 2009–10 (2005–10):13(42)

[1]Students not yet committed to a research specialty are entered under non-specialized.
[2]Five-year totals in parentheses.

4. Graduate Degree Requirements

Master's: For the M.S. degree without thesis a minimum of 30 semester hour credits for graded courses is required. Course work must include a specified core curriculum. For the M.S. degree with thesis a minimum of 30 credits is required of which 18 must be for graded courses. A minimum 3.0 GPA must be maintained. One academic year of residence is required. No foreign language is required. Must pass oral final examination. Students making normal progress toward Ph.D. satisfy requirements for non-thesis M.S. degree at the end of the second year of study.

Doctorate: For the Ph.D. degree a minimum of 72 semester hour credits is required of which 36 credits must be for graduate level graded course work in physics, astrophysics, or related fields. Course work must include a specified core curriculum. A minimum 3.0 GPA must be maintained. Minimum period of study is three years of which two years must be in residence including a minimum of two continuous semesters in an academic year. No foreign language is required. Qualifying examination, preliminary examinations, and final examination are required. Thesis is required.

Other Programs: Interdepartmental doctoral program is available in Materials Science.

Thesis: Thesis may be written *in absentia.*

Special Equipment, Facilities, or Programs: The department occupies a modern 96,000 square foot building. Surface and Solid State Physics Laboratories equipped for STM, LEED, Auger, molecular beam, optical and resonance spectroscopy, laser studies and x-ray correlation spectroscopy. The optical physics laboratories are equipped with several high-power ultrafast femto-, pico-, nanosec, and continuous wave lasers, as well as a wide assortment of detection systems. The physical acoustics research laboratory is equipped with a 6,000 gallon water tank for scattering and nonlinear acoustics experiments. Several computer systems are used in research and teaching. There is local access to a nuclear reactor, ESR and NMR spectrometers, x-ray spectrometers, and electron microscopes. The Institute for Shock Physics (ISP) is housed in a separate new state of the art 37,787 sq ft building completed in 2003. The ISP is equipped with high pressure gas guns, intense lasers, and pulsed-power facility for shock wave propagation studies and studies of material properties under high pressure.

Table B—Appointments to Graduate Students, 2009–10

Title of Appointee	Appointments Total	Appointments First year	Academic Load Allowed in Credit Hours	Hours of Service Per Week	Stipend for Academic Year ($)
			Semester		
Teaching Assistant	25	12	14 SCH	20	15,075.00[1]/15,858.00[2]
Research Assistant	39	1	18 SCH	20	15,381.00[1]/16,168.50[2]
Other (specify)					
Hourly Appointments			14 SCH	20	9.00/hr–15.30/hr[3]
Total	64	13			

[1]Pre-Master's rate for half-time appointment. Salary reduced by 14% if lack of progression to RA by 3rd year.
[2]Post-Master's rate for half-time appointment. Salary reduced by 14% if lack of progression to RA by 3rd year.
[3]Students with half-time appointments may be eligible for in-state tuition rate.

5. Personnel Engaged in Separately Budgeted Research, 7/09–6/10

Professorial faculty	16
Postdoctoral appointments	3
Graduate students	39
Undergraduate students	16
Other faculty	2
Total	60

6. Separately Budgeted Research Expenditures by Source of Support

	Departmental Research	Physics-related Research Outside Department
Federal government	$2,457,000	$5,097,000
Business and industry	$19,000	0
Other	$47,000	$42,000
Total	$2,523,000	$5,139,000

Table C—Separately Budgeted Research Expenditures

Research Specialty	No. of Grants	Expenditures ($)
Acoustics	10	$403,000
Astronomy & Astrophysics	11	$411,000
Atomic, Molecular & Optics	11	$496,000
Cluster Physics	4	$200,000
Laser Surface Interactions	1	$139,000
Materials Physics	6	$270,000
Nuclear Solid State Physics	2	—
Solid State Physics	4	$314,000
Total	49	$2,522,000

Table D—Physics-related Research Outside Department

Field and Unit Outside Department	No. of Grants	Expenditures ($)
Center for Materials Research	10	$1,328,241
Institute for Shock Physics	24	$3,834,001
Total	34	$5,162,242

FACULTY

Professors

Collins, Gary S., Ph.D., Rutgers, 1976. Condensed matter physics using nuclear hyperfine interactions.

Dickinson, J. Thomas, Ph.D., Michigan, 1968. Solid state physics; surface physics; chemical physics.

Gupta, Yogendra M., Ph.D., Washington State, 1973. Shock wave physics; condensed matter physics; spectroscopy. Director, Institute for Shock Physics.

Kuzyk, Mark G., Ph.D., Pennsylvania, 1985. Nonlinear optics.

Lynn, Kelvin G., Ph.D., Utah, 1974. Materials science. Director, Center for Materials Research.

Marston, Philip L., Stanford, 1976. Wave propagation and scattering; acoustics; optics; fluid mechanisms; microgravity.

McCluskey, Matthew D., Ph.D., University of California, Berkeley, 1997. Static high pressure; wide band gap semiconductors.

Miller, Michael D., Ph.D., Northwestern, 1974. Theoretical physics.

Tomsovic, Steven L., Ph.D., Rochester, 1987. Chair of the Department. Theoretical physics.

Associate Professors

Blume, Doerte, Ph.D., Georg-August-Universität zu Göttingen, Germany, 1998. Theoretical/Quantum Physics.

Bose, Sukanta, Ph.D., University of Wisconsin-Milwaukee, 1996. Astrophysics/Theoretical.

Dexheimer, Susan, Ph.D., California, Berkeley, 1990. Ultrafast optical physics.

Engels, Peter, Ph.D., University of Hannover, Germany, 2000. Experimental atomic/molecular.

Worthey, Guy, Ph.D., University of California-Santa Cruz, 1992. Astronomy.

Assistant Professors

Duez, Mathew, Ph.D., University of Illinois, Urbana, 2005. Theoretical Astrophysics and Relativity.

Gu, Yi, Ph.D., Columbia University, NY, 2004. Experimental, nanomaterials.

Zhang, Chuanwei, Ph.D., University of Texas, Austin, 2005. Theoretical, atomic.

Adjunct Professors

Anderson, Roger H., Ph.D., University of Washington, 1961. Condensed matter theoretical physics.

Baer, Donald R., Ph.D., Cornell, 1974. Experimental solid state physics.

Blakeslee, John, Ph.D., Massachusetts Institute of Technology, Cambridge, 1997. Astronomy and cosmology.

Clays, Koen, J., Ph.D., University of Leuven, Belgium, 1989. Chemistry/non-linear optics.

Exarhos, Gregory J., Ph.D., Brown, 1974. Physical chemistry.

Kouzes, Richard T., Ph.D., Princeton University, 1974. Nuclear physics.

Lytel, Rick, Ph.D., Stanford University, 1980. Theoretical high energy physics.

Raab, Frederick J., Ph.D., SUNY, Stony Brook, 1980. Experimental physics.

Wang, Lai-Sheng, Ph.D., California, Berkeley, 1989. Metal clusters; photoelectron spectroscopy.

Zacate, Matthew O., Ph.D., Oregon State University, 1997. Computational materials science, experimental physics.

Affiliated Professor

Asay, James R., Ph.D., Washington State University, 1971. Shock physics.

Professors Emeriti

Donaldson, Edward E., Ph.D., Washington State, 1953. Surface physics.

Dresser, Miles J., Ph.D., Iowa State, 1964. Solid state physics; surface physics.

Park, James L., Ph.D., Yale, 1967. Theoretical physics.

RESEARCH SPECIALTIES AND STAFF

Theoretical

Acoustics, Optics. Marston.

Astronomy and Astrophysics. Stellar populations. Worthey. Gravational waves. Bose.

Theoretical Astrophysics and Relativity. Duez.

Atomic. Quantum clusters, Bose-Einstein Condensates. Blume.

Condensed Matter Physics. Phase transitions in liquid mixtures. Nonlinear dynamics. Miller, Zhang. Mesoscopic systems. Tomsovic. Low temperature and many-body physics. Miller, Zhang.

Nonlinear dynamics. Quantum chaos and semi-classical theory. Tomsovic.

Quantum Information and Computation. Zhang.

Shock Wave and High Pressure Physics. Equations of state; finite amplitude wave propagation; material models of mechanical and thermal behavior. Gupta, Marston.

Experimental

Acoustics. Nonlinear acoustics; radiation pressure and scattering. Marston.

Chemical Physics. Molecular interactions on surfaces; problems in catalysis. Dickinson.

Fundamental properties. Vortex lattices and aggregates of Bose–Einstein condensates. Engels.

Materials Physics. Low energy positron beam studies of interfaces and layered structures. Lynn.

Nanomaterials. Synthesis and device design. Gu.

Nuclear Solid State Physics. Local atomic environments in solids; point defects in metals; perturbed angular correlation and Mössbauer spectroscopy. Collins.

Optics. Production and study of laser-induced plasmas and subpicosecond x-ray pulses; dynamics of excited states in solids and molecules; nonlinear optical properties of doped polymers; light scattering and Fourier optics; clusters; optomechanical effects; all-optical devices; time-resolved optical spectroscopy; atomic spectroscopy. Dexheimer, Gupta, Kuzyk, Marston, Wang.

Shock Wave and High Pressure Physics. Structural and chemical changes in condensed materials; time-resolved optical spectroscopy; nonlinear wave propagation; inelastic deformation of ceramics. Gupta, McCluskey.

Solid State Physics. Fracture of solids. Dickinson, Defects in semiconductor materials. Lynn. Photoelectron spectroscopy in clusters. Wang. Wide band gap semiconductors. Lynn, McCluskey.

Surface Physics. Molecular and atomic interactions and characterization of surfaces; reactive etching of surfaces; transmission electron microscopy on small metal clusters; photoelectric and thermal emission microscopy. Dickinson.

WEST VIRGINIA UNIVERSITY

DEPARTMENT OF PHYSICS

Morgantown, West Virginia 26506

Students Accepted For Degree	FIELDS		
	Physics	Astronomy	Related Fields
Doctorate	X		
Master's	X		

1. General

President: Jim Clements
Provost and V.P. for Academic Affairs & Research: Michelle Wheatly
Department Chairman: Earl E. Scime
Department Telephone Number: (304) 293-3422
Type of Institution: University
Control: Public
Setting: Small town
Total Faculty: 1,989
Total Graduate Faculty: 1,418
Total Students: 28,840
Total Graduate Students: 6,910
Annual Graduate Tuition:
In-state residents: Full-time—$5,612
Part-time-$346/credit
Out-of-state residents: Full-time—$16,270
Part-time-$1,004/credit
Tuition rates for: 2009–10
Deferred tuition plan: No
Other Fees: $0
Term: Semester

2. Number of Faculty in Department

The combined total of full-time faculty in the three professorial ranks is 18. The combined total of full-time, part-time, and other faculty at all ranks is 30.

3. Admission, Financial Aid, and Housing

Address admission inquiries to: Admissions Committee, Dept. of Physics, P.O. Box 6315
Internet Address: http://www.as.wvu.edu/phys/
Graduate application fee required: $50
Admission deadline (Fall admission): 2/15
Admission information: For fall admission, 2009–10, 2 students were accepted from 100 applicants.
Admission requirements: For admission to the graduate programs, a Bachelor's degree in physics is required with no minimum undergraduate GPA specified. The GRE is required. The GRE Advanced is strongly urged. No minimum score is specified. Students from non-English speaking countries are required to demonstrate proficiency in English via the TOEFL exam. Minimum acceptable score for admission is 213.
Undergraduate preparation assumed: Intermediate mechanics, electricity and magnetism, atomic and quantum physics, thermodynamics, and mathematics through partial differential equations. Typical physics texts include Davis (mechanics), Wangsness (electricity and magnetism), Saxon (quantum mechanics), and Sears and Salinger (thermodynamics).
Address financial aid inquiries to: Graduate Program Committee, Dept. of Physics
GAPSFAS application required: No
Financial aid deadline: 2/15
Loans available: Yes
Address housing inquiries to: Housing Office, Evansdale Campus
On-campus, single student housing available: Yes
Cost/month: $526–635 (incl. utils.)
On-campus, married student housing available: Yes
Cost/month: $526–635 (incl. utils.)

Table A—Faculty, Enrollments, and Degrees Granted

Research Specialty	2009–10 Faculty	Enrollment[1] Fall 2009		No. of Degrees Granted[2] 2009–10 (2005–10)			Median No. of Years for 2009–10 Ph.D.'s
		Master's	Doctorate	Master's	Terminal Master's	Doctorate	
Applied Physics	6	0	0	0(0)	0(0)	0(0)	–
Astrophysics	5	0	11	2(2)	0(0)	1(1)	5
Atmos./Space Phys., Cosmic Rays	4	0	1	0(0)	0(0)	0(0)	–
Biophysics	1	0	0	0(1)	0(0)	0(1)	5
Condensed Matter Physics	9	1	24	1(15)	0(4)	5(22)	7
Fluids & Rheology	2	0	3	1(2)	0(0)	0(0)	–
Materials Sci./ Metallurgy	3	0	0	0(0)	0(0)	0(0)	–
Medical Physics	0	0	3	1(1)	0(0)	0(0)	–
Particles & Fields	1	0	0	0(1)	0(0)	0(0)	–
Physics Education	1	0	0	0(0)	0(0)	0(0)	–
Plasma Physics & Fusion	6	0	15	0(7)	0(1)	3(10)	6
Statistical & Thermal	2	0	2	0(1)	0(0)	0(1)	5
Other Theoretical/ Math.	1	0	1	0(0)	0(3)	0(0)	–
Nonlinear dynamics	1	0	0	0(0)	0(0)	1(1)	5
Total		1	60	4(30)	0(5)	10(36)	
Full-time Grad. Stud.			60				
Part-time Grad. Stud.			0				
First-year Grad. Stud.		0	11				
Median Years in Grad. Study (2009–10 Degrees)				3	2	6	
Undergraduate Degrees, 20089-10 (2005-10):12(49)							

[1] Students not yet committed to a research specialty are entered under non-specialized.
[2] Five-year totals in parentheses.

4. Graduate Degree Requirements

Master's: Approved courses with minimum GPA of 3.0; no residence or language requirement. M.S. with thesis: 24 credits. M.S. without thesis: 30 credits.
Doctorate: Minimum of 36 hours of course work in approved program with minimum GPA of 3.0; written comprehensive exam, oral research exam, dissertation and oral dissertation defense.
Thesis: Thesis may be written *in absentia*.
Special Equipment, Facilities, or Programs: The department and associated instrument and electronics shops are housed in Hodges Hall, a six-story building located on the downtown campus. The building also houses a planetarium, a roof-top observatory, a small radio telescope, and eleven research laboratories. The facilities include a triple plasma source, a Q-machine for generating space-like plasmas and waves;

helicon plasma source, a space simulation chamber; a plasma processing test facitly, four molecular beam epitaxy (MBE) growth facilities; magnetic resonance laboratory (EPR, ENDOR); microimaging MRI; SQUID magnetometer with magneto-resistance probe; rotating anode x-ray source; x-ray diffractometers; e-beam writer, scanning probe microscope, atomic force microscope, Hall effect apparatus; optical spectrophotometer; FTIR spectrophotometer; high-temperature graphite furnace; ultrasonic, thermogravimetry, and differential scanning calorimetry; characterization capabilities for thermoluminescence, optical absorption, photoreflectance, photoconductance and photoluminescence of materials; and sputtering system for thin-film deposition. Laser facilities include cw argon ion lasers (4), dye lasers (2), tunable diode lasers (3), cw and Q-switched Nd:YAG lasers (3). Departmental computing facilities include 2 cluster computers. Cooperative research programs with National Energy Technology Laboratory and Pittsburgh Supercomputing Center are possible. Joint facility with Chem. Eng. Dept. for Auger and XPS studies; joint facility with CSEE Dept. for materials and device processing.

Table B—Appointments to Graduate Students, 2009–10

Title of Appointee	Appointments Total	Appointments First year	Academic Load Allowed in Credit Hours	Hours of Service Per Week	Stipend for Academic Year ($)
			Semester		
Teaching Assistant	21	11	12	20	14,500[1,2]
Research Assistant	39	0	12	20	19,000[1]
Total	60	11			

[1] Tuition waived for academic year and summer.
[2] Summer assistantships available, $4,500 more.

5. Personnel Engaged in Separately Budgeted Research, 7/08–6/10

Professorial faculty	21
Postdoctoral appointments	6
Graduate students	60
Undergraduate students	20
Total	107

6. Separately Budgeted Research Expenditures by Source of Support

	Departmental Research	Physics-related Research Outside Department
Federal government	$2,375,000	$
Other	80,000	
Total	$2,455,040	$

Table C—Separately Budgeted Research Expenditures

Research Specialty	No. of Grants	Expenditures ($)
Astrophysics	4	316,561
Condensed Matter Physics	14	1,318,779
Plasma Physics & Fusion	8	819,708
Total	26	2,455,040

FACULTY

Distinguished Professors

Edwards, Boyd F., Ph.D., Stanford, 1985. Theoretical statistical physics and fluid dynamics.
Koepke, Mark E., Ph.D., Maryland, 1984. Experimental plasma physics; nonlinear dynamics.
Lederman, David, Ph.D., California, Santa Barbara, 1992. Experimental solid state physics; magnetic materials; superconductors.
Scime, Earl E., Ph.D., Wisconsin-Madison, 1992. Chair of the Department. Experimental plasma physics.

Professors

Abdul-Razzaq, Wathiq, Ph.D., Illinois, Chicago, 1986. Experimental solid state; magnetism of nanoparticles; particulate matter in the environment.
Cooper, Bernard R., Ph.D., California, Berkeley, 1961. Professor Emeritus; Condensed matter and materials theory.
Ferer, Martin V., Ph.D., Illinois, 1971. Theory; statistical physics; applied physics; critical phenomena.
Ganguli, Gurudas, Ph.D., Boston College, 1980. Adjunct Professor. Plasma physics theory.
Golubovic, Leonardo, Ph.D., Belgrade, 1987. Condensed matter theory and statistical physics.
Halliburton, Larry E., Ph.D., Missouri-Columbia, 1971. Optical and magnetic properties of point defects.
Littleton, John E., Ph.D., Rochester, 1972. Professor Emeritus. Theoretical astrophysics; space plasma physics.
Smith, Duane, Ph.D., Univ. of Chicago, 1970. Adjunct Professor. Statistical and applied physics; fluids.
Pavlovic, Arthur S., Ph.D., Penn State, 1966. Professor Emeritus. Solid state experiment.
Treat, Richard P., Ph.D., California, Riverside, 1967. Professor Emeritus. Quantum field theory.
Weldon, H. Arthur, Ph.D., MIT, 1974. Particle theory.

Associate Professors

Lewis, James, Ph.D., Arizona State Univ., 1996. Computational physics.
Lorimer, Duncan R., Ph.D., Univ. Manchester, UK, 1994. Radio astronomy; astrophysics.
Raylman, Raymond R., Ph.D., Univ. Michigan, 1991. Adjunct. Medical physics, radiology, imaging.

Assistant Professors

Cassak, Paul, Ph.D., Univ. of Maryland, 2006. Theoretical plasma physics.
Ganikhanov, Feruz, Ph.D., Moscow State Univ., USSR, 1993. Nonlinear optics.
Holcomb, Mickel, Ph.D., Univ. of California, Berkeley, 2009. Experimental condensed matter physics.
McLaughlin, Maura A., Ph.D., Cornell University, 2001. Radio astronomy; astrophysics.
Pisano, Daniel J., Ph.D., Univ. of Wisconsin - Madison, 2001. Radio astronomy, astrophysics.
Stanescu, Tudor, Ph.D., U. of Illinois, Urbana-Champaign, 2002. Theoretical condensed matter physics.
Urazdhin, Sergei, Ph.D., Michigan State University, 2002. Experimental condensed matter physics; magnetic materials, spintronics.

Research Professors

Demidov, Vladimir, Ph.D., St. Petersburg State University, USSR, 1981. Experimental plasma physics.
Seehra, Mohindar S., Ph.D., Rochester, 1969. Solid state experiment; x-ray scattering; applied physics; magnetism.

Research Associate Professor

Vasiliadis, Dimitris, Ph.D., Univ. of Maryland, 1992. Space plasma physics.

Research Assistant Professor

Simien, Clayton, Ph.D., Rice Univ. 2008. Atomic Physics.

RESEARCH SPECIALTIES AND STAFF

Theoretical

Applied Physics. Aquifer remediation, fragmentation and coagulation kinetics, aerosol physics. Edwards, Ferer, Smith.
Astrophysics. Interstellar medium; galactic structure; stellar evolution; compact objects; general relativity. Littleton, Lorimer, McLaughlin, Pisano. 3 postdoctoral fellows.
Condensed Matter and Materials. Surface and interface phenomena; lattice stability and relaxation; molecular dynamics; properties of disordered materials; biomaterials; complex fluids and membranes; fracture; transport in random media; thin film growth. Ferer, Golubovic, Lewis, Stanescu. 5 postdoctoral fellow.
Elementary Particles and Fields. High-temperature quantum field theory; quark-gluon plasma; relativistic heavy-ion collisions. Weldon.
Fluids. Convective instabilities and nonlinear dynamics. Edwards, Smith.
Plasma. Plasma instabilities; simulations applicable to space and laboratory plasmas. Cassak, Ganguli, Vasiliadis.
Statistical Physics. Fractals; percolation theory; chaos; phase transitions and critical phenomena; nonequilibrium growth and pattern formation. Edwards, Ferer, Golubovic, Smith.

Experimental

Applied Physics. Preparation and characterization of nano-particles; iron-based catalysts; properties of air-borne particulate matter; coal-based high purity carbons and carbon fibers; electrochemical detection of Hg and other trace metals using boron-doped diamond films; visible and UV light emitters and sensors; nonlinear optical and photorefractive materials. Abdul-Razzaq, Ganikhanov, Halliburton, Seehra. 1 postdoctoral fellow.
Astrophysics. Radio Astronomy; X-ray astronomy; pulsars; tests of strong-field gravity; digital signal processing; computational astrophysics. Littleton, Lorimer, McLaughlin, Pisano. 2 postdoctoral fellows.
Condensed Matter and Materials. Electronic structure and magnetic properties of artificially grown surfaces and superlattices and nanoscale particles; spin transport; properties of magnetic ions and clusters; elementary excitations in antiferromagnets; magnetic susceptibility; magnetostriction; electrical, structural, and electro-optic properties of semiconductors; optical and magnetic resonance characterization of point defects. Ganikhanov, Halliburton, Holcomb, Lederman, Seehra, Urazhdin. 5 postdoctoral fellows.
Plasma. Plasma waves and instabilities; nonlinear interactions; turbulence and chaos; space plasma instrument design; space plasma data analysis and instrument (sensor) development; magnetic reconnection; plasma processing. Demidov, Koepke, Scime, Simien, Vassiliadis. 3 postdoctoral fellows.
Surface and Interface Physics. X-ray scattering from disordered systems; Auger and X-ray photoelectron spectroscopy deposition physics; molecular beam epitaxy; properties of monolayer and multilayer thin films; optical properties of quantum confined systems and semiconductors. Lederman, Seehra. 2 postdoctoral fellows.

UNIVERSITY OF WISCONSIN, MADISON

DEPARTMENT OF PHYSICS

Madison, Wisconsin 53706

Students Accepted For Degree	FIELDS		
	Physics	Astronomy	Related Fields
Doctorate	X		
Master's	X		

1. General

President, Univ. of Wisconsin System: Kevin Reilly
Chancellor, Univ. of Wisconsin, Madison: Carolyn A. Martin
Dean of Graduate School: M. Cadwallader
Department Chair: Baha Balantekin
Department Telephone Number: (608) 262-4526 (C)
Type of Institution: University
Control: Public
Setting: Urban
Total Faculty: 2,175
Total Graduate Faculty: Not separated
Total Students: 42,030
Total Graduate Students: 8,817
Annual Graduate Tuition:
In-state residents: Full-time—$5,258.88/semester
Part-time—$659.11/credit
Out-of-state residents: Full-time—$12,536.24/semester
Part-time—$1,568.78/credit
Tuition rates for: 2009–10
Deferred tuition plan: Yes[1]
Term: Semester

[1]Payroll deduction plan available to students holding graduate appointments.

2. Number of Faculty in Department

The combined total of full-time faculty in the three professorial ranks is 50.

3. Admission, Financial Aid, and Housing

Address admission inquiries to: Graduate Coordinator, Dept. of Physics, 1150 University Ave., Univ. of Wisconsin-Madison, WI 53706
Graduate application fee required: $56
Admission deadline (Fall admission): 1/1
Admission information: For fall admission, 2009–10, 46 students matriculated from 481 applicants.
Admission requirements: For admission to the graduate programs, a Bachelor's degree in science is required with a minimum undergraduate GPA of 3.00 specified. The GRE is required. The GRE Advanced is required. The average GRE scores for admissions were verbal-580; quantitative-780; total-1,395. The average GRE Subject score for admissions was 794. Students from non-English speaking countries are required to demonstrate proficiency in English via the TOEFL and/or Michigan English exam. Minimum acceptable TOEFL score for admission is 580 (237 computer-based or 92 internet-based) or 89 for MELAB. All international students who are admitted as teaching assistants will be required to take and pass the SPEAK test when they arrive on campus as well as participate in the 6-week Summer Orientation Program prior to the fall semester for which they have been admitted. An admitted applicant may be required to take the English Placement exam upon arrival and register for the recommended English as a second language (ESL) course.
Undergraduate preparation assumed: Classical mechanics; electromagnetic fields; electric circuits and elementary electronics; waves and optics; thermal physics; quantum physics; laboratory experience in classical and atomic physics.
Address financial aid inquiries to: Graduate Coordinator, Department of Physics, UW-Madison, 1150 University Avenue
GAPSFAS application required: No
Financial aid deadline: 1/1
Loans available: www.finaid.wisc.edu
Address housing inquiries to: www.housing.wisc.edu
On-campus, single student housing available: Yes[1]
On-campus, married student housing available: Yes[1]

[1]Demand heavy: apply early

Table A—Faculty, Enrollments, and Degrees Granted

Research Specialty	2008–09 Faculty	Enrollment[1] Fall 2009		No. of Degrees Granted[2] 2009–10 (2005–10)			Median No. of Years for 2009–10 Ph.D.'s
		Master's	Doctorate	Master's	Terminal Master's	Doctorate	
Astrophysics	5	0	15	0(2)	0(4)	4(14)	–
Atmos./Space Phys., Cosmic Rays	2	1	7	0(6)	1(3)	0(6)	–
Atomic, Molecular, & Optical Physics	5	1	20	1(5)	4(11)	3(18)	–
Condensed Matter Physics	13	0	41	2(4)	0(2)	5(27)	–
Engineering Physics/ Science	–	0	3	1(3)	0(3)	0(1)	–
Materials Sci./ Metallurgy	1	0	1	0(0)	0(2)	2(8)	–
Medical & Health Physics	–	0	0	0(1)	0(0)	1(5)	–
Nuclear Physics	3	1	5	1(2)	0(1)	0(2)	–
Particles & Fields	16	0	53	3(9)	0(0)	3(43)	–
Plasma Physics & Fusion	6	1	30	0(5)	0(2)	0(16)	–
Other Theoretical/ Math.	0	0	–	0(2)	0(1)	0(3)	–
Non-specialized	0	0	1	0(0)	1(3)	0(0)	–
Total		4	177	8(39)	6(32)	18(143)	
Full-time Grad. Stud.		4	177				
Part-time Grad. Stud.		–	–				
First-year Grad. Stud.		2	44				
Median Years in Grad. Study (2007–08 Degrees)					2.2	6.0	5.8

Undergraduate Degrees, 2008–09 (2007–08): 26(163)
2009–10 degrees include 12/09 and anticipated 5/10 only

[1]Students not yet committed to a research specialty are entered under non-specialized.
[2]Five-year totals in parentheses.

4. Graduate Degree Requirements

Master of Science: The department requires that at least 12 of the 18 must be in physics courses (other than research) numbered above 500. A Master of Science Degree is awarded to a student who has (1) satisfied the graduate-level credit and course requirements; (2) passed the Qualifying Exam; and (3) completed a Master's project, including a thesis.
Master of Arts: The department requires that at least 18 of the 24

credits must be in physics courses (other than research) numbered above 500. A Master of Arts Degree is awarded to a student who has (1) satisfied the graduate-level credit and course requirements; and (2) passed th Qualifying Exam.

Doctorate: Qualifying exam; preliminary exam; complete minor requirements; 32 credits (equivalent to four semesters), including required course work, with grades of B or better; Ph.D. thesis.

Thesis: Thesis work may not be done *in absentia*.

Table B—Appointments to Graduate Students, 2009–10

Title of Appointee	Appointments Total	Appointments First year	Academic Load Allowed in Credit Hours	Hours of Service Per Week	Stipend for Academic Year ($)
			Semester		
Teaching Assistant	54	26	6 credit	20	Reg. $14,087 Senior $16,264
Fellows	10	5			$16,605
Research Assistant	114	1			$15,724
Total	178	32			

5. Personnel Engaged in Separately Budgeted Research, 7/09–6/10

Professorial faculty	50
Postdoctoral appointments	80
Graduate students	178
Undergraduate students	30
Total	338

6. Separately Budgeted Research Expenditures by Source of Support

	Departmental Research	Physics-related Research Outside Department
Federal government	$20,806,550	$N/A
Private, non-profit organizations	2,279,660	
Total	$23,086,210	$N/A

Table C—Separately Budgeted Research Expenditures

Research Specialty	No. of Grants	Expenditures ($)
Astrophysics & Atmos./Space Phys., Cosmic Rays	18	1,238,475
Atomic, Molecular, & Optical Physics	19	1,666,987
Condensed Matter Physics	27	2,319,043
Nuclear Physics	8	650,000
Particles & Fields	66	7,674,101
Plasma Physics & Fusion	15	8,970,000
String Theory	3	280,000
Total	157	23,086,210

FACULTY

Professors

Balantekin, A. Baha, Ph.D., Yale, 1982. Theoretical physics at the interface of nuclear physics, particle physics, and astrophysics; mathematical physics; neutrino physics and astrophysics; fundamental symmetries; nuclear structure physics.

Barger, Vernon, Ph.D., Penn State, 1963. Theory and phenomenology of elementary particle physics; neutrino physics; electroweak gauge models; heavy quarks; supersymmetry; cosmology.

Bruch, L. W., Ph.D., California, San Diego, 1964. Chemical and theoretical physics; many-particle problems and statistical mechanics.

Carlsmith, Duncan L., Ph.D., Chicago, 1984. High-energy and fundamental particle physics at the Tevatron and LHC.

Chubukov, Andrey, Ph.D., Moscow State University, 1985. Condensed matter theory; low-D magnetism; frustrated antiferromagnets; fermi liquid theory; high Tc superconductivity.

Coppersmith, Susan N., Ph.D., Cornell, 1983. Theoretical condensed matter physics, nonlinear dynamics, quantum computation and information, biomineralization.

Dasu, S., Ph.D., Rochester, 1988. Experimental high energy and elementary particle physics; electro-weak symmetry breaking and search for new physics phenomena using the CMS experiment at the Large Hadron Collider; tigger and computing systems for high energy physics.

Eriksson, Mark, Ph.D., Harvard, 1997. Condensed matter physics, nanoscience, semiconductor membranes, semiconductor nanostructures, quantum dots, quantum computing, thermoelectric materials.

Forest, Cary B., Ph.D., Princeton, 1992. Experimental plasma physics, and liquid metal magnetohydrodynamics, with applications to astrophysics and magnetic confinement of fusion plasmas.

Gilbert, Pupa, Ph.D., First University of Rome "La Sapienza," 1987. Biophysics, specialized in biomineralization, nanobiology, synchrotron spectromicroscopy.

Halzen, Francis, Ph.D., Louvain, 1969. Theory and phenomenology of particle physics; particle astrophysics; neutrino astronomy.

Han, Tao, Ph.D., Wisconsin, 1990. Theoretical high energy physics, collider phenomenology; strong, electroweak and gravitational interactions; physics with extra dimensions.

Himpsel, Franz, J., Ph.D., Munich, 1977. Experimental condensed matter physics; synchroton radiation techniques; nanoscience.

Joynt, R. J., Ph.D., Maryland, 1982. Theory of superconductivity and heavy fermion systems; quantum Hall effect; magnetism, high-T_c; quantum computing.

Karle, A., Ph.D., Munich, 1994. Experimental particle astrophysics; high energy neutrino astronomy, neutrino physics, cosmic rays.

Knutson, Lynn, Ph.D., Wisconsin, 1973. Nuclear physics with polarized particles; properties of few-nucleon systems; medium energy physics.

Lagally, M. G., Ph.D., Wisconsin, 1968. Surface physics; structure and disorder; electronic materials; thin-film growth.[1]

Lawler, J. E., Ph.D., Wisconsin, 1978. Experimental atomic physics; laser spectroscopy; gas discharges, laboratory Astrophysics.

Lin, C. C., Ph.D., Harvard, 1955. Atomic and molecular physics; atomic collisions.

McCammon, D., Ph.D., Wisconsin, 1971. Astrophysics; x-ray astronomy; interstellar and intergalactic medium, x-ray detectors.

Ögelman, H., Ph.D., Cornell, 1966. Astrophysics; compact object observation in optical and x-ray; compact object theory.

Onellion, Marshall, Ph.D., Rice, 1984. Experimental solid state; synchroton radiation and ultra-fast optical techniques, nanomaterials.

Pondrom, L. G., Ph.D., Chicago, 1958. High-energy and fundamental particle physics; hadronic interactions.

Ramsey-Musolf, M., Ph.D., Princeton University, 1989. Nuclear and elementary particle theory; physics beyond the standard model; electroweak baryogenesis; low-energy tests of fundamental symmetries; neutrino properties and interactions; effective field theories of strong and electroweak interactions; extended Higgs sector models.

Rzchowski, Mark S., Ph.D., Stanford, 1988. Experimental condensed matter physics; magnetic heterostructures and nanostructures; low-temperature scanning tunneling spectroscopy; superconductivity in novel materials; thin film growth and fabrication.

Saffman, M. Ph.D. Colorado, 1994. Atomic physics; quantum computing with neutral atoms; quantum optics; entanglement; non-linear optics; solitons; pattern formation.

Sarff, John S., Ph.D., UW-Madison, 1988. Plasma physics; magnetic confinement; instabilities and turbulence.

Schnack, Dalton, Ph.D., University of California, Davis, 1977. Computational plasma physics.

Shiu, Gary, Ph.D., Cornell, 1998. String theory, Theoretical physics; elementary particle physics; cosmology.

Smith, Wesley H., Ph.D., California, Berkeley, 1981. High-energy and fundamental experimental particle physcis;/ep/ collisions at the LHC, CERN, Geneva, Switzerland.

Terry, P. W., Ph.D., Texas, 1981. Theory of turbulent plasmas and neutral fluids; plasma theory; anomalous transport and turbulence in fusion plasmas; plasma astrophysics.

Timbie, Peter T., Ph.D., Princeton 1985. Observational astrophysics and cosmology, measurements of the 2.7 K cosmic microwave background radiation; 21-cm hydrogen tomography; microwave detectors and cryogenics.

Walker, T., Ph.D., Princeton, 1988. Laser trapping of atoms; collisions between ultra-cold atoms, neutral atom quantum computing, spin-exchange optical pumping, biomagnetometry.

Winokur, Michael J., Ph.D., Michigan, 1985. Condensed matter physics; structure of novel materials; phase transitions.

Wu, Sau Lan, Ph.D., Harvard, 1970. High-energy and elementary particle physics; weak, electromagnetic, and strong interactions, Higgs Boson, CERN, Geneva, Switzerland.

Zweibel, Ellen, Ph.D., Princeton University, 1977. Theoretical astrophysics, plasma astrophysics; origin and evolution of astrophysical magnetic fields.[2]

Associate Professors

Boldyrev, S., Ph.D., Princeton, 1999. Plasma theory.

Chung, Daniel J.H., Ph.D., Chicago, 1998. Theoretical cosmology, high energy physics; quantum field theory in curved spacetime.

Hashimoto, Akikazu, Ph.D., Princeton, 1997. String theory, black hole physics, quantum field theory, theoretical physics.

Heeger, Karsten, Ph.D., University of Washington, Seattle, 2002. Neutrino physics and astroparticle physics.

Herndon, M., Ph.D., Maryland, 1998. Fundamental particle physics involving high energy hadron collisions with the CDF experiment at the Tevatron and the CMS experiment at the LHC. Research topics include rare decay of B hadrons, diboson physics, Higgs physics, and searches for fundamental new particles. Detector and algorithm development involving muon triggers and tracking detectors.

McDermott, Robert, Ph.D., University of California, Berkeley, 2002. Experimental condensed matter physics, quantum computing.

Montaruli, T., Ph.D., University of Bari, Italy, 1998. High-energy and elementary particle physics; neutrino physics, neutrino oscillations and astrophysics; cosmic ray physics.

Pan, Yibin, Ph.D., University of Wisconsin-Madison, 1991. High energy experimental particle physics.

Petriello, Frank, Ph.D., Stanford, 2003. Theoretical high-energy physics and collider phenomenology; perturbative QCD; techniques for higher-order calculations in quantum field theory; models of electroweak-scale physics.

Westerhoff, Stefan, Ph.D., University of Wuppertal, Germany, 1996. Experimental particle astrophysics, high energy neutrino astronomy, ultra high energy cosmic rays.

Assistant Professors

Everett, Lisa, Ph.D., University of Pennsylvania, 1998. Theoretical elementary particle physics, superstring phenomenology and supersymmetry.

Mellado, Bruce, Ph.D., Columbia University, 2001. Experimental high energy physics, supersymmetry and Higgs boson.

Perkins, Natalia, Ph.D., Moscow State University, 1997. Condensed matter theory; strongly correlated electron systems, orbital physics, frustrated magnetism, Kondo physics.

Vavilov, Maxim, Ph.D., Cornell University, 2001. Condensed matter theory: nanoscale and low dimensional systems.

Yavuz, Deniz, Ph.D., Stanford University, 2003. Experimental atomic, molecular, and optical physics.

Professors Emeriti

Anderson, L. W., Ph.D., Harvard, 1960. Atomic and molecular physics; atomic collisions; lasers.

Bincer, A.M., Ph.D., MIT, 1956. Theoretical physics, quantum field theory; group theory.

Callen, J. D., Ph.D., MIT, 1968. Plasma physics; theory of confinement and heating of magnetically confined plasmas, primarily for controlled thermonuclear fusion.[2]

Camerini, Ugo, B.S., São Paulo, 1946. High-energy and fundamental particle physics; cosmic rays; astrophysics.

Cox, Donald P., Ph.D., California, San Diego, 1970. Astrophysics and space physics; theoretical studies of interstellar matter; cosmic-ray acceleration.

Dexter, R. N., Ph.D., Wisconsin, 1955. Plasma physics; diagnostics of high-temperature plasma physics; fluctuation and turbulence studies.

Durand, Bernice, Ph.D., Iowa State, 1971. Theoretical high-energy physics; use of algebras in theoretical physics; also Associate Vice Provost.

Durand, L., Ph.D., Yale, 1957. Theoretical physics; elementary particle physics; electro-weak interactions; scattering processes; mathematical physics; special functions and group theory.

Ebel, Marvin E., Ph.D., Iowa State, 1953. Theoretical physics; high-energy physics and interactions of elementary particles.

Erwin, A. R., Jr., Ph.D., Harvard, 1959. High-energy and fundamental particle physics; strong interactions, neutrino Kaon and hyperon physics and neutrino oscillations.

Friedman, William A., Ph.D., MIT, 1966. Nuclear physics;

nuclear theory including reaction theory and collective effects in nuclear models; heavy-ion reactions.

Fry, W. F., Ph.D., Iowa State, 1951. High-energy and fundamental particle physics; weak interactions; high-energy astrophysics; physics of music.

Goebel, C. J., Ph.D., Chicago, 1954. Theoretical physics; quantum field theory, including high-energy interactions, elementary particles, and dispersion theory; general relativity.

Haeberli, Willy, Ph.D., Basel, 1952. Nuclear physics; polarized particle physics; polarized ion sources; polarized gas targets; tests of fundamental symmetries in hadronic and weak interactions.

Huber, David L., Ph.D., Harvard, 1964. Condensed matter theory; magnetic and optical properties of solids.

March, R. H., Ph.D., Chicago, 1960. High-energy and fundamental particle physics; high-energy astrophysics.[3]

Morse, R. Ph.D. Wisconsin, 1969. High-energy particle astrophysics, gamma-ray and neutrino astronomy.

Olsson, M. G., Ph.D., Maryland, 1964. Theory and phenomenology of fundamental particle physics, non-perturbative QCD, quark confinement, chiral dynamics, B-meson.

Prager, Stewart, Ph.D., Columbia, 1975. Plasma physics; magnetic confinement; instabilities and turbulence.

Prepost, R., Ph.D., Columbia, 1961. High-energy and fundamental particle physics; weak and electromagnetic interactions.

Quin, Paul A., Ph.D., Notre Dame, 1969. Nuclear reactions with polarized particles; weak and electro magnetic interactions, fundamental symmetries.

Reeder, D. D., Ph.D., Wisconsin, 1966. High-energy and fundamental particle physics; weak and electromagnetic interactions; *ep* colliders; cosmic rays.

Roesler, F. L., Ph.D., Wisconsin, 1962. Astrophysics; aeronomy; optical spectroscopy; interference spectroscopy.

Scherb, Frank, Ph.D., MIT, 1958. Astrophysics and space physics; space plasma physics; high-resolution astrophysical spectroscopy.

Sprott, J. C., Ph.D., Wisconsin, 1969. Plasma physics, computational nonlinear dynamics, chaos, complex systems.

Symon, K. R., Ph.D., Harvard, 1948. Particle accelerators; plasma physics; non-linear mechanics; particle orbit theory.

Webb, M. B., Ph.D., Wisconsin, 1956. Solid state physics; surface studies.

[1]Joint appointment in Materials Science and Engineering.
[2]Joint appointment in Astronomy.

RESEARCH SPECIALTIES AND STAFF

Theoretical

Astrophysics/Cosmology. Early universe cosmology; dark matter and energy; baryogenesis; cosmic microwave background radiation; modified gravity; string cosmology; interstellar medium; supernova remnants; gas dynamics and radiation; cosmic-ray acceleration; neutrino and gamma-ray astronomy. Balantekin, Chung, Halzen, Ögelman, Ramsey-Musolf, Zweibel.

Atomic and Molecular Physics. Scattering theory; electron-electron and electron-atom collisions; atomic collisions; molecular Rydberg states. Lin. Scientist/postdoctoral staff.

Biophysics. Modeling of a variety of complex biological systems. Coppersmith.

Chemical Physics. Variational methods; many-body problems; statistical mechanics. Bruch.

Condensed Matter. Magnetism; optical properties; energy band structure; many-body problems; superconductivity; heavy fermion systems; quantum Hall effect. Bruch, Chubukov, Coppersmith, Joynt, Lin, Perkins, Vavilov. Postdoctoral staff.

Elementary Particles. Quantum field theory; particle astrophysics; string theory; mathematical physics; phenomenology of particle physics; collider physics; standard model and extensions; cosmology. Balantekin, Barger, Chung, Everett, Halzen, Han, Hashimoto, Petriello, Ramsey-Musolf, Shiu.

Nuclear Physics. Reaction theory; scattering theory; nuclear structure; many-body theory; symmetry principles; heavy ions and intermediate energies; high-energy nuclear physics; nuclear astrophysics. Balantekin, Ramsey-Musolf. Postdoctoral staff.

Plasmas and Fusion. Stability theory, plasma confinement; turbulence theory; anomalous transport; heating theory; computer simulation. Boldyrev, Schnack, Terry, Zweibel. Scientist/postdoctoral staff.

Quantum Computing. Quantum algorithms, studies of decoherence, studies of novel experimental architectures. Coppersmith, Joynt. Scientist/postdoctoral staff.

Statistical and Thermal. Many-body problems; disordered systems; thin films. Bruch, Coppersmith, Joynt.

Experimental

Astrophysics/Cosmology. X-ray, gamma-ray and neutrino astronomy, observational cosmology, cosmic background radiation and spectroscopy. Karle, McCammon, Montaruli (Astronomy), Timbie, Westerhoff. Scientist/postdoctoral staff.

Atomic and Molecular. Atomic collisions; lasers; atomic oscillator strengths; high-resolution spectroscopy; trapped atoms; weakly ionized plasmas. Lawler, Lin, Saffman, Walker, Yavuz. Scientist/postdoctoral staff.

Biophysics: Photoelectron spectromicroscopy of biological systems, cancer therapy. Eriksson, Gilbert.

Condensed Matter Physics. Mesocopic systems; scanning force microscopies; strongly correlated magnetic materials; high-temperature superconductivity; magnetic nanostructures and heterostructures; structural properties of polymers; synchrotron radiation studies of strongly correlated systems. Eriksson, Himpsel, Lagally, McDermott, Onellion, Rzchowski, Winokur.

Elementary Particles. Weak, electromagnetic, and strong interactions; search for Higgs Bosons and new physics phenomena; study of B meson and neutrino physics. Study of proton structure. Carlsmith, Dasu, Heeger, Herndon, Mellado, Pan, Pondrom, Smith, Wu. Scientist/postdoctoral staff.

Nuclear Physics. Polarization phenomena at low and intermediate energies; properties of few-nuclear systems; electromagnetic and weak interactions; tests of fundamental symmetries; development of polarized-ion sources and polarized targets. Heeger, Knutson. Scientist/postdoctoral staff.

Plasma Physics and Magnetic Fusion. Toroidal confinement. Instabilities, turbulence, anomalous transport. Reversed field pinch, tokamak. Forest, Sarff, Terry. Scientist/postdoctoral staff.

Quantum Computing. Rydberg atom, flux qubit, and semiconductoring architectures. Eriksson, McDermott, Safman, Walker. Scientist/postdoctoral staff.

Spectroscopy. Very high-resolution studies of atomic, molecular and astrophysical phenomena; spectral line-strength determinations. Lawler. Postdoctoral staff.

UNIVERSITY OF WISCONSIN, MADISON

DEPARTMENT OF ASTRONOMY

Madison, Wisconsin 53706

Students Accepted For Degree	FIELDS		
	Physics	Astronomy	Related Fields
Doctorate		X	
Master's			

1. General

President: University of Wisconsin System: Kevin Reilly
Dean of Graduate School: Martin T. Cadwallader
Department Chairman: Robert D. Mathieu
Department Telephone Number: (608) 262–3071 (C)
Type of Institution: University
Control: Public
Setting: Urban
Total Faculty: 2,023
Total Graduate Faculty: (not separated)
Total Students: 42,099
Total Graduate Students: 9,116
Annual Graduate Tuition:
In-state residents: Full-time—$10,518.00
Out-of-state residents: Full-time—$25,072.00
Tuition rates for: 2009–10
Deferred tuition plan: Yes
Other Fees: None
Term: Semester

2. Number of Faculty in Department

The combined total of full-time faculty in the three professorial ranks is 13. The combined total of full-time, part-time, and other faculty at all ranks is 16.

3. Admission, Financial Aid, and Housing

Address admission inquiries to: Graduate Coordinator, Department of Astronomy, 475 N. Charter St., Madison, WI 53706.
E-mail address: gradinq@astro.wisc.edu
Graduate application fee required: $52
Admission deadline (Fall admission): 1/15
Admission information: For fall admission, 2010–11, 7 students were accepted from 95 applicants.
Admission requirements: To enter as a graduate student, an applicant must have undergraduate preparation that includes at least three years of college physics and mathematics through differential equations. Applicants are judged on the basis of: previous academic record, letters of recommendation, personal statement, research experience, and Graduate Record Exam (GRE) scores. Admission is competitive as it is for the fall only. Applicants must submit the following: Official transcripts of all undergraduate work, reasons for graduate study, at least three letters of recommendation from people well acquainted with past academic work (no special form), GRE scores (general test and subject test-Physics). International students must submit the Test Of English As A Foreign Language (TOEFL) scores. Financial support is provided through university fellowships or departmental assistantships. Unless a student is admitted conditionally, the admission includes a **four-year** guarantee of support as long as the student progresses satisfactorily through our program.
Address financial aid inquiries to: Graduate Coordinator, Department of Astronomy, 475 N. Charter St., Madison, WI 53706.
E-mail address: gradinq@astro.wisc.edu.
GAPSFAS application required: No
Financial aid deadline: 1/1
Loans available: Yes
Address housing inquiries to: UW Housing, Slichter Hall, 625 Babcock Drive, Campus Assistance, 716 Langdon St., universityapartments@housing.wisc.edu
On-campus, single student housing available: Yes
Cost/month: $605–785
On-campus, married student housing available: Yes
Cost/month: $605–$1005

Table A—Faculty, Enrollments, and Degrees Granted

Research Specialty	2009–10 Faculty	Enrollment[1] Fall 2009		No. of Degrees Granted[2] 2009–10 (2005–10)			Median No. of Years for 2005–06 Ph.D.'s
		Mas-ter's	Doc-torate	Mas-ter's	Terminal Master's	Doc-torate	
Astronomy	13	0	25	0(0)	0(0)	5(21)	6.0
Total		0	27			5(21)	
Full-time Grad. Stud.		0	24				
Part-time Grad. Stud.		0	0				
First-year Grad. Stud.		0	4				
Median Years in Grad. Study (2005–06 Degrees)			3.5				6.0
Undergraduate Degrees, 2005–06 (1999–04): 12(64)							

[1]Students not yet committed to a research specialty are entered under non-specialized.
[2]Five-year totals in parentheses.

4. Graduate Degree Requirements

Doctorate: Preliminary exam, complete minor requirements, 32 credits (equal to 4 semesters) including required course work, all with grades of "B" or better; Ph.D. thesis.
Theses: Theses may be written *in absentia*.

Table B—Appointments to Graduate Students, 2009–10

Title of Appointee	Appointments Total	Appointments First year	Academic Load Allowed in Credit Hours	Hours of Service Per Week	Stipend for Academic Year ($)
			Semester		
Teaching Assistant	2	1	6–12	20	14,088–16,264
Research Assistant	22	3	6–12	20	22,606–28,451[1]
Fellowship	1	0	6–12		
Total	25	3			

[1]Research Assistant appointments range from 50%–70%, and are on annual basis.

5. Personnel Engaged in Separately Budgeted Research, 7/09–6/10

Professorial faculty	12
Other faculty	5
Postdoctoral appointments	4
Graduate students	25
Undergraduate students	25
Nonteaching research personnel	9
Scientists	6
Total	86

6. Separately Budgeted Research Expenditures by Source of Support

	Departmental Research
Federal government	$9,546,600
Private, non-profit organizations	380,900
Other	2,800,400
Total	$16,156,000

Table C—Separately Budgeted Research Expenditures

Research Specialty	No. of Grants	Expenditures ($)
Astronomy	101	16,156,000
Total	101	16,156,000

FACULTY

Professors

Barger, Amy, Ph.D., Cambridge, 1997. Extragalactic astronomy, observations of high redshift galaxies.

Bershady, Matthew A., Ph.D., Chicago, 1994. Galaxy kinematics, stellar populations, galaxy and quasar evolution; optical and IR spectra and instrumentation.

Gallagher, John S. III, Ph.D., Wisconsin, 1972. Multi-wavelength observational investigations of evolutionary processes in galaxies, stellar populations, classical novae.

Lazarian, Alex, Ph.D., Cambridge, 1995. Interstellar turbulence, magnetohydrodynamics, dynamo, dust, and molecular clouds, star formation, polarizations from dust.

Mathieu, Robert D., Ph.D., California-Berkeley, 1983. Department Chair. Observational study of star formation, binary stars, and open star clusters.

Sparke, Linda S., Ph.D., California-Berkeley, 1981. Structure and dynamics of galaxies, dynamical constraints on dark matter, galactic warps, bars, and polar rings.

Wilcots, Eric M., Ph.D., Washington, 1992. Department Chair. The structure and evolution of galaxies through radio and optical observations; extended gas around galaxies; distributions and kinematics of the interstellar medium.

Zweibel, Ellen, Ph.D., Princeton, 1977. Theoretical plasma astrophysics. Generation and evolution of astrophysical magnetic fields, interstellar astrophysics, star formation, stellar physics.

Assistant Professors

Heinz, Sebastian, Ph.D., Colorado, 2000. Relativistic jets, black holes, AGNs. X-ray binaries, galaxy clusters, gamma ray bursts, interstellar and intergalactic medium. Numerical methods.

Sheinis, Andrew I., Ph.D., California-Santa Cruz, 2002. Astronomical instrumentation, Adaptive Optics, AGN, QSO's, black hole/galaxy formation and evolution.

Stanimirovic, Snezana, Ph.D., Western Sydney Nepean, 1999. Galactic disk/halos, dust properties in low-metallicity environments, physics of the interstellar medium.

Townsend, Richard, Ph.D., University College, London, 1997. Stellar astrophysics, magnetic fields, stellar winds, massive stars.

Tremonti, Christy, Ph.D., Johns Hopkins, 2003. Dwarf stars, quasar evolution, star formation.

Professors Emeriti

Cassinelli, Joseph P., Ph.D., Washington, 1970. Theory of stellar winds, stellar X-ray, and infrared observations; the effects of rotation and magnetic fields on the atmospheres and envelopes of hot stars.

Churchwell, Edward B., Ph.D., Indiana, 1970. Star formation, hot molecular cores, UC HII regions, atomic abundances, radio and infrared astronomy.

Nordsieck, Kenneth H., Ph.D., California-Santa Cruz, 1972. Stellar and extragalactic optical/ultraviolet spectropolarimetry, ground-based instrument control, space astronomy.

Reynolds, Ronald J., Ph.D., Wisconsin, 1971. High-resolution spectroscopy of diffuse sources, development of high-throughput spectrometers, physics of the interstellar medium.

Savage, Blair D., Ph.D., Princeton, 1968. Physical properties of the interstellar medium, gas in galactic halos and the intergalactic medium; high-resolution ultraviolet spectroscopy.

Scientists

Orio, Marins, Ph.D., Technion, Haifa, Israel, 1987. Close binary stars, classical and recurrent novae, low-mass X-ray binaries, supersoft X-ray sources, cataclysmic variables, ionization nebulae.

Haffner, Matt, Ph.D., Wisconsin, 1999. Observation studies of the interstellar medium: in particular, the distribution and physics conditions of the Warm Ionized Medium in the Milky Way.

Wakker, Bastiaan, Ph.D., Gronigen, 1990. The gaseous halo of the Milky Way, high- and intermediate-velocity gas, the interstellar medium of the Large Magellanic Cloud.

Percival, Jeffrey, Ph.D., Wisconsin, 1979. Using computers, computer software, and new computer algorithms to perform astronomical research and place instrumentation at the forefront.

RESEARCH SPECIALTIES AND STAFF

Stellar star formation, young stars: Cassinelli, Churchwell, Mathieu, Nordsieck, Orio, Townsend.

Interstellar and intergalactic media: Haffner, Heinz, Savage, Reynolds, Stanimirovic, Wakker.

Extragalactic astronomy and cosmology: Barger, Bershady, Gallagher, Sparke, Tremonti, Wilcots.

Instrumentation: Anderson, Bershady, Nordsieck, Percival, Reynolds, Sheinis.

Plasma astrophysics and magnetic fields: Cassinelli, Lazarian, Nordsieck, Zweibel.

UNIVERSITY OF WISCONSIN, MILWAUKEE

DEPARTMENT OF PHYSICS

Milwaukee, Wisconsin 53201

Students Accepted For Degree	FIELDS		
	Physics	Astronomy	Related Fields
Doctorate	X		
Master's	X		

1. General

President, Univ. of Wisconsin System: Kevin P. Reilly
Chancellor, Univ. of Wisconsin, Milwaukee: Carlos E. Santiago
Dean of Graduate School: Colin G. Scanes
Department Chairman: Alan G. Wiseman
Department Telephone Number: (414) 229-4474
Type of Institution: University
Control: Public
Setting: Urban
Total Faculty: 824
Total Graduate Faculty: 824
Total Students: 30,455
Total Graduate Students: 5,216
Annual Graduate Tuition:
In-state residents: Full-time—$9,184, Part-time—Varies per no. of credits
Out-of-state residents: Full-time—$22,852 Part-time—Varies per no. of credits
Tuition rates for: 2009–10
Deferred tuition plan: Yes
Other Fees: $816*
Term: Semester

*Segregated fee not included in amount quoted for full-time students; pro-rated for part-time students.

2. Number of Faculty in Department

The combined total of full-time faculty in the three professorial ranks is 22. The combined total of full-time, part-time, and other faculty at all ranks is 23.

3. Admission, Financial Aid, and Housing

Address admission inquiries to: Department of Physics, P.O. Box 413, Milwaukee, WI 53202
Graduate application fee required: $56, domestic; $96 (includes processing fee), foreign
Admission deadline (Fall admission): 2/15
Admission information: For fall admission, 2009–10, 8 students were admitted and enrolled from 60 applicants.
Admission requirements: For admission to the graduate programs, a Bachelor's degree in physics or a related field is required with a minimum undergraduate GPA of 2.75 specified. Applicants are encouraged to take the GRE General Test and the GRE Subject Test. Students from non-English speaking countries are required to demonstrate proficiency in English via the TOEFL exam. Minimum acceptable score for admissions is 550 (213 for computer based test). For financial support the minimum acceptable score is 580 (237 for computer based test).
Undergraduate preparation assumed: *Mechanics*, Barger, Olsson, and Marion; *Thermodynamics*, Zemansky and Dittman; *E.M.*, Corson, Lorrain, and Griffiths; *Quantum Physics*, Eisberg, Resnick, and Gasiorowitz.

Address financial aid inquiries to: UW-Milwaukee, Dept. of Financial Aid, P. O. Box 469, Milwaukee, WI 53201
GAPSFAS application required: No
Financial aid deadline: 1/1
Loans available: Yes
Address housing inquiries to: UW-Milwaukee, Attn.—Housing/Sandburg Hall, 3400 N. Maryland Ave., Milwaukee, WI 53211.
On-campus, single student housing available: Yes
Cost/term: $4,870 (AY)(Single room only)
On-campus, married student housing available: No

Table A—Faculty, Enrollments, and Degrees Granted

Research Specialty	2009–10 Faculty	Enrollment Fall 2009 Mas-ter's	Enrollment Fall 2009 Doc-torate	No. of Degrees Granted[1] 2009–10 (2005–09) Mas-ter's	Terminal Master's	Doc-torate	Median No. of Years for 2009–10 Ph.D.'s
Astrophysics	1	0	0	0(0)	0(0)	0(1)	0
Biophysics	4	0	4	0(0)	0(0)	0(1)	–
Condensed Matter Physics & Surface Physics	9	0	12	0(6)	2(3)	1(4)	6
Medical Physics	1	1	0	0(0)	2(0)	0(0)	–
Optics	1	0	2	0(0)	0(1)	0(1)	–
Particles & Fields	0	0	0	0(1)	0(1)	0(0)	–
Relativity & Gravitation	6	0	8	0(0)	1(2)	1(8)	6
Total		1	26	0(7)	5(7)	2(15)	
Full-time Grad. Stud.		1	26				
Part-time Grad. Stud.		0	0				
First-year Grad. Stud.		1	8				
Median Years in Grad. Study (2009–10 Degrees)				0	3	6	

[1]Five-year totals in parentheses.

4. Graduate Degree Requirements

Master's: The Master's graduate must complete 24 graduate credits (12 credits must be earned at UWM) and pass a comprehensive examination. The student may choose to engage in research and present a thesis for 6 credits of the total needed. The Graduate School requires an average of at least 3.0 (4.0 basis) for all graduate work.
Doctorate: The Ph.D. graduate is required to complete 54 graduate credits (27 credits must be earned in residence at UWM) beyond the Bachelor's degree and pass a written comprehensive examination. A dissertation reporting the results of original and independent research investigation must be written and defended. The Graduate School requires an average of at least 3.0 (4.0 basis) for all graduate work.
Thesis: Thesis may be written *in absentia*.
Special Equipment, Facilities, or Programs: Certain faculty members of the department also participate in Laboratory for Surface Studies, an interdepartmental unit (with members from physics, chemistry, material science, etc.) dedicated to the study of surfaces on a microscopic scale. Laboratory equipment in the Physics Building includes high and ultra high vacuum systems with facilities for the following ana-

lytical techniques: high- and low-energy-electron diffraction (RHEED & LEED), Auger spectroscopy, x-ray energy dispersive spectrometry (EDS), electron spectroscopy for chemical analysis (ESCA), infrared spectroscopy (IR), scanning tunneling microscopy (STM), atomic force microscopy (AFM), high resolution transmission electron microscopy (HRTEM) and molecular beam epitaxy (MBE). Experiments using synchrotron radiation are carried out at the nearby Wisconsin Synchrotron Radiation center and at Brookhaven National Laboratory and Argonne National Laboratory. Additional experimental condensed matter research facilities include a ^{3}He-^{4}He dilution refrigerator, several superconducting magnets (8–9.5T), pulsed-ultrasonic equipment (1 MHz–4 GHz), and a thin film deposition facility-sputtering, evaporation, microphotolithography and clean-room high temperature furnaces. Oxide synthesis in infrared image furnace and E-beam furnace for single crystal growth. Physical property measurement equipment. Ultra high power femtosecond pulsed laser is central part of the optics lab. Biophysics laboratories are equipped with all necessary tools to grow bacteria, to purify and crystallize proteins, to collect x-ray diffraction data and to analyze them for structure determination; also equipment for Förster Resonance Energy Transfer (FRET) experiments. Experimental collaborations exist with, among others, Argonne National Laboratory, Univ. of Illinois, Lucent Labs., Univ. of Hamburg, Illinois Institute of Technology, Univ. of Chicago, Science Univ. of Tokyo, Northwestern University and the National High Magnetic field Laboratory at Tallahasee and at Los Alamos.

Table B—Appointments to Graduate Students, 2009–10

Title of Appointee	Appointments Total	Appointments First year	Academic Load Allowed in Credit Hours	Hours of Service Per Week	Stipend for Academic Year ($)
			Semester		
Teaching Assistant (teaching assistants do not pay tuition fees)	19	8	6–9	20	23,209[1,2,3,4]
Research Assistant (research assistants do not pay tuition fees)	13	0	8–12	20	34,000[1,3,4]
Graduate School Fellowships	1	0	8–12		11,332[1,3]
Dissertation Fellowship	2	0	3	0	14,000[1,3]
Total	35	8			

[1]Additional Chancellor's and Physics Graduate Student Trust Fund awards are given competitively, and total up to $8,000.
[2]Non-doctoral rates.
[3]Non-resident portion of tuition waived.
[4]AY is 9 months; awards rarely exceed 50% of AY rate.

5. Personnel Engaged in Separately Budgeted Research, 7/09–6/10

Professorial faculty	22
Postdoctoral appointments	23
Graduate students	35
Total	80

6. Separately Budgeted Research Expenditures by Source of Support

	Departmental Research	Physics-related Research Outside Department
Federal government	$3,938,000	$475,000
Total	$3,938,000	$475,000

Table C—Separately Budgeted Research Expenditures

Research Specialty	No. of Grants	Expenditures ($)
Condensed Matter Physics & Surface Physics	6	886,000
Relativity & Gravitation	5	2,937,000
Research Education	1	115,000
Total	12	3,938,000

Table D—Physics-related Research Outside Department

Field and Unit Outside Department	No. of Grants	Expenditures ($)
Argonne National Laboratory	1	475,000
Total	1	475,000

FACULTY

Professors

Agterberg, Daniel, Ph.D., University of Toronto, 1996. Theoretical condensed matter physics (superconductivity and magnetism).

Brady, Patrick, Ph.D., University of Alberta, 1993. Gravitational wave detection; detection of periodic sources, coalescing binary search. Gravitational wave astronomy. Numerical relativity; flux conservative formalisms; intermediate black hole coalescence. Dynamics of gravitational collapse, quantum effects in gravitational collapse.

Friedman, John, Ph.D., University of Chicago, 1973. Gravitational physics; quantum gravity; spacetime topology; mathematical relativity; relativistic astrophysics.

Gajdardziska-Josifovska, Marija, Ph.D., Arizona State University, 1991. Experimental solid state physics; transmission and reflection electron microscopy of solid surfaces and interfaces; electron holography; reflection high energy electron diffraction; nanodiffraction; x-ray and electron-energy-loss spectroscopy; growth and optical properties of thin films.

Hirschmugl, Carol J., Ph.D., Yale University, 1994. Experimental surface physics. Infrared synchrotron radiation studies of surfaces; photoelectron diffraction studies of hydrocarbons on metals; reflection-absorption spectroscopy.

Li, Lian, Ph.D., Arizona State University, 1995. Experimental surface physics. Molecular beam epitaxy, scanning tunneling microscopy.

Lyman, Paul, Ph.D., University of Pennsylvania, 1991. Experimental surface science. Semiconductor adsorbates and thin film growth using x-ray scattering, x-ray standing waves, photoemission, and reflection high energy electron diffraction.

Ourmazd, Abbas, D.Phil., Oxford University, 1980. Semiconductor physics, electron microscopy and holography, protein crystallography.

Parker, Leonard, Ph.D., Harvard University, 1967. General relativity; cosmology and quantum field theory in curved spacetime; particle production by gravitational fields; renormalization of quantized fields interacting with gravity; use of computer symbolic manipulation in mathematical physics.

Saldin, Dilano K., D.Phil., Oxford University, 1975. Theoretical surface science; crystallography of ordered and disordered surfaces; protein crystallography; theory of electron diffraction and microscopy; x-ray absorption; electron energy-loss spectroscopy.

Sarma, Bimal, Ph.D., Northwestern University, 1980. Low-temperature physics; heavy-fermion superconductivity; high-T_c superconductivity; ultrasonics; neutron diffraction, and physical phenomena at high magnetic fields.

Sorbello, Richard, Ph.D., Stanford University, 1970. Theoretical condensed matter physics; electron and atom transport in solids and low-dimensional systems; electromigration; optical response of nanoparticle systems.

Weinert, Michael, Ph.D., Northwestern University, 1982. Theoretical condensed matter physics, magnetism and first-principles electronic structure.

Yakovlev, Vladislav, V., Ph.D., Moscow State University, 1990. Optical investigations of the interactions between atoms and molecules on the mesoscopic scale and the dynamics of biological molecules and processes in living cells.

Professors Emeriti

Beck, Donald, Ph.D.
Chow, Yutze, Ph.D.
Dittman, Richard, Ph.D.
Greenler, Robert, Ph.D.
Levy, Moises, Ph.D.
Lubkin, Elihu, Ph.D.
McQuistan, Richmond, Ph.D.
Schmieg, Glenn, Ph.D.
Shurman, Michael, Ph.D.
Snider, Dale, Ph.D.
Suchy, Raymond, Ph.M.
Walters, William, Ph.D.

Associate Professors

Creighton, Jolien, Ph.D., University of Waterloo, Canada, 1996. General relativistic theory of gravity. Physics of black holes and gravitational radiation.

Guptasarma, Prasenjit, Ph.D., Tata Institute of Fundamental Research, University of Bombay, 1993. Experimental condensed matter physics. Highly correlated electronic materials, unconventional superconductivity, floating-zone single crystal growth, physical properties of materials at low temperatures.

Patch, Sarah K., Ph.D., UC Berkeley, 1994. Image reconstruction, primarily for medical imaging. Current focus: thermo/opto-acoustic tomography, requiring inversion of the spherical Radon transform.

Raicu, Valerica, Ph.D., University of Bucharest, 1997. Studies of biological systems by optical methods.

Wiseman, Alan, Ph.D., Washington University in St. Louis, 1992. Gravitational wave source modeling. Computer analysis of gravitational-wave data.

Assistant Professors

Anchordoqui, Luis, Ph.D., Universidad Nacional de La Plata, 1998. Cosmic ray astrophysics and neutrino astrophysics with applications to particle physics and cosmology.

Schmidt, Marius, Ph.D., Technical University of Munich, 1996. Biophysics.

Siemens, Xavier, Ph.D., Tufts University, 2002. Gravitational Physics.

RESEARCH SPECIALTIES AND STAFF

Theoretical

Biophysics. Single molecule crystallography, direct methods. Ourmazd, Saldin. 1 senior scientist. 2 postdoctoral fellows.

Condensed Matter Physics. Transport phenomena; electron and atom transport in solids and low-dimensional systems, magnetism, phase stability, electronic structure, electromigration and superconductivity. Agterberg, Sorbello, Weinert. 1 postdoctoral fellow.

Gravitational Physics and Cosmology. Gravitational-wave astronomy; collaboration on the construction for LIGO (Laser-Interferometric Gravitational-wave Observatory) of templates to detect the gravity-wave background of the early universe and the waves emitted in the coalescence of binary systems of neutron stars (and of black holes). Early cosmology; use of renormalization-group methods to study the physics of inflationary universe models; gravitational waves from a network of cosmic strings. Quantum field theory on curved spacetime; blackhole evaporation, particle production in the early universe; field theory on spacetimes that are not globally hyperbolic. Aspects of quantum gravity associated with small-scale noneuclidean topology. Relativistic astrophysics; structure and stability of relativistic stars; limits on the mass and rotation of pulsars, and implied constraints on the equation of state of matter above nuclear density. Brady, Creighton, Friedman, Parker, Siemens, Wiseman. 8 postdoctoral fellows. 3 senior scientists.

Medical Physics. Image reconstruction and data corrections/preprocessing. Work relies heavily on fourier analysis, and inversion of Radon transforms. Patch.

Particle physics. Cosmic ray astrophysics and neutrino astronomy with applications to particle physics and cosmology. Anchordoqui.

Statistical Physics. Entropy and quantum coherence. Black-hole thermodynamics and information loss. Parker.

Surface Physics. Crystallography of ordered and disordered surfaces, theory of electron diffraction and microscopy, x-ray absorption, electron-energy-loss spectroscopy, electron holography, dynamical processes at surfaces and interfaces, surface phonons, atom-solid bonding, photoemission, inelastic low-energy electron diffraction, and surface electronic structure of materials reconstruction. Saldin and Weinert. 1 postdoctoral fellow.

Experimental

Biophysics. Protein crystallography. Development of infrared microspectroscopy, confocal microscopy, fluorescence, and Raman spectroscopy methods for characterization of living cells. Iron and toxic metal storage in organisms; atomic structure of botanical nanocrystalline iron biominerals; biomagnetism; biogenic self assembly; life in extreme environments. Gajdardziska-Josifovska, Hirschmugl, Ourmazd, Raicu, Schmidt, Yakovlev. 3 postdoctoral fellows.

Condensed Matter Physics. Studies of the electron-phonon interaction in superconductors, surface wave interaction with superconducting and magnetic films and spin-phonon interaction. The principal technique used in these studies is ultrasonic attenuation, and other physical property measurements at low temperatures. Areas of interest are high-T_c superconductivity, heavy fermion superconductivity, milliKelvin temperature physics and neutron diffraction. Single crystal growth by floating-zone and other techniques, oxide and intermetallic compounds, manganite and cuprate oxides, ferroelectric oxides. Sarma, Guptasarma. 2 postdoctoral fellows.

Optics. Femtosecond laser spectroscopy and optical investigations of the interactions between atoms and molecules on the mesoscopic scale and the dynamics of biological molecules and processes in living cells. Yakovlev. 1 senior scientist.

Surface Physics. Physical methods for synthesis of advanced materials with reduced dimensionality, including thin films and nanocrystals of wide band gap semiconductors and high k-dielectrics. Development and applications of electron and photon based imaging, diffraction and spectroscopy methods for determination of atomic structures, with emphasis on solid surfaces, interfaces and nanocrystals. Effects of surface structure on controlled epitaxial growth of single crystal films; Stabilization mechanisms for polar oxide surfaces and interfaces; Low energy dynamics and structure of water-oxide interfaces, electronic friction at adsorbate-superconductor interfaces. Gajdardziska-Josifovska, Hirschmugl, Li, Lyman. 1 postdoctoral fellow.

UNIVERSITY OF WYOMING

DEPARTMENT OF PHYSICS AND ASTRONOMY

Laramie, Wyoming 82071-3905

Students Accepted For Degree	FIELDS		
	Physics	Astronomy	Related Fields
Doctorate	X		
Master's	X		

1. General

President: Tom Buchanan
Grad School closed
Department Chairman: Daniel Dale
Department Telephone Number: (307) 766–6150 (C)
Type of Institution: University
Control: Public
Setting: Small town
Total Faculty: 730
Total Students: 13,476
Total Graduate Students: 3,240
Annual Graduate Tuition:
In-state residents: Full-time—$5298/sem
Out-of-state residents: Full-time—$13,458/sem
Tuition rates for: 2009–10
Term: Semester

2. Number of Faculty in Department

The combined total of full-time faculty in the three professorial ranks is 11. The combined total of full-time, part-time, and other faculty at all ranks is 15.

3. Admission, Financial Aid, and Housing

Address admission inquiries to: www.uwyo.edu/
Graduate application fee required: Yes ($50)
Admission deadline (Fall admission): 1/20
Admission information: For fall admission, 2010–11, 10 students were accepted from 55 applicants.
Admission requirements: For admission to the graduate programs, a Bachelor's degree in physics is required with no minimum undergraduate GPA specified. The GRE is required. The minimum acceptable score for admission is total–1000. The average GRE scores for admissions were verbal–561; quantitative–751; total–1312. The average GRE advanced score for admissions was 689. Students from non-English speaking countries are required to demonstrate proficiency in English via the TOEFL exam. Minimum acceptable score for admission is 540 (76 internet based).
Undergraduate preparation assumed: Undergraduate preparation in physics and mathematics equivalent to that specified for a physics major.
Address financial aid inquiries to: www.uwyo.edu/sfa/
GAPSFAS application required: No
Financial aid deadline: 3/1
Loans available: Yes
Address housing inquiries to: www.uwyo.edu/reslife-dining/
On-campus, single student housing available: Yes
Cost/term: $3,582–5,306
On-campus, married student housing available: Yes
Cost/term: $517–904

Table A—Faculty, Enrollments, and Degrees Granted

Research Specialty	2009–10 Faculty	Enrollment[1] Fall 2008		No. of Degrees Granted[2] 2008–09 (2004–09)			Median No. of Years for 2008–09 Ph.D.'s
		Mas-ter's	Doc-torate	Mas-ter's	Terminal Master's	Doc-torate	
Astrophysics	5	2	11	2(3)	0(2)	3(5)	6.5
Biophysics	1	0	0	0(0)	0(0)	0(0)	
Condensed Matter	5	1	6	1(1)	1(1)	0(1)	–
Plasma Physics & Fusion	1	0	0	0(0)	0(0)	0(0)	–
Total		3	14	0(3)	0(6)	2(4)	
Full-time Grad. Stud.			17				
Part-time Grad. Stud.			0				
First-year Grad. Stud.		3	5				
Median Years in Grad. Study (2009–10 Degrees)			5.5				

Undergraduate Degrees, 2009–10 (2005–10): 9(33)

[1]Students not yet committed to a research specialty are entered under non-specialized.
[2]Five-year totals in parentheses.

4. Graduate Degree Requirements

Master's: 30 hours of graduate course work. Thesis planning, development, and production guided by the committee chair and graduate committee. 3.0 GPA (on 4.0 scale) required.
Doctorate: 42 hours of course work at the graduate level, 30 hours of research. Dissertation planning, development, and production guided by the committee chair and graduate committee. 3.0 GPA (on 4.0 scale) required. Comprehensive exam required.

Table B—Appointments to Graduate Students, 2009–10

Title of Appointee	Appointments Total	Appointments First year	Academic Load Allowed in Credit Hours	Hours of Service Per Week	Stipend for Academic Year ($)
			Semester		
Teaching Assistant	8	4	9–12 credits	18	$15,795
Research Assistant	11	3	9–12 credits	18	$15,795
Total	19	7			

5. Personnel Engaged in Separately Budgeted Research, 7/09–6/10

Professorial faculty	9
Postdoctoral appointments	4
Graduate students	19
Nonteaching research personnel	3
Total	35

6. Separately Budgeted Research Expenditures by Source of Support

	Departmental Research	Physics-related Research Outside Department
Federal government	$8,424.031	$
State and local government	227,000	
Business and industry	448,991	
Other	111,596	$
Total	$8,984,618	$

Table D—Physics-related Research Outside Department

Research Specialty	No. of Grants	Expenditures ($)
National Aeronautics & Space Admin.		$4,568,956
NSF		3,855,075
State of Wyoming		227,000
Industry		448,991
Misc.		111,596
Total		$8,984,618

FACULTY

Professor

Dahnovsky, Yuri, Ph.D., Russian Academy of Sciences, 1983. Computational and theoretical physics: molecular electronic devices, solar cells, electronic properties of surfaces, non-equilibrium Green functions, photon-assisted tunneling and electron transfer reactions.

Dale, Daniel, Ph.D., Cornell, 1998. Ground- and space-based multi-wavelength studies of galaxies; clusters of galaxies; observational cosmology.

Johnson, Paul, Ph.D., Washington, 1979. Biophysics; detection of pathogenic micro-organisms.

Tang, Jinke, Ph.D., Iowa State, 1989. Experimental condensed matter physics and materials science: spintronics and opto-electronics; magnetic semiconductors; half-metals; magnetic, optical and thermoelectric properties of nanomaterials; tunneling magnetoresistance; thin films.

Associate Professor

Brotherton, Michael, Ph.D., Texas, 1996. Multi-wavelength observations of quasars and active galaxies; quasar/galaxy mutual evolution.

Kobulnicky, Henry, Ph.D., Minnesota, 1997. Ground and space-based studies of dynamics & chemical abundances in galaxies; radio, optical, and infrared spectroscopy; young star clusters; massive star formation; astronomical instrumentation.

Pierce, Michael, Ph.D., Hawaii, 1988. Galaxies, clusters of galaxies, large-scale structure of the universe, observational cosmology, astronomical instrumentation.

Thayer, David, Ph.D., MIT, 1983. Theoretical studies of plasmas, fusion, turbulence, nonlinear dynamics, global change, and quantum mechanics.

Assistant Professors

Michalak, Rudy, Ph.D., Physics, Ruhr-Universität Bochum, 1993. Experimental condensed matter physics; nuclear magnetic resonance; science education.

Feiguin, Adrian, Ph.D., Universidad Nacional de Rosario, 2000. Computational studies of strongly correlated systems in condensed matter. Density matrix renormalization groups, exact diagonalization, quantum Monte Carlo and analytical methods.

Wang, Wenyong, Ph.D., Yale, 2004. Experimental condensed matter physics; nanotechnology.

FACULTY PUBLICATIONS

Brotherton, Michael

A. Georgakakis *et al.*, "Are red 2MASS QSOs young?" MNRAS **394**, 533 (2009).

Dahnovsky, Yuri

Yu. Dahnovsky, "Electron-electron correlations in molecular tunnel junctions: A diagrammatic approach," Phys. Rev. B **80**, 5305 (2009).

Dale, Daniel

M. Boquien *et al.*, "Star-Forming or Starbursting: The Ultraviolet Conundrum," ApJ **706**, 553 (2009).

J. Lee *et al.*, "Comparison of Halpha and UV Star Formation Rates in the Local Volume: Systematic Discrepancies for Dwarf Galaxies," ApJ **706**, 599 (2009).

B. Bertincourt *et al.*, "A Spitzer Unbiased, Ultradeep Spectroscopic Survey," ApJ **705**, 68 (2009).

J. C. Munoz-Mateos *et al.*, "Radial Distribution of Stars, Gas, and Dust in SINGS Galaxies. I. Surface Photometry and Morphology," ApJ **703**, 1672 (2009).

P. Kennicutt *et al.*, "Extinction-Corrected Star Formation Rates of Galaxies. I. Combinations of Halpha and Infrared Tracers," ApJ **703**, 1672 (2009).

D. Dale , "The Spitzer Local Volume Legacy: Survey Description and Infrared Photometry," ApJ **703**, 517 (2009).

J. C. Munoz-Mateos *et al.*, "Radial Distribution of Stars, Gas, and Dust in SINGS Galaxies. II. Derived Dust Properties," ApJ **701**, 1965 (2009).

D. Dale *et al.*, "The Spitzer Infrared Nearby Galaxies Survey: A High-Resolution Spectroscopy Anthology," ApJ **693**, 1821 (2009).

Feiguin

F. Heidrich-Meisner *et al.*, "Quantum distillation: Dynamical generation of low-entropy states of strongly correlated fermions in an optical lattice," Phys. Rev. A **80**, 1603 (2009).

K. A. Al-Hassanieh *et al.*, "Robust pairing mechanism from repulsive interactions," Phys. Rev. B **80k**, 5116 (2009).

A. E. Feiguin and M. P. A. Fisher, "Exotic Paired States with Anisoltropic Spin-Dependent Fermi Surfaces," Phys. Rev. Lett. **103b**, 5303 (2009).

V. V. Dobrovitski *et al.*, "Decay of Rabi Oscillations by Dipolar-Coupled Dynamical Spin Environments," Phys. Rev. Lett. **102**, 7601 (2009).

F. Heidrich-Meisner *et al.*, "Real-time simulations of non-equilibrium transport in the single-impurity Anderson model," Phys. Rev. B **79**, 5336 (2009).

A. E. Feiguin *et al.*, "Spin polarization of the ν=5/2 quantum Hall state," Phys. Rev. B **79**, 5322 (2009).

A. E. Feiguin and D. A. Huse, "Spectral properties of a partially spin-polarized one-dimensional Hubbard/Luttinger superfield," Phys. Rev. B **102**, 6403 (2009).

F. Heidrich-Meisner *et al.*, "Transport through quantum dots: a combined DMRG and embedded-cluster approximation study," EPJB **67**, 527 (2009).

Kobulnicky, Henry

Michael J. Alexander *et al.*, "The Discovery of a Massive CLuster of Red Supergiants with GLIMPSE," AJ **137**, 4824 (2009).*et al.*, "Five More Massive Binaries in the

Cygnus OB2 Association," AJ **137**, 4608 (2009).

Toru Misawa *et al.*, "The Magellanic Bridge as a Damped Lyman Alpha System: Physical Properties of Cold Gas Toward PKS 0312-770," ApJ **695**, 1382 (2009).

N. Lehner *et al.*, "The Connection Between an Lyman Limit System, a Very Strong O VI Absorber, and Galaxies at z~0.203," ApJ **694**, 734 (2009).

B. Uzpen *et al.*, "The Frequency of Warm Debris Disks and Transistion Disks in a Complete Sample of Intermediate-Mass Glimpse Stars: Placing Constraints on Disk Lifetimes," AJ **137**, 3329 (2009).

Pierce, Michael

Andrew J. Monson and Michael J. Pierce, "BIRCM: A Near-Infrared Camera for The University of Wyoming Red Buttes Observatory," PASP **121**, 728M (2009).

Thayer

Diana F. Spears *et al.*, "Formation of swarm robotic chemical plume tracing from a fluid dynamics perspective," IJICC **745–785**, 2-4 (2009).

Tang, Jinke

J. Tang *et al.*, "Colossal Positive Seebeck Coefficient and Low Thermal Conductivity in Reduced TiO_2," J. Phys.: Cond. Matt. **21**, 205703 (2009).

X. Zhang *et al.*, "Influence of Pb doping on the crystal structure and multiferroic properties of the $BiFeO_3$ perovskite," J. Appl. Phys. **105**, 07D918 (2009).*et al.*, "Seebeck coefficient and thermal conductivity in doped C60," J. Renewable Sustainable Energy **1**, 023104 (2009).

T. Komesu *et al.*, "4f hybridization and band dispersion in gadolinium thin films and compounds," Physica Status Solidi B **246**, 975 (2009).

Q. Goa *et al.*, "Comparison of n-type Gd_2O_3 and Gd-doped HfO_2," J. Phys.: Condens. Matter **21**, 045602 (2009).

W. Wang *et al.*, "Enhanced tunneling magnetoresistance of Fe_3O_4 in a Fe_3O_4-hexabromobenzene (C_6Br_6) composite system," J. Appl. Phys. **105**, 07B105 (2009).

X. Wang *et al.*, "Amplification of magnetoresistance and Hall effect of Fe_3O_4/x{FF0D]SiO_2/x{FF0D}Si structure," J. Appl. Phys. **105**, 07B101 (2009).

G. X. Du *et al.*, "Tunneling magnetoresistance in (Ga,MN)As/Al-O/CoFeB hybrid structures," J. Appl. Phys. **105**, 07C707 (2009).

Wang, Wenyong

W Wang and C. A. Richter, "Magnetic Tunnel Junctions with Self-Assembled Molecules," J. Nanosci. Nanotechnol. **1008**, 9 (2009).

H. D. Xiong *et al.*, "Noise in ZnO Nanowire Field Effect Transistors," J. Nanosci. Nanotechnol. **1041**, 9 (2009).

PART II

MEXICO

Geographic Listing of Graduate Programs

BENEMÉRITA UNIVERSIDAD AUTÓNOMA DE PUEBLA

INSTITUTO DE FISICA

72570 Puebla, Puebla, México
http://www.ifuap.buap.mx

Students Accepted For Degree	FIELDS		
	Physics	Astronomy	Related Fields
Doctorate	X		X
Master's	X		X

1. General

President: Enrique Agüera Ibañez
Dean of Graduate School: Roberto Cartas Fuentevilla
Department Chairman: Juan Francisco Rivas Silva
Department Telephone Number: (+52-222) 229-5610
Type of Institution: University
Control: Public
Setting: Urban
Total Students: 66
Total Graduate Students: 266
Annual Graduate Tuition:
Masters Programs: $8,000*
Doctorals Programs: $10,000*
Tuition rates for: 2010–11
Deferred tuition plan: No
Other Fees: None
Term: 4-month

*Mexican pesos.

2. Number of Faculty in Department

The total of full time faculty members is 32.

3. Admission, Financial Aid, and Housing

Address admission inquiries to: Secretario Académico, Instituto de Física, UAP, Apdo. Postal J-48, 72570 Puebla, Puebla, México. Phone (+52-222) 229-5610; Fax (+52-222) 229-5611

Graduate application fee required: No

Admission deadline (Fall period): 8/30

Admission information: For the fall period, 2010–11, 19 students were accepted out of 57 applicants.

Admission requirements: For admission to the graduate programs, a Bachelor's degree in physics or related field is required with no minimum undergraduate GPA specified. The GRE or the GRE Advanced is not required. Students from non-Spanish speaking countries are required to demonstrate proficiency in Spanish. Passing an admission exam and/or attending and passing prerequisite courses are required for the Master's degree. Passing an admission exam is required for the Doctorate degree.

Physics–Undergraduate preparation assumed: Reif, *Statistical and Thermal Physics*; Marion, *Classical Dynamics of Particles and Systems*; Reitz, Milford, and Christy, *Foundations of Electromagnetic Theory*; Arfken, *Mathematical Methods for Physicists*, 2nd ed.; Merzbacher, *Quantum Mechanics*.

Materials Science–Undergraduate preparation assumed: M.R. Spiege, *Teoría y Problemas de Mateméticas Superiores para Ingenieros y Científicos* (McGraw Hill, 1971); F. Reif, *Fundamentos de la Física Estadística y Térmica* (McGraw-Hill, 1968); Berkeley Physics Course, *Mechanics* Vol. 1 (Reverté), Berkeley Physics Course, *Electromagnetism*, Vol. 2 (Reverté), G.M. Bonder, *Chemistry and Experimental Science*, 2nd Edition (John Wiley, 1995).

Address financial aid inquiries to: Secretario Académico, Instituto de Física, UAP, Apdo. Postal J-48, 72570 Puebla, Puebla, México.

GAPSFAS application required: No

Address housing inquiries to: Secretario Académico, Instituto de Física, UAP, Apdo. Postal J-48, 72570 Puebla, Puebla, México.

On-campus, single student housing available: No

On-campus, married student housing available: No

4. Graduate Degree Requirements

Physics–Spring and Summer Programs: consist of (B.S.-level) intensive courses on Modern Physics, Classical Mechanics, Electromagnetism and Mathematical Methods of Physics are offered as a prerequisite to enter the M.S. program.

Materials Science–Spring and Summer Programs: consist of (B.S.-level) intensive courses on Mathematical Methods, General Physics, General Chemistry and Thermal Physics are offered as a prerequisite to enter the M.S. program.

Master's in Physics: The Master's degree in Physics requires 9 courses (with a minimum combined grade average of 8, scale from 0 to 10) are required. During the second year the student must pass a qualifying examination. Altematively, to write a thesis under the supervision of a faculty member. The program must be completed within 2 years. The M.Sc. is a terminal program.

Master's in Material Science: The Master's in Material Science requires 10 courses (with a minimum combined grade average of 8, scale from 0 to 10) are required. During the second year the student must pass a qualifying examination. Altematively, to write a thesis under the supervision of a faculty member. The program must be completed within 2 years. The M.Sc. is a terminal program.

Doctorate in Physics: 3 advanced courses are required. During the second year of the program the student must pass a preliminary examination. The student must write a dissertation based on original research and must have at least one paper accepted for publication in a recognized journal prior to the final examination.

Doctorate in Materials Science: 3 advanced courses. During the second year of the program the student must pass a preliminary examination. The student must write a dissertation based on original research and must have a paper accepted for publication in a recognized journal prior to the final examination.

Thesis: Thesis may be written *in absentia.*

5. Special Equipment, Facilities, or Programs

Library: Monographic material: 5,826 items; 5708 books; 134 research magazines titles and, 250 thesis exemplars.

Services; data-base, books catalogues and electronic magazines, through page web: www.bibliocatalogo.buap.mx, copies, agreements with libraries within the country, like UNAM, UDLAP, CINVESTAV, INAOE, etc.

Laboratories: 17 specialized laboratories, 1 electronic workshop and, 1 mechanic workshop.

Computer Center: Workstations: 1WS SGI 02, CPU MIPS R12000 300 MHz, WS Microway, CPU ALPHA 21264 600 MHz, 1 WS HP Visualize J5600, 2 CPU parisc 8500, 1 ws sgi Octane/SE, CPU MIPS R12000 400 MHz, 1 Server SGI 1400l, 4 CPU Xeon 500 MHz, 1 SGI 2200, 8 CPU MIPS R12000 400 MHz, 1 Cluster Beowulf, 16 alpha nodes 21164 600 MHz, 5PC Linux Fedora Core, Pentim 4 HT 3 GHz, etc.

FACULTY

Arriaga-Rodríguez, J. Jesús, Ph.D., U.C. Madrid, Spain, 1992. Semiconductor physics.

Calixto Rodríguez, Ma. Estela, Ph.D., UNAM, 2001.

Carrillo-Estrada, José Luis, Ph.D., UNAM, México City, México, 1984, Semiconductor physics; statistical physics.

Cartas Fuentevilla, Roberto, Ph.D., IFUAP, Puebla, México, 1999. Mathematical Physics.

Castañeda Aviña, Luis, Ph.D., II-UNAM, 2003.

Dossetti Romero, Victor, Ph.D., IFUAP, BUAP, Puebla, México, 2005.

Escalante Hernández, Alberto, Ph.D., IFUAP, BUAP, Puebla, Mexico, 2005.

Flores-Riveros, Antonio, Ph.D., Uppsala, Sweden, 1986. Quantum chemistry.

García-Vázquez, Valentín, Ph.D., Arizona, USA, 1992. Artificially structured materials.

González Melchor, Minerva, Ph.D., CINVESTAV-IPN, México City, México, 2002. Molecular simulations, structural properties, dynamics and thermodynamics.

Gracia y Jiménez, Justo Miguel, Ph.D., UAP, Puebla, México, 1993. 0ptical properties of solids.

Hernández-Cocoletzi, Gregorio, Ph.D., UNAM, México City, México, 1991. Condensed matter physics; optical properties of solids.

Hernández-Tejeda, Pedro Hugo, Ph.D., Michigan, USA, 1987. Kinetics of phase transitions; X-ray diffraction.

Izrailev, Felix, Ph.D., Novosibirsk, URSS, 1969. Complex systems, quantum chaos, nonlinear dynamics.

López-Crúz, José Elías, Ph.D., CINVESTAV, México City, México, 1979. Electrical and optical properties of insulators and semiconductors.

Luna-Acosta, Germán Aurelio, Ph.D., New Mexico University, USA, 1984. Theoretical physics; nonlinear dynamics; classical-quantum correspondence.

Márquez Beltrán, César, Ph.D., Univ. De Paris-XI, Orsay, Francia, 2004.

Martínez-Montes, Gerardo, Ph.D., Arizona, USA, 1985. Optical properties of semiconductors'.

Méndez Bermúdez, José Antonio, Ph.D., IFUAP, Puebla, México, 2003. Chaotic systems, quantum chaos, random matrices.

Méndez Blas, Antonio, Ph.D., UAM, Madrid, Spain, 2003. Optical characterizations, UV-VIS and infrared.

Mendoza-Álvarez, María Eugenia, Ph.D., Geneva, Switzerland, 1985. Crystal growth of ferroic materials; structural characterization.

Meza-Montes, Lilia, Ph.D., UAP, Puebla, México, 1993. Semiconductor physics; statistical mechanics.

Pal, Umapada, Ph.D., 1. T. Karagpur, India, 1991. Condensed matter physics; optical and electrical properties of solids.

Palma-Almendra, Alejandro, Ph.D., Uppsala, Sweden, 1976. Quantum chemistry.

Pando-Lambruschini, Carlos Leopoldo, Ph.D., Lomonosov Univ., Moscow, USSR, 1990. Quantum and nonlinear optics.

Pérez-Rodríguez, Felipe, Ph.D., Kharkov, USSR, 1989. Metal physics; acoustic and optical properties of solids.

Rivas-Silva, Juan Francisco, Ph.D., UAP, Puebla, México, 1991. Atomic and molecular physics.

Rosado-Sánchez, Alfonso, Ph.D., CINVESTAV, México City, México, 1984. Elementary particle physics.

Saldaña-Saldaña, Xóchitl, Ph.D., BUAP, Puebla, México. Optical properties of solids.

Sánchez Mora, Enrique, Ph.D., UAM-Iztapalapa, México City, México, 2000. Thin film growth, optical properties of thin films.

Silva-González, Nicolás Rutilo, Dr.rer.nat., TU Dresden, GRD, 1988. Electro- and thermotransport in thin semiconductor films.

Soto-Manríquez, José, Ph.D., Arizona, 1983. Optical sciences.

RESEARCH SPECIALITIES AND STAFF

APPLIED PHYSICS

(Mexican Research Level nomination)

J. Jesús Arriaga Rodríguez,	S.N.I. II
Antonio Méndez Blas,	S.N.I. I (Leader)
José Elías López Crúz,	S.N.I. II
Ma. Estela Calixto Rodríguez	S.N.I. I

Fields of research:

—Surface and interface physics

—Optical and Acoustic Properties of Periodic Artificial systems

—Photonic and Phononic Crystals

COMPLEX SYSTEMS

Felix M. Izrailev,	S.N.I. III
Carlos L. Pando Lambruschini,	S.N.I. I
José Antonio Méndez Bermúdez,	S.N.I. I
Germán Aurelio Luna Acosta,	S.N.I. II (Leader)

Fields of research:

—Complex systems

COMPLEX, INTELLIGENT AND NANOSTRUCTERED MATERIALS

José Luis Carrillo Estrada,	S.N.I. II
María Eugenia Mendoza Álvarez,	S.N.I. II
Umapada Pal,	S.N.I. II
Lilia Meza Montes,	S.N.I. I (Leader)
César Márquez Beltrán	S.N.I. I
Victor Dossetti Romero	S.N.I. C

Fields of research:

—Complex and intelligent materials

—Nanoparticles and nanocomposites

PHOTOCATALITIC AND PHOTOCONDUCTIVE MATERIALS

Justo Miguel Gracia y Jiménez-IFUAP,	S.N.I. II
Estela Gómez Barojas-CIDS-BUAP,	S.N.I. I
Nicolás Rutilo Silva González-IFUAP,	S.N.I. II (Leader)
Luis Castañeda Aviña	S.N.I. I

Fields of research:

—Morphological and chemical properties of materials

—Photocatalitic, luminescent and photoelectric properties of materials.

COMPUTATIONAL PHYSICS OF THE CONDENSED MATTER

Minerva González Melchor,	S.N.I. I
Gregorio Hernáncez Cocoletzi,	S.N.I. III
Pedro Hugo Hernández Tejeda,	S.N.I. I
Gerardo Martínez Montes,	S.N.I. I
Juan Francisco Rivas Silva,	S.N.I. II
Antonio Flores Riveros,	S.N.I. II (Leader)
Alejandro Palma Almendra	S.N.I. III

Fields of research:

—Ab initio calculations of the electronic structure of atoms, molecules and Solids.

—Variational methods and their applications to confined systems

—Optical properties.

—Molecular simulation of liquids.

ADVANCED MATERIALS

Valentin García Vázquez,	S.N.I. I
Estela de Lourdes Juarez Ruíz,	
Felipe Pérez Rodríguez,	S.N.I. II (Leader)
Xóchitl Inés Saldaña Saldaña,	
José Soto Manriquez,	
Enrique Sánchez Mora,	S.N.I. I

Fields of Research:

—Superconductivity and Magnetism.

—Physics properties of advanced materials.

APPENDIX I
Geographic Listing of Departments

UNITED STATES

Nebraska

Nevada

New Hampshire

New Jersey

New Mexico

New York

North Carolina

North Dakota

Ohio

Oklahoma

Oregon

Pennsylvania

Rhode Island

South Carolina

South Dakota

Tennessee

Texas

Utah

Vermont

Virginia

Washington

West Virginia

Wisconsin

Wyoming

Mexico

APPENDIX II
Alphabetical Listing of Departments

APPENDIX III
Alphabetical Listing of Departments by Degree Granted

PHYSICS DOCTORAL PROGRAMS

PHYSICS MASTERS PROGRAMS

ASTRONOMY DOCTORAL PROGRAMS

ASTRONOMY MASTERS PROGRAMS

RELATED FIELDS DOCTORAL PROGRAMS

RELATED FIELDS MASTER'S PROGRAMS

APPENDIX IV
RESEARCH SPECIALTIES OF DOCTORAL PROGRAMS IN PHYSICS, ASTRONOMY, AND RELATED FIELDS

The following pages are summarized program offerings of all physics, astronomy, and physics-related master's-granting departments contacted by AIP which were willing to provide information about their programs. Research Specialties of doctoral Programs is geographically organized so that users can identify institutions in their area that offer research and/or courses in their field of interest. An **X** in the column means that the department offers only experimental work in that particular research specialty. A **T** means that work in that field is entirely theoretical, while **B** denotes that the program is both experimental and theoretical.

X–Experimental T–Theoretical B–Both	Accoustics	Applied Physics	Astronomy	Astrophysics	Atmosphere/Space Physics, Cosmic Rays	Atomic, Molecular & Optical Physics	Biophysics	Chemical Physics	Computer Science	Condensed Matter Physics	Electromagnetism	Electronics	Energy Sources & Environment	Engineering Physics/Science	Fluids, Rheology	Geophysics	History & Philosophy of Physics/Science	Low Temperature Physics	Marine Science/Oceanography	Materials Science, Metallurgy	Mechanics	Medical/Health Physics	Nuclear Engineering	Nuclear Physics	Optics	Particles and Fields	Physics Education	Physics of Beams	Plasma Physics & Fusion	Polymer Physics/Science	Relativity & Gravitation	Statistical & Thermal Physics	Systems Science/Engineering	Other see www.GradschoolShopper.com
UNITED STATES																																		
ALABAMA																																		
Alabama A&M (Physics)		X			X															X					X									
Alabama, U. of, Birmingham (Physics)				B			B			B															B									
Alabama, U. of, Huntsville (Physics)			B	B	B																				B									
Alabama, U. of, Tuscaloosa (Physics & Astronomy)		B	B	B	T					B										B				X		B								
Auburn U. (Physics)					B					B																								
ARIZONA																																		
Arizona State U. (Physics & Astronomy)		X	B	B			B			B										B			B		X									
Arizona, U. of																																		
(Astronomy)			B	B																														
(Optical Sciences)							B							B						B					B									
(Physics)	B	B	B	B	B		B	B		B	B	B		B	B	B		B	B	B	B	B	B	B	B	B	B	B		B			B	
(Planetary Sciences)			B																															
ARKANSAS																																		
Arkansas, U. of (Physics)				T			X			B															B		T							
CALIFORNIA																																		
California, U. of, Berkeley (Physics)				B	B		B	B		B								X						X		B								
California, U. of, Davis (Physics)				B			B			B								X		B				B		B	B							
California, U. of, Davis (Applied Science)		B		B	B		B	B		B	B			B						B					B									
California, U. of, Irvine (Physics)			B	B	B		B			B												X				B								
California, U. of, Los Angeles (Physics & Astronomy)	B		B	B	B		B	X		B					X	T		B						B		B		B		T				
California, U. of, Riverside (Physics & Astronomy)		X		X				X		B								X		X					X	B								
California, U. of, San Diego (Physics)	T	T	B	B	B		B			B					B			X		X				T	X	B	X	X		B				
California, U. of, Santa Barbara (Physics)			B	B			B			B																B				B				
California, U. of, Santa Cruz																																		
(Astronomy & Astrophysics)			B	B																														
(Physics)	B		B	B	B		B	X	T	B		X			T	T		B	T	B		X			X	B	T	X		B				

X–Experimental
T–Theoretical
B–Both

	Accoustics	Applied Physics	Astronomy	Astrophysics	Atmophere/Space Physics, Cosmic Rays	Atomic, Molecular & Optical Physics	Biophysics	Chemical Physics	Computer Science	Condensed Matter Physics	Electromagnetism	Electronics	Energy Sources & Environment	Engineering Physics/Science	Fluids, Rheology	Geophysics	History & Philosophy of Physics/Science	Low Temperature Physics	Marine Science/Oceanography	Materials Science, Metallurgy	Mechanics	Medical/Health Physics	Nuclear Engineering	Nuclear Physics	Optics	Particles and Fields	Physics Education	Physics of Beams	Plasma Physics & Fusion	Polymer Physics/Science	Relativity & Gravitation	Statistical & Thermal Physics	Systems Science/Engineering	Other see www.GradschoolShopper.com
Southern California, U. of (Physics & Astronomy)			B		B				B	B								X							B	T								
Stanford U.																																		
(Applied Physics)		B		B			B			B				X				B		B					X									
(Physics)			B	B			X			B								X						X	X	B								
COLORADO																																		
Colorado School of Mines (Physics)	B	B		X	X		X			B	B			B						B			X	X	X									
Colorado State U. (Physics)	X									B																X								
Colorado, U. of, Boulder																																		
(Astrophysical and Planetary Sciences)			B	B											T										X		B							
(Physics)		X		B			B	B	T	B				X		B		B						B	B	B	B			B				
Denver, U. of (Physics)		X	X	B	B		B			B				X				X		X							B							
CONNECTICUT																																		
Connecticut, U. of (Physics)		X						B		B				X		T		X		B				B	X	T				B				
Wesleyan U. (Physics)								B							X			X			T							T						
Yale U. (Physics)				B						B														B		B								
DELAWARE																																		
Delaware, U. of (Physics & Astronomy)	X	X	B	B	B					B								X		B				T	X	B								
DISTRICT OF COLUMBIA																																		
Catholic U. of America (Physics)		X		B	B		X			B										X				X										
George Washington U. (Physics)				T			B			X														B		T	X							
Georgetown U. (Physics)		X					B			B		X		X	B			B		B		X			X	T				X			X	
Howard U. (Physics & Astronomy)				T	B		B			B										B						T								
FLORIDA																																		
Central Florida, U. of (Physics)			X		X		B			B										B					B		B	X						
Florida Atlantic U. (Physics)		B			T		B			B	B				T			B		B	T				B	T	B	T						
Florida Institute of Technology (Physics & Space Sciences)		B	B	B	B					B	B					B				B						B	B			B				
Florida International U. (Physics)			B				B			B														B		B								
Florida State U. (Physics)				B			B			B								B		B				B		B								
Florida, U. of (Physics)	X	B		B	T		B	B		B	B	X						B		B		X		X	B	B				B				

X–Experimental T–Theoretical B–Both	Accoustics	Applied Physics	Astronomy	Astrophysics	Atmosphere/Space Physics, Cosmic Rays	Atomic, Molecular & Optical Physics	Biophysics	Chemical Physics	Computer Science	Condensed Matter Physics	Electromagnetism	Electronics	Energy Sources & Environment	Engineering Physics/Science	Fluids, Rheology	Geophysics	History & Philosophy of Physics/Science	Low Temperature Physics	Marine Science/Oceanography	Materials Science, Metallurgy	Mechanics	Medical/Health Physics	Nuclear Engineering	Nuclear Physics	Optics	Particles and Fields	Physics Education	Physics of Beams	Plasma Physics & Fusion	Polymer Physics/Science	Relativity & Gravitation	Statistical & Thermal Physics	Systems Science/Engineering	Other see www.GradschoolShopper.com
Miami, U. of (Physics)		T		B			T			B								B							B	T								
South Florida, U. of (Physics)		B					B			B										B		X			B									
GEORGIA																																		
Emory U. (Physics)							B			B																				X				
Georgia Institute of Technology (Physics)							B			B														B	B									
Georgia State U. (Physics & Astronomy)		X	X	X			X			X														X										
IDAHO																																		
Idaho, U. of (Physics)		X								B								B		X				B	X	T								
ILLINOIS																																		
Chicago, U. of																																		
(Astronomy & Astrophysics)			B	B																														
(Physics)		X	B	B	B		B	B		B					B			B		B				X	B	B		B		B				
Illinois Institute of Technology (Physics)							B			B																B		X						
Illinois, U. of, Chicago (Physics)		B					B			B										B		B		B	B	B								
Illinois, U. of, Urbana Champaign																																		
(Astronomy)			B	B																														
(Physics)			B	B			B	T		B						B		B						B	B	B	B	X						
Northern Illinois U. (Physics)				T						B										B						B		B						
Northwestern U. (Physics & Astronomy)				B			B			B								B		B				X		B								
Southern Illinois U., Carbondale (Physics)		B							T	B								B		B										B				
INDIANA																																		
Ball State U. (Physics & Astronomy)		B	B	B					B	B												B		B		B	B							
Indiana State U. (Physics & Astronomy)			B	B																														
Indiana U. - Purdue U. (Physics)							B			X										B					B		B							
Indiana U., Bloomington																																		
(Astronomy)			B	B																														
(Physics)				B			B	B		B								B						B		B		B						
Notre Dame, U. of (Physics)			B	B			T	B		B								B		B				B		B								
Purdue U. (Physics)		X	B	B	X		B			B						X		B						B	B	B	B							
IOWA																																		
Iowa State U. (Physics and Astronomy)			B	B	B		X			B								B		B				B	B	B	B			B				

X–Experimental T–Theoretical B–Both	Accoustics	Applied Physics	Astronomy	Astrophysics	Atmophere/Space Physics, Cosmic Rays	Atomic, Molecular & Optical Physics	Biophysics	Chemical Physics	Computer Science	Condensed Matter Physics	Electromagnetism	Electronics	Energy Sources & Environment	Engineering Physics/Science	Fluids, Rheology	Geophysics	History & Philosophy of Physics/Science	Low Temperature Physics	Marine Science/Oceanography	Materials Science, Metallurgy	Mechanics	Medical/Health Physics	Nuclear Engineering	Nuclear Physics	Optics	Particles and Fields	Physics Education	Physics of Beams	Plasma Physics & Fusion	Polymer Physics/Science	Relativity & Gravitation	Statistical & Thermal Physics	Systems Science/Engineering	Other see www.GradschoolShopper.com
Iowa, U. of (Physics & Astronomy)			B	B	B			X		B	B			B								X		B	B	B				X				
KANSAS																																		
Kansas, U. of (Physics & Astronomy)		X	X	B	T		X			B								X						X		B		T						
KENTUCKY																																		
Kentucky, U. of (Physics & Astronomy)			B	B						B	T													B		T	X							
LOUISIANA																																		
Louisiana State U. (Physics & Astronomy)		B	B	B	X			B	T	B	T							B		B		X		B		B								
New Orleans, U. of (Physics)	B	B						T		B	T							X		X					B									
Tulane U. (Physics)		B					X	T		B								X		X				T	T					X				
MAINE																																		
Maine, U. of (Physics and Astronomy)		X	B	B			B	B		B	T			X		B			X	B		B					B			B				
MARYLAND																																		
Johns Hopkins U. (Physics & Astronomy)			B	B						B															X	B								
Maryland, U. of																																		
(Astronomy)			B																															
(Chemical Physics Prog.)								B																										
(Physics Program)		B		B	X					B	B	B						B		B				B	B	B	B	B						
Maryland, U. of, Baltimore County (Physics)		B	B	X	B		X			B										X					B					B				
U. of Maryland (Baltimore) (Atmospheric Physics)					B																													
MASSACHUSETTS																																		
Boston College (Physics)								X		B								X																
Boston U.																																		
(Astronomy)			B	B	B																													
(Physics)							X			X																B								
Brandeis U. (Physics)				B			B			B																B								
Clark U. (Physics)							T			B																								
Harvard U. (Applied Sciences)	B	B			B				B	B	T	B		B	B			B	B	B	B				B					B			B	
Massachusetts Institute of Technology																																		
(Earth, Atmospheric & Planetary Sciences)			B		B										B	B			B		B													

X–Experimental T–Theoretical B–Both	Accoustics	Applied Physics	Astronomy	Astrophysics	Atmosphere/Space Physics, Cosmic Rays	Atomic, Molecular & Optical Physics	Biophysics	Chemical Physics	Computer Science	Condensed Matter Physics	Electromagnetism	Electronics	Energy Sources & Environment	Engineering Physics/Science	Fluids, Rheology	Geophysics	History & Philosophy of Physics/Science	Low Temperature Physics	Marine Science/Oceanography	Materials Science, Metallurgy	Mechanics	Medical/Health Physics	Nuclear Engineering	Nuclear Physics	Optics	Particles and Fields	Physics Education	Physics of Beams	Plasma Physics & Fusion	Polymer Physics/Science	Relativity & Gravitation	Statistical & Thermal Physics	Systems Science/Engineering	Other see www.GradschoolShopper.com
(Physics)				B			B			B	T													B	X	B								
Massachusetts, U. of, Amherst																																		
(5-college Astronomy)			B																															
(Physics & Astronomy)				B			B			B					B			B		X		X		B	X	B				B				
Massachusetts, U. of, Lowell (Physics & Applied Physics)		B			B					B								X			B	B		X	B	T				B				
Northeastern U. (Physics)							B			B												B				B								
Tufts U. (Physics & Astronomy)				T						B										X						B				X				
Worcester Polytechnic Institute (Physics)		B			T			X		B	T				T			X			X			T	B	T	B			B				
MEXICO, Puebla																																		
Benemérita U. Autónoma de Puebla (Physics)		B								B	B				B			B		B					B	B								
MICHIGAN																																		
Michigan State U. (Physics and Astronomy)			X	B				B		B								X		X				B		B		T						
Michigan Technological U. (Physics)		B	B	B	B		T			B	X			B	X					X	B				X									
Michigan, U. of (Astronomy)			B	B																														
Oakland U. (Physics)																						B												
Wayne State U. (Physics and Astronomy)		X		B						B				X				X		X				B	X	B		X						
Western Michigan U. (Physics)										B														B										
MINNESOTA																																		
Minnesota, U. of, Minneapolis																																		
(Astronomy)			B	B																														
(Physics)				X			X			X														T		X								
MISSISSIPPI																																		
Mississippi State U. (Physics & Astronomy)		X		B										B						B				B	B		B							
Mississippi, U. of (Physics & Astronomy)	B																									B								
MISSOURI																																		
Missouri, U. of, Columbia (Physics & Astronomy)			B	B			B			B										B		X			X					X				
Missouri, U. of, Kansas City (Physics)		X						B		B						X		X		B										X				
Missouri, U. of, Rolla (Physics)			B				B	B		B															B	B				B				

X–Experimental
T–Theoretical
B–Both

	Accoustics	Applied Physics	Astronomy	Astrophysics	Atmosphere/Space Physics, Cosmic Rays	Atomic, Molecular & Optical Physics	Biophysics	Chemical Physics	Computer Science	Condensed Matter Physics	Electromagnetism	Electronics	Energy Sources & Environment	Engineering Physics/Science	Fluids, Rheology	Geophysics	History & Philosophy of Physics/Science	Low Temperature Physics	Marine Science/Oceanography	Materials Science, Metallurgy	Mechanics	Medical/Health Physics	Nuclear Engineering	Nuclear Physics	Optics	Particles and Fields	Physics Education	Physics of Beams	Plasma Physics & Fusion	Polymer Physics/Science	Relativity & Gravitation	Statistical & Thermal Physics	Systems Science/Engineering	Other see www.GradschoolShopper.com
Missouri, U. of, Saint Louis (Physics & Astronomy)			B	B	T		B			B										B														
Washington U. (Physics)	X	X	X	B	X		B			B								X		B		B		B		T								
MONTANA																																		
Montana State U. (Physics)				T	B					B								X							X		B							
NEBRASKA																																		
Nebraska, U. of, Lincoln																																		
(Nebraska Center for Materials & Nanoscience)										B				B				B		B										B				
(Physics & Astronomy)								X		B										B					X	X				X				
NEVADA																																		
Nevada, U. of, Las Vegas (Physics)		B	B	B						B										B					X					X				
Nevada, U. of, Reno (Physics)					B			B		T															X									
NEW HAMPSHIRE																																		
Dartmouth College (Physics & Astronomy)			B	B						B								B								T								
New Hampshire, U. of (Physics)				B	B					B					T									B										
NEW JERSEY																																		
New Jersey Institute of Technology (Physics)	B	B	B				B	B		B	B	B		B						B	B	B			B					B			B	
Princeton U.																																		
(Astrophy. Sci.- Plasma Phys.)																																		
(Astrophysical Sciences)			B	B																														
(Physics)			T	X			B			B																B								
Rutgers, SUNJ, Busch Campus (Piscataway) (Physics & Astronomy)			B				T			B								X						B	X	B	X							
Stevens Institute of Technology (Physics and Engineering Physics)		X			B					B															B									
NEW MEXICO																																		
New Mexico Institute of Mining & Technology (Physics)		B		B	B																													
New Mexico State U. (Astronomy)			B	B	B																													

X–Experimental
T–Theoretical
B–Both

	Accoustics	Applied Physics	Astronomy	Astrophysics	Atmosphere/Space Physics, Cosmic Rays	Atomic, Molecular & Optical Physics	Biophysics	Chemical Physics	Computer Science	Condensed Matter Physics	Electromagnetism	Electronics	Energy Sources & Environment	Engineering Physics/Science	Fluids, Rheology	Geophysics	History & Philosophy of Physics/Science	Low Temperature Physics	Marine Science/Oceanography	Materials Science, Metallurgy	Mechanics	Medical/Health Physics	Nuclear Engineering	Nuclear Physics	Optics	Particles and Fields	Physics Education	Physics of Beams	Plasma Physics & Fusion	Polymer Physics/Science	Relativity & Gravitation	Statistical & Thermal Physics	Systems Science/Engineering	Other see www.GradschoolShopper.com
(Physics)					B										T	B				B				B			X							
New Mexico, U. of																																		
(Chemical & Nuclear Engineering)							B							T	B					B		B	B					B		B				
(Physics & Astronomy)		X	X	B	B		B			B	B			X				X		T		X		B	B	B		B						
NEW YORK																																		
Clarkson U. (Physics)							X	X		X					X					X					X									
Columbia U. (Applied Physics)		B			T			B		B	B			B	T	T		B	T	B		B			B					B				
Cornell U.																																		
(Applied & Engineering Physics)		B		B	B		B	B		B	B	B				B				B				B	B									
(Physics)		X		T			B		T	B		X			B			B		B	B				X	B		X					X	
Queens College of the CUNY (Physics)										B	B									B					B					X				
Rensselaer Polytechnic Institute (Physics, Applied Physics, Astronomy)		B	B	B			T			B										B				X	X	B								
Rochester, U. of																																		
(Institute of Optics)		B					B			B	B			B						X		B			B			B					B	
(Physics & Astronomy)				B			B	B		B				B								B		B		B				B				
Rockefeller U. (Physics)							X			B																B								
SUNY at Albany (Physics)		B		T	X		B	B		B										B		X			X	B	B			T				
SUNY at Stony Brook (Physics & Astronomy)		B	B	B	B		B			B								B				B		B	B	B	T	B						
Syracuse U. (Physics)				B			B		T	B					T			X				X		X		B								
University at Buffalo, SUNY (Physics)		X		T	X		B			B				X				X			T			T		B	T							
NORTH CAROLINA																																		
Duke U. (Physics)		X					B			B												X		B	X	B		X						
East Carolina U. (Physics)	X						B															X			B									
North Carolina State U., Raleigh (Physics)		B		B			B			B					B					B				B	B	B	B			B				
North Carolina, U. of, Chapel Hill (Physics & Astronomy)			B	B			T			B										X				B		T								
Wake Forest U. (Physics)		B	B	B			B			B	B	B								B	B	B			B	B		B		B				
NORTH DAKOTA																																		
North Dakota, U. of (Physics)				B						B	X									X		X												

X–Experimental **T**–Theoretical **B**–Both	Accoustics	Applied Physics	Astronomy	Astrophysics	Atmophere/Space Physics, Cosmic Rays	Atomic, Molecular & Optical Physics	Biophysics	Chemical Physics	Computer Science	Condensed Matter Physics	Electromagnetism	Electronics	Energy Sources & Environment	Engineering Physics/Science	Fluids, Rheology	Geophysics	History & Philosophy of Physics/Science	Low Temperature Physics	Marine Science/Oceanography	Materials Science, Metallurgy	Mechanics	Medical/Health Physics	Nuclear Engineering	Nuclear Physics	Optics	Particles and Fields	Physics Education	Physics of Beams	Plasma Physics & Fusion	Polymer Physics/Science	Relativity & Gravitation	Statistical & Thermal Physics	Systems Science/Engineering	Other see www.GradschoolShopper.com
OHIO																																		
Air Force Institute of Technology (Engineering Physics)		X			B			B		X				X						X			B	X	X									
Akron, U. of (Physics)								B																			X							
Case Western Reserve U.																																		
(Astronomy)			X	X																														
(Physics)		B		B			B			B				B	X			X							B	B				B				
Cincinnati, U. of (Physics)			B	B	T		T	T		B										B		X				B	B							
Dayton, U. of (Electro-Optics Program)		B			B						B	B													B			B						
Kent State U. (Physics)										B								B						B	B		B							
Ohio Univ. (Physics and Astronomy)	X		B	B						B						X		X						B										
The Ohio State U.																																		
(Astronomy)			B	B																														
(Physics)				B			B	B		B					X			B		B				B	X	B	B			X				
Toledo, U. of (Physics & Astronomy)		B	B	B	B		X			B										B		X			B									
OKLAHOMA																																		
Oklahoma State U. (Physics)				T			B	B		B					X					B					B	T				X				
Oklahoma, U. of (Physics & Astronomy)		B	B	B				B						X				B								B								
OREGON																																		
Oregon State U. (Physics)		B					X	X	T	B	T							X		B					X		B							
Oregon, U. of (Physics)			X	B			B			B								B							B	B	X							
Portland State U. (Physics)		B			B		B	B		B										B														
PENNSYLVANIA																																		
Bryn Mawr College (Physics)								X		X										X						T								
Carnegie Mellon U. (Physics)		B	B	B			B	B		B					B					B				B		B								
Drexel U. (Physics)			B	B			B			B								B		B			B			B								
Lehigh U. (Physics)			B	B			B			B								X							B					X				
Pennsylvania State U.																																		
(Astronomy & Astrophysics)			B	B																														
(Physics)	B	B		B	B		B		B	B				B				B		B				B	B	B				B				
Pittsburgh, U. of (Physics & Astronomy)		X	B	B	X		X	B		B				B				B				X			B	B	T	B						
Temple U. (Physics)	B	B	X	X	B			X	T	B	B	X			B			B		B	B	X			X	B	B	X		T				

X–Experimental
T–Theoretical
B–Both

	Accoustics	Applied Physics	Astronomy	Astrophysics	Atmosphere/Space Physics, Cosmic Rays	Atomic, Molecular & Optical Physics	Biophysics	Chemical Physics	Computer Science	Condensed Matter Physics	Electromagnetism	Electronics	Energy Sources & Environment	Engineering Physics/Science	Fluids, Rheology	Geophysics	History & Philosophy of Physics/Science	Low Temperature Physics	Marine Science/Oceanography	Materials Science, Metallurgy	Mechanics	Medical/Health Physics	Nuclear Engineering	Nuclear Physics	Optics	Particles and Fields	Physics Education	Physics of Beams	Plasma Physics & Fusion	Polymer Physics/Science	Relativity & Gravitation	Statistical & Thermal Physics	Systems Science/Engineering	Other see www.GradschoolShopper.com
RHODE ISLAND																																		
Brown U. (Physics)				B			B			B								X								B								
SOUTH CAROLINA																																		
Clemson U. (Physics & Astronomy)			B	B	B		B			B										B														
South Carolina, U. of (Physics & Astronomy)		X		B				B										B					B	B		B								
SOUTH DAKOTA																																		
South Dakota School of Mines & Technology (Physics)										B																								
TENNESSEE																																		
Tennessee, U. of, Knoxville (Physics & Astronomy)		B		B	B		B	B		B								B		B				B		B								
Vanderbilt U. (Physics & Astronomy)		X	B	B						B														B		B								
TEXAS																																		
Texas, U. of, Arlington (Physics)		B	B	B	B		B		B	B	B							B		B		B			B	B								
Baylor U. (Physics)				T	B			B		B																B								
North Texas, U. of (Physics)		X					T			B	T			B						B				X	B			X		X				
Rice U. (Physics & Astronomy)			B	B	B		B	T		B																B								
Southern Methodist U.																																		
(Physics)				T																						B								
Texas Christian U. (Physics and Astronomy)			X	X			X			X										X					X									
Texas Tech U. (Physics)		X	X	X			B	X		B										B				T	X	B	B							
Texas, U. of Brownsville (Physics & Astronomy)				B			B																		X		T							
Texas, U. of, Austin																																		
(Astronomy)			B																															
(Physics)	B	T		B						B														B		T								
Texas, U. of, Dallas (Physics Program)				B	B			B		B																B								
Texas, U. of, El Paso (Physics)																				B														
Texas, U. of, Health Science Center at San Antonio (Radiological Sciences)							B															B												

X–Experimental T–Theoretical B–Both	Accoustics	Applied Physics	Astronomy	Astrophysics	Atmosphere/Space Physics, Cosmic Rays	Atomic, Molecular & Optical Physics	Biophysics	Chemical Physics	Computer Science	Condensed Matter Physics	Electromagnetism	Electronics	Energy Sources & Environment	Engineering Physics/Science	Fluids, Rheology	Geophysics	History & Philosophy of Physics/Science	Low Temperature Physics	Marine Science/Oceanography	Materials Science, Metallurgy	Mechanics	Medical/Health Physics	Nuclear Engineering	Nuclear Physics	Optics	Particles and Fields	Physics Education	Physics of Beams	Plasma Physics & Fusion	Polymer Physics/Science	Relativity & Gravitation	Statistical & Thermal Physics	Systems Science/Engineering	Other see www.GradschoolShopper.com
UTAH																																		
Brigham Young U. (Physics & Astronomy)	X		X	X						B										X				X	B									
Utah State U. (Physics)				T	B		T			B	T									X					X	T								
Utah, U. of (Physics)	X			B	X		X	T		B								X								B	X							
VERMONT																																		
Vermont, U. of (Physics)																				B														
VIRGINIA																																		
George Mason U. (School of Computational Sci.)				T	T			T		T					T					T														
George Mason University (Physical Sciences)			B	B	B		B			B				B						B				B	B	B	B							
Virginia Commonwealth U. (Physics)																						B												
Virginia, U. of																																		
(Astronomy)			B	B																														
(Physics)		B					X	X		B	B							B				X		B	B	B								
William and Mary, College of (Physics)		X					B	X		B								X		X		X			X	B		X						
WASHINGTON																																		
Washington State U. (Physics)	B	B	B	B			B	B		B					B			B		B				B	B		B	B		B				
Washington, U. of																																		
(Astronomy)			B	B																														
(Physics)		X		B	X		B	B		B						B			X	X				B		B	B			B				
WEST VIRGINIA																																		
West Virginia U. (Physics)		B		B	X		T			B					T										X	T								
WISCONSIN																																		
Wisconsin, U. of, Madison																																		
(Astronomy)			B																															
(Physics)	X			B	B		B	X		B	B	X						X		B		X		B	X	B	X	X		X				
Wisconsin, U. of, Milwaukee (Physics)				T	T		B			B								B						T	X	T	T							
WYOMING																																		
Wyoming, U. of (Physics & Astronomy)				B																														

APPENDIX V
AREAS OF CONCENTRATION OF MASTER'S PROGRAMS
in
PHYSICS, ASTRONOMY, AND RELATED FIELDS

The following pages are summarized program offerings of all physics, astronomy, and physics-related master's-granting departments contacted by AIP which were willing to provide information about their programs. Areas of Concentration of Master's Programs is georgraphically organized so that users can identify institutions in their region that offer programs in their field of interest. A **G** in the column means that the degree is part of a program leading toward a Ph.D. An **M** means that this is a terminal master's degree, while a **B** denotes that the program can go either on to a Ph.D. or terminate at the master's level if the student chooses to do so.

M–Terminal Masters
G–Further Graduate Study
B–Both

	Accoustics	Applied Physics	Astronomy	Astrophysics	Atmosphere/Space Physics, Cosmic Rays	Atomic, Molecular & Optical Physics	Biophysics	Chemical Physics	Computer Science, Eng. Programming	Condensed Matter Physics	Electromagnetism	Electronics	Energy Sources & Environment	Engineering Physics/Science	Fluids, Rheology	Geophysics	History & Philosophy of Physics/Science	Low Temperature Physics	Marine Science/Oceanography	Materials Science, Metallurgy	Mechanics	Medical/Health Physics	Nuclear Engineering	Nuclear Physics	Optics	Particles and Fields	Physics Educ. (a) Jr. College/Secondary School	(b) Elementary School	Other Science Education	Physics of Beams	Plasma Physics & Fusion	Polymer Physics/Science	Relativity & Gravitation	Statistical Thermal Physics	Systems Science/Engineering	Other see www.GradschoolShopper.com
UNITED STATES																																				
ALABAMA																																				
Alabama A&M (Physics)		B			B															B					B											
Alabama, U. of, Birmingham (Physics)			B				B			B															B											
Alabama, U. of, Huntsville (Physics)		B	B	B	B																				B											+
Alabama, U. of, Tuscaloosa (Physics & Astronomy)		B	B	B	B					B										B				B		B										+
Auburn U. (Physics)					B					B																										+
ARIZONA																																				
Arizona State U. (Physics & Astronomy)		B		B			B		B	B	B	B		B						B	B		B	B	B	B	B		M							+
Arizona, U. of																																				
(Astronomy)			M																																	
(Optical Sciences)							B							B						B					B											+
(Physics)	B	B	B	B	B		B	B	B	B	B	B		B	B	B		B	B	B	B	B		B	B	B	B	B	B	B		B			B	+
(Planetary Sciences)																																				+
Northern Arizona U. (Physics)		B	B	B				B		B										B							M					B				
ARKANSAS																																				
Arkansas, U. of (Physics)			B	B			B			B															B		M									+
CALIFORNIA																																				
California State U., Fullerton (Physics)			G							G															B											+
California State U., Northridge (Physics & Astronomy)		M		M			M			M															M											+
California, U. of, Berkeley (Astronomy)				G																																+
California, U. of, Davis (Physics)				B			B			B								B		B				B		B			B							+
California, U. of, Davis (Applied Science)		B			B		B	B		B	B			B						B					B											+
California, U. of, Irvine (Physics)			B	B	B		B			B												B				B										+
California, U. of, Los Angeles (Physics and Astronomy)	G		G	G			G			G								G						G		G										+
California, U. of, Riverside (Physics & Astronomy)		B		B				B		B								B		B					B	B										+

M–Terminal Masters
G–Further Graduate Study
B–Both

	Accoustics	Applied Physics	Astronomy	Astrophysics	Atmosphere/Space Physics, Cosmic Rays	Atomic, Molecular & Optical Physics	Biophysics	Chemical Physics	Computer Science, Eng. Programming	Condensed Matter Physics	Electromagnetism	Electronics	Energy Sources & Environment	Engineering Physics/Science	Fluids, Rheology	Geophysics	History & Philosophy of Physics/Science	Low Temperature Physics	Marine Science/Oceanography	Materials Science, Metallurgy	Mechanics	Medical/Health Physics	Nuclear Engineering	Nuclear Physics	Optics	Particles and Fields	Physics Educ. (a) Jr. College/Secondary School	(b) Elementary School	Other Science Education	Physics of Beams	Plasma Physics & Fusion	Polymer Physics/Science	Relativity & Gravitation	Statistical Thermal Physics	Systems Science/Engineering	Other see www.GradschoolShopper.com
California, U. of, San Diego (Physics)																				M																
California, U. of, Santa Cruz (Physics)	B		B	B	B		B	B	B	B		B			B	B		B	B	B		B			B	B	M			B		B				+
San Diego State U.																																				
(Astronomy)			M																																	
(Physics)				M																		M		M	M		M					M				+
San Francisco State U. (Physics & Astronomy)		M	M	M			M			M	M			M	M			M							M	M	M		M							+
San Jose State U. (Physics)										M																										
Southern California, U. of (Physics & Astronomy)			B		B					B								B							B	B										+
Stanford U. (Applied Physics)		B																																		
COLORADO																																				
Colorado School of Mines (Physics)	B	B								B	B									B				B	B											
Colorado State U. (Physics)	B		M							B																B										+
Colorado, U. of, Boulder																																				
(Astrophysical and Planetary Sciences)			B	B											B										G		B									
(Physics)		B	B	B				B		B						B		B						B	B	B										+
Denver, U. of (Physics)		B	B	B	B		B			B				B				B		B																+
CONNECTICUT																																				
Connecticut, U. of (Physics)		B						B		B				B		B		B		B				B	B	B						B				+
Wesleyan U. (Physics)	B						B	B		B					B			B			B									B						+
DELAWARE																																				
Delaware, U. of (Physics & Astronomy)	B	B	B	B	B					B								B		B				B	B	B										+
DISTRICT OF COLUMBIA																																				
Catholic U. of America (Physics)		G		G	G		G			B														G												+
George Washington U. (Physics)				B			B			B														B		B	M									
Georgetown U. (Physics)		B					B	B		B		B													B											+
Howard U. (Physics & Astronomy)					B		B	B		B										B						B										+
FLORIDA																																				
Central Florida, U. of (Physics)							B			B										B					B		B									+

M–Terminal Masters
G–Further Graduate Study
B–Both

	Accoustics	Applied Physics	Astronomy	Astrophysics	Atmosphere/Space Physics, Cosmic Rays	Atomic, Molecular & Optical Physics	Biophysics	Chemical Physics	Computer Science, Eng. Programming	Condensed Matter Physics	Electromagnetism	Electronics	Energy Sources & Environment	Engineering Physics/Science	Fluids, Rheology	Geophysics	History & Philosophy of Physics/Science	Low Temperature Physics	Marine Science/Oceanography	Materials Science, Metallurgy	Mechanics	Medical/Health Physics	Nuclear Engineering	Nuclear Physics	Optics	Particles and Fields	Physics Educ. (a) Jr. College/Secondary School	(b) Elementary School	Other Science Education	Physics of Beams	Plasma Physics & Fusion	Polymer Physics/Science	Relativity & Gravitation	Statistical Thermal Physics	Systems Science/Engineering	Other see www.GradschoolShopper.com
Florida Atlantic U. (Physics)		B		B	B		G			G	G				B			B		B	B				G	B	M									+
Florida Institute of Technology (Physics & Space Sciences)		B	B	B	B					B	B					B				B						B	B					B				+
Florida International U. (Physics)			B				B			B														B		B										+
Florida State U. (Physics)				B			B			B								B		B				B		B										+
Florida, U. of (Physics)	B	B		B	B		B	B		B	B	B						B		B		B		B	B	B						B				+
Miami, U. of (Physics)		B		B			B			B								B							B	B										+
South Florida, U. of (Physics)		B					B			B										B		B			B											+
GEORGIA																																				
Emory U. (Physics)							G			G																										
Georgia Institute of Technology (Physics)							G	G		G										G				G	G										G	+
Georgia State U. (Physics & Astronomy)		B	B	B			B	B		B														B	B											+
IDAHO																																				
Idaho, U. of (Physics)		B								B								B		B				B	B	B	M	M	M							+
ILLINOIS																																				
Chicago, U. of																																				
(Astronomy & Astrophysics)			G	G																																
(Physics)		G	G	G	G		G			G					G			G		G				G	G	G				G		G				+
DePaul U. (Physics)		M		M						M														M	M											+
Illinois Institute of Technology (Physics)							B			B												M				B				B						
Illinois, U. of, Chicago (Physics)		B					B			B										B				B	B	B	M									+
Illinois, U. of, Urbana Champaign																																				
(Astronomy)			B	B																																+
(Physics)			B	B			B	B		B						B		B					B		B	B				B						+
Northern Illinois U. (Physics)		M		B						B										B						B	M			B						
Southern Illinois U., Carbondale (Physics)		B					B		B	B	B				B			B		B		B		B	B					B						+
Southern Illinois U., Edwardsville (Physics)										B															B											+
INDIANA																																				
Ball State U. (Physics & Astronomy)			M	M			M			M												M		M		M	M									+
Indiana State U. (Physics & Astronomy)			B	B																																

M–Terminal Masters G–Further Graduate Study B–Both	Accoustics	Applied Physics	Astronomy	Astrophysics	Atmosphere/Space Physics, Cosmic Rays	Atomic, Molecular & Optical Physics	Biophysics	Chemical Physics	Computer Science, Eng. Programming	Condensed Matter Physics	Electromagnetism	Electronics	Energy Sources & Environment	Engineering Physics/Science	Fluids, Rheology	Geophysics	History & Philosophy of Physics/Science	Low Temperature Physics	Marine Science/Oceanography	Materials Science, Metallurgy	Mechanics	Medical/Health Physics	Nuclear Engineering	Nuclear Physics	Optics	Particles and Fields	Physics Educ. (a) Jr. College/Secondary School	(b) Elementary School	Other Science Education	Physics of Beams	Plasma Physics & Fusion	Polymer Physics/Science	Relativity & Gravitation	Statistical Thermal Physics	Systems Science/Engineering	Other see www.GradschoolShopper.com
Indiana U. - Purdue U. (Physics)							B			B										B					B		B									+
Indiana U., Bloomington																																				
(Astronomy)			B	B																																
(Physics)				B			B	B		B								B						B		B				B						
Notre Dame, U. of (Physics)			G	G			G	G		G								G		G				G		G										+
Purdue U. (Physics)		B	B	B	B		B			B						B		B						B	B		B									+
IOWA																																				
Iowa State U. (Physics and Astronomy)			B	B	B		B			B								B						B	B	B	B					B				+
Iowa, U. of (Physics & Astronomy)			B	B	B			B		B	B			B								B		B	B	B						B				+
KANSAS																																				
Kansas, U. of (Physics & Astronomy)		B	B	B	B		B			B								B						B		B				B						+
Pittsburg State U. (Physics)		M								M								M						M	M		M					M				+
KENTUCKY																																				
Kentucky, U. of (Physics & Astronomy)			B	B						B	B													B		B										+
LOUISIANA																																				
Louisiana State U. (Physics & Astronomy)		B	B	B	B			B	B	B								B		B		M		B		B			M							+
Louisiana, U. of, Lafayette (Physics)	M	M		M																M																+
New Orleans, U. of (Physics)	B	B	B							B	B					B		B	B	B					B											+
Tulane U. (Physics)		B	B	B			B	B		B								B		B				B		B						B				+
MAINE																																				
Maine, U. of (Physics and Astronomy)		B	B	B			B	B		B	B			M		B			B	B		B					B		B			B				+
MARYLAND																																				
Maryland, U. of																																				
(Astronomy)			B																																	
(Chemical Physics Prog.)								B																												
(Physics Program)		B		B	B					B	B	B						B		B				B	B	B	B			B						+
Maryland, U. of, Baltimore County (Physics)		B	B	B	B		B			B										B					B							B				+

M–Terminal Masters
G–Further Graduate Study
B–Both

Institution	Accoustics	Applied Physics	Astronomy	Astrophysics	Atmosphere/Space Physics, Cosmic Rays	Atomic, Molecular & Optical Physics	Biophysics	Chemical Physics	Computer Science, Eng. Programming	Condensed Matter Physics	Electromagnetism	Electronics	Energy Sources & Environment	Engineering Physics/Science	Fluids, Rheology	Geophysics	History & Philosophy of Physics/Science	Low Temperature Physics	Marine Science/Oceanography	Materials Science, Metallurgy	Mechanics	Medical/Health Physics	Nuclear Engineering	Nuclear Physics	Optics	Particles and Fields	Physics Educ. (a) Jr. College/Secondary School	(b) Elementary School	Other Science Education	Physics of Beams	Plasma Physics & Fusion	Polymer Physics/Science	Relativity & Gravitation	Statistical Thermal Physics	Systems Science/Engineering	Other see www.GradschoolShopper.com
U. of Maryland (Baltimore) (Atmospheric Physics)					B																															
MASSACHUSETTS																																				
Boston College (Physics)										G																	M		M							
Boston U.																																				
(Astronomy)			B	B	B																															
(Physics)							G			B																G										
Clark U. (Physics)							B			B																										+
Harvard U. (Applied Sciences)	B	B			B				B	B	B	B		B	B			B	B	B	B				B							B			B	+
Massachusetts Institute of Technology																																				
(Earth, Atmospheric & Planetary Sciences)			B													B			M																	
(Physics)				B			B			B												B		B		B										+
Massachusetts, U. of, Amherst (5-college Astronomy)			B																																	
Massachusetts, U. of, Dartmouth (Physics)			M	M						M					M				M		M			M		M										+
Massachusetts, U. of, Lowell (Physics & Applied Physics)		B			B					B								B		B	B	B		B	B	B						B				+
Northeastern U. (Physics)				B			B			B								B		B	B	B			M	B	M		M							+
Tufts U. (Physics & Astronomy)			B	B			B			B								B							B							B				+
Worcester Polytechnic Institute (Physics)		B						B		B								B			B				B				B			B				+
MEXICO, Puebla																																				
Benemérita U. Autónoma de Puebla (Physics)		B								B	B				B			B		B					B	B										+
MICHIGAN																																				
Central Michigan U. (Physics)			M	M				M		M					M					M				M	M							M				+
Eastern Michigan U. (Physics and Astronomy)	M	M	M	M	M				M	M	M				M					M	M				M		M									+
Michigan State U. (Physics and Astronomy)			B	B				B		B								B		B				B		B	B			B						
Michigan Technological U. (Physics)		B	B	B	B		B			B	B			B	B					B	B				B											+
Michigan, U. of																																				
(Nuclear Engineering and Radiological Sciences)		B							B	G				B						B		B	B							B						+
(Physics)	B	B		B			B			B														B	B											+

M–Terminal Masters G–Further Graduate Study B–Both	Accoustics	Applied Physics	Astronomy	Astrophysics	Atmosphere/Space Physics, Cosmic Rays	Atomic, Molecular & Optical Physics	Biophysics	Chemical Physics	Computer Science, Eng. Programming	Condensed Matter Physics	Electromagnetism	Electronics	Energy Sources & Environment	Engineering Physics/Science	Fluids, Rheology	Geophysics	History & Philosophy of Physics/Science	Low Temperature Physics	Marine Science/Oceanography	Materials Science, Metallurgy	Mechanics	Medical/Health Physics	Nuclear Engineering	Nuclear Physics	Optics	Particles and Fields	Physics Educ. (a) Jr. College/Secondary School	(b) Elementary School	Other Science Education	Physics of Beams	Plasma Physics & Fusion	Polymer Physics/Science	Relativity & Gravitation	Statistical Thermal Physics	Systems Science/Engineering	Other see www.GradschoolShopper.com
Oakland U. (Physics)		B								B	B									B		B														+
Wayne State U. (Physics and Astronomy)		B		B						B				B				B		B				B	B	B				B						+
Western Michigan U. (Physics)			M							B														B												+
MINNESOTA																																				
Minnesota State U., Mankato (Physics and Astronomy)			M	M	M					M		M						M		M		M	M	M			M									+
Minnesota, U. of, Duluth (Physics)					M					M		M						M	M						M	M										+
Minnesota, U. of, Minneapolis																																				
(Astronomy)			B	B																																
(Physics)							M			B																B										
MISSISSIPPI																																				
Mississippi State U. (Physics & Astronomy)		B		B						B				B						B				B	B		B									+
Mississippi, U. of (Physics & Astronomy)	B																									B										+
MISSOURI																																				
Missouri, U. of, Columbia (Physics & Astronomy)			B	B			B			B										B		B			B							B				+
Missouri, U. of, Kansas City (Physics)		B						B		B						B		B		B												B				+
Missouri, U. of, Rolla (Physics)								B		B																						B				+
Missouri, U. of, Saint Louis (Physics & Astronomy)			B	B			B			B										B																+
Washington U. (Physics)	G	G	G	G	G		G	G		G								G		G		G		G		G									G	+
MONTANA																																				
Montana State U. (Physics)				B	B					B								B							B		B	B								+
NEBRASKA																																				
Creighton U. (Physics)					M		M			M														M		M										+
Nebraska, U. of, Lincoln																																				
(Nebraska Center for Materials & Nanoscience)										B				B				B		B												B				+
(Physics & Astronomy)								B		B										B					B	B						B				+

M–Terminal Masters
G–Further Graduate Study
B–Both

	Accoustics	Applied Physics	Astronomy	Astrophysics	Atmosphere/Space Physics, Cosmic Rays	Atomic, Molecular & Optical Physics	Biophysics	Chemical Physics	Computer Science, Eng. Programming	Condensed Matter Physics	Electromagnetism	Electronics	Energy Sources & Environment	Engineering Physics/Science	Fluids, Rheology	Geophysics	History & Philosophy of Physics/Science	Low Temperature Physics	Marine Science/Oceanography	Materials Science, Metallurgy	Mechanics	Medical/Health Physics	Nuclear Engineering	Nuclear Physics	Optics	Particles and Fields	Physics Educ. (a) Jr. College/Secondary School	(b) Elementary School	Other Science Education	Physics of Beams	Plasma Physics & Fusion	Polymer Physics/Science	Relativity & Gravitation	Statistical Thermal Physics	Systems Science/Engineering	Other see www.GradschoolShopper.com
NEVADA																																				
Nevada, U. of, Las Vegas (Physics)		B	B	B				B		B										B					B							B				+
Nevada, U. of, Reno (Physics)					B			B		B																										+
NEW HAMPSHIRE																																				
Dartmouth College (Physics & Astronomy)			B	B						B								B								B										+
New Hampshire, U. of (Physics)				B	B					B					B									B												+
NEW JERSEY																																				
New Jersey Institute of Technology (Physics)	B	B	B				B	B		B	B	B		B						B	B	B			B							B			B	+
Princeton U. (Astrophy. Sci.- Plasma Phys.)																																				+
Rutgers, SUNJ, Busch Campus (Piscataway) (Physics & Astronomy)			B	B			B			B								B						B		B										+
Stevens Institute of Technology (Physics and Engineering Physics)		B			B					B				B											B											+
NEW MEXICO																																				
New Mexico Institute of Mining & Technology (Physics)		B		B	B																															+
New Mexico State U.																																				
(Astronomy)			M	M	M																															
(Physics)					B					B					B	B				B				B	B	B	B									+
New Mexico, U. of																																				
(Chemical & Nuclear Engineering)							B	B						B	B					B		M	B							B		B				+
(Physics & Astronomy)		B	B	B	B		B			B	B			B				B		B		B		B	B	B				B						+
NEW YORK																																				
City College of the CUNY (Physics)		B					B	B		B	B			B	B			B				B			B	B	B									+
Clarkson U. (Physics)							B	B		B					B					B					B		B									+
Columbia U. (Applied Physics)		B			B			B		B	B			B	B	B		B	B	B		B			B							B				+
Cornell U. (Applied & Engineering Physics)														M																						
Queens College of the CUNY (Physics)										G	G									G					G		M		M			G				+
Rensselaer Polytechnic Institute (Physics, Applied Physics, Astronomy)		B	B	B			B			B										B				B	B	B										+

M–Terminal Masters
G–Further Graduate Study
B–Both

	Accoustics	Applied Physics	Astronomy	Astrophysics	Atmophere/Space Physics, Cosmic Rays	Atomic, Molecular & Optical Physics	Biophysics	Chemical Physics	Computer Science, Eng. Programming	Condensed Matter Physics	Electromagnetism	Electronics	Energy Sources & Environment	Engineering Physics/Science	Fluids, Rheology	Geophysics	History & Philosophy of Physics/Science	Low Temperature Physics	Marine Science/Oceanography	Materials Science, Metallurgy	Mechanics	Medical/Health Physics	Nuclear Engineering	Nuclear Physics	Optics	Particles and Fields	Physics Educ. (a) Jr. College/Secondary School	(b) Elementary School	Other Science Education	Physics of Beams	Plasma Physics & Fusion	Polymer Physics/Science	Relativity & Gravitation	Statistical Thermal Physics	Systems Science/Engineering	Other see www.GradschoolShopper.com
Rochester, U. of																																				
(Institute of Optics)											B			B						B					B											+
(Physics & Astronomy)							G	G		G				G										G		G						G				+
Rockefeller U. (Physics)							G			G				G				G							B	B				G						+
SUNY at Albany (Physics)		B		M			B	B		B										B		B		B	B	B	M									+
SUNY at Stony Brook (Physics & Astronomy)		M																									M	M							M	
Syracuse U. (Physics)				B			B		B	B					B			B				B		B		B										+
University at Buffalo, SUNY (Physics)				B	B		B			B				B				B			B			B		B	B									+
NORTH CAROLINA																																				
Appalachian State U. (Physics and Astronomy)			M	M	M			M		M		M			M	M									M										M	+
East Carolina U. (Physics)	B						B															B			B											+
North Carolina Agricultural and Technical State U. (Physics)								M		M														M			M									
North Carolina State U., Raleigh (Physics)		B		B			B			B					B					B				B	B	B	B					B				+
North Carolina, U. of, Chapel Hill (Physics & Astronomy)		B	B	B			B			B										B				B		B										+
Wake Forest U. (Physics)		B	B	B			B			B	B	B								B	B	B			B	B				B		B				+
NORTH DAKOTA																																				
North Dakota, U. of (Physics)				B						B	B									B		B														
OHIO																																				
Air Force Institute of Technology (Engineering Physics)		B			B			B		B				B						B			B	B	B											+
Akron, U. of (Physics)								M																								M				+
Bowling Green State U. (Physics & Astronomy)	M	M	M	M						M					M			M		M					M	M	M									+
Case Western Reserve U.																																				
(Astronomy)			B	B																																
(Physics)		B		B			B			B				B	B			B							B	B						B				+
Cincinnati, U. of (Physics)			B	B	B		B	B		B										B		B				B	M									+
Dayton, U. of (Electro-Optics Program)		B			B						B														B					B						
Kent State U. (Physics)										B								B						B	B											+

M–Terminal Masters
G–Further Graduate Study
B–Both

	Accoustics	Applied Physics	Astronomy	Astrophysics	Atmosphere/Space Physics, Cosmic Rays	Atomic, Molecular & Optical Physics	Biophysics	Chemical Physics	Computer Science, Eng. Programming	Condensed Matter Physics	Electromagnetism	Electronics	Energy Sources & Environment	Engineering Physics/Science	Fluids, Rheology	Geophysics	History & Philosophy of Physics/Science	Low Temperature Physics	Marine Science/Oceanography	Materials Science, Metallurgy	Mechanics	Medical/Health Physics	Nuclear Engineering	Nuclear Physics	Optics	Particles and Fields	Physics Educ. (a) Jr. College/Secondary School	(b) Elementary School	Other Science Education	Physics of Beams	Plasma Physics & Fusion	Polymer Physics/Science	Relativity & Gravitation	Statistical Thermal Physics	Systems Science/Engineering	Other see www.GradschoolShopper.com
Miami U. (Physics)				M	M		M			M										M	M				M		M		M							+
Ohio Univ. (Physics and Astronomy)	B		B	B						B						B		B						B												+
The Ohio State U.																																				
(Astronomy)			G	G																																
(Physics)				B			B	B		B					B			B		B				B	B	B			B			B				+
Toledo, U. of (Physics & Astronomy)		B	B	B	B		B			B										B		B			B											+
OKLAHOMA																																				
Oklahoma State U. (Physics)				B			B	B		B					B					B		B			B	B						B				+
Oklahoma, U. of (Physics & Astronomy)		B	B	B				B		B				B				B								B										+
OREGON																																				
Oregon State U. (Physics)		B					B	B	B	B	B							B		B					B	B	B									+
Oregon, U. of (Physics)		M		B			B	B		B		B						B		B	B				B	B										+
Portland State U. (Physics)		B			B		B			B										B					B											
PENNSYLVANIA																																				
Bryn Mawr College (Physics)								B		B																										+
Carnegie Mellon U. (Physics)		G	G	G			G	G		G					G					G				G		G										+
Drexel U. (Physics)			B	B			B			B								B		B						B	G									+
Indiana U. of Pennsylvania (Physics)		M								M										M			M		M		M									+
Lehigh U. (Physics)			B	B			B			B								B							B							B				+
Pennsylvania State U. (Physics)	B	B		B	B		B		B	B		B		B				B		B				B	B	B						B				+
Pittsburgh, U. of (Physics & Astronomy)																																				+
Temple U. (Physics)										B																B										+
RHODE ISLAND																																				
Brown U. (Physics)				G			G	G		G								G		G						G										+
SOUTH CAROLINA																																				
Clemson U. (Physics & Astronomy)			B	B	B		B			B										B																+
South Carolina, U. of (Physics & Astronomy)		B		B				B		B														B		B										+
SOUTH DAKOTA																																				

M–Terminal Masters G–Further Graduate Study B–Both	Accoustics	Applied Physics	Astronomy	Astrophysics	Atmosphere/Space Physics, Cosmic Rays	Atomic, Molecular & Optical Physics	Biophysics	Chemical Physics	Computer Science, Eng. Programming	Condensed Matter Physics	Electromagnetism	Electronics	Energy Sources & Environment	Engineering Physics/Science	Fluids, Rheology	Geophysics	History & Philosophy of Physics/Science	Low Temperature Physics	Marine Science/Oceanography	Materials Science, Metallurgy	Mechanics	Medical/Health Physics	Nuclear Engineering	Nuclear Physics	Optics	Particles and Fields	Physics Educ. (a) Jr. College/Secondary School	(b) Elementary School	Other Science Education	Physics of Beams	Plasma Physics & Fusion	Polymer Physics/Science	Relativity & Gravitation	Statistical Thermal Physics	Systems Science/Engineering	Other see www.GradschoolShopper.com
South Dakota School of Mines & Technology (Physics)		B								B										B																+
TENNESSEE																																				
Tennessee, U. of, Knoxville (Physics & Astronomy)		B		B	B		B			B						M		B		B		M		B	M	B	M									+
Vanderbilt U. (Physics & Astronomy)			B	B						B												B		B		B										+
TEXAS																																				
Texas, U. of, Arlington (Physics)			B	B	B		B	B		B	B							B		B		B			B	B										+
Baylor U. (Physics)				B	B			B		B																B										+
North Texas, U. of (Physics)										B				B						B					B											+
Rice U.																																				
(Physics & Astronomy)			G	G	G		G	G		G																G	M									+
(Professional Masters - Nanoscale)																																				+
Southern Methodist U. (Physics)				G																						G										
Texas Christian U. (Physics and Astronomy)																				M																+
Texas State University - San Marcos (Physics)										M										M					M				M							
Texas Tech U. (Physics)		B	B	B			B	B		G				B						B				B	B	B	B	B								+
Texas, U. of Brownsville (Physics & Astronomy)				B			B																		B											+
Texas, U. of, Austin																																				
(Astronomy)			B																																	
(Physics)	B	M		B						B														B		B										+
Texas, U. of, Dallas (Physics Program)		M		B	B			B		B																B										+
Texas, U. of, El Paso (Physics)			M							M										B				M												+
Texas, U. of, Health Science Center at San Antonio (Radiological Sciences)																						B														
UTAH																																				
Brigham Young U. (Physics & Astronomy)	B	B	B	B						B										B				B	B											+
Utah State U. (Physics)				B	B		B			B	B									B					B	B										+
Utah, U. of (Physics)	B			B	B		B			B								B								B										+

M–Terminal Masters
G–Further Graduate Study
B–Both

	Accoustics	Applied Physics	Astronomy	Astrophysics	Atmosphere/Space Physics, Cosmic Rays	Atomic, Molecular & Optical Physics	Biophysics	Chemical Physics	Computer Science, Eng. Programming	Condensed Matter Physics	Electromagnetism	Electronics	Energy Sources & Environment	Engineering Physics/Science	Fluids, Rheology	Geophysics	History & Philosophy of Physics/Science	Low Temperature Physics	Marine Science/Oceanography	Materials Science, Metallurgy	Mechanics	Medical/Health Physics	Nuclear Engineering	Nuclear Physics	Optics	Particles and Fields	Physics Educ. (a) Jr. College/Secondary School	(b) Elementary School	Other Science Education	Physics of Beams	Plasma Physics & Fusion	Polymer Physics/Science	Relativity & Gravitation	Statistical Thermal Physics	Systems Science/Engineering	Other see www.GradschoolShopper.com
VERMONT																																				
Vermont, U. of (Physics)	M			M			M			M				M													M					M				
VIRGINIA																																				
George Mason U. (School of Computational Sci.)				B	B			B		B					B					B																+
George Mason University (Physical Sciences)			B	B	B		B			B				B						B				B	B	B										+
Virginia Commonwealth U. (Physics)								B		M												B														+
Virginia, U. of																																				
(Astronomy)			B	B																																
(Physics)		B					B	B		B	B							B			M	B		B	B	B			M							+
William And Mary, College of (Physics)		G					G	G		G								G		G		G			G	G				G						+
WASHINGTON																																				
Washington State U. (Physics)	B	B	B	B			B	B		B					B			B		B				B	B		B		B	B		B				+
Washington, U. of																																				
(Astronomy)			G	G																																+
(Physics)	M	M		G	B		B	G		B	M	M				G			B	B		M	G	B	M	B	B					G				+
WEST VIRGINIA																																				
West Virginia U. (Physics)		B		B	B		B			B					B										B											+
WISCONSIN																																				
Wisconsin, U. of, Madison																																				
(Astronomy)			M																																	
(Physics)				B	B		B			B	B							B		B				B		B	B									+
Wisconsin, U. of, Milwaukee (Physics)				B	B		B			B								B		B				B	B	B										+
WYOMING																																				
Wyoming, U. of (Physics & Astronomy)				B																																

Use these cards to order additional copies of

2011 Graduate Programs in Physics, Astronomy, and Related Fields.

❑ **Yes!** Please rush me ______ copies of **2011 Graduate Programs in Physics, Astronomy, and Related Fields.** 978-0-7354-0840-1

List Price: **$79.00** *Shipping: $8.00 for the first book ($12.75 foreign) $1.00 each additional book ($6.00 foreign).*

METHOD OF PAYMENT

❑ Check enclosed (payable in U.S. dollars to the American Institute of Physics and drawn on a bank in the United States).

❑ Charge my credit card:

❑ AMEX ❑ VISA ❑ MasterCard ❑ Discover

Cardholder name (please print) ____________

Acct. No. ____________ Exp. date ______

Signature ____________
(Credit card orders not valid without signature)

Name ____________

Institution ____________

Address ____________

City/State/Zip ____________

mail orders to:

American Institute of Physics
c/o Springer
P.O. Box 2485
Secaucus, NJ 07096-2485
or fax (201) 348-4505

SATISFACTION GUARANTEED

Examine this book for 30 days without obligation. If not completely satisfied, return the book(s) in saleable condition with your invoice within 30 days for a full, prompt refund.

AMERICAN INSTITUTE OF PHYSICS

009

❑ **Yes!** Please rush me ______ copies of **2011 Graduate Programs in Physics, Astronomy, and Related Fields.** 978-0-7354-0840-1

List Price: **$79.00** *Shipping: $8.00 for the first book ($12.75 foreign) $1.00 each additional book ($6.00 foreign).*

METHOD OF PAYMENT

❑ Check enclosed (payable in U.S. dollars to the American Institute of Physics and drawn on a bank in the United States).

❑ Charge my credit card:

❑ AMEX ❑ VISA ❑ MasterCard ❑ Discover

Cardholder name (please print) ____________

Acct. No. ____________ Exp. date ______

Signature ____________
(Credit card orders not valid without signature)

Name ____________

Institution ____________

Address ____________

City/State/Zip ____________

mail orders to:

American Institute of Physics
c/o Springer
P.O. Box 2485
Secaucus, NJ 07096-2485
or fax (201) 348-4505

SATISFACTION GUARANTEED

Examine this book for 30 days without obligation. If not completely satisfied, return the book(s) in saleable condition with your invoice within 30 days for a full, prompt refund.

AMERICAN INSTITUTE OF PHYSICS

009

❑ **Yes!** Please rush me ______ copies of **2011 Graduate Programs in Physics, Astronomy, and Related Fields.** 978-0-7354-0840-1

List Price: **$79.00** *Shipping: $8.00 for the first book ($12.75 foreign) $1.00 each additional book ($6.00 foreign).*

METHOD OF PAYMENT

❑ Check enclosed (payable in U.S. dollars to the American Institute of Physics and drawn on a bank in the United States).

❑ Charge my credit card:

❑ AMEX ❑ VISA ❑ MasterCard ❑ Discover

Cardholder name (please print) ____________

Acct. No. ____________ Exp. date ______

Signature ____________
(Credit card orders not valid without signature)

Name ____________

Institution ____________

Address ____________

City/State/Zip ____________

mail orders to:

American Institute of Physics
c/o Springer
P.O. Box 2485
Secaucus, NJ 07096-2485
or fax (201) 348-4505

SATISFACTION GUARANTEED

Examine this book for 30 days without obligation. If not completely satisfied, return the book(s) in saleable condition with your invoice within 30 days for a full, prompt refund.

AMERICAN INSTITUTE OF PHYSICS

009

❑ **Yes!** Please rush me ______ copies of **2011 Graduate Programs in Physics, Astronomy, and Related Fields.** 978-0-7354-0840-1

List Price: **$79.00** *Shipping: $8.00 for the first book ($12.75foreign) $1.00 each additional book ($6.00 foreign).*

METHOD OF PAYMENT

❑ Check enclosed (payable in U.S. dollars to the American Institute of Physics and drawn on a bank in the United States).

❑ Charge my credit card:

❑ AMEX ❑ VISA ❑ MasterCard ❑ Discover

Cardholder name (please print) ____________

Acct. No. ____________ Exp. date ______

Signature ____________
(Credit card orders not valid without signature)

Name ____________

Institution ____________

Address ____________

City/State/Zip ____________

mail orders to:

American Institute of Physics
c/o Springer
P.O. Box 2485
Secaucus, NJ 07096-2485
or fax (201) 348-4505

SATISFACTION GUARANTEED

Examine this book for 30 days without obligation. If not completely satisfied, return the book(s) in saleable condition with your invoice within 30 days for a full, prompt refund.

AMERICAN INSTITUTE OF PHYSICS

009

➡ Use this card to order additional copies of

2011
Graduate Programs
in Physics, Astronomy,
and Related Fields

➡ Use this card to order additional copies of

2011
Graduate Programs
in Physics, Astronomy,
and Related Fields

➡ Use this card to order additional copies of

2011
Graduate Programs
in Physics, Astronomy,
and Related Fields

(PLEASE PRINT)

TO REQUEST APPLICATION FORMS AND FURTHER INFORMATION ABOUT GRADUATE DEPARTMENTS LISTED IN THIS BOOK, COMPLETE AND MAIL ONE OF THE ATTACHED POSTCARDS TO THE INDIVIDUAL DEPARTMENT.

Please send application forms and additional information concerning graduate study in your department. I saw your listing in the **2011 Graduate Programs in Physics, Astronomy, and Related Fields.**

Name ______________________________

Address ______________________________

City, State, Zip ______________________________

Highest Degree Received/Expected : ______________________________

Degree ____________ Field ____________ Date ____________

Institution ______________________________

Expected Entry Date ______________________________

My special area of interest (if any) : ______________________________

____

Other information requested (if any) : ______________________________

____

Please send application forms and additional information concerning graduate study in your department. I saw your listing in the **2011 Graduate Programs in Physics, Astronomy, and Related Fields.**

Name ______________________________

Address ______________________________

City, State, Zip ______________________________

Highest Degree Received/Expected : ______________________________

Degree ____________ Field ____________ Date ____________

Institution ______________________________

Expected Entry Date ______________________________

My special area of interest (if any) : ______________________________

____

Other information requested (if any) : ______________________________

____

Please send application forms and additional information concerning graduate study in your department. I saw your listing in the **2011 Graduate Programs in Physics, Astronomy, and Related Fields.**

Name ______________________________

Address ______________________________

City, State, Zip ______________________________

Highest Degree Received/Expected : ______________________________

Degree ____________ Field ____________ Date ____________

Institution ______________________________

Expected Entry Date ______________________________

My special area of interest (if any) : ______________________________

____

Other information requested (if any) : ______________________________

____

Please send application forms and additional information concerning graduate study in your department. I saw your listing in the **2011 Graduate Programs in Physics, Astronomy, and Related Fields.**

Name ______________________________

Address ______________________________

City, State, Zip ______________________________

Highest Degree Received/Expected : ______________________________

Degree ____________ Field ____________ Date ____________

Institution ______________________________

Expected Entry Date ______________________________

My special area of interest (if any) : ______________________________

____

Other information requested (if any) : ______________________________

____

STAMP

TO: ______________________________

____

____

____

STAMP

TO: ______________________________

____

____

____

STAMP

TO: ______________________________

____

____

____

STAMP

TO: ______________________________

____

____

____